GATE PREP SERIES

GATE

Graduate Aptitude Test in Engineering

2022

Topic-wise

Previous Solved Papers

35 Years' Solved Papers

Electronics & Communication Engineering

G. K. PUBLICATIONS (P) LTD.

Title : GATE 2022 : Electronics & Communication Engineering 35 Years Topic-wise Previous Solved Papers

Language : English

Editor's Name : Vinit Garg

Typeset & Published by :

Career Launcher Infrastructure (P) Ltd.

A-45, Mohan Cooperative Industrial Area, Near Mohan Estate Metro Station, New Delhi - 110044

Marketed by :

G.K. Publications (P) Ltd.

Plot No. 9A, Sector-27A, Mathura Road, Faridabad, Haryana-121003

ISBN : **978-93-90820-98-6**

Printer's Details : Made in India, New Delhi.

For product information :

Visit ***www.gkpublications.com*** or email to ***gkp@gkpublications.com***

GATE
Electronics & Communication Engineering

Contents

Solved Papers (Topic-Wise)

Unit III Controls System

Unit IV Electronic Device and Circuit

GATE

Graduate Aptitude Test in Engineering

IIT Institutes

GATE 2021 will be conducted by
Indian Institute of Technology, Mumbai

Indian Institute of Technology, Kharagpur

GATE 2020 conducted by
Indian Institute of Technology, Delhi

Indian Institute of Technology, Chennai

Indian Institute of Technology, Guwahati

Indian Institute of Technology, Roorkee

Indian Institute of Technology, Kanpur

Preface

Whether you want to work with esteemed PSUs including the Maharatna and Navratna companies or do your Masters from IITs and IISc, GATE is your entry to career and academic avenues post your graduation. Thousands of students write the GATE paper annually. The level of competition is fierce, owing to the increasing competition every year for a limited number of seats. If you are a serious aspirant, it is advisable to start preparing for GATE from the second year of your Graduation with the right books.

GK Publications has been the guiding force for GATE aspirants for the last 3 decades. Better known as the **"Publisher of Choice"** to students preparing for GATE and other technical test prep examinations in the country, we published the first set of books in 1994 when GATE exam, both objective and conventional, was conducted in the paper and pencil environment and used as a checkpoint for entry to postgraduate courses in IITs and IISCs. At that time, students had little access to technology and relied mainly on instructor-led learning followed by practice with books available for these examinations.

A lot has changed since then!

Today, GATE is conducted in an online-only mode with multiple choice and numerical based questions.The exam pattern too has evolved over the years. Instead of theory, the questions today test a student's basic understanding and his capacity to put fundamentals to practical applications.The exam only rewards those with a concrete foundational clarity of basics.

We, at GKP, have also embraced change. Today, our preparatory manuals are not mere books but torchbearers in your GATE preparation journey. We understand that tackling the syllabus one chapter at a time is not only easy but also allows students to cope with the looming exam pressure. A major game changer is the habit to practice and solve previous year questions and this is why, our **GATE 2022 Topic-wise Solved Papers of Electronics and Communication Engineering** is your best bet to be GATE READY by 2022!

Each chapter carries detailed analysis of 5 previous years' solved papers (all sets) so that you are well-versed with the latest exam pattern and how it has evolved over time. Both MCQs and Numerical have been explained in a lucid and detailed fashion, so that you take your preparation forward - one question at a time.

With each question, its marks' weightage has also been given so that you can gauge the level of question and make judicious use of your time.

Today's students have easy access to technology and the concept of a monologue within the classroom has changed to dialogue where students come prepared with concepts and then discuss topics. We have launched a web portal and mobile app to help you take your preparation on the go - so that you learn, revise and practice mock tests online.

The mobile app provides access to video lectures, short tests and provides regular updates about the exam. The web portal in addition to what is available on the app provides full-length mock tests to mimic the actual exam to help you gauge your level of preparedness. The combination of practice content in print, video lectures, and short and full-length tests on mobile and web make this product complete courseware for GATE preparation.

We also know that improvement is a never-ending process and hence we welcome your suggestions and feedback or spelling and technical errors if any. Please write to us at gkp@gkpublications.com

We hope that our small effort will help you prepare well for the examination.

We wish you all the best!

Team GKP

About GATE

The Graduate Aptitude Test in Engineering (GATE) conducted by IISc and IITs has emerged as one of the bench mark tests for engineering and science aptitude in facilitating admissions for higher education (M.Tech./Ph.D.) in IITs, IISc and various other Institutes/Universities/Laboratories in India. With the standard and high quality of the GATE examination in 27 disciplines of engineering and science, Humanities and Social Sciences subjects, it identifies the candidate's understanding of a subject and aptitude and eligibility for higher studies. During the last few years, GATE score is also being used as one of the criteria for recruitment in Government Organizations such as Cabinet Secretariat, and National/State Public Sector Undertakings in India. Because of the importance of the GATE examination, the number of candidates taking up GATE exams has increased tremendously. GATE exams are conducted by the IITs and IISc as a computer based test having multiple choice questions and numerical answer type questions. The questions are mostly fundamental, concept based and thought provoking. From 2017 onwards GATE Exam is being held in Bangladesh, Ethiopia, Nepal, Singapore, Sri Lanka and United Arab Emirates. An Institute with various nationalities in its campus widens the horizons of an academic environment. A foreign student brings with him/her a great diversity, culture and wisdom to share. Many GATE qualified candidates are paid scholarships/assistantship, especially funded by Ministry of Human Resources Development, Government of India and by other Ministries. Indian Institute of Technology Bombay is the Organizing Institute for GATE 2021.

Why GATE?

Admission to Post Graduate and Doctoral Programmes

Admission to postgraduate programmes with MHRD and some other government scholarships/ assistantships in engineering colleges/institutes is open to those who qualify through GATE. GATE qualified candidates with Bachelor's degree in Engineering/Technology/Architecture or Master's degree in any branch of Science/Mathematics/Statistics/Computer Applications are eligible for admission to Master/Doctoral programmes in Engineering/Technology/Architecture as well as for Doctoral programmes in relevant branches of Science with MHRD or other government scholarships/ assistantships. Candidates with Master's degree in Engineering/Technology/Architecture may seek admission to relevant Ph.D programmes with scholarship/assistantship without appearing in the GATE examination.

Financial Assistance

A valid GATE score is essential for obtaining financial assistance during Master's programs and direct Doctoral programs in Engineering/Technology/Architecture, and Doctoral programs in relevant branches of Science in Institutes supported by the MHRD or other Government agencies. As per the directives of the MHRD, the following procedure is to be adopted for admission to the post-graduate programs (Master's and Doctoral) with MHRD scholarship/assistantship. Depending upon the norms adopted by a specific institute or department of the Institute, a candidate may be admitted directly into a course based on his/her performance in GATE only **or** based on his/her performance in GATE **and** an admission test/interview conducted by the department to which he/she has applied **and/or** the candidate's academic record. If the candidate is to be selected through test/interview for post-graduate programs, a minimum of 70% weightage will be given to the performance in GATE and the remaining 30% weightage will be given to the candidate's performance in test/interview and/or academic record, as per MHRD guidelines. The admitting institutes could however prescribe a minimum passing percentage of marks in the test/interview. Some colleges/institutes specify GATE qualification as the mandatory requirement even for admission without MHRD scholarship/assistantship.

To avail of the financial assistance (scholarship), the candidate must first secure admission to a program in these Institutes, by a procedure that could vary from institute to institute. Qualification in GATE is also a minimum requirement to apply for various fellowships awarded by many Government organizations. Candidates are advised to seek complete details of admission procedures and availability of MHRD scholarship/assistantship from the concerned admitting institution. The criteria for postgraduate admission with scholarship/assistantship could be different for different institutions. The management of the post-graduate scholarship/assistantship is also the responsibility of the admitting institution. Similarly, reservation of seats under different categories is as per the policies and norms prevailing at the admitting institution and Government of India rules. *GATE offices will usually not entertain any enquiry about admission, reservation of seats and / or award of scholarship / assistantship.*

PSU Recruitments

As many as 37 PSUs are using GATE score for recruitment. It is likely that more number of PSUs may start doing so by next year. Below is the list of PSUs:

MDL, BPCL, GAIL, NLC LTD, CEL, Indian Oil, HPCL, NBPC, NECC, BHEL, WBSEDCL, NTPC, ONGC, Oil India, Power Grid, Cabinet Secretariat, Govt. of India, BAARC, NFL, IPR, PSPCL, PSTCL, DRDO, OPGC Ltd., THDC India Ltd., BBNL, RITES, IRCON, GHECL, NHAI, KRIBHCO, Mumbai Railway Vikas Corporation Ltd. (MRVC Ltd.), National Textile Corporation, Coal India Ltd., BNPM, AAI, NALCO, EdCIL India.

Important :

1. Admissions in IITs/IISc or other Institutes for M.Tech./Ph.D. through GATE scores shall be advertised separately by the Institutes and GATE does not take the responsibility of admissions.
2. Cabinet Secretariat has decided to recruit officers for the post of Senior Field Officer (Tele) (From GATE papers of EC, CS, PH), Senior Research Officer (Crypto) (From GATE papers of EC, CS, MA), Senior Research Officer (S&T) (From GATE papers EC, CS, CY, PH, AE, BT) in the Telecommunication Cadre, Cryptographic Cadre and Science & Technology Unit respectively of Cabinet Secretariat. The details of the scheme of recruitment shall be published in National Newspaper/Employment News by the concerned authority.
3. Some PSUs in India have expressed their interest to utilize GATE scores for their recruitment purpose. The Organizations who intend to utilize GATE scores shall make separate advertisement for this purpose in Newspapers etc.

Who Can Appear for GATE?

Eligibility for GATE

Before starting the application process, the candidate must ensure that he/she meets the eligibility criteria of GATE given in Table.

Eligibility Criteria for GATE 2021

Degree/Program	Qualifying Degree/Examination	Description of Eligible Candidates	Expected Year of Completion
B.E. / B.Tech. / B.Pharm.	Bachelor's degree in Engineering / Technology (4 years after 10+2 or 3 years after B.Sc. / Diploma in Engineering / Technology)	Currently in the 3rd year or higher grade or already completed	2022
B. Arch.	Bachelor's degree of Architecture (5- year course) / Naval Architecture (4- year course) / Planning (4- year course)	Currently in the 3rd year or higher grade or already completed	2023 (for 5-year program), 2022 (for 4-year program)
B.Sc. (Research) / B.S.	Bachelor's degree in Science (Post-Diploma/4 years after 10+2)	Currently in the 3rd year or higher grade or already completed	2022
Pharm. D. (after 10+2)	6 years degree program, consisting of internship or residency training, during third year onwards	Currently in the 3rd/4th/5th/6th year or already completed	2024
M.B.B.S.	Degree holders of M.B.B.S. and those who are in the 5th/6th/7th semester or higher semester of such programme.	5th, 6th, 7th or higher semester or already completed	2022
M. Sc. / M.A. / MCA or equivalent	Master's degree in any branch of Arts/Science/Mathematics/Statistics/ Computer Applications or equivalent	Currently in the first year or higher or already Completed	2022
Int. M.E./ M.Tech. (Post-B.Sc.)	Post-B.Sc Integrated Master's degree programs in Engineering/ Technology (4-year program)	Currently in the 1st / 2nd/3rd/4th year or already completed	2024
Int. M.E./ M.Tech. or Dual Degree (after Diploma or 10+2)	Integrated Master's degree program or Dual Degree program in Engineering/Technology (5-year program)	Currently in the 3rd /4th/5th year or alreadycompleted	2023
B.Sc. / B.A. / B.Com.	Bachelor degree in any branch of Science / Arts / Commerce (3 years program)	Currently in the 3rd year or already completed	2021
Int. M.Sc. / Int. B.S. / M.S.	Integrated M.Sc. or 5-year integrated B.S.-M.S. program	Currently in the 3rd year or higher or already completed	2022
Professional Society Examinations (equivalent to B.E. / B.Tech. / B.Arch.)	B.E./B.Tech./B.Arch. equivalent examinations of Professional Societies, recognized by MoE/UPSC/AICTE (e.g. AMIE by Institution of Engineers-India, AMICE by the Institute of Civil Engineers-India and so on)	Completed Section A or equivalent of such professional courses	NA

In case a candidate has passed one of the qualifying examinations as mentioned above in 2020 or earlier, the candidate has to submit the degree certificate / provisional certificate / course completion certificate / professional certificate / membership certificate issued by the society or institute. In case, the candidate is expected to complete one of the qualifying criteria in 2021 or later as mentioned above, he/she has to submit a certificate from Principal or a copy of marks card for section A of AMIE.

Certificate From Principal

Candidates who have to submit a certificate from their college Principal have to obtain one from his/her institution beforehand and upload the same during the online submission of the application form.

Candidates With Backlogs

Candidates, who have appeared in the final semester/year exam in 2021, but with a backlog (arrears/failed subjects) in any of the papers in their qualifying degree should upload a copy of any one of the mark sheets of the final year,

OR

obtain a declaration from their Principal along with the signature and seal beforehand and upload the same during the online submission of the application form.

GATE Structure

Structure of GATE

GATE 2021 will be conducted on 27 subjects (papers). Table below shows the list of papers and paper codes for GATE 2021. A candidate is allowed to appear in ANY ONE or UP TO TWO papers of the GATE examination. However, note that the combination of TWO papers in which a candidate can appear MUST be from the pre-defined list as given in Table. Also note that for a paper running in multiple sessions, a candidate will be mapped to appear for the examination in one of the sessions ONLY.

List of GATE Papers and Corresponding Codes

GATE Paper	Code	GATE Paper	Code
Aerospace Engineering	AE	Instrumentation Engineering	IN
Agricultural Engineering	AG	Mathematics	MA
Architecture and Planning	AR	Mechanical Engineering	ME
Bio-medical Engineering	BM	Mining Engineering	MN
Biotechnology	BT	Metallurgical Engineering	MT
Civil Engineering	CE	Petroleum Engineering	PE
Chemical Engineering	CH	Physics	PH
Computer Science and Information Technology	CS	Production and Industrial Engineering	PI
Chemistry	CY	Statistics	ST
Electronics and Communication Engineering	EC	Textile Engineering and Fibre Science	TF
Electrical Engineering	EE	Engineering Sciences	XE*
Environmental Science & Engineering	ES	Humanities & Social Sciences	XH**
Ecology and Evolution	EY	Life Sciences	XL***
Geology and Geophysics	GG		

Note: *Environmental Science and Engineering (ES) and Humanities and Social Sciences (XH) are two new papers introduced in GATE 2021.*

*XE Paper Sections	Code	**XH Paper Sections	Code	***XL Paper Sections	Code
Engineering Mathematics (Compulsory) (15 marks)	A	Reasoning and Comprehension (Compulsory) (25 marks)	B1	Chemistry (Compulsory) (25 marks)	P
Any TWO optional Sections (2x35 = 70 marks)		Any ONE optional Section (60 marks)		Any TWO optional Sections (2x30 = 60 marks)	
Fluid Mechanics	B	Economics	C1	Biochemistry	Q
Materials Science	C	English	C2	Botany	R
Solid Mechanics	D	Linguistics	C3	Microbiology	S
Thermodynamics	E	Philosophy	C4	Zoology	T
Polymer Science and Engineering	F	Psychology	C5	Food Technology	U
Food Technology	G	Sociology	C6		
Atmospheric and Oceanic Sciences	H				

*XE (Engineering Sciences), **XH (Humanities & Social Sciences), ***XL (Life Sciences), papers are of general nature and will be comprised of Sections listed in the above table

Note: *Each subject/paper is of total 100 marks. General Aptitude (GA) section of 15 marks is common for all papers. Hence remaining 85 marks are for the respective subject/paper code.*

Combination of Two Papers Allowed to Appear in GATE 2021 (subject to availability of infrastructure and schedule)

Code of The First (Primary) Paper	Codes of Papers Allowed as The Second Paper
AE	XE
AG	ES
AR	CE
BM	BT / XL
BT	BM / XL
CE	AR / ES
CH	CY / PE / XE
CS	MA
CY	CH / XL
EC	IN / PH
EE	IN
ES	AG / CE
EY	XL
GG	MN / PE / PH
IN	EC / EE / PH
MA	CS / PH / ST
ME	XE
MN	GG / XE
MT	PH / XE
PE	CH / GG / XE
PH	EC / GG / IN / MA / MT / ST
PI	XE
ST	MA / PH
TF	XE
XE	AE / CH / ME / MN / MT / PE / PI / TF
XH	----
XL	BM / BT / CY / EY

General Aptitude Questions

All the papers will have a few questions that test the General Aptitude (Language and Analytical Skills), apart from the core subject of the paper.

Duration and Examination Type

All the papers of the GATE 2021 examination will be for 3 hours duration and they consist of 65 questions for a total of 100 marks. Since the examination is an ONLINE computer based test (CBT), at the end of the stipulated time (3-hours), the computer screen will automatically close the examination inhibiting any further action.

Candidates will be permitted to occupy their allotted seats 40 minutes before the scheduled start of the examination. Candidates can login and start reading the instructions 20 minutes before the start of examination. The late login time (if any) recorded by the computer system MUST NOT be beyond 30 minutes from the actual starting time of the examination. Under NO circumstances, will a candidate be permitted to login after 30 minutes from the actual examination starting time. Candidates will NOT be permitted to leave the examination hall before the end of the examination.

Pattern of Question Papers

GATE 2021 may contain questions of THREE different types in all the papers:

(i) **Multiple Choice Questions (MCQ)** carrying 1 or 2 marks each, in all the papers and sections. These questions are objective in nature, and each will have choice of four answers, out of which ONLY ONE choice is correct.

Negative Marking for Wrong Answers: For a wrong answer chosen in a MCQ, there will be negative marking. For 1-mark MCQ, 1/3 mark will be deducted for a wrong answer. Likewise, for 2-mark MCQ, 2/3 mark will be deducted for a wrong answer.

(ii) **Multiple Select Questions (MSQ)** carrying 1 or 2 marks each in all the papers and sections. These questions are objective in nature, and each will have choice of four answers, out of which ONE or MORE than ONE choice(s) is / are correct.

Note: There is NO negative marking for a wrong answer in MSQ questions. However, there is NO partial credit for choosing partially correct combinations of choices or any single wrong choice.

(iii) **Numerical Answer Type (NAT)** Questions carrying 1 or 2 marks each in most of the papers and sections. For these questions, the answer is a signed real number, which needs to be entered by the candidate using the virtual numeric keypad on the monitor (keyboard of the computer will be disabled). No choices will be shown for these types of questions. The answer can be a number such as 10 or –10 (an integer only). The answer may be in decimals as well, for example, 10.1 (one decimal) or 10.01 (two decimals) or –10.001 (three decimals). These questions will be mentioned with, up to which decimal places, the candidates need to present the answer. Also, for some NAT type problems an appropriate range will be considered while evaluating these questions so that the candidate is not unduly penalized due to the usual round-off errors. Candidates are advised to do the rounding off at the end of the calculation (not in between steps). Wherever required and possible, it is better to give NAT answer up to a maximum of three decimal places.

Note: *There is NO negative marking for a wrong answer in NAT questions.*
Also, there is NO partial credit in NAT questions.

Marking Scheme – Distribution of Marks and Questions

General Aptitude (GA) Questions

In all papers, GA questions carry a total of 15 marks. The GA section includes 5 questions carrying 1-mark each (sub-total 5 marks) and 5 questions carrying 2-marks each (sub-total 10 marks).

Question Papers other than GG, XE, XH and XL

These papers would contain 25 questions carrying 1-mark each (sub-total 25 marks) and 30 questions carrying 2-marks each (sub-total 60 marks) consisting of some MCQ type questions, while the remaining may be MSQ and / or NAT questions.

GG (Geology and Geophysics) Paper

Apart from the General Aptitude (GA) section, the GG question paper consists of two parts: Part A and Part B. Part A is compulsory for all the candidates. Part B contains two sections: Section 1 (Geology) and Section 2 (Geophysics). Candidates will have to attempt questions in Part A and questions in either Section 1 or Section 2 of Part B. The choice of Section 1 OR Section 2 of Part B has to be made at the time of online registration in GOAPS. At the examination hall, candidate cannot request for change of section.

Part A consists of 25 questions carrying 1-mark each (sub-total 25 marks and some of these will be MSQ and/or numerical answer type questions while remaining questions will be MCQ type). Either section of Part B (Section 1 and Section 2) consists of 30 questions carrying 2-marks each (sub-total 60 marks and some of these will be MSQ and/or numerical answer type questions while remaining questions will be MCQ type).

XE Paper (Engineering Sciences)

A candidate appearing in the XE paper has to answer the following:

- GA – General Aptitude carrying a total of 15 marks.
- Section A – Engineering Mathematics (Compulsory): This section contains 11 questions carrying a total of 15 marks: 7 questions carrying 1-mark each (sub-total 7 marks), and 4 questions carrying 2-marks each (sub-total 8 marks). Some questions will be of numerical answer type while remaining questions will be MCQ type.
- Any two of XE Sections B to H: The choice of two sections from B to H can be made during the examination after viewing the questions. Only TWO optional sections can be answered at a time. A candidate wishing to change midway of the examination to another optional section must first choose to deselect one of the previously chosen optional sections (B to H). Each of the optional sections of the XE paper (Sections B through H) contains 22 questions carrying a total of 35 marks: 9 questions carrying 1-mark each (sub-total 9 marks) and 13 questions carrying 2-marks each (sub-total 26 marks). Some questions will be of MSQ and/or numerical answer type while remaining questions will be MCQ type.

XH Paper (Humanities and Social Sciences)

A candidate appearing in the XH paper has to answer the following:

- **GA – General Aptitude** carrying a total of 15 marks.
- **Section B1 – Reasoning and Comprehension (Compulsory):** This section contains 15 questions carrying a total of 25 marks: 5 questions carrying 1-mark each (sub-total 5 marks) and 10 questions carrying 2-marks each (sub-total 20 marks). Some questions will be of MSQ and/or numerical answer type while remaining questions will be MCQ type.
- **Any ONE of XH Sections C1 to C6:** The ONE choice of section from C1 to C6 has to be made at the time of online registration in GOAPS. At the examination hall, candidate cannot request for change of section. Each of the optional sections of the XH paper (Sections C1 through C6) contains 40 questions carrying a total of 60 marks: 20 questions carrying 1-mark each (sub-total 20 marks) and 20 questions carrying 2-marks each (sub-total 40 marks). Some questions will be of MSQ and/or numerical answer type while remaining questions will be MCQ type.

XL Paper (Life Sciences)

A candidate appearing in the XL paper has to answer the following:

- **GA – General Aptitude** carrying a total of 15 marks.
- **Section P–Chemistry (Compulsory):** This section contains 15 questions carrying a total of 25 marks: 5 questions carrying 1-mark each (sub-total 5 marks) and 10 questions carrying 2-marks each (sub-total 20 marks). Some questions will be of MSQ and/or numerical answer type while remaining questions will be MCQ type.
- **Any two of XL Sections Q to U:** The choice of two sections from Q to U can be made during the examination after viewing the questions. Only TWO optional sections can be answered at a time. A candidate wishing to change midway of the examination to another optional section must first choose to deselect one of the previously chosen optional sections (Q to U). Each of the optional sections of the XL paper (Sections Q through U) contains 20 questions carrying a total of 30 marks: 10 questions carrying 1-mark each (sub-total 10 marks) and 10 questions carrying 2-marks each (sub-total 20 marks). Some questions will be of MSQ and/or numerical answer type while remaining questions will be MCQ type.

GATE Score

After the evaluation of the answers, the actual (raw) marks obtained by a candidate will be considered for computing the GATE Score. For multi-session papers (subjects), raw marks obtained by the candidates in different sessions will be converted to Normalized marks for that particular subject. Thus, raw marks (for single session papers) or normalized marks (for multi-session papers) will be used for computing the GATE Score, based on the qualifying marks.

Calculation of Normalized Marks for Multi-Session Papers

In GATE 2021 examination, some papers may be conducted in multi-sessions. Hence, for these papers, a suitable normalization is applied to take into account any variation in the difficulty levels of the question papers across different sessions. The normalization is done based on the fundamental assumption that "in all multi-session GATE papers, the distribution of abilities of candidates is the same across all the sessions". This assumption is justified since the number of candidates appearing in multi-session papers in GATE 2021 is large and the procedure for allocation of session to candidates is random. Further, it is also ensured that for the same multi-session paper, the number of candidates allotted in each session is of the same order of magnitude.

Based on the above, and considering various normalization methods, the committee arrived at the following formula for calculating the normalized marks for the multi-session papers.

Nsormalization mark of jth candidate in the ith session $\widehat{M_{ij}}$ is given by

$$\widehat{M_{ij}} = \frac{\bar{M}_t^g - M_q^g}{\bar{M}_{ti} - M_{iq}}(M_{ij} - M_{iq}) + M_q^g$$

where

M_{ij} : is the actual marks obtained by the jth candidate in i^{th} session

$\bar{M}_t^g$: is the average marks of the top 0.1% of the candidates considering all sessions

M_q^g : is the sum of mean and standard deviation marks of the candidates in the paper considering all sessions

$\bar{M}_{ti}$: is the average marks of the top 0.1% of the candidates in the ith session

M_{iq} : is the sum of the mean marks and standard deviation of the ith session

Calculation of GATE Score for All Papers

For all papers for which there is only one session, actual marks obtained by the candidates will be used for calculating the GATE 2021 Score. For papers in multi-sessions, normalized marks will be calculated corresponding to the raw marks obtained by a candidate and the GATE 2021 Score will be calculated based on the normalized marks.

The GATE 2021 score will be computed using the formula given below.

$$\text{GATE Score} = S_q + (S_t - S_q)\frac{(M - M_q)}{(\bar{M}_t - M_q)}$$

where

- M : marks obtained by the candidate (actual marks for single session papers and normalized marks for multi-session papers)
- M_q : is the qualifying marks for general category candidate in the paper
- $\bar{M}_t$: is the mean of marks of top 0.1% or top 10 (whichever is larger) of the candidates who appeared in the paper (in case of multi-session papers including all sessions)
- S_q : 350, is the score assigned to M_q
- S_t : 900, is the score assigned to $\bar{M}_t$

In the GATE 2021 the qualifying marks (M_q) for general category student in each subject will be 25 marks (out of 100) or $\mu + \sigma$, whichever is larger. Here μ is the mean and σ is the standard deviation of marks of all the candidates who appeared in the paper.

After the declaration of results, GATE Scorecards can be downloaded by the GATE qualified candidates ONLY.

The GATE 2021 Committee has the authority to decide the qualifying mark/score for each GATE paper. In case of any claim or dispute with respect to GATE 2021 examination or score, the Courts and Tribunals in Mumbai alone will have the exclusive jurisdiction to entertain and settle them.

GATE Syllabus

Section 1: Engineering Mathematics

Linear Algebra: Vector space, basis, linear dependence and independence, matrix algebra, eigenvalues and eigenvectors, rank, solution of linear equations- existence and uniqueness.

Calculus: Mean value theorems, theorems of integral calculus, evaluation of definite and improper integrals, partial derivatives, maxima and minima, multiple integrals, line, surface and volume integrals, Taylor series.

Differential Equations: First order equations (linear and nonlinear), higher order linear differential equations, Cauchy's and Euler's equations, methods of solution using variation of parameters, complementary function and particular integral, partial differential equations, variable separable method, initial and boundary value problems.

Vector Analysis: Vectors in plane and space, vector operations, gradient, divergence and curl, Gauss's, Green's and Stokes' theorems.

Complex Analysis: Analytic functions, Cauchy's integral theorem, Cauchy's integral formula, sequences, series, convergence tests, Taylor and Laurent series, residue theorem.

Probability and Statistics: Mean, median, mode, standard deviation, combinatorial probability, probability distributions, binomial distribution, Poisson distribution, exponential distribution, normal distribution, joint and conditional probability.

Section 2: Networks, Signals and Systems

Circuit analysis: Node and mesh analysis, superposition, Thevenin's theorem, Norton's theorem, reciprocity.

Sinusoidal steady state analysis: phasors, complex power, maximum power transfer.

Time and frequency domain analysis of linear circuits: RL, RC and RLC circuits, solution of network equations using Laplace transform.

Linear 2-port network parameters, wye-delta transformation.

Continuous-time signals: Fourier series and Fourier transform, sampling theorem and applications.

Discrete-time signals: DTFT, DFT, z-transform, discrete-time processing of continuous-time signals. LTI systems: definition and properties, causality, stability, impulse response, convolution, poles and zeroes, frequency response, group delay, phase delay.

Section 3: Electronic Devices

Energy bands in intrinsic and extrinsic semiconductors, equilibrium carrier concentration, direct and indirect band-gap semiconductors.

Carrier transport: diffusion current, drift current, mobility and resistivity, generation and recombination of carriers, Poisson and continuity equations.

P-N junction, Zener diode, BJT, MOS capacitor, MOSFET, LED, photo diode and solar cell.

Section 4: Analog Circuits

Diode circuits: clipping, clamping and rectifiers.

BJT and MOSFET amplifiers: biasing, ac coupling, small signal analysis, frequency response.

Current mirrors and differential amplifiers.

Op-amp circuits: Amplifiers, summers, differentiators, integrators, active filters, Schmitt triggers and oscillators.

Section 5: Digital Circuits

Number representations: binary, integer and floating-point- numbers.

Combinatorial circuits: Boolean algebra, minimization of functions using Boolean identities and Karnaugh map, logic gates and their static CMOS implementations, arithmetic circuits, code converters, multiplexers, decoders.

Sequential circuits: latches and flip-flops, counters, shift-registers, finite state machines, propagation delay, setup and hold time, critical path delay.

Data converters: sample and hold circuits, ADCs and DACs.

Semiconductor memories: ROM, SRAM, DRAM.

Computer organization: Machine instructions and addressing modes, ALU, data-path and control unit, instruction pipelining.

Section 6: Control Systems

Basic control system components; Feedback principle; Transfer function; Block diagram representation; Signal flow graph; Transient and steady-state analysis of LTI systems; Frequency response; Routh-Hurwitz and Nyquist stability criteria; Bode and root-locus plots; Lag, lead and lag-lead compensation; State variable model and solution of state equation of LTI systems.

Section 7: Communications

Random processes: autocorrelation and power spectral density, properties of white noise, filtering of random signals through LTI systems.

Analog communications: amplitude modulation and demodulation, angle modulation and demodulation, spectra of AM and FM, superheterodyne receivers.

Information theory: entropy, mutual information and channel capacity theorem.

Digital communications: PCM, DPCM, digital modulation schemes (ASK, PSK, FSK, QAM), bandwidth, inter-symbol interference, MAP, ML detection, matched filter receiver, SNR and BER.

Fundamentals of error correction, Hamming codes, CRC.

Section 8: Electromagnetics

Maxwell's equations: differential and integral forms and their interpretation, boundary conditions,wave equation, Poynting vector.

Plane waves and properties: reflection and refraction, polarization, phase and group velocity, propagation through various media, skin depth.

Transmission lines: equations, characteristic impedance, impedance matching, impedance transformation, S-parameters, Smith chart.

Rectangular and circular waveguides, light propagation in optical fibers, dipole and monopole antennas, linear antenna arrays.

Chapter-Wise Analysis

GATE PAPERS (Electronics & Communication Engineering)

Subject	2011	2012	2013	2014	2015	2016	2017	2018	2019	2020	2021
1. Engineering Maths											
1 Mark Questions	3	4	4	3	4	5	4	5	6	4	5
2 Marks Questions	4	7	6	4	3	4	5	3	3	3	3
Total Marks	**11**	**18**	**16**	**11**	**10**	**13**	**14**	**11**	**12**	**10**	**11**
2. Network Theory											
1 Mark Questions	2	1	3	3	3	5	1	1	1	3	1
2 Marks Questions	3	4	2	2	3	3	2	3	4	2	5
Total Marks	**8**	**9**	**7**	**7**	**9**	**11**	**5**	**7**	**9**	**7**	**11**
3. Signals & Systems											
1 Mark Questions	1	2	0	4	2	1	3	2	3	2	2
2 Marks Questions	4	3	1	4	4	2	3	2	1	2	3
Total Marks	**9**	**8**	**2**	**12**	**10**	**5**	**9**	**6**	**5**	**6**	**8**
4. Electronic Devices & Circuits											
1 Mark Questions	2	4	6	4	4	3	5	3	3	2	3
2 Marks Questions	4	4	4	2	3	3	3	2	2	4	2
Total Marks	**10**	**12**	**14**	**8**	**10**	**9**	**11**	**7**	**7**	**10**	**7**
5. Analog Circuits											
1 Mark Questions	3	0	1	2	2	2	1	2	3	3	3
2 Marks Questions	3	2	6	5	3	3	4	5	5	5	5
Total Marks	**9**	**4**	**13**	**12**	**8**	**8**	**9**	**12**	**13**	**13**	**13**
6. Digital Circuits & Microprocessor											
1 Mark Questions	2	5	3	3	4	2	3	3	2	3	3
2 Marks Questions	3	1	3	5	4	3	4	4	1	3	2
Total Marks	**8**	**7**	**9**	**13**	**12**	**8**	**11**	**11**	**4**	**9**	**7**
7. Control Systems											
1 Mark Questions	6	1	4	3	2	2	3	3	1	2	2
2 Marks Questions	4	3	4	4	3	3	4	5	4	4	1
Total Marks	**14**	**7**	**12**	**11**	**8**	**8**	**11**	**13**	**9**	**10**	**4**

Subject	2011	2012	2013	2014	2015	2016	2017	2018	2019	2020	2021
8. Communication systems											
1 Mark Questions	1	2	3	2	2	2	4	4	3	3	4
2 Marks Questions	1	2	0	2	4	4	2	3	6	4	5
Total Marks	**3**	**6**	**3**	**6**	**10**	**10**	**8**	**10**	**15**	**11**	**14**
9. Electromagnetic Theory											
1 Mark Questions	3	6	1	1	2	3	1	2	3	3	2
2 Marks Questions	4	4	3	2	3	5	3	3	4	3	3
Total Marks	**11**	**14**	**7**	**5**	**8**	**13**	**7**	**8**	**11**	**9**	**8**
10. Computer-organization											
1 Mark Questions	–	–	–	–	–	–	–	–	–	–	0
2 Marks Questions	–	–	–	–	–	–	–	–	–	–	1
Total Marks	–	–	–	–	–	–	–	–	–	–	**2**
11. General Aptitude											
1 Mark Questions	5	5	5	5	5	5	5	5	5	5	5
2 Marks Questions	5	5	5	5	5	5	5	5	5	5	5
Total Marks	**15**	**15**	**15**	**15**	**15**	**15**	**15**	**15**	**15**	**15**	**15**

SOLVED PAPER - 2021

INSTRUCTIONS

1. Total of 65 questions carrying 100 marks, out of which 10 questions carrying a total of 15 marks are in General Aptitude (GA)
2. The Engineering Mathematics will carry around **15% of the total marks,** the General Aptitude section will carry **15% of the total marks** and the **remaining 70% of the total marks.**
3. **Types of Questions**
   (*a*) **Multiple Choice Questions (MCQ)** carrying 1 or 2 marks each in all papers and sections. These questions are objective in nature, and each will have a choice of four options, out of which the candidate has to mark the correct answer(s).
   (*b*) **Numerical Answer Questions** of 1 or 2 marks each in all papers and sections. For these questions the answer is a real number, to be entered by the candidate using the virtual keypad. No choices will be shown for these type of questions.
   (*c*) **Multiple Select Questions (MSQ)** carrying 1 or 2 marks each in all the papers and sections. These questions are objective in nature, and each will have choice of four answers, out of which ONE or MORE than ONE choice(s) is / are correct.
4. For **1-mark** multiple-choice questions, **1/3 marks** will be deducted for a wrong answer. Likewise, for **2-marks** multiple-choice questions, 2/3 marks will be deducted for a wrong answer. There is no negative marking for numerical answer type questions & multiple selection questions.

Chapter-Wise Analysis

#	Chapters	Marks		
		1	2	Total
1	General Aptitude	5	5	15
2	Engineering Mathematics	5	3	11
3	Network Theory	1	5	11
4	Signals and Systems	2	3	8
5	Electronic Devices & Circuits	3	2	7
6	Analog Circuits	3	5	13

#	Chapters	Marks		
		1	2	Total
7	Digital Circuits & Microprocessor	3	2	7
8	Control Systems	2	1	4
9	Communication Systems	4	5	14
10	Electromagnetic Theory	2	3	8
11	Computer-organization	-	1	2

GENERAL APTITUDE (GA)

Q.1 to Q.5: Multiple Choice Questions (MCQ), carry ONE mark each.

1. The current population of a city is 11,02,500. If it has been increasing at the rate of 5% per annum, what was its population 2 years ago?

(a) 9,92,500 (b) 9,95,006
(c) 10,00,000 (d) 12,51,506

2. p and q are positive integers and $\frac{p}{q}+\frac{q}{p}=3$, then, $\frac{p^2}{q^2}+\frac{q^2}{p^2}=$

(a) 3 (b) 7
(c) 9 (d) 11

3.

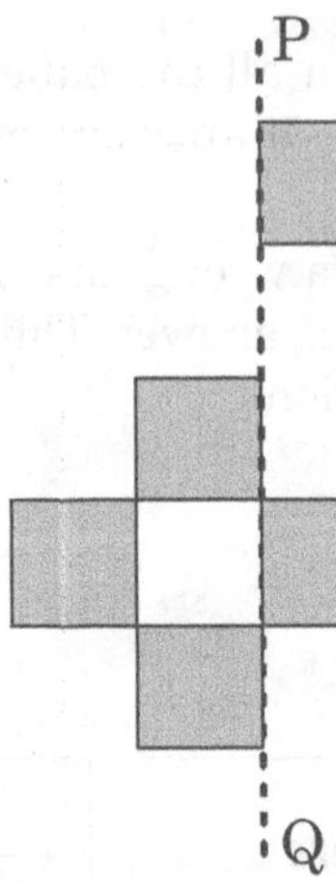

The least number of squares that must be added so that the line P-Q becomes the line of symmetry is ______

(a) 4 (b) 3
(c) 6 (d) 7

4. *Nostalgia* is to *anticipation* as _____ is to _____
Which one of the following options maintains a similar logical relation in the above sentence?

(a) Present, past (b) Future, past
(c) Past, future (d) Future, present

5. Consider the following sentences:

(i) I woke up from sleep.
(ii) I woked up from sleep.
(iii) I was woken up from sleep.
(iv) I was wokened up from sleep.

Which of the above sentences are grammatically CORRECT?

(a) (i) and (ii) (b) (i) and (iii)
(c) (ii) and (iii) (d) (i) and (iv)

Q.6 to Q.10: Multiple Choice Question (MCQ), carry TWO marks each.

6. Given below are two statements and two conclusions.

Statement 1 : All purple are green.
Statement 2 : All black are green.
Conclusion I : Some black are purple.
Conclusion II : No black is purple.

Based on the above statements and conclusions, which one of the following options is logically CORRECT?

(a) Only conclusion I is correct.
(b) Only conclusion II is correct.
(c) Either conclusion I or II is correct.
(d) Both conclusion I and II are correct.

7. Computers are ubiquitous. They are used to improve efficiency in almost all fields from agriculture to space exploration. Artificial intelligence (AI) is currently a hot topic. AI enables computers to learn, given enough training data. For humans, sitting in front of a computer for long hours can lead to health issues.

Which of the following can be deduced from the above passage?

(i) Nowadays, computers are present in almost all places.
(ii) Computers cannot be used for solving problems in engineering.
(iii) For humans, there are both positive and negative effects of using computers.
(iv) Artificial intelligence can be done without data.

(a) (ii) and (iii)
(b) (ii) and (iv)
(c) (i), (iii) and (iv)
(d) (i) and (iii)

8. Consider a square sheet of side 1 unit. In the first step, it is cut along the main diagonal to get two triangles. In the next step, one of the cut triangles is revolved about its short edge to form a solid cone. The volume of the resulting cone, in cubic units, is ______

(a) $\frac{\pi}{3}$

(b) $\frac{2\pi}{3}$

(c) $\frac{3\pi}{2}$

(d) 3π

9.

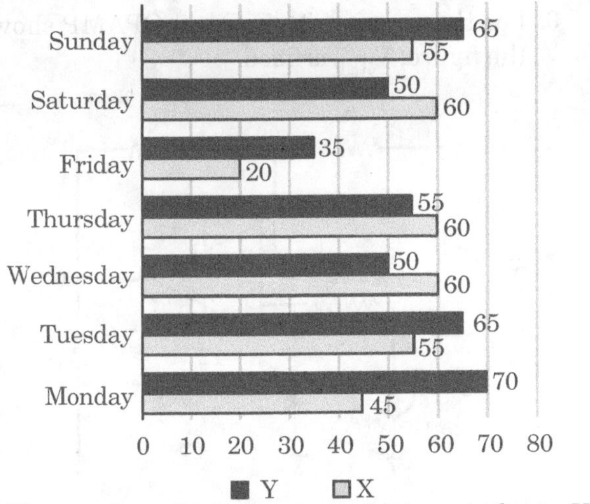

The number of minutes spent by two students, X and Y, exercising every day in a given week are shown in the bar chart above.

The number of days in the given week in which one of the students spent a minimum of 10% more than the other student, on a given day, is

(a) 4 (b) 5

(c) 6 (d) 7

10.

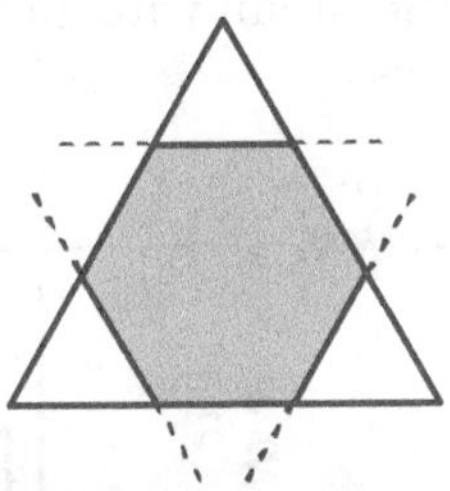

Corners are cut from an equilateral triangle to produce a regular convex hexagon as shown in the figure above.

The ratio of the area of the regular convex hexagon to the area of the original equilateral triangle is

(a) 2 : 3 (b) 3 : 4

(c) 4 : 5 (d) 5 : 6

ELECTRONICS & COMMUNICATION ENGINEERING

Q.1 - Q-15: Multiple Choice Questions (MCQ), carry ONE mark each.

1. The vector function $F(r) = -x\hat{i} + y\hat{j}$ is defined over a circular arc C shown in the figure.

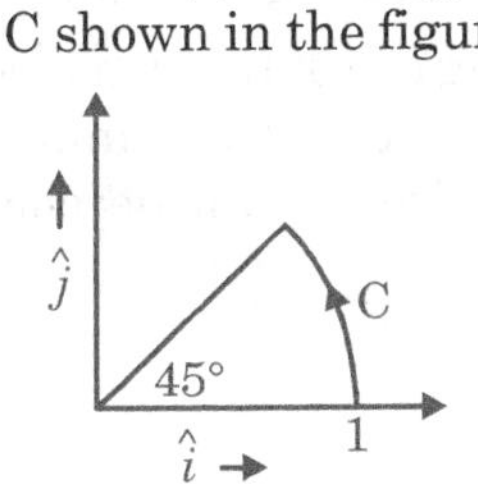

The line integral of $\int_C F(r).\,dr$ is

(a) $\frac{1}{2}$ (b) $\frac{1}{4}$

(c) $\frac{1}{6}$ (d) $\frac{1}{3}$

2. Consider the differential equation given below.

$$\frac{dy}{dx} + \frac{x}{1-x^2}y = x\sqrt{y}$$

The integrating factor of the differential equation is

(a) $\left(1-x^2\right)^{-\frac{3}{4}}$ (b) $\left(1-x^2\right)^{-\frac{1}{4}}$

(c) $\left(1-x^2\right)^{-\frac{3}{2}}$ (d) $\left(1-x^2\right)^{-\frac{1}{2}}$

3. Two continuous random variables X and Y are related as

$$Y = 2X + 3$$

Let σ_X^2 and σ_Y^2 denote the variances of X and Y, respectively. The variances are related as

(a) $\sigma_Y^2 = 2\sigma_X^2$ (b) $\sigma_Y^2 = 4\sigma_X^2$

(c) $\sigma_Y^2 = 5\sigma_X^2$ (d) $\sigma_Y^2 = 25\sigma_X^2$

4. Consider a real-valued base-band signal $x(t)$, band limited to 10 kHz. The Nyquist rate for the signal $y(t) = x(t)\,x\left(1+\frac{t}{2}\right)$ is

(a) 15 kHz (b) 30 kHz

(c) 60 kHz (d) 20 kHz

5. Consider two 16-point sequences $x[n]$ and $h[n]$. Let the linear convolution of $x[n]$ and $h[n]$ be denoted by $y[n]$, while $z[n]$ denotes the 16-point inverse discrete Fourier transform (IDFT) of the product of the 16-point DFTs of $x[n]$ and $h[n]$. The value(s) of k for which $z[k] = y[k]$ is/are

(a) $k = 0, 1, 2, ..., 15$ (b) $k = 0$

(c) $k = 15$ (d) $k = 0$ and $k = 15$

6. A bar of silicon is doped with boron concentration of 10^{16} cm^{-3} and assumed to be fully ionized. It is exposed to light such that electron-hole pairs are generated throughout the volume of the bar at the rate of 10^{20} cm^{-3}s^{-1}. If the recombination lifetime is 100 μs, intrinsic carrier concentration of silicon is 10^{10} cm^{-3} and assuming 100% ionization of boron, then the approximate product of steady-state electron and hole concentrations due to this light exposure is

(a) 10^{20} cm^{-6} (b) 2×10^{20} cm^{-6}

(c) 10^{32} cm^{-6} (d) 2×10^{32} cm^{-6}

7. The energy band diagram of a p-type semiconductor bar of length L under equilibrium condition (*i.e.*, the Fermi energy level E_F is constant) is shown in the figure. The valance band E_V is sloped since doping is non-uniform along the bar. The difference between the energy levels of the valence band at the two edges of the bar is Δ.

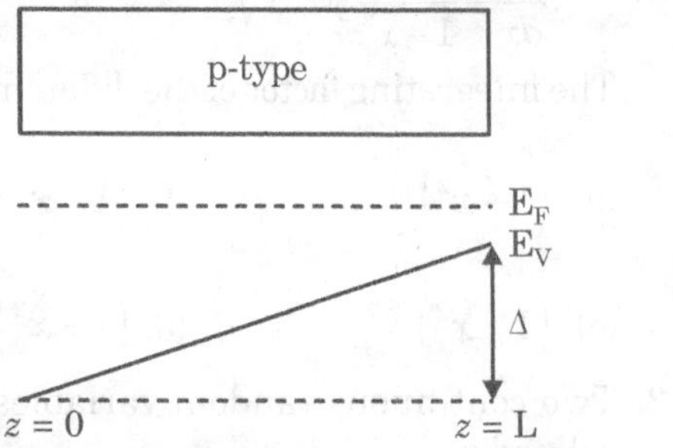

If the charge of an electron is q, then the magnitude of the electric field developed inside this semiconductor bar is

(a) $\frac{\Delta}{qL}$ (b) $\frac{2\Delta}{qL}$

(c) $\frac{\Delta}{2qL}$ (d) $\frac{3\Delta}{2qL}$

8. In the circuit shown in the figure, the transistors M_1 and M_2 are operating in saturation. The channel length modulation coefficients of both the transistors are non-zero. The transconductance of the MOSFETs M_1 and M_2 are g_{m1} and g_{m2}, respectively, and the internal resistance of the MOSFETs M_1 and M_2 are r_{01} and r_{02}, respectively.

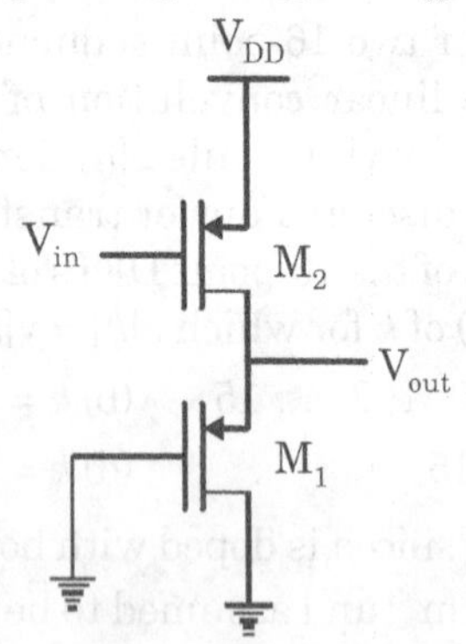

Ignoring the body effect, the ac small signal voltage gain $\left(\frac{\partial V_{out}}{\partial V_{in}}\right)$ of the circuit is

(a) $-g_{m2}(r_{01} \| r_{02})$

(b) $-g_{m2}\left(\frac{1}{g_{m1}} \| r_{02}\right)$

(c) $-g_{m1}\left(\frac{1}{g_{m2}} \| r_{01} \| r_{02}\right)$

(d) $-g_{m2}\left(\frac{1}{g_{m1}} \| r_{01} \| r_{02}\right)$

9. For the circuit with an ideal OPAMP shown in the figure, V_{REF} is fixed.

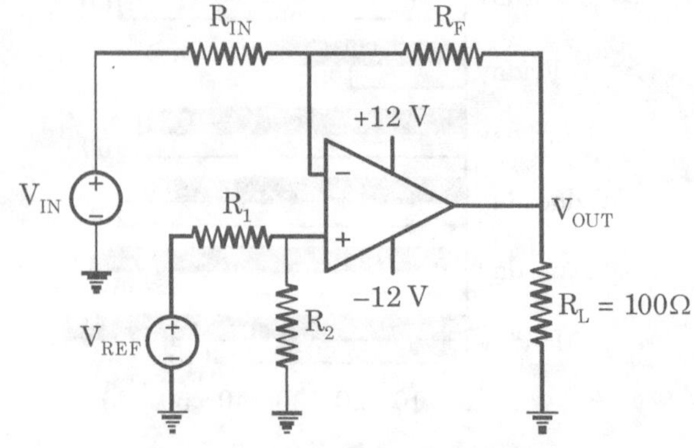

If $V_{OUT} = 1$ volt for $V_{IN} = 0.1$ volt and $V_{OUT} = 6$ volt for $V_{IN} = 1$ volt, where V_{OUT} is measured across R_L connected at the output of this OPAMP, the value of $\frac{R_F}{R_{IN}}$ is

(a) 3.285 (b) 2.860

(c) 3.825 (d) 5.555

10. Consider the circuit with an ideal OPAMP shown in the figure.

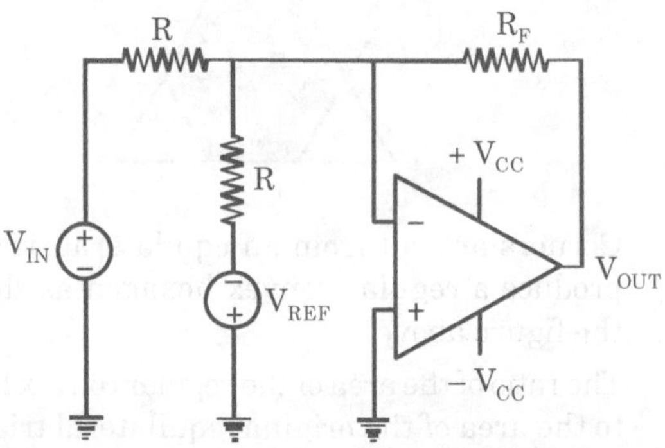

Assuming $|V_{IN}| << |V_{CC}|$ and $|V_{REF}| << |V_{CC}|$, the condition at which V_{OUT} equals to zero is

(a) $V_{IN} = V_{REF}$ (b) $V_{IN} = 0.5\,V_{REF}$

(c) $V_{IN} = 2\,V_{REF}$ (d) $V_{IN} = 2 + V_{REF}$

11. If $(1235)_x = (3033)_y$, where x and y indicate the bases of the corresponding numbers, then

(a) $x = 7$ and $y = 5$ (b) $x = 8$ and $y = 6$

(c) $x = 6$ and $y = 4$ (d) $x = 9$ and $y = 7$

12. Addressing of a 32K × 16 memory is realized using a single decoder. The minimum number of AND gates required for the decoder is

(a) 2^8

(b) 2^{32}

(c) 2^{15}

(d) 2^{19}

13. The block diagram of a feedback control system is shown in the figure.

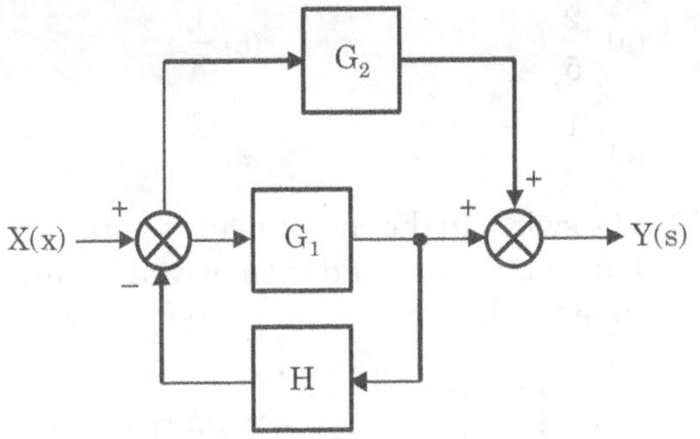

The transfer function $\frac{Y(s)}{X(s)}$ of the system is

(a) $\frac{G_1 + G_2 + G_1G_2H}{1 + G_1H}$ (b) $\frac{G_1 + G_2}{1 + G_1H + G_2H}$

(c) $\frac{G_1 + G_2}{1 + G_1H}$ (d) $\frac{G_1 + G_2 + G_1G_2H}{1 + G_1H + G_2H}$

14. The complete Nyquist plot of the open-loop transfer function G(*s*)H(*s*) of a feedback control system is shown in the figure.

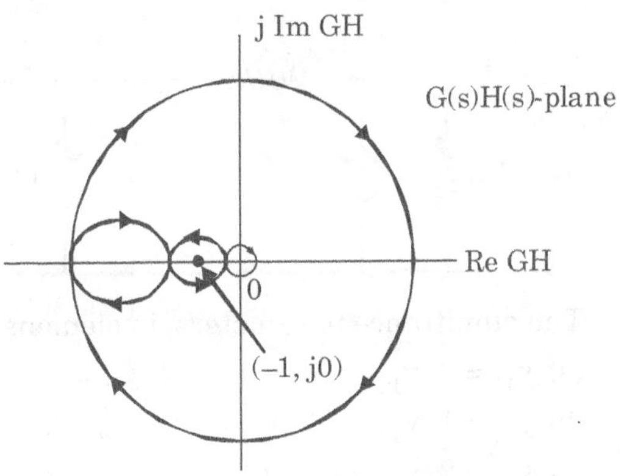

If G(*s*)H(*s*) has one zero in the right-half of the *s*-plane, the number of poles that the closed-loop system will have in the right-half of the s-plane is

(a) 0 (b) 1

(c) 4 (d) 3

15. Consider a rectangular coordinate system (*x*, *y*, *z*) with unit vectors a_x, a_y, and a_z. A plane wave traveling in the region $z \geq 0$ with electric field vector $E = 10 \cos(2 \times 10^8 t + \beta z)a_y$ is incident normally on the plane at $z = 0$, where β is the phase constant. The region $z \geq 0$ is in free space and the region $z < 0$ is filled with a lossless medium (permittivity $\varepsilon = \varepsilon_0$, permeability $\mu = 4\mu_0$, where $\varepsilon_0 = 8.85 \times 10^{-12}$ F/m and $\mu_0 = 4\pi \times 10^{-7}$ H/m). The value of the reflection coefficient is

(a) $\frac{1}{3}$ (b) $\frac{3}{5}$

(c) $\frac{2}{5}$ (d) $\frac{2}{3}$

Q. 16 to Q. 25 : Numerical Answer Type (NAT), carry ONE mark each.

16. If the vectors (1. 0, –1. 0, 2. 0), (7. 0, 3. 0, *x*) and (2. 0, 3. 0, 1. 0) in $\mathbb{R}^3$ are linearly dependent, the value of *x* is ________

17. Consider the vector field $F = a_x(4y - c_1z) + a_y(4x + 2z) + a_z(2y + z)$ in a rectangular coordinate system (*x*, *y*, *z*) with unit vectors a_x, a_y, and a_z. If the field F is irrotational (conservative), then the constant c_1 (in integer) is ________

18. Consider the circuit shown in the figure.

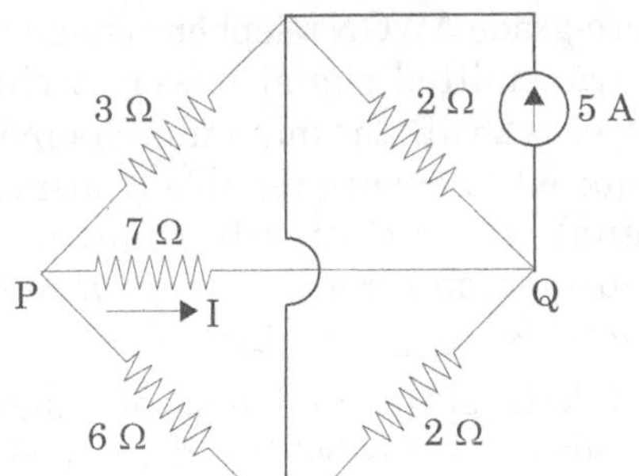

The current I flowing through the 7Ω resistor between P and Q (*rounded off to one decimal place*) is ________ A.

19. Consider the circuit shown in the figure.

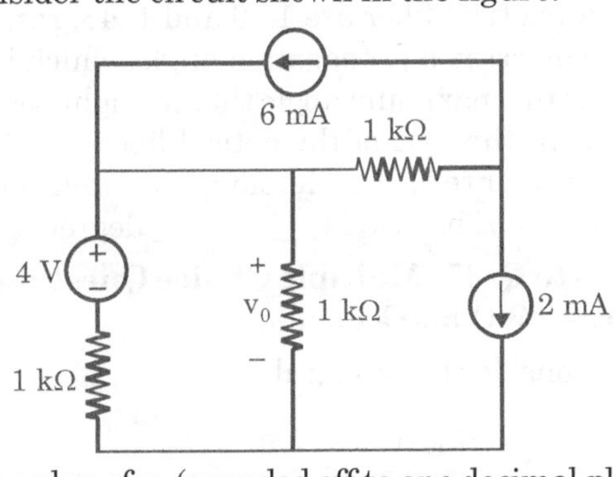

The value of v_0 (rounded off to one decimal place) is ________ V.

20. An 8-bit unipolar (all analog output values are positive) digital-to-analog converter (DAC) has a full-scale voltage range from 0 V to 7. 68 V. If the digital input code is 10010110 (the leftmost bit is MSB), then the analog output voltage of the DAC (*rounded off to one decimal place*) is ________V.

21. The autocorrelation function $R_X(\tau)$ of a wide-sense stationary random process X(*t*) is shown in the figure.

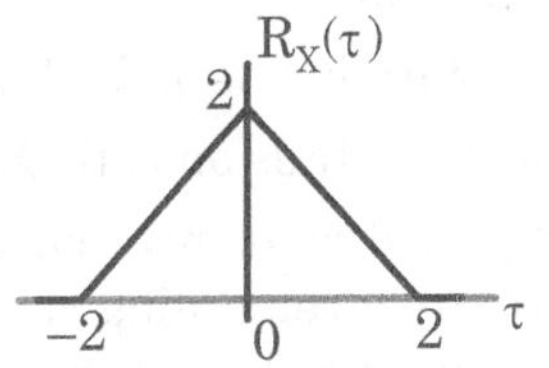

The average power of X(*t*) is ________

22. Consider a carrier signal which is amplitude modulated by a single-tone sinusoidal message signal with a modulation index of 50%. If the carrier and one of the sidebands are suppressed in the modulated signal, the percentage of power saved (*rounded off to one decimal place*) is ______

23. A speech signal, band limited to 4 kHz, is sampled at 1.25 times the Nyquist rate. The speech samples, assumed to be statistically independent and uniformly distributed in the range – 5 V to + 5 V, are subsequently quantized in an 8-bit uniform quantizer and then transmitted over a voice-grade AWGN telephone channel. If the ratio of transmitted signal power to channel noise power is 26 dB, the minimum channel bandwidth required to ensure reliable transmission of the signal with arbitrarily small probability of transmission error (*rounded off to two decimal places*) is ________ kHz.

24. A 4 kHz sinusoidal message signal having amplitude 4 V is fed to a delta modulator (DM) operating at a sampling rate of 32 kHz. The minimum step size required to avoid slope overload noise in the DM (*rounded off to two decimal places*) is ________ V.

25. The refractive indices of the core and cladding of an optical fiber are 1.50 and 1. 48, respectively. The critical propagation angle, which is defined as the maximum angle that the light beam makes with the axis of the optical fiber to achieve the total internal reflection, (*rounded off to two decimal places*) is ________ degree.

Q. 26 to Q. 35 : Multiple Choice Question (MCQ), carry TWO marks each.

26. Consider the integral

$$\oint_C \frac{\sin(x)}{x^2(x^2+4)}dx$$

where C is a counter-clockwise oriented circle defined as $|x - i| = 2$. The value of the integral is

(a) $-\frac{\pi}{8}\sin(2i)$ (b) $\frac{\pi}{8}\sin(2i)$

(c) $-\frac{\pi}{4}\sin(2i)$ (d) $\frac{\pi}{4}\sin(2i)$

27. A box contains the following three coins.

I. A fair coin with head on one face and tail on the other face.

II. A coin with heads on both the faces.

III. A coin with tails on both the faces.

A coin is picked randomly from the box and tossed. Out of the two remaining coins in the box, one coin is then picked randomly and tossed. If the first toss results in a head, the probability of getting a head in the second toss is

(a) $\frac{2}{5}$ (b) $\frac{1}{3}$

(c) $\frac{1}{2}$ (d) $\frac{2}{3}$

28. The switch in the circuit in the figure is in position P for a long time and then moved to position Q at time t = 0.

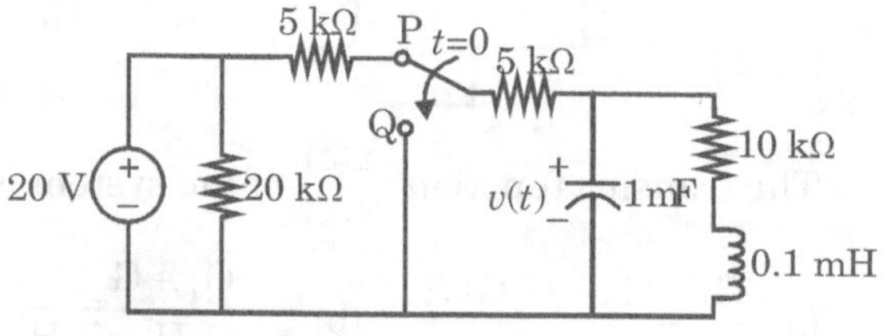

The value of $\frac{dv(t)}{dt}$ at $t = 0^+$ is

(a) 0 V/s (b) 3 V/s

(c) – 3 V/s (d) – 5 V/s

29. Consider the two-port network shown in the figure.

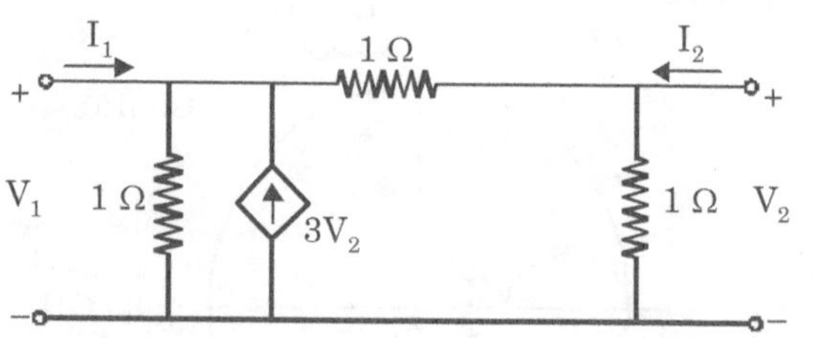

The admittance parameters, in siemens, are

(a) $y_{11} = 2, y_{12} = -4, y_{21} = -4, y_{22} = 2$

(b) $y_{11} = 1, y_{12} = -2, y_{21} = -1, y_{22} = 3$

(c) $y_{11} = 2, y_{12} = -4, y_{21} = -1, y_{22} = 2$

(d) $y_{11} = 2, y_{12} = -4, y_{21} = -4, y_{22} = 3$

30. For an n-channel silicon MOSFET with 10 nm gate oxide thickness, the substrate sensitivity $\frac{\partial V_T}{\partial |V_{BS}|}$ is found to be 50 mV/V at a substrate voltage $|V_{BS}| = 2$ V, where V_T is the threshold voltage of the MOSFET. Assume that, $|V_{BS}| >> 2\Phi_B$, where $q\Phi_B$ is the separation between the Fermi energy level E_F and the intrinsic level E_i in the bulk. Parameters given are

Electron charge (q) = 1.6×10^{-19} C

Vacuum permittivity (ε_0) = 8.85×10^{-12} F/m

Relative permittivity of silicon (ε_{Si}) = 12

Relative permittivity of oxide (ε_{ox}) = 4

The doping concentration of the substrate is

(a) 7.37×10^{15} cm^{-3} (b) 4.37×10^{15} cm^{-3}

(c) 2.37×10^{15} cm^{-3} (d) 9.37×10^{15} cm^{-3}

31. The propagation delays of the XOR gate, AND gate and multiplexer (MUX) in the circuit shown in the figure are 4 ns, 2 ns and 1 ns, respectively.

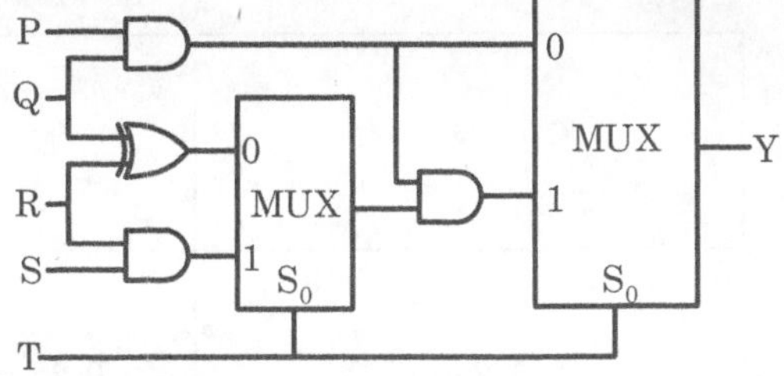

If all the inputs P, Q, R, S and T are applied simultaneously and held constant, the maximum propagation delay of the circuit is

(a) 3 ns (b) 5 ns
(c) 6 ns (d) 7 ns

32. The content of the registers are R_1 = 25 H, R_2 = 30 H and R_3 = 40 H. The following machine instructions are executed.

PUSH{R_1}
PUSH{R_2}
PUSH{R_3}
POP{R_1}
POP{R_2}
POP{R_3}

After execution, the content of registers R_1, R_2, R_3 are

(a) R_1 = 40H, R_2 = 30H, R_3 = 25H
(b) R_1 = 25H, R_2 = 30H, R_3 = 40H
(c) R_1 = 30H, R_2 = 40H, R_3 = 25H
(d) R_1 = 40H, R_2 = 25H, R_3 = 30H

33. The electrical system shown in the figure converts input source current $i_s(t)$ to output voltage $v_0(t)$.

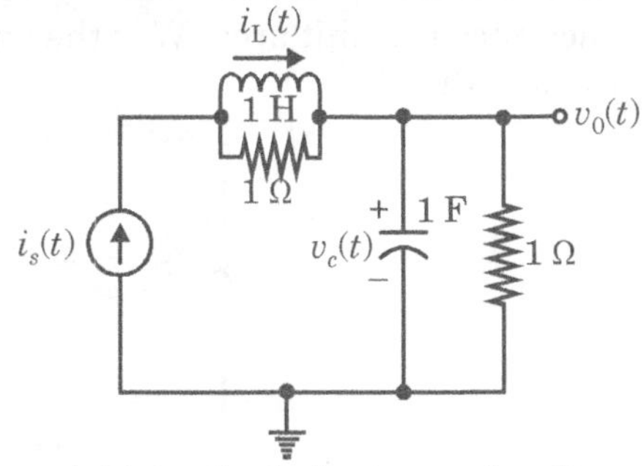

Current $i_L(t)$ in the inductor and voltage $v_C(t)$ across the capacitor are taken as the state variables, both assumed to be initially equal to zero, i.e., $i_L(0) = 0$ and $v_C(0) = 0$. The system is

(a) completely state controllable as well as completely observable
(b) completely state controllable but not observable
(c) completely observable but not state controllable
(d) neither state controllable nor observable

34. A digital transmission system uses a (7, 4) systematic linear Hamming code for transmitting data over a noisy channel. If three of the message-codeword pairs in this code $(m_i; c_i)$, where c_i is the codeword corresponding to the ith message m_i, are known to be (1 1 0 0 ; 0 1 0 1 1 0 0), (1 1 1 0 ; 0 0 1 1 1 1 0) and (0 1 1 0 ; 1 0 0 0 1 1 0), then which of the following is a valid codeword in this code?

(a) 1 1 0 1 0 0 1
(b) 1 0 1 1 0 1 0
(c) 0 0 0 1 0 1 1
(d) 0 1 1 0 1 0 0

35. The impedance matching network shown in the figure is to match a lossless line having characteristic impedance Z_0 = 50 Ω with a load impedance Z_L. A quarter–wave line having a characteristic impedance Z_1 = 75 Ω is connected to Z_L. Two stubs having characteristic impedance of 75 Ω each are connected to this quarter–wave line. One is a short-circuited (S.C.) stub of length 0.25λ connected across PQ and the other one is an open–circuited (O.C.) stub of length 0.5λ connected across RS.

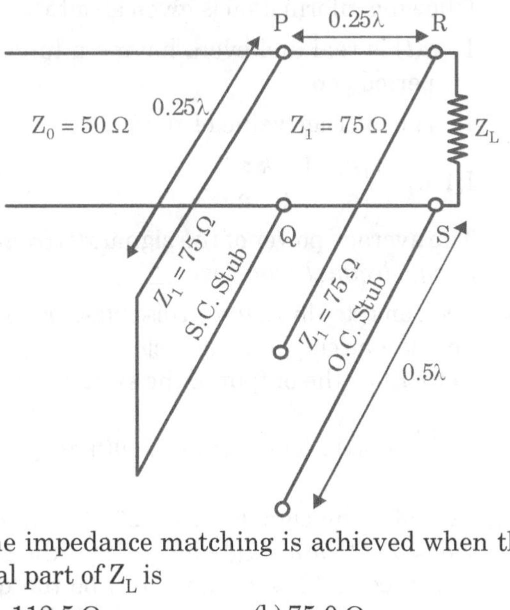

The impedance matching is achieved when the real part of Z_L is

(a) 112.5 Ω (b) 75.0 Ω
(c) 50.0 Ω (d) 33.3 Ω

Q. 36 to Q. 55 Numerical Answer Type (NAT), carry TWO marks each.

36. A real 2 × 2 non-singular matrix A with repeated eigenvalue is given as

$$A = \begin{bmatrix} x & -3.0 \\ 3.0 & 4.0 \end{bmatrix}$$

where x is a real positive number. The value of x (*rounded off to one decimal place*) is ________

37. For a vector field $D = \rho \cos^2 \varphi a_\rho + z^2 \sin^2 \varphi a_\varphi$ in a cylindrical coordinate system (ρ, φ, z) with unit vectors a_ρ, a_φ and a_z, the net flux of D leaving the closed surface of the cylinder $(\rho = 3, 0 \le z \le 2)$ *(rounded off to two decimal places)* is ________

38. In the circuit shown in the figure, the switch is closed at time t = 0, while the capacitor is initially charged to – 5V *(i.e.,* $v_c(0) = -5$ V).

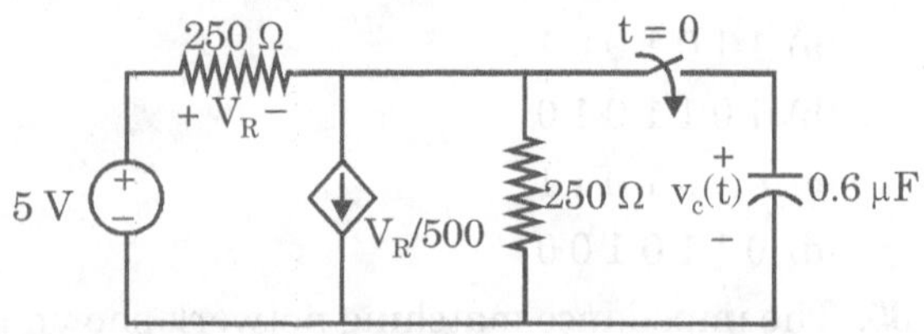

The time after which the voltage across the capacitor becomes zero *(rounded off to three decimal places)* is ________ ms.

39. The exponential Fourier series representation of a continuous-time periodic signal $x(t)$ is defined as

$$x(t) = \sum_{k=-\infty}^{\infty} a_k e^{jk\omega_0 t}$$

where ω_0 is the fundamental angular frequency of $x(t)$ and the coefficients of the series are *ak*. The following information is given about $x(t)$ and a_k.

I. $x(t)$ is real and even, having a fundamental period of 6

II. The average value of $x(t)$ is 2

III. $a_k = \begin{cases} k, & 1 \le k \le 3 \\ 0, & k > 3 \end{cases}$

The average power of the signal $x(t)$ *(rounded off to one decimal place)* is ________

40. For a unit step input $u[n]$, a discrete-time LTI system produces an output signal $(2\delta[n+1] + \delta[n] + \delta[n-1])$. Let $y[n]$ be the output of the system for an input $\left(\left(\frac{1}{2}\right)^n u(n)\right)$. The value of $y[0]$ is ________

41. Consider the signals $x[n] = 2^{n-1} u[-n+2]$ and $y[n] = 2^{-n+2} u[n+1]$, where $u[n]$ is the unit step sequence. Let $X(e^{j\omega})$ and $Y(e^{j\omega})$ be the discrete-time Fourier transform of $x[n]$ and $y[n]$, respectively. The value of the integral

$$\frac{1}{2\pi} \int_0^{2\pi} X(e^{j\omega}) Y(e^{-j\omega}) d\omega$$

(rounded off to one decimal place) is ________

42. A silicon P-N junction is shown in the figure. The doping in the P region is 5×10^{16} cm^{-3} and doping in the N region is 10×10^{16} cm^{-3}. The parameters given are

Built-in voltage $(\Phi_{bi}) = 0.8$ V

Electron charge $(q) = 1.6 \times 10^{-19}$ C

Vacuum permittivity $(\varepsilon_0) = 8.85 \times 10^{-12}$ F/m

Relative permittivity of silicon $(\varepsilon_{Si}) = 12$

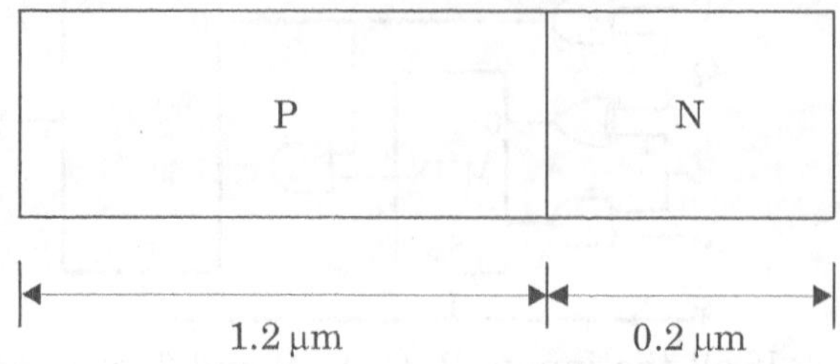

The magnitude of reverse bias voltage that would completely deplete one of the two regions (P or N) prior to the other *(rounded off to one decimal place)* is ________ V.

43. An asymmetrical periodic pulse train v_{in} of 10 V amplitude with on-time $T_{ON} = 1$ ms and off-time $T_{OFF} = 1$ μs is applied to the circuit shown in the figure. The diode D_1 is ideal.

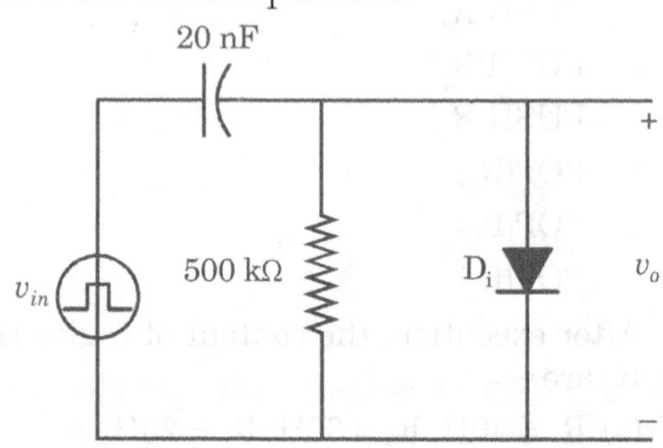

The difference between the maximum voltage and minimum voltage of the output waveform v_o *(in integer)* is ________ V.

44. For the transistor M_1 in the circuit shown in the figure, $\mu_n C_{ox} = 100$ μA/V^2 and (W/L) = 10, where μ_n is the mobility of electron, C_{ox} is the oxide capacitance per unit area, W is the width and L is the length.

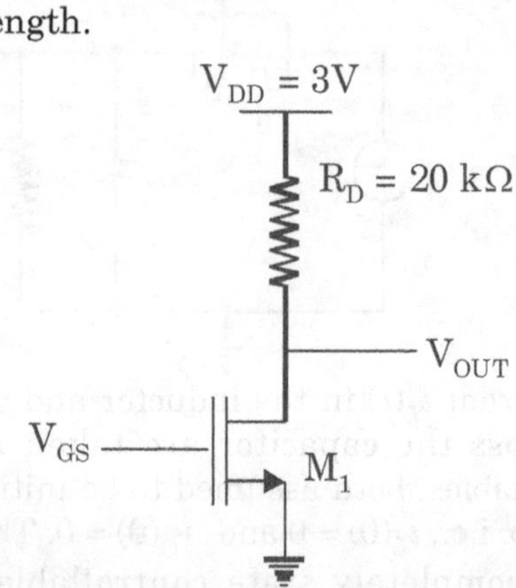

The channel length modulation coefficient is ignored. If the gate-to-source voltage V_{GS} is 1 V to keep the transistor at the edge of saturation, then the threshold voltage of the transistor *(rounded off to one decimal place)* is ________ V.

45. A circuit with an ideal OPAMP is shown in the figure. A pulse V_{IN} of 20 ms duration is applied to the input. The capacitors are initially uncharged.

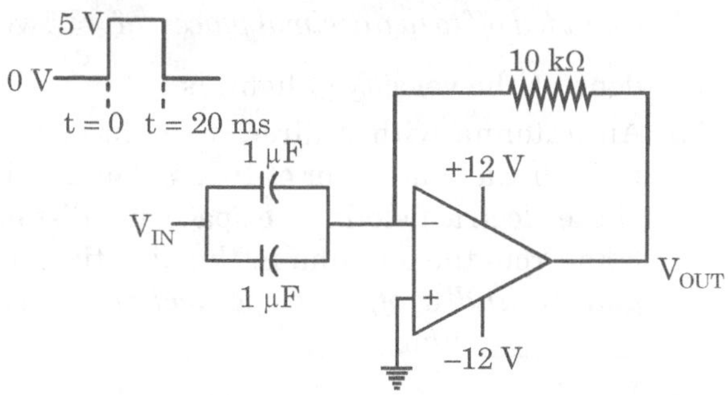

The output voltage V_{OUT} of this circuit at $t = 0^+$ (*in integer*) is ________ V.

46. The propagation delay of the exclusive-OR (XOR) gate in the circuit in the figure is 3 ns. The propagation delay of all the flip-flops is assumed to be zero. The clock (Clk) frequency provided to the circuit is 500 MHz.

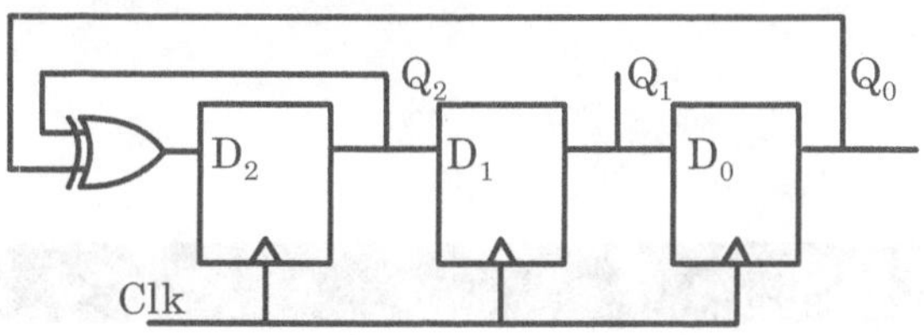

Starting from the initial value of the flip-flop outputs $Q_2Q_1Q_0 = 1\,1\,1$ with $D_2 = 1$, the minimum number of triggering clock edges after which the flip-flop outputs $Q_2Q_1Q_0$ becomes 1 0 0 (*in integer*) is ________

47. The circuit in the figure contains a current source driving a load having an inductor and a resistor in series, with a shunt capacitor across the load. The ammeter is assumed to have zero resistance. The switch is closed at time t = 0.

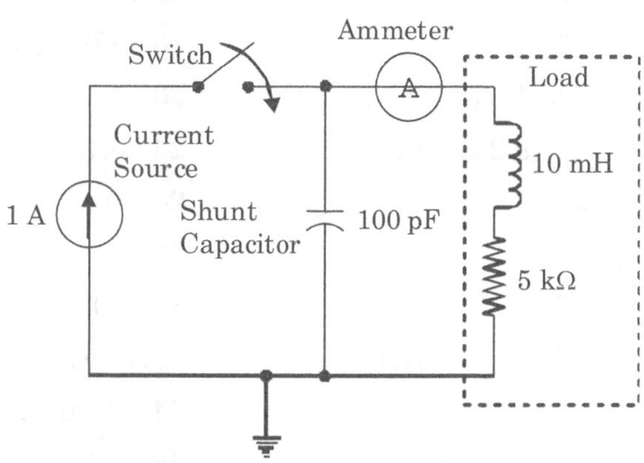

Initially, when the switch is open, the capacitor is discharged and the ammeter reads zero ampere. After the switch is closed, the ammeter reading keeps fluctuating for some time till it settles to a final steady value. The maximum ammeter reading that one will observe after the switch is closed (*rounded off to two decimal places*) is ________ A.

48. A unity feedback system that uses proportional-integral (PI) control is shown in the figure.

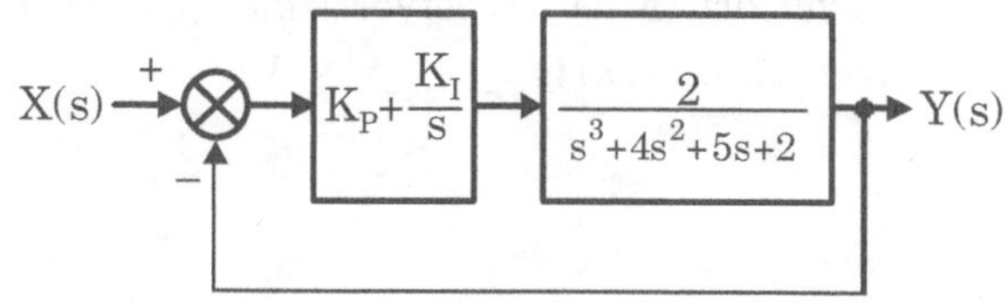

The stability of the overall system is controlled by tuning the PI control parameters K_P and K_I. The maximum value of K_I that can be chosen so as to keep the overall system stable or, in the worst case, marginally stable (*rounded off to three decimal places*) is ________

49. A sinusoidal message signal having root mean square value of 4 V and frequency of 1 kHz is fed to a phase modulator with phase deviation constant 2 rad/volt. If the carrier signal is $c(t) = 2\cos(2\pi 10^6 t)$, the maximum instantaneous frequency of the phase modulated signal (*rounded off to one decimal place*) is ________ Hz.

50. Consider a superheterodyne receiver tuned to 600 kHz. If the local oscillator feeds a 1000 kHz signal to the mixer, the image frequency (*in integer*) is ________ kHz.

51. In a high school having equal number of boy students and girl students, 75% of the students study Science and the remaining 25% students study Commerce. Commerce students are two times more likely to be a boy than are Science students. The amount of information gained in knowing that a randomly selected girl student studies Commerce (*rounded off to three decimal places*) is ________ bits.

52. A message signal having peak-to-peak value of 2 V, root mean square value of 0.1 V and bandwidth of 5 kHz is sampled and fed to a pulse code modulation (PCM) system that uses a uniform quantizer. The PCM output is transmitted over a channel that can support a maximum transmission rate of 50 kbps. Assuming that the quantization error is uniformly distributed, the maximum signal to quantization noise ratio that can be obtained by the PCM system (*rounded off to two decimal places*) is ________

53. Consider a polar non-return to zero (NRZ) waveform, using + 2V and – 2 V for representing binary '1' and '0' respectively, is transmitted in the presence of additive zero–mean white Gaussian noise with variance 0. 4 V^2. If the *a priori* probability of transmission of a binary '1' is 0. 4, the optimum threshold voltage for a maximum *a posteriori* (MAP) receiver (*rounded off to two decimal places*) is ________ V.

54. A standard air–filled rectangular waveguide with dimensions a = 8 cm, b = 4 cm, operates at 3.4 GHz. For the dominant mode of wave propagation, the phase velocity of the signal is v_p. The value (*rounded off to two decimal places*) of $\frac{v_p}{c}$, where c denotes the velocity of light, is ________

55. An antenna with a directive gain of 6 dB is radiating a total power of 16 kW. The amplitude of the electric field in free space at a distance of 8 km from the antenna in the direction of 6 dB gain (*rounded off to three decimal places*) is ________ V/m.

ANSWERS

General Aptitude (GA)

1. (c)	**2.** (b)	**3.** (c)	**4.** (c)
5. (b)	**6.** (c)	**7.** (d)	**8.** (a)
9. (c)	**10.** (a)		

Electronics & Communication Engineering

1. (a)	**2.** (b)	**3.** (b)	**4.** (b)
5. (c)	**6.** (d)	**7.** (a)	**8.** (d)
9. (d)	**10.** (a)	**11.** (b)	**12.** (c)
13. (c)	**14.** (d)	**15.** (a)	**16.** (8.0 to 8.0)
17. (0 to 0)	**18.** (0.5 to 0.5)	**19.** (1.0 to 1.0)	**20.** (4.5 to 4.5)
21. (2.0 to 2.0)	**22.** (94.2 to 94.6)	**23.** (9.24 to 9.28)	**24.** (3.12 to 3.16)
25. (9.30 to 9.44)	**26.** (a)	**27.** (b)	**28.** (c)
29. (c)	**30.** (a)	**31.** (c)	**32.** (a)
33. (d)	**34.** (d)	**35.** (a)	**36.** (10.0 to 10.0)
37. (56.50 to 56.60)	**38.** (0.132 to 0.146)	**39.** (31.9 to 32.1)	**40.** (0.0 to 0.0)
41. (7.9 to 8.1)	**42.** (8.1 to 8.4)	**43.** (10 to 10)	**44.** (0.5 to 0.5)
45. (–12 to –12)	**46.** (5 to 5)	**47.** (1.40 to 1.50)	**48.** (3.125 to 3.125)
49. (1011313.5 to 1011313.9)	**50.** (1400 to 1400)	**51.** (3.320 to 3.325)	**52.** (30.70 to 30.74)
53. (0.03 to 0.05)	**54.** (1.15 to 1.25)	**55.** (0.224 to 0.264)	

EXPLANATIONS

GENERAL APTITUDE

1. As given that,

Current population of a city = 11, 02, 500

Increasing rate = 5%/Annum

The population 2 years ago = ?

Let us Consider, the population 2 years ago = x

Using compound interest formula, we have

$$A = P\left(1+\frac{R}{100}\right)^n$$

$$\Rightarrow \quad 11{,}02{,}500 = x\left(1+\frac{5}{100}\right)^2$$

$$\Rightarrow \quad 11{,}02{,}500 \times \frac{100\times 100}{105\times 105} = x$$

$$\Rightarrow \quad \boxed{x = 10{,}00{,}000}$$

Hence, the population 2 years ago = 10,00,000.

2. As Given that, $\frac{p}{q}+\frac{q}{p}=3$

By Squaring on both sides;

$$\left(\frac{p}{q}+\frac{q}{p}\right)^2 = 9$$

$$\Rightarrow \frac{p^2}{q^2}+\frac{q^2}{p^2}+2 = 9$$

Hence, $\boxed{\frac{p^2}{q^2}+\frac{q^2}{p^2}=7}$

3.

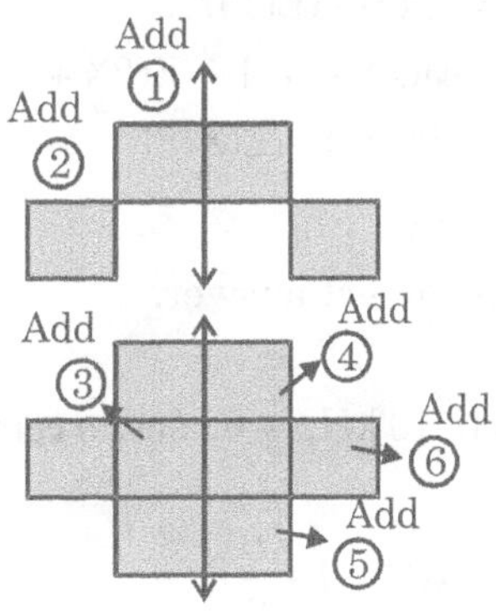

Hence, the least number of squares added = 6.

4. Nostalgia (means excessively sentimental yearning for return to or of some past period) is to anticipation. (means visualization of a future event or state)

5. Wake pastense is woke and third form is woken so, the verb forms are:

V1	V2	V3
Wake	Woke	Woken

6.

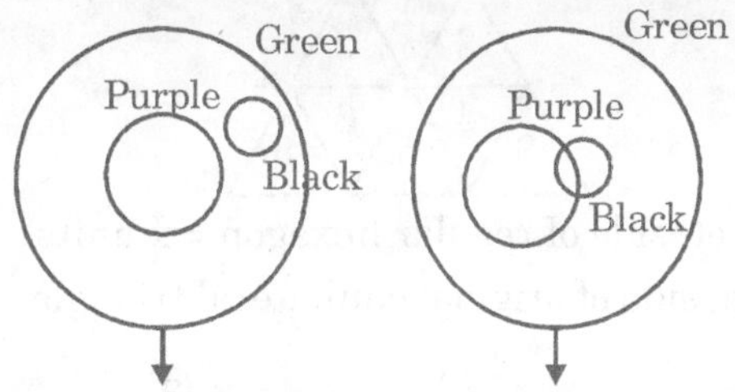

No black is purple   Some black are purple

7. The passage deduces that computers are present every where and they have both positive and negative effects on humans.

8. Square:

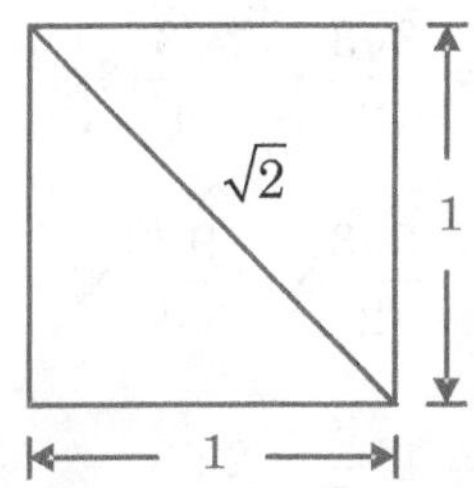

Solid core formed by revolving tringle about its short edge.

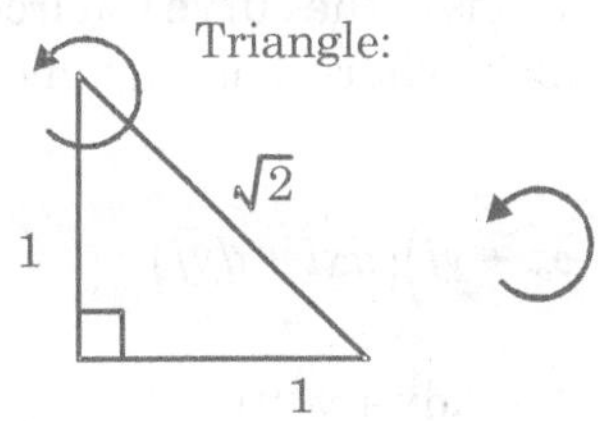

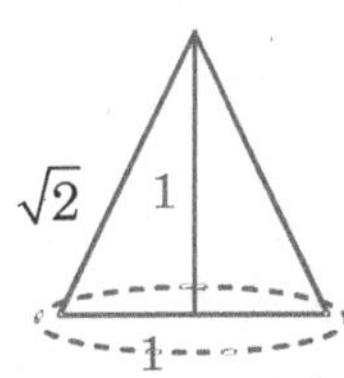

Radius of base circle(r) = 1 unit

Height of cone(h) = 1 unit

$$\text{Volume of cone} = \frac{\pi}{3}\times r^2h = \frac{\pi}{3}\times 1\times 1 = \frac{\pi}{3}\ \text{unit}^3$$

9. From the bar graph, it is clear that except Thursday, total 6 days one of the students spent a minimum of 10% more than the other student.

10. In regular hexagon, each interior angle = 120°,

So, if we cut corners of an equilateral triangle, then the removed triangles are also equilateral triangles.

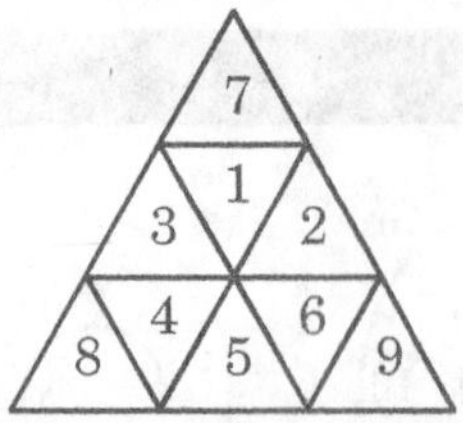

$\therefore$ Let, side of regular hexagon = x units

Then, side of original equilateral triangle = 3x

$\therefore$ Area of regular hexagon $= \frac{3\sqrt{3}}{2}x^2$

Area of equilateral triangle $= \frac{\sqrt{3}}{4}(3x)^2$

$\therefore$ Required ratio $= \frac{3\sqrt{3}}{2}x^2 : \frac{\sqrt{3}}{4}(3x)^2$

$= \frac{3\sqrt{3}}{2}x^2 : \frac{9\sqrt{3}}{4}x^2$

$= 1 : \frac{3}{2} = 2 : 3$

ELECTRONICS & COMMUNICATION ENGINEERING

1. Parametric equations of the curve C (circular arc) x = cos t; y = sin t where t is parameter, and radius = 1

$\int_C F(r)\,dr = \int(-x\hat{i} + y\hat{j})(dx\hat{i} + dy\hat{j})$

$\therefore \int_C F(r).dr = \int_C(-xdx + ydy)$

$= \int_C\{(-\cos t)(-\sin t\, dt) + (\sin t)(\cos t\, dt)\}$

where $(t \to 0 \text{ to } \frac{\pi}{4})$

$$= \int_0^{\frac{\pi}{4}} 2\sin t \cos t\, dt$$

$$= \int_0^{\frac{\pi}{4}} \sin 2t\, dt$$

$$= \left(\frac{-\cos 2t}{2}\right)_0^{\frac{\pi}{4}}$$

$$= \frac{-1}{2}(0-1) = \frac{1}{2}$$

Hence option (a) is correct answer.

2. As given differential equation $x(t) \times x(\frac{t}{2})$

$$\frac{dy}{dx} + \underbrace{\left(\frac{x}{1-x^2}\right)}_{P}y = \underbrace{(x)}_{Q}y^{1/2} \quad ...(i)$$

is a Bernoulli's equation in y.

Dividing both sides by $y^{1/2}$

$$y^{-1/2}\frac{dy}{dx} + \left(\frac{x}{1-x^2}\right)y^{1/2} = x \quad ...(ii)$$

Let $y^{1/2} = t$

$\Rightarrow \frac{1}{2}y^{-\frac{1}{2}}\frac{dy}{dx} = \frac{dt}{dx}$

$\therefore$ (2) becomes $2\frac{dt}{dx} + \left(\frac{x}{1+x^2}\right)t = x$

$$\Rightarrow \frac{dt}{dx} + \left(\frac{x}{2(1+x^2)}\right)t = \frac{x}{2} \quad ...(iii)$$

is a linear equation in t.

$\therefore$ I.F $= e^{\int \frac{x}{2(1-x^2)}dx}$

$= e^{-\frac{1}{4}\int \frac{-2x}{1-x^2}dx}$

$= e^{\ell n\left((1-x^2)^{-\frac{1}{4}}\right)}$

$\Rightarrow \boxed{I \cdot F = (1-x^2)^{-\frac{1}{4}}}$

Hence, option (b) is correct answer.

3. As Given that, Y = 2X + 3

$\text{Var}(Y) = E(Y^2) - (E(Y))^2$

$= E[(2X+3)^2] - [E(2X+3)]^2$

$= 4[E(X^2) - (E(X))^2]$

$\sigma_Y^2 = 4\sigma_X^2$

Hence, option (b) is correct answer.

4. As is given that

x(t) is band limited to 10 kHz, we need to compute Nyquist rate of

$$y(t) = x(t)x\left(1 + \frac{t}{2}\right)$$

x(t) having maximum frequency $f_{m_1} = 10$ kHz

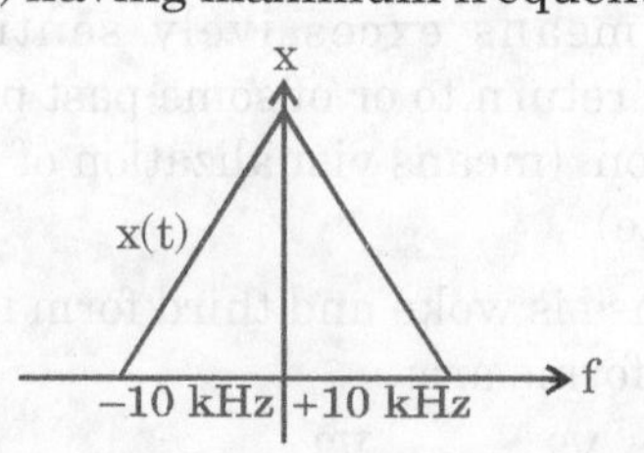

$x\left(\frac{t}{2}\right)$ having maximum frequency $f_{m_2} = 5$ kHz

$x\left(\frac{t+2}{2}\right) = x\left(\frac{1}{2}+1\right) = x\left(1+\frac{t}{2}\right)$ will have

maximum frequency same as $x\left(\frac{t}{2}\right)$ i.e., 5 kHz,

since time shifting does not affect the frequency limit of original signal.

$x(t).x\left(1+\frac{t}{2}\right)$ will have maximum frequency.

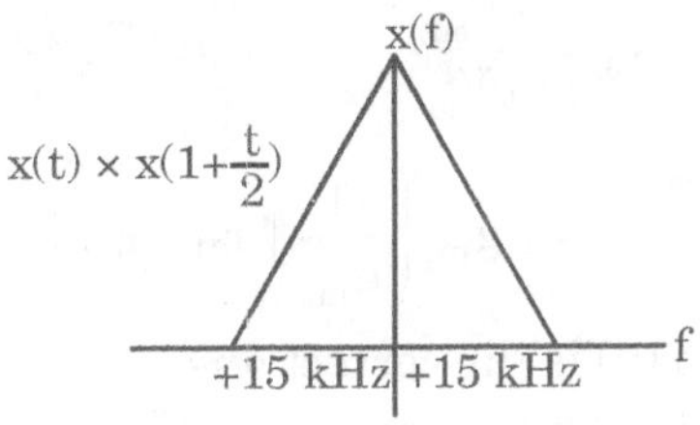

$$f_{m_1} + f_{m_2} = 10 + 5 = 15 \text{ kHz}$$

Since in frequency domain it will lead to convolution and in convolution upper limit is sum of upper limit.

So, hence, Nyquist rate = $2\left(f_{m_1} + f_{m_2}\right) = 2 \times 15$

= 30 kHz

Hence, option (b) is correct answer.

5. As given that

$y(n) = x(n) * h(n)$

where, * represents linear convolution

$Z(n) = \text{IDFT}\,[X(k).\,H(k)]$

16 point DFT is considered

We need to find for what value of k, y(k) = z(k)

$\text{IDFT}[X(k).\,H(k)] = x(n) \circledast h(n)$ [$\circledast$: circular convolution]

where,

x(n) have samples

h(n) have 16 samples

As we know that the length of linear convolution is = $N_1 + N_2 - 1$

Then x(n) * h(n) wil have 16 + 16 – 1 = 31 number of samples x(n) $\circledast$ h(n) will have 16 number of samples.

While computing circular convolution through liner convolution, we used to add last (k – 1) number of samples of linear convolution in beginning without disturbing the kth sample of linear and circular convolution. So the kth samples of linear convolution and circular convolution same.

Now, take an example, $x[n] = \left[\underset{\uparrow}{1}, 2, 3\right]$

$$h(n) = \left[\underset{\uparrow}{4}, 5, 6\right]$$

x(n) * h(n)	1	2	3		
	4	5	6		
	4	8	12		
		5	10	15	
			6	12	18
	4	13	28	27	18
	27	28			
	31	41	28		

So, $y(n) = x(n) * h(n) = \left[\underset{\uparrow}{4}, 13, 28, 27, 18\right]$

$z(n) = x(n) \circledast h(n) = [31, 31, 28]$

Here we have taken 3 samples in both x(n), h(n), and can see the 3rd samples of y(n), z(n) are same i.e., 28. [3rd sample is equivalent to n = 2 or k = 2]

Instead of 3 samples, if we take 16 samples then 16th samples will appear same in both y(n), z(n) [16th samples is equivalent to n = 15 or k = 15 sincestarting value of n/k is 0 not 1].

Hence, option (c) is correct answer.

6. As Given that: $P_o = 10^{16}$ cm^{-3}, $\tau = 100\ \mu$ sec, $n_i = 10^{10}$ cm^{-3}, $g_o = 10^{20}$ cm^{-3}s^{-1}

Now using mass action law

$$n_o = \frac{n_i^2}{p_o} = \frac{10^{20}}{10^{16}} = 10^4 \text{ cm}^{-3}$$

As we know that

Excess charge carrier:

$$\Delta P = \Delta n$$
$$= 10^{20} \text{ cm}^{-3}\text{s}^{-1} \times 100 \times 10^{-6}\text{s}$$
$$= 10^{16} \text{ cm}^{-3}$$
$$\text{Product} = (P_o + \Delta P)(n_o + \Delta n)$$
$$= (10^{16} + 10^{16}) \times (10^4 + 10^{16})$$
$$= 2 \times 10^{32} \text{ cm}^{-6}$$

Hence option (d) is correct answer.

7. This is the case of non-uniform doping.

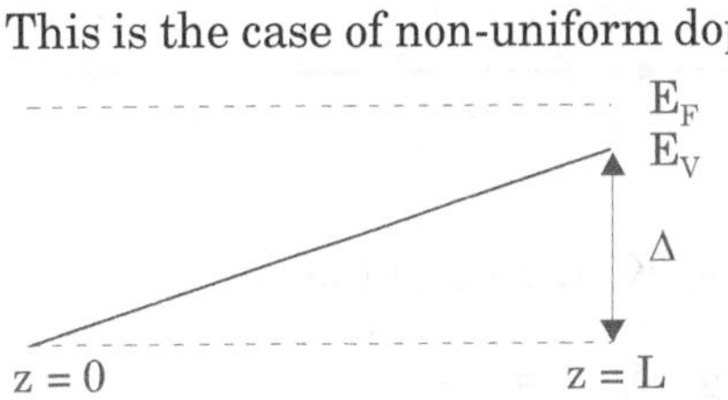

In a semi conductor result in built in electric filed so the relationship between the electric field and slope of energy-level is given by

$$E_x = \frac{-d\psi}{dx}$$

$$E_x = \frac{-d}{dx}\left[\frac{-E_x}{q}\right]$$

$$|E_x| = \left|\frac{1}{q}\frac{dE_v}{dx}\right|$$

$$\Rightarrow \quad E_x = \frac{1}{q}\left(\frac{\Delta - 0}{L - 0}\right)$$

$$\therefore\ E = \frac{1}{q}.\frac{\Delta}{L}$$

Hence, option (a) is corrct answer.

8. A.C equivalent of M_1 transistor ($\lambda \neq 0$)

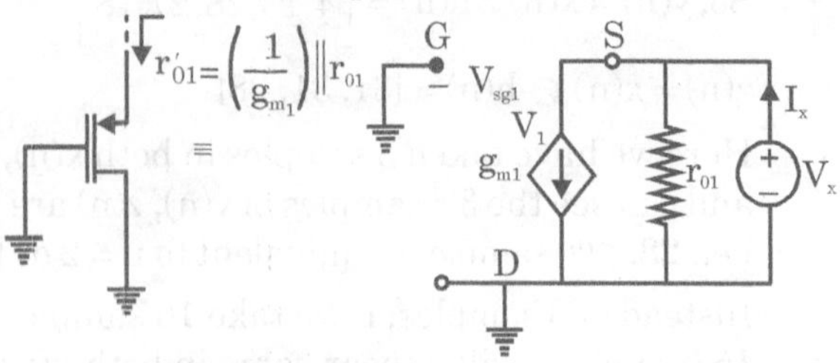

Apply KCL at source node

$$i_x = \frac{V_x}{r_{01}} + g_{m1}V_{sg1}$$

$$= \frac{V_x}{r_{01}} + g_{m1}V_x \qquad (\because V_{sg1} = V_x)$$

$$= \left(\frac{1}{r_{01}} + g_{m1}\right)V_x$$

$$r_{01}' = \frac{V_x}{i_x} = \frac{1}{\frac{1}{r_{01}} + g_{m1}} = \frac{1}{g_{m1}} \parallel r_{01}$$

A.c equivalent of complete amplifier circuit

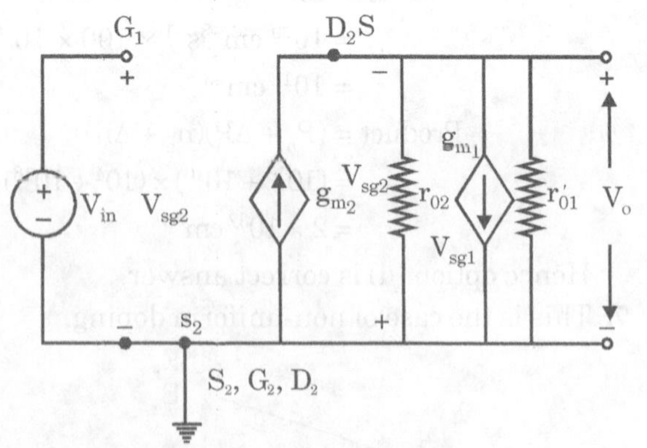

Now, Apply KCL at output node

$$g_{m_2}V_{sg2} = \frac{V_0}{r_{02}} + \frac{V_0}{r_{01}'}$$

$$g_{m_2}V_{sg2} = V_0\left(\frac{1}{r_{02}} + \frac{1}{r_{01}'}\right)$$

$$g_{m_2}V_{sg2} = V_0\frac{1}{r_{eq}}$$

$$V_{sg2} = V_{s2} - V_{g2}$$

$$= 0 - V_{g2} - V_{in} \qquad (\because V_{in} = V_{g2})$$

$$-g_{m_2}V_{in} = V_{out}r_{eq}$$

$$\frac{V_{out}}{V_{in}} = -g_{m_2}r_{eq}$$

$$= -g_{m_2}\left(r_{02} \parallel r_{01}'\right)$$

$$= -g_{m_2}\left(r_{01}' \parallel r_{02}\right)$$

$$A_g = \frac{\partial V_{out}}{\partial V_{in}}$$

$$= -g_{m_2}\left(\frac{1}{g_{m_1}} \parallel r_{01} \parallel r_{02}\right)$$

Hence, option (d) is the correct answer.

9. According to the Question:

For ideal op-amp and virtual short concept

$$V_+ = V_-$$

$$= \frac{R_2}{R_1 + R_2}.V_{REF}$$

Apply KCL at node V_-, then

$I_1 = I_2$ ($\because$ Input current of ideal op-amp is zero)

$$\frac{V_{IN} - V_-}{R_{IN}} = \frac{V_- - V_{OUT}}{R_F}$$

$$\frac{R_F}{R_{IN}} = \frac{V_- - V_{OUT}}{V_{IN} - V_-} \qquad ...(i)$$

If $V_{OUT} = 1$ volt for $V_{IN} = 0.1$ volt then equation (i) becomes

$$\frac{R_F}{R_{IN}} = \frac{V_- - 1}{0.1 - V_-} \qquad ...(ii)$$

For $V_{OUT} = 6V$ for $V_{IN} = 1$ volt, then equation (i) becomes

$$\frac{R_F}{R_{IN}} = \frac{V_- - 6}{1 - V_-} \qquad ...(iii)$$

By Compare equation (i) and (iii)

$$\frac{V_- - 1}{0.1 - V_-} = \frac{V_- - 6}{1 - V_-}$$

$$V_- = V_+ = -0.097 \text{ volt}$$

Substitute the value of V_- in equation (iii), then

$$\frac{R_F}{R_{IN}} = \frac{-0.097 - 6}{1 - (-0.097)} = 5.55$$

$$\therefore \frac{R_F}{R_{IN}} = 5.55$$

10. According to the question given circuit is:

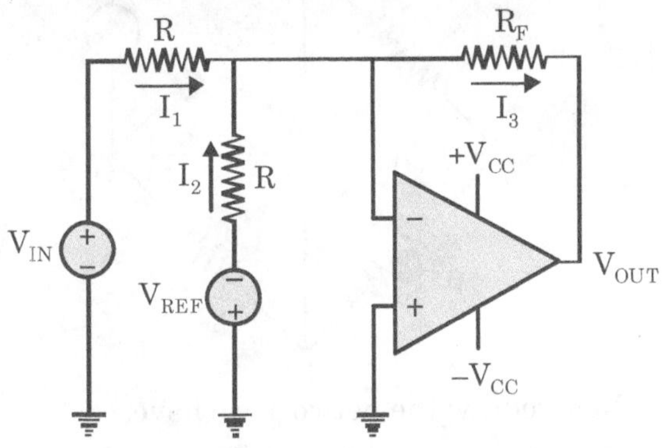

As per the virtual short concept of an ideal op-amp

$$V_+ = V_- = 0V$$

Apply KCL at (c)

$$I_1 + I_2 = I_3$$

$$\frac{V_{IN} - 0}{R} + \left(\frac{-V_{REF} - 0}{R}\right) = \frac{0 - V_{OUT}}{R_f}$$

$[\therefore V_{OUT} = 0V]$

$$\frac{V_{IN}}{R} - \frac{V_{REF}}{R} = 0$$

$$\boxed{\therefore V_{IN} = V_{REF}}$$

Hence, option (a) is correct answer.

11. As given that

$(1235)_x = (3033)_y$

By converting both side into base 10.

$(1 \times x^3) + (2 \times x^2) + (3 \times x^1) + (5 \times x^0)$

$= (3 \times y^3) + (0 \times y^2) + (3 \times y^1) + (3 \times y^0)$

$\Rightarrow x^3 + 2x^2 + 3x + 5 = 3y^3 + 3y + 3$

Let Checking with option B i.e., x = 8, y = 6

$8^3 + (2 \times 8^2) + (3 \times 8) + 5 = (3 \times 6^3) + (3 \times 6) + 3$

$\Rightarrow 669 = 699$ (valid)

12. Total of memory:

$32k \times 16 = 2^5 \times 2^{10} \times 16 = 2^{15} \times 16 = 2^P \times n$

Number of address line = p = 15

15 address line can generate 2^{15} number of different address, and each of these address will be accessed through output of decoder, so the size of decoder needed is 15 to 2^{15} which needs 2^{15} number of AND gates.

In a 2 × 4 Decoder the internal structure is

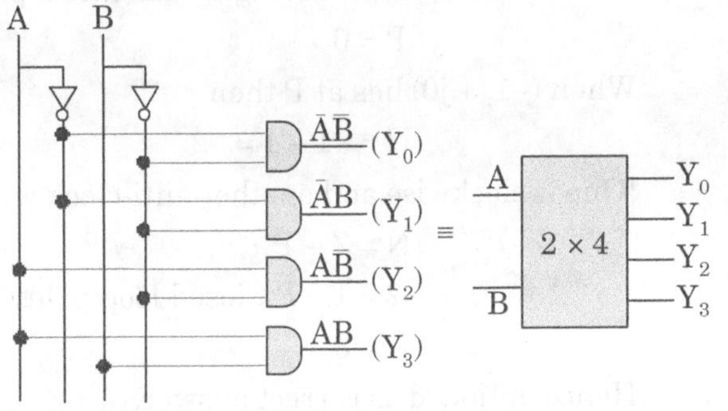

So, a 2 × 4 decoder need 4 AND gates.

2×2^{15} decoders need 2^{15} AND gates.

Basically, number of AND gate needed for a decoder is equal to number of output lines of decoder.

Hence, option (c) is correct answer.

13.

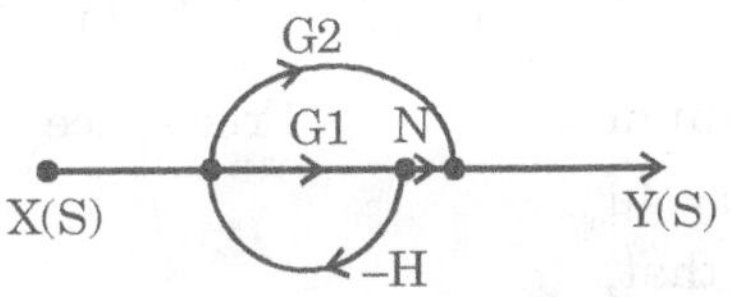

$$Y(s) = [X(s) - NH]G_2 + N$$

Where $N = [X(s) - NH]G_1$

$$N = X(s)\frac{G_1}{1+G_1H}$$

$$Y(s) = G_2X(s) - NG_2H + N$$

$$Y(s) = G_2X(s) - \frac{X(s)G_1G_2H}{1+G_1H} \times X(s)\frac{G_1}{1+G_1H}$$

$$Y(s) = X(s)\left[\left[\frac{G_2 + G_1G_2H - G_1G_2H}{1+G_1H}\right] + \frac{G_1}{1+G_1H}\right]$$

So, Transfer function $= \dfrac{Y(s)}{X(s)} = \dfrac{G_1 + G_2}{1+G_1H}$

Hence, option (c) is correct answer.

14. As given figure,

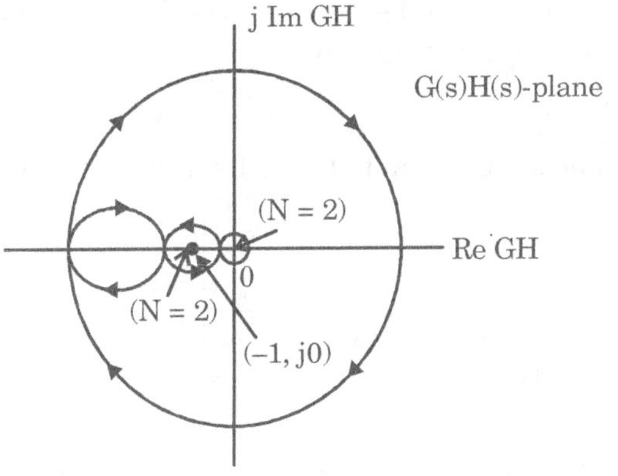

W.K. T, N = Z – P

GH have no poles on the right half(R.H)s-plane

$$P = 0$$

When $(-1, +j0)$ lies at P then

$$N = -1 - 1 = -2$$

[One is clockwise and another anti-clock wise]

$$N = Z - P$$

$$-2 = 1 - P \text{ Closed loop poles}$$

$$P = 3$$

Hence option 'd' is correct answer.

15. Considering two medium

$Z < 0$ $\quad Z = 0 \quad$ $Z > 0$

impedance$(\eta_1) = 120\pi\sqrt{\frac{\mu}{\epsilon}}$	impedance$(\eta_1) = 120\pi\sqrt{\frac{\mu}{e}}$
$= 120\pi\sqrt{\frac{4}{1}} = 240\pi$	$= 120\pi\sqrt{\frac{1}{1}} = 120\pi$
Lossless medium	Free space

with $\varepsilon = \varepsilon_o$, $\mu = 4\mu_o$

As we know that,

$$\Rightarrow \text{Reflection coefficient } \frac{\eta_2 - \eta_1}{\eta_2 + \eta_1} = \frac{240\pi - 120\pi}{240\pi + 120\pi}$$

$$= \frac{120\pi}{360\pi} = \frac{1}{3}$$

Hence, option (a) is the correct answer.

16. As given that, $X_1 = (1.0, -1.0, 2.0)$, $X_2 = (7.0, 3.0, x)$ and $X_3 = (2.0, 3.0, 1.0)$ be three linearly dependent vectors. Consider these vectors as rows in a matrix A. For linearly dependent vectors, we have

$$|A| = 0$$

$$\Rightarrow \begin{vmatrix} 1 & -1 & 2 \\ 7 & 3 & x \\ 2 & 3 & 1 \end{vmatrix} = 0$$

$$\Rightarrow 1 \times (3 - 3x) + 1 \times (7 - 2x) + 2 \times (21 - 6) = 0$$

$$\Rightarrow -5x + 10 + 30 = 0$$

$$\Rightarrow x = 8$$

17. For irrotational field, we have curl $(\nabla \times \vec{F}) = 0$

$$\Rightarrow \begin{vmatrix} \hat{i}_x & \hat{j}_y & \hat{k}_z \\ \frac{\partial}{\partial x} & \frac{\partial}{\partial y} & \frac{\partial}{\partial z} \\ 4y - c_1 z & 4x + 2z & 2y + z \end{vmatrix} = 0$$

$$\Rightarrow \hat{i}_x(2 - z) - \hat{j}_y(0 + c_1) + \hat{k}_z(4 - 4) = 0$$

$$\Rightarrow c_1 = 0$$

18. From the given circuit

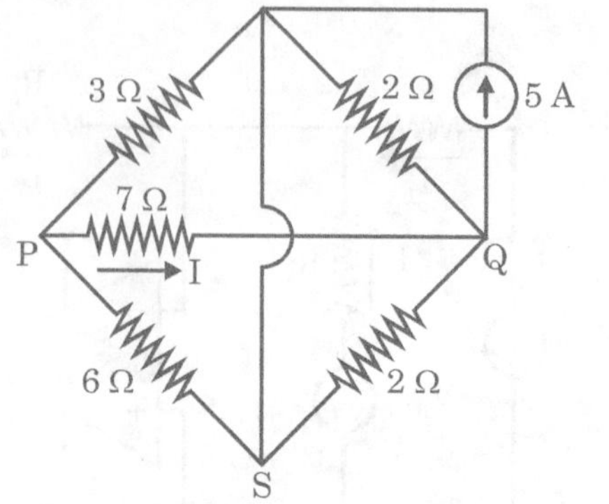

Now redraw the network we have,

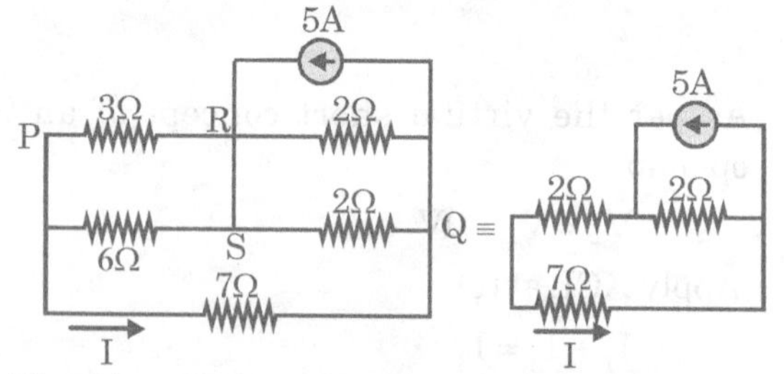

Now, by current division rule (CDR):

$$(I) = \frac{5 \times 1}{1 + 2 + 7} = \frac{5}{10} = 0.5 \text{ Amps}$$

19. From the given circuit

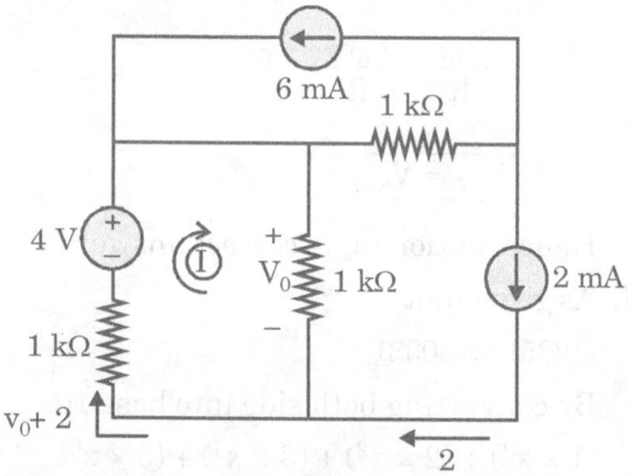

By KVL

$$4 - V_0 - (V_o + 2) = 0$$

$$V_0 + V_o = 4 - 2$$

$$2V_0 = 2$$

$$\Rightarrow V_0 = 1 \text{ Volt}$$

20. As given that

8-bit DAC have F.S.O = 7.68V

We need to obtain analog output voltage for input (10010110).

$$\text{Resolution} = \frac{V_{F.S.O}}{2^n - 1} = \frac{7.68}{2^8 - 1} = 0.03011$$

So, Analog output Voltage

= Resolution × Decimal equivalent of $(10010110)_2$

$$= 0.030 \times 150 = 4.5\ 17 \text{ V}$$

21. As we know that,
Average power of x(t) = $E[x^2(t)]$
$= R_x(0) = 2$

22. As we know that,
% power saving

$$= \frac{\text{Power Saved}}{\text{Total power}} \times 100$$

Total power in amplitude modulated wave by a single tune sinusoidal signal:

$$= \underbrace{P_C}_{\text{carrier}} + \underbrace{P_C\frac{\mu^2}{4}}_{\text{sideband}} + \underbrace{P_C\frac{\mu^2}{4}}_{\text{sideband}}$$

After suppressing carrier and one sideband

$$\text{Power} = \frac{P_C\mu^2}{4}$$

$$\Rightarrow \text{Savings in power} = P_C + P_C\frac{\mu^2}{4}$$

$$\%\text{ Power saved} = \frac{P_C + P_C\frac{\mu^2}{4}}{P_C + P_C\frac{\mu^2}{4} + P_C\frac{\mu^2}{4}} \times 100$$

$$= \frac{4+\mu^2}{2(2+\mu^2)} \times 100$$

For 50% modulation this value is $(\because \mu = 0.5)$

$$\frac{4+(0.5)^2 \times 100}{2(2+(0.5)^2)} = \frac{4.25}{4.5} \times 100 = 94.4\%$$

23. As we know that, Channel capacity relation for an AWGN channel

$$C = B\log_2[1 + \text{SNR}]$$

Data rate = $n.F_s$
$= 8 \times 1.25 \times 2 \times 4000$
$= 80 \times 10^3$ bits per record

$\Rightarrow$ Data rate (R) $\leq$ C

$$R \leq B\log_2[1+\text{SNR}]$$

$$\Rightarrow \quad 80 \times 10^3 \leq B\log_2[1+400]$$

$$\Rightarrow \quad B \geq \frac{80\times 10^3}{\log_2[401]}$$

$$\geq \frac{80\times 10^3}{8.64} \geq 9.259 \text{ kHz}$$

$B \geq 9.26$ KHz

24. As given that
$F_S = 32$ KHz Δ to avoid slope-overload error.
To avoid slope overload in DM

$$\Delta.F_S \geq = \left|\frac{d}{dt}m(t)\right|_{max}$$

$\therefore m(t) = 4\sin(2\pi \times 4 \times 10^3 t)$

$$\Rightarrow \quad \Delta.32 \times 10^3 \geq 4 \times 4 \times 2\pi \times 10^3 t$$

$$\Rightarrow \quad \Delta \geq \frac{32\pi}{32}$$

$$\Rightarrow \quad \Delta \geq 3.14$$

25. As we know that
The total interval neglection angle of incidence

$$(\theta_C) = \sin^{-1}\left[\frac{\eta_2}{\eta_1}\right]$$

$$\theta_c = \sin^{-1}\left(\frac{1.48}{1.5}\right) = \sin^{-1}(0.9866) = 80.63$$

$\Rightarrow$ Angle it should make with axis is
$\theta_1 = 90° - \theta_c$
$\theta_1 = 90 - 80.63 = 9.36°$

26. C(curve) is $|x - i \Rightarrow = 2$.
$x^2(x^2 + 4) = 0$ gives x = 0,2i, 2i are the singular points.
Only x = 0, 2i lies inside C.

$$\text{Let } f(x) = \frac{\sin x}{x^2(x^2+4)} = \left[\frac{\sin x}{4x^2}\right] - \left[\frac{\sin x}{4(x^2+4)}\right]$$

$$\therefore \oint_C \frac{\sin x}{x^2(x^2+4)}dx = \frac{1}{4}\oint_C \frac{\sin x}{(x-0)^2}dx - \frac{1}{4}\oint_C \frac{\left(\frac{\sin x}{x+2i}\right)}{x-2i}dx$$

$$\int_C f(z)\,dz = 2\pi i\begin{bmatrix}\text{Res} & +\text{Res}\\ x=0 & x=2i\end{bmatrix}$$

$$= \frac{1}{4}\times\frac{2\pi i}{1!}\left[\frac{d}{dx}(\sin x)\right]_{x=0} - \frac{1}{4}\times 2\pi i\left[\frac{\sin x}{x+2i}\right]_{x=2i}$$

($\because$ Using Cauchy's integral formula)

$$= 2\pi i\left[\frac{1}{4} - \frac{\sin 2i}{16i}\right]$$

$$= \frac{\pi i}{2} - \frac{\pi}{8}\sin(2i)$$

Hence, no option matched.

27. As givne that,
A box contains the three coins such as,(i) HT (ii) HH (iii) TT.

So, probability of head in first toss $P(H_1) = \frac{1}{2}$

As we know that $P[H_2/H_1] = P[H_2 \cap H_1]/P(H_1)$

$$\text{So, } P(H_2 \cap H_1) = \left(\frac{1}{12} + \frac{1}{12}\right) = \frac{1}{6}$$

$$\text{Therefore, } P\left(\frac{H_2}{H_1}\right) = \frac{1}{6} \times \frac{2}{1} = \frac{1}{3}$$

Hence, option (b) is correct answer.

28. As given network is

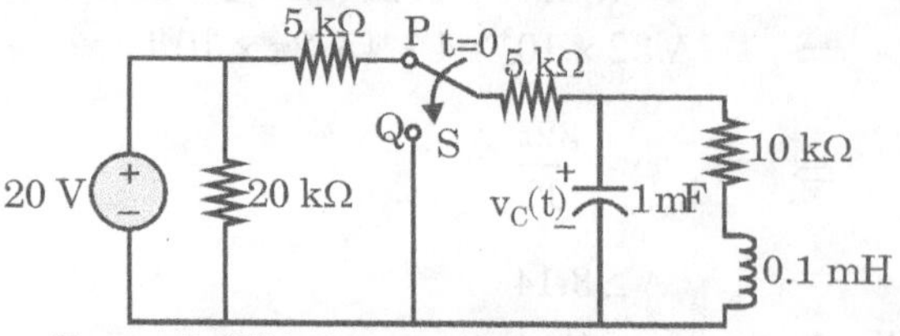

→ For t<0, 'S' is in position (P)

At, t = 0 – (S.S), L → S.C, C → O.C

We need to obtain $\left.\frac{dV_C(t)}{dt}\right|_{t=0^+}$

at $t = 0^-$:
- → Switch at P
- → Capacitor open circuit
- → Inductor short circuit
- → Network in steady state

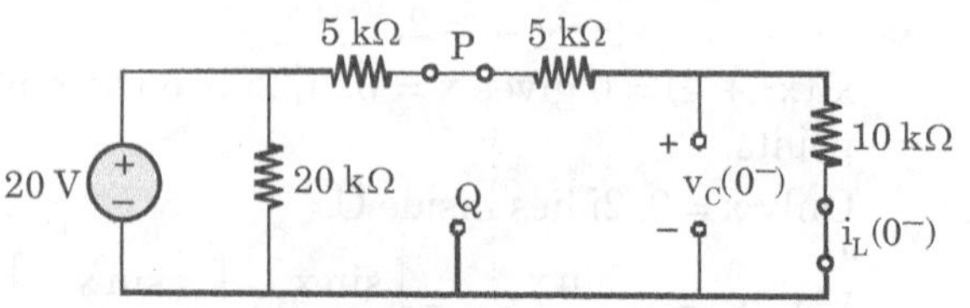

at t = 0⁻

$$i_L(0^-) = \frac{20V}{(5+5+10)k\Omega}$$

$$= \frac{20V}{20k\Omega} = 1 \text{ mA}$$

$$V_C(0^-) = (10\text{ k}\Omega)[i_L(0^-)]$$

$$= [10\text{ k}\Omega][1\text{ mA}] = 10\text{ V}$$

at $t = 0^+$:
- → Switch at Q
- → Capacitor replaced by ideal voltage source of value $V_C(0^-)$
- → Inductor replaced by ideal current source of value $i_L(0^-)$

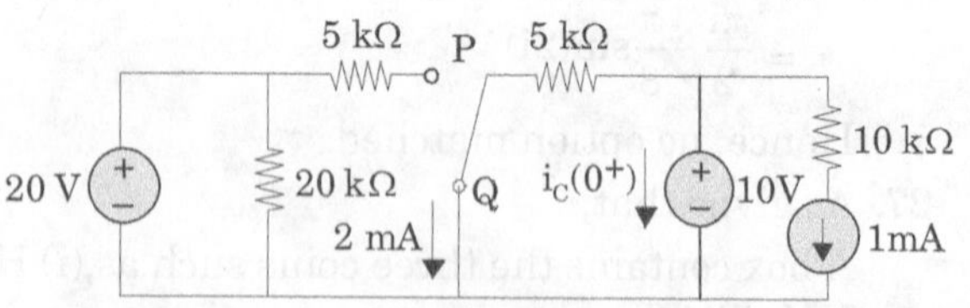

at t = 0⁻

Now By KCL $i_C(0^+) = -3$ mA

$$i_C(t) = C\frac{dV_C(t)}{dt}$$

$$\Rightarrow \frac{dV_C(t)}{dt} = \frac{i_C(t)}{C}$$

$$\Rightarrow \left.\frac{dV_C(t)}{dt}\right|_{t=0^+} = \frac{i_C(0^+)}{C}$$

$$= \frac{-3\text{ mA}}{1\text{ mF}} = -3\text{ V/sec}$$

Hence, option (c) is correct answer.

29. As given network

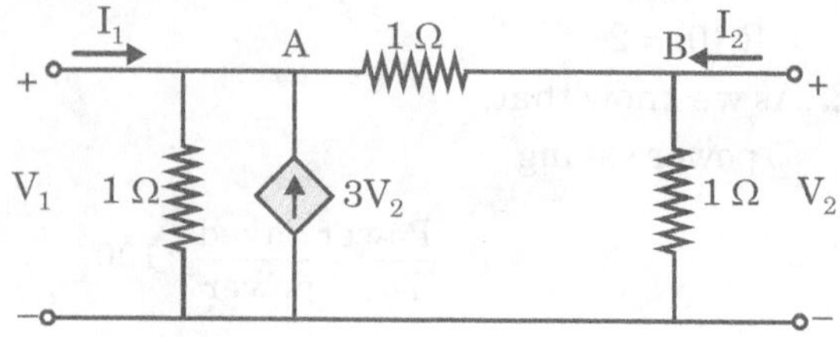

By nodal analysis at node A!

Node A	Node B
$-I_1 - 3V_2 + \frac{V_1}{1} + \frac{V_1 - V_1}{1} = 0$	$-I_2 + \frac{V_1}{1} + \frac{V_2 - V_1}{1} = 0$
$\Rightarrow I_1 = V_1 + V_1 - V_3 - 3V_2$	$\Rightarrow I_2 = -V_1 + V_2 + V_2$
$\Rightarrow I_1 = 2V_1 - 4V_2$ ....(i)	$\Rightarrow I_2 = -V_1 + 2V_2$ ....(ii)
$\Rightarrow I_1 = Y_{11}V_1 + Y_{12}V_2$	$\Rightarrow I_2 = Y_{21}V_1 + Y_{22}V_2$
So $Y_{11} = 2$, $Y_{21} = -4$	So $Y_{21} = -1$, $Y_{22} = 2$

From, equation (i) & (ii)

$$Y = \begin{bmatrix} 2 & -4 \\ -1 & 2 \end{bmatrix}$$

Hence, option (c) is the correct answer.

30. As we know that

For N-channel MOSFET their threshold voltage (V_T) is given by

$$V_T = V_{T_0} + \gamma\left(\sqrt{|V_{BS}| + |2\phi_B|} - \sqrt{|2\phi_B|}\right) \quad ....(i)$$

Where, V_{T_0} threshold voltage no body-bias i.e., $|V_{BS}| = 0V$

$$\gamma = \frac{\sqrt{2\epsilon_s qN_a}}{\epsilon_{ox}} \quad ...(ii)$$

(∵ N-channel MOSFET is having P-type susbstrate)

Differentiate equation (i) w.r.t V_{BS}

$$\frac{\partial V_T}{\partial |V_{BS}|} = \frac{\gamma}{2\sqrt{|V_{BS}| + |2\phi_B|}}$$

$$= 50\text{ mV/V} \quad ...(iii)$$

$$50 \times 10^{-3}\text{ v/v} = \frac{\gamma}{2\sqrt{2}} \quad (\because |V_{BS}| >> |2\phi_B|)$$

$$\gamma = 2\sqrt{2} \times 50 \times 10^{-3} = 0.141$$

From equation (iii)

$$\gamma = \frac{\sqrt{2\epsilon_s qN_a}}{\frac{\epsilon_{ox}}{t_{ox}}}$$

$$= \frac{t_{ox}\sqrt{2\epsilon_s qN_a}}{\epsilon_{ox}}$$

$$\gamma = \frac{t_{ox}\sqrt{2\epsilon_s qN_a}}{\epsilon_{ox}} \quad ...(iv)$$

Now, Square on both the sides

$$\gamma^2 = \left(\frac{t_{ox}}{\epsilon_{ox}}\right)^2 \times 2\epsilon_s q N_a$$

So, $$N_a = \gamma^2 \left(\frac{\epsilon_{ox}}{t_{ox}}\right)^2 \times \frac{1}{2\epsilon_s q}$$

By putting respective values

$$= (0.14)^2 \left(\frac{4 \times 8.854 \times 10^{-14}}{10 \times 10^{-9} \times 10^2}\right) \times \frac{1}{2 \times 12 \times 8.854 \times 10^{-4} \times 1.6 \times 10^{-19}}$$

$$= \frac{2.491 \times 10^{-27}}{3.3984 \times 10^{-43}} = 7.329 \times 10^{15}$$

$$\boxed{\text{So, } N_a = 7.37 \times 10^{15}\, cm^{-3}}$$

Hence, option (a) is the correct answer.

31. As given combination circuit:

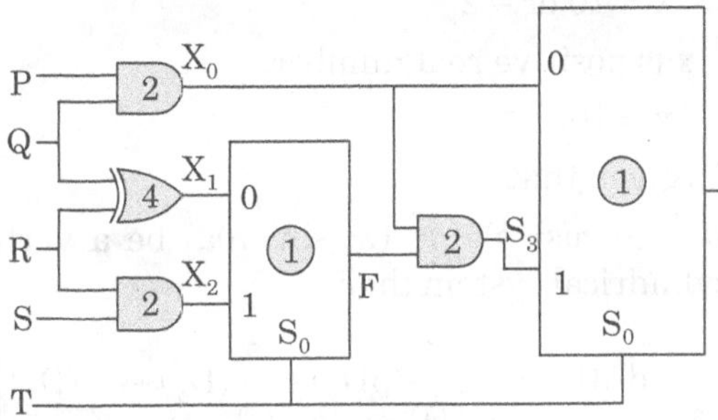

When T = 0, to get output only signal x_0 is essential for second MUX, so the delay involved in this case is 2n sec for x_0, MUX delay 1 nsec, total 2 + 1 = 3 nsec.

When T = 1, to get output Y we need X_2 and X_3, X_2 will be available at 2n sec (delay of AND gate)

F will be available after 1 n-sec of X_2 2 + 1 = 3 n sec (i.e., delay of MUX with X_2),

X_3 will available after 2 n sec of F i.e., 3 + 2 = 5 n sec

Y will be available after 1 n sec of X_3 i.e., 5 + 1 = 6 n sec

When T = 0, propagation delay (Y) at 3 n sec

T = 1, propagation delay Y at 6 n sec

Hence, Worst case delay is 6 n sec.

32. As given that

The content of Registers

$$R_1 = 25\text{ H}, R_2 = 30\text{ H}, R_3 = 40\text{ H}$$

PUSH $\{R_1\} \rightarrow$ content of R_1 will be pushed to top 2 bytes of stack

PUSH $\{R_2\} \rightarrow$ content of R_2 will be pushed to top 2 bytes of stack

PUSH $\{R_3\} \rightarrow$ content of R_3 will be pushed to top 2 bytes of stack

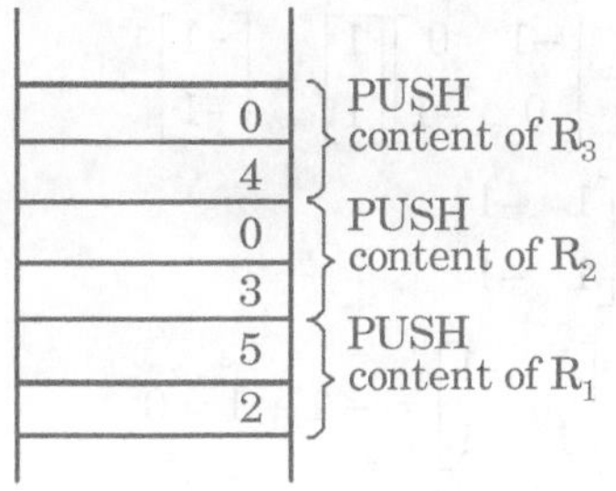

POP $R_1 \rightarrow$ Top 2 types of stack will be popped to R_1, So R_1 = 40H

POP $R_2 \rightarrow$ Top 2 bytes of stack will be popped to R_2, So R_2 = 30 H

POP $R_3 \rightarrow$ Top 2 bytes of stack will be popped R_3, So R_3 = 25H

Hence, after execution of above 6 instructions

R_1 = 40H, R_2 = 30H, R_3 = 25H.

Hence, option (a) is the correct answer.

33. As given that:

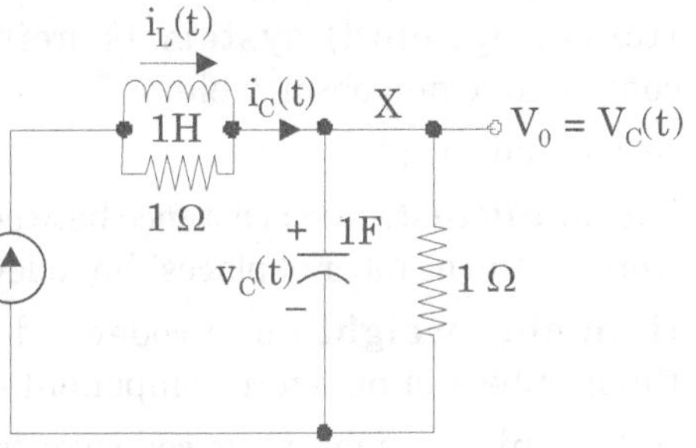

Here, $i_L(t)$ and $V_C(t)$ are state variables

At node X by KCL

$$-i_S(t) + 1\frac{dV_C(t)}{dt} + \frac{V_C(t)}{1} = 0$$

$$V_C'(t) = \frac{dV_C(t)}{dt}$$

$$= i_S(t) - V_C(t)$$

Now, by current-division principle

$$i_L(s) = \frac{i(s) \times 1}{s+1}$$

$$i_S(s) = \frac{di_L(t)}{dt} + i_L(t)$$

$$i_L(s) = \frac{di_L(t)}{d(t)} = i_S(t) - i_L(t)$$

So, state matrix A can be obtained by

$$\begin{bmatrix} i_L(t) \\ V_C(t) \end{bmatrix} = \begin{bmatrix} -1 & 0 \\ 0 & -1 \end{bmatrix} \begin{bmatrix} i_L(t) \\ V_C(t) \end{bmatrix} + \begin{bmatrix} 1 \\ 1 \end{bmatrix} i_S(t)$$

$$A = \begin{bmatrix} -1 & 0 \\ 0 & -1 \end{bmatrix} \quad B = \begin{bmatrix} 1 \\ 1 \end{bmatrix}$$

C = controllability = [B AB]

$$AB = \begin{bmatrix} -1 & 0 \\ 0 & -1 \end{bmatrix}\begin{bmatrix} 1 \\ 1 \end{bmatrix} = \begin{bmatrix} -1 \\ -1 \end{bmatrix}$$

$$C = \begin{bmatrix} 1 & -1 \\ 1 & -1 \end{bmatrix}$$

$$|C| = \begin{bmatrix} 1 & -1 \\ 1 & -1 \end{bmatrix} = -1 + 1 = 0$$

$$\text{Observability} = 0 = \begin{bmatrix} C \\ CA \end{bmatrix}$$

$$\boxed{V_0(t) = V_C(t) + i_S(t)}$$

$$O = \begin{bmatrix} 0 & 1 \\ 0 & -1 \end{bmatrix}$$

$|O| = 0$

C = [0 1]

$$C_A = [0\ \ 1]\begin{bmatrix} -1 & 0 \\ 0 & -1 \end{bmatrix} = [0\ \ 1]$$

Hence, option(d) system is neither state controllable non observable.

34. As we know that

Hamming distance : Distance between two code words is the number of places the codeword differ.

Hamming weight : of a codeword is equal to the numbers of non-zero components.

[If hamming weight of the code words formed by module-2]

As Given that, $[m_1, c_1]$ = [1100; 0101100]

$[m_2, c_2]$ = [1110,0011110]

$[m_3, c_3]$ = [0110; 1000110]

c_1 = 0101100

∴ c_2 = 0011110

c_3 = 1000110

Now by taking option 4 code word 0110 100 = c_4

$d(c_1, c_2) = 3 = W(c_1 \oplus c_2) = W(0110010) = 3$

$d(c_1, c_3) = 4 = W(c_1 \oplus c_3) = W(1101010) = 4$

$d(c_1, c_4) = 2 = W(c_1 \oplus c_4) = W(0011000) = 2$

Hence option (d) is valid code word.

35.

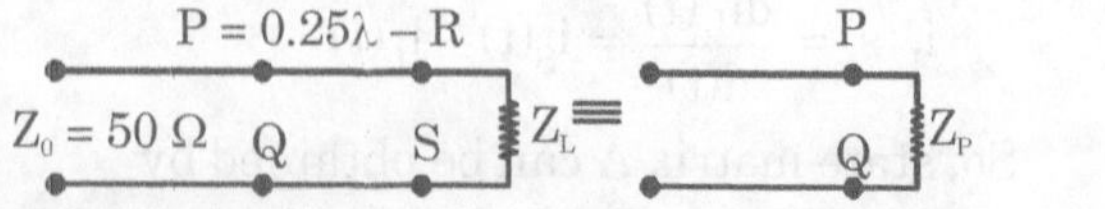

Impedance across RS = $Z_L \parallel \infty = Z_L$

Load impedance across PQ

$$= \frac{Z_0^2}{0} \parallel \frac{(75)^2}{Z_L} = \infty \parallel \frac{(75)^2}{Z_L} \parallel \frac{(75)^2}{Z_L}$$

For matching $50\ \Omega = \dfrac{(75)^2}{Z_L} \Rightarrow Z_L = 112.5$

Hence, option (a) is the correct answer.

36. As we know that

The characteristic equation is $|A - \lambda I| = 0$

$$\begin{bmatrix} x-\lambda & -3 \\ 3 & 4-\lambda \end{bmatrix} = 0$$

$\Rightarrow (x - \lambda)(4 - \lambda) + 9 = 0$

$\Rightarrow \lambda^2 + (-4 - x)\lambda + (4_x + 9) = 0$, a quadratic equatin in λ

For repeated eigen values, we have

$B^2 - 4AC = 0$, where A = 1, B = – 4 – x, C = 4x + 9

$(-4 - x)^2 - 4(1)(4x + 9) = 0$

$16 + x^2 + 8x - 16x - 36 = 0$

$x^2 - 8x - 20 = 0$

$(x - 10)(x + 2) = 0$

$\Rightarrow$ x = 10 or – 2

∵ x is positive real number.

∴ x = 10

37. As given that

$D = (\rho \cos^2 \phi)a_\rho + (z^2 \sin^2 \phi)a_\phi$ be a vector in cylindrical system then

$$\text{div}(D) = \frac{1}{\rho}\left\{\frac{\partial}{\partial \rho}(\rho D_\rho) + \frac{\partial}{\partial \phi}(D_\phi) + \frac{\partial}{\partial z}(D_z)\right\}$$

Where $D_\rho = \rho \cos^2\phi$; $D_\phi = z^2 \sin^2\phi$; $D_z = 0$

$$\Rightarrow \quad \text{div}(D) = \frac{1}{\rho}\{2\rho \cos^2\phi;\ 2z^2 \sin\phi \cos\phi\}$$

$$= (1 + \cos 2\phi) + \frac{z^2 \sin 2\phi}{\rho}$$

∴ The net Electric flux of D leaving the closed surface of the cylinder is

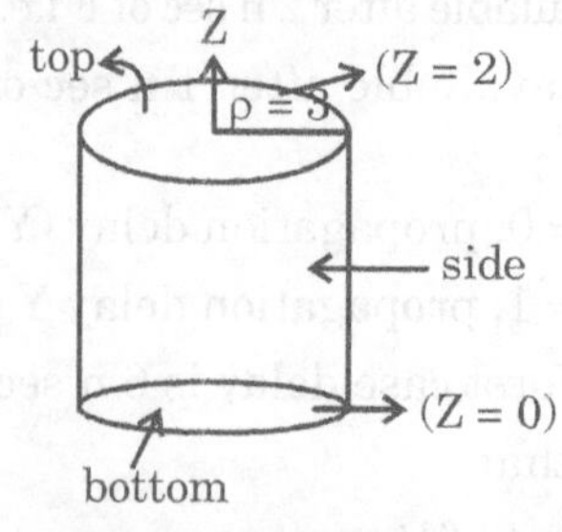

$$\Psi_{net} = \Psi_{bottom} + \Psi_{top} + \Psi_{side}$$

$$\Psi_{net} = \oiint_S D.d\vec{s} = \oiiint_V \text{div}(D)dV$$

(Using Gauss divergence theorem)

$$\Psi_{net} = \int_{\rho=0}^{3}\int_{\phi=0}^{2\pi}\int_{z=0}^{2}(1 + \cos 2\phi)\rho d\rho d\phi dz$$

$$+\int_{\rho=0}^{3}\int_{\phi=0}^{2\pi}\int_{z=0}^{2}\left(\frac{z^2\sin 2\phi}{\rho}\right)$$

$$\left(\because dV = \rho d\rho d\phi dz \text{ and } '\rho' \text{ limits; 0 to 3, } \phi \text{ limits; 0 to } 2\pi, \text{ z limits; 0 to 2}\right)$$

$$= \left(\frac{\rho^2}{2}\right)_0^3\left(\phi+\frac{\sin 2\phi}{2}\right)_0^{2\pi}\times(z)_0^2+(\ell n\rho)_0^3$$

$$\times\left(\frac{-\cos 2\phi}{2}\right)_0^{2\pi}\times\left(\frac{z^3}{3}\right)_0^2=\left(\frac{9}{2}\right)\times(2\pi+0)\times 2+0$$

$$\Psi_{net} = 18\pi \approx 56.55$$

38. As given Circuit:

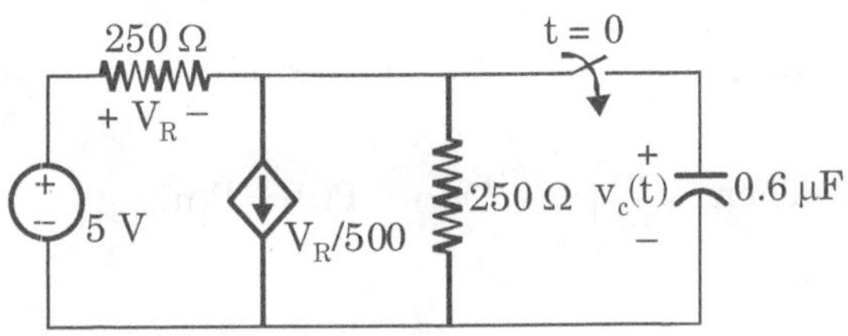

$V(0) = -5V = V_0$
for t < 0, is closed
For final value at t

= ∞ → Switch is closed
→ Capacitor is open circuit
→ Network in steady state

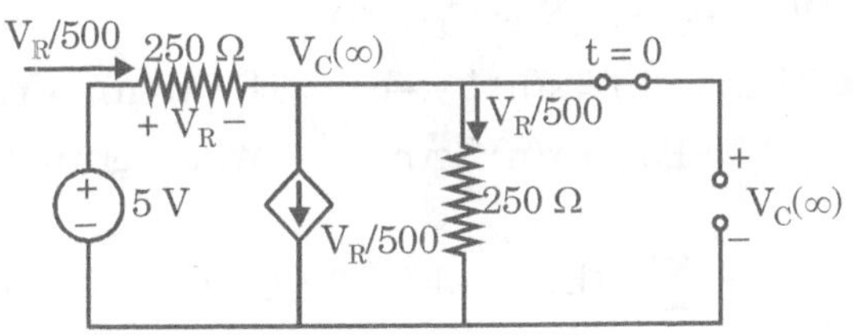

By KVL in outer loop

$$\Rightarrow 5 - V_R\left(\frac{V_R}{500}250\right) = 0$$

$$\Rightarrow 5 - V_R - \frac{V_R}{2} = 0$$

$$\Rightarrow \frac{3V_R}{2} = 0 \Rightarrow V_R = \frac{10}{3}V$$

$$\Rightarrow V_C(\infty) = \frac{V_R}{500}\times 250 = \frac{V_R}{2} = \frac{10}{6} = \frac{5}{3}V = 1.66\text{ V}$$

Time constant $\tau = R_{th}C$

$$R_{th} = \frac{V_{th}}{I_N}$$

- In the above circuit V_{th} and $V_C(\infty)$ are same

So $V_{th} = \frac{5}{3}V$

- Computation of I_N

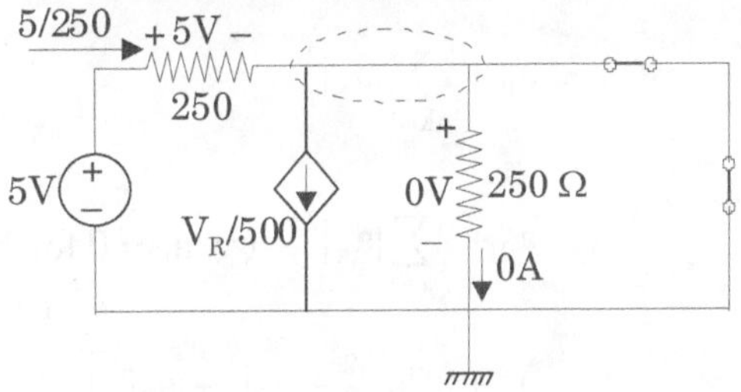

$$\text{By KCL } \frac{5}{250} = \frac{5}{500} + 0 + I_N$$

$$\Rightarrow I_N = \frac{5}{250}-\frac{5}{500} = \frac{1}{50}-\frac{1}{100} = \frac{2-1}{100} = \frac{1}{100}A$$

$$R_{th} = \frac{V_{th}}{I_N} = \frac{\frac{5}{3}}{\frac{1}{100}} = \frac{500}{3}\Omega$$

$$\tau = R_{th}C = \frac{500}{3}\times 0.6\times 10^{-6} = 1\times 10^{-4}\text{ sec}$$

$$V_C(t) = V_C(\infty)+\left[V_C(0^-)-V_C(\infty)\right]e^{-\frac{t}{\tau}}$$

$$= 1.66+[-5-1.66]e^{-10^4 t}$$

$$\Rightarrow 0 = 1.66-6.66\,e^{-10^4 t}$$

$$\Rightarrow e^{-10^4 t} = \frac{1.66}{6.66}$$

$$\Rightarrow -10^4 t = \ell n\left(\frac{1.66}{6.66}\right)$$

$$\Rightarrow t = -10^{-4}\ell n\left(\frac{1.66}{6.66}\right)$$

$$= (-10^4)(1.389)$$

$$= 1.389\times 10^{-4}\text{ sec} = 0.138\text{ msec}$$

39. As given that

For a periodic signal x(t) is

- Its period is 6, it is real, even.
- Average value of x(t) is 2 i.e., $a_0 = 2$.
- $a_k = \begin{cases} k; & k \le k \le 3 \\ 0; & k > 3 \end{cases} \Rightarrow \begin{matrix} a_1 = 1 \\ a_2 = 2 \\ a_3 = 3 \end{matrix}$

If $|a_1| = 1$ then $|a_{-1}| = 1$
$|a_2| = 2$ then $|a_{-2}| = 2$ etc.
$|a_3| = 3$ then $|a_{-3}| = 3$ etc.
We need to compute its average power.

Now, by Parseval's theorem

$$P = \frac{1}{T}\int_0^T |x(t)|^2\, dt = \sum_{k=-\infty}^{\infty} |a_k|^2$$

$$= a_0^2 + 2\sum_{k=1}^{\infty} |a_k|^2$$

$$= a_0^2 + 2\sum_{k=1}^{3} |a_k|^2 \quad (\because\ a_k = 0 \text{ for } k > 3)$$

$$= a_0^2 + 2\left[|a_1|^2 + |a_2|^2 + |a_3|^3\right]$$

$$= 2^2 + 2[1^2 + 2^2 + 3^2]$$

$$P = 4 + 28 = 32 \text{ watt}$$

40. As given that

For unit step input u(n), the output is $2\delta(n + 1) + \delta(n) + \delta(n - 1)$. We need to obtain the system response for input $\left(\frac{1}{2}\right)^n u(n)$.

$$s(n) = 2\delta(n + 1) + \delta(n) + \delta(n - 1)$$

[s(n) : step response]

$$h(n) = s(n) - s(n - 1)$$

(h(n) : impulse response)

$$= 2\delta(n + 1) + \delta(n) + \delta(n - 1)$$
$$-[2\delta[(n-1)+1] + \delta(n-1) + \delta(n-1) - 1]$$
$$= 2\delta(n + 1) + \delta(n) + \delta(n - 1) - 2\delta(n) - \delta(n - 1) - \delta(n - 2)$$
$$= 2\delta(n + 1) - \delta(n) - \delta(n - 2)$$

If the input is $x(n) = \left(\frac{1}{2}\right)^n u(n)$ then its response is: $y(n) = x(n)*h(n)$

$$= \left[\left(\frac{1}{2}\right)^n u(n)\right] * [2\delta(n + 1) - \delta(n) - \delta(n - 2)]$$

$$= \left[\left(\frac{1}{2}\right)^n u(n) * 2\delta(n+1)\right] - \left[\left(\frac{1}{2}\right)^n u(n) * \delta(n)\right] - \left[\left(\frac{1}{2}\right)^n u(n) * \delta(n-2)\right]$$

$$y(n) = 2\left(\frac{1}{2}\right)^{n+1} u(n+1) - \left(\frac{1}{2}\right)^n u(n) - \left(\frac{1}{2}\right)^{n-2} u(n-2)$$

As we want Y(0), it is due to first 3 terms of y(n)

$$\text{So, } y(0) = 2\left(\frac{1}{2}\right)^1 u(1) - \left(\frac{1}{2}\right)^0 u(0) - \left(\frac{1}{2}\right)^2 u(-2)$$

$$= \left(2 \times \frac{1}{2} \times 1\right) - (1 \times 1) - \left[\left(\frac{1}{2}\right)^{-2} \times 0\right]$$

$$= 1 - 1 - 0 = 0$$

41. As given that

$$x(n) = 2^{n-1} u(-n + 2)$$
$$y(n) = 2^{-n+2} u(n + 1)$$

$$I = \frac{1}{2\pi}\int_0^{2\pi} X(e^{j\omega})Y(e^{-j\omega})d\omega = ?$$

Let, $P(e^{j\omega}) = X(e^{j\omega})Y(e^{-j\omega})$

Then $\frac{1}{2\pi}\int_0^{2\pi} P(e^{j\omega})d\omega = P(0) = P(n)|_{n=0}$

$$P(n) = \frac{1}{2\pi}\int_0^{2\pi} P(e^{j\omega})e^{-j\omega n}d\omega$$

If we make n = 0 in above equation then

$$P(0) = \frac{1}{2\pi}\int_0^{2\pi} P(e^{j\omega})d\omega$$

If $P(e^{j\omega}) = X(e^{j\omega})Y(e^{-j\omega})$

Then $P(n) = x(n) * y(-n)$ $[\because\ y(-n) \leftrightarrow Y(e^{-j\omega})]$

$P(n) = x(n) * g(n)$ [Where $g(n) = y(-n)$]

$$= \sum_{k=-\infty}^{\infty} x(k)g(n - k) \qquad [\because g(n) = 2^{n+2}u(-n + 1)]$$

$$= \sum_{k=-\infty}^{\infty} \left[2^{k-1}u(-k+2)\right]\left[2^{n-k+2}u[-(n-k)+1]\right]$$

$$= \sum_{k=-\infty}^{\infty} (2^{k-1})(2^{n-k+2})u(-k+2)u(k-n+1)$$

$$\Rightarrow P(0) = \sum_{k=-\infty}^{\infty} 2^{k-1}.2^{-k+2}\left[u(-k+2).u(k+1)\right]$$

$$\left[\begin{array}{l} u(-k+2) = 1; -\infty \le k \le 2 \\ u(k+1) = 1; -1 \le k \le \infty \\ u(-k+2),\ u(k+1) = \begin{cases} 1; & -1 \le k \le 2 \\ 0; & \text{else} \end{cases} \end{array}\right]$$

$$\Rightarrow \quad P(0) = \sum_{k=-1}^{2} 2^{k-1+(-k+2)}$$

$$= \sum_{k=-1}^{2} 2$$

$$= \sum_{k=-1}^{2} 2(1)^k$$

$$= 2(1)^{-1} + 2(1)^0 + 2(1)^1 + 2(1)^2$$

$$= (2 \times 1) + (2 \times 1) + (2 \times 1) + (2 \times 1)$$

$$= 2 + 2 + 2 + 2 = 8$$

42. According to the question

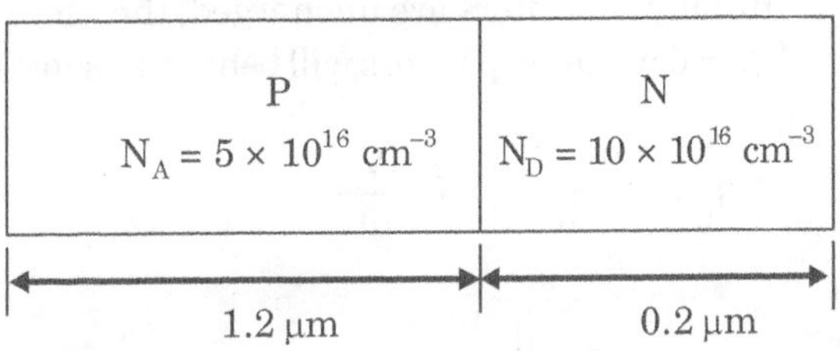

According to the junction law

$$\frac{W_P}{W_n} = \frac{N_D}{N_A} = \frac{10 \times 10^{16}}{5 \times 10^{16}} = 2$$

$$W_P = 2W_n$$

This indicates that, N-region will depleted entirely at first;

$$W = W_P + W_n$$

$$= 2W_n + W_n$$

$$W = 3W_n = 3 \times 0.2 \ \mu m$$

$$\therefore \quad W = 0.6 \ \mu m = 0.6 \times 10^{-4} \text{ cm} \qquad (\because \in - \in_0 \in_r)$$

$$W = \sqrt{\left(\frac{2\in}{q}\right)\left(\frac{1}{N_A} + \frac{1}{N_D}\right).V_J}$$

(where $V_J = V_{bi} + V_{RB}$)

$$= \sqrt{\left(\frac{2\in}{q}\right)\left(\frac{1}{N_A} + \frac{1}{N_D}\right)(V_{bi} + V_{RB})}$$

$$V_{RB} = \frac{W_q^2 \times N_A.N_D}{2\in(N_A + N_D)} - V_{bi}$$

$$V_{RB} = \frac{(0.6 \times 10^{-4})^2 \times 1.6 \times 10^{-19} \times 5 \times 10^{16} \times 10 \times 10^{16}}{2 \times 12 \times 8.85 \times 10^{-14}(5 \times 10^{16} + 10 \times 10^{16})} - 0.8\text{V}$$

$$= 9.095 \text{ V} - 0.8 \text{ V} = 8.24 \text{ V}$$

$$\boxed{\text{Hence, } V_{RB} = 8.24 \text{ V}}$$

43. As given that,

$V_{in} = 10$V

$T_{on} = 1$m sec,

$T_{off} = 1$ μ sec

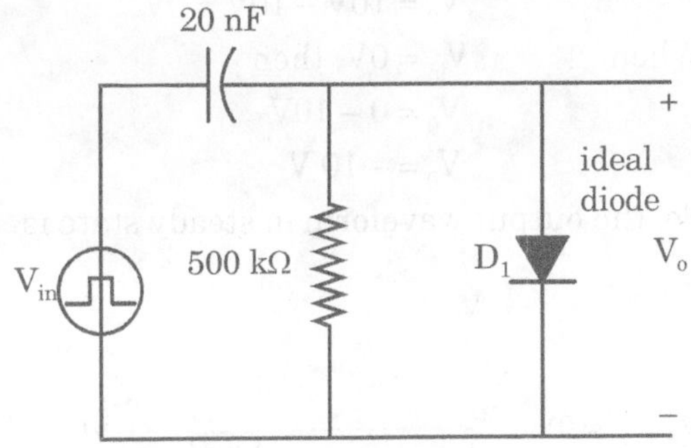

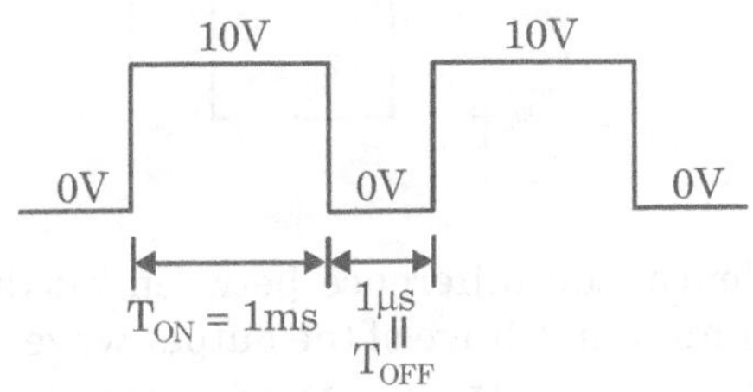

$$T = T_{ON} + T_{OFF} = 1.001 \text{ m sec.}$$

$$\tau = RC = 20 \times 10^{-9} \times 500 \times 10^3 = 10 \text{ ms}$$

So, $\quad \tau >> T$

In T_{ON} duration diode be in conduction and their equivalent circuit is

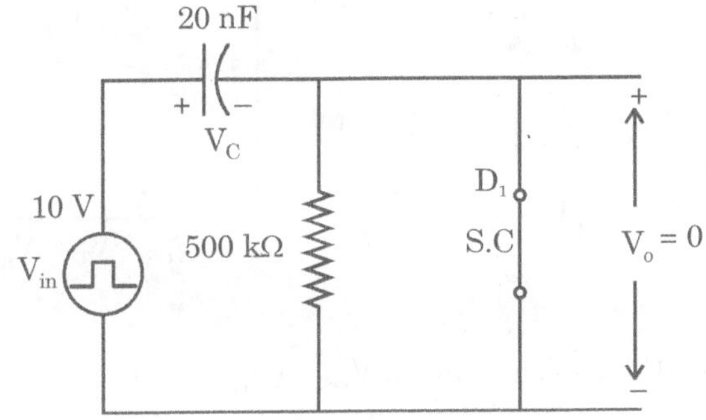

$$+V_{in} - V_C = 0\text{V}$$

$$\Rightarrow \quad \boxed{V_C = V_{in} = +10\text{V}}$$

$$\boxed{V_0 = 0\text{V}} \qquad \text{....(i)}$$

Once the capacitor in fully charge and $\tau >> T$; so diode will not conduct further and their equivalent circuit is

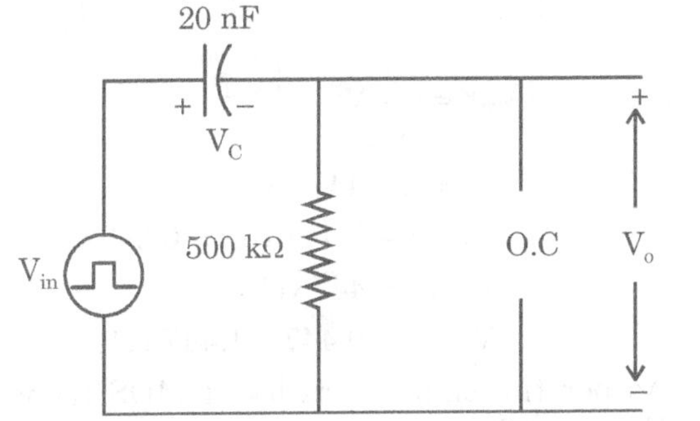

$$+V_{in} - V_C - V_0 = 0\text{V}$$

$$V_0 = V_{in} - V_C$$

In steady state: when $V_{in} = 10V$

$$V_0 = 10V - 10V = 0V$$

When $V_{in} = 0V$, then

$$V_0 = 0 - 10V$$

$$V_0 = -10\ V \quad ....(ii)$$

So, the output waveform in steady state is

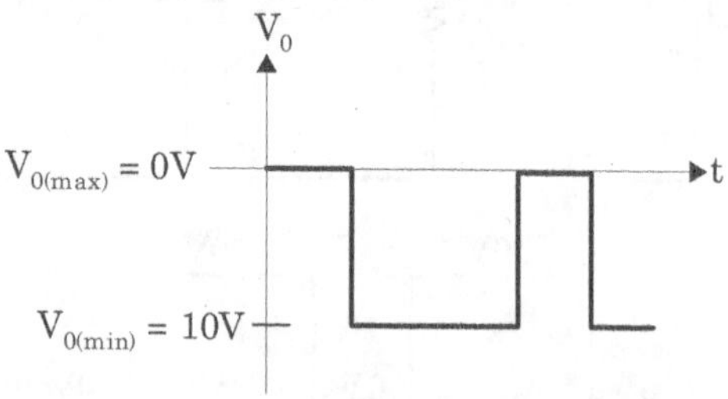

Hence, the difference between maximum and minimum voltage of the output wave form.

$$V_{0(differential)} = V_{0(max)} - V_{0(min)} = 0V - (-10V) = 10V$$

44. As given circuit:

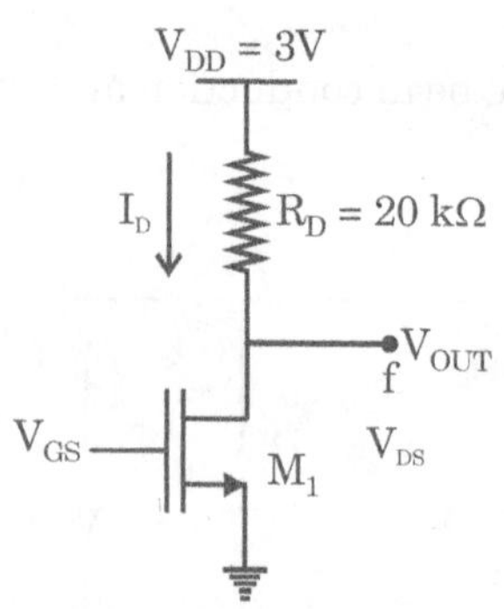

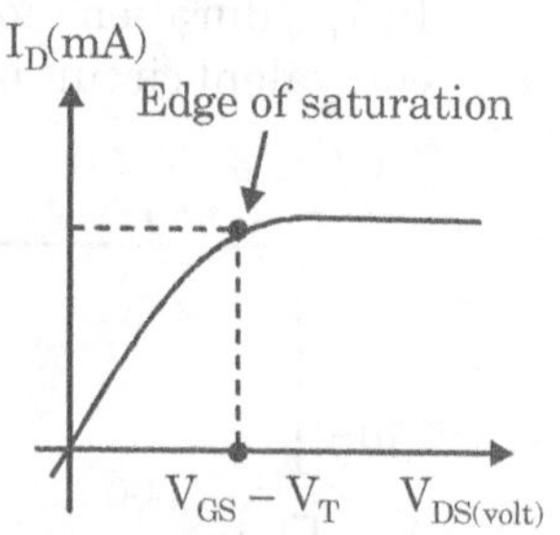

Edge of saturation $V_{DS} = (V_{GS} - V_T)$

$$I_{D(sat)} = \frac{1}{2}\mu_n C_{ox}\frac{W}{L}(V_{DS})^2$$

$$\frac{3-1}{20}\text{mA} = \frac{1}{2}\times 100\ \mu A/v^2 \times 10(V_{GS} - V_T)^2$$

$$\frac{1}{10}\text{mA} = \frac{1}{2}\times\frac{\text{mA}}{V^2}\times(1-V_T)^2$$

$$(1-V_T)^2 = \frac{2}{10}V^2 = \frac{1}{5}V^2$$

$$1 - V_T = \pm\ 0.447\ \text{volt}$$

$$V_T = 1 - 0.447\ \text{volt} = 0.552\ \text{volt}$$

$$1 - V_T = -0.447\ \text{volt}$$

$$V_T = 1 + 0.447 = 1.447\ \text{volt}$$

As per the characteristics of MOS transition $V_G - V_T > 0V$

Hence, $\boxed{V_T = 0.55\ \text{volt}}$ is accepted value.

45. As given circuit:

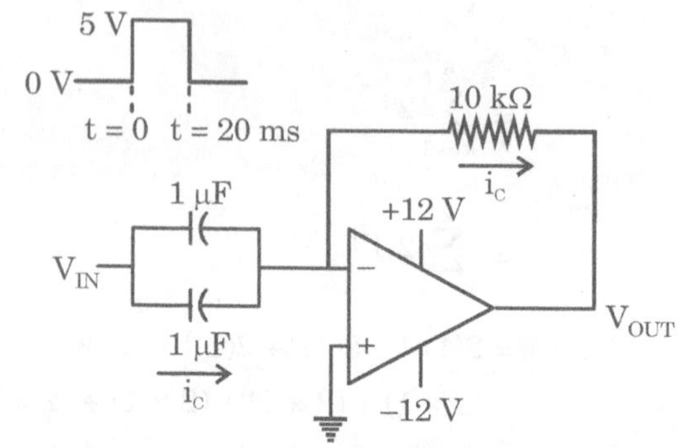

Initially capacitors are uncharged, therefore $V_{C1} = V_{C2} = 0$ means capacitors will behave as short cirict.

$$i_C = \left[\frac{0 - V_0}{R}\right] = C\frac{dV_{in}}{dt}$$

$$V_0 = -RC\frac{d}{dt}V_{in}$$

$= -RC$ [Slope of input V_{in}]

$= -RC(\infty)$

Capacitors doesn't allow sudden change in voltage, so at $t = 0^+$, $V_{c1} = V_{c2} = 0V$

So, the voltage at $V_- = 5V$ att $= 0$ and $V_+ = +V$

$V_- > V_+ \Rightarrow V_+ - V_- < 0V$, so op-amp will produce, saturated negative output voltage.

Hence $\boxed{V_{out} = -12V}$

46. As given circuit:

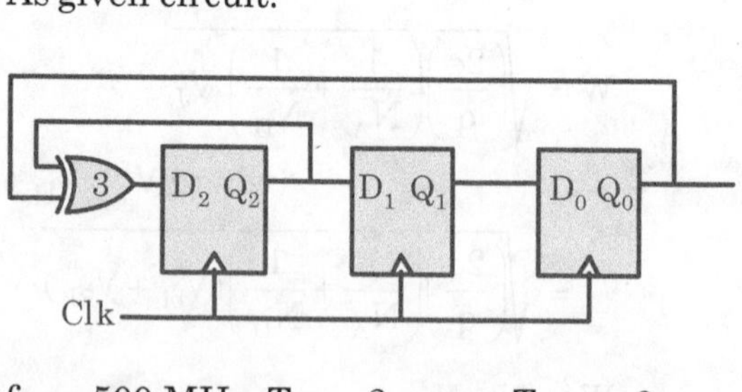

$f_{clk} = 500$ MHz, $T_{clk} = 2n$ secm $T_{XOR} = 3ns$,

Initially $Q_2\ Q_1\ Q_0 = 111$, $D_2 = 1$

Delay of XOR gate is 3 n sec.

We need to obtain after how many clock contents of $Q_2\ Q_1\ Q_0$ becomes100

Its clear from above circuit that

$$Q_2^+ = D_2,\ Q_1^+ = D_1,\ Q_0^+ = D_0$$

Now, draw the waveform's which will give a clear picture.

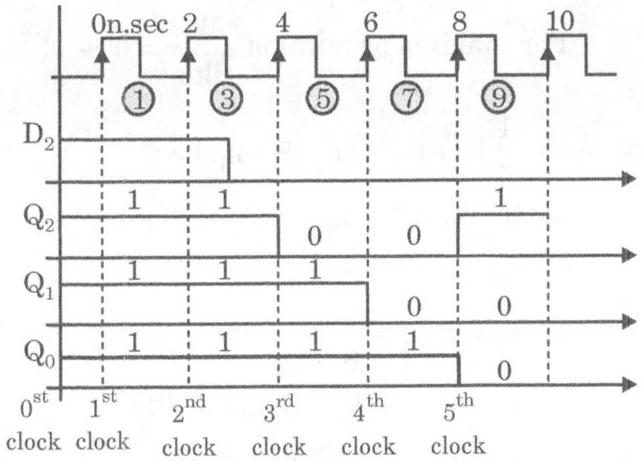

$Q_2Q_1Q_0$: 111 111 011 001 100

We can see after 5th clock content $Q_2\ Q_1\ Q_0$ is100

Here, XOR gate will change its output after 3n sec of its new input, which is happening here at 3nsec and 7n sec

47. As given circuit:

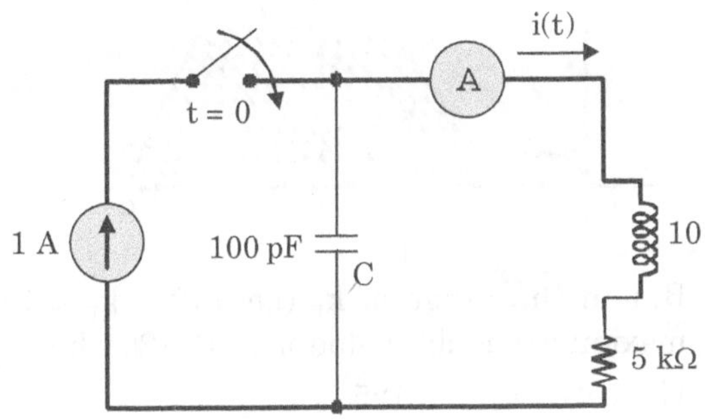

As given that $V_C(0^-)= 0V$

$$i_L(0^-) = 0A$$

In Laplace domain the above circuit for t > 0 becomes

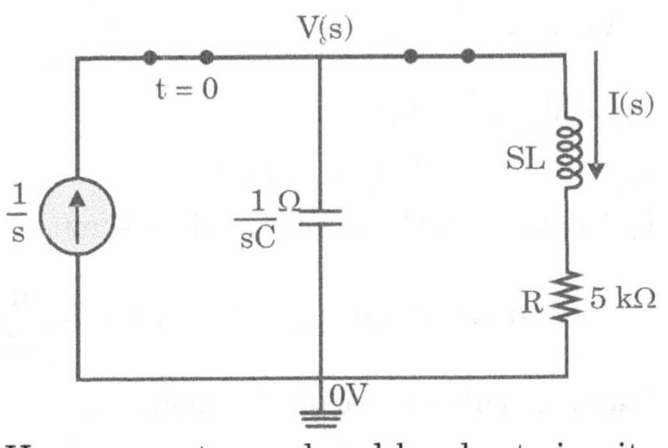

Here, ammeter replaced by short circuit as its internal resistance given is 0Ω.

Now, by current division rule

$$I(s) = \frac{1}{s}\frac{\frac{1}{sC}}{R + SL + \frac{1}{sC}}$$

$$= \frac{1}{s(s^2LC + SCR + 1)}$$

$$= \frac{\frac{1}{LC}}{s\left(s^2 + \frac{R}{L}s + \frac{1}{LC}\right)}$$

$$i(\infty) = \lim_{t\to\infty} SI(s) = 1$$

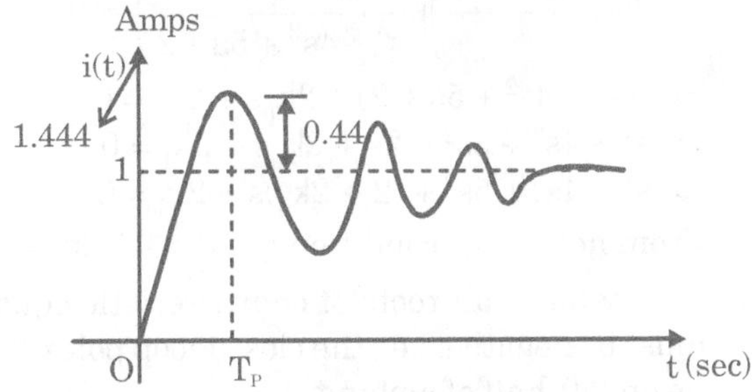

The expression of I(S) is identical to unit step response of a standard 2nd order closed loop system i.e.,

$$Y(s) = \frac{\omega_n^2}{s\left(s^2 + 2\xi w_n s + \omega_n^2\right)}$$

Now, by comparison $\omega_n = \frac{1}{\sqrt{LC}}$

$$= \frac{1}{\sqrt{10\times10^{-3}\times100\times10^{-12}}}$$

$$= 10^6 \text{ r/sec}$$

$$2\xi\omega_n = \frac{R}{L}$$

$$\Rightarrow \quad \xi = \frac{R}{2L\omega_n}$$

$$= \frac{5\times10^3}{2\times10\times10^{-3}\times10^6}$$

$$\xi = \frac{1}{4} = 0.25$$

$$\xi = 0.25$$

underdamped response:

$$\text{Overshoot} = e^{\frac{-\pi\xi}{\sqrt{1-\xi^2}}} = e^{\frac{-0.25A}{\sqrt{1-0.25^2}}} = 0.4443$$

Hence, the maximum ammeter reading just after the switch closed is:

$$\frac{i_L(t)}{\max} = i(\infty) + 0.444 = 1 + 0.444 = 1.444 \text{ A}$$

48. As given block diagram

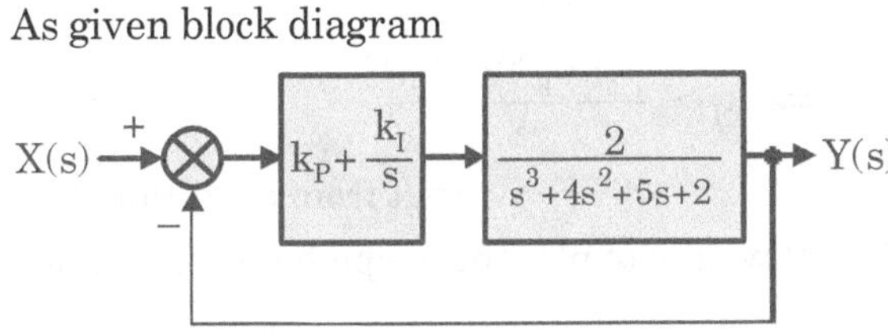

The characteristic equation of the system is

$$1 + G(s) = 0$$

$$1+\left(k_p+\frac{k_I}{s}\right)\left(\frac{2}{s^3+4s^2+5s+2}\right)=0$$

$$\Rightarrow 1+\left(\frac{k_p s+k_I}{s}\right)\left(\frac{2}{s^3+4s^2+5a+2}\right)=0$$

$$\Rightarrow s(s^3 + 4s^2 + 5s + 2) + 2k_p s + 2k_I = 0$$

$$\Rightarrow s^4 + 4s^3 + 5s^2 + 2s + 2k_p s + 2k_I = 0$$

$$\Rightarrow s^4 + 4s^3 + 5s^2 + (2 + 2k_p)s + 2k_I = 0$$

From necessary condition $\Rightarrow K_p > -1, k_1 > 0$

For stability all roots of characteristic equation must be negative i.e., the closed loop poles should lie on left half of s-plane.

By R-H criterion.

s^4	1	5	$2k_I$
s^3	4	$2+2k_p$	0
s^2	$\frac{18-2k_p}{4}$	$2k_I$	0
s^1	$\frac{\left(\frac{18-2k_p}{4}\right)(2+2k_p)-8k_p}{\frac{18-2k_p}{4}}$		0
s^0	$2k_I$		

For stability all element of 1st column must be positive.

Condition 1 : $2k_I > 0 \Rightarrow k_I > 0$

Condition 2 : $\left[\frac{18-2k_p}{4}\right] > 0 \Rightarrow k_p < 9$

Condition 3 : $\left[\frac{\left(\frac{18-2k_p}{4}\right)(2+2k_p)-8k_1}{\frac{18-2k_p}{4}}\right] > 0$

$$\Rightarrow (2+2k_p)-\left(\frac{8k_I\times 4}{18-2k_p}\right)>0$$

$$\Rightarrow 1+k_p>\frac{16k_I}{18-2k_p}$$

$$\Rightarrow 1+k_p>\frac{8k_I}{9-k_p}$$

$$\Rightarrow k_I>\frac{(1+k_p)(9-k_p)}{8}$$

[but from condition 2 : $k_p < 9$]

If we try to plot the graph between k_I and k_P k_I

$$=\frac{(1+k_p)(9-k_p)}{8}$$

For maxima or minima $\frac{dk_I}{dk_P} = 0$

$$\Rightarrow \frac{1}{8}[1 + k_P(-1) + (9 - k_P)] = 0$$

$$\Rightarrow 9 - k_P = 1 + k_P$$

$$\Rightarrow k_P = 4$$

If, $k_P = 0$, $k_I = \frac{9}{8} = 1.125$

$k_P = 4$, $k_I = 3.125$

$k_P = 9$, $k_I = 0$

So, the plot between k_P and k_I will be

The standard part is satisfying the condition $(k_I > 0)$ and $(k_P < 9)$

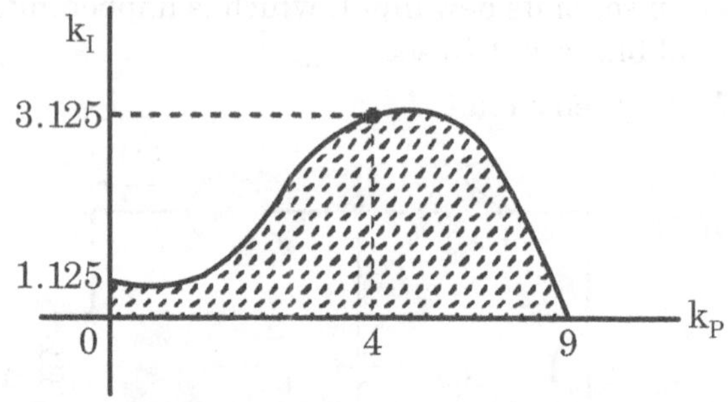

But in the range of k_P (i.e.,) $(0 < k_P < 9)$, the maximum possible value of k_I is 3.125 when $k_P = 4$.

Hence, $(k_I)_{max}$: 3.125

49. As given that, $V_{rms} = 4$

$f_m = 1$ KHz

$K_P = 2$ rad/Volt

$$\therefore Vrms = \frac{V_m}{\sqrt{2}}$$

$$V_m = V_{rms}\sqrt{2} = 4\sqrt{2}$$

$$x_{pm}(t) = A_C \cos(2\pi f_c t + k_p m(t))$$

Instantaneous phase $\phi_1(t) = 2\pi f_c t + k_p m(t)$

Instantaneous frequency $\omega_1(t) = 2\pi f_c + k_p \frac{d}{dt} m(t)$

Maximum instantaneous frequency (in Hz)

$$f_{i\,max} = f_C + \frac{1}{2\pi} k_P \left(\frac{d}{d+} m(t)\right)_{max}$$

$$= 1000 \times 10^3 + \frac{1}{2\pi} \times 2 \times 4\sqrt{2} \times 10^3$$

$$= 1000 \times 10^3 + 11.312 \times 10^3$$

$$m(t) = 4\sqrt{2}\sin\left(2\pi \times 10^3 t\right)$$

Hence maximum instantaneous frequency is 1011312 kHz.

50. As given that

$f_S = 600$ kHz

$f_{LO} = 1000$ kHz

f_{SI} = Image frequency

$F_{SI} = F_S + 2f_{IF}$

$\therefore f_{IF} = 1000k - 600k = 400$ k

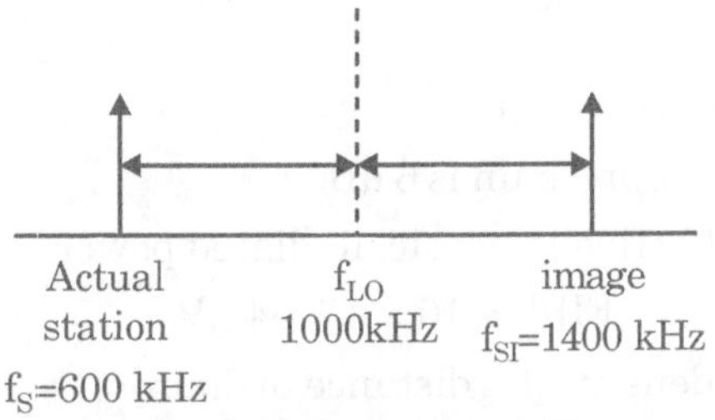

$f_{SI} = 600 + 2 \times 400 = 1400$ kHz

51. As given probabilities:

Students → Boys 50%, Girls 50%

Subjects → Science 75%, Commerce 25%

$P(B) = \frac{1}{2}$ $P(G) = \frac{1}{2}$ $P(S) = \frac{3}{4}$ $P(C) = \frac{1}{4}$

It is given that conditional probability of Boy being science student is half for boy being commerce student.

Let us consider

$$P\left\{\frac{\text{Boy}}{\text{Science}}\right\} = x$$

$$\Rightarrow P\left\{\frac{\text{Boy}}{\text{Commerce}}\right\} = 2x$$

$$\Rightarrow P\{\text{Boy}\} = P\left\{\frac{\text{Boy}}{\text{Science}}\right\}P\{\text{Science}\} + P\left\{\frac{\text{Boy}}{\text{Commerece}}\right\}P\{\text{Commerce}\}$$

$$\Rightarrow \frac{1}{2} = x\frac{3}{4} + 2x.\frac{1}{4}$$

$$\Rightarrow x = \frac{2}{5}$$

$$\Rightarrow P\left\{\frac{\text{Boy}}{\text{Commerece}}\right\} = \frac{4}{5}$$

$$\Rightarrow P\left\{\frac{\text{Girl}}{\text{Commerce}}\right\} = 1 - P\left\{\frac{\text{Boy}}{\text{Commerece}}\right\} = \frac{1}{5}$$

Now,

$$P\left\{\frac{\text{Commerce}}{\text{Girl}}\right\} = \left[\frac{P\left\{\frac{\text{Girl}}{\text{Commerce}}\right\}.P\{\text{Commerce}\}}{P\{\text{Girl}\}}\right]$$

$$= \frac{\frac{1}{5}\times\frac{1}{4}}{\frac{1}{2}} = \frac{1}{10}$$

$$\Rightarrow I\left(\frac{\text{Commerce}}{\text{Girl}}\right) = \log_2\left(\frac{1}{\frac{1}{110}}\right)$$

$$= \log_2 10 = 3.32 \text{ bits}$$

52. As we know that

Signal to quantization noise ratio $(SNR)_q$

$$= \frac{\text{signal power}}{\text{Quantization noise power}}$$

Signal power = $(RMS)^2 = (0.1)^2 = 0.01$

Quantization step size Δ

$$= \frac{x(t)_{max} - x(t)_{min}}{L} = \frac{2}{L}$$

Data rate = n.Fs

Where 'n' in number of encoder bits.

F_S is sampling frequency.

As given that

$n.F_S \leq 50 \times 10^3$ bits per second

Minimum $F_S = 2 \times 5$ kHz = 10 kHz

$\Rightarrow n \times 10 \times 10^3 \leq 50 \times 10^3$

$\Rightarrow n \leq 5$, taking maximum n = 5

$$\text{Quantization step size } \Delta = \frac{\text{Peak to peak Voltage}}{2^n}$$

$= \frac{2}{2^5} = \frac{1}{2^4}$, quantization noise power $= \frac{\Delta^2}{12}$ to maximum $(SNR)_q$, quantization noise power should be minimized.

$$\Rightarrow (SNR)_{q\,max} = \frac{\text{Signal power}}{\text{Noise power}} = \frac{(0.01)}{\frac{(1/24)^2}{12}}$$

$$= \frac{0.01 \times 12}{\left(\frac{1}{16}\right)^2} = 30.72$$

53. Optimum threshold for a maximum a posteriori (MAP) receiver under AWGN channel is

$$V_{th} = \left(\frac{a_1 + a_2}{2}\right) + \frac{\sigma_n^2}{a_1 - a_2}\ell_n\left[\frac{P_0}{P_1}\right]$$

$(\because a_1 = 2, a_2 = -2)$

$$= 0 + \frac{0.4}{2-(-2)}\ell_n\left(\frac{0.6}{0.4}\right)$$

$$= 0.1 \times 0.4 = 0.04 \text{ V}$$

54. As we know that

$$V_P = \frac{C}{\sqrt{1-\left(\frac{f_c}{f}\right)^2}},$$

Since a > b, dominant mode $f_C\,(TE)_{10}$:

$$f_C\left(TE_{10}\right) = \frac{C}{2a}$$

$$= \frac{3\times10^8}{2\times8\times10^{-2}} = 1.875\text{ GHz}$$

$$\Rightarrow \frac{V_P}{C} = \frac{1}{\sqrt{1-\left(\frac{1.875}{3.4}\right)^2}}$$

$$= \frac{1}{\sqrt{0.695}} = 1.198$$

Hence, the factor $\left[\frac{VP}{C}\right]$ is 1.198.

55. As given that

For directive gain:

10 log D = 6

$D = 10^{0.6}$

= 3.98

P_{rad} = 16 kw

Distance (r) = 8 km

Since antenna gain is 6 dB

Total effective Isotropic, Radiated power

EIRP = 16 × 4 = 64 kW

Power density at a distance of 8 km

$$= \frac{64\times10^3}{4\pi\times(8000)^2}$$

$$\Rightarrow \frac{|E_0|^2}{2\times120\pi} = \frac{64\times10^3}{4\pi\times(8000)^2}$$

$$\Rightarrow E_0 = \sqrt{\frac{64\times60\times10^3}{8000\times8000}} = 0.245\text{ V/m}$$

SOLVED PAPER - 2020

INSTRUCTIONS

1. Total of 65 questions carrying 100 marks, out of which 10 questions carrying a total of 15 marks are in General Aptitude (GA)
2. The Engineering Mathematics will carry around **15% of the total marks,** the General Aptitude section will carry **15% of the total marks** and the **remaining 70% of the total marks.**
3. **Types of Questions**
   (*a*) **Multiple Choice Questions (MCQ)** carrying 1 or 2 marks each in all papers and sections. These questions are objective in nature, and each will have a choice of four options, out of which the candidate has to mark the correct answer(s).
   (*b*) **Numerical Answer Questions** of 1 or 2 marks each in all papers and sections. For these questions the answer is a real number, to be entered by the candidate using the virtual keypad. No choices will be shown for these type of questions.
4. For **1-mark** multiple-choice questions, **1/3 marks** will be deducted for a wrong answer. Likewise, for **2-marks** multiple-choice questions, **2/3** marks will be deducted for a wrong answer. There is no negative marking for numerical answer type questions.

Chapter-Wise Analysis

#	Chapters	Marks		
		1	2	Total
1	General Aptitude	5	5	15
2	Engineering Mathematics	4	3	10
3	Network Theory	3	2	7
4	Signals and Systems	2	2	6
5	Electronic Devices & Circuits	2	4	10

#	Chapters	Marks		
		1	2	Total
6	Analog Circuits	3	5	13
7	Digital Circuits & Microprocessor	3	3	9
8	Control Systems	2	4	10
9	Communication Systems	3	4	11
10	Electromagnetic Theory	3	3	9

GENERAL APTITUDE (GA)

Q. 1 to Q. 5 : Carry One Mark Each.

1. The untimely loss of life is a cause of serious global concern as thousands of people get killed ____ accidents every year while many other die ____ diseases like cardio vascular disease, cancer, etc.

(a) in, of (b) from, of
(c) during, from (d) from, from

2. He was not only accused of theft ___ of conspiracy.

(a) rather (b) but also
(c) but even (d) rather than

3. Select the word that fits the analogy:
Explicit : Implicit :: Express : _____

(a) Impress (b) Repress
(c) Compress (d) Suppress

4. The Canadian constitution requires that equal importance be given to English and French. Last year, Air Canada lost a lawsuit, and had to pay a six- figure fine to a French-speaking couple after they filed complaints about formal in-flight announcements in English lasting 15 seconds, as opposed to informal 5 second messages in French.

The French-speaking couple were upset at ____.

(a) the in-flight announcements being made in English.
(b) the English announcements being clearer than the French ones.
(c) the English announcements being longer than the French ones.
(d) equal importance being given to English and French.

5. A superadditive function f(.) satisfies the following property

$$f(x_1 + x_2) \geq f(x_1) + f(x_2)$$

Which of the following functions is a superadditive function for x > 1 ?

(a) e^x (b) $\sqrt{x}$
(c) $\frac{1}{x}$ (d) e^{-x}

Q. 6 to Q. 10 : Carry Two Marks Each.

6. The global financial crisis in 2008 is considered to be the most serious world-wide financial crisis, which started with the sub-prime lending crisis in USA in 2007. The sub-prime lending crisis led to the banking crisis in 2008 with the collapse of Lehman Brothers in 2008. The sub-prime lending refers to the provision of loans to those borrowers who may have difficulties in repaying loans, and it arises because of excess liquidity following the East Asian crisis.
Which one of the following sequences shows the correct precedence as per the given passage?

(a) East Asian crisis — Sub-prime lending crisis — Banking crisis — Global financial crisis
(b) Sub-prime lending crisis — Global financial crisis — Banking crisis — East Asian crisis
(c) Banking crisis — Sub-prime lending crisis — Global financial crisis — East Asian crisis
(d) Global financial crisis — East Asian crisis — Banking crisis — Sub-prime lending crisis

7. It is quarter past three in your watch. The angle between the hour hand and the minute hand is ___

(a) 0° (b) 7.5°
(c) 15° (d) 22.5°

8. A circle with centre O is shown in the figure. A rectangle PQRS of maximum possible area is inscribed in the circle. If the radius of the circle is a, then the area of the shaded portion is _____.

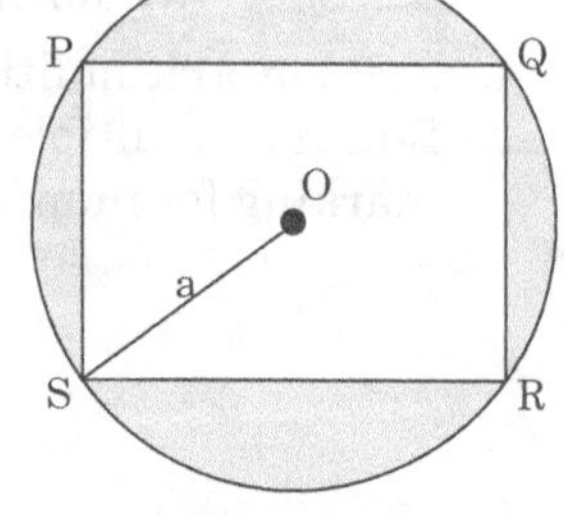

(a) $\pi a^2 - a^2$
(b) $\pi a^2 - \sqrt{2}a^2$
(c) $\pi a^2 - 2a^2$
(d) $\pi a^2 - 3a^2$

9. a, b, c are real numbers. The quadratic equation $ax^2 - bx + c = 0$ has equal roots, which is β, then

(a) $\beta = b/a$ (b) $\beta^2 = ac$
(c) $\beta^3 = bc/(2a^2)$ (d) $b^2 \neq 4ac$

10. The following figure shows the data of students enrolled in 5 years (2014 to 2018) for two schools P and Q. During this period, the ratio of the average number of the students enrolled in school P to the average of the difference of the number of students enrolled in schools P and Q is ____

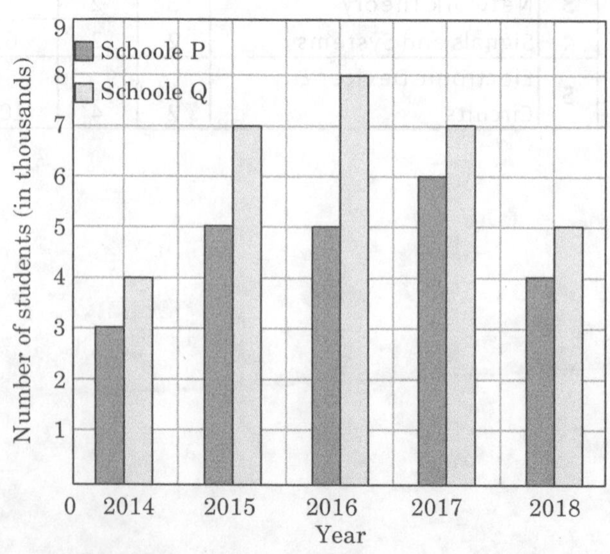

(a) 8 : 23 (b) 23 : 8
(c) 23 : 31 (d) 31 : 23

ELECTRONICS & COMMUNICATION ENGINEERING

Q. 1 to Q. 25 : Carry One Mark Each.

1. If $v_1, v_2, \ldots, v_6$ are six vectors in $\mathbb{R}^4$, which one of the following statements is FALSE ?

(a) It is not necessary that these vectors span $\mathbb{R}^4$

(b) These vectors are not linearly independent

(c) Any four of these vectors form a basis for $\mathbb{R}^4$

(d) If $\{v_1, v_3, v_5, v_6\}$ spans $\mathbb{R}^4$, then it forms a basis for $\mathbb{R}^4$

2. For a vector field $\vec{A}$, which one of the following is FALSE ?

(a) $\vec{A}$ is solenoidal if $\nabla . \vec{A} = 0$

(b) $\nabla \times \vec{A}$ is another vector field.

(c) $\vec{A}$ is irrotational if $\nabla^2 \vec{A} = 0$.

(d) $\nabla \times (\nabla \times \vec{A}) = \nabla(\nabla . \vec{A}) - \nabla^2 \vec{A}$

3. The partial derivative of the function

$$f(x, y, z) = e^{1-x\cos y} + xze^{-\frac{1}{(1+y^2)}}$$

with respect to x at the point (1, 0, e) is

(a) – 1 (b) 0

(c) 1 (d) $\frac{1}{e}$

4. The general solution of $\frac{d^2y}{dx^2} - 6\frac{dy}{dx} + 9y = 0$ is

(a) $y = C_1e^{3x} + C_2e^{-3x}$

(b) $y = (C_1 + C_2x)e^{-3x}$

(c) $y = (C_1 + C_2x)e^{3x}$

(d) $y = C_1e^{3x}$

5. The output y[n] of a discrete-time system for an input x[n] is

$$y[n] = \max_{-\infty \le k \le n} |x[k]|$$

The unit impulse response of the system is

(a) 0 for all n

(b) 1 for all n

(c) unit step signal u[n]

(d) unit impulse signal δ[n]

6. A single crystal intrinsic semiconductor is at a temperature of 300K with effective density of states for holes twice that of electrons. The thermal voltage is 26mV. The intrinsic Fermi level is shifted from mid-bandgap energy level by

(a) 18.02 meV (b) 9.01 meV

(c) 13.45 meV (d) 26.90 meV

7. Consider the recombination process via bulk traps in a forward biased pn homojunction diode. The maximum recombination rate is U_{max}. If the electron and the hole capture cross-sections are equal, which one of the following is FALSE ?

(a) With all other parameters unchanged, U_{max} decreases if the intrinsic carrier density is reduced.

(b) U_{max} occurs at the edges of the depletion region in the device

(c) U_{max} depends exponentially on the applied bias

(d) With all other parameters unchanged, U_{max} increases if the thermal velocity of the carriers increases.

8. The components in the circuit shown below are ideal. If the op-amp is in positive feedback and the input voltage V_i is a sine wave of amplitude 1 V, the output voltage V_0 is

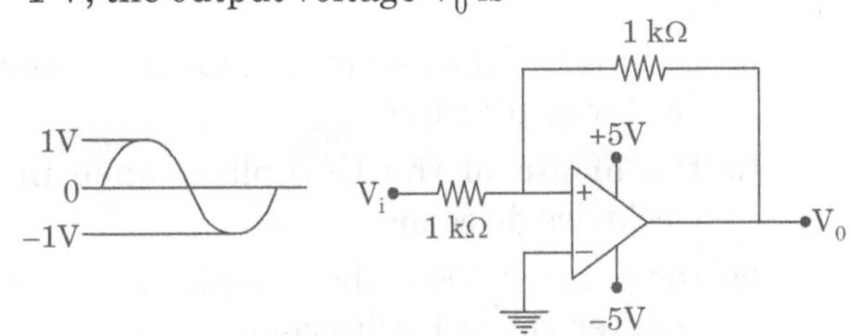

(a) a non-inverted sine wave of 2V amplitude.

(b) an inverted sine wave of 1V amplitude.

(c) a square wave of 5V amplitude.

(d) a constant of either +5 V or – 5V.

9. In the circuit shown below, the Thevenin voltage V_{TH} is

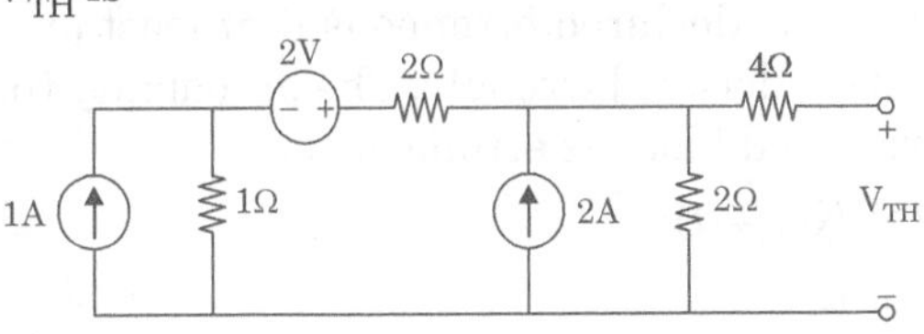

(a) 2.4 V (b) 2.8 V

(c) 3.6 V (d) 4.5 V

10. The figure below shows a multiplexer where S_1 and S_0 are the select lines, I_0 to I_3 are the input data lines, EN is the enable line, and F(P, Q, R) is the output. F is

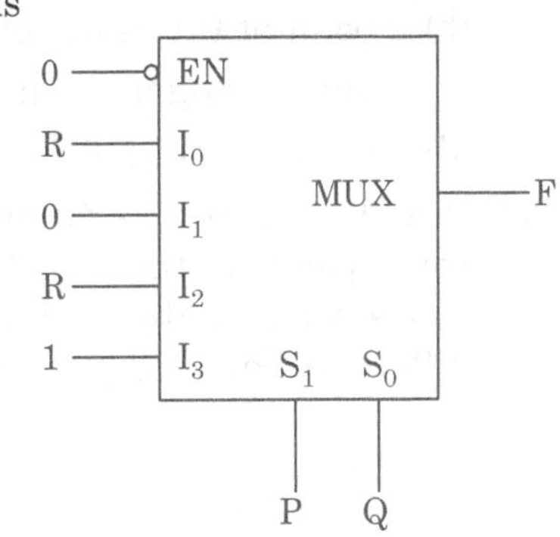

(a) $PQ + \bar{Q}R$

(b) $P + Q\bar{R}$

(c) $P\bar{Q}R + \bar{P}Q$

(d) $\bar{Q} + PR$

11. The pole-zero map of a rational function G(s) is shown below. When the closed contour Γ is mapped into the G(s)-plane, then the mapping encircles.

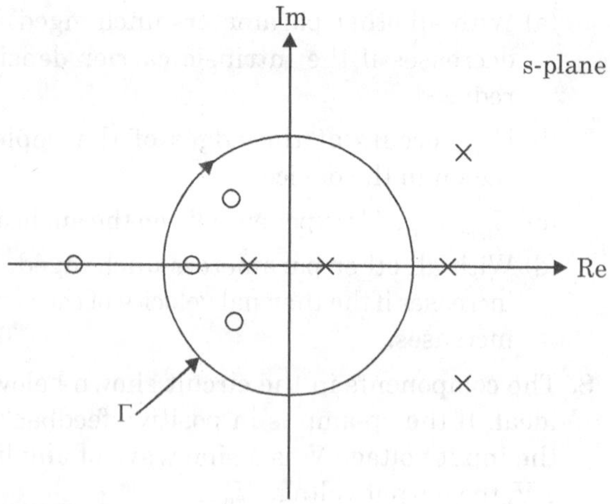

(a) the origin of the G(s)-plane once in the counter-clockwise direction.

(b) the origin of the G(s)-plane once in the clockwise direction.

(c) the point $-1 + j0$ of the G(s)-plane once in the counter-clockwise direction.

(d) the point $-1 + j0$ of the G(s)-plane once in the clockwise direction.

12. A digital communication system transmits a block of N bits. The probability of error in decoding a bit is α. The error event of each bit is independent of the error events of the other bits. The received block is declared erroneous if at least one of its bits is decoded wrongly. The probability that the received block is erroneous is

(a) $N(1-\alpha)$

(b) α^N

(c) $1-\alpha^N$

(d) $1-(1-\alpha)^N$

13. The impedances $Z = jX$, for all X in the range $(-\infty, \infty)$, map to the Smith chart as

(a) a circle of radius 1 with centre at (0, 0)

(b) a point at the centre of the chart.

(c) a line passing through the centre of the chart.

(d) a circle of radius 0.5 with centre at (0.5, 0)

14. Which one of the following pole-zero plots corresponds to the transfer function of an LTI system characterized by the input-output difference equation given below?

$$y[n] = \sum_{k=0}^{3} (-1)^k x[n-k]$$

(a)

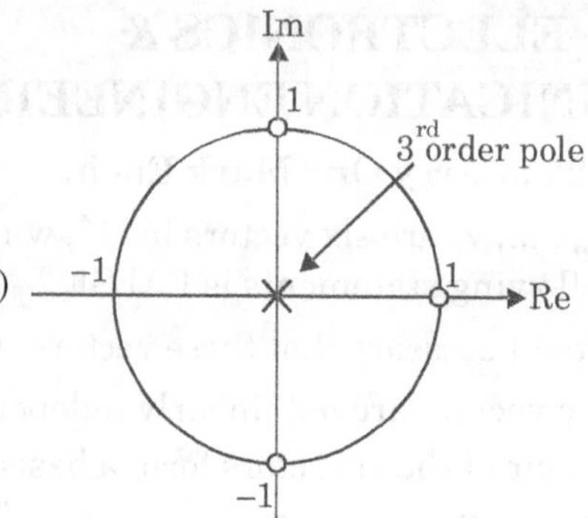

(b)

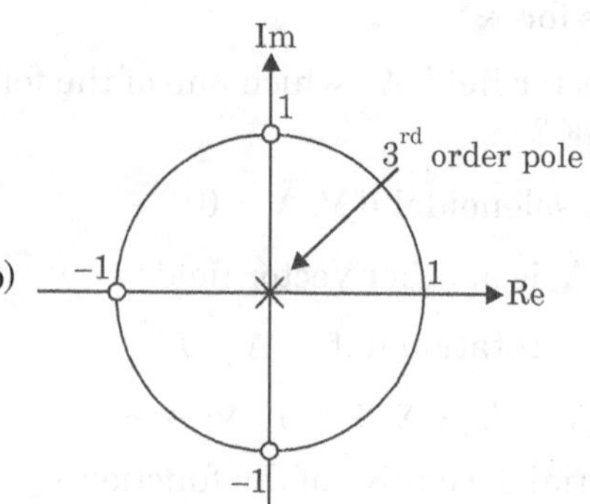

(c)

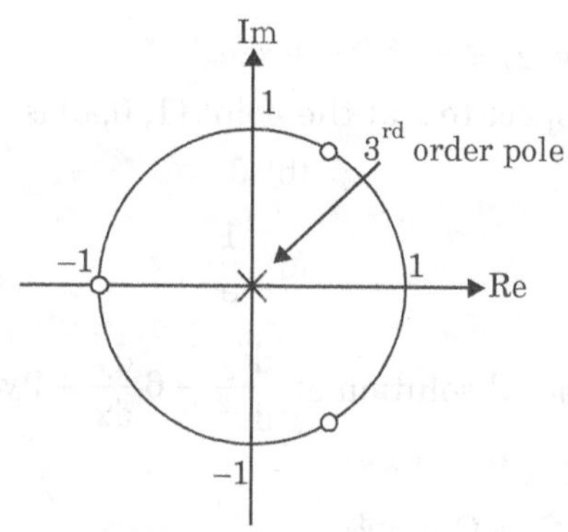

(d)

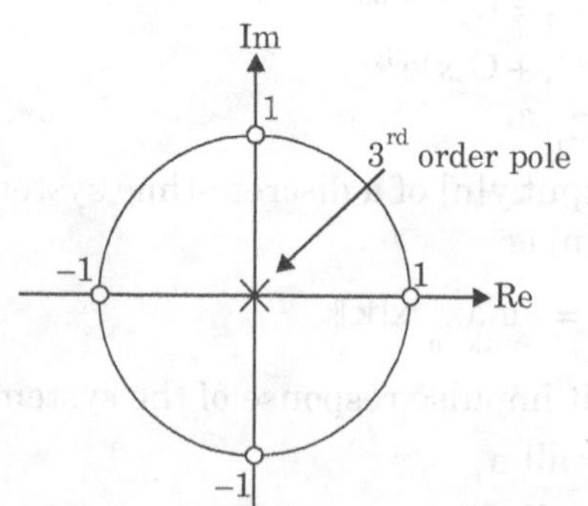

15. In the given circuit, the two-port network has the impedance matrix $Z = \begin{bmatrix} 40 & 60 \\ 60 & 120 \end{bmatrix}$. The value of Z_L for which maximum power is transferred to the load is ______ Ω.

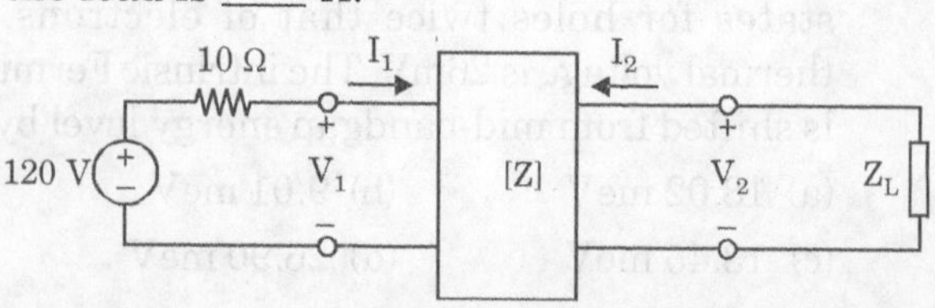

16. The current in the RL-circuit shown below is $i(t) = 10\cos(5t - \pi/4)$A.
The value of the inductor **(rounded off to two decimal places)** is ____ H.

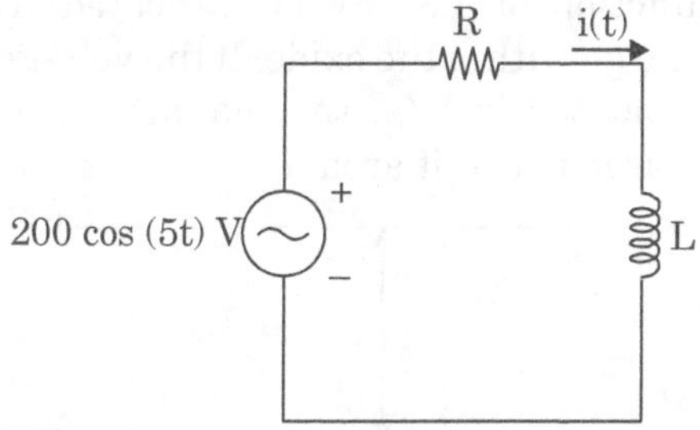

17. In the circuit shown below, all the components are ideal and the input voltage is sinusoidal. The magnitude of the steady-state output V_0 **(rounded off to two decimal places)** is ______ V.

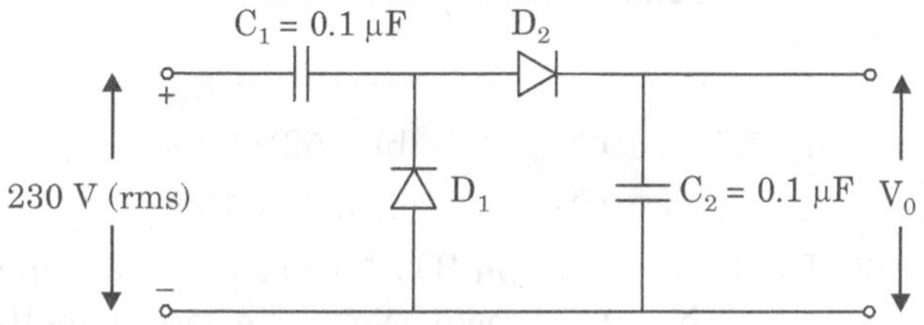

18. In the circuit shown below, all the components are ideal. If V_i is +2V, the current I_0 sourced by the op-amp is _____ mA.

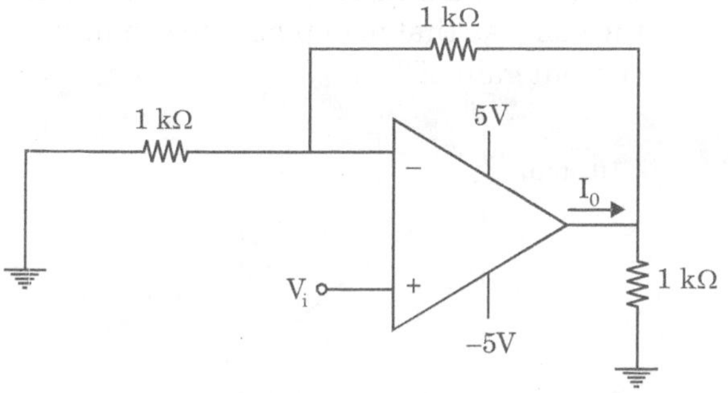

19. In an 8085 microprocessor, the number of address lines required to access a 16K byte memory bank is ______.

20. A 10-bit D/A converter is calibrated over the full range from 0 to 10V. If the input to the D/A converter is 13A (in hex), the output (**rounded off to three decimal places**) is ____ V.

21. A transmission line of length $3\lambda/4$ and having a characteristic impedance of 50Ω is terminated with a load of 400 Ω. The impedance (**rounded off to two decimal places**) seen at the input end of the transmission line is _____ Ω.

22. A binary random variable X takes the value +2 or –2. The probability $P(X = +2) = \alpha$. The value of α (**rounded off to one decimal place**), for which the entropy of X is maximum, is ____.

23. The loop transfer function of a negative feedback system is

$$G(s)H(s) = \frac{K(s+11)}{s(s+2)(s+8)}$$

The value of K, for which the system is marginally stable, is ______.

24. The random variable

$$Y = \int_{-\infty}^{\infty} W(t)\,\phi(t)dt$$

Where $\phi(t) = \begin{cases} 1; & 5 \le t \le 7 \\ 0; & \text{otherwise} \end{cases}$

and W(t) is a real white Gaussian noise process with two-sided power spectral density $S_W(f) = 3$ W/Hz, for all f. The variance of Y is ____.

25. The two sides of a fair coin are labelled as 0 and 1. The coin is tossed two times independently. Let M and N denote the labels corresponding to the outcomes of those tosses. For a random variable X, defined as X = min(M, N), the expected value E(X) (**rounded off to two decimal places**) is _____.

Q. 26 to Q. 55 : Carry Two Marks Each.

26. Consider the following system of linear equations.
$x_1 + 2x_2 = b_1$; $2x_1 + 4x_2 = b_2$; $3x_1 + 7x_2 = b_3$; $3x_1 + 9x_2 = b_4$
Which one of the following conditions ensures that a solution exists for the above system?
(a) $b_2 = 2b_1$ and $6b_1 - 3b_3 + b_4 = 0$
(b) $b_3 = 2b_1$ and $6b_1 - 3b_3 + b_4 = 0$
(c) $b_2 = 2b_1$ and $3b_1 - 6b_3 + b_4 = 0$
(d) $b_3 = 2b_1$ and $3b_1 - 6b_3 + b_4 = 0$

27. Which one of the following options contains two solutions of the differential equation $\frac{dy}{dx} = (y-1)x$?
(a) $\ln|y-1| = 0.5x^2 + C$ and $y = 1$
(b) $\ln|y-1| = 2x^2 + C$ and $y = 1$
(c) $\ln|y-1| = 0.5x^2 + C$ and $y = -1$
(d) $\ln|y-1| = 2x^2 + C$ and $y = -1$

28. The current I in the given network is

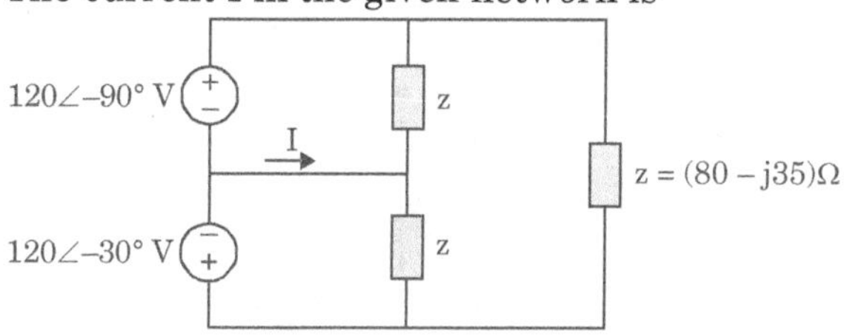

(a) 0A (b) 2.38 ∠–96.37° A
(c) 2.38 ∠143.63° A (d) 2.38 ∠– 23.63° A

29. A finite duration discrete time signals x[n] is obtained by sampling the continuous-time signal $x(t) = \cos(200\pi t)$ of sampling instants $t = \frac{n}{400}$, $n = 0, 1, 2, \ldots 7$. The 8-point discrete Fourier transform (DFT) of x[n] is defined as

$$X[k] = \sum_{n=0}^{7} x[n]e^{-j\frac{\pi kn}{4}}, \quad k = 0, 1, \ldots, 7$$

Which one of the following statements is TRUE?

(a) All X[k] are non-zero

(b) Only X[4] is non-zero

(c) Only X[2] and X[6] are non-zero

(d) Only X[3] and X[5] are non-zero

30. For the given circuit, which one of the following is the correct state equation?

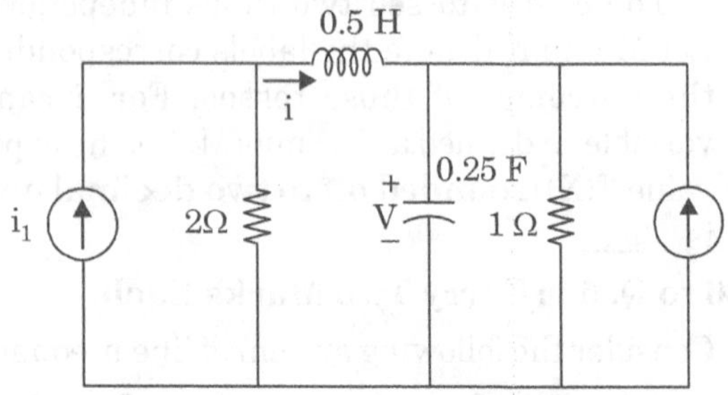

(a) $\frac{d}{dt}\begin{bmatrix} v \\ i \end{bmatrix} = \begin{bmatrix} -4 & 4 \\ -2 & -4 \end{bmatrix}\begin{bmatrix} v \\ i \end{bmatrix} + \begin{bmatrix} 0 & 4 \\ 4 & 0 \end{bmatrix}\begin{bmatrix} i_1 \\ i_2 \end{bmatrix}$

(b) $\frac{d}{dt}\begin{bmatrix} v \\ i \end{bmatrix} = \begin{bmatrix} -4 & -4 \\ -2 & 4 \end{bmatrix}\begin{bmatrix} v \\ i \end{bmatrix} + \begin{bmatrix} 4 & 4 \\ 4 & 0 \end{bmatrix}\begin{bmatrix} i_1 \\ i_2 \end{bmatrix}$

(c) $\frac{d}{dt}\begin{bmatrix} v \\ i \end{bmatrix} = \begin{bmatrix} 4 & -4 \\ -2 & -4 \end{bmatrix}\begin{bmatrix} v \\ i \end{bmatrix} + \begin{bmatrix} 0 & 4 \\ 4 & 4 \end{bmatrix}\begin{bmatrix} i_1 \\ i_2 \end{bmatrix}$

(d) $\frac{d}{dt}\begin{bmatrix} v \\ i \end{bmatrix} = \begin{bmatrix} -4 & -4 \\ -2 & -4 \end{bmatrix}\begin{bmatrix} v \\ i \end{bmatrix} + \begin{bmatrix} 4 & 0 \\ 0 & 4 \end{bmatrix}\begin{bmatrix} i_1 \\ i_2 \end{bmatrix}$

31. A one-sided abrupt pn junction diode has a depletion capacitance C_D of 50 pF at a reverse bias of 0.2V. The plot of $1/C_D^2$ versus the applied voltage V for this diode is a straight line as shown in the figure below. The slope of the plot is ______ $\times 10^{20}\ F^{-2}V^{-1}$.

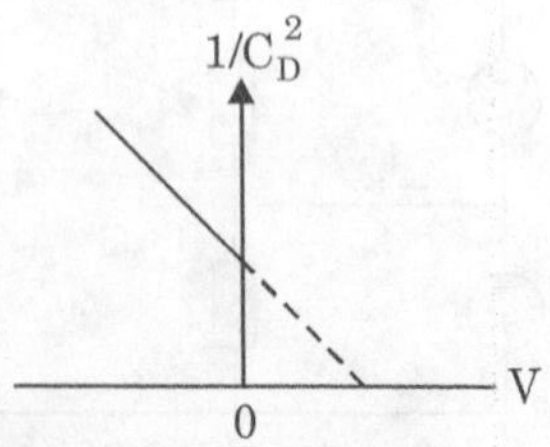

(a) − 5.7 (b) − 3.8

(c) − 1.2 (d) − 0.4

32. The band diagram of a p-type semiconductor with a band-gap of 1eV is shown. Using this semiconductor, a MOS capacitor having V_{TH} of − 0.16 V, C'_{ox} of 100 nF/cm² and a metal work function of 3.87 eV is fabricated. There is no charge within the oxide. If the voltage across the capacitor is V_{TH}, the magnitude of depletion charge per unit area (in C/cm²) is

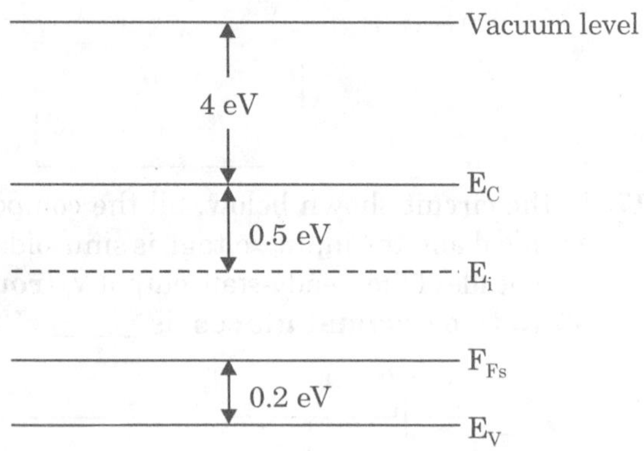

(a) 1.70×10^{-8} (b) 0.52×10^{-8}

(c) 1.41×10^{-8} (d) 0.93×10^{-8}

33. The base of an npn BJT T1 has a linear doping profile $N_B(x)$ as shown below. The base of another npn BJT T2 has a uniform doping N_B of 10^{17} cm^{-3}. All other parameters are identical for both the devices. Assuming that the hole density profile is the same as that of doping, the common-emitter current gain of T2 is

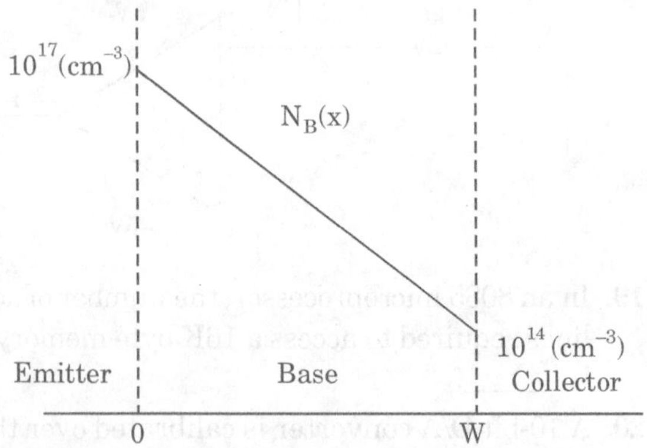

(a) approximately 2.0 times that of T1.

(b) approximately 0.3 times that of T1.

(c) approximately 2.5 times that of T1.

(d) approximately 0.7 times that of T1.

34. A pn junction solar cell of area 1.0 cm², illuminated uniformly with 100 mW cm^{-2}, has the following parameters: Efficiency = 15%, open circuit voltage = 0.7V, fill factor = 0.8, and thickness = 200 μm. The charge of an electron is 1.6×10^{-19} C. The average optical generation rate (in cm^{-3} s^{-1}) is

(a) 0.84×10^{19} (b) 5.57×10^{19}

(c) 1.04×10^{19} (d) 83.60×10^{19}

35. For the BJT in the amplifier shown below, V_{BE} = 0.7 V, kT/q = 26 mV. Assume that BJT output resistance (r_o) is very high and the base current is negligible. The capacitors are also assumed to be short circuited at signal frequencies. The input v_i is direct coupled. The low frequency voltage gain v_o/v_i of the amplifier is

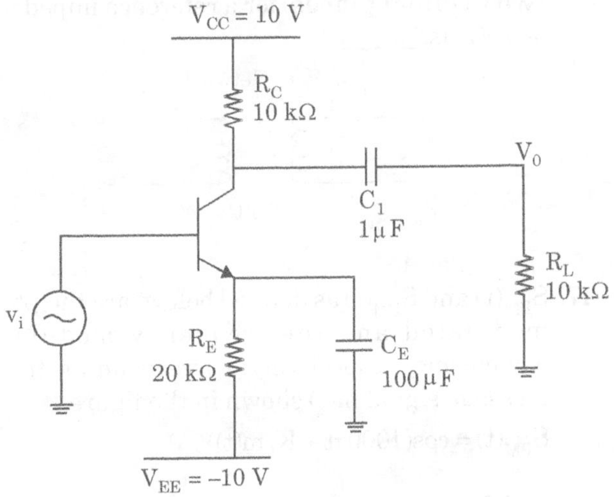

(a) – 89.42 (b) – 128.21

(c) – 178.85 (d) – 256.42

36. An ehhancement MOSFET of threshold voltage 3 V is being used in the sample and hold circuit given below. Assume that the substrate of the MOS device is connected to –10 V. If the input voltage V_1 lies between ±10 V, the minimum and the maximum values of V_G required for proper sampling and holding respectively, are

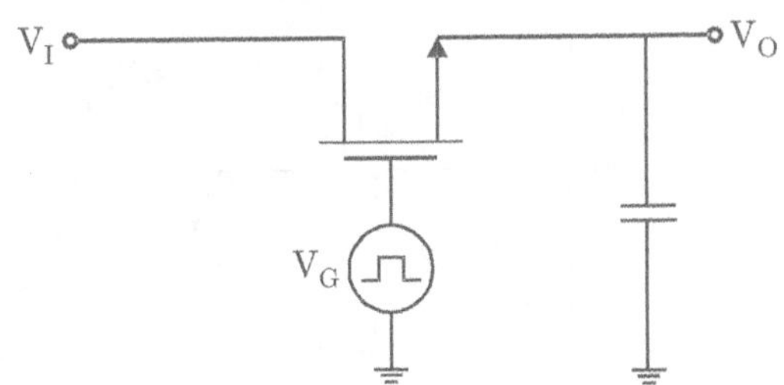

(a) 3 V and –3 V. (b) 10 V and –10 V.

(c) 13 V and –7V. (d) 10 V and –13 V.

37. Using the incremental low frequency small-signal model of the MOS device, the Norton equivalent resistance of the following circuit is

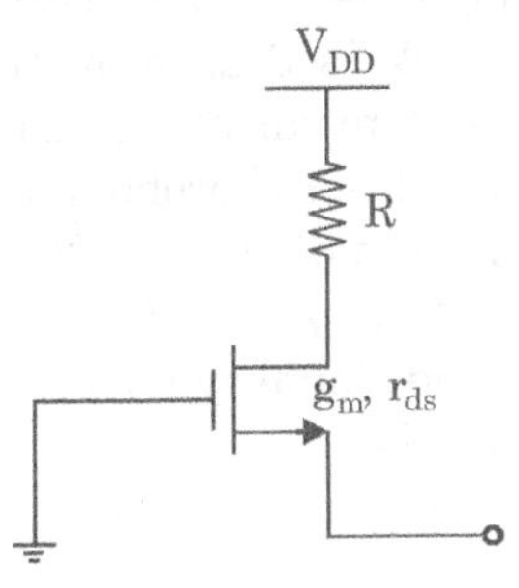

(a) $r_{ds} + R + g_m r_{ds} R$ (b) $\frac{r_{ds} + R}{1 + g_m r_{ds}}$

(c) $r_{ds} + \frac{1}{g_m} + R$ (d) $r_{ds} + R$

38. P, Q, and R are the decimal integers corresponding to the 4-bit binary number 1100 considered in signed magnitude, 1's complement and 2's complement representations, respectively. The 6-bit 2's complement representation of (P + Q + R) is

(a) 110101 (b) 110010

(c) 111101 (d) 111001

39. The state diagram of a sequence detector is shown below. State S_0 is the initial state of the sequence detector. If the output is 1, then

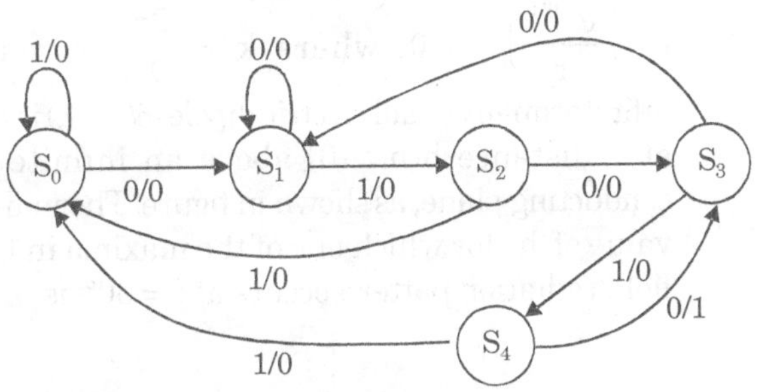

(b) the sequence 01010 is detected.

(d) the sequence 01011 is detected.

(a) the sequence 01110 is detected.

(c) the sequence 01001 is detected.

40. The characteristic equation of a system is

$$s^3 + 3s^2 + (K + 2)s + 3K = 0.$$

In the root locus plot for the given system, as K varies from 0 to ∞, the break-away or break-in point(s) lie within

(a) (–1, 0) (b) (– 2, – 1)

(c) (– 3, – 2) (d) (– ∞, – 3)

41. The components in the circuit given below are ideal. If R = 2 kΩ and C = 1 μF, the –3 dB cut-off frequency of the circuit in Hz is

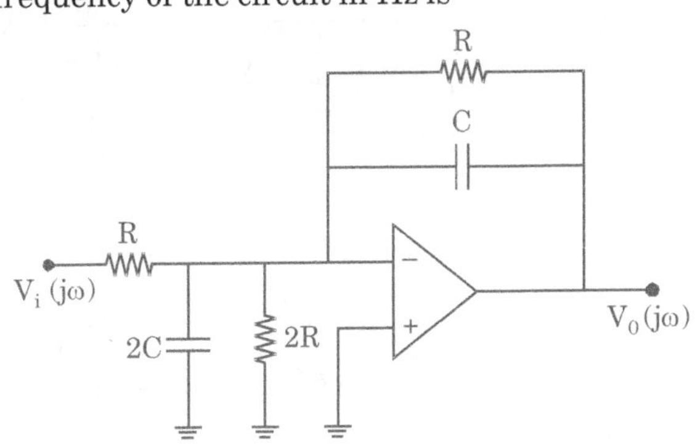

(a) 14.92 (b) 34.46

(c) 59.68 (d) 79.58

42. For the modulated signal $x(t) = m(t) \cos(2\pi f_c t)$, the message signal $m(t) = 4\cos(1000\pi t)$ and the carrier frequency f_c is 1 MHz. The signal x(t) is passed through a demodulator, as shown in the figure below. The output y(t) of the demodulator is

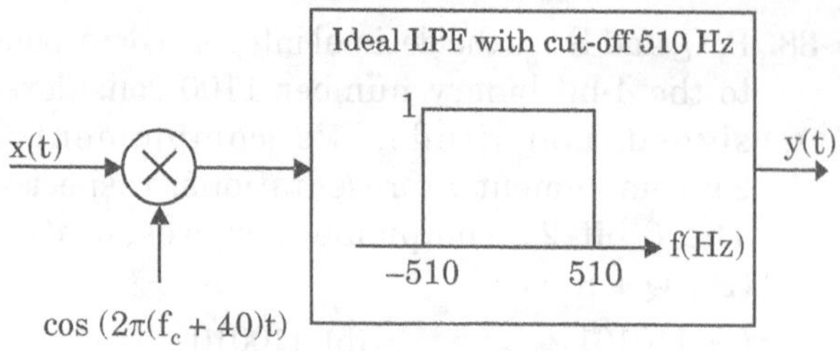

(a) $\cos(460\pi t)$ (b) $\cos(920\pi t)$

(c) $\cos(1000\pi t)$ (d) $\cos(540\pi t)$

43. For an infinitesimally small dipole in free space, the electric field E_θ in the far field is proportional to $\left(\frac{e^{-jkr}}{r}\right) \sin\theta$, where $k = \frac{2\pi}{\lambda}$. A vertical infinitesimally small electric dipole ($\delta l << \lambda$) is placed at a distance $h (h > 0)$ above an infinite ideal conducting plane, as shown in figure. The minimum value of 'h', for which one of the maxima in the far field radiation pattern occurs at $\theta = 60°$, is

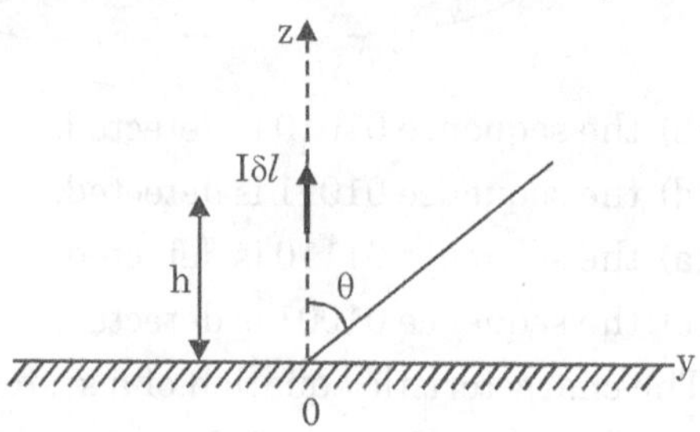

Infinite conducting plane

(a) λ (b) 0.5λ

(c) 0.25λ (d) 0.75λ

44. In the voltage regulator shown below, V_I is the unregulated input at 15 V. Assume $V_{BE} = 0.7$ V and the base current is negligible for both the BJTs. If the regulated output V_0 is 9 V, the value of R_2 is ______ Ω.

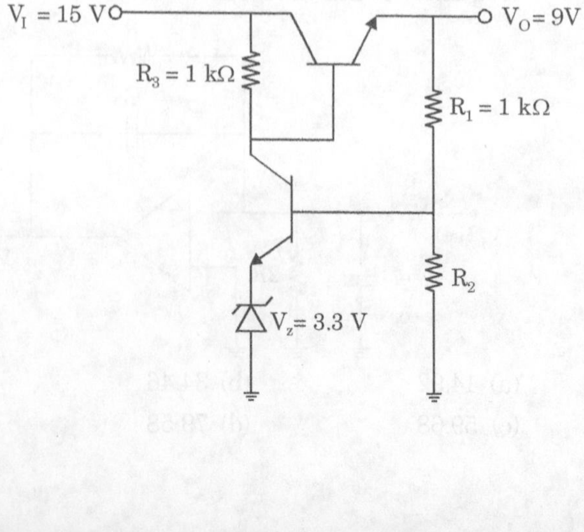

45. The magnetic field of a uniform plane wave in vacuum is given by

$$\vec{H}(x, y, z, t) = (\hat{a}_x + 2\hat{a}_y + b\hat{a}_z) \cos(\omega t + 3x - y - z)$$

The value of b is _____.

46. For a 2-port network consisting of an ideal lossless transformer, the parameter S_{21} **(rounded off to two decimal places)** for a reference impedance of 10Ω, is _______.

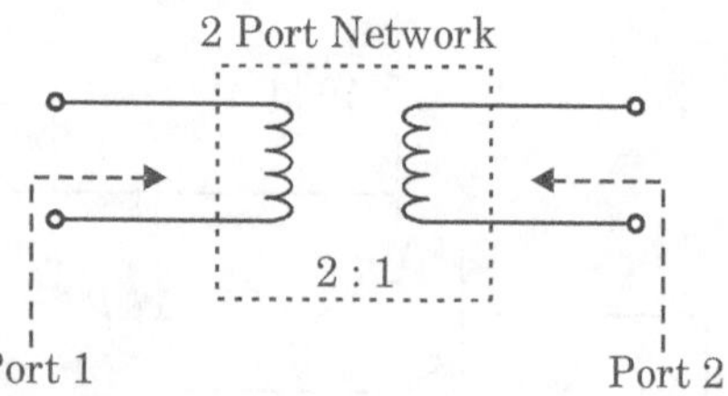

47. $S_{PM}(t)$ and $S_{FM}(t)$ as defined below, are the phase modulated and the frequency modulated waveforms, respectively, corresponding to the message signal m(t) shown in the figure.

$S_{PM}(t) = \cos(1000\pi t + K_P m(t))$

and $S_{PM}(t) = \cos\left(100\pi t + K_f \int_{-\infty}^{t} m(\tau) d\tau\right)$

where K_p is the phase deviation constant in radians/volt and K_f is the frequency deviation constant in radians/second/volt. If the highest instantaneous frequencies of $S_{PM}(t)$ and $S_{FM}(t)$ are same, then the value of the ratio $\frac{K_p}{K_f}$ is ___ seconds.

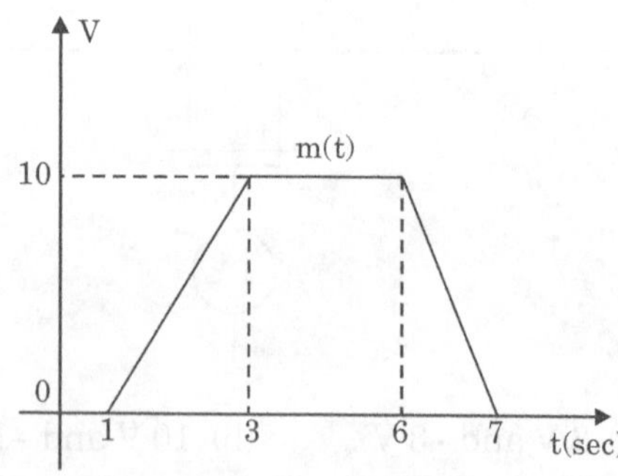

48. In a digital communication system, a symbol S randomly chosen from the set $\{s_1, s_2, s_3, s_4\}$ is transmitted. It is given that $s_1 = -3$, $s_2 = -1$, $s_3 = +1$ and $s_4 = +2$. The received symbol is $Y = S + W$. W is a zero-mean unit-variance Gaussian random variable and is independent of S. P_i is the conditional probability of symbol error for the maximum likelihood (ML) decoding when the transmitted symbol $S = s_i$. The index i for which the conditional symbol error probability P_i is the highest is _____.

49. A system with transfer function $G(s) = \dfrac{1}{(s+1)(s+a)}$, $a > 0$ is subjected to an input $5\cos 3t$. The steady state output of the system is $\dfrac{1}{\sqrt{10}}\cos(3t - 1.892)$. The value of a is _____.

50. For the components in sequential circuit shown below, t_{pd} is the propagation delay, t_{setup} is the setup time, and t_{hold} is the hold time. The maximum clock frequency (**rounded off to the nearest integer**), at which the given circuit can operate reliably is _____ MHz.

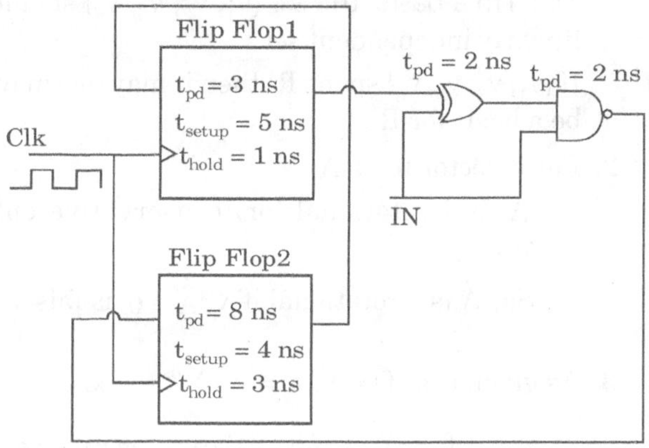

51. For the solid S shown below, the value of $\iiint_S x\,dx\,dy\,dz$ (**rounded off to two decimal places**) is

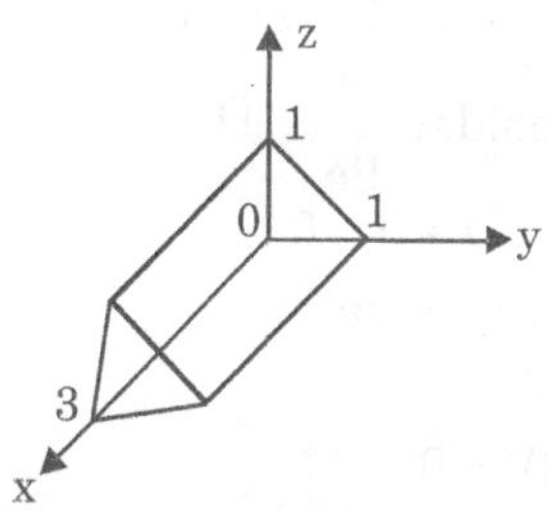

52. $X(\omega)$ is Fourier transform of x(t) shown below. The value of $\int_{-\infty}^{\infty} |X(\omega)|^2$ (**rounded off to two decimal places**) is _____.

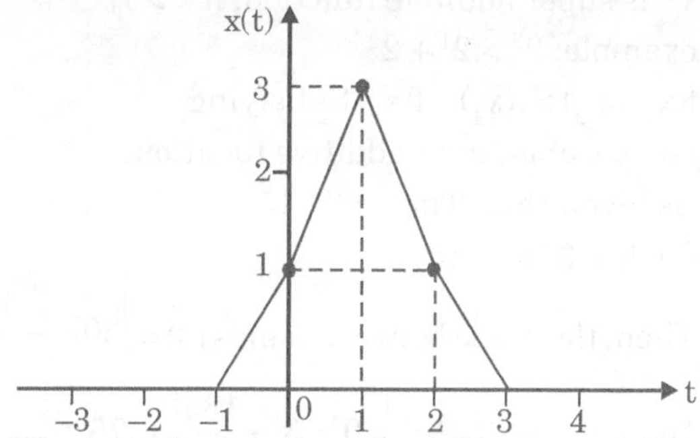

53. The transfer function of a stable discrete-time LTI system is $H(z) = \dfrac{K(z-\alpha)}{(z+0.5)}$, where K and α are real numbers. The value of α (**rounded off to one decimal place**) with $|\alpha| > 1$, for which the magnitude response of the system is constant over all frequencies, is _____.

54. X is a random variable with uniform probability density function in the interval [– 2, 10]. For $Y = 2X - 6$, the conditional probability $P(Y \le 7 \mid X \ge 5)$ (**rounded off to three decimal places**) is _____.

55. Consider the following closed loop control system

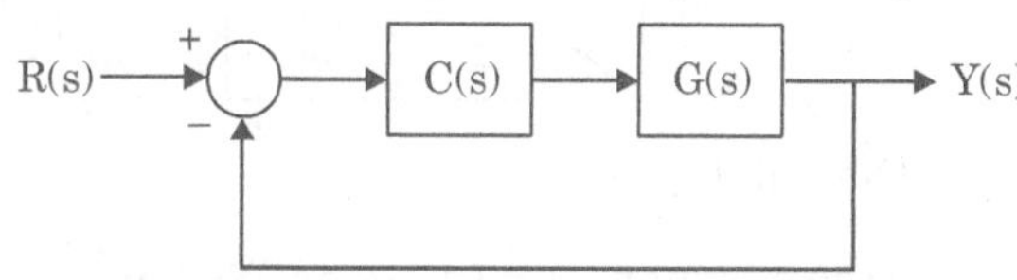

Where $G(s) = \dfrac{1}{s(s+1)}$ and $C(s) = K\dfrac{s+1}{s(s+3)}$

If the steady state error for a unit ramp input is 0.1, then the value of K is _____.

ANSWERS

General Aptitude (GA)

1. (a)	**2.** (b)	**3.** (b)	**4.** (c)	**5.** (a)	**6.** (a)
7. (b)	**8.** (c)	**9.** (c)	**10.** (b)		

Electronics & Communication Engineering

1. (c)	**2.** (c)	**3.** (b)	**4.** (c)	**5.** (c)	**6.** (b)
7. (b)	**8.** (d)	**9.** (c)	**10.** (a)	**11.** (b)	**12.** (d)
13. (a)	**14.** (a)	**15.** (48 to 48)	**16.** (2.80 to 2.85)	**17.** (644 to 657)	**18.** (6 to 6)
19. (14 to 14)	**20.** (3.050 to 3.080)	**21.** (6.25 to 6.25)	**22.** (0.5 to 0.5)	**23.** (160 to 160)	**24.** (6 to 6)
25. (0.25 to 0.25)	**26.** (a)	**27.** (a)	**28.** (c)	**29.** (c)	**30.** (a)
31. (a)	**32.** (a)	**33.** (a)	**34.** (a)	**35.** (a)	**36.** (c)
37. (b)	**38.** (a)	**39.** (a)	**40.** (a)	**41.** (d)	**42.** (b)
43. (a)	**44.** (800 to 800)	**45.** (1 to 1)	**46.** (0.8 to 0.8)	**47.** (2 to 2)	**48.** (3 to 3)
49. (4 to 4)	**50.** (76 to 78)	**51.** (2.25 to 2.25)	**52.** (58.50 to 58.80)	**53.** (–2 to –2)	**54.** (0.3 to 0.3)
55. (30 to 30)					

EXPLANATIONS

GENERAL APTITUDE

5. x^m is super additive function, if $x > 1$.

example: $2^3 > 2^1 + 2^2$

$f(x_1 + x_2) > f(x_1) + f(x_2)$ Satisfying

Hence, e^x is super additive fucntion.

7. As given that, Time: 03 : 15

So, h = 3, m = 15

Then, the angle between hands $(\theta) = \left|30h - \dfrac{11m}{2}\right|$

Hence, $\theta = \left|30 \times 3 - \dfrac{165}{2}\right| = 75°$

8. From given figure:

Area of shaded portion

= Area of circle – Area of rectangle

As we know that,

Area of the circle = πa^2 And

The maximum possible, area of rectangle inscribed in the circle, when it becomes square.

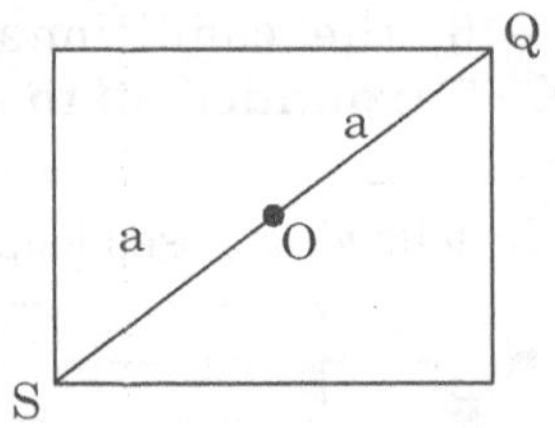

Diagonal = 2a

Area of square = $\dfrac{1}{2}$(diagonal)$^2 = 2a^2$

So, required shaded Area.

= Area of circle – Area of square = $\pi a^2 - 2a^2$

9. As given that, the quadratic equation:

$ax^2 - bx + c = 0$

According to question it has equal roots (β).

So, sum of roots = $\beta + \beta = \dfrac{b}{a} \Rightarrow 2\beta = \dfrac{b}{a}$ ...(1)

And, product of roots = $\beta * \beta = \dfrac{c}{a} \Rightarrow \beta^2 = \dfrac{c}{a}$ ...(2)

By multiplying (1) & (2) equation:

$\Rightarrow (2\beta)(\beta^2) = (b/a)(c/a)$

$(2b^3) = bc/a^2$

$b^3 = bc/(2a^2)$

10. Average number of students in school (P) in 5 years

$= \dfrac{3000 + 5000 + 5000 + 6000 + 4000}{5} = 4600$

Average number of students in school (Q) in 5 years

$= \dfrac{4000 + 7000 + 8000 + 7000 + 5000}{5} = 6200$

The difference of the number of students enrolled in school P and Q = (6200 – 4600) = 1600

The ratio of the average number of the student enrolled in school P to the average of the diffrence of the number of student enrolled in school P and Q then,

Required ratio $= \dfrac{4600}{1600} = \dfrac{23}{8} = 23 : 8$

ELECTRONICS & COMMUNICATION ENGINEERING

1. Option (d) is a wrong statement.

To form a basis, the set $\{v_1, v_3, v_5, v_6\}$ should be linearly independent set.

If $\{v_1, v_3, v_5, v_6\}$ spans R^4 then it may (or) may not be a basis for R^4.

2. For a vector field $\vec{A}$

$\vec{A}$ is irrotational (or) conservative only if $\nabla \times \vec{A} = 0$

So, $\vec{A}$ is irrotational if $\nabla^2\vec{A} = 0$ is false.

3. As given that $f(x, y, z) = e^{1-x\cos y} + xze^{\left(\frac{-1}{1+y^2}\right)}$

$$f_x = \frac{\partial f}{\partial x} = e^{1-x\cos y}(0 - \cos y) + z.e^{\frac{-1}{1+y^2}}$$

$$\therefore \left(\frac{\delta f}{\delta x}\right)_{(1,0),e} = e^{1-1}(-\cos 0) + (e).e^{\left(\frac{-1}{1+0}\right)}$$

$$= (-1 + 1) = 0$$

4. Let us Consider $\dfrac{d}{dx} = D$

$f(D) = D^2 - 6D + 9 = 0$

$\Rightarrow$ Auxiliary equation

$\Rightarrow$ $D^2 - 6D + 9 = 0$

$\Rightarrow$ $(D - 3)^2 = 0$

$\Rightarrow$ D = 3, 3 are real and equal roots

So, the general solution of given differential equation is $y = (C_1 + C_2x)e^{3x}$.

Hence, option (d) is correct.

5. The output of a discrete time system

$y(n) = \max |x(k)|; \quad -\infty \le k \le n$

The unit impulse response is

Now, apply $x(n) = \delta(n) \Rightarrow X(k) = \delta(k)$

So, $y(n) = \max|\delta(k)| = 1, n \ge 0 = u(n)$

6. As given that

$T = 300K, N_V = 2 N_C \; V_T = 26$ mV

As we know that,

$$E_F = \frac{E_C + E_V}{2} - \frac{kT}{2}\ln\left(\frac{N_C}{N_V}\right)$$

If $N_C = N_V$ then $E_{fi} = \dfrac{E_C + E_V}{2}$

$\Rightarrow$ Midband gap energy $(\because N_V = 2N_C)$

$$E_F = E_{fi} - \frac{kT}{2}\ln\left(\frac{N_C}{2N_C}\right)$$

$$(E_{fi} - E_F) = \frac{kT}{2}\ln\left(\frac{1}{2}\right) = \frac{0.026}{2}\ln\left(\frac{1}{2}\right)$$

$$= -9.01 \times 10^{-3} = -9.01 \text{ meV}$$

7. With all other parameters unchanges U_{max}, decreases if the intrinsic carrier density reduced is false because intrinsic carrier density will not affect generation and recombination rate. so, statement (c) is false.

8.

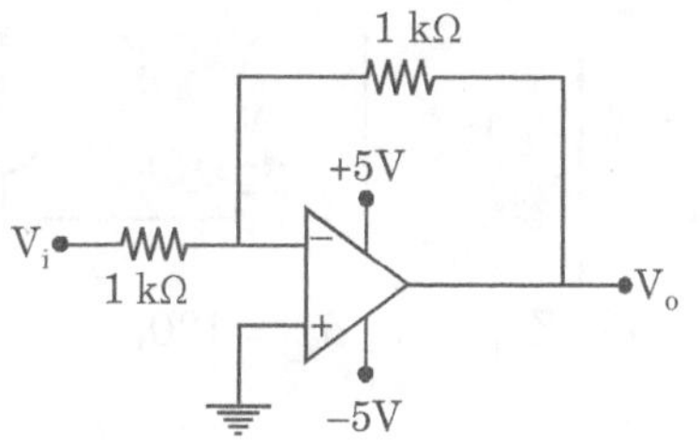

The given circuit is a schmittrigger of non-inverting type $V_0 = \pm 5V$

$$V^+ = \left[\frac{V_o \times 1 + V_i \times 1}{1+1}\right] = \left[\frac{V_o + V_i}{2}\right]$$

Consider $V_o = \pm 5V$, $V^+ = \left[\frac{\pm 5 + V_i}{2}\right]$

If $V_{in} > +5V$ then V_o changes from –5V to +5V

If $V_{in} < -5V$ then V_o changes from +5V to –5V

But the given input has maximum peak value $V_{in} = \pm 1V$, therefore the input cannot flip the output.

Hence, output stays or cannot change from +5V (or) – 5V

Transfer characteristics

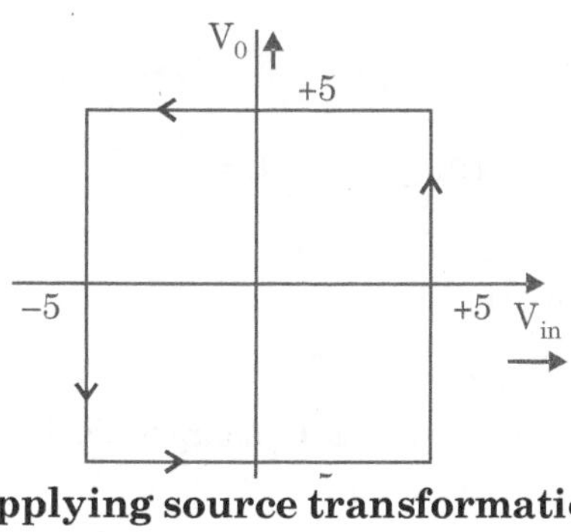

9. By Applying source transformation:

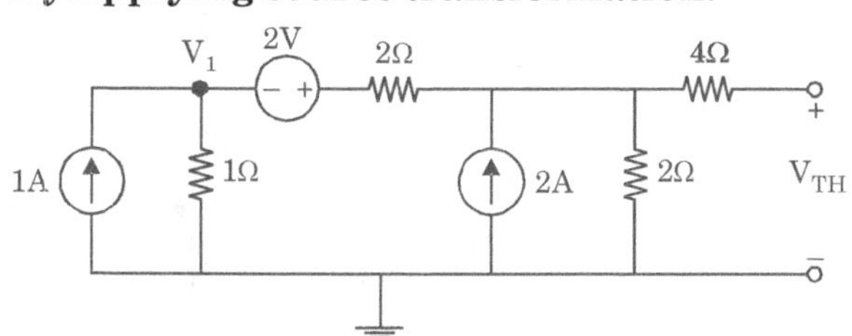

Now, By KCL:

$$-1 + \frac{V_1}{1} + \frac{(V_1 - V_{TH} + 2)}{2} = 0$$

$$3V_1 - V_{TH} = 0 \quad ...(1)$$

$$\frac{V_{TH}}{2} - 2 + \frac{(V_{TH} - V_1 - 2)}{2} = 0$$

$$-V_1 + 2V_{TH} = 6 \quad ...(2)$$

Now, equation (2) multiply by 3,

$$-3V + 6V_{TH} = 18 \quad ...(3)$$

And, By adding equation (1) with equation (3),

$$5V_{TH} = 18$$

$$V_{TH} = \frac{18}{5} = 3.6V$$

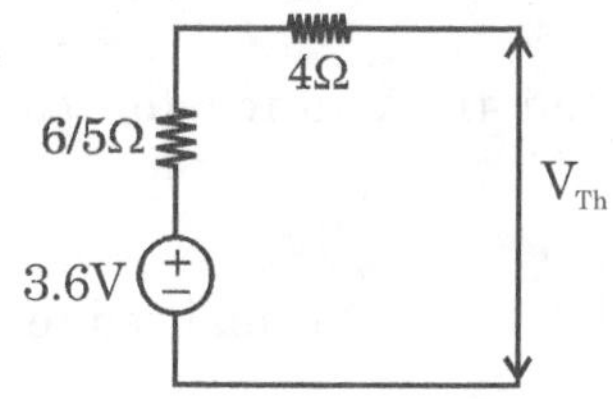

$V_{Th} = 3.6V$

10. If EN = 0 then the multiplexer is Enabled.

P \ QR	00	01	11	10
0	0	1	0	0
1	0	1	1	1

So, the output

$$F = \bar{P}\bar{Q}.R + P.\bar{Q}.R + P.Q.1 = \bar{Q}R + PQ.$$

11. Here, S-plane contour is encircling 2-poles and 3-zeros in clockwise direction. So, the corresponding G(s) plane contour encircle origin 2 times in anticlockwise direction and 3 times in clockwise direction.

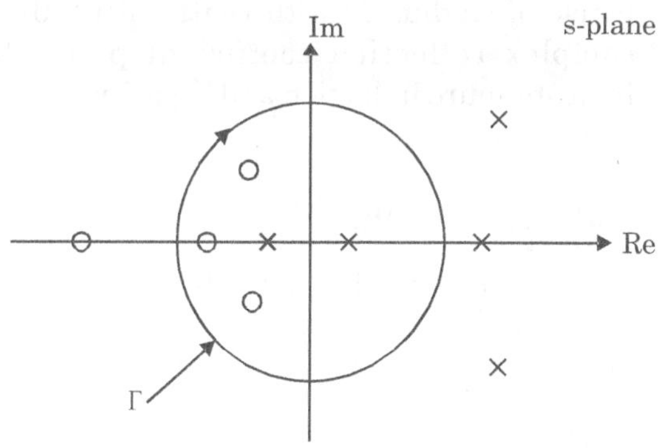

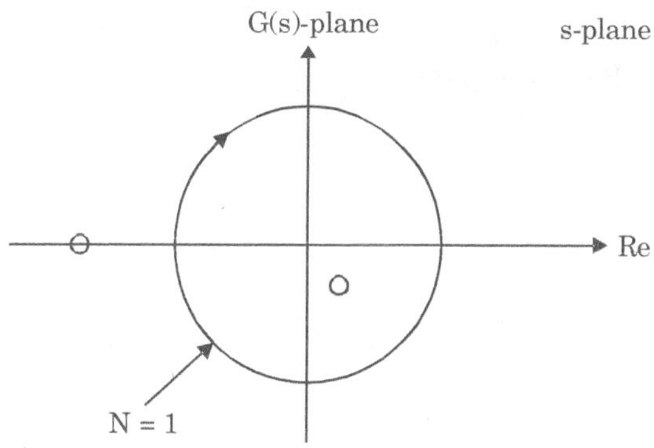

Effectively once encircles the origin in the clock wise direction.

12. The probability of error:

For; P(1 bit error) = α,

P(1 bit correct or Noerror) = $(1-\alpha)$

P(N bit error) = α^N,

P(N bit correct or No error in received block)

$$= (1-\alpha)^N$$

The received block declared erroneous if atleast one of bits decoding wrongly (i.e correct probability means all bits should be correct)

P(No error in received block is erroneous)

$$= 1-(1-\alpha)^N$$

13. The normalized impedance for the given impedance

$$\Rightarrow \left[\frac{Z}{Z_0}\right] = \left[\frac{jx}{Z_0}\right] \qquad [\because \text{Normalized resistance}]$$

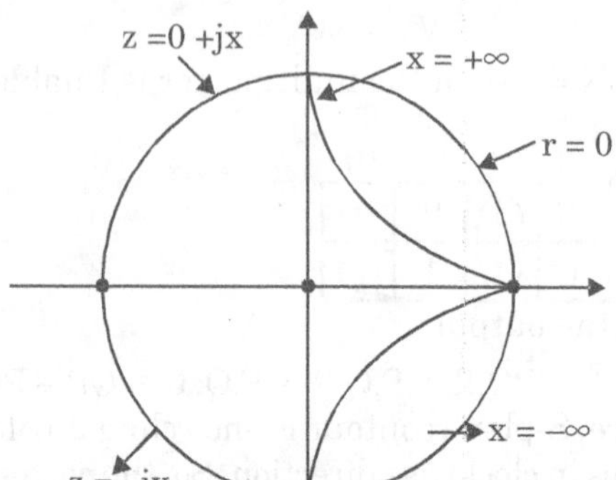

When r = 0 and x lies from $-\infty$ to ∞ is a unit Circle of (radius 1) with centre at (0, 0), on a complex reflection coefficient plane. That indicates pure induction and capacitive load.

14. As given that,

$$y(n) = \sum_{k=0}^{3} (-1)^k x(n-k)$$

$$y(n) = x(n) - x(n-1) + x(n-2) - x(n-3)$$

$$H(z) = \frac{Y(z)}{X(z)} = 1 - z^{-1} + z^{-2} + z^{-3}$$

$$= \frac{z^3 - z^2 + z - 1}{z^3} = \frac{(z-1)(z^2+1)}{z^3}$$

Pole zero plot:

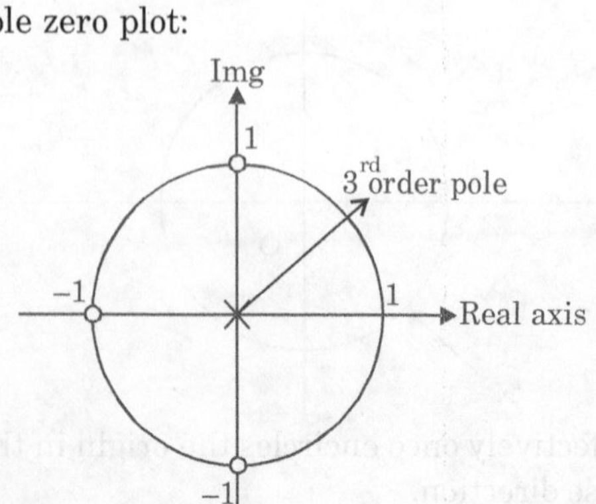

one zero at z = 1 & 2 zeros at z = ±j

3 poles at z = 0

Hence, option (c) is correct.

15. $V_1 = 40I_1 + 60I_2$ ...(1)

$V_2 = 60I_1 + 120I_2$ ...(2)

By compairing eq(i) and (ii) by standard Z-transform equation

$V_1 = Z_{11}I_1 + Z_{12}I_2$

$V_2 = Z_{21}I_1 + Z_{22}I_2$

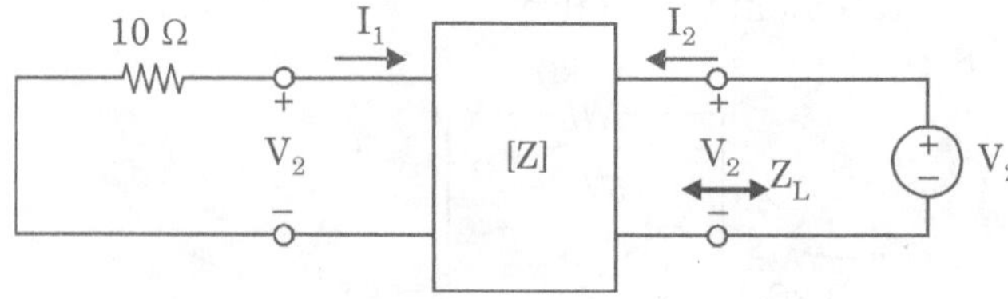

$Z_{11} = 40, Z_{12} = 60, Z_{21} = 60, Z_{22} = 120,$

$R_S = 10\Omega$(given)

From maximum power transfer theorem:-

$$Z_L = Z_{th} = Z_{22} - \left[\frac{Z_{12} \times Z_{21}}{R_S + Z_{11}}\right] = 120 - \left[\frac{60 \times 60}{10+40}\right]$$

$Z_L = 48\Omega$

16. As given that

$\bar{V} = 200\angle 0°$

$\bar{I} = 10\angle -45°$

$$Z = \frac{V}{I} = \frac{200\sqrt{0°}}{10\sqrt{-45°}} = 20\angle + 45°$$

$$Z = \left(10\sqrt{2} + j10\sqrt{2}\right)$$

$$Z = 10\sqrt{2} + j\,10\sqrt{2}$$

$$\downarrow$$

$$X_L$$

$$X_L = \omega_L = 10\sqrt{2} \Rightarrow L = \frac{10\sqrt{2}}{\omega} = \frac{10\sqrt{2}}{5}$$

So, L = 2.828H

17. As given that

$V_{in} = 230 \times \sqrt{2} = V_m$

During negative cycle C_1 charges [D_1 forward biased]

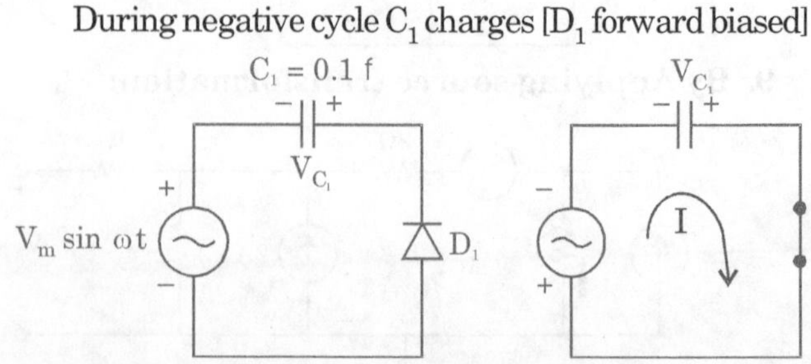

By KVL, $-V_m + V_{C_1} = 0$

$$V_{C_1} = V_m$$

During positive cycle D_2 forward biased

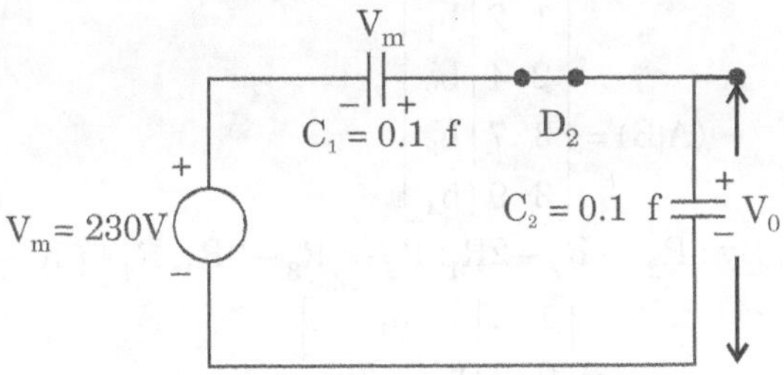

By KVL; $-V_m - V_m + V_o = 0$

$\because$ Voltage doubles $(V_o) = 2Vm$

$$= 2 \times 230 \times \sqrt{2} = 650.440V$$

$$V_0 = 650.44V$$

18.

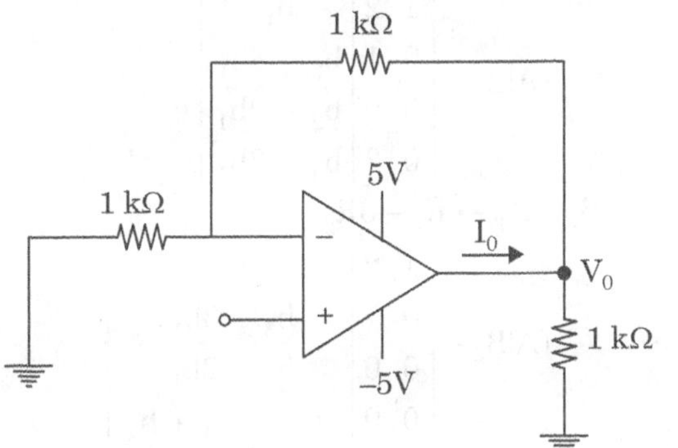

Given op-amp is in non-inverting mode, then output

$$V_0 = \left(1 + \frac{R_f}{R_i}\right) V_{in} = \left(1 + \frac{1k}{1k}\right) \times 2V = 4V$$

By appling KCL at node V_0

$$\frac{V_i - V_0}{1k\Omega} + \frac{0 - V_0}{2k\Omega} + I_0 = 0$$

$$\frac{2-4}{1k\Omega} + \frac{0-4}{1k\Omega} + I_0 = 0$$

$$-2 + I_0 - 4 = 0$$

$$I_o = 6mA$$

19. In 8085 microprocessor, the number of address lines(n) required to a access a 16KB number of memory locations(N) is $2^n = 16KB = 2^4 \times 2^{10}$

$$(\because 1KB = 2^{10})$$

$$\Rightarrow \quad 2^n = 2^{14}$$

Hence, the number of adder lines required are n = 14.

20. The Number of steps for digital input is $(13A)_{16}$

$$\Rightarrow \quad 1 \times 16^2 + 3 \times 16^1 + 10 \times 16^0 = 314$$

$$\text{step size} = \frac{10}{2^{10-1}} = \frac{10}{2^9}$$

As we know that

DAC output

= Number of steps for given digital input × step size

$$= 314 \times \frac{10}{2^9} = \frac{3140}{512} = 6.132 \text{ volts.}$$

21. As given that :

Length of tranmission line :

$$(\ell) = \frac{3\lambda}{4}$$

$$Z_o = 50\Omega$$

$$Z_R = 400\Omega \qquad \left(\because l = \frac{\lambda}{4}\right)$$

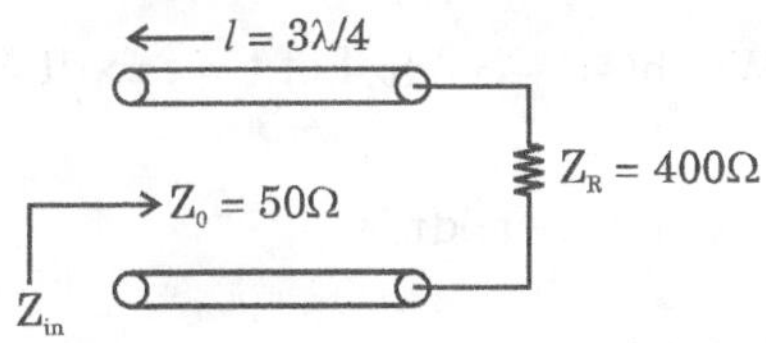

As length of the transmission line is $\frac{3\lambda}{4}$, (odd multiples of quarter wavelengths like $\frac{\lambda}{4}, \frac{3\lambda}{4}, \frac{5\lambda}{4}$....) and hence this line must be a quarter wave transmission line.

Assume this quarter wave transmission line is lossless then $Z_{in} = \frac{Z_o^2}{Z_R} = \frac{50^2}{400} = 6.25\Omega$

22. As given $p(x = 2) = \alpha$

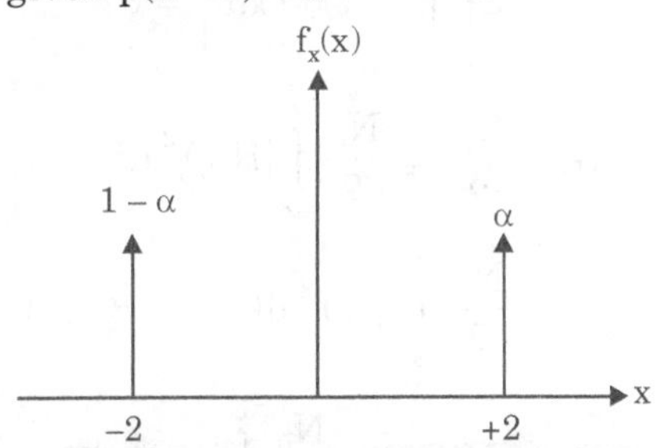

Entropy will be maximum when the provided probabilities are equal

i.e. $P(X = +2) = P(x = -2) = \alpha = 1/2 = 0.5$

23. Characteristic equation q(s) for the given open loop system will be: $1 + G(s)H(s) = 0$

$$q(s) = s(s + 2)(s + 8) + K(s + 11) = 0$$

$$q(s) = s^3 + 10s^2 + 16s + Ks + 11K = 0$$

$$q(s) = s^3 + 10s^2 + (16 + K)s + 11K = 0$$

Now, By using R–H criteria:

s^3	1	16 + K
s^2	10	11 K
s^1	$\frac{10(16+K) - 11K}{10}$,	0
s^0	11 K	

For system to be marginal stability,

$$\left(\frac{10(K + 16) - 11K}{10}\right) = 0$$

$$\text{or, } 10K + 160 - 11K = 0$$

Hence, K = 160

24. As given that

$$Y = \int_{-\infty}^{\infty} W(t)\phi(t)dt$$

where, $\phi(t) = \begin{cases} 1; 5 \le t \le 7, \\ 0; \text{otherwise} \end{cases}$

let $\phi(t) = h(-(t - \tau)) \Rightarrow$ where τ is the symbol duration

$$Y = \int_{-\infty}^{\infty} W(t)h(-(t-\tau))dt \quad \text{Let } t = \tau \Rightarrow dt = d\tau$$

$$Y = \int_{-\infty}^{\infty} W(\tau)h((\tau - t))d\tau$$

$$Y(t) = W(t) * h(t)$$

$$\Rightarrow \quad S_Y(f) = |H(f)|^2 s_w(f)$$

$$\delta_Y^2 = E[Y^2] - E[Y]^2$$

$$E[Y^2] = \int_{-\infty}^{\infty} S_Y(f)df = \int_{-\infty}^{\infty} |H(f)|^2 S_W(f)df$$

$$E[Y^2] = E\left[\int_{-\infty}^{\infty} W(t)\phi(t)\right]dt$$

$$= E\left[\int_{-\infty}^{\infty} W(t)\phi(t)dt\right] = 0$$

$$S_w(f) = \frac{N_0}{2} = \frac{N_0}{2}\int_{-\infty}^{\infty} |H(f)|^2 df$$

$$= \frac{N_0}{2}\int_{-\infty}^{\infty} |\phi(f)|^2 df \quad (\because \phi(t) = h(-(t-\tau)))$$

$$|\phi(f)^2| = |H(f)|2 = \frac{N_0}{2}\int_{-\infty}^{\infty} |\phi(t)|^2 dt$$

(from the Rayleigh's theorem)

$$E[y^2] = 3 \times 2 = 6, \quad (\because E_\phi = 2)$$

25. Coin tossed two times:

S = {(H, H), (H, T), (T, H), (T, T)}

M Outcomes		X = min (M, N)
0(T),	0(T)	0(T)
0(T),	1(H)	0(T)
1(H),	0(T)	0(T)
1(H),	1(H)	1(H)

P(X = 1) = 1/4, P(X = 0) = 3/4

As we know that,

$$E(X) = \sum_i x_i P(X = x_i) \quad (\because i = 0, 1)$$

So, $E(X) = 0 \times \frac{3}{4} + 1 \times \frac{1}{4} = 0.25$

26. As given that

$X_1 + 2X_2 = b_1;$

$2X_1 + 42X_2 = b_2;$

$3X_1 + 7X_2 = b_3;$

$3X_1 + 9X_2 = b_4;$

So, Consider augmented matrix

$$(A|B) = \left[\begin{array}{cc|c} 1 & 2 & b_1 \\ 2 & 4 & b_2 \\ 3 & 7 & b_3 \\ 3 & 9 & b_4 \end{array}\right]$$

By : $R_2 \to R_2 - 2R_1; R_3 \to R_3 - 3R_1; R_4 \to R_4 - 3R_1$

$$(A|B) \sim \left[\begin{array}{cc|c} 1 & 2 & b_1 \\ 0 & 0 & b_2 - 2b_1 \\ 0 & 1 & b_3 - 3b_1 \\ 0 & 3 & b_4 - 3b_1 \end{array}\right]$$

By : $R_2 \leftrightarrow R_3$

$$(A|B) \sim \left[\begin{array}{cc|c} 1 & 2 & b_1 \\ 0 & 1 & b_3 - 3b_1 \\ 0 & 0 & b_2 - 2b_1 \\ 0 & 3 & b_4 - 3b_1 \end{array}\right]$$

By : $R_4 \to R_4 - 3R_2$

$$(A|B) \sim \left[\begin{array}{cc|c} 1 & 2 & b_1 \\ 0 & 1 & b_3 - 3b_1 \\ 0 & 0 & b_2 - 2b_1 \\ 0 & 0 & 6b_1 - 3b_1 + b_4 \end{array}\right]$$

Here, $\rho(A) = 2$

To have a solution, $\rho(A) = 2 = \rho(A|B)$

if $b_2 - 2b_1 = 0$ and $6b_1 - 3b_3 + b_4 = 0$ then

$\rho(A|B) = 2$

$\therefore$ solution exists if $b_2 = 2b_1$

and $6b_1 - 3b_3 + b_4 = 0$

27. As given that

$$\frac{dy}{dx} = (y - 1)x$$

By variable separable method:

$$\Rightarrow \int \frac{1}{y-1}dy = \int x\,dx + c$$

$$\Rightarrow \log|(y-1)| = \frac{x^2}{2} + C \quad (\text{where, } y \ne 1)$$

$$\left(\because \int \frac{f'(x)dx}{f(x)}dx = \log|f(x)| + C\right)$$

So, $\log|y - 1| = (0.5)x^2 + C$ for $y \ne 1$ & $y = -1$ is a general solution.

Hence, option (a) is correct.

28. By applying super-position theorem

Step-I

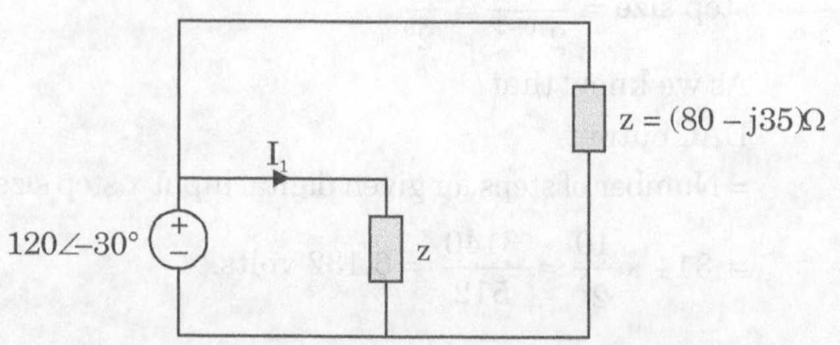

$$I_1 = -\frac{[120\angle -30°]}{Z} = \frac{-[120\angle -30°]}{80 - j35}$$

Step-II

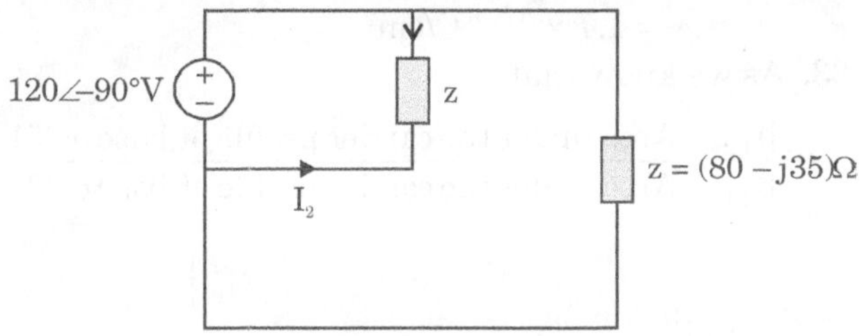

$$I_2 = -\frac{[120\angle -90°]}{Z} = -\frac{[-j120]}{80 - j35} = \frac{j120}{80 - j35}$$

$$I = I_1 + I_2 = \frac{1}{(80 - j35)}[-103.92 + j60 + j120]$$

$$I = \frac{(-103.92 + j180)}{(80 - j35)} = \left[\frac{207.34\angle 120°}{87.32\angle 23.63°}\right]$$

$$I = 2.380\angle 143.63° A$$

29. As given that

$$x(t) = \cos(200\pi t)$$

$$t = \frac{n}{400}$$

$$x(n) = \cos\left(\frac{200\pi n}{400}\right) \Rightarrow x(n) = \cos\left(\frac{\pi n}{2}\right)$$

$$(\because n = 0, 1, ..., 7)$$

$$x(n) = \{1, 0, -1, 0, 1, 0, -1, 0\}$$

When we look at x(n), it contains only even samples.

Let us consider $z(n) = \{1 -1\ 1 -1\} \xleftrightarrow{DFT} Z(k)$

$$Z(k) = Wn|_{(n=4)} Z(n) = \begin{bmatrix} Z(0) \\ Z(1) \\ Z(2) \\ Z(3) \end{bmatrix}$$

$$= \begin{bmatrix} 1 & 1 & 1 & 1 \\ 1 & -j & 1 & j \\ 1 & -1 & 1 & -1 \\ 1 & j & -1 & -j \end{bmatrix} \begin{bmatrix} 1 \\ -1 \\ 1 \\ -1 \end{bmatrix} = \begin{bmatrix} 0 \\ 0 \\ 4 \\ 0 \end{bmatrix}$$

N = 4

We know zero interpolation in time domain corresponds to replication of DFT spectrum

$$x(n) = z\left(\frac{n}{2}\right) \xleftrightarrow{DFT} X(k) = \{Z(k), Z(k)\}$$

$$= \{0,0,4,0,0,0,4,0\}$$

(or)

$$\cos\left(\frac{2\pi}{N}k_0 n\right) \leftrightarrow \frac{N}{2}[\delta(k - k_0) + \delta(k + k_0)]$$

$$\cos\left(\frac{n\pi}{2}\right) = \cos\left(\frac{2\pi}{8}2n\right) \leftrightarrow \frac{8}{2}[\delta(k-2) + \delta(k+2)]$$

So, non zero samples are –2, 2. From periodicity property non zero samples are –2 + 8, 2 = 6, 2.

30. From source transformation:

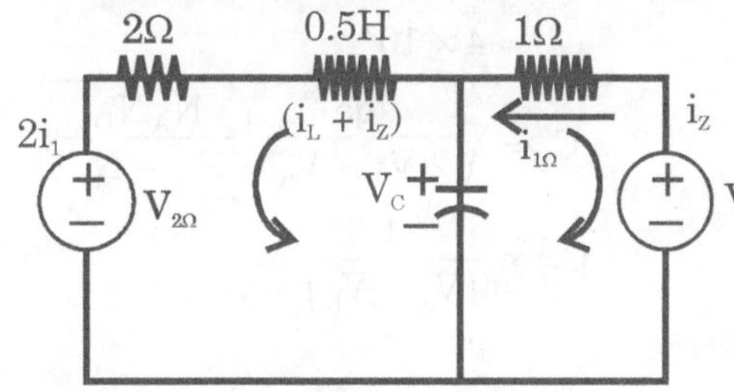

By KCL,

$$i_L + i_Z = C\frac{dV_C}{dt} + i_{1\Omega} \quad ...(1)$$

By KVL,

$$-V_{2\Omega} + L\frac{di_L}{dt} + V_C = 0 \quad ...(2)$$

Now, voltage across 2Ω, current theory 1Ω

$$i_1 = \frac{V_{2\Omega}}{2} + i_L \Rightarrow V_{2\Omega} = 2[i_1 - 1_L] \quad ...(3)$$

$$1.\ i_{1\Omega} - V_C = 0 \Rightarrow i_{1\Omega} = V_C \quad ...(4)$$

$$i_L + i_2 = \frac{1}{4}\left[\frac{dV_C}{dt}\right] + V_C$$

$$-[2(i_1 - i_L)] + \frac{1}{2}\frac{di_L}{dt} + V_C = 0$$

$$\frac{1}{4}\frac{dV_C}{dt} = -V_C + i_L + i_2$$

$$\frac{1}{2}\frac{di_L}{dt} = -V_C + i_L + i_2$$

$$\frac{dV_C}{dt} = -4V_C + 4i_L + 4i_2$$

$$\frac{di_L}{dt} = -2V_C - 4i_L + 4i_1$$

$$\begin{bmatrix} \frac{dV_C}{dt} \\ \frac{di_L}{dt} \end{bmatrix} = \frac{d}{dt}\begin{bmatrix} V_c \\ i_L \end{bmatrix} = \begin{bmatrix} -4 & 4 \\ -2 & -4 \end{bmatrix}\begin{bmatrix} V_C \\ i_L \end{bmatrix} + \begin{bmatrix} 0 & 4 \\ 4 & 0 \end{bmatrix}\begin{bmatrix} i_1 \\ i_2 \end{bmatrix}$$

31. If we assume $V_0 = 0.5$ V

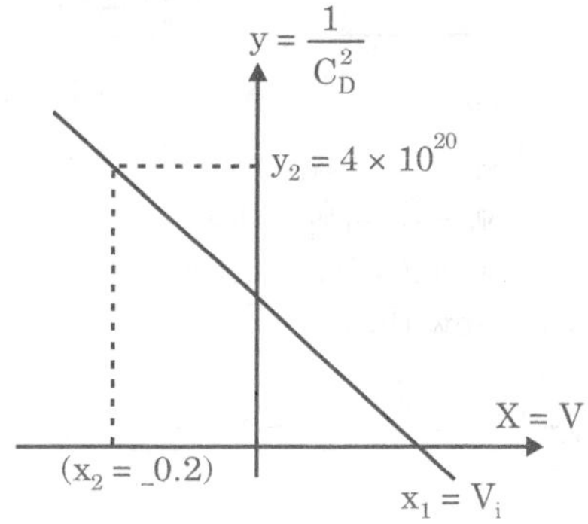

$$y_2 = \frac{1}{C_D^2}$$

$$= \left[\frac{1}{(50\times10^{-12})^2}\right]$$

$$= 4 \times 10^{20}$$

$$\because \quad C_D = \sqrt{\frac{q\varepsilon}{2(V_0 + V_R)}\left(\frac{N_A N_D}{N_A + N_D}\right)}$$

$$C_D^2 \propto \frac{1}{(V_o + V_R)}$$

or

$$\frac{1}{C_D^2} \propto (V_0 + V_R)$$

$$k = \frac{\frac{1}{C_D^2}}{V_0 + V_R} = \text{slope}$$

$$= \frac{\left(\frac{1}{50\text{pF}}\right)^2}{0.5 + 0.2}$$

$$= \frac{1}{(50\text{pF})^2 \times 0.7}$$

$$= 5.71 \times 10^{20}\ F^{-2}\ V^{-1}$$

Hence, option (a) is correct.

32. For Mos Capacitance:

As given that

$$f_S = 4eV + 0.5eV + 0.3eV = 4.8eV$$

$$V_{TH} = -0.16V,\ C_{ox} = 100nF/cm^2$$

$$\phi_m = 3.87\ eV$$

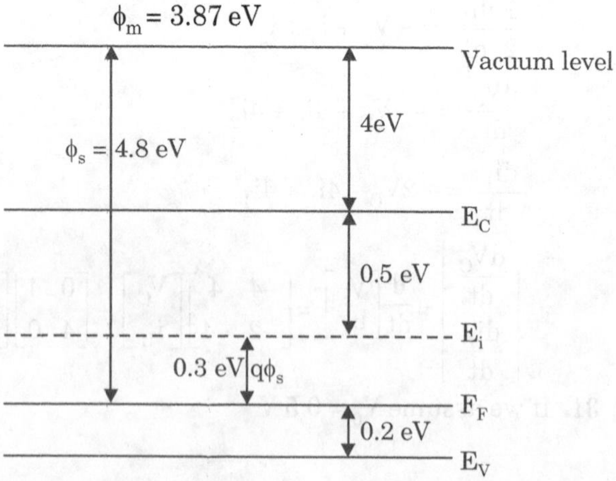

$$\therefore \quad V_{FB} = \phi_{ms} \qquad (\because \text{No oxide charge})$$

$$\therefore \quad V_{FB} = \phi_{ms} = \phi_m - \phi_s = 3.87 - 4.8 = -0.93$$

$$\phi_t = 2\,\phi_F,\ \phi_F = 0.3V$$

$$\phi_t = 2 \times 0.3 = 0.6V$$

As we know that,

$$V_T - V_{FB} = \frac{|Q_{dinv}|}{C_{ox}} + \phi_t$$

$$-0.16 - (-0.93) = \frac{|Q_{dinv}|}{100n} + 0.6$$

$$|Q_{dinv}| = 0.17 \times 100 \times 10^{-9} = 17 \times 10^{-9}$$

$$= 1.7 \times 10^{-8}\ C/cm^2$$

33. As we know that

$$\frac{\beta_{T2}}{\beta_{T1}} = \frac{\text{Area under the carrier profile of base of T1}}{\text{Area under the carrier profile of base of T2}}$$

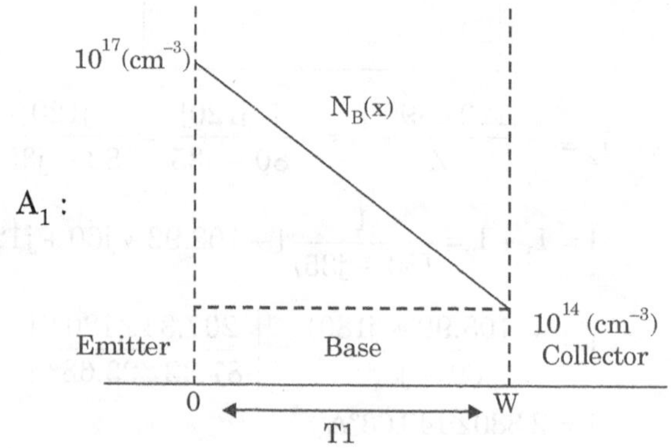

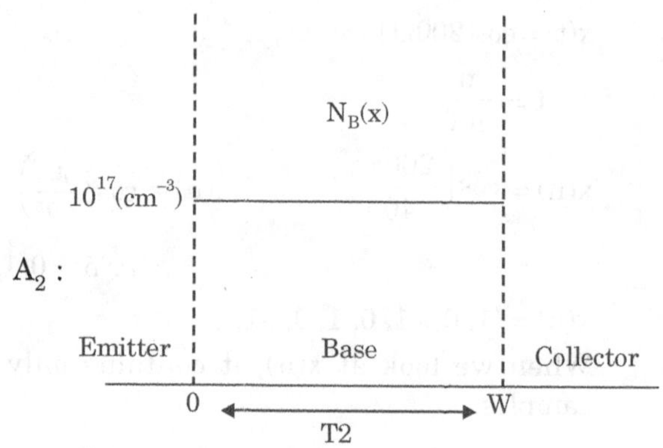

$$\frac{\beta_{T2}}{\beta_{T1}} = \frac{\int_0^W N_{A_2}(X)dx}{\int_0^W N_{A_1}(X)dx}$$

$$= \frac{10^{17} \times W}{\frac{1}{2}(10^{17} - 10^{14})W + 10^{14}W}$$

$$= \frac{10^{17}}{\frac{1}{2}(10^{17} - 10^{14}) + 10^{14}}$$

$$\Rightarrow \qquad \beta_{T2} = 2\beta_{T1}$$

34. As given that,

$$A = 1\ cm^2$$

$$P_{in}\ (/cm^2) = 100\ mW/cm^2$$

$$\eta = 0.15,\ V_{oc} = 0.7,\ \text{Fill factor}(F) = 0.8$$

$$\text{Thickness} = 200\ \mu m$$

$$\text{Power}\ (/cm^2) = \frac{100mW/cm^2}{200 \times 10^{-4}} = 5W/cm^3$$

As we know that,

$$\eta = \frac{V_m I_m}{P_{in}}, \; F = \frac{V_m I_m}{V_{OC} I_{SC}}$$

$$V_m I_m = \eta \, P_{in} = 0.15 \times 100\text{mw}$$

$$I_{SC} = \frac{V_m I_m}{V_{OC} \times F} = \frac{0.15 \times 100\,\text{mw}}{0.7 \times 0.8}$$

$$= \frac{15}{0.56}\ \text{mA}$$

$$G_L = \frac{I_{SC}}{q \times \text{Area} \times \text{thickness}}$$

$$= \frac{15 \times 10^{-3}}{0.56 \times 1.6 \times 10^{-19} \times 200 \times 10^{-4}}$$

$$= \frac{15 \times 10^{19}}{0.56 \times 32}$$

$$G_L = 0.84 \times 10^{19}/\text{cm}^3$$

Hence, Optical generation rate = $0.84 \times 10^{19}/\text{cm}^3$-s

35. DC Analysis: $I_E = \dfrac{10 - 0.7}{20} = 0.465\text{mA} = I_{C_{DC}}$

$$g_m = \frac{I_{C_{DC}}}{V_t} = \frac{0.465}{26}\ \text{A/V} = 0.0178$$

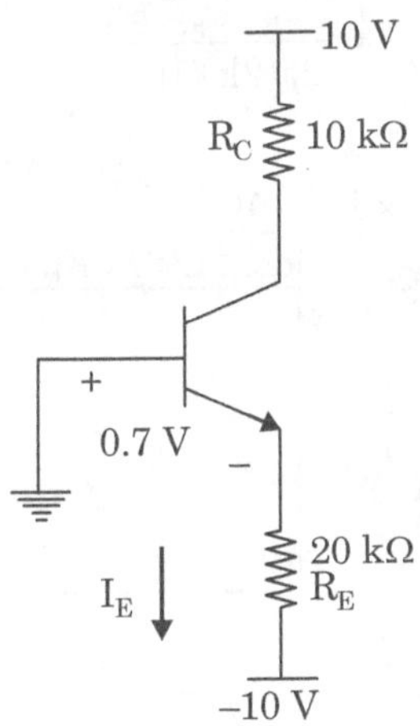

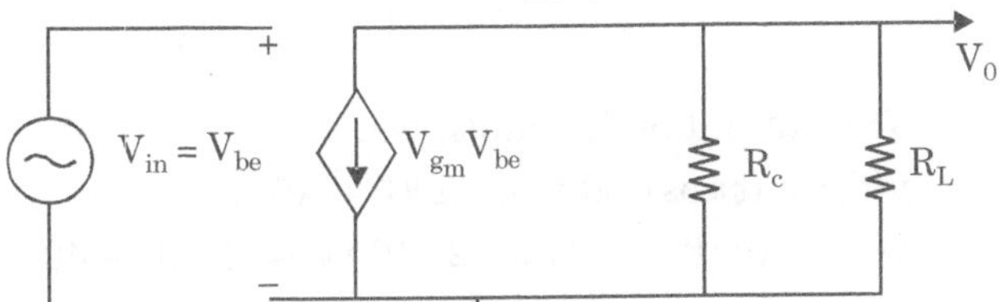

$$V_{out} = -g_m (R_C \parallel R_L)$$

$$V_{in} = V_{be}$$

$$\frac{V_{out}}{V_{in}} = \frac{-g_m (R_C \| R_L) V_{be}}{V_{be}}$$

$$\frac{V_{out}}{V_{in}} = -g_m (R_C \parallel R_L)$$

$$= -0.0178(5000) = -89.423$$

36.

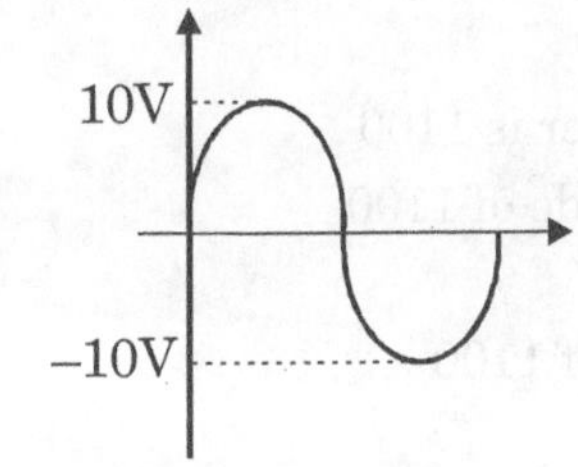

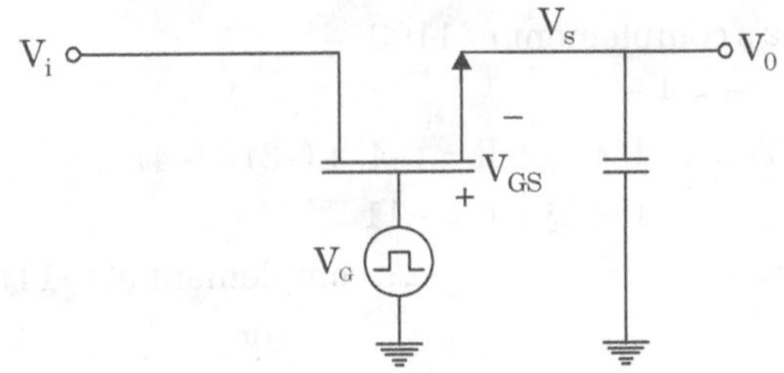

$$V_T = 3V$$

$$V_i = \pm 10V$$

During sampling (i.e when input ±10V) MOSFET must be ON:

when $V_i = 10V \;\Rightarrow\; V_G > 10V + 3V$

[when, $V_{i\,max} = 10V$]

$$V_G > 13\ V$$

For holding MOSFET should be off

when $V_i = -10V$

$\Rightarrow \quad V_G > V_i + V_T$ [when $V_{i\,min} = -10V$]

$$V_G > -10 + 3V$$

$$V_G > -7V$$

37.

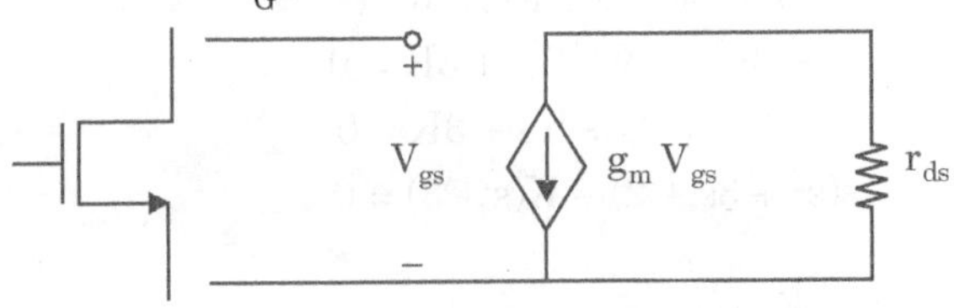

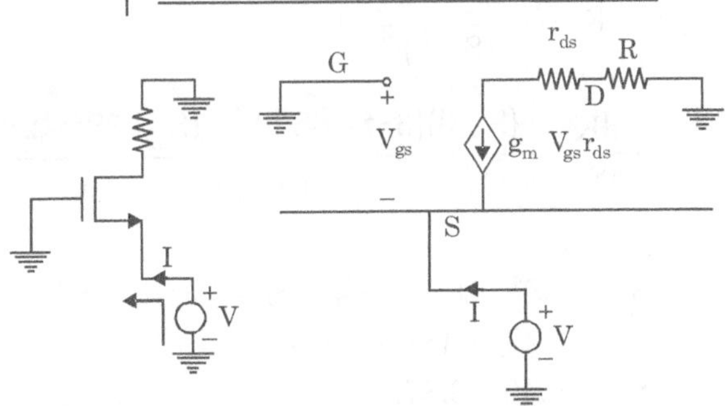

By Applying KVL:

$$V_{gs} + V = 0 \;\Rightarrow\; V = -V_{gs}$$

$$I_x = \frac{V_x}{r_{ds} + R} = \frac{-g_m r_{ds} V_{gs} + V}{r_{ds} + R}$$

$$I_x = \frac{g_m r_{ds} V + V}{r_{ds} + R}$$

$$= (r_{ds} + R) I_x = V(1 + g_m r_{ds})$$

$$R_N = \frac{V_x}{I_x} = \frac{r_{ds} + R}{1 + g_m r_{ds}}$$

38. As given that:

the binary number is 1100

So, sign magnitude of 1100

$= -4 = P$

1's complement of 1100

$= -3 = Q$

2's complement of 1100

$= -4 = R$

Thus, $P + Q + R = (-4) + (-3) + (-4)$

$P + Q + R = -11$

= 2's complement of (+11)

or

= 2's complement of (001011)

= (110101)

Hence, P + Q + R in 6-digit 2's complement representation is 110101.

39. As Given that S_0 is initial state of the sequence detector

If the output to be 1, the sequence of states traversed are

$S_0,\ S_1,\ S_2,\ S_3,\ S_4,\ S_3$

So, the sequence detected is → 0 1 0 1 0

40. As given that, the characteristic Equation of a system is:

$Q(s) = 1 + G(s).H(s) = 0$

$s^3 + 3s^2 + (K + 2)s + 3K = 0$

$s^3 + 3s^2 + 2s + Ks + 3K = 0$

$s(s^2 + 3s + 2) + K(s + 3) = 0$

$$\left[-K = \frac{S^3 + 3S^2 + 2S}{(S+3)}\right]$$

$$\frac{-dK}{dS} = \left[\frac{(S+3)(3S^2 + 6S + 2) - (S^3 + 3S^2 + 2S)}{(S+3)^2}\right] = 0$$

$3S^3 + 6S^2 + 2S + 9S^2 + 18S + 6 - S^3 - 3S^2 - 2S = 0$

$2S^3 + 12S^2 + 18S + 6 = 0$

$S = (-0.46, -3.87, -1.65)$

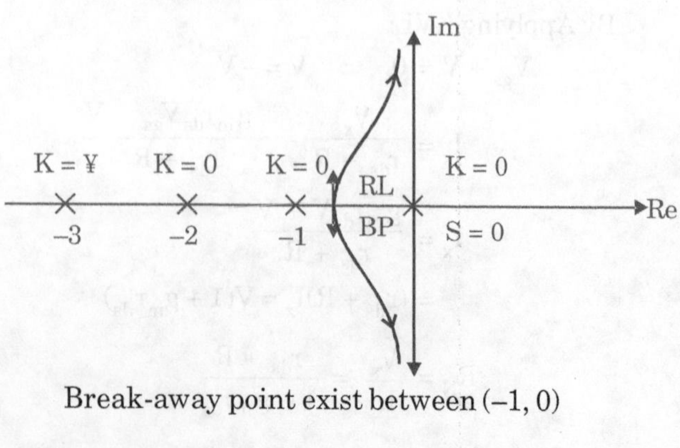

Break-away point exist between (–1, 0)

41.

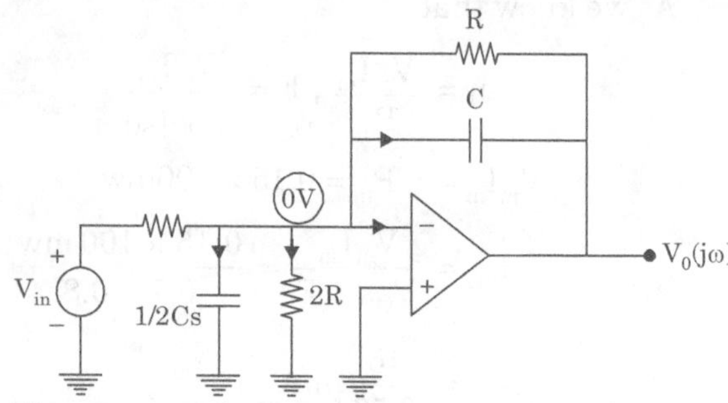

If no current flows through 2C and 2R in the given above Ckt.

Now, By using KCL

$$\frac{V_{in} - 0}{R} = \frac{0-0}{1/2sC} + \frac{0-0}{2R} + 0 + \frac{0 - V_o}{1/sC} + \frac{0 - V_o}{R}$$

$$\frac{V_{in}}{R} = -V_o\left[\frac{1}{R} + sC\right]$$

$$\left|\frac{V_o}{V_{in}}\right| = \frac{1}{1 + sCR} = \frac{1}{1 + s\tau}$$

$$\omega_{3dB} = \frac{1}{\tau} = \frac{1}{RC}$$

So, op-amp active filter (LPF) inverting type 3dB cut-off frequency.

$$f_{3dB} = \frac{1}{2\pi RC} = \frac{1}{2\pi(2k)(1\mu)}$$

$$= \frac{1}{2\pi \times 2 \times 10^3 \times 10^{-6}} = 79.58 \text{ Hz}$$

42.

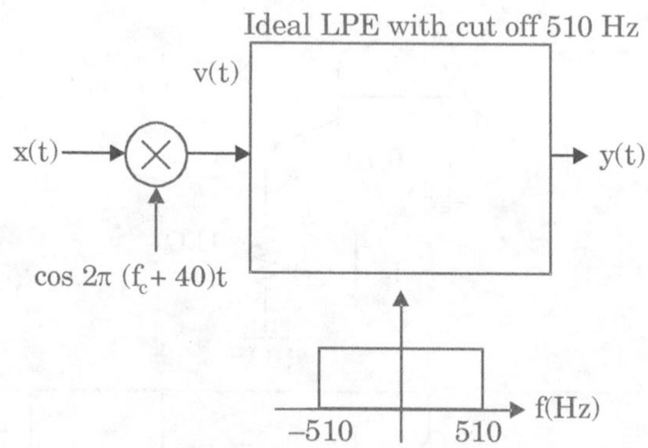

The out put of the multiplex:

$v(t) = m(t)\cos(2\pi f_c t)\cos(2\pi(f_c + 40)t)$

$v(t) = 4\cos(2\pi 500t)\cos(2\pi.10^6 t)\cos(2\pi(10^6 + 40)t)$

$v(t) = 4\cos(2\pi 500t)\frac{1}{2}\cos(4\pi 10^6 + 2\pi 40)t + \cos(2\pi 40t)$

$v(t) = 2\cos(2\pi.500t)[\cos(2\pi.40t) + \cos(2\pi(2.10^6 + 40)t]$

$v(t) = \cos(2\pi.540t) + \cos(2\pi.460t) + \cos(2\pi(2.10^6 + 540)t) + \cos(2\pi(2.10^6 + 460t))$

So, output of low pass filter:

$y(t) = \cos(2\pi 460t)$

$y(t) = \cos(920\pi t)$

43. Consider an infinitesimal electric dipole ($dl << \lambda$), which is vertically placed at height 'h' above an infinite grounded conductor.

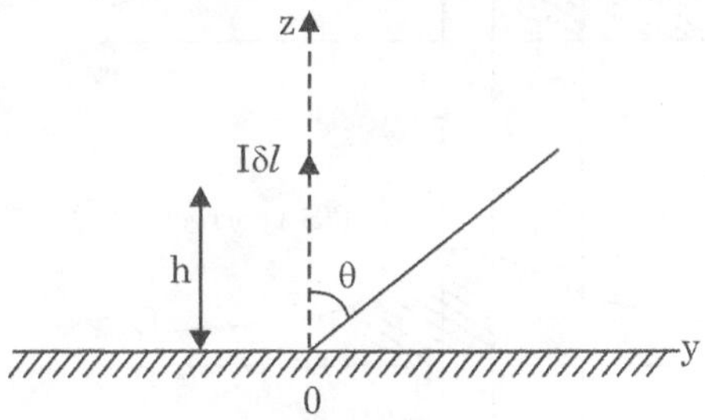

Infinite conduction plate

The resultant electric field intensity is given by

$$E_\theta = \frac{j\eta I_0 \beta dl \sin\theta}{4\pi r} e^{-j\beta r} 2\cos(\beta h\cos\theta)\ ;\ z \geq 0$$

$$= 0 \qquad ;\ z < 0$$

(or)

$$E_\theta = \frac{j\eta I_0 k dl \sin\theta}{4\pi r} e^{-j\beta r} 2\cos(kh\cos\theta)\ ;\ z \geq 0$$

$$= 0 \qquad ;\ z < 0$$

where k (or) β is phase shift constant (or) wave number.

$$\beta = \frac{2\pi}{\lambda}$$

Given $\theta_{max} = 60°$

Field will have maxima,

when $|\cos(kh\cos\theta_{max})| = 1$

$$\Rightarrow \quad kh\cos\theta_{max} = \pi$$

$$\frac{2\pi}{\lambda} h\cos 60° = \pi$$

$$\frac{2\pi}{\lambda}.h.\frac{1}{2} = \pi$$

$$\therefore \quad h = \lambda$$

(or) $$\frac{\partial E_0}{\partial h} = 0$$

Consider, $$\frac{j\eta k I_0 dl e^{-jkr}\sin\theta}{4\pi r} = P$$

$$\frac{\partial}{\partial h}\{2P\cos(kh\cos\theta_{max})\} = 0$$

$$-2pk\cos\theta_{max}\sin(kh\cos\theta_{max}) = 0$$

$$\sin(kh\cos\theta_{max}) = 0$$

$$kh\cos\theta_{max} = n\pi$$

For, $n = 1$

$$kh\cos 60 = \pi$$

$$\frac{2\pi}{\lambda} \times h.\frac{1}{2} = \pi$$

$\therefore$ So, minimum $h = \lambda$

44. $$V_{R2} = 0.7 + 3.3 = 4V$$

$$V_o = IR_1 + IR_2$$

$$9 = I(R_1 + R_2) \quad [\text{Given Vo} = 9V]$$

Now, By Neglecting base current

$$\text{So,} \quad 9 = \frac{4}{R_2}(1k + R_2)$$

$$\frac{9 \times R_2}{R_2 + 1k\Omega} = 4$$

$$5R_2 = 4k$$

$$R_2 = \frac{4000}{5}$$

$$R_2 = 800\ \Omega$$

45. As given that,

$$\bar{H} = (a_{\hat{x}} + 2a_{\hat{y}} + ba_{\hat{z}})\cos(\omega t + 3x - y - z)$$

$$\bar{H} = \bar{H}_0\cos(\omega t - \bar{k}.\bar{r})$$

$$= \bar{H}_0\cos\left[\omega t - (-3a_{\hat{x}} + a_{\hat{y}} + a_{\hat{z}})\cdot(xa_{\hat{x}} + ya_{\hat{y}} + za_{\hat{z}})\right]$$

So $\bar{H}_0 = a_{\hat{x}} + 2a_{\hat{y}} + ba_{\hat{z}}$

$$\hat{H}_0 = \left[\frac{(1a_{\hat{x}} + 2a_{\hat{y}} + ba_{\hat{z}})}{\sqrt{1^2 + 2^2 + b^2}}\right]$$

$$\bar{k} = -3a_{\hat{x}} + a_{\hat{y}} + a_{\hat{z}}$$

$$\hat{k} = \frac{-3a_{\hat{x}} + a_{\hat{y}} + a_{\hat{z}}}{\sqrt{(3)^2 + 1^2 + 1^2}}$$

For uniform plane wave $\hat{H}_0.\hat{k} = 0$

$$\frac{a\hat{x} + 2\hat{y} + b\hat{z}}{\sqrt{1^2 + 2^2 + b^2}}.\frac{(-3a_{\hat{x}} + a_{\hat{y}} + a_{\hat{z}})}{\sqrt{9 + 1 + 1}} = 0$$

$$-3 + 2 + b = 0$$

$$b = 1$$

46. For ideal Tranformer of n : 1, the scattering matrix is:

$$\begin{bmatrix} S_{11} & S_{12} \\ S_{21} & S_{22} \end{bmatrix} = \begin{bmatrix} \frac{n^2 - 1}{n^2 + 1} & \frac{2n}{n^2 + 1} \\ \frac{2n}{n^2 + 1} & \frac{1 - n^2}{n^2 + 1} \end{bmatrix} \qquad ...(1)$$

$$\therefore \quad n = 2$$

$$\Rightarrow S_{12} = S_{21} = \left[\frac{2n}{n^2 + 1}\right]$$

$$= \frac{2 \times 2}{2^2 + 1}$$

$$= \frac{4}{5} = 0.8$$

47. As given that

$$S(t)_{pm} = A_c \cos(2\pi f_c t + k_p\, m(t)$$

$$S(t)_{fm} = A_c \cos[2\pi f_c t + k_f \int_0^{\infty} m(t)\,dt\,]$$

Highest instantaneous frequencies are same.

$$f_i = \frac{1}{2\pi}\frac{d}{dt}\theta(t)$$

$$\because\ f_{i,\,PM} = f_{i,\,FM}$$

$$f_c + \frac{K_p}{2\pi}\frac{dm(t)}{dt}\bigg|_{max} = f_c + \frac{K_f}{2\pi}m(t)\bigg|_{max}$$

$$K_p \frac{dm(t)}{dt}\bigg|_{max} = K_f m(t)\big|_{max}$$

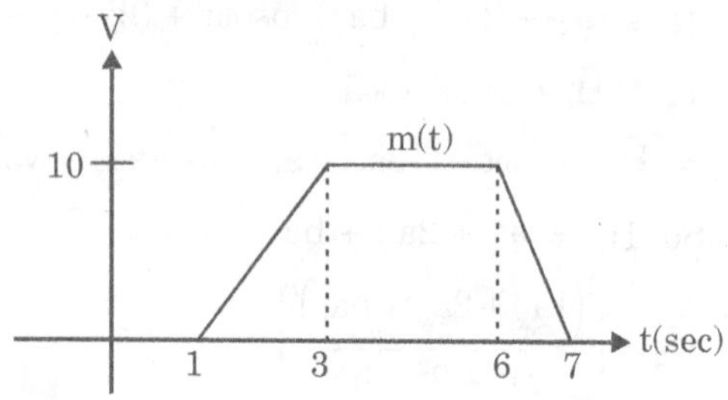

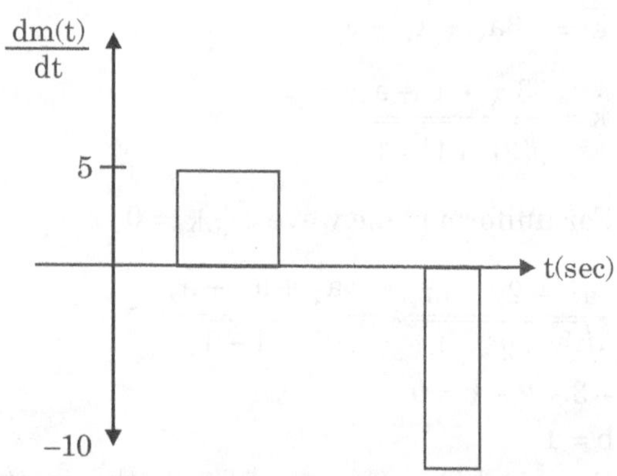

$$K_p.5 = k_f.10$$

$$\frac{K_p}{K_f} = \frac{10}{5} = 2 \text{ sec}$$

48. $S_1 = -3,\ S_2 = -1,\ S_3 = +1,\ S_4 = +2$

$Y = S + W$

$$W = N(0, 1) = \frac{1}{\sqrt{2\pi}}e^{-\omega^2/2}$$

$f_{s1}(s_1) = -3 + W = S_1$

Since the noise variable is gaussian with zero mean and ML decoding is used, the decision boundary between two adjacent signal points will be their arithmetic mean.

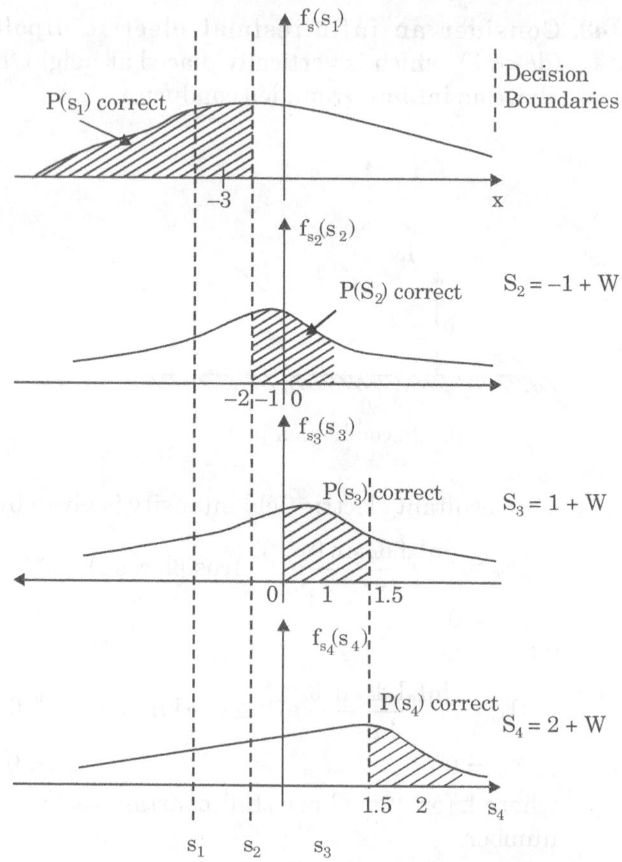

In all above graphs, the shaded area indicates the conditional probability of decoding a symbol correctly when it is transmited.

Since all symmbols are equally likely, decision boundaries are mid points.

i.e., for S_1, decision boundary – 2 V i.e., anything below – 2 V considered as S_1.

For S_2 decision boundaries are – 2 & 0V, i.e., any thing – 2V to 0V considered as S_2.

For S_3 decision boundaries are 0V & 1.5V.

For S_4 decision boundaries is 1.5V i.e., 1.5 V to ∞ considered as S_4.

P(Correct symbol receiveing is less for S_3 only therefore S_3 has highest conditional symbol error probability).

49. As given that,

$$G(s) = \frac{1}{(s+1)(s+a)} \quad a > 0$$

$$x(t) = 5 \cos 3t$$

$$y_{ss}(t) = \frac{1}{\sqrt{10}}\cos(3t - 1.892)$$

$$x(t) = A\cos(\omega_0 t + \theta)$$

$$H(s)\big|_{s\,j\omega_0} = |H(\omega_0)|\angle H(\omega_0)$$

$Y^{ss}(t) = A\,|H(\omega_0)|\cos(\omega_0 t + \theta + \angle H(\omega_0))$

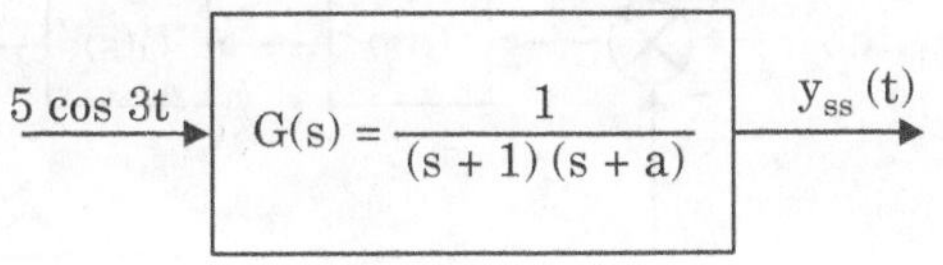

$$(\because S = j\omega_0)$$

$$G(S) = \frac{1}{(1 + j\omega_0)(a + j\omega_0)}$$

$$|G(j\omega_0)| = \frac{1}{\sqrt{(\omega_0^2 + 1)(\omega_0^2 + a^2)}} \qquad (\because \omega_0 = 3)$$

So, $\quad |G(j3)| = \dfrac{1}{\sqrt{a^2 + 9}\sqrt{1 + 9}}$

From the above expression

$$A|G(j3)| = \frac{1}{\sqrt{10}} \qquad (\because A = 5)$$

$$\frac{5}{\sqrt{10}\sqrt{a^2 + 9}} = \frac{1}{\sqrt{10}} \Rightarrow 5 = \sqrt{a^2 + 9}$$

$$25 - a^2 + 9 = 0$$

So, $\quad a = 4$

50. In any sequential circuit, the condition for proper operation is

Clock period (T) $\geq (t_{pd} + t_{comb\ logic} + t_{setup})$

So, total propagation delay:

For Flipflop 1 $\Rightarrow T \geq (8 + 5) \Rightarrow T \geq 13$ns

For Flipflop 2 $\Rightarrow T \geq (3 + 2 + 2 + 4) \Rightarrow T \geq 11$ns

Thus, $T \geq 13$ns

$\Rightarrow$ frequency of operation (f) = $\dfrac{1000}{13}$ MHz

$f_{max} = 76.92$ MHz.

51. From the given figures, the limits of te integralare given by

$$\text{limits} : \begin{cases} x : 0 \text{ to } 3 \\ y : 0 \text{ to } 1 \\ z : 0 \text{ to } 1 - y \end{cases}$$

$$I = \int_{y=0}^{1}\int_{z=0}^{1-y} xdz \int_{x=0}^{3} x\,dx\,dy\,dz$$

$$I = \int_{x=0}^{3}\left[\int_{y=0}^{1}\left\{\int_{z=0}^{1-y} xdz\right\}dy\right]dx$$

$$= \int_{x=0}^{3}\left[\int_{y=0}^{1} x(z)_0^{1-y}\,dy\right]dx$$

$$= \int_{x=0}^{3}\left[\int_{y=0}^{1} x.(1 - y)dy\right]dx$$

$$= \left(\frac{x^2}{2}\right)_0^3 . \left(y - \frac{y^2}{2}\right)_0^1$$

$$= \frac{9}{2}\left(1 - \frac{1}{2}\right) = \frac{9}{4} = 2.25$$

52. As given figure:

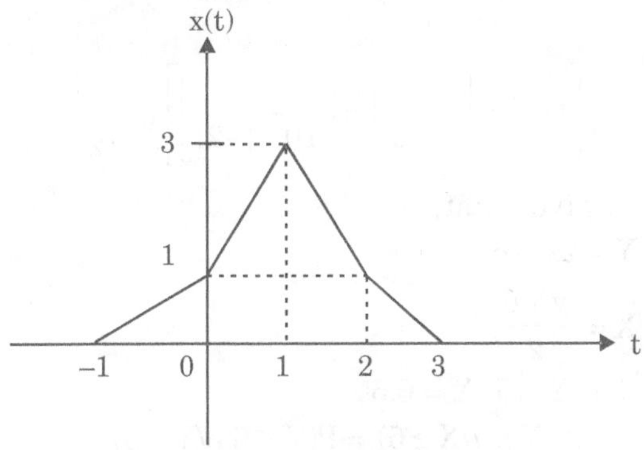

Because of symmetry of the signal w.r.t to t = 1

From praseval's theorem

$$\int_{-\infty}^{\infty} |X(\omega)|^2 d\omega = 2\pi\int_{-\infty}^{\infty} |X(t)|^2 dt$$

$$\int_{-\infty}^{\infty} |X(\omega)|^2 d\omega = 2\pi\left[\int_{-1}^{3} |X(t)|^2 dt\right]$$

$$= 2(2\pi)\left[\int_{-1}^{0} (t + 1)|^2 dt + \int_{0}^{1} (2t + 1)^2 dt\right]$$

$$2(2\pi)\left[\int_{-1}^{0} (t + 1)|^2 dt + \int_{0}^{1} (4t^2 + 4t + 1)dt\right]$$

$$= 4\pi\left[\left[\frac{(t + 1)^3}{3}\right]_{-1}^{0} + \left[\frac{4t^3}{3} + \frac{4t^2}{2} + t\right]_0^1\right]$$

$$= 4\pi\left[\frac{1}{3} + \frac{4}{3} + 2 + 1\right] = 4\pi\left[\frac{5}{3} + 2\right]$$

$$= \frac{56\pi}{3}$$

$$= 58.64$$

53. As given that T/f of LTI system is:

$$H(z) = \frac{K(z - \alpha)}{z + 0.5}, |\alpha| > 1$$

$$|H(e^{j\omega})| = 1\ \forall\ \omega$$

For an all pass filter poles & zeros are reciprocal to each other.

To get constant magnitude for all frequency pole located at z = – 0.5

$\therefore$ zero is located at = $\dfrac{1}{\text{Pole}} = \dfrac{1}{-0.5} = -2$

$\therefore$ zero = $\alpha = -2$

So, $H(z) = \dfrac{k(z + 2)}{z + 0.5}$

54. X follows uniform distribution over [–2, 10]

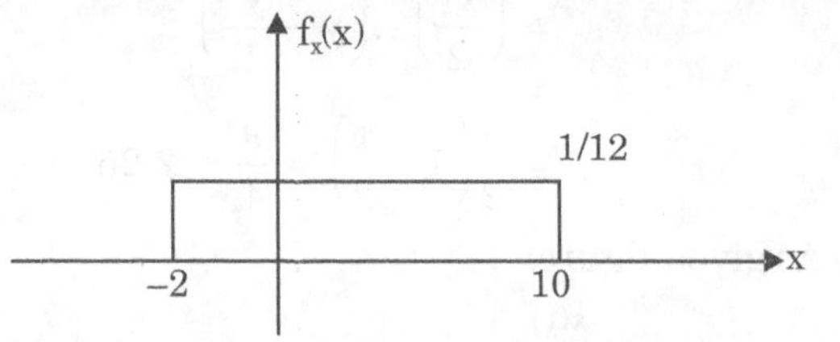

$$\left[f(x)=\left[\frac{1}{b-a}\right]=\left[\frac{1}{10-(-2)}\right]\right]=\frac{1}{12}$$

As given that,

$Y = 2x - 6$

$X = \frac{y+6}{2}$

for, Y = 7, X = 6.5

$P(Y \le 7/X \ge 5) = P(Y \le 6.5/Y \ge 5)$

$$\frac{P[X > 5 \text{ and } X < 6.5]}{P[X > 5]}$$

$$= \frac{\int_5^{6.5} 1/12\, dy}{\int_5^{10} 1/12\, dy} = \frac{1.5}{5} = 0.3$$

55.

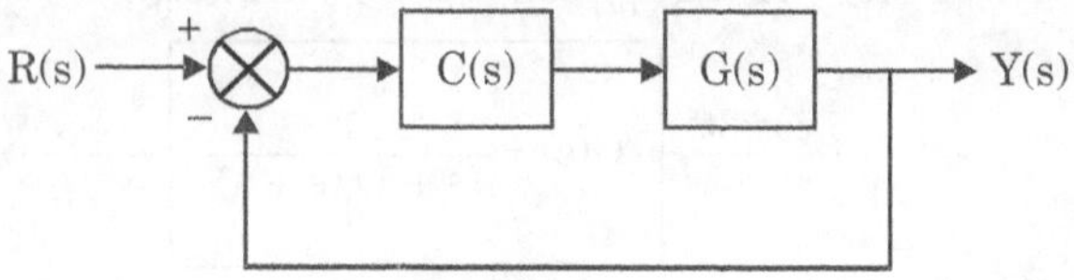

The open loop transfer function for the

$$\text{System} = C(s)G(s) = \frac{k(s+1)}{s(s+1)(s+3)} = \frac{k}{s(s+3)}$$

Since the system is type–1, so far a given Unit ramp input steady state.

$$e_{ss}\,(\text{ramp}) = \frac{1}{k_v} = 0.1$$

where, $k_v = \lim_{s\to 0} C(s)G(s) = \lim_{s\to 0} s.\frac{k}{s(s+3)} = \frac{k}{3}$

$$e_{ss} = \frac{1}{k_V} = \frac{3}{k} = 0.1$$

$$k = \frac{3}{0.1}$$

$$k = 30$$

Unit - I

Engineering Mathematics

Syllabus

Linear Algebra: Vector space, basis, linear dependence and independence, matrix algebra, eigen values and eigen vectors, rank, solution of linear equations – existence and uniqueness.

Calculus: Mean value theorems, theorems of integral calculus, evaluation of definite and improper integrals, partial derivatives, maxima and minima, multiple integrals, line, surface and volume integrals, Taylor series.

Differential Equations: First order equations (linear and nonlinear), higher order linear differential equations, Cauchy's and Euler's equations, methods of solution using variation of parameters, complementary function and particular integral, partial differential equations, variable separable method, initial and boundary value problems.

Vector Analysis: Vectors in plane and space, vector operations, gradient, divergence and curl, Gauss's, Green's and Stoke's theorems.

Complex Analysis: Analytic functions, Cauchy's integral theorem, Cauchy's integral formula; Taylor's and Laurent's series, residue theorem.

Numerical Methods: Solution of nonlinear equations, single and multi-step methods for differential equations, convergence criteria.

Probability and Statistics: Mean, median, mode and standard deviation; combinatorial probability, probability distribution functions - binomial, Poisson, exponential and normal; Joint and conditional probability; Correlation and regression analysis.

Contents

CHAPTER 1

Linear Algebra

Analysis of Previous GATE Papers												
	Year → Topics ↓		2019	2018	2017 Set 1	2017 Set 2	2016 Set 1	2016 Set 2	2016 Set 3	2015 Set 1	2015 Set 2	2015 Set 3
MATRIX ALGEBRA, VECTOR, SPACE, RANK:	**1 Mark**	MCQ Type			1		1					1
		Numerical Type		1		1						
	2 Marks	MCQ Type										
		Numerical Type				1						
		Total		**1**	**1**	**3**	**1**					**1**
SOLUTION OF LINEAR SYSTEM OF EQUATIONS	**1 Mark**	MCQ Type										
		Numerical Type										
	2 Marks	MCQ Type										
		Numerical Type										
		Total										
EIGEN VALUES & EIGEN VECTORS	**1 Mark**	MCQ Type		1	1				1		1	
		Numerical Type	1		1					1		
	2 Marks	MCQ Type										
		Numerical Type					1					
		Total	**1**	**1**	**2**		**2**		**1**	**1**	**1**	

MATRIX ALGEBRA, VECTOR, SPACE, RANK:

1. The rank of (m × n) matrix (where m < n) cannot be more than

(a) m (b) n
(c) mn (d) none

[1994 : 1 Mark]

2. Let, $A = \begin{bmatrix} 2 & -0.1 \\ 0 & 3 \end{bmatrix}$ and $A^{-1} = \begin{bmatrix} \frac{1}{2} & a \\ 0 & b \end{bmatrix}$. Then (a + b) =

(a) $\frac{7}{20}$ (b) $\frac{3}{20}$
(c) $\frac{19}{60}$ (d) $\frac{11}{20}$

[2005 : 2 Marks]

3. Given an orthogonal matrix

$$A = \begin{bmatrix} 1 & 1 & 1 & 1 \\ 1 & 1 & -1 & -1 \\ 1 & -1 & 0 & 0 \\ 0 & 0 & 1 & -1 \end{bmatrix} \quad [AA^T]^{-1} \text{ is}$$

(a) $\begin{bmatrix} \frac{1}{4} & 0 & 0 & 0 \\ 0 & \frac{1}{4} & 0 & 0 \\ 0 & 0 & \frac{1}{2} & 0 \\ 0 & 0 & 0 & \frac{1}{2} \end{bmatrix}$ (b) $\begin{bmatrix} \frac{1}{2} & 0 & 0 & 0 \\ 0 & \frac{1}{2} & 0 & 0 \\ 0 & 0 & \frac{1}{2} & 0 \\ 0 & 0 & 0 & \frac{1}{2} \end{bmatrix}$

(c) $\begin{bmatrix} 1 & 0 & 0 & 0 \\ 0 & 1 & 0 & 0 \\ 0 & 0 & 1 & 0 \\ 0 & 0 & 0 & 1 \end{bmatrix}$ (d) $\begin{bmatrix} \frac{1}{4} & 0 & 0 & 0 \\ 0 & \frac{1}{4} & 0 & 0 \\ 0 & 0 & \frac{1}{4} & 0 \\ 0 & 0 & 0 & \frac{1}{4} \end{bmatrix}$

[2005 : 2 Marks]

4. The rank of the matrix $\begin{bmatrix} 1 & 1 & 1 \\ 1 & -1 & 0 \\ 1 & 1 & 1 \end{bmatrix}$ is

(a) 0 (b) 1
(c) 2 (d) 3

[2006 : 1 Mark]

5. It is given that $X_1, X_2, \ldots X_M$ are M non-zero, orthogonal vectors. The dimension of the vector space spanned by the 2 M vectors $X_1, X_2, \ldots X_M, -X_1, -X_2, \ldots, -X_M$ is

(a) 2M
(b) M + 1
(c) M
(d) dependent on the choice of $X_1, X_2, \ldots, X_M$

[2007 : 2 Marks]

6. Given that $A = \begin{bmatrix} -5 & -3 \\ 2 & 0 \end{bmatrix}$ and $I = \begin{bmatrix} 1 & 0 \\ 0 & 1 \end{bmatrix}$, the value of A^3 is

(a) 15A + 12I (b) 19A + 30I
(c) 17A + 15I (d) 17A + 21I

[2011 : 2 Marks]

7. Let A be an m × n matrix and B an n × m matrix. It is given that determinant $(I_m + AB)$ = determinant $(I_n + BA)$, where I_k is the k × k identity matrix. Using the above property, the determinant of the matrix given below is

$$\begin{bmatrix} 2 & 1 & 1 & 1 \\ 1 & 2 & 1 & 1 \\ 1 & 1 & 2 & 1 \\ 1 & 1 & 1 & 2 \end{bmatrix}$$

(a) 2 (b) 5
(c) 8 (d) 16

[2013 : 2 Marks]

8. For matrices of same dimension M, N and scalar c, which one of these properties DOES NOT ALWAYS hold?

(a) $(M^T)^T = M$
(b) $(cM)^T = c(M)^T$
(c) $(M + N)^T = M^T + N^T$
(d) MN= NM

[2014 : 1 Mark, Set-1]

9. Consider the matrix

$$J_y = \begin{bmatrix} 0 & 0 & 0 & 0 & 0 & 1 \\ 0 & 0 & 0 & 0 & 1 & 0 \\ 0 & 0 & 0 & 1 & 0 & 0 \\ 0 & 0 & 1 & 0 & 0 & 0 \\ 0 & 1 & 0 & 0 & 0 & 0 \\ 1 & 0 & 0 & 0 & 0 & 0 \end{bmatrix}$$

which is obtained by reversing the order of the columns of the identity matrix I_6. Let $P = I_6 + \alpha J_6$, where α is a non-negativer real number. The value of a for which det(P) = 0 is ______.

[2014 : 2 Marks, Set-1]

10. The determinant of matrix A is 5 and the determinant of matrix B is 40. The determinant of matrix AB is _______.

[2014 : 1 Mark, Set-2]

11. The maximum value of the determinant among all 2 × 2 real symmetric matrices with trace 14 is _______.

[2014 : 2 Marks, Set-2]

12. For $A = \begin{bmatrix} 1 & \tan x \\ -\tan x & 1 \end{bmatrix}$, the determinant of $A^T A^{-1}$ is

(a) $\sec^2 x$ (b) $\cos 4x$
(c) 1 (d) 0

[2015 : 1 Mark, Set-3]

13. Let $M^4 = I$, (where I denotes the identity matrix) and $M \neq I$, $M^2 \neq I$ and $M^3 \neq I$. Then, for any natural number k, M^{-1} equals:

(a) M^{4k+1} (b) M^{4k+2}
(c) M^{4k+3} (d) M^{4k}

[2016 : 1 Mark, Set-1]

14. The matrix $A = \begin{bmatrix} a & 0 & 3 & 7 \\ 2 & 5 & 1 & 3 \\ 0 & 0 & 2 & 4 \\ 0 & 0 & 0 & b \end{bmatrix}$ has det(A) = 100 and trace [A] = 14. The value of $|a - b|$ is _______.

[2016 : 2 Marks, Set-2]

15. The rank of the matrix $M = \begin{bmatrix} 5 & 10 & 10 \\ 1 & 0 & 2 \\ 3 & 6 & 6 \end{bmatrix}$ is

(a) 0 (b) 1
(c) 2 (d) 3

[2017 : 1 Mark, Set-1]

16. The rank of the matrix

$$\begin{bmatrix} 1 & -1 & 0 & 0 & 0 \\ 0 & 0 & 1 & -1 & 0 \\ 0 & 1 & -1 & 0 & 0 \\ -1 & 0 & 0 & 0 & 1 \\ 0 & 0 & 0 & 1 & -1 \end{bmatrix}$$

is _______.

[2017 : 1 Mark, Set-2]

SOLUTION OF LINEAR SYSTEM OF EQUATIONS

17. Solve the following system of equations

$x_1 + x_2 + x_3 = 3$
$x_1 - x_2 = 0$
$x_1 - x_2 + x_3 = 1$

(a) Unique solution
(b) No solution
(c) Infinite number of solutions
(d) Only one solution

[1994 : 2 Marks]

18. The eigenvalues and the corresponding eigenvectors of a 2 × 2 matrix are given by

Eigenvalue Eigenvector

$\lambda_1 = 8$ $v_1 = \begin{bmatrix} 1 \\ 1 \end{bmatrix}$

$\lambda_2 = 4$ $v_2 = \begin{bmatrix} 1 \\ -1 \end{bmatrix}$

The matrix is

(a) $\begin{bmatrix} 6 & 2 \\ 2 & 6 \end{bmatrix}$ (b) $\begin{bmatrix} 4 & 6 \\ 6 & 4 \end{bmatrix}$
(c) $\begin{bmatrix} 2 & 4 \\ 4 & 2 \end{bmatrix}$ (d) $\begin{bmatrix} 4 & 8 \\ 8 & 4 \end{bmatrix}$

[2006 : 2 Marks]

19. The system of linear equations

$4x + 2y = 7$
$2x + y = 6$

has

(a) a unique solution.
(b) no solution.
(c) an infinite number of solutions.
(d) exactly two distinct solutions.

[2008 : 1 Mark]

20. The system of equations

$x + y + z = 6$
$x + 4y + 6z = 20$
$x + 4y + \lambda z = \mu$

has NO solution for values of λ and μ given by

(a) $\lambda = 6, \mu = 20$ (b) $\lambda - 6, \mu \neq 20$
(c) $\lambda \neq 6, \mu = 20$ (d) $\lambda \neq 6, \mu \neq 20$

21. The system of linear equations

$$\begin{bmatrix} 2 & 1 & 3 \\ 3 & 0 & 1 \\ 1 & 2 & 5 \end{bmatrix} \begin{bmatrix} a \\ b \\ c \end{bmatrix} = \begin{bmatrix} 5 \\ -4 \\ 14 \end{bmatrix}$$ has

(a) a unique solution
(b) infinitely many solutions
(c) no solution
(d) exactly two solutions

[2014 : 2 Marks, Set-2]

22. Consider a system of linear equations-

$x - 2y + 3z = -1$,

$x - 3y + 4z = 1$, and

$-2x + 4y - 6z = k$

The value of k for which the system has infinitely many solution is ________.

23. Consider matrix $A = \begin{bmatrix} k & 2k \\ k^2 - k & k^2 \end{bmatrix}$ and vector $X = \begin{bmatrix} x_1 \\ x_2 \end{bmatrix}$. The number of distinct real values of k for which the equation AX = 0 has infinitely many solutions is ______.

[2018 : 1 Mark]

EIGEN VALUES & EIGEN VECTORS

24. The eigen values of the matrix $\begin{bmatrix} 2 & -1 & 0 & 0 \\ 0 & 3 & 0 & 0 \\ 0 & 0 & -2 & 0 \\ 0 & 0 & -1 & 4 \end{bmatrix}$ are

(a) 2, –2, 1, –1

(b) 2, 3, –2, 4

(c) 2, 3, 1, 4

(d) None

[2000 : 1 Mark]

25. Given the matrix $\begin{bmatrix} -4 & 2 \\ 4 & 3 \end{bmatrix}$, the eigenvector is

(a) $\begin{bmatrix} 3 \\ 2 \end{bmatrix}$ (b) $\begin{bmatrix} 4 \\ 3 \end{bmatrix}$

(c) $\begin{bmatrix} 2 \\ -1 \end{bmatrix}$ (d) $\begin{bmatrix} -1 \\ 2 \end{bmatrix}$

[2005 : 2 Marks]

26. For the matrix $\begin{bmatrix} 4 & 2 \\ 2 & 4 \end{bmatrix}$ the eigenvalue corresponding to the eigenvector $\begin{bmatrix} 101 \\ 101 \end{bmatrix}$ is

(a) 2

(b) 4

(c) 6

(d) 8

[2006 : 2 Marks]

27. All the four entries of the 2 × 2 matrix $P = \begin{bmatrix} P_{11} & P_{12} \\ P_{21} & P_{22} \end{bmatrix}$ are nonzero, and one of its eigenvalue is zero. Which of the following statements is true?

(a) $P_{11}P_{22} - P_{12}P_{21} = 1$

(b) $P_{11}P_{22} - P_{12}P_{21} = -1$

(c) $P_{11}P_{22} - P_{12}P_{21} = 0$

(d) $P_{11}P_{22} + P_{12}P_{21} = 0$

[2008 : 1 Mark]

28. The eigen values of the following matrix are

$$\begin{bmatrix} -1 & 3 & 5 \\ -3 & -1 & 6 \\ 0 & 0 & 3 \end{bmatrix}$$

(a) 3, 3 + 5j, 6 – j

(b) –6 + 5j, 3 + j, 3 – j

(c) 3 + j, 3 – j, 5 + j

(d) 3, –1 + 3j, –1 – 3j

[2009 : 2 Marks]

29. The eigen values of a skew-symmetric matrix are

(a) always zero

(b) always pure imaginary

(c) either zero or pure imaginary

(d) always real

[2010 : 1 Mark]

30. The minimum eigen value of the following matrix is

$$\begin{bmatrix} 3 & 5 & 2 \\ 5 & 12 & 7 \\ 2 & 7 & 5 \end{bmatrix}$$

(a) 0 (b) 1

(c) 2 (d) 3

[2013 : 1 Mark]

31. A real (4 × 4) matrix A satisfies the equation $A^2 = I$, where I is the (4 × 4) identity matrix. The positive eigen value of A is ______.

[2014 : 1 Mark, Set-1]

32. Which one of the following statements is NOT true for a square matrix A?

(a) If A is upper triangular, the eigenvalues of A are the diagonal elements of it

(b) If A is real symmetric, the eigenvalues of A are always real and positive

(c) If A is real, the eigenvalues of A and A^T are always the same

(d) If all the principal minors of A are positive, all the eigenvalues of A are also positive

[2014 : 2 Marks, Set-3]

33. The value of p such that the vector $\begin{bmatrix}1\\2\\3\end{bmatrix}$ is an eigenvector of the matrix $\begin{bmatrix}4 & 1 & 2\\ p & 2 & 1\\ 14 & -4 & 10\end{bmatrix}$ is

[2015 : 1 Mark, Set-1]

34. The value of x for which all the eigen-values of the matrix given below are real is ______.

$$\begin{bmatrix}10 & 5+j & 4\\ x & 20 & 2\\ 4 & 2 & -10\end{bmatrix}$$

(a) $5 + j$
(b) $5 - j$
(c) $1 - 5j$
(d) $1 + 5j$

[2015 : 1 Mark, Set-2]

35. A sequence x[n] is specified as

$$\begin{bmatrix}x[n]\\ x(n-1)\end{bmatrix} = \begin{bmatrix}1 & 1\\ 1 & 0\end{bmatrix}\begin{bmatrix}1\\ 0\end{bmatrix}, \text{ for } n \geq 2.$$

The initial conditions are $x[0] = 1$, $x[1] = 1$, and $x[n] = 0$ for $n < 0$. The value of x[12] is ______.

[2016 : 2 Marks, Set-1]

36. The value of x for which the matrix

$$A = \begin{bmatrix}3 & 2 & 4\\ 9 & 7 & 13\\ -6 & -4 & -9+x\end{bmatrix}$$ has zero as an eigen value is ______.

[2017 : 1 Mark, Set-1]

37. Consider a 2 × 2 square matrix

$$A = \begin{bmatrix}\sigma & x\\ \omega & \sigma\end{bmatrix}$$

where x is unknown. If the eigenvalues of the matrix A are $(\sigma + j\omega)$ and $(\sigma - j\omega)$, then x is equal to

(a) $+j\omega$
(b) $-y\omega$
(c) $+\omega$
(d) $-\omega$

[2016 : 1 Mark, Set-3]

38. Consider the 5 × 5 matrix

$$A = \begin{bmatrix}1 & 2 & 3 & 4 & 5\\ 5 & 1 & 2 & 3 & 4\\ 4 & 5 & 1 & 2 & 3\\ 3 & 4 & 5 & 1 & 2\\ 2 & 3 & 4 & 5 & 1\end{bmatrix}$$

It is given that A has only one real eigenvalue. Then the real eigenvalue of A is

(a) –2.5
(b) 0
(c) 15
(d) 25

[2017 : 1 Mark, Set-1]

39. Let M be a real 4 × 4 matrix. Consider the following statements:

S_1 : M has 4 linearly independent eigenvectors.
S_2 : M has 4 distinct eigenvalues.
S_3 : M is non-singular(invertible).

Which one among the following is TRUE?

(a) S_1 implies S_2.
(b) S_1 implies S_3.
(c) S_2 implies S_1.
(d) S_3 implies S_2.

[2018 : 1 Mark]

40. The number of distinct eigen values of the matrix

$$A = \begin{bmatrix}2 & 2 & 3 & 3\\ 0 & 1 & 1 & 1\\ 0 & 0 & 3 & 3\\ 0 & 0 & 0 & 2\end{bmatrix}$$ is equal to ______.

[2019 : 1 Mark]

ANSWERS

1. (a)	2. (a)	3. (c)	4. (c)	5. (c)	6. (b)	7. (b)	8. (d)	9. (0· 99 to 1· 01)
10. (200)	11. (49)	12. (c)	13. (c)	14. (3)	15. (c)	16. (4)	17. (a)	18. (a)	19. (b)
20. (b)	21. (b)	22. (2)	23. (2)	24. (b)	25. (a)	26. (c)	27. (c)	28. (d)	29. (c)
30. (a)	31. (0· 99 to 1· 01)	32. (b)	33. (17)	34. (b)	35. (233)	36. (1)	37. (d)	38. (c)
39. (c)	40. (3)							

EXPLANATIONS

1. We know that $f(A_{max}) \leq$ main {m, n}

Now according to question m < n

$\therefore f(A_{max}) \leq m$

Hence $f(A_{max})$ cannot is more than 'm'.

$\therefore$ option (a) is correct

2. $A = \begin{bmatrix} 2 & -0.1 \\ 0 & 3 \end{bmatrix}$ and $A^{-1} = \begin{bmatrix} \frac{1}{2} & a \\ o & b \end{bmatrix}$

Now $A.A^{-1} = I$

$$\therefore \begin{bmatrix} 2 & -0.1 \\ 0 & 3 \end{bmatrix}\begin{bmatrix} \frac{1}{2} & a \\ 0 & b \end{bmatrix} = \begin{bmatrix} 1 & 0 \\ 0 & 1 \end{bmatrix}$$

$$\therefore \begin{bmatrix} 2\left(\frac{1}{2}\right) & 2a - 0.1b \\ 0 & 3b \end{bmatrix} = \begin{bmatrix} 1 & 0 \\ 0 & 1 \end{bmatrix}$$

$$\therefore \begin{bmatrix} 1 & 2a - 0.1b \\ 0 & 3b \end{bmatrix} = \begin{bmatrix} 1 & 0 \\ 0 & 1 \end{bmatrix}$$

$\therefore\ 2a - 0.1b = 0$

and $3b = 1 \Rightarrow b = \frac{1}{3}$

$\therefore 2a - 0.1\left(\frac{1}{3}\right) = 0$

$\therefore 2a = \frac{0.1}{3}$

$\therefore a = \frac{0.1}{6}$

3. $A = \begin{bmatrix} 1 & 1 & 1 & 1 \\ 1 & 1 & -1 & -1 \\ 1 & -1 & 0 & 0 \\ 0 & 0 & 1 & -1 \end{bmatrix}$

Here A is orthogonal matrix.l

$\therefore A.\ A^T = I.$

$$\therefore A.A^T = \begin{bmatrix} 1 & 0 & 0 & 0 \\ 0 & 1 & 0 & 0 \\ 0 & 0 & 1 & 0 \\ 0 & 0 & 0 & 1 \end{bmatrix} = I$$

Now inverse of identify matrix is identify matrix

$\therefore$ option (c) is correct.

4. $A = \begin{bmatrix} 1 & 1 & 1 \\ 1 & -1 & 0 \\ 1 & 1 & 1 \end{bmatrix}$

$|A| = 1(-1) - 1(1) + 1(1+1)$

$= -1 - 1 + 2$

$= 0$

Hence Lane is less than 3

Now $\begin{vmatrix} 1 & 1 \\ 1 & -1 \end{vmatrix} = -1 - 1 = -2 \neq 0$

Hence rank of above matri'x is '2'

5. There are M non-zero, orthogonal vectors, so there is required M dimension to represent them.

6. Characteristic equation of A is

$$|A - \lambda I| = 0$$

or, $\begin{vmatrix} -5-\lambda & -3 \\ 2 & 0-\lambda \end{vmatrix} = 0$

or, $(5+\lambda)\lambda + 6 = 0$

or, $\lambda^2 + 5\lambda + 6 = 0$ ...(1)

Now, CAYLEY-HAMILTON Theorm states that every square matrix satisfies its own characteristic equation; i.e. if the characteristic equation for the nth order square matrix A is,

$$|A - \lambda I| = (-1)^n \lambda^n + k_1\lambda^{n-1} + ... + k_n = 0$$

then, $(-1)^n A^n + k_1 A^{n-1} + ... + k_n = 0$

Then, by using this theorem to equation (1), we get

$$A^2 + 5A + 6I = 0$$

or, $$A^2 = -5A - 6I \quad ...(2)$$

Multiplying by A to equation (2) both sides, we get

$$A^3 = -5A^2 - 6A$$

or $$A^3 = -5[-5A - 6I] - 6A$$
$$= 19A + 30I$$

or, $$A^3 = 19A + 30I$$

7. Given A : $m \times n$ matrix

B : $n \times m$ matrix

det $(I_m + AB)$ = det $(I_n + BA)$

I_K : K × K identify matrix

To find:

$$\det\begin{bmatrix}2&1&1&1\\1&2&1&1\\1&1&2&1\\1&1&1&2\end{bmatrix} = \det(M) \qquad \{\because say\}$$

Analysis. We will break matrix m to match $(I_m + AB)$

Plan

1. As per analysis part we will break matrix m into sum of I_m and AB
2. Then use det $(I_m + AB)$ = det $(I_m + BA)$

Carrying out plan

$$\begin{bmatrix}2&1&1&1\\1&2&1&1\\1&1&2&1\\1&1&1&2\end{bmatrix} = \begin{bmatrix}1&0&0&0\\0&1&0&0\\0&0&1&0\\0&0&0&1\end{bmatrix} + \begin{bmatrix}1&1&1&1\\1&1&1&1\\1&1&1&1\\1&1&1&1\end{bmatrix} \quad ...(1)$$

Now we will break second matrix in RHS of above as follows

$$\begin{bmatrix}1&1&1&1\\1&1&1&1\\1&1&1&1\\1&1&1&1\end{bmatrix}_{4\times4} = \begin{bmatrix}1\\1\\1\\1\end{bmatrix}_{4\times1}\begin{bmatrix}1&1&1&1\end{bmatrix}_{1\times4} \quad ...(2)$$

Using (2) into (1), we get

$$\begin{bmatrix}2&1&1&1\\1&2&1&1\\1&1&2&1\\1&1&1&2\end{bmatrix} = I_4 + \begin{bmatrix}1\\1\\1\\1\end{bmatrix}\begin{bmatrix}1&1&1&1\end{bmatrix} \quad ...(3)$$

Let $A = \begin{bmatrix}1\\1\\1\\1\end{bmatrix}_{4\times1}$ and $B = \begin{bmatrix}1&1&1&1\end{bmatrix}_{1\times4}$

$$\therefore \begin{bmatrix}2&1&1&1\\1&2&1&1\\1&1&2&1\\1&1&1&2\end{bmatrix} = I_4 + A_{4\times1}B_{1\times4} \Rightarrow \boxed{\begin{matrix}m=4\\n=1\end{matrix}}$$

But we are given that det $(I_m + AB)$ = det $(I_n + BA)$

$$\therefore \det(I_4 + A_{4\times1}B_{1\times4}) = \det\left(I_1 + \begin{bmatrix}1&1&1&1\end{bmatrix}\begin{bmatrix}1\\1\\1\\1\end{bmatrix}\right)$$

$$= \det(1+4) \qquad \{\because I_1 = 1\}$$
$$= \det(5) = 5$$

{∵ determinant of a scalar is the same scalar}

8. Matrix multiplication is not commutative in general.

9. $$J_6 = \begin{bmatrix}0&0&0&0&0&1\\0&0&0&0&1&0\\0&0&0&1&0&0\\0&0&1&0&0&0\\0&1&0&0&0&0\\1&0&0&0&0&0\end{bmatrix}$$

$$I_6 = \begin{bmatrix}1&0&0&0&0&0\\0&1&0&0&0&0\\0&0&1&0&0&0\\0&0&0&1&0&0\\0&0&0&0&1&0\\0&0&0&0&0&1\end{bmatrix}$$

Now $P = I_6 + \alpha J_6$

$$= \begin{bmatrix}1&0&0&0&0&0\\0&1&0&0&0&0\\0&0&1&0&0&0\\0&0&0&1&0&0\\0&0&0&0&1&0\\0&0&0&0&0&1\end{bmatrix} + \begin{bmatrix}0&0&0&0&0&\alpha\\0&0&0&0&\alpha&0\\0&0&0&\alpha&0&0\\0&0&\alpha&0&0&0\\0&\alpha&0&0&0&0\\\alpha&0&0&0&0&0\end{bmatrix}$$

$$\therefore P = \begin{bmatrix}1&0&0&0&0&\alpha\\0&1&0&0&\alpha&0\\0&0&1&\alpha&0&0\\0&0&\alpha&1&0&0\\0&\alpha&0&0&1&0\\\alpha&0&0&0&0&1\end{bmatrix}$$

$$|P| = 1\begin{vmatrix}1&0&0&\alpha&0\\0&1&\alpha&0&0\\0&\alpha&1&0&0\\\alpha&0&0&1&0\\0&0&0&0&1\end{vmatrix} - \alpha\begin{vmatrix}0&1&0&0&\alpha\\0&0&1&\alpha&0\\0&0&\alpha&1&0\\0&\alpha&0&0&1\\\alpha&0&0&0&0\end{vmatrix}$$

$$= \begin{vmatrix} 1 & \alpha & 0 & 0 \\ \alpha & 1 & 0 & 0 \\ 0 & 0 & 1 & 0 \\ 0 & 0 & 0 & 1 \end{vmatrix} + \alpha \begin{vmatrix} 0 & 1 & \alpha & 0 \\ 0 & \alpha & 1 & 0 \\ 0 & 0 & 0 & 1 \\ \alpha & 0 & 0 & 0 \end{vmatrix}$$

$$= \begin{vmatrix} 1 & 0 & 0 \\ 0 & 1 & 0 \\ 0 & 0 & 1 \end{vmatrix} - \alpha \begin{vmatrix} \alpha & 0 & 0 \\ 0 & 1 & 0 \\ 0 & 0 & 1 \end{vmatrix} - \alpha \begin{vmatrix} 0 & 1 & 0 \\ 0 & 0 & 1 \\ \alpha & 0 & 0 \end{vmatrix}$$

$$+\alpha^2 \begin{vmatrix} 0 & \alpha & 0 \\ 0 & 0 & 1 \\ \alpha & 0 & 0 \end{vmatrix}$$

$$|P| = 1(1) - \alpha(\alpha) - \alpha((-1)(-\alpha)) + \alpha^2[(-\alpha)(-\alpha)] = 0$$

$\therefore \quad 1 - \alpha^2 - \alpha^2 + \alpha^4 = 0$

$\alpha^4 - 2\alpha^2 + 1 = 0$

$(\alpha^2 - 1)^2 = 0$

$\therefore \quad \alpha^2 = 1$

$\therefore \quad \alpha = \pm\sqrt{1}$

$\therefore \quad \alpha = 1$ [$\because$ α is non-negative rea

10. $|AB| = |A| \cdot |B| = (5) \cdot (40) = 200$

11. General 2 × 2 real symmetric matrix is $\begin{bmatrix} y & x \\ x & z \end{bmatrix}$

$\Rightarrow \quad \det = yz - x^2$

and trace is $y + z = 14$ (given)

$\Rightarrow \quad z = 14 - y$ ...(*)

Let $f = yz - x^2(\det)$

$= -x^2 - y^2 + 14\,y\,(u \sin g^*)$

Using maxima and minima of a function of two variables, we have f is maximum at $x = 0$, $y = 7$ and therefore, maximum value of the determinant is 49

12. $$A = \begin{bmatrix} 1 & \tan x \\ -\tan x & 1 \end{bmatrix}$$

$$A^T = \begin{bmatrix} 1 & -\tan x \\ \tan x & 1 \end{bmatrix}$$

Now $$A^{-1} = \frac{1}{|A|}(\text{adj}(A))$$

$$A = \frac{1}{\sec^2 x}\begin{bmatrix} 1 & -\tan x \\ \tan x & 1 \end{bmatrix}$$

$\therefore \quad |A^T.A^{-1}| = 1$

13. Given, $M^4 = I$

$\Rightarrow \quad M^8 = M^4 = I$

$\Rightarrow \quad M^7 = M^{-1}$

$\Rightarrow \quad M^{12} = M^8 = I$

$\Rightarrow \quad M^{11} = M^{-1}$

$\Rightarrow \quad M^{16} = M^{12} = I$

$\Rightarrow \quad M^{15} = M^{-1}$

....

$\therefore \quad M^{-1} = M^{4k+3}$

where k is a natural number.

14. Trace of A = 14

$a + 5 + 2 + b = 14$

(Taking the diagonal element and then adding)

$a + b = 7$ ...(*i*)

$\det(A) = 100$

$$5\begin{vmatrix} a & 3 & 7 \\ 0 & 2 & 4 \\ 0 & 0 & b \end{vmatrix} = 100$$

$5 \times 2 \times a \times b = 100$

$10\,ab = 100$

$ab = 10$ ... (*ii*)

From equation (*i*) and (*ii*)

either $a = 5, b = 2$

or $a = 2\ b = 5$

$|a - b| = |5 - 2| = 3$

15. $$M = \begin{bmatrix} 5 & 10 & 10 \\ 1 & 0 & 2 \\ 3 & 6 & 6 \end{bmatrix}$$

Now applying, $R_1 \leftrightarrow R_2$ in above matrix we get

$$\begin{bmatrix} 1 & 0 & 2 \\ 5 & 10 & 10 \\ 3 & 6 & 6 \end{bmatrix}$$

Applying $R_2 \leftarrow R_2 - 5R_1$ and $R_3 \leftarrow R_3 - 3R_1$, we get

$$\begin{bmatrix} 1 & 0 & 2 \\ 0 & 10 & 0 \\ 0 & 6 & 0 \end{bmatrix}$$

Now applying, $R_3 \leftarrow R_3 - \frac{6}{10}R_2$, we get

$$\begin{bmatrix} 1 & 0 & 2 \\ 0 & 10 & 0 \\ 0 & 0 & 0 \end{bmatrix}$$

The above matrix is in Echelon form

Rank of matrix M is given by

$\rho(M) = 2$

16. $A = \begin{bmatrix} 1 & -1 & 0 & 0 & 0 \\ 0 & 0 & 1 & -1 & 0 \\ 0 & 1 & -1 & 0 & 0 \\ -1 & 0 & 0 & 0 & 1 \\ 0 & 0 & 0 & 1 & -1 \end{bmatrix}$

$R_4 \to R_4 + R_1$

$$A = \begin{bmatrix} 1 & -1 & 0 & 0 & 0 \\ 0 & 0 & 1 & -1 & 0 \\ 0 & 1 & -1 & 0 & 0 \\ 0 & -1 & 0 & 0 & 1 \\ 0 & 0 & 0 & 1 & -1 \end{bmatrix}$$

Applying $R_2 \leftrightarrow R_3$

$$A = \begin{bmatrix} 1 & -1 & 0 & 0 & 0 \\ 0 & 1 & -1 & 0 & 0 \\ 0 & 0 & 1 & -1 & 0 \\ 0 & -1 & 0 & 0 & 1 \\ 0 & 0 & 0 & 1 & -1 \end{bmatrix}$$

$R_4 \to R_4 + R_2$

$$A = \begin{bmatrix} 1 & -1 & 0 & 0 & 0 \\ 0 & 1 & -1 & 0 & 0 \\ 0 & 0 & 1 & -1 & 0 \\ 0 & 0 & -1 & 0 & 1 \\ 0 & 0 & 0 & 1 & -1 \end{bmatrix}$$

$R_4 \to R_4 + R_3$

$$A = \begin{bmatrix} 1 & -1 & 0 & 0 & 0 \\ 0 & 1 & -1 & 0 & 0 \\ 0 & 0 & 1 & -1 & 0 \\ 0 & 0 & 0 & -1 & 1 \\ 0 & 0 & 0 & 1 & -1 \end{bmatrix}$$

$R_5 \to R_5 + R_4$

$$A = \begin{bmatrix} 1 & -1 & 0 & 0 & 0 \\ 0 & 1 & -1 & 0 & 0 \\ 0 & 0 & 1 & -1 & 0 \\ 0 & 0 & 0 & -1 & 1 \\ 0 & 0 & 0 & 0 & 0 \end{bmatrix}$$

17. $x_1 + x_2 + x_3 = 3 \to \dots (1)$

$x_1 - x_2 = 0 \to \dots (2)$

$x_1 - x_2 + x_3 = 1 \to \dots (3)$

Solving above eq we get.

$x_3 = 1$

putting value of x_3 in eq (1) we get

$x_1 + x_2 = 2$

$x_1 - x_2 = 0$

$2x_1 = 2$

$\therefore \boxed{x_1 = 1}$

$\Rightarrow \boxed{x_2 = 1}$

Hence $x_1 = 1$, $x_2 = 1$, and $x_3 = 1$

$\therefore$ option (a) is correct

18. For matrix, $A = \begin{bmatrix} 6 & 2 \\ 2 & 6 \end{bmatrix}$

We know $|\lambda I - A| = 0$

$$\Rightarrow \begin{vmatrix} \lambda - 6 & -2 \\ -2 & \lambda - 6 \end{vmatrix} = 0$$

$$\Rightarrow \lambda^2 - 12\lambda + 32 = 0,$$

$$\Rightarrow \lambda = 4, 8 \quad \text{(Eigen values)}$$

For $\lambda_1 = 4$,

$$(\lambda_1 I - A) = \begin{bmatrix} -2 & -2 \\ -2 & -2 \end{bmatrix}, v_1 = \begin{bmatrix} 1 \\ 1 \end{bmatrix}$$

For $\lambda_2 = 8$,

$$(\lambda_2 I - A) = \begin{bmatrix} 2 & -2 \\ -2 & 2 \end{bmatrix}, v_2 = \begin{bmatrix} 1 \\ -1 \end{bmatrix}$$

19. Writing equations in matrix form,

$$\begin{bmatrix} 4 & 2 \\ 2 & 1 \end{bmatrix}\begin{bmatrix} x \\ y \end{bmatrix} = \begin{bmatrix} 7 \\ 6 \end{bmatrix}$$

Where $[A] = \begin{bmatrix} 4 & 2 \\ 2 & 1 \end{bmatrix}, B = \begin{bmatrix} 7 \\ 6 \end{bmatrix}$

The augmented matrix is $[A | B]$ given as

$$\left[\begin{array}{cc|c} 4 & 2 & 7 \\ 2 & 1 & 6 \end{array}\right]$$

Using $R_2 - \dfrac{R_1}{2}$

We have $\left[\begin{array}{cc|c} 4 & 2 & 7 \\ 0 & 0 & 5/2 \end{array}\right]$

Now, rank of $[A | B] = 2$

and rank of $[A] = 1$

Since, Rank $[A| B] \neq$ Rank $[A]$, no solution exists

20. Given equations are

$$x + y + z = 6$$

$$x + 4y + 6z = 20$$

and $$x + 4y + \lambda z = \mu$$

If $\lambda = 6$ and $\mu = 20$, then

$$x + 4y + 6z = 20$$

$$x + 4y + 6z = 20 \text{ infinite solution}$$

If $\lambda = 6$ and $\mu \neq 20$, then

$$x + 4y + 6z = 20$$

$$(\mu \neq 20) \text{ no solution}$$

If $\lambda \neq 6$ and $\mu = 20$, then

$$x + 4y + 6z = 20 \text{ will have solution}$$

$$x + 4y + \lambda z = 20$$

$\lambda \neq 6$ and $\mu \neq 20$ will also give solution

21. $$\left[\frac{A}{B}\right] = \left[\begin{array}{ccc|c} 2 & 1 & 3 & 5 \\ 3 & 0 & 1 & -4 \\ 1 & 2 & 5 & 14 \end{array}\right]$$

$$\begin{array}{l} R_2 \to 2R_2 - 3R_1 \\ R_3 \to 2R_3 - R_1 \end{array} \left[\begin{array}{ccc|c} 2 & 1 & 3 & 5 \\ 0 & -3 & -7 & -23 \\ 0 & 3 & 7 & 23 \end{array}\right] \xrightarrow{R_3 + R_2}$$

$$\left[\begin{array}{ccc|c} 2 & 1 & 3 & 5 \\ 0 & -3 & -7 & -23 \\ 0 & 0 & 0 & 0 \end{array}\right]$$

Since, rank (A) = rank $\left(\frac{A}{B}\right)$ < number of unknowns

$\therefore$ Equations have infinitely many solutions.

22. $$x - 2y + 3z = -1$$

$$x - 3y + 4z = 1$$

$$-2x + 4y - 6z = k$$

Augmented matrix (A/B) is given by

$$(A/B) = \begin{bmatrix} 1 & -2 & 3 & -1 \\ 1 & -3 & 4 & 1 \\ -2 & 4 & -6 & k \end{bmatrix}$$

Now applying row reduction technique—

$R_2 \to R_2 - R_1$

$R_3 \to R_3 + 2R_1$

We get

$$\begin{bmatrix} 1 & -2 & 3 & -1 \\ 0 & -1 & 1 & 2 \\ 0 & 0 & 0 & k-2 \end{bmatrix}$$

For infinite solution

$\rho(A/B) = \rho(A) = r$

$\therefore \; k - 2 = 0$

23. AX = 0 has infinitely many solutions (given)

$\Rightarrow |A| = 0$

$$\therefore \begin{vmatrix} k & 2k \\ k^2 - k & k^2 \end{vmatrix} = 0$$

$\therefore k(k^2) - 2k(k^2 - k) = 0$

$\therefore \; k^3 - 2k^3 + 2k^2 = 0$

$\therefore \; k^2(2 - k) = 0$

$\therefore \; k = 0, 2$

$\Rightarrow$ "two" distinct values of k

24. $(\lambda - 2)(\lambda + 2)(\lambda - 4)(\lambda - 3) = 0$

$\lambda = 2, 3, -2, 4$

25. $A = \begin{bmatrix} -4 & 2 \\ 4 & 3 \end{bmatrix}$

first are have to find eigen value of matrix

$|A - \lambda I| = 0$

$$\begin{vmatrix} -4-\lambda & 2 \\ 4 & 3-\lambda \end{vmatrix} = 0$$

$\therefore (-4-\lambda)(3-\lambda) - 8 = 0$

$\therefore -12 + 4\lambda - 3\lambda + l^2 - \lambda 8 = 0$

$\therefore \lambda^2 + \lambda - 20 = 0$

$\therefore \lambda^2 + 5\lambda - 4\lambda - 20 = 0$

$\therefore (\lambda + 5) - 4(\lambda + 5) = 20$

$\therefore \lambda = 4$ or $\lambda = -5$

$$= \frac{0.1}{6} + \frac{1}{3}$$

$$= \frac{0.1 + 2}{6}$$

$$= \frac{2.1}{6}$$

$$= \frac{21}{6}$$

$$= \frac{7}{20}$$

$\therefore$ option (a) is correct

26. Given, Matrix $M = \begin{bmatrix} 4 & 2 \\ 2 & 4 \end{bmatrix}$

$$\therefore \quad M - \lambda I = \begin{bmatrix} 4-\lambda & 2 \\ 2 & 4-\lambda \end{bmatrix}$$

For the eigen vector $\begin{bmatrix} 101 \\ 101 \end{bmatrix}$,

$(4 - \lambda)(101) + 2(101) = 0$

$\Rightarrow \; 4 - \lambda + 2 = 0$

$\Rightarrow \; \lambda = 6$ (eigen value)

27. $$P - \lambda I = 0$$

For $\lambda = 0, P = 0$

$$\Rightarrow \begin{bmatrix} P_{11} & P_{12} \\ P_{21} & P_{22} \end{bmatrix} = 0$$

$$\Rightarrow P_{11} P_{22} - P_{12} P_{21} = 0$$

29. Skew symmetric matrix,

$$A = -A^T$$

Since magnitude of diagonal element is same only sign is changed, so eigen value must be either zero or pure imaginary to satisfy this relation.

30. $|A| = 3[60-49] - 5[25-14] + 2[35-24] = 0$

$|A| = a_1 \times a_2 \times a_3 = 0$

which, implies either of the above eigen values equal to zero. It may be one or two negative eigen values.

31. A real (4 × 4) matrix A satisfies the equation $A^2 = I$ where I is the (4 × 4) identity matrix. The positive eigen value of A is 1.

32. Consider,

$A\begin{bmatrix} -1 & 1 \\ 1 & -1 \end{bmatrix}$ which is real symmetric matrix

Characteristic equation is

$$|A - \lambda I| = 0$$

$$\Rightarrow (1 + \lambda)^2 - 1 = 0$$

$$\Rightarrow \lambda + 1 = \pm 1$$

$$\therefore \lambda = 0, -2 \text{ (not positive)}$$

(*b*) is not true

(*a*), (*c*), (*d*) are true using properties of eigen values

33. $$AX = \lambda X$$

$$\Rightarrow \begin{bmatrix} 4 & 1 & 2 \\ P & 2 & 1 \\ 14 & -4 & 10 \end{bmatrix} = \lambda \begin{bmatrix} 1 \\ 2 \\ 3 \end{bmatrix}$$

$$\Rightarrow \begin{bmatrix} 12 \\ P+7 \\ 36 \end{bmatrix} = \begin{bmatrix} \lambda \\ 2\lambda \\ 3\lambda \end{bmatrix}$$

$$\Rightarrow \lambda = 12 \quad ...(1)$$

$$2\lambda = P + 7 ...(2)$$

and $3\lambda = 36$

i.e., $\lambda = 12$

$\therefore$ Equation (2) gives $P + 7 = 24 \Rightarrow P = 17$

34. Let $A = \begin{bmatrix} 10 & 5+J & 4 \\ x & 20 & 2 \\ 4 & 2 & -10 \end{bmatrix}$

Given that all eigen values of A are real.

$\Rightarrow$ A is Hermitian

$\Rightarrow A^\theta = A$ i.e. $(\bar{A})^T = A$

$$\begin{bmatrix} 10 & \bar{x} & 4 \\ 5-j & 20 & 2 \\ 4 & 2 & -10 \end{bmatrix} = \begin{bmatrix} 10 & 5+J & 4 \\ x & 20 & 2 \\ 4 & 2 & -10 \end{bmatrix}$$

$\Rightarrow x = 5 - j$

35. Given sequence $x[n]$ is given by

$$\begin{bmatrix} x(n) \\ x(n-1) \end{bmatrix} = \begin{bmatrix} 1 & 1 \\ 1 & 0 \end{bmatrix}^n \begin{bmatrix} 1 \\ 0 \end{bmatrix}, n \ge 2$$

Put $n = 2$, we get

$$\begin{bmatrix} x(2) \\ x(1) \end{bmatrix} = \begin{bmatrix} 1 & 1 \\ 1 & 0 \end{bmatrix}^2 \begin{bmatrix} 1 \\ 0 \end{bmatrix}$$

$$= \begin{bmatrix} 2 & 1 \\ 1 & 1 \end{bmatrix} \begin{bmatrix} 1 \\ 0 \end{bmatrix} = \begin{bmatrix} 2 \\ 1 \end{bmatrix}$$

$x(2) = 2$

$x(1) = 1$

$n = 3$

$$\begin{bmatrix} x(3) \\ x(2) \end{bmatrix} = \begin{bmatrix} 1 & 1 \\ 1 & 0 \end{bmatrix}^3 \begin{bmatrix} 1 \\ 0 \end{bmatrix}$$

$$= \begin{bmatrix} 3 & 2 \\ 2 & 1 \end{bmatrix} \begin{bmatrix} 1 \\ 0 \end{bmatrix} = \begin{bmatrix} 3 \\ 2 \end{bmatrix}$$

$x(3) = 3$

$x(2) = 2$

From the above values we can write the recursive relation is as

$$x(n) = x(n-1) + x(n-2)$$

$$x(2) = x(1) + x(0)$$

$$= 1 + 1 = 2$$

$$x(3) = x(2) + x(1)$$

$$= 2 + 1 = 3$$

$$x(4) = x(3) + x(2)$$

$$= 3 + 2 = 5$$

$$x(5) = x(4) + x(3) = 5 + 3 = 8$$
$$x(6) = x(5) + x(4) = 8 + 5 = 13$$
$$x(7) = x(6) + x(5) = 13 + 8 = 21$$
$$x(8) = x(7) + x(6) = 21 + 13 = 34$$
$$x(9) = x(8) + x(7) = 34 + 21 = 55$$
$$x(10) = x(9) + x(8) = 55 + 34 = 89$$
$$x(11) = 89 + 55 = 144$$
$$x(12) = 144 + 89 = 233$$

So, the required value of $x(12)$ is 233.

36. From the given question,

A has an eigen value is zero

$$|A| = 0$$

$$\begin{vmatrix} 3 & 2 & 4 \\ 9 & 7 & 13 \\ -6 & -4 & -9 + x \end{vmatrix} = 0$$

$$3(-63 + 7x + 52) - 2(-81 + 9x + 78) + 4(-36 + 42) = 0$$
$$3(7x - 11) - 2(9x - 3) + 4(6) = 0$$
$$21x - 33 - 18x + 6 + 24 = 0$$
$$3x - 3 = 0$$
$$x = 1$$

37. From the given square matrix

$$\det(A) = \sigma^2 - \omega x$$
$$(\sigma + j\omega)(\sigma - j\omega) = \sigma^2 - \omega x$$
$$(\sigma)^2 - (j\omega)^2 = \sigma^2 - \omega x$$
$$\sigma^2 - (j^2)(\omega^2) = \sigma^2 - \omega x$$
$$\sigma^2 + \omega^2 = \sigma^2 - \omega x \qquad (\text{since } j^2 = -1)$$
$$\omega^2 = -\omega x$$
$$\therefore \quad x = -\omega$$

38.
$$|A - \lambda I| = 0$$

$$\therefore \begin{bmatrix} 1 & 2 & 3 & 4 & 5 \\ 5 & 1 & 2 & 3 & 4 \\ 4 & 5 & 1 & 2 & 3 \\ 3 & 4 & 5 & 1 & 2 \\ 2 & 3 & 4 & 5 & 1 \end{bmatrix} - \lambda \begin{bmatrix} 1 & 0 & 0 & 0 & 0 \\ 0 & 1 & 0 & 0 & 0 \\ 0 & 0 & 1 & 0 & 0 \\ 0 & 0 & 0 & 1 & 0 \\ 0 & 0 & 0 & 0 & 1 \end{bmatrix} = 0$$

$$\begin{bmatrix} 1-\lambda & 2 & 3 & 4 & 5 \\ 5 & 1-\lambda & 2 & 3 & 4 \\ 4 & 5 & 1-\lambda & 2 & 3 \\ 3 & 4 & 5 & 1-\lambda & 2 \\ 2 & 3 & 4 & 5 & 1-\lambda \end{bmatrix} = 0$$

Sum of all elements in any one row must be zero,

i.e., $\quad 15 - \lambda = 0$

$$\therefore \quad \lambda = 15$$

39. Eigen vectors corresponding to distinct eigen values are linearly independent.

40. Given,

$$\text{Matrix A} = \begin{bmatrix} 2 & 2 & 3 & 3 \\ 0 & 1 & 1 & 1 \\ 0 & 0 & 3 & 3 \\ 0 & 0 & 0 & 2 \end{bmatrix}$$

→ Upper triangular matrix

∴ Eigen values of A are $\lambda = 2, 1, 3, 2$ [diagonal elements]

∴ The number of distinct eigen values of the matrix = 3. i.e. 1, 2, 3.

2 CHAPTER

Calculus & Vector Analysis

Analysis of Previous GATE Papers

Year → Topics ↓			2019	2018	2017 Set 1	2017 Set 2	2016 Set 1	2016 Set 2	2016 Set 3	2015 Set 1	2015 Set 2	2015 Set 3
L IMI TS, CONTINUI TY & DI FFERENTIABI L I TY, MEAN VALUE THEOREM:	1 Mark	MCQ Type			1		1			1		
		Numerical Type										
	2 Marks	MCQ Type	1									
		Numerical Type										
	Total		2		1		1			1		
EVALUATION OF DEFINI TE & IMPROPER INTEGRAL	1 Mark	MCQ Type										
		Numerical Type	1						1			
	2 Marks	MCQ Type	1			1						
		Numerical Type			2		2					
	Total		3		4	2	4		1			
MAXIMA & MINIMA	1 Mark	MCQ Type						1				
		Numerical Type										
	2 Marks	MCQ Type								1		
		Numerical Type				1				1		
	Total					2		1		4		
PARTIAL DERIVATIVES & TAYOR SERIES:	1 Mark	MCQ Type										1
		Numerical Type										
	2 Marks	MCQ Type										
		Numerical Type										
	Total											1
VECTOR IN PLANE & SPACE, VECTOR OPERATINS:	1 Mark	MCQ Type										
		Numerical Type				1						
	2 Marks	MCQ Type										
		Numerical Type							1			
	Total					1			2			
CURL, DIVERGENCE, GRADIENT, GAURS'S GREEN'S & STOKE'S TEOREM	1 Mark	MCQ Type										
		Numerical Type						1				
	2 Marks	MCQ Type										
		Numerical Type										
	Total							1				

LIMITS, CONTINUITY & DIFFERENTIABILITY, MEAN VALUE THEOREM

1. The derivative of the symmetric function drawn in given figure will look like

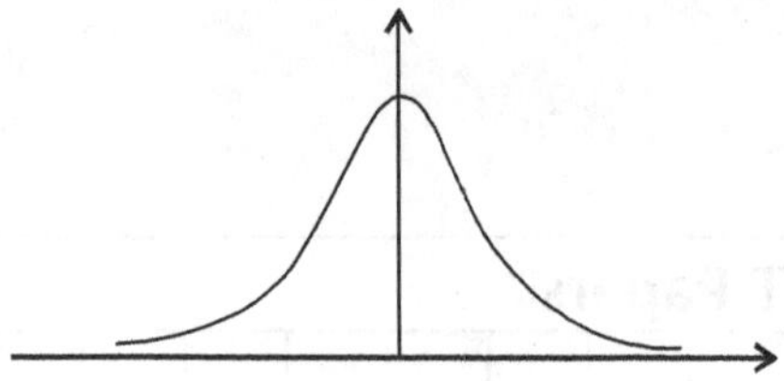

(a)
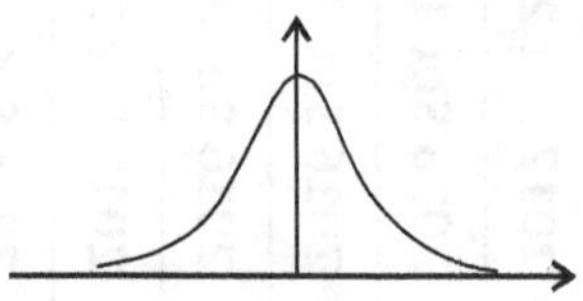

(b)
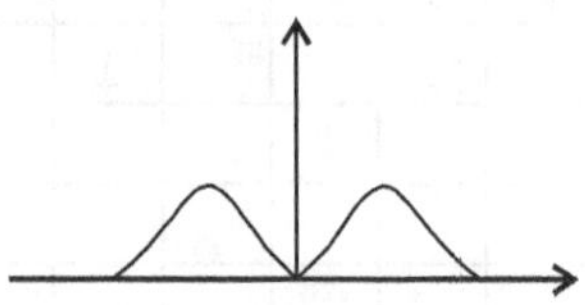

(c)
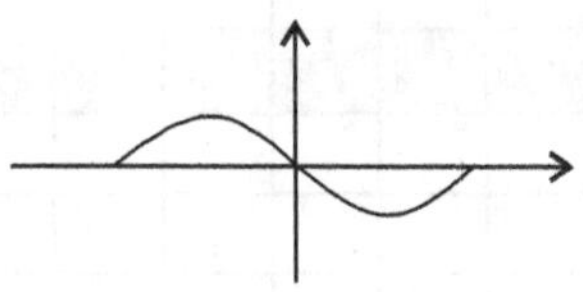

(d)
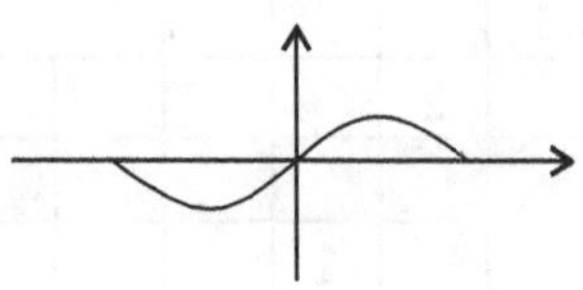

[2005 : 2 Marks]

2. For $|x| << 1$, cot h(x) can be approximated as

(a) x (b) x^2

(c) $\frac{1}{x}$ (d) $\frac{1}{x^2}$

[2007 : 1 Mark]

3. $\lim_{\theta \to 0} \frac{\sin\left(\frac{\theta}{2}\right)}{\theta}$ is

(a) 0.5 (b) 1

(c) 2 (d) not defined

[2007 : 1 Mark]

4. The value of $\lim_{x \to \infty} \left(1+\frac{1}{x}\right)^x$ is

(a) ln2 (b) 1.0

(c) e (d) ∞

[2014 : 1 Marks, Set-3]

5. A function $f(x) = 1 - x^2 + x^3$ is defined in the closed interval [–1, 1]. The value of x in the open interval (–1, 1) for which the mean value theorem is satisfied, is

(a) $-\frac{1}{2}$ (b) $-\frac{1}{3}$

(c) $\frac{1}{3}$ (d) $\frac{1}{2}$

[2015 : 1 Mark, Set-1]

6. Given the following statements about a function $f : R \to R$, select the right option:

P : If f(x) is continuous at $x = x_0$, then it is

Q : If f(x) is continuous at $x = x_0$, then it may not be differentiable at $x = x_0$.

R : If f(x) is differentiable at $x = x_0$, then it is also continuous at $x = x_0$.

(a) P is true, Q is false, R is false

(b) Pis false, Q is true, P is true

(c) P is false, Q is true, P is false

(d) Pis true, Q is false, P is true

[2016 : 1 Mark, Set-1]

7. Consider the following statements about the linear dependence of the real valued functions $y_1 = 1$, $y_2 = x$ and $y_3 = x^2$, over the field of real numbers.

I. y_1, y_2 and y_3 are linearly independent on $-1 \le x \le 0$.

II. y_1, y_2 and y_3 are linearly dependent on $-1 \le x \le 0$.

III. y_1, y_2 and y_3 are linearly independent on $1 \le x \le 0$.

IV. y_1, y_2 and y_3 are linearly dependent on $-1 \le x \le 0$.

Which one among the following is correct?

(a) Both I and II are true

(b) Both I and III are true

(c) Both II and IV are true

(d) Both III and IV are true

[2017 : 1 Mark, Set-1]

8. Consider a differentiable function f(x) on the set of real numbers such that $f(-1) = 0$ and $|f'(x)| \le 2$. Given these conditions, which one of the following inequalities is necessarily true for all $x \in [-2, 2]$?

(a) $f(x) \le 2|x+1|$ (b) $f(x) \le 2|x|$

(c) $f(x) \le \frac{1}{2}|x+1|$ (d) $f(x) \le \frac{1}{2}|x|$

[2019 : 2 Marks]

EVALUATION OF DEFINITE & IMPROPER INTEGRAL

9. The value of the integral $I = \frac{1}{\sqrt{2\pi}}\int_0^{\infty} \exp\left(\frac{x^2}{8}\right) dx$ is

(a) 1 (b) π

(c) 2 (d) 2π

[2005 : 2 Marks]

10. The integral $\int_0^{\pi} \sin^3 \theta d\theta$ is given by

(a) $\frac{1}{2}$ (b) $\frac{2}{3}$

(c) $\frac{4}{3}$ (d) $\frac{8}{3}$

[2006 : 2 Marks]

11. The following plot shows a function y which varies linearly with x. The value of the integral $I = \int_1^2 y dx$ is

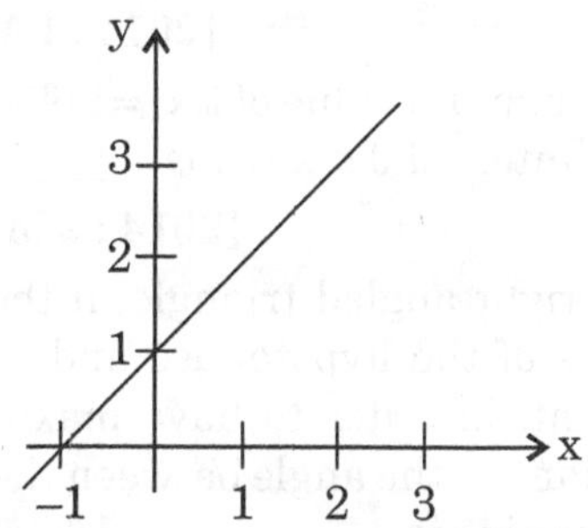

(a) 0.5 (b) 2.5

(c) 4.0 (d) 5.0

[2007 : 1 Mark]

12. The value of the integral of the function $g(x, y) \cong 4x^3 + 10y^4$ along the straight line segment from the point (0, 0) to the point (1, 2) in the x-y plane is

(a) 33 (b) 35

(c) 40 (d) 56

[2008 : 2 Marks]

13. Consider points P and Q in the x-y plane, with P = (1, 0) and Q = (0, 1). The line integral $2\int_P^Q (xdx + ydy)$ along the semicircle with the line segment PQ as its diameter

(a) – 1

(b) 0

(c) 1

(d) depends on the direction (clockwise or anti-clockwise) of the semicircle.

[2008 : 2 Marks]

14. The volume under the surface z(x, y) = x + y and above the triangle in the x-y plane defined by $\{0 \le y \le \text{ and } 0 \le x \le 12\}$ is ______.

[2014 : 2 Marks, Set-1]

15. The integral $\frac{1}{2\pi}\iint_D (x + y + 10)dx, dy$, where D denotes the disc: $x^2 + y^2 \le 4$, evaluates to ______.

[2016 : 2 Marks, Set-1]

16. The region specified by $\{(p, \phi, z) : 3 \le p \le 5, \frac{\pi}{8} \le \varphi \le \frac{\pi}{4}, 3 \le z \le 4.5\}$ in cylindrical coordinates has volume of ______.

[2016 : 2 Marks, Set-1]

17. The integral $\int_0^1 \frac{dx}{\sqrt{(1-x)}}$, is equal to ______.

[2016 : 1 Mark, Set-3]

18. Let $I = \int_C (2zdx + 2ydy + 2xdz)$ where x, y, z are real, and let C be the straight line segment from point A : (0, 2, 1) to point B : (4, 1, –1). The value of I is ______.

[2017 : 2 Marks, Set-1]

19. A three dimensional region R of finite volume is described by $x^2 + y^2 \le z^3; 0 \le z \le 1$, where x, y, z are real. The volume of R (up to two decimal places) is ______.

[2017 : 2 Marks, Set-1]

20. The values of the integrals

$$\int_0^1\left(\int_0^1 \frac{x-y}{(x+y)^3} dy\right)dx \text{ and } \int_0^1\left(\int_0^1 \frac{x-y}{(x+y)^3} dx\right)dy$$

are

(a) same and equal to 0.5.

(b) same and equal to –0.5.

(c) 0.5 and –0.5 respectively.

(d) –0.5 and 0.5 respectively.

[2017 : 2 Marks, Set-2]

21. The value of integral $\int_0^{\pi}\int_y^{\pi} \frac{\sin x}{x} dx\, dy$, is equal to ______.

[2019 : 1 Mark]

22. Consider the line integral $\int_C (xdy - ydx)$. The integral being taken in a counterclockwise direction over the closed curve C that forms the boundary of the region R shown in the figure below. The region R is the area enclosed by the union of a 2 × 3 rectangle and a semi-circle of radius 1. The line integral evaluates to

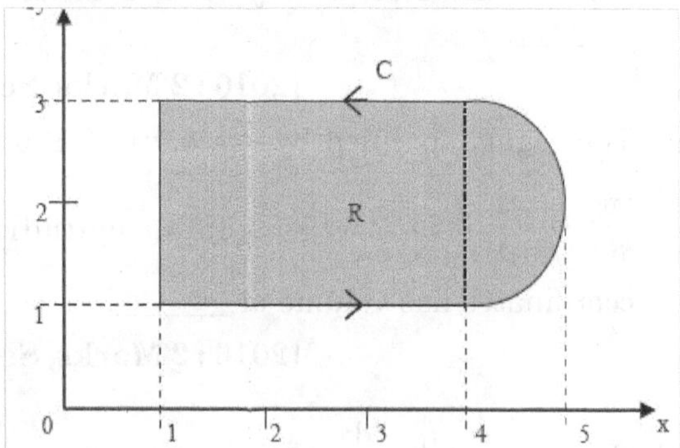

(a) $16 + 2\pi$ (b) $6 + \pi/2$
(c) $12 + \pi$ (d) $8 + \pi$

[2019 : 2 Mark]

MAXIMA & MINIMA

23. As x increased from $-\infty$ to ∞, the function

$$f(x) = \frac{e^x}{1+e^x}$$

(a) monotonically increases
(b) monotonically decreases
(c) increases to a maximum value and then decreases
(d) decreases to a minimum value and then increases

[2006 : 2 Marks]

24. Which one of the following functions is stricity bounded?

(a) $\frac{1}{x^2}$ (b) e^x
(c) x^2 (d) e^{-x^2}

[2007 : 1 Mark]

25. Consider the function $f(x) \cong x^2 - x - 2$. The maximum value of f(x) in the closed interval [–4, 4] is

(a) 18 (b) 10
(c) –2.25 (d) indeterminate

[2007 : 2 Marks]

26. For real values of x, the minimum value of the function f(x) = exp(x) + exp(–x) is

(a) 2 (b) 1
(c) 0.5 (d) 0

[2008 : 1 Mark]

27. If $e^y = x^{1/x}$, then y has a

(a) maximum at x = e
(b) minimum atx = e
(c) maximum at $x = e^{-1}$
(d) minimum atx = e^{-1}

[2010 : 2 Marks]

28. The maximum value of $f(x) = x^3 - 9x^2 + 24x + 5$ in the interval [1, 6] is

(a) 21 (b) 25
(c) 41 (d) 46

[2012 : 2 Marks]

29. For $0 \le t < \infty$, the maximum value of the function $f(t) = e^{-1} - 2e^{-2t}$ occurs at

(a) $t = \log_e 4$ (b) $t = \log_e 2$
(c) $t = 0$ (d) $t - \log_e 8$

[2014 : 1 Mark, Set-2]

30. The maximum value of the function $f(x) = \ln(1 + x) - x$ (where $x > -1$) occur at x = ______.

[2014 : 1 Mark, Set-3]

31. The maximum value of $t(x) = 2r^3 - 9x^2 + 12x - 3$ in the interval $0 \le x \le 3$ is ______.

[2014 : 2 Marks, Set-3]

32. For a right angled triangle, if the sum of the lengths of the hypotenuse and a side is kept constant, in order to have maximum area of the triangle, the angle between the hypotenuse and the side is

(a) 12° (b) 36°
(c) 60° (d) 45°

[2014 : 2 Marks, Set-4]

33. Which one of the following graphs describes the function $f(x) = e^{-x}(x^2 + x + 1)$?

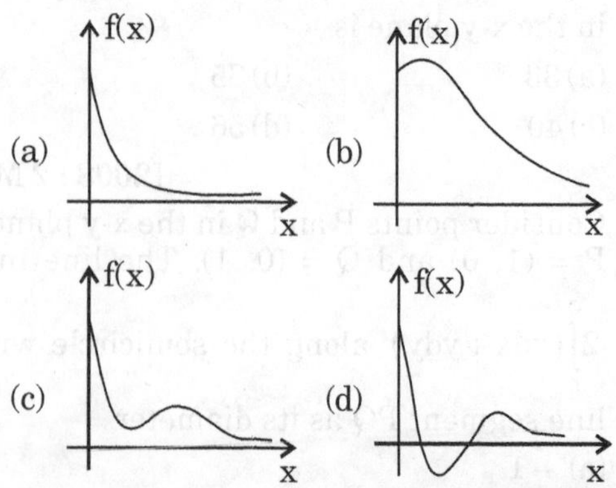

[2015 : 2 Marks, Set-1]

34. The maximum area (in square unit) of a rectangle whose vertices lies on the ellipse $x^2 + 4y^2 = 1$ is ______.

[2015 : 2 Marks, Set-1]

35. Consider the plot f(x) versusx as shown below.

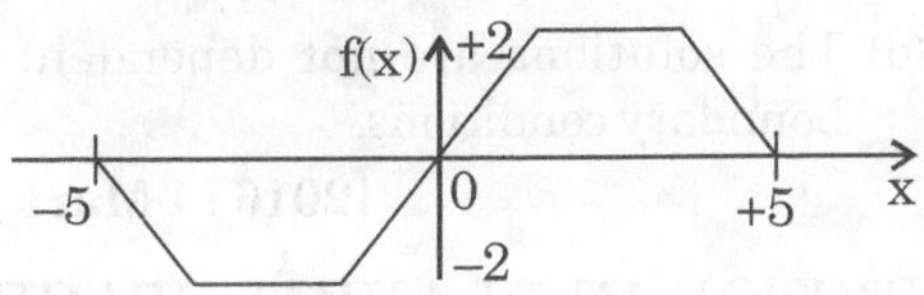

Suppose, $F(x) = \int_{-5}^{x} f(y)dy$. Which one of the following is a graph of F(x)?

(a)
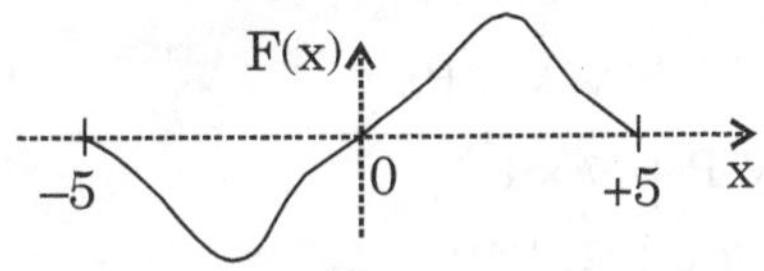

(b)
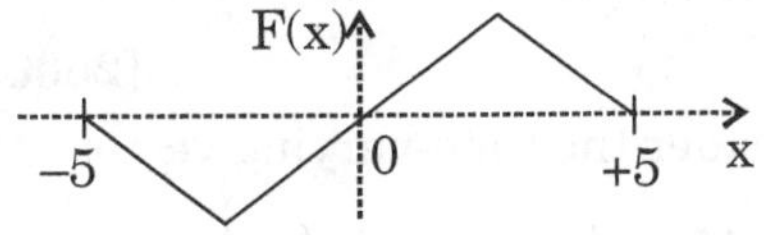

(c)
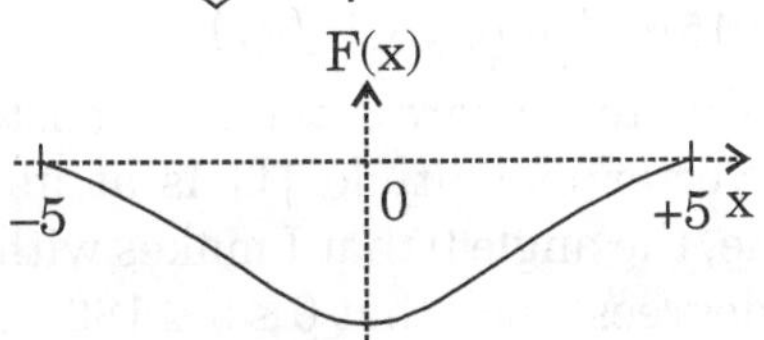

(d)
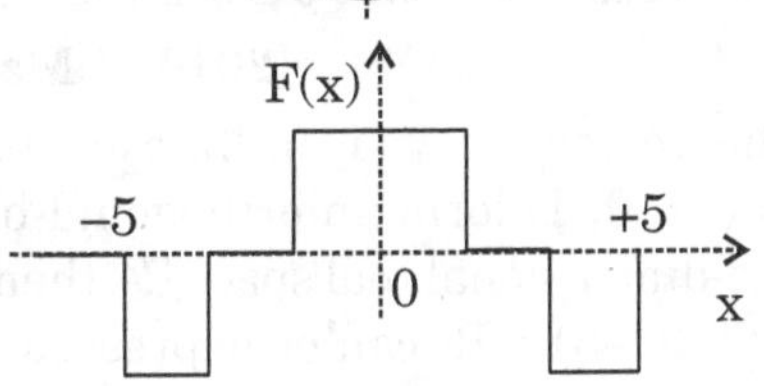

36. As x varies from −1 to +3, which one of the following describes the behaviour of the function $f(x) = x^3 - 3x^2 + 1$?

(a) f(x) increases monotonically.

(b) f(x) increases, then decreases and increases again.

(c) f(x) decreases, then increases and decreases again.

(d) f(x) increases and then decreases.

[2016 : 1 Mark, Set-2]

37. The minimum value of the function $f(x) = \frac{1}{3}x(x^2 - 3)$ in the interval $-100 \le x \le 100$ occurs at x = ________.

[2017 : 2 Marks, Set-2]

PARTIAL DERIVATIVES & TAYLOR SERIES

38. Which of the following functions would have only odd powers of x in its Taylor series expansion about the point x = 0?

(a) $\sin(x^3)$ (b) $\sin(x^2)$

(c) $\cos(x^3)$ (d) $\cos(x^2)$

[2008 : 1 Mark]

39. In the Taylor series expansion of exp(x) + sin(x) about the point x = n, the coefficient of $(x - \pi)^2$ is

(a) $\exp(\pi)$ (b) $0.5\exp(\pi)$

(c) $\exp(\pi) + 1$ (d) $\exp(\pi) - 1$

[2008 : 2 Marks]

40. The Taylor series expansion of $\frac{\sin x}{x - \pi}$ at $x = \pi$ is given by

(a) $1 + \frac{(x-\pi)^2}{3!} + ...$ (b) $-1 - \frac{(x-\pi)^2}{3!} + ...$

(c) $1 - \frac{(x-\pi)^2}{3!} + ...$ (d) $-1 + \frac{(x-\pi)^2}{3!} + ...$

[2009 : 2 Marks]

41. The Taylor series expansion of 3sinx + 2cos x is ________.

(a) $2 + 3x - x^2 - \frac{x^3}{2} + ...$

(b) $2 - 3x + x^2 - \frac{x^3}{2} + ...$

(c) $2 + 3x + x^2 - \frac{x^3}{2} + ...$

(d) $2 - 3x - x^2 + \frac{x^3}{2} + ...$

[2014 : 2 Marks, Set-1]

42. The series $\sum_{n=0}^{\infty} \frac{1}{n!}$ converges to

(a) 2l 2 (b) $\sqrt{2}$

(c) 2 (d) e

[2014 : 1 Mark, Set-4]

43. If z = xyln(xy), then

(a) $x\frac{\partial z}{\partial x} + y\frac{\partial z}{\partial y} = 0$ (b) $y\frac{\partial z}{\partial x} = x\frac{\partial z}{\partial y}$

(c) $x\frac{\partial z}{\partial x} = y\frac{\partial z}{\partial y}$ (d) $y\frac{\partial z}{\partial x} + x\frac{\partial z}{\partial y} = 0$

[2014 : 1 Mark, Set-3]

44. The contour on the x-y plane, where the partial derivative of $x^2 + y^2$ with respect to y is equal to the partial derivative of 6y + 4x with respect to x, is

(a) y = 2 (b) x = 2

(c) x = y = 4 (d) x − y = 0

[2015 : 1 Mark, Set-3]

45. A triangle in the xy-plaie is bounded by the straight lines $2x = 3y$, $y = 3$ and $x = 3$. The volume above the triangle and under the plane $x + y + z = 6$ is ______.

[2016 : 2 Marks, Set-3]

46. Let $f(x) = e^{x+x^2}$ for real x. From among the following, choose the Taylor series approximation of f(x) around x = 0, which includes all powers of x less than or equal to 3.

(a) $1 + x + x^2 + x^3$

(b) $1 + x + \frac{3}{2}x^2 + x^3$

(c) $1 + x + \frac{3}{2}x^2 + \frac{7}{6}x^3$

(d) $1 + x + 3x^2 + 7x^3$

[2017 : 2 Marks, Set-1]

47. Taylor series expansion of $f(x) = \int_0^x e^{-\left(\frac{t^2}{2}\right)} dt$ aront x = 0 has the form

$$f(x) = a_0 + a_1x + a_2x^2 + ...$$

The coefficient a_2 (correct to two decimal placees) is equal to ______.

[2018 : 1 Mark]

48. Let $f(x, y) = \frac{ax^2 + by^2}{xy}$, where a and b are constants. If $\frac{\partial f}{\partial x} = \frac{\partial f}{\partial y}$ at $x = 1$ and $y = 2$, then the relation between a and b is

(a) $a = \frac{b}{4}$ (b) $a = \frac{b}{2}$

(c) $a = 2b$ (d) $a = 4b$

[2018 : 1 Mark]

49. Let $r = x^2 + y - z$ and $z^3 - xy + yz + y^3 = 1$. Assume that x and yare independent variables. At $(x, y, z) = (2, -1, 1)$, the value (correct to two decimal places) of is ______.

[2018 : 2 Marks]

50. Which one of the following is a property of the solutions to the Laplace equation: $\nabla^2 f = 0$?

(a) The solutions have neither maxima nor minima anywhere except at the boundaries.

(b) The solutions are not separable in the coordinates.

(c) The solutions are not continuous.

(d) The solutions are not dependent on the boundary conditions.

[2016 : 1 Mark, Set-1]

VECTOR IN PLANE & SPACE, VECTOR OPERATIONS

51. $\nabla \times \nabla \times P$, where P is a vector is equal to

(a) $P \times \nabla \times P - \nabla^2 P$.

(b) $\nabla^2 P + \nabla(\nabla \times P)$.

(c) $\nabla^2 P + \nabla \times P$.

(d) $\nabla(\nabla . P) - \nabla^2 P$.

[2006 : 1 Mark]

52. Consider the time-varying vector $I = \hat{x}15\cos(\omega t) + \hat{y}5\sin(\omega t)$ in Cartesian coordinates, where $\omega > 0$ is a constant. When the vector magnitude $|I|$ is at its minimum value, the angle 0 that I makes with the x axis (in degrees, such that $0 \le \theta \le 180$) is ______.

[2016 : 1 Mark, Set-2]

53. If the vectors $e_1 = (1, 0, 2)$, $e_2 = (0, 1, 0)$ and $e_3 = (-2, 0, 1)$ form an orthogonal basis of the three-dimensional real space R^3, then the vector $u = (4, 3, -3) \in R^3$ can be expressed as

(a) $u = -\frac{2}{3}e_1 - 3e_2 - \frac{11}{5}e_3$

(b) $u = -\frac{2}{5}e_1 - 3e_2 + \frac{11}{5}e_3$

(c) $u = -\frac{2}{5}e_1 + 3e_2 + \frac{11}{5}e_3$

(d) $u = -\frac{2}{5}e_1 + 3e_2 - \frac{11}{5}e_3$

[2016 : 2 Marks, Set-3]

54. The smaller angle (in degrees) between the planes $x + y + z = 1$ and $2x - y + 2z = 0$ is ______.

[2017 : 1 Mark, Set-2]

CURL, DIVERGENCE, GRADIENT, GAUS'S GREEN'S & STOKE'S THEOREM

55. $\iint (\nabla \times P) \cdot ds$, where P is a vector, is equal to

(a) $\oint P \cdot dl$ (b) $\oint \nabla \times \nabla \times P \cdot dl$

(c) $\oint \nabla \times P \cdot dl$ (d) $\iiint \nabla \cdot P dv$

[2006 : 1 Mark]

56. If $\vec{A} = xy\hat{a}_x + x^2\hat{a}_y$, $\oint_C \vec{A}\cdot d\vec{l}$ over the path shown in the figure is

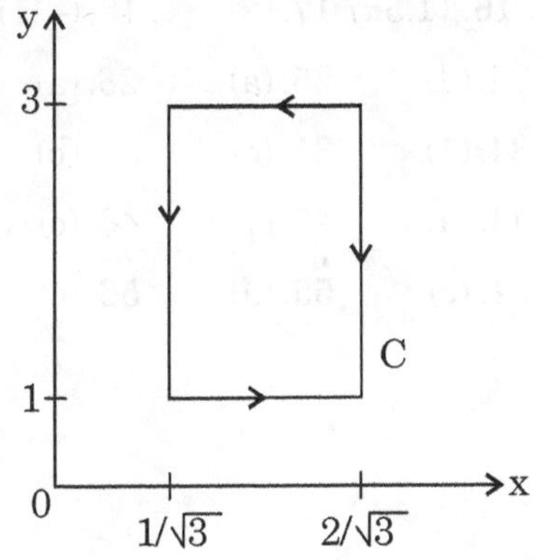

(a) 0 (b) $\frac{2}{\sqrt{3}}$

(c) 1 (d) $2\sqrt{3}$

[2010 : 2 Marks]

57. Consider a vector field $\vec{A}(\vec{r})$. The closed loop line integral $\oint \vec{A}\cdot\vec{dl}$ can be expressed as

(a) $\oiint(\nabla\times\vec{A})\cdot\vec{ds}$ over the closed surface bounded by the loop.

(b) $\oiiint(\nabla\cdot\vec{A})dv$ over the closed volume bounded by the top.

(c) $\oiiint(\nabla\cdot\vec{A})dv$ over the open volume bounded by the loop.

(d) $\iint(\nabla\times\vec{A})\cdot\vec{ds}$ over the open surface bounded by the loop.

[2013 : 1 Mark]

58. The divergence of the vector field

$\vec{A} = x\hat{a}_x + ya_y + z\hat{a}_z$ is

(a) 0 (b) $\frac{1}{3}$

(c) 1 (d) 3

[2013 : 1 Mark]

59. If $\vec{R} = x\hat{a}_x + y\hat{a}_y + z\hat{a}_z$ and $|\bar{r}| = r$, then $\text{div}(r^2\nabla(\ln r))$______.

[2014 : 2 Marks, Set-2]

60. The magnitude of the gradient for the function $f(x, y, z) = x^2 + 3y^2 + z^3$ at the point (1,1,1) is ______.

[2014 : 1 Mark, Set-4]

61. The directional derivative of $f(x,y) = \frac{xy}{\sqrt{2}}(x+y)$ at (1, 1) in the direction of the unit vector an angle of $\frac{\pi}{4}$ with y-axis, is given by ______.

[2014 : 1 Mark, Set-2]

62. Given $\vec{F} - z\hat{a}_x + x\hat{a}_y + y\hat{a}_z$. If S represents the portion of the sphere $x^2 + y^2 + z^2 = 1$ for $z \geq 0$, then $\oint_s \nabla\times\vec{F}\cdot\vec{ds}$ ______.

[2014 : 2 Marks, Set-4]

63. Suppose C is the closed curve defined as the circle $x^2 + y^2 = 1$ with C oriented anti clockwise. The value of $\oint(xy^2dx + x^2ydy)$ over the curve C equals ______.

[2016 : 2 Marks, Set-2]

ANSWERS

1. (c)	2. (c)	3. (a)	4. (c)	5. (b)	6. (b)	7. (b)	8. (a)	9. (a)	10. (c)
11. (b)	12. (a)	13. (b)	14. (862 to 866)		15. (20)	16. (1.5π)	17. (2)	18. (–11)	
19. (0· 7853)		20. (c)	21. (2)	22. (c)	23. (a)	24. (d)	25. (a)	26. (a)	27. (a)
28. (b)	29. (a)	30. (0)	31. (6)	32. (c)	33. (b)	34. (1)	35. (c)	36. (b)	37. (b)
38. (a)	39. (b)	40. (d)	41. (a)	42. (d)	43. (c)	44. (a)	45. (10)	46. (c)	47. (0)
48. (d)	49. (4· 50)	50. (c)	51. (d)	52. (90°)	53. (b)	54. (c)	55. (a)	56. (c)	57. (d)
58. (d)	59. (3)	60. (7)	61. (3)	62. (6.28)	63. (0)				

EXPLANATIONS

1. The graph of given function indicator –ve slope in +ve half and + v2 slope in – ve half.Hence option (c) satisfies the above condition.

4. $$\lim_{x\to\infty}\left(1+\frac{1}{x}\right)^x = e \text{(standard limit)}$$

5. The given function is

$$F(x) = 1 - x^2 + x^3,$$

where $x \in [-1, 1]$

$\Rightarrow \quad F'(x) = -2x + 3x^2$

By mean value theorem

$$F'(x) = \frac{F(1)-F(-1)}{1-(-1)}$$

Now $\quad F(1) = 1 - (1)^2 + (1)^3 = 1$

and $\quad F(-1) = 1 - (-1)^2 + (-1)^3 = 1 - 1 - 1 = -1$

$\therefore \quad F'(x) = \dfrac{1-(-1)}{1-(-1)} = \dfrac{2}{2} = 1$

$\therefore \quad F'(x) = 1$

$\therefore \quad -2x + 3x^2 = 1$

$\therefore \quad 3x^2 - 2x - 1 = 0$

$\therefore \quad 3x^2 - 3x + x - 1 = 0$

$\therefore \quad 3x(x-1) + 1(x-1) = 0$

$\therefore \quad (3x + 1) = 0$

or $\quad x - 1 = 0$

$\therefore \quad x = -\dfrac{1}{3}$

or $\quad x = 1$

Now $-\dfrac{1}{3}$ lies between $(-1, 1)$

$\therefore \quad x = -\dfrac{1}{3}$

6. Since continuous function may not be differentiable at $x = x_0$, but differentiable function is always continuous.

7. Any of the given three functions cannot be written as the linear combination of other two functions. Hence, the statements I and III are correct.

8. Satisfy the given condition

Given $|f'(x)| \le 2; f(-1) = 0$

$\Rightarrow -2 \le f'(x) \le 2$; where $x \in [-2, 2]$

Now applying Lagrange's Mean Value Theorem over interval [–1, 2]

$$\Rightarrow -2 \le f'(x) \le 2 \Rightarrow -2 \le \frac{f(2)-f(-1)}{2-(-1)} \le 2$$

$$\left[\because f'(c) = \frac{f(b)-f(a)}{b-a}\right]$$

$$\Rightarrow -2 \le \frac{f(2)-0}{3} \le 2$$

$\Rightarrow -2 \le f(2) \le 6$ ...(2)

Hence option (a) satisfies the equation.

9. $I = \frac{1}{\sqrt{2\pi}} \int_0^{\infty} \exp\left(\frac{-x^2}{8}\right) dx$

$dt \frac{-x^2}{8} = t$

$\therefore \frac{2x}{8} dx = dt$

$\therefore dx = \frac{4dt}{x}$

$\therefore dx = \frac{\sqrt{2}}{\sqrt{E}} dt$

$\therefore I = \frac{1}{\sqrt{2\pi}} \int_0^{\infty} e^{-\frac{2}{8}} dx$

$= \frac{1}{\sqrt{2\pi}} \int_0^{\infty} e^{-t} \frac{\sqrt{2}}{\sqrt{t}} dt$

$\therefore I = \frac{1}{\sqrt{\pi}} \times \sqrt{\pi} = 1$

$\therefore$ option (a) is correct.

10. $\text{Cot}\, ht = \frac{\text{Cos}\, hx}{\text{Sin}\, hx} = \frac{1}{x}$

11. Two points on line are (–1, 0) and (0, 1)

Hence line equation is,

$$y = \left(\frac{y_2 - y_1}{x_2 - x_1}\right) x + c$$

$$y = x + c$$

$$y = x + 1 \quad \text{....(i)}$$

$$I = \int_1^2 y\, dx$$

$$= \int_1^2 (x+1) dx$$

$$= \frac{5}{2} = 2.5$$

(since at $x = 1$, $y = 2$)

12. Here $g(x,y) = 4x^3 + 10y^4$

Now the given points on strainger line are (0,0)and (1,2)

$\therefore$ Equation of line is given by

$(y - y_1) = \frac{y, y_1}{x_2 - x_1}(x - x_1)$

$\therefore (y - o) = \frac{2}{1}(x - o)$

$\therefore y = 2x$

Now according to question,

$\int_O^1 g(x,y)dx = \int_o^1 (4x^3 + 10y^4) dx$

$= \int_o^1 (4x^3 + 10(2x)^4) dx$

$= \int_o^1 (4x^3 + 10 \times 16x^4) dx$

$= \int_o^1 (4x^3 + 160x^4) dx$

$= \left[\frac{4x^4}{4} + \frac{160x^5}{5}\right]_0^1 = \left[x^4 + 32x^5\right]_0^1$

$= 1 + 32 = 33$

$\therefore$ option (a) is correct

13. Here the given points are

P (1, 0) and Q (0, 1)

Equation of line is given by

$(y - y_1) = \frac{y_2 - y_1}{x_2 x_1}(x = x_1)$

$(y - 0) = \frac{1}{-1}(x = 1)$

$\therefore y = -x + 1$

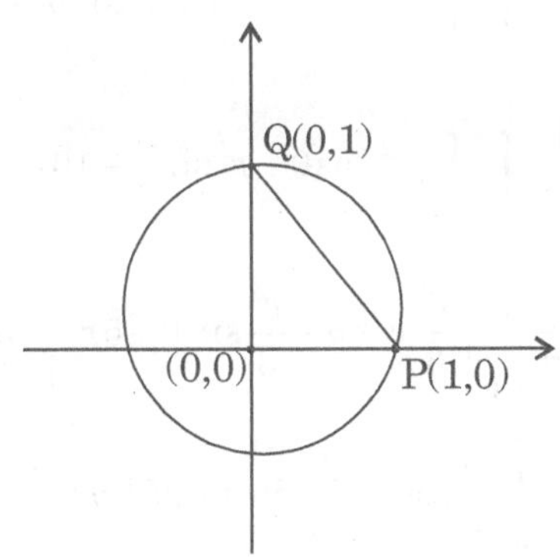

Now

$2\int_P^Q (xdx + ydy) = 2\int_P^Q xdx + 2\int_P^Q ydy$

$= 2\int_1^0 xdx + 2\int_0^1 ydy$

$= 2\left[\frac{x^2}{2}\right]_1^0 + 2\left[\frac{y^2}{2}\right]_0^1$

$= -1 + 1$

$= 0$

$\therefore$ Option (b) is correct

14. The volume under the surface $z(x, y) = x + y$ and above the triangle in the $x - y$ plane is given

$$\text{Volume} = \int_0^{12}\left[\int_0^x (x+y)\,dy\right]dx$$

$$= \int_0^{12}\left[xy + \frac{y^2}{2}\right]_0^x dx$$

$$= \int_0^{12}\left[x^2 + \frac{x^2}{2}\right]dx$$

$$\therefore \quad \text{Volume} = \left[\frac{x^3}{3} + \frac{x^3}{6}\right]_0^{12}$$

$$= \left[\frac{2x^3 + x^3}{6}\right]_0^{12}$$

$$= \left[\frac{x^3}{2}\right]_0^{12} = \frac{1}{2}(12)^3 = 864.$$

l number]

15. Given integral is converted into polar coordinates, we get

$$\frac{1}{2\pi}\iint_D (x + y + 10)\,dxdy$$

$$= \frac{1}{2\pi}\int_{r=0}^{2}\int_{\theta=0}^{2\pi}(r\cos\theta + r\sin\theta + 10)\,pdrd\theta$$

$$= \frac{1}{2\pi}\int_{r=0}^{2}\int_{\theta=0}^{2\pi}(r^2\cos\theta + r^2\sin\theta + 10r)\,dr\,d\theta$$

$$= \frac{1}{2\pi}\int_{0=0}^{2\pi}\left\{\frac{r^3}{3}\cos\theta + \frac{r^3}{3}\sin\theta + 5r^2\right\}_0^2 d\theta$$

$$= \frac{1}{2\pi}\int_0^{\pi}\left\{\frac{8}{3}\cos\theta + \frac{8}{3}\sin\theta + 20\right\}d\theta$$

$$= \frac{1}{2\pi}\left[\frac{8}{3}\sin\theta - \frac{8}{3}\cos\theta + 20\,\theta\right]_0^{2\pi}$$

$$= \frac{1}{2\pi}\left[\left(\frac{8}{3}\sin 2\pi - \frac{8}{3}\cos 2\pi + 40\pi\right) - \left(\frac{8}{3}\sin 0 - \frac{8}{3}\cos 0\right)\right]$$

$$= \frac{1}{2\pi}\left[\left(0 - \frac{8}{3} + 40\pi\right) - \left(-\frac{8}{3}\right)\right]$$

$$= \frac{1}{2\pi}\left\{\left(-\frac{8}{3} + 40\pi\right) - \left(-\frac{8}{3}\right)\right\}$$

$$= \frac{1}{2\pi}\left(-\frac{8}{3} + 40\pi + \frac{8}{3}\right) = \frac{1}{2\pi}\times 40\pi = 20$$

16. Required Volume $= \int_{P=3}^{P=5}\int_{\phi=\frac{\pi}{8}}^{\phi=\frac{\pi}{4}}\int_{z=3}^{4.5} P\,dp.\,d\phi dz$

$$= \int_3^{4.5}\int_{\frac{\pi}{8}}^{\frac{\pi}{4}}\left[\frac{P^2}{2}\right]_3^5 d\phi dz = \int_3^{4.5}\int_{\frac{\pi}{8}}^{\frac{\pi}{4}}\frac{1}{2}(25-9)\,d\phi dz$$

$$= 8\int_3^{4.5}\int_{\frac{\pi}{8}}^{\frac{\pi}{4}} d\phi dz = 8\int_3^{4.5}[\phi]_{\frac{\pi}{8}}^{\frac{\pi}{4}} dz = 8\int_3^{4.5}\left[\frac{\pi}{4} - \frac{\pi}{8}\right]dz$$

$$= 8\int_3^{4.5}\left(\frac{\pi}{8}\right)dz = 8\pi\int_3^{4.5} dz = \pi[Z]_3$$

$$= \pi\,(4.5 - 3) = 1.5\,\pi$$

17. Let $\quad I = \int_0^1 \frac{dx}{\sqrt{1-x}}$

$$= \left[-2\sqrt{1-x}\right]_0^1 = -2\,[(0) - 1]$$

$$\therefore \quad I = 2$$

18. A(0, 2, 1) and 6(4, 1, –1)

The equation of the line AB is

$$\frac{x-0}{4-0} = \frac{y-2}{1-2} = \frac{z-1}{-1-1} = t(\text{say})$$

$\therefore \quad x = 4t; \quad y = -t + 2; \quad z = -2t + 1$

$\therefore \quad dx = 4dt; \quad dy = -dt; \quad dz = -2dt$

Also t varies from 0 to 1

$$\therefore I = \int_0^1 2(-2t+1)4dt + 2(-t+2)(-dt) + 2(4t)(-2\,dt)$$

$$= \int_0^1 (-16t + 8 + 2t - 4 - 16t)\,dt = \int_0^1 (-30t + 4)\,dt$$

$$= \left(-30\frac{t^2}{2} + 4t\right)\Bigg|_0^1 = (-15 + 4) = -11$$

19. Let $\quad x^2 + y^2 = t^2$

and $\quad t^2 = z^3$

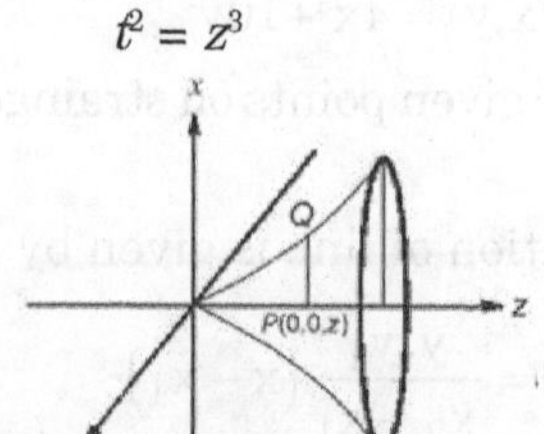

Here revolution is about z axis

Now volume of region R $= \int_0^1 \pi(PQ)^2\,dz$

Where PQ is radius of circle at some z.

Now $PQ = \sqrt{x^2 + y^2}$

$\therefore \quad (PQ)^2 = x^2 + y^2 = z^3$

$\therefore$ Volume of region R = $\int_0^1 \pi t^2\,dz = \int_0^1 \pi z^3\,dz$

$= \frac{\pi}{4} = 0.7853$

21. 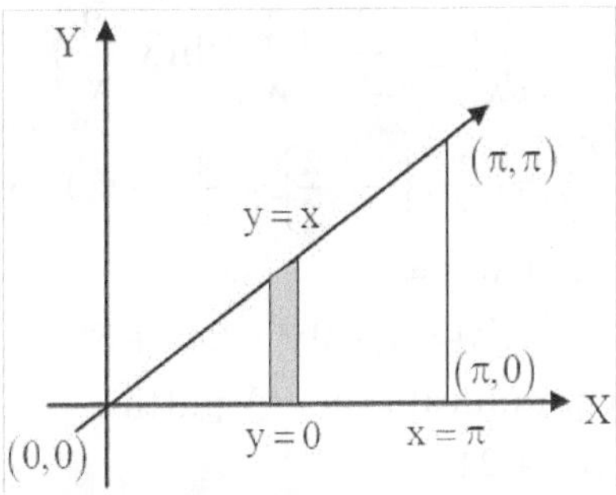

$\int_0^\pi\int_0^\pi \frac{\sin x}{x}\,dx\,dy$

Given limits are x = y → x = π & y = 0 → y = π.

For change of order of integration.

Consider a strip parallel to Y-axis. Then

Limits of y are : y = 0 → y = x and

Limits of x : x = 0 → x = π

$\therefore \int_0^\pi\int_0^\pi \frac{\sin x}{x}\,dx\,dy = \int_{x=0}^{\pi}\left[\int_{y=0}^{\pi} \frac{\sin x}{x}.dy\right]dx$

$= \int_{x=0}^{\pi} \frac{\sin x}{x}[y]_0^x .\,dx$

$= \int_{x=0}^{\pi} \sin x.dx$

$= [-\cos x]_0^\pi$

$= -[-1-1] = 2$

$\Rightarrow \int_0^\pi\int_y^\pi \frac{\sin y}{x}\,dx\,dy = 2$

22. 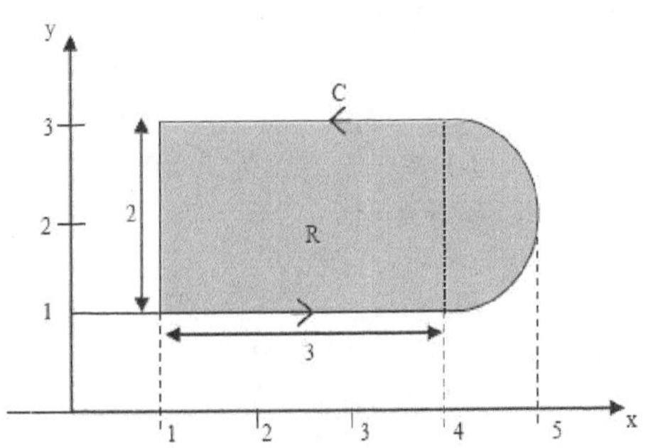

$\int_C (x\,dy - y\,dx) = \int_C (-y\,dx + x\,dy)$; where C is a Closed figure formed by rectangle and a semi circle having radius 1.

To countert line integral to area integral.

Now Using Green's theorem, we have

$\oint_C \left(-\underset{M}{y}\,dx + \underset{N}{y}\,dy\right) = \int_R (1+1)\,dx\,dy$

$= 2\int_R\int dx\,dy \rightarrow$ Area of the region R.

= 2[Area of rectangle + Area of semi-circle]

$= 2\left[(3\times 2) + \frac{1}{2}\pi(1^2)\right]$

$= 2\left[6 + \frac{\pi}{2}\right] = (12 + \pi)$

MAXIMA & MINIMA

23. Given, $f(x) = \frac{e^x}{1+e^x}$

$\therefore \quad f'(x) = \frac{(1+e^x).e^x - e^{2x}}{(1+e^x)^2} = \frac{e^x}{(1+e^x)^2}$

which is always +ve, so monotonically increases

24. graph of $y = \frac{1}{x^2}$

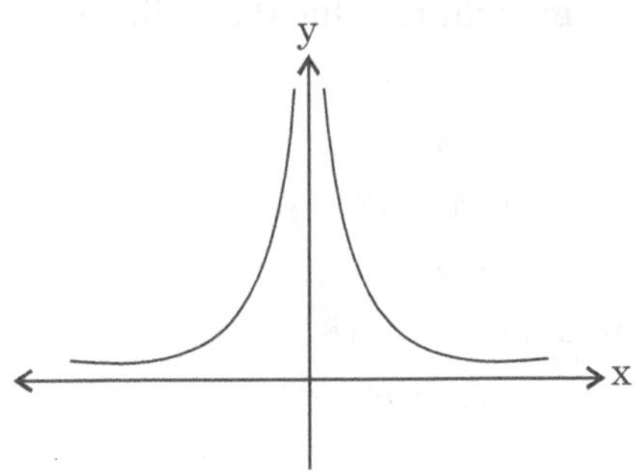

graph of y = e^x

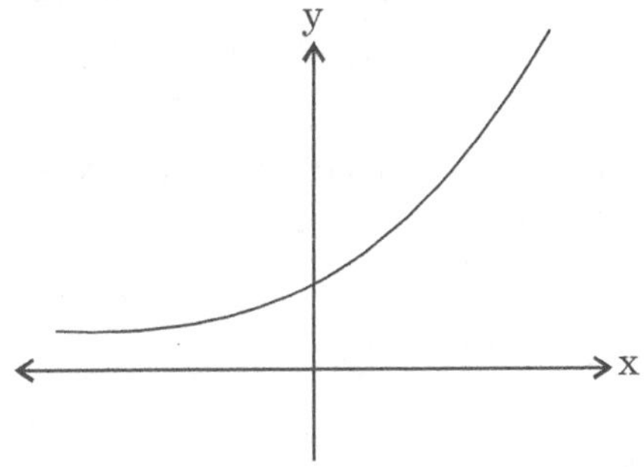

graph of y = e^x

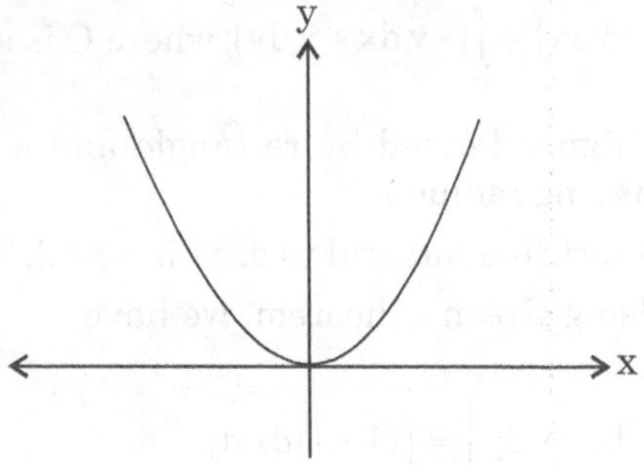

graph of y = e^{-x^2}

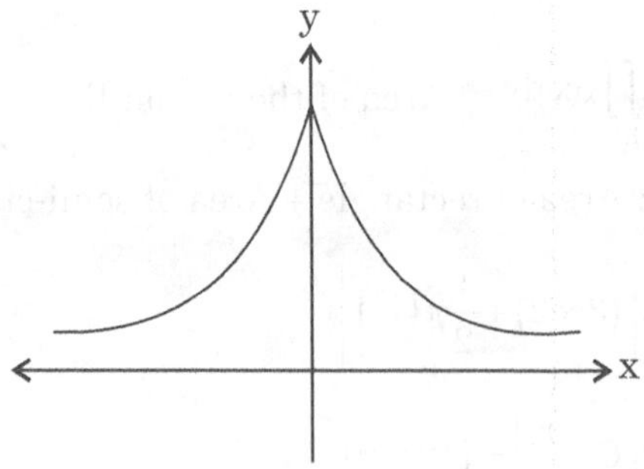

25. Given, $f(x) = x^2 - x - 2$

$$\frac{df(x)}{dx} = 0$$

$\Rightarrow$ $2x - 1 = 0$

$\Rightarrow$ $x = \frac{1}{2}$

$\therefore$ $\frac{d^2 f(x)}{dx^2} = 2$ (+ ve),

So it shows only minima for interval [–4, 4], it contains a maximum value that will be at

$x = -4$

or $x = 4$

$\therefore$ $f(-4) = 18$

and $f(+4) = 10$

So, maximum value is 18

26. $f(x) = e^x + e^{-x}$

$$= e^x + \frac{1}{e^x} = \left(e^{\frac{x}{2}} - e^{\frac{1}{\frac{x}{2}}}\right)^2 + 2$$

The minimum value of $\left(e^{\frac{x}{2}} - e^{\frac{1}{\frac{x}{2}}}\right)$ is zero at $x = 0$

then min $|f(x)| = 2$

27. Given : $e^y = x^{\frac{1}{x}}$

Taking log, $\ln e^y = \ln x^{\frac{1}{x}}$

$\Rightarrow$ $y = \frac{1}{x} \ln x$

Differentiating

$$\frac{dy}{dx} = \frac{-1}{x^2}(\ln x) + \left(\frac{1}{x^2}\right) = 0 \quad ...(i)$$

$\Rightarrow$ $\frac{1}{x^2}\{-\ln x + 1\} = 0$

$\Rightarrow$ $\frac{1}{x^2} = 0$ or $\ln x = 1$

$\Rightarrow$ $x = \infty$ or $x = e^1 = e$

Differentiating equation (*i*), we have

$$\frac{d^2 y}{dx^2} = \frac{-2}{x^3} - \left\{\frac{1}{x^3} - \ln x \cdot \frac{2}{x^3}\right\}$$

For $x = e$, $\frac{d^2 y}{dx^2} = -\frac{2}{e^3} - \left\{\frac{1}{e^3} - \frac{2}{e^3}\right\} = -\frac{1}{e^3}$ (– ve)

Hence maxima at $x = e$.

28. $f(x) = x^3 - 9x^2 + 24x + 5$

To find maximum value of $f(x)$, putting $f'(x) = 0$

or, $3x^2 - 18x + 24 = 0$

or, $x^2 - 6x + 8 = 0$

$x = 2,4$

Now, at $x = 2$, $f''(x) = (2x - 6)|_{x=2} = -2 < 0$

$x = 4$, $f''(x) = (2x - 6)|_{x=4} = 2 > 0$

Hence, a maxima occurs at $x = 2$ and a minima occurs at $x = 4$.

But at the end of interval, $x = 6$,

$f(x) = 6^3 - 9 \times 6^2 + 24 \times 6 + 5 = 41$

Which infers that maximum value of a function is simply the greatest value in the immediate neighbourhood of the maxima point ($x = 2$, in this case). In fact there may be several maximum and minimum values of a function is an interval same is also true for the minimum value of a function.

29. $f'(t) = -e^{-t} + 4e^{-2t} = 0$

$\Rightarrow$ $e^{-t}\left[4e^{-t} - 1\right] \Rightarrow e^{-t} = \frac{1}{4}$

$\Rightarrow$ $t = \log_e^4$

and $f''(t) < 0$

at $t = \log_e^4$

30. $f(x) = 0$

$\Rightarrow$ $\frac{1}{1+x} - 1 = 0$

$\Rightarrow$ $\frac{-x}{1+x} = 0$

$\Rightarrow$ $x = 0$

and $f^1(x) = \frac{-1}{(1+x)^2} < 0$ at $x = 0$

31. $f(x) = 6x^2 - 18x + 12 = 0$

$\Rightarrow$ $x = 1, 2 \in [0, 3]$

Now $f(0) = -3$;

$f(3) = 6$

and $f(1) = 2$;

$f(2) = 1$

Hence, $f(x)$ is maximum at $x = 3$ and the maximum value is 6.

32. 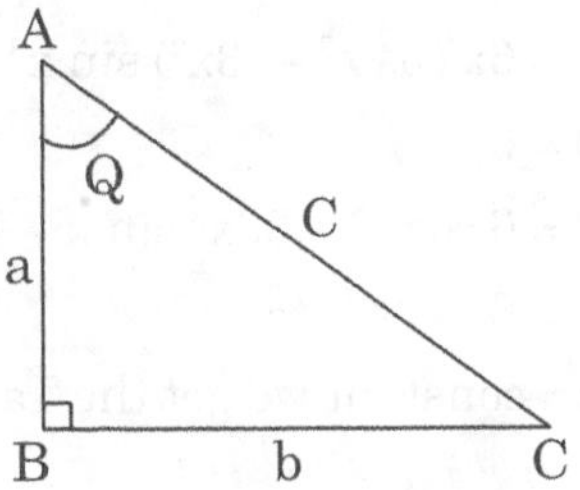

Here $C^2 = a^2+b^2$

Now, a + c = Constant

$\therefore$ Let a+ c = K

$\therefore \cos\theta = \frac{1}{2}$

$\therefore \theta = 60$.

33. $F(x) = e^{-x}(x^2+x+1)$

Now differentiating f (x) w.r.t (x) will get

$$F'(x) = e^{-x}\frac{d}{dx}(x^2+x+1) + (x^2+x+1)\frac{d}{dx}e^{-x}$$

$$= e^{-x}(2x+1)\bar{0}(x^2+x+1)e^{-x}$$

$$= e^{-x}(2x+1\bar{0}x^2\bar{0}x\bar{0}1)$$

$$= e^{-x}(x-x^2)$$

$$F'(x) = e^{-x}x(1-x)$$

Equation F' (x)=0 we get

$e^{-x}.x(1-x) = 0$

$\therefore$ x = 0 or x = 1

Now again differentiating F'(x) w.o.r.t (x) will get

$$F''(x) = \frac{d}{dx}e^{-x}(x-x^2)$$

$$= e^{-x}\left[\frac{d}{dx}(x-x^0)\right] + (x-x^2)\frac{d}{dx}e^{-x}$$

$$= e^{-x}(1-2x) + (x-x^2)e^{-x}(+)$$

$$= e^{-x}\left[1-2x-x+x^2\right]$$

$$F''(x) = e^{-x}\left[x^2 = 3x+1\right]$$

For x = 0

F" (0) 1 (1) > 0

$\therefore$ at x = 0 F (x) is minimum

and for x = 1

$F''(1) = e^{-1}(1-3+1)$

$= -\frac{1}{e} < 0$

$\therefore$ at x = 1, F(x) is maximum.

$\therefore$ Option (b) is correct

34. 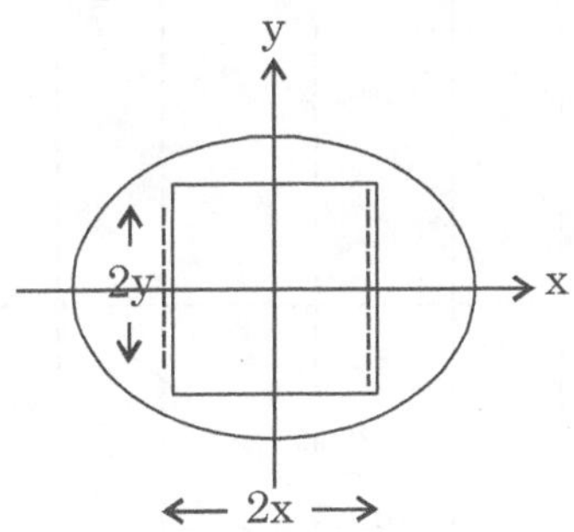

Area of rectangle = 2x× 2y

= 4 xy

$(\text{Area})^2 = 16x^2y^2$

$= 4x^2(1-x^2)$

$\therefore F = 4x^2(1-x^2)$

Now difference 'F' w.r.t 'x' we get

$$F'(x) = 4\frac{xd^2}{dn}(1-x^2) + (1-x^2)\frac{d}{dx}4x^2$$

$$= 4x^2(-2x) + (1-x^2)8x$$

$$= -8x^3 + 8x - 8x^3$$

$= -16x^3 + 8x$

Now Equaling

F' (x) = 0

$\therefore 8x(1-2x^2) = 0$

$\therefore x = 0 \text{ and } x^2 = \frac{1}{2}$

$\therefore x = \pm\frac{1}{\sqrt{2}}$

$\therefore y = \pm\frac{1}{\sqrt{8}}$

Now area of rectangle = 4xy

$$= 4\left(\frac{1}{\sqrt{2}}\right)\left(\frac{1}{\sqrt{8}}\right) = 4\times\frac{1}{\sqrt{16}} = 4\times\frac{1}{4}$$

= 1

35. From the given question integration of ramp is parabolic, integration of step is ramp.

So, the required graph of $f(x)$ is shown below.

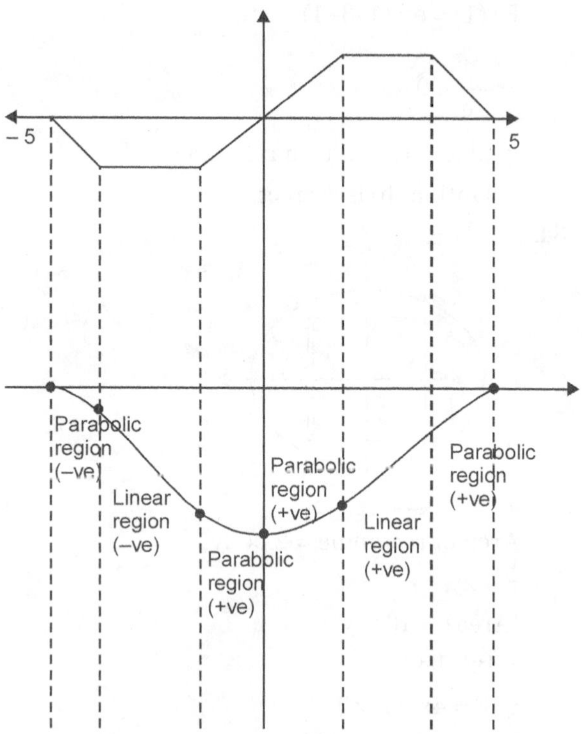

36. Given

$$f(x) = x^3 - 3x^2 + 1$$
$$f'(x) = 3x^2 - 6x$$
$$f'(x) = 0$$

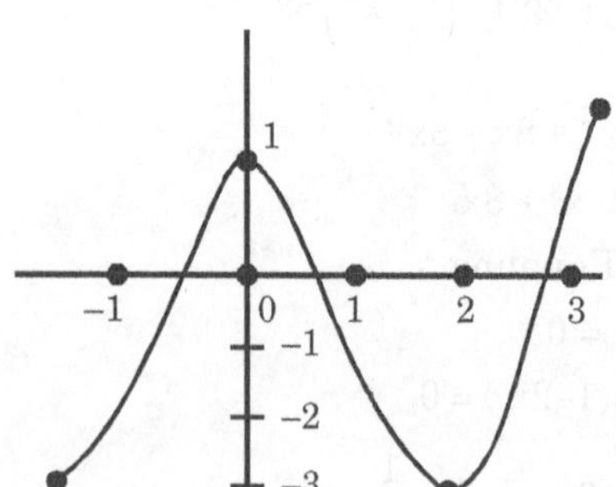

$$3x^2 - 6x = 0$$
$$3x(x-2) = 0$$
$$x = 0, 2$$
$$f''(x) = 6x - 6$$

At $x = 0$ $\quad f''(0) = -6$ maxima

$x = 2$ $\quad f''(2) = 6$ minima

Then the given function $f(x)$ increases, then decreases and increase again.

38. Taylor's series about x = 0 is

$$f(x) = f(0) + xf'(0) + \frac{x^2}{2!}f''(0) + \ldots\ldots$$

For $\quad f(x) = \sin x^3$

$\Rightarrow \quad t(0) = 0$

$f'(x) = 3x^2 \cos x^3$

$\Rightarrow \quad f'(0) = 0$

$f''(x) = 6x \cos x^3 - (3x^2) \sin x^3$

$\Rightarrow \quad f''(0) = 0$

$f'''(x) = 6 \cos x^3 - 36x^3 \sin x^3 - (3x^2)^2$

$\Rightarrow \quad f'''(0) = 6$

Similarly for $f^{1x}(x)$ = constant we get the Taylor's series

$$f(x) = \frac{x^3}{3!}.6 + k.\frac{x^9}{9!} + \ldots\ldots$$

39. Let $\quad f(x) = e^x + \sin x$

Taylor's series is

$$f(x) = f(a) + (x-a)f'(a) + \frac{(x-a)^2}{2!}f''(a)$$

where, $\quad a = \pi$

$$\therefore \quad f(x) = f(\pi) + (x-\pi)f'(\pi) + \frac{(x-\pi)^2}{2!}f''(\pi)$$

Coefficient of $(x-\pi)^2$ is $\dfrac{f''(\pi)}{2}$

$$f''(\pi) = e^x - \sin x \Big|_{at\, x=\pi} = e^\pi$$

$\therefore$ Coefficient of $(x-\pi)^2 = 0.5 \exp(\pi)$

40. Since $\sin x = x - \dfrac{x^3}{3!} + \dfrac{x^5}{5!} + \text{----}$

$$\therefore \quad \sin(x-\pi) = (x-\pi) - \frac{(x-\pi)^3}{3!} + \frac{(x-\pi)^5}{5!} + \ldots$$

or $$-\sin x = (x-\pi) - \frac{(x-\pi)^3}{3!} + \frac{(x-\pi)^5}{5!} + \ldots$$

or $$-\frac{\sin x}{(x-\pi)} = 1 - \frac{(x-\pi)^2}{3!} + \frac{(x-\pi)^4}{5!} + \ldots..$$

or $$\frac{\sin x}{(x-\pi)} = -1 + \frac{(x-\pi)^2}{3!} - \frac{(x-\pi)^4}{5!} + \ldots$$

41. Taylor series expansion of $3 \sin x + 2 \cos x$

Taylor series is given by

$$f(x) = f(x_0) + f'(x_0)(x-x_0) + \frac{f''(x_0)}{2!}(x-x_0)^2 + \frac{f'''(x_0)}{3!}(x-x_0)^3 + \ldots$$

Here $x_0 = 0$

$$\therefore\ f(x) = f(0) + f'(0)(x) + \frac{f''(0)}{2}x^2 + \frac{f'''(0)}{6}(x)^3 + \ldots \quad \ldots(i)$$

Now $\quad f(x) = 3\sin x + 2\cos x$

$\Rightarrow \quad f(0) = 2$

Now $\quad f'(x) = 3\cos x + 2(-\sin x)$

$= 3\cos x - 2\sin x$

$\Rightarrow \quad f'(0) = 3$

$f''(x) = -3\sin x - 2\cos x$

$\Rightarrow \quad f''(0) = -2$

$f'''(x) = -3\cos x + 2\sin x$

$\Rightarrow \quad f'''(0) = -3$

From equation (i),

$$f(x) = f(0) + f'(0)\,x + \frac{f''(0)}{2}x^2 + \frac{f'''(0)}{6}(x)^3 + \ldots$$

$$= 2 + 3x + \frac{(-2)}{2}x^2 + \left(\frac{-3}{6}\right)x^3 + \ldots$$

$$= 2 + 3x - x^2 - \frac{1}{2}x^3 + \ldots$$

42. $\displaystyle\sum_{n=0}^{\infty}\frac{1}{n!} = \frac{1}{0!} + \frac{1}{1!} + \frac{1}{2!} + \frac{1}{3!} = 1 + 1 + \frac{1}{2} + \frac{1}{6} + \ldots\ldots$

Now expansion of e^x

$$e^x = 1 + x + \frac{x^2}{2} + \frac{x^3}{6} + \ldots\ldots$$

For $x = 1$

$$e^1 = 1 + 1 + \frac{1}{2} + \frac{1}{6} + \ldots\ldots \sim$$

$$\therefore \sum_{n=0}^{\infty}\frac{1}{n!} = e$$

43. $$\frac{\partial z}{\partial x} = y\left[x \times \frac{1}{xy} \times y + \ln xy\right]$$

$$= y[1 + \ln xy]$$

and $$\frac{\partial z}{\partial y} = x(1 + \ln xy)$$

$$\Rightarrow \quad x\frac{\partial z}{\partial x} = y\frac{\partial z}{\partial y}$$

44. The partial derivative of x^2y^2 with respect to y is $0 + 2y \Rightarrow 2y$. The partial derivative of $6y + 4x$ with respect x is $0 + 4 = 4$. Green that both are equal.

$\Rightarrow \quad 2y = 4$

$\Rightarrow \quad y = 2$

46. At $\quad x = 0,$

$$f(x) = e^x e^{x^2}$$

$$= \left(1 + x + \frac{x^2}{2!} + \frac{x^3}{3!} + \ldots\right)\left(1 + x^2 + \frac{x^4}{2!} + \frac{x^6}{3!} + \ldots\right)$$

$$= 1 + x^2 + \frac{x^4}{2} + \frac{x^6}{6} + \ldots + x + x^3 + \frac{x^5}{2} + \frac{x^7}{6} + \ldots$$

$$+ \frac{x^2}{2} + \frac{x^4}{2} + \frac{x^6}{4} + \frac{x^8}{12} \ldots + \frac{x^3}{6} + \frac{x^5}{6} + \frac{x^7}{12} + \frac{x^9}{36} + \ldots$$

$$= 1 + x + \frac{3}{2}x^2 + \frac{7}{6}x^3$$

47. $$f(x) = \int_0^x e^{-t^2/2}dt$$

Differentiating $f(x)$ w.r.t. 'x' we get

$$f'(x) = e^{-x^2/2} \text{ and } f''(x) = e^{-x^2/2}(-x)$$

Now $\quad f''(0) = 0$

$$a_2 = \frac{f''(0)}{2!} = 0$$

48. Here,

$$f(x,y) = \frac{ax^2 + by^2}{xy} = a\left(\frac{x}{y}\right) + b\left(\frac{y}{x}\right)$$

Now $\displaystyle\frac{\partial f}{\partial x} = \frac{a}{y} - \frac{by}{x^2}$

Now $\displaystyle\frac{\partial f}{\partial x}$ at $x = 1$ and $y = 2$

$$\left.\frac{\partial f}{\partial x}\right|_{(1,2)} = \left.\left[\frac{a}{y} - \frac{by}{x^2}\right]\right|_{(1,2)} \left[\frac{a}{2} - \frac{b(2)}{1}\right] = \frac{a}{2} - 2b$$

Now $\displaystyle\frac{\partial f}{\partial y} = \left(-\frac{ax}{y^2} + \frac{b}{x}\right)$

Now $\displaystyle\frac{\partial f}{\partial y}$ at $x = 1$ and $y = 2$

$$\therefore \left.\frac{\partial f}{\partial y}\right|_{(1,2)} = \left.\left[-\frac{ax}{y^2} + \frac{b}{x}\right]\right|_{(1,2)} = -\frac{a}{4} + b$$

Now according to question,

$$\frac{\partial f}{\partial x} = \frac{\partial f}{\partial y}$$

$$\therefore \frac{a}{2} - 2b = -\frac{a}{4} + b$$

$$\therefore \frac{3a}{4} = 3b$$

$$\therefore a = 4b$$

49. $r = x^2 + y - z$ ...(i)

$\therefore \quad \frac{\partial r}{\partial x} = 2x - \frac{\partial z}{\partial x}$ ...(ii)

Now, $z^3 - xy + yz + y^3 = 1$ ...(iii)

$\therefore \quad 3z^2 \frac{\partial z}{\partial x} - y + y\frac{\partial z}{\partial x} = 0$

$\therefore \quad \frac{\partial z}{\partial x} = \frac{y}{3z^2 + y}$

By substituting the value of $\frac{\partial z}{\partial x}$ in equation (ii), we get,

$$\frac{\partial r}{\partial x} = 2x - \frac{y}{3z^2 + y}$$

Now $\left.\frac{\partial r}{\partial x}\right|_{(2,-1,1)} = 2(2) - \frac{(-1)}{3(1)^2 + (-1)}$

$$= 4 + \frac{1}{3-1} = 4.50$$

51. From triple product of vector

$A\times(B\times C) = B\,(A,C) - C\,(A, B)$

Now Here $A = \nabla$, $B = \nabla$ and $C = P$

$\therefore \nabla \times \nabla \times P = \nabla(J.P) - P\,(\nabla.\nabla)$

$= \nabla\,(\nabla.P) - \nabla^2 P$

$\therefore$ option (d) is correct

55. According to stokes theorem

$$\iint(\nabla \times p).ds = \oint \bar{p}.dl$$

56. $\bar{A} = xy\hat{a}_2 + x^2\hat{a}y,$

$\oint_0 \bar{A}.d\bar{l} = ?$

Here $\bar{A} = xy\hat{a}x + x^2 a\hat{y}$

and $\bar{l} = x\hat{a}x + y\hat{a}y$

Now $d\bar{l} = dxa\hat{x} + dy\hat{a}y$

Now $\bar{A}.d\bar{l} = \left(xya_{\hat{x}} + x^2 a_{\hat{y}}\right),(dxa\hat{x} + dy\hat{a}y)$

$\bar{A}.d\bar{l} = xydx + x^2dy$

$\therefore \oint_C \bar{A}.d\bar{l} = \oint_C xydx \times x^2dy$

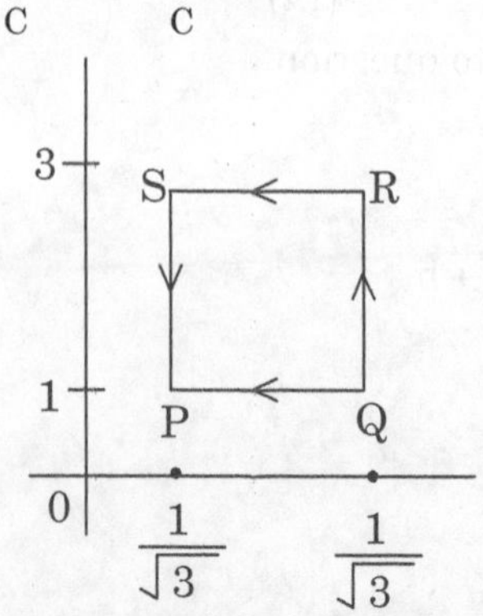

For P→ Q

y = 1 ad dy = o

$$\int_P^Q \bar{A}.d\bar{l} = \int_{\frac{1}{\sqrt{3}}}^{\frac{2}{\sqrt{3}}} xdx = \left[\frac{x^2}{2}\right]_{\frac{1}{\sqrt{3}}}^{\frac{2}{\sqrt{3}}} = \frac{1}{2}\left[\left(\frac{2}{\sqrt{3}}\right)^2 - \left(\frac{1}{\sqrt{3}}\right)^2\right]$$

$$= \frac{1}{2}\left[\frac{4}{3} - \frac{1}{3}\right] = \frac{1}{2}$$

Similary for Q → R

$x = \frac{2}{\sqrt{3}} dx = 0$

$$\therefore \int_Q^R \bar{A}.d\bar{l} = \int_1^3 \left(\frac{2}{\sqrt{3}}\right)^2 dy$$

$$= \frac{4}{3}[y]_1^3 = \frac{4}{3} \times 2 = \frac{8}{3}$$

Similarly for R → S

y = 3, dy = 0

$$\int_R^S \bar{A}.d\bar{l} = \int_{\frac{2}{\sqrt{3}}}^{\frac{1}{\sqrt{3}}} 3xdx = 3\left[\frac{x^2}{2}\right]_{\frac{2}{\sqrt{3}}}^{\frac{1}{\sqrt{3}}}$$

$$= \frac{3}{2}\left(\frac{1}{3} - \frac{4}{3}\right) = \frac{3}{2}(-1) = -\frac{3}{2}$$

Similarly for S → P

$x = \frac{1}{\sqrt{3}}$ ad dx = 0

$$\int_P^S \bar{A}.d\bar{l} = \int_2^1 \left(\frac{1}{\sqrt{3}}\right)^2 dy = \frac{1}{3}\int_2^1 dy = \frac{1}{3}\int_2^1 dy$$

$$= \frac{1}{3}[y]_2^1 = \frac{1}{3}[1-2] = -\frac{1}{3}$$

$$\therefore \oint_C \bar{A}.dl = \int_P^Q \bar{A}.d\bar{l} + \int_J^R \bar{A}.d\bar{l} + \int_R^S \bar{A}.d\bar{l} + \int_S^P \bar{A}.dl$$

$$= \left(\frac{1}{2} + \frac{8}{3} - \frac{3}{2} - \frac{2}{3}\right) = \left(\frac{3+16-9-4}{6}\right) = 1$$

$\therefore$ Option (c) is correct

57. According to stoke theorem.

$$\boxed{\oint_C \bar{A}.d\bar{l} = \iint_s \left(\nabla \times \bar{A}\right).d\bar{s}}$$

58. $\vec{A} = xa\hat{x} + ya\hat{y} + Za\hat{z}$

Difference of $\vec{A}$ is given by

$$\nabla.\bar{A} = \left(\frac{\partial}{\partial x}a\hat{x} + \frac{\partial}{\partial y}a\hat{y} + \frac{\partial}{\partial z}a\hat{z}\right).(xa\hat{x} + ya\hat{y} + za\hat{z})$$

$$= \frac{\partial}{\partial x}x + \frac{\partial}{\partial y}y + \frac{\partial}{\partial z}z$$

$= 1 + (+)$

$= 3.$

Option (d) is correct

59. $\bar{r} = xa\hat{x} + ya\hat{y} + za\hat{z}$ and $1\bar{r}1 = r$

div ($r^2 \nabla$ lnr)

$$= \nabla\left(r^2 \frac{1}{\partial}\left(\frac{\partial r}{\partial x}a\hat{r} + \frac{\partial r}{\partial y}a\hat{y} + \frac{\partial r}{\partial z}a\hat{z}\right)\right.$$

$$= \nabla\left(r\frac{\partial r}{\partial x}a\hat{x} + r\frac{\partial r}{\partial y}a\hat{y} + r\frac{\partial r}{\partial z}a\hat{z}\right)$$

Now $\bar{r} = xa\bar{x} + ya\hat{y} + za\hat{z}$

$$= \nabla\left[r\left(a\hat{x} + a\hat{y} + a\hat{z}\right)\right]$$

$$= \left(\frac{\partial}{\partial x}r\hat{x} + \frac{\partial}{ry}a\hat{y} + \frac{\partial}{\partial \hat{z}}a\hat{z}\right).\left[r\left(ax + a\hat{y} + a\hat{z}\right)\right]$$

$$= \frac{\partial r}{\partial x} + \frac{\partial r}{\partial y} + \frac{\partial r}{\partial z} \qquad \left(\because \frac{\partial r}{\partial x} = \frac{\partial r}{\partial y} = \frac{\partial r}{\partial z} = 1\right)$$

$=1 + 1 + 1$

$= 3$

60. $f(x,y,z) = x^2 + 3y^2 + z^3$

$$\nabla.f = \left(\frac{\partial}{\partial x}\hat{i} + \frac{\partial}{\partial y}\hat{j} + \frac{\partial}{\partial z}\hat{k}\right).F$$

$$= \frac{\partial f}{\partial x}\hat{i} + \frac{\partial F}{\partial y}\hat{j} + \frac{\partial F}{\partial z}\hat{k}$$

$$= \frac{\partial}{\partial x}\left(x^2 + 3y^2 + z^3\right)\hat{i} + \frac{\partial}{\partial y}\left(x^2 + 3y^2 + z^3\right)\hat{j} +$$

$$= \frac{\partial}{\partial z}\left(x^2 + 3y^2 + z^3\right)\hat{k}$$

$$= 2\times\hat{i} + 6y\hat{j} + 3z^2\hat{k}$$

$$\nabla.F|(1,1,1) = 2\hat{i} + 6\hat{j} + 3\hat{k}$$

$$\text{Now}|\nabla.F|\text{at}(1,1,1) = \sqrt{(2)^2 + (6)^2 + (3)^2}$$

$$= \sqrt{4 + 36 + 9}$$

$= 7$

61. $F(x,y) = \frac{xy}{\sqrt{2}}(x + y)$

$$\nabla F = \frac{\partial}{\partial x}\left(\frac{xy}{\sqrt{2}}(x + y)\right)\hat{i} + \frac{\partial}{\partial y}\left(\frac{xy}{\sqrt{2}}(x + y)\right)\hat{i}$$

$$= \left(\frac{2xy + y^2}{\sqrt{2}}\right)\hat{i} + \left(\frac{x^2 + 2xy}{\sqrt{2}}\right)\hat{i}$$

Now the given direction is

$$\hat{h} = \frac{1}{\sqrt{2}}\hat{i} + \frac{1}{\sqrt{2}}\hat{j}$$

Now directional derivative in direction of $\hat{h}$

$$= \left(\frac{2xy + y^2}{2} + \frac{x^2 + 2xy}{2}\right) = \left(\frac{x^2 + y^2 + 4xy}{2}\right)$$

∇ F in direction of $\hat{h}$ at (111)

$$\text{Now} = \left(\frac{1 + 1 + 4}{2}\right) = 3.$$

62.
$$\bar{\nabla} \times \bar{F} = \begin{vmatrix} a\hat{x} & z\hat{y} & a\hat{z} \\ \frac{\partial}{\partial x} & \frac{\partial}{ry} & \frac{\partial}{\partial z} \\ z & x & y \end{vmatrix}$$

$$= a\hat{x}\left(\frac{\partial y}{\partial y} - \frac{\partial x}{\partial z}\right) - a\hat{y}\left(\frac{\partial y}{\partial x} - \frac{\partial z}{\partial z}\right) + a\hat{z}\left(\frac{\partial x}{\partial x} - \frac{\partial z}{\partial y}\right)$$

$$+ a\hat{x}\left(1 - \frac{\partial x}{\partial z} - \frac{\partial z}{\partial y}\right) - a\hat{y}\left(\frac{\partial y}{\partial x} - 1\right) + a\hat{z}\left(1 - \frac{\partial x}{\partial y}\right)$$

$$= a\hat{x} + a\hat{y} + a\hat{z}$$

Now,

$$\int \nabla \times \vec{F}.ds = \int_s (a\hat{x} + a\hat{y} + a\hat{z}).(a\hat{z} - dxdy)$$

$$= \int_s dndy = 2\,p\,(1)^2 = 2\times 3.14 = 6.28$$

63. By Green's theorem

$$C = \oint xy^2dx + x^2ydy \text{ (where C = Closed curve)}$$

$$= \iint_R \left(\frac{d}{dx}(x^2y) - \frac{d}{dy}(xy^2)\right)dxdy$$

$$= \iint_R (2xy - 2xy) = 0$$

So, the value of closed curve C is 0.

3
CHAPTER

Differential Equations

Analysis of Previous GATE Papers

Topics ↓ / Year →			2019	2018	2017 Set 1	2017 Set 2	2016 Set 1	2016 Set 2	2016 Set 3	2015 Set 1	2015 Set 2	2015 Set 3
TYPE & ORDER, VARIATION OF PARAMETER, VARIABLE SEPARATION	**1 Mark**	MCQ Type									1	
		Numerical Type										
	2 Marks	MCQ Type		1	1							
		Numerical Type										
		Total		2	2						1	
COMPLEMENTARY FUNCTION & PARTICULAR INTEGRAL	**1 Mark**	MCQ Type	1			1						
		Numerical Type										
	2 Marks	MCQ Type										
		Numerical Type										
		Total	1			1						
INITIAL & BOUNDARY CONDITIONS	**1 Mark**	MCQ Type										
		Numerical Type										
	2 Marks	MCQ Type							1	1		
		Numerical Type	1									1
		Total	2						2	2		2

TYPE & ORDER, VARIATION OF PARAMETER, VARIABLE SEPARATION

1. A curve passes through the point (x = 1, y = 0) and satisfies the differential equation $\frac{dy}{dx} = \frac{x^2 + y^2}{2y} + \frac{y}{x}$. The equation that describes the curve is

(a) $\ln\left(1 + \frac{y^2}{x^2}\right) = x - 1$

(b) $\frac{1}{2}\ln\left(1 + \frac{y^2}{x^2}\right) = x - 1$

(c) $\ln\left(1 + \frac{y}{x}\right) = x - 1$

(d) $\frac{1}{2}\ln\left(1 + \frac{y}{x}\right) = x - 1$

[2003 : 2 Marks]

2. The following differential equation has

$$3\left(\frac{d^2y}{dt^2}\right) + 4\left(\frac{dy}{dt}\right)^3 + y^2 + 2 = x$$

(a) degree = 2, order = 1.
(b) degree = 1, order = 2.
(c) degree = 4, order = 3.
(d) degree = 2, order = 3.

[2005 : 1 Mark]

3. The order of the differential equation $\frac{d^2y}{dt^2} + \left(\frac{dy}{dt}\right)^3 + y^4 = e^{-t}$ is

(a) 1
(c) 3
(c) 3
(d) 4

4. Match each differential equation in Group I to its family of solution curves from Group II

Group I	Group II
A. $\frac{dy}{dx} = \frac{y}{x}$	1. Circles
B. $\frac{dy}{dx} = -\frac{y}{x}$	2. Straight lines
C. $\frac{dy}{dx} = \frac{x}{y}$	3. Hyperbolas
D. $\frac{dy}{dx} = -\frac{x}{y}$	

(a) A - 2, B - 3, C - 3, D - 1
(b) A - 1, B - 3, C - 2, D - 1
(c) A - 2, B - 1, C - 3, D - 3
(d) A - 3, B - 2, C - 1, D - 2

[2009 : 2 Marks]

5. The general solution of the differential equation $\frac{dy}{dx} = \frac{1 + \cos 2y}{1 - \cos 2x}$ is

(a) tan y – cot x = c (c is a constant).
(b) tan x – cot y = c (c is a constant).
(c) tan y + cot x = c (c is a constant).
(d) tan x + cot y = c (c is a constant).

[2015 : 1 Mark, Set-2]

6. Which one of the following is the general solution of the first order differential equation

$$\frac{dy}{dx} = (x + y - 1)^2$$

where x, y are real?

(a) $y = 1 + x + \tan^{-1}(x + c)$, where c is a constant.
(b) $y = 1 + x + \tan(x + c)$, where c is a constant.
(c) $y = 1 - x + \tan^{-1}(x + c)$, where cis a constant.
(d) $y = 1 - x + \tan(x + c)$, where c is a constant.

[2017 : 2 Marks, Set-1]

COMPLEMENTARY FUNCTION & PARTICULAR INTEGRAL

7. $y = e^{-2x}$ is a soltuion of the differential equaiton $y'' + y' - 2y = 0$. (True/False).

[1994 : 1 Mark]

8. A solution of the following differential equation is given by $\frac{d^2y}{dx^2} - 5\frac{dy}{dx} + 6y = 0$

(a) $y = e^{2x} + 3^{-3x}$
(b) $y = e^{2x} + e^{3x}$
(c) $y = e^{-2x} + e^{3x}$
(d) $y = e^{-2x} + e^{-3x}$

[2005 : 1 Mark]

9. Which of the following is a solution to the differential equation $\frac{dx(t)}{dt} + 3x(t) = 0$?

(a) $x(t) = 3e^{-t}$ (b) $x(t) = 2e^{-3t}$

(c) $x(t) = -\frac{3}{2}t^2$ (d) $x(t) = 3t^2$

10. If the characteristic equation of the differential equation

$$\frac{d^2y}{dx^2} + 2\alpha\frac{dy}{dx} + y = 0$$

has two equal roots, then the values of a are

(a) ±1 (b) 0, 0

(c) ±j (d) $\pm\frac{1}{2}$

[2014 : 1 Mark, Set-2]

11. Which ONE of the following is a linear non-homogeneous differential equation, where x and y are the independent and dependent variables respectively?

(a) $\frac{dy}{dx} + xy = e^{-x}$ (b) $\frac{dy}{dx} + xy = 0$

(c) $\frac{dy}{dx} + xy = e^{-y}$ (d) $\frac{dy}{dx} + e^{-y} = 0$

[2014 : 1 Mark, Set-3]

12. If a and b are constants, the most general solution of the differential equation

$\frac{d^2x}{dt^2} + 2\frac{dx}{dt} + x = 0$ is

(a) ae^{-t} (b) $ae^{-1} + bte^{-t}$

(c) $ae^{t} + bte^{-t}$ (d) ae^{-2t}

[2014 : 1 Mark, Set-4]

13. The general solution of the differential equation

$$\frac{d^2y}{dx^2} + 2\frac{dy}{dx} - 5y = 0$$

in terms of arbitrary constants K_1 and K_2 is

(a) $K_1e^{(-1+\sqrt{6})x} + K_2e^{(-1-\sqrt{8})x}$.

(b) $K_1e^{(-1+\sqrt{8})x} + K_2e^{(-1-\sqrt{8})x}$.

(c) $K_1e^{(-2+\sqrt{6})x} + K_2e^{(-2-\sqrt{6})x}$.

(d) $K_1e^{(-2+\sqrt{8})x} + K_2e^{(-2-\sqrt{8})x}$.

[2017 : 1 Mark, Set-2]

14. The families of curves represented by the solution of the equation $\frac{dy}{dx} = -\left(\frac{x}{y}\right)^n$

For n = – 1 and n = + 1, respectively, are

(*a*) Hyperbolas and Circles

(*b*) Circles and Hyperbolas

(*c*) Hyperbolas and Parabolas

(*d*) Parabolas and Circles

[2019 : 1 Mark]

INITIAL & BOUNDARY CONDITIONS

15. A solution for the differential equation

$\dot{x}(t) + 2x = \partial(t)$

with initial condition $x(0^-) = 0$ is

(a) $e^{-2t}u(t)$ (b) $e^{2t}u(t)$

(c) $e^{-1}u(t)$ (d) $e^{t}u(t)$

[2006 : 1 Mark]

16. For the differential equation $\frac{d^2y}{dx^2} + K^2y = 0$ the boundary conditions are

(i) y = 0 for x = 0 and

(ii) y = 0 for x = a

The form of non-zero solutions of y(where m varies overall integers) are

(a) $y = \sum_m A_m \sin\frac{m\pi x}{a}$.

(b) $y = \sum_m A_m \cos\frac{m\pi x}{a}$.

(c) $y = \sum_m A_m x^{\frac{m\pi}{a}}$.

(d) $y = \sum_m A_m e^{-\frac{m\pi x}{a}}$.

[2006 : 2 Marks]

17. The solution of the differential equation

$k^2\frac{d^2y}{dx^2} = y - y_2$ under the boundary conditions

(i) y = y, at x = 0 and

(ii) $y = y_2$ at $x = \infty$, where k, y_1 and y_2 are constants, is

(a) $y = (y_1 - y_2)\exp\left(\frac{-x}{K^2}\right) + y_2$.

(b) $y = (y_2 - y_1)\exp\left(\frac{-x}{K}\right) + y_1$.

(c) $y = (y_1 - y_2)\sinh\left(\frac{x}{K}\right) + y_1$.

(d) $y = (y_1 - y_2)\exp\left(\frac{-x}{K}\right) + y_2$.

[2007 : 2 Marks]

18. A function n(x) satisfies the differential equation $\frac{d^2n(x)}{dx^2} - \frac{n(x)}{L^2} = 0$ where L is a constant. The boundary conditions are : $n(0) = K$ and $n(\infty) = 0$. The solution to this equation is

(a) $n(x) = K\exp\frac{x}{L}$.

(b) $n(x) = K\exp\left(\frac{-x}{\sqrt{L}}\right)$.

(c) $n(x) = K^2\exp\left(\frac{-x}{L}\right)$.

(d) $n(x) = K\exp\left(\frac{-x}{L}\right)$.

[2010 : 1 Mark]

19. The solution of the differential equation $\frac{dy}{dx} = ky$, $y(0) = c$ is

(a) $x = ce^{-ky}$ (b) $x = ke^{cy}$

(c) $y = ce^{kx}$ (d) $y = ce^{-kx}$

[2011 : 1 Mark]

20. With initial condition $x(1) = 0.5$, the solution of the differential equation, $t\frac{dx}{dt} + x = t$ is

(a) $x = t - \frac{1}{2}$ (b) $x = t^2 - \frac{1}{2}$

(c) $x = \frac{t^2}{2}$ (d) $x = \frac{t}{2}$

[2012 : 1 Mark]

21. With initial values $y(0) = y'(0) = 1$, the solution of the differential equation

$$\frac{d^2y}{dx^2} + 4\frac{dy}{dx} + 4y = 0$$

at $x = 1$ is _______.

[2014 : 2 Marks, Set-4]

22. The solution of the differential equation $\frac{d^2y}{dt^2} + 2\frac{dy}{dt} + y = 0$ with $y(0) = y'(0) = 1$ is

(a) $(2 - t)e^t$ (b) $(1 + 2t)e^{-t}$

(c) $(2 + t)e^{-t}$ (d) $(1 - 2t)e^t$

[2015 : 2 Marks, Set-1]

23. Consider the differential equation

$\frac{d^2x(t)}{dt^2} + 3\frac{dx(t)}{dt} + 2x(t) = 0$. Given $x(0) = 20$ and $x(1) = \frac{10}{e}$, where $e = 2.718$, the value of $x(2)$ is _______.

[2015 : 2 Marks, Set-3]

24. The particular solution of the initial value problem given below is

$\frac{d^2y}{dx^2}12\frac{dy}{dx} + 36y = 0$ with $y(0) = 3$ and

$\left.\frac{dy}{dx}\right|_{x=0} = -36$

(a) $(3 - 18x)e^{-6x}$ (b) $(3 + 25x)e^{-6x}$

(c) $(3 + 20x)e^{-5x}$ (d) $(3 - 12x)e^{-6x}$

[2016 : 2 Marks, Set-3]

25. Consider the homogeneous oridinary differential equation

$x^2\frac{d^2y}{dx^2} - 3x\frac{dy}{dx} + 3y = 0,\ x > 0$ with $y(x)$ as a general solution. Given that $y(1) = 1$ and $y(2) = 14$ the value of $y(1.5)$, (rounded off of to two decimal places), is ______.

[2019 : 2 Marks]

ANSWERS

1. (a)	2. (b)	3. (b)	4. (a)	5. (d)	6. (d)	7. (true)	8. (b)	9. (b)	10. (d)
11. (a)	12. (b)	13. (*)	14. (a)	15. (a)	16. (a)	17. (d)	18. (d)	19. (c)	20. (d)
21. (0· 54)	22. (b)	23. (0.85)	24. (a)	25. (5.25)					

EXPLANATIONS

1. $\frac{dy}{dx} = \frac{x^2}{2y} + \frac{y}{2} + \frac{y}{x}$

Let $y = xt$

$\therefore \frac{dy}{dx} = t + x\frac{dt}{dx}$

$\therefore t + x\frac{dt}{dx} = \frac{x}{2t} + \frac{tx}{2} + t$

$\therefore x\frac{dt}{dx} = x\left(\frac{1}{2t} + \frac{t}{2}\right)$

$\therefore x\frac{dt}{dx} = x\left(\frac{1+t^2}{2t}\right)$

Now $\int \frac{2t}{1+t^2}dt = \int dx + C$

$In(1 + t^2) = x + C$

$\therefore t = \frac{y}{x}$

So, $In\left(1 + \frac{y^2}{x^2}\right) = x + C$

At $x = 1, y = 0$

$In\left(1 + \frac{0}{1}\right) = In(1) = 0 = 1 + C$

$C = -1$

So, $In\left(1 + \frac{y^2}{x^2}\right) = x - 1$

2. Order is highest derivative and degree is poulee of height derivative

$\therefore$ Order = 2 and degree = 1

3. The order of a differential equation is the order of the highest derivative involving in equation, so answer is 2.

4. A – 2, B – 3, C – 3, D – 1

(*i*) $\frac{dy}{dx} = \frac{y}{x}$

$\therefore \log y = \log x + \log c$

or $y = cx$, which is straight line equation

(*ii*) $\frac{dy}{dx} = \frac{-y}{x}$

$\therefore \log y + \log x = \log c$

or $c = xy$, which is hyperbola equation

(*iii*) $\frac{dy}{dx} = \frac{x}{y}$

$\therefore ydy = xdx$

or $y^2 = x^2 + c$, which is hyperbola equaiton

(*iv*) $\frac{dy}{dx} = \frac{-x}{y}$

$\therefore y^2 + x^2 = c$, which is equation of a circle

5. Given $\frac{dy}{dx} = \frac{1-\cos 2y}{1+\cos 2x}$

$\Rightarrow \frac{dy}{1-\cos 2y} = \frac{dx}{1+\cos 2x}$ Variable - Separable

$\Rightarrow \frac{dy}{2\sin^2 y} = \frac{dx}{2\cos^2 x}$

$\Rightarrow \int \operatorname{cosec}^2 y dy = \int \sec^2 x dx$

$\Rightarrow -\cot y = \tan x + k$

$\Rightarrow -\tan x - \cot y = k$

$\Rightarrow \tan x + \cot y = c$ where $c = -k$

6. $$\frac{dy}{dx} = (x + y - 1)^2 \quad ...(i)$$

Let $x + y - 1 = t$ ...(*ii*)

Now differiating equation (*ii*) w.r.t to '*x*' we get

$$1 + \frac{dy}{dx} = \frac{dt}{dx}$$

$$\therefore \quad \frac{dy}{dx} = \frac{dt}{dx} - 1$$

Substituting equations (*ii*) and (*iii*) in equation (*i*) we get

$$\frac{dt}{dx} - 1 = t^2$$

$$\frac{dt}{dx} = t^2 + 1$$

$$\frac{dt}{t^2+1} = dx$$

Now integrating on both side we get,

$$\int \frac{1}{t^2+1}\, dt = \int dx$$

$$\therefore \quad \tan^{-1} t = x + c$$

$$\therefore \quad x + y - 1 = \tan(x + c) \quad (\because\ t = x + y - 1)$$

$$\therefore \quad y = 1 - x + \tan(x + c)$$

7. $y = e^{-2x}$

$y^1 = -2e^{-2x}$

$y'^1 = (-2)(-2)e^{-2x}$

$= 4e^{-2x}$

Now $y'^1 + y^1 - 2y$

$= 4e^{-2x} + (-2)e^{-2x} - 2(e^{-2x})$

$= 4e^{-2x} - 2e^{-2x} - 2e^{-2x}$

$= 0$

y = e–2x is a solution of the given differential equation.

8. Auxillary Equation of given differential lauaton is

$D^2 - 5D + 6 = 0$

$\therefore$ (D – 3) (D–2) = 0

$\therefore$ D = 3, 2

$\therefore$ $y = e^2x + e^3x$ is solution of given differential equation.

9. $\therefore \quad \frac{dx(t)}{dt} + 3x(t) = 0$

$$\Rightarrow \quad \frac{dx(t)}{x(t)} = -3dt$$

$$\Rightarrow \quad \ln[x(t)] = -3t + c$$

$$\Rightarrow \quad x(t) = ke^{-3t}$$

10. $$\frac{d^2y(t)}{dt^2} + \frac{2dy(t)}{dt} + y(t) = \delta(t)$$

Converting to s-domain,

$$s^2y(s) - sy(0) - y'(0) + 2[sy(s) - y(0)] + y(s) = 1$$

$$[s^2 + 2s + 1]y(s) + 2s + 4 = 1$$

$$y(s) = \frac{-3 - 2s}{(s^2 + 2s + 1)}$$

Find inverse Lapalce transform

$$y(t) = [-2e^{-t} - te^{-t}]u(t)$$

$$\frac{dy(t)}{dt} = 2e^{-t} + te^{-t} - e^{-t}$$

$$\left.\frac{dy(t)}{dt}\right|_{t=0^+} = 2 - 1 = 1$$

11. (*a*) $\frac{dy}{dx} + xy = e^{-x}$ is a first order linear equation (non-homogeneous)

(*b*) $\frac{dy}{dx} + xy = 0$ is a first order linear equation (homogeneous)

(*c*), (*d*) are non linear equations

12. A.E. : $-m^2 + 2m + 1 = 0 \Rightarrow m = -1, -1$

$\therefore$ general solution is $x = (a + bt)e^{-t}$

13. The given differential equation is

$$\frac{d^2y}{dx^2} + \frac{2dy}{dx} - 5y = 0$$

Now auxillary equation for above differential equation is

$D^2 + 2D - 5y = 0$

ie $m^2 + 2m - 5 = 0$

$$\therefore m = \frac{-2 \pm \sqrt{(2)^2 - 4(1)(-5)}}{2(1)}$$

$$\therefore m = -1 \pm \sqrt{6}$$

Hence solution of above differential equation is

$$y = K_1 e^{(-1+\sqrt{6})x} + K_2 e^{(-1\sqrt{6})x}$$

14. Given differential equation is

$$\frac{dy}{dx} = -\left(\frac{x}{y}\right)^n$$

$$\therefore \quad \frac{dy}{dx} = -\frac{x^n}{y^n} \quad ...(1)$$

For n = –1; we have

$$\frac{dy}{dx} = -\frac{x^{-1}}{y^{-1}}$$

$$\Rightarrow \frac{dy}{dx} = -\frac{y}{x}$$

By method of variable separable form.

$$\Rightarrow \frac{1}{y}dy = \frac{-1}{x}dx$$

Integrative on both sides we get,

$$\ell ny = -\ell nx + \ell nc \Rightarrow \ell n(xy) = \ell nc$$

$$\Rightarrow xy = c^2$$

Which represents family of hyperbolas.

For n = 1; we have

$$\frac{dy}{dx} = -\frac{x}{y} \qquad [\because (1)]$$

By method of variable separable

$$\Rightarrow ydy = -xdx$$

Integrating on bothsides get

$$\frac{y^2}{2} = -\frac{x^2}{2} + c \Rightarrow \frac{x^2}{2} + \frac{y^2}{2} = c$$

$$\Rightarrow x^2 + y^2 = c$$

$\Rightarrow x^2 + y^2 = \left(\sqrt{c}\right)^2$ which represents family of circles.

16. Given, Differential equation,

$$\frac{d^2y}{dx^2} + k^2y = 0$$

Let $y = A\cos kx + jB\sin kx$

At $x = 0$, $y = 0$

$\Rightarrow A = 0$

$y = jB\sin kx$

At $x = a$, $y = 0$

$\Rightarrow B\sin ka = 0$

or $B \neq 0$ otherwise $y = 0$ always

$\Rightarrow \sin ka = 0$

$$\Rightarrow k = \frac{m\pi x}{a}$$

$$\therefore \quad y = \sum_m A_m \sin\left(\frac{m\pi x}{a}\right)$$

17.

$$k^2D^2y = y - y_2$$

$$\left(D^2 - \frac{1}{k^2}\right)y = \frac{-y_2}{k^2}$$

$$m_1 = \pm\frac{1}{k}$$

C.F. $\quad C_1e^{x/k} + C_2e^{-x/k}$

P. F. $\quad \dfrac{1}{D^2 - (1/k)^2}\left(\dfrac{-y^2}{k^2}\right)$

Let $\quad \lambda \to \dfrac{1}{k} \quad \dfrac{1}{D^2 - \lambda^2}$

$$= \frac{1}{2\lambda}\left[\frac{-1}{\lambda}\left(1 - \frac{D}{\lambda}\right)^{-1} - \frac{1}{\lambda}\left(1 + \frac{D}{\lambda}\right)^{-2}\right]$$

$\therefore$ P.F. $\quad \dfrac{-1}{2\lambda^2} \times \dfrac{-y_2}{k^2}$

$$\left[1 - \frac{D}{\lambda} + \frac{D^2}{\lambda^2} - - - + 1 - \frac{D}{\lambda} - \frac{D^2}{\lambda^2} - -\right]$$

$$= \frac{y_2}{k^2\lambda^2}\left(1 - \frac{D}{\lambda} + - -\right)$$

$$P.F. = y_2$$

$$y = C_1e^{x/k} + C_2e^{-x/k} + y_2$$

At $y = y_1$, $\quad x = 0$

$\therefore \quad y_1 = C_1 + C_2 + y_2 \quad ....(i)$

At $y = y_2$, $\quad x = \infty$ Hence C_1 must be zero

$\therefore \quad y_1 = C_2 + y_2$

$\Rightarrow \quad C_2 = y_1 - y_2$

$$\therefore \quad y = (y_1 - y_2)\exp\left(-\frac{x}{k}\right) + y_2$$

18. Given : $\dfrac{d^2n(x)}{dx^2} - \dfrac{n(x)}{L^2} = 0,$

where, $\quad L = \text{constant}$

Boundary conditions are : $n(0) = K$

and $\quad n(\infty) = 0$

$$\therefore \quad \left(p^2 - \frac{1}{L^2}\right) = 0$$

$$\Rightarrow \quad p = +\frac{1}{L}$$

$$\therefore \quad n(x) = A_1e^{-\frac{1}{L}x} + A_2e^{\frac{1}{L}x}$$

Applying *Boundary conditions*:

$$n(0) = Ae^0 = A_1 + A_2 = k$$

and $\quad n(\infty) = 0$

$\therefore \quad n(\infty) = A_1 \cdot 0 + A_2 e^{\infty} = 0$

$\Rightarrow \quad A_2 = 0$

$\therefore \quad A_1 = k$

Hence solution is $n(x) = ke^{-\frac{1}{L}x}$

19. Given : $y(0) = C$

and $\dfrac{dy}{dx} = ky \Rightarrow \dfrac{dy}{y} = kdx$

Taking log, $\ln y = kx + c$

$\Rightarrow \quad y = e^{kx} e^c$

When $y(0) = C$, $y = k_1 e^0$

$\therefore \quad y = ce^{kx} \quad ...(\because k_1 = C)$

20. Given differential equation is

$$t\frac{dx}{dt} + x = t \Rightarrow \frac{dx}{dt} + \frac{x}{t} = 1$$

$\text{IF} = e^{\int \frac{1}{t} dt} = e^{\log t} = t;$

solution is $x(\text{IF}) = \int (\text{IF})\, dt$

$xt = \int t\, dt$

$\Rightarrow \quad xt = \dfrac{t^2}{2} + c;$

Given: $x(1) = 0.5$

$\Rightarrow \quad 0.5 = \dfrac{1}{2} + c$

$\Rightarrow \quad c = 0$

Hence required solution $xt = \dfrac{t^2}{2}$

$\Rightarrow \quad x = \dfrac{t}{2}$

21. A.E : $m^2 + 4m + 4 = 0$

$\Rightarrow m = -2, -2$

$\therefore$ solutions is $y = (a + bx)e^{-2x} \quad ...(1)$

$y = (a + bx)(-2e^{-2x}) + e^{-2x}(b) \quad ...(2)$

using $y(0) = 1$; $y(0) = 1$, (1) and (2) gives

$a = 1$ and $b = 3$

$\therefore \quad y = (1 + 3x)e^{-2x}$

at $x = 1$, $y = 4e^{-2} = 0.541 \approx 0.54$

22. The given differential equation is of the form

$(D^2 + 2D + 1)y = 0$

$\therefore \quad D^2 + 2D + 1 = 0$

$\therefore \quad (D + 1)^2 = 0$

$\therefore \quad D = -1, -1$

$\therefore$ Solution of above differential equation is

$y(t) = (c_1 + c_2 t)e^t$

(Complimentary function)

$\therefore \quad y'(t) = c_2 e^{-t} + (c_1 + c_2 t)(-e^{-t})$

$\therefore \quad y(0) = 1, y'(0) = 1$

gives $c_1 = 1$

and $c_2 + c_1(-1) = 1$

$\therefore \quad c_2 = 2$

Hence $y(t) = (1 + 2t)e^{-t}$

23. The given diggerential equation is

$$\frac{d2x(t)}{dt^2} + \frac{3dx(t)}{dt} + 2x(t) = 0$$

Now Auxiliary eqation of above differential equation is

$D^2 + 3D + 2 = 0$

$\therefore$ $D^2 + 2D + D + 2 = 0$

$\therefore$ $D(D+2) + 1(D+2) = 0$

$\therefore$ $(D+1)(D+2) = 0$

$\therefore$ $D = -1$ and $D = -2$.

$\therefore$ $x(y) = C_1\, e{+}x\, C_2\, e^{-2t}$

Now according to question,

$X(0) = 20$

$\therefore$ $20 = C_1 + C_2 \rightarrow (1)$

$\text{ad} \times (1) = \dfrac{10}{e}$

$\therefore \dfrac{10}{e} = C_1 e^{-1} + C_2{}^{e-2}$

$\therefore \dfrac{10}{e} = \dfrac{C_1}{e} + \dfrac{C^2}{e^2}$

$\therefore \dfrac{10}{e} = \dfrac{eC_1 + C_2}{e^2}$

$\therefore$ $10e = eC_1 + C_2 \rightarrow \quad (2)$

Solving Eq (1) and (2) we get

$C_1 = \dfrac{10e - 20}{e - 1}$ and $C_2 = \dfrac{10e}{e - 1}$

$\therefore x(t) = \left(\dfrac{10e - 20}{e - 1}\right)e^{-t} + \left(\dfrac{10e}{e - 1}\right)e^{-2t}$

$\therefore x(2) = \left(\dfrac{10e - 20}{e - 1}\right)e^{-2} + \left(\dfrac{10e}{e - 1}\right)e^{-4}$

$= 0.85$

24. The differential equation is given by

$$\frac{d^2y}{dx^2}+12\frac{dy}{dx}+36y=0$$

$\Rightarrow \quad D^2 + 12D + 36 = 0$

$(D + 6)^2 = 0$

$\therefore \quad D = -6, -6$

The solution is given by

$$y = C_1e^{-6x}+C_2xe^{-6x} \quad ...(i)$$

$y(0) = 3$

$\Rightarrow \quad 3 = C_1$

$C_1 = 3$

$(i) \Rightarrow \quad y = 3e^{-6x}+C_2xe^{-6x}$

$$\frac{dy}{dx} = -18e^{-6x}+C_2\left\{-6xe^{-6x}+e^{-6x}\right\}$$

$\Rightarrow \quad \left.\frac{dy}{dx}\right|_{x=0} = -18 + C_2$

$\Rightarrow \quad -36 = -18 + C_2$

$\Rightarrow \quad C_2 = -18$

$\therefore$ The solution of $y = 3\,e^{-6x} - 18\,x\,e^{-6x}$

(Since $C_1 = 3$ & $C_2 = -18$)

25. Given differential equation,

$$x^2\frac{d^2y}{dx^2}-3x\frac{dy}{dx}+3y=0, \quad ...(i)$$

$x > 0$ & $y(1) = 1$ and $y(2) = 14$

(Cauchy-Euler Linear D. E.)

Equation (i) can be written as

$[x^2D^2 - 3xD + 3]y = 0$

Let $xD = \theta$; $x^2D^2 = \theta(\theta - 1)$;

where $\theta = \frac{d}{dz}$ and $x = e^z$

$\therefore [\theta(\theta-1)-3\theta+3]y = 0 \Rightarrow [\theta^2-4\theta+3]y = 0$

Consider the Auxillary equation as

$\theta^2 - 4\theta + 3 = 0 \Rightarrow (\theta - 3)(\theta - 1) = 0$

$\Rightarrow \theta = 1, 3 \rightarrow$ (Roots are real and distinct)

$\therefore$ The solution is,

$y = C_1e^{1.z} + C_2e^{3.z} \Rightarrow y = C_1x + C_2x^3\ [\because\ x = e^z]$ ...(ii)

Given y = 1at x = 1

$\therefore\ C_1 + C_2$ ...(iii)

and y = 14 at x = 2

$\therefore\ 14 = 2C_1 + 8C_2$

$\therefore\ C_1 + 4C_2 = 7$ ...(iv)

Solving eq. (iii) and (iv) we get

$\boxed{C_2 = 2 \text{ and } C_1 = 1}$

Now from eq. (ii),

$y = (-1)x + (2)x^3$

$\therefore\ y(1.5) = (-1)(1.5) + 2(1.5)^3$

$\therefore\ y(1.5) = 5.25$

4 CHAPTER

Complex Analysis

Analysis of Previous GATE Papers												
	Year → Topics ↓		2019	2018	2017 Set 1	2017 Set 2	2016 Set 1	2016 Set 2	2016 Set 3	2015 Set 1	2015 Set 2	2015 Set 3
ANALYTIC FUNCTION	1 Mark	MCQ Type	1									
		Numerical Type									1	
	2 Marks	MCQ Type										
		Numerical Type										
	Total		1								1	
COUNCHY'S INTEGRAL FORMULA & RESIDUE THEOREM	1 Mark	MCQ Type				1				1		1
		Numerical Type	1					1	1			
	2 Marks	MCQ Type				1			1			
		Numerical Type					1			1		
	Total		1			3	2	1	3	3		1

ANALYTIC FUNCTION

1. For the function of a complex variable W = lnZ (where, W = u + jv and Z = x + jy, the u = constant lines get mapped in z-plane as
   (a) set of radial straight lines.
   (b) set of concentric circles.
   (c) set of confocal hyperbolas.
   (d) set of confocal ellipses.

[2006 : 2 Marks]

2. The equation sin(z) = 10 has
   (a) no real or complex solution.
   (b) exactly two distinct complex solutions.
   (c) a unique solution.
   (d) an infinite number of complex solutions.

[2008 : 1 Mark]

3. If $x = \sqrt{-1}$. then the value of x^{-x} is
   (a) $e^{-\pi/2}$ (b) $e^{\pi/2}$
   (c) x (d) 1

[2012 : 1 Mark]

4. The real part of an analytic function f(z) where z = x + jy is given by $e^{-y}\cos(x)$. The imaginary part of f(z) is
   (a) $e^{y}\cos(x)$ (b) $e^{-y}\sin(x)$
   (c) $-e^{y}\sin(x)$ (d) $-e^{y}\sin(x)$

[2014 : 2 Marks, Set-2]

5. Let $f(z) = \frac{az+b}{cz+d}$. If $f(z_1) = f(z_2)$ for all $z_1 \neq z_2$, a = 2, b = 4 and c = 5, then d should be equal to ______.

[2015 : 1 Mark, Set-2]

6. Which one of the following functions is analytic over the entire complex plane?
   (a) $\ell n(z)$ (b) cos (z)
   (c) $e^{1/z}$ (d) $\frac{1}{1-z}$

[2019 : 1 Mark]

COUNCHY'S INTEGRAL FORMULA & RESIDUE THEOREM

7. The value of the contour integral $\oint_{|z-j|=2} \frac{1}{z^2+4}dz$ in positive sense is
   (a) $\frac{j\pi}{2}$ (b) $\frac{-\pi}{2}$
   (c) $\frac{-j\pi}{2}$ (d) $\frac{\pi}{2}$

[2006 : 2 Marks]

8. If the semicircular contour Dof radius 2 is as shown in the figure, then the value of the integral $\oint_D \frac{1}{(s^2-1)}dx$ is

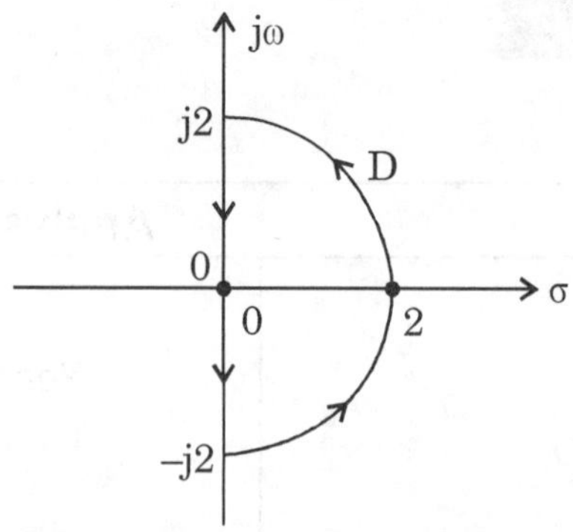

   (a) jπ (c) –jπ
   (c) –π (d) π

[2007 : 2 Marks]

9. The residue of the function $f(z) = \frac{1}{(z+2)^2(z-2)^2}$ at z = 2 is
   (a) $-\frac{1}{32}$ (b) $-\frac{1}{16}$
   (c) $\frac{1}{16}$ (d) $\frac{1}{32}$

10. If $f(z) = c_0 + c_1z^{-1}$, then $\oint_{\text{unit circle}} \frac{1+f(z)}{z}dz$ is given by
   (a) $2\pi c_1$ (b) $2\pi(1 + c_0)$
   (c) $2\pi j c_1$ (d) $2\pi j(1 + c_0)$

[2009 : 1 Mark]

11. The residues of a complex function $X(z) = \frac{1-2z}{z(z-1)(z-2)}$ at its poles are
   (a) $\frac{1}{2}, -\frac{1}{2}$ and 1 (b) $\frac{1}{2}, \frac{1}{2}$ and –1
   (c) $\frac{1}{2}$, 1 and $-\frac{3}{2}$ (d) $\frac{1}{2}$, –1 and $\frac{3}{2}$

[2010 : 2 Marks]

12. The value of the integral $\oint_C \frac{-3z+4}{(z^2+4z+5)}dz$ where c is the circle |z| = 1 is given by
   (a) 0 (b) $\frac{1}{10}$
   (c) $\frac{4}{5}$ (d) 1

[2011 : 1 Mark]

13. Given $f(z)=\frac{1}{z+1}-\frac{2}{z+3}$

If C is a countierclockwise path in the z-plane such that $|z+1|=1$, the value of $\frac{1}{2\pi j}\oint_C f(z)dz$ is

(a) –2 (b) –1

(c) 1 (d) 2

[2012 : 1 Mark]

14. Let z = x + iy be a complex variable. Consider that contour integration is performed along the unit circle in anticlockwise direction. Which one of the following statements is NOT TRUE?

(a) The residue of $\frac{z}{z^2-1}$ at z = 1 is $\frac{1}{2}$

(b) $\oint_C z^2 dz = 0$

(c) $\frac{1}{2\pi j}\oint_C \frac{1}{z}dz = 1$

(d) $\bar{z}$ (complex conjugate of z) is analytical function.

[2015 : 1 Mark, Set-1]

15. If Cdenotesthe counterclockwise unit circle, the value of the contour integral

$\frac{1}{2\pi j}\oint_C \text{Re}\{z\}dz$ is ______.

[2015 : 2 Marks, Set-2]

16. If C is a circle of radius r with centre z0, in the complex z-plane and if n is a non-zero integer, then $\oint \frac{dz}{(z-z_0)^{n+1}}$ equals

(a) 2πnj (b) 0

(c) $\frac{\pi j}{2\pi}$ (d) 2πn

[2015 1 Mark, Set-3]

17. In the following integral, the contour C encloses the points 2πj and –2πj

$$-\frac{1}{2\pi}\oint_C \frac{\sin z}{(z-2\pi j)^3}dz.$$

The value of the integral is ______.

[2016 : 2 Marks, Set-1]

18. Consider the complex valued function

$f(z)=2z^3+b|z|^3$ where z is a complex variable. The value of bfor which the function f(z) is analytic is ______.

[2016 : 1 Mark, Set-2]

19. The values of the integral $\frac{1}{2\pi}\oint_C \frac{e^z}{z-2}dz$ along a closed contour c in anti-clockwise direction for

(i) the point $z_0 = 2$ inside the contour c, and

(ii) the point $z_0 = 2$ outside the contour c, respectively, are

(a) (i) 2.72, (ii) 0 (b) (i) 7.39, (ii) 0

(c) (i) 0, (ii) 2.72 (d) (i) 0, (ii) 7.39

[2016 : 2 Marks, Set-3]

20. For $f(z)=\frac{\sin(z)}{z^2}$, the residue of the pole at z = 0 is ______.

[2016 : 1 Mark, Set-3]

21. The residues of a function $f(z)=\frac{1}{(z-4)(z+1)^3}$ is

(a) $\frac{-1}{27}$ and $\frac{-1}{125}$ (b) $\frac{1}{125}$ and $\frac{-1}{125}$

(c) $\frac{-1}{27}$ and $\frac{1}{5}$ (d) $\frac{1}{125}$ and $\frac{-1}{5}$

[2017 : 1 Mark, Set-2]

22. An integral I over a counter-clockwise circle C is given by

$$I=\oint_C \frac{z^2-1}{z^2+1}e^z\,dz.$$

If C is defined as $|z| = 3$, then the value of I is

(a) –πisin(I) (b) –2πi/sin(I)

(c) –3πisin(I) (d) –4πisin(I)

[2017 : 2 Marks, Set-2]

23. The contour Cgiven below is on the complex plane z = x + jy, where $j=\sqrt{-1}$

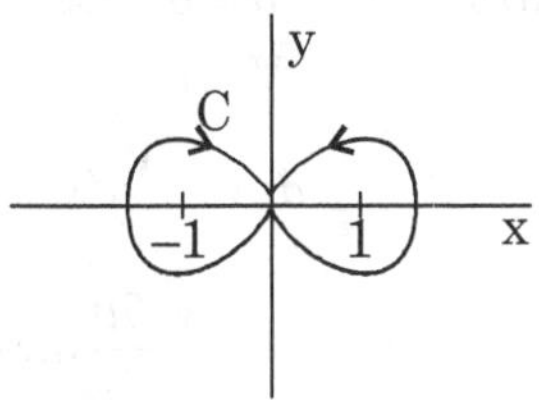

The value of the integral $\frac{1}{\pi j}\oint_C \frac{dz}{z^2-1}$ is ______.

[2018 : 2 Marks]

24. The value of the contour integral

$$\frac{1}{2\pi j}\oint\left(z+\frac{1}{z}\right)^2 dz$$

Evaluated over the unit circle |Z| = 1 is ______.

[2019 : 1 Mark]

ANSWERS

1. (b)	2. (a)	3. (b)	4. (b)	5. (10)	6. (b)	7. (d)	8. (a)	9. (a)	10. (d)
11. (c)	12. (a)	13. (c)	14. (d)	15. (c)	16. (b)	17. (–133.8)	18. (0)	19. (0)	20. (*)
21. (*)	22. (*)	23. (2)	24. (0)						

EXPLANATIONS

1. Given, $W = \log_e z$

$$\Rightarrow u + iv = \log_e(x+iy) = \frac{1}{2}\log(x^2+y^2) + i\tan^{-1}\left(\frac{y}{x}\right)$$

Since, u is constant, therefore

$$\frac{1}{2}\log(x^2+y^2) = c \quad \Rightarrow x^2 + y^2 = c'$$

which is represented set of concentric circles.

2. Since $-1 \le \sin\theta \le 1$

therefore, it has no real or complex solution exists.

3. $x = \sqrt{-1}$

Now $i = \sqrt{-1}$

$\therefore x = i$

Now representing in polar co-ordinates

x = Cos π/2 fi Sin π/2

= e π/2 i

$$\text{Now } x^{-x} = \left(e\frac{\pi}{2}i\right) - i$$

$$e^{\pi/2} = \left(\because i^2 = -1\right)$$

$\therefore$ option (b) is correct

4. Real part $\quad u = e^{-y}\cos x$

and $\quad V = ?$

$$dv = \frac{\partial v}{\partial x}dx + \frac{\partial v}{\partial y}dy$$

$$= -\frac{\partial u}{\partial y}dx + \frac{\partial u}{\partial x}dy$$

(Using C – R equations) $= e^{-y}\cos x\,dx - e^{-y}\sin x\,dy$

$$= d\left[e^{-y}\sin x\right]$$

Integrating, we get $V = e^{-y}\sin x$

5. $F(Z) = \dfrac{aZ+b}{cZ+d}$

Now $F(Z_1) = F(Z_2)$

$$\therefore \frac{aZ_1+b}{CZ_1+d} = \frac{aZ_2eb}{CZ_2+d.}$$

Now putting values of a = 2, b = 4, and c = 5, we get

$$\frac{2Z_1+4}{5Z_1+d} = \frac{2Z_2+4}{5Z_2+d.}$$

$$\Rightarrow 20(Z_2 - Z_1) = 2d(Z_2 - Z_1)$$

$$\therefore \boxed{d = 10}$$

6. $\ell n(z)$ is not analytic at z = 0;

and $e^{\frac{1}{z}}$ is also not analytic at z = 0 and

$\dfrac{1}{1-z}$ is also not analytic at z = 0

But cos (z) is analytic over the entire complex plane,

Since cos z = cos (x + iy) = cos x cos (iy) – sin x sin (iy)

= cos x cos h y – i sin x sin h y;

Now comparing with u + iv we get

Where $\quad u(x, y) = \cos x \cosh y;$

$\quad v(x,y) = -\sin x \sinh y$

$u_x = -\sin x.\cosh y$	$v_x = -\cos x \sinh y$
$u_y = \cos x \sinh y$	$v_y = -\sin x \cosh y$

7. Given, $\dfrac{1}{z^2+4} = \dfrac{1}{(z+2j)(z-2j)}$

Pole (0, 2) lies inside the circle $|z - j| = 2$

By Cauchy's Integral formula,

$$I = \frac{2\pi j}{(2j+2j)} = \frac{\pi}{2}$$

8. $\oint_D \dfrac{1}{(s^2-1)}ds = 2\pi j$ (sum of Residue)

Singular points are s = ± 1

Only s = + 1 lies inside the given contour

Residue $s = +1 = \lim_{S\to 1} (s-1) f(s)$

$$= \lim_{S\to 1} (s-1)\frac{1}{(s^2-1)}$$

$$= \frac{1}{2} = 2\pi j\left(\frac{1}{2}\right) = \pi j$$

9. $$f(z) = \frac{1}{(z+2)^2(z-2)^2}$$

Residue at $z = a$ of order n is, defined as,

$$= \lim_{z\to 0}\frac{1}{\lfloor n-1}\left[\frac{d^{n-1}}{dz^{n-1}}(z-a)^n f(z)\right]_{z=a}$$

$$= \text{Res}[f(z)]_{z=2}$$

$$= \left[\frac{d}{dz}(z-2)^2 f(z)\right]_{z=2}$$

$$= \left[\frac{d}{dz}\frac{1}{(z+2)^2}\right]_{z=2} = \left.\frac{-2}{(z+2)^3}\right|_{z=2}$$

$$= -\frac{1}{32}$$

10. $$f(z) = c_0 + c_1 z^{-1}$$

$$f(z) = \oint \frac{1+c_0+c_1z^{-1}}{z}dz$$

$$= \oint\left(\frac{(1+c_0)z+c_1}{z^2}\right)dz$$

∴ $f(z) = 2\pi j(\text{ residue of } f(z))$

Residue at $z = 0$ (2 – order)

$$= \frac{1}{L_1}\frac{d}{dz}\left(z^2 - \frac{(1+c_0)z+c_1}{z^2}\right)$$

$$= \frac{1}{L_1}(1+c_0).1 = (1+c_0)$$

∴ $f(z) = 2\pi j(1+c_0)$

11. $$X(Z) = \frac{1-22}{Z(Z-1)(Z-2)}$$

The ploes of the given function are at 0, 1, 2 now Residue at a point 'a' is given by

$$= \frac{\lim}{2a}\left(\frac{1-2Z}{Z(Z-1)(Z-2)}\right)$$

∴ Residue at 0 $= \lim_{Z\to 0}\left(\frac{1-2Z}{(Z-1)(Z-2)}\right)$

$$= \frac{1}{(-1)(-2)} = \frac{1}{2}$$

Residue at 1 $= \lim_{Z\to 1}\frac{1-2Z}{Z(Z-2)}$

$$= \frac{1-2}{1(1-2)}$$

$$= \frac{+1}{+1}$$

$= 1$

Residue at 2 $= \lim_{Z\to 2}\frac{1-2Z}{Z(Z-1)}$

$$= \frac{1-2(2)}{2(2-1)} = \frac{-3}{2}$$

12. $$\oint_c \frac{-3z+4}{z^2+4z+5}dz = 0$$

Since $z^2 + 4z + 5 = (z+2)^2 + 1 = 0$

$z = -2 \pm j$ will be outside the unit circle.

Since the poles lies outside the circle $|z| = 1$, $f(z)$ is analytic every where and $\oint_c f(z).dz = 2\pi i(0) = 0$.

13. $$f(z) = \frac{1}{z+1} - \frac{2}{z+3};\ f(z) = \frac{-z+1}{(z+1)(z+3)}$$

Pole at $z = -1$ lies within the circle $|z+1| = 1$ and Pole at $z = -3$ lies outside the circle $|z+1| = 1$

$\oint_c f(z).dz = 2\pi i$(Sum of the residues at the Singular points within C)

or, $$\frac{1}{2\pi i}\oint_c f(z)\cdot dz = \text{Res}\, f(-1) = \underset{z\to -1}{Lt}\frac{-z+1}{z+3} = 1$$

14. $F(Z) = \bar{Z}$

Now Z = x + i y

$\Rightarrow \bar{Z} = x - iy$

4 =x, V = – y

4x = 1, Vx = 0

uy = 0, Vy = –1

∴ 4x ≠ Vy ∴ $\bar{Z}$ is not analytic function.

15. $\frac{1}{2\pi i}\oint De\{Z\}$

Now $Re(Z) = \frac{Z+\bar{Z}}{2}$

and in Re (Z), there is no pole

∴ Reoduce at poles is zero.

$\therefore \frac{1}{2\pi i}\oint Re\{Z\}dZ = 0$

16. $\oint \frac{dZ}{(Z-Z_0)^{n+1}}$

Now using cauchy integral formulas use get

$\oint \frac{dZ}{(Z-Z_0)^{n+1}} = \frac{2\pi iF'(Z_0)}{n!}$

$= \frac{2\pi i}{n!} \times (0)$

$= 0$

17. $I = -\frac{1}{2\pi}\oint \frac{\sin Z}{(Z-2\pi j)3}dZ$

Applying Cauchygntcgral formula

$I = \frac{-1}{2\pi} \times \frac{2\pi jF''(2\pi j)}{2!}$

Now F (Z) = SinZ

∴ F'(Z) = – Sin Z

$\therefore I = \frac{-1}{2\pi} \times 2\pi j\frac{(-\sin(2\pi j))}{2}$

$= \frac{-1}{2}Sin(h2\pi)$

= -133.8

18. From the question, the complex variable function is:

$$f(z) = 2z^3 + b_1 |z|^3$$

Given $f(z)$ is analytic.

which is possible only when $b = 0$

since $|z^3|$ is differentiable at the origin but not analytic.

$2z^3$ is analytic everywhere

∴ $f(z) = 2z^3 + b|z^3|$ is analytic

only when $b = 0$

19. $\frac{1}{2\pi}\oint_c \frac{e^z}{z-2}dz$

For $Z_0 = 2$ (inside a circle)

$Res\ (Z0 = 2) = \lim_{Z\to 2}(Z-2)\frac{e^z}{z-2}$

$\lim_{Z\to 2} e^Z$

$= e^2 = 7.39$

For $Z_0 = 2$ li??? outside circle

⇒ Residue F(Z) = 0

$\therefore \oint_c \frac{e^z}{z-2}dZ = 2\pi i \times \frac{1}{2\pi i} \times (0)$

$= 0$

20. $F(Z) = \frac{\sin(Z)}{Z^2}$

Now Residue at Z = 0

$=$ wefficint of $\frac{1}{Z}$ in $\left(\frac{Z - \frac{Z^3}{Z!} + \frac{Z^5}{5!} + ...}{Z^2}\right)$

$=$ coefficent of $\frac{1}{2}$ in $\left(\frac{1}{Z} - \frac{Z}{3!} + + \frac{Z^3}{5!} - \frac{Z^5}{7!} + \right)$

$=1$

21. $F(Z) = \frac{1}{(Z-4)(Z+1)^3}$

Here poles are at z = 4 and z = –1

Residance of F(Z) at Z = 4

$\lim_{Z-4}(Z-4) \times \frac{1}{(Z-4)(Z+1)^3}$

$= \frac{1}{(4+1)^3} = \frac{1}{125}$

and residue of F (Z) at Z = –1 is

$\lim_{2\to -1}\frac{4}{2!}\frac{d^2}{dZ^2}\left((Z-1)^3\frac{\times 1}{(Z-4)(Z+1)^3}\right)$

$$\lim_{Z \to -1} \frac{1}{2!}\left(\frac{2}{(Z-4)^3}\right)$$

$$= \frac{1}{(-1-4)^3}$$

$$= \frac{-1}{125}$$

22. $I = \oint_c \frac{Z^2-1}{Z^2+1} e^Z dZ$

Poles are at $Z^2+1 = 0$

$\therefore\ Z^2 = -1$

$\therefore\ Z^2 = 0\ i2$

$\therefore\ Z = \pm i$

Now poth the pole at –i and +i lies injide $|Z| = 3$ Now reoidue at Z = –i

$$= \lim_{zj-i} (z+i)\frac{Z^2-1}{Z^2+1}e^z$$

$$= \lim_{z-s-i} (z+i)\frac{(Z^2-1)}{(Z^2-i^2)}d^z$$

$$= \lim_{z \to -i} (z+i)\frac{(Z^2-1)}{(Z-i)(Z+i)}e^Z$$

$$= \frac{(-i)^2-1}{(-i\bar{0}i)}e^{-i}$$

$= -i\ e^{-i}$

and residue at Z = i

$$= \lim_{Z \to i} (Z-i)\frac{Z^2-1}{Z^2+1}e^Z$$

$$= \lim_{Z \to i} (Z-i)\frac{(Z^2-1)}{(Z^2+i^2)}e^Z$$

$$= \lim_{Z \to i} (Z-i)\frac{(Z^2-1)}{(Z-i)(Z+i)}e^Z$$

$$= \frac{i^2-1}{i+i}e^i$$

$$= \frac{-2}{2i}e^i$$

$$= \frac{-1}{i}e^i$$

$= ie^i$

23.

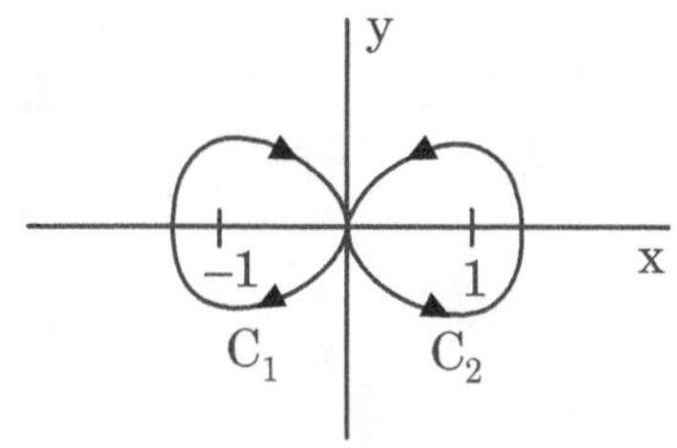

$$\frac{1}{\pi i}\oint_C \frac{dz}{z^2-1}$$

$$= 2\left[\frac{1}{2\pi i}\oint_{C_1}\frac{dz}{(z+1)(z-1)} + \frac{1}{2\pi i}\oint_{C_2}\frac{dz}{(z+1)(z-1)}\right]$$

$$= 2\left[-\left(\frac{1}{z-1}\right)\bigg|_{z=-1} + \left(\frac{1}{z+1}\right)\bigg|_{z=1}\right]$$

$$= 2\left[-\left(-\frac{1}{2}\right) + \left(\frac{1}{2}\right)\right] = 2$$

24. $\frac{1}{2\pi j}\oint\left(z+\frac{1}{z}\right)^2 dz = \frac{1}{2\pi j}\oint\frac{(z^2+1)^2}{z^2}dz$ ...(i)

$\therefore$ Singular point z = 0; which lies inside unit circle $|z| = 1$

$\therefore$ Using Cauchy's generalization integration formula, we get.

$$\oint_c \frac{f(z)}{(z-z_0)^n}dz = 2\pi j\frac{[f^{n-1}(z_0)]}{(n-1)!}$$

$$\Rightarrow \therefore \oint_C \frac{(z^2+1)^2}{(z-0)^2}dz = 2\pi j\frac{\left[(z^2+1)^2\right]'}{1!}\Big/ z = 0;$$

$= 2\pi j[2(z^2+1)(2z)]_{z=0}$

$= 2\pi j[0] = 0.$

From eq. (i) we get,

$$\frac{1}{2\pi j}\oint\left(z+\frac{1}{z}\right)^2 dz = \frac{1}{2\pi j}\oint\frac{(z^2+1)^2}{(z-0)^2} = \frac{1}{2\pi j}(0) = 0$$

5

CHAPTER

Numerical Methods

Analysis of Previous GATE Papers												
	Year → Topics ↓		2019	2018	2017 Set 1	2017 Set 2	2016 Set 1	2016 Set 2	2016 Set 3	2015 Set 1	2015 Set 2	2015 Set 3
SOLUTIONS OF NON-L INEAR EQUATIONS	1 Mark	MCQ Type						1				
		Numerical Type										
	2 Marks	MCQ Type										
		Numerical Type										
	Total							1				
SINGLE & MULTI -STEP METHODS FOR DI FFERENT EQUATION	1 Mark	MCQ Type		1								
		Numerical Type							1			
	2 Marks	MCQ Type										
		Numerical Type			1			1				1
	Total			1	2			2	1			2

SOLUTIONS OF NON-LINEAR EQUATIONS

1. How many distinct values of* satisfy the equation sin(x) = $\frac{x}{2}$, where x is in radians?

(a) 1 (b) 2
(c) 3 (d) 4 or more

[2006 : 1 Mark, Set-2]

2. For the function e^{-x}, the linear approximation around x = 2 is

(a) $(3 - x)e^{-2}$
(b) $1 - x$
(c) $\left[3+2\sqrt{2}-(1+\sqrt{2})\right]e^{-2}$
(d) e^{-2}

[2007 : 1 Mark]

3. The maximum value of θ until which the approximation sinθ = 0 holds to within 10% error is

(a) 10° (b) 18°
(c) 50° (d) 90°

[2013 : 1 Mark]

4. A polynomial $f(x) = a_4x^4 + a_3x^3 + a_2x^2 + a_1x - a_0$ with all coefficients positive has

(a) no real roots.
(b) no negative real root.
(c) odd number of real roots.
(d) at least one positive and one negative real root.

[2013 : 1 Mark]

SINGLE & MULTI-STEP METHODS FOR DIFFERENT EQUATION

5. The equation $x^3 - x^2 + 4x - 4 = 0$ is to be solved using the Newton-Raphson method. If x = 2 is taken as the initial approximation of the solution, then the next approximation using this method will be

(a) $\frac{2}{3}$ (b) $\frac{4}{3}$
(c) 1 (d) $\frac{3}{2}$

[2007 : 2 Marks]

6. The recursion relation to solve $x = e^{-x}$ using Newton-Raphson method is

(a) $x_{n+1} = e^{-x_n}$
(b) $x_{n+1} = x_n - e^{-x_n}$
(c) $x_{n+1} = (1+x_n)\frac{e^{-x_n}}{1+e^{-x_n}}$
(d) $x_{n+1} = \frac{x_n^2 - e^{-x_n}(1+x_n)-1}{x_n - e^{-x_n}}$

[2008 : 2 Marks]

7. Consider a differential equation $\frac{dy(x)}{dx} - y(x) = x$ with the initial condition y(0) = 0. Using Euler's first order method with a step size of 0.1, the value of y(0.3) is

(a) 0.01 (b) 0.031
(c) 0.0631 (d) 0.1

[2010 : 2 Marks]

8. A numerical solution of the equation $f(x) = x\sqrt{x} - 3 = 0$ can be obtained using Newton-Raphson method. If the starting value is x = 2 for the iteration, the value of x that is to be used in the next step is

(a) 0.306 (b) 0.739
(c) 1.694 (d) 2.306

[2011 : 2 Marks]

9. Match the application to appropriate numerical method.

Application

P1: Numerical integration
P2: Solution to a transcendental equation
P3: Solution to a system of linear equations
P4: Solution to a differential equation

Numerical Methods

M1: Newton-Raphson Method
M2: Runge-Kutta Method
M3: Simpson's $\frac{1}{3}$-rule
M4: Gauss Elimination Method

(a) P1— M3, P2—M2, P3—M4, P4—M1
(b) P1— M3, P2—M1, P3—M4, P4—M2
(c) P1—M4, P2—M1, P3—M3, P4—M2
(d) P1— M2; P2—M1, P3—M3, P4—M4

[2014 : 1 Mark, Set-3]

10. The Newton-Raphson method is used to solve the equation $f(x) = x^3 - 5x^2 + 6x - 8 = 0$. Taking the initial guess as x = 5, the solution obtained at the end of the first iteration is ______.

[2015 : 2 Marks, Set-3]

11. Consider the first order initial value problem

$y' = y + 2x - x^2, y(0) = 1, (0 \le x \le \infty)$

with exact solution $y(x) = x^2 + e^x$. For x = 0.1, the percentage difference between the exact solution and the solution obtained using a single iteration of the second-order Runge-Kutta method with step-size h = 0.1 is ______.

[2016 : 1 Mark, Set-3]

12. The ordinary differential equation

$$\frac{dx}{dt} = -3x + 2, \text{ with } x(0) = 1;$$

is to be solved using the forward Euler method. The largest time step that can be used to solve the equation without making the numerical solution unstable is ______.

[2016 : 2 Marks, Set-2]

13. Starting with x = 1, the solution of the equation $x^3 + x = 1$, after two iterations of Newton-Raphson's method (up to two decimal places) is ______.

[2017 : 2 Marks, Set-1]

14. Consider $p(s) = s^3 + a^2s^2 + a_1s + a_0$ with all real coefficients. It is known that its derivative p'(s) has no real roots. The number of real roots of p(s) is

(a) 0 (b) 1

(c) 2 (d) 3

[2018 : 1 Mark]

ANSWERS

1. (c) 2. (a) 3. (b) 4. (d) 5. (c) 6. (c) 7. (b) 8. (c) 9. (b) 10. (4· 2903)

11. (0· 06) 12. (0.66) 13. (0.68) 14. (b)

EXPLANATIONS

1. Here the given equation is

$$\sin(x) = \frac{x}{2} \text{ (where } x \text{ is in radians)}$$

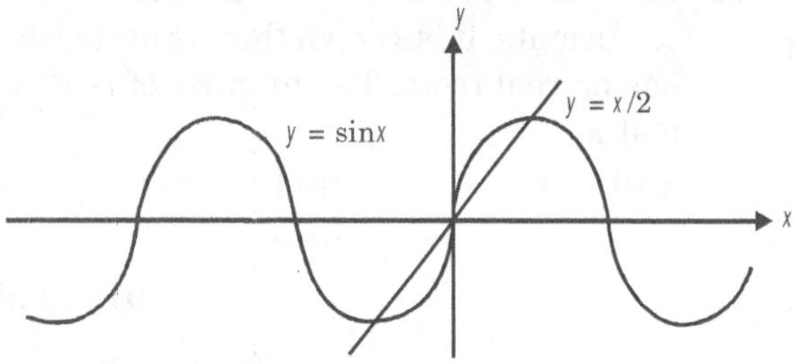

Hence there are 3 distinct values.

2. f(x) = e^{-x}

Lenear approximation at x = 2

L(x) = F(a) + F′(a) (x – a) at x = a

∴ L(x) = F(2) + F′(2) (x – 2)

Now F(2) = e^{-2}

ad F′(x) = – e^{-x}

∴ F′(2) = – e^{-2}

∴ L(x) = e^{-2} + (– e^{-2}) (x – 2)

∴ L(x) = e^{-2}[1 – (x – 2)]

∴ = e^{-2} (3 – x)

∴option (a) is correct

3. π radian = 180°

∴ ? 10°

$$\therefore \frac{10\pi}{180°} = 0.1745$$

Now sin 10° = 0.1736

∴ for 10°, $\sin\theta \cong \theta$ holds with in 10 % error

for 18°

$$18° - \frac{18\pi}{180\pi} = 0.3142$$

sin 18° = 6.3090

∴ for 18° → $\sin\theta \approx \sigma$ holds with in 10 % error

for 50°

$$50° - \frac{50\pi}{180°} = 0.8727$$

Now sin 50° = 0.766

∴ for 50° → $\sin\theta \approx \theta$ doesnot hold with in 10% error for 90°

$$90° = \frac{90 \times \pi}{180} = \frac{\pi}{2} = 1.571$$

Now sin 90 = 1

∴ for 90 → $\sin\theta \cong \theta$ doesnot hold with in 10 % error

Hence max[m] value of θ for approximation within 18 % error is 18°

4. According to Routh ??? criterea,

x^4 a_4 ↘ a_2 $-a_0$

x^3 a_3 a_1

x^2 x

x^1 a_1

x^0 $-a_0$

$$\text{Now } x = \frac{a_4a_1 - a_2a_3}{a_3}$$

from above table, there is atlest one sign change

∴ There will be atleast one positive and negative real roots.

5. $y(t) = x^3 - x^2 + 4x - 4 = 0;\ x_0 = 2$

Next approximation

$$x_1 = x_0 - \frac{f(x_o)}{f'(x_o)}$$

$$x_1 = x_0 - \frac{x_o^3 - x_o^2 + 4x_o - 4}{3x_o^2 - 2x_o + 4}$$

$$x_1 = 2 - \frac{8}{12} = \frac{4}{3}$$

6. Given: f(x) = x – e^{-x}

By Newton Raphson method,

$$x_{n+1} = x_n - \frac{f(x_n)}{f'(x_n)}$$

$$= x_n - \frac{x_n - e^{-x_n}}{1 + e^{-x_n}}$$

$$= (1 + x_n)\frac{e^{-x_n}}{1 + e^{-x_n}}$$

7. Given : $\frac{dy(x)}{dx} - y(x) = x;\ y(0) = 0$

$h = 0.1 \quad y(0.3) = ?$

$x_1 = 0; \quad y_1 = 0$

$$\therefore \left(\frac{dy}{dx}\right)_1 = 0 - 0 = 0$$

$$y_2 = y_1 + h\left(\frac{dy}{dx}\right)_1$$

$= 0 + 0.1 \times 0 = 0$

$x_2 = x_1 + h = 0.1$

$y_2 = 0 = x_2 + y_2(x)$

$\therefore \quad \left(\frac{dy}{dx}\right)_2 = 0.1 + 0 = 0.1$

$$y_3 = y_2 + h\left(\frac{dy}{dx}\right)_2$$

$= 0 + 0.1 \times 0.1 = 0.01$

$x_3 = x_2 + \Delta = 0.2$

$y_3 = 0.01$

$\therefore \quad \left(\frac{dy}{dx}\right)_3 = x_3 + y_3(x) = 0.2 + 0.01 = 0.21$

$\therefore \quad y_4 = y_3 + h\cdot\left(\frac{dy}{dy}\right)_3$

$= 0.01 + 0.1 * 0.21$

$= 0.01 + 0.021 = 0.031$

$x_4 = x_3 + \Delta = 0.3,$

$y_4 = 0.31$

$\therefore \quad y(0.3) = 0.031$

8. $$X_{n+1} = x_n - \frac{f(x_n)}{f'(x_n)}$$

$f(x_0) = f(2) = \left(2+\sqrt{2}-3\right) = \sqrt{2}-1$

and, $f'(x_0) = f'(2) = 1+\frac{1}{2\sqrt{2}} = \frac{2\sqrt{2}+1}{2\sqrt{2}}$,

$$X_1 = X_0 - \frac{f(X_0)}{f'(X_0)}$$

$\Rightarrow \quad X_1 = 2 - \dfrac{\left(\sqrt{2}-1\right)}{\dfrac{2\sqrt{2}+1}{2\sqrt{2}}} = 1.694$

10. $f(x) = x^3 - 5x^2 + 6x - 8$

$x_0 = 5$

$f'(x) = 3x^2 - 10x + 6$

By Newton-Raphson method.

$$x_1 = x_0 - \frac{f(x_0)}{f'(x_0)} = 5 - \frac{f(5)}{f'(5)}$$

$$= 5 - \frac{22}{31} = 5 - 0.7097 = 4.2903$$

11. Given the first order initial value problem is

$$\frac{dy}{dx} = y + 2x - x^2$$

$y(0) = 1$

$0 \le x < \infty$

Given $\quad f(x, y) = y + 2x - x^2,$

$x_o = 0,$

$y_o = 1,$

$h = 0.1$

$k_1 = hf(x_o, y_o)$

$= 0.1\,(1 + 2(0) - 0^2) = 0.1$

$k_2 = hg(x_0 + h, y_0 + k_1)$

$= 0.1\,((y_0 + k_1) + 2(x_0 + h) - (x_0 + h)^2)$

(since $h = 0.1$)

$= 0.1\,((1 + 0.1) + 2(0.1) - (0.1)^2)$

$= 0.1\,(1.1 + 0.2 - 0.01) = 0.129$

$\therefore \quad y_1 = y_0 + \frac{1}{2}(k_1 + k_2)$

$= 1 + \frac{1}{2}(0.1 + 0.129)$

(since $k_1 = 0.1$ and $k_2 = 0.129$)

$= 1 + 0.1145 = 1.1145$

For exact solution,

$y(x) = x^2 + e^x$

$y(0.1) = (0.1)^2 + e^{0.1}$

$= 0.01 + 1.1050 = 1.1152$

error $= 1.1152 - 1.1145 = 0.00069$

$$\text{Relative error} = \frac{\text{error}}{y(x)} = \frac{0.00069}{1.1152} = 0.00062$$

Percentage Error $= (0.00062 \times 100)\% = 0.06\%$

13. $F(x) = r^3 + x - 1$

Now $F(1) = 1 + 1 - 1$

$= 1$

ad $F'(x) = 3x^2 + 1$

$F'(1) = 3 + 1$

$= 4$

Now according to Newton – raphson method

$$x_{i+1} = xi - \frac{F(xi)}{F'(x1)}$$

$\therefore x1 = x_0 - \dfrac{F(x_0)}{F'(x_0)}$

$\therefore x1 = 1 - \frac{1}{4} = \frac{3}{4}$

Now for $x_1 = \frac{3}{4}$

$$x_2 = x_1 - \frac{F(x_1)}{F'(x_1)} = \frac{3}{4} - \frac{F\left(\frac{3}{4}\right)}{F'\left(\frac{3}{4}\right)} = 0.68$$

14. Here degree of p(s) is '3', so it will have three roots.

$p'(s) = 3s^2 + 2a_2s + a_1$ will have two roots and has no real root

$\therefore$ p(s) will have one real root.

6
CHAPTER

Probability & Statistics

Analysis of Previous GATE Papers												
	Year → Topics ↓		2019	2018	2017 Set 1	2017 Set 2	2016 Set 1	2016 Set 2	2016 Set 3	2015 Set 1	2015 Set 2	2015 Set 3
PROBABILITY & STATISTICS	1 Mark	MCQ Type								1		
		Numerical Type		1	1		1		1			
	2 Marks	MCQ Type										
		Numerical Type				1		1				1
	Total			1	1	2	1	2		1		2

1. Events A and B are mutually exclusive and have nonzero probability. Which of the following statement(s) are true?
(a) $P(A \cup B) = P(A) + P(B)$
(b) $P(B^C) > P(A)$
(c) $P(A \cap B) = P(A) P(B)$
(d) $P(B^C) < P(A)$

[1988 : 2 Marks]

2. A fair dice is rolled twice. The probability that an odd number will follow an even number is
(a) $\frac{1}{2}$ (b) $\frac{1}{6}$
(c) $\frac{1}{3}$ (d) $\frac{1}{4}$

[2005 : 1 Mark]

3. A probability density function is of the form $p(x) = Ke^{-a|x|}, x \in (-\infty, \infty)$
The value of K is
(a) 0.5 (b) 1
(c) 0.5 a (d) a

[2006 : 1 Mark]

4. Three companies X, Y and Z supply computers to a university. The percentage of computers supplied by them and the probability of those being defective are tabulated below

Company	% of computers supplied	Probability of being defective
X	60%	0.01
Y	30%	0.02
Z	10%	0.03

Given that a computer is defective, the probability that it was supplied by Y is
(a) 0.1 (b) 0.2
(c) 0.3 (d) 0.4

[2006 : 2 Marks]

5. An examination consists of two papers, Paper 1 and Paper 2. The probability of failing in Paper 1 is 0.3 and that in Paper 2 is 0.2. Given that a student has failed in Paper 2, the probability of failing in Paper 1 is 0.6. The probability of a student failing in both the papers is
(a) 0.5 (b) 0.18
(c) 0.12 (d) 0.06

[2007 : 2 Marks

6. A fair coin is tossed 10 times. What is the probability that ONLY the first two tosses will yield heads?
(a) $\left(\frac{1}{2}\right)^2$ (b) $^{10}C_2\left(\frac{1}{2}\right)^2$
(c) $\left(\frac{1}{2}\right)^{10}$ (d) $^{10}C_2\left(\frac{1}{2}\right)^{10}$

[2009 : 1 Mark]

7. Consider two independent random variables X and Y with identical distributions. The variables Xand /take values 0,1 and 2 with probabilities $\frac{1}{2}, \frac{1}{4}$ and $\frac{1}{4}$, and respectively.
What is the conditional probability $P(X + Y = 2 | X - Y = 0)$?
(a) 0 (b) $\frac{1}{16}$
(c) $\frac{1}{6}$ (d) 1

[2009 : 2 Marks]

8. A fair coin is tossed independently four times. The probability of the event "the number of times heads show up is more than the number of times tails show up is
(a) $\frac{1}{16}$ (b) $\frac{1}{8}$
(c) $\frac{1}{4}$ (d) $\frac{5}{16}$

[2010 : 2 Marks]

9. A fair dice is tossed two times. The probability that the second toss results in a value that is higher than the first toss is
(a) $\frac{2}{36}$ (b) $\frac{2}{6}$
(c) $\frac{5}{12}$ (d) $\frac{1}{2}$

[2011 : 2 Marks]

10. A fair coin is tossed till a head appears for the first time. The probability that the number of required tosses is odd, is
(a) 1/3 (b) 1/2
(c) 2/3 (d) 3/4

[2012 : 2 Marks]

11. In a housing society, half of the families have a single child per family, while the remaining half have two children per family. The probability that a child picked at random, has a sibling is ______.

[2014 : 1 Mark, Set-1]

12. An unbiased coin is tossed an infinite number of times. The probability that the fourth head appears at the tenth toss is

(a) 0.067 (b) 0.073
(c) 0.082 (d) 0.091

[2014 : 1 Mark, Set-3]

13. A fair coin is tossed repeatedly till both head and tail appear at least once. The average number of tosses required is ______.

[2014 : 2 Marks, Set-3]

14. Parcels from sender S to receiver R pass sequentially through two post-offices. Each post-office has a probability $\frac{1}{5}$ of osing an incoming parcel, independently of all other parcels. Given that a parcel is lost, the probaoility that it was lost by the second post-office is ______.

[2014 : 2 Marks, Set-4]

15. If calls arrive at a telephone exchange such that the time of arrival of any call is independent of the time of arrival of earlier or future calls, the probability distribution function of the total number of calls in a fixed time interval will be

(a) Poisson (b) Gaussian
(c) Exponential (d) Gamma

[2014 : 1 Mark, Set-4]

16. Suppose A and B are two independent events with probabilities P(A) ≠ 0 and P(B) ≠ 0. Let A and B be their complements. Which one of the following statements is FALSE?

(a) $P(A \cap B) = P(A) P\{B)$

(b) $P\left(\frac{A}{B}\right) = P(A)$

(c) $P(A \cup B) = P(A) + P(B)$

(d) $P(\bar{A} \cap \bar{B}) = P(\bar{A})P(\bar{B})$

[2015 : 1 Mark, Set-1]

17. A fair die with faces {1, 2, 3, 4, 5, 6} is thrown repeatedly till '3' is observed for the first time. Let X denote the number of times the die is thrown. The expected value of X is ______.

[2015 : 2 Marks, Set-3]

18. Two random variables X and Y are distributed according to

$$f_{X,Y}(x, y) = \begin{cases} (x+y) & 0 \le x \le 1 \quad 0 \le y \le 1 \\ 0 & \text{otherwise} \end{cases}$$

The probability P(X+ Y≤ 1) is ______.

[2016 : 2 Marks, Set-2]

19. The second moment of a Poisson-distributed random variable is 2. The mean of the random variable is ______.

[2016 : 1 Mark, Set-1]

20. The probability of getting a "head" in a single toss of a biased coin is 0.3. The coin is tossed repeatedly till a "head" is obtained. If the tosses are independent, then the probability of getting "head" for the first time in the fifth toss is ______.

[2016 : 1 Mark, Set-3]

21. Three fair cubical dice are thrown simultaneously. The probability that all three dice have the same number of dots on the faces showing up is (up to third decimal place) ______.

[2017 : 1 Mark, Set-1]

22. Passengers try repeatedly to get a seat reservation in any train running between two stations until they are successful. If there is 40% chance of getting reservation in any attempt by a passenger, ther the average number of attempts that passengers need to make to get a seat reserved is ______.

[2017 : 2 Marks, Set-2]

23. Let X_1, X_2, X_3 and X_4 be independent norma random variables with zero mean and unit variance. The probability that X_4 is the smallest among the four is ______.

[2018 : 1 Mark]

ANSWERS

1. (a)	2. (d)	3. (c)	4. (d)	5. (b)	6. (c)	7. (c)	8. (d)	9. (c)	10. (c)
11. (0· 65 to 0· 68)	12. (c)	13. (1· 99 to 2· 01)	14. (0.44)	15. (a)	16. (c)	17. (6)	18. (0· 33)		
19. (1)	20. (0· 07203)	21. (0· 028)	22. (2.5)	23. (0· 25)					

EXPLANATIONS

1. For mutually exclusive event
   P (A0B) = P (A) + P (B)

2. Required Probalility

   $= \frac{1}{2} \times \frac{1}{2} = \frac{1}{4}$ (Independent events)

3. Here P(x) is problility density function.

   $\int_{-\infty}^{\infty} P(x)dx = 1$

   Now $P(x) = Ke^{-a(x)}$

   $\Rightarrow P(x) = \begin{cases} Ke^{-ax}, & x \geq 0 \\ Ke^{-ax}, & x < 0 \end{cases}$

   $\therefore \int_{-\infty}^{0} Ke^{ax}dx + \int_{0}^{\infty} Ke^{-ax}dx = 1$

   $\therefore K \int_{-\infty}^{0} e^{ax}dx + K\int_{0}^{\infty} e^{-ax}dx = 1$

   $\therefore \frac{K}{a}\left[e^{ax}\right]_{-\infty}^{0} + \frac{K}{-a}\left[e^{-ax}\right]_{0}^{\infty} = 1$

   $\therefore K = 0.5\ a$

4. Probability that product supplied by 'y' and was found dfechre

   $P(s/d) = \frac{P(snd)}{P(d)}$ where' s dentes supply by 'y'

   and 'd' demoter defective.

   Now P (snd) = 0.3 × 0.02

   = 0.006

   and P (d) = 0.6×0.01 + 0.3 + 0.02 + 0.1×0.03

   = 0.015

   $\text{Now } P(s/d) = \frac{P(snd)}{P(d)}$

   $= \frac{0.006}{0.015}$

   $= \frac{6}{15}$

   $= \frac{2}{5}$

   = 0.4

5. Probability of failing in paper –1 is P (A) = 0.2 and Probalility of failing in paper –2 is P(B) = 0.3 and P (A/B) = 0.6

   $\text{Now } P(A/B) = \frac{P(AnB)}{P(B)}$

   Required Probalility P(AnB) = P (A/B) × P(B)

   = 0.6 × 0.3

   = 0.18

6. Desired outcomes = {H, H, T,T,T,T,T,T,T,T}

   Probalility that the first toss will yield head $= \frac{1}{2}$

   Probalility that second toss will also yield head $= \frac{1}{2}$ and similarly from 3rd toss till 10th toss will yield tail $= \frac{1}{2}$

   Required Probalility =

   $\left(\frac{1}{2}\right)^2 \times \left(\frac{1}{2}\right)^8 = \left(\frac{1}{2}\right)^{10} = \frac{1}{1024}$

8. Desired Outcomes

   {HHHH, HHHT, HHTH, HTHH, THHH}

   Now Probalility of each event

   $\left(\frac{1}{2}\right) \times \frac{1}{2} \times \frac{1}{2} \times \frac{1}{2} = \frac{1}{16}$

   Now as all events are mutually exclusive

   ∴ total probality

   $= \frac{1}{16} + \frac{1}{16} + \frac{1}{16} + \frac{1}{16} + \frac{1}{16} = \frac{5}{16}$

9. Total number of cause = 36

   Total number of favorable causes

   = 5 + 4 + 3 + 2 + 1 = 15

   $\therefore$ Probability $= \frac{15}{36} = \frac{5}{12}$

   (1, 1) (2, 1) (3, 1) (4, 1) (5, 1) (6, 1)

10. P(odd tosses) = P(H) + P(TTH) + P(TTTTH) + ...

    $= \frac{1}{2} + \left(\frac{1}{2}\right)^3 + \left(\frac{1}{2}\right)^5 + \ldots$

    $= \frac{1}{2}\left(1 + \left(\frac{1}{2}\right)^2 + \left(\frac{1}{2}\right)^4 + \ldots\right)$

$$= \frac{1}{2}\left[1+\left(\frac{1}{4}\right)+\left(\frac{1}{4}\right)^2+...\right]$$

$$= \frac{1}{2}\left(\frac{1}{1-\frac{1}{4}}\right)$$

$$= \frac{1}{2}\times\frac{4}{3}$$

$$= \frac{2}{3}$$

11. Let the number of families in housing society be x.

$\therefore$ $\frac{x}{2}$ family have single child.

$\therefore$ Total $\frac{x}{2}$ children

Now remaining $\frac{x}{2}$ families have 2 children.

$\therefore$ Total $\frac{x}{2}\times 2 = x$, sibling children

Probability that a child picked at random has a

$$\text{sibling} = \frac{x}{x+\frac{x}{2}}$$

$$= \frac{x\times 2}{3x}$$

$$= \frac{2}{3} = 0.667$$

12. P [fourth head appears at the tenth toss] = P [getting 3 heads in the first 9 tosses and one head at tenth toss]

$$= \left[{}^9C_3\cdot\left(\frac{1}{2}\right)^9\right]\times\left[\frac{1}{2}\right]$$

$$= \frac{21}{256}$$

$$= 0.082$$

13. In this problem random variable is L

L can be 1, 2, ...

$$P\{L = 1\} = \frac{1}{2}$$

$$P\{L = 2\} = \frac{1}{4}$$

$$P\{L = 3\} = \frac{1}{8}$$

$$H\{L\} = \frac{1}{2}\log_2\frac{1}{1/2}+\frac{1}{4}\log_2\frac{1}{1/4}+\frac{1}{8}\log_2\frac{1}{1/8}+....$$

$$= 0+1.\frac{1}{2}+2.\frac{1}{4}+3.\frac{1}{8}+....$$

[Arithmatic geometric series summation]

$$= \frac{\frac{1}{2}}{1-\frac{1}{2}}+\frac{\frac{1}{2}\cdot 1}{\left(1-\frac{1}{2}\right)^2} = 2$$

14. Probalility that parcel was lost at post office –1 $= \frac{1}{5}$ and probalility that parcel was lost at post office–2 $= \frac{4}{5}\times\frac{1}{5}$

Total Prbalility that parcel is lost $= \frac{4/25}{9/25} = 0.44$

18. The probability $P(X + Y \le 1)$

$$= \int_{x=0}^{1}\int_{y=0}^{(1-x)} f_{xy}(x,y)\,dx\,dy$$

$$= \int_{x=0}^{1}\int_{y=0}^{(1-x)} (x+y)\,dx\,dy$$

$$= \int_{x=0}^{1}\left(xy+\frac{y^2}{2}\right)_0^{1-x}$$

$$= \int_{x=0}^{1}\left(x(1-x)+\frac{(1-x)^2}{2}\right)dx$$

$$= \int_{x=0}^{1}(x-x^2)\,dx+\int_{x=0}^{1}\frac{(1-x)^2}{2}\,dx$$

$$= \int_{x=0}^{1}\left(\frac{1}{2}-\frac{x^2}{2}\right)dx$$

$$= \left(\frac{x}{2}-\frac{x^3}{6}\right)_0^1$$

$$= \frac{1}{2}-\frac{1}{6}$$

$$= \frac{1}{3} = 0.33$$

19. Here, $E(x^2) = 2$

$\therefore$ $V(x) = E(x^2) - (E(x))^2$

Let mean of the poisson random variable be x.

$$x = 2 - x^2$$

$$\left(\text{Since } V(x) = x \text{ and } (E(x))^2 = x^2\right)$$

$$x^2 + x - 2 = 0$$

$$x^2 + 2x - x - 2 = 0$$

$$x(x + 2) - 1\,(x + 2) = 0$$

$$(x - 1)\,(x + 2) = 0$$

$\therefore$ $x = 1, -2$

Mean of random variable $(\lambda) = 1$

20. The probability of getting "head" for the first time in fifth toss

$$(P) = (0.7)^4\,(0.3)$$

$$= 0.07203$$

21. When three dice are thrown

Total number of possible cases

$$= (6)^3 = 216$$

Outcomes for all three dice having same number are,

$$\begin{Bmatrix} (1,1,1) & (2,2,2) & (3,3,3) \\ (4,4,4) & (5,5,5) & (6,6,6) \end{Bmatrix}$$

Hence number of favourable cases = 6

$\therefore$ Required probability

$$= \frac{\text{Number of favourable cases}}{\text{Total number of outcomes}}$$

$$= \frac{6}{216} = \frac{1}{36} = 0.028$$

22. There is 40% Chance of getting reservation

$\therefore$ probalility of getting reservation $= \frac{40}{100} = \frac{2}{5}$

$\Rightarrow$ Probalility of not getting reservation $= 1 - \frac{2}{5} = \frac{3}{5}$

Now $E\,[X] = \sum x_i P(x_i)$

$$= 1 \times \frac{2}{5} + 2 \times \frac{2}{5} \times \frac{3}{5} + 3 \times \frac{2}{5}\left(\frac{3}{5}\right)^2 + \ldots\ldots$$

$$\therefore E[X] = \frac{10}{4} = 2.5$$

23. $P(X_4 \text{ is smallest}) = \frac{3!}{4!} = \frac{3 \times 2}{4 \times 3 \times 2} = \frac{1}{4} = 0.25$

7

CHAPTER

Transform Theory

Analysis of Previous GATE Papers

Topics ↓		Year →	2019	2018	2017 Set 1	2017 Set 2	2016 Set 1	2016 Set 2	2015 Set 1	2015 Set 2	2015 Set 3
CTFT	1 Mark	MCQ Type									
		Numerical Type									
	2 Marks	MCQ Type								1	
		Numerical Type									
		Total								**2**	
BI LATERAL LAPLACE TRANSFORM	1 Mark	MCQ Type									
		Numerical Type									
	2 Marks	MCQ Type									
		Numerical Type									
		Total									
UNI LATERAL LAPLACE TRANSFORM	1 Mark	MCQ Type									
		Numerical Type									
	2 Marks	MCQ Type									
		Numerical Type		1							
		Total		**2**							
FI LTER (CTFT)	1 Mark	MCQ Type									
		Numerical Type									
	2 Marks	MCQ Type									
		Numerical Type									
		Total									
C.T. FOURIER SERIES	1 Mark	MCQ Type									
		Numerical Type									
	2 Marks	MCQ Type									
		Numerical Type									
		Total									

CTFT

1. The value of the integral

$$\int_{-\infty}^{\infty} 12\cos(2\pi t)\frac{\sin(4\pi t)}{4\pi t}dt \text{ is } \_\_\_\_\_\_.$$

[2015 : 2 Marks, Set-2]

BILATERAL LAPLACE TRANSFORM

2. If $L\{f(c)\} = \frac{2(s+1)}{s^2+2s+s}$ then $f(0^+)$ and are $f(\propto)$ are given by ______.

(a) 0, 2 respectively (b) 2, 0 respectively

(c) 0, 1 respectively (d) $\frac{2}{5}$, 0 respectively

[1995 : 2 Marks]

3. If $L\{f(t)\} = \frac{\omega}{s^2+\omega^2}$ then the value of $\lim_{t\to\infty} f(t) =$ ______.

(a) can not be determined

(b) zero

(c) unity

(d) Infinite

[1998 : 1 Mark]

4. The Laplace transform of i(t) is given by $I(s) = \frac{2}{s(1+s)}$. As $t \to \infty$, the value of i(t)

(a) 0 (b) 1

(c) 2 (d) ∞

[2003 : 1 Mark]

5. In what range should Re(s) remain so that the Laplace transform of the function $g^{(a+2)t+5}$ exits?

(a) Re(s) > a + 2

(b) Re(s) > a + 7

(c) Re(s) < 2

(d) Re(s) > a + 5

[2005 : 2 Marks]

6. Given $L^{-1}\left[\frac{3s+1}{s^3+4s^2+(K-3)s}\right]$. If $\lim_{t\to\infty} f(t) = 1$ then the value of K is

(a) 1

(b) 2

(c) 3

(d) 4

[2007 : 2 Marks]

UNILATERAL LAPLACE TRANSFORM

7. The position of a particle y(t) is described by the differential equation:

$$\frac{d^2y}{dt^2} = -\frac{dy}{dt} - \frac{5y}{4}$$

The initial conditions are y(0) = 1 and $\left.\frac{dy}{dt}\right|_{t=0} = 0$. The position (accurate to two decimal places) of the particle at $t = \pi$ is ______.

[1996 : 2 Marks]

8. The inverse Laplace transform of the function $\frac{s+5}{(s+1)(s+3)}$ is ______.

(a) $2e^{-t} - e^{-3t}$ (b) $2e^{-t} + e^{-3t}$

(c) $e^{-t} - 2e^{-3t}$ (d) $e^{-t} + 2e^{-3t}$

[1996 : 1 Mark]

9. The Laplace transform of $e^{at}\cos\alpha t$ is equal to ______.

(a) $\frac{s-\alpha}{(s-\alpha)^2+\alpha^2}$ (b) $\frac{s-\alpha}{(s+\alpha)^2+\alpha^2}$

(c) $\frac{1}{(s-\alpha)^2}$ (d) n`one

[1997 : 1 Mark]

10. If $L\{f(t)\} = F(s)$ then $L\{f(t-T)\}$ is equal to

(a) $e^{sT}F(s)$ (b) $e^{-sT}F(s)$

(c) $\frac{F(s)}{1-e^{sT}}$ (d) $\frac{F(s)}{1-e^{-sT}}$

[1999 : 1 Mark]

11. If $L\{f(t)\} = \frac{s+2}{s^2+1}$, $L\{g(t)\} = \frac{s^2+1}{(s+3)(s+2)}$

$h(t) = \int_0^t f(T)g(t-T)dT$ then $L\{h(t)\}$ is ______.

(a) $\frac{s^2+1}{s+3}$

(b) $\frac{1}{s+3}$

(c) $\frac{s^2+1}{(s+3)(s+2)} + \frac{s+2}{s^2+1}$

(d) None of these

12. Consider the differential equation

$$\frac{d^2y(t)}{dt^2}+2\frac{dy(t)}{dt}+y(t)=\delta(t)$$

$$y(t)\Big|_{1=0}=-2 \text{ and } \frac{dy}{dt}\Big|_{t=0^-}=0$$

The numerical value of $\frac{dy}{dt}\Big|_{t=0^+}$ is

(a) –2 (b) –1

(c) 0 (d) 1

[2012 : 2 Marks]

FILTER (CTFT)

13. A low-pass filter having a frequency response $H(j\omega) = A(\omega)e^{j\phi(\omega)}$ does not produce any phase distortion if

(a) $A(\omega) = C\omega^2, \phi(\omega) = k\omega^3$.

(b) $A(\omega) = C\omega^2, \phi(\omega) = k\omega$.

(c) $A(\omega) = C\omega, \phi(\omega) = k\omega^2$.

(d) $A(\omega) = C, \phi(\omega) = k\omega^{-1}$.

[2006 : 1 Mark]

C.T. FOURIER SERIES

14. Three functions $f_1(t)$, $f_2(t)$ and $f_3(t)$, which are zero outside the interval [0, T], are shown in the figure. Which of the following statements is correct?

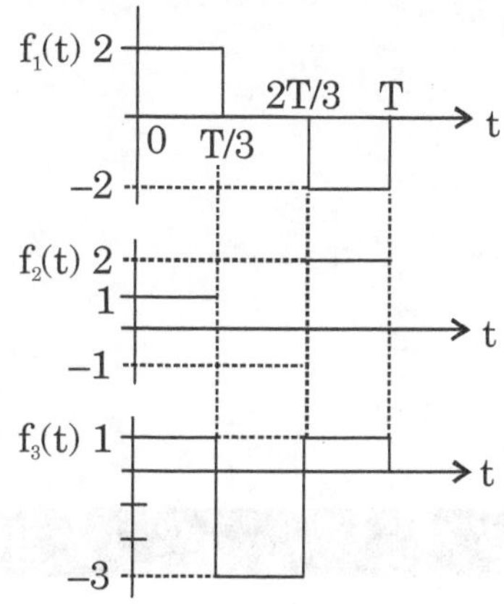

(a) $f_1(t)$ and $f_2(t)$ are orthogonal.

(b) $f_1(t)$ and $f_3(t)$ are orthogonal.

(c) $f_2(t)$ and $f_3(t)$ are orthogonal.

(d) $f_1(5)$ and $f_2(t)$ are orthonormal.

[2007 : 2 Marks]

15. The trigonometric Fourier series for the waveform f(t) shown below contains

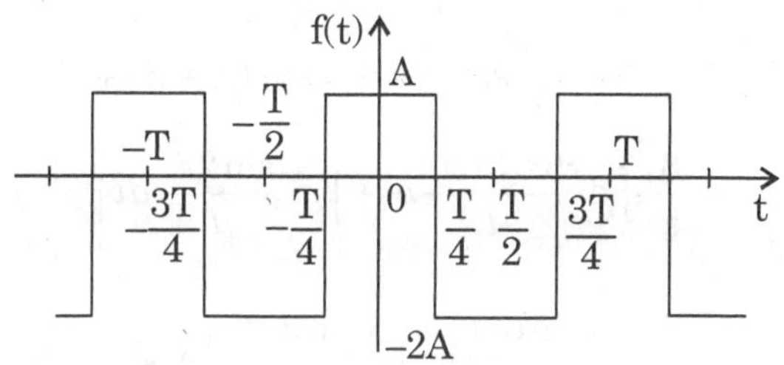

(a) only cosine terms and zero value for the dc component

(b) only cosine terms and a positive value for the dc component

(c) only cosine terms and a negative value for the dc component

(d) only sine terms and a negative value for the dc component

[2010 : 1 Mark]

ANSWERS

1. (b)	2. (b)	3. (a)	4. (c)	5. (a)	6. (d)	7. (0.21)	8. (a)	9. (a)	10. (b)
11. (b)	12. (d)	13. (b)	14. (c)	15. (c)					

EXPLANATIONS

1. $\int_{-\infty}^{\infty} 12\cos 2\pi t \frac{\sin 4\pi t}{4\pi t} dt$

$\frac{12}{4\pi}\int_{0}^{\infty} \frac{2\cos 2\pi t \sin 4\pi t}{t} dt$

$\frac{3}{\pi}\left[\int_0^{\infty}\frac{\sin 6\pi t\, dt}{t} + \int_0^{\infty}\frac{\sin 2\pi t\, dt}{t}\right]$

($\because$ sin A – cos B = sin (A + B) + sin (A – B))

$= \frac{3}{\pi}\left[\int_0^{\infty} e^{\theta t}\frac{6\sin 6\pi t}{t} dt + \int_0^{\infty} e^{\theta t}\frac{\sin 2\pi t}{t} dt\right]$

$= \frac{3}{\pi}\left[L\left\{\frac{\sin 6\pi t}{t}\right\} + L\left\{\frac{\sin 2\pi t}{t}\right\}\right]$ with $s = 0$

$= \frac{3}{\pi}\left[\int_s^{\infty}\frac{6\pi}{s^2+36\pi^2} ds + \int_s^{\infty}\frac{2\pi}{s^2+4\pi^2} ds\right]$ with $s = 0$

$= \frac{3}{\pi}\left[6\pi.\frac{1}{6\pi}\tan^{-1}\left(\frac{s}{6\pi}\right) + 2\pi.\frac{1}{2\pi}\tan^{-1}\left(\frac{s}{2\pi}\right)\right]_s^{\infty}$ with $s = 0$

$= \frac{3}{\pi}\left[\tan^{-1}\infty - \tan^{-1}\left(\frac{s}{6\pi}\right) + \tan^{-1}(\infty) - \tan^{-1}\left(\frac{s}{2\pi}\right)\right]$

$\Rightarrow \frac{3}{\pi}\left[\frac{\pi}{2} - \tan^{-1} 0 + \frac{\pi}{2} - \tan^{-1} 0\right]$

$\Rightarrow \frac{3}{\pi}\left[\frac{\pi}{2} - 0 + \frac{\pi}{2} - 0\right] = \frac{3}{\pi}\times\pi = 3$

2. $F(s) = \frac{2(s+1)}{s^2+2s+5}$

$= \frac{2(s+1)}{(s+1)^2+4} = \frac{2(s+1)}{(s+1)^2+2^2}$

By first shifting theorem

$f(t) = 2e^{-t}\cos 2t$

$f(0^+) = \lim_{t\to 0} f(t) = \lim_{t\to 0} 2e^{-t}\cos 2t = 2$

$f(\infty) = \lim_{t\to 0} f(t) = \lim_{t\to 0} 2e^{-t}\cos 2t = 0$

3. $L(f(t)) = \frac{\omega}{s^2+\omega^2}$

$\therefore f(t) = \sin\omega t$

$\therefore \mathop{Lt}_{t\to\infty} f(t) = \mathop{Lt}_{t\to\infty} \sin\omega t$

= lies between –1 and 1

4. $I(s) = \frac{2}{s(1+s)} = \frac{2}{s} - \frac{2}{1+s}$

$i(t) = 2 - 2e^{-t}$

$\lim_{t\to\infty} i(t) = \lim_{t\to\infty}(2 - 2e^{-t})$

$= 2$

5. $f(t) = e^{(a+2)t+5} = e^5.e^{(a+2)t}$

$F(s) = \left[\frac{1}{s-(a+2)}\right]e^5$

$\therefore$ For LT to exist, $Re(s) > a + 2$

6. $\lim_{t\to\infty} f(t) = \lim_{s\to 0} sF(s)$

Given that,

$$F(s) = \left[\frac{3s+1}{s^3+4s^2+(K-3)s}\right]$$

$$\lim_{t\to\infty} f(t) = 1$$

$$\Rightarrow \lim_{s\to 0}\left[\frac{3s+1}{s^3+4s^2+(K-3)s}\right] = 1$$

$$\Rightarrow \lim_{s\to 0}\left[\frac{3s+1}{s^2+4s+(K-3)}\right] = 1$$

$$\Rightarrow \frac{1}{K-3} = 1$$

$\Rightarrow K - 3 = 1$

$\Rightarrow K = 4$

7. $$\frac{d^2y}{dt^2} + \frac{dy}{dt} + \frac{5y}{4} = 0$$

$y(0) = 1$

$y'(0) = 0$

By applying Laplace transform,

$$s^2Y(s) - s(1) + sY(s) - 1 + \frac{5}{4}Y(s) = 0$$

$$Y(s) = \frac{s+1}{s^2+s+\frac{5}{4}} = \frac{s+1}{\left(s+\frac{1}{2}\right)+1}$$

$$= \frac{\left(s+\frac{1}{2}\right)}{\left(s+\frac{1}{2}\right)^2+1} + \frac{\frac{1}{2}}{\left(s+\frac{1}{2}\right)^2+1}$$

By taking inverse Laplace transform,

$$y(t) = e^{-t/2}\left[\cos(t) + \frac{1}{2}\sin(t)\right]; t > 0$$

At $t = \pi$,

$y(t = \pi) = e^{-\pi/2}[(-1) + (0)]$

$= -e^{-\pi/2} = -0.2075 \simeq -0.21$

8. $$L^{-1}\left[\frac{s+5}{(s+1)(s+3)}\right] = L^{-1}\left[\frac{2}{s+1} - \frac{1}{s+3}\right]$$

$= 2e^{-t} - e^{-3t}$

9. $$L(\cos \alpha t) = \frac{s}{s^2+\alpha^2}$$

By first shifting theorem

$$L(e^{\alpha t}\cos\alpha t) = \frac{(s-\alpha)}{(s-\alpha)^2+\alpha^2}$$

10. By second shifting theorem

If $L(f(t)) = F(s)$

then $L(f(t - T)) = e^{-sT} F(s)$

11. $$h(t) = \int_0^t f(T)g(t-T)dT$$

$= f(t) * g(t)$

$L(h(t)) = L(f(t) * g(t))$

$= F(s).G(s)$

$$= \frac{s+2}{s^2+1}\cdot\frac{s^2+1}{(s+3)(s+2)} = \frac{1}{s+3}$$

12. $$\frac{d^2y}{dt^2} + 2\frac{dy}{dt} + y(t) = \delta(t)$$

taking Laplace transform on both the sides we have

$[s^2Y(s) - sy(0^-) - y'(0)] + 2[Y(s) - y(0^-)] + Y(s) = 1$

$s^2Y(s) + 2s + 2sY(s) + 4 + Y(s) = 1$

$(s^2 + 2s + 1)\,Y(s) = -(2s + 3)$

$$Y(s) = \frac{-(2s+3)}{(s+1)^2}$$

$$Y(s) = -\left[\frac{2}{(s+1)} + \frac{1}{(s+1)^2}\right]$$

$\Rightarrow Y(t) = -[2e^{-t} + te^{-t}]u(t)$

$$\frac{dy}{dt} = -[-2e^{-t} + e^{-t} - te^{-t}]u(t)$$

$$\left.\frac{dy}{dt}\right|_{at\ t=0^+} = -[-2 + 1 - 0]$$

$$\left.\frac{dy}{dt}\right|_{at\ t=0^+} = 1$$

13. For distortionless transmission,

$$\frac{d\phi(\omega)}{d\omega} = \text{constant}$$

Phase response should be linear

$$\phi(\omega) = k\omega$$

14. Two functions f(x) & g(x) are said to be orthogonal

if $\int_{-\infty}^{\infty} f(x)g(x)dx = 0$

$$\int_{-\infty}^{\infty} f_2(t)f_3(t)dt = \int_0^T f_2(t)f_3(t)dt$$

$$= \frac{T}{3} - T + \frac{2T}{3}$$

$$= 0$$

15. Since f(t) is an even function, its trigonometric Fourier series contains only cosine terms.

D.C. component,

$$A_0 = \frac{1}{T}\int_{-T/2}^{T/2} f(t)dt = \frac{2}{T}\int_0^{T/2} f(t)dt$$

$$= \frac{2}{T}\left[\int_0^{T/A} A\,dt + \int_{T/4}^{T/2} (-2A)dt\right]$$

$$= \frac{2}{T}\left[\frac{AT}{4} - 2A\left(\frac{T}{2} - \frac{T}{4}\right)\right]$$

$$= \frac{2}{T}\left[-\frac{AT}{4}\right] = -\frac{A}{2}$$

Therefore, the trigonometric Fourier series for the waveform f(t) contains only cosine terms and a negative value for the dc component.

Unit - II

Electromagnetics

Syllabus

Electrostatics; Maxwell's equations: differential and integral forms and their interpretation, boundary conditions, wave equation, Poynting vector; Plane waves and properties: reflection and refraction, polarization, phase and group velocity, propagation through various media, skin depth; Transmission lines: equations, characteristic impedance, impedance matching, impedance transformation, S-parameters, Smith chart; Waveguides: modes, boundary conditions, cut-off frequencies, dispersion relations; Antennas: antenna types, radiation pattern, gain and directivity, return loss, antenna arrays; Basics of radar; Light propagation in optical fibers.

Contents

1
CHAPTER

Introduction to Vectors

Analysis of Previous GATE Papers												
Year → Topics ↓			2019	2018	2017 Set 1	2017 Set 2	2016 Set 1	2016 Set 2	2016 Set 3	2015 Set 1	2015 Set 2	2015 Set 3
VECTORS (Curl, Divergence, Gradient)	1 Mark	MCQ Type										
		Numerical Type										
	2 Marks	MCQ Type				1				1		
		Numerical Type										
	Total					2				2		
INTEGRATION (Line, Surface, Volume)	1 Mark	MCQ Type										
		Numerical Type										
	2 Marks	MCQ Type							1			
		Numerical Type										1
	Total											2

VECTORS (CURL, DIVERGENCE, GRADIENT)

1. Vector potential is a vector
(a) whose curl is equal to the magnetic flux density
(b) whose curl is equal to the electric field intensity
(c) whose divergence is equal to the electric potential
(d) which is equal to the vector product E×H

2. The direction of vector A is radially outward from the origin, with |A| = kr^n where $r^2 = x^2 + y^2 + z^2$ and k is a constant. The value of n for which $\nabla \cdot A = 0$ is
(a) –2 (b) 2
(c) 1 (d) 0

[2012 : 2 Marks]

3. A vector $\vec{P}$ is given by

$\vec{P} = x^3y\vec{a}_x - x^2y^2\vec{a}_y - x^2yz\vec{a}_z$. Which of the following statements is TRUE?

(a) $\vec{P}$ is solenoidal, but not irrotational
(b) $\vec{P}$ is irrotational, but not solensoidal
(c) $\vec{P}$ is neither solenoidal nor irrotational
(d) $\vec{P}$ is both solenoidal and irrotational

[2015 : 2 Marks, Set-1]

4. If the vector function

$\vec{F} = \hat{a}_x(3y - k_1z) + \hat{a}_y(k_2x - 2z) - \hat{a}_z(k_3y + z)$ is irrotational, then the values of the constants k_1, k_2 and k_3, respectively are
(a) 0.3, –2.5, 0.5 (b) 0.0, 3.0, 2.0
(c) 0.3, 0.33, 0.5 (d) 4.0, 3.0, 2.0

[2017 : 2 Marks, Set-2]

INTEGRATION (LINE, SURFACE, VOLUME)

5. $\vec{V} = x\cos^2 y\hat{i} + x^2e^z\hat{j} + z\sin^2 y\hat{k}$ and S is the surface of unit cube with one corner at the origin and edges parallel to the coordinate axis, the value of the integral $\iint_C \vec{V} \cdot \hat{n}dS$ is ______.

[1993 : 2 Marks]

6. $\oint_C \vec{A} \cdot d\vec{l} = \int_S \_\_\_ \cdot d\vec{s}$

[1994 : 1 Mark]

7. If a vector field $\vec{V}$ is related to another vector field $\vec{A}$ through $\vec{V} = \nabla\vec{A}$, which of the following is true? Note: C and S_C refer to any closed contour and any surface whose boundary is C.

(a) $\oint_C \vec{V} \cdot d\vec{l} = \iint_{S_C} \vec{A} \cdot d\vec{S}$

(b) $\oint_C \vec{A} \cdot d\vec{l} = \iint_{S_C} \vec{V} \cdot d\vec{S}$

(c) $\oint_C \nabla \times \vec{V} \cdot d\vec{l} = \iint_{S_C} \nabla \times \vec{A} \cdot d\vec{S}$

(d) $\oint_C \nabla \times \vec{A} \cdot d\vec{l} = \iint_{S_C} \vec{V} \cdot d\vec{S}$

[2009 : 2 Marks]

8. Consider a closed surface S surrounding a volume V. If $\vec{r}$ is the position vector of a point inside S, with $\hat{n}$ the unit normal on S, the value of the integral $\oiint_S 5\vec{r} \cdot \hat{n}\, dS$ is
(a) 3 V (b) 5 V
(c) 10 V (d) 15 V

[2011 : 1 Mark]

9. A vector field $\vec{D} = 2\rho^2\hat{a}_\rho + z\hat{a}_z$ exists inside a cylindrical region enclosed by the surfaces $\rho = 1$, $z = 0$ and $z = 5$. Let S be the surface bounding this cylindrical region. The surface integral of this field on S $\oiint_S \vec{D} \cdot d\vec{s}$ is ______.

[2015 : 2 Marks, Set-3]

10. Consider the charge profile shown in the figure. The resultant potential distribution is best described by

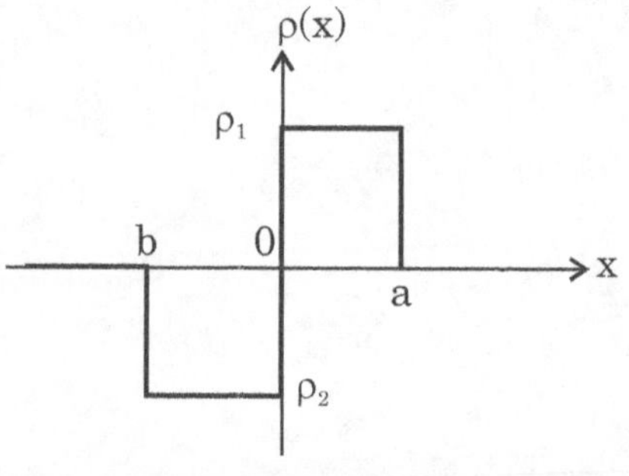

(a)

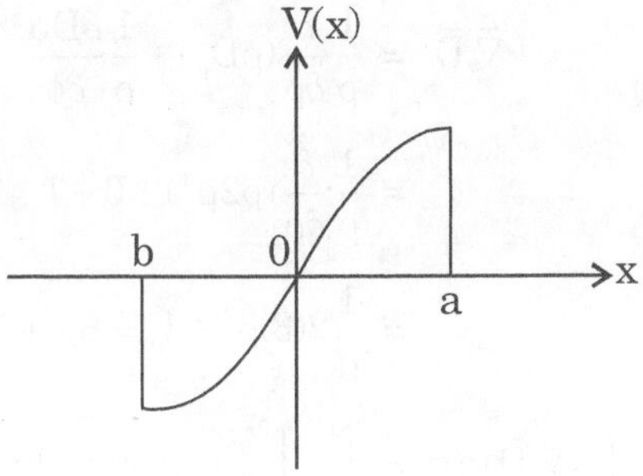

(b)

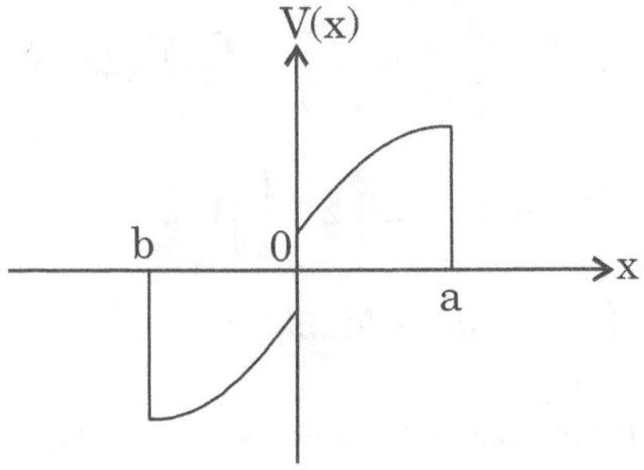

(c)

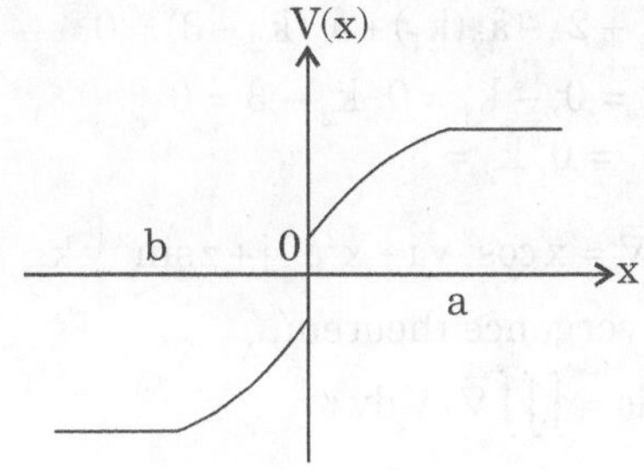

(d)

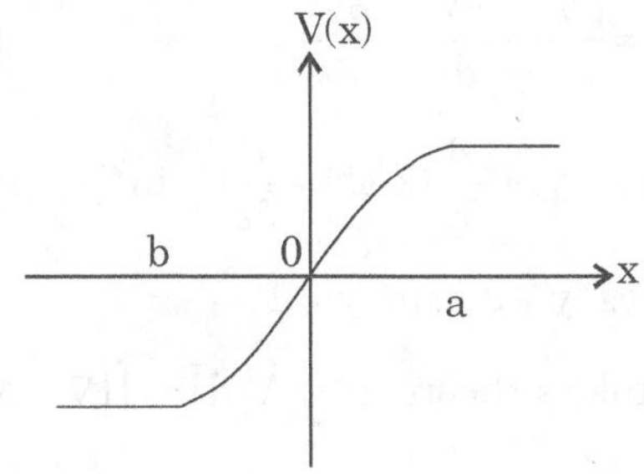

[2016 : 2 Marks, Set-3]

ANSWERS

1. (a) **2.** (a) **3.** (a) **4.** (b) **5.** (1) **6.** (*) **7.** (b) **8.** (d) **9.** (78.53) **10.** (d)

EXPLANATIONS

1. Vector magnetic potential $\bar{V}_B$ is a vector whose curl is eual to the magnetic flux density $\bar{B}$, i.e.,

$$\bar{\nabla} \times \bar{V}_B = \bar{B} = \mu\bar{H}$$

2. Given : $|A| = k \cdot r^n \Rightarrow \bar{A} = Kr^n \cdot \hat{a}_r$

(radially outward)

$$\bar{\nabla} \cdot \bar{A} = \frac{1}{r^2}\frac{\partial}{\partial r}(r^2 A_r) + \frac{1}{r\sin\theta}\frac{\partial}{\partial r}(A_\theta \sin\theta) + \frac{1}{r\sin\theta}\frac{\partial}{\partial \phi}(A_\phi)$$

$$\bar{\nabla} \cdot \bar{A} = \frac{1}{r^2} \cdot \frac{\partial}{\partial r}(r^2 \cdot kr^n) + 0 + 0$$

$$\bar{\nabla} \cdot \bar{A} = \frac{k}{r^2} \cdot \frac{\partial}{\partial r}(r^{n+2}) = \frac{k}{r^2}(n+2) \cdot r^{n+1}$$

$\Rightarrow \bar{\nabla} \cdot \bar{A}$ will be zero if $n + 2 = 0$

$\therefore\ n = -2$

3. Trace of A = 14

$a + 5 + 2 + b = 14$

(Taking the diagonal element and then adding)

$$a + b = 7 \qquad ...(i)$$

$$\det(A) = 100$$

$$5\begin{vmatrix} a & 3 & 7 \\ 0 & 2 & 4 \\ 0 & 0 & b \end{vmatrix} = 100$$

$$5 \times 2 \times a \times b = 100$$

$$10\,ab = 100$$

$$ab = 10 \qquad ...(ii)$$

From equation (*i*) and (*ii*)

either $a = 5, b = 2$

or $a = 2\ b = 5$

$$|a - b| = |5 - 2| = 3$$

4. If $\bar{F} = (3y - k_1 z)\hat{a}_x + (k_2 x - 2z)\hat{a}_y - (k_3 y + z)\hat{a}_z$

irrotational, then $\bar{\nabla} \times \bar{F} = 0$.

$$\bar{\nabla} \times \bar{F} = \begin{vmatrix} \hat{a}_x & \hat{a}_y & \hat{a}_z \\ \partial/\partial x & \partial/\partial y & \partial/\partial z \\ (3y - k_1 z) & (k_2 x - 2z) & (k_3 y + z) \end{vmatrix}$$

$= \hat{a}_z(-k_3 + 2) - \hat{a}_y(k_1) + \hat{a}_z(k_2 - 3) = 0$

$\Rightarrow k_3 + 2 = 0; - k_1 = 0; k_2 - 3 = 0$

$k_3 = 2, k_1 = 0, k_2 = 3.$

5. Given : $\bar{V} = x\cos^2 y\,\hat{i} + x^2e^z\,\hat{j} + z\sin^2 y\,\hat{k}.$

Using divergence theorem,

$$\oiint_s \bar{V}\cdot\hat{n}\,ds = \iiint_v \bar{\nabla}\cdot\bar{V}\,dv$$

$$\Rightarrow \bar{\nabla}\cdot\bar{V} = \frac{\partial V_x}{\partial x} + \frac{\partial V_y}{dy} + \frac{\partial V_z}{\partial z}$$

$$= \frac{\partial}{\partial x}(x\cos^2 y) + \frac{\partial}{\partial y}(x^2e^z) + \frac{\partial}{\partial z}(z\sin^2 y)$$

$$\bar{\nabla}\cdot\bar{V} = \cos^2 y + 0 + \sin^2 y = 1$$

6. Using stoke's theorem, $\oint_c \bar{A}\cdot d\bar{l} = \int_s (\bar{\nabla}\times\bar{A})\cdot d\bar{s}$

7. Given $\bar{V} = \bar{\nabla}\times\bar{A}$

Taking surface integral in both the sides,

$$\iint_s \bar{V}\cdot d\bar{s} = \iint_s (\bar{\nabla}\times\bar{A})\cdot d\bar{s}$$

Using stoke's theorem, $\iint_s (\bar{\nabla}\times\bar{A})\cdot d\bar{s} = \oint_c \bar{A}\cdot d\bar{\ell}$

Now, $\iint_s \bar{V}\cdot d\bar{s} = \oint_c \bar{A}\cdot d\bar{\ell}$

8. Using divergence theorem,

$$\iiint(\bar{\nabla}\cdot\bar{A})dv = \oint_s \bar{A}\cdot\hat{n}\,ds$$

Position vector, $\bar{r} = x\hat{a}_x + y\hat{a}_y + z\hat{a}_z$

$$\Rightarrow \bar{A} = 5\bar{r}$$

$$\bar{\nabla}\cdot\bar{A} = \left(\hat{a}_x\frac{\partial}{\partial x} + \hat{a}_y\frac{\partial}{\partial y} + \hat{a}_z\frac{\partial}{\partial z}\right)\cdot 5(x\hat{a}_x + y\hat{a}_y + z\hat{a}_z)$$

$= (1 + 1 + 1)\,5 = 3 \times 5 = 15.$

$$\Rightarrow \oiint_s 5\bar{r}\cdot\hat{n}\,ds = \iiint_v 15\,dv = 15v.$$

9. $D = 2\rho^2 a\rho + za_z$

$$\oint_z \bar{D}.d\bar{s} = \int_V (\bar{\nabla}.\bar{D})dv$$

$$\bar{\nabla}.\bar{D} = \frac{1}{\rho}\frac{\partial}{\partial\rho}(\rho D_\rho) + \frac{1}{\rho}\frac{\partial D\phi}{\partial\phi} + \frac{\partial D_z}{\partial z}$$

$$= \frac{1}{\rho}\frac{\partial}{\partial\rho}(\rho 2\rho^2) + 0 + 1$$

$$= \frac{1}{\rho}2(3)\rho^2 + 1 = 6\rho + 1$$

$$\int_V(\bar{\nabla}.\bar{D})dv = \int_{\rho=0}^{1}\int_{\phi=0}^{2\pi}\int_{z=0}^{5}(6\rho+1)\rho\,d\rho\,d\phi\,dz$$

$$= \left(\frac{6\rho^3}{3} + \frac{\rho^2}{2}\right)\Bigg|_0^1 (2\pi)(5)$$

$$= \left(2 + \frac{1}{2}\right)10\pi$$

$$\int_V(\bar{\nabla}.\bar{D})dv = 78.53$$

10. From poisson's equation, $\bar{\nabla}\cdot\bar{V} = -\rho(x)/\in$

For one dimensional charge density,

$$\frac{d^2v}{dx^2} = -\rho(x)/\in$$

For $x < 0 : \rho(x) = -\rho_2$

$$\frac{d^2y}{dx^2} = \rho_2/\in$$

On Solving, we get

$$V(\bar{x}) = \frac{\rho_2}{\in}x^2 + C_1x + C_2 \quad \forall\, x < 0.$$

Where C_1 and C_2 are arbitary constants.

thus, V(x) is an upward parabola.

for $x > 0$: $\rho(x) = \rho_2$; $\dfrac{d^2v}{dx^2} = \dfrac{-\rho_2}{\in}$

On solving, $V^+(x) = -\dfrac{\rho_1}{\in}x^2 + c_3x + c_4 \quad \forall x < 0.$

Where c_3 and c_4 are arbitary constant.

For $x = 0$; $V^-(x) = V^+(x) = 0.$

V(x) will be constant for x < b and x > a. There is no discontinuity.

2 CHAPTER Basics of Electromagnetics

Analysis of Previous GATE Papers												
	Year → Topics ↓		2019	2018	2017 Set 1	2017 Set 2	2016 Set 1	2016 Set 2	2016 Set 3	2015 Set 1	2015 Set 2	2015 Set 3
ELECTROSTATICS	1 Mark	MCQ Type	1			1						
		Numerical Type					1				1	
	2 Marks	MCQ Type						2				
		Numerical Type				1						
		Total	1			3	1	4			1	
MAGNETOSTATICS	1 Mark	MCQ Type						1				
		Numerical Type										
	2 Marks	MCQ Type	1				1					
		Numerical Type										
		Total	2				2	1				
MAXWELL'S EQUATION (Differential & Integral Form) & Their INTERPRETATION	1 Mark	MCQ Type	1						1			
		Numerical Type										
	2 Marks	MCQ Type										
		Numerical Type										
		Total	1						1			

ELECTROSTATICS

1. An electrostatic field is said to be conservative when
   (a) the divergence of the field is equal to zero
   (b) the curl of the field is equal to zero
   (c) the curl of the field is equal to $\frac{\partial E}{\partial t^2}$
   (d) the Laplacian of the field is equal to $\frac{\partial^2 E}{\partial t^2}$
   [1987 : 2 Marks]

2. On either side of a charge free interface between two media
   (a) the normal components of the electric field are equal
   (b) the tangential component of the electric field are equal
   (c) the normal components of the electric flux density are equal
   (d) the tangential components of the electric flux density are equal
   [1988 : 2 Marks]

3. The electric field strength at a far off point P due to a point charge, +q located at the origin, O is 100 millivolts/metre. Tie point charge is now enclosed by a perfectly conducting hollow metal sphere with its centre at the origin, O. The electric field strength at the point, P
   (a) remains unchanged in its magnitude and direction.
   (b) remains unchanged in its magnitude but reverse in direction.
   (c) would be that due to a dipole formed by the charge, +q, at O and -q induced.
   (d) would be zero.
   [1989 : 2 Marks]

4. For a uniformly charged sphere of radius R and charge density p, the ratio of magnitude of electric fields at distances R/2 and 2R from the centre, i.e.,
   $\frac{E\left(r=\frac{R}{2}\right)}{E(r=2R)}$ is ______. **[1993 : 2 Marks]**

5. The electric field strength at distant point, P, due to a point charge, +q, located at the origin, is 100 µV/m. If the point charge is now enclosed by a perfectly conducting metal sheet sphere whose center is at the origin, then the electric field strength at the point, P, outside the sphere, becomes
   (a) zero (b) 100 µV/m
   (c) –100 µV/m (d) 50 µV/m
   [1995 : 1 Mark]

6. In the infinite plane, y = 6 m, there exists a uniform surface charge density of $\left(\frac{1}{6000\pi}\right)\mu C/m^2$. The associated electric field strength is
   (a) $30\hat{i}$ V/m (b) $3\hat{i}$ V/m
   (c) $30\hat{k}$ V/m (d) $60\hat{j}$ V/m
   [1995 : 1 Mark]

7. A metal sphere with 1 m radius and a surface charge density of 10 Coulombs/m² is enclosed in a cube of 10 m side. The total outward electric displacement normal to the surface of the cube is
   (a) 40π Coulombs (b) 10π Coulombs
   (c) 5π Coulombs (d) None of the above
   [1995 : 1 Mark]

8. An electric field on a plane is described by its potential $V = 20(r^{-1} + r^{-2})$ where r is the distance from the source. The field is due to
   (a) a monopole
   (b) a dipole
   (c) both a monopole and adipole
   (d) a quadrupole
   [1999 : 1 Mark]

9. The electric field on the surface of a perfect conductor is 2 V/m. The conductor is immersed in water with $\epsilon = 80\,\epsilon_0$. The surface charge density on the conductor is
   (a) 0 C/m² (b) 2 C/m²
   (c) 1.8×10^{-11} C/m² (d) 1.41×10^{-9} C/m²
   [2002 : 2 Marks]

10. If the electric field intensity is given by $\vec{E} = (xu_x + yu_y + zu_z)$ Volt/m the potential difference between X(2,0,0) and Y(1,2, 3) is
    (a) +1 volt (b) – 1 volt
    (c) + 5 volt (d) + 6 volt
    [2003 : 2 Marks]

11. The force on a point charge +q kept at a distance dfrom the surface of an infinite grounded metal plate in a medium of permittivity ∊ is
    (a) 0
    (b) $\frac{q^2}{16\pi \epsilon d^2}$ away from the plate

(c) $\dfrac{q^2}{16\pi \in d^2}$ towards the plate

(d) $\dfrac{q^2}{4\pi \in d^2}$ towards the plate

[2014 : 1 Mark, Set-1]

12. The electric field (assumed to be one-dimensional) between two points A and B is shown. Let ψ_A and ψ_B be the electrostatic potentials at A and B, respectively. The value of $\psi_A - \psi_B$ in Volts is ______.

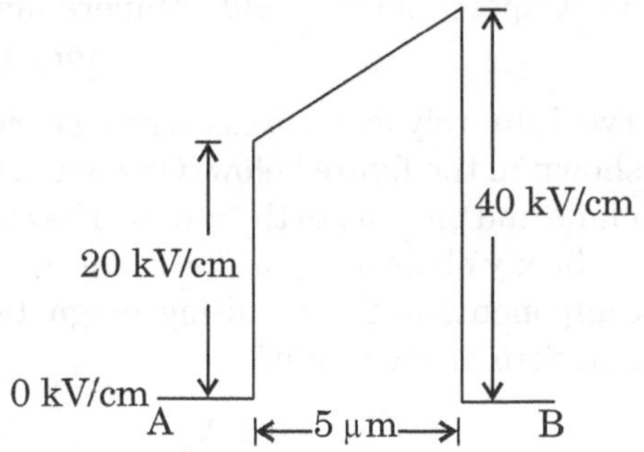

[2014 : 2 Marks, Set-4]

13. If $\vec{E} = -(2y^3 - 3y^2)\hat{x} - (6xy^2 - 3xz^2)\hat{y} + (6xyz)\hat{z}$ is the electric field in a source free region, a valid expression for the electrostatic potential is

(a) $xy^3 - yz^2$ (b) $2xy^3 - xyz^2$

(c) $y^3 + xyz^2$ (d) $2xy^3 - 3xyz^2$

[2014 : 2 Marks, Set-4]

14. In a source free region in vaccum, if the electrostatic potential $\varphi = 2x^2 + y^2 + cz^2$, the value of constant c must be ______.

[2015 : 1 Mark, Set-2]

15. Concentric spherical shells of radii 2 m, 4 m, and 8 m carry uniform surface charge densities of 20 nC/m^2, –4 nC/m^2 and ρ_s, respectively. The value of ρ_s(nC/m^2) required to ensure that the electric flux density $\vec{D} = \vec{0}$ at radius 10 m is ______.

[2016 : 1 Mark, Set-1]

16. The parallel-plate capacitor shown in the figure has movable plates. The capacitor is charged so that the energy stored in it is E when the plate separation is d. The capacitor is then isolated electrically and the plates are moved such that the plate separation becomes 2d.

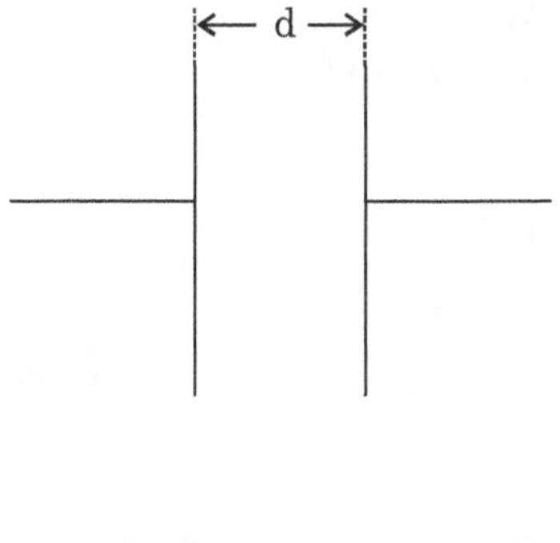

At this new plate separation, what is the energy stored in the capacitor, neglecting fringing effects?

(a) 2E (b) $\sqrt{2}\,E$

(c) E (d) $\dfrac{E}{2}$

[2016 : 2 Marks, Set-2]

17. Two conducting spheres S1 and S2 of radii a and b (b > a) respectively, are placed far apart and connected by a long, thin conducting wire, as shown in the figure.

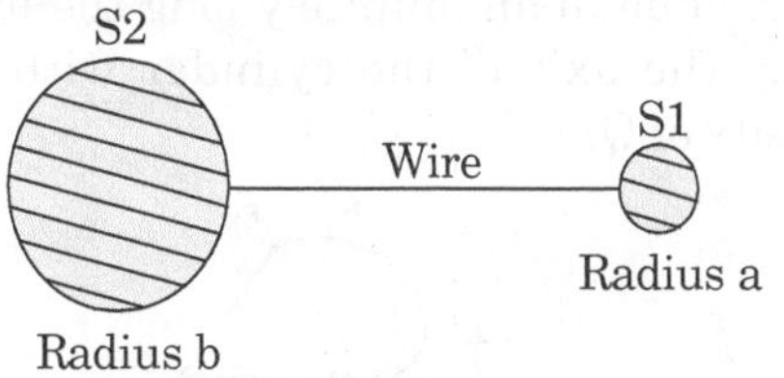

For some charge placed on this structure, the potential and surface electric field on S_1 are V_a and E_a, and that on S_2 are V_b and E_b respectively. Then, which of the following is CORRECT?

(a) $V_a = V_b$ and $E_a < E_b$

(b) $V_a > V_b$ and $E_a > E_b$

(c) $V_a = V_b$ and $E_a > E_b$

(d) $V_a > V_b$ and $E_a = E_b$

[2017 : 1 Mark, Set-2]

18. An electron (q_1) is moving in free space with velocity 10^5 m/s towards a stationary electron (q_2) far away. The closest distance that this moving electron gets to the stationary electron before the repulsive force diverts its path is ______ $\times 10^{-8}$ m. [Given, mass of electron m = 9.11×10^{-31} kg, charge of electron e = -1.6×10^{-19} C, and permittivity $\varepsilon_0 = \left(\dfrac{1}{36\pi}\right) \times 10^{-9}$ F/m].

[2017 : 2 Marks, Set-2]

19. A positive charge q is placed at x = 0 between two infinite metal plates placed at x = –d and at x = +d respectively. The metal plates lie in the yz plane.

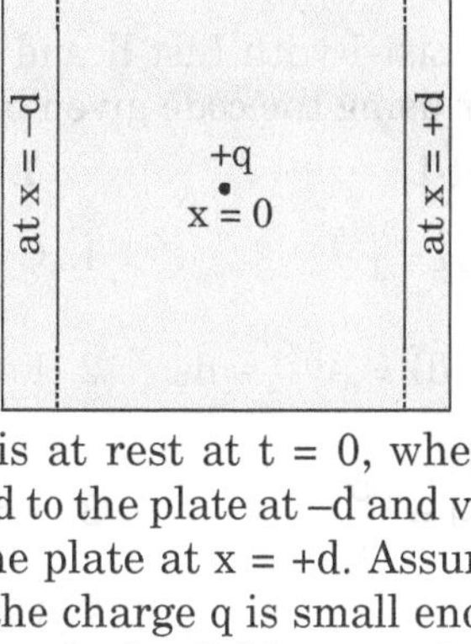

The charge is at rest at t = 0, when a voltage + V is applied to the plate at –d and voltage –V is applied to the plate at x = +d. Assume that the quantity of the charge q is small enough that it does not perturb the field set up by the metal plates. The time that the charge q takes to reach the right plate is proportional to

(a) $\frac{d}{V}$ (b) $\frac{\sqrt{d}}{V}$

(c) $\frac{d}{\sqrt{V}}$ (d) $\sqrt{\frac{d}{V}}$

[2017 : 2 Marks, Set-2]

20. What is the electric flux $\left(\int \vec{E}.d\hat{a}\right)$ through a quarter-cylinder of height H (as shown in the figure) due to an infinitely long the line charge along the axis of the cylinder with a charge density of Q?

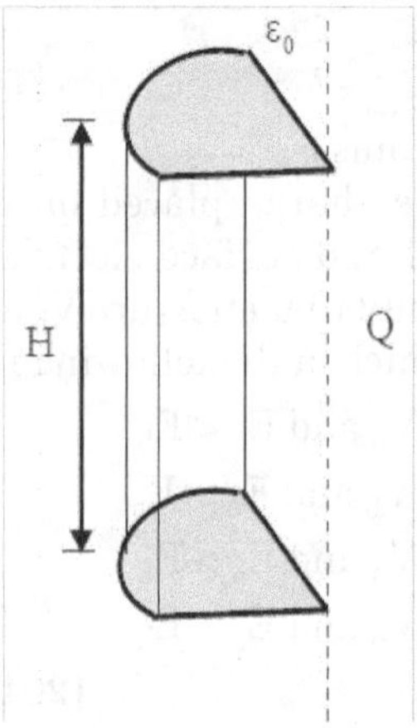

(a) $\frac{4H}{Q\varepsilon_0}$ (b) $\frac{HQ}{4\varepsilon_0}$

(c) $\frac{HQ}{\varepsilon_0}$ (d) $\frac{H\varepsilon_0}{4Q}$

[2019 : 1 Mark]

MAGNETOSTATICS

21. Which of the following field equations indicate that the free magnetic charge do not exist?

(a) $\vec{H} = \frac{1}{\mu}\nabla \times \vec{A}$ (b) $\vec{H} = \oint \frac{d\vec{l} \times \vec{R}}{4\pi R^2}$

(c) $\nabla \cdot \vec{H} = 0$ (d) $\nabla \times \vec{H} = \vec{j}$

[1990 : 2 Marks]

22. Match List-I with List-II and select the correct answer using the code given below the Lists:

List-I	List-II
A. $\nabla \times \vec{H} = \vec{j}$	1. Continuity equation
B. $\oint_C \vec{E} \cdot d\vec{l} = -\oint_S \frac{\partial B}{\partial t} \cdot d\vec{s}$	2. Faraday's Law
C. $\nabla \cdot \vec{J} = \frac{\partial \rho}{\partial t}$	3. Ampere's Law
	4. Gauss's Law
	5. Biot-Savart Law

Codes :

	A	B	C
(a)	3	2	1
(b)	2	1	3
(c)	4	3	1
(d)	1	2	3

[1994 : 2 Marks]

23. The unit of $\nabla \times \vec{H}$ is

(a) Ampere (b) Ampere/meter

(c) Ampere/meter^2 (d) Ampere-meter

[2003 : 1 Mark]

24. Two infinitely long wires carrying current are as shown in the figure below. One wire is in the y-z plane and parallel to the y-axis. The other wire is in the x-y plane and parallel to the .v-axis. Which components of the resulting magnetic field are non-zero at the origin?

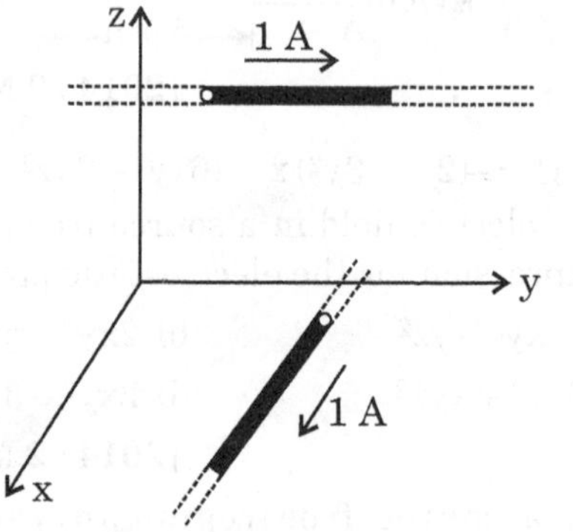

(a) x, y, z components

(b) x, y components

(c) y, z components

(d) x, z components

[2009 : 1 Marks]

25. A magnetic field in air is measured to be

$$\vec{B} = B_0\left(\frac{x}{x^2+y^2}\hat{y} - \frac{y}{x^2+y^2}\hat{x}\right)$$

What current distribution leads to this field? [Hint: the algebra is trivial in cylindrical coordinates].

(a) $\vec{J} = \frac{B_0\hat{z}}{\mu_0}\left(\frac{1}{x^2+y^2}\right), r \neq 0$

(b) $\vec{J} = \frac{B_0\hat{z}}{\mu_0}\left(\frac{2}{x^2+y^2}\right), r \neq 0$

(c) $\vec{J} = 0, r \neq 0$

(d) $\vec{J} = \frac{B_0\hat{z}}{\mu_0}\left(\frac{1}{x^2+y^2}\right), r \neq 0$

[2009 : 2 Marks]

Statement for Linked Answer Questions 26 and 27 : An infinitely long uniform solid wire of radius a carries a uniform dc current of density $\vec{J}$.

26. The magnetic field at a distance rfrom the center of the wire is proportional to

(a) r for r < a and $\frac{1}{r^2}$ for r > a

(b) 0 for r < a and $\frac{1}{r}$ for r > a

(c) r for r < a and $\frac{1}{r}$ for r > a

(d) 0 for r < a and $\frac{1}{r^2}$ for r > a

[2012 : 2 Marks]

27. A hole of radius b (b < a) is now drilled along the length of the wire at a distance d from the center of the wire as shown below.

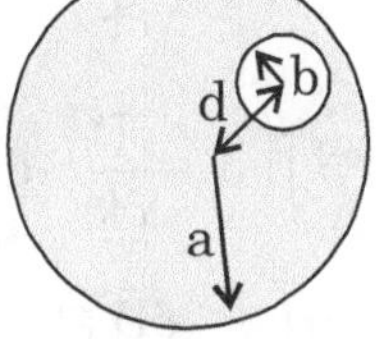

The magnetic field inside the hole is

(a) uniform and depends only on d.

(b) uniform and depends only on b.

(c) uniform and depends on both band d.

(d) non-uniform.

[2012 : 2 Marks]

28. A region shown below contains a perfect conducting half-space and air. The surface current $\vec{K}_s$ on the surface of the perfect conductor is $\vec{K}_s = \hat{x}2$ amperes per meter. The tangential $\vec{H}$ field in the air just above the perfect conductor is

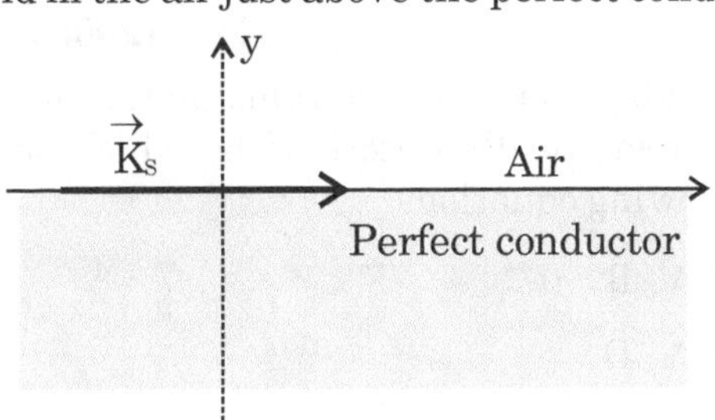

(a) $(\hat{x}+\hat{z})$ amperes per meter

(b) $\hat{x}2$ amperes per meter

(c) $-\hat{z}2$ amperese per meter

(d) $\hat{z}2$ amperes per meter

[2014 : 2 Marks, Set-3]

29. Consider a straight, infinitely long, current carrying conductor lying on the z-axis. Which one of the following plots (in linear scale) qualitatively represents the dependence of H_ϕ on r, where H_ϕ is the magnitude of the azimuthal component of magnetic field outside the conductor and ris the radial distance from the conductor?

(a)

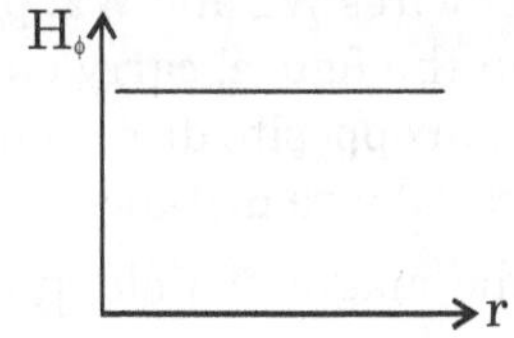

(b)

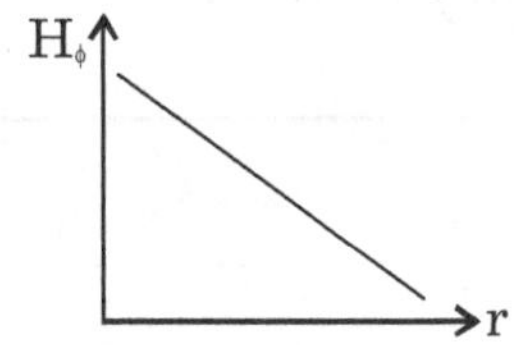

(c)

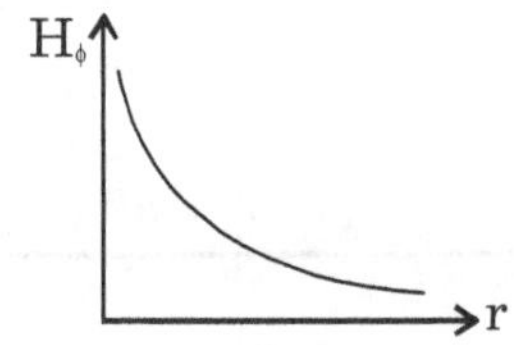

(d)

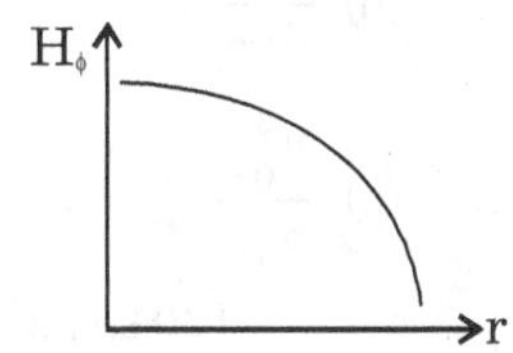

30. The current density in a medium is given by

$$\vec{J} = \frac{400\sin\theta}{2\pi(r^2+4)}\hat{a}_r \text{ A}-\text{m}^{-2}$$

The total current and the average current density flowing through the portion of a spherical surface $r = 0.8$ m, $\frac{\pi}{12} \le \theta \le \frac{\pi}{4}$, $0 \le \phi \le 2\pi$ are given respectively, by

(a) 15.09 A, 12.86 Am^{-2}

(b) 18.73 A, 13.65 Am^{-2}

(c) 12.86 A, 9.23 Am^{-2}

(d) 10.28 A, 7.56 Am^{-2}

[2016 : 2 Marks, Set-1]

31. A uniform and constant magnetic field $B = \hat{z}B$ exists in the $\hat{z}$ direction in vacuum. A particle of mass m with a small charge q is introduced into this region with an initial velocity $v = \hat{x}v_x + \hat{z}v_z$. Given that B, m, q, v_x and v_z are all non-zero, which one of the following describes the eventual trajectory of the particle?

(a) Helical motion in the $\hat{z}$ direction.
(b) Circular motion in the xy plane.
(c) Linear motion in the $\hat{z}$ direction.
(d) Linear motion in the $\hat{x}$ direction.

[2016 : 1 Mark, Set-2]

32. Two identical copper wires W1 and W2, placed in parallel as shown in the figure, carry currents I and 2I, respectively, in opposite directions. If the two wires are separated by a distance of 4r, then the magnitude of the magnetic field $\vec{B}$ between the wires at a distance r form W1 is

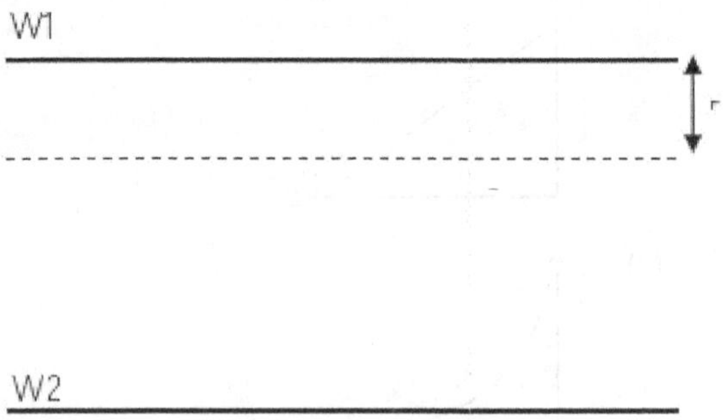

(a) $\dfrac{5\mu_0 I}{6\pi r}$ (b) $\dfrac{\mu_0 I}{6\pi r}$

(c) $\dfrac{6\mu_0 I}{5\pi r}$ (d) $\dfrac{\mu_0^2 I^2}{2\pi r^2}$

[2019 : 2 Marks]

MAXWELL'S EQUATIONS (DIFFERENTIAL & INTEGRAL FORM) & THEIR INTERPRETATION

33. A long solenoid of radius R, and having N turns power unit length carries a time dependent current $i(t) = I_0 \cos(at)$. The magnitude of induced electric field at a distance R/2 radially from the axis of the solenoid is p

(a) $\frac{R}{2}\mu_0 N I_0 \omega \sin(\omega t)$ (a) $\frac{R}{4}\mu_0 N I_0 \omega \cos(\omega t)$

(c) $\frac{R}{4}\mu_0 N I_0 \omega \sin(\omega t)$ (d) $R\mu_0 N I_0 \omega \sin(\omega t)$

[1993 : 2 Marks]

34. The Maxwell's equation, $\nabla \times \vec{H} = \vec{j} + \frac{\partial \vec{D}}{\partial t}$ is based on

(a) Ampere's law
(b) Gauss's law
(c) Faraday's law
(d) Coulomb's law

[1998 : 1 Mark]

35. A parallel plate air-filled capacitor has plate area of 10^{-4} m^2 and plate separation of 10^{-3} m. It is connected to a 0.5 V, 3.6 GHz source. The magnitude of the displacement current is ($\varepsilon_0 = 1/36\pi \times 10^{-9}$ F/m)

(a) 10 mA
(b) 100 mA
(c) 10 A
(d) 1.59 mA

[2004 : 2 Marks]

36. If C is a closed curve enclosing a surface S, then the magnetic field intensity $\vec{H}$, the current density $\vec{J}$. and the electric flux density $\vec{D}$ are related by

(a) $\iint_S \vec{H} \cdot d\vec{s} = \oint_C \left(\vec{J} + \frac{\partial \vec{D}}{\partial t}\right) \cdot d\vec{t}$

(b) $\int_C \vec{H} \cdot d\vec{l} = \oiint_S \left(\vec{J} + \frac{\partial \vec{D}}{\partial t}\right) \cdot d\vec{s}$

(c) $\oiint_S \vec{H} \cdot d\vec{s} = \int_C \left(\vec{J} + \frac{\partial \vec{D}}{\partial t}\right) \cdot d\vec{\ell}$

(d) $\oint_C \vec{H} . d\vec{l} = \iint_S \left(\vec{J} + \frac{\partial \vec{D}}{\partial t}\right) \cdot d\vec{s}$

[2007 : 1 Mark]

37. For static electric and magnetic fields in an inhomogeneous source-free medium, which of the following represents the correct form of two of Maxwell's equations?

(a) $\nabla \cdot \vec{E} = 0$, $\nabla \times \vec{B} = 0$ (b) $\nabla \times \vec{E} = 0$, $\nabla \cdot \vec{B} = 0$

(c) $\nabla \times \vec{E} = 0$, $\nabla \times \vec{B} = 0$ (d) $\nabla \times \vec{E} = 0$, $\nabla \cdot \vec{B} = 0$

[2008 : 1 Mark]

38. Faraday's law of electromagnetic induction is mathematically described by which one of the following equations?

(a) $\nabla \cdot \vec{B} = 0$

(b) $\nabla \cdot \vec{D} = \rho_v$

(c) $\nabla \times \vec{E} = \frac{\partial \vec{B}}{\partial t}$

(d) $\nabla \times \vec{H} = \sigma \vec{E} + \frac{\partial \vec{D}}{\partial t}$

[2016 : 1 Mark, Set-3]

39. A loop is rotating about the y-axis in a magnetic field $\vec{B} = B_0\cos(\omega t + \phi)\hat{a}_x T$. The voltage in the loop is

(a) zero.

(b) due to rotation only.

(c) due to transformer action only.

(d) due to both rotation and transformer action.

[1998 : 1 Mark]

40. In the table shown, List-I and List-II, respectively, contain terms appearing on the left-hand side and the right-hand side of Maxwell's equations (in their standard form). Match the left-hand side with the corresponding right-hand side.

List - I	List - II
1. $\nabla.D$	(P) 0
2. $\nabla \times E$	(Q) ρ
3. $\nabla.B$	(R) $-\frac{\partial B}{\partial t}$
4. $\nabla \times H$	(S) $J + \frac{\partial D}{\partial t}$

(*a*) 1-Q, 2-R, 3-P, 4-S

(*b*) 1-Q, 2-S, 3-P, 4-R

(*c*) 1-P, 2-R, 3-Q, 4-S

(*d*) 1-R, 2-Q, 3-S, 4-P

[2019 : 1 Mark]

ANSWERS

1. (b)	**2.** (b,c)	**3.** (d)	**4.** (8)	**5.** (b)	**6.** (b)	**7.** (a)	**8.** (c)	**9.** (d)	**10.** (c)
11. (b)	**12.** (–15)	**13.** (d)	**14.** (–3)	**15.** (0.1)	**16.** (a)	**17.** (c)	**18.** (5.063)	**19.** (c)	**20.** (b)
21. (c)	**22.** (b)	**23.** (c)	**24.** (d)	**25.** (c)	**26.** (c)	**27.** (a)	**28.** (a)	**29.** (b)	**30.** (b)
31. (a)	**32.** (a)	**33.** (c)	**34.** (a)	**35.** (a)	**36.** (d)	**37.** (b)	**38.** (c)	**39.** (d)	**40.** (a)

EXPLANATIONS

1. An electrostatic field $\vec{E}$ is said to be conservative if the closed line integral of the field is zero, i.e.,

$$\oint_s \bar{E}\cdot\overline{d\ell} = 0 \quad ...(i)$$

Using stoke's theorem, $\oint_s \bar{E}\cdot\overline{d\ell} = \int_s (\bar{\nabla}\times\bar{E}).\overline{ds}$.

Equation (i) becomes $\bar{\nabla}\times\bar{E} = 0$. i.e, the curl of the field $\bar{E}$ is equal to zero.

2. For a charge free interface between two media with dielectric constant ϵ_1 and ϵ_2,

(i) the tangential components of the electric field are equal,

(ii) the normal components of electric flux density are equal.

3. Given:

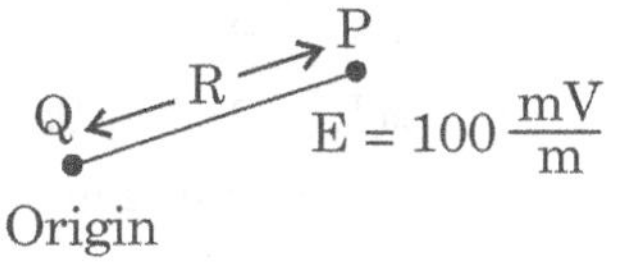

Electric field due to a point charge q is 100 $\frac{mV}{m}$.

According to Faraday's law, the point charge +q will induce – q charge on hollow conducting sphere.

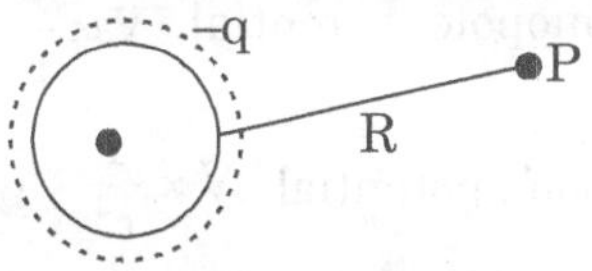

As point is very far, conducting sphere will act as a point charge of –q for point P.

Electric field intensity due to point charge –q = – 100 mV/m.

Net electric field strength at point P = 100 – 100 = 0

6. Given: Surface charge density

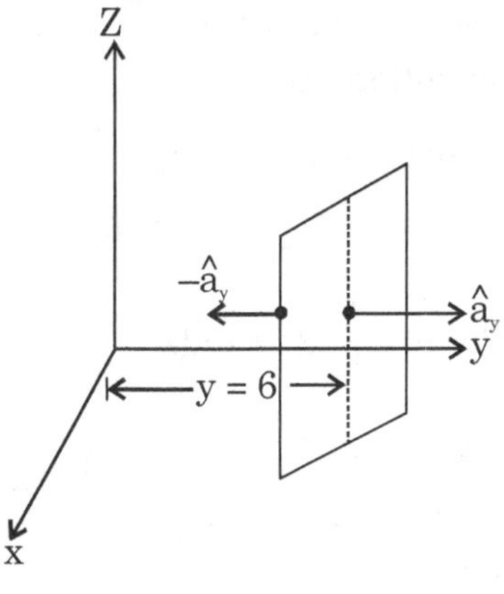

$$\rho_s = \left(\frac{1}{6000\pi}\right)\mu C/m^2$$

The electric field strength is

$$E = \frac{\rho_s}{2\in}\hat{a}_n$$

Where $\hat{a}_n$ is the unit vector normal to the plane

y = 6m.

$\hat{a}_n = \hat{a}_y$ or $-\hat{a}_y$.

Electric field, $\bar{E} = \frac{\rho_s}{2\in}\cdot\hat{a}_n$

$$= \frac{1}{6000\pi}\times 10^{-6}\times\frac{36\pi\times 10^9}{2}\hat{a}_n$$

$$\bar{E} = 3\hat{a}\ \text{V/m}$$

$$\Rightarrow \bar{E} = 3\hat{a}_y \text{ or } -3\hat{a}_y\ \text{V/m}$$

$$\bar{E} = 3j\ \text{V/m or } -3j\ \text{V/m}$$

7. Given: Sphere radius = 1 m, surface charge density ρ_s = 10 c/m², side of cube = 10 m

⇒ The total charge on the sphere

$Q = \rho_s\cdot 4\pi r^2 = 10\times 4\pi\times 1^2$, $Q = 40\pi$ C.

The sphere is enclosed in a cube. From Gauss's law, $\oint_s \bar{D}\cdot d\bar{s} = Q_{enclosed}$

Total outward electric flux or displacement normal to the surface of cube = 40π C.

8. For a monopole, potential, $V \propto \frac{1}{r}$

For a dipole, potential, $V \propto \frac{1}{r^2}$

9. Given: E = 2V/m, $\in = 80\in_0$

Total electric field of a perfect conductor is given by only normal component.

$$E_n = \frac{\rho_s}{\in_0\cdot\in_r} \Rightarrow 2 = \frac{\rho_s}{80\times 8\cdot 85\times 10^{-12}}$$

$\rho_s = 1.41\times 10^{-9}$ C/m².

10. Given: $E = x\hat{a}_x + y\hat{a}_y + z\hat{a}_z$

Potential difference,

$$V = -\int\bar{E}\cdot d\bar{\ell} = -\int\left(x\hat{a}_x + y\hat{a}_y + z\hat{a}_z\right)\cdot\left(\hat{a}_x dx + \hat{a}_y\cdot dy + \hat{a}_z\cdot dz\right)$$

$$V = -\left[\int_1^2 x\,dx + \int_2^0 y\,dy + \int_3^0 z\,dz\right]$$

$$= -\left[\frac{x^2}{2}\right]_1^2 - \left[\frac{y^2}{2}\right]_2^0 - \left[\frac{z^2}{2}\right]_3^0 = -\frac{1}{2}[3-4-9]$$

V = 5 volt.

11.

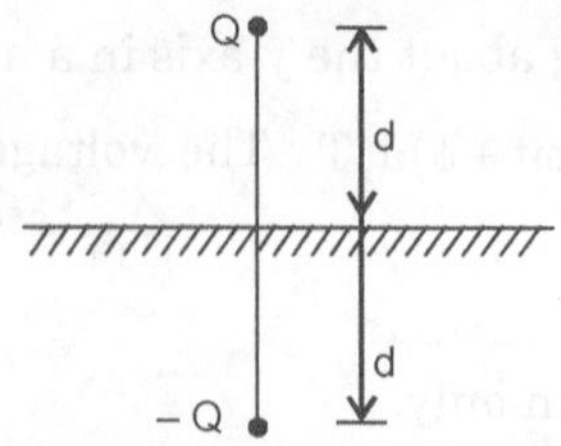

$$F = \frac{Q_1 Q_2}{4\pi\in R^2} = \frac{q^2}{4\pi\in(2d)^2} = \frac{q^2}{16\pi\in d^2}$$

12. A (0kV/cm, 20kV/cm) B (5×10^{-4}kV/cm, 40kV/cm)

$$E - 20 = \frac{40-20}{5\times 10^{-4}}(x-0)$$

$$\Rightarrow \quad E = 4\times 10^4 x + 20$$

$$V_{AB} = -\int_A^B E.dl$$

$$= -\int_0^{5\times 10^{-4}/cm}(4\times 10^4 x + 20)dx$$

$$= -\left(4\times 10^4\frac{x^2}{2} + 20x\right)\Bigg|_0^{5\times 10^{-4}}$$

$$= -(2\times 10^4\times 25\times 10^{-8} + 20\times 5\times 10^{-4})$$

$$= -(50\times 10^{-4} + 100\times 10^{-4})$$

$$= -150\times 10^{-4}\ \text{kV}$$

$$\Rightarrow \quad V_{AB} = -15\ \text{V}$$

13. Given $E = -\left(2y^3 - 3yz^2\right)a_x - \left(6xy^2 - 3xz^2\right)a_y + 6xyza_z$

By verification option (*d*) satisfy

$$\bar{E} = -\bar{\nabla}V$$

15. Given,

Propagation constant of a lossy transmission line (P) = (2 + *j*5) m⁻¹,

characteristic impedance (z_0) = 50 Ω,

angular frequency (ω) = 10⁶ rad/sec,

Now $P = \sqrt{(R + j\omega L)(G + j\omega C)}$

$$z_0 = \sqrt{\frac{(R + j\omega L)}{(G + jwC)}}$$

$$Pz_0 = R + j\omega L$$

$$\Rightarrow \quad R + j\omega L = (100 + j250)$$

$$\therefore \quad R = 100\ \Omega/\text{m}$$

$$L = \frac{250}{\omega}$$

$$L = \frac{250}{10^6} = 250\ \mu\text{H}/m$$

(Since ω = 10⁶ rad/s)

$$\therefore \quad \frac{P}{z_0} = G + j\omega C$$

$$G + j\omega C = \left(\frac{2}{50} + j\frac{5}{50}\right)$$

$$\therefore \quad G = \frac{2}{50} = \frac{1}{25} = 0.04 \text{ s/m}$$

and $$C = \frac{5}{50 \times \omega} = \frac{5}{50 \times 10^6} \text{ F/m}$$

$$= \frac{1}{10} \times 10^{-6} = 0.1 \ \mu \text{ F/m}$$

Hence line constant L, C, R & G are respectively

$$L = 250 \ \mu\text{H/m}$$
$$C = 0.1 \ \mu\text{F/m}$$
$$R = 100 \ \Omega\text{/m}$$

and $$G = 0.04 \text{ S/m}$$

16. Let $E = E_1$,

$$\therefore \quad \text{Energy } E_1 = \frac{Q_1^2}{2C_1}$$

(Here Q = charge across capacitor)

For electrically isolated

$$\Rightarrow \quad Q_2 = Q_1$$
$$d_2 = 2d_1$$
$$\Rightarrow \quad C_2 = \frac{C_1}{2}$$

(Here $d_1 = d_2$ = plate separation)

$$E_2 = \frac{Q_2^2}{2C_2} = \frac{Q_1^2}{\frac{2C_1}{2}} = 2\left(\frac{Q_1^2}{2C_1}\right)$$

$$= 2E_1 = 2E$$

Hence the energy stored in the capacitor is 2E.

17. Potential on sphere S_1, $V_a = \dfrac{q_a}{4\pi \in r_a}$

Potential on sphere S_2, $V_b = \dfrac{q_b}{4\pi \in r_b}$

When two sphere are connectd through conducting wire, the charge flows from higher potential to lower potential until both spheres attains same potential, i.e, $V_a = V_b$

$$\Rightarrow \frac{q_a}{r_a} = \frac{q_b}{r_b} \qquad \text{...(i)}$$

Electric field outside sphere A, $E_a = \dfrac{Kq}{r_a^2}$

Electric field outside sphere B, $E_b = \dfrac{Kq}{r_b^2}$

$$\text{Given : } r_b > r_a \Rightarrow \frac{1}{r_a} > \frac{1}{r_b} \qquad \text{...(ii)}$$

Using eq. (ii) in eq. (i), we get

$$\frac{Kq_a}{r_a^2} > \frac{Kq_b}{r_b^2} \Rightarrow E_a > E_b$$

18. Given $m_e = 9.11 \times 10^{-31}$ kg,
$q_e = 1.6 \times 10^{-19}$ C, $v_e = 10^5$ m/s.
From energy conservation principle,

$$(KE)_{initial} + (PE)_{initial} = (KE)_{final} + (PE)_{final}$$

$$\frac{1}{2}m_e v_e^2 + 0 = 0 + \frac{1}{4\pi \epsilon_0} \cdot \frac{q_e \cdot q_e}{r}$$

$$r = \frac{2qe^2}{4\pi \epsilon_0 \cdot m_e v_e^2} = \frac{2 \times 9 \times 10^9 \times (-1.6 \times 10^{-19})^2}{9.1 \times 10^{-31} \times (10^5)^2}$$

$r = 5.063 \times 10^{-8}$ m.

19. For free velocity, KE = Work done

$$\frac{1}{2}mv^2 = qv \quad \therefore v = \frac{d}{t}$$

$$v = \sqrt{\frac{2qv}{m}} \Rightarrow \frac{d}{t} = \sqrt{2q} \text{ v/m}$$

$$\therefore \ t \propto \frac{d}{\sqrt{v}}$$

20. The total electric flux leaving the cylinder of height "H" is

$$= \oiint \bar{E}.d\bar{A} = \frac{QH}{\epsilon_0}$$

$\therefore$ The electric flux through a quarter cylinder of height "H" is

$$= \frac{\oiint \bar{E}.d\bar{A}}{4} = \frac{QH}{4 \epsilon_0}$$

$\therefore$ Option (b) is correct.

21. According to Gauss's law for magnetic fields, the magnetic flux flowing through the closed surface is zero, because free magnetic charges do not exit and magnetic flux travels in a closed path, i.e.,

$$\bar{\nabla} \cdot \bar{B} = 0 \qquad \text{...(1)}$$

Using divergence theorem, $\oint_s \bar{B} \cdot \bar{ds} = \int_v (\bar{\nabla} \cdot \bar{B}) dv$

From eq (1), $\oint_s \bar{B} \cdot d\bar{s} = \int_v 0.dv = 0$

$$\Rightarrow \bar{\nabla} \cdot \bar{B} = 0 \Rightarrow \nabla \cdot (\mu \bar{H}) = 0$$

For homogeneous medium, μ does not depend on position, then, $\bar{\nabla} \cdot \bar{H} = 0$

22.

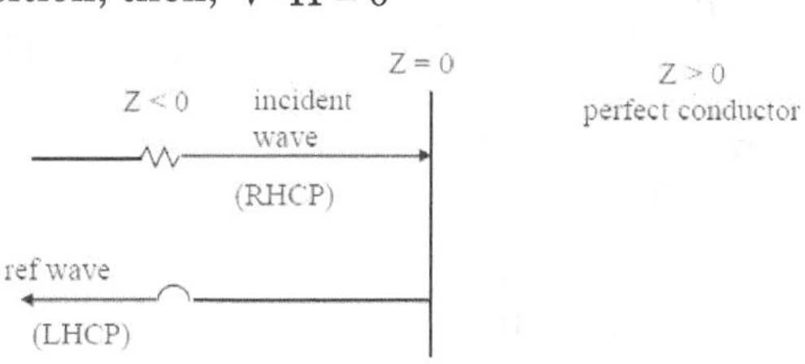

Here RHCP → right handed circularly polarized

LHCP → Left-handed circularly polarized

If the wave is incident on perfect conductor then reflection coefficient is given by

$$\Gamma \equiv \frac{E_{r_0}}{E_{i_0}} = -1$$

$$E_{r_0} = E_{i_0} \angle 180°$$

If incident wave is traveling along + Z direction then the reflected wave will be traveling along – Z direction.

Thus, the reflected wave is left hand circularly polarized (LHCP).

23. From Maxwell's equation, $\bar{\nabla} \times \bar{H} = \bar{J}_C + \bar{J}_D$

Where $\bar{J}_C$ and $\bar{J}_D$ are conduction and displacement current densities. Hence, Unit of $\bar{\nabla} \times \bar{H}$ is A/m².

24. By using right hand thumb rule, it is found that field due to wire in y-z plane is in negative x-direction and field due to wire in x-y plane is in negative z-direction. Therefore, x and z components of the resulting magnetic field are non-zero at the oxigin.

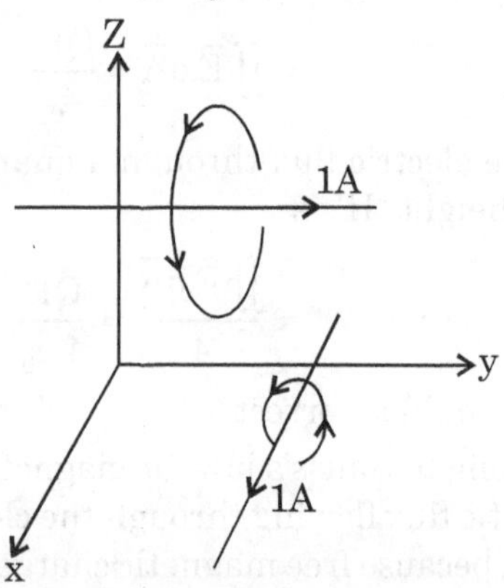

25. Given: $\bar{B} = B_0\left(\frac{x}{a^2+x^2}\hat{a}_y - \frac{y}{x^2+y^2}\hat{a}_x\right)$

Using cylindrical co-ordinates,

$x = r\cos\phi,\ \hat{a}_x = \cos\phi \cdot \hat{a}_r - \sin\phi \cdot \hat{a}_\phi$

$y = r\sin\phi,\ \hat{a}_y = \sin\phi \cdot \hat{a}_r + \cos\phi \cdot \hat{a}_\phi$

$\bar{B}(r,\phi,z)$

$$= B_0\left[\frac{r\cos\phi}{r^2}(\sin\phi\,\hat{a}_r + \cos\phi\,\hat{a}_\phi) - \frac{r\sin\phi}{r^2}(\cos\phi\,\hat{a}_r - \sin\phi\,\hat{a}_\phi)\right]$$

$$= \frac{B_0}{r}\left[(\cos\phi\sin\phi - \sin\phi\cos\phi)\hat{a}_r + \hat{a}_\phi(\cos^2\phi + \sin^2\phi)\right]$$

$$\bar{B} = \frac{B_0}{r}\cdot\hat{a}_\phi$$

$$\Rightarrow \bar{H} = \frac{\bar{B}}{\mu} = \frac{B_0}{\mu r}\hat{a}_\phi$$

$\Rightarrow \bar{\nabla} \times \bar{H} = \bar{J}$; where $H_r = H_z = 0$ and $H_\phi = \frac{B_0}{\mu \cdot r}$

$$\frac{1}{r}\begin{vmatrix} \hat{a}_r & r\hat{a}_\phi & \hat{a}_z \\ \frac{\partial}{\partial r} & \frac{\partial}{\partial \phi} & \frac{\partial}{\partial_z} \\ 0 & B_0/\mu r & 0 \end{vmatrix} = \bar{J}$$

$$\bar{J} = \frac{1}{r}\left[-\frac{\partial}{\partial_z}\left(r \times \frac{B_0}{\mu r}\right)\hat{a}_r + \frac{\partial}{\partial r}\left(r \times \frac{B_0}{\mu r}\right)\cdot\hat{a}_z\right]$$

$\bar{J} = 0$ for $r \neq 0$.

26. According to Ampere's law,

$$\oint \bar{H}\cdot d\bar{l} = \int \bar{J}\cdot \bar{ds} = I_{endosed}$$

Case 1 : $r < a$.

$\Rightarrow H\cdot(2\pi r) = J.\pi r^2$

$$H = \frac{J}{2}r$$

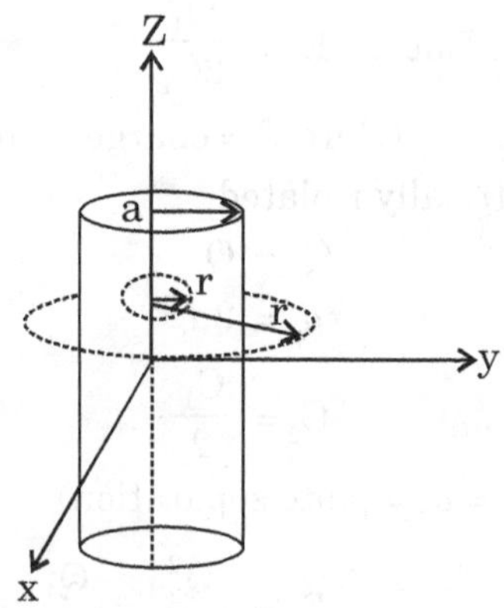

$H \propto r$

Case 2 : $r > a$

$H \cdot 2\pi r = J \times \pi a^2$

$$H = \frac{J}{2r}a^2$$

$$H \propto \frac{1}{r}$$

27. Assuming the cross-section of the wire on x-y plane as shown in figure.

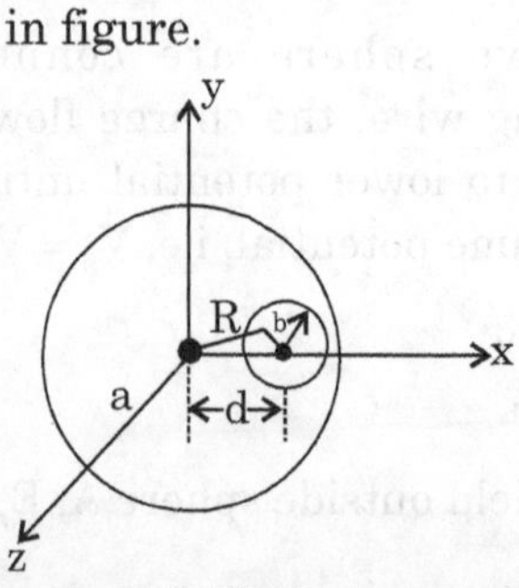

Since, the hole is drilled along the length of wire. So, it can be assumed that the drilled portion carries current density of – J.

Now, for the wire without hole, magnetic field intensity at point P is $H_{\phi_1} \cdot (2\pi R) = J(\pi R^2)$

$H_{\phi_1} = JR/2$.

Since, point O is at origin.

$$\bar{H}_1 = \frac{J}{2}\left(x\hat{a}_x + y\hat{a}_y\right)$$

Magnetic field intensity only due to the hole, $H_{\phi_2}(2\pi r) = -j \cdot \pi r^2$ $H\phi_2 = -Jr/2$.

If we take 0′ at origin then, $\bar{H}_2 = -\frac{J}{2}(x'\hat{a}_x + y'\hat{a}_y)$

where x′ and y′ denotes point 'p' is new co-ordinate system.

$$x = x' + d,\ y = y'.\ \bar{H}_2 = -\frac{J}{2}\left[(x-d)\hat{a}_x + y\hat{a}_y\right]$$

Total magnetic field intensity

$$= \bar{H}_1 + \bar{H}_2 = \frac{J}{2}\cdot d\ \hat{a}_x.$$

So, the magnetic field intensity inside the hole depends only on d.

28. Given medium (1) is perfect conductor Medium (2) is air

$$\therefore \qquad \bar{H}_1 = 0$$

From boundary conditions

$$\left(\bar{H}_1 - \bar{H}_2\right) \times \hat{a}_n = K_S$$

$$\left.\begin{array}{l}\bar{H}_1 = 0\\ a_n = a_y\end{array}\right| K_S = 2\hat{a}_x$$

$$-\bar{H}_2 \times \hat{a}_y = 2\hat{a}_x$$

$$-\left(H_x a_x + H_y a_y + H_z a_z\right) \times a_y = 2a_x$$

$$-H_x a_z + H_z a_x = 2a_x$$

$$\therefore \qquad H_z = 2$$

$$\boxed{H = 2a_z}$$

29. 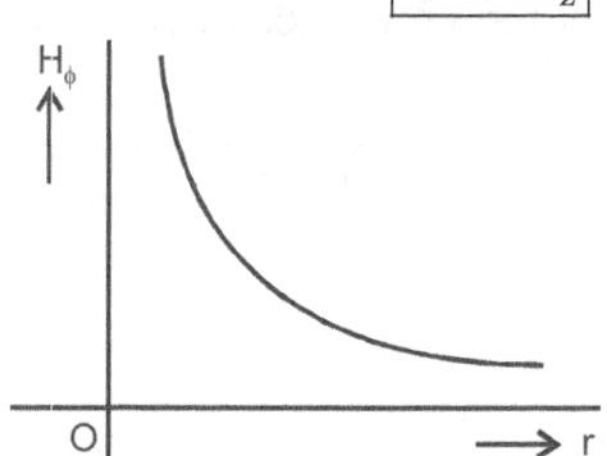

We know that magnetic field around a current carrying conductor

$$H_\phi = \frac{I}{2\pi r} i\phi$$

$$|H_\phi| = \frac{I}{2\pi r}$$

$$|H_\phi|\ \alpha\ \frac{1}{r}$$

30. Correct option is not given

Current density in a medium is given by

$$\vec{J} = \frac{400\sin\theta}{2\pi(r^2+4)}\hat{a}_r\ \text{A/m}^{-2}$$

Now current passing through the portion of sphere of radius $(r) = 0.8$ m is given by

$$I = \int_s \vec{J}.d\vec{s} \quad (r = \text{constant})$$

$$d\vec{s} = r^2\sin\theta\, d\theta\, d\phi\, \hat{a}rd \quad (\because r = 0.8\text{ m})$$

$$I = \int_{\theta=\frac{\pi}{2}}^{\frac{\pi}{4}}\int_{\phi=0}^{2\pi}\frac{400\sin\theta}{2\pi(r^2+4)}r^2\sin\theta\, d\theta\, d\phi$$

$$= \frac{400(0.8)^2}{2\pi(0.8^2+4)}\left[\left(\frac{\pi}{4}-\frac{\pi}{12}\right) - \left(\sin\left(\frac{\pi}{2}\right)-\sin\left(\frac{\pi}{6}\right)\right)\right]\times(2\pi)$$

$\therefore$ I = 7.45 A

The average current density through the given sphere surface is given by

$$J = \frac{1}{\text{Area of } r = 0.8\text{m sphere}}$$

$$= \frac{7.45}{(0.8)^2\int_{\theta=\frac{\pi}{2}}^{\frac{\pi}{4}}\int_{\phi=0}^{2\pi}\sin\theta\, d\theta\, d\phi} = \frac{7.45}{1.04}$$

$\therefore$ J = 7.15 A/m^2

Hence the total current and average current density are 7.45 A and 7.15 A/m^2.

31. Ba_z magnetic field (magnetic field exists in the $\hat{z}$ direction in vacuum).

$v_x a_x + v_z a_z$ velocity

$$F = Q(v \times B) \text{ by Lorent's law}$$

$$= Q(v_x a_x + v_z a_z) \times Ba_z$$

$$F_y = Q\, v_x . B(-a_y)$$

This results in a circular path in the XY plane with $v_z a_z$ component causing a linear path.

Both result in a helical path in Z axis.

32. 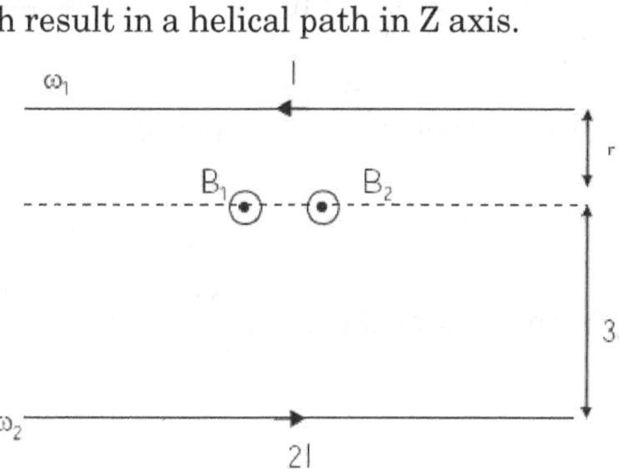

Between the wires ω_1 and ω_2 the $\overline{B}$ – fields due to I and 2I will gets added up.

∴ The B – field due to ω_1 at a distance 'r' from the ω_1 is

∴ The B – field due to ω_1 at a distance'r' from the ω_1 is

$$B_1 = \frac{\mu_0 I}{2\pi r} \rightarrow (1)$$

The B – field dut to ω_2 at a distance "3r" from the ω_2 is given by

$$B_2 = \frac{\mu_0 (2I)}{2\pi (3r)} \rightarrow (2)$$

∴ Now, total magnetic field (B)

$$= B_1 + B_2 = \frac{\mu_0 I}{2\pi}\left[\frac{1}{r} + \frac{2}{3r}\right]$$

$$B = \frac{5\mu_0 I}{6\pi r} \text{Wb/m}^2.$$

34. From Ampere's law, $\overline{\nabla} \times \overline{H} = \overline{J} + \overline{J}_D = \overline{J} + \frac{\partial \overline{D}}{\partial t}$

35. Given : $A = 10^{-4}$ m², $d = 10^{-3} m$, $v = 0.5$ V, $f = 3.6 \times 10^9$ Hz

Displacement current density $= \frac{I_d}{A} = \varepsilon \cdot \frac{\partial E}{\partial t}$,

$$E = E_o\, e^{j(\omega t + z\beta)}$$

$$E = \frac{-V}{d}\, e^{j(\omega t + z\beta)};$$

$$E_o = \frac{-V}{d}$$

$$\frac{\partial E}{\partial t} = +j\omega\, \frac{V}{d}\, e^{j(\omega t + z\beta)}$$

At $z = 0$, $\left|\frac{I_d}{A}\right| = \varepsilon_o \cdot \frac{V}{d}\omega$

$$I_d = \frac{\varepsilon_o V\omega}{d} A$$

$$= \frac{1}{36\pi} \times 10^{-9} \frac{\times 0.5 \times 3.6 \times 10^9 \times 2\pi \times 10^{-4}}{10^{-3}}$$

= **1.59 mA.**

36. From Ampere's law (or Maxwell's second equation)

$$\vec{\nabla} \times \vec{H} = \vec{J} + \frac{\partial \vec{D}}{\partial t}$$

From Stoke's theorem, $\iint_S \overline{\nabla} \times \overline{H}.\overline{ds} = \oint_C \vec{H}.\vec{dl}$

$$\therefore \quad \oint_C \vec{H}.\vec{dl} = \iint_S \left(\vec{J} + \frac{\partial \vec{D}}{\partial t}\right).\overline{ds}$$

37. For static electric and magnetic field, Maxwell's equations are

$$\overline{\nabla} \times \overline{H} = \overline{J}_C,\ \overline{\nabla} \times \overline{E} = 0,\ \overline{\nabla} \cdot \overline{D} = \rho_v, \overline{\nabla} \cdot \overline{B} = 0$$

For an inhomogeneous medium, $\in$, μ and σ are function of position or space. Now,

$$\overline{\nabla} \times \overline{B} \neq 0,\ \overline{\nabla} \times \overline{D} \neq 0, \overline{\nabla} \cdot \overline{E} \neq 0, \overline{\nabla} \cdot \overline{H} \neq 0.$$

Hence, for static electric and magnetic field in an inhomogeneous source free medium,

$$\overline{\nabla} \times \overline{E} = 0$$

$$\overline{\nabla} \cdot \overline{D} = 0$$

$$\overline{\nabla} \cdot \overline{B} = 0$$

$$\overline{\nabla} \times \overline{H} = 0$$

38. Faraday's law of electromagnetic induction is mathematically described by the equation is given by

$$\nabla \times \vec{E} = -\frac{\partial \vec{B}}{\partial t}$$

or

$$\nabla \times \vec{E} = -\mu \frac{\partial \vec{H}}{\partial t}$$

39. The voltage induced due to motion of conductor in static magnetic field,

$$V_m = \oint_l (\overline{v} \times \overline{B}) \cdot d\overline{l} \text{ (Generator action.)}$$

The voltage induced due to time-changing magnetic field,

$$V_t = -\int_s \frac{\partial \overline{B}}{\partial t} \cdot d\overline{s} \text{ (Transormer action.)}$$

The total induced voltage = $V_m + V_t$.

40. $\overline{\nabla}.\overline{D} = \rho$

$$\overline{\nabla} \times \overline{E} = \frac{-\partial \overline{B}}{\partial t}$$

$$\overline{\nabla}.\overline{B} = 0$$

$$\overline{\nabla} \times \overline{H} = \overline{J} + \frac{\partial \overline{D}}{\partial t}$$

∴ 1 – Q, 2 – R, 3 – P, 4 – S

Option (a) is correct.

3

CHAPTER

Uniform Plane Waves

Analysis of Previous GATE Papers												
Topics ↓		Year →	2019	2018	2017 Set 1	2017 Set 2	2016 Set 1	2016 Set 2	2016 Set 3	2015 Set 1	2015 Set 2	2015 Set 3
BOUNDARY CONDITIONS	1 Mark	MCQ Type										
		Numerical Type								1		
	2 Marks	MCQ Type										
		Numerical Type										
	Total									1		
WAVE EQUATION	1 Mark	MCQ Type										
		Numerical Type										
	2 Marks	MCQ Type										
		Numerical Type										
	Total											
PROPERTIES (Reflection, Refraction, Polarization)	1 Mark	MCQ Type							1		1	
		Numerical Type										
	2 Marks	MCQ Type			1			1			1	
		Numerical Type		1								
	Total			2	2			2	1		3	
PHASE AND GROUP VELOCITY	1 Mark	MCQ Type						1				
		Numerical Type										
	2 Marks	MCQ Type	1									
		Numerical Type										
	Total		2					1				
PROPAGATION THROUGH VARIOUS MEDIA & SKIN DEPTH	1 Mark	MCQ Type										
		Numerical Type										
	2 Marks	MCQ Type		1								
		Numerical Type								1		
	Total			2						2		

BOUNDARY CONDITIONS

1. A uniform plane wave in air impinges at 45° angle on a lossless dielectric material with dielectric constant ϵ_r. The transmitted wave propagates in a 30° direction with respect to the normal. The value of ϵ_r is

(a) 1.5 (b) $\sqrt{1.5}$

(c) 2 (d) $\sqrt{2}$

[2000 : 2 Marks]

2. Two coaxial cables 1 and 2 are filled with different dielectric constants ϵ_{r_1} and ϵ_{r_2} respectively. The ratio of the wavelengths in the two cables, $\left(\frac{\lambda_1}{\lambda_2}\right)$ is

(a) $\sqrt{\frac{\epsilon_{r_1}}{\epsilon_{r_2}}}$ (b) $\sqrt{\frac{\epsilon_{r_2}}{\epsilon_{r_1}}}$

(c) $\frac{\epsilon_{r_1}}{\epsilon_{r_2}}$ (d) $\frac{\epsilon_{r_2}}{\epsilon_{r_1}}$

[2000 : 2 Marks]

3. Medium 1 has the electrical permittivity $\varepsilon_1 = 1.5\varepsilon_0$ Farad/m and occupies the region to the left of x = 0 plane. Medium 2 has the electrical permittivity $\varepsilon_2 = 2.5\varepsilon_0$ Farad/m and occupies the region to the right of x = 0 plane. If E_1 in medium 1 is $E_1 = (2u_x - 3u_y + 1u_z)$ Voltm, then E_2 in medium 2 is

(a) $(2.0u_x - 7.5u_y + 2.5u_z)$ Volt/m

(b) $(2.0u_x - 2.0u_y + 0.6u_j)$ Volt/m

(c) $(1.2u_x - 3.0u_y + 1.0u_z)$ Volt/m

(d) $(1.2u_x - 2.0u_y + 0.6u_z)$ Volt/m

[2003 : 2 Marks]

4. A medium of relative permittivity $\epsilon_{r_2} = 2$ forms an interface with free-space. A point source of electromagnetic energy is located in the medium at a depth of 1 meter from the interface. Due to the total internal reflection, the transmitted beam has a circular cross-section over the interface. The area of the beam cross-section at the interface is given by

(a) 2π m^2 (b) π^2 m^2

(c) $\frac{\pi}{2}$ m^2 (d) π m^2

[2006 : 2 Marks]

5. A medium is divided into regions I and II about x = 0 plane, as shown in the figure below. An electromagnetic wave with electric field $\vec{E}_1 = 4\hat{a}_x + 3\hat{a}_y + 5\hat{a}_z$ is incident normally on the interface from region-I. The electric field $\vec{E}_2$ in region-II at the interface is

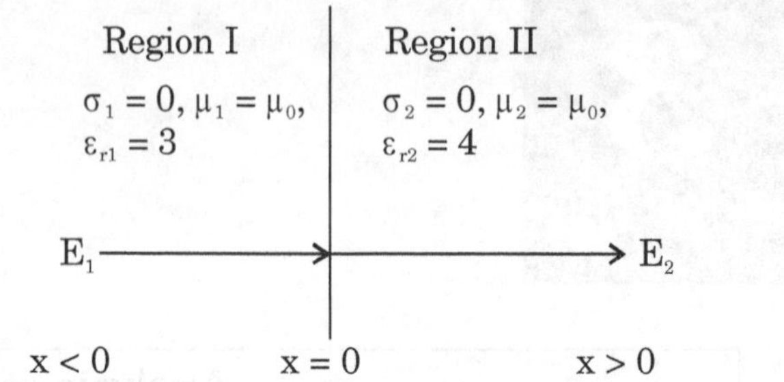

(a) $\vec{E}_2 = \vec{E}_1$ (b) $4\hat{a}_x + 0.75\hat{a}_y - 1.25\hat{a}_z$

(c) $3\hat{a}_x + 3\hat{a}_y + 5\hat{a}_z$ (d) $-3\hat{a}_x + 3\hat{a}_y + 5\hat{a}_z$

[2006 : 2 Marks]

6. A current sheet $\vec{J} = 10\hat{u}_y$ A/m lies on the dielectric interface x = 0 between two dielectric media with $\varepsilon_{r_1} = 1, \mu_{r_1} = 1$ in Region-1 (x < 0) and $\varepsilon_{r_2} = 2, \mu_{r_2} = 2$ in Region-2 (x > 0). If the magnetic field in Region-1 at x = 0$^-$ is $\vec{H}_1 = 3\hat{u}_x + 30\hat{u}_y$ A/m, the magnetic field in Region-2 at x = 0$^+$ is

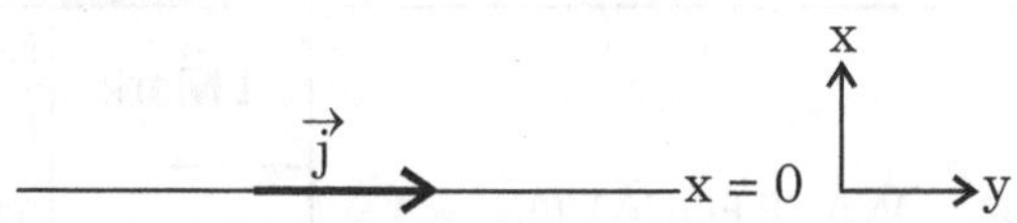

(a) $\vec{H}_2 = 1.5\hat{u}_x + 30\hat{u}_y - 10\hat{u}_z$ A/m

(b) $\vec{H}_2 = 3\hat{u}_x + 30\hat{u}_y - 10\hat{u}_z$ A/m

(c) $\vec{H}_2 = 1.5\hat{u}_x + 40\hat{u}_y$ A/m

(d) $\vec{H}_2 = 3\hat{u}_x + 30\hat{u}_y + 10\hat{u}_z$ A/m

[2011 : 2 Marks]

7. Consider a uniform plane wave with amplitude (E_0) of 10 V/m and 1.1 GHz frequency travelling an air, and incident normally on a dielectric medium with complex relative permittivity (ϵ_r) and permeability (μ_r) as shown in the figure.

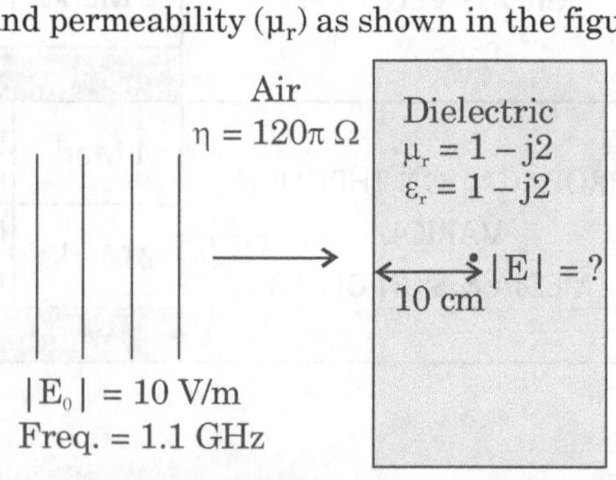

The magnitude of the transmitted electric field component (in V/m) after it has travelled a distance of 10 cm inside the dielectric region is _______.

[2015 : 2 Marks, Set-1]

WAVE EQUATION

8. The wavelength of a wave with propagation constant $(0.1\pi + j0.2\pi)\ m^{-1}$ is

(a) $\frac{2}{\sqrt{0.05}}$ m (b) 10 m

(c) 20 m (d) 30 m

[1998 : 1 Mark]

9. Identify which one of the following will NOT satisfy the wave equation.

(a) $50e^{j(\omega t - 3z)}$ (b) $\sin[\omega(10z + 5t)]$

(c) $\cos(y^2 + 5t)$ (d) $\sin(x)\cos(t)$

[1999 : 1 Mark]

10. If the electric field intensity associated with a uniform plane electromagnetic wave travelling in a perfect dielectric medium is given by $\vec{E}(z, t) = 10\cos(2\pi \times 10^7 t - 0.1\pi z)$ Volt/m, the velocity of the travelling wave is

(a) 3.00×10^8 m/sec (b) 2.00×10^8 m/sec

(c) 6.28×10^7 m/sec (d) 2.00×10^7 m/sec

[2003 : 2 Marks]

11. A plane wave of wavelength λ is travelling in a direction making an angle 30° with positive x-axis and 90° with positive y-axis. The $\vec{E}$ field of the plane wave can be represented as (E_0 is a constant)

(a) $\vec{E} = \hat{y}E_0 e^{j\left(\omega t - \frac{\sqrt{3}\pi}{\lambda}x - \frac{\pi}{\lambda}z\right)}$

(b) $\vec{E} = \hat{y}E_0 e^{j\left(\omega t - \frac{\pi}{\lambda}x + \frac{\pi}{\lambda}z\right)}$

(c) $\vec{E} = \hat{y}E_0 e^{j\left(\omega t - \frac{\sqrt{3}\pi}{\lambda}x + \frac{\pi}{\lambda}z\right)}$

(d) $\vec{E} = \hat{y}E_0 e^{j\left(\omega t - \frac{\pi}{\lambda}x + \frac{\sqrt{3}\pi}{\lambda}z\right)}$

[2007 : 1 Mark]

Statement for Linked Answer Questions 49 and 50: A monochromatic plane wave of wavelength λ = 600 μm is propagating in the direction as shown in the figure below. $\vec{E}_i$, $\vec{E}_r$ and $\vec{E}_t$, denote incident, reflected, and transmitted electric field vectors associated with the wave.

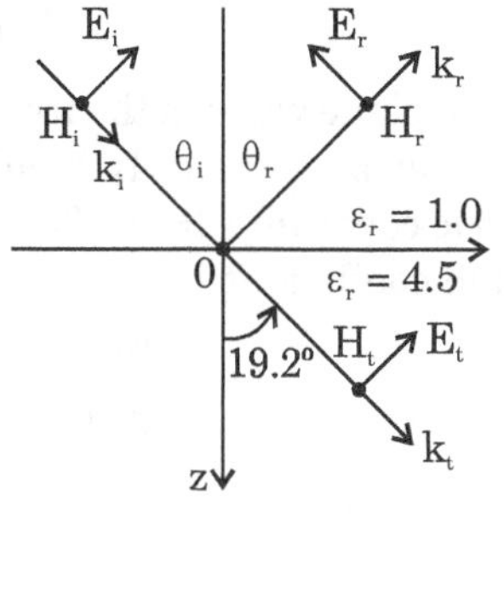

12. The angle of incidence 0(. and the expression for $\vec{E}_i$ are

(a) 60° and $\frac{E_0}{\sqrt{2}}(\hat{a}_x - \hat{a}_z)e^{-j\frac{\pi\times10^4(x+z)}{3\sqrt{2}}}$ V/m

(b) 45° and $\frac{E_0}{\sqrt{2}}(\hat{a}_x - \hat{a}_z)e^{-j\frac{\pi\times10^4 z}{3}}$ V/m

(c) 45° and $\frac{E_0}{\sqrt{2}}(\hat{a}_x - \hat{a}_z)e^{-j\frac{\pi\times10^4(z+z)}{3\sqrt{2}}}$ V/m

(d) 60° and $\frac{E_0}{\sqrt{2}}(\hat{a}_x - \hat{a}_z)e^{-j\frac{\pi\times10^4 z}{3}}$ V/m

[2013 : 2 Marks]

13. The expression for $\vec{E}_r$ is

(a) $0.23\frac{E_0}{\sqrt{2}}(\hat{a}_x + \hat{a}_z)e^{-j\frac{\pi\times10^4(x-z)}{3\sqrt{2}}}$ V/m

(b) $-\frac{E_0}{\sqrt{2}}(\hat{a}_x + \hat{a}_z)e^{j\frac{\pi\times10^4 z}{3}}$ V/m

(c) $0.44\frac{E_0}{\sqrt{2}}(\hat{a}_x + \hat{a}_z)e^{-j\frac{\pi\times10^4(x-z)}{3}}$ V/m

(d) $\frac{E_0}{\sqrt{2}}(\hat{a}_x + \hat{a}_z)e^{-j\frac{\pi\times10^4(r+z)}{3}}$ V/m

[2013: 2 Marks]

14. The electric field component of a plane wave travelling in a lossless dielectric medium is given by $\vec{E}(z,t) = \hat{a}_y 2\cos\left(10^8 t - \frac{z}{\sqrt{2}}\right)$ V/m. Wavelength (in m) for the wave is _______.

[2015 : 1 Mark, Set-1]

PROPERTIES (REFLECTION, REFRACTION, POLARIZATION)

15. For an electromagnetic wave incident from one medium to a second medium, total reflection takes place when

(a) the angle of incidence is equal to the Brewster angle with E field perpendicular to the plane of incidence

(b) the angle of incidence is equal to the Brewster angle with E field parallel to the plane of incidence

(c) the angle of incidence is equal to the critical angle with the wave moving from the denser medium to a rarer medium

(d) the angle of incidence is equal to the critical angle with the wave moving from a rarer medium to a denser medium

[1987 : 2 Marks]

16. A plane wave is incident normally on a perfect conductor as shown in figure. Here E_x^i, H_y^i and $\vec{P}$ are electric field, magnetic field and Poynting vector, respectively, for the incident wave. The reflected wave should have

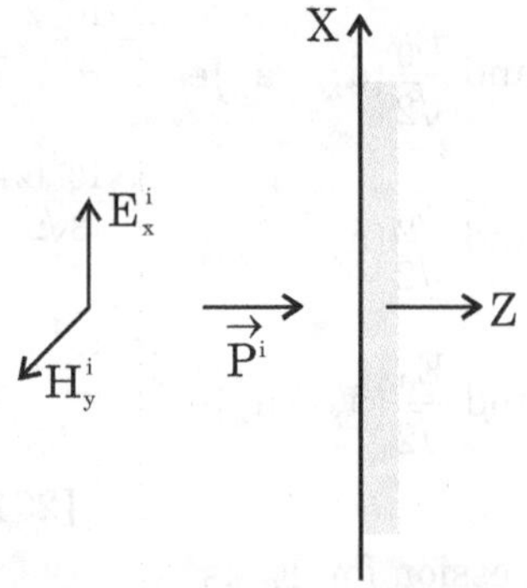

(a) $E_x^r = -E_x^i$ (b) $H_y^r = -H_y^i$
(c) $\vec{P}^r = -\vec{P}^i$ (d) $E_x^r = E_x^i$

[1993 : 2 Marks]

17. A plane electromagnetic wave traveling along the +z direction, has its electric field given by
$E_x = 2\cos(\omega t)$ and $E_y = 2\cos(\omega t + 90)°$
the wave is
(a) linearly polarized
(b) right circularly polarized
(c) left circularly polarized
(d) elliptically polarized

[1994 : 1 Mark]

18. A uniform plane wave in air is normally incident on an infinitely thick slab. If the refractive index of the glass slab is 1.5, then the percentage of the incident power that is reflected from the air-glass interface is
(a) 0% (b) 4%
(c) 20% (d) 100%

[1996 : 2 Marks]

19. The polarization of a wave with electric field vector $\vec{E} = E_0 e^{j(\omega t - \beta z)}(\vec{a}_x + \vec{a}_y)$ is
(a) linear (b) elliptical
(c) left hand circular (d) right hand circular

[1998 : 1 Mark]

20. A plane wave is characterized by $\vec{E} = (0.5\hat{x} + \hat{y}e^{j\pi/2})e^{j\omega t - jkz}$. This wave is
(a) linearly polarized
(b) circularly polarized
(c) elliptically polarized
(d) unpolarized

21. A uniform plane wave travelling in air is incident on the plane boundary between air and another dielectric medium with $\varepsilon_r = 4$. The reflection coefficient for the normal incidence, is
(a) zero (b) $0.5\angle 180°$
(c) $0.333\angle 0°$ (d) $0.333\angle 180°$

[2003 : 2 Marks]

22. The electric field of an electromagnetic wave propagating in the positive z-direction is given by

$$\vec{E} = \hat{a}_x \sin(\omega t - \beta z) + \hat{a}_y \sin\left(\omega t - \beta z + \frac{\pi}{2}\right).$$

The wave is
(a) linearly polarized in the z-direction
(b) elliptically polarized
(c) left-hand circularly polarized
(d) right-hand circularly polarized

[2006 : 1 Mark]

23. When a plane wave traveling in free-space is incident normally on a medium having $\epsilon_r = 4.0$, the fraction of power transmitted into the medium is given by
(a) $\frac{8}{9}$ (b) $\frac{1}{2}$
(c) $\frac{1}{3}$ (d) $\frac{5}{6}$

[2006 : 2 Marks]

24. A Right Circularly Polarized (RCP) plane wave is incident at an angle of 60° to the normal, on an air-dielectric interface. If the reflected wave is linearly polarized, the relative dielectric constant ϵ_{r_2} is

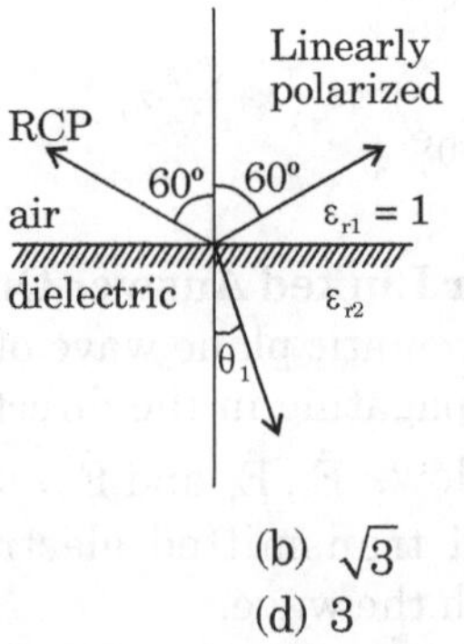

(a) $\sqrt{2}$ (b) $\sqrt{3}$
(c) 2 (d) 3

[2007 : 2 Marks]

25. A uniform plane wave in the free space is normally incident on an infinitely thick dielectric slab (dielectric constant $\varepsilon_r = 9$). The magnitude of the reflection coefficient is
(a) 0 (b) 0.3
(c) 0.5 (d) 0.8

[2008 : 2 Marks]

26. A plane wave having the electric field component $\vec{E}_i = 24\cos(3\times10^8 t - \beta y)\hat{a}_z$ V/m and traveling in free space is incident normally on a lossless medium with $\mu = \mu_0$ and $\epsilon = 9\epsilon_0$ which occupies the region $y \geq 0$. The reflected magnetic field component is given by

(a) $\frac{1}{10\pi}\cos(3\times10^8 t + \beta y)\hat{a}_z$ A/m

(b) $\frac{1}{20\pi}\cos(3\times10^8 t + \beta y)\hat{a}_x$ A/m

(c) $-\frac{1}{20\pi}\cos(3\times10^8 t + \beta y)\hat{a}_x$ A/m

(d) $-\frac{1}{10\pi}\cos(3\times10^8 t + \beta y)\hat{a}_x$ A/m

[2010 : 2 Marks]

27. The electric field of a uniform plane electromagnetic wave in free space, along the positive x–direction, is given by $\vec{E} = 10(\hat{a}_y + j\hat{a}_z)e^{-j25x}$. The frequency and polarization of the wave, respectively, are

(a) 1.2 GHz and left circular

(b) 4 Hz and left circular

(c) 1.2 GHz and right circular

(d) 4 GHz and right circular

[2012 : 1 Mark]

28. A plane wave propagating in air with $\vec{E} = (8\hat{a}_x + 6\hat{a}_y - 5\hat{a}_z)e^{j(\omega t+3x-4y)}$ V/m is incident on a perfectly conducting slab positioned at $x < 0$. the $\vec{E}$ field of the reflected waves is

(a) $(-8\hat{a}_x - 6\hat{a}_y - 5\hat{a}_z)e^{j(\omega t+3r+4y)}$ V/m

(b) $(-8\hat{a}_x - 6\hat{a}_y - 5\hat{a}_z)e^{j(\omega t+3x+4y)}$ V/m

(c) $(8\hat{a}_x - 6\hat{a}_y - 5\hat{a}_z)e^{j(\omega t-3r-4y)}$ V/m

(d) $(8\hat{a}_x + 6\hat{a}_y - 5\hat{a}_z)e^{j(\omega t-3x-4y)}$ V/m

[2012 : 1 Mark]

29. If the electric field of a plane wave is

$$\vec{E}(z,t) = \hat{x}3\cos(\omega t - kz + 30°) - \hat{y}3\cos(\omega t - kz + 45°)\ (\text{mV/m})$$

The polarization state of the plane wave is

(a) left elliptical (b) left circular

(c) right elliptical (d) right circular

[2014 : 2 Marks, Set-2]

30. Assume that a plane wave in air with an electric field $\vec{E} = 10\cos(\omega t - 3x - \sqrt{3}z)\hat{a}_y$ V/m is incident on a non-magnetic dielectric slab of relative permittivity 3 which covers the region $z > 0$. The angle of transmission in the dielectric slab is ________ degrees.

[2014 : 2 Marks, Set-3]

31. The electric field of a uniform plane electromagnetic wave is

$$\vec{E} = (\hat{a}_x + j2\hat{a}_y)\exp[j(2\pi\times10^7 t - 0.2z)]$$

The polarization of the wave is

(a) right handed circular

(b) right handed elliptical

(c) left handed circular

(d) left handed elliptical

[2015 : 1 Mark, Set-2]

32. The electric field of a plane wave propagating in a lossless non-magnetic medium is given by the following expression

$$\vec{E}(z,t) = \hat{a}_x(2\pi\times10^9 t + \beta z) + \hat{a}_y 3\cos\left(2\pi\times10^9 t + \beta z - \frac{\pi}{2}\right)$$

(a) right Hand Circular

(b) left Hand Elliptical

(c) right Hand Elliptical

(d) linear

[2015 : 2 Marks, Set-2]

33. A positive charge q is placed at x = 0 between two infinite metal plates placed at x = –d and at x = +d respectively. The metal plates lie in the yz plane.

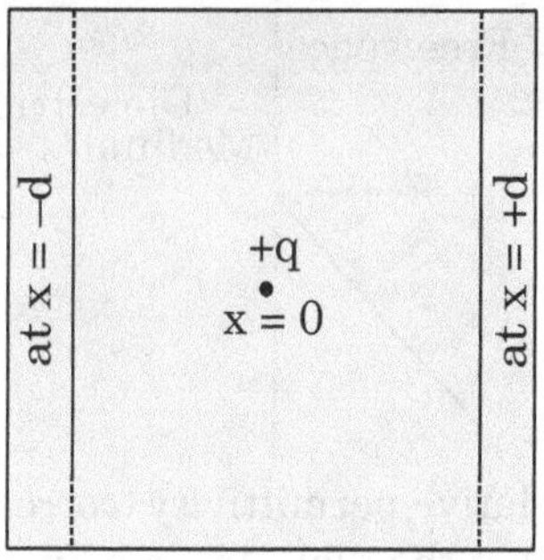

The charge is at rest at t = 0, when a voltage + V is applied to the plate at –d and voltage –V is applied to the plate at x = +d. Assume that the quantity of the charge q is small enough that it does not perturb the field set up by the metal plates. The time that the charge q takes to reach the right plate is proportional to

(a) $\frac{d}{V}$ (b) $\frac{\sqrt{d}}{V}$

(c) $\frac{d}{\sqrt{V}}$ (d) $\sqrt{\frac{d}{V}}$

[2016 : 2 Marks, Set-2]

34. If a right-handed circularly polarized wave is incident normally on a plane perfect conductor, then the reflected wave will be
(a) right-handed circularly polarized
(b) left-handed circularly polarized
(c) elliptically polarized with a tilt angle of 45°
(d) horizontally polarized

[2016 : 1 Mark, Set-3]

35. The expression for an electric field in free space is $\vec{E} = E_0(\hat{x} + y + j2\hat{z})e^{-j(\omega t - kx + ky)}$, where x, y, z represent the spatial coordinates, t represents time, and k are constants. This electric field
(a) does not represent a plane wave.
(b) represents a circularly polarized plane wave propagating normal to the z-axis.
(c) represents an elliptically polarized plane wave propagating along the y-plane.
(d) represents a linearly polarized plane wave.

[2017 : 2 Marks, Set-1]

36. A uniform plane wave traveling in free space and having the electric field

$\vec{E} = (\sqrt{2}\hat{a}_x - \hat{a}_z)\left[6\sqrt{3}\,\pi \times 10^8 t - 2\pi(x + \sqrt{2}z)\right]$ V/m is incident on a dielectric medium (relative permittivity > 1, relative permeability = 1) as shown in the figure and there is no reflected wave.

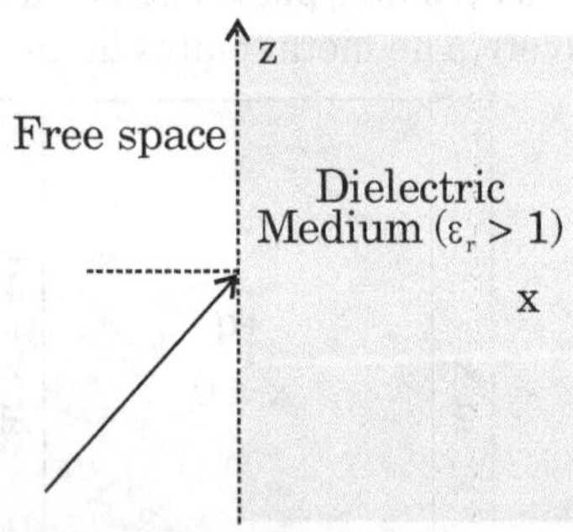

The relative permittivity (correct to two decimal places) of the dielectric medium is ______.

[2018 : 2 Marks]

PHASE AND GROUP VELOCITY

37. The magnetic field intensity vector of a plane wave is given by

$\bar{H}(x, t) = 10\sin(50000t + 0.004x + 30)\hat{a}_y$, where $\hat{a}_y$ denotes the unit vector in y-direction. The wave is propagating with a phase velocity
(a) 5×10^4 m/s
(b) 3×10^8 m/s
(c) 1.25×10^7 m/s
(d) 3×10^6 m/s

[2005 : 1 Mark]

38. Refractive index of glass is 1.5. Find the wavelength of a beam of light with frequency of 10^{14} Hz in glass. Assume velocity of light is 3×10^8 m/s in vacuum
(a) 3 pm
(b) 3 mm
(c) 2 mm
(d) 1 mm

[2005 : 1 Mark]

39. Let the electric field vector of a plane electromagnetic wave propagating in a homogenous medium be expressed as $\vec{E} = \hat{x}E_x e^{-j(\omega t - \beta z)}$, where the propagation constant β is a function of the angular frequency ω. Assume that $\beta(\omega)$ and E_x are known and are real. From the information available, which one of the following CANNOT be determined?
(a) The type of polarization of the wave.
(b) The group velocity of the wave.
(c) The phase velocity of the wave.
(d) The power flux through the z = 0 plane.

[2016 : 1 Mark, Set-2]

40. The dispersion equation of a waveguide, which relates the wave number k to the frequency ω, is

$k(\omega) = (1/c)\sqrt{\omega^2 - \omega_0^2}$

Where the speed of light $c = 3 \times 10^8$ m/s. and ω_0 is a constant. If the group velocity is 2×10^8 m/s, then the phase velocity is
(a) 2×10^8 m/s
(b) 1.5×10^8 m/s
(c) 3×10^8 m/s
(d) 4.5×10^8 m/s

[2019 : 2 Marks]

PROPAGATION THROUGH VARIOUS MEDIA & SKIN DEPTH

41. In a good conductor the phase relation between the tangential components of electric field E_t and the magnetic field H_t is as follows
(a) E_t and H_t are in phase
(b) E_t and H_t are out of phase
(c) H_t leads E_t by 90°
(d) E_t leads H_t by 45°

[1988 : 2 Marks]

42. The skin depth of copper at a frequency of 3 GHz is 1 micron (10^{-6} metre). At 12 GHz, for a non magnetic conductor whose conductivity is 1/9 times that of copper, the skin depth would be
(a) $\sqrt{9 \times 4}$ microns
(b) $\sqrt{\frac{9}{4}}$ microns
(c) $\sqrt{\frac{4}{9}}$ microns
(d) $\frac{1}{\sqrt{9 \times 4}}$ microns

[1989 : 2 Marks]

43. The incoming solar radiation at a place on the surface of the earth is 1.2 kW/m². The amplitude of the electric field corresponding to this incident power is nearly equal to
(a) 80 mV/m (b) 2.5 V/m
(c) 30 V/m (d) 950 V/m
[1990 : 2 Marks]

44. The electric field component of a uniform plane electromagnetic wave propagating in the Y-direction in a lossless medium will satisfy the equation

(a) $\frac{\partial^2 E^y}{\partial y^2} = \mu \in \frac{\partial^2 E_y}{\partial t^2}$ (b) $\frac{\partial^2 E^y}{\partial x^2} = \mu \in \frac{\partial^2 E_y}{\partial t^2}$

(c) $\frac{\partial^2 E_x}{\partial y^2} = \mu \in \frac{\partial^2 E_x}{\partial t^2}$ (d) $\frac{\sqrt{E_x^2 + E_z^2}}{\sqrt{H_x^2 + H_z^2}} = \sqrt{\frac{\mu}{\in}}$
[1991 : 2 Marks]

45. A material is described by the following electrical parameters at a frequency of 10 GHz, $\sigma = 10^6$ mho/m, $\mu = \mu_0$ and $\in/\in_0 = -10$. The material at this frequency is considered to be $\left(\in_0 = \frac{1}{36\pi} \times 10^{-6} \text{ F/m}\right)$
(a) a good conductor
(b) a good dielectric
(c) neither a good conductor, nor a good dielectric
(d) a good magnetic material
[1993 : 2 Marks]

46. The intrinsic impedance of a lossy dielectric medium is given by

(a) $\frac{j\omega\mu}{\sigma}$ (b) $\frac{j\omega \in}{\mu}$

(c) $\sqrt{\frac{j\omega\mu}{(\sigma + j\omega \in)}}$ (d) $\sqrt{\frac{\mu}{\in}}$
[1995 : 1 Mark]

47. Copper behaves as a
(a) conductor always
(b) conductor or dielectric depending on the applied electric field strength
(c) conductor or dielectric depending on the frequency
(d) conductor or dielectric depending on the electric current density
[1995 : 1 Mark]

48. Some unknown material has a conductivity of 106 mho/m and a permeability of $4\pi \times 10^{-7}$ H/m. The skin depth for the material at 1 GHz is
(a) 15.9 μm (b) 20.9 μm
(c) 25.9 μm (d) 30.9 μm
[1996 : 2 Marks]

49. The intrinsic impedance of copper at high frequency is
(a) purely resistive
(b) purely inductive
(c) complex with a capacitive component
(d) complex with an inductive component
[1998 : 1 Mark]

50. The depth of penetration of a wave in a lossy dielectric increases with increasing
(a) conductivity (b) permeability
(c) wavelength (d) permittivity
[1998 : 1 Mark]

51. The time averaged Poynting vector, in W/m2, for a wave with $\vec{E} = 24e^{j(\omega t + \beta z)}\vec{a}_y$ V/m in free space is

(a) $\frac{2.4}{\pi}\vec{a}_z$ (b) $-\frac{2.4}{\pi}\vec{a}_z$

(c) $\frac{4.8}{\pi}\vec{a}_z$ (d) $-\frac{4.8}{\pi}\vec{a}_z$
[1998 : 1 Mark]

52. A plane wave propagating through a medium $[\in_r = 8, \mu_r = 2, \text{ and } \sigma = 0]$ has its electric field given by $\vec{E} = 0.5 \sin(10^8 t - \beta z)$ V/m. The wave impedance, in ohms is
(a) 377 (b) 198.5∠180°
(c) 182.9∠140° (d) 188.5
[1999 : 2 Marks]

53. If a plane electromagnetic wave satisfies the equation $\frac{\partial^2 E_x}{\partial z^2} = \frac{1}{c^2}\frac{\partial^2 E_x}{\partial t^2}$, the wave propagates in the
(a) x-direction
(b) z-direction
(c) y-direction
(d) xy-plane at an angle of 45° between the x and z directions
[2001 : 1 Mark]

54. A material has conductivity of 10^{-2} mho/m and a relative permittivity of 4. The frequency at which the conduction current in the medium is equal to the displacement current is
(a) 45 MHz (b) 90 MHz
(c) 450 MHz (d) 900 MHz
[2001 : 2 Marks]

55. Distilled water at 25°C is characterized by $\sigma = 1.7 \times 10^{-4}$ mho/m and $\in = 78\in_0$ at a frequency of 3 GHz. Its loss tangent tan δ is
(a) 1.3×10^{-5} (b) 1.3×10^{-3}
(c) $1.7 \times 10^{-4}/78$ (d) $1.7 \times 10^{-4}/(78\in_0)$
$(\in = 10^{-9}/(36\pi)$ F/m)
[2002 : 2 Marks]

56. The depth of penetration of electromagnetic wave in a medium having conductivity a at a frequency of 1 MHz is 25 cm. The depth of penetration at a frequency of 4 MHz will be

(a) 6.25 cm (b) 12.50 cm
(c) 50.00 cm (d) 100.00 cm

[2003 : 1 Mark]

57. If $\vec{E} = (\hat{a}_x + j\hat{a}_y)e^{jkz-j\omega t}$ and $\vec{H} = \left(\frac{k}{\omega\mu}\right)(\hat{a}_y + j\hat{a}_x)e^{jkz-j\omega t}$, the time averaged Pointing vector is

(a) null vector (b) $\left(\frac{k}{\omega\mu}\right)\hat{a}_z$
(c) $\left(\frac{2k}{\omega\mu}\right)\hat{a}_z$ (d) $\left(\frac{k}{2\omega\mu}\right)\hat{a}_z$

[2004 : 2 Marks]

58. The $\vec{H}$ field (in A/m) of a plane wave propagating in free space is given by

$$\vec{H} = \hat{x}\frac{5\sqrt{3}}{\eta_0}\cos(\omega t - \beta z) + \hat{y}\frac{5}{\eta_0}\sin\left(\omega t - \beta z + \frac{\pi}{2}\right)$$

The time average power flow density in watt is

(a) $\frac{\eta_0}{100}$ (b) $\frac{100}{\eta_0}$
(c) $50\eta_0^2$ (d) $\frac{50}{\eta_0}$

[2007 : 2 Marks]

59. The electric field component of a time harmonic plane EM wave traveling in a nonmagnetic lossless dielectric medium has an amplitude of 1 V/m. If the relative permittivity of the medium is 4, the magnitude of the time average power density vector (in W/m²) is

(a) $\frac{1}{30\pi}$ (b) $\frac{1}{60\pi}$
(c) $\frac{1}{120\pi}$ (d) $\frac{1}{240\pi}$

[2010 : 1 Mark]

60. Consider the following statements regarding the complex Pointing vector $\vec{P}$ for the power radiated by a point source in an infinite homogenous and lossless medium. Re($\vec{P}$) denotes the real part of $\vec{P}$, S denotes a spherical surface whose centre is at the point source, and $\hat{n}$ denotes the unit surface normal on S. Which of the following statements is TRUE?

(a) Re($\vec{P}$) remains constant at any radial distance from the source

(b) Re($\vec{P}$) increases with increasing radial distance from the source

(c) $\oint\!\!\oint_S \text{Re}(\vec{P})\cdot\hat{n}\, dS$ remains constant at any radial distance from the source

(d) $\oint\!\!\oint_S \text{Re}(\vec{P})\cdot\hat{n}\, dS$ decreaseswith increasing radial distance from the source

[2011 : 1 Mark]

61. The electric field intensity of a plane wave traveling in free space is given by the following expression

$$\vec{E}(x, t) = \hat{a}_y 24\pi\cos(\omega t - k_0 x)\ (V/m)$$

In this field, consider a square area 10 cm × 10 cm on a plane x + y = 1. The total time-averaged power (in mW) passing through the square area is _______.

[2015 : 2 Marks, Set-1]

62. The distance (in meters) a wave has to propagate in a medium having a skin depth of 0.1 m so that the amplitude of the wave attenuates by 20 dB, is

(a) 0.12 (b) 0.23
(c) 0.46 (d) 2.3

[2018 : 2 Marks]

ANSWERS

1. (c)	2. (b)	3. (c)	4. (d)	5. (c)	6. (a)	7. (1)	8. (b)	9. (c)	10. (b)
11. (a)	12. (c)	13. (a)	14. (8.886)	15. (c)	16. (a,c)	17. (c)	18. (b)	19. (a)	20. (c)
21. (d)	22. (c)	23. (a)	24. (d)	25. (c)	26. (a)	27. (a)	28. (c)	29. (a)	30. (30°)
31. (d)	32. (b)	33. (c)	34. (b)	35. (c)	36. (2)	37. (c)	38. (c)	39. (d)	40. (d)
41. (d)	42. (b)	43. (d)	44. (c,d)	45. (a)	46. (c)	47. (a)	48. (a)	49. (d)	50. (d)
51. (a)	52. (a)	53. (b)	54. (a)	55. (a)	56. (b)	57. (a)	58. (d)	59. (c)	60. (d)
61. (0)	62. (b)								

EXPLANATIONS

1. $$\frac{\sin\theta_t}{\sin\theta_i} = \frac{\mu_1\varepsilon_1}{\mu_2\varepsilon_2}$$

$$\Rightarrow \quad \frac{\sin 30^\circ}{\sin 45^\circ} = \sqrt{\frac{1}{\varepsilon_r}}$$

$$\Rightarrow \quad \varepsilon_r = 2$$

2. For Cable 1, velocity $V_1 = \frac{1}{\sqrt{\mu_1 \cdot \epsilon_1}}$

For cable 2, velocity $V_2 = \frac{1}{\sqrt{\mu_2 \cdot \epsilon_2}}$

Since, frequency is same, $v \propto \lambda$

$$\frac{\lambda_1}{\lambda_2} = \sqrt{\frac{\epsilon_1}{\epsilon_2}} = \sqrt{\frac{\epsilon_0\epsilon_{r_1}}{\epsilon_0\epsilon_{r_3}}} = \sqrt{\frac{\epsilon_{r_1}}{\epsilon_{r_2}}}$$

3.

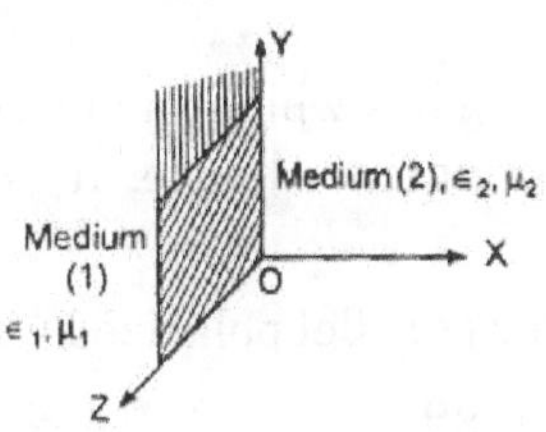

Since tangential components of electric field is same, therefore

$$\bar{E}_{x_1} = \bar{E}_{y_2} \quad \text{and} \quad \bar{E}_{z_1} = \bar{E}_{z_2}$$

and, as $\bar{D}_{x_1} = \bar{D}_{x_2}$

or $\bar{E}_{x_1}\epsilon_1 = \bar{E}_{x_2}\epsilon_2$

or $\bar{E}_{x_2} = \frac{(1.5)^2}{2.5}\bar{u}_x = 1.2\,\bar{u}_x$

thus $\bar{E}_2 = \bar{E}_{x_2} + \bar{E}_{y_2} + \bar{E}_{z_2} = \bar{E}_{x_2} + \bar{E}_{y_1} + \bar{E}_{z_1}$

$= (1.2\bar{u}_x - 3\bar{u}_y + 1\bar{u}_z)$ Volt/m.

4.

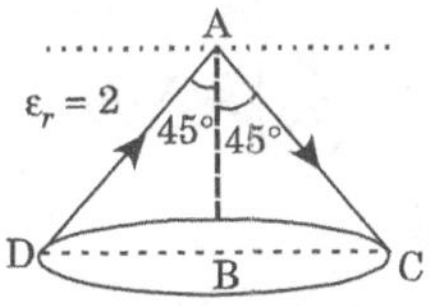

$$\sin\theta = \frac{1}{\sqrt{\varepsilon_r}} = \frac{1}{\sqrt{2}}$$

$\therefore$ $\theta = 45^\circ$ BD = AB = 1 m

Area = $\pi \times DB^2 = \pi$ m^2

5. From the figure, $\hat{a}_n = \hat{a}_x$

$\bar{E}_{n_1} = (E_1 \cdot \hat{a}_n)\hat{a}_n = [4\hat{a}_x + 3\hat{a}_y + 5\hat{a}_z \cdot \hat{a}_x]\hat{a}_x = 4\hat{a}_x$

$\bar{E}_{t_1} = \bar{E}_1 - \bar{E}_{n_1}$

$\bar{E}_{t_1} = 4\hat{a}_x + 3\hat{a}_y + 5\hat{a}_z - 4\hat{a}_x = 3\hat{a}_y + 5\hat{a}_z$

Assuming plane is charge free, i.e, $\rho_s = 0$

From Boundary conditions:

(i) $\bar{D}_{n_1} = \bar{D}_{n_2}$ (ii) $\bar{E}_{t_1} = \bar{E}_{t_2}$

$\epsilon_1\bar{E}_{n_1} = \epsilon_2\bar{E}_{n_2}$ $\quad$ $\bar{E}_{t_2} = 3\hat{a}_y + 5\hat{a}_z$

$En_2 = \frac{\epsilon_1}{\epsilon_2}\cdot\bar{E}n_1$ $\quad$ $\bar{E}_2 = \bar{E}_{t_2} + \bar{E}_{n_2}$

$= \frac{3\epsilon_0}{4\epsilon_0}\cdot 4\hat{a}_x$ $\quad$ $\bar{E}_2 = 3\hat{a}_x + 3\hat{a}_y + 5\hat{a}_z$

$\bar{E}_2 = 3\hat{a}_x$

6. $x > 0$ (Region 2) : $\epsilon_{r_2} = \mu_{r_2} = 2$

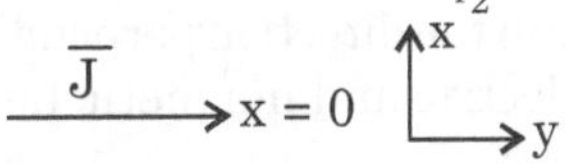

$\hat{a}_n = \hat{a}_x$

$\bar{H}_1 = 3\hat{a}_x + 30\hat{a}_y$

$\bar{H}_{n_1} = (H_1 \cdot \hat{a}_n)\hat{a}_n = [(3\hat{a}_x + 30\hat{a}_y)\cdot\hat{a}_x]\hat{a}_x$

$\bar{H}_{n_1} = 3\hat{a}_x$

$x < 0$ (Region 1) : $\epsilon_{r_1} = 5\ \mu_{r_1} = 1.$

$\bar{J} = 10\hat{a}_y \frac{A}{m}$

From Boundary wnditcons, (i) $\bar{B}_{n_1} = \bar{B}_{n_2}$

$\mu_1\bar{H}_{n_1} = \mu_2\bar{H}_{n_2}$

$\bar{H}_{n_2} = \frac{\mu_1}{\mu_2}\cdot\bar{H}_{n_1} = \frac{\mu_0}{2\mu 0}\cdot 3\hat{a}_x = 1.5\hat{a}_x$

(ii) $(\bar{H}_{t_1} - \bar{H}_{t_2}) = \hat{a}_n \times \bar{J}_s$

$(30\hat{a}_y - \bar{H}_{t_2}) = \hat{a}_x \times 10\hat{a}_y = 10\hat{a}_z$

$\bar{H}_{t_2} = 30\hat{a}_y - 10\hat{a}_z$

$\bar{H}_2 = \bar{H}_{t_2} + \bar{H}_{n_2} = 1.5\hat{a}_x + 30\hat{a}_y - 10\hat{a}_z$

8. Propagation constant,

$$\nu = 1\pi + j \times 2\pi$$

$$\Rightarrow \quad \beta = -2\pi$$

$$\therefore \quad \lambda = \frac{2\pi}{B} = 10 \text{ m}$$

9. $E = \cos(y^2 + 5t)$ does not satisfy the wave equation due to presence of non-linear term (y^2) in phase with distance.

10. $E(z, t) = 10\cos(2\pi \times 10^7\, t - 0.1\,\pi z)$

Now $2\pi \times 10^7\, t - 0.1\,\pi\, z$ = constant

Taking differentiation, we have

$$v = \frac{dz}{dt} = \frac{2\pi \times 10^7}{0.1\pi} = 2 \times 10^8 \text{ m/s}$$

11. The electric field in any arbitrary direction is given as $\bar{E} = E_0 e^{j(\omega t - \hat{n}\cdot\bar{r}\beta)}$

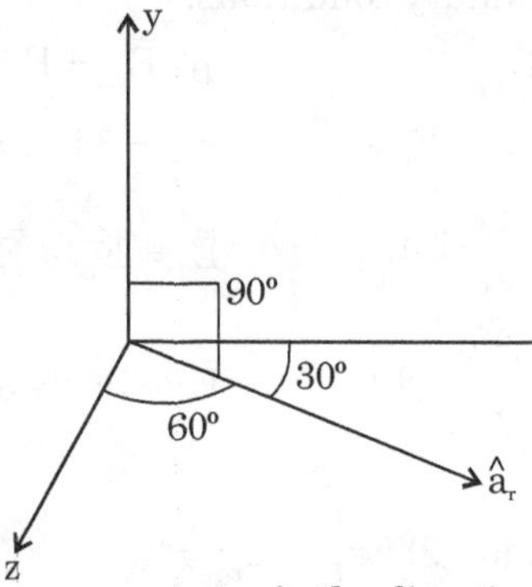

The wave propagates in the direction perpendicular to the direction of electric and magnetic field.

Now, $\bar{E} = \hat{y}\cdot E_0 e^{j(\omega t - \hat{n}\cdot\bar{r}\beta)}$

$$\hat{n}\cdot\bar{r} = x\cos 30° + y\cos 90° + z\cos 60°$$

$$\hat{n}\cdot\bar{r} = \frac{\sqrt{3}}{2}x + \frac{1}{2}z$$

$$\bar{E} = \hat{y}\,E_0 e^{j\left(\omega t - \frac{\sqrt{3}x}{2}\beta - \frac{1}{2}\beta z\right)} = \hat{y}\cdot E_0 e^{j\left(\omega t - \frac{\sqrt{3}}{2}\frac{2\pi}{\lambda}x - \frac{1}{2}\frac{2\pi}{\lambda}z\right)}$$

$$\bar{E} = \hat{y}\cdot E_0 e^{j\left(\omega t - \frac{\sqrt{3}\pi}{\lambda}x - \frac{\pi}{\lambda}z\right)}$$

12. From given figure, $\bar{E}_i$ is lying in the plane of incidence, thus, this is the case of parallel polarization for oblique incidence.

From Snell's law of refraction.

$$\frac{\sin\theta_t}{\sin\theta_i} = \sqrt{\frac{\epsilon_{r_2}}{\epsilon_{r_1}}} \quad (\mu_1 = \mu_2 = \mu_0 \text{ for non-magnetic medium})$$

$$\sin\theta_i = \sqrt{\frac{4.5}{1}}\cdot\sin 19.2° = 0.697.$$

$$\theta_i \cong 45°$$

$$E_{io}\sin\theta_i(-\hat{a}_z)$$

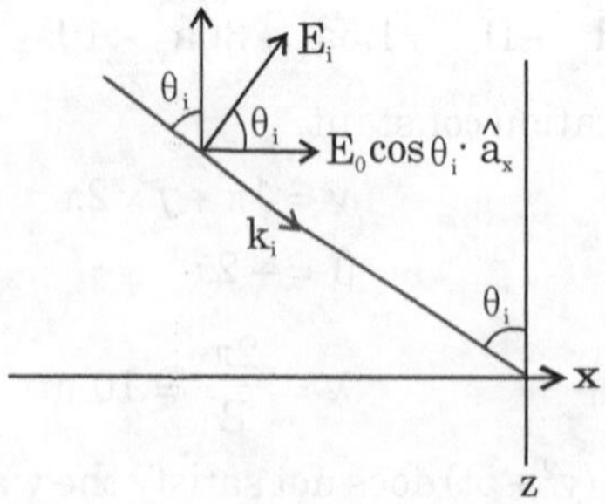

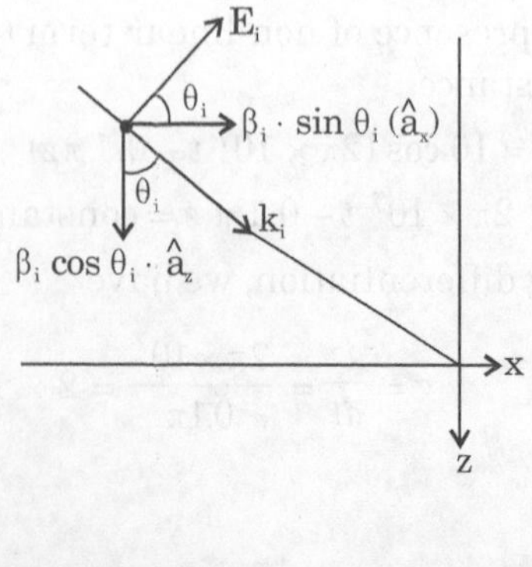

$$\Rightarrow \bar{E}_i = \left[E_0\cos\theta_i\cdot\hat{a}_x + E_0\sin\theta_i(-\hat{a}_z)\right]\cdot e^{-j(\bar{k}_i\cdot\bar{r})} \quad ...(1)$$

Propagating vector, $\bar{k}_i = \beta_1\sin\theta_i\cdot\hat{a}_x + \beta_1\cos\theta_i\cdot\hat{a}_z$

$$\beta_1 = \frac{2\pi}{\lambda} = \frac{2\pi}{600\times 10^{-16}} = \frac{\pi\times 10^4}{3} \text{ rad/m}$$

$$\bar{k}_i = \frac{\pi\times 10^4}{3}\left[\sin 45°\,\hat{a}_x + \cos 45°\,\hat{a}_y\right]$$

$$= \frac{\pi\times 10^4}{3\sqrt{2}}(\hat{a}_x + \hat{a}_z)$$

Position vector, $\bar{r} = x\hat{a}_x + y\hat{a}_y + z\hat{a}_z$

$$\bar{k}_i\cdot\bar{r} = \frac{\pi\times 10^4}{3\sqrt{2}}(x+z)$$

Incidence electric field intensity

$$\bar{E}_i = \frac{E_0}{\sqrt{2}}(\hat{a}_x - \hat{a}_z)\cdot e^{-j\frac{\pi\times 10^4(x+z)}{3\sqrt{2}}} \text{ V/m}$$

Wave is propagating in x-z plane and electric field has x and z components, hence, it is parallel polarization.

13. Reflection coeff for parallel polarized wave,

$$\tau_p = \frac{\eta_2\cos\theta_t - \eta_1\cos\theta_i}{\eta_2\cos\theta_i + \eta_1\cos\theta_i}$$

$$\therefore\ \theta_i = 45°,\ \theta_t = 19.2°,\ \eta_2 = \eta_0/\sqrt{4.5};\ \eta_1 = \eta_0$$

$$\tau_p = \frac{\cos 19.2° - \sqrt{4.5}\cos 45°}{\cos 19.2° + \sqrt{4.5}\cos 45°} = -0.23.$$

$$\bar{E}_r = \left[-E_{r_0}\cos\theta_r\hat{a}_x - E_{r_0}\sin\theta_r\hat{a}_z\right]e^{-j(\bar{k}\cdot\bar{r})}$$

$$\bar{E}_r = \left[0.23E_0\cos 45°\,\hat{a}_x + 0.23E_0\sin 45°\,\hat{a}_z\right]e^{-j\frac{\pi\times 10^4}{3\sqrt{2}}(x-z)}$$

$$\bar{E}_r = 0.23\frac{E_0}{\sqrt{2}}(\hat{a}_x + \hat{a}_z)e^{\frac{-j\pi\times 10^4}{3\sqrt{2}}(x-z)} \text{ V/m.}$$

14. $$\vec{E} = 2\cos\left(10^8 t - \frac{z}{\sqrt{2}}\right)\hat{a}_x \text{ V/m}$$

Compairing it with equation

$$\vec{E} = E_m\cos(\omega t - \beta z)\,\hat{a}_x \text{ V/m}$$

$$\omega = 10^8 \text{ rad/sec}$$

$$\beta = \frac{1}{\sqrt{2}} \text{ rad/m}$$

$$\lambda = \frac{2\pi}{\beta} \text{ m}$$

$$\lambda = \frac{2\pi}{1/\sqrt{2}} = 2\sqrt{2}\pi \text{ m}$$

$$\boxed{\lambda = 8.886 \text{ m}}$$

15. For total internal reflection, critical angle is

$$\theta_C = \sin^{-1}\sqrt{\frac{\epsilon_{r_2}}{\epsilon_{r_1}}}.$$

Total internal reflection takes place when $\theta_i \geq \theta_c$.

$$\theta_i \geq \sin^{-1}\left(\sqrt{\frac{\epsilon_{r_2}}{\epsilon_{r_1}}}\right)$$

The wave should move from a denser medium to a rarer medium $\left(\epsilon_{r_2} < \epsilon_{r_1}\right)$

16. Intrinsic impedance of medium 2, $\eta_2 = 0$ (perfect conductor). For normal incidence, reflection coeff. for electric field.

$$\tau = \frac{\eta_2 - \eta_1}{\eta_2 + \eta_1} = -\frac{\eta_1}{\eta_1} = -1$$

The tangential component of E must be continuous across the boundary. As E is zero within a perfect conductor.

$$E_x^i + E_x^r = 0. \text{ or } \frac{E_x^r}{E_x^i} = -1$$

$$E_x^r = -E_x^i$$

Since, direction of Poynting vector gives direction of propagation. After reflection, direction of propagation will be along z-axis.

$$\bar{P}_r = -\bar{P}_i$$

Direction of magnetic field will remains the same after reflection. $H_y^r = H_y^i$

17. Given: $E_x = 2\cos(\omega t)$, $E_y = 2\cos(\omega t + 90^\circ)$.

Direction of propagation, $\hat{a}_k = \hat{a}_z$

Phase difference, $\delta = 90^\circ$.

$\frac{E_{x_0}}{E_{y_0}} = 2$ V/m and E_y leads E_x by angle of 90°.

Therefore, the wave is left circularly polarized.

18. Refractive index, $\eta_2 = c\sqrt{\mu\,\epsilon} = c\sqrt{\mu_0\,\epsilon_0}\cdot\sqrt{\mu_r \epsilon_r}$

$$= \frac{c}{c}\sqrt{\mu_r \epsilon_r}$$

$$\eta_2 = \sqrt{\mu_r \epsilon_r}$$

Assuming non-magnetic medium, $\mu_r = 1$

$$n_2 = \sqrt{\epsilon_r} = 1.5$$

Intrinsic impedance of air, $\eta_{air} = \eta_0\sqrt{\frac{\mu}{\epsilon}} = \sqrt{\frac{\mu_0}{\epsilon_0}}$

$$\eta_{glass} = \eta_0 \cdot \sqrt{\frac{\mu_2}{\epsilon_2}} = \frac{\eta_0}{\sqrt{\epsilon_{r_2}}} = \frac{\eta_0}{1.5}$$

Reflection coeff.

$$\tau = \left(\eta_{glass} - \eta_{air}\right)/(\eta_{glass} + \eta_{air}) = \frac{1-1.5}{1+1.5} = \frac{-1}{5}$$

Reflected power, $P_r = |\tau|^2$. Incident power (P_i)

$$\frac{P_r}{P_i} = \left(\frac{1}{5}\right)^2 x = \frac{1}{25} \Rightarrow P_r = \frac{P_i}{25} \times 100\%$$

$$P_r = 4\% \text{ of } P_i$$

19. Given: $\bar{E} = E_0 e^{j(\omega t - \beta z)} \cdot (\hat{a}_x + \hat{a}_y)$

$$\bar{E} = E_0 e^{j(\omega t-\beta z)}\,\hat{a}_x + E_0 \cdot e^{j(\omega t-\beta z)}\hat{a}_y$$

$$E_x = E_0 \cdot e^{j(\omega t-\beta z)}, E_y = E_0 e^{j(\omega t-\beta z)}$$

Since $E_x = E_y$, the wave is linearly polarized.

20. $\bar{E}_X = (0.5).e^{j(wt-kz)}.\hat{x}$

$$\bar{E}_Y = j.\,e^{j(\omega t-kz)}.\,\hat{y}$$

then, $\frac{\bar{E}_x}{\bar{E}_y} = 0.5e^{-\frac{\pi}{2}} = \frac{1}{2}\angle -90^\circ$

And, wave is elliptically polarized, as $\left|\frac{\bar{E}_x}{\bar{E}_y}\right| \neq 1.$

21. Reflection coefficient,

$$\Gamma = \frac{\eta_2 - \eta_1}{\eta_2 + \eta_1} = \frac{\sqrt{\frac{\mu_0}{\varepsilon_o t_r}} - \sqrt{\frac{\mu_0}{\varepsilon_o}}}{\sqrt{\frac{\mu_0}{\varepsilon_o t_r}} + \sqrt{\frac{\mu_0}{\varepsilon_o}}} = -\frac{1}{3} = 0.333 \angle 180^\circ$$

22. Given, $E = \hat{a}_x \sin(\omega t - \beta z) + \hat{a}_y \sin\left(\omega t - \beta z + \frac{\pi}{2}\right)$

If the phase difference between E_x and E_y components is $\frac{\pi}{2}$, then wave is circularly polarised and when $\hat{E}_y$ leads the wave is left hand circularly polarised and when $\hat{E}_x$ leads the wave is right hand circularly polarised.

23. For free space, intrinsic impedance

$$\eta_1 = \sqrt{\frac{\mu_0}{\epsilon_0}} = 120\,\pi\,\Omega$$

and for medium having $\epsilon_r = 4$, $_r = 1$

$$\eta_2 = \sqrt{\frac{\mu_0\mu_r}{\epsilon_0\epsilon_r}} = \sqrt{\frac{\mu_0}{\epsilon_0}} \cdot \frac{1}{\sqrt{\epsilon_r}} = 60\pi = \frac{\eta_1}{2}$$

Power transmitted,

$$P_t = P_i \cdot \frac{\eta_1}{\eta_2}\left(\frac{2\eta_2}{\eta_1 + \eta_2}\right)^2$$

$$= P_i \cdot \frac{2\eta_2}{\eta_2}\left(\frac{2\eta_2}{2\eta_2 + \eta_2}\right)^2 = P_i \cdot \frac{8}{9}$$

24. When electromagnetic wave is incident at Brewster angle, the reflected wave is linearly polarized because reflection coefficient for parallel component is zero.

$$\tan(\theta_B) = \sqrt{\frac{\epsilon_{r_2}}{\epsilon_{r_1}}}$$

$$\tan 60^\circ = \sqrt{\frac{\epsilon_{r_2}}{1}}$$

$$\epsilon_{r_2} = 3$$

25. $$\frac{E_{reflected}}{E_{incident}} = \frac{E_r}{E_i} = \frac{\eta_2 - \eta_1}{\eta_2 + \eta_1}$$

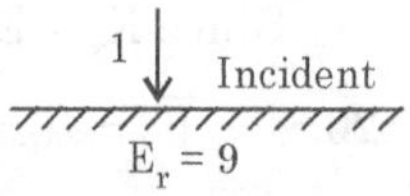

$$\eta = \sqrt{\frac{\mu}{E}}$$

But μ is same for both, therefore

$$K = \left|\frac{E_r}{E_i}\right| = \left|\frac{\sqrt{\frac{1}{E_r}} - 1}{\sqrt{\frac{1}{E_r}} + 1}\right| = \left|\frac{\frac{1}{3} - 1}{\frac{1}{3} + 1}\right| = \frac{2}{4} = 0.5$$

Alternately

Reflection coefficient,

$$T_L = \frac{\eta_2 - \eta_1}{\eta_2 + \eta_1}$$

$$T_L = \frac{\frac{\mu_2}{\epsilon_2} - \frac{\sqrt{\mu_1}}{\epsilon_1}}{\sqrt{\frac{\mu^2}{\epsilon_2}} + \sqrt{\frac{\mu_1}{\epsilon_1}}}$$

$$= \frac{\frac{1}{\sqrt{\epsilon_r}} - 1}{\frac{1}{\sqrt{\epsilon_r}} + 1} = \frac{\frac{1}{3} - 1}{\frac{1}{3} + 1} = -0.5$$

$$|T_L| = 0.5$$

26. $\bar{E}_i = 24\cos(3 \times 10^8 t - \beta y)\bar{a}_z$ V/m

Comparing $\bar{E}_i$ with $\bar{E} = E_m \cos(wt - \beta y) \cdot \bar{a}_z$ V/m it can be inferred that the plane wave given by $\bar{E} \times \bar{H}$ is travelling in the y direction having electric field along the $+\bar{a}_z$ direction and magnetic field along $+\bar{a}_x$ direction.

Now,

as, $$\frac{|\bar{E}_i|}{|\bar{H}_i|} = \eta$$

$$\bar{H}_i = \frac{\bar{E}_i}{\eta}$$

$$= \frac{1}{5\pi}\cos(3 \times 10^8 t - \beta y) \cdot \bar{a}_x \text{A/m}$$

Also, $$\frac{\bar{H}_r}{\bar{H}_i} = \frac{\eta_1 - \eta_2}{\eta_1 + \eta_2}$$

Given, $u_{r_1} = u_{r_2} = 1$

$E_{r_1} = E_{r_2} = 9$

$$\frac{\sqrt{\frac{\mu_0 \mu_{r_1}}{E_{r_1} E_0}} - \sqrt{\frac{\mu_0}{E_0} \cdot \frac{\mu_{r_2}}{E_{r_2}}}}{\sqrt{\frac{\mu_0}{E_0} \cdot \frac{\mu_{r_1}}{E_{r_1}}} \sqrt{\frac{\mu_0}{E_0} \cdot \frac{\mu_{r_2}}{E_{r_2}}}} = \frac{\frac{1}{\sqrt{E_{r_1}}} - \frac{1}{\sqrt{E_{r_2}}}}{\frac{1}{\sqrt{E_{r_1}}} + \frac{1}{\sqrt{E_{r_2}}}}$$

$$= \frac{\sqrt{9} - \sqrt{1}}{\sqrt{9} + \sqrt{1}} = \frac{1}{2}$$

then, $$\bar{H}_r = \frac{1}{10\pi}\cos(3 \times 10^8 t - \beta y)\bar{a}_x \text{ A/m}$$

27. Given : $\bar{E} = 10(\hat{a}_y + j\hat{a}_z)e^{-j25x}$ V/m

$\beta = 25$ rad/m, $\omega = \beta v_p = \beta.c$

$$f = \frac{25 \times 3 \times 10^8}{2\pi} = 1.2 \times 10^9 \text{ Hz} = 1.2\text{GHZ}.$$

$E_{y0} = E_{z0} = 10$ v/m.

Direction of propagation, $\hat{a}_k = \hat{a}_x$

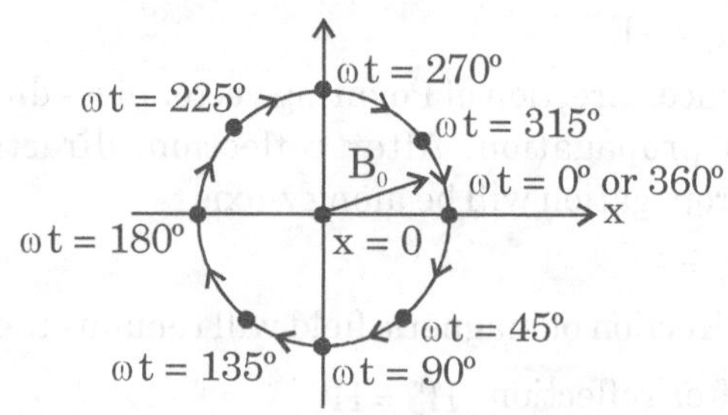

Left hand clockwise circlar polarixation

Phase difference, δ = 90°

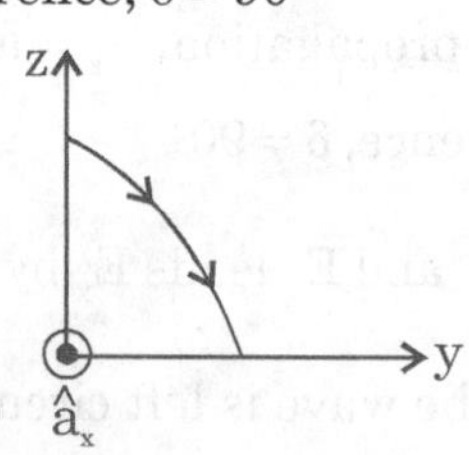

In the given wave, $E_{y0} = E_{z0} = 10$ V/m. and E_x leads E_y by an angle of 90°.

28. Propagation vector of reflected wave

$$\bar{k}_r = \beta_1 \cos\theta_r \cdot \hat{a}_x + \beta_1 \sin\theta_r \cdot \hat{a}_y$$

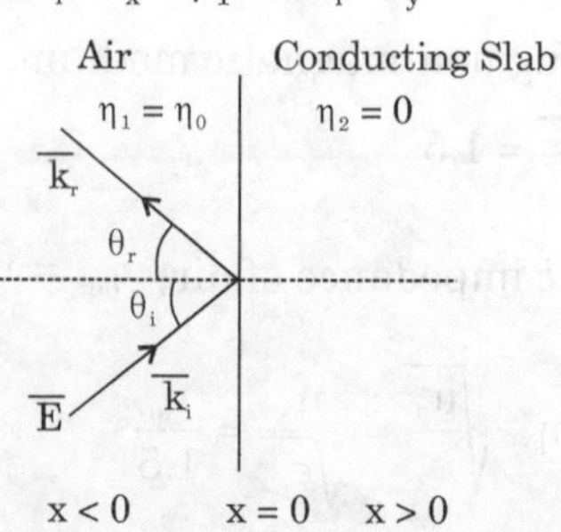

Phase constant, $\beta_1 = \sqrt{3^2 + 4^2} = 5$

Incident angle, $\theta_i = \tan^{-1}\left(\frac{4}{3}\right) = 53.13°$

Reflected angle, $\theta_r = \theta_i = 53.13°$

$\Rightarrow \quad \bar{k}_r = 3\hat{a}_x + 4\hat{a}_y$

$\bar{r} = x\hat{a}_x + y\hat{a}_y$

$\Rightarrow \bar{k}_r \cdot \bar{r} = 3x + 4y$

Reflected wave,

$$\bar{E}_r = \left(E_{0_x}\hat{a}_x + E_{0_y}\hat{a}_y + E_{0_z}\hat{a}_z\right)e^{j(\omega t - \bar{k}_r \cdot \bar{r})}$$
$$= \left(E_{0x}\hat{a}_x + E_{0y}\hat{a}_y + E_{0z}\hat{a}_z\right)e^{j(\omega t - 3x - 4y)}$$

Unsing Maxwell's eq.,

$$\bar{\nabla}\cdot\bar{E}_r = 0 \Rightarrow \bar{k}_r \cdot \bar{E}_r = 0 \Rightarrow 3E_{0x} + 4E_{0y} = 0$$

From Boundary condition (at x = 0), $E_{t_1} = E_{t_2} = 0$ (perfect conductor).

$\bar{E}_{t_1} = \bar{E}_{i_1} + \bar{E}_{r_1} = 6\hat{a}_y + 5\hat{a}_z + E_{oy}\hat{a}_y + E_{0z}\hat{a}_z = 0$

$6\hat{a}_y + 5\hat{a}_z = E_{0y}\hat{a}_y - E_{0z}\hat{a}_z \Rightarrow E_{0y} = -6$ and $E_{0z} = -5$

From eq. (i), $3E_{0x} + 4(-6) = 0 \Rightarrow E_{0x} = 8$

Reflected wave,

$$E_r = \left(8\hat{a}_x - 6\hat{a}_y - 5\hat{a}_z\right)e^{j(\omega t - 3x - 4y)}\ \text{V/m}$$

No option is correct.

29. $E(z,t) = 3\cos(cot - kz + 30°)\hat{a}_x - 4$

$-\sin(\omega t - kz + 45°)\hat{a}_y$

$E_x = 3\cos(\omega t - kz + 30°)$

$E_y = -4\cos(\omega t - kz + 45°)$

At $\quad z = 0$

$E_x = 3\cos(\omega t + 30°)$

$E_y = -4\sin(\omega t + 45°)$

$|E_x| \neq |E_y|$

→ So Elliptical polarization,

$Q = 30° - 135°$

$= -105°$

∴ left hand elliptical (LEP)

30. Given $\quad E = 10\cos\left(\omega t - 3x - \sqrt{3}z\right)a_y$

$E = E_0 e^{-J\beta(x\cos\theta_x + y\cos\theta_y + z\cos\theta_z)}$

So $\quad \beta_x = \beta\cos\theta_x = 3$

$\beta_y = \beta\cos\theta_y = 0$

$\beta_z = \beta\cos\theta_z = \sqrt{3}$

$\beta_x^2 + \beta_y^2 + \beta_z^2 = \beta^2$

$9 + 3 = \beta^2$

$\Rightarrow \quad \beta = \sqrt{13}$

$\beta\cos\theta_z = \sqrt{3}$

$\Rightarrow \quad \cos\theta_z = \sqrt{\frac{3}{13}}$

$\Rightarrow \quad \theta_z = 61.28 = \theta_i$

$$\frac{\sin\theta_i}{\sin\theta_t} = \sqrt{\frac{E_2}{E_1}}$$

$$\Rightarrow \quad \frac{\sin 61.28}{\sin\theta_t} = \sqrt{\frac{3}{1}}$$

$$\Rightarrow \quad \frac{0.8769}{\sqrt{3}} = \sin\theta_t$$

$\theta_t = 30.4$

$\Rightarrow \quad \theta_t \cong 30°$

31. $E = \left(a_x + 4ja_y\right)e^{j(2\pi\times 10^7 t - 0.2z)}$, $\quad \omega = 2\pi \times 10^7$

$E_z = \cos\omega t \quad , \quad \beta = 0.2$

$E_y = 4\cos(\omega + \pi/2) = -4\sin\omega t$

So, it left hand elliptical polarization

32. Given: $\bar{E}(z,t) = \hat{a}_x 5\cos(2\pi\times 10^9 t + \beta z)$

$$+\hat{a}_y \cdot 3\cos(2\pi\times 10^9 t + \beta z - \frac{\pi}{2}).$$

Wave is travelling is $-\hat{a}_z$ direction. It has orthogonal components with unequal amplitudes and $\hat{a}_y$ component lags $\hat{a}_y$ component.

Hence, wave is left hand elliptically polarized.

33. For velocity being free,

Since $\quad K.E. = \frac{1}{2}mV^2$

$V = \frac{E}{q}$

E = Energy

V = Applied voltage

q = electric charge of a metal plate

$$\frac{1}{2}mV^2 = qV$$

$$v = \frac{d}{t} = \sqrt{\frac{2qV}{m}}$$

$$\frac{d}{t} = \sqrt{\frac{2qV}{m}}$$

$$t = \frac{d}{\sqrt{\frac{2qV}{m}}}$$

$$\Rightarrow \quad t \propto \frac{d}{\sqrt{V}}$$

34. The reflected wave will be left circularly polarized because.

(i) 180° phase difference between reflected and incident wave,

(ii) Change in direction after reflection from conductor.

36.

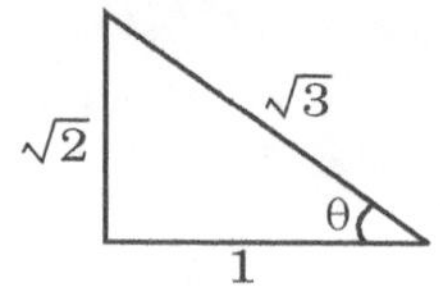

$$\vec{K}i = 2\pi\left(\hat{x} + \sqrt{2}\hat{z}\right) = 2\pi\sqrt{3}\left(\frac{1}{\sqrt{3}}\hat{x} + \sqrt{\frac{2}{3}}\hat{z}\right)$$

$$\cos\theta_{ix} = \frac{1}{\sqrt{3}}$$

$$\Rightarrow \quad \tan\theta_{ix} = \sqrt{2}$$

Now as there is no reflected wave,

$\therefore \ \theta_{ix} = \theta_B$ = Brewester angle

As the wave is parallelly polarized,

$$\therefore \quad \tan\theta_B = \sqrt{\frac{\epsilon_r\epsilon_0}{\epsilon_0}} = \sqrt{\epsilon_r} = \sqrt{2}$$

So, $\qquad \epsilon_r = 2$

37. Given : ω = 50,000, β = 0.004

Phase velocity, $v_p = \frac{\omega}{\beta} = \frac{5\times10^4}{4\times10^{-3}}$

$= 1.25 \times 10^7$ m/s.

38. In vaccum, $c = f\lambda$

$$\lambda = \frac{c}{f} = \frac{3\times10^8}{10^{14}} = 3\times10^{-6}\text{ m.}$$

In glass, $\lambda_g = \frac{\lambda}{n} = \frac{3\times10^{-6}}{1.5} = 2\ \mu\text{m.}$

39. $v_p = \omega/\beta$ can be calculated. Polarization can be identified.

μ_r and ϵ_r cannot be found, due to which power flux cannot be calculated as power flux

$$P = \frac{1}{2}\frac{|E|^2}{\eta},\text{ where } \eta = 120\pi\times\sqrt{\frac{\mu_r}{\epsilon_r}}$$

Hence, the power flux through the z = 0 plane.

40. Given data $c = 3\times10^8$ m/sec

Group velocity $v_g = 2\times10^8$ m/sec

Now $v_p v_g = c^2$

$$\therefore \quad \boxed{P = \frac{c^2}{v_g}}$$

$$\therefore \text{ Phase velocity } (v_p) = \frac{c^2}{v_g}$$

$$= \frac{\left(3\times10^8\right)^2}{2\times10^8} = 4.5\times10^8\text{ m/sec}$$

41. For a good conductor, intrinsic impedance,

$$\eta = \sqrt{\frac{j\omega\mu}{\sigma}} = \frac{E_t}{H_t}$$

$$\eta = \sqrt{\frac{\omega\mu}{\sigma}}\angle 45^\circ. \left(\because \sqrt{j} = \frac{90^\circ}{2} = 45^\circ\right)$$

So, E_t leads H_t by an angle of 45°.

42. Given: For copper, (f_1 = 3 GHz), δ_1 = μm.

For non-magnetic conductor, f_2 = 12 GHz, $\sigma_2 = \frac{\sigma_1}{9}$.

For a good conductor, $\delta = \frac{1}{\sqrt{\pi f\mu\sigma}}$

$$\frac{\delta_2}{\delta_1} = \sqrt{\frac{f_2\sigma_2}{f_1\sigma_1}} = 1\times\sqrt{\frac{3}{12}\times9} = \sqrt{\frac{9}{4}}$$

$$\delta_2 = \sqrt{\frac{9}{4}}\ \mu\text{m.}$$

43. Intrinsic impedance of medium, $\eta = \eta_0 = 120\pi$

Incident power, $P = E_0^2/2\eta = 1.2\text{ kW/m}^2$

$$E_0 = \sqrt{2\times1.2\times10^3\times120\pi}$$

$$E_0 \cong 950\text{ V/m.}$$

44. For an electromagnetic wave propagating in y-direction, $E_y = 0$, $H_y = 0$.

Equation of uniform plane wave propagating in y-direction,

$$\frac{\partial^2 E_x}{\partial y^2} = \mu\in\frac{\partial^2 E_x}{\partial t^2}$$

Amplitude of resultant electric field,

$$E = \sqrt{E_x^2 + E_y^2}$$

Amplitude of resultant magnetic field,

$$H = \sqrt{H_x^2 + E_y^2}$$

Intrinsic impedance of medium,

$$\eta = \frac{E}{H} = \sqrt{\frac{\mu}{\in}}$$

$$\frac{E}{H} = \frac{\sqrt{E_x^2 + E_y^2}}{\sqrt{H_x^2 + H_y^2}} = \sqrt{\frac{\mu}{\in}}$$

45. Given : $\bar{\nabla} = x\cos^2 y\,\hat{i} + x^2 e^z\,\hat{j} + z\sin^2 y\,\hat{k}$.

Using divergence theorem,

$$\oiint_s \bar{V}\cdot\hat{n}\,ds = \iiint_v \bar{\nabla}\cdot\bar{V}\,dv$$

$$\Rightarrow \bar{\nabla}\cdot\bar{V} = \frac{\partial V_x}{\partial x} + \frac{\partial V_y}{dy} + \frac{\partial V_z}{\partial z}$$

$$= \frac{\partial}{\partial x}(x\cos^2 y) + \frac{\partial}{\partial y}(x^2 e^z) + \frac{\partial}{\partial z}(z\sin^2 y)$$

$$\bar{\nabla}\cdot\bar{V} = \cos^2 y + 0 + \sin^2 y = 1$$

46. The intrinsic impedance of a lossy dielectric is

$$\eta = \sqrt{j\omega\mu/(\sigma + j\omega\varepsilon)}$$

47. A material behaves as a conductor, if

$$\frac{\sigma}{\omega\varepsilon} >> 1 \Rightarrow \sigma >> \omega\varepsilon$$

For copper, $\sigma = 5.8 \times 10^7$ S/m, $\in = 8.856 \times 10^{-12}$ F/m.

At large frequencies, $\sigma >> \omega\in$

48. Given : f = 1 GHz, $\sigma = 10^6$ mho/m.

For a material to be good conductor, $\frac{\sigma}{\omega\in} >> 1$.

$$\frac{\sigma}{\omega\in} = \frac{10^6}{2\pi\times10^9\times8.86\times10^{-12}} \cong 10^7 >> 1.$$

Skin depth, $\delta = \frac{1}{\sqrt{\pi f\mu\sigma}} = \frac{1}{\sqrt{\pi\times10^9\times10^6\times4\pi\times10^{-7}}}$

$$\delta = \frac{10^{-4}}{2\pi} = 15.9\ \mu\text{m}.$$

49. Given: Sphere radius = 1 m, surface charge density $\rho_s = 10$ c/m², Side of cube = 10 m

⇒ The total charge on the sphere

$Q = \rho_s\cdot 4\pi r^2 = 10\times4\pi\times1^2$ $Q = 40\pi$ C.

the sphere is enclosed in a cube. From Gauss's law, $\oint_s \bar{D}\cdot d\bar{s} = Q_{enclosed}$

total outward electric flux or displacement normal to the surface of cube = 40π C.

50. In a lossy dielectric, depth of penetration, $\delta = \frac{1}{\alpha}$

$$\therefore \alpha = \frac{\sigma}{2}\sqrt{\frac{\mu}{\in}}$$

$$\delta = \frac{2}{\sigma}\sqrt{\frac{\in}{\mu}} \Rightarrow \sigma \propto \sqrt{\in}$$

51. Given: $\bar{E} = 24\cdot e^{j(\omega t+\beta z)}\hat{a}_y$ V/m

$\hat{a}_k = -\hat{a}_z$, $E_0 = 24$ v / m,

For free space, $\eta = 120\pi\ \Omega$.

Time - average Poynting vector, $\bar{P}_{av} = \frac{E_0^2}{2\eta}\hat{a}_k$

$$\bar{P}_{av} = \frac{24^2}{2\times120\pi}\cdot(-\hat{a}_z) = -\frac{2.4}{\pi}\hat{a}_z\ \text{W/m}^2$$

52. Given: $\in_r = 8$, $\mu_r = 2$ and $\sigma = 0$.

Wave impedance, $\eta = \sqrt{\frac{j\omega\mu}{\sigma + j\omega\in}} = \sqrt{\frac{j\omega\mu}{j\omega\in}} = \sqrt{\frac{\mu}{\in}}\angle 0°$

$$\eta = \sqrt{\frac{\mu_0}{\in_0}\cdot\frac{\mu_r}{\in_r}} = 377\sqrt{\frac{2}{8}} = 188.5$$

53. Equations for electric and magnetic field for a plane EM have propagating in the z-direction

$$\frac{\partial^2 E_x}{\partial z^2} = C^2\,\frac{\partial^2 E_x}{\partial t^2}$$

$$\frac{\partial^2 E_y}{\partial z^2} = C^2\,.\,\frac{\partial^2 E_y}{\partial t^2}$$

54. $J = \sigma E = \omega D = \omega\in E$

or $\frac{\sigma E}{D} = 2\pi f$

or $2\pi f = \frac{10^{-2}}{8.85\times10^{-12}}$

or $f = 45$ MHz.

55. Loss tangent = $\frac{\sigma}{\omega\in}$

$$= \frac{1.7\times10^{-4}}{2\pi\times3\times10^9\times78\times8.86\times10^{-12}}$$

$\tan\delta = 0.13 \times 10^{-4} = 1.3 \times 10^{-15}$.

56. Depth of penetration, $\delta = \frac{1}{\sqrt{\pi f\mu\sigma}}$

or $\delta \propto \frac{1}{\sqrt{f}}$

or $\frac{\delta_1}{\delta_2} = \sqrt{\frac{f_2}{f_1}} = \sqrt{\frac{4}{1}}$

or $\delta_2 = \frac{\delta_1}{2} = 12.50$ cm

57. Time - averaged poynting vector,

$$P_{avg} = \frac{1}{2}\text{Re}\left[\bar{E}\times\bar{H}^*\right]$$

$$= \frac{1}{2}\text{Re}\left[\left(\bar{a}_x - j\bar{a}_y\right).e^{jkz-j\omega t} \times\left(\frac{k}{\omega\mu}\right)\left(\bar{a}_y - j\bar{a}_x\right)e^{-jkz+j\omega t}\right]$$

$$= \frac{k}{\omega\mu}\text{Re}\left[\left(\bar{a}_x + j\bar{a}_y\right)\times\left(\bar{a}_y - j\bar{a}_x\right)\right] = 0$$

= null vector.

58. Given, $\vec{H} = H_x \hat{x} + H_y \hat{y}$

So, plane wave propagating in free space will be more in z direction

We know that, pointing vector,

$$\vec{P} = \vec{E} \times \vec{H} = [\eta_0 H^2]\, \hat{k}$$

That shows the power flow in z-direction

Instantaneous power,

$$P = |\vec{P}| = \eta_0\, H^2$$

Average power over an interval $(0, 2\pi)$ will be

$$P_{av} = \frac{1}{2\pi}\int_0^{2\pi} P d\theta = \frac{1}{T}\int_0^T P dt = \frac{1}{T}\int_0^T \eta H^2 dt$$

$$= \frac{\eta_0}{T}\int_0^T (H_x^2 + H_y^2)\, dt$$

$$= \frac{\eta_0}{T}\int_0^T \left[\frac{25 \times 3}{\eta_0^2}\cos^2(\omega t - \beta z) + \frac{25}{\eta_0^2}\sin^2(\omega t - \beta z + \frac{\pi}{2})\right] dt$$

$$= \frac{100}{T\eta_0}\int_0^T \cos^2(\omega t - \beta z)\, dt = \frac{100}{T\eta_0}\left(\frac{T}{2}\right) = \left(\frac{50}{\eta_0}\right)$$

59. Magnitude of time-average power density vector

$$|\bar{P}_{av}| = \frac{E^2}{2\eta} \text{ where } \eta = \sqrt{\frac{\mu}{\epsilon}} \text{ for nonmagnetic lossless}$$

dielectric.

$$= \frac{(1)^2}{2\sqrt{\frac{\mu_0}{4\epsilon_0}}} = \frac{1}{1} = \frac{\frac{1}{120\pi}}{2} \times 2$$

$$= \frac{1}{120\pi}\ \text{W/m}^2$$

60. $\oiint_s \text{Re}(\bar{P}) \cdot \hat{n} ds$ gives average power and it decreases with increasing radial distance from the source.

61. $E = 24\pi \cos(\omega t - \beta z)\, \hat{a}_x \text{V/m}$

$$\vec{P}_{avg} = \text{average Poynting vector} = \frac{(24\pi)^2}{2\eta}\bar{a}_z$$

Surface over the plane

$$X + Y = 1\ (x + y + 0z = 1)$$

Normal to this plane is

$$\hat{n} = (1, 1, 0)$$

$$\text{Power crossing} = \int_S \bar{P}_{avg} \cdot d\bar{S}$$

$$= \int_S \frac{(24\pi)^2}{2\eta}\bar{a}_z \cdot (1,1,0)\, dS$$

$$= 0 \text{ mW}$$

62. Attenuation constant,

$$(\alpha) = \frac{1}{\text{skin depth}} = 10 \text{ Np/m}$$

$$\therefore\ 20\log_{10}\left(\frac{E_0}{E_x}\right) = 20 \text{ dB}$$

$$\therefore\ \frac{E_0}{E_x} = 10^1$$

$$\therefore\ \frac{E_0}{E_x} = 10$$

$$\therefore\ E_x = \frac{E_0}{10}$$

$$\text{Now } E_x = E_0 e^{-\alpha x} = E_0 e^{-10x} = \frac{E_0}{10}$$

$$\therefore\ e^{-10x} = \frac{1}{10}$$

$$x = \frac{1}{10}\text{In}(10) = 0.23 \text{ m}$$

4

CHAPTER

Transmission Lines

Analysis of Previous GATE Papers

Topics ↓		Year →	2019	2018	2017 Set 1	2017 Set 2	2016 Set 1	2016 Set 2	2015 Set 1	2015 Set 2	2015 Set 3
EQUATIONS	1 Mark	MCQ Type									
		Numerical Type				1					1
	2 Marks	MCQ Type									1
		Numerical Type						1			1
	Total					1		2			5
CHARACTERISTIC IMPEDANCE & IMPEDANCE MATHCING	1 Mark	MCQ Type					1				
		Numerical Type		1	1						
	2 Marks	MCQ Type						1			
		Numerical Type									1
	Total			1	1		1	2			2
TRANSFORMATION & S-PARAMETERS	1 Mark	MCQ Type									
		Numerical Type									
	2 Marks	MCQ Type									
		Numerical Type									
	Total										
SMI TH CHARTS	1 Mark	MCQ Type		1							
		Numerical Type									
	2 Marks	MCQ Type									
		Numerical Type									
	Total			1							

EQUATIONS

1. A transmission line of real characteristic impedance is terminated with an unknown load. The measured value of VSWR on the line is equal to 2 and a voltage minimum point is found to be at the load. The load impedance is then

(a) complex (b) purely capacitive
(c) purely resistive (d) purely inductive

[1987 : 2 Marks]

2. A 50 ohm lossless transmission line has a pure reactance of (j100)ohms as its load. The VSWR in the line is

(a) $\frac{1}{2}$ (Half) (b) 2(Two)
(c) 4 (Four) (d) ∞ (infinity)

[1989 : 2 Marks]

3. The input impedance of a short circuited lossless transmission line quarter wave long is

(a) purely reactive
(b) purely resistive
(c) infinite
(d) dependent on the characteristic impedance of the line

[1991 : 2 Marks]

4. If a pure resistance load, when connected to a lossless 75 ohm line, produces a VSWR of 3 on the line, then the load impedance can only be 25 ohms. True/False (Give Reason)

[1994 : 2 Marks]

5. In a twin-wire transmission line in air, the adjacent voltage maxima are at 12.5 cm and 27.5 cm. The operating frequency is

(a) 300 MHz (b) 1 GHz
(c) 2 GHz (d) 6.28 GHz

[1999 : 2 Marks]

6. In air, a lossless transmission line of length 50 cm with L = 10μ H/m, C = 40 pF/m is operated at 25 MHz. Its electrical path length is

(a) 0.5 meters (b) λ meters
(c) $\frac{\pi}{n}$ radians (d) 180 degrees

[1999 : 2 Marks]

7. A transmission line is distortionless if

(a) $RL = \frac{1}{GC}$ (b) $RL = GC$
(c) $LG = RC$ (d) $RG = LC$

[2001 : 1 Mark]

8. A uniform plane electromagnetic wave incident normally on a plane surface of a dielectric material is reflected with a VSWR of 3. What is the percentage of incident power that is reflected?

(a) 10% (b) 25%
(c) 50% (d) 75%

[2001 : 2 Marks]

9. The VSWR can have any value between

(a) 0 and 1 (b) –1 and +1
(c) 0 and ∞ (d) 1 and ∞

[2002 : 1 Mark]

10. Consider a 300 Ω, quarter-wave long (at 1 GHz) transmission line as shown in the figure. It is connected to a 10 V, 50 Ω source at one end and is left open circuited at the other end. The magnitude of the voltage at the open circuit end of the line is

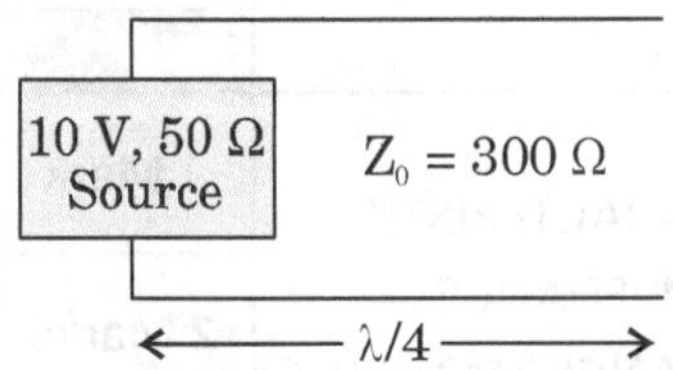

(a) 10 V (b) 5 V
(c) 60 V (d) $\frac{60}{7}$ V

[2004 : 2 Marks]

11. A plane electromagnetic wave propagating in free space is incident normally on a large slab of loss-less, non-magnetic, dielectric material with $\epsilon > \epsilon_0$. Maxima and minima are observed when the electric field is measured in front of the slab. The maximum electric field is found to be 5 times the minimum field. The intrinsic impedance of the medium should be

(a) 120π Ω
(b) 60π Ω
(c) 600π Ω
(d) 24π Ω

[2004 : 2 Marks]

12. A lossless transmission line is terminated in a load which reflects a part of the incident power. The measured VSWR is 2. The percentage of the power that is reflected back is

(a) 57.73
(b) 33.33
(c) 0.11
(d) 11.11

[2004 : 2 Marks]

Common data for Questions 13 and 14:

Voltage standing wave pattern in a lossless transmission line with characteristic impedance 50 Ω and a resistive load is shown in the figure.

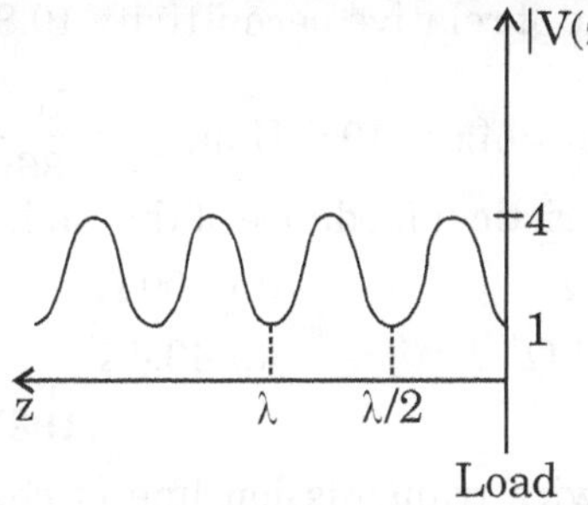

13. The value of the load resistance is

(a) 50 Ω (b) 200 Ω

(c) 12.5 Ω (d) 0 Ω

[2005 : 2 Marks]

14. The reflection coefficient is given by

(a) –0.6 (b) –1

(c) 0.6 (d) 0

[2005 : 2 Marks]

Common data for Questions 3.15 & 3.16: A 30 Volts battery with zero source resistance is connected to a coaxial line of characteristic impedance of 50 Ohms at t = 0 second and terminated in an unknown resistive load. The line length is such that it takes 400 μs for an electromagnetic wave to travel from source end to load end and vice-versa. At t = 400 μs, the voltage at the load end is found to be 40 Volts.

15. The load resistance is

(a) 25 Ohms (b) 50 Ohms

(c) 75 Ohms (d) 100 Ohms

[2006 : 2 Marks]

16. The steady-state current through the load resistance is

(a) 1.2 A (b) 0.3 A

(c) 0.6 A (d) 0.4 A

[2006 : 2 Marks]

17. In the design of a single mode step index optical fiber close to upper cut-off, the single-mode operation is NOT preserved if

(a) radius as well as operating wavelength are halved

(b) radius as well as operating wavelength are doubled

(c) radius is halved and operating wavelength is doubled

(d) radius is doubled and operating wavelength is halved

[2008 : 2 Marks]

18. The return loss of a device is found to be 20 dB. The voltage standing wave ratio (VSWR) and magnitude of reflection coefficient are respectively

(a) 1.22 and 0.1 (b) 0.81 and 0.1

(c) –1.22 and 0.1 (d) 2.44 and 0.2

[2013 : 1 Mark]

19. A coaxial cable is made of two brass conductors. The spacing between the conductors is filled with Teflon (ε_r = 2.1, tanδ = 0). Which one of the following circuits can represent the lumped element model of a small piece of this cable having length Δz?

(a)

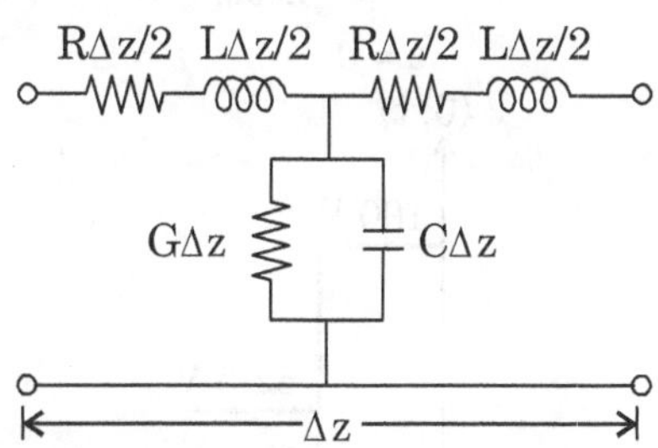

(b)

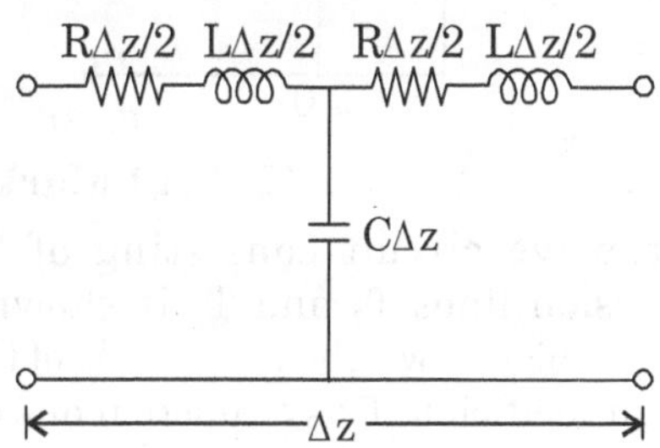

(c)

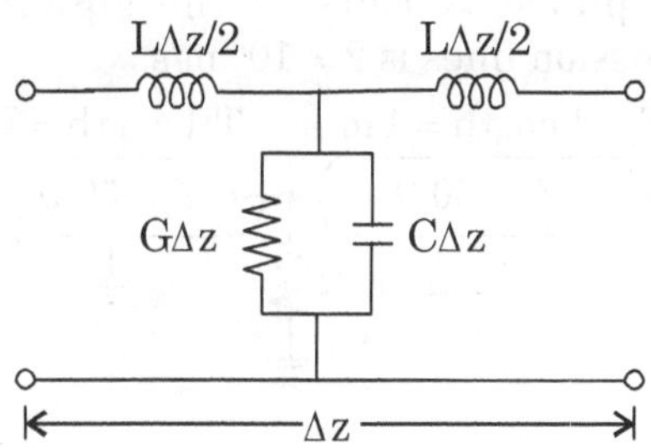

(d)

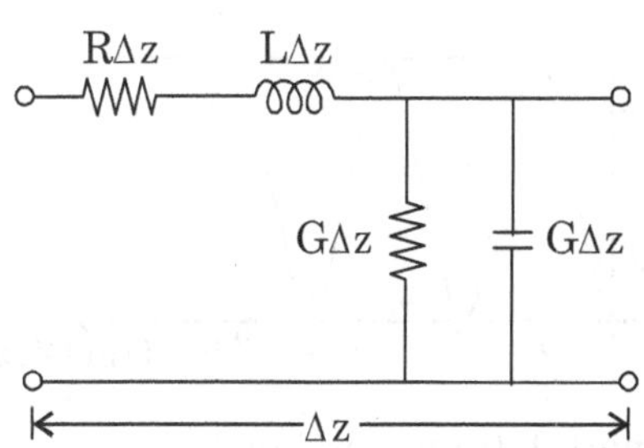

[2015 : 2 Marks, Set-3]

20. A coaxial capacitor of inner radius 1 mm and outer radius 5 mm has a capacitance a per unit length of 172 pF/m. If the ratio of outer radius to inner is doubled, the capacitance per unit length (in pF/m) is_______.

[2015 : 2 Marks, Set-3]

21. A 200 m long transmission line having parameters shown in the figure is terminated into a load R_L. The line is connected to a 400 V source having source resistance R_S through a switch which is closed at t = 0. The transient response of the circuit at the input of the line (z = 0) is also drawn in the figure. The value of R_L (in Ω) is________.

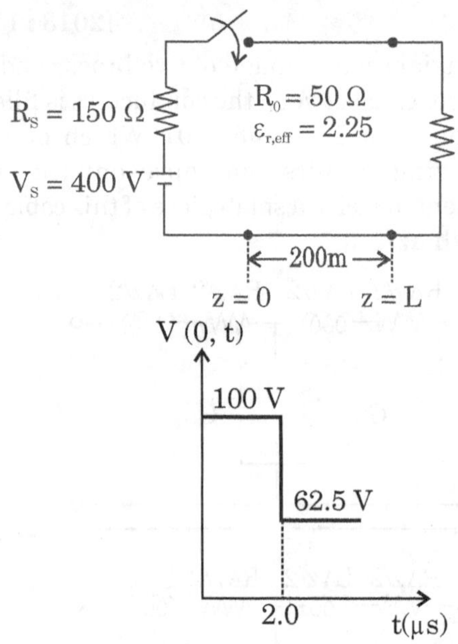

[2015 : 1 Mark, Set-3]

22. A microwave circuit consisting of lossless transmission lines T_1 and T_2 is shown in the figure. The plot shows the magnitude of the input reflection coefficient Γ as a function of frequency f. The phase velocity of the signal in the transmission lines is 2×10^8 m/s.

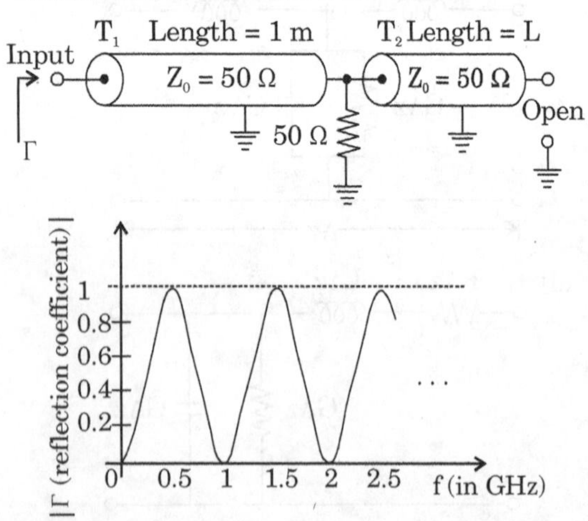

The length L (in meters) of T_2 is ______.

[2016 : 2 Marks, Set-2]

23. A two-wire transmission line terminates in a television set. The VSWR measured on the line is 5.8. The percentage of power that is reflected from the television set is ______.

[2017 : 1 Mark, Set-2]

CHARACTERISTIC IMPEDANCE & IMPEDANCE MATCHING

24. A coaxial-cable with an inner diameter of 1 mm and outer diameter of 2.4 mm is filled with a dielectric of relative permittivity 10.89.

Given $\mu_0 = 4\pi \times 10^{-7}$ H/m, $\varepsilon_0 = \frac{10^{-9}}{36\pi}$ F/m, the characteristic impedance of the cable as

(a) 330 Ω (b) 100 Ω
(c) 143.3 Ω (d) 43.4 Ω

[1987 : 1 Mark]

25. A Two-wire transmission line of characteristic impedance Z_0 is connected to a load of impedance $Z_L (Z_L \neq Z_0)$. Impedance matching cannot be achieved with

(a) a quarter-wavelength transformer
(b) a half-wavelength transformer
(c) an open-circuited parallel stub
(d) a short-circuited parallel stub

[1988 : 2 Marks]

26. A transmission line whose characteristic impedance is a real,

(a) must be a lossless line
(b) must be a distortionless line
(c) may not be a lossless line
(d) may not be a distortionless line

[1992 : 2 Marks]

27. Consider a transmission line of characteristic impedance 50 ohm. Let it be terminated at one end by (+j50)ohm. The VSWR produced by it in the transmission line will be

(a) +1 (b) 0
(c) ∞ (d) +j

[1993 : 2 Marks]

28. A load impedance, (200 + j0) Ω is to be matched to a 50 Ω lossless transmission line by using a quarter wave line transformer (QWT). The characteristic impedance of the QWT required is______.

[1994 : 1 Mark]

29. A lossless transmission line having 50 Ω characteristic impedance and length $\frac{\lambda}{4}$ is short circuited at one end and connected to an ideal voltage source of 1 V at the other end. The current drawn from the voltage sources is

(a) 0 (b) 0.02 A
(c) ∞ (d) None of the these

[1996 : 1 Mark]

30. The capacitance per unit length and the characteristic impedance of a lossless transmission line are C and Z_0 respectively. The velocity of a travelling wave on the transmission line is

(a) Z_0C
(b) $\frac{1}{(Z_0C)}$
(c) Z_0/C
(d) $\frac{C}{Z_0}$

[1996 : 1 Mark]

31. A transmission line of 50 characteristic impedance is terminated with a 100 Ω resistance. The minimum impedance measured on the line is equal to

(a) 0 Ω
(b) 25 Ω
(c) 50 Ω
(d) 100 Ω

[1997 : 1 Mark]

32. All transmission line section in Figure, have a characteristic impedance $R_0 + j0$. The input impedance Z_{in} equals

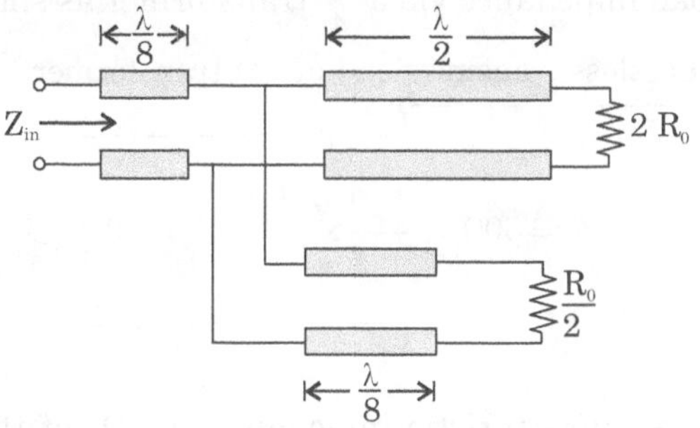

(a) $\frac{2}{3}R_0$
(b) R_0
(c) $\frac{3}{2}R_0$
(d) $2R_0$

[1998 : 1 Mark]

33. The magnitudes of the open-circuit and short-circuit input impedances of a transmission line are 100 Ω and 25 Ω respectively. The characteristic impedance of the line is

(a) 25 Ω
(b) 50 Ω
(c) 75 Ω
(d) 100 Ω

[2000 : 1 Mark]

34. Characteristic impedance of a transmission line is 50 Ω. Input impedance of the open-circuited line is $Z_{OC} = 100 + j150$ Ω. When the transmission line is short-circuited, then value of the input impedance will be

(a) 50 Ω
(b) 100 + j150 Ω
(c) 7.69 + j11.54 Ω
(d) 7.69 – j11.54 Ω

[2005 : 2 Marks]

35. The parallel branches of a 2-wire transmission line are terminated in 100 Ω and 200 Ω resistors as shown in the figure. The characteristic impedance of the line is Z_0 = 50 Ω and each section has a length of $\frac{\lambda}{4}$. The voltage reflection coefficient Γ at the input is

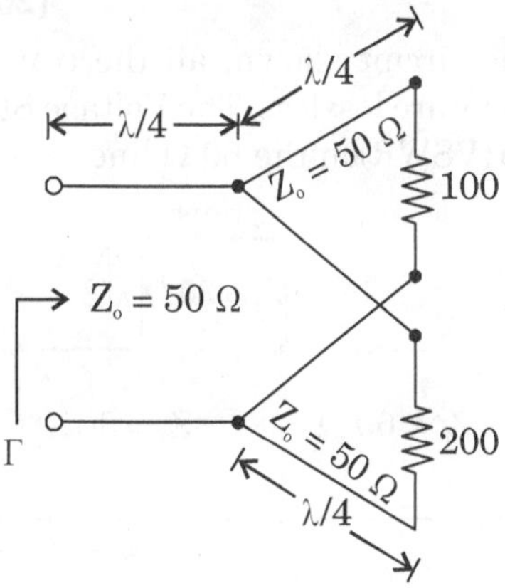

(a) $-j\frac{7}{5}$
(b) $\frac{-5}{7}$
(c) $j\frac{5}{7}$
(d) $\frac{5}{7}$

[2007 : 2 Marks]

36. One end of a loss-less transmission line having the characteristic impedance of 75 Ω and length of 1 cm is short-circuited. At 3 GHz, the input impedance at the other end of the transmission line is

(a) 0
(b) resistive
(c) capacitive
(d) inductive

[2008 : 2 Marks]

37. A transmission line terminates in two branches, each of length $\frac{\lambda}{4}$, as shown. The branches are terminated by 50 Ω loads. The lines are lossless and have the characte-ristic impedances shown. Determine the impedance Z_i as seen by the source.

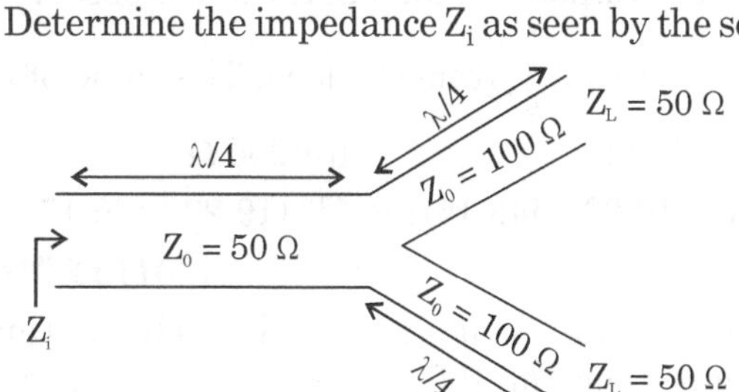

(a) 200 Ω
(b) 100 Ω
(c) 50 Ω
(d) 25 Ω

[2009 : 2 Marks]

38. A transmission line has a characteristic impedance of 50 Ω and a resistance of 0.1 Ω/m. If the line is distortionless, the attenuation constant (in Np/m) is

(a) 500 (b) 5
(c) 0.014 (d) 0.002

[2010 : 1 Mark]

39. In the circuit shown, all the transmission line sections are lossless. The Voltage Standing Wave Ratio (VSWR)on the 60 Ω line is

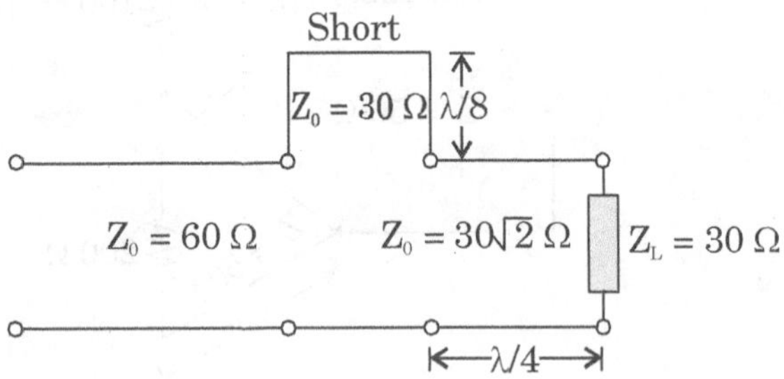

(a) 1.00 (b) 1.64
(c) 2.50 (d) 3.00

[2010 : 2 Marks]

40. A transmission line of characteristic impedance 50 Ω is terminated by a 50 load. When excited by a sinusoidal voltage source at 10 GHz, the phase difference between two points spaced 2 mm apart on the line is found to be $\frac{\pi}{4}$ radians. The phase velocity of the wave along the line is

(a) 0.8×10^8 m/s
(b) 1.2×10^8 m/s
(c) 1.6×10^8 m/s
(d) 3×10^8 m/s

[2011 : 1 Mark]

41. A transmission line of characteristic impedance 50 Ω is terminated in a load impedance Z_L. The VSWR of the line is measured as 5 and the first of the voltage maxima in the line is observed at a distance of $\frac{\lambda}{4}$ from the load. The value of Z_L is

(a) 10 Ω (b) 250 Ω
(c) (19.23 + j46.15) Ω (d) (19.23 – j46.15) Ω

[2011 : 2 Marks]

42. A transmission line with a characteristic impedance of 100 Ω is used to match a 50 Ω section to a 200 Ω section. If the matching is to be done both at 429 MHz and 1 GHz, the length of the transmission line can be aporoximately

(a) 82.5 cm (b) 1.05 m
(c) 1.58 m (d) 1.75 m

[2012 : 2 Marks]

43. For a parallel plate transmission line, let v be the speed of propagation and Z be the characteristic impedance. Neglecting fringe effects, a reduction of the spacing between the plates by a factor of two results in

(a) halving of v and no change in Z
(b) no changes in v and halving of Z
(c) no change in both v and z
(d) halving of both v and z

[2014 : 2 Marks, Set-1]

44. The input impedance of a $\frac{\lambda}{8}$ section of a lossless transmission line of characteristic impedance 50 Ω is found to be real when the other end is terminated by a load Z_L = (R + jX) Ω. If X is 30 Ω, the value of R (in Ω)is _______.

[2014 : 2 Marks, Set-1]

45. To maximize power transfer, a lossless transmission line is to be matched to a resistive load impedance via a $\frac{\lambda}{4}$ transformer as shown.

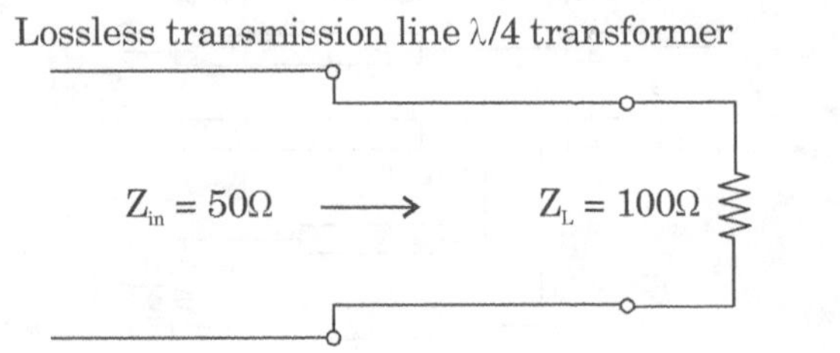

The characteristic impedance (in Ω) of the $\frac{\lambda}{4}$ transformer is_______.

[2014 : 1 Mark, Set-2]

46. In the transmission line shown, the impedance Z_{in} (in ohms) between node A and the ground is_______.

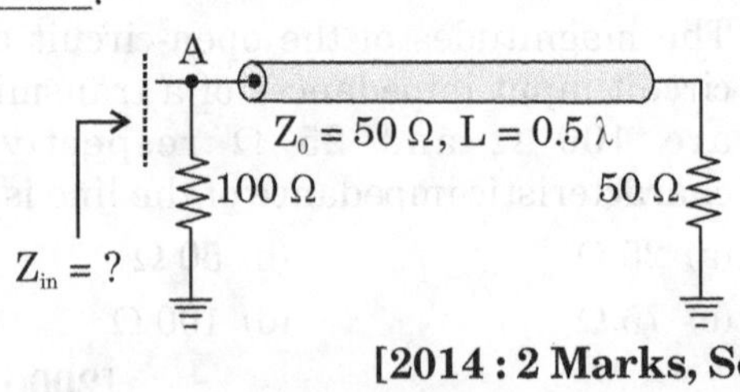

[2014 : 2 Marks, Set-2]

47. In the following figure, the transmitter T_x sends a wideband modulated RFsignal via a coaxial cable to the receiver R_x. The output impedance Z_r of T_x, the characteristic impedance Z_0 of the cable and the input impedance Z_R of R_x are all real.

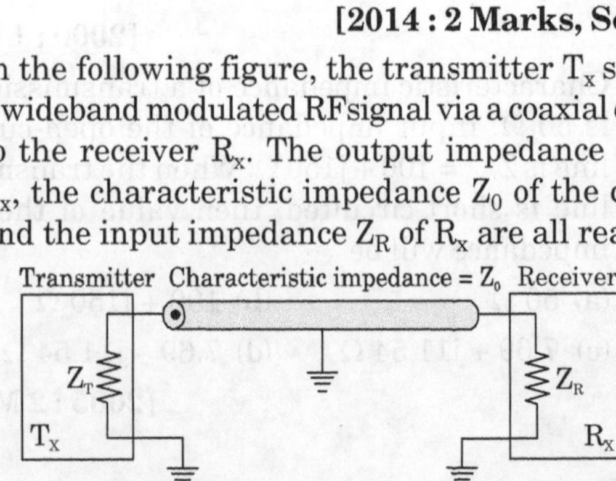

Which one of the following statements is TRUE about the distortion of the received signal due to impedance mismatch?

(a) The signal gets distorted if $Z_R \neq Z_0$, irrespective of the value of Z_T

(b) The signal gets distorted if $Z_T \neq Z_0$, irrespective of the value of Z_R.

(c) Signal distortion implies impedance mismatch at both ends: $Z_T \neq Z_0$ and $Z_R \neq Z_0$.

(d) Impedance mismatches do NOT result in signal distortion but reduce power transfer efficiency.

[2014 : 1 Mark, Set-3]

48. Consider the 3 m long lossless air-filled transmission line shown in the figure. It has a characteristic impedance of $120\pi\ \Omega$, is terminated by a short circuit, and is excited with a frequency of 37.5 MHz. What is the nature of the input impedance (Z_{in})?

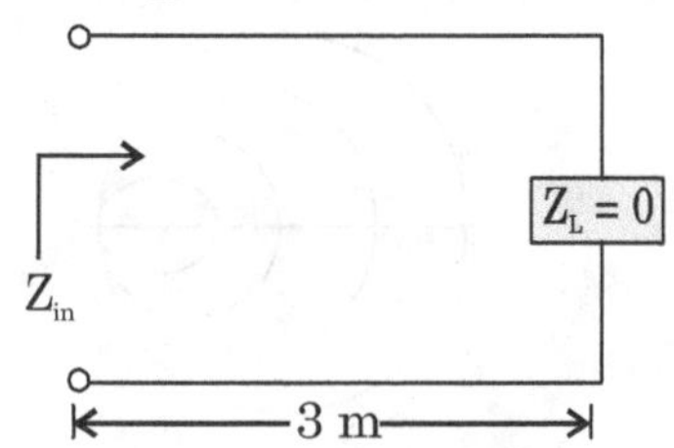

[2015 : 2 Marks, Set-3]

49. The propagation constant of a lossy transmission line is $(2 + j5)\ m^{-1}$ and its characteristic impedance is $(50 + j0)\ \Omega$ at $\omega = 10^6\ rad\text{-}s^{-1}$. The value of the line constants L, C, R, G are respectively,

(a) $L = 200\ \mu H/m$, $C = 0.1\ \mu F/m$, $R = 50\ \Omega/m$, $G = 0.02\ S/m$

(b) $L = 250\ \mu H/m$, $C = 0.1\ \mu F/m$, $R = 100\ \Omega/m$, $G = 0.04\ S/m$

(c) $L = 200\ \mu H/m$, $C = 0.2\ \mu F/m$, $R = 100\ \Omega/m$, $G = 0.02\ S/m$

(d) $L = 250\ \mu H/m$, $C = 0.2\ \mu F/m$, $R = 50\ \Omega/m$, $G = 0.04\ S/m$

[2016 : 1 Mark, Set-1]

50. A lossless microstrip transmission line consists of a trace of width W. It is drawn over a practically infinite ground plane and is separated by a dielectric slab of thickness t and relative permittivity $\varepsilon_r > 1$. The inductance per unit length and the characteristic impedance of this line are L and Z_0, respectively.

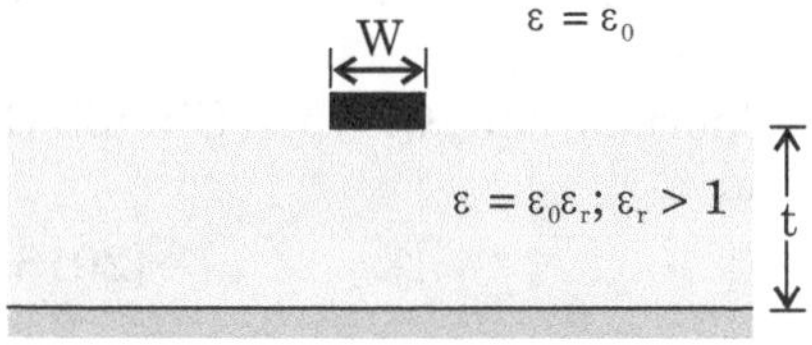

Which one of the following inequalities is always satisfied?

(a) $Z_0 > \sqrt{\dfrac{Lt}{\varepsilon_0 \varepsilon_r W}}$ (b) $Z_0 < \sqrt{\dfrac{Lt}{\varepsilon_0 \varepsilon_r W}}$

(c) $Z_0 > \sqrt{\dfrac{Lw}{\varepsilon_0 \varepsilon_r t}}$ (d) $Z_0 < \sqrt{\dfrac{LW}{\varepsilon_0 \varepsilon_r t}}$

[2016 : 2 Marks, Set-2]

51. The voltage of an electromagnetic wave propagating in a coaxial cable with uniform characteristic impedance is $V(l) = e^{-\gamma l + j\omega l}$ Volts, where l is the distance along the length of the cable in metres, $\gamma = (0.1 + j40)\ m^{-1}$ is the complex propagation constant, and $\omega = 2\pi \times 10^9$ rad/s is the angular frequency. The absolute value of the attenuation in the cable in dB/metre is ______.

[2017 : 1 Mark, Set-1]

52. A lossy transmission line has resistance per unit length $R = 0.05\ \Omega/m$. The line is distortionless and has characteristic impedance of $50\ \Omega$. The attenuation constant (in Np/m correct to three decimal places)of the line is ______.

[2018 : 1 Mark]

TRANSFORMATION & S-PARAMETERS

53. A short-circuited stub is shunt connected to a transmission line as shown in the figure. If $Z_0 = 50\ \Omega$, the admittance /seen at the junction of the stub and the transmission line is

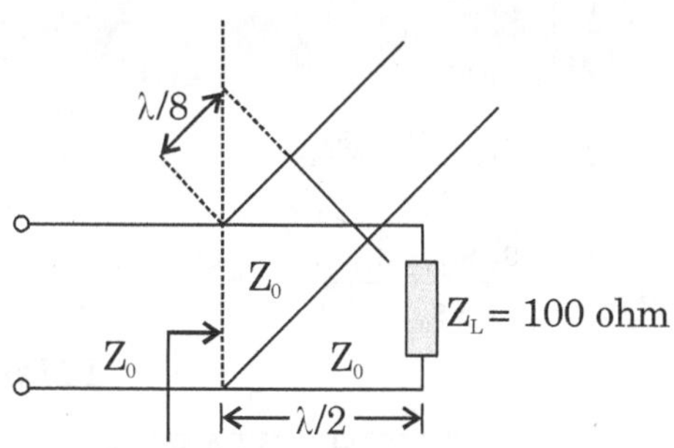

(a) (0.01 – j0.02) mho (b) (0.02 – j0.01) mho

(c) (0.04 – j0.02) mho (d) (0.02 + j0) mho

[2003 : 2 Marks]

54. A load of $50\ \Omega$ is connected in shunt in a 2-wire transmission line of $Z_0 = 50\ \Omega$ as shown in the figure. The 2-port scattering parameter matrix (S-matrix) of the shunt element is

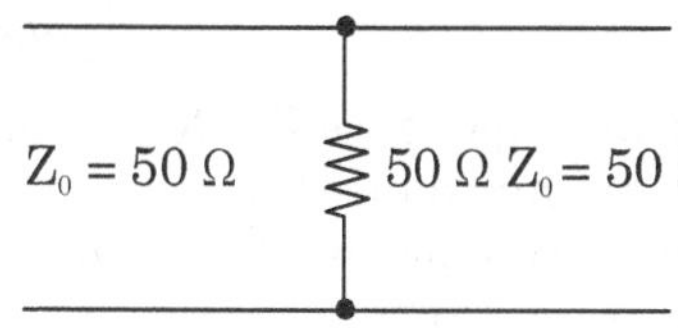

(a) $\begin{bmatrix} -\frac{1}{2} & \frac{1}{2} \\ \frac{1}{2} & -\frac{1}{2} \end{bmatrix}$ (b) $\begin{bmatrix} 0 & 1 \\ 1 & 0 \end{bmatrix}$

(c) $\begin{bmatrix} -\frac{1}{3} & \frac{2}{3} \\ \frac{2}{3} & -\frac{1}{3} \end{bmatrix}$ (d) $\begin{bmatrix} \frac{1}{4} & -\frac{3}{4} \\ -\frac{3}{4} & \frac{1}{4} \end{bmatrix}$

[2007 : 2 Marks]

55. If the scattering matrix [S] of a two port network is

$$[S] = \begin{bmatrix} 0.2\angle 0° & 0.9\angle 90° \\ 0.9\angle 90° & 0.1\angle 90° \end{bmatrix},$$

then the network is

(a) lossless and reciprocal
(b) lossless but not reciprocal
(c) not lossless but reciprocal
(d) neither lossless nor reciprocal

[2010 : 1 Mark]

56. A two-port network has scattering parameters given by $[S] = \begin{bmatrix} S_{11} & S_{12} \\ S_{21} & S_{22} \end{bmatrix}$. If the port-2 of the two-port is short circuited, the s_{11} parameter for the resultant one-port network is

(a) $\dfrac{s_{11} - s_{11}s_{22} + s_{12}s_{21}}{1 + s_{22}}$

(b) $\left.\dfrac{b_1}{a_1}\right|_{a_2 = 0}$

(c) $\dfrac{s_{11} + s_{11}s_{22} + s_{12}s_{21}}{1 - s_{22}}$

(d) $\dfrac{s_{11} - s_{11}s_{22} + s_{12}s_{21}}{1 - s_{22}}$

[2014 : 1 Mark, Set-1]

SMITH CHARTS

57. In an impedance Smith chart, a clockwise movement along a constant resistance circle gives rise to

(a) a decrease in the value of reactance
(b) an increase in the value of reactance
(c) no change in the reactance value
(d) no change in the impedance value

[2002 : 1 Mark]

58. Consider an impedance Z = R + jX marked with point P in an impedance Smith chart as shown in the figure. The movement from point P along a constant resistance circle in the clockwise direction by an angle 45° is equivalent to

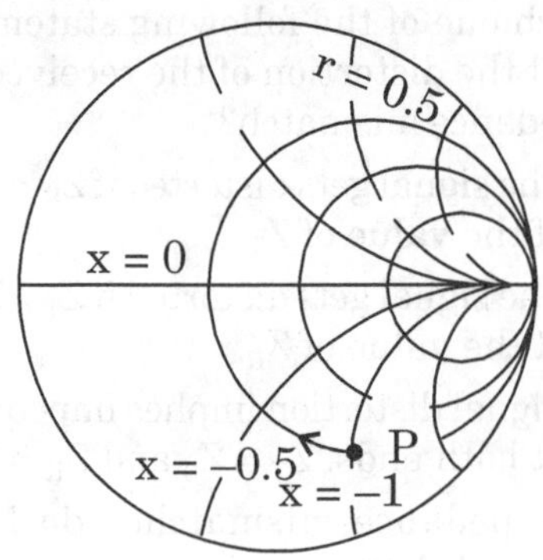

(a) adding an inductance in series with Z
(b) adding a capacitance in series with Z
(c) adding an inductance in shunt across Z
(d) adding a capacitance in shunt across Z

[2004 : 2 Marks]

59. Many circles are drawn in a Smith chart used for transmission line calculations. The circles shown in the figure represent

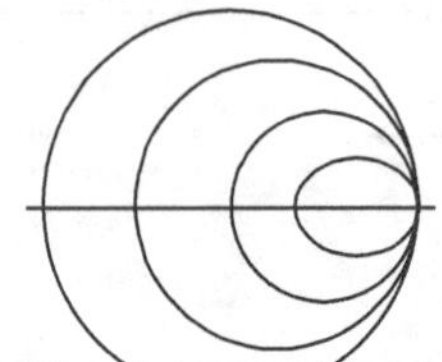

(a) unit circles
(b) constant resistance circles
(c) constant reactance circles
(d) constant reflection coefficient circles

[2005 : 2 Marks]

60. The points P, Q and R shows on the Smith chart (normalized impedance chart) in the following figure represent :

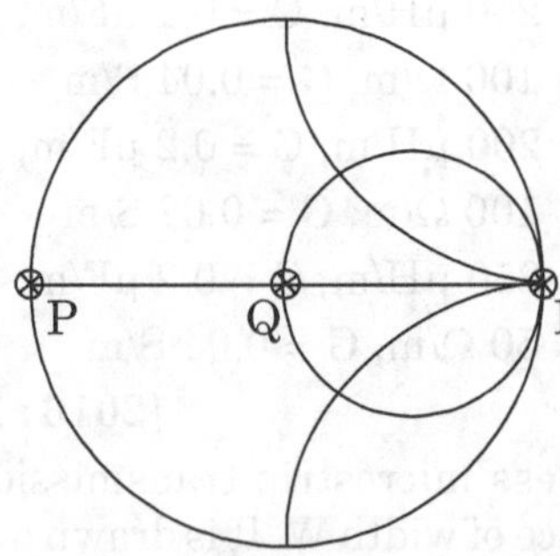

(a) P: Open Circuit, Q: Short Circuit, R: Matched Load
(b) P: Open Circuit, Q: Matched Load, R: Short Circuit
(c) P: Short Circuit, Q: Matched Load, R: Open Circuit
(d) P: Short Circuit, Q: Open Circuit, R: Matched Load

[2018 : 1 Mark]

ANSWERS

1. (c) **2.** (d) **3.** (c) **4.** (256) **5.** (b) **6.** (c) **7.** (c) **8.** (b) **9.** (d) **10.** (a)
11. (d) **12.** (d) **13.** (c) **14.** (a) **15.** (d) **16.** (b) **17.** (d) **18.** (*a*) **19.** (*b*) **20.** (*d*)
21. (30) **22.** (0.1) **23.** (49.83) **24.** (b) **25.** (b) **26.** (a,b) **27.** (c) **28.** (26) **29.** (a) **30.** (b)
31. (a) **32.** (b) **33.** (b) **34.** (d) **35.** (d) **36.** (d) **37.** (d) **38.** (d) **39.** (b) **40.** (c)
41. (*a*) **42.** (*b*) **43.** (b) **44.** (39 to 41) **45.** (70.7Ω) **46.** (33.33Ω) **47.** (*c*) **48.** (*d*)
49. (b) **50.** (b) **51.** (0.868) **52.** (0.001) **53.** (a) **54.** (b) **55.** (c) **56.** (*b*) **57.** (b) **58.** (a)
59. (b) **60.** (c)

EXPLANATIONS

1. Given : Characteristic impedance (Z_b) is pure real. Voltage minima point is found to be at load.

$$\text{VSWR} = 2 = \frac{1+|\Gamma|}{1-|\Gamma|}$$

$$|\Gamma| = \frac{1}{3} \Rightarrow \Gamma \pm \frac{1}{3}$$

$$\text{Reflection coeff } \Gamma = \frac{Z_L - Z_o}{Z_L + Z_o}$$

$$Z_o = \frac{1-\Gamma}{1+\Gamma} \times Z_L.$$

Because reflection coefficient is real, hence for Z_o to be pure resistive, Z_L must be pure resistive.

2. Given $Z_o = 50\ \Omega$, $Z_L = j100\ \Omega$.

$$\text{Reflection coeff, } \Gamma = \frac{Z_L - Z_o}{Z_L + Z_o} = \frac{j100-50}{j100+50} = \frac{j2-1}{j2+1}$$

$$|\Gamma| = \sqrt{\frac{4+1}{4+1}} = 1$$

$$\text{VSWR} = \frac{1+|\Gamma|}{1-|\Gamma|} = \frac{1+1}{1-1} = \infty.$$

3. Given : $l = \frac{\lambda}{4}$, $Z_L = 0\ \Omega$ (short circuit)

For a lossless transmission line, input impedance is

$$Z_{in} = Z_o\left[\frac{Z_L + jZ_o \tan\beta l}{Z_o + jZ_L \tan\beta l}\right]$$

$$\beta = \frac{2\pi}{\lambda} : \text{Bl} = \frac{2\pi}{\lambda} \times \frac{l}{4} = \frac{\pi}{2}$$

$$Z_{in} = Z_o\left[\frac{0 + jZ_o \tan\frac{\pi}{2}}{Z_o + 0}\right] = \infty.$$

4. For a lossless line connected with pure sensitive load

$$\text{VSWR} = \frac{R_L}{R_o} \text{ if } R_L > R_o$$

$$\text{VSWR} = \frac{R_o}{R_L} \text{ if } R_o > R_L. \text{ (True)}$$

$\therefore$ VSWR = 3

$\Rightarrow R_L = \text{VSWR} \times R_o = 3 \times 75 = 225\ \Omega.$

OR

$$\Rightarrow R_L = \frac{R_o}{\text{VSWR}} = \frac{75}{3} = 25\ \Omega$$

5. Adjacent volt maximum occurs at $\frac{\lambda}{2}$ distance,

i.e, $27.5 - 12.5 = \frac{\lambda}{2}$

$\Rightarrow l = 30$ cm $= 0.3$ m.

$$f = \frac{c}{\lambda} = \frac{3\times10^8}{0.3} = 1 \text{ GHz}$$

6. Electrical path length = βl.

$$v = \frac{1}{\sqrt{LC}} = \frac{1}{\sqrt{10\times10^{-6}\times40\times10^{-12}}} = 0.5\times10^8$$

$$\lambda = \frac{v}{f} = \frac{0.5\times10^8}{25\times10^6} = 2.$$

$$\beta l = \frac{2\pi}{\lambda}.l = \frac{2\pi}{2}\times0.5 = \frac{\pi}{2} \text{ radian.}$$

7. Characteristic equation of a transmission line

$$Z_0 = \sqrt{\frac{R + j\omega L}{G + j\omega C}}$$

[At microwave frequencies, $R << \omega L, G << \omega C$]

$$Z_0 \approx \sqrt{\frac{L}{C}}\left(1 + \frac{R}{2j\omega L}\right)\left(1 - \frac{G}{2j\omega C}\right)$$

$$\approx \sqrt{\frac{L}{C}}\left[1 + \frac{1}{2}\left(\frac{R}{j\omega L} - \frac{G}{j\omega C}\right)\right]$$

$$\approx \sqrt{\frac{L}{C}}\left[1 + \frac{1}{2j\omega LC}(RC - GL)\right]$$

Transmission line will be lossless when imaginary term is equal to zero.

i.e. RC = GL.

8. $\text{VSWR} = \frac{1+|\Gamma|}{1-|\Gamma|} \Rightarrow 3 = \frac{1+|\Gamma|}{1-|\Gamma|}$

$|\Gamma| = 0.5$

$\frac{\text{Reflected power, } P_r}{\text{Incident power, } P_t} = [\Gamma]^2 = 0.25$

$\Rightarrow$ 25% of total incident power is reflected.

9. $\text{VSWR} = \frac{1+|\tau_l|}{1-|\tau_l|}$

Since, $|\tau_l| \le 1$.

$\therefore \quad 0 < \text{VSWR} < \infty$

10. Moving in clock wise direction in the constant resistance circle gives rise to inductive effect, which is same as adding inductance in series. Similarly, moving is anticlockwise direction in the constant resistance circle gives rise to capacitive effect, which is same as adding capacitor in series.

11. $\frac{\eta_1}{\eta_2} = \frac{E_{min}}{E_{max}}$

$\frac{\eta_2}{\eta_0} = \frac{E_0}{5E_0}$

$\eta_2 = \frac{\eta_0}{5} = 24\pi$

12. VSWR, $\rho = \frac{1+\Gamma}{1-\Gamma}$

or, $2 = \frac{1+\Gamma}{1-\Gamma}$

or, $\Gamma = \frac{1}{3}$

Then, percentage of the power that is reflected back

$\frac{P_{ref}}{P_{inc}} = \Gamma^2 = \frac{1}{9} \times 100$

$= 11.11.$

13. From figure, $V_{max} = 4\text{V}\left(\text{at} \frac{\lambda}{4}\right)$, $V_{min} = 1\text{ V (at load)}$.

It is possible when $Z_L < Z_o$ with Z_L being purely resistive.

$\text{VSWR, S} = \frac{V_{max}}{V_{min}} = \frac{4}{1} = 4.$

$\Rightarrow S = \frac{Z_o}{Z_L} \Rightarrow Z_L = 12.5\ \Omega$

14. Reflection ceff, $\Gamma = \frac{Z_L - Z_o}{Z_L + Z_o} = \frac{12.5-50}{12.5+50} = -0.6$

15.

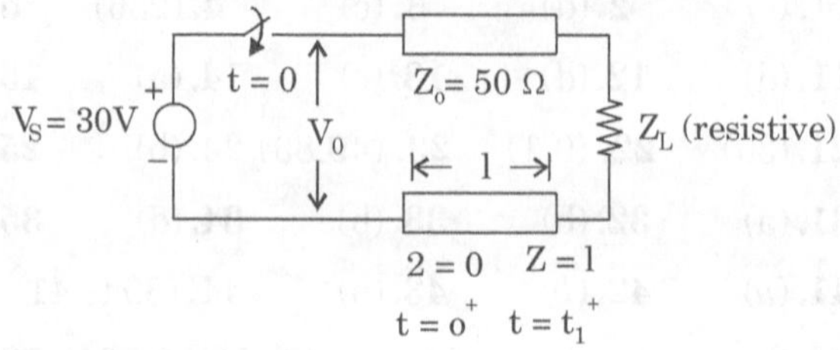

$V_S = 30\text{ V}, Z_S = 0\ \Omega,$

$Z_0 = 50\ \Omega.$

$t_1^+ = 400\ \mu s; V(l_2 t_1^+) = 40\text{V}$

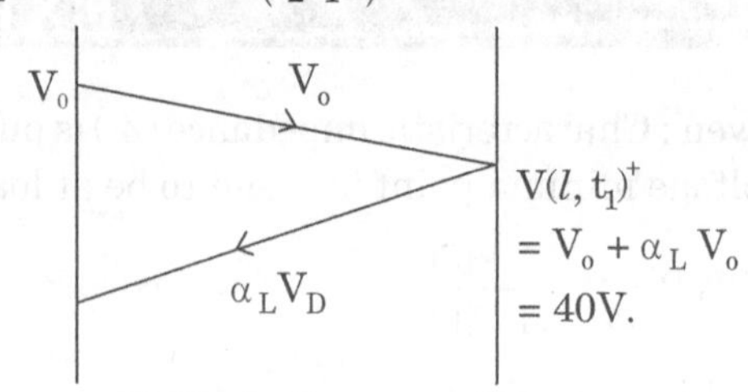

Voltage-bounce diagram

$40\ V_0 + \alpha_L\ Y_o.$

$\alpha_L = \frac{Z_L - Z_o}{Z_L + Z_o}$

At t = 0+ :

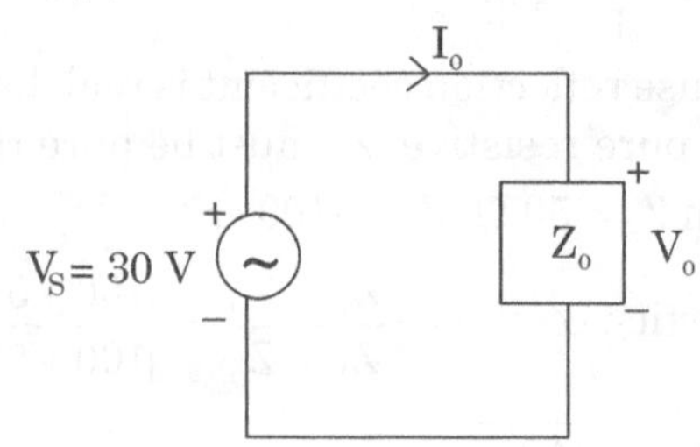

$V_o = V_S = 30\text{ V}$

$40 = V_S(1+\alpha_L) = 30(1+\alpha_L)$

$\frac{4}{3} = 1 + \frac{Z_L - 50}{Z_L + 50} \Rightarrow \frac{Z_L - 50}{Z_L + 50} = \frac{1}{3}$

$Z_L = 100\Omega.$

16. From the current bounce diagram, steady-state current through load resistance is

$I_{SS} = \frac{I_o(1-\alpha_L)}{1-\alpha_L-\alpha_S}$

$\alpha_S = \frac{Z_S - Z_o}{Z_S + Z_o} = \frac{0-50}{0+50} = -1$

$\alpha_L = \frac{1}{3}.$

$I_o = \frac{V_o}{Z_o} = \frac{30}{50} = 0.6$

$I_{SS} = \frac{0.6 \times \left(1-\frac{1}{3}\right)}{1-\frac{1}{3}\times(-1)} = \frac{0.2\times 2}{\frac{4}{3}} = 0.3\text{ A.}$

17. In optical fiber, the mode volume is given by,

$$V = \frac{nd}{\lambda}\sqrt{n_1^2 - n_2^2}$$

d = diameter of fiber, λ = wavelength

When V < 2.405, then fiber supports single mode

for V > 2.405, then fiber supports multi mode

So single mode is NOT preserved if radius is doubled and wavelength is halved became is that case

V > 2.405 and fibre will support multi mode.

18. The reflection co-efficient is $-20\log\Gamma = 20$ dB

$\Rightarrow \log\Gamma = -1$ dB; $\quad \Rightarrow \Gamma = 10^{-1} \Rightarrow \Gamma = 0.1$

Relation between Γ and VSWR is

$$S = \frac{1+|\Gamma|}{1-|\Gamma|} = \frac{1+0.1}{1-0.1} = \frac{1.1}{0.9} = 1.22$$

18. Loss tangent $\tan\delta = 0 = \dfrac{\sigma}{\omega\in}$

$$\sigma = 0$$

$G \to$ Conductivity of the dielectric material

So, $\quad \sigma = 0 = G$

20. Entropy is maximum when all symbols are equiprobable. If the probability of symbols are different then entropy is going to decrease.

21. Phase velocity,

$$v_p = \frac{1}{\sqrt{\mu\in}} = \frac{C}{\sqrt{\mu_r \in_r}} = \frac{3\times10^8}{\sqrt{2.25\times1}}$$

$v_p = 2\times10^8$ m/s ($\mu_r = 1 \Rightarrow$ Non-magnetic material)

The transient response of the circuit at the input or source end of line (Z = 0) is shown in the figure.

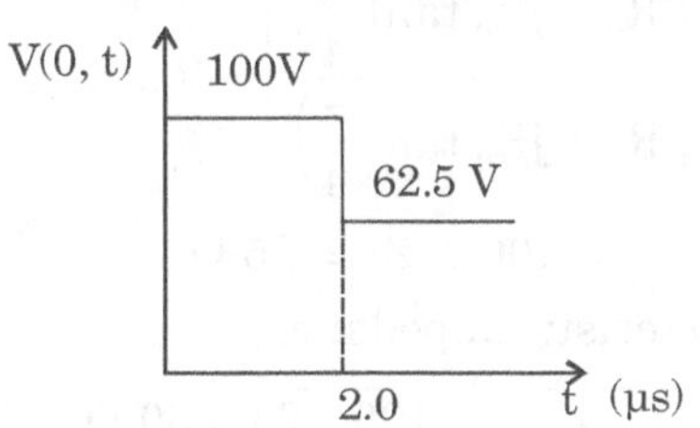

At t = 2 μs, the input voltage at transmission line becomes 62.5 V.

Time taken by wave to travel from source to load,

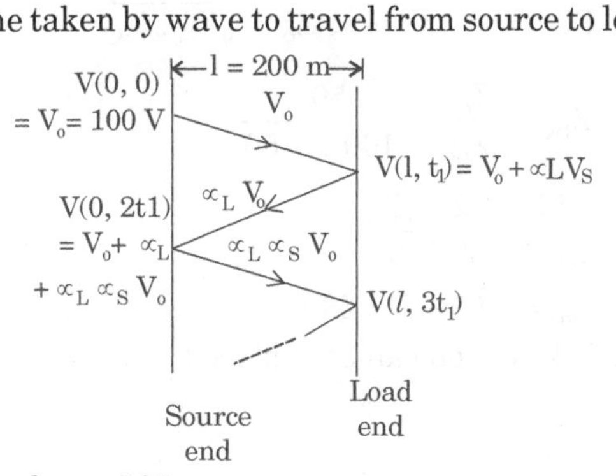

$$T = \frac{l}{v_p} = \frac{200}{2\times10^8} = 1\ \mu s.$$

From voltage bounce diagram,

$$V_o + \alpha_L V_o + \alpha_L \alpha_S V_o = 62.5$$

$$V_o = 100V$$

$$\alpha_L = \frac{R_L - R_0}{R_2 + R_0} = \frac{R_L - 50}{R_L + 50}$$

$$\propto_S = \frac{R_S - R_o}{R_S + R_o} = \frac{150-50}{150+50} = 0.5$$

$$100.(1+\alpha_L + 0.5\,\alpha_L) = 62.5$$

$$1.5\,\alpha_L = -0.375$$

$$\alpha_L = -0.25$$

$$\frac{R_L - 50}{R_L + 50} = -0.25$$

$$R_L = 30\ \Omega.$$

22. Phase velocity = 2×10^8 m/s

Input impedance at T_2

$$Z_{in_2} = -jZ_o \cot\beta l, (Z_L = \infty)$$

Load impedance at T_1,

$$Z_{L_1} = 50 \| (-jZ_o \cot\beta l)$$

$$Z_{L_1} = 50 \| (-j50 \cot\beta l)$$

From graph,

$\Gamma = 0$, at f = 0, 1, 2 GHz

$\Rightarrow \Gamma = 0$ only when $Z_{L_1} = Z_o$

$$-j50\omega t(\beta l) \| 50 = Z_o = 50$$

$$-j50\cot(\beta l) = \infty$$

$$\beta l = n\pi \Rightarrow l = \sigma\frac{\pi}{\beta}$$

$$l = n\frac{\pi}{\left(2\frac{\pi}{\lambda}\right)} = \frac{n\lambda}{2} = \frac{nv}{2f}$$

For l to be minimum, n = 1.

$$l = \frac{1}{2}\times\frac{2\times10^8}{1\times10^9} = 0.1\text{ m}.$$

23. VSWR, S = 5.8

$$= \frac{1+|\rho|}{1-|\rho|} \Rightarrow |\rho| = \frac{S-1}{S+1} = \frac{4.8}{6.8} = 0.7058$$

$$\frac{P_{reflected}}{P_{incident}} = \rho^2 = 0.4983 \text{ or } 49.83\%$$

24. Characteristic impedance of Cable is

$$Z_0 = \sqrt{\frac{\mu_0}{\epsilon_0 \epsilon_r}} \cdot \frac{1}{2\pi} 1n\left(\frac{b}{a}\right)$$

$$= \sqrt{\frac{4\pi\times10^{-7}}{\frac{1}{36\pi}\times10^{-9}\times10.89}} \cdot \frac{1}{2\pi} 1n\left(\frac{2.4}{1}\right) = 15.19\ \Omega$$

No option is correct.

25. A transmission line is said to be impedance matched, if $Z_L = Z_o$. If $Z_L \neq Z_o$, reflection occurs and standing waves are formed, as a result maximum power cannot be transferred.
Impedance matching can be achieved by

(i) connecting a $\frac{\lambda}{4}$ (quarter wave) line between mismatched line and load, or

(ii) connecting a stub (either short-circuit or open-circuit) across mismatched line.

However, $\frac{\lambda}{2}$ (half wave) line cannot be used for impedance matching.

26. Given: Characteristic impedance, $Z_o = R_o$.

Char. impedance, $Z_o = \sqrt{\frac{R + j\omega L}{G + j\omega C}}$

(∴ Lossless line, R = ω = 0)

$$Z_o\big|_{lossless} = \sqrt{\frac{L}{C}}.$$

For a lossless line, Z_o is purely resistive.
For a distortionless line,

$$\frac{R}{G} = \frac{L}{C} \Rightarrow \frac{L}{R} = \frac{C}{G}$$

$$Z_o\big|_{distortionless} = \sqrt{\frac{R}{G}\frac{\left(1 + j\omega\frac{L}{R}\right)}{\left(1 + j\omega\frac{C}{a}\right)}}$$

$$= \sqrt{\frac{L}{C}} = \sqrt{\frac{R}{G}} \text{ (Real).}$$

Therefore, a lossless line is distortion less, but a distortion less line not always lossless.

27. Reflection coeficient, $\rho = \frac{Z_L - Z_0}{Z_L + Z_0} = \frac{j50 - 50}{j50 + 50}$

Hence $|\rho| = \frac{Z_L - Z_0}{Z_L + Z_0} = \frac{\sqrt{50^2 + 50^2}}{\sqrt{50^2 + 50^2}} = 1,$

and the VSWR $\frac{1 + |\rho|}{1 - |\rho|} = \frac{2}{0} = \infty$

28. For quarter wave line transformer,

$$Z_{oq} = \sqrt{Z_{in}.Z_L} = \sqrt{50 \times 200} = 100\ \Omega.$$

29. A short circuited line of length λ/4 shows infinite impedance at the other end, hence the current is zero.

30. $$\omega = \frac{1}{\sqrt{LC}},$$

Now $Z_o = \sqrt{\frac{L}{C}}$

$$\Rightarrow \sqrt{L} = Z_o\sqrt{C}$$

$$\therefore \quad \omega = \frac{1}{Z_o\sqrt{C}\sqrt{C}} = \frac{1}{Z_oC}$$

31. Given: $R_L = 100\ \Omega$, $R_o = 50\ \Omega$.

$$SWR = \frac{R_L}{R_o} = 2$$

$$Z_{min} = \frac{Z_o}{SWR} = \frac{50}{2} = 25\Omega.$$

32. For a lossless transmission line, the input impedance is

$$Z_{in} = Z_o\left[\frac{Z_L + jZ_o\tan(\beta l)}{Z_L + jZ_L\tan(\beta l)}\right]; \beta = \frac{2\pi}{\lambda}$$

For $\frac{\lambda}{4}$ line,

$$Z_{in_1} = \frac{Z_o^2}{Z_L} = \frac{R_o^2}{\frac{R_o}{2}} = 2R_o.$$

For $\frac{\lambda}{2}$ line, $\beta l = \frac{2\pi}{\lambda} \times \frac{\lambda}{2} = \pi$

$$Z_{in_2} = R_o\left[\frac{2R_o + j.R_o\tan(\pi)}{R_o + j(2R_o)\tan(\pi)}\right] = 2R_o$$

$$Z_{in} = Z_{in_1}\|Z_{in_2} = \frac{2R_o \times 2R_o}{2R_o + 2R_o} = R_o.$$

For $\frac{\lambda}{8}$ line, $\beta l = \frac{2\pi}{\lambda} \times \frac{\lambda}{8} = \frac{\pi}{4}$

$$Z_{in} = \left[\frac{R_o + jR_o\tan\left(\frac{\pi}{4}\right)}{R_o + jR_o\tan\left(\frac{\pi}{4}\right)}\right] = R_o.$$

33. Given $Z_{oc} = 100\Omega$, $Z_{sc} = 25\ \Omega$.
characteristic impedance,

$$Z_o = \sqrt{Z_{oc}.Z_{sc}} = \sqrt{100 \times 25} = 50\ \Omega$$

34. Given $Z_o = 50\ \Omega$, $Z_{oC} = 100 + j50\ \Omega$.

Char. impedance, $Z_o = \sqrt{Z_{oc}.Z_{SC}}$

$$Z_{SC} = \frac{Z_o^2}{Z_{oc}} = \frac{50^2}{100 + j150}$$

$$\frac{50}{2 + j3} \times \frac{2 - j3}{2 - j3}$$

$$Z_{SC} = 7.69 - j11.54\ \Omega.$$

35. λ/4 section can be replaced by a

$$Z_L' = \frac{Z_0^2}{Z_L}$$

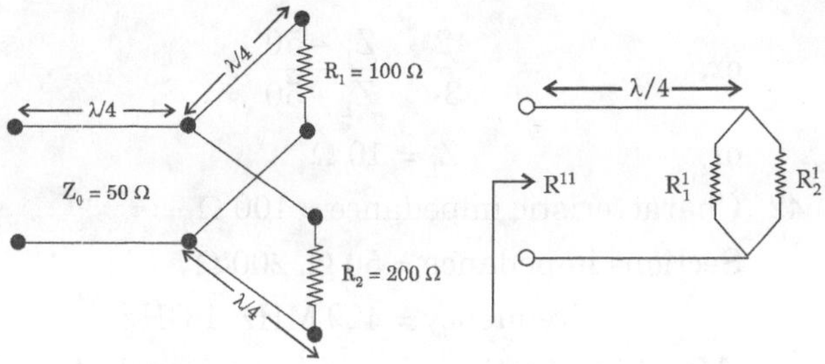

$$\therefore \quad R_1' = \frac{Z_0^2}{R_1} = \frac{50 \times 50}{100} = 25\ \Omega$$

$$\therefore \quad R_2' = \frac{50 \times 50}{200} = \frac{25}{2}\ \Omega$$

$$\therefore \quad R' = R_1' \parallel R_2' = 25 \parallel \frac{25}{2}$$

$$R' = \frac{25}{3}\ \Omega$$

$$\therefore \quad R'' = \frac{Z_0^2}{R_1} = \frac{50 \times 50 \times 3}{25} = 300\ \Omega$$

Voltage reflection coefficient,

$$\Gamma = \frac{Z_L - Z_0}{Z_L + Z_0} = \frac{300 - 50}{300 + 50} = \frac{5}{7}$$

36. For a lossless line,

$$Z_{in} = Z_o \left[\frac{Z_L + jZ_o \tan(\beta l)}{Z_o + jZ_L \tan(\beta l)}\right]$$

$$= 75\left[\frac{0 + j75\tan\left(\frac{\pi}{5}\right)}{75 + 0}\right]$$

$$= j75\tan\left(\frac{\pi}{5}\right) = j54.5\ \Omega \text{ (Inductive)}$$

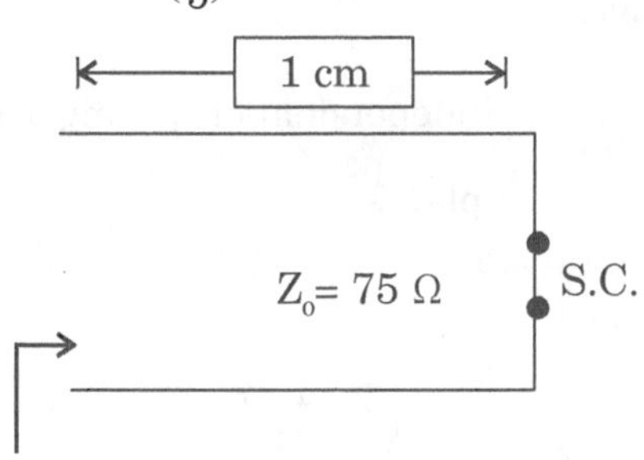

$$Z_{in}; \lambda = \frac{c}{f} = \frac{3 \times 10^8}{3 \times 10^9} = 0.1\ \text{m}$$

$$\beta l = \frac{2\pi}{10} \times 1 = \frac{\pi}{5}\ \text{rad}$$

37. For a quater wave transformer input impedance is

$$Z_{in} = \frac{Z_o^{\,2}}{Z_L}.$$

$$\Rightarrow Z_{in_1} = \frac{Z_o^{\,2}}{Z_L} = \frac{(100)^2}{50} = 200\ \Omega$$

$$Z_{in_2} = \frac{(100)^2}{50} = 200\Omega$$

For the front Z_o = 50 Ω line,

$$Z_L = Z_{in_1} \| Z_{in_2} = \frac{200 \times 200}{200 + 200} = 100\ \Omega$$

$$Z_{in} = \frac{(50)^2}{100} = 25\ \Omega$$

38. For a distortionless line, characteristic impedance is

$$Z_o = \sqrt{\frac{L}{C}} = \sqrt{\frac{R}{G}}.$$

Attenuation constant,

$$\alpha = \sqrt{RG} = \frac{R}{Z_o} = \frac{0.1}{50} = 0.002\ \text{Np/m}$$

39.

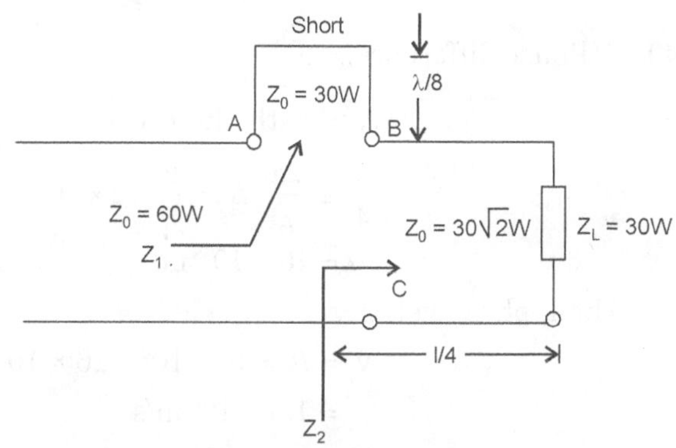

In input impedance is given by,

$$Z_{in} = Z_0\left(\frac{Z_L + jZ_0 + \tan\beta l}{Z_0 + jZ_L \tan\beta l}\right)$$

Where Z_0 = characteristics impedance,

Z_L = load impedance

l = length of the line,

$$\beta = \frac{2\pi}{\lambda}$$

then input impedance looking into terminal AB,

$$Z_1 = 30\left[\frac{0 + 30j\tan\left(\frac{2\pi}{\lambda}\cdot\frac{\lambda}{8}\right)}{30 + 0}\right] = 30j$$

Input impedance looking into terminal BC,

$$Z_2 = 30\sqrt{2}\left[\frac{30 + j30\sqrt{2}\tan\left(\frac{2\pi}{\lambda}\cdot\frac{\lambda}{4}\right)}{30\sqrt{2} + 30j\tan\left(\frac{2\pi}{\lambda}\cdot\frac{\lambda}{4}\right)}\right] = 60$$

$$= 30\sqrt{2}\left[\frac{\frac{30}{\tan\frac{\pi}{2}} + j30\sqrt{2}}{\frac{30\sqrt{2}}{\tan\frac{\pi}{2}} + 30j}\right] = 60$$

Total load impedance looking into terminal AC,

$$Z_L = Z_1 + Z_2 = 60 + 30j$$

Reflection coefficient,

$$T_e = \frac{Z_L - Z_0}{Z_L + Z_0}$$

$$= \left[\frac{60+30j-60}{60+30j+60}\right]$$

$$= \left[\frac{j}{4+j}\right] = \frac{1}{\sqrt{17}}$$

$$\text{Hence, VSWR, of} = \frac{1+|j_e|}{1-|j_e|} = \frac{1+\frac{1}{\sqrt{17}}}{1-\frac{1}{\sqrt{17}}} = 1.64$$

40. $\phi\text{(Phase difference)} = \frac{2\pi}{\lambda}\cdot\Delta x$

$$\Delta x = \text{path difference}$$

or, $$\frac{\pi}{4} = \frac{2\pi}{\lambda}\cdot\Delta x = \frac{2\pi}{\lambda}[2\times10^{-3}]$$

or, $$\lambda = 16\times10^{-3}\,\text{m}$$

Then, phase velocity

$$v_p = f\lambda = 10\times10^9\times16\times10^{-3}$$

$$= 1.6\times10^8\ \text{m/s}$$

41. The reflection coefficient at the receiving end,

$$J_l = \frac{Z_L - Z_0}{Z_L + Z_0},\ J_l = |J_l|\cdot e^{J\phi_l}$$

Q_l is the phase-angle between incident & reflected wave.

and standing-wave ratio, $P = \frac{1+|J_l|}{1-|J_l|}$

or $$5 = \frac{1+|J_l|}{1-|J_l|}, |J_x| = \frac{2}{3}$$

Now, voltage expression at distance d from the load,

$$V(d) = V_T e^{\beta l}\left[1+|J_l|\,e^{j(\phi_l - 2\beta d)}\right]$$

Given, $V(d)$ is maximum at $d = \frac{\lambda}{4}$

then, $$\phi_l - 2\beta\cdot\frac{\lambda}{4} = 0 \text{ for voltage maximum}$$

or, $$\phi_l = \frac{\beta\lambda}{2} = \frac{2\pi}{\lambda}\cdot\frac{\lambda}{2} = \pi$$

then, $$J_l = |J_l|\cdot e^{j\pi} = \frac{-2}{3}$$

and, $$\frac{-2}{3} = \frac{Z_L - Z_0}{Z_L + Z_0}$$

or, $$\frac{-2}{3} = \frac{Z_L - 50}{Z_L + 50}$$

or, $$Z_L = 10\ \Omega$$

42. Characteristic impedance = 100 Ω

Sections impedance = 50 Ω, 200 Ω

Frequency = 429 MHz, 1 GHz

(Matching section should have length l = odd multiple of λ/4 where λ is operating wavelength)

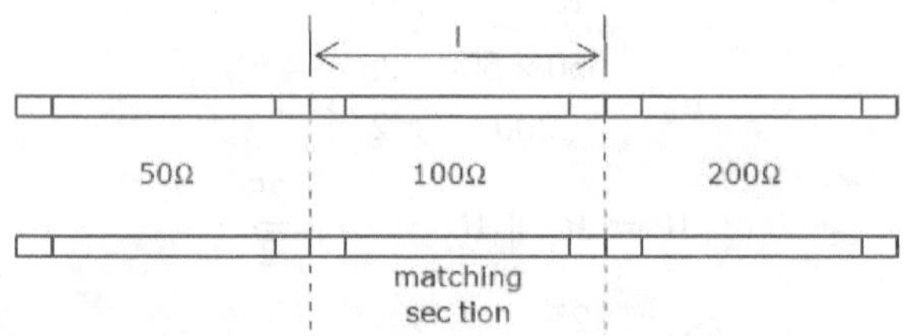

For 429 MHz, $I_1 = \frac{\lambda_1}{4} = \frac{C}{4f_1} = 0.175$ m;

For 1 GHz, $I_2 = \frac{\lambda_2}{4} = \frac{C}{4f_2} = 0.075$ m

Length I should be integral multiples of both I_1 and I_2.

∴ I = multiple of LCM of I_1 and I_2

= multiple of 0.525 m

Hence, 1.05 m is the appropriate solution

43. $Z_0 = \frac{276}{\sqrt{\epsilon_r}}\log\left(\frac{d}{r}\right)$

d → distance between the two plates

so, Z_0 – changes, if the spacing between the plates changes.

$v = \frac{1}{\sqrt{LC}}$ → independent of spacing between the plates

44. Given, $\ell = \lambda/8$

$Z_0 = 50\Omega$

$$Z_{in}\left(\ell = \lambda/8\right) = Z_0\left[\frac{Z_L + JZ_0}{Z_0 + KZ_L}\right]$$

$$Z_{in} = 50\left[\frac{Z_L + J50}{50 + JZ_L}\right] = 50\left[\frac{Z_L + J50}{50 + JZ_L}\times\frac{50 - JZ_L}{50 - JZ_L}\right]$$

$$Z_{in} = 50\left[\frac{50Z_L + 50Z_L + J(50^2 - Z_L^2)}{50^2 + Z_L^2}\right]$$

Given, Z_{in} → Real

So, $$I_{mg}(Z_{in}) = 0$$

$$50^2 - Z_L^2 = 0$$

$$Z_L^2 = 50^2$$

$$R^2 + X^2 = 50^2$$
$$R^2 = 50^2 - X^2 = 50^2 - 30^2$$
$$R = 40\ \Omega$$

45. Here impedance is matched by using QWT $\left(\frac{\lambda}{4}\right)$

$$\therefore \quad Z_0' = \sqrt{Z_L Z_{in}}$$
$$= \sqrt{100 \times 50}$$
$$= 50\sqrt{2}$$
$$Z_0' = 70.7\ \Omega$$

46. Here $\ell = \frac{\lambda}{2}$

$$Z_{in}\left(\ell = \frac{\lambda}{2}\right) = Z_L = 50\ \Omega$$
$$\therefore \quad Z_{in} = (100 \| 50)$$
$$= \frac{100}{3} = 33.33\ \Omega$$

47. Signal distortion implies impedance mismatch at both ends. i.e.,

$$Z_T \neq Z_0$$
$$Z_R \neq Z_0$$

48. $Z_{in} = JZ_0 \tan \beta l$

$$\beta l = \frac{2\pi}{\lambda}.l \qquad \lambda = \frac{3 \times 10^8}{37.5 \times 10^6}$$
$$= \frac{2\pi}{8}(3) \qquad = 8 \text{ unit}$$
$$= \frac{3\pi}{4}$$

Short circuited line

$$0 < \beta l < \frac{\pi}{2} \rightarrow \text{Inductor}$$
$$\frac{\pi}{2} < \beta l < \pi \rightarrow \text{Capacitor}$$

49. Propagation constant, $\gamma = 2 + j5\ m^{-1}$.
char. inpedance, $Z_o = 50 + j0\ \Omega$, $\omega = 10^6$ rad/s.

$$\gamma = \sqrt{(R + j\omega L)(G + j\omega C)};$$
$$Z_0 = \sqrt{\frac{R + j\omega L}{G + j\omega C}}$$

$\gamma Z_0 = R + j\omega L = (2 + j5).(50 + j0) = 100 + j250$
Hence, $R = 100\ \Omega/m$, $\omega L = 250$

$$L = \frac{250}{\omega} = \frac{250}{10^6} = 250\ \mu H/m$$
$$\frac{\gamma}{Z_0} = \frac{2 + j5}{50} \Rightarrow G + j\omega C = 0.04 + j0.1$$
$$\omega C = 0.1 \Rightarrow C = \frac{0.1}{10^6} = 0.1\ \mu F/m;$$
$$G = 0.04\ S/m$$

50. Char. impedance of transmission lines is

$$Z_o = \sqrt{\frac{L}{C}}.$$
$$\Rightarrow C = \frac{\in A}{d}; d = t;$$
$$A \cong W \text{ and } \in = \in_{eff}.$$
$$C = \frac{\in_{eff} W}{t}$$
$$Z_o = \sqrt{\frac{L}{\in_{eff}\frac{W}{t}}} = \sqrt{\frac{Lt}{\in_{eff} W}} \qquad (\because \in_{eff} > \in_o \in_r)$$
$$Z_o < \sqrt{\frac{Lt}{\in_o \in_r W}}$$

51. $V(l) = e^{\gamma l + j\omega l} = V_o.e^{-\gamma l}.e^{j\omega l}$

$$= V_o e^{-\alpha l}.e^{-j\beta l}.e^{j\omega l}$$
$$\text{Attenuation} = \frac{|\text{Input}|}{|\text{Output}|} = \frac{|V_o(0)|}{|V_o(l)|}$$
$$\text{Attenuation per meter} = \frac{|V_o|}{|V_o(lm)|} = e^{\alpha}$$

Attenuation in dB/m $= 20 \log_{10}(e^{\alpha})$
$= 20 \times 0.1. \log_{40} e$
$= 0.868$ dB/m.

52. For distortion less transmission line,

$$\frac{L}{R} = \frac{C}{G}.$$

Propagation constant,

$$\gamma = \alpha + j\beta = \sqrt{(R + j\omega L)(G + j\omega C)}$$
$$= \sqrt{RG}\left(1 + j\omega\frac{L}{R}\right)$$

Attnuation constant, $\alpha = \sqrt{RG}$
Characteristic impediance

$$Z_o = \sqrt{\frac{(R + j\omega L)}{(G + j\omega C)}} = \sqrt{\frac{R}{G}}$$
$$\sqrt{a} = \frac{\sqrt{R}}{Z_o}$$
$$\alpha = \sqrt{R}.\frac{\sqrt{R}}{Z_o} = \frac{R}{Z_o}$$
$$\alpha = \frac{0.05}{50} = \frac{0.01}{10} = 0.001\ Np/m.$$

53. $\beta = \frac{2\pi}{\lambda}$

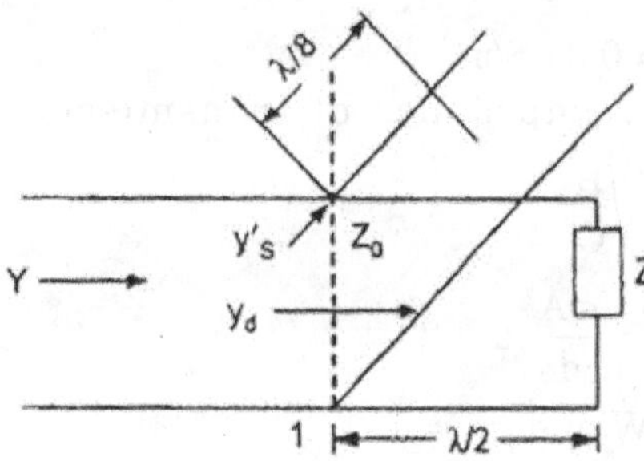

For y_d, $\beta l = \left(\frac{2\pi}{\lambda}\right)\frac{\lambda}{2} = \pi$

$$y_d = Z_0\left(\frac{Z_0 + jZ_L \tan \beta d}{Z_L + iZ_0 \tan \beta d}\right)$$

$$= \frac{50 + j100 \tan \pi}{50(100 + j50 \tan \pi)} = 0.01$$

For y_s, $\beta l = \frac{2\pi}{\lambda}\frac{\lambda}{8} = \frac{\pi}{4}$, $Z_L \to \infty$

then $y_s = \frac{Z_0 + jZ_L \tan \pi/4}{Z_0(Z_L + jZ_0 \tan \pi/4)} = \frac{j}{Z_0} = 0.021$

Thus, $Y = y_d - y_s = (0.01 - 0.02\,j)$ mho

54. S – matrix for 2 – port network

$$\begin{bmatrix} b_1 \\ b_2 \end{bmatrix} = \begin{bmatrix} S_{11} & S_{12} \\ S_{21} & S_{22} \end{bmatrix}\begin{bmatrix} a_1 \\ a_2 \end{bmatrix}$$

$$\overline{B} = \overline{S}\cdot\overline{A}$$

where, $\overline{S}$ = scattering matrix

$\overline{B}$ = scattered case matrix

$\overline{A}$ = incident case matrix

There is a shunt (R = 50 Ω) is equal to the characteristic impedance ; so perfectly power condition occurs at both parts

$\therefore\ S_{11} = S_{22} = 0$ (no reflection)

and $S_{12} = S_{21} = 1$ (complete power transfer)

$$\therefore\ \overline{S} = \begin{bmatrix} 0 & 1 \\ 1 & 0 \end{bmatrix}$$

55. Given : $[S] = \begin{bmatrix} 0.2\angle 0° & 0.9\angle 90° \\ 0.9\angle 90° & 0.1\angle 90° \end{bmatrix} = \begin{bmatrix} S_{11} & S_{12} \\ S_{21} & S_{22} \end{bmatrix}$

For lossless reciprocal networks,

$$|S_{11}| = |S_{22}|$$

and $|S_{11}|^2 + |S_{22}|^2 = 1$

For reciprocal, $S_{12} = S_{22}$, so system is reciprocal.

56.

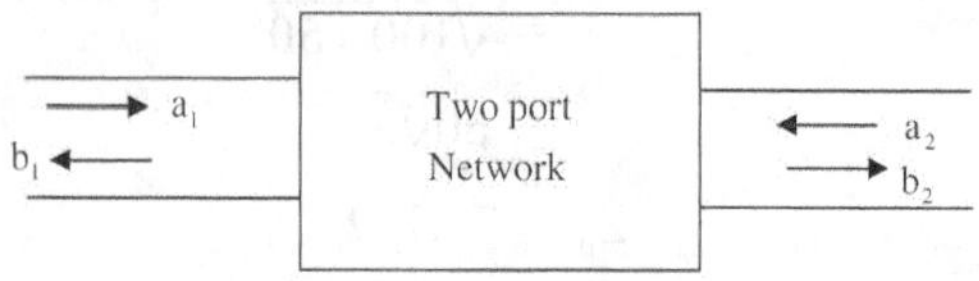

$b_1 = s_{11}a_1 + s_{12}a_2$

$b_2 = s_{21}a_1 + s_{22}a_2$

$$\begin{bmatrix} b_1 \\ b_2 \end{bmatrix} = \begin{bmatrix} s_{11} & s_{12} \\ s_{21} & s_{22} \end{bmatrix}\begin{bmatrix} a_1 \\ a_2 \end{bmatrix};\ s_1 = \left.\frac{b_1}{a_1}\right|_{a_2=0}$$

By verification answer (b) satisfies.

57. Moving clockwise in a constant resistance circle gives increases in the value of reactance.

58. Inductance is added in series when moving in clockwise direction.

59. Here, $\mu = 1.5$, $\therefore f = 10^{14}$ Hz

$$\mu = \frac{C_0}{C_\mu}$$

Now, $f\gamma = C\mu$

$$\Rightarrow\ f\gamma = \frac{C_0}{\mu}$$

$$\therefore\ \lambda = \frac{C_0}{\mu f}$$

$$= \frac{3\times 10^8}{1.5\times 10^{14}} = 2\times 10^{-6} = 2\ \mu\text{m}$$

60. For short-circuit : r = x = 0 ⇒ Point "P"

For open-circuit : r = x = ∞ ⇒ Point "R"

For matched load : r = 1, x = 0 ⇒ Point "Q"

5 CHAPTER

Wave Guides

Analysis of Previous GATE Papers												
	Year → Topics ↓		2019	2018	2017 Set 1	2017 Set 2	2016 Set 1	2016 Set 2	2016 Set 3	2015 Set 1	2015 Set 2	2015 Set 3
MODES	1 Mark	MCQ Type										
		Numerical Type						1				
	2 Marks	MCQ Type				1			1	1		
		Numerical Type	1		1	1			1			
	Total		2		2	4		1		2		
BOUNDARY CONDI TIONS	1 Mark	MCQ Type										
		Numerical Type										
	2 Marks	MCQ Type										
		Numerical Type										
	Total											
CUT-OFF FREQUENCIES	1 Mark	MCQ Type										
		Numerical Type										
	2 Marks	MCQ Type										
		Numerical Type		1							1	
	Total			2							2	
DISPERSION RELATIONS	1 Mark	MCQ Type										
		Numerical Type										
	2 Marks	MCQ Type										
		Numerical Type										
	Total											

MODES

1. The electric and magnetic fields for a TEM wave of frequency 14 GHz in a homogeneous medium of relative permittivity e^ and relative permeability $\mu_r = 1$ are given by

$$\vec{E} = E_p e^{j(\omega t - 280 g p y)} \hat{u}_z \text{ V/m}$$

$$\vec{H} = 3 e^{j(\omega t - 280 g p y)} \hat{u}_x \text{ V/m}$$

Assuming the speed of light in free space to be 3×10^8 m/s, intrinsic impedance of free space to be 120π, the relative permittivity ε_r of the medium and the electric field amplitude E_p are

(a) $\varepsilon_r = 3, E_p = 120\pi$ (b) $\varepsilon_r = 3, E_p = 360\pi$
(c) $\varepsilon_r = 9, E_p = 360\pi$ (d) $\varepsilon_r = 9, E_p = 120\pi$

[1990 : 2 Marks]

2. Choose the correct statements

For a wave propagating in an air filled rectangular waveguide

(a) guided wavelength is never less than the free space wavelength.
(b) wave impedance is never less than the free space impedance.
(c) phase velocity is never less than the free space velocity.
(d) TEM mode is possible if the dimensions of the wave guide are properly chosen.

[1990 : 2 Marks]

3. The interior of a $\frac{20}{3}$ cm $\times \frac{20}{4}$ cm rectangular waveguide is completely filled with a dielectric of $\epsilon_r = 4$. Waves of free space wave lengths shorter than ____ can be propagated in the TE_{11} mode.

[1994 : 1 Mark]

4. Assuming perfect conductors of a transmission line, pure TEM propagation is NOT possible in

(a) coaxial cable
(b) air-filled cyclindrical waveguide
(c) parallel twin-wire line in air
(d) semi-infinite parallel plate waveguide

[1999 : 1 Mark]

5. The dominant mode in a rectangular waveguide is TE_{10}, because this mode has

(a) no attenuation
(b) no cut-off
(c) no magnetic field component
(d) the highest cut-off wavelength

[2001 : 1 Mark]

6. The phase velocity for the TE_{10} mode in an air-filled rectangular waveguide is

(a) less than c (b) equal to c
(c) greater than c (d) None of the above

Note: c is the velocity of plane waves in free space)

[2002 : 1 Mark]

7. The phase velocity of an electromagnetic wave propagating in a hollow metallic rectangular waveguide in the TE_{10} mode is

(a) equal to its group velocity
(b) less than the velocity of light in free space
(c) equal to the velocity of light n free space
(d) greater than the velocity of light in free space

[2004 : 1 Mark]

8. Which one of the following does represent the electric field lines for the TE_{02} mode in the cross-section of a hollow rectangular metallic waveguide?

(a) 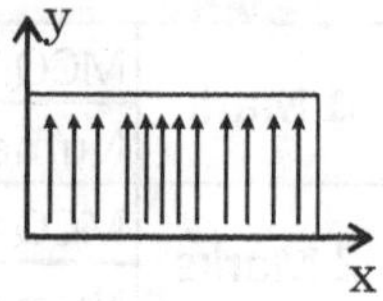

(b) 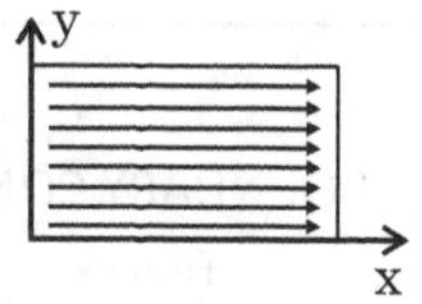

(c) 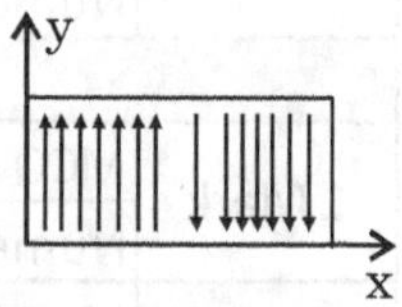

(d) 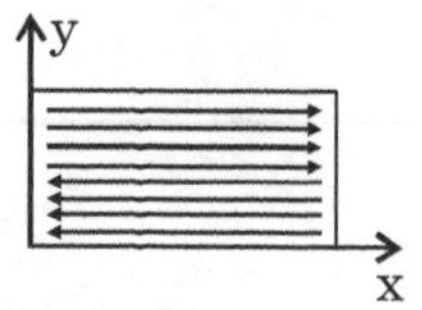

9. A rectangular waveguide having TE_{10} mode as dominant mode is having a cutoff frequency of 18-GHz for the TE_{30} mode. The inner broad-wall dimension of the rectangular waveguide is

(a) $\frac{5}{3}$ cm (b) 5 cm
(c) $\frac{5}{2}$ cm (d) 10 cm

[2006 : 2 Marks]

10. An air-filled rectangular waveguide has inner dimensions of 3 cm × 2 cm. The wave impedance of the TE_{20} mode of propagation in the waveguide at a frequency of 30 GHz is (free space impedance $\eta_0 = 377\ \Omega$)

(a) 308 Ω (b) 355 Ω
(c) 400 Ω (d) 461 Ω

[2007 : 2 Marks]

11. The $\vec{E}$ field in a rectangular waveguide of inner dimensions a × b is given by

$$\vec{E} = \frac{\omega\mu}{h^2}\left(\frac{\pi}{a}\right) H_0 \sin\left(\frac{2\pi x}{a}\right) \sin(\omega t - \beta z)\hat{y}$$

Where H_0 is a constant, and a and b are the dimensions along the x-axis and the y-axis respectively. The mode of propagation in the waveguide is

(a) TE_{20} (b) TM_{11}
(c) TM_{20} (d) TE_{10}

[2007 : 2 Marks]

12. A rectangular waveguide of internal dimensions (a = 4 cm and b = 3 cm) is to be operated in TE_{11} mode. The minimum operating frequency is

(a) 6.25 GHz (b) 6.0 GHz
(c) 5.0 GHz (d) 3.75 GHz

[2008 : 2 Marks]

13. The modes in a rectangular waveguide are denoted by TE_{mn}/TM_{mn} where m and n are the eigen numbers along the larger and smaller dimensions of the waveguide respectively. Which one of the following statements is TRUE?

(a) The TM_{10} mode of the waveguide does not exist
(b) The TE_{10} mode of the waveguide does not exist
(c) The TM_{10} and the TE_{10} modes both exist and have the same cut-off frequencies
(d) The TM_{10} and the TM_{01} modes both exist and have the same cut-off frequencies

[2011 : 1 Mark]

14. Which one of the following field patterns represents a TEM wave travelling in the positive x direction?

(a) E = +8y, H = –4z (b) E = –2y, H = –3z
(c) E = +2z, H = +2y (d) E = –3y, H = +4z

[2014 : 1 Mark, Set-2]

15. The longitudinal component of the magnetic field inside an air-filled rectangular waveguide made of a perfect electric conductor is given by the following expression

$H_z(x, y, z, t) = 0.1 \cos(25\pi x) \cos(30.3\pi y) \cos(12\pi \times 10^9 t - \beta z)$ (A/m)_______?

(a) TM_{12} (b) TM_{21}
(c) TE_{21} (d) TE_{12}

[2015 : 2 Marks, Set-1]

16. Light from free space is incident at an angle θ_i to the normal of the facet of a step-index large core optical fibre. The core and cladding refractive indices are n_1 = 1.5 and n_2 = 1.4, respectively.

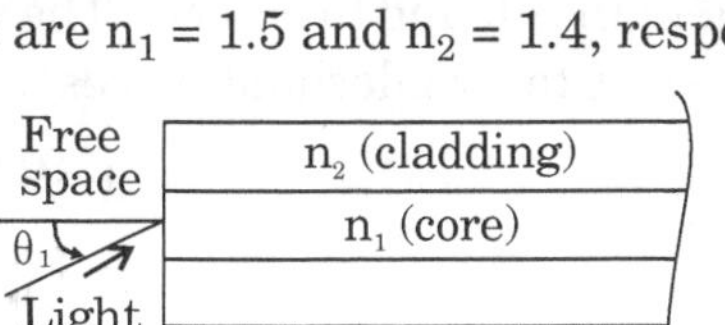

The maximum value of θ_i (in degrees) for which the incident light will be guided in the core of the fibre is _______.

[2016 : 1 Mark, Set-2]

17. Consider an air-filled rectangular waveguide with dimensions a = 2.286 cm and b= 1.016 cm. At 10 GHz operating frequency, the value of the propagation constant (per meter) of the corresponding propagating mode is _______.

[2016 : 2 Marks, Set-3]

18. Consider an air-filled rectangular waveguide with dimensions a = 2.286 cm and b = 1.016 cm. The increasing order of the cut-off frequencies for different modes is

(a) $TE_{01} < TE_{10} < TE_{11} < TE_{20}$
(b) $TE_{20} < TE_{11} < TE_{10} < TE_{01}$
(c) $TE_{10} < TE_{20} < TE_{01} < TE_{11}$
(d) $TE_{10} < TE_{11} < TE_{20} < TE_{01}$

[2016 : 2 Marks, Set-3]

19. An optical fiber is kept along the $\hat{z}$ direction. The refractive indices for the electric fields along $\hat{x}$ and $\hat{y}$ directions in the fiber are n_x = 1.5000 and n_y = 1.5001, respectively ($n_x \neq n_y$ due to the imperfection in the fiber cross-section). The free space wavelength of a light wave propagating in the fiber is 1.5 μm. If the lightwave is circularly polarized at the input of the fiber, the minimum propagation distance after which it becomes linearly polarized, in centimetres, is ______.

[2017 : 2 Marks, Set-1]

20. Standard air-filled rectangular waveguides of dimensions a = 2.29 cm and b = 1.02 cm are designed for radar applications. It is desired that these waveguides operate only in the dominant TE_{10} mode with the operating frequency at least 25% above the cutoff frequency of the TE_{10} mode but not higher than 95% of the next higher cutoff frequency. The range of the allowable operating frequency f is _______.

(a) 8.19 GHz ≤ f ≤ 13.1 GHz
(b) 8.19 GHz ≤ f ≤ 12.45 GHz
(c) 6.55 GHz ≤ f ≤ 13.1 GHz
(d) 1.64 GHz ≤ f ≤ 10.24 GHz

[2017 : 2 Marks, Set-2]

21. A rectangular waveguide of width W and height h has cut-off frequencies for TE_{10} and TE_{11} modes in the ratio 1 : 2. The aspect ratio w/h, rounded off to two decimal places, is ______.

[2019 : 2 Marks]

BOUNDARY CONDITIONS

22. A TEM wave is incident normally upon a perfect conductor. The E and H fields at the boundary will be, respectively,
(a) minimum and minimum
(b) maximum and maximum
(c) minimum and maximum
(d) maximum and minimum

[2000 : 1 Mark]

23. The permittivity of water at optical frequencies is $1.75\varepsilon_0$. It is found that an isotropic light source at a distance d under water forms an illuminated circular area of radius 5 m as shown in the figure. The critical angle is θ_c.

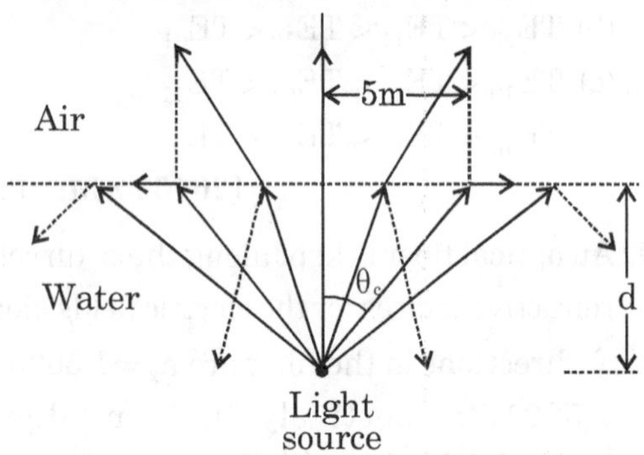

The value of d (in meter) is _______.

[2017 : 2 Marks, Set-2]

CUT-OFF FREQUENCIES

24. The cut off frequency of a waveguide depends upon
(a) the dimensions of waveguide
(b) the dielectric property of the medium in the waveguide
(c) the characteristic impedance of the waveguide
(d) the transverse and axial components of the fields

[1987 : 2 Marks]

25. A rectangular air filled waveguide has a cross section of 4 cm × 10 cm. The minimum frequency which can propagate in the waveguide is
(a) 1.5 GHz (b) 2.0 GHz
(c) 2.5 GHz (d) 3.0 GHz

[1997 : 1 Mark]

26. A rectangular waveguide has dimensions 1 cm × 0.5 cm. Its cut-off frequency is
(a) 5 GHz (b) 10 GHz
(c) 15 GHz (d) 12 GHz

[2000 : 2 Marks]

27. A rectangular metal wave guide filled with a dielectric material of relative permittivity $\epsilon_r = 4$ has the inside dimensions 3.0 cm × 1.2 cm. The cut-off frequency for the dominant mode is
(a) 2.5 GHz
(b) 5.0 GHz
(c) 10.0 GHz
(d) 12.5 GHz

[2003 : 2 Marks]

28. Which of the following statements is true regarding the fundamental mode of the metallic waveguides shown?

P: Coaxial   Q : Cylindrical   R : Rectangular

(a) Only P has no cutoff-frequency
(b) Only Q has no cutoff-frequency
(c) Only R has no cutoff-frequency
(d) All three have cutoff-frequencies

[2009 : 1 Mark]

29. For a rectangular waveguide of internal dimensions a × b (a > o), the cut-off frequency for the TE_{11} mode is the arithmetic mean of the cut-off frequencies for TE_{10} mode and TE_{20} mode. If a = $\sqrt{5}$ cm the value of b (in cm) is _______.

[2014 : 2 Marks, Set-2]

30. Consider an air filled rectangular waveguide with a cross-section of 5 cm × 3 cm. For this waveguide, the cut-off frequency (in MHz) of TE_{21} mode is

[2014 : 1 Mark, Set-3]

31. An air-filled rectangular waveguide of interval dimensions a cm × b cm (a > b) has a cutoff frequency of 6 GHz for the dominant TE_{10} mode. For the same waveguide, if the cutoff frequency of the TM_{11} mode is 15 GHz, the cutoff frequency of the TE_{01} mode in GHz is _______.

[2015 : 2 Marks, Set-2]

32. The cut-off frequency of TE_{01} mode of an air filled rectangular waveguide having inner dimensions a cm x b cm (a > b) is twice that of the dominant TE_{10} mode. When the waveguide is operated at a frequency which is 25% higher than the cut-off frequency of the dominant mode, the guide wavelength is found to be 4 cm. The value of b (in cm, correct to two decimal places) is _______.

[2018 : 2 Marks]

DISPERSION RELATIONS

33. For a normal mode EM wave propagating in a hollow rectangular wave guide,

(a) the phase velocity is greater than the group velocity.

(b) the phase velocity is greater than velocity of light in free space.

(c) the phase velocity is less than the velocity of light in free space.

(d) the phase velocity may be either greater than or less than the group velocity.

[1988 : 2 Marks]

34. The phase velocity of waves propagating in a hollow metal waveguide is

(a) greater than the velocity of light in free space.

(b) less than the velocity of light in free space

(c) equal to the velocity of light in free space

(d) equal to the group velocity

[2001 : 1 Mark]

35. In a microwave test bench, why s the microwave signal amplitude modulated at 1 kHz?

(a) To increase the sensitivity of measurement

(b) To transmit the signal to a far-off place

(c) To study amplitude modulation

(d) Because crystal detector fails at microwave frequencies

[2006 : 2 Marks]

36. The magnetic field along the propagation direction inside a rectangular waveguide with the cross-section shown in figure is

$H_z = 3\cos(2.094 \times 10^2 x)\cos(2.618 \times 10^2 y)\cos(6.283 \times 10^{10} t - \beta z)$

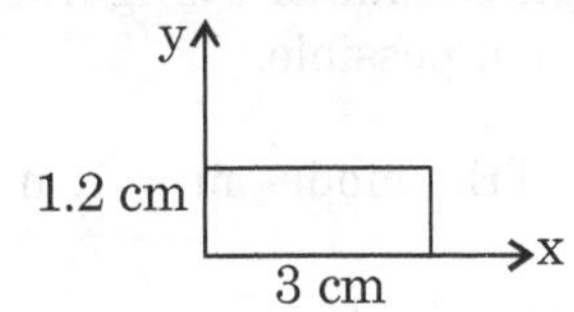

The phase velocity v of the wave inside the waveguide satisfies

(a) $v_p > c$ (b) $v_p = c$

(c) $0 < v_p < c$ (d) $v_p = 0$

[2012 : 2 Marks]

ANSWERS

1. (d)	**2.** (a, c)	**3.** (1)	**4.** (b)	**5.** (d)	**6.** (c)	**7.** (d)	**8.** (d)	**9.** (*c*)	**10.** (c)
11. (a)	**12.** (a)	**13.** (a)	**14.** (b)	**15.** (c)	**16.** (32.58)	**17.** (158.07)	**18.** (c)	**19.** (0.375)	**20.** (b)
21. (1.732)	**22.** (a)	**23.** (4.33)	**24.** (a,b)	**25.** (a)	**26.** (c)	**27.** (a)	**28.** (a)	**29.** (2)	
30. (7810 MHz)	**31.** (13.7)	**32.** (0.75)	**33.** (a)	**34.** (a)	**36.** (*d*)	**36.** (d)			

EXPLANATIONS

1. From the given expressions for $\bar{E}$ and $\bar{H}$

$$\beta = 280\pi$$

$$\Rightarrow \quad \frac{2\pi}{\lambda} = 280\pi, \ \lambda = \frac{1}{140} m.$$

$$v = f\lambda = 14 \times 10^9 \times \frac{1}{140}$$

$$= 1 \times 10^8 \text{ m/s}$$

Now, $$v = \frac{C}{\sqrt{\mu_r t_r}} = \frac{C}{\sqrt{t_r}}$$

or, $$10^8 = \frac{3 \times 10^8}{\sqrt{t_e}}$$

$$\Rightarrow \quad t_r = 9$$

and, $$\frac{E_p}{H_p} = \eta = \sqrt{\frac{\mu}{t}} = \sqrt{\frac{\mu_0 \mu_2}{t_0 t_r}} = \frac{1}{3}\sqrt{\frac{\mu_0}{t_0}}$$

$$\Rightarrow \quad E_p = 3 \times \frac{1}{3} \times 120\pi = 120\pi$$

2. Guide wavelength, $$\lambda_g = \frac{\lambda}{\sqrt{1 - \left(\frac{f_c}{f}\right)^2}}$$

$\therefore$ For waveguide, $\frac{f}{f_c} > 1.$

$$\boxed{\lambda_g > \lambda}$$

wave impedance,

$$\eta_{TE} = \frac{\eta_o}{\sqrt{1-\left(\frac{f_c}{f}\right)}} \Rightarrow \eta_{TE} > \eta_o$$

$$\eta_{TM} = \eta_o\sqrt{1-\left(\frac{f_c}{f}\right)^2} \Rightarrow \eta_{TM} < n_o$$

$\Rightarrow$ Phase velocity, $v_p = \frac{c}{\sqrt{1-\left(\frac{f_c}{f}\right)^2}}$(for air)

$v_P > c$.

$\Rightarrow$ For TEM mode to exist, presence of two conductors is necessary and because waveguide is a single conductor configuration. Hence, TEM mode is not possible.

3. Given : TE_{11} mode- m = 1, n = 1, $a = \frac{20}{3}$ cm,

$b = \frac{20}{4}$ cm, $\epsilon_r = 4$

cut-off wavelength,

$$\lambda_c = \frac{2}{\sqrt{\left(\frac{m}{a}\right)^2+\left(\frac{n}{b}\right)^2}} = \frac{2}{\sqrt{\left(\frac{1}{\frac{20}{3}}\right)^2+\left(\frac{1}{\frac{20}{4}}\right)^2}}$$

$\lambda_c = 8$ cm.

Cut-off wavelength is the maximum value of wavelength below which wave propagation can take place. Hence, waves of free space wavelengths shorter than 8 cm can be propagate in the TE_{11} mode.

4. A pure TEM mode is possible in two or more conductor waveguides not is single conductor waveguides such as rectangular or cylindrical waveguides.

5. The dominant mode in a particular waveguide is the mode having lowest cut-off frequency, below this cut-off frequency, EM will be attenuated to a negligible value. Lowest-cut-off frequency corresponds to highest cut-off wavelength.

7. For TE_{10} mode, phase velocity,

$$V_g = \frac{v_p}{\left[1-(f_c/f)^2\right]^{1/2}}$$

and $f_c = \frac{1}{2\sqrt{\mu\varepsilon}}\sqrt{\frac{m^2}{a^2}+\frac{n^2}{b^2}} = \left(\frac{1}{2\sqrt{\mu\varepsilon}}\right)\cdot\frac{1}{a}$

$m = 1$; $n = 0$ for TE_{10} mode]

$$f_c = \frac{c}{2a}$$

Now, $\frac{f_c}{f} < 1$, then $v_g > v_p$ (velocity of light).

8. x-component of electric field for TE_{mn} nodes is given as

$$E_x = \frac{j\omega\mu}{h^2}\left(\frac{n\pi}{b}\right)H_0\cos\left(\frac{m\pi x}{a}\right)\sin\left(\frac{n\pi y}{b}\right)e^{-xz}$$

Where a, b are the dimensions of the rectangular waveguide.

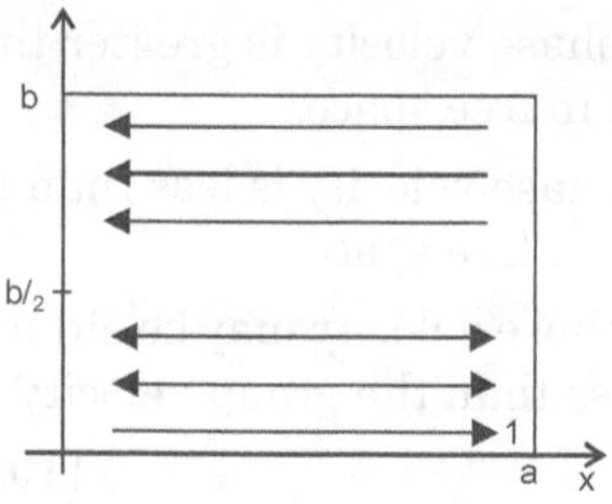

For TE_{02} mode, $m = 0$, $n = 2$

then, $E_x \propto \sin\left(\frac{2\pi y}{b}\right)$

At, $y = 0$, E_x is zero

At $y = \frac{b}{4}$,

E_x is maximum positive.

At $y = \frac{b}{2}$, E_x is zero again.

When $y > \frac{b}{2}$, then direction of E_x gets reversed

and at $y = \frac{b}{4}$, E_x is negative maximum

At $y = b$, E_x is zero.

9. Cut-off frequency, f_c for TE_{mn} mode is given by,

$$f_c = \frac{C}{2}\sqrt{\left(\frac{m}{a}\right)^2+\left(\frac{n}{b}\right)^2}$$

For, TE_{30} mode, $m = 3$, $n = 0$, $C = 3 \times 10^8$ m/s

$$f_c = 18 \times 10^9 = \frac{3\times10^8}{2}\sqrt{\left(\frac{3}{a}\right)^2}$$

$$\Rightarrow \frac{3}{a}\cdot\frac{3\times10^8}{2} = 18\times10^9$$

$$\Rightarrow a = \frac{5}{2} \text{ cm}$$

10. Given, $f_0 = 30$ GHz

$$\therefore \lambda_0 = \frac{c}{f_o} = \frac{3\times10^8}{30\times10^9} = 1 \text{ cm}$$

$$\frac{1}{\lambda_c^2} = \left(\frac{m}{2a}\right)^2 + \left(\frac{n}{2b}\right)^2$$

For TE_{20} mode, $m = 2$ and $n = 0$

$\therefore$ $\lambda_c = a = 3$ cm

$$\eta = \frac{\eta_0}{\sqrt{1-\left(\frac{\lambda_0}{\lambda_c}\right)^2}}$$

$$= \frac{377}{\sqrt{1-\left(\frac{1}{3}\right)^2}} = 400\ \Omega$$

11. For TE_{mn} mode, magnetic field in z - direction is expended as,

$$\bar{H}_z \simeq H_{20} \cos\frac{m\pi x}{a} \cos\frac{m\pi y}{b} \cdot e^{-j\beta mnZ} \quad ...(1)$$

then, $\bar{E}_y = J\frac{\omega\mu}{k^2}\frac{\partial \bar{H}2}{\partial x}$

$$= \frac{-j\left(\frac{m\pi\omega\mu}{axz}H_{20}\right)\sin\left(\frac{m\pi x}{a}\right)}{}$$

constant term $\cos\left(\frac{n\pi y}{b}\right)\cdot e^{-j\beta_{mn}Z}$

then for TE_{20} mode $m = 2, n = 0$

We have

$$\bar{E}_y = -j\left(\frac{2\pi\omega\mu}{akz}H_0\right)\sin\left(\frac{2\pi x}{a}\right)\cdot e^{-j\beta_{mn}Z}$$

Comparing with given equation for $\bar{E}_y$, it can be inferred that the mode of propagation is TE_{20} mode.

12. For TE_{11} mode,

$$f_c = \frac{1}{2\sqrt{\mu\epsilon}}\sqrt{\left(\frac{m}{a}\right)^2 + \left(\frac{n}{b}\right)^2};$$

where, a and b in cm.

and $$c = \frac{1}{\sqrt{\mu\epsilon}}$$

$$f_c = 1.5 \times 10^{10}\sqrt{\left(\frac{1}{4}\right)^2 + \left(\frac{1}{3}\right)^2}$$

$$= 6.25 \text{ GHz}$$

13. In case of rectangular waveguide TE_{mm} exists for all values of m and n except m = 0 and n = 0. For TM_{mm} to exit both values of m and n must be non-zero.

14. For TEM wave

Electric field (E), Magnetic field (H) and

Direction of propagation (P) are orthogonal to each other.

Here $P = +a_x$

By verification

$$E = -2a_y,$$

$$H = -3a_z$$

$$E \times H = -a_y \times -a_z$$

$$= +a_x \rightarrow P$$

15. $E_x = 5\cos(\omega t + \beta z)$

$$E_y = 3\cos\left(\omega t + \beta z - \frac{\pi}{2}\right)$$

$$\phi = -\frac{\pi}{2}$$

But the wave is propagating along negative z-direction

So, left hand elliptical

16. $\sin\alpha_{max} = \sin\theta_i = \sqrt{n_1^2 - n_2^2} = \sqrt{1.5^2 - 1.4^2}$

(Here θ_i be the incident angle)

$\Rightarrow \alpha_{max} = \sin\theta_i = \sin^{-1}(0.5385) = 32.58°$

$\therefore\ \theta_i = 32.58°$

Hence, the maximum value of θ_i is 32.58°.

17. Given

Air filled rectangular waveguide

a = 2.286 cm, b = 1.016 cm, f = 10 GHz

Assume dominant mode (TE_{10}) is propagating in the waveguide,

So cut-off frequency of TE_{10} mode is given by

$$f_c(TE_{10}) = \frac{c}{2a} = \frac{3\times10^{10}}{2\times2.286}$$

$$f_c = 6.56 \text{ GHz}$$

propagation constant $\bar{\gamma}$ is given by

$$\bar{\gamma} = j\vec{\beta}$$

$$= j\omega\sqrt{\mu_0\epsilon_0}\sqrt{1-\left(\frac{f_c}{f}\right)^2}$$

$$= j2\pi\times10\times10^9\times\frac{1}{3\times10^8}\sqrt{1-\left(\frac{6.56}{10}\right)^2}$$

$$\bar{\gamma} = j158.07 \text{ m}^{-1}$$

Therefore the value of propagation constant is given by

$$|\bar{\gamma}| = 158.07 \text{ m}^{-1}$$

18. Given : a = 2.286 cm, b = 1.016 cm

Air filled rectangular waveguide

$$f_{c(TE_{11})} = \frac{c}{2}\sqrt{\frac{1}{a^2}+\frac{1}{b^2}} \quad (\text{Since } m = 1, n = 1)$$

$$= \frac{3\times10^{10}}{2}\sqrt{\left(\frac{1}{(2.216)^2}+\frac{1}{1.016^2}\right)}$$

$$f_{c(TE_{11})} = 16.15 \text{ GHz}$$

$$f_{c(TE_{01})} = \frac{c}{2b} = \frac{3\times10^{10}}{2\times1.016} = 14.76 \text{ GHz}$$

$$f_{c(TE_{20})} = \frac{c}{a} = \frac{3\times10^{10}}{2.286} = 13.12 \text{ GHz}$$

$$f_{c(TE_{10})} = \frac{c}{2a} = \frac{3\times10^{10}}{2\times2.286} = 6.56 \text{ GHz}$$

∴ Increasing order of the cut-off frequency is given by

$TE_{10} < TE_{20} < TE_{01} < TE_{11}$

19. Initially the wave is circularly polarized. So, the initial phase difference between field components in $\hat{a}_x$ direction and $\hat{a}_y$ direction is $\frac{\pi}{2}$.

To become linearly polarized, the wave must travel a minimum distance, such that, the phase difference at that point between the field components in $\hat{a}_x$ direction and $\hat{a}_y$ direction is π (i.e., the travel of this minimum distance should provide an additional phase difference of $\frac{\pi}{2}$ between $\hat{a}_x$ and $\hat{a}_y$ field components).

$$\therefore \quad z_{min}k_x \sim z_{min}k_y = \frac{\pi}{2}$$

$$\therefore \quad z_{min}\left[\frac{\omega}{v_{px}} \sim \frac{\omega}{v_{py}}\right] = \frac{\pi}{2}$$

$$\left[\because k_x = \frac{\omega}{v_{px}} \text{ and } k_y = \frac{\omega}{v_{py}}\right]$$

$$\therefore \quad 2\pi z_{min}\left[\frac{f}{c}\sqrt{\varepsilon_{rx}} \sim \frac{f}{c}\sqrt{\varepsilon_{rx}}\right] = \frac{\pi}{2}$$

$$\therefore \quad \frac{4z_{min}}{\lambda_0}[n_x \sim n_y] = 1$$

$$\therefore \quad z_{min} = \frac{\lambda_0}{4(n_x \sim n_y)}$$

$$= \frac{1.5}{4(1.5 \sim 1.5001)}\ \mu m$$

$$= \frac{1.5}{4(0.0001)}\ \mu m = \frac{1.5}{4} \text{ cm}$$

$$\therefore \quad z_{min} = 0.375 \text{ cm}$$

20. Given : a = 2.29 cm, b = 1.02 cm.

Case 1: $f \geq f_c + 0.25 f_c \Rightarrow f \geq \frac{5}{4}f_c$

Cut-off frequency,

$$f_c = \frac{c}{2}\sqrt{\left(\frac{m}{a}\right)^2 + \left(\frac{n}{b}\right)^2}$$

For TE_{10} mode, m = 1, n = 0

$$f_c = \frac{3\times10^8}{2}\sqrt{\left(\frac{1\times10^2}{2.29}\right)^2} = 6.55 \text{ GHz}$$

$$\Rightarrow f \geq \frac{5}{4}f_c \Rightarrow f \geq 819 \text{ GHz.}$$

Case 2: $f \leq 95\%$ of $f_c \Rightarrow f \leq 0.95 f_c$.

Next higher cut-off frequency (for TE_{20} mode), m = 2, n = 2

$$f_c = \frac{c}{a} = \frac{3\times10^8}{2.29\times10^{-2}} = 13.1 \text{ GHz}$$

$f \leq 0.95 \times 13.1 \times 10^9$.

Range of operating frequency, 8.19 GHz $\leq f \leq$ 12.44 GHz.

21. Cutoff frequency of TE_{10} mode is $f_{C_{10}} = \frac{C}{2\omega}$

Cutoff frequency of TE_{11} mode

$$f_{C_{11}} = \frac{C}{2\omega}\sqrt{1^2 + \left(\frac{\omega}{h}\right)^2}$$

$$\text{Given } \frac{f_{C_{10}}}{f_{C_{11}}} = \frac{1}{2} \Rightarrow \frac{(C/2\omega)}{\left(\frac{C}{2\omega}\right)\sqrt{1^2 + \left(\frac{\omega}{h}\right)^3}} = \frac{1}{2}$$

$$\frac{1}{\sqrt{1 + \frac{\omega^2}{h^2}}} = \frac{1}{2}$$

Now squaring both the side,

$$\therefore \quad 1^2 + \left(\frac{\omega}{h}\right)^2 = 4$$

$$\therefore \quad \frac{\omega}{h} = \sqrt{3}$$

∴ Required ratio $= \frac{\omega}{h} = \sqrt{3} = 1.732$.

22. For good conductors,

$$\overline{E} = e^{\frac{-z}{\delta}}.\cos\left(\omega t - \frac{z}{\delta}\right)\hat{a}_x; \ \delta = \frac{1}{\sqrt{\pi f\sigma\mu}}$$

We know that δ is the measure of damping (exponential) of the wave as it travels through the conductor. Electric field $\overline{E}$ and magnetic field $\overline{H}$ can hardly propagate through good conductors.

23. Critical angle,

$$\theta_c = \sin^{-1}\sqrt{\frac{\epsilon_2}{\epsilon_1}} = \sin^{-1}\sqrt{\frac{\epsilon_0}{1.75\,\epsilon_0}} = 49.1°$$

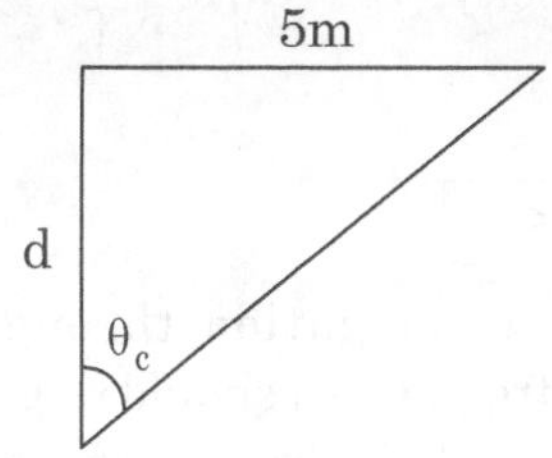

$$\Rightarrow \tan\theta_c = \frac{5}{d}$$

$$d = \frac{5}{\tan 49.1°} = \frac{5}{1.154}$$

d = 4.33 m.

24. The cut-off frequency of the waveguide depends upon the dimension of the cross-section and the dielectric property of the medium inside. For rectangular waveguide cut-off frequency is

$$f_c = \frac{c}{2\sqrt{\mu_r\,\epsilon_r}}\sqrt{\left(\frac{m}{a}\right)^2 + \left(\frac{n}{b}\right)^2}$$

where a, b = cross-sectional dimensions,

m,n = mode numbers, m, n $\in$ I

μ_r, ϵ_r = permeability and permittivity of the medium.

25. Given : b = 4 cm, a = 10 cm, $v_p = c = 3 \times 10^8$ cm. Assuming dominant mode (TE_{10} mode); m = 1, n = 0.

$$\text{Cut-off frequency, } f_c = \frac{v_p}{2}\sqrt{\left(\frac{m}{a}\right)^2 + \left(\frac{n}{b}\right)^2}$$

$$= \frac{3 \times 10^8}{2}\sqrt{\left(\frac{1}{10 \times 10^{-2}}\right)^2}$$

f_c = 1.5 GHz.

26. Given a = 1 cm, b = 0.5 cm, $v_p = c = 3 \times 10^8$ m/s (for air).

Assuming dominant (TE_{10}) mode, m = 1, n = 0

$$\text{cut-off frequency, } (f_c) = \frac{v_p}{2}\sqrt{\left(\frac{m}{a}\right)^2 + \left(\frac{n}{b}\right)^2}$$

$$= \frac{3 \times 10^8}{2}\sqrt{\left(\frac{1}{1}\right)^2 + 0 \times 10^{+1}}$$

f_c = 15 GHz.

27. Given: a = 3 cm, b = 1.2 cm, ϵ_r = 4.

$$v = \frac{1}{\sqrt{\mu\,\epsilon}} = \frac{c}{\sqrt{\epsilon_r \times \mu_r}} = \frac{c}{\sqrt{\epsilon_r}}$$

(for non-magnetic material μ_r =1)

$$\frac{3}{2} \times 10^8\,\text{m/s}.$$

Assuming dominant mode (m = 1, n = 1),

$$\text{cut-off freq, } f_c = \frac{v}{2}\sqrt{\left(\frac{m}{a}\right)^2 + \left(\frac{n}{b}\right)^2}$$

$$= \frac{3 \times 10^8}{4}\sqrt{\left(\frac{1}{3}\right)^2 + 0 \times 10^{+1}} = 2.5 \text{ GHz.}$$

28. P is coaxial line (Two conductors - inner and outer conductors) and supports TEM wave. Hence, P has no cut-off frequency.

Q and R are single conductor systems (cylindrical and rectangular respectively) and have cut-off frequency which depends on their dimensions.

29.

$$t_{c10} = \frac{C}{2}\sqrt{\left(\frac{1}{a}\right)^2}$$

$$t_{c10} = K\left(\frac{1}{a}\right);$$

$$t_{c20} = K\left(\frac{2}{a}\right)$$

$$t_{c11} = K\sqrt{\frac{1}{a^2} + \frac{1}{b^2}}$$

Given
$$t_{c11} = \frac{f_{c10} + f_{c20}}{2}$$

$$K\sqrt{\frac{1}{a^2} + \frac{1}{b^2}} = \frac{K}{2}\left[\frac{1}{a} + \frac{2}{a}\right]$$

$$\sqrt{\frac{1}{a^2} + \frac{1}{b^2}} = \frac{3}{2a}$$

$$\frac{1}{5} + \frac{1}{b^2} = \frac{9}{4(5)}$$

$$\Rightarrow \quad -\frac{1}{5} + \frac{9}{20} = \frac{1}{b^2}$$

$$-0.2 + 0.45 = \frac{1}{b^2}$$

$$\therefore \quad \frac{1}{b^2} = \frac{1}{2^2}$$

$$\Rightarrow \quad b = 2 \text{ cm}$$

30.

$$f_c(TE_{21}) = \frac{C}{2}\sqrt{\left(\frac{2}{9}\right)^2 + \left(\frac{1}{b}\right)^2}$$

$$= \frac{3 \times 10^{10}}{2}\sqrt{\left(\frac{2}{5}\right)^2 + \left(\frac{1}{3}\right)^2}$$

$$= 1.5 \times 10^{10}\sqrt{0.16 + 0.111}$$

$$= 0.52 \times 1.5 \times 10^{10}$$

= 7.81 GHz

= 7810 MHz

31. **1st case :** $V_{wx1} = 100\text{ V}$

So, $Vy'z_1 = \frac{M_2}{M_1} Vwx_1 = 1.25 \times 100 = 125V$

$\therefore$ $Vy'z_1 = Vy'z_1 \times x$

$= 125 \times 0.8 = 100\ v.$

$\therefore$ $Vyz_1/Vwx_1 = \frac{100}{100}.$

2nd case: $Vyz_2 = 100\text{ V}$

$\therefore$ $Vy't_2 = \frac{100}{\alpha} = \frac{100}{0.8} = 125v$

Now, $Vwx_2 = \frac{M_1}{M_2} Vy'z_2 = \frac{1}{1.25} \times 125 = 100v$

$\therefore$ $Vwx_2/Vyz_2 = \frac{100}{100}$

32. $f_{c(01)} = 2 f_{c(10)} = \frac{2c}{2a} = \frac{c}{a}$

Now $\frac{c}{2b} = \frac{c}{a} \Rightarrow a = 2b$

$\therefore$ $b = \frac{a}{2}$

Operating frequency, $f = 1.25 f_{c(10)}$

$f_{c(10)} < 1.25 f_{c(10)} < [f_{c(01)} = 2 f_{c(10)}]$

So, for the given frequency, the waveguide will work in TE_{10} mode.

$$\text{So, } \lambda_g = \frac{\lambda_0}{\sqrt{1 - \left(\frac{f_{c(10)}}{f}\right)^2}} = \frac{c/f}{\sqrt{1 - \left(\frac{1}{1.25}\right)^{2}}} = \frac{c/f}{0.6}$$

$$\frac{C}{(1.25) f_{c(10)} (0.6)} = \lambda_g = 4 \text{ cm}$$

$$\frac{C}{f_{c(10)}} = 3 \times 10^{-2} = 2a$$

$\therefore$ a = 1.5 cm

$\Rightarrow b = \frac{a}{2} = 0.75$ cm

33. For wave propagation through the waveguide operating frequency should be greater than f_c, i.e,

$f_c > f_c \Rightarrow \frac{f_c}{f} < 1.$

Phase velocity, $v_p = \frac{c}{\sqrt{1 - \left(\frac{f_c}{f}\right)^2}}$ (for free space)

$\boxed{v_p > c}$

Group velocity, $v_g = c.\sqrt{1 - \left(\frac{f_c}{f}\right)^2}$ (for free space)

$v_g < c.$

$\Rightarrow \boxed{v_g < c < v_p}$

34. For wave propagation through waveguide, operating frequency f should be greater than cut-off frequency, f_c for free space, phase velocity

$$v_p = \frac{c}{\sqrt{1 - \left(\frac{f_c}{f}\right)^2}},$$

$f_c < f$

$v_p > v_c.$

35. Microwave signals are amplitude modulate because crystal detector fails of microwave frequencies.

36. Given a = 3, b = 1.2 cm, $H_z = 3\cos(2.094 \times 10^2 x)$ $\cos(2.618 \times 10^2 y) \cos(6.283 \times 10^{10} t - \beta z)$...(i)

For TE_{mn} mode $E_{zs} = 0$.

$$H_{zs} = H_o \cos\left(\frac{m\pi x}{a}\right).\cos\left(\frac{n\pi y}{b}\right) \cos(cot - \beta x) \text{ ...(ii)}$$

On comparing eq. (i) and (ii),

$\omega = 6.283 \times 10^{10}$ rad/s, $f = \frac{\omega}{2\pi} = 10$ GHz.

$$\frac{m\pi}{a} = 2.094 \times 10^2 \Rightarrow \frac{m}{a} = 66.65/\text{m}$$

$$\frac{n\pi}{b} = 2.618 \times 10^2 \Rightarrow \frac{n}{b} = 83.33/\text{m}$$

$$\text{cut-off freq, } f_c = \frac{c}{2}\sqrt{\left(\frac{m}{a}\right)^2 + \left(\frac{n}{b}\right)^2}$$

$$= \sqrt{66.65^2 + 83.33^2} \times \frac{3 \times 10^8}{2}$$

$f_c = 16$ GHz.

The wave with frequency, f = 10 GHz ($<f_c$) will not propagate through the quide.

Hence, $v_p = 0$.

6 CHAPTER

Basics of Antenna & Radars

Analysis of Previous GATE Papers

Topics ↓		Year →	2019	2018	2017 Set 1	2017 Set 2	2016 Set 1	2016 Set 2	2016 Set 3	2015 Set 1	2015 Set 2	2015 Set 3
TYPES OF ANTENNA & RADIATION PATTERN	1 Mark	MCQ Type										
		Numerical Type	1									
	2 Marks	MCQ Type			1							
		Numerical Type									1	
	Total		1		2						2	
GAIN & DIRECTIVI TY	1 Mark	MCQ Type			1							1
		Numerical Type										
	2 Marks	MCQ Type					1		1			
		Numerical Type					1					
	Total				1		4		2			1
RETURN LOSS & ANTENNA ARRAYS	1 Mark	MCQ Type										
		Numerical Type										
	2 Marks	MCQ Type					1					
		Numerical Type										
	Total						2					

TYPES OF ANTENNA & RADIATION PATTERN

1. The electric field E and the magnetic field H of a short dipole antenna satisfy the condition
(a) the r component of E is equal to zero
(b) both r and θ components of H are equal to zero
(c) the θ component of E dominates the r component in the far - field region
(d) the θ and ϕ components of Hare of the same order of magnitude in the near field region

[1988 : 2 Marks]

2. Two isotropic antennas are separated by a distance of two wavelengths. If both the antennas are fed with currents of equal phase and magnitude, the number of lobes in the radiation pattern in the horizontal plane are
(a) 2 (b) 4
(c) 6 (d) 8

[1990 : 2 Marks]

3. In a board side array of 20 isotropic radiators, equally spaced at distance of $\frac{\lambda}{2}$, the beam width between first nulls is
(a) 51.3 degrees (b) 11.46 degrees
(c) 22.9 degrees (d) 102.6 degrees

[1991 : 2 Marks]

4. The beamwidth between first nulls of a uniform linear array of Nequally spaced (element spacing = d), equally excited antennas, is determine by
(a) N alone and not by d
(b) d alone and not by N
(c) the ratio, (N/d)
(d) the product, (Net)

5. For a dipole antenna,
(a) the radiation intensity is maximum along the normal to the dipole axis
(b) the current distribution along its length is uniform irrespective of the length
(c) the effective length equals its physical length
(d) the input impedance is independent of the location of the feed-point

[1994 : 1 Mark]

6. An antenna, when radiating, has a highly directional radiation pattern. When the antenna is receiving, its radiation pattern
(a) is more directive
(b) is less directive
(c) is the same
(d) exhibits no directivity at all

[1995 : 1 Mark]

7. Two half-wave dipole antennas placed as shown in the figure are excited with sinusoidally varying currents of frequency 3 MHz and phase shift of $\frac{\pi}{2}$ between them (the element at the origin leads in phase). If the maximum radiated E-field at the point P in the x-y plane cccurs at an azimuthal angle of 60°, the distance d (in meters) between the antennas is ________.

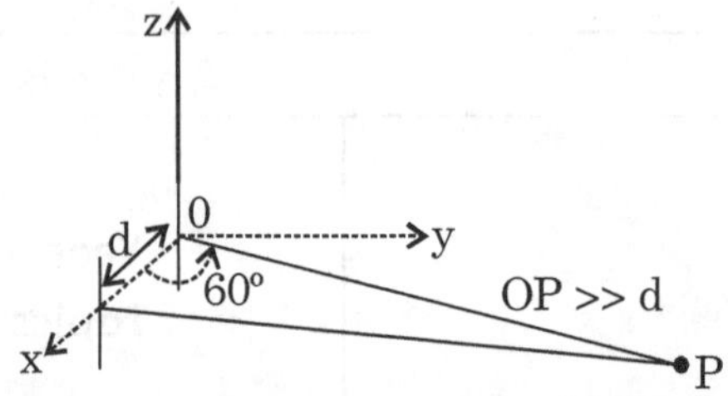

[1996 : 2 Marks]

8. A half wavelength dipole is kept in the x-y plane and oriented along 45° from the x-axis. Determine the direction of null in the radiation pattern for $0 \le \phi \le \pi$. Here the angle $\theta(0 \le \theta \le \pi)$ is measured from the z-axis, and the angle $\phi(0 \le \phi \le 2\pi)$ is measured from the x-axis in the x- y plane
(a) θ = 90°, ϕ = 45° (b) θ = 45°, ϕ = 90°
(c) θ = 90°, ϕ = 135° (d) θ = 45°, ϕ = 135°

[1997 : 2 Marks]

9. The vector $\vec{H}$ in the far field of an antenna satisfies
(a) $\nabla \cdot \vec{H} = 0$ and $\nabla \times \vec{H} = 0$
(b) $\nabla \cdot \vec{H} \neq 0$ and $\nabla \times \vec{H} \neq 0$
(c) $\nabla \cdot \vec{H} = 0$ and $\nabla \times \vec{H} \neq 0$
(d) $\nabla \cdot \vec{H} \neq 0$ and $\nabla \times \vec{H} = 0$

[1998 : 1 Mark]

10. The radiation resistance of a circular loop of one turn is 0.01 Ω. The radiation resistance of five turns of such a loop will be
(a) 0.002 Ω (b) 0.01 Ω
(c) 0.05 Ω (d) 0.25 Ω

[1998 : 1 Mark]

11. An antenna in free space receives 2 μW of power when the incident electric field is 20 mV/m rms. The effective aperture of the antenna is
(a) 0.005 m^2 (b) 0.05 m^2
(c) 1.885 m^2 (d) 3.77 m^2

[1998 : 1 Mark]

12. The far field of an antenna varies with distance r as
(a) $\frac{1}{r}$ (b) $\frac{1}{r^2}$
(c) $\frac{1}{r^3}$ (d) $\frac{1}{\sqrt{r}}$

1998 : 1 Mark]

13. The frequency range for satellite communication is

(a) 1 kHz to 100 kHz (b) 100 kHz to 10 kHz

(c) 10 MHz to 30 MHz (d) 1 GHz to 30 GHz

[2000 : 1 Mark]

14. If the diameter of a $\lambda/2$ dipole antenna is increased from $\frac{\lambda}{100}$ to $\frac{\lambda}{50}$, then its

(a) bandwidth increases

(b) bandwidth decreases

(c) gain increases

(d) gain decreases

[2000 : 1 Mark]

15. For an 8 feet (2.4 m) parabolic dish antenna operating at 4 GHz, the minimum distance required for far field measurement is closest to

(a) 7.5 cm (b) 15 cm

(c) 15 m (d) 150 m

[2000 : 2 Marks]

16. In a uniform linear array, four isotropic radiating elements are spaced $\frac{\lambda}{4}$ apart. The progressive phase shift between the elements required for forming the main beam at 60° off the end-fire is

(a) $-\pi$ radians (b) $\frac{-\pi}{2}$ radians

(c) $\frac{-\pi}{4}$ radians (d) $\frac{-\pi}{8}$ radians

[2001 : 2 Marks]

17. The line-of-sight communication requires the transmit and receive antennas to face each other. If the transmit antenna is vertically polarized, for best reception the receiver antenna should be

(a) horizontally polarized

(b) vertically polarized

(c) at 45° with respect to horizontal polarization

(d) at 45° with respect to vertical polarization

[2002 : 1 Mark]

18. Two identical and parallel dipole antennas are kept apart by a distance of $\frac{\lambda}{4}$ in the H-plane. They are fed with equal currents but the right most antenna has a phase shift of +90°. The radiation pattern is given as

(a)

(b)

(c)
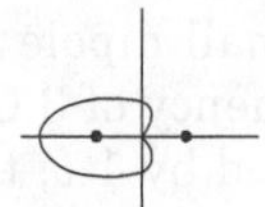

(d)
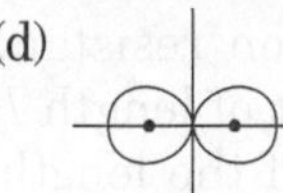

[2006 : 2 Marks]

19. A mast antenna consisting of a 50 meter long vertical conductor operates over a perfectly conducting ground plane. It is base-fed at a frequency of 600 kHz. The radiation resistance of the antenna in Ohms is

(a) $\frac{2\pi^2}{5}$ (b) $\frac{\pi^2}{5}$

(c) $\frac{4\pi^2}{5}$ (d) $20\pi^2$

[2006 : 2 Marks]

20. A $\frac{\lambda}{2}$ dipole is kept horizontally at a height of $\frac{\lambda_0}{2}$ above a perfectly conducting infinite ground plane. The radiation pattern in the plane of the dipole ($\vec{E}$ plane) looks approximately as

(a)
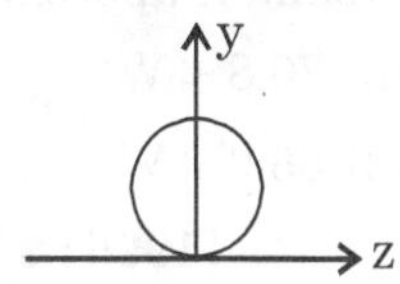

(b)
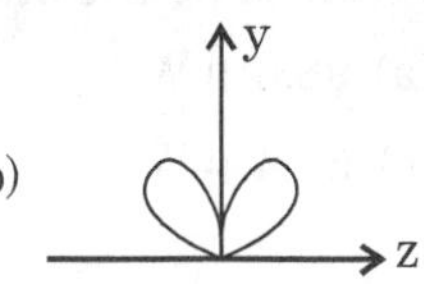

(c)
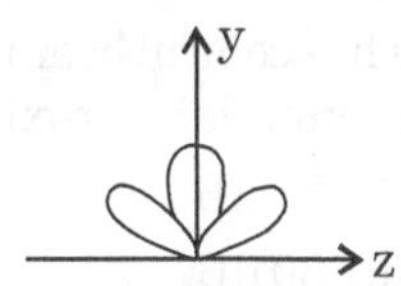

(d)
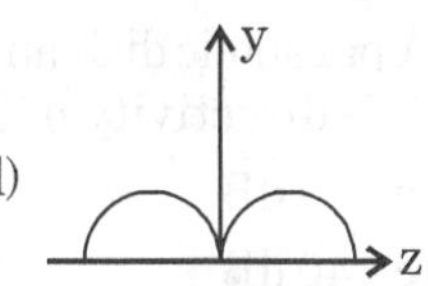

[2007 : 2 Marks]

21. For a Hertz dipole antenna, the Half Power Beam Width (HPBW) in the E-plane is

(a) 360° (b) 180°

(c) 90° (d) 45°

[2008 : 1 Mark]

22. Match Column A with Column B.

Column-A

1. Point electromagnetic source
2. Dish antenna
3. Yagi-Uda antenna

Column-B

P. Highly directional

Q. End fire

R. Isotropic

	P	Q	R
(a)	1	2	3
(b)	2	3	1
(c)	2	1	3
(d)	3	2	1

[2014: 1 Mark, Set-4]

23. Radiation resistance of a small dipole current element of length l at a frequency of 3 GHz is 3 ohms. If the length is changed by 1%, then the percentage change in the radiation resistance, rounded off two decimal places, is ________%.

[2019 : 1 Marks]

GAIN & DIRECTIVITY

24. Two dissimilar antennas having their maximum directivities equal,
   (a) must have their beam widths also equal
   (b) cannot have their beam widths equal because they are dissimilar antenna
   (c) may not necessarily have their maximum power gains equal
   (d) must have their effective aperture areas (capture areas) also equal

[1992 : 2 Marks]

25. A 1 km long microwave link uses two antennas each having 30 dB gain. If the power transmitted by one antenna is 1 W at 3 GHz, the power received by the other antenna is approximately
   (a) 98.6 μW (b) 76.8 μW
   (c) 63.4 μW (d) 55.2 μW

[1996 : 2 Marks]

26. A parabolic dish antenna has conical beam 2° wide. The directivity of the antenna is approximately
   (a) 20 dB (b) 30 dB
   (c) 40 dB (d) 50 dB

[1997 : 2 Marks]

27. A transmitting antenna radiates 251 W isotropically. A receiving antenna, located 100 m away from the transmitting antenna, has an effective aperture of 500 cm^2. The total power received by the antenna is
   (a) 10 μW (b) 1 μW
   (c) 20 μW (d) 100 μW

[1999 : 2 Marks]

28. A person with a receiver is 5 km away from the transmitter. What is the distance that this person must move further to detect a 3-dB decrease in signal strength?
   (a) 942 m (b) 2070 m
   (c) 4978 m (d) 5320 m

[2002 : 2 Marks]

29. Consider a lossless antenna with a directive gain of +6 dB. If 1 mW of power is fed to it the total power radiated by the antenna will be
   (a) 4 mW (b) 1 mW
   (c) 7 mW (d) 1/4 mW

[2004 : 1 Mark]

30. A transmission line is feeding 1 Watt of power to a horn antenna having a gain of 10 dB. The antenna is matched to the transmission line. The total power radiated
   (a) 10 Watts (b) 1 Watt
   (c) 0.1 Watt (d) 0.01 Watt

[2006 : 1 Mark]

31. At 20 GHz, the gain of a parabolic dish antenna of 1 meter diameter and 70% efficiency is
   (a) 15 dB (b) 25 dB
   (c) 35 dB (d) 45 dB

[2008 : 2 Marks]

32. The radiation pattern of an antenna in spherical co-ordiantes is given by

$$F(\theta) = \cos^4\theta,\ 0 \le \theta \le \frac{\pi}{2}$$

The directivity of the antenna is
   (a) 10 dB (b) 12.6 dB
   (c) 11.5dB (d) 18 dB

[2012 : 1 Mark]

33. In spherical coordinates, let $\hat{a}_\theta, \hat{a}_\phi$ denote unit vectors along the θ, φ directions.

$$\vec{E} = \frac{100}{r}\sin\theta\cos(\omega t - \beta r)\hat{a}_\theta \text{ V/m and}$$

$$\vec{H} = \frac{0.265}{r}\sin\theta\cos(\omega t - \beta r)\hat{a}_\phi \text{ V/m}$$

represent the electric and magnetic field components of the EM wave at large distances r from a dipole antenna, in free space. The average power (W) crossing the hemispherical shell located at r = 1 km, $0 < \theta \le \frac{\pi}{2}$ is ________.

[2014 : 2 Marks, Set-1]

34. For an antenna radiating in free space, the electric field at a distance of 1 km is found to be 12 mV/m. Given that intrinsic impedance of the free space is 120π Ω, the magnitude of average power density due to this antenna at a distance of 2 km from the antenna (in nW/m^2) is ________.

[2014 : 1 Mark, Set-4]

35. The directivity of an antenna array can be increased by adding more antenna elements, as a larger number of elements
   (a) improves the radiation efficiency
   (b) increases the effective area of the antenna
   (c) results in a better impedance matching
   (d) allows more power to be transmitted by the antenna

[2015 : 1 Mark, Set-3]

36. Two lossless X- band horn antennas are separated by a distance of 200λ. The amplitude reflection coefficients at the terminals of the transmitting and receiving antennas are 0.15 and 0.18, respectively. The maximum directivities of the transmitting and receiving antennas (over the isotropic antenna) are 18 dB and 22 dB, respectively. Assuming that the input power in the lossless transmission line connected to the antenna is 2 W, and that the antennas are perfectly aligned and polarization matched, the power (in mW) delivered to the load at the receiver is _______.

[2016 : 2 Marks, Set-1]

37. The far-zone power density radiated by a helical antenna is approximated as:

$$\vec{W}_{rad} = \vec{W}_{average} = \hat{a}_r C_0 \frac{1}{r^2} \cos^4\theta$$

The radiated power density is symmetrical with respect to ϕ and exists only in the upper hemisphere $0 \le \theta \le \frac{\pi}{2}; 0 \le \phi \le 2\pi; C_0$ is a constant. The power radiated by the antenna (in watts) and the maximum directivity of antenna, respectively, are

(a) $1.5C_0$, 10 dB (b) $1.256C_0$, 10 dB
(c) $1.256C_0$, 12dB (d) $1.5C_0$, 12dB

[2016 : 2 Marks, Set-1]

38. A radar operating at 5 GHz uses a common antenna for transmission and reception. The antenna has a gain of 150 and is aligned for maximum directional radiation and reception to a target 1 km away having radar cross-section of 3 m^2. If it transmits 100 kW, then the received power (in μW) is ______.

[2016 : 2 Marks, Set-3]

39. Consider a wireless communication link between a transmitter and a receiver located in free space, with finite and strictly positive capacity. If the effective areas of the transmitter and the receiver antennas, and the distance between them are all doubled, and everything else remains unchanged, the maximum capacity of the wireless link

(a) increases by a factor of 2
(b) decreases by a factor of 2
(c) remains unchanged
(d) decreases by a factor of $\sqrt{2}$

[2017 : 1 Mark, Set-1]

RETURN LOSS & ANTENNA ARRAYS

40. The electric field of a uniform plane wave travelling along the negative z direction is given by the following equation:

$$\vec{E}_w^i = (\hat{a}_x + j\hat{a}_y)E_0 e^{jrz}$$

This wave is incident upon a receiving antenna placed at the origin and whose radiated electric field towards the incident wave is given by the following equation:

$$\vec{E}_w^i = (\hat{a}_x + 2\hat{a}_y)E_t \frac{1}{4} e^{-jkr}$$

The polarization of the incident wave, the polarization of the antenna and losses due to the polarization mismatch are, respectively,

(a) linear, circular (clockwise), –5 dB
(b) circular (clockwise), linear, –5 dB
(c) circular (clockwise), linear, –3 dB
(d) circular (anticlockwise), linear, –3 dB

[2016 : 2 Marks, Set-1]

41. A transverse electromagnetic wave with circular polarization is received by a dipole antenna. Due to polarization mismatch, the power transfer efficiency from the wave to the antenna is reduced to about

(a) 50% (b) 35.3%
(c) 25% (d) 0%

[1996 : 2 Marks]

42. A medium wave radio transmitter operating at a wavelength of 492 m has a tower antenna of height 124 m. What is the radiation resistance of the antenna?

(a) 25 Ω (b) 36.5 Ω
(c) 50 Ω (d) 73 Ω

[2001 : 2 Marks]

43. Two identical antennas are placed in the $\phi = \frac{\pi}{2}$ plane as shown in the figure. The elements have equal amplitude excitation with 180° polarity difference, operating at wavelength λ. The correct value of the magnitude of the far-zone resultant electric field strength normalized with that of a single element, both computed for $\phi = 0$, is

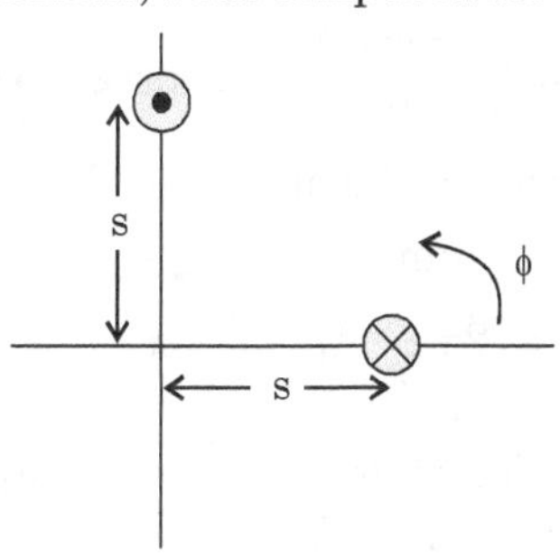

(a) $2\cos\left(\frac{2\pi s}{\lambda}\right)$ (b) $2\sin\left(\frac{2\pi s}{\lambda}\right)$
(c) $2\cos\left(\frac{\pi s}{\lambda}\right)$ (d) $2\sin\left(\frac{\pi s}{\lambda}\right)$

[2003 : 2 Marks]

ANSWERS

1. (b,c)	**2.** (d)	**3.** (b)	**4.** (d)	**5.** (a)	**6.** (c)	**7.** (50)	**8.** (a)	**9.** (c)	**10.** (d)
11. (d)	**12.** (a)	**13.** (d)	**14.** (c)	**15.** (b)	**16.** (c)	**17.** (b)	**18.** (a)	**19.** (a)	**20.** (c)
21. (c)	**22.** (c)	**23.** (2)	**24.** (a,d)	**25.** (c)	**26.** (c)	**27.** (d)	**28.** (b)	**29.** (b)	**30.** (a)
31. (d)	**32.** (a)	**33.** (55.5)	**34.** (47.7)	**35.** (b)	**36.** (2.99)	**37.** (b)	**38.** (0.0122)	**39.** (c)	**40.** (c,d)
41. (a)	**42.** (b)	**43.** (d)							

EXPLANATIONS

1. For a short-dipole antenna, magnetic field is $H_{r_s} = 0, H_{\theta_s} = 0,$

$$H_{\phi_s} = \frac{I_0 dl \sin\theta}{G\pi}\left[\frac{j\beta}{r} + \frac{1}{r^2}\right]e^{-j\beta r}$$

Electric field: $E_{\phi_s} = 0.$

$$E_{\theta_s} = \frac{\eta I_0 dl \sin\theta}{4\pi}\left[\left(\frac{j\beta}{r}\right) + \frac{1}{r} - \frac{1}{\beta r^2}\right]e^{-j\beta r}$$

$$E_{r_s} = \frac{\beta}{\omega \in}\frac{I_0 dl \sin\theta}{2\pi}\left[\frac{1}{r^2} - \frac{j}{\beta r^3}\right]e^{-j\beta r}$$

In the far field region $(\beta r > 1)$, E_{r_s} will be smaller than E_{θ_s}.

So, θ components of E dominates the r component of E in far-field region.

2. Given : $d = 2\lambda, \propto = 0$ (currents are of equal phase).

Array factor, $AF = 2\cos\left(\frac{\Psi}{2}\right); \Psi \propto +\beta d \sin\theta$

$$\Psi = \frac{2\pi}{\lambda}.2\lambda\cos\theta = 4\pi\cos\theta$$

$$AF = 2\cos\left(\frac{4\pi\cos\theta}{2}\right) = 2\cos(2\pi\cos\theta)$$

In the horizontal plane, $\theta \in [0, 2\pi]$.

AF is maximum at

$$\theta = 0, \frac{\pi}{3}, \frac{\pi}{2}, \frac{2\pi}{3}, \pi, \frac{4p}{3}, \frac{3\pi}{2}, \frac{5\pi}{3}.$$

Hence, No. of lobes in radiation patter in horizontal plane = 8

3. Given : $N = 20, d = \frac{\lambda}{2}$

$$BWFN = \frac{2\lambda}{Nd}$$

$$= \frac{2\lambda}{20 \times \frac{\lambda}{2}} \times \frac{180}{\pi} \text{ deg ree}$$

$$BWFN = \frac{36}{\pi} = 11.46°$$

4. For linear arrays, $Nd >> \lambda$.

For Broadside array, $BWFN = \frac{2\pi}{Nd}$ rad.

For End fire array, $BWFN = \frac{2}{3}\sqrt{\frac{2\lambda}{Nd}}$.

Thus, the BWFN in both arrays is determined by the product Nd.

5. In a dipole antenna, radiation is almost nil along its axis and maximum in a direction perpedicular to its axis.

6. Antenna is a passive and reciprocal device which characteristics are the same when it is transmitting and receiving. It is transducer.

7. Given: $f = 3$ MHz, $\propto = -90°, \theta = 60°$.

The wave makes an angle 30° with + x-axis and 90° with +y–axis. Hence, te electric field is directed along y-axis.

Array factor, $AF = \cos\left(\frac{\Psi}{2}\right)$

For principle,

$$\Psi = \beta d \cos\theta + \propto = 0°$$

$$\therefore \beta = \frac{2\pi}{\lambda} = \frac{2\pi \times 3 \times 10^6}{3 \times 10^8} = \frac{2\pi}{100}$$

$$0 = \frac{2\pi}{100} d \cos 60° - 90°$$

$$\frac{\pi}{2} = \frac{2\pi}{100} \times d \times \frac{1}{2}$$

d = 50 m.

8. For half wave dipole,

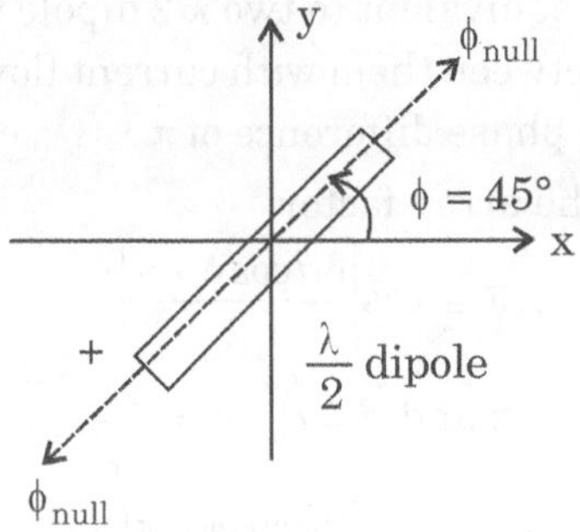

$$E_{\theta_s} = jI_o\eta e^{-j\beta r} \frac{\times \cos\left(\frac{\pi}{2}\cos(90° - \theta)\right)}{2\pi r \sin\theta(90° - \theta)}$$

---- in x – y plane

For $\theta = 90°$

$$E_{\theta_s} = \frac{\cos\left(\frac{\pi}{2}\sin 90°\right)}{\cos 90°} = \frac{0}{0}$$

Applying L – hospital rule,

$$E_{\theta_s} = \lim_{\theta \to 90°} \frac{-\frac{\pi}{2}\sin\left(\frac{\pi}{2}\sin\theta\right)\cos\theta}{-\sin\theta}$$

$$E_{\theta_s} = \frac{0}{1} = 0.$$

9. $\bar{H} = \bar{\nabla} \times \bar{A}$

$\therefore$ $\nabla \times H = \nabla \times (\nabla \times A) \neq 0$

$\nabla . H = \nabla . (\nabla \times A) = 0$

10. Consider it as new solution

Given : $R_{rad} = 0.01\ \Omega$.

For loop antenna, radiation resistance,

$$R_{rad} = \frac{320\pi^4 s^2}{\lambda^4}$$

$R_{rad} \propto s^2$.

For circular loop with N turns, $s = N\pi a^2$.

$R_{rad} \propto N^2$.

$$R_{rad} = \left(\frac{N_2}{N_1}\right)^2 .(R_{rad})_1 = \left(\frac{5}{1}\right)^2 \times 0.01 = 0.25\Omega$$

11. $P_r = \frac{|E_{rs}|^2}{2\eta_o} A_e$.

$$\therefore \quad A_e = \frac{2 \times 10^{-6} \times 2 \times 120\pi}{20 \times 20 \times 10^{-6}} = 3.77\ m^2.$$

12. The fields around an antenna may be divided into near field (or fresnel field) and far-field (fraunhofer field) $\frac{1}{r}$ is the only term that remain at the far zone.

13. The frequency range for satellite communication is 1 GHz to 30 GHz.

14. Radiation efficiency,

$$\eta = \frac{R_{rad}}{R_{rad} + R_L} = \frac{G}{D}$$

where, R_{rad} = radiation resistance,

R_L = loss resistance $= \frac{l}{\pi d}.R_s$

where, l = length, d = diameter, R_s = surface resistance.

Therefore, $A_s d \uparrow, R_L \downarrow$ and $\eta \uparrow, G \uparrow$.

15. $$R \geq \frac{2d^2}{\lambda} \quad d = 2.4\ m$$

$$\lambda = \frac{3 \times 10^8}{4 \times 10^9} = 7.5\ cm$$

$$R = \frac{2 \times (2.4)^2}{7.5 \times 10^{-2}} = 153\ m$$

16. Given q = 60°, $d = \frac{\lambda}{4}$.

An end fire array has its maximum radiation directed along the axis of array.

In the direction of main beam, $\Psi = 0$.

$$\propto = -\beta d \cos\theta = \frac{-2\pi}{\lambda}.\frac{\lambda}{4}.\cos 60° = \frac{-\pi}{4}\ \text{rad.}$$

17. If the transmit antenna is vertically polarized, then for best reception the receiver antenna should be vertically polarized.

18. Given : $d = \frac{\lambda}{4}, \propto = 90°$

The overall radiation pattern can be obtained by simply multiplying the element factor and array factor

For Hertzian dipole placed along z-axis, electric field component is $E_\theta \propto \sin\theta$.

$\theta = 0°$	$\frac{\pi}{4}$	$\frac{\pi}{2}$	$\frac{3\pi}{4}$	π	$\frac{5\pi}{4}$	$\frac{3\pi}{2}$	$\frac{7\pi}{4}$	2π
$\|AF\| = 0$	0.707	1	0.707	0	0.707	1	0.707	0

$$\text{Array factor, } AF = \cos\left[\frac{1}{2}(\beta d.\cos\theta + \alpha)\right]$$

$$= \cos\left[\frac{1}{2}\left(\frac{2\pi}{\lambda}\right) \times \frac{\lambda}{4}\cos\theta + 90°\right]$$

$$AF = \cos(45° \cos\theta + 45°).$$

$\theta = 0°$	$\frac{\pi}{4}$	$\frac{\pi}{2}$	$\frac{3\pi}{4}$	π	$\frac{5\pi}{4}$	$\frac{3\pi}{2}$	$\frac{7\pi}{4}$	2π
$\|AF\| = 0$	0.22	0.7	0.97	1	0.97	0.7	0.22	0

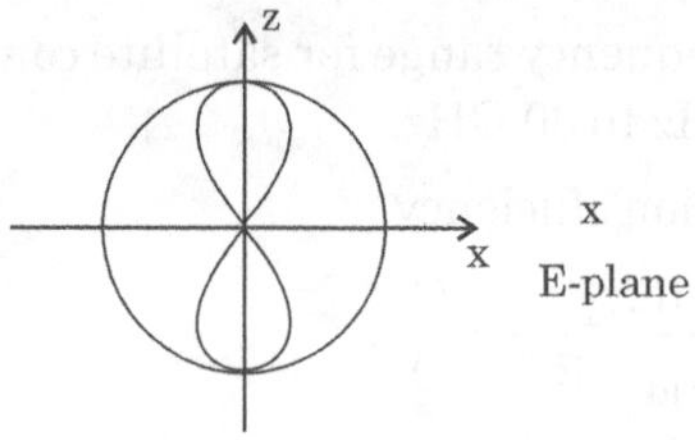

Unit pattern

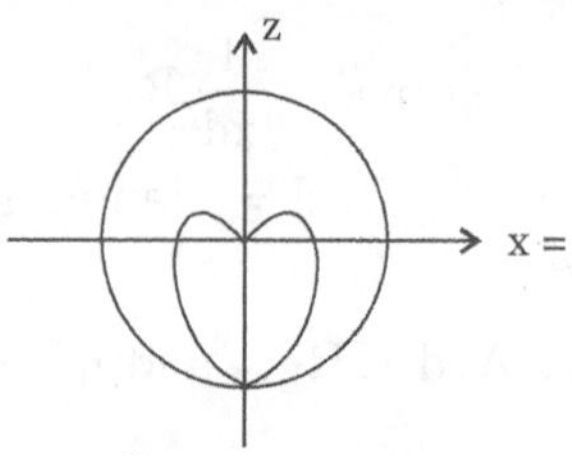

Group pattern

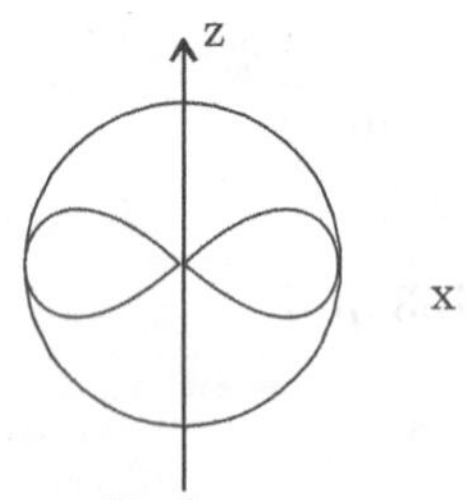

E-plane
overall radiation
patter

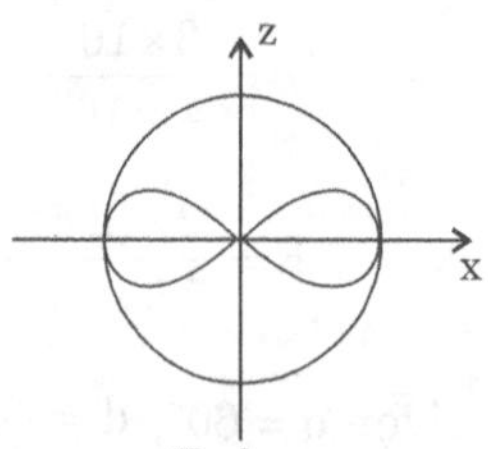

E-plane
overall radiation
patter

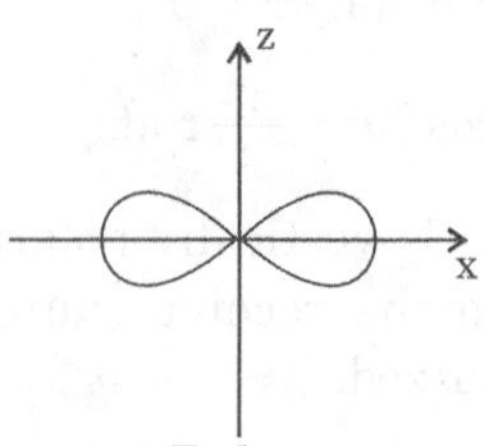

E-plane

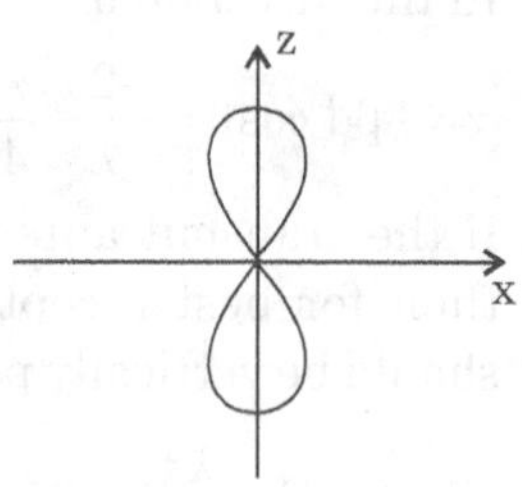

19. Given: length of vertical radiator, $dl = 50$ m, f = 600 kHz.

$$\lambda = \frac{c}{f} = \frac{3\times 10^8}{600\times 10^3} = 500 \text{ m.}$$

Radiation resistance of base fed Hertz dipole over a perfectly conducting ground is

$$R_{rad} = 40\pi^2\left(\frac{dl}{\lambda}\right)^2$$

$$= 40\pi^2 \times \left(\frac{50}{500}\right)^2 = \frac{2\pi^2}{5}\Omega.$$

20.

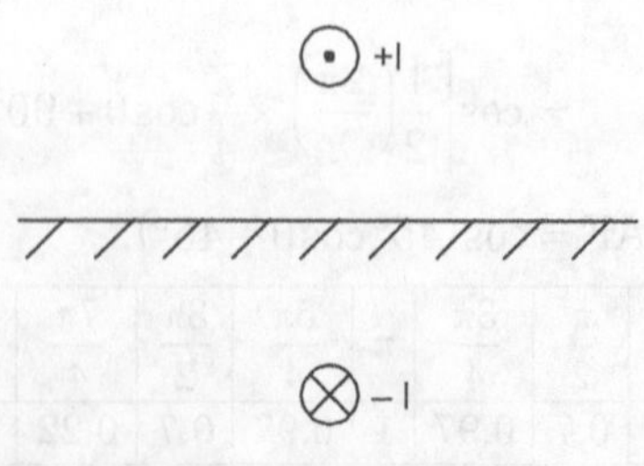

A $\lambda/2$ dipole kept $\lambda/2$ at a distance $\lambda_0/2$ above the grand equivalent to two $\lambda/2$ dipole at a distance of λ_0 between them with current flowing in them with a phase difference of π.

Now the array factor

$$AF = \cos\left|\frac{\beta d\cos\phi + \alpha}{2}\right|$$

Here $\alpha = \pi$ and $d = \lambda_0$, $\beta = \dfrac{2\pi}{\lambda_0}$

$$AF = \cos\left|\frac{2\pi\cos\phi + \pi}{2}\right| = |\sin(\pi\cos\phi)|$$

At $\phi = \dfrac{\pi}{2}$, array factor is zero and field strength is 0

At $\phi = 0$, field strength is also zero.

Hence, radiation field pattern will be as shown below

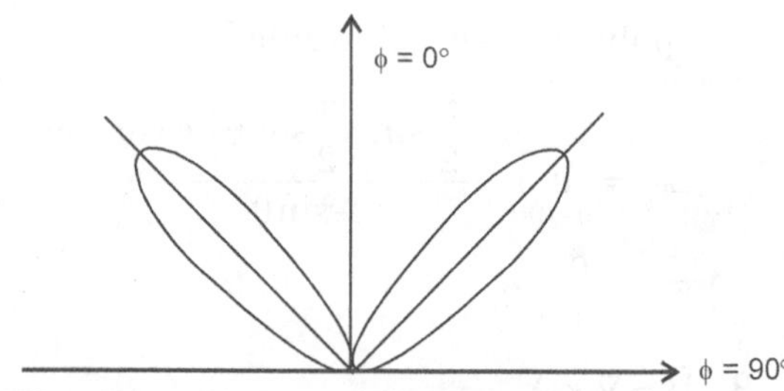

21. For Hertizian dipole, radiation field $E_\theta \propto \sin\theta$.

$\theta = 0°$	$\frac{\pi}{4}$	$\frac{\pi}{2}$	$\frac{3\pi}{4}$	π
|AF|= 0	0.707	1	0.707	0

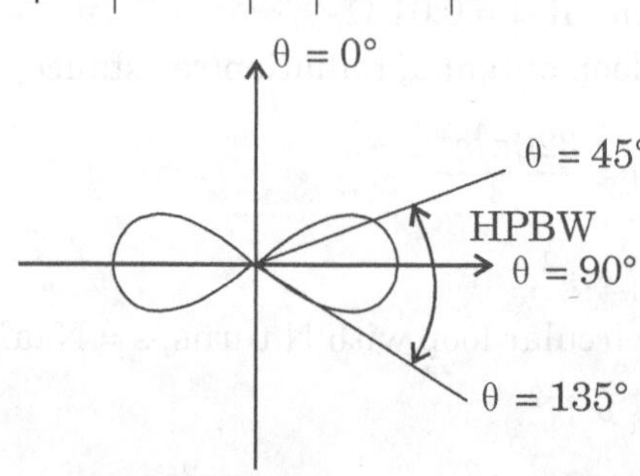

The half power bandwidth = 135° – 45° = 90°.

22. Point electromagnetic source radiates in all direction ⇒ Isotropic

* Dish antenna radiates any electromagnetic energy is any particular direction with narrow beamwidth and high directivity ⇒ highly direction.

* Yogi-Uda antenna is high bandwidth antenna used for T.V. reception ⇒ End-fire.

23. Given, $R_{rad} = 3\ \Omega$, and f = 3 GHz

$$\% \text{ change in length } \frac{d\ell}{\ell}\times 100\%$$

$$\frac{dR_{rad}}{R_{rad}} \times 100\% = ?$$

$$R_{rad} = \frac{80\pi^2}{\lambda^2}\ell^2 \quad ...(i)$$

Change in resistance with respect to change in length is

$$\frac{dR_{rad}}{d\ell} = 2\ell\left(\frac{80\pi^2}{\lambda^2}\right) \quad ...(ii)$$

dividing eq. (i) and (ii) we get

$$= \frac{\frac{dR_{rad}}{d\ell}}{R_{rad}} = \frac{2\ell}{\ell^2} = \frac{2}{\ell}$$

$$\therefore \frac{dR_{rad}}{R_{rad}} = 2 \times \frac{d\ell}{\ell}$$

$$\therefore \frac{dR_{rad}}{R_{rad}} \times 100\% = 2 \times \left(\frac{\partial \ell}{\ell} \times 100\%\right)$$

$$\left[\because \frac{d\ell}{\ell} \times 100\% = 1\%\right]$$

$\therefore$ Percent change in radiation resistance = 2%

24. Directivity of an antenna,

$$D \propto \frac{1}{\text{Beam width}} \quad \left(\therefore D = \frac{4\pi}{\text{Beam width}}\right)$$

and also $D = \frac{4p}{\lambda^2}.A_e.$

If two dissimilar antennas, having their maximum directivities equal, if their beam widths or effective operate areas are equal.

25. Given : d = 1 km = 10^3 m.

P_t = 1 W, f = 3 GHz.

$G_t = G_r$ = 30 dB

30 = 10 log G_t.

$G_t = G_r = 10^3 = 1000.$

$$\lambda = \frac{c}{f} = \frac{3 \times 10^8}{3 \times 10^9} = 0.1 \text{ m.}$$

Using friss equation,

$$P_r = P_t G_t G_r \left(\frac{\lambda}{4\pi d}\right)^2$$

$$= 1 \times 10^3 \times 10^3 \left(\frac{0.1}{4\pi \times 10^3}\right)^2$$

$$= \frac{1 \times 0.1 \times 0.1}{(4\pi)^2} = 63.4\ \mu\text{W.}$$

26. Directivity, $D = \frac{41,253}{\theta_{3dB} \cdot \phi_{3dB}} = \frac{41253}{2^\circ \times 2^\circ}$

$$D \cong \frac{4000}{4} = 1000$$

$D \cong$ 40dB.

27. Give: P_t = 251 W, d = 100 m, $(A_e)_r$ = 500 cm², G_t = 1 (isotropic)

$$\Rightarrow G = \frac{4\pi}{\lambda^2}.A_e$$

From friss formula, $P_t = P_t G_t G_r \left(\frac{\lambda}{4\pi d}\right)^2$

$$P_t G_t = \frac{4\pi}{\lambda^2}.A_e.\left(\frac{\lambda}{4\pi d}\right)^2$$

$$P_r = \frac{251 \times 1 \times 500 \times 10^{-4}}{4\pi \times (100)^2}$$

P_r = 100 μW.

28. Signal strength $= \frac{P}{4\pi R^2}$. For 3dB decrease,

$$\frac{P_2}{P_1} = \frac{1}{2}.$$

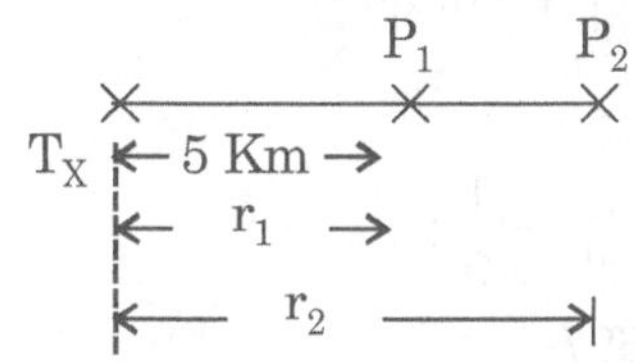

$$P_1 = \frac{P}{4\pi}(5000)^2, P_2 \frac{P}{4\pi.r_2^2}$$

$$\Rightarrow r_2 = 5000\sqrt{2} \text{ m.}$$

Required distance $= 5000\sqrt{2} - 5000 = 2071$ m.

29. Given: Directive gain of lossless antenna (W_l = 0) = 6 dB = 4.

Input power to antenna = 1 mW

$P_{in} = P_r \quad P_l = P_r + O$

$P_r = P_{in}$

Power radiated by antenna = 1 mW.

30. Input Power, P_{in} = 1W

Gain, G = 10 dB = 10^1 = 10

Radiated power, $P_{rad} = P_{in} \times G = 1 \times 10$ W.

31. Given: d = 1 m, η = 0.7, f = 20 GHz.

$$\lambda = \frac{c}{f} = \frac{3 \times 10^8}{20 \times 10^9} = 1.5 \text{ cm} = 15 \times 10^{-3} \text{ m}$$

Gain of parabolic dish antenna,

$$G_D = \pi^2\left(\frac{d}{\lambda}\right)^2 = \pi^2\left(\frac{1}{15 \times 10^{-3}}\right)^2$$

G_D = 43684.9.

$\Rightarrow G_P = \eta.\ G_D = 0.7 \times 43684.9 = 30705.44$

(G_P) in dB = 10log G_P = 10log G_P = 10 log(30705.44) = 44.87

32. Given: $F(\theta) = \cos^4 \theta;\ 0 \le \theta \le \frac{\pi}{2}.$

Assuming the given quantity is field strength.

Radiation intensity,

$$U(\theta, \phi) \propto [F(\theta)]^2$$

$$U(\theta, \phi) = \cos^2 \theta$$

$$|U(\theta, \phi)|_{max} = 1$$

Radiated power,

$$P_{rad} = \int_{\theta}\int_{\phi} U(\theta, \phi) \sin\theta\, d\theta\, d\phi$$

$$= \int_{\theta=0}^{\frac{\pi}{2}} \int_{\phi=0}^{2\pi} \cos^2 \theta.\ \sin\theta\, d\theta\, d\phi$$

$\therefore$ Let $\cos\theta = t \Rightarrow -\sin\theta\, d\theta = dt$

$$P_{rad} = \int_{1}^{0} \int_{\phi=0}^{2\pi} t^8.\ dt\, d\phi$$

$$\left[\frac{t^9}{9}\right]_0^1 .[\phi]_0^{2\pi} = \frac{1}{9}.2\pi$$

$P_{rad} = 0.697.$

$$\text{Directivity, } D = 4\pi.\frac{|U(\theta,\phi)|_{max}}{P_{rad}} = \frac{4\pi \times 1}{0.697} = 18.02$$

$D(\text{in dB}) = 10 \log_{10}(18.02) = 12.55$

33. Given: $\overline{E} = \frac{100}{r} \sin\theta.\cos(\omega r - \beta r)\hat{a}_\theta \text{ V/m},$

$\overline{H} = \frac{0.265}{r} \sin\theta.\cos(\cot - \beta r)\hat{a}_\phi \text{ A/m}.$

$$\text{Average power, } Pav = \frac{1}{2}\int_s (\overline{E}_\theta \times \overline{H}_\phi).\overline{ds} = \frac{1}{2}$$

$$.\overline{ds} = \frac{1}{2}\int_s \frac{100 \times 0.265}{r^2} \sin^3 \theta d\theta d\phi$$

$$= 13.25 \int_{\theta=0}^{\frac{\pi}{2}} \sin^3 \theta d\theta \int_{\phi=0}^{\frac{\pi}{2}} d\phi$$

$$\therefore \int_0^{\frac{\pi}{2}} \sin^n \theta \frac{(n-1)(n-2)(n-5)...}{n(n-2)(n-4)}$$

$$P_{av} = 13.25 \times \frac{(3-1)}{3} \times 2\pi = 55.5 \text{ W}.$$

34. Given E_1 = 12 mV/m, r_1 = 1 km, r_2 = 2 km.

Electric field of an antenna

$$E_\theta = \frac{\eta I_o dl}{4\pi} \sin\left[\frac{j\beta}{r} + \frac{1}{r^2} - \frac{j}{\beta r^3}\right]$$

At far field, $E \propto \frac{1}{r}$

$$\frac{E_2}{E_1} = \frac{r_2}{r_1}$$

$$\Rightarrow E_2 = \frac{E_1 r_2}{r_1} = \frac{12 \times 10^{-13} \times 1 \times 10^3}{2 \times 10^3} = 6 \text{ mV/m}.$$

$$P_{av} = \frac{1}{2}\frac{E_2^2}{\eta} = \frac{1}{2} \times \frac{(6 \times 10^{-3})^2}{120\pi} = 47.7 \text{ nW/m}^2.$$

35. Directly, $D = \frac{4\pi}{\lambda^2}.A_e$

Directly is increased by adding more antenna eternents because this increases the effective area of antenna which result into in directivity.

36. For lossless horn antennas

$$\eta_T = \eta_R = 1$$

Power Gain = Directivity

Directivity of Transmitting antenna,

$$D_T = 18 \text{ dB}$$

$$10 \log D_T = 18$$

$$G_T \text{ (or) } D_T = 63.09$$

Directivity of Receiving antenna,

$$D_R = 22 \text{ dB}$$

$$10 \log D_R = 22$$

$$G_R \text{ (or) } D_R = 158.48$$

input power P_{in} = 2 W

Spacing, $r = 200\ \lambda$

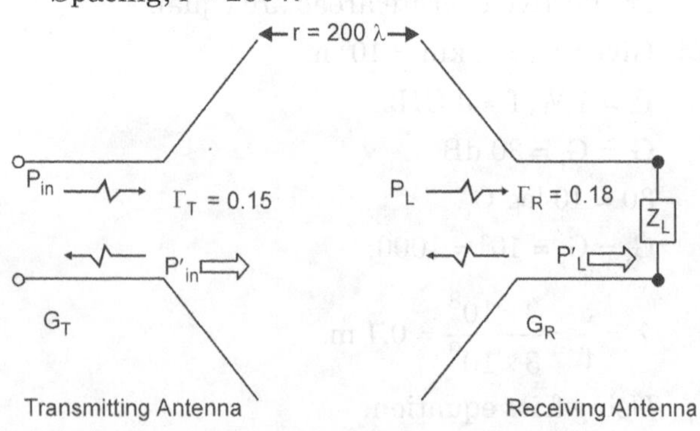

Friis transmission formula in given by

$$P_L = G_T G_R \left[\frac{\lambda}{4\pi r}\right]^2 P'_{in}$$

where,

P'_{in} : Input power (prime indicates power due to reflection)

$$P'_{in} = \left|1 - \Gamma_T^2\right| P_{in}$$

$$= \left|1 - (0.15)^2\right| \times 2$$

$$P'_{in} = 1.955\ W$$

$$P_L = 63.09 \times 158.48 \left[\frac{\lambda}{4\pi \times 200\lambda}\right]^2 \times 1.955$$

$$= 3.1 \times 10^{-3}$$

As there is a reflection at the terminals of receiving antenna power delivered to the load in given by

$$P'_L = \{1 - |\Gamma_R^2|\} \times P_L$$

$$= \{|1 - (0.18)^2|\} \times 3.1 \times 10^{-3}$$

$$\therefore \quad P'_L = 2.99\ mW$$

Hence, the power delivered to the load at the receiver is 2.99 mW.

37. Given : $\overline{W}_{rad} = \overline{W}_{average} = co\frac{1}{r^2}\cos^4\theta.\, \hat{a}_r.$

Power radiated by antenna, $P_{rad} = \int_s \overline{W_{rad}}.\overline{ds}$

$$P_{rad} = \int_{\theta=0}^{\frac{\pi}{2}} \int_{\phi=0}^{2\pi} co.\frac{1}{r^2}\cos^4\theta.\, r^2 \sin q\, d\theta\, d\phi$$

$$= 2\pi co \int_{\theta=0}^{\pi^4} \cos^4\theta\, d\theta.\sin\theta$$

$$\therefore d(-\cos\theta) = \sin\theta d\theta$$

$$P_{rad} = 2\pi c_o\left(-\frac{\cos^5\theta}{5}\right)\Bigg|_0^{\frac{\pi}{2}}$$

$$\frac{2\pi}{5}co = 1.256\ co$$

$$\text{Directivity, } D = \frac{4\pi V}{\int W_{rad} d\Omega}$$

$$\frac{4\pi\, co^4\,\theta}{\int_{\theta=0}^{\frac{\pi}{2}}\int_{\phi=0}^{2\pi} \cos^4\theta \sin\theta d\theta d\phi}$$

$$D = \frac{2\cos^4\theta}{\frac{1}{5}} = 10\cos^4\theta$$

$$D_{max} = 10$$

D_{max} in dB $= 10 \log_{10} 10 = 10$ dB.

38. Given

frequency, $f = 5$ GHz $= 5 \times 10^9$ Hz

$$\text{wave length, } \lambda = \frac{c}{f} = \frac{3 \times 10^8}{5 \times 10^9} = 0.06\ m$$

gain of antenna, G = 150

Range of target, $R_{max} = 1$ km $= 10^3$ m,

radar cross-section, $\sigma = 3m^2$,

transmitted power, $P_t = 100$ kW

Radar range equation is given by

$$R_{max} = \left[\frac{P_t \times G \times \frac{\lambda^2}{4\pi} \times G \times \sigma}{(4\pi)^2 \times P_R}\right]^{\frac{1}{4}} \quad \left(\text{Since } A_e = \frac{\lambda^2}{4\pi}G\right)$$

The received power, P_R is given by

$$P_R = \frac{100 \times 10^3 \times 150 \times 150 \times (0.06)^2 \times 3}{(4\pi)^3 \times (10^3)^4}$$

$$= 1.22 \times 10^{-8} = 0.0122\ \mu W$$

Hence, the received power is 0.0122 μW.

39. From friis free space propagation

$$\text{eq, } W_r = \frac{W_t A_{er}\, A_{et}}{(\lambda d)^2}.$$

If A_{e_r} and A_{et} are doubled with doubled d, W_r is same. Hence, capacity is also same.

40. Given

Electric field of incident wave is given by

$$E_W^i = (\hat{a}_x + j\hat{a}_y)E_0 e^{jkz}$$

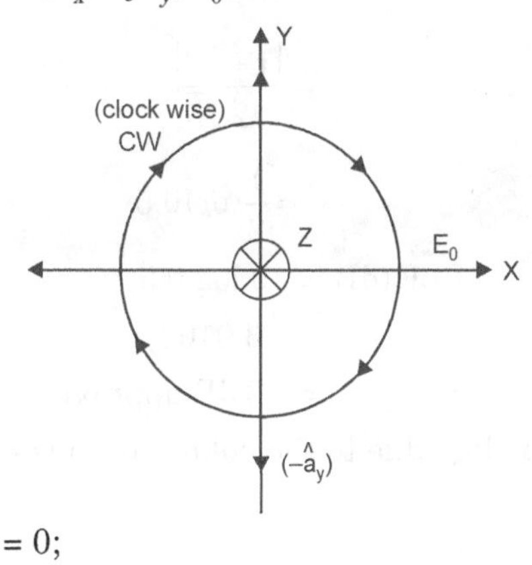

at $z = 0$;

$$\vec{E}_w^i = E_0 \cos\omega t\hat{a}_x - E_0 \sin\omega t\hat{a}_y$$

(in time varying form)

at $\omega t = 0$

$$\vec{E}_w^i = E_0\,\hat{a}_x$$

$$\text{at} \quad \omega t = \frac{\pi}{2}$$

$$\vec{E}_w^i = E_0(-\hat{a}_y)$$

As a tip of electric field intensity is tracing a circle when time varies, hence the wave is said to be circularly polarized in clockwise direction (or) RHCP. Polarizing vector of incident wave is given by,

$$\hat{P}_i = \frac{\hat{a}_x + j\hat{a}_y}{\sqrt{2}}$$

radiated electric field from the antenna is

$$\vec{E}_a = (\hat{a}_x + 2\hat{a}_y)E_I \frac{1}{\gamma} e^{-jk\gamma}$$

at $r = 0$

$$\vec{E}_a = E_I \cos\omega t\, \hat{a}_x + 2E_I \cos\omega t\, \hat{a}_y$$

(in time varying form)

As both x and y components are in-phase hence the wave is said to be linear polarized. Polarizing vector of radiated field is $\hat{P}_a = \frac{(\hat{a}_x + 2\hat{a}_y)}{\sqrt{5}}$ polarizing mismatch; The polarizing mismatch is said to have, if the polarization of receiving antenna is not same on the polarization of the incident wave. The polarization loss factor (PLF) characterizes the loss of EM power due to polarization mismatch.

$$\text{PLF} = |\hat{P}_i \cdot \hat{P}_a|^2$$

in dB; PLF (dB) = 10 log (PLF)

$$\text{PLF} = \left|\left(\frac{\hat{a}_x + j\hat{a}_y}{2}\right) \cdot \left(\frac{\hat{a}_x + 2\hat{a}_y}{\sqrt{5}}\right)\right|$$

$$= \left|\frac{1 + j2}{\sqrt{2}\sqrt{5}}\right|^2$$

$$= \frac{1}{2} \text{(or)} 0.5$$

$$\text{PLF (dB)} = 10 \log 0.5$$

$$= -3.0102$$

$$= -3 \text{ dB(approx)}$$

So, the loss due to the polarization is – 3 dB.

41. TEM wave with circular polarization $(\hat{a}x + ja\hat{y})$ is received by a dipole, which is linearly polarized $(\hat{a}x)$. Due to polarization mismatch, there will be polarization loss.

$$\text{Polarization loss factor} = |P_w \cdot Pla|^2$$

$$= \left|\hat{a}_x \frac{\hat{a}_x + j\hat{a}_y}{\sqrt{2}}\right|^2 = \frac{1}{2}.$$

Hence, power efficiency is 50%.

42. Given: $\lambda = 492$ m, $h_t = 124\text{m} \cong \frac{\lambda}{4}$.

Tower antenna is like quarter wave monopole, so $R_{rad} = 36.5\ \Omega$.

43. Normalized array factor $= 2\left|\cos\frac{\Psi}{2}\right|$

where $\Psi = \beta d \sin\theta \cos\phi + \delta$

Here $\theta = 90°$, $d = \sqrt{2}s$, $\phi = 45°$, $r = 180°$.

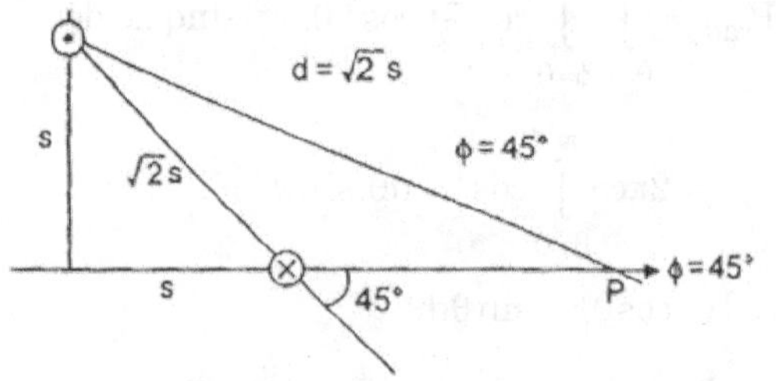

$$\text{Then, } 2\left|\cos\frac{\Psi}{2}\right| = 2\cos\left(\frac{\beta d \sin\theta\cos\phi + \delta}{2}\right)$$

$$= 2\cos\left[\frac{2\pi}{\lambda.2}\sqrt{2}s\cos 45° + \left(\frac{180°}{2}\right)\right]$$

$$= 2\cos\left(\frac{\pi s}{\lambda} + 90°\right) = 2\sin\left(\frac{\pi s}{\lambda}\right)$$

Unit - III

Controls System

Syllabus

Basic control system components; Feedback principle; Transfer function; Block diagram representation; Signal flow graph; Transient and steady-state analysis of LTI systems; Frequency response; Routh-Hurwitz and Nyquist stability criteria; Bode and root-locus plots; Lag, lead and lag-lead compensation; State variable model and solution of state equation of LTI systems.

Contents

1 CHAPTER

Modeling & Transfer Function

Analysis of Previous GATE Papers												
Year → Topics ↓			2019	2018	2017 Set 1	2017 Set 2	2016 Set 1	2016 Set 2	2016 Set 3	2015 Set 1	2015 Set 2	2015 Set 3
TRANSFER FUNCTION OF LINEAR SYSTEM	1 Mark	MCQ Type										
		Numerical Type										
	2 Marks	MCQ Type										
		Numerical Type										
	Total											
FEEDBACK PRINCIPLE	1 Mark	MCQ Type										
		Numerical Type										
	2 Marks	MCQ Type										
		Numerical Type										
	Total											

TRANSFER FUNCTION OF LINEAR SYSTEM

1. The transfer function of a linear system is the

(a) ratio of the output, $v_0(t)$ and input $v_i(t)$

(b) ratio of the derivatives of the output and the input

(c) ratio of the Lapalace transform of the output and that of the input with all initial conditions zeros.

(d) None of these.

[1995 : 1 Mark]

2. The transfer function of a tachometer is of the form

(a) Ks (b) $\frac{K}{s}$

(c) $\frac{K}{(s+1)}$ (d) $\frac{K}{s(s+1)}$

[1995 : 1 Mark]

3. The unit-step response of a system starting from rest is given by $c(t) = 1 - e^{-2t}$ for $t \geq 0$.
The transfer function of the system is

(a) $\frac{1}{1+2s}$ (b) $\frac{2}{2+s}$

(c) $\frac{1}{2+s}$ (d) $\frac{2s}{1+2s}$

[2006 : 2 Marks]

4. A system with the transfer function $\frac{Y(s)}{X(s)} = \frac{s}{s+p}$ has an output $y(t) = \cos\left(2t - \frac{\pi}{3}\right)$, Then, the system parameter 'p' is

(a) $\sqrt{3}$ (b) $\frac{2}{\sqrt{3}}$

(c) 1 (d) $\frac{\sqrt{3}}{2}$

[2010 : 1 Mark]

FEEDBACK PRINCIPLE

5. Tachometer feedback in a d.c. position control system enhances stability (T/F)

[1994 : 1 Mark]

6. The open-loop DC gain of a unity negative feedback system with closed-loop transfere function $\frac{s+4}{s^2+7s+13}$ is

(a) $\frac{4}{13}$

(b) $\frac{4}{9}$

(c) 4

(d) 13

[2001 : 2 Marks]

7. Despite the presence of negative feedback, control systems still have problems of instability because the

(a) components used have nonlinearities

(b) dynamic equations of the systems are not known exactly

(c) mathematical analysis involves approximations

(d) system has large negative phase angle at high frequencies

[2005 : 1 Mark]

ANSWERS

1. (c)	2. (a)	3. (b)	4. (b)	5. (True)	6. (b)	7. (a)

EXPLANATIONS

1. The transfer function of a linear, time-invariant system is defined as the ratio of the Laplace transform of the output and that of the input with all initial conditions are zero.

2. Transfer function of a tachometer = $\frac{E(s)}{Q(s)} = k.s$

3. Input, r(t) = 4(t)

$$R(s) = \frac{1}{s}$$

Response, $e(t) = 1 - e^{-2t} \;\forall\; t \geq 0$

$$C(s) = \frac{1}{s} - \frac{1}{s+2} = \frac{2}{s(s+2)}$$

$$\text{Transfer function, } T(s) = \frac{C(s)}{R(s)} = \frac{2}{s(s+2) \times \frac{1}{s}}$$

$$T(s) = \frac{2}{(s+2)}$$

4. Phase difference between input and output = 30°

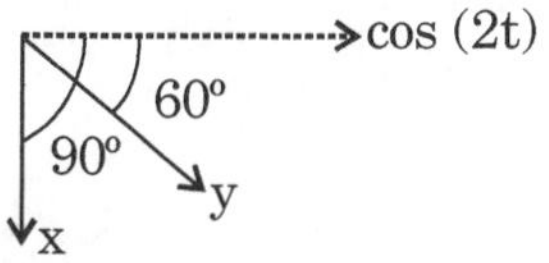

Given: ω = 2 rad/s.

$$\frac{Y(s)}{X(s)} = \frac{S}{S+p}$$

$$\angle Y(j\omega)/X(j\omega) = 90° - \tan^{-1}\left(\frac{\omega}{P}\right) = 30°$$

$$\tan^{-1}\left(\frac{2}{p}\right) = 60°$$

$$\frac{2}{p} = \sqrt{3}$$

$$P = \frac{2}{\sqrt{3}}$$

5. Tranfer function of tachometer = KS

The tachometer feedback is a derivative feedback. It adds zero at origins. Hence, type of system decreases and stability improves.

6.

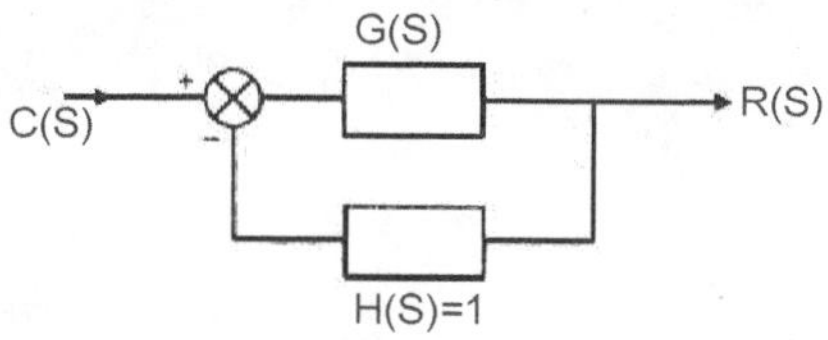

$$\frac{C(s)}{R(s)} = \frac{G(s)}{1 + G(s)\,H(s)}$$

$$\Rightarrow \quad \frac{G(s)}{1+G(s)} = \frac{s+4}{s^2+7s+13}$$

$$\Rightarrow \quad \frac{1+G(s)}{G(s)} = \frac{s^2+7s+13}{s+4}$$

$$\Rightarrow \quad \frac{1}{G(s)} = \frac{s^2+6s+9}{s+4}$$

$$\Rightarrow \quad G(s) = \frac{s+4}{s^2+6s+9}$$

DC gain will occur at ω = 0

$$|G|_{DC} = \frac{4}{9}.$$

7. In general, the components used (like amplifier controllers, etc.) are assumed to behave linearly. But non-lineaities (saturation) are inhently present which make the system unstable.

2

CHAPTER

Block Diagram Reduction

Analysis of Previous GATE Papers												
	Year → Topics ↓		2019	2018	2017 Set 1	2017 Set 2	2016 Set 1	2016 Set 2	2016 Set 3	2015 Set 1	2015 Set 2	2015 Set 3
BLOCK DIAGRAM ALGEBRA	1 Mark	MCQ Type							1			
		Numerical Type				1						
	2 Marks	MCQ Type	1								1	
		Numerical Type										
	Total		2			1					2	
SIGNAL FLOW GRAPH	1 Mark	MCQ Type									1	
		Numerical Type										
	2 Marks	MCQ Type										
		Numerical Type										
	Total										1	

BLOCK DIAGRAM ALGEBRA

1. For the system shown in the figure, $Y(s)/X(s) =$ ________.

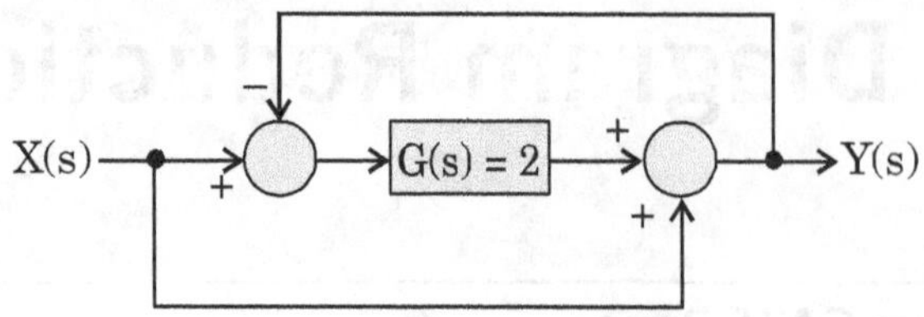

[1986 : 1 Mark]

2. For the system shown in figure the transfer function $\frac{C(s)}{R(s)}$ is equal to

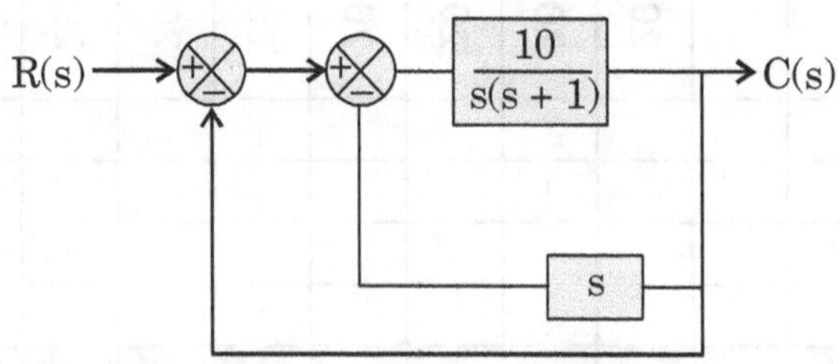

(a) $\frac{10}{s^2+s+10}$ (b) $\frac{10}{s^2+11s+10}$

(c) $\frac{10}{s^2+9s+10}$ (d) $\frac{10}{s^2+2s+10}$

[1987 : 2 Marks]

3. The equivalent of the block diagram in the figure is given as

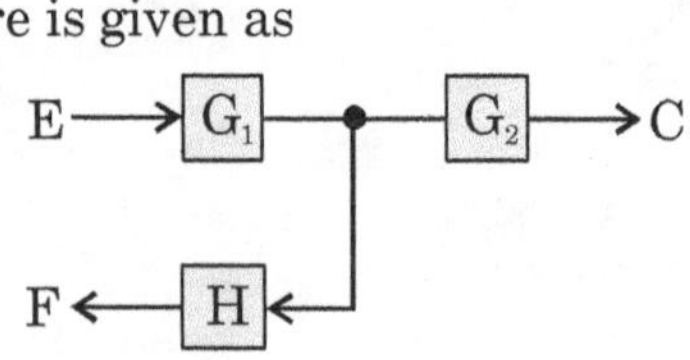

(a)

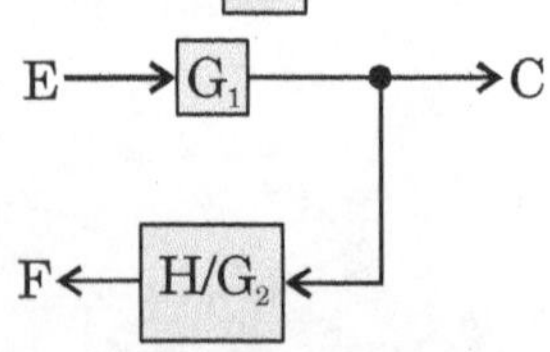

(b)

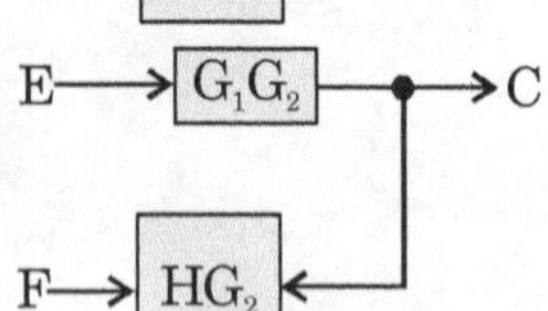

(c)

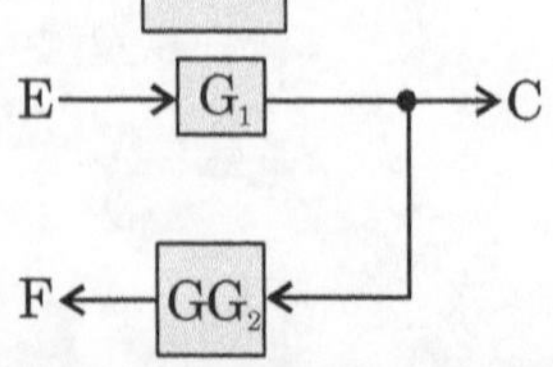

(d)

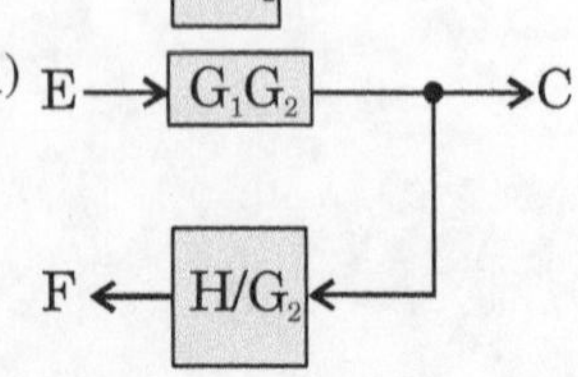

[2001 : 1 Mark]

4. The transfer function Y(s)/R(s) of the system shown is

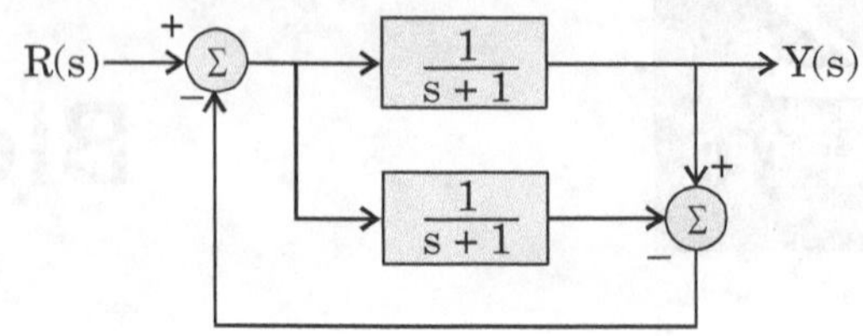

(a) 0 (b) $\frac{1}{s+1}$

(c) $\frac{2}{s+1}$ (d) $\frac{2}{s+3}$

[2010 : 1 Mark]

5. For the following system,

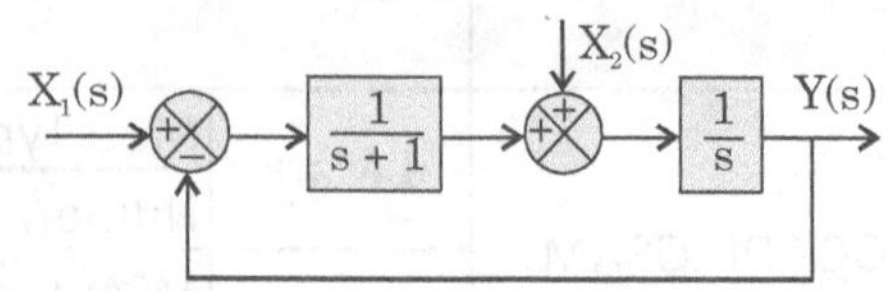

when $X_1(s) = 0$, the transfer function $\frac{Y(s)}{X_2(s)}$ is

(a) $\frac{s+1}{s^2}$ (b) $\frac{1}{s+1}$

(c) $\frac{s+2}{s(s+1)}$ (d) $\frac{s+1}{s(s+2)}$

[2014 : 1 Mark, Set-2]

6. Consider the following block diagram in the figure.

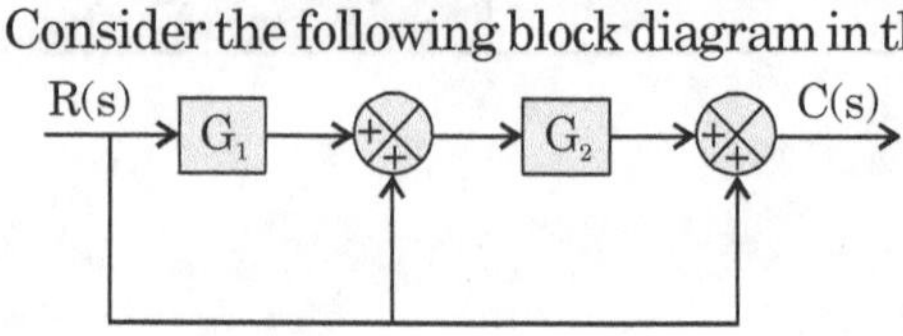

The transfer function $\frac{C(s)}{R(s)}$ is

(a) $\frac{G_1G_2}{1+G_1+G_2}$ (b) $G_1G_2 + G_1 + 1$

(c) $G_1G_2 + G_2 + 1$ (d) $\frac{G_1}{1+G_1G_2}$

[2014 : 1 Mark]

7. By performing cascading and/or summing/ differencing operations using transfer function blocke $G_1(s)$ and $G_2(s)$, one CANNOT realizede a transfer function of the form

(a) $G_1(s)G_2(s)$

(b) $\frac{G_1(s)}{G_2(s)}$

(c) $G_1(s)\left(\frac{1}{G_1(s)}+G_2(s)\right)$

(d) $G_1(s)\left(\frac{1}{G_1(s)}-G_2(s)\right)$

[2015 : 2 Marks, Set-2]

8. The block diagram of a feedback control system is shown in the figure. The overall closed-loop gain G of the system is

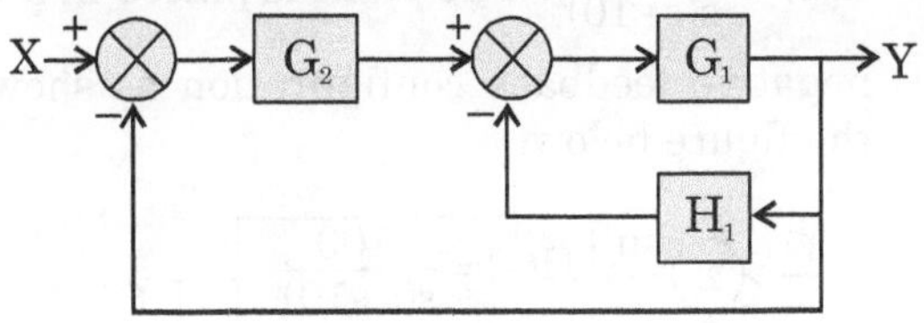

(a) $G = \dfrac{G_1G_2}{1+G_1H_1}$

(b) $G = \dfrac{G_1G_2}{1+G_1G_2+G_1H_1}$

(c) $G = \dfrac{G_1G_2}{1+G_1G_2H_1}$

(d) $G = \dfrac{G_1G_2}{1+G_1G_2+G_1G_2H_1}$

[2016 : 1 Mark, Set-3]

9. The block diagram of a system is illustrated in the figure shown, where X(s) is the input and Y(s) is the output. The transfter function $H(s) = \dfrac{Y(s)}{X(s)}$ is

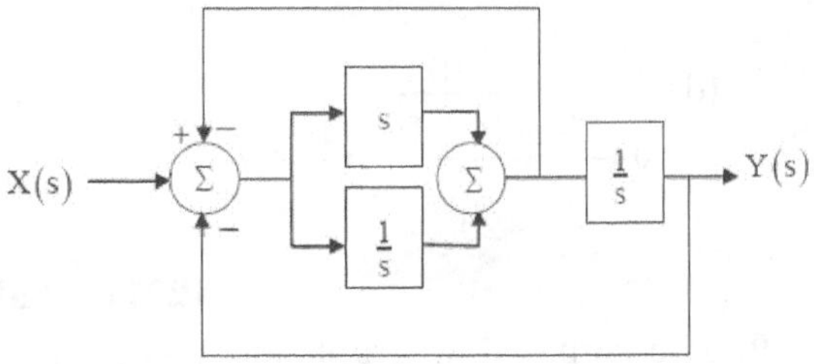

(a) $H(s) = \dfrac{s^2+1}{2s^2+1}$ (b) $H(s) = \dfrac{s^2+1}{s^3+2s^2+s+1}$

(c) $H(s) = \dfrac{s+1}{s^2+s+1}$ (d) $H(s) = \dfrac{s^2+1}{s^3+s^2+s+1}$

[2019 : 2 Marks]

SIGNAL FLOW GRAPH

10. An electrical system and its signal-flow graph representations are shown in the figure (a) and (b) respectively. the values of G_2 and H, respectively. are

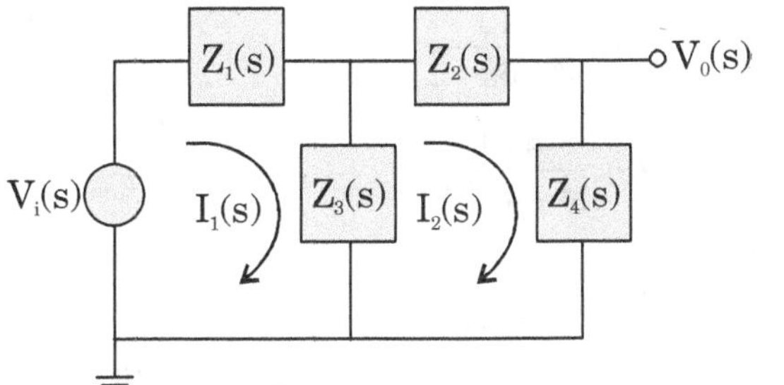

Figure (a)

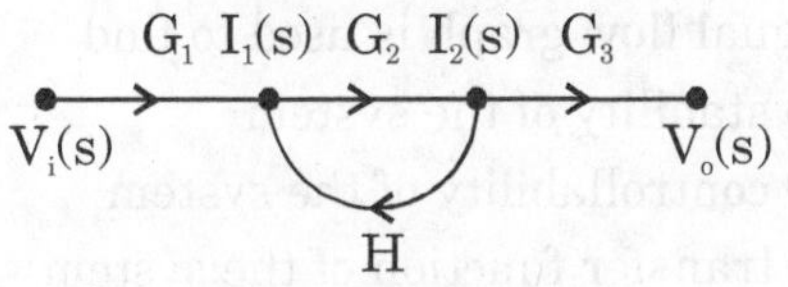

Figure (b)

(a) $\dfrac{Z_3(s)}{Z_2(s)+Z_3(s)+Z_4(s)}, \dfrac{-Z_3(s)}{Z_1(s)+Z_3(s)}$

(b) $\dfrac{-Z_3(s)}{Z_2(s)-Z_3(s)+Z_4(s)}, \dfrac{-Z_3(s)}{Z_1(s)+Z_3(s)}$

(c) $\dfrac{Z_3(s)}{Z_2(s)+Z_3(s)+Z_4(s)}, \dfrac{-Z_3(s)}{Z_1(s)+Z_3(s)}$

(d) $\dfrac{-Z_3(s)}{Z_2(s)-Z_3(s)+Z_4(s)}, \dfrac{-Z_3(s)}{Z_1(s)+Z_3(s)}$

[1986 : 2 Marks]

11. In the signal flow graph shown in figure $X_2 = TX_1$ where T, is equal to

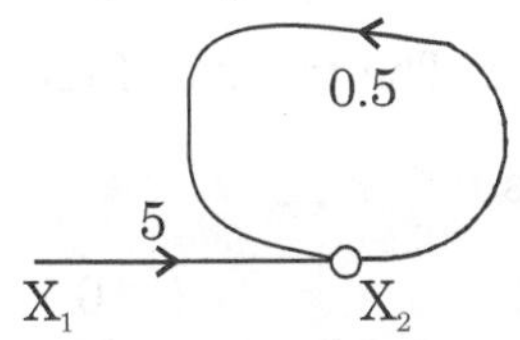

(a) 2.5 (b) 5

(c) 5.5 (d) 10

[1987 : 2 Marks]

12. The C/R for the signal flow graph in figure is:

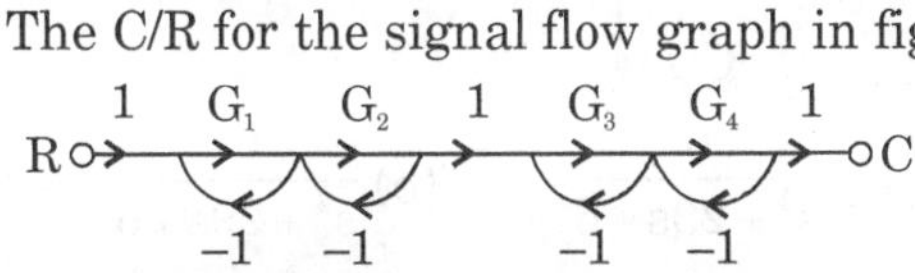

(a) $\dfrac{G_1G_2G_3G_4}{(1+G_1G_2)(1+G_3G_4)}$

(b) $\dfrac{G_1G_2G_3G_4}{(1+G_1+G_2+G_1G_2)(1+G_3+G_4+G_3G_4)}$

(c) $\dfrac{G_1G_2G_3G_4}{(1+G_1+G_2)(1+G_3+G_4)}$

(d) $\dfrac{G_1G_2G_3G_4}{(1+G_1+G_2+G_3+G_4)}$ **[1989 : 2 Marks]**

13. In the signal flow graph of figure the gain c/r will be

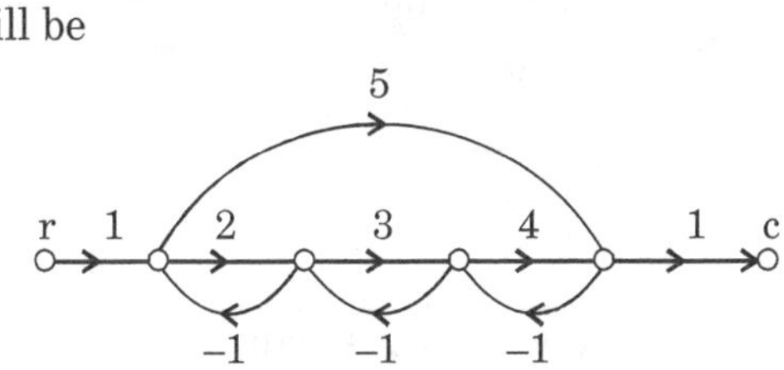

(a) $\dfrac{11}{9}$ (b) $\dfrac{22}{15}$

(c) $\dfrac{24}{23}$ (d) $\dfrac{44}{23}$

[1991 : 2 Marks]

14. Signal flow graph is used to find
(a) stability of the system
(b) controllability of the system
(c) transfer function of the system
(d) poles of the system

[1995 : 1 Mark]

15. In the signal flow graph of figure, y/x equals

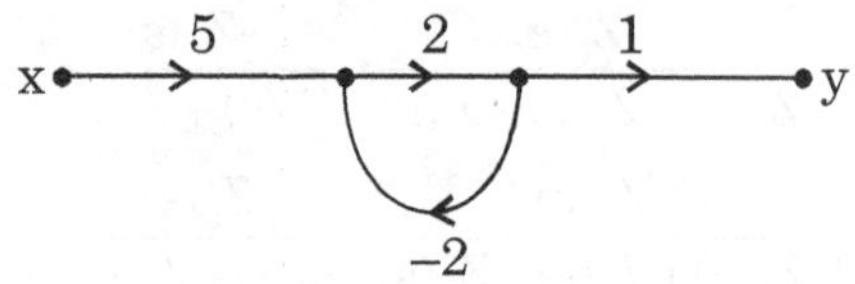

(a) 3 (b) $\frac{5}{2}$
(c) 2 (d) None of the above

[1997 : 2 Marks]

16. The signal flow graph of a system is shown in the figure. The transfer function $\frac{C(s)}{R(s)}$ of the system is

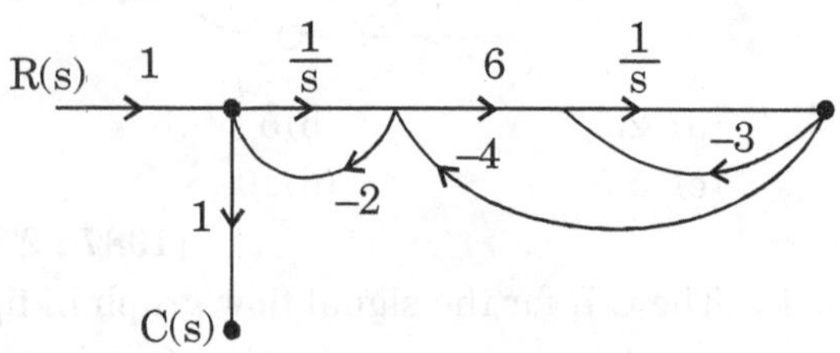

(a) $\frac{6}{s^2+29s+6}$ (b) $\frac{6s}{s^2+29s+6}$

(c) $\frac{s(s+2)}{s^2+29s+6}$ (d) $\frac{s(s+27)}{s^2+29s+6}$

[2003 : 2 Marks]

17. Consider the signal flow graph shown in the figure. The gain x_5/x_1 is

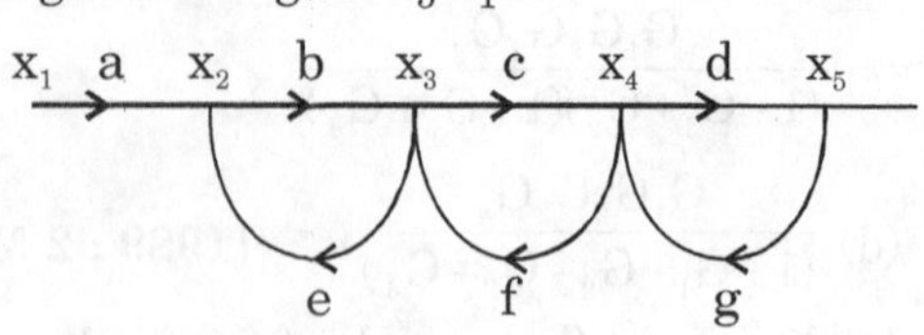

(a) $\frac{1-(be+cf+dg)}{abc}$

(b) $\frac{bcdg}{1-(be+cf+dg)}$

(c) $\frac{abcd}{1-(be+cf+dg)+bedg}$

(d) $\frac{1-(be+cf+dg)+bedg}{abcd}$

[2004 : 2 Marks]

18. The input-output transfer function of a plant $H(s) = \frac{100}{s(s+10)^2}$ The plant is placed in a unity negative feedback configuration as shown in the figure below.

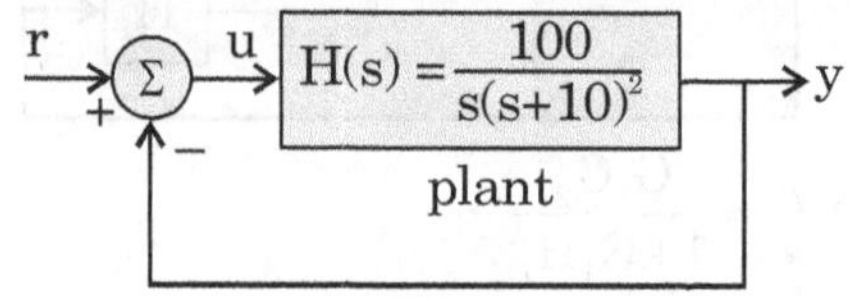

The signal flow graph that DOES NOT model the plant transfer function H(s)

(a)
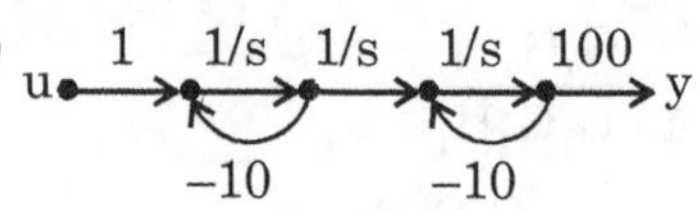

(b)
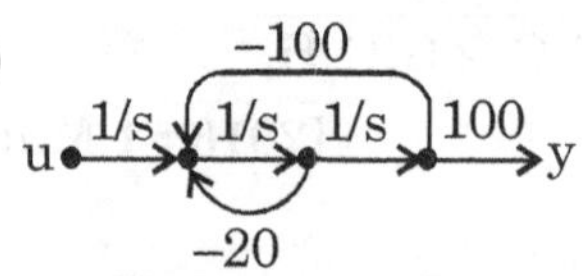

(c)
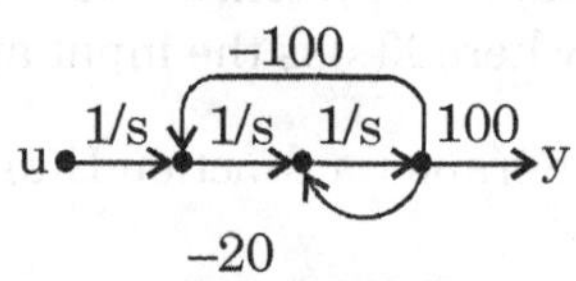

(d)
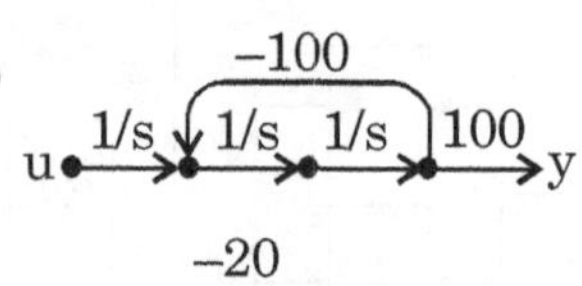

[2011 : 2 Marks]

19. The signal flow graph for a system is given below. The transfer function $\frac{Y(s)}{U(s)}$ for this system is

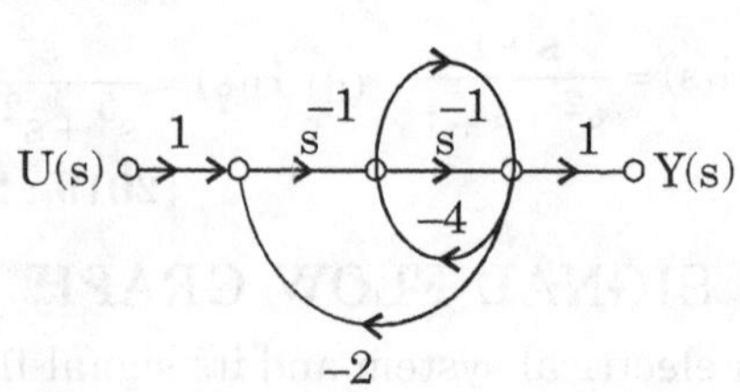

(a) $\frac{s+1}{5s^2+6s+2}$

(b) $\frac{s+1}{s^2+6s+2}$

(c) $\frac{s+1}{s^2+4s+2}$

(d) $\frac{1}{5s^2+6s+2}$

[2013 : 2 Marks]

20. For the signal flow graph shown in the figure, the value of $\frac{C(s)}{R(s)}$ is

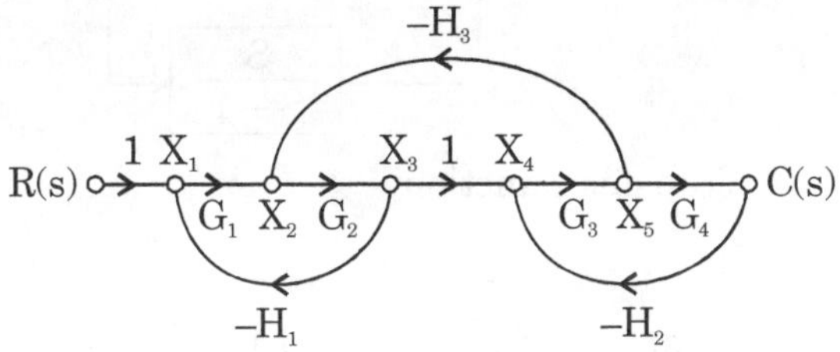

(a) $\dfrac{G_1G_2G_3G_4}{1-G_1G_2H_1-G_3G_4H_2-G_2G_3H_3+G_1G_2G_3H_1H_2}$

(b) $\dfrac{G_1G_2G_3G_4}{1+G_1G_2H_1+G_3G_4H_2+G_2G_3H_3+G_1G_2G_3H_1H_2}$

(c) $\dfrac{1}{1+G_1G_2H_1+G_3G_4H_2+G_2G_3H_3+G_1G_2G_3G_4H_1H_2}$

(d) $\dfrac{1}{1-G_1G_2H_1+G_3G_4H_2+G_2G_3H_3+G_1G_2G_3G_4H_1H_2}$

[2015 : 1 Mark, Set-2]

ANSWERS

1. (1)	2. (b)	3. (d)	4. (b)	5. (d)	6. (c)	7. (b)	8. (b)	9. (b)	10. (c)
11. (d)	12. (c)	13. (d)	14. (c)	15. (c)	16. (d)	17. (c)	18. (d)	19. (a)	20. (c)

EXPLANATIONS

1.

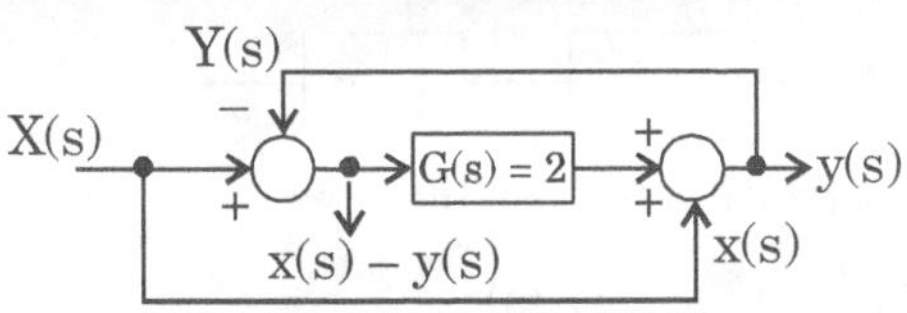

From figure,

$Y(s) = [X(s) - Y(s)]\,G(s) + X(s)$

$[1 + G(s)]\,Y(s) = [1 + G(s)]\,X(s)$

$$\frac{Y(s)}{X(s)} = 1$$

2. Redrawing the figure

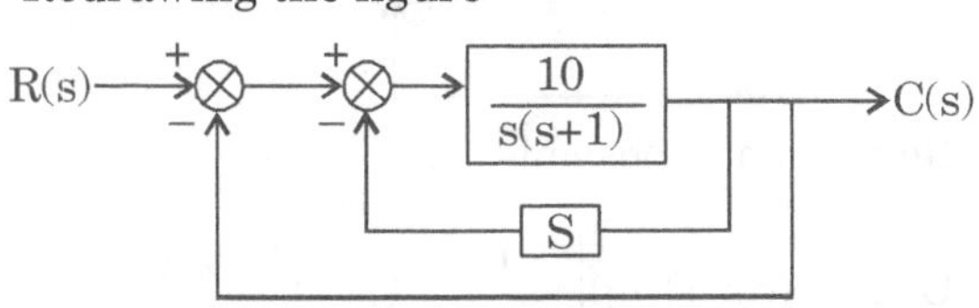

$$\therefore \frac{\dfrac{10}{s(s+1)}}{1+\dfrac{10}{s(s+1)}\cdot s} = \frac{10}{s^2+s+10s} = \frac{10}{s^2+11s}$$

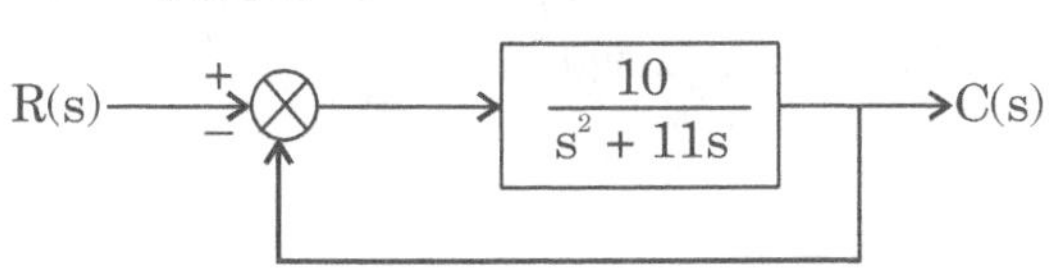

$$\frac{C(s)}{R(s)} = \frac{10/(s+11)s}{1+\dfrac{10}{s^2+11s}}\cdot 1 = \frac{10}{s^2+11s+10}$$

3.

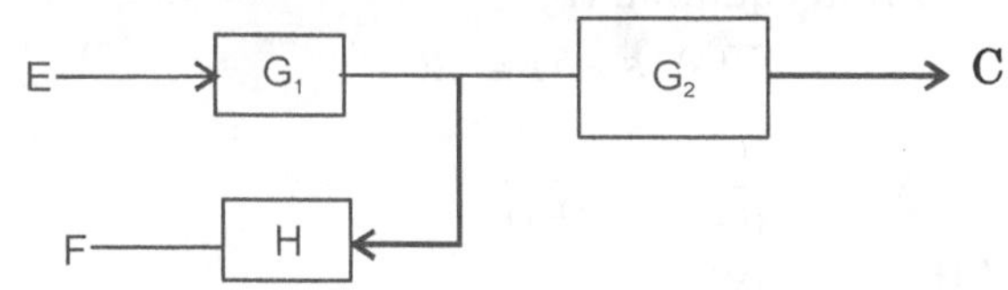

$F = EG_1H$

(*a*)

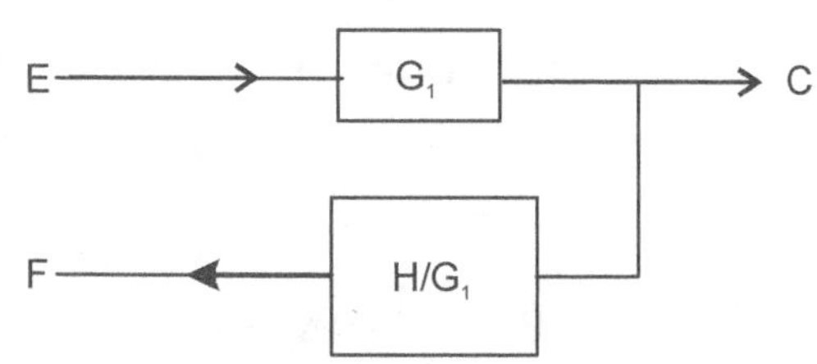

$F = EH$

(*b*)

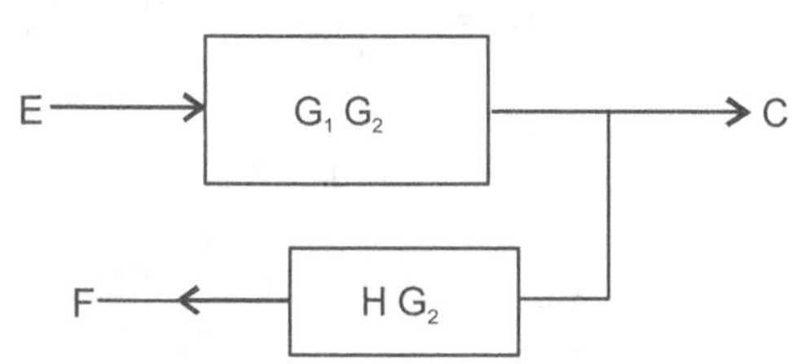

$F = G_1G_2^2H$

(*c*)

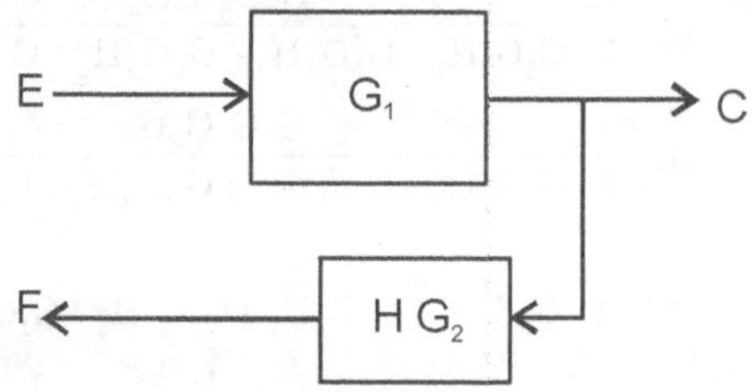

$$F = EG_1G_2H$$

(*d*)

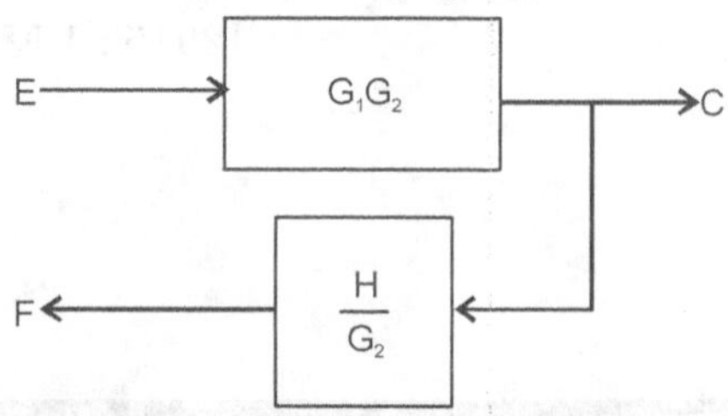

$$F = EG_1H$$

Hence the right option is (*d*).

4.

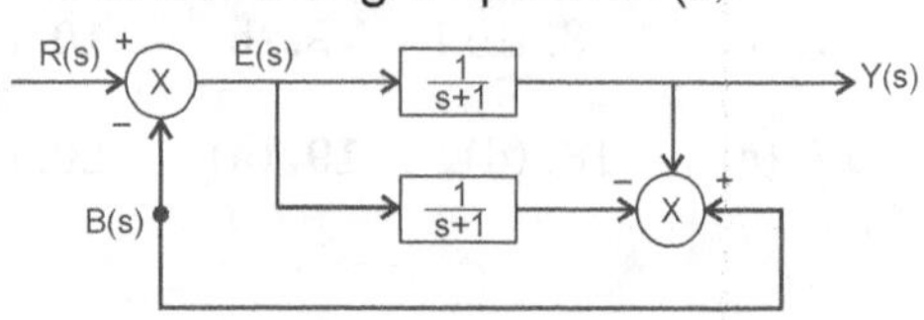

$$Y(s) = E(s)\cdot\frac{1}{(s+1)} \quad \ldots(i)$$

$$E(s) = R(s) - B(s)$$

$$= R(s) - \left\{Y(s) - \frac{1}{(s+1)}\cdot E(s)\right\}$$

$$E(s) = R(s) - \{Y(s) - Y(s)\}$$

$$E(s) = R(s)$$

From equation (*i*),

$$Y(s)\,(s+1) = R(s)$$

$$\therefore \quad \frac{Y(s)}{R(s)} = \frac{1}{(s+1)}$$

Alternately

Signal flow graph of given system

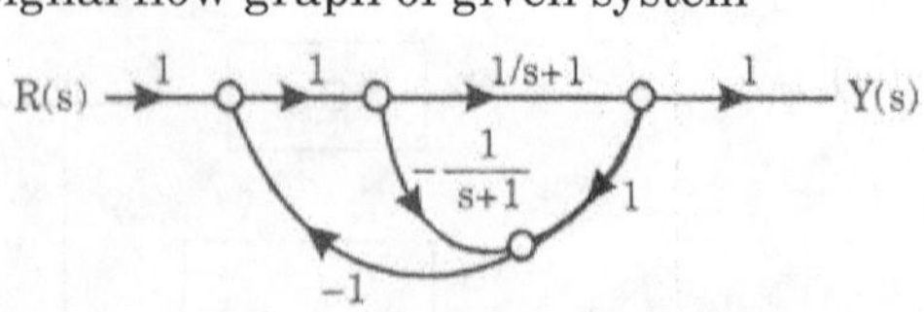

$$P_1 = \frac{1}{s+1};\ \Delta_1 = 1$$

$$L_1 = \frac{1}{s+1};\ \Delta = 1 - \{L_1 + L_2\}$$

$$L_2 = -\frac{1}{s+1}$$

$$\therefore \quad \frac{Y(s)}{R(s)} = \frac{P_1\Delta_1}{\Delta} = \frac{\left(\frac{1}{s+1}\right)(1)}{(1)} = \frac{1}{s+1}$$

5. Redrawing the block diagram with $X_1(s) = 0$

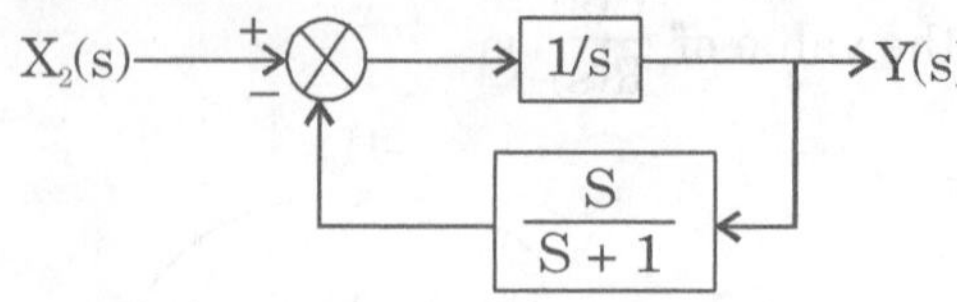

Transfer Function,

$$\frac{Y(s)}{X_2(s)} = \frac{\frac{1}{s}}{1+\frac{1}{s}\cdot\frac{s}{s+1}} = \frac{s+1}{s(s+2)}$$

6. Redrawing the block diagram,

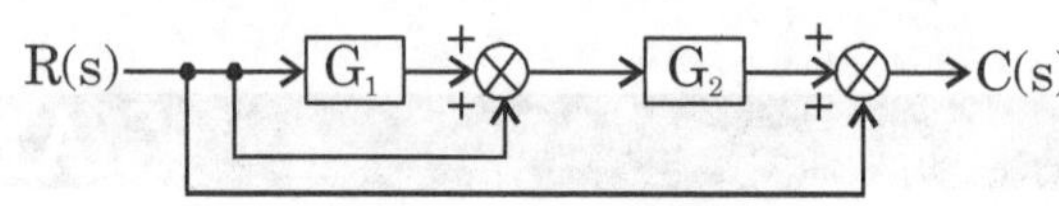

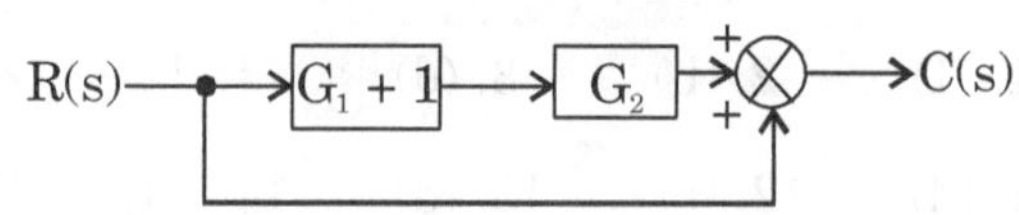

$$\frac{C(s)}{R(s)} = (G_1 + 1)\,G_2 + 1 = G_1\,G_2 + G_2 + 1.$$

7. In cascade connection : $G_1(s) + G_2(s)$.

In parallel connection : $G_1(s) \pm G_2(s)$.

Option (a) : $G_1(s)\cdot G_2(s)$

$\Rightarrow$

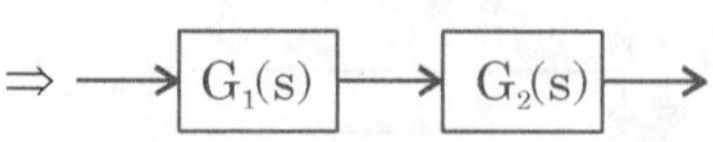

Relization is possible.

Option (b) : $\frac{G_1(s)}{G_2(s)} \Rightarrow$ Relization is not possible.

Option (c) : $G_1(s)\left[\frac{1}{G_1(s)} + G_2(s)\right]$

$\Rightarrow 1 + G_1(s)\cdot G_2(s)$

$\Rightarrow$

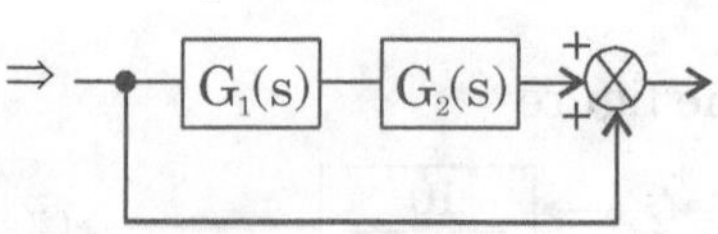

Relization is possible.

Option (d): $G_1(s)\left[\frac{1}{G_1(s)} - G_2(s)\right]$

$\Rightarrow [1 - G_1(s)\cdot G_2(s)]$

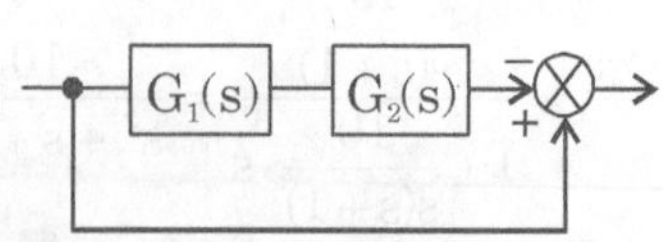

Realization is possible.

8. Redrawing the block diagram,

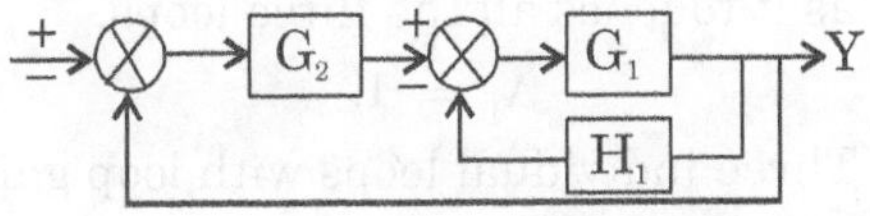

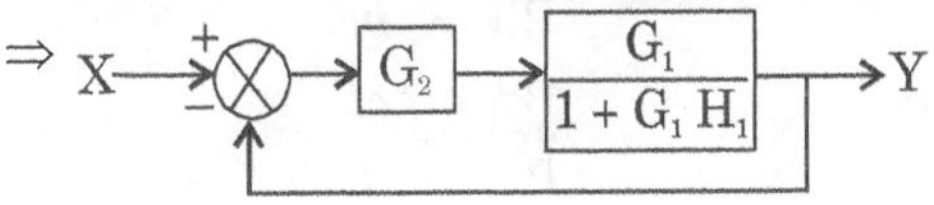

$$\frac{Y(s)}{X(s)} = \frac{\dfrac{G_1G_2}{(1+G_1H_1)}}{1+\dfrac{G_2G_1}{1+G_1H_1}} = \frac{G_1G_2}{1+G_1G_2+G_1H_1}$$

9. Given block diagram

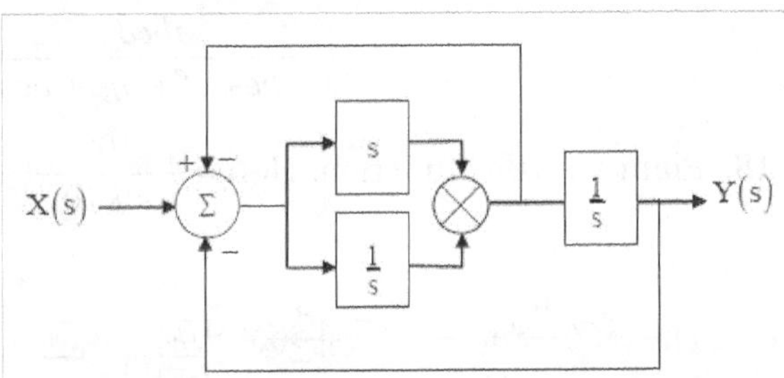

It can be reduced to

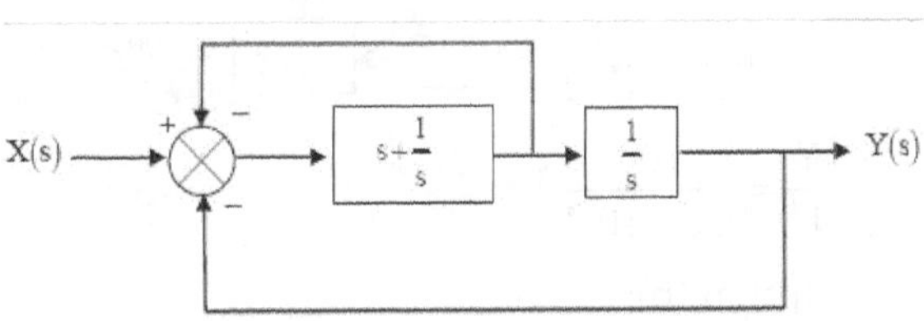

$$H(S) = \frac{Y(s)}{X(s)} = \frac{\dfrac{(s^2+1)}{s}\times\dfrac{1}{s}}{1+\dfrac{s^2+1}{s}+\dfrac{s^2+1}{s^2}}$$

$$= \frac{\dfrac{(s^2+1)}{s^2}}{\dfrac{s^2+(s^2+1)s+s^2+1}{s^2}}$$

Hence, the transfer function

$$H(s) = \frac{Y(s)}{X(s)} = \frac{(s^2+1)}{(s^3+2s^2+s+1)}$$

10. Applying KVL in both loops,

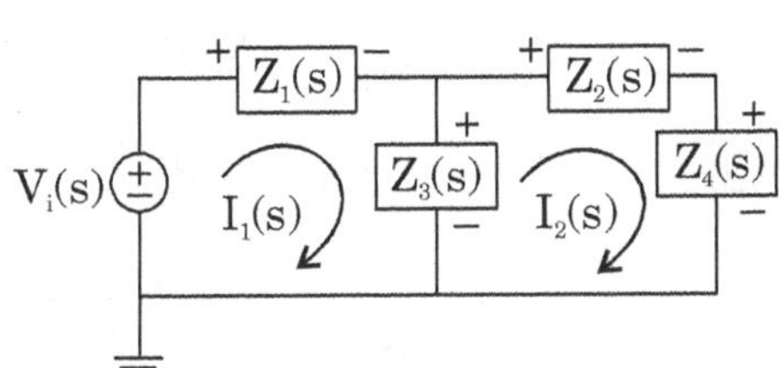

$$-V_i(s)+Z_1(s)I_1(s)+Z_3(s)[I_1(s)-I_2(s)] = 0$$

$$V_i(s) = I_1(s)[Z_1(s)+Z_3(s)]-I_2(s)Z_3(s)$$

$$I_1(s) = \frac{V_i(s)}{Z_1(s)+Z_3(s)}+\frac{I_2(s)\cdot Z_3(s)}{Z_1(s)+Z_3(s)}$$

$$-\ Z_3(s)[I_1(s)-I_2(s)]+Z_2(s)I_2+(s)+Z_4(s)I_2(s)$$

$$I_2(s)[Z_2(s)+Z_3(s)+Z_4(s)] = I_1(s)\cdot Z_3(s)$$

$$\text{Now } G_2 = \frac{I_2(s)}{I_1(s)} = \frac{Z_3(s)}{Z_2(s)+Z_3(s)+Z_4(s)}$$

[Using eq. (2)]

$I_1(s) = V_i(s)\cdot G_1(s)+I_2(s)\cdot H(s)$ [From eq. (1)]

$$H(s) = \frac{Z_3(s)}{Z_1(s)+Z_3(s)}$$

11.

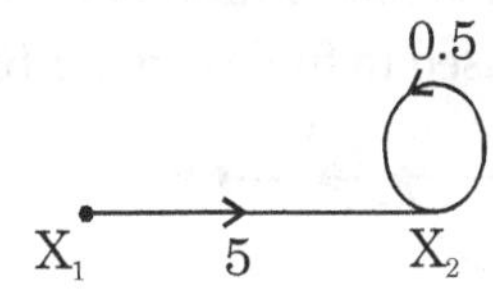

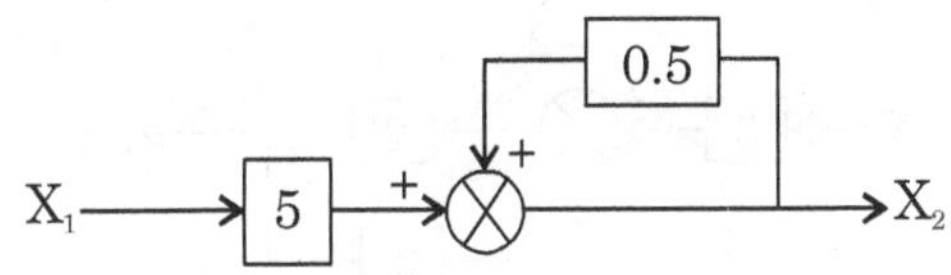

$\Rightarrow X_2 = 5x_1 + 0.5\ X_2$

$$\frac{X_2}{X_1} = \frac{5}{1-0.5} = 10$$

12. Using Mason's Gain formula, $\frac{C}{R} = \sum\frac{P_R\Delta_k}{\Delta}$

Foroard path gain(s): $P_1 = G_1\ G_2\ G_3\ G_4$

Individual loop gain(s):

$L_1 = -\ G_1,\ L_2 = -\ G_2$

$L_3 = -G_3,\ L_4 = -\ G_4$

Gain (s) of pair of non-touching loops: $G_1\ G_3$, $G_1\ G_4$, G_2, G_3, G_2, G_4

$$\Rightarrow \frac{C}{R} = \frac{G_1G_2G_3G_4}{1-[-G_1-G_2-G_3-G_4]+[G_1G_3+G_1G_4+G_2G_3+G_2G_4]}$$

$$\frac{C}{R} = \frac{G_1G_2G_3G_4}{1+G_1+G_2+G_3+G_4+G_1G_3+G_2G_3+G_1G_4+G_2G_4}$$

$$\frac{C}{R} = \frac{G_1G_2G_3G_4}{(1+G_1+G_2)(1+G_3+G_4)}$$

13. Using Mason's gain formula, $\frac{C}{R} = \frac{\sum P_k\Delta_k}{\Delta}$

Forward Path gain(s) : $P_1 = 1\times 2\times 3\times 4\times 1 = 24$

$P_2 = 1\times 5\times 1 = 5$

Individual loop gains: $L_1 = 2 \times (-1) = -2$,

$L_2 = 3 \times (-1) = -3$,

$L_3 = 4 \times (-1) = -4$,

$L_5 = 5 \times (-1) \times (-1) \times (-1) = -5$.

Gain(s) of pair of non-touching loops:

$L_1 L_3 = (-2) \times (-4) = 8$

$\Delta_2 = (1 + 3) = 4$.

$$\Rightarrow \frac{C}{R} = \frac{P_1\Delta_1 + P_2\Delta_2}{\Delta} = \frac{24 + 5 \times 4}{1 - (-2 - 3 - 4 - 5) + 8} = \frac{24 + 20}{1 + 14 + 8}$$

$$\therefore \frac{C}{R} = \frac{44}{23}$$

14. The signal flow graph cannot be used to find stability, controllability, observability or pole-zero locations. It is only used to find transfer function.

15. 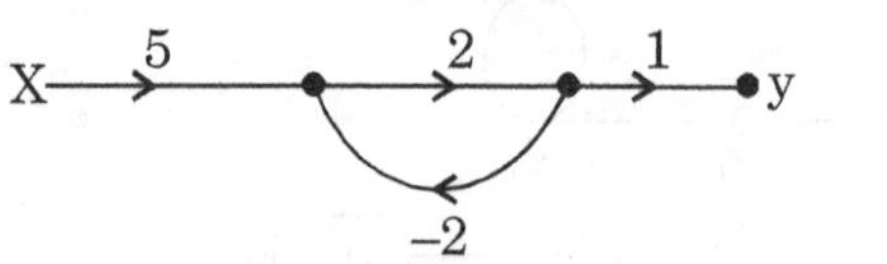

$\Rightarrow$ 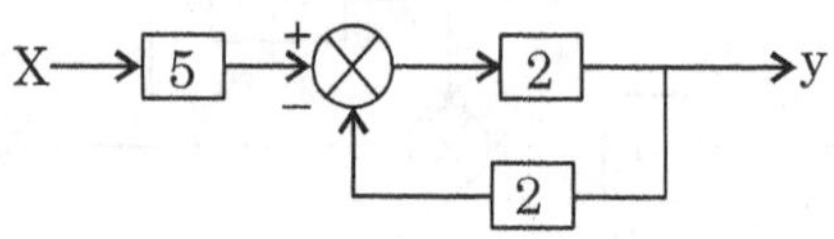

$$\frac{Y}{X} = 5 \times \frac{2}{5} = 2.$$

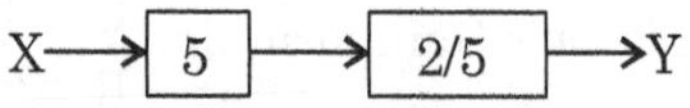

16. Forward path gain(s) : $P_1 = 1$.

Individual loop gain(s):

$$L_1 = \frac{-3}{s}, L_2 = \frac{-24}{s}, L_3 = \frac{-2}{s}$$

Gain(s) of pair of non-touching loops

$$= L_1 L_3 = \frac{6}{s^2}$$

$$\Delta_1 = 1 + \frac{3}{s} + \frac{24}{s} = \frac{s + 27}{s}$$

Mason's Gain formula, $\frac{C}{R} = \frac{P_1\Delta_1}{\Delta}$

$$\frac{C}{R} = \frac{(s + 27)s}{1 - \left(-\frac{3}{s} - \frac{24}{s} - \frac{2}{s}\right) + \left(\frac{6}{s^2}\right)} = \frac{\frac{(s + 27)}{s}}{1 + \frac{29}{s} + \frac{6}{s^2}}$$

$$\therefore \frac{C}{R} = \frac{s(s + 27)}{s^2 + 29s + 6}$$

17. Forward path gain, $P_1 = abcd$

and, as it touches all the three loops,

$$\Delta_1 = 1.$$

Now, Three individual loops with loop gains.

$$P_{11} = be,$$
$$P_{21} = cf,$$
$$P_{31} = dg$$

and combination of two non-touching loops with loop gains products,

$$P_{12} = bedg$$

Then,
$$T = \frac{P_1\Delta_1}{1 - P_{11} - P_{21} - P_{31} + P_{12}}$$

$$= \frac{abcd}{1 - (be + cf + dg) + bedg}$$

18. Plant transfer function. $H(s) = \frac{100}{s(s + 10)^2}$

Option (a):

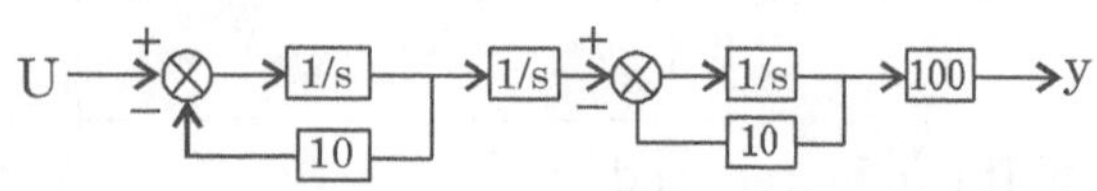

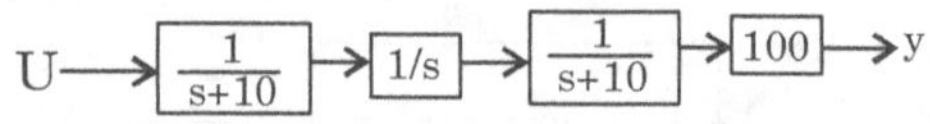

$$\frac{Y}{U} = \frac{100}{s(s + 10)^2}$$

Option (b):

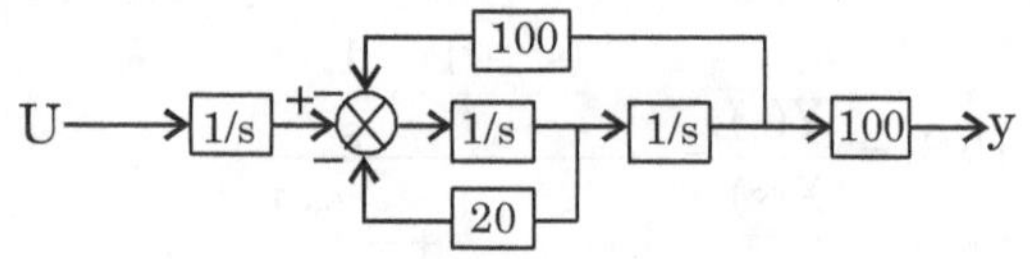

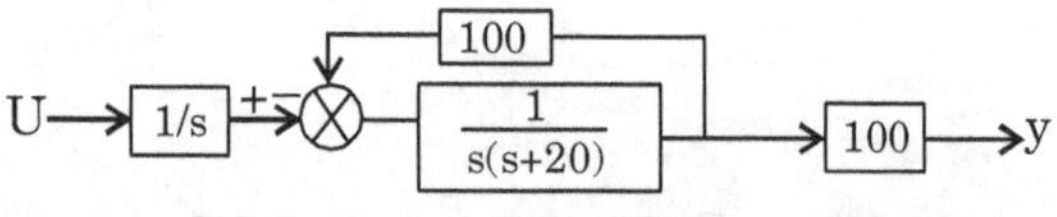

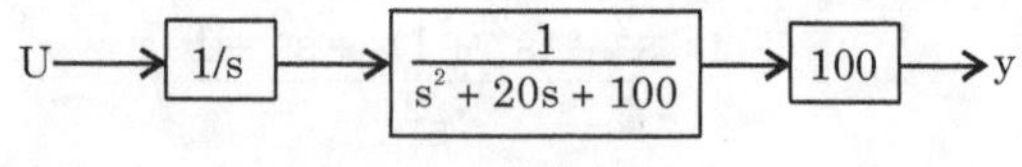

$$\frac{Y}{U} = \frac{100}{s(s + 10)^2}$$

Option (c) :

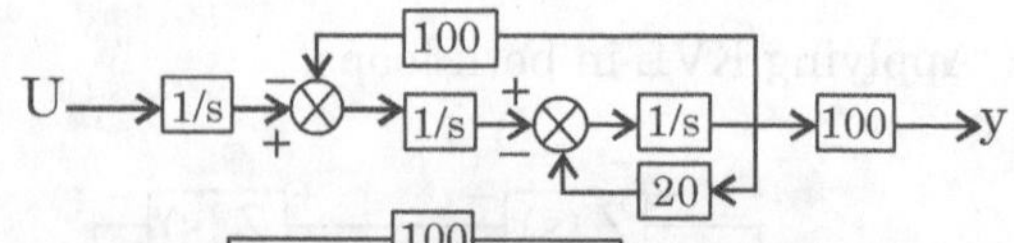

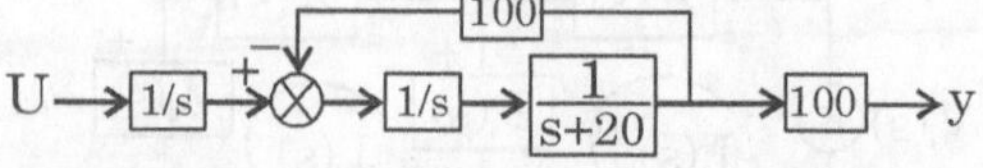

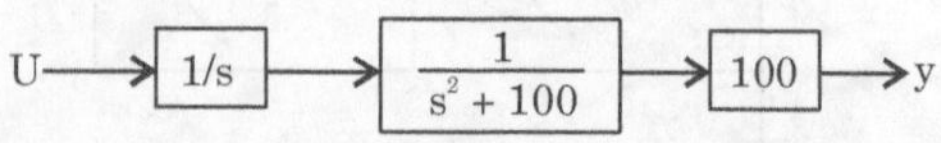

$$\frac{Y}{U} = \frac{100}{s(s+10)^2}.$$

Option (D):

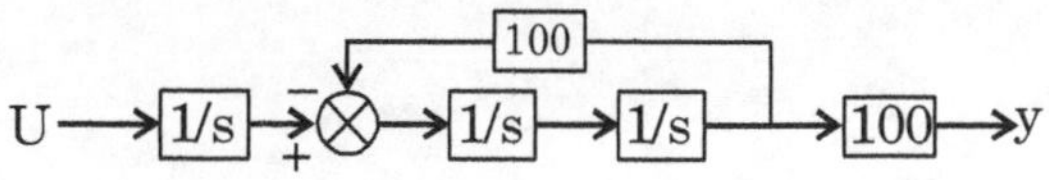

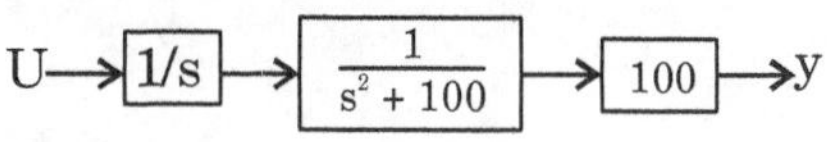

$$\frac{Y}{U} = \frac{100}{s(s^2+100)}.$$

Option (d) is the correct answer.

19. Forward path gain (s):

$$P_1 = \frac{1}{s}\cdot\frac{1}{s} = \frac{1}{s^2}, P_2 = \frac{1}{s}\cdot 1 = \frac{1}{s}$$

Individual loop gain(s):

$$L_1 = \frac{1}{s}(-2) = -\frac{2}{s},\ L_2 = \frac{1}{s}.\frac{1}{s}(-2) = -\frac{2}{s^2}$$

$$L_3 = \frac{1}{s}.(-4) = -\frac{4}{s}, L_4 = (4)\times(-1) = -4$$

$\Delta_1 = 1$, $\Delta_2 = 2$.

Using Mason's gain formula,

$$\frac{Y(s)}{U(s)} = \frac{P_1\Delta_1 + P_2\Delta_2}{\Delta}$$

$$= \left[\frac{\frac{1}{s^2}\times 1 + \frac{1}{s}\times 1}{1 - \frac{2}{s} - \frac{2}{s^2} - \frac{4}{s} - 4}\right] = \frac{s+1}{5s^2+6s+2}$$

20. Forward path gain(s):

$P_1 = G_1\, G_2 \cdot 1 \cdot G_3\, G_4 = G_1G_2G_3G_4$

Individual loop gain(s):

$L_1 = G_1G_2H_1$, $L_2 = -\ G_3G_4H_2$

$L_3 = -\ G_2G_3H_3$

Gain(s) of Pair of non-touching loops:

$L_1\ L_2 = G_1G_2G_3G_4H_1H_2$

$\Delta_1 = 1.$

From Mason's gain formula,

$$\frac{C(s)}{R(s)} = \frac{P_1\Delta_1}{\Delta}$$

$$= \frac{G_1G_2G_3G_4}{1-(-G_1G_2H_1 - G_2G_4H_2 - G_2G_3H_3) + G_1G_2G_3G_4H_1H_2}$$

$$\frac{C(s)}{R(s)} = \frac{G_1G_2G_3G_4}{1+G_1G_2H_1+G_2G_4H_2+G_2G_3H_3+G_1G_2G_3G_4H_1H_2}$$

3

CHAPTER

Time Response Analysis

Analysis of Previous GATE Papers											
	Year → Topics ↓		2019	2018	2017 Set 1	2017 Set 2	2016 Set 1	2016 Set 2	2015 Set 1	2015 Set 2	2015 Set 3
TYPE OF ORDER OF A SYSTEM	1 Mark	MCQ Type									
		Numerical Type									
	2 Marks	MCQ Type									
		Numerical Type									
	Total										
TIME RESPONSE OF L INEAR SYSTEMS	1 Mark	MCQ Type									
		Numerical Type			1			1		1	
	2 Marks	MCQ Type	1							1	
		Numerical Type						2			1
	Total		2		1			5		3	2
STEADY-STATE ERRORS	1 Mark	MCQ Type									
		Numerical Type									
	2 Marks	MCQ Type									1
		Numerical Type									
	Total										2

TYPE OF ORDER OF A SYSTEM

1. Consider a feedback control system with loop transfer function

$$G(s)H(s) = \frac{K(1+0.5s)}{s(1+s)(1+2s)}$$

The type of the closed loop system is

(a) zero (b) one

(c) two (d) three

[1998 : 1 Mark]

2. The transfer function of a plant is

$T(s) = \dfrac{5}{(s+5)(s^2+s+1)}$. The second order approximation of T(s)using dominant pole concept is

(a) $\dfrac{1}{(s+5)(s+1)}$ (b) $\dfrac{5}{(s+5)(s+1)}$

(c) $\dfrac{5}{s^2+s+1}$ (d) $\dfrac{1}{s^2+s+1}$

[2007 : 2 Marks]

TIME RESPONSE OF LINEAR SYSTEMS

3. A system described by the following differential equation $\dfrac{d^3y}{dt^2}+3\dfrac{dy}{dt}+2y = x(t)$ is inistially at rest.

For input x(t) = 2u(t), the output y(t) is

(a) $(1 - 2e^{-t}+e^{-2t})u(t)$ (b) $(1 + 2e^{-t}+e^{-2t})u(t)$

(c) $(0.5 + e^{-t}+1.5e^{-2t})u(t)$ (d) $(0.5 + 2e^{-t}+2e^{-2t})u(t)$

[1990 : 2 Marks]

4. The unit impulse response of a system is $h(t) = e^{-t}, t \geq 0$

For this system, the steady-state value of teh output for unit step input is equal to

(a) –1 (b) 0

(c) 1 (d) ∞

[1990 : 2 Marks]

5. A linear, time-invariant, causal continuous time system has a rational transfer function with simple poles at s = -2 and s = -4, and one simple zero at s = -1. A unit step u(t)is applied at the input of the system. At steady state, the output has constant value of 1. The impulse response of this system is

(a) [exp(–2t) + exp(–4t)u(t)]

(b) [–4exp(–2t) + 12exp(–4t)–expu(–t)]u(t)

(c) [–4exp(–2t) + 12exp(–4t)u(t)]

(d) [–0.5exp(–2t) + 1.5exp(–4t)u(t)]

[1991 : 2 Marks]

6. The position control of a DC servo-motor is given in the figure. The values of the parameters are K_T = 1 N-m A, R_a = 1 Ω, L_a = 0.1 H. J = 5 kg-m^2, B = 1 N-m rad/sec and K_b = 1 V/(rad/sec). The steady-state position response (in radians) due to unit impulse dsturbance torque T_d is ________.

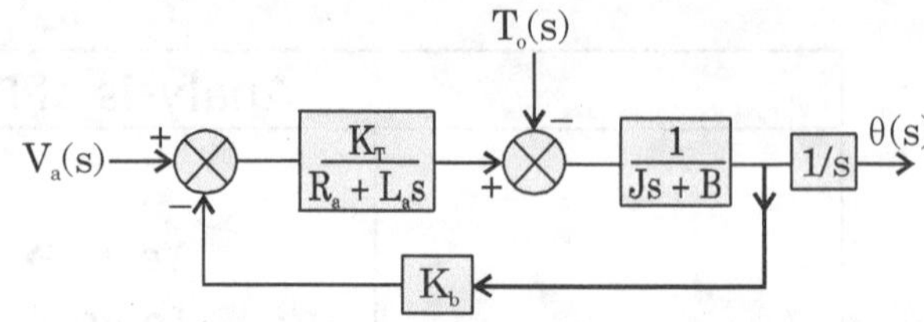

[1991 : 2 Marks]

7. A second-order system has a transfer function given by $G(s) = \dfrac{25}{s^2+8s+25}$

If the system, initially at rest, is subjected to a unit step input at t = 0, the second peak in the response will occur at

(a) π sec (b) π/3 sec

(c) 2π/3 sec (d) π/2 sec

[1991 : 2 Marks]

8. The poles of a continuous time oscillator are _____.

[1994 : 1 Mark]

9. The response of an LCR circuit to a step input is

(a) over damped,

(b) critically damped,

(c) oscillatory

If the transfer function has

(1) poles on the negative real axis,

(2) poles on the imaginary axis

(3) multiple poles on the positive real axis,

(4) poles on the positive real axis,

(5) multiple poles on the negative real axis.

[1994 : 2 Marks]

10. Match the following codes with List-I with List- II:

List-I

(a) Very low response at very high frequencies

(b) Over shoot

(c) Synchro-control transformer output

List-II

(1) Low pass systems

(2) Velocity damping

(3) Natural frequency

(4) Phase-sensitive modulation

(5) Damping ratio.

[1994 : 2 Marks]

11. For a second order system, damping ratio is $0 < \xi < 1$, then the roots of the characteristic polynomial are

(a) real but not equal

(b) real and equal

(c) complex conjugates

(d) imaginary

[1996 : 1 Mark]

12. If $L[f(t)] = \dfrac{2(s+1)}{s^2+2s+5}$, then $f(0^+)$ and $f(\infty)$ are given by

(a) 0, 2 respectively

(b) 2, 0 respectively

(c) 0, 1 respectively

(d) 2, 5, 0 respectively.

[Note : 'L' stands for 'Laplace Transform of]

[1997 : 1 Mark]

13. The final value theorem is used to find the

(a) steady state value of the system output

(b) initial value of the system output

(c) transient behavior of the system output

(d) none of these.

[1997 : 1 Mark]

14. If $F(s) = \dfrac{\omega}{s^2+\omega^2}$, then the value of $\lim_{t\to\infty} f(t)$, {where F(s) is the L[f(t)]}

(a) cannot be determined

(b) is zero

(c) is unity

(d) is infinite

[1998 : 1 Mark]

15. The unit impulse response of a linear time invariant system is the unit step function u(t). For t > 0, the response of the system to an excitation $e^{-at}u(t)$, a> 0 will be

(a) ae^{-at} (b) $(1/a)(1 - e^{-at})$

(c) $a(1 - e^{-at})$ (d) $1 - e^{-at}$

[1998 : 1 Mark]

16. For a second-order system with the closed- loop transfer function $T(s) = \dfrac{9}{s^2+4s+9}$, the settling time for 2-percent band, in seconds, is

(a) 1.5 (b) 2.0

(c) 3.0 (d) 4.0

[1999 : 1 Mark]

17. If the characteristic equation of a closed-loop system is $s^2 + 2s + 2 = 0$, then the system is

(a) overdamped (b) critically damped

(c) underdamped (d) undamped

[2001 : 1 Mark]

18. Consider a system with the transfer function $G(s) = \dfrac{s+6}{Ks^2+s+6}$. Its damping ratio will be 0.5 when the value of K is

(a) 2/6 (b) 3

(c) 1/6 (d) 6

[2002 : 1 Mark]

19. The transfer function of a system is $G(s) = \dfrac{100}{(s+1)(s+100)}$. For a unit-step input to the system the approximate settling time for 2% criterion is

(a) 100 sec (b) 4 sec

(c) 1 sec (d) 0.01 sec

[2002 : 2 Marks]

20. A second-order system has the transfer function $\dfrac{C(s)}{R(s)} = \dfrac{4}{s^2+4s+4}$. With r(t) as the unit-step function, the response c(t) of the system is represented by

(a)

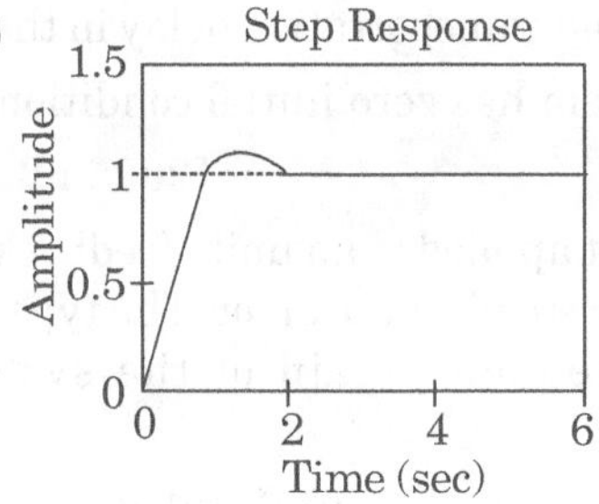

(b)

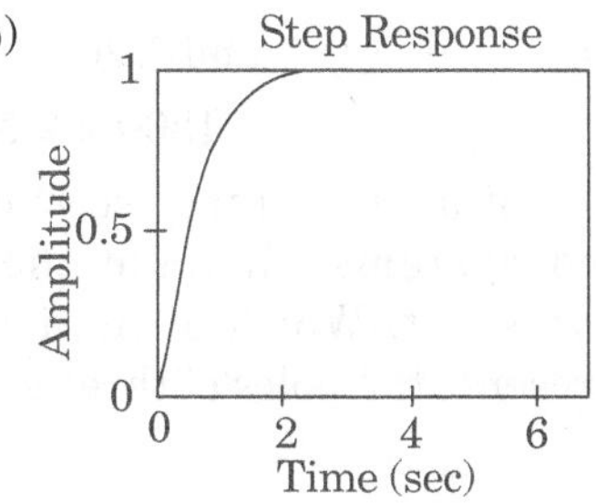

(c)

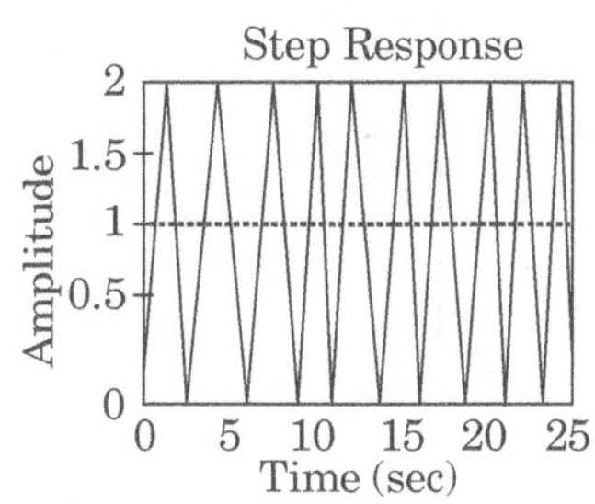

(d) 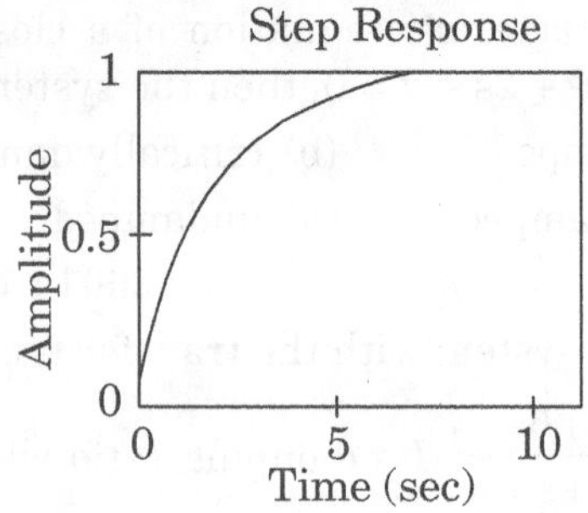

[2003 : 2 Mark]

21. A casual system having the transfer function H(s) $= \frac{1}{s+2}$ is excited with 10u(t). The time at which the output reaches 99% of its steady state value is

(a) 2.7 sec (b) 2.5 sec
(c) 2.3 sec (d) 2.1 sec

[2004 : 2 Marks]

22. In the derivation of expression for peak percent overshoot, $M_p = \exp\left(\frac{-\pi\xi}{\sqrt{1-\xi^2}}\right) \times 100$ which one of the following conditions is NOT required?

(a) System is linear and time invariant
(b) The system transfer function has a pair of complex conjugate poles and no zeroes
(c) There is no transportation delay in the system
(d) The system has zero initial conditions

[2005 : 2 Marks]

23. A ramp input applied to an unity feedback system results in 5% steady state error. The type number and zero frequency gain of the system are respectively

(a) 1 and 20 (b) 0 and 20
(c) 0 and 1/20 (d) 1 and 1/20

[2005: 2 Marks]

24. Step responses of a set of three second-order underdamped systems all have the same percentage overshoot. Which of the following diagrams represents the poles of three systems?

(a) 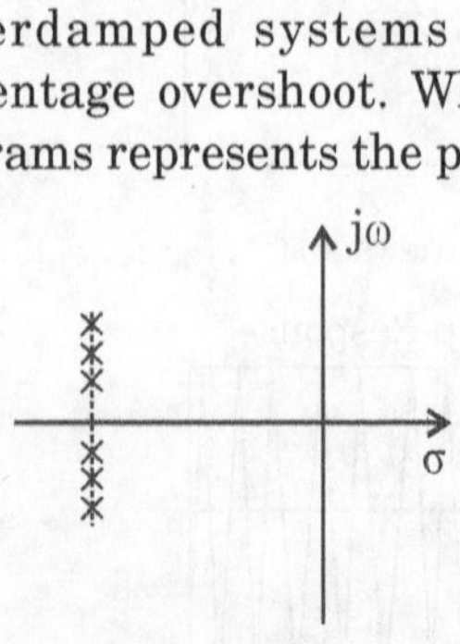

(b) 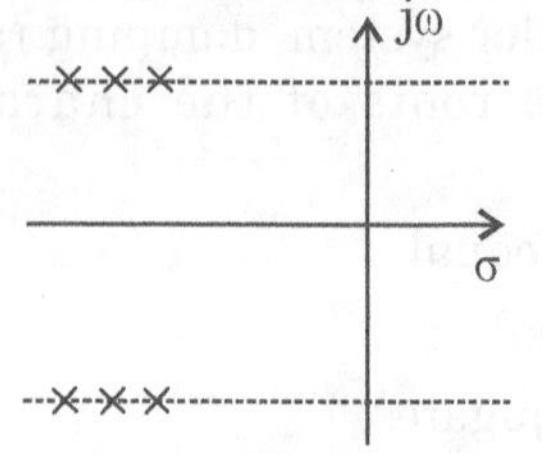

(c) 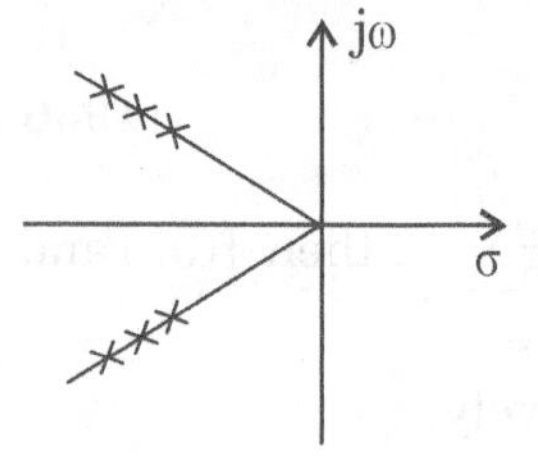

(d) 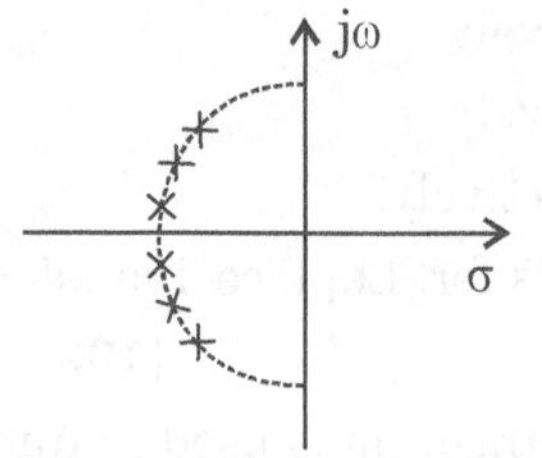

[2008 : 1 Mark]

25. Group I lists a set of four transfer functions. Group II gives a list of possible step responses y(t). Match the step responses with the corresponding transfer functions.

Group I

$P = \frac{25}{s^2+25}$:$Q = \frac{36}{s^2+20s+36}$

$R = \frac{36}{s^2+12s+36}$:$T = \frac{36}{s^2+7s+49}$

Group II

1. 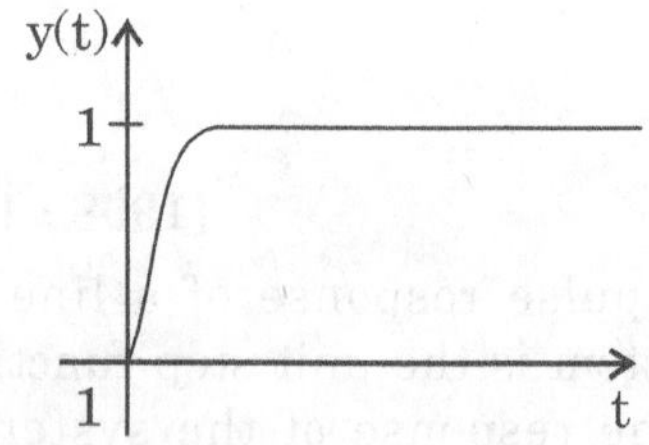

2. 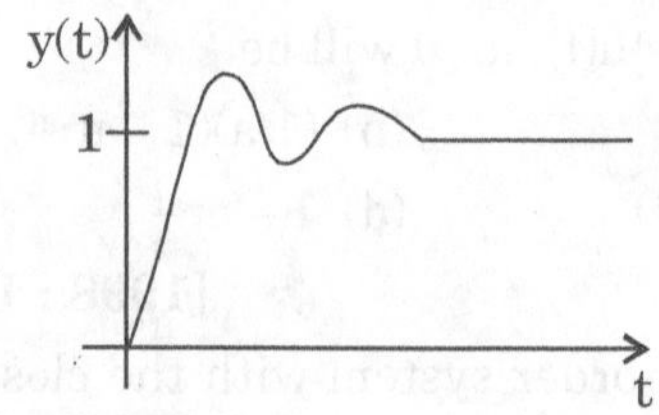

3. 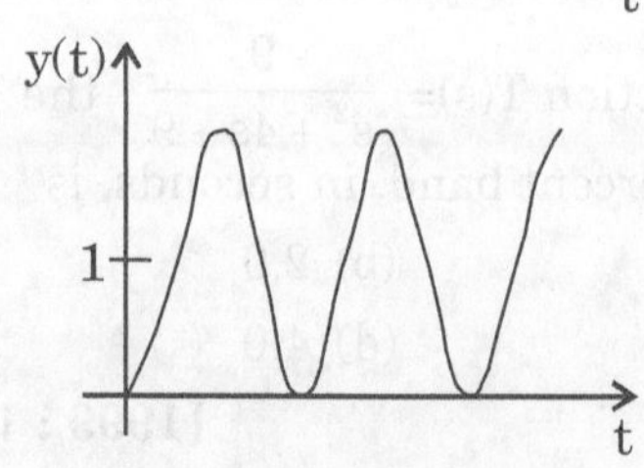

4. 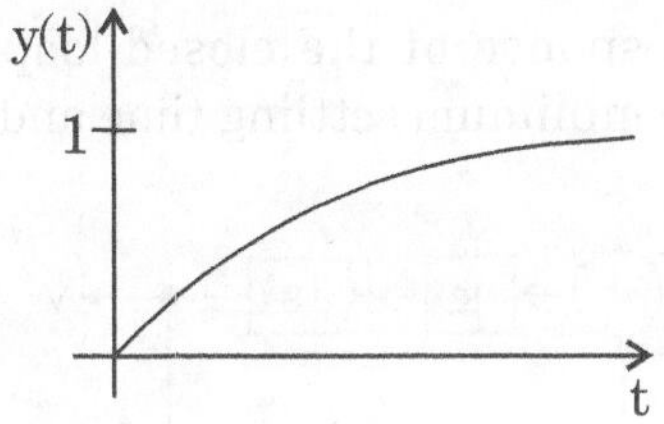

(a) P-3, 0-1, R-4, S-2
(b) P-3, 0-2, R-4, S-1
(c) P-2, 0-1, R-4, S-3
(d) P-3, 0-4, R-1, S-2

[2008 : 2 Marks]

26. The unit step response of an under-damped second order system has steady state value of –2. Which one of the following transfer functions has these properties?

(a) $\dfrac{-2.24}{s^2+2.59s+1.12}$ (b) $\dfrac{-3.82}{s^2+1.91s+1.91}$

(c) $\dfrac{-2.24}{s^2+2.59s+1.12}$ (d) $\dfrac{-3.82}{s^2-1.91s+1.91}$

[2009 : 2 Marks]

27. The diferential equation

$$100\frac{d^2y}{dt^2}-20\frac{dy}{dt}+y=x(t)$$

describes a system with an input x(t) and an output y(t). The system, which is initially relaxed, is excited by a unit step input. The output y(t) can be represented by the waveform

(a) 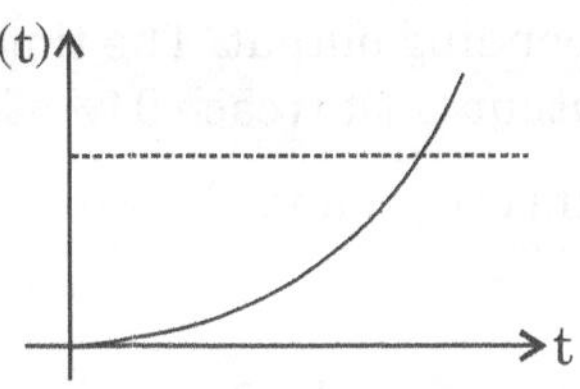

(b) 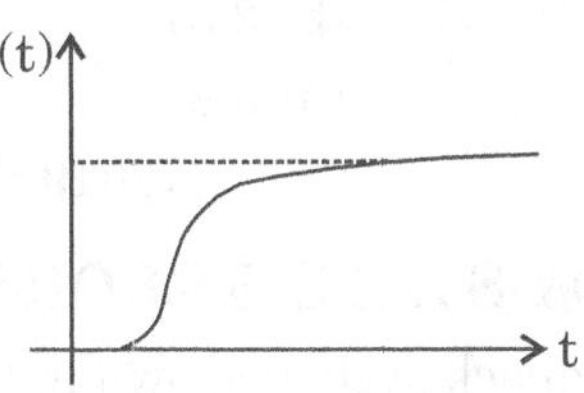

(c) 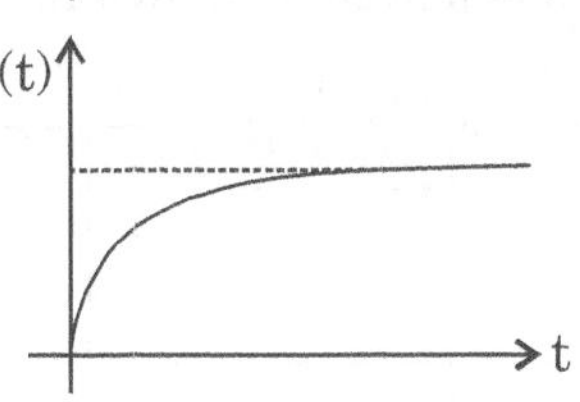

(d) 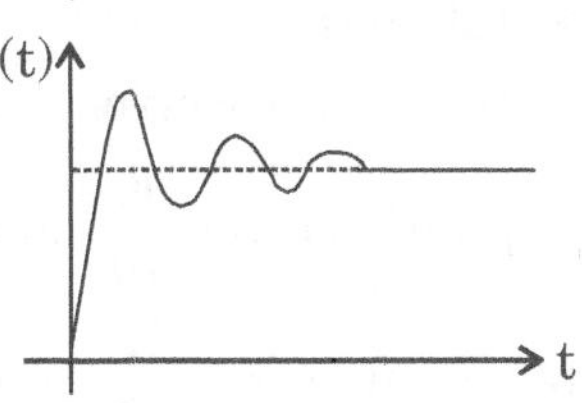

[2011 : 1 Mark]

28. A system described by a linear, constant coefficient, ordinary, first order differential equation has an exact solution given by y(t) for t > 0, when the forcing function is x(t) and the initial condition is y(0). If one wishes to modify the system so that the solution becomes –2y(t) for t > 0, we need to

(a) change the initial condition to -y(0) and the forcing function to at(t)

(b) change the initial condition to 2y(0) and the forcing function to –x(t)

(c) change the initial condition to $j\sqrt{2}\,y(0)$ and the forcing function to $j\sqrt{2}\,x(t)$

(d) change the initial condition to –2y(0)and the forcing function to –2x(t)

[2013 : 2 Marks]

29. The open-loop transfer function of a dc motor is given as $\dfrac{\omega(s)}{V_a(a)}=\dfrac{10}{1+10s}$. When connected in feedback as shown below, the approximate value of K_a that will reduce the time constant of the closed loop system by one hundred times as compared to that of the open-loop system is

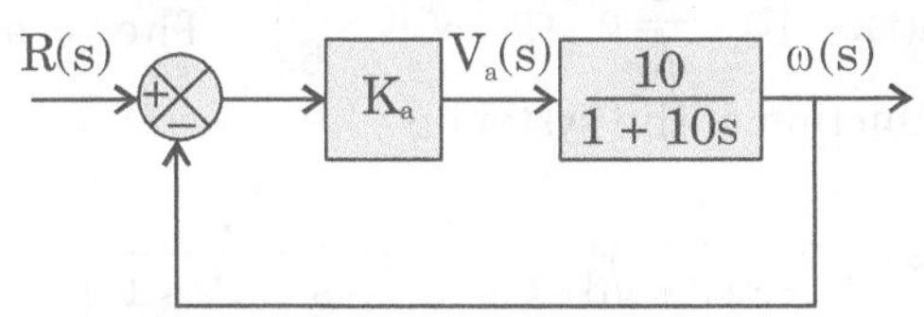

(a) 1 (b) 5

(c) 10 (d) 100

[2013 : 2 Marks]

30. For the following feedback system

$G(s)=\dfrac{1}{(s+1)(s+2)}$. The 2%-settling time of the step response is required to be less than 2 seconds.

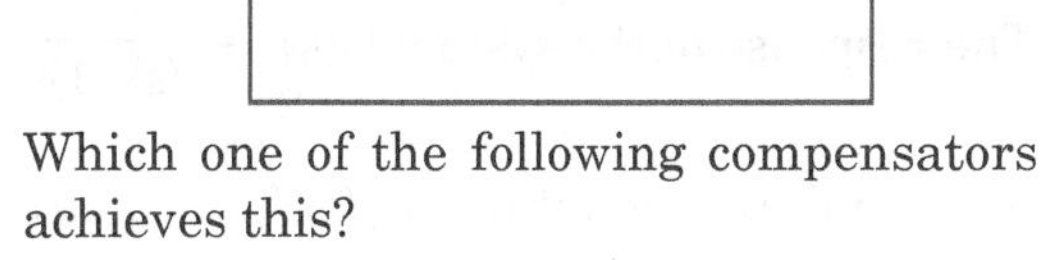

Which one of the following compensators C(s) achieves this?

(a) $3\left(\dfrac{1}{s+5}\right)$ (b) $5\left(\dfrac{0.03}{s}+1\right)$

(c) 2(s + 4) (d) $4\left(\dfrac{s+8}{s+3}\right)$

[2014 : 2 Marks, Set-1]

31. The natural frequency of an undamped second-order system is 40 rad/s. If the system is damped with a damping ratio 0.3, the damped natural frequency in radis is ______.

[2014: 1 Mark, Set-2]

32. For the second order closed-loop system shown in the figure, the natural frequency (in rad/s) is

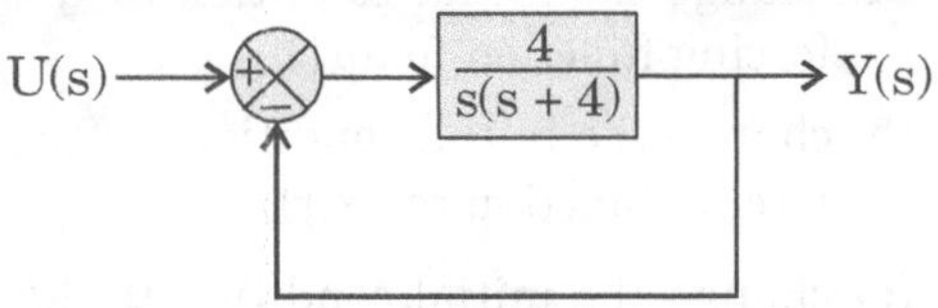

(a) 16 (b) 4

(c) 2 (d) 1

[2014 : 1 Mark, Set-4]

33. A unity negative feedback system has an open-loop transfer function $G(s) = \frac{K}{s(s+10)}$. The gain K for the system to have a damping ratio of 0.25 is ______.

[2015 : 1 Mark, Set-2]

34. The output of a standard second-order system for a unit step input is given as

$y(t) = 1 - \frac{2}{\sqrt{3}} e^{-t} \cos\left(\sqrt{3}t - \frac{\pi}{6}\right)$. The transfer function of the system is

(a) $\frac{2}{(s+2)(s+\sqrt{3})}$ (b) $\frac{1}{s^2+2s+1}$

(c) $\frac{3}{s^2+2s+3}$ (d) $\frac{4}{s^2+2s+4}$

[2015 : 2 Marks, Set-2]

35. The open-loop transfer function of unity-feedback control system is given by

$G(s) = \frac{K}{s(s+2)}$. For the peak overshoot of the closed-loop system to a unit step input to be 10%, the value of K is ______.

[2016 : 2 Marks, Set-2]

36. The response of the system $G(s) = \frac{s-2}{(s+1)(s+3)}$ to the unit step input u(t)is y(t). The value of $\frac{dy}{dt}$ at $t = 0^+$ is ______.

[2016 : 1 Mark, Set-2]

37. In the feedback system shown below

$$G(s) = \frac{1}{(s^2+2s)}$$

The step response of the closed-loop system should have minimum settling time and have no overshoot.

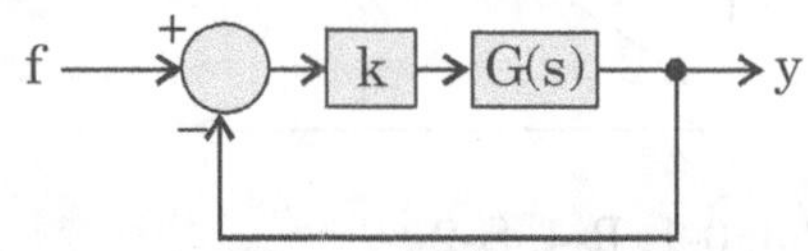

The required value of gain k to achieve this is ____.

[2016 : 2 Marks, Set-2]

38. The open loop transfer function

$$G(s) = \frac{(s+1)}{s^p(s+2)(s+3)}$$

where p is an integer, is connected in unity feedbackk configuration as shown in the figure.

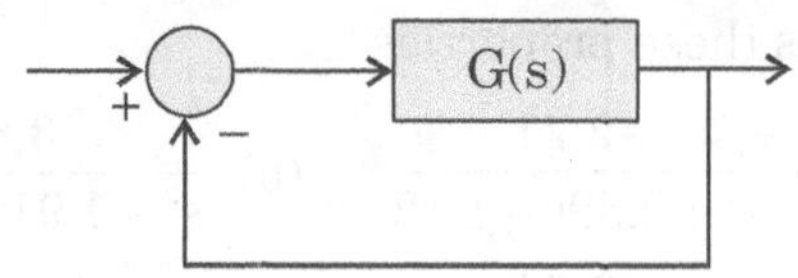

Given that the steady state error is zero for unit step input and is 6 for unit ramp input, the value of the parameter p is ______.

[2017: 1 Mark, Set-1]

39. Consider a causal second-order system with the transfer function $G(s) = \frac{1}{1+2s+s^2}$

With a until-step $R(s) = \frac{1}{s}$ as an input. Let C(s) be the corresponding output. The time taken by the system output C(t)to reach 94% of its steady-state value $\lim_{t\to\infty} c(t)$ rounded off to two decimal places, is

(a) 5.25 (b) 2.81

(c) 4.50 (b) 3.89

[2019 : 2 Marks]

STEADY-STATE ERRORS

40. The unity feedback system shown in fig. has

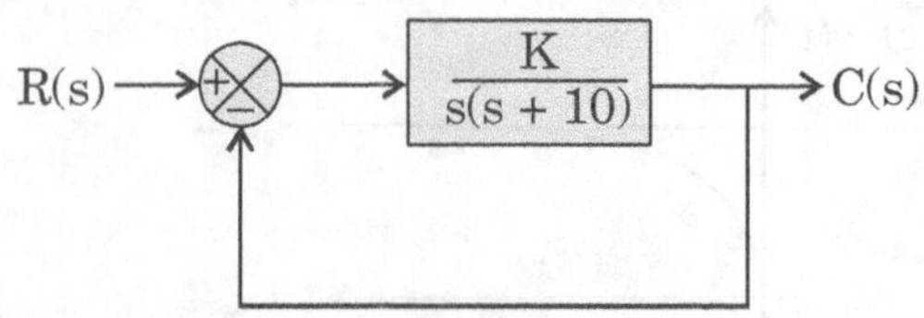

(a) zero steady state position error

(b) zero steady state velocity error

(c) steady state position error $\frac{K}{10}$ units

(d) steady state velocity error $\frac{K}{10}$ units

[1987 : 2 Marks]

41. The steady state error of a stable 'type-0' unity feedback system for a unit step function is

(a) 0 (b) $\frac{1}{1+K_P}$

(c) ∞ (d) $\frac{1}{K_P}$

[1990 : 2 Marks]

42. A unity-feedback control system has the open-loop transfer function G(s) $\frac{4(1+2s)}{s^2(s+2)}$, if the input to the system is a unit ramp, the steady-state error will be

(a) 0 (b) 0.5

(0) 2 (d) infinity

[1991 : 2 Marks]

43. The step error coefficient of a system

$G(s)=\frac{1}{(s+6)(s+1)}$ with unity feedback is

(a) 1/6 (b) ∞

(c) 0 (d) 1

[1995 : 1 Mark]

44. Consider a unity feedback control system with open-loop transfer function $G(s) = \frac{K}{s(s+1)}$ the steady state error of the system due to a unit step input is

(a) zero (b) K

(c) 1/K (d) infinite

[1998 : 1 Mark]

45. If the closed-loop transfer function T(s) of a unity negative feedback system is given by

$$T(s) = \frac{a_{n-1}s + a_n}{s^n + a_1 s^{n-1} + ... + a_{n-1}s + a_n}$$

Then the steady state error for a unit ramp input is

(a) $\frac{a_n}{a_{n-1}}$ (b) $\frac{a_n}{a_{n-2}}$

(c) $\frac{a_n - 2}{a_{n-1}}$ (d) zero

[1999 : 2 Marks]

46. The steady state error of the system shown in the figure for a unit step input is ______.

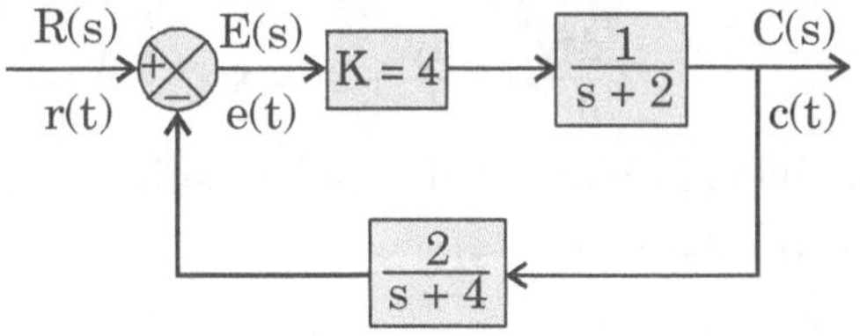

[2014 : 2 Marks, Set-3]

47. For the unity feedback control system shown in the figure, the open-Coop transfer function G(s) is given as $G(s) = \frac{2}{s(s+1)}$. The steady state error e_{ss} due to unit step input is

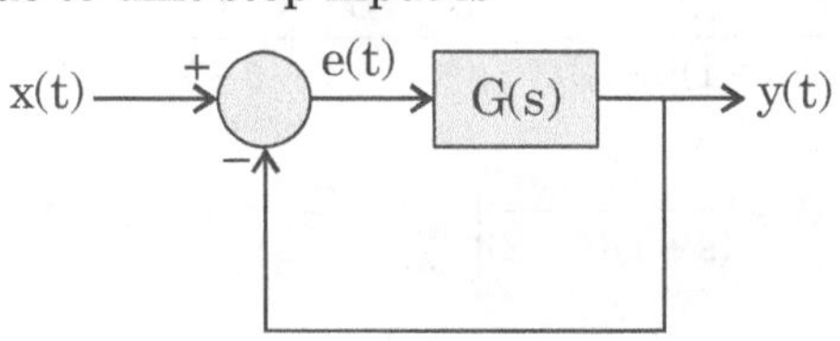

(a) 0 (b) 0.5

(c) 1.0 (d) ∞

[2015 : 2 Mark, Set-3]

ANSWERS

1. (b)	**2.** (d)	**3.** (a)	**4.** (c)	**5.** (c)	**6.** (–0.5)	**7.** (a)	**8.** (Imaginary)	**9.** (*)	
10. (*)	**11.** (c)	**12.** (b)	**13.** (a)	**14.** (a)	**15.** (b)	**16.** (b)	**17.** (c)	**18.** (c)	**19.** (b)
20. (b)	**21.** (c)	**22.** (c)	**23.** (a)	**24.** (c)	**25.** (d)	**26.** (b)	**27.** (a)	**28.** (d)	**29.** (c)
30. (c)	**31.** (38.16)	**32.** (c)	**33.** (400)	**34.** (d)	**35.** (2·87)	**36.** (1)	**37.** (1)	**38.** (1)	
39. (c)	**40.** (a)	**41.** (b)	**42.** (a)	**43.** (a)	**44.** (a)	**45.** (d)	**46.** (1)	**47.** (a)	

EXPLANATIONS

1. Type of closed loop system having a feedback control system as

$$G(s)\,H(s) = \frac{K(1+0.5s)}{s(1+s)(1+2s)} \text{ is one.}$$

2. In dominant pole approximation, poles which are near to origin considered.

$$T(s) = \frac{5}{5\left(\frac{s}{5}+1\right)(s^2+s+1)}$$

$$T(s) \cong \frac{1}{s^2+s+1}$$

3. $$\frac{d^2y}{dt^2} + 3\frac{dy}{dt} + 2y = x(t)$$

Applying Laplace transform on both sides,

$s^2y(s) + 3sy(s) + 2y(s) = x(s)$

$$\therefore\ x(t) = 2u(t) \longleftrightarrow X(s) = \frac{2}{s}$$

$(s^2 + 3s + 2)\, y(s) = X(s)$

$$Y(s) = \frac{2}{s(s+1)(s+2)}$$

$$\frac{2}{s(s+1)(s+2)} = \frac{A}{s} + \frac{B}{s+1} + \frac{c}{s+2}$$

$$A = \left.\frac{2}{(s+1)(s+2)}\right|_{s=0} = \frac{2}{1\times 2} = 1$$

$$B = \left.\frac{2}{s(s+2)}\right|_{s=-1} = \frac{2}{-1\times 1} = -2$$

$$C = \left.\frac{2}{s(s+1)}\right|_{s=-2} = \frac{2}{-2\times(-1)} = 1$$

$$Y(s) = \frac{1}{s} + \left(\frac{-2}{s+1}\right) + \frac{1}{s+2}$$

$Y(t) = 1 - 2e^{-t} + e^{-2t}\ t > 0.$

4. Input, $r(t) = u(t) \Rightarrow R(s) = \frac{1}{s}$.

Laplace [Impulse resonse] = Transfer function.

$$\text{Transfer function} = H(s) = \frac{1}{s+1}$$

$$\text{Output, } C(s) = H(s)\cdot R(s) = \frac{1}{s(s+1)} = \frac{1}{s} - \frac{1}{s+1}$$

$c(t) = 1 - e^{-t}\ \forall\ t \geq 0$

Steady state value, $c(t \to \infty) = 1 - e^{-\infty} = 1$.

5. Simple poles: $p_1 = -2$, $p_2 = -4$

Simple zero : $z_1 = -1$

Transfer function, T(s)

$$= \frac{K(s-z_1)}{(s-p_1)(s-p_2)} = \frac{K(s+1)}{(s+2)(s+4)}$$

Input, $r(t) = u(t) \Rightarrow R(s) = \frac{1}{s}$.

$$\text{Output, } C(s) = T(s)\cdot R(s) = \frac{K(s+1)}{S(s+2)(s+4)}$$

Given: $\lim\limits_{s\to 0} sC(s) = 1$

$$\lim_{s\to 0} \frac{sk(s+1)}{s(s+2)(s+4)} = 1$$

$$\therefore\ \frac{K}{8} = 1$$

$\therefore\ K = 8$

$$\text{Transfer function, } T(s) = \frac{8(s+1)}{(s+2)(s+4)}$$

Impulse response = L^{-1} [T(s)]

$$= L^{-1}\left[\frac{8(s+1)}{(s+2)(s+4)}\right]$$

$$\therefore\ \frac{8(s+1)}{(s+2)(s+4)} = \frac{-4}{s+2} + \frac{12}{s+4}$$

$$\text{Impulse response} = \left[-4e^{-2t} + 12\cdot e^{-4t}\right] u(t).$$

6. Redrawing the block diagram for $V_a(s) = 0$.

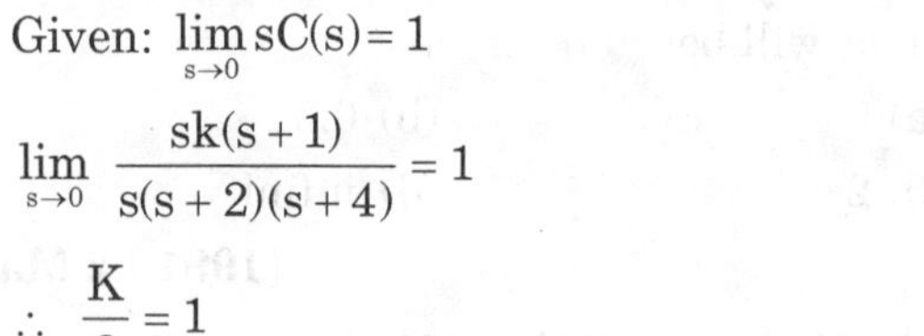

$$\frac{Q(s)}{T_d(s)} = \frac{-\frac{1}{J_s+B}\cdot\frac{1}{s}}{1 + \frac{1}{(J_s+B)}\cdot\left(\frac{K_b\cdot K_T}{R_a+sL_a}\right)}$$

$$= \frac{\frac{-(R_a+sL_a)}{s}}{(R_a+L_a\cdot s)(J_s+B)+K_bK_T}$$

The steady-state value of response for unit impulse input [$T_d(s) = 1$]

$$Q(\infty) = \lim_{s\to 0} s\theta(s) = \lim_{s\to 0}\frac{\frac{-s(R_a+sL_a)}{s}}{(R_a+sL_a)(J_s+B)+K_bK_T}$$

$$= -\frac{R_a}{R_aB + K_TK_b}$$

Given : $K_T = \frac{N-m}{A}$, $R_a = 1\Omega$.

$B = \frac{N-m}{rad/s}$, $K_b = 1 \frac{V}{\frac{rad}{s}}$.

$Q(\infty) = -\frac{1}{1\times1+1\times1} = -0.5$ radians.

7. $G(s) = \frac{25}{s^2+8s+25}$

Comparing with standard second-order transfer function,

$\omega_n^2 = 25 \Rightarrow \omega_n = 5$ rads/s.

$2\xi\omega n = 8 \Rightarrow \xi = \frac{8}{2\times5} = 0.8$

For second peak of response (n = 3),

$t_p = \frac{n\pi}{\omega_d} = \frac{3\pi}{3} = \pi\,s.$

8. The poles of a continuous time oscillator (undamped response) are pure imaginary.

9. Characteristic equation of 2nd order system.

$\Delta(s) = s^2 + 2\xi\omega_n s + \omega_n^2 = 0$

$S_{1,2} = -\xi\omega_n \pm \omega_n\sqrt{\xi^2-1}$

(a) Overdamped : $\xi > 1$

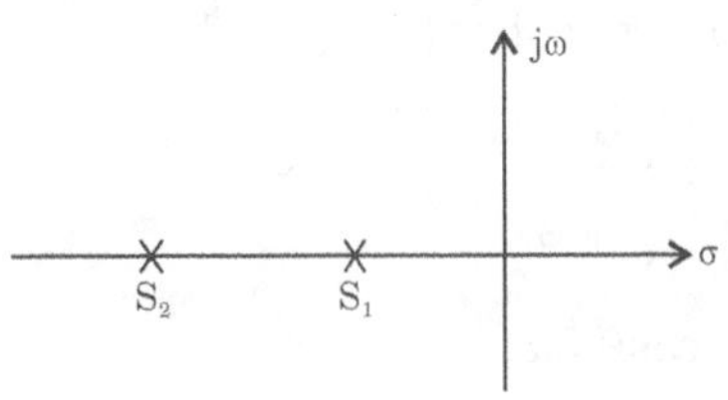

$S_{1,2} = -\xi\omega_n \pm \sqrt{\xi^2-1}$

(b) Critically damped : $\xi = 1$

$S_{1,2} = -\omega_n$

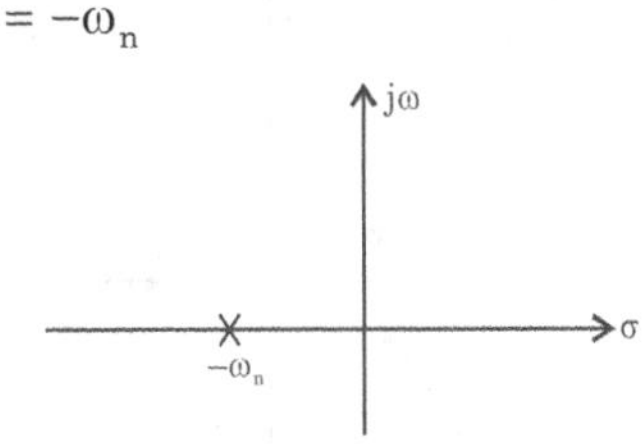

(c) Oscillatory : $\xi = 0$

$S = \pm j\omega_n$

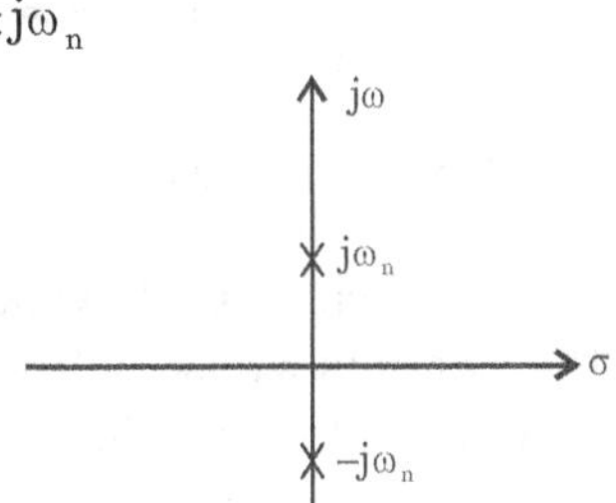

(a) → (1), (b) → (5), (c) → (2)

10. Low pass systems (eq. LPF) produce very low (or zero) response at very high frequencies.

$M_P = e^{-\pi\xi/\sqrt{1-\xi^2}}$

$\Rightarrow M_P \;\alpha\; \xi$

Synchro-control transformer output → phase sensitive modulation.

(a) → (1), (b) → (5), (c) → (4)

11. For underdamped ($\xi < 1$) second-order system, the roots of characteristic polynomial are complex conjugates

$\left(S_{1,2} = -\xi\omega_n \pm j\omega_n\sqrt{1-\xi^2}\right)$

12. $$\frac{2(s+1)}{(s^2+2s+5)} = \frac{2(s+1)}{(s^2+1)^2+4}$$

$$= 2\left[\frac{s+1}{(s+1)^2+4}\right]$$

This is standard Laplace Transform for the function

$$L\,|\exp(-at)\cos\omega t| = \frac{(s+a)}{(s+a)^2+\omega^2}$$

where $a = 1,\quad \omega = 2$

Hence $f(t) = 2e^{-t}\cos 2t$

$f(0) = f(0+) = 2e^{-0}\cos 2\times0 = 2$

$f(\infty) = 2e^{-\infty}\,t = 0$

13. The final value theorem is used to find the steady state value of the system output.

14. $F(s) = \frac{\omega}{S^2+\omega^2} \xleftrightarrow{L.T} f(t) = \sin(\omega t)$

Hence, final value of f(t) cannot be determined.

15. Given : [Impulse response] = Transfer function.

Tranfer function $H(s) = L[u(t)] = \frac{1}{s}$

$\Rightarrow$ Excitation, $r(t) = e^{-at}u(t) \Rightarrow R(S) = \frac{1}{s+a}$

Response, $C(s) = H(s)\cdot R(s)$

$C(s) = \frac{1}{s+a}\cdot\frac{1}{s} = \frac{1}{a}\left[\frac{1}{s} - \frac{1}{s+a}\right]$

$C(t) = \frac{1}{a}[1-e^{-at}]\;\forall\; t>0$

16. $T(s) = \frac{9}{s^2+4s+9}$

$\Rightarrow 2\xi\omega_n = 4$

$\omega_n\xi = \frac{4}{2} = 2$

Setling – time, $t_s = \frac{4}{\xi\omega_n}$ (for ± 2% error)

$t_s = \frac{4}{2} = 2s$

17. Characteristic equation of a closed-loop system is

$$s^2 + 2s + 2 = 0$$

$$\omega_n = \sqrt{2} = 0$$

$$\delta = \frac{1}{\sqrt{2}} < 1, \text{ Hence underdamped system}$$

18. Transfer function,

$$G(s) = \frac{s+6}{Ks^2+s+6} = \frac{s+6}{K\left(s^2+\frac{S}{K}+\frac{6}{K}\right)}$$

Comparing with general eq. $(s^2 + 2\xi\omega_n \cdot s + \omega_n^2)$

$$\omega_n = \sqrt{6/K}$$

$$2\xi\omega_n = \frac{1}{K}$$

$$2 \times 0.5 \times \sqrt{\frac{6}{K}} = \frac{1}{K}$$

$$\frac{6}{K} = \frac{1}{K^2}$$

$$K = \frac{1}{6}$$

19. Taking dominant pole consideration (i.e, neglecting the pole(s) which is too far from origin.

$$G(s) = \frac{100}{\left(\frac{s}{1}+1\right)100\left(\frac{s}{100}+1\right)} \cong \frac{1}{s+1}$$

Now, it is a first-order system with time-constnat, τ = 1s.

Setting time, $t_s = 4\tau$ (for ± 2% error)

t_s = 4s

20. Given $\frac{C(s)}{R(s)} = \frac{4}{s^2+4s+4}$

Comparing with standard 2nd order transfer function, $2\xi\omega_n = 4$ and $\omega_n^2 = 4 \Rightarrow \omega_n = 2$ rad/s

$$\xi\omega_n = \frac{4}{2} = 2$$

$\Rightarrow \xi = 1$ (critically damped system)

Setling time, $t_s = \frac{4}{\xi\omega_n} = \frac{4}{1\times 2} = 2s$

21. Given : $r(t) = 10\,u(t), H(s) = \frac{1}{s+2}$

$$R(s) = \frac{10}{s}$$

Output $C(s) = H(s)\cdot R(s) = \frac{10}{s(s+2)} = \frac{5}{s} - \frac{5}{s+2}$

$$C(t) = 5 - 5e^{-2t} = 5(1-e^{-2t}) = 5(1-e^{-2t})\ \forall t > 0.$$

Steady – state value of c(t) = 5.

$$5[1-e^{-2t}] = 5 \times 0.99$$

$$1 - e^{-2t} = 0.99$$

$$e^{-2t} = 0.1$$

$$-2t = \ln(0.1)$$

$$t = 2.3 \text{ s.}$$

22. Even if there is some transportation delay in the system, various time parameters, *e.g.*, t_r, t_s, t_d, t_p are affected but M_p remains same.

23. Given : steady-state error, $e_{ss} = 5\% = \frac{1}{20}$.

We know that steady-state error for ramp input and unity feedback is finite only when system type is 1.

Zero frequency gain or velocity error coefficient

$$K_v = \lim_{s\to 0} sG(s) = \frac{1}{e_{ss}} = \frac{1}{\frac{1}{20}}$$

$$K_v = 20.$$

24. Peak overshoot, $M_P = e^{-\xi\pi/\sqrt{1-\xi^2}}$

$$M_P = e^{-\pi\cdot\xi\omega_n/\omega_n\sqrt{1-\xi^2}} = e^{-\pi\omega_d/\sigma_d}$$

$$M_P = e^{-\pi\cdot\tan\theta} \quad \text{... (1)}$$

where $\cos\theta = \xi$

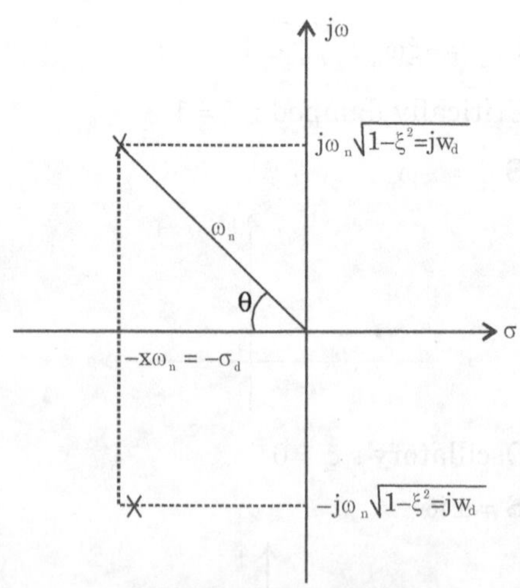

Pole plot for an underdamp second order system.

From eq. (1), it is clear that peak overshoot is same for the systems whose pole(s) are lying on the same line with slope tan θ.

25. Standard tansfer function of a second-order system

$$= \frac{\omega_n^2}{s^2 + 2\xi\omega_n s + \omega_n^2}$$

$$\Rightarrow P = \frac{25}{s^2+25}; \quad 2\xi\omega_n = 0$$

$$\Rightarrow \xi = 0, \ \omega_n = 5 \qquad \text{(undamped)}$$

$$\Rightarrow Q = \frac{36}{s^2+20s+36}; \ \omega_n = 6, 2\xi\omega_n = 20$$

$$\xi = \frac{20}{2\times 6} = 1.67 \qquad \text{(overdamped)}$$

$$\Rightarrow R = \frac{36}{s^2+12s+36}; \ \omega_n = 6; \ 2\xi\omega_n = 12$$

$$\xi = \frac{12}{2\times 6} = 1 \qquad \text{(critically damped)}$$

$$\Rightarrow T = \frac{36}{s^2+7s+49}; \ \omega_n = 7; 2\xi\omega_n = 7$$

$$\xi = \frac{7}{2\times 7} = 0.5 \qquad \text{(Underdamped)}$$

$$\Rightarrow (P)\to(3), (Q)\to(4), (R)\to(1), (S)\to(2)$$

26. Given : $c(\infty) = (2), \xi < 1$ (Underdamped)

All options except (d) have steady-state value of −2. Now, looking for damping.

Option (a) and (c) have damping ratio more than 1. Option (b) has damping ratio less than one.

27. $100\frac{d^2y}{dt^2} - 20\frac{dy}{dt} + y = x(t); x(t) = u(t)$

$$\Rightarrow X(s) = \frac{1}{s}$$

Taking laplace transform,

$$100s^2Y(s) - 20sY(s) + Y(s) = X(s)$$

$$\frac{Y(S)}{X(S)} = \frac{1}{100s^2 - 20s + 1} = \frac{1}{(10s-1)^2}$$

Transfer function has two repeated poles $\left(S = \frac{1}{10}, \frac{1}{10}\right)$ at right side of s-plane, so given system is unstable. Hence, option (a) is correct answer.

28. Linear, constant coefficient, ordinary first-order differential equation is given as :

$$\frac{dy(t)}{dt} + ky(t) = x(t)$$

Taking laplace transform,

$$s^2Y(s) - y(0) + kY(s) = X(s)$$

$$Y(s) = \frac{X(s)}{S+K} + \frac{y(0)}{S+K}$$

$$y(t) = e^{-kt}\cdot x(t) + y(0)e^{-kt}$$

(taking inverse laplace transform)

If we want-2y(t) as a solution, both x(t) and y(0) has to be multiplied by-2. Hence, change x(t) diagram by-2x(t) and y(0) by-2y(0).

29. From block diagram, $\frac{\omega(s)}{v_a(s)} = \frac{10}{1+10s}$

It is a first order system with time constant

$\tau = 10s$

$$\text{Given } \tau' = \frac{\tau}{100} = \frac{10}{100} = \frac{1}{10}s$$

$$\frac{\omega(s)}{R(s)} = \frac{K_a\cdot\frac{10}{1+10s}}{1+K_a\cdot\frac{10}{1+10s}} = \frac{10K_a}{(1+10K_a)+10s}$$

$$\frac{\omega(s)}{R(s)} = \frac{10K_a}{(1+10K_a)\left[1+\frac{10}{1+10K_a}\cdot s\right]}$$

$$\tau' = \frac{10}{1+10K_a} = \frac{1}{10}$$

$$1 + 10K_a = 100$$

$$K_a = 9.9 \cong 10$$

30. Given : open-loop transfer function,

$$G(s) = \frac{1}{(S+1)(s+2)}$$

Closed-loop transfer function, $T(s) = \frac{G(s)}{1+G(s)}$

$$T(s) = \frac{1/(s+1)(s+2)}{1+1/(s+1)(s+2)} = \frac{1}{s^2+3s+3}$$

Comparing with standard transfer function equation $\left(\frac{\omega_n^2}{s^2+2\xi\omega_n s+\omega_n^2}\right)$

$$\xi\omega_n = \frac{3}{2} = 1.5$$

Setling time, $t_s = \frac{4}{\xi\omega_n}$ (for ± 2% error)

$$= \frac{4}{1.5} = 2.67 > 2s.$$

Thus, to reduce setling time lower than 2s, PD controller should be used. Hence option (c) is correct answer.

Now, $C(s) = 2(s+4)$

New, closed loop transfer function

$$T'(s) = \frac{C(s)G(s)}{1+C(s)G(s)} = \frac{2(s+4)}{s^2+3s+2+2s+8}$$

$$= \frac{2(s+4)}{s^2+5s+10}$$

$$t'_s = \frac{4}{\xi\omega_n} = \frac{4}{\frac{5}{2}} = \frac{4\times 2}{5} = 1.6\text{ s} < 2\text{s}$$

31. Given : $\omega_n = 40$ rad/s, $\xi = 0.3$

Damped natural frequency,

$$\omega_d = \omega_n\sqrt{1-\xi^2} = 40\sqrt{1-0.9^2} = 38.16\text{ s}$$

32. Closed loop transfer function,

$$\frac{Y(s)}{U(s)} = \frac{\frac{4}{s(s+4)}}{1+\frac{4}{s(s+4)}\cdot 1} = \frac{4}{s^2+4s+4}$$

Comparing with standard equation

$$\left(\frac{\omega_n^2}{s^2+2\xi\omega_n s+\omega_n^2}\right)$$

$$\omega_n^2 = 4$$

$$\omega_n = 2\ \text{rad/s}$$

33. Closed-loop transfer function

$$T(s) = \frac{G(s)}{1+G(s)} = \frac{\frac{k}{s(s+10)}}{1+\frac{k}{s(s+10)}} = \frac{k}{s^2+10s+k}$$

Comparing above with standard transfer function

$$\left(\frac{\omega_n^2}{s^2+2\xi\omega_n s+\omega_n^2}\right),\ \text{we get}$$

$$\Rightarrow \quad \omega_n^2 = k \Rightarrow \omega_n = \sqrt{k}$$

$$\Rightarrow 2\omega_n\xi = 10$$

$$\therefore \quad \omega_n = \frac{10}{2\xi} = \frac{10}{2\times 0.25} = 20$$

$$k = \omega_n^2 = 400.$$

34. For unit-step input, output of second-order system is

$$y(t) = 1-\frac{2}{\sqrt{3}}e^{-t}\cdot\cos\left(\sqrt{3}t-\frac{\pi}{6}\right)$$

$$= 1-\frac{2}{\sqrt{3}}e^{-t}\cdot\sin\left(\sqrt{3}t-\frac{\pi}{6}-\frac{\pi}{2}\right)$$

Comparing with standard equation

$$y(t) = 1-\frac{e^{-\xi\omega_n t}}{\sqrt{1-\xi^2}}\sin(\omega_d t-\theta)$$

$$\Rightarrow \frac{1}{\sqrt{1-\xi^2}} = \frac{2}{\sqrt{3}}$$

$$1-\xi^2 = \frac{3}{4} \Rightarrow \xi^2 = \frac{1}{4}$$

$$\therefore \quad \xi = \frac{1}{2}$$

$$\Rightarrow \xi\omega n = 1 \Rightarrow \omega_n = 2\text{ rad/s.}$$

Transfer function, T(s)

$$= \frac{\omega_n^2}{s^2+2\xi\omega_n s+\omega_n^2} = \frac{4}{s^2+2s+4}$$

35. Given the peak overshoot of the closed-loop system to a unit step input

$$\%M_p = 10\%$$

$$M_p = 0.1$$

$$\Rightarrow \quad M_p = e^{-\pi\xi/\sqrt{1-\xi^2}}$$

$$0.1 = e^{-\pi\xi/\sqrt{1-\xi^2}}$$

$$\Rightarrow \quad \ln(0.1) = \frac{-\pi\xi}{\sqrt{1-\xi^2}}$$

$$\Rightarrow \quad 2.3 = \frac{\pi\xi}{\sqrt{1-\xi^2}}$$

$$\xi = 0.59$$

Also given $\quad G(s) = \dfrac{K}{s(s+2)}$

CE:-

$$1 + G(s) = 0$$

$$\Rightarrow \quad s^2 + 2s + K = 0$$

$$2\,\varepsilon\omega_n = 2$$

$$2\times 0.59\times \omega_n = 2$$

$$\omega_n = 1.69\text{ r/sec}$$

$$K = \omega_n^2 = 1.69\times 1.69$$

$$= 2.87\text{ r/sec}$$

Hence the value of K is 2.87 r/sec.

36. Given the response of the system

$$G(s) = \frac{s-2}{(s+1)(s+3)}$$

Now $\quad L\left(\dfrac{dy}{dt}\right) = (sY(s)-y(0))$

$$Y(s) = G(s)\times\frac{1}{s} = \frac{s-2}{s(s+1)(s+3)}$$

$$y(0) = \underset{s\to\infty}{\text{Lt}}\ sY(s)$$

(Applying initial value theorem)

$$= \underset{s\to\infty}{\text{Lt}} \frac{s-2}{(s+1)(s+3)}$$

$$= \frac{\left(1-\frac{2}{s}\right)}{s\left(1+\frac{1}{s}\right)\left(1+\frac{3}{s}\right)}$$

$$y(0) = 0$$

$$L\left(\frac{dy}{dt}\right) = sY(s) = \frac{s\times(s-2)}{s(s+1)(s+3)}$$

$$= \frac{s-2}{(s+1)(s+3)}$$

$$\left.\frac{dy}{dt}\right|_{t=0} = \underset{s\to\infty}{\text{Lt}}\ sL\left(\frac{dy}{dt}\right)$$

$$= \underset{s\to\infty}{\text{Lt}} \frac{s\times(s-2)}{(s+1)(s+3)}$$

$$= \frac{s\times s\left(1-\frac{2}{s}\right)}{s\left(1+\frac{1}{s}\right)s\left(1+\frac{3}{s}\right)}$$

$$= \frac{1-0}{(1+0)(1+0)} = 1$$

37. $$G(s) = \frac{1}{s^2+2s}$$

$$\frac{Y(s)}{R(s)} = \frac{\frac{K}{s^2+2s}}{1+\frac{K}{s^2+2s}} = \frac{K}{s^2+2s+K}$$

Minimum settling time and no overshoot implies

$$\xi = 1$$

$$\omega_n = \sqrt{K}$$

$$2\times\xi\times\omega_n = 2$$

$$\omega_n = 1$$

$\Rightarrow$ $\sqrt{K} = 1$ or $K = 1$

Hence, the required value of gain K to achieve this is 1.

38. We know that with unity feedback and step input to a type-1 system, steady-state error is zero and also for ramp input is finite $\left(\frac{1}{k_v}\right)$ and for parabolic input is infinity.

Hence, the given system must be type-1 (p = 1).

39. Given, $G(s) = \frac{1}{1+2s+s^2}$

$$\therefore C(s) = G(s)R(s) = \frac{1}{s(s+1)^2}$$

Taking increase Laplace transform,

$C(t) = 1 - e^{-1} - te^{-1}$

Now checking from options,

From Option (A)

$0.94 = 1 - e^{-5.25} - 5.25\, e^{-5.25} = 0.89$

option A is wrong

From option (C)

$\therefore\ 0.94 = 1 - e^{-4.50} - 4.50\, e^{-4.50} = 0.94$

40. Position error coefficient, $k_P = \lim_{s\to 0} G(s)\cdot H(s)$

$$= \lim_{s\to 0}\frac{k}{s(s+10)} = \infty$$

Steady-state error, $l_{ss} = \frac{A}{1+k_p} = \frac{1}{\infty} = 0.$

41. Steady-state error for a stable 'type – 0' system with unity feedback $= \frac{1}{1+k_p}$.

42. Velocity error coefficient, $K_v = \lim_{s\to 0} SG(s)H(s)$

$$= \lim_{s\to 0}\frac{s\cdot 4(1+2s)}{s^2(s+2)} = \infty$$

Steady-state error, $e_{ss} = \frac{A}{K_v} = \frac{A}{\infty} = 0.$

43. $K_p = lt\ G(S)\ H(s)$

$$= \underset{s\to 0}{lt}\frac{1}{(s+6)(s+1)} = \frac{1}{6}$$

44. Position error coefficient,

$$K_P = \lim_{s\to 0} G(s)\cdot H(s) = \lim_{s\to 0}\frac{K}{S(s+1)} = \infty$$

Steady-state error, $e_{ss} = \frac{A}{1+K_P} = \frac{A}{1+\infty} = 0.$

45. Closed loop transfer function,

$$T(s) = \frac{a_{n-1}\cdot s + a_n}{s^n + a_1 s^{n-1} + \ldots + a_{n-1}s + a_n} \times \frac{s^n + a_1 s^{n-1} + \ldots + a_{n-2}\cdot s^2}{s^n + a_1 s^{n-1} + \ldots + a_{n-2}\cdot s^2}$$

$$T(s) = \frac{(a_{n-1}\cdot s + a_n)/(s^n + a_1\cdot s^{n-1} + \ldots + a_{n-2}\cdot s^2)}{1 + \frac{a_{n-1}\cdot s + a_n}{s^n + a_1 s^{n-1} + \ldots + a_{n-2}\cdot s^2}}$$

On comparing with standard eq.

$$T(s) = \frac{G(s)}{1+G(s)}$$

$$\Rightarrow G(s) = \frac{a_{n-1} \cdot s + a_n}{s^n + a_1 \cdot s^{n-1} + ... + a_{n-2} \cdot s^2}$$

For type-2, ramp input,

Velocity error coefficient,

$$K_v = \lim_{s \to 0} s \cdot G(s) = \infty$$

Steady-state error, $e_{ss} = \frac{1}{K_v} = 0$

46. In general, $E(s) = \frac{R(s)}{1+G(s)H(s)}$

From block diagram, $G(s) = \frac{4}{(s+2)}, H(s) = \frac{2}{s+4}$

Given : $R(s) = \frac{1}{s}$

Steady-state error,

$$e_{ss} = \lim_{s \to 0} sE(s) = \lim_{s \to 0} \frac{s \times \frac{1}{s}}{1 + \frac{4}{s+2} \cdot \frac{2}{s+4}}$$

$$= \frac{1}{1 + \frac{4}{2} \cdot \frac{2}{4}} = 1.$$

47. Given: $R(s) = \frac{1}{s}$, $G(s) = \frac{2}{s(s+1)}$, $H(s) = 1$.

Error, $E(s) = \frac{R(s)}{1 + G(s) \cdot H(s)}$

$$= \frac{\frac{1}{s}}{1 + \frac{2}{s(s+1)} \cdot 1} = \frac{s+1}{s^2 + s + 2}$$

Steady-state error, $e_{ss} = \lim_{s \to 0} S \cdot E(s)$

$$\lim_{s \to 0} \frac{s(s+1)}{s^2 + s + 2} = \frac{0}{2} = 0.$$

4 CHAPTER

Stability & Root Locus

Analysis of Previous GATE Papers												
	Year → Topics ↓		2019	2018	2017 Set 1	2017 Set 2	2016 Set 1	2016 Set 2	2016 Set 3	2015 Set 1	2015 Set 2	2015 Set 3
CONCEPT OF STABI L I TY	1 Mark	MCQ Type										
		Numerical Type										
	2 Marks	MCQ Type										
		Numerical Type										
	Total											
ROUTH-HURWI TZ CRI TERION	1 Mark	MCQ Type						1				
		Numerical Type										
	2 Marks	MCQ Type			1					1		
		Numerical Type				1		1	1			1
	Total				2	2		3	2	2		2
ROOT LOCUS PLOT	1 Mark	MCQ Type										
		Numerical Type								1		
	2 Marks	MCQ Type			1							
		Numerical Type					1		1	1		1
	Total				2		2		2	3		2

CONCEPT OF STABILITY

1. If G(s) is a stable transfer function, then $F(s) = \frac{1}{G(s)}$ is always a stable transfer function, (T/F).

[1994 : 2 Marks]

2. If the closed-loop transfer function of a control system is given as

$T(s) = \frac{s-5}{(s+2)(s+3)}$, then it is

(a) an unstable system
(b) an uncontrollable system
(c) a minimum phase system
(d) a non-minimum phase system

[2007 : 1 Mark]

ROUTH-HURWITZ CRITERION

3. Consider a characteristic equation given by + $s^4 + 3s^4 + 5s^2 + 6s + K + 10$

The condition for stability is

(a) $K > 5$ (b) $-10 < K$
(c) $K > -4$ (d) $-10 < K - 4$

[1988 : 2 Marks]

4. In order to stabilize the system shown in figure T_i should satisfy,

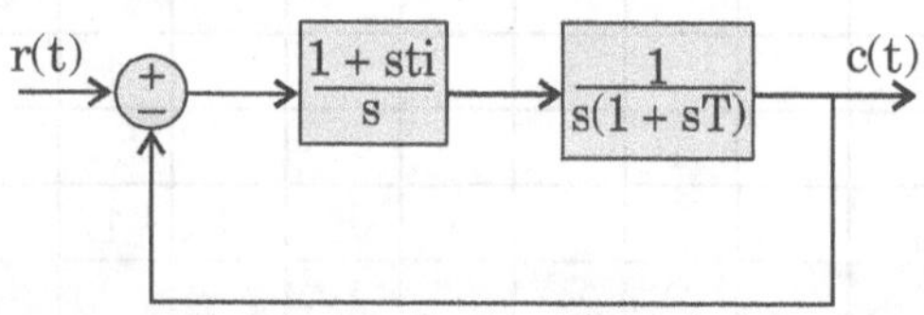

(a) $T_i = -T$ (b) $T_i = T$
(c) $T_i < T$ (d) $T_i > T$

[1989 : 2 Marks]

5. An electromechanical closed-loop control system has the following characteristic equation; $s^3 + 6K s^2 + (K + 2)s + 8 = 0$. Where K is the forward gain of the system. The condition for closed loop stability is

(a) K = 0.528 (b) K = 2
(c) K = 0 (d) K = -2.258

[1990 : 2 Marks]

6. If $s^3 + 3s^2 + 4s + A = 0$, then all the roots of this equation are in the left half plane provided that

(a) $A > 12$
(b) $-3 < A < 4$
(c) $0 < A < 12$
(d) $5 < A < 12$

[1993 : 2 Marks]

7. The number of roots of $s^3 + 5s^2 + 7s + 3 = 0$ in the left half of the s-plane is

(a) zero (b) one
(c) two (d) three

[1998 : 1 Mark]

8. The open loop transfer function of an unity feedback open loop system is $\frac{2s^2+6s+5}{(s+1)^2(s+2)}$. The characteristic equation of the closed loop system is

(a) $2s^2 + 6s + 5 = 0$
(b) $(s + 1)^2(s + 2) = 0$
(c) $2s^2 + 6s + 5 + (s + 1)^2 (s + 2) = 0$
(d) $2s^2 + 6s + 5 - (s + 1)^2(s + 2) = 0$

[1998 : 1 Mark]

9. An amplifier with resistive negative feedback has two left half plane poles in its open-loop transfer function. The amplifier

(a) will always be unstable at high frequency
(b) will be stable for all frequency
(c) may be unstable, depending on the feedback factor
(d) will oscillate at low frequency

[2000 : 1 Mark]

10. A system described by the transfer function

$H(s) = \frac{1}{s^3 + \alpha s^2 + Ks + 3}$ is stable.

The constraints on a and K are,

(a) $\alpha > 0, \alpha K < 3$ (b) $\alpha > 0, \alpha K > 3$
(c) $\alpha < 0, \alpha K > 3$ (d) $\alpha < 0, \alpha K < 3$

[2000 : 2 Marks]

11. The feedback control system in the figure is stable

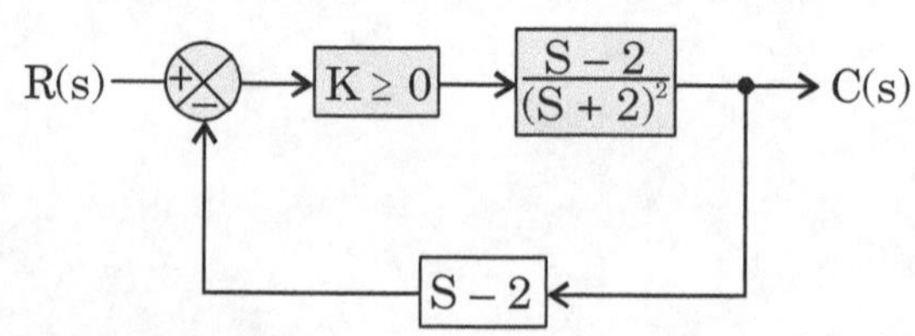

(a) for all $K \geq 0$
(b) only is $K \geq 0$
(c) only if $0 \leq k < 1$
(d) only if $0 \leq K \leq 2$

[2001 : 2 Marks]

12. The system shown in the figure remains stable when

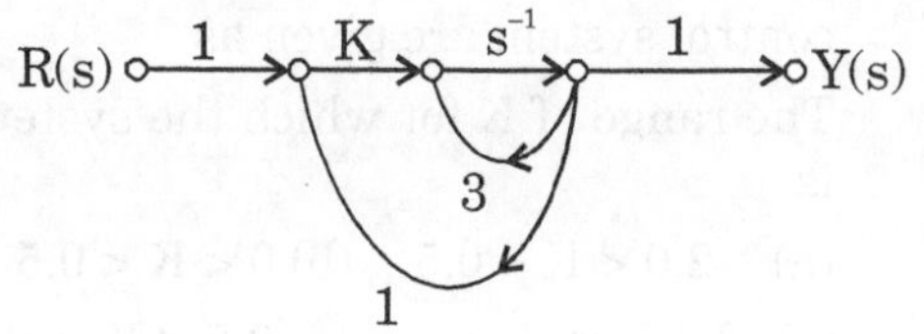

(a) $K < -1$ (b) $-1 < K < 1$
(c) $1 < K < 3$ (d) $K < -3$

[2002 : 2 Marks]

13. The characteristic polynomial of a system is $q(s) = 2s^5 + S^4 + 4s^3 + 2s^2 + 2s + 1$. The system is

(a) stable (b) marginally stable
(c) unstable (d) oscillatory

[2002 : 2 Marks]

14. The open-loop transfer function of a unity feedback system is

$$G(s) = \frac{K}{s(s^2+s+2)(s+3)}$$

The range of K forwhich the system is stable is

(a) $\frac{21}{4} > K > 0$ (b) $13 > K > 0$
(c) $\frac{21}{4} > K > \infty$ (d) $-6 < K < \infty$

[2004 : 2 Mark]

15. For the polynomial

$P(s) = s^5 + s^4 + 2s^3 + 2s^2 + 3s + 15$, the number of roots which lie in the right half of the s-plane is

(a) 4 (b) 2
(c) 3 (d) 1

[2004 : 2 Marks]

16. The positive values of "K" and "a" so that the system shown in the figure below oscillates at a frequency of 2 rad/sec respectively are

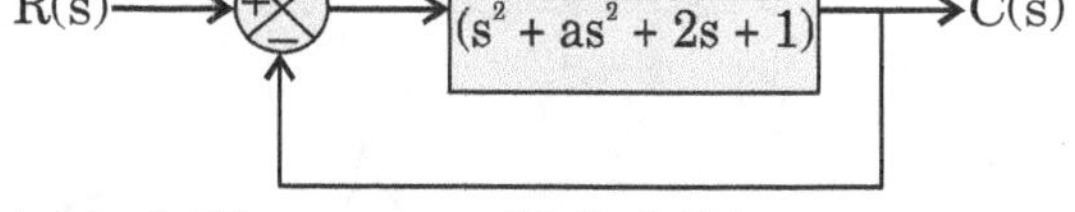

(a) 1, 0.75 (b) 2, 0.75
(c) 1, 1 (d) 2, 2

[2006 : 2 Marks]

Common Data for Questions 17 & 18:

Consider a unity-gain feedback control system whose open-loop transfer function is $G(s) = \frac{as+1}{s^2}$

17. A certain system has transfer function

$G(s) = \frac{s+8}{s^2 + \alpha s - 4}$, where a is a parameter.

Consider the standard negative unity feedback configuration as shown below.

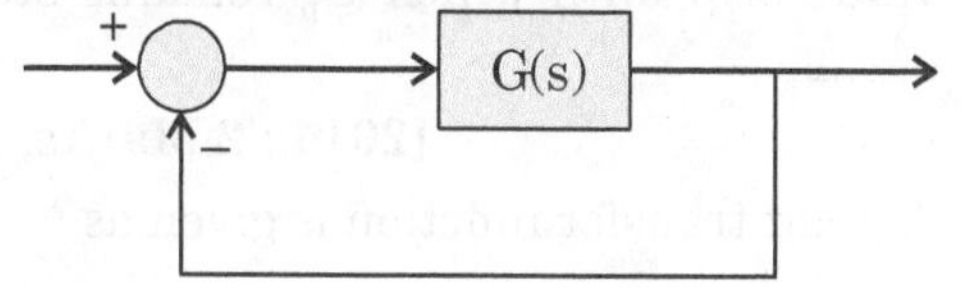

Which of the following statements is true?

(a) The closed loop system is never stable for any value of a

(b) For some positive values of a, the closed loop system is stable, but not for all positive values

(c) For all positive values of a, the closed loop system is stable

(d) The closed loop system is stable for all values of a, both positive and negative

[2008 : 2 Marks]

18. The number of open right half plane poles of

$G(s) = \frac{10}{s^5 + 2s^4 + 3s^3 + 6s^2 + 5s + 3}$ is

(a) 0
(b) 1
(c) 2
(d) 3

[2008 : 2 Marks]

19. The feedback system shown below oscillates at 2 rad/s when

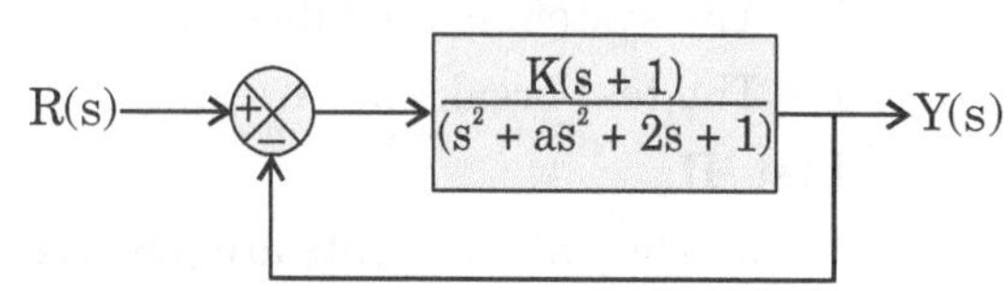

(a) K = 2 and a = 0.75
(b) K = 3 and a = 0.75
(c) K = 4 and a = 0.5
(d) K = 2 and a = 0.5

[2012 : 2 Marks]

20. The forward path transfer function of a unity negative feedback system is given by

$G(s) = \frac{K}{(s+2)(s-1)}$,

The value of K which will place both the poles of the closed-loop system at the same location, is ______.

[2014 : 1 Mark, set-1]

21. Consider a transfer function

$$G_p(s) = \frac{ps^2 + 3ps - 2}{s^2 + (3+p)s + (2-p)}$$

with p a positive real parameter. The maximum value of p until which G_p remains stable is ______.

[2014 : 2 Marks, Set-4]

22. A plant transfer function is given as

$$G(s) = \left(K_p + \frac{K_1}{s}\right)\frac{1}{s(s+2)}$$

When the plant operates in a unity feedback configuration, the condition for the stability of the closed loop system is

(a) $K_P > \frac{K_1}{2} > 0$ (b) $2K_1 > K_P > 0$

(c) $2K_1 < K_P$ (d) $2K_1 > K_P$

[2015 : 2 Marks, Set-1]

23. The characteristic equation of an LTI system is given by

$F(s) = s^5 + 2s^4 + 3s^3 + 6s^2 - 4s - 8$. The number of roots that lie strictly in the left half s-plane is ______.

[2015 : 2 Marks, Set-3]

24. Match the inferences X, Y and Z about a system, to the corresponding properties of the elements of first column in Routh's Table of the system characteristic equation.

List – I

X. The system is stable.....

Y. The system is unstable .....

Z. The test breaks down .....

List -II

P. ... when all elements are positive

Q. ...when any one element is zero

R. ...when there is a change in sign of coefficients

(a) X –P; Y– Q; Z – R

(b) X – C); Y – P; Z – R

(c) X – R; Y– Q; Z – P

(d) X –P; Y – R: Z – Q

[2016 : 1 Mark, Set-1]

25. The transfer function of a linear time invariant system is given by

$H(s) = 2s^4 - 5s^3 + 5s - 2$

The number of zeroes in the right half of the s-plane is ....

[2016 : 2 Marks, Set-1]

26. The first two rows in the Routh table for the characteristic equation of a certain closed-loop control system are given as

The range of K for which the system is stable is

(a) $-2.0 < K < 0.5$ (b) $0 < K < 0.5$

(c) $0 < K <$ (d) $0.5 < K <$

[2016: 2 Marks, Set-3]

27. Which one of the following options correctly describes the locations of the roots of the equation $s^4 + s^2 + 1 = 0$ on the complex plane?

(a) Four left half plane (LHP) roots.

(b) One right half plane (RHP) root, one LHP root and two roots on the imaginary axis.

(c) Two RHP roots and two LHP roots.

(d) All four roots are on the imaginary axis.

[2017 : 2 Marks, Set-1]

28. A unity feedback control system is characterized by the open-loop transfer function

$$G(s) = \frac{2(s+1)}{s^3 + Ks^2 + 2s + 1}.$$

The value of K for which the system oscillates at 2 rad/s is ______.

[2017 : 2 Marks, Set-2]

ROOT LOCUS PLOT

29. The characteristic equation of a unity negative feedback system is 1 + KG(s) = 0. The open loop transfer function G(s) has one pole at 0 and two poles at –1. The root locus of the system for varying K is shown in the figure.

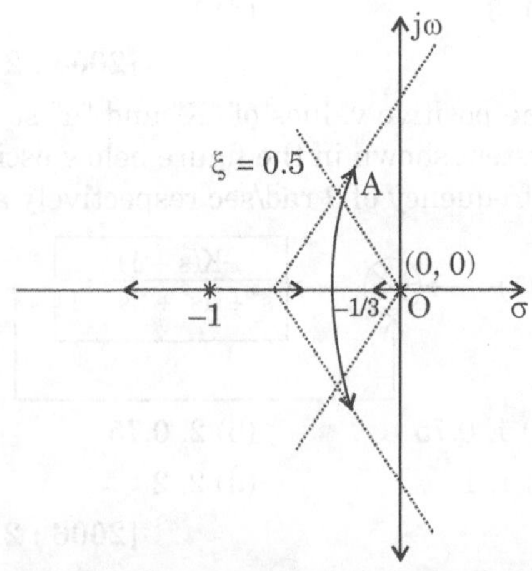

The constant damping ratio line, for ξ = 0.5, intersects the root locus at point A. The distance from the origin to point A is given as 0.5. The value of K at point A is ______.

[2014 : 2 Mark]

30. Consider a closed-loop system shown in Figure (A) below. The root locus for it is shown in Figure (B). the closed loop transfer function for the system is

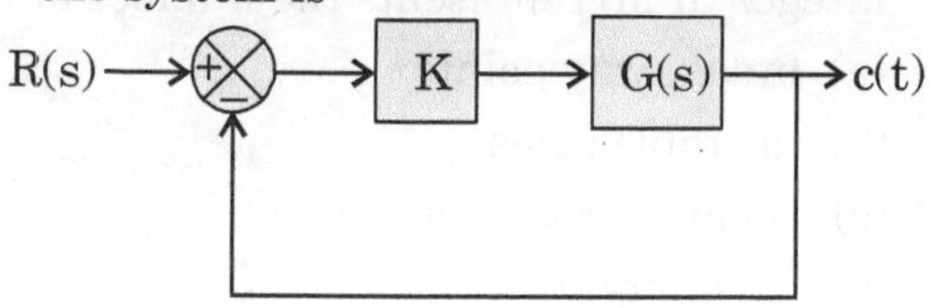

(A)

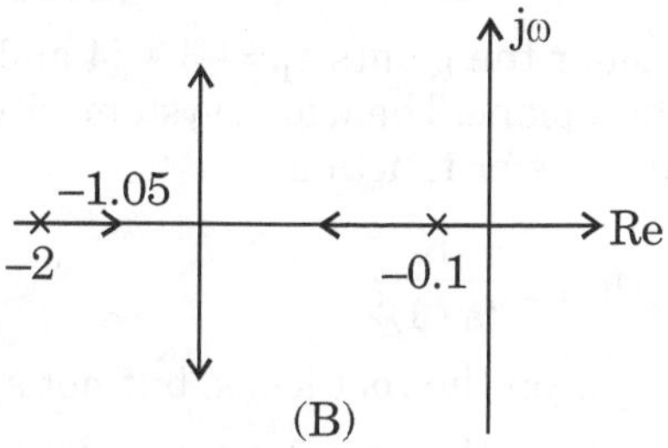

(B)

(a) $\dfrac{K}{1+(0.5s+1)(10s+1)}$

(b) $\dfrac{K}{(s+2)(s+0.1)}$

(c) $\dfrac{K}{1+K(0.5s+1)(10s+1)}$

(d) $\dfrac{K}{K+0.2(0.5s+1)(10s+1)}$

[1988 : 2 Marks]

31. The OLTF of a feedback system is

$$G(s)H(s) = \frac{K(s+1)(s+3)}{s^2+4s+8}.$$

The root locus for the same is

(a)

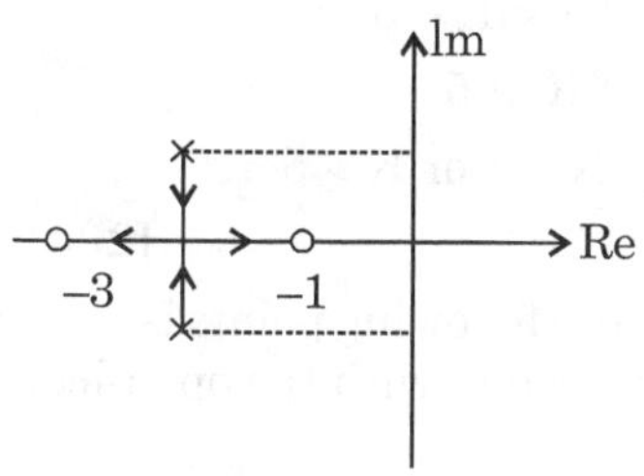

(b)

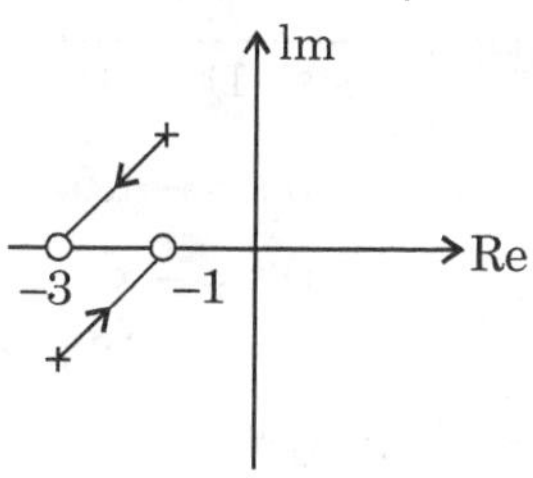

(c)

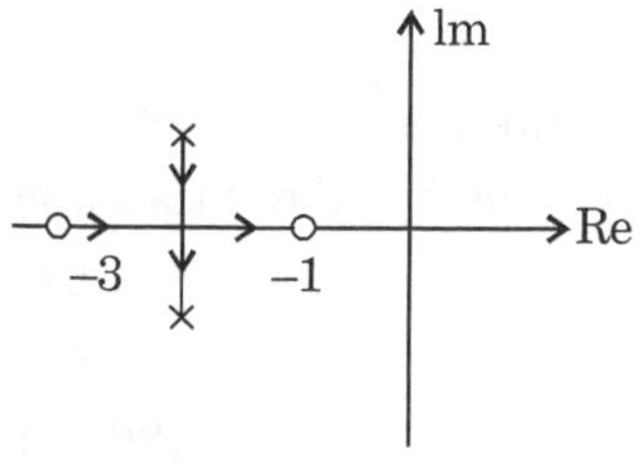

(d)

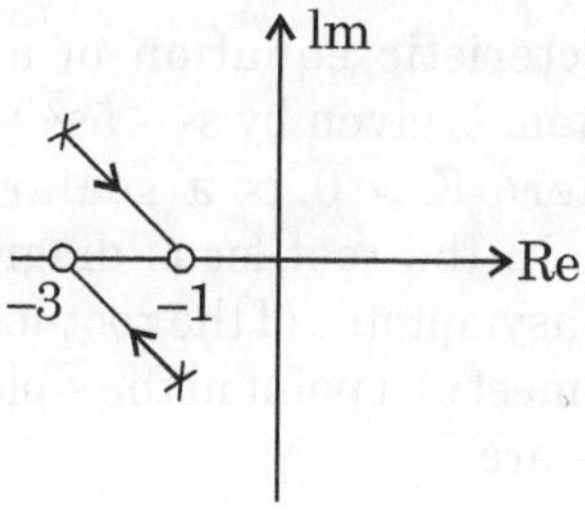

[1989 : 2 Marks]

32. The transfer function of a closed loop system is:

$$T(s) = \frac{K}{s^2+(3-K)(s+1)}$$

Where K is the forward path gain. The root locus plot of the system is:

(a)

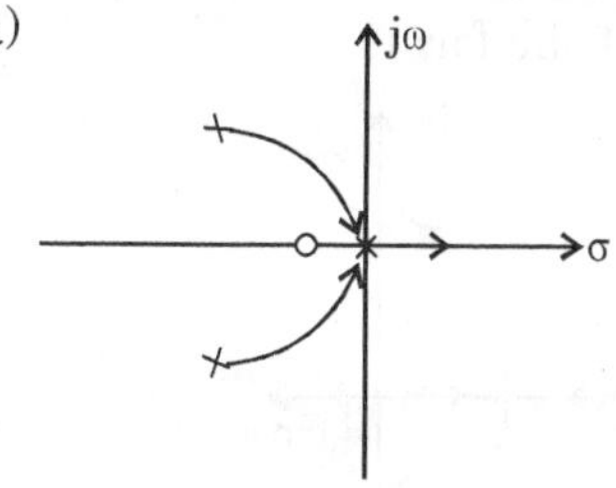

(b)

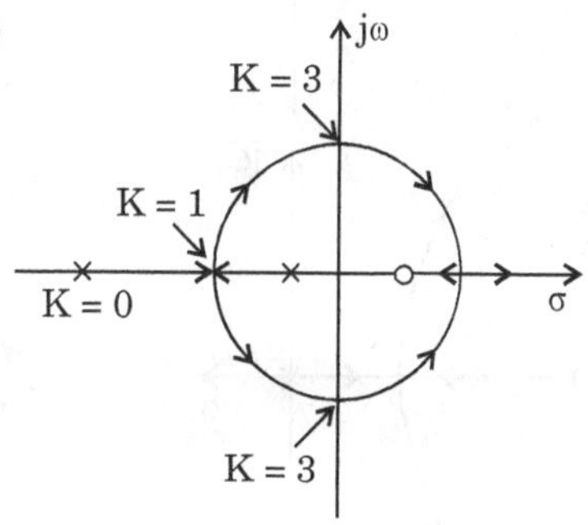

(c)

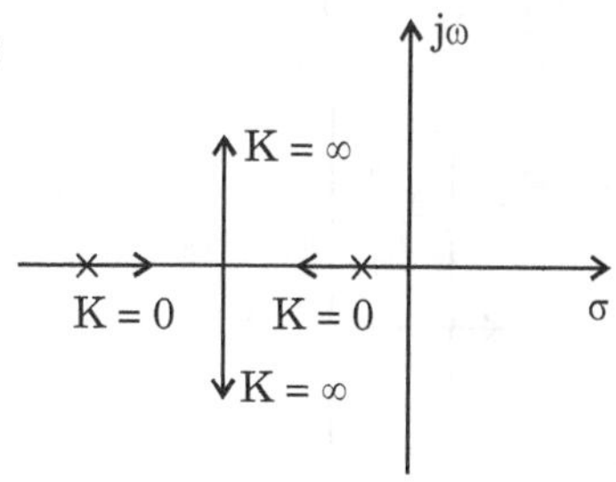

(d)

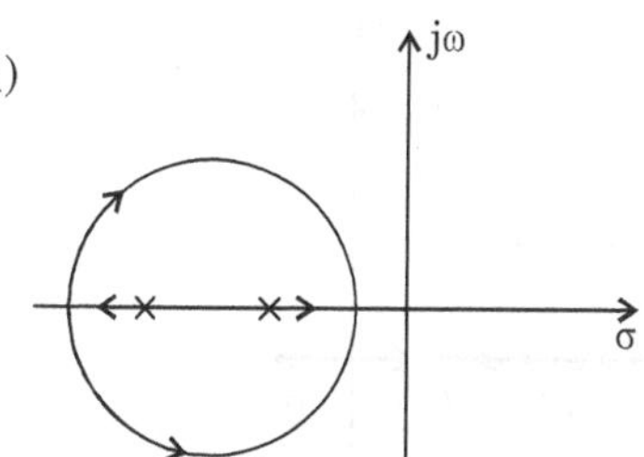

[1990 : 2 Marks]

33. The characteristic equation of a feedback control system is given by $s^3 + 5s^2 + (K + 6)s + K = 0$ Where $K > 0$ is a scalar variable parameter. In the root-locus diagram of the system the asymptotes of the root-loci for large values of K meet at a point in the s-plane, whose coordinates are

(a) (–3, 0) (b) (–2, 0)
(c) (–1, 0) (d) (2, 0)

[1991 : 2 Marks]

34. Given a unity feedback system with open-loop transfer function.

$G(s) = \frac{K}{s(s+1)(s+2)}$. The root locus plot of the system is of the form.

(a)

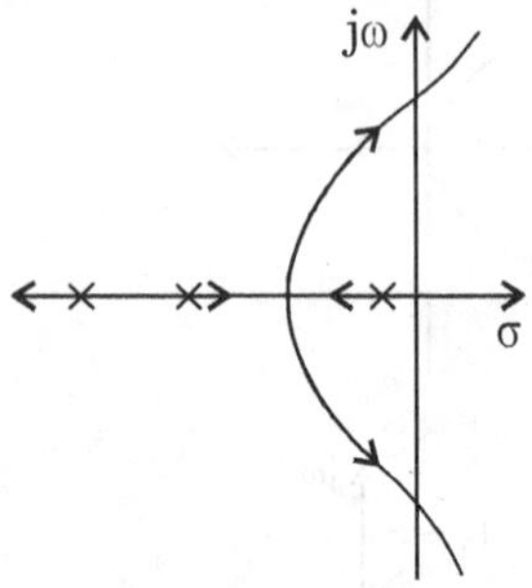

(b)

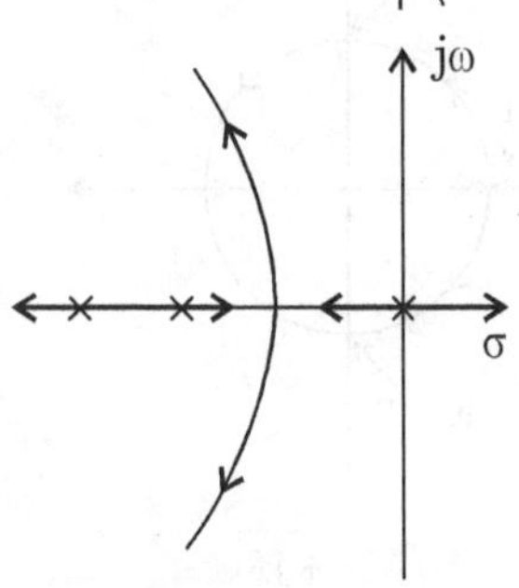

(c)

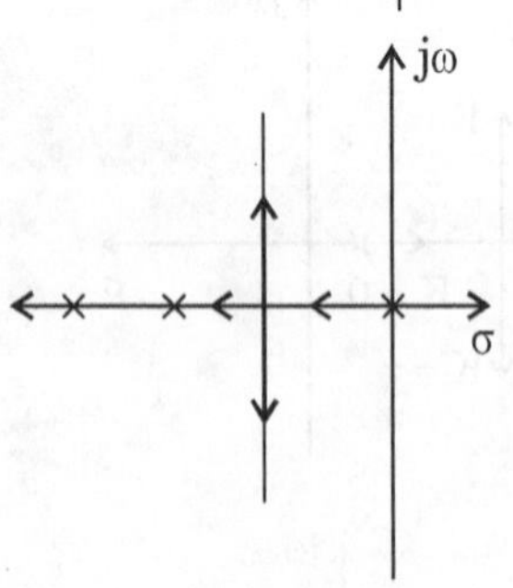

(d)

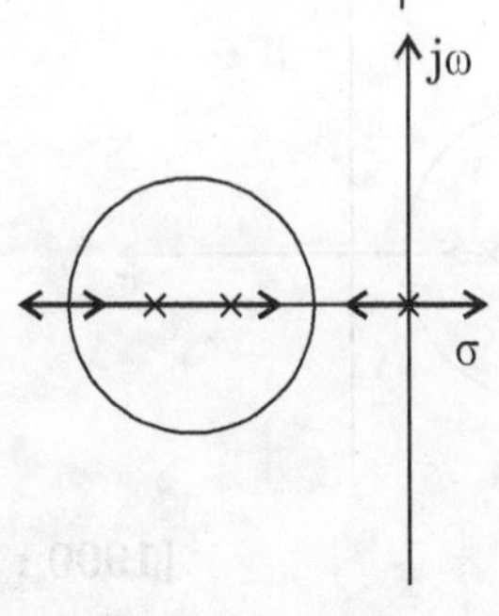

[1992 : 2 Marks]

35. If the open-loop transfer function is a ratio of a numerator polynomial of degree 'm' and a denominator polynomial of degree 'n' then the integer (n-m) represents the number of

(a) breakaway points
(b) unstable poles
(c) separate root loci
(d) asymptotes

[1994 : 1 Mark]

36. Consider the points $s_1 = -3 + j4$ and $s_2 = -3$ j2 in the s-plane. Then, for a system with the open-loop transfer function

$$G(s)H(s) = \frac{K}{(s+1)^4}$$

(a) s_1 is on the root locus, but not s_2
(b) s_2 is on the root locus, but not s_1
(c) both si and s_2 are on the root locus
(d) neither s_1 nor s_2 is on the root locus

[1999 : 2 Marks]

37. The root-locus diagram for a closed-loop feedback system is shown in the figure. The system is overdamped

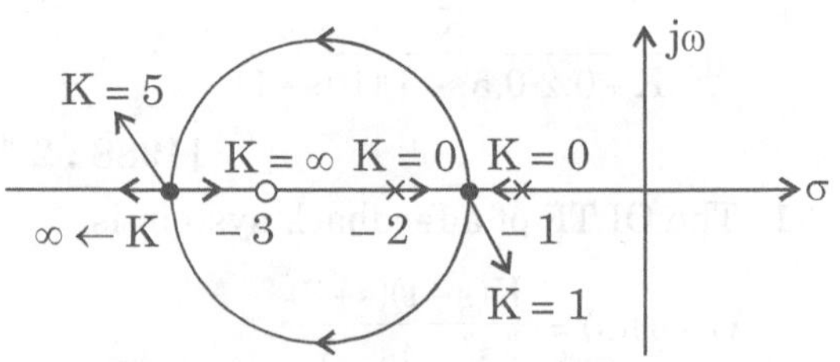

(a) only if $0 \leq K \leq 1$
(b) only if $1 < K < 5$
(c) only if $K > 5$
(d) if $0 \leq K < 1$ or $K > 5$

[2001 : 1 Mark]

38. Which of the following points is NOT on the root locus of a system with the open-loop transfer function $G(s)H(s) = \frac{K}{s(s+1)(s+3)}$?

(a) $s = -j\sqrt{3}$ (b) $s = -1.5$
(c) $s = -3$ (d) $s = -\infty$

[2002 : 1 Mark]

39. The root locus of the system

$$G(s)H(s) = \frac{K}{s(s+2)(s+3)},$$

has the break-away point located at

(a) (–0.5, 0) (b) (–2.548, 0)
(c) (–4, 0) (d) (–0.784, 0)

[2003 : 2 Marks]

40. Given $G(s)H(s) = \dfrac{K}{s(s+1)(s+3)}$, the point of intersection of the asymptotes of the root loci with the real axis is

(a) –4 (b) 1.33

(c) –1.33 (d) 4

[2004 : 1 Mark]

41. An unity feedback system is given as,

$G(s) = \dfrac{K(1-s)}{s(s+3)}$.

Indicate the correct root locus diagram

(a)

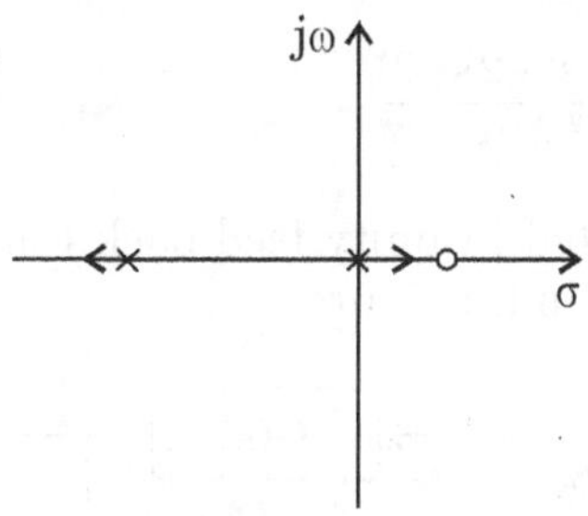

(b)

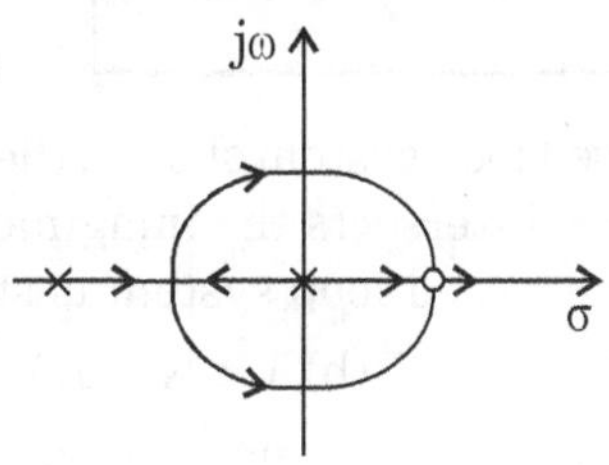

(c)

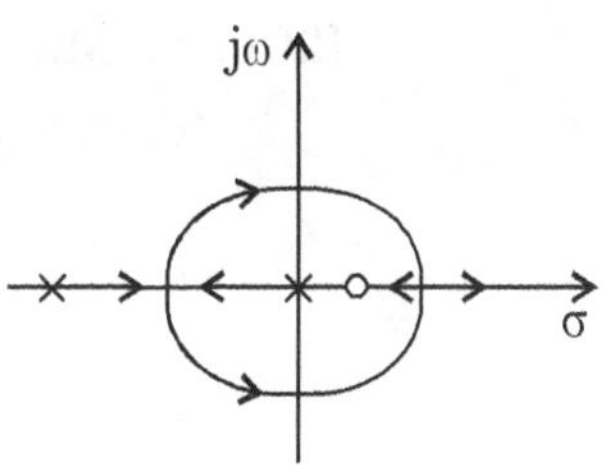

(d)

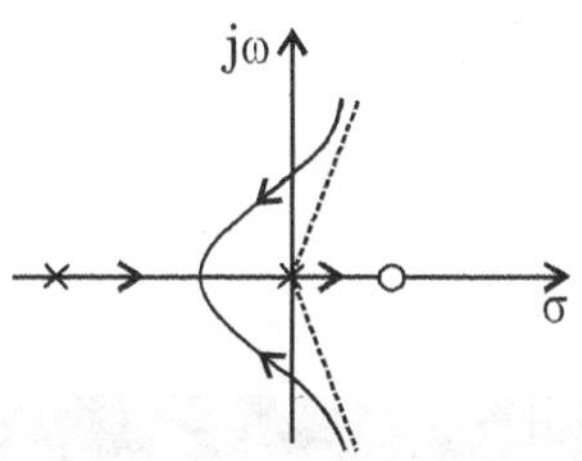

[2005 : 2 Marks]

42. A unity feedback control system has an openloop transfer function $G(s) = \dfrac{K}{s(s^2+7s+12)}$.

The gain K for which s = –1 + j1 will lie on the root locus of this system is

(a) 4 (b) 5.5

(c) 6.5 (d) 10

[2007 : 2 Marks]

43. The feedback configuration and the pole-zero locations of $G(s) = \dfrac{s^2-2s+2}{s^2+2s+2}$ are shown below. The root locus for Negative values of K, i.e., for $-\infty < K < 0$, has breakaway/break-in points and angle of departure at pole P (with respect to the positive real axis) equal to

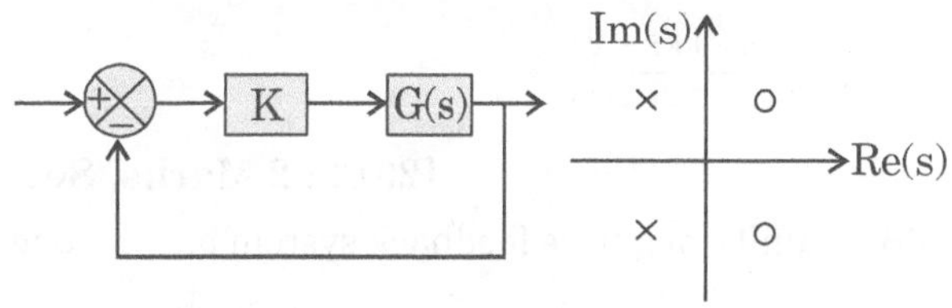

(a) $\pm\sqrt{2}$ and 0° (b) $\pm\sqrt{2}$ and 45°

(c) $\pm\sqrt{3}$ and 0° (d) $\pm\sqrt{3}$ and 45°

[2009 : 2 Marks]

44. The root locus plot for a system is given below. The open loop transfer function corresponding to this plot is given by

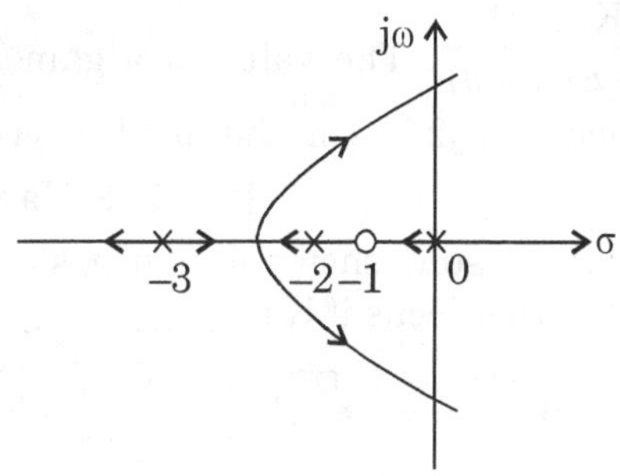

(a) $G(s)H(s) = K\dfrac{s(s+1)}{(s+2)(s+3)}$

(b) $G(s)H(s) = K\dfrac{(s+1)}{s(s+2)(s+3)^2}$

(c) $G(s)H(s) = K\dfrac{1}{s(s-1)(s+2)(s+3)}$

(d) $G(s)H(s) = K\dfrac{(s+1)}{s(s+2)(s+3)}$

[2011 : 1 Mark]

45. In the root locus plot shown in the figure, the pole/ zero marks and the arrows have been removed. Which one of the following transfer functions has this root locus?

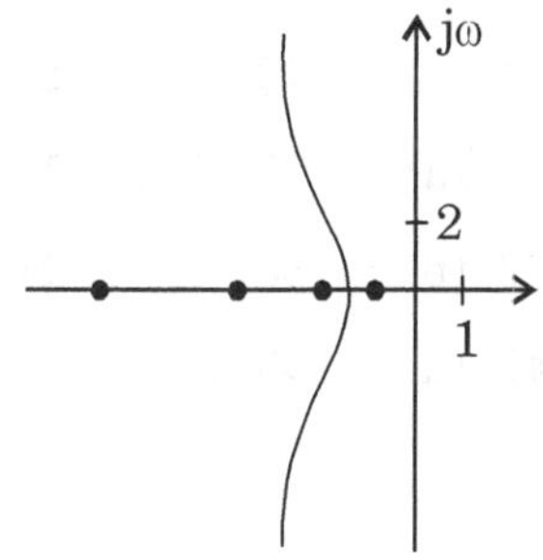

(a) $\frac{s+1}{(s+2)(s+4)(s+7)}$

(b) $\frac{s+4}{(s+1)(s+2)(s+7)}$

(c) $\frac{s+7}{(s+1)(s+2)(s+4)}$

(d) $\frac{(s+1)(s+2)}{(s+7)(s+4)}$

[2014 : 2 Marks, Set-3]

46. A unity negative feedback system has the open-loop transfer function $G(s) = \frac{K}{(s+1)(s+3)}$. The value of the gain K(>0) at which the root locus crosses the imaginary axis is _____.

[2015 : 1 Mark, Set-1]

47. The open-loop transfer function of a unity feedback configuration is given as $G(s) = \frac{K(s+4)}{(s+8)(s^2+9)}$. The value of a gain K(>0) for which −1 + j2 lies on the root locus is _____.

[2015 : 2 Mark, Set-1]

48. For the system shown in figure, s = −2.75 lies on the root locus if K is _____.

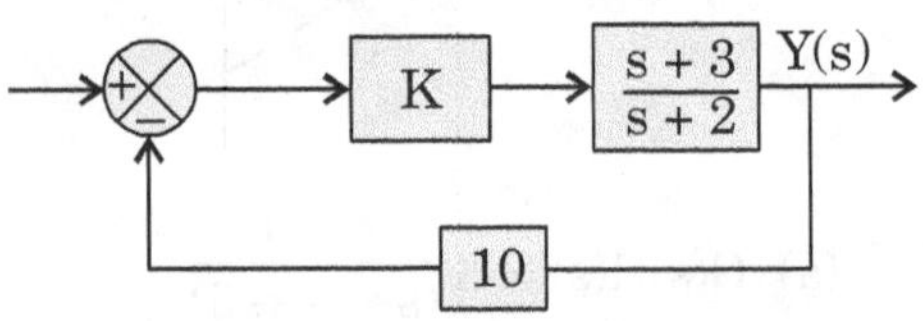

[2015 : 2 Marks, Set-3]

49. The open-loop transfer function of a unity-feedback control system is

$$G(s) = \frac{K}{s^2+5s+5}.$$

The value of K at the breakaway point of the feedback control system's root-locus plot is ......

[2016 : 2 Marks, Set-1]

50. The forward-path transfer function and the feedback-path transfer function of a single loop negative feedback control system are given as

$$G(s) = \frac{K(s+2)}{s^2+2s+2} \text{ and } H(s) = 1,$$

respectively. If the variable parameter K is real positive, then the location of the breakaway point on the root locus diagram of the system is _____.

[2016 : 2 Marks, Set-3]

51. A linear time invariant (LTI) system with the transfer function

$$G(s) = \frac{K(s^2+2s+2)}{(s^2-3s+2)}$$

is connected in unity feed back configuration as shown in the figure.

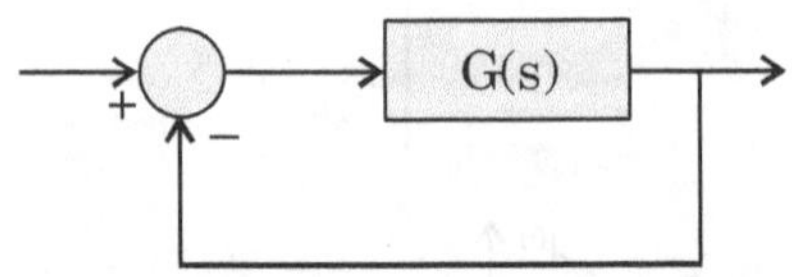

For the closed loop system shown, the root locus for $0 < K < \infty$ intersects the imaginary axis for K = 1.5. The closed loop system is stable for

(a) K > 1.5
(b) 1 < K < 1.5
(c) 0 < K < 1.5
(d) no positive value of K

[2017 : 2 Marks, Set-1]

ANSWERS

1. (False)	2. (d)	3. (d)	4. (d)	5. (b)	6. (c)	7. (a)	8. (c)	9. (b)	10. (b)
11. (c)	12. (d)	13. (c)	14. (a)	15. (b)	16. (b)	17. (c)	18. (c)	19. (a)	20. (2.25)
21. (2)	22. (a)	23. (2)	24. (d)	25. (3)	26. (d)	27. (c)	28. (0.75)	29. (0.37)	30. (d)
31. (a)	32. (b)	33. (b)	34. (a)	35. (d)	36. (d)	37. (d)	38. (b)	39. (d)	40. (c)
41. (c)	42. (d)	43. (b)	44. (b)	45. (b,c)	46. (12)	47. (25.24)	48. (0.3)	49. (1.25)	50. (−3· 41)
51. (a)									

EXPLANATIONS

1. The condition for stability is that no poles of G (*s*) should be on the R.H.S. of the *s*–plane. If G (*s*) has a zero in the R.H.S. of *s*–plane, with all poles in the L.H.S. or *s*–plane, then F (*s*) would have a pole in the R.H.S. of *s*–plane and hence is unstable.

2. A minimum phase system is one which do not have any poles or zeros on right half s-plane where T(s) has on zero, s = 5 at right half s-place, so it is non-minimum phase system

3. Using Rout-Hurwitz Criterion

S^4	1	5	K + 10
S^3	3	6	
S^2	3	K + 10	
S^1	$\frac{-12-3k}{3}$		
S^0	K + 10		

For a stable system, all the coefficient of first column should be positive, i.e.,

$$-\frac{12-3k}{3} > 0 \text{ and } k + 10 > 0$$

$- 12 - 3\,k > 0$ and $k > - 10$

$3k < - 12$

$k < - 4$

the condition for stability is $- 10 < k < - 4$.

4. Given: $G(s) = \frac{(1+STi)}{S} \cdot \frac{1}{s(1+sT)}$;

$H(s) = 1$

Characteristic equation :

$\Delta(s) = 1 + G(s)\,H(s) = 0$

$$1 + \frac{(1+ST_i)}{s^2(1+ST)} \cdot 1 = 0$$

$s^2\,(1 + sT) + (1 + sT_i) = 0$

$\Rightarrow s^3\,T + s^2 + sT_i + 1 = 0$

Using Routh – Hurwitz Criterion

S^3	T	T_i
S^2	1	1
S^1	$T_i - T$	
S^0	T_i	

For a stable system, all the coefficient of first column should be positive, i.e.,

$T_i - T > 0$ and $T_i > 0\ T_i > T$.

5. Characteristic equation :

$$\Delta(S) = S^3 + 6KS^2 + (K + 2)S + 8 = 0$$

Using Routh – Hurwitz Criterion,

S^3	1	K + 2
S^2	6K	8
S^1	$\frac{6k^2 + 12\,k - 8}{6k}$	
S^0	8	

For an stable system, first column element should be positive, i.e.,

$6k > 0$

$k > 0$ ...(1)

and $\frac{6k^2 + 12k - 8}{6k} > 0$

$6k^2 + 12k - 8 > 0$

$3k^2 + 6k - 4 > 0$ .. (ii)

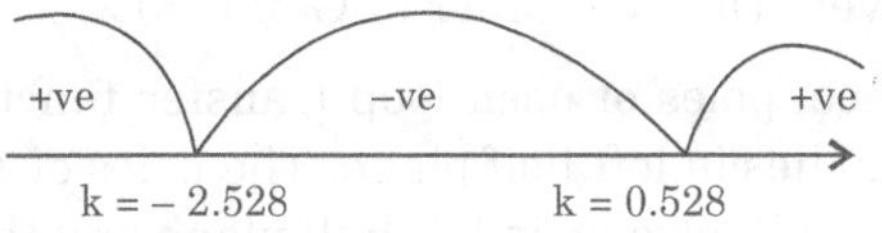

$$k = -\frac{6 \pm \sqrt{36+48}}{6}$$

$k = - 2.528, + 0.528$

For eq. (ii),

Range of k

$- 2.528 > k > 0.528$

For eq. (i),

Range of k

$k > 0.528$.

For given options, k = 2 is the right answer.

6. $\Delta(s) : s^3 + 3s^2 + 4s + A = 0$

Using Routh – Hurwitz criterion,

S^3	1	A
S^2	–3	A
S^1	$\frac{12-A}{3}$	
S^0	A	

All the roots of the eq. $\Delta(s) = 0$ to be lie on left half plane, first column element should be positive, i.e.,

$\frac{12-A}{3} > 0$ and $A > 0$

$12 - A > 0$

$A < 12$

Now, $0 < A < 12$.

7. $\Delta(s) = s^3 + 5s^2 + 7s + 3 = 0$

Using Routh–Hurwitz Criterion,

S^3	1	7
S^2	5	3
S^1	6.4	
S^0	3	

Since, there is no sign change in first column of R-H array so no root will lie on right half of s-plane.

The order of $\Delta(s)$ is 3. Hence, these 3 roots will be in the left half of s-plane.

8. Given: $G(s) = \dfrac{2s^2 + 6s + 5}{(s+1)^2(s+5)}$, $H(s) = 1$.

Characteristic equation $\Delta(s) : 1 + G(s)\,H(s) = 0$

$(s + 1)^2 (s + 5) + 2s^2 + 6s + 5 = 0$

$(s^2 + 2s + 1)(s + 5) + 2s^2 + 6s + 5 = 0$

9. Given: $H(s) = 1$, order of $G(s)\,H(s) = 2$.

Since, poles of open loop transfer function G(s) H(s) lies in left half plane. The poles of $1 + G(s) \cdot H(s)$ will also lie in left half plane and the closed loop system will be stable.

10. Transfer function,

$$H(s) = \frac{\sum P_k \Delta_k}{\Delta(s)} = \frac{1}{s^3 \alpha\, s^2 + ks + 3}$$

Char. eq.: $\Delta(s) = s^3 + \alpha\, s^2 + ks + 3 = 0$

Using Routh-Hurwitz criteion,

S^3	1	k
S^2	5	3
S^1	$\frac{\alpha k - 3}{\alpha}$	
S^0	3	

For a system to be stable, all elements of first column should be positive, i.e.,

$\alpha > 0$ and $\dfrac{\alpha\, k - 3}{\alpha} > 0$

$\alpha\, k - 3 > 0$

$\alpha\, k > 3$

11. Characteristic equation is

$\Rightarrow \quad (k + 1)\, s^2 + s(4 - 4K) + 4 + 4K = 0$

Routh array drawn below,

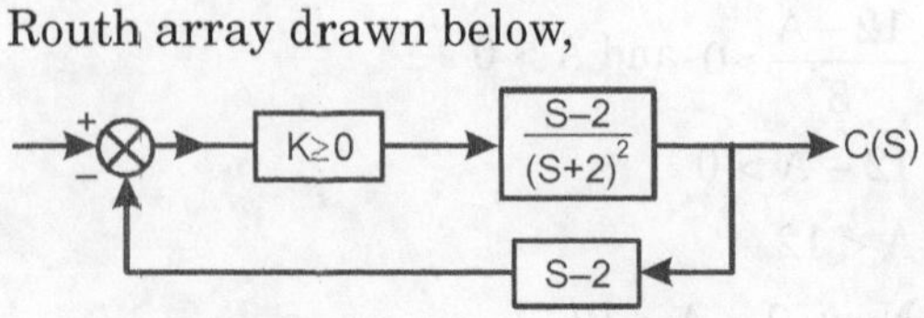

s^2	K+1	4 (1+K)
s	4 (1 – K)	0

For stability, co-efficients in the first column is greater than zero, *i.e.* + ve

$K + 1 > 0$

$1 > K$ and $K \geq 0$.

or $\quad 0 \leq K < 1$

12.

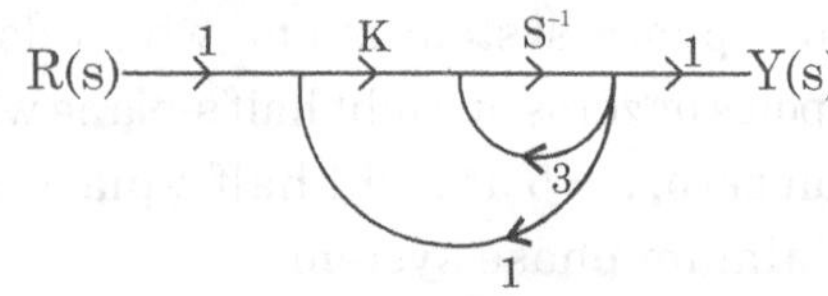

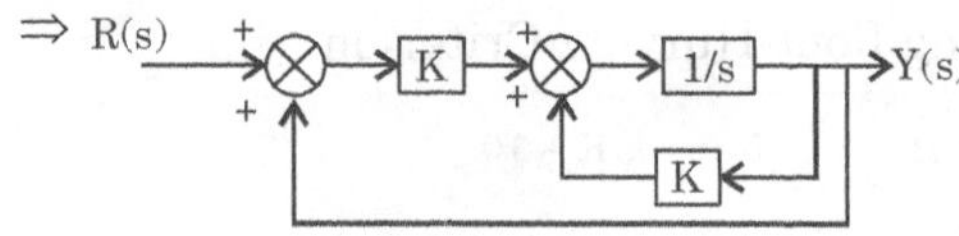

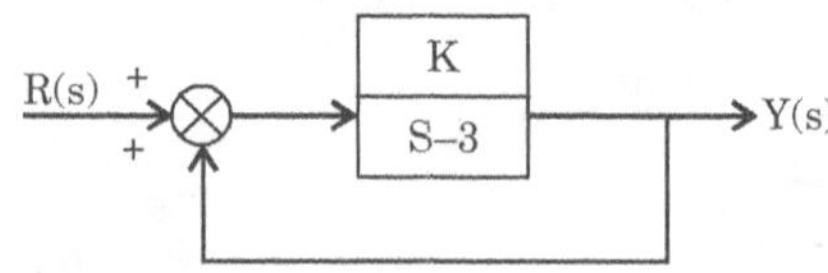

Transfer function,

$$\frac{Y(s)}{R(s)} = \frac{k}{s - 3 - k} = \frac{k}{s - (3 + k)}$$

For a stable system, its pole must lie in left half of s-plane,

$3 + k < 0$

$k < -3$.

13. Characteristic polynomial :

$q(s) = 2s^5 + s^4 + 4s^3 + 2s^2 + 2s + 1$

Using Routh-Hurwitz criterion,

S^5	2	4	2
S^4	1	2	1
S^3	0	0	0
S^2	1	1	
S^1	0	0	
S^0			

Since, all the entries of s^3 row are zero it means characteristic polynomial is completely divisible by auxillary polynomial.

for Auxilliary equation :

$$\frac{d}{ds}(s^4 + 2s^2 + 1) = 0$$

$4s^3 + 4s = 0$

$s(s^2 + 1) = 0$

$s = \pm j, 0$

$$
\begin{array}{r}
2s^3+s^2+2s+1 \\
s^2+1 \overline{)\, 2s^5+s^4+4s^3+2s^2+2s+1} \\
-2s^5 \quad \pm 2s^3 \qquad\qquad \\
\hline
s^4+2s^3+2s^2 \qquad \\
-s^4 \quad \pm s^2 \qquad \\
\hline
2s^3+s^2+2s \qquad \\
-2s^3 + \; -2s \qquad \\
\hline
s^2+1 \qquad \\
s^2+1 \qquad \\
\hline
\times \qquad
\end{array}
$$

Again For Auxilliary eq.

$$\frac{d}{ds}(s^2+1)=0$$

$$2s = 0$$

$$
\begin{array}{r}
s^2+1 \,)\, 2s^3+s^2+2s+1\,(\,2s+1 \\
\underline{2s^3 \qquad \pm 2s} \qquad\qquad \\
\hline
s^2+1 \qquad\qquad \\
s^2+1 \qquad\qquad \\
\hline
\times \qquad\qquad
\end{array}
$$

$q(s) = (s^2 + 1)^2 \cdot (2s + 1)$

q(s) has double poles on imaginary axis, hence, the system is unstable.

14. The characteristic equation is, $1 + G(s) = 0$

or $s^4 + 4s^3 + 5s^2 + 6s + K = 0$

Constructing Routh – Hurwitz array, we have

s^4	1	5	K
s^3	4	6	0
s^2	$\frac{7}{2}$	K	
s^1	$(21-4K)\frac{2}{7}$		0
s^0	K		

For the system to be stable, we have,

$$21 - 4K > 0 \quad \text{and} \quad K > 0$$

Combining on we get, $\frac{21}{4} > K > 0.$

15. Constructing Routh array,

s^5	1	2	3
s^4	1	2	15
s^3	ε	– 12	0
s^2	$\frac{2\varepsilon+12}{\varepsilon}$	15	0
s^1	$\frac{-12(2\varepsilon+12)}{\varepsilon} - 15\varepsilon$	0	0
s^0	15	$\frac{2\varepsilon+12}{\varepsilon}$	

Here ε is small positive quantity

Now, as number of sign changes are 2.

Therefore number of roots in RHS-plane is also 2.

16. Given: $G(s) = \frac{k(s+1)}{s^3+as^2+2s+1}, H(s) = 1.$

Characteristic equation : $\Delta(s) = 1 + G(s)\cdot H(s) = 0$

$s^3 + as^2 + (k + 2)s + (k + 1) = 0$

Using Routh – Hurwitz criterion,

S^3	1	2 + k
S^2	a	k + 1
S^1	$\frac{a(2+k)-(k+1)}{a}$	0
S^0	k + 1	

For system to oscillate at ω = 2 rad/s one row elements of Routh array should be zero, i.e.,

$$\frac{a(2+k)-(k+1)}{a} = 0$$

$$a = \frac{k+1}{k+2}$$

Auxillary equation : $as^2 + k + 1 = 0$;

Puting $s = j\omega \Rightarrow s = j2$

$-4 + k + 1 = 0$

$$a = \frac{k+1}{4}$$

$$\Rightarrow \frac{k+1}{4} = \frac{k+1}{K+2}$$

$\Rightarrow k = 4 - 2 = 2$

$\Rightarrow a = 0.75$

17. Closed loop transfer function

$$= \frac{G(s)}{1+a(s)} = \frac{s+8}{s^2+\alpha\, s-4+s+8}$$

$$= \frac{s+8}{s^2+(\alpha+1)s+4}$$

For closed loop system to be stable, coefficients of characteristic polynomial $(s^2 + (\alpha + 1)\, s + 4)$ must be positive, i.e.,

$\alpha + 1 > 0.$

$\alpha > -1.$

Therefore, all positive values of α, system is stable.

18. $1 + G(s) = 0$

or $s^5 + 2s^4 + 3s^3 + 6s^2 + 5s + 13 = 0$

Constructing Routh -array, we have

s^5	1	3	5
s^4	2	6	13
s^3	$0 \to \in$	$\frac{-3}{2}$	0
s^2	$\frac{6\in+3}{\in}$	13	0
s	$\frac{-3}{2}$	0	
s^0	13		

As $\in \to 0^+$, $\lim_{\in \to 0^+}\left(\frac{6\in+3}{\in}\right) = +ve$

hence two sign changes, so two positive poles in RHS plane

19. Given: $G(s) = \frac{k(s+1)}{s^3+as^2+2s+1}, H(s) = 1.$

Characteristic equation : $\Delta(s) = 1 + G(s)\,H(s) = 0$

$$1+\frac{k(s+1)}{s^3+as^2+2s+1} = 0$$

$s^3 + as^2 + (k + 2)s + (k + 1) = 0$

Using Routh-Hurwitz Criterion,

S^3	1	2 + k
S^2	6K	1 + k
S^1	$\frac{a(2+k)-(k+1)}{a}$	0
S^0	1 + k	

For system to oscillate at $\omega = 2$ rad/s.

One row elements should be zero, i.e.,

$$\frac{a(2+k)-(1+k)}{a} = 0$$

$$a = \frac{1+k}{2+k}$$

$\Rightarrow as^2 + (1 + k) = 0$; putting $s = j2$

$-4a + (1 + k) = 0$

$$a = \frac{1+k}{4}$$

$\Rightarrow k = 4 - 2 = 2$ and $a = 0.75$.

20. Given: $G(s) = \frac{k}{(s+2)(s-1)}, H(s) = 1.$

Characteristic equation : $1 + G(s)\,H(s) = 0$

$$\Rightarrow 1+\frac{k}{(s+2)(s-1)}\cdot 1 = 0$$

$\therefore\ s^2 + s + (k - 2) = 0$

Given system is a 2nd order system. Both poles lie at the same location when damping ratio (ξ) is unity, i.e., critically damping.

Now, $\omega_n^2 = (k - 2)$

$2\xi\omega_n = 1$

$$2 \times 1 \times \omega_n = 1 \Rightarrow \omega_n^2 = \frac{1}{4}$$

$$k - 2 = \frac{1}{4}$$

$\Rightarrow k = 2.25.$

21. Transfer function,

$$G_p(s) = \frac{ps^2+3ps-2}{s^2+(3+p)s+(2-p)} \text{ and } p > 0.$$

Characteristic equation :

$\Delta(s) = s^2 + (3 + p)\,s + (2 - p) = 0$

Using Routh-Hurwitz Criterion.

S^2	1	(2 – p)
S^1	3 + p	0
S^0	2 – p	

For given system to be stable, all the first column elements should be positive, i.e.,

$2 - p > 0$

$p < 2$

For marginally stable,

$2 - p_{max} = 0$

$p_{max} = 2$

22. Given:

$$G(s) = \left(k_p+\frac{k_i}{s}\right)\cdot\frac{1}{s(s+2)} = \frac{k_i+sk_p}{s^2(s+2)}, H(s) = 1.$$

Characteristic equation : $\Delta(s) = 1 + G(s)\,H(s) = 0$

$s^2(s + 2) + k_i + sk_p = 0$

$s^3 + 2s^2 + sk_p + k_i = 0$

Using Routh - Hurwitz Criterion,

S^3	1	k_p
S^2	2	k_i
S^1	$\frac{2k_p-k_i}{2}$	
S^0	k_p	

For system to be stable, all the first column elements should be positive.

$$\frac{2k_p-k_i}{2} > 0 \text{ and } k_p > 0$$

$2k_p - k_i > 0$

$$k_p > \frac{k_i}{2} > 0.$$

23. Using Routh-Hurwitz Criterion,

s^5	1	3	–4	
s^4	2	6	–8	–4
s^3	04	06	0	
s^2	3/2	–4		
s^1	25/3			
s^0	–4			

One entire row is zero.

Auxillary eq. $s^4 + 3s^2 - 4 = 0$

$$\frac{d}{ds}(s^4 + 3s^2 - 4) = 0$$

$4s^3 + 6s = 0$

Total no poles = 5

No. of poles on $j\omega$ – axis = 2.

No. of poles on R.H. side = 1

No. of poles on L.H. side = 5 – 2 – 1 = 2.

24. For first column elements in Routh's table

⇒ When all elements are positive → the system is stable

⇒ When there is a change in sign of coefficients → the system is unstable.

⇒ When any one element is zero → the test beaks down.

25. Transfer function,

$H(s) = 2s^4 - 5s^3 + 5s - 2.$

Using Routh – Hurwitz criterion,

s^4	2	0	–2
s^3	–5	5	
s^2	2	–2	
s^1	0		
s^0	–2		

Auxillary eq. $s^2 - 1 = 0$

$$\frac{d}{ds}(s^2 - 1) = 0$$

$2s = 0$

Number of zeros in right half of s-plane = no. of sign change = 3.

26.

s^3	1	2k + 3
s^2	2K	4
s^1	$\frac{4k^2+6k-1}{2k}$	0
s^0	4	

For stability, first column elements must be positive, i.e.,

$2k > 0$

$k > 0$

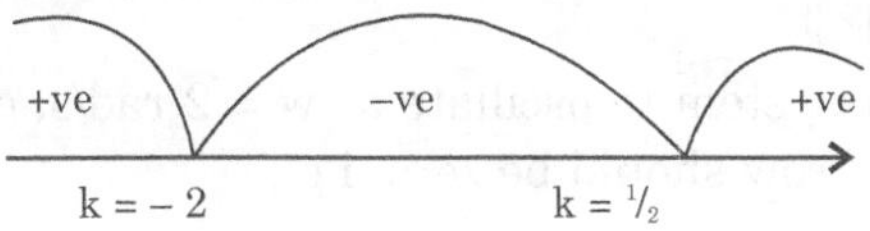

and $\frac{4k^2 + 6k - 4}{2k} > 0$

$2k^2 + 3k - 2 > 0$

$(2k - 1)(k + 2) > 0$

$k < -2$ and $k > \frac{1}{2}$

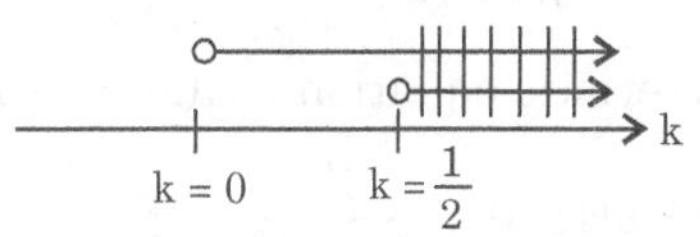

$\boxed{0.5 < k < \infty}$

27. Given equation : $q(s) = s^4 + s^2 + 1 = 0$

using Routh-Hurwitz criterion,

s^4	1	1	1
s^3	0	0	0
s^2	0.5	1	
s^1	–3	0	
s^0	1		

Sign Change ($s^2 \to s^1$, $s^1 \to s^0$)

All elements of s^3 row are zero.

Auxillary eq. $s^4 + s^2 + 1 = 0$

$$\frac{d}{ds}(s^4 + s^2 + 1) = 0$$

$4s^3 + 2s = 0$

$2s^3 + s = 0$

No. of roots on R.H. of s-plane = no of sign change = 2

No. of roots = order = 4

No. of roots on L.H. of s-plane = 4 – 2 = 2.

28. Given : $G(s) = \frac{2(s+1)}{s^3 + ks^2 + 2s + 1}$, $H(s) = 1$.

Char. eq. $\Delta(s) = 1 + G(s)H(s) = 0$

$$1 + \frac{2(s+1}{s^3 + ks^2 + 2s + 1} = 0$$

$s^3 + ks^2 + 4s + 3 = 0$

Using Routh-Hurvitz criterion,

$$\begin{array}{c|cc} S^3 & 1 & 4 \\ S^2 & k & 3 \\ S^1 & \frac{4k-3}{k} & 0 \\ S^0 & 3 & \end{array}$$

For system to oscillate at w = 2 rad/s, elements of s' row should be zero, i.e.,

$$\frac{4k-3}{k} = 0$$

$$4k = 3$$

$$k = 0.75$$

29. Given : H(s) = 1, $\xi = 0.5$, OA = 0.5

and $G(s) = \dfrac{1}{s(s+1)^2}$

Characteristic equation : $\Delta(s) = 1 + KG(s) = 0$

$$K \cdot G(s)H(s) = \frac{K}{s(s+1)^2} \quad \text{... (i)}$$

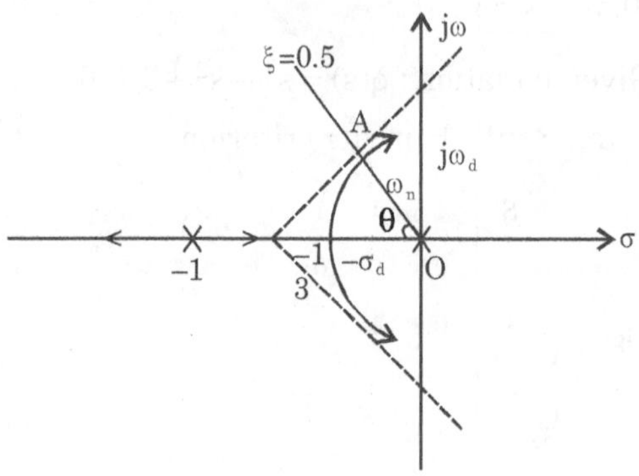

where $\sigma_d = \xi\omega_n$

$$\omega_d = \omega_n\sqrt{1-\xi^2}$$

$$\tan\theta = \frac{\sqrt{1-\xi^2}}{\xi}$$

$\Rightarrow$ damping ratio, $\xi = \cos\theta$

$$\theta = \cos^{-1}(0.5) = 60°$$

$\Rightarrow$ OA = ω_n = 0.5

$$\omega_d = 0.5\sqrt{1-(0.5)^2} = 0.433$$

$$\sigma_d = \xi\omega_n = 0.5 \times 0.5 = 0.25$$

The root locus cuts the $\xi = 0.5$ line at $s = -\sigma \pm j\omega_d$ = –0.5 ±j0.433. The value of system gain at this point can be obtained as :

$$K|G(s)H(s)| = 1 \quad \text{at} \quad s = -0.5 + j\,0.433$$

We get, K = 0.37.

30. Given: H(s) = 1 and

Open loop transfer function,

G(s) : Poles at s = – 2, – 0.1

$$G(s) = \frac{1}{(s+0.1)(s+2)}$$

Closed loop transfer function,

$$T(s) = \frac{kG(s)}{1+G(s)\cdot H(s)\cdot k}$$

$$= \frac{\dfrac{k}{(s+0.1)(s+2)}}{1+\dfrac{k}{(s+0.1)(s+2)}\cdot 1}$$

$$T(s) = \frac{k}{k+(s+0.1)(s+2)}$$

$$= \frac{k}{k+0.2(1+10s)(1+0.5s)}$$

31. We know that the root locus

(i) is symmetrical about real or σ-axis,

(ii) traverses from poles to zero.

32. Given: $T(s) = \dfrac{k}{s^2+(3-k)s+1}$

$$T(s) = \frac{\dfrac{k}{(s^2+3s+1)}}{1-\dfrac{ks}{s^2+3s+1}}$$

$$G(s) = \frac{k}{s^2+3s+1}; H(s) = s$$

$$G(s) = \frac{k}{s^2+3s+1}; \; H(s) = s.$$

Open loop transfer function

$$G(s)\,H(s) = \frac{KS}{s^2+3s+1} = \frac{KS}{(s+0.381)(s+2.618)}$$

$\Rightarrow$ System has 2 real poles and one real zero at origin.

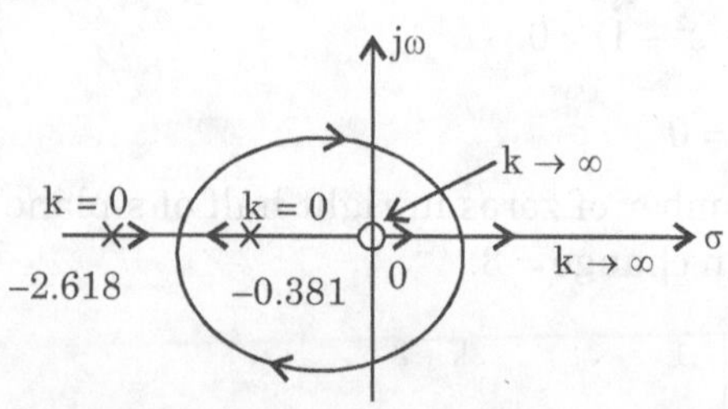

Above figure is an approximate root locus plot. From given options, only option (b) is correct answer.

33. Characteristic equation :

$\Delta(s) = s^3 + 5s^2 + (k + 6)\, s + k = 0$

$\Delta(s) = s^3 + 5s^2 + 6s + k(s + 1) = 0$

Open loop transfer function:

$$G(s)\,H(s) = \frac{k(s+1)}{s^3+5s^2+6s} = \frac{k(s+1)}{s(s^2+5s+6)}$$

$$= \frac{k(s+1)}{s(s+2)(s+3)}$$

No. of poles, P = 3 i.e. (0, – 2, – 3)

No. of zeros, z = 1 i.e. (–1).

The asymptotes of the root loci meet at the centroid (anywhere in real axis) in the s-plane.

$$\text{Centroid, } \sigma = \frac{\sum P_i - \sum Z_i}{P - Z} = \frac{(-2-3-0)-(-1)}{-1}$$

$$= -2.$$

Co-ordinates ≡ (–2, 0)

34. Given : $G(s) = \dfrac{k}{s(s+1)(s+2)}, H(s) = 1.$

Open loop tranfer junction

$$G(s)\,H(s) = \frac{k}{s(s+1)(s+2)}$$

In all the given option, location of poles, segment on real axis and centroid are same. Now, looking for angle of asymptotes for correct answer.

⇒ P = 3 i.e. (0, – 1, – 2); z = 0

$$\text{Angle of asymptotes} = \frac{(2n+1)180^\circ}{P-Z};$$

n = 0, 1, 2, ..., P – Z – 1

n = 0, 1, 2

= 60°, 180°, 300

35. Root locus branch starts from a pole and teminates at zero of open-loop transfer function. If number of zeros (m) are less than no of poles (n) then remaining poles (n – m) traverse in s-plane using (n–m) asymptotes to terminate at zero (s) at infinity.

36. If a point s = σ + jω lies on root locus then angle or phase of open-loop transfer function is odd multiple of 180°, i.e.,

$$\angle G(s)H(s) = \pm n\pi, \text{ where } n = 1, 3, 5, ...$$

$$\Rightarrow G(s)\,H(s) = \frac{k}{(s+1)^4};$$

$$G(s)H(s)\big|_{s=s_1=-3+j4} = \frac{k}{(-3+j4+1)^4} = \frac{k}{(-2+j4)^4}$$

$$\angle G(s)H(s)\big|_{s=s_1} = -4\left[\pi - \tan^{-1}\left(\frac{4}{2}\right)\right]$$

$= -466^\circ \neq n\pi.$

Hence, S_1 does not lie on root locus

$$G(s)\cdot H(s)\big|_{s=s_2=-3+j4} = \frac{k}{(-3-j^2+4)^2} = \frac{k}{(2-2j)^4}$$

$$\angle G(s)H(s)\big|_{s=s_2} = -4\left[\tan^{-1}\left(\frac{2}{2}\right)\right] = -4\times\frac{\pi}{4} = -\pi$$

Hence, s_2 lies on root locus.

37. For 0 < K < 1, and K < 5, both the roots lie on the negative real axis which corresponds to overdamped system.

For, K = 1, system is critically damped.

For 1 < K < 5, the system is underdamped.

38. Poles, P = 3 i.e. (0, – 2, – 3)

Zeros, Z = 0

Hence, S = – 1.5 does not lie on root locus.

Real axis-segment

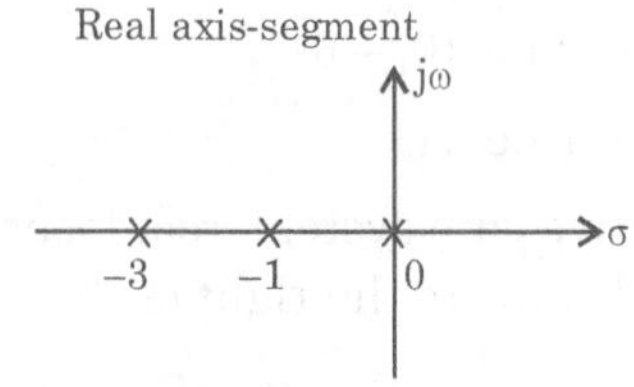

39. The characteristic equations is

$1 + G(s)\,H(s) = 0$

or $\quad 1 + \dfrac{k}{s(s+2)\,(s+3)} = 0$

or $\quad k = -(s^3 + 5s^2 + 6s)$

For Break-away points,

$$\frac{dk}{ds} = 0 = -(3s^2 + 10s + 6)$$

or $\quad 3s^2 + 10s + 6 = 0$

or $\quad s = -0.784, -2.54$

The root-locus is drawn below

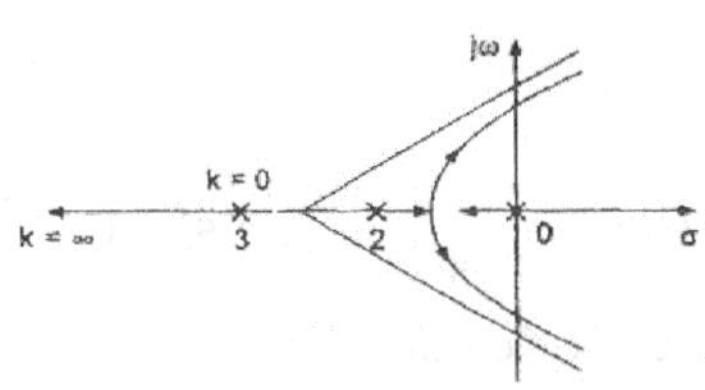

Since the point $s = -2.54$, doesn't lie on the root-locus, it can never be breakway point.

Hence $\quad s = -0.784.$

40. Centroid, $(-\sigma_A) = \dfrac{\sum_{i=1}^{m} P_i - \sum_{j=1}^{n} Z_j}{m-n}$

where m and n are the number of poles and zero respectively.

$\therefore$ Centroid $= \dfrac{-4-0}{3} = -1.33$

41. Given: $G(s) = \dfrac{k(1-s)}{s(s+3)} = \dfrac{-k(s-1)}{s(s+3)}$; $H(s) = 1$

Open loop transfer function:

$G(s)\,H(s) = -\dfrac{k(s-1)}{s(s+3)}$ (Negative gain)

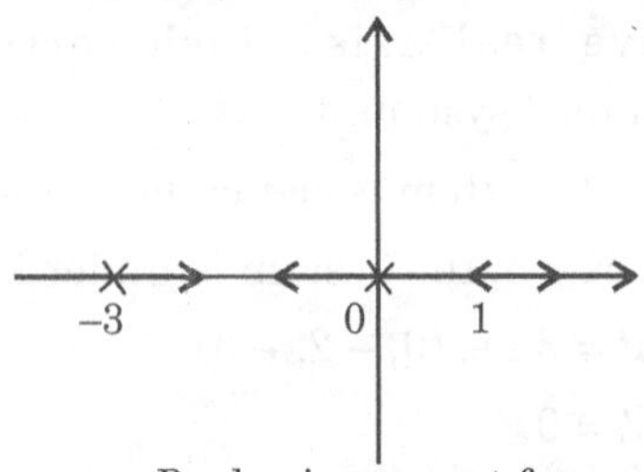

Real axis segment for G(s) H(s)

Poles, P = 2 i.e. (0, – 3)

Zeros, Z = 1 i.e. (1)

For negative gain system, root locus where no. of poles and zeros to the right of the pole or zero is even.

Only option (c) satisfies above.

42. Given: $G(s) = \dfrac{k}{s(s^2+7s+12)}, H(s) = 1$

Characteristic equations :

$\Delta(s) = 1 + G(s)\cdot H(s) = 0$

$$1 + \frac{k}{s(s^2+7s+12)}\cdot 1 = 0$$

$s^3 + 7s^2 + 12s + k = 0$

Point $s = -1 + j1$ lies on root locus,

$-1 + 2j + 1 + 7 - 7j - 12 - j + 2 - j - 7j - 7 + 12j + k = 0$

$\therefore\ k = +10$

43. Given $G(s) = \dfrac{s^2-2s+2}{s^2+2s+2}, H(s) = 1.$

Characteristic equations :

$\Delta(s) = 1 + G(s)\cdot H(s) = 0$

$$1 + \frac{k(s^2-2s+2)}{s^2+2s+2}\cdot 1 = 0$$

$$k = -\frac{s^2+2s+2}{s^2-2s+2}$$

For break points, solving $\dfrac{\partial k}{\partial s} = 0$

$(s^2 - 2s + 2)(2s + 2) - (s^2 + 2s + 2)(2s - 2) = 0$

$(s^2 - 2s + 2)(s + 1) - (s^2 + 2s + 2)(s - 1) = 0$

$s^3 + s^2 - 2s^2 - 2s + 2s + 2 - s^3 + s^2 - 2s^2 + 2s - 2s + 2 = 0$

$\therefore\ -2s^2 + 4 = 0$

$\therefore\ \boxed{s = \pm\sqrt{2}}$

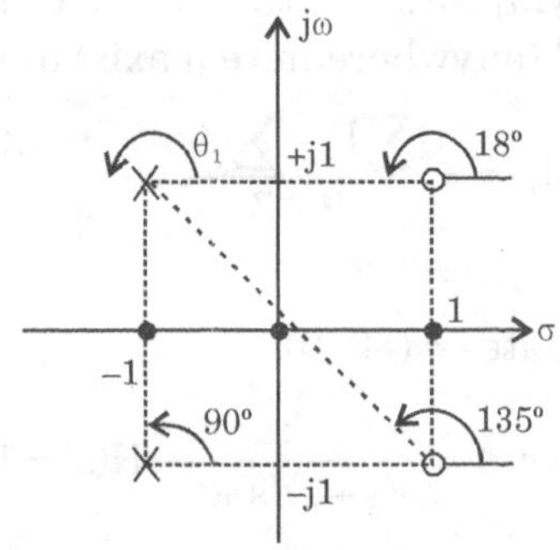

For angle of departure,

$-\theta_1 - 90° + 100° + 135° = 180°$

$-\theta_1 = -45°$

$\theta_1 = 45°.$

Pole-zero plot for G(s) · H(s)

44. For given plot, root locus exists from – 3 to – ∞, so there must be odd number of poles and zeros. There is double pole at s = – 3.

No. of poles, P = 4 i.e. (0, – 2, – 3, – 3)

No. of zeros, z = 1 i.e. (–1).

The open-loop transfer function is

$$G(s)\cdot H(s) = \frac{k(s+1)}{s(s+2)(s+3)^2}$$

45. We know that root locus laves the real-axis when their is a breakway point or two real poles ($-\sigma_1$ and σ_2) converge to each other and split symmetrically in s-plane. Here, $-\sigma_1$ and $-\sigma_2$ are near to origin.

From options, $-\sigma_1 = -1$ and $-\sigma_2 = -2$ is correct.

We also know that when a root locus segment lies only on real axis then starting point will be a real pole ($-\sigma_3$ or $-\sigma_4$) and terminating point ($-\sigma_4$ or $-\sigma_3$).

Hence, Combinations of $-\sigma_3$ and $-\sigma_4$ are:

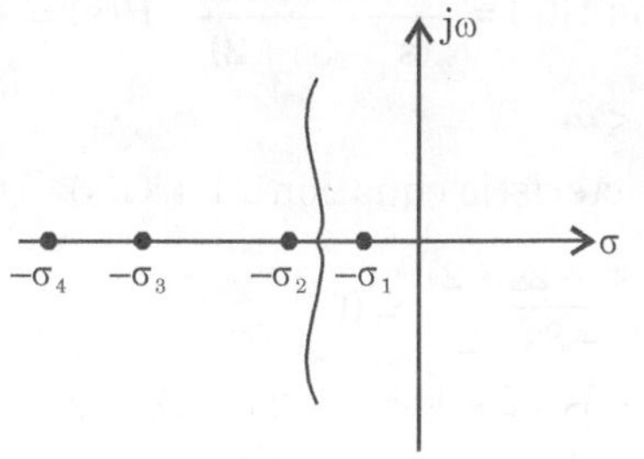

If $-\sigma_3$ is zero $\rightarrow -\sigma_4$ will be pole.

If $-\sigma_3$ is pole $\rightarrow -\sigma_4$ will be zero.

Now, considering above drawn conclusions, possible transfer function(s) are:

$$\frac{s+4}{(s+1)(s+2)(s+7)} \text{ or } \frac{s+7}{(s+1)(s+2)(s+4)}$$

46. Given: $G(s) = \frac{k}{s(s+1)(s+3)}, H(s) = 1.$

Characteristic equations :

$\Delta(s) = 1 + G(s) \cdot H(s) = 0$

$s(s + 1)(s + 3) + k = 0$

$s^3 + 4s^2 + 3s + k = 0$

Using Routh-Hurwitz crietrion,

S^3	1	3
S^2	4	k
S^1	(12-k)/4	0
S^0	k	

In order to cross the imaginary axis, system should be marginally stable or elements of s' row will be zero.

$$\frac{12-k}{4} = 0$$

$k = 12$

47. Given:

$$G(s) = \frac{k(s+4)}{(s+8)(s^2-9)} = \frac{k(s+4)}{(s+8)(s+3)(s-3)}; H(s) = 1$$

For the point $s = -1 + j2$ to lie on root locus, then it must satisfy,

$$|G(s)H(s)|_{s=-1+2j} = 1.$$

$$\left|\frac{k(-1+2j+4)}{(-1+2j+8)(-1+2j+3)(-1+2j-3)}\right| = 1$$

$$\frac{k\cdot\sqrt{(-1+4)^2+2^2}}{\sqrt{(-1+8)^2+2^2}\cdot\sqrt{(-1+3)^2+2^2}\cdot\sqrt{(-1-3)^2+2^2}} = 1$$

$$\frac{k\sqrt{13}}{\sqrt{53}\sqrt{8}\sqrt{20}} = 1.$$

$$k = \sqrt{\frac{53\times 8\times 20}{13}} = 25.24.$$

48. Given: $G(s) = \left(\frac{s+3}{s+2}\right)$, $H(s) = 10.$

For the point $\delta = -2.75$ to lie on root locus, then it must satisfy $|G(s)\cdot H(s)|_{s=-2.75} = 1.$

$$\left|\frac{k(s+3)}{(s+2)}\times 10\right|_{s=-2.75} = 1$$

$$\Rightarrow \frac{10k\sqrt{(3-2.75)^2}}{\sqrt{(2-2.75)^2}} = 1.$$

$$\frac{0.25\times 10k}{0.75} = 1 \Rightarrow k = 0.3$$

49. Given: $G(s) = \frac{k}{s^2+5s+5}$, $H(s) = 1.$

Char. eq.: $1 + G(s) H(s) = 0$

$s^2 + 2s + 5 + k = 0$

$k = -s^2 - 2s - 5$

For break away point, $dk/ds = 0$

$-2s - 5 = 0 \Rightarrow s = -2.5$

Magnitude of k at $s = -2.5$ is given by

$$|G(s)H(s)|_{s=-2.5} = 1.$$

$$\frac{|k|}{|(-2.5)^2+5\times(-2.5)+5|} = 1$$

$|k| = 1.25$

$k = \pm 1.25.$

50. Given:

$$G(s) = \frac{k(s+2)}{s^2+2s+2}$$

$$= \frac{s(s+2)}{(s+1+j1)(s+1-j1)}, H(s) = 1; k > 0$$

Open loop transfer function,

$$G(s)\cdot H(s) = \frac{k(s+2)}{(s+1+j1)(s+1-j1)}$$

Characteristic equation $1 + G(s)\cdot H(s) = 0$

$s^2 + 2s + 2 + k(s + 2) = 0$

$$k = -\frac{s^2+2s+2}{s+2}$$

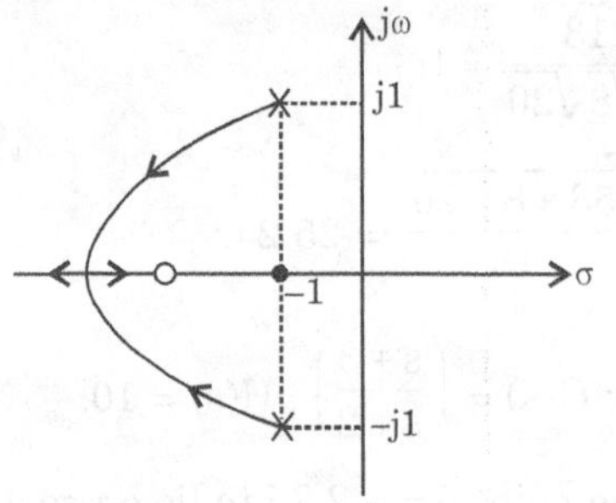

For breakaway points,

$$\frac{dk}{ds} = 0$$

$$(2s + 2)(s + 2) - (s^2 + 2s + 2) = 0$$

$$2s^2 + 6s + 4 - s^2 - 2s - 2 = 0$$

$$s^2 + 4s + 2 = 0$$

$$s_{1,2} = \frac{-4 \pm \sqrt{16-8}}{2}$$

$$s_{1,2} = -2 \pm \sqrt{2}$$

$$s_{1,2} = -0.58, -3.41$$

But $\delta = -3.41$ lie on root locus.

51. Given : $G(s) = \dfrac{k(s^2+2s+2)}{(s^2-3s+2)}$, $H(s) = 1$;

$0 < k < \infty$

Characteristic equation : $1 + G(s) \cdot H(s) = 0$

$$1 + \frac{k(s^2+2s+2)}{s^2-3s+2} = 0$$

$$\Rightarrow s^2 - 3s + 2 + ks^2 + k \cdot 2s + 2k = 0$$

$$(1 + k)s^2 + (2k - 3)s + 2(k + 1) = 0$$

Using Routh-Hurwitz criterion,

S^2	$(1 + k)$	$2(k + 1)$
S^1	$(2k - 3)$	0
S^0	$\dfrac{(2k-3)(k+1)\cdot 2 - 0}{(2k-3)} = 2(k+1)$	

For closed loop system to stable, all elements of of first column should be positve, i.e.,

$$2k - 3 > 0$$

$$k > 1.5.$$

5 CHAPTER Frequency Response Analysis

Analysis of Previous GATE Papers											
	Year → Topics ↓		2019	2018	2017 Set 1	2017 Set 2	2016 Set 1	2016 Set 2	2015 Set 1	2015 Set 2	2015 Set 3
FREQUENCY RESPONSE	1 Mark	MCQ Type		1	1		1	1	1		
		Numerical Type	1								1
	2 Marks	MCQ Type			1	1					
		Numerical Type	1	1				1		1	
		Total	**3**	**3**	**3**	**2**	**1**	**3**	**1**	**2**	**1**
POLAR, NYQUIST & BODE PLOT	1 Mark	MCQ Type									
		Numerical Type									
	2 Marks	MCQ Type									
		Numerical Type									
		Total									
GAIN & PHASE MARGIN	1 Mark	MCQ Type									
		Numerical Type									1
	2 Marks	MCQ Type									
		Numerical Type						1			
		Total						**2**			**1**

FREQUENCY RESPONSE

1. In the system shown below, x(t) = (sint)u(t). In steady-state, the response y(t) will be

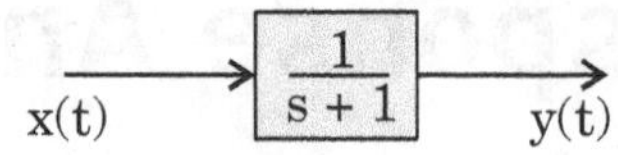

(a) $\frac{1}{\sqrt{2}}\sin\left(t-\frac{\pi}{4}\right)$ (b) $\frac{1}{\sqrt{2}}\sin\left(t+\frac{\pi}{4}\right)$

(c) $\frac{1}{\sqrt{2}}e^{-t}\sin t$ (d) sint – cost

[2006 : 1 Mark]

2. The frequency response of a linear, time-invariant system is given by $H(f) = \frac{5}{1+j10\pi f}$.

The step response of the system is

(a) $5(1-e^{-5t})\,u(t)$ (b) 5(1 –e–t/5)u(t)

(c) $\frac{1}{5}(1-e^{-5t})u(t)$ (d) $\frac{1}{(s+5)(s+1)}$

[2007 : 2 Marks]

3. A system with transfer function

$$G(s) = \frac{(s^2+9)(s+2)}{(s+1)(s+3)(s+4)}$$

is excited by sin(ω t). The steady-state output of the system is zero at

(a) ω = 1 rad/s (b) ω = 2 rad/s

(c) ω = 3 rad/s (d) ω = 4 rad/s

[2012 : 1 Mark]

4. The 3-dB bandwidth of a typical second-order system with the transfer function

$\frac{C(s)}{R(s)} = \frac{\omega_n^2}{s^2+2\xi\,\omega_n s+\omega_n^2}$ is given by

(a) $\omega_n\sqrt{1-2\xi^2}$

(b) $\omega_n\sqrt{(1-\xi^2)+\sqrt{\xi^4-\xi^2+1}}$

(c) $\omega_n\sqrt{(1-2\xi^2)+\sqrt{4\xi^4-4\xi^2+2}}$

(d) $\omega_n\sqrt{(1-2\xi^2)-\sqrt{4\xi^4-4\xi^2+2}}$

[1994 : 1 Mark]

5. The open loop frequency response of a system at two particular frequencies are given by :1.2 Z-180° and 1.0∠190°. The closed loop unity feedback control is then

[1994 : 1 Mark]

6. Non-minimum phase transfer function is defined as the transfer function

(a) which has zeros in the right-half s-plane

(b) which has zeros only in the left-half s--plane

(c) which has poles in the right-half s-plane

(d) which has poles in the left-half s-plane

[1995 : 1 Mark]

7. A system has poles at 0.01 Hz, 1 Hz and 80 Hz; zeros at 5 Hz, 100 Hz and 200 Hz. The approximate phase of the system response at 20 Hz is

(a) –90° (b) 0°

(c) 90° (d) –180°

[2004 : 2 Marks]

8. The magnitude of frequency response of an underdamped second order system is 5 at 0 rad/sec and peaks to $\frac{10}{\sqrt{3}}$ at $5\sqrt{2}$ red/sec. The transfer function of the system is

(a) $\frac{500}{s^2+10s+100}$ (b) $\frac{375}{s^2+5s+75}$

(c) $\frac{720}{s^2+12s+144}$ (d) $\frac{1125}{s^2+25s+225}$

[2006 : 2 Marks]

9. The transfer function of a mass-spring-damper system is given by

$$G(s) = \frac{1}{Ms^2+Bs+K}$$

The frequency response data for the system are given in the following table.

ω in rad/s	|G(jω)| in dB	arg (G(jω)) in deg
0.01	– 18.5	–0.2
0.1	–18.5	–1.3
0.2	–18.4	–2.6
1	– 16	–16.9
2	–11.4	–89.4
3	–21.5	–151
5	–32.8	–167
10	–45.3	–174.5

The unit step response of the system approaches a steady state value of

[2015 : 2 Marks, Set-2]

10. For a unity feedback control system with the forward path transfer function $G(s) = \frac{K}{s(s+2)}$.

The peak resonant magnitude M_t of the closed-loop frequency response is 2. The corresponding value of the gain K (correct to two decimal places) is.

[2018 : 2 Marks]

POLAR, NYQUIST & BODE PLOT

11. A system has fourteen poles and two zeros. Its high frequency asymptote in its magnitude plot having a slope of

(a) –40 dB/decade (b) –240 dB/decade
(c) –280 dB/decade (d) –320 dB/decade

[1987 : 2 Marks]

12. The polar plot of $G(s) = \frac{10}{s(s+1)^2}$ intercepts real axis at $\omega = \omega_0$ Then, the real part and ω_0 are respectively given by:

(a) –2.5, 1 (b) –5, 0.5
(c) –5, 1 (d) –5, 2

[1987 : 2 Marks]

13. Bode plot of a stable system is shown in figure The transfer function of the system is

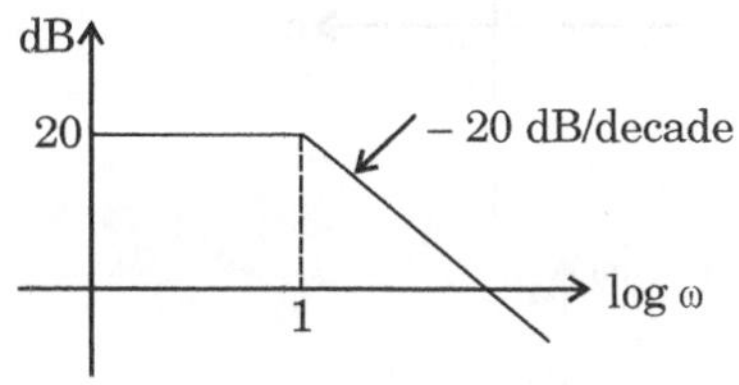

[1992 : 2 Marks]

14. The Nyquist plot for the open-loop transfer function G(s) of a unity negative feedback system is shown in the figure, if G(s) has no pole in the right-half of s-plane, the number of roots of the system characteristic equation in the right-half of s-plane is

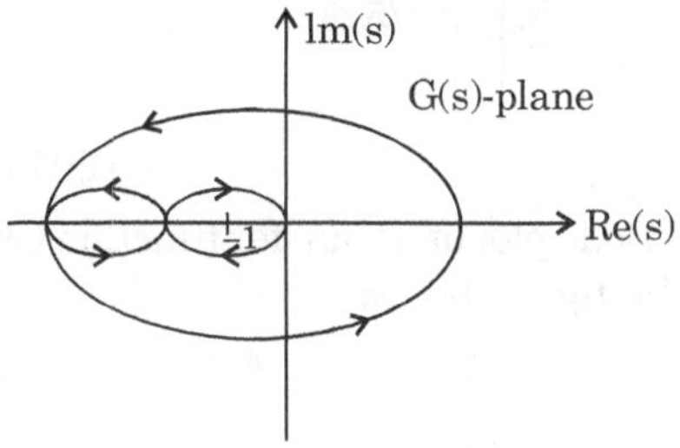

(a) 0 (b) 1
(c) 2 (d) 3

[2001 : 1 Mark]

15. In the figure, the Nyquist plot of the open-loop transfer function G(s)H(s) of a system is shown. If G(s)H(s) has one right-hand pole, the closed-loop system is

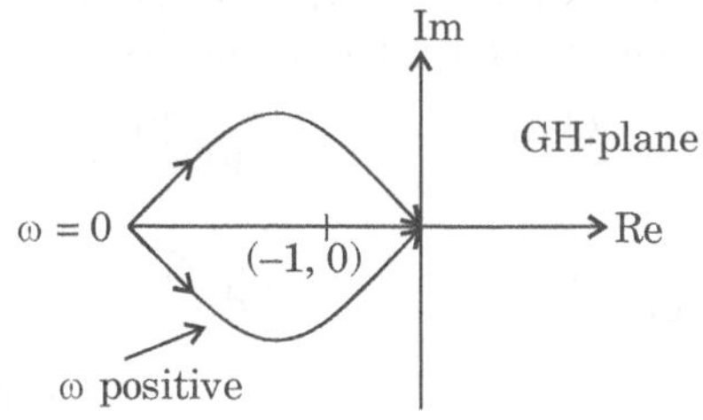

(a) always stable
(b) unstable with one closed-loop right hand pole
(c) unstable with two closed-loop right hand poles
(d) unstable with three closed-loop right hand poles

[2003 : 1 Mark]

16. The approximate Bode magnitude plot of a minimum-phase system is shown in the figure The transfer function of the system is

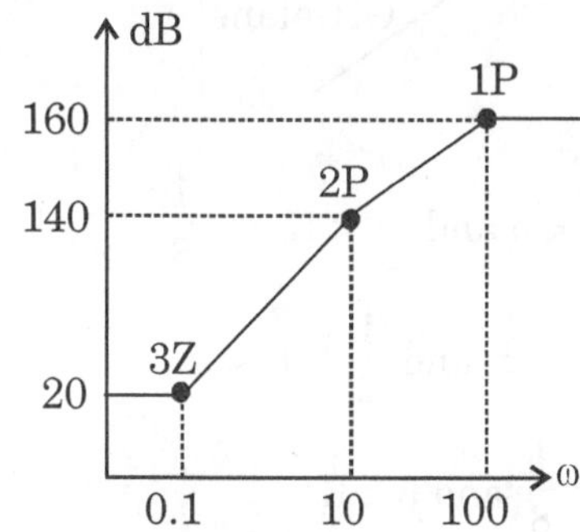

(a) $10^8\frac{(s+0.1)^3}{(s+10)^2(s+100)}$.

(b) $10^7\frac{(s+0.1)^3}{(s+10)(s+100)}$.

(c) $10^8\frac{(s+0.1)^2}{(s+10)^2(s+100)}$.

(d) $10^9\frac{(s+0.1)^3}{(s+10)(s+100)^2}$.

[2003 : 2 Marks]

17. Consider the Bode magnitude plot shown in the figure. The transfer function H(s) is

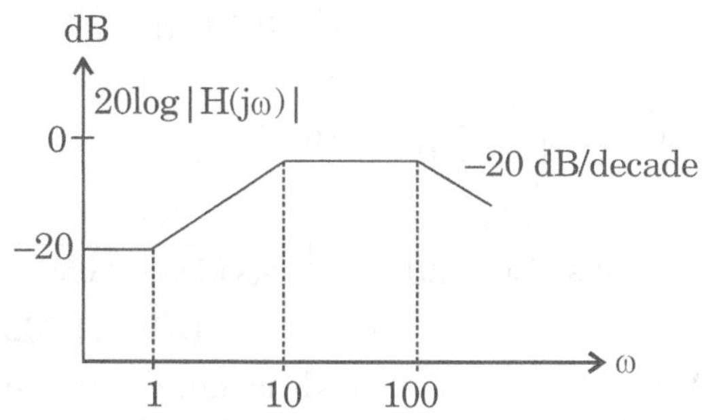

(a) $\frac{(s+10)}{(s+1)(s+100)}$. (b) $\frac{10(s+1)}{(s+10)(s+100)}$.

(c) $\frac{10^2(s+1)}{(s+10)(s+100)}$. (d) $\frac{10^3(s+100)}{(s+1)(s+10)}$.

[2004 : 2 Marks]

18. The polar diagram of a conditionally stable system for open loop gain K = 1 is shown in the figure. The open loop transfer function of the system is known to be stable. The closed loop system is stable for

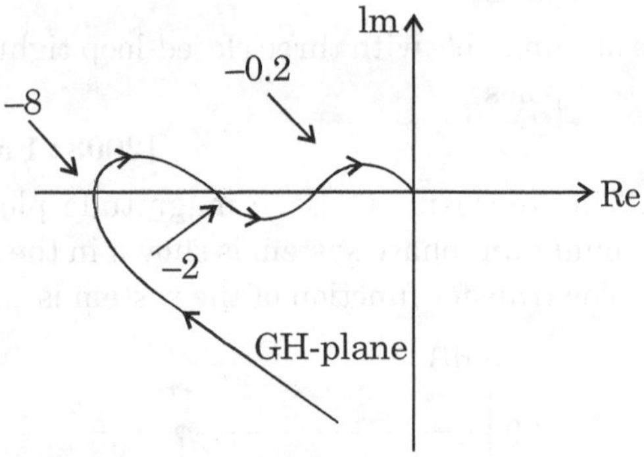

(a) $K < 5$ and $\frac{1}{2} < K < -\frac{1}{8}$

(b) $K < \frac{1}{8}$ and $\frac{1}{2} < K < 5$

(c) $K < \frac{1}{8}$ and $5 < K$

(d) $K < \frac{1}{8}$ and $K < 5$

[2005 : 2 Marks]

19. The asymptotic Bode plot of a transfer function is as shown in the figure. The transfer function G(s) corresponding to this Bode plot is

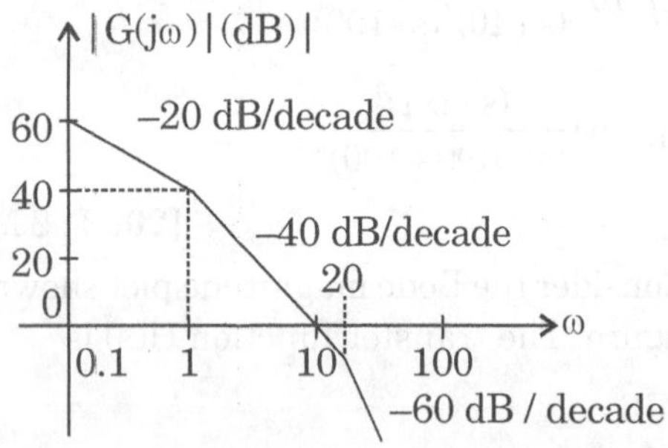

(a) $\frac{1}{(s+1)(s+20)}$ (b) $\frac{1}{s(s+1)(s+20)}$

(c) $\frac{100}{s(s+1)(s+20)}$ (d) $\frac{100}{s(s+1)(1+0.05s)}$

[2007 : 2 Marks]

20. For the asymptotic Bode magnitude plot shown below, the system transfer function can be

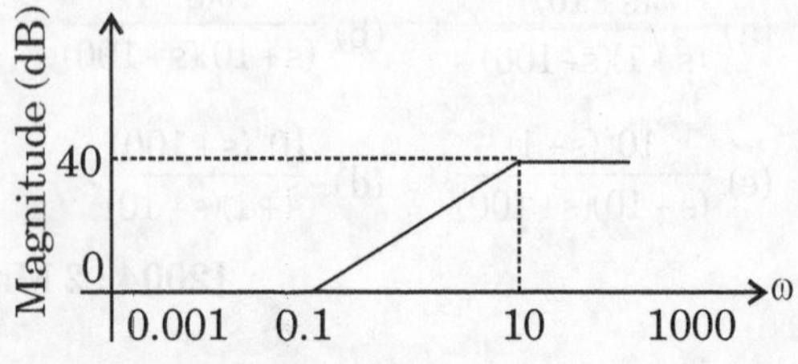

(a) $\frac{10s+1}{0.1s+1}$ (b) $\frac{100s+1}{0.1s+1}$

(c) $\frac{100s}{10s+1}$ (d) $\frac{0.1s+1}{10s+1}$

[2010 : 1 Mark]

21. TheFor the transfer function $G(j\omega) = 5 + j\omega$, the corresponding Nyquist plot for positive frequency has the form

(a)

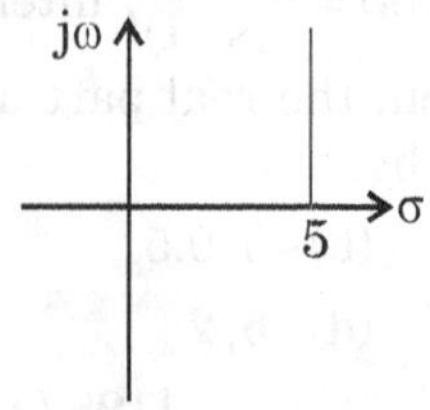

(b)

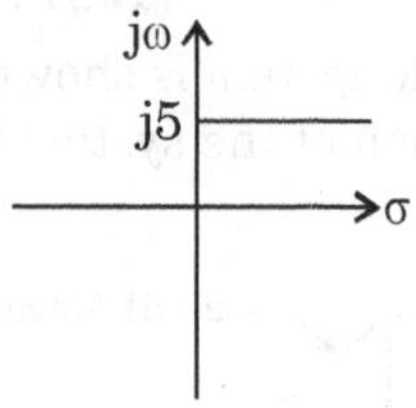

(c)

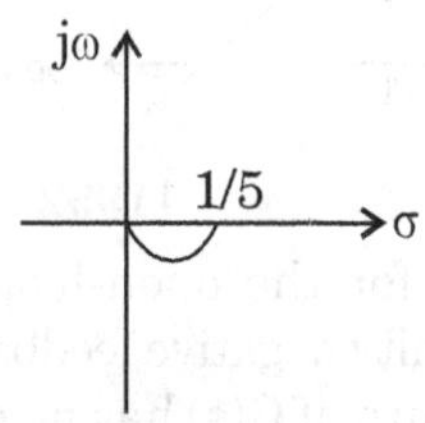

(d)

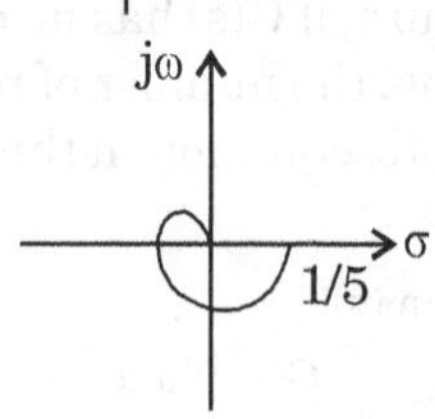

[2011 : 1 Mark]

22. The Bode plot of a transfer function G(s) is shown in the figure below:

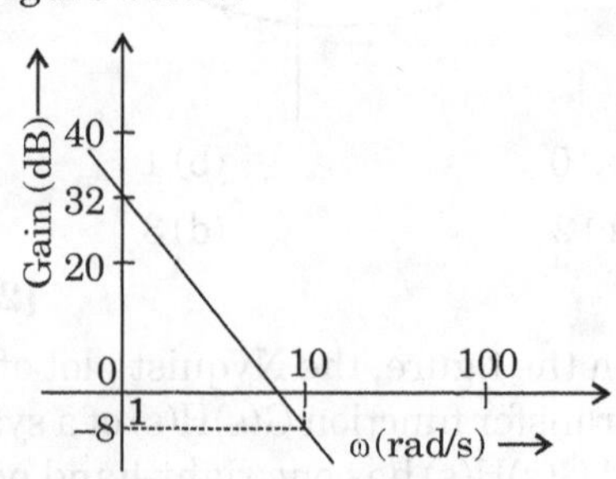

The gain ($20\log|G(s)|$) is 32 dB and –8 dB at 1 rad/s and 10 rad/s respectively. The phase is negative for all ω. Then G(s) is

(a) $\frac{39.8}{s}$ (b) $\frac{39.8}{s^2}$

(c) $\frac{32}{s}$ (d) $\frac{32}{s^2}$

[2013 : 1 Mark]

23. Consider the feedback system shown in the figure. The Nyquist plot of G(s) is also shown. which one of the following conclusions is correct?

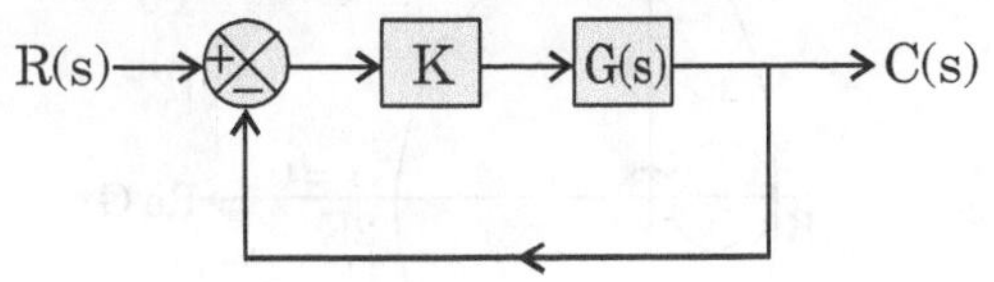

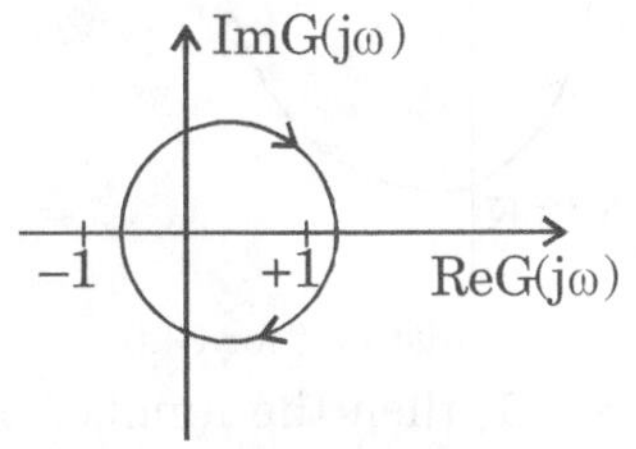

(a) G(s) is an all-pass filter

(b) G(s) is a strictly proper transfer function

(c) G(s) is a stable and minimum-phase transfer function

(d) The closed-loop system is unstable for sufficiently large and positive K

[2014 : 1 Mark, Set-1]

24. The bode asymptotic magnitude plot of a minimum phase system is shown in the figure.

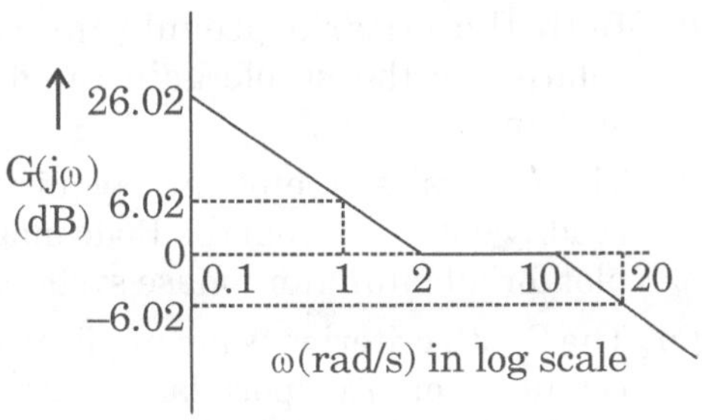

It the system is connected in a unity negative feedback configuration, the steady state error of the closed loop system, to a unit ramp input, is ______.

[2014 : 2 Marks, Set-2]

25. In a Bode magnitude plot, which one of the following slopes would be exhibited at high frequencies by a 4th order all-pole system?

(a) –80 dB/decade (b) –40 dB/decade

(c) +40 dB/decade (d) +80 dB/decade

[2014 : 1 Mark, Set-4]

26. The polar plot of the transfer function

$$G(s) = \frac{10(s+1)}{s+10} \text{ for } \omega < \infty \text{ will be in the}$$

(a) first quadrant (b) second quadrant

(c) third quadrant (d) fouth quadrant

[2015 : 1 Mark, Set-1]

27. Consider the Bode plot shown in figure, Assume that all the poles and zeros are real valued.

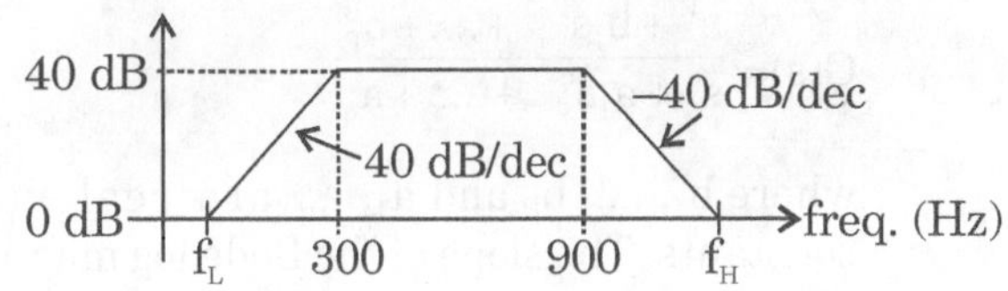

The value of $f_H - f_L$ (in Hz) is ______.

[2015 : 1 Mark, Set-3]

28. A closed-loop control system is stable if the Nyquist plot of the corresponding open-loop transfer function

(a) encircles the s-plane point (–1 + j0) in the counterclockwise direction as many times as the number of right-half s-plane poles.

(b) encircles the s-plane point (0 – j1) in the clockwise direction as many times as the number of right-half s-plane poles.

(c) encircles the s-plane point (–1 + j0) in the counterclockwise direction as many times as the number of left-half s-plane poles.

(d) encircles the s-plane point (–1 + j0) in the counterclockwise direction as many times as the number of right-half s-plane zeros.

[2016: 1 Mark, Set-1]

29. The number and direction of encirclements around the point -1 + j0 in the complex plane by the Nyquist plot of $G(s) = \frac{1-s}{4+2s}$ is

(a) zero. (b) one, anti-clockwise.

(c) one, clockwise. (d) two, clockwise.

[2016 : 1 Mark, Set-2]

30. The asymptotic Bode phase plot of $G(s) = \frac{k}{(s+0.1)(s+10)(s+p_1)}$ with k and p_1 both positive, is shown below.

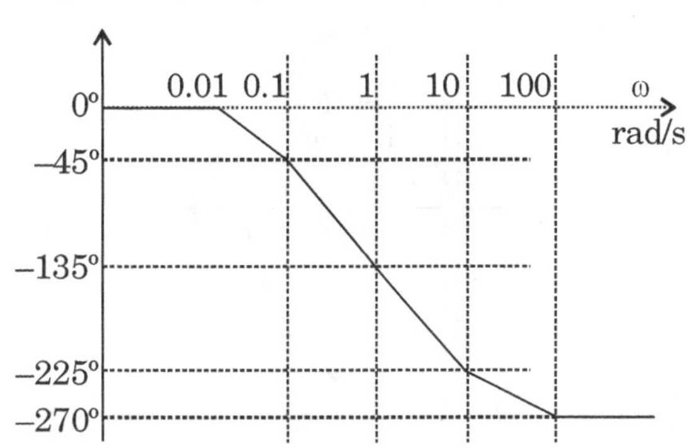

The value of p1 is ______.

[2016 : 2 Marks, Set-2]

31. Consider a stable system with transfer function

$$G(s)=\frac{s^P+b_1s^{P-1}+....+b_P}{s^q+a_1s^{q-1}+.....+a_q},$$

where b_1, ..., b_P and a_1, a_q are real valued constants. The slope of the Bode log magnitude curve of G(s) converges to -60 dB/decade as $\omega\to\infty$. A possible pair of values for p and q is

(a) p = 0 and q = 3 (b) p = 1 and q = 7

(c) p = 2 and q = 3 (d) p = 3 and q = 5

[2017 : 1 Mark, Set-1]

32. The Nyquist plot of the transfer function

$$G(s)\ \frac{K}{(s^2+2s+2)(s+2)},$$

does not encircle the point (–1 + j0) for K = 10 but does encircle the point (–1 +j0) for K = 100. Then the closed loop system (having unity gain feedback) is

(a) stable for K= 10 and stable for K = 100

(b) stable for K = 10 and unstable for K = 100

(c) unstable for K = 10 and stable for K = 100

(d) unstable for K = 10 and unstable for K = 100

[2017 : 2 Marks, Set-1]

33. A unity feedback control system is characterized by the open-loop transfer function

$$G(s)=\frac{10K(s+2)}{s^3+3s^2+10}.$$

The Nyquist path and the corresponding Nyquist plot of G(s) are shown in the figures below.

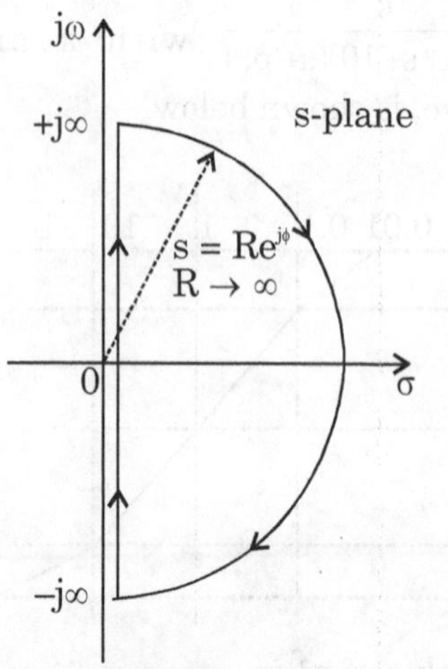

Nyquist path for G(s)

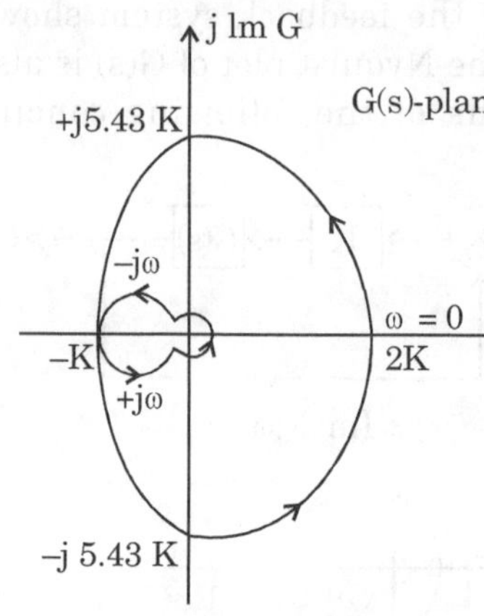

Nyquist plot for G(s)

If 0 < K < 1, then the number of poles of the closed-loop transfer function that lie in the right half of the s-plance is

(a) 0 (b) 1

(c) 2 (d) 3

[2017 : 2 Marks, Set-2]

34. The Nyquist stability crit erion and the Routh criterion both are powerful analysis tools for determining the stability of feedback controllers. Identify which of the following statements is FALSE.

(a) Both the criteria provide information relative to the stable gain range of the system.

(b) The general shape of the Nyquist plot is readily obtained from the Bode magnitude plot for all minimum-phase systems.

(c) The Routh criterion is not applicable in the condition of transport lag, which can be readily handled by the Nyquist criterion.

(d) The closed-loop frequency response for a unity feedback system cannot be obtained from the Nyquist plot.

[2018 : 1 Mark]

35. The figure below shows the Bode magnitude and phase plots of a stable transfer function

$$G(s)=\frac{n_0}{s^3+d_2s^2+d_1s+d_0}$$

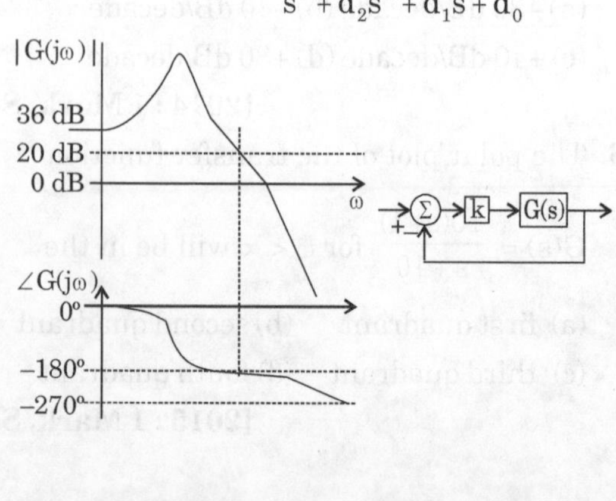

Consider the negative unity feedback configuration with gain kin the feedforward path. The closed loop is stable for k< k_0. The maximum value of k_0 is

[2018 : 2 Marks]

36. For an LTI system, the Bode plot for its gain is as illustrated in the figure shown. The number of system poles N_p and the number of system zeros N_z in the frequency range 1Hz ≤ f ≤ 10^7 Hz is

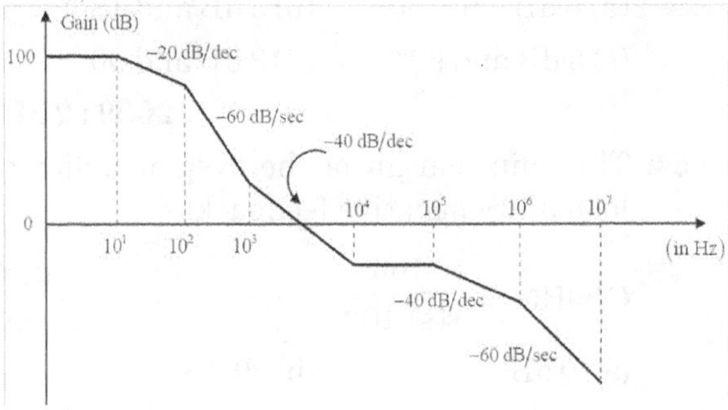

(a) $N_p = 4, N_z = 2$ (b) $N_p = 7, N_z = 4$

(c) $N_p = 6, N_z = 3$ (d) $N_p = 5, N_z = 2$

[2019 : 1 Mark]

37. Consider a unity feedback system, as in the figure shown, with an integral compensator $\frac{k}{s}$ and open-loop transfer function $G(s) = \frac{1}{s^2 + 3s + 2}$

Where K > 0. The positive value of K for which there are exactly two poles of the unity feedback system on the jω axis is equal to ______ (rounded off to two decimal places).

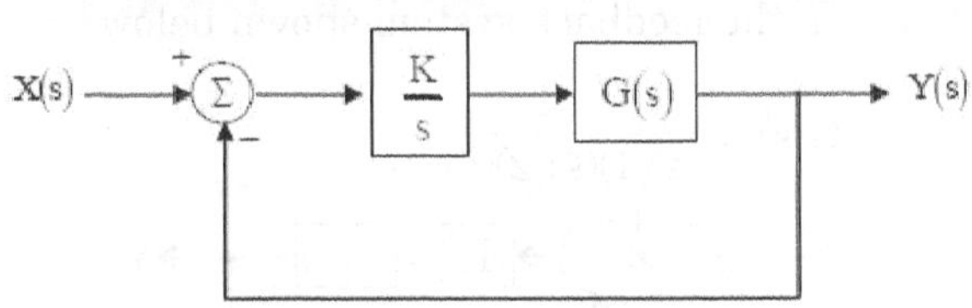

[2019 : 2 Marks]

GAIN & PHASE MARGIN

38. The gain margin for the system with open-loop transfer function $G(s)H(s) = \frac{2(1+s)}{s^2}$, is

(a) ∞ (b) 0

(c) 1 (d) −∞

[2004 : 1 Mark]

39. The value of "a" so that the system has a phase-margin equal to π/4 is approximately equal to

(a) 2.40 (b) 1.40

(c) 0.84 (d) 0.74

[2006 : 2 Marks]

40. With the value of "a" set for phase-margin of $\pi/2$, the value of unit-impulse response of the open-loop system at t 1 second is equal to

(a) 3.40 (b) 2.40

(c) 1.84 (d) 1.74

[2006 : 2 Marks]

41. The open-loop transfer function of a feedback control system is

$$G(s).\,H(s) = \frac{1}{(s+1)^3}$$

The gain margin of the system is

(a) 2 (b) 4

(c) 8 (d) 16

[1991 : 2 Marks]

42. The Nyquist plot of a loop transfer function G(jω).H(jω) of a system encloses the (–1, 0) point. The gain margin of the system is

(a) less than zero (b) zero

(c) greater than zero (d) infinity

[1998 : 1 Mark]

43. The gain margin (in dB) of a system having the loop transfer function $G(s)H(s) = \frac{\sqrt{2}}{s(s+1)}$ is

(a) 0 (b) 3

(c) 6 (d) ∞

[1999 : 1 Mark]

44. The phase margin (in degrees) of a system having the loop transfer function

$$G(s)H(s) = \frac{2\sqrt{3}}{s(s+1)} \text{ is}$$

(a) 45° (b) -30°

(c) 60° (d) 30°

[1999 : 1 Mark]

45. The system with the open loop transfer function

$G(s)H(s) = \frac{1}{s(s^2+s+1)}$ has a gain margin of

(a) –6 dB (b) 0 dB

(c) 3.5 dB (d) 6 dB

[2002 : 2 Marks]

46. The phase margin of a system with the open-loop transfer function

$$G(s)H(s) = \frac{(1-s)}{(s+1)(2+s)} \text{ is}$$

(a) 0° (b) 63.4°

(c) 90° (d) ∞

[2002 : 1 Mark]

47. The gain margin and the phase margin of a feedback system with

$$G(s)H(s) = \frac{s}{(s+100)^3} \text{ are}$$

(a) 0 dB, 0° (b) ∞, ∞
(c) ∞, 0° (d) 88.5 dB.

[2003 : 2 Marks]

Common Data for Questions 48 and 49.

The open loop transfer function of a unity feedback system is given by $G(s) = \frac{3e^{-2s}}{s(s+2)}$.

48. The gain and phase crossover frequencies in rad/sec are, respectively

(a) 0.032 and 1.26 (b) 0.632 and 0.485
(c) 0.485 and 0.632 (d) 1.26 and 0.632

[2005 : 2 Marks]

49. Based on the above results, the gain and phase margins of the system will be

(a) –7.09 and 87.5°
(b) 7.09 and 87.5°
(c) 7.09 dB and –87.5°
(d) –7.09 dB and –87.5°

[2005 : 2 Marks]

50. The open-loop transfer function of a unity-gain feedback control system is given by

$$G(s) = \frac{K}{(s+1)(s+2)}.$$

The gain margin of the system in dB is given by

(a) 0 (b) 1
(c) 20 (d) ∞

[2006 : 1 Mark]

51. The Nyquist plot of $G(j\omega)H(j\omega)$ for a closed loop control system, passed through (–1, j0) point in the GH plane. The gain margin of the system in dB is equal to

(a) infinite (b) greater than zero
(c) less than zero (d) zero

[2008 : 2 Marks]

Common Data for Question 52 and 53:

The Nyquist plot of a stable transfer function G(s) is shown in the figure. We are interested in the stability of the closed loop system in the feedback configuration shown

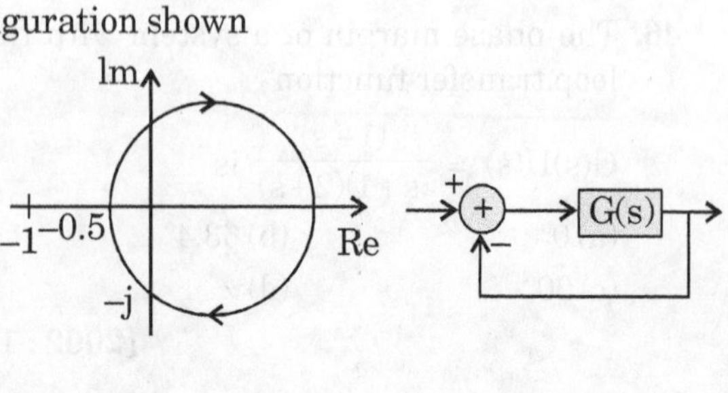

52. Which of the following statement is true?

(a) G(s) is an all-pass filter
(b) G(s) has a zero in the right-half plane
(c) G(s) is the impedance of a passive network
(d) G(s) is marginally stable

[2009 : 2 Marks]

53. The gain and phase margins of G(s) for closed loop stability are

(a) 6 dB and 180° (b) 3 dB and 180°
(c) 6 dB and 90° (d) 3 dB and 90°

[2009 : 2 Marks]

54. The gain margin of the system under closed loop unity negative feedback is

$$G(s)H(s) = \frac{100}{s(s+10)^2}$$

(a) 0 dB (b) 20 dB
(c) 26 dB (d) 46 dB

[2011 : 2 Marks]

55. The phase margin in degrees of $G(s) = \frac{10}{(s+0.1)(s+1)(s+10)}$ calculated using the asymptotic Bode plot is.

[2014 : 2 Marks, Set-1]

56. The phase margin (in degrees) of the system $G(s) = \frac{10}{s(s+10)}$ is ______.

[2015 : 1 Mark, Set-3]

57. In the feedback system shown below

$$G(s) = \frac{1}{(s+1)(s+2)(s+3)}$$

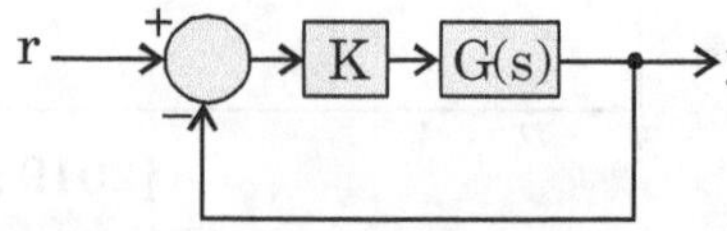

The positive value of k for which the gain margin of the loop is exactly 0 dB and the phase margin of the loop is exactly zero degree is

[2016 : 2 Marks, Set-2]

EXTRA

58. In the Bode-plot of a unity feedback control system, the value of phase of G(jw) at the gain cross over frequency is -125°. The phase margin of the system is

(a) –125° (b) –55°
(c) 55° (d) 125'

[1998 : 1 Mark]

ANSWERS

1. (a)	2. (b)	3. (c)	4. (c)	5. (Unstable)	6. (a,c)	7. (a)	8. (a)	9. (0.12)	
10. (14.92)	11. (b)	12. (c)	13. ($\frac{100}{S+10}$)	14. (a)	15. (a)	16. (a)	17. (c)	18. (b)	
19. (d)	20. (a)	21. (a)	22. (b)	23. (d)	24. (0.5)	25. (a)	26. (a)	27. (8970)	28. (a)
29. (a)	30. (1)	31. (a)	32. (b)	33. (c)	34. (d)	35. (0.1)	36. (c)	37. (6)	38. (b)
39. (c)	40. (c)	41. (c)	42. (a)	43. (d)	44. (d)	45. (b)	46. (d)	47. (b)	48. (d)
49. (d)	50. (d)	51. (d)	52. (b)	53. (c)	54. (c)	55. (45°)	56. (84.36)	57. (60)	58. (c)

EXPLANATIONS

1. $H(s) = \frac{1}{s+1} \Rightarrow H(j\omega) = \frac{1}{j\omega+1} = \frac{1}{\sqrt{2}}\angle -45^\circ$

$x(t) = \sin t;\ t > 0. \Rightarrow x(jw) = 1\angle 0^\circ$

The response, Y(s) = X(s) · H(s) [Convolution]

$Y(j\omega) = -(j\omega) \cdot H(j\omega)$

$$Y(j\omega) = 1\angle 0^\circ \cdot \frac{1}{\sqrt{2}}\angle -45^\circ = \frac{1}{\sqrt{2}}\angle -45^\circ$$

$$Y(t) = \frac{1}{\sqrt{2}}\sin\left(\omega t - \frac{\pi}{4}\right)\ \forall t > 0.$$

2. Frequency response, $H(f) = \frac{5}{1 + j10\pi f}$

Putting $S = j\omega = jz\pi f$

$$\text{Transfer function, } H(s) = \frac{5}{1+5s} = \frac{1}{s+\frac{1}{5}}$$

$$\text{Step-response} = H(s) \times \frac{1}{s} = \frac{1}{s\left(s+\frac{1}{5}\right)}$$

$$\frac{1}{s\left(s+\frac{1}{5}\right)} = \frac{A}{s} + \frac{B}{s+\frac{1}{5}}$$

$$A = \left.\frac{1}{s+\frac{1}{5}}\right|_{s=0} = 5$$

$$B = \left.\frac{1}{s}\right|_{s=-\frac{1}{5}} = -5$$

$$\text{Step-response} = \frac{5}{s} - \frac{5}{s+\frac{1}{5}} = 5\left[1 - e^{-\frac{t}{5}}\right]\ \forall\ t \geq 0.$$

3. Steady state output of system is

$$y(t) = |G(j\omega)| \sin\left(\omega t + \angle G(j\omega)\right)$$

for $y(t)$ to be zero.

$|G(j\omega)|$ can be zero.

$$|G(j\omega)| = \frac{(-\omega^2+9)\sqrt{\omega^2-4}}{\sqrt{\omega^2+1}\sqrt{\omega^2+9}\sqrt{\omega^2+16}}$$

$\Rightarrow$ at $\omega = 3$ rad/sec

$|G(j\omega)| = 0$, thus $y(t) = 0$

4.
$$H(s) = \frac{\omega_n^2}{s^2 + 2\zeta\omega_n s + \omega_n^2}$$

$$H(j\omega) = \frac{\omega_n^2}{-\omega^2 + 2\zeta\omega_n j\omega + \omega_n^2}$$

$$|H(jw)| = \frac{\omega_n^2}{\sqrt{\left(\omega_n^2 - \omega^2\right)^2 + (2\zeta w_n \omega)^2}}$$

$|H(j0)| = 1$

If ω_c is the 3 – dB frequency, then

$$[H(j\omega_c)] = \frac{\omega_n^2}{\sqrt{(\omega_n^2 - \omega_c^2)^2 + (2\zeta w_n \omega_c)^2}}$$

$$= 0.707$$

or $\omega_n^4 = 0.5\left[\omega_n^4 + \omega_c^4 - 2\omega_c^2\omega_n^2 + 4\zeta^2\omega_n^2\omega_c^2\right]$

Rearranging,

$$0.5\,\omega_c^4 - \omega_c^2\left(\omega_c^2 - 2\zeta^2\omega_n^2\right) - 0.5\,\omega_n^4 = 0$$

or $\omega_c^2 = \dfrac{(\omega_n^2 - 2\xi^2\omega_n^2) \pm \sqrt{(\omega_n^2 - 2\zeta^2\omega_n^2)^2 + \omega_n^4}}{1}$

$$= \omega_n^2\left[1 - 2 \pm \sqrt{2 + 4\zeta^4 - 4\zeta^2}\right]$$

or $\omega_c = \omega_n\sqrt{(1 - 2\zeta^2) \pm \sqrt{4\zeta^4 - 4\zeta^2 + 2}}$

since ω_c is real.

5. Given : $GH(\omega_1) = 1.2\angle{-180°}$, $GH(\omega_2) = 1.0\angle{-190°}$

At ω_1, $|GH| = 1.2$ and $\angle GH = -180°$

$$\text{Gain-margin} = 20\log\left[\frac{1}{|GH|}\right] = 20\log\left(\frac{1}{1.2}\right)$$

$$= -1.6 \text{ dB}$$

At ω_2, $|GH| = 1$ and $\angle GH = -190°$

Phase-margin $= 180° + \phi = 180° - 190° = -10°$

Since, both gain-margin and phase-margin are negative. The system is unstable.

6. Non-minimum phase transfer function has atleast one zero or pole in right side of s-plane.

7. Transfer function $= K\dfrac{\left(\frac{s}{5}+1\right)\left(\frac{s}{100}+1\right)\left(\frac{s}{200}+1\right)}{\left(\frac{s}{.01}+1\right)\left(\frac{s}{1}+1\right)\left(\frac{s}{80}+1\right)}$

$$= \tan^{-1}\left(\frac{\omega}{5}\right) + \tan^{-1}\left(\frac{\omega}{100}\right) + \tan^{-1}\left(\frac{\omega}{200}\right)$$

$$- \tan^{-1}\left(\frac{\omega}{.01}\right) - \tan^{-1}\omega - \tan^{-1}\left(\frac{\omega}{80}\right)$$

At $\omega = 20$, transfer function.

$$= \tan^{-1}(4) + \tan^{-1}\left(\frac{1}{5}\right) + \tan^{-1}\left(\frac{1}{10}\right) - \tan^{-1}(2000)$$

$$- \tan^{-1} 20 - \tan^{-1}\left(\frac{1}{4}\right) = -90°.$$

8. Given $M_r = \dfrac{10}{\sqrt{3}}$ rad/s, $\omega_r = 5\sqrt{2}$ rad/s; k = 5

$$M_r = \frac{k}{2\xi\sqrt{1-\xi^2}} = \frac{10}{\sqrt{3}}$$

$$\frac{5}{2\xi\sqrt{1-\xi^2}} = \frac{10}{\sqrt{3}}$$

$$\Rightarrow \xi = \frac{1}{2}$$

$$\omega_r = \omega_n\sqrt{1 - 2\xi^2}$$

$$\Rightarrow 5\sqrt{2} = \omega_n\sqrt{1 - 2\times\frac{1}{4}}$$

$$\omega_n = 10 \text{ rad/s.}$$

Only option (a) satisfies the conditions.

9. Transfer function,

$$\frac{C(s)}{R(s)} = G(s) = \frac{1}{Ms^2 + Bs + k};\ R(s) = 1/s.$$

Steady-state value of unit step response

$$C(\infty) = \lim_{s\to 0} S \cdot R(s)G(s) = \lim_{s\to 0}\frac{s}{Ms^2 + Bs + k}\cdot\frac{1}{s}$$

$$C(\infty) = \frac{1}{k}$$

From Table, $G(\omega \cong 0) = -18.5$ dB

$$\frac{1}{k} = -18.5 \text{ dB} = 10^{-18.5/20} = 0.12$$

$$\boxed{c(\infty) = 0.12}$$

10. Given : $G(s) = \dfrac{K}{s(s+2)}$, $H(s) = 1$, $M_r = 2$

Closed loop transfer function,

$$T(s) = \frac{G(s)}{1 + G(s)H(s)} = \frac{\dfrac{K}{(s+2)s}}{1 + \dfrac{K}{s(s+2)}\cdot 1}$$

$$= \frac{K}{s^2 + 2s + K}$$

D.C. Gain = $T(j\omega = jo) = 1$

$$\Rightarrow M_r = \frac{\text{DC gain}}{2\xi\sqrt{1-\xi^2}} = 2$$

$$\Rightarrow 4\xi\sqrt{1-\xi^2} = 1$$

$$16\xi^4 - 16\xi^2 + 1 = 0$$

$$\Rightarrow \boxed{\xi = 0.258}$$

Characteristic equation :

$$\Delta(s) = s^2 + 2s + K = 0$$

$$\omega_n^2 = K \Rightarrow \omega_n = \sqrt{K}$$

$$2\xi\omega_n = 2$$

$$\xi = \frac{1}{\omega_n} = \frac{1}{\sqrt{K}}$$

$$K = \frac{1}{\xi^2} = 14.92$$

11. Slope produced by 1 pole = –20 dB/decade

Slope produced by 1 zero = +20 dB.decade

Net slope (for high frequency asymptote)

$$= -20\,(14-2)$$

$$= -240 \text{ dB/decade.}$$

12. Given : $G(s) = \dfrac{10}{s(s+1)^2} = \dfrac{10}{\omega(1+\omega^2)}\ \underline{|-90° - 2\tan^{-1}(\omega)}$

For intercept at real axis, $\underline{|G(s)} = -180°$

$$-90° - 2\tan^{-1}(\omega) = -180°$$

$$\tan^{-1}(\omega) = 45°$$

$$\therefore\ \omega = \omega_{pc} = 1 \text{ rad/s}$$

$$|G(\omega = \omega_{pc})| = \frac{10}{1(1+1^2)} = 5$$

$\Rightarrow$ At $\omega = \omega_{pc} = 1$ rad/s, then plot crosses the negative real axis at –5.

13. From bode plot, corner frequency (ω_c) = 1 rad/s.

$$\Rightarrow 20 \text{ dB} = 20.\log(K)$$

$$\log(K) = 1$$

$$K = 10$$

Transfer function, $T(s) = \dfrac{K}{\left(\dfrac{s}{\omega_c}+1\right)} = \dfrac{10}{s+1}$

14. Pole in right half P = 0

Encirclement of $-1 + j0$

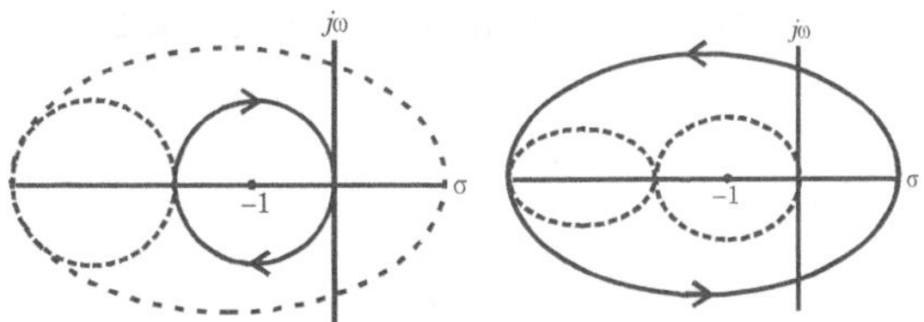

Clockwise *Anticlockwise*

N = (No. of anticlockwise encirclement – No. of clockwise encirclement)

$$N = (1-1) = 0$$

Thus, $N = P - Z = 0$

$\therefore$ Zero roots on RH of s-plane.

15. The encirclement of point (1, 0) is only once in anti-clockwise direction, then N = 1 and as P = 1 therefore $N = P - Z = 0$

and, the system is stable as there are no zeroes in right-hand s-plane.

16. From $\omega = 0.1$ to 10, i.e. in two decade dB change is + 120 dB which infers that there are three zeroes at $\omega = 0.1$.

Now from $\omega = 10$ to $\omega = 100$, i.e. in one decade dB change is – 20 dB, thus having a pole at $\omega = 10$.

Likewise at $\omega = 100$, slope changes to 0 dB/decade which infers a pole at $\omega = 100$.

Thus, $$T(s) = \frac{k(s+0.1)^3}{(s+10)^2\,(s+100)}$$

Given, $20 \log [T(s)]|_{\omega = 0.1} = 20 = 20 \log (k \times 10^{-7})$

or $k = 10^8$

17. At $\omega = 1$, slope changes from OdB/decade to 20 *d*B/decade, and, at $\omega = 10$, slope changes from 20*d*B/decade to 0*d*B/decade and at $\omega = 100$, slope changes from 0*d*B/decade to – 20 *d*B/decade.

Hence, there is a zero at $\omega = 1$ and poles at $\omega = 10, 100$.

Now, $$T(s) = \frac{k(s+1)}{(s+10)\,(s+100)}$$

$$T(j\omega) = \frac{k(j\omega+1)}{(j\omega+10)\,(j\omega+100)}$$

$$= \frac{k(j\omega+1)\cdot 10^{-3}}{\left(\dfrac{j\omega}{10}+1\right)\left(\dfrac{j\omega}{100}+1\right)}$$

Then, $20 \log (10^{-3}k) = -\,20$

or, $k = 10^2$

Hence, $$T(s) = \frac{10^2(s+1)}{(s+10)\,(s+100)}.$$

18. As the open loop transfer function is stable, therefore number of poles in RHS of s-plane is zero. Now, according to Nyquist criteria, the closed-loop system is stable if number of encirclements of $(-1 + j0)$ point in anti-clockwise direction is equal to P-Z, where P and Z are the number of poles and zeros in the RHS of s-plane.

Here P = 0, therefore N = –Z

Hence, for the closed-loop system to be stable number of encirclements in anti-clockwise direction should be zero which will make number of zeros in RH-side of s-plane is zero.

Now, Drawing Nyquist plot.

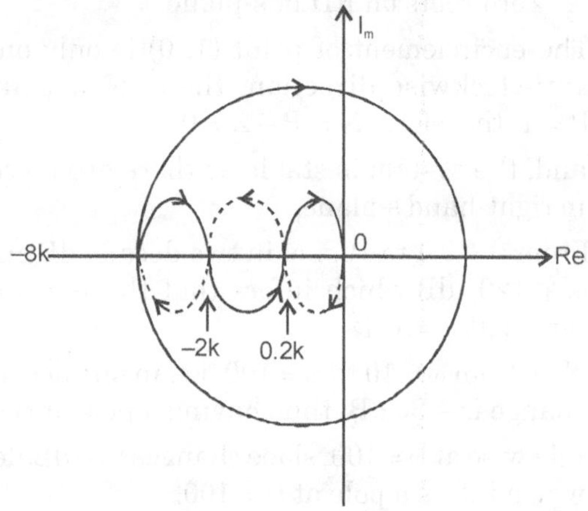

If $(-1 + j0)$ point lie between $-0.2k$ and $-2k$, then number of encirclements is zero. i.e. $2k > 1$ and $0.2k < 1$. Aslo, if $-8k > -1$, then there is no encirclements of $(-1 + j0)$.

Hence, the condition for stability is

$$k < \frac{1}{8} \text{ and } \frac{1}{2} < k < 5$$

19. From given figure, corner frequencies at

$$\omega_1 = 0, \quad \omega_2 = 1, \quad \omega_3 = 20$$

$$\therefore \quad G(s) = \frac{k}{s(1+s\tau_1)(1+s\tau_2)}$$

$$\therefore \quad \tau_1 = \frac{1}{\omega_2} = 1, \quad \tau_2 = \frac{1}{\omega_3} = 0.05$$

$$\therefore \quad G(s) = \frac{k}{s(s+1)(1+0.05\,s)}$$

From graph $|G z(j\omega)|$ dB = 60 at $\omega = 0.1$

$$20 \log_{10} |G(j\omega)| = 60$$

$$\Rightarrow \quad 20 \log_{10} \left|\frac{k}{j\omega(1+j\omega)(1+0.00\,5j\omega)}\right| = 60$$

$$\Rightarrow \quad k = 100$$

$$\therefore \quad G(s) = \frac{100}{s(s+1)(1+0.05\,s)}$$

20. At $\omega_1 = 0.1$, slope changes from 0 dB to + 20 dB

$$\therefore \quad \text{Factor} = \left(1+\frac{s}{0.1}\right)$$

At $\omega_2 = 10$, slope changes from + 20 dB to 0 dB

$$\therefore \quad \text{Factor} = \frac{1}{\left(1+\frac{s}{10}\right)}$$

$$\text{So,} \quad T(s) = \frac{k\left(1+\frac{s}{0.1}\right)}{\left(1+\frac{s}{10}\right)} = \frac{k(10s+1)}{0.1s+1}$$

At $\omega = 0.1$, magnitude = 0 dB,

$$\therefore \quad 20 \log k = 0$$

$$\Rightarrow \quad k = 1$$

$$\therefore \quad T(s) = \frac{1+10s}{1+0.1s}$$

21. $|G(jw)|_{w=0} = 5, \quad \angle G(jw)|_{w=0} = 0$

$|G(jw)|_{w=\infty} = \infty, \quad \angle G(jw)|_{w=\infty} = \frac{\pi}{2}$

Hence, the correct option is (*a*)

22. Change is magnitude from 1 rad/s to 10 rad/s

$$= 32 - (-8) = 40 \text{ dB}.$$

$$\text{Slope} = \frac{\text{Change is magnitude}}{\text{Change is frequency}}$$

$$= -\frac{40}{1} = -40 \text{ dB/decade}$$

Initial slope of –40 bB/decade means there are two poles at origin. It means either option (b) or option (d) is correct.

Given: $|G(s)|_{\omega=1}$ rad/s = 32 dB.

$$\text{Option (b)}: 20\log\left|\frac{39.8}{(1)^2}\right| = 32\,\text{dB}$$

$$\text{Option (c)}: 20\log\left|\frac{32}{1^2}\right| = 30.1 \text{ dB}$$

23. Nyquist plot does not encircle the critical point $(-1, 0)$,

So, $N = P^+ - Z^+$, $N = 0$.

If dc gain is increased to a large value, it will lead to instability. Hence, option (d) is correct.

24. We know that 6.02 dB/octave = 20 dB/decade

Redrawing the Bode plot,

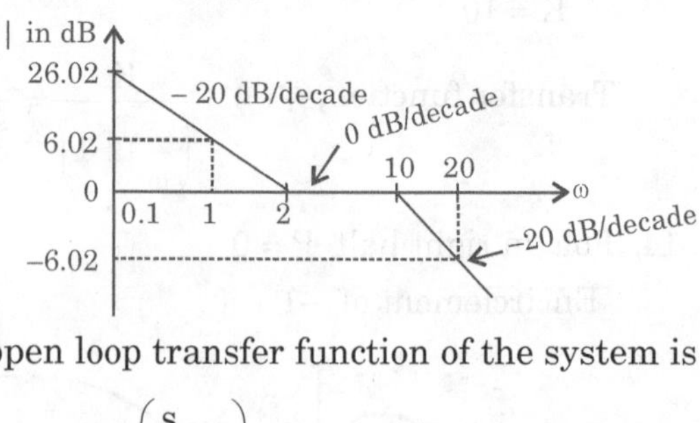

The open loop transfer function of the system is

$$G(s)\,H(s) = \frac{k\left(\frac{s}{2}+1\right)}{s\left(\frac{s}{10}+1\right)}$$

$\Rightarrow 26.02 = 20 \log k - \log(\omega)$ for $\omega = 0.1$

$20 \log k = 26.02 + \log (0.1)$

k = 1.99 ≅ 2.

$$\Rightarrow G(s)\cdot H(s) = \frac{2\left(\frac{s}{2}+1\right)}{s\left(\frac{s}{10}+1\right)} = \frac{20(s+2)}{2s(s+10)}$$

Steady - state error for ramp unit $e_{ss} = \frac{1}{k_v}$

$$k_v = \lim_{s\to0} S.G(s) \lim_{s\to0} \frac{s.20(s+2)}{2s(s+10)} = 2$$

$$e_{ss} = \frac{1}{2} = 0.5$$

25. Slope produced by one pole = – 20 dB/decade
For all system, resultant slope produced by 4 poles = – 20 × 4 = – 80 dB/dec.

26. Given: $G(s)\,H(s) = \frac{10(s+1)}{s+10}$

$$\Rightarrow G(j\omega)H(j\omega) = \frac{10\sqrt{\omega^2+1}}{\sqrt{\omega^2+100}} \underline{\left|\tan^{-1}(\omega) - \tan^{t}\frac{\omega}{10}\right.}$$

For ω = 0, $G(j0)\,H(j0) = 1\underline{|0°}$

For ω = ∞, $G(j\infty)\,H(j\infty) = 10\underline{|0°}$

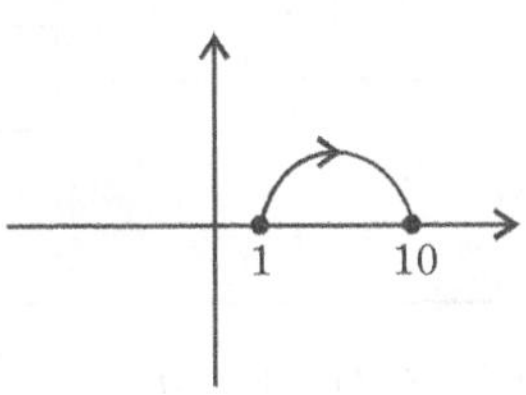

27. Using slope formula, slope = $\frac{y_2 - y_1}{x_2 - x_1}$

$$40 \text{ db/dec} = \frac{40-0}{\log 300 - \log f_L} = \frac{40}{\log\left(\frac{300}{f_L}\right)}$$

$$\log\left(\frac{300}{f_L}\right) = 1 \therefore f_L = 30 \text{ Hz.}$$

$$\Rightarrow -40 \text{ dB/decade} = \frac{40-0}{\log 900 - \log f_H} = \frac{40}{\log\left(\frac{900}{f_H}\right)}$$

$$\log\left(\frac{900}{f_H}\right) = -1$$

$$f_H = 9000 \text{ Hz.}$$

$$\Rightarrow f_H - f_L = 9000 - 30 = 8970 \text{ Hz.}$$

28. Here the Nyquist plot of the corresponding open-loop transfer function is

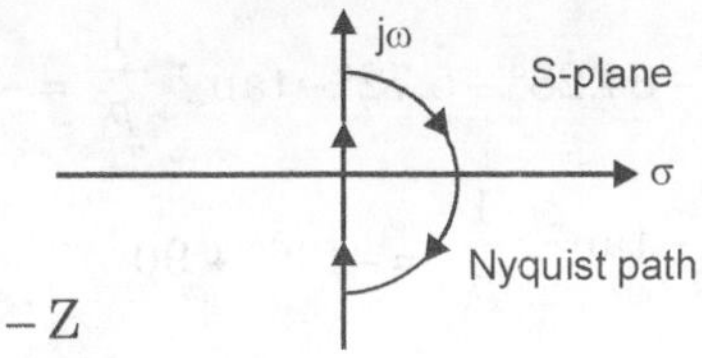

$$N = P - Z$$

For closed loop stability

$$Z = 0,\ N = P$$

So, the point (– 1, j0) encircles the s-plane is the counter clockwise direction equaling P poles of the right-half s-plane.

29. Here the Nyquist plot is

$$G(s) = \frac{1-s}{4+2s}$$

and $\angle G(s)| = -\tan^{-1}\omega - \tan^{-1}\frac{\omega}{2}$

at $s = 0$, $|G(s)| = \frac{1-0}{4+0} = \frac{1}{4} = 0.25$

and $\angle G(s)|_{\omega=0} = -\tan^{-1}0° - \tan^{-1}\frac{0°}{2} = 0°$

at $s = \infty$, $|G(s)| = \frac{s\left(\frac{1}{s}-1\right)}{s\left(\frac{4}{s}+2\right)} = \frac{\left(\frac{1}{s}-1\right)}{\left(\frac{4}{s}+2\right)}$

$$\frac{(0+1)}{(0+2)} = -\frac{1}{2} = -0.5$$

and $\angle G(s)|_{\omega=\infty} = -\tan^{-1}180° - \tan^{-1}90° = -180°$

Nyquist plot is

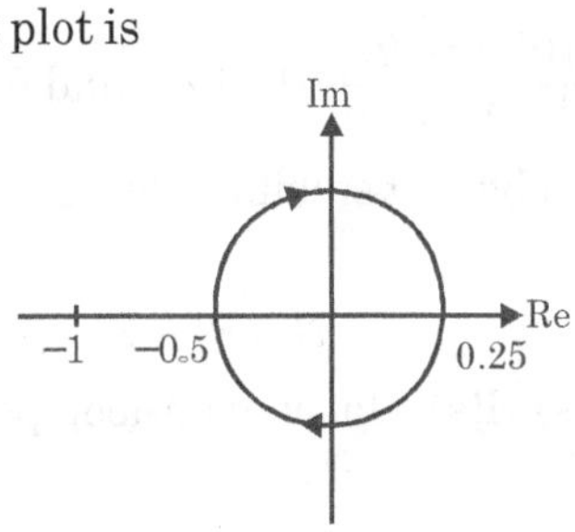

Hence the number of encirclements of $(-1 + j0) = 0$

30. $$G(s) = \frac{K}{(s+0.1)(s+10)(s+p_1)}$$

$$\angle G(s) = -\tan^{-1}\frac{\omega}{0.1} - \tan^{-1}\frac{\omega}{10} - \tan^{-1}\frac{\omega}{p_1}$$

$$\angle G(s)|_{\omega=1} = -\tan^{-1}\frac{1}{0.1} - \tan^{-1}\frac{1}{10} - \tan^{-1}\frac{1}{p_1}$$

$$= -135°$$

$$-\tan^{-1} 10 - \tan^{-1} 0.1 - \tan^{-1}\frac{1}{p_1} = -135°$$

$$-84.28° - 5.72° - \tan^{-1}\frac{1}{p_1} = -135°$$

$$-\tan^{-1}\frac{1}{p_1} = -135° + 90°$$

$$\tan^{-1}\frac{1}{p_1} = 45°$$

$$\frac{1}{p_1} = 1 \Rightarrow p_1 = 1$$

Hence, the value of p_1 is 1.

31. No. of poles = q

No. of zeroes = p

Resultant slope = – (q – p) × 20 db/decade

= – 60 dB/decade

q – p = 3.

Only option (a) satisfies above condition.

32. According to Nyquist stability criterion,

$N = P^+ - Z^+$

Where P^+ = Open-loop poles in RHS of s-plane

= 0 (given)

Z^+ = closed - loop poles in RHS of s-plane

N = No. of encirclement about the point (–1, 0)

$\Rightarrow$ For k = 10, $N = P^+ - Z^+ = 0$

$\Rightarrow Z^+ = 0$: Stable system

$\Rightarrow$ For K = 100, $N = P^+ - Z^+ = 1$

$\Rightarrow Z^+ \neq 0$: Unstable system.

33. Given :

$$G(s) = \frac{10K(s+2)}{s^3 + 3s^2 + 10}, H(s) = 1 \text{ and } 0 < k < 1$$

Applying Nyquist stability criterion for given plot, N = 0.

$$N = P^+ - Z^+$$

From G(s) H(s), No. of open-loop poles is RHS of s-plane

$$P^+ = 2$$

$\Rightarrow 0 = 2 - Z^+$

$Z^+ = 2$ (Closed loop poles in RHS of s-plane).

34. Open loop tranfer function = G(s) H(s)

where G(s) = forward path transfer function,

$$\text{Closed loop tranfer function} : T(s) = \frac{G(s)}{1 + G(s)H(s)}$$

H(s) = Feedback path transfer function.

We can draw its Nyquist plot by simply putting s = jω and obtain polar co-ordinates of T(jω) for different vlaues of ω.

35. Let $G'(s) = K \cdot G(s)$

For the closed loop system to be stable,

Gain-margin > 1

$$\frac{1}{|G'(j\omega)|_{\omega=\omega_{pc}}} > 1$$

$$|G'(j\omega)|_{\omega=\omega_{pc}} < 1$$

$$20\log K + |G(j\omega_{pc})| < 0 \text{ dB}$$

$$20.\log K + 20 \text{ dB} < 0 \text{ dB}$$

$$20\log K < -20 \text{ dB}$$

$$\log K < -1$$

$$K < 0.1$$

Hence, maximum value of K, i.e., K_0, for which system is stable is $K_{max} = K_0 = 0.1$

36. From the given bode plot

$$N_p = 6, N_z = 3$$

37. Given, A unit feedback system as shown in figure

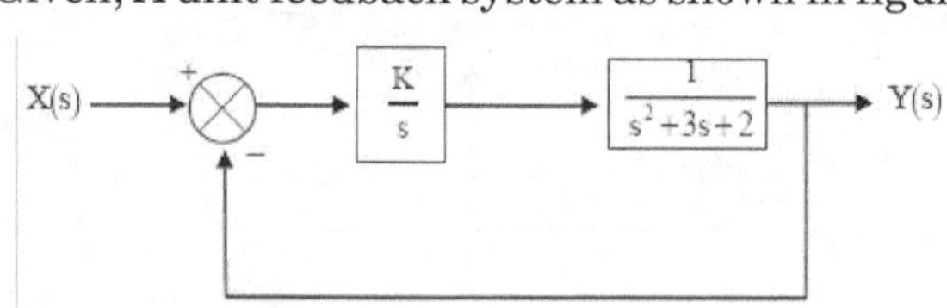

$CE = s(s^2 + 3s + 2) + k = 0$

$\Rightarrow s^3 + 3s^2 + 5 + k = 0$

Rowth Hurwitz criteria.

s^3	1	2
s^2	3	k
s^1	$\frac{6-k}{3}$	0
s^0	k	0

s^1 row → If this row will be zero then two poles will be on jω axis

$\therefore$ 6 – K = 0 $\Rightarrow$ K = 6

Lets cross check : $3s^2 + 6 = 0 \Rightarrow s^2 + 2 = 0$

$s = \pm j2$ → Two poles on jω

38. Given: $G(s)H(s) = \frac{2(1+5)}{s^2}$

$\angle G(j\omega) \cdot H(j\omega) = -180° + \tan^{-1}\omega$

For gain margin, $\angle G(j\omega_{pc}) \cdot H(j\omega_{pc}) = -180°$

$-180° + \tan^{-1}(\omega_{pc}) = -180°$

$\omega_{pc} = 0$ rad/s.

$$|G(s)H(s)|\ \omega = \omega_{pc} = \left.\frac{2\sqrt{1+\omega^2}}{\omega^2}\right|_{\omega=\omega_{pc}} = \infty.$$

$$\text{Gain margin} = \frac{1}{\infty} = 0.$$

bB (G.M.) = ∞.

39. Given: $G(s) = \dfrac{as+1}{s^2}$, $H(s) = 1$,

Phase margin or PM = $\dfrac{\pi}{4}$.

$\underline{|G(j\omega)H(j\omega)}$ = $180° + \tan^{-1}(a\omega)$

For phase margin, $180° + \tan^{-1}(a\omega_{gc}) = 180° + \dfrac{\pi}{4}$

$$a\omega_{gc} = \tan\frac{\pi}{4} = 1.$$

For gain cross-over frequency

$$|G(j\omega_{gc})\cdot H(j\omega_{gc})| = 1$$

$$\frac{\sqrt{1+a^2\omega_{gc}^2}}{\omega_{gc}^2} = 1 \qquad (\because a\omega_{gc} = 1)$$

$$\frac{\sqrt{1+1}}{\omega_{gc}^2} = 1$$

$$\therefore\ \omega_{gc} = (2)^{\frac{1}{4}}$$

$$a = \frac{1}{2^{\frac{1}{4}}} = 0.84.$$

40. $G(s) = \dfrac{as+1}{s^2} = \dfrac{0.84s+1}{s^2}$ for PM = $\dfrac{\pi}{4}$.

Unit impulse resonse = L^{-1} [Open - loop transfer function]

$$C(t) = L^{-1}[G(s)H(s)] = L^{-1}\left[\frac{0.84s+1}{s^2}\right]$$

$$= L^{-1}\left[\frac{0.84}{s} + \frac{1}{s^2}\right]$$

$c(t) = 0.84 + t$; $\forall\ t > 0$.

At t = 1, c(1) = 1 + 0.84 = 1.84

41. Given :

$$G(s)\cdot H(s) = \frac{1}{(s+1)^3} = \frac{1}{(1+\omega^2)^{3/2}}\ \underline{|-3\tan^{-1}(\omega)}$$

For phase – crossover frequency,

$$\underline{|G(s)H(s)} = -180°$$

$$-3\tan^{-1}(\omega_{pc}) = -180°$$

$$\tan^{-1}(\omega_{pc}) = 60°$$

$$\omega_{pc} = \sqrt{3}\ \text{rad/s}$$

$$|G(s)\cdot H(s)|_{\omega=\omega_{pc}} = \frac{1}{\left[1+(\sqrt{3})^2\right]^{3/2}}$$

$$= \frac{1}{2^3} = \frac{1}{8}$$

$$\text{Gain – margin} = \frac{1}{|G(s)H(s)|_{\omega=\omega_{pc}}} = \frac{1}{\frac{1}{8}} = 8$$

42. Gain margin of the system for which Nyquist plot of a loop transfer function $G(j\omega)\ H(j\omega)$ encloses $(-1, +j0)$ point is less than zero.

43.
$$G(j\omega)H(j\omega) = \frac{\sqrt{2}}{j\omega H(j\omega+1)}$$

$$= \frac{\sqrt{2}}{\omega(j-\omega)}\cdot\frac{(j+\omega)}{(j+\omega)}$$

$$= \frac{-\sqrt{2}\omega}{\omega(1+\omega^2)} - \frac{-\sqrt{2}j}{\omega(1-\omega^2)}$$

Equating imaginary part of 0, we have

$$\frac{-\sqrt{2}j}{\omega(1+\omega^2)} = 0$$

$$\Rightarrow \quad \omega = \infty$$

$$\text{G.M.} = -20\log\left|\frac{-\sqrt{2}}{1+\omega^2}\right|_{\omega=\infty}$$

$$= -20\log 0$$

$$= \infty$$

44. Given : $G(s)H(s) = \dfrac{2\sqrt{3}}{s\cdot(s+1)}$

$$= \frac{2\sqrt{3}}{\omega\sqrt{1+\omega^2}}\ \underline{|-90° - \tan^{-1}\omega}$$

For gain crossover frequency ω_{gc}, $|GH| = 1$

$$\frac{2\sqrt{3}}{\omega_{gc}\sqrt{1+\omega_{gc}^2}} = 1$$

$$2\sqrt{3} = \omega_{gc}\sqrt{1+\omega_{gc}^2}$$

$$\Rightarrow \omega_{gc} = \sqrt{3}\ \text{rad/s}$$

$$\Rightarrow \underline{|G(s)H(s)}\ |\ \omega = \omega_{gc} = -90° - \tan^{-1}(\sqrt{3})$$

$$= -90° - 60° = -150°$$

Phase-margin = 180° – 150° = 30°

45. Given : $G(s)H(s) = \dfrac{1}{s(s^2+s+1)}$

$$= \frac{1}{\omega\sqrt{(1-\omega^2)^2+\omega^2}} \angle\left(-90° - \tan^{-1}\left(\frac{\omega}{1-\omega^2}\right)\right)$$

For phase-crossover frequency ω_{pc},

$$\angle G(s)H(s) = -180°$$

$$-90° - \tan^{-1}\frac{\omega}{1-\omega^2} = 180°$$

$$-\tan^{-1}\left(\frac{\omega}{1-\omega^2}\right) = -90°$$

$$\frac{\omega}{1-\omega^2} = \frac{1}{0}$$

$$1-\omega^2 = 0$$

$$\omega_{pc} = \omega = 1 \text{ rad/s}$$

$$|G(s)H(s)|_{\omega=\omega_{pc}} = \frac{1}{1\sqrt{(1-1^2)^2+1^2}} = 1$$

$$\text{Gain - margin} = 20\log\left(\frac{1}{GH|_{\omega=\omega_{pc}}}\right)$$

$= -20 \log 1 = 0$ dB.

46. For gain-crossover frequency ω_{gc},

$$|G(s)\cdot H(s)|_{\omega=\omega_{gc}} = 1$$

$$\frac{\sqrt{1+\omega_{gc}^2}}{\sqrt{1+\omega_{gc}^2}\sqrt{4+\omega_{gc}^2}} = 1$$

$$\sqrt{4+\omega_{gc}^2} = 1$$

$\omega_{gc}^2 = -3$ (No gain crassover frequency exists)

$\Rightarrow$ Phase-margin $= \infty$.

47. Given: $G(s)H(s) = \dfrac{s}{(s+100)^3}$

For ω_{pc}, $\angle G(s)H(s) = -180°$

$$90° - 3\cdot\tan^{-1}\left(\frac{\omega_{pc}}{100}\right) = -180°$$

$$\tan^{-1}\left(\frac{\omega_{pc}}{100}\right) = 90°$$

$\omega_{pc} = \infty$.

$\Rightarrow$ Hence, gain margin cannot be determined.

For ω_{gc}, $G(s)\cdot H(s) = 1$

$$\frac{\omega_{gc}}{\left(\sqrt{100+\omega^2}\right)^3} = 1$$

$\omega_{gc} < 0$.

Hence, G.M. and P.M. of the system cannot be determined.

48. Given: $G(s) = \dfrac{3.e^{-2s}}{s(s+2)}$, $H(s) = 1$.

Open loop tranfer function :

$$G(j\omega)H(j\omega) = \frac{3.e^{-j2\omega}}{j\omega(j\omega+2)}$$

For gain crossover frequency ω_{gc},

$|G(s)H(s)| = 1$.

$$\frac{3\times 1}{\omega_{gc}\cdot\sqrt{\omega_{gc}^2+4}} = 1$$

$$\frac{9}{\omega_{gc}^2(\omega_{gc}^2+4)} = 1$$

$$\omega_{gc}^4 + 4\omega_{gc}^2 - 9 = 0$$

$\omega_{gc} = 1.26$ rad/s.

For phase crossover frequency, ω_{pc}, $\angle GH = -180°$

$$-2\omega_{pc} - 90° - \tan^{-1}\left(\frac{\omega_{pc}}{2}\right) = -180°$$

$$2\omega_{pc} + \tan^{-1}\left(\frac{\omega_{pc}}{2}\right) = \frac{\pi}{2}$$

$$2\omega_{pc} + \left[\frac{\omega_{pc}}{2} - \frac{1}{3}\left(\frac{\omega_{pc}}{2}\right)^3\right] = \frac{\pi}{2}$$

$$\therefore \quad \frac{5\omega_{pc}}{2} - \frac{\omega_{pc}^3}{24} = \frac{\pi}{2}$$

$$\frac{5\omega_{pc}}{2} \cong \frac{\pi}{2}$$

$\omega_{pc} = 0.63$ rad/s.

49. Gain-margin $= -20\log|GH(\omega = \omega_{pc})|$

$$= -20\log\left|\frac{3}{0.63(0.63^2+4)^{\frac{1}{2}}}\right|$$

$= -20\log|2.27|$

$= -7.08$ dB

Since, gain-margin is negative, system is unstable.

Phase – margin $= 180° + \angle GH(\omega = \omega_{gc})$

$$= 180° - 2\omega_{gc} - \frac{\pi}{2} - \tan^{-1}\left(\frac{\omega_{pc}}{2}\right)$$

$$= -2\times 1.26 + \frac{\pi}{2} - \tan^{-1}\left(\frac{1.26}{2}\right)$$

$= -4.65$ rad or $-86.5°$

50. Given: $G(s) = \frac{k}{(s+1)(s+2)}$, $H(s) = 1$.

Open loop transfer function :

$$G(s)H(s) = \frac{k}{(s+1)(s+2)} = \frac{k}{\sqrt{\omega^2+1}\sqrt{\omega^2+4}} \angle -\tan^{-1}(\omega) - \tan^{-1}\left(\frac{\omega}{2}\right)$$

For phase crossover frequency ω_{pc},

$$\angle GH(\omega_{pc}) = -180^\circ$$

$$-\tan^{-1}(\omega) - \tan^{-1}\left(\frac{\omega}{2}\right) = -180^\circ$$

$$\tan^{-1}\left(\frac{\omega + \frac{\omega}{2}}{1 - \frac{\omega^2}{2}}\right) = 180^\circ$$

$$\frac{\omega + \frac{\omega}{2}}{1-\omega^2} = 0$$

$\therefore\ \omega = 0$

Hence, Gain margin is infinite.

51. Given: $G(j\omega) \cdot H(j\omega)|_{\omega=\omega_{pc}} = 1\angle -180^\circ$

Gain-margin $= -20 \log |GH(\omega_{pc})|$

$= -20 \log 1 = 0$ dB.

52. According to Nyquist stability criterion,

$N = p^+ - z^+$

For Nyquist plot, $N = -1$, $P = 0$ (given)

$z^+ = 1$.

Thus, one open loop zero lies is right half of s-plane

53. Magnitude at negative real axis

$= |GH(\omega = \omega_{pc}) = 0.5$

Gain-margin $= -20 \log |GH(\omega_{pc})|$

$= -20 \log 0.5$

$= 6$ dB.

Phase (when magnitude is unity), $\phi = -90^\circ$

Phase margin $= 180^\circ - 90^\circ = 90^\circ$.

54. Given $G(s) \cdot H(s) = \frac{100}{s(s+10)^2}$

$$= \frac{100}{\omega(\omega^2+100)} \angle -90^\circ - 2\tan^{-1}\left(\frac{\omega}{10}\right)$$

For phase crossover frequency ω_{pc}, $\angle GH = -180^\circ$

$$-90^\circ - 2 \cdot \tan^{-1}\left(\frac{\omega_{pc}}{10}\right) = -180^\circ$$

$$\tan^{-1}\left(\frac{\omega_{pc}}{10}\right) = 45^\circ$$

$\Rightarrow \omega_{pc} = 10$ rad/s.

Gain-margin $= -20 \log |GH(\omega_{pc})|$

$$= -20 \log \left|\frac{100}{10(10^2+100)}\right|$$

$= 20 \log 20$

$= 26$ dB.

55. Given: $(G(s) = \frac{10}{(s+0.1)(s+1)(s+10)}$

$$= \frac{10}{0.1 \times 1 \times 10\left(\frac{s}{0.1}+1\right)\left(\frac{s}{10}+1\right)\left(\frac{s}{1}+1\right)}$$

$$G(s) = \frac{10}{\left(\frac{s}{0.1}+1\right)\left(\frac{s}{1}+1\right)\left(\frac{s}{10}+1\right)}$$

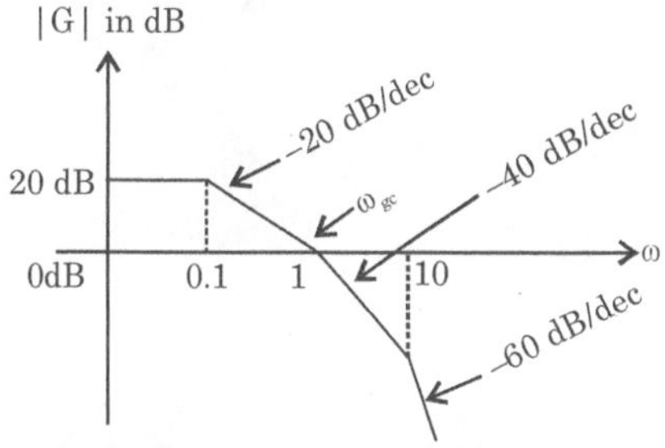

From Bode plot,

$\omega_{gc} = 1$ rad/s.

Phase - margin $= 180^\circ + \angle G(\omega_{pc})$

$$= 180^\circ - \tan^{-1}\left(\frac{\omega_{gc}}{0.1}\right) - \tan^{-1}(\omega_{gc}) - \tan^{-1}\left(\frac{\omega_{gc}}{10}\right)$$

$= 180^\circ - 84 \cdot 289^\circ - 45^\circ - 5.711^\circ$

$= 180^\circ - 135^\circ$

$= 45^\circ$.

56. Given:

$$G(s) = \frac{10}{s(s+10)} = \frac{10}{\omega\sqrt{\omega^2+100}} \angle -\frac{\pi}{2} - \tan^{-1}\frac{\omega}{10}$$

For gain cross-over frequency ω_{gc}, $|G(j\omega_{gc})| = 1$.

$$\frac{10}{\omega_{gc} \cdot \sqrt{\omega_{gc}^2 + 100}} = 1$$

$$\omega_{gc}^4 + 100\omega_{gc}^2 - 100 = 0$$

$\omega_{gc} = 0.99$ rad/s.

Phase - margin = $180° + \underline{|G(j\omega_{gc})}$

$$= 180° + \left[-\frac{\pi}{2} - \tan^{-1}\left(\frac{0.99}{10}\right)\right]$$

$\cong 84.364°$

57. $1 + G(s)H(s) = 1 + \frac{K}{(s+1)(s+2)(s+3)} = 0$

$(s+1)(s^2+5s+6) + K = 0$

$s^3 + 5s^2 + 6s + s^2 + 5s + 6 + K = 0$

$\Rightarrow s^3 + 6s^2 + 11s + 6 + K = 0$

Gain margin = 0 dB and phase margin = 0°

It implies marginal stable system

By Routh Array

s^3	1	11
s^2	6	$(6+K)$
s	$\frac{66-6-K}{6}$	0
s^0	$6+K$	

For marginal stable system,

$$60 - K = 0$$

$$\Rightarrow \quad K = 60$$

Hence, the positive value of K is 60.

58. Given that in a Bode-plot

$$G_{\phi}(j\omega)|_{\omega = w_{ge}} = -125°$$

Phase margin of system,

$$\phi_M = -125° + 180°$$

$$= 55°$$

6
CHAPTER

State Space Analysis

Analysis of Previous GATE Papers												
	Year → Topics ↓		2019	2018	2017 Set 1	2017 Set 2	2016 Set 1	2016 Set 2	2016 Set 3	2015 Set 1	2015 Set 2	2015 Set 3
STATE VARIABLE MODEL OF LINEAR SYSTEMS	1 Mark	MCQ Type										1
		Numerical Type										
	2 Marks	MCQ Type	1	1		1						
		Numerical Type										
	Total		2	2		2						1
SOLUTION OF STATE EQUATIONS OF LTI SYSTEMS	1 Mark	MCQ Type										
		Numerical Type				1						
	2 Marks	MCQ Type									1	
		Numerical Type										
	Total					1					2	
CONTROLLABILITY & OBSERVABILITY	1 Mark	MCQ Type										
		Numerical Type										
	2 Marks	MCQ Type							1			
		Numerical Type										
	Total								2			

STATE VARIABLE MODEL OF LINEAR SYSTEMS

1. a linear system is equivalently represented by two sets of state equations.

$\dot{X} = AX + BU$ and $W = CW + DU$

The eigen values of the representations are also computed as $[\lambda]$ and $[\mu]$. Which one of the following statement is true?

(a) $[\lambda] = [\mu]$ and $X = W$

(b) $[\lambda] = [\mu]$ and $X \neq W$

(c) $[\lambda] \neq [\mu]$ and $X = W$

(d) $[\lambda] \neq [\mu]$ and $X \neq W$

[2005 : 1 Mark]

2. The state space representation of a separately excited DC servo motor dynamics is given as

$$\begin{bmatrix} \frac{d\omega}{dt} \\ \frac{di_a}{dt} \end{bmatrix} = \begin{bmatrix} -1 & 1 \\ -1 & -10 \end{bmatrix} \begin{bmatrix} \omega \\ i_a \end{bmatrix} + \begin{bmatrix} 0 \\ 10 \end{bmatrix} u$$

where ω is the speed of the motor i_a is the armature current and u is the armature voltage.

The transfer function $\frac{\omega(s)}{U(s)}$ of the motor is

(a) $\frac{10}{s^2 + 11s + 11}$ (b) $\frac{1}{s^2 + 11s + 11}$

(c) $\frac{10s + 10}{s^2 + 11s + 11}$ (d) $\frac{1}{s^2 + s + 1}$

[2007 : 2 Marks]

3. A signal flow graph of a system is given below.

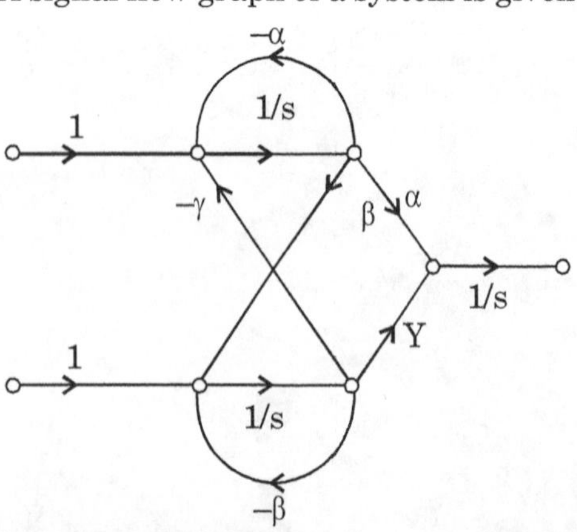

The set of equations that correspond to this signal flow graph is

(a) $$\frac{d}{dt}\begin{pmatrix} x_1 \\ x_2 \\ x_3 \end{pmatrix} = \begin{bmatrix} \beta & -\gamma & 0 \\ \gamma & \alpha & 0 \\ -\alpha & -\beta & 0 \end{bmatrix}\begin{pmatrix} x_1 \\ x_2 \\ x_3 \end{pmatrix} + \begin{bmatrix} 1 & 0 \\ 0 & 0 \\ 0 & 1 \end{bmatrix}\begin{pmatrix} u_1 \\ u_2 \end{pmatrix}$$

(b) $$\frac{d}{dt}\begin{pmatrix} x_1 \\ x_2 \\ x_3 \end{pmatrix} = \begin{bmatrix} 0 & \alpha & \gamma \\ 0 & -\alpha & -\gamma \\ 0 & \beta & -\beta \end{bmatrix}\begin{pmatrix} x_1 \\ x_2 \\ x_3 \end{pmatrix} + \begin{bmatrix} 0 & 0 \\ 0 & 1 \\ 1 & 0 \end{bmatrix}\begin{pmatrix} u_1 \\ u_2 \end{pmatrix}$$

(c) $$\frac{d}{dt}\begin{pmatrix} x_1 \\ x_2 \\ x_3 \end{pmatrix} = \begin{bmatrix} -\alpha & \beta & 0 \\ -\beta & -\gamma & 0 \\ \alpha & \gamma & 0 \end{bmatrix}\begin{pmatrix} x_1 \\ x_2 \\ x_3 \end{pmatrix} + \begin{bmatrix} 1 & 0 \\ 0 & 1 \\ 0 & 0 \end{bmatrix}\begin{pmatrix} u_1 \\ u_2 \end{pmatrix}$$

(d) $$\frac{d}{dt}\begin{pmatrix} x_1 \\ x_2 \\ x_3 \end{pmatrix} = \begin{bmatrix} -\gamma & \beta & \beta \\ \gamma & -\gamma & \alpha \\ -\beta & \gamma & -\alpha \end{bmatrix}\begin{pmatrix} x_1 \\ x_2 \\ x_3 \end{pmatrix} + \begin{bmatrix} 0 & 1 \\ 0 & 0 \\ 1 & 0 \end{bmatrix}\begin{pmatrix} u_1 \\ u_2 \end{pmatrix}$$

[2008 : 2 Marks]

Common Data for Questions 4 and 5:

The signal flow graph of a system is shown below:

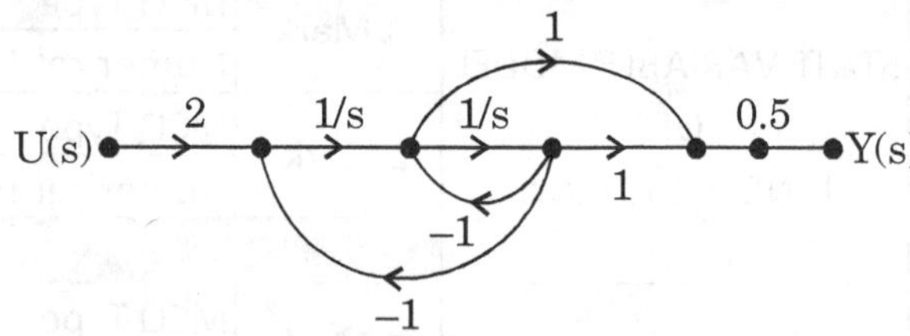

4. The state variable representation of the system can be

(a) $$\dot{x} = \begin{bmatrix} 1 & 1 \\ -1 & 0 \end{bmatrix} x + \begin{bmatrix} 0 \\ 2 \end{bmatrix} u$$
$$y = [0 \quad 0.5] x$$

(b) $$\dot{x} = \begin{bmatrix} -1 & 1 \\ -1 & 0 \end{bmatrix} x + \begin{bmatrix} 0 \\ 2 \end{bmatrix} u$$
$$y = [0 \quad 0.5] x$$

(c) $$\dot{x} = \begin{bmatrix} 1 & 1 \\ -1 & 0 \end{bmatrix} x + \begin{bmatrix} 0 \\ 2 \end{bmatrix} u$$
$$y = [0.5 \quad 0.5] x$$

(d) $$\dot{x} = \begin{bmatrix} -1 & 1 \\ -1 & 0 \end{bmatrix} x + \begin{bmatrix} 0 \\ 2 \end{bmatrix} u$$
$$y = [0.5 \quad 0.5] x$$

[2010 : 2 Marks]

5. The transfer function of the system is

(a) $\frac{s+1}{s^2+1}$

(b) $\frac{s-1}{s^2-1}$

(c) $\frac{s+1}{s^2+s+1}$

(d) $\frac{s-1}{s^2+s+1}$

[2010 : 2 Marks]

6. The block diagram of a system with one input u and two outputs y_1 and y_2 is given below.

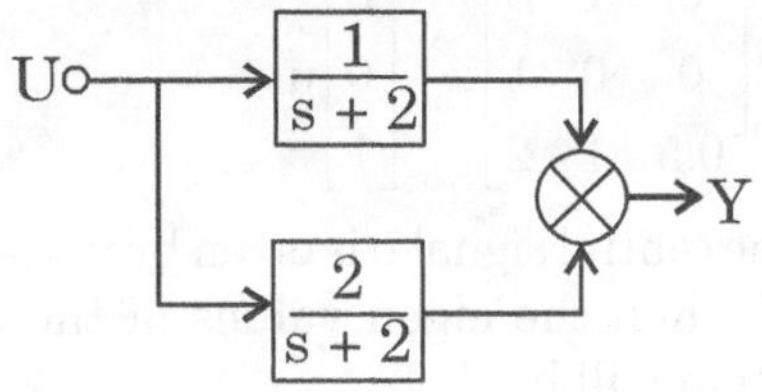

A state space model of the above system in terms of the state vector x and the output vector $y = [y_1 \ \ y_2]^t$ is

(a) $\dot{x} = [2]x + [1]u; \quad y = [1 \ \ 2]x$

(b) $\dot{x} = [-2]x + [1]u; \quad y = \begin{bmatrix} 1 \\ 2 \end{bmatrix} x$

(c) $\dot{x} = \begin{bmatrix} -2 & 0 \\ 0 & -2 \end{bmatrix} x + \begin{bmatrix} 1 \\ 1 \end{bmatrix} u; \quad y = [1 \ \ 2]x$

(d) $\dot{x} = \begin{bmatrix} 2 & 0 \\ 0 & 2 \end{bmatrix} x + \begin{bmatrix} 1 \\ 2 \end{bmatrix} u; \quad y = \begin{bmatrix} 1 \\ 2 \end{bmatrix} x$

[2011 : 2 Marks]

Statement for Linked Answer Questions 7 and 8 : The state diagram of a system is shown below is described by the state-variable equations:

$\underline{\dot{X}}(t) = AX + Bu, \quad y = CX + Du.$

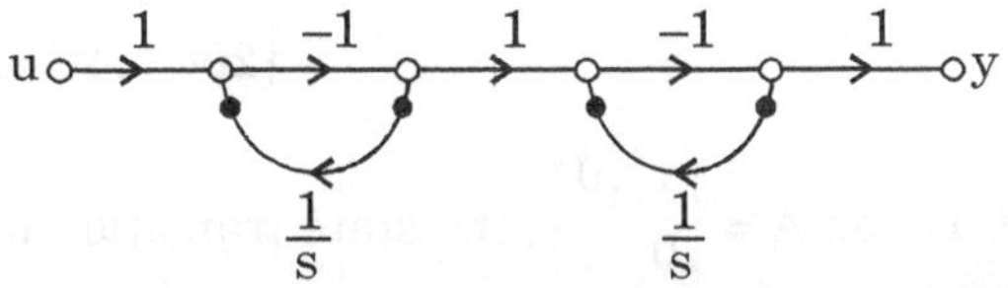

7. The state-variable equations of the system in the figure above are

(a) $\dot{X} = \begin{bmatrix} -1 & 0 \\ 1 & -1 \end{bmatrix} X + \begin{bmatrix} -1 \\ 1 \end{bmatrix} u,$
$y = [1 \ \ -1]X + u$

(b) $\dot{X} = \begin{bmatrix} -1 & 0 \\ -1 & -1 \end{bmatrix} X + \begin{bmatrix} -1 \\ 1 \end{bmatrix} u,$
$y = [-1 \ \ -1]X + u$

(c) $\dot{X} = \begin{bmatrix} -1 & 0 \\ -1 & -1 \end{bmatrix} X + \begin{bmatrix} -1 \\ 1 \end{bmatrix} u,$
$y = [-1 \ \ -1]X - u$

(d) $\dot{X} = \begin{bmatrix} -1 & -1 \\ 0 & -1 \end{bmatrix} X + \begin{bmatrix} -1 \\ 1 \end{bmatrix} u,$
$y = [1 \ \ -1]X - u$

[2013 : 2 Marks]

8. The state transition matrix e^{At} of the system shown in figure above is

(a) $\begin{bmatrix} e^{-t} & 0 \\ te^{-t} & e^{-t} \end{bmatrix}$ (b) $\begin{bmatrix} e^{-t} & 0 \\ -te^{-t} & e^{-t} \end{bmatrix}$

(c) $\begin{bmatrix} e^{-t} & 0 \\ e^{-t} & e^{-t} \end{bmatrix}$ (d) $\begin{bmatrix} e^{-t} & -te^{-t} \\ 0 & e^{-t} \end{bmatrix}$

[2013 : 2 Marks]

9. A network is described by the state model as

$x_1 = 2x_1 - x_2 + 3u,$

$x_2 = -4x_2 - u,$

$y = 3x_1 - 2x_2.$

The transfer function $H(s) \left(= \dfrac{Y(s)}{U(s)} \right)$ is

(a) $\dfrac{11s + 35}{(s-2)(s+4)}$ (b) $\dfrac{11s + 35}{(s-2)(s+4)}$

(c) $\dfrac{11s + 38}{(s-2)(s+4)}$ (d) $\dfrac{11s + 38}{(s-2)(s+4)}$

[2015 : 1 Marks, Set-3]

10. A second order LTI system is described by the following state equations

$$\frac{d}{dt}x_1(t) - x_2(t) = 0$$

$$\frac{d}{dt}x_2(t) + 2x_1(t) + 3x_2(t) = r(t);$$

where $x_1(t)$ and $x_2(t)$ are the two state variables and r(t) denotes the input. The output c(t) $= x_1(t)$.

The system is

(a) undamped (oscilatory).

(b) underdamped.

(c) critically damped.

(d) overdamped.

[2017 : 2 Marks, Set-2]

11. The state equation and the output equation of a control system are given below:

$$\dot{x} = \begin{bmatrix} -4 & -1.5 \\ 4 & 0 \end{bmatrix} x + \begin{bmatrix} 2 \\ 0 \end{bmatrix} u$$

$y = [1.5 \ \ 0.625]x$

The transfer function representation of the system is

(a) $\dfrac{3s + 5}{s^2 + 4s + 6}$ (b) $\dfrac{3s - 1.875}{s^2 + 4s + 6}$

(c) $\dfrac{4s - 1.5}{s^2 + 4s + 6}$ (d) $\dfrac{6s + 5}{s^2 + 4s + 6}$

[2018 : 2 Marks]

12. Let the state-space representation of an LT1 system be $\dot{x}(t) = A_x(t) + Bu(t)$, $y(t) = Cx(t) + du(t)$ where A,B,C are matrices, d is a scalar, u(t) is the input to the system, and y (t) is its output. Let $B = [0\ 0\ 1]^T$ and d = 0. Which one of the following options for A and C will ensure that the transfer function of this LTI system is

$$H(s) = \frac{1}{S^3 + 3s^2 + 2s + 1}?$$

(a) $A = \begin{bmatrix} 0 & 1 & 0 \\ 0 & 0 & 1 \\ -1 & -2 & -3 \end{bmatrix}$ and $C = [1\ \ 0\ \ 0]$

(b) $A = \begin{bmatrix} 0 & 1 & 0 \\ 0 & 0 & 1 \\ -1 & -2 & -3 \end{bmatrix}$ and $C = [0\ \ 0\ \ 1]$

(c) $A = \begin{bmatrix} 0 & 1 & 0 \\ 0 & 0 & 1 \\ -3 & -2 & -1 \end{bmatrix}$ and $C = [0\ \ 0\ \ 1]$

(d) $A = \begin{bmatrix} 0 & 1 & 0 \\ 0 & 0 & 1 \\ -3 & -2 & -1 \end{bmatrix}$ and $C = [1\ \ 0\ \ 0]$

[2019 : 2 Marks]

SOLUTION OF STATE EQUATIONS OF LTI SYSTEMS

13. Give Given the folloing state-space description of a system

$$\begin{bmatrix} \dot{x}_1 \\ \dot{x}_2 \end{bmatrix} = \begin{bmatrix} -2 & 0 \\ 0 & -4 \end{bmatrix}\begin{bmatrix} x_1 \\ x_2 \end{bmatrix} + \begin{bmatrix} 0 \\ 1 \end{bmatrix}u,$$

$$y = [1\ 0]\begin{bmatrix} x_1 \\ x_2 \end{bmatrix}.$$

Find the state-transition matrix.

[1988 : 2 Marks]

14. A certain linear time invariant system has the sate and the output equations given below:

$$\begin{bmatrix} \dot{x}_1 \\ \dot{x}_2 \end{bmatrix} = \begin{bmatrix} -1 & -1 \\ 0 & 1 \end{bmatrix}\begin{bmatrix} x_1 \\ x_2 \end{bmatrix} + \begin{bmatrix} 0 \\ 1 \end{bmatrix}u,$$

$$y = [1\ \ 1]\begin{bmatrix} x_1 \\ x_2 \end{bmatrix} \text{ is}$$

$x_1(0) = 1$, $x_2(0) = -1$, $u(0) = 0$ then $\left.\frac{dy}{dt}\right|_{t=0}$ is

(a) 1 (b) –1

(c) 0 (d) None of the above

[1997 : 2 Marks]

15. For the system described byf the state equation

$$\dot{x} = \begin{bmatrix} 0 & 1 & 0 \\ 0 & 0 & 1 \\ 0.5 & 1 & 2 \end{bmatrix}x + \begin{bmatrix} 0 \\ 0 \\ 1 \end{bmatrix}u$$

If the control signal u is given by $u = [-0.5 - 3 - 5]x + V$, then the eigen values of the closed-loop system will be

(a) 0, –1, –2 (b) 0, –1, –3

(c) –1, –1, –2 (d) 0, –1, –1

[1999 : 2 Marks]

16. The transfer function Y(s)/U(s) of a system described by the sate equations $\dot{x}(t) = -2x(t) + 2u(t)$ and $y(t) = 0.5x(t)$ is

(a) 0.5/(s – 2) (b) 1/(s – 2)

(c) 0.5/(s + 2) (d) 1/(s + 2)

[2002 : 1 Mark]

17. The zero-input response of a system given by the state-space equation

(a) $\begin{bmatrix} te^t \\ t \end{bmatrix}$ (b) $\begin{bmatrix} e^t \\ t \end{bmatrix}$

(c) $\begin{bmatrix} e^t \\ te^t \end{bmatrix}$ (d) $\begin{bmatrix} e^t \\ te^t \end{bmatrix}$

[2003 : 2 Marks]

18. Given $A = \begin{bmatrix} 1 & 0 \\ 0 & 1 \end{bmatrix}$, the state transition matrix e^{At} is given by

(a) $\begin{bmatrix} 0 & e^{-t} \\ e^{-t} & 0 \end{bmatrix}$ (b) $\begin{bmatrix} e^t & 0 \\ 0 & e^t \end{bmatrix}$

(c) $\begin{bmatrix} e^{-t} & 0 \\ 0 & e^{-t} \end{bmatrix}$ (d) $\begin{bmatrix} 0 & e^t \\ e^t & 0 \end{bmatrix}$

[2004 : 2 Marks]

19. A linear system is described by the following state equation

$$\dot{X}(t) = AX(t) + BU(t),\ A = \begin{bmatrix} 0 & 1 \\ -1 & 0 \end{bmatrix}$$

The state-transition matrix of the system is

(a) $\begin{bmatrix} \cos t & \sin t \\ -\sin t & \cos t \end{bmatrix}$ (b) $\begin{bmatrix} -\cos t & \sin t \\ -\sin t & -\cos t \end{bmatrix}$

(c) $\begin{bmatrix} -\cos t & -\sin t \\ -\sin t & \cos t \end{bmatrix}$ (d) $\begin{bmatrix} \cos t & -\sin t \\ \sin t & \cos t \end{bmatrix}$

[2006 : 2 Marks]

Statement for Linked answer Questions 20 & 21:
Consider a linear system whose state space representation is $\dot{\underline{x}}(t) = Ax(t)$. If the initial state vector of the system is $x(0) = \begin{bmatrix} 1 \\ -2 \end{bmatrix}$, then the system response is $x(t) = \begin{bmatrix} e^{-2t} \\ -2e^{-2t} \end{bmatrix}$. If the initial state vector of the system changes to $x(0) = \begin{bmatrix} 1 \\ -1 \end{bmatrix}$ then the system response becomes $x(t) = \begin{bmatrix} e^{-t} \\ -e^{-t} \end{bmatrix}$.

20. The eigen-value and eigen-vector pairs (λ_i, v_i) for the system are

(a) $\left(-1, \begin{bmatrix} 1 \\ 1 \end{bmatrix}\right)$ and $\left(-2, \begin{bmatrix} 1 \\ 2 \end{bmatrix}\right)$

(b) $\left(-2, \begin{bmatrix} 1 \\ -1 \end{bmatrix}\right)$ and $\left(-1, \begin{bmatrix} 1 \\ -2 \end{bmatrix}\right)$

(c) $\left(-1, \begin{bmatrix} 1 \\ -1 \end{bmatrix}\right)$ and $\left(-2, \begin{bmatrix} 1 \\ -2 \end{bmatrix}\right)$

(d) $\left(-2, \begin{bmatrix} 1 \\ -1 \end{bmatrix}\right)$ and $\left(-2, \begin{bmatrix} 1 \\ -2 \end{bmatrix}\right)$

[2007 : 2 Marks]

21. The system matrix A is

(a) $\begin{bmatrix} 0 & 1 \\ -1 & 1 \end{bmatrix}$ (b) $\begin{bmatrix} 1 & 1 \\ -1 & -2 \end{bmatrix}$

(c) $\begin{bmatrix} 1 & 1 \\ -1 & -2 \end{bmatrix}$ (d) $\begin{bmatrix} 0 & 1 \\ -2 & -3 \end{bmatrix}$

[2007 : 2 Marks]

22. Consider the matrix $P = \begin{bmatrix} 0 & 1 \\ -2 & -3 \end{bmatrix}$. The value of e^P is

(a) $\begin{bmatrix} 2e^{-2} - 3e^{-1} & e^{-1} - e^{-2} \\ 2e^{-2} - 2e^{-1} & 5e^{-2} - e^{-1} \end{bmatrix}$

(b) $\begin{bmatrix} e^{-2} + e^{-2} & 2e^{-2} - e^{-1} \\ 2e^{-1} - 4e^{2} & 3e^{-1} - 2e^{-2} \end{bmatrix}$

(c) $\begin{bmatrix} 5e^{-2} - e^{-1} & 3e^{-1} - e^{-2} \\ 2e^{-2} - 6e^{-1} & 4e^{-2} - e^{-1} \end{bmatrix}$

(d) $\begin{bmatrix} 2e^{-1} - e^{-2} & e^{-1} - e^{-2} \\ -2e^{-1} + 2e^{-2} & -e^{-1} + 2e^{-2} \end{bmatrix}$

[2008 : 2 Marks]

23. An unforced linear time invariant (LTI) system is represented by $\begin{bmatrix} \dot{x}_1 \\ \dot{x}_2 \end{bmatrix} = \begin{bmatrix} -1 & 0 \\ 0 & 2 \end{bmatrix} \begin{bmatrix} x_1 \\ x_2 \end{bmatrix}$.

If the initial conditions are $x_1(0) = 1$ and $x_2(0) = -1$, the solution of the state equation is

(a) $x_1(t) = -1, x_2(t) = 2$

(b) $x_1(t) = -e^{-t}, x_2(t) = 2e^{-t}$

(c) $x_1(t) = e^{-1}, x_2(t) = -e^{-2t}$

(d) $x_1(t) = -e^{-t}, x_2(t) = -2e^{-t}$

[2014 : 2 Marks, Set-2]

24. The state equation of a second-order linear system is given by,

$\dot{x}(t) = Ax(t), x(0) = x_0$

For $x_0 = \begin{bmatrix} 1 \\ -2 \end{bmatrix}$, $x(t) = \begin{bmatrix} e^{-t} \\ -e^{-t} \end{bmatrix}$ and for $x_0 = \begin{bmatrix} 0 \\ 1 \end{bmatrix}$,

$x(t) = \begin{bmatrix} e^{-t} - e^{-2t} \\ -e^{-t} + 2e^{-2t} \end{bmatrix}$

When $x_0 = \begin{bmatrix} 3 \\ 5 \end{bmatrix}$, x(t) is

(a) $\begin{bmatrix} -8e^{-t} + 11e^{-2t} \\ 8e^{-t} - 22e^{-2t} \end{bmatrix}$ (b) $\begin{bmatrix} 11e^{-t} - 8e^{-2t} \\ -11e^{-t} - 16e^{-2t} \end{bmatrix}$

(c) $\begin{bmatrix} 3e^{-t} - 5e^{-2t} \\ -3e^{-t} - 10e^{-2t} \end{bmatrix}$ (d) $\begin{bmatrix} -5e^{-t} + 6e^{-2t} \\ -3e^{-t} + 6e^{-2t} \end{bmatrix}$

[2014 : 2 Marks, Set-3]

25. The state transition matrix $\phi(t)$ of a system

$\begin{bmatrix} \dot{x}_1 \\ \dot{x}_2 \end{bmatrix} = \begin{bmatrix} 0 & 1 \\ 0 & 0 \end{bmatrix} \begin{bmatrix} x_1 \\ x_2 \end{bmatrix}$ is

(a) $\begin{bmatrix} t & 1 \\ 1 & 0 \end{bmatrix}$ (b) $\begin{bmatrix} 1 & 0 \\ t & 1 \end{bmatrix}$

(c) $\begin{bmatrix} 0 & 1 \\ 1 & t \end{bmatrix}$ (d) $\begin{bmatrix} 1 & t \\ 0 & 1 \end{bmatrix}$

[2014 : 2 Marks, Set-4]

26. The state variable representation of a system is given as

$$\dot{x} = \begin{bmatrix} 0 & 1 \\ 0 & -1 \end{bmatrix} x, \quad x(0) = \begin{bmatrix} 1 \\ 0 \end{bmatrix},$$

$y = [0 \ \ 1]x$

The response y(t) is

(a) sin(t) (b) $1 - e^{t}$

(c) $1 - \cos(t)$ (d) 0

[2015 : 2 Marks, Set-2]

27. Consider the state space realization:

$$\begin{bmatrix}\dot{x}_1(t)\\ \dot{x}_2(t)\end{bmatrix}=\begin{bmatrix}0 & 0\\ 0 & -9\end{bmatrix}\begin{bmatrix}x_1(t)\\ x_2(t)\end{bmatrix}+\begin{bmatrix}0\\ 45\end{bmatrix}u(t), \text{ with the}$$

initial condition $\begin{bmatrix}x_1(0)\\ x_2(0)\end{bmatrix}+\begin{bmatrix}0\\ 0\end{bmatrix}$;

where u(t) denotes the unit step function the value of $\lim_{x\to\infty}\left|\sqrt{x_1^2(t)+x_2^2(t)}\right|$ is ______.

[2017 : 1 Mark, Set-2]

CONTROLLABILITY & OBSERVABILITY

28. A liner second-order single-input continuous-time system is described by following set of differential equations

$x_1(t) = -2x_1(t) + 4x_2(t)$;

$x_2(t) = 2x_1(t) - x_2(t) + u(t)$.

Where $x_1(t)$ and $x_2(t)$ are the state variables and u(t) is the control variable. The system is

(a) controllable and stable.

(b) controllable but unstable.

(c) uncontrollable and unstable.

(d) uncontrollable but stable.

[1991 : 2 Marks]

29. A linear time-invariant system is described by the state variable model

$$\begin{bmatrix}\dot{x}_1\\ \dot{x}_2\end{bmatrix}\begin{bmatrix}-1 & 0\\ 0 & -2\end{bmatrix}\begin{bmatrix}x_1\\ x_2\end{bmatrix}+\begin{bmatrix}0\\ 1\end{bmatrix}u,$$

$$y=[1\ \ 2]\begin{bmatrix}x_1\\ x_2\end{bmatrix}$$

(a) The system is completely controllable

(b) The system is not completely controllable

(c) The system is completely observable

(d) The system is not completely observable

[1992 : 2 Marks]

30. The system mode described by the state equations

$$\begin{bmatrix}\dot{x}_1\\ \dot{x}_2\end{bmatrix}x=\begin{bmatrix}0 & 1\\ 2 & -3\end{bmatrix}x+\begin{bmatrix}0\\ 1\end{bmatrix}u,\ y=[1\ \ 1]x;$$

(a) controllable and observable.

(b) controllable, but not observable.

(c) observable, but not controllable.

(d) neither controllable nor observable.

[1999 : 1 Mark]

31. The state variable equations of a system are:

1. $\dot{x}_1 = -3x_1 - x_2 + u$,
2. $\dot{x}_2 = 2x_1$; $y = x_1 + u$.

The system is

(a) controllable but not observable.

(b) observable but not controllable.

(c) neither controllable nor observable.

(d) controllable and observable.

[2004 : 2 Marks]

32. Consider the system $\frac{dx}{dt} = Ax + Bu$ with $A = \begin{bmatrix}1 & 0\\ 0 & 1\end{bmatrix}$ and $B=\begin{bmatrix}p\\ q\end{bmatrix}$ where p and q are arbitrary real numbers. Which of the following statements about the controllability of the system is true?

(a) The system is completely state controllable for any nonzero values of p and q.

(b) Only p = 0 and q = 0 result in controllability.

(c) The system is uncontrollable for all values of p and q.

(d) Cannot conclude about controllability from the given data.

[2009 : 1 Mark]

33. The state variable description of an LTI system is given by

$$\begin{pmatrix}\dot{x}_1\\ \dot{x}_1\\ \dot{x}_1\end{pmatrix}=\begin{pmatrix}0 & a_1 & 0\\ 0 & 0 & a_2\\ a_3 & 0 & 0\end{pmatrix}\begin{pmatrix}x_1\\ x_2\\ x_3\end{pmatrix}+\begin{pmatrix}0\\ 0\\ 1\end{pmatrix}u,$$

$$y=(1\ \ 0\ \ 0)\begin{pmatrix}x_1\\ x_2\\ x_3\end{pmatrix};$$

where y is the output and u is the input. The system is controllable for

(a) $a_1 \neq 0, a_2 = 0, a_3 \neq 0$

(b) $a_1 = 0, a_2 \neq 0, a_3 \neq 0$

(c) $a_1 = 0, a_2 \neq 0, a_3 = 0$

(d) $a_1 \neq 0, a_2 \neq 0, a_3 = 0$

[2012 : 2 Marks]

34. Consider the state space model of a system, as given below:

$$\begin{bmatrix}\dot{x}_1\\ \dot{x}_2\\ \dot{x}_3\end{bmatrix}=\begin{bmatrix}-1 & 1 & 0\\ 0 & -1 & 0\\ 0 & 0 & -2\end{bmatrix}\begin{bmatrix}x_1\\ x_2\\ x_3\end{bmatrix}+\begin{bmatrix}0\\ 4\\ 0\end{bmatrix}u;\ y=[1\ \ 1\ \ 1]\begin{bmatrix}x_1\\ x_2\\ x_3\end{bmatrix}$$

The system is

(a) controllable and observable

(b) un controllable and observable

(c) less than zero

(d) controllable and unobservable

[2014 : 2 Marks, Set-1]

35. Consider the state space system expressed by the signal flow diagram shown in the figure.

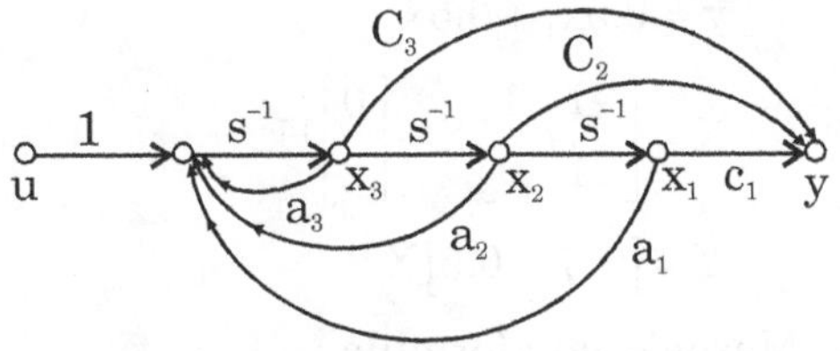

The corresponding system is

(a) always controllable

(b) always observable

(c) always stable

(d) always unstable

[2014 : 2 Marks, Set-2]

36. A second-order linear time-invariant system is described by the following state equations

$$\frac{d}{dt}x_1(t) + 2x_1(t) = 3u(t),$$

$$\frac{d}{dt}x_1(t) + x_2(t) = u(t);$$

where $x_1(t)$ and $x_2(t)$ are the two state variables and u(t) denotes the input if the output c(t) = $x_1(t)$, then the system is

(a) controllable but not observable.

(b) observable but not controllable.

(c) both controllable nad observable.

(d) neither controllable nor observable.

[2016 : 2 Marks, Set-3]

ANSWERS

1. (c)	2. (a)	3. (d)	4. (d)	5. (c)	6. (b)	7. (a)	8. (a)	9. (a)	10. (d)
11. (a)	12. (b)	13. (*)	14. (a)	15. (a)	16. (d)	17. (c)	18. (b)	19. (a)	20. (c)
21. (d)	22. (d)	23. (c)	24. (b)	25. (d)	26. (d)	27. (5)	28. (b)	29. (b,c)	30. (a)
31. (d)	32. (c)	33. (d)	34. (b)	35. (a)	36. (a)				

EXPLANATIONS

1. $\dot{x} = AX + BU$ with [λ] be the set of eigen values.

$\dot{W} = CW + DU$ with [μ] be the set of eigen values.

If a linear system is represented by two sets of state equations, then for both sets, states will be same but their set of eigen values will not be smae, i.e.,

$$X = W \text{ but } \lambda \neq \mu$$

2. $\frac{d\omega}{dt} = -\omega + i_a$

Taking laplace transform

$$s \cdot \omega(s) = -\omega(s) + I_a(s)$$

$$(s+1) \cdot \omega(s) = I_a(s)$$

$$\omega(s) = \frac{1}{s+1} \cdot I_a(s)$$

$$\Rightarrow \frac{di_a}{dt} = -\omega - 10i_a + 10u$$

$$s \cdot I_a(s) = -\omega(s) - 10I_a(s) + 10U(s)$$

$$\omega(s) = -(s+10)I_a(s) + 10U(s)$$

$$= -(s+10)(s+1)\omega(s) + 10U(s)$$

$$\therefore \omega(s)[1 + s^2 + 11s + 10] = 10 \cdot U(s)$$

$$\therefore \frac{\omega(s)}{U(s)} = \frac{10}{s^2 + 11s + 11}$$

3. We get

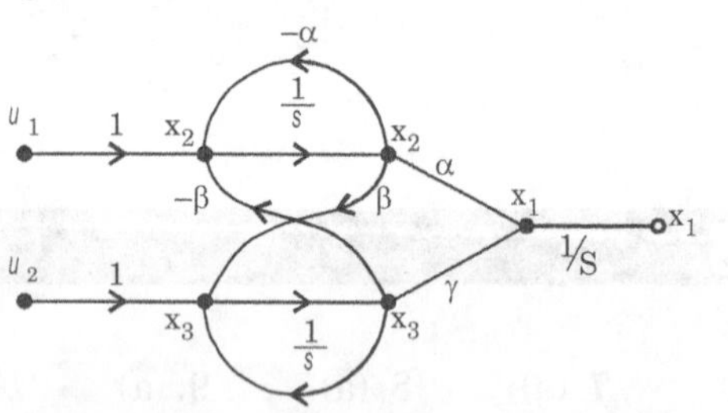

$$\dot{x}_1 = \alpha x_2 + \gamma x_3$$

$$\dot{x}_2 = \mu_1 - \alpha x_2 - \beta x_3$$

$$\dot{x}_3 = \mu_2 - \gamma x_3 + \beta x_2$$

$$\Rightarrow \begin{bmatrix} \dot{x}_1 \\ \dot{x}_2 \\ \dot{x}_3 \end{bmatrix} = \begin{bmatrix} 0 & \alpha & \gamma \\ 0 & -\alpha & -\beta \\ 0 & \beta & -\gamma \end{bmatrix} \begin{bmatrix} x_1 \\ x_2 \\ x_3 \end{bmatrix} + \begin{bmatrix} 0 & 0 \\ 1 & 0 \\ 0 & 1 \end{bmatrix} \begin{pmatrix} u_1 \\ u_2 \end{pmatrix}$$

Depending upon position of x_1, x_2 and x_3 and u_1 and u_2, the rows and columns may interchange but the corresponding entries will remain the same, hence the solution is (d).

4.

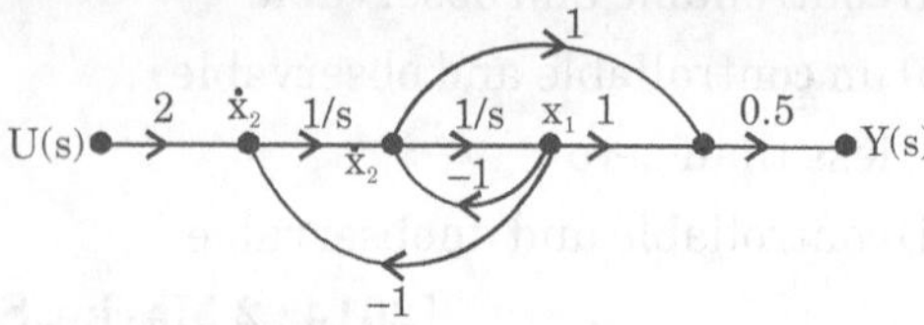

$$\Rightarrow \dot{x}_1 = -x_1 + x_2$$

$$\dot{x}_2 = -x_1 + 2u$$

$$y = 0.5x_1 + 0.5x_2$$

$$\Rightarrow \dot{X} = \begin{bmatrix} -1 & 1 \\ -1 & 0 \end{bmatrix} X + \begin{bmatrix} 0 \\ 2 \end{bmatrix} U$$

$$\dot{Y} = \begin{bmatrix} 0.5 & 0.5 \end{bmatrix} X$$

5. Mason's gain formula :

$$T(s) = \frac{Y(s)}{U(s)} = \frac{\Sigma P_K \Delta_K}{\Delta}$$

Forward path gain(s) :

$$P_1 = 2 \times \frac{1}{5} \times \frac{1}{5} \times 0.5 = \frac{1}{5^2}$$

$$P_2 = 2 \times \frac{1}{s} \times 1 \times 0.5 = \frac{1}{s}$$

Loop gain(s) : $\alpha_1 = \frac{-1}{s}$, $\alpha_2 = \frac{1}{s} \times \frac{1}{s} \times (-1) = \frac{-1}{s^2}$

$$\Delta = 1 - [\alpha_1 + \alpha_2] = 1 - \left[-\frac{1}{s} - \frac{1}{s^2}\right] = 1 + \frac{1}{s} + \frac{1}{s^2}$$

$$\Delta_1 = 1, \quad \Delta_2 = 2$$

$$T(s) = \frac{P_1\Delta_1 + P_2\Delta_2}{\Delta} = \frac{\frac{1}{s^2} \times 1 + \frac{1}{s} \times 1}{1 + \frac{1}{s} + \frac{1}{s^2}} = \frac{(1+s)}{(s^2 + s + 1)}$$

6. $x = y_1$ and $\dot{x} = \frac{dy_1}{dx}$

$$y = \begin{bmatrix} y_1 \\ y_2 \end{bmatrix} = \begin{bmatrix} x \\ 2x \end{bmatrix} = \begin{bmatrix} 1 \\ 2 \end{bmatrix} x$$

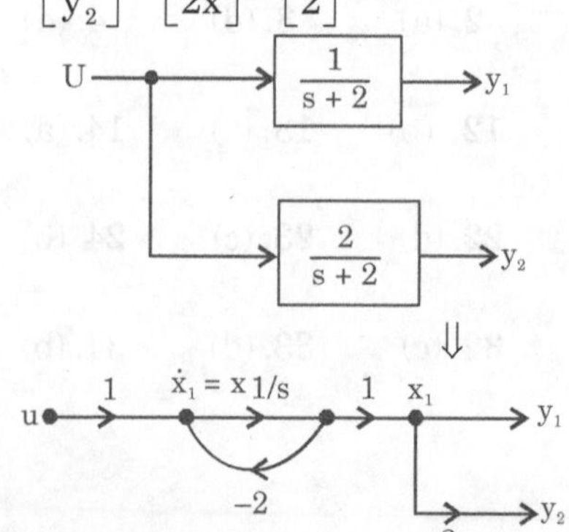

$$y_1(s+2) = u$$

$$\dot{y}_1 + 2y_1 = u$$

$$\dot{x} + 2x = u$$

$$\dot{x} = -2x + u$$

$$\dot{x} = -2x + 1.u$$

$$y_1 = x_1; y_2 = 2x_1$$

$$y = \begin{bmatrix} y_1 \\ y_2 \end{bmatrix} = \begin{bmatrix} 1 \\ 2 \end{bmatrix} x$$

7. 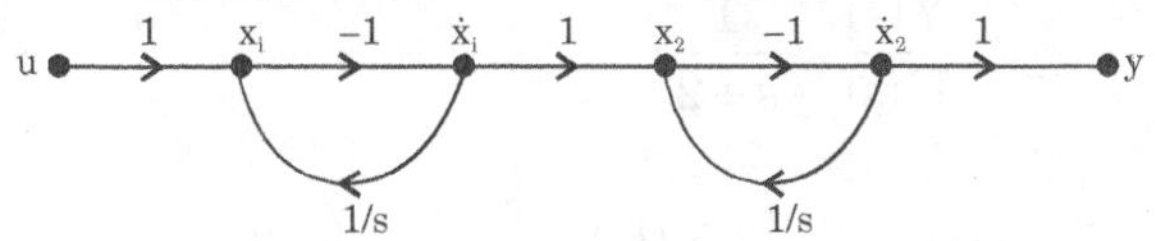

$$\Rightarrow \dot{x}_1 = -x_1 - u; \quad \dot{x}_2 = -(x_2 + \dot{x}_1)$$

$$= -x_2 + x_1 + u$$

$$y = \dot{x}_2 = -x_2 + x_1 + u$$

$$\begin{bmatrix} \dot{x}_1 \\ \dot{x}_2 \end{bmatrix} = \begin{bmatrix} -1 & 0 \\ 1 & -1 \end{bmatrix}\begin{bmatrix} x_1 \\ x_2 \end{bmatrix} + \begin{bmatrix} -1 \\ 1 \end{bmatrix} u$$

$$\dot{x} = \begin{bmatrix} -1 & 0 \\ 1 & -1 \end{bmatrix} X + \begin{bmatrix} -1 \\ 1 \end{bmatrix} u$$

8. $A = \begin{bmatrix} -1 & 0 \\ 1 & -1 \end{bmatrix}$

$$[sI - A] = \begin{bmatrix} s & 0 \\ 0 & s \end{bmatrix} - \begin{bmatrix} -1 & 0 \\ 1 & -1 \end{bmatrix} = \begin{bmatrix} s+1 & 0 \\ -1 & s+1 \end{bmatrix}$$

$$[sI - A]^{-1} = \frac{1}{(s+1)(s+1)}\begin{bmatrix} s+1 & 0 \\ 1 & s+1 \end{bmatrix}$$

$$= \begin{bmatrix} \frac{1}{s+1} & 0 \\ \frac{1}{(s+1)^2} & \frac{1}{s+1} \end{bmatrix}$$

State transition matrix, $\phi(t) = e^{At} = L^{-1}[sI - A]^{-1}$

$$= \begin{bmatrix} e^{-t} & 0 \\ te^{-t} & e^{-t} \end{bmatrix}$$

9. $\dot{x} = \begin{bmatrix} 2 & -1 \\ 0 & -4 \end{bmatrix} x + \begin{bmatrix} 3 \\ -1 \end{bmatrix} u; \; y = [3 \;\; -2] x.$

Transfer function, $T(s) = C[SI - A]^{-1} B + D$

$\Rightarrow [SI - A]^{-1}$

$$= \begin{bmatrix} s-2 & 1 \\ 0 & s+4 \end{bmatrix}^{-1} = \frac{1}{(s-2)(s+4)}\begin{bmatrix} s+4 & -1 \\ 0 & s-2 \end{bmatrix}$$

$\Rightarrow C\,[SI - A]^{-1} B$

$$= \frac{1}{(s-2)(s+4)} \cdot [3 \;\; -2]\begin{bmatrix} s+4 & -1 \\ 0 & s-2 \end{bmatrix}\begin{bmatrix} 3 \\ -1 \end{bmatrix}$$

$$= \frac{1}{(s-2)(s+4)} \cdot [3 \;\; -2] \cdot \begin{bmatrix} 3s+4\times 3+1 \\ -s+2 \end{bmatrix}$$

$$= \frac{1}{(s-2)(s+4)}[9s + 39 + 2s - 4]$$

$$C\,[SI - A]^{-1} B = \frac{11s+35}{(s-2)(s+4)}$$

10. $\frac{d}{dt}x_1(t) - x_2(t) = 0$

Applying Laplace transform

$$sx_1(s) = x_2(s)$$

$$\frac{d}{dt}x_2(t) + 2x_1(t) + 3x_2(t) = r(t)$$

$$sx_2(s) + 2x_1(s) + 3x_2(s) = R(s)$$

$$(s+3)x_2(s) + 2x_1(s) = R(s)$$

$$s(s+3)x_1(s) + 2x_1(s) = R(s)$$

$$(s^2 + 3s + 2) \cdot x_1(s) = R(s)$$

$$\therefore c(t) = x_1(t) \Rightarrow c(s) = x_1(s)$$

Transfer function,

$$\frac{C(s)}{R(s)} = \frac{X_1(s)}{R(s)} = \frac{1}{s^2 + 3s + 2}$$

$$\Rightarrow \omega_n^2 = 2$$

$$2\xi\omega_n = 3 \Rightarrow \xi = \frac{3}{2\omega_n} = \frac{3}{2\sqrt{2}} = 1.06 > 1$$

As $\xi > 1$, the system is overdamped.

12. Given,

C matrix start from right to left same sign.

$$H(S)\frac{Y(s)}{V(s)} = \frac{1}{s^3 + 3s^2 + 2s + 1}$$

A matrix element start from right to left and take sopposite sign.

Controllable.

$$A \begin{bmatrix} 0 & 1 & 0 \\ 0 & 0 & 1 \\ -1 & -2 & -3 \end{bmatrix} \text{and } C = [0 \;\; 0 \;\; 1]$$

13. Given : $A = \begin{bmatrix} -2 & 0 \\ 0 & 4 \end{bmatrix}$

$$sI - A = \begin{bmatrix} s+2 & 0 \\ 0 & s+4 \end{bmatrix}$$

$$[sI - A]^{-1} = \frac{1}{(s+2)(s+4)}\begin{bmatrix} s+4 & 0 \\ 0 & s+2 \end{bmatrix}$$

$$= \begin{bmatrix} \frac{1}{(s+2)} & 0 \\ 0 & \frac{1}{(s+4)} \end{bmatrix}$$

State transition matrix, $\phi(t) = L^{-1}[sI - A]^{-1}$

$$\phi(t) = \begin{bmatrix} e^{-2t} & 0 \\ 0 & e^{-4t} \end{bmatrix}$$

14. Given : $x_1(0) = 1,\ x_2(0) = -1, 3(0) = 0$

$$\Rightarrow \dot{x}_1 = x_1 - x_2$$

$$\dot{x}_1(0) = x_1(0) - x_2(0) = 1-(-1) = 2$$

$$\Rightarrow \dot{x}_2 = x_2$$

$$\dot{x}_2(0) = x_2(0) = -1$$

$$\Rightarrow y = x_1 + x_2$$

$$\frac{dy}{dt} = \dot{x}_1 + \dot{x}_2$$

$$\left.\frac{dy}{dt}\right|_{t=0} = \dot{x}_1(0) + \dot{x}_2(0)$$

$$= 2 - 1$$

$$= 1$$

15. Given : $\dot{x} = \begin{bmatrix} 0 & 1 & 0 \\ 0 & 0 & 1 \\ 0.5 & 1 & 2 \end{bmatrix} x + \begin{bmatrix} 0 \\ 0 \\ 1 \end{bmatrix} u,$

$$u = [-0.5 \quad -3 \quad -5]x + V$$

$$\dot{x} = \begin{bmatrix} 0 & 1 & 0 \\ 0 & 0 & 1 \\ 0.5 & 1 & 2 \end{bmatrix} x + \begin{bmatrix} 0 \\ 0 \\ 1 \end{bmatrix} [-0.5 \quad -3 \quad -5]x + V$$

$$\dot{x} = \begin{bmatrix} 0 & 1 & 0 \\ 0 & 0 & 1 \\ 0 & -2 & -3 \end{bmatrix} x + V$$

$$A = \begin{bmatrix} 0 & 1 & 0 \\ 0 & 0 & 1 \\ 0 & -2 & -3 \end{bmatrix}$$

Characteristic equation : $\Delta(s) = |sI - A|$

$$\begin{vmatrix} s & -1 & 0 \\ 0 & s & -1 \\ 0 & 2 & s+3 \end{vmatrix} = 0$$

$$\Rightarrow s^3 + 3s^2 + 2s + 0 = 0$$

$$s = 0, -1, -2$$

16. $\dot{x}(t) = -2x(t) + 2u(t)$

Taking laplace transfrom

$$sX(s) = -2 \cdot X(s) + 2U(s)$$

$$X(s) \cdot (s+2) = 2U(s)$$

$$X(s) = \frac{2}{s+2} \cdot U(s)$$

$$\Rightarrow y(t) = 0.5\, x(t)$$

$$Y(s) = 0.5 \times (s)$$

$$Y(s) = 0.5 \times \frac{2}{s+2} \times U(s)$$

$$\frac{Y(s)}{U(s)} = \frac{1}{s+2}$$

17. $x(t) = e^{A(t-t_0)} X(t_0) + \int_{t_0}^{t} e^{A(t-s)} B\, U(\tau) ds$

$$= e^{A(t-t_0)} X(t_0) + 0 \quad (\because B = 0)$$

$$(sI - A) = \begin{bmatrix} s & 0 \\ 0 & s \end{bmatrix} - \begin{bmatrix} 1 & 0 \\ 1 & 1 \end{bmatrix}$$

where, $A = \begin{bmatrix} 1 & 0 \\ 1 & 1 \end{bmatrix}$ and $X(t_0) = \begin{bmatrix} 1 \\ 0 \end{bmatrix}$

$$\therefore \quad (sI - A) = \begin{bmatrix} s-1 & 0 \\ -1 & s-1 \end{bmatrix}$$

$$(sI - A)^{-1} = \frac{\begin{bmatrix} (s-1) & 0 \\ 1 & (s-1) \end{bmatrix}}{(s-1)^2} = \begin{bmatrix} \frac{1}{s-1} & 0 \\ \frac{1}{(s-1)^2} & \frac{1}{(s-1)} \end{bmatrix}$$

$$L^{-1}[(sI - A)^{-1}] = e^{At} = \begin{bmatrix} e^t & 0 \\ te^t & e^t \end{bmatrix}$$

$$\Rightarrow \quad x(t) = e^{At} \times x[t_0] = \begin{bmatrix} e^t & 0 \\ te^t & e^t \end{bmatrix} \begin{bmatrix} 1 \\ 0 \end{bmatrix} = \begin{bmatrix} e^t \\ te^t \end{bmatrix}$$

18. Here $\quad [sI - A]^{-1} = \begin{bmatrix} s-1 & 0 \\ 0 & s-1 \end{bmatrix}^{-1}$

Then, $\mathcal{L}^{-1}[sI - A]^{-1} = \mathcal{L}^{-1}\left\{\begin{bmatrix} \frac{1}{s-1} & 0 \\ 0 & \frac{1}{s-1} \end{bmatrix}\right\}$

$$= \begin{bmatrix} e^t & 0 \\ 0 & e^t \end{bmatrix}$$

19. $$\phi(t) = L^{-1}[SI - A]^{-1}$$

$$= L^{-1}\begin{bmatrix} s & -1 \\ 1 & s \end{bmatrix}^{-1}$$

$$= L^{-1}\begin{bmatrix} \frac{s}{s^2+1} & \frac{1}{s^2+1} \\ \frac{-1}{s^2+1} & \frac{s}{s^2+1} \end{bmatrix}$$

$$= \begin{bmatrix} \cos t & \sin t \\ -\sin t & \cos t \end{bmatrix}$$

20. We know that

sum of the eigen value = Trace of the principle diagonal matrix.

Sum = –3

Only option (a) satisfies both conditions.

21. Multiplication of eigen-value = determinant of matrix

From option, it seems determinant should be ±2.

Only option (d) satisfies as determinant = 2.

22. $e^{P} = L^{-1}[sI - P]^{-1} = L^{-1}\left(\begin{bmatrix} s & 0 \\ 0 & s \end{bmatrix} - \begin{bmatrix} 0 & 1 \\ -2 & -3 \end{bmatrix}\right)^{-1}$

$$= L^{-1}\left(\begin{bmatrix} s & -1 \\ 2 & s+3 \end{bmatrix}\right)^{-1}$$

$$= L^{-1}\left(\begin{bmatrix} \frac{(s+3)}{(s+1)(s+2)} & \frac{1}{(s+1)(s+2)} \\ \frac{-2}{(s+1)(s+2)} & \frac{s}{(s+1)(s+2)} \end{bmatrix}\right)$$

$$= \begin{bmatrix} 2e^{-1} - e^{-2} & e^{-1} - e^{-2} \\ -2e^{-1} + 2e^{-2} & -e^{-1} + 2e^{-2} \end{bmatrix}$$

23. $\dot{x}_1 = -x_1 + 0.x_2 = -x_1$

$\dot{x}_2 = 0.x_1 - 2x_2 = -2x_2$

Applying Laplace transform,

$sx_1(s) - x_1(0) = -x_1(s)$

$x_1(s) = 1/(s+1)$

$x_1(t) = e^{-t}$

$\Rightarrow sx_2(s) - x_2(0) = -2x_2(s)$

$(s + 2)\, x_2(s) = x_2(0)$

$x_2(s) = \dfrac{-1}{s+2}$

$x_2(t) = -e^{-2t}$.

24. Given: $\dot{x} = Ax(t)$, $x(0) = x_0$;

Taking Laplace transform,

$sX(s) - x(0) = AX(s)$

$[SI - A]\, X(s) = x(0)$

$x(s) = [SI - A]^{-1}.\, x(0)$

$x(t) = L^{-1}[SI - A]^{-1}.x(0)$

for $x_0 = \begin{bmatrix} 1 \\ -1 \end{bmatrix}$, $x(t) = \begin{bmatrix} e^{-t} \\ -e^{-t} \end{bmatrix}$

For $x_0 = \begin{bmatrix} 0 \\ 1 \end{bmatrix}$, $x(t) = \begin{bmatrix} e^{-t} - e^{-2t} \\ -e^{-t} + 2e^{-2t} \end{bmatrix}$

Using additivity property

$k_1x_1(t) = L^{-1}[SI - A]^{-1} \cdot x_1(0) \cdot k_1$

$k_2x_2(t) = L^{-1}[SI - A]^{-1} \cdot x_2(0) \cdot k_2$

$k_1x_1(t) + k_2x_2(t)$

$= L^{-1}[SI - A]^{-1}\left[k_1x_1(0) + k_2x_2(0)\right]$

$x_3(s) = [SI - A]^{-1} \cdot x_3(0)$

$k_1\begin{bmatrix} 1 \\ -1 \end{bmatrix} + k_2\begin{bmatrix} 0 \\ 1 \end{bmatrix} = \begin{bmatrix} 3 \\ 5 \end{bmatrix}$

$\begin{bmatrix} k_1 \\ -k_1 + k_2 \end{bmatrix} = \begin{bmatrix} 3 \\ 5 \end{bmatrix} \Rightarrow \begin{matrix} k_1 = 3 \\ k_2 = 8 \end{matrix}$

$\Rightarrow x(t) = k_1x_1(t) + k_2x_2(t)$

$$x(t) = 3\begin{bmatrix} e^{-t} \\ -e^{-t} \end{bmatrix} + 8\begin{bmatrix} e^{-t} - e^{-2t} \\ -e^{-t} + 2e^{-2t} \end{bmatrix} = \begin{bmatrix} 11e^{-t} - 8e^{-2t} \\ -11e^{-t} + 16e^{-2t} \end{bmatrix}$$

25. $A = \begin{bmatrix} 0 & 1 \\ 0 & 0 \end{bmatrix}$

$[SI - A] = \begin{bmatrix} s & 0 \\ 0 & s \end{bmatrix} - \begin{bmatrix} 0 & 1 \\ 0 & 0 \end{bmatrix} = \begin{bmatrix} s & -1 \\ 0 & s \end{bmatrix}$

$|SI - A| = s^2$.

S.T.M : $\phi(t) = L^{-1}[SI - A]^{-1}$

$$= L^{-1}\begin{bmatrix} \frac{s}{s^2} & \frac{1}{s^2} \\ 0 & \frac{s}{s^2} \end{bmatrix} = L^{-1}\begin{bmatrix} \frac{1}{s} & \frac{1}{s^2} \\ 0 & \frac{1}{s} \end{bmatrix} = \begin{bmatrix} 1 & t \\ 0 & 1 \end{bmatrix}$$

26. $A = \begin{bmatrix} 0 & 1 \\ 0 & -1 \end{bmatrix}$, $B = 0$, $c = \begin{bmatrix} 0 & 1 \end{bmatrix}$; $x(0) = \begin{bmatrix} 1 \\ 0 \end{bmatrix}$

$\dot{x} = Ax$

$sx(s) - x(0) = A{\cdot}x(s)$

$[SI - A]^{-1}\, x(s) = x(0)$

$x(s) = [SI - A]^{-1} \cdot x(0)$

$$[SI - A]^{-1} = \begin{bmatrix} s & -1 \\ 0 & s+1 \end{bmatrix}^{-1} = \frac{1}{s(s+1)}\begin{bmatrix} s+1 & 1 \\ 0 & s \end{bmatrix}$$

$$\Rightarrow x(s) = \frac{1}{s(s+1)}\begin{bmatrix} s+1 & 1 \\ 0 & s \end{bmatrix}\begin{bmatrix} 1 \\ 0 \end{bmatrix} = \frac{1}{s(s+1)}\begin{bmatrix} s+1 \\ 0 \end{bmatrix}$$

$$x(s) = \begin{bmatrix} \frac{1}{s} \\ 0 \end{bmatrix}$$

$$x(t) = \begin{bmatrix} 1 \\ 0 \end{bmatrix} u(t)$$

$\Rightarrow y(t) = [0 \ \ 1]\, x(t) = [0 \ \ 1]\begin{bmatrix}1\\0\end{bmatrix}$ y(t) = 0.

27. $\dot{x}_1(t) = 0$

Applying Laplace transform,

$SX(s) - x_1(0) = 0; \ \ x_1(0) = 0$

$x_1(s) = \frac{x_1(0)}{2} = 0$

$x_1(t) = 0.$

$\Rightarrow \dot{x}_2(t) - 9x_2(t) + 45.u(t)$

$sx_2(s) - x_2(0) = -9x_2(s) + 45/s$

$x_2(s) = \frac{45}{s(s+9)}$ $\qquad \therefore {}_x2(0) = 0$

$\lim_{t\to0} x_2(t) = \lim_{t\to0} sx_2(s) = \frac{45}{9} = 5$

$\lim_{t\to0} \sqrt{x_1^2(t) + x_2^2(t)}$

$\sqrt{\lim_{t\to\infty}\left[x_1^2(t)\right] + \lim_{t\to\infty}\left[x_2^2 + 1\right]}$

$= \sqrt{0 + 5^2}$

$= 5.$

28. Given : $\begin{bmatrix}x_1\\x_2\end{bmatrix} = \begin{bmatrix}-2 & 4\\2 & -1\end{bmatrix}\begin{bmatrix}x_1\\x_2\end{bmatrix} + \begin{bmatrix}0\\1\end{bmatrix}u(t)$

$A = \begin{bmatrix}-2 & 4\\2 & -1\end{bmatrix}, B = \begin{bmatrix}0\\1\end{bmatrix}$

$AB = \begin{bmatrix}-2 & 4\\2 & -1\end{bmatrix}\begin{bmatrix}0\\1\end{bmatrix} = \begin{bmatrix}0+4\\0-1\end{bmatrix} = \begin{bmatrix}4\\-1\end{bmatrix}$

$Q_C = [B \ \ AB] = \begin{bmatrix}0 & 4\\1 & -1\end{bmatrix}$

$|Q_C| = 0 - 4 = -4 \neq 0$

Hence, the given system is controllable.

Char. eq. :

$\Delta(s) = [sI - A] = 0$

$\begin{bmatrix}s+2 & -4\\-2 & s+1\end{bmatrix} = 0$

Since, 1 pole lie is RHS of S-plane. Hence, the given systme is stable.

$(s + 1)(s + 2) - 8 = 0$

$s^2 + 3s - 6 = 0$

$s = -4.37, 1.37$

29. Given $A = \begin{bmatrix}-1 & 0\\0 & -2\end{bmatrix}, B = \begin{bmatrix}0\\1\end{bmatrix}, C = [1\,2]$

$AB = \begin{bmatrix}-1 & 0\\0 & -2\end{bmatrix}\begin{bmatrix}0\\1\end{bmatrix} = \begin{bmatrix}0+0\\0-2\end{bmatrix} = \begin{bmatrix}0\\-2\end{bmatrix}$

$Q_C = [B \ \ AB] = \begin{bmatrix}0 & 0\\1 & -2\end{bmatrix}$

$|Q_C| = 0 - 0 = 0$

$\Rightarrow$ System is not controllable.

$A^T = \begin{bmatrix}-1 & 0\\0 & -2\end{bmatrix}, C^T = \begin{bmatrix}1\\2\end{bmatrix}$

$A^TC^T = \begin{bmatrix}-1 & 0\\0 & -2\end{bmatrix}\begin{bmatrix}1\\2\end{bmatrix} = \begin{bmatrix}-1+0\\0-4\end{bmatrix} = \begin{bmatrix}-1\\-4\end{bmatrix}$

$Q_0 = \left[C^T \ \ A^TC^T\right] = \begin{bmatrix}1 & -1\\2 & -4\end{bmatrix}$

$|Q_0| = -4 - (-2) = -2 \neq 0$

$\Rightarrow$ System is observable.

30. Given : $A = \begin{bmatrix}0 & 1\\2 & -3\end{bmatrix}, B = \begin{bmatrix}0\\1\end{bmatrix}, C = [1 \ 1]$

$\Rightarrow AB = \begin{bmatrix}0 & 1\\2 & -3\end{bmatrix}\begin{bmatrix}0\\1\end{bmatrix} = \begin{bmatrix}0+1\\0-3\end{bmatrix} = \begin{bmatrix}1\\-3\end{bmatrix}$

$Q_C = [B \ \ AB] = \begin{bmatrix}0 & 1\\1 & -3\end{bmatrix} = \begin{bmatrix}0 & 1\\1 & -3\end{bmatrix}$

$= 0 - 1 = -1 \neq 0$

$\Rightarrow$ Controllable

$\Rightarrow C^T = \begin{bmatrix}1\\1\end{bmatrix}, A^T = \begin{bmatrix}0 & 2\\1-3\end{bmatrix}$

$A^TC^T = \begin{bmatrix}0 & 2\\1 & -3\end{bmatrix}\begin{bmatrix}1\\1\end{bmatrix} = \begin{bmatrix}0+2\\1-3\end{bmatrix} = \begin{bmatrix}2\\-2\end{bmatrix}$

$Q_0 = \left[C^T \ \ A^TC^T\right] = \begin{bmatrix}1 & 2\\1 & -2\end{bmatrix}$

$= -2 - 2 \neq 0$

$\Rightarrow$ Observable.

31. Rewriting state equations in matrix form, we get

$$\begin{bmatrix}\dot{x}_1\\\dot{x}_2\end{bmatrix} = \begin{bmatrix}-3 & -1\\2 & 0\end{bmatrix}\begin{bmatrix}x_1\\x_2\end{bmatrix} + \begin{bmatrix}1\\0\end{bmatrix}u$$

$$y = [1 \ 0]\begin{bmatrix}x_1\\x_2\end{bmatrix} + \begin{bmatrix}1\\0\end{bmatrix}u$$

Then, $A = \begin{bmatrix} -3 & -1 \\ 2 & 0 \end{bmatrix}$,

$B = \begin{bmatrix} 1 \\ 0 \end{bmatrix}$,

$C = [1 \ 0]$

Now, $Q_c = [B : BA] = \begin{bmatrix} 1 & -3 \\ 0 & 2 \end{bmatrix}$,

$$\det Q_c = \begin{vmatrix} 1 & -3 \\ 0 & 2 \end{vmatrix} \neq 0$$

Also, $Q_0 = [C^T \quad C^TA^T] = \begin{bmatrix} 1 & -3 \\ 0 & -1 \end{bmatrix}$,

$$\det Q_c = \begin{bmatrix} 1 & -3 \\ 0 & -1 \end{bmatrix} \neq 0$$

Hence the system is controllable and observable.

32. $A = \begin{bmatrix} 1 & 0 \\ 0 & 1 \end{bmatrix}, B = \begin{bmatrix} p \\ q \end{bmatrix}$

$$AB = \begin{bmatrix} 1 & 0 \\ 0 & 1 \end{bmatrix}\begin{bmatrix} p \\ q \end{bmatrix} = \begin{bmatrix} p \\ q \end{bmatrix}$$

$$Q_C = [B \quad AB] = \begin{bmatrix} p & p \\ q & q \end{bmatrix} = 0$$

Hence, the system is uncontrollable for all values of p and q.

33. $A = \begin{bmatrix} 0 & a_1 & 0 \\ 0 & 0 & a_2 \\ a_3 & 0 & 0 \end{bmatrix}, B = \begin{bmatrix} 0 \\ 0 \\ 1 \end{bmatrix}$

$$AB = \begin{bmatrix} 0 & a_1 & 0 \\ 0 & 0 & a_2 \\ a_3 & 0 & 0 \end{bmatrix}\begin{bmatrix} 0 \\ 0 \\ 1 \end{bmatrix} = \begin{bmatrix} 0+0+0 \\ 0+0+a_2 \\ 0+0+0 \end{bmatrix} = \begin{bmatrix} 0 \\ a_2 \\ 0 \end{bmatrix}$$

$$A^2B = A \cdot AB = \begin{bmatrix} 0 & a_1 & 0 \\ 0 & 0 & a_2 \\ a_3 & 0 & 0 \end{bmatrix}\begin{bmatrix} 0 \\ a_2 \\ 0 \end{bmatrix}$$

$$= \begin{bmatrix} 0+a_1a_2+0 \\ 0+0+0 \\ 0+0+0 \end{bmatrix} = \begin{bmatrix} a_1a_2 \\ 0 \\ 0 \end{bmatrix}$$

$$\Rightarrow Q_C = [B \quad AB \quad A^2B] = \begin{bmatrix} 0 & 0 & a_1a_2 \\ 0 & a_2 & 0 \\ 1 & 0 & 0 \end{bmatrix}$$

$|Q_C| = 0 - a_1a_2^2 = -a_1a_2^2 \neq 0$ (for controllability)

$a_1 \neq 0, \ a_2 \neq 0$

34. $A = \begin{bmatrix} -1 & 1 & 0 \\ 0 & -1 & 0 \\ 0 & 0 & -2 \end{bmatrix}, B = \begin{bmatrix} 0 \\ 4 \\ 0 \end{bmatrix}$,

$C = [1 \ 1 \ 1] \Rightarrow C^T = \begin{bmatrix} 1 \\ 1 \\ 1 \end{bmatrix}$

$$AB = \begin{bmatrix} -1 & 1 & 0 \\ 0 & -1 & 0 \\ 0 & 0 & -2 \end{bmatrix}\begin{bmatrix} 0 \\ 4 \\ 0 \end{bmatrix} = \begin{bmatrix} 0+4+0 \\ 0-4+0 \\ 0+0+0 \end{bmatrix} = \begin{bmatrix} 4 \\ -4 \\ 0 \end{bmatrix}$$

$A^2B = A \cdot AB$

$$= \begin{bmatrix} -1 & 1 & 0 \\ 0 & -1 & 0 \\ 0 & 0 & -2 \end{bmatrix}\begin{bmatrix} 4 \\ -4 \\ 0 \end{bmatrix} = \begin{bmatrix} -4-4+0 \\ 0+4+0 \\ 0+0+0 \end{bmatrix} = \begin{bmatrix} -8 \\ 4 \\ 0 \end{bmatrix}$$

$$A^TC^T = \begin{bmatrix} -1 \\ 0 \\ -2 \end{bmatrix}; \ A^{2T}C^T = A^TA^TC^T = \begin{bmatrix} 1 \\ -1 \\ 4 \end{bmatrix}$$

For controllability, $|Q_C| \neq 0$

$$Q_C = [B \quad AB \quad A^2B] = \begin{bmatrix} 0 & 4 & -8 \\ 4 & -4 & 4 \\ 0 & 0 & 0 \end{bmatrix}$$

$|Q_C| = 4 \times 0 = 0 \Rightarrow$ Uncontrollable.

for observability, $|Q_0| \neq 0$

$$Q_0 = [C^T \quad A^TC^T \quad A^{2T}C^T] = \begin{bmatrix} 1 & -1 & -1 \\ 1 & 0 & -1 \\ 1 & -2 & -4 \end{bmatrix}$$

$|Q_0| = 1 \neq 0 \Rightarrow$ observable.

35. From signal flow graph,

$\dot{x}_1 = x_2$

$\dot{x}_2 = x_3$

$\dot{x}_3 = a_3x_3 + a_2x_2 + a_1x_1 + u$

$y = c_1x_1 + c_2x_2 + c_3x_3$

$$\begin{bmatrix} \dot{x}_1 \\ \dot{x}_2 \\ \dot{x}_3 \end{bmatrix} = \begin{bmatrix} 0 & 1 & 0 \\ 0 & 0 & 1 \\ a_1 & a_2 & a_3 \end{bmatrix}\begin{bmatrix} x_1 \\ x_2 \\ x_3 \end{bmatrix} + \begin{bmatrix} 0 \\ 0 \\ 1 \end{bmatrix}u$$

$$y = [c_1 \ c_2 \ c_3]\begin{bmatrix} x_1 \\ x_2 \\ x_3 \end{bmatrix}$$

For controllability, $|Q_C| \neq 0$.

$$Q_C = \begin{bmatrix} B & AB & A^2B \end{bmatrix} = \begin{bmatrix} 0 & 0 & 1 \\ 0 & 1 & a_3 \\ 1 & a_3 & a_2 + a_3^2 \end{bmatrix}$$

$|Q_C| \neq 0 \Rightarrow$ The system is controllable.

For observability, $|Q_0| \neq 0$

$$Q_0 = \begin{bmatrix} C^T & A^TC^T & A^{T^2}C^T \end{bmatrix}$$

$$= \begin{bmatrix} c_1 & c_2 & c_3 \\ c_3a_1 & c_1 + c_2a_2 & c_2 + c_2a_3 \\ c_1(c_2 + c_3a_3) & c_3a_1 + (c_2 + c_3a_3)a_2 & c_1 + c_2a_2 + (c_2 + c_3a_3)a_3 \end{bmatrix}$$

$\because |Q_0|$ depends on unknown parameters.

Hence, the system is not always observable.

36. $\dot{x}_1(t) + 2x_1(t) = 3u(t)$

$\dot{x}_2(t) = x_2(t) = u(t)$

$\dot{x}_1(t) = -2x_1(t) + 3u(t)$

$\dot{x}_2(t) = -x_2(t) + u(t)$

$$\begin{bmatrix} \dot{x}_1 \\ \dot{x}_2 \end{bmatrix} = \begin{bmatrix} -2 & 0 \\ 0 & -1 \end{bmatrix}\begin{bmatrix} x_1 \\ x_2 \end{bmatrix} + \begin{bmatrix} 3 \\ 1 \end{bmatrix}u$$

$$\Rightarrow c(t) = x_1(t) = \begin{bmatrix} 1 & 0 \end{bmatrix}\begin{bmatrix} x_1 \\ x_2 \end{bmatrix}$$

$$A = \begin{bmatrix} -2 & 0 \\ 0 & -1 \end{bmatrix}, B = \begin{bmatrix} 3 \\ 1 \end{bmatrix}$$

$$AB = \begin{bmatrix} -2 & 0 \\ 0 & -1 \end{bmatrix}\begin{bmatrix} 3 \\ 1 \end{bmatrix} = \begin{bmatrix} -6 \\ -1 \end{bmatrix}$$

For controllability, $Q_C = [B \quad AB] = \begin{bmatrix} 3 & -6 \\ 1 & -1 \end{bmatrix}$

$|Q_C| = -3 + 6 = 3 \neq 0 \Rightarrow$ Controllable.

For observability, $Q_0 = \begin{bmatrix} C \\ CA \end{bmatrix} = \begin{bmatrix} 1 & 0 \\ -2 & 0 \end{bmatrix}$

$|Q_0| = 0 \Rightarrow$ NOT observable.

7 CHAPTER

Compensators and Controllers

Analysis of Previous GATE Papers

Topics ↓		Year →	2019	2018	2017 Set 1	2017 Set 2	2016 Set 1	2016 Set 2	2015 Set 1	2015 Set 2	2015 Set 3
COMPENSATORS (LAG, LEAD, LAG-LEAD, LEAD-LAG)	1 Mark	MCQ Type			1	1					1
		Numerical Type									
	2 Marks	MCQ Type									
		Numerical Type							1		
	Total				1	1			2		1
CONTROL LERS (P, PI , PD, PID)	1 Mark	MCQ Type									
		Numerical Type									
	2 Marks	MCQ Type									
		Numerical Type									
	Total										

COMPENSATORS (LAG, LEAD, LAG-LEAD, LEAD-LAG)

1. The transfer function of a simple RC network functioning as a controller is

$$G_C(s) = \frac{s+z_1}{s+p_1}$$

Tje condition for the RC netwprl tp act as a phase lead controller is:

(a) $p_1 < z_1$ (b) $p_1 = 0$

(c) $p_1 = z_1$ (d) $p_1 > z_1$

[1990 : 2 Marks]

2. The transfer function of a phase lead controller is $\frac{1+3Ts}{1+Ts}$. The maximum value of phase provided by this controller is

(a) 90° (b) 60°

(c) 45° (d) 30°

[1998 : 1 Mark]

3. A double integrator plant, $G(s) = \frac{K}{s^2}$, $H(s) = 1$ is to be compensated to achieve the damping ratio = 0.5, and an undamped natural frequecny, $\omega_n = 5$ rad/s. Which one of the following compensator $G_e(s)$ will be suitable?

(a) $\frac{s+3}{s+9.9}$ (b) $\frac{s+9.9}{s+3}$

(c) $\frac{s-6}{s+8.33}$ (d) $\frac{s+6}{s}$

[2005 : 2 Marks]

4. The transfer function of a phase-lead compensator is given by

$$G_e(s) = \frac{1+3Ts}{1+Ts} \text{ where } T > 0$$

The maximum phase-shift provided by such a compensator is

(a) $\frac{\pi}{2}$ (b) $\frac{\pi}{3}$

(c) $\frac{\pi}{4}$ (d) $\frac{\pi}{6}$

[2006 : 2 Marks]

5. The open-loop transfer function of a plant is given as $G(s) = \frac{1}{s^2-1}$ If the plant is operated in a unity feedback configuration, then the lead compensator that can stabilize this control system is

(a) $\frac{10(s-1)}{(s+2)}$ (b) $\frac{10(s+4)}{(s+2)}$

(c) $\frac{10(s+2)}{(s+10)}$ (d) $\frac{10(s+4)}{(s+1)}$

[2007 : 2 Marks]

6. Group I gives two possible choices for the impedance Z in the diagram. The circuit elements in Z satisfy the condition $R_2C_2 > R_1 C_1$. The transfer function $\frac{V_0}{V_i}$ represents a kind of controller. Match the impedances in Group I with the types of controllers in Group II.

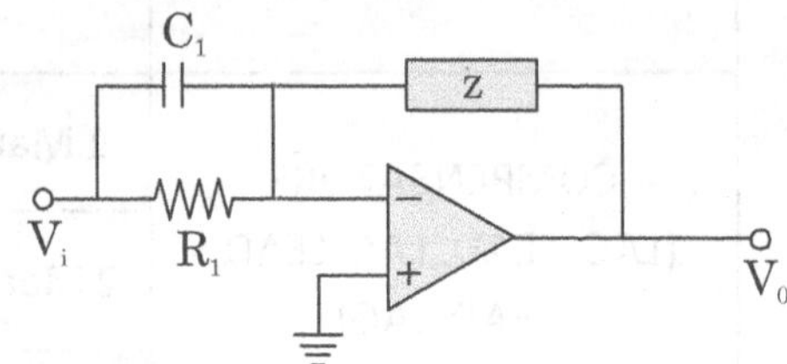

Group-I

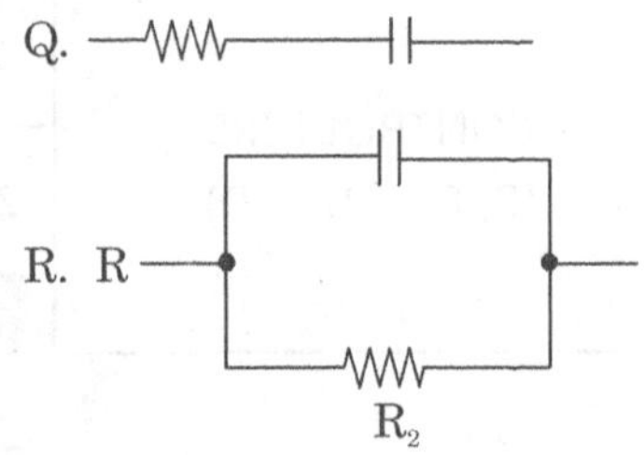

Group-II

1. PID controller
2. Lead compensator
3. Lag compensator

(a) Q - 1, R - 2 (b) Q - 1, R - 3

(c) Q - 2, R - 3 (d) Q - 3, R - 2

[2008 : 2 Marks]

7. The magnitude plot of a rational transfer function G(s) with real coefficients is shown below. Which of the following compensators has such a magnitude plot?

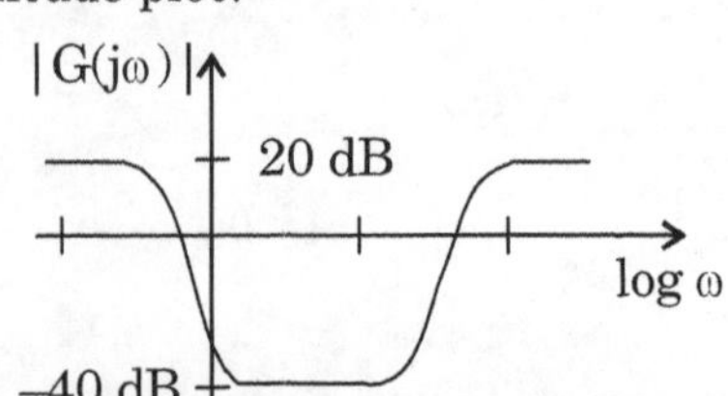

(a) Lead compensator

(b) Lag compensator

(c) PID compensator

(d) Lead-lag compensator

[2009 : 1 Mark]

8. $G_C(s)$ is a lead compensator if

(a) a = 1, b = 2 (b) a = 3, b = 2

(c) a = – 3, b = – 1 (d) a = 3, b = 1

[2012 : 2 Marks]

Linked Answer questions

Statement for Linked Answer questions 9 and 10 : The transfer function of a compensator is given as

$$G_C(s) = \frac{s+a}{s+b}$$

9. The phase of the above lead compensator is maximum at

(a) $\sqrt{2}$ rad/s (b) $\sqrt{3}$ rad/s

(c) $\sqrt{6}$ rad/s (d) $1/\sqrt{3}$ rad/s

[2015 : 2 Mark]

10. A lead compensator network includes a parallel combination of R and C in the feed-forward path. If the transfer function of the compensator is $G_C(s) = \frac{s+2}{s+4}$, the value of RC is ....

[2015 : 2 Marks, Set-1]

11. The transfer function of a first order controller is given as

$$G_C(s) = \frac{K(s+a)}{s+b},$$

where, K, a and bare positive real numbers. The condition for this controller to act as a phase lead compensator is

(a) a < b (b) a > b

(c) K < ab (d) K > ab

[2015 : 1 Mark, Set-3]

12. Which of the following can be the pole-zero configuration of a phase-lag controller (lag compensator)?

(a)

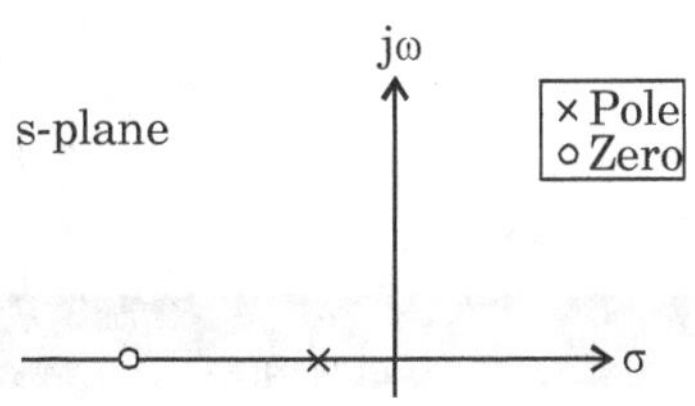

(b)

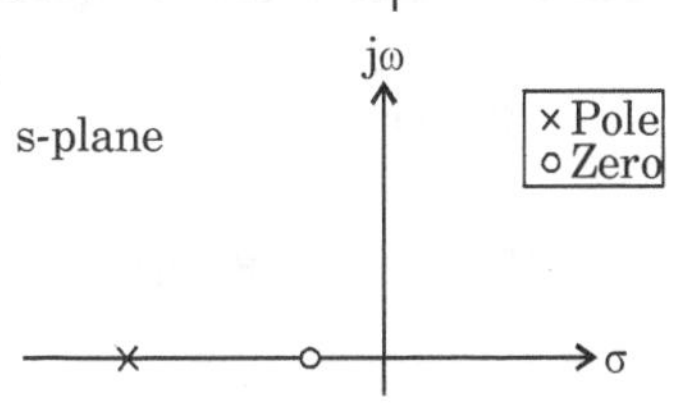

(c)

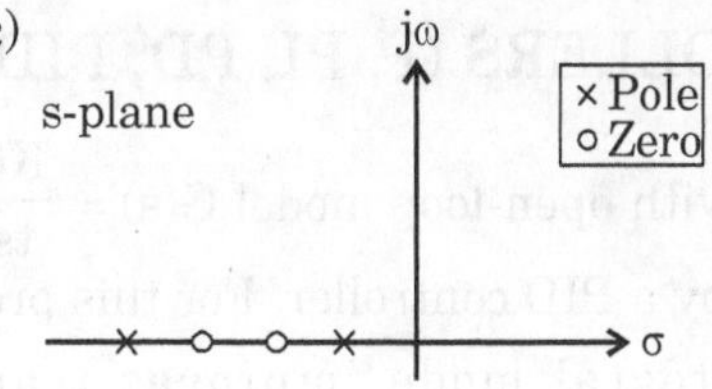

(d)

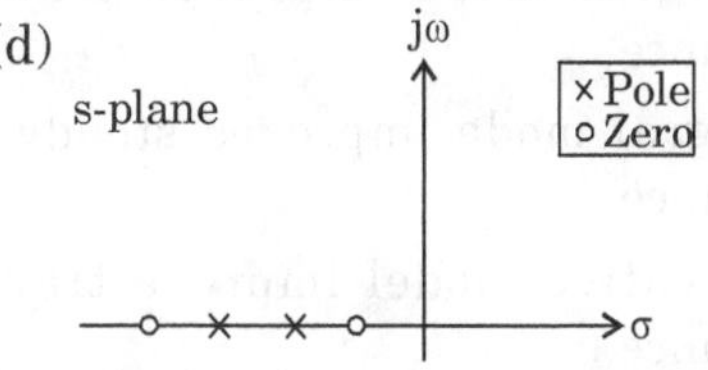

[2017 : 1 Mark, Set-1]

13. Which of the following statements is incorrect?

(a) Lead compensator is used to reduce the settling time.

(b) Lag compensator is used to reduce the steady state error.

(c) Lead compensator may increase the order of a system.

(d) Lag compensator always stabilizes an unstable system.

[2017 : 1 Mark, Set-2]

14. Which one of the following polar diagrams corresponds to a lag network?

(a)

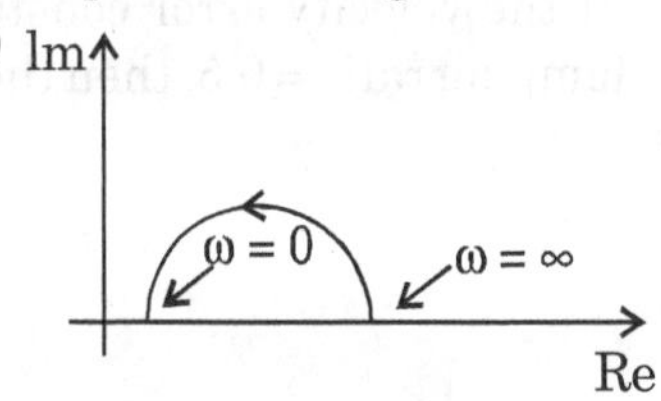

(b)

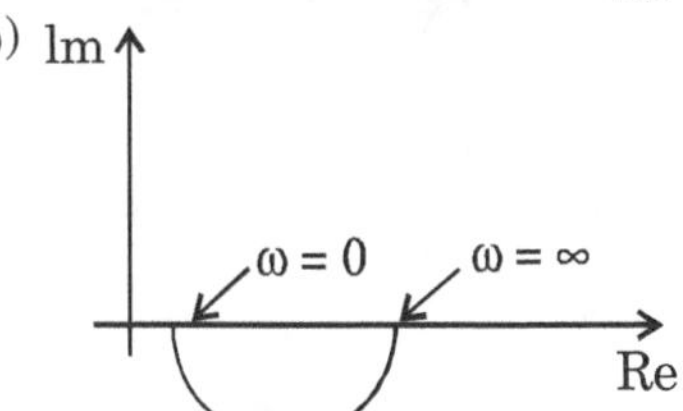

(c)

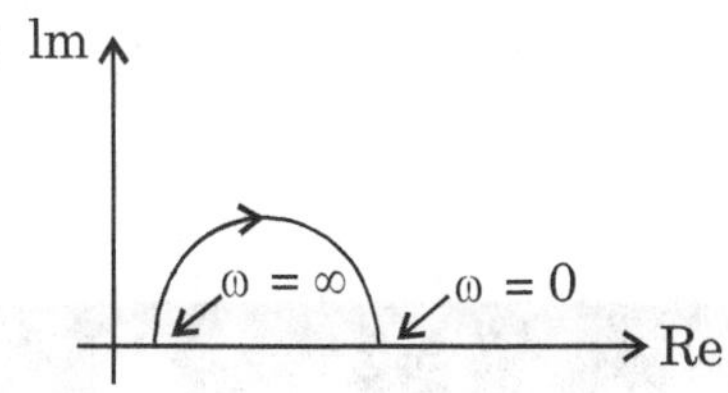

(d)

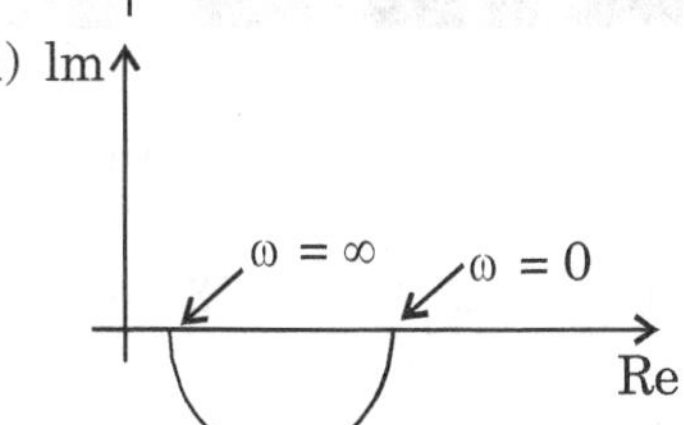

[2018 : 1 Mark]

CONTROLLERS (P, PI, PD, PID)

15. A process with open-loop model $G(s) = \frac{Ke^{-sI_D}}{ts+1}$ is controlled by a PID controller. For this process

(a) the integral mode improves transient performance

(b) the integral mode improves steady-state performance

(c) the derivative model improve transient performance

(d) the derivative mode improves steady-state performance.

[1992 : 2 Marks]

16. A PD controller is used to compensate a system. Compared to the uncompensated system, the compensated system has

(a) a higher type number

(b) reduced damping

(c) higher noise amplification

(d) larger transient overshoot

[2003 : 1 Mark]

17. A control system with a PD controller is shown in the figure. If the velocity error constant K_v= 1000 and the damping ratio = 0.5, then the values of K_p and K_D are

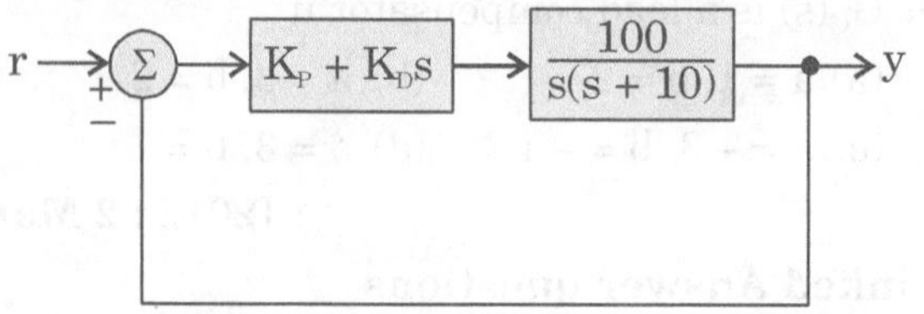

(a) $K_p = 100, K_D = 0.09$

(b) $K_p = 100, K_D = 0.9$

(c) $K_p = 10, K_D = 0.09$

(d) $K_p = 10, K_D = 0.9$

[2007 : 2 Mark]

18. A unity negative feedback closed loop system has a plant with the transfer function $G(s) = \frac{1}{s^2+2s+2}$ and a controller $G_C(s)$ in the feed forward path, For a unit step input, the transfer function of the controller that gives minimum steady state error is

(a) $G_C(s) = \frac{s+1}{s+2}$

(b) $G_C(s) = \frac{s+2}{s+1}$

(c) $G_C(s) = \frac{(s+1)(s+4)}{(s+2)(s+3)}$

(d) $G_C(s) = 1 + \frac{2}{s} + 3s$

[2010 : 2 Marks]

ANSWERS

1. (d)	2. (d)	3. (a)	4. (d)	5. (c)	6. (b)	7. (d)	8. (a)	9. (a)
10. (0.5)	11. (a)	12. (a)	13. (d)	14. (d)	15. (b,c)	16. (c)	17. (b)	18. (d)

EXPLANATIONS

1. $G_C(s) = \dfrac{s+z_1}{s+p_1}$

$\angle G_C(j\omega) = \tan^{-1}\left(\dfrac{\omega}{z_1}\right) - \tan^{-1}\left(\dfrac{\omega}{P_1}\right)$

For phase - lead controller $\angle G_C(j\omega) > 0$

$\tan^{-1}\left(\dfrac{\omega}{z_1}\right) - \tan^{-1}\left(\dfrac{\omega}{P_1}\right) > 0$

$\dfrac{\omega}{z_1} > \dfrac{\omega}{p_1}$ $\boxed{p_1 > z_1}$

2. Given that transfer function as of phase lead controller.

$$F = \frac{1+3T_S}{1+T_S}$$

$$\phi = \tan^{-1} 3T\omega - \tan^{-1} T\omega$$

$$= \tan^{-1} \frac{2T\omega}{1+3T^2\omega^2}$$

$$\frac{d\phi}{d\omega} = 2T(1+3T^2\omega^2) - (2T\omega)(3T^2)(2\omega)$$

$$= 1 + 3T^2\omega^2 - 6T^2\omega^2$$

$$= 0$$

or $1 = 3T^2\omega^2$

or $T\omega = \dfrac{1}{\sqrt{3}}$

$$\phi_{max} = \tan^{-1}\frac{1}{\sqrt{3}} = 30°$$

3. Damping ratio, $\xi = 0.5$

$\theta = \cos^{-1}(\xi) = 60°$

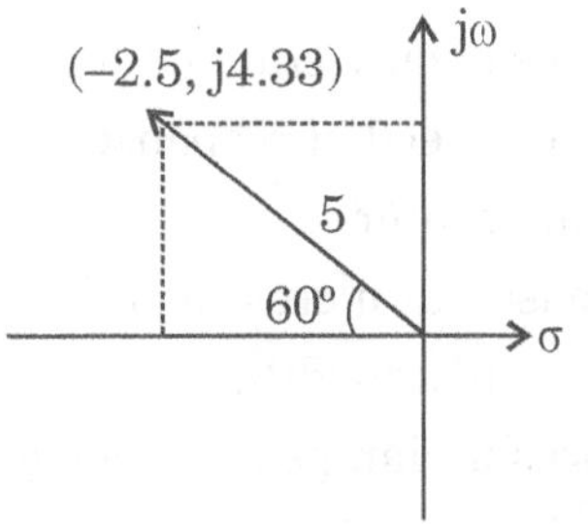

$\angle G(s) = \dfrac{K}{S^2}\Big|_{s=-2.5+j4.33} = -2\tan^{-1}\left(\dfrac{4.33}{-2.55}\right) = 120°$

∴ For compensated system, angle

$$= 180° - 120° = 60°.$$

(b) & (d) are lag network & for compensating lag $\dfrac{k}{s^2}$, a lead network is required.

∴ Putting $s = -2.5 + j4.33$ in option (a) gives

$$\frac{k(s+3)}{s^2(s+9.9)} = \frac{0.5+j4.33}{7.4+j4.33} = 53° \cong 60°$$

Hence, option (a) is the correct answer.

4. $\phi = \angle G_c(s)\big|_{s=j\omega} = -\tan^{-1} T\omega + \tan^{-1} 3T\omega$

For maximum phase-shift;

$$\frac{d\phi}{d\omega} = 0 = \frac{-T}{1+(T\omega)^2} + \frac{3T}{1+(3T\omega)^2}$$

$$\Rightarrow \frac{1}{1+(T\omega)^2} = \frac{3}{1+(3T\omega)^2}$$

$$\Rightarrow \omega T = \frac{1}{\sqrt{3}}$$

$$\therefore \phi_{max} = \tan^{-1}\sqrt{3} - \tan^{-1}\frac{1}{\sqrt{3}} = \frac{\pi}{6}$$

5. For a lead compensator, $\angle G_{lead}(j\omega) > 0°$.

Controllers gives in options (a), (b) & (d) are not lead compensator. Option (c) will stabilize this control system.

$$\text{Transfer function} = \frac{G(s)\cdot G_c(s)}{1+G(s)G_c(s)\cdot 1}$$

$$= \frac{\dfrac{1}{(s^2-1)}\cdot\dfrac{10(s+2)}{(s+10)}}{1+\dfrac{1}{s^2-1}\times\dfrac{10(s+2)}{s+10}\times 1}$$

$$= \frac{10(s+2)}{s^3+10s^2-s-10+10s+20}$$

$$= \frac{10(s+2)}{s^3+10s^2+9s+10}. \text{ [Stable System]}$$

6. Redrawing circuit(s) in s-domain,

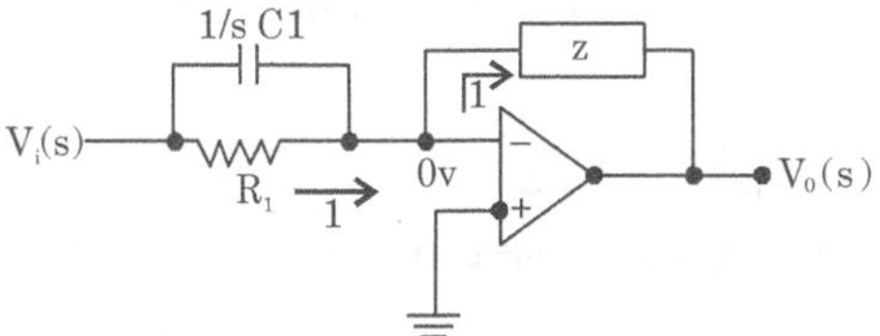

Applying nodal analysis,

$$\frac{\dfrac{V_i(s)-0}{R_1}}{SC_1R_1+1} = \frac{0-V_0(s)}{Z}$$

$$\frac{V_0(s)}{V_i(s)} = -\frac{Z\cdot(R_1C_1S+1)}{R_1}$$

For Q : $Z = R_2 + \dfrac{1}{SC_2} = \dfrac{R_2C_2s+1}{SC_2}$

For R : $Z = R_2 || \frac{1}{SC_2} = \frac{R_2}{R_2C_2s+1}$

Considering Q for z

$$\frac{V_o}{V_i} = -\frac{(R_2C_2S+1)(R_1C_1s+1)}{SC_2 \cdot R_1}$$

Considering R for z

$$\frac{V_o}{V_i} = -\frac{R_2(R_1C_1S+1)}{(R_2C_2S+1)\cdot R_1}$$

$\Rightarrow$ Given : $R_2 C_2 >> R_1 C_1$

$\rightarrow$ Considering R, controller is lag-compensator.

$\rightarrow$ Considering Q, controller is PID controller.

7. G (*s*) for lead compensator = $\frac{(s+z_1)(s+z_2)}{(s+p_1)(s+p_2)}$

and the Bode -plot for $G(j\omega)$ is shown below.

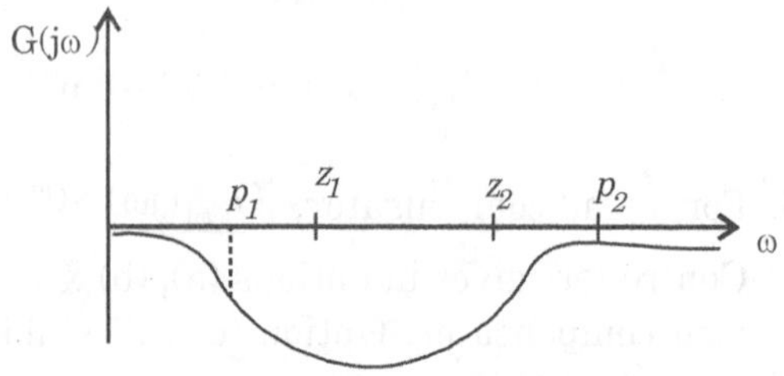

8. For lead compensator, $\angle G_C(j\omega) > 0°$.

$$\tan^{-1}\left(\frac{\omega}{a}\right) - \tan^{-1}\left(\frac{\omega}{b}\right) < 0$$

$$\frac{\omega}{a} > \frac{\omega}{b} \Rightarrow a < b.$$

Considering the options with above inequality, option (a) is correct.

9. For phase to be maximum,

$$\frac{\partial}{\partial\omega}\left[\angle G_C(j\omega)\right] = 0$$

$$\frac{\partial}{\partial\omega}\left[\tan^{-1}\left(\frac{\omega}{1}\right) - \tan^{-1}\left(\frac{\omega}{2}\right)\right] = 0$$

$$\frac{1}{1+\omega^2} - \frac{2}{\omega^2+4} = 0$$

$$\omega^2 + 4 - 2 - 2\omega^2 = 0$$

$$\therefore \omega^2 = 2$$

$$\therefore \omega = \sqrt{2}\ \frac{\text{rad}}{\text{s}}.$$

10. Transfer function = $\frac{1+ST}{1+\alpha\ ST}$

$$\tau = R_1C,\ \alpha = \frac{R_2}{R_1+R_2}$$

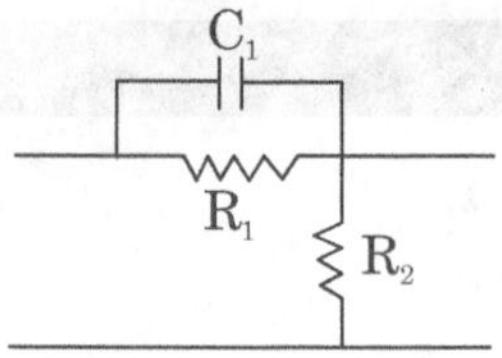

[Lead commentator]

Given : $G_C(s) = \frac{S+2}{S+4}$

Now comparing, $\tau = \frac{1}{2}$, $\alpha\ T = \frac{1}{4}$

$$\therefore \qquad \alpha = \frac{1}{2}$$

Time-constant τ = RC = 0.5S.

11. For phase lead compensator,

$$G_C(s) = \frac{1+sT}{1+\alpha\ sT};\ \alpha < 1.$$

Given : $G_C(s) = \frac{k(s+a)}{(s+b)}$

Now comparing, $\tau = \frac{1}{a}$ and $\alpha\ \tau = \frac{1}{b}$

$$\alpha = \frac{a}{b} < 1 \Rightarrow \boxed{a<b}$$

12. For phase lag controller, $G_C(s) = \frac{1+s\tau}{1+\beta\cdot\tau}$; $\beta>1$.

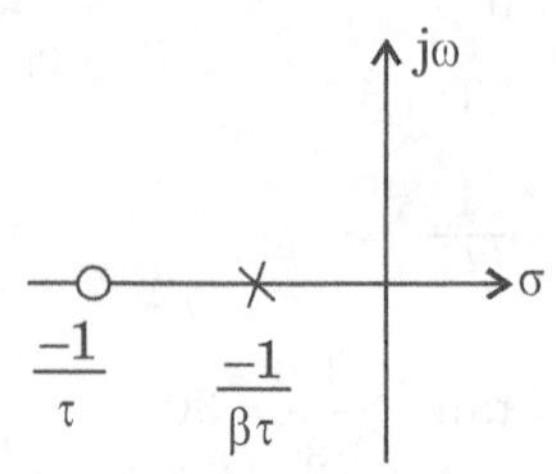

13. Lag compensator :

$\rightarrow$ Increases gain of original network without stability.

$\rightarrow$ reduces steady-state error.

$\rightarrow$ reduces speed of response.

Lead compensator :

$\rightarrow$ Increases gain crossover frequency.

$\rightarrow$ reduces bandwidth.

$\rightarrow$ reduces undamped frequency.

15. Controller Integral mode improves steady state response. Derivative mode improves transient response i.e. speeds up response.

16. Effect of PD controller,

(i) does not effect the type of the system,

(ii) reduces overshoot and increases damping,

(iii) increases the bandwidth, therefore, SNR decreases,

(iv) increases the noise.

17. $G = G_1G_2 = (K_P + K_Ds)\left(\frac{100}{s(s+10)}\right)$

$$K_v = \lim_{s\to 0} s\,G(s) = \lim_{s\to 0} \frac{s[K_P + K_Ds]100}{s(s+10)}$$

$$\therefore \quad 1000 = 10\, K_P$$

$$\Rightarrow \quad K_P = 100$$

$$T(s) = \frac{G(s)}{1+G(s)} = \frac{(K_P + K_Ds)\left(\frac{100}{s(s+10)}\right)}{1+(K_P + K_Ds)\left(\frac{100}{s(s+10)}\right)}$$

$$= \frac{100(K_P + K_Ds)}{s^2 + 10s + 100\,K_Ds + 100K_P}$$

$$= \frac{100(K_P + K_Ds)}{s^2 + s(10 + 100K_D) + 100K_P}$$

Compare characteristic equation with

$$s^2 + 2\xi\,\omega_n s + \omega_n^2$$

$$100\,K_D + 10 = 2\xi\omega_n \quad \text{and} \quad 100\,K_P = \omega_n^2$$

$$\Rightarrow \quad \omega_n = \sqrt{100 \times 100} = 100$$

$$\Rightarrow \quad 100\,K_D + 10 = 2 \times 0.5 \times 100$$

$$\Rightarrow \quad K_D = 0.90$$

18.

$$E(s) = R(s) - C(s)$$

$$E(s) = R(s) - E(s) \cdot C(s)\,G_c(s)$$

$$E(s) = \frac{R(s)}{1+G(s)G_c(s)}$$

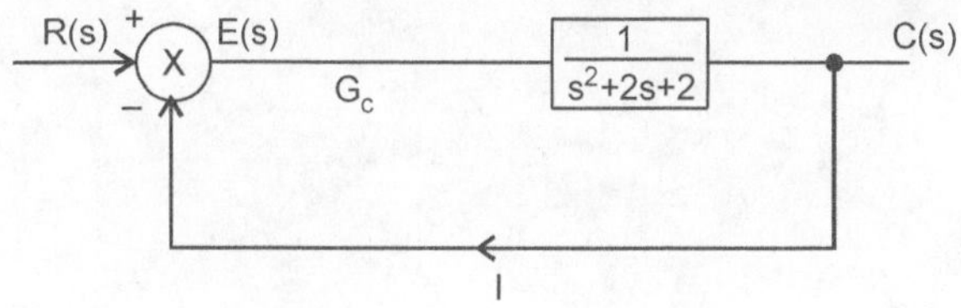

Here, $R(s) = \frac{1}{s}$

$$\lim_{s\to 0} G(s) = \frac{1}{0+0+2} = \frac{1}{2} = 0.5$$

$\therefore$ Steady state error, $e_{ss} = \lim_{s\to 0} sE(s)$

$$= \lim_{s\to 0} \frac{1}{1+G(s)G_c(s)}$$

For unit step input,

$$k(s) = \frac{1}{s}$$

$$\therefore \quad e_{ss} = \lim_{s\to 0} \frac{1}{1+G(s)G_c(s)}$$

$$= \frac{1}{1+\lim_{s\to 0} G(s) \lim_{s\to 0} G_c(s)}$$

$$= \frac{1}{1+0.5\lim_{s\to 0} G_c(s)}$$

$$e_{ss} = \text{minimum}$$

if $\quad \underset{s\to 0}{G_c(s)} = \text{maximum}$

Hence for option (*d*) : $\underset{s\to 0}{G_c(s)} = \infty$

$$\therefore \quad e_{ss} = 0$$

Unit - IV

Electronic Device and Circuit

Syllabus

Energy bands in intrinsic and extrinsic silicon; Carrier transport: diffusion current, drift current, mobility and resistivity; Generation and recombination of carriers; Poisson and continuity equations; P-N junction, Zener diode, BJT, MOS capacitor, MOSFET, LED, photo diode and solar cell; Integrated circuit fabrication process: oxidation, diffusion, ion implantation, photolithography and twin-tub CMOS process.

Contents

1 CHAPTER Basics of Semiconductor Physics

Analysis of Previous GATE Papers

Topics ↓	Year →		2019	2018	2017 Set 1	2017 Set 2	2016 Set 1	2016 Set 2	2016 Set 3	2015 Set 1	2015 Set 2	2015 Set 3
ENERGY BONDS	1 Mark	MCQ Type										
		Numerical Type		1								
	2 Marks	MCQ Type									1	
		Numerical Type										
		Total		**1**							**2**	
INTSINSIC & EXTXIURTC SEMICONDUCTOR MATERIALS	1 Mark	MCQ Type			1							
		Numerical Type										
	2 Marks	MCQ Type										
		Numerical Type										
		Total			**1**							
DRIFF & DIFFUSION CURRENTS	1 Mark	MCQ Type										
		Numerical Type			1							
	2 Marks	MCQ Type										
		Numerical Type						1				
		Total			**1**			**2**				
MOBILITY & RESISTIVITY	1 Mark	MCQ Type										
		Numerical Type								1	1	
	2 Marks	MCQ Type										
		Numerical Type									1	
		Total								**1**	**3**	
GENERATION & RECOMBINATION OF CARRIERS	1 Mark	MCQ Type										
		Numerical Type									1	
	2 Marks	MCQ Type					1					
		Numerical Type										
		Total					**2**				**1**	

ENERGY BONDS

1. The drift velocity of electrons, in silicon
   (a) is proportional to the electric field for all values of electric field
   (b) is independent of the electric field.
   (c) increases at low values of electric field and decreases at high values of electric field exhibiting negative differential resistance.
   (d) increases linearly with electric field at low values of electric field and gradually saturates at higher values of electric field.
   [1995 : 1 Mark]

2. The probability that an electron in a metal occupies the fermi level, at any temperature. (T > 0K)
   (a) 0 (b) 1
   (c) 0.5 (d) 1.0
   [1995 : 1 Mark]

3. The units of $\frac{q}{KT}$ are
   (a) V (b) V^{-1}
   (c) J (d) J/K
   [1997 : 1 Mark]

4. The unit of $\frac{q}{KT}$ are
   (a) V (b) V^{-1}
   (c) J (d) J/K
   [1998 : 1 Mark]

5. The bandgap of silicon at 300 K is
   (a) 1.36 eV (b) 1.10 eV
   (c) 0.80 eV (d) 0.67 eV
   [2003 : 1 Mark]

6. The longest wavelength that can be absorbed by silicon, which has the bandgap of 1.12 eV, is 1.1 μm. If the longest wavelength that can be absorbed by another material is 0.87 μm, then the bandgap of this material is
   (a) 1.416 eV (b) 0.886 eV
   (c) 0.854 eV (d) 0.706 eV
   [2004 : 2 Marks]

7. The bandgap of Silicon at room temperature is
   (a) 1.3 eV (b) 0.7 eV
   (c) 1.1 eV (d) 1.4 eV
   [2005 : 1 Mark]

8. Silicon is doped with boron to a concentration of 4×10^{17} atoms/cm^3. Assume the intrinsic carrier concentration of silicon to be 1.5×10^{10}/cm^3 and the value of kT/q to be 25 mV at 300 K. Compared to undoped silicon, the Fermi level of doped silicon
   (a) goes down by 0.13 eV
   (b) goes up by 0.13 eV
   (c) goes down by 0.427 eV
   (d) goes up by 0.427 eV
   [2008 : 2 Marks]

9. At T = 300 K, the band gap and the intrinsic carrier concentration of GaAs are 1.42 eV and 10^6 cm^{-3}, respectively. In order to generate electron hole pairs in GaAs, which one of the wavelength (λ_c) ranges of incident radiation, s most suitable? (Given that: Plank's constant is 6.62×10^{-34} J-s, velocity of light is 3×10^{10}cm/s and charge of electron is 1.6×10^{-19}C)
   (a) 0.42 μm < λ_c < 0.87 μm
   (b) 0.87 μm < λ_c < 1.42 μm
   (c) 1.42 μm < λ_c < 1.62 μm
   (d) 1.62 μm < λ_c < 6.62 μm
   [2014 : 1 Mark, Set-4]

10. In the figure, In (p_i) is plotted as a function of 1/T, where p_i is the intrinsic resistivity of silicon, T is the temperature, and the plot is almost linear.

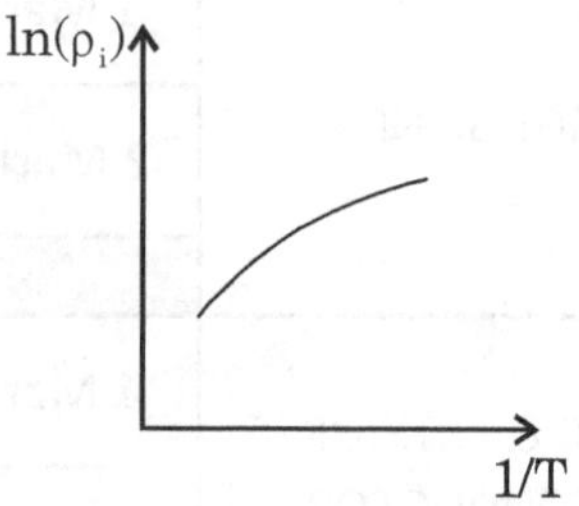

   The slope of the line can be used to estimate
   (a) band gap energy of silicon (E_g).
   (b) sum of electron and hole mobility in silicon ($\mu_n + \mu_p$).
   (c) reciprocal of the sum of electron and hole mobility in silicon $(\mu_n + \mu_p)^{-1}$.
   (d) intrinsic carrier concentration of silicon (n_i).
   [2014: 1 Mark, Set-4]

11. The cut-off wavelength (in μm) of light that can be used for intrinsic excitation of a semiconductor material of bandgap E_g = 1.1 eV is ______.
   [2014 : 1 Mark, Set-4]

12. An N-type semiconductor having uniform doping is biased as shown in the figure.

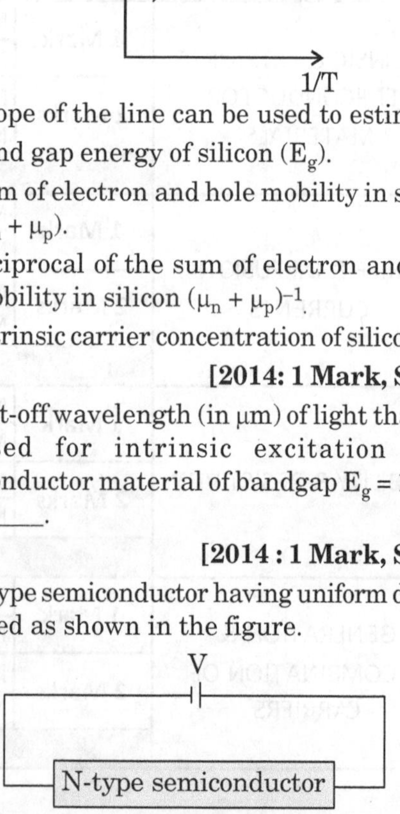

   If E_c is the lowest energy level of the conduction band, E_v is the highest energy level of the valance band and E_f is the Fermi level, which one of the following represents the energy band) diagram for the biased N-type semiconductor?

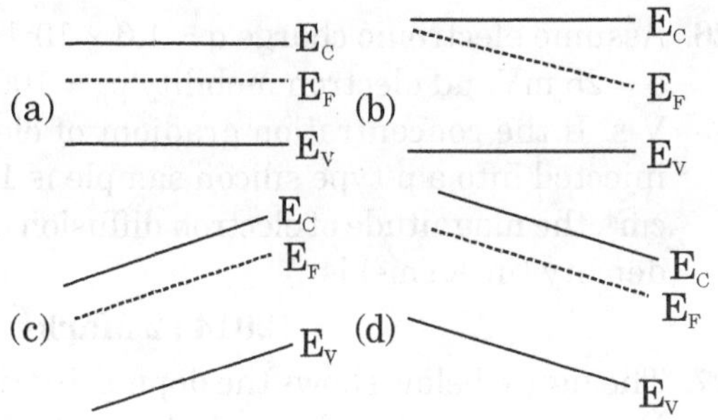

[2014 : 2 Marks, Set-4]

13. The energy band diagram and the' electron,n density profile n(x) in a semiconductor are shown in the figures.

Assume that $n(x) = 10^{15} e^{\left(\frac{q\alpha x}{kT}\right)}$ cm^3, with $\alpha = 0.1$ V/cm and x expressers in cm. Given $\frac{kT}{q} = 0.026$ V, $D_n = 36$ cm^2-s^{-1}, and $\frac{D}{\mu} = \frac{kT}{q}$. The electron-current density (in A/cm^2) at x = 0 is

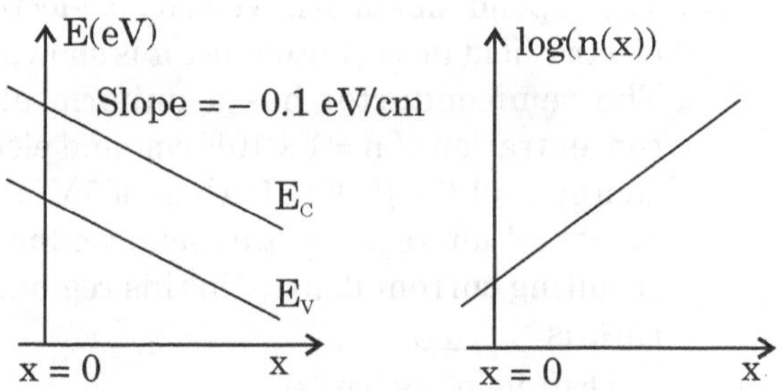

(a) -4.4×10^{-2}
(b) 0
(c) -2.2×10^{-2}
(d) 2.2×10^{-2}

[2015 : 2 Marks, Set-2]

14. A small percentage of impurity is added to an intrinsic semiconductor at 300 K. Which one of the following statements is true for the energy band diagram shown in the following figure?

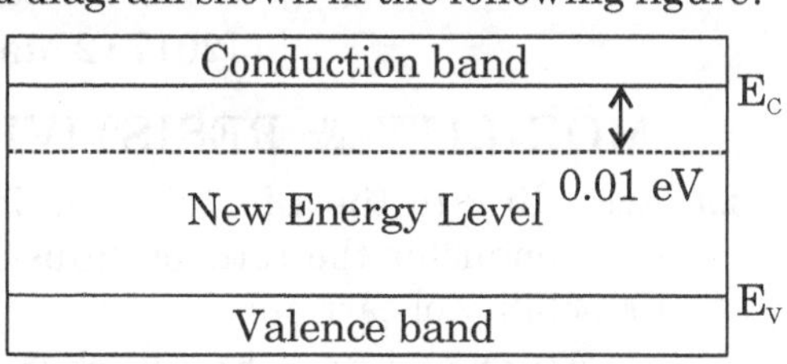

(a) Intrinsic semiconductor doped with pentavalent atoms to form n-type semiconductor
(b) Intrinsic semiconductor doped with trivalent atoms to form n-type semiconductor
(c) Intrinsic semiconductor doped with pentavalent atoms to form p-type semiconductor
(d) Intrinsic semiconductor doped with trivalent atoms to form p-type semiconductor

[2016 : 1 Mark, Set-1]

15. There are two photolithography systems: one with light source of wavelength λ_1, = 156 nm (System 1) and another with light source of wavelength λ_2 = 325 nm (System 2). Both photolithography systems are otherwise identical. If the minimum feature sizes that can be realized using System 1 and System2 are L_{min1} and L_{min2} respectively, the ratio L_{min1}/L_{min2} (correct to two decimal places) is ______.

[2018 : 1 Mark]

INTSINSIC & EXTXIURTC SEMICONDUCTOR MATERIALS

16. In an intrinsic semiconductor the free electron concentration depends on
(a) effective mass of electrons only.
(b) effective mass of holes only.
(c) temperature of the semiconductor.
(d) width of the forbidden energy band of the semiconductor.

[1987 : 2 Marks]

17. n-type silicon is obtained by doping silicon with
(a) Germanium
(b) Aluminium
(c) Boron
(d) Phosphorus

[2003 : 1 Mark]

18. The primary reason for the widespread use of Silicon in semiconductor device technology is
(a) abundance of Silicon on the surface of the Earth.
(b) larger bandgap of Silicon in comparison to Germanium
(c) favorable properties of Silicon-dioxide (SiO_2)
(d) lower melting point

[2005 : 1 Mark]

19. Which of the following is true?
(a) A silicon wafer heavily doped with boron is a p^+ substrate.
(b) A silicon wafer lightly doped with boron is a p^+ substrate.
(c) A silicon wafer heavily doped with arsenic is a p^+ substrate.
(d) A silicon wafer lightly doped with arsenic is a p^+ substrate.

[2008 : 1 Mark]

20. A silicon wafer has 100 nm of oxide on it and is inserted in a furnace at a temperature above 1000°C for further oxidation in dry oxygen. The oxidation rate
(a) is independent of current oxide thickness and temperature.
(b) is independent of current oxide thickness but depends on temperature.
(c) slows down as the oxide grows.
(d) is zero as the existing oxide prevents further oxidation.

[2008 : 1 Mark]

21. A bar of Gallium Arsenide (GaAs) is doped with Silicon such that the Silicon atoms occupy Gallium and Arsenic sites in the GaAs crystal. Which one of the following statements is true?
   (a) Silicon atoms act as p-type dopants in Arsenic sites and n-type dopants in Gallium sites.
   (b) Silicon atoms act as n-type dopants in Arsenic sites and p-type dopants in Gallium sites.
   (c) Silicon atoms act as p-type dopants in Arsenic as well as Gallium sites.
   (d) Silicon atoms act as n-type dopants in Arsenic as well as Gallium sites.

[2017 : 1 Mark, Set-1]

DRIFF & DIFFUSION CURRENTS

22. The electron concentration in a sample of non-uniformly doped n-type silicon at 300 K varies linearly from $10^{17}/cm^3$ at x = 0 to 6 $10^{16}/cm^3$ at x = 2 μm. Assume a situation that electrons are supplied to keep this concentration gradient constant with time. If electronic charge is 1.6×10^{-19} Coulomb and the diffusion constant $D_n = 35\ cm^2/s$, the current density in the silicon, if no electric field is present is
   (a) zero (b) 120 A/cm^2
   (c) +1120 A/cm^2 (d) -1120 A/cm^2

[1997 : 2 Marks]

23. Under low level injection assumption, the injected minority carrier current for an extrinsic semiconductor is essentially the
   (a) diffusion current
   (b) drift current
   (c) recombination current
   (d) induced current

[2006 : 1 Mark]

24. The majority carriers in an n-type semiconductor have an average drift velocity v in a direction perpendicular to a uniform magnetic field B. The electric field E induced due to Hall effect acts in the direction
   (a) v × B (b) B × v
   (c) along v (d) opposite to v

[2006 : 2 Marks]

25. Drift current in semiconductors depends upon
   (a) only the electric field.
   (b) only the carrier concentration gradient.
   (c) both the electric field and the carrier concentration.
   (d) both the electric field and the carrier concentration gradient.

[2011 : 1 Mark]

26. Assume electronic charge $q = 1.6 \times 10^{-19}$C, kT/q = 25 mVand electron mobility $\mu_n = 1000\ cm^2/V\text{-}s$. If the concentration gradient of electrons injected into a p-type silicon sample is $1 \times 10^{21}\ cm^4$, the magnitude of electron diffusion current density (in A/cm^2) is ______ .

[2014 : 2 Marks, Set-2]

27. The figure below shows the doping distribution in a p-type semiconductor in log scale.

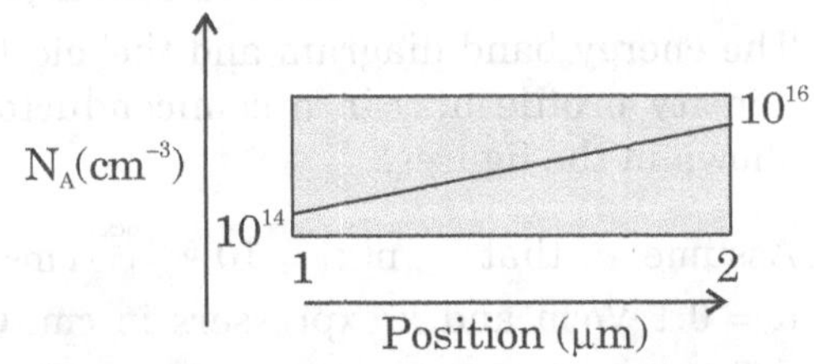

The magnitude of the electric field (in kV/cm) in the semiconductor due to non uniform doping is ___.

[2016 : 2 Marks, Set-1]

28. The dependence of drift velocity of electrons on electric field in a semiconductor is shown below. The semiconductor has a uniform electron concentration of $n = 1 \times 10^{16}\ cm^3$ and electronic charge $q = 1.6 \times 10^{-19}$ C. If a bias of 5 V is applied across a 1 μm region of this semiconductor, the resulting current density in this region, in kA/cm^2, is ______.

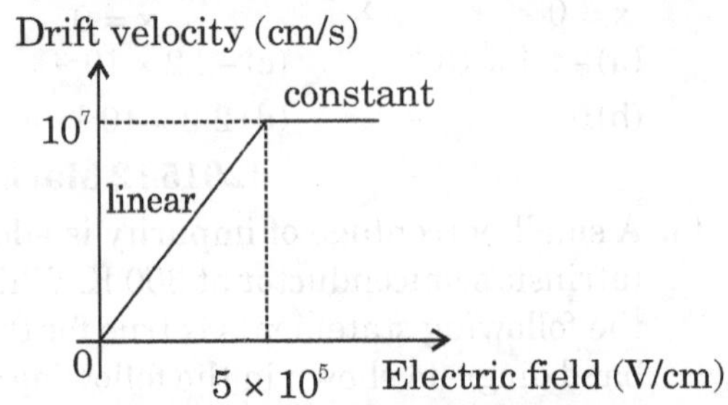

[2017 : 2 Marks, Set-1]

MOBILITY & RESISTIVITY

29. According to the Einstein relation, for any semiconductor the ratio of diffusion constant to mobility of carriers
   (a) depends upon the temperature of the semiconductor
   (b) depends upon the type of the semi conductor
   (c) varies with life time of the semi conductor
   (d) is a universal constant **[1987 : 2 Marks]**

30. Under high electric fields, in a semiconductor with increasing electric field,
   (a) the mobility of charge carriers decreases
   (b) the mobility of the carries increases
   (c) the velocity of the charge carriers saturates
   (d) the velocity of the charge carriers increases

[1990 : 2 Marks]

31. A small concentration of minority carries is injected into a homogeneous semiconductor crystal at one point. An electric field of 10 V/cm is applied across the crystal and this moves the minority carries a distance of 1 cm in 20 μsee. The mobility (in cm²/V-sec) will be

(a) 1,000 (b) 2,000

(c) 5,000 (d) 500,000

[1995 : 1 Mark]

32. An n-type silicon bar 0.1 cm long and 100 μm² in cross-sectional area has a majority carrier concentration of $5 \times 10^{20}/m^3$ and the carrier mobility is 0.13 m²/V-s at 300 K. If the charge of an electron is 1.6×10^{-19} Coulomb, then the resistance of the bar is

(a) $10^6\,\Omega$ (b) $10^4\,\Omega$

(c) $10^{-1}\,\Omega$ (d) $10^4\,\Omega$

[1997 : 2 Marks]

33. The resistivity of a uniformly doped n-type silicon sample is 0.5 Ω–cm. If the electron mobility (μ_n) is 1250 cm²/V-sec and the charge of an electron is 1.6×10^{-19} Coulomb, the donor impurity concentration (N_D) in the sample is

(a) $2 \times 10^{16}/cm^3$ (b) $1 \times 10^{16}/cm^3$

(c) $2.5 \times 10^{15}/cm^3$ (d) $2 \times 10^{15}/cm^3$

[2004 : 2 Marks]

34. A Silicon sampled is doped with 10^{18} atoms/cm³ of Boron. Another sample B of identical dimensions is doped with 10^{18} atoms/cm³ of Phosphorus. The ratio of electron to hole mobility is 3. The ratio of conductivity of the sample A to B is

(a) 3 (b) 1/3

(b) 2/3 (d) 3/2

[2005 : 2 Marks]

35. A heavily doped n-typed semiconductor has the following data:

Hole-electron mobility ratio: 0.4

Doping concentration : 4.2×10^8 atoms/m³

Intrinsic concentration : 1.5×10^4 atoms/m³

The ratio of conductance of the n-type semiconductor to that of the intrinsic semiconductor of same material and at the same temperature is given by

(a) 0.00005

(b) 2,000

(c) 10,000

(d) 20,000

[2006 : 2 Marks]

36. The ratio of the mobility to the diffusion coefficient in a semiconductor has the unit

(a) V^{-1} (b) $cm - V^{-1}$

(c) $V - cm^{-1}$ (d) $V - s$

[2009 : 1 Mark]

Linked Answer Questions 37 and 38: The silicon sample with unit cross-sectional areas shown below is in thermal equilibrium. The following information is given: T= 300 K, electronic charge s 1.6×10^{-19} C, thermal voltage = 26 mV and electron mobility = 1350 cm²/V-s.

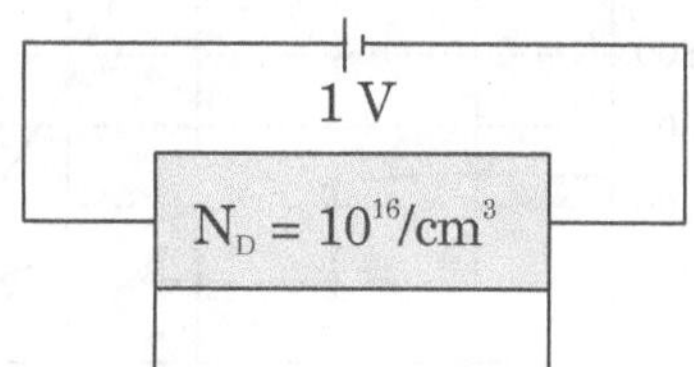

37. The magnitude of the electric field at x = 0.5 nm is

(a) 1 kV/cm

(b) 5 kV/cm

(c) 10 kV/cm

(d) 26 kV/cm

[2010 : 2 Marks]

38. The magnitude of the electron drift current density at x = 0.5 μm is

(a) 2.16×10^4 A/cm²

(b) 1.08 × 104 A/cm²

(c) 4.32×10^3 A/cm²

(d) 6.48×10^2 A/cm²

[2010 : 2 Marks]

39. At T = 300 K, the hole mobility of a semiconductor μp = 500 cm²/V-s and $\frac{kT}{q}$ = 26 mV. The hole diffusion constant D_p in cm²/s is_____.

[2014 : 1 Mark, Set-3]

40. A silicon sample is uniformly doped with donor type impurities with a concentration of $10^{16}/cm^3$. The electron and hole mobilities in the sample are 1200 cm² N-s and 400 cm² N-s respectively. Assume complete ionization of impurities. The charge of an electron is 1.6×10^{-19} C. The resistivity of the sample (in Ω–cm) is _______.

[2015 : 1 Mark, Set-1]

41. A piece of silicon is doped uniformly with phosphorous with a doping concentration of $10^{16}/cm^3$. The expected value of mobility versus doping versus doping concentration for silicon assuming full dopant ionization is shown below. The charge of an electron is 1.6×10^{-19} C. The conductivity (in $S\text{-}cm^{-1}$) of the silicon sample at 300 K is____.

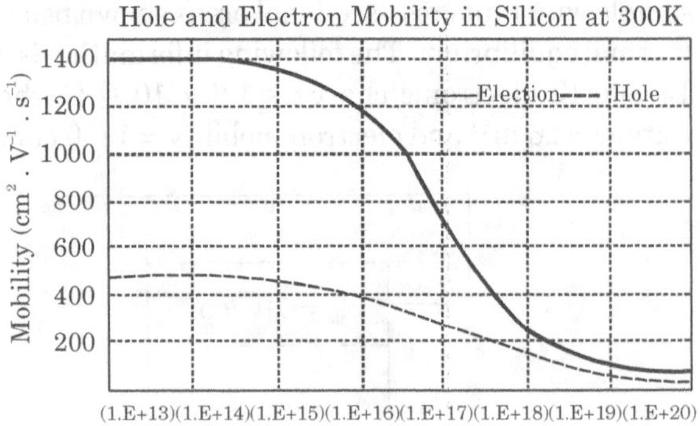

[2015 : 1 Mark, Set-2]

42. A dc voltage of 10 V is applied across an n-type silicon bar having a rectangular cross-section, and length of 1 cm as shown in figure. The donor doping concentration N_D and the mobility of electrons μ_n are 10^{16} cm^{-3} and 1000 $cm^2\text{-}V^{-1}-s^{-1}$, respectively. The average time (in μs) taken by the electrons to move from one end of the bar to other end is ____.

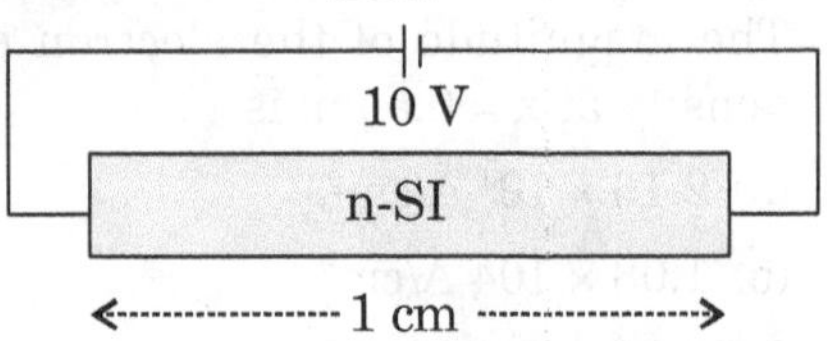

[2015 : 2 Marks, Set-2]

GENERATION & RECOMBINATION OF CARRIERS

43. Direct band gap semi conductors
(a) exhibit short carrier life time and they are used for fabricating BJT's.
(b) exhibit long carrier life time and they are used for fabricating BJT's.
(c) exhibit short carrier life time and they are used for fabricating lasers.
(d) exhibit long carrier life time and they are used for fabricating lasers.

[1987 : 2 Marks]

44. Due to illumination by light, the electron and hole concentrations in a heavily doped N type semi conductor increases by Δn and Δp respectively if n. is the intrinsic concentration then,
(a) $\Delta n < \Delta p$ (c) $\Delta n = \Delta p$
(b) $\Delta n > \Delta p$ (d) $\Delta n \times \Delta p = n^2_i$

[1989 : 2 Marks]

45. The concentration of ionized acceptors and donors in a semi conductor are N_A, N_D respectively. If $N_A > N_D$ and n_i is the intrinsic concentration, the position of the fermi level with respect to the intrinsic level depends on
(a) $N_A - N_D$ (b) $N_A + N_D$
(c) $\frac{N_A N_D}{n_i^2}$ (d) n_i

[1989 : 2 Marks]

46. A silicon sample is uniformly doped with 10^{16} phosphorus atoms/cm³ and 2×10^{16} boron atoms/cm³. If all the dopants are fully ionized, the material is
(a) n-type with carrier concentration of $10^{16}/cm^3$
(b) p-type with carrier concentration of $10^{16}/cm^3$
(c) p-type with carrier of $2 \times 10^{16}/cm^3$
(d) n-type with a carrier concentration of $2 \times 10^{16}/cm^3$

[1991 : 2 Marks]

47. A semi conductor is irradiated with light such that carriers are uniformly generated throughout its volume. The semiconductor is n-type with $N_D = 10^{19}/cm^3$. If the excess electron concentration in the steady state is $\Delta n = 10^{15}/cm^3$ and if t_p = 10 μsec, (minority carries life time) the generation rate due to irradiation
(a) is 10^{20} e-h pairs/cm³/s
(b) is 10^{24} e-h pairs/cm³/s
(c) is 10^{10} e-h pairs/cm³/s
(d) cannot be determined, the given data is insufficient

[1992 : 2 Marks]

48. A p-type silicon sample has a higher conductivity compared to an n-type silicon sample having the same dopant concentration. (True/False)

[1994 : 1 Mark]

49. In a p-type Si simple the hole concentration is $2.25 \times 10^{15}/cm^3$. The intrinsic carrier concentration is $1.5 \times 10^{10}/cm^3$ the electron concentration is
(a) zero (b) $10^{10}/cm^3$
(c) $10^5/cm^3$ (d) $1.5 \times 10^{25}/cm^3$

[1995 : 1 Mark]

50. The intrinsic carrier density at 300 K is $1.5 \times 10^{10}/cm^3$, in silicon for n-type silicon doped to 2.25×10^{15} atoms/cm³, the equilibrium electron and hole densities are
(a) $n = 1.5 \times 10^{15}$, $p = 1.5 \times 10^{10}/cm^3$
(b) $n = 1.5 \times 10^{10}$, $p = 2.25 \times 10^{15}/cm^3$
(c) $n = 2.25 \times 10^{15}$, $p = 1.0 \times 10^5/cm^3$
(d) $n = 1.5 \times 10^{10}$, $p = 1.5 \times 10^{10}/cm^3$

[1997 : 2 Marks]

51. The intrinsic carrier concentration of silicon sample at 300 K is $1.5 \times 10^{16}/m^3$. If after doping, the number of majority carriers is $5 \times 10^{20}/m^3$, the minority carrier density is

(a) $4.50 \times 10^{11}/m^3$ (b) $3.33 \times 104/m^3$

(c) $5.00 \times 10^{20}/m^3$ (d) $3.00 \times 10^{-5}/m^3$

[2003 : 1 Mark]

52. The concentration of minority carriers in an extrinsic semiconductor under equilibrium is

(a) directly proportional to the doping concentration

(b) inversely proportional to the doping concentration

(c) directly proportional to the intrinsic concentration

(d) inversely proportional to the intrinsic concentration

[2006 : 1 Mark]

53. The electron and hole concentrations in an intrinsic semiconductor are n_i, per cm^3 at 300 K. Now, if acceptor impurities are introduced with a concentration of N_A per cm^3 (where $N_A >> n_i$) the electron concentration per cm^3 at 300 K will be

(a) n_i (b) $n_i + N_A$

(c) $N_A - n_i$ (d) $\frac{n_i^2}{N_A}$

[2007 : 1 Mark]

54. In an n-type silicon crystal at room temperature, which of the following can have a concentration of $4 \times 10^{19}\ cm^{-3}$?

(a) Silicon atoms

(b) Holes

(c) Dopant atoms

(d) Valence electrons

[2009 : 1 Mark]

55. The doping concentrations on the p-side and n-side of a silicon diode are $1 \times 10^{16}\ cm^3$ and $1 \times 10^{17}\ cm^3$, respectively. A forward bias of 0.3 V is applied to the diode. At T = 300 K, the intrinsic carrier concentration of silicon $n_i = 1.5 \times 10^{10}\ cm^{-3}$ and $\frac{kT}{q} = 26$ mV. The electron concentration at the edge of the depletion region on the p-side is

(a) $2.3 \times 10^9\ cm^{-3}$

(b) $1 \times 10^{16}\ cm^3$

(c) $1 \times 10^{17}\ cm^{-3}$

(d) $2.25 \times 10^6\ cm^3$

[2014 : 2 Marks, Set-1]

56. A silicon bar is doped with donor impurities $N_D = 2.25 \times 10^{15}$ atoms/cm^3. Given the intrinsic carrier concentration of silicon at T= 300 K is $n_i = 1.5 \times 10^{10}\ cm^{-3}$. Assuming complete impurity ionization, the equilibrium electron and hole concentrations are

(a) $n_0 = 1.5 \times 10^{16}\ cm^{-3}$, $p_0 = 1.5 \times 10^5\ cm^3$

(b) $n_0 = 1.5 \times 10^{10}\ cm^{-3}$, $p_0 = 1.5 \times 10^{16}\ cm^3$

(c) $n_0 = 2.25 \times 10^{15}\ cm^{-3}$, $p_0 = 1.5 \times 10^{10}\ cm^3$

(d) $n_0 = 2.25 \times 10^{15}\ cm^{-3}$, $p_0 = 1 \times 10^5\ cm^3$

[2014 : 1 Mark, Set-2]

57. When a silicon diode having a doping concentration of $N_A = 9 \times 10^{16}\ cm^3$ on p-side and $N_D = 1 \times 10^{16}\ cm^{-3}$ on n-side is reverse biased, the total depletion width is found to be 3 μm. Given that the permittivity of silicon is 1.04×10^{-12} F/cm, the depletion width on the p-side and the maximum electric field in the depletion region, respectively are

(a) 2.7 μm and 2.3×10^5 V/cm

(b) 0.3 μm and 4.15×10^5 V/cm

(c) 0.3 μm and 0.42×10^5 V/cm

(d) 2.1 μm and 0.42×10^5 V/cm

[2014 : 2 Marks, Set-2]

58. A thin P-type silicon samole is uniformly illuminated with light which generstes excess carriers. The recombination rate is directly proportional to

(a) the minority carrier mobility.

(b) the minority carrier recombination lifetime.

(c) the majority carrier concentration.

(d) the excess minority carrier concentration.

[2014 : 1 Mark, Set-3]

59. Consider a silicon sample doped with $N_D = 1 \times 10^{15}/cm^3$ donor atoms. Assume that the intrinsic carrier concentration $N_D = 1.5 \times 10^{10}/cm^3$. If the sample is additionally doped with $N_A = 1 \times 10^{18}/cm^3$ accepter atoms, the approximate number of electrons/cm^3 in the sample, at T = 300 K, will be _____.

[2014: 2 Marks, Set-4]

60. An n-type silicon sample is uniformly illuminated with light which generates 10^{20} electron-hole pairs per cm^3 per second. The minority carrier lifetime in the sample is 1 μs. In the steady state, the hole concentration in the sample is approximately 10x, where x is an integer. The value of x is __________,

[2015 : 1 Mark, Set-2]

61. Consider a silicon sample at T = 300 K, with a uniform donor density $N_d = 5 \times 10^{16}$ cm^{-3}, illuminated uniformly such that the optical generation rate is $G_{opt} = 1.5 \times 10^{20}$ cm^{-3} S$_{-1}$ throughout the sample. The incident radiation is turned off at t = 0. Assume low-level injection to be valid and ignore surface effects. The carrier lifetimes are $t_{p0} = 0.1$ μs and $t_{n0} = 0.5$ μs

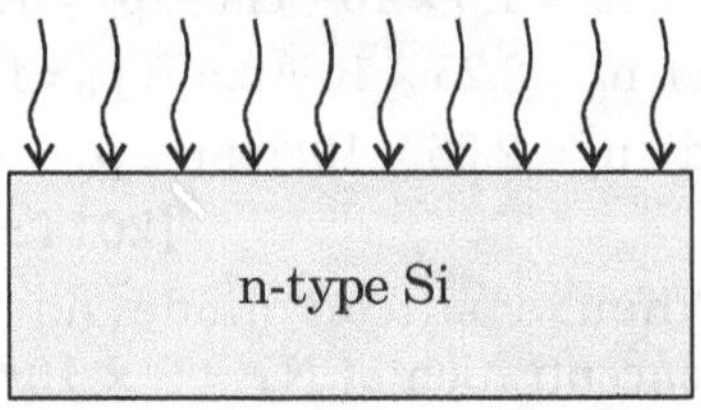

The hole concentration at t = 0 and the hole concentration at t = 0.3 μs, respectively, are

(a) 1.5×10^{13} cm^{-3} and 7.47×10^{11} cm^{-3}

(b) 1.5×10^{13} cm^{-3} and 8.23×10^{11} cm^{-3}

(c) 7.5×10^{13} cm^{-3} and 3.73×10^{11} cm^{-3}

(d) 7.5×10^{13} cm^{-3} and 4.12×10^{11} cm^{-3}

[2016 : 2 Marks, Set-1]

ANSWERS

1. (d)	**2.** (b)	**3.** (b)	**4.** (*)	**5.** (b)	**6.** (a)	**7.** (c)	**8.** (c)	**9.** (a)	**10.** (a)
11. 1.125	**12.** (d)	**13.** (b)	**14.** (a)	**15.** (0.48)	**16.** (c)	**17.** (d)	**18.** (c)	**19.** (a)	**20.** (c)
21. (d)	**22.** (d)	**23.** (a)	**24.** (b)	**25.** (c)	**26.** 4000	**27.** (0.0133)		**28.** (1.6)	**29.** (a)
30. (a & c)	**31.** (c)	**32.** (a)	**33.** (b)	**34.** (b)	**35.** (d)	**36.** (a)	**37.** (c)	**38.** (a)	
39. 13	**40.** 0.52	**41.** 1.92	**42.** (100)	**43.** (c)	**44.** (c)	**45.** (a)	**46.** (b)	**47.** (a)	**48.** (False)
49. (c)	**50.** (c)	**51.** (a)	**52.** (b)	**53.** (d)	**54.** (c)	**55.** (a)	**56.** (d)	**57.** (b)	**58.** (d)
59. 225.2	**60.** 14	**61.** (c)							

EXPLANATIONS

1. By theory, $\vec{v}_n = E\,\mu n$,

where $\vec{v}_n$ = drift velocity,

E = applied electric field

μn = electron mobility

However when the applied field is large the number of electrons in conduction band becomes very large and due to collisions the motion becomes erratic and linear relationship becomes invalid. For $E > 10^4$ V/ cm, μ is inversely proportional to E and so V_n.

2. Electrons are promoted from the valence band to the conduction band by thermal energy $f(E)$ is essentially 1 in the valence band and 0 throughout the conduction band.

3. Thermal voltage or volt–equivalent to temperature $V_T = \frac{K_T}{q}$ is volts. $\left[\frac{q}{K_T}\right] = \left[V^{-1}\right]$

5. $E_g(t) = 1.27 - 3.60 \times 10^{-4}$ T

For silicon, T = 300 K

6. We know $E = h\nu = \frac{hc}{\lambda}$

$$\therefore \quad \frac{E_2}{E_1} = \frac{\lambda_1}{\lambda_2}$$

$$\Rightarrow \quad E_2 = \frac{1\cdot1}{0.87} \times 1.12\,eV = \mathbf{1.41609}$$

7. Band gap of Si at any temperature T

$= 1.21 - 3.6 \times 10^{-4}$T

For T = 300° K

Bandgap $= 1.21 - 3.6 \times 10^{-4} \times 300 = 1.1$ eV

8. $N_A = 4 \times 10^{17}$ atom/cm³

$n_i = 1.5 \times 10^{10}$/cm³

For P-type semiconductor,

$$E_f - E_v = KT \ln \frac{N_A}{N_i}$$

$$= 0.025 \ln\left(\frac{4 \times 10^{17}}{1.5 \times 10^{10}}\right) ev$$

$= 0.427$ ev

9. $E = \frac{hC}{\lambda} \Rightarrow \lambda = \frac{6.62 \times 10^{-34} \times 3 \times 10^8}{1.42 \times 1.6 \times 10^{-19}} = 0.87\ \mu m$

10. $n_i \alpha T^{3/2} e^{-Eg/kT}$ and $\rho_i \,\alpha \frac{1}{\eta_i}$

$\therefore$ From the graph, Energy graph of S_i can be estimated

11. $E = \frac{hC}{\lambda}$

$$\Rightarrow \lambda = \frac{6.6 \times 10^{-34} \times 3 \times 10^8}{1.1 \times 1.6 \times 10^{-19}} = 1.125\ \mu m$$

12. Terminal A will be less reverse and terminal B will be more reverse biased because $V_A < V_B$.

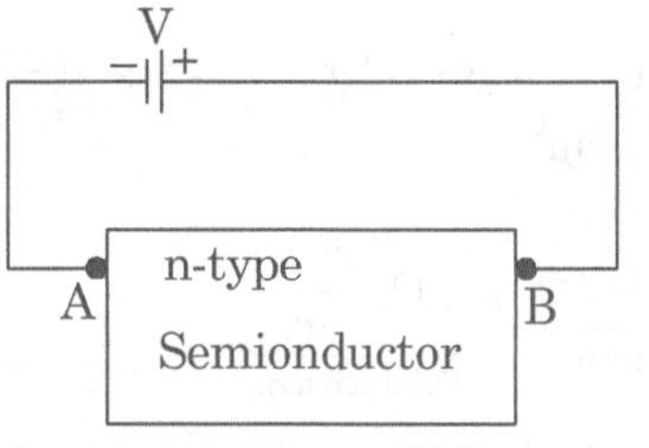

As reverse bias will increase, energy levels will move down. Therefore, height of energy level towards terminal A will be higher and height of energy level toward terminal B will be lower.

14. Donor energy level close to conduction band.

15. $\therefore$ $L = \frac{k\lambda}{NA}$

$L_{min} \propto \lambda$

$$\frac{L_{min\,1}}{L_{min\,2}} = \frac{\lambda_1}{\lambda_2} = \frac{156\text{ nm}}{325\text{ nm}} = 0.48$$

16. From Mass - action law, $n.p = ni^2$

$ni^2 = A_0 T^3 e^{-E}90/kT$

$ni^2 \propto T^3$

For intrinsic semiconductor, $n = P = n_i$

$n \propto T^{3/2}$

17. Phosphorus is a pentavalent atom, therefore one free electron after making four bonds with Si atoms, thus making doping n–type

18. Abundance of silicon on the surface of the earth makes use of it in semiconductor technology.

19. Boron is an acceptor impurity high concentration of which makes p^+ substrate.

20. The oxidation rate is zero as the existing oxide prevents further oxidation.

22. General formula is, $J_n = q\, u_n\, n\varepsilon + q.D_n \dfrac{dn}{dx}$

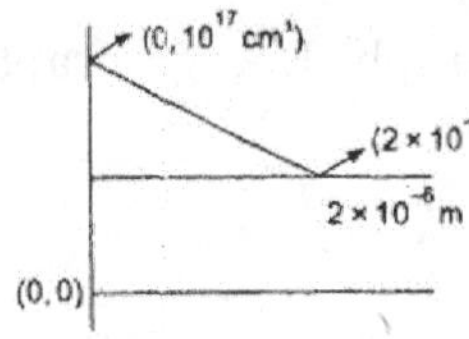

Here, $\varepsilon = 0$

$$\Rightarrow J_n = qD_n \frac{dn}{dx}$$

where $\dfrac{dn}{dx} = \dfrac{-6\times10^{+16}}{2\times10^{-6}} = -3 \times 10^{22}$

$$\Rightarrow J_n = 1.6 \times 10^{-19} \times 35 \times 10^{-4} \times (-3 \times 10^{22})$$
$$= -1120\ \text{A}/\text{cm}^2$$

23. $J_n(x) = \underbrace{q\, u_n(x)\, E(x)}_{\text{Drift term}} + \underbrace{q\, D_n \dfrac{dn(x)}{dx}}_{\text{Diffusion term}}$

For minority carrier current, low level injection, $n(x)$ is very less.

i.e. negligible but $\dfrac{d}{dx}$ of it is significant

Hence, the current is essentially due to diffusion term.

24. Lorentz force, $F = q(\overline{V} \times \overline{B})$

$$= -e(VB)\bar{a}_x \times \bar{a}_z$$

The *n*-type carriers will be carried towards back face. Thus clearly an electric field along negative *y*-axis, i.e. in the direction of $\overline{B} \times \overline{V}$.

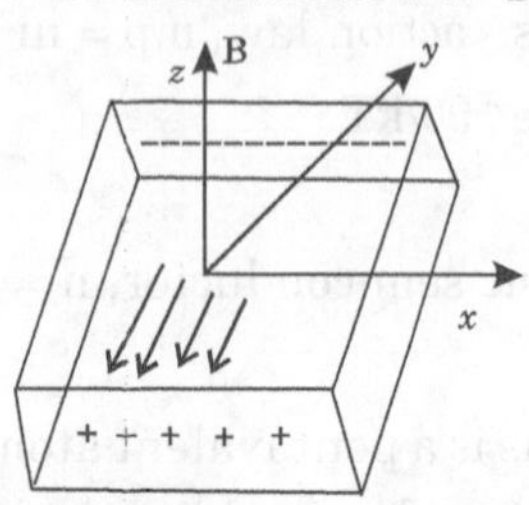

25. Drift current, $J = \sigma E$

$$J = (n\mu_n + p\mu_p)\, qE$$

Hence it depends on carrier concentration and electric field.

26. Given $q = 1.6 \times 10^{-19}$;

$$\frac{kJ}{q} = 2.5\ \text{mV},$$
$$\mu_n = 1000\ \text{cm}^2/\text{v} - \text{s}$$

From Einstein relation,

$$\frac{D_n}{\mu_n} = \frac{kJ}{q}$$

$$\Rightarrow D_n = 25\ \text{mV} \times 1000\ \text{cm}^2/\text{v} - \text{S}$$
$$\Rightarrow \quad = 25\ \text{cm}^2/\text{s}$$

Diffusion current Density

$$J = qD_n \frac{dn}{dx}$$
$$= 1.6 \times 10^{-19} \times 25 \times 1 \times 10^{21}$$
$$= 4000\ \text{A/cm}^2$$

27. The doping distribution in P-type semiconductor (in log scale) is shown below in the figures.

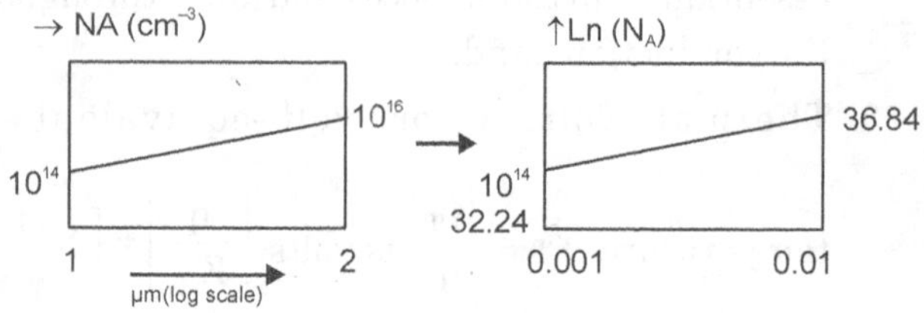

$$qD_p \frac{dp}{dx} = q\mu_p\, p\varepsilon$$

$$\mu_p V_T \frac{dp}{dx} = \mu_p\, p\varepsilon$$

$$\varepsilon = \frac{V_T}{p}\frac{dp}{dx}$$

$$p \cong N_A$$

$$\varepsilon = \frac{V_T}{N_A}\frac{dN_A}{dx}$$

$$\Rightarrow \varepsilon = V_T \frac{d}{dx}\ln\left[N_A(x)\right]$$

$$\log_{10} x_1 = 1\ \mu\text{m}$$

$$\Rightarrow x_1 = 10^1\ \mu\text{m}$$
$$= 0.001\ \text{cm}$$

$$\log_{10} x_2 = 2\mu\text{m}$$

$$\Rightarrow x_2 = 10^2\ \mu\text{m} = 0.01\ \text{cm}$$

$$\ln(10^{14}) = 32.23$$
$$\ln(10^{16}) = 36.84$$

$$\varepsilon = 0.026\left[\frac{36.84 - 32.23}{0.01 - 0.001}\right]$$
$$= 0.0133\ /\text{cm}$$

Hence, the magnitude of electric field in the semiconductor due to non-uniform doping is 0.0133 / cm.

28. $E = \frac{V}{d} = \frac{5}{10^{-4}} = 5 \times 10^4$ V/cm

slope of the curve, (m)

$$= \frac{10^7 - 0}{5 \times 10^5} = 20 \quad \begin{pmatrix} \because y = mx \\ \Rightarrow m = \frac{y}{x} \end{pmatrix}$$

Now $v_d = 20 \times E = 20 \times 5 \times 10^4 = 10^6$ V/cm

and $J = nev_d = 1 \times 10^{16} \times 1.6 \times 10^{-19} \times 10^6$

$= 1.6 \times 10^3$ A/cm^2 = 1.6 kA/cm^2

29. Einstein's relation is

$$\frac{Dn}{\mu n} = \frac{DP}{\mu p} = V_T = \frac{T}{11600} \quad -(1)$$

Ratio of diffusion constant to mobility of carrier will depend upon temperature of semiconductor.

30. Drift velocity, $V_d = \mu E$.

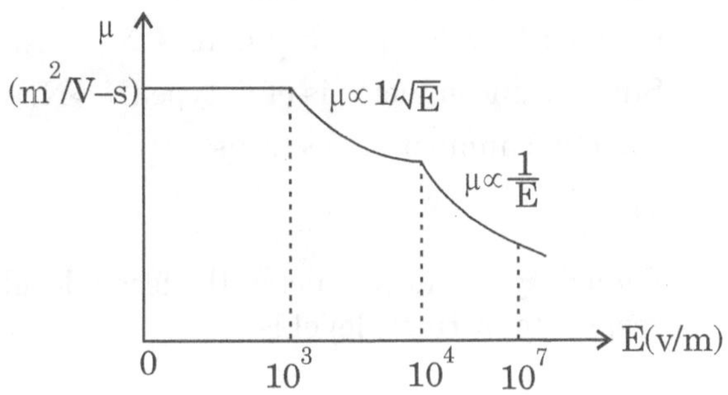

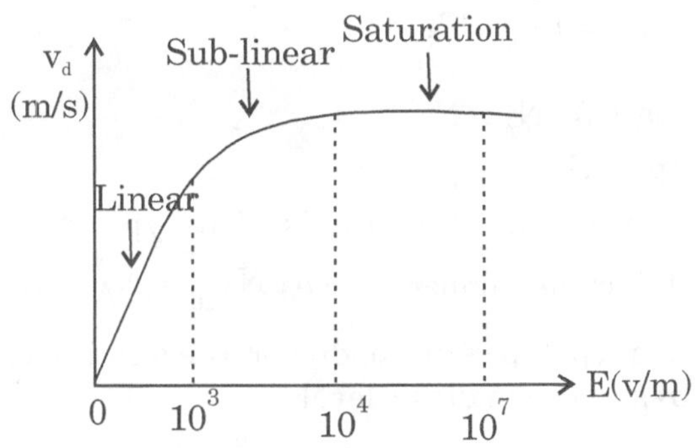

From above graphs, under high electric field (with increasing electric field), (i) the mobility of charge carriers decreases as electric field increases.
(ii) The drift velocity of charge carriers saturates.

31. Mobility $\mu = \frac{v}{E}$

where, Drift velocity $= \frac{1}{20 \times 10^{-6}}$

$$\therefore \mu = \frac{1}{20 \times 10^{-6}} \times \frac{1}{10} = 0.5 \times 10^4$$

$= \mathbf{5000\ cm^2/volt\text{- }sec}$

32. Conductivity, $\sigma = ne\mu_n + pe\mu_p$

As, silicon-bar is n–type, $\sigma \approx ne\mu_n$

or Resistivity, $\rho = \frac{1}{ne\mu_n}$

and,resistance of the bar $= \frac{\rho l}{A} = \frac{l}{ne\mu_n A}$

or $R = \frac{10^{-3}}{5 \times 10^{20} \times 1.6 \times 10^{-19} \times 0.13 \times 100 \times 10^{-12}}$

$= 0.96 \times 10^6\ \Omega$

33. Conductivity, $\sigma = e(N_D \mu_n + N_A \mu_p)$

For, n – type silicon sample, $N_D >> N_A$

Then, $\sigma \approx eN_D\mu_n$

or, $N_D = \frac{\sigma}{e\mu_n} = \frac{1}{\rho e\mu_n}$

$$= \frac{1}{0.5 \times 1.6 \times 10^{-19} \times 1250} = 1 \times 10^{16}/$$

34. Since, $\frac{\mu_n}{\mu_p} = 3$

and $N_A = N_D$

$$\therefore \frac{\sigma_A}{\sigma_B} = \frac{N_A \mu_P}{N_D \mu_n} = \frac{\mu_p}{\mu_e} = \frac{1}{3}$$

35. Ratio $= \frac{qn\mu_u}{qn_i(\mu_u + \mu_p)} = \frac{n}{n_i\left(1 + \frac{\mu_p}{\mu_u}\right)}$

$$= \frac{4.2 \times 10^8}{1.5 \times 10^4 (1 + 0.4)} = 20{,}000$$

36. Unit for D : m^2/s

Unit for μ : m^2/V– s

$\therefore$ Ratio = V^{-1}

Alternately

$J = e\mu nE$

$J = eD\frac{dn}{dx}$

$$\frac{\mu}{D} = \frac{\frac{dn}{dx}}{nE}$$

$$= \frac{\text{number of } e^-}{cm} \times \frac{1}{\text{number of } e^-} \times \frac{\text{meter}}{\text{volt}}$$

$= V^{-1}$

37. Given : T = 300 k, V_T = 26 mV, μn = 1350 cm^2/V–s, $N_D = 10^{16}$ cm^{-3} . The sample is inthermal equilibrium. Hence, at any distance from x = 0 dielectric field will be constant.

$$E = \frac{V}{d} = \frac{1}{1 \times 10^{-6}} = 10^6 v/m.$$

38. Electron drift current density, Jn = N_D. q.μn.E Jn = $10^{16} \times 1.6 \times 10^{-19} \times 1350 \times 10^6 = 2.16 \times 10^4$ A/cm^2.

39. From Einstein relation,

$$\frac{D_P}{\mu_P} = \frac{kJ}{q}$$

$$\Rightarrow \quad D_P = 26 \text{ mV} \times 500 \text{ cm}^2/v-s = 13 \text{ cm}^2/s$$

40. For donor type impurities,

Conductivity

$$\sigma_N = N_D q\mu_n = 10^{16} \times 1.6 \times 10^{-19} \times 1200 = 1.92 \text{ ℧/cm}$$

Resistivity

$$\rho = \frac{1}{\sigma_N} = \frac{1}{1.92}$$

$$\rho = 0.52 \, \Omega - \text{cm}$$

41. As per the graph mobility of electrons at the concentration $10^{16}/cm^3$ is 1200 $\frac{cm^2}{V-s}$

So, $\quad _n = 1200 \frac{cm^2}{V-s}$

$$\sigma_N = N_D q \;_n = 10^{16} \times 1.6 \times 10^{-19} \times 1200 = 1.92 \text{ S cm}^{-1}$$

43. Direct band gap semiconductors exhibit short carrier life time they are used for fabricating LASERS. In direct band gap semiconductor during the recombination, the energy is released in the form of light.

44. Due to illumination by light, electron and hole pair generation will occur. Number of generated hole and generated electrons are some because holes are absence of electrons.

There fore, Δn = ΔP

45. Given: (i) The concentration of ionized acceptors and donors in a semiconductors are N_A, N_D respectively (ii) $N_A > N_D$ and (iii) n_i = intrinsic concentration. For intrinsic semiconductor, $n_o = p_o$.

When donor atom with concentration N_D is added in an intrinsic semiconductor then semiconductor becomes n-type semiconductor. Hence concentration of electron in this n–type semiconductor will increase by N_D and concentration of hole will not change

Number of electrons = $N_O + N_D$

Number of holes = p_o.

When above obtained n–type semiconductor is doped with acceptor atom having concentration N_A then concentration of hole will increase by N_A and concentration of electron will not change.

Hence, number of holes = po + N_A.

Number of electrons = $n_O + N_D$.

Case 1 : $N_A = N_D$ Electron concentration = $n_o + N_D$

Hole concentration = $p_O + N_A$

∴ $N_A = N_D$ and no = po.

Hence, semiconductor is intrinsic.

Case 2 : $N_D > N_A$ $(p_o + N_A) < (n_O + N_D)$

Since, semiconductor is of n–type (N_D>N_A).

Effective number of electrons

$$N_{D_{eff}} = N_D - N_A$$

For a n–type semiconductor, the fermi–level with respect to intrinsic level is

$$E_{F_n} = E_{F_i} + KT \ln\left(\frac{N_{D_{ett}}}{ni}\right)$$

Case 3 : $N_A > N_D$

$(p_O + N_A) > (n_O + N_D)$.

Since semiconductor is of p–type (N_A> N_D).

Effective number of holes $N_{A_{eff}} = N_A - N_D$

For a p–type semiconductor, the fermi level with respect to intrinsic level

$$E_{FP} = E_{Fi} - KT\ln\left(\frac{N_{A_{eff}}}{n_i}\right).$$

46. Given : Donor concentration (phosphorus), $N_D = 10^{16}$ cm^{-3}

Acceptor concentration (Boron), $N_A = 2 \times 10^{16}$ cm^{-3}

Since N_A>> N_D, hence it is p–type semiconductor. concentration of here will be equal to the effective concentration of acceptor atom.

p= $N_{Aeff} = N_A - N_D = 2 \times 10^{16} - 10^{16} = 10^{16}$ cm^{-3}

47. Generation rate of minority carrier

$$= \frac{\text{excess hole conc}}{\text{hole life time}} = \frac{10^{15}}{10x\, 10^{-6}}$$

$$= 10^{20} \text{ electron-hole pairs per second per cm}^3$$

48. False. The conductivity of a p–type sample $\sigma = qu_p \cdot p$ and a n–type sample $\sigma = qu_n n$.

For the same dopant concentration, as the mobility of electron is about 3 times that of holes in silicon, the conductivity of p–type sample is less than that of n–type.

49. $$n = \frac{n_1^2}{p} = \frac{(1.5\times 10^{-10})^2}{2.25\times 10^{15}}$$

$$= 10^5/\text{cm}^3$$

50. Given : $n_i = 1.5 \times 10^{10}\ \text{Cm}^{-3}$, $N_D = 2.25 \times 10^{15}\ \text{cm}^{-3}$. Since to crate n–type semiconductor, intrinsic semiconductor is doped with donor atom. Donor atom concentration, $N_D = 2.25 \times 10^{15}\ \text{cm}^{-3}$ since, to create n–type semiconductor, intrinsic semiconductor is doped with donor atom.

Donor atom concentration, $N_D = 2.25 \times 10^{15}\ \text{cm}^{-3}$ Concentration of majority carrier (electron), $n = N_D = 2.25 \times 10^{15}\ \text{cm}^{-3}$ From mass action law, $np = n_i^2$

$$p = \frac{n_i^2}{n} = \frac{\left(1.5\times 10^{10}\right)^2}{2.25\times 10^{15}} = 10^{15}\,\text{cm}^{-3}$$

51. Let majority carriers be electrons,

$$n_i = 1.5 \times 10^{16},\ n_e = 5 \times 10^{20}$$

According to mass action law, $np = n_i^2$

or $$p = \frac{n_i^2}{n} = \frac{(1.5\times 10^{16})^2}{5\times 10^{20}}$$

$$= 4.50 \times 10^{11}\ /\text{m}^3$$

52. $$np = n_i^2$$

where n_i = intrinsic concentration,

N_d = doping concentration for a n-type material

$$n \cong N_d$$

$$\Rightarrow \quad p = \frac{n_i^2}{N_d}$$

$$\Rightarrow \quad p \propto \frac{1}{N_d}$$

where p = Minority carrier concentration

53. We know that at a particular temperature

$$np = (n_i)^2 \text{ (constant)}$$

where, n = electron concentration

p = hole concentration

By doping acceptor impurities with a concentration of N_A per cm^2

$$p \simeq N_A$$

$$\therefore \quad N_A \cdot n = n_i^2$$

$$n = \frac{n_i^2}{N_A}$$

54. The concentration of Dopand atoms in n–type silicon is $4 \times 10^{19}\ \text{cm}^{-3}$.

55. Electron concentration,

$$n \simeq \frac{n_i^2}{N_A} e^{V_{in}/V_T}$$

$$= \frac{(1.5\times 10^{10})^2}{1\times 10^{16}} e^{0.3/26\text{mV}}$$

$$= 2.3 \times 10^9/\ \text{cm}^3$$

56. $$N_D = 2.25 \times 10^{15}\ \text{Atom/cm}^3$$

$$h_i = 1.5 \times 10^{10}\ /\text{cm}^3$$

Since complete ionization taken place,

$$h_0 = N_D = 2.25 \times 10^{15}\ /\text{cm}^3$$

$$P_0 = \frac{n_i^2}{n_0} = \frac{\left(1.5\times 10^{10}\right)^2}{2.25\times 10^{15}}$$

$$= 1 \times 10^5\ /\text{cm}^3$$

57. Given $$N_A = 9 \times 10^{16}/\text{cm}^3;$$

$$N_D = 1 \times 10^{16}/\text{cm}^3$$

Total depletion width

$$x = x_n + x_p = 3\ \mu\text{m}$$

$$\in = 1.04 \times 10^{-12}\ \text{F/cm}$$

$$\frac{x_n}{x_p} = \frac{N_A}{N_D}$$

$$= \frac{9\times 10^{16}}{1\times 10^{16}}$$

$$x_n = 9x_p \quad \text{...(1)}$$

Total Depletion width,

$$x_n + x_p = 3\ \mu\text{m}$$

$$9x_p + x_p = 3\ \mu\text{m}$$

$$x_p = 0.3\ \mu\text{m}$$

Maximum Electric field,

$$E = \frac{qN_A x_p}{\in}$$

$$= \frac{1.6\times 10^{-19}\times 9\times 10^{16}\times 0.3\ \mu\text{m}}{1.04\times 10^{-12}}$$

$$= 4.15 \times 10^5\ \text{V/cm}$$

58. Recombination rate,

$$R = B\left(n_{n_o} + n'_n\right)\left[P_{n_o} + P'_n\right]$$

n_{n_o} & P_{n_o} = Electron and hole concentrations respectively under thermal equilibrium

n'_n & p'_n = Excess elements and hole concentrations respectively

59. $P = N_A - N_D = 1 \times 10^{18} - 1 \times 10^{13}$
$= 9.99 \times 1017$

$$\eta = \frac{\eta_i^2}{P} = \frac{\left(1.5\times10^{10}\right)}{9.99\times10^{17}} = 225.2/\text{cm}^3$$

60. The concentration of hole-electron pair in 1 sec $= 10^{20} \times 10^{-6} = 10^{14}/\text{cm}^3$

So, the power of 10 is 14.

$x = 14$

61. The hole Concentration

$$P_n(t) = P_{n_0} + P_n(0)\, e^{-t/\tau_p}$$

at low level injuction

$\Rightarrow P_{n_0}$ (neglective)

$$\therefore \quad GR = \frac{P_n(0)}{\tau_{n_0}}$$

$$\Rightarrow \quad P_{n_0}(0) = GR \times \tau_{n_0}$$
$$= 1.5 \times 10^{20} \times 0.5 \times 10^{-6}$$
$$= 7.5 \times 10^{13}/\text{cm}^3$$

At $t = 0$

$$\Rightarrow \quad P(t) = P_n(0) \cdot e^0$$
$$= 7.5 \times 10^{13}/\text{cm}^3$$

At $t = 0.3$ μs

$$\Rightarrow \quad P(t) = P_n(0)\, e^{\frac{0.3}{0.1}}$$
$$= 3.73 \times 10^{11}/\text{cm}^3$$

So, the hole concentration at $t = 0$ and at $t = 0.3$ μs are 7.5×10^{13} cm^{-3} and 3.73×10^4 cm^{-3} respectively.

CHAPTER 2

Diodes

Analysis of Previous GATE Papers												
	Year → Topics ↓		2019	2018	2017 Set 1	2017 Set 2	2016 Set 1	2016 Set 2	2016 Set 3	2015 Set 1	2015 Set 2	2015 Set 3
DIODES	1 Mark	MCQ Type	1	1								
		Numerical Type		1								
	2 Marks	MCQ Type		1		1						
		Numerical Type	1		1			1				
	Total		3	4	2	2		2				
P-N JUNCTION DIODE	1 Mark	MCQ Type							1	1		
		Numerical Type										
	2 Marks	MCQ Type	1					1				1
		Numerical Type					1			2		
	Total		2				2	2	1	5		2
ZENER DIODE	1 Mark	MCQ Type										
		Numerical Type										
	2 Marks	MCQ Type										
		Numerical Type										
	Total											
LED & SOLAR CELL	1 Mark	MCQ Type										
		Numerical Type										
	2 Marks	MCQ Type	1									
		Numerical Type		1								
	Total		2	2								
TUNNEL DIODE, LASER & PHOTODIODE	1 Mark	MCQ Type			1							
		Numerical Type							1			
	2 Marks	MCQ Type										
		Numerical Type				1						
	Total				1	2			1			

DIODES

1. Consider a region of silicon devoid of electrons and holes, with an ionized donor density of $N_d^+ = 10^{17}$ cm^{-3}. The electric field at x = 0 is 0 V/cm, and the electric field at x = L is 50 kV/cm in the positive x direction. Assume that the electric field is zero in the y and z-directions at all points.

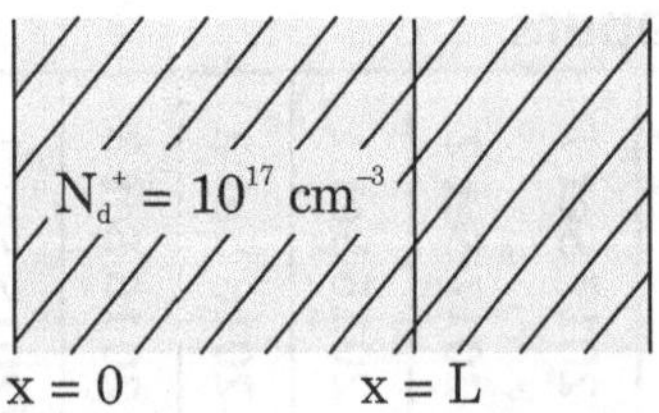

Given g =1.6 × 10^{-19} Coulomb,

ε_0 = 8.85 × 10^{-14} F/cm, ε_r= 11.7 for silicon, the value of L in nm is ______.

[2016 : 2 Marks, Set-2]

2. As shown, two Silicon (Si) abrupt p-n junction diodes are fabricated with uniform donor doping concentrations of $N_{D1} = 10^{14}$ cm^{-3} and $N_{D2} = 10^{16}$ cm^3 in the n-regions of the diodes, and uniform acceptor doping concentrations of $N_{A1} = 10^{14}$ cm^{-3} and $N_{A2} = 10^{16}$ cm^{-3} in the p-regions of the diodes, respectvely. Assuming that the reverse bias voltage is >> built-in potentials of the diodes, the ratio C_2/C_1 of their reverse bias capacitances for the same applied reverse bias, is ________.

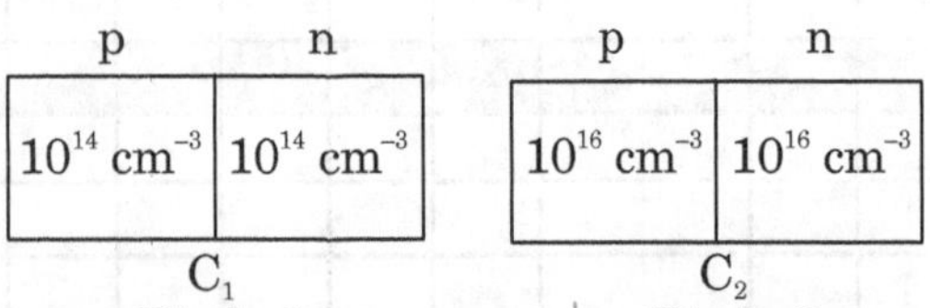

Diode 1 Diode 2

[2017: 2 Marks, Set-1]

3. An abrupt pn junction (located at x = 0) is uniformly doped on both p and n sides. The width of the depletion region is W and the electric field variation in the x-direction is E(x). Which of the following figures represents the electric field profile near the pn junction?

(a)

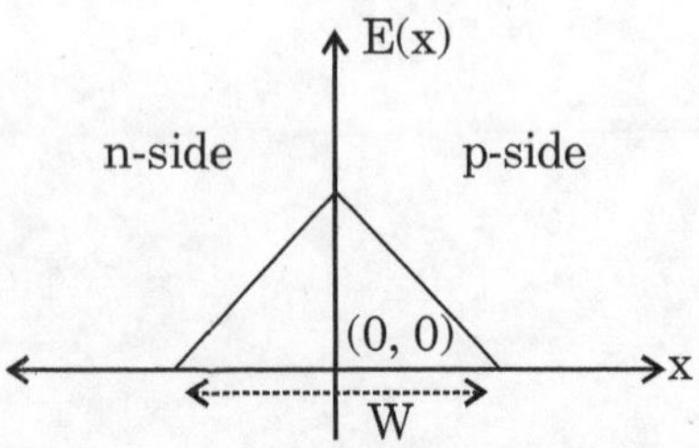

(b)

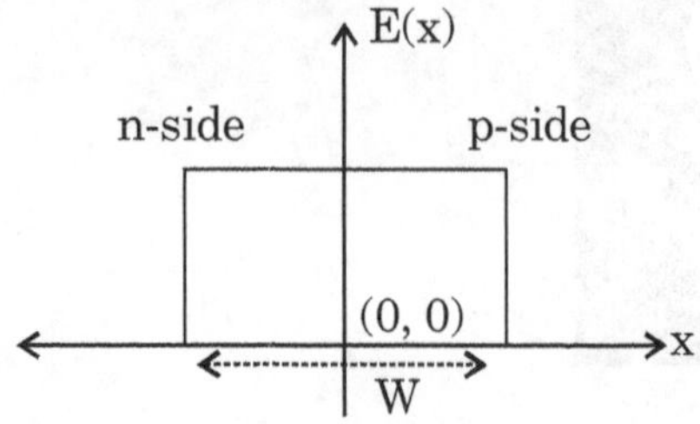

(c)

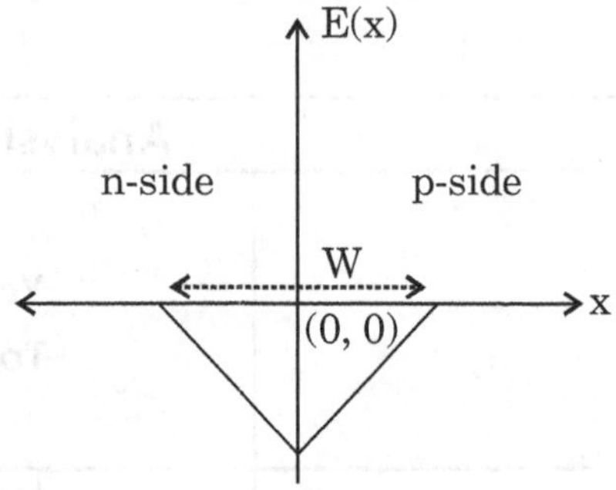

(d)

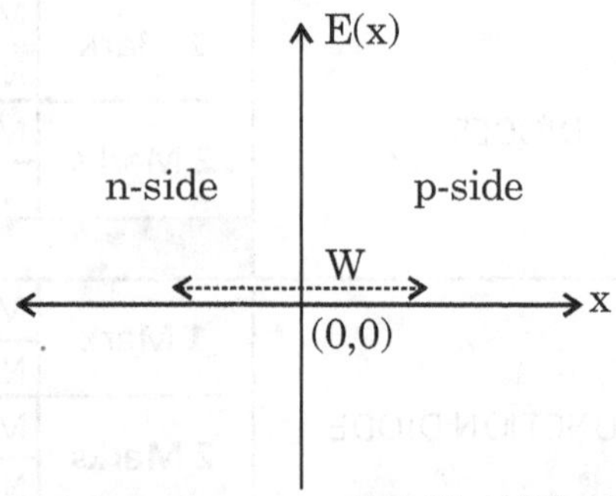

[2017 : 2 Marks, Set-2]

4. In a p-n junction diode at equilibrium, which one of the following statements is NOT TRUE?
   (a) The hole and electron diffusion current components are in the same direction.
   (b) The hole and electron drift current components are in the same direction.
   (c) On an average, holes and electrons drift in opposite direction.
   (d) On an average, electrons drift and diffuse in the same direction.

[2018 : 1 Mark]

5. A p-n step junction diode with a contact potential of 0.65 V has a depletion width of 1 μm at equilibrium. The forward voltage (in volts, correct to two decimal places) at which this width reduces to 0.6 μm is ______.

[2018 : 1 Mark]

6. A junction is made between p$^-$ Si with doping density $N_{A1} = 10^{15}$ cm^{-3} and p Si with doping density $N_{A2} = 10^{17}$ cm^{-3}.

Given: Boltzmannconstant k = 1.38 × 10^{-23} J. K^{-1}, electronic charge q = 1.6 × 10^{-19} C. Assume 100% acceptor ionization.

At room temperature (T = 300 K), the magnitude of the built-in potential (in volts, correct to two decimal places) across this junction will be _____.

[2018: 2 Marks]

7. Which one of the following options describes correctly the equilibrium band diagram at T = 300 K of a Silicon pnn$^+$p^{++} configuration shown in the figure?

p	n	n$^+$	p^{++}

(A) 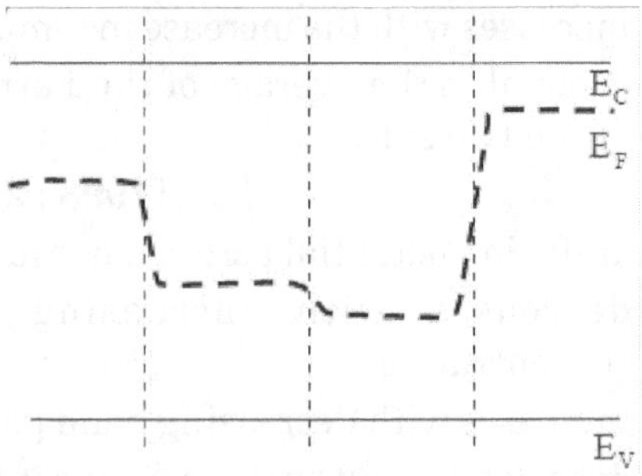

(B) 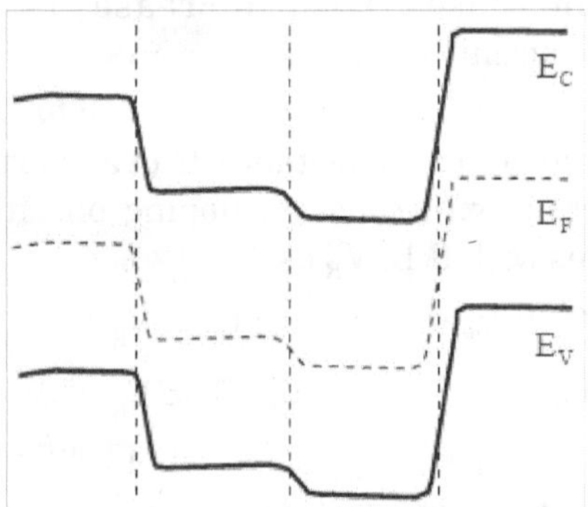

(C) 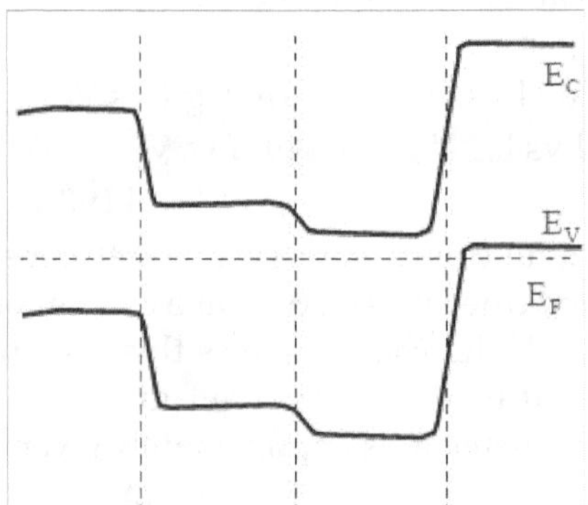

(D) 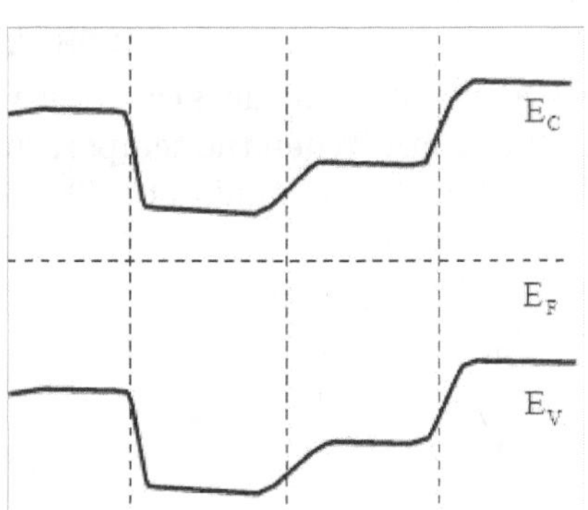

[2019 : 1 Mark]

8. In an ideal p-n junction with an ideality factor of 1 at T = 300 K, the magnitude of the reverse-bias voltage required to reach 75% of its reverse saturation current, rounded off to 2 decimal places, is ______mV. [k = 1.38×10^{-23} JK^{-1}, h = 6.625×10^{-34} J – s, q = 1.602×10^{-19}C]

[2019 : 2 Marks]

P-N JUNCTION DIODE

9. The diffusion capacitance of a p-n junction
   (a) decreases with increasing current and increasing temperature
   (b) decreases with decreasing current and increasing temperature
   (c) increases with increasing current and increasing temperature
   (d) does not depend on current and temperature

[1987 : 2 Marks]

10. For a p-n-junction match the type of breakdovvn with phenomenon
    1. Avalanche breakdown
    2. Zener breakdown
    3. Punch through
    A. Collision of carriers with crystal ions
    B. Earlyeffect
    C. Rupture of covalent bond due to strong electric field.

    (a) 1-B, 2-A, 3-C (b) 1-C, 2-A, 3-B
    (c) 1-A, 2-B, 3-C (d) 1-A, 2-C, 3-B

[1988 : 2 Marks]

11. In the circuit shown below the current voltage relationship when D_1 and D_2 are identical is given by (Assume Ge diodes)

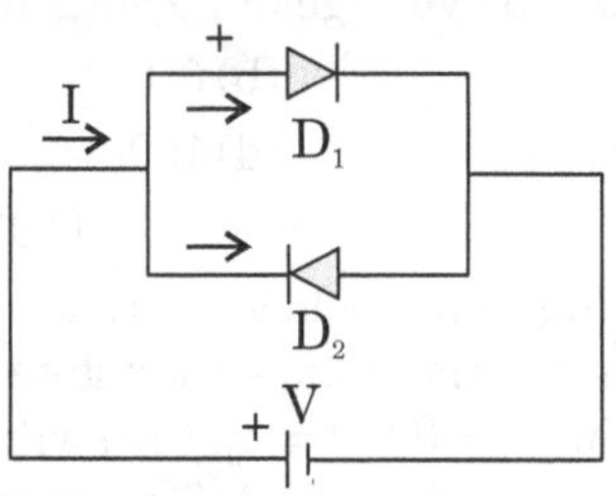

(a) $V = \frac{\bar{K}T}{q}\sinh\left(\frac{I}{2}\right)$

(b) $V = \frac{\bar{K}T}{q}\ln\left(\frac{I}{I_0}\right)$

(c) $V = \frac{\bar{K}T}{q}\sinh^{-1}\left(\frac{I}{2}\right)$

(d) $V = \frac{\bar{K}T}{q}[\text{Exp}(-I) - 1]$

[1988 : 2 Marks]

12. The switching speed of $P^+ - N$ junction (having a heavily doped Pregion) depends primarily on

(a) the mobility of minority carriers in the P^+-region.

(b) the lifetime of minority carriers in the P^+-region

(c) the mobility of majority carriers in the N-region

(d) the lifetime of majority carriers in the N-region

[1989 : 2 Marks]

13. In a junction diode

(a) the depletion capacitance increases with increase in the reverse bias.

(b) the depletion capacitance decreases with increase in the reverse bias.

(c) the depletion capacitance increases with increase in the forward bias.

(d) the depletion capacitance is much higher than the depletion capacitance when it is forward biased

[1990 : 1 Mark]

14. In a uniformly doped abrupt p-n junction the doping level of the n-side is four times the doping level of the p-side the ratio of the depletion layer width of n-side verses p-side is

(a) 0.25 (b) 0.5

(c) 1.0 (d) 2.0

[1990 : 2 Marks]

15. The small signal capacitance of an abrupt $P^+ - n$ junction is 1 nF/cm^2 at zero bias. If the built-in voltage is 1 volt, the capacitance at a reverse bias voltage of 99 volts in

(a) 10 (b) 0.1

(c) 0.01 (d) 100

[1991 : 2 Marks]

16. Referring to the below figure the switch S is in position 1 initially and steady state condition exist from time t = 0 to t = t_0, the switch is suddenly thrown into position 2. The current I through the 10 K resistor as a function of time t, from f = 0 is? (give the sketch showing the magnitudes of the current at t = 0, t = t_0 and t = ∞).

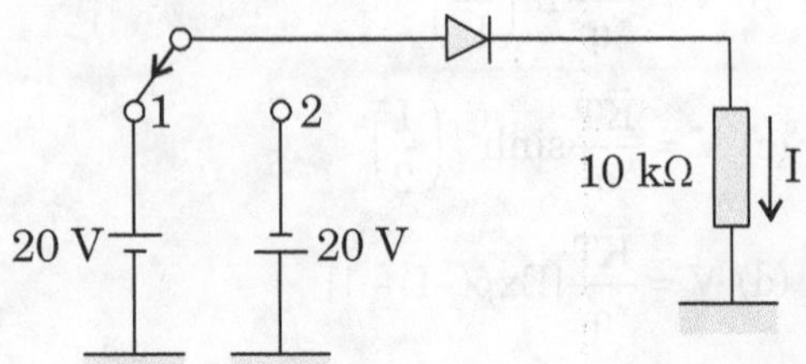

[1991 : 2 Marks]

17. The built-in potential (diffusion potential) in a p-n junction

(a) is equal to the difference in the Fermi-level of the two sides, expressed in volts.

(b) increases with the increase in the doping levels of the two sides.

(c) increases with the increase in temperature.

(d) is equal to the average of the Fermi levels of the two sides.

[1993 : 2 Marks]

18. The diffusion potential across a p-n junction

(a) decreases with increasing doping concentration.

(b) increases with decreasing band gap.

(c) does not depend on doping concentrations.

(d) increases with increase in doping concentration.

[1995 : 1 Mark]

19. The depletion capacitance, C_j of an abruptly p-n junction with constant doping on either side varies with R.B. V_R as

(a) $C_j \propto V_R$ (b) $C_j \propto V_R^{-1}$

(c) $C_j \propto V_R^{-1/2}$ (d) $C_j \propto V_R^{-1/3}$

[1995 : 1 Mark]

20. The static characteristic of an adequately forward biased p-n junction is a straight line, if the plot is of

(a) log I vs.log V (b) log I vs V

(c) I vs log V (d) I vs V

[1998 : 1 Mark]

21. A p-n-junction in series with a 100 ohms resistor, is forwarded biased. So that a current of 100 mA flows. If the voltage across this combination is instantaneously reversed at t = 0, current through diodes is approximately given by

(a) 0 mA (b) 100 mA

(c) –100 mA (d) 50 mA

[1998 : 2 Marks]

22. In the figure, silicon diode is carrying a constant current of 1 mA. When the temperature of the diode is 20°C, V_D is found to be 700 mV. If the temperature rises to 40 °C, V_D becomes approximately equal to

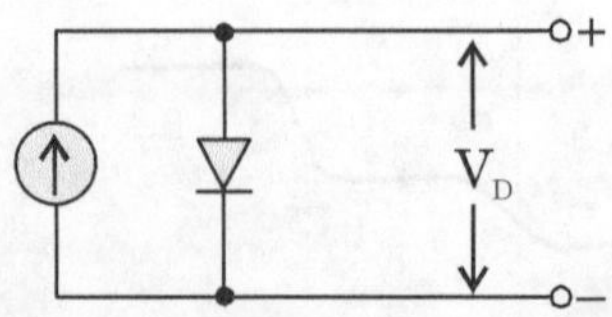

(a) 740 mV (b) 660 mV

(c) 680 mV (d) 700 mV

[2002 : 1 Mark]

23. At 300 K, for a diode current of 1 mA, a certain germanium diode requires a forward bias of 0.1435 V, whereas a certain Silicon diode requires a forward bias of 0.713 V. Under the conditions stated above, the closest approximation of the ratio of reverse saturation current in germanium diode to that in silicon diode is

(a) 1 (b) 5

(c) 4×10^3 (d) 8×10^3

[2003 : 2 Marks]

24. In an abrupt p-n junction, the doping concentrations on the p-side and n-side are $N_A = 9 \times 5\ 10^{16}/cm^3$ and $N_D = 1 \times 10^{16}/cm^3$ respectively. The p-n junction is reverse biased and the total depletion width is 3 nm. The depletion width on the p-side is

(a) 2.7 nm

(b) 0.3 nm

(c) 2.25 nm

(d) 0.75 nm

[2004 : 2 Marks]

25. Consider an abrupt p-junction. Let V_{bi} the built-in potential of this junction and V_R be the applied reverse bias. If the junction capacitance (C_j) is 1 pF for $V_{bi} + V_R = 1V$, then for $V_{bi} + V_R = 4$ V, C_j will be

(a) 4 pF (b) 2 pF

(c) 0.25 pF (d) 0.5 pF

[2004 : 2 Marks]

26. A Silicon PN junction at a temperature of 20°C has a reverse saturation current of 10 pico-Amperes (pA). The reverse saturation current at 40°C for the same bias is approximately

(a) 30 pA (b) 40 pA

(c) 50 pA (d) 60 pA

[2005 : 1 Mark]

27. A Silicon PN junction diode under reverse bias has depletion region of width 10 μm. The relative permittivity of Silicon, $\varepsilon_r = 11.7$ and the permittivity of free space $\varepsilon_0 = 8.85 \times 10^{-12}$ F/m. The depletion capacitance of the diode per square meter is

(a) 100 μ (b) 10 μ

(c) 1 μ (d) 20 μ

[2005 : 2 Marks]

28. In the circuit shown below, the svvitch was connected to position 1 at f < 0 and at t = 0, it is changed to position 2. Assume that the diode has zero voltage drop and a Storage time t_s. For $0 < t \leq t_s$ V_R is given by (all in Volts)

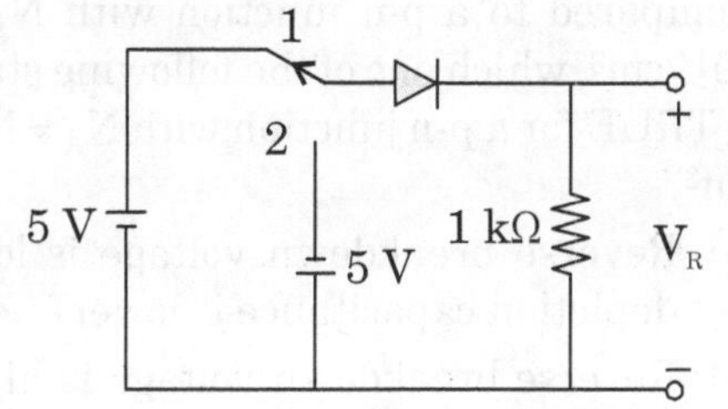

(a) $V_R = -5$ (b) $V_R < +5$

(a) $0 \leq VR < 5$ (d) $-5 < V_R < 0$

[2006 : 2 Marks]

29. In a $p^+ - n$ junction diode under reverse bias, the magnitude of electric field is maximum at

(a) the edge of the depletion region on the p-side

(b) the edge of the depletion region on the n-side

(c) the $p^+ - n$ junction

(d) the centre of the depletion region on the n-side

[2007 : 1 Mark]

30. A $p^+ - n$ junction has a built-in potential of 0.8 V. The depletion layer width at a reverse bias of 1.2 V is 2 nm. For a reverse bias of 7.2 V, the depletion layer widthwill be

(a) 4 μm (b) 4.9 μm

(c) 8 μm (d) 12 μm

[2007 : 2 Marks]

31. Which of the follovving is NOT associated with a p-n junction?

(a) Junction Capacitance

(b) Charge Storage Capacitance

(c) Depletion Capacitance

(d) Channel Length Modulation

[2008 : 1 Mark]

32. The built-in potential of the junction

(a) is 0.70 V

(b) is 0.76 V

(c) is 0.82 V

(d) cannot be estimated from the data given

[2009 : 2 Marks]

33. The peak electric field in the device is

(a) 0.15 MV-cm^{-1}, directed from p-region to n-region

(b) 0.15 MV-cm^{-1}, directed from n-region to p-region

(c) 1.80 MV-cm^{-1}, directed from p-region to n-region

(d) 1.80 MV-cm^{-1}, directed from n-region to p-region

[2009 : 2 Marks]

34. Compared to a p-n junction with $N_A = N_D = 10^{14}/cm^3$, which one of the following statements is TRUE for a p-n junction with $N_A = N_D = 10^{20}/cm^3$?

(a) Reverse breakdown voltage is lower and depletion capacitance is lower.

(b) Reverse breakdown voltage is higher and depletion capacitance is lower.

(c) Reverse breakdown voltage is lovver and depletion capacitance is higher.

(d) Reverse breakdown voltage is higher and depletion capacitance is higher.

[2010 : 2 Marks]

35. A silicon PN junction is forward biased with a constant current at room temperature. When the temperature is increased by 10°C, the forward bias voltage across the PN junction

(a) increases by 60 mV

(b) decreases by 60 mV

(c) increases by 25 mV

(d) decreases by 25 mV

[2011 : 1 Mark]

36. The i-v characteristics of the diode in the circuit given below are

$$i = \begin{cases} \frac{v - 0.7}{500} A, v \geq 0.7V \\ 0A, \quad v < 0.7V. \end{cases}$$

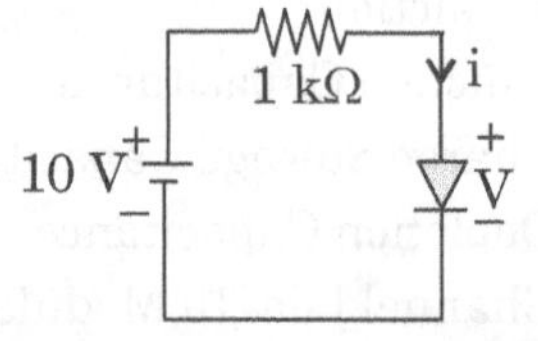

The current in the circuit is

(a) 10 mA (b) 9.3 mA

(c) 6.67 mA (d) 6.2 mA

[2012 : 1 Mark]

37. In a forward biased pn junction, the sequence of events that best describes the mechanism of current flow is

(a) injection, and subsequent diffusion and recombination of minority carriers.

(b) injection, and subsequent drift and generation of minority carriers.

(c) extraction, and subsequent diffusion and generation of minority carriers.

(d) extraction, and subsequent drift and recombination of minority carriers.

[2013: 1 Mark]

38. Consider an abrupt PN junction (at T = 300 K) shown in the figure. The depletion region width x_n on the N-side of the junction is 0.2 μm and the permittivity of Silicon (ε_{si}) is 1.044 × 10^{-12} m. At the junction, the approximate value of the peak electric field (in kV/cm) is________.

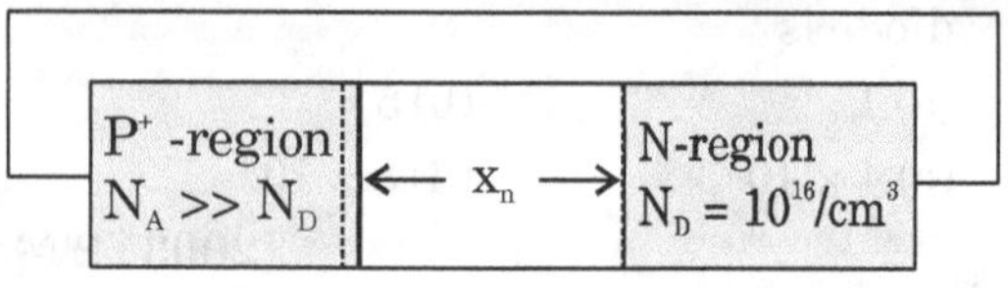

[2014 : 2 Marks, Set-2]

39. The donor and acceptor impurities in an abrupt junction silicon diode are 1×10^{16} cm^{-3} and 5×10^{18} cm^{-3}, respectively. Assume that the intrinsic carrier concentration in silicon $n_i = 1.5 \times 10^{10}$ cm^{-3} at 300K, $\frac{kT}{q}$ = 26 mV and the permittivity of silicon $\varepsilon_{si} = 1.04 \times 10^{-12}$ F/cm. The built-in potential and the depletion width of the diode under thermal equilibrium conditions, respectively, are

(a) 0.7 V and 1×10^{-4} cm

(b) 0.86 V and 1×10^{-4} cm

(c) 0.7 V and 3.3×10^{-5} cm

(d) 0.86 Vand 3.3×10^{-5} cm

[2014 : 2 Marks, Set-3]

40. A region of negative differential resistance is observed in the current voltage characteristics of a silicon PN junction if

(a) both the P-region and the N-region are heavily doped

(b) the N- region is heavily doped compared to the P-region

(c) the P-region is heavily doped compared to the N-region

(d) an intrinsic silicon region is inserted between the P-region and the N-region

[2015 : 1 Mark, Set-1]

41. The built-in potential of an abrupt p-n junction is 0.75 V. If its junction capacitance (C_j) at a reverse bias (V_R) of 1.25 V is 5 pF, the value of C_j,(in pF) when V_R = 7.25 V is______.

[2015 : 2 Marks, Set-1]

42. For a silicon diode with long Pand N regions, the accepter and donor impurity concentrations are 1×10^{17} cm^3 and 1×10^{15} cm^3, respectively. The lifetimes of electrons in P region and holes in N region are both 100 us. The electron and hole diffusion coefficients are 49 cm^2/s and 36 cm^2/s, respectively. Assume kT/q = 26 mV, the intrinsic carrier concentration is 1×10^{10} cm^3, and $q = 1.6 \times 10^{19}$ C. When a forward voltage of 208 mV is applied across the diode, the hole current density (in nA/cm^2) injected from P region to N region is _______.

[2015 : 2 Marks, Set-1]

43. The electric field profile in the depletion region of a p-n junction in equilibrium is shown in the figure. Which one of the following statements is NOT TRUE?

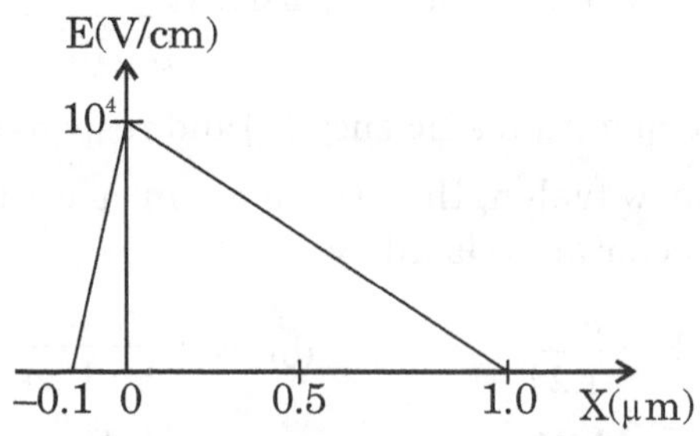

(a) The left side of the junction is n-type and the right side is p-type

(b) Both the n-type and p-type depletion regions are uniformly doped

(c) The potential difference across the depletion region is 700 mV

(d) If the p-type region has a doping concentration of 10^{15} cm^{-3}, then the doping concentration in the n-type region will be 10^{16} cm^{-3}

[2015 : 2 Marks, Set-3]

44. Consider a silicon p-n junction with a uniform acceptor doping concentration of 10^{17} cm^{-3} on the p-side and a uniform donor doping concentration of 10^{16} cm^{-3} on the n-side. No external voltage is applied to the diode.

Given: kT/q = 26 mV, $n_i = 1.5 \times 10^{10}$ cm^3, $\varepsilon_{si} = 12\,\varepsilon_0$, $\varepsilon_0 = 8.85 \times 10^{-14}$ F/m, and

$q = 1.6 \times 10^{-19}$C.

The charge per unit junction area (nC cm^{-2}) in the depletion region on the p-side is ______.

[2016 :2 Marks, Set-1]

45. Consider avalanche breakdown in a silicon p+n junction. The n-region is uniformly doped with a donor density N_D. Assume that breakdown occurs when the magnitude of the electric field at any point in the device becomes equal to the critical field E_{crit}. Assume E_{crit} to be independent of N_D. If the built-in voltage of the p$^+$ n junction is much smaller than the breakdown voltage, V_{BR}, the relationship between V_{BR} and N_D is given by

(a) $V_{BR} \times \sqrt{N_D}$ = constant.

(b) $N_D \times \sqrt{V_{BR}}$ = constant.

(c) $N_D \times V_{BR}$ = constant.

(d) N_D/V_{BR} = constant.

[2016 : 2 Marks, Set-2]

46. The I – V characteristics of three types of diodes at the room temperature, made of semiconductors X, Y and Z, are shown in the figure. Assume that the diodes are uniformly doped and identical in all respects except their materials. If E_{gX}, E_{gY}, and E_{gZ} are the band gaps of X, Yand Z, respectively, then

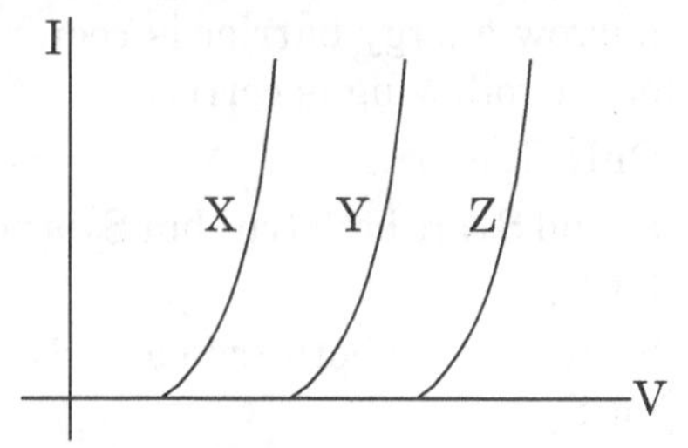

(a) $E_{gX} > E_{gY} > E_{gZ}$

(b) $E_{gX} = E_{gY} = E_{gZ}$

(c) $E_{gX} < E_{gY} < E_{gZ}$

(d) no relationship among these band gaps exists

[2016 : 1 Mark, Set-3]

47. In the circuit shown, the breakdown voltage and the maximum current of the Zener diod are 20 V and 60 mA, respectively. The values of R_1 and R_L are 200 Ω and 1k Ω, respectively. What is the range of V_i that will maintain the Zener diode in the 'n' state?

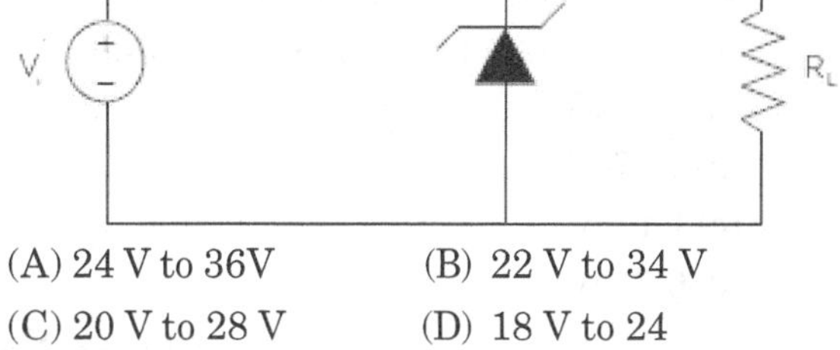

(A) 24 V to 36V (B) 22 V to 34 V

(C) 20 V to 28 V (D) 18 V to 24

[2019 : 2 Marks]

ZENER DIODE

48. In a Zener diode

(a) only the P-region is heavily doped.

(b) only the N-region is heavily doped.

(c) both P and N-regions are heavily doped.

(d) both P and P-regions are lightly doped.

[1989 : 2 Marks]

49. A Zener diode works on the principle of

(a) tunneling of charge carriers across the junction.

(b) thermionic emission.

(c) diffusion of charge carriers across the junction.

(d) doping of charge carriers across the junction.

[1995 : 1 Mark]

50. Consider the following assertions.

S_1: For Zener effect to occur, a very abrupt junction is required

S_2: For quantum tunneling to occur, a very narrow energy barrier is required. Which of the following is correct?

(a) Only S_2 is true.

(b) S_1 and S_2 are both true but S_2 is not a reason for S_1.

(c) S_1 and S_2 are both true and S_2 is a reason for S_1.

(d) Both S_1 and S_2 are false.

[2008 : 2 Marks]

51. A Zener diode, when used in voltage stabilization circuits, is biased in

(a) reverse bias region below the breakdown voltage.

(b) reverse breakdown region.

(c) forward bias region.

(d) forward bias constant current mode.

[2011 : 1 Mark]

LED & SOLAR CELL

52. A particular green LED emits light of wavelength 5490 Å. The energy bandgap of the semiconductor material used there is

(Plank's constant = 6.626×10^{-34} J-s)

(a) 2.26 eV

(b) 1.98eV

(c) 1.17eV

(d) 0.74eV

[2003 : 2 Marks]

53. Red (R), Green (G) and Blue (B) Light Emitting Diodes (LEDs) were fabricated using p-n junctions of three different inorganic semiconductors having different band-gaps. The built-in voltages of red, green and blue diodes are V_R, V_G and V_B, respectively. Assume donor and acceptor doping to be the same (N_A and N_D respectively) in the p and n sides of ali the three diodes.

Which one of the following relationships about the built-in voltages is TRUE?

(a) $V_R > V_G > V_B$ (b) $V_R < V_G < V_B$

(c) $V_R = V_G = V_B$ (d) $V_R > V_G < V_B$

54. A solar cell of area 1.0 cm², operating at 1.0 sun intensity, has a short circuit current of 20 mA, and an open circuit voltage of 0.65 V. Assuming room temperature operation and thermal equivalent voltage of 26 mV, the open circuit voltage (in volts, correct to two decimal places) at 0.2 sun intensity is ______.

[2018 : 2 Marks]

55. The quantum efficiency (η) and responsivity (R) at a wavelength λ (in μ m) in a p-i-n photo detector are related by

(a) $R = \dfrac{\eta \times \lambda}{1.24}$ (b) $R = \dfrac{\lambda}{\eta \times 1.24}$

(c) $R = \dfrac{1.24 \times \lambda}{\eta}$ (d) $R = \dfrac{1.24}{\eta \times \lambda}$

[2019 : 2 Marks]

TUNNEL DIODE, LASER & PHOTODIODE

56. Choose proper substitutes for X and Y to make the following statement correct. Tunnel diode and Avalanche photodiode are operated in X bias and Y bias respectively.

(a) X : reverse, Y : reverse

(b) X : reverse, V : forward

(c) X : forward, Y : reverse

(d) X : forward, Y : forward

[2003 : 1 Mark]

57. Match items in **Group-1** with items in **Group-2**, most suitably.

Group-1	Group-2
P. LED	1. Heavy doping
Q. Avalanche Photodiode	2. choerent radiation
R. Tunnel diode	3. Spontaneous emission
S. LASER	4. Current gain

(a) P – 1; Q – 2; R – 4; S–3

(b) P – 2; Q – 3; R – 1; S–4

(c) P – 3; Q – 4; R – 1; S–2

(d) P – 2; Q – 1; R – 4; S–3

[2003 : 2 Marks]

58. The values of voltage(V_D) across a tunnel-diode corresponding to peak and valley currents are V and V_v respectively. The range of tunnel-diode voltage V_D for which the slope of its I - V_D characteristics is negative would be

(a) $V_D < 0$ (b) $0 \leq V_D < V_p$
(c) $V_p \leq V_D < V_v$ (d) $V_D \geq V_v$

[2006 : 1 Mark]

59. Find the correct match between **Group 1** and **Group 2**

Group 1	Group 2
E. Varactor diode	1. Voltage reference
F. PIN diode	2. Hign-frequency switch
G. Zener diode	3. Tuned circuits
H. Schottky diode	4. Current controlled attenuator

(a) E-4, F-2, G-1, H-3
(b) E-2, F-4, G-1, H-3
(c) E-3, F-4, G-1, H-2
(d) E-1, F-3, G-2, H-4

[2006 : 2 Marks]

60. **Group I** lists four types of p-n junction diodes. Match each device in **Group I** with one of the options in **Group II** to indicate the bias condition of that device in its normal mode of operation.

Group I	Group II
R . Zener Diode	1. Forward bias
Q. Solarcell	2. Reverse bias
R. LASER diode	
S. Avalanche Photodiode	

(a) P-1, Q-2, R-1, S-2
(b) P-2, Q-1, R-1, S-2
(c) P-2, Q-2, R-2, S-1
(d) P-2, Q-1, R-2, S-2

[2007 : 2 Marks]

61. When the optical power incident on a photodiode is 10 μW and the responsivity is 0.8 A/W, the photocurrent generated (in μA) is ______.

[2014 : 1 Mark, Set-1]

62. The figure shows the I – V characteristics of a solar cell illuminated uniformly with solar light of power 100 mW/cm². The solar cell has an area of 3 cm² and a fill factor of 0.7. The maximum efficiency (in %) of the device is____.

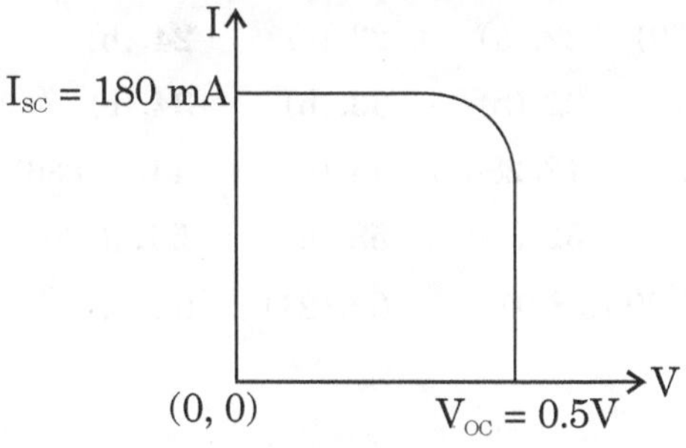

[2016 : 1 Mark, Set-3]

63. An n^+ –n Silicon device is fabricated with uniform and non-degenerate donor doping concentrations of $N_{D1} = 1 \times 10^{18}$ cm³ and $N_{D2} = 1 \times 10^{15}$ cm^{-3} corresponding to the n^+ and n – regions respectively. At the operational temperature T, assume complete impurity ionization, kT/q = 25 mV, and intrinsic carrier concentration to be n. = 1×10^{10} cm^{-3}. What is the magnitude of the built-in potential of this device?

(a) 0.748 V (b) 0.460 V
(c) 0.288 V (d) 0.173V

[2017 : 1 Mark, Set-1]

64. For a particular intensity of incident light on a silicon pn junction solar cell, the photocurrent density (J_L) is 2.5 mA/cm² and the open-circuit voltage (V_{oc}) is 0.451 V. Consider thermal voltage (V_T) to be 25 mV. If the intensity of the incident light is increased by 20 times, assuming that the temperature remains unchanged, V_{oc} (in volts) will be ________.

[2017: 2 Marks, Set-2]

ANSWERS

1.(32.36)	**2.** (10)	**3.** (a)	**4.** (d)	**5.** (0.41)	**6.** (0.119)	**7.** (c)	**8.** (35.87)	**9.** (b)	**10.** (d)
11. (b)	**12.** (d)	**13.** (b)	**14.** (a)	**15.** (b)	**16.** (5)	**17.** (a,b,c)	**18.** (d)	**19.** (c)	**20.** (b)
21. (100)	**22.** (a)	**23.** (c)	**24.** (b)	**25.** (d)	**26.** (b)	**27.** (b)	**28.** (a)	**29.** (c)	**30.** (a)
31. (d)	**32.** (b)	**33.** (b)	**34.** (c)	**35.** (d)	**36.** (d)	**37.** (a)	**38.** 30.66	**39.** (d)	**40.** (a)
41. (2.5)	**42.** 28-30	**43.** (c)	**44.** (4.836)	**45.** (c)	**46.** (c)	**47.** (A)	**48.** (c)	**49.** (a)	**50.** (b)
51. (b)	**52.** (a)	**53.** (b)	**54.** (0.608)	**55.** (a)	**56.** (c)	**57.** (c)	**58.** (c)	**59.** (c)	**60.** (b)
61. 7.99 to 8.01		**62.** (21)	**63.** (d)	**64.** (0.526)					

EXPLANATIONS

1. $|\varepsilon| = \dfrac{q}{\varepsilon_0} N_D X_N$

$$50 \times 10^3 = \frac{1.6\times10^{-19}}{8.85\times10^{-14}\times11.7}\times10^{17}\times X_N$$

(Since $\in = \in_0 \in_r$)

On solving $X_N = 3.2356 \times 10^{-6}$ cm

$= 3.236 \times 10^{-8}$ m

$\therefore$ $X_N = L = 32.36$ nm

Hence, the value of L is 32.36 nm.

3.

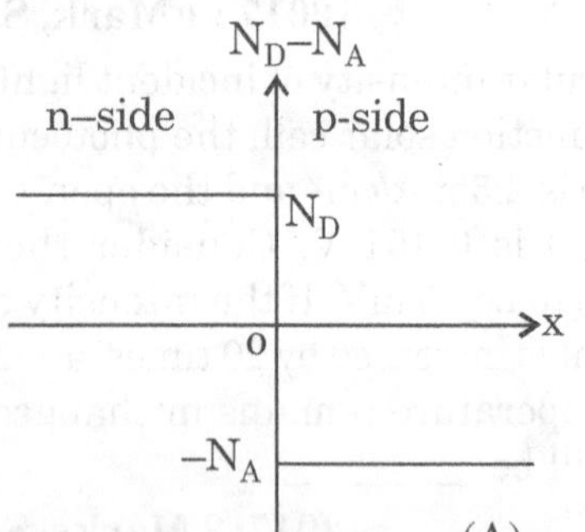

(A)

Doping Profile

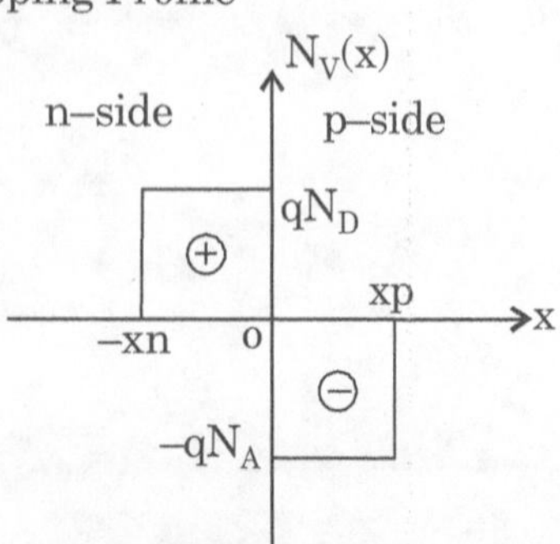

Charge Density

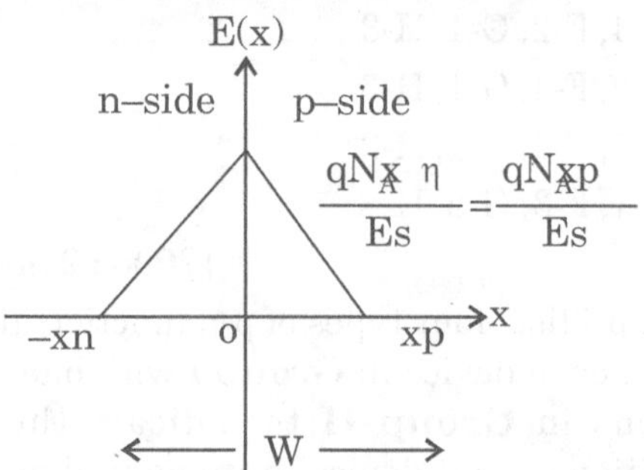

Electric Field

$$E(x) = \int_{-\infty}^{x} \rho v(x)/E_s \, dx$$

6. Given: Doping density at p side, $N_{A1} = 10^{15}$ cm^{-3}

Doping density at P– side, $N_{A2} = 10^{17}$ cm^{-3}

$V_T = 26$ mV.

Built in potential

$$= V_T \ln\left(\frac{N_{A2}}{N_{A1}}\right) = 26\times10^{-3}\ln\left(\frac{10^{17}}{10^{15}}\right) = 26\times10^{-3}\times4.605$$

$= 0.119$ V

7.

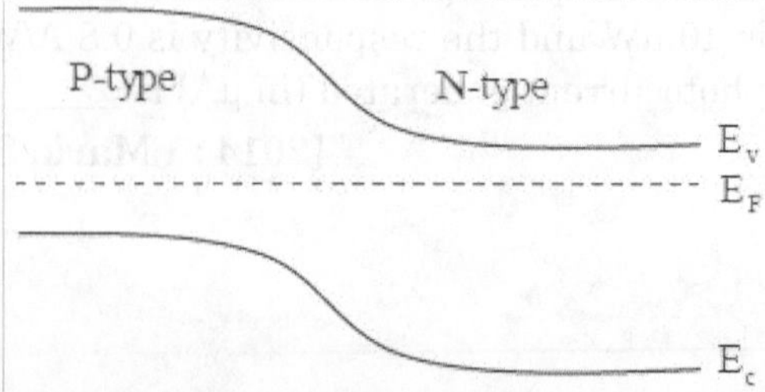

Energy 'band diagram of PN junction diode.

- For pnn$^+$p^{++} configuration, Fermi-energy level (E_F) is more closer to E_V compare to E_C.
- For P-type of semiconductor

$$E_F = E_V - kT\ell n\left(\frac{N_A}{N_V}\right) \quad [\text{For } N_A < N_V]$$

8. $I = I_0\left[\exp\left(-\frac{V_R}{V_T}\right) - 1\right]$

$$\therefore \quad \frac{-3}{4}I_0 = I_0\left[\exp\left(\frac{V_R}{V_T}\right) - 1\right]$$

$$\therefore \quad \exp\left[\frac{-V_R}{V_T}\right] = \frac{-3}{4} + 1 = \frac{1}{4}$$

$$\therefore \quad \frac{-V_R}{V_T} = \ell n\left(\frac{1}{4}\right)$$

$$\therefore \quad -V_R = V_T\ell n\left(\frac{1}{4}\right) = 25.9 \times \ell n\left(\frac{1}{4}\right) \text{mV}$$

$$\therefore \quad |-V_R| = 35.87 \text{ mV}$$

9. (b) For a p–n junction diode, diffusion capacitance

is $C_D = \tau.g = \frac{\tau}{r} = \tau.\frac{I_D}{\eta V\tau} = \frac{\tau I_D}{\eta\tau} \times 11600.$

$$C_D \propto I_D \text{and} C_D \propto \frac{1}{T}.$$

10. Avalanche breakdown : collision of carriers with crystal ions zener break down : Rupture of covalent bond due to strong electric field.
Punch through : Early effect
1→A, 2→C, 3→B

11. (b) In the figure, diode D_1 is in forward bias and diode D_2 is in reverse bias. I_D will be forward bias current and current I_O will be reverse saturation current. Hence, total current

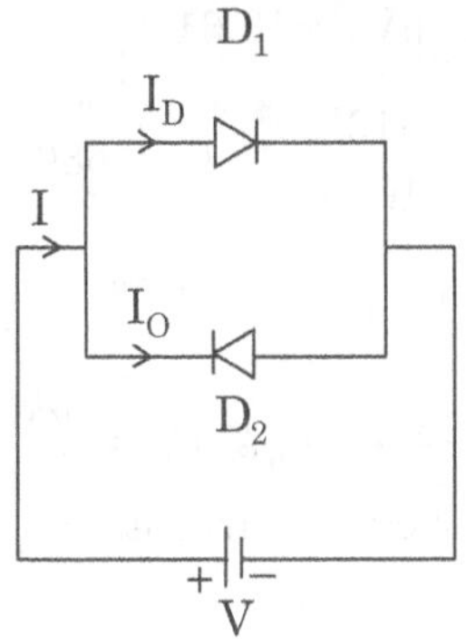

$I = I_O + I_D$
$I = I_O + I_O\,(e^{V/\eta v\tau} - 1)$
$I = I_O\, e^{V/\eta v\tau}$
$V = \eta V\tau\, \text{In}\,(I/I_O)$

$$V = \frac{kT}{q}.\text{In}(I/I_O)[\because \eta = 1 \text{ for Ge}]$$

12. (d) The lifetime of majority carriers in the N-vegion. The switching speed of a $P^+ n$ (heavily doped p – region) Junction depends on the lifetime (t) of majority carriers (electrons) in the N-region (lightly doped region.)

13. (b) Depletion capacitance,

$$C_T = \frac{C_{TO}}{(1 + V_{RB}/V_{bi})^{1/n}}$$

where n = 2 for abrupt p–n junction
V_{bi} = built in potential
V_{RB} = Reverse bias voltage

$$C_T \propto \frac{1}{\sqrt{V_{RB} + V_{bi}}} \propto \frac{1}{\sqrt{V_{RB}}}$$

14. (a) Given : $N_D = 4\,N_A$
According to charge neutrality equation, charge density in the depletion width of both side should be equal. Hence

$$\frac{W_n}{W_P} = \frac{N_A}{N_D} = \frac{N_A}{4N_A} = \frac{1}{4} = 0.25$$

15. (b) For abrupt p–n junction, $C\tau \propto \frac{1}{\sqrt{V_{RB} + V_{bi}}}$

$$\frac{C\tau_2}{C\tau_1} = \sqrt{\frac{V_{RB_1} + V_{bi_1}}{V_{RB_2} + V_{bi_2}}} = \sqrt{\frac{1+0}{1+99}} = \sqrt{\frac{1}{100}} = \frac{1}{10}$$

$$C\tau_2 = \frac{C\tau_1}{10} = \frac{1}{10} = 0.1\eta F/cm^2$$

16. Case 1 : When switch is at position 1. Diode will be in forward bias and will act as short–circuit and therefore current I is

$$I = \frac{20}{10} = 2\text{mA}.$$

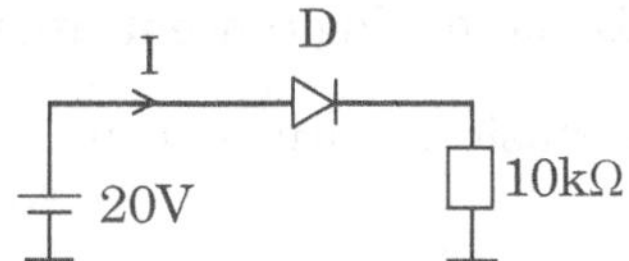

Case 2 : When switch is at position 2. Diode is in reverse bias but due to stored excess minority charge carriers, current flows in opposite direction till storage time is.

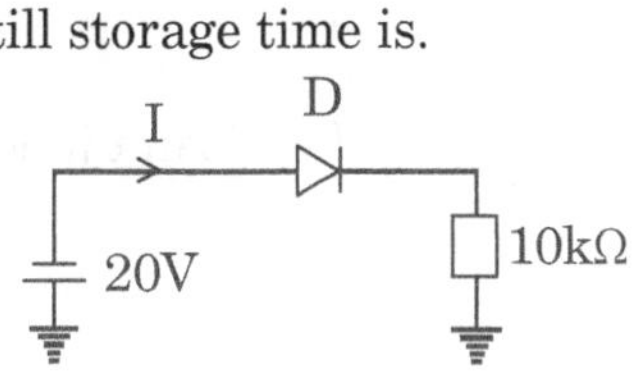

After storage time, current will reduce to reverse saturation current saturation current is due to minority charge carriers.

Reverse recovery time

$t_{rr} = t_s + t_r$.

Hence the current through the 10 KΩ resistor at $t = t_0$ is $I = -2$ mA

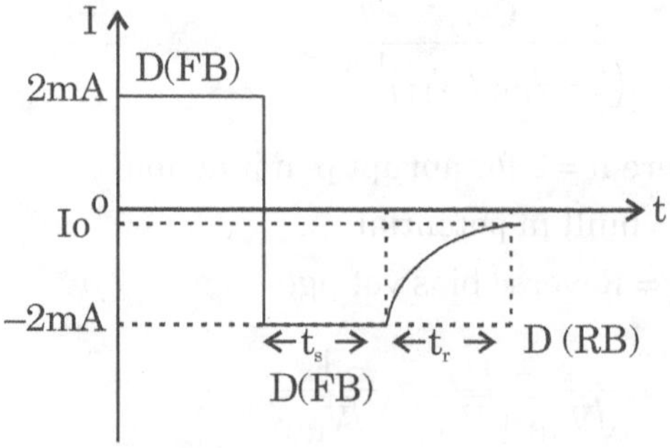

17.

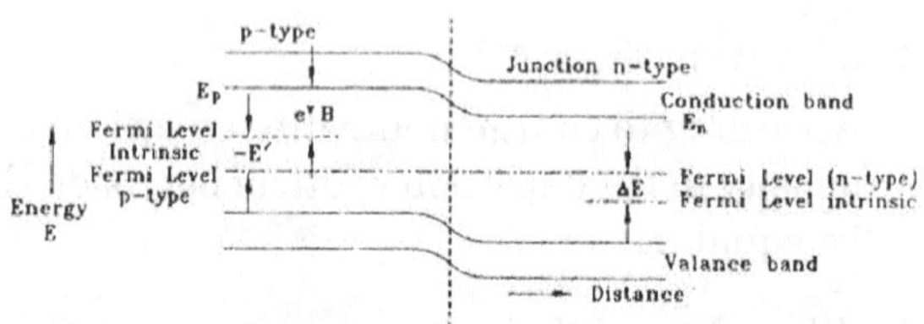

Fig. 6.6. Energy band diagram for p-n junction

Energy band diagram for p-n junction

$$E_p - E_n = e^V$$

B, barrier potential = ΔE′ + ΔE,

$$V_B = \frac{kT}{e}\ln\left(\frac{N_a N_d}{n_i^2}\right)$$

18. The diffusion or barrier or junction is given by

$$V_o \frac{KT}{q}\log\left(\frac{N_A N_D}{n_i^2}\right)$$

where V_o = Diffusion potential,

q = Election charge,

N_A = Acceptor doping concentration in p material

N_D = Donor doping concentration in n material

n_i = Intrinsic impurity density

19. $$C_J = \frac{C_0}{\left|1+\frac{V_R}{V_T}\right|^n}$$

where, $n = \frac{1}{2}$ for abrupt junction

20. Static characteristic of a forward biased p-n junction

$$I = I_S\,(e^{Va/Vr} - 1)$$

$$V_T \log\left(\frac{I}{I_s}+1\right) = V_a$$

Straight line is log I *vs* V.

22. Diode current, $I = I_0\left(e^{V_{D/n}V_{T\_1}}\right)$

I_0 is the saturation current, For silicon, η = 2,

Now, At $T_1 = 273 + 20 = 293$ K, $V_{D1} = 0.70$ volts

$$I_1 = I_0\left(e^{\frac{V_D}{\eta V_{T_1}}} - 1\right) \quad \text{... (A)}$$

$$V_{T1} = \frac{293.}{11600} = 0.025\text{V}$$

At $T_2 = 273 + 40 = 313$ K

and $I_{0_2} = I_{0_1}\cdot 2^{\Delta T/10} = I_{0_1}\cdot 2^2 = 4I_{0_1}, V_{T_2} = \dfrac{373}{11600}$

$$I_2 = I_{02}\left(e^{V_{D2}/\eta V_{T2}} - 1\right) \quad \text{... (B)}$$

As diode current remains unchanged,

$I_1 = I_2$

or, $I_{0_1}\left(e^{\frac{V_{D1}}{\eta V_{T1}}} - 1\right) = 4I_{0_1}\left(e^{\frac{V_{D_2}}{\eta V_{T2}}} - 1\right)$

or, $3 = 4e^{\frac{V_{D2}}{\eta V_{T2}}} - e^{\frac{V_{D1}}{\eta V_{T2}}}$

$$3 = 4e^{\frac{V_{D2}}{\eta V_{T2}}} - e^{\frac{0.70}{2\times 0.025}}$$

or, $V_{D_2} = \eta V_{T_2} \times 12.61$

$$= 2\times\frac{313}{11600}\times 12.61 = 680 \text{ mV}$$

23. For Si, $I = I_{0,Si}\,(1 - e^{0.718/\eta V_T})$

and for Ge, $I = I_{0,}\,ge(1 - e^{0.1435/\eta V_T})$

Hence for equal currents, $\eta_{Si} = 2$; $\eta_{Ge} = 1$

$$\frac{I_{0,Si}}{I_{0,Ge}} = \frac{1 - e^{-0.1435/V_T}}{1 - e^{-0.359/V_T}} \approx e^{-0.2155/V_T}$$

or $$\frac{I_{0,Ge}}{I_{0,Si}} \approx e^{(0.2155/26\times 10^{-3})} = 4\times 10^3$$

24. Since the net change must be zero, then

$$N_A W_p = N_D W_n \quad \dots (i)$$

and $$W_p + W_n = 3\ \mu m \quad \dots (ii)$$

Solving equations (i) and (ii) with given values, we have

$$W_p = 0.3\ \mu m$$

25. Junction capacitance

$$C_j = \frac{tA}{W}$$

and $$W = \sqrt{\frac{2t}{qN_D} Vj}$$

N_D= donor density

and V_j= function potential

where

Now, $$C_j = V_j^{-½}$$

or, $$\frac{C_{j1}}{C_{j2}} = \sqrt{\frac{V_{j2}}{V_{j1}}} \Rightarrow C_{j2} = C_{j1}\sqrt{\frac{V_{j1}}{V_{j2}}} = 0.5PF$$

26. Reverse saturation current, at temperature, T_2,

$$I(T_2) = I(T_1) \times 2^{\frac{T_2 - T_1}{10}}$$

or, $I(40°C) = I(20°C) \times 2^{20/10} = 10 \times 4 = 40$ pA

27. $$C_T = \frac{\varepsilon_r \varepsilon_0 A}{d}$$

$$\therefore \quad \frac{C_T}{A} = \frac{\varepsilon_r \varepsilon_0}{d} = \frac{11.7 \times 8.85 \times 10^{-12}}{10 \times 10^{-6}} = 10\mu F$$

28.

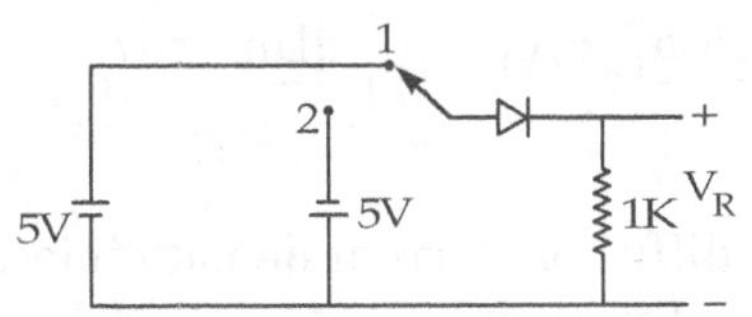

The injected charge carriers across the junction can not be removed instantaneously, but takes certain time (t_s). Thus for time t_s, diode will remain forward biased and will conduct.

Since, voltage drop across diode is zero, hence

$$V_R = -5V \text{ for } 0 \le t < t_s.$$

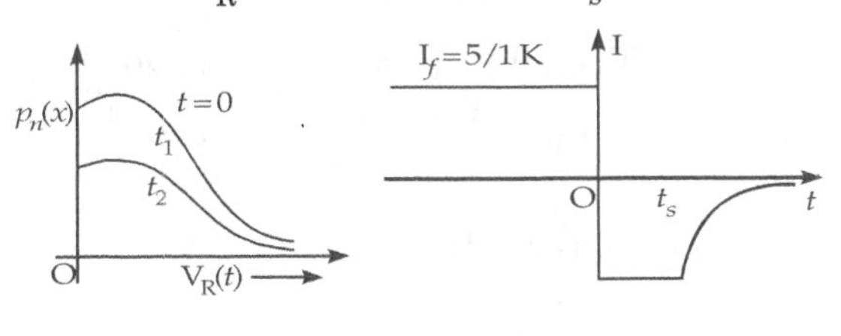

29. For PN junction

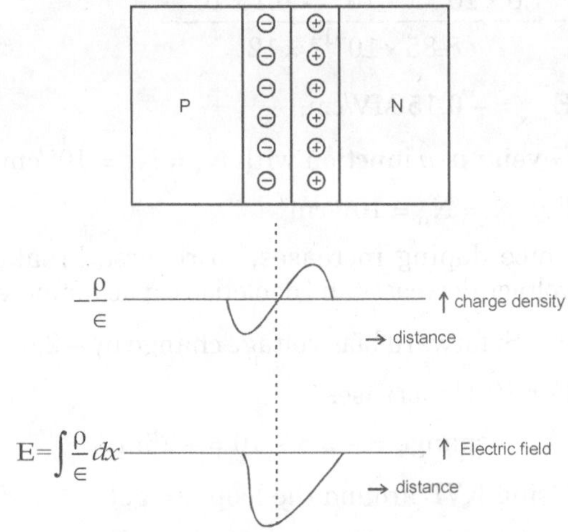

So maximum E field will be at the junction.

30. Depletion capacitance,

$$C_d \propto V^{-1/2}$$

$$\frac{\varepsilon A}{d} \propto V^{-1/2}$$

$$\therefore \quad d \propto V^{+1/2}$$

$$\therefore \quad \frac{d_2}{d_1} = \left(\frac{V_2}{V_1}\right)^{1/2}$$

$$\Rightarrow \quad d_2 = d_1\left(\frac{V_2}{V_1}\right)^{1/2} = (2\mu m)\left(\frac{7.2+0.8}{1.2+0.8}\right)^{1/2} = 4\mu m$$

31. Channel length modulation is related to FET.

32. (b) From change neutrality equation , $N_A W_P = N_D W_N$.

$$N_A = \frac{N_D W\eta}{W_P} = \frac{10^{17} \times 0.1}{1} = 10^{16}\ cm^{-3}$$

Built in potential,

$$V_{bi} = V_T \, l\eta\left(\frac{N_A N_D}{\eta_i^2}\right) = 26 \times 10^{-3}.\ln\left[\frac{10^{16} \times 10^{17}}{\left(1.4 \times 10^{10}\right)^2}\right]$$

Vbi = 0.7 V

33. (b) The electric field will be directed from n to p. The maximum electric field occurs at the junction. Hence,

$$E_{max} = \frac{qN_D W\eta}{\epsilon} = -\frac{qN_A W_P}{\epsilon} =$$

$$-\frac{1.6\times10^{-19}\times10^{17}\times0.1\times10^{-4}}{8.85\times10^{-14}\times12}$$

$E_{max} = -0.15$ MV/cm.

34. Given : p–n junction with $N_A = N_D = 10^4/cm^3$

For $N_A = N_D = 10^{20}/cm^3$

Since doping increases, so reverse breakdown voltage decreases and depletion capacitor increases.

35. For Si forward bias voltage change by – 2.5mV/°C

For 10°C increases,

change = – 2.5 × 10 = – 25mV

36. Using KVL around the loop, we get

$10 - 1000i - V = 0$ ...(1)

Given $i = \frac{V-0.7}{500}$

or, $V = 500i + 0.7$ ...(2)

Putting equation (2) in equation (1), we have

$10 - 1000i - (500i + 0.7) = 0$

or, $i = \frac{10-0.7}{1500} = 6.2\times10^{-3}$ A = 6.2mA

37. Due to application of voltage (forward bias minority carrier are injected from either side of diode one subsequent diffusion takes place and finally recombination.

Injection and subsequent diffusion and recombination of minority carriers

38. Given $x_n = 0.2\ \mu m$,

$\epsilon_{Si} = 1.044\times10^{-12}$ F/μ_n

$N_D = 10^{16}/cm^3$

Peak Electric field,

$$E = \frac{qN_D x_n}{\epsilon}$$

$$= \frac{1.6\times10^{-19}\times10^{16}\times0.00002}{1.044\times10^{-12}}$$

= 30.66 KV/cm

39. $V_{bi} = V_T \ln\frac{N_A N_D}{n_i^2}$

$$= 26 \text{ mv } \ln\left[\frac{5\times10^{18}\times1\times10^{16}}{(1.5\times10^{10})^2}\right]$$

= 0.859 V

$$W = \sqrt{\frac{2\varepsilon_S V_{bi}}{q}\left[\frac{N_A+N_D}{N_A N_D}\right]}$$

$= 3.34\times10^{-5}$ cm

40. n Tunnel diode, both the P region and N region are heavily doped. It shows the negative differential resistance in the current (–) voltage characterstics.

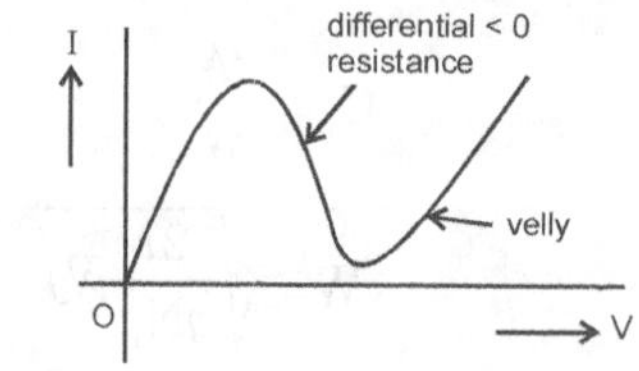

41. $C_j \propto \frac{1}{\sqrt{V_{b1}+V_R}}$

$$\frac{C_{2j}}{C_{1j}} = \sqrt{\frac{V_{b1}+V_{R_1}}{V_{b1}+V_{R_2}}}$$

$$C_{2j} = C_{1j}\sqrt{\frac{2}{8}} = \frac{C_{1j}}{2} = 2.5\ pF$$

42. Diode current, $I_D = I_O\,(e^{V/V\tau} - 1)$------(i)

where, $I_O = \frac{AqDpP\eta o}{Lp} + \frac{AqDn\eta po}{L\eta}$..............(ii)

Current density, $j = \frac{I_D}{A}$...........(iii)

Using eq. (i), (ii) and (iii),

$$J = \frac{qDpPno}{L_p}\left(e^{V/V\tau}-1\right) + \frac{qDn.\eta po}{L_n}\left(e^{V/V\tau}-1\right)$$

Electron diffusion current density injected from n– region ot p–region is

$$J_n = \frac{qDn.\eta po}{Ln}\left(e^{V/V\tau}-1\right)$$

Hole diffusion current density injected from p–segion or n-region is

$$J_p = \frac{qD_p.pno}{L_p}\left(e^{V/V\tau}-1\right) = \frac{q\eta_i^2.Dp}{N_D L_P}\left(e^{V/V\tau}-1\right)$$

$\because p\eta_o = n_1^2 / NdD.$

Diffusion length,

$$L_p = \sqrt{\tau_P D_P} = \sqrt{100 \times 10^{-6} \times 36} = 0.06\text{cm}.$$

$\because V\tau = 26\text{mv}/1$

$$Jp = \frac{1.6 \times 10^{-19} \times (10^{10})^2 \times 36}{10^{15} \times 6 \times 10^{-2}}.$$

$$\left(e^{208/26} - 1\right) = 28.6 \times 10^{-9}\,\text{A}/\text{cm}^2$$

$Jp = 28.6\ \text{nA/cm}^2$.

43. (c) Given $E_{max} = 10^4$ V/cm $= 10^6$ V/m.

$W = 1 - (-0.1)\ \mu\text{m} = 1.1 \times 10^{-6}$ m.

From poisson's equation,

$$\frac{d2V}{dx_2} = -\rho v/E \text{ and } \underline{|T} = -\frac{dv}{dx}$$

$$\frac{dE}{dx} = sv/E$$

$$E = \int (sv/\epsilon)\,dv$$

$E \propto \int sv$, where sv = change density given, electric field profile is of p-n junction.

⇒ option (A) is true because left side of junction is n–type and right side is p–type.

⇒ option (B) is true because sv = constant, i.e., uniformly doped.

⇒ Potential difference,

$$V = -\int E dx = -(\text{Area under E} - \text{x curve})$$

$$V = -\frac{1}{2} \times 1.1 \times 10^{-4} \times 10^4 = -550\text{mV}.$$

option (c) is Not true.

⇒ From change neutrality equation, $N_A W_P = N_D W_n$

⇒ $N_D = 10^{16}$ cm^{-3}. option (d) is also true.

44. From given data

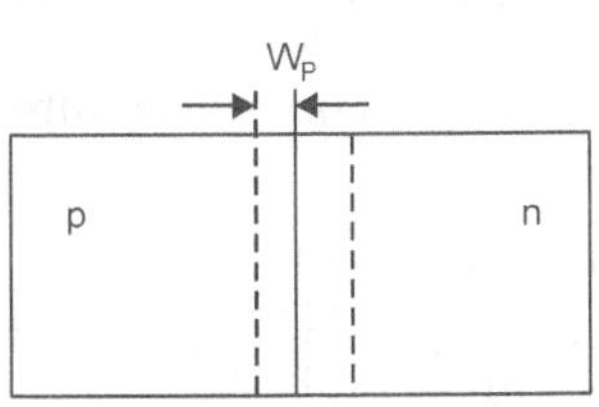

$\varepsilon_{si} = 12\,\varepsilon_0$

$= 12 \times 8.85 \times 10^{-14}$ F/m

$N_D = 10^{16}$ cm^{-3}

$= 10^{22}$ m^{-3}

$N_A = 10^{17}$ cm^{-3}

$= 10^{23}$ m^{-3}

$$V_0 = \frac{kT}{q}\ell n\left[\frac{N_A N_D}{ni^2}\right]$$

$$= 0.026\ell n\left[\frac{10^{23} \times 10^{22}}{\left(1.5 \times 10^{16}\right)^2}\right]$$

$= 0.757$ V

$$W = \sqrt{\frac{2\varepsilon}{q} V_0 \left(\frac{1}{N_A} + \frac{1}{N_D}\right)}$$

$$= \sqrt{\frac{2 \times 12 \times 8.85 \times 10^{-14}}{1.6 \times 10^{-19}} \times 0.757 \left(\frac{1}{10^{23}} + \frac{1}{10^{22}}\right)}$$

$= 3.325 \times 10^{-8}$ m

$= 3.325 \times 10^{-6}$ cm

$$W_p = \frac{N_D}{N_A + N_D}\omega$$

$$= \frac{10^{22}}{10^{22} + 10^{23}} \times 3.325 \times 10^{-8}$$

$= 3.023 \times 10^{-9}$ m

$= 3.023 \times 10^{-7}$ cm

$Q = W_P N_A eA$

$$\Rightarrow \quad \frac{Q}{A} = W_P N_A$$

$e = 3.023 \times 10^{-7} \times 10^{17} \times 1.6 \times 10^{-19}$

$= 4.836 \times 10^{-9}$ cm^{-2}

$= 4.836$ nC-cm^{-2}

Hence, the charge per unit function area (nC-cm^{-2}) in the depletion region on the *p*-side is 4.836 nC-cm^{-2}.

45. In any type of PN junction

$$V_{BR} \propto \frac{1}{\text{Doping Concentration}}$$

$$\text{i.e. } V_{BR} \propto \frac{1}{N_D} \text{ (or) } V_{BR} = \frac{\varepsilon E^2}{2qN_D}$$

(Here V_{BR} = break down voltage, N_D = donor density)

$\therefore\ V_{BR} \times N_D \rightarrow$ is a constant

46. From the given fig :

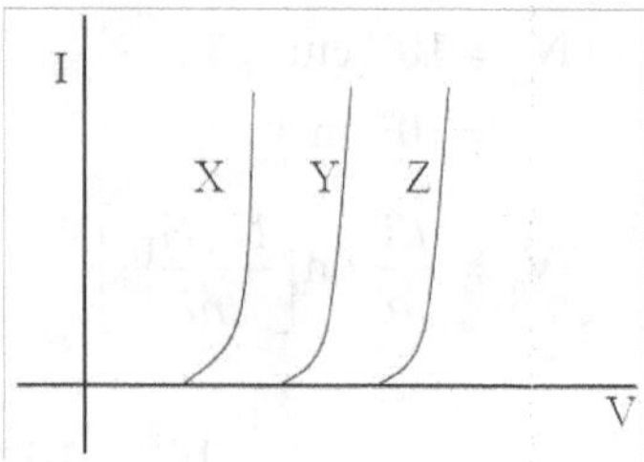

$$V_{r_3} > V_{r_2} > V_{r_1}$$
$$V_r \propto E_g$$

Where V_r is cut-in voltage

So, $\quad E_{gZ} > E_{gY} > E_{gX}$

47.

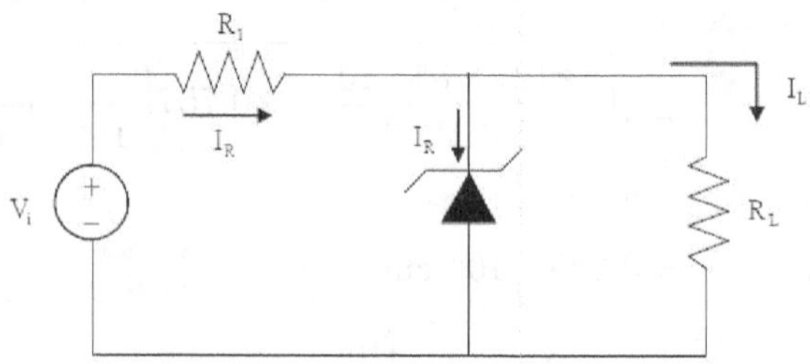

$$I_L = \frac{20-0}{1} mA = 20mA$$

At node A

Now, $I_R = I_Z + I_L$

Now, $V_{i\,min} = I_{R_{min}} R_1 + V_Z$

$$I_{R_{min}} = 0 + I_2 = 20mA.$$

$$V_{i\,min} = (20 \times 10^{-3} \times 200) + 20 = 4V + 20v = 24v$$

and $V_{i\,max} = I_{R_{max}} + R_1 + V_Z$

Now, $I_{R_{max}} = I_{z\,max} + I_L = 60mA + 20mA = 80mA$

$\therefore\ V_{i\,max} = 80 \times 10^{-3} \times 200 + 20 = 16V + 20V = 36V$

$\therefore\ V_{i\,max} = 24V$ and $V_{i\,max} = 36V$

48. (c) Both P and N regions are heavily doped. Doping level of zener diode is $1 : 10^5$

49. In Zener diode the carriers are accelerated by electric field. When they collide with atoms they ionise the atoms due to their kinetic energy. Hence an avalanche break down occurs due to the large increase in the number of carriers.

50. S1 $\rightarrow$ True, in linearly graded function avalanche breakdown occurs.

S2 $\rightarrow$ For tunneling phenomenon to occur narrow energy barrier is required to charge tunnel from filled state in one side to empty state in other side.

Also, tunneling is independent on the doping profile of junction.

51. Zener Diode I-V Characteristics

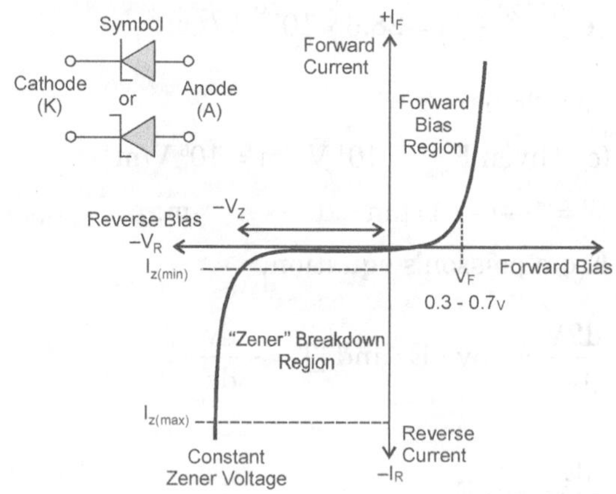

From the I-V characteristics curve above, we can see that the zener diode has a region in its reverse bias characteristics of almost a constant negative voltage regardless of the value of the current flowing through the diode and remains nearly constant even with large changes in current as long as the zener diodes current remains between the breakdown current $I_{z(min)}$ and the maxium current rating $I_{z(max)}$.

This ability to control itself can be used to great effect to regulate or stabilise a voltage source against supply or load variations and this is accomplished when using zener diode in reverse-bias region below the breakdown voltage.

52. $E = \frac{hc}{\lambda} = \frac{6.626 \times 10^{-34} \times 3 \times 10^8}{5490 \times 10^{-10}}$

$= 3.62 \times 10^{-19}$ Joules

In eV, the energy Band gap,

$$E_g = \frac{E}{e} = \frac{3.62 \times 10^{-19}}{1.6 \times 10^{-19}} = 2.26 \text{ eV}$$

54. Given Voc = 0.65 at operating internity x = 1 son, n = 1

$I_{SC} = 20$ mA, A = 1 cm^2

$$Vo_c^1 = \eta \frac{kT}{q} \ln\left[\frac{Isc}{Io}.x\right]$$

when x = 1, $V_{oc}^{1} = V_{oc} = \frac{kT}{q}.\ln\left(\frac{Isc}{Io}\right) = 0.65V.$

When x = 0.2 $V_{oc}^{1} = \eta \times \frac{kT}{q}\ln\left(\frac{Isc}{Io} \times 0.2\right) = \frac{kT}{Q}$

$$\ln\left(\frac{Isc}{Io}\right) + \frac{kT}{q}\ln(0.2) = 0.65 + 0.026 \times (-1.6)$$

$= 0.65 - 0.0416$

$V_{oc}^{1} = 0.608$ V.

55. Responsivity $(R) = \frac{e\eta}{hv} = \frac{e\eta s}{hc} = \frac{\eta \times \lambda}{1.24}(A/W)$

56. (c)

Diode	LED	Photodiode	Tunnel Diode	Zener diode	Pin Diode	Solar Cell	Laser Diode
Biasing	Forward Bias	Reverse bias	Forward bias	Reverse bias	Reverse bias	Forward bais	Forward bias

57. (c) LED : When an atom absorb extra energy and goes in excited state, then to return to its normal or ground state it emits extra energy (photon) at an undetermined time. This unpredictable release of photon energy by an atom is called spontaneous emission. Avalanche Photodiode : Current gain, $M = \frac{I_M}{I_P} > 1.$

Tunnel diode : Heavily doped (1: 10^3) or 10^{20} cm^{-3} LASER : coherent radiation takes place.

58.

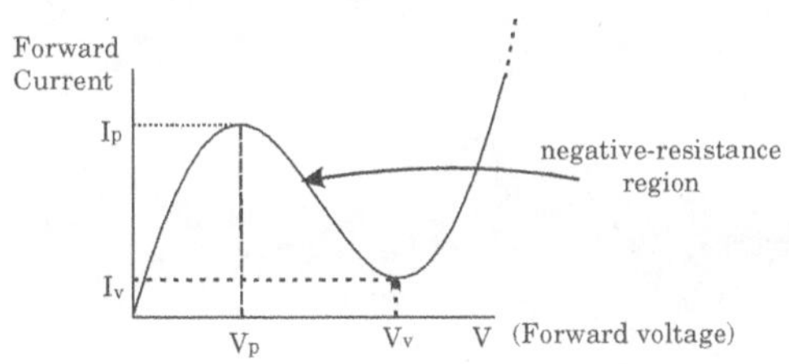

59. (c) Varactor diode : used in tuned circuit.

PIN diode : due to sandwitched intrinsic semiconductor between two extrinsic semiconductor, it can be used in current controlled attenuator.

Zener diode : Used as voltage regulator in voltage reference circuit.

Schottky diode : is the fastest device and hence used for high frequency switching.

60. P. Zener diode is used in reverse bias to give fixed zener voltage across it.

Q. Solar cell operates in forward bias.

R. Laser diode operate in very high voltage forward bias to gives population insertion

S. Photo diode operate in reverse bias in avalanche region.

61. Responsivity $(R) = \frac{I_P}{P_0}$

where I_P = photo current

P_0 = Incident power

$I_P = R \times P_0$

$= 8$ A

62. Fill factor $= \dfrac{P_{max}}{P_T} = \dfrac{P_{max}}{I_{SC} \cdot V_{OC}}$

where P_{max} = Maximum power

P_T = Theoretical power

$= I_{SC} \cdot V_{OC}$

$$0.7 = \frac{P_{max}}{180 \times 10^{-3} \times 0.5}$$

$\therefore \quad P_{max} = 63 \times 10^{-3}$ W

Now Maximum efficiency

$$(\eta_{max}) = \frac{P_{max}}{P_{in}}$$

$$= \left(\frac{63 \times 10^{-3}\,\text{W}}{100 \times 10^{-3} \dfrac{\text{W}}{\text{cm}^2} \times 3\text{cm}^2} \times 100 \right)\%$$

$= 21\%$

Thus the maximum efficiency (in %) of the device is 21%.

64. $Voc = V_T . \ln(J_L / J_5); J_L \propto$ light intensity

$$Voc_2 - Voc_1 = V_T . \ln\left(\frac{J_{L2}}{J_{L1}}\right) = 25\ln(20) \cong 75\text{mv} = 0.075\text{V}.$$

$Voc_2 = 0.451\text{V} + 0.075\text{V} = 0.526$ V.

3 CHAPTER

Bipolar Junction Transistor

Analysis of Previous GATE Papers												
	Year → Topics ↓		2019	2018	2017 Set 1	2017 Set 2	2016 Set 1	2016 Set 2	2016 Set 3	2015 Set 1	2015 Set 2	2015 Set 3
FABRICATION	1 Mark	MCQ Type	1									
		Numerical Type										
	2 Marks	MCQ Type										
		Numerical Type	1									
	Total		3									
CONFIGURATION & BIASING	1 Mark	MCQ Type			1	1						1
		Numerical Type										
	2 Marks	MCQ Type										
		Numerical Type										1
	Total				1	1						3

FABRICATION

1. The impurity commonly used for realizing the base region of a silicon n-p-n transistor is
(a) Gallium
(b) Indium
(c) Boron
(d) Phosphorus

[2004 : 1 Mark]

2. The neutral base width of a bipolar transistor, biased in the active region, is 0.5 μm. The maximum electron concentration and the diffusion constant in the base are $10^{14}/cm^3$ and $D_n = 25\ cm^2/sec$ respectively. Assuming negligible recombination in the base, the collector current density is (the electron charge is 1.6×10^{-19} Coulomb)
(a) 800 A/cm^2
(b) 8 A/cm^2
(c) 200 A/cm^2
(d) 2 A/cm^2

[2004 : 2 Marks]

3. The correct circuit representation of the structure shown in the figure is

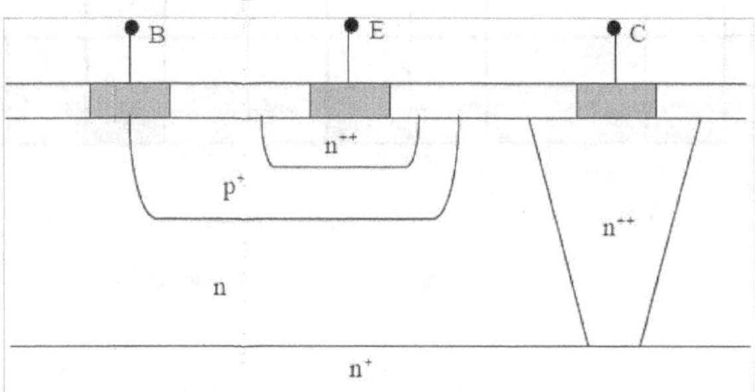

(A) 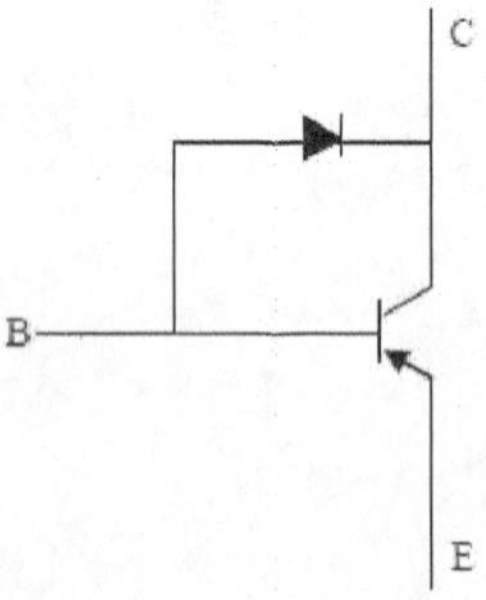

(B) 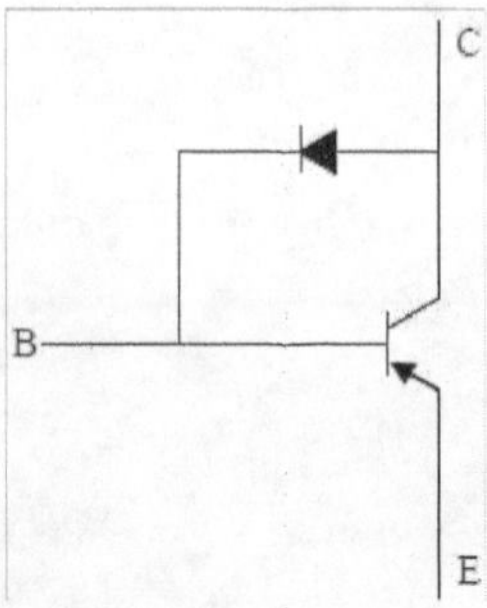

(C) 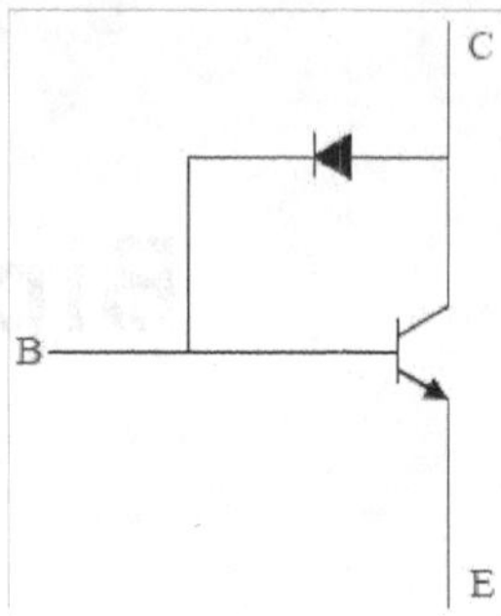

(D) 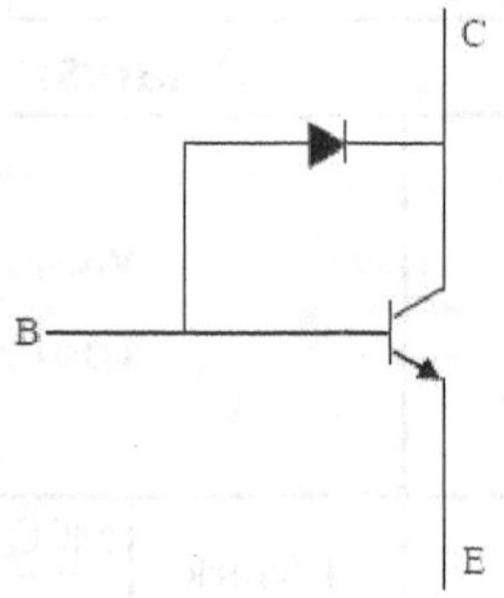

[2019 : 1 Mark]

4. A Germanium sample of dimension 1 cm × 1 cm is illuminated with a 20 mW, 600 nm laser light source as shown in the figure. The illuminated sample surface has a 100 nm of loss-less Silicon dioxide layer that reflects one-fourth of the incident light. From the remaining light, one-third of the power is reflected form the silicon dioxide-Germanium interface, one-third is absorbed in the Germanium layer, and one-third is transmitted through the other side of the sample. If the absorption coefficient of Germanium at 600 nm is $3 \times 10^4\ cm^{-1}$ and the bandgap is 0.66 eV, the thickness of the Germanium layer, rounded off to 3 decimal places, is _____ μm.

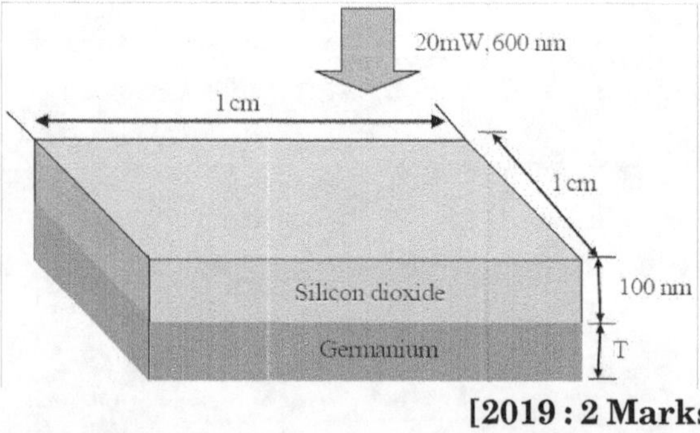

[2019 : 2 Marks]

CONFIGURATION & BIASING

5. The break down voltage of a transistor with its base open is BV_{CEO} and that with emitter open is $BV/_{CBO}$, then
(a) $BV_{CEO} = BV_{CBO}$
(b) $BV_{CBO} > BV_{CBC}$
(c) $BV_{CEO} < BV_{CBO}$
(d) BV_{CEO} is not related to BV_{CBO}

[1995 : 1 Mark]

6. A BJT is said to be operating in the saturation region if

(a) both junctions are reverse biased.
(b) base-emitter junction is reverse biased and base collector junction is forward biased.
(c) base-emitter junction is forward biased and base-collector junction reverse biased.
(d) both the junctions are forward biased.

[1995 : 1 Mark]

7. The early-effect in a bipolar junction transistor is caused by

(a) fast-turn-on
(b) fast-turn-off
(c) large collector-base reverse bias
(d) large emitter-base forward bias

[1995 & 1999 : 1 Mark]

8. If a transistor is operating with both of its junctions forward biased, but with the collector base forward bias greater than the emitter-base forward bias, then it is operating in the

(a) forward active mode
(b) reverse saturation mode
(c) reverse active mode
(d) forward saturation anode

[1996 : 2 Marks]

9. In a bipolar transistor at room temperature, if the emitter current is doubled, the voltage across its base-emitter junction

(a) doubles
(b) halves
(c) increases by about 20 mV
(d) decreases by about 20 mV

[1997 : 2 Marks]

10. If for a silicon n-p-n transistor, the base-to-emitter voltage (V_{BE}) is 0.7 V and the collector-to-base voltage (V_{CB}) is 0.2 V, then the transistor is operating in the

(a) normal active mode
(b) saturation mode
(c) inverse active mode
(d) cut use-off mode

[2004 : 1 Mark]

11. Consider the following statements S_1 and S_2.

S_1: The β of a bipolar transistor reduces if the base width is increased.

S_2: The β of a bipolar transistor increases if the doping concentration in the base is increased.

Which one of the following is correct?

(a) S_1 is FALSE and S_2 is TRUE
(b) Both S_1 and S_2 are TRUE
(c) Both S_1 and S_2 are FALSE
(d) S_1 and TRUE and S_2 is FALSE

[2004 : 1 Mark]

12. The phenomenon known as "Early Effect" in a bipolar transistor refers to a reduction of the effective base-width caused by

(a) electron-hole recombination at the base
(b) the reverse biasing of the base-collector junction
(c) the forward biasing of emitter-base junction
(d) the early removal of stored base charge during saturation-to-cut-off switching

[2006 : 1 Mark]

13. In a uniformly doped BJT, assume that N_E, N_B and N_C are the emitter, base and collector dopings in atoms/cm^3, respectively. If the emitter injection efficiency of the BJT is close to unity, which one of the following conditions is TRUE?

(a) $N_E = N_B = N_C$
(b) $N_E >> N_B$ and $N_B > N_C$
(c) $N_E = N_B$ and $N_B < N_C$
(d) $N_E < N_B < N_C$

[2010 : 2 Marks]

14. For a BJT, thecommon-base current gain α = 0.98 and the collector base junction reverse bias saturation current I_{co} = 0.6 μA. This BJT is connected in the common emitter mode and operated in the active region with a base drive current I_B = 20 μA. The collector current I_c for this mode of operation is

(a) 0.98 mA (b) 0.99 mA
(c) 1.0 mA (d) 1.01 mA

[2011 : 2 Marks]

15. A BJT is biased in forward active mode. Assume V_{BE} = 0.7 V, kT/q = 25 mV and reverse saturation current $I_s = 10^{-13}$ mA. The transconductance of the BJT (in mA/V) is.

[2014 : 2 Marks, Set-1]

16. Consider two BJTs biased at the same collector current with area, A_1 = 0.2 μm × 0.2 μm and A_2 = 300 μm. Assuming that all other device parameters are identical kT/q = 26 mV, the intrinsic carrier concentrations is 1×10^{10} cm^{-3}, and q = 1.6×10^{-19} C, the difference betweenthe base- emitter voltages (in mV) of the two BJTs (i.e., $V_{BE1} - V_{BE2}$) is ______.

[2014 : 2 Marks, Set-4]

17. If the base width in a bipolar junction transistor is doubled, vvriieh one of the follovving statements will be TRUE?

(a) Current gain will increase.
(b) Unity gain frequency will increase.
(c) Emitter base junction capacitance will increase.
(d) Early voltage will increase.

[2015 : 1 Mark, Set-3]

18. An npn BJT having reverse saturation current $I_s = 10^{-15}$A is biased in the forward active region with $V_{BE} = 700$ mV. The thermal voltage (V_T) is 25 mV and the current gain (β) may vary from 50 to 150 due to manufacturing variations. The maximum emitter current (in μA) is ______.

[2015 : 2 Marks, Set-3]

19. For a narrow base PNP BJT, the excess minority carrier concentrations (Δn_E for emitter, Δp_B for base, Δn_c for collector) normalized to equilibrium minority carrier concentrations (n_{E0} for emitter, p_{B0} for base, n_{c0} for collector) in the quasi-neutral emitter, base and collector regions are shown below. Which one of the following biasing modes is the transistor operating in?

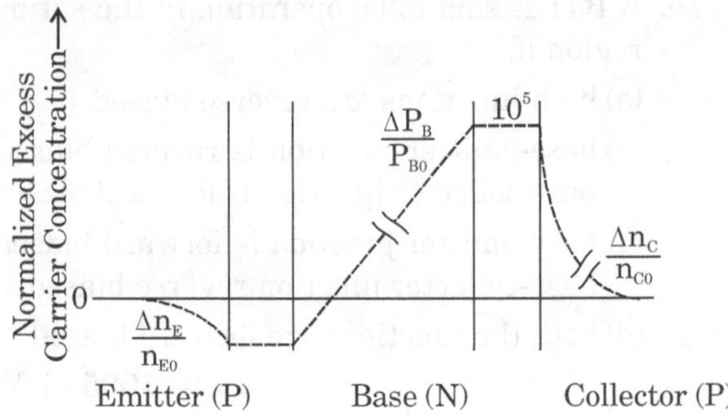

(a) Forward active
(b) Saturation
(c) Inverse active
(d) Cut-off

[2017: 1 Mark, Set-1]

20. An npn bipolar junction transistor (BJT) is operating in the active region. If the reverse bias across the base-collector junction is increased, then

(a) the effective base width increases and common-emitter current gain increases
(b) the effective base width increases and common-emitter current gain decreases
(c) the effective base width decreases and common-emitter current gain increases
(d) the effective base width decreases and common-emitter current gain decreases

[2017 : 1 Mark, Set-2]

ANSWERS

1. (c)	**2.** (b)	**3.** (d)	**4.** (0.115)	**5.** (c)	**6.** (d)	**7.** (c)	**8.** (c)	**9.** (b)	**10.** (c)
11. (b)	**12.** (b)	**13.** (b)	**14.** (d)	**15.** 5.7 to 5.9		**16.** 381	**17.** (d)	**18.** 1475	**19.** (c)
20. (c)									

EXPLANATIONS

1. Boron is trivalent atom. So it acts as an acceptor type impurity.

2. $J_p = D_n q\frac{\partial N}{\partial x} = \frac{25\times1.6\times10^{-19}\times10^{14}}{0.5\times10^{-6}\times10^{2}} = 8\ A/cm^2$

3. Aluminium metal in contact with a lightly doped n type silicon forms a non ohmic rectifying contact because the trivalent aluminium easily dissolves into silicon and convert the part of n type semiconductor into p type.

4.

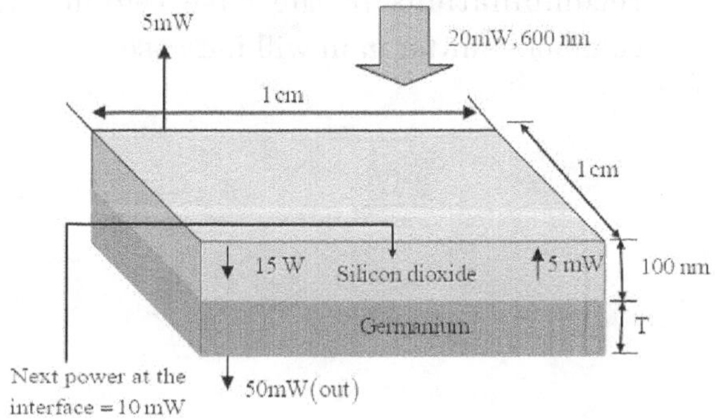

Given

$$\alpha = 3\times10^4\ \text{cm}$$

$$2\alpha = 6\times10^4\ \text{cm}$$

Now, $P(L) = P(0)e^{-2\alpha 1.}$

$$L = -\frac{1}{2\alpha}\ln\frac{P(2)}{P(0)}$$

$$L = -\frac{1}{6\times10^4}\ln\frac{5}{10}$$

$$= 0.11\times10^{-6}\ \text{m}$$

$$= 0.115\ \mu_m$$

5. Because the emitter is the most heavily doped region in the transistor.

   It is about 40 to 50% of B_{CBO}.

6. (d) In saturation region, collector - base junction is forward bias and emitter - base junction is forward bias.

7. (c) The process where the effective base width of the transistor is altered by varying the collector junction voltage is called base width modulation or early - effect. In a transistor, by changing the collector junction voltage, it is seen that the base width of the transistor is altered and this property is called as early - effect.

8. $V_{CE} = V_{BE} + V_{CB}$.

   As V_{CB} is increased, V_{CE} is pushed to the negative region, I_C start even from the negative region and transistor operates in the reverse active mode.

10. $V_{BE} = 0.7\ V$ [forward - biased]

    and $V_{BC} = -V_{CB} = -0.2V$ [Reverse - biased]

    Hence the transistor operates in normal active mode.

11. Reducing base width causes decreasing recombination base current and therefore I_C increases as well as β, whereare, if doping concentration increases the result is vice-versa.

12. In a bipolar transistor, the increase in the depletion region of the reverse biased junction, tends to decrease the width of the base with increased reverse voltage. This in turn causes the current gain to increase with increased voltage resulting in a nearly linear increase in current even with the device in the saturation.

13. N_E, N_B, N_C

    Since emitter injection efficiency of the BJT is close to unity.

    ∴ $N_E > N_B > N_C$

    and $N_E >> N_B$

14. We know $I_C = \beta I_B + (1+\beta)\,I_{CBO}$

    where, $\beta = \frac{\alpha}{1-\alpha} = \frac{0.98}{1-0.98} = 49$

    $I_B = 20\mu A,\ I_{CBO} = 0.6\mu A$

    ∴ $I_C = 1.01\ mA$

15. $$g_m = \frac{I_C}{V_T}$$

$$I_C = I_S\left[e^{\frac{V_{BE}}{V_T}} - 1\right]$$

$$I_C = 10^{-13}\times\left[e^{\frac{0.7}{0.025}} - 1\right]$$

$$I_C = 144.6257\ mA$$

$$g_m = \frac{0.144}{0.025} = 5.78\ \frac{mA}{V}$$

16. $I_{C_1} = I_{C_2}$ (Given)

$$I_{S_1} e^{\frac{V_{BE_1}}{V_T}} = I_{S_2} e^{\frac{V_{BE_2}}{V_T}}$$

$$e^{\frac{(V_{BE_1}-V_{BE_2})}{V_T}} = \frac{I_{S_2}}{I_{S_1}}$$

$$V_{BE_1} - V_{BE_2} = V_T \ln \frac{I_{S_2}}{I_{S_1}}$$

$$= 26\times10^{-3} \ln\left[\frac{300\times300}{0.2\times0.2}\right] \because I_S \alpha A$$

$$\left(V_{BE_1} - V_{BE_2}\right) = 381 \text{ mV}$$

17. If base width is doubled or increase, early effect is still present but its effect is less severe relative to previous base width. Hence, scope of I_c versurs V_{CE} decreases.

18. $$I_B = \frac{I_C}{\beta} = \frac{I_S}{\beta} e^{V_{BE}/V_T}$$

$$I_E = (\beta + 1)I_B = \frac{\beta+1}{\beta} I_S . e^{V_{BE}/V_T}$$

$$= (1.02)(10^{-9} \times 10^{-6}) e^{\frac{700\times10^{-3}}{25\times10^{-3}}}$$

$$= 1475\ \mu A$$

20. (c) when a BJT is in active region, as the reverse bias voltage across collector-base junction increased, the width of deplation region increases, which result in decrease of effective base width. This decrease in effective base width reduces the recombinations in the base region. Hence, common-emitter gain will increase.

4

CHAPTER

Field Effect Transistor

Analysis of Previous GATE Papers												
	Year → Topics ↓		2019	2018	2017 Set 1	2017 Set 2	2016 Set 1	2016 Set 2	2016 Set 3	2015 Set 1	2015 Set 2	2015 Set 3
FABRICATION	1 Mark	MCQ Type						1	1			1
		Numerical Type				1						
	2 Marks	MCQ Type										
		Numerical Type				1	1		1	1	1	1
	Total					3	2	1		2	2	3
CHARACTERISTICS & BIANING	1 Mark	MCQ Type					1					
		Numerical Type										
	2 Marks	MCQ Type				1		1		1		
		Numerical Type	1					1				
	Total		2			2	1	4		2		

FABRICATION

1. If P is Passivation, Q is n-well implant, R is metallization and S is source/drain diffusion, then the order in which they are carried out in a standard n-well CMOS fabrication process, is
(a) P-Q-R-S (b) Q-S-R-P
(c) R-P-S-Q (d) S-R-Q-P
[2003 : 2 Marks]

2. Thin gate oxide in a CMOS process is preferably grown using
(a) wet oxidation (b) dry oxidation
(c) epitaxial deposition (d) ion implantation
[2010 : 1 Mark]

3. In IC technology, dry oxidation (using dry oxygen) as compared to wet oxidation (using stem or water vapour) produces
(a) superior quality oxide with a higher growth rate.
(b) inferior quality oxide with a higher growth rate.
(c) inferior quality oxide with a lower growth rate.
(d) superior quality oxide with a lower growth rate.
[2013 : 1 Mark]

4. Consider the following statements S_1 and S_2.
S_1: The threshold voltage (V_T) of a MOS capacitor decreases with increase in gate oxide thickness
S_2: The threshold voltage (V_T) of a MOS capacitor decreases with increase in substrate doping concentration.
Which one of the following is correct?
(a) S_1 is FALSE and S_2 is TRUE
(b) Both S_1 and S_2 are TRUE
(c) Both S_1 and S_2 are FALSE
(d) S_1 is TRUE and S_2 is FALSE
[2004 : 2 Marks]

5. A MOS capacitor made using p-type substrate is in the accumulation mode. The dominant charge in the Channel is due to the presence of
(a) holes
(b) electrons
(c) positively charged ions
(d) negatively charged ions
[2005 : 2 Marks]

6. The gate oxide thickness in the MOS capacitor is
(a) 50 nm (b) 143 nm
(c) 350 nm (d) 1 μm
[2007 : 2 Marks]

7. The maximum depletion layer width in Silicon is
(a) 0.143 nm (b) 0.857 nm
(c) 1 nm (d) 1.143 nm
[2007 : 2 Marks]

8. Consider the following statements about the C–V characteristics plot:
S_1: The MOS capacitor has as n-type substrate.
S_2: If positive charges are introduced in the oxide, the C–V plot will shift to the left. Then which of the following is true?
(a) Both S_1 and S_2 are true
(b) S_1 is true and S_2 is false
(c) S_1 is false and S_2 is true
(d) Both S_1 and S_2 are false
[2007 : 2 Marks]

9. The cross section of a JFET is shown in the following figure. Let V_G be -2 V and let V_p be the initial pinch-off voltage. If the width W is doubled (with other geometrical parameters and doping levels remaining the same), then the ratio between the mutual transconductances of the initial and the modified JFET is

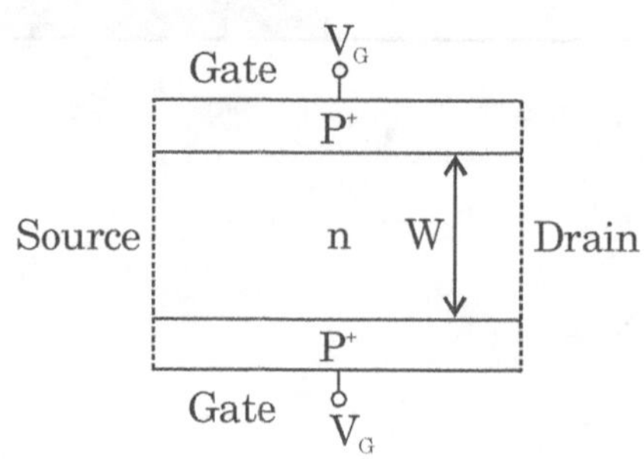

(a) 4 (b) $\frac{1}{2}\left(\frac{1-\sqrt{2/V_p}}{1-\sqrt{1/(2V_p)}}\right)$

(a) $\frac{1-\sqrt{2/V_p}}{1-\sqrt{1/(2V_p)}}$ (d) $\frac{1-\left(2/\sqrt{V_p}\right)}{1-\left(\sqrt{1/\left(2\sqrt{V_p}\right)}\right)}$

[2008 : 2 Marks]

Common Data for Questions 39 and 40:

The channel resistance of an N-channel JFET shown in the figure below is 600 Ω when the full Channel thickness (t_{ch}) of 10 pm is available for conduction. The built-in voltage of the gate P+ – N-junction (V_{bi}) is -1. When the gate to source voltage (V_{GS}) is 0 V, the Channel is depleted by 1 pm on each side due to the built-in voltage and hence the thickness available for conduction is only 8 μm.

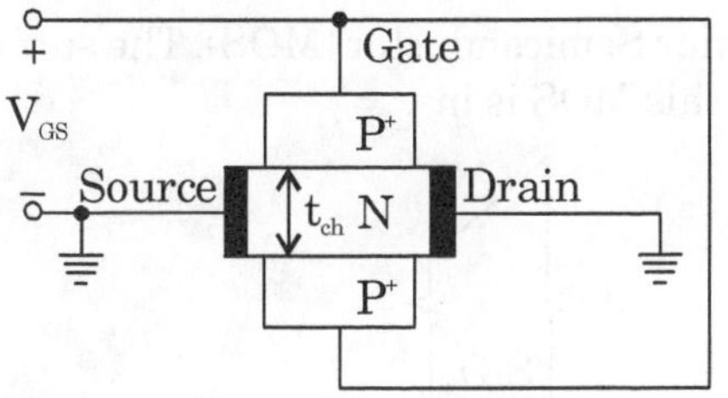

10. The Channel resistance when $V_{GS} = 0$ Vis

(a) 480 Ω (c) 600 Ω

(b) 750 Ω (d) 1000 Ω

[2011 : 2 Marks]

11. The Channel resistance when $V_{GS} = -3$ V is

(a) 360 Ω (c) 917 Ω

(b) 1000 Ω (d) 3000 Ω

[2011 : 2 Marks]

12. The source of a silicon ($n_i = 10^{10}$ per cm^3) n-channel MOS transistor has an area of 1 sq μm and a depth of 1 μm . If the dopant density in the source is $10^{19}/cm^3$, the number of holes in the source region with the above volume is approximately

(a) 10^7 (b) 100

(c) 10 (d) 0

[2012 : 2 Marks]

Common Data for Questions 42 and 43: In the three dimensional view of a silicon n channel MOS transistor shown below, δ = 20 nm . The transistor is of width 1 μm . The depletion width formed at every p-n junction is 10 nm. The relative permittivities of Si and SiO_2, respectively, are 11.7 and 3.9, and $\varepsilon_0 = 8.9 \times 10^{-12}$ F/m.

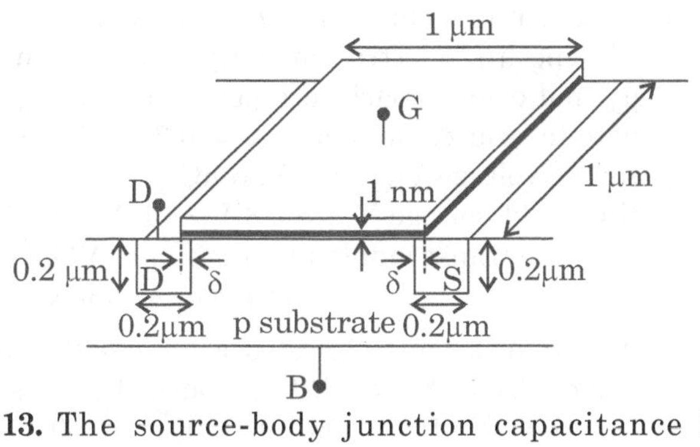

13. The source-body junction capacitance is approximately

(a) 2 fF (b) 7 fF

(c) 2 pF (d) 7 pF

[2012 : 2 Marks]

14. The gate-source overlap capacitance is approximately

(a) 0.7 fF (b) 0.7 pF

(c) 0.35 fF (d) 0.24 pF

[2012: 2 Marks]

15. If fixed positive charges are present in the gate oxide of an n-channel enhancement type MOSFET. it will lead to

(a) a decrease in the threshold voltage.

(b) channel length modulation.

(c) an increase in substrate leakage current.

(d) an increase in accumulation capacitance.

[2014 : 1 Mark, Set-1]

16. In CMOS technology, shallow P-well or N-well regions can be formed using

(a) low pressure chemical vapour deposition.

(b) low energy sputtering.

(c) low temperature dry oxidation.

(d) low energy ion-implantation.

[2014 : 1 Mark, Set-2]

17. In MOSFET fabrication, the channel length is defined during the process of

(a) isolation oxide growth.

(b) channel stop implantation.

(c) poly-silicon gate patterning.

(d) lithography step leading to the contact pads.

[2014 : 1 Mark, Set-3]

18. An ideal MOS capacitor has boron doping-concentration of 10^{15} cm^{-3} in the substrate. When a gate voltage is applied, a depletion region of width 0.5 μm is formed with a surface (channel) potential of 0.2 V. Given that $\varepsilon_0 = 8.854 \times 10^{-14}$ F/cm and the relative permittivities of silicon and silicon dioxide are 12 and 4, respectively, the peak electric field (in V/μm) in the oxide region is _____.

[2014 : 2 Marks, Set-3]

19. A MOSFET in saturation has a drain current of 1 mA for $V_{DS} = 0.5$ V. If the channel length modulation coefficient is 0.05 V^{-1}, the output resistance (in kΩ) of the MOSFET is _____.

[2015 : 2 Marks, Set-1]

20. In MOS capacitor with an oxide layer thickness of 10 mm. The maximum depletion layer thickness is 100 mm. The primitivities of the semiconductor and the oxide layer are ε_s and ε_{ax} respectively.

Assuming $\varepsilon_s \varepsilon_{ox} = 3$ the ratio of the maximum capacitance to the minimum capacitance of this MOS capacitor is ______.

[2015 : 2 Marks, Set-2]

21. Which one of the following processes is preferred to form the gate dielectric (SiO_2) of MOSFETs?

(a) Sputtering

(b) Molecular beam epitaxy

(c) Wet oxidation

(d) Dry oxidation

[2015 : 1 Mark, Set-3]

22. The current in an enhancement mode NMOS transistor biased in saturation mode was measured to be 1 mA at a drain-source voltage of 5 V. When the drain-source voltage was increased to 6 V while keeping gate-source voltage same, the drain current increased to 1.02 mA. Assume that drain to source saturation voltages is much smaller than the applied drain-source voltage. The Channel length modulation parameter λ (in V^{-1}) is .

[2015 : 2 Marks, Set-3]

23. Consider an n-channel metal oxide semiconductor field effect transistor (MOSFET) with a gate-to- source voltage of 1.8 V. Assume that $\frac{W}{L} = 4$, $\mu_n C_x = 70 \times 10^{-6}$ A-V^2, the threshold voltage is 0.3 V and the Channel length modulation parameter is 0.09 V^{-1}. In the saturation region, the drain conductance (in micro-siemens) is

[2016 : 2 Marks, Set-1]

24. A long-channel NMOS transistor is biased in the linear region with V_{DS} = 50 mV and is used as a resistance. Which one of the following statements is NOT correct?

(a) If the device width W is increased, the resistance decreases.

(b) If the threshold voltage is reduced, the resistance decreases.

(c) If the device length L is increased, the resistance increase.

(d) If V_{GS} is increased, the resistance increases.

[2016 : 1 Mark, Set-2]

25. The figure shows the band diagram of a Metal Oxide Semiconductor (MOS). The surface region of this MOS is in

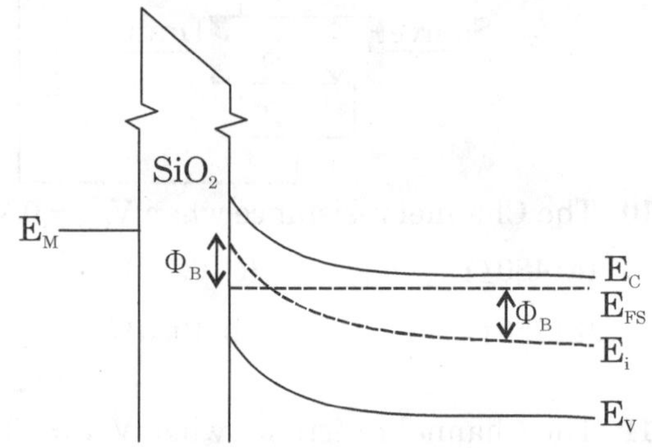

(a) inversion (b) accumulation

(c) depletion (d) flat band

[2016 : 1 Mark, Set-3]

26. Figures I and II show two MOS capacitors of unit area. The capacitor in Figure I has insulator materials X (of thickness t_1 = 1 nm and dielectric constant ε_1 = 4) and /(of thickness t_2 = 3 nm and dielectric constant ε_2 = 20). The capacitor in Figure II has only insulator material X of thickness t_{Eq} . If the capacitors are of equal capacitance, then the value of t_{Eq} (in nm) is_____.

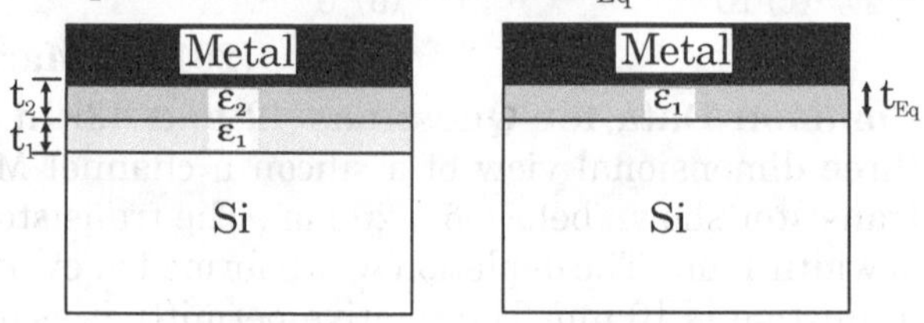

Figure I Figure II

[2016 : 2 Marks, Set-3]

27. Consider an n-channel MOSFET having vvidth W, length L, electron mobility in the Channel μ_{in} and oxide capacitance per unit area C_{ox}. If gate-to- source voltage V_{GS} = 0.7 V, drain-to-source voltage V_{DS} = 0.1 V. (μnC_{ox} = 100 $\mu A/V^2$, threshold voltage V_{Th} = 0.3 V and (W/L) = 50, then the transconductance g_m (in mA/V) is __.

[2017 : 1 Mark, Set-2]

28. A MOS capacitor is fabricated on p-type Si(Silicon) where the metal work function is 4.1 eV and electron affinity of Si is 4.0 eV. E_c - E_F = 0.9 eV, where E_c and E_F are the conduction band minimum and the Fermi energy levels of Si respectively. Oxide ε_r = 3.9, $\varepsilon_0 = 8.85 \times 10^{-14}$ F/cm, oxide thickness t_{ox} = 0.1 µm and electronic charge q = 1.6×10^{-19} C. If The measured flat band voltage of this capacitor is-1 V, then the magnitude of the fixed charge at the oxide-semiconductor interface, in nC/cm^2 is _______.

[2017 : 2 Marks, Set-2]

CHARACTERISTICS & BIANING

29. The pinch off voltage for a n-channel JFET is 4 V, when V_{GS} = 1 V, the pinch-off occurs for V_{DS} equal to

(a) 3 V (b) 5 V
(c) 4 V (d) 1 V

[1987 : 2 Marks]

30. In an n-channel JFET, V_{GS} held constant. V_{DS} is less than the breakdown voltage. As V_{DS} is increased,

(a) conducting cross-sectional area of the channel 'S' and the channel current density 'J' both increase
(b) 'S' decrease and 'J' decreases
(c) 'S' decreases and 'J' increases
(d) 'S' increases and 'J' decreases

[1988 : 2 Marks]

31. In MOSFET devices the n-channel type is better than the P-channel type in the following respects,

(a) it has better noise immunity.
(b) it is faster.
(c) it is TTL compatible.
(d) it has better drive capability.

[1988 : 2 Marks]

32. In a MOSFET, the polarity of the inversion layer is the same as that of the

(a) charge on the GATE-EC-electrode.
(b) minority carries in the drain.
(c) majority carries in the substrate.
(d) majority carries in the source.

[1989 : 2 Marks]

33. The 'Pinch-off' voltage of a JFET is 5.0 volts, Its 'cut-off voltage is

(a) $(5.0)^{1/2}$ V (b) 2.5 V
(c) 5.0 V (d) $(5.0)^{3/2}$ V

[1990 : 2 Marks]

34. Which of the following effects can be caused by a rise in the temperature?

(a) Increase in MOSFET current (I_{DS})
(b) Increase in BJT current (I_C)
(c) Decrease in MOSFET current (I_{DS})
(d) Decrease in BJT current (I_c)

[1990 : 2 Marks]

35. In a transistor having finite β, forward bias across the base emitter junction is kept constant and the reverse bias across the collector base junction is increased. Neglecting the leakage across the collector base junction and the depletion region generations current, the base current will (increase/decrease/remains constant).

[1992 : 2 Marks]

36. An n-channel JFET has a pinch-off voltage V_p = –5 V, V_{DS}(max) = 20 V, and gm = 2 mA/V. The min 'ON' resistance is achieved in the JFET for

(a) V_{GS} = –7 V and V_{DS} = 0 V.
(b) V_{GS} = 0 V and V_{DS} = 0 V.
(c) V_{GS} = 0 V and V_{DS} = 20 V.
(d) V_{GS} = –7 V and V_{DS} = 20 V.

[1992 : 2 Marks]

37. The threshold voltage of an n-channel MOSFET can be increased by

(a) increasing the channel dopant concentration.
(b) reducing the channel dopant concentration.
(c) reducing the GATE oxide thickness.
(d) reducing the cnannel length.

[1994 : 1 Mark]

38. The transit time of the current carries through the channel of a JFET decides its characteristic

(a) source (b) drain
(c) GATE (d) source and drain

[1994 : 1 Mark]

39. Channel current is reduced on application of a more positive voltage to the GATE of the depletion mode n-channel MOSFET. (True/False)

[1994 : 1 Mark]

40. In a JFET

List-I

A. The pinch-off voltage decreases
B. The transconductance increases
C. The transit time of the carriers in the Channel is reduced

List–II

1. The channel doping is reduced
2. The channel length is increases
3. The conductivity of the channel is increased
4. The channel length is reduced
5. The GATE area is reduced

[1995 : 2 Marks]

41. An n-channel JFET has I_{DSS} = 2 mA and V_p = –4 V. It's transconductance g_m(in mA/V). An applied GATE to source voltage V_{GS} of –2 V is

(a) 0.25 (b) 0.5
(c) 0.75 (d) 1.0

[1999 : 2 Marks]

42. MOSFET can be used as a

(a) current controlled capacitor
(b) voltage controlled capacitor
(c) current controlled inductor
(d) voltage controlled inductor

[2001 : 1 Mark]

43. The effective channel length of a MOSFET in saturation decreases with increase in

(a) gate voltage

(b) drain voltage

(c) source voltage

(d) body voltage

[2001 : 1 Mark]

44. For an n-channel enhancement type MOSFET, if the source is connected at a higher potential than that of the bulk (i.e. $V_{SB} > 0$), the threshold voltage V_T of the MOSFET will

(a) remain unchanged

(b) decrease

(c) change polarity

(d) increase

[2003 : 1 Mark]

45. When the gate-to-source voltage (V_{GS}) of a MOSFET with threshold voltage of 400 mV, working in saturation is 900 mV, the drain current is observed to be 1 mA. Neglecting the channel width modulation effect and assuming that the MOSFET is operating at saturation, the drain current for an applied V_{GS} of 1400 mV is

(a) 0.5 mA (b) 2.0 mA

(c) 3.5 mA (d) 4.0 mA

[2003 : 2 Marks]

46. The drain of an n-channel MOSFET is shorted to the gate so that $V_{GS} = V_{DS}$. The threshold voltage (V_T) of MOSFET is 1 V. If the drain current (I_D) is 1 mA for $V_{GS} = 2$ V, then for $V_{GS} = 3$ V, I_D is

(a) 2 mA

(b) 3 mA

(c) 9 mA

(d) 4 mA

[2004 : 2 Marks]

47. The drain current of a MOSFET in saturation is given by $I_D = K(V_{GS} - V_T)^2$ where K is a constant. The magnitude of the trans-conductance g_m is

(a) $\frac{K(V_{GS} - V_T)^2}{V_{DS}}$

(b) $2K(V_{GS} - V_T)$

(c) $\frac{I_D}{V_{GS} - V_{DS}}$

(d) $\frac{K(V_{GS} - V_T)^2}{V_{GS}}$

[2008 : 1 Mark]

48. The measured trans-conductance g_m of an NMOS transistor operating in the linear region is plotted against the gate voltage V_G at a constant drain voltage V_D. Which of the following figures represents the expected dependence g_m on V_G?

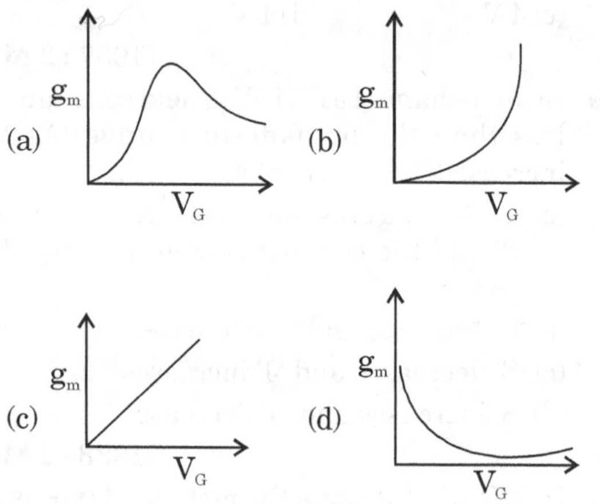

[2008 : 2 Marks]

49. Consider the following two statements about the internal conditions in an n-channel MOSFET operating in the active region:

S_1: The inversion charge decreases from source to drain.

S_2: The Channel potential increases from source to drain.

Which of the following is correct?

(a) Only S_2 is true.

(b) Both S_1 and S_2 are false.

(c) Both S_1 and S_2 are true, but S_2 is not a reason for S_1.

(d) Both S_1 and S_2 are true, and S_2 is a reason for S_1.

[2009 : 2 Marks]

50. At room temperature, a possible value for the mobility of electrons in the inversion layer of a silicon n-channel MOSFET is

(a) 450 cm²/V-s (b) 1350 cm²/V-s

(c) 1800 cm²/V-s (d) 3600 cm²/V-s

[2010 : 1 Mark]

51. A depletion type N-channel MOSFET is biased in its linear region for use as a voltage controlled resistor. Assume threshold voltage $V_{TH} = -0.5$ V, $V_{GS} = 2.0$ V, $V_{DS} = 5$ V, W/L = 100, $C_{ox} = 10^{-8}$ F/cm² and $\mu n = 800$ cm²/V–s. The value of the resistance of the voltage controlled resistor (in Ω) is ______.

[2014: 2 Marks, Set-1]

52. The slope of the I_D vs V_{GS} curve of an n-channel MOSFET in linear region is 10^{-3} Ω the same device, neglecting channel length modulation, the slope of the $\sqrt{I_D}$ vs. V_{GS} (in $\sqrt{A}/V$) under saturation region is approximately _____.

[2014: 2 Marks, Set-3]

53. For the N-MOSFET in the circuit shown. the threshold voltage is V_{th}, where $V_{th} > 0$. The source voltage V_{ss} is varied from 0 to V_{DD}. Neglecting the channel length modulation, the drain current I_D as a function of V_{ss} is represented by

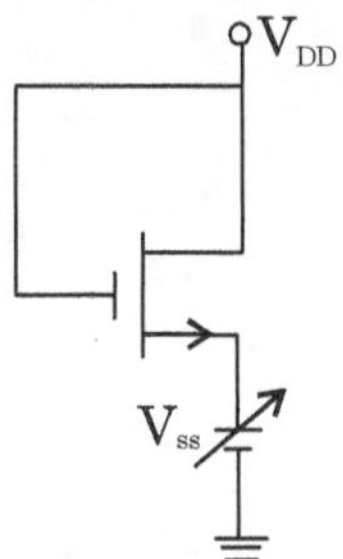

(a)

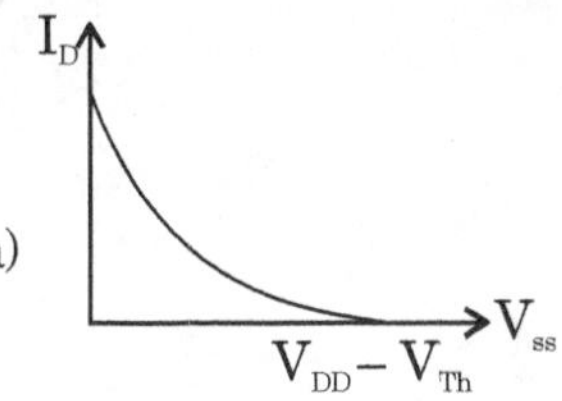

(b)

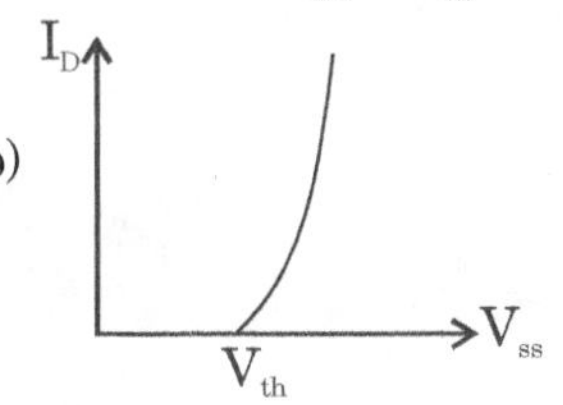

(c)

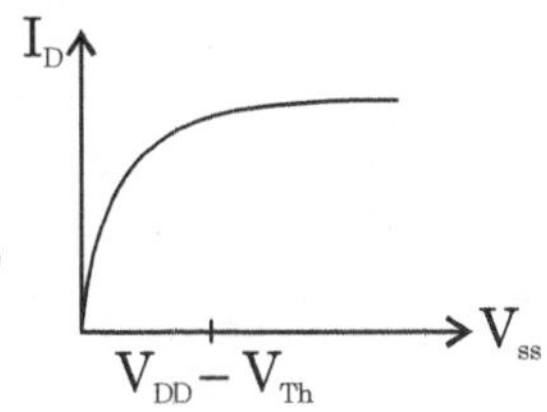

(d)

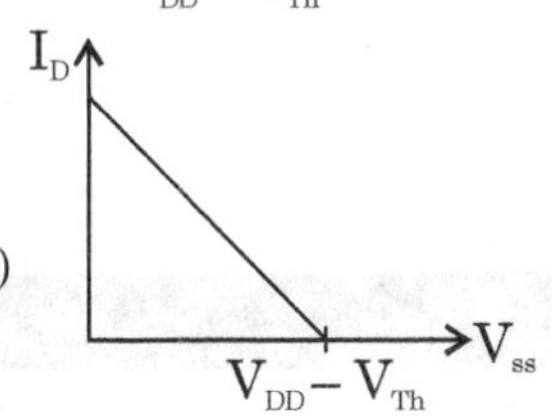

[2015 : 2 Marks, Set-1]

54. Consider the following statements for a metal oxide semiconductor field effect transistor (MOSFET):

P: As channel length reduces, OFF-state current increases.

Q: As channel length reduces, output resistance increases.

R: As Channel length reduces, threshold voltage remains constant.

S: As channel length reduces, ON current increases.

Which of the above statements are INCORRECT?

(a) Pand Q (b) P and S
(c) Q and R (d) R and S

[2016 :1 Mark, Set-1]

55. A voltage V_G is applied across a MOS capacitor with metal gate and p-type Silicon substrate at T = 300 K. The inversion carrier density (in number of carriers per unit area) for $V_G = 0.8$ V is 2×10^{11} cm^{-2}. For $V_G = 1.3$ V, the inversion carrier density is 4×10^{11} cm^{-2}. What is the value of the inversion carrier density for $V_G = 1.8$ V?

(a) 4.5×10^{11} cm^2 (b) 6.0×10^{11} cm^2
(c) 7.2×10^{11} cm^2 (d) 8.4×10^{11} cm^{-2}

[2016 : 2 Marks, Set-2]

56. Consider a long-channel NMOS transistor with source and body connected together. Assume that the electron mobility is independent of V_{GS} and V_{DS}. Given,

$g_m = 0.5$ μA/V for $V_{DS} = 50$ mV and $V_{GS} = 2$ V,
$g_d = 8$μA/V for $V_{GS} = 2$ Vand $V_{DS} = 0$ V.

where $g_m = \frac{\partial I_D}{\partial V_{GS}}$ and $g_d = \frac{\partial I_D}{\partial V_{DS}}$

The threshold voltage (in volts) of the transitor is ______ .

[2016 : 2 Marks, Set-2]

57. The injected excess electron concentration profile in the base region of an npn BJT, biased in the active region, is linear, as shown in the figure. If the area of the emitter-base junction is 0.001 cm^2, $\mu_n = 800$ cm^2/(V-s) in the base region and depletion layer widths are negligible, then the collector current I_c (in mA) at room temperature is (Given: thermal voltage $V_T = 26$ mV at room temperature, electronic charge q= 1.6×10^{-19}C)

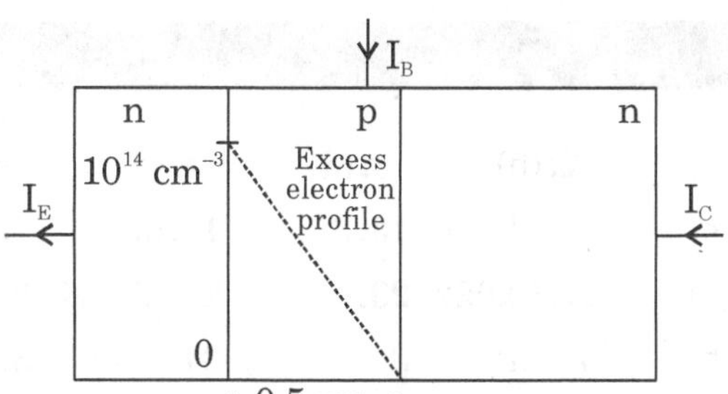

[2016 : 2 Marks, Set-3]

58. Two n-channel MOSFETs, T1 and T2, are identical in ali respects except that the width of T2 is double that of T1. Both the transistors are biased in the saturation region of operation, but the gateoverdrive voltage ($V_{GS} - V_{TH}$) of T2 is double that of T1, where V_{GS} and V_{TH} are the gate-to-source voltage and threshold voltage of the transistors, respectively. If the drain current and transconductance of T1 are I_{D1} and g_{m1} respectively, the corresponding values of these two parameters for T2 are

(a) $8I_{D1}$ and $2g_{m1}$ (b) $8\ I_{D1}$ and $4g_{m1}$
(c) $4I_{D1}$ and $4g_{m1}$ (d) $4I_{D1}$ and $2g_{m1}$

[2017 : 2 Marks, Set-2]

59. Consider a long-channel MOSFET with a channel length 1 μ m and width 10μm. The device parameters are acceptor concentration $N_A = 5 \times 10^{16}$ cm^{-3}, electron mobility μ_n = 800 cm^2/V–s, oxide capacitance/area C_{ox} = 3.45 ×10^{-7}F/cm^2, threshold voltage V_T = 0.7V. The drain saturation current $\left(I_{D_{sat}}\right)$ for a gate voltage of 5V is ______ mA (rounded off to two decimal places).

$$\left[\varepsilon_0 = 8.854 \times 10^{-14}\,\text{F/cm}, \varepsilon_{Si,} = 11.9\right]$$

[2019 : 2 Marks]

ANSWERS

1. (b)	**2.** (b)	**3.** (d)	**4.** (d)	**5.** (b)	**6.** (a)	**7.** (b)	**8.** (c)	**9.** (c)	**10.** (c)
11. (c)	**12.** (d)	**13.** (b)	**14.** (a)	**15.** (a)	**16.** (d)	**17.** (c)	**18.** 2.4	**19.** 20	**20.** 4.33
21. (d)	**22.** (0.022)	**23.** (*)	**24.** (d)	**25.** (a)	**26.** (1.6)	**27.** (6.5)	**28.** (6.903)	**29.** (a)	**30.** (c)
31. (b)	**32.** (d)	**33.** (c)	**34.** (b,c)	**35.** (Decrease)	**36.** (b)	**37.** (b)	**38.** (b)	**39.** (False)	
40. (A-3, B-3, C-4)	**41.** (b)	**42.** (b)	**43.** (b)	**44.** (d)	**45.** (d)	**46.** (d)	**47.** (b)	**48.** (c)	
49. (d)	**50.** (b)	**51.** 499 to 501	**52.** 0.07	**53.** (a)	**54.** (c)	**55.** (b)	**56.** (1.2)	**57.** (6.656)	
58. (b)	**59.** (25.51)								

EXPLANATIONS

2. To achieve high quality oxide growth (i.e., uniform film with good dielectric properties), dry oxide alone is employed when gate oxide is grown. Wet oxidation is used to grow field oxide, since the quality of the dielectric properties of the field oxide are not as critical as they are for the gate oxide. Dry oxidation is slower than the wet oxidation.

3. To achieve high quality oxide growth, dry oxidation alone is employed when gate oxide is grown. Dry oxidation is slower then the wet oxidation.

4. S1: $C_{ox} = \frac{E_{ox}}{t_{ox}}$, t_{ox} = Gate oxide thickness

$$V_T = \phi_{GC} - 2\phi_F - \frac{Q_B}{C_{ox}} - \frac{Q_{ox}}{C_{ox}}$$

if $C_{ox}\uparrow$, then $V_T\uparrow$, $\frac{Q_B}{C_{ox}}\downarrow$, $\frac{Q_{ox}}{C_{ox}}\downarrow$

If $C_{ox}\downarrow$, then t_{ox} increases.
Hence, S1 is true.
S 2 : V_T of MOS increases by increase in substrate doping concentration.
Hence, S2 is false.

5. Accumulation occurs when we set $V_{GB} < V_{FB}$, pulling the gate to a potential below the substrate. From the graph of potential accross the MOS structure (see figure 1), we know that $\frac{dV}{dx}$ is positive, meaning E(x), the electric field, is negative, or pointing in the negative x direction. This causes holes to accumulate at the top of the substrate and electrons to accumulate at the bottom of the n + gate.

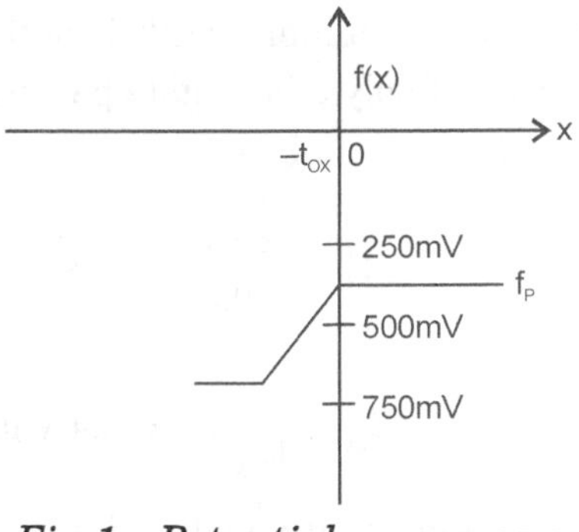

Fig 1 : Potential across as a MOS capacitor in accumulation

6. capacitance,

$$C = \frac{\in A}{d}$$

$$d = \frac{\in A}{d} = \frac{3.5\times10^{-13}\times1\times10^{-4}\times10^{-4}}{7\times10^{-12}\times10^{-2}}$$

$= 5 \times 10^{-8}$m.

d = 50 nm.

7. $\Rightarrow \in si = 10^{-12}$ F/cm $= 10^{-10}$ F/m.

$$\Rightarrow C_{min} = \frac{C_{max}.C_d}{C_{max}+C_d}$$

$$C_d = \frac{C_{max}.C_{min}}{C_{max}-C_{min}}$$

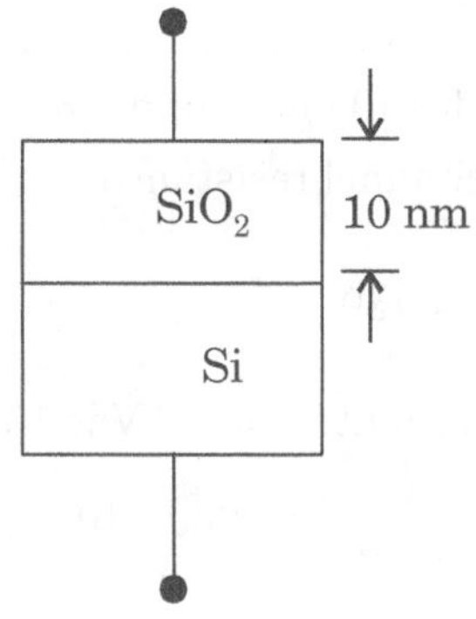

$$C_d = \frac{7\times1}{7-1} = 1.1.66\text{pf}.$$

$$C_d = \frac{\in_{si}.A}{w}$$

$$w = \frac{10^{-10}\times10^{-4}\times10^{-4}}{1.166\times10^{-12}} = 0.857\mu\text{m}.$$

8. (1) Since, as shown in the given c-v plot, V_{th} is positive, hence,

(i) The MOSFET is n-channel.

(ii) The substrate of the MOSFET is p-type.

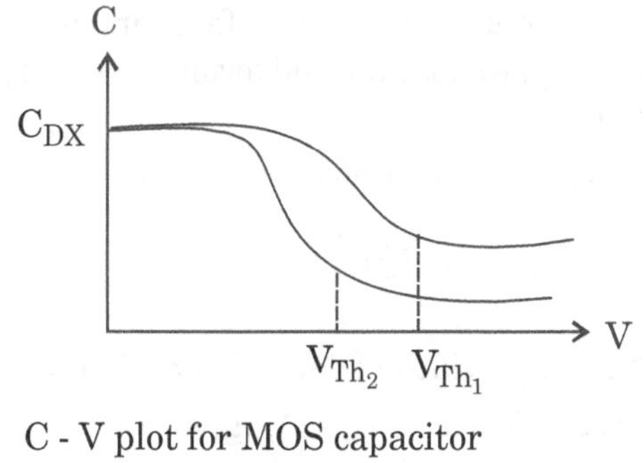

C - V plot for MOS capacitor

s1 is false.

(2) If positive charges are introduced in the oxide, the accumulation mode will be obtained for lesser value of V_a, i.e., threshold will occur for lower value of V_a. Hence, C - V plot will move towards left.

S_2 is true.

10. Depletion layer width,

$$w \propto \sqrt{V_{as} + V_{bi}}\,;\because V_{bi} = -1V.$$

for V_{as} = OV, w_1 = 1 μm.

$$\frac{w^2}{w_1} = \sqrt{\frac{V_{as2} + V_{bi}}{V_{as1} + V_{bi}}}$$

$$\Rightarrow w^2 = \sqrt{\frac{-3-1}{0-1}} \times 1\mu m$$

w_2 = 2 μm,

As channel width increase, channel resistance decreases. for 10 μm, channel resistance = 600 Ω.

for 8 μm, channel resistance

$$= \frac{600 \times 10}{8} = 750\Omega.$$

11. channel width at V_{as} = – 3 V is 10 – 2_{w2} = 6 μm.

$$\text{channel resistance} = \frac{600 \times 10}{6} = 1000\Omega..$$

12. $$p = \frac{n_i^2}{10^{19}} = 10 \text{ cm}^{-3}$$

Volume = $10^{-6} \times 10^{-6}$ m³ = 10^{-12} cm³

∴ Total holes = $10^{-11} \approx 0$

13. Given : δ = 20 nm, width of transistor = 1 μm.

Depletion width at every p - n junction = 10 nm.

$(\epsilon_r)_{si}$ = 11.7, $(\epsilon_r)_{sio2}$ = 3.9,

ϵ_o = 8.85 × 10^{-12} F/m.

It has rectangular shape in 3 – D, except for the top face (out of six faces), all faces are in connection with p-type substrate and forming the p-n junction diode.

Source body junction capacitance,

$$c = \frac{\epsilon A}{d} = \frac{\epsilon o(\epsilon r)_{si} A}{d}$$

where, d = width of depletion region = 10 nm.

A = $A_1 + A_2 + A_3 + A_4 + A_5$ (given)

= $2A_1 + 2A_2 + A_5$

A = 2 × (0.2 × 1) + 2 (0.2 × 0.2) + 1 (0.2 ×1)

A = 0.68 μm².

$$\Rightarrow c = \frac{11.7 \times 8.85 \times 10^{-12} \times 0.68 \times 10^{-6}}{10^{-8}}$$

= 7 × 10^{-15} F = 7 fF.

14. $CO_v = WLO_v\, c_{Ox}$

$$\because C_{ox} = \frac{\epsilon_0 (\epsilon_r)_{si02}}{t_{ox}}$$

$$C_{ov} = WL_{ov} \cdot \frac{\epsilon_o (\epsilon_r)_{sio2}}{t_{ox}}$$

$$= \frac{10^{-6} \times 20 \times 10^{-9} \times 8.9 \times 10^{-12} \times 3.9}{10^{-9}}$$

C_{ov} = 694.2 × 10^{-18}F = 0.6942 × 10^{-15} F

$C_{ov} \simeq 0.7$fF.

15. when positive gate voltage is applied then electrons of p-type substrate are attracted and they form an inversion layer. If already fixed positive charges are present on gate, then less gate voltage is required to induce n-channel. Hence, it will reduce the threshold voltage.

16. In CMOS, an n-well is used to fabricate PMOS if the substrate is p-type. similarly, p-well is used to fabricate NMOS if the substrate is n-type for n-well CMOS process, an initial thick oxide layer (5000 Å) is grown on the entire surface. The first lithographic mark defines the n-well region. Donor atoms usually phosphorus are implanted through this window in the oxide.

17. Though MOSFET stands for metal oxide semiconductor FET. But, practically a polycrystalline si (or pdy si) is deposited above the gate. The channel length is defined during the process of poly-silicon gate patterning.

18. $$E_s = \frac{2 \times 0.2}{0.5} = 0.8 \text{ v/}\mu\text{m}$$

$$E_{ox} = \frac{E_s}{E_{ox}} E_s = 2.4 \text{ v/}\mu\text{m}$$

19. Under channel length modulation

$$I_D = I_{Dsat}\,(1 + \lambda V_{DS})$$

$$\frac{dI_D}{dV_{DS}} = \frac{I}{r_0} = \lambda I_{Dsat}$$

$$r_0 = \frac{1}{\lambda I_{Dsat}} = \frac{1}{0.05\times10^{-3}} = 20k\Omega$$

20. $$\frac{C_{max}}{C_{min}} = \frac{\frac{\epsilon_{ox}}{t_{ox}}}{\frac{\frac{\epsilon_{ox}}{t_{ox}}\times\frac{\epsilon_s}{X_{dmax}}}{\frac{\epsilon_{ox}}{t_{ox}}+\frac{\epsilon_s}{X_{dmax}}}}$$

$$= \left[1+\frac{X_{dmax}}{t_{ox}}\times\frac{\epsilon_{ox}}{\epsilon_s}\right]$$

$$= \left[1+\frac{100}{10}\times\frac{1}{3}\right] = 4.33$$

21. $$y(t) = A_c\cos[2\pi f_c t + m(t)]$$

$$m(t) = A_m\sin(2\pi f_m t)$$

Since $y(t)$ is phase modulated signal,

$$\phi(t) = 2\pi f_c t + m(t)$$

$$\text{Bandwidth} = 2[\Delta f + f_m]$$

$$\Delta f = \frac{1}{2\pi}\frac{d}{dt}m(t)$$

⇒ Δf depends on A_m as well as f_m. Thus Bandwidth depends on both A_m and f_m.

23. In saturation region,

$$g_d = \lambda.I_{DS} = \lambda\left[\frac{1}{2}\mu nC_{ox}\frac{W}{L}(V_{as}-V_T)^2\right]$$

$$g_d = 0.09\left[\frac{1}{2}\times70\times10^{-6}\times4(1.8-0.3)^2\right]$$

$g_d = 28.35\ \mu s$

24. Here $$r_{ds} = \frac{1}{\mu_n C_{OX}\frac{W}{L}(V_{GS}-V_T)}$$

r_{ds} = channel resistance

W ↑ r ↓ A → correct

V_T ↓ r ↓ B → correct

L ↑ r ↑ C → correct

V_{GS} ↑ r ↓

∴ D → Wrong statement

So, statement (*d*) is not correct.

25. (a) Conclusion from Energy band diagram:

(i) Fami-energy level E_{FS} is close E_c. Therefore, MOS has n-type substrate. Hence it is PMOS transistors.

(ii) Band bending is in upward direction, the direction of electric field is upward. Gate voltage is negative then MOS capacitor will be either in deputation region or strong invension region.

(iii) Surface potential, $\phi s = 2\phi f_{Fn}$.

Therefore, MOS capacitor is in invension region.

26. We known that

Now $C = \frac{\varepsilon A}{d}$

Where C = Capacitance

A = Area of a capacitor

ε = dielectric constant

Now $C = \frac{C_1C_2}{C_1+C_2}$

$$= \left[\frac{\frac{4}{1\times10^{-9}}\times\frac{20}{3\times10^{-9}}}{\frac{4}{1\times10^{-9}}+\frac{10}{3\times10^{-9}}}\right]8.8521\times10^{-12}$$

$= 2.5\times10^9\ \varepsilon_0$

$$C = \frac{\varepsilon_r\varepsilon_0}{t_{eq}}$$

$$t_{eq} = \frac{\varepsilon_r\varepsilon_0}{2.5\times10^9\ \varepsilon_0}$$

$$= \frac{4\times\varepsilon_0}{2.5\times10^9\ \varepsilon_0}$$

$= 1.6\times10^{-9}$ m $= 1.6$ nm

So, the value of thickness (t_{eq}) is 1.6 nm.

27. Given : $V_{as} = 0.7$ V, $V_{th} = 0.3$ V, $V_{DS} = 0.1$ V.

$V_{as} - V_{th} = 0.7 - 0.3 = 0.4\text{ V} > V_{as}$

Hence, MOSFET is in linear region. In linear region,

$$\text{Transconductance, } g_m = \frac{\partial I_D}{\partial V_{as}} = 2.k_n V_{DS}$$

$$\frac{\mu_{ncox}C}{2}\left(\frac{W}{L}\right).V_{DS}$$

$= 100\times50\times0.1$

$= 0.5$ mA/V.

28. Given : $\epsilon_r = 0.9$, $\epsilon o = 8.86\times10^{-14}$ F/cm,

$t_{ox} = 0.1\ \mu$m, $q = 1.6\times10^{-19}$ c.

metal work function $e\phi m = 4.1$ eV

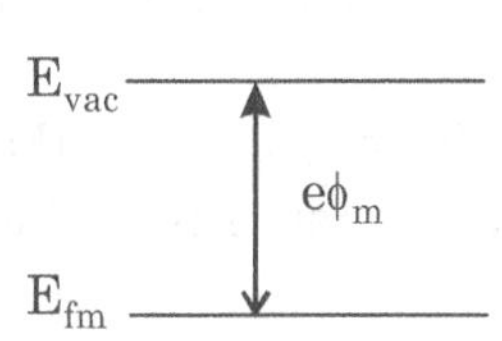

Metal

Electron affinity ex = 4 eV

Flat band gap $V_{FP} = -1$ V.

Vaccum energy level = E_{vac}

$\Rightarrow E_c - E_{pp} = 0.9$ eV

$e\phi_s = ex + 0.9 = 4.9$eV

$\phi s = 4.9$ V, $e\phi_m$ 4.1 eV, $\phi_m = 4.1$ V.

$\Rightarrow \phi_{ms} = \phi_m - \phi_s = 4.1 - 4.9 = -0.8$ V

$$\Rightarrow C_{ox} = \frac{\epsilon_{ox}}{t_{ox}} = \frac{3.9 \times 8.85 \times 10^{-14}}{0.1 \times 10^{-4}}$$

= 34.515 nf/cm^2

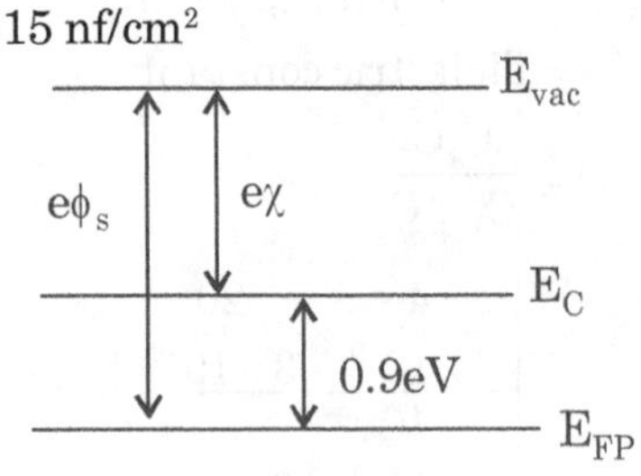

Semiconductor

$$\Rightarrow V_{FB} = \phi_{ms} - \frac{Q_{SS}}{C_{ox}}$$

$$\frac{Q_{SS}}{C_{ox}} = -0.8 + 1 = 0.2V$$

$Q_{ss} = 0.2 \times Cox = 0.2 \times 34.515 \times 10^{-19}$ c/cm^2

$Q_{ss} =$ 6.903 nc/cm^2.

29. For $V_{as} = -1$ V, pinch - off voltage, $V_{p1} = 4$ V.

If V_{as} will increase negatively then the value of V_{DS} at which drain current becomes constant will decrease and hence, pinch - off voltage decrease. The new volume of pinch - off voltage is

$V_{p2} = V_{DS} = V_{p1} - |V_{as}| = 4 - 1 - 3$ V.

30. V_{as} is held constant and V_{DS} is increase then the depletion width increases, so the cross-sectional area of the channel 's'.

$\because$ current density, $J = \frac{I}{A} = \frac{I}{S}$.

As S decrease, J will decrease. So, the channel current density increase.

31. n - channel MOSFET is faster than the p - channel MOSFET. Mobility of electrons is always higher than the mobility of holes. $\mu_n > \mu_p$.

In n - channel the charge carriers are electrons whereas in p - channel MOSFET, the carriers are hores.

32. In a MOSFET, the polarity of the inversion layer is the same as that of the minority carriers in the source.

33. Pinch - off voltage = cut - off voltage = 5.0 v.

34. (i) Effect of temperature on BJT :If temperature will increase then more number of covalent bonds will break and number of minority carrier will increase. Due to increase in number of minority carrier reverse saturation currect I_{co} will increase and I_c will increase.

(ii) Effect of temperature on MOSFET : If temperature will increase, then mobility ($\mu \propto T^{-m}$) of carriers will decrease and drain current I_{DS} will decrease.

35. As the collector junction reverse bias is increased in magnitude, the width of collector junction depletion layer increases. Hence the effective width of base region decreases.

36. For n - channel JFET, JERT offers minimum 'ON' resistance when V_{as} is positive and large and voltage V_{DS} is very small.

Ideally, V_{DS} = OV, V_{as} = OV.

37. For n - channel MOSFET, threshold voltage is

$$V_{th} = V_{To} = +\gamma\left[\sqrt{2\phi_{FP} + |V_{SB}|} - \sqrt{2\phi_{FP}}\right]$$

$$\text{Body - effect parameter } \gamma = \frac{\sqrt{2qN_A E_S}}{C_{ox}}$$

$$= \frac{t_{ox}\sqrt{2qN_A E_S}}{3.45 \times 10^{-11}}.$$

Therefore, threshold voltage increase when source to body voltage is applied, i.e., when source is connected at a higher potential than bulk.

38. The transit - time of the current carries through the channel of a JFET decides its drain characterstic.

39. **False.** An *n* channel depletion mode MOSFET can be operated both in the depletion (V_{GS} negative) and enhancement (V_{GS} positive) mode. So for application of positive voltage the drain current is more.

40. $\Rightarrow$ As the channel doping is reduced then the pinch - off voltage decreases.

$\Rightarrow g_m = \frac{I_D}{V_{as}}$. So, the transconductances increases as drain current increases and the drain current increases as the conductivity of the channel is increased.

$\Rightarrow$ If the channel length is reduced then the transit time of the carriers in the channel is reduced.

A $\rightarrow$ 1, B $\rightarrow$ 3, C $\rightarrow$ 4

44. $V_T = V_{To} + \gamma \sqrt{|V_{SB} - 2\phi_F|} - \sqrt{|2\phi_F|}$

where, V_T = Threshold voltage

γ = substrate bias coefficient

V_{SB} = substrate bias voltage

46. Given : $V_{th} = 1$ V, $I_{D1} = 1$ mA for $V_{as1} = 2$ V.

$V_{as} = V_{DS}$.

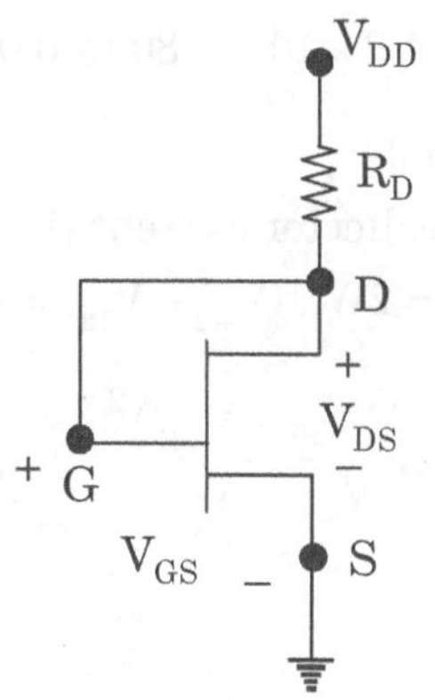

$\Rightarrow V_{DS} > V_{as} - V_{th}$.

Therefore, n - MOSFET is in saturation region.

Drain current of a n - MOSFET working in saturation region is

$$I_D = \frac{1}{2}k_n (V_{as} - V_{th})^2$$

$$I_D \propto (V_{as} - V_{th})^2.$$

$$\Rightarrow I_D = I_{D1} \times \left(\frac{V_{as_2} - V_{th}}{V_{as_1} - V_{th}}\right)^2$$

$$= 1 \times \frac{(3-1)^2}{(2-1)^2} = 4\text{mA}.$$

48. In linear region of NMOS transistor, diode current is given by,

$$I_D = \frac{\mu_n C_{ox} W}{L}\left(V_{GS} - V_T - \frac{1}{2}V_{DS}\right)V_{DS}$$

and $V_{DS} \le V_{GS} - V_T$

Now, $g_m = \left.\frac{\partial I_n}{\partial v_{gs}}\right|_{V_{DS=const.}}$

$$\frac{\partial I_D}{\partial v_{gs}} = \frac{\mu_n C_{ox} W}{L} \cdot V_{DS} = K \text{ [Constant]}.$$

Hence, it can be infered that $\frac{\partial I_n}{\partial V_{gs}}$ follows a constant Linear variation.

49.

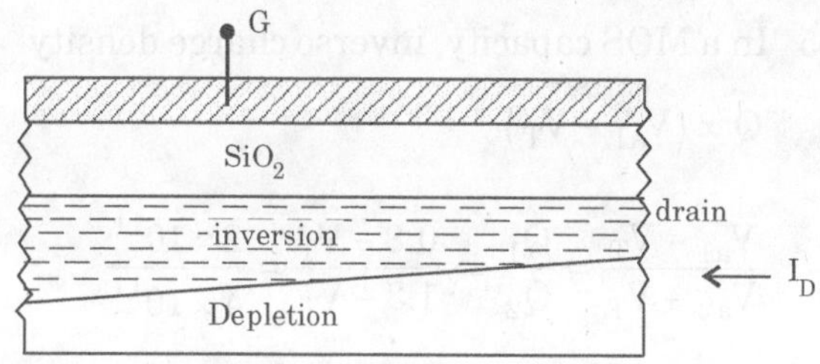

Both S_1 and S_2 are true, and S_2 is reason for S_1

50. For *n*-channel MOSFET,

Mobility of inversion layer = 1350 cm²/V-s

51.

$$r_{DS} = \frac{1}{(\mu_n C_{on})\left(\frac{W}{L}\right)(V_{GS} - V_T)}$$

$$r_{DS} = \frac{1}{800 \times 10^{-4} \times 10^{8} \times 100 \times (20 - 0.5)}$$

$$r_{DS} = 500\ \Omega$$

52. In linear region,

$$I_D = k\left[(V_{GS} - V_T)V_{DS} - \frac{V_{DS}^2}{2}\right]$$

$$\frac{\partial I_D}{\partial V_{GS}} = 10^{-3} = kV_{DS}$$

$\because V_{DS}$ is small, $\frac{V_{DS}^2}{2}$ is neglected

$$\Rightarrow \quad K = \frac{10^{-3}}{0.1}$$

$$= 0.01$$

In saturation region,

$$I_D = \frac{1}{2}k(V_{GS} - V_T)^2$$

$$\sqrt{I_D} = \sqrt{\frac{k}{2}}(V_{GS} - V_T)$$

$$\frac{\partial \sqrt{I_D}}{\partial V_{GS}} = \sqrt{\frac{k}{2}} = \sqrt{\frac{0.01}{2}} = 0.07$$

54. As per the given statement

P : TRUE

Q : FALSE, As channel length reduces, output resistance reduces.

R : FALSE, As channel length reduces, threshold voltage reduces

S : TRUE

So, the given statement Q and R are incorrect.

55. In a MOS capacity, inverse charge density

$Q \propto (V_{a1} - V_T)$

$$\frac{V_{a1} - V_T}{V_{a2} - V_T} = \frac{Q_1}{Q_2} \Rightarrow \frac{0.8 - V_T}{1.3 - V_T} = \frac{2 \times 10^{11}}{4 \times 10^{11}}$$

$V_T = 0.3$ V.

$$\Rightarrow \frac{V_{a2} - V^T}{V_{a3} - V_T} = \frac{Q_2}{Q_3} \Rightarrow Q_3 = \frac{V_{a3} - V_T}{V_{a2} - V_T} \times Q_2$$

56. Since $V_{GS} > V_{DS}$, MOSFET is in linear operation

$$I_D = K'_N [V_{GS} - V_T] V_{DS}$$

$$\frac{dI_D}{dV_{GS}} = K'_N V_{DS}$$

$$g_m = K'_N V_{DS}$$

$$0.5 \times 10^{-6} = K'_N [50 \times 10^{-3}]$$

$$\therefore \quad K'_N = 10^{-5}$$

$$I_D = K'_N [V_{GS} - V_T] V_{DS}$$

$$\frac{dI_D}{dV_{dS}} = K'_N [V_{GS} - V_T]$$

$$g_d = K'_N [V_{GS} - V_T]$$

$$8 \times 10^{-6} = 10^{-5} [2 - V_T]$$

$$\frac{8 \times 10^{-6}}{10^{-5}} = 2 - V_T$$

$$8 \times 10^{-1} = 2 - V_T$$

$$0.8 = 2 - V_T$$

$$\therefore \quad V_T = 1.2 \text{ V}$$

Hence threshold voltage of the transistor is 1.2 V.

57. We know that

Collector current $(I_C) = AeD_n \frac{dn}{dx} = Ae\mu_n V_T \frac{dn}{dx}$

Where

A = Area for the emitter-base function

e = Electronic charge

V_T = Thermal voltage

$$I_C = 0.001 \times 1.6 \times 10^{-19} \times 800 \times 0.026 \times \left(\frac{10^4 - 0}{0.5 \times 10^{-4}}\right)$$

$I_C = 6.656$ mA

Hence the collector current (I_C) is 6.656 mA.

58. Given : $W_2 = 2W_1$, $(V_{as2} - V_{T2}) = 2(V_{as1} - V_{T1})$

$$\frac{I_{D2}}{I_{D1}} = \frac{W_2}{W_1} \times \left(\frac{V_{as2} - V_{T2}}{V_{as1} - V_{T1}}\right)^2$$

$= 2 \times 2^2 = 8$

$I_{D2} = 8\, I_{D1}$

$$\Rightarrow \frac{g_{m2}}{g_{m1}} = \frac{W_2}{W_1} \cdot \frac{V_{as2} - V_{T2}}{V_{as1} - V_{T1}}$$

$= 2 \times 2 = 4$

$g_{m2} = 4.\, g_{m1}$.

59. Given data; L = 1 μm μm, W = 10 μm

$NA = 5 \times 10^{16}\ \text{cm}^{-3}$, $\mu_n = 800 \frac{\text{cm}^2}{\text{v - sec}}$

$C_{ox} = 3.45 \times 10^{-7}$ F/cm, $V_T = 0.7$V

$\varepsilon_0 = 8.854 \times 10^{-14}$ F/cm, $\varepsilon_{si} = 11.9$ and $V_G = 5$V

$$\text{Now, } I_{D_{sat}} = \frac{1}{2} \mu_n C_{ox} \frac{W}{L} [V_G - V_T]^2$$

$$= \frac{1}{2} \times 800 \times 3.45 \times 10^{-7} \times \frac{10}{1} [5 - 0.7]^2 = 0.0255\text{A}$$

≈ 25.51mA

Unit - V

Analog Circuits

Syllabus

Small signal equivalent circuits of diodes, BJTs and MOSFETs; Simple diode circuits: clipping, clamping and rectifiers; Single-stage BJT and MOSFET amplifiers: biasing, bias stability, mid-frequency small signal analysis and frequency response; BJT and MOSFET amplifiers: multi-stage, differential, feedback, power and operational; Simple op-amp circuits; Active filters; Sinusoidal oscillators: criterion for oscillation, single-transistor and op-amp configurations; Function generators, wave-shaping circuits and 555 timers; Voltage reference circuits; Power supplies: ripple removal and regulation.

Contents

1 CHAPTER

Diode Applications

Analysis of Previous GATE Papers												
Topics ↓ / Year →			2019	2018	2017 Set 1	2017 Set 2	2016 Set 1	2016 Set 2	2016 Set 3	2015 Set 1	2015 Set 2	2015 Set 3
SMALL SIGNAL EQUIVALENT CIRCUITS OF DIODE	1 Mark	MCQ Type										
		Numerical Type										
	2 Marks	MCQ Type										
		Numerical Type										
	Total											
SIMPLE DIODE CIRCUITS	1 Mark	MCQ Type				1				1	1	
		Numerical Type				1		1	2	1		1
	2 Marks	MCQ Type		1				2	1			
		Numerical Type		1							1	
	Total			4		2		5	4	2	3	1
RECTIFIERS	1 Mark	MCQ Type										
		Numerical Type	1									
	2 Marks	MCQ Type										
		Numerical Type	1									
	Total		3									

SMALL SIGNAL EQUIVALENT CIRCUITS OF DIODE

1. For small signal ac operation, a practical forward biased diode can be modeled as

(a) a resistance and a capacitance.

(b) an ideal diode and resistance in parallel.

(c) a resistance and an ideal diode in series.

(d) a resistance. **[1998 : 1 Mark]**

2. The bias current I_{DC} through the diodes is

(a) 1 mA (b) 1.28 mA

(c) 1.5 mA (d) 2 mA **[2011 : 2 Marks]**

3. The ac output voltage v_{ac} is

(a) 0.25cos(ωt) mV (b) 1cos(ωt) mV

(c) 2cos(ωt) mV (d) 22cos(ωt) mV

[2011 : 2 Marks]

SIMPLE DIODE CIRCUITS

4. The 6 V Zener diode shown below has zero Zener resistance and a knee current of 5 mA. The minimum value of R so that the voltage across it does not fall below 6 V is

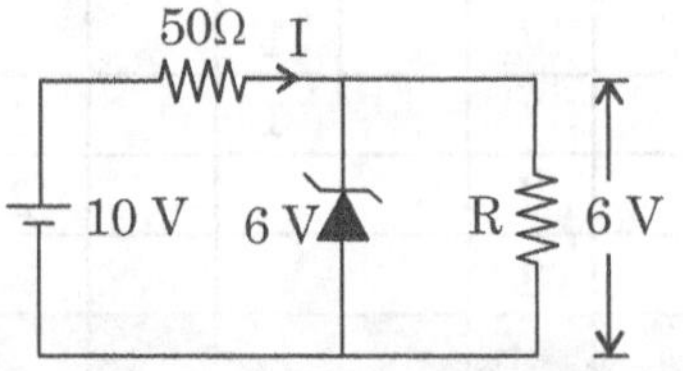

(a) 1.2 kΩ

(b) 50 Ω

(c) 80 Ω

(d) 0 Ω

[1992 : 2 Marks]

5. The wave shape of V_0 in figure is

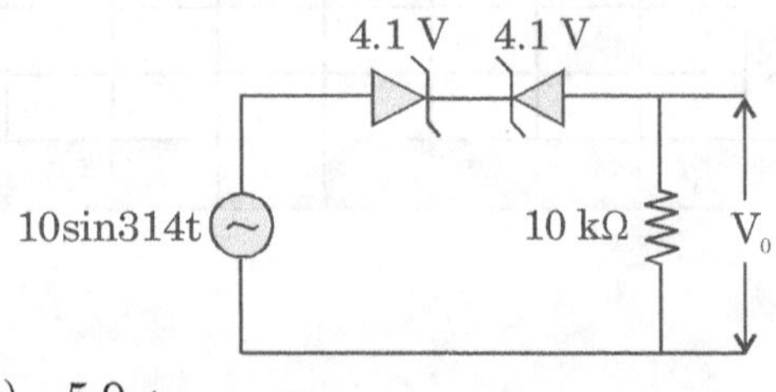

(a)
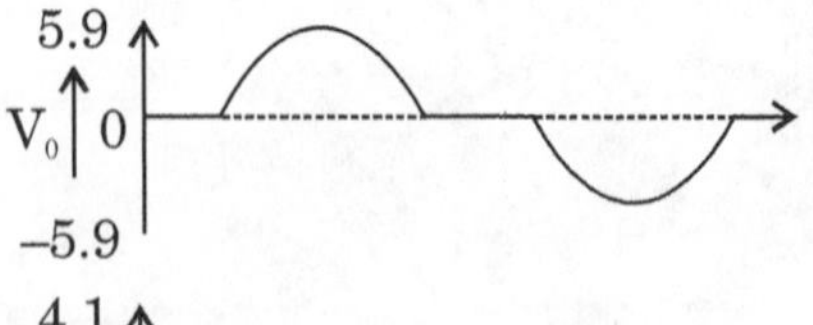

(b)
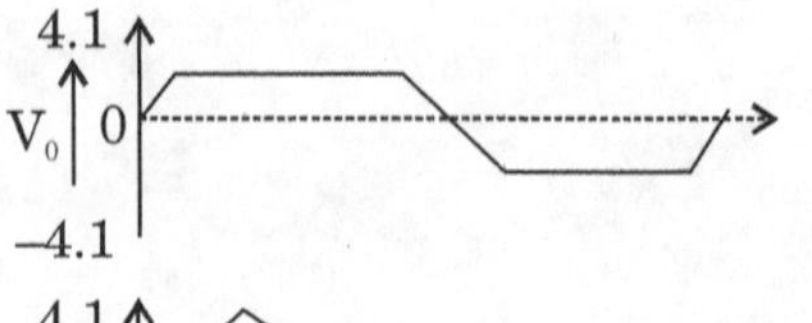

(c)
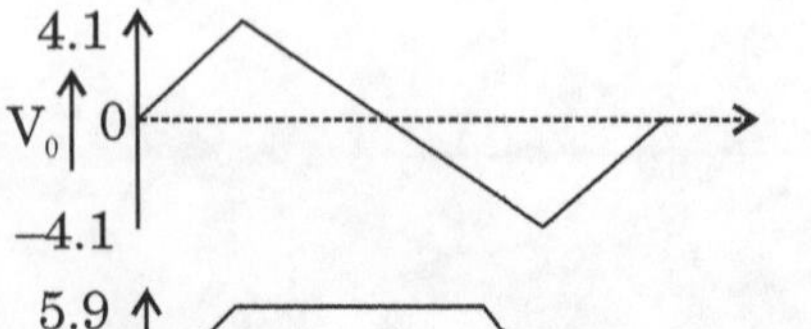

(d)
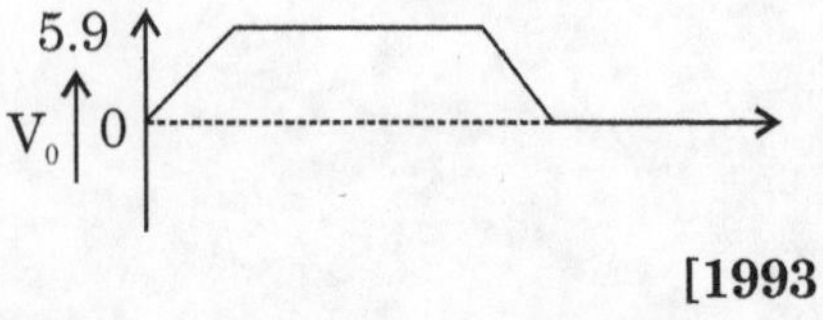

[1993 : 1 Mark]

6. A Zener diode in the circuit shown in below figure has a knee current of 5 mA, and a maximum allowed power dissipation of 300 mW. What are the minimum and maximum load currents that can be drawn safely from the circuit, keeping the output voltage V_0 constant at 6 V?

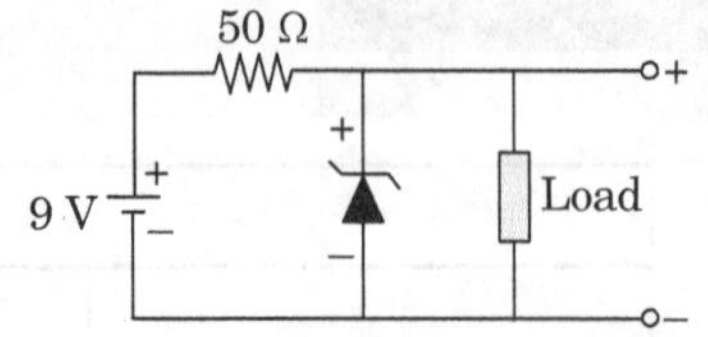

(a) 0 mA, 180 mA

(b) 5 mA, 110 mA

(c) 10 mA, 55 mA

(d) 60 mA, 180 mA

[1996 : 2 Marks]

7. A zener diode regulator in the figure is to be designed to meet the specifications: $I_L = 10$ mA, $V_0 = 10$ V and V_{in} varies from 30 V to 50 V. The zener diode has $V_Z = 10$ V and I_{zk} (knee current) = 1 mA. For satisfactory operation

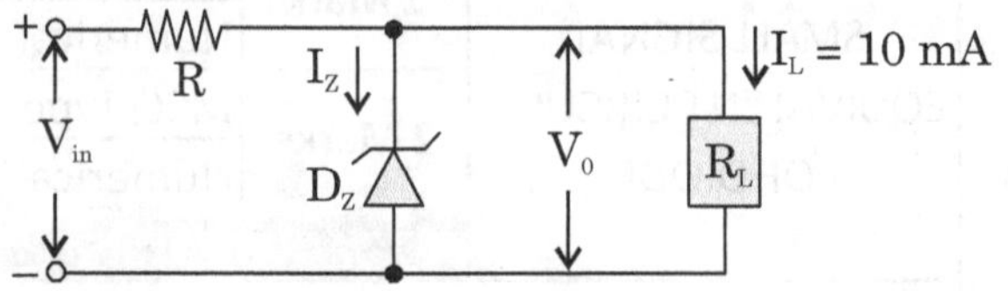

(a) R ≤ 1800 Ω

(b) 2000 Ω ≤ R ≤ 2200 Ω

(c) 3700 Ω ≤ R ≤ 4000 Ω

(d) R > 4000 Ω **[2002 : 2 Marks]**

8. The circuit shown in the figure is best described as a

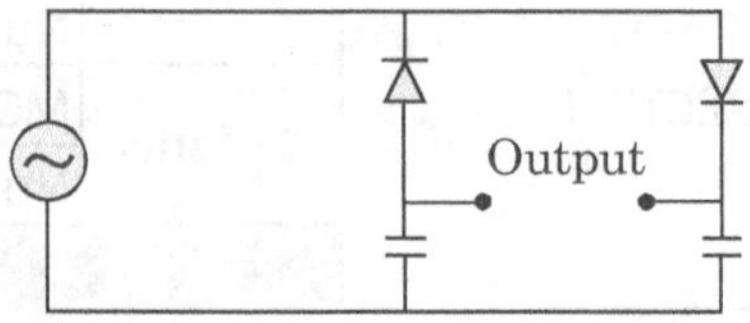

(a) bridge rectifier.

(b) ring modulator.

(c) frequency discriminatory.

(d) voltage doubler. **[2003 : 1 Mark]**

9. In the voltage regulator shown in the figure, the load current can vary from 100 mA to 500 mA. Assuming that the Zener diode is ideal (i.e., the Zener knee current is negligibly small and Zener resistance is zero in the breakdown region), the value of R is

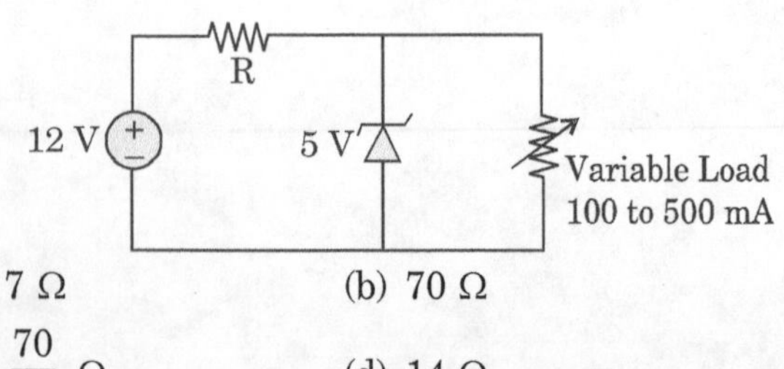

(a) 7 Ω (b) 70 Ω

(c) $\frac{70}{3}$ Ω (d) 14 Ω

[2004 : 2 Marks]

10. The Zener diode in the regulator circuit shown in the figure has a Zener voltage of 5.8 volts and a Zener knee current of 0.5 mA. The maximum load current drawn from this circuit ensuring proper functioning over the input voltage range between 20 and 30 volts, is

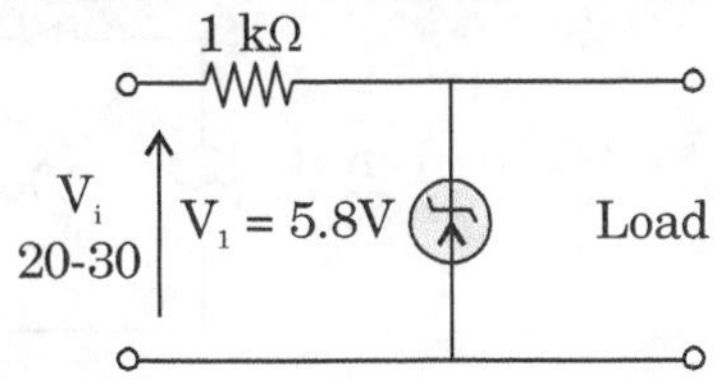

(a) 23.7 mA
(b) 14.2 mA
(c) 13.7 mA
(d) 24.2 mA

[2005 : 2 Marks]

11. For the circuit shown below, assume that the zener diode is ideal with a breakdown voltage of 6 volts. The waveform observed across R is

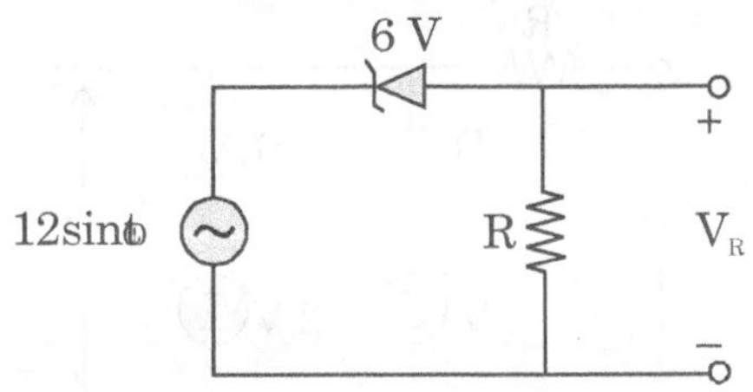

(a) 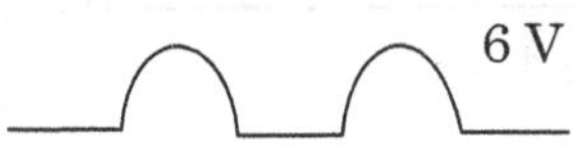

(b) 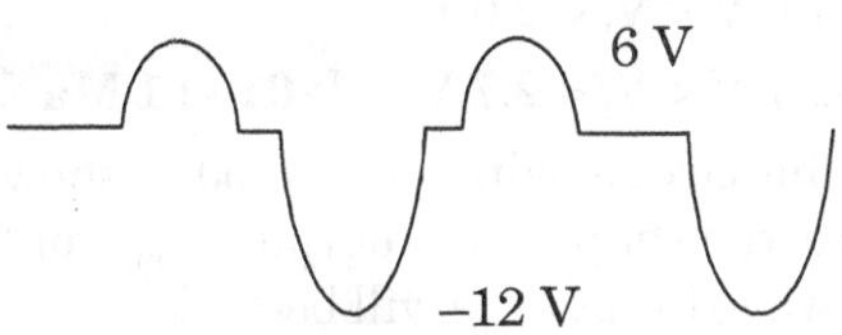

(c) 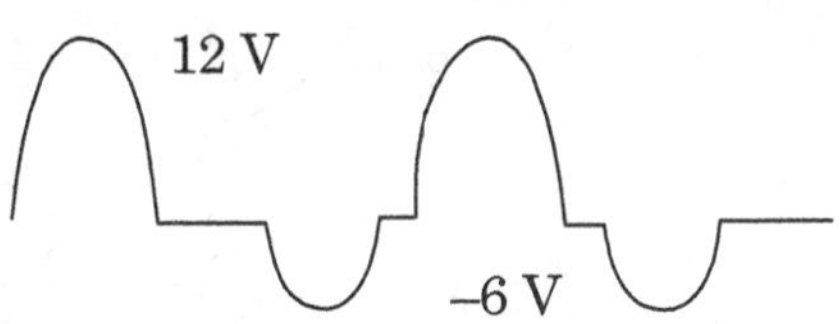

(d) 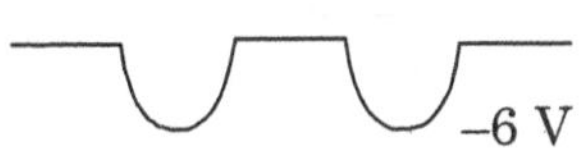

[2006 : 2 Marks]

12. For the Zener diode shown in the figure, the Zener voltage at knee is 7 V, the knee current is negligible and the Zener dynamic resistance is 10 Ω. If the input voltage (V_i) range is from 10 to 16 V, the output voltage (V_0) ranges from

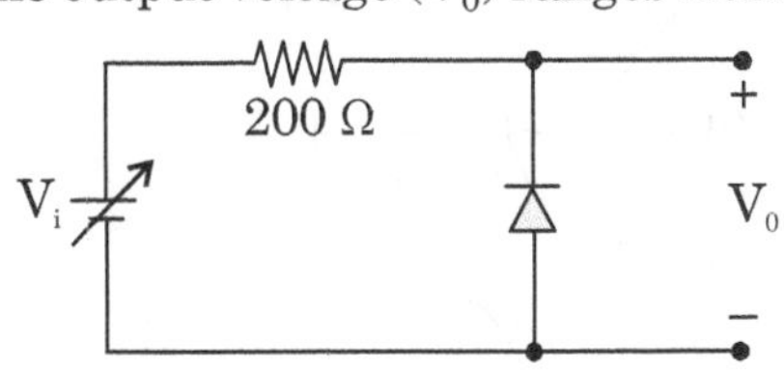

(a) 7.00 to 7.29 V (b) 7.14 to 7.29 V
(c) 7.14 to 7.43 V (d) 7.29 to 7.43 V

[2007 : 2 Marks]

13. In the following limiter circuit, an input voltage $V_i = 10\sin 100\pi V$ is applied. Assume that the diode drop is 0.7 V when it is forward biased. The Zener breakdown voltage is 6.8 V.

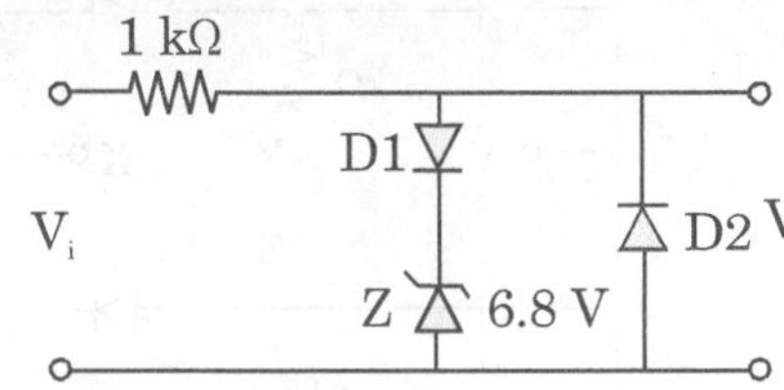

The maximum and minimum values of the output voltage respectively are

(a) 6.1 V, –0.7 V (b) 0.7 V, –7.5 V
(c) 7.5 V, –0.7 V (d) 7.5 V, –7.5 V

[2008 : 1 Mark]

14. In the circuit below, the diode is ideal. The voltage V is given by

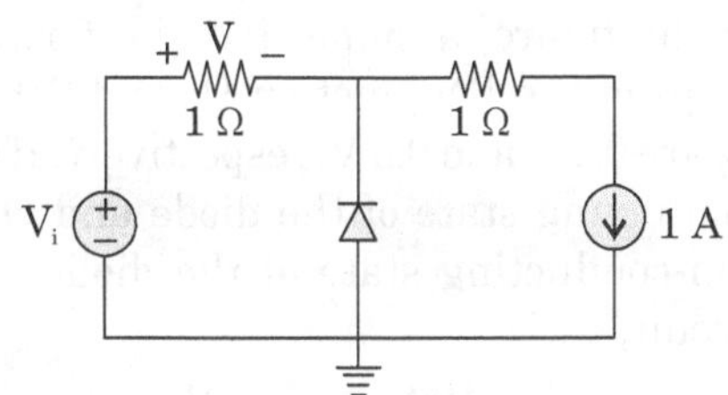

(a) $\min(V_i, 1)$ (b) $\max(V_i, 1)$
(c) $\min(-V_i, 1)$ (d) $\max(-V_i, 1)$

[2009 : 2 Marks]

15. The diodes and capacitors in the circuit shown are ideal. The voltage v(t) across the diode D1 is

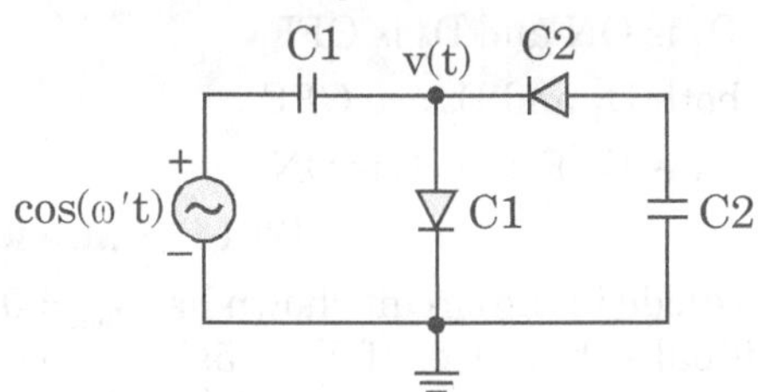

(a) $\cos(\omega t)-1$ (b) $\sin(\omega t)$
(c) $1-\cos(\omega t)$ (d) $1-\sin(\omega t)$

[2012 : 1 Mark]

16. In the circuit shown below, the knee current of the ideal Zener diode is 10 mA. To maintain 5 V across R_L, the minimum value of R_L in Ω and the minimum power rating of the Zener diode in mW, respectively, are

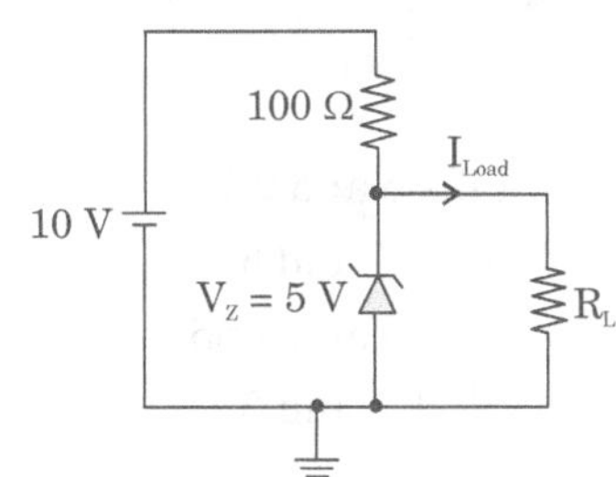

(a) 125 and 125
(b) 125 and 250
(c) 250 and 125
(d) 250 and 250

[2013 : 2 Marks]

17. A voltage 1000sinωt volts is applied across YZ. Assuming ideal diodes, the voltage measured across WX in volts, is

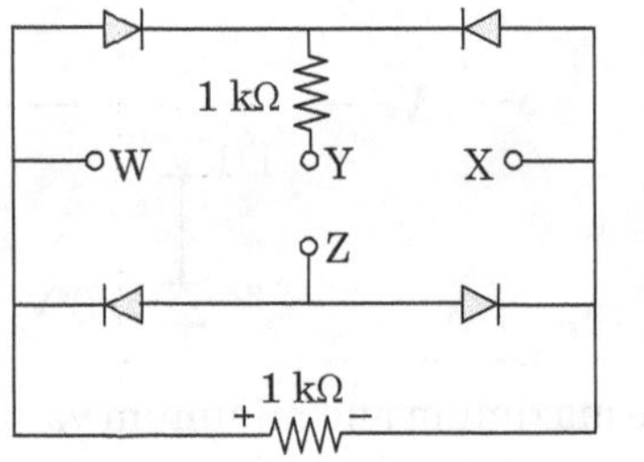

(a) sinωt
(b) $\frac{(\sin\omega t + |\sin\omega t|)}{2}$
(c) $\frac{(\sin\omega t - |\sin\omega t|)}{2}$
(d) 0 for all t

[2013 : 2 Marks]

18. In the figure, assume that the forward voltage drops of the PN junction D_1 and Schottky diode D_2 are 0.7 V and 0.3 V, respectively. If ON denotes conducting state of the diode and OFF denotes non-conducting state of the diode, then in the circuit,

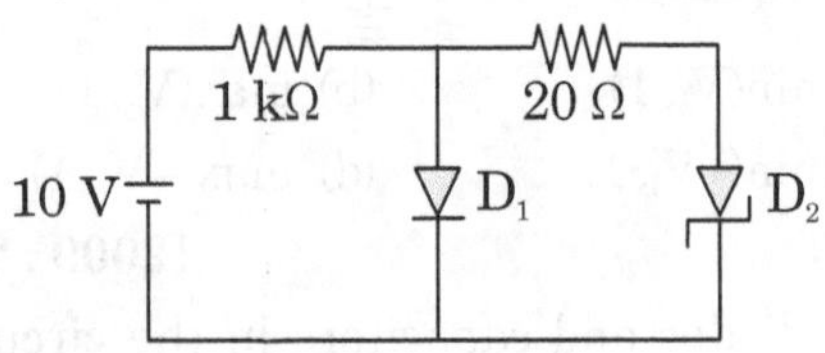

(a) both D_1 and D_2 are ON
(b) D_1 is ON and D_2 is OFF
(c) both D_1 and D_2 are OFF
(d) D_1 is OFF and D_2 is ON

[2014 : 1 Mark, Set-1]

19. The diode in the circuit shown has V_{on} = 0.7 V but is ideal otherwise. If V_i = 5sin(ωt) volts, the minimum and maximum values of V_0 (in Volts) are, respectively,

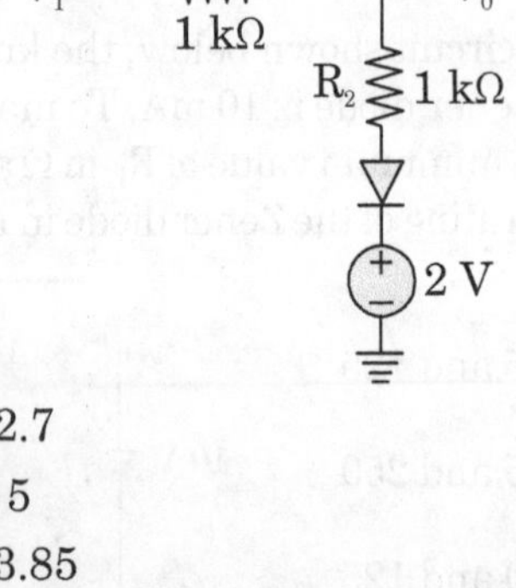

(a) –5 and 2.7
(b) 2.7 and 5
(c) –5 and 3.85
(d) 1.3 and 5

[2014 : 2 Mark, Set-2]

20. The figure shows a half-wave rectifier. The diode D is ideal. The average steady-state current (in Amperes) through the diode is approximately ______.

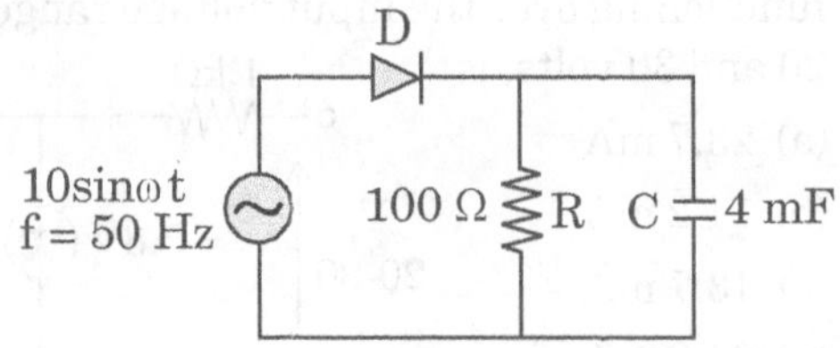

[2014 : 1 Marks, Set-3]

21. Two silicon diodes, with a forward voltage drop of 0.7 V, are used in the circuit shown in the figure. The range of input voltage V_i for which the output voltage $V_0 = V_i$ is

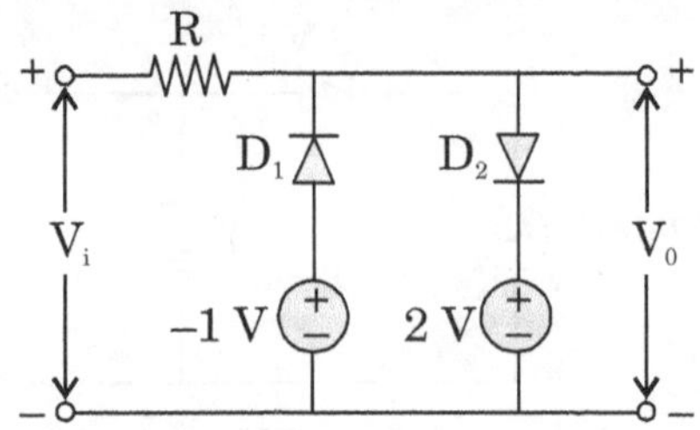

(a) $-0.3\text{ V} < V_i < 1.3\text{ V}$
(b) $-0.3\text{ V} < V_i < 2\text{ V}$
(c) $-1.0\text{ V} < V_i < 2.0\text{ V}$
(d) $-1.7\text{ V} < V_i < 2.7\text{ V}$ **[2014 : 1 Mark, Set-4]**

22. For the circuit with ideal diodes shown in the figure, the shape of the output (V_{out}) for the given sine wave input (V_{in}) will be

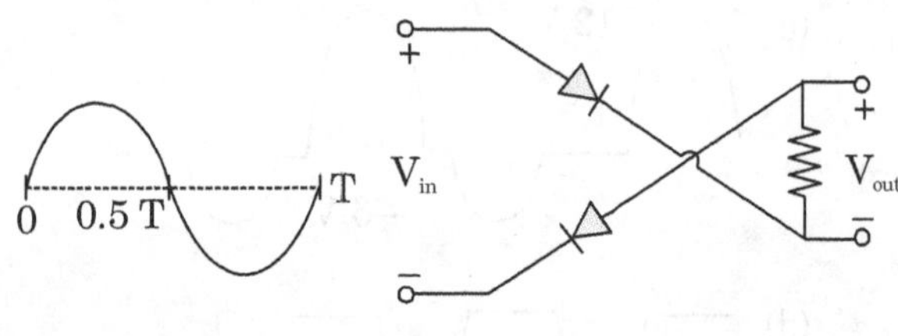

(a)
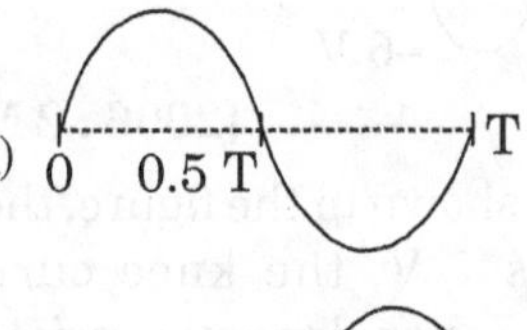

(b)
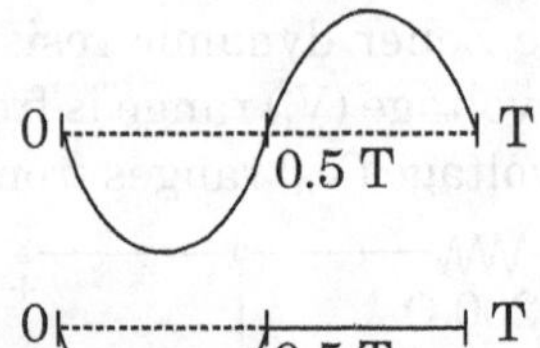

(c)
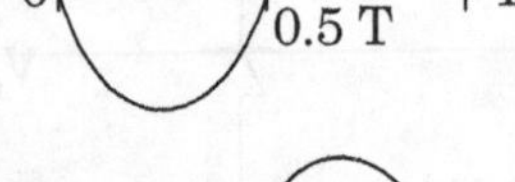

(d)
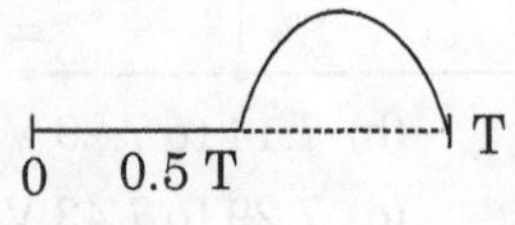

[2015 : 1 Mark, Set-1]

23. In the circuit shown below, the Zener diode is ideal and the Zener voltage is 6 V. The output voltage V_0 (in volts) is ______.

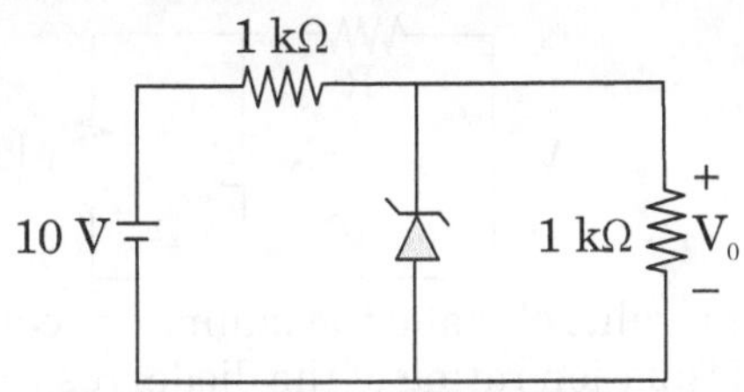

[2015 : 1 Mark, Set-1]

24. If the circuit shown has to function as a clamping circuit, then which one of the following conditions should be satisfied for the sinusoidal signal of period T?

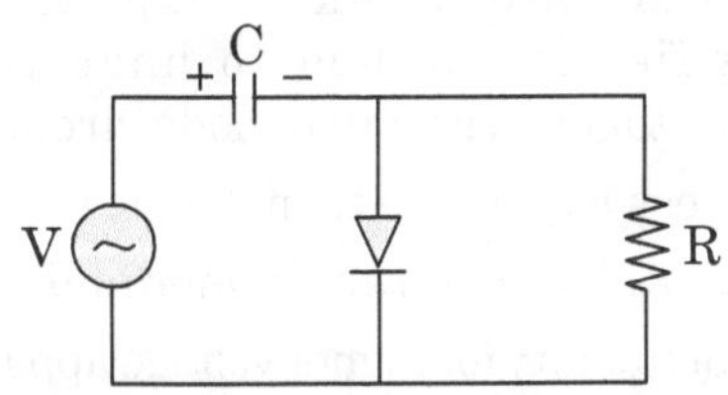

(a) RC << T

(b) RC = 0.35T

(c) RC = T

(d) RC >> T

[2015 : 1 Mark, Set-2]

25. The diode in the circuit given below has V_{ON} = 0.7 V but is ideal otherwise. The current (in mA) in the 4 kΩ resistor is ______.

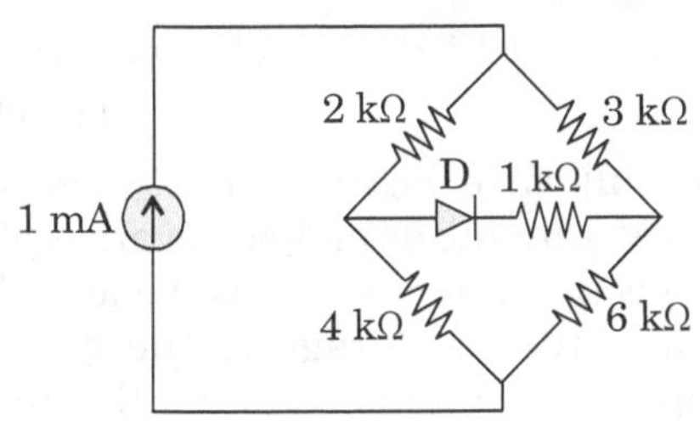

[2015 : 2 Marks, Set-2]

26. In the circuit shown, assume that diodes D_1 and D_2 are ideal. In the steady-state condition the average voltage V_{ab} (in Volts) across the 0.5 μF capacitor is ______.

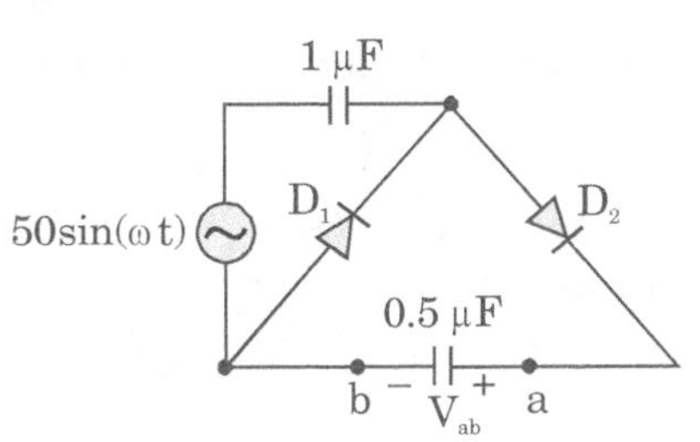

[2015 : 1 Marks, Set-3]

27. The switch S in the circuit shown has been closed for a long time. It is opened at time t = 0 and remains open after that. Assume that the diode has zero reverse current and zero forward voltage drop.

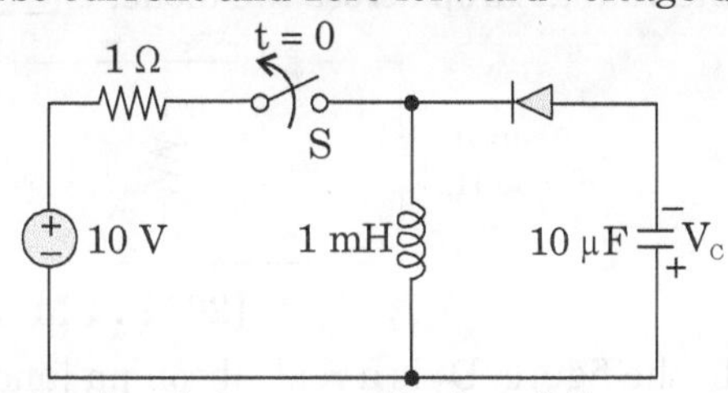

The steady state magnitude of the capacitor voltage V_C (in volts) is ______.

[2016 : 2 Marks, Set-2]

28. The diodes D_1 and D_2 in the figure are ideal and the capacitors are identical. The product RC is very large compared to the time period of the ac voltage. Assuming that the diodes do not breakdown in the reverse bias, the output voltage V_0 (in volt) at the steady state is ______.

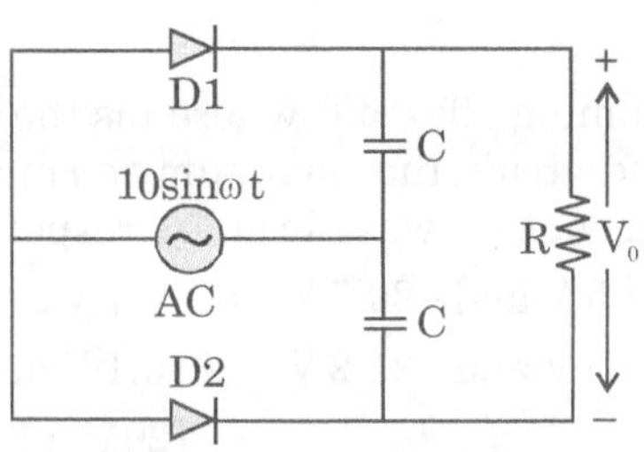

[2016 : 1 Mark, Set-3]

29. Assume that the diode in the figures has V_{on} = 0.7 V, but is otherwise ideal.

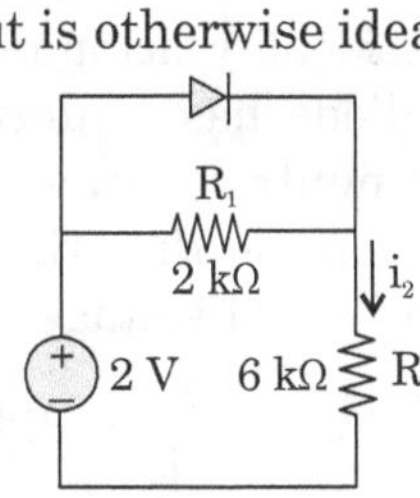

The magnitude of the current i_2 (in mA) is equal to ______. **[2016 : 1 Mark, Set-2]**

30. The I-V characteristics of the zener diodes D_1 and D_2 are shown in Figure I. These diodes are used in the circuit given in Figure II. If the supply voltage is varied from 0 to 100 V, then breakdown occurs in

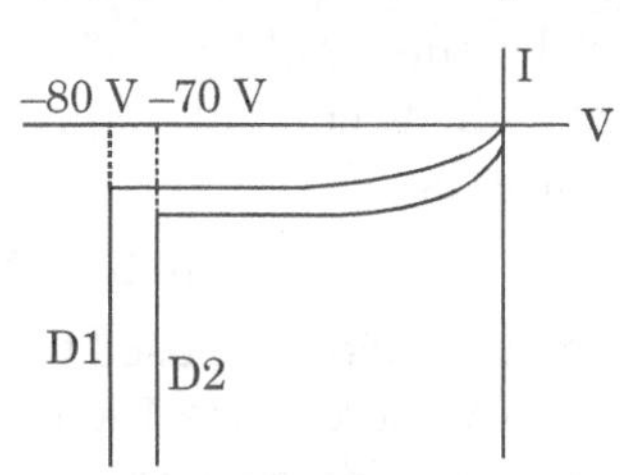

Figure I

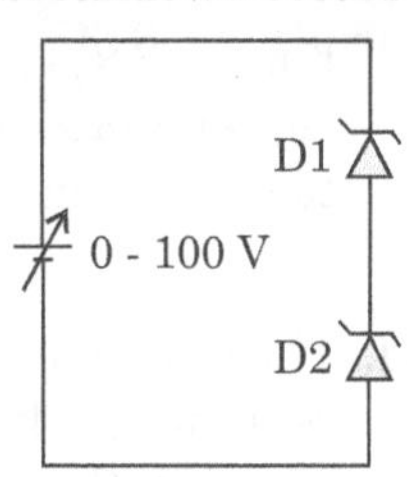

Figure II

(a) D1 only

(b) D2 only

(c) both D1 and D2

(d) none of D1 and D2

[2016 : 2 Marks, Set-3]

31. The output V_0 of the diode circuit shown in the figure is connected to an averaging DC voltmeter. The reading on the DC voltmeter in Volts, neglecting the voltage drop across the diode, is ______.

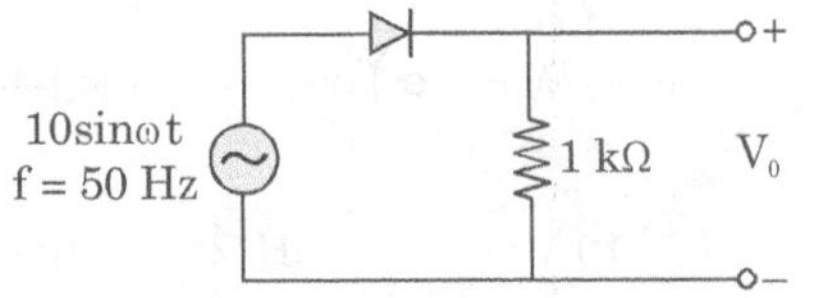

[2017 : 1 Mark, Set-2]

32. In the figure, D_1 is a real silicon pn junction diode with a drop of 0.7 V under forward bias condition and D_2 is a Zener diode with breakdown voltage of –6.8 V. The input $V_{in}(t)$ is a periodic square wave of period T, whose one period is shown in the figure.

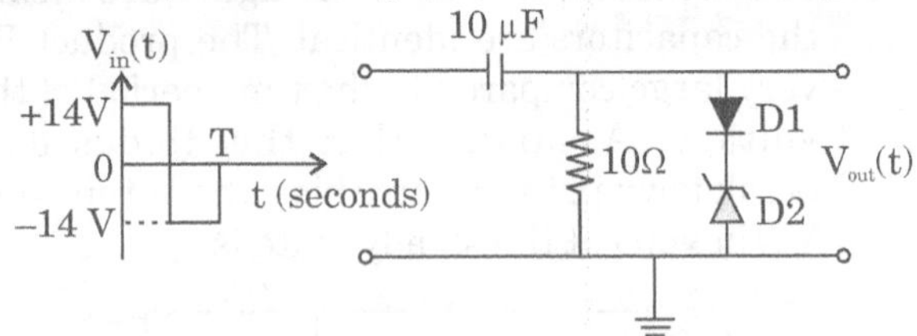

Assuming 10τ << T, where τ is the time constant of the circuit, the maximum and minimum values of the output waveform are respectively,

(a) 7.5 V and –20.5 V (b) 6.1 V and –21.9 V

(c) 7.5 V and –21.2 V (d) 6.1 V and –22.6 V

[2017 : 1 Mark, Set-2]

33. A DC current of 26 μA flows through the circuit shown. The diode in the circuit is forward biased and it has an ideality factor of one. At the quiescent point, the diode has a junction capacitance of 0.5 nF. Its neutral region resistances can be neglected. Assume that the room temperature thermal equivalent voltage is 26 mV.

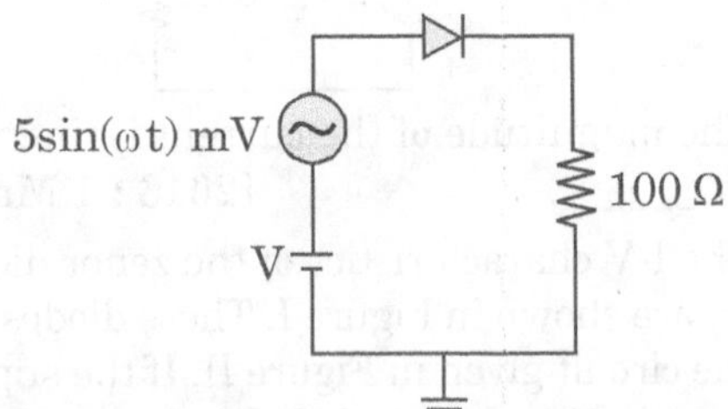

For ω = 2 × 10^6 rad/s, the amplitude of the small-signal component of diode current (in μA) correct to one decimal place) is ______.

[2018 : 2 Marks]

34. The circuit shown in the figure is used to provide regulated voltage (5 V) across the 1 kΩ resistor. Assume that the Zener diode has a constant reverse breakdown voltage for a current range, starting from a minimum required Zener current, I_{Zmin} = 2 mA to its maximum allowable current. The input voltage V_1 may vary by 5% from its nominal value of 6 V. The resistance of the diode in the breakdown region is negligible.

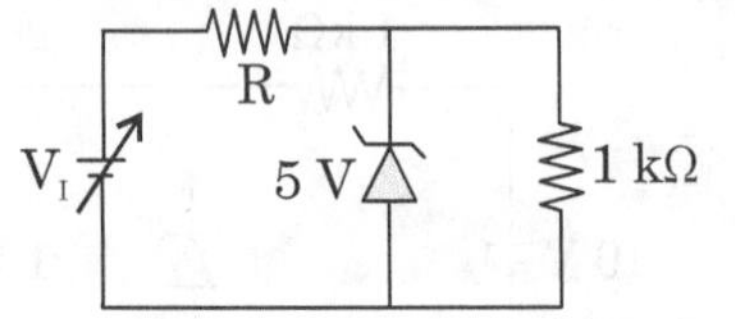

The value of R and the minimum required power dissipation rating of the diode, respectively, are

(a) 186 Ω and 10 mW (b) 100 Ω and 40 mW

(c) 100 Ω and 10 mW (d) 186 W and 40 mW

[2018 : 2 Marks]

RECTIFIERS

35. For full wave rectification, a four diode bridge rectifier is claimed to have the following advantages over a two diode circuit.

1. Less expensive transformer,
2. Smaller size transformer and,
3. Suitability for higher voltage application of these.

(a) Only (1) and (2) are true.

(b) Only (1) and (3) are true.

(c) Only (2) and (3) are true.

(d) (1), (2) as well as (3) are true. **[1998 : 1 Mark]**

36. A dc power supply has a no-load voltage of 30 V, and a full-load voltage of 25 V at a full-load current of 1 A. Its output resistance and load regulation, respectively, are

(a) 5 Ω and 20% (b) 25 Ω and 20%

(c) 5 Ω and 16.7% (d) 25 Ω and 16.7%

[1999 : 2 Marks]

37. In a full-wave rectifier using two ideal diodes, V_{dc} and V_m are the dc and peak values of the voltage respectively across a resistive load. If PIV is the peak inverse voltage of the diode, then the appropriate relationships for this rectifier are

(a) $V_{dc} = \frac{V_m}{\pi}$, PIV = $2V_m$ (b) $V_{dc} = 2\frac{V_m}{\pi}$, PIV = $2V_m$

(c) $V_{dc} = 2\frac{V_m}{\pi}$, PIV = V_m (d) $V_{dc} = \frac{V_m}{\pi}$, PIV = V_m

[2004 : 2 Marks]

38. The correct full wave rectifier circuit is

(a)

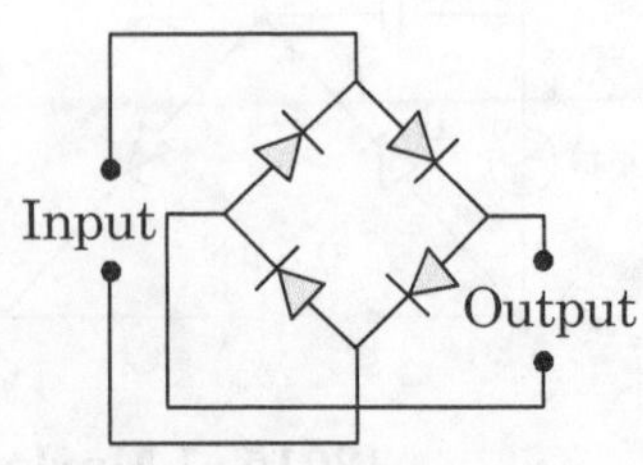

(b)

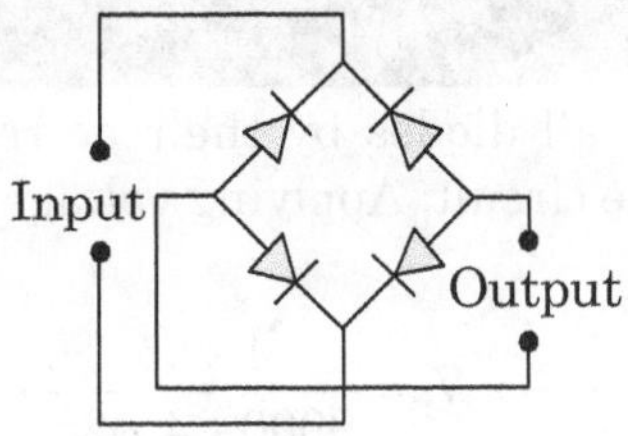

(c)

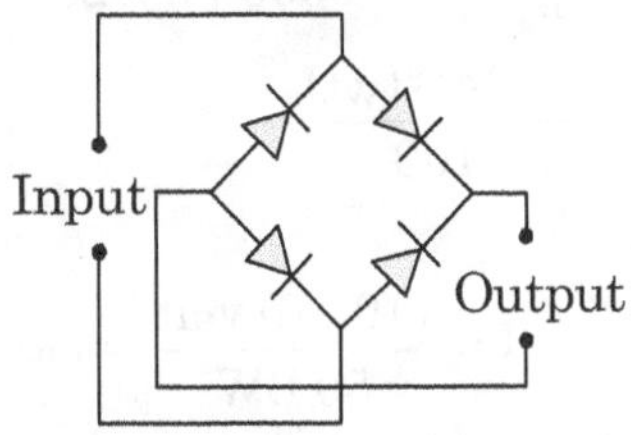

(d)

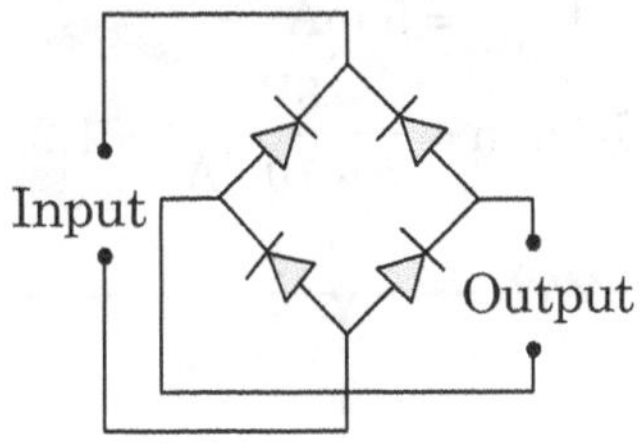

[2007 : 1 Mark]

39. The figure shows a half-wave rectifier with a 475 μF filter capacitor. The load draws a constant current I_0 = 1 A from the rectifier. The figure also shows the input voltage V_i the output voltage V_C and the peak-to-peak voltage ripple u on V_C. The input voltage V_i is a triangle-wave with an amplitude of 10 V and a period of 1 ms.

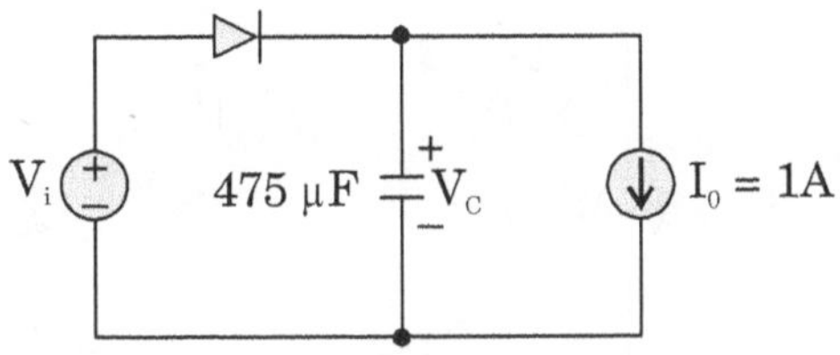

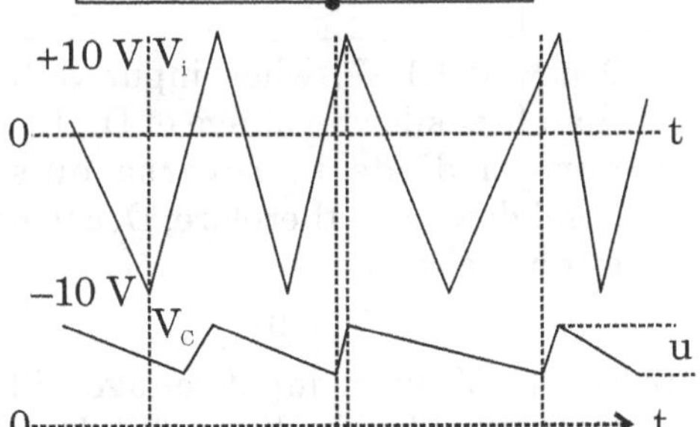

The value of the ripple u (in volts) is _____.

[2016 : 2 Marks, Set-2]

40. In the circuit shown, V_S is a square wave of period T with maximum and minimum values of 8V and – 10 V, respectively. Assume that the diode is ideal and R_1 = R_2 = 50 Ω. The average value of V_L is ________ volts (rounded off to 1 decimal place).

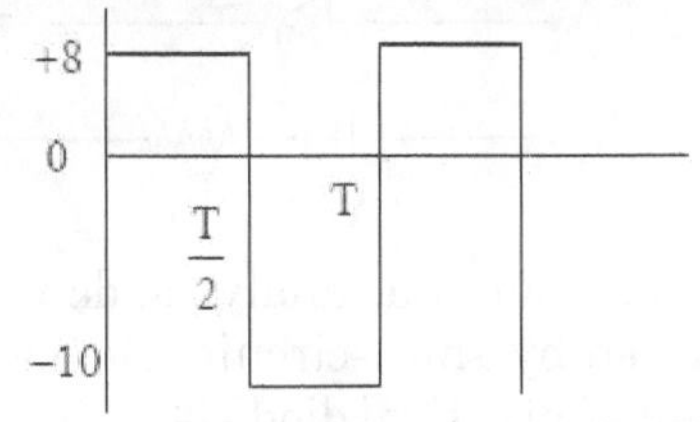

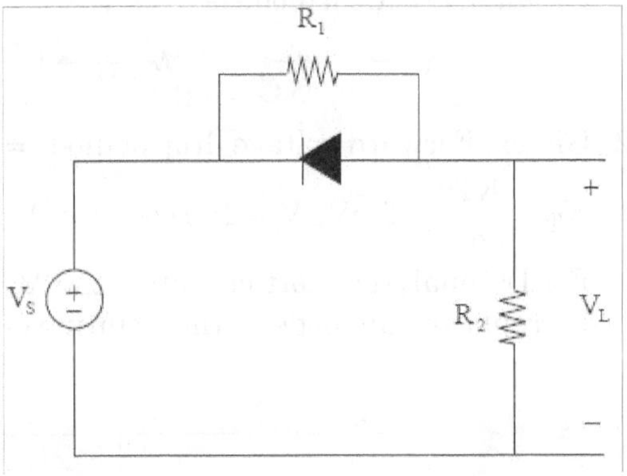

[2019 : 1 Mark]

41. In the circuit shown, V_S is a 10 V square wave of period, T = 4 ms with R = 500 Ω and C = 10 μF.The capacitor is initially unchanged at t = 0, and the diode is assumed to be ideal. The voltage across the capacitor (V_C) at 3 ms is equal to ______ volts (rounded off to one decimal place).

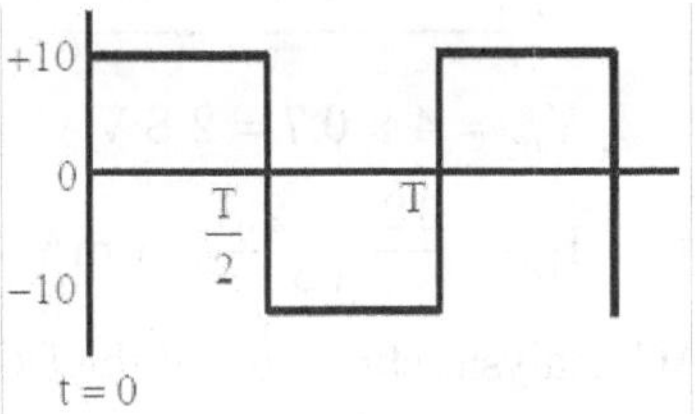

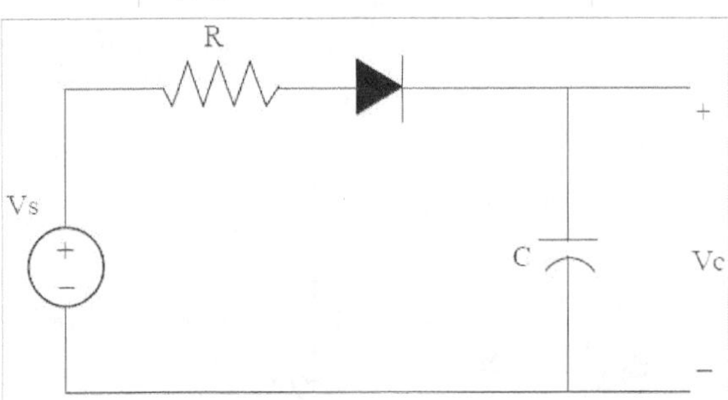

[2019 : 2 Marks]

ANSWERS

1. (d)	**2.** (a)	**3.** (b)	**4.** (a)	**5.** (a)	**6.** (c)	**7.** (a)	**8.** (d)	**9.** (d)	**10.** (a)
11. (b)	**12.** (c)	**13.** (c)	**14.** (a)	**15.** (a)	**16.** (b)	**17.** (d)	**18.** (d)	**19.** (c)	**20.** (0· 09)
21. (d)	**22.** (c)	**23.** (5)	**24.** (d)	**25.** (b)	**26.** (100)	**27.** (100)	**28.** (0)	**29.** (b)	**30.** (a)
31. (3.183)	**32.** (a)	**33.** 6.40	**34.** (b)	**35.** (d)	**36.** (d)	**37.** (b)	**38.** (a)	**39.** (2· 105)	**40.** (– 3)
41. (3.31)									

EXPLANATIONS

1. Equivalent circuit representation of practical diode is series combination of a forward resistance and cut-in voltage.

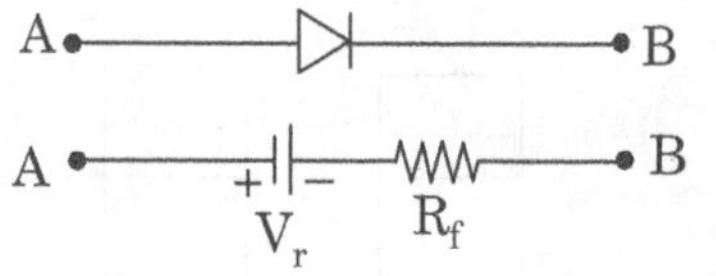

In small signal ac analysis, dc voltage source is replaced by short-circuit. Hence, ac equivalent circuit of practical diode is

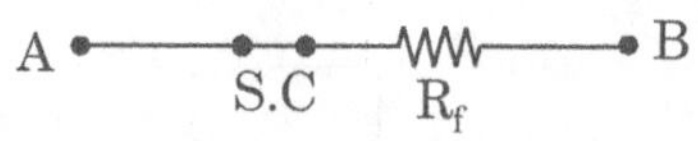

2. Given : Forward voltage drop of diode = 0.7 V,

$V_T = \frac{KT}{q} = 25\text{mV},\ V_i = 100\cos(\omega t)\text{ mV}.$

For DC analysis, short-circuiting the AC voltage. By replacing all diode by their cut-in voltage,

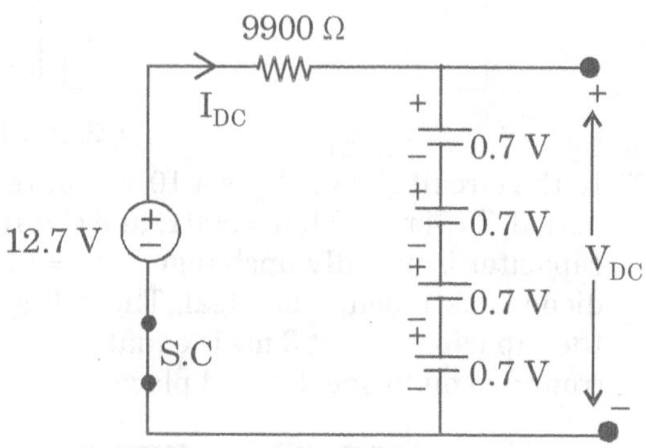

$$V_{DC} = 4 \times 0.7 = 2.8\text{ V}$$

$$\text{currect, } I_{DC} = \frac{12.7 - 2.8}{9.9} = 1\text{ mA}.$$

3. For AC analysis, short-circuit the DC voltage source.

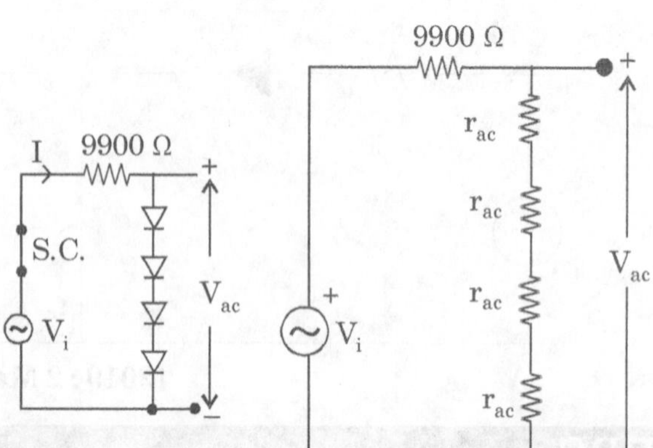

Dynamic resistance or ac resistance of diode is

$$r_{ac} = \frac{\eta V_T}{I} = \frac{V_T}{I} = \frac{25\text{ mV}}{1\text{ mA}} = 25\ \Omega$$

(Assume, $\eta = 1$).

By replacing all diodes by their ac resistance redrawing the circuit; Applying voltage division rule,

$$V_{ac} = \frac{4r_{ac}}{9900 + 4r_{ac}} \times V_i = \frac{4 \times 25}{9900 + 4 \times 25} \times V_i$$

$$V_{ac} = \frac{1}{100}.V_i = \frac{100\cos(wt)}{100}$$

$$V_{ac} = \cos(\omega t)\text{ mV}.$$

4. $$I = \frac{(10-6)\text{ volt}}{50\text{ BW}} = 80\text{ mA}$$

$$I_{L\,max} = 5\text{ mA}$$

$$R_L\text{ min} = \frac{6\text{V}}{5 \times 10^{-3}\text{A}} = 1.2.k\Omega$$

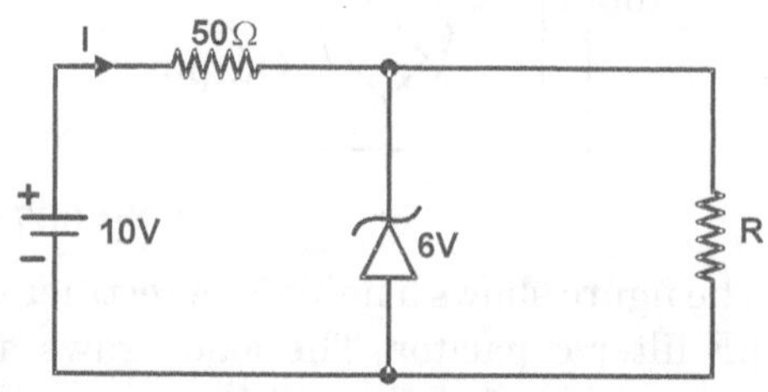

5. Source voltage is sinusoidal, hence, consider two cases for positive and negative half-cycles.

Case 1 : During positive half cycle : Diode D_1, will be in forward bias. hence it acts as short-circuit.

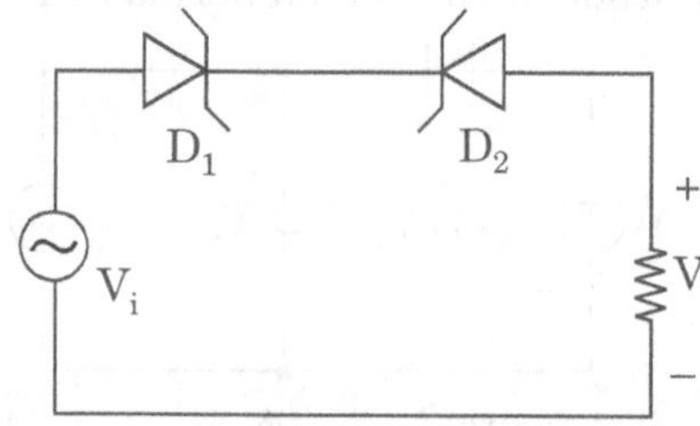

(a) $0 < V_i < 4.1$ V : when input voltage is less than breakdown voltage of D_2, D_2 will act as normal diode in reverse bias without breakdown and therefore, D_2 can be replaced by open circuit.

$$V_0 = 0\text{V}$$

(b) $V_i > 4.1$ V: when input voltage will be higher than breakdown voltage of D_2, D_2 will work in reverse bias breakdown. Therefore, D_2 can be replaced by its breakdown voltage. Now, applying KVL,

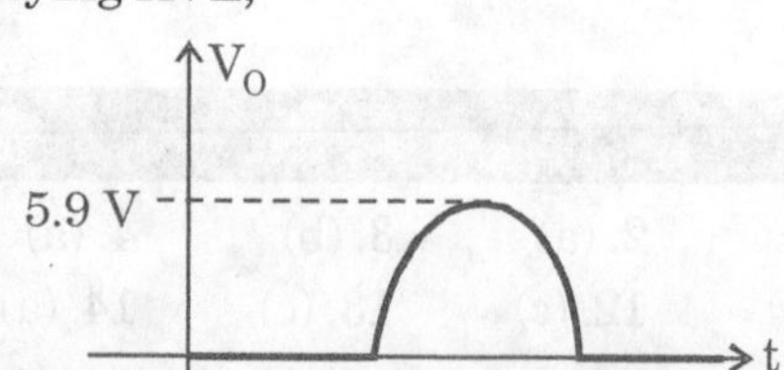

$$-V_i + 4.1 + V_0 = 0$$

$$\therefore\quad V_0 = V_i - 4.1 = 10\sin(\omega t) - 4.1$$

Case 2 : During negative half cycle : Diode D_2 will be in forward bias, hence acts as a short-circuit

(a) $-4.1\ V < V_i < 0\ V$: when input voltage will be less than breakdown voltage of D_1, D_1 will act as a normal diode in reverse bias. Therefore, D_1 can be replaced by open-circut.

$$V_0 = 0 \text{ volt}$$

(b) $V_i < -4.1\ V$: when input voltage will be higher than the breakdown voltage of D_1, D_1 will work in reverse bias breakdown region. Therefore, D_1 can be replaced by its breakdown voltage. Applying KVL,

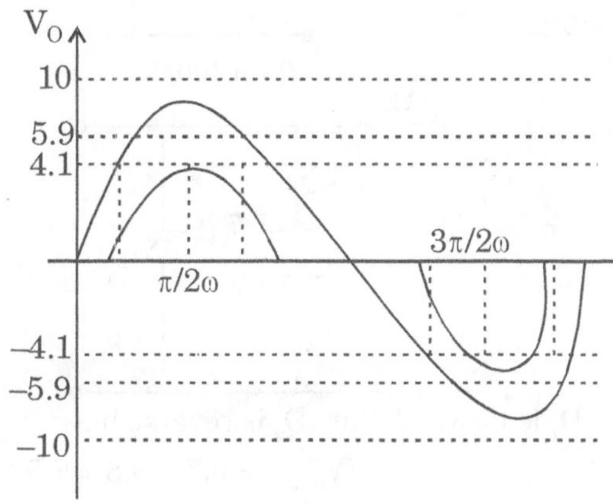

$$-V_i - 4.1 + V_0 = 0$$

$$V0 = V_i + 4.1$$

$$\text{Output voltage, } V_o = \begin{cases} 0\ V; & 0\ V < V_i < 4.1\ V \\ V_i - 4.1; & V_i > 4.1\ V \\ 0\ V; & -4.1\ V < V_i < 0\ V \\ V_i + 4.1; & V_i < -4.1\ V \end{cases}$$

6. $$I_{in} = \frac{9-6}{50} = 0.06\ A = 60\ mA$$

Minimum current through zener diode

$$= \frac{300 \times 10^{-3}}{6} = 50\ mA$$

$\therefore$ Minimum current through load

$$= 60 - 50 = 10\ mA$$

and maximum current through load

$$= 60 - 5 = 55\ mA.$$

7. $$\frac{V_{in} - V_z}{R} \geq I_{2K} + I_L$$

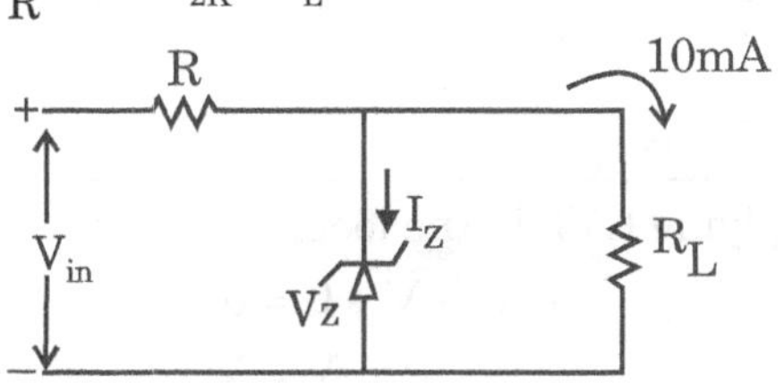

When $V_{in} = 20$ V, $\dfrac{20}{R} \geq 11 \times 10^{-3}$

or $R \leq 1818\ \Omega$ ...(*i*)

and when $V_i = 50$ V

then $\dfrac{40}{R} \geq L \times 10^{-3}$

or $R \leq 3636\Omega$ ...(*ii*)

Combining equations (*i*) and (*ii*), we get

8. 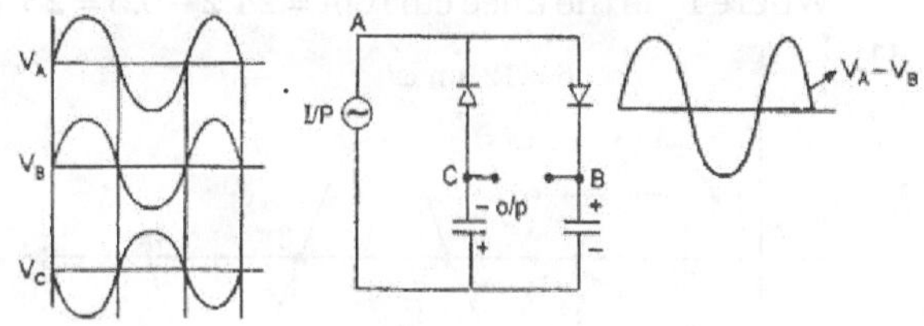

9. From the circuit given, we have

$$\frac{12 - V_z}{R} = I_L + I_z$$

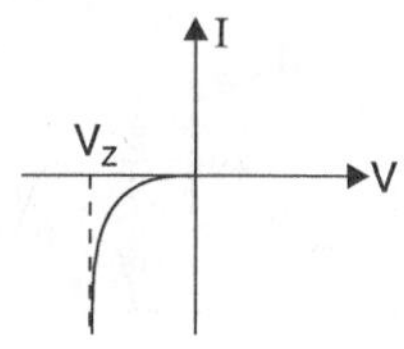

or, $$I_L = \left(\frac{12 - V_z}{R}\right) - I_z$$

Given : $100\ mA < I_L < 500m\ A$,

then, $$100 \times 10^{-3} < \left(\frac{7}{R} - I_z\right) < 500 \times 10^{-3}$$

or $$100 \times 10^{-3} < \frac{7}{R} < 500 \times 10^{-3}$$

{Zener knee current is negligibly small }

or $14\ \Omega < R < 70\ \Omega$

Hence required value of $R = 14\ \Omega$

[choosing minimum value]

10. 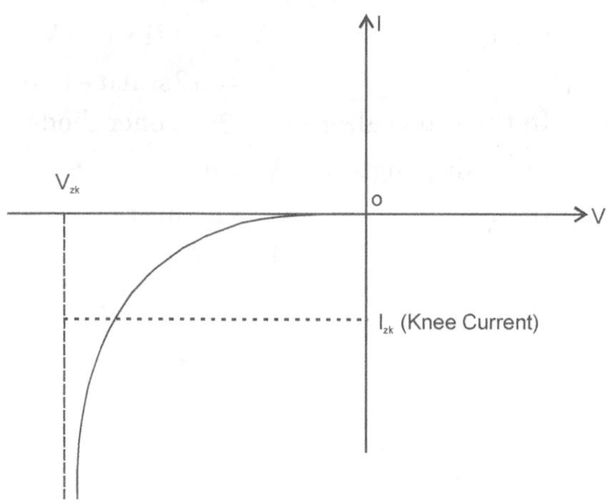

V–I characteristics of zener diode

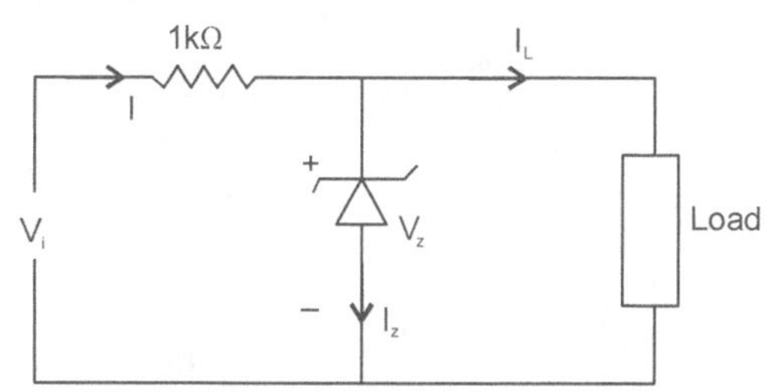

For $V_i = 20$ volts, minimum value of current I

$$I_{min} = \frac{20 - 58}{1k} = 14.2\ mA$$

For V_i = 30 V, and maximum value of current I

$$I_{max} = \frac{30-5.8}{1k}\ 24.2 \text{ mA}$$

Maximum value of load current

$$I_{L,max} = I_{max} - I_{zk}$$

Where I_{zk} is the knee current = 24.2 – 0.5 = 23.7 mA

11. 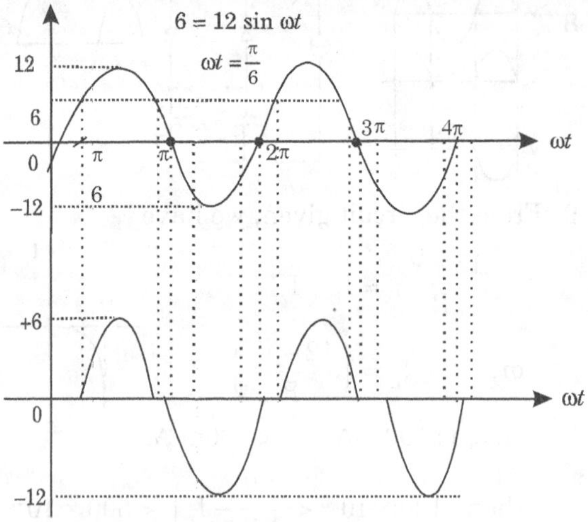

During $0<\omega t<\frac{\pi}{6}$

Diode is OFF, No conduction

then, $V_R = 0$ volts

In the interval, $\frac{\pi}{6}<\omega t<\pi$,

Zener diode voltage, $V_Z = 6$ volts

then, $V_0 = 12\sin\omega t - V_Z$

$= 12\sin\omega t - 6$

In the interval, $\pi<\omega t<2\pi$, zener diode is in conducting state, $V_2 = 0$

then, $V_0 = 12\sin\omega t$

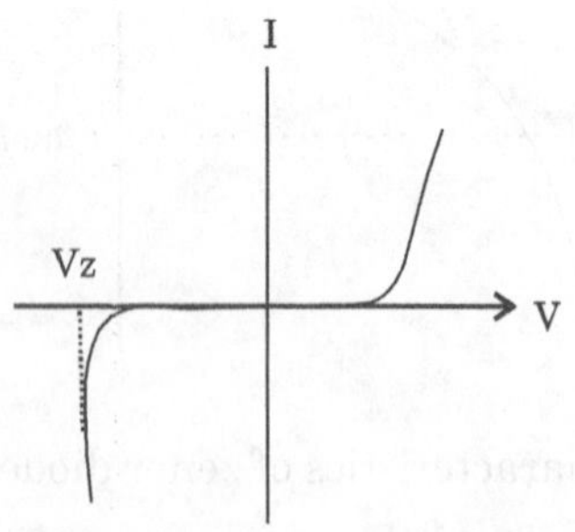

12. 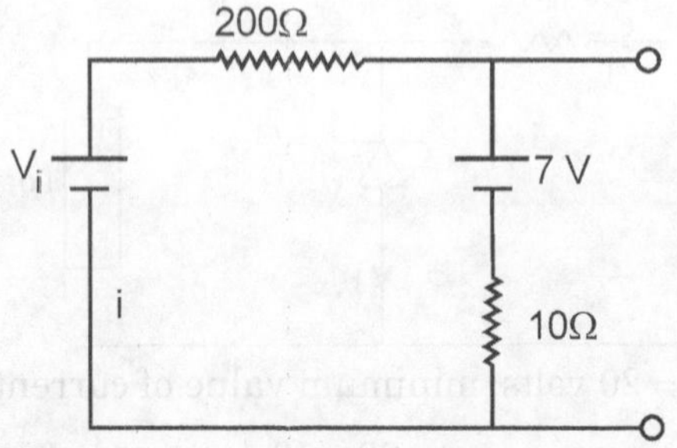

When V_i = 10 V, $i = \frac{10-7}{210} = \frac{3}{210}$ A

$$V_{01} = 7 + 10i = 7 + 10\ \frac{3}{210} = 7.14 \text{ volts}$$

When V_i = 16 V, $i = \frac{9}{210}$ amp

$$V_{02} = 7 + 10\left(\frac{9}{210}\right) = 7.43 \text{ volts}$$

13. In positive half cycle

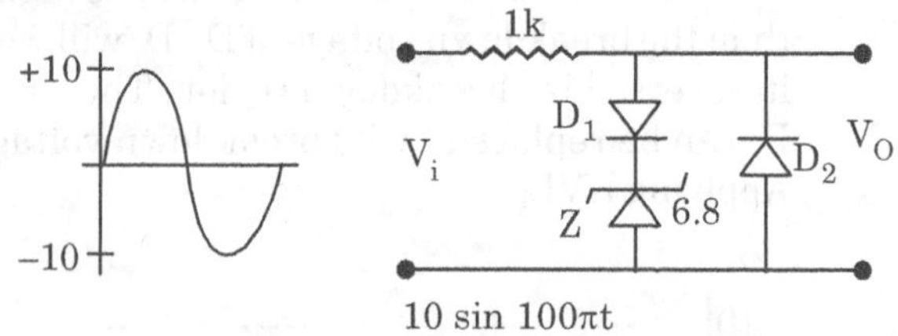

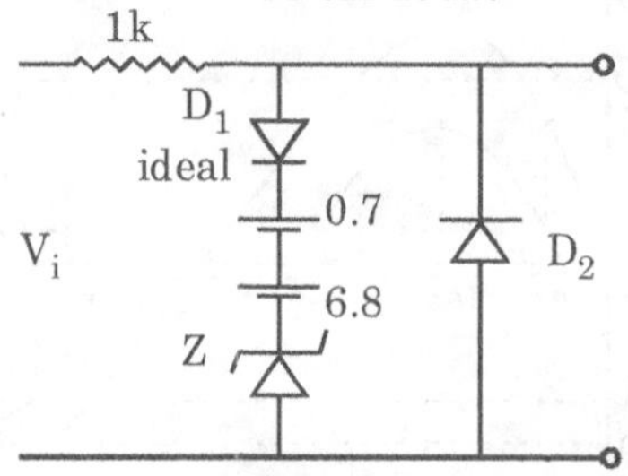

D_1 is forward bias, D_2 is reverse bias

$\therefore$ $(V_0)_{max} = 0.7 + 6.8 = 7.5$ V

In negative half cycle, D_2 is Forward bias and D_1 is Reverse bias

$\therefore$ $(V_0)_{min} = V_{D2} = -0.7$ Volt

Alternately

During positive cycle, D_1 is ON and D_2 is OFF,

$$V_0 = V_Z + V_{D_1} = 6.8 + 0.7 = 7.5\text{V}$$

During negative cycle,

D_1 is OFF and D_2 is ON

$$V_0 = -V_{D_2} = -0.7\text{V}$$

14. **Case 1 :** Assuming diode is ON.

Hence, diode can be replaced by short-circuit.

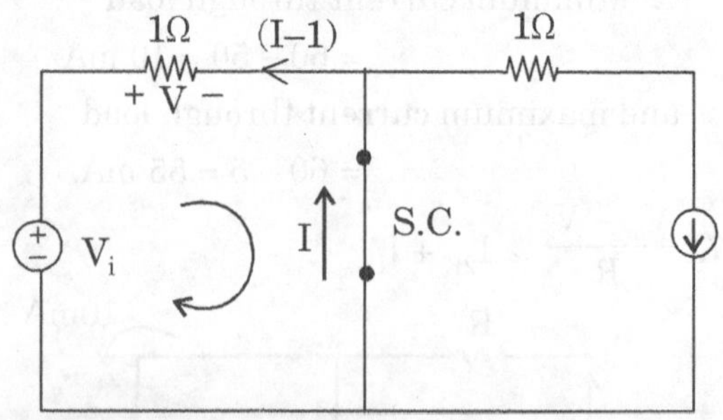

Applying KVL in the loop,

$$-V_i + V + 0 = 0$$

$$V = V_i$$

$$\because \quad V = -(I-1)\times 1 = 1 - I$$

$$I = 1 - V_i$$

For diode to be ON, currect I must be positive, i.e., $I > 0$.

$1 - V_i > 0 \Rightarrow V_i < 1$ V.

Hence, for diode to be ON, the voltage V_i must be less then 1V.

If $V_i = 0.5$V, then V = 0.5 volt.

option	V	Status
(A) $\min(V_i, 1)$	0.5 V	Correct
(B) $\max(V_i, 1)$	1 V	Wrong
(C) $\min(-V_i, 1)$	-0.5 V	Wrong
(D) $\max(-V_i, 1)$	1 V	Wrong

Case 2 : Assume diode is OFF.

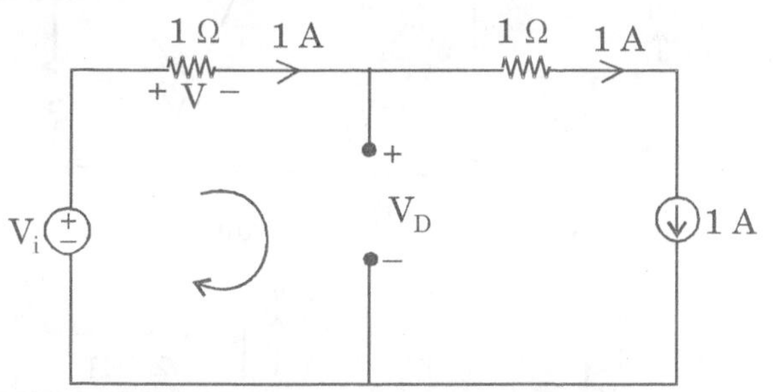

Hence, diode can be replaced by open-circuit.

$\Rightarrow \quad V = 1 \times 1 = 1$ volt.

Applying KVL in the loop

$$-V_i + V + V_D = 0$$

$$V_D = V_i - V = V_i - 1.$$

For diode to be OFF, V_D must be positive,

i.e., $\quad V_D > 0$V.

$$V_i - 1 > 0.$$

$$V_i > 1 \text{ V}.$$

Hence, for diode to be OFF, V_i must be greater than 1 V and V must be equal to 1 volt.

If $V_i = 1.5$ volt and V = 1 volt

Option	V	Status
(A) $\min(V_i, 1)$	1 V	Correct
(B) $\max(V_i, 1)$	1.5 V	Wrong
(C) $\min(-V_i, 1)$	-1.5 V	Wrong
(D) $\max(-V_i, 1)$	1 V	Correct

$\Rightarrow$ Hence, from the Case 1 and Case 2, it is clear that option (a) is common for both cases.

15. It is voltage doubler circuit in which C_1 will be charged to maximum value of input that is 1 V. So, $v(t) = (\cos\omega t - 1)$.

16.

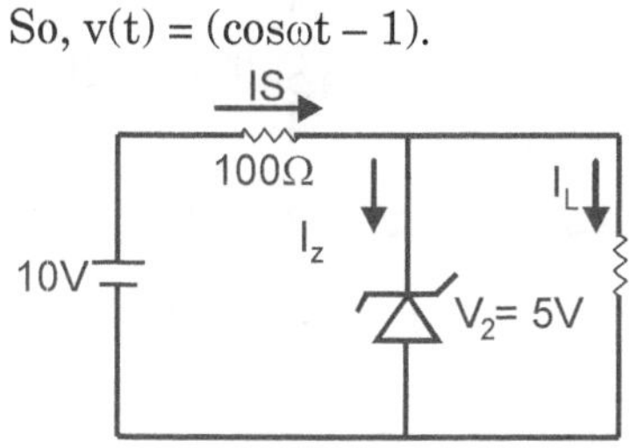

$$I_s = I_z + I_L$$

$$I_s - I_z = I_L$$

Two extreme condition:

If I_z(min), then I_L (max)

If I_z(max) then I_L (min) = 0

$$I_z(\text{max}) = I_s = \frac{10-5}{10} = 50 \text{ mA}$$

$$I_z(\text{min}) = I_s - I_L(\text{max})$$

$$I_L(\text{max}) = I_s - I_z(\text{min}) = I_s - I_z = (50 - 10) = 40 \text{ mA}$$

$$R_L(\text{min}) = \frac{V}{I_L(\text{max})} = \frac{5}{40}\text{k} = 125\ \Omega$$

$$P_z = V_z \times I_z(\text{max}) = 5 \times 50 \text{ mA} = 250 \text{ mw}$$

17. 'D' 0 for all +

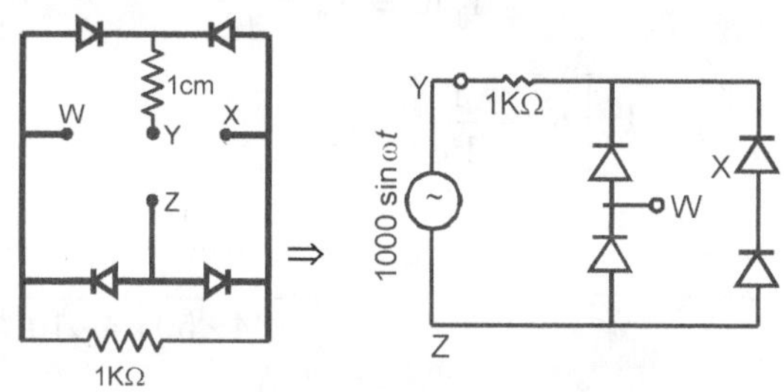

Note:

$\Rightarrow$ All diode conducts only in negative half cycle.

$\Rightarrow$ XW is at symmetrical point so voltage across XW is zero for all time.

18. Assume both the diode ON.

Then circuit will be as per figure (2)

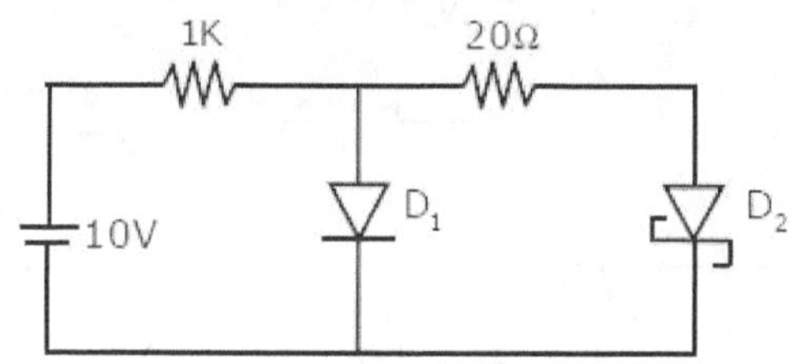

Figure (1)

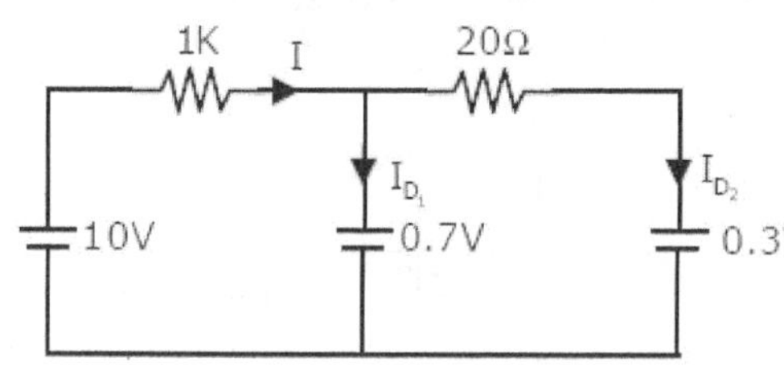

Figure (2)

$$\therefore \quad I = \frac{10-0.7}{1\text{k}} = 9.3 \text{ mA}$$

$$I_{D2} = \frac{0.7-0.3}{20} = 20 \text{ mA}$$

Now, $\quad I_{D1} = I - I_{D2}$

$= -10.7$ mA (Not Possible)

$\therefore$ D_1 is OFF and hence D_2 – ON

19. When V_i makes Diode 'D' OFF,

$$V_0 = V_i$$

$\therefore$ $$V_0\,(\text{min}) = -5\text{ V}$$

When V_i makes diode 'D' ON,

$$V_0 = \frac{(V_i - 0.7 - 2)}{R_1 + R_2} + V_{on} + 2\text{ V}$$

$\therefore$ $$V_0\,(\text{max}) = \frac{(5 - 0.7 - 2)1k}{1k + 1k} + 0.7 + 2\text{ V}$$

$$= 3.85\text{ V}$$

20. $$V_{dc} = V_m - \frac{I_{dc}}{4fc}$$

$$I_{dc}R_L = V_m - \frac{I_{dc}}{4fc}$$

$$I_{dc}\left[R_L + \frac{1}{4fc}\right] = V_m$$

$\Rightarrow$ $$I_{dc} = \frac{10}{100 + \dfrac{1}{4 \times 50 \times 4 \times 10^{-3}}}$$

$$= 0.09\text{ A}$$

21. When $V_i < -1.7$ V ; D_1 – ON and D_2 – OFF

$\therefore$ $$V_0 = -1.7\text{ V}$$

When $V_i > 2.7$ V; D_1 – OFF & D_2 – ON

$\therefore$ $$V_0 = 2.7\text{V}$$

When $-1.7 < V_i < 2.7$ V, Both D_1 & D_2 OFF

$\therefore$ $$V_0 = V_i$$

22.

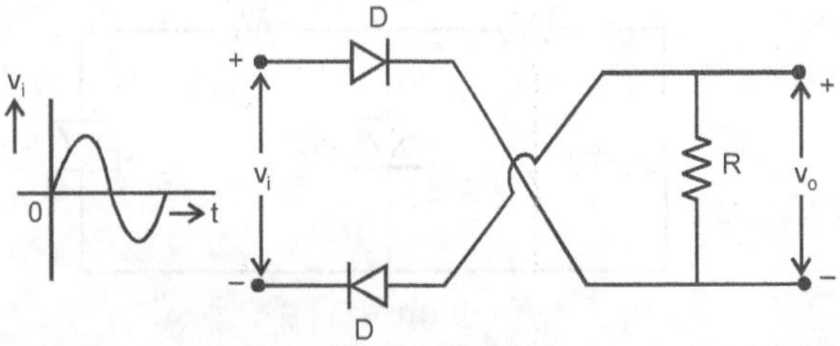

Consider + ve half cycle,

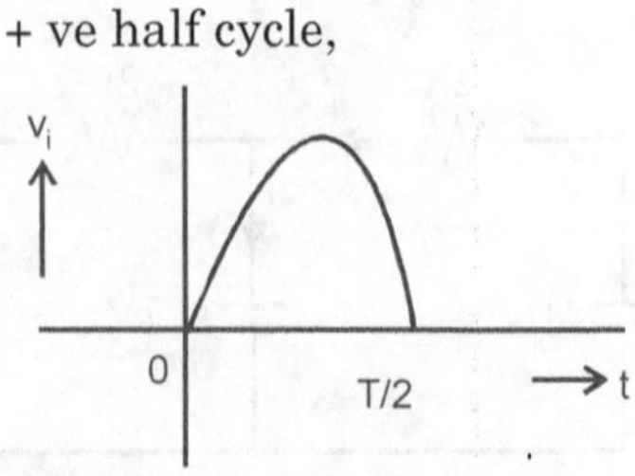

Both the diodes are forward bias.

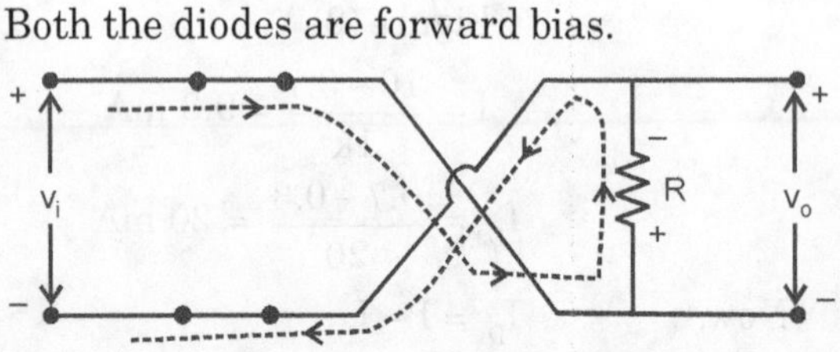

$\boxed{V_o = -V_i}$

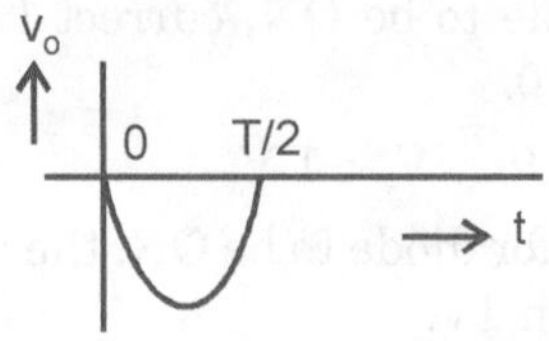

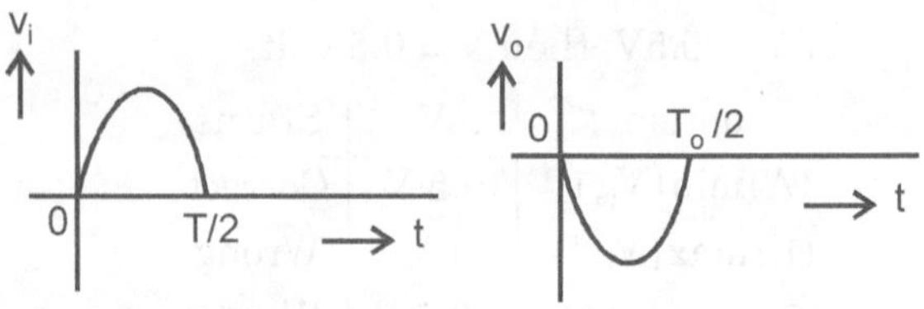

Consider –ve half cycle

Both the diodes are reverse biased.

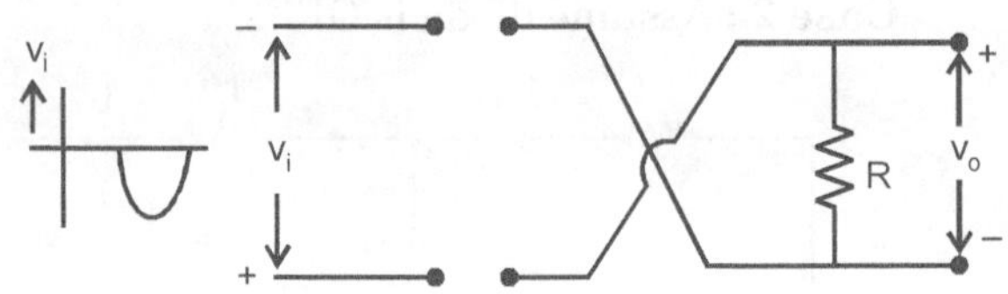

$\boxed{V_o = 0}$ volt

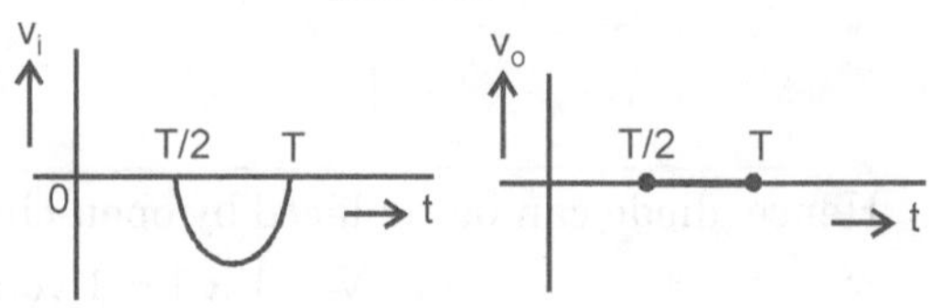

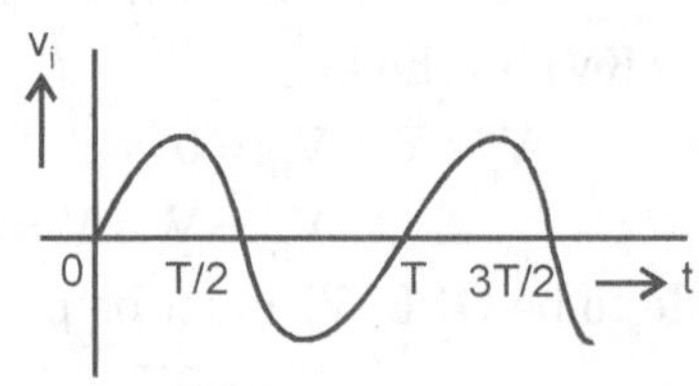

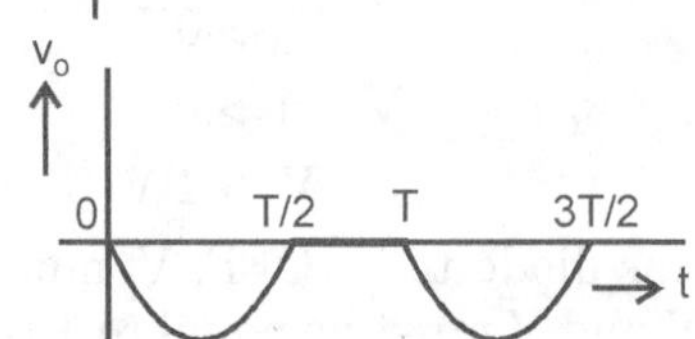

23.

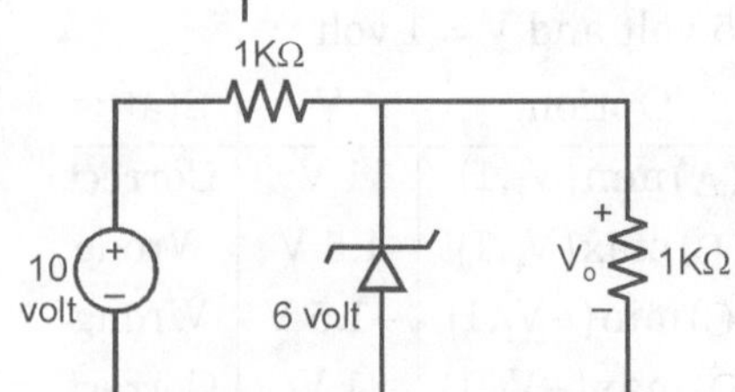

Using voltage divider Rule,

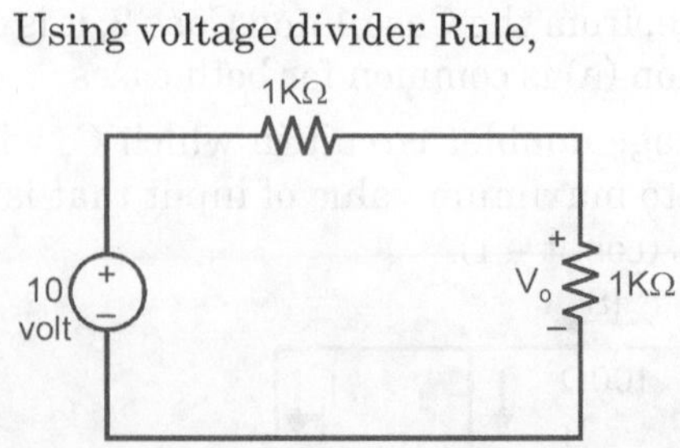

$$V_o = 10 \times \frac{1}{1+1} = 5\text{ volt}$$

since $V_o < V_z$

Zener wouldn't breakdown

Output voltage

$$V_o = 5 \text{ volt}$$

24. **Case (i)** : If any input is logic 0 (i.e., 0V) then corresponding diode is 'ON' and due to ideal diode output voltage $V_0 = 0$ as well as if there is any input logic 1 (i.e., 10V) corresponding diode will be OFF.

Case (ii) : If all the inputs are high (i.e., 10V) then all the diodes are R.B (OFF) and output voltage $V_0 = 10V$.

So, it is a positive logic 3-inputs AND gate.

25.

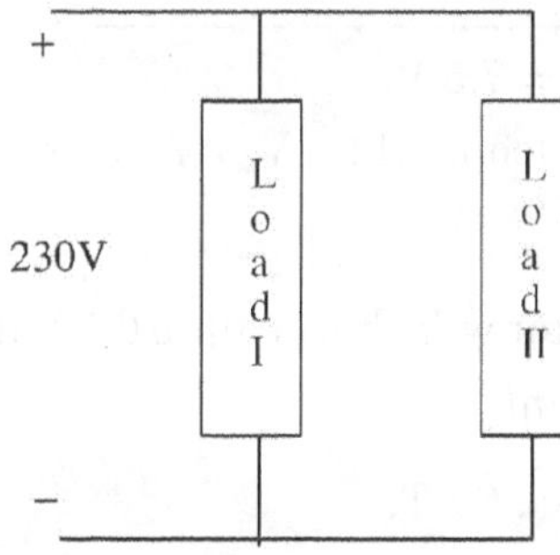

Load I:

$P = 10$ kw

$\cos\phi = 0.8$

$Q = P\tan\phi = 7.5\,\text{kVAR}$ $\quad S_I = P - jQ = 10 - j7.5$ KVA

Load II : $S_{II} = 10$ kVA

$$\cos\phi = 0.8, \sin\phi = \frac{Q}{S}$$

$$\cos\phi = \frac{P}{S_{II}}$$

$$0.8 = \frac{P}{10} \rightarrow P = 8\,\text{kW} \quad Q = 6\ \text{KVAR}$$

$S_1 = P + jQ = 8 + j6$

Complex power delivered by the source is $S_I + S_{II} = 18 - j1.5$ KVA

26. The circuit works as a voltage doubler.

$V_{ab} = 2 \times 50 = 100$ Volt.

27. At $t = 0^-$

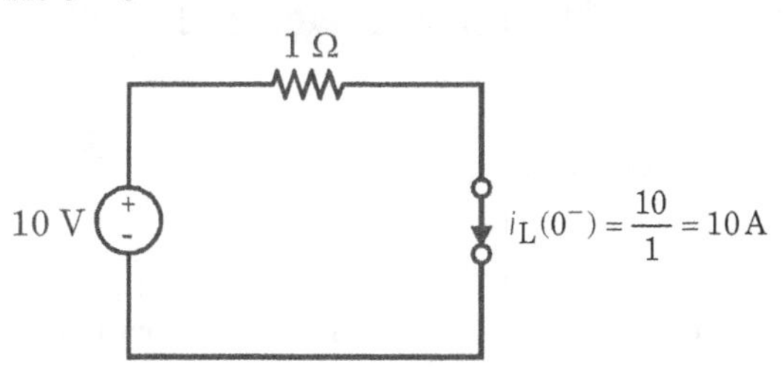

$$i_L(0^-) = \frac{10}{1} = 10 \text{ A.}$$

For $t > 0$ (using Laplace transform)

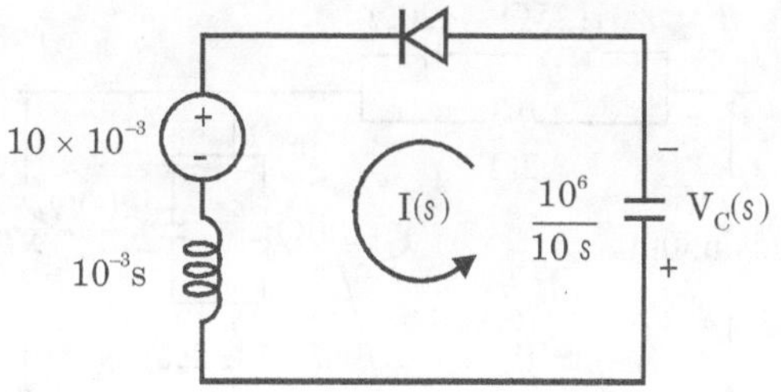

$$I(s) = \frac{10 \times 10^{-3}}{10^{-3}s + \dfrac{10^6}{10s}}$$

$$V_c(s) = I(s) \times \frac{10^6}{10s}$$

$$V_c(s) = \frac{10^6}{s^2 + 10^8}$$

Taking inverse Laplace transform, we get

$V_c(t) = 100\sin 10^4 t\,\text{V}$ $\quad (\because V_c(t) = V_0 \sin\omega t)$

$\therefore$ Steady state magnitude voltage across capacitor is 100 V.

28.

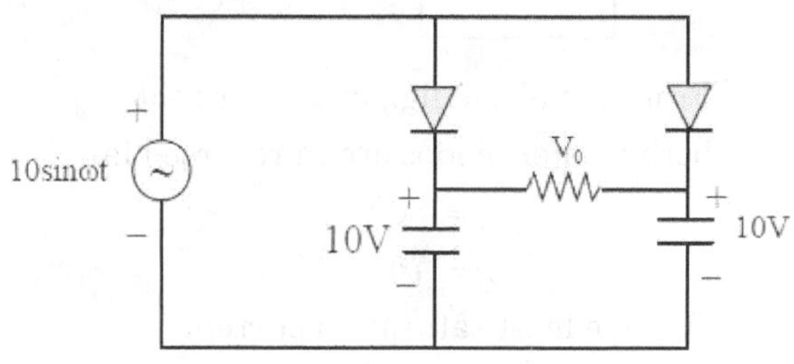

Diodes are ideal therefore during Positive cycle of output voltage

$$(V_0) = 10 - 10 = 0 \text{ V}$$

During Negative cycle, the diodes are Reverse biased so output

$$(V_0) = 0 \text{ V}$$

$$\therefore \quad V_0 = 0 \text{ V (always)}$$

So, the output voltage V_0 (in volt) at steady state is 0 V.

29. The input voltage is given by

$$V_i(t) = 2\cos 200t + 4\sin 500t$$

Let us apply Superposition theorem only consider $2\cos 200t$, then circuit becomes

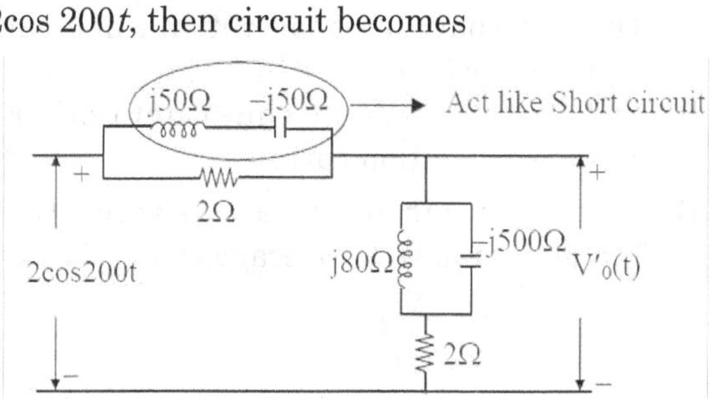

So, $\quad V_0'(t) = 2\cos 200t$

Now consider only $4\sin 500t$, then circuit becomes

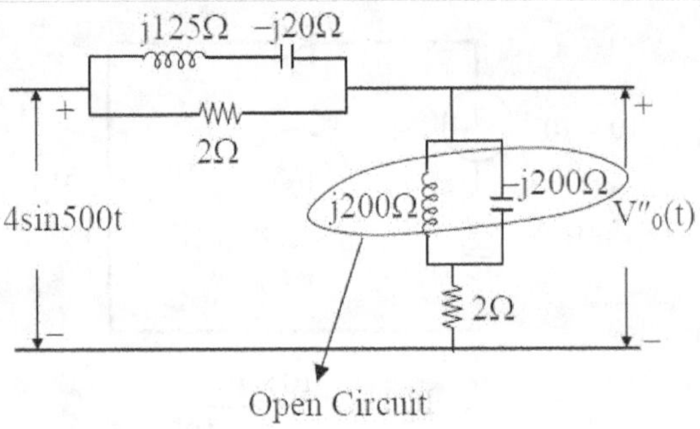

So, again $V_0''(t) = 4 \sin 500\, t$

Finally (according to superposition theorem)

$$V_0(t) = V_0'(t) + V_0''(t)$$

$$\therefore \quad V_0(t) = 2\cos(200t) + 4\sin(500\, t)$$

30.

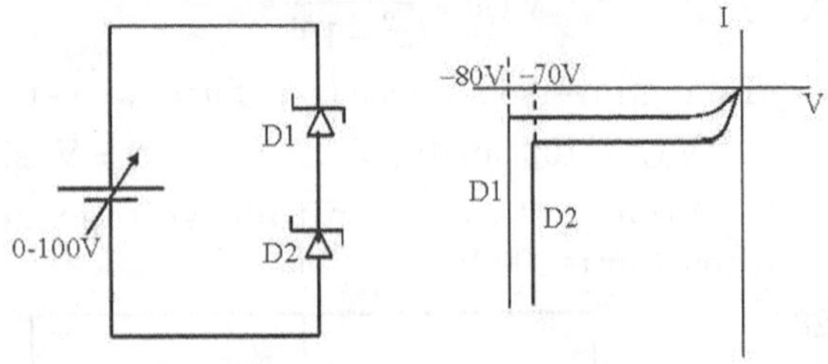

From the above fig., it is clear that

both zZener diodes are in reverse biased

$$V_{BZ1} = 80 \text{ V}$$

$$V_{BZ2} = 70 \text{ V}$$

D_1 have least saturation current.

When we will vary the voltage above 80 V

D_1 get breaks down & will replaced by 80 V. & through it 'infinite' current can flow through it.

But because of D_2 we will take minimum current i.e. net current equals to reverse saturation current of D_2 as we know.

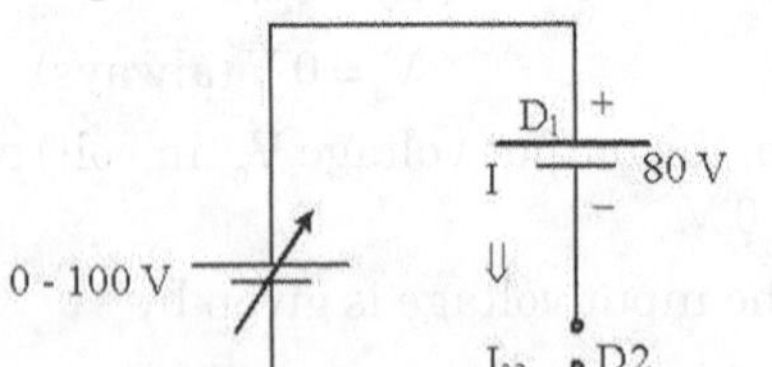

The diode have least saturation will break down first and it will replaced by its break down voltage and the net current equal upto other diode reverse saturation current.

31. The given circuit is a halfwave rectifier. Voltmeter reads the average value of V_0. Average value of $V_0 = \frac{V_m}{\pi}$.

Peak value of the applied sine-wave, $V_m = 10$ V.

Reading of meter, $V_0 = \frac{V_m}{\pi} = \frac{10}{\pi} = 3.183$ V.

32. Given : $V_r = 0.7$ V, $V_z = -6.8$V, $10\tau << T$.

Case 1 : when $t < 0 < \frac{T}{2}$.

Diode is in forward bias and hence it can be replaced by 0.7 V. Zener diode is in reverse bias and because voltage across zener diode is more than breakdown region. Therefore, it can be replaced by its breakdown voltage.

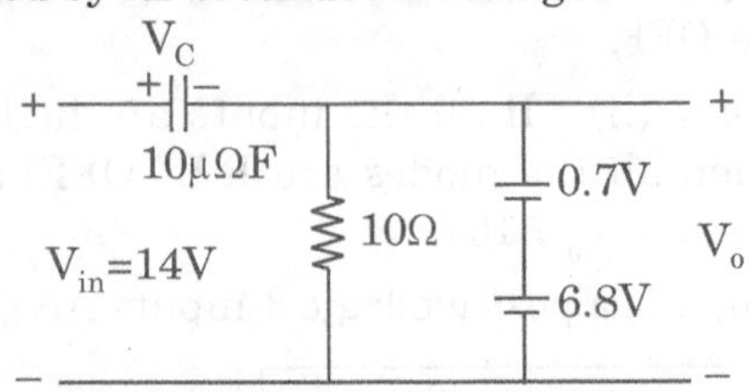

$(V_o)_{max} = 0.7 + 6.8 = 7.5$ V.

Applying KVL in loop, $-14 + V_c + 7.5 = 0$

$\Rightarrow V_c = 6.5$ V.

At $t = \frac{T}{2}$, capacitor will charge upto 6.5 V due to small time constant.

Case 2 : when $\frac{T}{2} < t < T$.

Diode is in reverse bias and hence it can be replaced by open-circuit. Zener diode is forward biased and therefore it will act as normal diode. Hence, it can be replaced by short-circuit. Applying KVL,

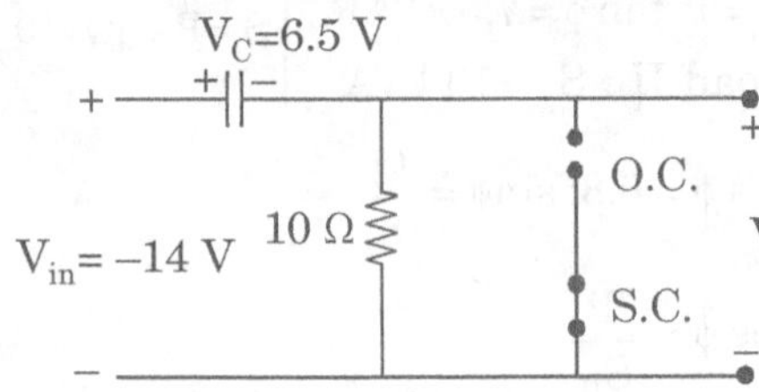

$$-V_{in} + V_c + V_0 = 0$$

$$14 + 6.5 + V_0 = 0$$

$$\therefore \quad (V_0)_{min} = -20.5 \text{ V}.$$

33. The small-signal equivalent model of the given circuit is given below.

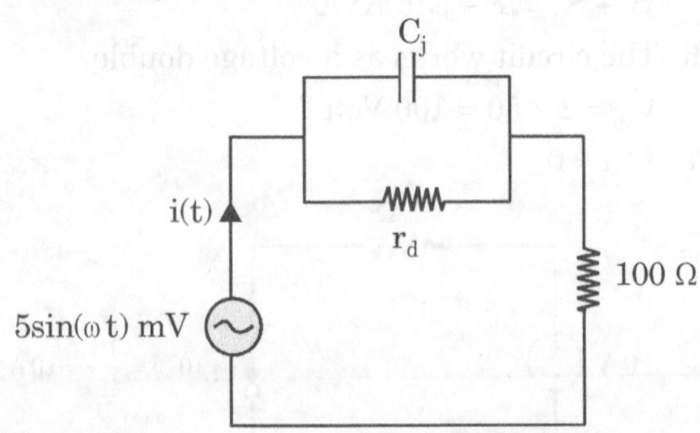

Given data,

$\omega = 2 \times 10^6$ rad/sec

$C_j = 0.5$ nF

$I_{DC} = 26\ \mu$A

$V_T = 26$ mV

$\eta = 1$

Now, $r_d = \frac{\eta V_T}{I_{DC}} = \frac{26 \text{ mV}}{26 \ \mu\text{A}} = 1 \text{ k}\Omega$

$$\frac{1}{\omega Cj} = \frac{1}{2 \times 10^6 \times 0.5 \times 10^{-9}} \Omega = 1 \text{ k}\Omega$$

$\therefore$ Total impedance of the circuit

$$Z = \left(r_d \parallel \frac{1}{j\omega Cj} \right) + 100\Omega$$

$$\left(r_d \parallel \frac{1}{j\omega Cj} \right) = \frac{(1000)(-j1000)}{1000 - j1000} \Omega = \frac{-j(1+j)}{2} \text{k}\Omega$$

$$= \frac{1}{2}(1-j)\text{k}\Omega = (500 - j500)\Omega$$

$\therefore \ Z = \left[(500 + j500) + 100\right]\Omega$

$\therefore \ Z = 600 - j\ 500\ \Omega$

$\sqrt{(600)^2 + (500)^2}$

Now magnitude is

$$|Z| = 100\sqrt{36+25} = 100\sqrt{61}\ \Omega$$

Now, $I_m = \frac{V_m}{|Z|} = \frac{5 \text{ mV}}{100\sqrt{61}\Omega} = \frac{50}{\sqrt{61}} \mu\text{A} = 6.40\ \mu\text{A}$

34. Given : $V_z = V_0 = 5\text{V}$, $I_{z_k} = 2$ mA,

$V_i = 6 \pm 5\%$ of 6 V

$V_i = 6 \pm 0.3$ V

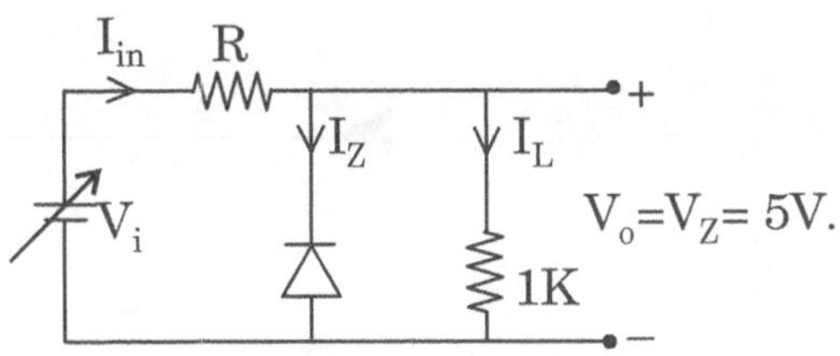

Voltage regulated output = 5 V.

Load currect, $I_L = \frac{V}{R_L} = \frac{5}{1} = 5$ mA.

Input voltage, $V_i \in [5.7, 6.3]$ V

$$\Rightarrow I_{in} = \frac{V_{in} - V_z}{R}$$

$$(I_{in})_{min} = \frac{(V_{in})_{min} - V_z}{R_{min}},$$

$$(I_{in})_{max} = \frac{(V_{in})_{max} - V_z}{R_{max}},$$

$\Rightarrow (I_{in})_{min} = (I_z)_{min} + (I_L)_{fixed} = 2 + 5 = 7$ mA.

$$R_{min} = \frac{5.7 - 5}{7} = 100\ \Omega.$$

$$\Rightarrow (I_{in})_{max} = (I_z)_{max} + (I_L)_{fixed}$$

$$= \frac{(V_{in})_{max} - V_z}{R} = \frac{6.3-5}{100} = 13\text{mA}.$$

$(I_z)_{Max} + 5 = 13$

$(I_z)_{max} = 8$ mA.

Therefore, minimum power rating of diode, i.e., maximum power dissipation

$$= V_z \times (I_z)_{max} = 5 \times 8 = 40 \text{ mW}.$$

35. In a four diode bridge rectifier, the transformer is used with no centre tap and small size.

36. Load regulation $= \frac{V_{no\ load} - V_{full\ load}}{V_{no\ load}} \times 10$

$$= \frac{30-25}{30} \times 100 = 16.7\%$$

Output resistance $= \frac{V}{I} = \frac{25}{1} = 25\ \Omega$

37. In full wave rectifier, $V_{DC} = \frac{2V_m}{\pi}$

38. Full wave rectifier is

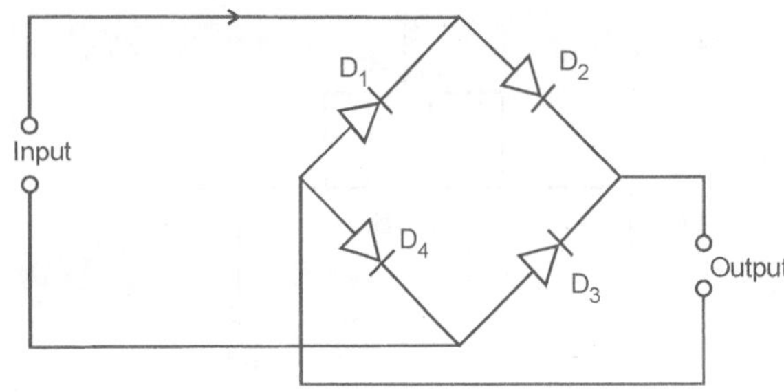

because

(1) during +ve cycle of input

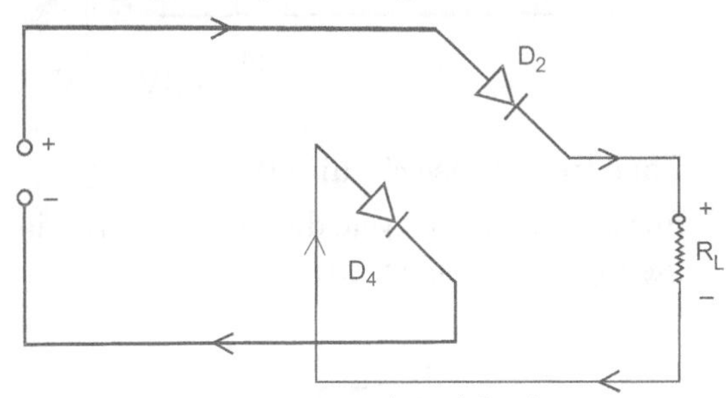

A complete path from input to output

(2) during – ve cycle of input

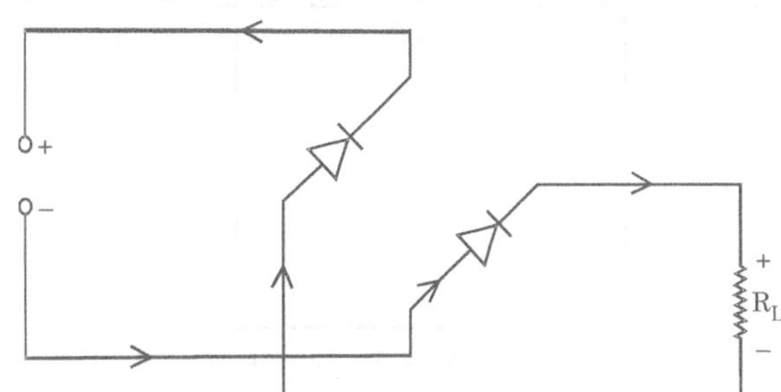

A complete path from input to output

In both cases current passes through R_L is in same direction (rectification)

39. $$V_{ripple} = \frac{I_{DC} \cdot T}{C}$$

Here T = Time period

C = Filter capacitor

$$\therefore \quad V_{ripple} = \frac{1 \times 1 \times 10^{-3}}{475 \times 10^{-6}} = 2.105 \text{ Volts}$$

40.

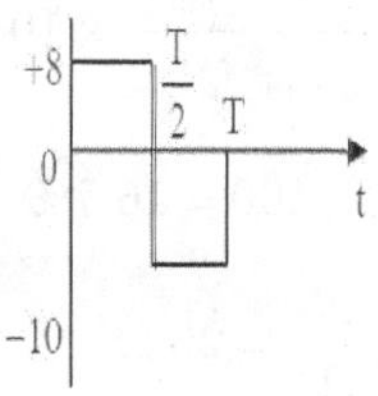

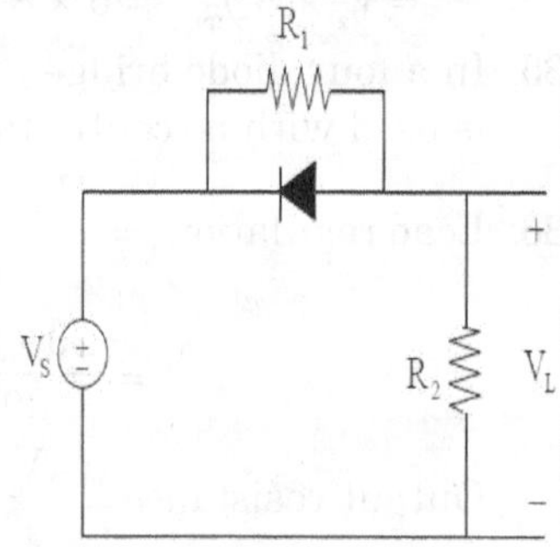

Case (i) :

For +ve half cycle diode is reverse bias : (open circuited)

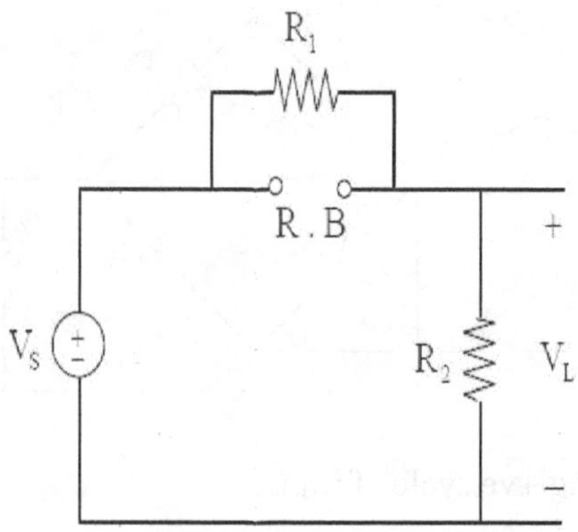

$$V_L = \frac{R_2}{R_1 + R_2} \cdot V_S = \frac{50}{100} \times 8V = 4V$$

Case (ii) : (Close circuited)

For negative half cycle diode is forward bias and ideally acts as short circuit.

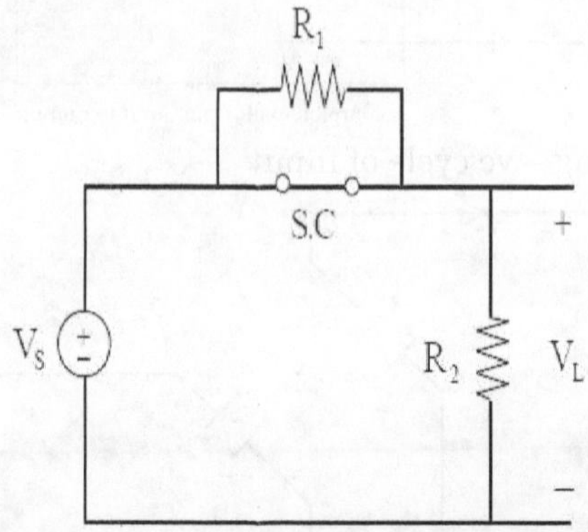

$$V_L = V_S = -10 \text{ V.}$$

Output Waveform

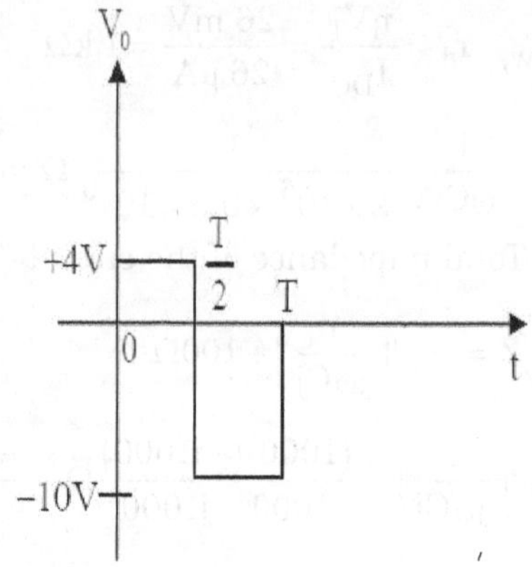

$$V_{avg} = \frac{4V \times \frac{T}{2} + \left(-10V \times \frac{T}{2}\right)}{T} = 2\text{ V} - 5\text{ V} = -3\text{ V}$$

$$\therefore \quad V_{avg} = -3 \text{ V}$$

41. Given: T = 4 ms, R = 500 Ω, C = 10 μF

Now, $\tau = RC = 500 \times 10 \times 10^{-6}$ s = 5 ms

and $\frac{T}{2} = \frac{4 \text{ ms}}{2} = 2$ ms

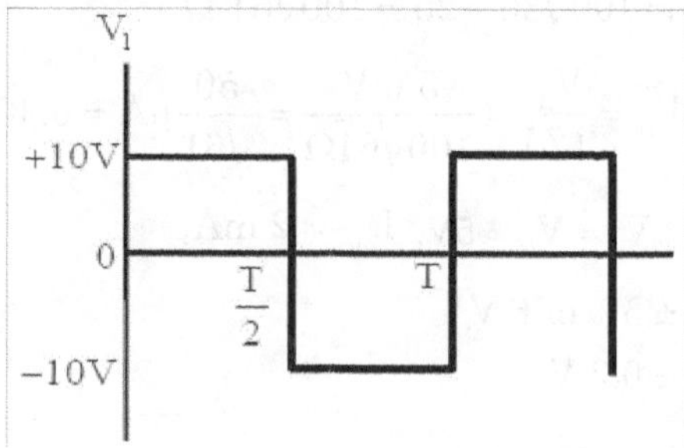

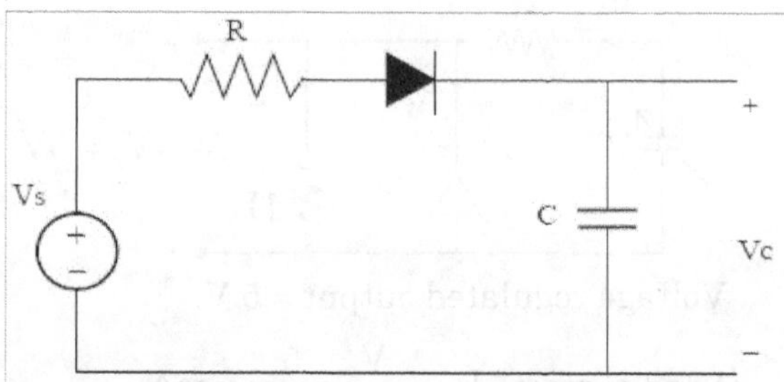

For positive half cycle, diode will be is forward biased and capacitor start to charge

$$\therefore \quad V_C = V_C(\infty)\left[1 - e^{\frac{-t}{RC}}\right] = 10\left[1 - e^{\frac{2ms}{5ms}}\right]$$

$$= 10\left[1 - e^{\frac{2}{5}}\right] = 4.51 \text{ V}$$

$$\therefore \quad V_C = 3.31\text{V}$$

2

CHAPTER

BJT & FET Analysis

Analysis of Previous GATE Papers												
	Year → Topics ↓		2019	2018	2017 Set 1	2017 Set 2	2016 Set 1	2016 Set 2	2016 Set 3	2015 Set 1	2015 Set 2	2015 Set 3
SMALL SIGNAL EQUIVALENT CIRCUITS	1 Mark	MCQ Type		1		1						
		Numerical Type										
	2 Marks	MCQ Type							1	1	1	
		Numerical Type							1			
	Total			1		1			4	2	2	
ANALYSIS & BIASING	1 Mark	MCQ Type					1	1				
		Numerical Type							2			
	2 Marks	MCQ Type				1						
		Numerical Type	3	1	2					1		1
	Total		6	2	4	2	1	1	2	2		2

SMALL SIGNAL EQUIVALENT CIRCUIT

1. The Ebers-Moll model is applicable to
   (a) Bipolar junction transistors.
   (b) NMOS transistors.
   (c) Unipolar junction transistors.
   (d) Junction field-effect.

[1987 : 1 Mark]

2. Each transistor in the Darlington pair (see Fig. below) has h_{FE} = 100. The overall h_{FE} of the composite transistor neglecting the leakage currents is

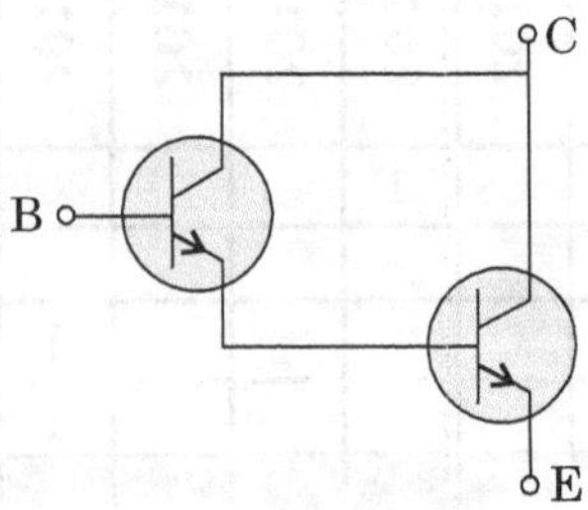

   (a) 10000 (b) 10001
   (c) 10100 (d) 10200

3. The amplifier circuit shown below uses a composite transistor of a MOSFET and BIPOLAR in cascade. All capacitances are large. g_m of the MOSFET = 2 mA/V and h_{fe} of the BIPOLAR = 99. The overall Transconductance g_m of the composite transistor is

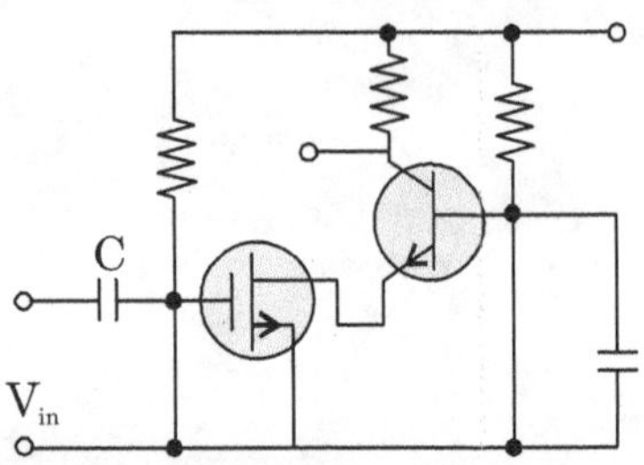

   (a) 198 mA/V (c) 4.95 mA/V
   (b) 9.9 mA/V (d) 1.98 mA/V

[1988 : 2 Marks]

4. The transistor in the amplifier shown below has following parameters:
   h_{le} = 100, h_{ie} = 2 kΩ, h_{re} = 0, h_{oe} = 0.05 mhos. C is very large. The output impedance is

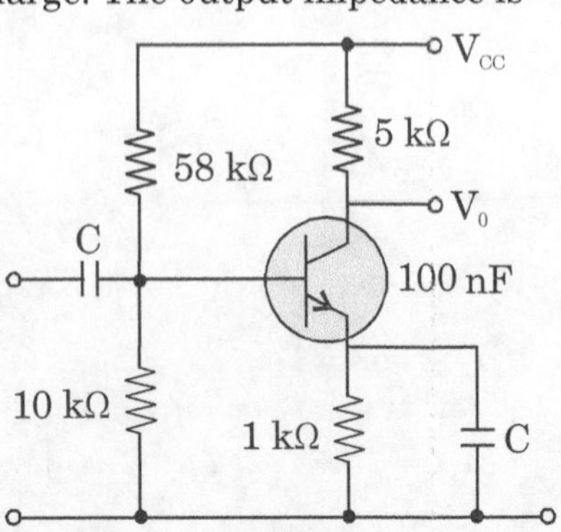

   (a) 20 kΩ (b) 16 kΩ
   (c) 5 kΩ (d) 4 kΩ

[1988 : 2 Marks]

5. In figure all transistors are identical and have a high value of beta. The voltage V_{DC} is equal to ____.

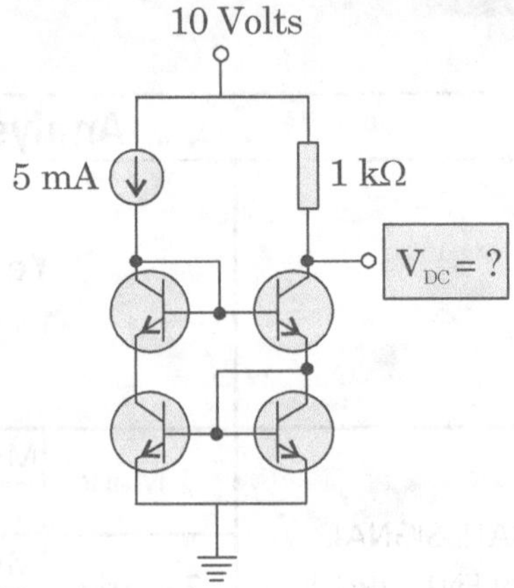

[1991 : 2 Marks]

6. If the transistor in fig., has high value of β and V_{BE} of 0.65y, the current I flowing through the 2 kilo-ohms resistance will be ______.

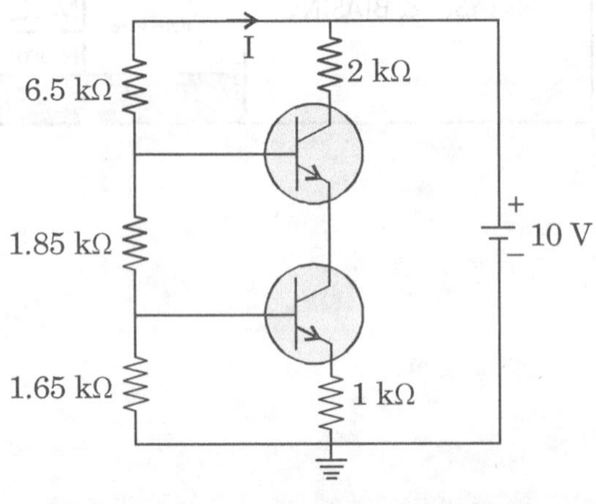

[1992 : 2 Marks]

7. A transistor having A = 0.99 and V_{BE} = 0.7 V, is used in the circuit of the figure is the value of the collector current will be ________.

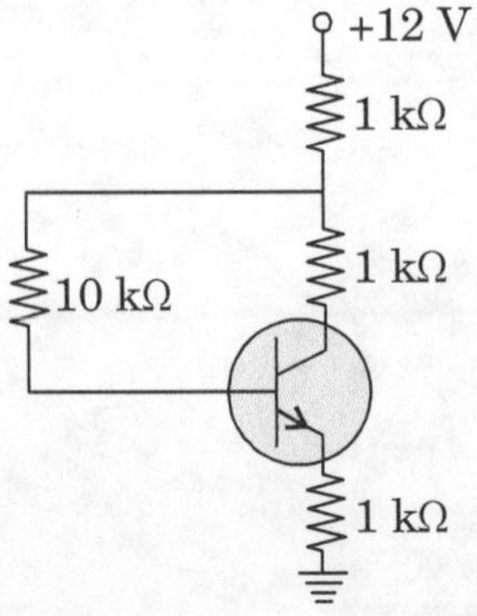

[1995 : 1 Mark]

8. A Darlington stage is shown in the figure is, if the transconductance of Q_1 is g_{m1} and Q_2 is g_{m2}, then the overall transconductance is given by g_m is given by

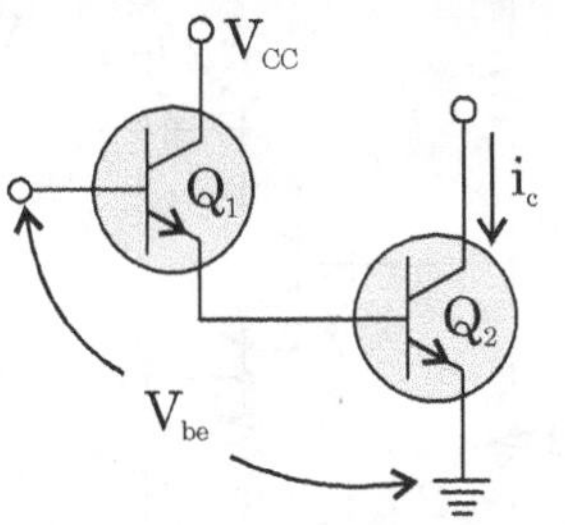

(a) g_{m1} (b) $0.5g_{m1}$

(c) g_{m2} (d) $0.5g_{m2}$

[1996 : 2 Marks]

9. In the BJT amplifier shown in the figure is the transistor is biased in the forward active region putting a capacitor across R_E will

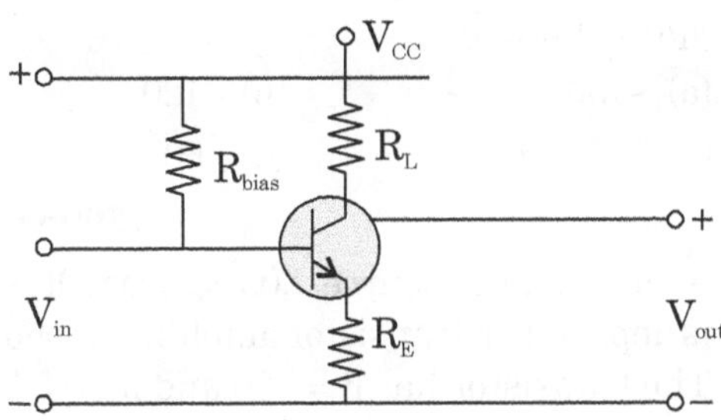

(a) decrease the voltage gain and decrease the input impedance

(b) increase the voltage gain and decrease the input impedance

(c) decrease the voltage gain and increase the input impedance

(d) increase the voltage gain and increase the input impedance

[1997 : 1 Mark]

10. The emitter coupled pair of BJT's gives a linear transfer relation between the differential o/p voltage and the differential input voltage V_{id} only. When the magnitude of V_{id} is less 'a' times the thermal voltage, where 'a' is

(a) 4 (b) 3

(c) 2 (d) 1

[1998 : 1 Mark]

11. In a series regulated power supply circuit the voltage gain A_v of the "pass", transistor satisfies the condition

(a) $A_v \to \infty$ (b) $1 << A_v << \infty$

(c) $A_v = 1$ (d) $A_v << 1$

[1998 : 1 Mark]

12. In the cascode amplifier shown in the figure, if the common-emitter stage (Q_1) has a transconductance g_{m1}, and the common base stage (Q_2)has a transconductance g_{m2}, then the overall transconductance $g(= i_0/V_i)$ of the cascode amplifier is

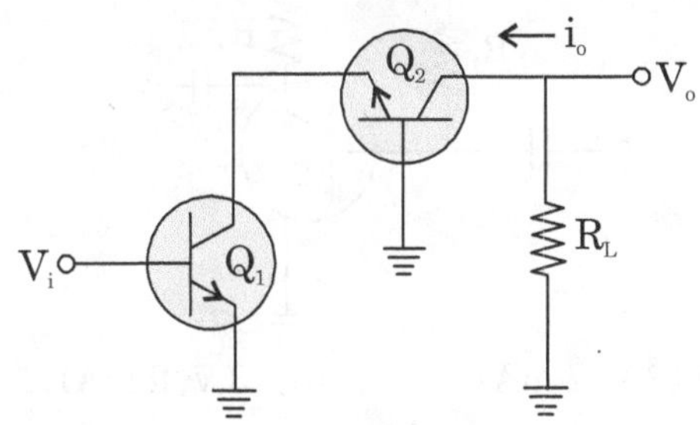

(a) g_{m1} (b) g_{m2}

(c) $\frac{g_{m1}}{2}$ (d) $\frac{g_{m2}}{2}$

[1999 : 1 Mark]

13. The current gain of a bipolar transistor drops at high frequencies because of

(a) transistor capacitances

(b) high current effects in the base

(c) parasitic inductive elements

(d) the Early effect **[2000 : 1 Mark]**

14. In the circuit of the figure, assume that the transistor is in the active region. It has a large β and its base-emitter voltage is 0.7 V. The value of I_C is

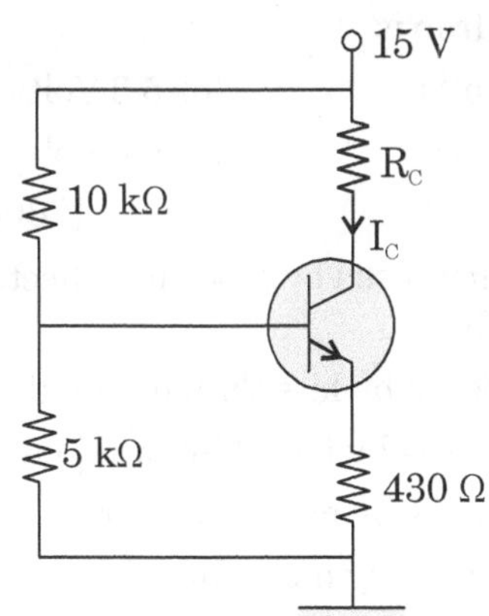

(a) indeterminate since R_c is not given

(b) 1 mA

(c) 5 mA

(d) 10 mA **[2000 : 2 Marks]**

15. The current gain of a BJT is :

(a) $g_m r_0$ (b) $\frac{g_m}{r_0}$

(c) $g_m r_\pi$ (d) $\frac{g_m}{r_\pi}$

[2002 : 1 Mark]

16. In the amplifier circuit shown in the figure, the values of R_1 and R_2 are such that the transistor is operating at $V_{CE} = 3$ V and $I_C = 1.5$ mA when its β is 150. For a transistor with β of 200, the operating point (V_{CE}, I_C) is :

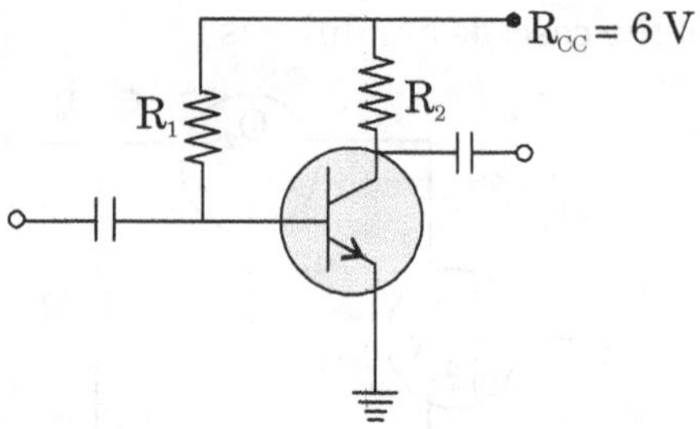

(a) (2 V, 2 mA) (b) (3 V, 2 mA)

(c) (4 V, 2 mA) (d) (4 V, 1 mA)

[2003 : 2 Marks]

17. Assuming $V_{CEsat} = 0.2$ V and β = 50, the minimum base current (I_B) required to drive the transistor in the figure to saturation is

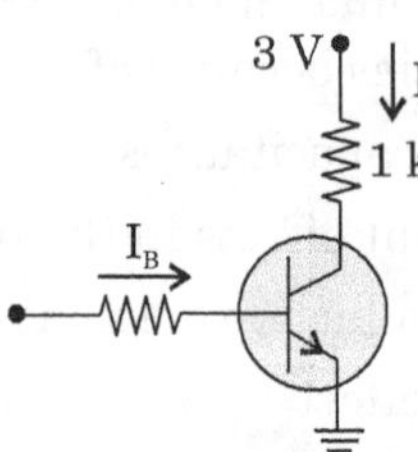

(a) 56 μA (b) 140 μA

(c) 60 μA (d) 3 μA **[2004 : 1 Mark]**

18. Under the DC conditions, the collector-to-emitter voltage drop is

(a) 4.8 Volts (b) 5.3 Volts

(c) 6.0 Volts (d) 6.6 Volts

[2006 : 2 Marks]

19. If β_{ac} is increased by 10%, the collector-to-emitter voltage drop

(a) increases by less than or equal to 10%

(b) decreases by less than or equal to 10%

(c) increases by more than 10%

(d) decreases by more than 10%

[2006 : 2 Marks]

20. The small-signal gain of the amplifier V_c/V_s is

(a) –10 (b) –5.3

(c) 5.3 (d) 10 **[2006 : 2 Marks]**

21. The DC current gain (β) of a BJT is 50. Assuming that the emitter injection efficiency is 0.995, the base transport factor is

(a) 0.980 (b) 0.985

(c) 0.990 (d) 0.995

[2007 : 2 Marks]

Statement for linked Answer Questions 22 and 23 : In the following transistor circuit, $V_{BE} = 0.7$ V, $r_e = 25\text{ mV}/I_E$, and β and all the capacitances are very large.

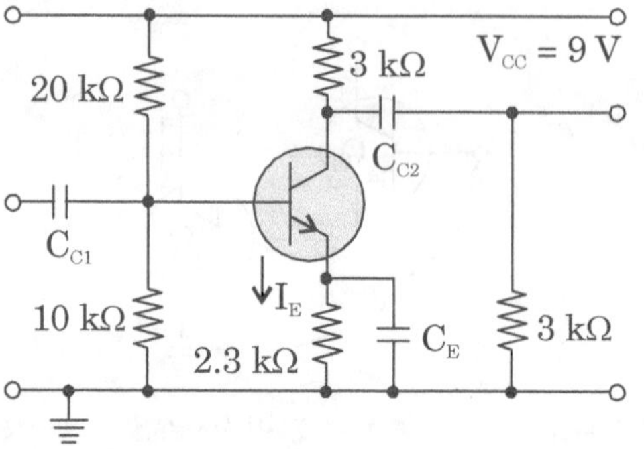

22. The Value of DC current I_E is

(a) 1 mA (b) 2 mA

(c) 5 mA (d) 10 mA

[2008 : 2 Marks]

23. The mid-band voltage gain of the amplifier is approximately

(a) –180 (b) –120

(c) –90 (d) –60

[2008 : 2 Marks]

24. A small signal source $v_i(t) = A\cos 20t + B\sin 10^6 t$ is applied to a transistor amplifier as shown below. The transistor has β = 150 and $h_{ie} = 3$ kΩ. Which expression best approximates $v_0(t)$?

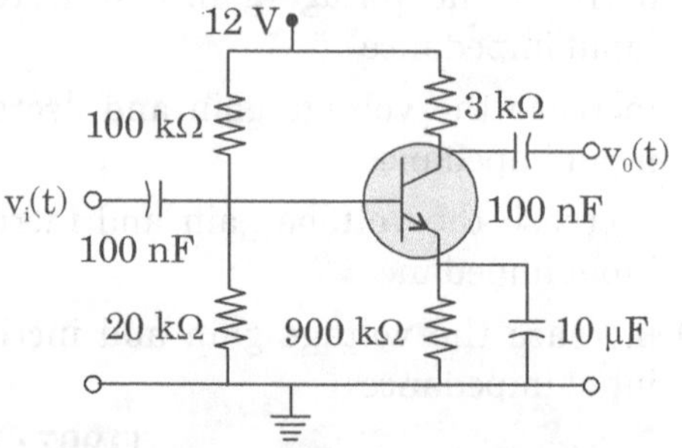

(a) $v_0(t) = -1500(A\cos 20t + B\sin 10^6 t)$

(b) $v_0(t) = -150(A\cos 20t + B\sin 10^6)$

(c) $v_0(t) = -1500B\sin 10^6 t$

(d) $v_0(t) = -150B\sin 10^6 t$ **[2009 : 2 Marks]**

25. In the silicon BJT circuit shown below, assume that the emitter area of transistor Q1 is half that of transistor Q2.

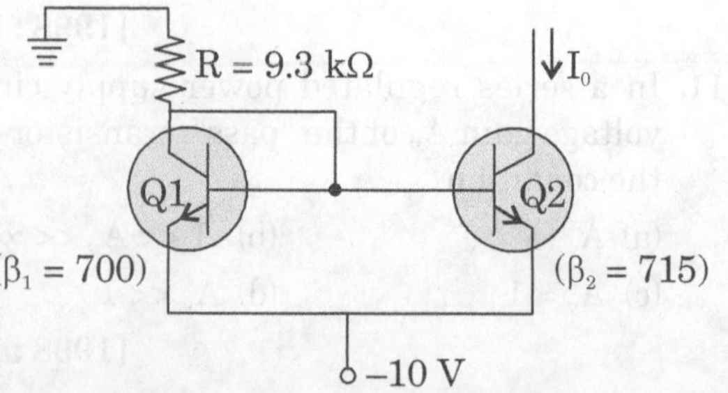

The value of current I_0 is approximately

(a) 0.5 mA (b) 2 mA

(c) 9.3 mA (d) 15 mA

[2010 : 1 Mark]

26. The amplifier circuit shown below uses a silicon transistor. The capacitors C_C and C_E can be assumed to be short at signal frequency and the effect of output resistance r_0 can be ignored. If C_E is disconnected from the circuit, which one of the following statements is TRUE?

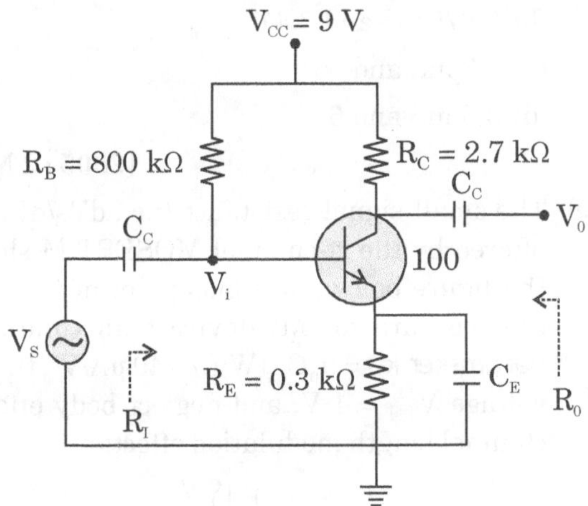

(a) The input resistance R_i increases and the magnitude of voltage gain A_V decreases

(b) The input resistance R_t decreases and the magnitude of voltage gain A_V increases

(c) Both input resistance R_i and the magnitude of voltage gain A_V decrease

(d) Both input resistance R_i and the magnitude of voltage gain A_V increase

[2010 : 1 Mark]

Common Data for Questions 27 and 28:

Consider the common emitter amplifier shown below with the following circuit parameters:

$\beta = 100$, $g_m = 0.3861$ A/V, $r_0 = \infty$, $r_\pi = 259\ \Omega$

$R_S = 1\ k\Omega$, $R_B = 93\ k\Omega$, $R_C = 250\ \Omega$, $R_L = 1\ k\Omega$, $C_1 = \infty$ and $C_2 = 4.7\ \mu F$.

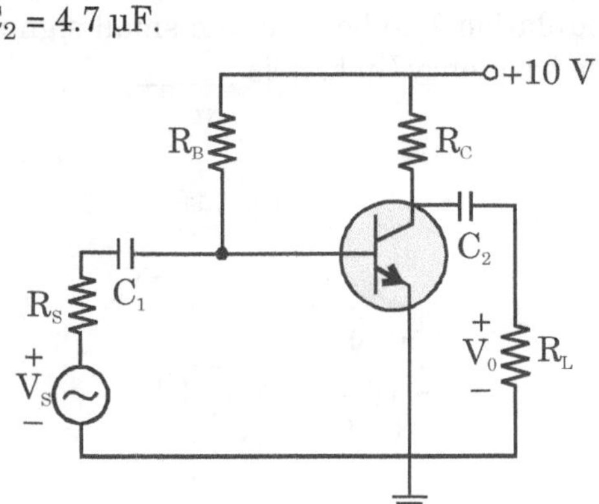

27. The resistance seen by the source V_s is

(a) 250 Ω (b) 1258 Ω

(c) 93 kΩ (d) ∞

[2010 : 2 Marks]

28. The lower cut-off frequency due to C_2 is

(a) 33.9 Hz (b) 27.1 Hz

(c) 13.6 Hz (d) 16.9 Hz

[2010 : 2 Marks]

29. The current i_b through the base of a silicon npn transistor is $1 + 0.1\cos(10000\pi t)$ mA. At 300 K, the r_π in the small signal model of the transistor is

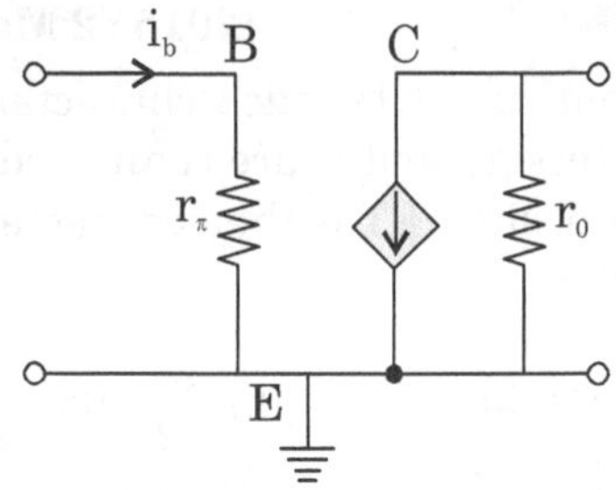

(a) 250 Ω

(b) 27.5 Ω

(c) 25 Ω

(d) 22.5 Ω

[2012 : 1 Mark]

30. An increase in the base recombination of a BJT increase

(a) the common emitter dc current gain β

(b) the breakdown voltage V_{CEO}

(c) the unity-gain cut-off frequency f_T

(d) the transconductance g_m

[2012 : 2 Marks]

31. In the circuit shown, $I_1 = 80$ mA and $I_2 = 4$ mA. Transistors T_1 and T_2 are identical. Assume that the thermal voltage V_T is 26 mV at 27 °C. At 50 °C, the value of voltage $V_{12} = V_1 - V_2$ (in mV) is

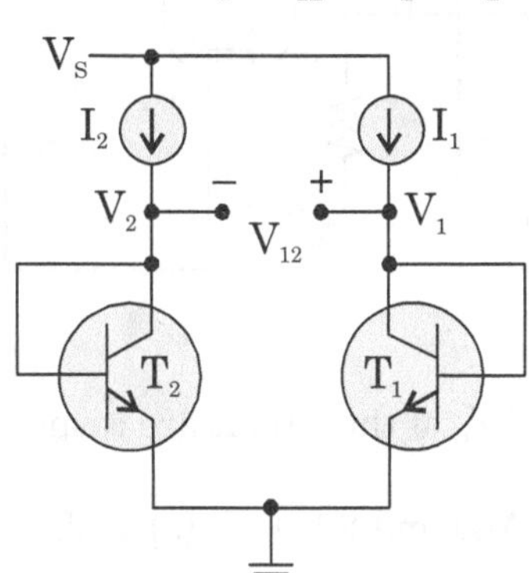

[2014 : 1 Mark, Set-2]

32. In the ac equivalent circuit shown, the two BJTs are biased in active region and have identical parameters with $\pi >> 1$. The open circuit small signal voltage gain is approximately ______.

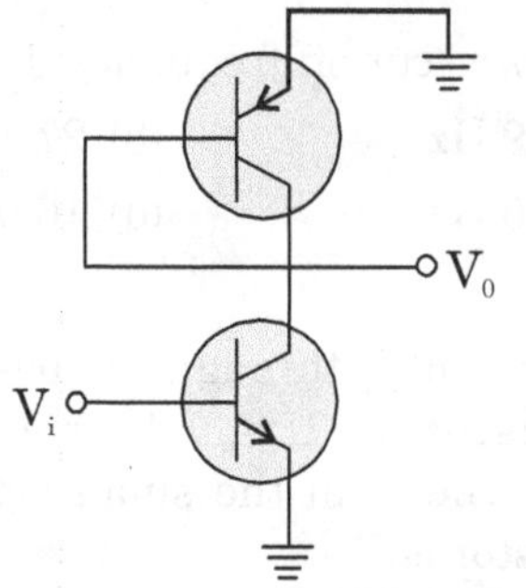

[2015 : 2 Marks, Set-1]

33. Two identical FETs, each characterized by the parameters g_m and r_d are connected in parallel. The composite FET is then characterized by the parameters

(a) $\frac{g_m}{2}$ and $2r_d$ (b) $\frac{g_m}{2}$ and $\frac{r_d}{2}$

(c) $2g_m$ and $\frac{r_d}{2}$ (d) $2g_m$ and $2r_d$

34. The action of a JFET in its equivalent circuit can best be represented as a

(a) Current Controlled Current Source.

(b) Current Controlled Voltage Source.

(c) Voltage Controlled Voltage Source.

(d) Voltage Controlled Current Source.

[2015 : 2 Marks, Set-2]

Common Data Questions 35, 36 & 37.

Given, $r_d = 20\ k\Omega$, $I_{DSS} = 10$ mA, $V_p = -8$ V

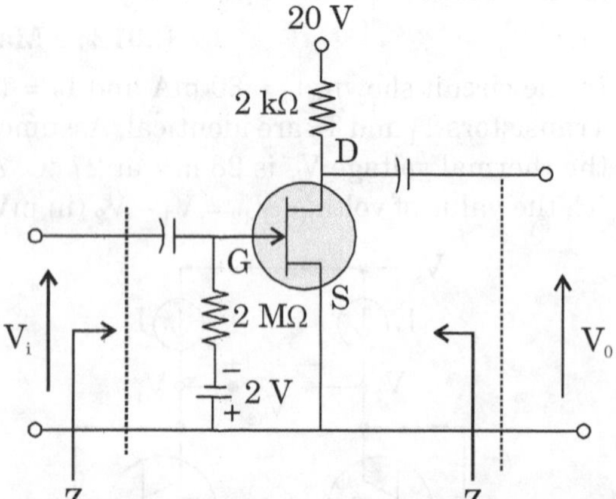

35. Z_i and Z_0 of the circuit are respectively

(a) 2 MΩ and 2 kΩ (b) 2 MΩ and $\frac{20}{11}$ kΩ

(c) infinity and 2 MΩ (d) infinity and $\frac{20}{11}$ kΩ

[2005 : 2 Marks]

36. I_0 and V_{DS} under DC conditions are respectively

(a) 5.625 mA and 8.75 V

(b) 7.500 mA and 5.00 V

(c) 4.500 mA and 11.00 V

(d) 6.250 mA and 7.50 V

[2005 : 2 Marks]

37. Transconductance in milli-Siemens (mS) and voltage gain of the amplifier are respectively

(a) 1.875 mS and 3.41

(b) 1.875 mS and-3.41

(c) 3.3 mS and –6

(d) 3.3 mS and 6

[2005 : 2 Marks]

38. The small-signal resistance (i.e., dV_B/dI_D) in kΩ offered by the rf-channel MOSFET M shown in the figure below, at a bias point of $V_B = 2$ V is (device data for M_1 device transconductance parameter $k_N = \mu_n C_{ox}(W/L) = 40\ \mu A/V^2$, threshold voltage $V_{TN} = 1$ V, and neglect body effect and channel length modulation effects)

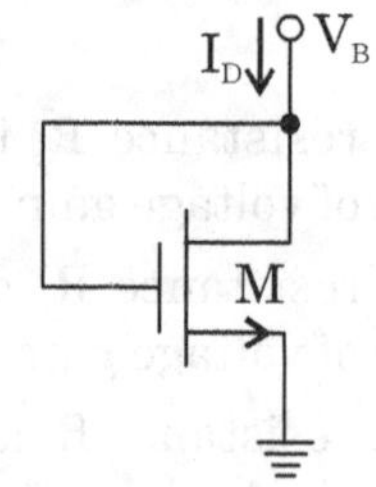

(a) 12.5

(b) 25

(c) 50

(d) 100

[2013 : 2 Marks]

39. In the circuit shown in the figure, transistor M_1 is in saturation and has transconductance $g_m = 0.01$ Siemens. Ignoring internal parasitic capacitances and assuming the channel length modulation λ to be zero, the small signal input pole frequency (in kHz)is ______.

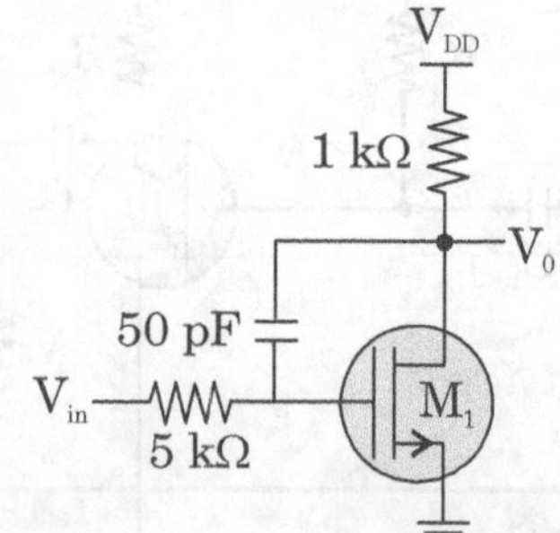

[2016 : 2 Marks, Set-3]

40. In the circuit shown in the figure, the channel length modulation of all transistors is non-zero ($\lambda \neq 0$). Also, all transistors operate in saturation and have negligible body effect. The ac small signal voltage gain (V_0/V_{in}) of the circuit is

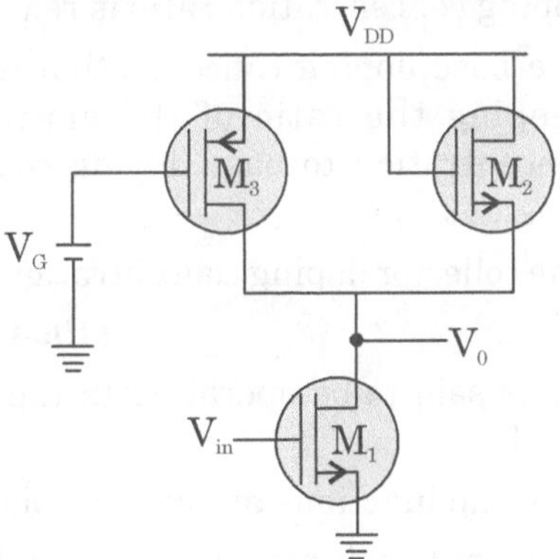

(a) $-g_{m1}\left(r_{01} \| r_{02} \| r_{03}\right)$

(b) $-g_{m1}\left(r_{01} \left\| \frac{1}{g_{m3}} \right\| r_{03}\right)$

(c) $-g_{m1}\left(r_{01} \left\| \frac{1}{g_{m2}} \right\| r_{02} \| r_{03}\right)$

(d) $-g_{m1}\left(r_{01} \left\| \frac{1}{g_{m3}} \right\| r_{03} \| r_{02}\right)$

[2016 : 2 Marks, Set-3]

41. An n-channel enhancement mode MOSFET is biased at $V_{GS} > V_{TH}$ and $V_{DS} > (V_{GS} - V_{TH})$, where V_{GS} is the gate-to-source voltage. V_{DS} is the drain-to-source voltage and V_{TH} is the threshold voltage. Considering channel length modulation effect to be significant, the MOSFET behaves as a

(a) voltage source with zero output impedance

(b) voltage source with non-zero output impedance

(c) current source with finite output impedance

(d) current source with infinite output impedance

[2017 : 1 Mark, Set-2]

42. Two identical nMOS transistors M_1 and M_2 are connected as shown below. The circuit is used as an amplifier with the input connected between G and S terminals and the output taken between D ans S terminals, V_{bias} and V_D are so adjusted that both transistors are in saturation. The transconductance of this combination is defined as $g_m = \frac{\partial i_D}{\partial v_{GS}}$ while the output resistance is drain of M_2. Let g_{m1}, g_{m2} be the transcondcutances and r_{01}, r_{02} be the output resistance of transistors M_1 and M_2, respectively

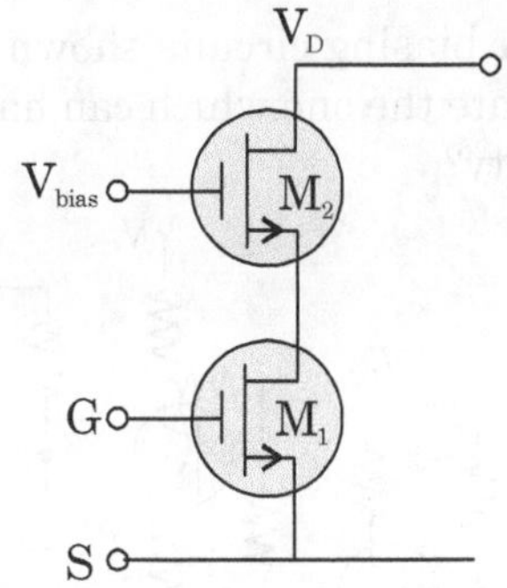

Which of the following statements about estimates for g_m and r_0 is correct?

(a) $g_m = g_{m1} \cdot g_{m2} \cdot r_{02}$ and $r_0 = r_{01} + r_{02}$

(b) $g_m = g_{m1} + g_{m2}$ and $r_0 = + r_{02}$

(c) $g_m = g_{m1}$ and $r_0 = r_{01} \cdot g_{m2} \cdot r_{02}$

(d) $g_m = g_{m1}$ and $r_0 = r_{02}$

[2018 : 1 Mark]

ANALYSIS & BIASING

43. The transistor shunt regulator shown in the figure has a regulated output voltage of 10 V, when the input varies from 20 V to 30 V. The relevant parameters for the zener diode and the transistor are: $V_Z = 9.5$, $V_{BE} = 0.5$ V, $\beta = 99$. Neglect the current through RB. Then the maximum power dissipated in the zener diode (P_Z) and the transistor (P_T) are

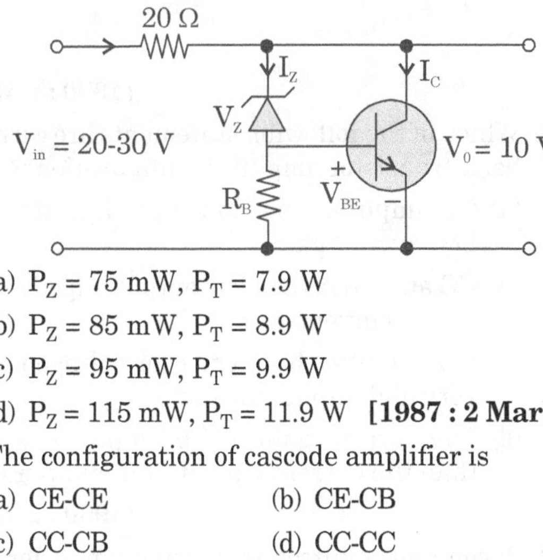

(a) $P_Z = 75$ mW, $P_T = 7.9$ W

(b) $P_Z = 85$ mW, $P_T = 8.9$ W

(c) $P_Z = 95$ mW, $P_T = 9.9$ W

(d) $P_Z = 115$ mW, $P_T = 11.9$ W **[1987 : 2 Marks]**

44. The configuration of cascode amplifier is

(a) CE-CE (b) CE-CB

(c) CC-CB (d) CC-CC

[1987 : 2 Marks]

45. The quiescent collector current I_C, of a transistor is increased by changing resistances. As a result

(a) g_m will not be affected

(b) g_m will decrease

(c) g_m will increase

(d) g_m will increase or decrease depending upon bias stability.

[1988 : 2 Marks]

46. Of the four biasing circuits shown in Fig. For a BJT, indicate the one which can have maximum bias stability?

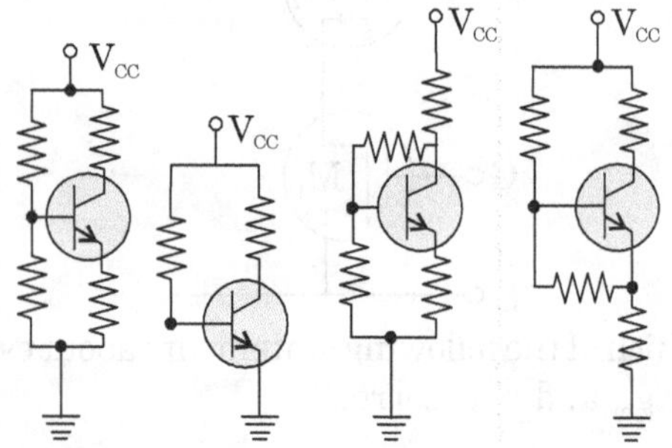

(a) (b) (c) (d)

[1989 : 2 Marks]

47. For good stabilized biasing of the transistor of the CE Amplifier of fig; we should have

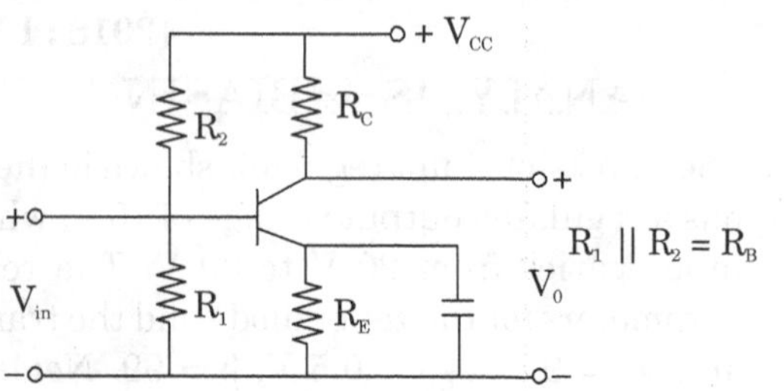

(a) $\frac{R_E}{R_B} << 1$ (b) $\frac{R_E}{R_B} >> 1$

(c) $\frac{R_E}{R_B} << h_{FE}$ (d) $\frac{R_E}{R_B} >> h_{FE}$

[1990 : 2 Marks]

48. Which of the following statements are correct for basic transistor amplifier configurations?

(a) CB amplifiers has low input impedance and low current gain

(b) CC amplifiers has low output impedance and a high current gain.

(c) CE amplifier has very poor voltage gain but very high input impedance

(d) The current gain of CB amplifier is higher than the current gain of CC amplifiers

[1990 : 2 Marks]

49. A common emitter transistor amplifier has a collector current of 1.0 mA when its base current is 25 μA at the room temperature. It's input resistance is approximately equal to ______.

[1994 : 1 Mark]

50. Match the following:

List-I

A. The current gain of a BJT will be increased.

B. The current gain of a BJT will be reduced.

C. The break-down voltage of a BJT will be reduced .

List-II

1. The collector doping concentration is increased.
2. The base width is reduced.
3. The emitter doping concentration to base doping concentration ratio is reduced.
4. The base doping concentration is increased keeping the ratio of the emitter doping concentration to base doping concentration constant.
5. The collector doping concentration is reduced.

[1994 : 2 Marks]

51. A BJT is said to be operating in the saturation region if

(a) both the junctions are reverse biased.

(b) base-emitter junction is reverse biased and base-collector junction is forward biased.

(c) base-emitter junction is forward biased and base-collector junction is reverse-biased.

(d) both the junctions are forward biased.

[1995 : 1 Mark]

52. Match the following.

(a) Cascode Amplifier

(b) Differential Amplifier

(c) Darlington pair common-collector Amplifier

(1) does not provide current gain

(2) is a wide band Amplifier

(3) has very low input impedance emitter amplifier and very high current gain

(4) has very high input impedance and very high current gain

(5) provides high common mode voltage rejection

[1996 : 2 Marks]

53. A cascode Amplifier stage is equivalent to

(a) a common emitter stage followed by a common base stage

(b) a common base stage followed by an emitter follower

(c) an emitter follower stage followed by a common base stage

(d) a common base stage followed by a common emitter stage

[1997 : 1 Mark]

54. Introducing a resistor in the emitter of a common emitter amplifier stabilizes the dc operating point against variations in

(a) only the temperature

(b) only the β of the transistor

(c) both temperature and β

(d) None of the above

[2000 : 1 Mark]

55. If the transistor in the figure is in saturation, then

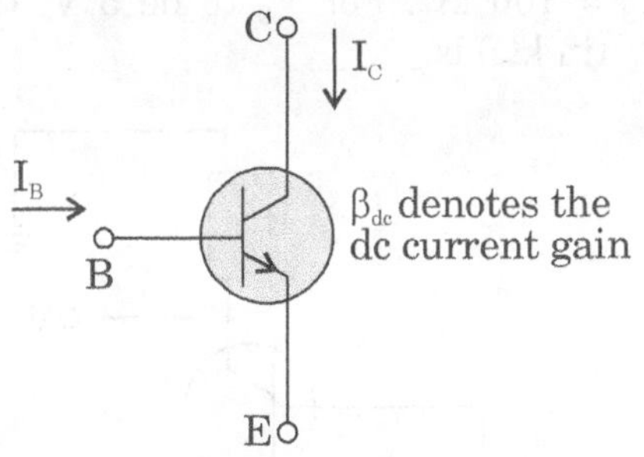

(a) I_C is always equal to $\beta_{dc}I_B$
(b) I_C is always equal to $-\beta_{dc}I_B$
(c) I_C is greater than or equal to $\beta_{dc}I_B$
(d) I_C is less than or equal to $\beta_{dc}I_B$

[2002 : 1 Mark]

56. Choose the correct match for input resistance of various amplifier configurations shown below

Configuration

CB: Common Base,

CC: Common Collector,

CE: Common Emitter,

Input resistance

LO: Low MO: Moderate, HI: High

(a) CB-LO, CC-MO, CE-HI
(b) CB-LO, CC-HI.CE-MO
(c) CB-MO, CC-HI, CE-LO
(d) CB-HI, CC-LO, CE-MO

[2003 : 1 Mark]

57. Assuming that the β of the transistor is extremely large and V_{BE} = 0.7 V, I_C and V_{CE} in the circuit shown in the figure are

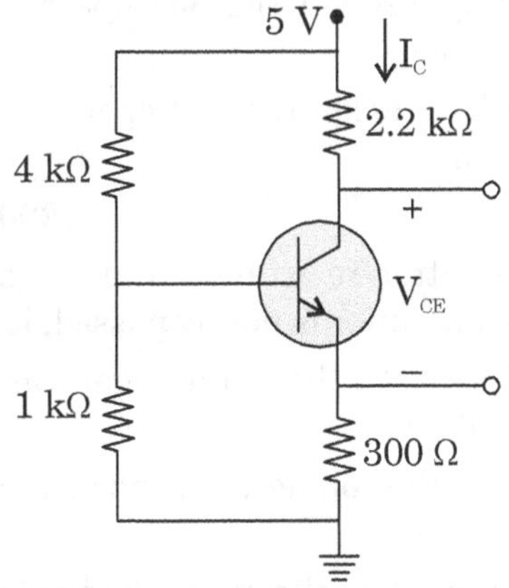

(a) I_C = 1mA, V_{EC} = 4.7 V
(c) I_C = 0.5 mA, V_{CE} = 3.75 V
(c) I_C = 1 mA, V_{CE} = 2.5V
(d) I_C = 0.5 mA, V_{CE} = 3.9 V

[2004 : 2 Marks]

58. For an npn transistor connected as shown in the figure, V_{BE} = 0.7 volts. Given that reverse saturation current of the junction at room temperature 300 K

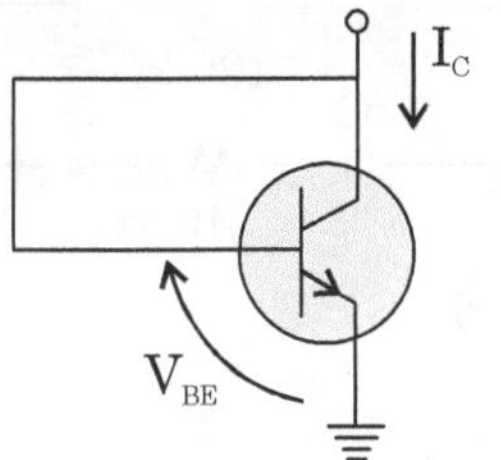

(a) 30 mA (b) 39 mA
(c) 49 mA (d) 20 mA

[2005 : 2 Marks]

59. The circuit using a BJT with β = 50 and V_{BE} = 0.7 V is shown in the figure. The base current I_B and collector voltage Vc are respectively

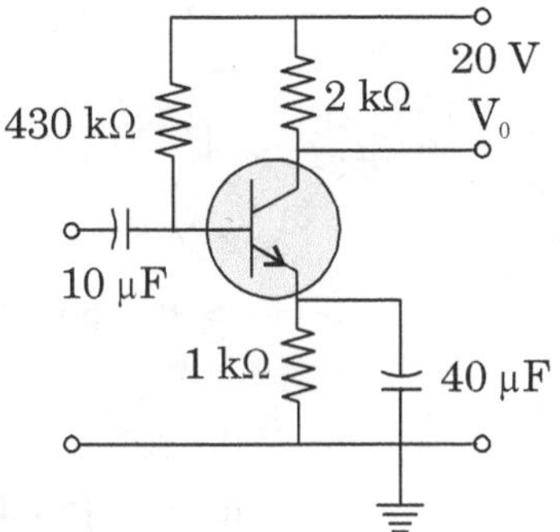

(a) 43 μA and 11.4 Volts
(b) 40 μA and 16 Volts
(c) 45 μA and 11 Volts
(d) 50 μA and 10 Volts

[2005 : 2 Marks]

60. For the BJT Q_1 in the circuit shown below, β = ∞, V_{BEon} = 07 V_{CEsat} = 0.7 V. The switch is initially closed. At time t = 0, the switch is opened. The time t at which Q_1 leaves the active region is

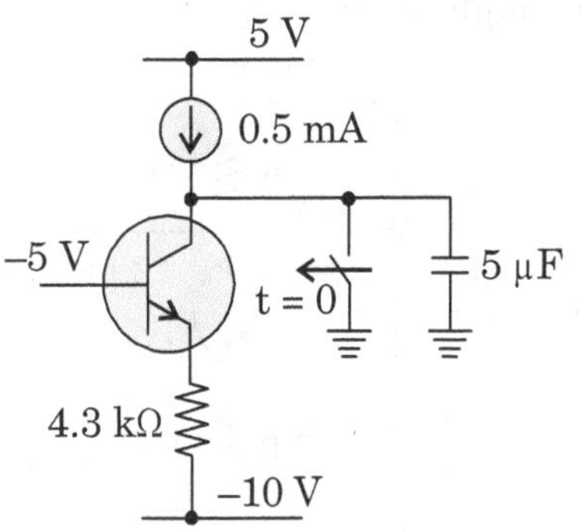

(a) 10 ms (b) 25 ms
(c) 50 ms (d) 100 ms

[2005 : 1 Mark]

61. The voltage gain A_v, of the circuit shown below is

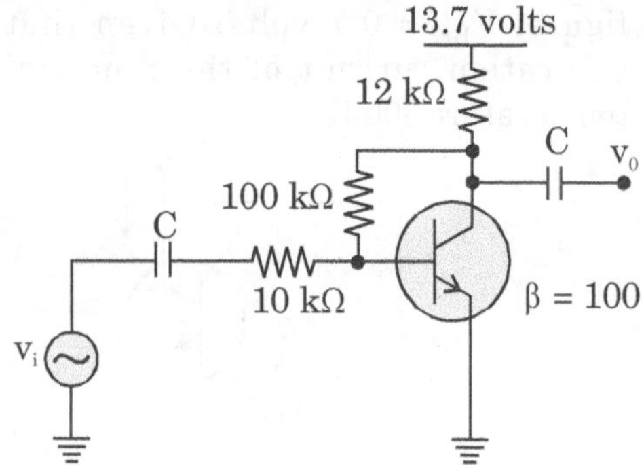

(a) $|A_v| = 200$ (b) $|A_v| = 100$
(c) $|A_v| = 20$ (d) $|A_v| = 10$

[2005 : 2 Marks]

62. In the circuit shown below, the silicon npn transistor Q has a very high value of β. The required value of R_2 in kΩ to produce $I_C = 1$ mA is

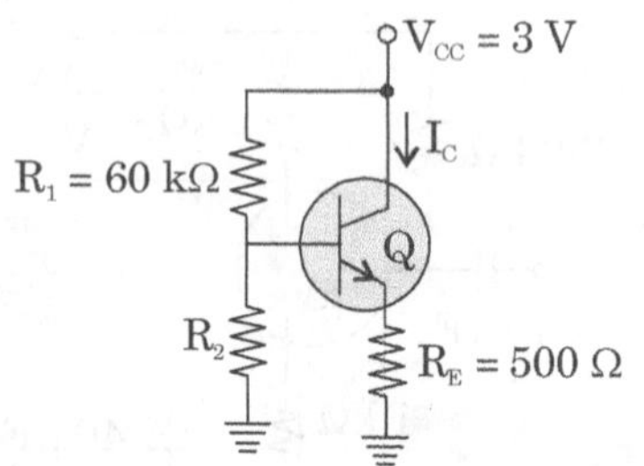

(a) 20 (b) 30
(c) 40 (d) 50 **[2005 : 2 Marks]**

63. A good current buffer has

(a) low input impedance and low output impedance
(b) low input impedance and high output impedance
(c) high input impedance and low output impedance
(d) high input impedance and high output impedance

[2006 : 2 Marks]

64. For the amplifier shown in the figure, the BJT parameters are $V_{BE} = 0.7$ V, β = 200, and thermal voltage $V_T = 25$ mV. The voltage gain (v_o/v_i) of the amplifier is ______.

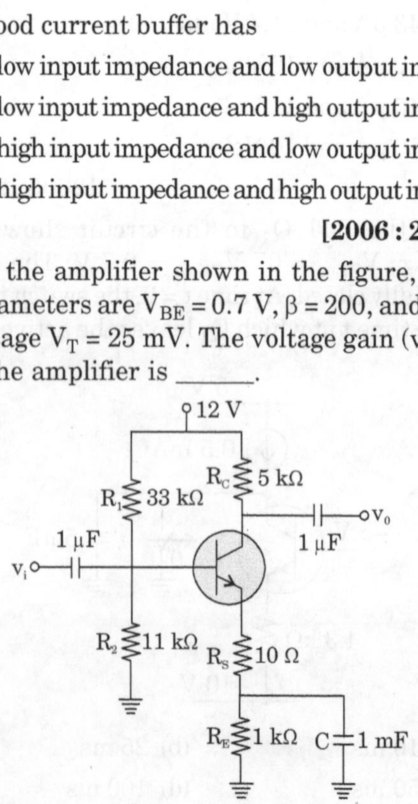

65. In the circuit shown, the PNP transistor has $|V_{BE}| = 0.7$ V and β = 50. Assume that that $R_B = 100$ kΩ. For V_0 to be 5 V, the value of R_C (in kΩ) is ______.

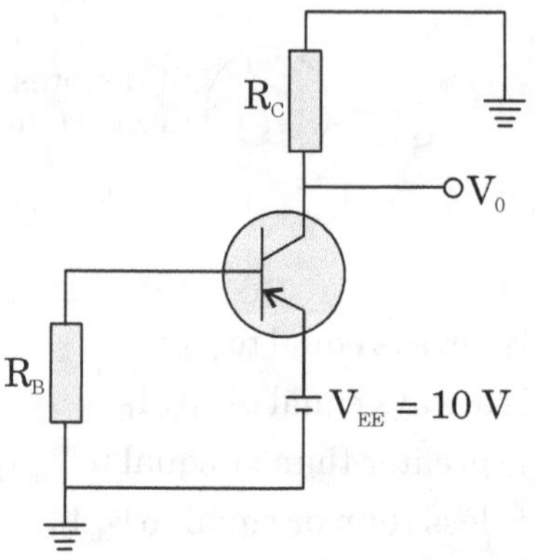

[2006 : 2 Marks]

66. In the circuit shown, the silicon BJT has β = 50. Assume $V_{BE} = 0.7$ V and $V_{CE(sat)} = 0.2$ V. Which one of the following statements is correct?

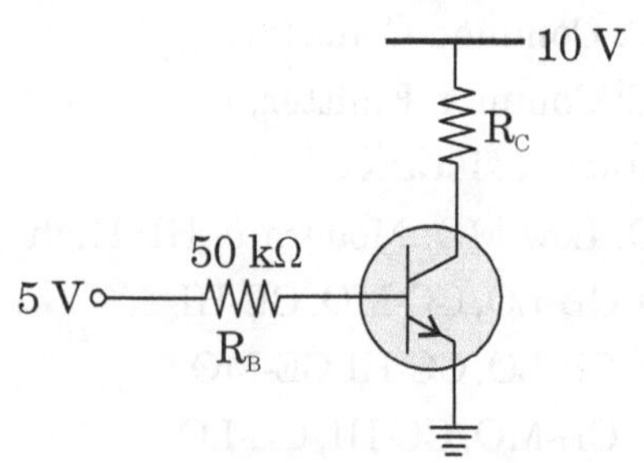

(a) For RC= 1 kQ, the BJT operates in the saturation region
(b) For $R_C = 1$ kΩ, the BJT operates in the saturation region
(c) For $R_C = 20$ kΩ, the BJT operates in the cut-off region
(d) For $R_C = 20$ kΩ, the BJT operates in the linear region

[2006 : 1 Mark]

67. If the emitter resistance in a common-emitter voltage amplifier is not by passed, it will

(a) reduce both the voltage gain and the input impedance
(b) reduce the voltage gain and increase the input impedance
(c) increase the voltage gain and reduce the input impedance
(d) increase both the voltage gain and the input impedance

[2007 : 2 Marks]

68. Consider the common-collector amplifier in the figure (bias circuitry ensures that the transistor operates in forward active region, but has been omitted for simplicity). Let I_C be the collector current, V_{BE} be the base-emitter voltage and V_T be the thermal voltage. Also, g_m and r_0 are the small-signal transconductance and output resistance of the transistor, respectively. Which one of the following conditions ensures a nearly constant small signal voltage gain for a wide range of values of R_E?

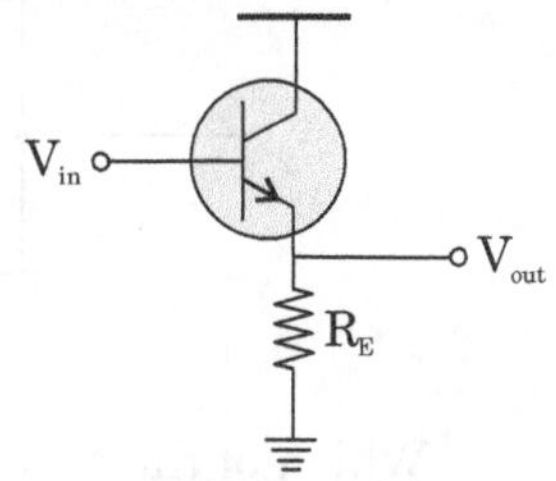

(a) $g_m R_E << 1$

(b) $I_C R_E >> V_T$

(c) $g_m r_0 >> 1$

(d) $V_{BE} >> V_T$

[2008 : 2 Marks]

69. For the common collector amplifier shown in the figure, the BJT has high β, negligible $V_{CE(sat)}$, and V_{BE} = 0.7 V. The maximum undistorted peak-to-peak output voltage v_0 (in Volts) is ______.

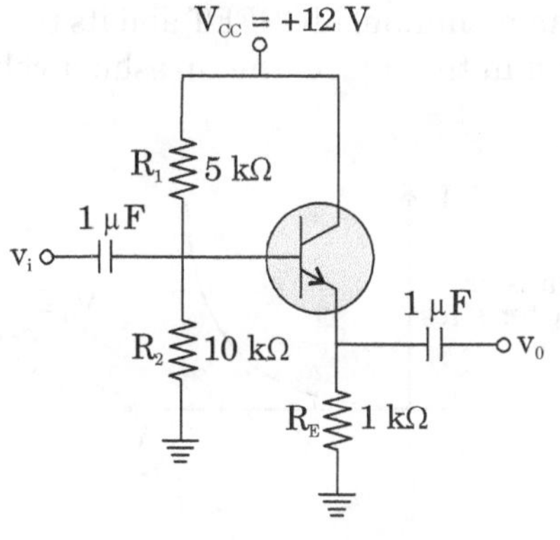

70. A BJT in a common-base configuration is used to amplify a signal received by a 50 Ω antenna. Assume kT/q = 25 mV. The value of the collector bias current (in mA) required to match the input impedance of the amplifier to the impedance of the antenna is ______.

[2009 : 2 Marks]

71. In the circuit shown in the figure, the BJT has a current gain (β) of 50. For an emitter base voltage V_{EB} = 600 mV, the emitter collector voltage V_{EC} (in Volts) is ______.

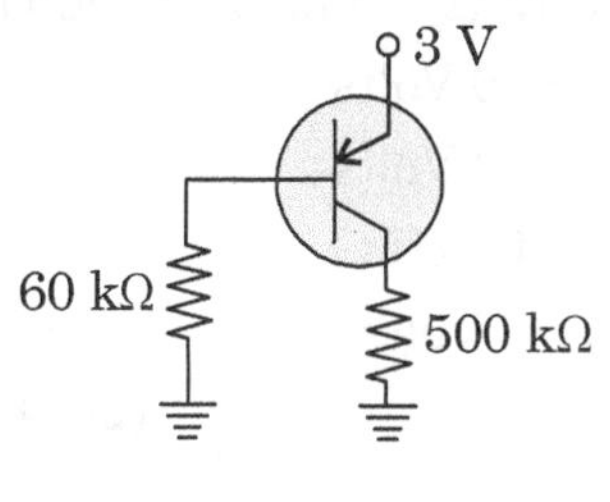

72. The Ebers-Moll model of a BJT is valid

(a) only in a active mode

(b) only in active and saturation modes

(c) only in active and cut-off modes

(d) in active, saturation and cut-off modes

[2009 : 2 Marks]

73. Resistor R_1 in the circuit below has been adjusted so that I_1 = 1 mA. The bipolar transistors Q_1 and Q_2 are perfectly matched and have very high current gain, so their base currents are negligible. The supply voltage V_{CC} is 6 V. The thermal voltage kT/q is 26 mV.

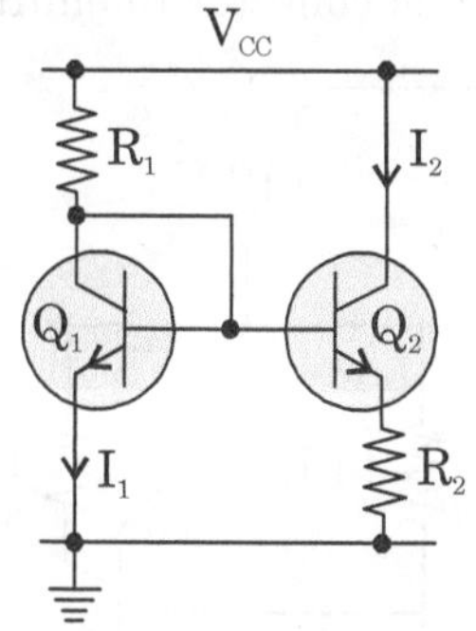

The value of R_2 (in Ω) for which I_2 = 100 μA is _.

[2011 : 2 Marks]

74. Consider the circuit shown in the figure. Assuming $V_{BE1} = V_{EB2}$ = 0.7 volt, the value of the dc voltage V_{C2} (in volt) is ______.

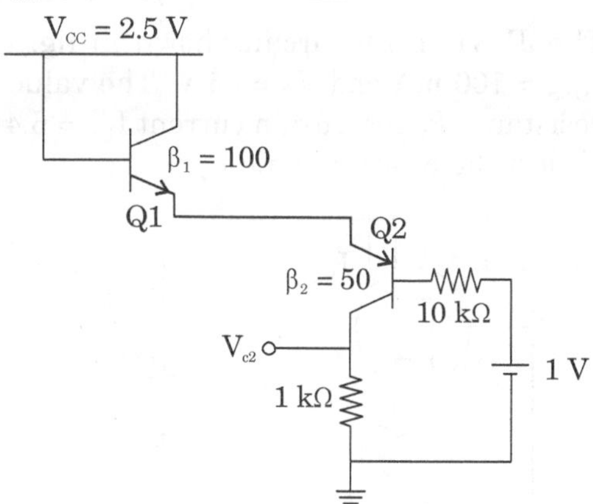

[2011 : 2 Marks]

75. In the figure shown, the npn transistor acts as a switch

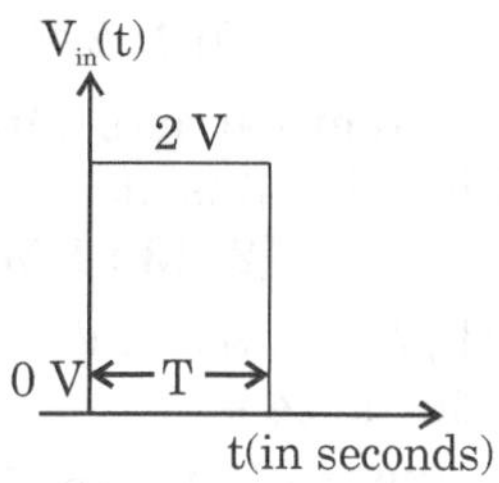

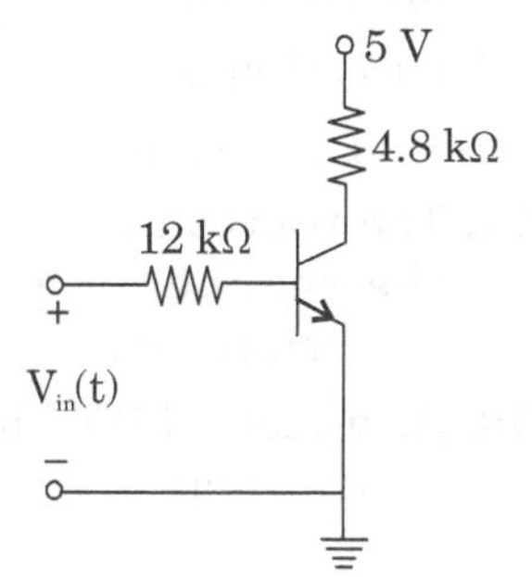

For the input $V_{in}(t)$ as shown in the figure, the transistor switches between the cut-off and saturation regions of operation, when T is large. Assume collector-to-emitter voltage at saturation $V_{CE(sat)}$= 0 2 V and base-to-emitter voltage $V_{BE} = 0.7$ V. The minimum value of the common-base current gain (α) of the transistor for the switching should be ______.

[2012 : 2 Marks]

76. Consider the circuit shown in the figure. Assume base-to-emitter voltage $V_{BE} = 0.8$ V and common-base current gain (α) of the transistor is unity. The value of collector-to-emitter voltage V_{CE} (in Volt) is ______.

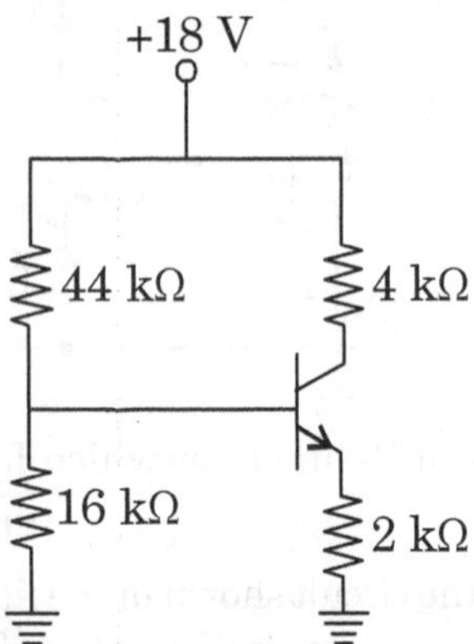

[2012 : 2 Marks]

77. The JFET in the circuit shown in fig. has an $I_{DSS} = 100$ mA and $V_P = -5$ V. The value of the resistance R_S for a drain current $I_{DS} = 6.4$ mA is (Select the Nearest value.)

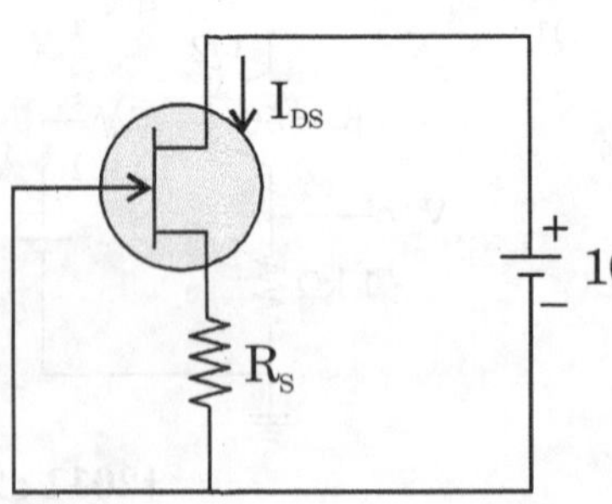

(a) 150 ohms
(b) 470 ohms
(c) 560 ohms
(d) 1 Kilo ohm **[2012 : 2 Marks]**

78. The transit time of a current carriers through the channel of an FET decides it ______ characteristics. **[2013 : 1 Mark]**

79. An n-channel JFET has $I_{DSS} = 1$ mA and $V_P = -5$ V. Its maximum transconductance is ______.

[2014 : 2 Marks, Set-1]

80. Consider the following statements in connection with the CMOS inverter in the figure, where both the MOSFETs are of enhancement type and both have a threshold voltage of 2 V.

Statement 1 : T_1 conducts when $V_i \geq 2$ V.

Statement 2 : T_1 is always in saturation when $V_0 = 0$ V.

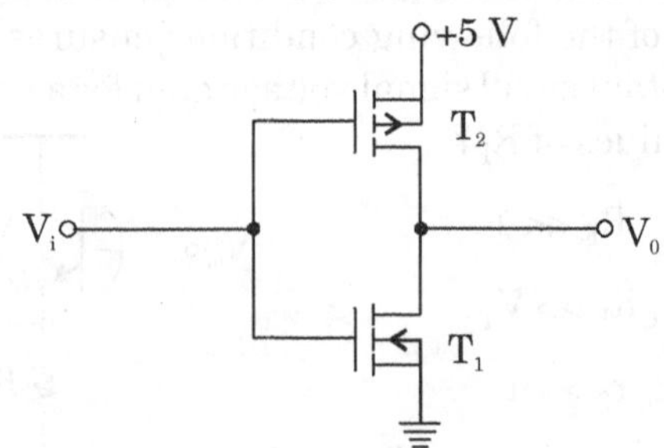

Which of the following is correct?

(a) Only Statement 1 is TRUE
(b) Only Statement 2 is TRUE
(c) Both the statements are TRUE
(d) Both the statements are FALSE

[2014 : 1 Mark, Set-1]

81. For an n-channel MOSFET and its transfer curve shown in the figure, the threshold voltage is

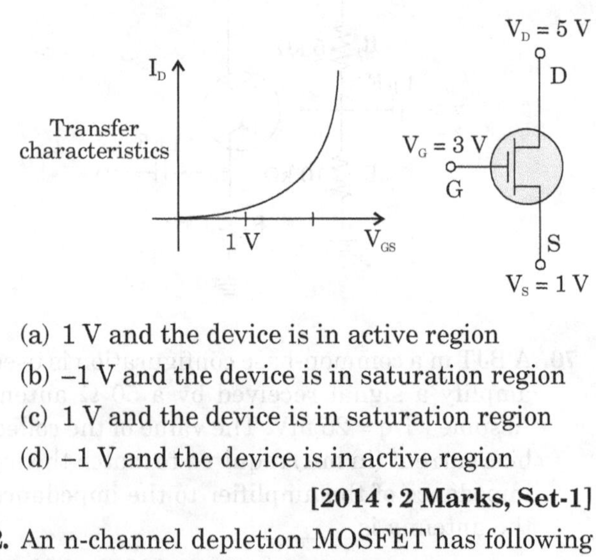

(a) 1 V and the device is in active region
(b) –1 V and the device is in saturation region
(c) 1 V and the device is in saturation region
(d) –1 V and the device is in active region

[2014 : 2 Marks, Set-1]

82. An n-channel depletion MOSFET has following two points on its $I_D - V_{GS}$ curve:

(i) $V_{GS} = 0$ at $I_D = 12$ mA and
(ii) $V_{GS} = -6$ Volts at $Z_0 = \infty$

Which of the following Q-points will give the highest trans-conductance gain for small signals?

(a) $V_{GS} = -6$ Volts
(b) $V_{GS} = -3$ Volts
(c) $V_{GS} = 0$ Volts
(d) $V_{GS} = 3$ Volts

[2014 : 1 Mark, Set-1]

83. In the CMOS inverter circuit shown, if the transconductance parameters of the NMOS and PMOS transistors are

$$K_n = K_p = \mu_n C_{ox} \frac{W_n}{L_n} = \mu_p C_{ox} \frac{W_P}{L_P} = 40\ \mu A/V^2$$

and their voltages are $V_{THn} = |V_{THp}| = 1$ V, the current I is

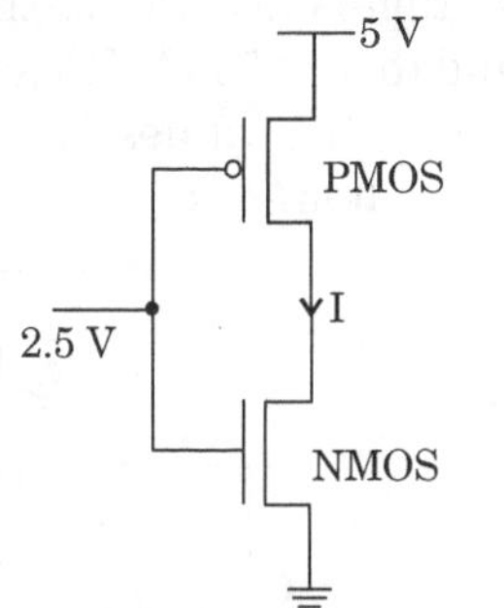

(a) 0 A
(b) 25 μA
(c) 45 μA
(d) 90 μA

[2014 : 2 Marks, Set-2]

84. Two identical NMOS transistors M_1 and M_2 are connected as shown below. V_{bias} is chosen so that both transistors are in saturation. The equivalent g_m of the pair is defined to be $\frac{\partial I_{out}}{\partial V_i}$ at constant V_{out}

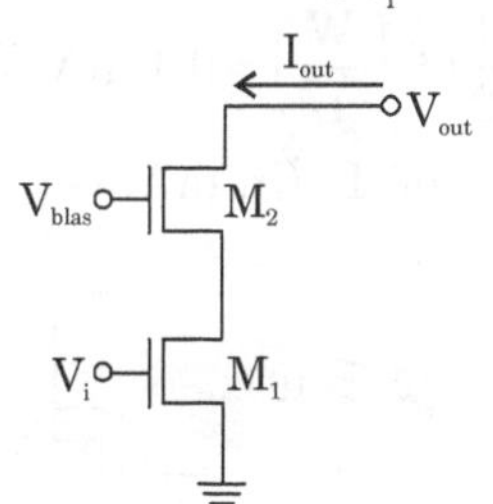

The equivalent g_m of the pair is

(a) the sum of individual g_m's of the transistors
(b) the product of individual g_m's of the transistors
(c) nearly equal to the g_m of M_1
(d) nearly equal to g_m/g_0 of M_2

[2014 : 2 Marks, Set-2]

85. For the circuit shown in the following figure, transistors M_1 and M_2 are identical NMOS transistors. Assume that M_2 is in saturation and the output is unloaded.

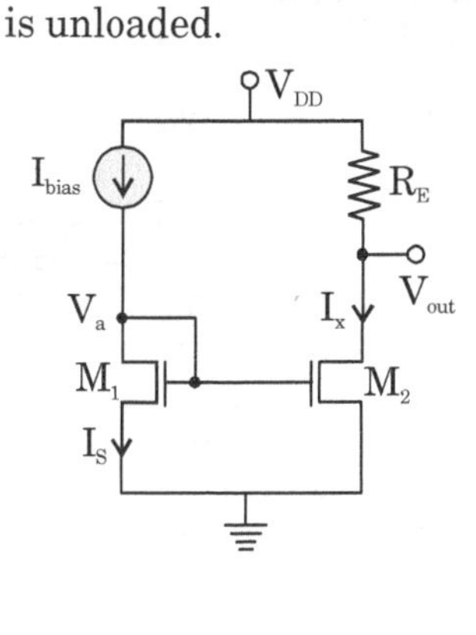

The equivalent g_m of the pair is

(a) $I_x = I_{bias} + I_s$
(b) $I_x = I_{bias}$
(c) $I_x = I_{bias} - I_s$
(d) $I_x = I_{bias} - \left(V_{DD} - \frac{V_{out}}{R_E}\right)$

[2014 : 1 Mark, Set-3]

Statement for Linked Answer Question 86 and 87.

Consider the CMOS circuit shown, where the gate voltage V_G of the n-MOSFET is increased from zero, while the gate voltage of the p-MOSFET is kept constant at 3 V. Assume that, for both transistors, the magnitude of the threshold voltage is 1 V and the product of the transconductance parameter and the (W/L) ratio, i.e. the quantity $nC_{ox}(W/L)$, is 1 mA – V^{-2}.

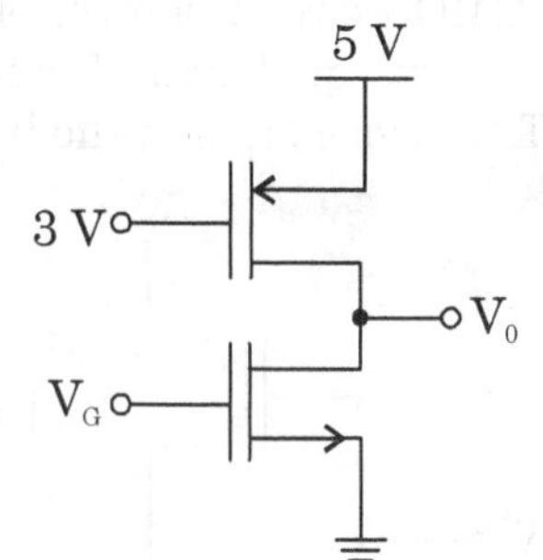

86. For small increase in V_G beyond 1 V, which of the following gives the correct description of the region of operation of each MOSFET?

(a) Both the MOSFETs are in saturation region
(b) Both the MOSFETs are in triode region
(c) n-MOSFET is in triode and p-MOSFET is in saturation region
(d) n-MOSFET is in saturation and p-MOSFET is in triode region

[2014 : 2 Marks, Set-3]

87. Estimate the output voltage V_0 for $V_G = 1.5$ V.

(a) $\left(4 - \frac{1}{\sqrt{2}}\right)$ V
(b) $\left(4 + \frac{1}{\sqrt{2}}\right)$ V
(c) $\left(4 - \frac{\sqrt{3}}{2}\right)$ V
(d) $\left(4 + \frac{\sqrt{3}}{2}\right)$ V

[2014 : 2 Marks, Set-3]

88. In the circuit shown below, for the MOS transistors, $\mu_0 C_{ox} = 100\ \mu A/V^2$ and the threshold voltage $V_T = 1$ V. The voltage V_x at the source of the upper transistor is

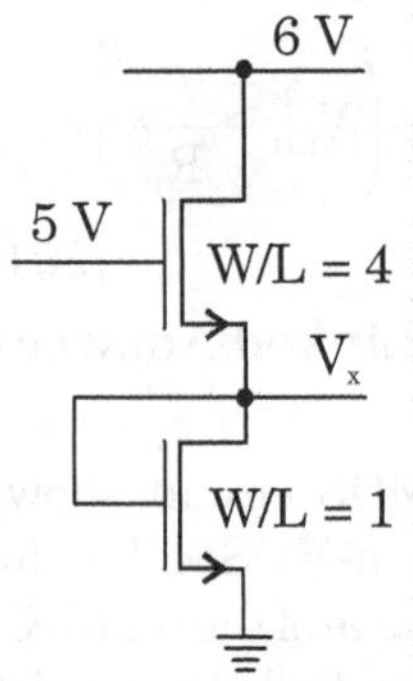

(a) 1 V (b) 2 V
(c) 3 V (d) 3.67 V

[2014 : 2 Marks, Set-4]

89. In the CMOS circuit shown, electron and hole mobilities are equal, and M_1 and M_2 are equally sized. The device M_1 is in the linear region if

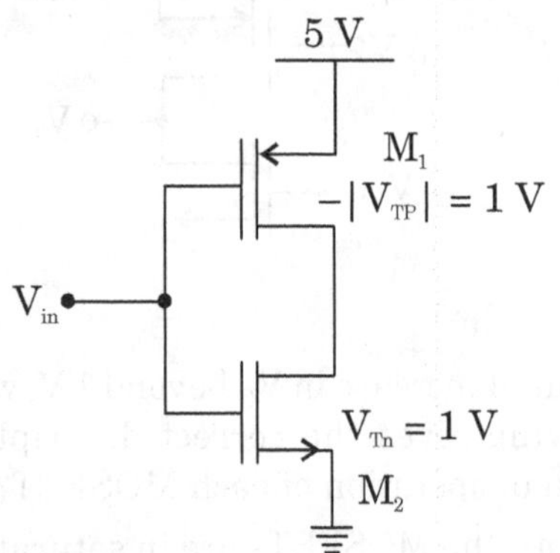

(a) $V_{in} < 1.875$ V (b) $1.875\text{ V} < V_{in} < 3.125$ V
(c) $V_{in} > 3.125$ V (d) $0 < V_{in} < 5$ V

[2014 : 2 Marks, Set-4]

90. In a MOSFET operating in the saturation region, the channel length modulation effect causes

(a) an increase in the gate-source capacitance
(b) a decrease in the transconductance
(c) a decrease in the unity-gain cut-off frequency
(d) a decrease in the output resistance

[2014 : 1 Mark, Set-4]

91. In the following circuit employing pass transistor logic, all NMOS transistors are identical with a threshold voltage of 1 V. Ignoring the body-effect, the output voltages at P, Q and R are,

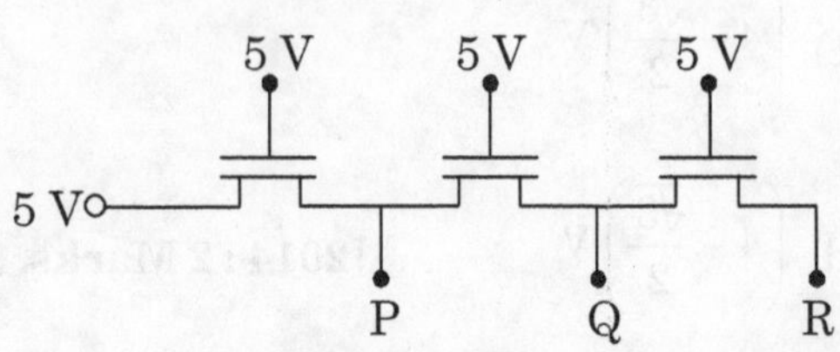

(a) 4 V, 3 V, 2 V (b) 5 V, 5 V, 5 V
(c) 4 V, 4 V, 4 V (d) 5 V, 4 V, 3 V

[2015 : 2 Marks, Set-1]

92. For the n-channel MOS transistor shown in figure, the threshold voltage V_{Th} is 0.8 V. Neglect channel length modulation effects. When the drain voltage $V_D = 1.6$ V, the drain current I_D was found to be 0.5 mA. If V_D is adjusted to be 2 V by changing the values of R and V_{DD}, the new value of I_D (in mA) is

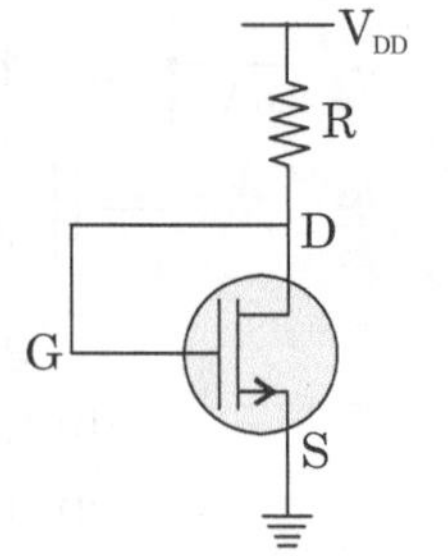

(a) 0.625 (b) 0.75
(c) 1.125 (d) 1.5

[2016 : 1 Mark, Set-2]

93. For the MOSFET shown in the figure, the threshold voltage $|V_t| = 2$ V and

$$K = \frac{1}{2}\mu C\left(\frac{W}{L}\right) = 0.1\ mA/V^2 .$$

The value of I_D (in mA) is _______.

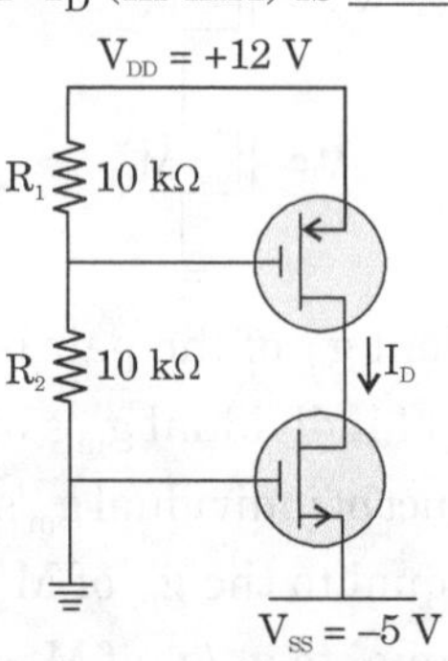

[2016 : 1 Marks, Set-3]

94. For the MOSFET M_1 shown in the figure, assume W/L = 2, $V_{DD} = 2.0$ V, $\mu_n C_{ox} = 100\ \mu A/V^2$ and $V_{TH} = 0.5$ V. The transistor M_1 switches from saturation region to linear region when V_{in} (in volts) is _______.

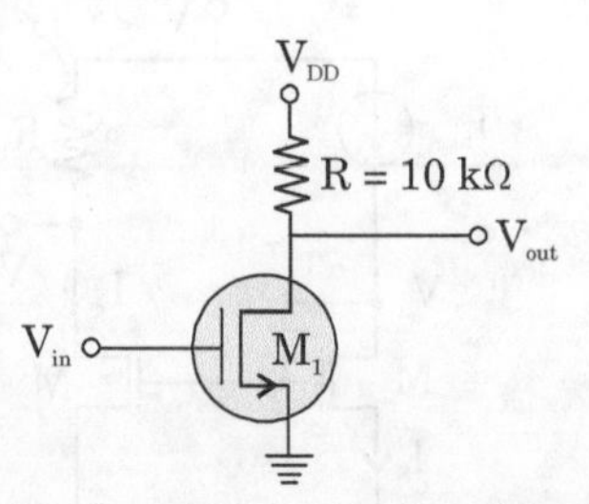

[2016 : 1 Mark, Set-1]

95. In the circuit shown, both the enhancement mode NMOS transistors have the following characteristics: $K_n = \mu_n C_{ox}(W/L) = 1\ mA/V^2$ $V_{TN} = 1$ V. Assume that the channel length modulation parameter λ is zero and body is shorted to source. The minimum supply voltage V_{DD} (in volts) needed to ensure that transistor M_1 operates in saturation mode of operation is _______.

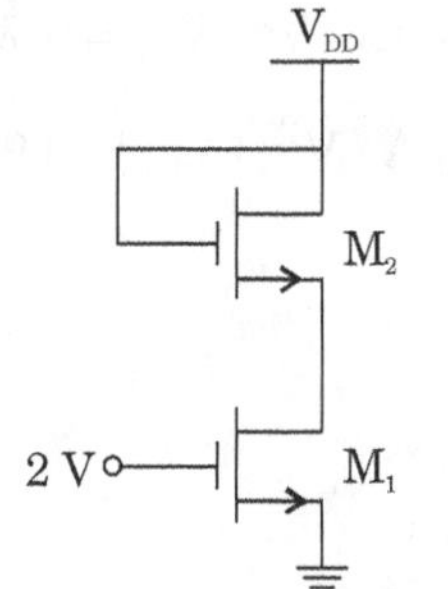

[2016 : 1 Mark, Set-3]

96. What is the voltage V_{out} in the following circuit?

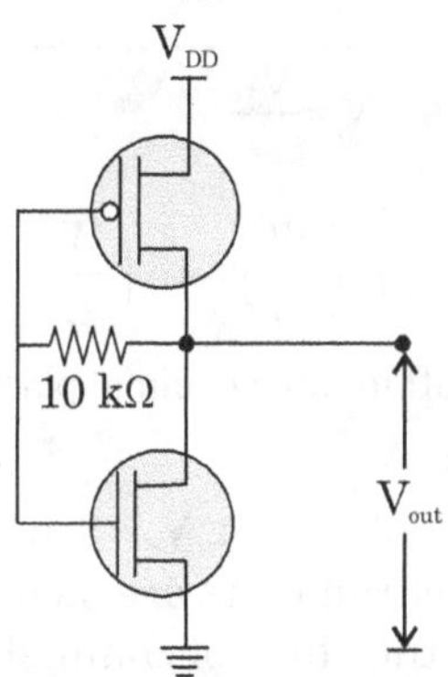

(a) 0 V

(b) (|T of PMOS| + V_T of NMOS)/2

(c) Switching threshold of inverter

(d) V_{DD}

[2017 : 2 Marks, Set-1]

97. For the circuit shown, assume that the NMOS transistor is in saturation. Its threshold voltage V_{in} = 1 V and its transconductance parameter $\mu_n C_{ox}\left(\frac{W}{L}\right) = 1\ mA/V^2$. Neglect channel length modulation and body bias effects. Under these conditions, the drain current I_D in mA is _______.

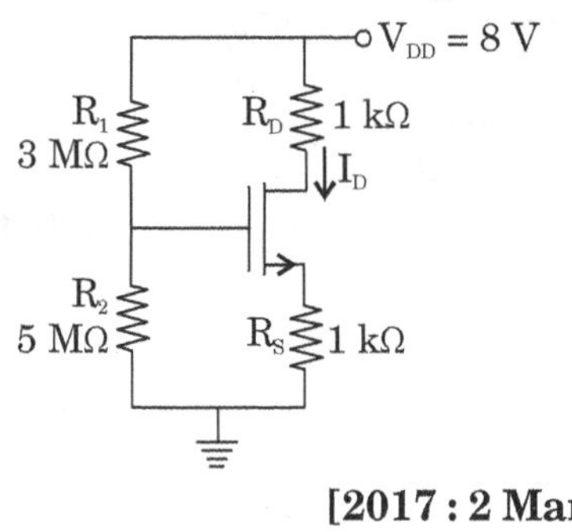

[2017 : 2 Marks, Set-1]

98. Assuming that transistors M_1 and M_2 are identical and have a threshold voltage of 1 V, the state of transistors M_1 and M_2 are respectively

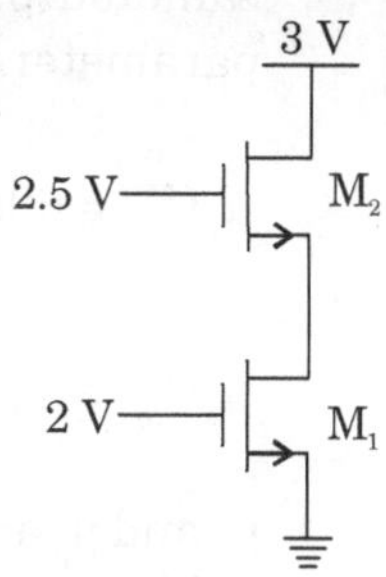

(a) Saturation, Saturation

(b) Linear, Linear

(c) Linear, Saturation

(d) Saturation, Linear

[2017 : 2 Marks, Set-2]

99. In the circuit shown below, the (W/L) value for M_2 is twice that for M_1. The two nMOS transistors are otherwise identical. The threshold voltage for both transistors is 1.0 V. Note that V_{GS} for M_2 must be > 1.0 V.

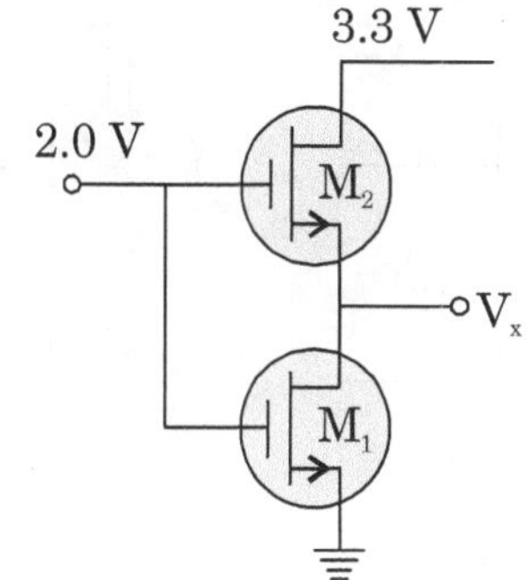

Current through the nMOS transistors can be modeled as

$$I_{DS} = \mu C_{ox}\left(\frac{W}{L}\right)\left((V_{GS} - V_T)V_{DS} - \frac{1}{2}V_{DS}^2\right)$$

for $V_{DS} \le V_{GS} - V_T$

$$I_{DS} = \mu C_{ox}\left(\frac{W}{L}\right)(V_{GS} - V_T)^2/2 \text{ for } V_{DS} \ge V_{GS} - V_T$$

The voltage (in volts, accurate to two decimal places) at V_x is _______.

[2018 : 2 Marks]

100. The figure shows the high-frequency C-V curve of a MOS capacitor (at T = 300 K) with Φ_{ms} = 0 V and no oxide charges. The flat-band, inversion, and accumulation conditions are represented, respectively, by the points.

(A) R, P, Q

(B) Q, P, R

(C) P, Q, R

(D) Q, R, P

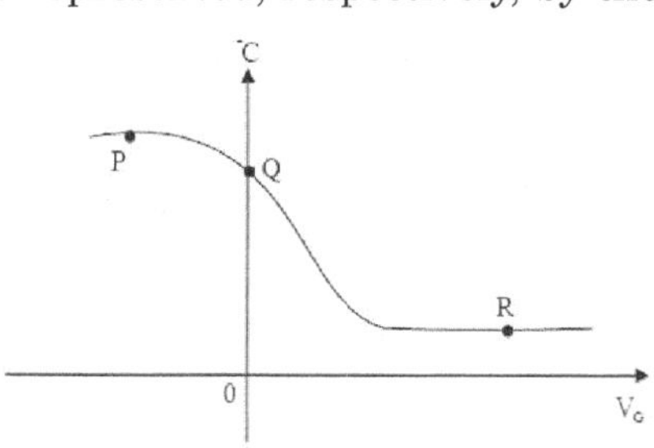

101. In the circuit shown, the threshold voltages of the pMOS($|V_{tp}|$) and nMOS (V_{tn}) transistors are both equal to 1 V. All the transistors have the same output resistance r_{ds} of 6 MΩ. The other parameters are listed below:

$$\mu_n C_{ox} = 60\mu A / V^2; \left(\frac{W}{L}\right)_{nMOS} = 5$$

$$\mu_p C_{ox} = 30\mu A / V^2; \left(\frac{W}{L}\right)_{pMOS} = 10$$

μ_n and μ_p are the carrier mobilities, and C_{ox} is the oxide capacitance per unit area. Ignoring the effect of channel length modulation and body bias, the gain of the circuit is ________ (rounded off to 1 decimal place).

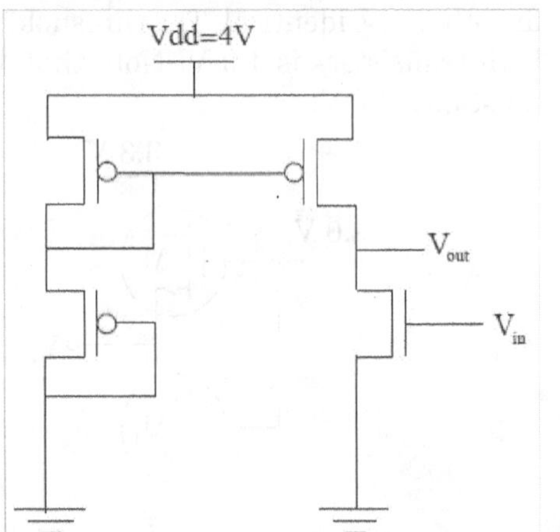

[2019 : 2 Marks]

102. In the circuits shown, the threshold voltage of each nMOS transistor is 0.6V. Ignoring the effect of channel length modulation and body bias, the values of V_{out1} and V_{out2}, respectively, in volts, are

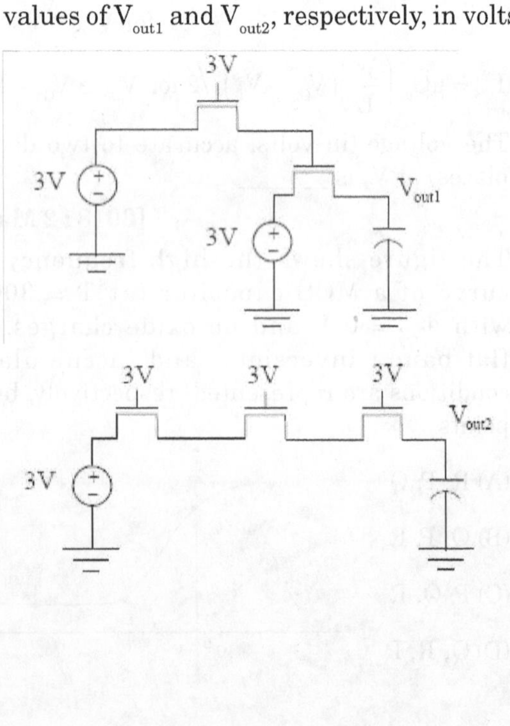

(a) 2.4 and 1.2
(b) 2.4 and 2.4
(c) 1.8 and 1.2
(d) 1.8 and 2.4

[2019 : 2 Marks]

103. A CMOS inverter, designed to have a mid-point voltage V_1 equal to half of V_{dd}, as shown in the figure, has the following parameters:

$V_{dd} = 3V$

$$\mu_n C_{ox} = 100\mu A / V^2; \; V_{tn} = 0.7V \text{ for nMOS}$$

$$\mu_p C_{ox} = 40\mu A / V^2; \; |V_{tp}| = 0.9V \text{ for pMOS}$$

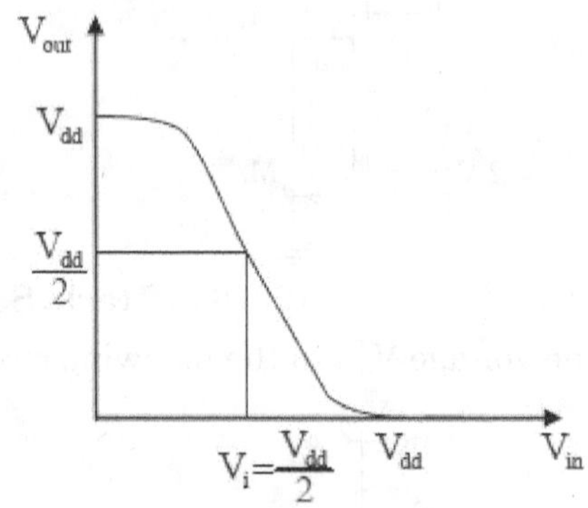

The ratio of $\left(\frac{W}{L}\right)_n$ to $\left(\frac{W}{L}\right)_p$ is equal to ______. (rounded off to three decimal places).

[2019 : 2 Marks]

104. In the circuit shown, $V_1 = 0$ and $V_2 = V_{dd}$. The other relevant parameters are mentioned in the figure. Ignoring the effect of channel length modulation and the body effect, the value of I_{out} is ____mA (rounded off to one decimal place).

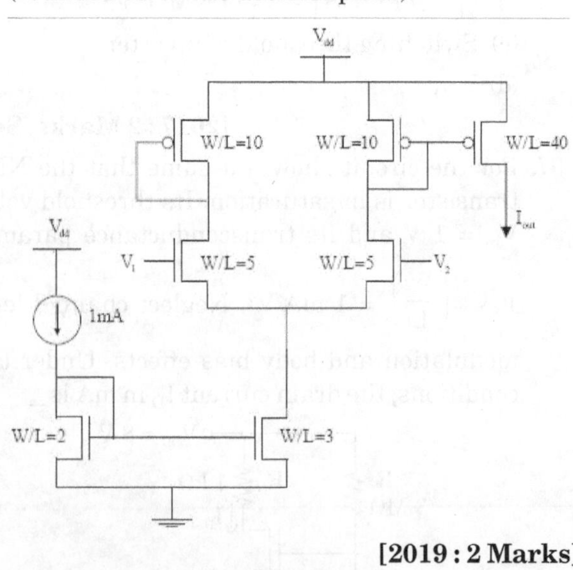

[2019 : 2 Marks]

ANSWERS

1. (a) 2. (d) 3. (d) 4. (d) 5. (5) 6. (1) 7. (5·33 mA) 8. (a) 9. (b)
10. (d) 11. (c) 12. (a) 13. (a) 14. (d) 15. (c) 16. (a) 17. (a) 18. (c) 19. (b)
20. (a) 21. (b) 22. (a) 23. (d) 24. (b) 25. (b) 26. (a) 27. (b) 28. (b) 29. (c)
30. (d) 31. (83·5 – 84·0) 32. (–1) 33. (c) 34. (d) 35. (b) 36. (a) 37. (b) 38. (b)
39. (57·8745) 40. (c) 41. (c) 42. (c) 43. (c) 44. (b) 45. (c) 46. (a) 47. (b)
48. (a,b) 49. (1000 W) 50. (*) 51. (d) 52. (*) 53. (b) 54. (c) 55. (d) 56. (b)
57. (c) 58. (c) 59. (b) 60. (c) 61. (d) 62. (c) 63. (b) 64. (– 240 to – 230) 65. (1·075)
66. (b) 67. (b) 68. (b) 69. (9·4) 70. (0.50) 71. (2) 72. (d) 73. (598.67) 74. (0·5) 75. (0.902)
76. (6) 77. (a) 78. (Switching) 79. (0.4) 80. (a) 81. (c) 82. (d) 83. (d) 84. (c)
85. (b) 86. (d) 87. (d) 88. (c) 89. (a) 90. (d) 91. (c) 92. (c) 93. (0.9) 94. (1.5)
95. (3) 96. (b) 97. (2) 98. (c) 99. (0.4226) 100. (d) 101. (–900) 102. (d)
103. (0.225) 104. (6)

EXPLANATIONS

1. Ebers-Moll model is a composite model and is used to predict the operation of BJT in all of its possible modes.

2.

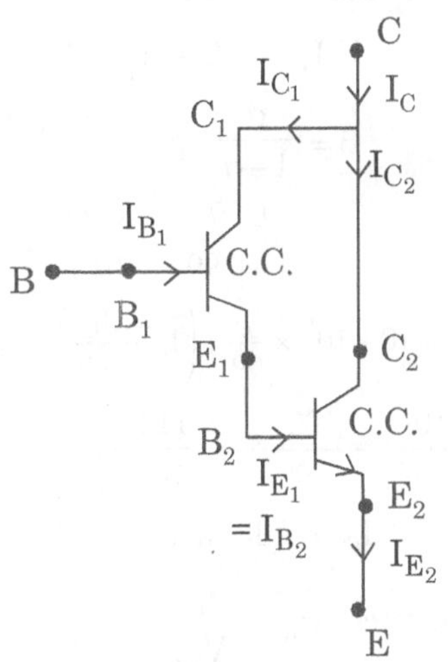

Current gain for each transistor

$h_{fe} = \beta = 100$

$\Rightarrow I_C = I_{C1} + I_{C2} = \beta_1 I_{B1} + \beta_2 I_{B2}$

$\because \quad I_{B_2} = I_{E_1}$

$I_C = \beta_1 I_{B1} + \beta_2 I_{E1}$

$\because \quad I_{E_1} = (1+\beta) I_{B_1}$

$I_C = \beta_1 IB_1 + \beta_2 (\beta_1 + 1) I_{B1} = [\beta_1 + \beta_2 (\beta_1 + 1)] I_{B1}$

overall current gain of darlington pair,

$$h'_{fe} = \beta = \frac{\text{output current}}{\text{Input current}} = \frac{I_c}{I_{B_1}}$$

$$\beta' = \frac{\beta_1 + \beta_2(\beta_1 + 1)}{I_{B_1}} . I_{B_1} = \beta_1 + \beta_2(\beta_1 + 1) = \beta + \beta^2 + \beta$$

$\beta' = 100^2 + 2 \times 100$

$\beta' = 10200.$

3.

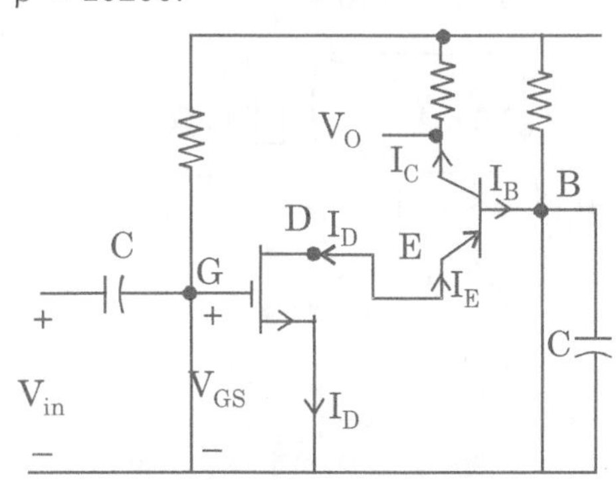

Given : g_m of MOSFET = 2mA/V

h_{fe} of BJT = β = 99.

Input stage is common source MOSFET amplifier (CS). Output stage is common base BJT amplifier (CB). Composite transistor is a cascade amplifier. Overall transconductance of cascade amplifier is

$$g_m \text{ (cascade)} = \left|\frac{\partial I_C}{\partial V_{GS}}\right| = \left|\frac{\partial I_D}{\partial V_{GS}}\right| \times \left|\frac{\partial I_C}{\partial I_D}\right|$$

$\therefore$ From figure, $I_D = -I_E$, i.e., $|I_D| = |I_E|$

$$g_m(\text{cascade}) = g_m(\text{cs}) \times \alpha[\text{CB}] = g_m[\text{CS}] \times \frac{\beta}{1+\beta}[\text{CB}]$$

$$= 2\frac{\text{mA}}{\text{V}} \times \frac{99}{1+99} = 1.98 \text{ mA/V}$$

4. Given $h_{fe} = 200$, h_{ie} 2 kΩ, $h_{re} = 0$, $h_{oe} = 0.05 \times 10^{-3}$ ℧

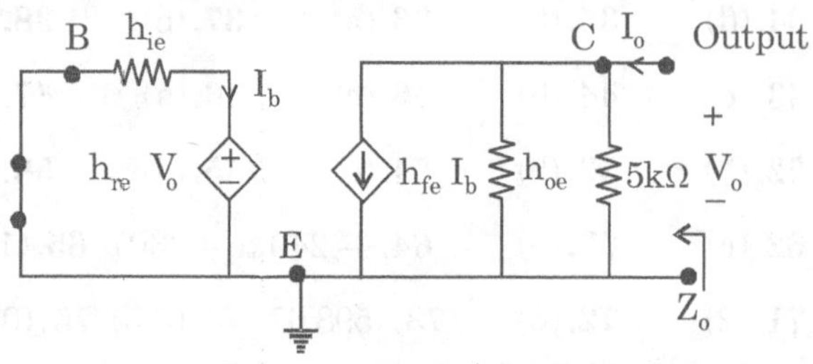

Output impedance, $Z_0 = \left.\frac{V_0}{I_0}\right|_{v_i=0}$

Applying KCL at collector terminal,

$$I_o = \frac{V_0}{(5 \times 1000)} + (V_0 \times h_{oe}) + h_{fe} \times I_b \qquad \text{(i)}$$

Applying KVL in input circuit,

$$I_b = -\frac{h_{re} \times V_0}{h_{ie}} = 0$$

$$\Rightarrow I_0 = \frac{V_0}{5000} + V_0 \times h_{oe} = \left(0.2 \times 10^{-3} + 0.05 \times 10^{-3}\right) V_0$$

$I_o = 0.25 \times 10^{-3} V_o$

$$\Rightarrow Z_0 = \frac{V_0}{I_0} = 4\text{k}\Omega$$

5. Given: (i) All transistors are identical therefore, $I_{C1} = I_{C2}$ (ii) β is very large (β = ∞) so $I_B = 0$A and $I_C = I_E$.

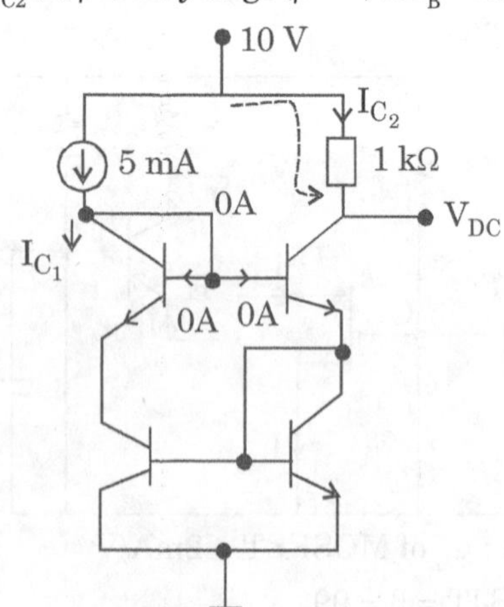

From figure, $I_4 = 5$ mA.

From the concept of current mirror circuit current through 1 KΩ resistor is

$I_{C1} = I_{C2} = 5$ mA

Applying KVL in loop,

$-10 + I_{C2} \times 1 + V_{DC} = 0$

$V_{DC} = 5$ V.

6. Given : (i) β is very large (β = ∞). So, $I_B = 0$A and $I_C = I_E$ Hence R_1, R_2 and R_3 are in series.

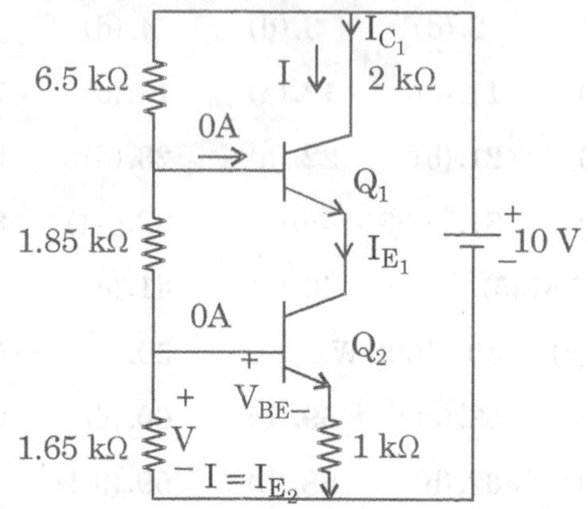

Base current for both transistors are zero. Therefore $I_{C1} = I_{E1} = I_{C2} = I_{E2} = I$.

(ii) $V_{BE} = 0.65$ V.

$$\text{Voltage } V = \frac{1.65}{1.65 + 1.85 + 6.5} \times 10 \text{(by voltage division rule)}$$

V = 1.65 Volt.

Applying KVL in loop, $-V + V_{BE} + I \times 1 = 0$

$-1.65 + 0.65 + I = 0$

I = 1 mA

7. Taking KVL for the base circuit,

$(I_C + I_B)\,1\text{K} + 10\,\text{K}\,I_B + 0.7 + (I_B + I_C)\,1\text{K} = 12$

$$\because \qquad \beta = \frac{\alpha}{1-\alpha}$$

$$= \frac{0.99}{1-0.99} = \frac{0.99}{0.01} = 99$$

$$\left(I_C + \frac{I_C}{99}\right)10^3 + 10 \times 10^3 \times \frac{I_C}{99} + \left(I_C + \frac{I_C}{99}\right) \times 10^3 = 12$$

$$\text{or} \quad \frac{100 I_C}{99} + \frac{10 I_C}{99} + \frac{100 I_C}{99} = \frac{11.3}{10^3}$$

$$\therefore \qquad I_C = 5.33 \text{ mA}$$

8. Transconductance, $g_m = \frac{I_C}{V_{be}}$

From figure $V_{be} = V_{be1} + V_{be2}$

Assuming that transistors are identical.

$V_{be1} = V_{be2}$

$$\text{overall transconductance, } g_m = \frac{I_C}{V_{be1} + V_{be2}}$$

$\because I_C = I_{C1} + I_{C2}$ and $I_{C2} >> I_{C1}$

$I_C \cong I_{C2}$

$$g_m = \frac{I_{C2}}{V_{be_2} + V_{be_2}} = \frac{g_m}{2} = 0.5\, g_{m2}$$

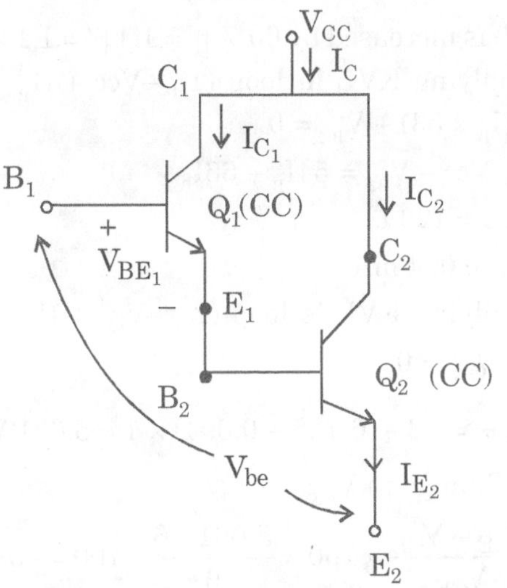

9. The bypass caption C across R_E will act as short – circuit for ac signal. So, the bypass capacitor C will increase the voltage gain and decrease the input impedance.

10. Linear transfer relation between differential output voltage and differential input voltage only when V_{id} is less than thermal voltage.

11. The pass transistor in a series voltage regulator is in common collector configuration (emitter follower) having voltage gain close to unity.

12.

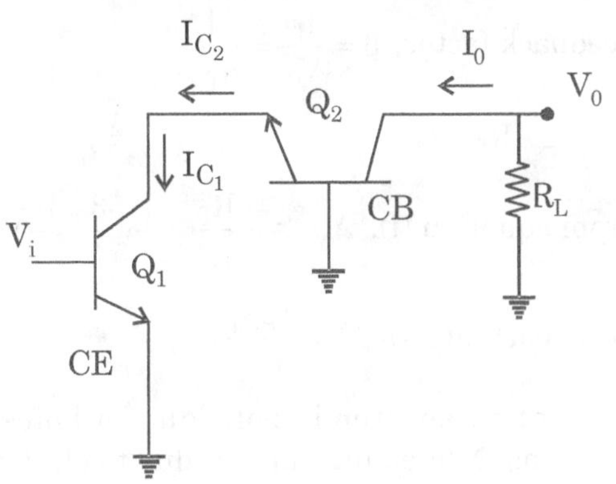

Transconductance of Q_1 is $g_{m1} = \dfrac{I_{C_1}}{V_i}$

Overall tansconductance is $g_m = \dfrac{I_{C_2}}{V_i} = \dfrac{I_O}{V_i}$

$$g_m = \frac{I_O}{I_{e_2}} \times \frac{Ie_2}{I_{e_1}} \times \frac{I_{C_1}}{V_i}$$

$\because$ From figure, $I_{C2} = I_{C1}$ and

$\alpha = I_0/I_{e2}$ (common base configuration)

$\alpha \simeq 1$

$g_m = \alpha \times 1 \times g_{m1} = 1 \times \alpha g_{m1}$

$\therefore\ g_m = g_{m1}$

13. Current gain of BJT, $A_i = \dfrac{-9m}{9_{b'e} + jw(c_e + c_c)}$

Where, Ce and Cc are the transistor capacitances. At high frequencies the current gain of BJT drops due to the transistor capacitors.

14.

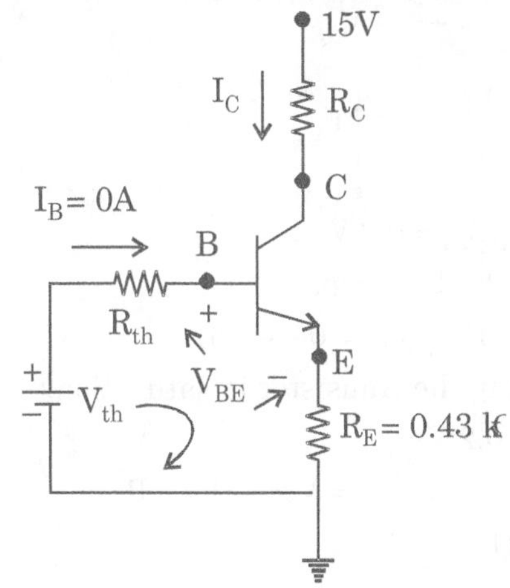

Thevenin equivalent circuit

Given: $R_1 = 10\ k\Omega$, $R_2 = 5\ k\Omega$, $R_E = 0.43\ k\Omega$

$V_{BE} = 0.7$ V

$\Rightarrow \beta$ is very large ($\beta = \infty$). $I_B = 0A$, $I_C = I_E$

$\Rightarrow$ Operating region of transistor is active region.

$$\Rightarrow\ V_{th} = \frac{R_2 \times 15}{R_1 + R_2} = \frac{5 \times 15}{10 + 5} = 5\ V$$

$$\Rightarrow\ R_{th} = \frac{R_1 \times R_2}{R_1 + R_2} = \frac{10 \times 5}{10 + 5} = 3.34\ k\Omega$$

Applying KVL in the input loop,

$-V_{th} + (I_B \times R_{th}) + V_{BE} + (I_E \times R_E) = 0$

$-5 + 0 \times R_{th} + 0.7 + I_C \times R_E = 0$

$$I_C = \frac{5 - 0.7}{0.43} = 10\ mA$$

15. Current gain of BJT, $h_{fe} = g_m \cdot r_\pi$

16. $V_{CC} - I_C R_2 = V_{CE}$

or $R_2 = \dfrac{3\ V}{1.5\ mA}$

$= 2\ k\Omega.$

$I_B = \dfrac{I_c}{\beta} = 0.01$ mA

With change of β equal to 200,

$I_C' = 0.01 \times 200$

$= 2$ mA

and $V_{CE} = V_{CC} - I_C' R_2$

$= 6 - 2 \times 2$

$= 2$ volts.

17.

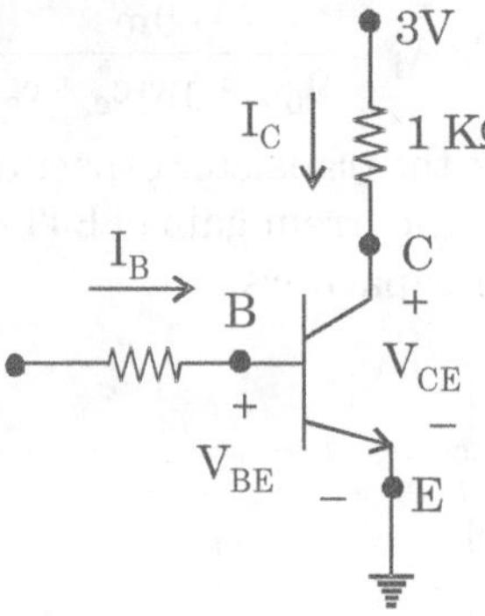

Given $(V_{CE})_{sat} = 0.2$ V, $\beta = 50$.

Applying KVL in loop,

$- V_{CC} + I_C R_C + V_{CE} = 0$ ——(i)

For driving the transistor in saturation,

$(V_{CE})_{sat} = 0.2$ V.

From eq (i), $(V_{CE})_{sat} = Vcc - (I_C)_{sat} R_C$

$0.2 = 3- (I_C)_{sat} \times 1$

$(I_C)_{sat} = 2.8$ mA.

$$(I_B)_{min} = \frac{(I_C)sat}{\beta} = \frac{2.8}{50} = 56\mu A.$$

18. Given : $\beta_{dc} = 60$, $V_{BE} = 0.7$ V, $h_{ie} \to \infty$, $h_{fe} \to \infty$.

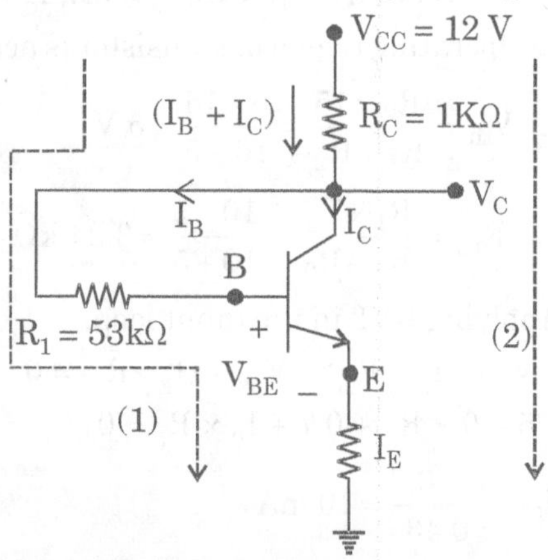

DC equivalent circuit

Applying KVL in loop (1),

$-V_{CC} + (I_B + I_C) \times 1 + I_B \times 53 + V_{BE} = 0$

$V_{CC} - V_{BE} = 54I_B + I_C$ ....(i)

$\because I_C = \beta I_B = 60\ I_B$

$V_{CC} - V_{BE} = 54\ I_B + 60\ I_B = 114\ I_B$

$$I_B = \frac{V_{CC} - V_{BE}}{114} = \frac{12 - 0.7}{114} \cong 0.1 \text{ mA}.$$

$\Rightarrow I_C = \beta_{IB} = 60 \times 0.1$ mA $= 6$ mA

Applying KVL in loop (2),

$-V_{CC} + (I_C + I_B) R_C + V_{CE} = 0$

$V_{CE} = 12 - (I_C + I_B) R_C = 12- (6 + 0.1) \times 1$

$V_{CE} = 6.0$ V.

19. If β is increased by 60 % $\beta' = 1.1\ \beta' = 1.1 \times 60 = 66$.

Applying KVL in loop (1), $-Vcc + (I_B + I_C) \times 1 + (I_B \times 53) + V_{BE} = 0$

$\therefore\ Vcc - V_{BE} = 54I_B + 66I_B$

$11.3 = 120\ I_B$

$I_B = 0.094$ mA

Applying KVL is loop (2), $-V_{CC} + (I_C + R_B).\ R_C + V'_{CE} = 0$

$V'_{CE} = 12 - (6.125 + 0.094) \times 1 = 5.691$ V.

% Change in V_{CE}

$$\frac{V'_{CE} - V_{CE}}{V_{CE}} \times 100 = \frac{5.691 - 6}{6} \times 100 = -5.15\%$$

20. The given circuit is a voltage shunt feedback amplifier and voltage gain of this voltage shunt feedback amplifier is

$$A_V = A_{V_f} = \frac{V_0}{V_i} = \frac{R_{Mf}}{R_s} \quad \text{.....................(i)}$$

Where, R_S = Source resistance,

R_{Mf} = Transconductance gain of feedback amplifier $\cong \frac{1}{\beta}$

Feedback factor, $\beta = \frac{I_f}{V_0} = \frac{-1}{R_f}$

$R_{Mf} \cong -R_f$

From equation (i), $A_{V_f} \cong \frac{-R_f}{R_s} = -\frac{53}{5.3} = -10.$

21. Transport factors $\beta^* = \frac{I_{PC1}}{I_{PE}}$

Current in emitter is both due to holes and electrons. Neglecting current due to electrons,

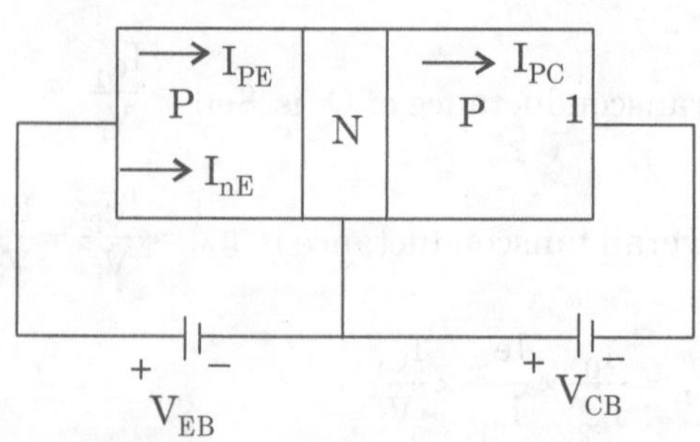

$$\alpha = \frac{I_{PC1}}{I_E} = \frac{I_{PC1}}{I_{PE}} \times \frac{I_{PE}}{I_E} = \beta^* \times \eta$$

$$\beta^* = \frac{\alpha}{\eta} = \frac{50}{51 \times 0.995} = 0.9853.$$

22.

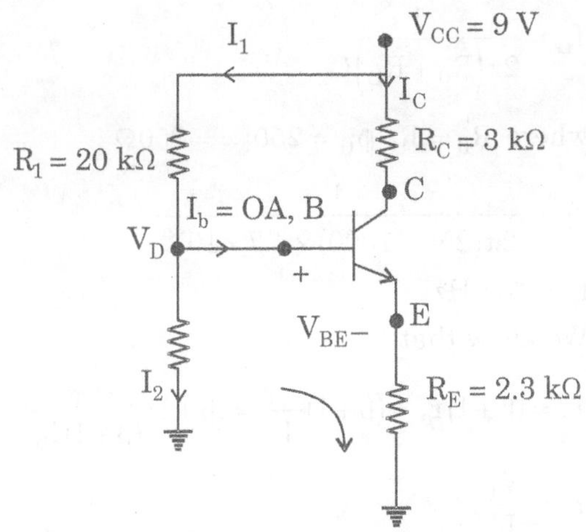

Given : $V_{BE} \cong 0.7$ V, $R_E = \dfrac{25 \text{ mA}}{I_E}$

$\beta = \infty, I_B \cong 0, I_C = I_E.$

From figure, $I_2 = I_1 - I_B = I_1$

$$I_1 = \frac{V_{CC}}{R_1 + R_2} = \frac{9}{20+10} = 0.3 \text{ mA}$$

$V_b = R_2 I_2 = 0.3 \times 10 = 3$ V.

Applying KVL in input loop,

$-V_b + V_{BE} + I_E R_E = 0$

$-3 + 0.7 + I_E \times 2.3 = 0$

$I_E = 1$ mA

23. The midband voltage gain for common – emitter amplifier without R_E is

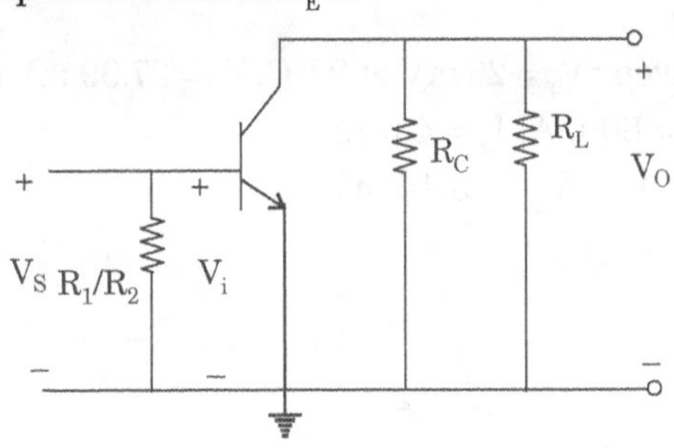

$$A_V = -g_m R_L' = -\frac{R_L'}{r_e}$$

$$\because R_L' = R_C \| R_L = \frac{3 \times 3}{3+3} = 1.5 \text{ k}\Omega$$

$g_m = I_C/V_T = 1\text{mA}/25\text{mV} = 40$ mA/V

$A_V = -40 \times 1.5 = -60.$

24. Drawing AC equivalent circuit, for AC sources, capacitors are short circuited

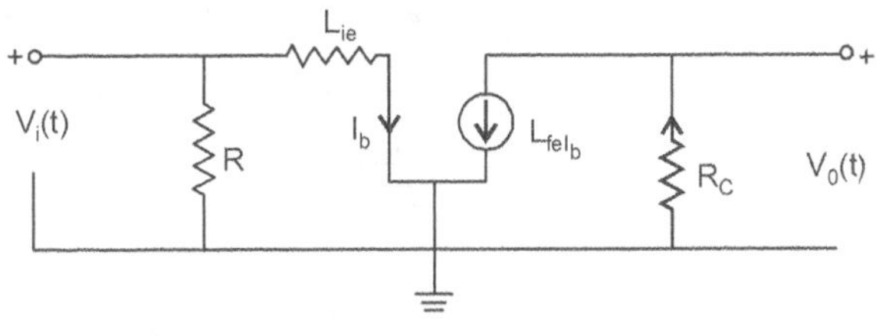

$R = 100$ kΩ ∥ 20 kΩ

$= \dfrac{50}{3}$ kΩ from the circuit

$V_i(t) = +h_{ie} I_b$

and, $V_o(t) = -h_{fe} I_b R_C$

$$= \frac{-h_{fe} R_C}{h_{ie}} v_i(t)$$

$$= \frac{-150(3K\Omega)}{(3K\Omega)}(A\cos 20t + B\sin 10^6 t)$$

$= -150\,[A\cos 20t + B\sin 10^6 t]$

25.

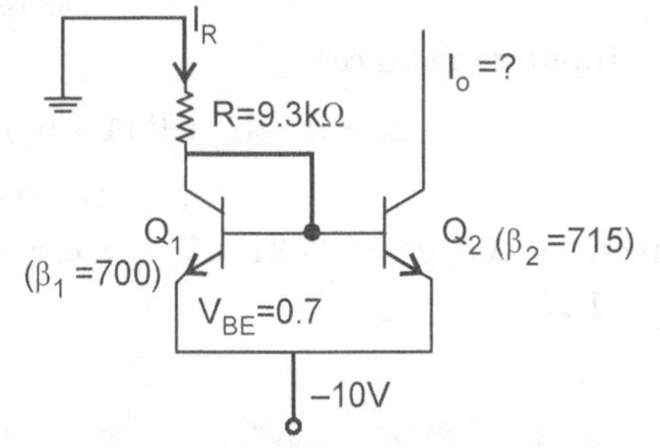

Let both transistors are in active region, therefore voltage at Q_1 base

$(V_{Base})_{Q1} = 0.7 - 10 = -9.3$V

Current through R,

$$I_R = \frac{9.3\text{V}}{9.3\,k\Omega} = 1 \text{ mA} = I_C$$

Since emitter area of $Q_1 = \dfrac{1}{2}$ [Emitter area of Q_2]

i.e. $A_{Q1} = \dfrac{A_{Q2}}{2}$

$\therefore$ $(\beta_2)_{effective} = 2 \times \beta_2 = 1430$

Since effective β of Q_2 is double of Q_1, so collector current also will be double nearly.

$I_0 = I_{C2} \simeq 2 \times I_{C1} \simeq 2$ mA

26. Drawing low-frequency simplified *h*-Parameter model of the given circuit,

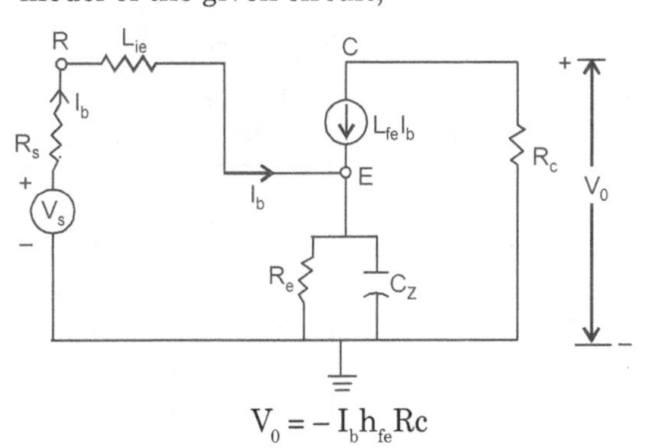

$V_0 = -I_b h_{fe} Rc$

$$= -h_{fe} Rc \frac{V_s}{R_s + h_{ie} + Z_e'}$$

or,
$$A_v = \frac{-v_0}{v_s} = \frac{-h_{fe}Rc}{R_s + h_{ie} + Z'_e}$$

$$Z'_e = \frac{(1+h_{fe})R_e}{1+fwc_2R_e}$$

Input impedance,
$$Z'_i = R_s + h_{ie} + Z'_e$$

When C_2 is disconnected,
$$A'_v = \frac{V_0}{V_s} = \frac{-h_{ef}Rc}{R_s + h_{ie} + R_e(1+h_{fe})}$$
[Decrease]

and Input impedance,
$$Z'_i = R_j + h_{ie} + R_e(1 + h_{fe})$$
[Increase]

27. Given $\beta = 100$, $g_m = 0.3681$ A/V, $r_o = \infty$, $r_\pi = 259\ \Omega$, $R_S = 1$ KΩ

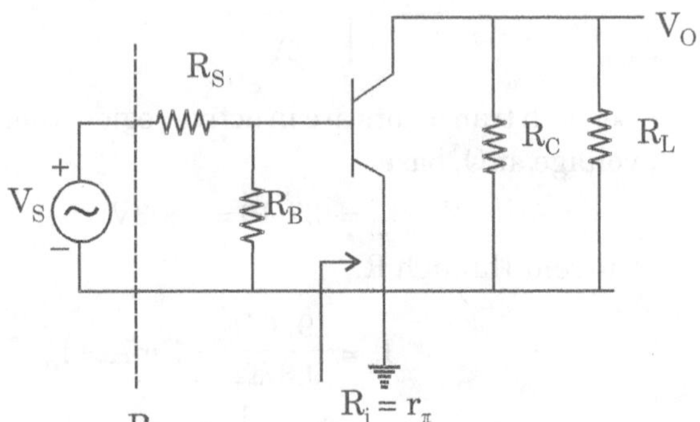

$R_B = 93$ kΩ, $R_C = 250\ \Omega$, $R_L = 1$ kΩ,
$C_1 = \infty$, $C_2 = 4.7\ \mu$F.
AC analysis is used for calculation of resistance (input and output) and gain (voltage and current)
The resistance seen by source,
$$R'_L = R_s + R_B \| r_\pi = 1000 + 93000 \| 259$$
$= 1000 + 258.28 = 1258.28\ \Omega$.

28. Lower cut –off frequency due to C_2

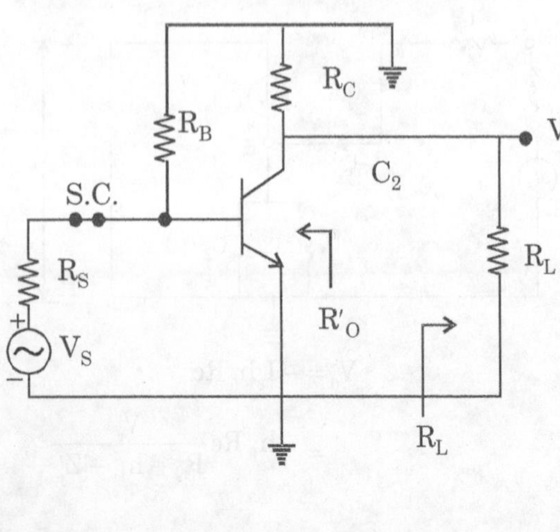

$$f_L = \frac{1}{2\pi(R'_0 + R_L)C_2}$$

where $R'_0 = R_C \| r_0 = 250 \| \infty = 250\Omega$

$$f_L = \frac{1}{2\pi(250+1000)\times 4.7\times 10^{-6}}$$

$f_L = 27.1$ Hz

29. We know that
$$r_\pi = (\beta+1)r_e = (\beta+1)\frac{V_T}{I_e} = (\beta+1)\frac{V_T}{(\beta+1)I_b}$$
$$r_\pi = \frac{V_T}{I_b}.$$
Where I_b is dc current through base so $I_b = 1$ mA.
$V_T = 25$ mV at room temperature.
$$\text{So, } r_\pi = \frac{25\times 10^{-3}}{1\times 10^{-3}} = 25\Omega$$

30. An increase in the base recombination of a BJT, increases the transconductance g_m.

31. $\Rightarrow I_2 = I_s e^{\frac{V_{BE_2}}{\eta V_T}}$, where $V_{BE_2} = V_2$.

$\Rightarrow I_1 = I_s e^{\frac{V_{BE_1}}{\eta V_T}}$, where $V_{BE_1} = V_1$.

$$\frac{I_1}{I_2} = e^{\frac{(V_1 - V_2)}{\eta V_T}}$$

Given : $V_T = 26$ mV at 27°C, $V_T = 27.99$ mV at 50°C.
$I_1 = 80$ mA, $I_2 = 4$ mA.
$\Rightarrow V_1 - V_2 = 83.15$ mV

32.

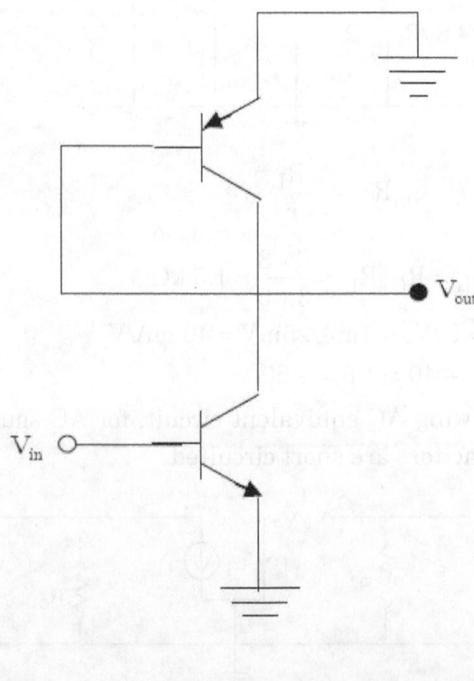

When Base and collector is shorted, it act as a diode.

So, $V_0 = -0.7\text{ V}$.

Gain, $\dfrac{V_0}{V_{in}} = -\dfrac{0.7V}{0.7V} = -1$

33. Composite FET having 2 identical FET's connected in parallel with g_m and r_d is characterised by 2 g_m and $r_d/2$.

34.

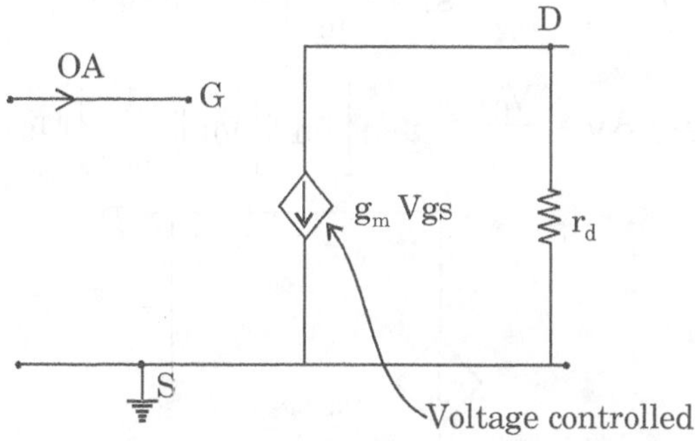

JFET equivalent circuit

35. Drawing small signal model for the given circuit,

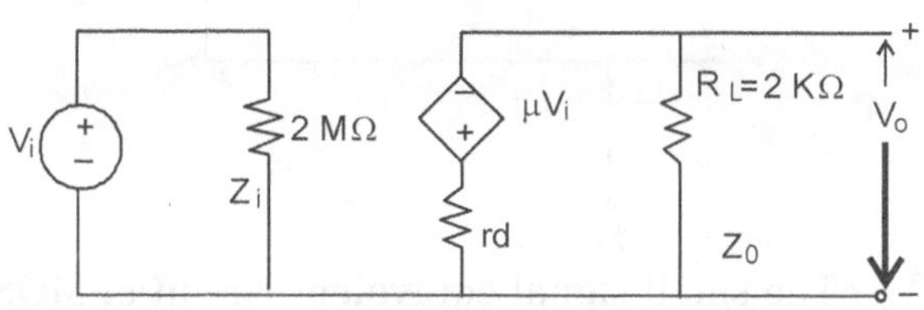

Fig (A)

$$Z_i = 2\text{ m}\Omega$$

$$Z_o = R_L \| rd = 20\text{ k}\Omega \,\|\, 2\text{k}\Omega$$

$$= \frac{(20\text{ k}\Omega)(2\text{ k}\Omega)}{22\text{ k}\Omega} = \frac{20}{11}\text{k}\Omega$$

36. Under DC conditions, capacitors gets open-circuited

$$V_{GS} = -2\text{ volts}$$
$$V_P = -8\text{ volts}$$

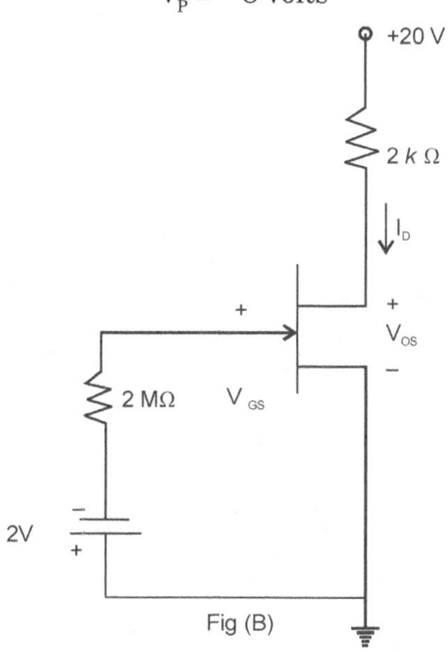

Fig (B)

N-channel MOSFET requires zero or negative gate bias and +ve drain voltage for to operate in active region.

Now, drain current

$$I_D = I_{DSS}\left(1 - \frac{V_{GS}}{V_P}\right)^2$$

$$\Rightarrow \quad I_D = 10 \times 10^{-3}\left(1 - \frac{2}{8}\right)^2 = 5.625\text{ mA}$$

37. Trans conductance

$$g_m = g_{m0}\left(1 - \frac{V_{GS}}{V_P}\right) \quad \text{...(A)}$$

Drain to source current

$$I_{DS} = I_{DSS}\left(1 - \frac{V_{GS}}{V_P}\right)^2 \quad \text{...(B)}$$

$$g_m = \frac{\partial I_{DS}}{\partial V_{GS}} = \frac{-2I_{DSS}}{V_P}\left(1 - \frac{V_{GS}}{V_P}\right) = \frac{-2I_{DSS}}{V_P}\sqrt{\frac{I_{DS}}{I_{DSS}}}$$

$$g_m = \frac{2\sqrt{I_{DSS}\cdot I_{DS}}}{V_P}$$

$$\Rightarrow \quad g_m = \frac{2\sqrt{I_{DSS}\cdot I_{DS}}}{|V_P|} = \frac{2}{8}\sqrt{10\times10^{-3}\times 5.625\times10^{-3}}$$

$$= 1.875\text{ mS}$$

From figure (A),

$$V_o = \frac{-\mu V_i R_L}{R_L + r_d} = \frac{-g_m r_d V_i R_L}{R_L + r_d}$$

$$\Rightarrow \quad \frac{V_o}{V_i} = \frac{-g_m r_d R_L}{R_L + r_d} = -1.875\times10^{-3}\times\frac{20}{11}\times10^3 = -3.41$$

38. From figure, $V_B = V_a$ and $V_S = 0V$.

$V_{GS} = V_G - V_S = V_G - O \Rightarrow V_G = V_B$

$V_D = V_a \Rightarrow V_D - V_S = V_a - V_S$

$V_{DS} = V_{GS}$

$V_{DS} > V_{GS} - V_T$

Hence, MOSFET is in saturation region

$$g_m = \mu_{ncox}\left(\frac{W}{L}\right)(V_{as} - V_T)$$

$$g_m = 10 \times 10^{-6} \times (2 - 1) = 4 \times 10^{-6}\text{ S}$$

$$\frac{\partial V_B}{\partial I_D} = \frac{1}{g_m} = 25\text{ k}\Omega.$$

39.

$$C_{M1} = 50\text{ PF}[1 - A_V]$$
$$A_V = -g_m R_D$$
$$= -0.01 \times 1$$
$$A_V = -10$$

$$C_{Mi} = 50\text{ PF }[1 + 10]$$
$$= 0.55 \times 10^{-9}\text{ F}$$
$$= 0.55\text{ nF}$$

$$f_p = \frac{1}{2\pi R_i C_{mi}}$$

$$f_p = \frac{1}{2\pi \times 5\text{ k} \times 0.55\text{ mF}}$$

$$= \frac{1}{2\pi \times 5 \times 10^3 \times 0.55 \times 10^{-9}}$$

$$= 57.8745\text{ kHz}$$

So, the small signal input pole frequency is 57.8745 kHz.

40. Internal output resistance of MOSFET is

$$r_0 = \left|\frac{1}{\lambda I_D}\right|$$

If $\lambda = 0$, then $r_0 = \infty$.

If $\lambda = 0$, then $r_0 \neq \infty$ (given).

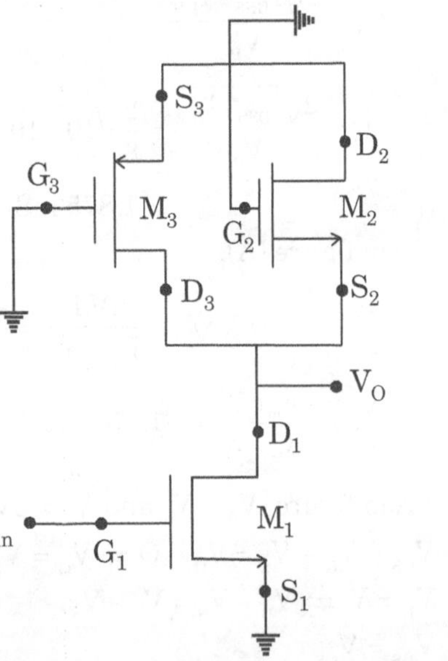

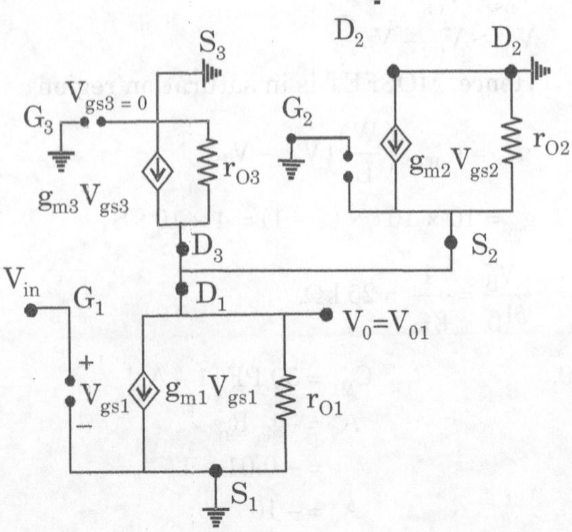

From figure, $V_{gs3} = V_{g3} - V_{S3} = 0 - 0 = 0V$

$V_{in} = V_{gs}$; $V_{01} = V_0 = -V_{gs2}$

Applying KCL at D_1 ($\cong D_3$),

$$gm_1 Vgs_1 + \frac{V_O}{r_{01}} + \frac{V_O}{r_{02}} - g_{m2}V_{gs2} - g_{m3}V_{gs3} + \frac{V_O}{r_{o3}} = 0.$$

$$V_O\left(\frac{1}{r_{01}} + \frac{1}{r_{02}} + g_{m3} + \frac{1}{r_{03}}\right) = -g_{m1}V_{in}$$

$$\Rightarrow A_V = \frac{V_O}{V_{in}} = -g_{m1}\left[r_{01} \parallel \left[r_{01} \parallel \frac{1}{g_{m2}}\right] \parallel r_{03}\right]$$

41.

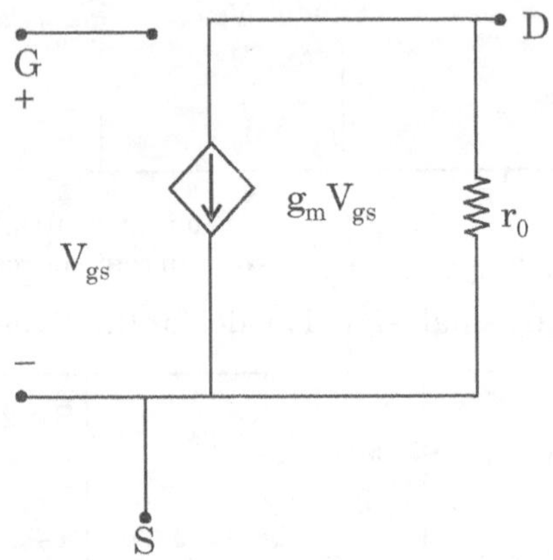

The small signal equivalent circuit of MOSFET in saturation is shown in figure. So, when the channel length modulation effect is significant, the MOSFET can be modelled as a current source with finite output impedance.

42.

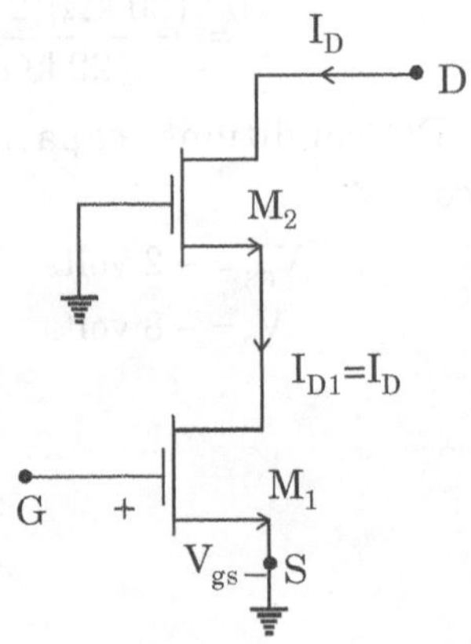

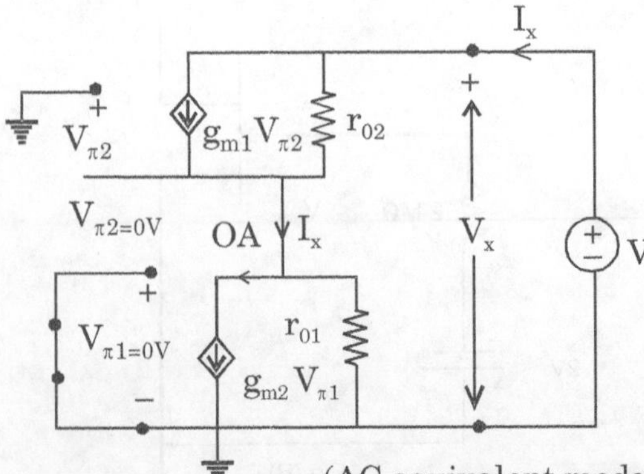

(AC equivalent model)

$\Rightarrow$ From figure, $V\pi_2 = -I_x \cdot r_{01}$

$$I_x = g_{m2} V\pi_2 + \frac{(V_x - I_x . r_{01})}{r_{02}}$$

$$= -g_{m2} r_{01} I_x + \frac{V_x}{r_{02}} - I_x \frac{r_{01}}{r_{02}}$$

$$V_x = r_{02}\left[1 + r_{01}\, g_{m2} + \frac{r_{01}}{r_{02}}\right] I_x$$

$$r_0 = \frac{V_x}{I_x} = r_{01} + r_{02} + r_{01} \cdot r_{02}\, g_{m2}$$

$$r_o = r_{01} \cdot r_{02}\, gm_2$$

43.

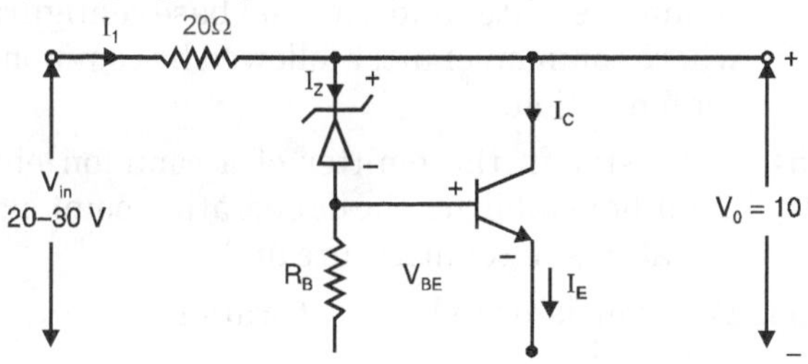

As, $I_1 \max = \frac{V_{in\,max} - V_0}{20} = \frac{30 - 10}{20} = IA$

Also, $I_E = I_Z + I_C$ ...(*i*)

And $I_B = I_Z$

$$\frac{I_C}{I_B} = 99$$

From equation (*i*), $I_E = 100\, I_Z$

From the figure, $I_1 = I_Z + I_C = I_E = 100\, I_Z$

$$\Rightarrow \quad I_Z = \frac{1}{100} = 0.01 \text{ A}$$

$$I_C = 99\, I_Z = 0.99 \text{ A}$$

$$\therefore \quad P_Z = V_Z\, I_Z = 9.5 \times 0.01 \text{ W}$$

$$= 95 \text{ mW}$$

$$P_T = V_C\, I_C = 10 \times 0.99 = \mathbf{9.9\ W}$$

44.

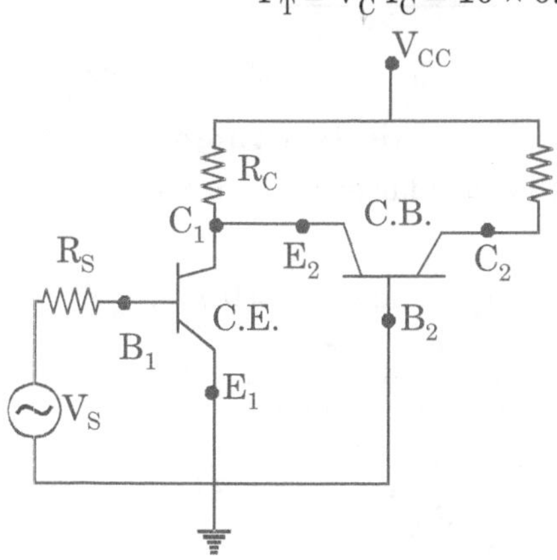

Cascade amplifier is the cascade connection of common emitter and common base configuration where common emitter followed by common base configuration.

45. Transconductance of BJT is $g_m = \left|\frac{I_C}{V_T}\right| . gm\ \alpha Ic.$

Hence if quiescent collector current I_c increases then the transconductance g_m also increases.

46. Option (b) represents a fixed bias circuit without emitter resistance R_E and option (c) represents collector to base bias circuit with emitter resistance R_E.

In BJT, self bias circuit or voltage (potential) divider circuit provides the maximum bias stability because it provides stable I_C, irrespective of variation in temperature and β.

47. Given circuit is self - bias circuit or voltage divider bias circuit . (i) All capacitors are open – circuited (ii) AC voltage source is replaced by short–circuit.

D.C. equivalent circuit

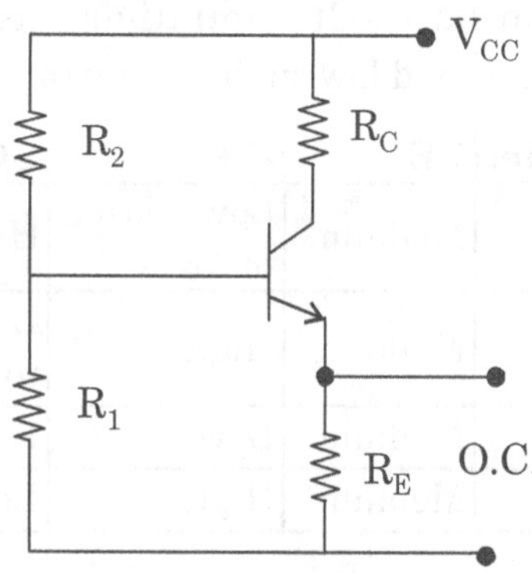

Thevenin equivalent circuit

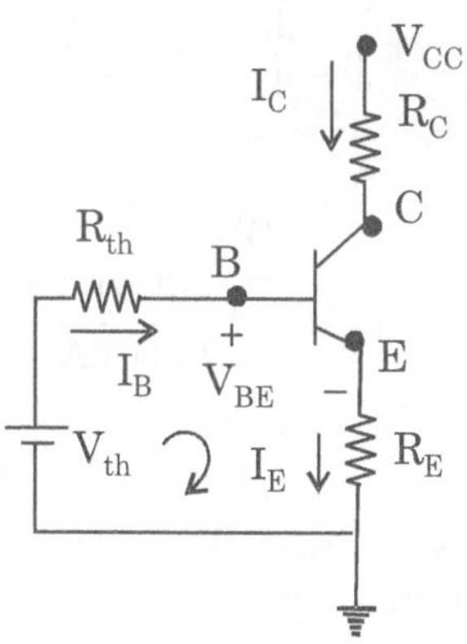

By voltage division rule,

$$V_{th} = Vcc. \frac{R_1}{R_1 + R_2}$$

$$R_{th} = R_1 \| R_2 = \frac{R_1 R_2}{R_1 + R_2} = R_B$$

Applying KVL in input loop,

$$-V_{th} + I_B R_{th} + V_{BE} + I_E R_E = 0$$

$$\because I_B + I_C = I_E$$

$$V_{th} = (R_{th} + R_E)\, I_B + I_C R_E + V_{BE}$$

Differentiating w.r.t to I_C (keeping β and V_{BE} constant)

$$0 = (R_{th} + R_E)\frac{\partial I_B}{\partial I_C} + R_E + 0$$

$$\frac{\partial I_B}{\partial I_C} = -\frac{R_E}{R_E + R_B}.$$

Stability factor's

$$= \frac{(1+\beta)}{1 - \frac{\partial I_B}{\partial I_C}} = \frac{1+\beta}{1-\beta\left(\frac{-R_E}{R_E+R_B}\right)} = \frac{(1+\beta)\left(1+\frac{R_B}{R_E}\right)}{\frac{R_B}{R_E}+(1+\beta)}$$

For better stability, s≅1. Hence,

$$\frac{R_B}{R_E} \ll 1 \Rightarrow \frac{R_E}{R_B} \gg 1.$$

48. Common base (CB) amplifier has low input impedance and low current gain (α)

Paramet	CE	CB	CC
A_I	Medium	Low (closed to unity)	High
A_V	Medium	High	Low (closed to unity)
R_i	Medium	Low	High
R_o	Medium	High	Low

49. $I_I = I_B + I_C = 0.025 + 1.0 = 1.025$ mA

$$\beta = \frac{I_C}{I_B} = \frac{1}{0.025} = 40$$

For CE, $$r_m = \frac{V_T(mV)(b+1)}{1.025(mA)} = \frac{26}{1.025} = 1000\ \Omega$$

50. ⇒ As the base width of the BJT is reduced then the recombination current (base current I_B) decreases as a result collector current (I_C) increases so, the current gain of the BJT ($\alpha = I_C/I_E$) increases.

⇒ If the emitter doping concentration to base doping concentration ratio is reduced then the emitter injection efficiency decreases, so the current gain (α) of the BJT reduces.

⇒ If the collector dropping concentration is increased then the breakdown (V_{BR}) of a BJT will be reduced.

(A)→2, (B)→ 3, (C)→ 1.

51. When both the emitter junction and collector junction are in forward biased then BJT is said to be operating in saturation.

52. Cascade amplifier : is used when large bandwidth is required It is wide band amplifier.

Differential amplifier : provides high common mode voltage rejection

Darlington pair : has very high input impedance and very high current gain.

(A)→2, (B)→ 5, (C)→ 4.

53. Cascade amplifier is the cascade connection of common emitter and common base configuration where common emitter followed by common base configuration.

54. A resistor in the emitter of a common emitter amplifier stabilizes the dc operating point against variations in temperature and β.

55. For transistor to be in saturation mode,

$(I_B)_{min} \le I_B$

$\therefore \frac{I_C}{\beta dc} \le I_B$

$I_C \le \beta dc.\ I_B$

56. CB < CE < CE (order of input resistance)

57.

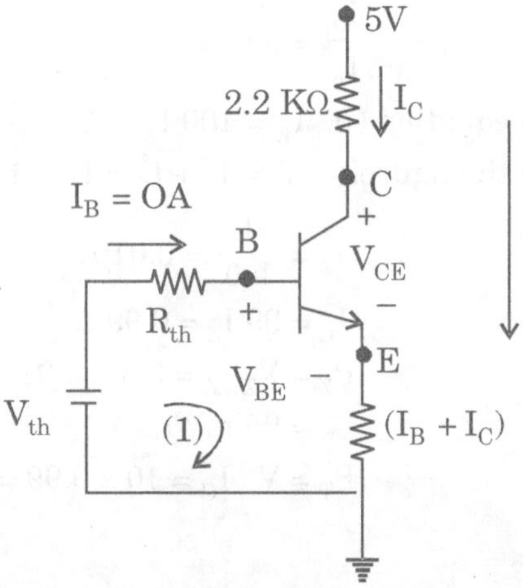

Thevenin equivalent circuit

Given: b is vary large, (β = ∞). So, I_B = OA, $I_C = I_B$.

$R_1 = 4$ kΩ, $R_2 = 1$ kΩ, $V_{BE} = 0.7$ V

$$\Rightarrow V_{th} = \frac{R_2}{R_1+R_2}.5 = \frac{5\times 1}{4+1} = 1\text{ V}.$$

$$\Rightarrow R_{th} = \frac{R_1 \times R_2}{R_1+R_2} = \frac{4\times 1}{4+1} = 0.8\text{ k}\Omega$$

Applying KVL in loop (1).

$-V_{th} + (0 \times R_{th}) + V_{BE} + [R_E \times (0 + I_C)] = 0$

$V_{th} = V_{BE} + (R_E\ I_C)$

$$I_C = \frac{V_{th} - V_{BE}}{R_E} = \frac{1-0.7}{0.3} = 1\text{ mA}.$$

Applying KVL in loop (2),

$-5 + (I_C \times 2.2) + V_{CE} + 0.3 \times (0 + I_C) = 0$

$5 = I_C \times 2.2 + V_{CE} + 0.3 \times I_C$

$V_{CE} = 5 - (2.2 + 0.3) = 2.5$ V

58. When two terminals of a transistor are shorted if acts as diode.

$$I = I_0\left(e^{\frac{VD}{\eta VT}} - 1\right) = 10^{-13}\left(\frac{0.7}{e^{1\times 26\times 10^{-3}}} - 1\right) = 49 \text{ mA.}$$

59.

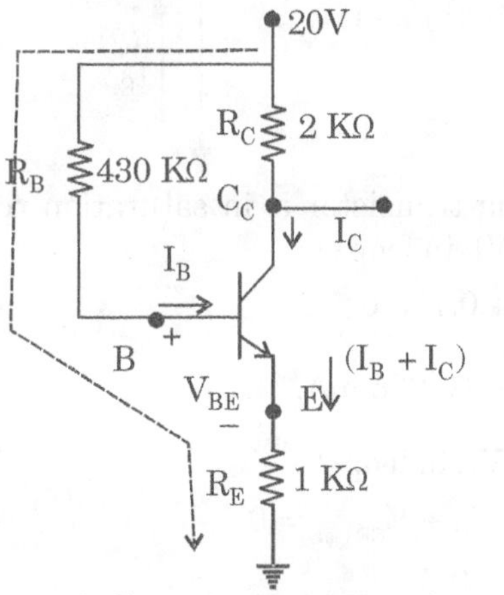

DC equivalent circuit

Given: $\beta = 50$, $V_{BE} = 0.7$ V.

Applying KVL in loop,

$-20 + (I_BR_B) + V_{BE} + R_E(I_C + I_B) = 0$

$20 = 430\, I_B + 0.7 + 1 \times (\beta I_B + I_B)$

$$I_B = \frac{20 - 0.7}{430 + (\beta + 1) \times 1} = \frac{19.3}{430 + 51}$$

$I_B = 0.04$ mA $= 40\ \mu$A

$I_C = \beta I_B = 50 \times 40\ \mu\text{A} = 2$ mA.

From the output circuit,

$V_C = 20 - I_CR_C = 20 - 2 \times 2 = 16$ V.

60. Apply KVL at the BE junction

$$I_E = \frac{-5 - 0.7 + 10}{4.3\text{ k}\Omega}$$

$$= \frac{4.3}{4.3\text{ k}\Omega} = 1 \text{ mA}$$

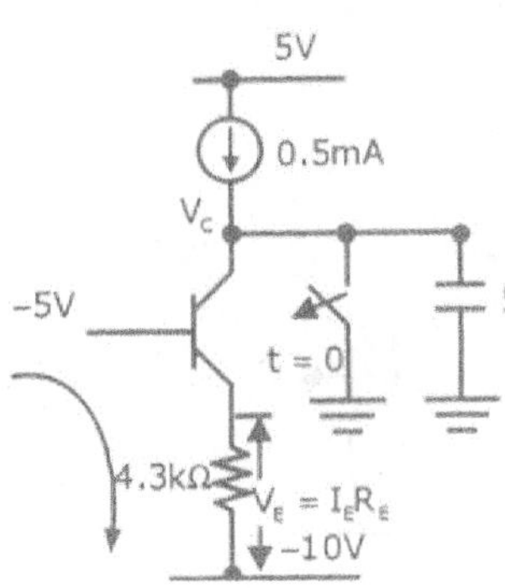

Always $I_E = 1$ mA;

At collector junction

$I_{cap} + (0.5 \text{ mA}) = 1$ mA (since $\beta = \infty$; $I_E = I_C$)

$I_{cap} = 1 - 0.5$

$= 0.5$ mA (always constant)

$V_{CE} = V_C - V_E$

$\Rightarrow$ $V_C = V_{CE} + V_E$

$= 0.7 + (4.3) \times 10^3 \times 10^{-3}$

$= 0.7 + 4.3$ (since $V_E = I_ER_E$)

$V_C = 5\text{V} = V_{cap}$

$$V_{cap} = I_{cap}\frac{t}{C}$$

or $$t = \frac{V_{cap}(C)}{I_{cap}} = \frac{(5) \times 5 \times 10^{-6}}{0.5 \times 10^{-3}}$$

$= 50$ ms

61.

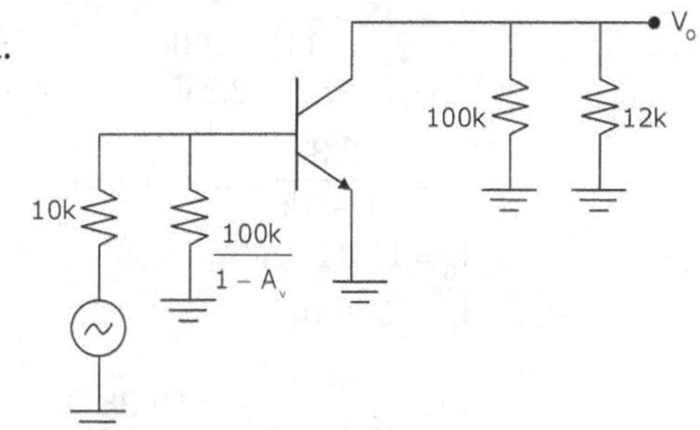

KVL in put loop,

$13.7 - (I_C + I_B)12\text{k} - 100\text{k}(I_B) - 0.7 = 0$

$\Rightarrow$ $I_B = 9.9\ \mu$A;

$I_C = \beta I_B = 0.99$ mA;

$I_E = 1$ mA

$\therefore$ $$r_e = \frac{26 \text{ mA}}{I_E} = 26\ \Omega;$$

$z_i = \beta r_e = 2.6$ kΩ;

$\therefore$ $$A_V = \frac{(100\text{k} \,\|\, 12\text{k})}{26} = 412$$

$$z_i' = z_i \,\|\left(\frac{100\text{k}}{1 + 412}\right) = 221\ \Omega;$$

$$A_{vs} = A_v\frac{z_i'}{z_i' + R_s}$$

$$= (412)\left(\frac{221}{221 + 10\text{k}}\right)$$

$|A_{vs}| \approx 10$

62. Given : $\beta = \infty$, $I_B = 0$, $I_C = I_E = 1$ mA, $V_{BE} = 0.7$ V.

$V_E = R_EI_E = 1 \times 10^{-3} \times 500 = 0.5$ V.

$V_{R2} = V_{BE} + V_E = 0.7 + 0.5 = 1.2$ V.

It is a self – bias circuit. So, $V_{R_2} = \dfrac{R_2}{R_1 + R_2}.V_{CC}$

$$1.2 = \frac{R_2}{60 + R_2} \times 3$$

$72 + 1.2R_2 = 3R_2$

$R_2 = 40$ kΩ.

63.

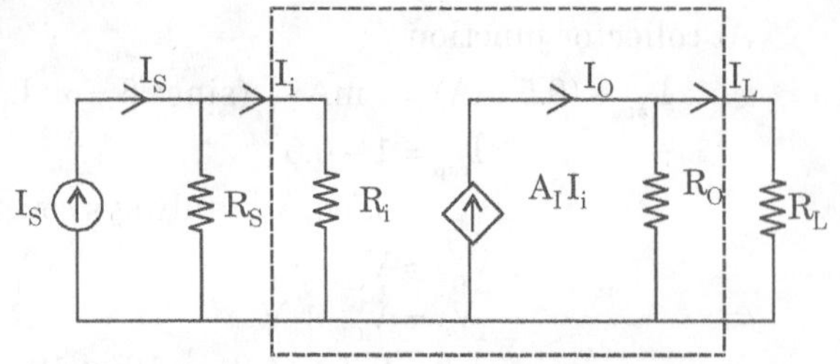

Current buffer or current amplifier circuit is shown in figure. for ideal current amplifier, $I_i = I_s$ [$\because R_s >> R_i$]

So value of R_i should be low, and $I_L = I_O$

[$\because R_O >> R_L$]

64. $V_{BE} = 0.7$ V, $\beta = 200$, $V_T = 25$ mV

DC Analysis:

$$V_B = 12 \times \frac{11k}{11k + 33k} = 3V$$

$$V_E = 3 - 0.7 = 2.3V$$

$$I_E = \frac{2.3}{10 + 1k} = 2.277 \text{ mA}$$

$$I_B = 11.34 \text{ A}$$

$$I_C = 2.26 \text{ mA}$$

$$r_e = \frac{25\,\text{mV}}{2.277\,\text{mA}} = 10.98\Omega$$

$$A_V = \frac{V_0}{V_i}$$

$$= \frac{-\beta R_C}{\beta r_e + (1+\beta)(R_s)}$$

$$= \frac{-200 \times 5K}{200 \times 10.98 + (201)10}$$

$$A_V = -237.76$$

65. KVL in base loop gives,

$$I_B = \frac{10 - 0.7}{100K}$$

$$= 93\ \mu A$$

$\Rightarrow$ $I_C = \beta I_B = 50 \times 93\ \mu A$

$= 4.65$ mA

from figure, $V_0 = I_C R_C$

$\Rightarrow$ $R_C = \dfrac{V_0}{I_C}$

$$= \frac{5V}{4.65mA}$$

$$= 1.075\ \Omega$$

66. Given : $\beta = 50$ $V_{CE\ (Sat)} = 0.2$ V

$V_{BE} = 0.7$ V, this value can be used for both active and saturation region.

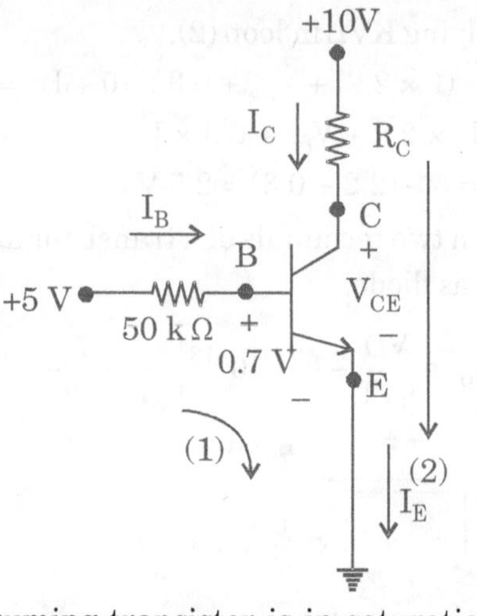

$\Rightarrow$ Assuming transistor is in saturation region, applying KVL in loop (1)

$- 5 + 50\ I_B + 0.7 = 0$

$$I_B = \frac{5 - 0.7}{50} = 0.086 \text{ mA.}$$

Applying KVL in loop (2),

$- 10 + I_{C(sat)} R_C + V_{CE\ (sat)} = 0$

$$I_{C(sat)} = \frac{10 - 0.2}{R_C}$$

Condition of saturation is $I_{B(min)} \le I_B$, where $I_{B\ (min)}$ $= \dfrac{I_{C\ (sat)}}{\beta}$

$$\frac{I_{C(sat)}}{\beta} \le I_B$$

$$\frac{10 - 0.2}{50R_C} \le I_B \Rightarrow \frac{9.8}{50R_C} \le 0.086$$

$R_C \ge 2.279$ KΩ.

Above condition of R_C is satisfied only by option (B), i.e., for $R_C = 3$ KΩ, the BJT operates in saturation region.

For $R_C = 1$ kΩ, BJT will operate in active region.

Fro $R_C = 20$ kΩ, BJT will still be saturation region.

67. For unbypassed R_E, $R_i = \beta r_e + (1+ \beta) R_E$ and

$$A_V = \frac{A_I R_L}{R_i}$$

68.

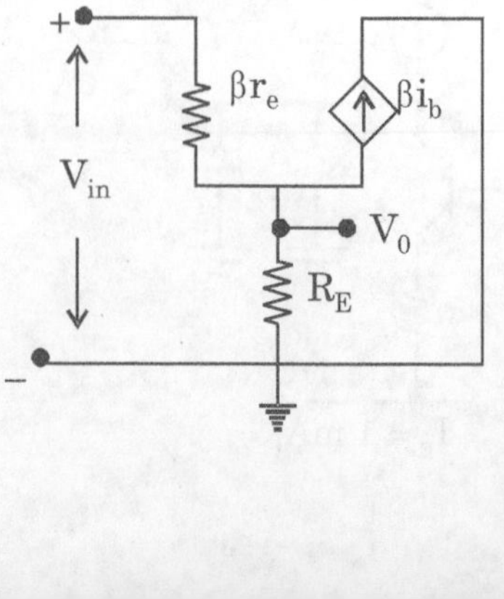

$V_{in} = \beta_{re} i_b + (1+\beta) i_b R_E$

$V_0 = (1+\beta) R_E i_b$

Then, $\dfrac{V_0}{V_{in}} = \dfrac{(1+\beta)R_E}{\beta_{re} + (1+\beta)R_E}$

$\dfrac{V_0}{V_{in}} = \dfrac{\beta R_E}{\beta r_e + \beta R_E} = \dfrac{R_E}{r_e + R_E}$

The condition for small signal voltage gain to be nearly constant is

$R_E >> r_e$

$R_E >> V_T / I_C$

$I_C R_E >> V_T$

69. $\because$ β = high, I_B is neglected

$\therefore \quad V_B = 12 \times \dfrac{10k}{10k + 5k} = 8V$

$V_E = V_B - 0.7 = 7.3V$

$\therefore \quad V_{CE} = 12 - 7.3 = 4.7V$

$\therefore$ Maximum undistorted

$V_0(p-p) = 2 \times 4.7\ V = 9.4\ V$

70. The input impedance of CB amplifier is $Z_i = r_e = 50\Omega$.

$\dfrac{V_T}{I_C} = 50$

$\therefore \quad I_C = \dfrac{25\ mV}{50\ \Omega} = 0.50\ mA.$

71. $V_{EB} = 0.7\ V$

$I_B = 0.0383\ mA$

$I_C = 1.916\ mA$

72. Ebers-Moll model is valid for all the region of operation of BJT.

73. $\Rightarrow R_2 = \dfrac{V_T}{I_2} \ln\left(\dfrac{I_1}{I_2}\right)$

$= \dfrac{26 \times 10^{-3}}{100 \times 10^{-6}} \ln\left(\dfrac{1 \times 10^{-3}}{100 \times 10^{-6}}\right) = 598.67\Omega$

74. Here $V_{E_1} = (2.5 - 0.7) = 1.8\ V$

$V_{B_2} = V_{E1} - V_{EB2}$

$= (1.8 - 0.7) = 1.1\ V$

$I_{B_2} = \dfrac{V_{B_2} - 1}{10k} = \dfrac{1.1 - 1}{10k} = \dfrac{0.1}{10k}$

$I_{C2} = \beta I_{B2} = 50\left[\dfrac{0.1}{10k}\right]$

$\therefore \quad V_{C2} = I_{C2}(1k)$

$= \dfrac{50(0.1)}{10k}(1k) = 0.5\ V$

Hence, the value of the dc voltage V_{C2} is 0.5V.

75.

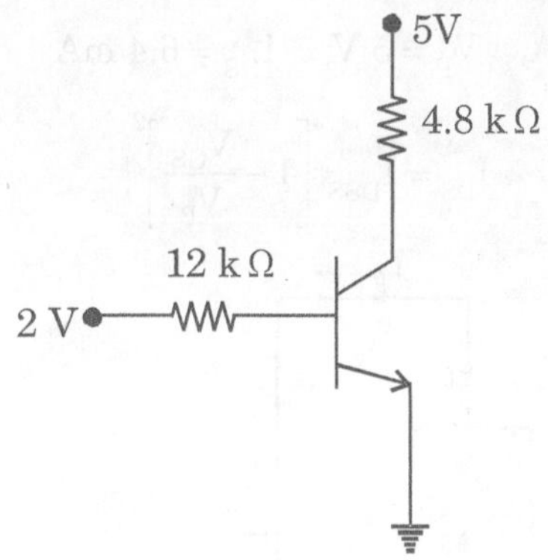

$I_B = \dfrac{2 - 0.7}{12} = 0.10833\ mA.$

$I_{c(sat)} = \dfrac{5 - 0.2}{4.8} = 1\ mA$

$I_B \geq I_{B(min)} = \dfrac{I_{c(sat)}}{\beta}$

$I_{B\geq} \dfrac{1mA}{\beta} \Rightarrow \beta \geq \dfrac{1}{0.10833}; \beta_{min} = 9.23.$

$\alpha_{min} = \dfrac{\beta_{min}}{1 + \beta_{min}} = 0.902$

76.

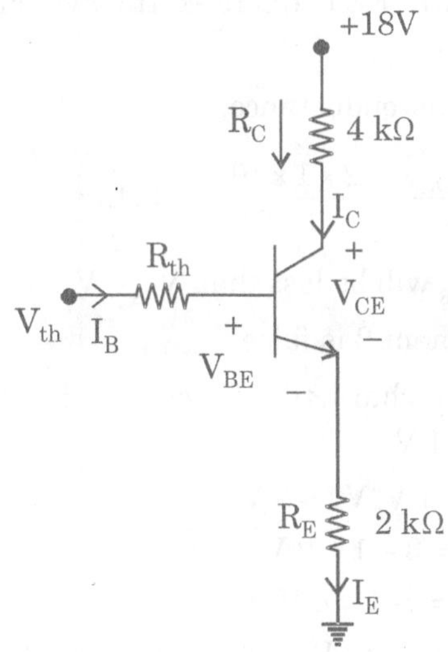

Thevenin's equivalent circuit

$V_{th} = \dfrac{16}{16 + 44} \times 18 = 4.8V.$

$I_E R_E = V_{th} - V_{BE} - I_B R_{th} \qquad \because \alpha = 1$

$\because I_B = 0A$

$I_E R_E = 4.8 - 0.8 = 4\ V.$

$I_E = \dfrac{4}{2}\ mA = 2\ mA.$

$I_C = I_E = 2\ mA.$

$V_{CE} = V_{CC} - I_C R_C - I_E R_E = 18 - 2 \times 4 - 2 \times 2$

$V_{CE} = 6\ V.$

77. $I_{DSS} = 10$ mA, $V_P = 5$ V; $I_{DS} = 6.4$ mA

$$I_{DS} = I_{DSS}\left[1 - \frac{V_{GS}}{V_P}\right]^2$$

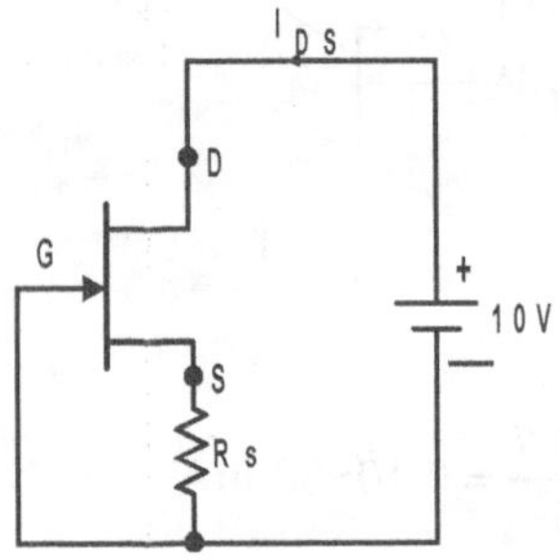

or $$6.4 = 10\left[1 - \frac{V_{GS}}{V_P}\right]^2$$

or $$V_{GS} = 0.2V_P = 0.2 \times 5 = 1 \text{ volt}$$

$\therefore$ $$I_{DS} \cdot R_s = 1 \text{ volt}$$

Hence $$R_s = \frac{1}{6.4 \times 10^{-3}} = 156\,\Omega$$

78. The transit time of a current carries through the channel of an FET decides its switching characteristics.

79. Maximum transconductance,

$$g_{m\,(max)} = \frac{2I_{DSS}}{|V_P|} = \frac{2 \times 1 \times 10^{-3}}{1-51} = 0.4\text{ms.}$$

80. If $V_1 = 0$V, V_{DS} will be less than $V_{GS} - V_T$.
Hence, Statement 2 is false.

81. From transfer characteristic curve, threshold voltage, $V_{th} = 1$ V.
$V_G = 3$ V, $V_S = 1$ V, $V_D = 5$ V.
$V_{GS} = V_a - V_S = 3 - 1 = 2$ V
$V_{DS} = V_D - V_S = 5 - 1 = 4$ V
Over – drive voltage, $V_{OV} = V_{GS} - V_{th} = 2 - 1 = 1$ V
Hence, $V_{GS} > V_{th}$ and $V_{DS} > V_{GS} - V_{th}$.
Therefore, MOSFET is in saturation region.

82. $$g_m = \left.\frac{\partial i_D}{\partial V_{GS}}\right|_{V_{DS}}$$

and, $$I_D = I_{DSS}\left(1 - \frac{V_{GS}}{V_P}\right)^2$$

$$g_m = \frac{\partial I_D}{\partial V_{GS}} = -\frac{2I_{DSS}}{V_p}\left(1 - \frac{V_{GS}}{V_P}\right)$$

g_m will be maximum when $V_{GS} = 0$ and given by,

$$g_{m0} = -\frac{2I_{DSS}}{V_p}$$

83. Assuiming both PMOS and NMOS are in saturation

Then, $$I_{D1} = \frac{\mu_{P_1} C_0 \times W}{2L}(V_{GS1} - V_{Tn})^2$$
[For NMOS transistor]

$$= \frac{40}{2}(2.5 - 1)^2$$

$$= 20 \times 1.5^2 = 45\,\mu\text{A}$$

Now, for PMOS transistor

$$I_{D2} = \frac{\mu_p C_0 \times W}{2L}(V_{GS2} - V_{TP})^2$$

$$= \frac{40}{2}(5 - 2.5 - 1)^2 = 45\,\mu\text{A}$$

As $I_{D1} = I_{D2}$, both transistors are in saturation and

$$I = I_{D1} = I_{D2} = 45 \text{ A}$$

84. $$\frac{1}{g_m} = \frac{1}{g_{m1}} + \frac{1}{g_{m2}}$$

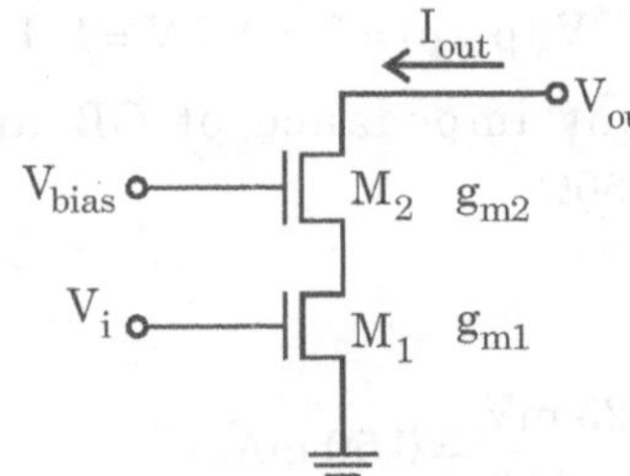

$\therefore$ M_2 is always saturated due to bias but g_{m1} changes according to V_i

$$g_m = \frac{g_{m1}\,g_{m2}}{g_{m1} + g_{m2}}$$

$$= \frac{g_{m1}\,g_{m2}}{g_{m2}\left(1 + \frac{g_{m_1}}{g_{m_2}}\right)}$$

But $$g_{m2} >> g_{m1}$$

$\therefore$ $$g_m = g_{m1}$$

85. In situation region, drain current for M1

$$I_{D1} = \frac{\mu_n C_{ox} W}{2L}(V_{GS_1} - V_T)^2 = I_{Bias}$$

and for M_2,

$$I_{D2} = \frac{\mu_n C_{ox} W}{2L}(V_{GS_2} - V_T)^2$$

As $V_{GS1} = V_{GS2}$ for this circuit and the two transistors are identical,

$$I_{D1} = I_{D2} = I_{bias}$$

or, $$I_x = I_{bias}$$

86. When V_G is little higher than 1 V,

$\Rightarrow$ For n–MOSFET :

$V_{GSn} = V_a = 1 + hv$; h is small positive quantity

$V_{GSn} - V_{Tn} = h$

$\Rightarrow$ For p – MOSFET:

$V_{sGp} = V_{DD} - 3 = 2$ V; $V_{sG} - |V_{TP}| = 1$ V

$(V_{GSn} - V_{Tn}) < (V_{sGp} - |V_{TP}|)$

So, n–MOSFET will be in saturation region and p–MOSFET will be in triode region.

87. When $V_G = 1.5$ V:

$\Rightarrow$ For n–MOSFET : $V_{aGn} = V_G = 1.5$ V and

$V_{aGn} - V_{th} = 0.5$ V.

$\Rightarrow$ For p-MOSFET : $V_{sGp} = V_{DD} - 3 = 2$ V and

$V_{sap} - |VTp| = 2 - 1 = 1$V.

$(V_{GSn} - V_{Tn}) < (V_{sGP} - |V_{TP}|)$

So, n – MOSFET will be in saturation region and p–MOSFET will be in triode region.

For V_o: $I_{DSn} = I_{DSp}$

$k_n (V_{Gsn} - V_{Tn})^2 = k_p[2(V_{sGp} - |V_{Tp}|)V_{SDD} - V^2_{SDP}]$

Given that, $k_n = k_p$

$(0.5)^2 = 2\times1 (V_{DD} - V_0) - (V_{DD} - V_0)^2$

$0.25 = 10–2V_0 - 25 + 10 V_0 - V_0^2$

$V_0^2 - 8V_0 + 15.25 = 0$

$$V_0 = 4 \pm \sqrt{\frac{3}{2}}\text{V.} = 4.866 \text{ V or } 3.134 \text{ V}$$

Check for valid V_0 :

$\Rightarrow$ We know that n–MOSFET is in saturation region and p–MOSFET is in triode region so, $V_0 \geq V_{Gsn} - V_{Th}$

Here, both possibilities of V_o satisfies this

$\Rightarrow$ For $V_o = 4.866$ V.

$[V_{SDp} = 5–V_0 = 0.314 \text{ V}] < [V_{saGp} - |V_{Tp}| = 1 \text{ V}]$

$\Rightarrow$ For $V_0 = 3.14$ V

$[V_{sDp} = 5 - V_0 = 0.866\text{V}] > [V_{sGp} - |V_{Tp}| = 1\text{V}]$

The valid answer of V_0 is 4.866 or $4 + \frac{\sqrt{3}}{2}$V.

88. For upper MOS, $V_{DS} = 6 - V_x$

$V_{Gs} - V_T = 5 - V_x - 1 = 4 - V_x$

Upper MOS will be in saturation because $V_{DS} > VGs - V_T$

For lower MOS, $V_{DS} = V_x$ and $V_{as} - V_T = V_x - 1$

$V_{DS} > V_{Gs} - V_T$

Lower MOS will be in saturation.

$$\Rightarrow I_{D_1} = \mu.c_{ox}.\left(\frac{W}{L}\right)^2 (V_{Gs} - V_T)^2 = \mu n.c_{ox} 2^2 (4 - Vx)^2$$

$I_{D2} = \mu n.\ c_{ox}\ 1^2 (V_X–1)^2$

But $I_{D1} = I_{D2}$

$4 (4 - V_X)^2 = (V_X - 1)^2$

$V_X = 3$ Volt

89. For p– MOS, $V_{sG} = V_s - V_a = 5 - V_{in}$

For p–MOS to be ON, $V_{sG} > |V_{Tp}| \rightarrow 5 - V_{in} > 1$

$V_{in} < 4$ Volt

So, V_{in} must be less than 4 V for MOS to be in linear agion, hence option (c) and (d) are incorrect We know that for small V_{in} output is high and pMOS is in linear agion and NMOS is in cut off region. Similarly, for high V_{in}, PMOS is in cut off and NMOS is in linear region and for V_{in} in between both are in saturation.

So, PMOS will be in linear region for $V_{in} < 1.875$ V.

90. In a MOSFET operating in the saturation region the channel length modulation effect cuses a decrease in the output resistance.

91. For proper operation, $V_{DS} = V_{Gs} - V_T$

$V_D - V_S = V_G - V_s - V_T \Rightarrow V_D = V_G - V_T$.

At P : $V_D = V_G - V_T = 5 - 1 = 4$ V

At Q : $V_D = V_G - V_T = 5 - 1 = 4$ V

At R : $V_D = V_G - V_T = 5 - 1 = 4$ V

$\because$ Va = 5 V and $V_T = 1$ V For all transistors.

According to the given options, the only condition for the voltage at P,Q and R is only in option (c)

92. Given $V_{Th} = 0.8$ V

When $V_D = 1.6$ V,

$I_D = 0.5$ mA

$$= \frac{1}{2}\mu_n \cos\frac{W}{L}(V_{DS} - V_{Th})^2$$

[$\because$ Device is in saturation]

$$\Rightarrow \quad \frac{1}{2}\mu_n \cos\frac{W}{L} = 0.78125 \times 10^{-3}\ \text{A/V}^2$$

When $V_D = 2$V

$$I_D = \frac{1}{2}\mu_n \cos\frac{\omega}{L}(V_{DS} - V_{Th})^2$$

$= 078125 \times 10^{-3}(2–0.8)1.125$ mA

93. Given : $|V_T| = 2$ V and $K_n = \frac{1}{2}\mu n C_{ox}\left(\frac{W}{L}\right)$

$= 0.1\text{mA}/\text{V}^2$

$\Rightarrow I_{D1} = I_{D2} = Kn\ (V_{as} - V_T)^2 = 0.1\ (5–2)^2 = 0.9$ mA.

94. Transistor m_1 switch from saturation to linear

$$\Rightarrow \quad V_{DS} = V_{GS} - V_T;$$

where $V_{DS} = V_0$

and $V_{GS} = V_i$

$$\therefore \quad V_{DS} = V_0 = V_i - V_T$$

Drain current $I_D = \frac{1}{2}\mu_n \cos\frac{W}{L}(V_{GS} - V_T)^2$

$$\frac{V_{DD} - V_o}{10\text{ K}} = \frac{1}{2}\times 100\times 10^{-6}\times 2(V_{GS} - 0.5)^2$$

$$\frac{2-(V_i - 0.5)}{10\text{ K}} = 100\times 10^{-6}(V_i - 0.5)^2$$

$$\Rightarrow \quad V_i = 1.5\text{V}$$

95. Lower transistor (M_1) to work in saturation

$$V_{DS1} \geq V_{GS1} - V_+$$

So, for minimum V_{DD}

$$V_{DS1} = V_{GS1} - V_+$$

$$V_{DS1} = 2 - 1 = 1\text{V}$$

$$V_{DS1} = V_{D1} - V_{S1}$$

$$1\text{V} = V_{D1} - 0$$

$$\therefore \quad V_{D1} = 1\text{V}$$

and $I_{D1} = K'(V_{GS1} - V_+)^2$

$$V_{D1} = \frac{1\text{ mA}}{V^2}\times(2-1)^2 = 1\text{ mA}$$

Now transistor M_2, $V_{DG} = 0V$

So, it will work into saturation region and same current will flow

$$I_{D2} = I_{D1} = K'(V_{GS2} - V_+)^2$$

$$1\text{ mA} = \frac{1\text{ mA}}{V^2}\times(V_{DD}-1-1)^2 \quad (\because V_{S2} = V_{D1})$$

$$\therefore \quad V_{DD} = 3\text{V}$$

96.

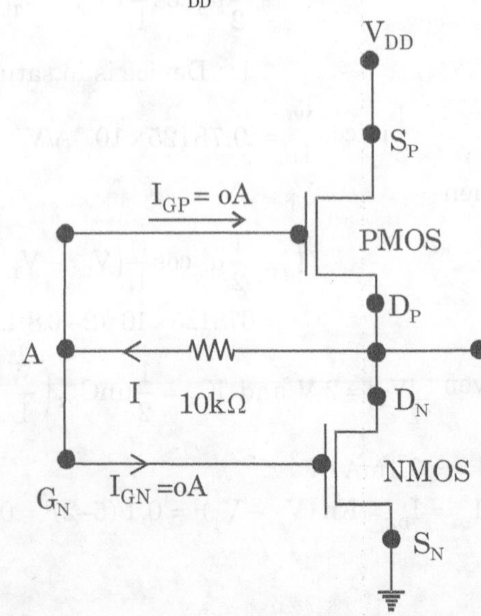

Due to very large input impedance gate current of both PMDS will be zero, i.e., $Ia_p = Ia_N = 0A$.

From circuit, $I = Ia_p + Ia_N = 0A$.

Voltage across 10 kΩ resistor

$= V_{out} - V_A = 10\,I = 0.$

$\Rightarrow V_A = V_{out}, V_A = Va_p = Va_N = V_{out}.$

⇒ Since, drain of PMOS is shouted to gate of PMOS, hence PMOS is operating in saturation region.

⇒ Since drain of NMOS is short region to gate of NMOS, hence NMOS is operating in saturation region.

⇒ It is clear that when PMOS and NMOS both are is saturation region then output will be switching threshold of the inverter because of output voltage is switching from

$$\frac{V_{DD}}{2} + V_{th} \text{ to } \frac{V_{DD}}{2} - V_{th}$$

97. $V_{Gs} = \frac{8\times 5}{8} - 1\times I_D = 5 - I_D$

$I_D = 5 - V_{Gs}$ ....(i)

$$I_D = \frac{\mu_n c_{ox}}{2}\left(\frac{W}{L}\right)(V_{Gs} - V_T)^2$$

$$5 - V_{Gs} = \frac{1}{2}(V_{Gs} - 1)^2$$

$$10 - 2V_{Gs} = V_{Gs}^2 + 1 - 2V_{Gs}$$

$$V_{Gs} = 3\text{ V}.$$

$$I_D = 5 - 3 = 2\text{ mA}.$$

98.

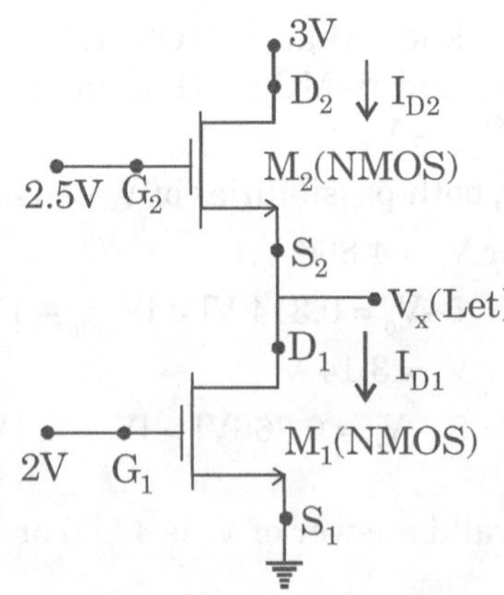

For transistor M_2 : $V_{DS2} = V_{D2} - V_{S2} = 3 - V_x$

$V_{GS2} = V_{G2} - V_{s2} = 2.5 - V_x$

Overdrive voltage, $V_{oV2} = V_{Gs2} - V_{th} = 2.5 - V_x - 1$

$V_0V_2 = 1.5 - V_x$

Since, $V_{Gs2} > V_{th}$ and $V_{DS2} \geq V_{OV2}$

So transistor M_2 is in saturation.

For transistor M_1 : Assuming, M_1 is also in saturation.

$V_{DS1} = V_{D1} - V_{S1} = V_X - 0 = V_X$

$V_{Gs1} = V_{G1} - V_{S1} = 2 - 0 = 2$ V ....(i)

$V_0V_1 = V_{GS1} - V_{th} = 1$ V ....(ii)

Since, transistors are connected in sales. Hence,

I_{D2} (sat) = I_{D1} (sat)

$K_n (V_{Gs2} - V_{th})^2 = K_n (V_{GS1} - V_{th})^2$

$(2.5 - V_x - 1)^2 = (2 - 1)^2 \Rightarrow 1.5 - V_x = \pm 1$

$V_x = 0.5$ V or 2.5 V.

Case 1 : For $V_x = 2.5$ V

$V_{Gs2} = V_{G2} - V_{s2} = V_{G2} - V_x = 2.5 - 2.5$

$V_{s2} = 0V < V_{th}$

$\Rightarrow M_2$ is in cut off region, $V_X \neq 2.5$ V

Case 2 : For $V_x = 0.5$ V

(i) For transistor M_1:

$V_{Gs2} = V_{G2} - V_{S2} = V_{G2} - V_x$

$V_{Gs2} = 2.5 - 0.5 = 2V.$

$V_{DS2} = V_{D2} - V_{S2} = 3 - 0.5 = 2.5$ V

$V_{OV2} = V_{Gs2} - V_{th} = 2 - 1 = 1V$

$V_{s2} > V_{th}$ and $V_{DS2} > V_{OV2}$. M2 $\rightarrow$ saturation

(i) For transitor M_1 : $V_{Gs1} = 2 - 0 = 2V$

$V_{DS1} = V_x - 0 = 0.5$ V ....(iii)

From eq. (i), (ii) & (iii), $V_{Gs1} > V_{th}$ & $V_{DS1} < V_{OV1}$

Our assumption is wrong and hence, M_1 is in linear region.

99. Given : $\left(\frac{W}{L}\right)_2 = 2\left(\frac{W}{L}\right)_1 \Rightarrow k_{n2} = 2k_{n1}$

For M_1, $V_{Gs1} - V_T = 2 - 1 = 1V.$

For M_2, $V_{Gs2} - V_T = 2-1 - V_x = 1-V_x < 1$ Volt

$V_{DS2} = (3.3 - V_x) > (V_{Gs2} - V_T)$

So, M_1 will be in linear region an M_2 will be in saturation region.

$I_{D1} = I_{D2}$

$$k_{n1}\left[2(V_{Gs1} - V_T)V_{DS1} - V_{DS1}^2\right] = K_{\eta 2}(V_{as2} - V_T)^2$$

$$k_{n1}\left[2(2-1)V_x - V_x^2\right] = 2K_{n1}(2 - Vx - 1)^2$$

$$2V_x - V_x^2 = 2\left(1 + V_x^2 - 2V_x\right) = 2V_x^2 - 4V_x + 2$$

$$3V_x^2 - 6V_x + 2 = 0$$

$$V_x = 1 \pm \sqrt{\frac{4 - 8/3}{4}} = 1 \pm \sqrt{\frac{1}{3}}V.$$

$\Rightarrow V_{Gs2} = (2 - V_X) \geq V_T$

$(1 - V_x) \geq 0$

So, Valid answer is $V_X = 1 - \sqrt{\frac{1}{3}} = 0.4226$ Volt

100.

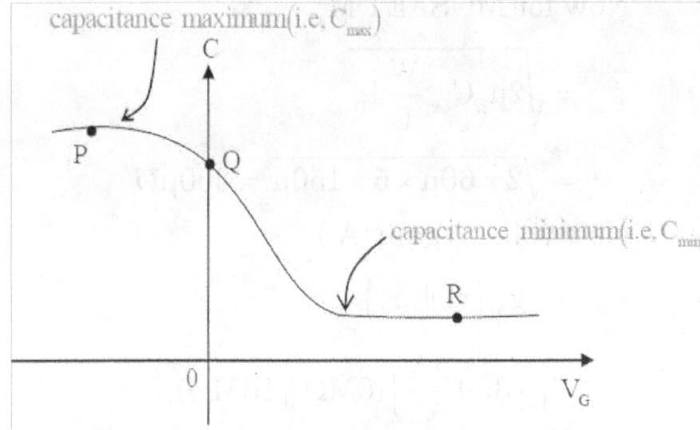

At higher frequency C_{min} minimum capacitance obtained in inversion-regime so point R-belongs to inversion regime.

Maximum capacitance obtained in accumulation regime so point p-line is accumulation regime.

For $\phi_m = 0V$, flat band occurs at $V_G = 0V$, so point ϕ lies at flat-band regime.

101.

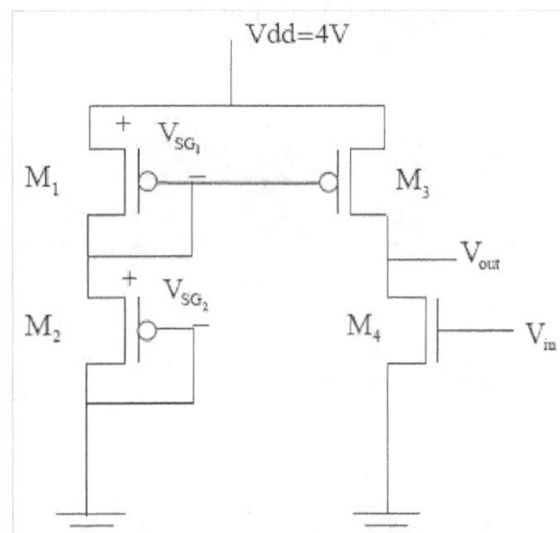

M_1 and M_2 will have equal current flowing also since they are identical $\left(\frac{W}{L}\right)_1 = \left(\frac{W}{L}\right)_2$

$\therefore \quad V_{SG_1} = V_{SG_2}$ also, By KVL in loop

$4 = V_{SG_1} + V_{SG_2}$

$\therefore V_{SG_1} = V_{SG_2} = 2$ V

$\therefore$ Current through M_1 and M_2 is given by

$$I = \left(\frac{1}{2}\right)(\mu_p C_{ox})\left(\frac{W}{L}\right)\left[V_{SG} - |V_T|\right]^2$$

$$= \left(\frac{1}{2}\right)(30)(10)(2-1)^2$$

$$= 15(10)(1)^2 = 0 = 150\ \mu A$$

M_1 and M_3 are matched with same $\left(\frac{W}{L}\right)$ and same V_{SG}

$\therefore \quad I_{M_3} = I_{M_4} = 150\ \mu A$

Now for MOSFET M_4,

$$g_m = \sqrt{2\mu_n C_{ox}\frac{W}{L}I_D}$$

$$= \sqrt{2\times 60\mu\times 5\times 150\mu} = 300\mu℧$$

Now voltage gain (A_v)

$$= -g_m\left(r_d \parallel r_d\right)$$

$$= \left(-300\frac{\mu A}{V}\right)(6M\Omega \parallel 16M\Omega)$$

$$= \left(-300\frac{\mu A}{V}\right)(3M\Omega) = -900V/V$$

102.

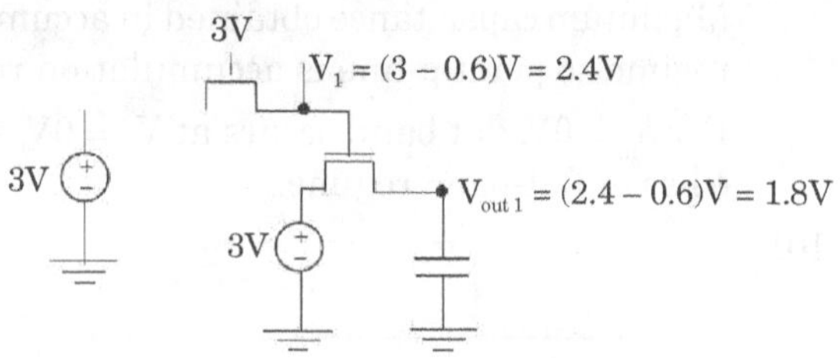

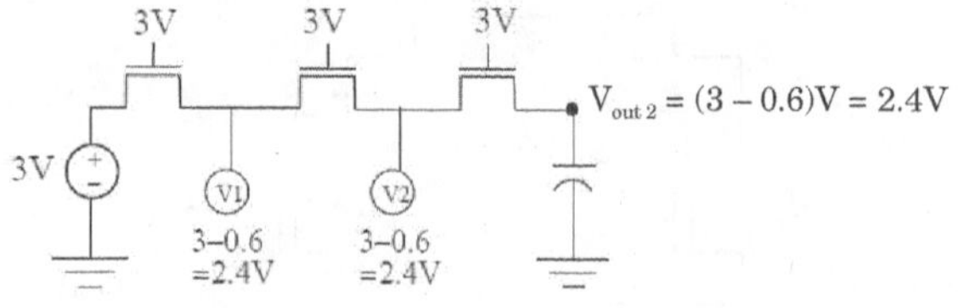

103.

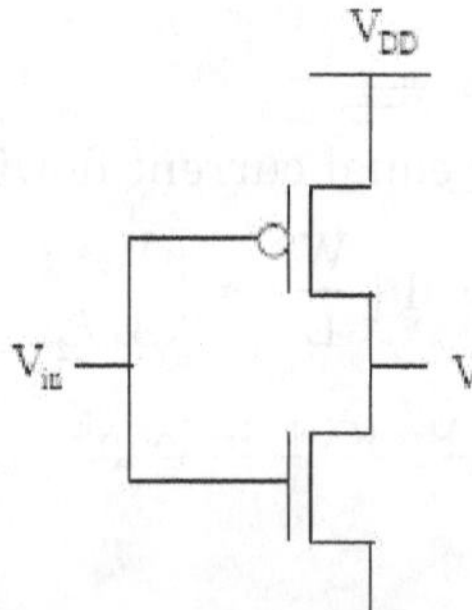

At $V_i = \frac{V_{DD}}{2}$, both the MOSFETs are in saturation and both MOSFETs have the same current.

$$\left(\frac{V_{dd}}{2}\right)$$
$$\uparrow$$

$$\therefore \left(\frac{1}{2}\right)(100\mu A/V^2)\left(\frac{W}{L}\right)_n [1.5-0.7]^2$$

$$= \left(\frac{1}{2}\right)(40\mu A/V^2)\left(\frac{W}{L}\right)_P (3-1.5-0.9)^2$$

$$\frac{\left(\frac{W}{L}\right)_n}{\left(\frac{W}{L}\right)_p} = \frac{(40)(0.6)^2}{(100)(0.8)^2} = \frac{40\times 0.36}{100\times 0.64} = 0.225$$

104.

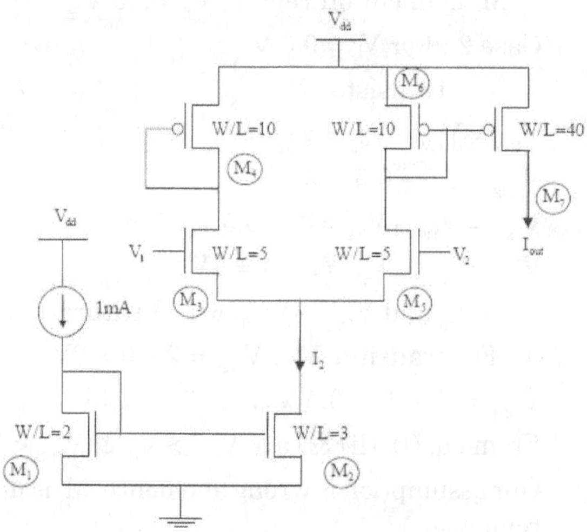

M_1 and M_2 have the same V_{GS}

$\therefore$ Current flows in the ratio of W/L

$$\therefore I_2 = \left(\frac{3}{2}\right)^{(1\,mA)} = 1.5\text{ mA}$$

$V_1 = 0$; Therefore M_3 is in cut off and entire I_2 current flows through M_5 branch.

$$\therefore I_5 = 1.5\text{ mA}$$

$$I_{out} = \left(\frac{40}{10}\right)^{[I_5]}$$

$$\therefore I_{out} = 4\times 1.5\text{ mA} = 6\text{ mA}$$

3
CHAPTER

Frequency Response of Amplifiers

Analysis of Previous GATE Papers

Topics ↓ / Year →			2019	2018	2017 Set 1	2017 Set 2	2016 Set 1	2016 Set 2	2016 Set 3	2015 Set 1	2015 Set 2	2015 Set 3
LOW & HIGH FREQUENCY RESPONSE OF AMPLIFIERS USING BJT & FET	1 Mark	MCQ Type						1				
		Numerical Type										
	2 Marks	MCQ Type			1	1						
		Numerical Type										
		Total			2	2		1				
MULTISTAGE AMPLIFIERS	1 Mark	MCQ Type										
		Numerical Type										
	2 Marks	MCQ Type										
		Numerical Type										
		Total										
MILLER-EFFECT	1 Mark	MCQ Type			1							
		Numerical Type										
	2 Marks	MCQ Type										
		Numerical Type										
		Total			1							

LOW & HIGH FREQUENCY RESPONSE OF AMPLIFIERS USING BJT & FET

1. For the amplifier circuit of fig., the transistor has a β of 800. The mid-band voltage gain V_o/V_T of the circuit will be

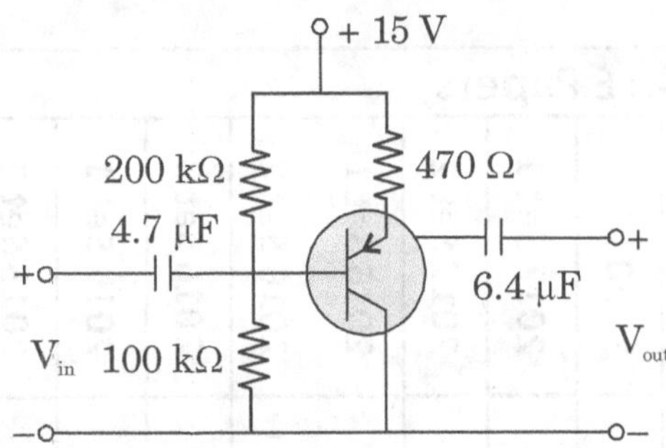

(a) 0 (b) < 1

(c) ≈1 (d) 800

[1993 : 1 Mark]

2. α cut-off frequency of a bipolar junction transistor

(a) increase with the increase in base width

(b) increase with the increase in emitter width

(c) increases with the increase in collector width

(d) increase with decrease in the base width

[1993 : 2 Marks]

3. In order to reduce the harmonic distortion in an amplifier its dynamic range has to be _____.

[1994 : 1 Mark]

4. From measurement of the rise time of the output pulse of an Amplifier whose input is a small amplitude square wave, one can estimate the following parameter of the amplifier

(a) gain-bandwidth product

(b) slew Rate

(c) upper 3-dB frequency

(d) lower 3-dB frequency

[1995 : 1 Mark]

5. A distorted sinusoid has the Amplitude, A_1, A_2, A_3, ... of the fundamental, second harmonic, third harmonic, distortion is _____ respectively. The total harmonic

(a) $\frac{A_2 + A_3 + ...}{A_1}$ (b) $\frac{\sqrt{A_2^2 + A_3^2 + ...}}{A_1}$

(c) $\frac{\sqrt{A_2^2 + A_3^2 + ...}}{\sqrt{A_1^2 + A_2^2 + A_3^2 ...}}$ (d) $\frac{(A_2^2 + A_3^2 + ...)}{A_1}$

[1995 : 1 Mark]

6. In the differential amplifier of the figure, if the source resistance of the current source I_{EE} is infinite, then the common-mode gain is

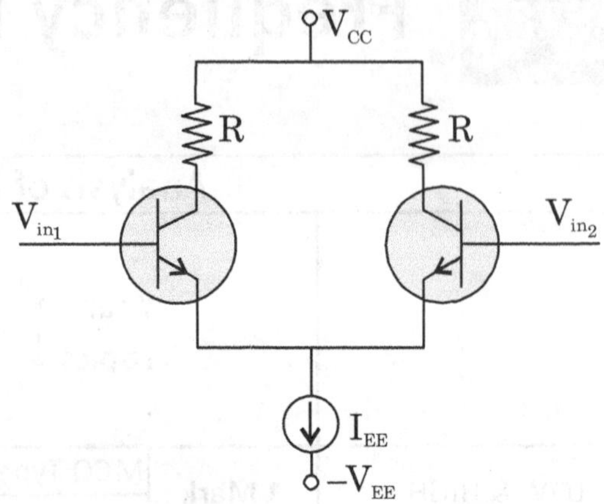

(a) zero

(b) infinite

(c) indeterminate

(d) $\frac{V_{in_1} + V_{in_2}}{2V_T}$

[1996 : 1 Mark]

7. Generally, the gain of a transistor amplifier falls at high frequencies due to the

(a) internal capacitances of the device.

(b) coupling capacitor at the input.

(c) skin effect.

(d) coupling capacitor at the output.

[1996 : 1 Mark]

8. In an ideal differential amplifier shown in the figure, a large value of (R_E).

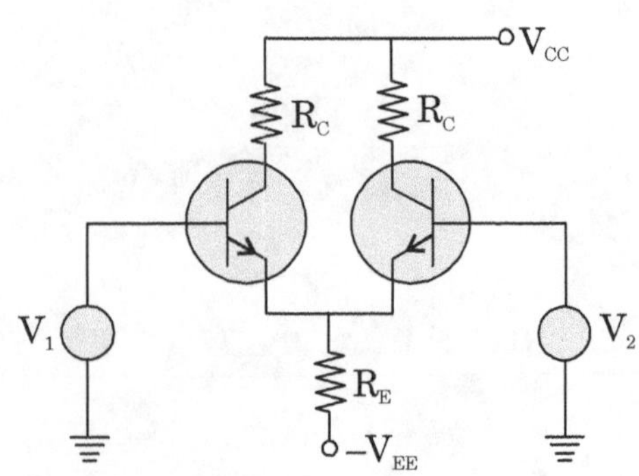

(a) increases both the differential and common- mode gains

(b) increases the common-mode gain only

(c) decreases the differential-mode gain only

(d) decreases the common-mode gain only

[1997 : 1 Mark]

9. In the circuit shown below, capacitors C_1 and C_2 are very large and shorts at the input frequency. v_i is a small input. The gain magnitude $|v_o/v_i|$ at 10 Mrad/s is

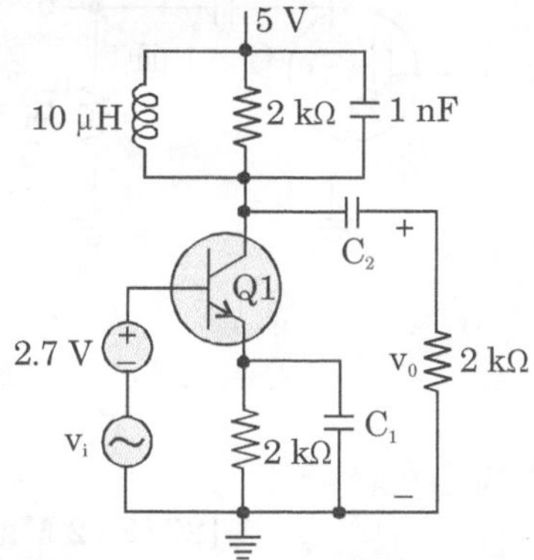

(a) maximum (b) minimum

(c) unity (d) zero **[1998 : 1 Mark]**

10. For the DC analysis of the Common-Emitter amplifier shown, neglect the base current and assume that the emitter and collector currents are equal. Given that $V_T = 25$ mV, $V_{BE} = 0.7$ V, and the BJT output resistance r_0 is practically infinite. Under these conditions, the midband voltage gain magnitude, $A_v = |v_0/v_i|$ V/V, is ______.

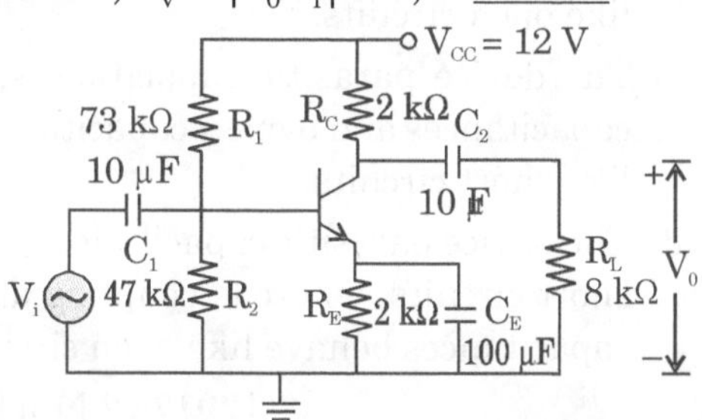

[1999 : 1 Mark]

11. In the circuit shown, transistors Q_1 and Q_2 are biased at a collector current of 2.6 mA. Assuming that transistor current gains are sufficiently large to assume collector current equal to emitter current and thermal voltage of 26 mV, the magnitude of voltage gain V_0/V_s in the mid-band frequency range is ______ (up to second decimal place).

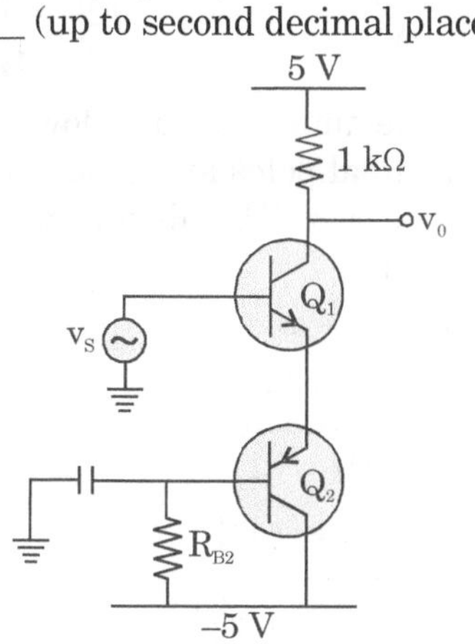

12. In the MOSFET amplifier of the figure is the signal output V_1 and V_2 obey the relationship

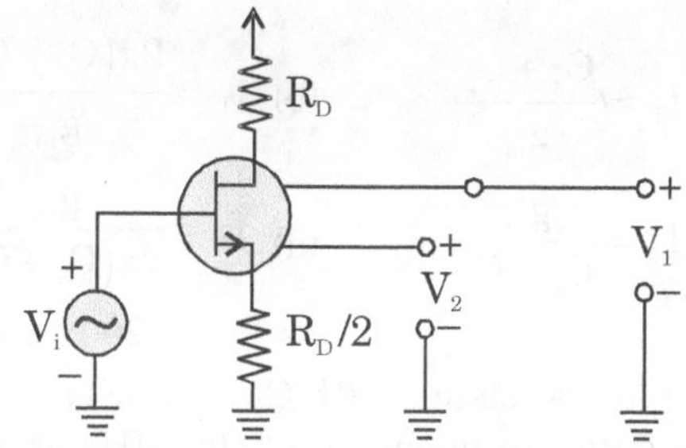

(a) $V_1 = \frac{V_2}{2}$ (b) $V_1 = -\frac{V_2}{2}$

(c) $V_1 = 2V_2$ (d) $V_1 = -2V_2$

[1999 : 1 Mark]

13. The voltage gain $A_v = \frac{V_0}{V_i}$ of the JFET amplifier shown in the figure is

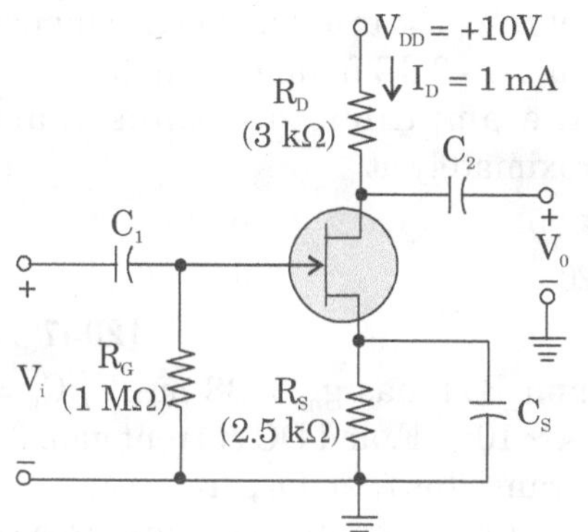

(Assume C_1, C_2 and C_3 to be very large.)

(a) +18 (b) –18

(c) +6 (d) –6

[2000 : 2 Marks]

14. An RC-Coupled Amplifier is assumed to have a single-pole low frequency transfer function. The maximum lower-cutoff frequency allowed for the Amplifier to pass 50 Hz. Square wave with no more than 10% tilt is ______. **[2000 : 1 Mark]**

15. An Amplifier has an open-loop gain of 100 and its lower and upper-cut-off frequency of 100 Hz and 100 kHz respectively, a feedback network with a feedback factor of 0.99 is connected to the amplifier. The new lower-and upper-cut-off frequencies are at ______ and ______.

[2001 : 1 Mark]

16. An npn transistor has a beta cut-off frequency f_β of 1 MHz and Common-emitter short-circuit low frequency current gain β_0 of 200 at unity gain frequency f_T and the alpha cut-off frequency f_α respectively are

(a) 200 MHz, 201 MHz (b) 200 MHz, 1999 MHz

(c) 199 MHz, 200 MHz (d) 201 MHz, 200 MHz

[2002 : 1 Mark]

17. The f_T of a BJT is related to its g_m, C_n and C_μ as follows

(a) $f_T = \dfrac{C_\pi + C_\mu}{g_m}$ (b) $f_T = \dfrac{2\pi(C_\pi + C_\mu)}{g_m}$

(c) $f_T = \dfrac{g_m}{C_\pi + C_\mu}$ (d) $f_T = \dfrac{g_m}{2\pi(C_\pi + C_\mu)}$

[2004 : 1 Mark]

18. An npn transistor (with C = 0.3 pF) has a unity-gain cutoff frequency f_T of 400 MHz at a dc bias current I_c = 1 mA. The value of its C_μ (in pF) is approximately (V_T = 26 mV)

(a) 15 (b) 30

(c) 50 (d) 96

[2005 : 2 Marks]

19. An amplifier is assumed to have a single-pole high-frequency transfer function. The rise time of its output response to a step function input is 35 nsec. The upper –3 dB frequency (in MHz) for the amplifier to a sinusoidal input is approximately at

(a) 4.55 (b) 10

(c) 20 (d) 28.6

[2007 : 2 Marks]

20. An npn BJT has g_m = 38 mA/V, $C_\mu = 10^{-14}$ F, $C_\pi = 4 \times 10^{-13}$ F, and DC current gain β_0 = 90. For this transistor f_T and f_β, are

(a) $f_T = 1.64 \times 10^8$ Hz and 1.47×10^{10} Hz

(b) $f_T = 1.47 \times 10^{10}$ Hz and $f_\beta = 1.64 \times 10^8$ Hz

(c) $f_T = 1.33 \times 10^{12}$ Hz and $f_\beta = 1.47 \times 10^{10}$ Hz

(d) $f_T = 1.47 \times 10^{10}$ Hz and $f_\beta = 1.33 \times 10^{12}$ Hz

[2013 : 2 Marks]

21. A bipolar transistor is operating in the active region with a collector current of 1 mA. Assuming that the p of the transistor is 100 and the thermal voltage (V_T)is 25 mV, the transconductance (g_m) and the input resistance (r_π) of the transistor in the common emitter configuration, are

(a) g_m = 25 mA/ V and r_π = 15.625 kΩ

(b) g_m = 40 mA/ V and r_π = 4.0 kΩ

(c) g_m = 25 mA/ V and r_π = 2.5 kΩ

(d) g_m = 40 mA/ V and r_π = 2.5 kΩ

[2016 : 1 Mark, Set-2]

22. The ac schematic of an NMOS common-source stage is shown in the figure below, where part of the biasing circuits has been omitted for simplicity. For the n-channel MOSFET M, the transconductance g_m = 1 mA/V, and body effect and channel length modulation effect are to be neglected. The lower cut-off frequency in Hz of the circuit is approximately at

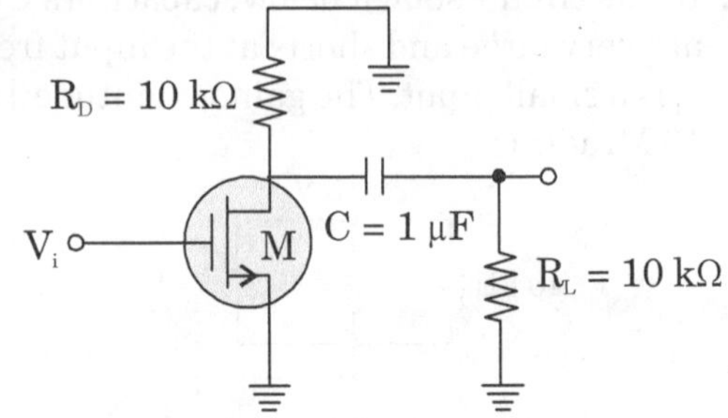

(a) 8

(b) 32

(c) 50

(d) 200

[2017 : 2 Marks, Set-2]

23. Which one of the following statements is correct about an ac-coupled common-emitter amplifier operating in the mid-band region?

(a) The device parasitic capacitances behave like open circuits, whereas coupling and bypass capacitances behave like short circuits.

(b) The device parasitic capacitances, coupling capacitances and bypass capacitances behave like open circuits.

(c) The device parasitic capacitances, coupling capacitances and bypass capacitances behave like short circuits.

(d) The device parasitic capacitances behave like short circuits, whereas coupling and bypass capacitances behave like open circuits.

[2017 : 2 Marks, Set-1]

MULTISTAGE AMPLIFIERS

24. The bandwidth of an n-stage tuned Amplifier, with each stage having a bandwidth of B, is given by

(a) B/n (b) $B/\sqrt{n}$

(c) $B\sqrt{2^{1/n} - 1}$ (d) $\dfrac{B}{\sqrt{2^{1/n} - 1}}$

[1993 : 1 Mark]

25. A multistage amplifier has a low-pass Response with three Real poles at S = ω_1, $-\omega_2$ and ω_3. The approximate overall bandwidth B of the amplifier will be given by

(a) $B = \omega_1 + \omega_2 + \omega_3$

(b) $\dfrac{1}{B} = \dfrac{1}{\omega_1} + \dfrac{1}{\omega_2} + \dfrac{1}{\omega_2}$

(c) $B = (\omega_1 + \omega_2 + \omega_3)^{1/3}$

(d) $B = \sqrt{\left(\omega_1^2 + \omega_2^2 + \omega_3^2\right)}$

[1998 : 1 Mark]

26. Three identical RC-coupled transistor amplifiers are cascaded. If each of the amplifiers has a frequency response as shown in the figure, the overall frequency response is as given in

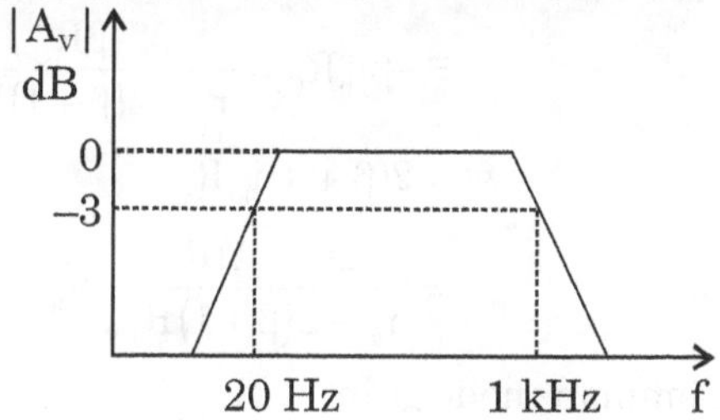

(a)
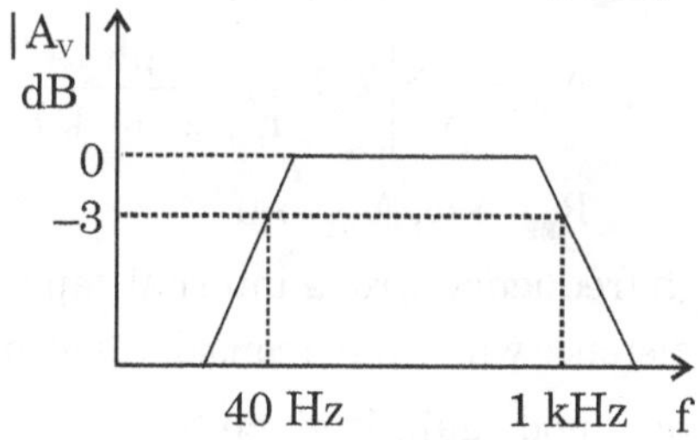

(b)
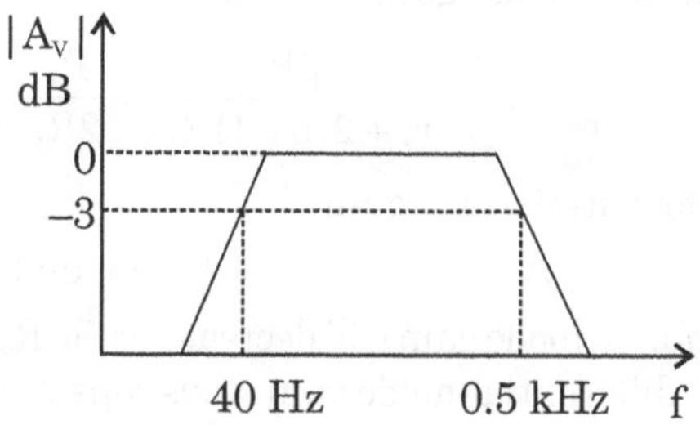

(c)
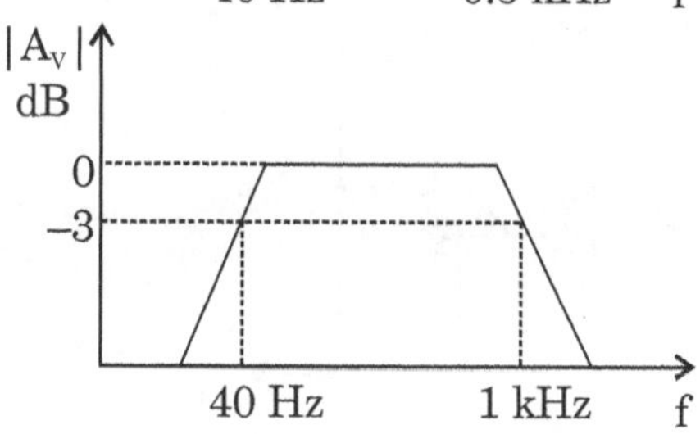

(d)
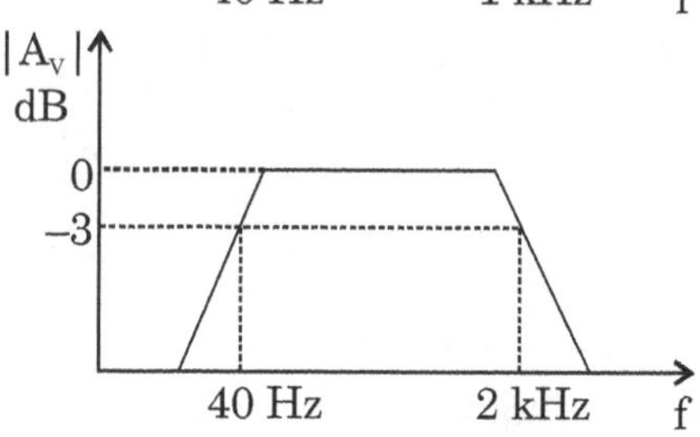

[2001 : 1 Mark]

27. The cascode amplifier is a multistage configuration of

(a) CC-CB
(b) CE-CB
(c) CB-CC
(d) CE-CC

[1993 : 1 Mark]

28. A cascade connection of two voltage amplifiers A, and A_2 is shown in the figure. The open-loop gain A_{VO}, input resistance R_{in}, and output resistance R_0 for A_1 and A_2 are as follows:

$A_1 : A_{v_0} = 10$, $R_{in} = 10\ k\Omega$, $R_0 = 1\ k\Omega$

$A_1 : A_{v_0} = 5$, $R_{in} = 5\ k\Omega$, $R_0 = 200\ k\Omega$

The approximate overall voltage gain vout/vin is

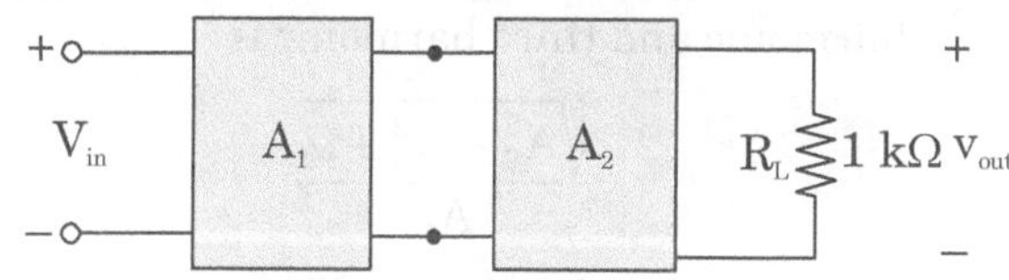

[2005 : 1 Mark]

29. In a multi-stage RC-Coupled Amplifier the coupling capacitor

(a) limits the low frequency response
(b) limits the high frequency response
(c) does not effect the frequency response
(d) blocks the d.c components without effecting the frequency response.

[2014 : 1 Mark, Set-2]

MILLER-EFFECT

30. The Miller effect in the context of a common emitter amplifier explains

(a) an increase in the low-frequency cutoff frequency
(b) an increase in the high-frequency cutoff frequency
(c) a decrease in the low-frequency cutoff frequency
(d) a decrease in the high-frequency cutoff frequency

[2017 : 1 Mark, Set-1]

ANSWERS

1. (c) 2. (d) 3. (Compressed) 4. (c) 5. (b) 6. (a) 7. (a) 8. (d) 9. (a) 10. (10^7) 11. (50) 12. (a) 13. (d) 14. (2 πF) 15. (1 Hz) 16. (a) 17. (d) 18. (c) 19. (d) 20. (b) 21. (d) 22. (a) 23. (a) 24. (c) 25. (a) 26. (a) 27. (b) 28. (34· 722) 29. (a) 30. (d)

EXPLANATIONS

2. As base width decreases, recombination decreases so collector current I_C increases.

$$\alpha = \frac{I_C}{I_E}.$$

So, α also increases.

3. In order to reduce the harmonic distortion in an amplifier its dynamic range has to be large.

4. Rise time, $t_r = \frac{0.35}{f_H}$

5. Total harmonic distortion in a distorted harmonic having A_1, A_2, A_3 of fundamental, second harmonic and third harmonic is

$$\frac{\sqrt{A_2^2 + A_3^2 + ...}}{A_1}$$

6. Both arms of differential amplifier are symmetrical. So terminal currents and collector voltages are equal. Characteristics of differential pair with common-mode input are similar to those of a C-E amplifier with large emitter resistor.

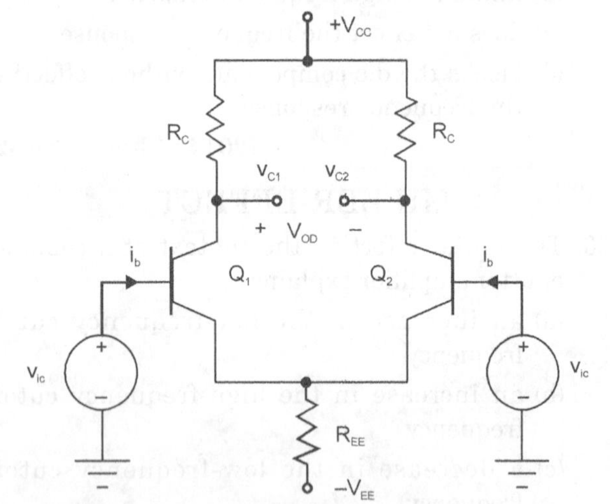

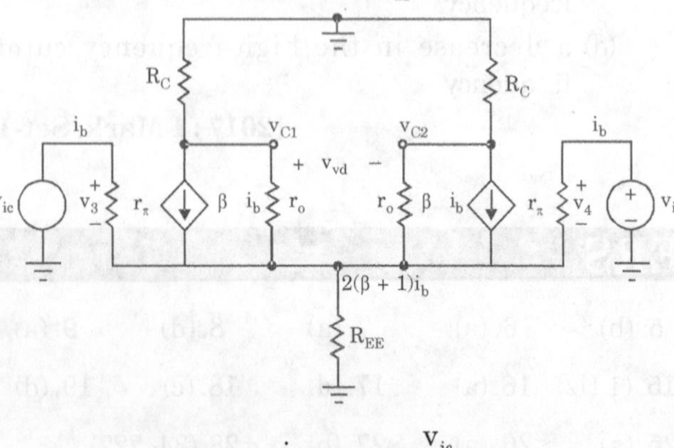

$$i_b = \frac{V_{ic}}{r_\pi + 2(\beta + 1)R_{EE}}$$

Output voltages are:

$$v_{c1} = v_{c2}$$

$$= -\beta i_b R_C = \frac{-\beta R_C}{r_\pi + 2(\beta+1)R_{EE}} V_{ic}$$

$$v_e = 2(\beta + 1) i_b R_{EE}$$

$$= \frac{2(\beta+1)R_{EE}}{r_\pi + 2(\beta+1)R_{EE}} V_{ic} \cong V_{ic}$$

Common-mode gain,

$$A_{CC} = \left.\frac{V_{0c}}{V_{ic}}\right|_{V_{i=0}} = \frac{-\beta R_C}{r_\pi + 2(\beta+1)R_{EE}}$$

For $R_{EE} \to \infty, A_{CC} \to 0$

7. At high frequency strong internal capacitor come into existence which have considerable impedance.

8. Common mode gain is given by,

$$A_{cc} = \frac{-\beta R_C}{r_k + 2(\beta+1)R_E} \approx \frac{R_C}{2R_E}$$

Differential-mode gain,

$$A_{dd} = -g_m R_C \qquad \text{[Independent of } R_E\text{]}$$

Common mode gain will decrease while R_E increases and differential-mode gain stays constant.

9.

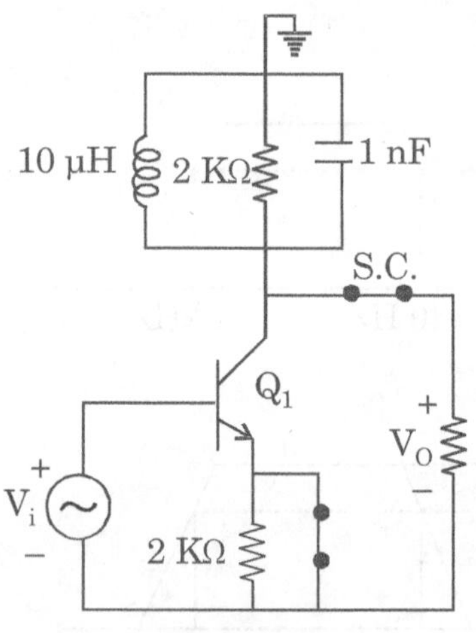

AC equivalent circuit

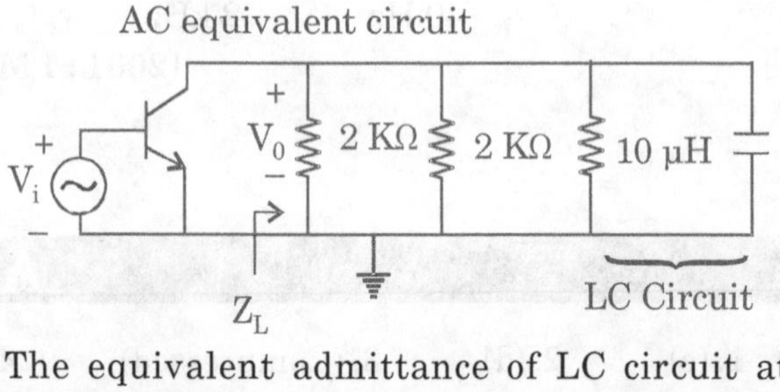

The equivalent admittance of LC circuit at $\omega = 10^7$ rad/s

$$y = \frac{1}{j\omega L} + j\omega C = 0.$$

Hence, $Z = \infty$ (maximum).

The circuit is in resonance for parallel LC circuit. The total impedance,

$Z_L = 2\|2\|\infty = 1\ k\Omega$ (maximum)

since, $A_V = A_I . Z_L / R_L$

So, voltage gain is maximum at 10^7 rad/s.

11. In AC equivalent circuit, Q_2 becomes diode connected transistor because collector and base get shorted.

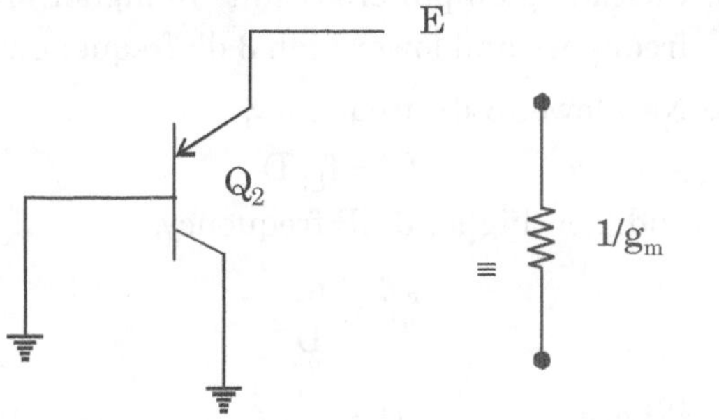

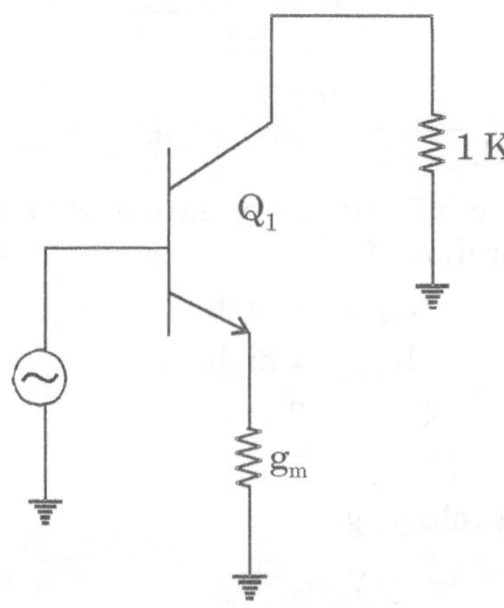

Complete AC equivalent circuit is as shown below:
It is a CE amplifier with unbypassed R_E.

$$A_v = \frac{-g_m R_L'}{1 + g_m R_E}$$

$$\because g_m = \frac{I_C}{V_T} = \frac{2.6}{26} = 100\ m\mho$$

$$R_L' = 11\Omega, \text{ and } R_E = \frac{1}{9m}$$

$$A_v = \frac{-100 \times 1}{1+1} = -50.$$

$$|A_v| = 50$$

13.
$$g_m = .\, g_{m_0}\left(1 - \frac{V_{GS}}{V_p}\right) = \frac{2I_{DSS}}{[V_p]} . \sqrt{\frac{I_D}{I_{DSS}}}$$

$V_G = 0,$

$V_s = I_D . R_s$

$= 1\ mA \times 2.5\ k$

$= 2.5\ V$

14. The emitter base capacitance is called diffusion capacitance C_D or G = 12 pF.

Depletion capacitance is across collector-base.

$$C_D = \frac{\tau I}{\eta V_T}$$

Here, $\eta = 1$

$$\therefore \quad C_D = \frac{260 \times 10^{-12} \times 10^3}{26 \times 10^{-3}}$$

$$= 10\ pF$$

$\therefore$ Deplection capacitance = 12 – 10 = 2 pF

15.
$$Af = \frac{A_{CL}}{1 + A_{OL} b}$$

$$= \frac{100}{1 + 100 \times 0.99} = 1$$

$$f_H = f_h(1 + \beta A_{OL})$$

$$= 100 \times 10^3 (1 + 0.99 \times 100)$$

$$= 10 \times 10^6\ Hz = 10\ MHz$$

$$f_I = \frac{f_I}{1 + \beta A_{OL}}$$

$$= \frac{100}{100} = 1\ Hz$$

16. Given $f_\beta = 1$ MHz, $\beta = 200$.

unity gain frequency or gain band width of a BJT is $F_T = h_{fe} . f_\beta = \beta f_\beta = 200 \times 1\ MHz = 200\ MHz. = 20\ MHz.$

Alpha - cut off frequency ,

$$f_\alpha = \frac{f_\beta}{(1-\alpha)} = (1+\beta) f_\beta = (1+200) \times 1$$

$$f_\alpha = 201\ MHz.$$

17. Unity gain bandwidth or gain bandwidth product of a BJT is $f_T = h_{fe} f_{\beta} = \frac{g_m}{2\pi(c_\mu + c_\pi)}$

where, $g_m = \frac{|I_C|}{V_T}$ = trans conduc tan ce,

$C\mu$ = transition capacitance and $c\pi$ = diffusion capacitance

18. NPN transistor has a unity gain cut off frequency, $f_T = 400$ MHz

dc bias current $I_c = 1$ mA.

$$C_\mu = 0.3\ \mu F$$

$$g_m = \frac{I_C(mA)}{26} = \frac{1}{26}\ A/V$$

$$C_\pi = \frac{g_m}{2\pi f_T} = \frac{\frac{1/26}{2\pi \times 400 \times 10^6}}{1}$$

$$= \frac{1}{26 \times 2\pi \times 4} \times 10^{-8}$$

$$= \frac{1}{2.6 \times 8\pi} \ P_F$$

20. $$f_T = \frac{g_m}{2\pi(C_\mu + C_\pi)}$$

$$f_\beta \cdot h_{fe} = f_T$$

Putting the given values,

$$f_T = 1.47 \times 10^{10} \text{ Hz},$$

$$f_\beta = 1.64 \times 10^8 \text{ Hz}.$$

21. $$g_m = \frac{|I_C|}{V_T} = \frac{1\text{m A}}{25\text{mV}} = 0.04 \text{ A/V} = 40 \text{ mA/V}$$

$$r_\pi = \frac{\beta}{g_m} = \frac{100}{40\,mA/V} = 2.5 \text{ k}\Omega$$

22.

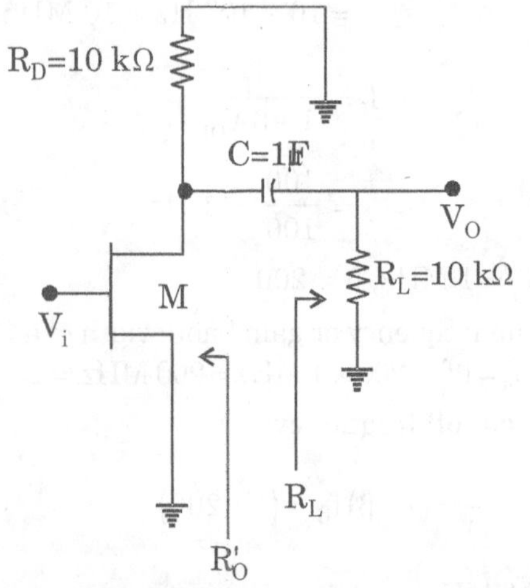

Given: $\lambda = 0$, $g_m = 1$ mA/V.

Lower cut - off frequency or lower 3–dB frequency

$$f_L = \frac{1}{2\pi\left(R_0' + R_L\right)C}; \quad r_0 = \left|\frac{1}{\lambda I_0}\right| = \infty$$

$\because$ From figure, $R_0' = R_D \parallel r_0 = 10 \parallel \infty = 10 \text{ k}\Omega$

$R_L = 10 \text{ k}\Omega$, $C = 1\ \mu\text{F}$

$$f_L = \frac{1}{2\pi\left(10 \times 10^3 + 10 \times 10^3\right) \times 1 \times 10^{-6}} = \frac{100}{4\pi}$$

$f_L = 7.96$ Hz.

23. Parasitic capacitors are small capacitor, where as bypass capacitor and coupling capacitors are large capacitor. In mid–band region, small capacitor acts as an open–circuit and large capacitor will act as a short – circuit.

24. As tuned amplifier is specified, high frequency response of *n*-stage is

$$B' = B\sqrt{2^{1/n} - 1}$$

25. Cascading amplifiers results in higher low 3 *d*B frequency and lower high 3 db frequency.

26. New lower 3-dB frequency,

$$f_L' = f_L \cdot D$$

and, new higher 3-dB frequency,

$$f_H' = \frac{f_H}{D}.$$

Where $$D = \frac{1}{\sqrt{2^{1/n} - 1}}$$

$$n = 3$$

27. CE – CC,

Because of the impedance match of the configuration

$R_{(out)CE}$ = high,

and $R_{(in)CC}$ = high

4 3

28. Overall voltage gain,

$$A_v = \frac{V_0}{V_i}$$

$$= A_{V_1} A_{V_2} \left[\frac{Z_{i_2}}{Z_{i_2} + Z_{0_1}}\right]\left[\frac{R_L}{R_L + Z_{0_2}}\right]$$

$$= 10 \times 5\left[\frac{5k}{5k + 1k}\right]\left[\frac{1k}{1k + 200}\right]$$

$$A_v = 34.722$$

29. Coupling capacitance comes in series with the circuit and hence affects low frequency response as at low frequency its impedance becomes high. It also blocks dc but affect the frequency response.

30. Miller's theorem is used for analysis of feedback element which is connected between input and output terminal. Miller effect increases input capacitance and there by decreases the higher cut – off frequency.

4

CHAPTER

Operational Amplifiers

Analysis of Previous GATE Papers												
	Year → Topics ↓		2019	2018	2017 Set 1	2017 Set 2	2016 Set 1	2016 Set 2	2016 Set 3	2015 Set 1	2015 Set 2	2015 Set 3
IDEAL & PRACTICAL OP-AMP	1 Mark	MCQ Type		1	1							
		Numerical Type										
	2 Marks	MCQ Type		1	1	1	2	2	2			
		Numerical Type										
	Total			3	3	2	4	4	4			
DIFFERENT OP-AMP CONFIGURATIONS	1 Mark	MCQ Type					1					
		Numerical Type					1				2	1
	2 Marks	MCQ Type										
		Numerical Type								1	2	1
	Total						2			2	6	3

IDEAL & PRACTICAL OP-AMP

1. In figure shown below, if the CMRR of the operational amplifier is 60 dB, then the magnitude of the output voltage is ______.

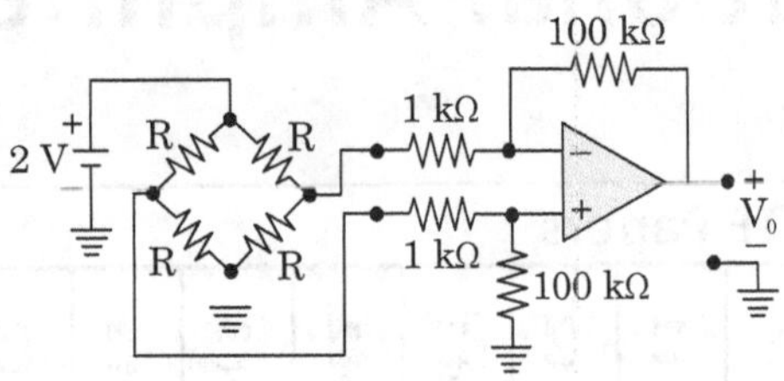

[1987 : 2 Marks]

2. The Op-Amp of figure shown below has a very poor open loop voltage gain of 45 but is otherwise ideal. The gain of the Amplifier equals

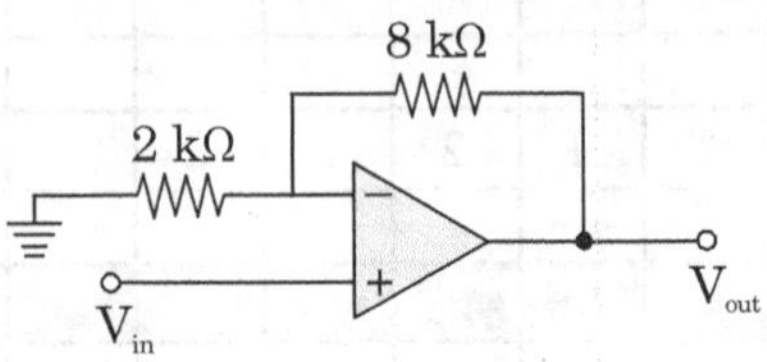

(a) 5 (b) 20

(c) 4 (d) 4.5

[1990 : 2 Marks]

3. The CMRR of the differential Amplifier of the figure shown below is equal to

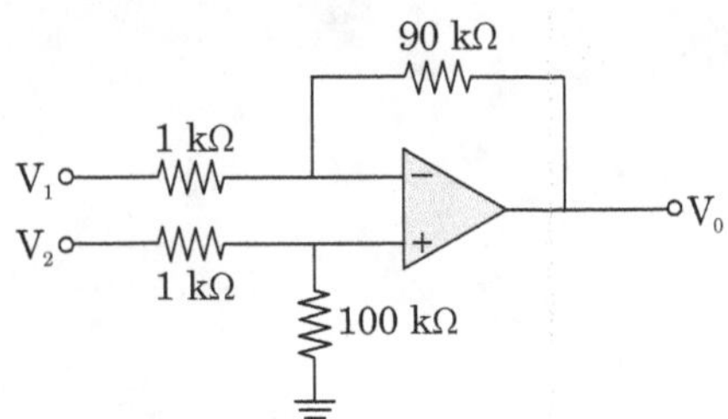

(a) ∞

(b) 0

(c) 1000

(d) 1800

4. An Op-Amp has an offset voltage of 1 mV and is ideal in all other respects, if this Op-Amp is used in the circuit shown in fig. The output voltage will be (Select the nearest value)

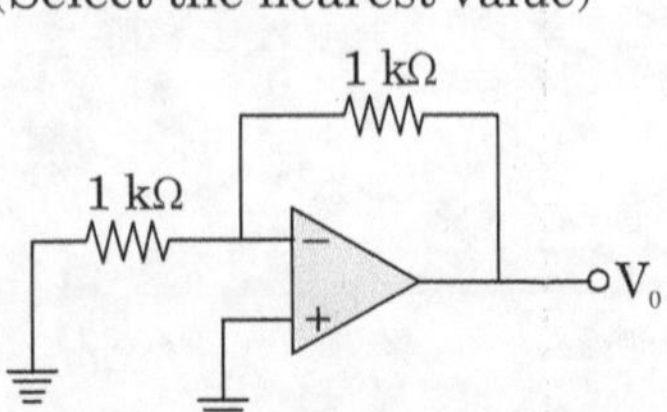

(a) 1 mV (b) 1 V

(c) ±1 V (d) 0 V

[1992 : 2 Marks]

5. The frequency compensation is used in OP-Amps to increase its ______.

[1994 : 1 Mark]

6. In the given circuit figure, if the voltage inputs V_- and V_+ are to be amplified by the same amplification factor, the value of 'R' should be ______.

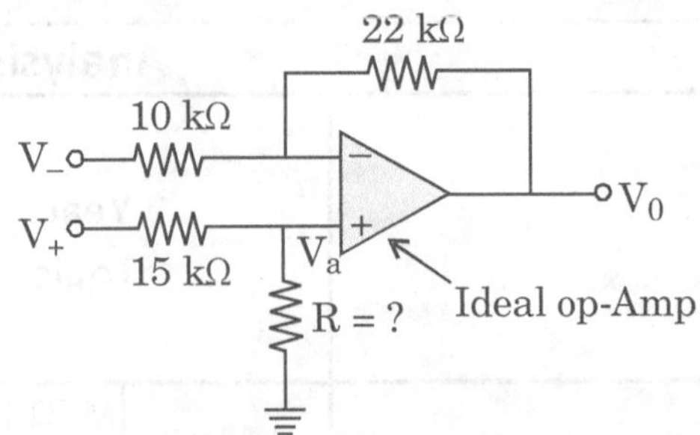

[1995 : 1 Marks]

7. If the op-amp in the figure has an input offset voltage of 5 mV and an open-loop voltage gain of 10,000; then V_0 will be

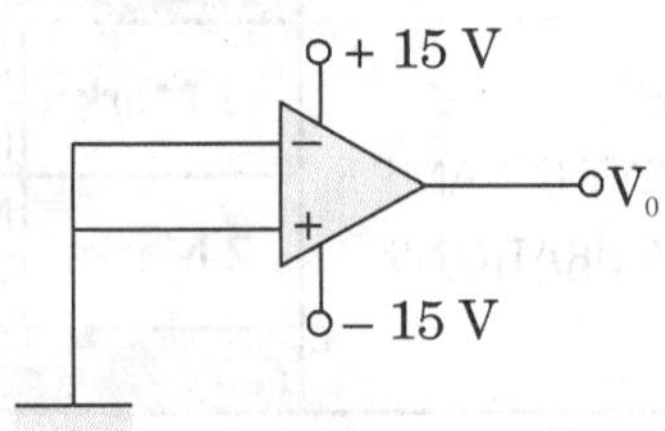

(a) 0 V

(b) 5 mV

(c) +15 V or – 15 V

(d) + 50 V or – 50 V

[2000 : 2 Marks]

8. The ideal OP-AMP has the following characteristics.

(a) $R_i = \infty, A = \infty, R_0 = 0$

(b) $R_i = 0, A = \infty, R_0 = 0$

(c) $R_i = \infty, A = \infty, R_0 = \infty$

(d) $R_i = 0, A = \infty, R_0 = \infty$

[2001 : 1 Mark]

9. An amplifier using an op-amp with a slew-rate SR = 1 V/μ sec has a gain of 40 dB. If this amplifier has to faithfully amplify sinusoidal signals from dc to 20 kHz without introducing any slew-rate induced distortion, then the input signal level must not exceed

(a) 795 mV

(b) 395 mV

(c) 79.5 mV

(d) 39.5 mV

[2002 : 2 Marks]

10. An ideal op-amp is an ideal

(a) voltage controlled current source.

(b) voltage controlled voltage source.

(c) current controlled current source.

(d) current controlled voltage source.

[2004 : 1 Mark]

11. An ideal op-amp has voltage sources, V_1, V_3, V_5, ... V_{N-1} connected to the non-inverting input and V_2, V_4, V_6, ..., V_N connected to the inverting input as shown in the figure below ($+V_{CC}$ = 15 volt, $-V_{CC}$ = −15 volt). The voltages V_1, V_2, V_3, V_4, V_5, V_6, ... are 1, $-\frac{1}{2}$, $\frac{1}{3}$, $-\frac{1}{4}$, $\frac{1}{5}$, $-\frac{1}{6}$, ... Volt, respectively. As N approaches infinity, the output voltage (in volt) is ______.

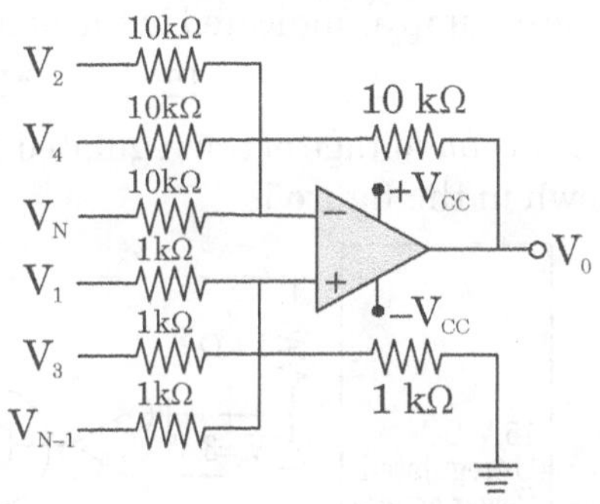

[2005 : 1 Mark]

12. A p-i-n photodiode of responsivity 0.8 A/W is connected to the inverting input of an ideal op-amp as shown in the figure, $+V_{CC}$ = 15 V, $-V_{CC}$ = −15 V, load resistor R_L = 10 kΩ. If 10 μW of power is incident on the photodiode, then the value of the photocurrent (in μA) through the load is ______.

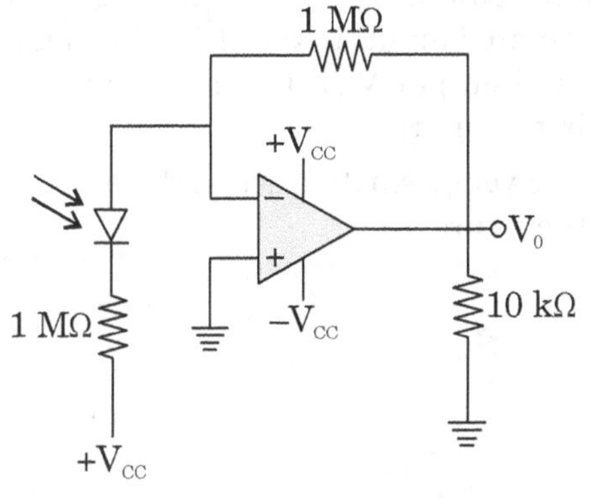

[2007 : 2 Mark]

13. In the op-amp circuit shown, the Zener diodes Z_1 and Z_2 clamp the output voltage V_0 to + 5 V or −5 V. The switch S is initially closed and is opened at time t= 0.

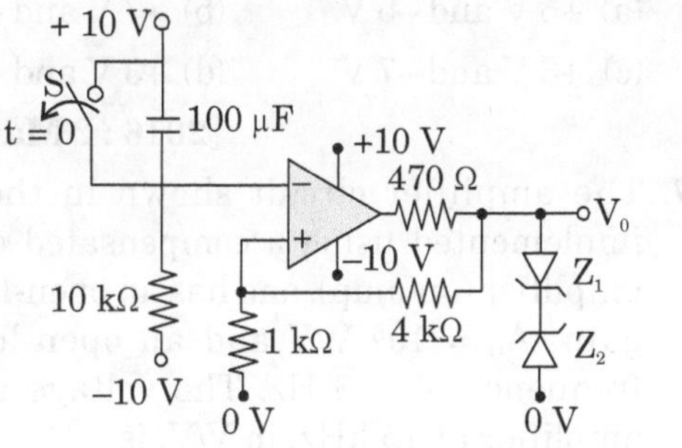

The time t = t_1 (in seconds) at which V_0 changes state is ______.

[2011 : 1 Mark]

14. An op-amp has a finite open loop voltage gain of 100. Its input offset voltage V_{ios}(= + 5 mV) is modeled as shown in the circuit below. The amplifier is ideal in all other respects. V_{input} is 25 mV.

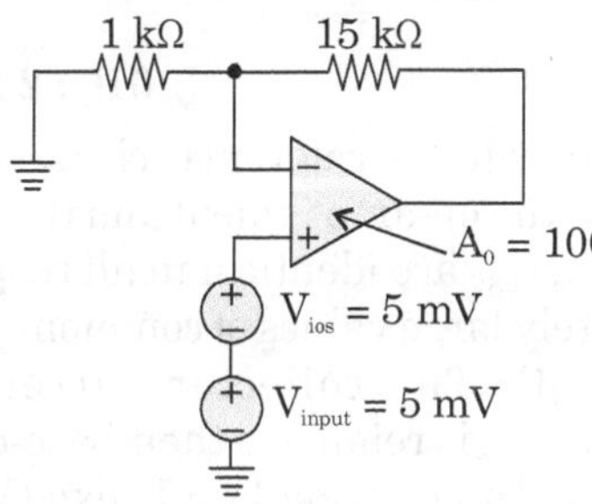

The output voltage (in millivolts) is ______.

[2016 : 2 Marks, Set-1]

15. For the circuit shown in the figure, $R_1 = R_2 = R_3$ = 1 Ω, L = 1 μH and C = 1 μF. If the input V_{in} = cos(10^6 t), then the overall voltage gain (V_{out}/V_{in}) of the circuit is ______.

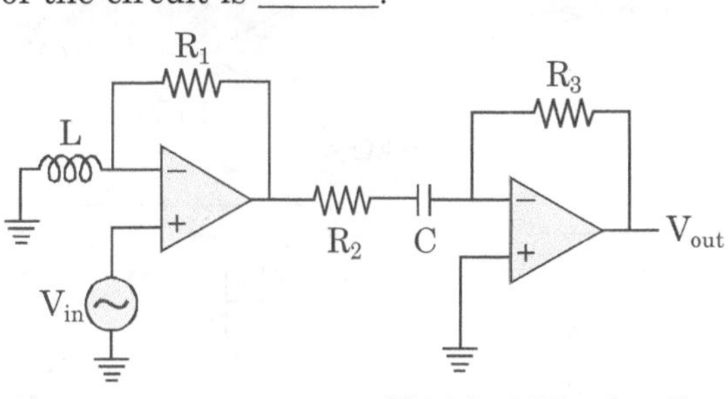

[2016 : 2 Marks, Set-1]

16. For the operational amplifier circuit shown, the output saturation voltages are ±15 V. The upper and lower threshold voltages for the circuit are, respectively,

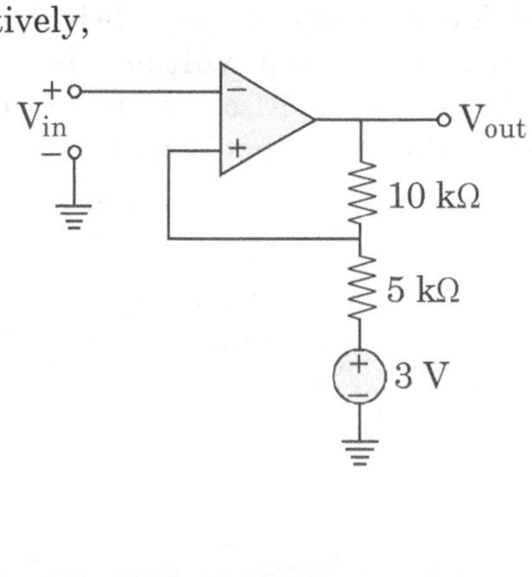

(a) +5 V and –5 V (b) +7 V and –3 V
(c) +3 V and –7 V (d) +3 V and –3 V

[2016 : 2 Marks, Set-2]

17. The amplifier circuit shown in the figure is implemented using a compensated operational amplifier (op-amp), and has an open-loop voltage gain, $A_0 = 10^5$ V/V and an open-loop cut-off frequency, $f_c = 8$ Hz. The voltage gain of the amplifier at 15 kHz, in V/V, is ______.

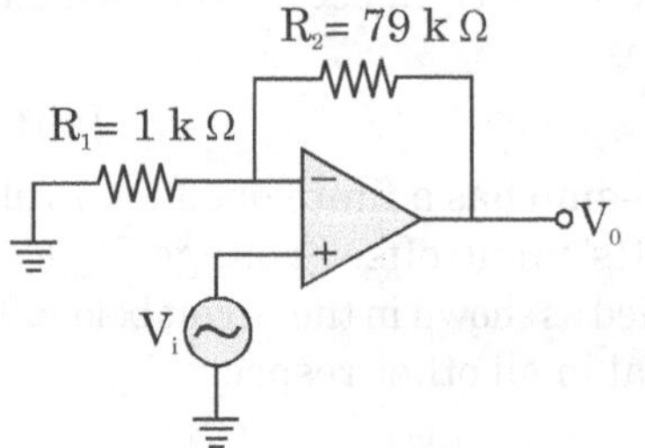

[2016 : 2 Marks, Set-2]

18. In the voltage reference circuit shown in the figure, the op-amp is ideal and the transistors Q_1, Q_2, ..., Q_{32} are identical in all respects and have infinitely large values of common-emitter current gain (β). The collector current (I_C) of the transistors is related to their base-emitter voltage (V_{BE}) by the relation $I_C = I_S \exp(V_{BE}/V_T)$, where I_S is the saturation current. Assume that the voltage V_P shown in the figure is 0.7 V and the thermal voltage $V_T = 26$ mV.

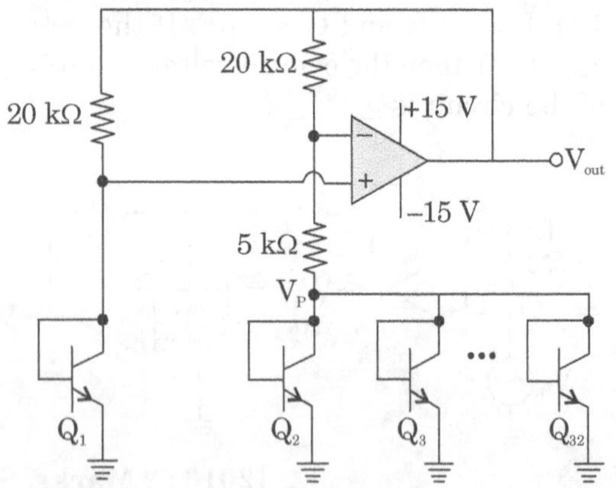

The output voltage V_{out} (in volts) is ______.

[2016 : 2 Marks, Set-3]

19. In the circuit shown below, the op-amp is ideal and Zener voltage of the diode is 2.5 volts. At the input, unit step voltage is applied i.e. $v_{IN}(t) = u(t)$ volts. Also at t = 0, the voltage across each of the capacitors is zero

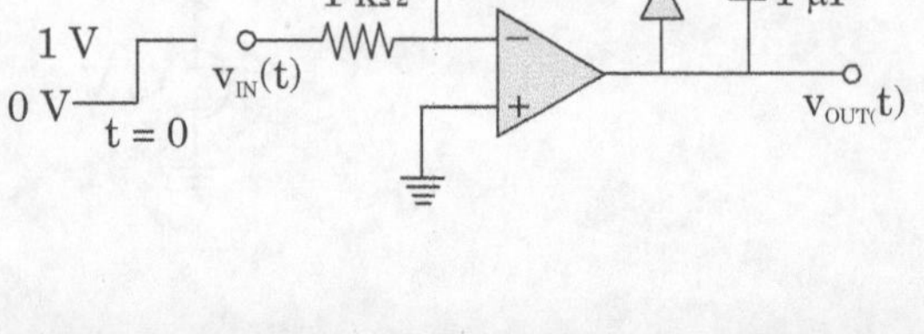

The time t, in milliseconds, at which the output voltage v_{OUT} crosses –10 V is

(a) 2.5 (b) 5
(c) 7.5 (d) 10

[2017 : 1 Mark, Set-1]

20. An op-amp based circuit is implemented as shown below.

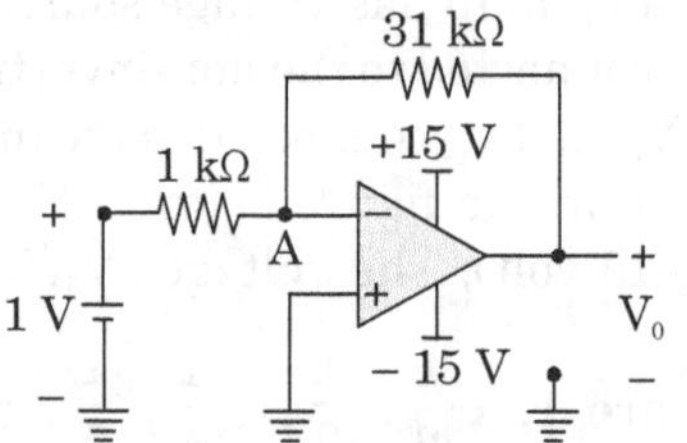

In the above circuit, assume the op-amp to be ideal. The voltage (in volts, correct to one decimal place) at node A, connected to the negative input of the op-amp as indicated in the figure is ______.

[2017 : 2 Marks, Set-1]

21. The output voltage of the regulated power supply shown in the figure is

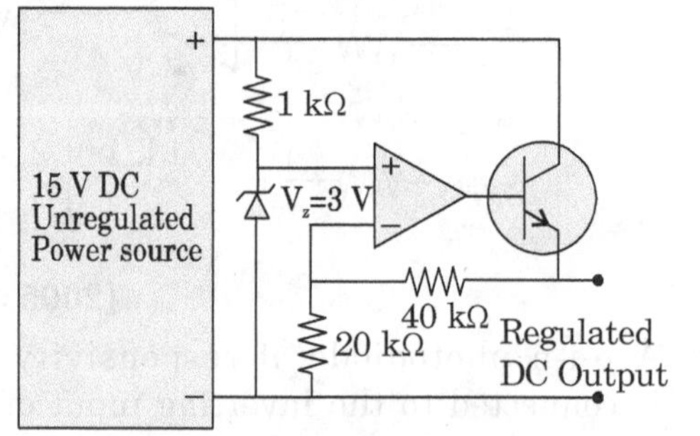

(a) 3 V (b) 6 V
(c) 9 V (d) 12 V

[2017 : 2 Marks, Set-2]

Common Data for Questions 22 & 23 :

A regulated power supply, shown in figure below, has an unregulated input (UR) of 15 Volts and generates a regulated output V_{out}. Use the component values shown in the figure

22. The power dissipation across the transistor shown in the figure is

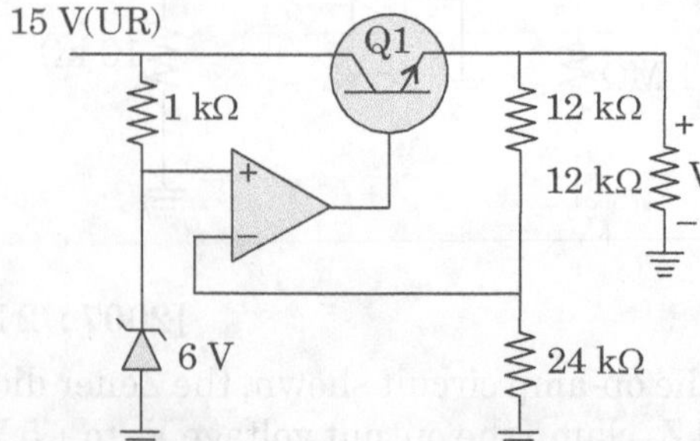

(a) 4.8 Watts (b) 5.0 Watts
(c) 5.4 Watts (d) 6.0 Watts

[2018 : 1 Mark]

23. If the unregulated voltage increases by 20%, the power dissipation across the transistor Q_1

(a) increases by 20%

(b) increases by 50%

(c) remains unchanged

(d) decreases by 20%

[2018 : 2 Marks]

DIFFERENT OP-AMP CONFIGURATIONS

24. The Op-Amp shown in figure below is ideal, $R = \sqrt{\frac{L}{C}}$. The phase angle between V_0 and V_i, at $\omega = \frac{1}{\sqrt{LC}}$ is

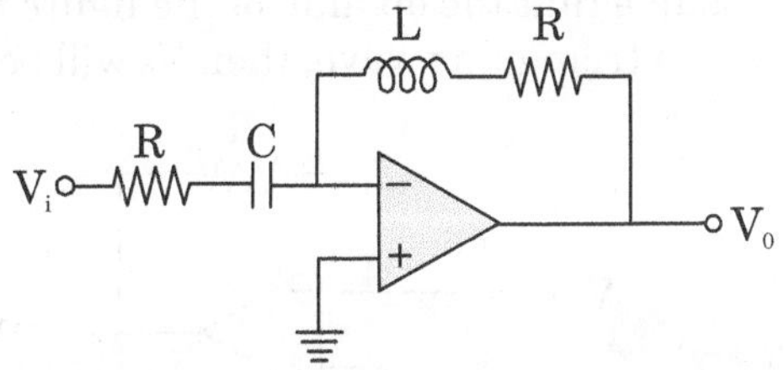

(a) $\frac{\pi}{2}$ (b) π

(c) $\frac{3\pi}{2}$ (d) 2π

[1988 : 2 Marks]

25. Refer to figure shown below :

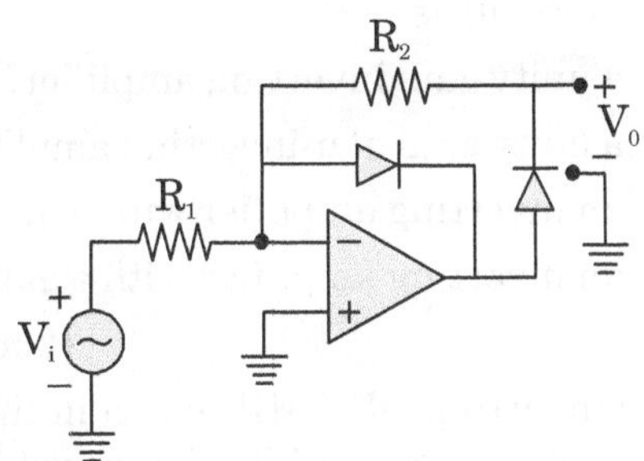

(a) For $V_i > 0$, $V_0 = \frac{R_2}{R_1} V_i$

(b) For $V_i > 0$, $V_0 = 0$

(c) For $V_i < 0$, $V_0 = \frac{R_2}{R_1} V_i$

(d) For $V_i < 0$, $V_0 = 0$

[1989 : 2 Marks]

26. If the input to the circuit of figure is a sine wave the output will be

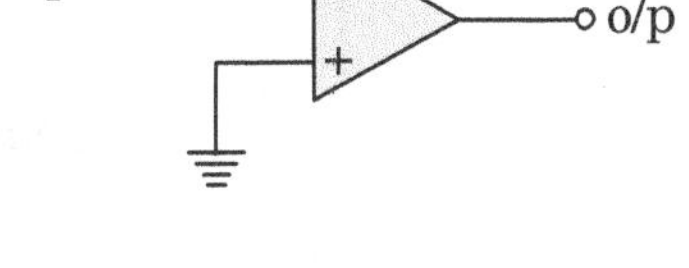

(a) a half-wave rectified sine wave.

(b) a full-wave rectified sine wave.

(c) a triangular wave.

(d) a square wave.

[1990 : 2 Marks]

27. The circuit of fig. uses on ideal Op-Amp for small positive values of V_{in}, the circuit works as

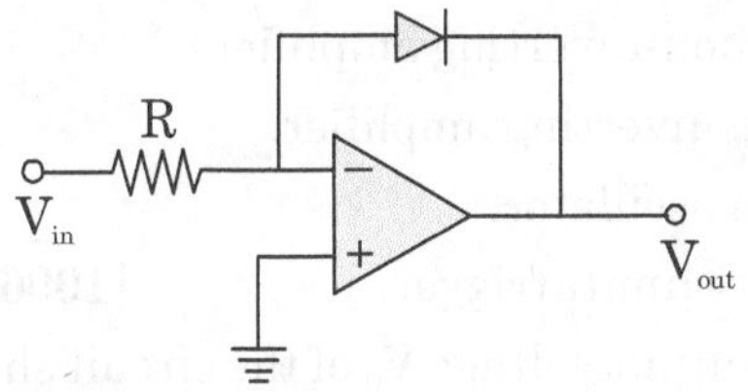

(a) a half wave rectifier.

(b) a differentiator.

(c) a logarithmic amplifier.

(d) an exponential amplifier.

[1992 : 2 Marks]

28. Assume that the operational amplifier in figure is ideal the current I through the 1 KΩ resistor is

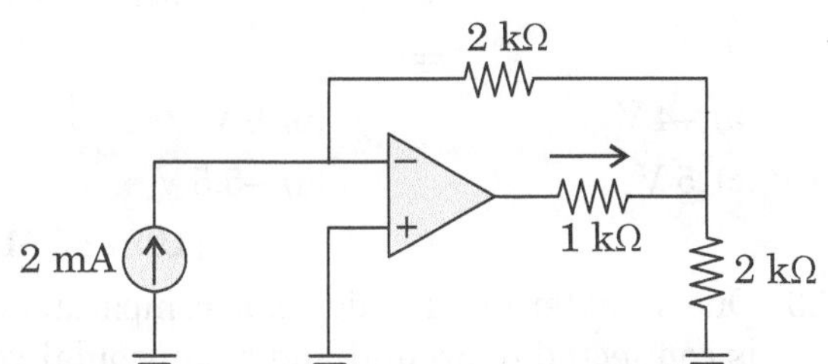

[1992 : 2 Mark]

29. For the ideal Op-Amp circuit of fig. Determine the output voltage V_0.

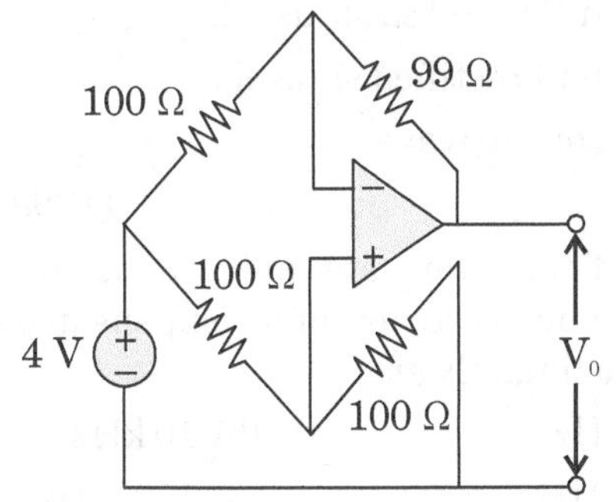

[1993 : 2 Marks]

30. An Op-Amp is used as a zero-crossing detector. If maximum output available from the Op-Amp is ±12 V_{p-p} and the slew rate of the Op-Amp is 12 V/μ sec then the maximum frequency of the input signal that can be applied without causing a reduction in the P- P output is _______.

[1995 : 1 Mark]

31. The circuit shown in the figure is that of

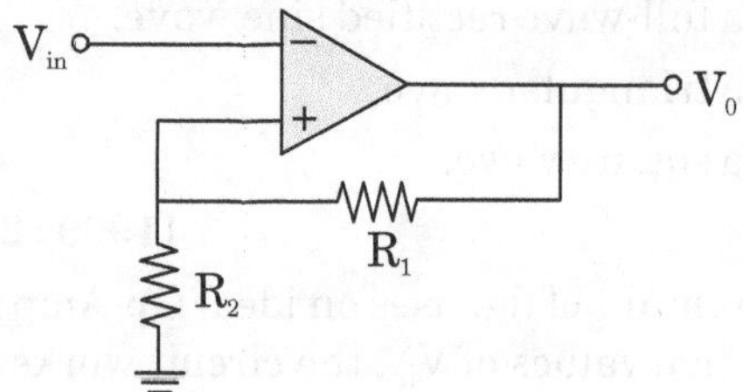

(a) a non-inverting amplifiers.
(b) an inverting amplifier.
(c) an oscillator.
(d) a Schmitt trigger. **[1996 : 1 Mark]**

32. The output voltage V_0 of the circuit shown in the figure is

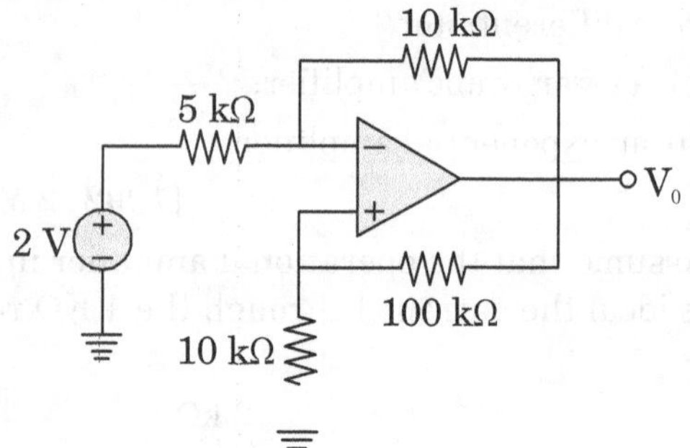

(a) −4 V (b) 6 V
(c) 5 V (d) −5.5 V

[1997 : 2 Marks]

33. One input terminal of high gain comparator circuit is connected to ground and a sinusoidal voltage is applied to the other input. The output of comparator will be
(a) a sinusoid.
(b) a full rectified sinusoid.
(c) a half rectified sinusoid.
(d) a square wave.

[1998 : 1 Mark]

34. The first dominant pole encountered in the frequency response of a compensated op-amp is approximately at
(a) 5 Hz (b) 10 kHz
(c) 1 MHz (d) 100 MHz

[1999 : 1 Mark]

35. In the circuit of the figure, V_0 is

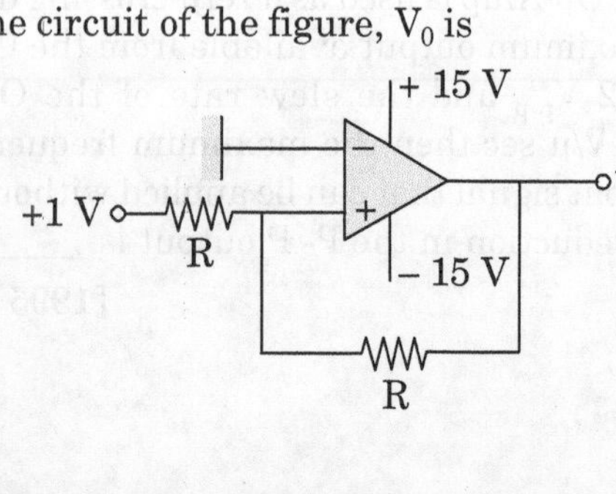

(a) −1 V (b) 2 V
(c) +1 V (d) +15 V

[2000 : 1 Mark]

36. If the op-amp in the figure, is ideal, then V_0 is

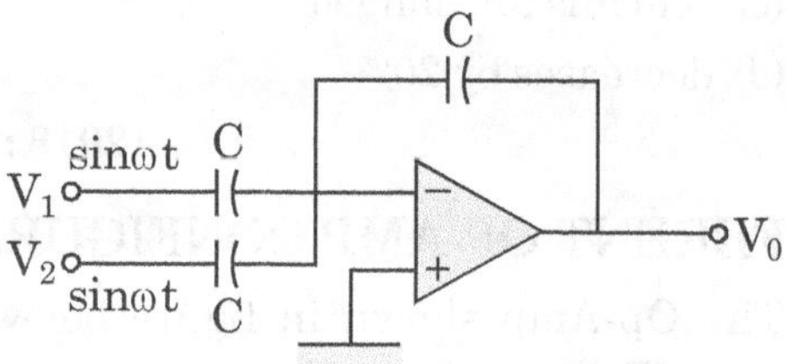

(a) Zero (b) $(V_1 - V_2)\sin\omega t$
(c) $-(V_1 + V_2)\sin\omega t$ (d) $(V_1 + V_2)\sin\omega t$

[2000 : 1 Mark]

37. Assume that the op-amp of the figure is ideal. If V_i is a triangular wave, then V_0 will be

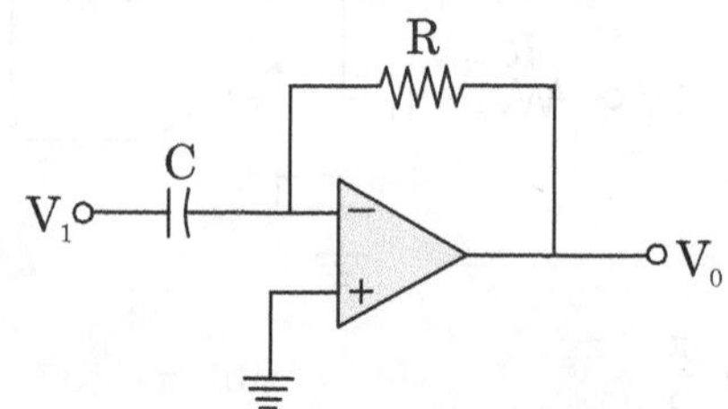

(a) square wave (b) triangular wave
(c) parabolic wave (d) sine wave

[2000 : 1 Mark]

38. The most commonly used amplifier in sample and hold circuits is
(a) a unity gain inverting amplifier.
(b) a unity gain non-inverting amplifier.
(c) an inverting amplifier with a gain of 10.
(d) an inverting amplifier with a gain of 100.

[2000 : 1 Mark]

39. The inverting OP-AMP shown in the figure has an open-loop gain of 100. The closed-loop gain V_0/V_s is

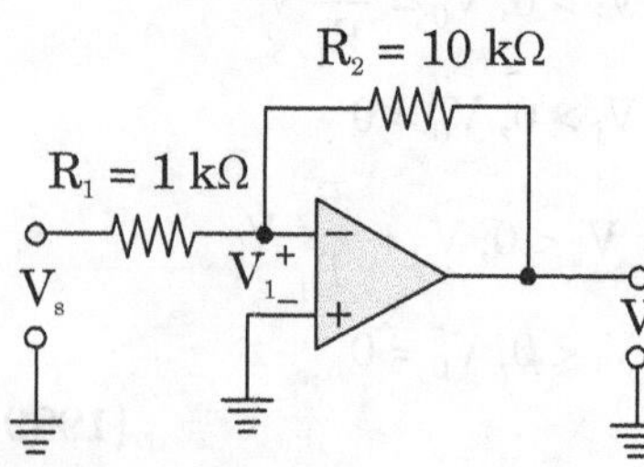

(a) −8
(b) −9
(c) −10
(d) −11

[2001 : 2 Marks]

40. In the figure assume the Op-Amps to be ideal. The output V_0 of the circuit is :

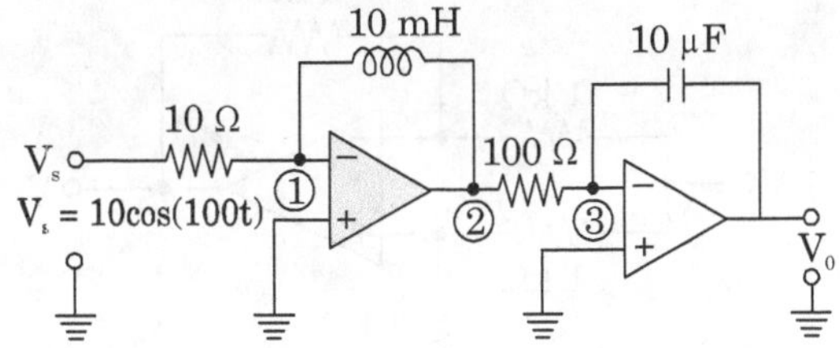

(a) 10 cos(100t)
(b) $10\int_0^t \cos(100\tau)d\tau$

(c) $10^{-4}\int_0^t \cos(100\tau)d\tau$
(d) $10^{-4}\frac{d}{dt}\cos(100t)$

[2001 : 2 Marks]

41. A 741 -type op-amp has a gain-bandwidth product of 1 MHz. A non-inverting amplifier using this op-amp and having a voltage gain of 20 dB will exhibit a -3 dB bandwidth of

(a) 50 kHz
(b) 100 kHz
(c) $\frac{1000}{17}$ kHz
(d) $\frac{1000}{7.07}$ kHz

[2002 : 1 Mark]

42. If the input to the ideal comparator shown in the figure is a sinusoidal signal of 8 V (peak to peak) without any DC component, then the output of the comparator has a duty cycle of

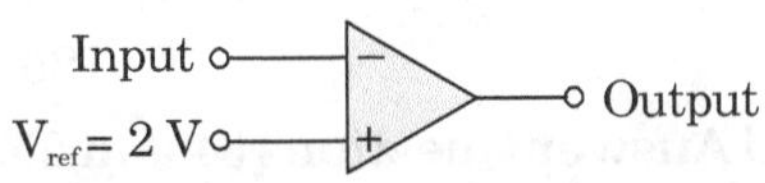

(a) $\frac{1}{2}$
(b) $\frac{1}{3}$
(c) $\frac{1}{6}$
(d) $\frac{1}{12}$

[2003 : 1 Mark]

43. If the differential voltage gain and the common mode voltage gain of a differential amplifier are 48 dB and 2 dB respectively, then its common mode rejection ratio is

(a) 23 dB
(b) 25 dB
(c) 46 dB
(d) 50 dB

[2003 : 1 Mark]

44. If the op-amp in the figure is ideal, the output voltage V_{out} will be equal to

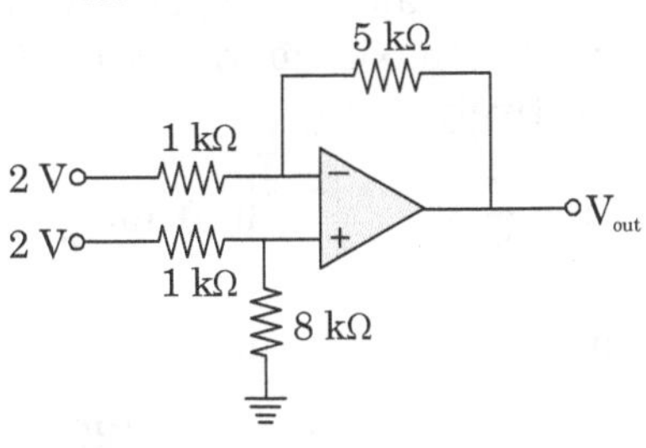

(a) 1 V
(b) 6 V
(c) 14 V
(d) 17 V

[2003 : 2 Marks]

45. Three identical amplifiers with each one having a voltage gain of 50, input resistance of 1 kΩ and output resistance of 250 Ω, are cascaded. The open circuit voltage gain of the combined amplifier is

(a) 49 dB
(b) 51 dB
(c) 98 dB
(d) 102 dB

[2003 : 2 Marks]

46. The circuit in the figure is

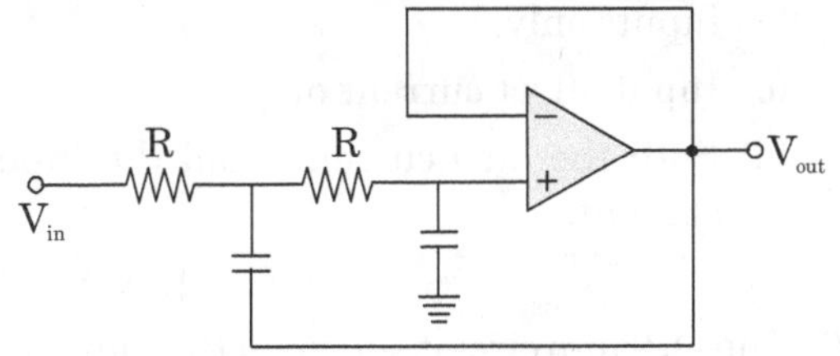

(a) low-pass filter
(b) high-pass filter
(c) band-pass filter
(d) band-reject filter

[2004 : 1 Mark]

47. In the op-amp circuit given in the figure, the load current i_L is

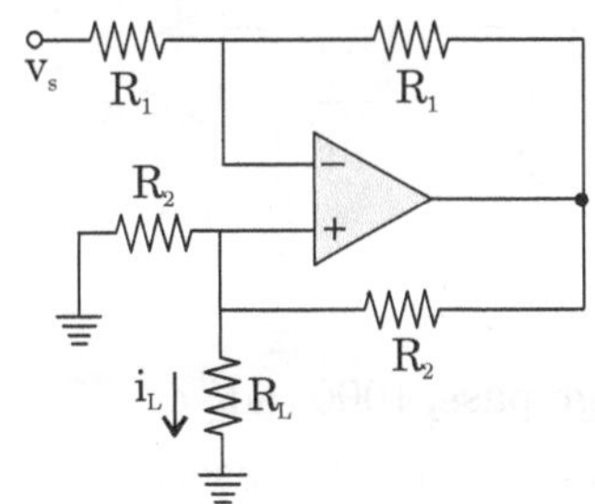

(a) $-\frac{V_s}{R_2}$
(b) $\frac{V_s}{R_2}$
(c) $-\frac{V_s}{R_L}$
(d) $\frac{V_s}{R_1}$

[2004 : 2 Marks]

48. The input resistance R_i of the amplifier shown in the figure is

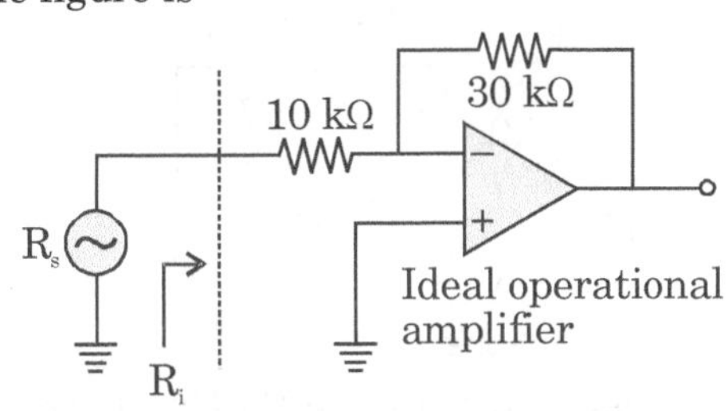

(a) $\frac{30}{4}$ kΩ
(b) 10 kΩ
(c) 40 kΩ
(d) infinite

[2005 : 1 Marks]

49. The voltage e_0 indicated in the figure has been measured by an ideal voltmeter. Which of the following can be calculated?

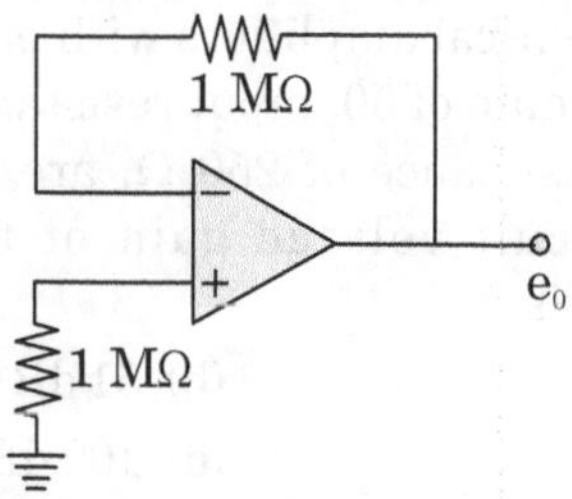

(a) Bias current of the inverting input only.
(b) Bias current of the inverting and non-inverting inputs only.
(c) Input offset current only.
(d) Both the bias currents and the input offset current.

[2005 : 2 Marks]

50. The OP-amp circuit shown in the figure is a filter. The type of filter and its cut-off frequency are respectively

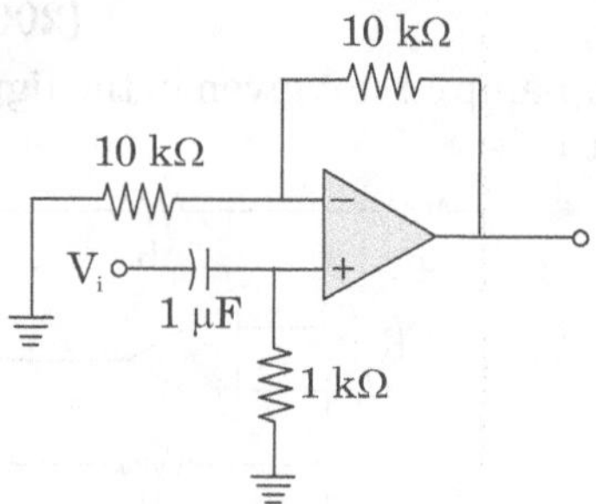

(a) high pass, 1000 rad/sec.
(b) low pass, 1000 rad/sec.
(c) high pass, 10000 rad/sec.
(d) low pass, 10000 rad/sec.

[2005 : 2 Marks]

51. For the circuit shown in the following figure, the capacitor C is initially uncharged. At t = 0, the switch S is closed. The voltage V_C across the capacitor at t = 1 millisecond is

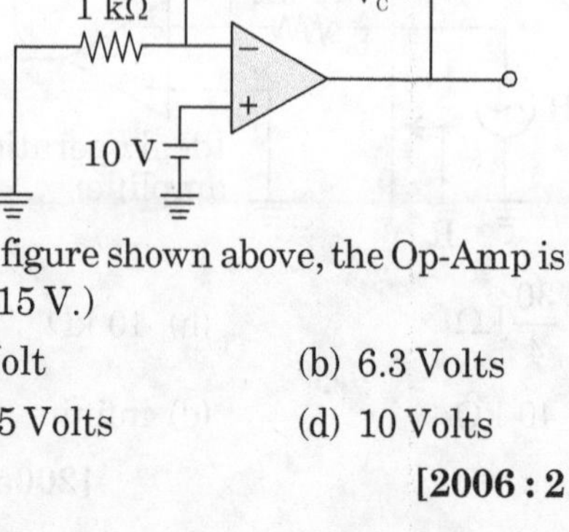

(In the figure shown above, the Op-Amp is supplied with ±15 V.)

(a) 0 Volt (b) 6.3 Volts
(c) 9.45 Volts (d) 10 Volts

[2006 : 2 Marks]

52. For the Op-Amp circuit shown in the figure, V_0 is

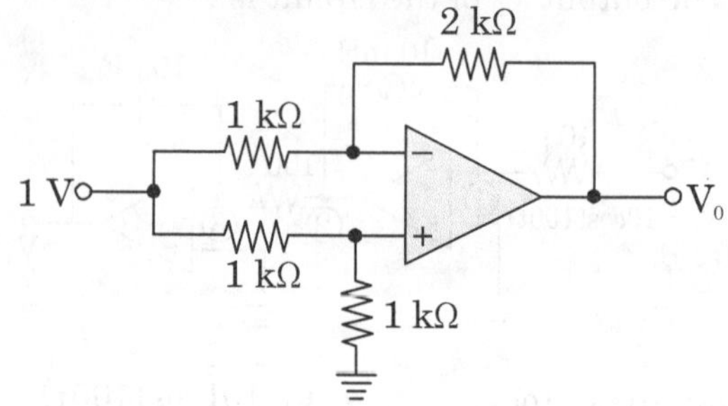

(a) –2 V (b) –1 V
(c) –0.5 V (d) 0.5 V

[2007 : 2 Marks]

53. In the OP-Amp circuit shown, assume that the diode current follows the equation $I = I_s \exp\left(\frac{V}{V_T}\right)$. For $V_i = 2$ V, $V_0 = V_{01}$, and for $V_i = 4$ V, $V_0 = V_{02}$. The relationship between V_{01} and V_{02} is

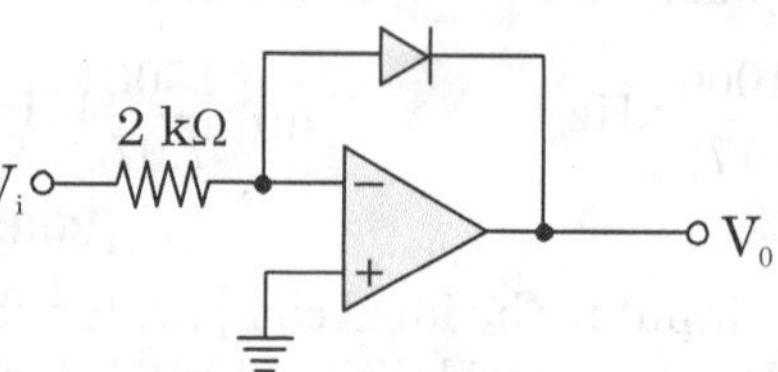

(b) $V_{02} = \sqrt{2}\, V_{01}$ (b) $V_{02} = e^2 V_{01}$
(c) $V_{02} = V_{01} \ln 2$ (d) $V_{01} - V_{02} = V_T \ln 2$

[2007 : 2 Marks]

Linked Answer Questions 54 and 55: Consider the Op-Amp circuit shown in the figure.

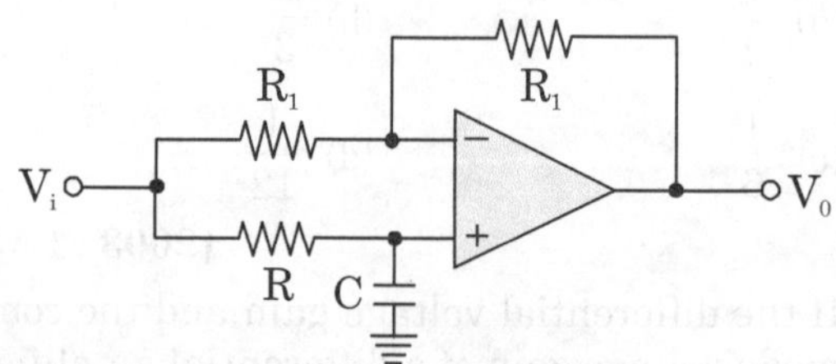

54. The transfer function $V_0(s)/V_i(s)$ is

(a) $\frac{1-sRC}{1+sRC}$ (b) $\frac{1+sRC}{1-sRC}$

(c) $\frac{1}{1-sRC}$ (d) $\frac{1}{1+sRC}$

[2007 : 2 Marks]

55. If $V_i = V_1 \sin(\omega t)$ and $V_0 = V_2 \sin(\omega t + \phi)$, then the minimum and maximum values of ϕ (in radians) are respectively

(a) $\frac{-\pi}{2}$ and $\frac{\pi}{2}$ (b) 0 and $\frac{\pi}{2}$

(c) $-\pi$ and 0 (d) $\frac{-\pi}{2}$ and 0

[2007 : 2 Marks]

56. Consider the following circuit using an ideal OP-Amp. The I-V characteristics of the diode is described by the relation $I = I_0\left(e^{\frac{V}{V_T}} - 1\right)$ where $V_T = 25$ mV, $I_0 = 1$ μA and V is the voltage across the diode (taken as positive for forward bias).

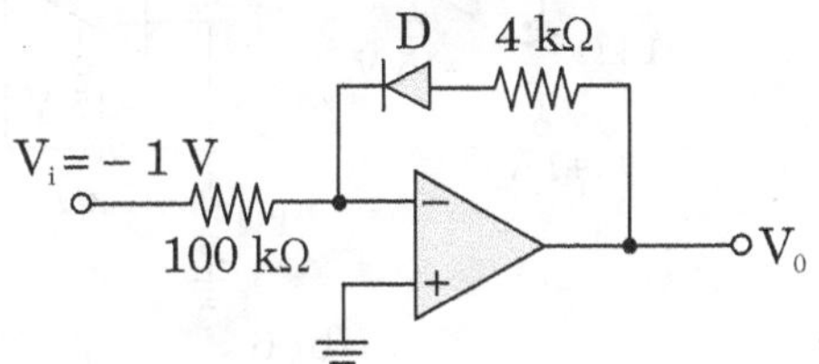

For an input voltage $V_i = -1$ V, the output voltage V_0 is

(a) 0 V (b) 0.1 V

(c) 0.7 V (d) 1.1 V

[2008 : 2 Marks]

57. The OP-Amp circuit shown below represents a

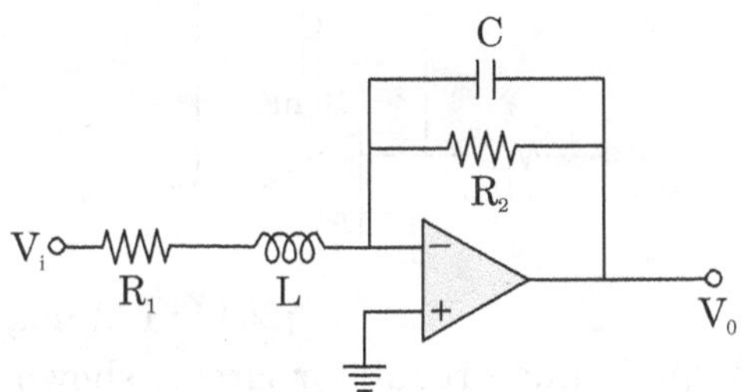

(a) high pass filter

(b) low pass filter

(c) band pass filter

(d) band reject filter

[2008 : 2 Marks]

58. In the circuit shown below, the op-amp is ideal, the transistor has $V_{BE} = 0.6$ V and $\beta = 150$. Decide whether the feedback in the circuit is positive or negative and determine the voltage V at the output of the op-amp.

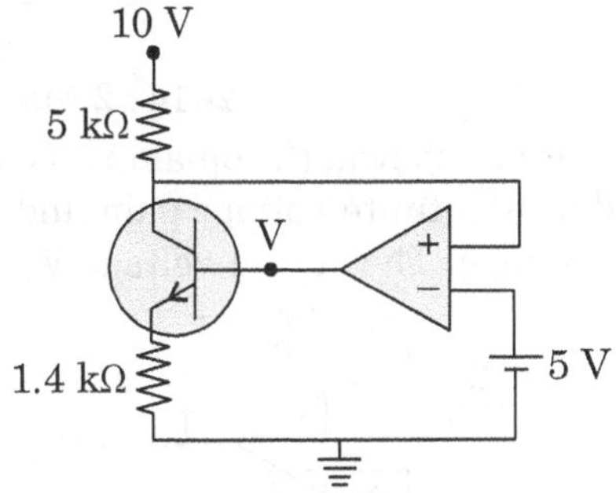

(a) Positive feedback, V = 10 V

(b) Positive feedback, V = 0 V

(c) Negative feedback, V = 5 V

(d) Negative feedback, V = 2 V

[2009 : 2 Marks]

59. Assuming the OP-AMP to be ideal, the voltage gain of the amplifier shown below is

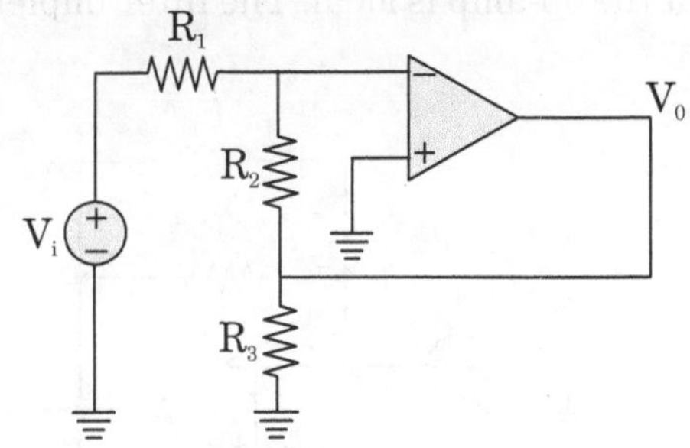

(a) $-\frac{R_2}{R_1}$ (b) $-\frac{R_3}{R_1}$

(c) $-\left(\frac{R_2 \| R_3}{R_1}\right)$ (d) $-\left(\frac{R_2 + R_3}{R_1}\right)$

[2010 : 1 Mark]

60. The transfer characteristic for the precision rectifier circuit shown below is (assume ideal OP-AMP and practical diodes)

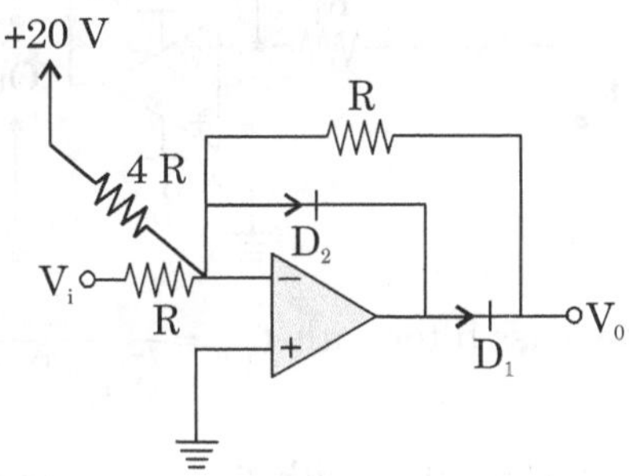

(a)

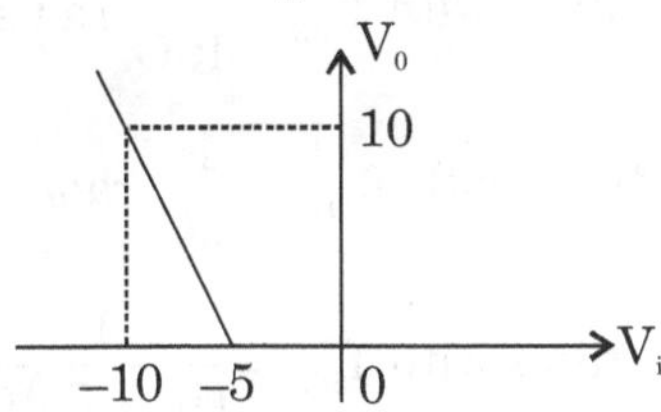

(b)

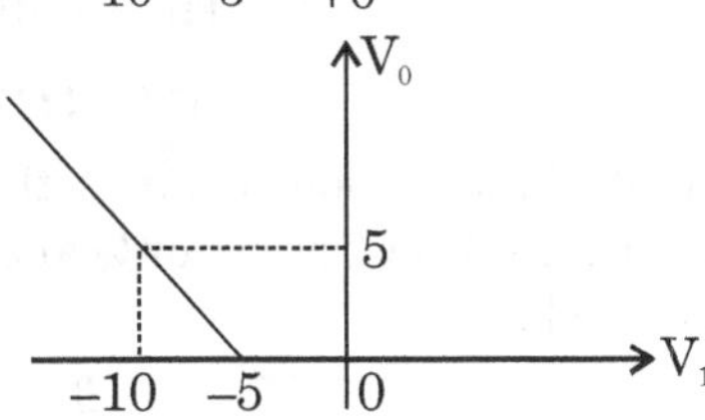

(c)

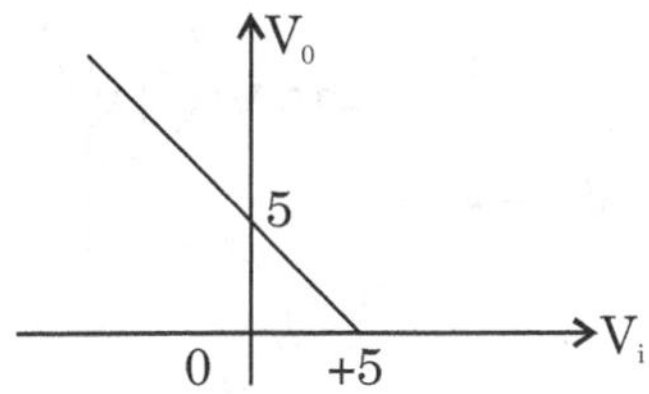

(d)

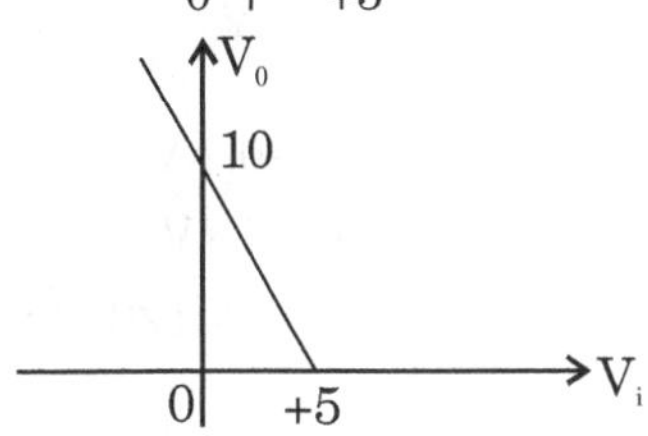

[2010 : 2 Marks]

61. The circuit below implement a filter between the input current i_i and output voltage v_0. Assume that the op-amp is ideal. The filter implemented is a

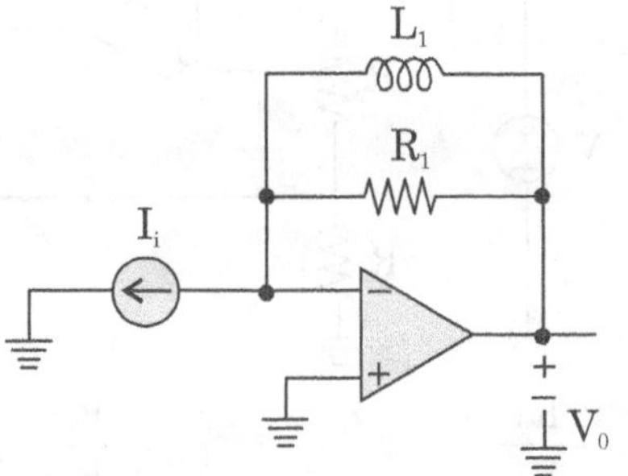

(a) low pass filter (b) band pass filter
(c) band stop filter (d) high pass filter

[2011 : 1 Mark]

62. The circuit shown is a

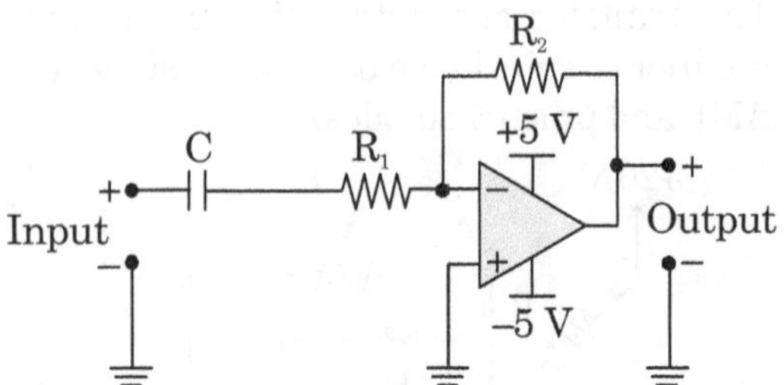

(a) low pass filter with $f_{3dB} = \frac{1}{(R_1 + R_2)C}$ rad/s

(b) high pass filter with $f_{3dB} = \frac{1}{R_1C}$ rad/s

(c) low pass filter with $f_{3dB} = \frac{1}{R_1C}$ rad/s

(d) high pass filter with $f_{3dB} = \frac{1}{(R_1 + R_2)C}$ rad/s

[2012 : 2 Marks]

63. In the circuit shown below what is the output voltage (V_{out}) if a silicon transistor Q and an ideal op-amp are used?

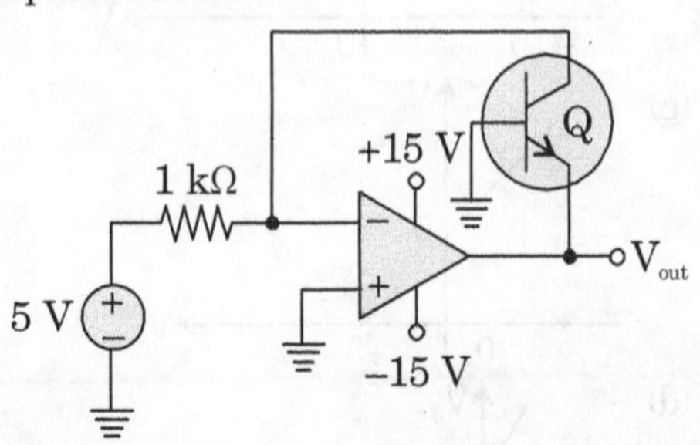

(a) –15 V (b) –0.7 V
(c) +0.7 V (d) +15 V

[2013 : 1 Mark]

64. In the circuit shown below the op-amps are ideal. Then V_{out} in volts is

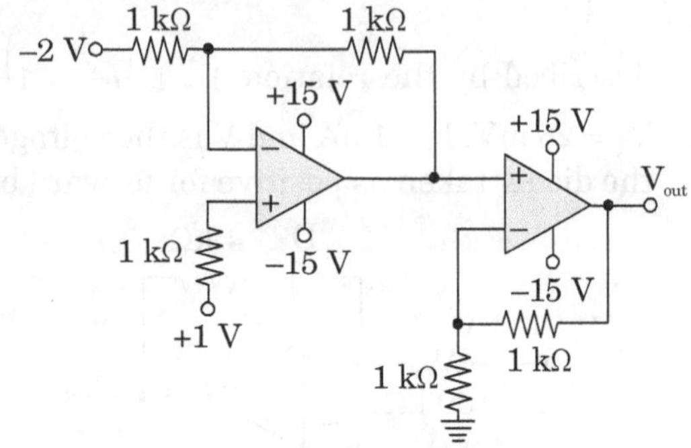

(a) 4 (b) 6
(c) 8 (d) 10

[2013 : 2 Marks]

65. In the low-pass filter shown in the figure, for a cut-off frequency of 5 kHz, the value of R_2 (in kΩ) is _______.

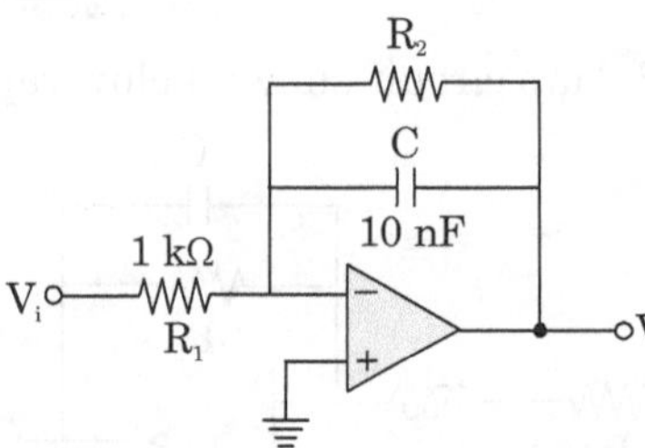

[2014 : 1 Mark, Set-1]

66. In the voltage regulator circuit shown in the figure, the op-amp is ideal. The BJT has $V_{BE} = 0.7$ V and $\beta = 100$, and the zener voltage is 4.7 V. For a regulated output of 9 V, the value of R (in Ω) is _______.

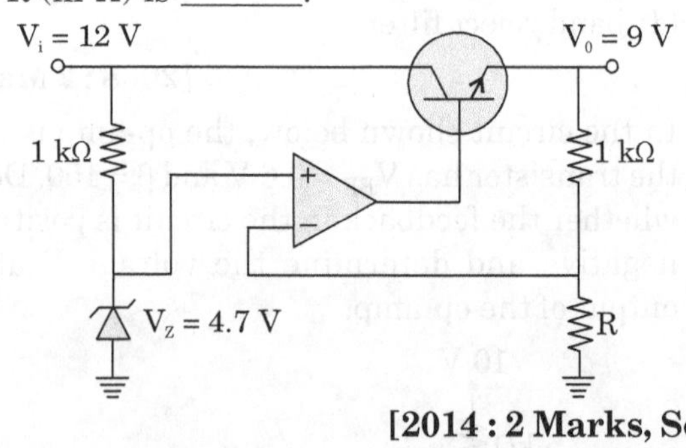

[2014 : 2 Marks, Set-1]

67. In the circuit shown, the op-amp has finite input impedance, infinite voltage gain and zero input offset voltage. The output voltage V_{out} is

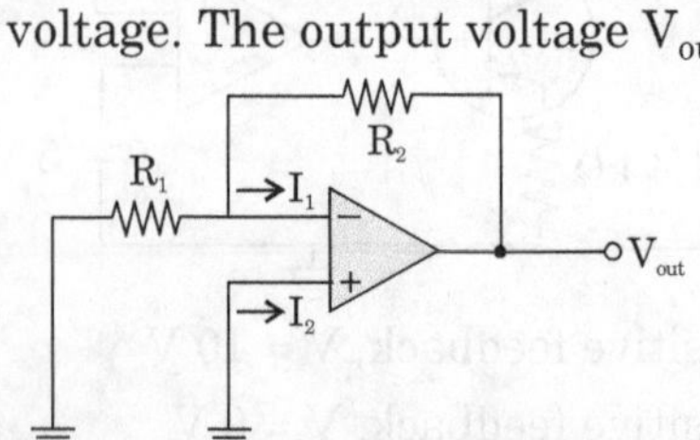

(a) $-I_2(R_1 + R_2)$ (b) I_2R_2
(c) I_1R_2 (d) $-I_1(R_1 + R_2)$

[2014 : 2 Marks, Set-1]

68. In the differential amplifier shown in the figure, the magnitudes of the common-mode and differential-mode gains are A_{CM} and A_D, respectively. If the resistance R_E is increased, then

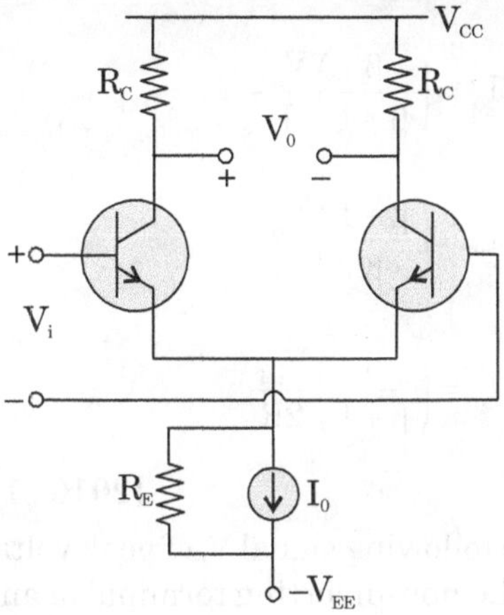

(a) A_{CM} increase

(b) common-mode rejection ratio increases

(c) A_d increase

(d) common-mode rejection ratio decreases

[2014 : 1 Mark, Set-2]

69. Assuming that the Op-amp in the circuit shown is ideal, V_0 is given by

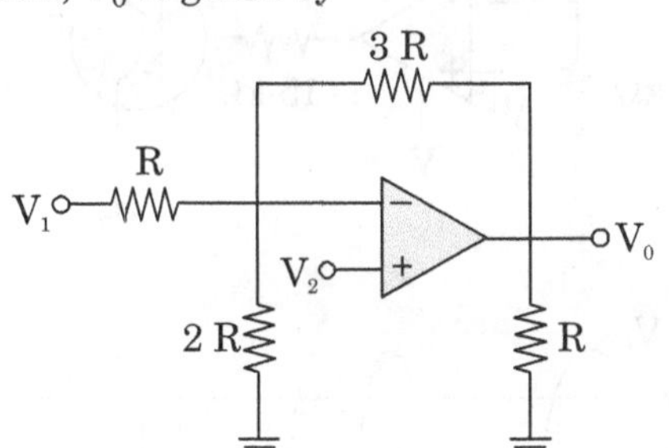

(a) $\frac{5}{2}V_1 - 3V_2$ (b) $2V_1 - \frac{5}{2}V_2$

(c) $-\frac{3}{2}V_1 + \frac{7}{2}V_2$ (d) $-3V_1 + \frac{11}{2}V_2$

[2014 : 2 Marks, Set-3]

70. The circuit shown represents

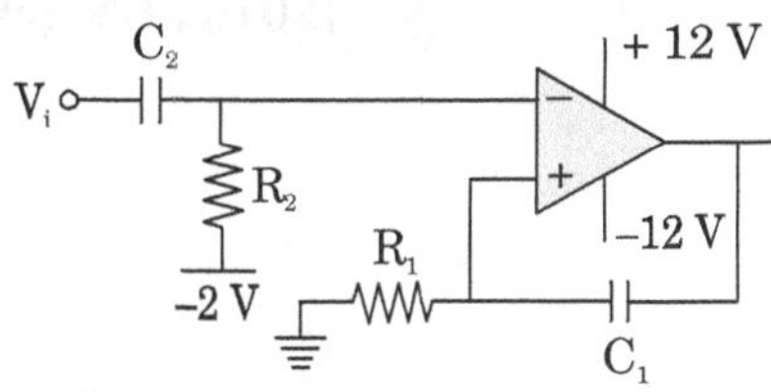

(a) a bandpass filter

(b) a voltage controlled oscillator

(c) an amplitude modulator

(d) a monostable multivibrator

[2014 : 1 Mark, Set-4]

71. In the circuit shown, assume that the op-amp is ideal. The bridge output voltage V_0 (in mV) for $\delta = 0.05$ is _______.

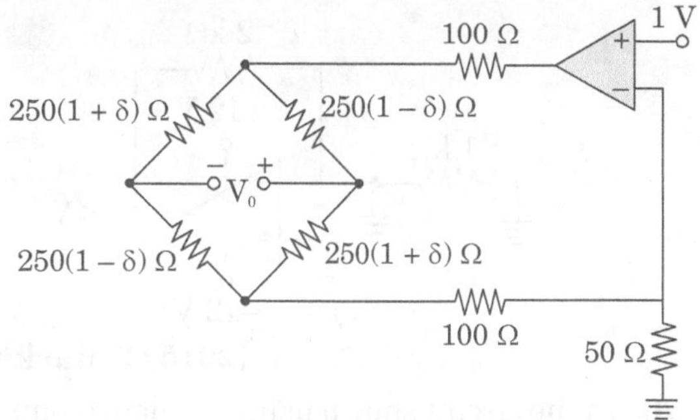

[2015 : 2 Marks, Set-1]

72. In the circuit shown, $V_0 = V_{0A}$ for switch SW in position A and $V_0 = V_{0B}$ for SW in position B. Assume that the op-amp is deal. The value of $\frac{V_{0B}}{V_{0A}}$ is _______.

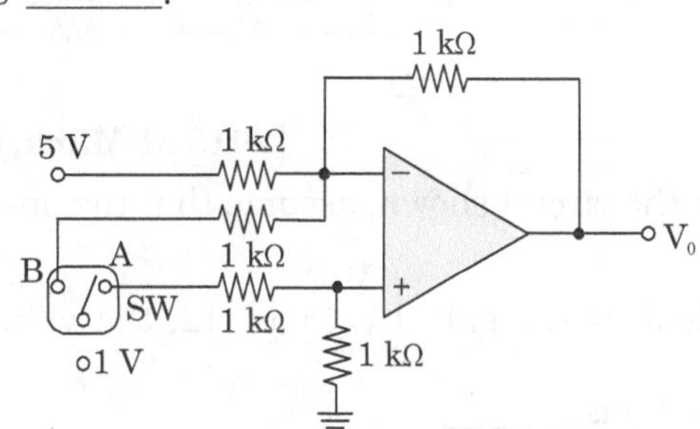

[2015 : 1 Mark, Set-2]

73. In the bistable circuit shown, the ideal op-amp has saturation levels of ±5 V. The value of R_1 (in kΩ) that gives a hysteresis width of 500 mV is _______.

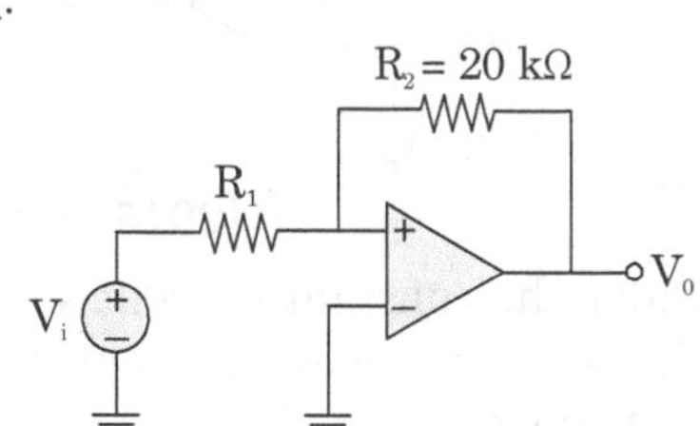

[2015 : 1 Mark, Set-2]

74. For the voltage regulator circuit shown, the input voltage (V_{in}) is 20 V ±20% and the regulated output voltage (V_{out}) is 10 V. Assume the op-amp to be ideal. For a load R_L drawing 200 mA, the maximum power dissipation in Q_1 (in Watts) is ____.

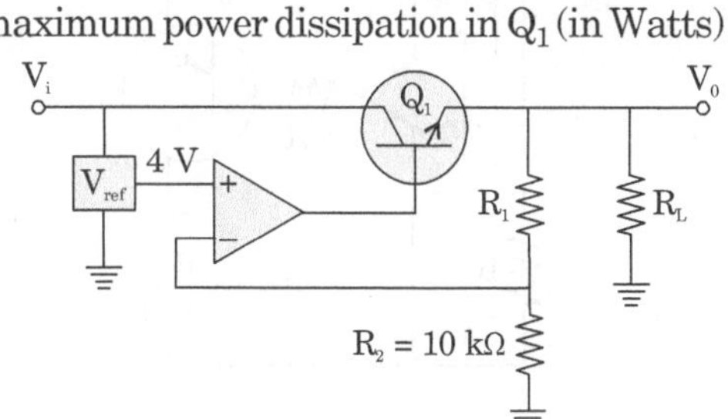

[2015 : 2 Marks, Set-2]

75. Assuming that the op-amp in the circuit shown below is deal, the output voltage V_0 (in volts) is ________.

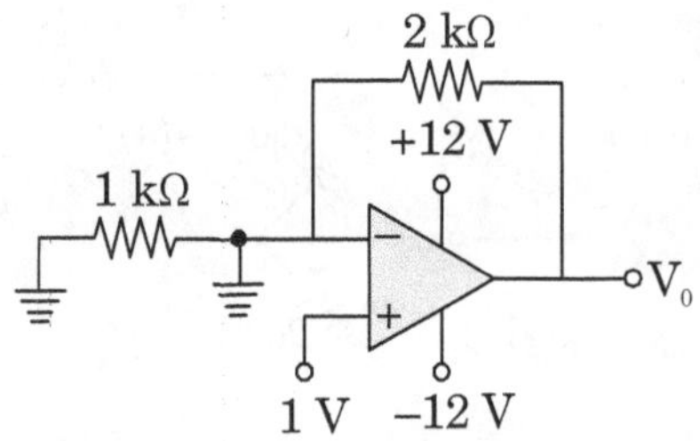

[2015 : 2 Marks, Set-2]

76. In the circuit shown using an ideal op-amp, the 3-dB cut-off frequency (in Hz) is ________.

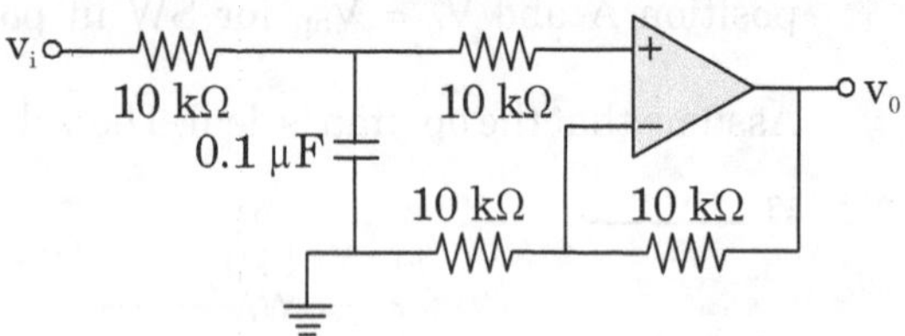

[2015 : 1 Mark, Set-3]

77. In the circuit shown, assume that the op-amp is ideal. If the gain $\left(\frac{V_0}{V_{in}}\right)$ is -12, the value of R (in Ω) is ________.

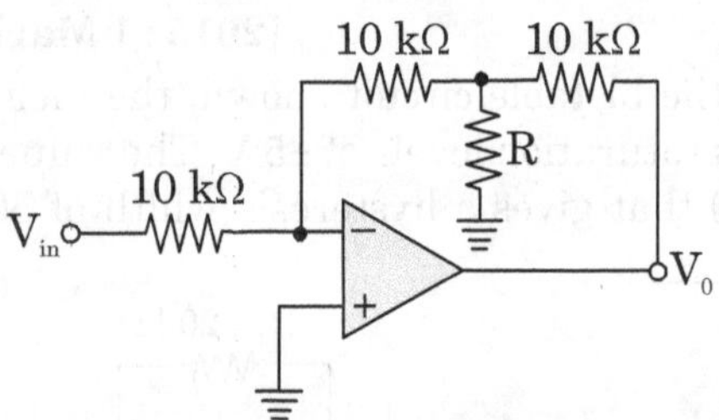

[2015 : 2 Marks, Set-3]

78. Consider the constant current source shown in the figure below. Let β represent the current gain of the transistor.

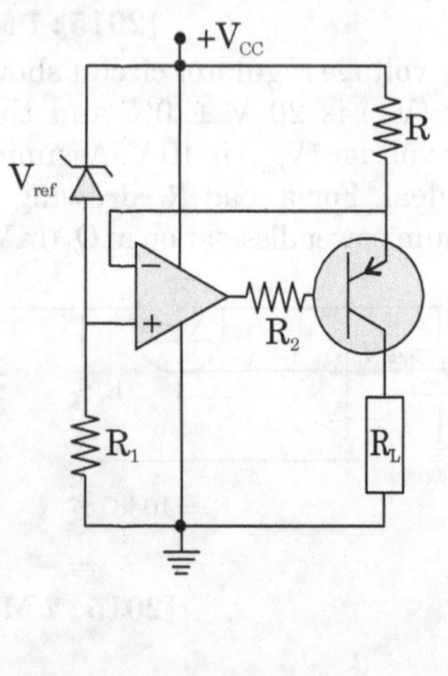

The load current I_0 through R_L is

(a) $I_0 = \left(\frac{\beta+1}{\beta}\right)\frac{V_{ref}}{R}$

(b) $I_0 = \left(\frac{\beta}{\beta+1}\right)\frac{V_{ref}}{R}$

(c) $I_0 = \left(\frac{\beta+1}{\beta}\right)\frac{V_{ref}}{2R}$

(d) $I_0 = \left(\frac{\beta}{\beta+1}\right)\frac{V_{ref}}{2R}$

[2016 : 1 Mark, Set-1]

79. The following signal V_i of peak voltage 8 V applied to the non-inverting terminal of an ideal op-amp. The transistor has V_{BE} = 0.7 V, β = 100; V_{LED} = 1.5 V, V_{CC} = 10 V and $-V_{CC}$ = −10 V

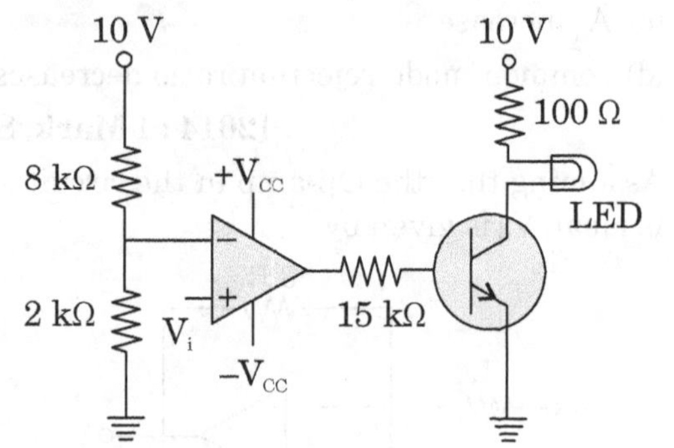

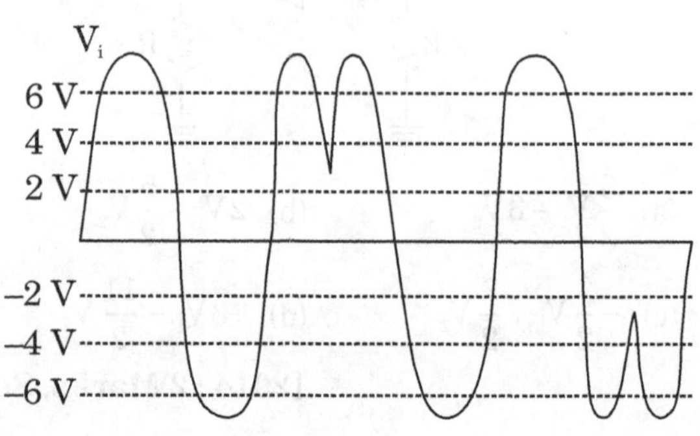

The number of times the LED glows is ________.

[2016 : 1 Mark, Set-1]

ANSWERS

1. (100) 2. (d) 3. (c) 4. (c) 5. (Stability) 6. (3.3) 7. (d) 8. (a) 9. (c) 10. (b)

11. (15) 12. (–8) 13. (0· 7985) 14. (413· 8) 15. (–1) 16. (b) 17. (44.4) 18. (1.145) 19. (c) 20. (0.5)

21. (c) 22. (d) 23. (b) 24. (c) 25. (b,c) 26. (d) 27. (c) 28. (–4) 29. (0.02)

30. ($12 \cdot 10^6$) 31. (d) 32. (d) 33. (d) 34. (a) 35. (d) 36. (c) 37. (a) 38. (b) 39. (d)

40. (a) 41. (b) 42. (b) 43. (c) 44. (b) 45. (c) 46. (a) 47. (a) 48. (b) 49. (c)

50. (a) 51. (c) 52. (c) 53. (d) 54. (a) 55. (c) 56. (b) 57. (b) 58. (d) 59. (a)

60. (b) 61. (d) 62. (b) 63. (b) 64. (c) 65. (3· 1 to 3· 26) 66. (1092 to 1094) 67. (c)

68. (b) 69. (d) 70. (d) 71. (c) 72. (11–12) 73. (1) 74. (2· 8056) 75. (11–12)

76. (159· 15) 77. (1) 78. (b) 79. (3)

EXPLANATIONS

1. Given : CMRR of given op-Amp = 60 dB.

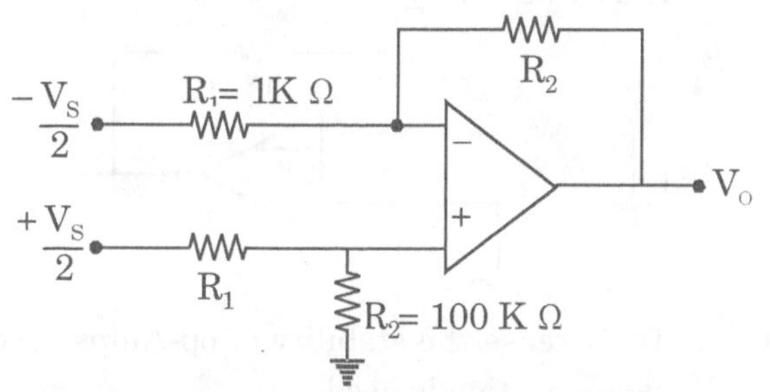

Op-amp circuit without bridge is shown in figure.

Differential gain, $|A_d| = \frac{R_2}{R_1} = 100.$

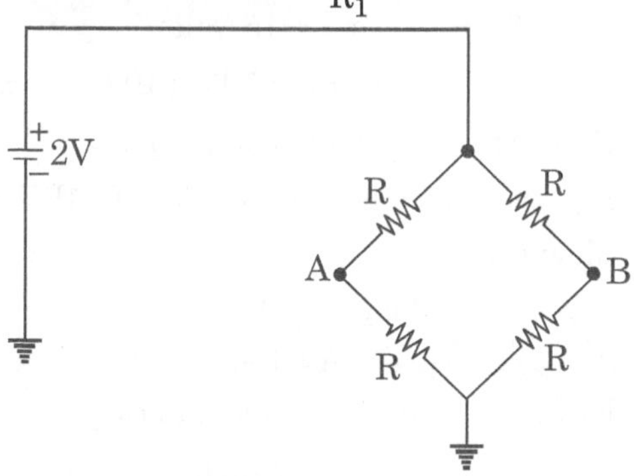

Bridge circuit of given op-amp is shown in figure. In this circuit, bridge is balanced, hence,

$$V_A = \frac{R}{R+R} \times 2 = 1V$$

$$V_B = \frac{R}{R+R} \times 2 = 1V.$$

⇒ Differential input voltage, $V_d = V_A - V_B = 0V$.

Output voltage when CMRR ≠ ∞,

is $V_0 = A_d V_d + A_c V_c = A_c V_c$.

⇒ Common mode input voltage,

$$V_c = \frac{V_A + V_B}{2} = 1 \text{ V}.$$

⇒ CMRR in dB = 20log (CMRR)

60 = 20log (CMRR)

CMRR = 10^3 = 1000.

$$\left|\frac{A_d}{A_c}\right| = 1000.$$

$$|A_c| = \frac{100}{1000} = 0.1$$

⇒ $V_o = A_c V_c = 0.1 \times 1 = 0.1$ V

= 100 mV

2. Given : $A_{OL} = 45$.

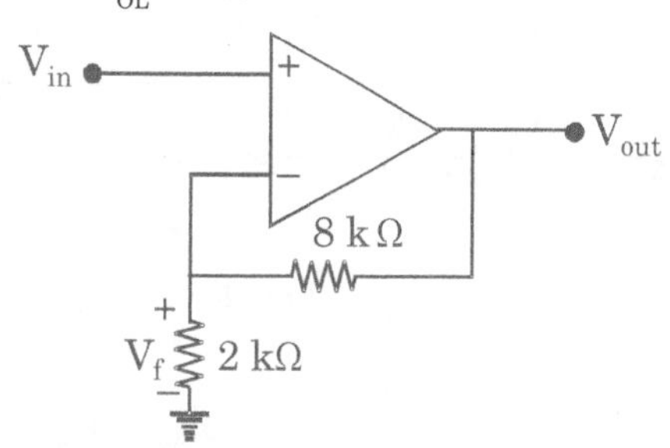

It is a non-inverting op-Amp. Closed loop gain of non-inverting op-Amp

$$A_F = \frac{A_{OL}}{1 + A_{OL} \cdot \beta}$$

Feedback gain, $\beta = \frac{V_f}{V_{out}}$

$\because$ By voltage division rule,

$$V_f = \frac{2}{2+8}.V_{out} = \frac{1}{5}V_{out}$$

$$\beta = \frac{1}{5}.$$

$$\Rightarrow A_f = \frac{45}{1+45\times\frac{1}{5}} = 4.5$$

3. CMRR $= \left|\frac{A_d}{A_c}\right|$;

where $A_d = \frac{A_1 - A_2}{2}, A_c = A_1 + A_2.$

Calculation of A_1 :

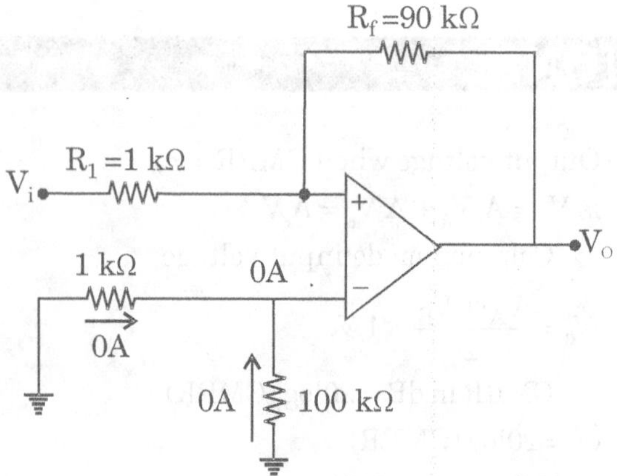

$$A_1 = \left.\frac{V_o}{V_1}\right|_{V_2=0} = \frac{-R_f}{R_1} = \frac{-90}{1} = -90.$$

calculation of A_2 :

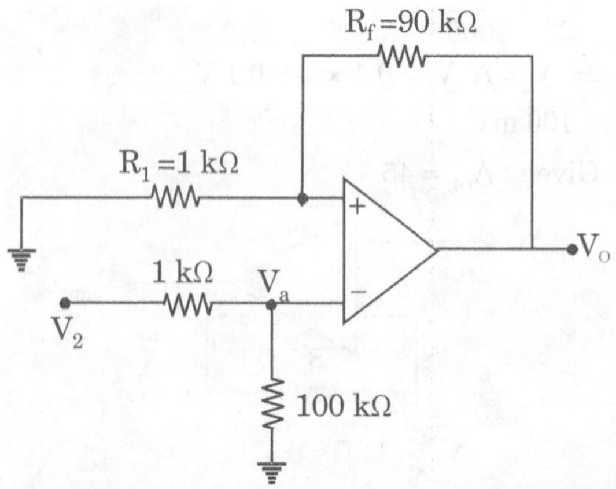

$$\Rightarrow A_2 = \left.\frac{V_o}{V_2}\right|_{V_1=0}$$

By voltage division rule,

$$V_a = \frac{100}{100+1}.V_2$$

Output of non-inverting amplifier,

$$V_o = \left(1+\frac{R_f}{R_1}\right)V_a$$

$$V_o = \left(1+\frac{90}{1}\right)\left(\frac{100}{100+1}.V_2\right) = \frac{91\times100}{101}.V_2$$

$$A_2 = \frac{V_o}{V_2} = \frac{91\times100}{101} = 90.1$$

$$\Rightarrow \text{CMRR} = \left|\frac{A_d}{A_c}\right| = \left|\frac{A_1 - A_2}{2(A_1+A_2)}\right|$$

$$= \left|\frac{-90-90.1}{2(90.1-90)}\right| = \left|\frac{-180.1}{2\times0.1}\right|$$

CMRR = 900.5

4. $V_o = V_{os} \times$ Gain

$$= 1\times10^{-3}\times\frac{10^6}{10_3}$$

$$= 1\text{ V}$$

It may be – i.e. $\pm$ 1V

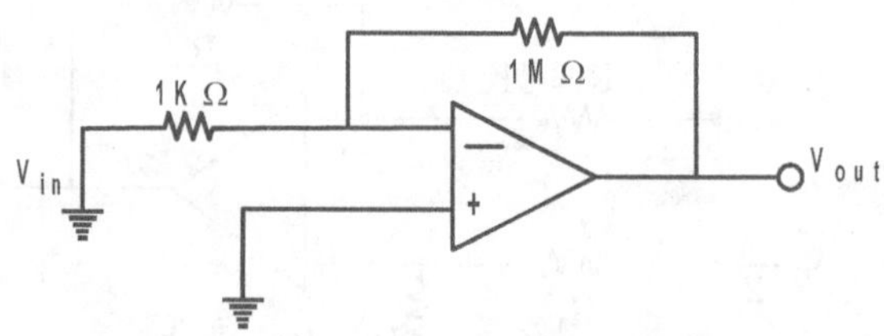

5. To increase the stability of op-Amps, frequency compensation is used.

7. $V_o = V_i . A_v$

But V_o oscillates behei + V_{sat} to $-V_{sat}$

Therefore, . Vo = $\pm$ 15 Volts

$V_o = 5\times10^{-3}\times10{,}000 = \pm 50$V

8. Characteristic of ideal op-Amp :

$A_{oL} = \infty$ $R_i = \infty$, $R_o = \infty$, BW = ∞, CMRR = ∞, Slew rate = ∞.

9. As, $20\log A = 40$

$\therefore$ $A = 100$

Now, $v_0 = A\,v_m \sin\omega t$

or $v_0 = \left|\frac{dv_0}{dt}\right|_{max} = AV_m\,2\pi f$

$\therefore$ $v_m = \left|\frac{dv_0}{dt}\right|_{max}\cdot\frac{1}{2\,\pi Af}$

$$= \frac{10^6}{2\pi\times100\times20\times10^3}$$

= 79.5 mV.

10. Id eal op-amp is an Ideal Voltage controlled voltage source.

11. Here an ideal opamp has voltage source V_1, V_3, V_5, ..., V_{N-1} connected to the non-inverting input and V_2, V_4, V_6, ..., V_N connected to the inverting input as shown in the fig. below.

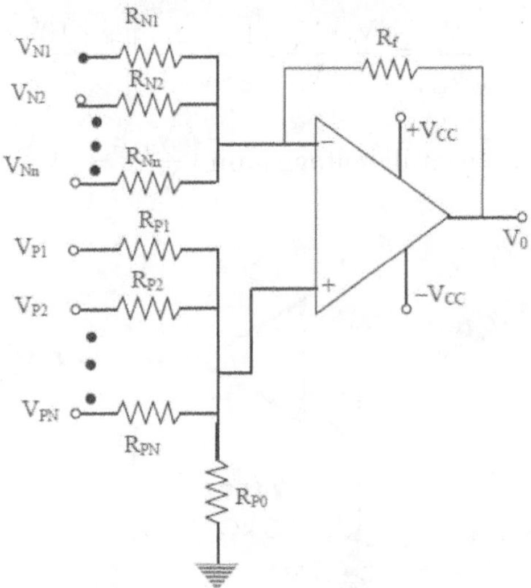

Using superposition, the output can be shown by

$$V_0 = \left[1 + \frac{R_f}{R_N}\right]$$

$$\left[\frac{R_p}{R_{p1}}V_{p1} + \frac{R_p}{R_{p2}}V_{p2} + \ldots \frac{R_P}{R_{PN}}V_{pn}\right]$$

$$-\left[\frac{R_f}{R_{N1}}V_{N1} + \frac{R_f}{R_{N2}}V_{N2} + \ldots \frac{R_f}{R_{Nn}}V_{Nn}\right]$$

Where $\quad R_N = R_{N1} \| R_{N2} \| \| \ldots \| R_{Nn}$

and $\quad R_p = R_{p1} \| R_{p2} \ldots R_{PN} \| R_{PO}$

In the problem given

$$R_f = R_{N1} = R_{N2} = \ldots$$
$$= R_{Nn} = 10\ k\Omega$$
$$R_{p1} = R_{p2} = R_{p3} = \ldots$$
$$= R_{PN} = R_{PO} = 1\ k\Omega$$

$$\therefore \quad V_0 = \left[1 + \frac{10k}{\left(\frac{10k}{n}\right)}\right]$$

$$\left[\frac{\frac{1k}{(1+n)}}{1k}V_{P1} + \frac{\left(\frac{1k}{1+n}\right)}{1k}V_{P2} + \ldots\ldots\right]$$

$$-\left[\frac{10k}{10k}V_{N1} + \frac{10k}{10k}V_{N2} + \ldots\ldots\right]$$

$$\therefore \quad V_0 = (V_{P1} + V_{P2} + \ldots\ldots V_{Pn}) - (V_{N1} + V_{N2} + \ldots\ldots V_{Nn})$$

If the series approaches ∞ then

$$V_0 = \left(1 + \frac{1}{3} + \frac{1}{5} + \frac{1}{7} \ldots\ldots\right) - \left(\frac{-1}{2}\frac{-1}{4}\frac{-1}{6} - \ldots\ldots\right)$$

$$= 1 + \frac{1}{2} + \frac{1}{3} + \frac{1}{4} + \frac{1}{5} + \ldots\ldots = \infty$$

This series is called harmonic series which is a divergent infinite series

$$\therefore \quad V_0 = +\infty = +V_{sat}$$
$$= +V_{CC} = 15V$$

Hence, the output voltage is 15 V (positive).

12. From given data

The photo diode with responsivity 0.8 A/W

$$\therefore \quad \text{Diode current} = 0.8\ \text{A/W}[10\ \mu\text{W}]$$
$$= 8 \times 10^{-6}\ \text{A} = 8\ \mu\text{A}$$
$$V_0 = -8\mu\ (1\text{M})$$
$$= -8 \times 10^{-6} \times 10^{-6}\ \text{V} = -8\ \text{V}$$
$$I_L = \frac{-8}{10k} = \frac{-8}{10 \times 10^3}$$
$$= \frac{-8}{10^4} = -8 \times 10^{-4}\ \text{A}$$
$$= -800 \times 10^{-6}\ \text{A}$$
$$= -800\ \mu\text{A}$$

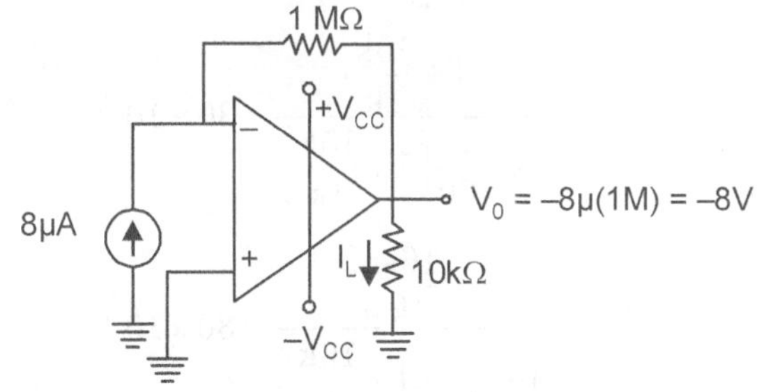

Hence, the value of photo current (throughput the load) is – 800 μA.

13. Initially switch is closed and $V_B = 10$ V

$$\Rightarrow V_{01} = -10\ \text{V}$$
$$\Rightarrow V_0 = -V_2 = -5\ \text{V}$$
$$\Rightarrow V_A = \frac{V_0}{4K + 1K} \times 1K = \frac{-5}{5K} \times K = -1\ \text{V}$$

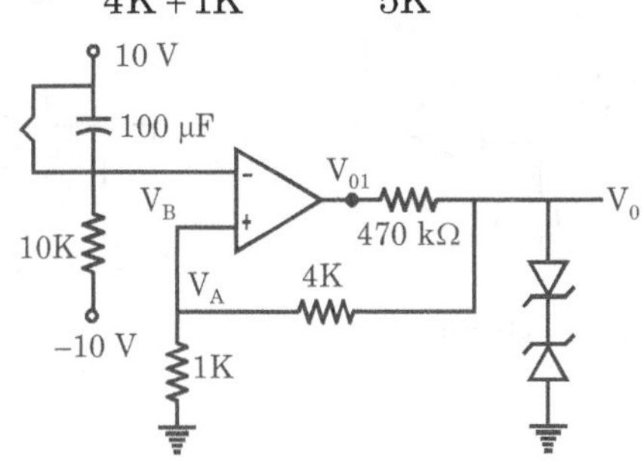

At t = 0;

The switch is opened and as $t \to \infty$, V_B approaches –10 V.

Let at $t = T_1$,

V_B exceeds V_A (–1 V) so that V_{01} changes from –10 V to 10 V

$\Rightarrow$ V_0 changes from –5 V to 5 V

$$V_B = V_f + (V_i - V_f)e^{-t/\tau}$$
$$= -10 + [10 - 1 - 10]e^{-t/RC}$$

(Since $\tau = RC$)

At $t = T_1$ $\quad V_B = -1$

$$-1\text{ V} = -10 + 20e^{-T_1/RC}$$

$$\Rightarrow \quad T_1 = RC \ln \frac{20}{9}$$
$$= 10 \times 10^3 \times 100 \times 10^{-6} \times 0.798$$
$$= 0.798 \text{ sec}$$

Hence, the required time at which V_0 changes state is 0.798 s.

14. Overall input = $V_{ios} + V_{input}$

(Here V_{ios} = input offset voltage)

$$= 5\text{ mV} + 25\text{ mV} = 30\text{ mV}$$

$$V_0 = \frac{\left(1+\frac{R_F}{R_1}\right)}{1+\frac{1}{A_{OL}}\left(1+\frac{R_F}{R_1}\right)} \times \text{Overall input}$$

$$= \frac{1+\frac{15k}{1k}}{1+\frac{1}{100}\left(1+\frac{15k}{1k}\right)} \times 30 \times 10^{-3}$$

$$= \frac{\frac{16k}{1k}}{1+\frac{1}{100}\left(1+\frac{16k}{k}\right)} \times 30 \times 10^{-3}$$

$$= \frac{16}{116} \times 30 \times 10^{-3} \times 100$$

$$= \frac{48000}{116} \times 10^{-3}\text{ V} = 413.79\text{ mV}$$

Hence, the output V_0 is 413.79 mV.

15.

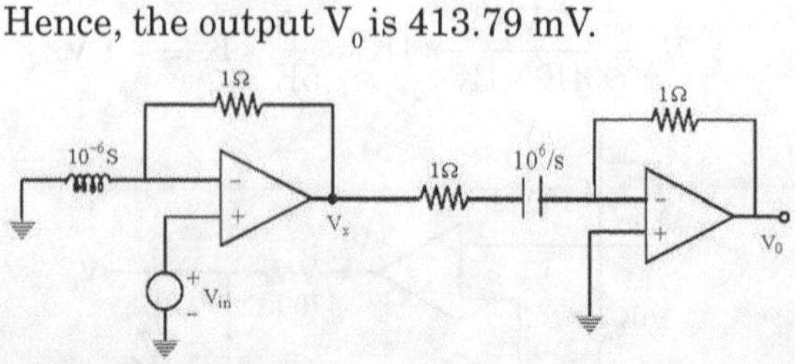

From above fig.

$$\frac{V_x}{V_{in}} = 1 - \frac{1}{10^{-6}s} = 1 + \frac{10^6}{s} = \frac{s+10^6}{s}$$

$$V_0 = \left[\frac{-1}{1+\frac{10^6}{s}}\right] V_x = \frac{V_0}{V_x} - \frac{-s}{s+10^6}$$

$$\therefore \quad \frac{V_0}{V_{in}} = \frac{V_0}{V_x} \cdot \frac{V_x}{V_{in}} = \left[\frac{-s}{s+10^6}\right]\left[\frac{s+10^6}{s}\right] = -1$$

$\therefore$ overall voltage gain $\left(\frac{V_0}{V_{in}}\right) = -1$

16.

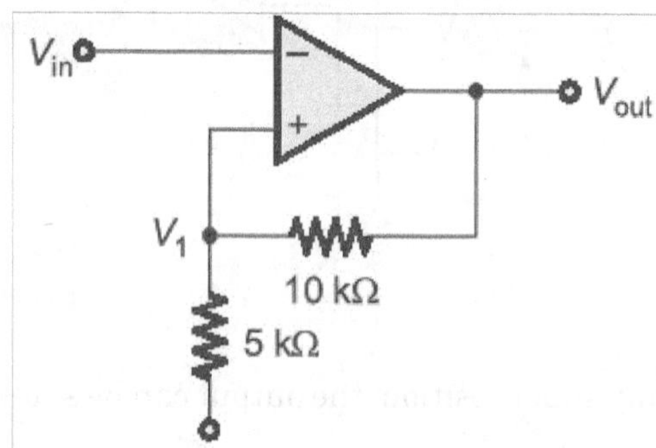

At node V_1,

$$V_1 = \frac{(3 \times 10) + (V_0 \times 5)}{15}$$
$$= \frac{30 + 5V_0}{15}$$
$$= \frac{5}{15}(6 + V_0) = \frac{(6 + V_0)}{3}$$

Now, $\quad V_{UT} = \frac{6+15}{3} = 7\text{V}$

and $\quad V_{LT} = \frac{6-15}{3} = -3\text{V}$

17. In the given circuit,

Feedback factor (β) = $\frac{R_1}{R_1 + R_2} = \frac{1}{80}$

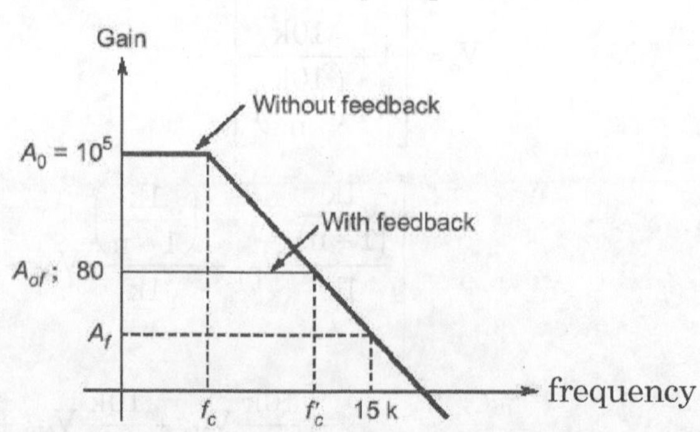

$\therefore \quad A_{nf} = \dfrac{A_0}{1 + A_0\beta} \simeq 80$

and $\quad f_c' = f_c(1 + A_0\beta)$

$$= 8\left(1 + \frac{10^5}{80}\right) \text{Hz} = 10{,}008 \text{ Hz}$$

Now $\quad$ Gain at f = 15 kHz given by

$$A_f = \frac{A_{0f}}{\sqrt{1 + \left(\frac{f}{f_c'}\right)}}$$

$$= \frac{80}{\sqrt{\left(1 + \left(\frac{15000}{10008}\right)^5\right)}} \simeq 44.4$$

19. For t > 0,

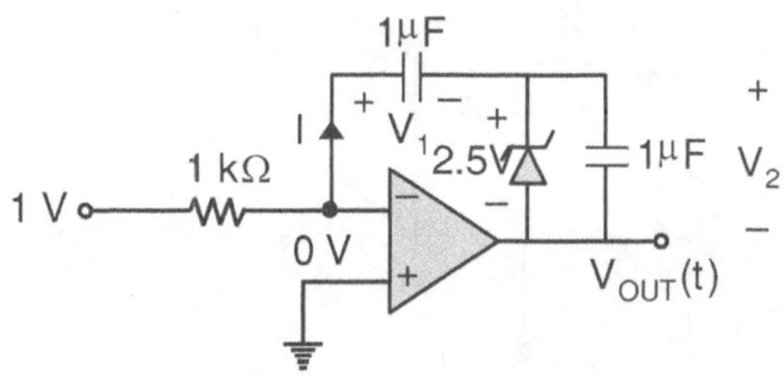

$$I = \frac{1\text{V}}{1\text{k}\Omega} = 1\text{mA}$$

Till t = 2.5 msec, both V_1 and V_2 will increase and after t = 2.5 msec, V_1 = 2.5 V and V_2 increases with time.

when $v_{out}(t) = -10$ V, $V_1 = 7.5$ V

So, $\quad \dfrac{1}{1\mu\text{F}} \displaystyle\int_0^t (1\text{mA})\,dt = 7.5\text{V}$

$$10^3 t = 7.5$$

$$t = 7.5 \text{ msec}$$

20. Applying the concept of virtual ground, we get,

$V_0 = \dfrac{R_2}{R_1} \cdot V_{in}$ $\quad$ [∵ non-inverting amplifier]

$\therefore \quad V_0 = \dfrac{31\text{k}\Omega}{1\text{k}\Omega} \times 1\text{V}$

$V_0 = -31 \text{ V} < -15 \text{ V}$ $\quad$ [Not possible]

Hence, the output voltage of the op-amp is equal to –15 V.

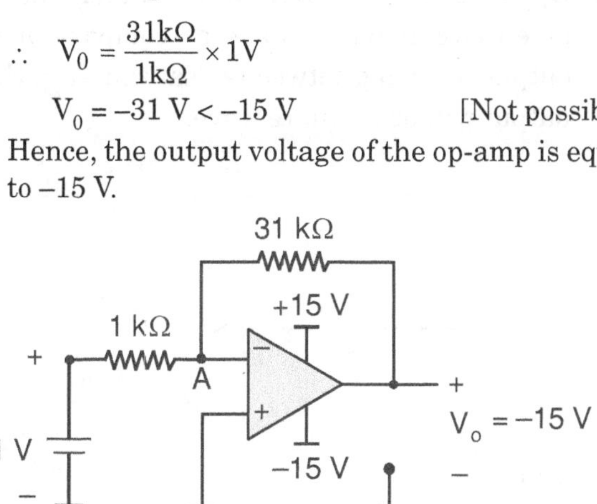

Now applying KCL at node 'A', we get,

$$\frac{V_A - 1}{1\text{k}\Omega} + \frac{V_A - (-15)}{31\text{k}\Omega} = 0$$

$$\frac{V_A}{1\text{k}\Omega} + \frac{V_A}{31\text{k}\Omega} = \frac{-15}{31\text{k}\Omega} + \frac{1}{1\text{k}\Omega}$$

$$\therefore \quad V_A\left[\frac{1}{1} + \frac{1}{31}\right] = -\frac{15}{31} + 1$$

$$V_A = 0.5 \text{ V}$$

21. The voltage at OPAMP inputs, $V_+ = V_- = V_2 = 3$ volts

then, by voltage division rule, $\dfrac{V_0}{20 + 40} \times 20 = 3$

or $\quad V_0 = 9$ V

22.

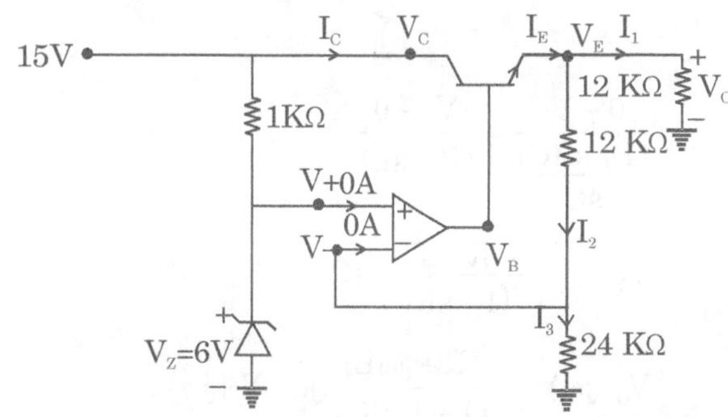

Since, supply voltage (15V) > V_z (6V). So, Zener diode will be in breakdown region.

$V_+ = V_z = 6$V.

There is a negative feedback through base-emitter junction of transistor. From virtual ground concept,

$V_+ = V_- = 6$V.

From figure, $I_3 = \dfrac{V_-}{24} = \dfrac{6}{24} = 0.25$mA.

$I_2 = I_3 = 0.25$ mA

$V_E = I_2 \times 12 + V_- = 0.25 \times 12 + 6 = 9$ V.

$I_1 = \dfrac{V_E}{12} = \dfrac{V_0}{12} = \dfrac{9}{12} = 0.75$mA

If there is no information of β of transistor then we assume β = ∞,

$I_B = 0$A. $I_c \cong I_E = I_2 + I_1 = 0.75 + 0.25 = 1$ mA

From figure, $V_c = 15$V.

$V_{cE} = 15 - 9 = 6$V.

Power dissipation across transistor, $P_D = V_{cE} \times I_c = 6 \times 1 = 6$ mW.

23. If unregulated power supply increases by 20%, i.e., new unregulated power supply is,

$$V_S' = 15 + \frac{20}{100} \times 15 = 18\text{V}.$$

$V'_{cE} = V_c - V_E = 18 - 9 = 9$ V.

$P'_D = V'_{cE} \times I_c = 9 \times 1 = 9$ mW.

Increase in power dissipation is

$$\frac{P_D' - P_D}{P_D} \times 100 = \frac{9-6}{6} \times 100 = 50\%.$$

24. Readrawing the given circuit in S-domain, Applying KLC at inverting terminal,

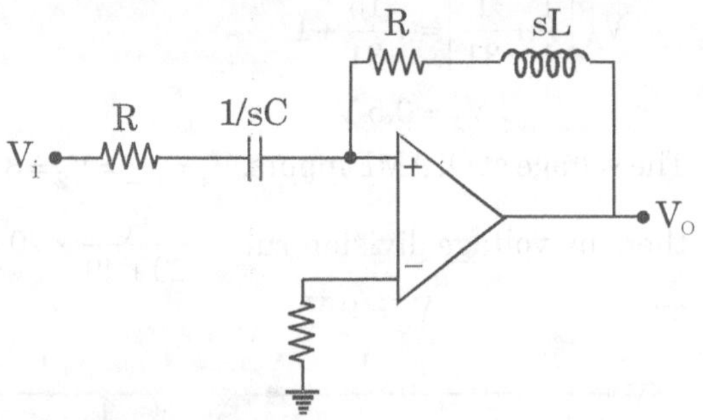

$$\frac{0 - V_i}{\frac{(1+sRC)}{sC}} = \frac{V_o - 0}{(R+sL)}$$

$$V_o(s) = -\frac{(R+sL)}{(1+sRC)}.sC.V_i(s)$$

$$V_o(j\omega) = -\frac{(R+j\omega L)}{(1+j\omega RC)}.j\omega c.V_i(j\omega)$$

Given : $R = \sqrt{\frac{L}{C}}, \omega = \frac{1}{\sqrt{LC}}$

$$V_o(j\omega) = \frac{-\left(\sqrt{\frac{L}{C}} + j\frac{1}{\sqrt{LC}} \times L\right)}{1 + j\frac{1}{\sqrt{LC}} \times \sqrt{\frac{L}{C}} \times C} \times j\frac{1}{\sqrt{LC}} \times C \times V_i(j\omega)$$

$$V_o(j\omega) = -jV_i(j\omega)$$

$$V_o(j\omega) = 1\underline{|-90^\circ} \times V_1\underline{|0^\circ}$$

$$V_o(j\omega) = V_1\underline{\left|-\frac{\pi}{2}\right.} = V_1\underline{\left|\frac{3\pi}{2}\right.}$$

25. Given circuit consists of two diodes and initially at t = o, the states of diode are unknown.

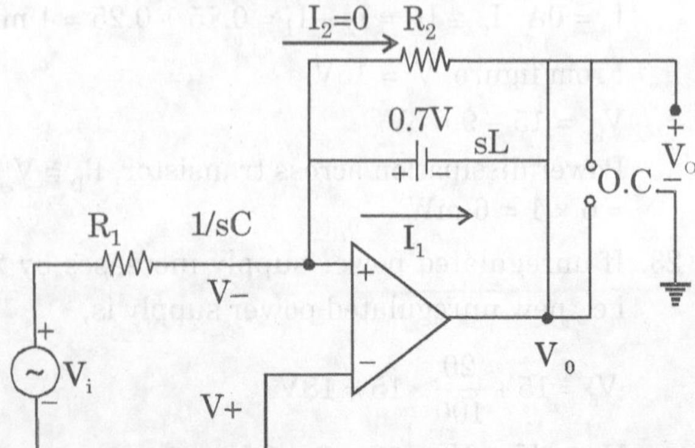

Case 1 : $V_i > 0 \Rightarrow V_- > V_+$

It means V_0' goes toward negative saturation.

when $V_0' = -0.7$ V, then diode (feedback) becomes forward biased. Other diode becomes reverse biased.

⇒ From virtual ground concept,

$V_- = V_+ = 0V_1$, and $I_2 = 0A$.

$$I_2 = \frac{0 - V_0}{R_2} = O \Rightarrow V_o = -0 \times R_2 = oV.$$

Case 2 : $V_i < O \Rightarrow V_- < V_+$

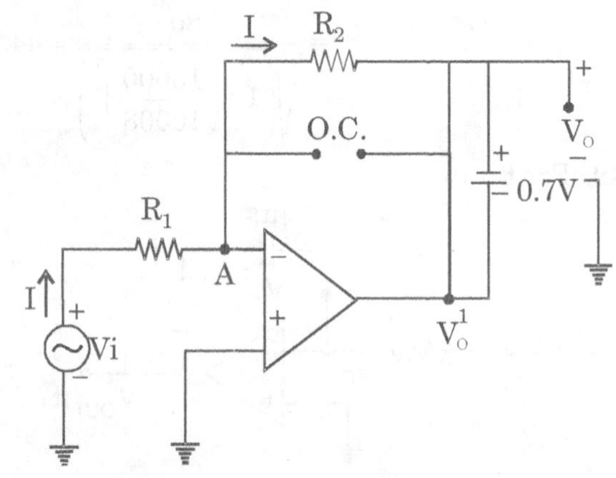

It means V_0' goes toward positive saturation. when $V_0' = 0.7$ V, then feedback diode becomes reverse biased. Another diode becomes forward biased.

From virtual ground concept,

$V_- = V_A = V_+ = 0V.$

Modified circuit represents inventing op-Amp circuit.

Output voltage, $V_o = -\frac{R_2}{R_1} \times V_i$

$$\text{Now, } V_o = \begin{cases} O, & V_i > 0V \\ \frac{-R_2}{R_1}V_i, & V_i < 0V \end{cases}$$

26. Open-loop gain of Op-Amp is very high, therefore given circuit will behave as comparator with output switching between +Vsat and $-V_{sat}$. Hence, output will be a square wave.

28.

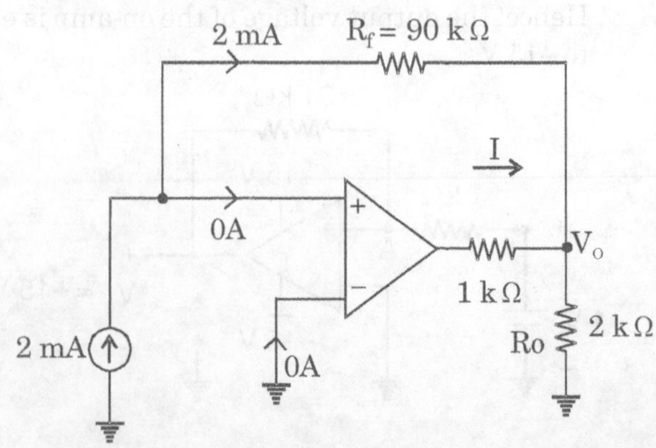

Applying KLC at inventing terminal,

$$\frac{0 - V_o}{R} - 2\text{ mA} = 0$$

$V_o = -2 \times 10^{-3} \times 2 \times 10^3 = -4V.$

Applying KCL at output node,

$$-2 \times 10^{-13} - I + \frac{V_o}{R_o} = 0$$

$$-2 - I + \left(\frac{-4}{2 \times 10^3}\right) = 0$$

$I = -4$ mA.

29.

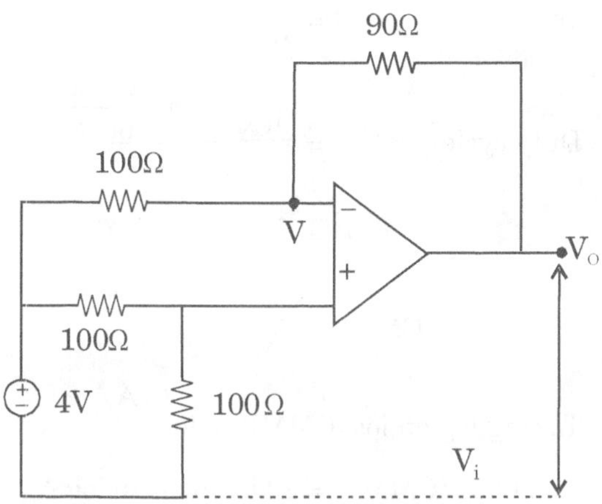

Applying KCL at non-inverting terminal,

$$\frac{V-4}{100} + \frac{V}{100} = 0 \Rightarrow 2V = 4 \Rightarrow V = 2 \text{ volt}$$

Applying KCL at inverting terminal,

$$\frac{V-4}{100} + \frac{V - V_o}{100} = 0 \Rightarrow \frac{2-4}{100} + \frac{2 - V_o}{100} = 0$$

$200 - 100\,V_o = 198 \Rightarrow V_o = 0.02$ volt.

31. Given circuit is Schmitt trigger (positive feedback circuit), which is used to convert any sinusoidal signal into square wave.

32. When the op-Amp circuit has both positive feedback as well as negative feedback then virtual ground concept will be applicable.

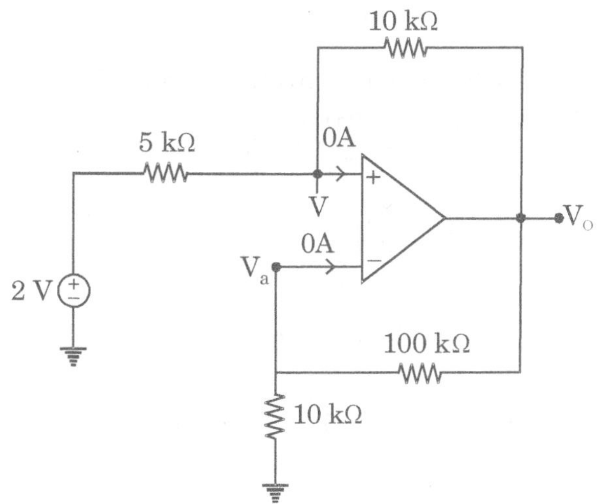

By using voltage division rule at non-inverting terminal,

$$V_a = \frac{10}{10 + 100} . V_o$$

$$V_a = \frac{1}{11} . V_o \qquad \text{...(i)}$$

Applying KCL at inverting terminal.

$$\frac{V_a - 2}{5} + \frac{V_a - V_o}{10} = 0$$

$-4 + 2V_a + V_a - V_0 = o$

$\therefore\ 3V_a - V_o = 4$

from equation (i),

$$\frac{3V_o}{11} - V_o = 4 \Rightarrow V_o = \frac{-44}{8} = -5.5.$$

33.

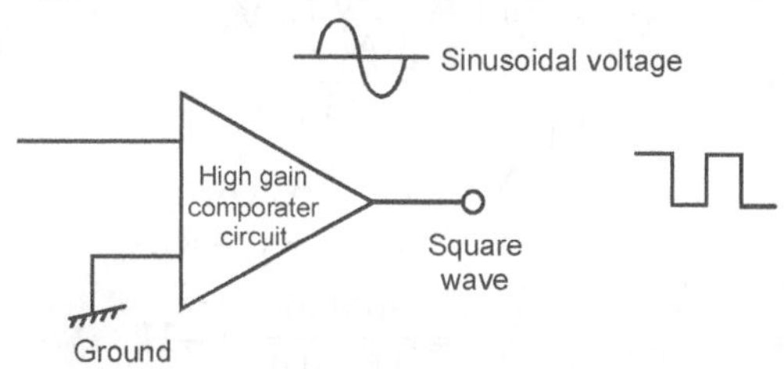

34. The pole closest to the origin is called dominant pole.The first dominant pole has to be at a very low value to ensure stability for an inverting amplifier.

35. Because of positive feedback

$$V_o = +V_{sat} = +15V$$

with positive feedback op-amp operates in saturation region $V_o = \pm V_{sat}$. Here +1 volt is applied at non inverting termal, hence

$$V_o = +V_{sat} = +15V$$

36.

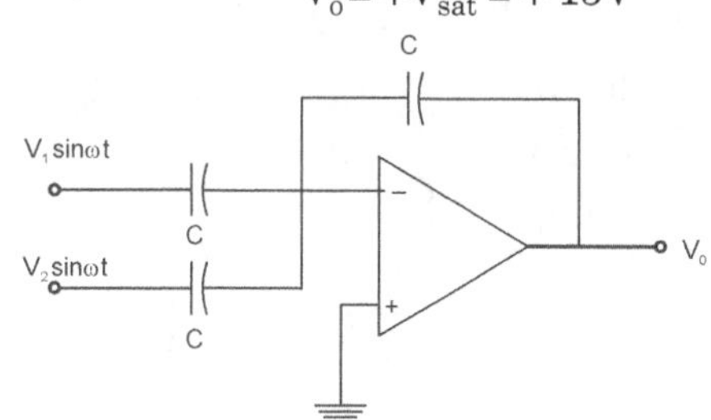

Using KCL at inverting terminal,

$$\frac{(V_1 \sin \omega t - 0)}{1/\omega C} + \frac{(V_2 \sin \omega t - 0)}{1/\omega C} = \frac{0 - V_0}{1/\omega C}$$

$$\Rightarrow \qquad V_0 = -(V_1 + V_2) \sin \omega t$$

37. $V_o = -R\omega C V_i = -RC\,\dfrac{dv_i}{dt}$

38.

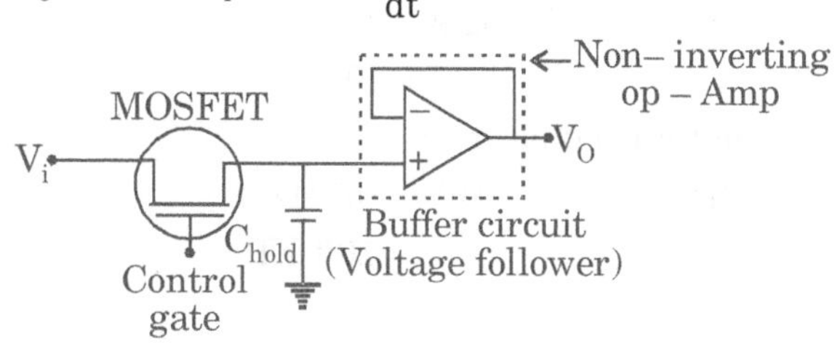

Buffer is used in sample and hold circuit. Generally, a buffer is a unity gain non-inverting amplifier.

39.

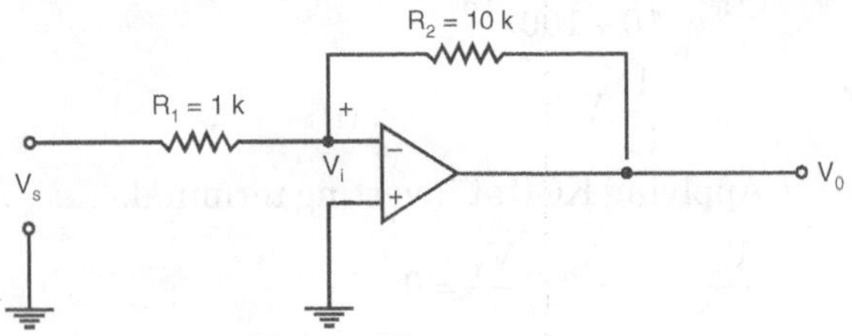

$$V_i A_v = V_0$$

KCL at inverting node,

$$\frac{V_s - V_i}{R_1} = \frac{V_i - V_0}{R_2}$$

$$\Rightarrow \quad \frac{V_s - \left(\frac{V_0}{A_v}\right)}{R_1} = \frac{\left(\frac{V_0}{A_v}\right) - V_0}{R_2}$$

$$\Rightarrow \quad \frac{V_0}{V_s} = \frac{A_v . R_2}{(R_1 + R_2) - R_1 A_v}$$

$$= \frac{100.10}{11 - 100} \approx \mathbf{-11.}$$

40.

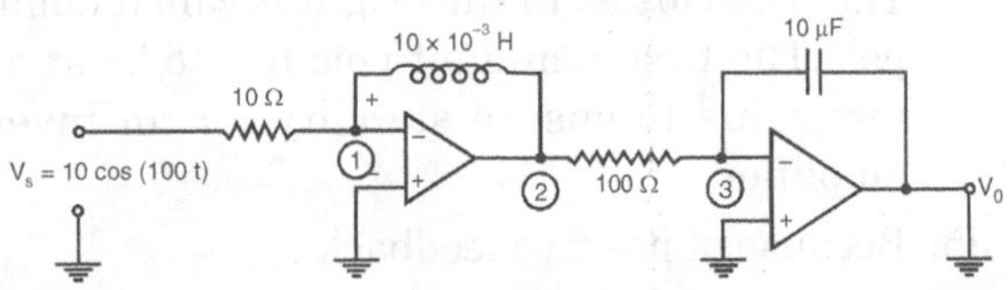

KCL at node (1),

$$\frac{V_s}{10} = -\frac{1}{L}\int V_2\, dt$$

$$\Rightarrow \quad V_s = \frac{-10}{L}\int V_2\, dt \qquad \text{...(i)}$$

KCL at node (3),

$$\Rightarrow \quad \frac{V_3 - V_2}{100} = C\,\frac{dV_0}{dt}$$

$$\Rightarrow \quad -V_2 = 100\,C\,\frac{dV_0}{dt} \quad [\because V_3 = 0]$$

$$\Rightarrow \quad V_0 = -\frac{1}{10^{-3}}\int V_2\, dt$$

From equation (i), we get

$$V_0 = 10 \cos 100\,t$$

41. Gain-bandwidth product = BW × A

$$\Rightarrow \quad 10^6 = BW \times A \qquad \text{...(i)}$$

Given, $20 \log A = 20$

$$\Rightarrow \quad A = 10$$

put A in equation (i)

then $BW = 10^5$ Hz

= **100 KHz**

42. The output voltage waveform is shown in the figure,

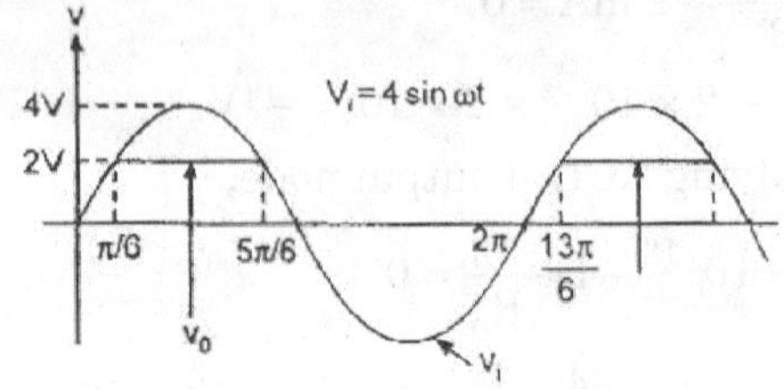

Here, at $V_i = 2$, $\sin \omega t = \frac{1}{2}$

or $\omega t = \frac{\pi}{6}$

$$\text{Duty cycle} = \frac{T_{ON}}{T} = \frac{\pi - \frac{\pi}{6} - \frac{\pi}{6}}{2\pi}$$

$$= \frac{1}{3}$$

43.

$$\text{CMMR} = \frac{A_d}{A_c}$$

Taking log on, $\log(\text{CMMR}) = \log \frac{A_d}{A_c}$

$$\Rightarrow \quad 20 \log(\text{CMMR}) = 20 \log A_d - 20 \log A_c$$

$$\therefore \quad = 48 - 2 = 46 \text{ dB}$$

44. By using voltage division rule,

$$V_a = \frac{3 \times 8}{1 + 8} = \frac{8}{3}\text{V}.$$

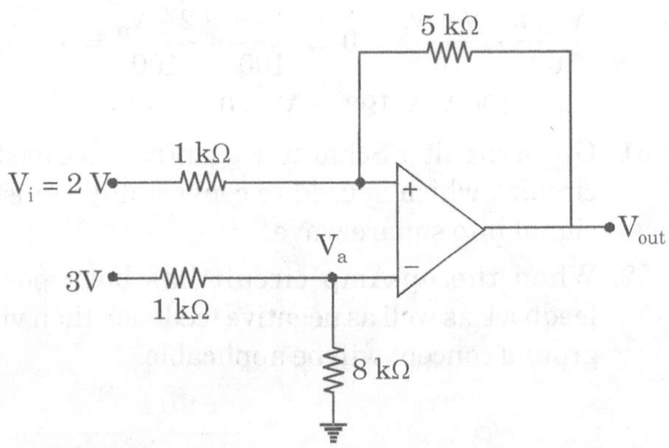

Applying KCL at inverting terminal,

$$\frac{V_a - V_1}{1} + \frac{V_a - V_{out}}{5} = 0$$

$$\frac{\left(\frac{8}{3}\right) - 2}{1} + \frac{\left(\frac{8}{3}\right) - V_{out}}{5} = 0$$

$$\frac{40}{3} - 10 + \frac{8}{3} - V_{out} = 0$$

$$\Rightarrow V_{out} = \frac{48}{3} - 10 = 16 - 10 = 6\text{V}.$$

45. The voltage amplifier can be represented as

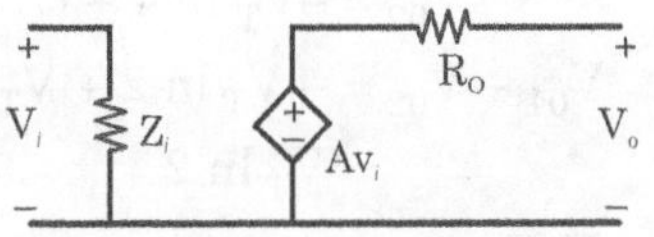

For three such cascaded amplifiers

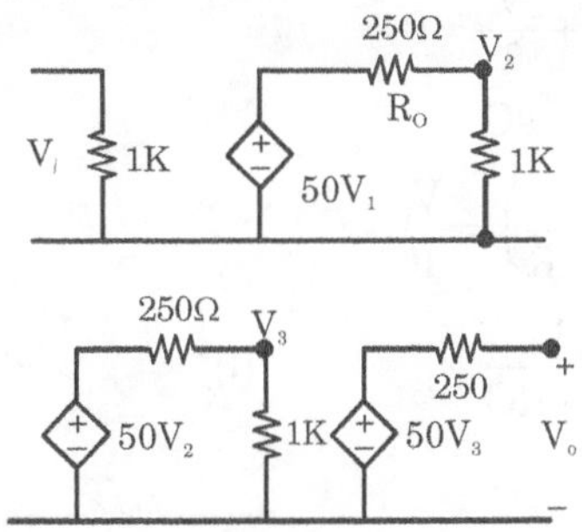

$$V_o = 50\ V_3$$

$$\downarrow$$

$$V_3 = \frac{1000}{1250} \times 50\ V_2 = 40\ V_2$$

$$\downarrow$$

$$V_2 = \frac{1000}{1250} \times 50\ V_1 = 40\ V_1$$

$$V_0 = 50 \times 40 \times 40\ V_1$$

$$A_v = \frac{V_o}{V_1}$$

$$= 50 \times 40 \times 40$$

$$= 8 \times 10^4$$

$\therefore$ A_v in dB = 20 log (8×10^4) = 98 dB

46. The circuit is second-order low-pass filter.

47.

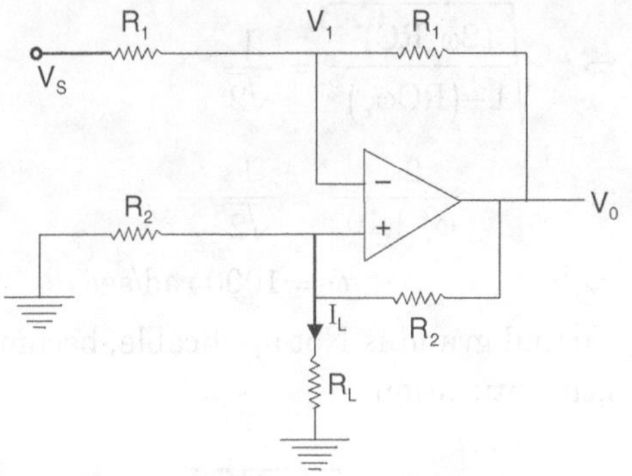

$$\frac{V_s - V_1}{R_1} = \frac{V_1 - V_O}{R_1}$$

$$\Rightarrow \quad \frac{V_s + V_O}{2} = V_1$$

$$I_L = \frac{V_1}{R_L}$$

$$\frac{V_O - V_1}{R_1} = \frac{V_1}{R_2} + \frac{V_1}{R_L}$$

$$V_o = \left(2 + \frac{R_2}{R_L}\right) V_1$$

$$= (2V_1 - V_s)$$

$$\frac{R_2 V_1}{R_L} = -V_s$$

$$\frac{V_1}{R_L} = \frac{-V_s}{R_2}.$$

48. For an ideal OPAMP, $R_{in} = \infty$, $R_o = 0$, $A_V = \infty$

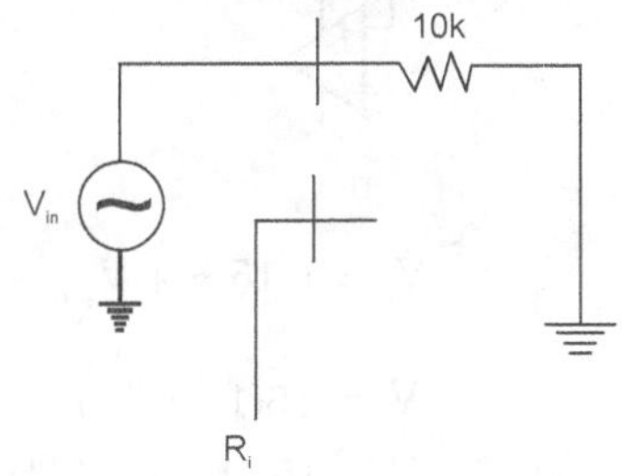

From the equivalent circuit, R_i = 10 KΩ

Equivalent circuit for an ideal OPAMP

49. Only input offset current can be measured.

50. Using KCL at riverting terminal,

$$\frac{V_- - V_o}{10k} + \frac{V_- - 0}{10k} = 0$$

$$\Rightarrow \qquad V_- = \frac{V_O}{2} \quad ...(A)$$

At the non-inverting terminal,

$$\frac{V_+}{R} + \frac{V_+ - V_i}{\frac{1}{CS}} = 0$$

Since $\quad V_+ = V_-$

then, $V_+\left[\frac{1}{R} + CS\right] = CS\ V_i$

$$\frac{V_O}{2}\left[\frac{1}{R} + CS\right] = CS\ V_i$$

$$\Rightarrow \quad \frac{V_O(s)}{V_i(s)} = \frac{2RCS}{RCS + 1}$$

$$\Rightarrow \quad \frac{V_O(j\omega)}{V_i(j\omega)} = T(j\omega) = \frac{2RCj\omega}{RCj\omega + 1}$$

When $\quad \omega \to 0,\ |T(j\omega)| \to 0$

When $\quad \omega \to \infty,\ |T(j\omega)| = 2$

Therefore this is high-pass filter. At cut-off frequency

$$|T(j\omega)|_{\omega_c = \omega_c} = \frac{1}{\sqrt{2}}|T(j\omega)|_{\omega=\infty}$$

$$\Rightarrow \quad \left|\frac{j2\omega_c RC}{jRC\omega_c + 1}\right| = \frac{1}{\sqrt{2}}$$

$$\Rightarrow \sqrt{\frac{(2\omega_c RC)^2}{1+(RC\omega_c)^2}} = \frac{1}{\sqrt{2}}$$

$$\Rightarrow \frac{2\omega_c}{\omega_c^2+10^6} = \frac{1}{\sqrt{2}}$$

$$\Rightarrow \omega_c = 1000 \text{ rad/sec}$$

51. Virtual grand is Not applicable, became op-amp is in saturation

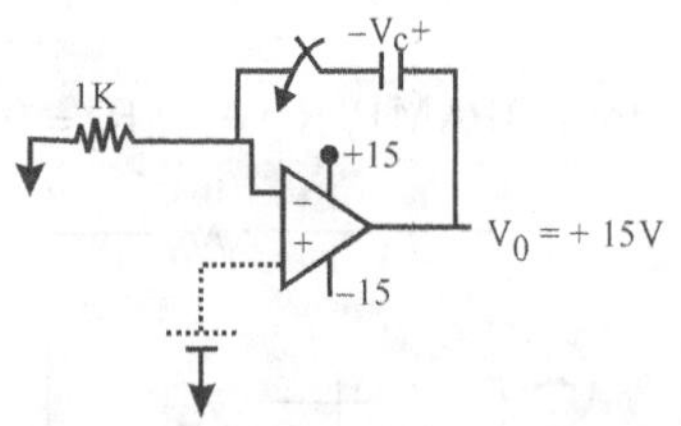

$$V_0 = +15 = +V_{sat}$$

$$V_c = 15(1-e^{-\frac{t}{Rc}})$$

$$V_c = 15\,(1-e^{-1000t})$$

at $t = 1$ ms

$$V_c = 9.45 \text{ Volts}$$

52. In Op-Amp

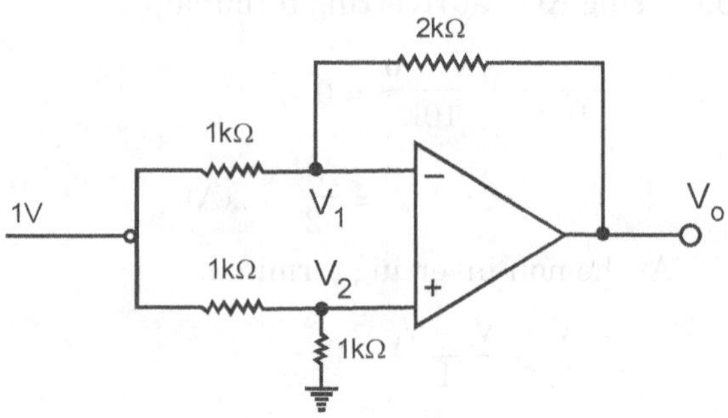

$$V_1 \simeq V_2$$

$$V_2 = 1 \times \left(\frac{1}{1+1}\right) = \frac{1}{2} = 0.5 \text{ volts}$$

Apply KCL at V_1,

$$\frac{V_0 - V_1}{2k\Omega} = \frac{V_1 - 1}{1\,k\Omega}$$

$$\Rightarrow 3V_1 - 2 = V_0$$

$$\Rightarrow 3 \times 0.5 - 2 = V_0$$

$$\therefore V_0 = -0.5 \text{ volts}$$

53.

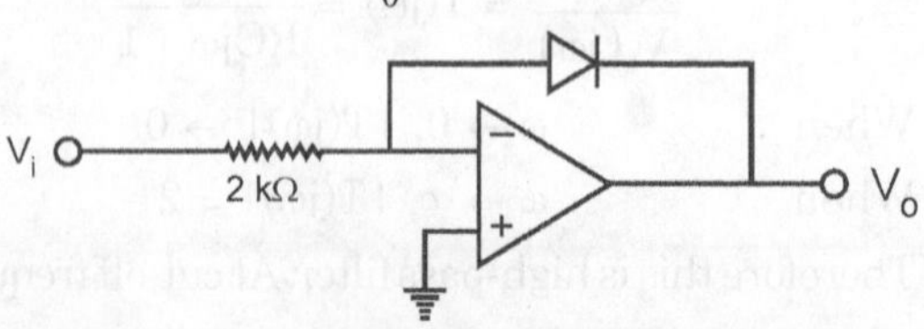

This is a circuit diagram of log amplifier

$$V_0 = -V_T \ln V_i + C$$

$$V_{01} = -V_T \ln V_{i1} + C$$

$$\Rightarrow V_{01} = -V_T \ln 2 + C$$

and

$$V_{02} = -V_T \ln V_{i2} + C$$

$$\Rightarrow V_{02} = -V_T \ln 4 + C$$

$$\therefore V_{01} - V_{02} = -V_T \ln 2 + V_T \ln 4$$

$$= V_T \ln 2$$

54. Applying KCL at non-inverting terminal,

$$\frac{V - V_i}{R} + \frac{V}{\frac{1}{sC}} = 0$$

$$V\left(sC + \frac{1}{R}\right) = \frac{V_i}{R}$$

$$V = \frac{1}{1+sRC}.V_i \qquad \text{(i)}$$

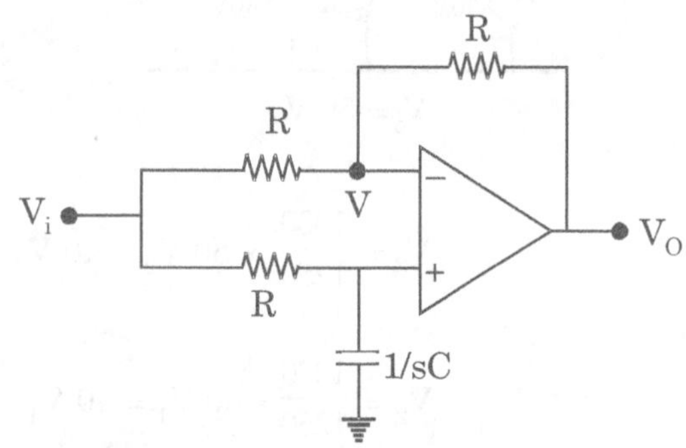

Applying kCL at inverting terminal,

$$\frac{V - V_i}{R} + \frac{V - V_o}{R} = o \Rightarrow 2V - V_i = V_o$$

from eq. (i), $\frac{2V_i}{1+sRC} - V_i = V_o$

$$\frac{V_i(1-sRC)}{1+sRC} = V_o$$

Transfer function, $T(S) = \frac{V_o(S)}{V_i(S)} = \frac{1-sRC}{1+sRC}.$

55. Given : $V_i = V_1 \sin(\omega t)$, $V_o = V_2 \sin(\omega t + \phi)$

$$\underline{|T(j\omega)} = -\tan^{-1}(\omega RC) - \tan^{-1}(\omega RC)$$

$$= -2\tan^{-1}(\omega RC)$$

$$\phi = \underline{|T(j\omega)} = -2\tan^{-1}(\omega RC).$$

$\omega = 0$, $\phi = 0° \to$ Maximum

$\omega \to \infty$, $\phi = -\pi$ Minimum.

56.

$$I = \frac{1}{100\text{ k}} = 10\ \mu A$$

$$I_o = 1\ \mu A$$

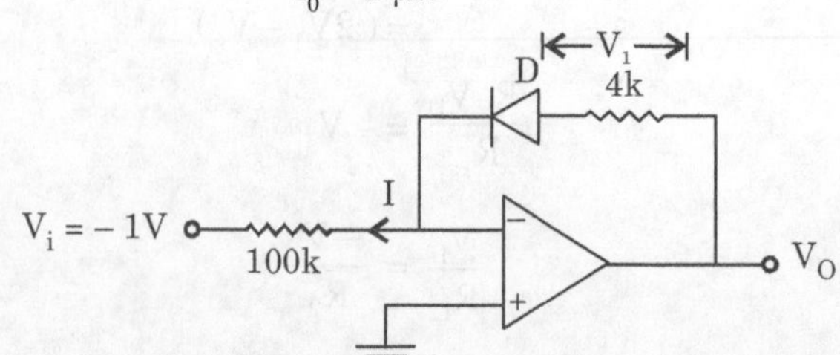

$$\Rightarrow \qquad 10\ \mu A = 1\ \mu A\left[e^{\frac{V_D}{25}} - 1\right]$$

$$\Rightarrow \qquad V_D = 60\ mV$$

Voltage drop across 4k resistance

$$V_1 = 4k \times 10\ \mu A = 40\ mV$$

Total voltage output $= V_1 + V_D = 1$

$$= (40 + 60)\ mV$$

$$= 0.1\ volts$$

57. $$\frac{V_o}{V_i} = -\frac{R_F}{R_i}, \quad R_F = R_2 \parallel \frac{1}{sC} = \frac{R_2}{R_2Cs + 1}$$

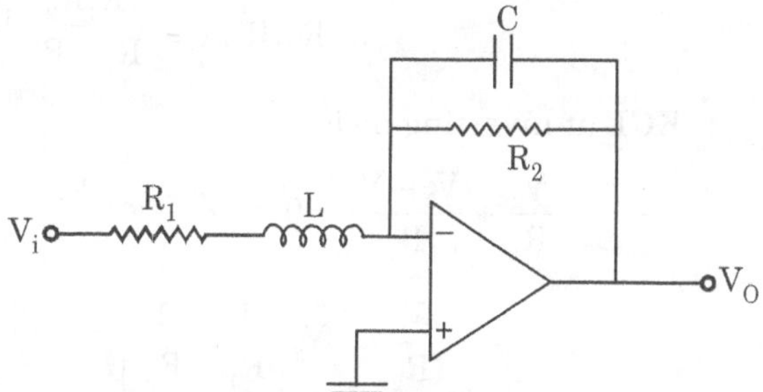

$$R_i = R_1 + sL$$

$$\frac{V_o}{V_i} = \frac{K}{(R_1 + sL)(R_2Cs + 1)}$$

$$= \frac{K}{\left(1 + j\frac{f}{f_H'}\right)\left(1 + j\left(\frac{f}{f_H''}\right)\right)}$$

As $\left.\frac{V_o}{V_i}\right|_{\omega \to 0} = 0$, it is a low pass filter.

58. For ideal op-amp, voltage at non-inverting point is equal to inverting point, so current from 5 kΩ resistor is

$$\frac{10 - 5}{5\ k} = 1\ mA$$

Ideal op-amp has no current in input, so

$$I_c = 1\ mA \approx I_E$$

$$V_o = 1.4 \times 1\ mA + 0.6 = 2\ volts$$

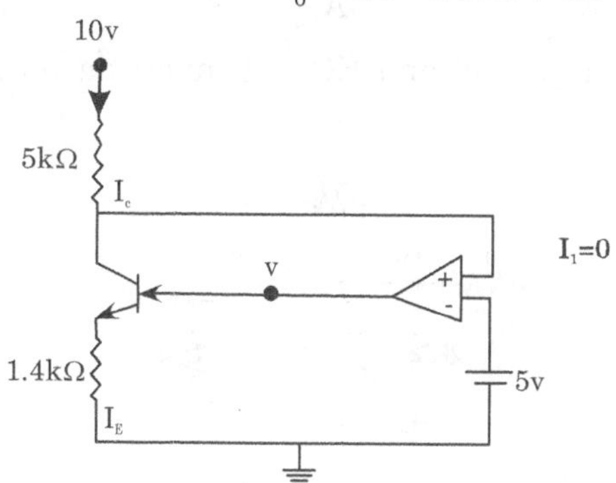

59.

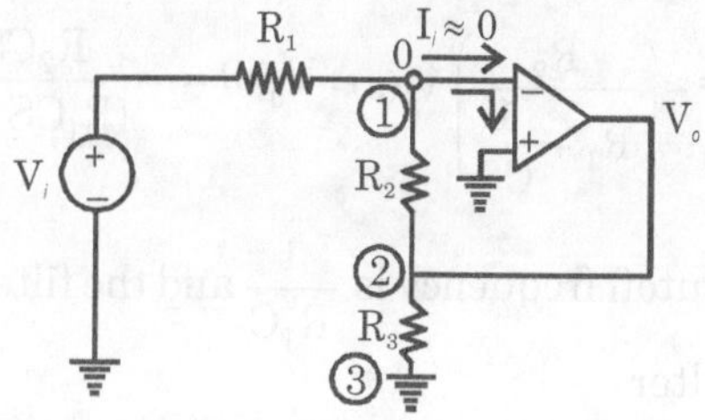

As $V_2 - V_1 = V_2 - V_3$

R_2 and R_3 in parallel.

Using KCL at mode (1), we have

$$\Rightarrow \quad \frac{0 - v_i}{R_1} + 0 - v_0\left[\frac{1}{R_2} + \frac{1}{R_3}\right] = 0$$

or, $$\frac{v_o}{v_i} = -\left(\frac{R_2 \parallel R_3}{R_1}\right)$$

60.

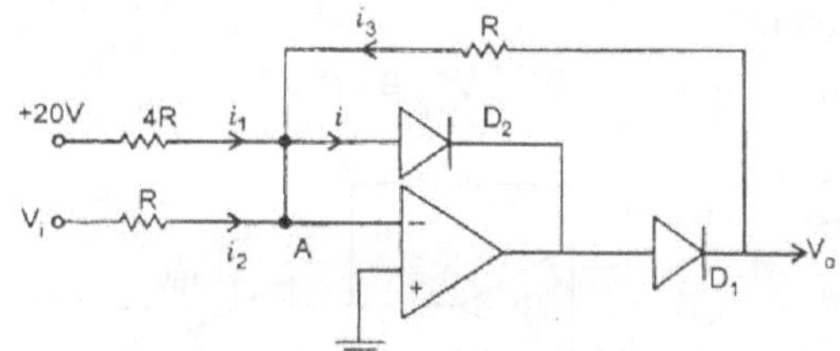

Since input is connected to negative terminal, so output always positive.

Since $\qquad v_A = 0$ (virtually grounded)

Case 1 : $i_1 + i_2 + i_3 = i$

$$\therefore \quad \frac{20}{4R} + \frac{v_i}{R} + \frac{v_0}{R} = i$$

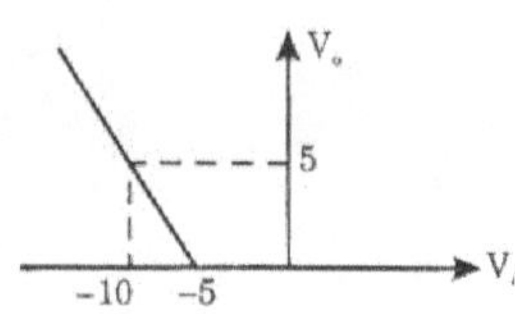

If v_0 = + ve, diode not conduct, so i = 0

$$\therefore \quad \frac{20}{4R} + \frac{v_i}{R} + \frac{v_0}{R} = 0$$

$$\Rightarrow \quad 5 + v_i + v_0 = 0$$

$$\Rightarrow \quad v_0 = -5 - v_i \geq 0$$

$$\Rightarrow \quad -v_i \geq 5$$

$$\Rightarrow \quad v_i \leq -5$$

Case 2 : Since diode will conduct, hence

For $v_i > -5, \quad v_0 = 0$

For input $v_i = -10$,

$$v_0 = -5 + 10 = +5\ volts$$

61. When ω = 0; inductor acts as a short circuit so, $V_0 = 0$

When ω = ∞, inductor acts as open circuit

$$\Rightarrow \quad V_0 = i_1R_1$$

So, it acts as a high pass filter.

62. $V_0(S) = -\left(\dfrac{R_2}{R_1 + \dfrac{1}{CS}}\right) V_i(s)\,;\, V_0(s) = -\dfrac{R_2CS}{(R_1CS+1)} V_i(s)$

Thus cutoff frequency is $\dfrac{1}{R_1C}$ and the filter is high pass filter

63. Using the concept of "virtual ground" in an operational amplifier, we can set the voltage at the point to zero volts since the non inverting terminal is grounded.

Once $V_A = 0$, V_C will also be zero.

We know that for a silicon n–p–n transistor,

$V_{BE} = V_B - V_E = 0.7$ V

Since, $V_B = 0 \Rightarrow V_E = -0.7$ V

Hence the output voltage is the same as the emitter voltage

so, $V_{out} = -0.7$ V

64.

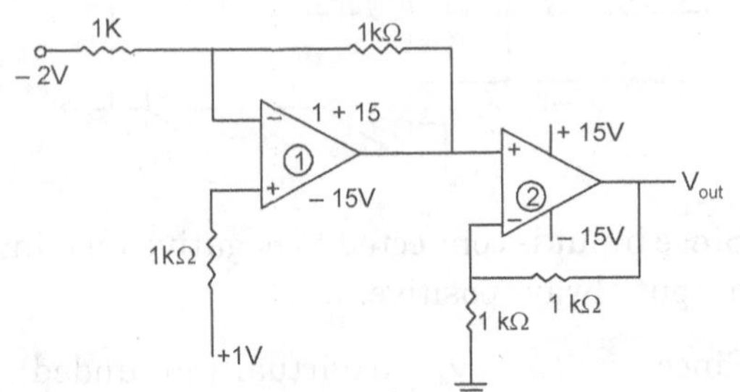

$$V_{out} = \left[+1v\left(1+\frac{1}{1}\right) - 2v\left(\frac{-1}{1}\right)\right]\left[1+\frac{1}{1}\right]$$

⇑ Gain of non inverty amp (1) ⇑ Gain of inverty amp (1) ⇑ Gain of non inverty amp (2)

$= [1 \times 2 + 2] \times 2$ v $= 8$v

65. f = 5KHz

$$\text{Cut off frequency (LPF)} = \frac{1}{2\pi R_2 C} = 5 \text{ KHz}$$

$$\Rightarrow R_2 = \frac{1}{2\pi \times 5 \times 10^3 \times 10 \times 10^{-9}} = 3.18 \text{ k}\Omega$$

66. Given circuit is an op-amp series regulator, V_0 is given by

$$V_0 = \left(1+\frac{R_1}{R_2}\right)V_z$$

$$9\text{ V} = \left(1+\frac{1\text{ k}\Omega}{R_2}\right)4.7$$

$$R_2 = 1093.02\ \Omega$$

67. Given, $Z_i = \infty$

$A_{0_L} = \infty$

$V_{i_0} = 0$

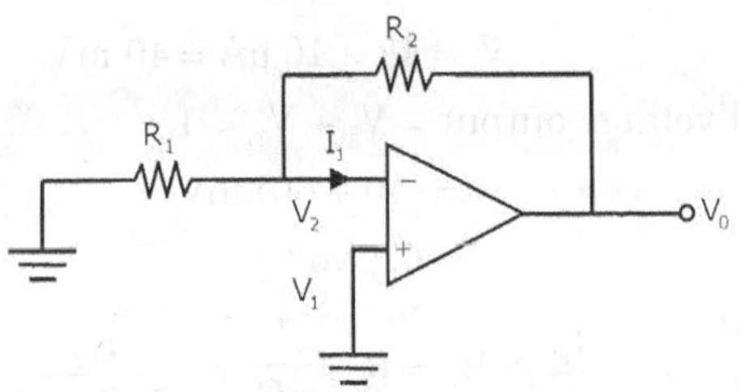

$$V_2 = (R_1//R_2)\, I_1 = \frac{R_1R_2}{R_1+R_2} I_1 \quad ...(1)$$

KCL at inverting node

$$\frac{V_2}{R_1} + \frac{V_2 - V_0}{R_2} = 0 \quad (\because Z_i = \infty)$$

$$\frac{V_0}{R_2} = V_2\left[\frac{1}{R_1} + \frac{1}{R_2}\right]$$

$$\frac{V_0}{R_2} = \left(\frac{R_1R_2}{R_1+R_2}\right)I_1\left[\frac{R_2+R_1}{R_1R_2}\right]$$

$$\Rightarrow \quad V_0 = I_1R_2$$

68.

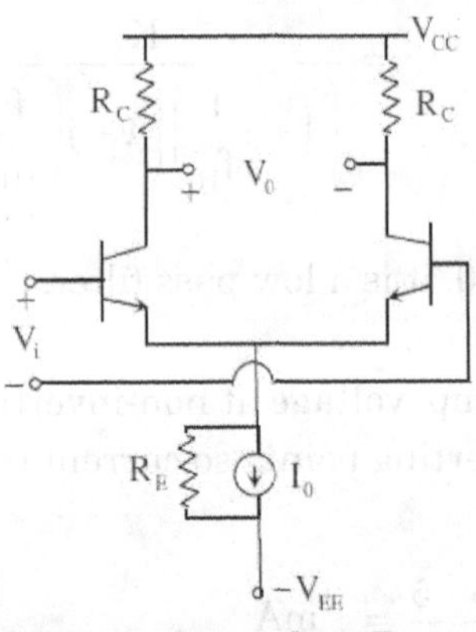

A_d does not depend on R_E

A_{cm} decreases as R_E is increased

$$\therefore \quad \text{CMRR} = \frac{A_d}{A_{cm}} = \text{Increases}$$

69. Virtual ground and KCL at inverting terminal gives

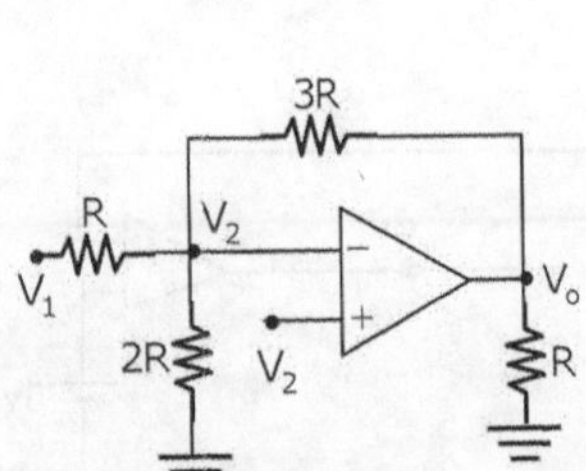

$$\frac{V_2 - V_1}{R} + \frac{V_2}{2R} + \frac{V_2 - V_0}{3R} = 0$$

$$\frac{V_0}{3R} = \frac{V_2}{R} + \frac{V_2}{3R} + \frac{V_2}{2R} - \frac{V_1}{R}$$

$$V_0 = -3V_1 + \frac{11}{2}V_2$$

70. The given circuit is a monostable multivibrator where v_i acts as a trigger to change output state, while charging of capacitor C_1 bring the output state to original state.

71. When V_i makes Diode 'D' OFF,

$$V_0 = V_i$$

$$\therefore \quad V_0\,(\text{min}) = -5\text{ V}$$

When V_i makes diode 'D' ON,

$$V_0 = \frac{(V_i - 0.7 - 2)}{R_1 + R_2} + V_{on} + 2\text{ V}$$

$$\therefore \quad V_0\,(\text{max}) = \frac{(5 - 0.7 - 2)1k}{1k + 1k} + 0.7 + 2\text{ V}$$

$$= 3.85\text{ V}$$

72. When switch is in position A :

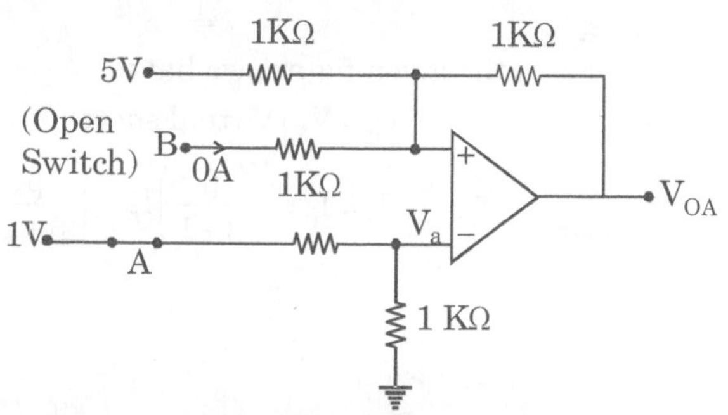

Applying KCL at inverting terminal,

$$\frac{V_a - 5}{1} + \frac{V_a - V_{OA}}{1} = 0$$

$$V_{OA} = 2V_a - 5$$

$$\because V_a = \frac{1}{1+1} \times 1 = 0.5\text{ V (voltage division rule)}$$

$$V_{OA} = 2 \times 0.5 - 5 = -4\text{ V.}$$

where switch is in position B :

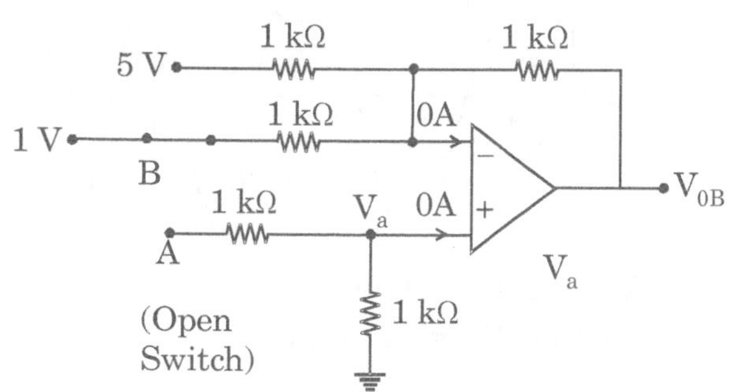

Non-inverting input, V_a = 0V.

Applying KCL at inventing terminal

$$\frac{0-1}{1} + \frac{0-5}{1} + \frac{0 - V_{OB}}{1} = 0$$

$$V_{OB} = -6V$$

$$\Rightarrow \frac{V_{OB}}{V_{OA}} = \frac{-6}{-4} = 1.5.$$

73.

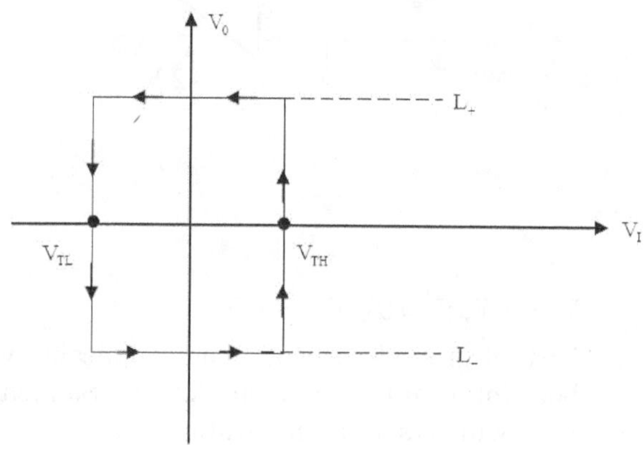

$$\text{Hysteresis} = V_{TH} - V_{TL}$$

$$= -L_-\left(\frac{R_1}{R_2}\right) + L_+\left(\frac{R_1}{R_2}\right)$$

$$500\text{ mV} = -(-5)\left(\frac{R_1}{20k}\right) + 5\left(\frac{R_1}{20k}\right)$$

$$= \frac{R_1}{2\text{ k}}$$

$$\therefore \quad R_1 = 500 \times 2 \times 10^3 \times 10^{-3}$$

$$= 1000\ \Omega = 1\text{ k}\Omega$$

74.

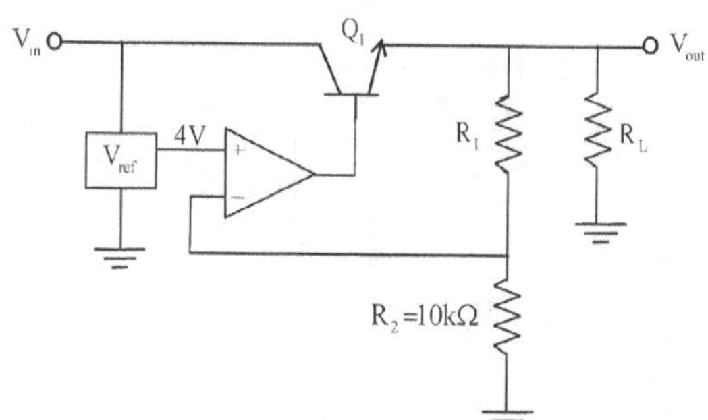

$$P_{Q_1(max)} = V_{CE(max)} \times I_{c\,max} \quad \text{...(i)}$$

$$V_{CE(max)} = (24 - 10)V = 14\text{ V}$$

$$I_{c\,max} = (200 + 0.4)\text{ mA}$$

$$I_E = I_c = 200\text{ mA} + 0.4\text{ mA}$$

$$= 200.4\text{ mA}$$

$$\left(\because I_{R_2} = I_{R_1} = \frac{4-0}{10}\text{mA}\right)$$

Put values in equation (i), we get

$$P_{Q_1(max)} = 14 \times 200.4 \times 10^{-3}\text{ Watt}$$

$$= 2.8056\text{ Watt}$$

75. From the given circuit,

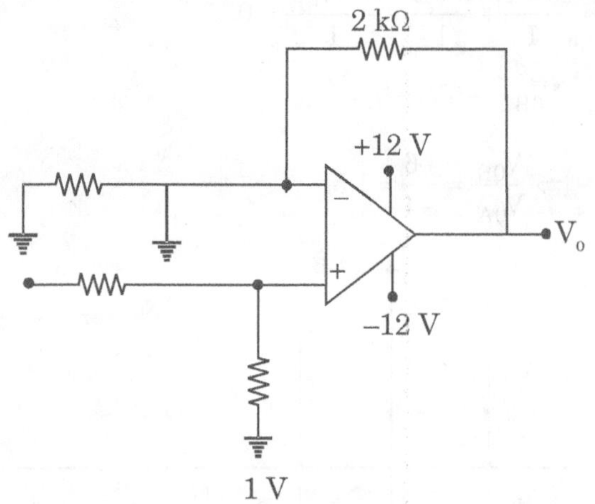

$V_+ = 1$ V, $V_- = 0$V. ($V_+ > V_-$).

Virtual ground concept is not applicable when both internal terminals are known. So ideal op-Amp will acts as a comparator circuit.

$\Rightarrow V_o = +V_{sat} = +12$V.

76. $$f_{3dB} = \frac{1}{2\pi RC}$$

$$= \frac{1}{2\pi \times 10 \times 10^3 \times 0.1 \times 10^{-6}}$$

$$= 159.15 \text{ Hz}$$

77. Applying KCL at inverting terminal,

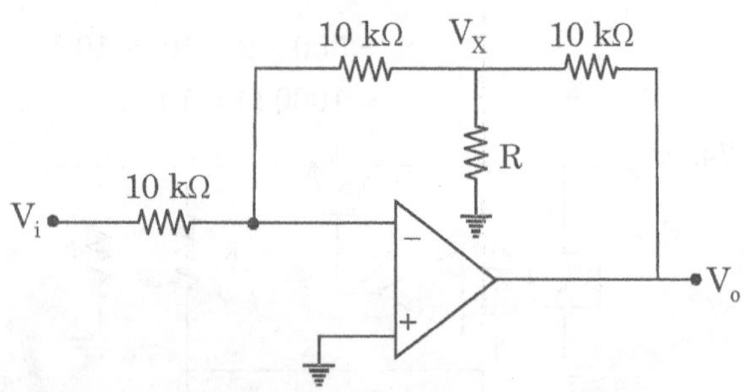

$$\frac{0 - V_i}{10} + \frac{0 - V_x}{10} = 0$$

$$V_x = -V_i \qquad \text{...(i)}$$

Applying KCL at V_x,

$$\frac{V_x}{10} + \frac{V_x}{R} + \frac{V_x - V_0}{10} = 0$$

$$\frac{-V_i}{10} - \frac{V_i}{R} - \frac{V_i}{10} = \frac{V_0}{10} \text{ (using eq.(i))}$$

$$-1 - \frac{10}{R} - 1 = \frac{V_0}{V_i}$$

$$\frac{-10}{R} - 2 = \frac{V_o}{V_i}$$

$$\because \frac{V_0}{V_i} = -12$$

$$\frac{-10}{R} - 2 = -12$$

$$\frac{10}{R} = 10$$

$R = 1$ kΩ.

78.

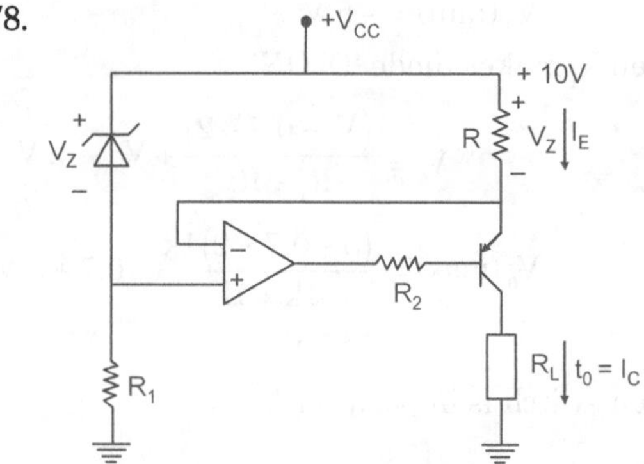

From the given figure, we have

$V_P = V_N$ (Virtual short)

$$I_0 = I_C = \left(\frac{\beta}{\beta+1}\right) I_E = \left(\frac{\beta}{\beta+1}\right)\frac{V_Z}{R}$$

$$= \left(\frac{\beta}{\beta+1}\right)\frac{V_{ref}}{R}$$

79.

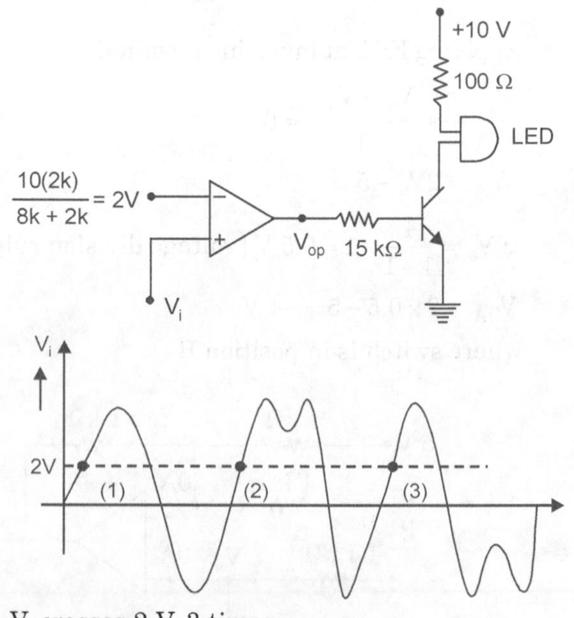

V_i crosses 2 V, 3 times

So, the LED glows 3 times.

5
CHAPTER

Feedback & Power Amplifier

Analysis of Previous GATE Papers

Topics ↓ / Year →			2019	2018	2017 Set 1	2017 Set 2	2016 Set 1	2016 Set 2	2016 Set 3	2015 Set 1	2015 Set 2	2015 Set 3
FEEDBACK (CONCEPT & CONNECTION TYPES)	1 Mark	MCQ Type		1	1							
		Numerical Type										
	2 Marks	MCQ Type										
		Numerical Type										
	Total			1	1							
POWER AMPLIFIER (DEFINITION & TYPES)	1 Mark	MCQ Type										
		Numerical Type										
	2 Marks	MCQ Type										
		Numerical Type										
	Total											

FEEDBACK (CONCEPT & CONNECTION TYPES)

1. The circuit of the figure is an example of feedback of the following type

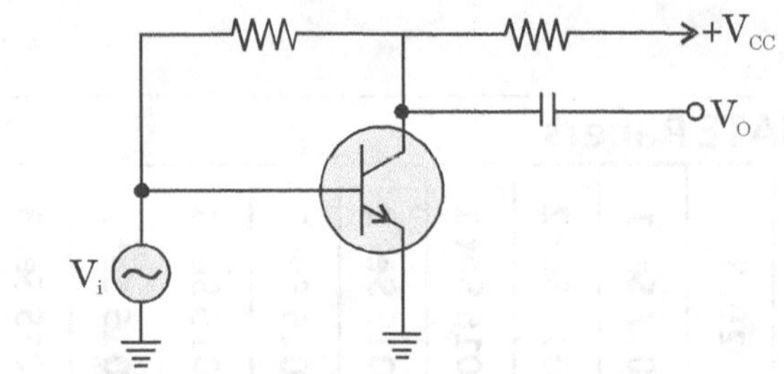

(a) current series (b) current shunt
(c) voltage series (d) voltage shunt

[1896 : 1 Mark]

2. In the circuit shown in fig. is a finite gain amplifier with a gain of K, a very large input impedance, and a very low output impedance. The input impedance of the feedback amplifier with the feedback impedance Z connected as shown will be

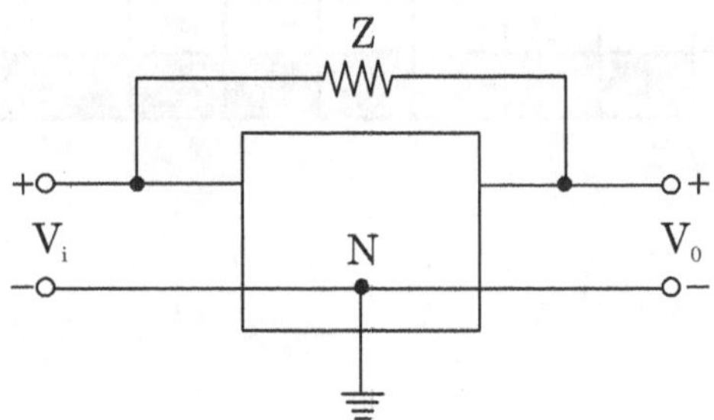

(a) $Z\left[1-\frac{1}{K}\right]$ (b) $Z(1-K)$
(c) $\left[\frac{Z}{K-1}\right]$ (d) $\left[\frac{Z}{1-K}\right]$

[1988 : 2 Marks]

3. The feedback amplifier shown in fig. has

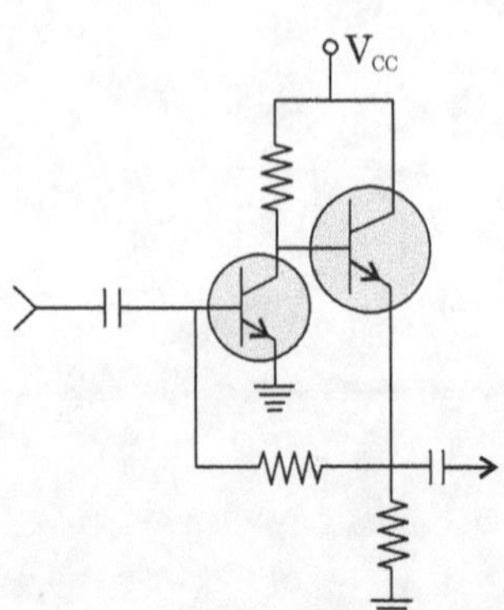

(a) current-series feedback with large input impedance and large output impedance.
(b) voltage-series feedback with large input impedance and low output impedance.
(c) voltage-shunt feedback with low input impedance and low output impedance.
(d) current-shunt feedback with low input impedance and output impedance.

[1989 : 2 Marks]

4. Two non-inverting amplifiers, one having a unity gain and the other having a gain of twenty, are made using identical operational amplifiers. As compared to the unity gain amplifier, the amplifier with gain twenty has
(a) less negative feedback
(b) greater input impedance
(c) less bandwidth
(d) None of the above.

[1991 : 2 Marks]

5. Negative feedback in Amplifiers
(a) improves the signal to noise ratio at the input
(b) improves the signal to noise ratio at the cutput
(c) does not effect the signal to noise ratio at the output
(d) reduces distortion

[1993 : 1 Mark]

6. To obtain very high input and output impedances in a feedback Amplifier, the mostly used is
(a) voltage-series
(b) current-series
(c) voltage-shunt
(d) current-shunt

[1995 : 1 Mark]

7. Negative feedback in
(1) voltage series configuration
(2) current shunt configuration
(a) increases input impedance
(b) decreases input impedance
(c) increases closed loop gain
(d) leads to oscillation

[1997 : 2 Marks]

8. In a shunt-shunt negative feedback Amplifier, as compared to the basic Amplifier.

(a) both input and output impedance decreases.

(b) input impedance decreases but output impedance increases.

(c) input impedance increases but output impedance decreases.

(d) both input and output impedance increases.

[1998 : 1 Mark]

9. Negative feedback in an amplifier

(a) reduces gain.

(b) increases frequency and phase distortions.

(c) reduces bandwidth.

(d) increases noise.

[1999 : 1 Mark]

10. An amplifier has an open-loop gain of 100, an input impedance of 1 kΩ, and an output impedance of 100 Ω. A feedback network with a feedback factor of 0.99 is connected to the amplifier in a voltage series feedback mode. The new input and output impedances, respectively, are

(a) 10 Ω and 1 Ω (b) 10 Ω and 10 Ω

(c) 100 kΩ and 1 Ω (d) 100 kΩ and 1 kΩ

[1999 : 2 Marks]

11. In a negative feedback amplifier using voltage-series (i.e. voltage-sampling, series mixing) feedback.

(a) R_i decreases and R_0 decreases

(b) R_i decreases and R_0 increases

(c) R_i increases and R_0 decreases

(d) R_i increases and R_0 increases

(R_i and R_0 denote the input and output resistances respectively.)

[2002 : 1 Mark]

12. An amplifier without feedback has a voltage gain of 50, input resistance of 1 kΩ and output resistance of 2.5 kΩ. The input resistance of the current-shunt negative feedback amplifier using the above amplifier with a feedback factor of 0.2 is

(a) $\frac{1}{11}$ kΩ (b) $\frac{1}{5}$ kΩ

(c) 5 kΩ (d) 11 kΩ

[2003 : 2 Marks]

13. Voltage series feedback (also called series-shunt feedback) results in

(a) increase in both input and output impedances

(b) decrease in both input and output impedances

(c) increase in input impedance and decrease in output impedance

(d) decrease in input impedance and increase in output impedance **[2004 : 1 Mark]**

14. The effect of current shunt feedback in an amplifier is to

(a) increase the input resistance and decrease the output resistance

(b) increase both input and output resistances

(c) decrease both input and output resistances.

(d) decrease the input resistance and increase the output resistance **[2005 : 1 Mark]**

15. The input impedance (Z_i) and the output impedance (Z_0) of an ideal trans-conductance (voltage controlled current source) amplifier are

(a) $Z_i = 0, Z_0 = 0$ (b) $Z_i = 0, Z_0 = \infty$

(c) $Zt = \infty, Z_0 = 0$ (d) $Z_i = \infty, Z_0 = \infty$

[2006 : 1 Mark]

16. In a transconductance amplifier, it is desirable to have

(a) a large input resistance and a large output resistance

(b) a large input resistance and a small output resistance

(c) a small input resistance and a large output resistance

(d) a small input resistance and a small output resistance

[2007 : 1 Mark]

17. In a voltage-voltage feedback as shown below, which one of the following statements is TRUE if the gain k is increased?

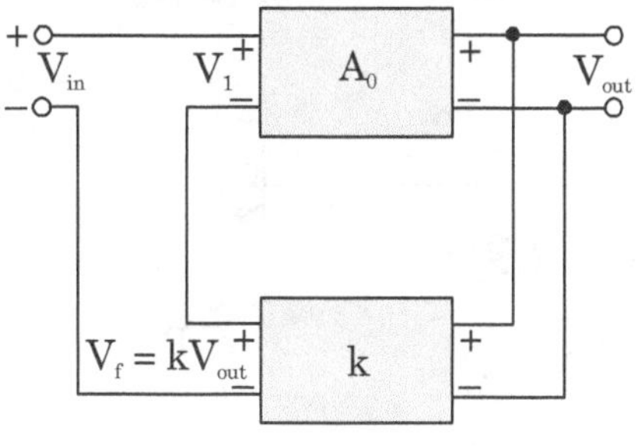

(a) The input impedance increases and output impedance decreases.

(b) The input impedance increases and output impedance also increases.

(c) The input impedance decreases and output impedance also decreases.

(d) The input impedance decreases and output impedance increases.

[2013 : 1 Mark]

18. In the ac equivalent circuit shown in the figure, if i_{in} is the input current and R_F is very large, the type of feedback is

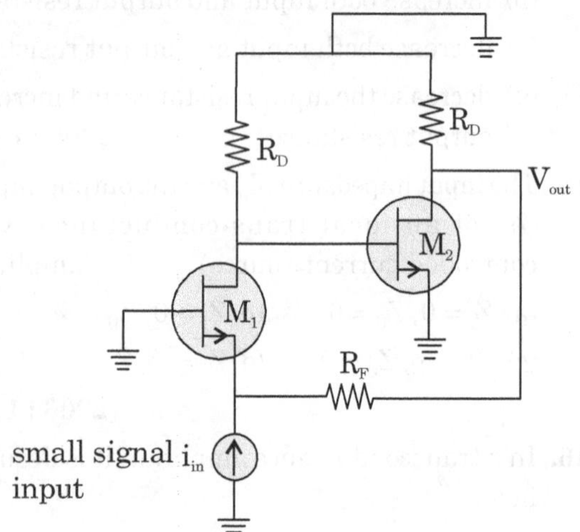

(a) voltage-voltage feedback

(b) voltage-current feedback

(c) current-voltage feedback

(d) current-current feedback

[2014 : 1 Mark, Set-1]

19. The feedback topology in the amplifier circuit (the base bias circuit is not shown for simplicity) in the figure is

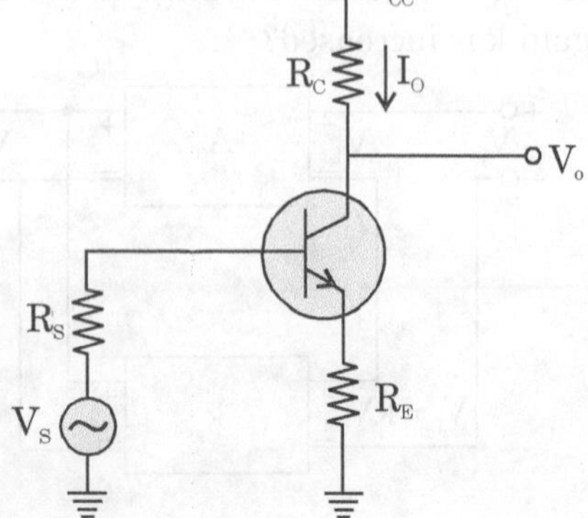

(a) voltage shunt feedback

(b) current series feedback

(c) current shunt feedback

(d) voltage series feedback

[2014 : 1 Mark, Set-2]

20. The desirable characteristics of a transconductance amplifier are

(a) high input resistance and high output resistance

(b) high input resistance and low output resistance

(c) low input resistance and high output resistance

(d) low input resistance and low output resistance

[2014 : 1 Mark, Set-3]

21. A good transconductance amplifier should have

(a) high input resistance and low output resistance

(b) low input resistance and high output resistance

(c) high input and output resistances

(d) low input and output resistances

[2017 : 1 Mark, Set-1]

22. A good transimpedance amplifier has

(a) low input impedance and high output impedance

(b) high input impedance and high output impedance

(c) high input impedance and low output impedance

(d) low input impedance and low output impedance

[2018 : 1 Mark]

POWER AMPLIFIER (DEFINITION & TYPES)

23. In case of class A amplifiers the ratio (efficiency of transformer coupled amplifier)/(efficiency of a transformer less amplifier) is

(a) 2.9 (b) 1.36

(c) 1.0 (d) 0.5

[1987 : 2 Marks]

24. In a transistor push-pull Amplifier

(a) there is no d.c present in the output

(b) there is no distortion in the output

(c) there is no even harmonics in the output

(d) there is no add harmonics in the output

[1993 : 1 Mark]

25. A class – A transformer coupled, transistor power Amplifier is required to deliver a power output of 10 watts. The maximum power Rating of the transistor should not be less than

(a) 5 W (b) 10 W

(c) 20 W (d) 40 W

[1999 : 1 Mark]

26. The circuit shown in the figure supplies power to an 8 Ω speaker, LS. The values of I_C and V_{CE} for this circuit will be I_C = ______ and V_{CE} = ______

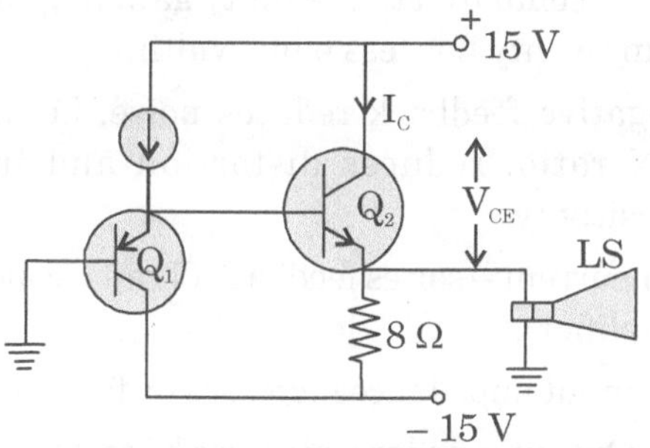

[1999 : 1 Mark]

27. A power amplifier delivers 50 W output at 50% efficiency. The ambient temperature is 25°C. If the maximum allowable junction temperature is 150° C, then the maximum thermal resistance θ_{gc} that can be tolerated is ______.

[2000 : 1 Mark]

28. Crossover distortion behaviour is characteristic of

(a) class A output stage

(b) class B output stage

(c) class AB output stage

(d) common-base output stage

[2004 : 1 Mark]

ANSWERS

1. (d)	**2.** (d)	**3.** (c)	**4.** (c)	**5.** (b, d)	**6.** (b)	**7.** (c)	**8.** (b)	**9.** (a)	**10.** (c)
11. (c)	**12.** (a)	**13.** (c)	**14.** (d)	**15.** (d)	**16.** (a)	**17.** (a)	**18.** (b)	**19.** (b)	**20.** (a)
21. (c)	**22.** (d)	**23.** (b)	**24.** (a,c)	**25.** (c)	**26.** (0.7 V)	**27.** (5° C/w)		**28.** (b)	

EXPLANATIONS

1.

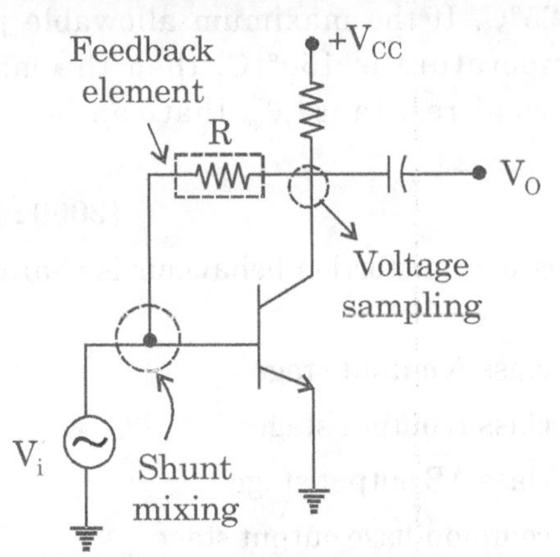

⇒ Feedback element is directly connected to output voltage, hence it is voltage sampling.

⇒ Feedback element is directly connected to the input voltage in shunt form, hence it is shunt mixing.

Therefore, feedback circuit is voltage– shunt feedback.

2.

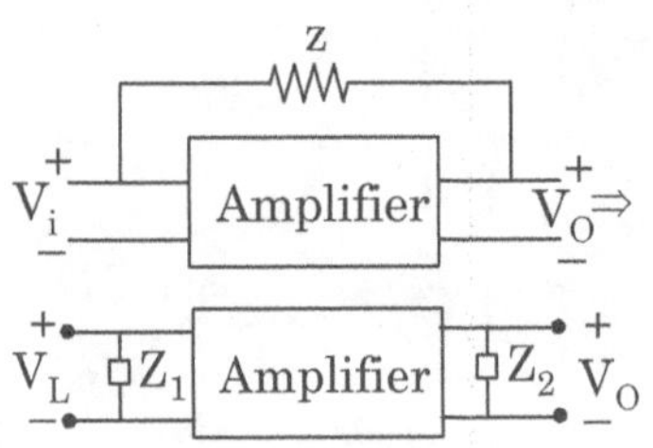

Given, $k = \frac{V_O}{V_i}$ = Voltage gain of the network/ amplifier using Miller's theorem,

$$Z_1 = \frac{Z}{1-k} \text{ and } Z_2 = \frac{Z}{1-1/k}$$

3.

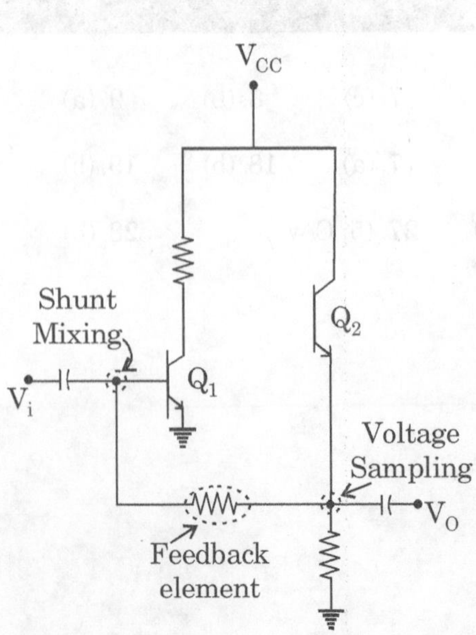

The feedback circuit is voltage –shunt feedback

For voltage shunt feedback,

(i) input inpedace decreases by factor $(1+R_m\beta)$,

(ii) Output impedance decreases by factor $(1+R_m\beta)$.

4. For indentical operational amplifiers, gain band width product is constant, i.e, $A_1 \times BW_1 = A_2 \times BW_2$

$$BW_2 = \frac{A_1 \times BW_1}{A_2} = \frac{1 \times BW_1}{20} = \frac{BW_2}{20}.$$

So, as compared to the unity gain amplifier with gain twenty has less band width.

5. Negative feedback reduces noise, i.e. improves S/N ratio, reduces distortion and improves sensitivity.

6. For current – series feedback (Trans–conductance amplifier),

(i) input impedance increases by factor $(1+G_m\beta)$,

(ii) Output inpedane increase by factor $(1+G_m\beta)$.

7. ⇒ In voltage – series configuration, input inpedance increases by a factor. $Ri_f = R_iD = R_i(1+A_V\beta)$.

⇒ In current – shunt configuration, input impedance decreases by a factor. of D.

$$R_{if} = \frac{R_i}{D} = \frac{R_i}{(1+A_I\beta)}$$

⇒ Negative feedback decreases the closed loop gain.

⇒ Negative feedback increases the stability (no oscillation)

8. In a shunt negative feedback amplifier, input impedance decreases but output impedance increases.

9. Negative feedback in an amplifier reduces gain stabilizing it. However, the reduction in gain can always be compensated by increasing the number of stages.

10. Given Open loop gain = 100

Input impedance = 1 k Ω

Output impedance = 100 Ω

Feedback factor, β = 0.99

For voltage series feedback.

$$Z_{if} = Z_i(1 + A\beta)$$
$$= 1\,K(1 + 100 \times 0.99)$$
$$= 1\,K \times 100 = 100\,K$$

$$Z_{of} = \frac{Z_0}{1+A\beta} = \frac{100}{1+99} = 1\Omega$$

11. $R_{if} = R_i\,(1 + \beta A)$

$$R_{of} = \frac{R_\delta}{1+\beta A}$$

12. In current-shunt, negative feedback amplifier, the input resistance,

$$R_i' = \frac{R_i}{1+\beta A} = \frac{1\,k\Omega}{1+50\times 0.2} = \left(\frac{1}{11}\right)k\Omega.$$

13. Voltage series feedback results in increase in i/p impedance and decrease in o/p impedance.

14.

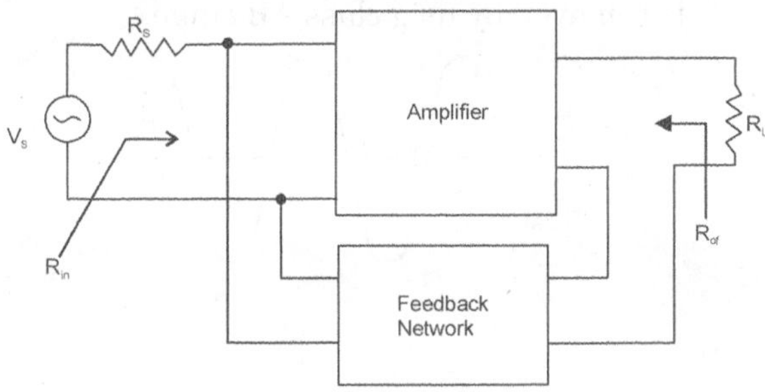

current shunt feedback network

As feedback network is connected in series, with output. Output resistance (R_{of}) increases, and the same is connected in shunt with the source resistance R_s, intput resistance decreases.

15.

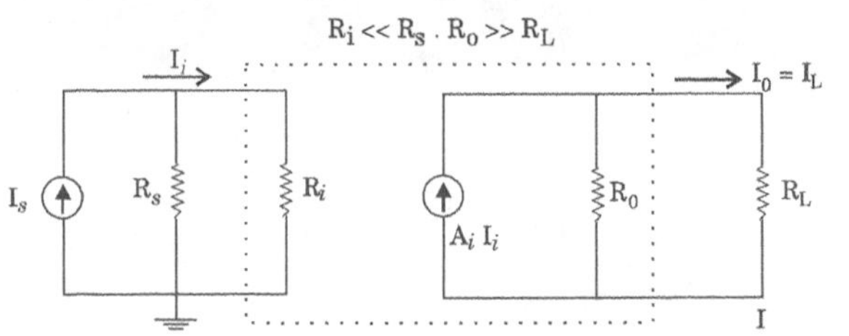

Noton's equivalent circuit of a current amplifier

For ideal transconductance amplifier, output current is proportional to the signal voltage independent of the magnitudes of R_S and R_L.

16.

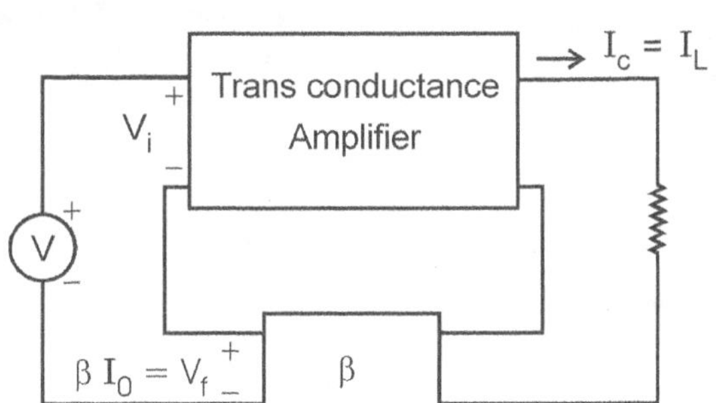

Input resistance, $R_{if} = R_i\,(1 + \beta\ Gm)$

So high input resistance

Output resistance, $R_{of} = Ro\,(1 + \beta\ Gm)$

So, Output resistance also will be high.

17.

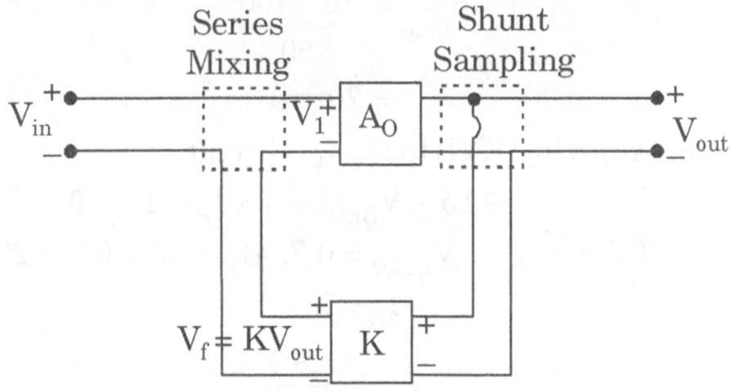

Given configuration is of voltage series feedback topology for voltage series feedback,

(i) input impedance increases by factor $(1+A_0K)$, i.e., $R_{if} = R_i\,(1+A_0k)$.

(ii) output impedance decreases by factor

$$(1+A_0K),\ \text{i.e., } Rof = \frac{Ro}{1+A_0k}$$

18. Output sample is voltage and is added at the input or current

∴ It is voltage-shunt negative feedback i.e., voltage-current negative feedback

20. A good transconductance amplifier should have very high input resistance and very high output resistance.

21. A good transconductance amplifier should have very high input resistance and very high output resistance.

22. A good trans-impedance amplifier should have low input impedance and low output impedance.

23. For class A amplifiers efficiency of transformer coupled amplifier, $\eta_1 = 50\%$

effciency of transformer less amplifier, $\eta_2 = 25\ \%$

$$\text{Ratio, } \frac{\eta_1}{\eta_2} = \frac{0.50}{0.25} = 2.$$

$$\left[\text{Taking closest approximation, } \frac{\eta_1}{\eta_2} \cong 1.36\right]$$

24. Push-pull amplifiers are operated in Class B. In Class B, there is no quiescent current as biasing is at cut–off and the output current

$$i_L = 2\,(A_1\cos\omega t + A_3\cos 3\,\omega t + ...)$$

Hence there is no dc current and even harmonics are cancelled out.

25. In transformer coupling efficiency is 50%

$$P_{ac} = 10 \text{ watts},$$

$$\therefore \quad P_{dc} = 20 \text{ watts}.$$

For class A, $\eta = 50$

$$\therefore \quad P_{Dmax} \geq \frac{10 \times 100}{50}$$

$$= 20 \text{ watts}$$

26. Applying KVL to Q_1, Q_2 Circuit,

$$-15 + V_{BEQ2} + 4 \times I_E - 15 = 0$$

Taking $V_{BEQ2} = 0.7$, $4I_E = 30 - 0.7 = 29.3$

$$I_C = I_F$$

$$= \frac{29.3}{4}$$

$$= 7.325 \text{ A}$$

In Q_2 circuit by KVL,

$$15 + V_{CE} + 29.3 - 15 = 0$$

or $V_C = 0.7$ V

tTranistor Q_2 is in saturation

27. P_D = Power dissipated

$$= 50 \times \frac{50}{100} = 25 \text{ W}$$

Given, $T_j = 150°C$,

$T_A = 25°C$

From
$$P_D = \frac{T_j - T_A}{\theta_{jc}}$$

$$\theta_{jc} = \frac{150 - 25}{26} = \frac{125}{25} = 5° \text{ C/W}$$

28. Cross over behaviour is characteristic of class B output stage as the amplifier is biased at cut-off. Here 2 transistors are operated one arranged to amplify the positive going portion and other for negative going portion. The corss-over distortion is removed by usig class AB stage.

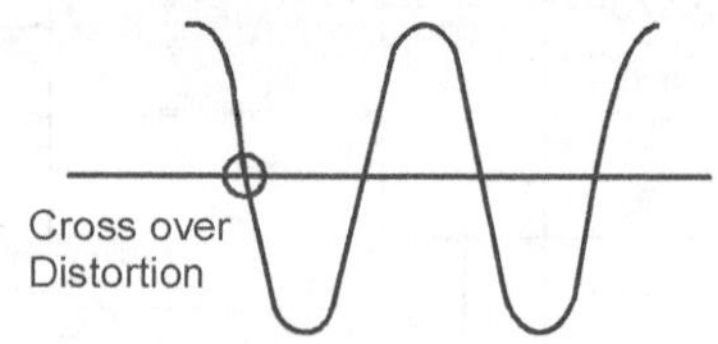

6 CHAPTER

Analog Circuit

Analysis of Previous GATE Papers												
	Year → Topics ↓		**2019**	**2018**	**2017 Set 1**	**2017 Set 2**	**2016 Set 1**	**2016 Set 2**	**2016 Set 3**	**2015 Set 1**	**2015 Set 2**	**2015 Set 3**
SINUSOIDAL OSCILLATORS	**1 Mark**	MCQ Type					1					
		Numerical Type										
	2 Marks	MCQ Type								1		
		Numerical Type										
	Total						1			2		
FUNCTION GENERATOR & WAVE-SHAPING CIRCUITS	**1 Mark**	MCQ Type										
		Numerical Type							1			
	2 Marks	MCQ Type										
		Numerical Type										
	Total								1			

SINUSOIDAL OSCILLATORS

1. Match the following

List-I	List-II
(a) Hartley	(1) Low frequency oscillator
(b) Wein-bridge	(2) High frequency oscillator
(c) Crystal	(3) Stable frequency oscillator
	(4) Relaxation oscillator
	(5) Negative Resistance oscillator

[1994 : 2 Marks]

2. Value of R in the oscillator shown in the given figure. So chosen that it just oscillates at an angular frequencies of 'ω'. The value of 'ω' and the required value of R will respectively be

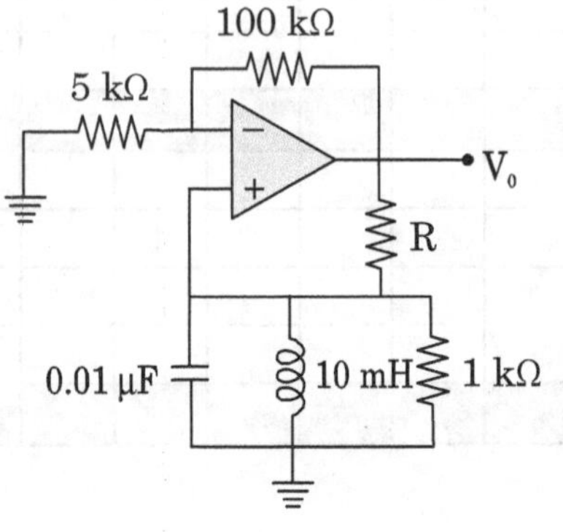

(a) 10^5 rad/sec, 2×10^4 Ω

(b) 2×10^4 rad/sec, 2×10^4 Ω

(c) 2×10^4 rad/sec, 10^5 Ω

(d) 10^5 rad/sec, 10^5 Ω

[1996 : 2 Marks]

3. The configuration of the figure is a

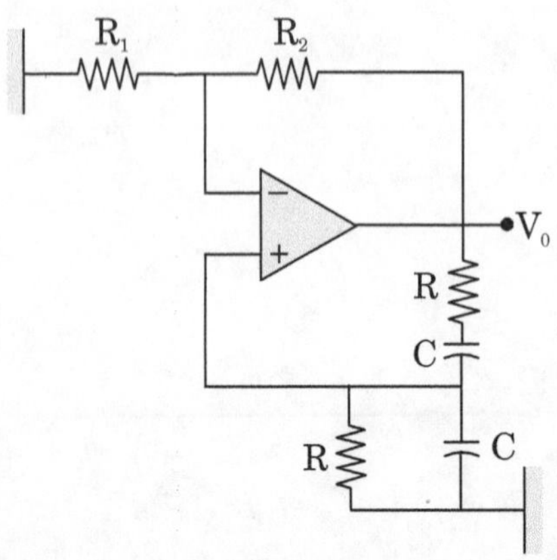

(a) Precision integrator

(b) Hartley oscillator

(c) Butterworth high pass filter

(d) Wien-bridge oscillator

[2000 : 1 Mark]

4. The oscillator circuit shown in the figure is

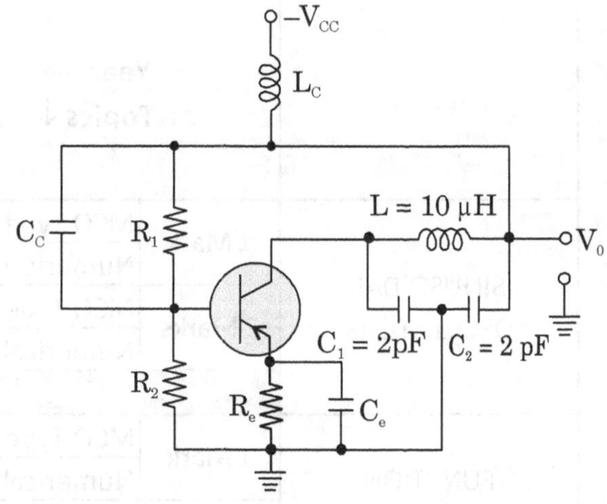

(a) Hartely oscillator with $f_{oscillation}$ = 79.6 MHz

(b) Colpitts oscillator with $f_{oscillation}$ = 50.3 MHz

(c) Hartley oscillator with $f_{oscillation}$ = 159.2 MHz

(d) Colpitts oscillator with $f_{oscillation}$ = 159.2 MHz

[2001 : 2 Marks]

5. The circuit in the figure employs positive feedback and is intended to generate sinusoidal oscillation.

If at a frequency f_0, $B(f) = \frac{V_t(f)}{V_0(f)} = \frac{1}{6}\angle 0°$, then to sustain oscillation at this frequency

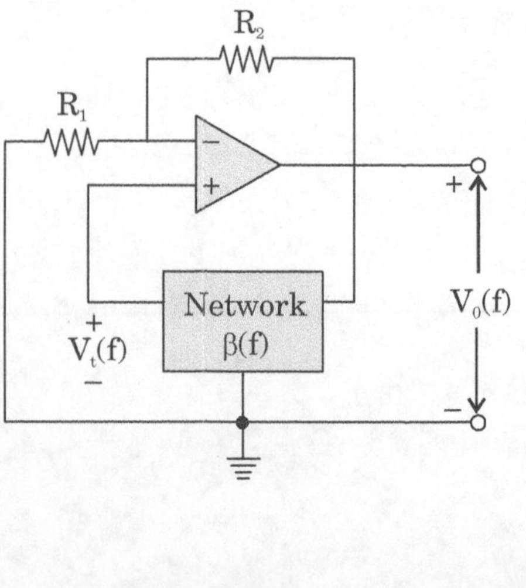

(a) $R_2 = 5R_1$ (b) $R_2 = 6R_1$

(c) $R_2 = \frac{R_1}{6}$ (d) $R_2 = \frac{R_1}{5}$

[2002 : 2 Marks]

6. The oscillator circuit shown in the figure has an ideal inverting amplifier. Its frequency of oscillation (in Hz) is

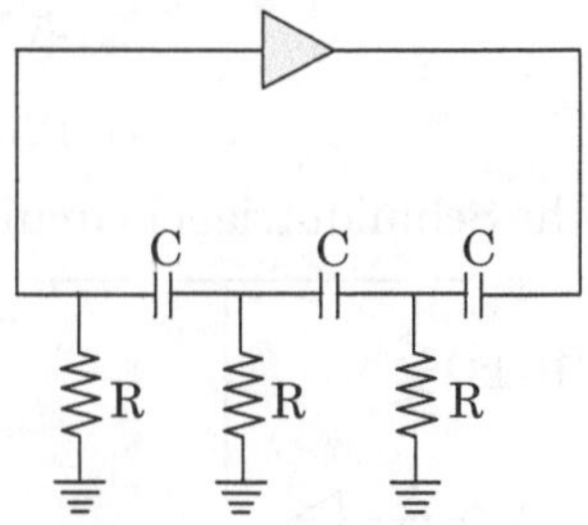

(a) $\frac{1}{(2\pi\sqrt{6}RC)}$ (b) $\frac{1}{(2\pi RC)}$

(c) $\frac{1}{(\sqrt{6}RC)}$ (d) $\frac{1}{\sqrt{6}(2\pi RC)}$

[2003 : 2 Marks]

7. The value of C required for sinusoidal oscillations of frequency 1 kHz in the circuit of the figure is

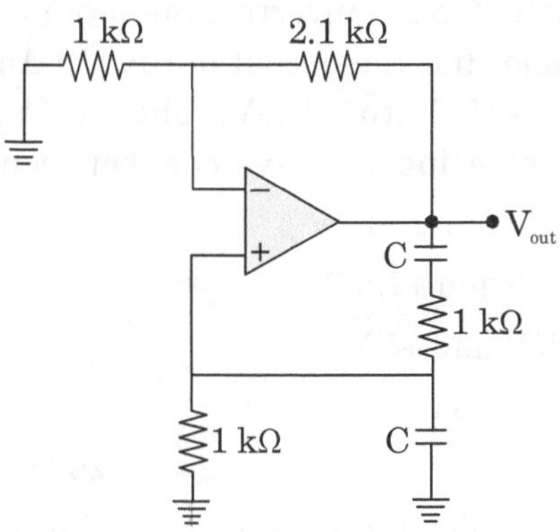

(a) $\frac{1}{2\pi}$ μF (b) 2π μF

(c) $\frac{1}{2\pi\sqrt{6}}$ μF (d) $2\pi\sqrt{6}$ μF

[2004 : 2 Marks]

8. The circuit shown in the figure has an ideal op-amp. The oscillation frequency and the condition to sustain the oscillations, respectively, are

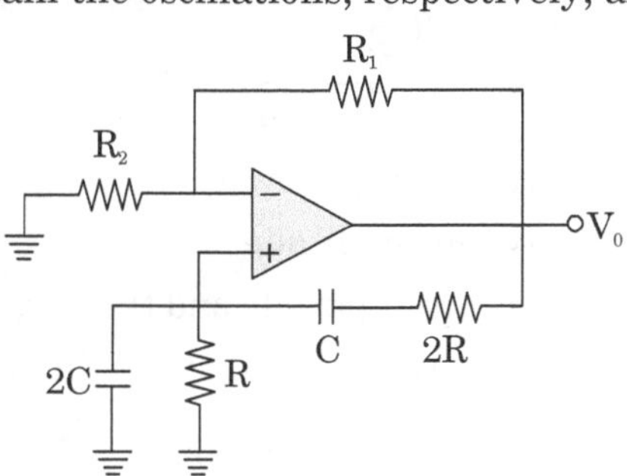

(a) $\frac{1}{CR}$ and $R_1 = R_2$ (b) $\frac{1}{CR}$ and $R_1 = 4R_2$

(c) $\frac{1}{2CR}$ and $R_1 = R_2$ (d) $\frac{1}{2CR}$ and $R_1 = 4R_2$

[2015 : 2 Marks, Set-1]

9. Consider the oscillator circuit shown in the figure. The function of the network (shown in dotted lines) consisting of the 100 kΩ resistor in series with the two diodes connected back-to-back is to

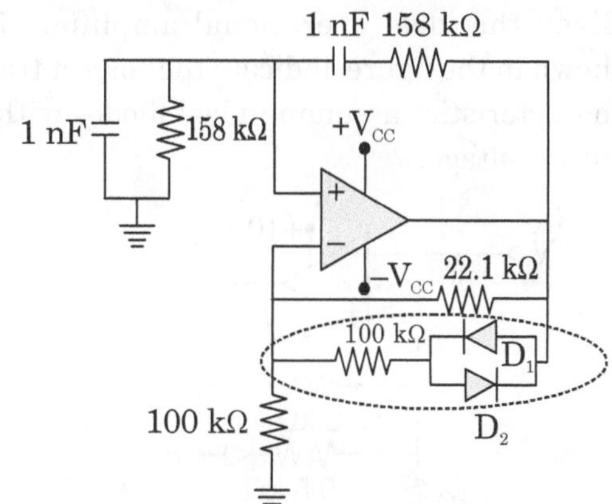

(a) introduce amplitude stabilization by preventing the op amp from saturating and thus producing sinusoidal oscillations of fixed amplitude.

(b) introduce amplitude stabilization by forcing the op-amp to swing between positive and negative saturation and thus producing square wave oscillations of fixed amplitude.

(c) introduce frequency stabilization by forcing the circuit to oscillate at a single frequency.

(d) enable the loop gain to take on a value that produces square wave oscillations.

[2016 : 1 Mark, Set-1]

FUNCTION GENERATOR & WAVE-SHAPING CIRCUITS

10. Consider the following two statements:

Statement 1: Astable multivibrator can be used for generating square wave.

Statement 2: Bistable multivibrator can be used for storing binary information.

(a) Only statement 1 is correct.

(b) Only statement 2 is correct.

(c) Both the statements 1 and 2 are correct.

(d) Both the statements 1 and 2 are incorrect.

[2001 : 1 Mark]

11. An ideal sawtooth voltage waveform of frequency 500 Hz and amplitude 3 V is generated by charging a capacitor of 2 μF in every cycle. The charging requires
   (a) constant voltage source of 3 V for 1 ms.
   (b) constant voltage source of 3 V for 2 ms.
   (c) constant current source of mA for 1 ms.
   (d) constant current source of 3 mA for 2 ms.

[2003 : 2 Marks]

12. Given the ideal operational amplifier circuit shown in the figure, indicate the correct transfer characteristics assuming ideal diodes with zero cut-in voltage.

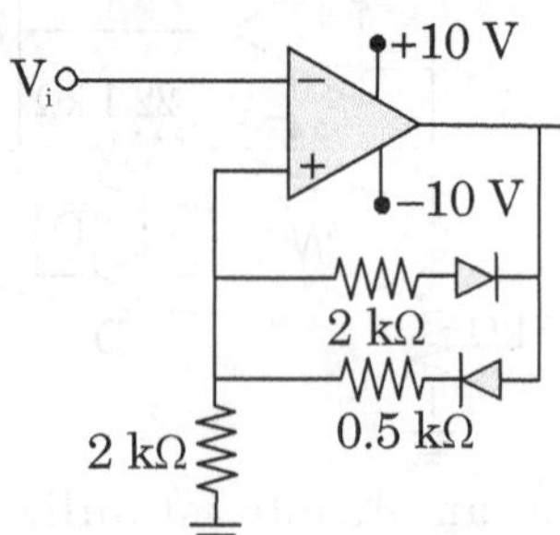

(a)
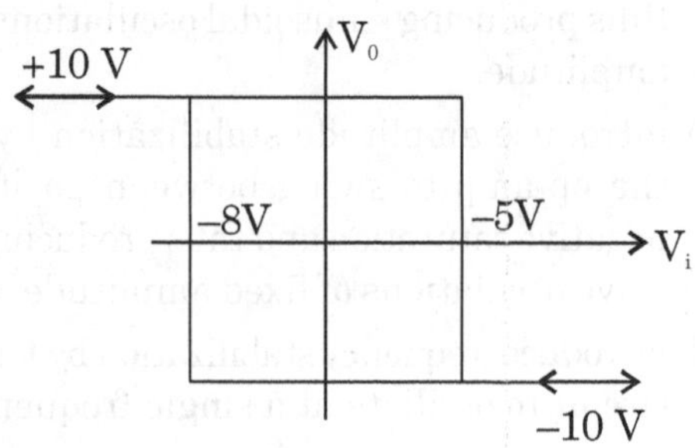

(b)
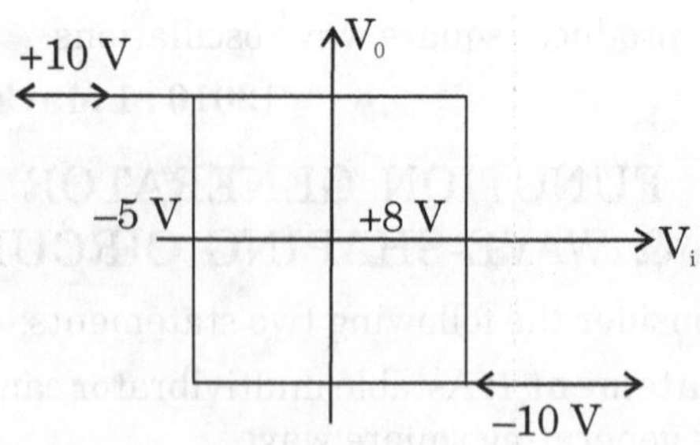

(c)
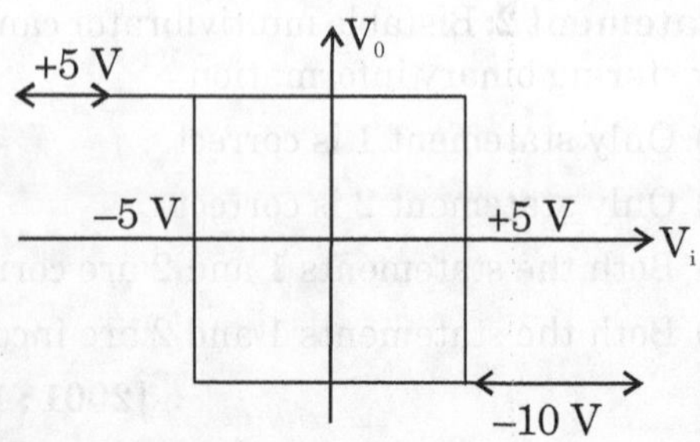

(d)
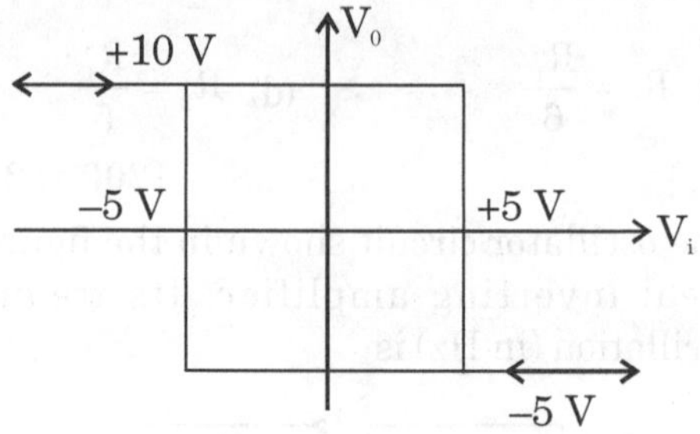

[2005 : 2 Marks]

13. Consider the Schmidt trigger circuit shown below:

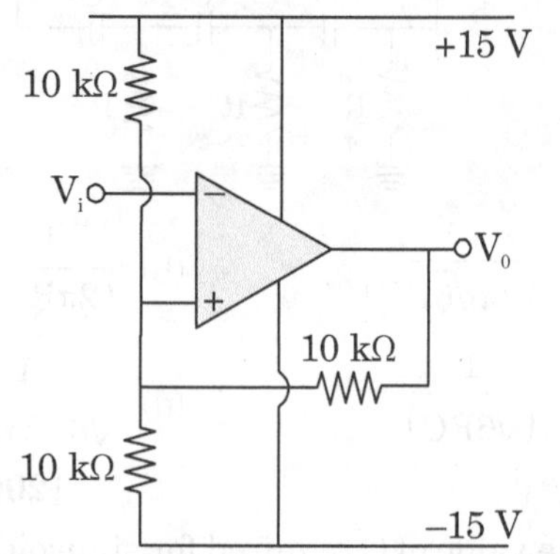

A triangular wave which goes from −12 V to 12 V is applied to the inverting input of the OP-Amp. Assume that the output of the OP-Amp swings from +15 V to −15 V. The voltage at the non- inverting input switches between
   (a) −12 V and +12 V
   (b) −7.5 and +7.5 V
   (c) −5 V and +5 V
   (d) 0 V and 5 V

[2008 : 2 Marks]

14. In the following astable multivibrator circuit, which properties of $v_0(t)$ depend on R_2?

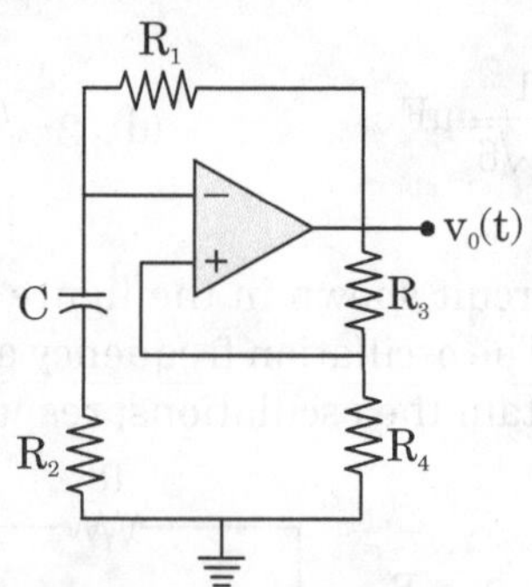

   (a) Only the frequency
   (b) Only the amplitude
   (c) Both the amplitude and the frequency
   (d) Neither the amplitude nor the frequency

[2009 : 2 Marks]

555 TIMER

15. An astable multi-vibrator circuit using IC 555 timer is shown below. Assume that the circuit is oscillating steadily.

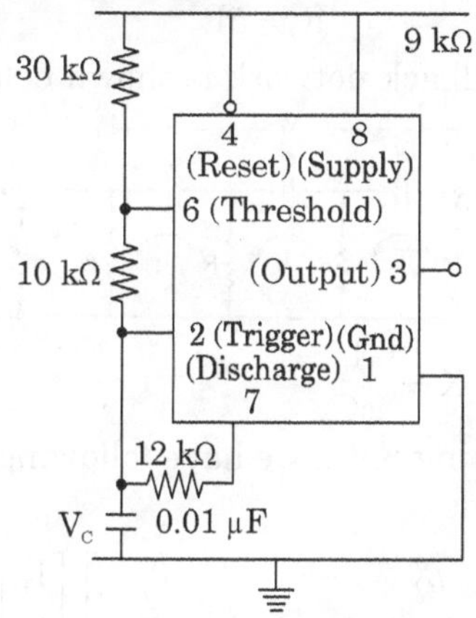

The voltage V_C across the capacitor varies between

(a) 3 V to 5 V
(b) 3 V to 6 V
(c) 3.6 V to 6 V
(d) 3.6 V to 5 V

[2008 : 2 Marks]

16. In the astable multivibrator circuit shown in the figure, the frequency of oscillat on (in kHz) at the output pin 3 is ______.

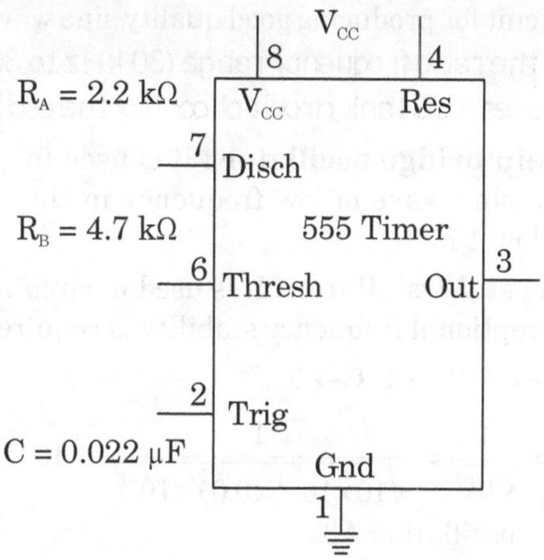

[2016: 1 Mark, Set-3]

ANSWERS

1. (*)	2. (a)	3. (d)	4. (b)	5. (a)	6. (a)	7. (a)	8. (d)	9. (a)	10. (c)
11. (d)	12. (b)	13. (c)	14. (a)	15. (b)	16. (5.65)				

EXPLANATIONS

1. **Hartley Oscillator :** It is a particular useful circuit for producing good quality sine wave signal in the ratio frequency range (30 kHz to 30 MHz). It uses a LC tank circuit ot control the oscillations.

   Wein bridge oscillator : It is used to generate the sine wave of low frequency in the range of below 1 MHz.

   Crystal Oscillator : It is used in circuits where exceptional frequency stability is required.

   $A \rightarrow 2, B \rightarrow 1, C \rightarrow 3.$

2. $$\omega = \frac{1}{\sqrt{LC}} = \frac{1}{\sqrt{10\times10^{-3}\times0.01\times10^{-6}}} = 10^5 \text{ rad/sec}$$

   For oscillation Aβ

   $$= 1\left(1+\frac{500K}{5K}\right)\frac{1000}{1000+R} = 1$$

   $$\therefore \quad R = 20000\ \Omega = 2\times10^4\ \Omega$$

3. 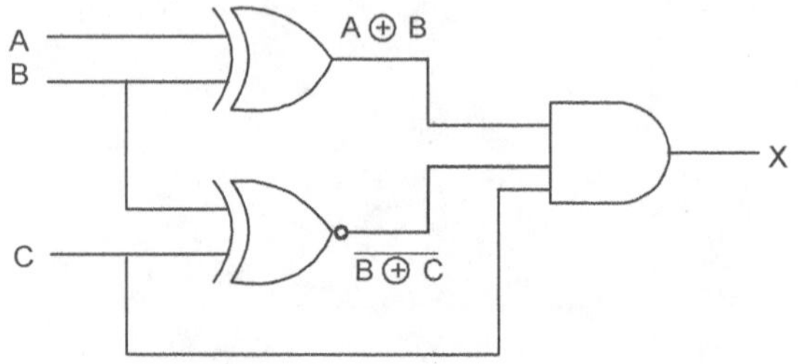

   $$X = (A\oplus B)(\overline{B\oplus C})\cdot C\ (A\bar{B}+\bar{A}B)\left(\overline{\bar{B}C+B\bar{C}}\right).C$$

   $$= (A\bar{B}+\bar{A}B)(B+\bar{C})\cdot(\bar{B}+C)C$$

   $$X = (A\bar{B}+\bar{A}B)(\bar{B}\bar{C}+BC)C = \bar{A}BC$$

   For X= 1, A = 0, B = 1, C = 1

4. This is Colpitts oscillator.

   $$f_{oscillate} = \frac{1}{2\pi\sqrt{L.C_{eq}}}$$

   where, $$C_{eq} = \frac{C_1C_2}{C_1+C_2} = \frac{2\times2}{4} = 1\text{ pF}$$

   $$L = 10\ \mu F$$

   $$f_{oscillate} = \frac{1}{2\pi\sqrt{10\times10^{-6}\times10^{-12}}}$$

   $$= 50.3\text{ MHz}$$

5. Using KCL at inverting terminal, we get

   $$\frac{\beta V_0 - 0}{R_1} + \frac{\beta V_0 - V_0}{R_2} = 0$$

   or $$\beta\left(\frac{R_1+R_2}{R_1R_2}\right) = \frac{1}{R_2}$$

   or $$\beta = \frac{R_1}{R_1+R_2} = \frac{1}{6}$$

   or $$R_2 = 5R_1$$

6. The feedback network is shown below.

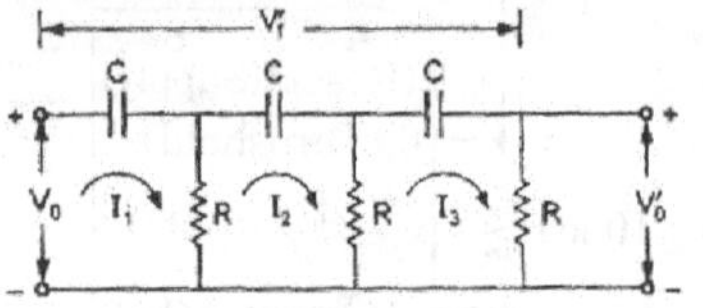

   Let $$X = \frac{1}{\omega C}$$

   then, using KVL, we have following set of linear equations

   $$\begin{bmatrix} R-jX & -R & 0 \\ -R & 2R-jX & -R \\ 0 & -R & 2R-jX \end{bmatrix}\begin{bmatrix} I_1 \\ I_2 \\ I_3 \end{bmatrix} = \begin{bmatrix} V_0 \\ 0 \\ 0 \end{bmatrix}$$

   Putting $\alpha = \frac{X}{R} = \frac{1}{\omega RC}$, we get

   or $$R^3\begin{bmatrix} 1-i\alpha & -1 & 0 \\ -1 & 2-i\alpha & -1 \\ 0 & -1 & 2-i\alpha \end{bmatrix}\begin{bmatrix} I_1 \\ I_2 \\ I_3 \end{bmatrix} = \begin{bmatrix} V_0 \\ 0 \\ 0 \end{bmatrix}$$

   Now, $$I_3 = \frac{\Delta_3}{\Delta}$$

   where $$\Delta_3 = R^3\begin{vmatrix} 1-i\alpha & -1 & V_0 \\ -1 & 2-i\alpha & 0 \\ 0 & -1 & 0 \end{vmatrix}$$

   and, $$\Delta = R^3\begin{vmatrix} 1-i\alpha & -1 & 0 \\ -1 & 2-i\alpha & -1 \\ 0 & -1 & 2-i\alpha \end{vmatrix}$$

   then, $$I_3 = \frac{\Delta_3}{\Delta} = \frac{V_0}{1-5\alpha^2+j(\alpha^3-6\alpha)}$$

   and, $$\frac{V_0'}{V_0} = \frac{-I_3R}{V_0}$$

   $$= \frac{-1}{(1-5\alpha^2)+j(\alpha^3-6\alpha)}$$

   For 180° phase shift, $\alpha^3 - 6\alpha = 0$

   or $$\alpha = \sqrt{6}$$

   or $$\frac{1}{\omega RC} = \sqrt{6}$$

   or $$f = \frac{1}{2\pi\ RC\sqrt{6}}$$

7. $$\frac{V_o - V_1}{X_C + R} = \frac{V_1}{X_C} + \frac{V_1}{R}$$

$$V_0 = (X_C + R)\left[\frac{1}{X_C} + \frac{1}{R}\right]V_1 + V_1$$

$$\frac{V_0}{V_1} = \left(\frac{(X_C + R)}{X_C R}\right)^2 + 1 = \text{feedback}$$

gain

$$X_C = -\frac{j}{\omega C}$$

and $R = 1 \times 10^3$

$$\frac{V_0}{V_1} = \frac{\frac{+1}{\omega^2 C^2} + R^2 - \frac{2jR}{\omega C}}{\frac{+R}{j\omega C}} + 1$$

$$= +j\omega C\left(\frac{\frac{+1}{\omega^2 C^2} + R^2 - \frac{2jR}{\omega C}}{R}\right) - 1$$

$$= \frac{+j\omega C\left(R^2 - \frac{1}{\omega^2 C^2}\right) + 2R}{R}$$

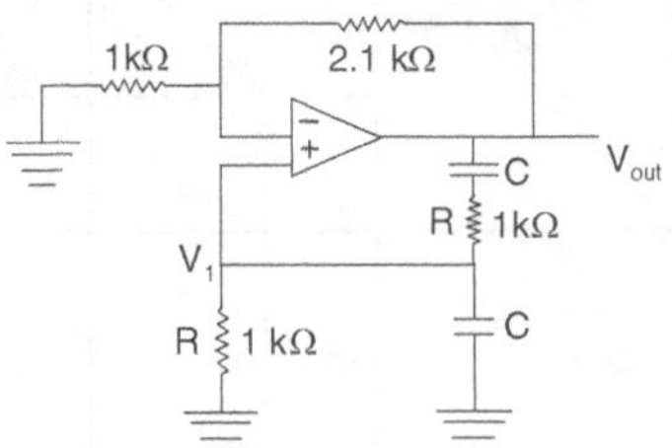

For oscillation imaginary part = 0

i.e. $R^2 - \frac{1}{\omega^2 C^2} = 0$

$$\Rightarrow \quad C = \frac{1}{\omega R} = \frac{1}{2\pi \times 1 \times 10^3 \times 1 \times 10^3} = \frac{1}{2\pi}\ \mu F$$

8. Given circuit is wein bridge oscillator, frequency of wein bridge oscillator is $\omega_0 = \frac{1}{RC}$.

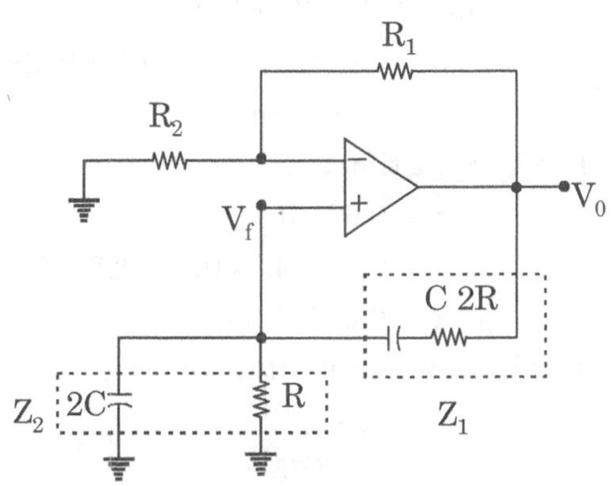

But in the question, time–constant is doubled so, frequency becomes half $\omega_0 = \frac{1}{2R_C}$.

$$\Rightarrow Z_1 = 2R + \frac{1}{j\omega C} = 2(R - jR)$$

$$\Rightarrow Z_2 = \frac{R \times 1/2j\omega C}{R + \frac{1}{2j\omega c}} = \frac{R^2/j}{R - jR}$$

$$\Rightarrow \beta = \frac{Z_2}{Z_1 + Z_2} = \frac{1}{5}$$

$$1 + \frac{R_1}{R_2} = 5 \Rightarrow \boxed{R_1 = 4R_2}$$

9. The circuit shown (in fig.) is a Wein-bridge oscillator. So the amplitude of oscillations can be determined and stabilized by using a nonlinear control network. As the oscillations grow, the diodes start to conducting that causes the effective resistance in the feedback to decrease. Therefore equilibrium will be reached at the output amplitude that causes the loop gain to be exactly unity.

10. In astable multivibrator, none of the states are stable. It is used to generate the square wave flip-flops are bistable multivibrator which are used to store 1 bit of information.

11. Ideal sawtooth voltage waveform is shown below

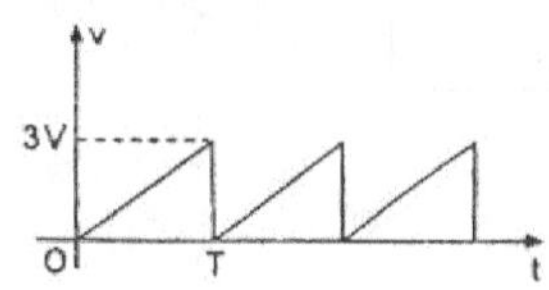

$$T = \frac{1}{f} = \frac{1}{500} = 2 \text{ ms}$$

and, $$I = C\frac{dV}{dt} = C\ \frac{3-0}{T}$$

$$= \frac{3 \times 2 \times 10^{-6}}{2 \times 10^{-3}} = 3 \text{ mA}$$

Thus charging requires a constant current source of 3 mA for 2 ms.

12.

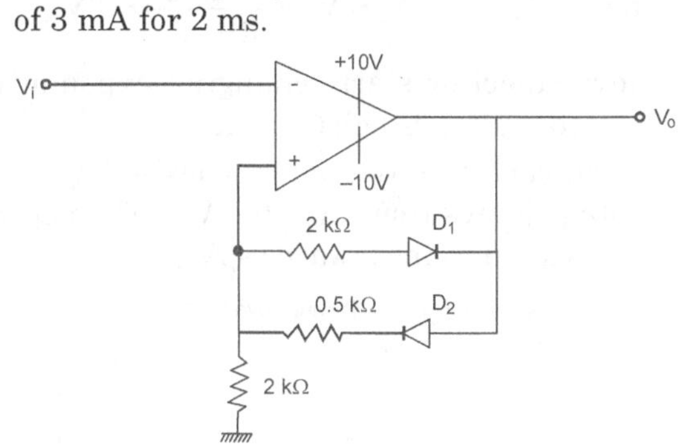

OPAMP output varies between $+V_{sat} = 10$ V to $-V_{sat} = -10$ V

When $V_o > 0$, D_2 is forward-biased and D_1 is reverse-biased.

Then, V_+ = (Upper threshold voltage)

$$= +V_{sat}\frac{2}{2 + 0.5} = 10 \times \frac{2}{2.5} = 12 \text{ volts}$$

When $V_o < 0$, D_2 is reverse-biased and D_1 is forward-biased.

Then $\quad V_- =$ lower-threshold voltage

$$= -V_{sat} \cdot \frac{2}{2+2} = -5 \text{ volts}$$

Hence, the right option is (b)

13. When output is + 15 V, let voltage at non inverting terminal is V_0

$$\therefore \quad \left(\frac{V_0 - 15}{10}\right) 2 + \left(\frac{V_0 + 15}{10}\right) = 0$$

$$\Rightarrow \quad V_0 = +5 \text{ volts}$$

When output is – 15V

$$\frac{V_0 - 15}{10} + 2\left(\frac{V_0 + 15}{10}\right) = 0$$

$$\Rightarrow \quad V_0 = -5 \text{ volts}$$

14.

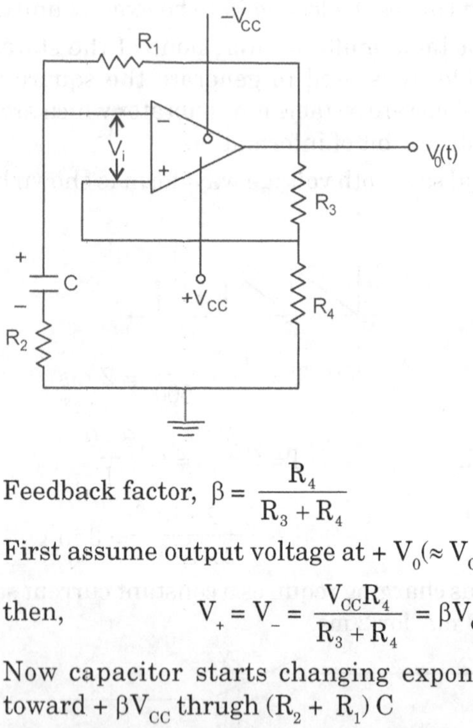

Feedback factor, $\beta = \frac{R_4}{R_3 + R_4}$

First assume output voltage at $+V_0 (\approx V_{CC})$

then, $\quad V_+ = V_- = \frac{V_{CC} R_4}{R_3 + R_4} = \beta V_{CC}$

Now capacitor starts changing exponentially toward $+\beta V_{CC}$ thrugh $(R_2 + R_1)$ C

When capacitor voltage reaches $+\beta V_{CC}$, output voltage jumps from $+V_{CC}$ to $-V_{CC}$. Then cpacitor starts discharging towards $-\beta V_{CC}$

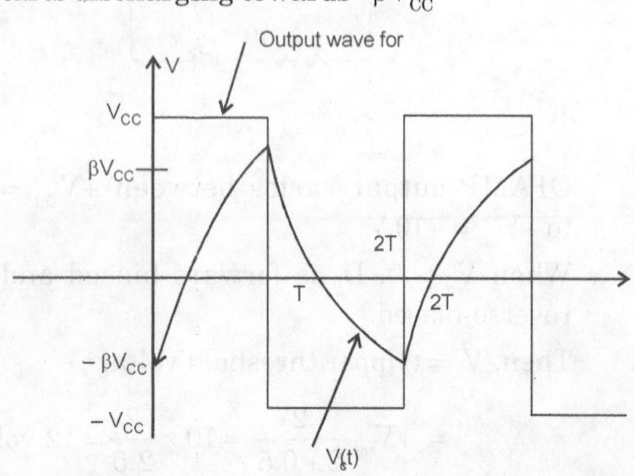

Now, $\quad V_C(t) = V_F + (V_1 - V_F)\, e^{-\frac{T}{RC}} \quad$ ...(A)

where $\quad R = R_2 + R_1$

$V_1 =$ (tuntial voltage) $= -\beta V_{CC}$

$V_F = V_{CC}$

at $\quad t = T$ (half-period)

$V_C(t) = \beta V_{CC}$

then, Equation (A) given

$$\beta V_{CC} = V_{CC} + (-\beta V_{CC} - V_{CC})\, e^{-\frac{T}{RC}}$$

or, $$T = RC \ln\left[\frac{1+\beta}{1-\beta}\right]$$

It can be seen that only time period frequency depends in R_2.

15. Capacitor voltage and output waveform are shown below :

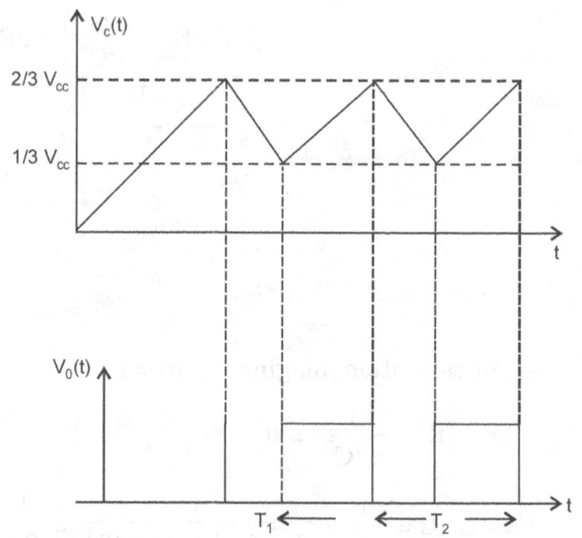

Hence $V_{cc} = 9$ volts

therefore, $V_c(t)$ varies between 3 volts to 6 volts.

16. Charging time,

$$\tau_C = 0.693\,(R_A + R_B).C$$

$$= 0.693 \times (2.2+4.7) \times 10^3 \times 0.022 \times 10^{-6}$$

$$\tau_C = 0.1052 \text{ ms.}$$

Discharging time,

$$\tau_d = 0.693\, R_B C$$

$$= 0.693 \times 4.7 \times 10^3 \times 0.022 \times 10^{-6}$$

$$\tau_d = 0.07165 \text{ ms.}$$

Total period of output waveform,

$$T = \tau_d + \tau_C$$

$$T = 0.17865 \text{ ms.}$$

Frequency of oscillation, $f = \frac{1}{T} = 5.65\text{kHz}.$

Unit - VI

Digital Circuits

Syllabus

Number systems; Combinatorial circuits: Boolean algebra, minimization of functions using Boolean identities and Karnaugh map, logic gates and their static CMOS implementations, arithmetic circuits, code converters, multiplexers, decoders and PLAs; Sequential circuits: latches and flip flops, counters, shift registers and finite state machines; Data converters: sample and hold circuits, ADCs and DACs; Semiconductor memories: ROM, SRAM, DRAM.

Contents

1 CHAPTER

Number System

Analysis of Previous GATE Papers													
	Year → Topics ↓		2019	2018	2017 Set 1	2017 Set 2	2016 Set 1	2016 Set 2	2016 Set 3	2015 Set 1	2015 Set 2	2015 Set 3
DIFFERENT NUMBER SYSTEMS & THEIR RANGE	1 Mark	MCQ Type										
		Numerical Type										
	2 Marks	MCQ Type										
		Numerical Type										
		Total										
REPRESENTATION OF NEGATIVE NUMBERS	1 Mark	MCQ Type										
		Numerical Type										
	2 Marks	MCQ Type										
		Numerical Type										
		Total										

DIFFERENT NUMBER SYSTEMS & THEIR RANGE

1. The range of signed decimal numbers that can be represented by 6-bit 1's complement number is

(a) –31 to +31 (b) –63 to 63
(c) –64 to + 63 (d) –32 to + 31

[2004 : 1 Mark]

2. Decimal 43 in Hexadecimal and BCD number system is respectively

(a) B2, 0100 0011 (b) 2B, 0100 0011
(c) 2B, 0011 0100 (d) B2, 0100 0100

[2005 : 1 Mark]

3. A new Binary Coded Pentary (BCP) number system is proposed in which every digit of a base-5 number is represented by its correspon ding 3-bit binary code. For example, the base-5 number 24 will be represented by its BOP code 010100, in this numbering system, the BCP code 100010011001 corresponds to the following number in base-5 system

(a) 423 (b) 1324
(c) 2201 (d) 4231

[2006 : 2 Marks]

4. The number of bytes required to represent the decimal number 1856357 in packed BCD (Binary Coded Decimal) form is ________.

[2014: 1 Mark, Set-2]

REPRESENTATION OF NEGATIVE NUMBERS

5. The subtraction of a binary number Y from another binary number X, done by adding the 2's complement of Y to X, results in a binary number without overflow. This implies that the result is

(a) negative and is in normal form
(b) negative and is in 2's complement form
(c) positive and is in normal form
(d) positive and is in 2's complement form

[1987 : 1 Mark]

6. 2's complement representation of a 16-bit number (one sign bit and 15 magnitude bits) if FFFF, its magnitude in decimal representation is

(a) 0 (b) 1
(c) 32,767 (d) 65,535

[1993 1 Mark]

7. A signed integer has been stored in a byte using the 2's complement format. We wish to store the same integer in a 16 bit word. We should

(a) copy the original byte to the less significant byte of the word and fill the more significant byte with zeros.
(b) copy the original byte to the more significant byte of the word and fill the less significant byte with zeros.
(c) copy the original byte to the less significant byte of the word and make each bit of the more significant byte equal to the most significant bit of the original byte.
(d) copy the original byte to the less significant bytes well as the more significant byte of the word.

[1997 : 1 Mark]

8. An equivalent 2's complement representation of the 2's complement number 1101 is

(a) 110100 (b) 001101
(c) 110111 (d) 111101

[1998 : 1 Mark]

9. The 2's complement representation of –17 is

(a) 101110 (b) 101111
(c) 111110 (d) 110001

[2001 : 1 Mark]

10. 4-bit 2's complement representation of a decimal number is 1000. The number is

(a) +8 (b) 0
(c) –7 (d) –8

[2002 : 1 Mark]

11. 11001, 1001 and 111001 correspond to the 2's complement representation of which one of the following sets of number?

(a) 25, 9 and 57 respectively
(b) –6, –6 and –6 respectively
(c) –7, –7 and –7 respectively
(d) –25, –9 and –57 respectively

[2004 : 2 Marks]

12. X = 01110 and Y = 11001 are two 5-bit binary numbers represented in 2's complement format. The sum of X and Y represented in 2's complement format using 6 bits is

(a) 100111 (b) 001000
(c) 000111 (d) 101001

[2007 : 1 Mark]

13. The two numbers represented in signed 2's complement form are

P = 11101101 and Q = 11100110. If Q is subtracted from P, the value obtained in signed 2's complement form is

(a) 100000111 (b) 00000111
(c) 11111001 (d) 111111001

[2008 : 2 Marks]

ANSWERS

1. (a)	**2.** (b)	**3.** (d)	**4.** (4)	**5.** (b)	**6.** (c)	**7.** (c)	**8.** (c)	**9.** (b)	**10.** (d)
11. (c)	**12.** (c)	**13.** (b)							

EXPLANATIONS

1. The first complement range of numbers is $-2^{n-1}+1$ to $2^{n-1}-1$.

   When $n = 6$, so -31 to $+31$

2. 43 becomes 0100 0011 in BCD

   When converted to binary,

   $$43 \equiv \underbrace{0010}_{2}\ \underbrace{1110}_{B}$$

   ∴ Hexadecimal code is 2 B

3. According to Binary Coded Pentary (BCP) number system,

   24 is represented by

   $$\overset{32\,16\,8}{\underbrace{010}_{2}}\ \overset{4\,2\,1}{\underbrace{100}_{4}}$$

   Thus, $\underbrace{100}_{4}\ \underbrace{010}_{2}\ \underbrace{011}_{3}\ \underbrace{001}_{1}$ corresponds to 4231

4. To represent decimal number into BCD number each decimal is represented in 4-bits white converting in BCD number are:

   $(1856357)_{10}$

   $= \overbrace{0001\ 1000}^{1\text{ Byte}}\ \overbrace{0101\ 0110}^{1\text{ Byte}}\ \overbrace{0011\ 0101}^{1\text{ Byte}}\ 0111.$

   = 0000 0001 1000 0101 0110 0011 0101 0111.

   ⇒ Hence, total 4 bytes are required.

5. If there is overflow bit then answer is positive and in normal form. If there is no overflow, then answer is negative and in 2's complement form.

6. Given: FFFFH = 1111 1111 1111 1111b (Number of negative).

   1's complement of FFFFH = 0000 0000 0000 0000b

   Normal form of FFFFH = 0000H + 1

   = 0001H. $= (1)_{10}$.

7. The MSB of the integer is 8-bit format should be repeated to expand the representation of 2's complement form of 16-bit.

8. Given number $(1101)_2$ is $(-3)_{10}$. According to sign bit extension, the sign bit can be extended towards left.

   $$(1101)_2 = \underbrace{11}_{\substack{\text{bit}\\\text{extension}}}1101.$$

9. $(17)_{10} = (10001)_2$

   2's complement of $(17)_{10}$ = 1's complement of (10001) + 1

   = 01110 + 1 = 01111b

   2's complement of $(-17)_{10}$ = 101111b.

10.
    ```
       1 0 0 0   |+8
       0 1 1 1   1's complement
    +        1
    ---------
       1 0 0 0   2's complement = –8
    ```

11. 11001 ⇒ – 00111 ⇒ – 7

    1001 ⇒ – 0111 ⇒ – 7

    111001 ⇒ – 000111 ⇒ – 7

12. Given, X = 01110 Y = 11001

    In 2′s complement format MSB shows the sign of number (0 for positive and 1 for negative) and negative number is represented in 2′s complement

X = 01110 Y = 11001

= + 14 00110

1

00111

Y = – 7

∴ X + Y = + 14 – 7 = + 7

S = + 7 in 6 bits representation

000111

It will remain same in 2′s complement because it is positive value

So answer is 000111

13. Signed 2's complement of

P = 11101101 is:

⇒ Number P = (1's complement of 11101101) + 1 = 00010011.

⇒ Signed 2's complement of Q = 1110010.

⇒ P – Q = P + (2's complement of Q)

= 11111001.

2's complement of P – Q

= 00000110 + 1 = 00000111.

2 CHAPTER Boolean Algebra & Logic Gates

Analysis of Previous GATE Papers												
	Year → Topics ↓		2019	2018	2017 Set 1	2017 Set 2	2016 Set 1	2016 Set 2	2016 Set 3	2015 Set 1	2015 Set 2	2015 Set 3
BOOLEAN ALGEBRA IDENTITIES & MINIMIZATION OF BOLLEAN FUNCTIONS	1 Mark	MCQ Type										
		Numerical Type										
	2 Marks	MCQ Type										
		Numerical Type										
	Total											
LOGIC GATES	1 Mark	MCQ Type					1		1			1
		Numerical Type										
	2 Marks	MCQ Type								1		1
		Numerical Type		1						1		
	Total			2			1		1	4		3
K-MAP	1 Mark	MCQ Type		1								
		Numerical Type										
	2 Marks	MCQ Type			1				1	1	1	
		Numerical Type										
	Total			1	2				2	2	2	

BOOLEAN ALGEBRA IDENTITIES & MINIMIZATION OF BOOLEAN FUNCTIONS

1. The number of Boolean functions that can be generated by n variables is equal to:

(a) 2^{2n-1} (b) 2^{2n}

(c) 2^{n-1} (d) 2^{n}

[1990 : 1 Mark]

2. Two 2's complement numbers having sign bits x and y are added and the sign bit of the result is z. Then, the occurrence of overflow is indicated by the Boolean function

(a) xyz (b) $\bar{x}\bar{y}\bar{z}$

(c) $\bar{x}\bar{y}z + xy\bar{z}$ (d) $xy + yz + zx$

[1998 : 1 Mark]

3. The logical expression $y = A + \bar{A}B$ is equivalent to

(a) $y = AB$ (b) $y = \bar{A}B$

(c) $y = \bar{A} + B$ (d) $y = A + B$

[1999 : 1 Mark]

4. The minimized form of the logical expression $\left(\bar{A}\bar{B}\bar{C} + \bar{A}B\bar{C} + \bar{A}BC + AB\bar{C}\right)$ is

(a) $\bar{A}\bar{C} + B\bar{C} + \bar{A}B$ (b) $A\bar{C} + \bar{B}\bar{C} + \bar{A}B$

(c) $\bar{A}C + \bar{B}C + \bar{A}B$ (d) $A\bar{C} + \bar{B}C + A\bar{B}$

[1999 : 2 Marks]

5. The number of distinct Boolean expressions of 4 variables is

(a) 16 (b) 256

(c) 1024 (d) 65536

[2003 : 1 Mark]

6. The Boolean expression for the truth table shown is

A	B	C	F
0	0	0	0
0	0	1	0
0	1	0	0
0	1	1	1
1	0	0	0
1	0	1	0
1	1	0	1
1	1	1	0

(a) $B(A + C)(\bar{A} + \bar{C})$ (b) $B(A + \bar{C})(\bar{A} + C)$

(c) $B(A + C)(\bar{A} + \bar{C})$ (d) $\bar{B}(A + C)(\bar{A} + \bar{C})$

[2005 : 2 Marks]

7. If X = 1 in the logic equation

$[X + Z\{\bar{Y} + (\bar{Z} + X\bar{Y})]\{\bar{X} + \bar{Z}(X + Y)\} = 1$, then

(a) $Y = Z$ (b) $Y = \bar{Z}$

(c) $Z = 1$ (d) $Z = 0$

[2009 : 2 Marks]

8. In the circuit shown below, Q_1 has negligible collector-to-emitter saturation voltage and the diode drops negligble voltage across it under forward bias. If V_{cc}, is +5 V X and Y are digital signals with DV as logic d and V_{cc} as logic 1, the Boolean expression for Z is

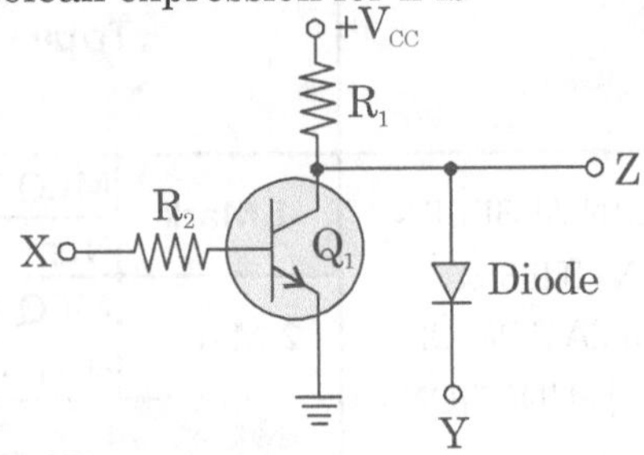

(a) XY (b) $\bar{X}Y$

(c) $X\bar{Y}$ (d) $\bar{X}\bar{Y}$

[2011 : 1 Mark]

9. The Boolean expression

$(X + Y)(X + \bar{Y}) + (\bar{X}\bar{Y}) + \bar{X}$ simplifies to

(a) X (b) Y

(c) XY (d) X + Y

[2013 : 2 Marks]

10. The Boolean function A + BC is a reduced form of

(a) AB + BC (b) (A + B). (A + C)

(c) $\bar{A}B + ABC$ (d) (A + C).B

[2014: 1 Mark, Set-1]

LOGIC GATES

11. The Boolean function Y= AB + CD is to be realized using only 2-input NAND gates. The minimum number of gates required is

(a) 2 (b) 3

(c) 4 (d) 5

[1988 : 1 Mark]

12. For the circuit shown below the output F is given by

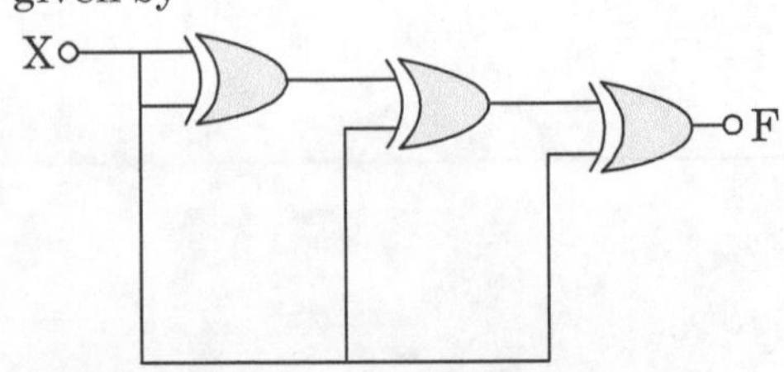

(a) F = 1 (b) F= 0

(c) F = X (d) F= $\bar{X}$

[1988 : 1 Mark]

13. Minimum number of 2-input NAND gates required to implement the function, F (X + Y) $F = (\overline{X} + \overline{Y})(Z + W)$ is

(a) 3 (b) 4
(c) 5 (d) 6

[1988: 1 Mark]

14. Indicate which of the following logic gates can be used to realized all possible combinational Logic functions?

(a) OR gates only (b) NAND gates only
(c) EX-OR gates only (d) NOR gates only

[1989 : 1 Mark]

15. Boolean expression for the output of XNOR (equivalence) logic gate with inputs A and B is

(a) $A\overline{B} + \overline{A}B$ (b) $\overline{A}\overline{B} + AB$
(c) $(\overline{A} + B)(A + \overline{B})$ (d) $(\overline{A} + \overline{B})(A + B)$

[1993 : 1 Mark]

16. For the logic circuit shown in figure, the output is equal to

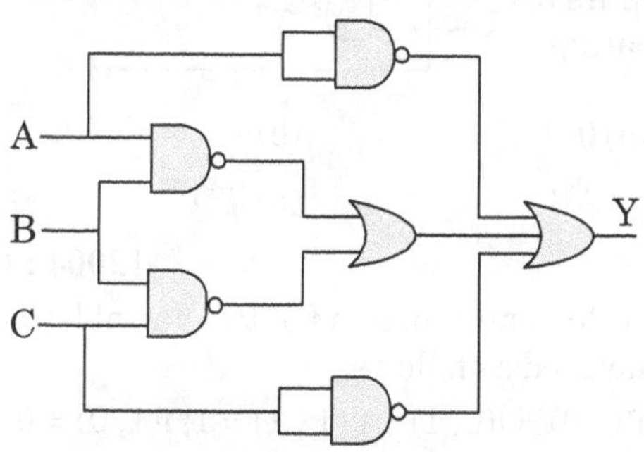

(a) $\overline{A}\overline{B}\overline{C}$ (b) $\overline{A} + \overline{B} + \overline{C}$
(c) $\overline{A}\overline{B} + \overline{B}\overline{C} + \overline{A} + \overline{C}$ (d) $\overline{A}\overline{B} + \overline{B}C$

[1993 : 1 Mark]

17. The output of a logic gate is '1' when all its a inputs are at logic '0'. The gate is either

(a) a NAND or an EX-OR gate
(b) a NOR or an EX-NOR gate
(c) an OR or an EX-NOR gate
(d) an AND or an EX-OR gate

[1994:1 Mark]

18. A ring oscillator consisting of 5 inverters is running at a frequency of 1.0 MHz. The propagation delay per gate ____ is n sec.

[1994: 1 Mark]

19. The minimum number of NAND gates required to implement the Boolean function $A + A\overline{B} + A\overline{B}C$ is equal to

(a) zero (b) 1
(c) 4 (d) 7

[1995 : 1 Mark]

20. The output of the logic gate in figure is

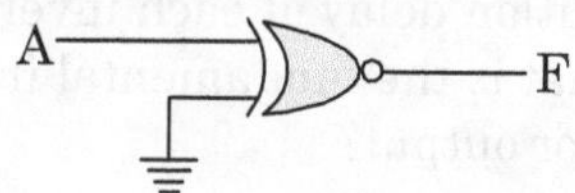

(a) 0 (b) 1
(c) $\overline{A}$ (d) A

[1997 : 1 Mark]

21. The minimum number of 2-input NAND gates required to implement the Boolean function $Z = A\overline{B}C$, assuming that A, B and C are available, is

(a) two (b) three
(c) five (d) six

[1998 : 1 Mark]

22. For the identity $AB + \overline{A}C + BC = AB + \overline{A}C$, the dual form is

(a) $(A + B)(\overline{A} + C)(B + C) = (A + B)(\overline{A} + C)$
(b) $(\overline{A} + \overline{B})(A + \overline{C})(\overline{B} + \overline{C}) = (\overline{A} + \overline{B})(A + \overline{C})$
(c) $(A + B)(\overline{A} + C)(B + C) = (\overline{A} + \overline{B})(A + C)$
(d) $\overline{A}\overline{B} + A\overline{C} + \overline{B}\overline{C} = \overline{A}\overline{B} + A\overline{C}$

[1998 : 1 Mark]

23. For the logic circuit shown in the figure, the required input condition (A, B, C) to make the output (X) = 1 is

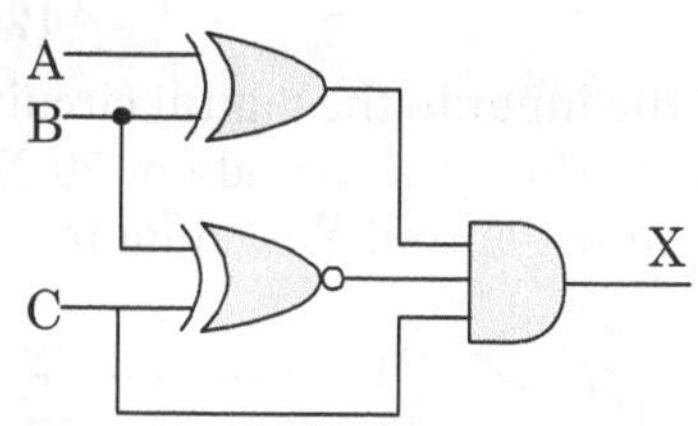

(a) 1, 0, 1 (b) 0, 0, 1
(c) 1, 1, 1 (d) 0, 1, 1

[2000 : 1 Mark]

24. For the logic circuit shown in the figure, the simplified Boolean expression for the output

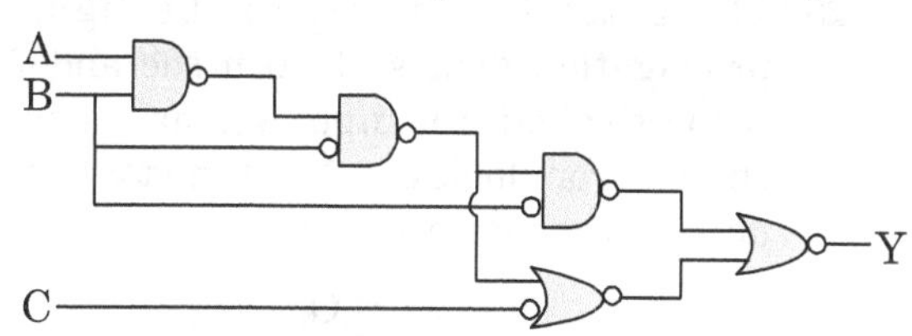

(a) A + B + C
(b) A
(c) B
(d) C

[2000 : 2 Marks]

25. For the ring oscillator shown in the figure, the propagation delay of each inverter is 100 pico sec. What is the fundamental frequency of the oscillator output?

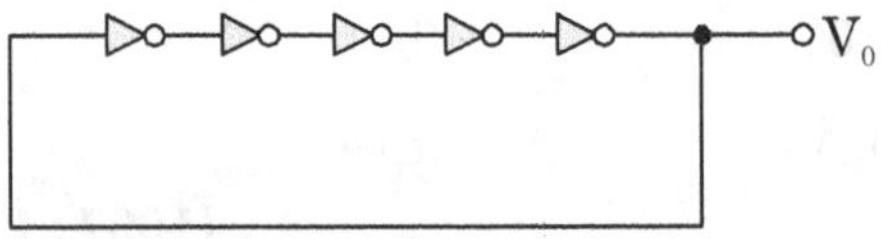

(a) 10 MHz
(b) 100 MHz
(c) 1 GHz
(d) 2 GHz

[2001 : 1 Mark]

26. in the figure, the LED

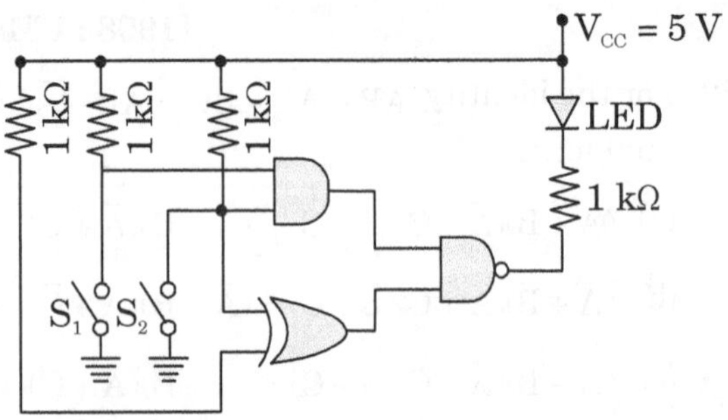

(a) emits light when both S_1 and S_2; are closed.
(b) emits light when both S_1 and S_2 are open.
(c) emits light when only S_1 and S_2 is closed.
(d) does not emit light, irrespective of the switch positions.

[2001 : 2 Marks]

27. If the input to the digital circuit (in the figure) consisting of a cascade of 20 XOR-gates is X, then the output Y is equal to

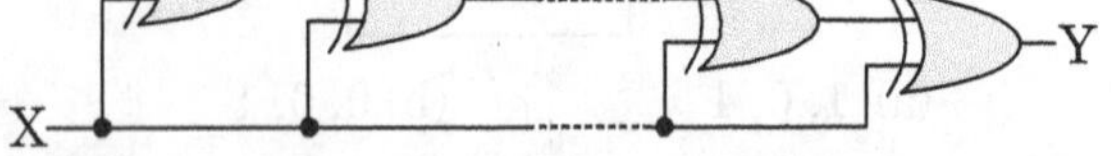

(a) 0 (b) 1
(c) $\overline{X}$ (d) X

[2002 : 1 Mark]

28. The gates G_1 and G_2 in the figure have propagation delays of 10 n sec and 20 nsec respectively. If the input V_1, makes an abrupt change from logic 0 to 1 at time t = t_o, then the output wave form V_o is

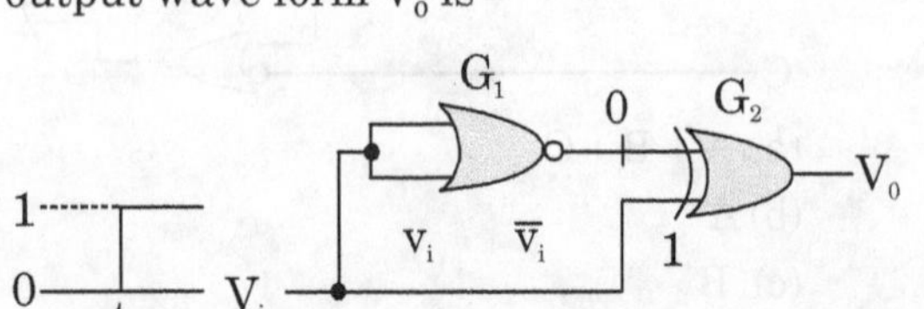

(a) 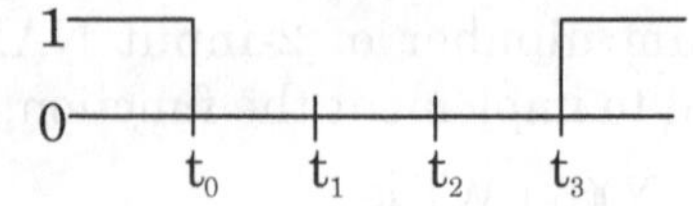

(b) 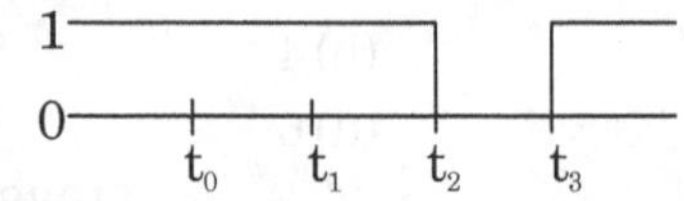

(c) 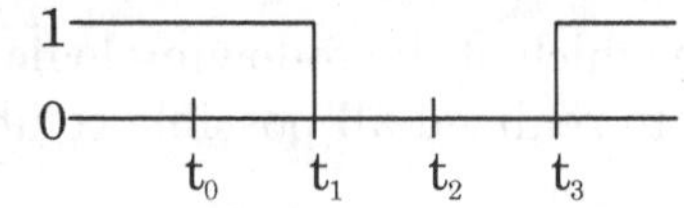

(d) 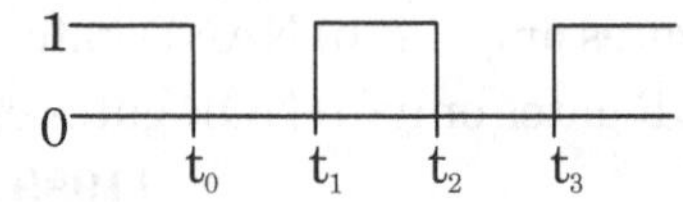

[2002 : 2 Marks]

29. The figure shows the internal schematic of a TTL AND-OR-Invert (AOI) gate. For the inputs shown in the figure, the output Y is

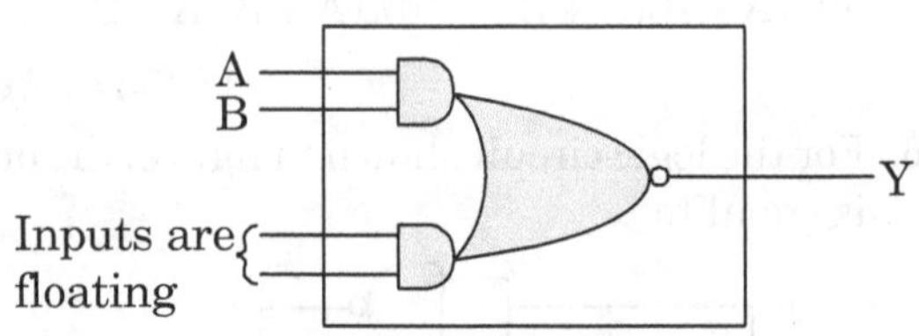

(a) 0 (b) 1
(c) AB (d) $\overline{A}\overline{B}$

[2004 : 1 Mark]

30. A Boolean function f of two variables x and y is defined as follows:

f(0, 0) = f(0, 1) = f (1 , 1) = 1; f(1, 0) = 0

Assuming complements of x and y are not available, a minimum cost solution for realizing f using only 2-input NOR gates and 2-input OR gates (each having unit cost) would have a total cost of

(a) 1 unit (b) 4 unit
(c) 3 unit (d) 2 unit

[2004 : 2 Marks]

31. The point P in the following figure is stuck-at-1. The output f will be

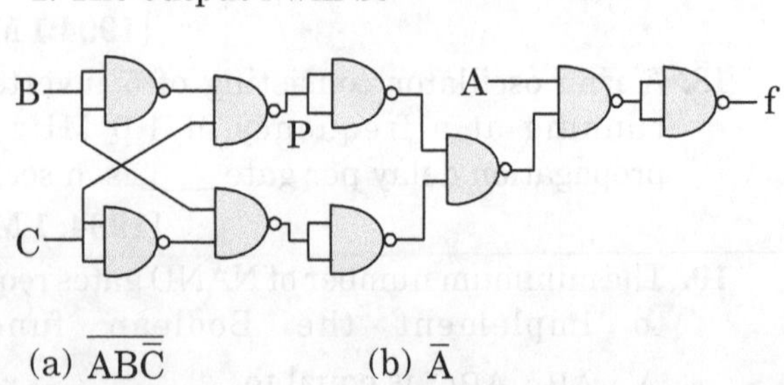

(a) $AB\overline{C}$ (b) $\overline{A}$
(c) ABC (d) A

[2006 : 2 Marks]

32. Which of the following Boolean expressions correctly represents the relation between P, Q, R and M_1?

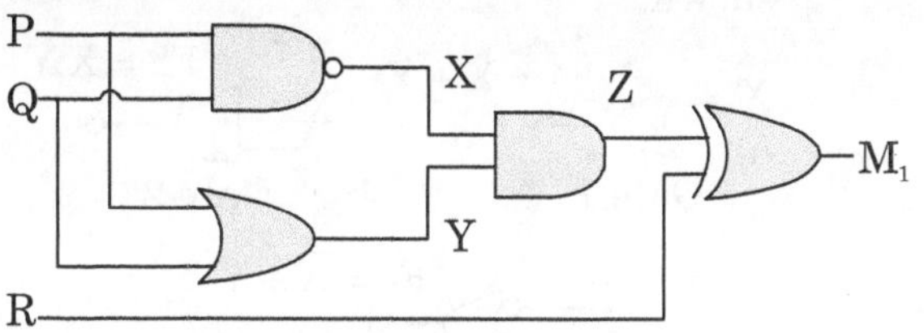

(a) M_1 = (P OR Q) XOR R.
(b) M_1 = (P AND Q) XOR R.
(c) M_1 = (P NOR Q) XOR R.
(d) M_1 = (P XOR Q) XOR R.

[2008 : 2 Marks]

33. Match the logic gates in Column A with their equivalents in Column B.

Column A **Column B**

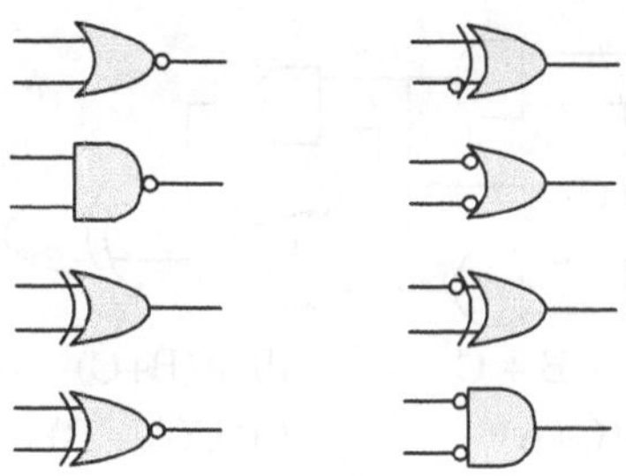

(a) P-2, Q-4, R-1, S-3
(b) P-4, Q-2, R-1, S-3
(c) P-2, Q-4, R-3, S-1
(d) P-4, Q-2, R-3, S-1

[2010 : 1 Mark]

34. For the output F to be 1 is the logic circuit shown, the input combination should be

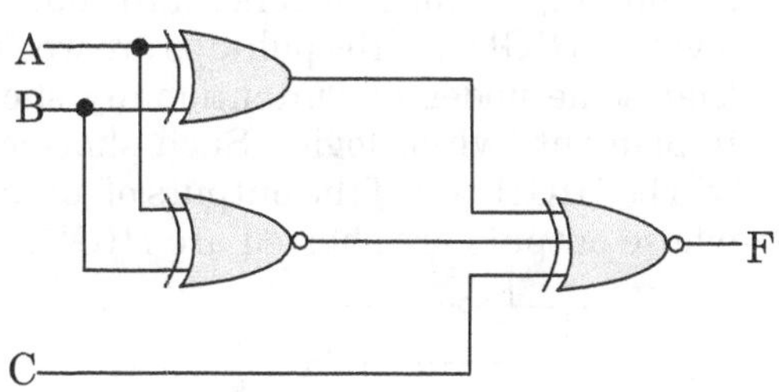

(a) A = 1, B = 1, C = 0
(b) A = 1, B = 0, C = 0
(c) A = 0, B = 1, C = 0
(d) A = 0, B = 0, C = 1

[2010 : 1 Mark]

35. The output Y in the circuit below is always "1" when

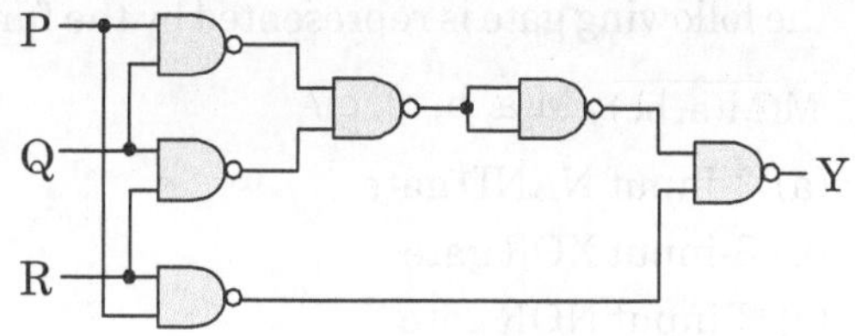

(a) two or more of the inputs P, Q, R are "0"
(b) two or more of the inputs P, Q, R are "1"
(c) any odd number of the inputs P, Q, R is "0'
(d) any odd number of the inputs P, Q, R is "1"

[2011 : 1 Mark]

36. A bulb in a staircase has two switches, one switch being at the ground floor and the other one at the first floor, The bulb can be turned ON and also can be turned OFF by any one of the switches irrespective of the state of the other switch. The logic of switching of the bulb resembles

(a) an AND gate (b) an OR gate
(c) an XOR gate (d) a NAND gate

[2012 : 1 Mark]

37. The output F in the digital logic circuit shown in the figure is

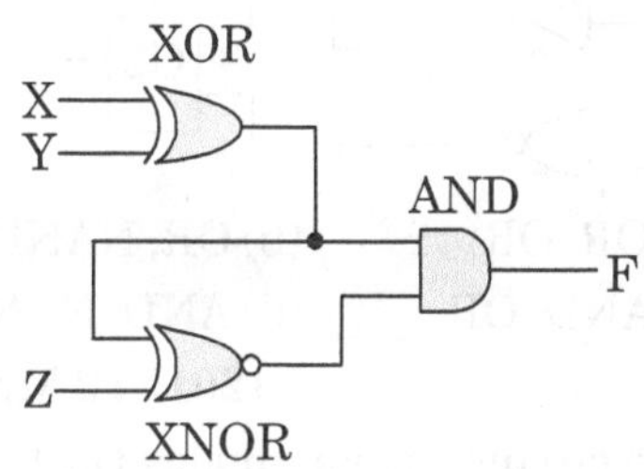

(a) $F = \bar{X}YZ + X\bar{Y}Z$ (b) $F = \bar{X}Y\bar{Z} + XY\bar{Z}$
(c) $F = \bar{X}\bar{Y}Z + XYZ$ (d) $F = \bar{X}\bar{Y}Z + XYZ$

[2013 : 1 Mark]

38. In the circuit shown in the figure, if C = 0, the expression for Y is

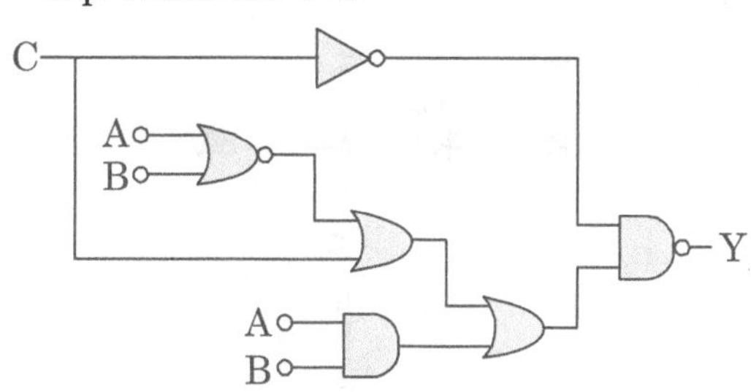

(a) $Y = A\bar{B} + \bar{A}B$ (b) Y = A + B
(c) $Y = \bar{A} + \bar{B}$ (d) Y = AB

[2013 : 1 Mark, Set-2]

39. A 3-input majority gate is defined by the logic function M(a, b, c) = ab + bc + ca. Which one of the following gate is represented by the function

$M(\overline{M(a, bc)}, M(a, b, c), c)$?

(a) 3-input NAND gate
(b) 3-input XOR gate
(c) 3-input NOR gate
(d) 3-input XNOR gate

[2014 : 2 Marks, Set-1]

40. All the logic gates shown in the figure have a propagation delay of 20 ns. Let A = C = 0 and B = 1 until time t = 0. At t = 0, all the inputs flip (i.e. A = C= 1 and B = 0) and remain in that state. For t > 0, output Z = 1 for a duration (in ns) of _____.

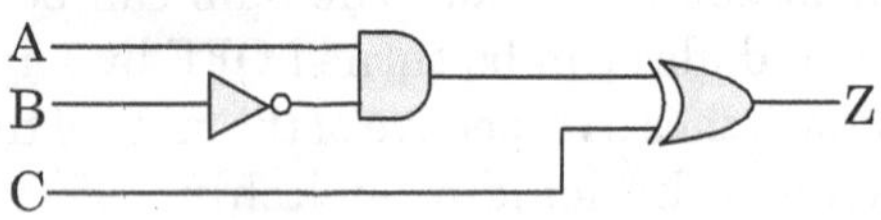

[2014 : 1 Mark, Set-4]

41. In the figure shown, the output Y is required to be $Y = AB + \overline{C}\overline{D}$. The gates G1 and G2 must be

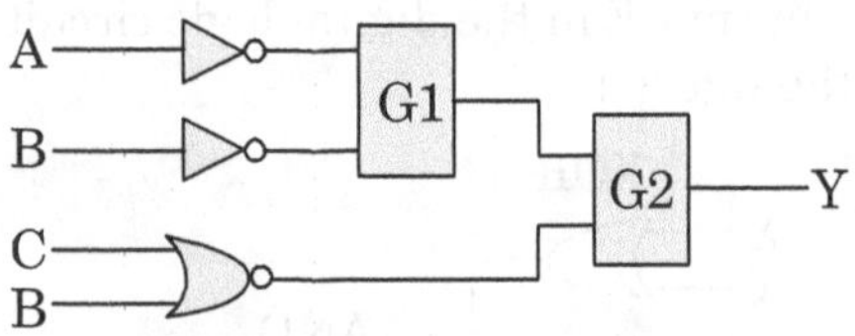

(a) NOR, OR
(b) OR, NAND
(c) NAND, OR
(d) AND, NAND

[2015 : 2 Marks, Set-1]

42. In the circuit shown, diodes D_1, D_2 and D_3 are ideal. and the inputs E_1, E_2, and E_3 are '0 V' for logic '0' and '10 V' for logic '1'. What logic gate does the circuit represent?

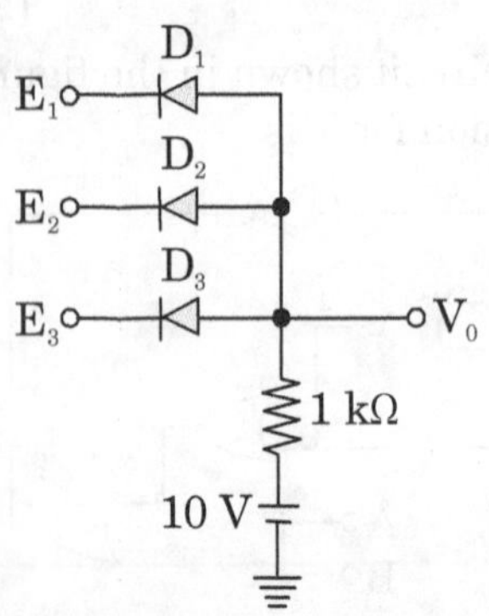

(a) 3 input OR gate
(b) 3 input NOR gate
(c) 3 input AND gate
(d) 3 input XOR gate

2015 : 2 Marks, Set-1]

43. A universal logic gate can mplement any Boolean function by connecting sufficient number of them appropriately. Three gates are shown.

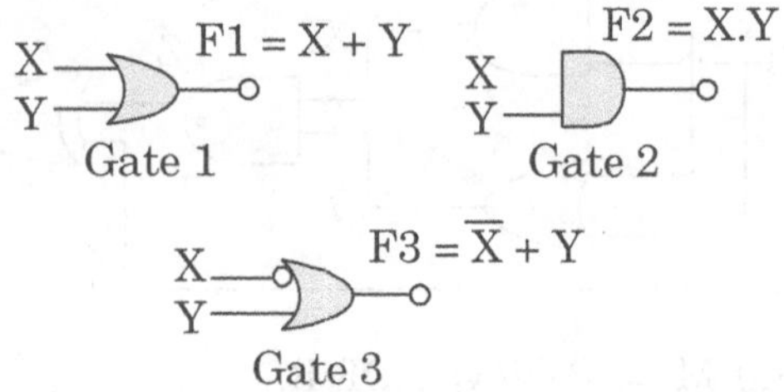

Which one of the following statements is TRUE?

(a) Gate 1 is a universal gate.
(b) Gate 2 is a universal gate.
(c) Gate 3 is a universal gate.
(d) None of the gates shown is a universal gate.

[2015 : 1 Mark, Set-3]

44. The output of the combinational circuit given below is

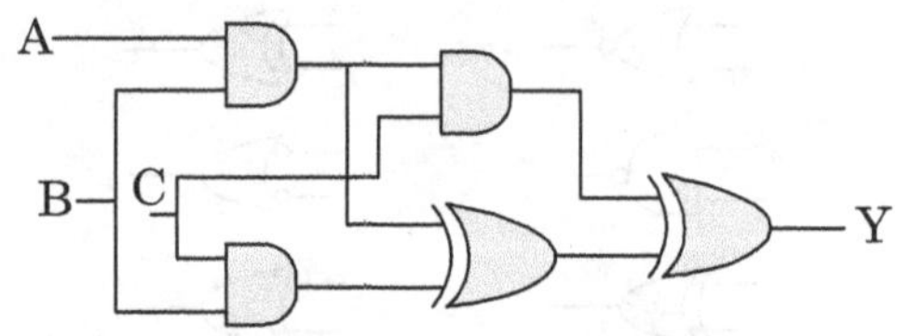

(a) A + B + C
(b) A(B+C)
(c) B(C+A)
(d) C(A + B)

[2015 : 2 Marks, Set-3]

45. The minimum number of 2-input NAND gates required to implement a 2-input XOR gate is

(a) 4
(b) 5
(c) 6
(d) 7

[2016 : 1 Mark, Set-1]

46. The logic gates shown in the digital circuit below use strong pull-down nMOS transistors for LOW logic level at the outputs. When the pull-downs are off, high-value resistors set the output logic levels to HIGH (i.e. the pull-ups are weak). Note that some nodes are intentionally shorted to implement "wired logic". Such shorted nodes will be HIGH only if the outputs of all the gates whose outputs are shorted are HIGH.

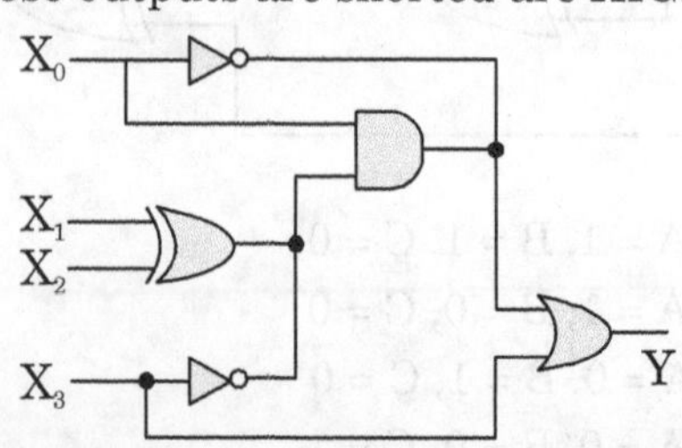

The number of distinct values of $X_3 X_2 X_1 X_0$ (out of the 16 possible values) that give Y = 1 is ____.

[2016 : 1 Mark, Set-3]

47. The output of the circuit shown in figure is equal to

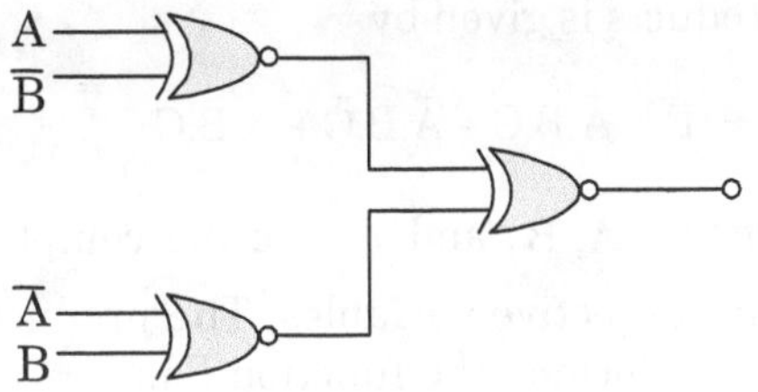

(a) 0 (b) 1

(c) $\bar{A}B + A\bar{B}$ (d) $(\overline{A \oplus B}) \oplus (\overline{A \oplus B})$

[2018 : 2 Marks]

K-MAP

48. The K-map for a Boolean function is shown in figure. The number of essential prime implicants for this function is

CD \ AB	00	01	11	10
00	1	1	0	1
01	0	0	0	1
11	1	0	0	0
10	1	0	0	1

(a) 4 (b) 5

(c) 6 (d) 8

[1998: 1 Mark]

49. If the functions W, X, Y and Z are as follows W $= R + \bar{P}Q + \bar{R}S$.

$X = PQ\bar{R}\bar{S} + \bar{P}\bar{Q}\bar{R}\bar{S} + P\bar{Q}\bar{R}\bar{S}$

$Y = RS + \overline{PR + P\bar{Q} + \bar{P}\bar{Q}}$

$Z = R + S + \overline{PQ + \bar{P}\bar{Q}\bar{R} + P\bar{Q}\bar{S}}$

Then

(a) $W = Z, X = \bar{Z}$ (b) $W = Z, X = Y$

(c) $W = Y$ (d) $W = Y = \bar{Z}$

[2003 : 2 Marks]

50. The Boolean expression $AC + B\bar{C}$ is equivalent to

(a) $\bar{A}C + B\bar{C} + AC$

(b) $\bar{B}C + AC + B\bar{C} + \bar{A}C\bar{B}$

(c) $AC + B\bar{C} + \bar{B}C + ABC$

(d) $ABC + \bar{A}B\bar{C} + AB\bar{C} + A\bar{B}C$

[2004 : 2 Marks]

51. The Boolean expression

$Y = \bar{A}\bar{B}\bar{C}D + \bar{A}BC\bar{D} + A\bar{B}\bar{C}D + AB\bar{C}D$ can be minimized to

(a) $Y = \bar{A}\bar{B}\bar{C}D + \bar{A}B\bar{C} + A\bar{C}D + A\bar{C}D$

(b) $Y = \bar{A}\bar{B}\bar{C}D + BC\bar{D} + A\bar{B}\bar{C}D + A\bar{C}D$

(c) $Y = \bar{A}BC\bar{D} + \bar{B}\bar{C}D + A\bar{B}\bar{C}D + A\bar{B}\bar{C}D$

(d) $Y = \bar{A}BC\bar{D} + \bar{B}\bar{C}D + A\bar{B}\bar{C}D + AB\bar{C}\bar{D}$

[2007 : 2 Marks]

52. In the sum of products function f(X. Y, Z) = $\sum(2,3,4,5)$, the prime implicates are

(a) $\bar{X}Y, X\bar{Y}$ (b) $\bar{X}Y, X\bar{Y}\bar{Z}, X\bar{Y}Z$

(c) $\bar{X}Y\bar{Z}, \bar{X}YZ, X\bar{Y}$ (d) $\bar{X}Y\bar{Z}, \bar{X}YZ, X\bar{Y}\bar{Z}, X\bar{Y}Z$

[2008 : 2 Marks]

53. Consider the Boolean function,

$F(w, x, y, z) = wy + xy + \bar{w}xyz + \bar{w}\bar{x}y + xz + \bar{x}\bar{y}\bar{z}$.

Which one of the following is the complete set of essential prime implicants?

(a) w, y, xz, $\bar{x}\bar{z}$

(b) w, y, xz

(c) y, $\bar{x}\bar{y}\bar{z}$

(d) y, $\bar{x}\bar{z}$, $\bar{x}\bar{z}$

[2012: 1 Mark]

54. For an n-variable Boolean function, the maximum number of prime implicants is

(a) 2(n–1) (b) n/2

(c) 2^n (d) $2^{(n-1)}$

[2014 : 2 Marks, Set-1]

55. The Boolean expression

$F(X, Y, Z) = \bar{X}Y\bar{Z} + X\bar{Y}\bar{Z} + XY\bar{Z} + XYZ$

converted into canonical product of sum (POS) form is

(a) $(X + Y + Z)(X + Y + \bar{Z})(X + \bar{Y} + \bar{Z})(\bar{X} + Y + \bar{Z})$

(b) $(X + \bar{Y} + Z)(\bar{X} + Y + \bar{Z})(\bar{X} + \bar{Y} + Z)(\bar{X} + \bar{Y} + \bar{Z})$

(c) $(X + Y + Z)(\bar{X} + Y + \bar{Z})(X + \bar{Y} + Z)(\bar{X} + \bar{Y} + \bar{Z})$

(d) $(X + \bar{Y} + \bar{Z})(\bar{X} + Y + Z)(\bar{X} + \bar{Y} + Z)(X + Y + Z)$

[2014 : 1 Mark, Set-2]

56. A function of Boolean variables, X, Y and Z is expressed in terms of the min-terms as

$$F(X, Y, Z) = \sum(1,2,5,6,7).$$

Which one of the product of sums given below is equal to the function F(X, Y, Z)?

(a) $(\bar{X}+\bar{Y}+\bar{Z}).(\bar{X}+Y+Z).(X+\bar{Y}+\bar{Z})$

(b) $(X+Y+Z).(X+\bar{Y}+\bar{Z}).(\bar{X}+Y+Z)$

(c) $(\bar{X}+\bar{Y}+Z).(\bar{X}+Y+\bar{Z}).(X+\bar{Y}+Z)$
$(X+Y+\bar{Z}).(X+Y+\bar{Z})$

(d) $(X+Y+\bar{Z}).(\bar{X}+Y+Z).(\bar{X}+Y+\bar{Z})$
$(\bar{X}+\bar{Y}+Z).(\bar{X}+\bar{Y}+\bar{Z})$

[2015 : 2 Marks, Set-1]

57. Following is the K-map of a Boolean function of five variables P, Q, R, S and X. The minimum sum of-product (SOP) expression for the function is

RS \ PQ	00	01	11	10
00	0	0	0	0
01	1	0	0	1
11	1	0	0	1
10	0	0	0	0

X = 0

RS \ PQ	00	01	11	10
00	0	1	1	0
01	0	0	0	0
11	0	0	0	0
10	0	1	1	0

X = 1

(a) $\bar{P}\bar{Q}S\bar{X} + P\bar{Q}S\bar{X} + Q\bar{R}\bar{S}X + QR\bar{S}X$

(b) $\bar{Q}S\bar{X} + Q\bar{S}\bar{X}$

(c) $\bar{Q}SX + Q\bar{S}\bar{X}$

(d) $\bar{Q}S + Q\bar{S}$

[2015 : 2 Marks, Set-2]

58. Which one of the following gives the simplified sum of products expression for the Boolean function $F = m_0 + m_2 + m_3 + m_5$, where m_0, m_2, m_3 and m_5 are minterms corresponding to the inputs A, B and C with A as the MSB and C as the LSB?

(a) $\bar{A}B + \bar{A}\bar{B}\bar{C} + A\bar{B}C$

(b) $\bar{A}\bar{C} + \bar{A}B + A\bar{B}C$

(c) $\bar{A}\bar{C} + A\bar{B} + A\bar{B}C$

(d) $\bar{A}BC + \bar{A}\bar{C} + A\bar{B}C$

[2016 : 2 Marks, Set-3]

59. A function F(A B, C) defined by three Boolean variables A, B and C when expressed as sum of products is given by

$$F = F = \bar{A}.B.C + \bar{A}.B.\bar{C} + A.\bar{B}.\bar{C}$$

where, $\bar{A}$, B, and $\bar{C}$ are the complements of the respective variables, The product of sums (POS) form of the function F is

(a) $F = (\bar{A}+\bar{B}+\bar{C}).(\bar{A}+B+\bar{C}).(A+\bar{B}+\bar{C})$

(b) $F = (\bar{A}+\bar{B}+\bar{C}).(\bar{A}+B+\bar{C}).(A+\bar{B}+\bar{C})$

(c) $F = (A+B+\bar{C}).(A+\bar{B}+\bar{C}).(\bar{A}+B+\bar{C})$
$.(\bar{A}+\bar{B}+C).(\bar{A}+\bar{B}+\bar{C})$

(d) $F = (\bar{A}+\bar{B}+C).(\bar{A}+B+C).(A+\bar{B}+C)$
$.(A+B+\bar{C}).(A+B+C)$

[2017: 2 Marks, Set-1]

60. The number of product terms in the minimized sumof-product expression obtained through the following K-map is (where, "d" denotes don't care states)

1	0	0	1
0	d	0	0
0	0	d	1
1	0	0	1

(a) 2

(b) 3

(c) 4

(d) 5 **[2018 : 1 Mark]**

ANSWERS

1. (b)	2. (d)	3. (d)	4. (a)	5. (d)	6. (a)	7. (d)	8. (b)	9. (a)	10. (b)
11. (b)	12. (b)	13. (b)	14. (b & d)	15. (c)	16. (b)	17. (b)	18. (c)	19. (a)	20. (c)
21. (c)	22. (a)	23. (d)	24. (c)	25. (c)	26. (d)	27. (d)	28. (b)	29. (a)	30. (d)
31. (d)	32. (d)	33. (d)	34. (d)	35. (b)	36. (c)	37. (a)	38. (a)	39. (b)	40. (–2)
41. (a)	42. (c)	43. (c)	44. (c)	45. (a)	46. (8)	47. (a)	48. (a)	49. (a)	50. (d)
51. (d)	52. (a)	53. (d)	54. (d)	55. (a)	56. (b)	57. (b)	58. (b)	59. (c)	60. (a)

EXPLANATIONS

1. Number of possible Boolean function = 2^{2^n}.

2. The condition for overflow to occur is $\bar{x}\bar{y}z + xy\bar{z}$

3. $$Y = A + \bar{A}B$$
$$= (A + A)(\bar{A} + B)$$
$$= 1(A + B) = A + B.$$

4. $\bar{A}\bar{B}\bar{C} + \bar{A}B\bar{C} + \bar{A}BC + AB\bar{C} + AB\bar{C} + \bar{A}B\bar{C}$
$= \bar{A}\bar{C}(B + \bar{B}) + \bar{A}B(C + \bar{C}) + B\bar{C}(A + \bar{A})$
$= \bar{A}\bar{C} + \bar{A}B + B\bar{C}$

5. Let ABCD be four variables.
$\Rightarrow$ Total number of variables
$= A, \bar{A}, B, \bar{B}\ C, \bar{C}, D, \bar{D}$ *i.e.* 8
Number of total elements in the 4 variable K Map
$= 2^4 = 16$
$\therefore$ Number of distinct Boolean expressions
$= 2^{16} = 65536$

6. $f = \bar{A}BC + AB\bar{C} = B(\bar{A}C + A\bar{C}) = B(A + C)(\bar{A} + \bar{C})$
$\{\because \text{XOR} = \bar{A}C + C\bar{A} \equiv (A + C)(\bar{A} + \bar{C})\}$

7. $$\left[X + Z\{\bar{Y} + (\bar{Z} + X\bar{Y})\}\right]\left[\{\bar{X} + \bar{Z}(X + Y)\}\right] = 1$$
$\Rightarrow \left[X\bar{Z}(X + Y) + \bar{X}Z\left[\bar{Y} + (\bar{Z} + X\bar{Y})\right]\right] = 1$
$\Rightarrow \left[X\bar{Z} + XY + \bar{X}Z\bar{Y}\right] = 1$
For $X = 1$, $\bar{Z} + Y = 1$, $Z = 0$
[since 1.1 = 1, and 0 + 1 = 1]

8. 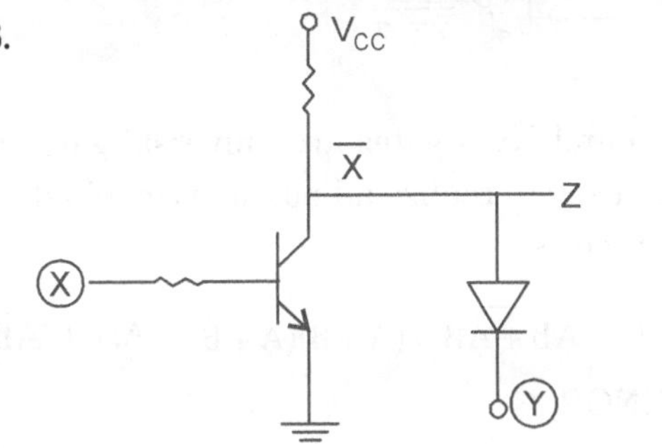

$z = \bar{x}.y$

9. $$f = (x + y)(x + \bar{y}) + \overline{(x\bar{y}) + \bar{x}}$$
$$= x + xy + x\bar{y} + \overline{(x\bar{y})} \cdot x$$
$$= x + xy + x\bar{y} + (\bar{x} + y) \cdot x$$
$$= x(1 + y + \bar{y})$$
$$= x$$

10. Distributive property:
$(A + B)(A + C) = A \cdot A + A \cdot C + A \cdot B + BC$
$= A + A \cdot C + A \cdot B + B \cdot C$
$= A(1 + C + B) + BC$
$= A + BC.$

11. $Y = AB + CD$

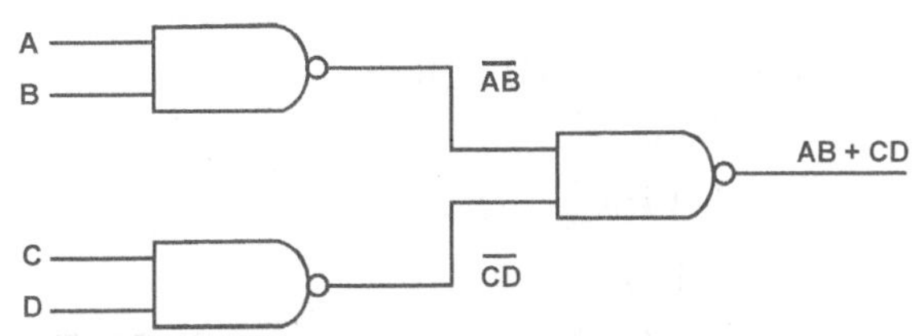

So, there is only 3 NAND gates will be required.

12. Output of 1st Ex-OR gate, $F_1 = X \oplus X = 0.$
Output of 2nd Ex-OR gate,
$F_2 = X \oplus F_1 = X \oplus 0 = X.$
Output of 3rd Ex-OR gate,
$F_3 = X \oplus F_2 = X \oplus X = 0.$

13. $F = (\bar{X}+\bar{Y})(Z+W)$

$= \overline{\overline{(\bar{X}+\bar{Y})}}(Z+W) = (\overline{XY})\cdot(Z+W)$

$= \overline{XY}\cdot Z + \overline{XY}\cdot W$

2- input NAND gate implementation is

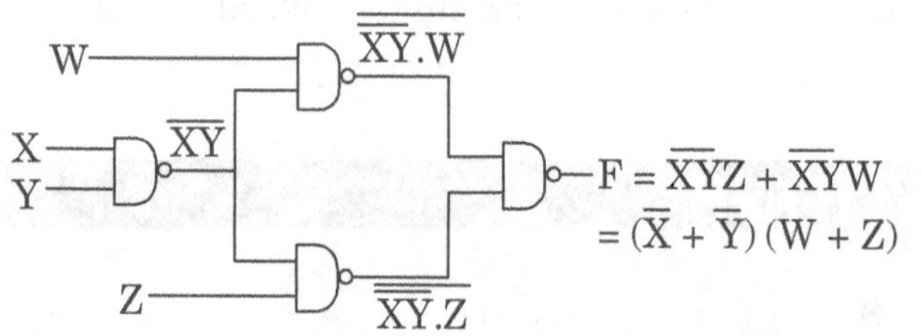

14. NAND and NOR gates are universal gates and can be used to realize all possible combinational logic circuits.

15. XNOR = $\overline{A\bar{B}+\bar{A}B} = (\bar{A}+B)(A+\bar{B}) = AB + \overline{AB}$ is also XNOR.

16.

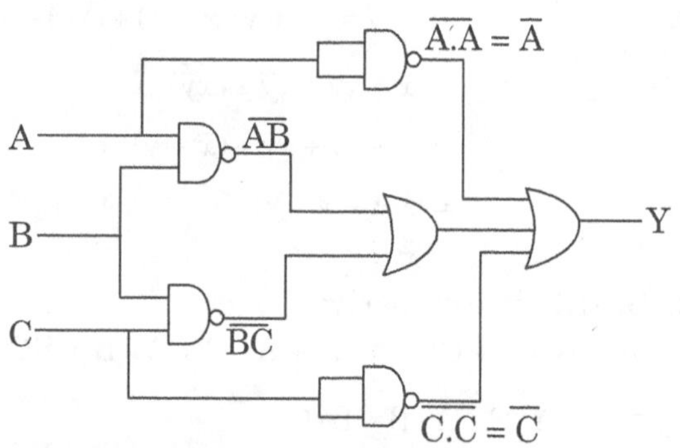

$\Rightarrow Y = \bar{A} + \overline{AB} + \overline{BC} + \bar{C}$

$Y = \bar{A}(1+\bar{B}) + (\bar{B}+1)\bar{C}$

$Y = \bar{A}+\bar{B}+\bar{C}.$

17. Truth Table

NOR			X- NOR		
INPUT		OUTPUT	INPUT		OUTPUT
0	0	1	0	0	1
0	1	0	0	1	0
1	0	0	1	0	0
1	1	0	1	1	1

18. Clock period = 1 μsec

Propagation delay per gate = $\frac{1}{5}$ = 0.2 μsec

= 200 nsec

19. $A + A\bar{B} + A\bar{B}C = A + A\bar{B}(1+C)$

$= A + A\bar{B}$

$= A(1+\bar{B})$

= A (The output is taken directly from A and hence 0)

20. Output of logic gate,

$F = A \odot 0 = \bar{A}\cdot\bar{0} + A\cdot 0 = \bar{A}\cdot 1 + 0$

$F = \bar{A}$.

21. 2-input NAND gate implementation of function Z $= A\bar{B}C$ is

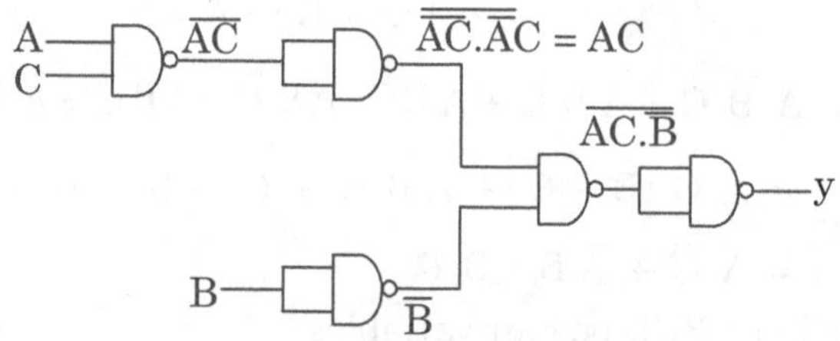

$y = \overline{(\overline{AC\cdot \bar{B}})\,(\overline{AC\cdot \bar{B}})} = AC\cdot\bar{B} = A\bar{B}C.$

The minimum number of NAND gate = 5.

22. For dual form, AND ↔ OR & OR ↔ AND.

$AB + \bar{A}C + BC = AB + \bar{A}C$

Dual form is

$(A+B)(\bar{A}+C)(B+C) = (A+B)(\bar{A}+C).$

23. For output X = 1, input of NAND gate (F_1, F_2 and C) must be 1. Now,

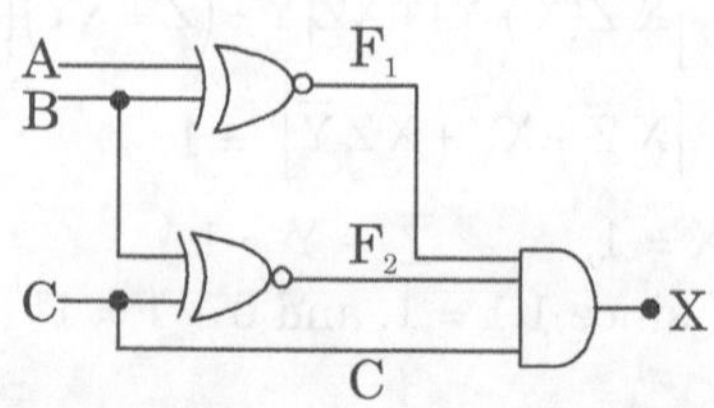

$\Rightarrow C = 1$

$\Rightarrow F_2 = B \odot C = BC + \bar{B}\bar{C} = 1.$

$B \cdot 1 + \bar{B} \cdot 0 = 1$

$B = 1$

$\Rightarrow F_1 = A\bar{B} + \bar{A}B = 1$

$A \cdot \bar{1} + \bar{A} \cdot 1 = 1$

$\bar{A} = 1$

$A = 0.$

$(A, B, C) \equiv (0, 1, 1).$

24.

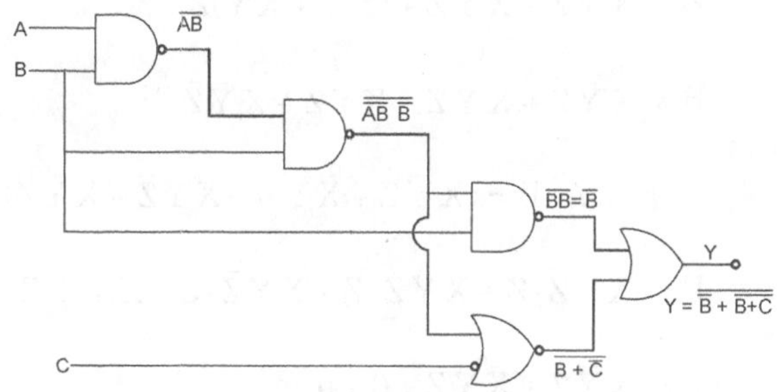

$$y = \overline{\overline{B} + \overline{B+C}}$$

$$= B(B + \bar{C})$$

$$= B + B\bar{C} = B$$

25. Propagation delay of all invertery t_{pd}

$= 5 \times 100$ ps $= 500$ ps

Fundamental frequecy of oscillator output V_0

$$= \frac{1}{2^{tpd}} = \frac{1}{2 \times 500 \times 10^{-9}} = 1\text{GHz}.$$

26. Output at AND gate $= S_1S_2$

Output at XOR gate $= S_1\bar{S}_2 + \bar{S}_1S_2$

Output at NAND gate $= \overline{(S_1S_2)(S_1\bar{S}_2 + \bar{S}_1S_2)}$

$$= \overline{S_1S_2\bar{S}_2 + S_1\bar{S}_1S_2} = 1$$

As output is 1 irrespective of switch positions of S_1 and S_2, LED is reversed-biased and will not emit light.

27. Output of first XOR gate

$= 1 \oplus X = \bar{X} \cdot 1 + 0 = \bar{X}$

Output of second XOR gate

$= \bar{X} \oplus X = (\bar{\bar{X}}) \cdot X + \bar{X} \cdot \bar{X} = X + \bar{X} = 1.$

Similarly, output of 20 XOR gates will be 1.

28.

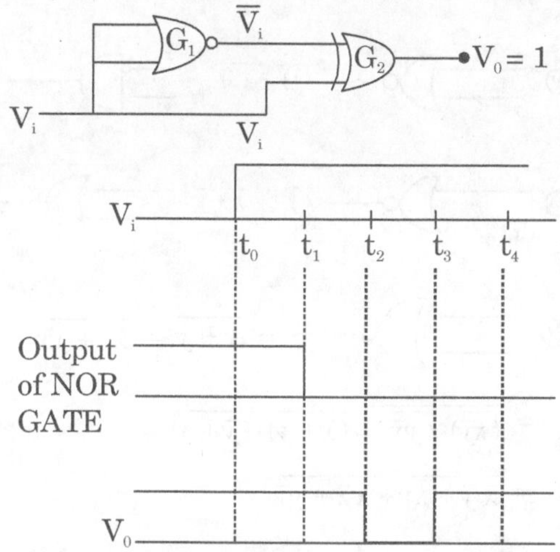

29. In TTL AND-OR invertor gate propagation delay of transistor depend on RC when *i/p* is floating then $R \to 0$, hence *o/p* is zero.

30.

x	*y*	F
0	0	1
0	1	1
1	0	0
1	1	1

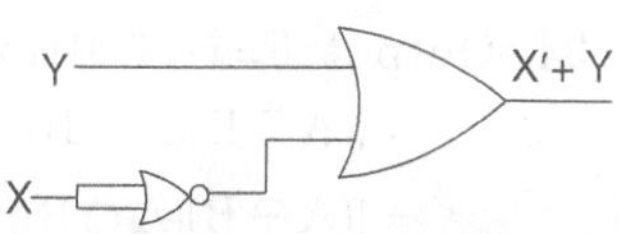

$F = \bar{X} + Y$

Hence 2 units are required.

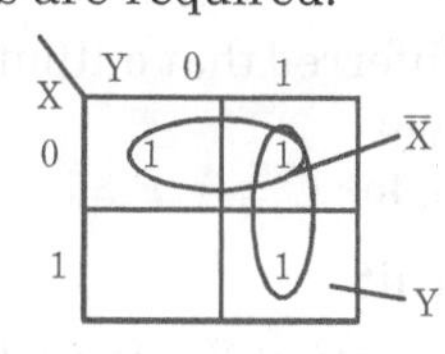

from k-map

$f = \bar{X} + Y$

31. Redrawing the logic circuit with P = 1.

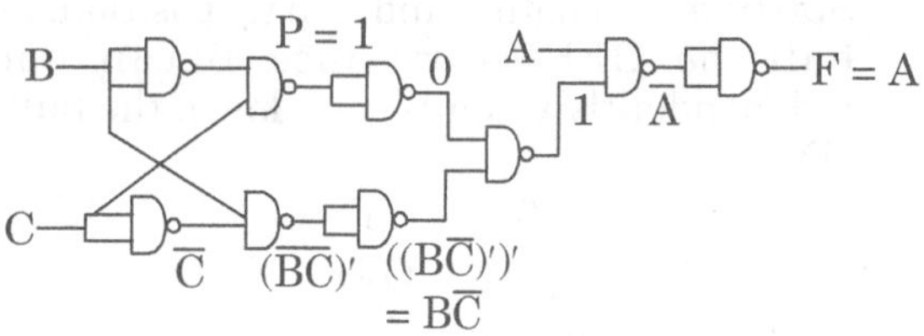

32. $\Rightarrow M_1 = \left[\overline{PQ}(P + Q)\right] \oplus R$

$= \left[(\bar{P} + \bar{Q})\,(P + Q)\right] \oplus R$

$M_1 = P \oplus Q \oplus R$

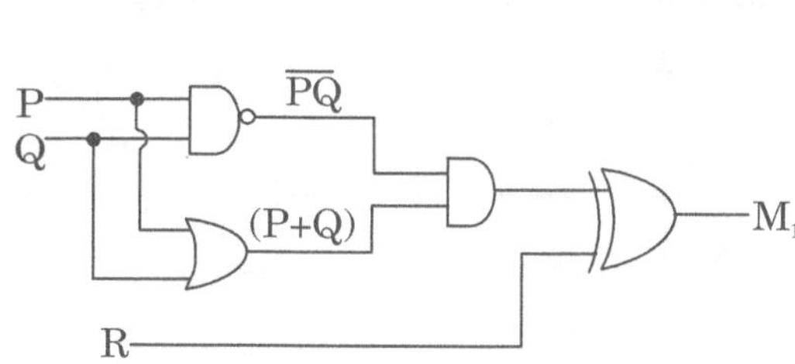

33. 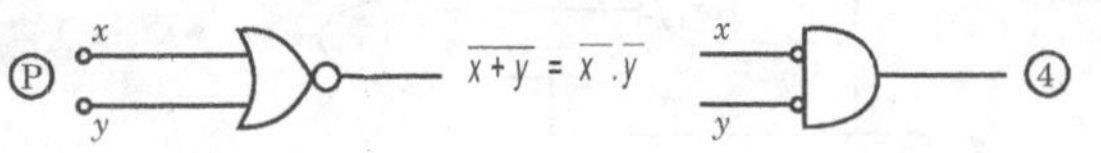

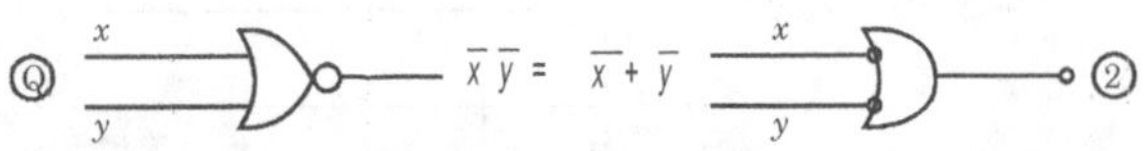

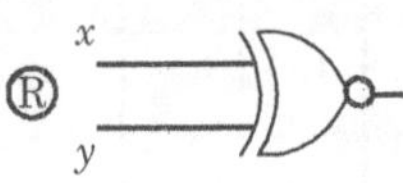

$= \overline{x \oplus y} = \overline{x\bar{y} + y\bar{x}}$

$= \overline{(\overline{x\bar{y}}) \cdot (\overline{y\bar{x}})} = \overline{(\bar{x} + y) \cdot (\bar{y} + x)}$

$= \overline{\bar{x}\bar{y} + xy + x\bar{x} + y\bar{y}}$

$= \overline{\bar{x}\bar{y} + xy} \equiv$ 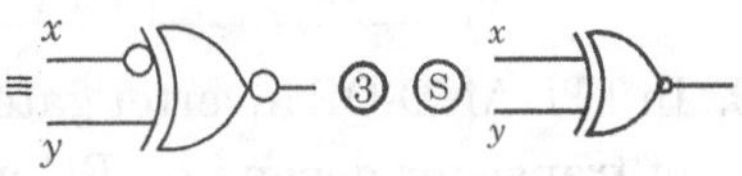

$= x \oplus y = \bar{x}\bar{y} + xy$ 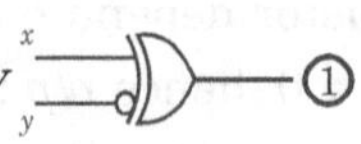

34. Output $F = (A \oplus B) \odot (A \odot B) \odot C$

$= \{(A \oplus B)(A \odot B) + (A \odot B)(A \oplus B)\} \odot C$

$= \{(A \oplus B)(A \odot B)\} \odot C$

$= 0 \odot C$

$= \bar{C}$

It can be inferred that output F is only completed of c function.

Therefore, for C = 0, F =1

35. In the circuit

output Y = PQ + PR + RQ

Hence, two or more inputs are '1', Y is always '1'.

36. Let us consider the switches A and B and bulb Y. Switches can be 2 positions up (0) or down (1)

Starting with both A and B in up position. Let the bulb be OFF. Now since B can operate independently when B goes down, the bulb goes ON

A	B	Y
up (0)	up (0)	OFF
up (0)	down (1)	ON

Now keeping A in down position when B goes down, the bulb will go OFF.

A	B	Y
down (1)	up (0)	ON
down (1)	down (1)	OFF

find truth table corresponds to XOR gate.

37. 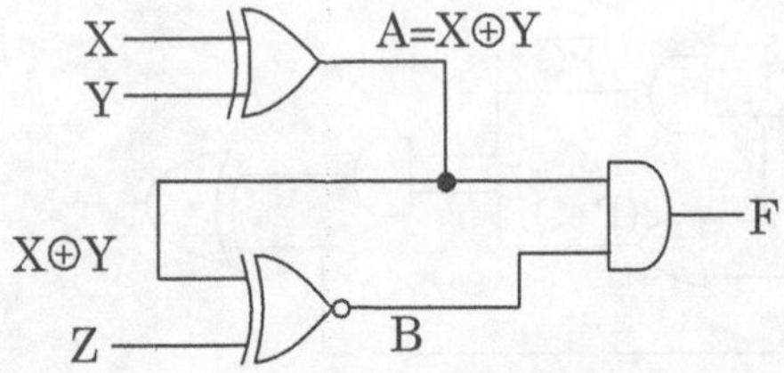

Assuming, $A = X \oplus Y = X\bar{Y} + \bar{X}Y$

$B = A \odot Z = AZ + \bar{A}\bar{Z}$

$B = (X\bar{Y} + \bar{X}Y)Z + \overline{(X\bar{Y} + \bar{X}Y)}\bar{Z}$

$B = X\bar{Y}Z + \bar{X}YZ + (\overline{X\bar{Y}} \cdot \overline{\bar{X}Y})\bar{Z}$

$B = X\bar{Y}Z + \bar{X}YZ + (\bar{X} + Y) \cdot (X + \bar{Y})\bar{Z}$

$B = X\bar{Y}Z + \bar{X}YZ + (XY + \bar{X}\bar{Y})\bar{Z}$

$B = X\bar{Y}Z + \bar{X}YZ + XY\bar{Z} + X\bar{Y}\bar{Z}$

$\Rightarrow F = A \cdot B = (X\bar{Y}Z + \bar{X}YZ + XY\bar{Z} + X\bar{Y}\bar{Z}) \cdot Z$

$F = X\bar{Y}Z \cdot Z + \bar{X}YZ \cdot Z + XY\bar{Z} \cdot Z + X\bar{Y}\bar{Z} \cdot Z$

$F = X\bar{Y}Z + \bar{X}YZ + 0 + 0$

$F = X\bar{Y}Z + \bar{X}YZ.$

38. Redrawing the logic circuit with outputs,

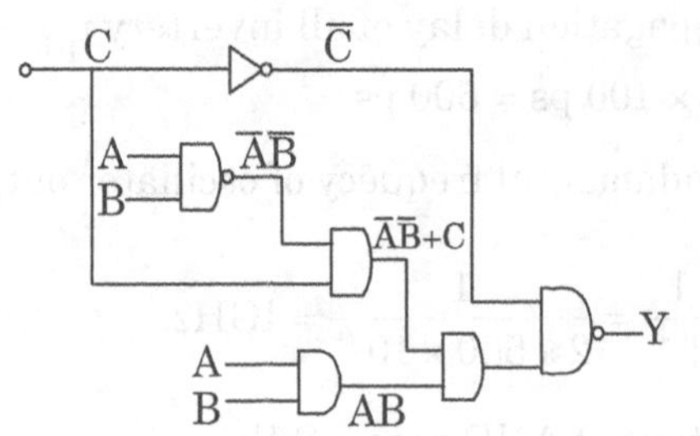

$\Rightarrow Y = \overline{(AB + \bar{A}\bar{B} + C) \cdot \bar{C}}$

Using De-morgan's theorem,

$Y = \overline{(AB + \bar{A}\bar{B} + C)} + \bar{\bar{C}}$

$Y = \overline{AB} \cdot \overline{\bar{A}\bar{B}} . \bar{C} + \bar{C}$

$Y = (\bar{A} + \bar{B}) \cdot (A + B) \cdot \bar{C} + \bar{C}$

Given: $C = 0,\ \bar{C} = 1.$

$Y = \bar{A}B + A\bar{B}.$

39. Given: M(a, b, c) = ab + bc + ca.

$\Rightarrow \overline{M(a,b,c)} = \overline{ab + bc + ca} = \overline{ab} \cdot \overline{bc} \cdot \overline{ca}$

$= (\bar{a} + \bar{b}) \cdot (\bar{b} + \bar{c}) \cdot (\bar{c} + \bar{a})$

$\Rightarrow M(a,b,\bar{c}) = ab + b\bar{c} + \bar{c}a$

$\Rightarrow M(\overline{M(a,b,c), M(a,b,\bar{c})}, c)$

$$= \begin{bmatrix} (\overline{ab}\cdot\overline{bc}\cdot\overline{ca})\cdot(ab + b\bar{c} + \bar{c}a) \\ +(ab + b\bar{c} + \bar{c}a)\cdot c + c(\overline{ab}\cdot\overline{bc}\cdot\overline{ca}) \end{bmatrix}$$

$= (\bar{a}+\bar{b})(\bar{b}+\bar{c})(\bar{c}+\bar{a})(ab + b\bar{c} + \bar{c}a) + abc + 0 + 0$

$+ (\bar{a}+\bar{b})(\bar{b}+\bar{c})(\bar{c}+\bar{a})c$

$= (\bar{a}+\bar{b})(\bar{b}+\bar{c})(\bar{c}+\bar{a})(c + ab + b\bar{c} + \bar{c}a) + abc$

$= (\bar{a}+\bar{b})(\bar{b}+\bar{c})(\bar{c}+\bar{a})(a+b+c) + abc$

$\therefore$ On solving, $c + ab + b\bar{c} + \bar{c}a = a + b + c.$

$= (\bar{a}\bar{b} + \bar{a}\bar{c} + \bar{b} + \bar{b}\bar{c})(\bar{c}+\bar{a})(a+b+c) + abc$

$= (\bar{a}\bar{b} + \bar{a}\bar{c} + \bar{b}\bar{c} + \bar{b})(\bar{c}a + b\bar{c} + 0 + 0 + \bar{a}b + \bar{a}c) + abc.$

$= (\bar{a}\bar{b} + \bar{a}\bar{c} + \bar{b}\bar{c} + \bar{b})(a\bar{c} + b\bar{c} + \bar{a}b + \bar{a}c) + abc.$

$= 0+0+0+\bar{a}\bar{b}c + 0 + \bar{a}b\bar{c} + \bar{a}b\bar{c} + 0 + a\bar{b}\bar{c} + 0$

$+ 0 + 0 + a\bar{b}\bar{c} + 0 + a + \bar{a}\bar{b}c$

$= a\bar{b}\bar{c} + b\bar{c}\bar{a} + c\bar{a}\bar{b} + abc$

$= a \oplus b \oplus c\cdot$

$\Rightarrow$ 3 input XOR gate.

41. Given expression is $Y = AB + \overline{C}\overline{D}$

The first term can be obtained by considering G_1 as NOR gate, and second term $(\overline{C}\overline{D})$ is obtained from another lower NOR-Gate. So, final expression can be implemented by considering G_2 as OR-Gate.

42. **Case (i)** : If any input is logic 0 (i.e., 0V) then corresponding diode is 'ON' and due to ideal diode output voltage $V_0 = 0$ as well as if there is any input logic 1 (i.e., 10V) corresponding diode will be OFF.

Case (ii) : If all the inputs are high (i.e., 10V) then all the diodes are R.B (OFF) and output voltage $V_0 = 10V$.

43. Universal gate is a gate by which every other gate can be realized.

Gate 1 and Gate 2 are basic gates and can not be used as universal gate. All the boolean functions can be implemented by using the gate 3. Hence it is a universal gate.

44. From the given combinational circuit output (Y)

$= ABC \oplus AB \oplus BC$

$= [\overline{ABC}\cdot AB + ABC\cdot\overline{AB}]$

$= AB\bar{C} \oplus BC$ (Since $C \oplus 1 = C\cdot\bar{1} + 1\cdot\bar{C}$)

$= C\cdot 0 + \bar{C}$

$= 0 + \bar{C} = \bar{C}$

$= B(A\bar{C} \oplus C)$

$= B\left[A\bar{C}\cdot\bar{C} + C\,\overline{A\bar{C}}\right]$

$= B\left[A\bar{C} + C\left(\bar{A} + \bar{\bar{C}}\right)\right]$

$= B\left[A\bar{C} + \bar{A}C + C\right]$ $\left(\text{Since } \bar{\bar{C}} = C\right)$

$= B\left[A\bar{C} + C\left(\bar{A} + 1\right)\right]$

$= B\left[C + A\bar{C}\right]$ $\left(\text{Since } \bar{A} + 1 = 1\right)$

$= B[C + A]$

$\therefore \quad Y = B(C + A)$

45. using 2-input NAND gate implementation of 2-input XOR

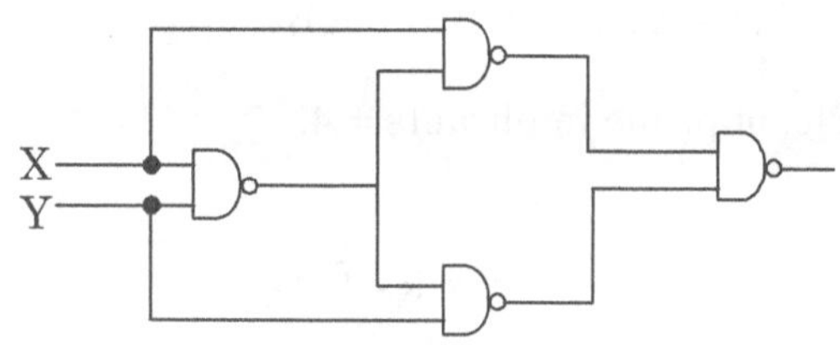

No. of NAND gates required = 4.

46.

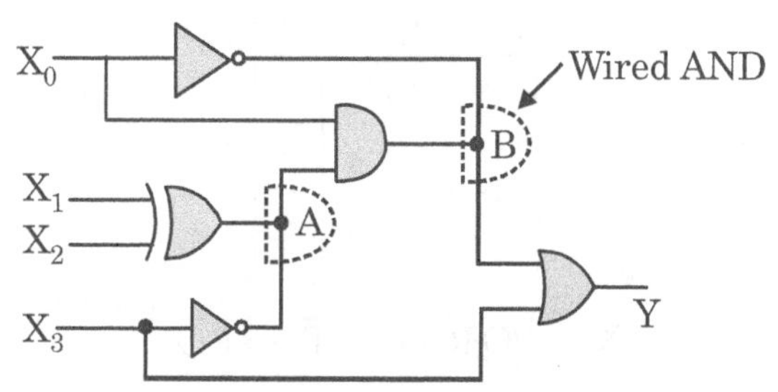

$A = (X_1 \oplus X_1)\,\bar{X}_3$

$B = \left[(X_1 \oplus X_2)\bar{X}_3\, X_0\right]\cdot\bar{X}_0 = 0$

$Y = B + X_3 = 0 + X_3 = X_3$

Out of 16 possible combinations of $X_3\, X_2\, X_1\, X_0$ Output will be high for 8 combinations.

$\therefore$ Y will be high for 8 combinations.

47.

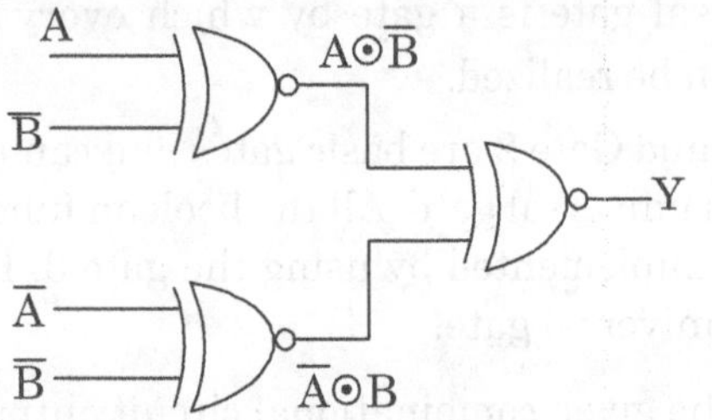

$\Rightarrow Y = (A \odot \bar{B}) \odot (\bar{A} \odot B)$

$Y = A \odot \bar{B} \odot \bar{A} \odot B$

$Y = (A \odot \bar{A}) \odot (\bar{B} \odot B)$

$Y = 0 \odot 0$

$Y = 1$

48.

CD \ AB	00	01	11	10
00	1	1		1
01				1
11	1			
10	1			1

Solution of k-Map is

$= \bar{B}\bar{D} + \bar{A}\bar{B}\bar{D} + A\bar{B}\bar{C} + \bar{A}\bar{B}C$

No. of prime implicants = 4.

49. $W = R + \bar{P}Q + \bar{R}S$

$\Rightarrow$

PQ \ RS	00	01	11	10
00		1	1	1
01	1	1	1	1
11		1	1	1
10		1	1	1

(i)

$X = PQ\bar{R}\bar{S} + \bar{P}\bar{Q}\bar{R}\bar{S} + P\bar{Q}\bar{R}\bar{S}$

$\Rightarrow$

PQ \ RS	00	01	11	10
00	1			
01				
11	1			
10	1			

(ii)

$Y = RS + \overline{PR + P\bar{Q} + \bar{P}\bar{Q}}$

$= RS + \overline{PR}\,.\,\overline{P\bar{Q}}\,.\,\overline{\bar{P}\bar{Q}}$

$= RS + (\bar{P}+\bar{R})(\bar{P}+Q)(P+Q)$

$= RS + (\bar{P}+\bar{R}Q)(P+Q)$

$= RS + \bar{P}Q + \bar{R}PQ + QR$

$\Rightarrow$

PQ \ RS	00	01	11	10
00			1	
01		1	1	1
11	1	1	1	
10			1	

(iii)

$Z = R + S + \overline{PQ + \bar{P}.\bar{Q}\bar{R} + P\bar{Q}\bar{S}}$

$= R + S + \overline{PQ}\,.\,\overline{\bar{P}\bar{Q}\bar{R}}\,.\,\overline{P\bar{Q}\bar{S}}$

$= R + S + (P + \bar{Q})(P + Q + R)(\bar{P} + Q + S)$

$= R + S + (\bar{P} + \bar{Q}(Q+S))(P+Q+R)$

$= R + S + (\bar{P} + \bar{Q}S)(P+Q+R)$

$= R + S + \bar{P}Q + \bar{P}R + \bar{Q}PS + \bar{Q}RS$

$\Rightarrow$

PQ \ RS	00	01	11	10
00		1	1	1
01	1	1	1	1
11		1	1	1
10		1	1	1

(iv)

From Fig (i) and Fig (iv), we see that W = Z

From Fig (ii) and Fig (iv), we see that $X = \bar{Z}$

50. AC + BC′

$\Rightarrow$

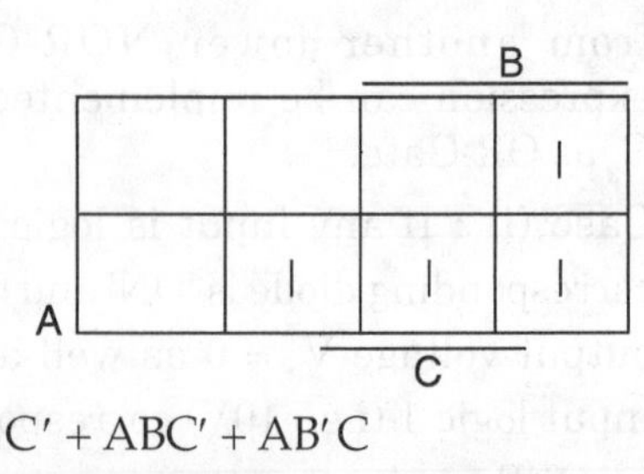

$ABC + \bar{A}BC' + ABC' + AB'C$

$\Rightarrow$

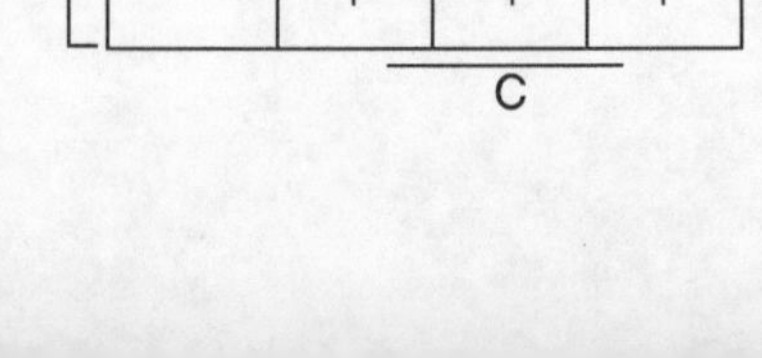

$AC + BC' + B'C + ABC$

$\Rightarrow$ 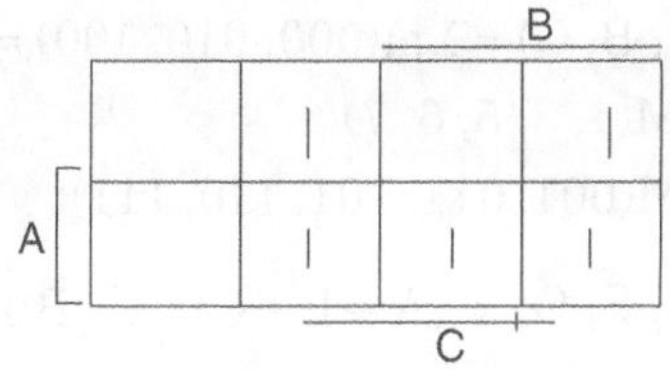

$B'C + AC + BC' + A'CB'$

$\Rightarrow$ 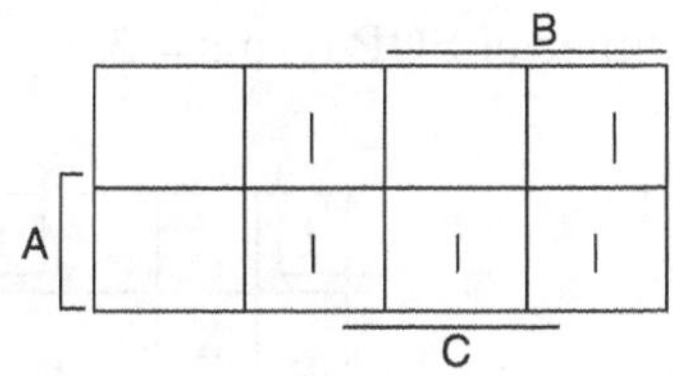

$A'C + BC' + AC$

$\Rightarrow$ 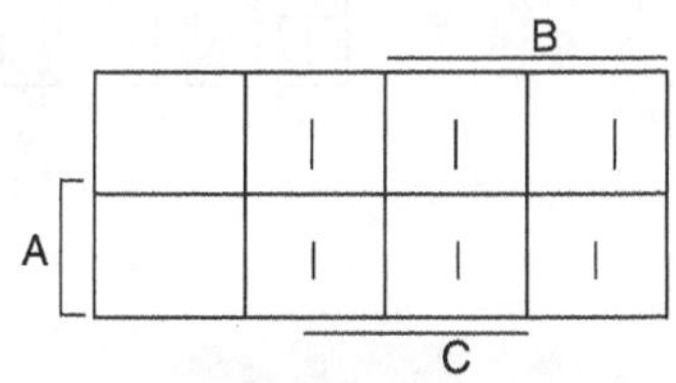

On comparing all, we find that ∞ is correct.

51. 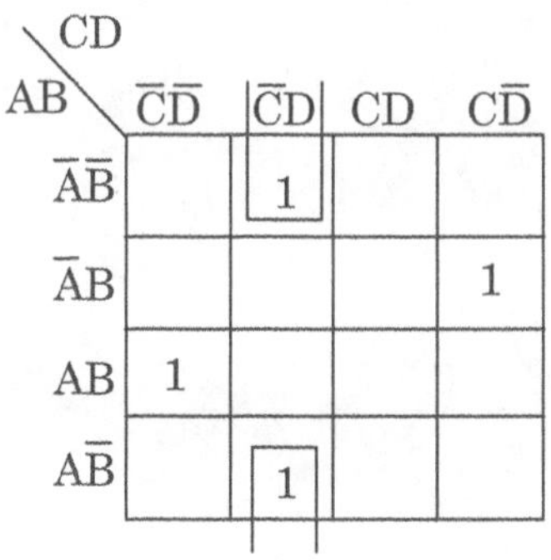

K-Map for Y

Simplified expression for K-map:

$Y = \bar{A}BC\bar{D} + AB\bar{C}\bar{D} + \bar{B}\bar{C}D.$

52. $f(x, y, z) = \Sigma(2, 3, 4, 5).$

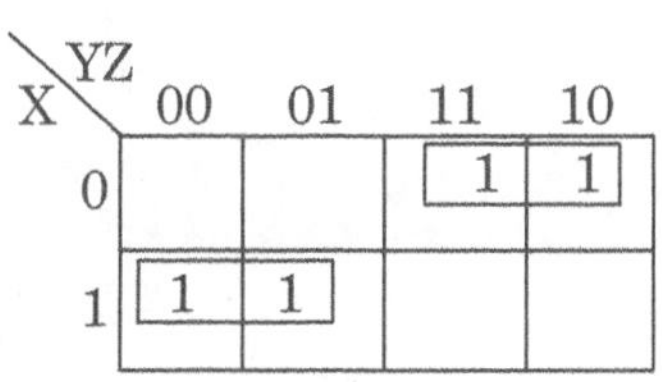

Simplified expression of

$f(x,y,z) = x\bar{y} + \bar{x}y.$

53. Given: $F(w, x, y, z) = wy + xy + \bar{w}xyz$

$+ \bar{w}\bar{x}y + xz + \bar{x}\bar{y}\bar{z}.$

Drawing the K-Map for above expression,

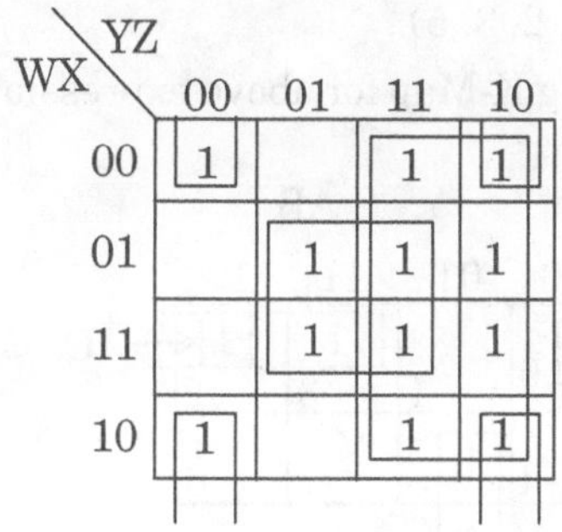

$\Rightarrow F(w, x, y, z) = y, xz, \bar{x}\bar{z}.$

Hence, the prime implicants are $y, xz, \bar{x}\bar{z}$.

54. For n-variable Boolean function, the maximum numbr of prime implicants = 2^{n-1}.

55. Boolean expression in SOP form

$$F(X, Y, Z) = \bar{X}Y\bar{Z} + X\bar{Y}\bar{Z} + XY\bar{Z} + XYZ$$

$$= \Sigma m\,(2, 4, 6, 7)$$

$$= \pi m\,(0, 1, 3, 5)$$

$$= (X + Y + Z)\cdot(X + Y + \bar{Z})$$

$$\cdot\,(X + \bar{Y} + \bar{Z})\cdot(\bar{X} + Y + \bar{Z})$$

56. Given minterm is :

$$F(X, Y, Z) = \Sigma(1, 2, 5, 6, 7)$$

So, maxterm is : $F(X, Y, Z) = \pi M(0, 3, 4)$

$$POS = (X + Y + Z)(X + \bar{Y} + \bar{Z})(\bar{X} + Y + Z)$$

57. 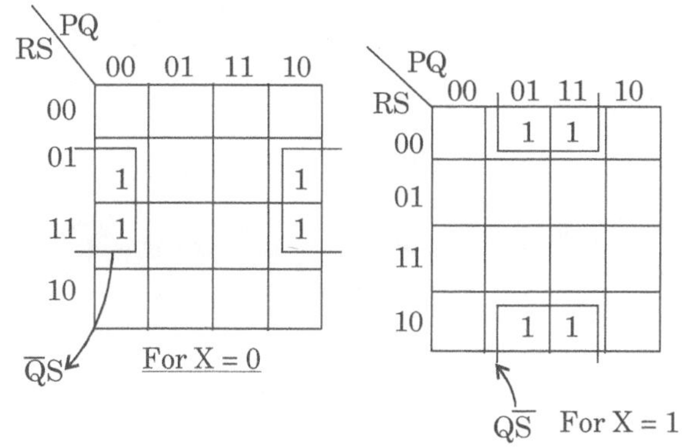

$\Rightarrow$ Minimized SOP expression for function is $\bar{Q}S\bar{X} + Q\bar{S}X$.

58. Given: $F = m_0 + m_2 + m_3 + m_5$

$= \Sigma m(0, 2, 3, 5)$.

Drawing K-Map for above expression,

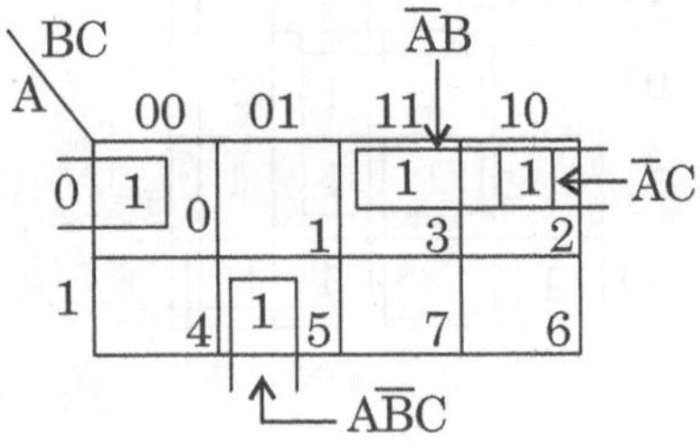

$F = \bar{A}\,\bar{C} + \bar{A}B + A\bar{B}C.$

59. $F(A, B, C) = \bar{A}\bar{B}\bar{C} + \bar{A}B\bar{C} + A\bar{B}\bar{C}$

$F(A, B, C) = \Sigma m(000, 010, 100) = \Sigma(0, 2, 4)$

$= \pi M(1, 3, 5, 6, 7)$

$= \pi M(001, 011, 101, 110, 111)$

$F(A, B, C) = (A + B + \bar{C})\,(A + \bar{B} + \bar{C}) \cdot (\bar{A} + B + \bar{C})$

$(\bar{A} + \bar{B} + C)\,(\bar{A} + \bar{B} + \bar{C})$

60. Number of SOP terms = 2.

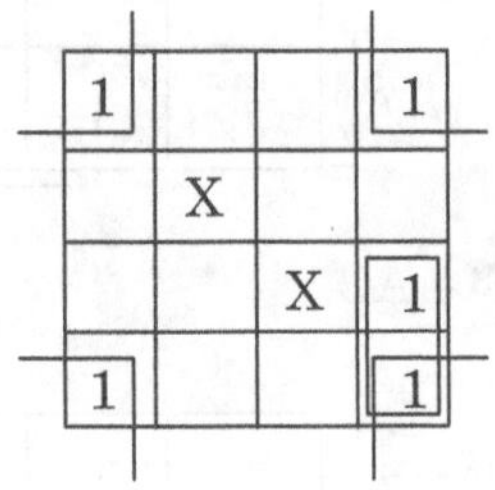

3

CHAPTER

Combinational Circuits

Analysis of Previous GATE Papers												
	Year → Topics ↓		2019	2018	2017 Set 1	2017 Set 2	2016 Set 1	2016 Set 2	2016 Set 3	2015 Set 1	2015 Set 2	2015 Set 3
ARITHMETIC CIRCUITS	1 Mark	MCQ Type										
		Numerical Type										
	2 Marks	MCQ Type				1						
		Numerical Type										
	Total					2						
MULTIPLEXERS	1 Mark	MCQ Type				1		1				
		Numerical Type										
	2 Marks	MCQ Type		1								
		Numerical Type							1			
	Total			2		1		1	2			
DECODERS & CODE CONVERTER	1 Mark	MCQ Type										
		Numerical Type										
	2 Marks	MCQ Type					2					
		Numerical Type										
	Total						4					
PROGRAMMABLE LOGIC ARRAY	1 Mark	MCQ Type										
		Numerical Type										
	2 Marks	MCQ Type				1						
		Numerical Type										
	Total					2						

ARITHMETIC CIRCUITS

1. A 2-bit binary multiplier can be implemented using
   (a) 2 input ANDs only.
   (b) 2 input X-ORs and 4-input AND gates only.
   (c) Two (2) input NORs and one XNOR gate.
   (d) XOR gates and shift registers.

[1997 : 1 Mark]

2. For a binary half-subtractor having two inputs A and B. the correct set of logical expressions for the outputs D(= A minus B) and X(= borrow) are
   (a) $D = AB + \bar{A}B, X = \bar{A}B$
   (b) $D = \bar{A}B + \bar{A}B + A\bar{B}, X = A\bar{B}$
   (c) $D = \bar{A}B + \bar{A}B, X = A\bar{B}$
   (d) $D = AB + \bar{A}\bar{B}, X = A\bar{B}$

[1999 : 2 Marks]

3. The circuit shown in the figure has 4 boxes each described by inputs P. 0, Rand outputs Y, Z with

$Y = P \oplus Q \oplus R$, $Z = RQ + \bar{P}R + Q\bar{P}$

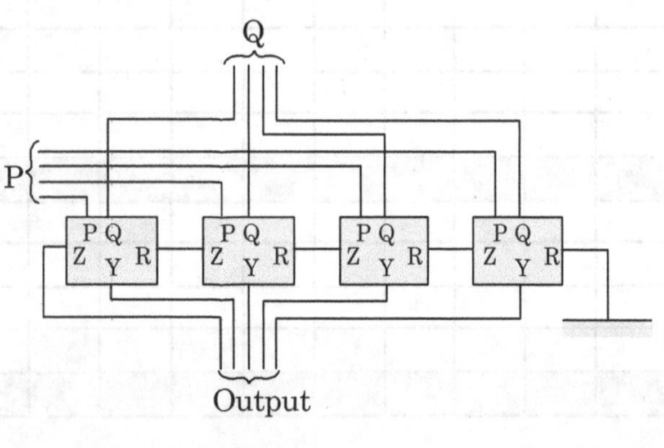

The circuit acts as a
   (a) 4 bit adder given P + Q
   (b) 4 bit subtractor given P – Q
   (c) 4 bit subtractor given Q – P
   (d) 4 bit adder given P + Q + R

[2003 : 2 Marks]

4. The output Y of a 2-bit comparator is logic 1 whenever the 2-bit input A is greater than the 2-bit input B. The number of combination for which the output is logic 1, is
   (a) 4 (b) 6
   (c) 8 (d) 10

[2012 : 1 Mark]

5. In a half-subtractor circuit with X and Y as inputs, the Borrow (M) and Difference (N = X – Y) are given by
   (a) $M = X \oplus Y, N = XY$
   (b) $M = XY, N = X \oplus Y$
   (c) $M = \bar{X}Y, N = X \oplus Y$
   (d) $M = \bar{X}Y, N = \overline{X \oplus Y}$

[2014 : 1 Mark, Set-2]

6. A 16-bit ripple carry adder is realized using 16 identical full adders (FA) as shown in the figure. The carry-propagation delay of each FA is 12 ns and the sum-propagation delay of each FA is 15 ns. The worst case delay (in ns) of this 16-bit adder will be _______ .

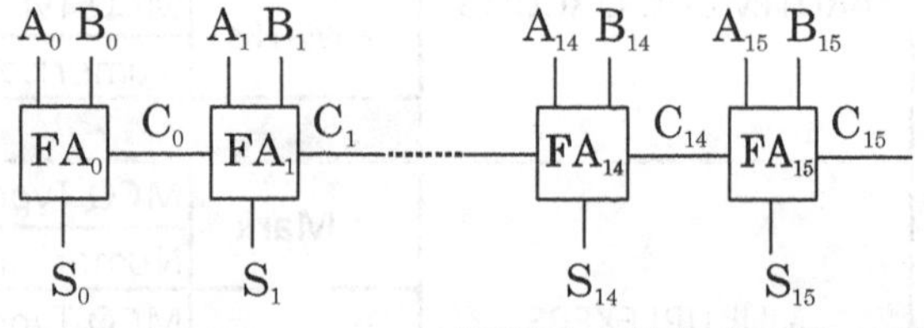

[2014 : 2 Marks, Set-4]

7. Figure I shows a 4-bit ripple carry adder realized using full adders and Figure II shows the circuit of a full-adder (FA). The propagation delay of the XOR. AND and OR gates in Figure II are 20 ns, 15 ns and 10 ns, respectively. Assume all the inputs to the 4-bit adder are initially reset to 0.

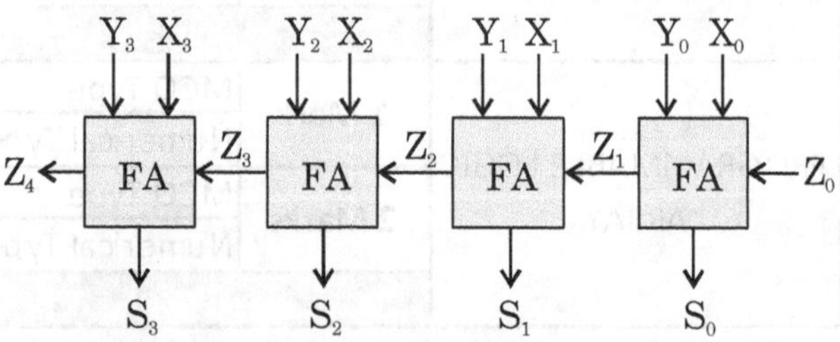

Figure I

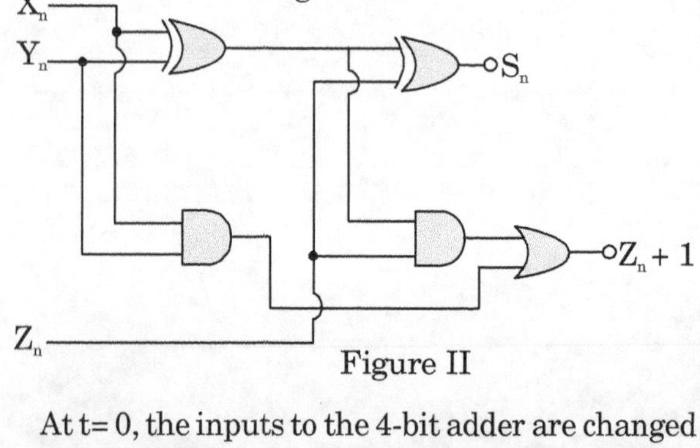

Figure II

At t= 0, the inputs to the 4-bit adder are changed to $X_3X_2X_1X_0 = 1100$, $Y_3Y_2Y_1Y_0 = 0100$ and $Z_0 = 1$. The output of the ripple carry adder will be stable at t (in ns) = _________.

[2017 : 2 Marks, Set-2]

MULTIPLEXERS

8. The logic realized by the circuit shown in figure is

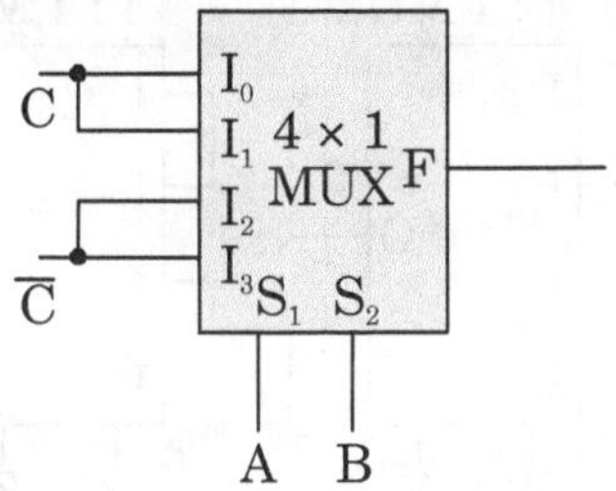

(a) F = AC

(b) F = AC

(c) F = BC

(d) F = BC

[1992 : 1 Mark]

9. In the TTL circuit in the figure, S_2 and S_0 are select lines and X_7 and X_0 are input lines. S_0 and X_0 are LSBs, The output Y is

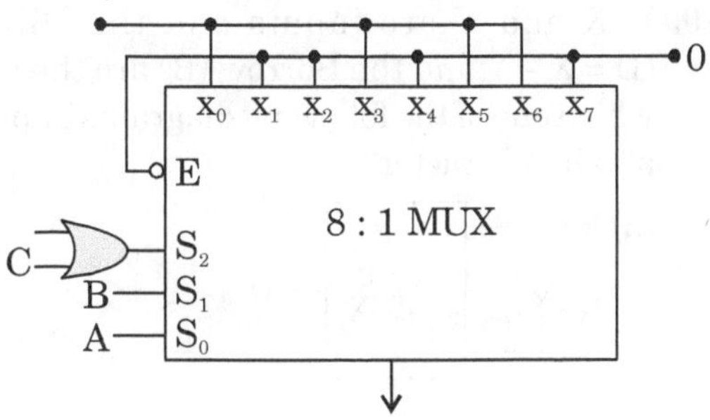

(a) indeterminate

(b) $A \oplus B$

(c) $\overline{A \oplus B}$

(d) $\bar{C}\left(\overline{A \oplus B}\right) + C\left(A \oplus B\right)$

[2001 : 2 Marks]

10. Without any additional circuitry, an 8 :1 MUX can be used to obtain

(a) some but not all Boolean functions of 3 variables.

(b) all functions of 3 variables but none of 4 variables.

(c) all functions of 3 variables and some but not all of 4 variables.

(d) all functions of 4 variables.

[2003 : 1 Mark]

11. The minimum number of 2-to-1 multiplexers required to realize a 4-to-1 multiplexer is

(a) 1 (b) 2

(c) 3 (d) 4

[2004 : 2 Marks]

12. The Boolean function f implemented in the figure using two input multiplexers is

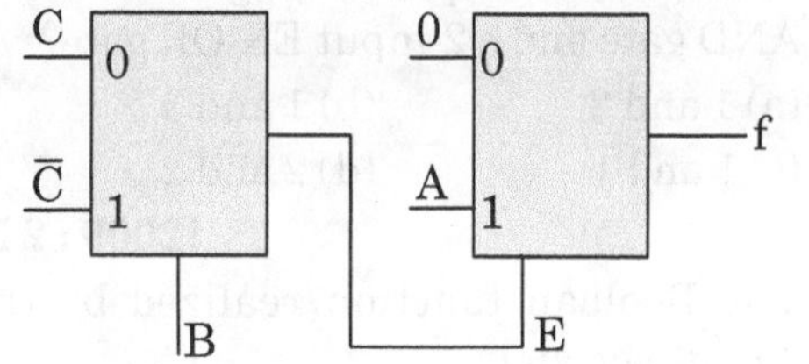

(a) $A\bar{B}C + AB\bar{C}$ (b) $ABC + A\bar{B}\bar{C}$

(c) $\bar{A}BC + \bar{A}\bar{B}\bar{C}$ (d) $\overline{AB}C + \bar{A}B\bar{C}$

[2005: 1 Mark]

13. In the following circuit, X is given by

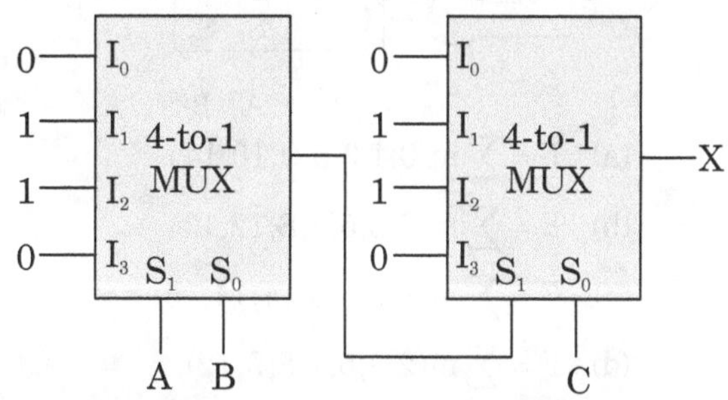

(a) $X = A\bar{B}\bar{C} + \bar{A}B\bar{C} + \bar{A}\bar{B}C + ABC$

(b) $X = \bar{A}BC + A\bar{B}C + AB\bar{C} + \bar{A}\bar{B}\bar{C}$

(c) X= AB + BC + AC

(d) $X = \bar{A}\bar{B} + \bar{B}\bar{C} + \bar{A}\bar{C}$

[2007 : 2 Marks]

14. For the circuit shown in the following figure, I_0,– I_3 are inputs to the 4 :1 multiplexer. R(MSB) and S are control bits.

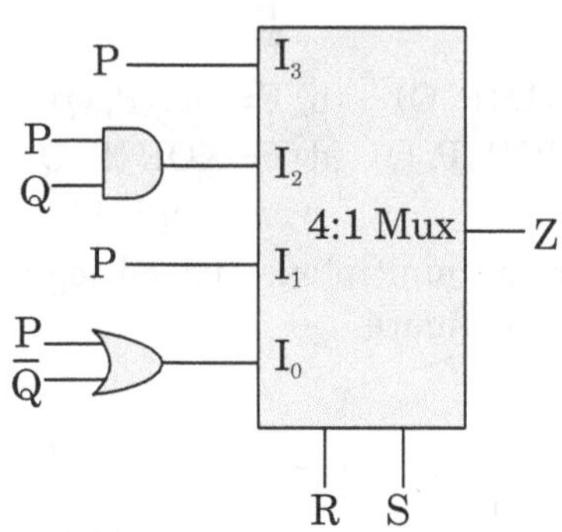

The output Z can be represented by

(a) $PQ + P\bar{Q}S + \bar{Q}\bar{R}\bar{S}$

(b) $P\bar{Q} + PQR + \bar{P}\bar{Q}\bar{S}$

(c) $P\bar{Q}\bar{R} + \bar{P}QR + PQRS + \bar{Q}\bar{R}\bar{S}$

(d) $PQ\bar{R} + PQR\bar{S} + P\bar{Q}\bar{R}S + \bar{Q}\bar{R}\bar{S}$

[2008 : 2 Marks]

15. What are the minimum number of 2-to-1 multiplexers required to generate a 2-input AND gate and a 2-input EX-OR gate?

(a) 1 and 2
(b) 1 and 3
(c) 1 and 1
(d) 2 and 2

[2009 : 2 Marks]

16. The Boolean function realized by the logic circuit shown is

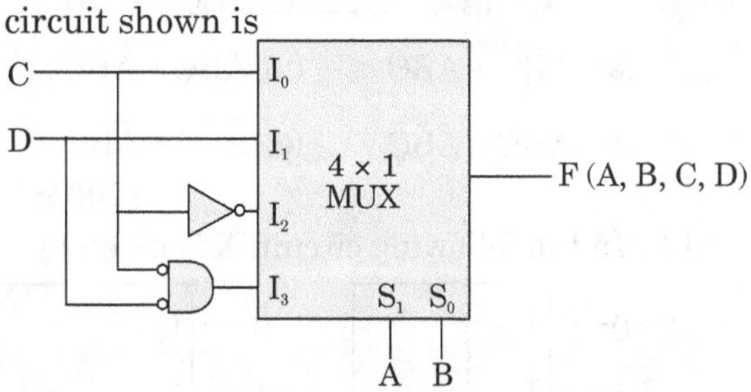

(a) $F = \sum m(0,1,3,5,9,10,14)$

(b) $F = \sum m(2,3,5,7,8,12,13)$

(c) $F = \sum m(1,2,4,5,11,14,15)$

(d) $F = \sum m(2,3,5,7,8,9,12)$

[2010 : 2 Marks]

17. The logic function implemented by the circuit below is (ground implies a logic '0')

4 × 1 MUX

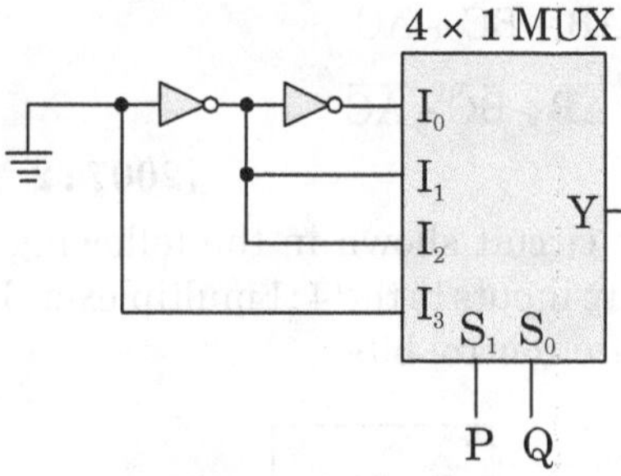

(a) F= AND (P, Q)
(b) F= OR (P, Q)
(c) F= XNOR (P, Q)
(d) F= XOR (P, Q)

[2011 : 1 Mark]

18. Consider the multiplexer based logic circuit shown in the figure.

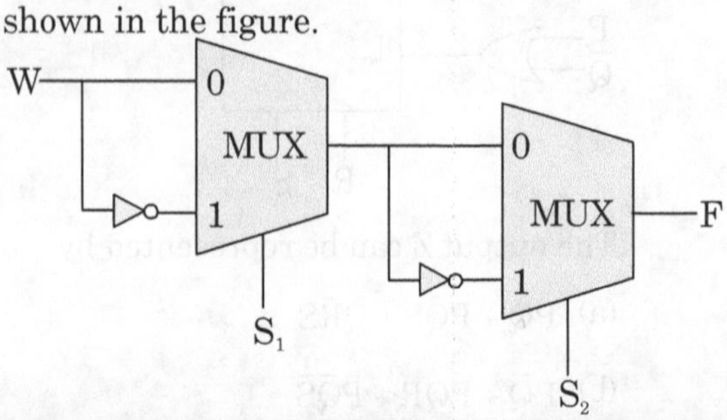

Which one of the following Boolean functions is realized by the circuit?

(a) $F = W\bar{S}_1\bar{S}_2$
(b) $F = WS_1 + WS_2 + S_1S_2$
(C) $F = \bar{W} + S_1 + S_2$
(d) $F = W \oplus S_1 \oplus S_2$

[2011 : 1 Mark]

19. In the circuit shown, W and Y are MSBs of the control inputs. The output MSBs is given by

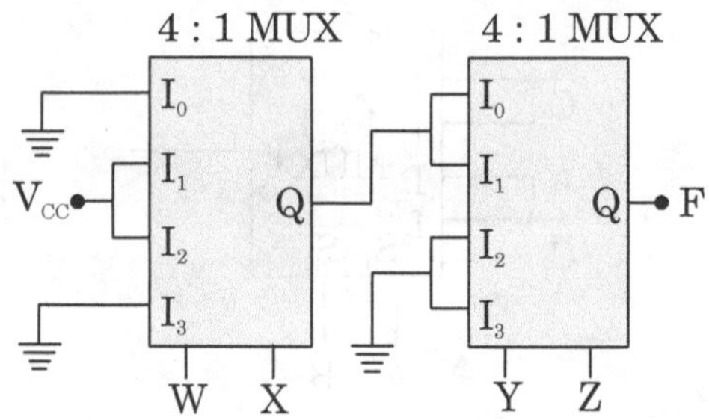

(a) $F = W\bar{X} + \bar{W}X + \bar{Y}\bar{Z}$

(b) $F = W\bar{X} + \bar{W}X + \bar{Y}Z$

(c) $F = W\bar{X}\bar{Y} + \bar{W}X\bar{Y}$

(d) $F = (\bar{W} + \bar{X})\bar{Y}\bar{Z}$

[2014 : 1 Mark, Set-3]

20. If X and Y are inputs and the Difference (D = X – Y and the Borrow (B) are the outputs, which one of the following diagrams implements a half-subtractor?

(a)

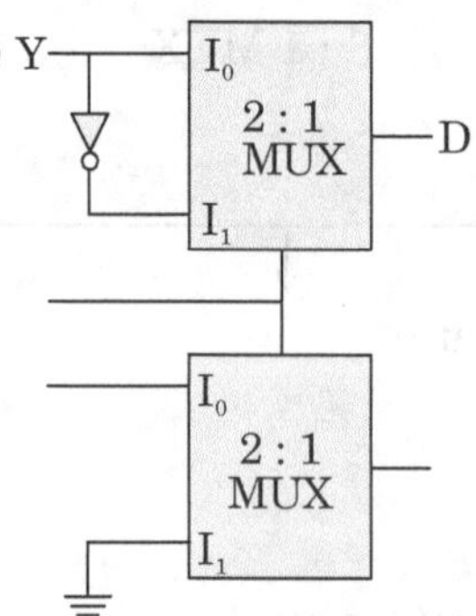

(b)

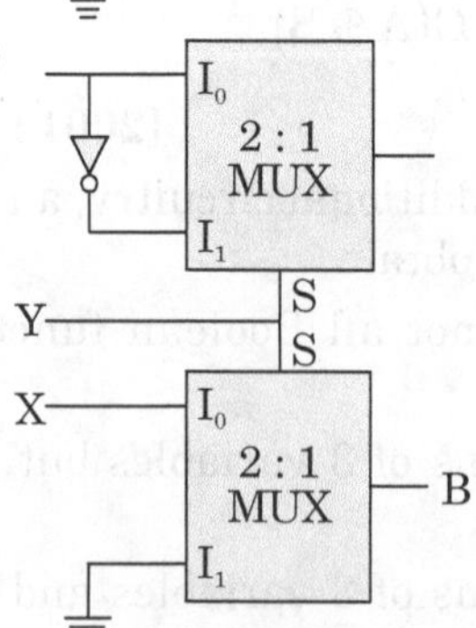

(c)

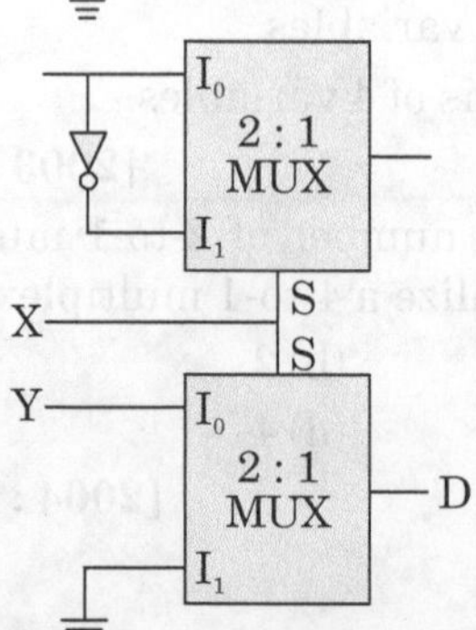

(d) 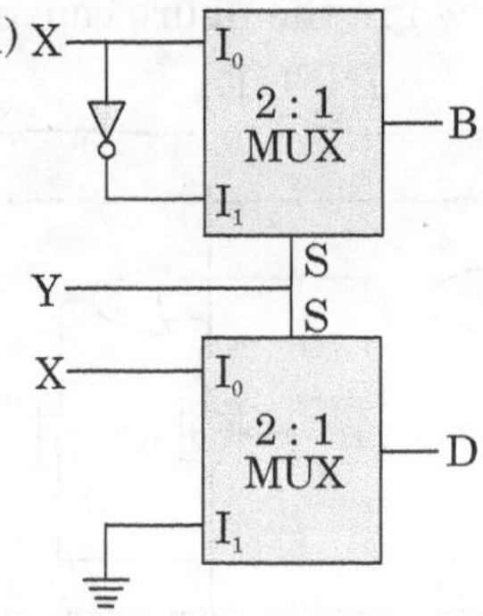

[2014 : 2 Marks, Set-3]

21. An 8-to-1 multiplexer is used to implement a logical function Y as shown in the figure. The output

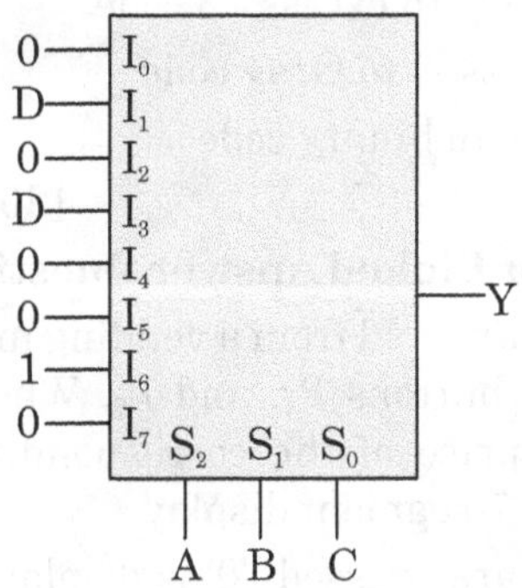

(a) $Y = A\bar{B}C + A\bar{C}D$ (b) $Y = \bar{A}BC + A\bar{B}D$

(c) $Y = AB\bar{C} + \bar{A}CD$ (d) $Y = \bar{A}\bar{B}D + \bar{A}BC$

2014 : 2 Marks, Set-3]

22. A 1-to-8 demultiplexer with data input D_n", address inputs S_o, S_1, S_2, (with S_o as the LSB) and $\bar{Y}_0$ to $\bar{Y}_1$ as the eight demultiplexed outputs, is to be designed using two 2-to-4 decoders (with enable input E and address inputs A_o and A_1) as shown in the figure D_n, S_o, S_1 and S_2, are to be connected to P, Q, R and S, but not necessarily in this order. The respective input connections to P, Q, R and S terminals should be

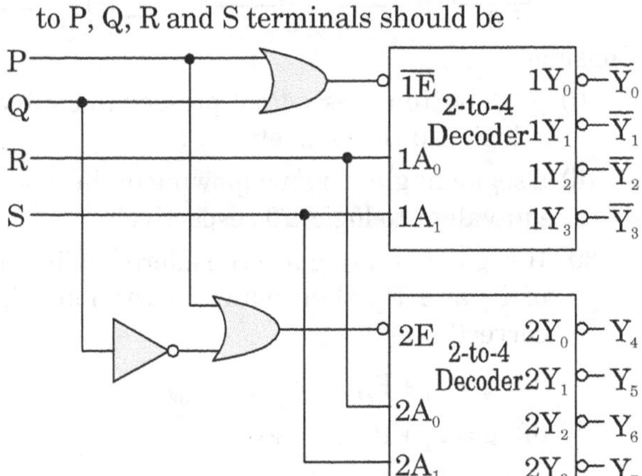

(a) S_2, D_n, S_0, S_1 (b) S_1, D_n, S_0, S_2

(C) S_n, D_0, S_1, S_2 (d) S_n, D_2, S_0, S_1

[2014 : 2 Marks, Set-4]

23. A 4:1 multiplexer is to be used for generating the output carry of a full adder. A and B are the bits to be added while C_{in} is the input carry and C_{out} is the output carry. A and B are to be used as the select bits with A being the more significant select bit.

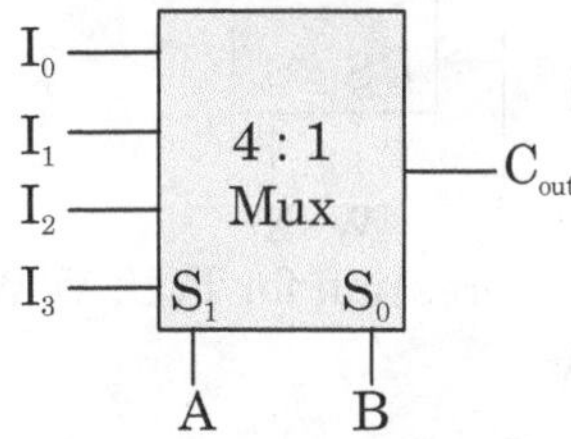

Which one of the following statements correctly describes the choice of signals to be connected to the inputs I_0, I_1, I_2 and I_3 so that the output is C_{out}?

(a) $I_0 = 0, I_1 = C_{in}, I_2 = C_{in}$ and $I_3 = 1$

(b) $I_0 = 1, I_1 = C_{in}, I_2 = C_{in}$ and $I_3 = 1$

(c) $I_0 = C_{in}, I_1 = 0, I_2 = C_{in}$ and $I_3 = 1$

(d) $I_0 = 0, I_1 = C_{in}, I_2 = 1$ and $I_3 = C_{in}$

[2016 : 1 Mark, Set-2]

24. For the circuit shown in the figure, the delays of NOR gates, multiplexers and inverters are 2 ns, 1.5 ns and 1 ns, respectively. If all the inputs P, Q, R, S and T are applied at the same time instant, the maximum propagation delay (in ns) of the circuit is ______.

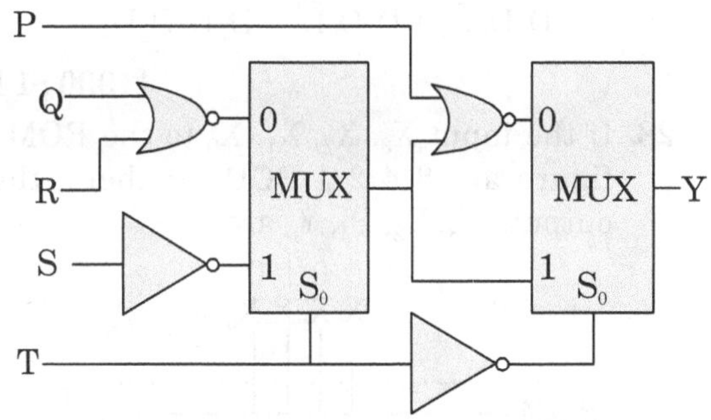

[2016 : 2 Marks, Set-3]

25. Consider the circuit shown in the figure.

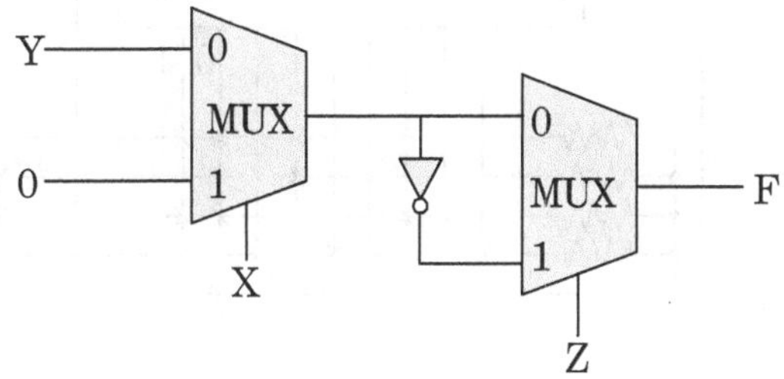

The Boolean expression F implemented by the circuit is

(a) $\bar{X}\bar{Y}\bar{Z} + XY + \bar{Y}Z$ (b) $\bar{X}Y\bar{Z} + XZ + \bar{Y}Z$

(c) $\bar{X}Y\bar{Z} + XY + \bar{Y}Z$ (d) $\bar{X}\bar{Y}\bar{Z} + XZ + \bar{Y}Z$

[2017 : 1 Mark, Set-2]

26. A four-variable Boolean function is realized using 4 × 1 multiplexers as shown in the figure.

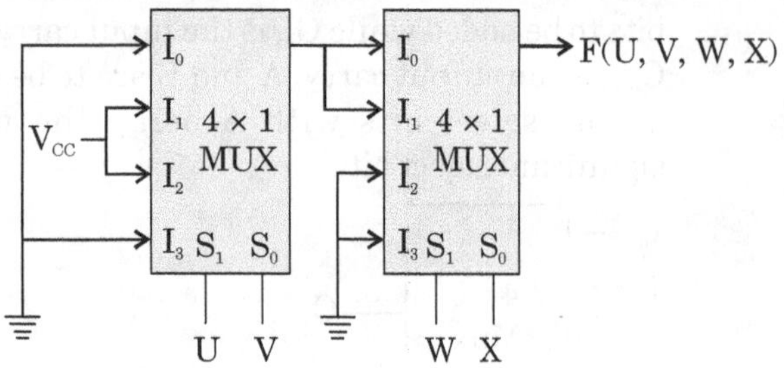

The minimized expression for F(U, V, W, X) is

(a) $(UV + \bar{U}\bar{V})\bar{W}$

(b) $(UV + \bar{U}\bar{V})(\bar{W}\bar{X} + \bar{W}X)$

(c) $(U\bar{V} + \bar{U}V)\bar{W}$

(d) $(U\bar{V} + \bar{U}V)(\bar{W}\bar{X} + \bar{W}X)$

[2018 : 2 Marks]

DECODERS & CODE CONVERTER

27. The minimal function that can detect a "divisible by 3" 8421 BCD code digit (representation is $D_8 D_4 D_2 D_1$) is given by

(a) $D_8D_1 + D_4D_2 + \bar{D}_8D_2D_1$

(b) $D_8D_1 + D_4D_2D_1 + \bar{D}_4D_2D_1 + \bar{D}_8\bar{D}_4\bar{D}_2\bar{D}_1$

(c) $D_8D_1 + D_4D_2 + \bar{D}_8\bar{D}_4\bar{D}_2\bar{D}_1$

(d) $D_4D_2\bar{D}_1 + D_4D_2D_1 + D_8\bar{D}_4D_2D_1$

[1990 : 1 Mark]

28. If the input X_3, X_2, X_1, X_0 to the ROM in the figure are 8 4 2 1 BCD numbers, then the outputs Y_3, Y_2, Y_1, Y_0 are

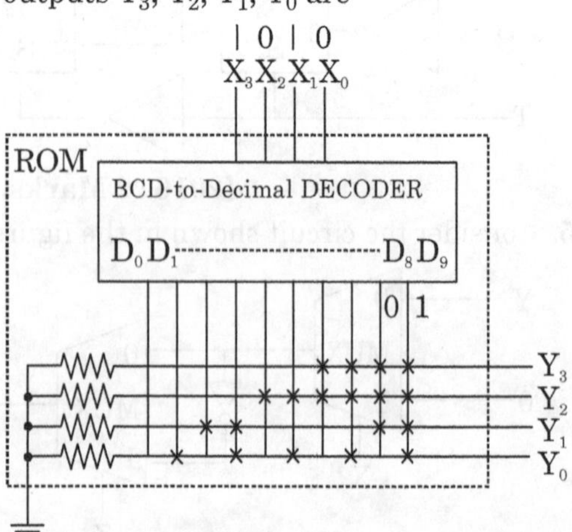

(a) gray code numbers

(b) 2 4 2 1 BCD numbers

(c) excess-3 code numbers

(d) None of the above

[2002 : 2 Marks]

29. The circuit shown in the figure converts

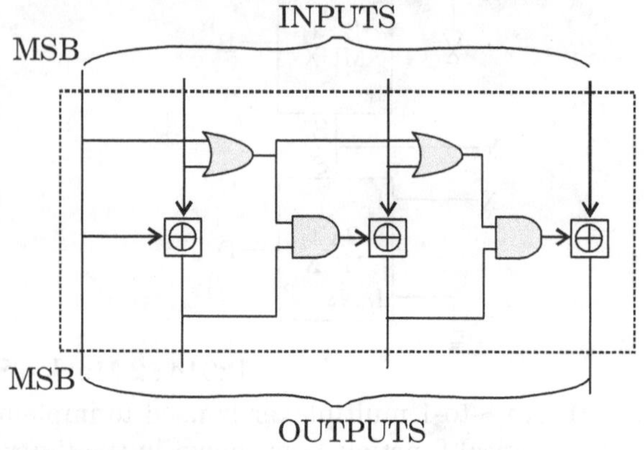

(a) BCD to binary code

(b) binary to excess - 3 code

(c) excess–3 to Gray code

(d) gray to binary code

[2003 : 2 Marks]

Statement for Linked Answer Question 30 and 31.

Two products are sold from a vending machine, which has two push buttons P_1, and P_2. When a button is pressed, the price of the corresponding product is displayed in a 7-segment display.

If no buttons are pressed, '0' is displayed. signifying 'Rs. 0'

If only P_1 is pressed, '2' is displayed, signifying 'Rs, 2'

If only P_2 is pressed, '5' is displayed. signifying 'Rs. 5'

If both P_1 and P_2 are pressed, 'E is displayed, signifying 'Error'

The names of the segments in the 7-segment display, and the glow of the display for '0, '2', '5' and `E. are shown below.

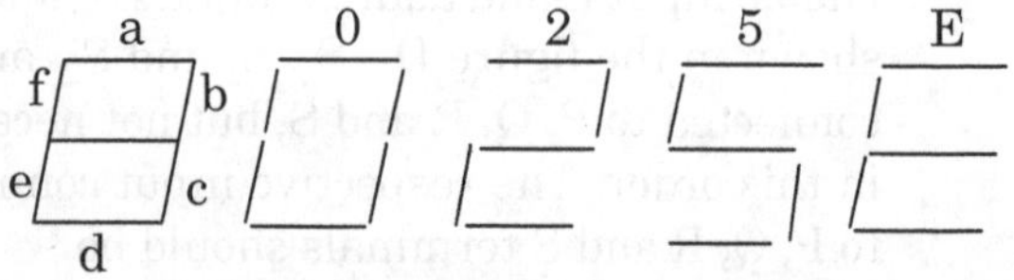

Consider

(i) push button pressed/not pressed in equivalent to logic 1/0 respectively,

(ii) a segment glowing/not glowing in the display is equivalent to logic 1/0 respectively

30. If segments a to g are considered as functions of P_1 and P_2, then which of the following is correct?

(a) $g = \bar{P}_1 + P_2, d = c + e$

(b) $g = P_1 + P_2, d = c + e$

(c) $g = \bar{P}_1 + P_2, d = b + c$

(d) $g = P_1 + P_2, d = b + c$

[2009 : 2 Marks]

31. What are the minimum numbers of NOT gates and 2-input OR gates required to design the logic of the driver for this 7-segment display?

(a) 3 NOT and 4 OR (b) 2 NOT and 4 OR

(c) 1 NOT and 3 OR (d) 2 NOT and 3 OR

[2009 : 2 Marks]

32. Identify the circuit below,

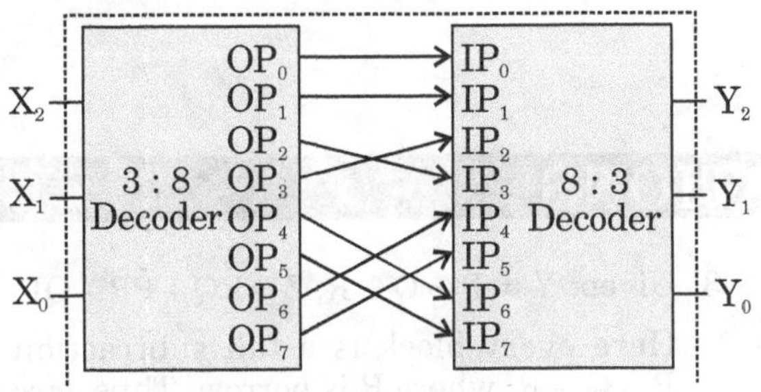

(a) binary to Gray code converter.

(b) binary to XS3 converter.

(c) gray to Binary converter.

(d) XS3 to Binary converter.

[2016: 2 Marks, Set-1]

33. The functionality implemented by the circuit below is

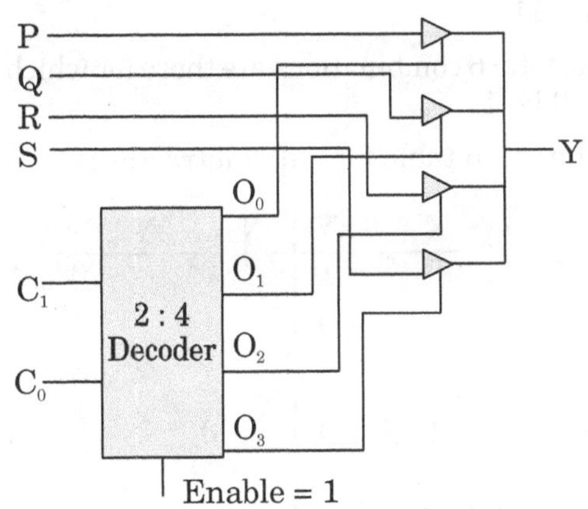

is a tristate buffer

(a) multiplexer

(b) 4-to-1 multiplexer

(c) 7-to-1 multiplexer

(d) 6-to-1 multiplexer

[2016 : 2 Marks, Set-1]

PROGRAMMABLE LOGIC ARRAY

34. A PLA can be

(a) as a microprocessor.

(b) as a dynamic memory.

(c) to realize a sequential logic.

(d) to realize a combinational logic.

[1994 : 1 Mark]

35. A programmable logic array (PLA) is shown in the figure.

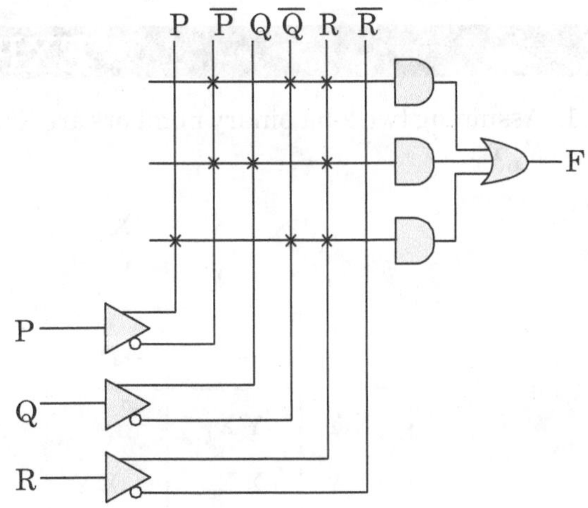

The Boolean function F implemented is

(a) $\bar{P}\bar{Q}R + \bar{P}QR + P\bar{Q}\bar{R}$

(b) $(\bar{P} + \bar{Q} + R)(\bar{P} + Q + R)(P + \bar{Q} + \bar{R})$

(c) $\bar{P}\bar{Q}R + \bar{P}QR + P\bar{Q}R$

(d) $(\bar{P} + \bar{Q} + R)(\bar{P} + Q + R)(P + \bar{Q} + R)$

[2017 : 2 Marks, Set-2)

ANSWERS

1. (b)	2. (c)	3. (b)	4. (b)	5. (c)	6. (195)	7. (50)	8. (b)	9. (d)	10. (d)
11. (b)	12. (a)	13. (a)	14. (a)	15. (a)	16. (d)	17. (d)	18. (d)	19. (c)	20. (a)
21. (c)	22. (d)	23. (a)	24. (7)	25. (b)	26. (c)	27. (b)	28. (b)	29. (a)	30. (b)
31. (d)	32. (a)	33. (b)	34. (d)	35. (c)					

EXPLANATIONS

1. Assuming two 2-bit binary numbers are $X_1 X_0$ and $Y_1 Y_0$.

		X_1	X_0
	×	Y_1	Y_0
		X_1Y_0	X_0Y_0
	X_1Y_1	Y_1X_0	
	X_1Y_1	$X_1Y_0 + Y_1X_0$	X_0Y_0
C_3	C_2	C_1	C_0

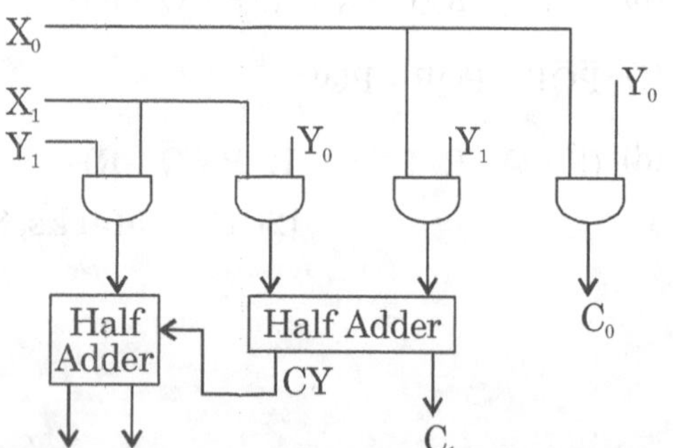

2-bit binary multiplier

2. Truth Table for half substractor having 2 inputs A and B, outputs D and X (borrow) is

A	B	D	X
0	0	0	0
0	1	1	1
1	0	1	1
1	1	0	0

Hence $D = \bar{A}B + A\bar{B}$ and $X = \bar{A}B$.

3. Given: $Y = P \oplus Q \oplus R$, $Z = RQ + \bar{P}R + Q\bar{P}$.

Here every block is a full subtraction giving $P - Q - R$, where R is borrow. Thus, circuit acts as a 4-bit subtractors giving $P - Q$.

4. Output will be 1 if $A > B$.

$\Rightarrow$ If $B = 00$, then there will be three combinations for which output will be 1, i.e., $A = 01$, 10 or 11.

$\Rightarrow$ If $B = 01$, there will be two conditions, i.e., $A = 10$ and 11.

$\Rightarrow$ If $B = 10$, there will be one condition, i.e., $A = 11$.

So, total 6 combinations are there for which output will be 1.

5. The truth table for half-subtractor is

X	Y	N = X – Y	M
0	0	0	0
0	1	1	1
1	0	1	0
1	1	0	0

$N = X \oplus Y$, $M = \bar{X}Y$.

6. 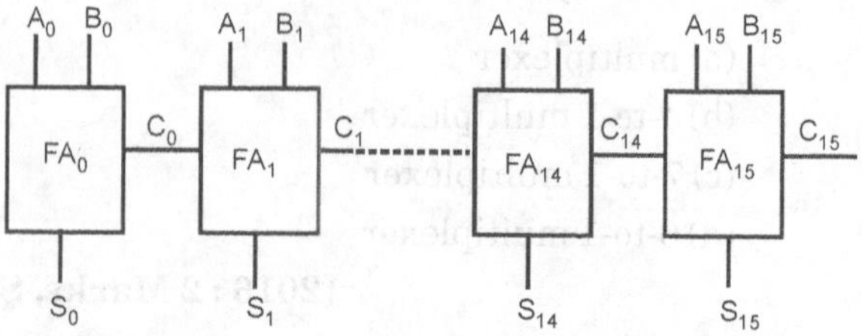

This is 16-bit ripple carry adder circuit, in their operation carry signal is propagating from 1st stage FA0 to last state FA15, so their propagation delay is added together but sum result is not propagating. We can say that next stage sum result depends upon previous carry.

So, last stage carry (C_{15}) will be produced after $16 \times 12\text{ns} = 192\text{ns}$

7. Assuming inputs to be added are
$X_3 X_2 X_1 X_0 = 1100, X_3 Y_2 Y_1 Y_0 = 0100$ and $Z_0 = 1$.
For this combination of addition, total minimum delays depends on the addition of most-significant two bits (since least significant two bits are zeros, they do not cause any change in Z_1 and Z_2). So, in the process of addition of given two digits waveforms at Z_1 and Z_2 become stable at t = 0 itself.

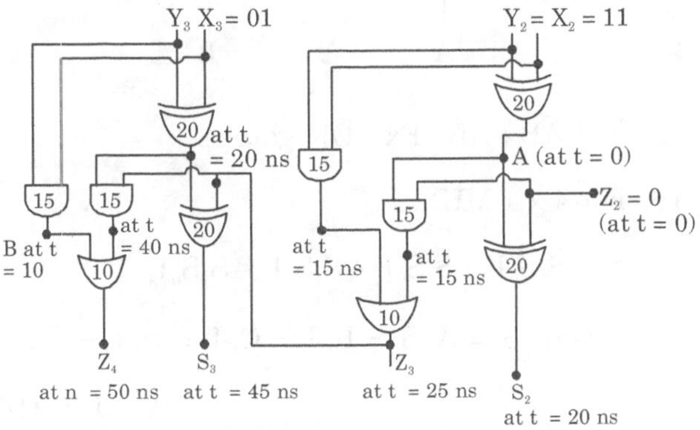

In above diagram, the waveform at A and B becomes stable at t = 0 itself, as the applied input combinations does not cause any change. So, for the given combination of inputs, output will settle at t = 50 ns.

8. Given: $I_0 = I_1 = C$,

$I_2 = I_3 = \bar{C}, S_1 = A, S_2 = B$.

Output of (4 × 1) MUX,

$$F = \bar{S}_1 \bar{S}_2 I_0 + \bar{S}_1 S_2 I_1 + S_1 \bar{S}_2 I_2 + S_1 S_2 I_3$$

$$= \bar{A}\bar{B}C + \bar{A}BC + A\bar{B}\bar{C} + AB\bar{C}$$

$$= \bar{A}C(\bar{B} + B) + A\bar{C}(\bar{B} + B) = \bar{A}C + A\bar{C}$$

$$= A \oplus C.$$

9. TTL logic gate accepts floating input as logic 1.

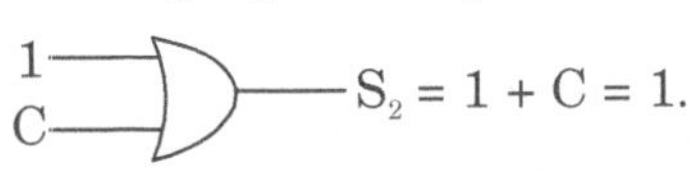

Inputs x_1, x_2, x_4, x_7 of MUX are zero. Hence, these term will vanish in output.

The output of (8 × 1) MUX is

$$F = \bar{S}_2 \bar{S}_1 \bar{S}_0 x_0 + \bar{S}_2 S_1 S_0 x_3 + S_2 \bar{S}_1 S_0 \cdot x_5 + S_2 S_1 \bar{S}_0 \cdot x_6$$

$$= 0 + 0 + 1 \cdot \bar{B} \cdot A \cdot 1 + 1 \cdot B, \bar{A}.0$$

$$F = A\bar{B} + \bar{A}B = A \oplus B.$$

10. A 2^n : 1 MUX can implement all logic functions of (n + 1) variables without any additional circuitry. Hence, n = 3, i.e., 8 × 1 MUX can implement all logic functions of 4 variable.

11.

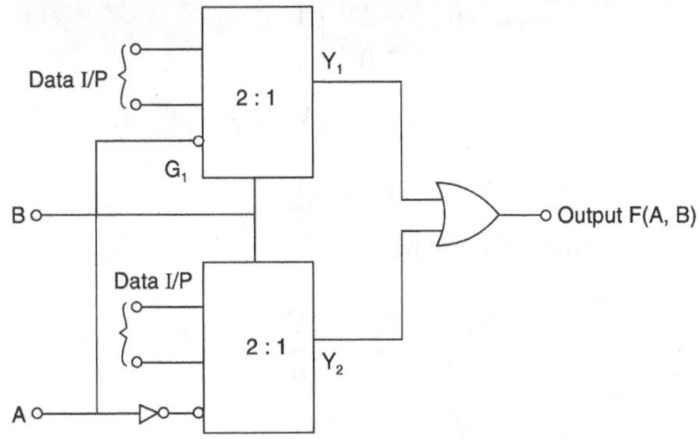

12. First multiplexer

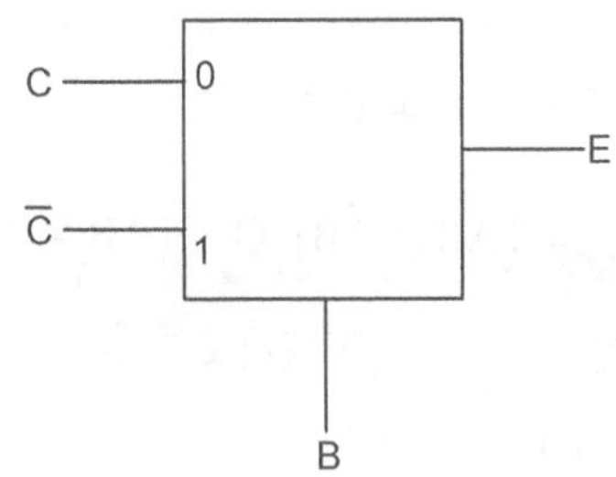

K-map for E

B \ C	0	1
0	0	1
1	1	0

$$E = \bar{B}C + \bar{C}B$$

For second multiplexer

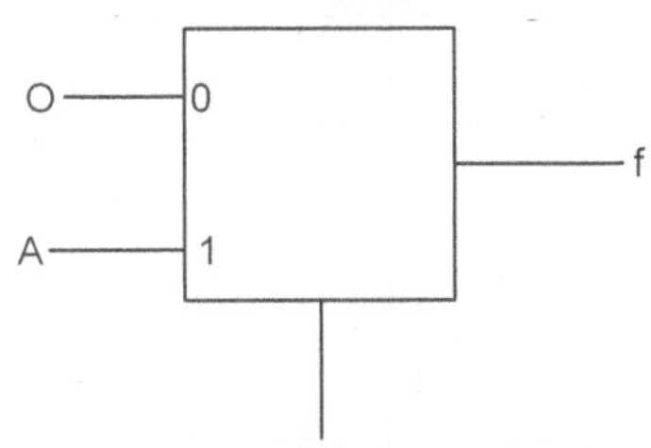

$$f = AE$$

$$= A\bar{B}C + AB\bar{C}$$

13.

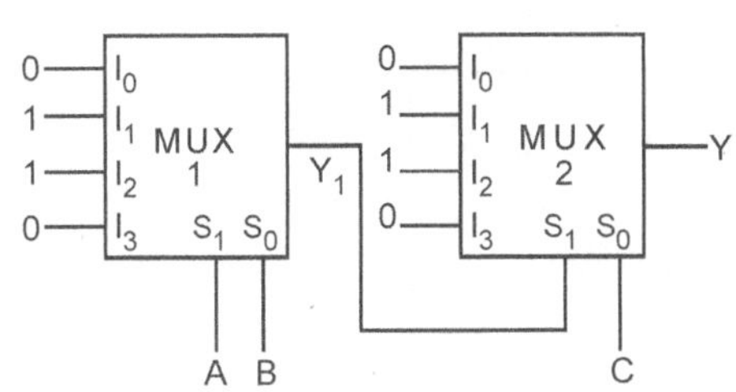

For MUX – 1

$A(S_1)$	$B(S_0)$	Y_1
0	0	0
0	1	1
1	0	1
1	1	0

$\therefore \quad Y_1 = \overline{A}B + A\overline{B}$

Similarly for MUX - 2

$Y_1(S_1)$	$C(S_0)$	Y
0	0	0
0	1	1
1	0	1
1	1	0

$$Y = \overline{Y_1}\,C + Y_1\overline{C}$$

$$= \overline{(\overline{A}B + A\overline{B})}\; C + (\overline{A}B + A\overline{B})\cdot\overline{C}$$

$$\therefore \quad Y = ABC + \overline{A}\,\overline{B}\,C + \overline{A}\,B\,\overline{C} + A\,\overline{B}\,\overline{C}$$

14. Output of 4 : 1 MUX,

$$Z = \overline{R}\overline{S}\cdot(P+\overline{Q}) + \overline{R}S\cdot P + R\overline{S}\cdot PQ + RS\cdot P$$

$$Z = P\overline{R}\overline{S} + \overline{Q}\overline{R}\overline{S} + P\overline{R}S + PQR\overline{S} + PRS.$$

Mapping above terms in K-map,

PQ \ RS	00	01	11	10
00	1			
01				
11	1	1	1	1
10	1	1	1	

$$Z = PQ + P\overline{Q}S + \overline{Q}\overline{R}\overline{S}$$

$$Z = PQ + P\overline{Q}S + \overline{Q}\overline{R}\overline{S}.$$

15.

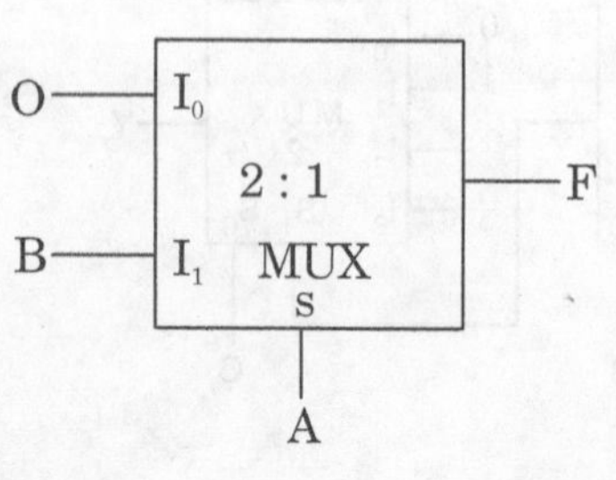

$$F = \overline{S}\cdot I_0 + S\cdot I_1$$

$$F = \overline{A}\cdot 0 + A\cdot B$$

$$F = AB \text{ (AND gate)}$$

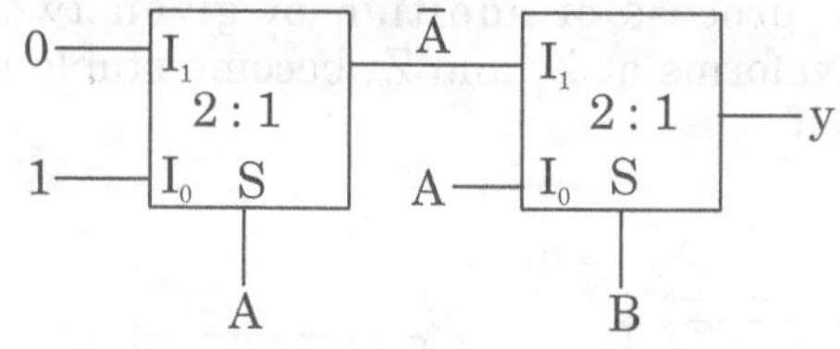

$$Y = \overline{A}B + A\overline{B}\cdot\text{(Ex – OR gate)}.$$

16. For 4 × 1 MUX,

$$F = \overline{S}_1\overline{S}_0 I_0 + \overline{S}_1 S_0 I_1 + S_1\overline{S}_0 I_2 + S_1 S_0 I_3$$

Given : $S_1 = A$, $S_0 = B$, $I_0 = C$, $I_1 = D$, $I_2 = \overline{C}$,

$I_3 = \overline{C}\cdot\overline{D}$

$\therefore$ F (A, B, C, D)

17.

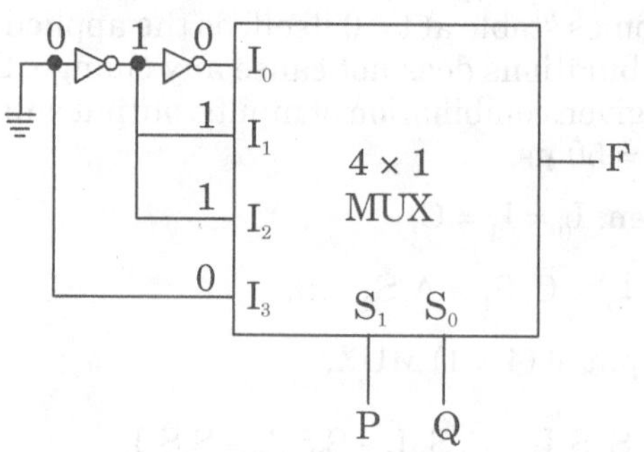

$$F = \overline{S}_1\overline{S}_0 I_0 + \overline{S}_1 S_0 I_1 + S_1\overline{S}_0 I_2 + S_1 S_0 I_3$$

$$F = \overline{P}\overline{Q}\cdot 0 + \overline{P}Q\cdot 1 + P\overline{Q}\cdot 1 + PQ\cdot 0$$

$$F = 0 + \overline{P}Q + P\overline{Q} + 0$$

$$F = P \oplus Q.$$

18.

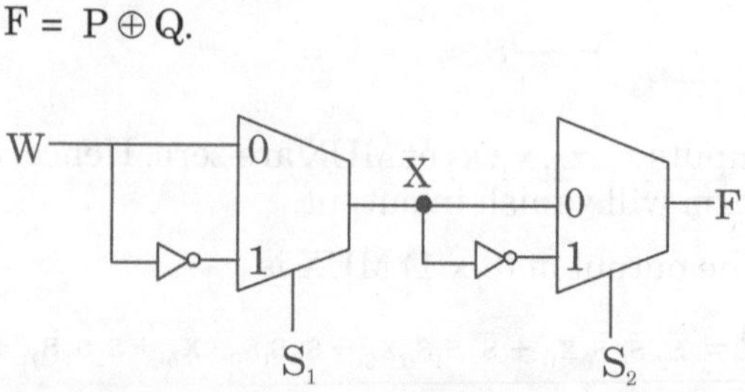

$$\Rightarrow X = \overline{S}_1 W + S_1\overline{W}$$

$$\Rightarrow F = \overline{S}_2\cdot X + S_2\cdot\overline{X}$$

$$= \bar{S}_2(\bar{S}_1 W + S_1\bar{W}) + S_2(\overline{\bar{S}_1 W + S_1 \bar{W}})$$

$$= \bar{S}_1\bar{S}_2 W + \bar{S}_2 S_1 \bar{W} + S_2(\bar{S}_1\bar{W} + S_1 W)$$

$$F = \bar{S}_1\bar{S}_2 W + \bar{S}_2 S_1 \bar{W} + S_2\bar{S}_1\bar{W} + S_2 S_1 W$$

$$F = W \oplus S_1 \oplus W_2.$$

19. $A = \bar{W}\bar{X}I_0 + \bar{W}XI_1 + W\bar{X}I_2 + WXI_3$

$$= \bar{W}X + W\bar{X}.$$

$$F = \bar{Y}\bar{Z}I_0 + \bar{Y}ZI_1 + Y\bar{Z}I_2 + YZI_3$$

$$= \bar{W}X\bar{Y}\bar{Z} + W\bar{X}\bar{Y}\bar{Z} + \bar{W}X\bar{Y}Z + W\bar{X}\bar{Y}Z$$

$$= \bar{W}X\bar{Y}(\bar{Z}+Z) + W\bar{X}\bar{Y}(\bar{Z}+Z)$$

$$F = \bar{W}X\bar{Y} + W\bar{X}\bar{Y}$$

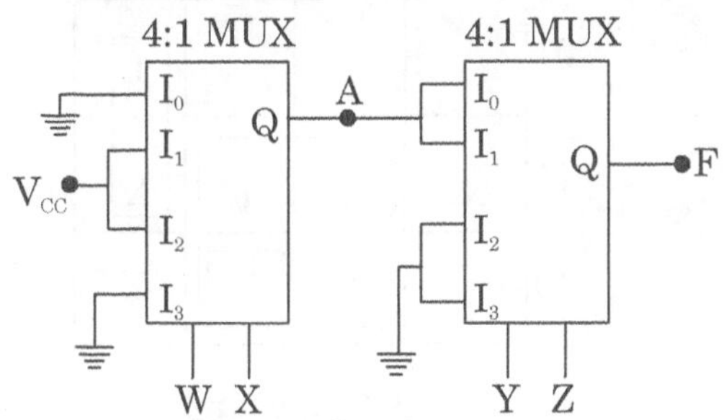

$V_{CC} = 1.$

20.

X	Y	D	B
0	0	0	0
0	1	1	1
1	0	1	0
1	1	0	0

So, $D = X \oplus Y$

$$= \bar{X}Y + X\bar{Y}$$

and $B = \bar{X}.Y$

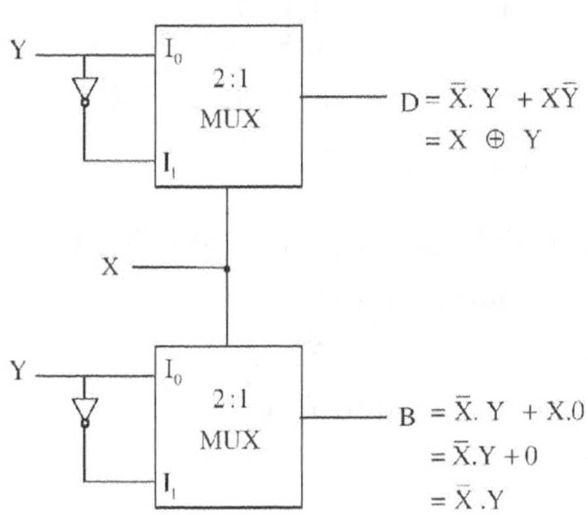

21.

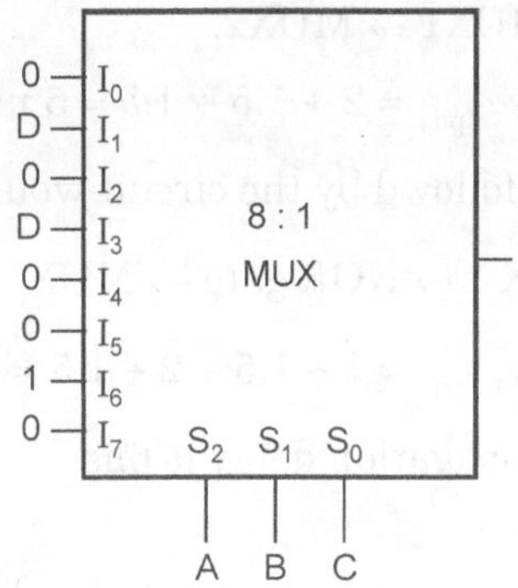

$$Y = \bar{A}\bar{B}CD + \bar{A}BCD + AB\bar{C}$$

Remaining combinations of the select lines will produce output 0.

So, $Y = \bar{A}CD\,(\bar{B} + B) + AB\bar{C}$

$$= \bar{A}CD + AB\bar{C}$$

$$= AB\bar{C} + \bar{A}CD$$

22. From the given circuit, considering as 1 × 8 DMUX,

$$\bar{Y}_1 = 1A_0 + 1A_1 + \bar{E} = (R+S+P+Q) = P+Q+R+S.$$

Similarly,

$$\bar{Y}_1 = (1\bar{A}_0 + 1A_1 + 1\bar{E}) = P+Q+\bar{R}+S.$$

$$\bar{Y}_4 = (2\bar{A}_0 + 2A_1 + 2\bar{E}) = R+S+P+\bar{Q}$$

$$= P+\bar{Q}+R+S.$$

$$\Rightarrow P = \bar{D}_{in}, Q = S_2, R = S_1, S = S_0.$$

23. In case of a full adder,

$C_{out}(A, B, C_{in}) = \Sigma(3, 5, 6, 7)$

(A, B) → Applied at select input of MUX; C_{in} → Applied at data input of MUX

	I_0	I_1	I_2	I_3
$\bar{C}_{in}$	0	2	4	(6)
C_{in}	1	(3)	(5)	(7)
	0	C_{in}	C_{in}	1

∴ $I_0 = 0$

$I_1 = C_{in}$

$I_2 = C_{in}$

$I_3 = 1$

24. When, T = 0, path followed by the circuit would be NOR gate → MUX1 → MUX2.

Propagation delay, $t_{pd_1} = 2 + 1.5 + 1.5 = 5$ ns.

When T = 1, path followd by the circuit would be

NOR gate → MUX 1 → NOR gate → MUX2

Propagation delay, $t_{pd_2} = 1 + 1.5 + 2 + 1.5 = 6$ ns.

The maximum propagation delay is 6ns.

25.

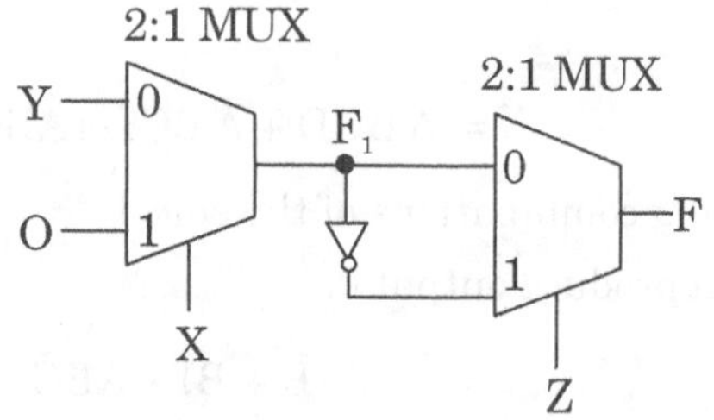

$$\Rightarrow F_1 = \bar{X}Y + 0 = \bar{X}Y$$

$$\Rightarrow F = \bar{Z}\cdot F_1 + Z\bar{F}_1$$

$$= (\bar{X}Y)\bar{Z} + (\overline{\bar{X}Y})Z$$

$$= \bar{X}Y\bar{Z} + (X + \bar{Y})Z$$

$$= \bar{X}Y\bar{Z} + XZ + X\bar{Y}Z$$

26.

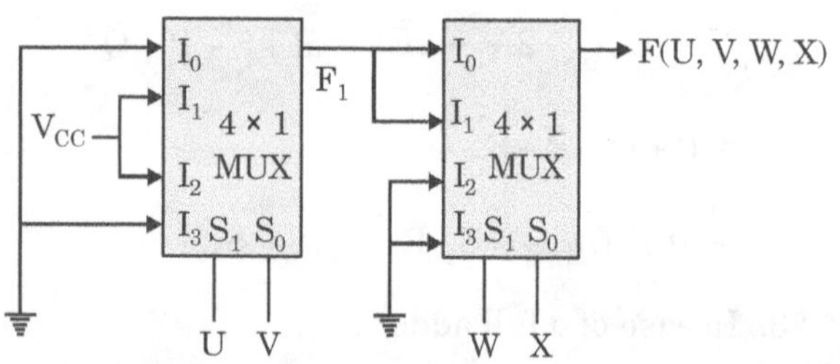

Output of first multiplexer is given by

$$F_1 = \bar{U}V + U\bar{V}$$

and Output of the second multiplexer is given by

$$F = \bar{W}\bar{X}F_1 + \bar{W}XF_1 = \bar{W}F_1 = (\bar{U}V + U\bar{V})\bar{W}$$

$$(\because F_1 = \bar{U}V + U\bar{V})$$

27. The output of function that can detect a "divisible by 3" 8421 BCD code digit is high or logic 1 when input (in decimal) is 0, 3, 6, 9. The truth table and K-Map for above function are

Decimal number	D_8	D_4	D_2	D_1	Y
0	0	0	0	0	1
1	0	0	0	1	0
2	0	0	1	0	0
3	0	0	1	1	1
4	0	1	0	0	0
5	0	1	0	1	0
6	0	1	1	0	1
7	0	1	1	1	0
8	1	0	0	0	0
9	1	0	0	1	1

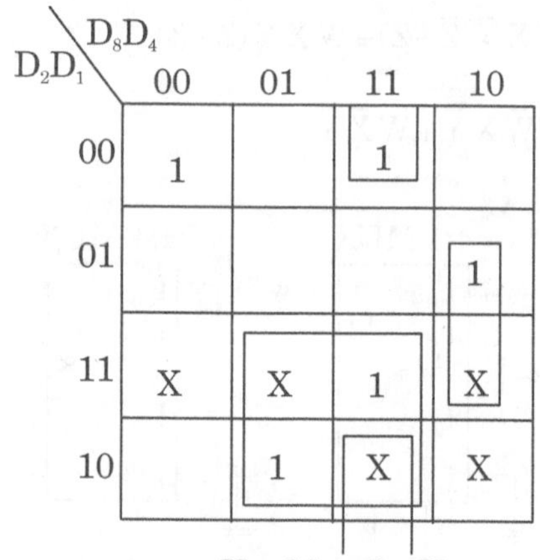

K – Map for Y

$$\Rightarrow Y = D_8D_1 + D_4D_2\bar{D}_1 + \bar{D}_4D_2D_1 + \bar{D}_8\bar{D}_4\bar{D}_2\bar{D}_1.$$

28. 8 – 4 2 – 1 BCD INPUT

X_3	X_2	X_1	X_0	Output
0	0	0	0	D_0
0	0	0	1	D_1
0	0	1	0	D_2
0	0	1	1	D_3
0	1	0	0	D_4
0	1	0	1	D_5
0	1	1	0	D_6
0	1	1	1	D_7
1	0	0	0	D_8
1	0	0	1	D_9

Now as D_0 is not connected to Y_3, Y_2, Y_1, Y_0

D_1 is connected to Y_0

D_2 is connected to Y_1

D_3 is connected to Y_1,Y_0

D_4 is connected to Y_2

D_5 is connected to Y_2, Y_0

D_6 is coneected to Y_3, Y_2

D_7 is coneected to Y_3, Y_2, Y_0

D_8 is coneected to Y_3, Y_2, Y_1

D_9 is coneected to Y_3, Y_2, Y_1, Y_0

Thus constructing truth-table for output Y,

	2 – 4 – 2 –1				
	Y_3	Y_2	Y_1	Y_0	Output
D_0	0	0	0	0	0
D_1	0	0	0	1	1
D_2	0	0	1	0	2
D_3	0	0	1	1	3
D_4	0	1	0	0	4
D_5	0	1	0	1	5
D_6	1	1	0	0	6
D_7	1	1	0	1	7
D_8	1	1	1	0	8
D_9	1	1	1	1	9

By the output shown

for $7 = 2Y_3 + 4Y_2 + Y_0$

$8 = 2Y_3 + 4Y_2 + 2Y_1$

$9 = 2Y_3 + 4Y_2 + 2Y_1 + 2Y_0$

Hence it can be inferred that out is 2 – 4 – 2 – 1 BCD code.

29. 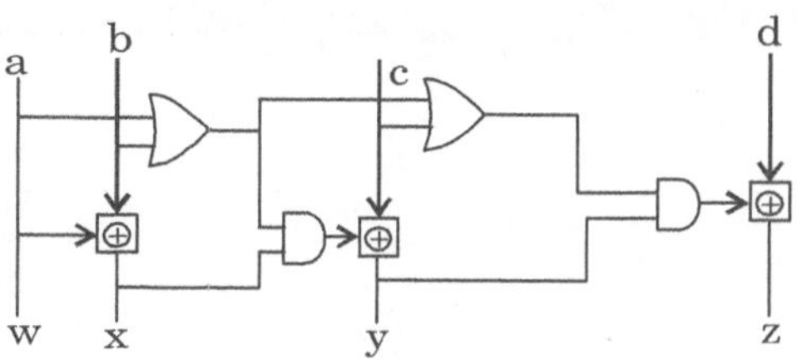

Assuming inputs are a, b, c & d and outputs are w, x, y & z.

$\Rightarrow w = a, x = a \oplus b, y = c \oplus x \cdot (a + b), z$

$= d \oplus y\,(a + b + c).$

Let input be 1010; output will be 1101.

Let input be 0110; output will be 0100.

Thus, it converts gray code number to binary code number.

30. With given conditions, the truth table is shown below.

P_1	P_2	a	b	c	d	e	f	g
0	0	1	1	1	1	1	1	0
0	1	1	0	1	1	0	1	1
1	0	1	1	0	1	1	0	1
1	1	1	0	0	1	1	1	1

$\Rightarrow a = 1$

$\Rightarrow b = \bar{P}_2 \rightarrow$ NOT

$\Rightarrow b = \bar{P}_1 \rightarrow$ NOT

$\Rightarrow d = 1$

$\Rightarrow e = P_1 + \bar{P}_2 \rightarrow$ OR

$\Rightarrow f = \bar{P}_1 + P_2 \rightarrow$ OR

$\Rightarrow g = P_1 + P_2 \rightarrow$ OR

$\Rightarrow d = c + e \;\&\; g = P_1 + P_2.$

31. 2 NOT gates and 3 OR gates are required.

32. Here the truth table for the given circuit is shown below.

X_2	X_1	X_0	OP_0	OP_1	OP_2	OP_3	OP_4	OP_5	OP_6	OP_7	IP_0	IP_1	IP_2	IP_3	IP_4	IP_5	IP_6	IP_7	Y_2	Y_1	Y_0
0	0	0	1	0	0	0	0	0	0	0	1	0	0	0	0	0	0	0	0	0	0
0	0	1	0	1	0	0	0	0	0	0	0	1	0	0	0	0	0	0	0	0	1
0	1	0	0	0	1	0	0	0	0	0	0	0	0	1	0	0	0	0	0	1	1
0	1	1	0	0	0	1	0	0	0	0	0	0	1	0	0	0	0	0	0	1	0
1	0	0	0	0	0	0	1	0	0	0	0	0	0	1	0	0	0	0	–	–	–
1	0	1	0	0	0	0	0	1	0	0	0	0	0	0	1	0	0	0	–	–	–
1	1	0	0	0	0	0	0	0	1	0	0	0	0	0	0	1	0	0	1	0	1
1	1	1	0	0	0	0	0	0	0	1	0	0	0	0	1	0	0	0	1	0	0

Hence it will be binary to Gray Code Converter.

33. When the output $(0_0, 0_1, 0_2, 0_3)$ of the decoder are at logic 1, the corresponding tristate luffer is activated. In that case, whatever data is applied at the input of a buffer, becomes its output. Hence, when

$\Rightarrow C_1C_0 \equiv OO$, then $O_0 = 1$; $Y = P$

$\Rightarrow C_1\ C_0 = 01$, then $O_1 = 1$; $Y = Q$

$\Rightarrow C_1C_0 = 10$, then $O_2 = 1$; $Y = R$

$\Rightarrow C_1\ C_0 = 11$, then $O_3 = 1$; $Y = S$.

Hence, the circuit behaves as a 4 : 1 multiplexer.

PROGRAMMABLE LOGIC ARRAY

34. PLA is a type of fixed architecture logic devices with programmable AND gates followed by programmable OR gates. The PLA can be used to implement a complex combinational circuits.

35. In given circuit, enable and output of any tristate buffer is not connected as inputs for same gate.

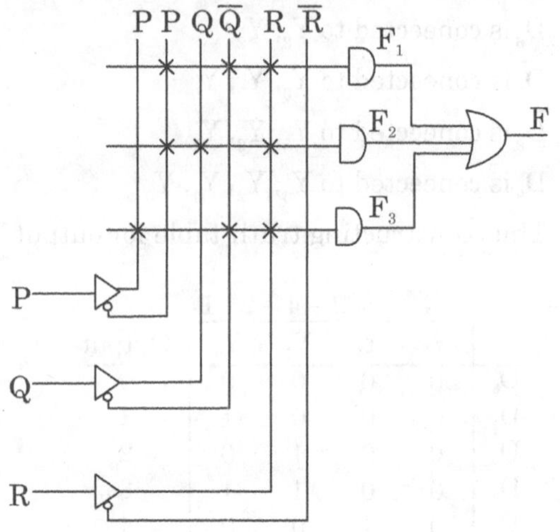

$\Rightarrow F_1 = \bar{P}\bar{Q}R$

$F_2 = \bar{P}QR$

$F_3 = P\bar{Q}R$

$\Rightarrow F = F_1 + F_2 + F_3$

$= \bar{P}\bar{Q}R + \bar{P}QR + P\bar{Q}R.$

4
CHAPTER

Sequential Circuits

Analysis of Previous GATE Papers

Topics ↓		Year →	2019	2018	2017 Set 1	2017 Set 2	2016 Set 1	2016 Set 2	2016 Set 3	2015 Set 1	2015 Set 2	2015 Set 3
LATCHES & FLIP-FLOPS	1 Mark	MCQ Type	1		1							
		Numerical Type			1			1				
	2 Marks	MCQ Type	1									1
		Numerical Type		1								
	Total		3	2	2			1				2
COUNTERS & SHIFT REGISTERS	1 Mark	MCQ Type									1	1
		Numerical Type	1		1						1	1
	2 Marks	MCQ Type							1			
		Numerical Type										
	Total		1		1				2		2	2
FINITE STATE MACHINE	1 Mark	MCQ Type										
		Numerical Type		1								
	2 Marks	MCQ Type			1			1				
		Numerical Type				1						
	Total			1	2	2		2				

LATCHES & FLIP-FLOPS

1. Choose the correct statements relating to the circuit of figure

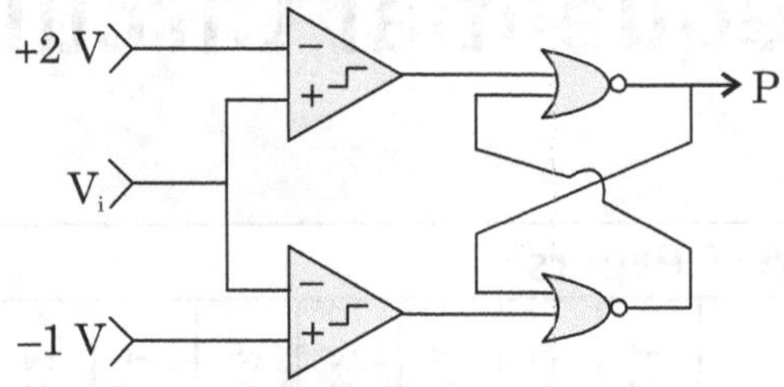

(a) For V_i = 2 V, P = 0
(b) For V_i = + 3 V, P = 0
(c) For V_i = 0 V, P = 0 always
(d) For V_i = 0 V, P can be either 0 or 1.

[1987 : 1 Mark]

2. The circuit given below is a

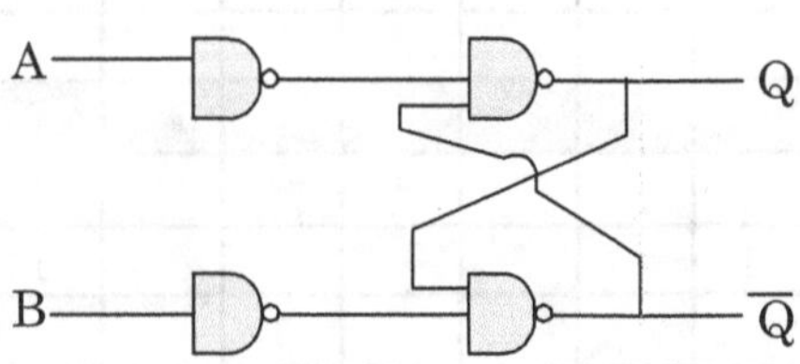

(a) J-K flip-flop (b) Johnson's counter
(c) R-S latch (d) None of above.

[1988 : 1 Mark]

3. An S-R FLIP-FLOP can be converted into a T flip–flop by connecting to ________ Q and ____ to $\overline{Q}$.

[1991 : 1 Mark]

4. An R-S latch is a
(a) combinat circuit.
(b) synchronous sequential circuit.
(c) one bit memory element.
(d) one clock delay element.

[1995 : 1 Mark]

5. In a J-K flip-flop we have J = $\overline{Q}$ and K = 1 (see figure) Assuming the flip-flop was initially cleared and then clocked for 6 pulses, the sequence at the Q output will be

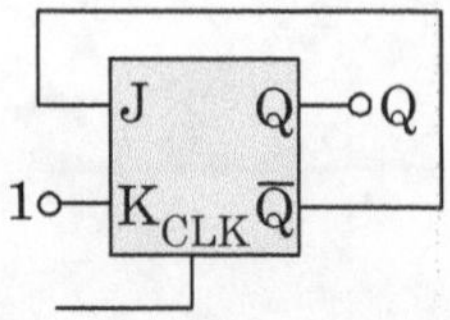

(a) 010000 (b) 011001
(c) 010010 (d) 010101

[1997 : 1 Mark]

6. A sequential circuit using D Flip-Flop and logic gates is shown in the figure, where X and Yare the inputs and Z is the output. The circuit is

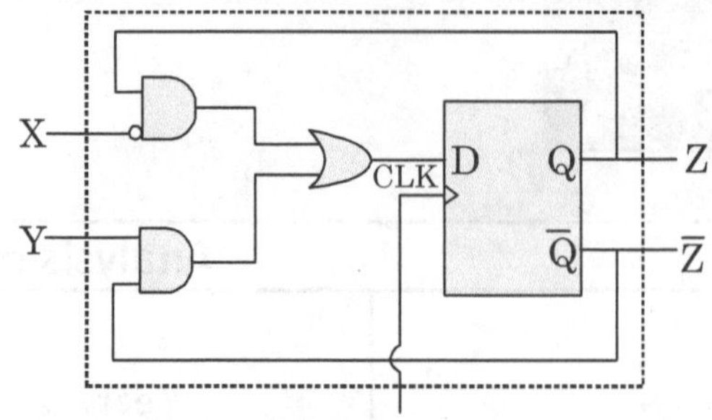

(a) S–R Flip-Flop with inputs X = R and Y= S.
(b) S–R Flip-Flop with inputs X = S and Y = R.
(c) J–K Flip-Flop with inputs X = J and Y = K.
(d) J–K Flip-Flop with inputs X = K and Y= J.

[2000 : 2 Marks]

7. The digital block in the figure is realized using two positive edge triggered D-flip-flops. Assume that for $t < t_0$, $Q_1 = Q_2 = 0$. The circuit in the digital block is given by:

(a)
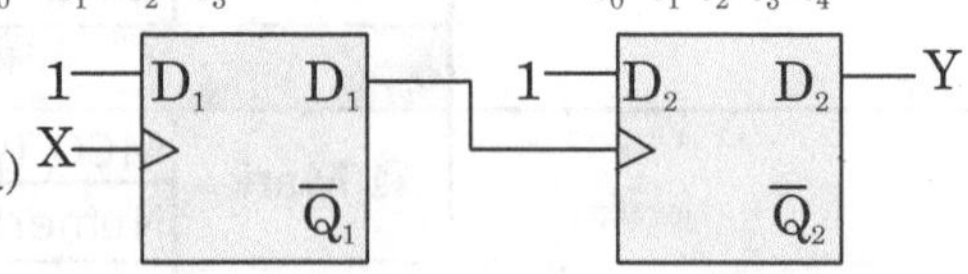

(b)
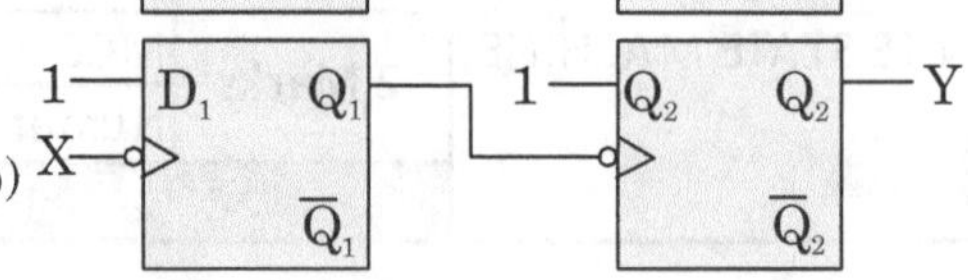

(c)
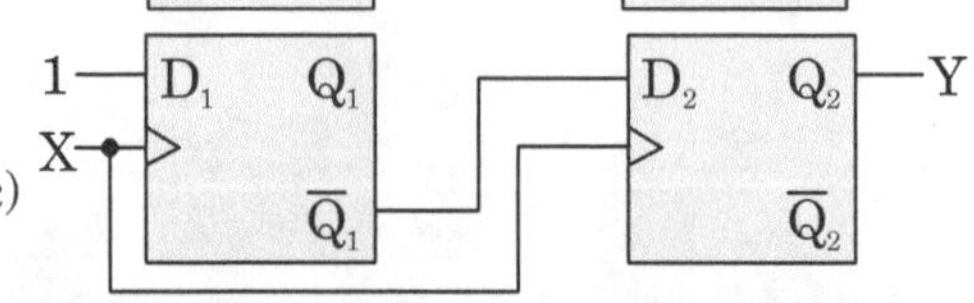

(d)
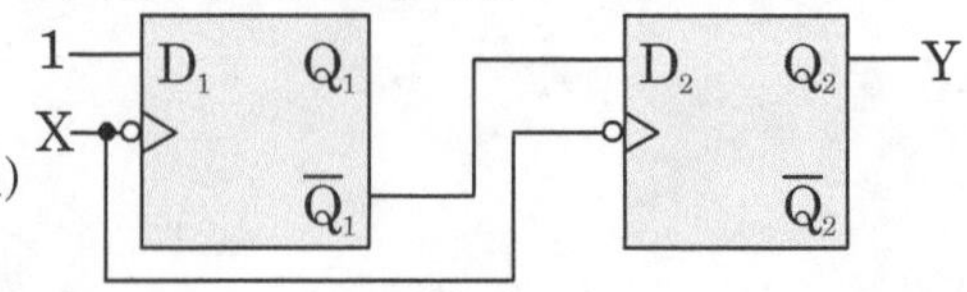

[2001 : 2 Marks]

8. A master-slave flip-flop has the characteristic that
(a) change in the input immediately reflected in the output.
(b) change in the output occurs when the state of the master is affected.
(c) change in the output occurs when the state of the slave is affected.
(d) both the master and the slave states are affected at the same time.

[2004 : 1 Mark]

9. The present output Q_n of an edge triggered JK flip-flop is logic 0. If J = 1, then Q_{n+1}

(a) cannot be determined.

(b) will be logic 0.

(c) will be logic 1.

(d) will race around.

[2005 : 2 Marks]

10. The following binary values were applied to the X and Y inputs of the NAND latch shown in the figure in the sequence indicated below:

X = 0, Y = 1;

X = 0, Y = O;

X = 1, Y = 1.

The corresponding stable P, Q outputs will be

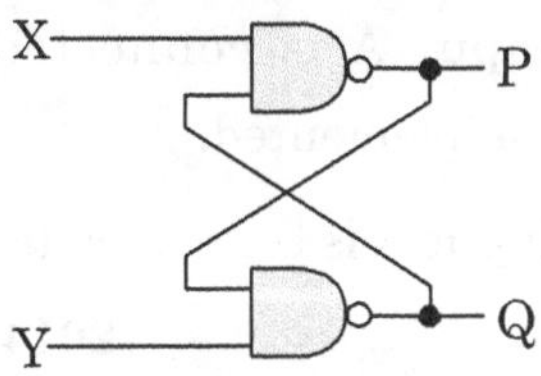

(a) P= 1, Q = 0; P= 1, Q= 0; P= 1, Q = 0 or P = 0, Q = 1.

(b) P= 1, Q = 0; P= 0, Q = 1 or P = 0, Q = 1; P = 0, = 1

(c) P = 1, Q = 0; P = 1, Q = 1; P = 1, Q = 0 or P = 0, Q = 1.

(d) P = 1, Q = 0; P = 1, Q = 1; P = 1, Q = 1.

[2007 : 2 Marks]

11. For each of the positive edge-triggered J-K flip flop used in the following figure, the propagation delay is ΔT

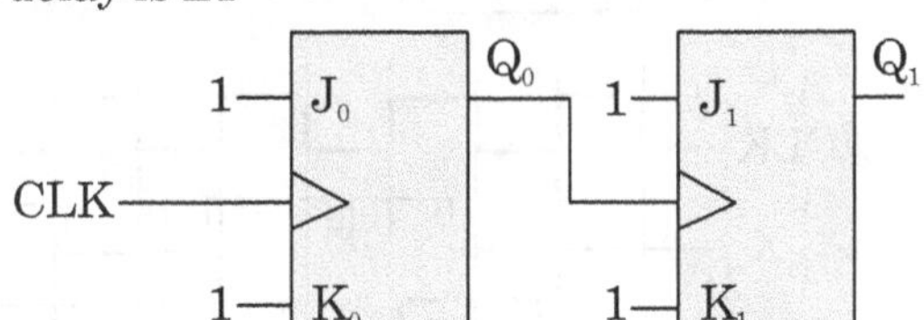

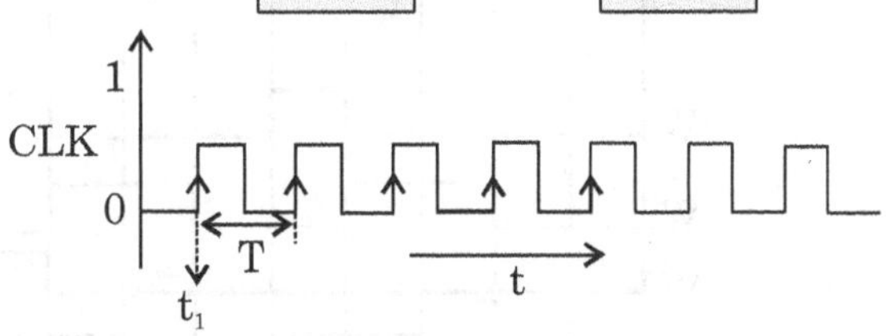

Which of the following waveforms correctly represents the output at Q_1?

(a)

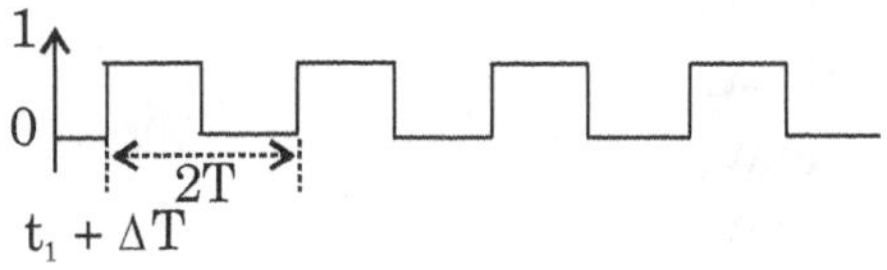

(b)

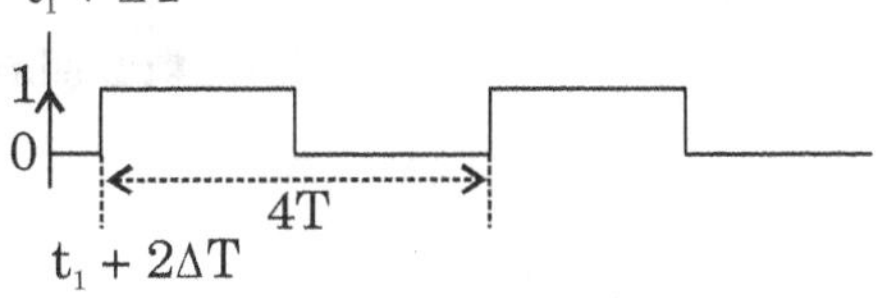

(c)

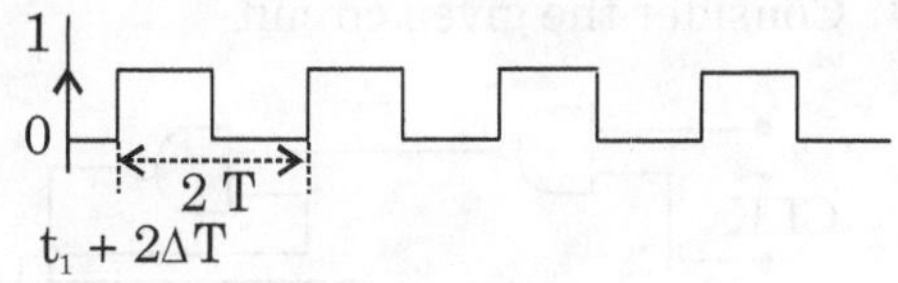

(d)

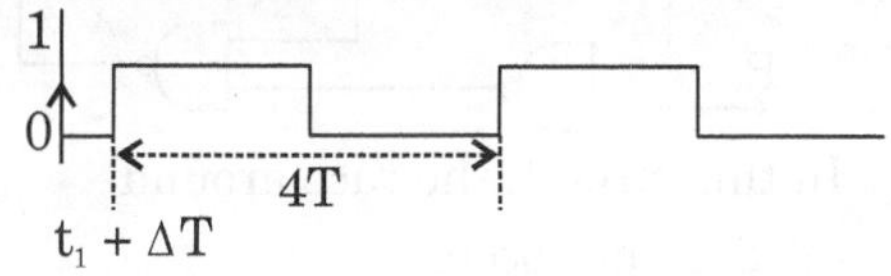

[2008 : 2 Marks]

12. For the circuit shown in the figure, 0 has a transition from 0 to 1 after CLK changes from 1 to 0. Assume gate delays to be negligible

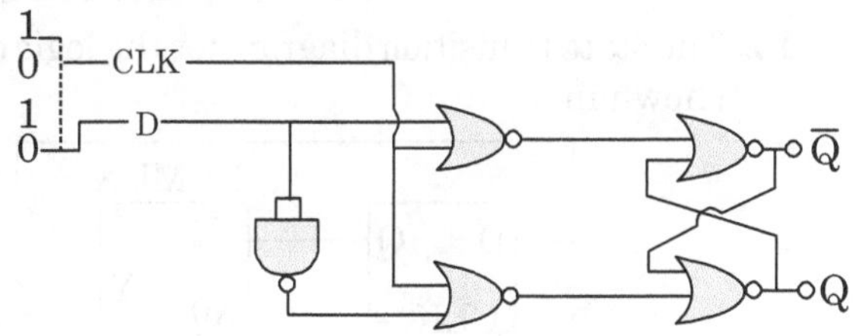

Which of the following statements is true?

(a) Q goes to 1 at the CLK transition and stays at 1.

(b) Q goes to 0 at the CLK transition and stays at 0.

(c) Q goes to 1 at the CLK transition and goes to 0 when D goes to 1.

(d) Q goes to 0 at the CLK transition and goes to 1 when D goes to 1.

[2008 : 2 Marks]

13. Refer to the NAND and NOR latches shown in the figure. The inputs (P_1. P_2) for both the latches are first made (0.1) and then, after a few seconds, made (1,1). The corresponding stable outputs (Q_1, Q_2) are

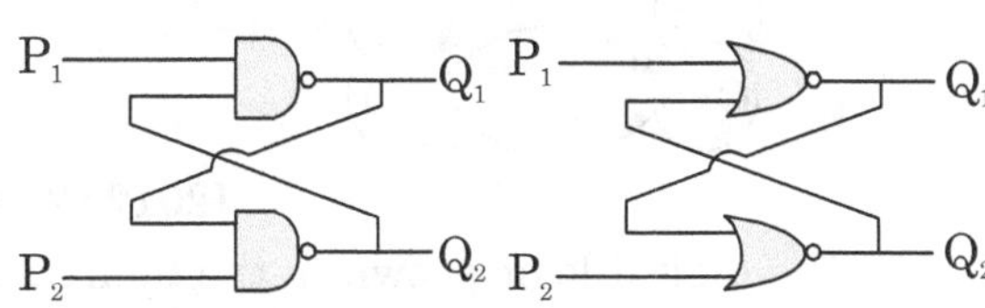

(a) NAND: first (0, 1) then (0, 1) NOR: first (1, 0) then (0, 0)

(b) NAND: first (1, 0) then (1, 0) NOR: first (1, 0) then (1, 0)

(c) NAND: first (1, 0) then (1, 0) NOR: first (1, 0) then (0, 0)

(d) NAND: first (1, 0) then (1, 1) NOR: first (0, 1) then (0, 1)

[2009 : 2 Marks]

14. Consider the given circuit

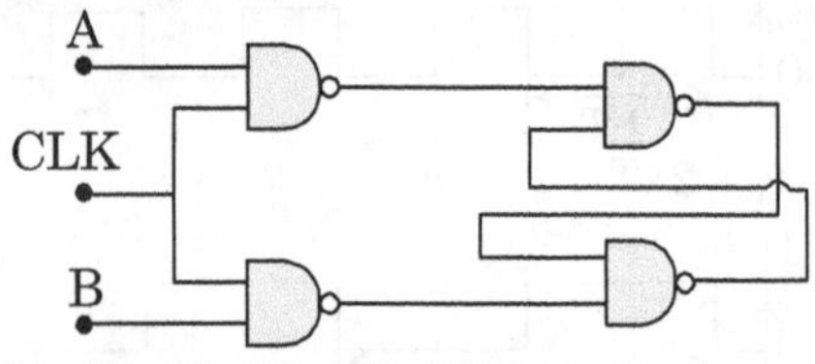

In this circuit, the race around

(a) does not occur.

(b) occurs when CLK = 0.

(c) occurs when CLK = 1 and A = B = 1.

(d) occurs when CLK = 1 and A = B = 0.

[2012 : 1 Mark]

15. The state transition diagram for the logic circuit shown in

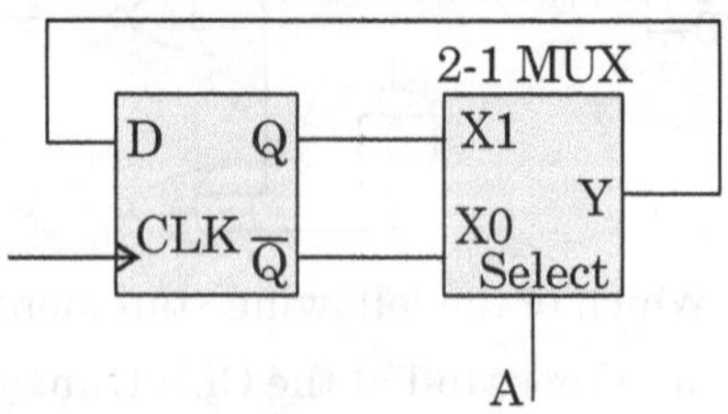

(a)
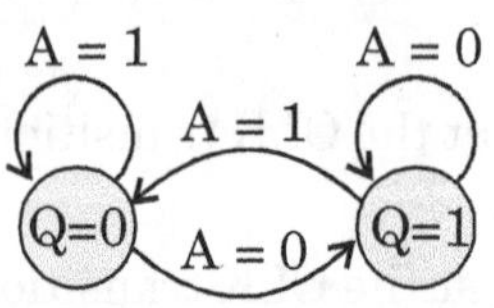

(b)
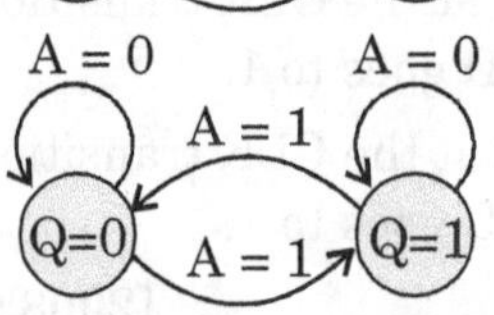

(c)
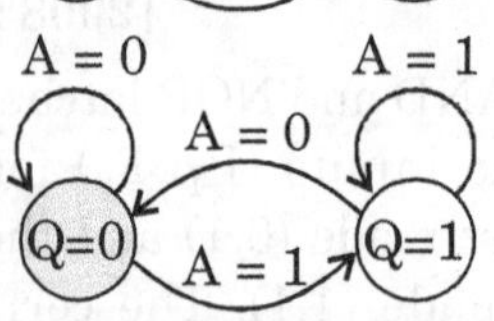

(d)
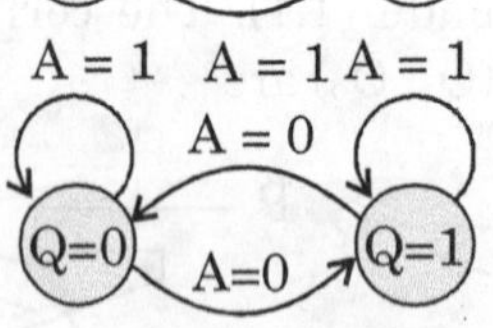

[2012 : 2 Marks]

16. The digital logic shown in the figure satisfies the given state diagram when Q_1 is connected to input A of the XOR gate.

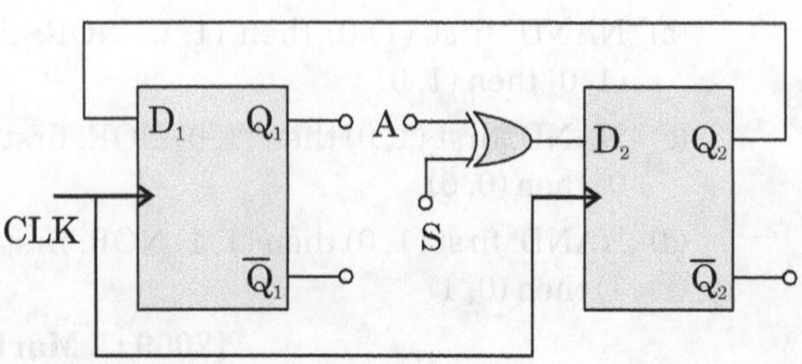

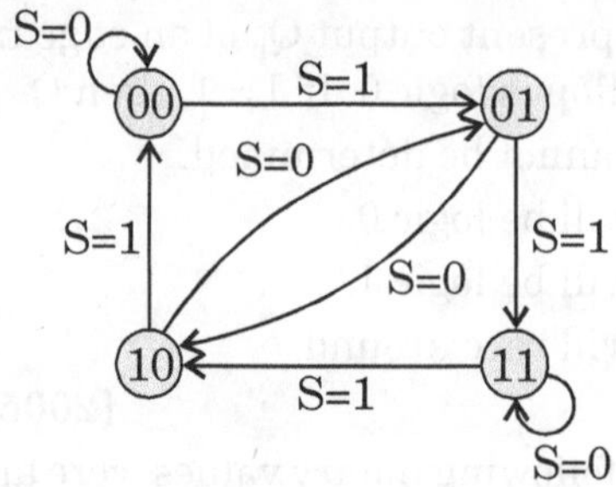

Suppose the XOR gate is replaced by an XNOR gate. Which one of the following options preserves the state diagram?

(a) Input A is connected to $\bar{Q}_2$.

(b) input A is connected to Q_2.

(c) Input A is connected to $\bar{Q}_1$ and S is complemented.

(d) Input A is connected to $\bar{Q}_1$.

[2014 : 2 Marks, Set-1]

17. In the circuit shown, choose the correct timing diagram of the output (Y) from the given waveforms W_1, W_2, W_3 and W_4.

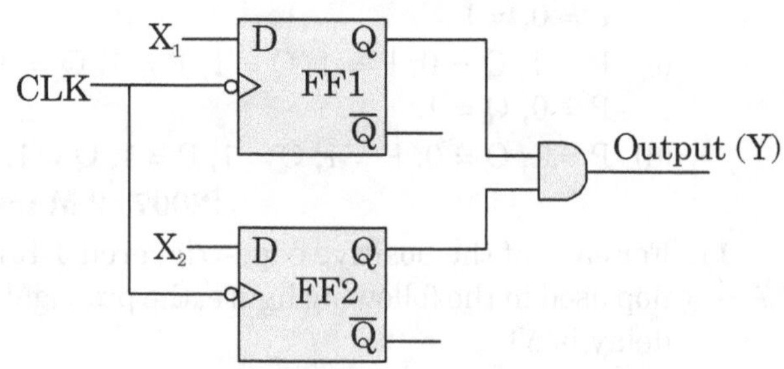

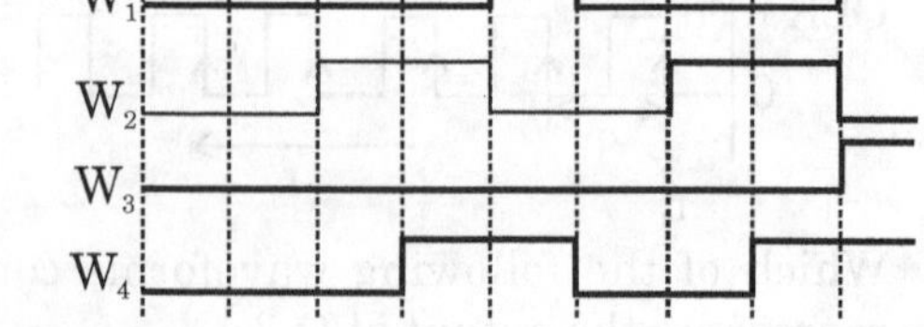

(a) W_1

(b) W_2

(c) W_3

(d) W_4

[2014 : 2 Marks, Set-2]

18. The circuit shown in the figure is a

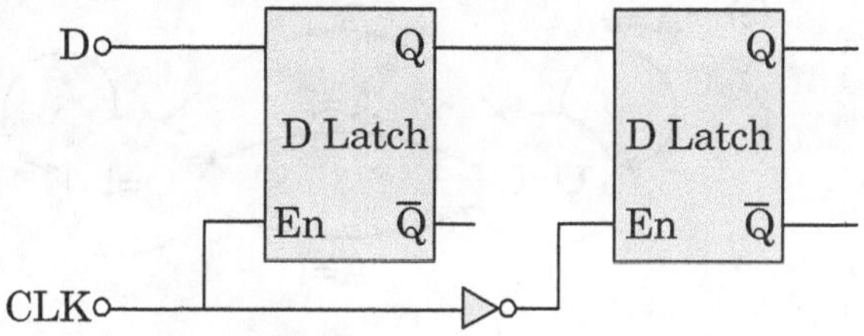

(a) Toggle flip-flop
(b) JK flip-flop
(c) SR Latch
(d) Master-Slave D flip-flop

[2014 : 1 Mark, Set-3]

19. An SR latch is implemented using TTL gates as shown in the figure. The set and reset pulse inputs are provided using the push-button switches. It is observed that the circuit fails to work as desired. The SR latch can be made functional by changing

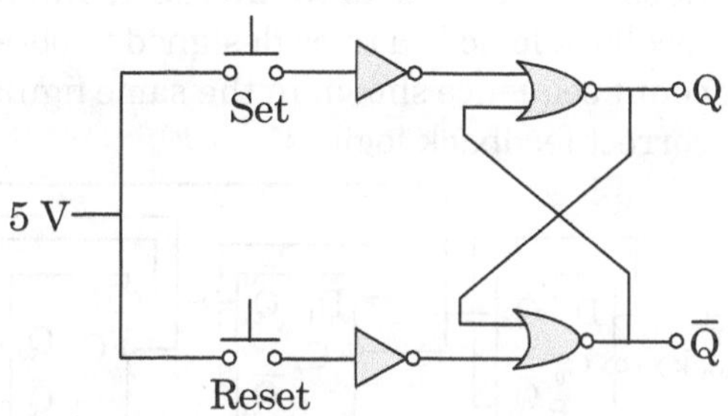

(a) NOR gates to NAND gates
(b) inverters to buffers
(c) NOR gates to NAND gates and inverters to buffers
(d) 5 V to ground

[2015 : 2 Marks, Set-3]

20. Assume that all the digital gates in the circuit shown in the figure are ideal, the resistor R = 10 kΩ and the supply voltage is 5 V. The D flip-flops D_1, D_2, D_3, D_4 and D_5 are initialized with logic values 0,1,0,1 and 0, respectively. The clock has a 30% duty cycle.

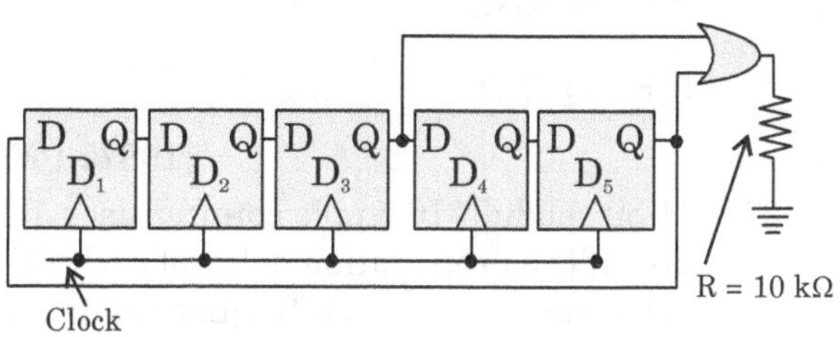

The average power dissipated (in mW) in the resistor R is ______.

[2016 : 1 Mark, Set-2]

21. In the latch circuit shown, the NAND gates have non-zero, but unequal propagation delays. The present input condition is: P= Q = '0'. if the input condition is changed simultaneously to P = Q = '1' , the outputs X and Y are

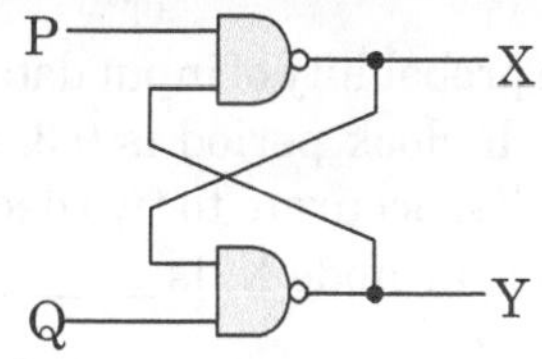

(a) X = '1', Y= 1.
(b) either X = '1', Y = '0' or X = '0', Y = '1'.
(c) either X = '1', Y = '1' or X = '0', Y = '0'.
(d) X = "0', = '0'.

[2017 : 1 Mark, Set-1]

22. Consider the D-Latch shown in the figure, which is transparent when its clock input CK is high and has zero propagation delay. In the figure, the clock signal CLK1 has a 50% duty cycle and CLK2 is a one-fifth period delayed version of CLK1. The duty cycle at the output of the latch in percentage is _____.

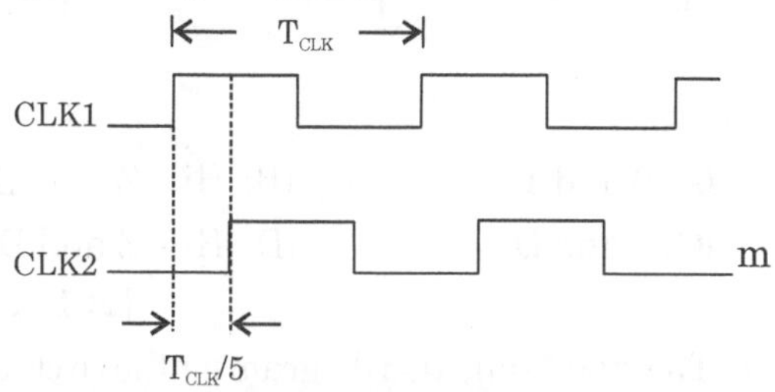

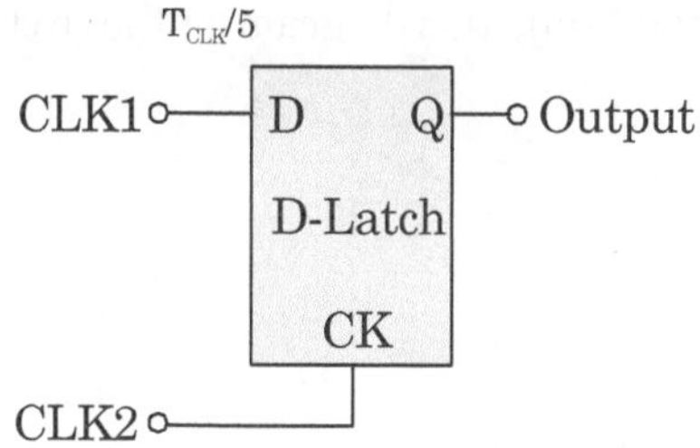

[2017 : 1 Mark, Set-1]

23. In the circuit shown below, a positive edge-triggered D flip-flop is used for sampling input data D_{IN} using clock CK, The XOR gate outputs 3.3 volts for logic HIGH and 0 volts for logic LOW levels. The data bit and clock periods are equal and the value of $\Delta T/T_{CK}$ = 0.15, where the parameters ΔT and T_{CK} are shown in the figure. Assume that the Flip-Flop and the XOR gate are ideal.

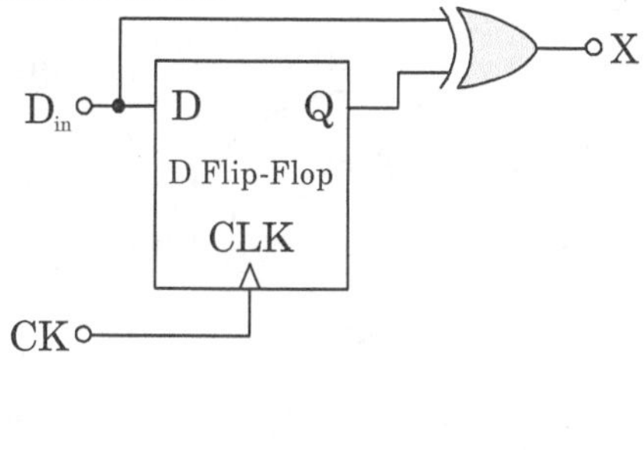

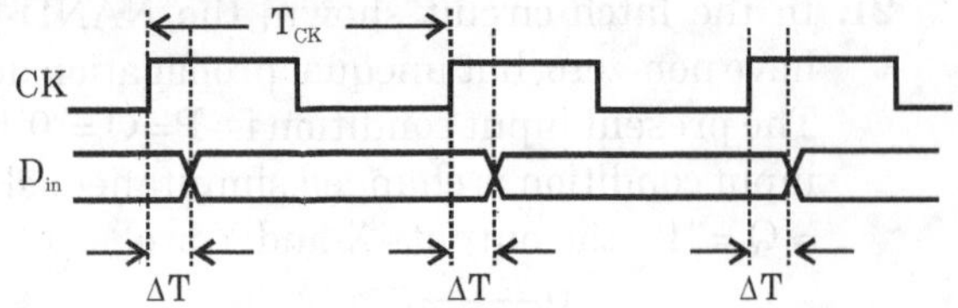

If the probability of input data bit (D_{in}) transition in each clock period is 0.3, the average value (in volts, accurate to two decimal places) of the voltage at node X, is _______ .

[2018 : 2 Marks]

24. In the circuit shown, what are the values of F for EN = 0 and EN = 1, respectively?

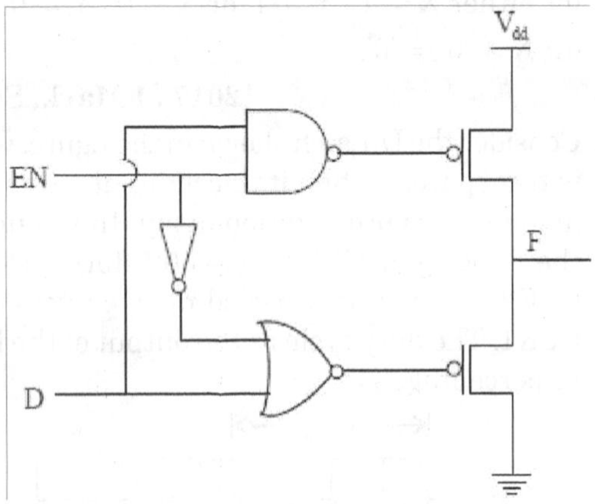

(A) 0 and 1 (B) Hi – Z and $\overline{D}$

(C) 0 and D (D) Hi – Z and D

[2019 : 1 Mark]

25. The state transition diagram for the circuit shown is

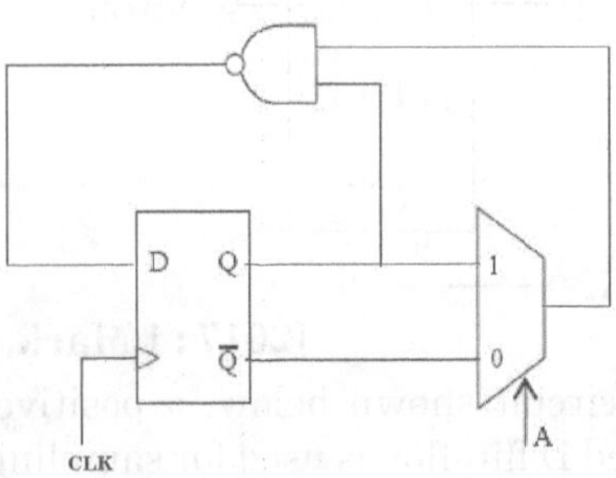

(a)

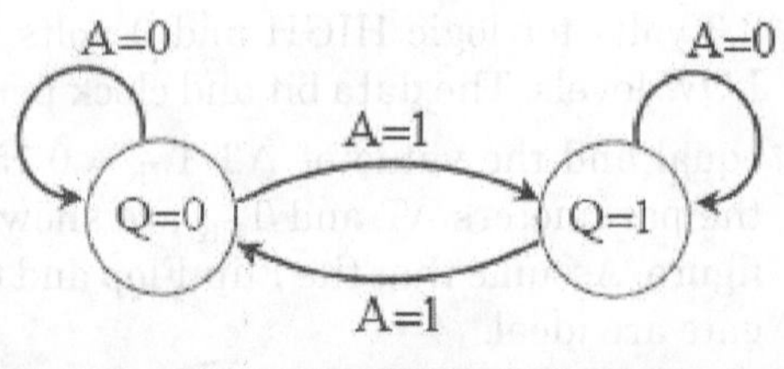

(b)

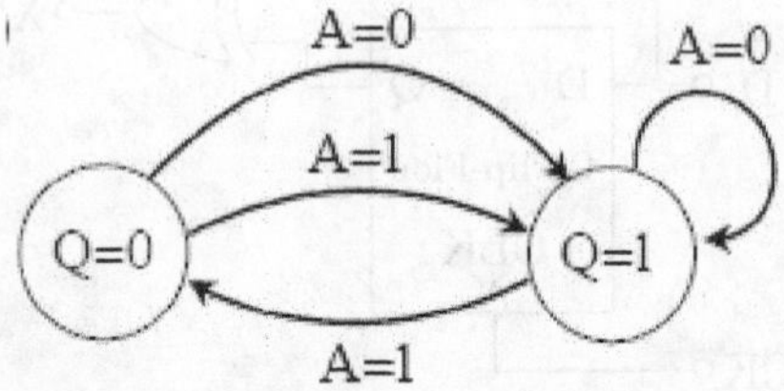

(c)

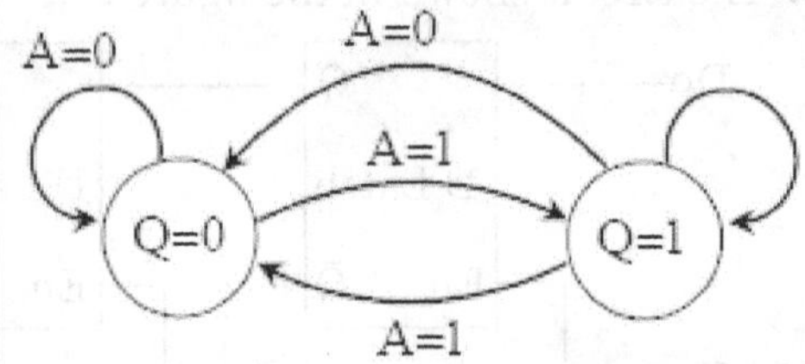

(d)

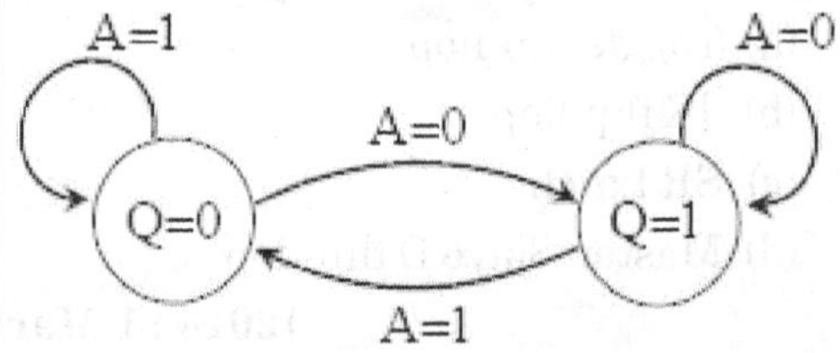

[2019 : 2 Marks]

COUNTERS & SHIFT REGISTERS

26. A ripple counter using negative edge-triggered D-flip flops is shown in figure below. The flip-flops are cleared to '0' at the R input. The feedback logic is a to be designed to obtain the count sequence shown in the same figure. The correct feedback logic is

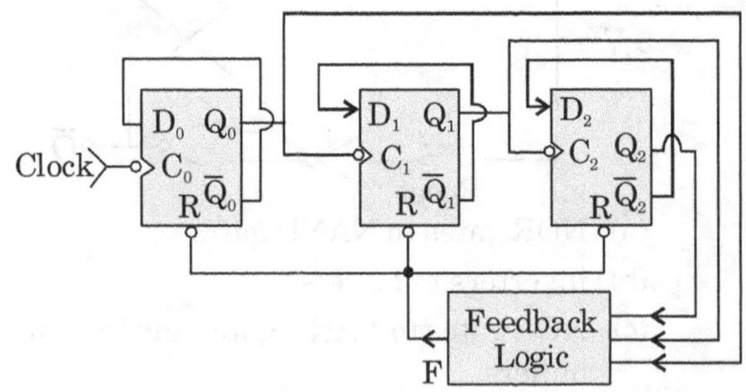

Count sequence in the order of $Q_2\ Q_1\ Q_0$:

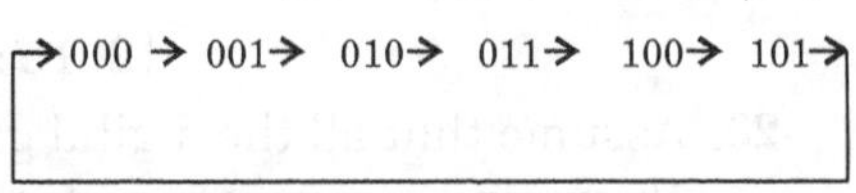

(a) $F = \overline{Q_2 Q_1 \overline{Q}_0}$

(b) $F = Q_2 \overline{Q}_1 Q_0$

(c) $F = \overline{Q}_2 \overline{Q}_1 Q_0$

(d) $F = \overline{Q}_2 \overline{Q}_1 \overline{Q}_0$

[1987 : 1 Mark]

27. A 4-bit modulo-16 ripple counter uses JK flip-flops. If the propagation delay of each FF is 50 ns, the maximum clock frequency that can be used is equal to

(a) 20 MHz (b) 10 MHz

(c) 5 MHz (d) 4 MHz

[1990 : 1 Mark]

28. The initial contents of the 4-bit serial-in-parallel-out, right-shift, Shift Register shown in the figure is 0110. After three clock pulses are applied, the contents of the Shift Register will be

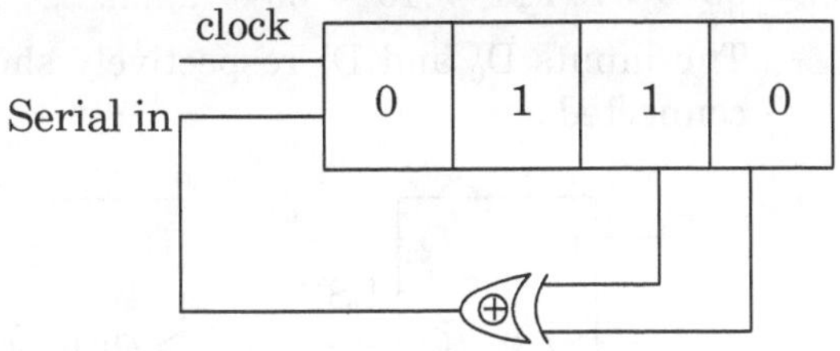

(a) 0000 (h) 0101
(c) 1010 (d) 1111

[1992 : 1 Mark]

29. A pulse train with a frequency of 1 MHz is counted using a modulo-1024 ripple-counter built with J-K flip flops. For proper operation of the counter. The maximum permissible propagation delay per flip flop stage is___n sec.

[1993 : 1 Mark]

30. Synchronous counters are ____ than the ripple counters.

[1994 : 1 Mark]

31. A switch-tail ring counter is made by using a single flip-flop. The resulting circuit is a

(a) SR flip-flop (b) JK flip-flop
(c) D flip-flop (d) T flip-flop

[1995 : 1 Mark]

32. Figure shows a mod-K counter, here K is equal to

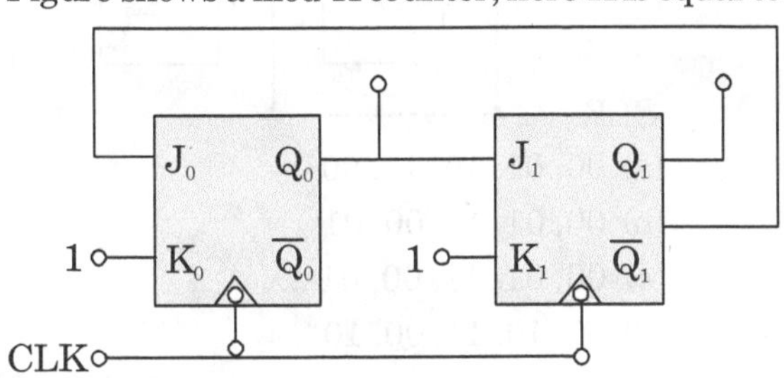

(a) 1 (b) 2
(c) 3 (d) 4

[1998: 1 Mark]

33. In fig. below, A = 1 and 8= 1 The input 8 is now replaced by a sequence 101010...., the outputs X and Y will be

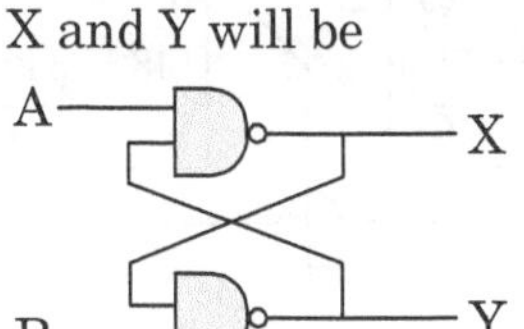

(a) fixed at 0 and 1, respectively.
(b) x = 1010....... while y= 0101 ....
(c) x = 1010 ...... and y = 1010 ....
(d) fixed at 1 and 0, respectively

[1998 : 2 Marks]

34. The ripple counter shown in the figure works as a

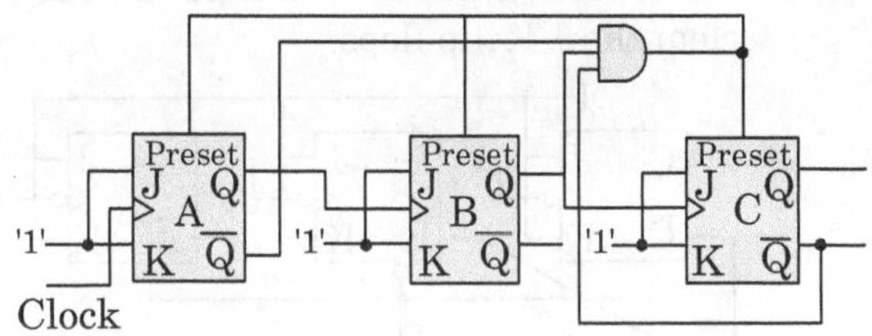

(a) mod-3 up counter
(b) mod-5 up counter
(c) mod-3 down counter
(d) mod-5 down counter **[1999 : 2 Marks]**

35. In the figure, the J and K inputs of all the four flip-flops are made high The frequency of the signal at output Y is

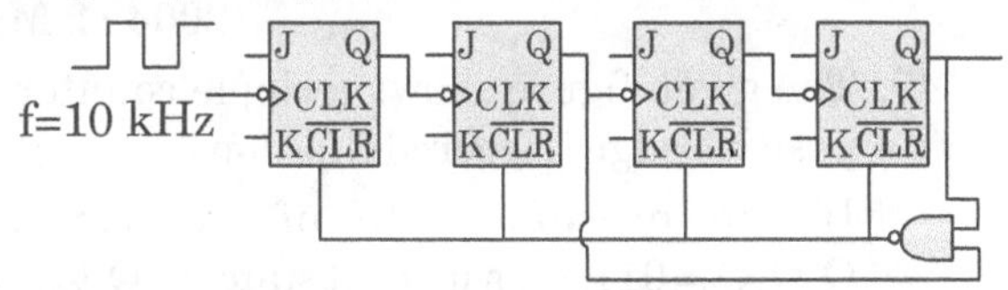

(a) 0.833 kHz (b) 1.0 kHz
(c) 0.91 kHz (d) 0.77 kHz

[2000 : 2 Marks]

36. A 0 to 6 counter consists of 3 flip flops and a combination circuit of 2 input gate(s). The combination circuit consists of

(a) one AND gate
(b) one OR gate
(c) one AND gate and one OR gate
(d) two AND gates **[2003 : 1 Mark]**

37. A 4 bit ripple counter and a 4 bit synchronous counter are made using flip-flops having a propagation delay of 10 ns each. If the worst case delay in the ripple counter and the synchronous counter be R and S respectively, then

(a) R= 10 ns, S = 40 ns
(b) R = 40 ns, S= 10 ns
(c) R =10 ns, S = 30 ns
(d) R = 30 ns, S =10 ns **[2003 : 2 Marks]**

38. A master-slave flip-flop has the characteristic that

(a) change in the input immediately reflected in the output.
(b) change in the output occurs when the state of the master is affected.
(c) change in the output occurs when the state of the slave is affected.
(d) both the master and the slave states are affected at the same time.

[2004 : 1 Mark]

39. In the modulo-6 ripple counter shown in the figure, the output of the 2-input gate is used to clear the J-K flip-flops.

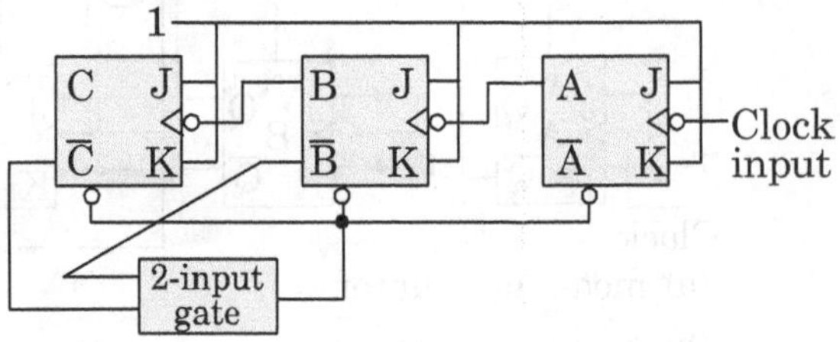

The 2-input gate is

(a) a NAND gate
(b) a NOR gate
(c) an OR gate
(d) an AND gate

[2004 : 2 Marks]

40. The given figure shows a ripple counter using positive edge triggered flip-flops.

If the present state of the counter is $Q_2Q_1Q_0$= 011, then its next state ($Q_2Q_1Q_0$) will be

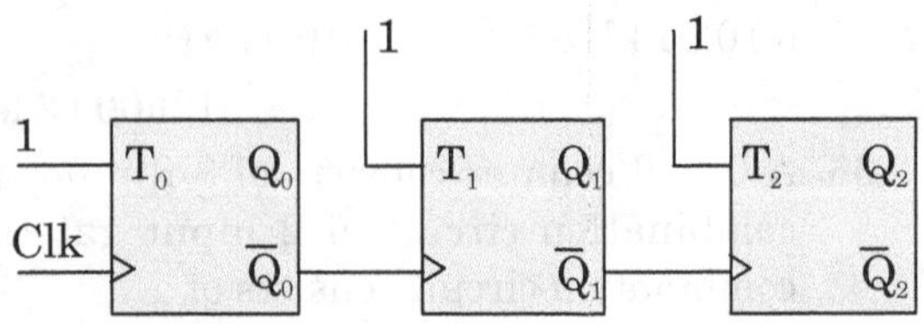

(a) 010 (b) 100
(c) 111 (d) 101

[2005 : 2 Marks]

41. For the circuit shown in the figure below, two 4-bit parallel-in serial-out shift registers loaded with the data shown are used to feed the data to a full adder. Initially, all the flip-flops are in clear state. After applying two clock pulses, the outputs of the full adder should be

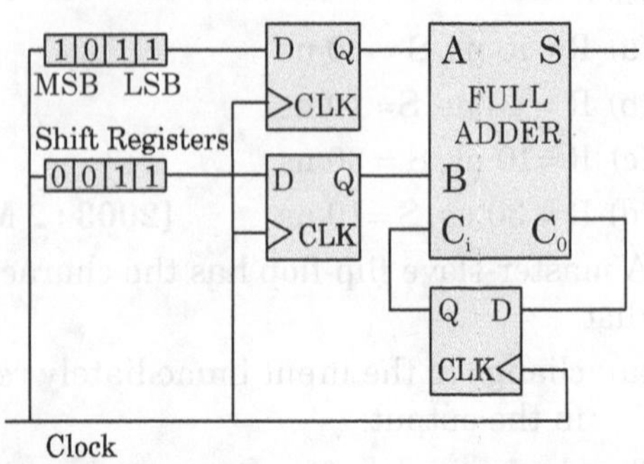

(a) S = 0 C_0 = 0
(b) S = 0 C_0 = 1
(c) S = 1 C_0 = 0
(d) S = 1 C_0 = 1

[2006 : 2 Marks]

42. Two D–flip-flops, as shown below, are to be connected as a synchronous counter that goes through the following Q_1Q_0 sequence

00→ 01→11→ 10→ 00→ ....

The inputs D_0 and D_1 respectively should be connected as

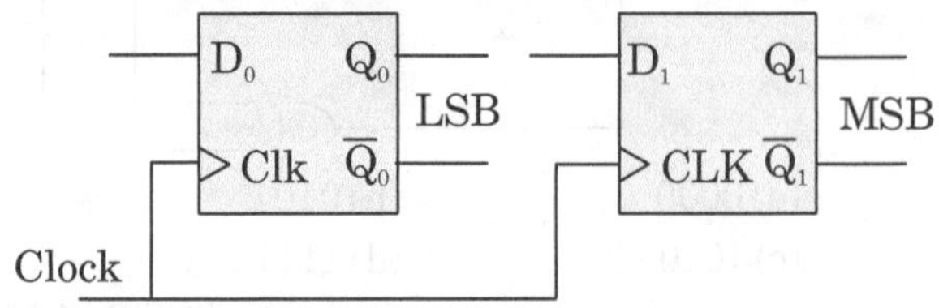

(a) $\bar{Q}_1$ and Q_0

(b) $\bar{Q}_0$ and Q_1

(c) $\bar{Q}_1Q_0$ and $\bar{Q}_1Q_0$

(d) $\bar{Q}_1\bar{Q}_0$ and Q_1Q_0

[2006 : 2 Marks]

43. For the circuit shown, the counter state (Q_1,Q_0) follows the sequence

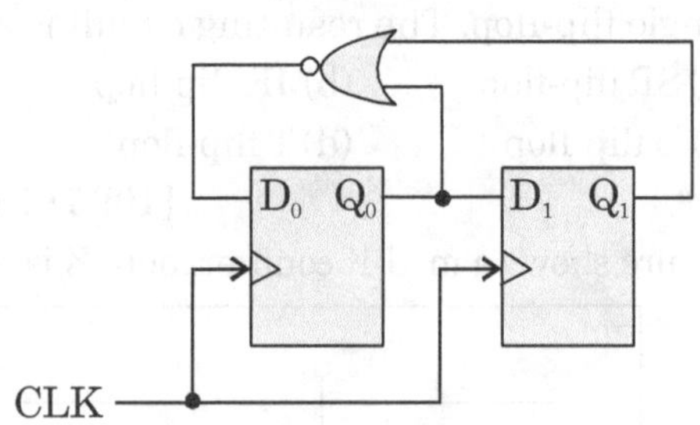

(a) 00, 01, 10, 11, 00 ..
(b) 00, 01. 10, 00, 01
(c) 00, 01, 11, 00, 01 ...
(d) 00. 10, 11, 00, 10

[2007 : 2 Marks]

44. What are the counting states (Q_1, Q_2) for the counter shown in the figure below?

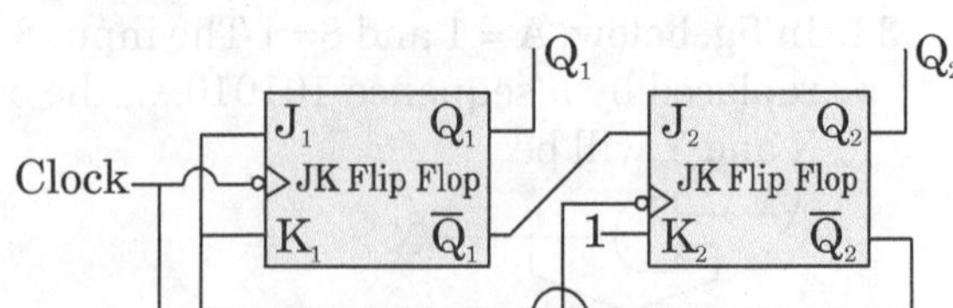

(a) 11, 10, 00, 11, 10, ...
(b) 01, 10, 11, 00, 01,........
(c) 00, 11, 01, 10, 00, ...
(d) 01, 10, 00, 01, 10, ...

[2009 : 2 Marks]

45. Assuming that all flip-flops are in reset conditions initially, the count sequence observed at Q_A in the circuit shown is

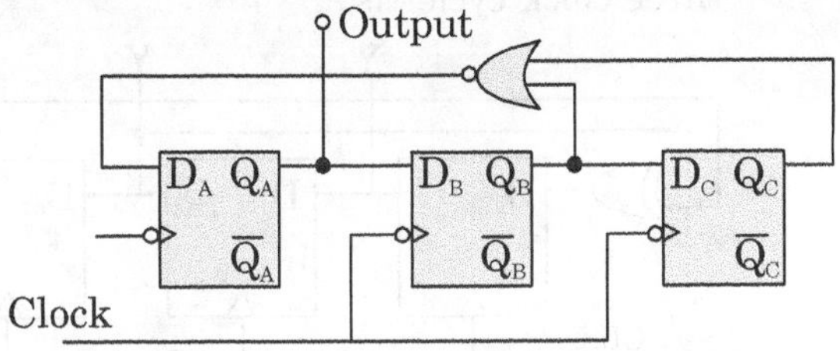

(a) 0010111... (b) 0001011 ..
(c) 0101111.. (d) 0110100...

[2010 : 2 Marks]

46. When the output Y in the circuit below is "1", it implies that data has

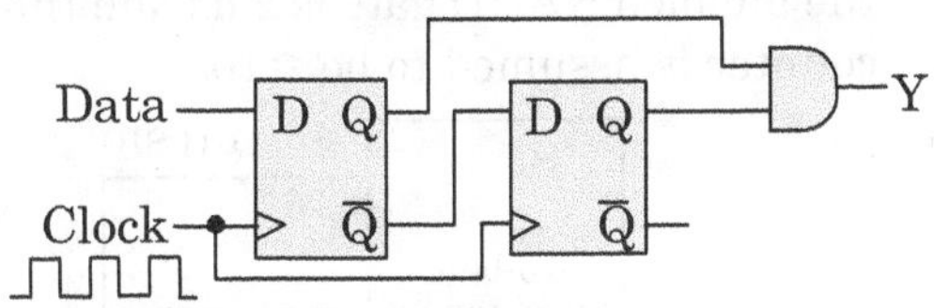

(a) changed from "0" to "1'.
(b) changed from '1" to "0".
(c) changed in either direction.
(d) not changed.

[2011 : 1 Mark]

47. The output of a 3-stage Johnson (twisted-ring) counter is fed to a digital-to-analog (D/A) converter as shown in the figure below. Assume all states of the counter to be unset initially. The waveform which represents the D/A converter output V_0 is

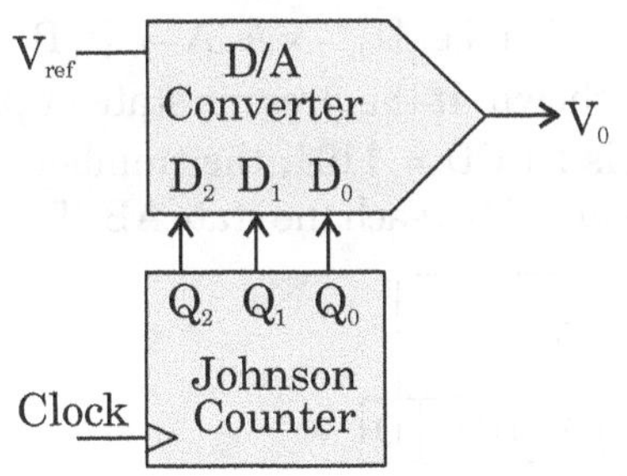

(a)
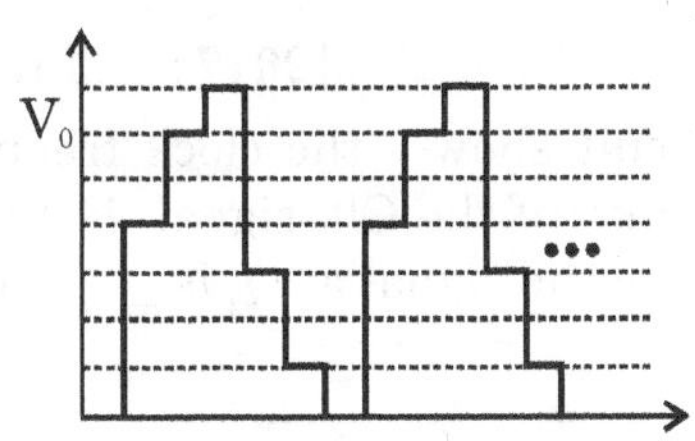

(b)
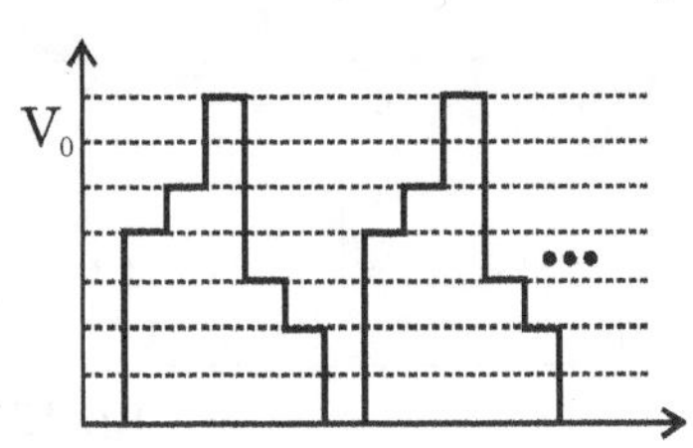

(c)
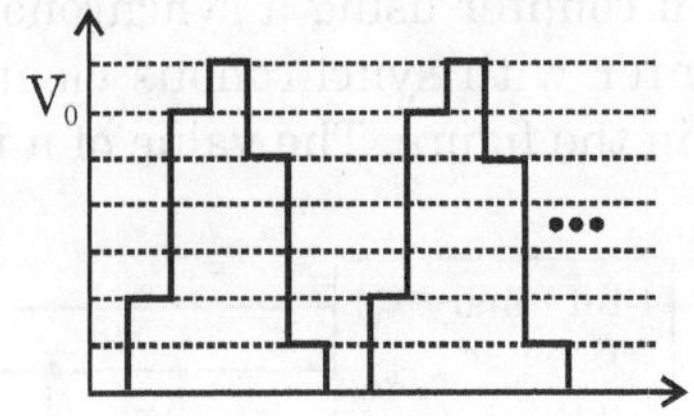

(d)
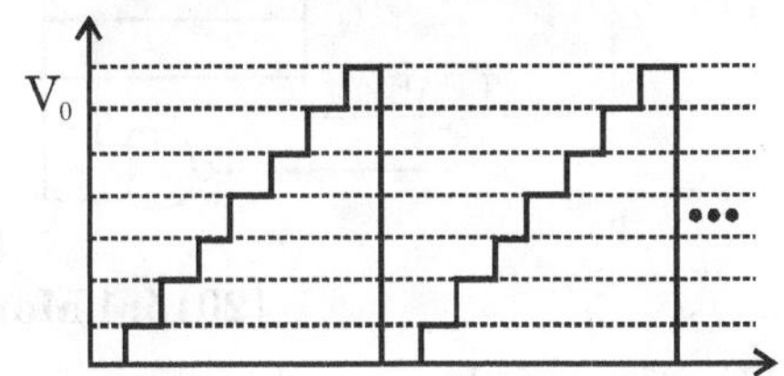

[2011 : 2 Marks]

48. Two D flip-flops are connected as a synchronous counter that goes through the following Q_B Q_A sequence $00 \rightarrow 11 \rightarrow 01 \rightarrow 10 \rightarrow 00 \rightarrow$.......

The connections to the inputs D_A and D_B are

(a) $D_A = Q_B$, $D_B = Q_4$.

(b) $D_A = \overline{Q}_A, D_B = \overline{Q}_B$.

(c) $D_A = (Q_A, \overline{D}_B + \overline{Q}_A Q_B), D_B = Q_A$.

(d) $D_A = (Q_A, D_B + \overline{Q}_A \overline{Q}_B), D_B = \overline{Q}_A$.

[2011 : 2 Marks]

49. Five JK flip-flops are cascaded to form the circuit shown in figure. Clock pulses at a frequency of 1 MHz are applied as shown. The frequency (in kHz) of the waveform at Q_3 is ______ .

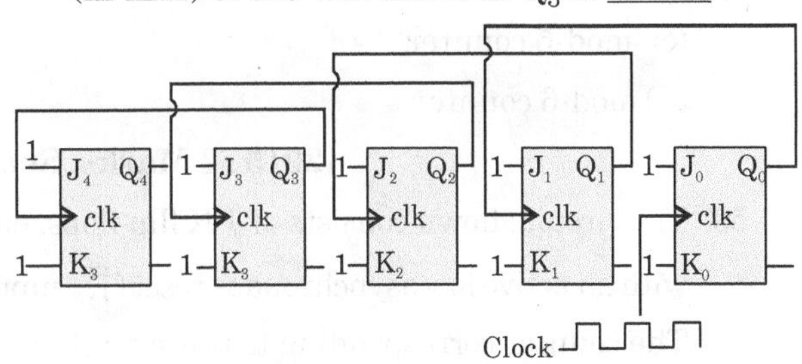

[2014 : 1 Mark, Set-11

50. The outputs of the two flip-flops Q_1, Q_2 in the figure shown are initialized to 0, 0. The sequence generated at Q_1 upon application of clock signal is

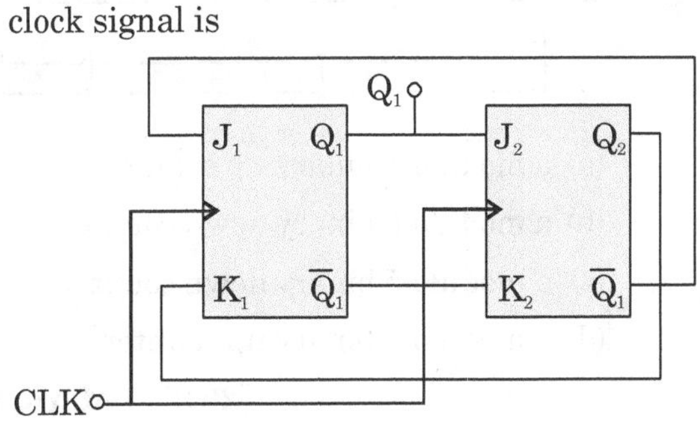

(a) 01110... (b) 01010...
(c) 00110... (d) 01100....

[2014 : 2 Marks, Set-2]

51. A mod-n counter using a synchronous binary up-counter with synchronous clear input is shown in the figure. The value of n is _____.

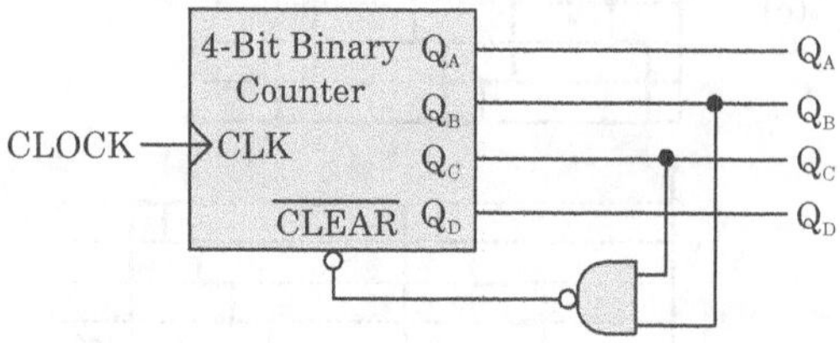

[2015: 1 Mark, Set-2]

52. The figure shows a binary counter with synchronous clear input. With the decoding logic shown, the counter works as a

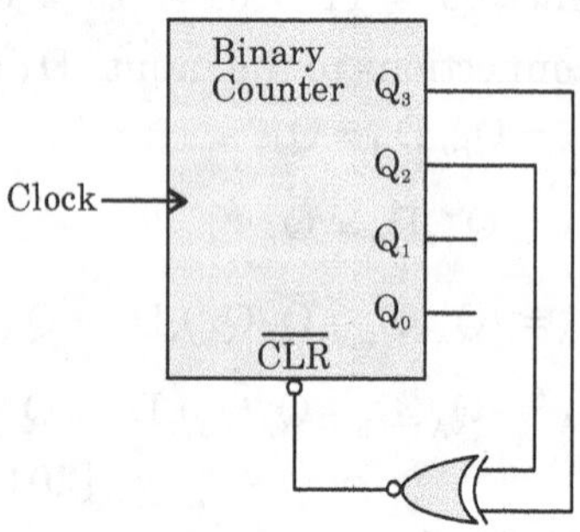

(a) mod-2 counter

(b) mod-4 counter

(c) mod-5 counter

(d) mod-6 counter

[2015 : 2 Marks, Set-2]

53. The circuit shown consists of J-K flip-flops, each with an active low asynchronous reset ($\overline{R}_d$ input). The counter corresponding to this circuit is

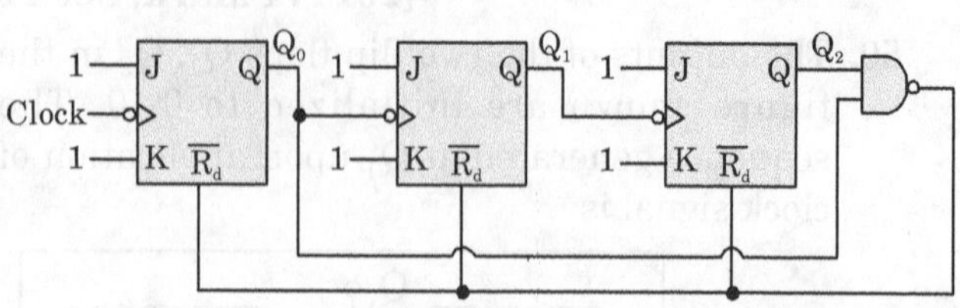

(a) a modulo-5 binary up counter

(b) a modulo-6 binary down counter

(c) a modulo-5 binary down counter

(d) a modulo-6 binary up counter

[2015 : 1 Mark, Set-31]

54. A three bq pseudo random number generator is shown. Initially the value of output Y = $Y_2 Y_1 Y_0$ is set to 111. The value of output Y after three clock cycles is

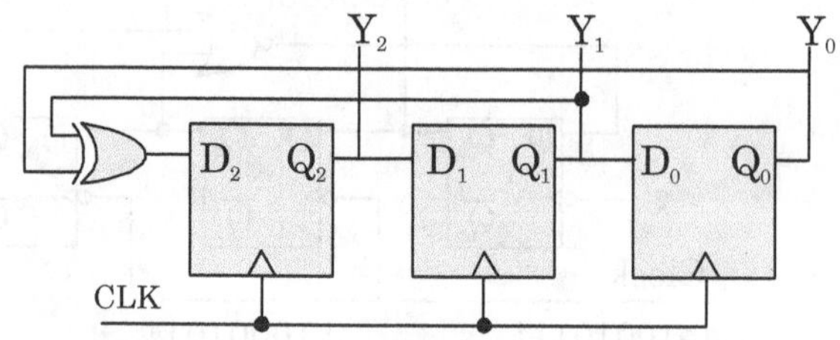

(a) 000 (b) 001

(c) 010 (d) 100

[2015 : 2 Marks, Set-3]

55. For the circuit shown in the figure, the delay of the bubbled NAND gate is 2 ns and that of the counter is assumed to be zero.

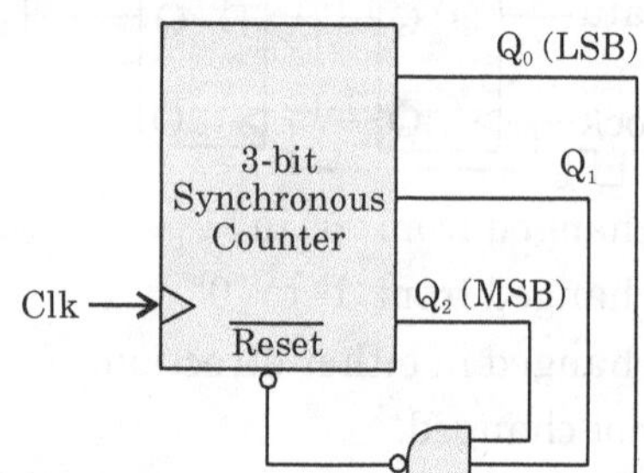

If the clock (Clk) frequency is 1 GHz, then the counter behaves as a

(a) mod-5 counter. (b) mod-6 counter.

(c) mod-7 counter. (d) mod-8 counter.

[2016 : 2 Marks, Set-3]

56. A 4-bit shift register circuit configured for right-shift operation, i.e. $D_{in} \rightarrow A$, $A \rightarrow B$, $B \rightarrow C$, $C \rightarrow D$, is shown. If the present state of the shift register is ABCD = 1101, the number of clock cycles required to reach the state ABCD = 1111 is

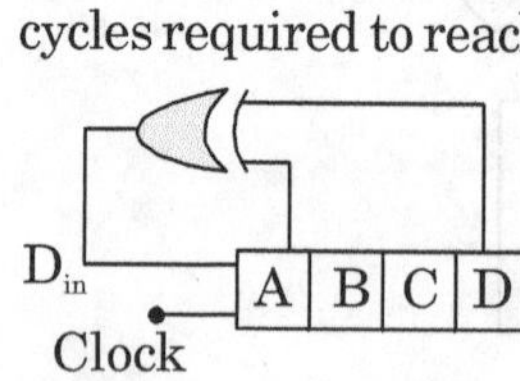

[2017 : 2 Marks, Set-1]

57. In the circuit shown, the clock frequency, i.e., the frequency of the Clk signal, is 12 kHz. The frequency of the signal at Q_2 is _____ kHz.

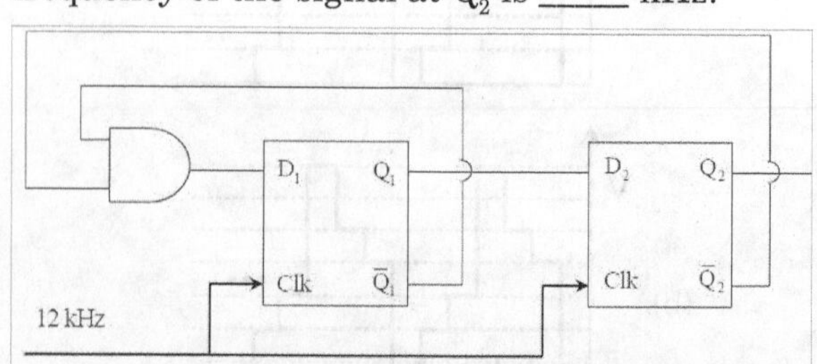

[2019 : 1 Mark]

FINITE STATE MACHINE

58. The state transition diagram for a finite state machine with states A. B and C, and binary inputs X, Yand 7, is shown in the figure.

Which one of the following statements is correct?

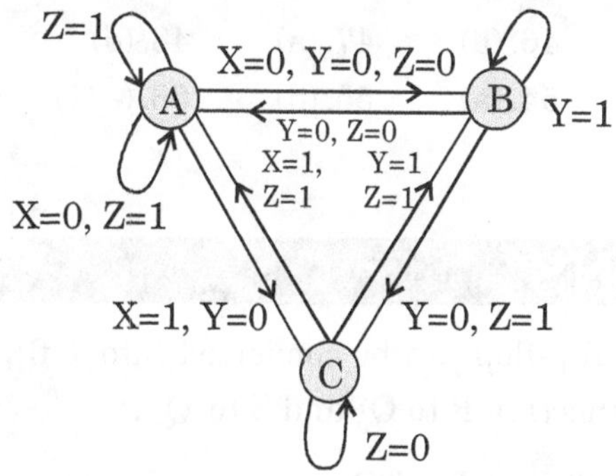

(a) Transitions from State A are ambiguously defined

(b) Transitions from State B are ambiguously defined.

(c) Transitions from State C are ambiguously defined.

(d) All of the state transitions are defined unambiguously.

[2016 : 2 Marks, Set-2]

59. A finite state machine (FSM) is implemented using the D flip-flops A and B, and logic gates, as shown in the figure below. The four possible states of the FSM are Q_AQ_B = 00, 01, 10, and 11.

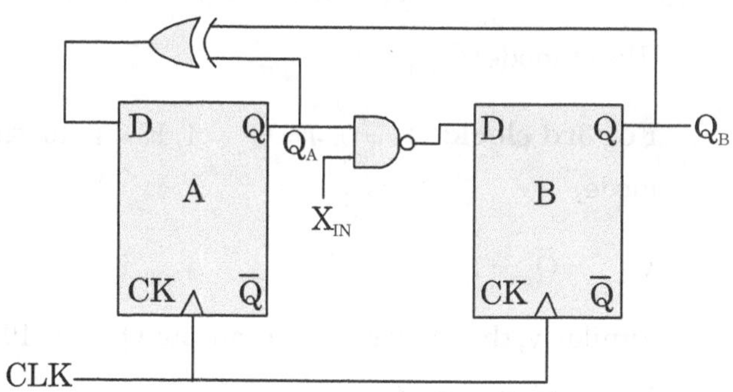

Assume that X_{IN} is held at constant logic level throughout the operation of the FSM. When the FSM is initialized to the Q_AQ_B = 00 and clocked, after a few clock cycles, it starts cycling through

(a) all of the four possible states if X_{IN} = 1.

(b) three of the four possible states if X_{IN} = 0.

(c) only two of the four possible states if X_{IN} =1.

(d) only two of the four possible states if X_{IN}, = 0.

[2017 : 2 Marks, Set-1]

60. The state diagram of a finite state machine (FSM) designed to detect an overlapping sequence of three bits is shown in the figure. The FSM has an input 'In' and an output 'Out'. The initial state of the FSM is S_0,

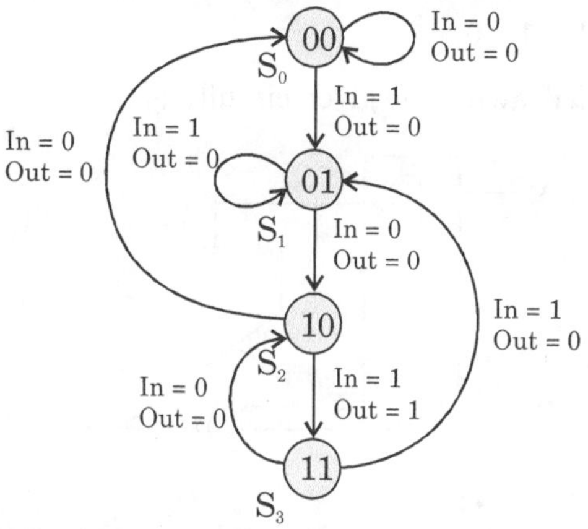

If the input sequence is 10101101001101, starting with the left-most bit, then the number of times 'Out' will be 1 is ______.

[2017 : 2 Marks, Set-2]

61. A traffic signal cycles from GREEN to YELLOW, YELLOW to RED and RED to GREEN, In each cycle, GREEN is turned on for 70 seconds, YELLOW is turned on for 5 seconds and the RED is turned on for 75 seconds. This traffic light has to be implemented using a finite state machine (FSM). The only input to this FSM is a clock of 5 second period. The minimum number of flip-flops required to implement this FSM is ______.

[2018 : 1 Mark]

ANSWERS

1. (b) 2. (c) 3. (*) 4. (c) 5. (d) 6. (d) 7. (c) 8. (b) 9. (c) 10. (c)
11. (b) 12. (c) 13. (c) 14. (a) 15. (d) 16. (d) 17. (c) 18. (d) 19. (d) 20. (1· 5)
21. (b) 22. (30) 23. (0· 8415) 24. (d) 25. (b) 26. (a) 27. (c) 28. (c) 29. (100 ns)
30. (Faster) 31. (d) 32. (d) 33. (a) 34. (d) 35. (b) 36. (b) 37. (b) 38. (b) 39. (c)
40. (b) 41. (d) 42. (a) 43. (c) 44. (a) 45. (d) 46. (a) 47. (a) 48. (d)
49. (62· 4 to 62· 6) 50. (d) 51. (7) 52. (c) 53. (a) 54. (a) 55. (d) 56. (–11) 57. (4)
58. (c) 59. (c) 60. (4) 61. (5)

EXPLANATIONS

1. When $V_i = +3V$, output of comparator 1 is logic 1 and output of comparator 2 is logic 0.

   Now, P = 1 + output of 2nd NOR gate

   $P = \bar{1} = 0.$

2. Redrawing the given circuit,

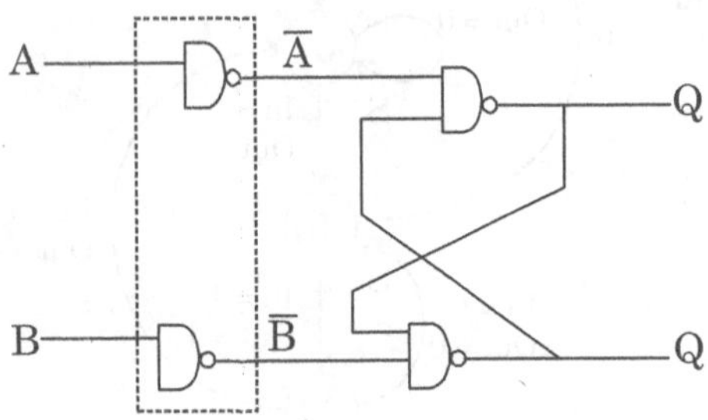

Inverter

Case 1: A = 0, B = 0

$$\left.\begin{aligned} Q_{n+1} &= \overline{1\cdot\bar{Q}_n} = \bar{\bar{Q}}_n QN \\ \bar{Q}_{n+1} &= \overline{1\cdot Q_n} = \bar{Q}_n \end{aligned}\right\}\text{Hold state}$$

Case 2: A = 0, B = 1

$$\left.\begin{aligned} Q_{n+1} &= \overline{1\cdot\bar{Q}_n} = Q_n = Q_n \equiv 0 \\ \bar{Q}_{n+1} &= \overline{0\cdot Q_n} = \bar{0} = 1 \end{aligned}\right\}\text{Reset state}$$

Case 3: A = 1, B = 0

$$\left.\begin{aligned} Q_{n+1} &= \overline{0\cdot\bar{Q}_n} = \bar{0} = 1 \\ \bar{Q}_{n+1} &= \overline{1\cdot Q_n} = \bar{Q}_n = 0 \end{aligned}\right\}\text{Set state}$$

Case 4: A = 1, B = 1

$$\left.\begin{aligned} Q_{n+1} &= \overline{0.\bar{Q}_n} = \bar{0} = 1 \\ \bar{Q}_{n+1} &= \overline{0.Q_n} = \bar{0} = 1 \end{aligned}\right\}\text{Invalid stats}$$

⇒ Hence, it is a RS latch.

3. SR flip-flop can be converted into T-flip flop by connecting R to Q_n and S to $\bar{Q}_n$.

   $S = T\bar{Q}_n, \quad R = TQ_n$

4. An R-S latch is a one-bit memory element. If does not have any input clock signal.

5. Given: $J = \bar{Q}, \; K = 1.$

   Flip-flop is initially cleared, i.e; Q = 0

   $\Rightarrow \bar{Q} = 1.$

   For 1st clock : $J = \bar{Q} = 1, \; K = 1$ (Toggle mode)

   $\bar{Q}_{n+1} = \bar{Q}_n = 1$

   For 2nd clock: $Q_n = 1, \; \bar{Q}_n = 0, K = 1, \; J = 0.$

   (Reset mode) $Q_{n+1} = 0$

   For 3rd clock: $Q_n = 0, J = \bar{Q}_n = 1, K = 1$ (to ggle mode)

   $Q_{n+1} = \bar{Q}_n = 1.$

   Similarly, the sequence of the output Q is 010101.

6. From given circuit,

   $D = \bar{X}Z + Y\bar{Z}$

X	Y	Z_n	D	Z_{n+1}
0	0	0	0	0
0	0	1	1	1
0	1	0	1	1
0	1	1	1	1
1	0	0	0	0
1	0	1	0	0
1	1	0	1	1
1	1	1	0	0

Truth table for X – Y

X	Y	Z_{n+1}
0	0	Z_n
0	1	1
1	0	0
1	1	Z_n

Comparing with J-K truth-table,

We have X = K, Y = J

J	K	Q_{n+1}
0	0	Q_n
0	1	0
1	0	1
1	1	$\bar{Q}_n$

7. It is given that clock is positive edge triggered in option (a) and (c).

Since, D flip-flop is transparent flip-flop means its input is transferred at output as it is, when clock is given.

In option (a), Since, $Q_1 = Q_2 = 0$,

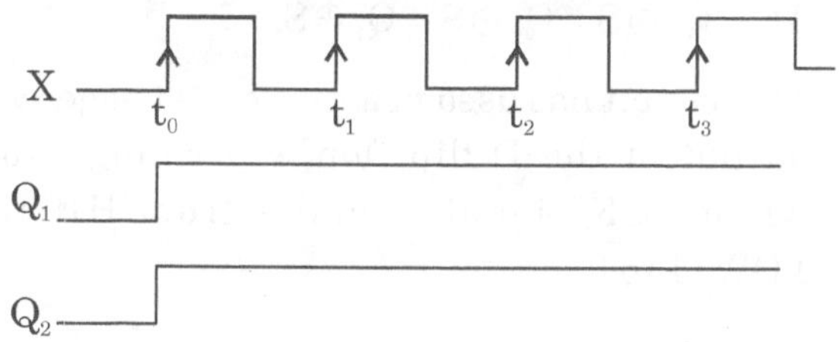

For option (c),

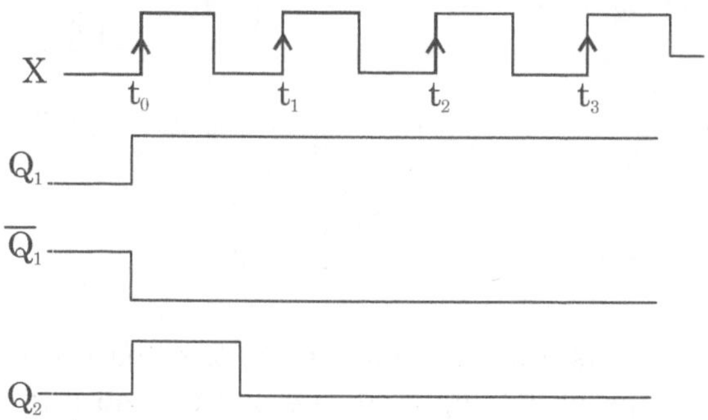

Hence, option (c) is correct.

8. Change in *o/p* occurs when the state of the master is affected

∵ State of slave is depend on state of master.

Hence, the output is also depend on master.

9. Given: For J-K flip-flop, $Q_n = 0$, J = 1.

If K = 0, $Q_{n+1} = 1$ (set state)

If K = 1, $Q_{n+1} = 1$ (toggle state)

10. 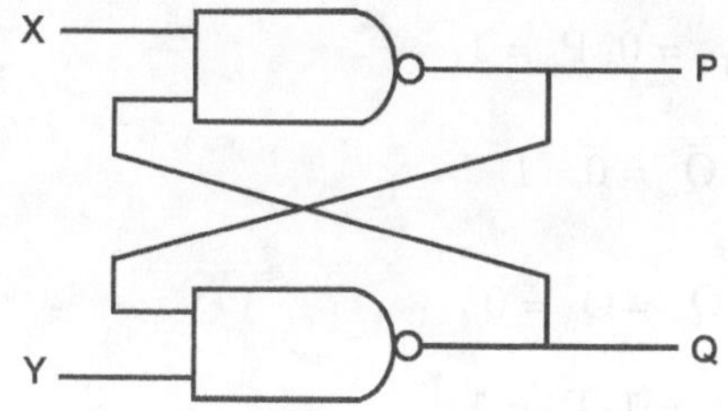

$$P_{t+1} = \overline{X.Q_t} = \bar{X} + \bar{Q}_t$$

$$Q_{t+1} = \overline{Y.P_t} = \bar{Y} + \bar{P}_t$$

X	Y	P_t	Q_t	P_{t+1}	Q_{t+1}
0	1	d	d	1	0
0	0	1	0	1	1
1	1	1	1	0	1
(Not stable)				1	0

Last condition is not stable, it continuously changes its state. Because changes of output again change its states.

11. For this logic circuit, in case of n flip-flops, time period of last output waveform = 2^n. T.

Where, T = Time period for clock pulse

ΔT = propagation delay of one flip-flop

n·DT = Delay time.

Time period of waveform of output at $Q_1 = 2 \times 2T$ = 4T

Time delay at output, $Q_1 = 2 \cdot \Delta T$.

12. 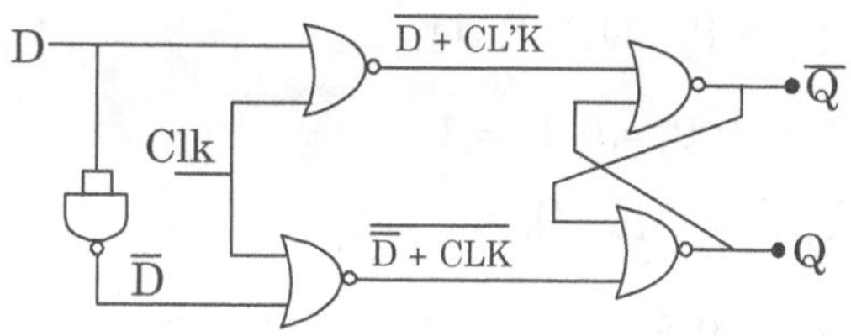

$$\Rightarrow \bar{Q}_{n+1} = \overline{(\overline{D+CLK} + Q_n)} = (D + CLK)\cdot \bar{Q}_n$$

$$\Rightarrow Q_{n+1} = \overline{(\overline{\bar{D}+CLK} + \bar{Q}_n)} = (\bar{D} + CLK)\cdot Q_n$$

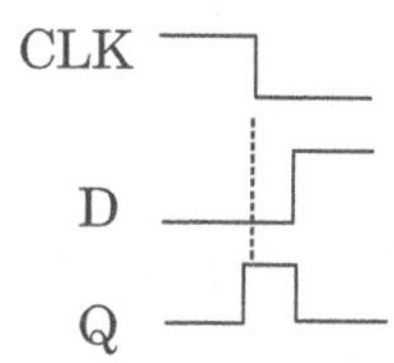

13. For NAND gate Latch:

$$Q_{n+1} = \overline{P_1 \cdot Q_n} \text{ and } \bar{Q}_{n+1} = \overline{P_2 \cdot Q_n}$$

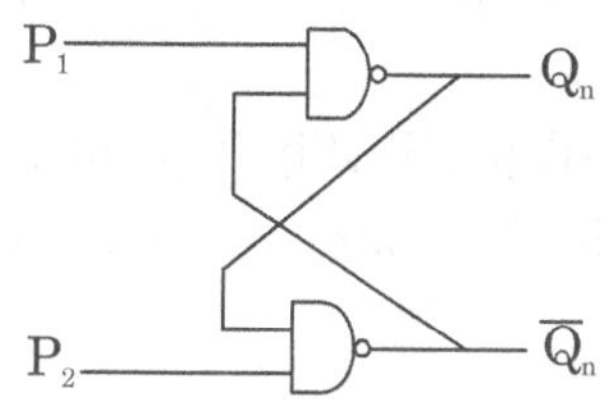

Case 1 : $_{P1}$ = 0, P_2 = 1.

$$Q_{n+1} = \overline{0\cdot\bar{Q}_n} = \bar{0} = 1$$

$$\overline{Q_{n+1}} + \overline{1\cdot Q_n} = \bar{Q}_n = 0.$$

Case 2: P_1 = 1, P_2 = 1.

$$\left.\begin{array}{l} Q_{n+1} = 1\cdot Q_n = Q_n = Q_n = 1 \\ \bar{Q}_{n+1} = \overline{1\cdot Q_n} = \bar{Q}_n = \bar{Q}_n = 1 \end{array}\right\} \text{memory state}$$

P_1	P_2	Q_1	Q_2
0	1	1	0
1	1	1	0

For NOR gate latch:

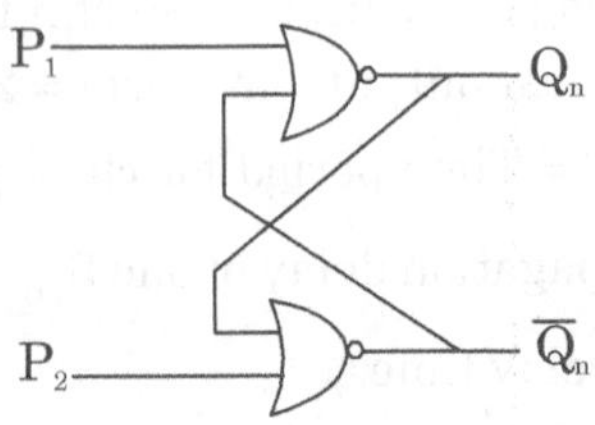

$$Q_{n+1} = \overline{P_1 + \bar{Q}_n} = \bar{P}_1 \cdot Q_n$$

$$\bar{Q}_{n+1} = \overline{P_2 + Q_n} = \bar{P}_2 \cdot \bar{Q}_n$$

Case 1: P_1 = 0, P_2 = 1.

$$Q_{n+1} = 1 \cdot Q_n = Q_n = 0$$

$$\bar{Q}_{n+1} = 0 \cdot \bar{Q}_n = 1$$

Case 2: P_1 = 1, P_2 = 1

$$Q_{n+1} = 0 \cdot Q_n = 0$$

$$\bar{Q}_{n+1} = 0 \cdot \bar{Q}_n = 0.$$

Clock	J_1	K_1	J_2	K_2	Q_1	Q_2
0	1	1	1	1	0	0
1	1	1	1	1	1	1
2	0	0	0	1	1	0
3	1	1	0	1	0	0

14. Given flip-flop is S-R flip-flop with A = S and B = R. In S-R flip-flop, race around condition does not occur.

15. When A = 1, Q will be selected by MUX and feedback to D-FF which gives output Q again. It is hold state.

When A = 0, $\bar{Q}$ will be selected by MUX and feedback to D-FF and output will be inverted. It is toggle state.
So, option (d) is correct.

16. Initially, when EX-OR gate is connected to the input of FF2, i.e., input of D_2. D_2

$$= A \oplus S = Q_1 \oplus S.$$

Now, Ex – OR is replaced by Ex–NOR, keeping state diagram to be unchanged, it means the input to D_2 should not get changed. Now,

$$D_2 = A \odot S.$$

$$\therefore \overline{A \oplus B} = \bar{A} \oplus B = A \odot B$$

We require output D_2 to be as $_{D2}$ = $A \oplus S$; A should be connected to $\bar{Q}_1$ as

$$D_2 = \bar{Q}_1 \odot S = \overline{Q_1 \odot S} = Q_1 \oplus S.$$

17. This circuit has used negative edge triggered, so output of the D-flip flop will changed only when CLK signal is going from HIGH to LOW (1 to 0)

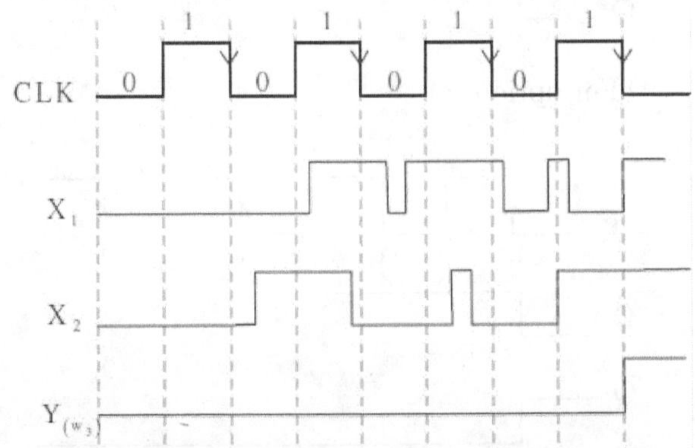

This is a synchronous circuit, so both the flip flops will trigger at the same time and will respond on falling edge of the Clock. So, the correct output (Y) waveform is associated to w_3 waveform.

18. Latches are used to construct Flip-Flop. Latches are level triggered, so if you use two latches in cascaded with inverted clock, then one latch will behave as master and another latch which is having inverted clock will be used as a slave and combined it will behave as a flip-flop. So given circuit is implementing Master-Slave D flip-flop

19. For TTL gate, open end acts as logic '1'.

If we connect SV battery to ground then pressing the switches allow logic '0', white opening the switches allow logic '1', In case of 5V battery both opening and closing of switches allow logic '1' and hence circiut can't act as SR latch.

20.

CLK	Q_1	Q_2	Q_3	Q_4	Q_5	$Y = Q_3 + Q_5$
0	0	1	0	1	0	0
1	0	0	1	0	1	1
2	1	0	0	1	0	0
3	0	1	0	0	1	1
4	1	0	1	0	0	1
5	0	1	0	1	0	0

The waveform of the gate output

$$Y = Q_3 + Q_5$$

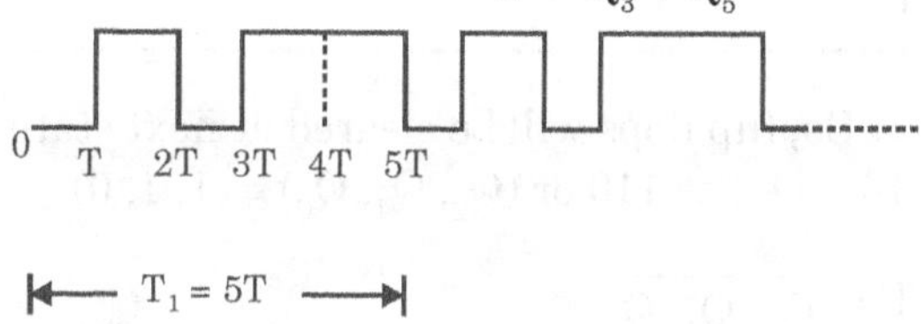

Average power dissipated

$$P = \frac{V^2}{R} \times \frac{T_{ON}}{T}$$

$$= \frac{5^2}{10} \times \frac{3T}{5T}$$

$$= 1.5 \text{ mW}$$

21.

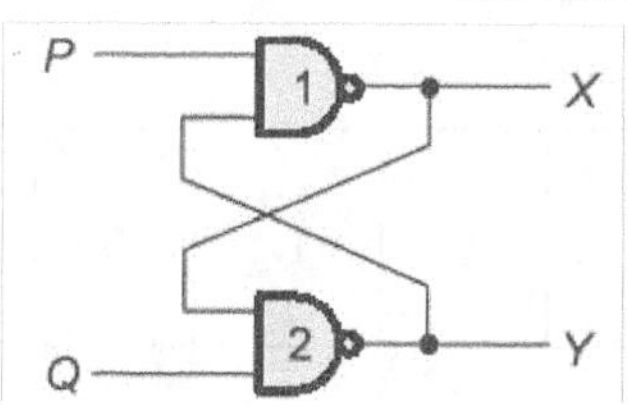

Present input condition : P = Q = 0

⇒ Corresponding outputs are X = Y = 1

When input condition is changed to

P = Q = 1 from P = Q = 0:

Possibility - 1 :

Let gate-2 is faster than gate-1, then the possible outputs are X = 1, Y = 0

Possibility - 2 :

Let gate-1 is faster than gate-2, then the possible outputs are X = 0, Y = 1

Hence, option (*b*) is correct.

22.

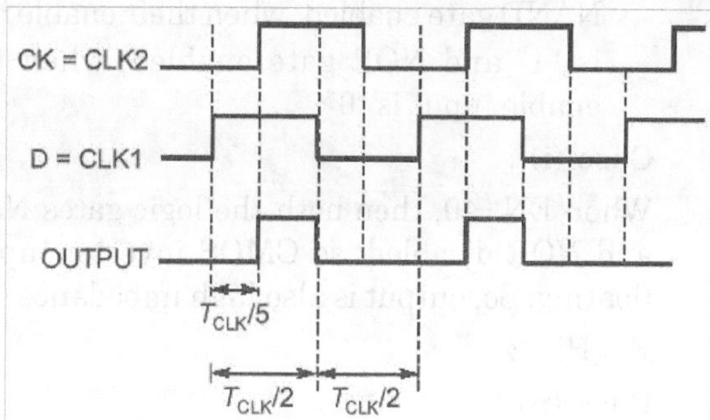

Duty cycle at the output

$$= \left[\frac{\frac{T_{CLK}}{2} - \frac{T_{CLK}}{5}}{T_{CLK}} \times 100\right] \%$$

$$= \left(\frac{3}{10} \times 100\right)\% = 30\%$$

23.

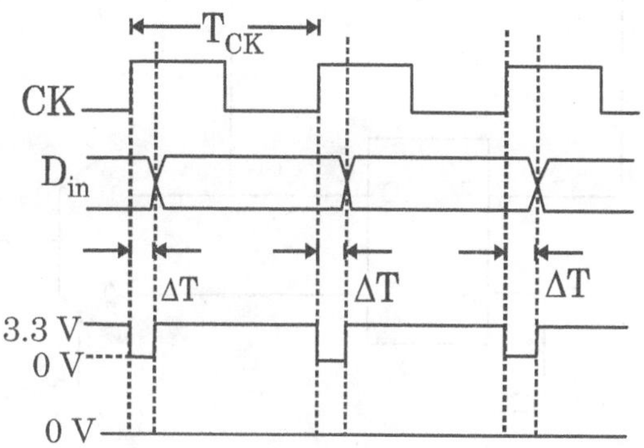

When input is changed during clock period (with probability 0.3)

When input is not changed during clock period (with probability 0.7)

$$V_{X(avg)} = \left[0.3 \times 3.3\left(1 - \frac{\Delta T}{T_{CK}}\right)\right] + [0.7 \times 0]\text{V}$$

$$\left(\because \frac{\Delta T}{T_{CK}} = 0.15\text{V}\right)$$

$$= 0.3 \times 3.3 \times (1 - 0.15) \text{ V}$$

$$= (0.3 \times 3.3 \times 0.85) \text{ V} = 0.8415 \text{ V}$$

24.

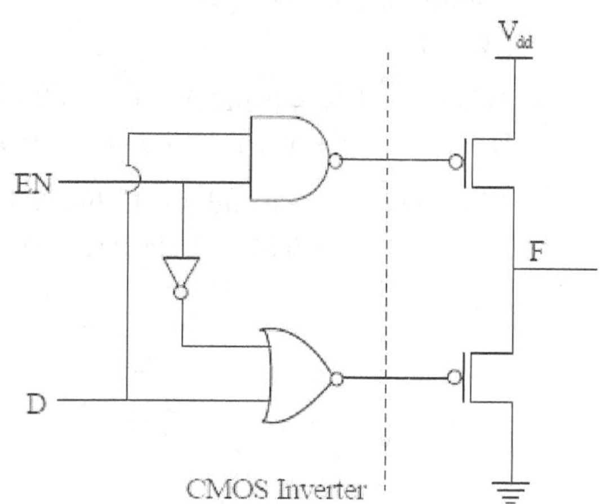

→ NAND gate enabled, when their enable input is "1" and NOR gate enabled, when their enable input is "0".

Case (i) :

When EN = 0, then both the logic gates NAND and NOR disabled, so CMOS inverter input is floating. So, output is also high impedance state.

F = Hi – z

Case (ii) :

When EN = 1, then both the logic gates NAND and NOR are enabled with output $\overline{D}$ that is input of CMOS inverter.

So, $F = \overline{\overline{D}} = D$

25. The given circuit is

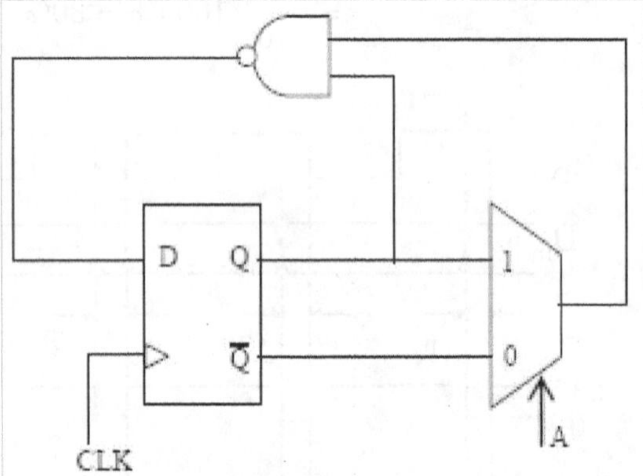

For Q = 0, state diagram

When $Q = 0, \overline{Q} = 1$ and A = 0, then $Y = \overline{Q} = 1$,

so $D = \overline{QY} = \overline{0.1} = \overline{0} = 1$ after 1 clock $Q^+ = 1$

When $Q = 0, \overline{Q} = 1$ and A = 1, then Y = Q = 1,

so $D = \overline{QY} = \overline{0.0} = 1$ after 1 clock $Q^+ = 1$

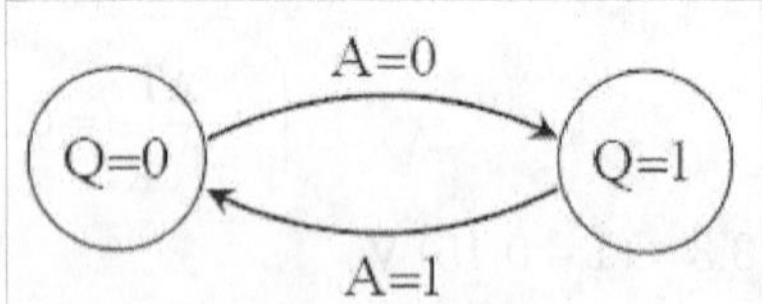

[State Diagram for Q = 0]

For Q = 1

→ When $Q = 1, \overline{Q} = 0$ and A = 0, then $Y = \overline{Q} = 0$,

so $D = \overline{QY} = \overline{1.0} = \overline{0} = 1$ after 1 clock $Q^+ = 1$

→ When $Q = 1, \overline{Q} = 0$ and A = 1, then Y = Q = 1,

so $D = \overline{QY} = \overline{1.1} = 0$ after 1 clock $Q^+ = 0$

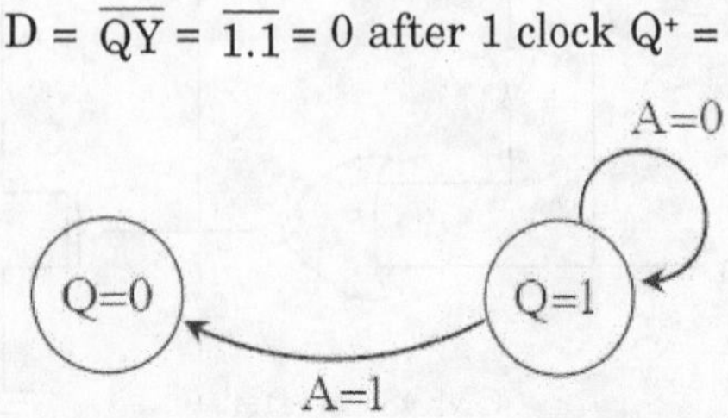

[State Diagram for Q = 1]

By combining both we can draw a single state diagram

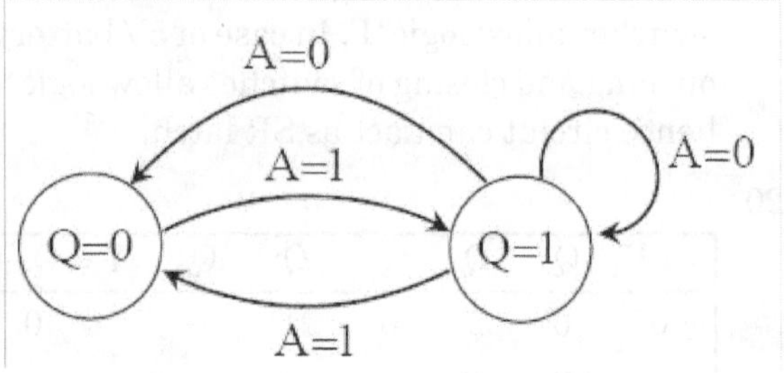

[Combined State Diagram]

26. Inputs to the feedback logic are Q_3, Q_1 and Q_0.

From logic circuit, $D_0 = \overline{Q}_0, D_1 = \overline{Q}_1, D_2 = \overline{Q}_2$.

Count sequence in the order of Q_2, Q_1, Q_0 is

→000→001→010→011→100→101 (back to 000)

⇒ So, flip flops will be cleared at next state after 101, i.e., at 110 or $(Q_2, Q_1, Q_0) \equiv (1, 1, 0)$

$F = \overline{Q_2 \cdot Q_1 \cdot \overline{Q}_0}$

27. Given: no. of bits, n = 4, modulus = ^{2n} = 16.

For ripple counter, max clock frequency

$$f = \frac{1}{n \cdot t_{pd_{FF}}} = \frac{1}{4 \times 50 \times 10^{-9}} = 5 \text{ MHz.}$$

28.

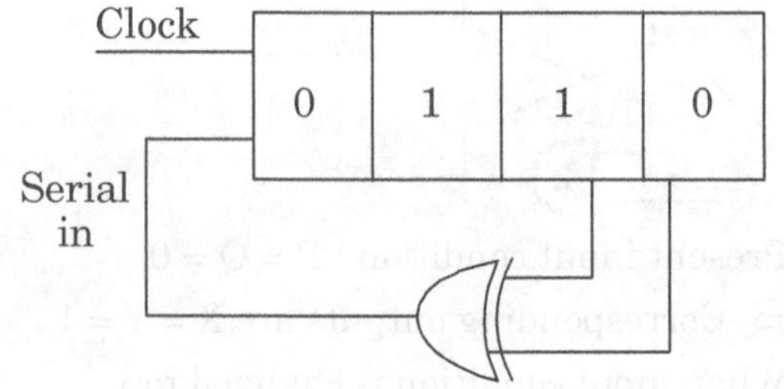

The output will be 1010.

Serial is $= D_1 \oplus D_0$

CLK	Serial is			
	0	1	1	0
1	1	0	1	1
2	0	1	0	1
3	1	0	1	0

29. For ripple counter, no. of states, n = log2 (1024) = 10.
Max propagation delay,

$$k\,t_{pd_{FF}} = \frac{1}{nf_{CLK}} = \frac{1}{10 \times 10^6} = 100 \text{ ns.}$$

30. Synchronous counters are faster than the ripple counters because all the flip-flops are triggered at same time in synchronous counters.

31. If a switch tail ring counter using single D flip-flop, the complementary output $\bar{Q}$ is connected to D. So, it becomes a T-flip-flop.

32. Given: $J_0 = \bar{Q}_1, J_1 = Q_0;\ K_0 = K_1 = 1.$

Assuming, $Q_1 = 0, Q_0 = 0.$

Clock	J_0	K_0	J_1	K_1	Q_1	Q_0
0					0	0
1	1	1	0	1	0	1
2	1	1	1	1	1	0
3	0	1	0	1	0	0
4	1	1	0	1	0	0

(Rows 1–3: Unique states = modulus (K) = 3)

33. 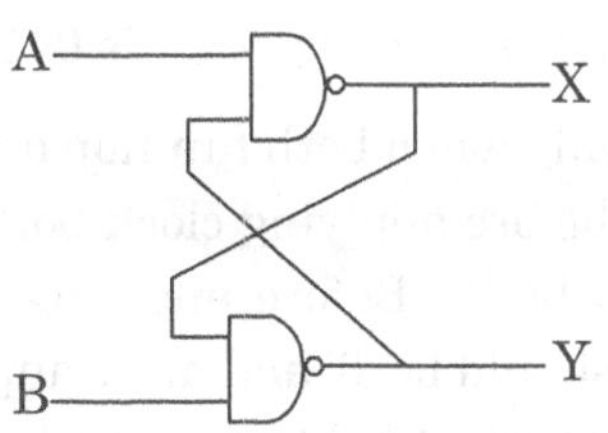

Given: A = 1 (fixed), B = 101010.

For A = 1, B = 1: $X = \overline{Y \cdot 1} = \bar{Y};\ Y = \overline{1 \cdot X} = \bar{X}$

For A = 1, B = 0: $X = \overline{1 \cdot Y} = \bar{Y} = 0.$

$Y = \overline{0 \cdot X} = 1.$

For A = 1, B = 1: $X = \overline{1 \cdot Y} = \bar{Y} = 0$

$Y = \overline{1 \cdot X} = \bar{X} = 1$

Similarly, the outputs X and Y will be fixed at 0 and 1 respectively.

34. 3 flip flops are used and it has –ve edge triggering. Hence it is mod-5 down counter.

35. Output of NAND is zero when $Q_3Q_2Q_1Q_0$ have state 1010 = $(10)_{Dec}$. Therefore, the given figure represents mod-10 counter. And, frequency of the signal will be

$$f = \frac{10\text{KHz}}{10} = 1\text{KHz}$$

36. Given: no of flip-flop, n = 3, of states required = 7 (0 to 6).

It is possible only with asynchronous counter.

Undesired state is 111.

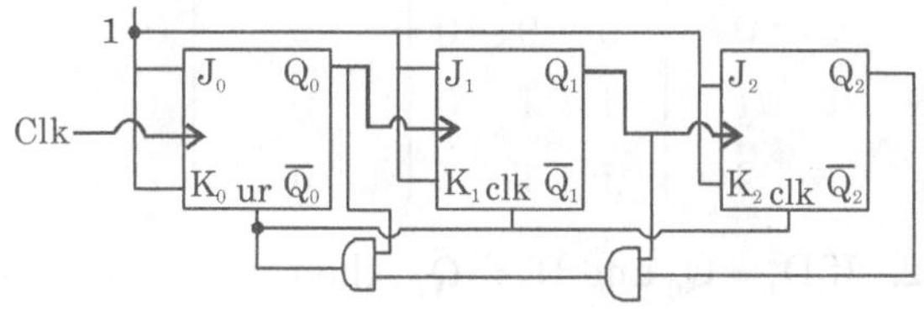

37. Propagation delay in 4-bit ripple counter = 4 T_d
= 40 ns

and, since all the flip-flops in the synchronous counter are clocked simultaneously, then its worst delay will be equal to 10 ns.

38. Change in *o/p* occurs when the state of the master is affected

∵ State of slave is depend on state of master.

Hence, the output is also depend on master.

39. 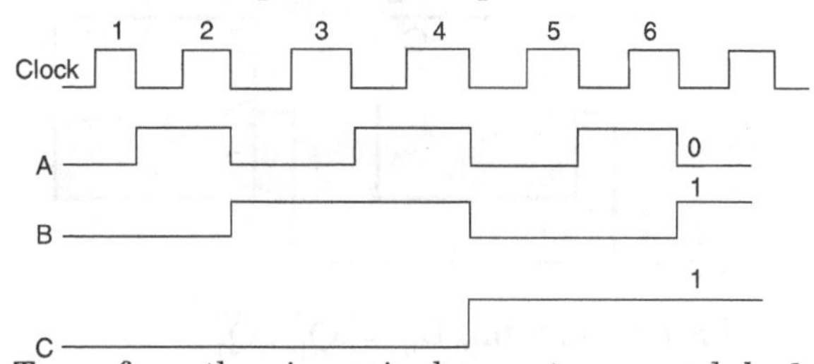

To perform the given ripple counter as module-6, the flip-flop should be cleared at the end of sixth pulse.

Since at 6th pulse,

$$B = C = 1,$$

then $\bar{B} = \bar{C} = 0.$

Hence, the 2 - input gate, should be OR gate so that CLR = 0 for all the three flip-flops.

40. Given: Present state of counter, $Q_2\ Q_1\ Q_3$ = 011.

Since, it is a ripple counter, $\bar{Q}_0$ triggers T_1FF and $\bar{Q}_1$ triggers T_2 FF. For toggle or T-FF,

T	Qn+1
0	Qn
1	$\bar{Q}_n$

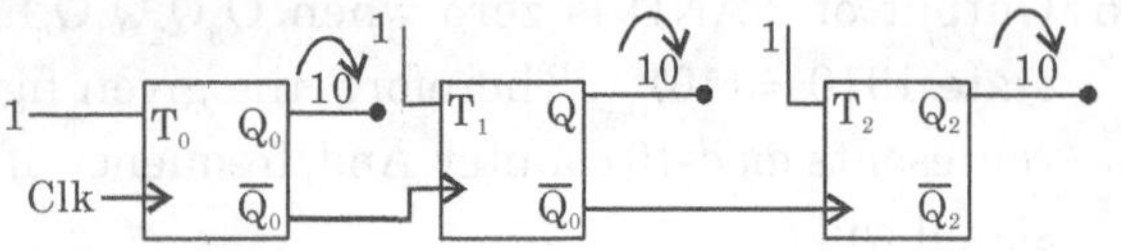

After 1st clock, $Q_2\, Q_1\, Q_0 = 100$.

41. Given: Initially, A = 0, B = 0, $C_i = 0$, $C_0 = 0$ (All FFs are close).

S is addition of bits at A, B and C_1 : C_0 is the next carry to be generated.

Clock	A	A	C_i	S = A + B	C_0
0	0	0	0	0	0
1	1	1	0	0	1
2	1	1	1	0	1

42. If $D_1 = Q_0$ and $D_0 = \overline{Q}_1$, then

Clock	Q_1	Q_0	$D_1 = Q_0$	$D_0 = \overline{Q}_1$
0	0	0	0	1
1	0	1	1	1
2	1	1	1	0
3	1	0	0	0

43.

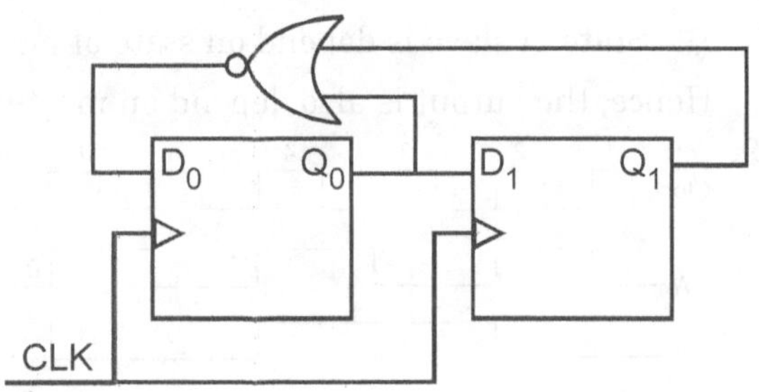

From diagram, $D_0 = \overline{Q_0 + Q_1}$

$= \overline{Q}_0 \cdot \overline{Q}_1$

and $D_1 = Q_0$

$Q_1(t)$	$Q_0(t)$	$Q_1(t+1) = D_1 = Q_0(t)$	$Q_0(t+1) = D_0 = \overline{Q_0+Q_1}$
0	0	0	1
0	1	1	0
1	0	0	0
0	0	0	1
⋮	⋮	⋮	⋮

44. From figure, $J_1 = K_1 = \overline{Q}_2$, $J_2 = \overline{Q}_1$, $K_2 = 1$.

Assuming initial outputs as 0, 0.

Clk	D_A	D_B	D_C	Q_A	Q_B	Q_C
0	1	0	0	0	0	0
1	1	0	0	0	0	0
2	1	1	0	1	0	0
3	1	0	1	1	1	0
4	0	1	0	1	0	1
5	0	0	1	0	1	0
6	0			0	0	1

45. From given figure,

$D_A = \overline{Q}_C \cdot \overline{D}_B$ $D_B = Q_A$, $D_C = Q_B$.

given: Initially, $Q_A = Q_B = Q_C = 0$.

$D_A = 1.1 = 1$, $D_B = 0$, $D_C = 0$.

Q_2	Q_1	Q_0	D_2	D_1	D_0	V_0
0	0	0	0	0	0	0
1	0	0	1	0	0	4
1	1	0	1	1	0	6
1	1	1	1	1	1	7
0	1	1	0	1	1	3
0	0	1	0	0	1	1
0	0	0	0	0	0	0

Therefore, the count sequency at Q_A is 0110100.

46. Y = 1 is possible only when both flip-flop outputs are '1'. It means before applying clock both flip-flop input should be '1'. Before applying clock, output of 1st FF should be '0' and after applying clock, output of 1st FF should be '1'. It depends only input data when changes from '0' to '1'.

47. Sequence of Johnson counter and output voltage is describe below:

Q_2	Q_1	Q_0	D_2	D_1	D_0	V_0
0	0	0	0	0	0	0
1	0	0	1	0	0	4
1	1	0	1	1	0	6
1	1	1	1	1	1	7
0	1	1	0	1	1	3
0	0	1	0	0	1	1
0	0	0	0	0	0	0

48.

Present State Q_B	Q_A	Next State Q_B	Q_A
0	0	1	1
1	1	0	1
0	1	1	0
1	0	0	0
0	0	1	1

Now, using excitation table of D-Ff.

$D_A = Q_A Q_B + \bar{Q}_A \bar{Q}_B;\ D_B = \bar{Q}_B$.

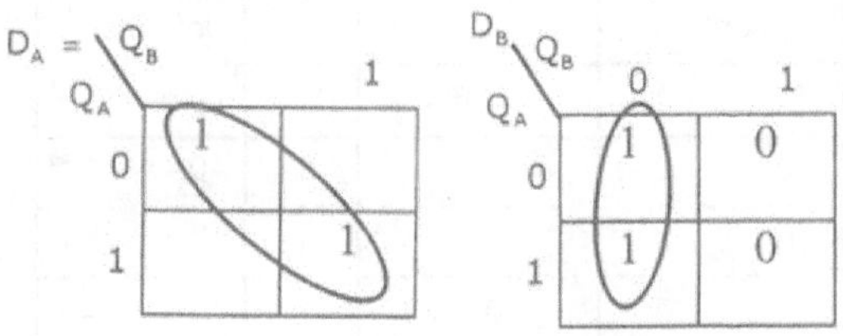

49.

$$f_{Q_3} = \frac{\text{input frequency}}{\text{modulus of counter}} = \frac{1\text{MHz}}{16}$$

$$f_{Q_3} = 62.5 \text{ kHz}.$$

50.

Clock	$J_1(\bar{Q}_2)$	$K_1(Q_2)$	$J_2(Q_1)$	$K_2(\bar{Q}_1)$	Q_1	Q_2
Initial →	–	–	–	–	0	0
1st CP →	1	0	0	1	1	0
2nd CP →	1	0	1	0	1	1
3rd CP →	0	1	1	0	0	1
4th CP →	0	1	0	1	0	0

So, the output sequence generated at Q_1 is 01100 .....

51. To find the modulus of the counter, consider the status of the inputs (Q_B, Q_C) as 1.

So, $Q_A Q_B Q_C Q_D = 0110$

So, it is a MOD-7 counter

52.

Q_3	Q_2	Q_1	Q_0
0	0	0	0
0	0	0	1
0	0	1	0
0	0	1	1
0	1	0	0

Once the output of Ex-NOR gate is 0 then counter will be RESET. So, Ex-NOR-gate will produce logic 0 for $Q_3 = 0$, $Q_2 = 1$. So, the counter will show the sequence like:

$$0 \rightarrow 1 \rightarrow 2 \rightarrow 3 \rightarrow 4$$

So, it is MOD-5 counter.

53.
1. Clock is taken from normal output and it is –ve edge triggering. So, it is UP-counter.
2. Input of the NAND-gate is taken from Q_2 and Q_0. So $Q_2 = 1$ and $Q_0 = 1$.
3. To find the modulus
   $(Q_2, Q_1, Q_0) = (1\ 0\ 1)$

So, it is MOD – 5 binary UP-counter.

54. Consider as new solution.

From given figure,

$$D_2 = Q_0 \oplus Y_1,\ D_1 = Q_2 = Y_2, D_0 = Q_1 = Y_1$$

	Y_2	Y_1	Y_0
Initial	1	1	1
1st Clock	0	1	1
2nd Clock	0	0	1
3rd Clock	1	0	0

⇒ After three clock

$Y = Y_2\ Y_1\ Y_1$ is 100.

55. Given: Clock frequency = 1GHz, Time period = 1 ns.

If the propagation delay of the NAND gate were 0 ns, the circuit would have behaved mod-6 counter.

However, the delay of NAND gate is 2 ns. During this time, two more clock pulses would reach the counter, and therefore it would count two more states. Hence, it act as mod-8 counter.

56.

Clock Number	$D_{in} = A \oplus D$ Just before clock	A	B	C	D
Initial	-	1	1	0	1
1	0 →	0	1	1	0
2	0 →	0	0	1	1
3	1 →	1	0	0	1
4	0 →	0	1	0	0
5	0 →	0	0	1	0
6	0 →	0	0	0	1
7	1 →	1	0	0	0
8	1 →	1	1	0	0
9	1 →	1	1	1	0
10	1 →	1	1	1	1

Hence inorder to reach the state ABCD = 1111, total 10 clock pulse is required.

57.

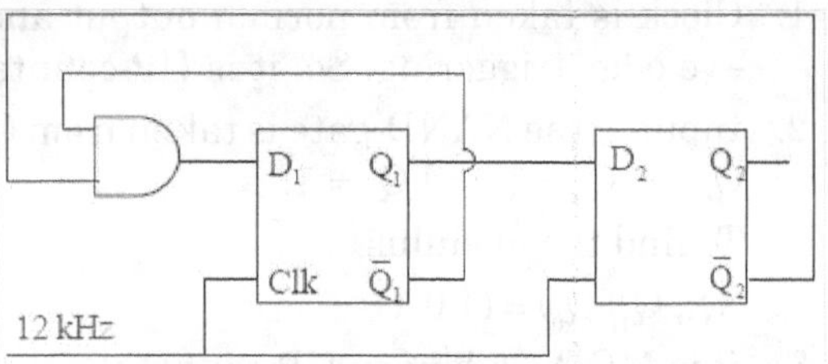

$D_2(Q)_1$	$D_1(\bar{Q}_1 \bar{Q})_2$	Q_2	Q_1	
–	–	0	0	← Initially
0	1	0	1	1st clock pulse
1	0	1	0	2nd clock pulse
0	0	0	0	3rd clock pulse

$$Q_2Q_1 \rightarrow 00 \rightarrow 01 \rightarrow 10 \rightarrow 00$$

It is a MOD-3 synchronous counter

$$\text{So, } f_{Q_2} = \frac{f_i}{3} = \frac{12\text{kHz}}{3} = 4 \text{ kHz}$$

58. For State A

Binary input			Present State	Next State
X	Y	Z		
0	0	0	A	B
0	0	1	A	A
0	1	0	A	A
0	1	1	A	A
1	0	0	A	C
1	0	1	A	C
1	1	0	A	A
1	1	1	A	A

Present State B

Binary input			Present State	Next State
X	Y	Z		
0	0	0	B	A
0	0	1	B	C
0	1	0	B	B
0	1	1	B	B
1	0	0	B	A
1	0	1	B	B
1	1	0	B	B
1	1	1	B	B

Present State C

Binary input			Present State	Next State
X	Y	Z		
0	0	0	C	C
0	0	1	C	C
0	1	0	C	C
0	1	1	C	B
1	0	0	C	C
1	0	1	C	A
1	1	0	C	C
1	1	1	C	AB

In state 'C' when XYZ = 111; the ambiguity occurs. Because from state 'C'

When, X = 1, Z = 1 ⇒ Next state = A

When, Y = 1, Z = 1 ⇒ Next state = B

So, transitions from state 'C' are ambiguous defined.

59. From the given diagram

$$D_A = Q_A \oplus Q_B$$

and $$D_B = \overline{Q_A X_{IN}}$$

For $$X_{IN} = 0$$

$$D_A = Q_A \oplus Q_B$$

and $$D_B = 1$$

Present State				Next State	
Q_A	Q_B	$D_A = Q_A \oplus Q_B$	D_B	Q_A^-	Q_B^-
0	0	0	1	0	1
0	1	1	1	1	1
1	1	0	1	0	1
0	0	1	1	1	1

So, for $X_{IN} = 0$,

Number of possible states = 2

For $X_{IN} = 1$:

$$D_A = Q_A \oplus Q_B$$

and $$D_B = \overline{Q_A}$$

Present State				Next State	
Q_A	Q_B	$D_A = Q_A \oplus Q_B$	D_B	Q_A^-	Q_B^-
0	0	0	1	0	1
0	1	1	1	1	1
1	1	0	0	0	0
0	0	0	1	0	1

So, for $X_{IN} = 1$,

Number of possible states = 3

So, the option (*d*) is correct.

60. By observing the given state diagram, it is clear that FSM can be used to detect the sequence '101'. It is given is the question that the FSM detects overlapping sequences also. The given input sequence is 10101101001101 So, output will be 1 for 4 times.

61. Duration for individual signal is given as

GREEN → 70 seconds

YELLOW → 5 seconds

RED → 75 second

Clock period → 5 seconds

Total number of unique states required

$$= \frac{70 + 5 + 75}{5} = 30.$$

Minimum number of flip-flops required,

$n = \log_2(30) = 4.91$

$n \cong 5.$

5 CHAPTER

Logic Families & Semiconductor Memories

Analysis of Previous GATE Papers												
	Year → Topics ↓		2019	2018	2017 Set 1	2017 Set 2	2016 Set 1	2016 Set 2	2016 Set 3	2015 Set 1	2015 Set 2	2015 Set 3
LOGIC FAMILIES & THEIR COMPARISON	1 Mark	MCQ Type	1									
		Numerical Type										
	2 Marks	MCQ Type										
		Numerical Type										
	Total		1									
STATIC CMOS IMPLEMENTATION OF LOGIC GATES	1 Mark	MCQ Type	1	1		1		1	1			
		Numerical Type										
	2 Marks	MCQ Type										
		Numerical Type										
	Total		1	1		1		1	1			
SEMI CONDUCTOR MEMORIES (SRAM, DRAM, ROM)	1 Mark	MCQ Type				1						
		Numerical Type										
	2 Marks	MCQ Type	1									
		Numerical Type										
	Total		2			1						

LOGIC FAMILIES & THEIR COMPARISON

1. Fill in the blanks of the statements below concerning the following Logic Families :

Standard TTL (74 XXLL), Low power TTL (74L XX) Low power schottky

TTL(74L SXX), schottky TTL(74 SXX), Emitter coupled Logic (ECL), CMOS

(a) Among the TTL Families, ________ family requires considerably less power khan the standard TTL (74 XX) and also has comparable propagation delay.

(b) Only the ________ family can operate over a wide range of power supply voltages

[1987 : 1 Mark]

2. Given that for a logic family,

V_{OH} is the minimum output high-level voltage.
V_{OL} is the maximum output-tow-level voltage.
V_{IH} is the minimum acceptable input high-fevel voltage and

V_{IL} is the maximum acceptable input low-level voltage,

The correct relationship is

(a) $V_{IH} > V_{OH} > V_{IL} > V_{OL}$

(b) $V_{OH} > V_{1H} > V_{IL} > V_{OL}$

(c) $V_{IH} > V_{OH} > V_{OL} > V_{IL}$

(d) $V_{OH} > V_{OL} > V_{OL} > V_{IL}$

[1987 : 2 Marks]

3. Among the digital IC-families-ECL, TTL and CMOS:

(a) ECL has the least propagation delay.

(b) TTL has the largest fan-out.

(c) COS has the biggest noise margin.

(d) TTL has the lowest power consumption.

[1989 : 1 Mark]

4. A logic family has threshold voltage $V_R = 2$ V, minimum guaranteed output high voltage

$V_{OH} = 4$ V. minimum accepted input high voltage $V_{IH} = 3$ V, maximum guaranteed output low voltage

$V_{OL} = 1$ V, and maximum accepted input low voltage $V_{IL} = 1.5$ V. Its noise margin is

(a) 2 V

(b) 1 V

(c) 1.5 V

(d) 0.5 V

[1989 : 1 Mark]

5. Figure shows the circuit of a gate in the Resistor Transistor Logic (RTL) family. The circuit represents a

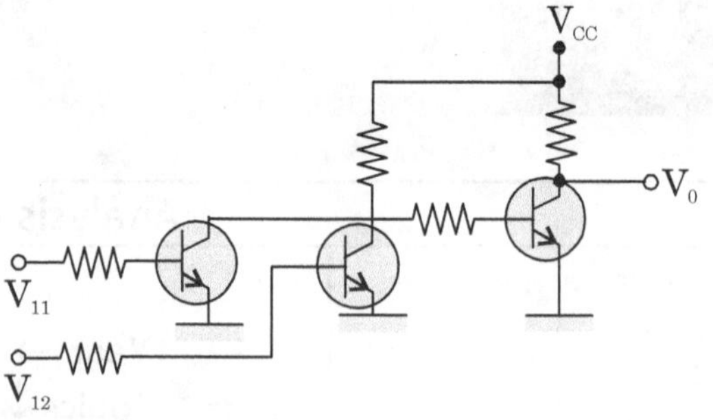

(a) NAND (b) AND

(c) NOR (d) OR

[1992 : 1 Mark]

6. In standard TTL the 'totem pole' stage refers to

(a) the multi-emitter input stage.

(b) the phase splitter.

(c) the output buffer.

(d) open collector output stage.

[1997 : 1 Mark]

7. The inverter 74 AL S01 has the following specifications:

$I_{OHmax} = 0.4$ mA, $I_{OLmax} = 8$ mA, $I_{IH\,max} = 20$ μA, $I_{ILmax} = 0.1$ mA

The fan-out based on the above will be

(a) 10 (b) 20

(c) 60 (d) 100

[1997 : 1 Mark]

8. The gate delay of an NMOS inverter is dominated by charge time rather than discharge time because

(a) the driver transistor has a larger threshold voltage than the load transistor

(b) the driver transistor has larger leakage currents compared to the load transistor

(c) the load transistor has a smaller WiL ratio compared to the driver transistor

(d) None of the above.

[1997 : 1 Mark]

9. The noise margin of a TTL gate is about

(a) 0.2 V

(b) 0.4 V

(c) 0.6 V

(d) 0.8 V

[1998 : 1 Mark]

10. The threshold voltage for each transistor in figure is 2 V. For this circuit to work as an inverter, V_i must take the values

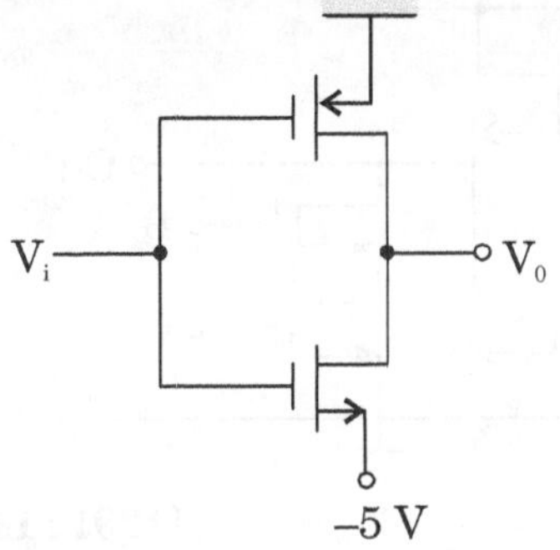

(a) –5 V and 0 V (b) – 5 V and 5 V

(c) 0 V and 5 V (d) – 3 V and 3 V

[1998 : 1 Mark]

11. A Darlington emitter-follower circuit is sometimes used in the output stage of a TTL gate in order to

(a) increase its I_{OL}.

(b) reduce its I_{OH}.

(c) increase its speed of operation.

(d) reduce power dissipation.

[1999 : 1 Mark]

12. Commercially available ECL Gates use two ground lines and one negative supply in order to

(a) reduce power dissipation.

(b) increase fan-out.

(c) reduce loading effect.

(d) eliminate the effect of power line glitches or the biasing circuit.

[1999 : 1 Mark]

13. The output of the 74 series GATE of TTL gates is taken from a BJT in

(a) totem pole and common collector configuration.

(b) either totem pole or open collector configuration.

(c) common base configuration.

(d) common collector configuration.

[2003 : 1 Mark]

14. The DTL, TTL, ECL and CMOS Tamil GATE of digital ICs are compared in the following 4 columns

	(P)	(Q)	(R)	(S)
Fanout is miniumut	DTL	DTL	TTL	CMOS
Power consumption is minimum.	TTL	CMOS	ECL	DTL
Propagation delay is minimum.	CMOS	ECL	TTL	TTL

The correct column is

(a) P (b) Q

(c) R (d) S

[2003 : 2 Marks]

15. Given figure is the voltage transfer characteristic of

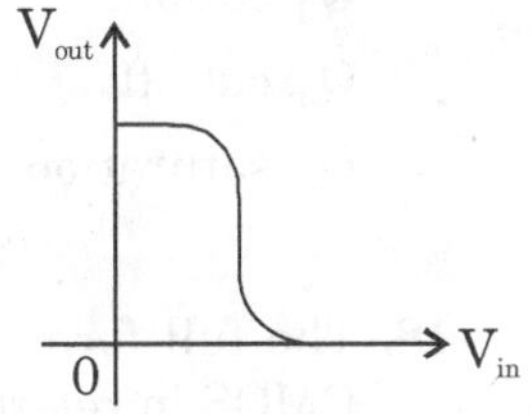

(a) an NMOS inverter with enhancement mode transistor as load

(b) an NMOS inverter with depletion mode transistor as load

(c) a CMOS inverter

(d) a BJT inverter

[2004 : 1 Mark]

16. The transistors used in a portion of the TTL gate shown in the figure have a β = 100. The base-emitter voltage of is 0.7 V for a transistor in active region and 0.75 V for a transistor in saturation. If the sink current I = 1 mA and the output is at logic 0, then the current I_R will be equal to

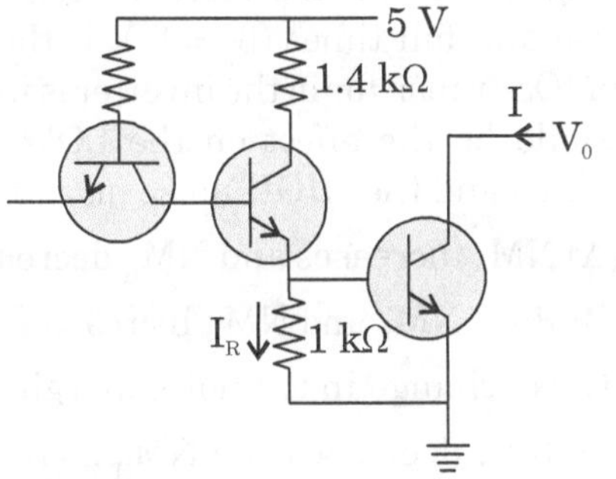

(a) 0.65 mA (b) 0.70 mA

(c) 0.75 mA (d) 1.00 mA

[2005 : 2 Marks]

17. The circuit diagram of a standard TTL NOT gate is shown in the figure. When V_i = 2.5 V the modes of operation of the transistors will be

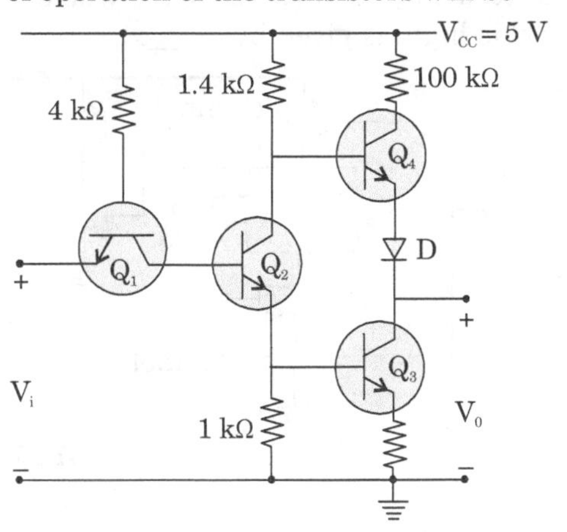

(a) Q_1: reverse active; Q_2: normal active; Q_3: saturation, Q_4: cut-off
(b) Q_1: reverse active; Q_2; saturation; Q_3: saturation; Q_4: cut-off
(c) Q_1: normal active: Q_2: cut-off; Q_3: cut-off; Q_4. saturation
(d) Q_1: saturation; Q_2: saturation; Q_3: saturation; Q_4: normal active

[2007 : 2 Marks]

18. The full forms of the abbreviations TTL and CMOS in reference to logic families are
(a) Triple Transistor Logic and Chip Metal Oxide Semiconductor.
(b) Tristate Transistor Logic and Chip Metal Oxide Semiconductor.
(c) Transistor Transistor Logic and Complementary Metal Oxide Semiconductor.
(d) Tristate Transistor Logic and Complementary Metal Oxide Silicon.

[2009 : 1 Mark]

19. A standard CMOS inverter is designed with equal rise and fall times ($\beta_n = \beta_p$). If the width of the pMOS transistor in the inverter is increased, what would be the effect on the LOW noise margin (NM_L) and the HIGH noise margin NM_H?
(A) NM_L increases and NM_H decrease
(B) Both NM_L and NM_H increase
(C) No change in the noise margins
(D) NM_L decreases and NM_H increases

[2019 : 1 Mark]

STATIC CMOS IMPLEMENTATION OF LOGIC GATES

20. In figure, the Boolean expression for the output in terms of inputs A, B and C when the clock 'ck' is high, is given by _____.

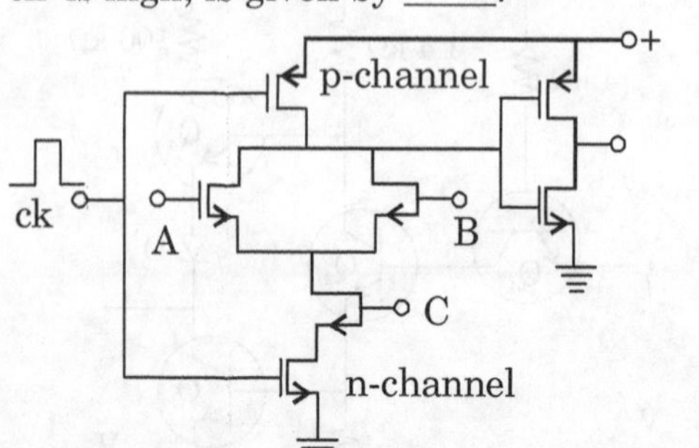

[1991 : 1 Mark]

21. The CMOS equivalent of the following n MOS gate (figure) is (draw the circuit).

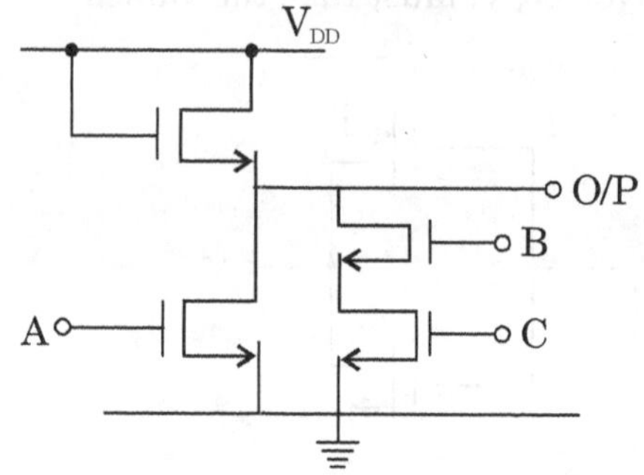

[1991 : 1 Mark]

22. In the output stage of a standard TTL, we have a diode between the emitter of the pull-up transistor and the collector of the pull-down transistor. The purpose of this diode is to isolate the output node from the power supply V_{CC}.

[1994 : 2 Marks]

23. For the NMOS logic gate shown in figure, the logic function implemented is

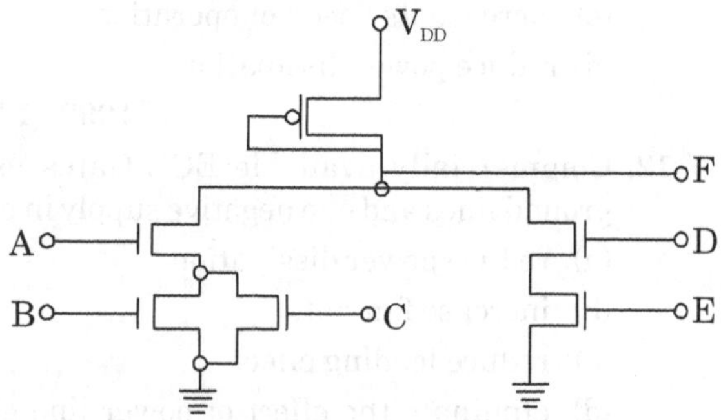

(a) $\overline{ABCDE}$
(h) $(AB + \bar{C}).(\overline{D + E})$
(c) $\overline{A.(B + C) + D.E}$
(d) $(\overline{A + B}).C + \bar{D}.\bar{E}$

24. The circuit in the figure has two CMOS NOR-gates. This circuit functions as a

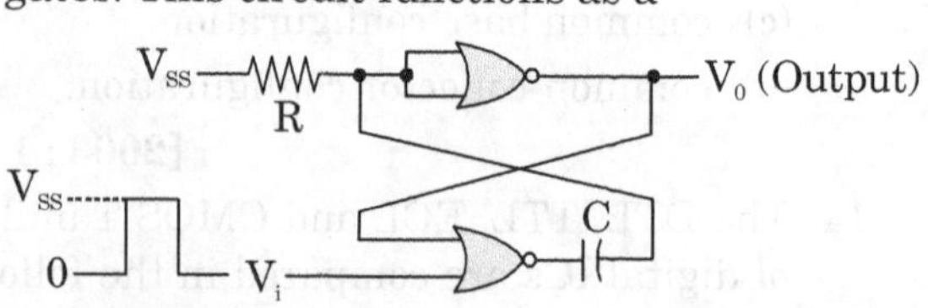

(a) flip-flop.
(b) Schmitt trigger.
(c) monostable multi-vibrator.
(d) astable multi-vibrator.

[2001 : 2 Marks]

25. Both transistors T_1 and T_2 shown in the figure, have a threshold voltage of 1 Volts. The device parameters K_1 and K_2 of T_1 and T_2 are, respectively, 36 $\mu A/V^2$ and 9 $\mu A/V^2$. The output voltage V_o is

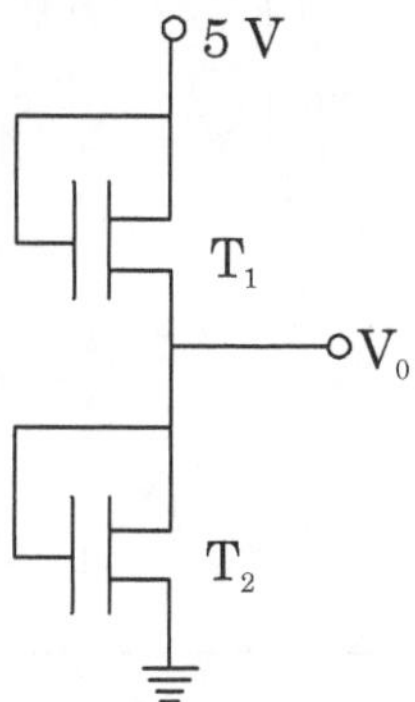

(a) 1 V (b) 2 V
(c) 3 V (d) 4 V

[2005 : 2 Marks]

26. The logic function implemented by the following circuit at the terminal OUT is

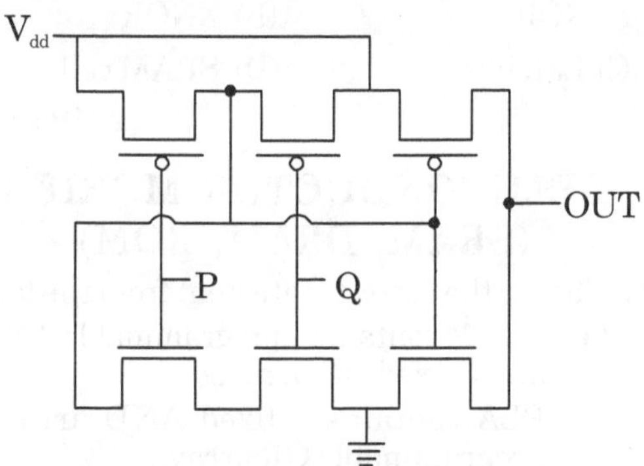

(a) P NOR Q (b) P NAND Q
(c) P OR Q (d) P AND Q

[2008 : 2 Marks]

27. In the circuit shown

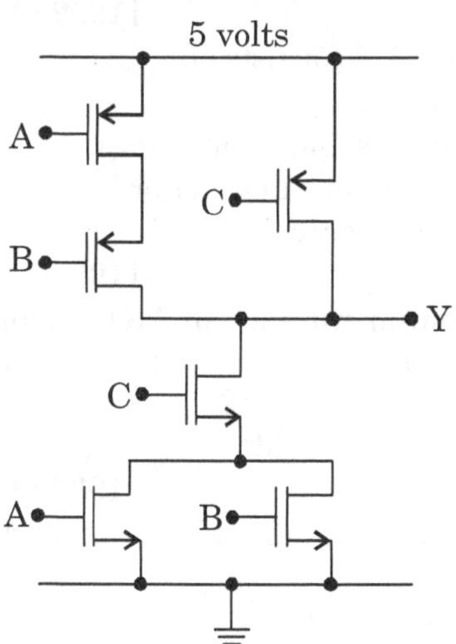

(a) $Y = \bar{A}\bar{B} + \bar{C}$ (b) $Y = (A + B)C$
(c) $Y = (\bar{A} + \bar{B})\bar{C}$ (d) $Y = AB + C$

[2012 : 1 Mark]

28. The output (Y) of the circuit shown in the figure is

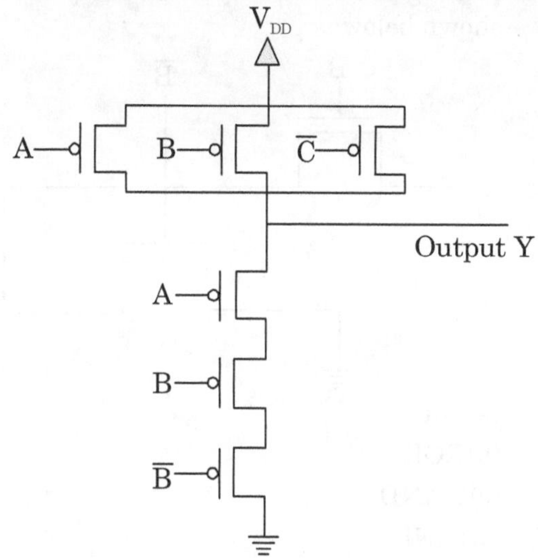

(a) $\bar{A} + \bar{B} + C$ (b) $A + \bar{B}.\bar{C} + A.\bar{C}$
(c) $\bar{A} + B + \bar{C}$ (d) $A.B.\bar{C}$

[2014 : 1 Mark, Set-4]

29. Transistor geometries in a CMOS inverter have been adjusted to meet the requirement for worst case charge and discharge times for driving a load capacitor C. This design is to be converted to that of a NOR circuit in the same technology, so that its worst case charge and discharge times while driving the same capacitor are similar, The channel lengths of all transistors are to be kept unchanged. Which one of the following statements is correct?

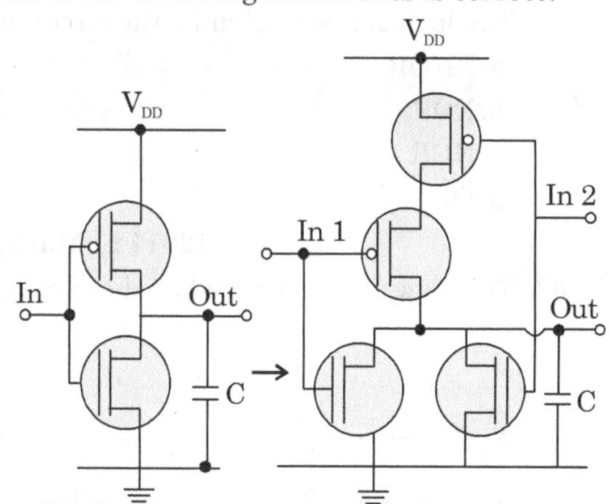

(a) widths of PMOS transistors should be doubled, while widths of NMOS transistors should be halved.
(b) widths of PMOS transistors should be doubled, while widths of NMOS transistors should not be changed.
(c) widths of PMOS transistors should be halved, while widths of NMOS transistors should not be changed.
(d) widths of PMOS transistors should be unchanged, while widths of NMOS transistors should be halved.

[2016 : 1 Mark, Set-2]

30. The logic functionality realized by the circuit shown below is

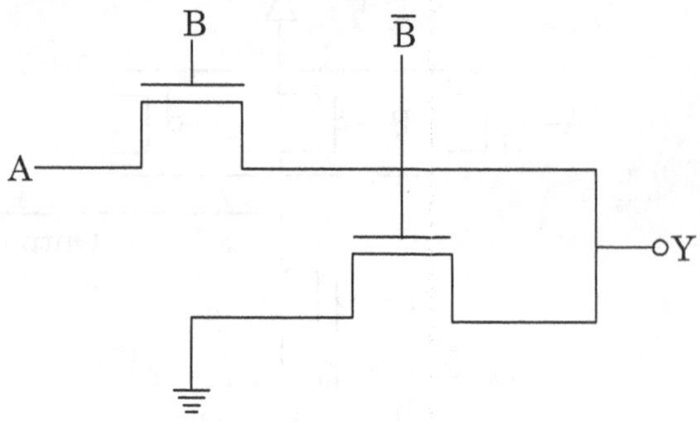

(a) OR
(b) XOR
(c) NAND
(d) AND

[2016 : 1 Mark, Set-3]

31. For the circuit shown in the figure. P and D are the inputs and Y is the output.

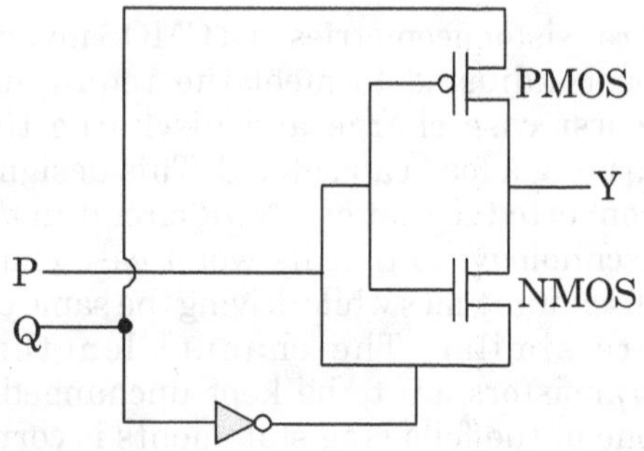

The logic implemented by the circuit is
(a) XNOR
(b) XOR
(c) NOR
(d) OR

[2017 : 1 Mark, Set-21

32. The logic f(X, Y) realized by the given circuit is

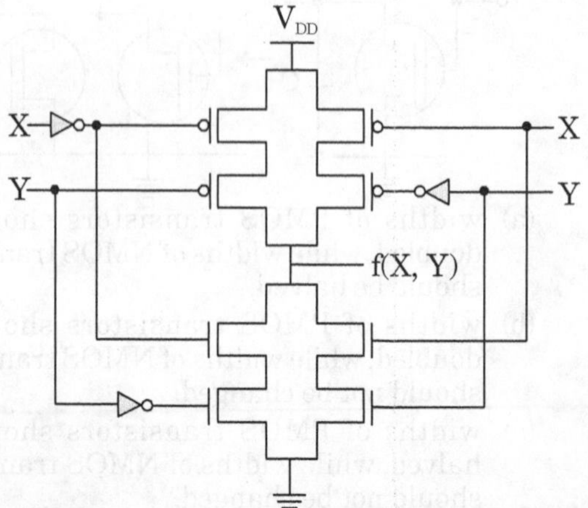

(a) NOR (b) AND
(c) NAND (d) XOR

[2018 : 1 Mark]

33. In the circuit shown, A and B are the inputs and F is the output. What is the functionality of the circuit?

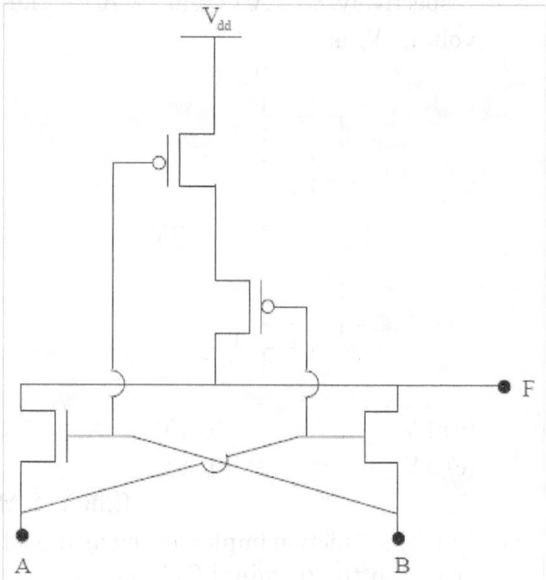

(A) XOR (B) XNOR
(C) Latch (D) SRAM cell

[2019 : 1 Mark]

SEMICONDUCTOR MORIES (SRAM, DRAM, ROM)

34. Choose the correct statement from the following:
(a) PROM contains a programmable AND array and a fixed OR array.
(b) PLA contains a fixed AND array and a programmable OR array.
(c) PROM contains a fixed AND array and a programmable OR array.
(d) PLA contains a programmable AND array and a programmable OR array.

[1992 : 1 Mark]

35. A dynamic RAM consists of
(a) 6 transistors.
(b) 2 transistors and 2 capacitors.
(c) 1 transistor and 1 capacitor.
(d) 2 capacitors only.

[1994 : 1 Mark]

36. The minimum number of MOS transistors required to make a dynamic RAM cell is
(a) 1 (b) 2
(C) 3 (d) 4

[1995 : 1 Mark]

37. Each cell of a static Random Access Memory contains
(a) 6 MOS transistors.
(b) 4 MOS transistors and 2 capacitors
(c) 2 MOS transistors and 4 capacitors
(d) 1 MOS transistor and 1 capacitor

[1996 : 1Mark]

38. In the DRAM cell in the figure, the V_t of the NMOSFET is 1 V. For the following three combinations of WL and BL voltages.

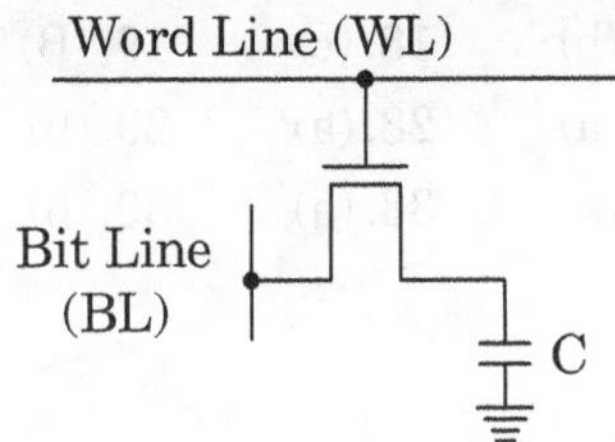

(a) 5 V; 3 V; 7 V (b) 4 V; 3 V: 4 V
(c) 5 V; 5 V; 5 V (d) 4 V; 4 V; 4 V

[2001 : 2 Marks]

39. If WL is the Word Line and BL the Bit Line, an SRAM cell is shown in

(a)

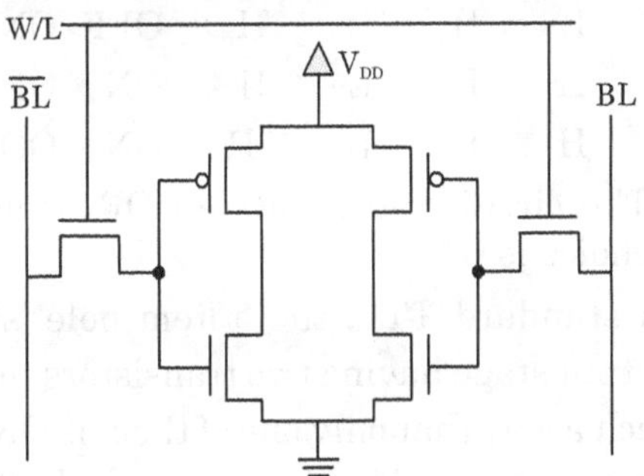

(b)

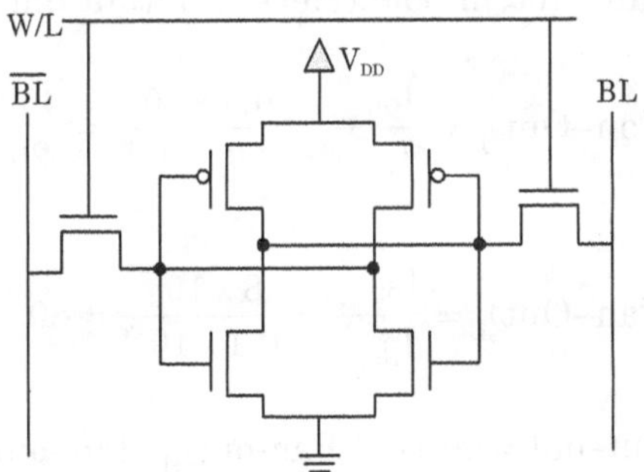

(c)

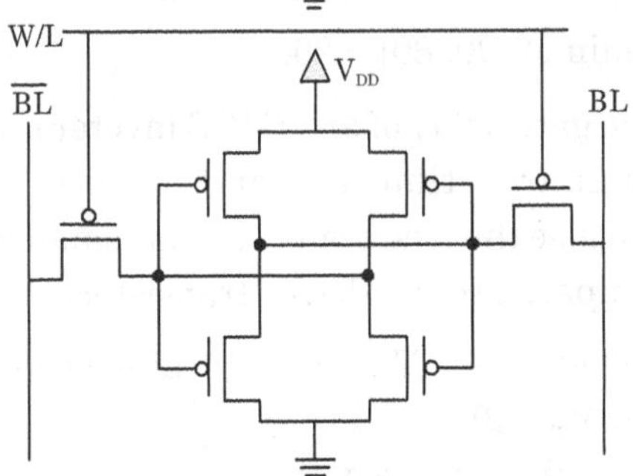

(d)

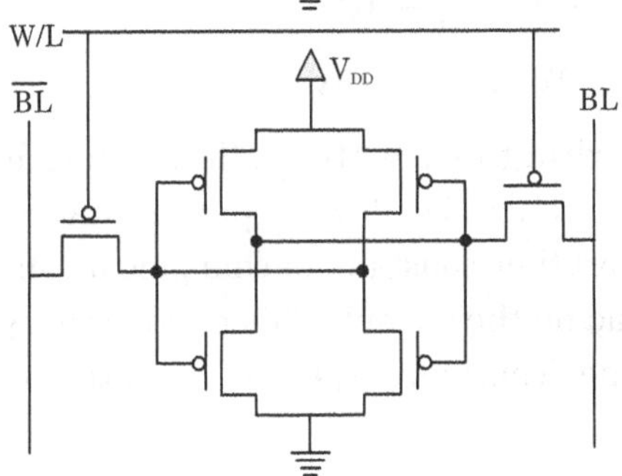

[2014 : 2 Marks, Set-3]

40. In a DRAM,

(a) periodic refreshing is not required.
(b) information is stored in a capacitor.
(c) information is stored in a latch.
(d) both read and write operations can be performed simultaneously.

[2017 : 1 Mark, Set-2]

41. A 2 × 2 ROM array is built with the help of diodes as shown in the circuit below Here W_0 and W_1, are signals that select the word lines and B_0 and B_1 are signals that are output of the sense amps based on the stored data corresponding to the bit lines during the read operation.

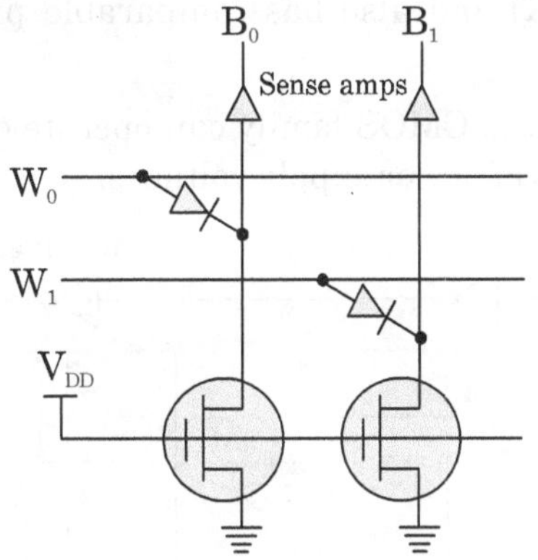

$$\begin{array}{c c} & \begin{array}{cc} B_0 & B_1 \end{array} \\ \begin{array}{c} W_0 \\ W_1 \end{array} & \begin{bmatrix} D_{00} & D_{01} \\ D_{10} & D_{11} \end{bmatrix} \end{array}$$ Bits stored in the ROM Array

During the read operation, the selected word line goes high and the other word line is in a high impedance state. As per the implementation shown in the circuit diagram above, what are the bits corresponding to D_{ij} (where i = 0 or 1 and j = 0 or 1) stored in the ROM?

(a) $\begin{bmatrix} 1 & 0 \\ 0 & 1 \end{bmatrix}$ (b) $\begin{bmatrix} 0 & 1 \\ 1 & 0 \end{bmatrix}$

(c) $\begin{bmatrix} 1 & 0 \\ 1 & 0 \end{bmatrix}$ (d) $\begin{bmatrix} 1 & 1 \\ 0 & 0 \end{bmatrix}$

[2018 : 2 Marks]

ANSWERS

1. (*)	2. (b)	3. (a)	4. (d)	5. (d)	6. (c)	7. (b)	8. (c)	9. (b)	10. (a)
11. (c)	12. (d)	13. (b)	14. (b)	15. (c)	16. (c)	17. (b)	18. (c)	19. (B)	20. (1)
21. (*)	22. (True)	23. (c)	24. (c)	25. (c)	26. (d)	27. (a)	28. (a)	29. (b)	30. (d)
31. (b)	32. (d)	33. (b)	34. (c,d)	35. (c)	36. (a)	37. (a)	38. (a)	39. (b)	40. (a)
41. (a)									

EXPLANATIONS

1. Among the TTL families, 74LS family requires considerably less power than the standard TTL (74XX) and also has comparable propagation delay.

   Only the CMOS family can operate over a wide range of power supply voltages.

2. 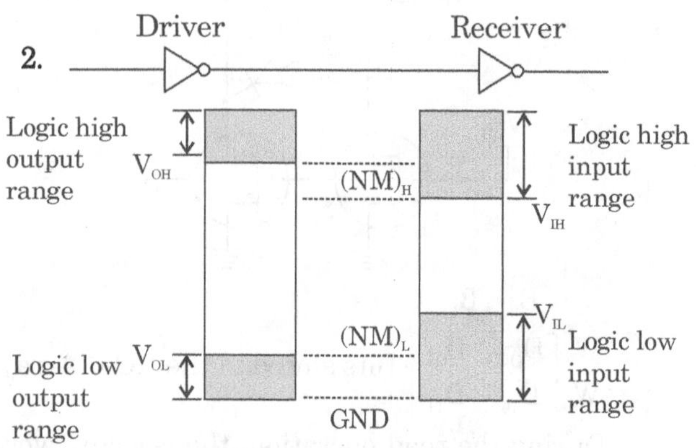

   $V_{OH} > V_{IH} > V_{IL} > V_{OL}$

3. Among given I_C families, ECL is the fastet logic family, hence, least propagation delay.

4. $(N.M.)_H = V_{OH} - V_{IL} = 4 - 3 = 1V.$

   $(NM)_L = V_{IL} - V_{OL} = 1.5 - 1 = 0.5V$

   Noise margin = min of $[(NM)_H, (NM)_L]$

   = 0.5 V.

5. 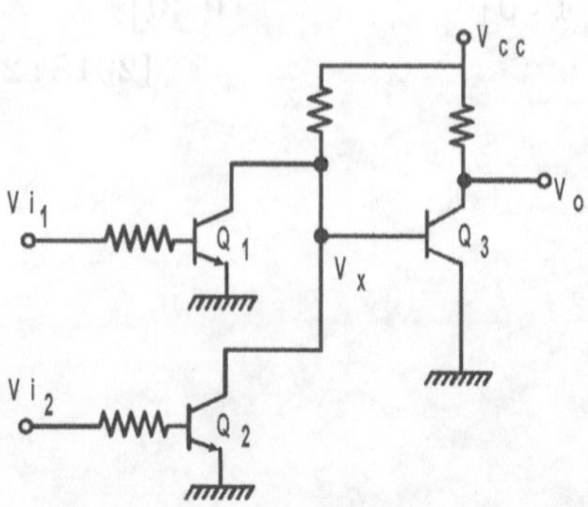

V_{i1}	V_{i2}	V_X	V_0	Q_1	Q_2	Q_3
L	L	H	L	OFF	OFF	ON
L	H	L	H	OFF	ON	OFF
H	L	L	H	ON	OFF	OFF
H	H	L	H	ON	OON	OFF

   The circuit represents an OR. Hence correct choice is (*d*).

6. In standard TTL, the 'totem pole' stage is an output stage having two transistors connected in such a way that only one of them is ON at a time. Thus, totem pole refers to output buffer.

7. $$(\text{Fan–Out})_H = \frac{I_{OH_{max}}}{I_{IH_{max}}} = \frac{0.4 \times 10^{-3}}{20 \times 10^{-6}} = \frac{400}{20} = 20$$

   $$(\text{Fan–Out})_L = \frac{I_{OL_{max}}}{I_{IL_{max}}} = \frac{8 \times 10^{-3}}{0.1 \times 10^{-3}} = 80.$$

   Fan–out = min of $[(\text{Fan-out})_H, (\text{fan–out})_L]$

   = min of [20, 80] = 20.

8. The gate delay of an NMOS inveter is dominated by charge time rather than discharge time because the load transistor has smaller W/L ratio compared to the driver transistor.

9. The noise margin of a TTL gate is about 0.04 V.

10. For $V_T = 2V$;

    $V_i = 5V \Rightarrow V_0 = 0V$

    $V_i = 0V \Rightarrow V_0 = -5V.$

11. Darlington emitter follower has low output resistance. The low output resistance provides a short time constant for changing up any capacitive load on the output. This provides very small rise time. Therefore, speed of operation is inreased.

12. ECL gates use two ground lines and one negative supply in order to eliminate the effect of power line glitches or the biasing circuit.
13. The output of the 74 series GATE of TTL gates is taken from a BJT in either totem pole or open collector configration.
14. DTL < TTL < ECL < CMOS

Comparison of Fan-out

CMOS < DTL < TTL < ECL

Comparision of Power consumption

ECL < TTL < DTL < CMOS

Comparison of propagation delay

15. Its CMOS invertor voltage transfer characteristic
16.

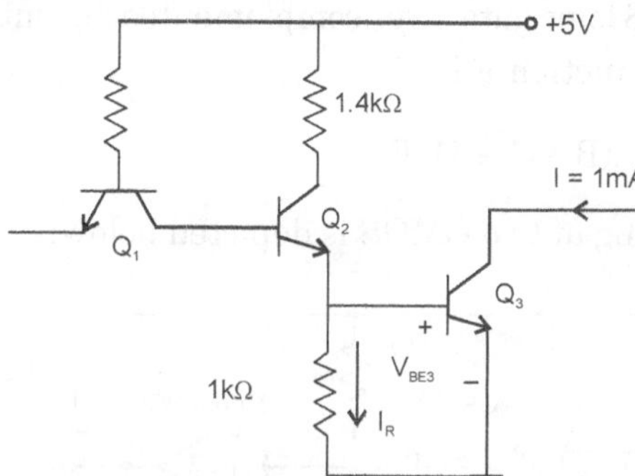

For output is at logic O, $V_o = 0$ V_oH

$$V_o = 0$$

only when Q_3(transistor) is in saturation.

then $V_{BE3} = 0.75$ volts

Using KVL in B – E loop of transistor Q_3,

$$I_R \times 1 \times 10^3 - V_{BE3} = 0$$

$$\Rightarrow I_R = \frac{V_{BE3}}{10^3} \Rightarrow I_R = 0.75 \times 10^{-3} = 75\text{mA}$$

17.

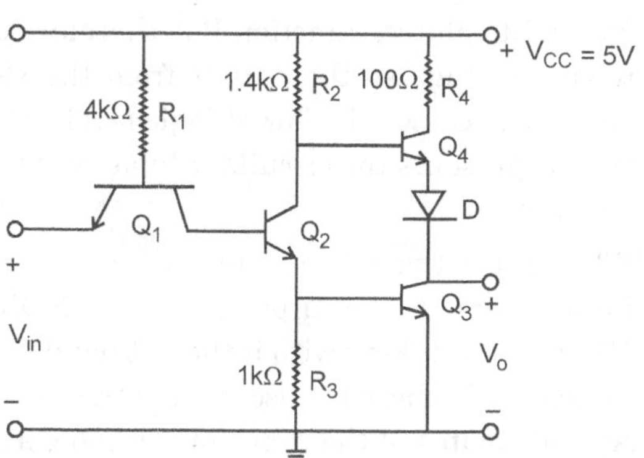

When V_{in} is at high voltage (2V – 5V), BE junction of Q_1 becomes reverse biased and current flows through R_1 and BC junction of Q_1 into the base of Q_2. So, Q_1 operate in reverse active.

Because of base current of Q_2 it drives into saturation because

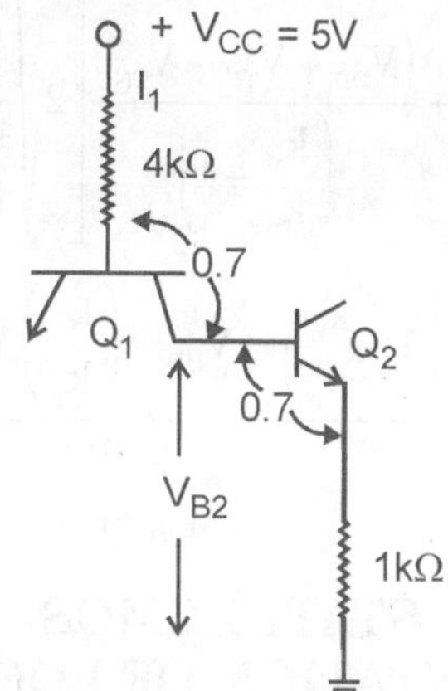

$$I_1 = \frac{5 - 0.7 - 0.7}{4 + 1} = 0.72 \text{ mA}$$

and $V_{B2} = 5 - 0.7 - I \times 4k$

$= 5 - 0.7 - 0.72 \times 4 = 1.42$ volts

$\therefore V_{B2} > 0.7$ volts so Q_2 operate in saturation region.

Because of saturation of Q_2, a voltage drop across R_3

$$I_2 = \frac{V_{CC}}{R_2 + R_3} = \frac{5}{1.4 + 1} = \frac{5}{2.4} \text{ mA} = 2.03 \text{ mA}$$

$V_{B3} = (I_1 + I_3)R_3 = (0.72 + 2.03)\ 1k = 2.75$ volts

Since $V_{B3} > 0.7$ volts, so Q_3 also operates in saturation region.

Q_3 and Q_4 together form a totem pole Output, one transistor operate at a time, so Q_4 will be in cut off.

18. TTL – Transistor - transistor logic.

CMOS – Complementary Metal oxide semiconductor.

19. B $NM_L = V_{IL} - V_{OLU}$ = Increase

$NM_H = V_{OHU} - V_{IH}$ = Increase

$$V_{IH} = V_{TN} + \frac{\left(V_{DD} + V_{TP} - V_{TN}\right)}{\left(\frac{k_n}{k_p} - 1\right)} \left[\frac{2\frac{k_n}{k_p}}{\sqrt{\frac{34n}{k_p} + 1}} - 1\right] > 1$$

$$k_p \alpha \left(\frac{W}{L}\right)_p, k_n \alpha \left(\frac{W}{L}\right)_n$$

Here, pMOS width is increased $\frac{k_p}{k_n} > 1 \Rightarrow \frac{k_n}{k_p} < 1$

$$V_{OH} = \frac{1}{2}\left[\left(1 + \frac{k_n}{k_p}\right)V_1 + V_{DD} - \left(\frac{k_n}{k_p}\right)V_{TN} - V_{TP}\right] < 1$$

Similarly,

$$V_{IL} = V_{TN} + \frac{\left(V_{DD} + V_{TP} - V_{TN}\right)}{\left(\frac{k_n}{k_p} - 1\right)}\left[2\sqrt{\frac{\frac{k_n}{k_p}}{\frac{k_n}{k_p}+3}} - 1\right] > 1$$

$$V_{OLU} = \frac{VI\left(1+\frac{k_n}{k_p}\right) - V_{DD} - \left(\frac{k_n}{k_p}\right)V_{TN} - V_{TP}}{2\left(\frac{k_n}{k_p}\right)} < 1$$

STATIC CMOS IMPLEMENTATION OF LOGIC GATES

20. When clock is high than p-channel is off, so the input to CMOS inverter is logic '0'. Then the output of the CMOS inverter is logic '1'.
21. The output is at logic '1', when pull-down network is not active. The output is at logic '0' when pull-down network is active.

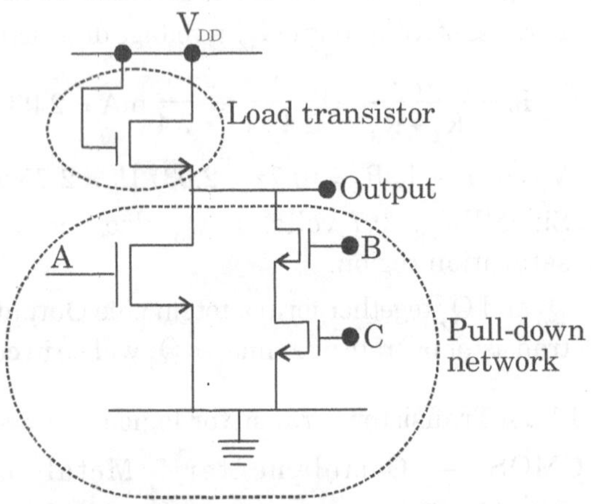

⇒ NMOS, CMOS, PMOS logic gates are used to get complement output.

Output = $\overline{A + BC}$

⇒ CMOS equivalent circuit consists both pull-up and pull-down network.

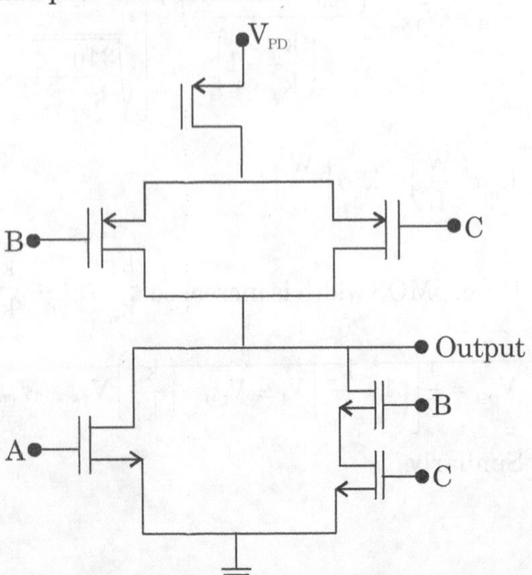

22. **True.** The circuit is

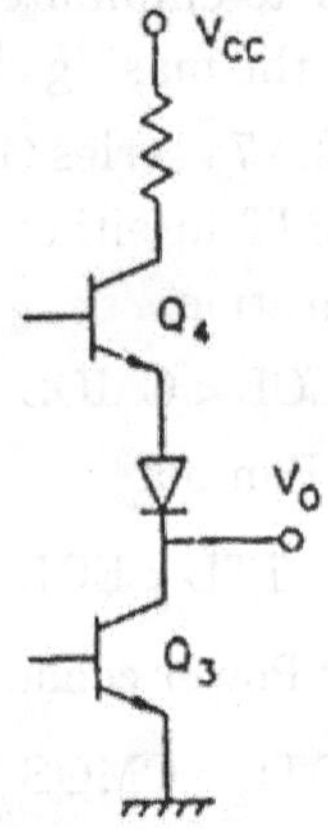

The diode switches off transistor Q_4 and V_0 is the voltage of Q_3. V_{CE3}.

23. NMOS logic gate gives complemented output. The logic function F is.

$$F = \overline{A \cdot (B + C) + D \cdot E}\,.$$

24. Working of two CMOS is depicted below:

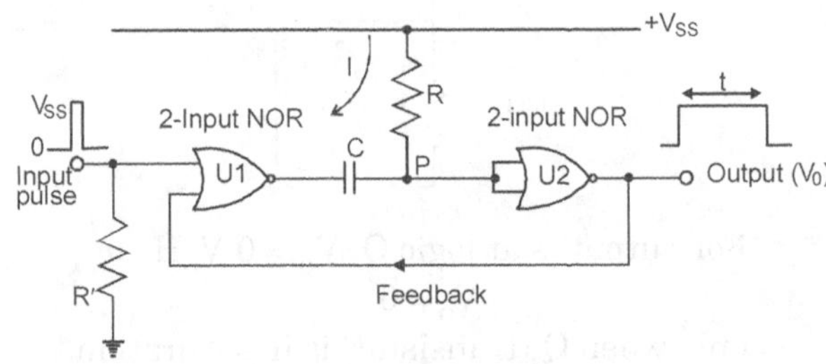

Suppose initially that the trigger input is LOW at a logic level "0" so that the output from the first NOR gate U1 is HIGH at logic level "1". The resistor, R is connected to the supply voltage so it is also equal to logic level "1", which means that the capacitor, C has the same charge on both of its plates. Junction P is therefore equal to this voltage so the output from the second NOR gate U2 will be low at logic level "0". This then represents the circuits "Stable State" with zero output.

When a positive trigger pulse is applied to the input at time t_0, the output of the first NOR gate U1 goes low taking with it the left hand plate of capacitor C1 there by discharging the capacitor. As both plates of the capacitor are now at logic level "0", so too is the input to the second NOR gate, U2 resulting in an output equal to logic level "1". This then represents the circuits second state, the "Unstable State" with an output voltage equal to $+V_{SS}$.

The second NOR gate, U2 will maintain this second unstable state until the timing capacitor now charging up through resistor, R reaches the minimum input Thershold voltage of U2 causing it to change state as a logic level "1" value has now appeared on its input. This causes the output to be rest to logic "0" which inturn is fed back (feed back loop) to one input of U2. This action automatically returns the monostable back to its original stable state and awaiting a second trigger pulse to restart the timing process once again.

25.

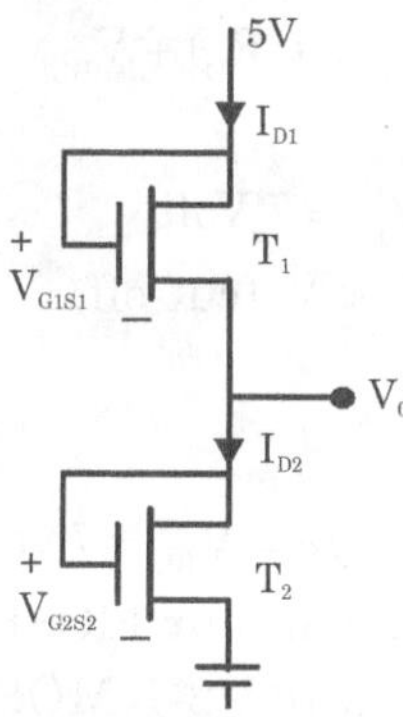

$$I_{01} = I_{02}$$

$$K_1 (V_{G1S1} - V_t)^2 = K_2 (V_{G2S2} - V_t)^2$$

$$\Rightarrow \quad 36 [(5 - V_0) - 1]^2 = 9[(V_0 - 0) - 1]^2$$

$$V_0 = 3V$$

26. The output Out is 1 only when the mode (S) is 0, since then the UPPER MOSFET is shorted and connected to V_{dd}.

Node S is zero when P = Q =1 as it shorts then lower MOSFET's

Hence Output logic Y = P AND Q

27. Series combination of N-MOS is equivalent to AND and parallel combination is equivalent to OR.

$$Y = \overline{C \cdot (A + B)} = \overline{C} + \overline{A + B} = \overline{C} + \overline{A} \cdot \overline{B}$$

28.

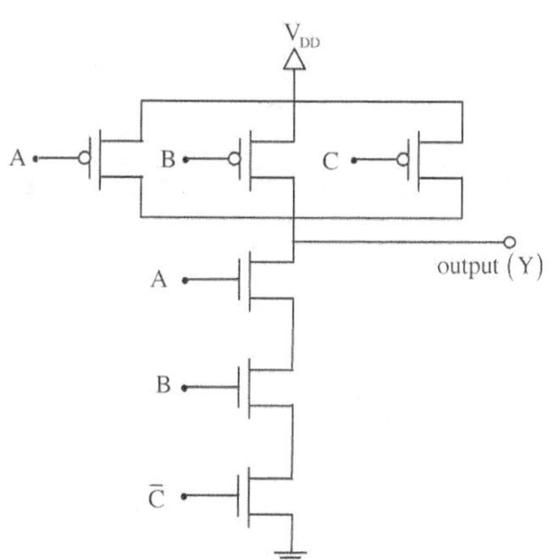

This circuit is CMOS implementation

If the NMOS is connected in series, then the output expression is product of each input with complement to the final product.

So, $\quad Y = \overline{A.B.\overline{C}} = \overline{A} + \overline{B} + C$

29. With of PMOS transistor should be halved while width of N-MOS transistors should not be changed, because.

NMOS transistors are in parallel. If any tranistor is ON, output goes low.

30. The output Y will be logic 1, when A = 1, B = 1 and $\overline{B} = 0$.

$\Rightarrow Y = A \cdot B \cdot B = A \cdot B.$

The logic implement by the ciracuit is AND operation.

31. If P = 1 ⇒ PMOS is OFF, NMOS is on } $Y = \overline{Q}$

If P = 0 ⇒ PMOS is ON, NMOS is OFF } $Y = Q$

P	Q	Y
0	0	0
0	1	1
1	0	1
1	1	0

⇒ Ex – OR gate.

32. From pull-down network,

$$\overline{f(X,Y)} = \overline{X}\overline{Y} + XY = X \odot Y$$

$$f(X,Y) = \overline{X \odot Y} = X \oplus Y$$

33. PMOS behaves as ON switch for logic 0.

and NMOS behaves as ON switch for logic 1.

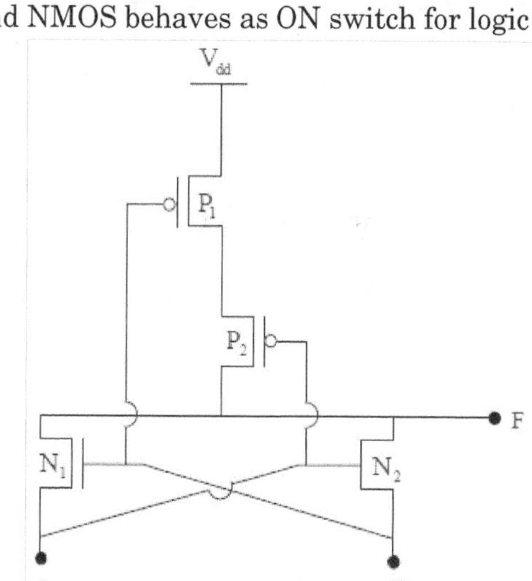

Truth table

A	B	P_1	N_1	P_2	N_2	F
0	0	0	0	0	0	1
0	1	1	1	0	0	0
1	0	0	0	1	1	0
1	1	1	1	1	1	1

So from the above truth table we can say $F = A \odot B$.

SEMICONDUCTOR MORIES (SRAM, DRAM, ROM)

34. PROM contains of fixed AND array and a programmable OR array.

    PLA contains a programmable AND array and a programmable OR array.

35. A dynamic RAM consists of 1 transistor and 1 capacitor.

36.

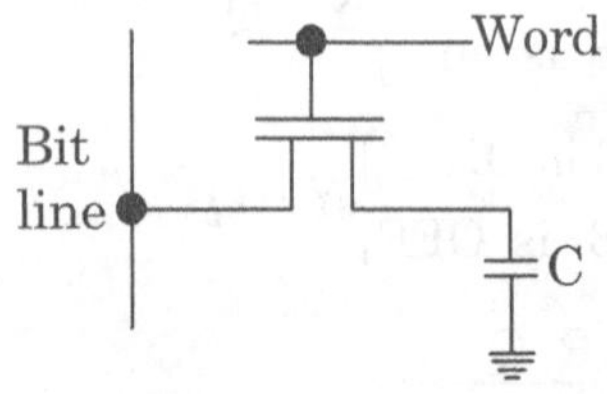

37. Each cell of a static RAM contains six MOS transistors. Each bit on static RAM is stored on 4 transistors out of which 2 are PMOS and 2 are NMOS that form cross-coupled inverters. Remaining two transistors are used to control reading from or writing into cell.

38.

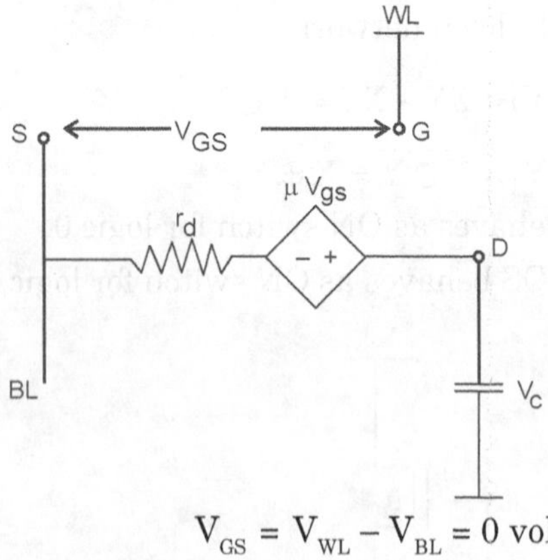

$$V_{GS} = V_{WL} - V_{BL} = 0 \text{ volt}$$

(A) $V_{WL} = 5V, V_{BL} = 5V$

For, $V_{gs} = 0V < V_T$ (cut-off),

$I_D = 0$

and $V_{gs} = 0,$

the capacitor voltage $= V_{BL} = 5V$

(B) $V_{WL} = 5V, V_{BL} = 3V,$

$V_{GS} = V_{WL} - V_{BL} = 2$ volt

$V_{GS} > V_T$, therefore current I_D from through capacitor

and $V_C = -(V_{gs} + I_D R_D) + V_{BL}$

or $V_C < V_{BL}$

(C) $V_{WL} = 5V, V_{BL} = 7$ Volt

$V_{GS} = -2$ volt $< V_T$ (cut-off)

Then, $I_D = 0$

and $V_{gs} = 0$

then capacitor voltage $= V_{BL} = 7$ volts

39. For an SRAM construction four MOSFETs are required (2-PMOS and 2-NMOS) with interchanged outputs connected to each CMOS inverter. So option (*b*) is correct.

40. In a DRAM, data is stored in the form of charge on capacitor and periodic refreshing is needed to restore the charge on capacitor.

41. When $W_0 = V_{DD}$, $B_0 = V_{DD}$; otherwise $B_0 = 0$.

    $\Rightarrow$ When $W_1 = V_{DD}$, $B_1 = V_{DD}$; otherwise $B_1 = 0$.

    So, $B_0 = W_0$ and $B_1 = W_1$.

$$\Rightarrow \begin{matrix} W_0 \\ W_1 \end{matrix} \begin{bmatrix} B_0 & B_1 \\ D_{00} & D_{01} \\ D_{10} & D_{11} \end{bmatrix} = \begin{bmatrix} 1 & 0 \\ 0 & 1 \end{bmatrix}$$

6 CHAPTER

Data Converters

Analysis of Previous GATE Papers												
	Year → Topics ↓		2019	2018	2017 Set 1	2017 Set 2	2016 Set 1	2016 Set 2	2016 Set 3	2015 Set 1	2015 Set 2	2015 Set 3
ADC	1 Mark	MCQ Type										
		Numerical Type										
	2 Marks	MCQ Type						1				
		Numerical Type										
	Total							2				
DAC	1 Mark	MCQ Type										
		Numerical Type										
	2 Marks	MCQ Type										
		Numerical Type										
	Total											

ADC

1. (a) Successive approximation.

(b) Dual-slope.

(b) Parallel Comparator.

Maximum conversion time for 8-bit ADC in clock cycles.

(1) 1 (2) 2 (3) 16

(4) 256 (5) 512

[1994 : 1 Mark]

2. For an ADC, match the following: if

List -I

(A) Flash converter

(B) Dual slope converter

(C) Successive Approximation Converter

List - II

(1) requires a conversion time of the order of a few seconds.

(2) requires a digital-to-analog converter.

(3) minimizes the effect of power supply interference.

(4) requires a very complex hardware.

(5) is a tracking A/D converter.

[1995 : 1 Mark]

3. The advantage of using a dual slope ADC in a digital voltmeter is that

(a) its conversion time is small.

(b) its accuracy is high.

(c) it gives output in BCD format.

(d) it does not require a comparator.

[1998 : 1 Mark]

4. The resolution of a 4-bit counting ADC is 0.5 Volts. For an analog input of 6.6 Volts, the digital output of the ADC will be

(a) 1011 (b) 1101

(c) 1100 (d) 1110

[1999 : 1 Mark]

5. An 8 bit successive approximation analog to digital converter has full scale reading of 2.55 V and its conversion time for an analog input of 1 V is 20 ps. The conversion time for a 2 V input will be

(a) 10 μs (b) 20 μs

(c) 40 μs (d) 50 μs

[2000 : 1 Mark]

6. The number of comparators in a 4-bit flash ADC is

(a) 4 (b) 5

(c) 15 (d) 16

[2000 : 1 Mark]

7. The number of comparators required in a 3-bit comparator type ADC is

(a) 2 (b) 3

(c) 7 (d) 8

[2002 : 1 Marc]

8. The minimum number of comparators required to build an 8-bit flash ADC is

(a) 8

(b) 63

(c) 255

(d) 256

[2003 : 1 Mark]

9. In an N-bit flash ADC, the analog voltage is fed simultaneously to $2^N - 1$ comparators. The output of the comparators is then encoded to a binary format using digital circuits. Assume that the analog voltage source V_{in}, (whose output is being converted to digital format) has a source resistance of 75 Ω as shown in the circuit diagram below and the input capacitance of each comparator is 8 pF.

The input must settle to an accuracy of 1/2 LSB even for a full scale input change for proper conversion. Assume that the time taken by the thermometer to binary encoder is negligible.

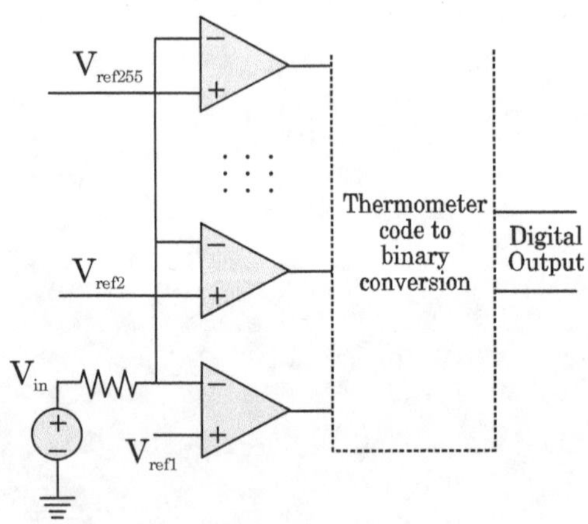

If the flash ADC has 8 bit resolution, which one of the following alternatives is closest to the maximum sampling rate ?

(a) 1 megasamples per second

(b) 6 megasamples per second

(c) 64 megasamples per second

(d) 256 megasamples per second

[2016 : 2 Marks, Set-2]

DAC

10. Which of the resistance networks of figure can be used as 3 bit R-2R ladder DAC? Assume V_0 corresponds to LSB.

(i)

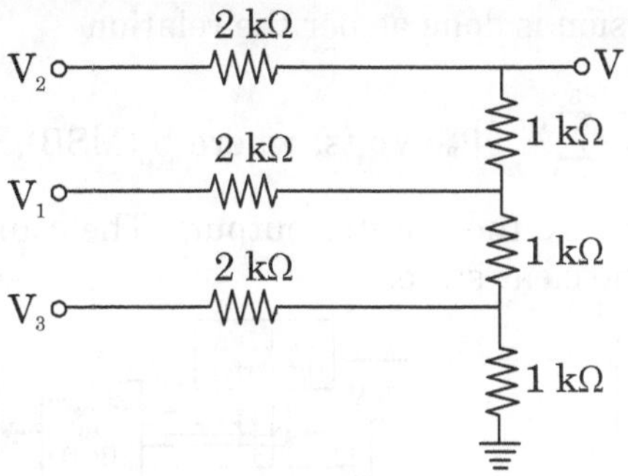

(ii)

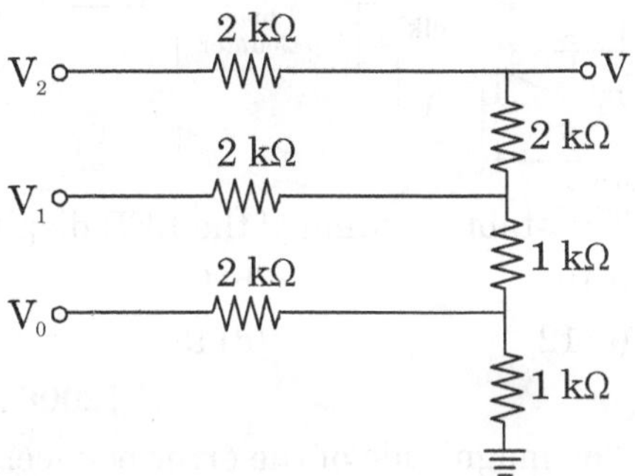

(iii)

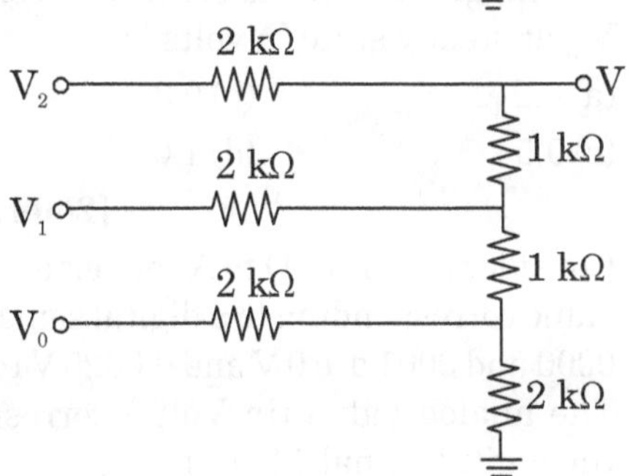

(a) Both (i) and (ii) (b) Both (i) and (ii)
(c) Only (iii) (d) Only (ii)

[1990 : 1 Mark]

11. For the 4-bit DAC shown in the figure, the output voltage V_0 is

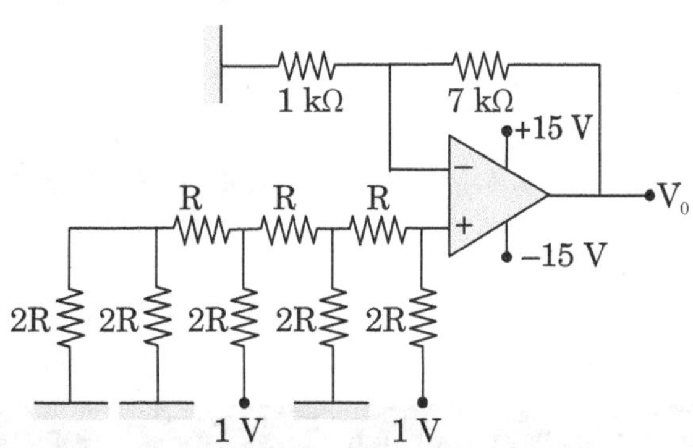

(a) 10 V (b) 5 V
(c) 4 V (d) 8 V

[2002 : 2 Marks]

12. The circuit shown in the figure is a 4 bit DAC

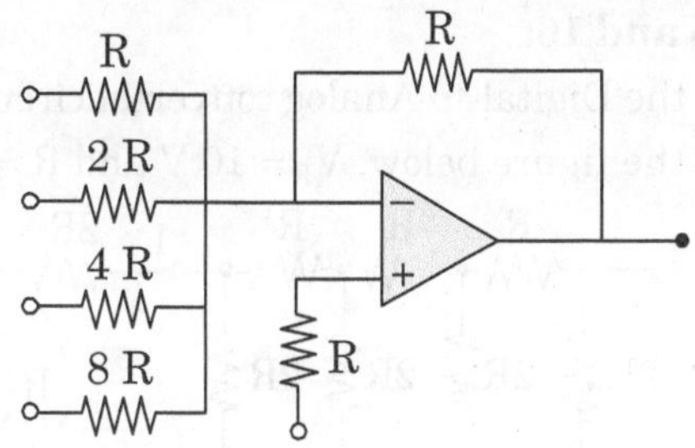

The input bits 0 and 1 are represented by 0 and 5 V respectively. The OpAmp is ideal, but all the resistances and the 5 V inputs have a tolerance of ±10%. The specification (rounded to the nearest multiple of 5%) for the tolerance of the DAC is

(a) ±35% (b) ±20%
(c) ±10% (d) ±5%

[2003 : 2 Marks]

13. A digital system is required to amplify a binary encoded audio signal. The user should be able to control the gain of the amplifier from a minimum to a maximum in 100 increments. The minimum number of bits required to encode, in straight binary, is

(a) 8 (b) 6
(c) 5 (d) 7

[2004 : 1 Mark]

14. A 4-bit D/A converter is connected to a free-running 3-bit UP counter, as shown in the following figure. Which of the following waveforms will be observed at V_0?

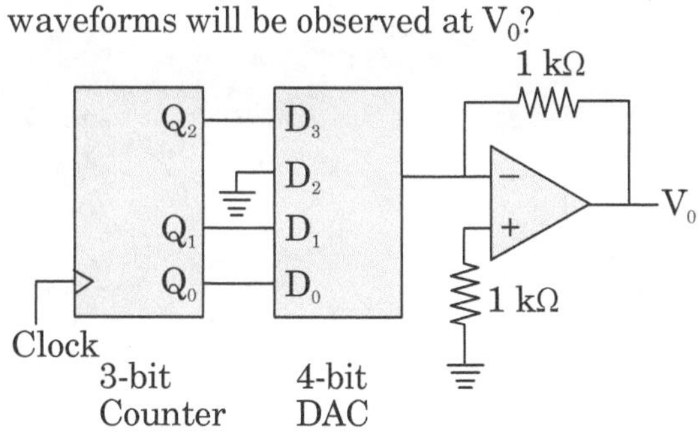

In the figure shown above, the ground has been shown by the symbol

(a) 

(b) 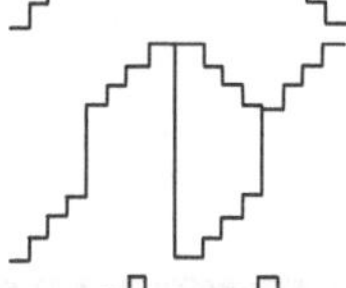

(c) 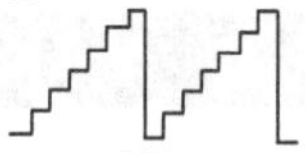

(d) 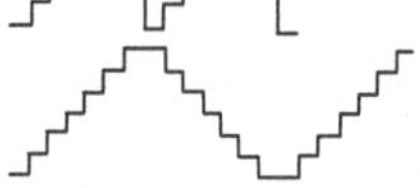

[2006 : 2 Marks]

Statement for Linked Answer Questions 15 and 16:

In the Digital-to-Analog converter circuit shown in the figure below. V_R = 10 V and R = 10 kΩ

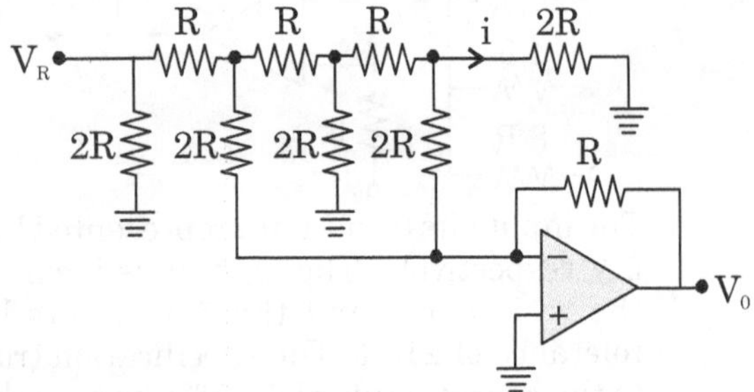

15. The current is
(a) 31.25 μA
(b) 62.5 μA
(c) 125 μA
(d) 250 μA

[2007 : 2 Marks]

16. The voltage V_o is
(a) –0.781 V
(b) –1.562 V
(c) –3,125 V
(d) –6.250 V

[2007 : 2 Marks]

Statement for linked Answer Questions 17 and 18:

In the following circuit, the comparator output is logic "1", if $V_1 > V_2$ and is logic "0" otherwise. The D/A conversion is done as per the relation

$$V_{DAC} = \sum_{n=0}^{3} 2^{n-1} b_n \text{ Volts, where } b_3\text{(MSB)}, b_1, b_2 \text{ and}$$

b_0(LSB) are the counter outputs. The counter starts from the clear state.

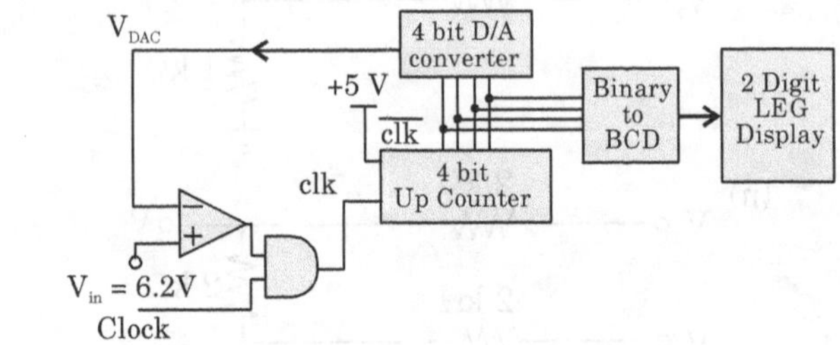

17. The stable reading of the LED displays is
(a) 06 (b) 07
(c) 12 (d) 13

[2008 : 2 Marks]

18. The magnitude of the error between V_{DAC} and V_{in} at steady state in volts is
(a) 0.2 (b) 0.3
(c) 0.5 (d) 1.0

[2008 : 2 Marks]

19. Consider a four bit D to A converter, The analog value corresponding to a digital signals of values 0000 and 0001 are 0 V and 0.0625 V respectively. The analog value (in Volts) corresponding to the digital signal 1111 is ______.

[2015 : 1 Mark, Set-1]

ANSWERS

1. (A-2, B-4, C-1) 2. (A-5, B-3, C-2) 3. (b) 4. (b) 5. (b) 6. (c) 7. (c) 8. (c)
9. (a) 10. (c) 11. (b) 12. (a) 13. (d) 14. (b) 15. (b) 16. (c) 17. (d) 18. (b)
19. (0· 93 - 0· 94)

EXPLANATIONS

1. (a) For successive approximation,

$T_{max} = nT_{CLK} = 8T_{CLK}$.

(b) For dual slope, $T_{max} = 2^{n+1} \cdot T_{CLK}$

$= 2^{8+1} \cdot T_{CLK} = 512\ T_{CLK}$.

(c) Parallel comparator, $T_{max} = 1.\ T_{CLK} = T_{CLK}$.

2. ⇒ Flash converter : requires a very complex hardware.

⇒ Dual slope converter : minimizes the effect of power supply interference.

⇒ Successive Approximation converter: requires a digital-to-analog converter.

3. The advantage of using a dual slope ADC in a digital voltmeter is its high accuracy.

4. Output voltage,

$$V_0 = \frac{\text{Input voltage, } (V_i)}{\text{Resoultion}} = \frac{6.6}{0.5} = 13.2 \cong 14.$$

Binary output = 1110.

5. Conversion time of successive approximate ADC depends upon the number of bits only.

6. Number of comparators

$= 2^n - 1$

$= 2^4 - 1 = 15$

7. Number of comparators $= 2^N - 1 = 2^3 - 1 = 7$

8. Minimum number of comparators $= 2^N - 1$

$= 2^8 - 1 = 255$

9. The total capacitance $= (2^N - 1) \times C$

$= (2^8 - 1) \times 8$ pF

$= 2.04$ nF

(Here the input capacitance of each comparator = 8 pF)

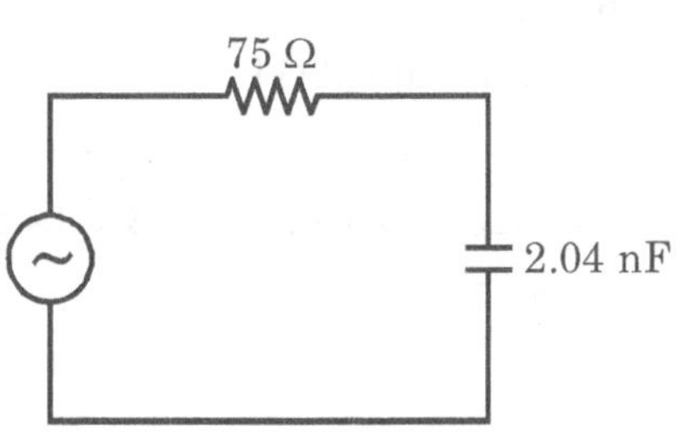

The time constant $= \tau = RC = 153$ ns

∵ Settling time $(T_s) = 5\tau = 5\ RC = 765$ ns

$$\therefore \text{ Sampling rate } = \frac{1}{\text{Settling time } (T_s)}$$

≈ 1 M samples/sec

Hence, the maximum sampling rate is 1 mega sample per second.

10. R-2R ladder DAC has R with ground and 2R with input.

11.

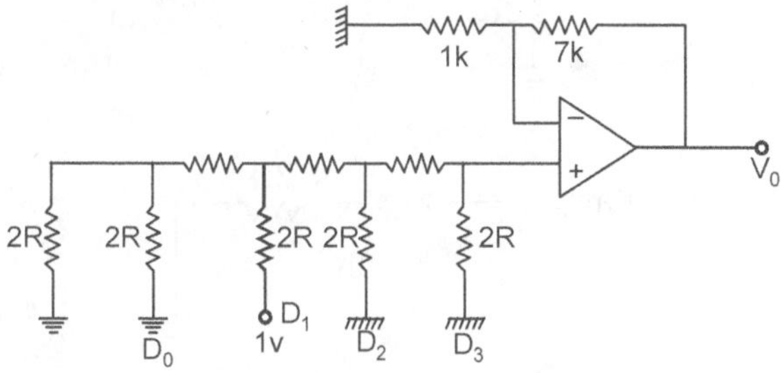

Voltage at non-inverting terminal,

$$V_T = \frac{V_R}{2^n}\left[D_0 2^0 + D_1 2^1 + D_2 2^2 + \cdots + D_{n-1} 2^{n-1}\right]$$

$$= \frac{V_R}{2^n}[0 + 2 \times 1 + 0 + 2^3 \times 1]$$

$$= \frac{5}{8} \text{volts}$$

Using KCL at inverting terminal, we have

$$\frac{V_- - 0}{1K} + \frac{V_- - V_0}{7} = 0$$

⇒ $V_0 = 8V_- = 8V_+ = 5$ volts

12. Output voltage,

$$V_0 = -V_R\left[d_3 \cdot \frac{R}{R} + d_2 \cdot \frac{R}{2R} + d_1 \cdot \frac{R}{4R} + d_o \cdot \frac{R}{8R}\right]$$

$$= -V_R \cdot \frac{R}{R} \text{ (constant).}$$

Worst case tolerant is $V_0 = \left[\frac{1.1 \times 1.1}{0.9}\right]\% = 35\%$

13. $2^7 = 128$

Hence for 100 increments 7 bits are required

14.

Number of clock pulse:	1	2	3	4	5	6	7
	↓	↓	↓	↓	↓	↓	↓
Output of counter:	001	010	011	100	101	110	1000
	↓	↓	↓	↓	↓	↓	↓
Output of decode :	0001	0010	0011	1000	1001	1010	0000
	1	2	3 ↷	8	9	10	0

On looking the DAC output, option (b) is correct.

15.

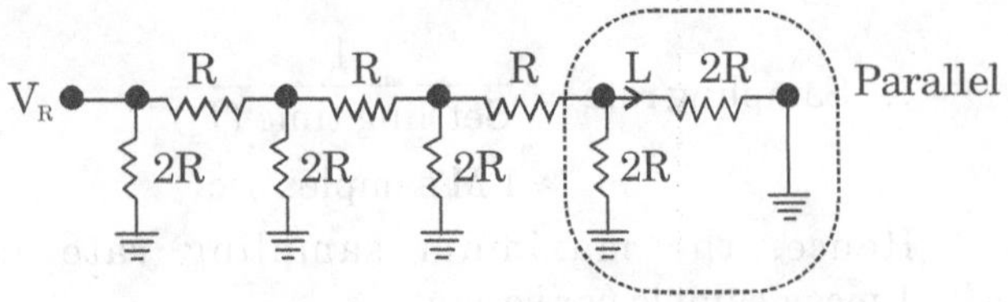

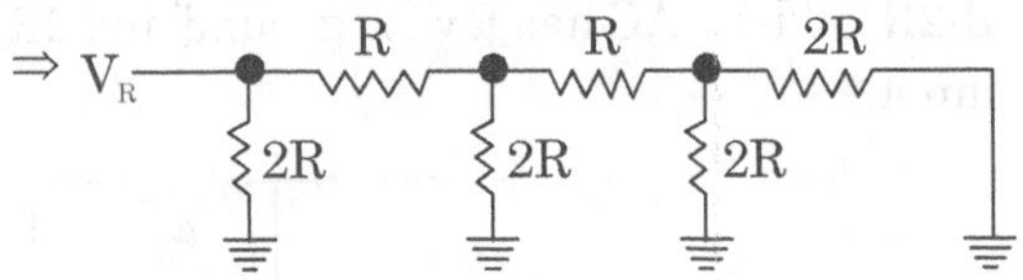

Similarly

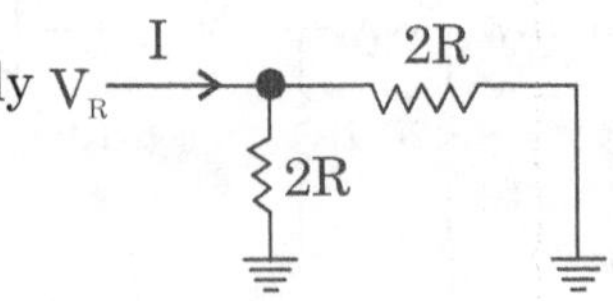

$$I = \frac{V_R}{QR} = \frac{10}{10k\Omega} = 1 \text{ mA}$$

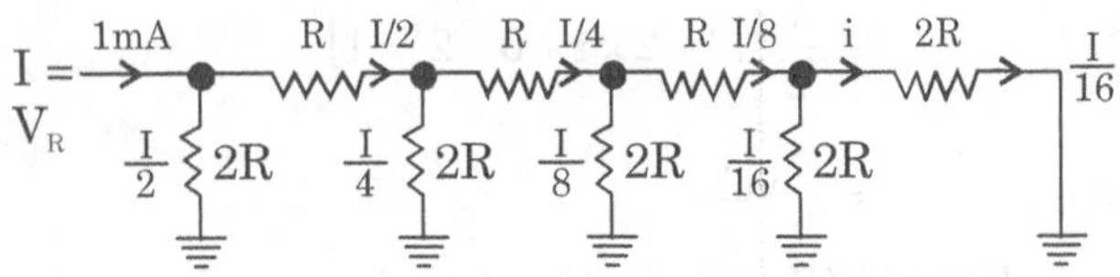

$$\Rightarrow i = \frac{I}{16} = \frac{1}{16} \text{ mA} = 62.5\ \mu\text{A}.$$

16. Total current in inverting terminal of 0p – Amp,

$$= \frac{I}{4} + \frac{I}{16} = \frac{5I}{16}$$

Output voltage,

$$V_0 = -R \times \frac{5I}{16} = \frac{-10 \times 10^3 \times 5 \times 1 \times 10^{-3}}{16}$$

$$= -3.125\text{V}$$

17. Output voltage of DAC,

$$V_0 = 2^{-1}.b_0 + 2^0.b_1 + 2^1.b_2 + 2^2.b_3$$

$$= 0.5\, b_0 + b_1 + 2b_2 + 4b_3.$$

Counter output will start from 0000 and will increase by 1 at every clock pulse.

b_3	b_2	b_1	b_0	V_0
0	0	0	0	0
0	0	0	1	0.5
0	0	1	0	1
0	0	1	1	1.5
0	1	0	0	2
0	1	0	1	2.5
0	1	1	0	3
0	1	1	1	3.5
1	0	0	0	4
1	0	0	1	4.5
1	0	1	0	5
1	0	1	1	5.5
1	1	0	0	6
1	1	0	1	6.5
1	1	1	0	7
1	1	1	1	7.5

Counter will increase till $V_{in} > V_0$. So, when $V_0 = 6.5$ V, the comparator output will be zero and the counter will be stable at that reading. The corresponding reading of LED display is 13.

18. Magnitude of error between V_0 and V_{in} at steady-state = 6.5 – 6.2 = 0.3V.

19. We know that for n bit D/A converter

$$V_0 = K\{2^{n-1} b_{n-1} + 2^{n-2} b_{n-2} + ... + 2^0 b_0\}$$

For 4 bit D/A converter

$$n = 4$$

$$V_0 = K\{2^3 b_3 + 2^2 b_2 + 2^1 b_1 + 2^0 b_0\}$$

when digital signal is 0001

$$V_0 = 0.0615 \text{ volt}$$

$$0.0615 = K\{2^3 \times 0 + 2^2 \times 0 + 2^1 \times 0 + 2^0 \times 1\}$$

$$K = 0.0615$$

when digital signal is 1111

$$V_0 = ?$$

$$V_0 = 0.0615\{2^3 \times 1 + 2^2 \times 1 + 2^1 \times 1 + 2^0 \times 1\}$$

$$= 0.0615 \times 15$$

Unit - VII

Microprocessor

Syllabus

8-bit microprocessor (8085): architecture, programming, memory and I/O interfacing.

Contents

1
CHAPTER

Basics of Microprocessor

Analysis of Previous GATE Papers

Year → / Topics ↓			2019	2018	2017 Set 1	2017 Set 2	2016 Set 1	2016 Set 2	2015 Set 1	2015 Set 2	2015 Set 3
COMPONENTS OF A MICROPROCESSOR	1 Mark	MCQ Type									
		Numerical Type									
	2 Marks	MCQ Type									
		Numerical Type									
	Total										
ARCHI TECTURE OF 8085 MICROPROCESSOR	1 Mark	MCQ Type									
		Numerical Type									
	2 Marks	MCQ Type									
		Numerical Type									
	Total										

COMPONENTS OF A MICROPROCESSOR

1. An 'Assembler' for a microprocessor is used for
   (a) assembly of processors in a production line.
   (b) creation of new programs using different modules.
   (c) translation of a program from assembly language to machine language.
   (d) translation of a higher level language into English text.

[1995 : 1 Mark]

2. An I/O processor controls the flow of information between
   (a) cache memory and I/O devices
   (b) main memory and I/O devices
   (c) two I/O devices
   (d) cache and main memories

[1995 : 1 Mark]

3. When a CPU is interrupted, it
   (a) stops execution of instructions.
   (b) acknowledges interrupt and branches to a subroutine.
   (c) acknowledge interrupt and continues.
   (d) acknowledges interrupt and waits for the next instruction from the interrupting device.

[1995 : 1 Mark]

ARCHITECTURE OF 8085 MICROPROCESSOR

4. In a microcomputer, WAIT states are used to
   (a) make the processor wait during a DMA operation.
   (b) make the processor wait during an interrupt processing.
   (c) make the processor wait during a power shutdown.
   (d) interface slow peripherals to the processor.

[1993 : 1 Mark]

5. In an 8085 microprocessor, the shift registers which store the result of an addition and the overflow bit are, respectively
   (a) B and F (b) A and F
   (c) H and F (d) A and C

[1993 : 1 Mark, Set-1

6. In a microprocessor system, the stack is used for
   (a) storing the program return address whenever a sub-routine jump instruction is executed.
   (b) transmitting and receiving input-output data.
   (c) storing all important CPU register contents whenever an interrupt is to be serviced.
   (d) storing program instructions for interrupt service routines.

[1993 : 1 Mark]

7. In a microprocessor, the register which holds the address of the next instruction to be fetched is
   (a) Accumulator (b) Program Counter
   (c) Stack Pointer (d) Instruction Register

[1993 : 1 Mark]

8. The number of hardware interrupts (which require an external signal to interrupt) present in an 8085 microprocessor are
   (a) 1 (b) 4
   (c) 5 (d) 13

[2000 : 1 Mark]

9. In a microprocessor, the service routine for a certain interrupt starts from a fixed location of memory which cannot be externally set, but the interrupt can be delayed or rejected. Such an interrupt is
   (a) non-maskable and non-vectored.
   (b) maskable and non-vectored.
   (c) non-maskable and vectored.
   (d) maskable and vectored.

[2009 : 1 Mark]

ANSWERS

1. (c) **2.** (b) **3.** (b) **4.** (c) **5.** (b) **6.** (a) **7.** (b) **8.** (c) **9.** (d)

EXPLANATIONS

2. An I/O processor controls the flow of data (or information) between main memory and I/O devices.
3. Microprocessor is also called as "CPU on a chip." When microprocessor is interrupted, it acknowledges interrupt and waits for the next instruction from the interrupting device.
4. In a microcomputer, WAIT states (with READY signal) are used to interface slow peripherals to the microprocessor.
5. In 8085 M.P.

   Result of addition → Accumulator overflow bit is carry which is stored in flag register

   Shift Registers are A (accumulator), F (flag register)
6. Stack is a temporary set of memory locations in the main memory (generally upper end). These memory locations are used to store binary information temporarily during execution of a program. Whenever a subroutine jump instruction (CALL) is executed then the stack is used for storing the program return address.
7. Program counter register is a special purpose register which stores the address of the next instruction to be executed.
8. In 8085 microprocessor, there are five hardware interrupts : TRAP, RST 7.5, RST 6.5, RST 5.5, INTR
9. If an interrupt has an address, then it is vectored interrupt, and if an interrupt can be rejected, then it is maskable interrupt.

2

CHAPTER

Programming in 8085 Microprocessor

Analysis of Previous GATE Papers

		Year → Topics ↓	2019	2018	2017 Set 1	2017 Set 2	2016 Set 1	2016 Set 2	2016 Set 3	2015 Set 1	2015 Set 2	2015 Set 3
PROGRAMMING IN 8085 MICROPROCESSOR	1 Mark	MCQ Type			1				1		1	1
		Numerical Type										
	2 Marks	MCQ Type			1		1					
		Numerical Type										
	Total				3		2		1		1	1

PROGRAMMING IN 8085 MICROPROCESSOR

1. The following program is run on an 8085 M.P. Consider the following assembly language microprocessor:

   Memory Address is in HEX Instruction

2000 H	LXI SP, 1000 H
2003 H	PUSH H
2004 H	PUSH D
2005 H	CALL 2050 H
2008 H	POP H
2009 H	HLT

   At the completion of execution of the program. the program counter of the 8085 contains ____ and the stack pointer contains ______.

   [1992 : 1 Mark]

2. The following sequence of instructions are executed by an 8085 microprocessor:

1000 H	LXI SP, 27FF H	
1003 H	CALL	1006 H
1006 H		POP H

   The contents of the stack pointer (SP) and the HL register pair on completion of execution of these instructions are,

   (a) SP = 27FF, HL = 1003

   (b) SP = 27FD, HL = 1003

   (c) SP =27FF, HL = 1006

   (d) SP = 27FD, HL = 1006 **[1996 : 1 Mark]**

3. The following instructions have been executed by an 8085 M.P.

ADDRESS (HEX)	INSTRUCTION
6010	LXI H, 8A79 H
6013	MOV A, L
6014	ADD H
6015	DAA
6016	MOV H, A
6017	PCHL

   From which address will the next instruction be fetched?

   (a) 6019

   (b) 6379

   (c) 6979

   (d) None of the above

   [1997 : 1 Mark]

4. In an 8085 microprocessor system, the RST instruction will cause an interrupt

   (a) only if an interrupt service routine is not being executed.

   (b) only if a bit in the interrupt mask is made 0.

   (c) only if interrupts have been enabled by an El instruction.

   (d) None of the above.

   [1997 : 1 Mark]

5. In register index addressing mode the effective address is given by

   (a) the index register value

   (b) the sum of the index register value and the operand.

   (c) the operand.

   (d) the difference of the index register value and the operand.

   [1988 : 1 Mark]

6. An instruction used to set the carry flag in a computer can be classified as

   (a) data transfer

   (b) arithmetic

   (c) logical

   (d) program control

   [1998 : 1 Mark]

7. The total number of memory accesses involved (inclusive of the op-code fetch) when an 8085 processor executes the instruction LDA 2003 H is

   (a) 1 (b) 2

   (c) 3 (d) 4

   [1999 : 1 Mark]

8. The contents of Register (B) and Accumulator (A) of 8085 microprocessor are 49 H and 3A H respectively. The contents of A and the status of carry flag (CY) and sign flag (S) after executing SUB B instruction are

   (a) A = F1 H, CY= 1, S = 1.

   (b) A = 0F H, CY = 1, S = 1.

   (c) A = F0 H, CY = 0, S = 0.

   (d) A= 1F H, CY = 1, S = 1.

   [2000 : 2 Marks]

9. In the 8085 microprocessor, the RST6 instruction transfers the program execution to the following location

   (a) 30 H (b) 24 H

   (c) 48 H (d) 60 H

   [2000 : 1 Mark]

10. Consider the following assembly language program.

```
          MVI B, 87H
          MOV A, B
START:    JMP NEXT
          MVI B, 00 H
          XRA    B
          OUT PORT 1
          HLT
NEXT:     XRA B
          JP  START
          OUT PORT 2
          HLT
```

The execution of the above program in an 8085 microprocessor will result in

(a) an output of 87H at PORT/1.

(b) an output of 87H at PORT/2.

(c) infinite looping of the program execution with accumulator data remaining at 00H.

(d) infinite looping of the program execution with accumulator data alternating between 00H and 87H.

[2002 : 2 Marks]

11. In an 8085 microprocessor, the instruction CMP B has been executed while the content of the accumulator is less than that of register B. As a result

(a) carry flag will be set but zero flag will be reset

(b) carry flag will be reset but zero flag will be set

(c) both carry flag and zero flag will be reset

(d) both carry flag and zero flag will be set

[2003 : 2 Marks]

12. The number of memory cycles required to execute the following 8085 instructions

(I) LDA 3000 H

(II) LXI D, F0F1 H

would be

(a) 2 for (1) and 2 for (II)

(b) 4 for (I) and 3 for II)

(c) 3 for (I) and 3 for (II)

(d) 3 for (I) and 4 for (II) **[2004 : 2 Marks]**

13. Consider the sequence of 8085 instructions given below.

LXI H, 9258 H, MOV A, M; CMA; MOV M, A.

Which one of the following is performed by this sequence?

(a) Contents of location 9258 H are moved to the accumulator.

(b) Contents of location 9258 H are compared with the contents of the accumulator.

(c) Contents of location 9258 H are complemented and stored in location 9268 H.

(d) Contents of location 5892 H are complemented and stored in location 5892 H.

[2004 : 2 Marks]

14. It is desired to multiply the numbers 0AH by OB H and store the result in the accumulator. The numbers are available in registers B and C respectively. A Part of the 8085 program for this purpose is given below:

```
MVI A, 00H
LOOP; ........
......................
......................
HLT END
```

The sequence of instructions to complete the program would be

(a) JNZ LOOP, ADD B, DCR C

(b) ADD B, JNZ LOOP, DCR C

(c) DCR C, JNZ LOOP, ADD B

(d) ADD B, DCR C, JNZ LOOP

[2004 : 2 Marks]

Common data for Questions 15 and 16. Consider an 8085 microprocessor system

15. The following program starts at location 0100 H.

```
LXI SP, 00FF H
LXI H, 0107 H
MVI A, 20 H
SUB M
```

The content of accumulator when the program counter reaches 0109 H is

(a) 20 H (b) 02 H

(c) 00 H (d) FF H

[2005 : 2 Marks]

16. Initially content of accumulator are 00 H and content of memory location are 20 H.

If in addition following code exists from 0109 H onwards,

ORI 40 H

ADD M

What will be the result in the accumulator after the last instruction is executed?

(a) 40 H (b) 20 H
(c) 60 H (d) 42 H

[2005 : 2 Marks]

17. Following is the segment of a 8085 assembly language program:

```
        LXI SP, EFFF H
        CALL 3000 H
3000 H : LXI H, 3CF4 H
        PUSH PSW
        SPHL
        POP PSW
        RET
```

On completion of RET execution, the contents of SP is

(a) 3CF0 H (b) 3CF8 H
(c) EFFD H (d) EFFF H

[2006 : 2 Marks]

Statement for Linked Answer Questions 18 and 19 : An 8085 assembly language program is given below.

```
Line 1 : MVI A, B5H
     2 : MVI B, 0EH
     3 : XRI 69 H
     4 : ADD B
     5 : ANI 9B H
     6 : CPI 9F H
     7 : STA 3010 H
     8 : HLT
```

18. The contents of the accumulator just after execution of the ADD instruction in line 4 will be

(a) C3 H (b) EA H
(c) DC H (d) 69 H

[2007 : 2 Marks]

19. After execution of line 7 of the program, the status of the CY and Z flags will be

(a) CY = 0, Z = 0 (b) CY = 0, Z= 1
(c) CY = 1, Z = 0 (d) CY = 1, Z = 1

[2007 : 2 Marks]

20. An 8085 executes the following instructions

```
2710 H  LXI    H,  30A  0H
2713 H  DAD    H
2714 H  PCHL
```

All addresses and constants are in Hex. Let PC be the contents of the program counter and HL be the contents of the HL register pair just after executing PCHL.

Which of the following statements is correct?

(a) PC = 2715 H, HL = 30A0 H
(b) PC = 30A0 H, HL = 2715 H
(c) PC = 6140 H, HL = 6140 H
(d) PC = 6140 H, HL = 2715 H

[2008 : 2 Marks]

21. For the 8085 assembly language program given below, the content of the accumulator after the execution of the program is

```
3000 H    MVI   A, 45 H
3002 H    MOV B, A
3003 H    STC
3004 H    CMC
3005 H    RAR
3006 H    XRA B
```

(a) 00 H (b) 45 H
(c) 67 H (d) E7 H

[2010 : 2 Marks]

22. An 8085 assembly language program is given below. Assume that the carry flag is initially tinsel. The content of the accumulator after the execution of the program is

```
MVI     A, 07 H
RLC
MOV     B, A
RLC
RLC
ADD     B
RRC
```

(a) 8C H (b) 64 H
(c) 23 H (d) 15 H

[2011 : 2 Marks]

23. For 8085 microprocessor, the following program is executed.

```
        MVI A, 05H;
        MVI B, 05H;
PTR:    ADD B;
        DCR B;
        JNZ PTR,
        ADI 03H;
        HLT;
```

At the end of program, accumulator contains

(a) 17 H (b) 20 H
(c) 23 H (d) 05 H

[2013 : 1 Mark]

24. An 8085 microprocessor executes "STA 1234 H" with starting address location 1-FFE H (STA copies the contents of the Accumulator to the 16-bit address location). While the instruction is fetched and executed, the sequence of values written at the address pins $A_{15} - A_8$ is

(a) 1F H, 1F H, 20 H, 12 H.

(b) 1F H, FE H, 1F H, FF H, 12 H.

(c) 1F H, 1F H, 12 H, 12 H.

(d) 1F H, 1F H, 12 H, 20 H, 12 H.

[2014 : 2 Marks, Set-4]

25. Which one of the following 8085 microprocessor programs correctly calculates the product of two 8-bit numbers stored in registers B and C?

(a)
```
        MVI A, 00 H
        JNZ LOOP
        CMP C
LOOP:   DCR B
        HLT
```
(b)
```
        MVI, A. 00 H
        CMP C
LOOP:   DCR B
        HLT
```
(c)
```
        MVI A, 00 H
LOOP:   ADD C
        DCR B
        JNZ LOOP
        HLT
```
(d)
```
        MVI A, 00H
        ADD C
        JNZ LOOP
        LOOP INR B
        HLT
```

[2015 : 1 Mark, Set-3]

26. In an 8085 microprocessor, which one of the following instructions changes the content of the accumulator?

(a) MOV B, M (b) PCHL

(c) RNZ (d) SBI BE H

[2015 : 1 Mark, Set-2]

27. In an 8085 system, a PUSH operation requires more clock cycles than a POP operation. Which one of the following options is the correct reason for this?

(a) For POP, the data transceivers remain in the same direction as for instruction fetch (memory to processor), whereas for PUSH their direction has to be reversed.

(b) Memory write operations are slower than memory read operations in an 8085 based system.

(c) The stack pointer needs to be pre-decremented before writing registers in a PUSH, whereas a POP operation uses the address already in the stack pointer.

(d) Order of registers has to be interchanged for a PUSH operation, whereas POP uses their natural order.

[2016 : 2 Marks, Set-1]

28. In an 8085 microprocessor, the contents of the accumulator and the carry flag are A7 (in hex) and 0, respectively. If the instruction RLC is executed, then the contents of the accumulator (in hex) and the carry flag, respectively, will be

(a) 4E and 0 (b) 4E and 1

(c) 4F and 0 (d) 4F and 1

[2016 : 1 Mark, Set-3]

29. The clock frequency of an 8085 microprocessor is 5 MHz, If the time required to execute an instruction is 1.4 μs, then the number of T- states needed for executing the instruction is

(a) 1 (b) 6

(c) 7 (d) 8

[2017 : 1 Mark, Set-11]

30. The following FIVE instructions were executed on an 8085 microprocessor.

```
MVI A, 33 H
MVI B, 78 H
ADD B
CMA
ANI 32 H
```

The accumulator value immediately after the execution of the fifth instruction is

(a) 00 H (b) 10 H

(c) 11 H (d) 32 H

[2017 : 2 Marks, Set-1]

ANSWERS

1. (*)	2. (c)	3. (c)	4. (c)	5. (b)	6. (c)	7. (d)	8. (a)	9. (a)	10. (b)
11. (a)	12. (b)	13. (c)	14. (d)	15. (a)	16. (c)	17. (b)	18. (b)	19. (c)	20. (c)
21. (c)	22. (c)	23. (a)	24. (b)	25. (c)	26. (d)	27. (c)	28. (d)	29. (c)	30. (b)

EXPLANATIONS

1. RST 6 is software vectored interrupt, which transfers the program execution to its vector address.
   Vector address = $(6 \times 8)_{10} = (48)_{10}$ = 0030 H.
2. Program counter register contains the address of the next instruction to be executed. If program contains instructions after HLT instruction (1 Byte), then content of program counter is 2009 H + 1 H = 200A H.
   LXI SP, 1000H $\Rightarrow$ (SP) = 1000H
   PUSH H $\Rightarrow$ (SP) $\leftarrow$ (SP) – 2 = 1000 – 2 = 0FFEH
   PUSH D $\Rightarrow$ (SP) $\leftarrow$ (SP) – 2 = OFFE – 2 = 0FFCH
   CALL 2050 H $\Rightarrow$ (SP) $\leftarrow$ (SP) – 2 = 0FFA H.
   Now, program execution is transferred to subroutine. Once RET instruction is executed at the end of subroutine, stack pointer content is incremented by 2.
   (SP) $\leftarrow$ (SP) + 2 = . 0FFA H + 2 H = 0FFCH.
   POP H $\Rightarrow$ (SP) $\leftarrow$ (SP) + 2 = 0FFE H.
3. 1000 H : LXI SP, 27FF H $\Rightarrow$ (SP) = 27FF H
   1003 H : CALL 1006 H $\Rightarrow$ (SP) $\leftarrow$ (SP) – 2 ; (SP) = 27FD H
   1006 H : POP H $\Rightarrow$ (SP) $\leftarrow$ (SP) + 2 ; (SP) = 27 FF H
   (HL) = 1006 H.
   6010 H : LXI H, 8A79 H $\Rightarrow$ (HL) = 8A79 H
   6013 H : MOV A, L $\Rightarrow$ (A) $\leftarrow$ (L) ; (A) = 79 H
   6014 H : ADD H
   $\Rightarrow$ (A) $\leftarrow$ (A) + (H) = 79 H + 8A $\Rightarrow$ 03H (CY) = 1b, (AC) = 1b.
   6015 H: DAA $\Rightarrow$ (A) $\leftarrow$ (A) + 66 = 3 + 66 H = 69 H
   6016 H: MOV H, A $\Rightarrow$ (H) = 69 H, (L) = 79 H
   6017 H: PCHL $\Rightarrow$ (PC) $\leftarrow$ (HL) ; (PC) = 6979 H.
4. From given data,
   $$f_{CLK} = 5 \text{ MHz}$$
   Execution time = 1.4 μs
   Execution time = n(T – state)
   where n is number of T-states required to execute the instruction
   $$\therefore \quad T_{CLK} = \frac{1}{f_{CLK}} = 0.2\ \mu s$$
   $$\text{Hence,} \quad n = \frac{1.4\ \mu s}{T_{CLK}} = \frac{1.4}{0.2} = 7$$
5. In register index addressing mode, the effective address is given by the sum of the index register value and the operand.
6. Logical instruction is used to set the carry flag in a processor. For example, in 8085 U.P. STC instruction sets the carry flag.
7. Machine cycle 1 : LDA $\Rightarrow$ Opcode fetch
   Machine cycle 2 : 03H $\Rightarrow$ Memory read (lower byte)
   Machine cycle 3 : 20H $\Rightarrow$ Memory read (higher byte)
   Machine cycle 4 : Copying the content of address 2003 H into accumulator.
8. Before: A [3A H], B [49 H], CY [x b], S [x b]
   During SUB B:
   A = 0011 1010
   –B = 0100 10001
   A = 1111 0001 = F 1 H
   After: A [3A H], B [49 H], CY [x b], S [x b]
   MOV A, M $\Rightarrow$
   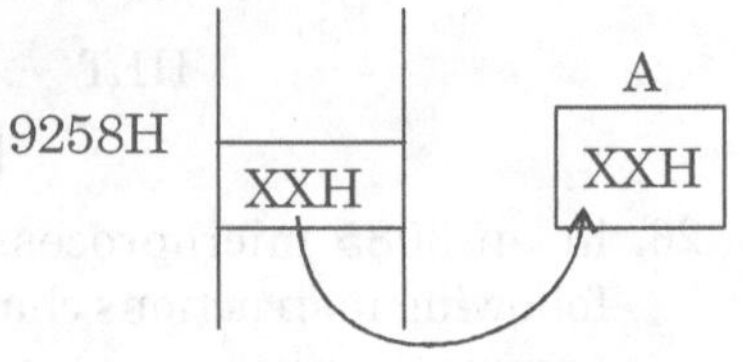

   CMA $\Rightarrow$ Comlement of content of accumulator.
   MOVM, A $\Rightarrow$ content of accumulator is stored in location 9258H
9. RST instruction will cause an interruption only if interrupts have been enabled by an interrupt enable instruction (EI) at the beginning of programm.
11. CMP B subtractes the contents of register B from the accumulator A. Since accumulator has content less than that of register B, there CY = 1 as it shows negative result. Also, the result is being non-zero the zero flag will be reset.

12. For LDA, 4 memory cycles
For LXI, 3 memory cycles

13. LXI H, 9258H ⟶ memory location 9258
MOV A, M ⟶ data moved from memory to accumulator
CMA ⟶ data complement in accumulator
MOV M, A ⟶ complemented data moved to A

Best option is *c* as data is moved from memory location 9058 H to the accumulator.

14. Procedure of Multiplication in 8085:

$0A\ H \times 0B\ H = \underbrace{0A + 0A + 0A + ... + 0A}_{0B\ times}$ or

$\underbrace{0B + 0B + ... + 0B}_{0A\ times.}$

Before: B [0A H], C [0B H], A [XXH]

Program: MVI A, 00H ⇒ A [00 H]

....

LOOP: ADD B ⇒ (A) = (A) + (B)

DCR C ⇒ (C) = (C) – 1

JNZ LOOP ⇒ Loop will continue till C = 00H i.e. loop. will execute 11 times.

....

HLT

15.

Location	Muemonics	Operation
0100 H	LXI SP, 00FF H	SP → [00FF]H
0103 H	LXIH,0107 H	H-L → [0107]H
0106 H	MVI A, 20 H	A → [20H]
0108 H	SUB M	A → [00]H
0109 H		

Address of M specified by contents of H-L memory and H-L → [0107]H, and 0107H → [20H]

When PC reaches 0109 H, contents of accumlater reduces to zero after having operation of instruction SUB M.

16. **Before:** A [00 H], M [20H]

Program: 0109H : ORI 40H ⇒

A ≡ 0000 0000b
(OR) 0100 0000b
A ≡ 01000000b = 40H

010BH: ADD M ⇒ (A) = (A) + ((M))

(A) = 40H + 20H = 60H

After : A [60H], M [20H]

17. MVI B, 87 H : copie 87 H into register B.

MOV A, B : copies the content of reg. (B) into reg. (A).

⇒ JMP NEXT : jumps the program sequence to location 'NEXT'.

⇒ XRA B : peforms EX – OR operation of contents of reg. A & B.

A ≡ 1000 0111b
B ≡ 1000 0111b
XOR ≡ 0000 0000b

Sign flag, (S) = 0b ⇒ positive.

(A) = 00H; (B) = 87 H

⇒ JP START : jumps the program sequence to location 'START'.

JMP NEXT : Again, program sequence is at location 'NEXT'.

⇒ XRA B : (A) = 87 H; (S) = 1 b (negative)

⇒ 0UT PORT 2 : 87 H at PORT2.

⇒ Halt the program.

18. ⇒ LXI SP, EFFF H : (SP) = EFFF H.

CALL 3000H : Jump to address 3000 H and (SP) = EFFF – 2 = EFFDH.

⇒ LXI H, 3CF4 H : (HL) = 3CF4H

PUSH PSW : (SP) = EFFD – 2 = EFFB H

SPHL : (SP) ← (HL) ; (SP) = 3CF4 H

POP PSW : (SP) ← (SP) + 2 ; (SP) = 3CF6 H.

RET : Return to the main program and (SP) ← (SP) + 2 ; (SP) = 3CF8 H.

19. Line 1 : MVI A, B5 H ⇒ (A) = B5 H = $(1011\,0101)_2$

Line 2 : MVI B, OEH ⇒ (B) = 0E H = $(00001110)_2$

Line 3 : XRI 69 H ⇒ A ≡ 101101016
⊕69H ≡ 011010016
(A) ≡ 11011100 = DCH

Line 4 : ADD B ⇒ A ≡ DC H
B ≡ 0E H
(A) = EAH

20. Program: 2710 H : LXI H, 30A0H ⇒ H [30] L [A0] H

2713H : DAD H ⇒ (HL) = (HL) + (HL)

= 30A0H + 30A0H

(HL) = 6140H

2714H : PCHL ⇒ (PC) = (HL)

PC [6140 H]

21. MVI A, 45H (A) = 45H
MOV B, A (B) = 45 H

STC	set carry ⇒ carry flag = 1
CMC	compliment carry ⇒ carry Flag = 0
RAR	Rotate with carry

0 1 0 0 0 1 0 0 1 0

⇒ carry Flag = 1

$$A = \underbrace{0010}_{2}\,\underbrace{0010}_{2} = 22H$$

XRA B → XOR with 45H

22H = 00100010
45 H = 01000101
Output 01100111

⇒ A = 67 H

23. (A) = EA H, (B)= 0E H

A ≡ 11101010
⊕ 9B ≡ 10011011
(A) = 8A H

Line 5 : ANI 9B H ⇒ (A) = 8 AH
Line 6 : CPI 9FH ⇒ (A) = 8A H
∴ (A) < 9FH
CY = 1, Z = 0.

24. MVI A, 05 H ⇒ (A) = 05H
MVI B, 05H ⇒ (B) = 05 H
PTR : ADD B
DCR B
JNZ PTR
⇒ (A) = 05 + 05 + 04 + 03 + 02 + 01
(A) = 20 = 14 H
ADI 03H ⇒ (A) ← (A) + 03 H
HLT (A) = 17 H.

25. Let the opcode of STA is XX H and content of accumulator is YY H.
Instruction: STA 1234 H
Starting address (given) = 1FFE H
So, the sequence of data and addresses is given below:

Address (in hex) : Data (in hex)

$A_{15} - A_8$	$A_7 - A_0$		
1F	FEH	→	XXH
1F	FFH	→	34H
20	00 H	→	12H
12	34 H	→	YYH

26. MVI A, 07 H ⇒ A • 07 H or A 0000 0111 b

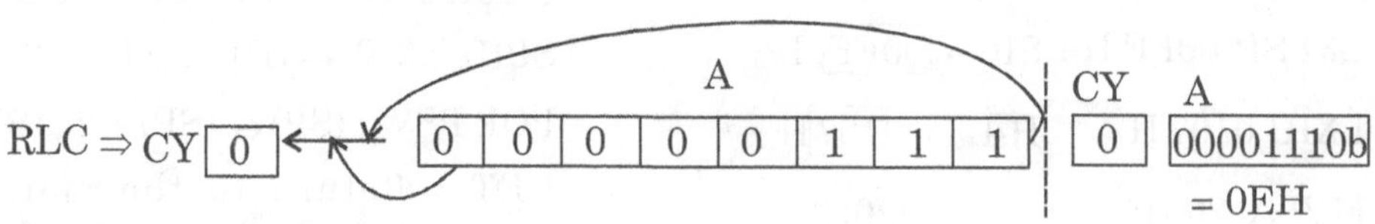

MOV B, A ⇒ (B) ← (A) ; B 0E H

CY
RLC ⇒ ob ; A 0 0 0 1 1 1 0 0 B = 1C H

RLC ⇒ CY ob ; A 0 0 0 1 1 1 0 0 B = 38 H

ADD B ⇒ (A) = (A) + (B) = 38H + 0EH = 46H = 01000110b

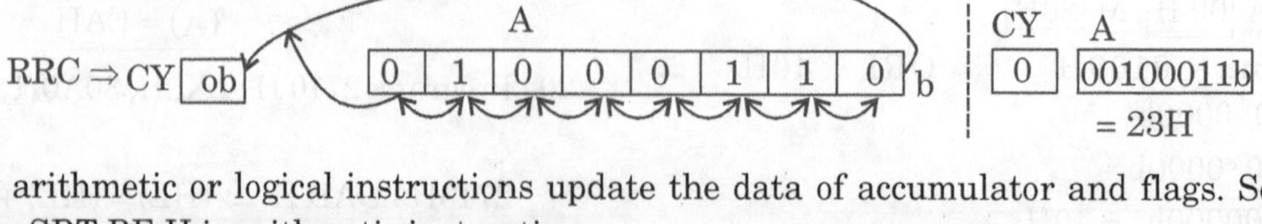

27. Generally arithmetic or logical instructions update the data of accumulator and flags. So, in the given option only SBT BE H is arithmetic instruction.
SBI BEH→ Add the content of accumulator with immediate data BE H and store the result in accumulator.

28. The stack pointer needs to be decremented before writing data into stack for reading data from stack, such pre-decrement or pre-increment operations are not needed, as already stack pointer indicates the address of stack top where the read operation takes place.

Hence, PUSH operation (12 T-state) requires more clock cycles than pop operation (10 T-states).

29. **Before :**

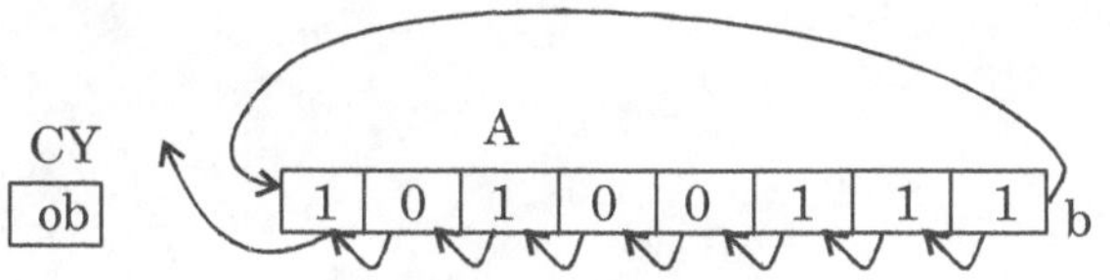

= A7 H

After :

A	0	1	0	0	1	1	1	1	b = 4F H

CY

1 b

30. 8085 microprocessor cannot perform multiplication directly The multiplication can be obtained by repeated addition, only (c) option satisfies addition repeatedly.

Here, register B is used as count to add register 'C' to accumulator, initially having 00H.

3 CHAPTER

Interfacing (Memory & I/O) with 8085 Microprocessor

Analysis of Previous GATE Papers												
	Year → Topics ↓		2019	2018	2017 Set 1	2017 Set 2	2016 Set 1	2016 Set 2	2016 Set 3	2015 Set 1	2015 Set 2	2015 Set 3
INTERFACING (MEMORY & I /O) WI TH 8085 MICROPROCESSOR	1 Mark	MCQ Type										
		Numerical Type								1		
	2 Marks	MCQ Type										
		Numerical Type										
	Total									1		

INTERFACING (MEMORY & I/O) WITH 8085 MICROPROCESSOR

1. An 8-bit microprocessor has 16-bit address bus A_0–A_{15}. The processor has a 1 kB memory chip as shown. The address range for the chip is

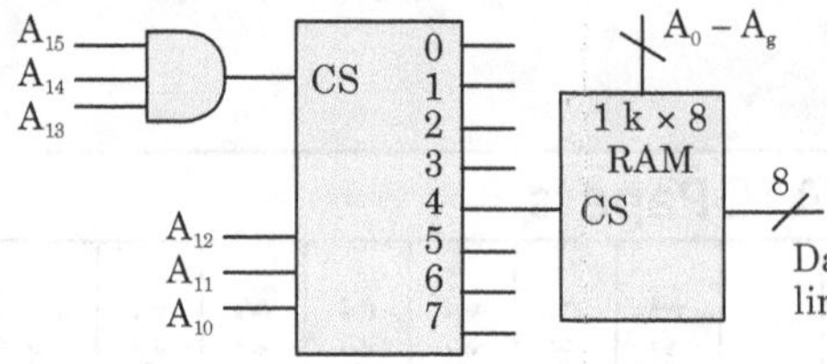

(a) F00F H to F40E H.
(b) F100 H to F4FF.
(c) P000 H to F3FF H.
(d) F700 H to FAFF H. **[1988 : 1 Mark]**

2. A microprocessor with a 16-bit address bus is used in a linear memory selection configuration (i.e. Address bus lines are directly used as chip selects of memory chips) with 4 memory chips. The maximum addressable memory space is

(a) 64 k (b) 16 k
(c) S k (d) 4 k **[1988 : 1 Mark]**

3. The minimal function that can detect a "divisible by 3" 8421 BCD code digit (representation is D_8 D_4 D_2 D_1) is given by

(a) $D_8D_1 + D_4D_2 + \bar{D}_8D_2D_1$

(b) $D_8D_1 + D_4D_2D_1 + \bar{D}_4D_2D_1 + \bar{D}_8\bar{D}_4\bar{D}_2\bar{D}_1$

(c) $D_8D_1 + D_4D_2 + \bar{D}_8\bar{D}_4\bar{D}_2\bar{D}_1$

(d) $D_4D_2\bar{D}_1 + D_4D_2D_1 + D_8\bar{D}_4D_2D_1$

[1990 : 1 Mark]

4. In an 8085 microprocessor system with memory mapped I/O,

(a) I/O devices have 16-bit addresses.
(b) I/O devices are accessed using IN and 0UT instructions.
(c) there can be a maximum of 256 input devices and 256 output devices.
(d) arithmetic and logic operations can be directly performed with the I/O data.

[1992 : 1 Mark]

5. The logic realized by the circuit shown in figure is

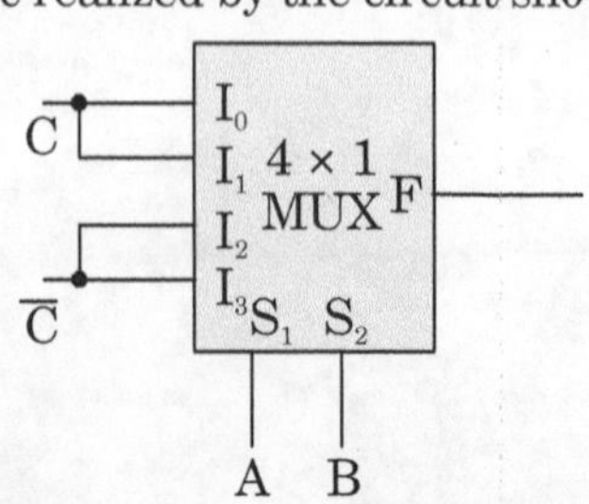

(a) F = AC (b) F = AC
(c) F = BC (d) F = BC

[1992 : 1 Mark]

6. The output of the circuit shown in figure is equal to

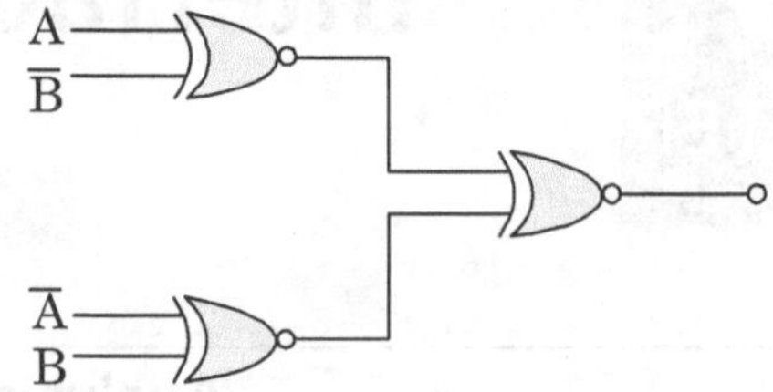

(a) 0 (b) 1

(c) $\bar{A}B + A\bar{B}$ (d) $(\overline{A \oplus B}) \oplus (\overline{A \oplus B})$

[1995 : 1 Mark]

7. A 2-bit binary multiplier can be implemented using

(a) 2 input ANDs only.
(b) 2 input XORs and 4-input AND gates only.
(c) Two (2) input NORs and one XNOR gate.
(d) XOR gates and shift registers.

[1997 : 1 Mark]

8. The decoding circuit shown in figure has been used to generate the active low chip select signal for a microprocessor peripheral. (The address lines are designated as At) to A7 for I/O addresses)

The peripheral will correspond to I/O addresses in the range

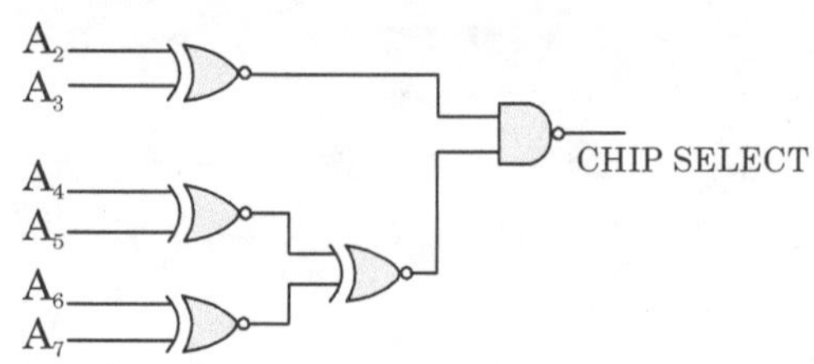

(a) 60 H to 63 H (b) A4 H to A7 H
(c) 50 H to AF H (d) 70 H to 73 H

[1997 : 1 Mark]

9. If CS = $A_{15}A_{14}A_{13}$ is used as the chip select logicof a 4K RAM in an 8085 system, then its memory range will be

(a) 3000 H - 3FFF H
(b) 7000 H - 7FFF H
(c) 5000 H - 5FFF H and 6000 H - 6FFF H
(d) 6000 H - 6FFF H and 7000 H - 7FFF H

[1999 : 2 Marks]

10. An 8085 microprocessor based system uses a 4 K × 8 bit RAM whose starting address is AA00 H. The address of the last byte in this RAM is

(a) 0FFF H (b) 1000 H
(c) B9FF H (d) BA00 H

[2001 : 1 Mark]

11. In the circuit shown in the figure, A is parallel-in. parallel-out 4 bit register, which loads at the rising edge of the clock G. The input lines are connected to a 4 bit bus, W. Its output acts as the input to a 16 × 4 ROM whose output is floating when the enable input E is 0. A partial table of the contents of the ROM is as follows:

Address	0	2	4	6	8	10	11	14
Data	0011	1111	0100	1010	1011	1000	0010	1000

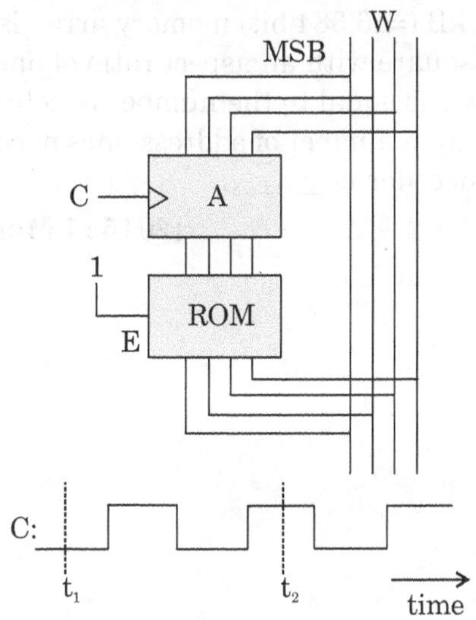

The clock to the register is shown, on the W bus at time t_1 is 0110. The bus at time t_2 is

(a) 1111 (b) 1011

(c) 1000 (d) 0010

[2003 : 2 Marks]

12. What memory address range is NOT represented by chip #1 and chip #2 in the figure. A_0 to A_{15} in this figure are the address lines and CS means Chip select.

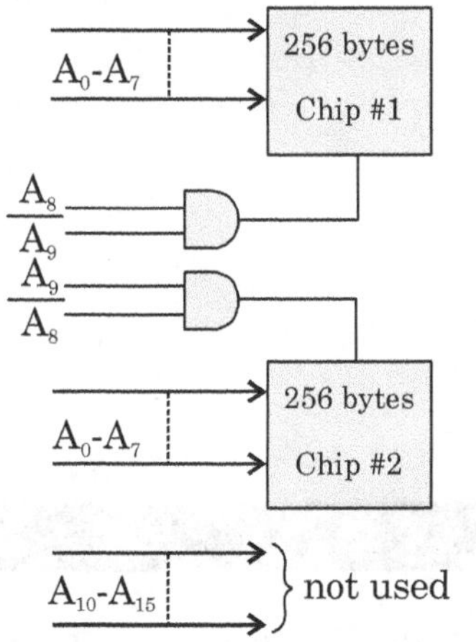

(a) 0100-02FF H (b) 1500-16FF H

(c) F900-FAFF H (d) F800-F9FF H

[2005 : 2 Marks]

13. An I/O peripheral device shown in the figure below is to be interfaced to an 8085 microprocessor. To select the I/O device in the I/O address range D4 H – D7H, its chip-select ($\overline{CS}$) should be connected to the figure

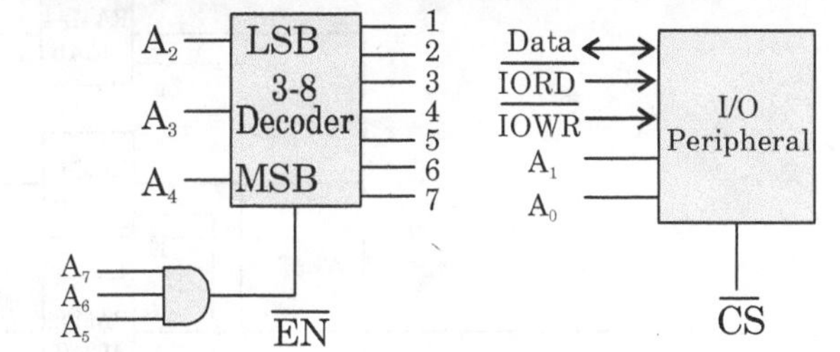

(a) output 7 (b) output 5

(c) output 2 (d) output 0

[2006 : 2 Marks]

14. An 8255 chip is interfaced to an 8085 microprocessor system as an I/O mapped I/O as shown in the figure. The address lines A_0 and A_1 of the 8085 are used by the 8255 chip to decode internally its three ports and the Control register. The address lines A_3 to A_7 as well as the IO/$\overline{M}$ signal are used for address decoding. The range of addresses for which the 8255 chip would get selected is

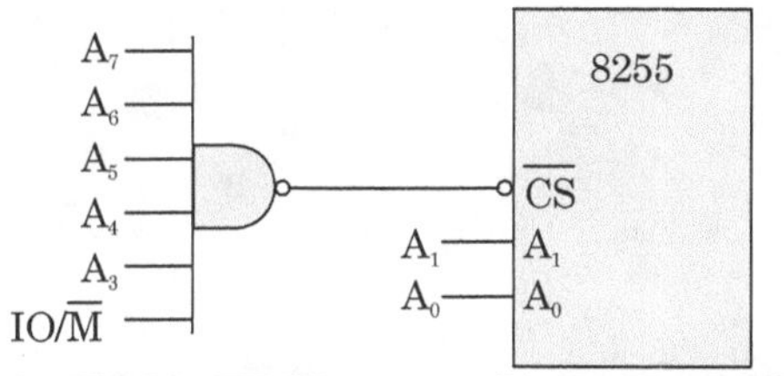

(a) F8 H –FB H (b) F8 H –FC H

(c) F8 H – FF H (d) FO H– F7 H

[2007 : 2 Marks]

15. In the circuit shown, the device connected to Y_5 can have address in the range

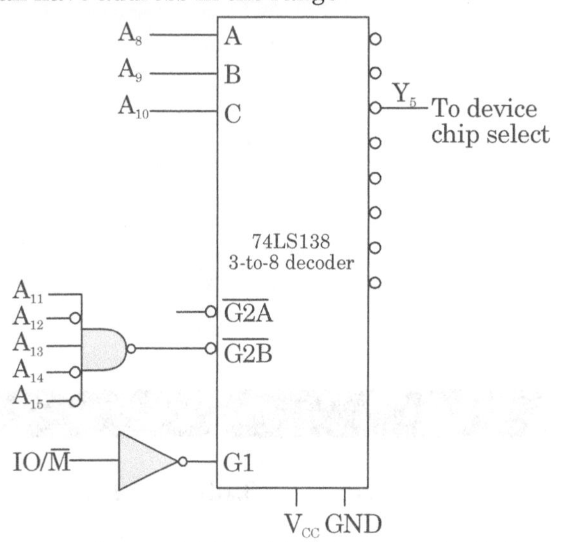

(a) 2000 H – 20FF H (b) 2D00 H – 2DFF H

(c) 2E00 H – 2EFF H (d) FD00 H – FDFF H

[2010 : 1 Mark]

16. There are four chips each of 1024 bytes connected to a 16 big address bus as shown in the figure below. RAMs 1, 2, 3 and 4 respectively are mapped to address

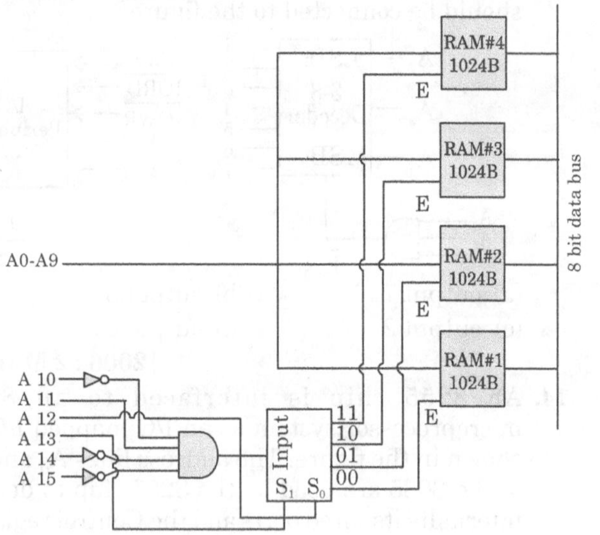

(a) 0C00 H-0FFF H, 1C000 H-1FFF H, 2000 H-2FFF H, 3000 H-3FFF H

(b) 1800 H-1FFF H, 2800 H-2FFF H, 3800 H-3FFF H, 4800 H-4FFF H

(c) 0500 H-08FF H, 1500 H-18FF H, 3500 H-38FF H, 5500 H-58FF H

(d) 0800 H-OBFF H, 18001-1BFF H, 2800 H-2BFF H, 3800 H-3BFF H

[2013 : 2 Marks]

17. A 16 kB (=16,384 bit) memory array is designed as a square with an aspect ratio of one (number of rows is equal to the number of columns). The minimum number of address lines needed for the row decoder is ______.

[2015 : 1 Mark, Set-1]

ANSWERS

1. (c)	2. (b)	3. (d)	4. (a,d)	5. (b)	6. (a)	7. (b)	8. (a)	9. (d)	10. (c)
11. (c)	12. (b)	13. (b)	14. (b)	15. (b)	16. (d)	17. (7)			

EXPLANATIONS

1. Size of memory = 1 kB = 2^{10} Bytes = 2^{10}×8–bit
   Address line = 10 ($A_2 \rightarrow A_0$)
   Data lines = 8 bit
   To enable the decorder (3×8), CS must be 1.
   Hence, $A_{15}\,A_{14}\,A_{13} \equiv 111$.
   To enable the chip through output 4 of decorder,
   $A_{12}\,A_{11}\,A_{10} \equiv 100$.

Decode enable			Chip enable				
A_{15}	A_{14}	A_{13}	A_{12}	A_{11}	A_{10}	A_9	A_8
1	1	1	1	0	0	0	0
1	1	1	1	0	0	1	1
F				3			

A_7	A_6	A_5	A_4	A_3	A_2	A_1	A_0
0	0	0	0	0	0	0	0
1	1	1	1	1	1	1	1
F				F			

Address range for chip ≡ F000 H to F3FF H.

2. Among 16 address lines, as there are 4 chips A_{15} $A_{14}\,A_{13}\,A_{12}$ are used as chip select lines. 12 Address lines (A_{11}–A_0) are left.
   Size of single chip = $2^{12} = 2^2 \times 2^{10}$ = 4 k
   for 4 chips, max addressable space = 4 × 4 k.
   = 16 k.

3. For I/O mapped I/O, number of device = 2^8 = 256.
   Memory size = 2^{16} × 8 – bit = $2^6 \times 2^{10}$ × 1B = 64 kB.

4. In memory mapped I/O, the I/O devices are also treated as memory locations and they will be given 16-bit addrees. Microprocessor uses related instructions it communicate with I/O devices Arithmetic and logic instructions are directly executed with I/O data.

5. Option are incorrect since part A can be operated as bidirectional part only in mode 2.
   ⇒ option (b) can be correct if it is mode 1 for (I) and mode 2 for (II).

6. $$F = \frac{\text{The gates are XNOR}}{(A \text{ O } \bar{B}) + (\bar{A} \text{ O } B)} = \overline{(A\bar{B} + \bar{A}B) + (\overline{AB} + \overline{AB})}$$
   $$= \overline{(A\bar{B} + \bar{A}B)\,(\bar{A}B\,.\,A\bar{B})} = \overline{A\bar{B} + \bar{A}B}$$
   $$= (A\bar{B})\,(\bar{A}\,B) = 0$$

7. For 8 kB ROM, data lines = 8 – bit (D_7 –D_0)
   Address lines = $\log_2$(8k) = $\log_2$ ($2^3 . 2^{10}$) = 13.
   ⇒ 13 address lines are required for memory chip. But the address range given is 1000 H to 2FFF H.

A_{15}	A_{14}	A_{13}	A_{12}	A_{11}	A_{10}	A_9	A_8	A_7	A_6	A_5	A_4	A_3	A_2	A_1	A_0
0	0	0	1	0	0	0	0	0	0	0	0	0	0	0	0
0	0	1	0	1	1	1	1	1	1	1	1	1	1	1	1

In order to get $\overline{CS}$ as low ('0'), the conditions of address lines are : A_{15} = 0, A_{14} = 0, A_{13} = 0 or1, A_{12} = 1 or 0.
The logic circuit is

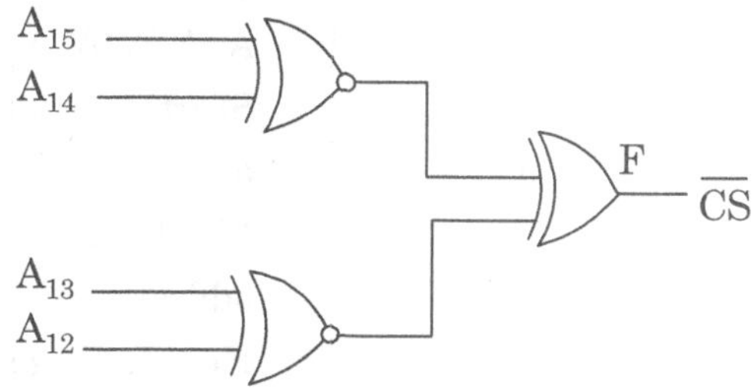

The logical expression is $F = A_{15} + A_{14} + A_{13} \odot A_{12} = A_{15} + A_{14} + A_{13}.\,A_{12} + \bar{A}13.\,\bar{A}12$.

8. For active low chip select, ($\overline{CS}$), NAND gate output should be '0' NAND gate output will be '0' when all the inputs are '1'.

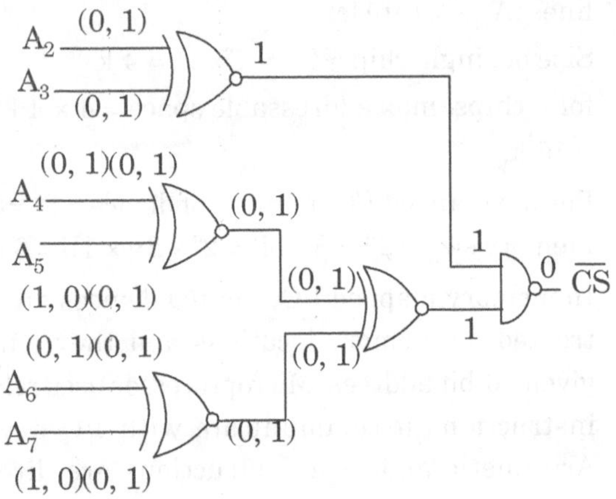

A_7	A_6	A_5	A_4	A_3	A_2	A_1	A_0	
0	1	1	0	0	0	0	0	60 H
0	1	1	0	0	0	1	1	63 H
1	0	1	0	0	1	0	0	A4 H
1	0	1	0	1	0	1	1	A7 H
	0	0	1	0	0	0	0	50 H
	1	1	0	1	1	1	1	AF H
0	1	1	1	0	0	0	0	70 H
0	1	1	1	0	1	1	1	73 H

A2 and A3 should be same. option (b) does not satisfy this.

Because all chip select signals are not same, option (c) is incorrect.

option (d) can't be correct because it gives logic '1' output Hence, option (a) is correct.

9. For starting address, $A_{12} \ldots A_0$ should be 0...0.

	A_{15}	A_{14}	A_{13}	A_{12}	A_0
Starting: Address	X	X	X	0	0
Final Address:	X	X	X	1	1

Checking the options for CS.

$A_{15}\,A_{14}\,A_{13}\,A_{12}$

$\Rightarrow$ X X X 0 $\rightarrow$ even Number

$\Rightarrow$ X X X 1 $\rightarrow$ odd Number.

Hence, the memory range is 6000 H – 7FFF H.

10. Size of RAM = 4k × 8 – bit

No. of address lines required = $\log_2$ (4K) = $\log_2(2^2 \times 2^{10})$ = 12.

For stating address, $A_{11} \ldots A_0$ should be 0...0, and for final address 1.....1.

AA00 H = 1010 1010 0000 00006

Last Address : 1010 1111 1111 11116 = BFFF H.

11. At the first rising edge of clock after $t = t_2$, output of shift register is 0110 = 6 in decimal. At the address 6 data 1010 is stored which is applied to the input of shift register and at the next rising edge, the output of register is 1010 = 10. At the address 10 of ROM, it contains 1000.

12. Consider it as a new solution.

For Chip 1: $A_{15}\,A_{14}\,A_{13}\,A_{12}\,A_{11}\,A_{10}\,A_9\,A_8\,A_7 \ldots A_0$

Stating address : X X X X X X 0 1 0 0

Final address: X X X X X X 0 1 1 ... 1

For Chip 2: $A_{15}\,A_{14}\,A_{13}\,A_{12}\,A_{11}\,A_{10}\,A_9\,A_8\,A_7 \ldots A_0$

Stating address: X X X X X X 1 0 0 ... 0

Final addres: X X X X X X 1 0 1 ... 1

$\Rightarrow$ F800 H – F9FF H cannot be the memory range for chip 1 cna chip 2.

13. To enable decode, $\overline{EN} = 0$

Hence, $(A_7, A_6, A_5) = (1, 1, 0)$.

	A_7	A_6	A_5	A_4	A_3	A_2	A_1	A_0
D4 H:	1	1	0	1	0	1	0	0
D7 H:	1	1	0	1	0	1	1	1

Input line to decode ($A_4\,A_3\,A_2$)

$\Rightarrow$ Output is taken from 5th line.

14. In figure, A_2 is not given, hence, can be considered as '0' or '1'.

output of NAND-gate is 0 if A_7 to A_3 & $IO/\overline{M}$ are '1'

$A_7\,A_6\,A_5\,A_4\,A_3\,A_2\,A_1\,A_0$

Starting address 1 1 1 1 1 0 0 0 = F8 H

Final address 1 1 1 1 1 1 1 1 = FF H.

Hence, range is F8 H – FF H.

15.

A_{15}	A_{14}	A_{13}	A_{12}	A_{11}	A_{10}	A_9	A_8
0	0	1	0	1	1	0	1
To enable chip					for y_5		

A_7	A_6	A_5	A_4	A_3	A_2	A_1	A_0
0	0	0	0	0	0	0	0
⋮	⋮	⋮	⋮	⋮	⋮	⋮	⋮
1	1	1	1	1	1	1	1

Hence range of the address is

2D00 – 2DFF H

16. Since, the RAM 1 range is different is all the four options. So, we need to check for RAM 1 only and then the same procedure can be followed for RAM 2, 3 & 4.

RAM 1 will be selected when

$S_0 = 0, S_1 = 0, S_0 = A_{12} = 0, S_1 = A_{13} = 0.$

Now, the RAM 1 will be enable when the input of MUX is 1, or the output of AND gate is 1.

	A_{15}	A_{14}	A_{13}	A_{12}	A_{11}	A_{10}	A_9	A_8	A_7	A_6	A_5	A_4	A_3	A_2	A_1	A_0
Starting address:	0	0	0	1	0	0	0	0	0	0	0	0	0	0	0	0
			0		8				0					0	H	
Final address:	0	0	0	0	1	0	1	1	1	1	1	1	1	1	1	1
			0			8				F			F	H		

$\Rightarrow$ Range of RAM 1 is 0800H – OBFFH.

17. Memory size = 16 kB = 2^{14} bits

Number of address lines = Number of data lines

$2^n \cdot 2^n = 2^{14} \Rightarrow \boxed{n = 2}$

Unit - VIII

Signals & System

Syllabus

Continuous-time Signals: Fourier series and Fourier transform representations, sampling theorem and applications; Discrete-time signals: discrete-time Fourier transform (DTFT), DFT, FFT, Z-transform, interpolation of discrete-time signals; LTI systems: definition and properties, causality, stability, impulse response, convolution, poles and zeros, parallel and cascade structure, frequency response, group delay, phase delay, digital filter design techniques.

Contents

CHAPTER 1

Basics of Signals & Systems

Analysis of Previous GATE Papers

Topics ↓		Year →	2019	2018	2017 Set 1	2017 Set 2	2016 Set 1	2016 Set 2	2016 Set 3	2015 Set 1	2015 Set 2	2015 Set 3
TRANSFORMATION & CLASSIFICATION OF CONTINUOUS & DISCRETE-TIME	1 Mark	MCQ Type		1		2						
		Numerical Type	1							1		
	2 Marks	MCQ Type										
		Numerical Type										1
	Total		1	1		2				1		2
CLASSIFICATION OF CONTINUOUS & DISCRETE-TIME SYSTEMS	1 Mark	MCQ Type			1							
		Numerical Type										
	2 Marks	MCQ Type										
		Numerical Type										
	Total				1							

TRANSFORMATION & CLASSIFICATION OF CONTINUOUS & DISCRETE-TIME

1. Let $\delta(t)$ denote the delta function. The value of the integral $\int_{-\infty}^{\infty} \delta(t)\cos\left(\frac{3t}{2}\right)dt$ is

(a) 1 (c) –1

(b) 0 (d) $\frac{\pi}{2}$

[2001 : 1 Mark]

2. If a signal f(t) has energy E, the energy of the signal f(2t) is equal to

(a) E (c) $\frac{E}{2}$

(c) 2E (d) 4E

[2001 : 1 Mark]

3. Let P be linearity, Q be time-invariance, R be causality and S be stability. A discrete-time system has the input-output relationship,

$$y(n) = \begin{cases} x(n), & n \geq 1 \\ 0, & n = 0 \\ x(n+1), & n \leq -1 \end{cases}$$

where x(n) is the input and y(n) is the output. The above system has the properties

(a) P, S but not Q, R.

(b) P. Q, S but not R.

(c) P, Q, R, S.

(d) O, R, S but not P.

[2003 : 2 Marks]

4. Consider the sequence

$x[n] = \{-4 - j5, 1 + j2, 4\}$

The conjugate antisymmetric part of the sequence is

(a) {–4 – j2.5, j2, 4 – j2.5}.

(b) {–j2.5, 1, j2.5}.

(c) {–j5, j2, 0}.

(d) {–4, 1, 4}.

[2005 : 1 Mark]

5. The function x(t) is shown in the figure. Even and odd parts of a unit-step function u(t) are respectively,

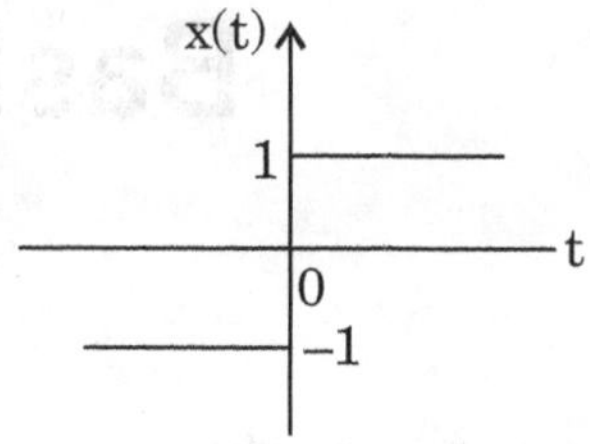

(a) $\frac{1}{2}, \frac{1}{2}x(t)$ (b) $-\frac{1}{2}, \frac{1}{2}x(t)$

(c) $\frac{1}{2}, -\frac{1}{2}x(t)$ (d) $-\frac{1}{2}, \frac{1}{2}x(t)$

6. The power in the signal

$s(t) = 8\cos\left(20\pi t - \frac{\pi}{2}\right) + 4\sin(15\pi t)$ is

(a) 40 (b) 41

(c) 42 (d) 82

[2005 : 1 Mark]

7. The Dirac-delta function $\delta(t)$ is defined as

(a) $\delta(t) = \begin{cases} 1, & t = 0 \\ 0, & \text{otherwise.} \end{cases}$

(b) $\delta(t) = \begin{cases} \infty, & t = 0 \\ 0, & \text{otherwise.} \end{cases}$

(c) $\delta(t) = \begin{cases} 1, & t = 0 \\ 0, & \text{otherwise} \end{cases}$ and $\int_{-\infty}^{\infty} \delta(t)dt = 1.$

(d) $\delta(t) = \begin{cases} \infty, & t = 0 \\ 0, & \text{otherwise} \end{cases}$ and $\int_{-\infty}^{\infty} \delta(t)dt = 1.$

[2006 : 1 Mark]

8. The waveform of a periodic signal jc(?) is shown in the figure.

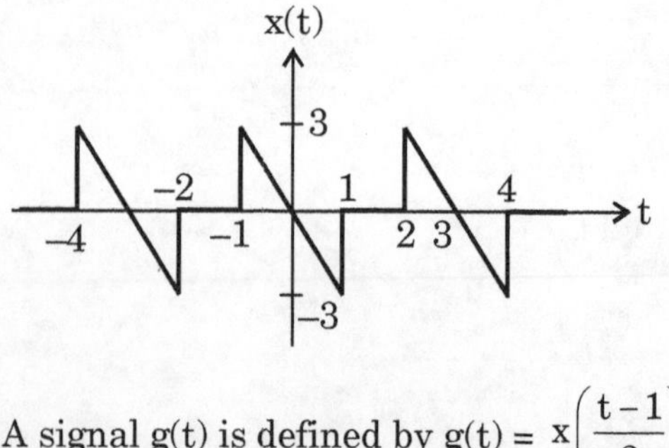

A signal g(t) is defined by $g(t) = x\left(\frac{t-1}{2}\right)$. The average power of g(t) is ______.

[2015 : 1 Mark, Set-1]

9. Two sequences $x_1[n]$ and $x_2[n]$ have the same energy. Suppose $x_1[n] = \alpha 0.5^n u[n]$, where a is a positive real number and u[n] is the unit step sequence. Assume

$$x_2[n] = \begin{cases} \sqrt{1.5}, & \text{for } n = 0,1 \\ 0, & \text{otherwise.} \end{cases}$$

Then the value of a is _______.

[2015 : 2 Marks, Set-3]

10. The input x(t) and the output y(t) of a continuous- time system are related as

$$y(t) = \int_{t-T}^{t} x(u)du$$

The system is

(a) linear and time-variant

(b) linear and time-invariant

(c) non-linear and time-variant

(d) non-linear and time-invariant

[2017 : 1 Mark, Set-2]

11. Let the input be u and the output be y of a system, and the other parameters are real constants. Identify which among the following systems is not a linear system :

(a) $\frac{d^3y}{dt^3} + a_1\frac{d^2y}{dt^2} + a_2\frac{dy}{dt} + a_3y = b_3u + b_2\frac{du}{dt} + b_1\frac{d^2u}{dt^2}$

(with initial rest conditions)

(b) $y(t) = \int_0^t e^{\alpha(t-\tau)}\beta u(\tau)d\tau$

(c) y = au + b, b ≠ 0

(d) y = au **[2018 : 1 Mark]**

12. Which of the following signals is/are periodic?

(a) $s(t) = \cos^2 t + \cos 3t + \cos 5t$

(b) $s(t) = \exp(j8\pi t)$

(c) $s(t) = \exp(-7t)\sin 10\pi t$

(d) $s(t) = \cos^2 t \cos 4t$ **[1992 : 2 Marks]**

13. The sequence

$$y(n) = \begin{cases} x\left(\frac{n}{2} - 1\right), & \text{for n even will be} \\ 0, & \text{for n odd} \end{cases}$$

(a)

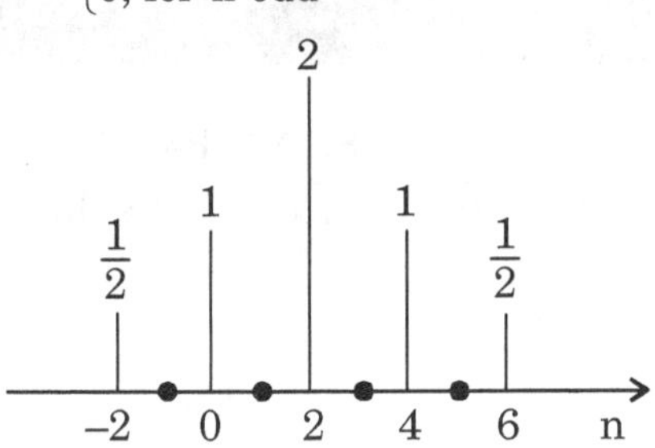

(b)

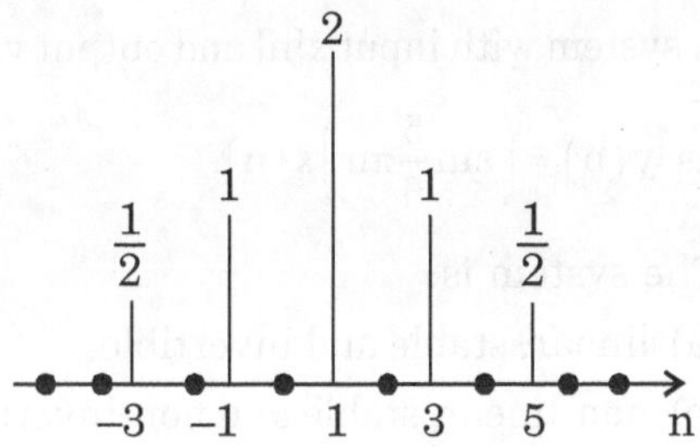

(c)

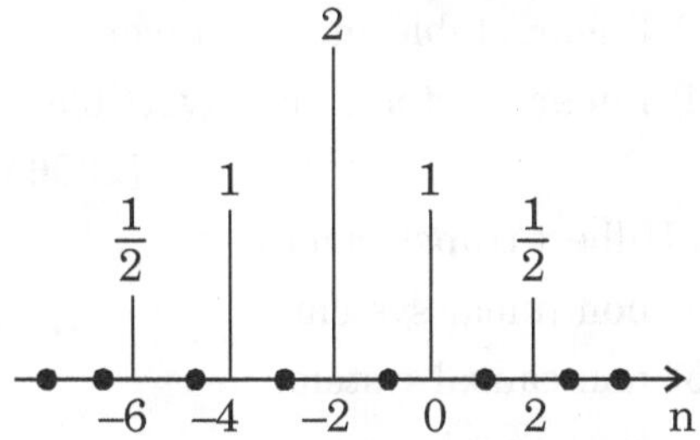

(d)

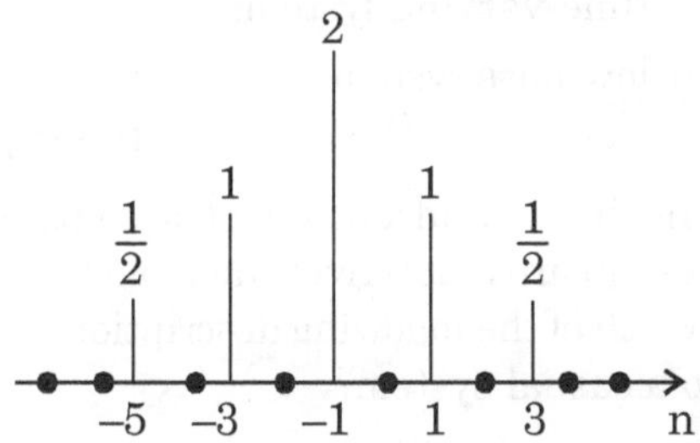

[2005 : 2 Marks]

14. Consider the signal $f(t) = 1 + 2\cos(\pi t) + 3\sin\left(\frac{2\pi}{3}t\right) + 4\cos\left(\frac{\pi}{2}t + \frac{\pi}{4}\right)$, where t is in seconds.

Its fundamental time period, in seconds, is _____.

[2019 : 1 Marks]

CLASSIFICATION OF CONTINUOUS & DISCRETE-TIME SYSTEMS

15. An excitation is applied to a system at t = land its response is zero for $-\infty < t < T$. Such a system is a

(a) non-causal system.

(b) stable system.

(c) causal system.

(d) unstable system.

[1991 : 2 Marks]

16. A system with an input x(t) and output y(t) is described by the relation: y(t) = tx(t). This system is

(a) linear and time-invariant

(b) linear and time varying

(c) non-linear & time-invariant

(d) non-linear and time-varying

[2000 : 1 Mark]

17. A system with input x[n] and output y[n] is given as $y(n) = \left(\sin\frac{5}{6}\pi n\right)x(n)$.

The system is

(a) linear, stable and invertible.
(b) non-linear, stable and non-invertible.
(c) linear, stable and non-invertible.
(d) linear, unstable and invertible.

[2006 : 2 Marks]

18. A Hilbert transformer is a

(a) non-linear system.
(b) non-causal system.
(c) time-varying system.
(d) low-pass system.

[2007 : 2 Marks]

19. The input and output of a continuous time system are respectively denoted by x(t) and y(t). Which of the following descriptions corresponds to a casual system?

(a) $y(t) = x(t - 2) + x(t + 4)$
(b) $y(t) = (t - 4)x(t + 1)$
(c) $y(t) = (t + 4)x(t - 1)$
(d) $y(t) = (t + 5)x(t + 5)$

[2008 : 1 Mark]

20. Letx(t) be the input and y(t) be the output of a continuous time system. Match the system properties P_1, P_2 and P_3 with system relations R_1, R_2, P_3, P_4.

Properties

P_1 : Linear but NOT time-invariant
P_2 : Time-invariant but NOT linear
P_3 : Linear and time-invariant

Relations

$R_1 : y(t) = t^2x(t)$
$R_2 : y(t) = t\,|x(t)|$
$R_3 : y(t) = |x(t)|$
$R4 : y(t) = x(t - 5)$

(a) $(P_1, R_1), (P_2, R_3), (P_3, R_4)$.
(b) $(P_1, R_2), (P_2, P_3), (P_3, R_4)$.
(c) $(P_1, R_3), (P_2, R_1), (P_3, R_2)$.
(d) $(P_1, R_1), (P_2, R_2), (P_3, R_3)$.

[2008 : 2 Marks]

21. The input x(t) and output y(t) of a system are related as $y(t) = \int_{-\infty}^{t} x(\tau)\cos(3\tau)d\tau$. The system is

(a) time-invariant and stable.
(b) stable and not time-invariant.
(c) time-invariant and not stable.
(d) not time-invariant and not stable.

[2012 : 2 Marks]

22. Consider a single input single output discrete-time system with x[n] as input and y[n] as output, where the two are related as

$$y[n] = \begin{cases} n|x[n]|, & \text{for } 0 \le n \le 10 \\ x[n] - x[n-1], & \text{otherwise} \end{cases}$$

Which one of the following statements is true about the system?

(a) It is causal and stable.
(b) It is causal but not stable.
(c) It is not causal but stable.
(d) It is neither causal nor stable.

[2017 : 1 Mark, Set-1]

ANSWERS

1. (a)	2. (b)	3. (a)	4. (a)	5. (a)	6. (a)	7. (d)	8. (A-2, B-4, C-1)
9. (1· 49-1· 51)		10. (a)	11. (c)	12. (a,b)	13. (a)	14. (12)	15. (c)	16. (b)
17. (c)	18. (a)	19. (c)	20. (b)	21. (d)	22. (a)		

EXPLANATIONS

1. As $\int_{-\infty}^{\infty} \delta(at)\phi(t)\,dt = \dfrac{\phi(0)}{|a|}$ for any $a < 0$.

Thus $\int_{-\infty}^{\infty} \delta(t)\cos\left(\dfrac{3t}{2}\right)dt = \dfrac{\cos 0}{1} = 1.$

2. Energy content of a signal $f(t)$,

$$E = \int_{-\infty}^{\infty} |f(t)|^2\,dt \quad \text{...(i)}$$

Now, $E' = \int_{-\infty}^{\infty} |f(2t)|^2\,dt$ for signal $f(2t)$

Putting, $2t = z$, we get

$$E' = \frac{1}{2}\int_{-\infty}^{\infty} |f(z)|^2\,dz$$

$$= \frac{1}{2}E \quad \text{(from equation (i))}$$

3. The equation is homogeneous

$\Rightarrow$ It is linear

$$y(n-n_0) \neq x(n-n_0)$$

$\Rightarrow$ not time invarient

$$y(n) = x(n+1),\ n \leq -1$$

not causal.

The system shows that not for bonded O/P but for bounded I/P

$\Rightarrow$ it is stable.

$\Rightarrow$ The system is linear and stable but not time invariant and causal.

4. $x[n] = [-4 - j5,\ \underset{\uparrow}{1 + 2j},\ 4]$

$$x[-n] = [4,\ \underset{\uparrow}{1+2j},\ -4-j5]$$

$$x^*[-n] = [4,\ \underset{\uparrow}{1-2j},\ -4+j5]$$

$$X_{CAS}[n] = \frac{x[n] - x^*[-n]}{2}$$

$$X_{CAS}[n] = [-4 - j2.5,\ j2,\ 4 - j2.5]$$

5. Even part $= \dfrac{u(t) + u(-t)}{2}$

Now $u(t) = 0\ ;\ t < 0$

$= 1,\ t \geq 0$

$\therefore$ $u(-t) = 0,\ -t < 0$

$= 1,\ -t \geq 0$

i.e., $u(-t) = 1,\quad t \leq 0$

$= 0,\quad t > 0$

$\therefore$ $\dfrac{u(t)+u(-t)}{2} = \dfrac{1}{2};\ t \leq 0 = \dfrac{1}{2},\quad t > 0$

$\therefore$ Even$[v(t)] = \dfrac{1}{2}$

Odd $(u(t)) = \dfrac{u(t)+u(-t)}{2}\begin{bmatrix} -\dfrac{1}{2}, & t \leq 0 \\ \dfrac{1}{2}, & t > 0 \end{bmatrix}$

$= \dfrac{x(t)}{2}$ (from given figure)

6. Power of $s(t) = 8\cos\left(20\pi t - \dfrac{\pi}{2}\right) + 4\sin(15\pi t)$

$$= \frac{8^2}{2} + \frac{16}{2} = \frac{64+16}{2} = 40$$

7. The heaviside function H(x) is defined.

$$H(x) = \begin{cases} 0 & \text{for } x < 0 \\ 1 & \text{for } x > 0 \end{cases}$$

The derivative of the Heaviside function is zero for $x \neq 0$. At $x = 0$ the derivative is undefined. The derivative of the Heaviside function is the Dirac delta function, $\delta(x)$. The delta function is zero for $x \neq 0$ and infinite at the point $x = 0$. Since the derivative of H(x) is undefined, $\delta(x)$ is not a function in the conventional sense of the word.

The Dirac delta function is defined by the properties.

$$\delta(x) = \begin{cases} 0 & \text{for } x \neq 0, \\ \infty & \text{for } x = 0, \end{cases} \quad \text{and} \int_{-\infty}^{\infty} \delta(x)\,dx = 1$$

The second property comes from the fact that $\delta(x)$ represents the derivative of H(x). The Dirac delta function is conceptually pictured in figure A.

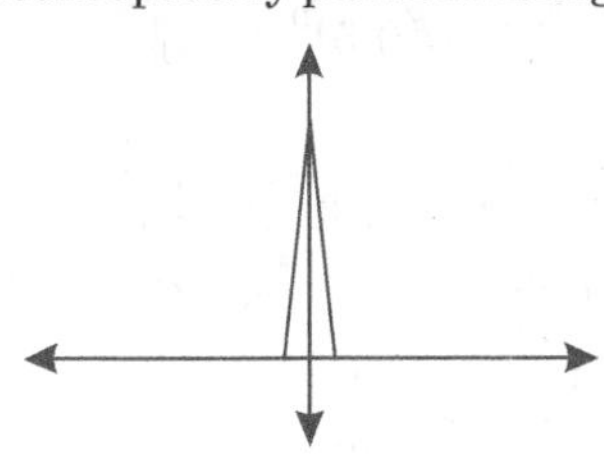

Figure A : The Dirac Delta Function.

8. From given figure,

$$x(+) = \begin{cases} -3t, & - < t < 1 \\ 0, & 1 < t < 2 \end{cases}$$

$$\therefore g(+) = x\left(\frac{t-1}{2}\right) = x\left(\frac{t}{2} - \frac{1}{2}\right)$$

$$x\left(t-\frac{1}{2}\right)=\begin{cases}-3\left(t-\frac{1}{2}\right), & -1<t-\frac{1}{2}<1\\ 0, & 1<t-\frac{1}{2}<2\end{cases}$$

$$=\begin{cases}-3\left(t-\frac{1}{2}\right), & -\frac{1}{2}<t\frac{3}{2}\\ 0, & \frac{3}{2}<t<\frac{5}{2}\end{cases}$$

$$x\left(\frac{t}{2}-\frac{1}{2}\right)=\begin{cases}-3(t-\frac{1}{2}, & -\frac{1}{2}<\frac{t}{2}<\frac{3}{2}\\ 0, & \frac{3}{2}<\frac{t}{2}<\frac{5}{2}\end{cases}$$

$$=\begin{cases}-\frac{3}{2}(t-1), & -1<t<3\\ 0, & 3<t<5\end{cases}$$

$$g(t)=\begin{cases}-\frac{3}{2}(t-1), & -1<t<3\\ 0, & 3<t<5\end{cases};$$

$\Rightarrow$ Time period, T = 6.

$$\text{Average power of } g(t)=\frac{1}{6}\int_{-1}^{3}\left[-\frac{3}{2}(t-t\right]^2 dt=\frac{1}{4}$$

$$=\frac{1}{4}\int_{-1}^{3}(t-1)^2\,dt$$

= 2.

9. $$\text{Energy of } x_1=\sum_{n=-\infty}^{\infty}|x_1[n]^2|$$

$$=x_1=\sum_{n=-\infty}^{\infty}\left|\alpha.(0.5)^n.u[n]\right|^2$$

$$=\alpha^2.\sum_{n=0}^{\infty}\left(\frac{1}{4}\right)^{2n}=\alpha^2.\sum_{n=0}^{\infty}\left(\frac{1}{4}\right)^{n}$$

$$=\alpha^2.\frac{1}{1-\frac{1}{4}}=\alpha^2.\frac{4}{3}$$

$$\text{Energy of } x_2[n]=\sum_{n=-\infty}^{\infty}|x_2[n]^2|$$

$$=x_2^2[0]+x_2^2[1]=1.5+1.5$$

= 3

$$\therefore \alpha^2\frac{4}{3}=3\Rightarrow\alpha=1.5$$

10. Given input – output relationship describes integration over a fundamental period T. The integration over one period is linea and time - invariout.

11. Option (a) consists of homogeneous differential equation with constant coeff. which constitutes a linear system. option (b) has running integration input - output relationship, hence, linear system. option (c) and (d) describe linear system if b = 0. in option (c), b ≠ 0, hence system is non–linear.

12. $S(t) = \cos 2t + \cos 3t + \cos 5t$

Period = 2π

$S(t+T)=e^{j8\pi(t+T)}=e^{j8\pi t+j8\pi T}$

T = 1/4 period

$S(t+T)=e^{-7(t+T)}\sin 10\pi(t+T)=e^{-7t}e^{7T}\sin 10\pi t$

Non periodic

$S(t+T)=\cos(2t+2T)\cos(4t+4T)$

T = π Periodic

13. $$y(x)=x\left(\frac{n}{2}-1\right);\quad n\text{ even}$$
$$=0;\quad n\text{ odd}$$

$\therefore$ $y(x)$ is shifted 1 towards left and axis expanded by 2

14. It is given that,

$$f(t)=1+2\cos\pi t+3\sin\frac{2\pi}{3}t+4\cos\left(\frac{\pi}{2}t+\frac{\pi}{4}\right)$$

Its fundamental frequency (ω_0)

$$\omega_0=\frac{\text{HCF of }(\pi,2\pi,\pi)}{\text{LCM of }(1,3,2)}=\frac{\pi}{6}$$

$$\text{Now, } T_0=\frac{2\pi}{\omega_0}=\frac{2\pi\times 6}{\pi}=12\text{ sec}$$

15. For the given system if the sesposes in zero prior to the (c)application of the excitation (or input), then such a system is called causal system.

16. For the system H to be linear, the condition to be satisfied is

Here, $ay_1(t)=$ at $x_1(t)$

$ay_2(t)=$ at $x_2(t)$H$(\alpha f_1(x)+\beta f_2(x)$

$=\alpha\,H(f_1(x)+\beta\,H(f_2(x))$ .....(A)

$a[y_1(t)+y_2(t))=at[x_1(t)+x_2(t)]$

Hence system is linear as condition (A) is satisfied.

If $g(x)=H[f(x)$ then for time-invariance system

$g(x+x_0)=H[f(x+x_0)$ .... (B)

Here $y(t-t_0)=(t-t_0)\,x(t-t_0)$

Therefore system is time-variant as condition (B) is not satisfied.

17. $y[n] = \left(\sin\frac{5}{6}\pi n\right) x(n)$

As $-1 \leq \sin\frac{5}{6}\pi n \leq 1$,

Then, for bounded $x[n]$, output $Y[n]$ is also bounded and the system is stable.

If any system to be invertible, input can be determined from the output of the system. For this to be true two different input signals should produce two different outputs. If some different input signals produce the same output signal then by processing output it can not be said which input produced the output.

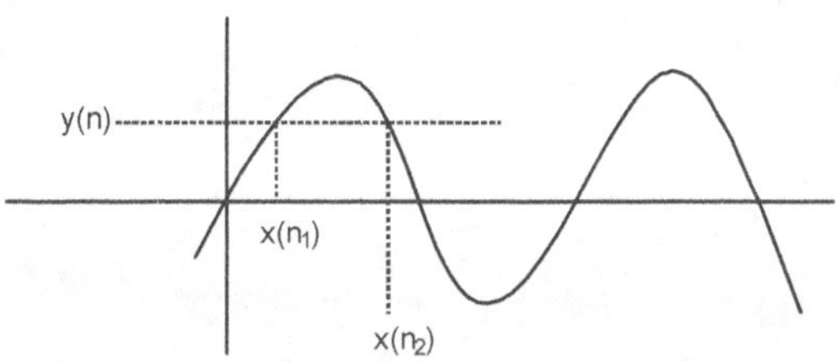

For different input $x(n_1)$ and $x(n_2)$, the output remains same. Therefore, it can not be ascentained which input produce the output and system is non-invertible.

18. Hilbert transformer

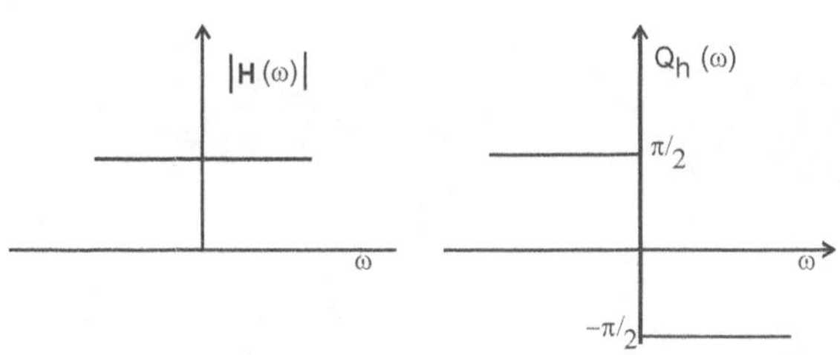

$H(\omega) = -j\text{sgn}(\omega)$

$$\begin{cases} -j = 1.e^{-\pi/2J} & \omega>0 \\ +j = 1.e^{+\pi/2J} & \omega<0 \end{cases}$$

For linear system, $Q_h(\omega) = -\omega t_d$

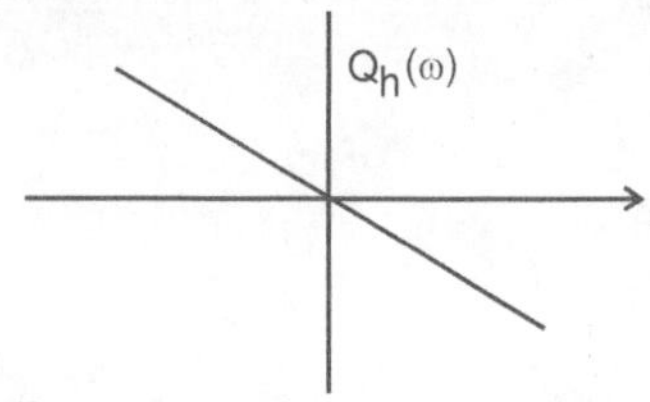

So, Hilbert transformer is non-linear and time invariant system.

19. A system is said to be causal if output depends only on present and past states only.

20. R_1 : $y(t) = t^2x(t) \Rightarrow$ Linear and time variant.
R_2 : $y(t) = t|x(t)| \Rightarrow$ Non linear and time variant.
R_3 : $y(t) = |x(t)| \Rightarrow$ Non linear and time invariant
R_4 : $y(t) = x(t-5) \Rightarrow$ Linear and time invariant

21. $$y(t) = \int_{-\infty}^{t} x(\tau)\cos(3\tau)\, d\tau$$

Since $y(t)$ and $x(t)$ are related with some function of time, so they are not time-invariant.

Let $x(t)$ be bounded to some finite value k.

$$y(t) = \int_{-\infty}^{t} K\cos(3\tau)\, d\tau < \infty$$

$y(t)$ is also bounded. Thus system is stable.

22. Since present output does not depend upon future values of input hence system is causal and also every bounded input produces bounded output, So we can say that system is stable.

2
CHAPTER

LTI Systems

Analysis of Previous GATE Papers												
	Year → Topics ↓		2019	2018	2017 Set 1	2017 Set 2	2016 Set 1	2016 Set 2	2016 Set 3	2015 Set 1	2015 Set 2	2015 Set 3
PROPERTIES	**1 Mark**	MCQ Type					1					
		Numerical Type										
	2 Marks	MCQ Type						2				
		Numerical Type										
	Total						1	4				
CONVOLUTION	**1 Mark**	MCQ Type								1		
		Numerical Type										
	2 Marks	MCQ Type										
		Numerical Type			1							
	Total				2					1		
INTERCONNECTION	**1 Mark**	MCQ Type										
		Numerical Type										
	2 Marks	MCQ Type										
		Numerical Type				1						
	Total					2						

PROPERTIES

1. The impulse response of a continuous time system is given by $h(t) = \delta(t-1) + \delta(t-3)$. The value of the step response at $t = 2$ is

(a) 0 (b) 1

(c) 2 (d) 3

[1989 : 2 Marks]

2. The impulse response and the excitation function of a linear time invariant causal system are shown in figure (a) and (b) respectively. The output of the system at $t = 2$ sec is equal to

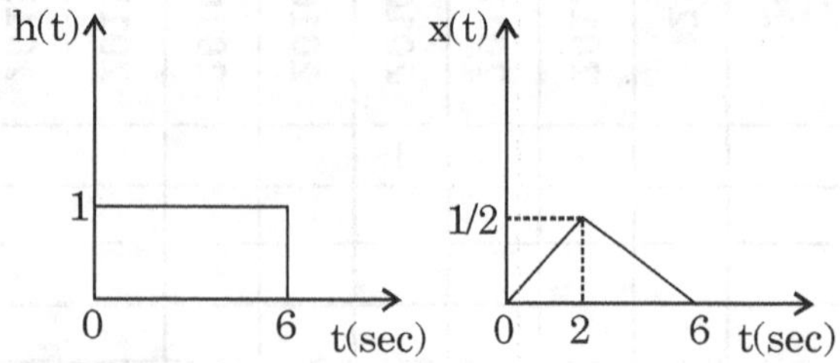

(a) 0 (b) $\frac{1}{2}$

(c) $\frac{3}{2}$ (d) 1 **[1990 : 2 Marks]**

3. The impulse response functions of four linear systems S_1, S_2, S_3, S_4 are given respectively by

$h_1(t) = 1$, $h_2(t) = u(t)$, $h_3(t) = \frac{u(t)}{t+1}$,
$h_4(t) = e^{-3t}u(t)$

Where u(t) is the unit step function. Which of these systems is time invariant, causal, and stable?

(a) S_1 (b) S_2

(c) S_3 (d) S_4 **[2001 : 2 Marks]**

4. The impulse response h[n] of a linear time-invariant system is given by

$h[n] = u[n+3] + u[n-2] - 2u[n-7]$,

where u[n] is the unit step sequence. The above system is

(a) stable but not causal.

(b) stable and causal.

(c) causal but unstable.

(d) unstable and not causal.

[2004 : 1 Mark]

5. Which of the following can be impulse response of a causal system?

(a)
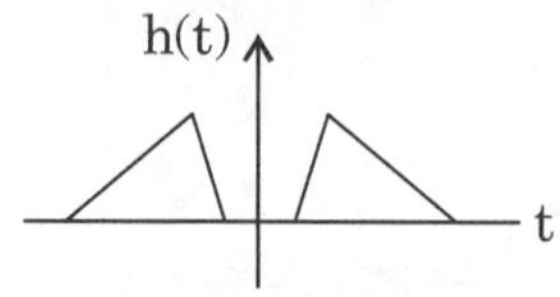

(b)
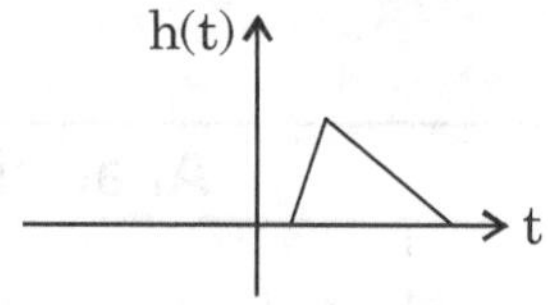

(c)
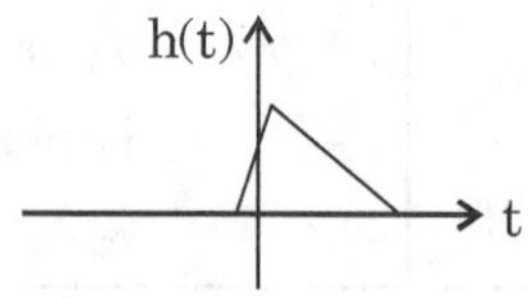

(d)
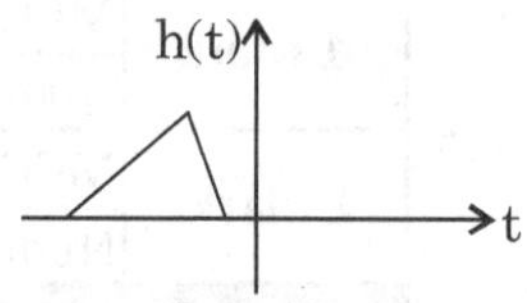

[2005 : 1 Mark]

6. The impulse response h(t) of a linear time-invariant continuous time system is described by $h(t) = \exp(\alpha t)u(t) + \exp(\beta t)u(-t)$, where u(t) denotes the unit step function, and α and β are real constants. This system is stable if

(a) α is positive and β is positive.

(b) α is negative and β is negative.

(c) α is positive and β is negative.

(d) α is negative and β is positive.

[2008 : 1 Mark]

7. A system is defined by its impulse response $h(n) = 2^n u(n-2)$. The system is

(a) stable and causal

(b) causal but not stable

(c) stable but not causal

(d) unstable and noncausal **[2011 : 1 Mark]**

8. The impulse response of an LTI system can be obtained by

(a) differentiating the unit ramp response.

(b) differentiating the unit step response.

(c) integrating the unit ramp response.

(d) integrating the unit step response.

[2015 : 1 Mark, Set-3]

9. Which one of the following is an eigen function of the class of all continuous- time, linear, time-invariant systems (u(t) denotes the unit-step function)?

(a) $e^{j\omega_0 t}u(t)$ (b) $\cos(\omega_0 t)$

(c) $e^{j\omega_0 t}$ (d) $\sin(\omega_0 t)$

[2016 : 1 Mark, Set-1]

10. A network consisting of a finite number of linear resistor (R), inducer (L), and capacitor (C) elements, connected all in series or all in parallel, is excited with a source of the form

$\sum_{k=1}^{3} a_x \cos(k\omega_0 t)$, were $a_k \neq 0, \omega_0 \neq 0$.

The source has nonzero impedance. Which one of the following is a possible form of the output measured across a resistor in the network?

(a) $\sum_{k=1}^{3} b_x \cos(k\omega_0 t + \phi_k)$, were $b_k \neq a_k, \forall K$

(b) $\sum_{k=1}^{3} b_x \cos(k\omega_0 t + \phi_k)$, were $b_k \neq 0, \forall K$

(c) $\sum_{k=1}^{3} a_x \cos(k\omega_0 t + \phi_k)$

(d) $\sum_{k=1}^{2} a_x \cos(k\omega_0 t + \phi_k)$

[2016 : 2 Marks, Set-1]

CONVOLUTION

11. Let h(t) be the impulse response of a linear time invariant system. Then the response of the system for any input u(t) is

(a) $\int_0^t h(\tau)u(t-\tau)d\tau$.

(b) $\frac{d}{dt}\int_0^t h(\tau)u(t-\tau)d\tau$.

(c) $\int_0^t\left[\int_0^t h(\tau)u(t-\tau)d\tau\right]dt$.

(d) $\int_0^t h^2(\tau)u(t-\tau)d\tau$.

[1995 : 1 Mark]

12. Convolution of x(t + 5) with impulse function δ(t – 7) is equal to

(a) x(t – 12) (b) x(t + 12)

(c) x(t – 2) (d) x(t + 2)

[2002 : 1 Mark]

13. The impulse response h[n] of a linear time invariant system is given as

$$h[n] = \begin{cases} -2\sqrt{2}, & n = 1, -1 \\ 4\sqrt{2}, & n = 2, -2 \\ 0, & \text{otherwise} \end{cases}$$

If the input to the above system is the sequence $e^{j\pi n/4}$, the output is

(a) $4\sqrt{2}e^{j\pi n/4}$ (b) $4\sqrt{2}e^{-j\pi n/4}$

(c) $4e^{j\pi n/4}$ (d) $-4e^{j\pi n/4}$

[2004 : 2 Marks]

14. A discrete time linear shift-invariant system has an impulse response h[n] with h[0] = 1, h[1] = –1, h[2] = 2, and zero otherwise. The system is given an input sequence x[n] with x[0] = x[2] = 1 and zero otherwise. The number of nonzero samples in the output sequence y[n], and the value of y[2] are, respectively

(a) 5, 2 (b) 6, 2

(c) 6, 1 (d) 5, 3

[2008 2 Marks]

15. Let y[n] denote the convolution of h[n] and g[n], where $h[n] = \left(\frac{1}{2}\right)^n u[n]$ and g[n] is a causal sequence. If y[0] = 1 and $y[1] = \frac{1}{2}$, then g[1] equals

(a) 0 (b) $\frac{1}{2}$

(c) 1 (d) $\frac{3}{2}$

[2012 : 2 Marks]

16. Consider a discrete-time signal

$$x[n] = \begin{cases} n \text{ for } & 0 \le 0 \le 10 \\ 0, & \text{otherwise.} \end{cases}$$

If y[n] is the convolution of x[n] with itself, the value of y[4] is _______.

[2014 : 2 Marks, Set-2]

17. The sequence $x[n] = 0.5^n u[n]$ is the unit step sequence, is convolved with itself to obtain y[n].

Then $\sum_{n=-\infty}^{+\infty} y[n]$ is _______.

[2014 : 1 Mark, Set-4]

18. The result of the convolution

$x(-t) * \delta(-t - t_0)$ is

(a) $x(t + t_0)$ (b) $x(t - t_0)$

(c) $x(-t + t_0)$ (d) $x(-t - t_0)$

[2015 : 1 Mark, Set-1]

19. Two discrete-time signals x[n] and h[n] are both non-zero only for n = 0, 1, 2 and are zero otherwise. It is given that

$x[0] = 1, x[1] = 2, x[2] = 1, h[0] = 1.$

Let y[n] be the linear convolution of x[n] and h[n]. Given that y[1] = 3 and y[2] = 4, the value of the expression (10y[3] + y[4]) is ______.

[2017 : 2 Marks, Set-1]

INTERCONNECTION

20. Two systems with impulse responses $h_1(t)$ and $h_2(t)$ are connected in cascade. Then the overall impulse response of the cascaded system is given by

(a) product of $h_1(t)$ and $h_2(t)$.

(b) sum of $h_1(t)$ and $h_2(t)$.

(c) convolution of $h_1(t)$ and $h_2(t)$.

(d) subtraction of $h_2(t)$ from $h_1(t)$.

[2013 : 1 Mark]

21. Consider the parallel combination of two LTI systems shown in the figure.

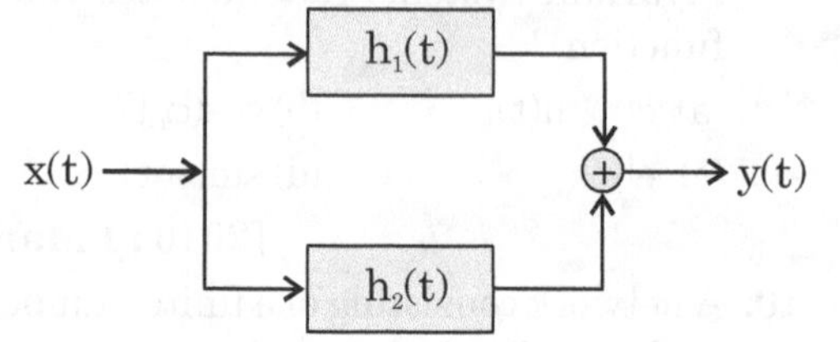

The impulse responses of the systems are

$h_1(t) = 2\delta(t+2) - 3\delta(t+1),$

$h_2(t) = 5(t-2).$

If the input x(t) is a unit step signal, then the energy of y(t) is ______.

[2017 : 2 Marks, Set-2]

ANSWERS

1. (b)	2. (b)	3. (d)	4. (a)	5. (b)	6. (d)	7. (b)	8. (b)	9. (a)	10. (c)
11. (a)	12. (c)	13. (d)	14. (a)	15. (a)	16. (10)	17. (4)	18. (d)	19. (31)	20. (c)
21. (c)									

EXPLANATIONS

1. As $h(t) = \delta(t-1) + \delta(t-3)$

$r(t) = h(t) \oplus u(t)$

$= [\delta(t-1) + \delta(t-3)] \oplus u(t)$

$= u(t-1) + u(t-3)$

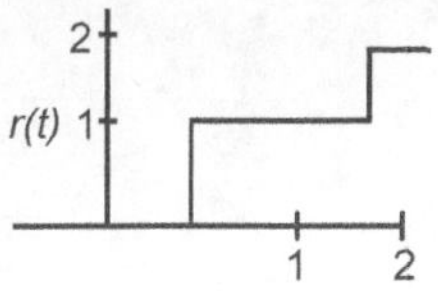

2. For a causal LTI system, $h(t) = 0 \forall t < 0$.

$$\Rightarrow \text{Output, } y(t) = x(t)h(t) = \sum_{-\infty}^{t} x(\tau).h(t-\tau)d\tau$$

$$\therefore x(t) = 0 \,|\, \forall \,|\, t < o \text{ and } h(t) = o \,|\, \forall \,|\, t < o$$

$$y(t) = \sum_{0}^{t} x(\tau).h(t-\tau)d\tau$$

$$y(2) = \sum_{0}^{t} x(\tau).h(2-\tau)d\tau$$

$$\Rightarrow y(2) = \frac{1}{2} \times (2 \times \frac{1}{2})$$

$$= \frac{1}{2}$$

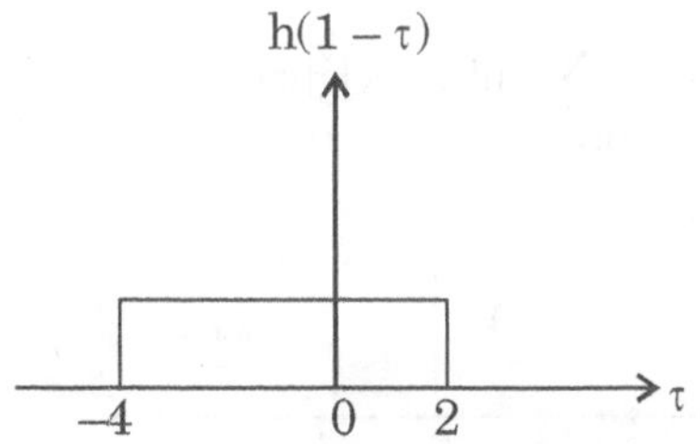

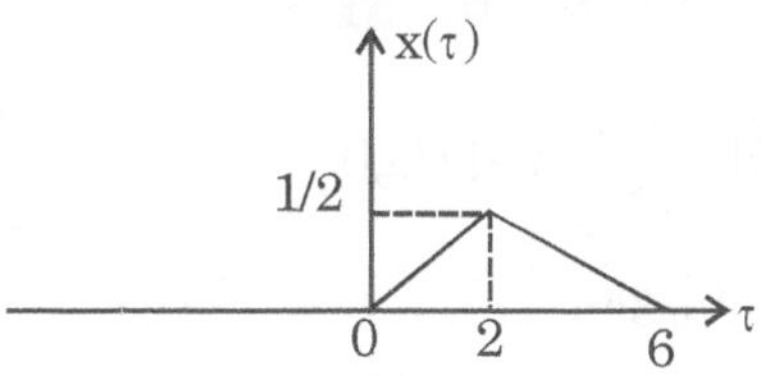

3. $\Rightarrow h_1(t) \neq o \forall t < o \Rightarrow S_1$ is non - causal.

$\Rightarrow h_2(t) = u(t)$

$$\int_{-8}^{8} n_2(t)dt = \int_{-\infty}^{8} u(t)dt \to \infty; \Rightarrow \; s_2 \text{ is unstable.}$$

$$\Rightarrow h_3(t) = \frac{u(t)}{S+1}$$

$t = -1, n_3(t) \to \infty \Rightarrow S_3$ is unstable.

$\Rightarrow h_\mu(t) = e^{-3t}u(t)$: Su is time - invariant, causel and stable.

4. For random variable X, the probability density function of the quantization error Q is given as,

$$f_X(x) = \begin{cases} 1, & 0 \le x \le 1 \\ 0, & \text{otherwise} \end{cases}$$

and $\int_0^1 f_X(x) = 1$

Now, Mean square value of qunatisation error is given by

$$\sigma_Q^2 = E[Q^2] = \int_0^1 q^2 f_x(x).dx$$

$$\sigma_Q^2 = \int_0^{0.3} x^2 f_x(x)\,dx + \int_{0.3}^{1} (x - x_q)^2 \cdot f_x(x) \cdot dx$$

Where $q = x - x_q = \left.\frac{x^3}{3}\right|_0^{0.3} + \int_{0.3}^{1} (x - 0.7)^2 \cdot dx$

$$\sigma_Q^2 = 0.039$$

Root-mean square value $= \sqrt{\sigma_Q^2} = 0.198$

5. $f(t) = 0$ for $t < 0$ for causality

6. $u(t) = e^{\alpha t}u(t) + e^{\beta t}u(-t)$

For the system to be stable,

$$\int_{-\infty}^{\infty} u(t) \cdot dt < \infty$$

or, $\int_{-\infty}^{\infty} \left(e^{\alpha t} \cdot u(t) + e^{\beta t} u(-t)\right) \cdot dt < \infty$

or, $\int_0^{\infty} e^{\alpha t} \cdot dt + \int_{-\infty}^{0} e^{\beta t} \cdot dt < \infty$ ...(A)

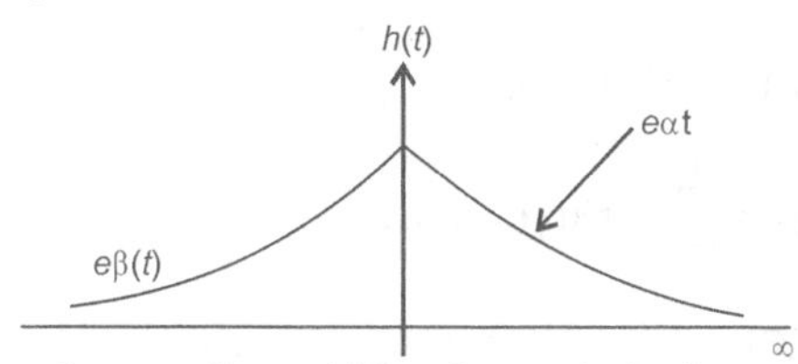

For, the conditon (A) to be satisfied,

$\alpha < 0$ and $\beta > 0$

7. $h(n) = 2^n u(n-2)$

$h(n)$ is existing for $n > 2$; thus

$h(n) = 0;\ n < 0$; hence causal

$$\sum_{n=-\infty}^{\infty} |h(n)| = \sum_{n=\infty}^{\infty} 2^n u(n-2)$$

$$\sum_{n=2}^{\infty} 2^n = \infty; \text{ hence system is unstable}$$

8. Let $h(t)$ be the impulse response of the system

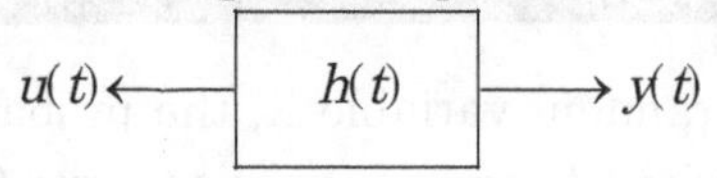

$y(t)$ is unit step response of the system

$$y(t) = \int_{-\infty}^{t} h(\tau)\, d\tau$$

If we need to get $h(t)$, then we have to differentiate $y(t)$.

Thus differentiating the unit-step response gives impulse response for LTI system.

9. From, z-transform property,

$$Z^{-m} \leftrightarrow \delta[n-m]$$

We have, $X(Z) \leftrightarrow 5\delta[n+2] + 4\delta[n-1] + 3\delta[n]$

10. Impulse response of the matched filter,

$$h(t) = s(T-t)$$

Given,

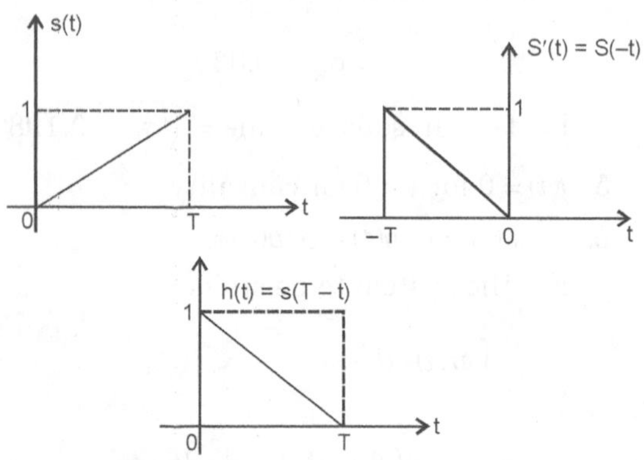

11. For LTI system, y(t) = u(t) * h(t)

$$y(t) = \sum_{-\infty}^{t} h(t)x(t-\tau)\, d\tau$$

$\therefore$ u(t) = 0 $\forall$ t < 0.

$$y(t) = \sum_{-\infty}^{t} h(\tau)u(t-\tau)\, d\tau$$

12. As, $g(t) * \delta(t-t_0) = g(t_0)$

then $x(t+5) * \delta(t-7) = x(t-2)$

13. $h[n] = [4\sqrt{2}, -2\sqrt{2}, 0, -2\sqrt{2}, 4\sqrt{2}]$
$\uparrow$ (under 0)

$$x[n] = e^{\frac{j\pi n}{4}}$$

$$y[n] = x[n] \otimes h[n]$$

Using property $x[n] \otimes \delta[n-n_0] = x[n-n_0]$

$$y[n] = e^{\frac{j\pi n}{4}} [0 - 2\sqrt{2} \times 2 \cos \pi/4]$$

$$y[n] = -4e^{\frac{j\pi n}{4}}$$

14. Here the convolution of two sinc pulses is sinc pulse.

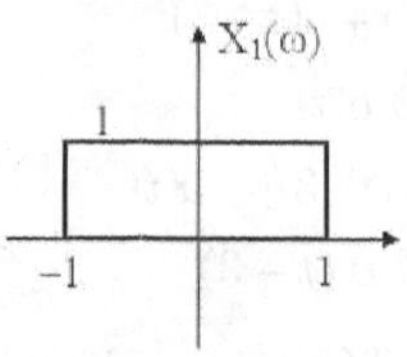

So, $x_1(t) = \dfrac{\sin t}{\pi t}$

Now $x(t) = x_1(t) * x_1(t)$

$$X(\omega) = X_1(\omega) \cdot X_1(\omega)$$
$$= X_1(\omega)$$

$\therefore$ $x(t) = x_1(t) = \dfrac{\sin t}{\pi t}$

So, the value of $x(t)$ is $\dfrac{\sin t}{\pi t}$

15. Given : $h[n] = \left(\dfrac{1}{2}\right)^n u[n]$,

$g[n] = 0 \forall n < 0, n < 0, y[0] = 1, y[1] = \dfrac{1}{2.}$

$$\Rightarrow y[n] = \sum_{k=-\infty} h[n-k]g[k]$$

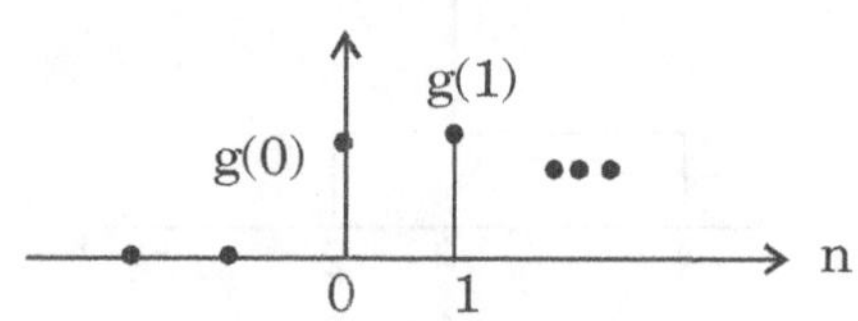

$$\Rightarrow y[0] = \sum_{k=-\infty} h-[k].g[k]$$

y[0]=n[0]. g[0]

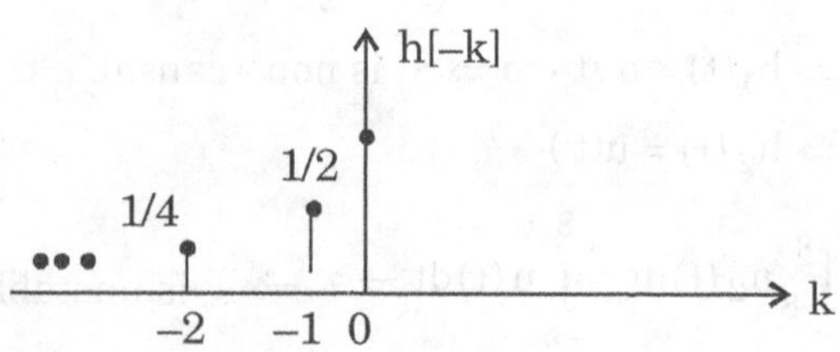

$$\Rightarrow y[1] = \sum_{k=-\infty} h[1-k].g[k]$$

= g[k] g[0] + h[0]. g[1]

h[1-k] will be zero for k > 1 and g[k] will be zero for k < 0 as it is causal sequence.

$$\frac{1}{2} = \frac{1}{2} \times 1 + 1 \times g[1] \Rightarrow g[1] = 0.$$

16. Given, $x[n] = \begin{cases} n & \text{for } 0 \le n \le 10 \\ 0 & \text{elsewhere} \end{cases}$

$$y[n] = x[n] * x[n]$$

$$y[n] = \sum_{k=0}^{n} x[k] \cdot x[n-k]$$

$$\Rightarrow \quad y[4] = \sum_{k=0}^{4} x[k] \cdot x[G-k]$$

17. $y[n] = x[n] * x[n]$

Let $Y(e^{j\Omega})$ is F.T. pair with $y[n]$

$\Rightarrow Y(e^{j\Omega}) = X(e^{j\Omega}) \,.\, X(e^{j\Omega})$

$$Y(e^{j\Omega}) = \frac{1}{1-0.5e^{-j\Omega}} \cdot \frac{1}{1-0.5e^{-j\Omega}}$$

also $$Y(e^{j\Omega}) = \sum_{h=-\infty}^{\infty} y[n].e^{-j\Omega n}$$

$$\Rightarrow \quad \sum_{h=-\infty}^{\infty} y[n] = Y(e^{j0}) = \frac{1}{0.5} \cdot \frac{1}{0.5} = 4$$

18. Using property of Impulse function,

$x(-t) * \delta(-t-t_o) = x(-t) * \delta(t+t_o) = x(-t-t_o)$

19.

$x[n] = \{1, 2, 1\}$

$h[n] = \{1, a, b\}$

$y[n] = [A, 3, 4, B, C\}$

x \ h	1	a	b
1	1	a	b
2	2	2a	2b
1	1	a	b

$y[0] = 1$; $y[1] = 2 + a$; $y[2] = 1 + 2a + b$; $y[3] = a + 2b$; $y[4] = b$

Given, $y[1] = 2 + a = 3 \Rightarrow a = 1$

and $y[2] = 1 + 2a + b = 4 \Rightarrow b = 1$

$\therefore$ $y[3] = a + 2b = 1 + 2(1) = 3$

and $y[4] = b = 1$

$\therefore$ $10y[3] + y[4] = (10 \times 3) + 1 = 31$

20.

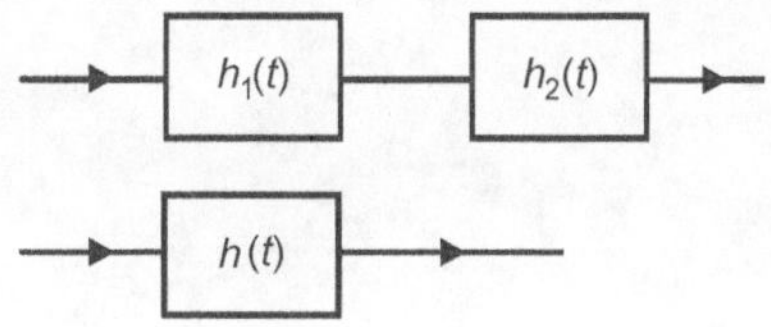

$h(t) = h_1(t) \oplus h_2(t)$

21. Given : $h_1(t) = 2\delta(t+2) - 3\delta(t+1)$, $h_2(t) = \delta(t-2)$, $x(t) = u(t)$.

Overall impulse response, $h(t) = h_1(t) + h_2(t)$

$= 2\delta(t+2) - 3\delta(t+1) + \delta(t-2)$.

$\Rightarrow u(t) = x(t) * h(t) = u(t) [2\delta(t+2) - 3\delta(t+1) + \delta(t-2)\}$

$= 2\,u(t+2) - 3\,u(t+1) + u(t-2)$

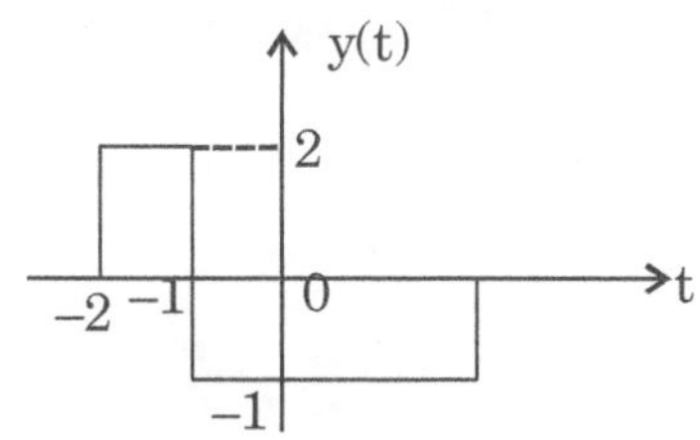

$\Rightarrow$ Energy of $y(t) = \int_{-\infty}^{\infty} y^2(t)\,dt$

= Area under the curve $y^2(t)$

$= (4 \times 1) + (3 \times 1) = 7$.

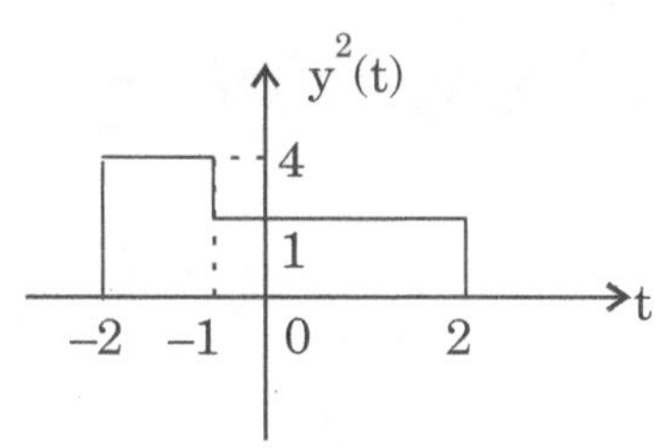

3 CHAPTER

Fourier Series

Analysis of Previous GATE Papers												
	Year → Topics ↓		2019	2018	2017 Set 1	2017 Set 2	2016 Set 1	2016 Set 2	2016 Set 3	2015 Set 1	2015 Set 2	2015 Set 3
CONTINUOUS TIME FOURIER SERIES	1 Mark	MCQ Type		1	1						1	
		Numerical Type										
	2 Marks	MCQ Type			1							
		Numerical Type										
	Total			1	3						1	
DISCRETE-TIME FOURIER SERIES	1 Mark	MCQ Type										
		Numerical Type										
	2 Marks	MCQ Type										
		Numerical Type										1
	Total											2

CONTINUOUS TIME FOURIER SERIES

1. A half-wave rectified sinusoidal waveform has a peak voltage of 10 V. Its average value and the peak value of the fundamental component are respectively given by

(a) $\frac{20}{\pi}$ V, $\frac{10}{\sqrt{2}}$ V (b) $\frac{10}{\pi}$ V, $\frac{10}{\sqrt{2}}$ V

(c) $\frac{10}{\pi}$ V, 5 V (d) $\frac{20}{\pi}$ V, 5 V

[1987 : 2 Marks]

2. Fourier series of the periodic function (period 2π) defined by

$$f(x) = \begin{cases} 0, & 0\pi < x0 \\ x, & 0 < x < \pi \end{cases}$$

is $\frac{\pi}{4} + \sum_{1}^{\infty}\left[\frac{1}{\pi n^2}[\cos(n\pi) - 1]\cos(nx) - \frac{1}{n}\cos(n\pi)\sin(nx)\right]$.

By putting $x = \pi$ in the above, one can deduce that the sum of the series $1 + \frac{1}{3^2} + \frac{1}{5^2} + \frac{1}{7^2} + \ldots$ is _______.

[1993 : 2 MArks]

3. The Fourier series of an odd periodic function, contains only

(a) odd harmonics

(b) even harmonics

(c) cosine terms

(d) sine terms

[1994 : 1 Mark]

4. The RMS value of a rectangular wave of period T, having a value of +V for a duration, $T_1 (<T)$ and $-V$ for the duration, $T - T_1 = T_2$ equals

(a) V (b) $\frac{T_1 - T_2}{T}$ V

(c) $\frac{V}{\sqrt{2}}$ (d) $\frac{T_1}{T_2}$ V

[1995 : 1 Mark]

5. The trigonometric Fourier series of an even function of time does not have

(a) the dc term

(b) cosine terms

(c) sine terms

(d) odd harmonic terms

[1996 : 1 Mark]

6. The trigonometric Fourier series of a periodic time function can have only

(a) cosine terms

(b) sine terms

(c) cosine and sine terms

(d) dc and cosine terms

[1998 : 1 Mark]

7. A periodic signal x(t) of period T_0 is given by

$$x(t) = \begin{cases} 1, & |t| < T_1 \\ 0, & T_1 < |t| < \frac{T_0}{2} \end{cases}$$

the dc component of x(t) is

(a) $\frac{T_1}{T_0}$ (b) $\frac{T_1}{(2T_0)}$

(c) $\frac{2T_1}{T_0}$ (d) $\frac{T_0}{T_1}$

[1998 : 1 Mark]

8. The Fourier series representation of an impulse train denoted by

$s(t) = \sum_{n=-\infty}^{\infty} \delta(t - nT_0)$ is given by

(a) $\frac{1}{T_0}\sum_{n=-\infty}^{\infty} \exp\left(-\frac{j2\pi nt}{T_0}\right)$

(b) $\frac{1}{T_0}\sum_{n=-\infty}^{\infty} \exp\left(-\frac{j\pi nt}{T_0}\right)$

(c) $\frac{1}{T_0}\sum_{n=-\infty}^{\infty} \exp\left(\frac{j\pi nt}{T_0}\right)$

(d) $\frac{1}{T_0}\sum_{n=-\infty}^{\infty} \exp\left(\frac{j2\pi nt}{T_0}\right)$

[1999 : 2 Marks]

9. One period (0, T) each of two periodic waveforms W_1 and W_2 are shown in the figure. The magnitudes of the nth Fourier series coefficients of W_1 and W_2, for $n \geq 1$, n is odd, are respectively proportional to

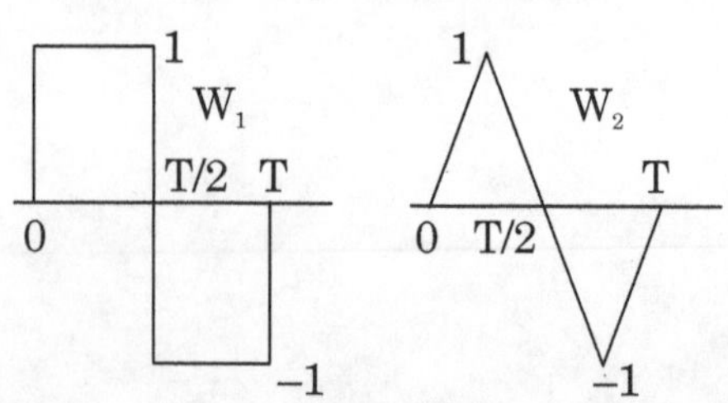

(a) $|n|^{-3}$ and $|n|^{-2}$ (b) $|n|^{-1}$ and $|n|^{-3}$

(c) $|n|^{-1}$ and $|n|^{-2}$ (d) $|n|^{-4}$ and $|n|^{-2}$

[2000 : 2 Marks]

10. Which of the following cannot be the Fourier series expansion of a periodic signal?

(a) $x(t) = 2\cos t + 3\cos 3t$

(b) $x(t) = 2\cos\pi t + 7\cos t$

(c) $x(t) = \cos t + 0.5$

(d) $x(t) = 2\cos 1.5\pi t + \sin 3.5\pi t$

[2002 : 1 Mark]

11. The Fourier series expansion of a real periodic signal with fundamental frequency f_0 is given by

$$g_p(t) = \sum_{n=-\infty}^{\infty} c_n e^{j2\pi f_0 t};$$

it is given that $c_3 = 3 + j5$. Then c_3 is

(a) $5 + j3$ (b) $-3 - j5$

(c) $-5 + j3$ (d) $3 - j5$

[2003 : 1 Mark]

12. Choose the function $f(t)$, $-\infty < t < \infty$, for which a Fourier series cannot be defined ?

(a) $3\sin(25t)$

(b) $4\cos(20t + 3) + 2\sin(710t)$

(c) $\exp(-|t|)\sin(25t)$

(d) 1 **[2005 : 1 Mark]**

13. The Fourier series of a real periodic function has only

P. Cosine terms if it is even

Q. Sine terms if it is even

R. Cosine terms if it is odd

S. Sine terms if it is odd

Which of the above statements are correct?

(a) P and S (b) P and R

(c) Q and S (d) Q and R

[2009 : 1 Mark]

14. The trigonometric Fourier series of an even function does not have the

(a) dc term

(b) cosine terms

(c) sine terms

(d) odd harmonic terms

[2011 : 1 Mark]

15. For a periodic signal

$$v(t) = 30\sin 100t + 10\cos 300t + 6\sin(500t + \frac{\pi}{4}),$$

the fundamental frequency in rad/s is

(a) 100 (b) 300

(c) 500 (d) 1500

[2013 : 1 Mark]

16. A discrete-time signal $x[n] = \sin(\pi^2 n)$, n being an integer, is

(a) periodic with period π.

(b) periodic with period π^2.

(c) periodic with period $\frac{\pi}{2}$.

(d) not periodic.

[2014 : 1 Mark, SeM]

17. Consider the periodic square wave in the figure shown.

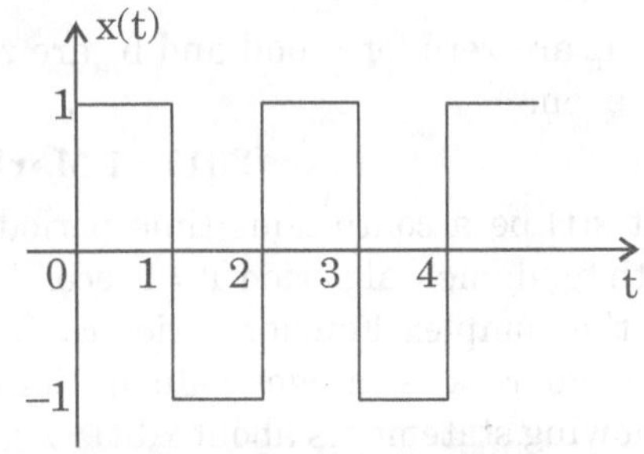

The ratio of the power in the 7th harmonic to the power in the 5th harmonic for this waveform is closest in value to ______.

[2014 : 1 Mark, Set-2]

18. The magnitude and phase of the complex Fourier series coefficient a_k of a periodic signal x(t) are shown in the figure. Choose the correct statement from the four choices given. Notation: C is the set of complex number, R is the set of purely real numbers, and P is the set of purely imaginary numbers.

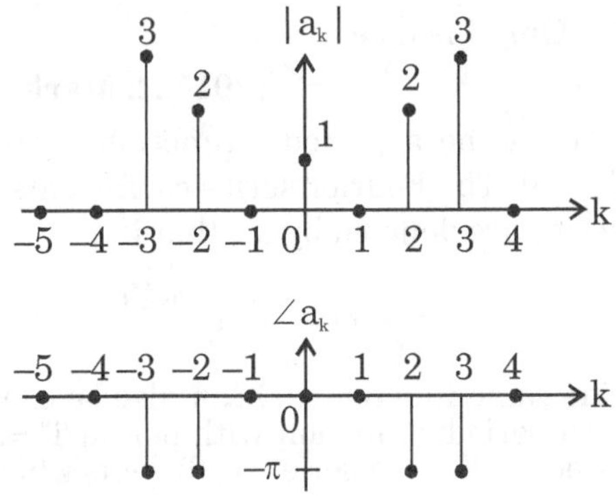

(a) $x(t) \in R$

(b) $x(t) \in P$

(c) $x(t) \in (C - R)$

(d) The information given is not sufficient to draw any conclusion about x(t)

[2015 : 1 Mark, Set-2]

19. A periodic signal x(t) has a trigonometric Fourier series expansion

$$x(t) = a_0 + \sum_{n=1}^{\infty}(a_n \cos n\omega_0 t + b_n \sin n\omega_0 t)$$

If $x(t) = -x(-t) = -x(t - \pi/\omega_0)$, we can conclude that

(a) a_n are zero for all n and b_n are zero for n even.

(b) a_n are zero for all n and b_n are zero for n odd.

(c) a_n are zero for n even and b_n are zero for n odd.

(d) a_n are zero for n odd and b_n are zero for n even.

[2017 : 1 Mark, Set-1]

20. Let x(t) be a continuous time periodic signal with fundamental period T = 1 seconds. Let $\{a_k\}$ be the complex Fourier series coefficients of x(t), where k is integer valued. Consider the following statements about x(3t) :

I. The complex Fourier series coefficients of x(3t) are $\{a_k\}$ where k is integer valued.

II. The complex Fourier series coefficients of x(3f) are $\{3a_k\}$ where k is integer valued.

III. The fundamental angular frequency of x(3t) is 6π rad/s.

For the three statements above, which one of the following is correct?

(a) Only II and III are true.

(b) Only I and III are true.

(c) Only III is true.

(d) Only I is true.

[2017 : 2 Marks, Set-1]

21. Let x(t) be a periodic function with period T = 10. The Fourier series coefficients for this series are denoted by a_k, that is

$$x(t) = \sum_{k=-\infty}^{\infty} a_k e^{jk\frac{2\pi}{T}t}.$$

The same function x(t) can also be considered as a periodic function with period T' = 40. Let b_k be the Fourier series coefficients when period is taken as T'. If $\sum_{k=-\infty}^{\infty} |a_k| = 16$, then $\sum_{k=-\infty}^{\infty} |b_k|$ is equal to

(a) 256 (b) 64

(c) 16 (d) 4 **[2018 : 1 Mark]**

22. A function is given by $f(t) = \sin^2 t + \cos 2t$. Which of the following is true?

(a) f has frequency components at 0 and $\frac{1}{2\pi}$ Hz

(b) f has frequency components at 0 and $\frac{1}{\pi}$ Hz

(c) f has frequency components at $\frac{1}{2\pi}$ and $\frac{1}{\pi}$ Hz

(d) f has frequency components at 0, $\frac{1}{2\pi}$ and $\frac{1}{\pi}$ HZ

[2009 : 1 Mark]

DISCRETE-TIME FOURIER SERIES

23. Let $\bar{x}[n] = 1 + \cos\left(\frac{\pi n}{8}\right)$ be a periodic signal with period 16. Its DFS coefficients are defined by $a_k = \frac{1}{16}\sum_{n=0}^{15} x[n]\exp\left(-j\frac{\pi}{8}kn\right)$ for all k. The value of the coefficient a_{31} is ______.

[2015 : 2 Marks, Set-3]

24. Consider a discrete time periodic signal $x[n] = \sin\left(\frac{\pi n}{5}\right)$. Let a_k be the complex Fourier series coefficients of x[n]. The coefficients $\{a_k\}$ are non-zero when $k = Bm \pm 1$, where m is any integer. The value of B is ______.

[2014 : 2 Marks, Set-1]

ANSWERS

1. (c)	**2.** (c)	**3.** (b)	**4.** (a)	**5.** (c)	**6.** (d)	**7.** (c)	**8.** (a)	**9.** (c)	**10.** (b)
11. (d)	**12.** (c)	**13.** (a)	**14.** (c)	**15.** (a)	**16.** (d)	**17.** (0· 5)	**18.** (a)	**19.** (a)	**20.** (b)
21. (c)	**22.** (b)	**23.** (0· 5)	**24.** (d)						

EXPLANATIONS

1. Assuming half wave rectified output as shown in figure.

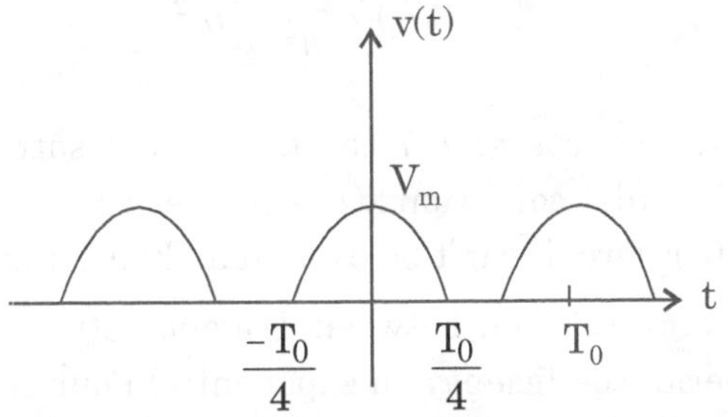

$u(t = V_m \cos(co_o t), |t| < \frac{To}{4}$

Timr period = T_0. Fundamental frequency or 1st

harmonic $= \frac{1}{T_o}$

Founier series of u(t) is given by

$x(t) = a_o + a_1 \cos(\omega_o t) + a_2 \cos(2\omega_o t) + ...$

where a_o = dc or average value $= \frac{1}{40} \int_{T_o} u(t)\,dt$

$$a_o \frac{1}{T_o} \int_{-\frac{T_o}{4}} v_m \cos(\omega_o t)\,dt = \frac{V_m}{T_o} \frac{[\sin(\omega_o t)]_{-\frac{T_o}{4}}^{\frac{T_o}{4}}}{\omega_o}$$

$$= \frac{V_m}{2\pi}\left[\sin\left(\frac{\pi}{2}\right) - \sin\left(-\frac{\pi}{2}\right)\right] = \frac{V_m}{2\pi} \times 2$$

$$a_o = \frac{V_m}{\pi} = \frac{10}{\pi} V.$$

$$\Rightarrow a_o \frac{2}{T_o} \int_{T_o} u(t) \cos(\omega_o t)\,dt$$

$$= \frac{2}{10} \int_{-\frac{To}{4}}^{To} V_m \cos(\omega_o t).\cos(\omega_o t)\,dt$$

$$= \frac{2}{10} \int_{-\frac{To}{4}}^{\frac{To}{4}} V_m \cos^2(\omega_o t)\,dt.$$

$$= \frac{2}{10} \int_{-\frac{To}{4}}^{\frac{To}{4}} \frac{(1 + \cos 2\omega_o t)}{2}\,dt.$$

$$= \frac{2V_m}{2T_o}\left[\int_{\frac{-To}{4}}^{\frac{To}{4}} 1.dt + \int_{\frac{-To}{4}}^{\frac{To}{4}} \cos(2w_o t)\,dt\right.$$

$$= \frac{V_m}{To}\left[\frac{To}{4} + \frac{To}{4}\right]$$

$$= \frac{V_m}{To} . \frac{To}{2}$$

$$a_1 = \frac{V_m}{To} . \frac{10}{2}$$

$a_1 = 5$ V.

2. Time period, T = 2π.

At the point of discontinuity, x = π

f(x) is exressed in Fourier series converges to the

middle Value $= \frac{\pi}{2}$

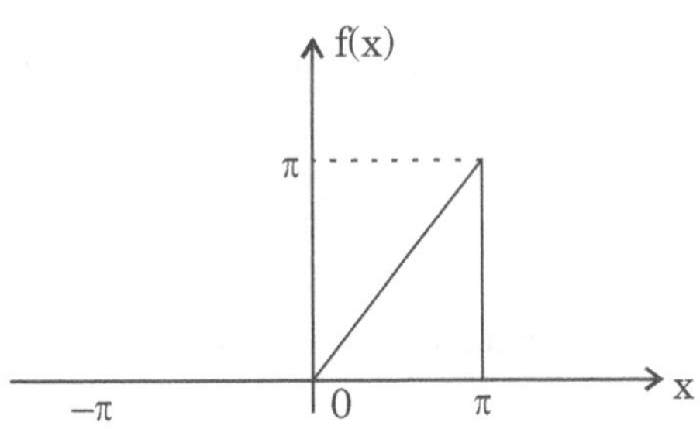

From the given trigonometric form of fourier series at x = π.

$$f(\pi) = \frac{\pi}{4} + \frac{2}{\pi}s = \frac{\pi}{2}$$

$$\frac{2}{\pi}s = \frac{\pi}{2} - \frac{\pi}{4} = \frac{\pi}{4}$$

$$S = \frac{\pi^2}{8}$$

4. Mean square value

$$= \frac{1}{T}\left[\int_0^{T_1} V^2 dt + \int_{T_1}^{T} V^2 dt\right] . T_1 + T_2 = T$$

$$= \frac{1}{T}[V^2 T_1 + V^2 T + V^2 T_1] = V^2$$

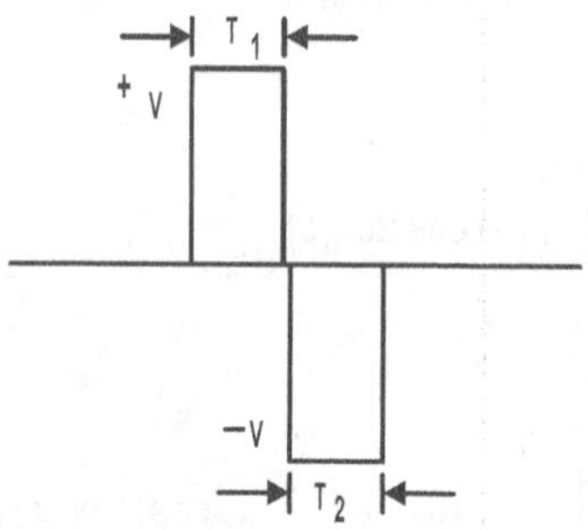

∴ RMS value = $\sqrt{V^2} = V$

5. $f(t)$ is even function, hence

where, b_k = coefficient of sine terms, $b_k = 0$

6. The trigonometric Fourier series of a periodic time function can have dc and cosine terms.

7. The periodic signal $X(t)$ of period T_o

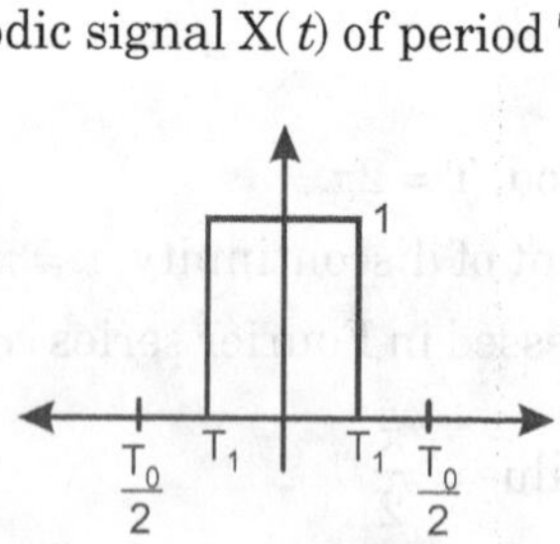

$X(t) = 1, \ |t| < T_1$

$X(t) = 0, \ T_1 < |t| < \frac{T_0}{2}$

$$A_o = \frac{1}{T_0}\int_{-T_1}^{T_1} x(t)\,dt = \frac{2T_1}{T_o}$$

8. (d) Complex form of fourier is

$$\delta_{T_o}(t) = \sum_{k=-\infty}^{\infty} C_k e^{jk\omega ot}$$

where $C_k = \frac{1}{T_o}\int_{-\frac{T_o}{2}}^{\frac{T_o}{2}} \delta(t).e^{-jk\omega_o t}dt$

$$= \int_{-\frac{T_o}{2}}^{\frac{T_o}{2}} \delta(t)e^o dt = \frac{1}{T_o}$$

$$\delta_{T_o}(t) = \frac{1}{T_o}\sum_{K=-\infty}^{\infty} e^{jkw_o t}$$

9. Magnitude $(W_1) \propto \frac{1}{n} \times n^{-1}$

$$(W_2) \propto \frac{1}{n^2} \times n^{-2}$$

10. $x(t) = 2\cos \pi t + 7\cos t$, does not satisfy the **Dirichlet condition** $x(t)$ is not periodic Function, therefore, it can't be expand as Fourier series.

11. Using relation between Trigonometric Fourier series coefficient and exponential Fourier series coefficient

$$c_n = a_n - jb_n \quad ..(i)$$

$$c_{-n} = a_n + jb_n \quad ...(A)$$

Given, $c_3 = 3 + j5$

$$c_3 = 3 - j(-5) \quad ...(ii)$$

Compare (*i*) with (*ii*)

$a_n = 3, \ b_n = -5$

Then using (A)

12. All other f's are periodic oconstant.

13. Fourier series for periodic function

$$f(t) = a_0 + \sum_{n=1}^{\infty} (a_n \cos n\omega_0 t + b_n \sin n\omega_0 t)$$

Here, $a_0 = \frac{1}{T}\int_{-T/2}^{T/3} f(t)dt$

$$b_0 = 0$$

$$a_n = \frac{2}{T}\int_{-T/2}^{T/2} f(t) \cos n\omega_0 t\, dt$$

$$b_n = \frac{2}{T}\int_{-T/2}^{T/2} f(t) \sin n\omega_0 t\, dt$$

For even function, $f(t) = f(-t)$

so $b_n = 0$

Hence even function has only cosine terms.

For odd function, $f(t) = -f(-t)$

so $a_n = 0$

Hence odd function has only sine terms.

14. Trigonometric Fouries series of an even function has dc and cosine terms only while for odd function, only sine exist.

16. Time period of a discrete signal,

$$\frac{w_0}{2\pi} = \frac{K}{N}$$

$$N = \frac{2\pi K}{w_0}$$

$$= \frac{2\pi K}{\pi^2} = \frac{2K}{\pi}$$

N is a irrational number so signal is not periodic.

17. For a periodic sequence wave, n^{th} harmonic component is $\alpha \frac{1}{n}$

$\Rightarrow$ Power in n^{th} harmonic component is $\alpha \frac{1}{n^2}$

$\Rightarrow$ Ratio of the power in 7th harmonic to power in 5th harmonic for given waveform is

$$\frac{\frac{1}{7^2}}{\frac{1}{5^2}} = \frac{25}{49} \approx 0.5$$

18. $\angle a_k = -\pi$ only changes the sign of the magnitude $|a_k|$. Since the magnitude spectrum $|a_k|$ is even the corresponding time-domain signal is real.

19. As signal has odd and half wave symmetries, hence, all a_n are zero and b_n are zero for n even.

20. Initially T = 1 sec, so $\omega_0 = 2\pi$ rad/sec. When $x(t)$ is compressed by 3, frequency will expand by same factor but there is no change in values of a_k, a_k remains constant.

So, both statement I and III are correct.

21. Change in only time period or frequency does not change in the value of Fourier series coefficients.

So, $\quad b_k = a_k$

$$\sum_{k=-\infty}^{\infty} |b_k| = \sum_{k=-\infty}^{\infty} |a_k| = 16$$

22. $$f(t) = \frac{1}{2} - \frac{\cos 2t}{2} + \cos 2t$$

$$f(t) = \frac{1}{2} + \frac{\cos(2t)}{2}$$

Thus $f(t)$ has frequency components

$$f_1\ (dc) = 0$$

$$f_2 = \frac{2}{2\pi} = \frac{1}{\pi}\ \text{Hz}$$

23. $$x[n] = 1 + \cos\left(\frac{\pi}{8}n\right)$$

$$N = 16$$

$$x[n] = 1 + \frac{1}{2}e^{j\frac{2\pi n}{16}} + \frac{1}{2}e^{-j\frac{2\pi n}{16}}$$

$$a_{-1} = \frac{1}{2},\ a_1 = \frac{1}{2},\ a_0 = 1$$

$$a_1 = a_{-1+16}$$

$$\Rightarrow \quad a_{-1} = a_{15} = \frac{1}{2}$$

$$\Rightarrow \quad a_0 = 1, a_1 = \frac{1}{L}, a_2 \text{ to } a_{14} = 0, a_{15} = \frac{1}{2}$$

DFS coefficients are also periodic with period 16.

$$a_{31} = a_{16+15}$$

$$a_{31} = a_{15}$$

$$\Rightarrow \quad a_{31} = \frac{1}{2}$$

CHAPTER 4

Fourier Transform & Correlation

Analysis of Previous GATE Papers												
Topics ↓		Year →	2019	2018	2017 Set 1	2017 Set 2	2016 Set 1	2016 Set 2	2016 Set 3	2015 Set 1	2015 Set 2	2015 Set 3
CONTINOUS-TIME FOURIER TRANSFORM	1 Mark	MCQ Type							1			
		Numerical Type						1				1
	2 Marks	MCQ Type										1
		Numerical Type		1		2						
		Total		2		4		1	1			3
CORRELATION	1 Mark	MCQ Type										
		Numerical Type										
	2 Marks	MCQ Type										
		Numerical Type										
		Total										
GROUP & PHASE DELAY	1 Mark	MCQ Type										
		Numerical Type										
	2 Marks	MCQ Type										
		Numerical Type										
		Total										

CONTINUOUS-TIME FOURIER TRANSFORM

1. If G(f) represents the Fourier transform of a signal g(t) which is real and odd symmetric in time, then

(a) G(f) is complex

(b) G(f) is imaginary

(c) G(f) is real

(d) G(f) is real and non-negative

[1992 : 2 Marks]

2. Sketch the waveform (with properly marked axes) at the output of a matched filter matched for a signal s(t), of duration T, given by

$$s(t) = \begin{cases} A \text{ for } & 0 \le t < \frac{2}{3}T \\ 0 \text{ for } & \frac{2}{3}T \le t < T \end{cases}$$

[1993 : 2 Marks]

3. Match each of the items, A, B and C with an appropriate item from 1, 2, 3, 4 and 5.

List-I

A. Fourier transform of a Gaussian function

B. Convolution of a rectangular pulse with itself

C. Current through an inductor for a step input voltage

List-II

1. Gaussian function
2. Rectangular pulse
3. Triangular pulse
4. Ramp function
5. Zero

[1995 : 2 Marks]

4. A rectangular pulse of duration T is applied to a filter matched to this input. The output of the filter is a

(a) rectangular pulse of duration T

(b) rectangular pulse of duration 2T

(c) triangular pulse

(d) sine function

[1996 : 1 Mark]

5. The Fourier transform of a real valued time signal has

(a) odd symmetry

(b) even symmetry

(c) conjugate symmetry

(d) no symmetry

[1996 : 1 Mark]

6. The function f(t) has the Fourier transform f(ω). The Fourier transform of

$$g(t) = \left(\int_{-\infty}^{\infty} g(t) e^{-j\omega} dt \right) \text{ is}$$

(a) $\frac{1}{2\pi}F(\omega)$ (b) $\frac{1}{2\pi}F(-\omega)$

(c) $\frac{1}{2\pi}F(-\omega)$ (d) None of these

[1997 : 1 Mark]

7. If the Fourier transform of a deterministic signal g(t) is G(f), then

1. The Fourier transform of g(t – 2) is

2. The Fourier transform of $g\left(\frac{t}{2}\right)$ is

(a) $G(f)e^{-j(4\pi f)}$ (b) G(2f)

(c) 2G(2f) (d) G(f – 2)

Match each of the items 1, 2 on the left with the most approptiate item A, B, C, or D on the right.

[1997 : 2 Marks]

8. The amplitude spectrum of a Gaussian pulse is

(a) uniform (b) a sine function

(c) Gaussian (d) an impulse function

[1998 : 1 Mark]

9. The Fourier Transform of a function x(t) is X(f). The Fourier transform of $\frac{dx(t)}{dt}$ will be

(a) $\frac{dx(t)}{dt}$ (b) $j2\pi fX(f)$

(c) jfX(f) (d) $\frac{X(f)}{jf}$

[1998 : 1 Mark]

10. The Fourier transform of a voltage signal x(t) is X(f). The unit of |X(f)| is

(a) Volt

(b) Volt-sec

(c) Volt/sec

(d) $Volt^2$

[1998 : 1 Mark]

11. A signal x(t) has a Fourier transform X(ω). If x(t) is a real and odd function of t, then X(ω) is

(a) a real and even function of ω

(b) an imaginary and odd function of ω

(c) an imaginary and even function of ω

(d) a real and odd function of ω

[1999 : 1 Mark]

12. The Fourier Transform of the signal $x(t) = e^{-3t^2}$ is of the following form, where A and B are constants

(a) Ae^{-Bf^2} (b) Ae^{-Bt^2}

(c) $A + B|f|^2$ (d) Ae^{-Bf}

[2000 : 1 Mark]

13. The Fourier transform F{e^{-t} u(t)} is equal to $\frac{1}{1+j2\pi f}$. Therefore, $F\left\{\frac{1}{1+j2\pi f}\right\}$ is

(a) $e^{f}u(f)$ (b) $e^{-f}u(f)$

(c) $e^{f}u(-f)$ (d) $e^{-f}u(-f)$

[2002 : 1 Mark]

14. Let x(t) be the input to a linear, time-invariant system. The required output is 4x(t – 2). The transfer function of the system should be

(a) $4e^{j4\pi f}$ (b) $2e^{-j8\pi f}$

(c) $4e^{-j4\pi f}$ (d) $2e^{j8\pi f}$

15. A rectangular pulse train s(t) as shown in the figure is convolved with the signal $\cos^2(4\pi \times 10^3 t)$. The convolved signal will be a

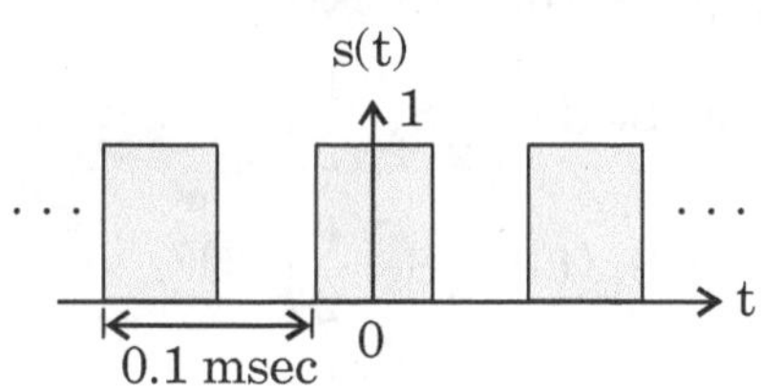

(a) DC (b) 12 kHz sinusoid

(c) 8 kHz sinusoid (d) 14 kHz sinusoid

[2004 : 2 Marks]

16. Let x(t) and y(t) (with Fourier transforms X(f) and Y(f) respectively) be related as shown in the figure. Then Y(f) is

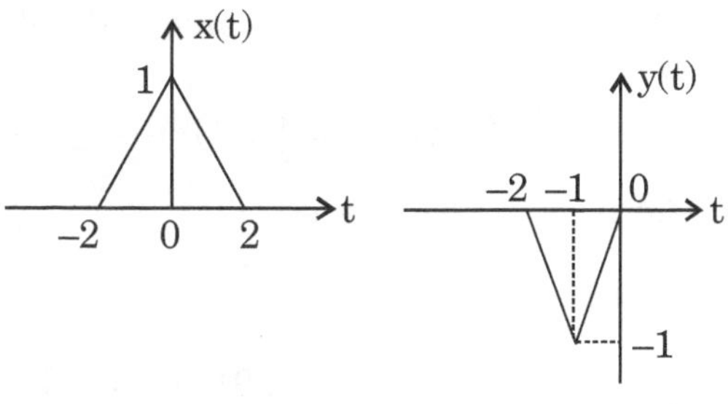

(a) $-\frac{1}{2}X\left(\frac{f}{2}\right)e^{-j2\pi f}$ (b) $-\frac{1}{2}X\left(\frac{f}{2}\right)e^{j2\pi f}$

(c) $-X\left(\frac{f}{2}\right)e^{j2\pi f}$ (d) $-X\left(\frac{f}{2}\right)e^{-j2\pi f}$

[2004 : 2 Marks]

17. Match the following and choose the correct combination.

Group 1

E. Continuous and aperiodic signal

F. Continuous and periodic signal

G. Discrete and aperiodic signal

H. Discrete and periodic signal

Group 2

1. Fourier representation is continuous and aperiodic.
2. Fourier representation is discrete and aperiodic.
3. Fourier representation is continuous and periodic.
4. Fourier representation is discrete and periodic.

(a) E-3, F-2, G-4, H-1

(b) E-1, F-3, G-2, H-4

(c) E-1, F-2, G-3, H-4

(d) E-2, F-1, G-4, H-3

[2005 : 2 Marks]

18. For a signal x(t) the Fourier transform is X(f). Then the inverse Fourier transform of X(3f + 2) is given by

(a) $\frac{1}{2}x\left(\frac{t}{2}e^{j3\pi t}\right)$ (b) $\frac{1}{2}x\left(\frac{t}{2}\right)e^{-j4\pi t/3}$

(c) $3x(3t)e^{-j4\pi t}$ (d) $x(3t+2)$

[2005 : 2 Marks]

19. The output y(t) of a linear time invariant system is related to its input x(t) by the following equation:

$y(t) = 0.5x(t - t_d + T) + x(t - t_d) + 0.5x(t - t_d - T)$.

The filter transfer function H(ω) of such a system is given by

(a) $(1 + \cos\omega T)e^{-j\omega t_d}$

(b) $(1 + 0.5\cos\omega T)e^{-j\omega t_d}$

(c) $(1 - \cos\omega T)e^{-j\omega t_d}$

(d) $(1 - 0.5\cos\omega T)e^{-j\omega t_d}$

[2005 : 2 Marks]

20. Let $x(t) \leftrightarrow X(j\omega)$ be Fourier Transform pair. The Fourier Transform of the signal $x(5t - 3)$ in terms of $X(j\omega)$ is given as

(a) $\frac{1}{5}e^{-\frac{j3\omega}{5}}X\left(\frac{j\omega}{5}\right)$ (b) $\frac{1}{5}e^{\frac{j3\omega}{5}}X\left(\frac{j\omega}{5}\right)$

(c) $\frac{1}{5}e^{-j3\omega}X\left(\frac{j\omega}{5}\right)$ (d) $\frac{1}{5}e^{j3\omega}X\left(\frac{j\omega}{5}\right)$

[2006 : 1 Mark]

21. The signal x(t) is described by

$$x(t) = \begin{cases} 1 \text{ for } & -1 \le t \le +1 \\ 0, & \text{otherwise} \end{cases}$$

Two of the angular frequencies at which its Fourier transform becomes zero are

(a) $\pi, 2\pi$ (b) $0.5\pi, 1.5\pi$

(c) $0, \pi$ (d) $2\pi, 2.5\pi$

[2008 : 2 Marks]

Statement for Linked Answer Questions 22 and 23 :

The impulse response h(t) of a linear time-invariant continuous time system is given by $h(t) = \exp(-2t)u(t)$, where u(t) denotes the unit step function.

22. The frequency response $H(\omega)$ of this system in terms of angular frequency ω, is given by, $H(\omega) =$

(a) $\frac{1}{1+j2\omega}$ (b) $\frac{\sin(\omega)}{\omega}$

(c) $\frac{1}{2+j\omega}$ (d) $\frac{j\omega}{2+j\omega}$

[2008 : 2 Marks]

23. The output of this system, to the sinusoidal input $x(t) = 2\cos(2t)$ for all time t, is

(a) 0

(b) $2^{-0.25}\cos(2t - 0.125\pi)$

(c) $2^{-0.5}\cos(2t - 0.125\pi)$

(d) $2^{-0.5}\cos(2t - 0.25\pi)$

[2008 : 2 Marks]

24. Consider a system whose input .r and output y are related by the equation

$$y(t) = \int_{-\infty}^{\infty} x(t-\tau)h(2\tau)d\tau$$

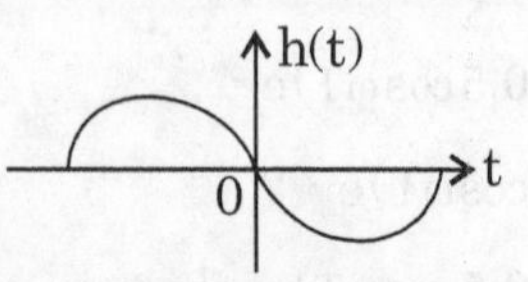

where h(t) is shown in the graph

Which of the following four properties are possessed by the system? BIBO: Bounded input gives a bounded output

Causal: The system is causal.

LP : The system is low pass.

LTI: The system is linear and time-invariant.

(a) Causal, LP.

(b) BIBO, LTI.

(c) BIBO, Causal, LTI .

(d) LP, LTI.

[2009 : 2 Marks]

25. The Fourier transform of a signal

h(t) is $H(j\omega) = \frac{(2\cos\omega)(\sin\omega)}{\omega}$

The value of h(0) is

(a) $\frac{1}{4}$ (b) $\frac{1}{2}$

(c) 1 (d) 2

[2012 : 2 Marks]

26. Let $g(t) = e^{-\pi t^2}$, and h(t) is a filter matched to g(t). If g(t) is applied as input to h(t), then the Fourier transform of the output is

(a) $e^{-\pi t^2}$ (b) $e^{-\pi t^2/2}$

(c) $e^{-\pi|f|}$ (d) $e^{-2\pi t^2}$

[2013 : 1 Mark]

27. Let x(t) be a wide sense stationary (WSS) random with power spectral density $S_x(f)$. If Y(t) is the process defined as $y(t) = x(2t - 1)$, the power spectral density $S_Y(f)$ is

(a) $S_Y(f) = \frac{1}{2}S_X\left(\frac{f}{2}\right)e^{-j\pi t}$..

(b) $S_Y(f) = \frac{1}{2}S_X\left(\frac{f}{2}\right)e^{-j\pi t/2}$.

(c) $S_Y(f) = \frac{1}{2}S_X\left(\frac{f}{2}\right)$.

(d) $S_Y(f) = \frac{1}{2}S_X\left(\frac{f}{2}\right)e^{j2\pi f}$.

[2014 : 2 Marks, Set-3]

28. For a function g(t), it is given that

$\int_{-\infty}^{+\infty} g(t)e^{-j\omega t}dt = \omega e^{-2\omega^2}$ for any real value ω. If

$y(t) = \int_{-\infty}^{t} g(\tau)d\tau$, then $\int_{-\infty}^{+\infty} y(t)dt$ is ______.

(a) 0 (b) $-j$

(c) $-\frac{j}{2}$ (d) $\frac{j}{2}$

[2014 : 2 Marks, Set-1]

29. The complex envelope of the bandpass signal $x(t) = \sqrt{2}\left(\frac{\sin\left(\frac{\pi t}{5}\right)}{\frac{\pi t}{5}}\right)\sin\left(\pi t - \frac{\pi}{4}\right)$, centered about $f = \frac{1}{2}$ Hz, is

(a) $\left(\frac{\sin\left(\frac{\pi t}{5}\right)}{\frac{\pi t}{5}}e^{j\frac{\pi}{4}}\right)$ (b) $\left(\frac{\sin\left(\frac{\pi t}{5}\right)}{\frac{\pi t}{5}}e^{-j\frac{\pi}{4}}\right)$

(c) $\sqrt{2}\left(\frac{\sin\left(\frac{\pi t}{5}\right)}{\frac{\pi t}{5}}e^{j\frac{\pi}{4}}\right)$ (d) $\sqrt{2}\left(\frac{\sin\left(\frac{\pi t}{5}\right)}{\frac{\pi t}{5}}e^{-j\frac{\pi}{4}}\right)$

[2015 : 2 Marks, Set-3]

30. If the signal $x(t) = \frac{\sin(t)}{\pi t} * \frac{\sin(t)}{\pi t}$ with $*$ denoting the convolution operation, then x(t) is equal to

(a) $\frac{\sin(t)}{\pi t}$ (b) $\frac{\sin(2t)}{2\pi t}$

(c) $\frac{2\sin(t)}{\pi t}$ (d) $\left(\frac{\sin(t)}{\pi t}\right)^2$

[2016 : 1 Mark, Set-3]

31. The value of the integral $\int_{-\infty}^{\infty} \text{sinc}^2(5t)dt$ is ______.

[2014 : 2 Marks, Set-2]

32. A real-valued signal x(t) limited to the frequency band $|f| \le \frac{W}{2}$ is passed through a linear time invariant system whose frequency response is

$$H(f) = \begin{cases} e^{-j4\pi f}, & |f| \le \frac{W}{2} \\ 0, & |f| > \frac{W}{2} \end{cases}$$

The output of the system is

(a) x(t + 4) (b) x(t – 4)

(c) x(t + 2) (d) x(t – 2)

[2014 : 1 Mark, Set-4]

33. Consider the function $g(t) = e^{-t}\sin(2\pi t)\,u(t)$ where u(t) is the unit step function. The area under g(t) is ______.

[2015 : 1 Mark, Set-3]

34. The energy of the signal $x(t) = \frac{\sin(4\pi t)}{4\pi t}$ is ______.

[2016 : 1 Mark, Set-2]

35. A continuous time signal x(t) = 4cos(200πt) + 8cos(400πt). where t is in seconds, is the input to a linear time invariant (LTI) filter with the impulse response

$$h(t) = \begin{cases} \frac{2\sin(300\pi t)}{\pi t} & t \ne 0 \\ 600, & t = 0 \end{cases}$$

Let y(t) be the output of this filter. The maximum value of |y(t)| is______.

[2017 : 2 Marks, Set-1]

36. Consider an LTI system with magnitude response

$$|H(f)| = \begin{cases} 1 - \frac{|f|}{20}, & |f| \le 20 \\ 0, & |f| > 20 \end{cases}$$

and phase response arg {H(f)} = –2f.

If the input to the system is

$$x(t) = 8\cos\left(20\pi t + \frac{\pi}{4}\right) + 16\sin\left(40\pi t + \frac{\pi}{8}\right) + 24\cos\left(80\pi t + \frac{\pi}{16}\right),$$

then the average power of the output signal y(t) is ______.

[2017 : 2 Marks, Set-2]

37. The input 4sinc(2t) is fed to a Hilbert transformer to obtain y(t), as shown in the figure below:

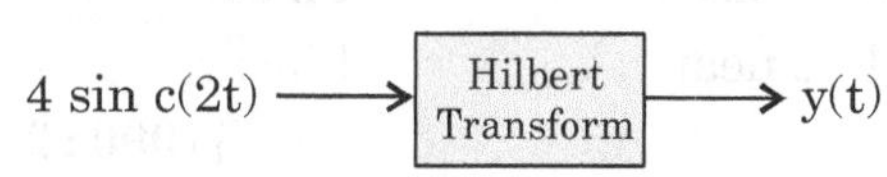

Here $\text{sinc}(x) = \frac{\sin(\pi x)}{\pi x}$. The value (accurate to two decimal places) of $\int_{-\infty}^{\infty} |y(t)^2| dt$ is ______.

[2018 : 2 Marks]

38. A 1 kHz sinusoidal signal is ideally sampled at 1500 samples/sec and the sampled signal is passed through an ideal low-pass filter with cut-off frequency 800 Hz. The output signal has the frequency

(a) zero Hz (b) 0.75 kHz

(c) 0.5 kHz (d) 0.25 kHz

[2018 : 2 Marks]

39. A signal m(t) with bandwidth 500 Hz is first multiplied by a signal g(t) where

$$g(t) = \sum_{k=-\infty}^{\infty} (-1)^k \delta(t - 0.5\times10^{-4}k)$$

The resulting signal is then passed through an ideal low pass filter with bandwidth 1 kHz. The output of the low pass filter would be

(a) δ(t) (b) m(t)

(c) 0 (d) m(t) δ(t)

[2018 : 2 Marks]

CORRELATION

40. The auto-correlation function of an energy signal has

(a) no symmetry.

(b) conjugate symmetry.

(c) odd symmetry.

(d) even symmetry.

[1996 : 2 Marks]

41. The ACF of a rectangular pulse of duration T is

(a) a rectangular pulse of duration T.

(b) a rectangular pulse of duration 2T.

(c) a triangular pulse of duration T.

(d) a triangular pulse of duration 2T.

[1998 : 1 Mark]

GROUP & PHASE DELAY

42. The magnitude and phase functions for a distortionless filter should respectively be

(Magnitude)	**(Phase)**
(a) Linear	Constant
(b) Constant	Constant
(c) Constant	Linear
(d) Linear	Linear

[1990 : 2 Marks]

43. The input to a channel is a bandpass signal. It is obtained by linearly modulating a sinusoidal carrier with a single-tone signal. The output of the channel due to this input is given by

$$y(t) = \left(\frac{1}{100}\right)\cos(100t - 10^{-6})\cos(10^6 t - 1.56)$$

The group delay (t_g) and the phase delay (t_p) in seconds, of the channel are

(a) $t_g = 10^{-6}$, $t_p = 1.56$

(b) $t_g = 1.56$, $t_p = 10^{-6}$

(c) $t_g = 10^{-8}$, $t_p = 1.56 \times 10^{-6}$

(d) $t_g = 10^8$, $t_p = 1.56$

[1999 : 1 Mark]

44. A linear phase channel with phase delay T_p and group delay T_g must have

(a) $T_p = T_g$ = constant.

(b) T_p α f and $T_g \propto$ f.

(c) T_p = constant and T_g α f.

(d) T_p α f and t_g = constant (f denotes frequency).

[2002 : 1 Mark]

Data given below for two questions. Solve the problem and choose the correct answer.

The system under consideration is an RC low-pass filter (RC-LPF) with R = 1.0 kΩ and C = 1.0 μF

45. Let H(f) denote the frequency response of the RC-LPF. Let f_1 be the highest frequency such that $0 \le |f| \le f_1$; $\frac{|H(f_1)|}{H(0)} \ge 0.95$. Then f_1 (in Hz) is

(a) 327.8 (b) 163.9

(c) 52.2 (d) 104.4

[2003 : 2 Marks]

46. Let $t_g(f)$ be the group delay function of the given RC-LPF and f_2 = 100 Hz. Then $t_g(f_2)$ in ms, is

(a) 0.717 (b) 7.17

(c) 71.7 (d) 4.505

[2003 : 2 Marks]

47. The phase response of a passband waveform at the receiver is given by

$$\varphi(f) = -2\pi\alpha(f - f_c) - 2\alpha\beta f_c,$$

where f_c is the centre frequency, and α and β are positive constants. The actual signal propagation delay from the transmittance to receiver is

(a) $\frac{\alpha-\beta}{\alpha+\beta}$ (b) $\frac{\alpha\beta}{\alpha+\beta}$

(c) α (d) β

[2014 : 1 Mark, Set-3]

ANSWERS

1. (a)	2. (*)	3. (A-1, B-3, C-4)	4. (c)	5. (c)	6. (c)	7. (*)	8. (c)	9. (c)
10. (a)	11. (a)	12. (b)	13. (c)	14. (c)	15. (a)	16. (b)	17. (c)	18. (b)
19. (a)	20. (a)	21. (a)	22. (c)	23. (d)	24. (b)	25. (c)	26. (d)	27. (c)
28. (b)	29. (c)	30. (a)	31. (Imaginary)	32. (d)	33. (0· 028)	34. (A-1, B-3, C-4)	35. (8)	36. (8)
37. (8)	38. (c)	39. (b)	40. (d)	41. (d)	42. (c)	43. (d)	44. (d)	45. (c)
46. (a)	47. (c)							

EXPLANATIONS

1.

2. 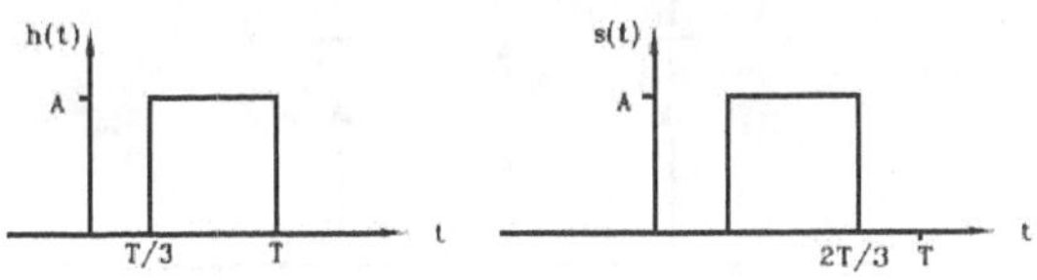

Matched filter response, $h(t) = s(T - t)$

For $t < \frac{T}{3}$, output $y(t) = 0$

For $\frac{T}{3} \leq t \leq T$, output

$$y(t) = \int_0^{t-\frac{T}{3}} A^2\, dt$$

$$= A^2\left(t - \frac{T}{1}\right)$$

For $T \leq t \leq \frac{2T}{3}$,

output $\int_{t-T}^{\frac{2}{3}T} A^3\, dt = A^2$

$$\left(\frac{2}{3}T - t + T\right) = A^2\left(\frac{5T}{3} - t\right)$$

and for $T > \frac{2T}{3} + T$, output $y(t) = 0$

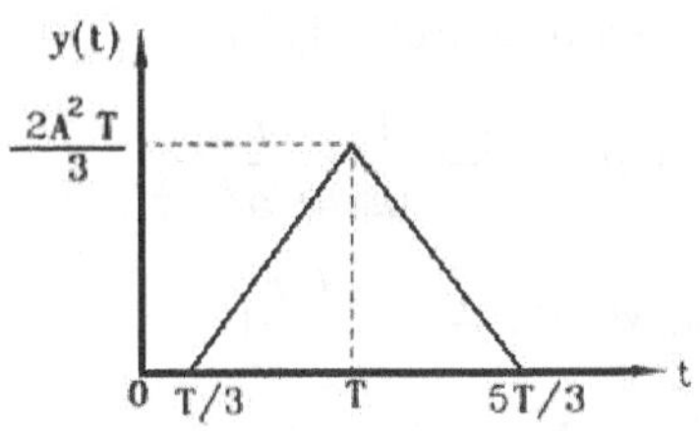

3. (a) Fourier transform of a Gaussion function is also a Gaussion function. $e^{-\pi t^2} \xleftrightarrow{\text{F.T.}} e^{-\pi f^2}$

(b) Convolution of a rectangular pulse with itself is a triangular pulse.

(c) Correct through an inductor for a step input voltage is ramp function.

$$i_L = \frac{1}{L}\int_{-\infty}^{t} v_L\, dt = \frac{1}{L}\int_{-\infty}^{t} u(t)\,dt = \frac{1}{L}r(t).$$

$(A) \rightarrow 1, (B) \rightarrow 3, (c) \rightarrow 4.$

4. 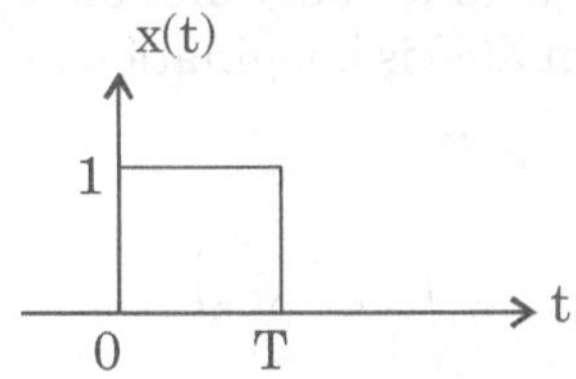

Given

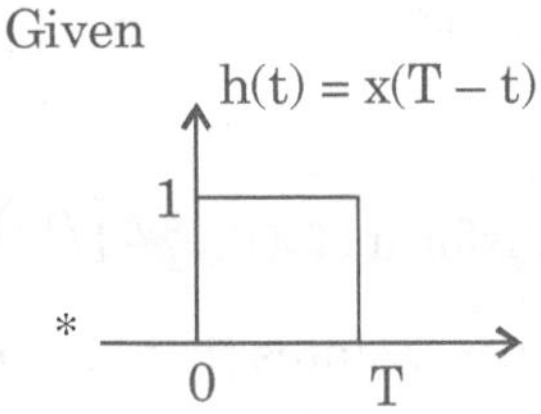

Matched filter

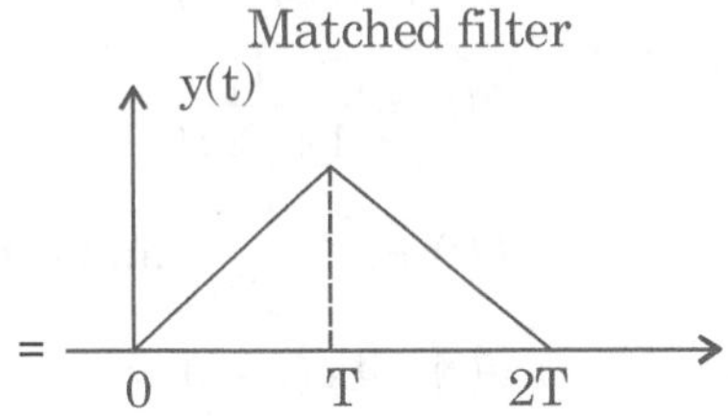

Triangular pulse

5. Fourier transform of real valued signal has conjugate symmetry, i.e., $X^*(j\omega) = X(-j\omega)$.

6. According to dual property of Fourier transform,

$$f(t) \xleftrightarrow{\text{F.T.}} g(w)$$

$$g(t) \xleftrightarrow{\text{F.T.}} 2\pi f(-w).$$

7. $g(t) \xleftrightarrow{\text{F.T.}} G(f).$

$$g(t) \xleftrightarrow{\text{F.T.}} e^{-j2\pi f x 2} G(f)$$

$$= e^{-ju\pi f} G(f)$$

$$g\left(\frac{t}{2}\right) \xleftrightarrow{\text{F.T.}} \frac{1}{\frac{1}{2}} G\left(\frac{f}{\frac{1}{2}}\right)$$

$= G(2f)$.

$(1) \rightarrow (A), (2) \rightarrow (C)$

8. $\frac{r_b}{B_o} = 1$ and $\frac{r_b}{B_p} = \frac{1}{22}$

$\therefore \quad B_p = 2\,B_o$

9. $\frac{r_b}{B_o} = 1$ and $\frac{r_b}{B_p} = \frac{1}{22}$

$\therefore \quad B_p = 2\,B_o$

10. By difinition of Fourier transform,

$$X(f) = \int_{-\infty}^{\infty} x(t).e^{-2\pi ft}dt$$

$[x(t)] \rightarrow$ volts, so $[X(t)] \rightarrow$ volt – sec.

11. For a real and odd function x(t), its Fourier transform X(ω) is imagination and odd.

12. $e^{-\pi t^2} \longleftrightarrow e^{-\pi f^2}$

$$e^{-\pi\left(\sqrt{\frac{3}{\pi}}t\right)^2} \longleftrightarrow \sqrt{\frac{\pi}{3}}\, e^{-\pi\left(\frac{f}{\sqrt{3/\pi}}\right)^2}$$

$e^{-3t2} \longleftrightarrow Ae^{-Bf^2}$

13. $f(t) = e^{-t}u(t)$

Fourier transform of $f(t) = \mathcal{L}[f(t)] = \frac{1}{1+j2\pi f}$

Duality property states that, if $g(t) \leftrightarrow G(f)$

then $\quad G(t) \leftrightarrow g(-f)$

Here $\quad F(f) = \frac{1}{1+j2\pi f}$

$F(t) = \frac{1}{1+j2\pi t}$ and $F(t) \leftrightarrow e^f u(-f)$

14. Output, $\quad y(t) = 4x(t-2)$

Its, fourier transform, $Y(f) = 4x(f).\, e^{-j4\pi f}$

and the transfer function, $\frac{Y(f)}{X(f)} = 4e^{-j4\pi f}$

15.

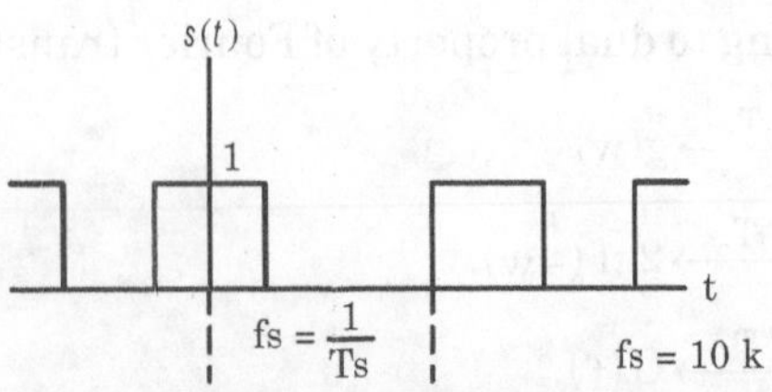

Using Fourier series expansion

$$s(t) = \frac{1}{2} + \frac{2}{\pi}\sum_{n=1}^{\infty}\frac{(-1)^{n-1}}{2n-1}\cos 2\pi(2n-1)f_s t$$

$$s(t) = \frac{1}{2} + \frac{2}{\pi}\cos 2\pi f_s t - \frac{2}{3\pi}\cos 2\pi \times 3 f_s t + \ldots$$

and let $h(t) = \cos^2(2\pi \times 2 \times 10^3)t$

$$= \frac{1}{2} + \frac{\cos(2\pi \times 4 \times 10^3)t}{2}$$

Then, $s(t) \otimes h(t) \longleftrightarrow S(f).H(f)$

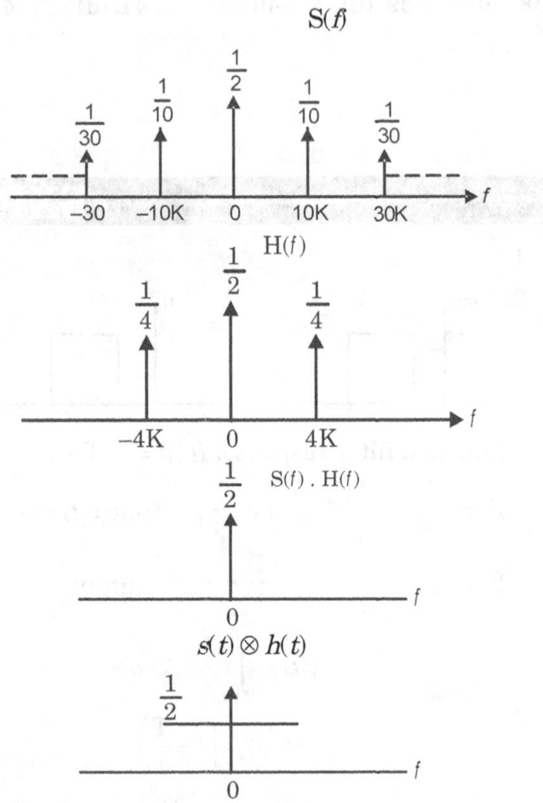

Hence, option (a) is correct.

16. $\quad y(t) = x[\,2(t+2)\,]$

Now, Using $x(t-t_o) \longleftrightarrow X(f).\, e^{-j\omega\pi ft_0}$

and $\quad x(at) \longleftrightarrow \frac{1}{|a|}X\left(\frac{f}{a}\right)$

We have $\quad y(t) \longleftrightarrow -\frac{1}{2}X\left(\frac{f}{2}\right).e^{j2\pi f}$

18. Time-scaling property

$$x(at) \xleftrightarrow{\text{FT}} \frac{1}{|a|}\times\left(\frac{f}{a}\right)$$

or, $|a|\,x(at) \xleftrightarrow{\text{FT}} \times\left(\frac{f}{a}\right)$ ...(A)

and frequency-shifting property

$e^{-j2\pi f_0 t}x(t) \xleftrightarrow{\text{FT}} \times(f+f_0)$ ...(B)

Using equation (B) in eq. (A), we get

$$|a|\,e^{-j2\pi f_0 at}x(at) \longleftrightarrow X\left\{\frac{1}{a}(f+af_0)\right\}$$

Putting $a = \frac{1}{3}$ and $f_0 = 2$, we get

$$\frac{1}{3}x\left(\frac{t}{3}\right)e^{-j4\pi t/3} \longleftrightarrow X(3f+2)$$

19. Taking Fourier transform,

$$Y(\omega) = 0.5\, e^{j\omega(-t_d + T)} x(\omega) + 0.\,0.5\, e^{-t_d j\omega} X(\omega) + 0.5X(\omega)e^{(-t_d - T)j\omega}$$

By time shifting property,

$$\frac{Y(\omega)}{X(\omega)} = H(\omega)$$

$$= (1 + \cos\omega t)e^{-j\omega t_d}$$

20.
$$X(\omega) = \int_{-\infty}^{\infty} x(5t-3).e^{-j\omega t}.dt$$

Let, $5t - 3 = p$

or, $dp = 5dt$

$$\therefore \quad X(\omega) = \frac{1}{5}\int_{-\infty}^{\infty} x(p).e^{-\frac{j(p+3)\omega}{5}}.dp$$

$$= \frac{1}{5}\int_{-\infty}^{\infty} x(p).e^{\frac{-jp\omega}{5}}.e^{\frac{-i3\omega}{5}}.dp$$

$$= \frac{1}{5}.e^{-\frac{j3\omega}{5}}\int_{-\infty}^{\infty} x(p).e^{-\frac{jp\omega}{5}}.dp$$

$$= \frac{1}{5}.e^{-\frac{j3\omega}{5}} X\left(\frac{j\omega}{5}\right)$$

21.
$$x(\omega) = \int_{-\infty}^{\infty} x(t)\cdot e^{-jwt}\cdot dt$$

$$= \int_{-1}^{1} e^{-jwt}\cdot dt = \frac{1}{jw}\left(e^{jw} - e^{-jw}\right)$$

$$X(\omega) = 0$$

$$\Rightarrow \quad e^{jw} - \frac{1}{e^{jw}} = 0$$

$$e^{2jw} - 1 = 0$$

$$e^{2jw} = 1$$

$$e^{jw} = \pm 1$$

For this equality to hold, $\omega = \pi, 2\pi$

22. Given : $h(t) = e^{-2t}u(t)$

$$H(\omega) = \int_{-\infty}^{\infty} h(t).e^{-j\omega t}dt$$

$$= \int_0^{\infty} e.^{-2t}.e^{-j\omega t}$$

$$= \int_0^{\infty} e^{-(2+j\omega)}dt$$

$$H(\omega) = -\frac{1}{2+j\omega}.e^{-(2+j\omega)t}\Big|_0^{\infty}$$

$$= \frac{1}{2+j\omega}$$

23. Given : $x(t) = 2\cos(2t)$.

$X(\omega) = 2\pi\,[\delta(\omega - 2) + \delta(\omega + 2)]$

$$H(\omega) = \frac{1}{2+j\omega}.$$

Output, $Y(\omega) = H(\omega).\,Y(\omega)$

$$= \frac{1}{2+j\omega}.2\pi\left[\delta(\omega-2)+\delta(\omega+2)\right]$$

$$Y(\omega) = \frac{2\pi}{2+j^2}.\delta(\omega-2) + \frac{2\pi}{2-j^2}.\delta(\omega+2)$$

$$= \frac{2\pi}{8}\left[\left(2-j^2\right)+\delta(\omega+2)+\left(2+j^2\right).\delta(\omega+2)\right]$$

$$Y(\omega) = \frac{\pi}{2}\left[\delta(\omega-2)+\delta(\omega+2)\right]$$

$$-j\frac{\pi}{2}\left[\delta(\omega-2)-\delta(\omega+2)\right]$$

$$y(t) = \frac{\cos 2t}{2} + \frac{\sin 2t}{2}$$

$$= \frac{1}{\sqrt{2}}\cos(2t - 0.25\pi)$$

$$= 2^{-0.5}.\cos(2t - 0.2\pi).$$

24. System is LTI and BIBO system.

25. $\frac{\sin 2\omega}{\omega} \longleftrightarrow$

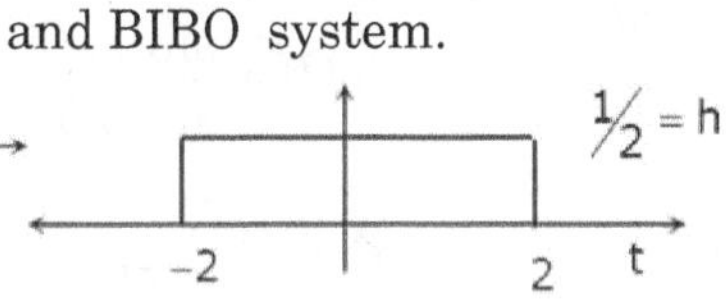

$$2\cos\omega\left(\frac{\sin 2\omega}{\omega}\right) \to h(t) = h'(t-1) + h'(t+1)$$

$$\left[e^{j\omega} + e^{-j\omega}\right]\left(\frac{\sin 2\omega}{\omega}\right)$$

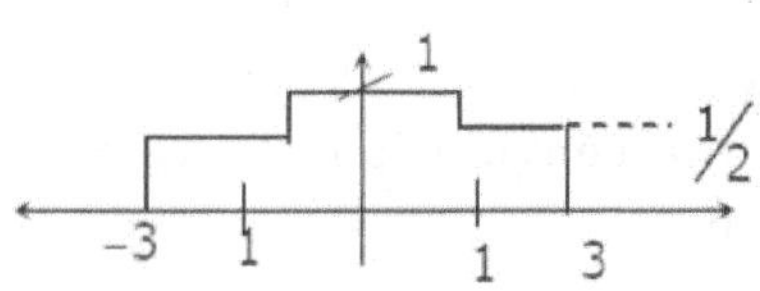

27. Shifting in time domain does not change PSD. Since PSD is Fourier transform of autocorrelation function of WSS process, autocorrelation function depends on time difference.

$$X(t) \leftrightarrow R_x(z) \leftrightarrow S_x(f)$$

$$Y(t) = X(2t-1) \leftrightarrow R_y(2\zeta) \leftrightarrow \frac{1}{2}S_x\left(\frac{f}{2}\right)$$

[time scaling property of Fourier transform]

28. $G(jw) = w.e^{-2\omega 2}$

$$\sum_{-8}^{\infty} g(t)dt = G(j0) = 0$$

$y(t) = g(t) * u(t)$

$$Y(j\omega) = G(j\omega)\,U(j\omega) = G(j\omega)\left[\frac{1}{j\omega} + \pi\delta(\omega)\right]$$

$$= \frac{\omega.e^{-2\omega 2}}{j\omega} + \omega.e^{-2\omega 2}\pi\delta(\omega)$$

$$\Rightarrow \int_{-\infty}^{\infty} y(t)dt = Y(j0) = \frac{1}{j} = -j$$

29. $$x(t) = \sqrt{2}\left(\frac{\sin(\pi t/5)}{\pi t/5}\right)\sin\left(\pi t - \frac{\pi}{4}\right)$$

we can write above expression as

$$x(t) = -\sqrt{2}\left(\frac{\sin(\pi t/5)}{\pi t/5}\right)\left[\cos\frac{\pi}{4}\sin\pi t - \sin\frac{\pi}{4}\cos\pi t\right]$$

$$= \left(\frac{\sin(\pi t/5)}{\pi t/5}\right)\cos\pi t - \frac{\sin(\pi t/5)}{(\pi t/5)}\sin(\pi t)$$

Also

$x(t) = x_c(t)\cos 2\pi f_c t - x_s(t)\sin(2\pi f_c t)$

[Low pass representation of Bandpass signals]

$$x_c(t) = \frac{\sin(\pi t/5)}{\pi t/5}$$

$$x_s(t) = \frac{\sin(\pi t/5)}{\pi t/5}$$

$x_{ce}(t)$ is the complex envelope of $x(t)$

$$x_{ce}(t) = x_c(t) + jx_s(t)$$

$$= \frac{\sin(\pi t/5)}{\pi t/5}[1+j]$$

$$= \frac{\sqrt{2}\sin(\pi t/5)}{(\pi t/5)}e^{j\pi/4}$$

30. Here the convolution of two sinc pulses is sinc pulse.

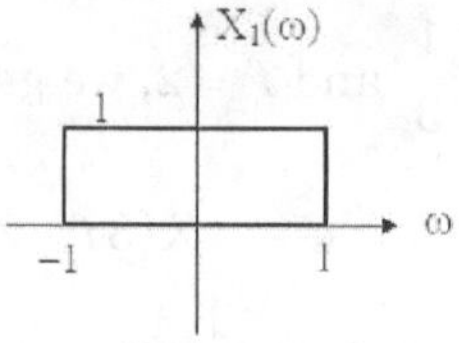

So, $x_1(t) = \dfrac{\sin t}{\pi t}$

Now $x(t) = x_1(t) * x_1(t)$

$$X(\omega) = X_1(\omega)\cdot X_1(\omega)$$
$$= X_1(\omega)$$

$\therefore$ $x(t) = x_1(t) = \dfrac{\sin t}{\pi t}$

So, the value of $x(t)$ is $\dfrac{\sin t}{\pi t}$

31. Assume, $f(t) = \text{sinc}(st) = S_a(5\pi t)$.

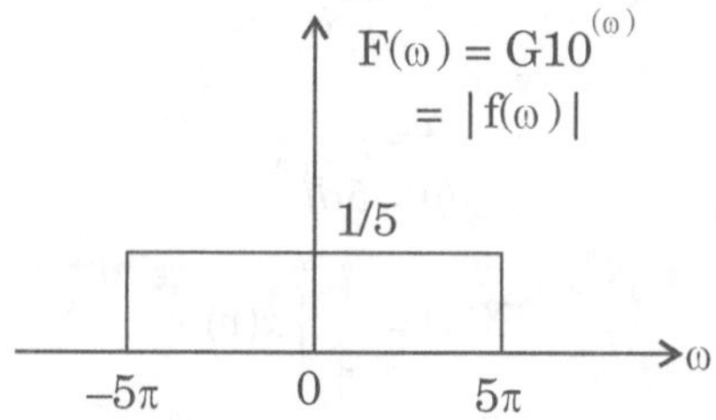

$$\therefore T.S_a\left(t.\frac{T}{2}\right) \longleftrightarrow 2\pi G_T(\omega); \frac{T}{2} = 5\pi$$

$$10p.S_a(5\pi t) \longleftrightarrow 2\pi G_{10\pi}^{(w)}$$

$$ESD_F = |F(\omega)|^2$$

$$E_f = \int_{-\infty}^{\infty} f^2(t)dt$$

$$= \frac{1}{2\pi}\int_{-\infty}^{\infty} ESD_f.d\omega$$

$$\int_{-\infty}^{\infty} \text{sinc}^2(st)dt$$

$$\frac{1}{2\pi}\int_{-\infty}^{\infty}\left(\frac{1}{5}\right)^2.d\omega$$

$$= \frac{1}{2\pi}\times\frac{1}{25}\times 10\pi = \frac{1}{5}.$$

32. Let $x(t)$ Fourier transform be $x(t)$

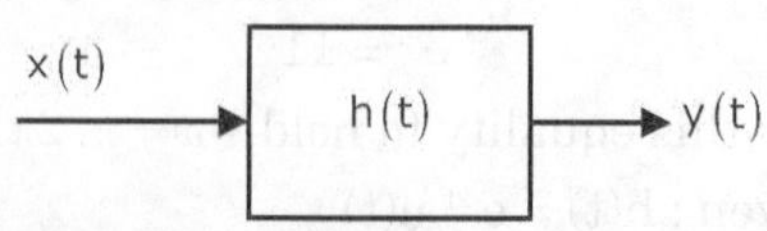

$y(t) = x(t) * h(t)$ (convolution)

$\Rightarrow Y(f) = X(f).H(f)$

$\Rightarrow Y(f) = e^{-j4\pi f}.X(f)$

$\Rightarrow y(t) = x(t-2)$

33. Probability of error

$$P_e = P^3 + 3P^2(1-P)$$
$$P = 0.1$$
$$P_e = (0.1)^3 + 3 \times (0.1)^2 (1-0.1)$$
$$= 0.001 + 3 \times 0.01 \times 0.9$$
$$= 0.001 + 0.027$$
$$= 0.028$$

34. Let $x(t) = \dfrac{\sin at}{\pi t} \xleftrightarrow{\text{F.T.}} X(\omega)$

$$x(t) = \frac{\sin(4\pi t)}{4\pi} = \frac{1}{4} \times \frac{\sin(4\pi t)}{\pi t}$$

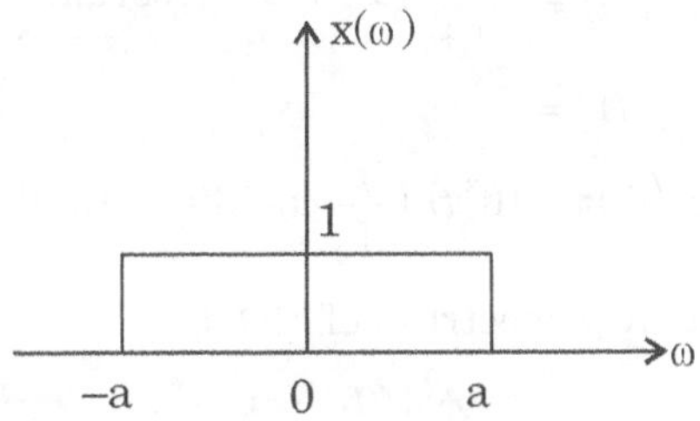

$$\text{Energy of } x(t) = \frac{1}{2\pi}\int_{-\infty}^{\infty} 1 \times |\omega|^2 \, d\omega$$

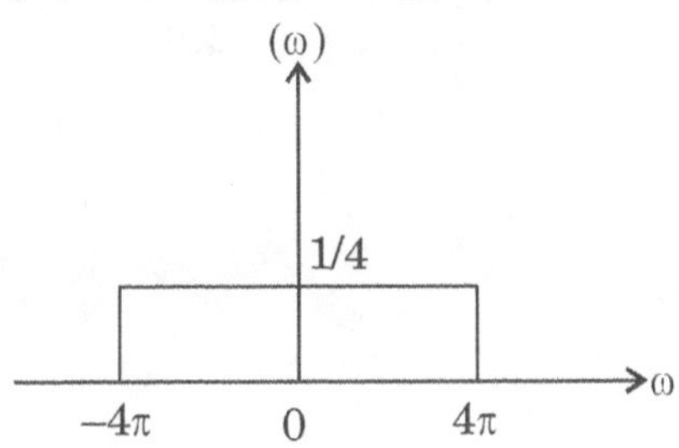

$$= \frac{1}{2\pi}\int_{-4\pi}^{4\pi}\left(\frac{1}{4}\right)^2 d\omega$$

$$= \frac{1}{2\pi} \times \frac{1}{10} \times 8\pi$$

$= 0.25.$

35. $x(t) = 4\cos 200\pi t + 8\cos 400\pi t$

Taking fourier transform of $h(t)$ we get

$$h(t) = \frac{2\sin 300\pi t}{\pi t} \xleftrightarrow{\text{F.T}}$$

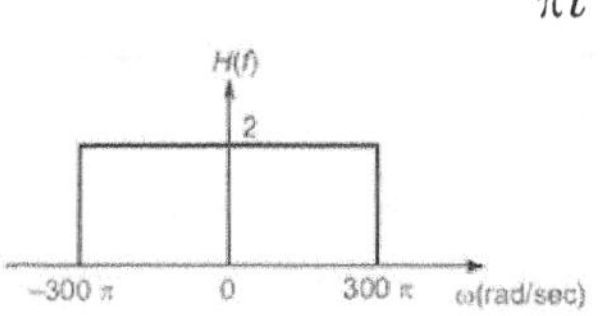

So, $y(t) = 8\cos 200\,\pi t$

Hence $|y(t)|_{max} = 8$

36. Given : $|H(f)|$

$$= \begin{cases} 1 - \dfrac{|f|}{20}, & |f| \le 20 \\ 0, & |f| < 20 \end{cases} : \underline{||H(f)| = -2f.}$$

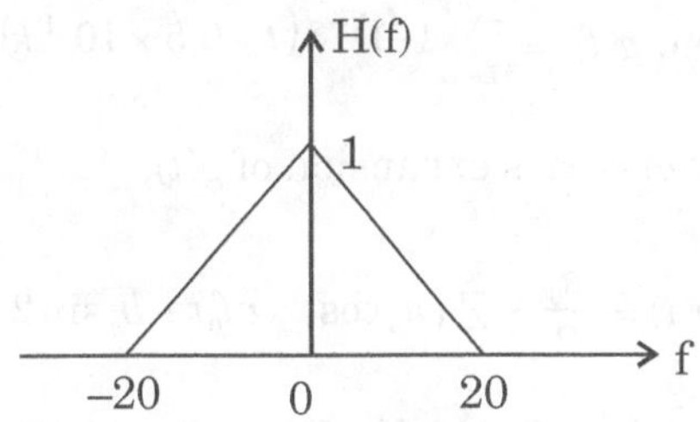

$$\Rightarrow y(t) = \frac{1}{2} \times 8\cos\left(20\pi t + \frac{\pi}{4}\right)$$

$= 4\cos(20\text{ pt} + \phi)$

Power of y(t)

$$= \left(\frac{4}{\sqrt{2}}\right)^2 = 8.$$

37. Hilbert transorm does not alter the amplitude spectrum of the signal.

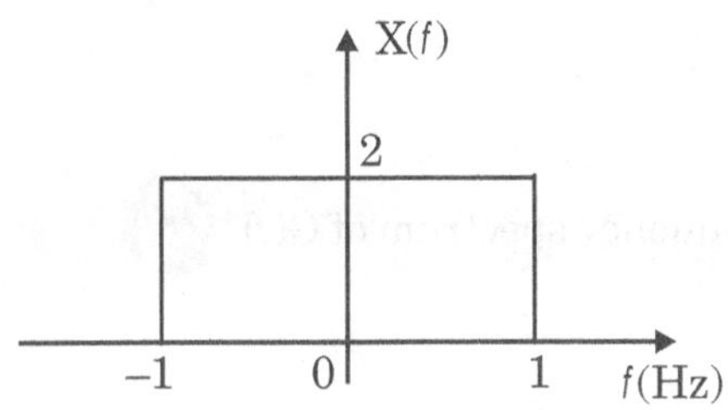

So, $\displaystyle\int_{-\infty}^{\infty} |y(t)|^2 dt = \int_{-\infty}^{\infty} |x(t)|^2 dt = \int_{-\infty}^{\infty} |X(f)|^2 df$

$x(t) = 4\text{sinc}(2t)$

$$\text{sinc}(t) \xleftrightarrow{\text{CTFT}} \text{rect}(f)$$

$$4\,\text{sinc}(2t) \xleftrightarrow{\text{CTFT}} \frac{4}{2}\text{rect}\left(\frac{f}{2}\right) = 2\,\text{rect}\left(\frac{f}{2}\right)$$

$$\int_{-\infty}^{\infty} |X(f)|^2 \, df = 2 \times (2)^2 = 8$$

So, $\displaystyle\int_{-\infty}^{\infty} |y(t)|^2 \, dt = 8$

38. Here, f_s = 1500 samples/sec, f_m = 1 kHz

The sampled frequency are 2.5 kHz, 0.5 kHz, Since LPF has cut-off frequency 800 Hz, then only output signal of frequency 0.5 kHz would pass through it.

39. Given, $g(t) = \sum_{k=-\infty}^{\infty} (-1)^k \delta(t - 0.5\times 10^{-4} k)$

Fourier series expansion of $g(t)$,

$$g(t) = \frac{a_0}{2} + \sum_{n=1}^{\infty}\left(a_n \cos 2\pi n f_0 t + b_n \sin 2\pi n f_0 t\right)$$

where $f_0 = 20 \times 10^3$Hz, $T_0 = 0.5 \times 10^{-4}$T

as $g(t)$ is even function of time, then $b_n = 0$

$$a_0 = \frac{2}{T}\int_{-T_0/2}^{T_0/2} g(t)\,dt - \frac{4}{T_s}\int_0^{T_0/2}\left[g(t) - g\left(t-\frac{T_0}{2}\right)\right]dt$$

$$= \frac{4}{T_s}\int_0^{T_0/2} g(t)\cdot dt - \frac{4}{T_s}\int_0^{T_0/2} g\left(t-\frac{T_0}{2}\right)dt = \frac{4}{T_0} - \frac{4}{T_0} = 0$$

$$\left[\int_{-\infty}^{\infty} g(t)\,dt = 1\right]$$

a_n =

$$[g(t)\,x(t) = x(0)\,g(t)]$$

$$\frac{4}{T_0}\int_0^{T_0/2} g(t)\cos\frac{2\pi n}{T_0}\cdot t\,dt - \frac{4}{T_s}\int_0^{T_s/2} g\left(t-\frac{T_0}{2}\right)\cos\frac{2\pi n}{T_0}\cdot t\,dt$$

=

$$\frac{4}{T_0}\int_0^{T_s/2} g(t)\,dt - \frac{4}{T_0}\cos n\pi\int_0^{T/2} g\left(t-\frac{T_0}{2}\right)dt = \frac{4}{T_0} - \frac{4}{T_0}\cos\pi n$$

$$\left[g(t-t_0)\,x(t) = x(t_0)\cdot g(t-t_0)\right]$$

then, $g(t) = \sum_{n=1}^{\infty}\frac{4}{T_0}\left[1-(-1)^n\right]\cos(2\pi f_0 n\, t)$

$g(t)$ =

$$\frac{8}{T_0}\cos(40\pi\times 10^3 t) + \frac{8}{T_0}\cos(60\pi\times 10^3 t) + \ldots$$

Frequency spectrum of M(f) =

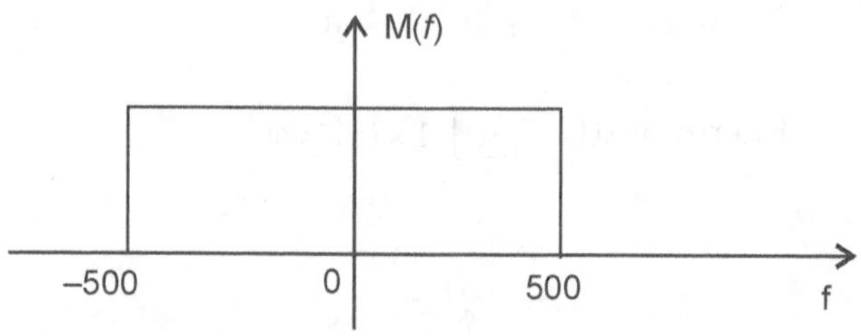

Frequency spectrum of G(f)

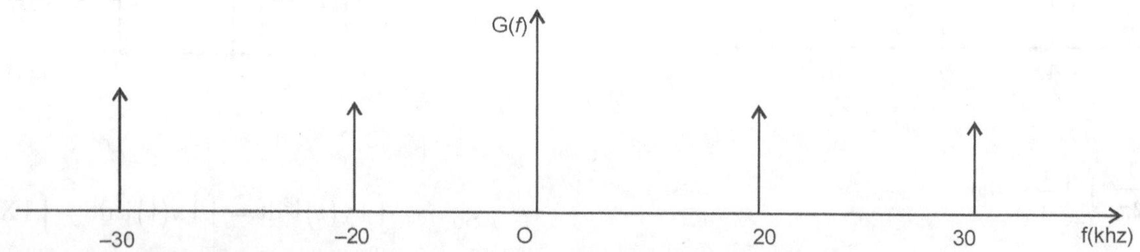

As $m(t)\,g(t) = M(f) \times G(f)$

Multiplication in time domain is equivalent to convolution in frequency domain

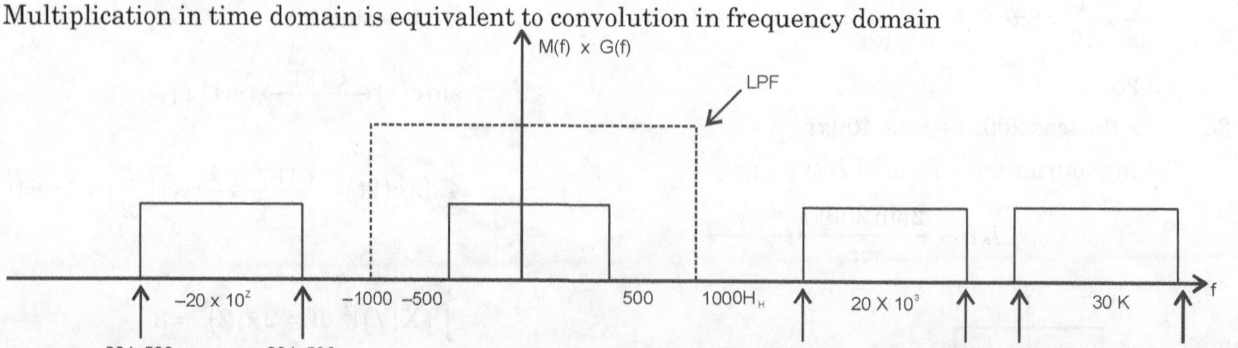

When the resulting signal passes through LPF having BW of 1 KHz, then only message signal $m(t)$ produces at the output.

40. (b) & (d) Auto - correlation function of energy signal has conjugate symmetric,

$R_x(\tau) = R_x^*(-\tau)$.

If the fuction is real, then the autocorrelation function has even symmetry

$R_x(\tau) = R_x(-\tau)$.

41. The autocorreclation function (ACF) of a rectangular pulpe af duration T is a triangular pulse of duration 2T.

42. For distortionless transmission, $|H(f)|$ = constant and $\underline{|H(f)} = -kf$, where K is a constant.

⇒ H(f) is the transfer function, so the magnitude function of a distortionless filter should be constant. The phase function of a distortionless filter be linear.

43. Given : $y(t) = \frac{1}{100}\cos(100t - 10^6).\cos(10^6t - 1.56)$

Companing with $y(t) = \cos[k|t\text{-}t_9)].\ \cos[\omega_c(t - t_p)]$

$t_p = 1.56 \times 10^{-6}$ s

$t_g = 10^{-8}$ s.

44. For linear phase channel,

$$Q(\omega) = -\omega t_0$$

$$T_P(\text{phase delay}) = \frac{-\theta(\omega)}{\omega} = t_0$$

$$\text{and } T_g(\text{group delay}) = \frac{-d\theta(\omega)}{d\omega} = t_0$$

Hence $T_P = T_g = t_0$ = constant

45. $H(f) = \frac{1}{j2\pi R f C + 1}$, $H(0) = 1$

Now, as $\frac{|H(f_1)|}{H(0)} \geq 0.95$

then, $\frac{1}{\sqrt{4\pi^2 R^2 f_1^2 C^2 + 1}} \geq 0.95$

or $4\pi^2 R^2 f_1^2 C^2 \leq 0.108$

or $(f_1)_{max} = \frac{\sqrt{0.108}}{2\pi RC} = \frac{\sqrt{0.108}}{2 \times 3.14 \times 10^{-3}} = 52.2$ Hz.

46. Group Delay $t_g(f) = -\frac{1}{2\pi}\frac{d\phi(f)}{df}$

here, $\phi(f) = -\tan^{-1}(2\pi RCf)$

then, $t_g(f) = \frac{RC}{1 + 4\pi^2R^2C^2f^2}$

$$tg(t)\big|_{f=f_2} = \frac{1 \times 10^3 \times 10^{-6}}{1 + 4\pi^2(10^3)^2 \times (10^{-6})^2 \times (100)^2}$$

$= 0.717$ ms.

47. Phase response of pass band waveform

$$\phi(f) = -2\pi\alpha(f - f_c) - 2\pi\beta f_c$$

Group delay $t_y = \frac{-d\phi(f)}{2\pi\ df} = \alpha$

Thus 'α' is actual signal propagation delay from transmitter to receiver.

5 CHAPTER

Laplace Transform

Analysis of Previous GATE Papers												
	Year → Topics ↓		2019	2018	2017 Set 1	2017 Set 2	2016 Set 1	2016 Set 2	2016 Set 3	2015 Set 1	2015 Set 2	2015 Set 3
BILATERAL LAPLACE TRANSFORM	1 Mark	MCQ Type										
		Numerical Type										
	2 Marks	MCQ Type										
		Numerical Type				1						
	Total					2						
UNILATERAL LAPLACE TRANSFORM	1 Mark	MCQ Type	1									
		Numerical Type										
	2 Marks	MCQ Type					1		1			
		Numerical Type										
	Total		1				2		2			
POLES & ZEROS	1 Mark	MCQ Type			1							
		Numerical Type										
	2 Marks	MCQ Type										
		Numerical Type										
	Total				1							

BILATERAL LAPLACE TRANSFORM

1. Which one of the following statements is NOT TRUE for a continuous time causal and stable LTI system?

(a) All the poles of the system must lie on the left side of the $j\omega$–axis.

(b) Zeros of the system can lie anywhere in the s-plane.

(c) All the poles must lie within $|s| = 1$.

(d) All the roots of the characteristic equation must be located on the left side of the $j\omega$–axis.

[1993 : 1 Mark]

2. If $F(s) = L|f(t)| = \frac{K}{(s+1)(s^2+4)}$, then $\lim_{t\to\infty} f(t)$ is given by

(a) $\frac{K}{4}$ (b) zero

(c) infinite (d) undefined

[1993 : 2 Marks]

3. If $L[f(t)] = \frac{2(s+1)}{s^2+2s+5}$, then $f(0^+)$ and $f(\infty)$ are given by

(a) 0, 2 respectively (b) 2, 0 respectively

(c) 0, 1 respectively (d) $\frac{2}{5}$, 0 respectively

[Note: 'L' stands for 'Laplace transform of]

[1995 : 1 Mark]

4. The final value theorem is used to find the

(a) steady state value of the system output.

(b) initial value of the system output.

(c) transient behaviour of the system output.

(d) None of these **[1995 : 1 Mark]**

5. If $sL[f(t)] = \frac{\omega}{(s^2+\omega^2)}$, then the value of $\lim_{t\to\infty} f(t)$

(a) cannot be determined

(b) is zero

(c) is unity

(d) is infinite **[1998 : 1 Mark]**

6. The Laplace transform of i(t) is given by

$$I(s) = \frac{2}{s(1+s)}$$

As $t \to \infty$, the value of i(t) tends to

(a) 0 (b) 1

(c) 2 (d) ∞ **[2003 : 1 Mark]**

7. Consider the function f(t) having Laplace transform

$$F(s) = \frac{\omega_0}{s^2+\omega_0^2} \quad \text{Re}|s| > 0$$

The final value of f(t) would be:

(a) 0 (b) 1

(c) $-1 \le f(\infty) \le 1$ (d) ∞ **[2006 : 2 Marks]**

8. If the Laplace transform of a signal y(t) is $Y(s) = \frac{1}{s(s-1)}$, then its final value is

(a) –1 (b) 0

(c) 1 (d) unbounded

[2007 : 1 Mark]

9. If $F(s) = L[f(t)] = \frac{2(s+1)}{s^2+4s+7}$ then the initial and final values of f(t) are respectively

(a) 0, 2 (b) 2, 0

(c) 0, $\frac{2}{7}$ (d) $\frac{2}{7}$, 0

[2011 : 2 Marks]

10. Input x(t) and output y(t) of an LTI system are related by the differential equation $y''(t) - y'(t) - 6y(t) = x(t)$. If the system is neither causal nor stable, the impulse response h(t) of the system is

(a) $\frac{1}{5}e^{3t}u(-t) + \frac{1}{5}e^{-2t}u(-t)$

(b) $-\frac{1}{5}e^{3t}u(-t) + \frac{1}{5}e^{-2t}u(-t)$

(c) $\frac{1}{5}e^{3t}u(-t) - \frac{1}{5}e^{-2t}u(t)$

(d) $-\frac{1}{5}e^{3t}u(-t) - \frac{1}{5}e^{-2t}u(t)$

[2015 : 2 Marks; Set-2]

11. Consider the differential equation

$\frac{dx}{dt} = 10 - 0.2x$ with initial conduction

x(0) = 1. The response x(t) for t > 0 is

(a) $2 - e^{-0.2t}$ (b) $2 - e^{-0.2t}$

(c) $50 - 49e^{-0.2t}$ (d) $50 - 49e^{0.2t}$

[2015 : 2 Marks, Set-2]

12. The bilateral Laplace transform of a function

$$f(t) = \begin{cases} 1, & \text{if } a \le t \le b \\ 0 & \text{otherwise} \end{cases} \text{ is}$$

(a) $\frac{a-b}{s}$ (b) $\frac{e^z(a-b)}{s}$

(c) $\frac{e^{-as} - e^{-bs}}{s}$ (d) $\frac{e^{-(a-b)}}{s}$

[2015 : 1 Mark, Set-2]

13. Let the signal i(t) = 0 outside the interval $[T_1, T_2]$, where T_1 and T_2 are finite. Furthermore, $|f(t)| < \infty$. The region of convergence (ROC) of the signal's bilateral Laplace transform F(s) is

(a) a parallel strip containing the $j\Omega$ axis

(b) a parallel strip not containing the $j\Omega$ axis

(c) the entire s-plane

(d) a half plane containing the $j\Omega$ axis

[2015 : 1 Mark, Set-2]

14. Let $x(t) = \alpha\beta s(t) + \beta s(-t)$ with $s(t) = e^{-4t}u(t)$, where u(t) is unit step function. If the bilateral Laplace transform of x(t) is

$$X(s) = \frac{16}{s^2 - 16} - 4 < \text{Re}\{s\} < 4;$$

then the value of β is _______.

[2015 : 2 Marks, Set-2]

15. The transfer function of a causal LTI system is $H(s) = \frac{1}{s}$. If the input to the system is

$x(t) = \left[\frac{\sin(t)}{\pi t}\right]u(t)$, where u(t) is a unit step function, the system output y(t) as $f \to \infty$ is _______.

[2017 : 2 Marks, Set-2]

UNILATERAL LAPLACE TRANSFORM

16. The unit impulse response of a linear time invariant system is the unit step function u(t). For t > 0, the response of the system to an excitation e^{-at} u(t), a > 0 will be

(a) ae^{-at} (b) $\left(\frac{1}{a}\right)(1 - e^{-at})$

(c) $a(1 - e^{-at})$ (d) $1 - e^{-at}$

[1998 : 1 Mark]

17. Let u(t) be the step function. Which of the waveforms in the figure corresponds to the convolution of u(t) – u(t – 1) with u(t) – u(t – 2)?

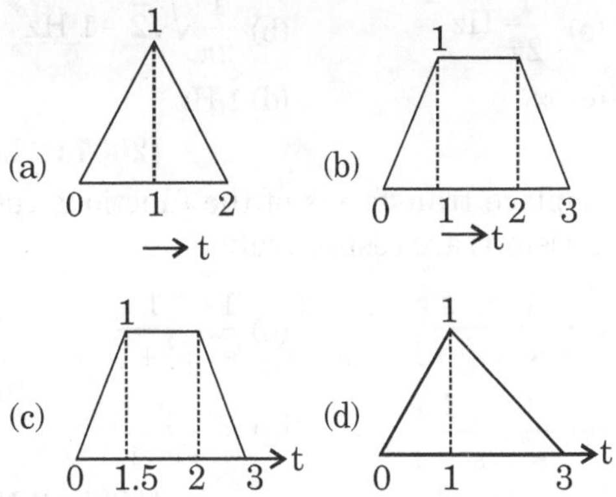

[2000 : 2 Marks]

18. Convolution of x(t + 5) with impulse function $\delta(t - 7)$ is equal to

(a) x(t – 12) (b) x(t + 12)

(c) x(t – 2) (d) x(t + 2)

[2002 : 1 Mark]

19. The input $-3e^{-2t}u(t)$, where u(t) is the unit step function, is applied to a system with transfer function $\frac{s-2}{s+3}$. If the initial value of the output is –2, then the vlaue of the output at steady state is _______.

[2014 : 1 Mark, Set-3]

20. A first-order low-pass filter of time constant T is excited with different input signals (with zero initial conditions up to t = 0). Match the excitation signals X, V, Z with the corresponding time responses for $t \ge 0$:

List-I	List-II
X. Impulse	P. $1 - e^{-t/T}$
Y. Unit step	Q. $t - T(1 - e^{-t/T})$
Z. Ramp	R. $e^{-t/T}$

(a) X – R; Y– Q; Z – P

(b) X – Q; Y – P; Z – R

(c) X – R; Y – P; Z – Q

(d) X – P; Y – R; Z – Q

21. The transfer function of a zero-order hold is

(a) $\frac{1 - \exp(-T_s)}{s}$ (b) $\frac{1}{s}$

(c) 1 (d) $\frac{1}{[1 - \exp(-T_s)]}$

[1988 : 2 Marks]

22. The 3-dB bandwidth of the low-pass signal $e^{-t}u(t)$, where u(t) is the unit step function, is given by

(a) $\frac{1}{2\pi}$ Hz (b) $\frac{1}{2\pi}\sqrt{\sqrt{2}-1}$ Hz

(c) ∞ (d) 1 Hz

[2007 : 2 Marks]

23. Laplace transforms of the functions tu(t) and u(t)sin(t) are respectively

(a) $\frac{1}{s^2}, \frac{s}{s^2+1}$ (b) $\frac{1}{s}, \frac{1}{s^2+1}$

(c) $\frac{1}{s^2}, \frac{1}{s^2+1}$ (d) $s, \frac{s}{s^2+1}$

[1987 : 2 Marks]

24. Specify the filter type if its voltage transfer function H(s) is given by

$$H(s) = \frac{K(s^2+1\omega_0^2)}{s^2+\left(\frac{\omega_0}{Q}\right)s+\omega_0^2}$$

(a) all pass filter (b) low pass filter

(c) band pass filter (d) notch filter

[1988 : 2 Marks]

25. The Laplace transform of a function f(t) u{t), where f(t) is periodic with period T, is A(s) times the Laplace transform of its first period. Then

(a) A(s) = s

(b) $A(s) = \frac{1}{(1-\exp(-T_s))}$

(c) $A(s) = \frac{1}{(1+\exp(-T_s))}$

(d) $A(s) = \exp(T_s)$

[1988 : 2 Marks]

26. The response of an initially relaxed linear constant parameter network to a unit impulse applied at t = 0 is $4e^{-2t}u(t)$. The response of this network to a unit step function will be

(a) $2[1-e^{-2t}]u(t)$ (b) $4[e^{-t}-e^{-2t}]u(t)$

(c) sin2t (d) $(1-4e^{-4t})u(t)$

[1990 : 2 Marks]

27. The voltage across an impedance in a network is V(s) = Z(s). I(s), where V(s), Z(s) and I(s) are the Laplace transform of the corresponding time functions v(t), z(t) and i(t). The voltage v(t) is

(a) v(t) = z(t).i(t)

(b) $v(t) = \int_0^t i(\tau)z(t-\tau)d\tau$

(c) $v(t) = \int_0^t i(\tau)z(t+\tau)d\tau$

(d) v(t) = z(t) + i(t)

[1991 : 2 Marks]

28. The Laplace transform of the periodic function f(t) described by the curve below, i.e.

$$f(t) = \begin{cases} \sin t, & \text{if } (2n-1)\pi \le t \le 2n\pi \ (n = 1,2,3,...) \\ 0, & \text{otherwise;} \end{cases}$$

is ______.

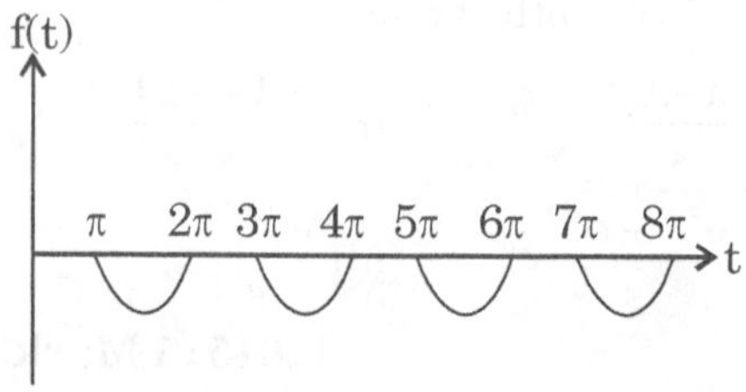

[1993 : 2 Marks]

29. The Laplace transform of a unit ramp function starting at t = a, is

(a) $\frac{1}{(s+a)^2}$ (b) $\frac{e^{-as}}{(s+a)^2}$

(c) $\frac{e^{-as}}{s^2}$ (d) $\frac{a}{s^2}$

[1994 : 1 Mark]

30. The inverse Laplace transform of the function $\frac{s+5}{(s+1)(s+3)}$ is

(a) $2e^{-t}-e^{-3t}$ (b) $2e^{-t}+e^{-3t}$

(c) $e^{-t}-2e^{-3t}$ (d) $e^{-t}+2e^{-3t}$

[1996 : 2 Marks]

31. The Laplace transform of $e^{\alpha t}\cos(\alpha t)$ is equal to

(a) $\frac{(s-\alpha)}{(s-\alpha)^2+\alpha^2}$ (b) $\frac{(s+\alpha)}{(s+\alpha)^2+\alpha^2}$

(c) $\frac{1}{(s-\alpha)^2}$ (d) None of these

[1997 : 1 Mark]

32. If L[f(t)] = F(s), then L[f(t – T)] is equal to

(a) $e^{sT}F(s)$ (b) $e^{-sT}F(s)$

(c) $\frac{F(s)}{1+e^{sT}}$ (d) $\frac{F(s)}{1-e^{-sT}}$

[1999 : 1 Mark]

33. Given that

$$L[f(t)] = \frac{s+2}{s^2+1}, L[g(t)] = \frac{s^2+1}{(s+3)(s+2)},$$

$$h(t) = \int_0^t f(\tau)g(t-\tau)d\tau$$

L[h(t)] is

(a) $\frac{s^2+1}{s+3}$ (b) $\frac{1}{s+3}$

(c) $\frac{s^2+1}{(s+3)(s+2)}+\frac{s+2}{s^2+1}$

(d) None of the above

[2000 : 1 Mark]

34. A linear time invariant system has an impulse response e^{2t}, for $t > 0$. If initial conditions are 0 and the input is e^{3t}, the output for $t > 0$ is

(a) $e^{3t} - e^{2t}$ (b) e^{5t}

(c) $e^{3t} + e^{2t}$ (d) None of these

[2000 : 2 Marks]

35. The transfer function of a system is given by $H(s)=\frac{1}{s^2(s-2)}$. The impulse response of the syste is

(a) $(t^2 * e^{-2t})u(t)$ (b) $(t * e^{2t})u(t)$

(c) $(te^{-2t})u(t)$ (d) $(te^{-2t})u(t)$

(* denotes convolution, and u(t) is unit step function)

[2001 : 1 Mark]

36. The Laplace transform of a continuous-time signal x(t) is $X(s)=\frac{5-s}{s^2-s-2}$. If the Fourier transform of this signal exists, then x(t) is

(a) $e^{2t}u(t) - 2e^{-t}u\{t)$

(b) $-e^{2t}u(-t) + 2e^{-t}u(t)$

(c) $-e^{2t}u(-t) - 2e^{-t}u(t)$

(d) $e^{2t}u(-t) - 2e^{-t}u(t)$

[2002 : 2 Marks]

37. Given that F(s) is the one-sided Laplace transform of f(t), the Laplace transform of $\int_0^t f(\tau)d\tau$ is

(a) $sF(s) - f(0)$ (b) $\frac{1}{s}F(s)$

(c) $\int_0^s F(\tau)d\tau$ (d) $\frac{1}{s}[F(s)-f(0)]$

[2009 : 2 Marks]

38. A continuous time LTI system is described by

$$\frac{d^2y(t)}{dt^2}+4\frac{dy(t)}{dt}+3y(t)=2\frac{dx(t)}{dt}+4x(t)$$

Assuming zero initial conditions, the response y(t) of the above system for the input $x(t) = e^{-2t}u(t)$ is given by

(a) $(e^{t} - e^{3t})u(t)$ (b) $(e^{-t} - e^{-3t})u(t)$

(c) $(e^{-t} + e^{-3t})u(t)$ (d) $(e^{t} + e^{3t})u(t)$

[2002 : 2 Marks]

39. The unilateral Laplace transform of f(t) is $\frac{1}{s^2+s+1}$. The unilateral Laplace transform of tf(t) is

(a) $-\frac{s}{(s^2+s+1)}$ (b) $-\frac{2s+1}{(s^2+s+1)^2}$

(c) $\frac{s}{(s^2+s+1)^2}$ (d) $\frac{2s+1}{(s^2+s+1)^2}$

[2012 : 1 Mark]

40. The impulse response of a system is $h(t) = tu(t)$. For an input $u(t-1)$, the output is

(a) $\frac{t^2}{2}u(t)$ (b) $\frac{t(t-1)}{2}u(t-1)$

(c) $\frac{(t-1)^2}{2}u(t-1)$ (d) $\frac{t^2-1}{2}u(t-1)$

[2013 : 1 Mark]

41. Assuming zero initial condition, the response y(t) of the system given below to a unit step input u(t) is

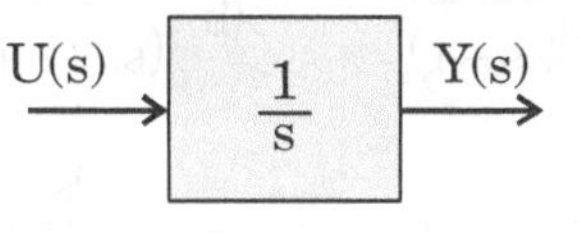

(a) u(t) (b) tu(t)

(c) $\frac{t^2}{2}u(t)$ (d) $e^{-t}u(t)$

[2013 : 1 Mark]

42. A system is described by the differential equation $\frac{d^2y}{dt^2}+5\frac{dy}{dt}+6y(t)=x(t)$. Let x(t) be a rectangular pulse given by

$$x(t)=\begin{cases}1, & 0<t<2\\ 0, & \text{otherwise.}\end{cases}$$

Assuming that $y(0) = 0$ and $\frac{dy}{dt}=0$ at $t = 0$, the Laplace transform of y(t) is

(a) $\frac{e^{-2s}}{s(s+2)(s+3)}$ (b) $\frac{1-e^{-2s}}{s(s+2)(s+3)}$

(c) $\frac{e^{-2s}}{(s+2)(s+3)}$ (d) $\frac{1-e^{-2s}}{(s+2)(s+3)}$

[2013 : 2 Marks]

43. Let h(t) denote the impulse response of a causal system with transfer function $\frac{1}{s+1}$. Consider the following three statements:

S_1 : The system is stable.

S_2 : $\frac{h(t+1)}{h(t)}$ independent of t for t > 0.

S_3 : A non-causal system with the same transfer function is stable.

For the above system,

(a) only S_1 and S_2 are true.
(b) only S_2 and S_3 are true.
(c) only S_1 and S_3 are true.
(d) S_1, S_2 and S_3 are true.

[2014 : 2 Marks, Set-3]

44. The unilateral Laplace transform of $f(t)$ is $\frac{1}{s^2+s+1}$. Which one of the followi9ng is the unilateral Laplace transform of g(t) = t.f(t)?

(a) $\frac{-s}{(s^2+s+1)^2}$ (b) $\frac{-(2s+1)}{(s^2+s+1)^2}$

(c) $\frac{s}{(s^2+s+1)^2}$ (d) $\frac{2s+1}{(s^2+s+1)^2}$

[2014 : 2 Marks, Set-4]

45. A stable linear time invariant (LTI) system has a transfer function H(s) = $\frac{1}{s^2+s-6}$. To make this system causal it needs to be cascaded with another LTI system having a transfer function $H_1(s)$. A correct choice for $H_1(s)$ among the following options is

(a) s + 3 (b) s – 2
(c) s – 6 (d) s + 1

[2014 : 2 Marks, Set-4]

46. A causal LTI system has zero initial conditions and impulse response h(t). Its input y(t) and output x(t) are related through the linear constant- coefficient differential equation

$$\frac{d^2y(t)}{dt^2}+\alpha\frac{dy(t)}{dt}+\alpha^2y(t)=x(t).$$

Let another signal g{t) be defined as

$$g(t)=\alpha^2\int_0^1 h(\tau)d\tau+\frac{dh(t)}{dt}+\alpha h(t).$$

If G(s) is the Laplace transform of g(t), then the number of poles of G(s) is ______.

[2014 : 2 Marks, Set-4]

47. A system is described by the following differential equation, where u(t) is the input to the system and y(t) is the output of the system.

y(t) + 5y(t) = u(t)

When y(0) = 1 and u(t) is a unit step function, y(t) is

(a) $0.2 + 0.8e^{-5t}$ (b) $0.2 - 0.2e^{-5t}$
(c) $0.8 + 0.2e^{-5t}$ (d) $0.8 - 0.8e^{-5t}$

[2014 : 2 Marks, Set-1]

48. The Laplace transform of the causal periodic square wave of period T shown in the figure below is It)

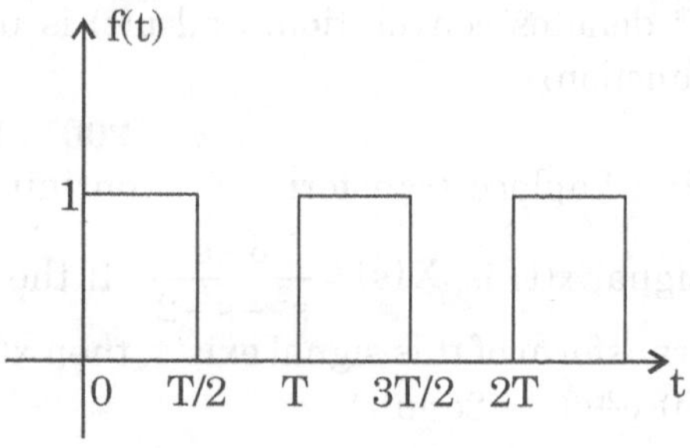

(a) $F(s)=\frac{1}{1+e^{-sT/2}}$

(b) $F(s)=\frac{1}{s(1+e^{-sT/2})}$

(c) $F(s)=\frac{1}{s(1-e^{-sT/2})}$

(d) $F(s)=\frac{1}{1-e^{-sT}}$

[2016 : 2 Marks, Set-1]

49. A signal $2\cos\left(\frac{2\pi}{3}t\right)-\cos(\pi t)$ is the input to an LTI system with the transfer function

$$H(s)=e^s+e^{-s}$$

If C_k denote the kth coefficient in the exponential Fourier series of the output signal, then C_3 is equal to

(a) 0 (b) 1
(c) 2 (d) 3

[2016 : 2 Marks, Set-3]

50. Let Y(s) be the unit-step response of a causal system having a transfer function

$$G(s) = \frac{3-s}{(s+1)(s+3)}$$

That is, $Y(s) = \frac{G(s)}{s}$. The forced response of the system is

(A) $u(t) - 2e^{-t}u(t) + e^{-3t}u(t)$

(B) $2u(t)$

(C) $u(t)$

(D) $2u(t) - 2e^{-t}u(t) + e^{-3t}u(t)$

[2019 : 1 Marks]

POLES & ZEROS

51. The pole-zero pattern of a certain filter is shown in figure. The filter must be of the following type

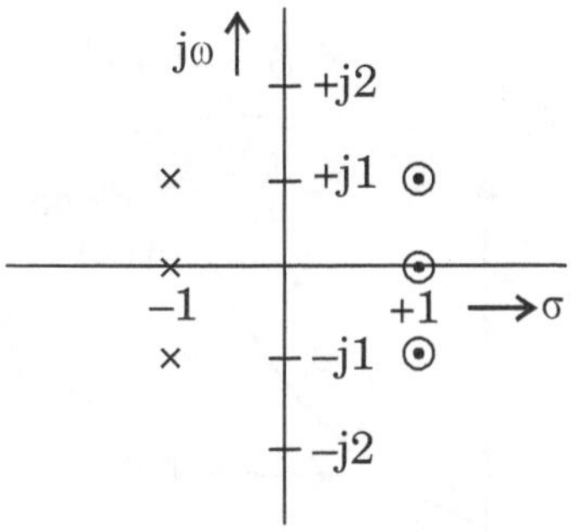

(a) low-pass (b) high-pass

(c) all-pass (d) band-pass

[1991 : 2 Marks]

52. Match each of the items 1, 2 on the left with most appropriate item A, B, C or D on the right. In the case of a linear time invariant system

1. Poles in the right half plane implies.
2. Impulse response zero for $t \le 0$ implies.

(a) Exponential decay of output.

(b) System is causal.

(c) No stored energy in the system.

(d) System is unstable.

[1997 : 2 Marks]

53. Consider the following statements for continuous–time linear time invariant (LTI) systems.

I. There is no bounded input bounded output (BIBO) stable system with a pole in the right half of the complex plane.

II. There is no causal and BIBO stable system with a pole in the right half of the complex plane.

Which one among the following is correct?

(a) Both I and II are true.

(b) Both I and II are not true.

(c) Only I is true.

(d) Only II is true.

[2017 : 1 Mark, Set-1]

ANSWERS

1. (c)	2. (c)	3. (b)	4. (a)	5. (b)	6. (c)	7. (c)	8. (d)	9. (b)	10. (b)
11. (c)	12. (c)	13. (c)	14. (–2)	15. (1/2)	16. (c)	17. (b)	18. (c)	19. (0)	20. (c)
21. (a)	22. (a)	23. (c)	24. (d)	25. (b)	26. (a)	27. (b)	28. (*)	29. (c)	30. (a)
31. (a)	32. (b)	33. (b)	34. (a)	35. (b)	36. (c)	37. (b)	38. (b)	39. (d)	40. (c)
41. (b)	42. (b)	43. (a)	44. (c)	45. (b)	46. (1)	47. (a)	48. (b)	49. (b)	50. (a)
51. (c)	52. (*)	53. (d)							

EXPLANATIONS

1. **Consider option A:** In which all the pols lie on the left of $j\omega$-axis which satisfy casual stable LT1 system.

   Option B: For a stable casual system, there are no restriction for the position of zeroes on s plane.

   Option C: text true.

   Option D: Roots of characteristic equation are all closed loop poles and they all line on the left side of the $j\omega$ axis.

2. From final value theorem, $\lim_{t\to\infty} f(t) = \lim_{s\to 0} sF(s)$.

   provided $s \to$ F(s) has poles with negative real parts, *i.e.* F (s) is a stable function exponentially decaying to 0 as $s \to 0\ (t \to \infty)$

   Here $s^2 + 4 = 0$, or $s = \pm$ 2j has poles on jω-axis. Hence the function is infinite.

3. $$\frac{2(s+1)}{(s^2+2s+5)} = \frac{2(s+1)}{(s^2+1)^2+4}$$

   $$= 2\left[\frac{s+1}{(s+1)^2+4}\right]$$

   This is standard Laplace Transform for the function

   $$L\,|\exp(-at)\cos\omega t| = \frac{(s+a)}{(s+a)^2+\omega^2}$$

   where $a = 1,\quad \omega = 2$

   Hence $f(t) = 2e^{-t}\cos 2t$

   $$f(0) = f(0+) = 2e^{-0}\cos 2\times 0 = 2$$

   $$f(\infty) = 2e^{-\infty}\ t = 0$$

4.

5. Given : $h(t) = u(t)$

   $$x(t) = e^{-at}\ u(t)$$

   $$y(s) = X(s)\ H(s) = \frac{1}{s+a}\times\frac{1}{s}$$

   $$= \frac{1}{s}$$

   $$= \frac{1}{a}\left(\frac{1}{s} - \frac{1}{s+a}\right)$$

   $$y(t) = \frac{1}{a}\ (1 - e^{-at})\ u(t)$$

6. From final value theorem,

   $$\underset{t\to\infty}{\text{Lt}}\ i(t) = \underset{s'\to 0}{Lt}\ sI(s) = \underset{s\to 0}{\text{Lt}}\ \frac{2}{(1+s)} = 2$$

7. $$f(t) = L^{-1}f(x)$$

   $$= \sin\ \omega_0 t$$

   As, $\quad -1 \le \sin\theta \le 1$

   Thus, $-1 \le f(\infty) \le 1$

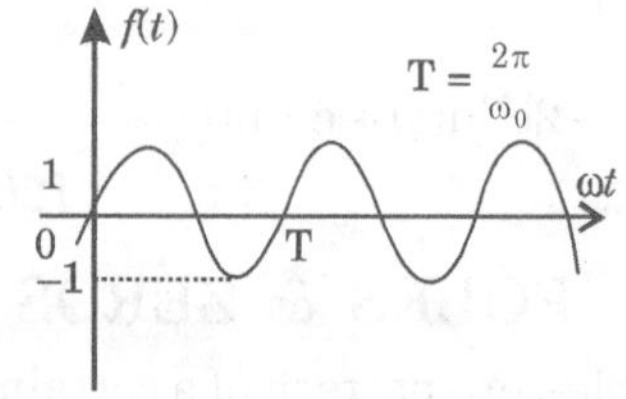

8. Given, $\quad Y(s) = \frac{1}{s(s-1)} = \frac{-1}{s} + \frac{1}{s-1}$

   $$Y(t) = -1 + e^t$$

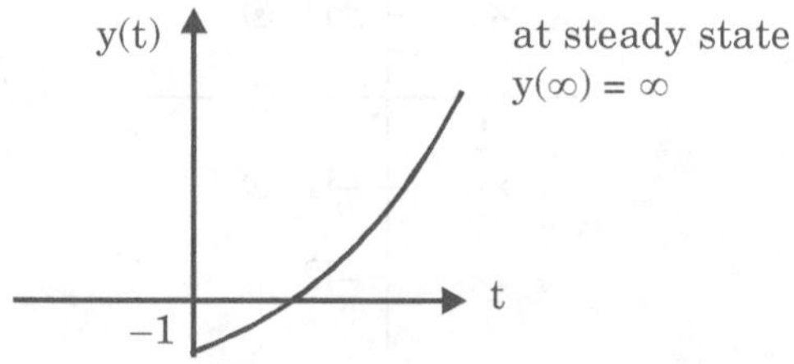

This graph show as $t \to \infty$, $y(t) \to \infty$

9. All poles of F(s) lie in left half of s-plane.

   $$F(s) = L[f(t)] = \frac{2(s+1)}{s^2+4s+7}.$$

   Initial value,

   $$\lim_{t\to 0} f(t) = \lim_{s\to\infty} sF(s) = \lim_{s\to\infty} s.\frac{2(s+1)}{s^2+4s+7}$$

   $$= \lim_{s\to\infty} s.\frac{2s^2\left(1+\frac{1}{s}\right)}{s^2+\left[\left(\frac{4}{5}\right)+\left(\frac{7}{s^2}\right)\right]}$$

   $$= \frac{2(1+0)}{1+0+0} = 2.$$

   Final value, $\lim_{t\to\infty} f(t) = \lim_{s\to 0} sF(s)$

   $$= \lim_{s\to 0} sf(s)\frac{s.(s+1)}{s^2+4s+7} = 0.$$

10. The given differential equation is,

$$y''(t) - y'(t) - 6y(t) = x(t)$$

On applying Laplace transform on both sides,

$$s^2y(s) - sy(0) - y(0) - [sy(s) - y(0)] - 6y(s) = x(s)$$

To calculate the transfer function all initial conditions are taken as '0'.

$$\therefore \quad (s^2 - s - 6)\,y(s) = x(s)$$

$$H(s) = \frac{1}{(s^2 - s - 6)}$$

$$= \frac{1}{(s-3)(s+2)}$$

$$= \frac{1}{5}\left[\frac{1}{s-3} - \frac{2}{s+2}\right]$$

It is given that $h(t)$ is non-casual and un-stable. To satisfy both the conditions ROC should be left of the left most pole.

Using the following standard pair

$$\frac{1}{s+a} \longleftrightarrow -e^{-at}u(-t);\ \sigma < -a$$

$$\frac{1}{s-a} \longleftrightarrow -e^{at}u(-t);\ \sigma < a$$

$$H(s) = \frac{1}{5}\left[\frac{1}{s-3} - \frac{1}{s+2}\right]$$

$$= \frac{1}{5}\left[-e^{3t}u(-t) + e^{-2t}u(-t)\right]$$

$$= \frac{-1}{5}e^{3t}u(-t) + \frac{1}{5}e^{-2t}u(-t)$$

So option (*b*) is correct.

11. Given D.E $\frac{dx}{dt} = 10 - 0.2\,x$

$$x(0) = 1$$

$$\Rightarrow \quad \frac{dx}{dt} + (0.2)x = 10$$

Auxiliary equation is $m + 0.2 = 0$

$$m = -0.2$$

Complementary solution

$$x_c = C\,e^{(-0.2)t}$$

$$x_p = \frac{1}{D+(0.2)}10\,e^{0t} = \frac{10e^{0t}}{0.2}$$

$$= 50e^{0t} = 50$$

$$x = x_c + x_p$$

$$= C\,e^{(-0.2)t} + 50$$

Given $x(0) = 1$

$\Rightarrow$ $C + 50 = 1$

$\Rightarrow$ $C = -49$

$$x = 50 - 49\,e^{(-0.2)t}$$

12. Given $f(t) = \begin{cases} 1 & a \le t \le b \\ 0 & \text{otherwise} \end{cases}$

$$L\{f(t)\} = \int_0^\infty e^{-st}f(t)dt$$

$$= \int_0^\infty e^{-st}f(t) + \int_a^\infty e^{-st}f(t)dt + \int_b^\infty e^{-st}f(t)dt$$

$$= 0 + \int_a^b e^{-st}dt + 0 = \left.\frac{e^{-st}}{-s}\right|_a^b$$

$$= \frac{-1}{s}\left[e^{-bs} - e^{-as}\right] = \frac{e^{-as} - e^{-bs}}{s}$$

13. For a finite duration time domain signal, ROC is entire s-plane.

14. $x(t) = \alpha s(t) + s(-t)$ & $s(t) = \beta e^{-4t}u(t)$

$x(t) = \alpha\beta e^{-4t}u(t) + \beta e^{4t}u(-t)$

$$\alpha\beta e^{-4t}u(t) \xrightarrow{L} \frac{\alpha\beta}{s+4}$$

$$\beta e^{4t}u(-t) \xrightarrow{L} \frac{\beta}{s-4}$$

$$\therefore \quad X(s) = \frac{\alpha\beta}{s+4} - \frac{\beta}{s-4}$$

$$\beta\left[\frac{\alpha(s-4) - (s+4)}{s^2 - 16}\right] = \frac{16}{s^2-16};\ -4 < \sigma < +4$$

On solving the numerator, $\beta = -2$

15. Given : $x(t) = \frac{\sin t.u(t)}{\pi t}$

Frequency Integration property :

$$\frac{x(t)}{t} \xleftrightarrow{L.T.} \int_{-\infty}^{t} X_1(u)du.$$

$$x_1(t) = \sin t.u(t) \xleftrightarrow{L.T.} \frac{1}{s^2+1} = X_1(s)$$

$$\because \int_{-\infty}^{t}\frac{1}{u^2+1}du = \frac{\pi}{2} - \tan^{-1}(s).$$

$$L[x(t)] = \frac{1}{\pi}\left[\frac{\pi}{2} - \tan^{-1}(s)\right] = X(s)$$

$$\therefore Y(s) = H(s).\times(s) = \frac{1}{2s} - \frac{1}{\pi s}\tan^{-1}(s)$$

By using final value theorem

$$\lim_{t\to\infty} y(t) = \lim_{s\to 0} F(s)$$

$$= \lim_{s\to 0} F(s), \left[\frac{1}{2} - \frac{\tan^{-1}(s)}{2}\right] = \frac{1}{2}.$$

16. As $h(t) \quad = t\,a(f)$

input response

$\delta(t) \to t\,a(t)$

$$u(t) \to \int_{-\infty}^{t} +a(I)dt = \int_{0}^{t} t dt = \frac{t^2}{2}a(t)$$

$$u(t-1) \to \frac{(t-1)^2}{2}a(t-1)$$

17. Given, Laplace transform of

$$f_1(t) = \mathcal{L}\, f_1(t) = \mathcal{L}\,[U(t) - U(t-1)]$$

$$F_1(S) = \frac{1}{S} - \frac{e^{-s}}{s}$$

and $\mathcal{L}\, f_2(t) = F_2(s) = \mathcal{L}\,[U(t) - U(t-2)]$

$$= \frac{1}{s}(1 - e^{-2s})$$

As convolution in time-domain is multiplication in s-domain *i.e.*

$$f_1(t) \otimes f_2(t) = F_1(1)F_2(1)$$

then, $F(S) = \frac{1}{s^2}(1 - e^{-3s} + e^{-s} + e^{-2s})$

Inverse laplace transform gives,

$$f(t) = \mathcal{L}^{-1}\, F(s)$$

$$= t - tu(t-2) - tu(t-1) + tu(t-3)$$

$$= t - t[u(t-1) - u(t-2) + tu(t-3)]$$

Which is dapicted in figure of option (*b*).

18. As, $g(t) * \delta(t - t_0) = g(t_0)$

then $x(t+5) * \delta(t-7) = x(t-2)$

19. $\frac{Y(s)}{X(s)} = \frac{S-2}{S+3}$

$\Rightarrow \quad sY(s) + 3Y(s) = S \times (s) - 2X(s)$

Due to initial condition, we can write above equation as

$$Sy(s) - y(0) + 3y(s) = sx(s) - x(0^-) - 2x(s)$$

$$y(0^-) = -2,$$

$$x(0^-) = 0$$

$$\left[x(t) = 3e^{2t}u(t)\right]$$

$$\Rightarrow Sy(s) + 2 + 3y(s) = (s-2)\left(\frac{-3}{s-2}\right)$$

$$(s+3)y(s) = -3 - 2$$

$$\Rightarrow \quad y(s) = \frac{-5}{5+3}$$

$$\Rightarrow \quad y(t) = -5e^{-3t}u(t)$$

$y(\infty)$ (steady sate) $= 0$

2. $H(s) = \frac{s-2}{s+3};$

$$X(t) = -3e^{2t}.u(t)$$

$$\therefore \quad X(s) = \frac{-3}{s-2}$$

$$\Rightarrow \quad Y(s) = \frac{-3}{s+3}$$

$$y(t)\big|_{\text{at } t=\infty} \Rightarrow y(\infty) = \lim_{s\to 0} S.\,y(s)$$

$$= \lim_{s\to 0} \frac{-3s}{s+3}$$

$$y(\infty) = 0$$

20. Transfer function

$$H(s) = \frac{1}{1+s\tau}$$

$$\therefore \quad V_0(s) = H(s) \cdot V_I(s)$$

(*a*) if $V_i(t) = \delta(t)$

$$V_1(s) = 1$$

$$V_0(s) = H(s) \cdot V_I(s) = \frac{1}{1+s\tau}$$

$$\upsilon_0(t) = \frac{1}{\tau}e^{-\frac{t}{\tau}} \quad \text{(impulse)}$$

(*b*) if $v_i(t) = u(t)$

$$V_I(s) = \frac{1}{s}$$

$$V_0(s) = \frac{1}{s(1+s\tau)} = \frac{1}{s} - \frac{1}{s+\frac{1}{\tau}}$$

$$\upsilon_0(t) = \left(1 - e^{-t/\tau}\right) \quad \text{(Unit step)}$$

(*c*) $v_i(t) = r(t)$

$$\Rightarrow \quad V_I(s) = \frac{1}{s^2}$$

$$V_0(s) = H(s) \cdot V_I(s)$$

$$= \frac{1}{s^2(1+s\tau)} = \frac{1}{s^2} - \frac{\tau}{s} + \frac{t}{s+\frac{1}{\tau}}$$

$$\upsilon_0(t) = t - \tau\left(1 - e^{-t/\tau}\right) \quad \text{(Ramp)}$$

21. A zero - order hold system holds the input signal value for a period of T, i.e., for an input of short duration (Δ) pulse, it produces an output pulse of duration T (sampling period).

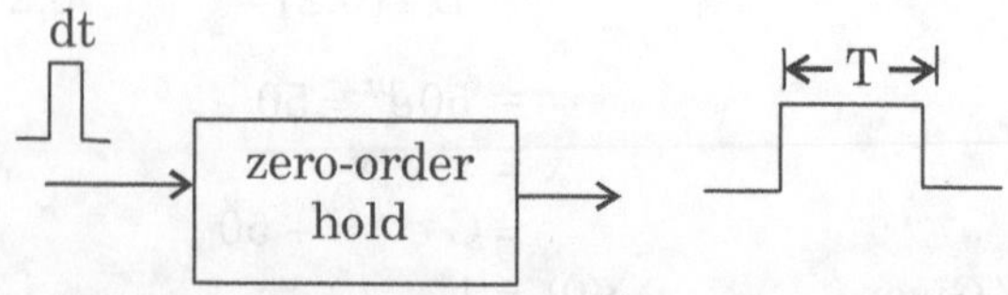

$\Rightarrow$ Input, $x(t) = \delta(t)$,

$X(s) = 1.$

$\Rightarrow$ output, $y(t) = u(t) - u(t - T)$; $Y(s)$

$$= \frac{1}{s} - \frac{e^{-st}}{s} = \frac{1 - s^{-ST}}{s}$$

$$\text{Transfer function} = \frac{Y(s)}{X(s)} = \frac{1 - e^{-ST}}{s}$$

22. $$f(t) = e^{-t}u(t)$$

$$F(s) = L[f(t)] = L[e^{-t}u(t)] = \frac{1}{s+1}$$

In frequency domain, $s = j\omega$

$$F(j\omega) = \frac{1}{(1 + j\omega)} = \frac{-1}{\sqrt{1+\omega^2}} \angle \tan^{-1}\omega$$

For 3 dB bandwidth,

$$\frac{1}{\sqrt{1+\omega^2}} = \frac{1}{\sqrt{2}}$$

$\therefore \quad \omega = +1$

$\omega = 2\pi f_c$

$$\Rightarrow f_c = \frac{1}{2\pi} \text{ Hz}$$

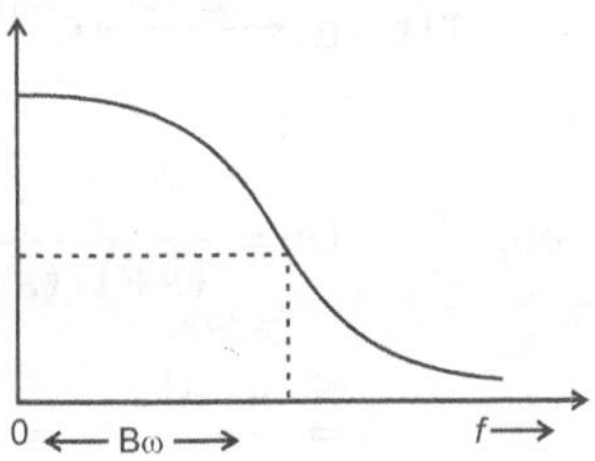

23. $\Rightarrow u(t) \xleftrightarrow{\text{L.T.}} \frac{1}{s}$

$$t.u(t) \xleftrightarrow{\text{L.T.}} \frac{1}{s^2}.$$

$$\sin at u(t) \xleftrightarrow{\text{L.T.}} \frac{a}{s^2 + a^2}$$

$$\sin .u(t) \xleftrightarrow{\text{L.T.}} \frac{a}{s^2 + 1^2} = \frac{1}{s^2 + 1}.$$

24. Given : Transfer function, $H(s)$

$$= \frac{k\left(s^2 + \omega_0^2\right)}{s^2 + \left(\frac{\omega_0}{Q}\right)s + \omega_0^2}$$

putting, $s = j\omega$

$$H(j\omega)\Big|_{\omega=0} = \frac{k\left(0 + \omega_0^2\right)}{0 + 0 + \omega_0^2}$$

$= k \Rightarrow H(0) \neq 0$

$$H(j\infty)\Big|_{\omega \to \infty}$$

$$= \lim_{\omega \to 0} \frac{k\left[1 + \left(\frac{\omega_0}{j\omega}\right)^2\right].(j\omega)^2}{(j\omega)^2\left[1 + \left(\frac{\omega_0}{Q}\right).\frac{1}{j\omega} + \left(\frac{\omega_0}{j\omega}\right)^2\right]}$$

$$= \frac{k(1+0)}{1+0+0}$$

$= k. \Rightarrow H(\infty) \neq 0.$

$\Rightarrow$ The given transfer function is a notch filter.

25. The given function represents a cousal periodic signal.

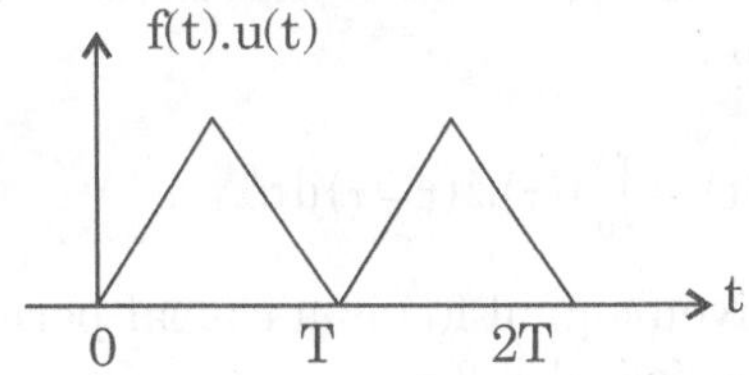

$f(t).\, u(t) = 0 \;\; \forall\, t < 0.$

Period of $F(t)$ for $t > 0$, is T.

Let $f_1(t) = f(t)u(t) \;\; \forall 0 \leq t \leq T$

$= 0$, otherwise.

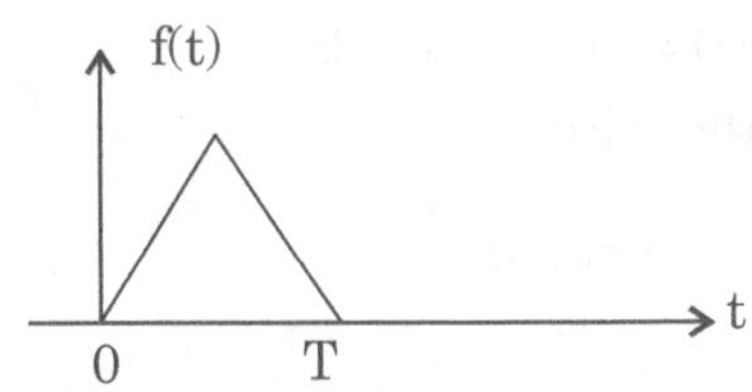

$$f(t)u(t) = \sum_{n=0}^{\infty} f_1(t - nT).$$

$$\text{Let } f_1(t) \xleftrightarrow{\text{L.T.}} F_1(s).$$

$$f_1(t - nT) \xleftrightarrow{\text{L.T.}} e^{-nTs}.F_1(s).$$

$$= \sum_{n=0}^{\infty} e^{-nT.s.}F_1(s)$$

$$= \frac{F_1(s)}{1 - e^{-sT}}$$

$$f(t)u(t) \xleftrightarrow{\text{L.T.}} \left[\frac{1}{1 - e^{-sT}}\right]$$

$$\times \left[\begin{array}{l}\text{Laplace transform of fisrt} \\ \text{part of } f(t)u(t).\end{array}\right.$$

$$A(s) = \frac{1}{1 - e^{-sT}}$$

26. Given : $h(t) = 4.e^{-2t}$; $x(t) = u(t)$

$$H(s) = \frac{4}{s+2};\ X(s) = \frac{1}{s}.$$

$$\Rightarrow \text{Output, } y(t) = L^{-1}[H(s).\times(s)]$$

$$= L^{-1}\left[\frac{4}{s+2}.\frac{1}{s}\right] = L^{-1}\left[2\left(\frac{4}{s+2}.-\frac{1}{s}\right)\right]$$

$y(t) = 2[1-e^{-2t}].u(t)$.

27. Multiplication of two function in frequency or s - domain is equivalent to the convolution in time - domain.

$$\text{So, } v(t) = \int_0^t i(\tau).z(t-\tau)d\tau.$$

28. The given signal f(t) is a causal periodic signal with period $T_o = 2\pi$.

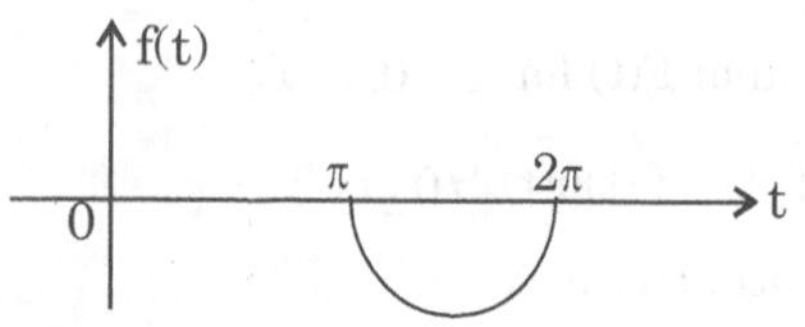

Let, $f_1(t) = f(t)$, $0 < t < 2\pi$,

= 0, otherwise.

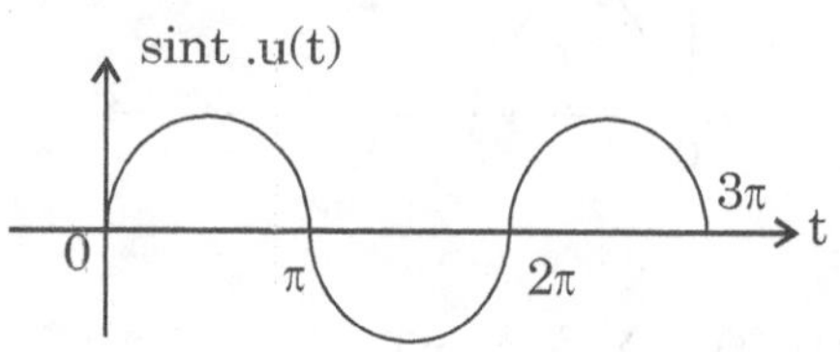

$$f_1(t) = -[\sin(t-\pi)\,u(t-\pi) + \sin(t-2\pi).u(t-2\pi)]$$

$$\because \sin t.u(t) \xleftrightarrow{L.T.} \frac{1}{s^2+1}$$

$$\sin(t-\pi).u(t-\pi) \xleftrightarrow{L.T.} \frac{e^{-\pi s}}{s^2+1}$$

$$\sin(t-2\pi).u(t-2\pi) \xleftrightarrow{L.T.} \frac{e^{-2\pi s}}{s^2+1}$$

$$\Rightarrow F_1(s) = -\left[\frac{e^{-\pi s}}{s^2+1} + \frac{e^{-2\pi s}}{s^2+1}\right]$$

$$= \frac{e^{-\pi s}}{s^2+1}(1+e^{-\pi s})$$

$$\because F(s) = \frac{F_1(s)}{1-e^{sT_o}}$$

$$= -e^{-\pi s}\left[\frac{1+e^{-\pi s}}{s^2+1}\right] \times \frac{1}{1-e^{-2\pi s}}$$

$$F(s) = \frac{-e^{-\pi s}.(1+e^{-\pi s})}{(s^2+1)(1+e^{-\pi s})(1-e^{-\pi s})}$$

$$F(s) = \frac{-e^{-\pi s}.}{(s^2+1)(1-e^{-\pi s})}$$

29. $r(t) \xleftrightarrow{L.T.} e^{-as}$

$$r(t-a) \xleftrightarrow{L.T.} e^{-as}.\frac{1}{s^2} = \frac{e^{-as}}{s^2}.$$

30. $$F(s) = \frac{s+5}{(s+1)(s+3)} = \frac{A}{s+1} + \frac{B}{s+3}$$

$$A = \left.\frac{s+5}{s+3}\right|_{s=-1} = 2,\quad B = \left.\frac{s+5}{s+1}\right|_{s=-3} = \frac{2}{-2} = -1$$

$\therefore$ $L^{-1} F(s) = 2e^{-t} - e^{-3t}$

31.

32. If $L[f(t)] = F(s)$ then $L[f(t-T)]$ is equal to $e^{sT} F(s)$.
We apply time shifting property of Laplace transform.

33. $$H(s) = F(s)\,G(s) = \frac{(s+2)}{(s^2+1)} \times \frac{(s^2+1)}{(s+3)(s+2)}$$

$$= \frac{1}{(s+3)}$$

34. $$H(s) = \frac{1}{(s-2)},\quad R(s) = \frac{1}{(s-3)}$$

$$\text{Output} = \frac{1}{(s-2)}\ \frac{1}{(s-3)} = \frac{1}{(s-3)} - \frac{1}{(s-2)}$$

$$c(t) = e^{3t} - e^{2t}$$

35. The impulse response,

$$C(s) = H(s)\,R(s)$$

$$\Rightarrow \quad C(s) = H(s)$$

For impulse signal

$$\mathcal{L}[\delta(t) = 1.$$

$$c(t) = \mathcal{L}^{-1}[H(s)] = \mathcal{L}^{-1}\left[\frac{1}{s^2}.\frac{1}{(s-2)}\right]$$

$$f_1(t) = \mathcal{L}^{-1}[F_1(s)] = tu(t)$$

$$f_2(t) = \mathcal{L}^{-1}[F_2(s)] = e^{2t}u(t)$$

then $e(t) = f_1(t) * f_2(t)$

$$= [e^{2t} * t]\, u(t)$$

36. $X(s) = \dfrac{5-s}{(s+1)(s-2)} = \dfrac{A}{(s+1)} + \dfrac{B}{s-2}$

$$A = \left.\frac{5-s}{s-2}\right|_{s=-1} = -2.$$

and $B = \left.\dfrac{5-s}{s+1}\right|_{s=2} = 1$

then, $x(t) = L^{-1}X(s) = L^{-1}\left[\dfrac{-2}{s+1} + \dfrac{1}{s-2}\right]$

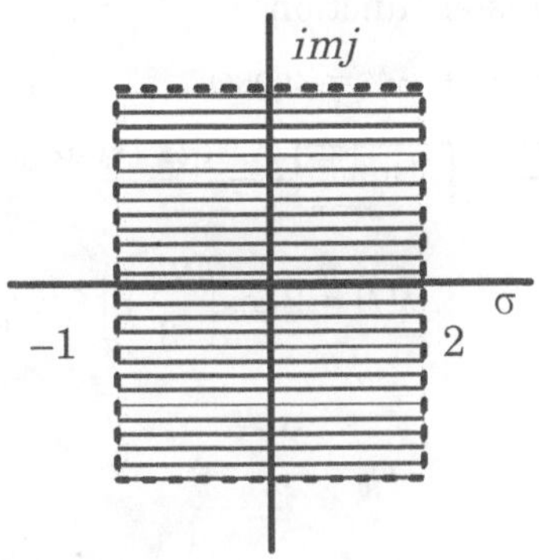

Given, ROC: $Re(s) < Re(a)$

$$\frac{1}{s-a} \longleftrightarrow -e^{at}u(-t)$$

$x(t)$ to be Fourier transformable

$x(t) = -2e^{-t}u(t) - e^{2t}u(-t)$

37. $\displaystyle\int_0^t f(\tau)\,d\tau = \frac{1}{s}f(s)$ ....(Laplace formulae)

38. Zero initial conditions : $y(t) = ?$ for $x(t) = e^{-2t}u(t)$

Given equation is

$$\ddot{y}(t) + 4\dot{y}(t) + 3y(t) = 2\dot{x}(t) + 4x(t)$$

Taking Laplace

$s^2Y(s) + 4sY(s) + 3Y(s) = 2sX(s) + 4X(s)$

$$\Rightarrow \quad \frac{Y(s)}{X(s)} = \frac{2s+4}{s^2+4s+3} \quad ...(i)$$

Given : $X(s) = \dfrac{1}{(s+2)}$

$\therefore$ $Y(s) =$

$$\frac{(2s+4)}{(s^2+4s+3)(s+2)} = \frac{2}{(s+1)(s+3)}$$

Taking inverse laplace transform

$$\Rightarrow \quad Y(s) = \left\{\frac{1}{s+1} - \frac{1}{s+3}\right\}$$

$$\therefore \quad y(t) = \left(e^{-t} - e^{-3t}\right)u(t)$$

39. If $f(t) \leftrightarrow F(s)$, then $tf(t) \leftrightarrow -\dfrac{d}{ds}F(s)$

Thus if $F(s) = \dfrac{1}{s^2+s+1}$

$$tf(t) \rightarrow -\frac{d}{ds}\left(\frac{1}{s^2+s+1}\right) = \frac{2s+1}{s^2+s+1}$$

40. As $h(t) = t\,a(f)$

input	response
$\delta(t) \rightarrow$	$t\,a(t)$

$$u(t) \rightarrow \int_{-\infty}^{t} +(I)dt = \int_0^t t\,dt = \frac{t^2}{2}a(t)$$

$$u(t-1) \rightarrow \frac{(t-1)^2}{2}a(t-1)$$

41. $H(s) = \dfrac{1}{s}$

$h(t) = u(t)$

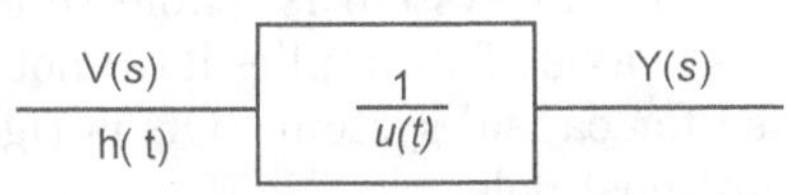

u(t) = input

output $= u(t) \oplus h(t) = u(t) \oplus u(t) = r(t) = +u(t)$

42. Since $n(t)$ can be written in function of t using step function $x(t) = 4(t) = 4(t-L)$

we need to $x(s)$ laplace transform $x(t)$

$$x(s) = \frac{1}{s} - \frac{1}{s}e^{-2s}$$

$$\therefore \quad L\left(\frac{d^2y(t)}{d(t)}\right) = s^2Y(s)$$

$$\therefore \quad L\left[\frac{d^2y(t)}{dt} + s\frac{dy}{dt} + 6y = x\right]$$

$$\Rightarrow \quad s^2Y(s) + sY(s) + 6Y(s) = X(s)$$

$$\therefore \quad Y(s) = \frac{X(s)}{s^2+5s+6} = \frac{1-e^{-2s}}{s(s+2)(s+3)}$$

43. $h(t) \leftrightarrow H(s) = \dfrac{1}{s+1}$

$$\Rightarrow \quad h(t) = e^{-t}u(t)$$

S_1 : System is stable (TRUE)

Because $h(t)$ absolutely integrable

S_2 : $\dfrac{h(t+1)}{h(t)}$ is independent of time (TRUE)

$$\frac{e^{-(t+1)}}{e^{-t}} \Rightarrow e^{-1} \text{ (independent of time)}$$

S_3 : A non-causal system with same transfer function is stable

$\frac{1}{s+1} \leftrightarrow -e^{-t}u(-t)$ (a non-causal system) but this is not absolutely integrable thus unstable.

Only S_1 and S_2 are TRUE

44. If $f(t) \leftrightarrow F(s)$, then $tf(t) \leftrightarrow -\frac{d}{ds}F(s)$

Thus if $\quad F(s) = \frac{1}{s^2+s+1}$

$$tf(t) \rightarrow -\frac{d}{ds}\left(\frac{1}{s^2+s+1}\right) = \frac{2s+1}{s^2+s+1}$$

45. Given, $\quad H(s) = \frac{1}{s^2+s-6}$

$$= \frac{1}{(s+3)(s-2)}$$

It is given that system is stable thus its ROC includes $j\omega$ axis. This implies it cannot be causal, because for causal system ROC is right side of the rightmost pole.

$\Rightarrow$ Poles at $s = 2$ must be removes so that it can be become causal and stable simultaneously.

$$\Rightarrow \frac{1}{(s+3)(s-2)}(s-2) = \frac{1}{s+3}$$

Thus $H_1(s) = s - 2$

46. Given differential equation

$$s^2 y(s) + \alpha\, sy(s) + a^2 y(s) = x(s)$$

$$\Rightarrow \frac{y(s)}{x(s)} = \frac{1}{s^2+\alpha s+\alpha^2} = H(s)$$

$$g(t) = \alpha^2\int_0^t h(z)dz + \frac{d}{dt}h(t) + \alpha h(t)$$

$$= \alpha^2\frac{H(s)}{s} + sH(s) + \alpha H(s)$$

$$= \alpha^2\frac{1}{s(s^2+\alpha s+\alpha^2)} + s\frac{1}{(s^2+2s+\alpha^2)} + \frac{\alpha}{s^2+\alpha s+\alpha^2}$$

$$= \frac{\alpha^2+\alpha s+s^2}{s(s^2+\alpha s+\alpha^2)} = \frac{1}{s}$$

48. $\quad \frac{dy}{dt} + 5y(t) = u(t)$

$$y(0) = 1$$

$$sY(s) - y(0) + 5Y(s) = \frac{1}{s}$$

$$5Y(s) - 1 + sY(s) = \frac{1}{s}$$

$$Y(s)[s+5] = \left[\frac{1}{s}+1\right]$$

$$Y(s) = \frac{s+1}{s(s+5)}$$

$$Y(s) = \frac{1}{5s} + \frac{4}{5(s+5)}$$

Applying inverse laplace transform

$$y(t) = \frac{1}{5}u(t) + \frac{4}{5}e^{-5t}u(t)$$

$$y(t) = (0.2 + 0.8\, e^{-5t})$$

49. The transfer function

$$H(e^{j\omega}) = e^{j\omega} + e^{-j\omega} = 2\cos\omega$$

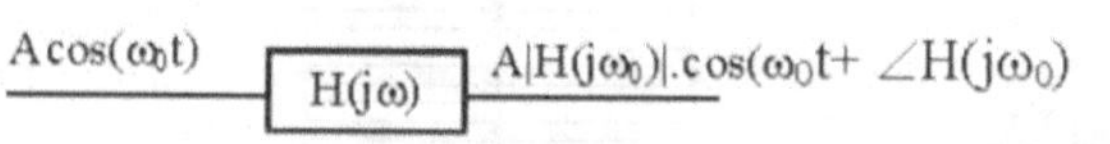

Here $\quad x(t) = 2\cos\left(\frac{2\pi}{3}t\right)$

$$\omega_0 = \frac{2\pi}{3}$$

$$H(j\omega_0) = 2\cos\left(\frac{2\pi}{3}\right) = 2\left(\frac{-1}{2}\right) = -1$$

$$y(t) = 2\cos\left(\frac{2\pi}{3}t + 180°\right)$$

$$x(t) = \cos \pi t$$

$$\omega_0 = \pi$$

$$H(j\omega_0) = 2\cos(\pi) = -2$$

$$y(t) = 2\cos(\pi t + 180°)$$

$$y(t) = 2\cos\left(\frac{2\pi}{3}t + \pi\right) - 2\cos(\pi t + \pi)$$

$$\omega_1 = \frac{2\pi}{3} \qquad \omega_2 = \pi$$

$$T_1 = 3, \qquad T_2 = 2$$

$$T_0 = 6,$$

$$\omega_0 = \frac{2\pi}{T_0} = \frac{\pi}{3}$$

$$y(t) = 2\cos(2\omega_0 t + \pi) - 2\cos(3\omega_0 t + \pi)$$

$$y(t) = e^{j(2\omega_0 t+\pi)} + e^{-j(2\omega_0 t+\pi)}$$

$$- e^{j(3\omega_0 t+\pi)} - e^{-j(3\omega_0 t+\pi)}$$

$$y(t) = -e^{j(2\omega_0 t)} - e^{-j(2\omega_0 t)}$$

$$+ e^{j(3\omega_0 t)} + e^{-j(3\omega_0 t)}$$

$$C_3 = 1$$

Thus the value of C_3 is 1.

50. Forced response is the response due to external input signal.

$$Y(s) = \frac{G(s)}{s} = \frac{3-s}{s(s+1)(s+3)}$$

By partial fraction,

$$= \frac{1}{s} + \frac{-2}{s+1} + \frac{1}{s+3}$$

Taking inverse laplace transform.

$$\Rightarrow y(t) = u(t) - 2e^{-t}u(t) + e^{-3t}u(t)$$

51. In the given pole - zero pattern, poles and zeros are symmetrical about imaginary axis. This is case of pass filter.

52. (1) poles in the right half plane implies $\rightarrow$ unstable system.

(2) If $h(t) = 0 \forall t \leq 0 \rightarrow$ causal system.

$(1) \rightarrow (D), (2) \rightarrow (B)$

53. A BIBO stable system can have poles in right half of complex plane, if it is a non-causal system. So, statement-I is wrong.

A causal and BIBO stable system should have all poles in the left half of complex plane. So, statement-II is correct.

$\therefore$ option (*d*) is correct.

6

CHAPTER

Z-transform

Analysis of Previous GATE Papers												
	Year → Topics ↓		2019	2018	2017 Set 1	2017 Set 2	2016 Set 1	2016 Set 2	2016 Set 3	2015 Set 1	2015 Set 2	2015 Set 3
Z-TRANSFORM OF DISCRETE SIGNALS	1 Mark	MCQ Type	1				1		1			
		Numerical Type									1	
	2 Marks	MCQ Type							1	1		1
		Numerical Type										
	Total		1				1		3	2	1	2
INTERCONNECTION	1 Mark	MCQ Type										
		Numerical Type										
	2 Marks	MCQ Type									1	1
		Numerical Type										
	Total										2	2
DIGITAL FILTER DESIGN	1 Mark	MCQ Type		1								1
		Numerical Type										
	2 Marks	MCQ Type	1						1			
		Numerical Type	1									
	Total		4	1					2			1

Z-TRANSFORM OF DISCRETE SIGNALS

1. The z-transform of the following real exponential sequence

$x(nT) = a^n, nT \geq 0$

$= 0, nT < 0, a > 0$

(a) $\frac{1}{1-z^{-1}}; |z| > 1$ (b) $\frac{1}{1-az^{-1}}; |z| > a$

(c) 1 for all z (d) $\frac{1}{1-az^{-1}}; |z| < a$

[1990 : 2 Marks]

2. A linear discrete-time system has the characteristics equation, $z^3 - 0.81z = 0$. The system

(a) is stable.

(b) is marginally stable.

(c) is unstable.

(d) stability cannot be assessed from the given information. **[1992 : 2 Marks]**

3. The z-transform of the time function $\sum_{k=0}^{\infty} \delta(n-K)$ is

(a) $\frac{z}{z-a^T}$ (b) $\frac{z}{z+a^T}$

(c) $\frac{z}{z-a^{-T}}$ (d) $\frac{z}{z+a^{-T}}$

[1998 : 1 Mark]

4. The z-transform F(z) of the function $f(nT) = a^{nT}$ is

(a) $\frac{z}{z-a^T}$ (b) $\frac{z}{z+a^T}$

(c) $\frac{z}{z-a^{-T}}$ (d) $\frac{z}{z+a^{-T}}$

[1999 : 1 Mark]

5. The z-transform of a signal is given by $C(z) = \frac{1}{4}\frac{z^{-1}(1-z^{-4})}{(1-z^{-1})^2}$. Its final value is

(a) $\frac{1}{4}$ (b) zero

(c) 1.0 (d) infinity

[1999 : 2 Marks]

6. The region of convergence of the z-transform of a unit step function is

(a) $|z| > 1$

(b) $|z| < 1$

(c) (Real part of z) > 0

(d) (Real part of z) < 0

[2001 : 1 Mark]

7. If the impulse response of a discrete-time system is $h[n] = -5^n u[-n-1]$, then the system function H(z) is equal to

(a) $\frac{-z}{z-5}$ and the system is stable

(b) $\frac{z}{z-5}$ and the system is stable

(c) $\frac{-z}{z-5}$ and the sVstem is unstable

(d) $\frac{z}{z-5}$ and the system is unstable

[2002 : 2 Marks]

8. A sequence x(n) with the z-transform $X(z) = z^4 + z^2 - 2z + 2 - 3z^{-4}$ is applied as an input to a linear, time-invariant system with the impulse response $h(n) = 2\delta(n-3)$ where

$$\delta(n) = \begin{cases} 1, & n = 0 \\ 0, & \text{otherwise} \end{cases}$$

The output at n = 4 is

(a) –6 (b) zero

(c) 2 (d) –4

[2003 : 1 Mark]

9. The z-transform of a system is $H(z) = \frac{z}{z-0.2}$.

If the ROC is $|z| < 0.2$, then the impulse response of the system is

(a) $(0.2)^n u[n]$

(b) $(0.2)^n u[-n-1]$

(c) $-(0.2)^n u[n]$

(d) $-(0.2)^n u[-n-1]$

[2004 : 1 Mark]

10. A causal LTI system is described by the difference equation
$2y[n] = \alpha y[n-2] - 2x|n| - \beta x[n-1]$.
The system is stable only if
(a) $|\alpha| = 2, |\beta| < 2$
(b) $|\alpha| > 2, |\pi| > 2$
(c) $|\alpha| < 2$, any value of β
(d) $|\beta| < 2$, any value of α

[2004 : 2 Marks]

11. The region of convergence of z-transform of the sequence

$\left(\frac{5}{6}\right)^n u(n) - \left(\frac{6}{5}\right)^n u(-n-1)$ must be

(a) $|z| < \frac{5}{6}$ (b) $|z| > \frac{5}{6}$

(c) $\frac{5}{6} < |z| < \frac{6}{5}$ (d) $\frac{6}{5} < |z| < \infty$

[2005 : 1 Mark]

12. If the region of convergence of $x_1[n] + x_2[n]$ is $\frac{1}{3} < |z| < \frac{2}{3}$, then the region of convergence of $x_1[n] - x_2[n]$ includes

(a) $\frac{1}{3} < |z| < 3$ (b) $\frac{2}{3} < |z| < 3$

(c) $\frac{3}{2} < |z| < 3$ (d) $\frac{1}{3} < |z| < \frac{2}{3}$

[2006 : 1 Mark]

13. The z-transform X[z] of a sequence x[n] is given by $X[z] = \frac{0.5}{1-2z^{-1}}$. It is given that the region of convergence of X[z] includes the unit circle. The value of x[0] is
(a) –0.5 (b) 0
(c) 0.25 (d) 0.5

[2007 : 2 Marks]

14. The ROC of z-transform of the discrete time sequence

$x(n) = \left(\frac{1}{3}\right)^n u(n) - \left(\frac{1}{2}\right)^n u(-n-1)$ is

(a) $\frac{1}{3} < |z| < \frac{1}{2}$ (b) $|z| > \frac{1}{2}$

(c) $|z| < \frac{1}{3}$ (d) $2 < |z| < 3$

[2009 : 1 Mark]

15. Consider the z-transform $X(z) = 5z^2 + 4z^{-1} + 3$; $0 < |z| < \infty$. The inverse z-transform x[n] is
(a) $5\delta[n+2] + 3\delta[n] + 4\delta[n-1]$.
(b) $5[n-2] + 3\delta[n] + 4\delta[n+1]$.
(c) $5u[n+2] + 3u[n] + 4u[n-1]$.
(d) $5u[n-2] + 3u[n] + 4u[n+1]$.

[2010 : 1 Mark]

16. The transfer function of a discrete time LTI system is given by

$$H(z) = \frac{2 - \frac{3}{4}z^{-1}}{1 - \frac{3}{4}z^{-1} + \frac{1}{8}z^{-2}}$$

Consider the following statements:
S_1 : The system is stable and causal for ROC : $|z| > \frac{1}{2}$
S_2 : The system is stable but not causal for ROC : $|z| < \frac{1}{4}$
S_3 : The system is neither stable nor causal for ROC : $\frac{1}{4} < |z| < \frac{1}{2}$
Which one of the following statements is valid?
(a) Both S_1 and S_2 are true
(b) Both S_2 and S_3 true
(c) Both S_1 and S_3 are true
(d) S_1, S_2 and S_3 are all true

[2010 : 2 Marks]

17. If $x[n] = \left(\frac{1}{3}\right)^{|n|} - \left(\frac{1}{2}\right)^n u[n]$, then the region of convergence (ROC) of its z-transform in the z-plane will be

(a) $\frac{1}{3} < |z| < 3$ (b) $\frac{1}{3} < |z| < \frac{1}{2}$

(c) $\frac{1}{2} < |z| < 3$ (d) $\frac{1}{3} < |z|$

[2012 : 1 Mark]

18. C is a closed path in the z-plane given by $|z| = 3$. The value of the integral $\oint_c \left(\frac{z^2 - z + 4j}{z + 2j}\right) dz$ is
(a) $-4\pi(1+j2)$ (b) $4\pi(3-j2)$
(c) $-4\pi(3+j2)$ (d) $4\pi(1-j2)$

[2014 : 1 Mark, Set-1]

19. Let $x[n]=\left(-\frac{1}{9}\right)^n u(n)-\left(-\frac{1}{3}\right)^n u(-n-1)$.

The Region of Convergence (ROC) of the z-transform of x[n]

(a) is $|z|>\frac{1}{9}$ (b) is $|z|<\frac{1}{3}$

(c) is $\frac{1}{3}>|z|>\frac{1}{9}$ (d) does not exist

[2014 : 2 Marks, Set-1]

20. Let x[n] = x[–n]. Let X(z) be the z-transform of x[n]. If 0.5 + j0.25 is a zero of X(z), which one of the following must also be a zero of X(z).

(a) 0.5 – j0.25 (b) $\frac{1}{(0.5+j0.25)}$

(c) $\frac{1}{(0.5-j0.25)}$ (d) 2 + j4

[2014 : 1 Mark, Set-2]

21. The input-output relationship of a causal stable LTI system is given as

$y[n]=\alpha y[n-1]+\beta x[n]$. If the impulse response h[n] of this system satisfies the condition

$\sum_{n=0}^{\infty} h[n]=2$, the relationship between α and β is

(a) $\alpha=1-\frac{\beta}{2}$ (b) $\alpha=1+\frac{\beta}{2}$

(c) $\alpha=2\beta$ (d) $\alpha=-2\beta$

[2014 : 2 Marks, Set-2]

22. The z-transform of the sequence x[n] is given by $X(z)=\frac{1}{(1-2z^{-1})^2}$, with the region of convergence |z| > 2. Then x[2] is ______.

[2014 : 2 Marks, Set-3]

23. The pole-zero diagram of a causal and stable discrete-time system is shown in the figure. The zero at the origin has multiplicity 4. The impulse response of the system is h[n]. If h[0] = 1, we can conclude

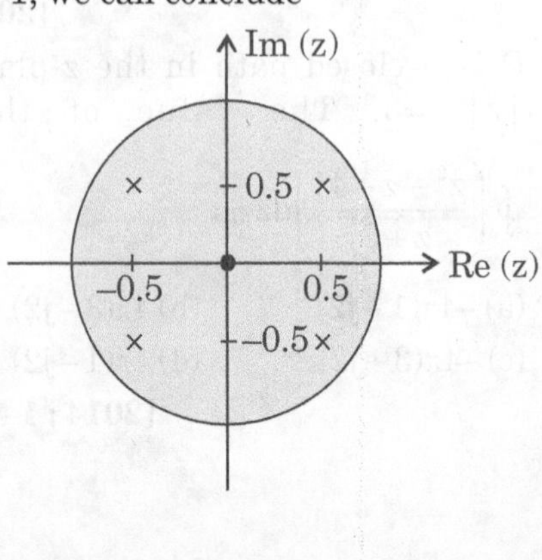

(a) h[n] is real for all n

(b) h[n] is purely imaginary for all n

(c) h[n] is real for only even n

(d) h[n] is purely imaginary for only odd n

[2015 : 2 Marks, Set-1]

24. Two causal discrete-time signalsx[n] and y[n] are related as $y[n]=\sum_{m=0}^{n} x[m]$. If the z-transform of y[n] is $\frac{2}{z(z-1)^2}$, the value of x[2] is ______.

[2015 : 1 Mark, Set-2]

25. Suppose x[n] is an absolutely summable discrete- time signal. Its z-transform is a rational function with two poles and two zeroes. The poles are at z = ±2j. Which one of the following statements is TRUE for the signal x[n]?

(a) It is a finite duration signal.

(b) It is a causal signal.

(c) It is a non-causal signal.

(d) It is a periodic signal.

[2015 : 2 Marks, Set-3]

26. Consider the sequence $x[n]=a^n u[n]+b^n u[n]$, where u[n] denotes the unit-step sequence and $0<|a|<|b|<1$. The region of convergence (ROC) of the z-transform of x[n] is

(a) |z| > |a| (b) |z| > |b|

(c) |z| < |a| (d) |a| < |z|< |b|

[2016 : 1 Mark, Set-1]

27. The ROC (region of convergence) of the z-transform of a discrete-time signal is represented by the shaded region in the z-plane. If the signal $x[n]=(2.0)^{|n|}, -\infty<n<+\infty$ then the ROC of its z-transform is represented by

(a)

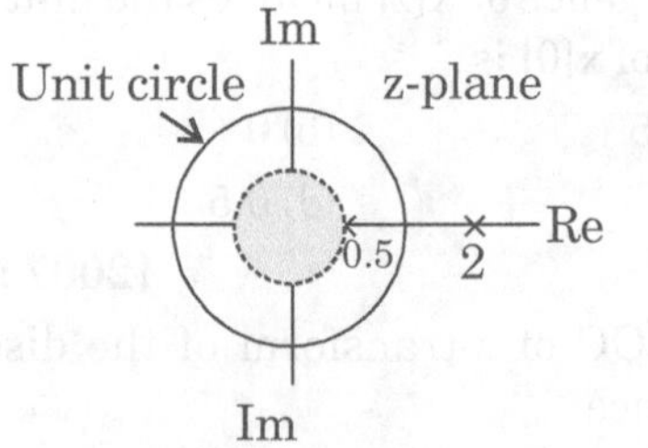

(b)

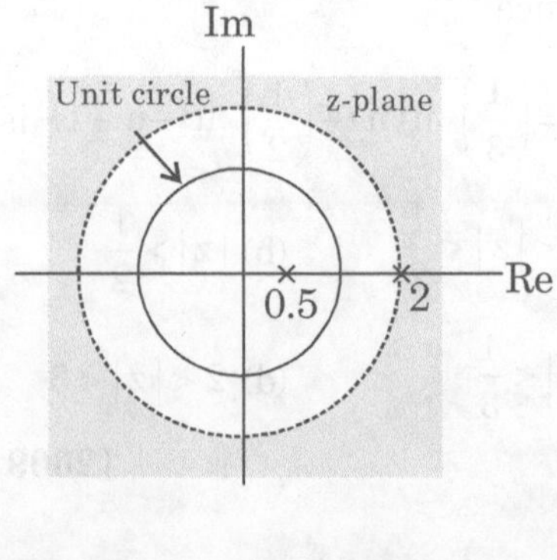

(c) 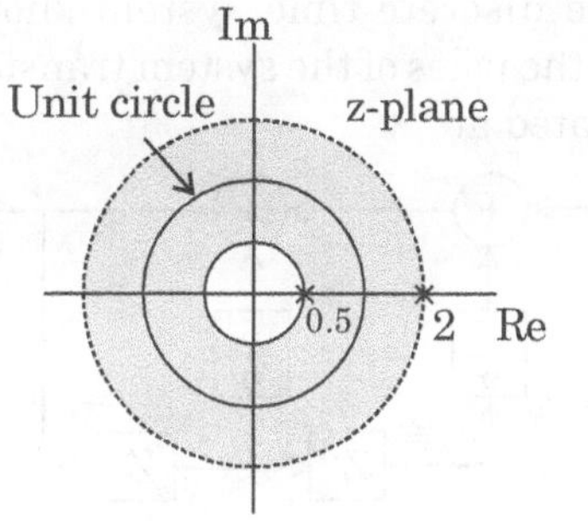

(d) 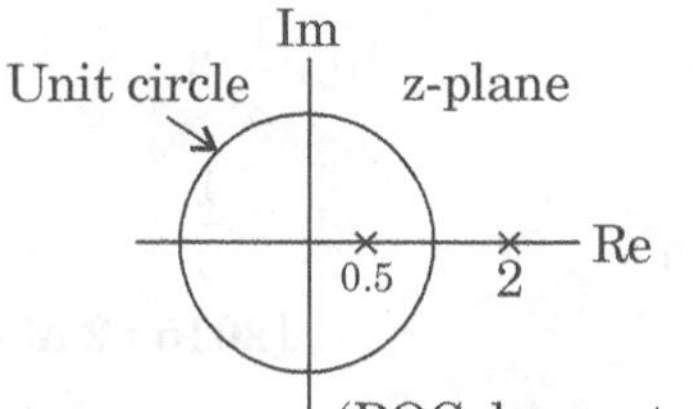

(ROC does not exist)

[2016 : 2 Marks, Set-3]

28. A discrete-time signal $x[n] = \delta[n-3] + 2\delta[n-5]$ has z-transform X(z). If $Y(z) = X(-z)$ is the z-transform of another signal y[n], then

(a) $y[n] = x[n]$ (b) $y[n] = x[-n]$
(c) $y[n] = -x[n]$ (d) $y[n] = -x[-n]$

[2016 : 1 Mark, Set-3]

29. Let H(z) be the z-transform of a real-valued discrete time signal h[n]. If $P(z) = H(z)H\left(\frac{1}{z}\right)$ has a zero $z = \frac{1}{2}+\frac{1}{2}j$, and P(z) has a total of four zeros, which one of the following plots represents all the zeros correctly?

(A) 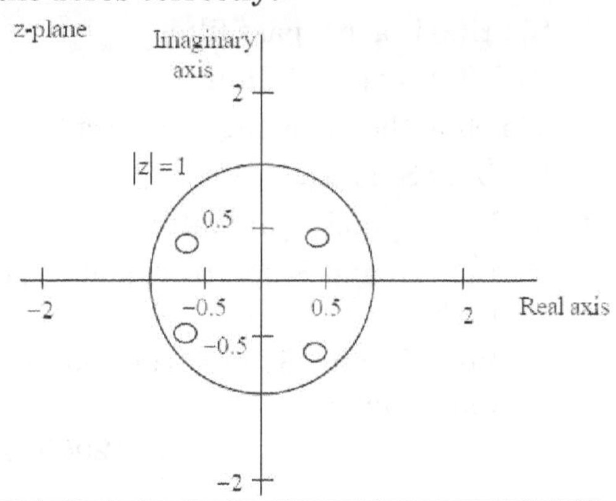

(B) 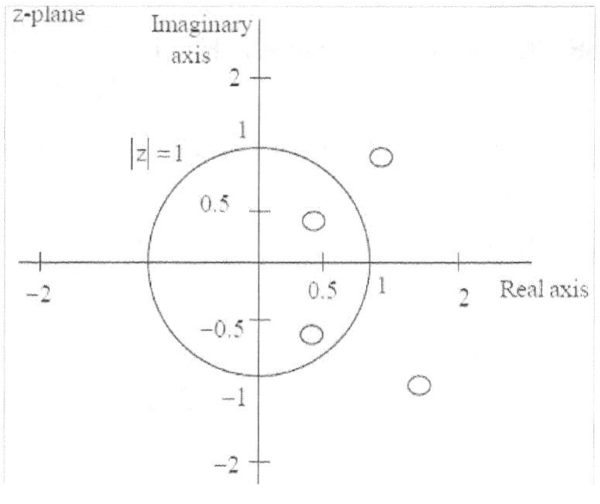

(C) 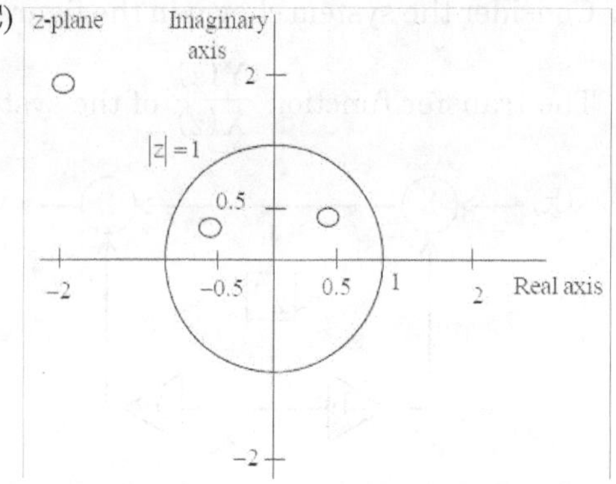

(D) 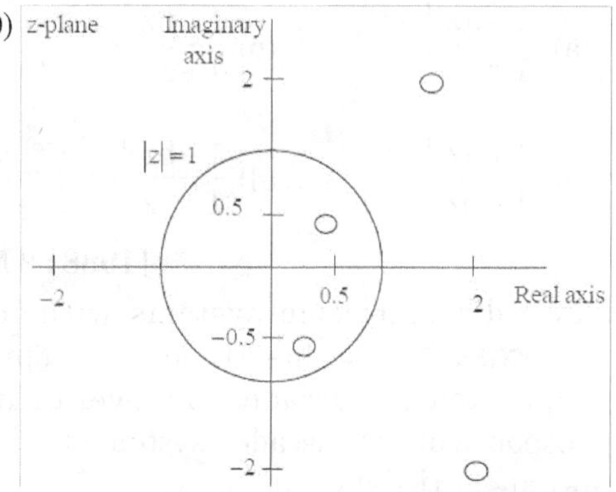

[2019 : 1 Mark]

INTERCONNECTION

30. A realization of a stable discrete time system is shown in figure. If the system is excited by a unit step sequence input x[n], the response y[n] is

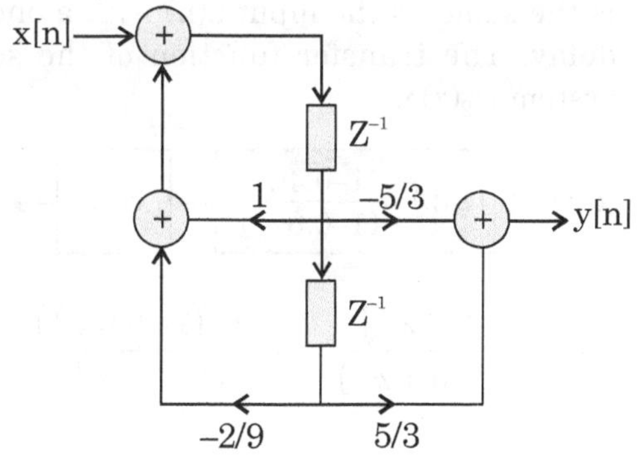

(a) $4\left(-\frac{1}{3}\right)^n u[n] - 5\left(-\frac{2}{3}\right)^n u[n]$

(b) $5\left(-\frac{2}{3}\right)^n u[n] - 3\left(-\frac{1}{3}\right)^n u[n]$

(c) $5\left(-\frac{1}{3}\right)^n u[n] - 3\left(-\frac{2}{3}\right)^n u[n]$

(d) $5\left(-\frac{2}{3}\right)^n u[n] - 5\left(-\frac{1}{3}\right)^n u[n]$

[1988 : 2 Marks]

31. Consider the system shown in the figure below.

The transfer function $\frac{Y(z)}{X(z)}$ of the system is

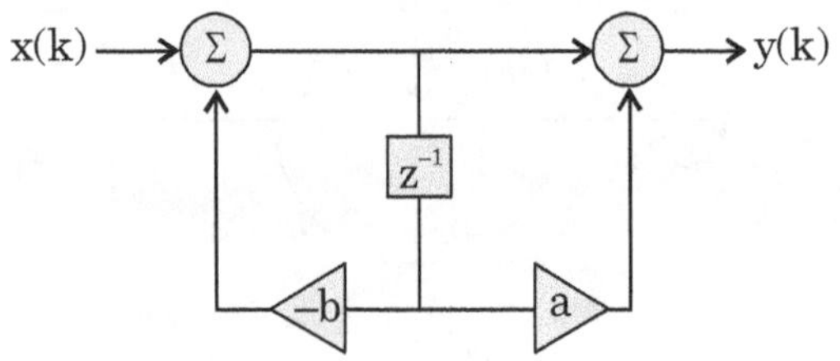

(a) $\frac{1+az^{-1}}{1+bz^{-1}}$ (b) $\frac{1+bz^{-1}}{1+az^{-1}}$

(c) $\frac{1+az^{-1}}{1-bz^{-1}}$ (d) $\frac{1-bz^{-1}}{1-az^{-1}}$

[1988 : 2 Marks]

32. Two discrete time systems with impulse responses $h_1[n] = \delta[n-1]$ and $h_2[n] = \delta[n-2]$ are connected in cascade. The overall impulse response of the cascaded system is

(a) $\delta[n-1] + \delta[n-2]$

(b) $\delta[n-4]$

(c) $\delta[n-3]$

(d) $\delta[n-1]\,\delta[n-2]$

[2010 : 1 Mark]

33. Two systems $H_1(z)$ and $H_2(z)$ are connected in cascade as shown below. The overall output y(n) is the same as the input x(n) with a one unit delay. The transfer function of the second system $H_2(z)$ is

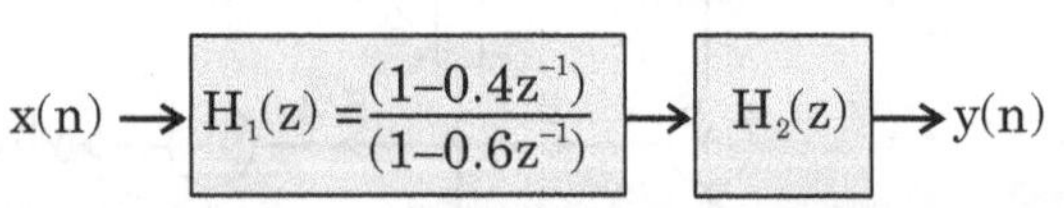

(a) $\frac{(1-0.6\,z^{-1})}{z^{-1}(1-0.4\,z^{-1})}$ (b) $\frac{z^{-1}(1-0.6\,z^{-1})}{(1-0.4\,z^{-1})}$

(c) $\frac{z^{-1}(1-0.4\,z^{-1})}{(1-0.6\,z^{-1})}$ (d) $\frac{(1-0.4\,z^{-1})}{z^{-1}(1-0.6\,z^{-1})}$

[2011 : 2 Marks]

34. Let $H_1(z) = (1 - pz^{-1})$, $H_2(z) = (1 - qz^{-1})^{-1}$, $H(z) - H_1(z) + rH_2(z)$. The quantities p, q, r are real numbers. Consider $p - \frac{1}{2}, q - \frac{1}{4}, |r| < 1$. If the zero of H(z) lies on the unit circle, then r = ________.

[2014 : 2 Marks, Set-3]

35. For the discrete-time system shown in the figure, the poles of the system transfer function are located at

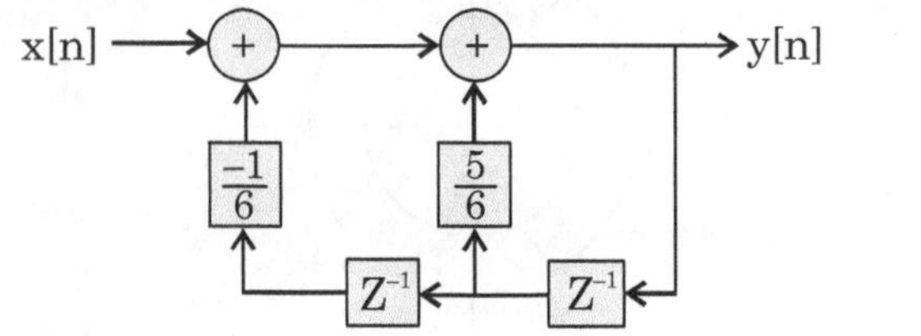

(a) 2, 3 (b) $\frac{1}{2}, 3$

(c) $\frac{1}{2}, \frac{1}{3}$ (d) $2, \frac{1}{3}$

[2015 : 2 Mark, Set-?]

DIGITAL FILTER DESIGN

36. An FIR system is described by the system function

$$H(z) = 1 + \frac{7}{2}z^{-1} + \frac{3}{z}z^{-2}$$

The system is

(a) maximum phase.

(b) minimum phase.

(c) mixed phase.

(d) zero phase. **[2008: 1 Mark]**

37. A system with transfer function H(z) has impulse response h(n) defined as h(2) = 1, h(3) = –1 and h(k) = 0 otherwise. Consider the following statements.

S_1 : H(z) is a low-pass filter.

S_2 : H(z) is an FIR filter.

Which of the following is correct?

(a) Only S_2 is true.

(b) Both S_1 and S_2 are false.

(c) Both S_1 and S_2 are true, and S_2 is a reason for S_1.

(d) Both S_1 and S_2 are true, but S_2 is not a reason for S_1.

[2009 : 2 Marks]

38. For an all-pass system $H(z) = \frac{(z^{-1} - b)}{(1 - az^{-1})}$, where $|H(e^{-j\omega})| = 1$, for all ω. If Re(a) ≠ 0, lm(a) ≠ 0, then b equals

(b) a (b) a^*

(c) $\frac{1}{a^*}$ (d) $\frac{1}{a}$

[2014 : 1 Mark, Set-3]

39. Consider a four point moving average filter defined by the equation $y[n] = \sum_{i=0}^{3} \alpha_i x[n-i]$. The condition on the filter coefficients that results in a null at zero frequency is

(a) $\alpha_1 = \alpha_2 = 0;\ \alpha_0 = -\alpha_3$.

(b) $\alpha_1 = \alpha_2 = 1;\ \alpha_0 = -\alpha_3$.

(c) $\alpha_0 = \alpha_3 = 0;\ \alpha_1 = \alpha_2$.

(d) $\alpha_1 = \alpha_2 = 0;\ \alpha_0 = \alpha_3$.

[2015 : 1 Mark, Set-3]

40. A discrete-time all-pass system has two of its poles at $0.25\angle 0°$ and $2\angle 30°$. Which one of the following statements about the system is TRUE?

(a) It has two more poles at $0.5\angle 30°$ and $4\angle 0°$.

(b) It is stable only when the impulse response is two-sided.

(c) It has constant phase response over all frequencies.

(d) It has constant phase response over the entire z-plane.

[2018 : 1 Mark]

41. The direct form structure of an FIR (finite impulse response) filter is shown in the figure.

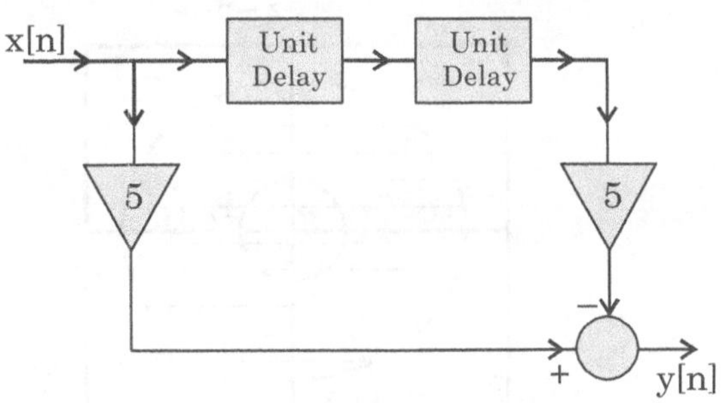

The filter can be used to approximate a

(a) low-pass filter (b) high-pass filter

(c) band-pass filter (d) band-stop filter

[2016 : 2 Marks, Set-3]

42. An LTI system with unit sample response $h[n] = 5\delta[n] - 7\delta[n-1] + 7\delta[n-3] - 5\delta[n-4]$ is a

(a) low-pass filter (b) high-pass filter

(c) band-pass filter (d) band-stop filter

[2017 : 1 Mark, Set-2]

43. It is desired to find three-tap causal filter which gives zero signal as an output to and input of the form

$$x[n] = c_1 \exp\left(-\frac{j\pi n}{2}\right) + c_2 \exp\left(\frac{j\pi n}{2}\right),$$

Where c_1 and c_2 are arbitrary real numbers. The desired three-tap filter is given by

$h[0] = 1,\ h[1] = a,\quad h[2] = b$ and

$h[n] = 0$ for $n < 0$ or $n > 2$.

What are the values of the filter taps a and b if the output is $y[n] = 0$ for all n, when x[n] is as given above?

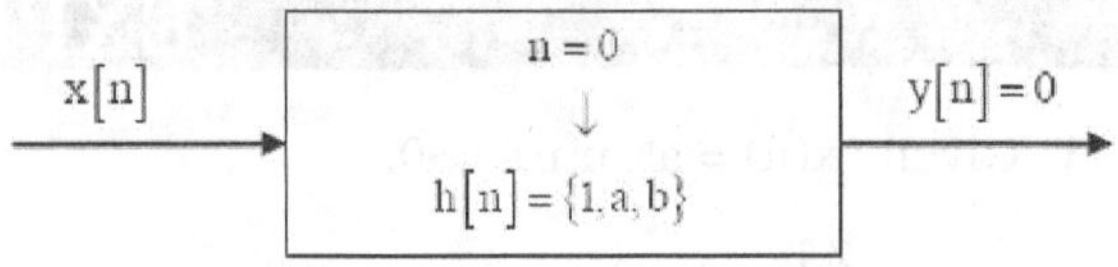

(A) a = −1, b = 1 (B) a = 0, b = 1

(C) a = 1, b = 1 (D) a = 0, b = − 1

[2019 : 2 Marks]

44. Let h[n] be length-7 discrete-time finite impulse response filter, given by

$h[0] = 4, h[1] = 3, h[2] = 2, h[3] = 1$

$h[-1] = -3, h[-2] = -2, h[-3] = -1,$

and h[n] is zero for $|n| \geq 4$. A length-3 finite impulse response approximation g[n] of h[n] has to be obtained such that

$E(h, g) = \int_{-\pi}^{\pi} \left|H(e^{j\omega}) - G(e^{j\omega})\right|^2 d\omega$ is minimized,

where $H(e^{j\omega})$ and $G(e^{j\omega})$ are the discrete-time Fourier transforms of h[n] and g[n], respectively. For the filter that minimizes E(h, g), the value of $10g[-1] + g[1]$, rounded off to 2 decimal places, is__________.

[2019 : 2 Marks]

ANSWERS

1. (b)	2. (a)	3. (c)	4. (a)	5. (b)	6. (a)	7. (b)	8. (b)	9. (d)	10. (c)
11. (c)	12. (d)	13. (b)	14. (a)	15. (a)	16. (c)	17. (c)	18. (c)	19. (c)	20. (b)
21. (a)	22. (12)	23. (a)	24. (0)	25. (c)	26. (b)	27. (d)	28. (c)	29. (b)	30. (c)
31. (a)	32. (c)	33. (b)	34. (–0. 5)	35. (c)	36. (c)	37. (a)	38. (b)	39. (a)	40. (b)
41. (2.10)	42. (c)	43. (b)	44. (–27)						

EXPLANATIONS

1. Given : $x(n) = a^n . u(n), a>0.$

$$x(n) \xleftrightarrow{\text{Z.T.}} X(z) = \frac{z}{z-a}, |z| > |a|$$

$$X(z) = \frac{1}{1-az^{-1}}, |z| > a.$$

2. Linear discrete time system has characterisitic equation.

$$z^3 - 0.81z = 0$$

$$\text{or } z(z^2 - 0.9^2) = 0$$

$$\text{or } z(z-0.9)(z+0.9) = 0$$

Hence $\quad z = 0, 0.9, -0.9$

All roots lie inside unit circle. Hence the system is stable.

3. Given time function as $\sum_{k=0}^{\infty} \delta(n-k)$

$$z_{transform} = \sum_{k=0}^{\infty} z^{-k} = \frac{1}{1-z^{-1}} = \frac{z}{z-1}$$

4. z transform $F(z)$ of the function of $f(nT) = a^nT$

$$x(n) = a^n u(n)$$

$$\text{or} \quad X(z) = \frac{1}{1-az^{-1}}$$

$$= \frac{z}{1-a}$$

5. Z transform of a signal is given by,

$$C(z) = \frac{1}{4} \times \frac{z^{-1}(1-z^{-4})}{(1-z^{-1})^2}$$

Final value theorem, for a casual sequence

$$f(n) \lim_{n\to\infty} = f(x) = \lim_{z\to 1} (z-1) F(z).$$

6. $$H(z) = \sum_{n=0}^{\infty} u(n) \cdot z^{-n} = \sum_{n=0}^{\infty} 1 \cdot z^{-n}$$

For convergence, $\left|\sum_{n=0}^{\infty} z^{-n}\right| < \infty$

or, $\quad |z^{-1}| < 1$

or, $\quad |z| > 1$

7. $\because \quad -a^n u[-n-1] \leftrightarrow \frac{z}{z-a} \qquad |z| < |a|$

$\therefore \quad -5^n u[-n-1] \leftrightarrow \frac{z}{z-5} \qquad |z| < |5|$

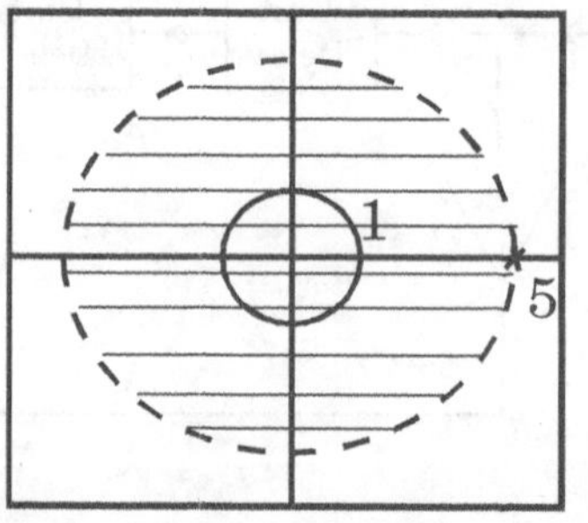

$\because$ ROC contains unit circle. Hence system is stable.

8. Output $\quad Y(z) = H(z)\,X(z)$

$$= 2(z^4 + z^2 - 2z + 2 - 3z^{-4})z^{-3}$$
$$= 2(z + z^{-1} - 2z^{-2} + 2z^{-3} - 3z^{-7})$$

Taking inverse z-transform, we have

$$y(n) = 2\,[\delta(n+1) + \delta(n-1) - 2\,\delta(n-2) + 2\,\delta(n-3) - 3\delta(n-7)]$$

At $n = 4$, $\quad y(4) = 0$

9. Using the following transform pair,

$$-A_k\,(d_k)^n\,u[-n-1] \xleftrightarrow{z} \frac{A_k}{1-d_k z^{-1}}$$

with ROC : $|z| < d_k$

We have $\frac{z}{z-0.2} \xleftrightarrow{Z} -(0.2)^n\,u[-n-1].$

10. $$2y(z) = \alpha z^{-2}y(z) + \beta z^{-1}x(z) - 2x(z)$$

$$(2-\alpha z^{-2})\,y(z) = (\beta z^{-1}-2)\,x(z)$$

$$\frac{y(z)}{x(z)} = \frac{(\beta z^{-1}-2)}{(2-\alpha z^{-2})}$$

For system to be stable, the ROC should include unit circle.

$\Rightarrow \quad 2-\alpha z^{-2} > 0$

$\Rightarrow \quad 2 > \alpha z^{-2}$

$\Rightarrow \quad z > \sqrt{\frac{\alpha}{2}}$

$\beta z^{-1} > 2$

$$\frac{\beta}{2} > z > \sqrt{\frac{\alpha}{2}}$$

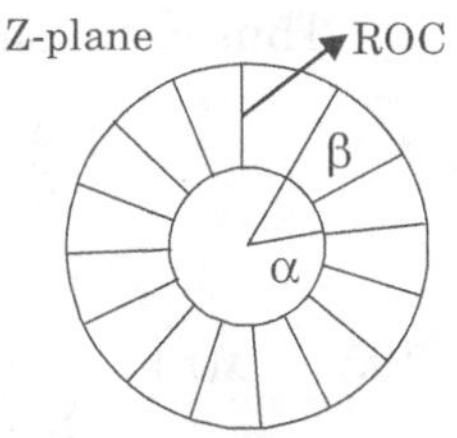

Hence $|\alpha| < 2$ and $|\beta|$ for any value.

11. $$x(n) = \left(\frac{5}{6}\right)^n u(n) - \left(\frac{6}{5}\right)^n u(-n-1)$$

$$X(z) = \sum_{n=-\infty}^{\infty}\left(\frac{5}{6}\right)^n u[n]\,z^{-n} - \sum_{n=-\infty}^{\infty}\left(\frac{6}{5}\right)^n u[-n-1]\cdot z^{-n}$$

$$= \sum_{n=0}^{\infty}\left(\frac{5}{6}\right)^n z^{-n} - \left(1+\sum_{n=-\infty}^{\infty}\left(\frac{6}{5}\right)^n u(-n)\,z^{-n}\right)$$

$$= \sum_{n=0}^{\infty}\left(\frac{5}{6}\right)^n z^{-n+1} + L - \sum_{n=0}^{\infty}\left(\frac{6}{5}\right)^{-n} z^n$$

The first term will converge when $\left|\frac{5}{6}z^{-1}\right| < 1$

or, $|z| > \frac{5}{6}$

The second term will converge, when, $\left|\frac{6}{5}z^{-1}\right| > 1$

or, $\quad |z| < \frac{6}{5}$

Thus, the region of convergence is

$$\frac{5}{6} < |z| < \frac{6}{5}$$

12. For **right sided** exponential sequence,

$$x_1[n] = a^n u[n]$$

$$X_1[z] = \sum_{n=-\infty}^{\infty} a^n u[n]\,z^{-n} = \sum_{n=0}^{\infty}(az^{-1})^n$$

Convergence requires that $\sum_{n=-0}^{\infty}|az^{-1}|^n < \infty$

This is possible only or, if $|az^{-1}| < 1$

or, $\quad |z| > |a|$

For **left-sided** exponential sequence

$$x_2[n] = b^{-n}\,u[-n-1]$$

$$X_2[z] = \sum_{n=-\infty}^{\infty} -b^n u[-n-1]\,z^{-n} = 1-\sum_{n=0}^{\infty}(b^{-1}z)^n$$

$X_2(n)$ converges only if $|b^{-1}z| < 1$ or, $|z| < |b|$

Hence, $x_1[n] + x_2[n]$ have ROC given by

$$\frac{1}{3} < |z| < \frac{2}{3},\ a = \frac{1}{3},\ b = \frac{2}{3}$$

For, $\quad x_1[n] - x_2[n];$

$$X_3[z] = X_1[z] - X_2[z] = \sum_{n=0}^{\infty}(az^{-1})^n - 1 + \sum_{n=0}^{\infty}(b^{-1}z)^n$$

$X_2[z]$ will be converge for $|az^{-1}| < 1$ and $|b^{-1}z| < 1$

or, $|z| < |b|$ and $|z| > |a|$

ROC remains same $\frac{1}{3} < |z| < \frac{2}{3}$

13. For left- handed signal,

$$X[z] = -\sum_{n=-\infty}^{\infty} a^n u(-n-1)z^{-n} = -\sum_{n=-\infty}^{-1} a^n z^{-n}$$

$$= -\sum_{n=1}^{\infty}(a^{-1}z)^n = 1-\sum_{n=0}^{\infty}(a^{-1}z)^n$$

as $\sum_{n=0}^{\infty}(a^{-1}z)^n = 1 + a^{-1}z + (a^{-1}z)^2 + \ldots$

$$= \frac{1}{1-a^{-1}z}$$

then, $\quad X[z] = 1-\frac{1}{1-a^{-1}z} = \frac{1}{1-az^{-1}}$

and ROC is given by $|z| < |a|$

Here $\quad a = 2,$

then, ROC: $|z| < 2$ includes unit circle

Hence, $\quad x[n] = -0.5[z]^n u[-n-1]$

and $\quad x[0] = 0$

14. $$x(n) = \left(\frac{1}{3}\right)^n u(n) - \left(\frac{1}{2}\right)^n u(-n-1)$$

Let, $\quad x_1(n) = \left(\frac{1}{3}\right)^n u[n]$

$$x_1[z] = \sum_{n=-\infty}^{\infty}\left(\frac{1}{3}\right)^n u(n)_z^{-n}$$

$$= -\sum_{n=0}^{\infty}\left(\frac{1}{3}z^{-1}\right)^n = \frac{1}{1-\frac{1}{3}z^{-1}}$$

$X_1[z]$ will converge where $\left|\frac{1}{3}z^{-1}\right| < 1$ or, $|z| > \frac{1}{3}$

and, $\quad x_2[n] = -\left(\frac{1}{2}\right)^n u(-n-1)$

$$X_2[z] = -\sum_{n=-\infty}^{\infty}\left(\frac{1}{2}\right)^n u(-n-1)_z^{-n} - \sum_{n=-\infty}^{\infty}\left(\frac{1}{2}\right)^n z^{-n}$$

$$= -\sum_{n=1}^{\infty}\left(\frac{1}{2}z^{-1}\right)^{-n} = -\sum_{n=1}^{\infty}(2z)^n$$

$$= 1-\sum_{n=0}^{\infty}(2z)^n$$

For convergence of $X_2(z)$, $|2z| < 1$ or $|z| < \frac{1}{2}$

Hence, ROC for $x(n)$ should be in the range of $\frac{1}{3} < |z| < \frac{1}{2}$

15. From, z-transform property of

$$z^{-m} \leftrightarrow \delta[n-m]$$

We has $\quad X(z) \leftrightarrow 5\delta[n+2]+4\delta[n-1]+3\delta[n]$

16. A discrete-time LT1 system is B1B0 (Bounded input bounded output) stable if and only if its impulse response $h[n]$ is absolutely summable, that is

$$\sum_{n=-\infty}^{\infty}|h[n]| < \infty$$

Now, $\quad H(z) = \sum_{n=-\infty}^{\infty} h[n]z^{-n}$

let z $= e^{j\Omega}$ so that

$$|z| = |e^{-j\Omega}| = 1 \text{ then}$$

$$|H(e^{j\Omega})| = \frac{N_0}{2} = \sum_{n=-\infty}^{\infty}|h[n]| < \infty$$

It can be inferred that if the system is stable, then $H(z)$ converges for $z = e^{j\Omega}$. hence, for a stable discrete-time LT1 system, the ROC of $H(z)$ must contain the unit circle $|z| = 1$.

Further, for the system having impulse response

$$h[n] = \alpha^n u[n],\ H(z) = \frac{z}{z-\alpha},\quad |z| > |\alpha|$$

For $|z| > |\alpha|$, ROC of $H(z)$ includes $z = \infty$, therefore $h(n)$ is a causal sequence and thus the system is causal.

Now, Given, $\quad H(z) = \dfrac{2-\frac{3}{4}z^{-1}}{1-\frac{3}{4}z^{-1}+\frac{1}{8}z^{-2}}$

$$= \frac{1}{1-\frac{1}{4}Z^{-1}} + \frac{1}{1-\frac{1}{2}Z^{-1}}$$

For, S1 : ROC, $|z| > \frac{1}{2}$, $H(z)$ contain unit circle and includes $z = \infty$, the system is stable and causal.

For, S2 : ROC $|z| < \frac{1}{4}$, $H(z)$ does not contain unit circle and excludes $z = \infty$, the system is neither stable nor causal.

For, S3 : ROC $\frac{1}{4} < |z| < \frac{1}{2}$, $H(z)$ does not contain unit circle and excludes $z = \infty$, the system is neither stable nor causal.

17. $\quad x[n] = (1/3)^{|n|} - \left(\frac{1}{2}\right)^n u[n]$

For $(1/3)^{|n|}$, $\quad$ ROC is $\frac{1}{3} < |z| < 3$

For $(1/2)^{|n|}u[n]$, $\quad$ ROC is $|z| > \frac{1}{2}$

Thus common ROC is $\frac{1}{2} < |z| < 3$

19. $\quad x(n) = \underbrace{\left[-\frac{1}{9}\right]4(x)}_{\text{Right side Signal}} - \underbrace{\left(\frac{1}{3}\right)^n 4(-n-1)}_{\text{Left side signal}}$

ROC is $|z| > \frac{1}{9}$ $\qquad$ ROC is $|z| < \frac{1}{3}$

So ROC is $\frac{1}{3} > |z| > \frac{1}{9}$

20. Given $\quad x[n] = x[-n]$

$\Rightarrow \quad x(z) = x(z^{-1})$

[Time reversal property in z – transform]

$\Rightarrow$ if one zero is $0.5 + j0.25$

then other zero will be $\dfrac{1}{0.5+j0.25}$

21. Given system equation as

$$y[n] = \alpha y[n-1] + \beta x[n]$$

$$\Rightarrow \quad \frac{y(z)}{x(z)} = \frac{\beta}{1-\alpha z^{-1}}$$

$$\Rightarrow \quad H(z) = \frac{\beta}{1-\alpha z^{-1}}$$

$h[n] = \beta(\alpha)^h u[n]$ [causal system]

Also given that $\sum_{h=0}^{\infty} h[n] = 2$

$$\beta\left[\frac{1}{1-\alpha}\right] = 2$$

$$1-\alpha = \frac{\beta}{2}$$

$$\alpha = 1-\frac{\beta}{2}$$

22. $X(z) = \dfrac{1}{(1-2z^{-1})^2} = \dfrac{1}{(1-2z^{-1})}\dfrac{1}{(1-2z^{-1})}$

$$x[n] = 2^n u[n] * 2^n u[n]$$

$$x[n] = \sum_{k=0}^{n} 2^k . 2^{(n-k)}$$

$$\Rightarrow x[2] = \sum_{k=0}^{2} 2^k . 2^{(2-k)} \; 2^0.2^2 + 2^1.2^1 + 2^2.2^0$$

$$= 2^0.2^2 + 2^1.2^1 + 2^2.2^0$$

$$= 4 + 4 + 4 = 12$$

24. $$y[n] = \sum_{m=0}^{n} x[m]$$

According to accumulation property of z-transform,

$$Y[z] = \frac{X(z)}{(1-z^{-1})}$$

$$\Rightarrow \quad \frac{2}{z(z-1)} = \frac{zX(z)}{(z-1)}$$

$$\therefore \quad X[z] = \frac{2z^{-2}}{(z-1)} = \frac{2z^{-3}}{(1-z^{-1})}$$

$$\therefore \quad x[n] = 2u[n-3]$$

thus $x[2] = 0$

25. Since $x[n]$ in absolutely summable thus its ROC must include unit circle.

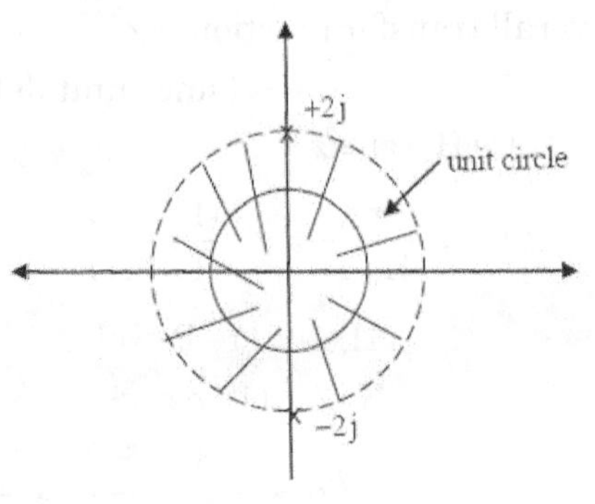

Thus ROC must be inside the circling radius 2. $x[n]$ must be a non-causal signal.

26. Given sequence

$$x(n) = (a)^n x(n) + (b)^n x(n),$$

Also given $0 < |a| < |b| < 1$

$\therefore$ The region of convergence (ROC)

$$= (|z| > |a|) \cap (|z| > |b|)$$

$$= |z| > |b|$$

27. $x(y) = (2)^n u(n) + \left(\frac{1}{2}\right)^n u(-n-1)$

ROC = $(|z| > 2) \cap (|z| > ½) = \phi$

So, the ROC of z-tansform is null.

28. Here $(a)^n x(n) \leftrightarrow X(z/a)$

$$a = -1$$

$$(-1)^n x(n) \leftrightarrow X(-z)$$

but $x(n) = \delta[n-3] + 2\delta[n-5]$

$$y(n) = (-1)^n x(n)$$

$$= (-1)^n [\delta(n-3) + 2\delta[n-5]$$

$$\therefore \quad y(n) = -\delta(n-3) - 2\delta(n-5)$$

$$= -x(n)$$

Hence, the value of signal $y(n)$ is $-x(n)$.

29. It is given that H(z) is z-transform of a real-valued signal h(n).

$P(z) = H(z)H\left(\frac{1}{z}\right)$ and P(z) has 4 zeros $\Rightarrow$ P(z) are,

sum of zeros of H(z) and zeros of $H\left(\frac{1}{z}\right)$.

If $z_1 = \frac{1}{2} + j\frac{1}{2}$ is one zero then there must be a zero at $z_1^* = \frac{1}{2} - j\frac{1}{2}$

Let z_1, z_1^* represent zeros of H(z) then the zeros of $H\left(\frac{1}{z}\right)$ will be $\frac{1}{z_1}$ and $\left(\frac{1}{z_1}\right)^*$

$$\frac{1}{z_1} = \frac{1}{\frac{1}{2} + j\frac{1}{2}} = 1 - i$$

$$\left(\frac{1}{z_1}\right)^* = 1 + j$$

Hence zeros of P(z) are $\left(\frac{1}{2} \pm j\frac{1}{2}\right)$ and $(1 \pm j)$ or

$[0.707 \angle \pm 45°]$ and $[\sqrt{2} \angle \pm 45°]$

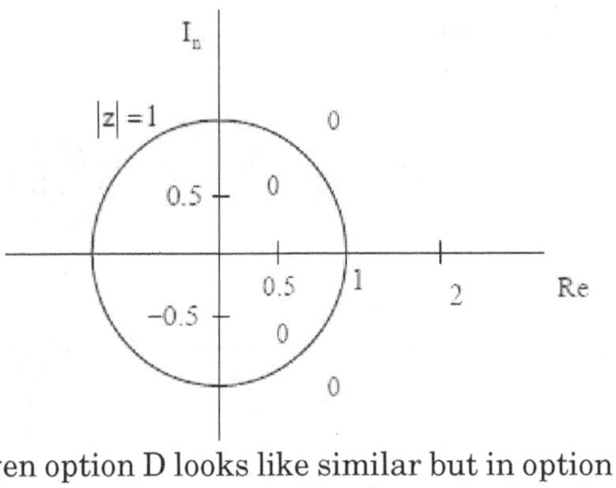

Even option D looks like similar but in option B, the zeros that are outside the unit circle have real part 2, but we need 1.

30.

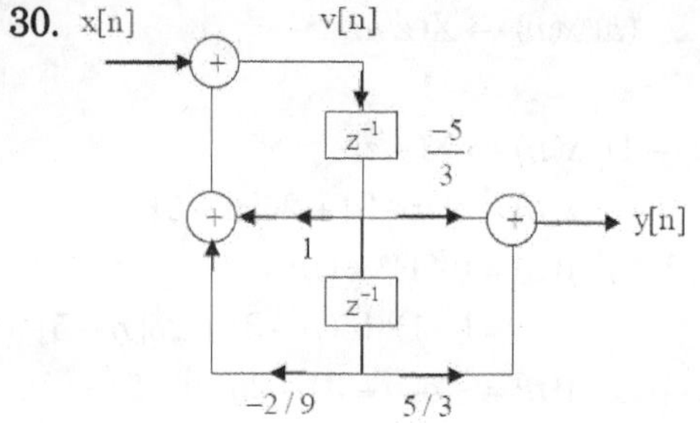

From the figure

$$v[n] = x[n] + v[n-1] - \frac{2}{9}v[n-2]$$

$$y[n] = -\frac{5}{3}v[n-1] + \frac{5}{3}v[n-2]$$

$$V(z) = \left[1 - z^{-1} + \frac{2}{9}z^{-2}\right] = X(z)$$

$$\Rightarrow \quad \frac{V(z)}{X(z)} = \frac{1}{1 - z^{-1} + \frac{2}{9}z^{-2}} \quad ...(i)$$

$$\frac{Y(z)}{V(z)} = \frac{-5}{3}z^{-1} + \frac{5}{3}z^{-2} \quad ...(ii)$$

Multiplying (*i*) and (*ii*) we get

$$\Rightarrow \quad \frac{Y(z)}{X(z)} = \frac{-\frac{5}{3}z^{-1}[1 - z^{-1}]}{1 - z^{-1} + \frac{2}{9}z^{-2}}$$

For unit step response,

$$X(z) = \frac{1}{1 - z^{-1}}$$

$$\Rightarrow \quad Y(z) = \frac{-\frac{5}{3}z^{-1}}{1 - z^{-1} + \frac{2}{9}z^{-2}}$$

$$= \frac{A}{1 - \frac{1}{3}z^{-1}} + \frac{B}{1 - \frac{2}{3}z^{-1}}$$

On solving,

A = 5; B = –5

$$\Rightarrow \quad y[n] = 5\left(\frac{1}{3}\right)^n u[n] - 5\left(\frac{2}{3}\right)^n u[n]$$

31.

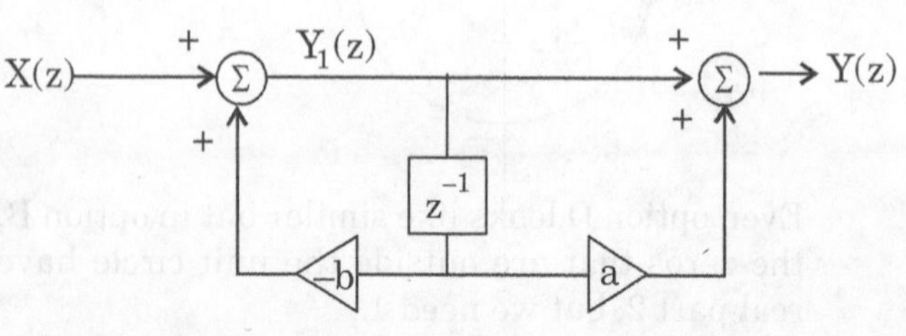

$$\Rightarrow y_1(z) = X(z) - bz^{-1}.y_1(z)$$

$$y_1(z)[1 + bz^{-1}] = x(z)$$

$$y1(z) = = \frac{x(z)}{1 + bz^{-1}}$$

$$\Rightarrow \quad y(z) = y_1(z) + az^{-1}.\, y_1(z)$$

$$y(z) = y_1(z)\,[1 + az^{-1}]$$

$$y(z) = \frac{x(z)}{[1 + bz_{-1}]}.[1 + az^{-1}]$$

Transfer function, H(z)

$$= \frac{y(z)}{x(z)} = \frac{1 + az^{-1}}{1 + bz^{-1}}$$

32.

$$h_1[n] = \delta(n-1) \qquad h_1(z) = z^{-1}$$

$$h_2(n) = \delta(n-2) \qquad h_2(z) = z^{-2}$$

Hence in cascade, overall z-transform of impulse response,

$$H(z) = h_1(z) \cdot h_2(z)$$

$$= z^{-1} \cdot z^{-2} = z^{-3}$$

$$\therefore \quad h(n) = \delta(n-3)$$

33. Overall transfer function = z^{-1}

(since unit delay T.F = z^{-1})

$$H_1(z)\, H_2(z) = z^{-1}$$

$$H_2(z) = \frac{z^{-1}}{H_1(z)} = z^{-1}\frac{(1 - 0.6z^{-1})}{(1 - 0.4z^{-1})}$$

34.

$$H_1(z) = (1 - Pz^{-1})^{-1}$$

$$H_2(z) = (1 - qz^{-1})^{-1}$$

$$H(z) = \frac{1}{1 - Pz^{-1}} + r\frac{1}{(1 - qz^{-1})}$$

$$= \frac{1 - qz^{-1} + r(1 - Pz^{-1})}{(1 - Pz^{-1})(1 - Pz^{-1})}$$

$$= \frac{(1 + r) - (q + rp)z^{-1}}{(1 - Pz^{-1})(1 - Pz^{-1})}$$

zero of $\quad H(z) = \dfrac{q + rp}{1 + r}$

Since zero is existing on unit circle

$$\Rightarrow \quad \frac{q + rp}{1 + r} = 1$$

or

$$\frac{q + rp}{1 + r} = -1$$

$$\frac{-\frac{1}{4} + \frac{r}{2}}{1 + r} = 1$$

or $$\frac{-\frac{1}{4}+\frac{r}{2}}{1+r}=-1$$

$$-\frac{1}{4}+\frac{r}{2}=1+r$$

or $$-\frac{1}{4}+\frac{r}{2}=-1-r$$

$\Rightarrow$ $$r=-\frac{5}{2}$$

$\Rightarrow$ $$\frac{r}{2}=-\frac{5}{4}$$

or $$\frac{3}{4}=\frac{-3r}{2}$$

$$r=-\frac{1}{2}$$

$\Rightarrow$ $$r=-0.5$$

$r=-\frac{5}{2}$ is not possible

35.

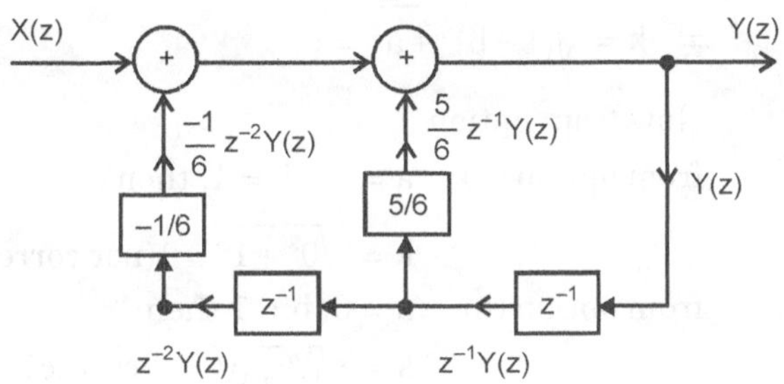

$$X(z)-\frac{1}{6}z^{-2}Y(z)+\frac{5}{6}z^{-1}Y(z)=Y(z)$$

$$Y(z)-\frac{5}{6}z^{-1}Y(z)+\frac{1}{6}z^{-2}Y(z)=X(z)$$

$$Y(z)\left\{1-\frac{5}{6}z^{-1}+\frac{1}{6}z^{-2}\right\}=X(z)$$

T.F. of the system, $$H(z)=\frac{Y(z)}{X(z)}=\frac{1}{\left(1-\frac{5}{6}z^{-1}+\frac{1}{6}z^{-2}\right)}$$

$$H(z)=\frac{z^2}{\left(z^2-\frac{5}{6}z+\frac{1}{6}\right)}$$

$$H(z)=\frac{z^2}{\left(z-\frac{1}{2}\right)\left(z-\frac{1}{3}\right)}$$

Pole location :

$$z-\frac{1}{2}=0 \quad \Rightarrow z=\frac{1}{2}$$

$$z-\frac{1}{3}=0 \quad \Rightarrow z=\frac{1}{3}$$

36. Minimum phase system has all zeros inside unit circle maximum phase system has all zeros outside unit circle mixed phase system has some zero outside unit circle and some zeros inside unit circle.

for $$H(s)=1+\frac{7}{2}z^{-1}+\frac{3}{2}z^{-2}$$

One zero is inside and one zero outside unit circle hence mixed phase system

37.

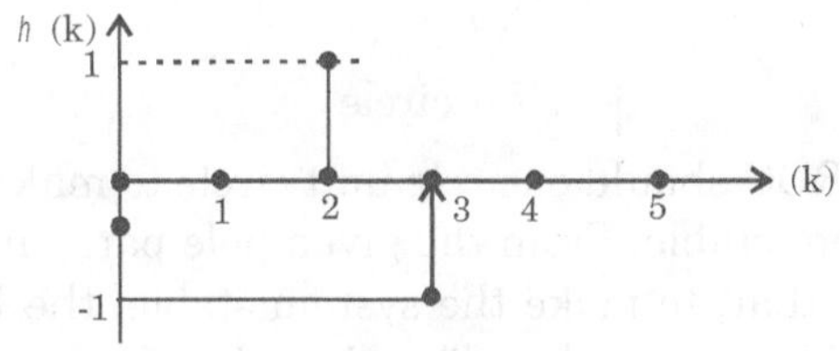

$\therefore$ $H(e^{j\omega})=e^{-j2\omega}-e^{-j3\omega}$

So, it is FIR high pass filter.

38. For an all pass system,

$$\text{pole}=\frac{1}{\text{zero}^*}$$

or $$\text{zero}=\frac{1}{\text{pole}^*}$$

$$\text{pole}=a$$

$$\text{zero}=\frac{1}{b}$$

$\Rightarrow$ $$\frac{1}{b}=\frac{1}{a^*}$$

or $$b=a^*$$

39. Given $y[n]=\sum_{i=0}^{2}\alpha_i x(n-i)$

$\Rightarrow$ $y[n]=\alpha_0 x[n]+\alpha_1 x[n-1]+\alpha_2 x[n-2]+\alpha_3 x[n-3]$

Getting a null at zero frequency implies that given filter can be high pass filter but it cannot be low pass filter.

High pass filter is possible if we have negative coefficients.

Let say, $\alpha_1=\alpha_2=0,\ \alpha_0=-\alpha_3$

$\Rightarrow$ $$y[n]=-\alpha_3 x[n]+a_3 x[n-3]$$

$$H(z)=-\alpha_3[1-z^{-3}]$$

$\Rightarrow$ $$H(e^{j\Omega})=-\alpha_3[1-e^{-j3\Omega}]$$

$$=-\alpha_3 e^{\frac{-j3\Omega}{2}}\left[\frac{e^{\frac{j3\Omega}{2}}-e^{-j\frac{3\Omega}{2}}}{2j}\right]\times 2j$$

$$=-\alpha_3 2j\sin\left(\frac{3\Omega}{2}\right).e^{-j\frac{3\Omega}{2}}$$

$$=-\alpha_3 2.\sin\frac{3\Omega}{2}.e^{-j\frac{3\Omega}{2}}.e^{j\frac{\pi}{2}}$$

$\Rightarrow$ $$H(e^{j\Omega})|_{\Omega=0}=0$$

In other cases it in not possible.

40.

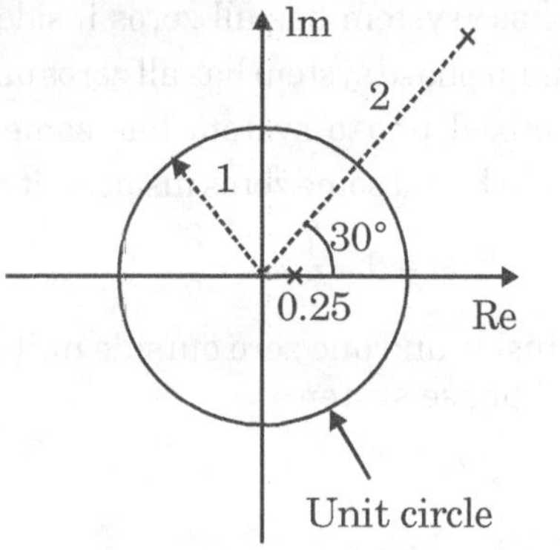

The ROC should encircle unit circle to make the system stable. From the given pole pattern it is clear that, to make the system stable, the ROC should be two-sided. Thus impulse response for the system should be two-sided.

41. Given : $h[n] = \frac{1}{3}\delta[n] + \frac{1}{3}\delta[n-1] + \frac{1}{3}.\delta[n-2]$.

$$H(e^{j\omega}) = \frac{1}{3}f^{j\omega}[1 + 2\cos\omega]$$

$$H(e^{j\omega}) = 0 \Rightarrow 1 + 2\cos\omega_0 = 0$$

$$\Rightarrow \cos\omega_0 = \frac{-1}{2}; \omega_0 = \frac{2\pi}{3} = 2.10\text{rad.}$$

42. Given : h [n] = 5 [n] – 7δ [n – 1] + 7δ [n – 3] – 5δ [n – 4].

$H(e^{j\omega}) = 5 - 7 \,.\, e^{-j\omega} + 7e^{-3j\omega} - 5e^{-4j\omega}$

Now, for ω = 0, $H(e^{j0}) = 5 - 7 \times 1 + 7 \times 1 - 5 \times 1 = 0$.

For ω = π, H(ejπ) = 5 – 7 (–1) + 7 (–1) + 5 (–1) = 5 + 7 – 7 – 5 = 0.

System is attenuating low and high frequencies whereas passing the mid frequencies. So, it is a bandpass filter.

43. It is given that

h(n) = [1, a, b]

and $x(n) = C_1e^{-j\frac{\pi}{2}n} + C_2e^{j\frac{\pi}{2}n}$

$\Rightarrow y(n) = 0$

Now If $h(n) = [1, a, b]$

$\therefore\ H(e_{j\omega}) = 1 + ae^{-j\omega} + be^{-j2\omega}$

→ When $x(n) = C_1e^{-j\frac{\pi}{2}n} + C_2e^{j\frac{\pi}{2}n}$ then expression of y(n)

Since the input x(n) contain 2 frequencies $\pm\frac{\pi}{2}$,

Let evaluate $|H(e^{j\omega})|$ at f nil 2 frequency

$$H(e^{j-\pi/2}) = 1 + ae^{-j\left(\frac{-\pi}{2}\right)} + be^{-j2\left(\frac{-\pi}{2}\right)}$$

$$= 1 + ae^{+j\pi/2} + be^{j\pi}$$

$$= 1 + [a(j)] + [b(-1)]$$

$$= (1 - b) + j(a)$$

$$\therefore\quad H(e^{j\pi/2}) = (1 - b) - ja$$

$$|H(e^{j\pi/2})| = \left|H\left(e^{-j\frac{\pi}{2}}\right)\right| = \sqrt{(1-b)^2 + a^2}$$

So the expression of y(n) is

$$y(n) = \left[(1-b)^2 + a^2\right]^{1/2} C_1e^{-j\left(\frac{1}{2}n+\phi_1\right)} + \left[(1-b)^2 + a^2\right]^{1/2} C_2e^{j\left(\frac{\pi}{2}n+\phi_2\right)}$$

for y(n) = 0

$$\Rightarrow k = \sqrt{(1-b)^2 + a^2} = 0$$

Now from option

from option (a) a = –1, b = 1, then

$k = \sqrt{0^2 + 1^2} = 1$(not correct)

from option (b) a = 0, b = 1 then

$k = \sqrt{0^2 + 0^2} = 0$(correct)

from option (c) a = 1, b = 1, then

$k = \sqrt{0^2 + 1^2} = 1$(not correct)

from option (d) a = 0, b = 1, then

$k = \sqrt{2^2 + 0^2} = 2$(not correct)

44. It is given that

$$h(n) = [-1, -2, -3, 4, 3, 2, 1] \quad \text{...(i)}$$

$$g(n) = [a, \underset{\uparrow}{b}, c] \quad \text{...(ii)}$$

It is mentioned that

$E(h,g)]\int_{-x}^{\pi}|H(e^{j\omega}) - G(e^{j\omega})|^2 d\omega$, is minimised,

If h(n) and g(n) represent IDTFT of $H(e^{j\omega})$, $G(e^{j\omega})$

then $E(h, g) = 2\pi\Sigma|h(n) - g(n)|^2$ (By Parseval theorem)

$h(n) \leftrightarrow H(e^{j\omega})$

$g(n) \leftrightarrow G(e^{j\omega})$

Now, $h(n) - g(n) \leftrightarrow H(e^{j\omega}) - G(e^{j\omega})$

Energy of

$$[h(n) - g(n)] = \frac{1}{2\pi}\int_{-\pi}^{\pi}|H(e^{j\omega}) - G(e^{j\omega})|^2 d\omega$$

⇒ 2π energy of

$$\left[h(n)-g(n)\right] = \int_{-\pi}^{\pi}\left|H\left(e^{j\omega}\right)-G\left(e^{j\omega}\right)\right|^2 d\omega$$

$$\Rightarrow E(h, g) = \int_{-x}^{\pi}\left|H\left(e^{j\omega}\right)-G\left(e^{j\omega}\right)\right|^2 d\omega$$

$$= 2\pi\sum_{n=\infty}^{\infty}\left|h(n)-g(n)\right|^2$$

We want to minimize E (h, g)

Using equation (i) and equation (ii) we get

h(n) – g(n) = [–1, –2, –3 – a, 4 – b, 3 – c, 2, 1]

$$\therefore\quad E(h,g) = 2\pi\Sigma\left|h(n)-g(n)\right|^2$$

$$= 2\pi\Sigma\left[h(n)-g(n)\right]^2$$

$$\therefore\quad E(h, g) = 2\pi[(-1)^2 + (-2)^2 + (-3 - a)^2 + (4 - b)^2 + (3 - c)^2 + 2^2 + 1^2]$$

$$= 2\pi[10 + (-3 - a)^2 + (4 - b)^2 + (3 - 1)^2]$$

To minimize the value of E(h,g)

Equate, –3 – a = 0 ⇒ a = –3

and 4 – b = 0 ⇒ b = 4

and 3 – c = 0 ⇒ c = 3

→ g(n) = [a, b, c] = [–3, 4, 3]

So 10g(–1) + g(1) = 10a + c

= (10(–3) + 3) = –30 + 3 = – 27

7 CHAPTER

DTFT, DFT & FFT

Analysis of Previous GATE Papers

Topics ↓		Year →	2019	2018	2017 Set 1	2017 Set 2	2016 Set 1	2016 Set 2	2016 Set 3	2015 Set 1	2015 Set 2	2015 Set 3
FOURIER TRANSFORM OF DISCRETE-TIME SIGNAL	1 Mark	MCQ Type										
		Numerical Type										
	2 Marks	MCQ Type										
		Numerical Type		1	1		1					
		Total		2	2		2					
DISCRETE-TIME FOURIER TRANSFORM	1 Mark	MCQ Type										
		Numerical Type										
	2 Marks	MCQ Type								1		
		Numerical Type						1			1	
		Total						2		2	2	
FAST FOURIER TRANSFORM	1 Mark	MCQ Type										
		Numerical Type										
	2 Marks	MCQ Type	1									
		Numerical Type							1			
		Total	2						2			

FOURIER TRANSFORM OF DISCRETE-TIME SIGNAL

1. A Fourier transform pair is given by

$$\left(\frac{2}{3}\right)^n u(n+3) \xleftrightarrow{\text{F.T.}} \frac{Ae^{+j6\pi t}}{1-\left(\frac{2}{3}\right)e^{-j2\pi f}},$$

when u[n] denotes the unit step sequence, the values of A is ______.

[2004 : 1 Mark]

2. Let $x(n) = \left(\frac{1}{2}\right)^n u(n)$, $y(n) = x^2(n)$ and $y(e^{j\omega})$ be the Fourier transform of y(n). Then $Y(e^{j0})$ is

(a) $\frac{1}{4}$ (b) 2

(c) 4 (d) $\frac{4}{3}$

[2005 : 1 Mark]

3. The Fourier transform of y(2n) will be

(a) $e^{-j2\omega} | \cos 4\omega + 2\cos 2\omega + 2]$

(b) $[\cos 2\omega + 2\cos\omega + 2]$

(c) $e^{-j\omega} [\cos 2\omega + 2\cos\omega + 2]$

(d) $e^{\frac{-j\omega}{2}} [\cos 2\omega + 2\cos\omega + 2]$

[2005 : 2 Marks]

4. A signal $x(n) = \sin(\omega_0 n + f)$ is the input to a linear time-invariant system having a frequency response $H(e^{j\omega})$. If ihe output of the system $Ax(n - n_0)$, then the most general form of $\angle H(e^{j\omega})$ will be

(a) $-n_0\omega_0 + \beta$ for any arbitrary real β

(b) $-n_0\omega_0 + 2\pi k$ for any arbitrary integer k.

(c) $n_0\omega_0 + 2\pi k$ for any arbitrary integer k.

(d) $-n_0\omega_0\phi$

[2005 : 2 Marks]

5. Consider the signal

$x[n] = 6\delta[n + 2] + 3\delta[n + 1] + 8\delta[n] + 7\delta[n - 1] + 4\delta[n - 2]$. If $X(e^{j\omega})$ is the discrete-time Fourier transform of x[n],

then $\frac{1}{\pi}\int_{-\pi}^{\pi} X(e^{j\omega})\sin^2(2\omega)d\omega$ is equal to ______

[2016 : 2 Marks, Set-1]

6. Let h[n] be the impulse response of a discrete-time linear time invariant (LTI) filter. The impulse response is given by

$$h[0] = \frac{1}{3};\ h[1] = \frac{1}{3};\ h[2] = \frac{1}{3}$$

and h[n] = 0 for n < 0 and n > 2.

Let $H(\omega)$ be the discrete-time Fourier transform (DTFT) of h[n], where to is the normalized angular frequency in radians. Given that $H(\omega_0) = 0$ and $0 < \omega_0 < \pi$, the value of ω_0 (in radians) is equal to ______.

[2017 : 2 Marks, Set-1]

7. Let $X[k] = k + 1, 0 \le k \le 7$ be 8-point DFT of a sequence x[n], where $X[k]= \sum_{n=0}^{N-1} x[n]e^{-j2\pi nk/N}$. The value (correct to two decimal places) of $\sum_{n=0}^{3} x[2n]$ is ______.

[2018 : 2 Marks]

DISCRETE-TIME FOURIER TRANSFORM

8. A 5-point sequence x[n] is given as

$x[-3] = 1, x[-2] = 1, x[-1] = 0, x[0] = 5, x[1] = 1$. Let $X(e^{j\omega})$ denote the discrete-time Fourier transform of x[n]. The value of $\int_{-\pi}^{\pi} X(e^{j\omega})d\omega$

(a) 5 (b) 10π

(c) 16π (d) $5 + j10\pi$

[2007 : 2 Marks]

9. {a(n)} is a real-valued periodic sequence with a period N. x(n) and X(k) form N-point Discrete Fourier Transform (DFT) pairs. The DFT Y(k) of the sequence

$y(n) = \frac{1}{N}\sum_{r=0}^{N-1} x(r)x(n+r)$ is

(a) $|X(k)|^2$

(b) $\frac{1}{N}\sum_{r=0}^{N-1} X(r)X'(k+r)$

(c) $\frac{1}{N}\sum_{r=0}^{N-1} X(r)X(k+r)$

(d) 0

[2008 : 2 Marks]

10. The 4-point discrete Fourier Transform (DFT) of a discrete time sequence {1, 0, 2, 3} is

(a) {0, –2 + 2j, 2, –2 – 2j}.

(b) {2, 2 + 2j, 6, 2 – 2j}.

(c) {6, 1 – 3j, 2, 1 + 3j}.

(d) {6, –1 + 3j, 0, –1 – 3j}.

[2009 : 2 Marks]

11. The first five points of the 8-point DFT of a real valued sequence are 5, 1 – j3, 0, 3 – j4 and 3 + j4. The last two points of the DFT are respectively

(a) 0, 1 – j3 (b) 0, 1 + j3

(c)1 + j3, 5 (d) 1 – j3, 5

[2011 : 2 Marks]

12. The DFT of a vector [a b c d] is the vector [α β γ δ]. Consider the product

$$[p\ q\ r\ s]=[a\ b\ c\ d]\begin{bmatrix} a & b & c & d\\ d & a & b & c\\ c & d & a & b\\ b & c & d & a\end{bmatrix}$$

The DFT of the vector [p q r s] is a scaled version of

(a) $\left[\alpha^2\ \beta^2\ \gamma^2\ \delta^2\right]$

(b) $\left[\sqrt{\alpha}\ \ \sqrt{\beta}\ \ \sqrt{\gamma}\ \ \sqrt{\delta}\right]$

(c) $\left[\alpha+\beta\ \ \beta+\delta\ \ \delta+\gamma\ \ \gamma+\alpha\right]$

(d) $\left[\alpha\ \ \beta\ \ \gamma\ \ \delta\right]$

[2013 : 2 Marks]

13. The N-point DFTof a sequence x[n], $0 \le n \le N-1$ is given by

$$X[K]=\frac{1}{\sqrt{N}}\sum_{n=0}^{N-1}x[n]e^{-j\frac{2\pi}{N}nK},\ 0\le K\le N-1$$

Denote this relation as X = DFT(x). For N = 4, which one of the following sequences satisfies DFT (DFT (x)) = x.

(a) x = [1 2 3 r] (b) x = [1 2 3 2]

(c) x = [1 3 2 2] (d) x = [1 2 2 3]

[2014 : 2 Marks, Set-4]

14. Two sequences [a, b, c] and [A, B, C] are related as,

$$\begin{bmatrix}A\\B\\C\end{bmatrix}=\begin{bmatrix}1 & 1 & 1\\ 1 & W_3^{-1} & W_3^{-2}\\ 1 & W_3^{-2} & W_3^{-4}\end{bmatrix}\begin{bmatrix}a\\b\\c\end{bmatrix}\ \text{where, } W_3=e^{i\frac{2\pi}{3}}.$$

If another sequence [p, q, r] is derived as,

$$\begin{bmatrix}a\\b\\c\end{bmatrix}=\begin{bmatrix}1 & 1 & 1\\ 1 & W_3^{1} & W_3^{2}\\ 1 & W_3^{2} & W_3^{4}\end{bmatrix}\begin{bmatrix}1 & 0 & 0\\ 0 & W_3^{2} & 0\\ 0 & 0 & W_3^{4}\end{bmatrix}\begin{bmatrix}A/3\\B/3\\C/3\end{bmatrix}$$

then the relationship between the sequences [p, q, r] and [a, b, c] is

(a) [p, q, r] = [b, a, c]

(b) [p, q, r] = [b, c, a]

(c) [p, q, r] = [c, a, b]

(d) [p, q, r] = [c, b, a]

[2015 : 2 Marks, Set-1]

15. Consider two real sequences with time-origin marked by the bold value,

$x_1[n]$ = {1, 2, 3, 0}, $x_2[n]$ = {1, 3, 2, 1}.

Let $X_1(k)$ and $X_2(k)$ be 4-point DFTs of $x_1[n]$ and $x_2[n]$, respectively.

Another sequence $x_3[n]$ is derived by taking 4-point inverse DFT of $X_3(k) = X_1(k)X_2(k)$. The value of $x_3[2]$ is _______.

[2015 : 2 Marks, Set-2]

16. The Discrete Fourier Transform (DFT) of the 4-point sequence

x[n] = {x[0], x[1], x[2], x[3]} = {3, 2, 3, 4) is

X[k] = {X[0], X[1], X[2], X[3]} = {12, 2j, 0, –2j}.

If $X_1[k]$ is the DFT of the 12-point sequence

$x_1[n]$ = {3, 0, 0, 2, 0, 0, 3, 0, 0, 4, 0, 0), the value of $\left|\frac{X_1(8)}{X_1[11]}\right|$ is _______.

[2016 : 2 Marks, Set-2]

FAST FOURISE TRANSFORM

17. For an N-point FFT algorithm with $N = 2^m$, which one of the following statements is TRUE?

(a) It is not possible to construct a signal flow graph with both input and output in normal order

(b) The number of butterflies in the mn state is $\frac{N}{m}$

(c) In-place computation requires storage of only 2N node data

(d) Computation of a butterfly requires only one complex multiplication

[2010 : 1 Mark]

18. A continuous-time speech signal $x_a(t)$ is sampled at a rate of 8 kHz and the samples are subsequently grouped in blocks, each of size N. The DFT of each block is to be computed in real time using the radix-2 decimation-in-frequency FFT algorithm. If the processor performs all operations sequentially, and takes 20 ps for computing each complex multiplication (including multiplications by 1 and -1) and the time required for addition/ subtraction is negligible, then the maximum value of N is _______.

[2016 : 2 Marks, Set-3]

19. Consider a six-point decimation-in-time Fast Fourier Transform (FFT) algorithm, for which the signal-flow graph corresponding to X[I] is shown in the figure. Let $W_6 = \exp\left(-\frac{j2\pi}{6}\right)$. In the figure, what should be the values of the coefficients a_1, a_2, a_3 in terms of W_6 so that X[I] is obtained correctly?

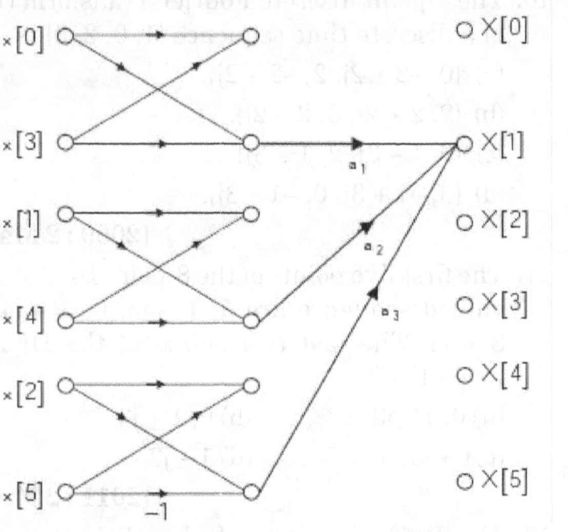

(A) $a_1 = 1, a_2 = W_6^2, a_3 = W_6$

(B) $a_1 = -1, a_2 = W_6^2, a_3 = W_6$

(C) $a_1 = -1, a_2 = W_6, a_3 = W_6^2$

(D) $a_1 = 1, a_2 = W_6, a_3 = W_6^2$

[2019 : 2 Marks]

ANSWERS

1. (3· 375)	2. (d)	3. (c)	4. (b)	5. (8)	6. (2· 10 radians)	7. (3)	8. (b)	9. (a)
10. (d)	11. (a)	12. (a)	13. (b)	14. (c)	15. (11)	16. (6)	17. (d)	18. (8)
19. (d)								

EXPLANATIONS

1. $x[n] = \left(\frac{2}{3}\right)^n u[n+3]$

$$X(e^{j\Omega}) = \sum_{n=-3}^{\infty}\left(\frac{2}{3}\right)^n .e^{-j\Omega n} = \frac{\left(\frac{2}{3}\right)^{-3} .e^{-j3\Omega}}{1-\frac{2}{3}e^{-j\Omega}}$$

$$\Rightarrow \quad A = \left(\frac{3}{2}\right)^3 = \frac{27}{8} = 3.375$$

2. $x(n) = \left(\frac{1}{2}\right)^n u(n)$

$$\therefore \quad y(n) = \left(\frac{1}{2}\right)^{2n} u^2(n)$$

$$y(n) = \left(\frac{1}{2}\right)^{2n} u(n) = \left(\left(\frac{1}{2}\right)^2\right)^n u(n)$$

$$\therefore \quad y(n) = \left(\frac{1}{4}\right)^n u(n)$$

$$\therefore \quad y(z) = \frac{1}{1-\frac{1}{4}z^{-1}}$$

$$y(e^{j\omega}) = \frac{1}{1-\frac{1}{4}e^{-iw}}$$

$$y(e^0) = \frac{4}{3}$$

3. From the known $y(n)$,

$$y(2n) = x(n-1)$$

$\therefore$ Fourier transform

$$= \frac{1}{2}e^{j\omega} + 1 + 2e^{-j\omega} + e^{-2j\omega} + \frac{1}{2}e^{-3j\omega}$$

$$= e^{-j\omega}\left[\frac{1}{2}\left(e^{2j\omega} + e^{-2j\omega}\right) + 2 + \left(e^{j\omega} + e^{-j\omega}\right)\right]$$

$$= e^{-j\omega}\left[\cos 2\omega + 2 + 2\cos\omega\right]$$

4. As, $x(n-n_0) \xleftrightarrow{FT} e^{-j\omega_0 n_0} \times (e^{j\omega})$

then, $\quad Y(e^{j\omega}) = Ae^{-j\omega_0 n_0} \times (e^{j\omega})$

$$\Rightarrow \quad H(f\omega) = \frac{Y(e^{j\omega})}{X(e^{j\omega})} = Ae^{-j\omega_0 n_0} \quad ...(i)$$

Given, frequency reforms

$$H(e^{j\omega}) = |H(e^{j\omega})|\, e^{j\angle H(e^{j\omega})} \quad ...(ii)$$

Comparing equation. (i) with (ii),

$$A = |H(e^{j\omega})|$$

and $\quad \angle H(e^{j\omega}) = -\omega_0 n_0$

For LTI system, phase and frequency reform are periodic with 2π. The general form of $\angle H(e^{j\omega})$ is $-\omega_0 n_0 + 2\pi k$.

5. Plancheral's relation is given by

$$\frac{1}{2\pi}\int_{-\pi}^{\pi} X(e^{j\omega})Y(e^{j\omega})\,d\omega = \sum_{n=-\infty}^{\infty} x(n)y(n)$$

$$Y(e^{j\omega}) = \sin^2(2\omega) = \frac{1-\cos(4\omega)}{2}$$

$$= \frac{1}{2} - \frac{1}{4}e^{j4\omega} - \frac{1}{4}e^{-j4\omega}$$

$$y(n) = \frac{1}{2}\delta(n) - \frac{1}{4}\delta(n+4) - \frac{1}{4}\delta(n-4)$$

$$y(n) = \left\{-\frac{1}{4}, 0, 0, 0, \underset{\uparrow}{\frac{1}{2}}, 0, 0, 0, -\frac{1}{4}\right\}$$

$$x(n) = \left\{6, 3, \underset{\uparrow}{8}, 7, 4\right\}$$

$$\frac{1}{\pi}\int_{-\pi}^{\pi} X(e^{j\omega})\cdot Y(e^{j\omega})\,d\omega = 2\sum_{n=-\infty}^{\infty} x(n)y(n)$$

$$= 2\sum_{n=-\infty}^{\infty} x(n)y(n) = 2\times 8\times\frac{1}{2} = 8$$

6. Since $\quad h[n] = \frac{1}{3}\delta[n] + \frac{1}{3}\delta[n-1] + \frac{1}{3}\delta[n-2]$

$$\therefore \quad H(e^{j\omega}) = \frac{1}{3}e^{j\omega}[1 + 2\cos\omega]$$

$$\therefore \quad H(e^{j\omega_0}) = 0,$$

$$\therefore \quad (1 + 2\cos\omega_0) = 0$$

$$\therefore \quad \cos\omega_0 = -\frac{1}{2}$$

$$\therefore \quad \omega_0 = \frac{2\pi}{3} = 2.10 \text{ radians}$$

7. $X(k) = \{1, 2, 3, 4, 5, 6, 7, 8\}$

$$\sum_{n=0}^{3} x[2n] = x[0] + x[2] + x[4] + x[6]$$

$$= 4.5 - 0.5 - 0.5j - 0.5 - 0.5 + 0.5j$$

$$= 4.5 - 1.5 = 3$$

8. For discrete time Fourier transform (DTFT)

$$x(n) = \sum_{k=n} \frac{1}{n} X(e^{(jk\omega_0)})\ e^{jk\omega_0 n}$$

When lim $N \to \infty$,

$$x[n] = \frac{1}{2\pi} \int_{-\pi}^{\pi} X(e^{j\omega})\ e^{j\omega n}\, d\omega$$

Putting $n = 0$,

$$x[0] = \frac{1}{2\pi} \int_{-\pi}^{\pi} X(e^{j\omega})\ e^{j\omega 0}\, d\omega$$

$$\therefore \quad \int_{-\pi}^{\pi} X(e^{j\omega})\, d\omega = 2\pi\, x[0] = 2\pi \times 5 = 10\pi$$

9. Given: $$y(n) = \frac{1}{N} \sum_{r=0}^{N-1} x(r)\, x(n+r)$$

It is auto correlation

Hence $y(n) = r_{xx}(n)$

$$\therefore \quad r_{xx}(n) \xrightarrow[N]{DFT} |X(k)|^2$$

Alternately

$$X[n-r] = \sum_{n=0}^{N-1} x[n].W_N(N-r)n$$

$$\sum_{n=0}^{N-1} x(n).e^{-1}\left(\frac{2\pi}{N}\right)(N-r)n$$

$$\text{Now, } e^{-1}\left(\frac{2\pi}{N}\right)(N-r) = e^{-12\pi n}\, e^{1}\left(\frac{2\pi}{N}\right)^{m} = e^{1}\left(\frac{2\pi}{N}\right)^{m}$$

If x[n] is real, then x[n] = x[n]

$$\text{and } x[N-r] = \sum_{n=0}^{N-1} x[n].e^{j}\left(\frac{2\pi}{N}\right)^{rn}$$

$$= \left[\sum_{n=0}^{N-1} x[n].e^{-j}\left(\frac{2\pi}{N}\right)^{rn}\right] = X(k)$$

Hence, $Y[k] = [x(k)\{x^*[k]\} = |X(k)|^2$

10. 4 – point DFT of sequence [1, 0, 2, 3] is

$$\begin{bmatrix} 1 & 1 & 1 & 1 \\ 1 & -j & -1 & j \\ 1 & -1 & 1 & -1 \\ 1 & j & -1 & -j \end{bmatrix} \begin{bmatrix} 1 \\ 0 \\ 2 \\ 3 \end{bmatrix} =$$

$$\begin{bmatrix} 1 + 2 + 3 \\ 1 - 2 + j3 \\ 1 + 2 - 3 \\ 1 - 2 - j3 \end{bmatrix} = \begin{bmatrix} 6 \\ -1 + j3 \\ 0 \\ -1 - j3 \end{bmatrix}$$

$$\begin{bmatrix} A \\ B \\ C \end{bmatrix} = \begin{bmatrix} a+b+c+ \\ a+bW_3^{-1}+cW_3^{-2} \\ a+bW_3^{-2}+cW_3-1 \end{bmatrix}$$

$$\begin{bmatrix} p \\ q \\ r \end{bmatrix} = \begin{bmatrix} 1 & W_3^{2} & W_3^{1} \\ 1 & W_3^{1}W_3^{2} & W_3^{2}W_3^{1} \\ 1 & W_3^{2}W_3^{2} & W_3^{1}W_3^{1} \end{bmatrix}$$

$$\begin{bmatrix} \frac{(a+b+c)}{3} \\ \frac{(a+bW_3^{-1}+cW_3^{2})}{3} \\ \frac{(a+bW_3^{-2}+cW_3^{-1})}{3} \end{bmatrix}$$

where, $W_3^1 = e^{\frac{j2\pi}{3}}, W_3^2 = e^{\frac{j4\pi}{3}}$,

$W_3^1 = e^{\frac{-j2\pi}{3}}, W_3^{-2} = e^{\frac{-j4\pi}{3}}$,

$W_3^{-4} =, W_3^{-1} = e^{\frac{-j2\pi}{3}}$.

$$\Rightarrow \begin{bmatrix} p \\ q \\ r \end{bmatrix} = \begin{bmatrix} c \\ a \\ b \end{bmatrix}$$

11. $$f(t) = e^{-t}u(t)$$

$$F(s) = L[f(t)] = L[e^{-t}u(t)] = \frac{1}{s+1}$$

In frequency domain, $s = j\omega$

$$F(j\omega) = \frac{1}{(1+j\omega)} = \frac{-1}{\sqrt{1+\omega^2}} \angle \tan^{-1}\omega$$

For 3 dB bandwidth,

$$\frac{1}{\sqrt{1+\omega^2}} = \frac{1}{\sqrt{2}}$$

$\therefore \quad \omega = +1$

$\omega = 2\pi f_c$

$$\Rightarrow f_c = \frac{1}{2\pi}\ \text{Hz}$$

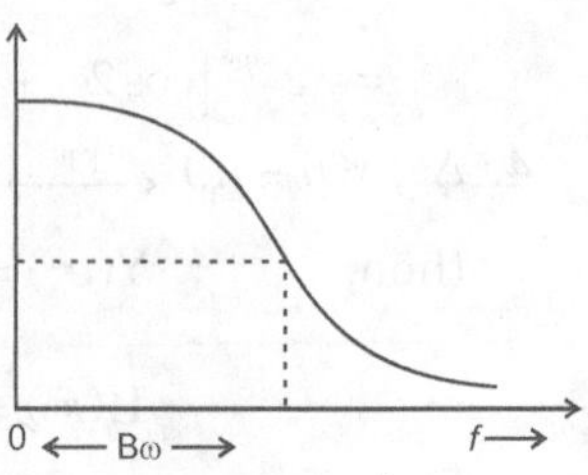

12. DFT of vector [a b c] is $[\propto \beta\gamma\delta]$.

$$\begin{bmatrix} \propto \\ \gamma \\ \beta \\ \delta \end{bmatrix} = \begin{bmatrix} 1 & 1 & 1 & 1 \\ 1 & -j & -1 & j \\ 1 & -1 & 1 & -1 \\ 1 & j & 1 & -j \end{bmatrix} \begin{bmatrix} a \\ b \\ c \\ d \end{bmatrix}$$

$$= \begin{bmatrix} a+b+c+d \\ a-jb-c+jd \\ a-b+c-d \\ a+jb-c-jd \end{bmatrix} \quad ...(i)$$

Given: [p q r s] = [a b c d]

$$\begin{bmatrix} a & b & c & d \\ b & a & b & c \\ c & c & a & b \\ d & d & d & a \end{bmatrix} \quad ...(ii)$$

$= [\, a_2 + bd + c_2bd \;\; ab + ab + cd+cd \;\; 2ac+b_2+d_2 \;\; 2ad + 2bc]$

DFT of [p q r s] is given as

$$[p\,q\,r\,s] \begin{bmatrix} 1 & 1 & 1 & 1 \\ 1 & -j & -1 & j \\ 1 & -1 & 1 & -1 \\ 1 & j & -1 & -j \end{bmatrix}$$

$$= \begin{bmatrix} (p+q+r+s)(p-jq-pr+js)(p-q+r-s) \\ (p+jq-r-js) \end{bmatrix}$$

$$= \begin{bmatrix} \propto^2 \beta^2 \gamma^2 \delta^2 \end{bmatrix}.$$

$\Rightarrow$ p + q + r+ s = $(a^2 + c^2 + 2bd) + (2ab + 2cd)$
$+ (b^2+ d^2 + 2ac) + (2ab + 2bc$ ... from eq(ii)
$= (a + b + c + d)^2$... from eq(i)
= a

13. This can be solved by directly using option and satisfying the condition given in question

X = DFT (x)

$$D_{FT}(D_{FT}(x)) = DFT(X) = \frac{1}{\sqrt{N}} \sum_{n=0}^{N-1} X[n] e^{-j\frac{2\pi}{N}nk}$$

DFT y[1 2 3 4]

$$X = \frac{1}{\sqrt{4}} \begin{bmatrix} 1 & 1 & 1 & 1 \\ 1 & -j & -1 & j \\ 1 & -1 & 1 & -1 \\ 1 & +j & -1 & -j \end{bmatrix} \begin{bmatrix} 1 \\ 2 \\ 3 \\ 4 \end{bmatrix}$$

$$= \frac{1}{\sqrt{4}} \begin{bmatrix} 10 \\ 2+2j \\ 2 \\ -2-j2 \end{bmatrix}$$

DFT of (x) will not result in [1 2 3 4]
Try with DFT of y[1 2 3 2]

$$X = \frac{1}{\sqrt{4}} \begin{bmatrix} 1 & 1 & 1 & 1 \\ 1 & -j & -1 & j \\ 1 & -1 & 1 & -1 \\ 1 & +j & -1 & -j \end{bmatrix} \begin{bmatrix} 1 \\ 2 \\ 3 \\ 2 \end{bmatrix}$$

$$= \frac{1}{\sqrt{4}} \begin{bmatrix} 8 \\ -2 \\ 0 \\ -2 \end{bmatrix} = \begin{bmatrix} 4 \\ -1 \\ 0 \\ -1 \end{bmatrix}$$

$$\text{DFT of} \begin{bmatrix} 4 \\ -1 \\ 0 \\ -1 \end{bmatrix} = \frac{1}{\sqrt{4}} \begin{bmatrix} 1 & 1 & 1 & 1 \\ 1 & -j & -1 & j \\ 1 & -1 & 1 & -1 \\ 1 & +j & -1 & -j \end{bmatrix} \begin{bmatrix} 4 \\ -1 \\ 0 \\ -1 \end{bmatrix}$$

$$= \frac{1}{2} \begin{bmatrix} 2 \\ 4 \\ 6 \\ 4 \end{bmatrix} = \begin{bmatrix} 1 \\ 2 \\ 3 \\ 2 \end{bmatrix}$$

Same as x
Then 'b' is right option

15. $x_1[n] = \{1, 2, 3, 0\}$, $x_2[n] = \{1, 3, 2, 1\}$
$X_3(k) = X_1(k) X_2(k)$
Based on the properties of DFT,
$x_1[n] \otimes x_2[n] = X_1(k) X_2(k) = x_3[n]$
Circular convolution between two 4-point signals is as follows :

$$\begin{bmatrix} 1 & 0 & 3 & 2 \\ 2 & 1 & 0 & 3 \\ 3 & 2 & 1 & 0 \\ 0 & 3 & 2 & 1 \end{bmatrix} \begin{bmatrix} 1 \\ 3 \\ 2 \\ 1 \end{bmatrix} = \begin{bmatrix} 9 \\ 8 \\ 11 \\ 14 \end{bmatrix}$$

$\therefore \; x_3[2] = 11$

16. Discrete Fourier Transform (DFT) of the 4-point sequence is

$$x_1[n] = x\left[\frac{n}{3}\right]$$

$X_1[K] = \{12, 2j, 0, -2j, 12, 2j, 0, -2j, 12, 2j, 0, -2j\}$
$X_1[8] = 12;$ $\quad X_1(11) = -2j$

$$\left|\frac{X_1[8]}{X_1[11]}\right| = \left|\frac{12}{-2j}\right| = 6$$

17. For N-point Fast Fourier Transform (FFT), with $N = 2^m$.

We need only one complex multiplication to compute a butterfly.

18. The number of complex multiplications required

for DIF-FFT = $\left(\frac{N}{2}\log_2 N\right)$

$\therefore \quad \left(\frac{N}{2}\log_2 N\right)(20\ \mu\text{ sec}) = 125\ \mu\text{sec}$

19. Using DIT algorithm are can obtain FFT coefficient [X(I)].

The given butterfly structure is a standard structure where ,

$$a_1 = W_6^0 = 1$$

$$a_2 = W_6^1 = W_6$$

$$a_3 = W_6^2$$

8
CHAPTER

Sampling

Analysis of Previous GATE Papers												
	Year → **Topics ↓**		**2019**	**2018**	**2017 Set 1**	**2017 Set 2**	**2016 Set 1**	**2016 Set 2**	**2016 Set 3**	**2015 Set 1**	**2015 Set 2**	**2015 Set 3**
SAMPLING THEOREM	**1 Mark**	MCQ Type					1		1		1	
		Numerical Type					1					
	2 Marks	MCQ Type				1						
		Numerical Type		1				1				1
	Total			**2**		**2**	**2**	**2**	**1**		**1**	**2**
APPLICATION	**1 Mark**	MCQ Type										
		Numerical Type										
	2 Marks	MCQ Type										
		Numerical Type										
	Total											

SAMPLING THEOREM

1. A signal containing only two frequency components (3 kHz and 6 kHz) is sampled at the rate of 8 kHz, and then passed through a low pass filter with a cut-off frequency of 8 kHz. The filter output

(a) is an undistorted version of the original signal.

(b) contains only the 3 kHz component.

(c) contains the 3 kHz component and a spurious component of 2 kHz.

(d) contains both the components of the original signal and two spurious components of 2 kHz and 5 kHz. **[1988 : 2 Marks]**

Statement for linked Answer Questions 2 and 3 :

In the following network, the switch is closed at $t = 0^-$ and the sampling starts from t = 0. The sampling frequency is 10 Hz.

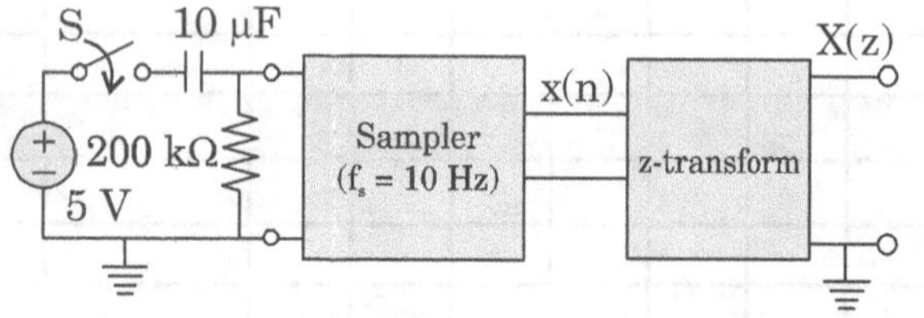

2. The sample x(n) (n = 0, 1, 2,....) are given by

(a) $5(1 - e^{-0.05n})$ (b) $5e^{-0.05n}$

(c) $5(1 - e^{-5n})$ (d) $5e^{-5n}$

[2008 : 2 Marks]

3. The expression and the region of convergence of the z-transform of the sampled signal are

(a) $\frac{5z}{z - e^5}, |z| < e^{-5}$

(b) $\frac{5z}{z - e^{-0.05}}, |z| < e^{-0.05}$

(c) $\frac{5z}{z - e^{-0.05}}, |z| > e^{-0.05}$

(d) $\frac{5z}{z - e^{-5}}, |z| > e^{-5}$ **[2008 : 2 Marks]**

4. An LTI system having transfer function $\frac{s^2 + 1}{s^2 + 2s + 1}$ and input $x(t) = \sin(t + 1)$ is in steady state. The output is sampled at a rate ω_s rad/s to obtain the final output {y(k)}. Which of the following is true?

(a) y is zero for all sampling frequencies ω_s.

(b) y is nonzero for all sampling frequencies ω_s.

(c) y is nonzero for $\omega_s > 2$ but zero for $\omega_s < 2$.

(d) y is zero for $\omega_s > 2$ but nonzero for $\omega_s < 2$.

[2009 : 2 Marks]

5. A band-limited signal with a maximum frequency of 5 kHz is to be sampled. According to the sampling theorem, the sampling frequency which is not valid is

(a) 5 kHz (b) 12 kHz

(c) 15 kHz (d) 20 kHz

[2013 : 1 Mark]

6. Consider two real valued signals, x(t) band-limited to [–500 Hz, 500 Hz] and y(t) band-limited to [–1 kHz, 1 kHz], For z(t) = x(t) y(t), the Nyquist sampling frequency (in kHz) is _______.

[2014 : 1 Mark, Set-1]

7. Let $x(t) = \cos(10\pi t) + \cos(30\pi t)$ be sampled at 20 Hz and reconstructed using an ideal low-pass filter with cut-off frequency of 20 Hz. The frequency/frequencies present in the reconstructed signal is/are

(a) 5 Hz and 15 Hz only

(b) 10 Hz and 15 Hz only

(c) 5 Hz, 10 Hz and 15 Hz only

(d) 5 Hz only

[2014 : 1 Mark, Set-3]

8. Consider a continuous time signal defined as

$$x(t) = \left(\frac{\sin\left(\frac{\pi t}{2}\right)}{\left(\frac{\pi t}{2}\right)}\right) * \sum_{n=-\infty}^{\infty} \delta(t - 10n)$$

Where '*' denotes the convolution operation and t is in seconds. The Nyquist sampling rate (in samples/sec) for x(t) is _______.

[2015 : 2 Marks, Set-3]

9. A continuous-time sinusoid of frequency 33 Hz is multiplied with a periodic Dirac-impulse train of frequency 46 Hz. The resulting signal is passed through an ideal analog low-pass filter with a cutoff frequency of 23 Hz. The fundamental frequency (in Hz) of the output is _______.

[2016 : 1 Mark, Set-1]

10. The signal $\cos\left(10\pi t + \frac{\pi}{4}\right)$ is ideally sampled at a sampling frequency of 15 Hz. The sampled signal is passed through a filter with impulse response $\left(\frac{\sin(\pi t)}{\pi t}\right)\cos\left(40\pi t - \frac{\pi}{2}\right)$. The filter output is

(a) $\frac{15}{2}\cos\left(40\pi t-\frac{\pi}{4}\right)$

(b) $\frac{15}{2}\left(\frac{\sin(\pi t)}{\pi t}\right)\cos\left(10\pi t+\frac{\pi}{4}\right)$

(c) $\frac{15}{2}\cos\left(10\pi t-\frac{\pi}{4}\right)$

(d) $\frac{15}{2}\left(\frac{\sin(\pi t)}{\pi t}\right)\cos\left(10\pi t-\frac{\pi}{2}\right)$

[2015 : 1 Mark, Set-2]

11. A continuous time function x(t) is periodic with period T. The function is sampled uniformly with a sampling period T_s. In which one of the following cases is the sampled signal periodic?

(a) $T = \sqrt{2}T_s$ (b) $T = 1.2T_s$

(c) Always (d) Never

[2016 : 1 Mark, Set-1]

12. A continuous-time fiher with transfer function $H(s)=\frac{2s+6}{s^2+6s+8}$ is converted to a discrete time filter with transfer function $G(s)=\frac{2z^2-0.5032z}{z^2-0.5032z+k}$, so that the impulse response of the continuous-time filter, sampled at 2 Hz, is identical at the sampling instants to the impulse response of the discrete time filter. The value of k is _______.

[2016 : 2 Marks, Set-2]

13. Consider the signal x(t) = cos(6πt) + sin(8πt), where t is in seconds. The Nyquist sampling rate (in samples/second) for the signal y(t) = x(2t + 5) is

(a) 8 (b) 12

(c) 16 (d) 32

[2016 : 1 Mark, Set-3]

14. The signal x(t) = sin(14000πt), where t is in seconds is sampled at a rate of 9000 samples per second. The sampled signal is the input to an ideal lowpass filter with frequency response H(t) as follows :

$$H(f)=\begin{cases}1, & |f|\le 12\text{ kHz}\\ 0, & |f|>12\text{ kHz}\end{cases}$$

What is the number of sinusoids in the output and their frequencies in kHz?

(a) Number = 1, frequency = 7

(b) Number = 3, frequencies = 2, 7, 11

(c) Number = 2, frequencies = 2, 7

(d) Number = 2, frequencies = 7, 11

[2017 : 2 Marks, Set-2]

15. A band limited low-pass signal x(t) of bandwidth 5 kHz is sampled at a sampling rate f_s. The signal x(t) is reconstructed using the reconstruction filter H(f) whose magnitude response is shown below:

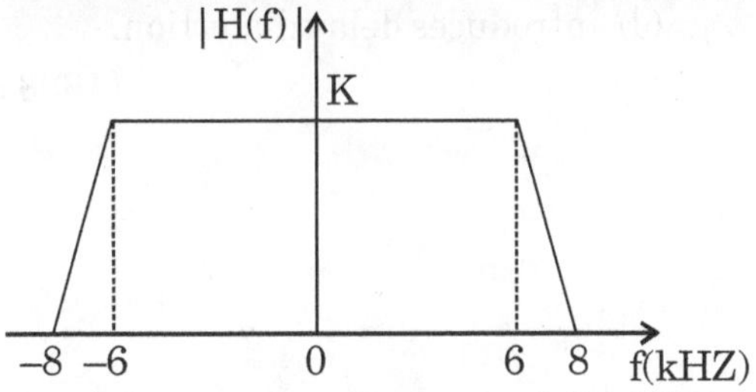

The minimum sampling rate f_s(in kHz) for perfect reconstruction of x(t) is _______.

[2018 : 2 Marks]

APPLICATION

16. Increased pulse-width in the flat-top sampling, leads to

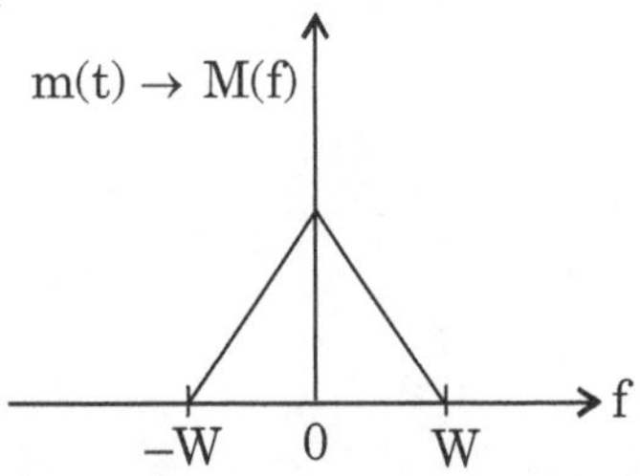

(a) attenuation of high frequencies in reproduction.

(b) attenuation of low frequencies in reproduction.

(c) greater aliasing errors in reproduction.

(d) no harmful effects in reproduction.

[1994 : 1 Mark]

17. A 1.0 kHz signal is flat-top sampled at the rate of 1800 samples/sec and the samples are applied to an ideal rectangular LPF with cut-off frequency of 1100 Hz, then the output of the filter contains

(a) only 800 Hz component.

(b) 800 Hz and 900 Hz components.

(c) 800 Hz and 1000 Hz components.

(d) 800 Hz, 900 Hz and 1000 Hz components.

[1995 : 1 Mark]

18. The transfer function of a zero-order-hold system is

(a) $\left(\frac{1}{s}\right)(1+e^{-sT})$ (b) $\left(\frac{1}{s}\right)(1-e^{-sT})$

(c) $1-\left(\frac{1}{s}\right)e^{-sT}$ (d) $1+\left(\frac{1}{s}\right)e^{-sT}$

[1998 : 1 Mark]

19. Flat top sampling of low pass signals

(a) gives rise to aperture effect.

(b) implies oversampling.

(c) leads to aliasing.

(d) introduces delay distortion.

[1998 : 1 Mark]

20. For a given sample-and-hold circuit, if the value of the hold capacitor is increased, then

(a) drop rate decreases and acquisition time decreases

(b) drop rate decreases and acquisition time increases

(c) drop rate increases and acquisition time decreases

(d) drop rate increases and acquisition time increases

[2014 : 1 Mark, Set-4]

ANSWERS

1. (d)	2. (b)	3. (c)	4. (a)	5. (a)	6. (2· 99 to 3· 01)	7. (a)	8. (0· 4)	9. (13)
10. (a)	11. (b)	12. (0· 049)	13. (c)	14. (b)	15. (13)	16. (a)	17. (b)	18. (b)
19. (a)	20. (b)							

EXPLANATIONS

1. Sampling frequency, fs = 8000 sanoes/s.

f_{m1} = 3 KHZ, f_{M2} = 6 KHZ.

The spectrum of sampled signal would have

$nf_s \pm f_m$.

$\Rightarrow$ 03 kHz ; 8 ± 3, 16 ± 3,

= 3 kHz, 5 kHz, 11 kHz,...

$\Rightarrow$ 6 kHz ; 8 ± 6, 16 ± 6,...

= 6 kHz, 14 kHz,...

Cut - off frequency of L.P.F. = 8 kHz

So, filter output would have 3 kHz, 6 kHz, 2 kHz and 5 kHz.

2. current through resistor (or capacitor) is

$$I = I(0^+).\ e^{\frac{-t}{RC}} :$$

$$\Rightarrow I(0^+) = \frac{V}{R} = \frac{5}{200K} = 25\ \mu A.$$

$$\Rightarrow RC = 200\text{ k} \times 10\mu = 25.$$

$$I = 25.e^{\frac{-t}{2}}\ \mu A$$

$$V_R = R.I = 200\times10^3\times25.e^{\frac{-t}{2}}\times10^{-6} = 5.e^{\frac{-t}{2}}\ V.$$

The voltage across the resistor is input to sampler at frequency of 10 Hz. Thus, $t = nT = \frac{n}{f} = \frac{n}{10}$

$$x(n) = 5e^{\frac{-n}{2\times10}}$$

$$x(n) = 5e^{-0.05n}\ \forall\ t > 0.$$

3. Since, $x(n) = 5.e^{-0.05n}\ u(n)$ is a cousal signal.

$$X(z) = \sum_{n=0}^{\infty} 5.e^{-0.05n} z^{-n} = \frac{5z}{z - e^{-0.05}}$$

Its ROC is $|e^{-0.05}z^{-1}| > | \Rightarrow |z| > e^{-0.05}$

4. Input $x(t) = \sin(\omega t+1) = \sin(t+1)$

$\Rightarrow$ $\omega = 1$

$$T(s) = \frac{s^2+1}{s^2+2s+1}$$

$$T(j\omega) = \left.\frac{-\omega^2+1}{-\omega^2+1+2j\omega}\right|_{\omega=1} = 0$$

So $y(0)$ is zero for all sampling frequency ω_s.

5. Here f_m = 5 kHz

$\therefore\ f_s \geq 2f_m$ = 10 kHz

B,C,D options are greater than 10 KHz

6. Multiplication in the time domain

= Convolution in frequency domain

$$x_1(t)\cdot x_2(t) = X_1(j\omega)X_2(j\omega)$$

Fundamental frequencies = $f_1,\ f_1 \pm f_2,\ f_1 \pm 2f_2$ ...

= 500, 1500....

Nyquist rate = 2 × 1500

= 3000 Hz

= 3 kHz

7. $x(t) = \cos(10\pi t) + \cos(30\pi t)$,

f_s = 20 Hz

Spectrum of $x(t)$

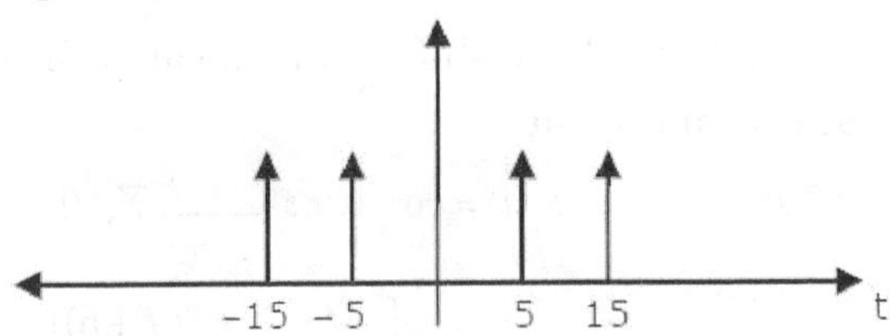

Spectrum of sampled version of $x(t)$

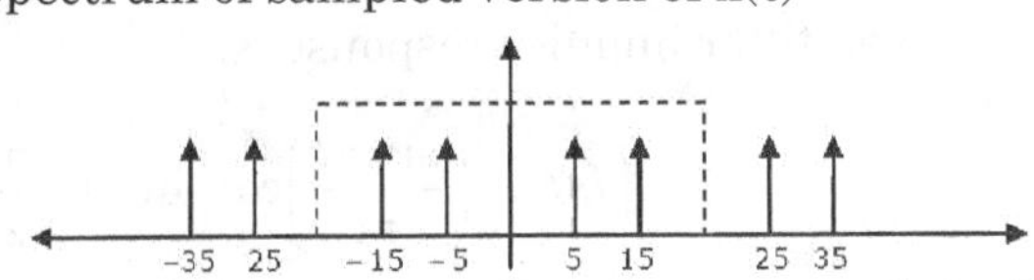

After LPF, signal will contain 5 and 15 Hz component only

8. $$x(t) = \frac{\sin(\pi t/2)}{(\pi t/2)} * \sum_{n=-\infty}^{\infty}\delta(t-10n)$$

Convolution in time domain becomes multiplication in frequency domain.

$$\frac{1}{10}\sum_{n=-\infty}^{\infty}\delta(f-kf_s)$$

$$f_s = \frac{1}{T_s} = 0.1$$

$$\frac{\sin(\pi t/2)}{(\pi t/2)} \longleftrightarrow$$

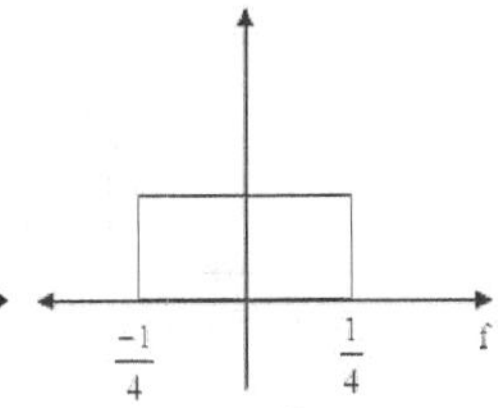

$$\sum_{n=-\infty}^{\infty}\delta(t-10n) \longleftrightarrow$$

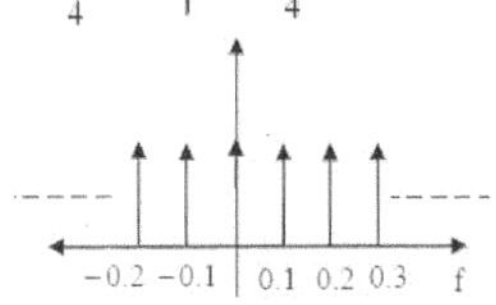

Multiplication in frequency domain will result maximum frequency is 0.2.

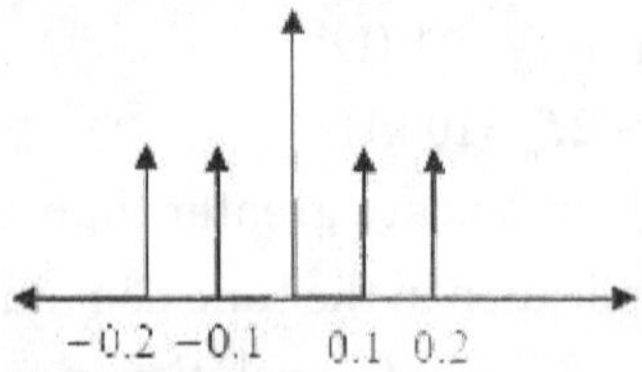

Thus Nyquist rate= 0.4 samples/sec

9. Given, $f_m = 33$ Hz,
$f_s = 46$ Hz

The frequency in sampled singal = ± 33, 13, 79, 59, 125. .... The above frequencies are passed to a LPF of cut-off frequency 23 Hz. The output frequency = 13 Hz.

10. Given signal is $x(t) = \cos\left(10\pi t + \frac{\pi}{4}\right)$

Neglect the phase-shift $\frac{\pi}{4}$ and it can be inserted at the end result.

∴ If $x_1(t) = \cos 10\pi t \xrightarrow{L} X_1(f)$

$$= \frac{1}{2}[\delta(f-5) + \delta(f+5)]$$

Given filter impulse response is,

$$h(t) = \left(\frac{\sin \pi t}{\pi t}\right)\cos\left(40\pi t - \frac{\pi}{2}\right)$$

$$= (\sin ct) \sin (40\, \pi t)$$

$$\therefore \text{H}(f) = \text{rect } f * \frac{1}{2j}[\delta(f-20) - \delta(f+20)]$$

$$= \frac{1}{2j}[\text{rect}(f-20) - \text{rect}(f+20)]$$

$X_1(f)$ repeats with a value $f_o = 15$Hz and each impulse value is $\frac{15}{2}$

Thus the sampled signal spectrum and the spectrum of the filter are as follows:

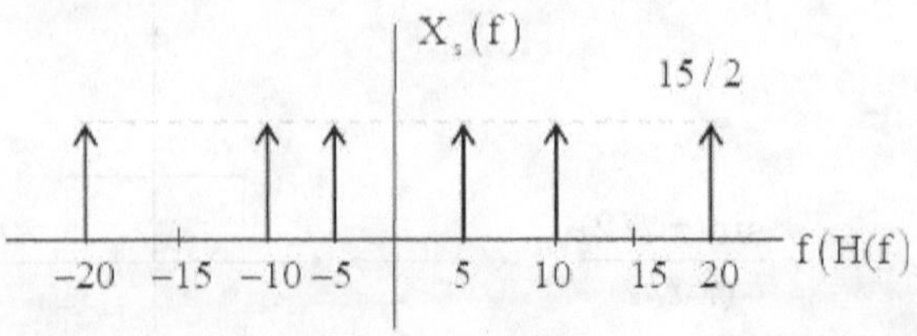

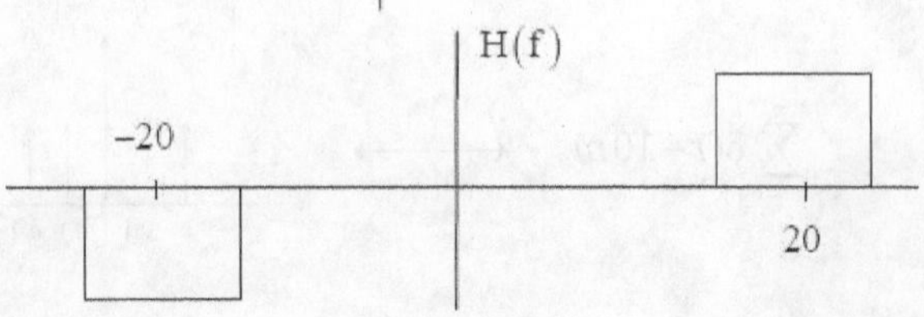

$$\therefore \quad X_s(f)\,H(f) = \frac{15}{4j}[\delta(f-20) - \delta(f+20)]$$

$$x_r(t) = \frac{15}{2}\sin(40\pi t) \rightarrow \text{recovered signal}$$

$$= \frac{15}{2}\cos\left(40\pi t - \frac{\pi}{2}\right)$$

Insert the neglected phase shift $\frac{\pi}{4}$

$$\therefore \quad x_r(t) = \frac{15}{2}\cos\left(40\pi t - \frac{\pi}{2} + \frac{\pi}{4}\right)$$

$$= \frac{15}{2}\cos\left(40\pi t - \frac{\pi}{4}\right)$$

11. If $\frac{\omega_0}{2\pi}$ is a rational number, then discrete time signal $x(n) = \cos(\omega_0 n)$ will be periodic.

12. Given a continuous-time filter with transfer function

$$\text{H}(s) = \frac{2s+6}{s^2+6s+8} = \frac{1}{s+2} + \frac{1}{s+4}$$

$$h(t) = e^{-2t}\,u(t) + e^{-4t}\,u(t)$$

Given sampling frequency $(f_s) = 2$ Hz

For discrete time,

$$t = n\text{T}_s = \frac{n}{2}$$

$$h[n] = (e^{-n} + e^{-2n})\,u[n]$$

$$\text{H}(z) = \frac{1}{1-e^{-1}z^{-1}} + \frac{1}{1-e^{-2}z^{-1}}$$

$$= \frac{z}{z-e^{-1}} + \frac{z}{z-e^{-2}}$$

$$= \frac{2z^2 - 0.5032z}{z^2 - 0.5032z + 0.049}$$

$$\therefore \quad k = 0.049$$

Hence the value of k is 0.049.

13. Given signal

$$x(t) = \cos(6\pi t) + \sin(8\pi t)$$

where t is in seconds,

and $y(t) = x(2t+5)$

$$y(t) = \cos(12\,\pi t + 30\pi) + \sin(16\,\pi t + 40\pi)$$

$$f_{m_1} = 6$$

$$f_{m_2} = 8$$

$$\therefore \quad f_m = 8 \text{ Hz}$$

$$(f_S)_{\min} = 2f_m = 16 \text{ Hz}$$

Thus the Nyquist sampling rate is 16 Hz.

14. $x(t) = \sin(14000\,\pi t)$; $f_m = 7$ kHz

$f_s = 9000$ samples/s.

$f_s = 9$ kHz

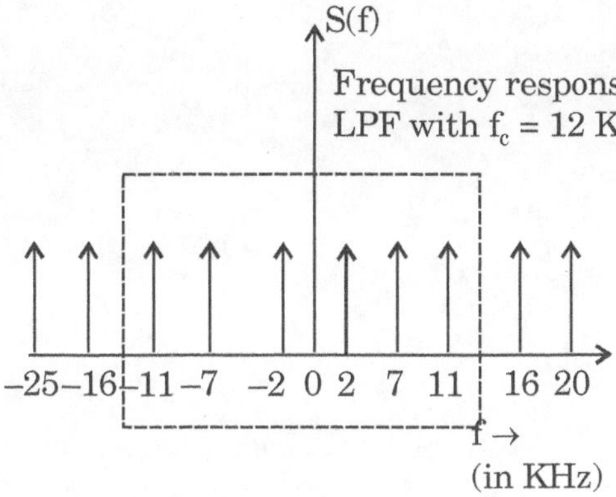

The spectrum of the sampled signal can be given as shown in figure.

So, three sinusoids will be there at the output of the LPF and the frequencies of those sinusoids are 2 kHz, 7 kHz and 11 kHz.

15. Let an arbitrary spectrum for x(t) as shown below:

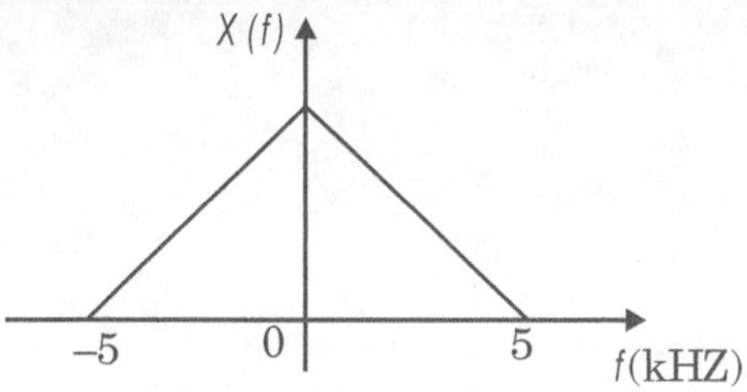

The spectrum of the sampled signal can be given as,

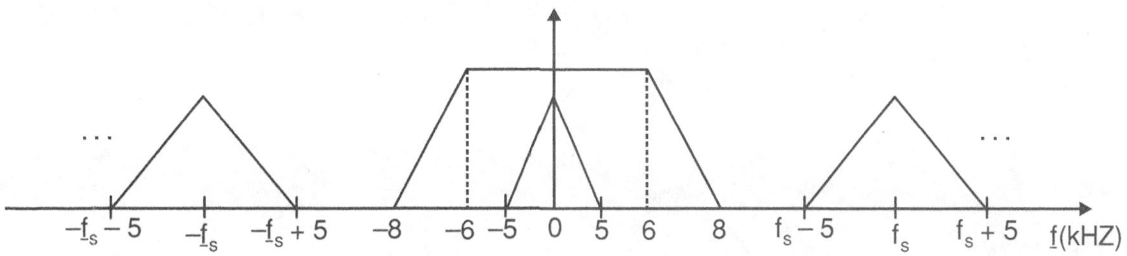

For proper reconstruction of the signal,

$f_s - 5 \geq 8$

$\therefore \quad f_s \geq 8 + 5 = 13$ kHz

$\therefore \quad f_{s(min)} = 13$ kHz

16.

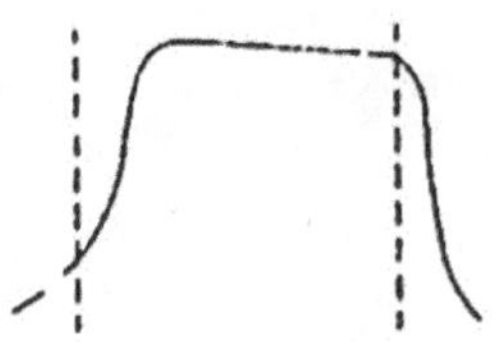

Increased rise and fall lines indicate inadequate high frequency response. Till in top indicates poor low frequency response.

17. Since the sampling rate is 1800 samples/sec, the highest frequency that can be recovered is 900 Hz (sampling frequency/2). Frequencies higher than 900 Hz will be rellected as low frequency due to well known aliasing effect.

18. Transfer function of zero order hold system as $\frac{1}{s}(1 - e^{-sT})$

19. Flat - top sampling of low pass signals gives rise to aperture effect.

20. Capacitor drop rate = $\frac{dv}{dt}$

For a capacitor, $\frac{dv}{dt} \propto \frac{1}{c}$

$\therefore$ Drop rate decreases as capacitor value is increased

For a capacitor, $Q = cv = i \times t \Rightarrow t \propto c$

$\therefore$ Acquisition time increases as capacitor value increased

Unit - IX

Communication

Syllabus

Random processes: autocorrelation and power spectral density, properties of white noise, filtering of random signals through LTI systems; Analog communications: amplitude modulation and demodulation, angle modulation and demodulation, spectra of AM and FM, superheterodyne receivers, circuits for analog communications; Information theory: entropy, mutual information and channel capacity theorem; Digital communications: PCM, DPCM, digital modulation schemes, amplitude, phase and frequency shift keying (ASK, PSK, FSK), QAM, MAP and ML decoding, matched filter receiver, calculation of bandwidth, SNR and BER for digital modulation; Fundamentals of error correction, Hamming codes; Timing and frequency synchronization, inter-symbol interference and its mitigation; Basics of TDMA, FDMA and CDMA.

Contents

1 CHAPTER

Random Signals & Noise

Analysis of Previous GATE Papers

Year → Topics ↓			2019	2018	2017 Set 1	2017 Set 2	2016 Set 1	2016 Set 2	2016 Set 3	2015 Set 1	2015 Set 2	2015 Set 3
VARIANCE & PROBABILITY DENSITY FUNCTION	1 Mark	MCQ Type										
		Numerical Type	2			1						
	2 Marks	MCQ Type									1	
		Numerical Type									1	2
	Total		2			1					4	4
AUTO-CORRELATION & POWER SPECTRAL DENSITY	1 Mark	MCQ Type										
		Numerical Type										
	2 Marks	MCQ Type		1					1			
		Numerical Type	1					2				
	Total		2	2				4	2			
PROPERTIES OF WITH NOISE	1 Mark	MCQ Type										
		Numerical Type										
	2 Marks	MCQ Type	1				1					
		Numerical Type										
	Total		2				2					
FILTERING OF RANDOM SIGNALS USING LTI SYSTEMS	1 Mark	MCQ Type										
		Numerical Type										
	2 Marks	MCQ Type			1							1
		Numerical Type	1									
	Total		2		2							2

VARIANCE & PROBABILITY DENSITY FUNCTION

1. The variance of a random variable X is σ_x^2. Then the variance of –kx (Where k is an positive constant) is

(a) σ_x^2 (b) $-k\,\sigma_x^2$

(c) $k\sigma_x^2$ (d) $k^2\sigma_x^2$

[1987 : 2 Marks]

2. Zero mean Gaussian noise of variance N is applied to a half wave rectifier. The mean squared value of the rectifier output will be

(a) zero (b) $\frac{N}{2}$

(c) $\frac{N}{\sqrt{2}}$ (d) N

[1989 : 2 Marks]

3. For a random variable 'X' following the probability density function, p(x), shown in figure, the mean and the variance are, respectively

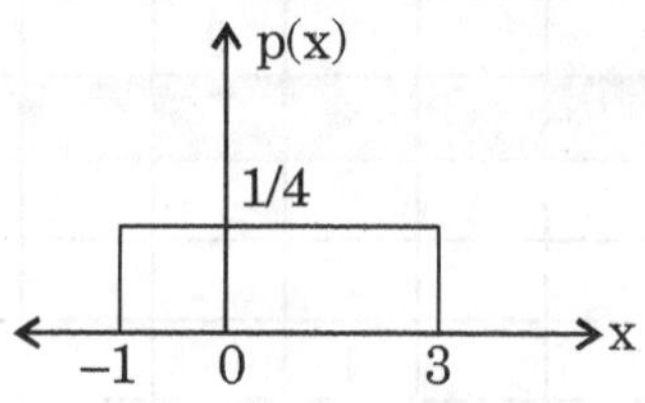

(a) $\frac{1}{2}$ and $\frac{2}{3}$ (b) 1 and $\frac{4}{3}$

(c) 1 and $\frac{2}{3}$ (d) 2 and $\frac{4}{3}$

[1992 : 2 Marks]

4. For a narrow band noise with Gaussian quadrature components, the probability density function of its envelope will be

(a) uniform

(b) Gaussian

(c) exponential

(d) Rayleigh

[1995 : 1 Mark]

5. A probability density function is given by

$$P(x) = K\exp\left(-\frac{x^2}{2}\right), -\infty < x < \infty$$

The value of K should be

(a) $\frac{1}{\sqrt{2\pi}}$ (b) $\sqrt{\frac{2}{\pi}}$

(c) $\frac{1}{2}\sqrt{\pi}$ (d) $\frac{1}{\pi\sqrt{2}}$

[1998 : 1 Mark]

6. The probability density function of the envelope of narrow band Gaussian noise is

(a) Poisson (b) Gaussian

(c) Rayleigh (d) Rician

[1998 : 1 Mark]

7. The PDF of a Gaussian random variable X is given by $P_x(x) = \frac{1}{3\sqrt{2\pi}}e^{\frac{-(x-4)^2}{18}}$. The probability of the event {X = 4} is

(a) $\frac{1}{2}$ (b) $\frac{1}{3\sqrt{2\pi}}$

(c) 0 (d) $\frac{1}{4}$

[2001 : 1 Mark]

8. The distribution function $F_x(x)$ of a random variable X is shown in the figure. The probability that X = 1 is

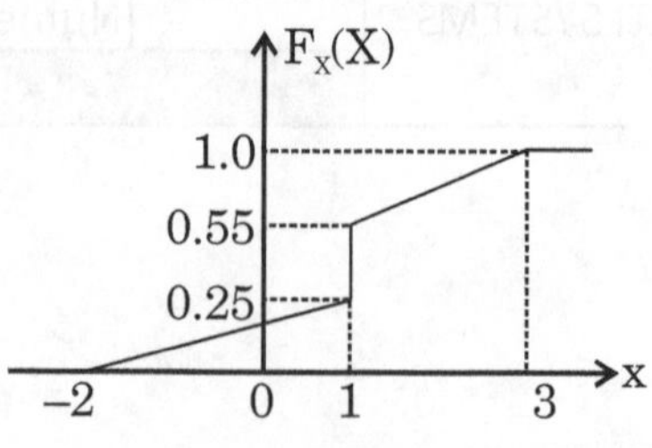

(a) zero (b) 0.25

(c) 0.55 (d) 0.30

[2004 : 1 Mark]

9. A random variable Xwith uniform density in the interval 0 to 1 is quantized as follows:
If $0 \le X \le 0.3$, $x_q = 0$
If $0.3 < X \le 1$, $x_q = 0.7$
where, x_q is the quantized value of X.
The root-mean square value of the quantization noise is
(a) 0.573 (b) 0.198
(c) 2.205 (d) 0.266

[2004 : 2 Marks]

10. An output of a communication channel is a random variable with the probability density function as shown in the figure. The mean square value of v is

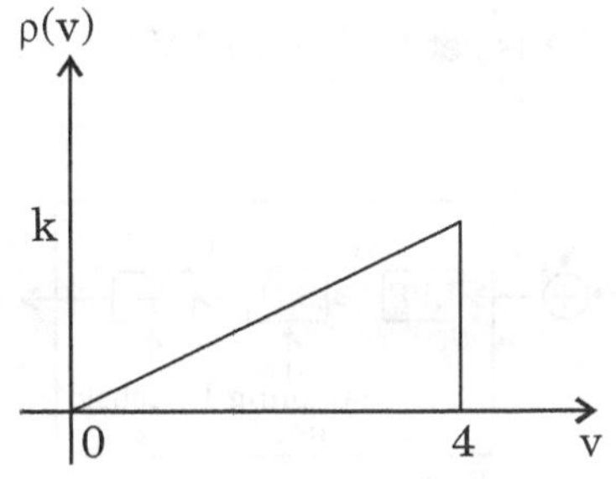

(a) 4 (b) 6
(c) 8 (d) 9

[2005 : 2 Marks]

Common Data for Questions 26 and 27.

Asymmetric three-level midtread quantizer is to be designed assuming equiprobable occurrence of all quantization levels.

11. If the probability density function is divided into three regions as shown in the figure, the value of a in the figure is

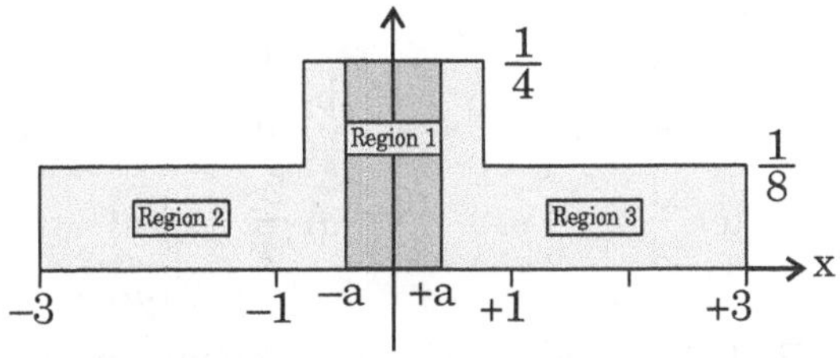

(a) $\frac{1}{3}$ (b) $\frac{2}{3}$

(c) $\frac{1}{2}$ (d) $\frac{1}{4}$

[2005 : 2 Marks]

12. The quantization noise power for the quantization region between -a and +a in the figure is

(a) $\frac{4}{81}$ (b) $\frac{1}{9}$

(c) $\frac{5}{81}$ (d) $\frac{2}{81}$

[2005 : 2 Marks]

13. A uniformly distributed random variable Xwith probability density function

$$f_x(x) = \frac{1}{10}(u(x+5) - u(x-5))$$

where u(.) is the unit step function is passed through a transformation given in the figure below. The probability density function of the transformed random variable y would be

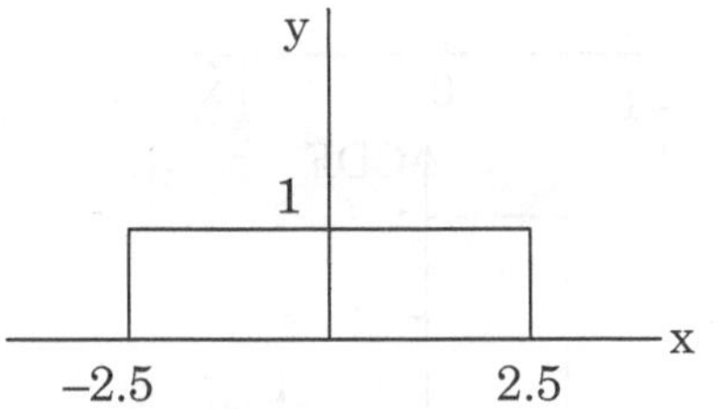

(a) $f_Y(y) = \frac{1}{5}(u(y+2.5) - u(y-2.5))$.

(b) $f_Y(y) = 0.5\delta(y) + 0.5\delta(y-1)$.

(c) $f_Y(y) = 0.25\delta(y+2.5) + 0.25\delta(y-2.5) + 0.5\delta(y)$.

(d) $f_Y(y) = 0.25\delta(y+2.5) + 0.25\delta(y-2.5) + \frac{1}{10}(u(y+2.5) - u(y-2.5))$

[2006 : 2 Marks]

14. If Edenotes expectation, the variance of a random variable X is given by
(a) $E[X^2] - E^2[X]$ (b) $E[X^2] + E^2[X]$
(c) $E[X^2]$ (d) $E^2[X]$

[2007 : 1 Mark]

15. The Probability Density Function (PDF) of a random variable X is as shown below.

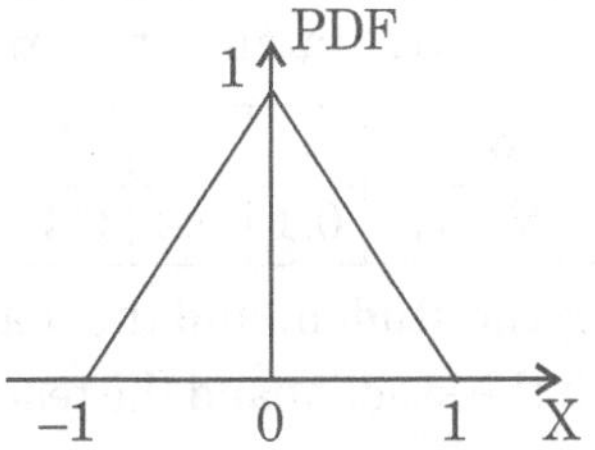

The corresponding Cumulative Distribution Function (CDF) has the form

(a) 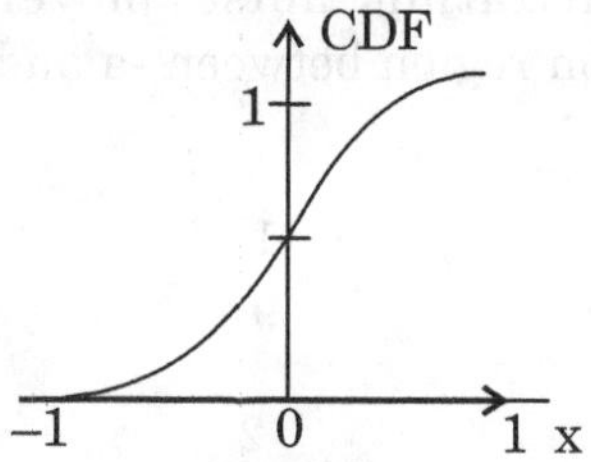

(b) 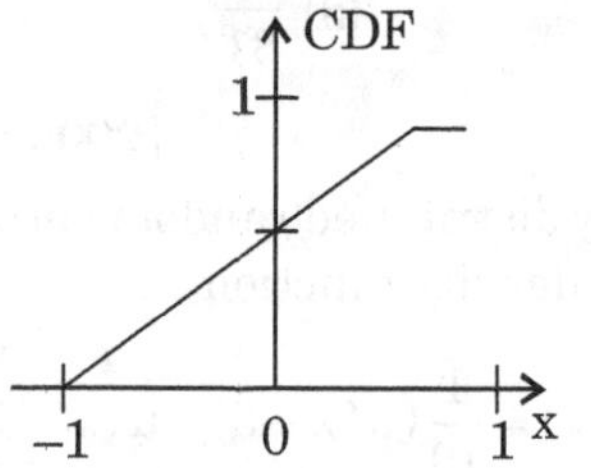

(c) 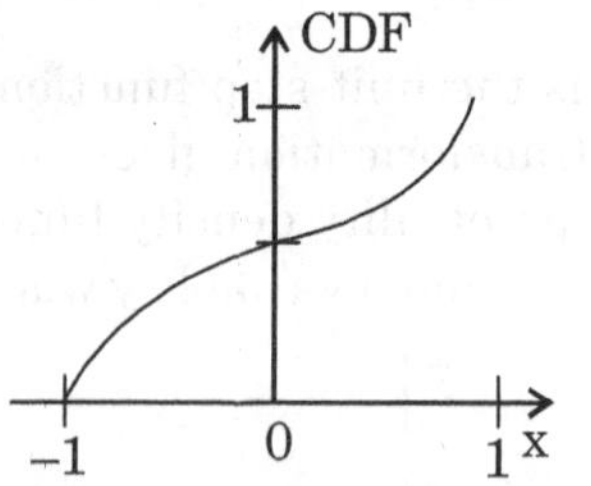

(d) 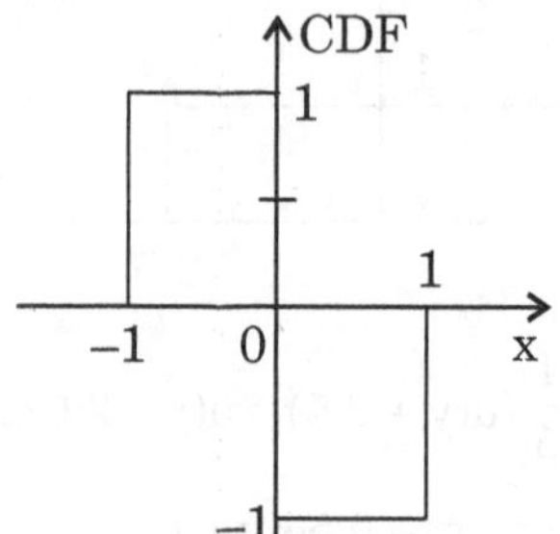

[2008 : 2 Marks]

16. $P_x(x) = M\exp(-2|x|) + N\exp(-3|x|)$ is the probability density function for the real random variable X, over the entire x axis. M and N are both positive real numbers. The equation relating M and N is

(a) $M + \frac{2}{3}N = 1$ (b) $M + \frac{2}{3}N = 1$

(c) $M + N = 1$ (d) $M + N = 3$

[2008 : 2 Marks]

17. A discrete random variable X takes values from 1 to 5 with probabilities as shown in the table. A student calculates the mean X as 3.5 and her teacher calculates the variance of X as 1.5. Which of the following statements is true?

K	1	2	3	4	5
P(X = K)	0.1	0.2	0.4	0.2	0.1

(a) Both the student and the teacher are right.

(b) Both the student and the teacher are wrong.

(c) The student is wrong but the teacher is right.

(d) The student is right but the teacher is wrong.

[2009 : 2 Marks]

Statement for Linked Answer Questions 18 and 19 :

Consider a baseband binary PAM receiver shown below. The additive channel noise n(f) is white with power spectral density $S_N(f) = \frac{N_0}{2} = 10^{-20}$ W/Hz. The low-pass filter is ideal with unity gain and cut-off frequency 1 MHz. Let Y_k represent the random variable $y(t_k)$.

$Y_k = N_k$ if transmitted bit $b_k = 0$

$Y_k = a + N_k$ if transmitted bit b = 1;

where N_k represents the noise sample value. The noise sample has a probability density function, $P_{Nk}(n) = 0.5\alpha e^{-\alpha|n|}$ (This has mean zero and variance $\frac{2}{\alpha^2}$). Assume transmitted bits to be equiprobable and threshold z is set to $\frac{a}{2} = 10^{-6}$ V.

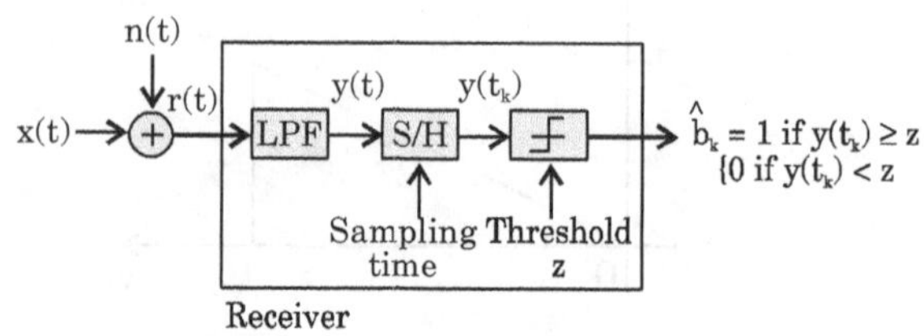

18. The value of the parameter a (in V^{-1}) is

(a) 10^{10} (b) 10^7

(c) 1.414×10^{-10} (d) 2×10^{-20}

[2010 : 2 Marks]

19. The probability of bit error is

(a) $0.5 \times e^{-3.5}$ (b) $0.5 \times e^{-5}$

(c) $0.5 \times e^{-7}$ (d) $0.5 \times e^{-10}$

[2010 : 2 Marks]

20. Two independent random variables X and Y are uniformly distributed in the interval [–1, 1]. The probability that max [X, Y] is less than $\frac{1}{2}$ is

(a) $\frac{3}{4}$ (b) $\frac{9}{16}$

(c) $\frac{1}{4}$ (d) $\frac{2}{3}$

[2012 : 1 Mark]

21. Let U and V be two independent zero mean Gaussian random variables of variances $\frac{1}{4}$ and $\frac{1}{9}$ respectively. The probability $P(3V \geq 2U)$ is

(a) $\frac{4}{9}$ (b) $\frac{1}{2}$

(c) $\frac{2}{3}$ (d) $\frac{2}{3}$ **[2013 : 2 Marks]**

22. Consider two identically zero-mean random variables U and V. Let the cumulative distribution functions of U and 2V be F(z) and G(x) respectively. Then, for all values of x

(a) $F(x) - G(x) \leq 0$

(b) $F(x) - G(x) > 0$

(c) $(F(x) - G(x)).x \leq 0$

(d) $(F(x) - G(x)).x \geq 0$

[2013 : 2 Marks]

23. Let, x_1, x_2, and x_3 be independent and identically distributed random variables with the uniform distribution on [0, 1], The probability P{x_1 is the largest} is _______.

[2014 : 1 Mark, Set-1]

24. Let X be a real-valued random variable with E[X] and E[X_2] denoting the mean values of X and X^2, respectively. The relation which always holds

(a) $(E[X]^2 > E[X^2])$ (b) $E[X^2] \geq (E[X])^2$

(c) $E[X^2] = (E[X])^2$ (d) $E[X^2] > (E[X])^2$

[2014 : 2 Marks, Set-1]

25. Let X be a random variable which is uniformly chosen from the set of positive odd numbers less than 100. The expectation, E[X] is _____.

[2014 : 1 Mark, Set-2]

26. The input to a 1-bit quantizer is a random variable X with PDF $f_X(x) = 2e^{-2x}$ for $x \geq 0$ and $f_X(x) = 0$ for $x < 0$. For outputs to be of equal probability, the quantizer threshold should be _______.

[2014 : 2 Marks, Set-2]

27. Let X_1, X_2 and X_3 be independent and identically distributed random variables with the uniform distribution on [0, 1]. The probability $P(X_1 + X_2 < X_3)$ is _______.

[2014 : 2 Marks, Set-3]

28. A binary random variable X takes the value of 1 with probability $\frac{1}{3}$. X is input to a cascade of 2 independent identical binary symmetric channels (BSCs) each with crossover probability $\frac{1}{2}$. The output of BSCs are the random variables Y_1 and Y_2 as shown in the figure.

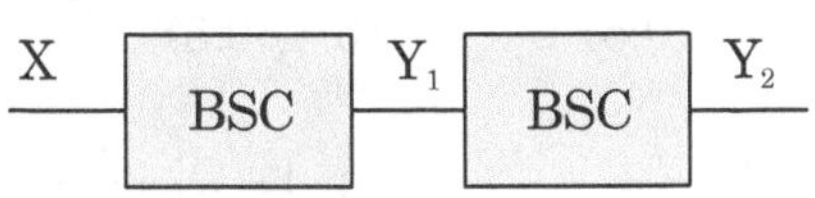

The value of $H(Y_1) + H(Y_2)$ in bits is _______.

[2014 : 2 Marks, Set-3]

29. Let, X be a zero mean unit variance Gaussian random variable. E[|X|] is equal to ______.

[2014 : 1 Mark, Set-4]

30. Consider a communication scheme where the binary values signal X satisfies P{X = 1} = 0.75 and P{X= −1} = 0.25. The received signal Y = X + Z, where Z is a Gaussian rardom variable with zero mean and variance σ^2. The received signal V is fed to the threshold detector. The output of the threshold detector $\hat{X}$ is :

$$\hat{X} = \begin{cases} +1, & Y > \tau \\ -1 & Y \leq \tau \end{cases}$$

To achieve a minimum probability of error $P|\hat{X} \neq X|$, the threshold t should be

(a) strictly positive.

(b) zero.

(c) strictly negative.

(d) strictly positive, zero, or strictly negative depending on the nonzero value of σ^2.

[2014 : 2 Marks, Set-4]

31. Consider the Z-channel given in the figure. The input is 0 or 1 with equal probability. 1.0.

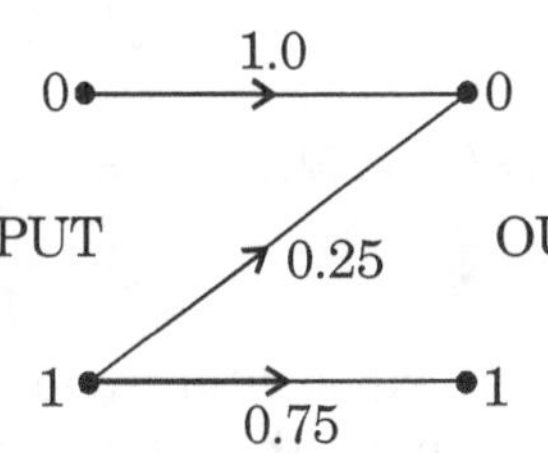

If the output is 0, the probability that the input is also 0 equals _______.

[2014 : 2 Marks, Set-4]

32. Let the random variable X represent the number of times a fair coin needs to be tossed till two consecutive heads appear for the first time. The expectation of X is ______.

[2015 : 2 Marks, Set-2]

33. Let $X \in \{0, 1\}$ and $Y \in \{0, 1\}$ be two independent binary random variables. If P(X = 0) = p and R(Y = 0) = q, then $P(X + Y \geq 1)$ is equal to

(a) $pq(1 - p)(1 - q)$

(b) pq

(c) $p(1 - q)$

(d) $1 - pq$

[2015 : 2 Marks, Set-2]

34. A random binary wave y(t) is given by

$$y(t)=\sum_{n=-\infty}^{\infty}X_n p(t-nT-\phi),$$

where p(t) = u(t) – u(t – T), u(t) is the unit step function and ϕ is an independent random variable with uniform distribution in [0, T], The sequence consists of independent and identically distributed binary valued random variables with $P[X_n = +1\} = P\{X_n = -1\} = 0.5$ for each n. The value of the autocorrelation

$R_{yy}\left(\frac{3T}{4}\right)\triangleq E\left[y(t)y\left(t-\frac{3T}{4}\right)\right]$ is equals______.

[2015 : 2 Marks, Set-3]

35. The variance of the random variable X with probability density function $f(x)=\frac{1}{2}|x|e^{-|x|}$ is _______.

[2015 : 2 Marks, Set-3]

36. Consider the random process

X(t) = U + Vt,

where, U is a zero-mean Gaussian random variable and Vis a random variable uniformly distributed between 0 and 2. Assume that U and V are statistically independent. The mean value of the random process at t = 2 is _____.

[2017 : 1 Mark, Set-2]

37. If X and Y are random variables such that E[2X + Y] = 0 and E[X + 2Y] = 33, the E[X] + E[Y] = _____.

[2019 : 1 Marks]

38. Let Z be an exponential random variable with mean 1. That is, the cumulative distribution function of Z is given by

$$F_z(x)=\begin{cases}1-e^{-x} & \text{if } x\ge 0\\ 0 & \text{if } x<0\end{cases}$$

The $\Pr(Z>2\,|\,Z>1)$, rounded off two decimal places, is equal to _____.

[2019 : 1 Marks]

AUTO-CORRELATION & POWER SPECTRAL DENSITY

39. The power spectral density of a deterministic signal is given by $[\sin(f)/f]^2$. where 'f' is frequency, the autocorrelation function of this signal in the time domain is

(a) a rectangular pulse

(b) a delta function

(c) a sine pulse

(d) a triangular pulse

[1997 : 2 Marks]

40. The amplitude spectrum of a Gaussian pulse is

(a) uniform.

(b) a sine function.

(c) Gaussian.

(d) an impulse function.

[1998 : 1 Mark]

41. The PSD and the power of a signal g(t) are, respectively, $S_g(\omega)$ and P_g. The PSD and the power of the signal ag(t) are, respectively,

(a) $a^2S_g(\omega)$ and a^2P_g (b) $a^2S_g(\omega)$ and aP_g

(c) $aS_g(\omega)$ and a^2P_g (d) $aS_g(\omega)$ and aP_g

[2001 : 2 Marks]

42. If the variance σ_x^2 of d(n) = x(n) – x(n – 1) is one-tenth the variance of a stationary zero-mean discrete-time signal x{n), then the normalized autocorrelation function $R_{XX}(k)/\sigma_x^2$ at k = 1 is

(a) 0.95 (b) 0.90

(c) 0.10 (d) 0.05

[2002 : 2 Marks]

43. The noise at the input to an ideal frequency detector is white. The power spectral density of the noise at the output is

(a) raised-cosine (b) flat

(c) parabolic (d) Gaussian

[2003 : 1 Mark]

44. Let X and Y be two statistically independent random variables uniformly distributed in the ranges (–1, 1) and (–2, 1) respectively. Let Z = X + Y. Then the probability that (Z ≤ –2) is

(a) zero (b) $\frac{1}{6}$

(c) $\frac{1}{3}$ (d) $\frac{1}{12}$

[2003 : 2 Marks]

Common Data for Questions 45 and 46 :

Let X be the Gaussian random variable obtained by sampling the process at t = t_1 and let

$$Q(\alpha)=\int_{\alpha}^{\infty}\frac{1}{\sqrt{2\pi}}e^{-\frac{x^2}{2}}dy$$

Auto correlation function $R_{XX}(\tau)=4(e^{-0.2|\tau|}+1)$ and mena = 0

45. The probability that [x ≤ 1] is

(a) 1 - Q(0.5) (b) Q(0.5)

(c) $Q\left(\frac{1}{2\sqrt{2}}\right)$ (d) $1-Q\left(\frac{1}{2\sqrt{2}}\right)$

[2003 : 2 Marks]

46. Let Vand Z be the random variables obtained by sampling X(t) at t = 2 and t = 4 respectively. Let W = Y – Z The variance of W is

(a) 13.36 (b) 9.36

(c) 2.64 (d) 8.00

[2003 : 2 Marks]

47. A 1 mW video signal having a bandwidth of 100 MHz is transmitted to a receiver through a cable that has 40 dB loss. If the effective one-sided noise spectral density at the receiver is 10^{-20} Watt/Hz, then the signal-to-noise ratio at the receiver is

(a) 50 dB (b) 30 dB

(c) 40 dB (d) 60 dB

[2004 : 2 Marks]

48. Noise with uniform power spectral density of N_0W/Hz is passed through a filter $H(\omega) = 2\exp(-j\omega t_d)$ followed by an ideal low pass filter of bandwidth B Hz. The output noise power in Watts is

(a) $2N_0B$ (b) $4N_0B$

(c) $8N_0B$ (d) $16N_0B$

[2005 : 2 Marks]

49. If S(f) is the power spectral dersityof a real, wide- sense stationary random process, then which of the following is ALWAYS true?

(a) $S(0) \geq S(f)$ (b) $S(f) \geq 0$

(c) $S(-f) = -S(f)$ (d) $\int_{-\infty}^{\infty} S(f)df = 0$

[2007 : 1 Mark]

50. If $R(\tau)$ is the auto-correlation function of a real, wide-sense stationary random process, then which of the following is NOT true?

(a) $R(\tau) = R(-\tau)$

(b) $|R(\tau)| \leq R(0)$

(c) $R(\tau) = -R(-\tau)$

(d) The mean square value of the process is R(0)

[2007 : 1 Mark]

51. If the power spectral density of stationary random process is a sinc-squared function of frequency, the shape of its auto-correlation is

(a)
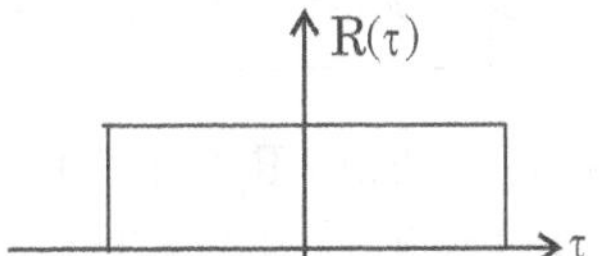

(b)
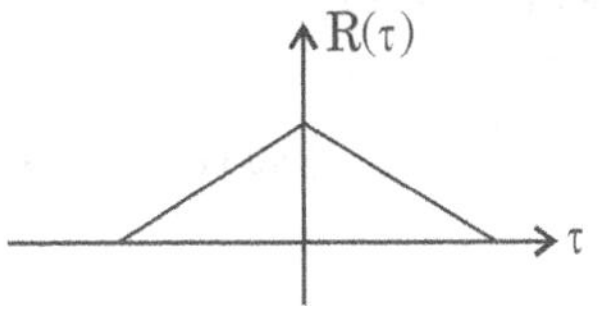

(c)
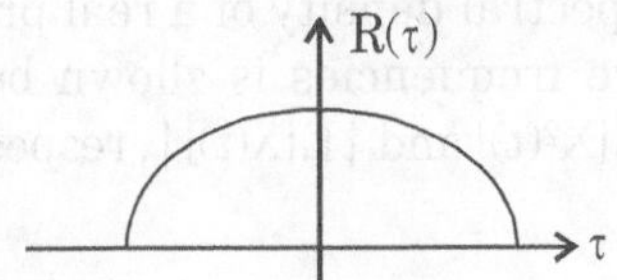

(d)
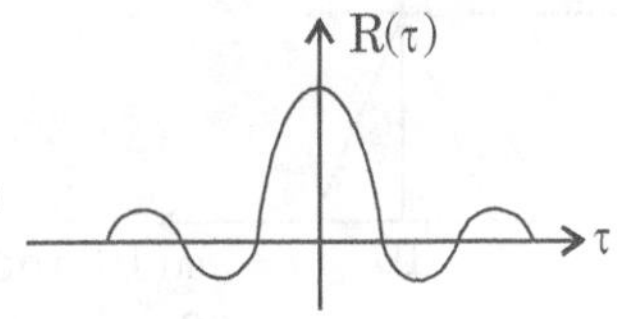

[2009 : 1 Mark]

52. X(t) is a stationary process with the power spectral density $S_X(f) > 0$ for all f. The process is passed through a system shown below.

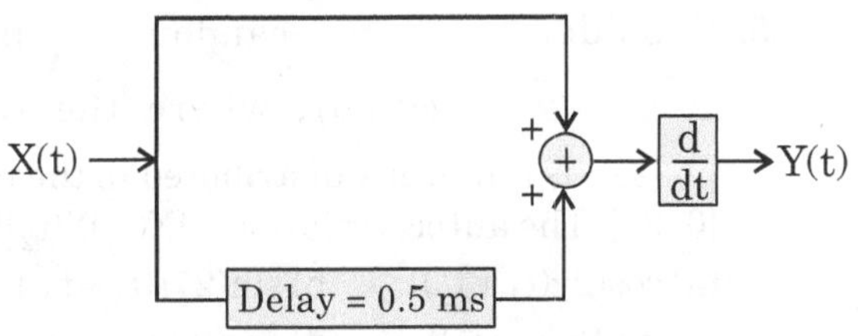

Let $S_Y(f)$ be the power spectral density of Y(t). Which one of the following statements is correct?

(a) $S_Y(f) > 0$ for all f

(b) $S_Y(f) = 0$ for $|f| > 1$ kHz

(c) $S_Y(f) = 0$ for $f = nf_0$, $f_0 = 2$ kHz, n any integer

(d) $S_Y(f) = 0$ for $f = (2n + 1)f_0$, $f_0 = 1$ kHz, n any integer

[2010 : 2 Marks]

53. X(t) is a stationary random process with autocorrelation function $R_x(\tau) = \exp(-\pi\tau^2)$. This process is passed through the system below. The power spectral density of the output process Y(t) is

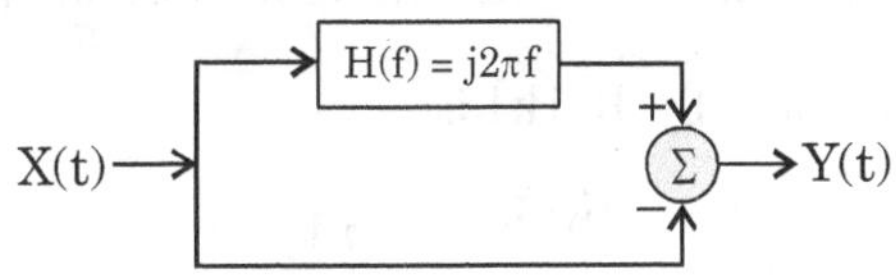

(a) $(4\pi^2 f^2 + 1)\exp(-\pi f^2)$

(b) $(4\pi^2 f^2 - 1)\exp(-\pi f^2)$

(c) $(4\pi^2 f^2 + 1)\exp(-\pi f)$

(d) $(4\pi^2 f^2 - 1)\exp(-\pi f)$

[2011 : 2 Marks]

54. A power spectral density of a real process X(t) for positive frequencies is shown below. The values of $E[X^2(t)]$ and $|E[X(t)]|$, respectively are

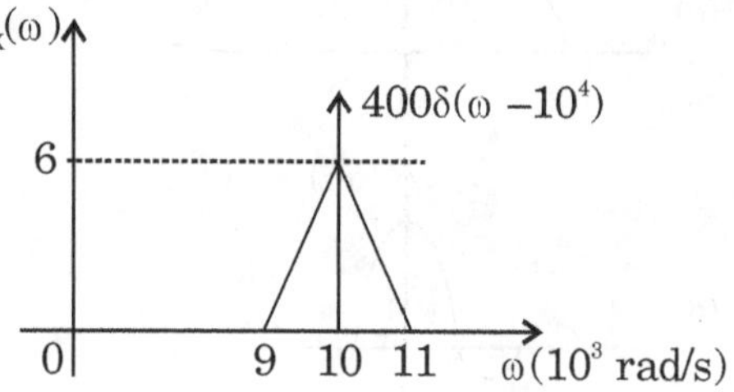

(a) $\frac{6000}{\pi}, 0$ (b) $\frac{6400}{\pi}, 0$

(c) $\frac{6400}{\pi}, \frac{20}{\left(\frac{\pi}{\sqrt{2}}\right)}$ (d) $\frac{6000}{\pi}, \frac{20}{\left(\frac{\pi}{\sqrt{2}}\right)}$

[2012 : 1 Mark]

55. Consider a random process $X(t) = \sqrt{2}\sin(2\pi t + \phi)$, where the random phase ϕ is uniformly distributed in the interval $[0, 2\pi]$. The autocorrelation $E[X(t_1)X(t_2)]$ is

(a) $\cos(2\pi(t_1 + t_2))$ (b) $\sin(27r(t_1 - t_2))$

(c) $\sin(2\pi(t_1 + t_2))$ (d) $\cos(2\pi(t_1 - t_2))$

[2014 : 2 Marks, Set-1]

56. The power spectral density of a real stationary random process X{t) is given by

$$S_x(f) = \begin{cases} \frac{1}{W}, & |f| \le W \\ 0 & |f| > W \end{cases}$$

The value of the expectation

$E\left[\pi x(t) \cdot x\left(t - \frac{1}{4W}\right)\right]$ is ________.

[2014 : 2 Marks, Set-2]

57. $\{X\}_{n=-\infty}^{n=\infty}$ is an independent and identically distributed (i.i.d.) random process X_n equally likely to be + 1 or −1. $\{Y_n\}_{n=-\infty}^{n=\infty}$ is another random process obtained as $Y_n = X_n + 0.5\, X_{n-1}$. The autocorrelation function of $\{Y_n\}_{n=-\infty}^{n=\infty}$ denoted by $R_y[k]$ is

(a)

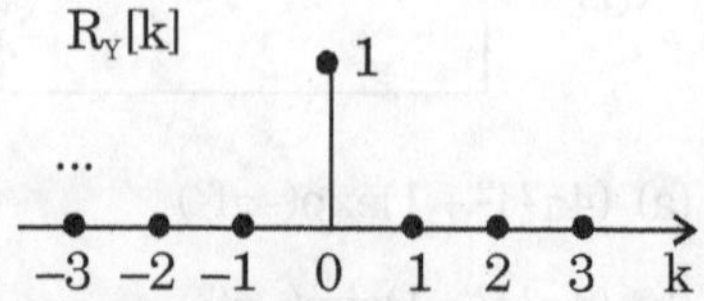

(b)

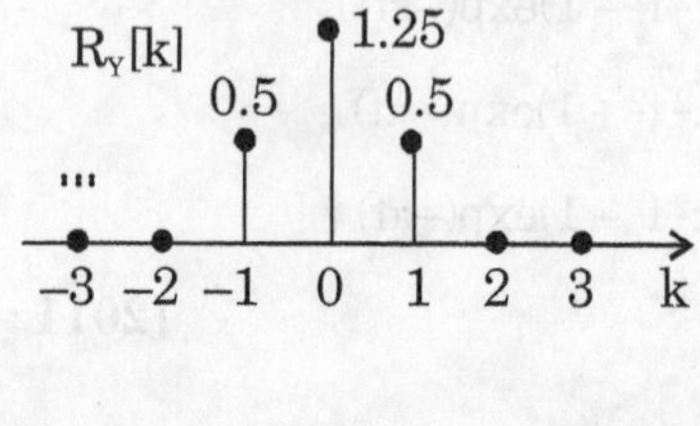

(c)

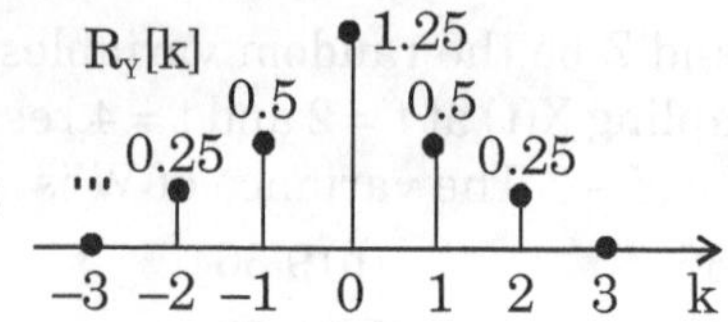

(d)

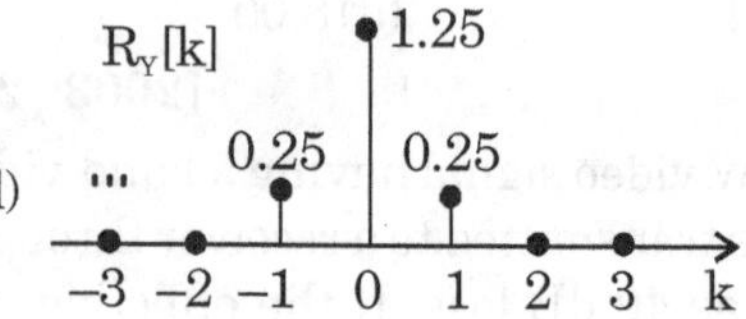

58. An information source generates a binary sequence $\{\alpha_n\}$.α_n can take one of the two possible values −1 and +1 with equal probability and are statistically independent and identically distributed. This sequence is precoded to obtain another sequence $\{\beta_n\}$, as $\beta_n = \alpha_n + k\alpha_{n-3}$. The sequence $\{\beta_n\}$ is used to modulate a pulse g(t) to generate the baseband signal

$$X(t) = \sum_{n=-\infty}^{\infty} \beta_n g(t - nT)$$

where $g(t) = \begin{cases} 1, & 0 \le t \le T \\ 0, & \text{otherwise} \end{cases}$

If there is a null at $f = \frac{1}{3T}$ in the power spectral density of X(t), then k is ________.

[2016 : 2 Marks, Set-2]

59. Consider a random process X(t) = 3V(f) − 8, where V(t) is a zero mean stationary random process with autocorrelation $R_v(\tau) = 4e^{-5|\tau|}$. The power in X(t) is ________.

[2016 : 2 Marks, Set-2]

60. A wide sense stationary random process X[t) passes through the LTI system shown in the figure. If the autocorrelation function of X(t) is $R_x(\tau)$, then the autocorrelation function $R_Y(\tau)$ of the output Y(t) is equal to

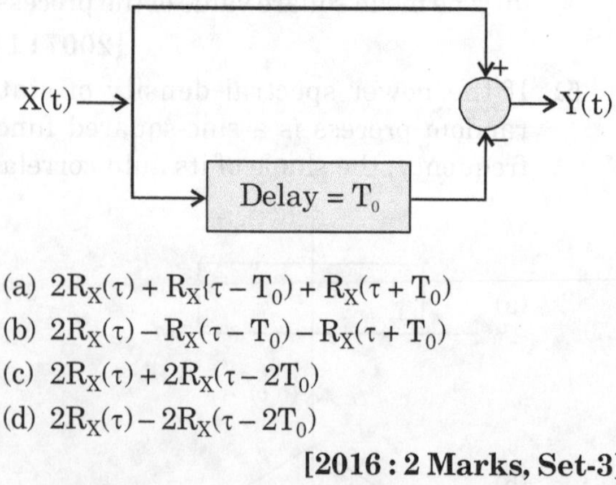

(a) $2R_X(\tau) + R_X(\tau - T_0) + R_X(\tau + T_0)$

(b) $2R_X(\tau) - R_X(\tau - T_0) - R_X(\tau + T_0)$

(c) $2R_X(\tau) + 2R_X(\tau - 2T_0)$

(d) $2R_X(\tau) - 2R_X(\tau - 2T_0)$

[2016 : 2 Marks, Set-3]

61. Consider a white Gaussian noise process N(t) with two-sided power spectral density $S_N(t) = 0.5$ W/Hz as input to a filter with impulse response $0.5e^{-t^2/2}$ (where t is in seconds) resulting in output Y(t). The power in Y(t) in watts is

(a) 0.11 (b) 0.22

(c) 0.33 (d) 0.44

[2018 : 2 Marks]

62. Let a random process Y(t) be described as Y(t) = h(t) × X(t) + Z(t), where X(t) is a white noise process with power spectral density $S_x(f)$ = 5W/Hz. The filter h(t) has a magnitude response given by |H(f)| = 0.5 for $-5 \le f \le 5$, and zero elsewhere. Z(t) is a stationary random process, uncorrelated with X(t), with power spectral density as shown in the figure.

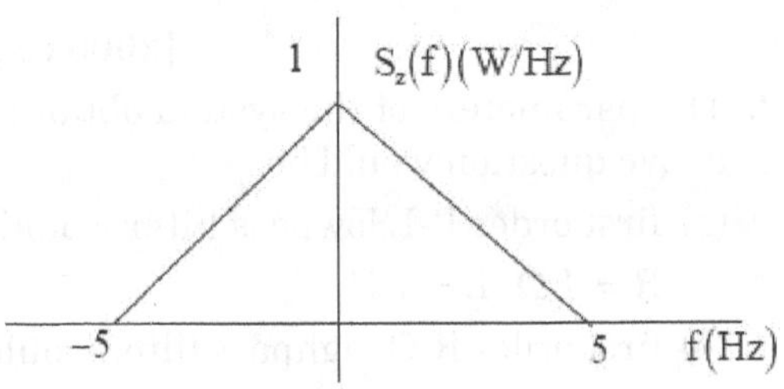

The power in Y(t), in watts, is equal to ______W. (rounded off to two decimal places).

[2019 : 2 Marks]

PROPERTIES OF WHITE NOISE

63. White Gaussian noise is passed through a linear narrow band filter. The probability density function of the envelope of the noise at the filter output is

(a) uniform

(b) Poisson

(c) Gaussian

(d) Rayleigh

[1987 : 2 Marks]

64. In a radar receiver the antenna is connected to the receiver through a waveguide. Placing the preamplifier on the antenna side of the waveguide rather than on the receivar side leads to

(a) a reduction in the overall noise figure

(b) a reduction in interference

(c) an improvement in selectivity characteristics

(d) an improvement in directional characteristics

[1987 : 2 Marks]

65. A part of a communication system consists of an amplifier of effective noise temperature, T_e = 21 K, and a gain of 13 dB, followed by a cable with a loss of 3 dB. Assuming the ambient temperature to be 300 K, we have for this part of the communication system,

(a) effective noise temperature = 30 K

(b) effective noise temperature = 36 K

(c) noise figure = 0.49 dB

(d) noise figure = 1.61 dB

[1991 : 2 Marks]

66. Two resistors R_1 and R_2 (in ohms) at temperatures T_1 K and T_2 K respectively, are connected in series. Their equivalent noise temperature is _______ K.

[1991 : 2 Marks]

67. An amplifier A has 6 dB gain and 50 Ω input and output impedances. The noise figure of this amplifier as shown in the figure is 3 dB. A cascade of two such amplifiers as in the figure will have a noise figure of

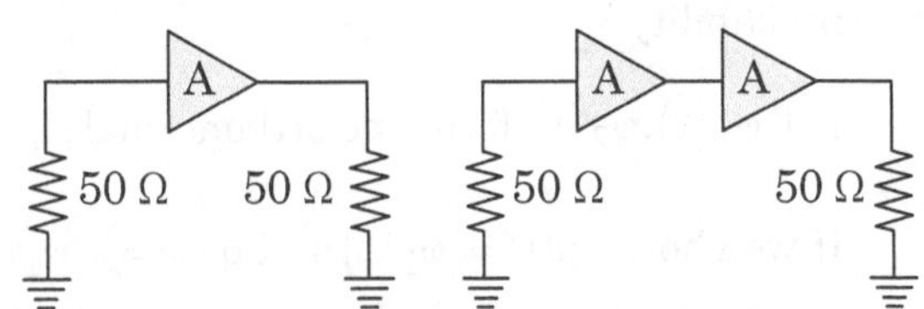

(a) 6 dB

(b) 8 dB

(c) 12 dB

(d) None of the above

[1997 : 1 Mark]

68. Consider a discrete-time channel Y = X + Z, where the additive noise Z is signal-dependent. In particular, given the transmitted symbol $X \in \{-a, +a\}$ at any instant, the noise sample Z is chosen independently from a Gaussian distribution with mean βX and unit variance. Assume a threshold detector with zero threshold at the receiver.

When, β = 0, the BER was found to be $Q(a) = 1 \times 10^{-8}$.

$$\left(Q(v) = \frac{1}{2\pi}\int_v^{\infty} e^{-u^2} du, \text{ and for } v > 1, \text{ use } Q(v) = e^{-v^2/2}\right)$$

When, β = –0.3, the BER is closest to

(a) 10^{-7} (b) 10^{-6}

(c) 10^{-4} (d) 10^{-2}

[2014 : 2 Marks, Set-4]

69. An antenna pointing in a certain direction has a noise temperature of 50 K. The ambient temperature is 290 K. The antenna is connected to a pre-amplifier that has a noise figure of 2 dB and an available gain of 40 dB over an effective bandwidth of 12 MHz. The effective input noise temperature T_e for the amplifier and the noise power P_{a0} at the output of the preamplifier, respectively, are

(a) $T_e = 169.36$ K and $P_{a0} = 3.73 \times 10^{-10}$ W

(b) $T_e = 170.8$ K and $P_{a0} = 4.56 \times 10^{-10}$ W

(c) $T_e = 182.5$ K and $P_{a0} = 3.85 \times 10^{-10}$ W

(d) $T_e = 160.62$ K and $P_{a0} = 4.6 \times 10^{-10}$ W

[2016 : 2 Marks, Set-1]

70. A single bit, equally likely to be 0 and 1, is to be sent across an additive white Gaussian noise (AWGN) channel with power spectral density $N_0/2$. Binary signaling, with $0 \rightarrow p(t)$ and $I \rightarrow q(t)$, is used for the transmission, along with an optimal receiver that minimizes the bit-error probability.

Let $\varphi_1(t), \varphi_2(t)$ form and orthonormal signal set.

If we choose $p(t) = \varphi_1(t)$ and $q(t) = -\varphi_1(t)$, we would obtain a certain bit-error probability P_b.

If we keep $p(t) = \varphi_1(t)$, but take $q(t) = \sqrt{E}\varphi_2(t)$, for what value of E would we obtain the same bit-error probability P_b?

(A) 3

(B) 1

(C) 2

(D) 0

[2019 : 2 Marks]

FILTERING OF RANDOM SIGNALS USING LTI SYSTEMS

71. A zero-mean white Gaussian noise is passed through an ideal lowpass filter of bandwidth 10 kHz. The output is then uniformly sampled with sampling period $t_s = 0.03$ msec. The samples so obtained would be

(a) correlated.

(b) statistically independent.

(c) uncorrelated.

(d) orthogonal.

[2006 : 2 Marks]

Common Data for Questions 29 & 30:

The following two questions refer to wide sense stationary stochastic processes

72. It is desired to generate a stochastic process (as voltage process) with power spectral density

$$S(\omega) = \frac{16}{16 + \omega^2}$$

by driving a Linear-Time-Invariant system by zero mean white noise (as voltage process) with power spectral density being constant equal to 1. The system which can perform the desired task could be

(a) first order lowpass R-L filter.

(b) first order highpass R-C filter.

(c) tuned L-C filter.

(d) series R-L-C filter.

[2006 : 2 Marks]

73. The parameters of the system obtained in the above question would be

(a) first order R-L lowpass filter would have $R = 4\ \Omega$, $L = 1$ H

(b) first order R-C highpass filter would have $R = 4\ \Omega$, $C = 0.25$ F

(c) tuned L-C filter would have $L = 4$ H, $C = 4$ F

(d) series R-L-C lowpass filter would have $R = 1\ \Omega$, $L = 4$ H, $C = 4$ F

[2006 : 2 Marks]

74. Noise with double-sided power spectral density of K over all frequencies is passed through a RC low pass filter with 3 dB cut-off frequency of f_c. The noise power at the filter output is

(a) K (b) Kf_c

(c) $K\pi f_c$ (d) ∞

[2008 : 2 Marks]

75. A white noise process x(t) with two-sided power spectral density 1×10^{-10} W/Hz is input to a filter whose magnitude squared response is shown below.

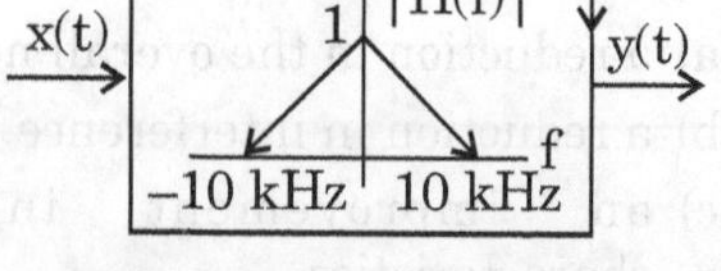

The power of the output process y(t) is given by

(a) 5×10^{-7} W (b) 1×10^{-6} W

(c) 2×10^{-6} W (d) 1×10^{-5} W

[2009 : 1 Mark]

76. A real band-limited random process X(t) has two-sided power spectral density

$$S_x(f) = \begin{cases} 10^{-6}1(300) - |f| \text{ Watts/Hz} & \text{for } |f| \le 3 \text{ kHz} \\ 0, & \end{cases}$$

where f is the frequency expressed in Hz. The signal x(t) modulates a carrier $\cos 16000\pi t$ and the resultant signal is passed through an ideal band-pass filter of unity gain with centre frequency of 8 kHz and band-width of 2 kHz. The output power (in Watts) is ______.

[2014 : 2 Marks, Set-3]

77. A zero mean white Gaussian noise having power spectral density $\frac{N_0}{2}$ is passed through an LTI filter whose impulse response h(t) is shown in the figure. The variance of the filtered noise at t = 4 is

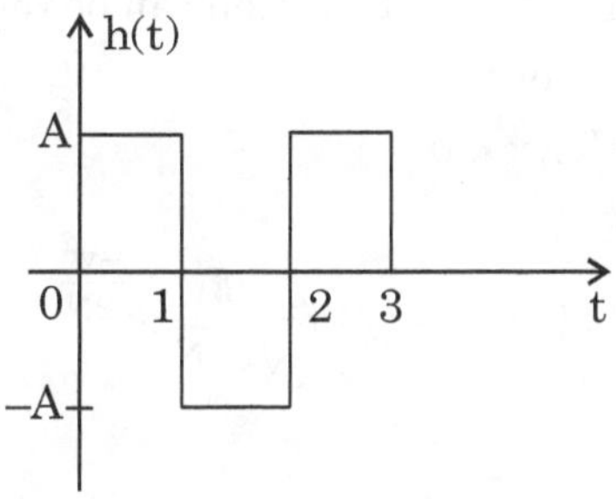

(a) $\frac{3}{2}A^2N_0$ (b) $\frac{3}{4}A^2N_0$

(c) A^2N_0 (d) $\frac{1}{2}A^2N_0$

[2015 : 2 Marks, Set-2]

78. Let X(t) be a wide sense stationary random process with the power spectral density $S_x(f)$ as shown in figure (a), where f is in Hertz (Hz). The random process X(t) is input to an ideal lowpass filter with the frequency response

$$H(f) = \begin{cases} 1, & |f| \le \frac{1}{2} \text{Hz} \\ 0, & |f| > \frac{1}{2} \text{Hz} \end{cases}$$

as shown in figure (b). The output of the lowpass filter is Y(t).

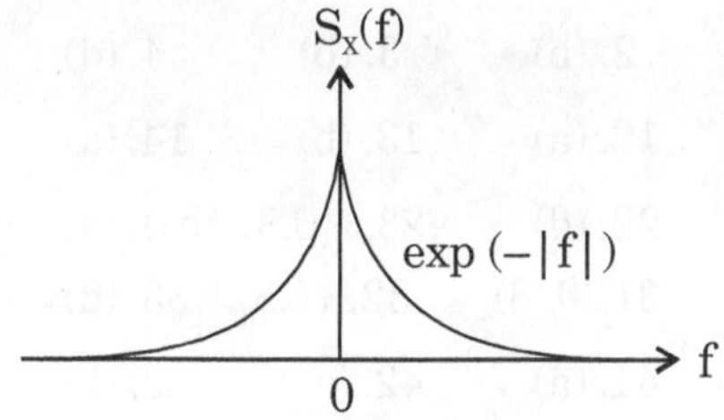

(a)

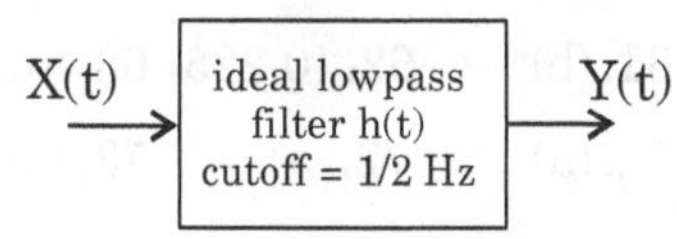

(b)

Let E be the expectation operator and consider the following statements:

I. E(X(t)) = E(Y(t))

II. $E(X^2(t)) = E(Y^2(t))$

III. $E(Y^2(t)) = 2$

Select the correct option:

(a) Only I is true.

(b) Only II and III are true.

(c) Only I and II are true.

(d) Only I and III are true.

[2017 : 2 Marks, Set-1]

79. A random variable X takes values – 1 and + 1 with probabilities 0.2 and 0.8, respectively. It is transmitted across a channel which adds noise N, so that the random variable at the channel output is Y = X + N. The noise N is independent of X, and is uniformly distributed over the interval [–2, 2]. The receiver makes a decision

$$X = \begin{cases} -1, & \text{if } Y \le \theta \\ +1, & \text{if } Y > \theta \end{cases}$$

Where the threshold $\theta \in [-1, 1]$ is chosen so as to minimize the probability of error, rounded off to 1 decimal place, is ______.

[2019 : 2 Marks]

ANSWERS

1. (d)	2. (b)	3. (b)	4. (d)	5. (a)	6. (a)	7. (c)	8. (a)	9. (b)	10. (c)
11. (b)	12. (a)	13. (b)	14. (a)	15. (a)	16. (a)	17. (b)	18. (b)	19. (d)	20. (c)
21. (c)	22. (d)	23. (0.32 to 0.34)	24. (b)	25. (50)	26. (0·35)	27. (0.16)	28. (2)	29. (31·503)	
30. (c)	31. (0·8)	32. (1.5)	33. (d)	34. (0·25)	35. (6)	36. (2)	37. (11)	38. (0.37)	39. (d)
40. (c)	41. (a)	42. (a)	43. (b)	44. (d)	45. (d)	46. (c)	47. (a)	48. (b)	49. (b)
50. (c)	51. (b)	52. (d)	53. (a)	54. (b)	55. (d)	56. (4)	57. (b)	58. (–1)	59. (100)
60. (b)	61. (b)	62. (0.225)	63. (d)	64. (a)	65. (b,c)	66. (4)	67. (a)	68. (c)	69. (a)
70. (a)	71. (a)	72. (a)	73. (a)	74. (c)	75. (b)	76. (2·5)	77. (d)	78. (a)	79. (0.1)

EXPLANATIONS

1. Variance of random X,

$$\sigma_x^2 = E[X^2] - (\overline{X})^2 \quad \text{...(i)}$$

Mean squere, $E[X^2] = \int_{-\infty}^{\infty} x^2 f_x(X)\,dx$

$$E[(-kx)^2] = \int_{-\infty}^{\infty} (-kx)^2 f_x(x)\,dx$$

$$= k^2 \int_{-\infty}^{\infty} x^2 f_x(x)\,dx$$

$$E[(-kx)^2] = K^2 E[x^2] \quad \text{...(ii)}$$

Mean Value, $\overline{x} = E[x] = \int_{-\infty}^{\infty} x f_x(x)\,dx$

$$E[-kx] = \int_{-\infty}^{\infty} (-kx) f_x(x)\,dx = -k\overline{x} \quad \text{...(iii)}$$

Variance, σ^2 = Mean-square value – (Mean)^2

$$\sigma^2 = k^2 E[X^2] - (-k\overline{X})^2$$

$$= k^2 E[X^2] - k^2 (\overline{X})^2$$

$$= k^2 \{E[X^2] - (\overline{X})^2\}$$

$$\sigma^2 = k^2 \sigma_x^2.$$

2. Half-wave rectification can be represented as

$y = x$ for $x \geq 0$

$= 0$ for $x < 0$.

So, $f(y) = \frac{1}{2}\delta(y) + \frac{1}{\sqrt{2\pi N}} . e^{\frac{-y^2}{2N}}$

$$E(y^2) = \int_0^{\infty} y^2 f(y)\,dy$$

$$= \int_0^{\infty} y^2 \left[\frac{1}{2}\delta(y) + \frac{1}{\sqrt{2\pi N}} e^{\frac{-y^2}{2N}}\right] dy$$

$$= E[y^2] = 0 + \int_0^{\infty} \frac{1}{\sqrt{2\pi N}} e^{\frac{-y^2}{2N}}\,dy$$

Let $\frac{y}{\sqrt{N}} = y \Rightarrow dy = \sqrt{N}.dt$

$$E[y^2] = \frac{1}{\sqrt{2\pi N}} . \int_0^{\infty} N t^2 . e^{\frac{-t^2}{2}} . \sqrt{N} dt$$

$$= \frac{N}{\sqrt{2\pi}} \int_0^{\infty} t^2 . e^{\frac{-t^2}{2}}\,dt$$

$$E[y^2] = \frac{N}{2}.$$

3. Mean $\bar{x} = \int_{-1}^{3} p(x)\,dx$

$$= \frac{1}{4}\int_{-1}^{3} x\,dx = \frac{1}{4}\left[\frac{9-1}{2}\right] = 1$$

Variance, $\sigma^2 = \int_{-1}^{3} x^2\,p(x)\,dx - \overline{(x)}^2$

$$= \frac{1}{4}\int_{-1}^{3} x^2\,dx. - (\bar{x})^2$$

$$= \frac{1}{12}[27+1] - 1 = \frac{4}{3}$$

4. The questions should be read as . . . . . . with Gaussian in phase and quadrature components. The actual function is of the form,

$$f_V(v) = ve^{v2}\ 2 \text{ for } v \geq 0$$

$$= 0 \text{ for } v\ 0$$

5. Given : Probability density function,

$$P(x)\frac{-x^2}{2}, \quad -\infty<x<\infty \qquad ...(i)$$

For Gaussian random variable, probability density function is

$$P(x) = \frac{1}{\sqrt{2\pi\sigma_x^2}}.e^{\frac{(x-\bar{x})^2}{2\sigma_x^2}} \quad (ii)$$

On companing eq (i) and eq (ii),

$$\sigma_x^2=1,\ \bar{x}=0,\ K=\frac{1}{\sqrt{2\pi}}.$$

6. For coherent BPSK synchronization required for detection and efficiency is $\frac{1}{2}$.

7. Since, $P_x(x)$ is continuous function

Hence, probability at point $[x = 4]$ tends to zero.

8. Probability distribution function of x is the function $F_X(x) = P(X \leq x)$ for every x from $-\infty$ to ∞.

and $P\{x_1 < X \leq x_2\} = F_X(x_2) - F_X(x_1)$

then, at $X = 1$,

$$P\{1^- < X \leq 1^+\} = F_X(1^+) - F_X(1^-)$$

$$= 0.55 - 0.25$$

$$= 0.30$$

9. $$m^2 = \int_{-\infty}^{\infty} x^2 f(x)\,dx = \int_{0}^{1} x^2\,fx\,dx$$

$$\int_{0.3}^{1} x^2\,dx = \left[\frac{x^3}{3}\right]_{0.3}^{1}$$

$$= 1 - 0.027$$

$$m^2 = 0.324$$

$$m = 0.5695 \simeq 0.57$$

10. $$\int_{-\infty}^{\infty} P(v)dv = 1$$

$$\Rightarrow \quad \frac{1}{2}4 \times k = 1$$

$$\Rightarrow \quad k = \frac{1}{2}$$

$$\therefore \quad P(v) = \frac{v}{8}$$

Now, mean square value

$$= \int_{-\infty}^{\infty} v^2P(v)\cdot dv = \int_{0}^{4} v^2P(v)\cdot dv$$

$$= \int_{0}^{4}\frac{v^2v}{8}\cdot dv = \frac{1}{8}\left[\frac{v^4}{4}\right]_0^4 = 8$$

11. $$\int_{-\infty}^{\infty} P(x)dx = 1$$

Since, equiprobable, hence each has same probability $= \frac{1}{3}$

$$\therefore \quad 2a \times \frac{1}{4} = \frac{1}{3}$$

$$\Rightarrow \quad a = \frac{2}{3}$$

12. Quantization noise power

$$= \int_{-a}^{a} f(x)x^2dx = \frac{1}{4}\frac{\left[x^3\right]_{-a}^{a}}{3}$$

$$= \frac{1}{12}\left[a^3 + a^3\right] = \frac{a^3}{6} = \frac{\left(\frac{2}{3}\right)^3}{6} = \frac{4}{81}$$

13.

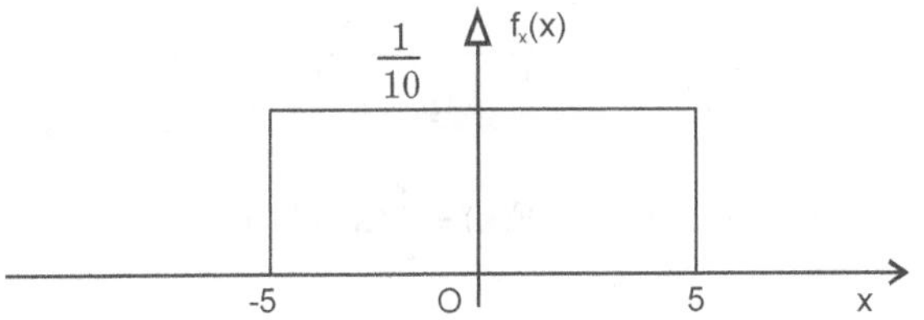

Now, $$f_Y(y) = \sum_{i=1}^{2} P(Y = y_i)\delta(y - y_i)$$

For $$i = 1,\ y_i = 0$$

$$P(Y = 0) = P(X < 2.5) + P(2.5 \leq X < \infty)$$

$$= \int_{-\infty}^{-2.5} f_X(x)\cdot dx + \int_{2.5}^{-\infty} f_X(x)\cdot dx$$

$$= \int_{-5}^{-2.5} \frac{1}{10} dx + \int_{2.5}^{5} \frac{1}{10} dx = 0.5$$

For $i = 2$, $Y_2 = 1$,

$$P(Y = 1) = P(-2.5 \le X \le 2.5)$$

$$= \int_{-2.5}^{2.5} f_x(x) \cdot dx = \int_{-2.5}^{2.5} \frac{1}{10} dx = 0.5$$

Hence, $$f_Y(y) = \sum_{i=1}^{2} P(Y = y_i)\delta(y - y_i)$$

$$= P(Y = 0)\,\delta(y - 0) + P(Y = 1)\,\delta(Y - 1)$$

$$= 0.5\,\delta(Y) + 0.5\,\delta(Y - 1)$$

14. $$\text{Var}[x] = \sigma_x^2 = E[(x - \mu_x)^2]$$

where, $\mu_x = E[x]$

μ_x = expected or mean value of X

Defining, $$E[X] = \int_{-\infty}^{\infty} x f_x(x) dx$$

$$= \int_{-\infty}^{\infty} x\left[\sum_i \rho(x_i)\delta(x - x_i)\right]dx$$

$$= \sum_i x_i \rho(x_i)$$

Variance σ_x^2 is a measure of the spread of the values of X from its mean μ_x.

Using relation, $E[X + Y] = E[X] + E[Y]$

and $E[CX] = C\,E[X]$

on $$\text{var}[X] = \sigma_x^2 = E[(X - \mu_x)^2]$$

$$\sigma_x^2 = E[X]^2 - \mu_x^2$$

$$= E[X^2] - E^2[X]$$

15. CDF : $$F(x) = \int_{-\infty}^{x} (\text{PDF})\, dx$$

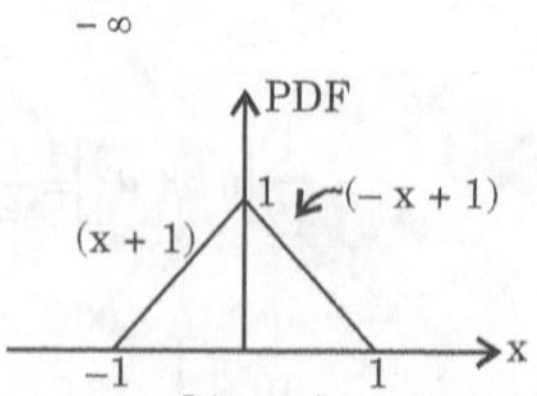

Fig. (a)

For $x < 0$, $$F(x) = \int_{-1}^{x} (x + 1)\, dx$$

$$= \frac{x^2}{2} + x + \frac{1}{2} \text{(concave upwards)}$$

$\because$ $$F(0) = \frac{1}{2}$$

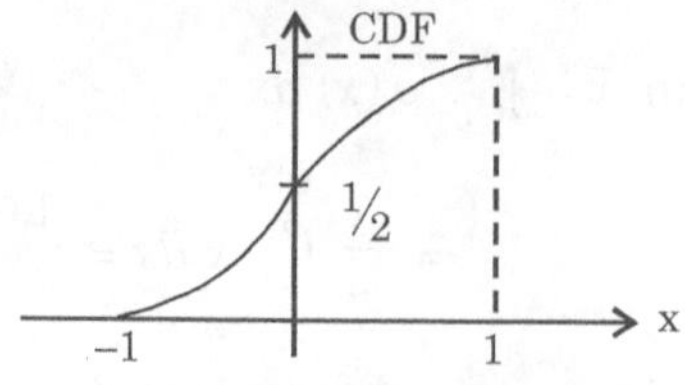

Fig. (b)

For $x > 0$, $$F(x) = F(0) + \int_{0}^{x} (-x + 1)\, dx$$

$$= \frac{1}{2} + \left(\frac{-x^2}{2} + x\right) \text{(concave downwards)}$$

Hence the CDF is shown in the figure (a).

16. Given: $P_x(x) = Me^{-2|x|} + Ne^{-3|x|}$

For PDF < $P_x(x)$,

$$\int_{-\infty}^{\infty} P_x(x)\, dx = 1$$

$$\int_{-\infty}^{\infty} (Me^{-2|x|} + Ne^{-3|x|})\, dx$$

$$= 2\int_{0}^{\infty} (Me^{-2|x|} + Ne^{-3|x|})\, dx = 1$$

$$\Rightarrow \quad M + \frac{2}{3}N = 1$$

17. Both the teacher and student are wrong

$$\overline{X} = (\text{Mean}) = \sum_{i=1}^{n} p_i k_i$$

$$= 0.1 + 0.4 + 1.2 + 0.8 + 0.5$$

$$= 3.0$$

$$\overline{X}^2 = \sum_{i=1}^{n} p_i k_i^{-2}$$

$$= 0.1 \times 1 + 0.2 \times 4 + 0.4 \times 9 + 0.2 \times 16 + 0.1 \times 25$$

$$= 0.1 + 0.8 + 3.6 + 3.2 + 2.5$$

$$= 10.2$$

18. Output noise power

$$|H(t)|^2 \cdot (\text{Input noise PSD})$$

$$S_{No}(f) = |H(f)|^2 \cdot S_{Ni}(f) = 10^{-20} \cdot |H(f)|^2.$$

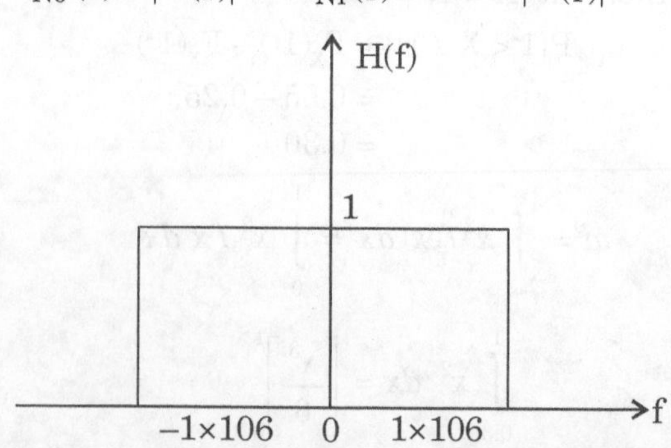

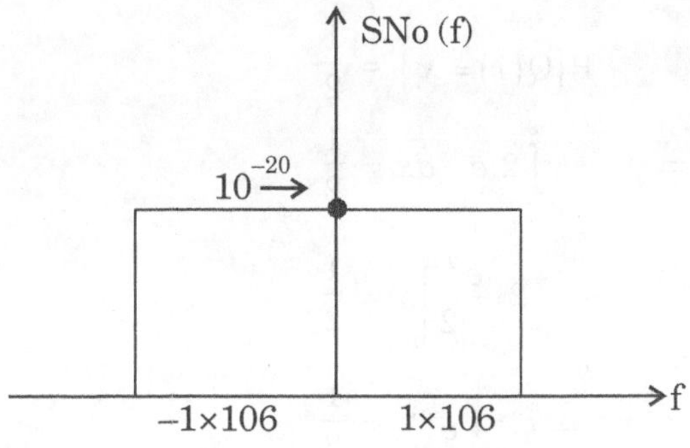

Output noise power = $\int^{\infty} S_{No}(f).df$

$= 10^{-20}(W/Hz) \times 2 \times 10^6 Hz$

$= 2 \times 10^{-14} W.$

⇒ Mean square value = power of signal.

$\frac{2}{\propto^2} = 2 \times 10^{-14} \Rightarrow \boxed{\propto = 10^7}$

19. When a '1' is transmitted : $Y_k = a + N_k$.

Threshold, $Z = \frac{a}{2} = 10^{-6} \Rightarrow a = 2 \times 10^{-6}$.

for error to occur, $Y_k < 10^{-16}$

$2 \times 10^{-6} + N_k < 10^{-6} \Rightarrow N_k < -10^{-6}$.

$$P\left(\frac{0}{1}\right) = \int_{-\infty}^{-10^{-6}} P_{NK}(n)dn$$

$$= \int_{-\infty}^{-10^{-6}} 0.5 \propto e^{\propto n} dn$$

$P\left(\frac{0}{1}\right) = 0.5e^{-10}$ with $\propto = 10^7$.

when a '0' is transmitted : $Y_k = N_k$.

for error to accur, $Y_k > 10^{-6}$

$$\therefore P\left(\frac{1}{0}\right) = \int_{10^{-6}}^{\infty} P_{NK}(n).dn$$

$0.5 \times e^{-10}$.

Since both bits are equiprobable,

$P(0) = P(1) = \frac{1}{2}$

The probability of error =

$$P(1).P\left(\frac{0}{1}\right) + P(0).P\left(\frac{1}{0}\right)$$

$$= \frac{P\left(\frac{0}{1}\right) + P\left(\frac{1}{0}\right)}{2} = 0.5 \times 10^{-10}.$$

20. $E_x = 5\cos(\omega t + \beta z)$

$E_y = 3\cos\left(\omega t + \beta z - \frac{\pi}{2}\right)$

$\phi = -\frac{\pi}{2}$

But the wave is propagating along negative z-direction

So, left hand elliptical (LED)

21.

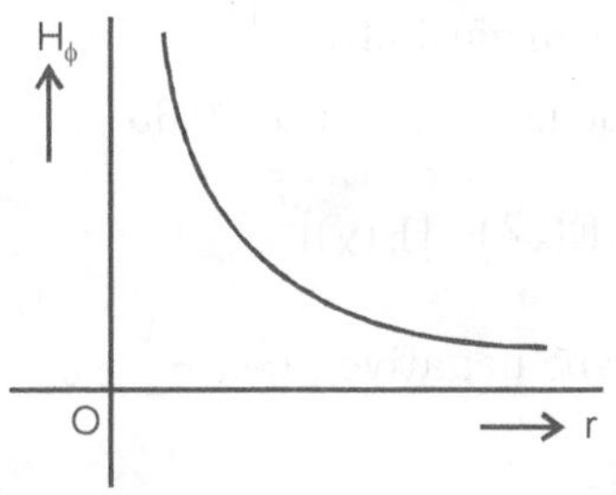

We know that magnetic field around a current carrying conductor

$H_\phi = \frac{I}{2\pi r} i\phi$

$|H_\phi| = \frac{I}{2\pi r}$

$|H_\phi| \ \alpha \ \frac{1}{r}$

22. Let distribution function U and V be $F_u(u)$ and $F_v(v)$ respectively.

$F_v(u) = F_v(v) = F(x)$

CDF of U and 2V are F (x) and G (x).

CDF is Fu (u) = $P(u \le u)$ =F (x).

$F_v(u) = P(v \le u)$

$F_{2v} = P(2V \le v) = P\left(V \le \frac{v}{2}\right) = G(x)$.

Since U and V are identical, u and v are also same, i.e., u = v = x.

for positive value of random variable U and V ; if $x_1 > x_2$ then

$F_x(x_1) > F_x(x_2)$.

$F(x) > G(x) \quad \because x > \frac{x}{2}$

$F(x) - G(x) > 0 \qquad ...(i)$

For negative value of random variable

U and V ; $\frac{v}{2} > u$.

$G(x) - F(x) > 0 \qquad ...(ii)$

for zero value of random variable

U and V, i.e., x = 0

$F(x) = G(x)$

$F(x) - G(x) = 0 \qquad ....(iii)$

Combining equations (i), (ii) and (iii),

$[F(x) - G(x)].x \geq o.$

23. X_1, X_2, X_3 are independent and identically distributed random variables.

So P{X_1 is the largest} = 1/3 = 0.33

24. Variance, $\sigma_x^2 = E(x^2) - [E(x)]^2$

$\because \sigma_x^2$ can never be negative ; i.e., $\sigma_x^2 \geq 0$

$E(x^2) \geq [E(x)]^2$.

25. $X = 1, 3, 5, ..., 99$

$\Rightarrow n = 50$ (number of observations)

$$\therefore \quad E(x) = \frac{1}{n}\sum_{i=1}^{n} x_i$$

$$= \frac{1}{50}[1+3+5+...+99]$$

$$= \frac{1}{50}(50)^2 = 50$$

26.

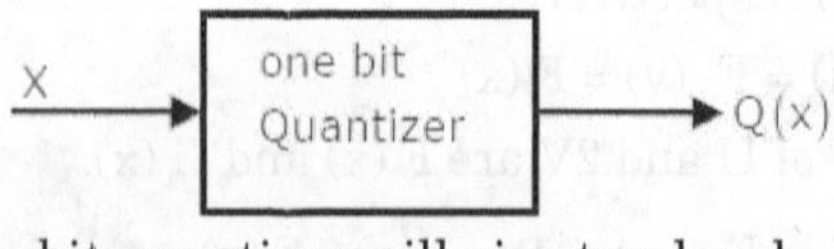

One bit quantizer will give two levels.

Both levels have probability of $\frac{1}{2}$

Pd of input X is

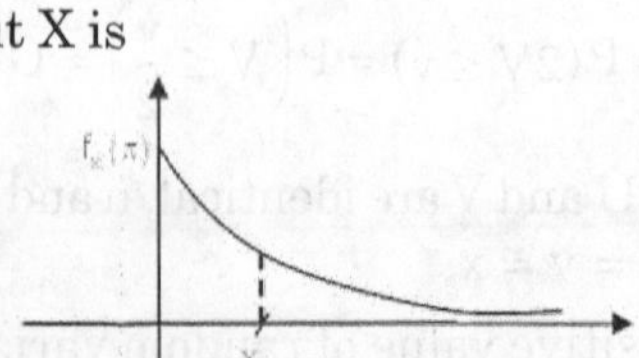

Let x_T be the threshold

$$Q(x) = \begin{bmatrix} x_1, & x \geq x_T \\ x_2, & x < x_T \end{bmatrix}$$

Where x_1 and x_2 are two levels

$$P\{Q(r) = x_1\} = \frac{1}{2}$$

$$\Rightarrow \int_{x_T}^{\infty} 2.e^{-2x}dx = \frac{1}{2}$$

$$2 \cdot \left.\frac{e^{-2x}}{-2}\right|_{x_T}^{\infty} = \frac{1}{2}$$

$$-e^{-2\infty} + e^{-2x_T} = \frac{1}{2}$$

$$e^{-2x_T} = \frac{1}{2}$$

$$-2x_T = \ln\frac{1}{2}$$

$$-2x_T = -0.693$$

$$x_T = 0.35$$

27. Given $x_1 x_2$ and x_3 be independent and identically distributed with uniform distribution on [0, 1]

Let $\qquad z = x_1 + x_2 - x_3$

$$\Rightarrow P\{x_1 + x_2 \leq x_3\} = P\{x_1 + x_2 - x_3 \leq 0\}$$

$$= P\{z \leq 0\}$$

Let us find probability density function of random variable z.

Since Z is summation of three random variable x_1, x_2 and $-x_3$

Overall pdf of z is convolution of the pdf of $x_1 x_2$ and $-x_3$

pdf of $\{x_1 + x_2\}$ is

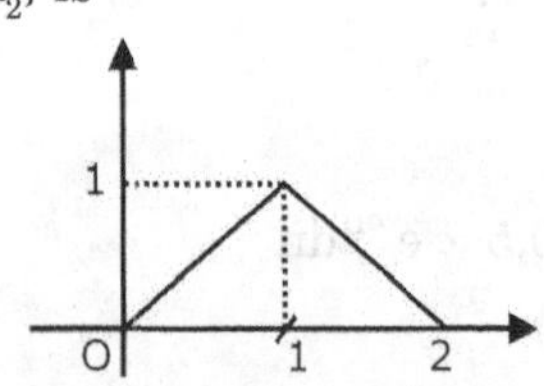

pdf of $-x_3$ is

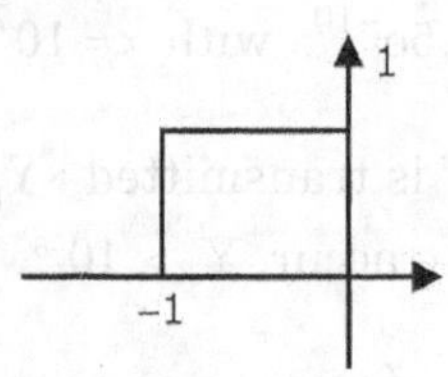

$$P\{z \leq 0\} = \int_{-1}^{0} \frac{(z+1)^2}{2}dz$$

$$= \left.\frac{(z+1)^3}{6}\right|_{-1}^{0}$$

$$= \frac{1}{6} = 0.16$$

28. Let ___________$P\{x = 2\}$________$= \frac{1}{3}$,

$$P\{x = 0\} = \frac{2}{3}$$

To find $H(Y_1)$ we need to know

$$P\{y_1 = 0\} \text{ and } P\{y_2 = 1\}$$

$$P\{Y_1 = 0\} = P\{Y_1 = 0/x_1 = 0\}$$

$$P\{x_1 = 0\} + P\{y_1 = 0/x_1 = 1\}\ P\{x_1 = 1\}$$

$$= \frac{1}{2}.\frac{1}{3} + \frac{1}{2}\times\frac{2}{3} = \frac{1}{2}$$

$$P\{y_1 = 1\} = \frac{1}{2}$$

$$\Rightarrow \quad H(y_1) = \frac{1}{2}\log_2^2 + \frac{1}{2}\log_2^2 = 1$$

Similarly, $P(y_2 = 0\} = \frac{1}{2}$

and $P\{y_2 = 1\} = \frac{1}{2}$

$$\Rightarrow \quad H\{y_2\} = 1$$

$$\Rightarrow \quad H\{y_1\} + H\{y_2\} = 2 \text{ bits}$$

29. From given data,

Channel transmission rate (C) = 52 kbps

Channel band width B = 4 kHz

$$\frac{\eta}{2} = 2.5 \times 10^{-5}$$

$$N = 4 \times 10^3 \times 2.5 \times 10^{-5} \times 2$$

$$C = B \log_2\left[1 + \frac{S}{N}\right]$$

$$S = 1638.2$$

$$E_b = \frac{S}{R_b} = \frac{\text{J/sec}}{\text{bits/sec}} = 31.503$$

$$\frac{C}{B} = \log_2(1 + S/N)$$

$$\Rightarrow \log_2(1 + S/N) = \frac{C}{B}$$

$$\Rightarrow (1 + S/N) = 2^{C/B} = 2^{52/4} = 2^{13} = 8192$$

$$\Rightarrow S/N = 8191$$

$$\therefore \quad S = 8191 \times N$$

$$\Rightarrow S = 8191 \times 4 \times 10^3 \times 2.5 \times 10^{-5} \times 2$$

$$= 819.1 \times 2$$

$$E_b = \frac{819.1 \times 2}{R_b} = 31.503$$

So the minimum bit energy (E_b) is 31.503 mJ/bit.

30. $H_1 : x = +1;\ H_0 : x = -1$

$P(H_1) = 0.75;\ P(H_2) = 0.25$

Received signal $\gamma = X + Z$

Where $Z \sim N(0, -2);\ f_z(z) = \frac{1}{\sigma\sqrt{2\pi}} e^{-z^2/2\sigma^2}$

$$\text{Received signal } \gamma = \begin{cases} 1 + Z & \text{if} \quad X = 1 \\ -1 + Z & \text{if} \quad X = -1 \end{cases}$$

$$f_y(y/H_1) = \frac{1}{\sigma\sqrt{2\pi}} e^{-\frac{1}{2\sigma^2}(\gamma-1)^2}$$

$$f_y(y/H_0) = \frac{1}{\sigma\sqrt{2\pi}} e^{-\frac{1}{2\sigma^2}(\gamma+1)^2}$$

At optimum threshold y_{opt}: for minimum probability of error

$$\left.\frac{f_y(y/H_1)}{f_y(y/H_0)}\right|_{y=y_{opt}} = \frac{P(H_0)}{P(H_1)}$$

$$\left. e^{-\frac{1}{2\sigma^2}\left[(\gamma-1)^2-(\gamma+1)^2\right]}\right|_{y_{opt}} = \frac{P(H_0)}{P(H_1)}$$

$$e^{+2y_{opt}/\sigma^2} = \frac{P(H_0)}{P(H_1)}$$

$$y_{opt} = \frac{\sigma^2}{2} 1_n \left(\frac{P(H_0)}{P(H_1)}\right) = \frac{-1.1\sigma^2}{2} = -0.55\sigma^2$$

y_{opt} = Optimum threshold

$y_{opt} < 0 \ \therefore$ Threshold is negative

31. Given channel

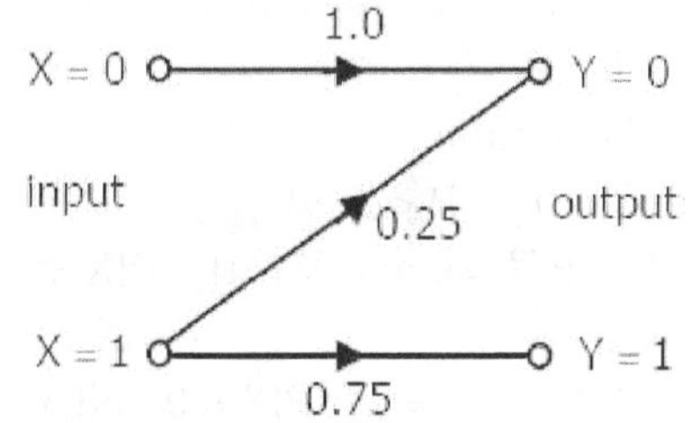

We have to determine, $P\{x = 0 / y = 0\}$

$$P\{x = 0 / y = 0\} = \frac{P\{y = 0/x = 0\}P\{x = 0\}}{P\{y = 0\}}$$

$$= \frac{1 \cdot 1/2}{1 \cdot 1/2 + 0.25 \times \frac{1}{2}}$$

$$= \frac{4}{5} = 0.8$$

32. Let x be a random variable which denotes number of tosses to get two heads.

$$P(x = 2) = HH = \frac{1}{2}\times\frac{1}{2}$$

$P(x = 3) = THH = \frac{1}{2}\times\frac{1}{2}\times\frac{1}{2}$

$P(x = 4) = TTHH = \frac{1}{2}\times\frac{1}{2}\times\frac{1}{2}\times\frac{1}{2}$

....

$$E(x) = 2\left(\frac{1}{2}\times\frac{1}{2}\right)+3\times\left(\frac{1}{2}\times\frac{1}{2}\times\frac{1}{2}\right)+4\left(\frac{1}{2}\times\frac{1}{2}\times\frac{1}{2}\times\frac{1}{2}\right)+.....$$

$$= 2\times\frac{1}{2^2}+3\times\frac{1}{2^3}+4\times\frac{1}{2^4}+.....$$

$$= \frac{1}{2}\left[2.\frac{1}{2^2}+3.\frac{1}{2^2}+4\times\frac{1}{2^3}+.....\right]$$

$$= \frac{1}{2}\left[\left(1+2.\frac{1}{2}+3\frac{1}{2^2}+.....\right)-1\right]$$

$$= \frac{1}{2}\left[\left(1-\frac{1}{2}\right)^{-2}-1\right]$$

$$= \frac{1}{2}[4-1]$$

$$= \frac{3}{2}$$

33. $P\{x = 0\} = P \Rightarrow P\{x = 1\} = 1 - p$

$P\{y = 0\} = q \Rightarrow P\{y = 1\} = 1 - q$

Let $Z = X + Y$

X	Y	Z
0	0	0
0	1	1
1	0	1
1	1	2

From above table,

$P\{X + Y + Z\} \Rightarrow P < Z \geq B$

$P\{Z \geq 1\} = P\{X = 0 \text{ and } Y = 1\} + P\{X = 1 \text{ and } Y = 0\} + P\{X = 1 \text{ and } Y = 1\}$

$= 1 - P\{X = 0 \text{ and } Y = 0\}$

$= 1 - pq$

34. $$y(t) = \sum_{n=-\infty}^{\infty} X_n P(t - nT - \phi)$$

$$R_{yy(z)} = \left[1-\frac{|\tau|}{T}\right]$$

Derivation of above autocorrelation function can be found in any book dealing with random process. [B.P. Lathi, Simon, Haykin, Schaum series].

$$R_{yy}\left(\frac{3T}{4}\right) = \left[1-\frac{3\pi/4}{\pi}\right]$$

$$= \frac{1}{4} = 0.25$$

35. Given : PDF $f(x) = \frac{1}{2}|x|.e^{-x}$.

$V(x) = E(x^2) - [E(x)]^2$

$$E(x) = \int_{-\infty}^{\infty} xf(x)\,dx$$

$$= \int_{-\infty}^{\infty} \frac{1}{2}|x|e^{x}dx = 0 \quad \because \text{ function is odd.}$$

$$E(x^2) = \int_{-\infty}^{\infty} x^2 f(x)\,dx$$

$$= \int_{-\infty}^{\infty} x^2\frac{1}{2}|x|e^{-x}dx = 6.$$

$V(x) = 6.$

36. Process, $x(t) = U + Vt$.

At $t = 2$, $x(2) = U + 2V$.

$E[x(t)] = E[U+2V]$

$E[U] = E[U] + 2E[V]$

$= 0.$

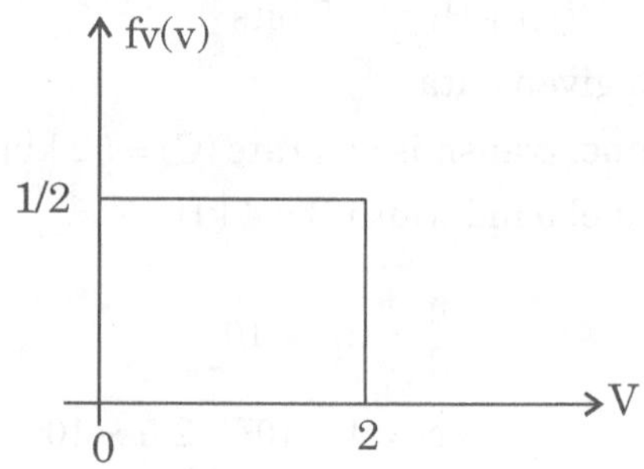

Also given that V is uniformly distributed between 0 and 2.

$$\Rightarrow E[V] = \int_{-\infty}^{\infty} f_v(v)\,dv = \int_0^2\left(\frac{1}{2}\right)dv = 1.$$

$\Rightarrow E[x(t) = 0 + 2 \times 1$

$= 2.$

37. Given, $E[2x + y] = 0$

and $E[x + 2y] = 33$

$\Rightarrow \quad 2E(x) + E(y) = 0 \quad$ ...(i)

$\Rightarrow \quad E(x) + 2E(y) = 33 \quad$ ...(ii)

Solving eq. (i) and (ii) we get

$E(y) = 22$ and $E(x) = -11$

$\therefore \quad E(x) + E(y) = -11 + 22 = 11$

$\Rightarrow \quad E(x) + E(y) = 11$

38. Given,

Cumulative distribution function,

$$F_z(x) = \begin{cases} 1 - e^{-x} & \text{if } x \geq 0 \\ 0 & \text{if } x < 0 \end{cases}$$

$\rightarrow$ probability density function.

$$F_z'(x) = f_z(x) = \begin{cases} e^{-x}; & x \geq 0 \\ 0; & x < 0 \end{cases}$$

Where Z be an exponential R.V with mean '1'

Using conditional probability

$$\therefore \Pr\left[\frac{z > 2}{z > 1}\right] = \frac{\Pr[z > 2 \cap z > 1]}{\Pr[z > 1]}$$

$$= \frac{\Pr[z > 2]}{\Pr[z > 1]} = \frac{\int_2^{\infty} f(x)dx}{\int_1^{\infty} f(x)dx} = \frac{\int_0^{\infty} e^{-x}dx}{\int_1^{\infty} e^{-x}dx}$$

$$\Rightarrow \Pr\left[\frac{z > 2}{z > 1}\right] = \frac{\left[-e^{-x}\right]_2^{\infty}}{\left[-e^{-x}\right]_1^{\infty}} = \frac{\left[0 + e^{-2}\right]}{\left[0 + e^{-1}\right]}$$

$$= \frac{e^{-2}}{e^{-1}} = \frac{1}{e} \approx 0.37$$

39. Auto correlation function and power spectral density makes the fourier transform pair.

$$R_X(\tau) \xleftrightarrow{\text{F.T.}} G_X(w).$$

$$R_X(\tau) = F^{-1}\left[\left(\frac{\sin f}{f}\right)^2\right]$$

$$= F^{-1}\left[\sin c^2\left(\frac{f}{\pi}\right)\right]$$

Inverse fourier transform of square of sinc function is always a triangular signal in time-domain.

40. $\frac{r_b}{B_o} = 1$ and $\frac{r_b}{B_p} = \frac{1}{22}$

$$\therefore \qquad B_p = 2\,B_o$$

41. $$(\text{PSD})\, S_g = \int_{-\infty}^{\infty} R_x(\tau)\, e^{-jW\tau}\, d\tau$$

$$R_x(\tau) = \underset{T\to\infty}{\text{Lt}} \frac{1}{T} \int_{-T/2}^{T/2} g(t)\, g(t+\tau)\, dt$$

$$P_g = \underset{T\to\infty}{\text{Lt}} \frac{1}{T} \int_{-T/2}^{T/2} |g(t)|^2\, dt$$

For a $g(t)$, $\quad S'_g = a^2 S_g$

$$P'_g = a^2 P_g$$

42. $V_{ar}[d(n)] = V_{ar}[x(n) - x(n-1)]$

$\Rightarrow E[\{x(n) - x(n-1)\}^2]$

$\Rightarrow E[x^2(n)] + E[x^2(n-1)] - 2E[x(n).\, x(n-1)]$

$\Rightarrow V_{ar}[x(n)] + V_{ar}[x(n-1)] - 2\,R_{xx}(-1)$

[shifting does not effect variance]

$\sigma_d^2 = \sigma_x^2 + \sigma_x^2 - 2\,R_{xx}$ (1)... even symmetric

$$\frac{\sigma_x^2}{10} = 2\sigma_x^2 - 2R_{xx}(1)$$

$$R_{xx}(1) = 0.95\,\sigma_x^2$$

$$\Rightarrow \qquad \frac{R_{XX}(1)}{\sigma_x^2} = 0.95$$

43.

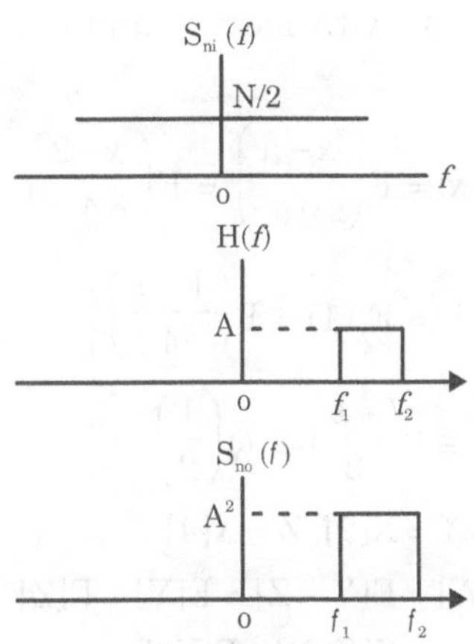

$S_{ni}(f)$= input white noise spectral density

$H(f)$= ideal frequency detector

= Output power spectral density

$$S_{no}(f) = |H(f)|^2 \cdot S_{ni}(f)$$

44.

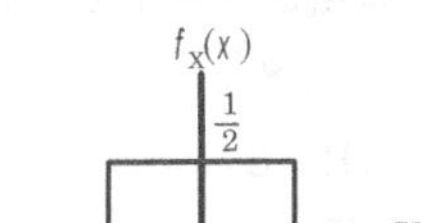

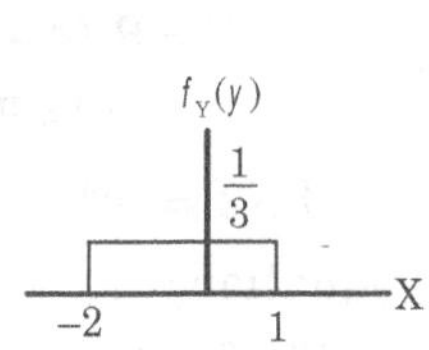

using central limit theorem

$$Z = X + Y$$

$$f_Z(z) = f_X(x) \otimes f_Y(y)$$

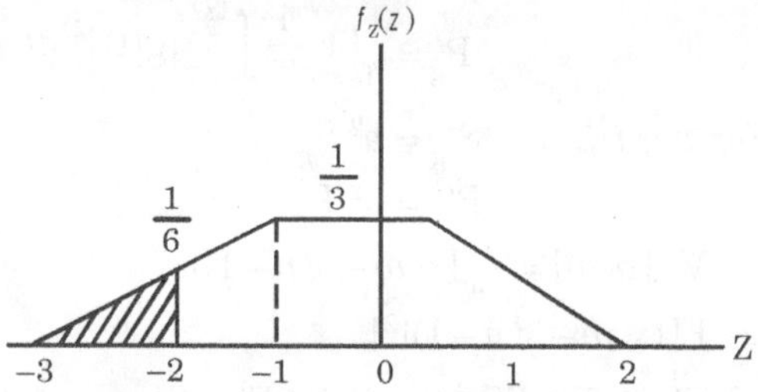

$$P(Z \leq -2) = \frac{1}{6} \times 1 \times \frac{1}{2} = \frac{1}{12}$$

45. Variance of $x(t)$, $\sigma x^2 = \overline{X}^2 - \mu^2$

μ is the mean value of $x(t)$ and $\mu = 2$

$\overline{X^2}$ ________$= R_x(0) = 8$

then $\sigma x^2 = 8 - 4 = 4$

Commulative distribution function,

Now, $F(X) = 1 - F(-X)$ and $F(-X) = Q(X)$

Now, $P(x \leq 1) = F_X(x)$

F(X) is expressed as with mean μ and variance σx^2

$$F_X(x) = F\left(\frac{x-\mu}{\sigma_x}\right) = F\left(\frac{x-2}{4}\right)$$

and $P(X \leq 1) = F_X(1) = F\left(\frac{1-2}{4}\right)$

$$= F\left(\frac{-1}{2}\right) = Q\left(\frac{1}{2}\right)$$

46. $Y = X[2], Z = X[4]$

$$E[W] = E[Y - Z] = E[Y] - E[Z]$$
$$= E[X(2)] - E[X(4)]$$
$$= 0$$

[As $x(t)$ is a random process having count at value of 2]

Now, $E[W^2] = E[Y^2] + E[Z^2] - 2E[YZ]$ ...(i)

$$E[X^2(t)] = R_x(0) = 4[e^{-0.2101} + 1] = 8$$

And, $E[Y^2] = E[X^2(2)] = 8$

$$E(Z^2) = E^2[X^2(4)] = 8$$
$$E[XZ] = E[X(Z)\,X(4)]$$
$$= R_X(4-2) = R_X(2)$$

[As $E[X(t)\,X(t+J) = R_X(J)]$

$$R_X(2) = [e^{-0.2|(4-2)|} + 1] = 6.68$$

then from equation (i),

$$E[W^2] = 8 + 8 - (2 \times 6.68) = 2.64$$

47. Total noise-power at the receiver is,

$$E[Y^2(t)] = \frac{1}{2\pi}\int_{-\infty}^{\infty} S_{YY}(\omega)\,d\omega$$

$$= N_0 W = 10^{-20} \times 100 \times 10^6$$
$$= 10^{-12} \text{ Watt.}$$

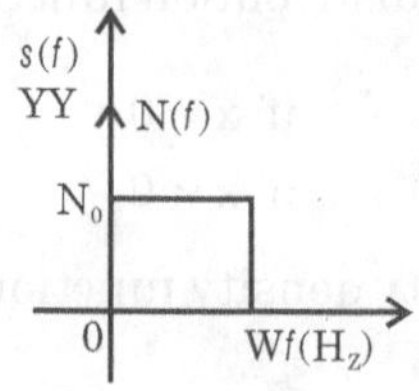

W = 100 MHz

$$\text{Signal - to - noise ratio} = \frac{1 \times 10^{-3}}{10^{-12}} = 10^9$$

In decibel, $10 \log_{10} 9 = 90$ dB

Considering loss of 40 dB, the SNR at the receiver is $(90 - 40)$ dB = 50 dB.

48. $H(\omega) = 2 \exp(-j\omega t_d)$

or, $H(f) = 2 \exp(-j2\pi f t_d)$

$$N_i(f) = N_o \frac{W}{Hz}$$

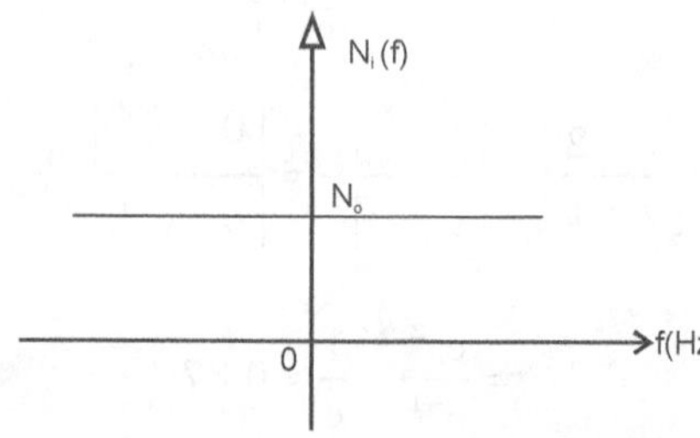

Ouput none power spectral density

$$N_O(f) = |H(f)|^2 Ni(f) = 4N_O \frac{W}{Hz}$$

PSD of LPF is shown below

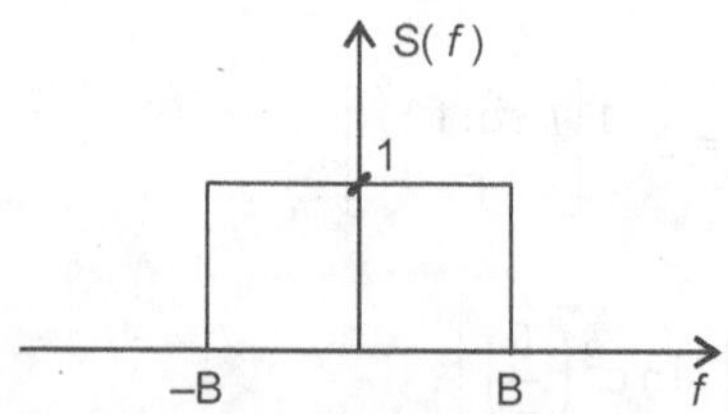

If output non passes through LPF, then PSD at the output of LPF will be as shown here

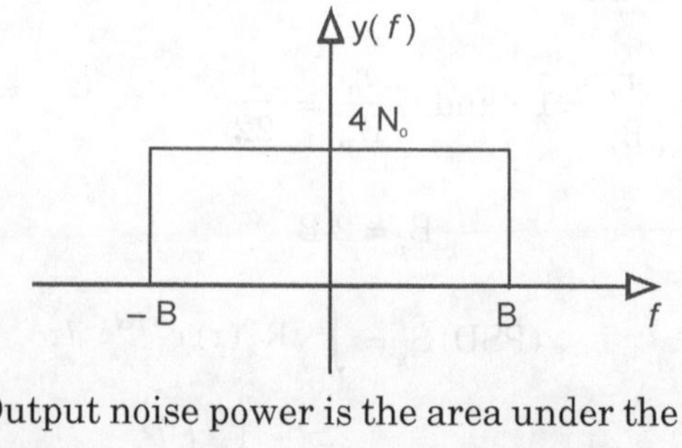

Output noise power is the area under the curved

$$= 4N_o \times 2B = 8N_o B$$

49. Power spectral density, $S(\omega) = \lim_{\tau\to\infty} \dfrac{|X_\tau(j\omega)|^2}{\tau}$

So for wide sense stationary random process, $S(f) \geq 0$

50. $R(\tau)$ is auto correlation function

$$R(\tau) = \frac{1}{T}\int_{-T/2}^{T/2} V(f)V(t+\tau)dt$$

and $$R(-\tau) = \frac{1}{T}\int_{-T/2}^{T/2} V(f)V(t-\tau)dt$$

Let $t-\tau = \sigma$

$$\therefore \quad R(-\tau) = \lim_{T\to\infty}\frac{1}{T}\int_{-T/2}^{T/2} V(\tau+\sigma)V(\sigma)d\sigma$$

$$R(-\tau) = R(\tau) \quad \text{(even function)}$$

51. Auto-correlation function and power spectral density form fourier transform pair,

$$R(\tau) \xleftrightarrow{\text{F.T.}} S(f).$$

Fourier transform of triangular pulse is sine squared function

$$A\left(1-\frac{|\tau|}{T}\right) \xleftrightarrow{\text{F.T.}} AT\sin^2[fT].$$

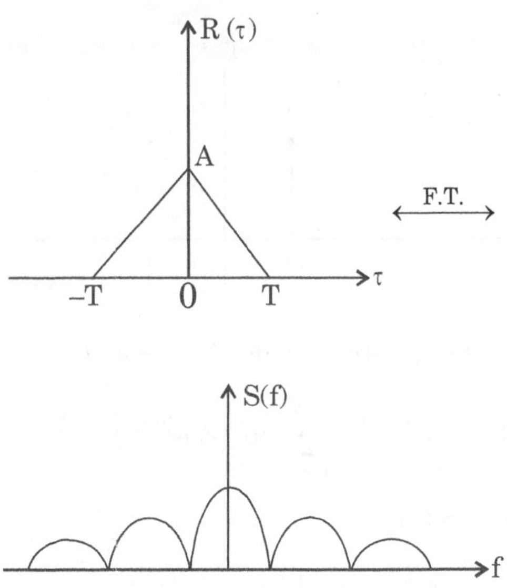

52.

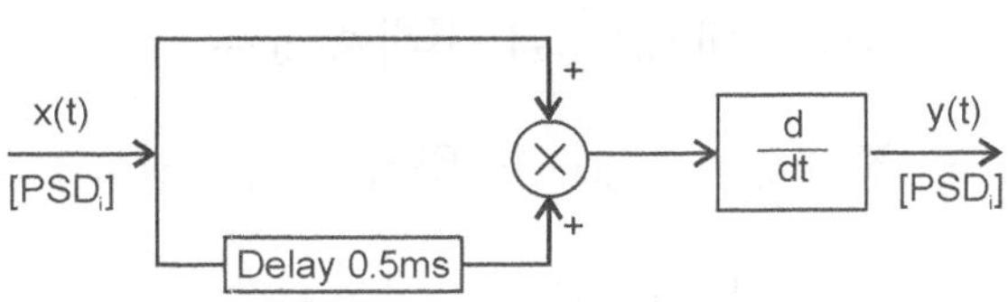

$$y'(t) = \frac{d}{dt}[x(t) + x(t-T)]$$

$$y'(f) = \left[x(f) + e^{-j2\pi fT} \times (f)\right]j\omega$$

$$H'(f) = j\omega\left[1 + e^{-j2\pi fT}\right]$$

$$= [1 + \cos 2\pi fT - j\sin 2\pi fT]j\omega$$

$$|H'(f)|^2 = \omega^2.2.(1 + \cos 2\pi fT)$$

$$\therefore \quad PSD_o = |H(f)|^2 PSD_i$$

$$= 2\omega^2[1 + \cos(\pi.2fT)]\, PSD_i$$

$$\therefore \quad [1 + \cos(\pi 2fT)] = 0;\ \text{for } f = \frac{(2n+1)}{2T}$$

53. In terms of fourier transform,

$$\mathcal{L}[x(t)] = x(f)$$

$$\mathcal{L}[y(t)] = y(f)$$

then, $$y(f) = x(f)\,H(f) - x(f)$$

$$y(f) = x(f)[H(f) - 1]$$

$$= x(f) - [j2\pi f - 1]$$

Power spectral density of $y(f)$,

$$S_y(f) = [j2\pi f - 1]^2 \cdot S_u(f)$$

[Using $S_y(f) = |H(f)^2| \cdot S_x f$]

Hence $H(f)$ is the transfer characteristic of the filter.

Given, $R_x(T) = e^{-(\pi J^2)}$

$$S_x(f) = e^{-(f^2)} \qquad \left[e^{-\pi t^2} \longleftrightarrow e^{-\pi f^2}\right]$$

as $e^{-\pi t^2}$ is a Gaussian function.

then, $$S_y(f) = e^{-\pi f^2} \cdot \left[1 + 4\pi^2 f^2\right]$$

54. PSD of $x(t)$

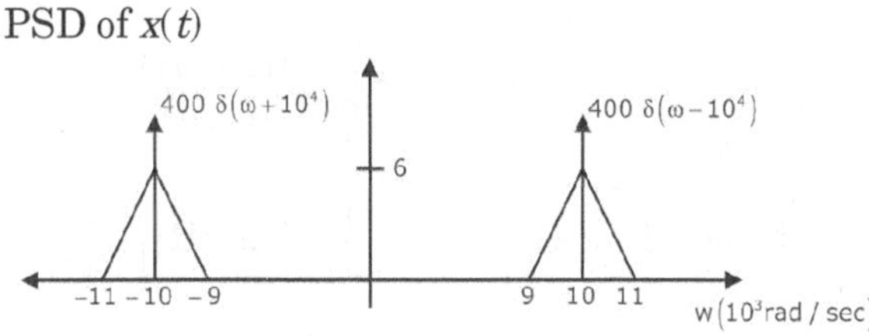

$$E[x^2(t)] = R_{xx}(0)$$

$$R_{xx}(0) = \frac{1}{2\pi}\int_{-\infty}^{\infty} S_{xx}(\omega)\, d\omega$$

$$R_{xx}(\tau) \leftrightarrow S_{xx}(\omega)$$

Fourier transform pair

$$= \frac{1}{2\pi}\left[\frac{1}{2}\times 2\times 10^3 \times 6 + \frac{1}{2}\times 2 \times 10^3 \times 6 + 400 + 400\right]$$

$$= \frac{6400}{\pi}$$

Since PSD of $x(t)$ does not contain any DC component, the mean value of $x(t)$ is zero.

55. Given $X(t) = \sqrt{2}\sin(2\pi t + \phi)$

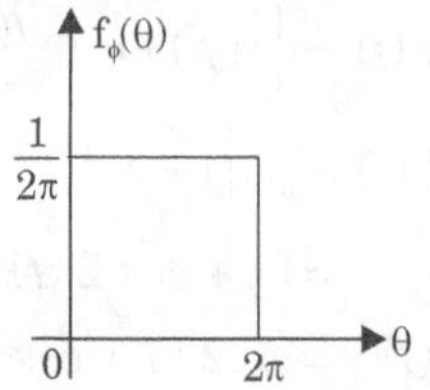

ϕ in uniformly distributed in the interval $[0, 2\pi]$

$E[X(t_1)X(t_2)]$

$$= \int_0^{2\pi} \sqrt{2}\sin(2\pi t_1 + \theta)\sqrt{2}\sin(2\pi t_2 + \theta) f_\phi(\theta)\, d\theta$$

$$= 2\int_0^{2\pi} \sin(2\pi t_1 + \theta)\sin(2\pi t_2 + \theta).\frac{1}{2\pi}.d\theta$$

$$= \frac{1}{2\pi}\int_0^{2\pi} \sin(2\pi(t_1 + t_2) + 2\theta)\, d\theta$$

$$+ \frac{1}{2\pi}\int_0^{2\pi} \cos(2\pi(t_1 - t_2))\, d\theta$$

First integral will result into zero as we are integrating from 0 to 2π.

Second integral result into $\cos[2\pi(t_1 - t_2)]$

$$\Rightarrow E[X(t_1)X(t_2)] = \cos(2\pi(t_1 - t_2))$$

56. Given $S_x(f) = \begin{cases} \frac{1}{\omega}, & |f| \le \omega \\ 0, & |f| \ge \omega \end{cases}$

$$R_x(t) = \int_{-\omega}^{\omega} \frac{1}{\omega} \cdot e^{j2\pi f t}\, df$$

$$= \frac{1}{\omega}\frac{e^{j2\pi\omega t} - e^{-j2\pi\omega t}}{j2\pi t}$$

$$= \frac{1}{\omega}\left(\frac{\sin(2\pi\omega t)}{\pi t}\right)$$

Now, $E\left[\pi \times (t) \cdot x\left(t - \frac{1}{4\omega}\right)\right] = \pi R_x\left(\frac{1}{4w}\right)$

$$\Rightarrow \quad \pi \cdot \frac{1}{w} \cdot \frac{\sin\left(2\pi w \cdot \frac{1}{4w}\right)}{\pi \cdot \frac{1}{4w}} = \frac{4}{1}$$

57.

$$R_Y(k) = R_y(n, n+k)$$

$$= E[Y(n).Y(n+k)]$$

$$Y(n) = x(n) + 0.5x(n-1)$$

$$R_Y(k) = E[(x[n]+0.5x[n-1])(x(n+k)+0.5x(n+k-1))]$$

$$= E[(x[n].x(n+k)+x(n)0.5x(n+k-1)+0.5x(n-1).x(n+k)$$
$$+0.25x(n-1)x(n+k-1)$$

$$= E[x[n].x(n+k)+0.5E[x(n)x(n+k-1)]$$
$$+0.5E[(x(n-1)x(n+k))]$$
$$+0.25E[x(n-1)x(n+k-1)]]$$

$$= R_X(k) + 0.5R_x(k-1) + 0.5R_x(k+1) + 0.25R_x(k)$$

$$R_y(k) = 1.25R_x(k) + 0.5R_x(k-1) + 0.5R_x(k+1)$$

$$R_x(k) = E[x(n).x(n+k)]$$

if $k = 0$

$$R_x(0) = E[x^2(n)]$$

$$= 1^2.\frac{1}{2} + (-1)^2 \times \frac{1}{2} = 1$$

if $k \ne 0$,

$$R_x(k) = E[x(n)].E[x(n+k)] = 0$$

$$\left\{\begin{array}{l} \because E[x(n)] = 0 \\ E[x(n+k)] = 0 \end{array}\right\}$$

$$\Rightarrow R_y(0) = 1.25R_x(0) + 0.5R_x(-1) + 0.5R_x(1)$$
$$= 1.25$$

$$R_y(1) = 1.25R_x(1) + 0.5R_x(0) + 0.5R_x(2) = 0.5$$

$$R_y(-1) = 1.25R_x(-1) + 0.5R_x(-2) + 0.5R_x(0)$$
$$= 0.5$$

$R_y(k)$ for k other than 0, 1 and $-1 = 0$

$\Rightarrow R_y(k)$

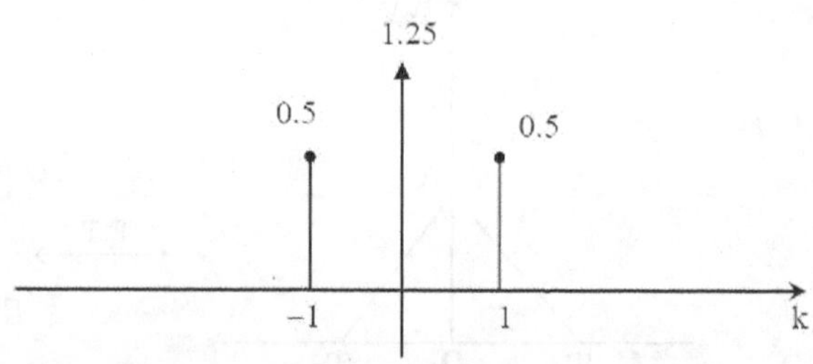

58. power spectral density of x(t) = s_x (f)

$$S_X(f) = \frac{|a(f)|^2}{T} \cdot \sum_{n=-\infty}^{\infty} R_b(\tau).e^{j2\pi f n\tau}$$

$$R_b(\tau) = E[\beta_n \beta_n - \tau]$$

$$= E\left[\propto_n + k \propto_{n-3} \left(\propto_{n-\tau} + k \propto_{n-\tau-3}\right)\right]$$

$$R_b(\tau) = E[\propto_n \propto_{n-\tau}] + KE[\propto_{n-3} \propto_{n-\tau}]$$

$$+ KE[\propto_n \propto_{n-\tau-3}] + K^3E[\propto_{n-3} \propto_{n-\tau-3}]$$

$$R_b(\tau) = E[\propto_n \propto_{n-\tau}] + KE[\propto_{n-3} \propto_{n-3+3-\tau}]$$

$$+KE[\propto_n \propto_{n-\tau-3}] + K^2E[\propto_{n-3}\propto_{n-\tau-3}]$$

$$R_b(\tau) = R(\tau) + KR(\tau-3) + KR(\tau-3) + K^2R(\tau).$$

$$R_b(\tau) = (1+K^2)R(\tau) + KR(\tau+3) + KR(\tau-3).$$

Auto-correlation function,

$$R_b(\tau) = \begin{cases} 1+k^2, & \tau = 0 \\ k, & \tau = \pm 3 \\ 0, & \text{otherwise} \end{cases}$$

Power spectral density,

$s_b(f) = 1 + k^2 + 2k\cos(2\pi f.3T)$

Null will occur at $f = \frac{1}{3T} \Rightarrow f = \frac{1}{3T}$.

$$s_b\left(f = \frac{1}{3T}\right) = 1 + k^2 + 2k.\cos 2\pi = 0$$

$1 + k^2 + 2k = 0$

$(k + 1)^2 = 0$

$k = -1$

59. Given random process

$$X(t) = 3V(t) - 8 \quad \text{and} \quad E[V(t)] = 0$$

$$R_v(\tau) = 4e^{-5|\tau|}$$

Power of X(t) $= E[X^2(t)]$

$= E[9V^2(t)] + 64 - 48\,E[V(t)]$

$= 9E[V^2(t)] + 64 - 48\,E[V(t)]$

$E[V^2(t)] = R_v(0) = 4$

Power of $X(t) = ((9 \times 4) + 64)$

$= 100$

Hence the power in $X(t)$ is 100.

60. $Y(t) = X(t) - X(t - T_o)$

Autocorrelation function for o/p

$= R_y(\tau) = E[y(t)\,Y(t+\tau)]$

$R_y(\tau) = E\,[(X(t) - X(t-T_o)]\,[X(t+\tau) - X(t+\tau-T_o)]$

$R_y(\tau) = E\,[(X(t)\,X(t+\tau) - X(t)\,X(t+\tau-T_o) - X(t-T_o)\,X(t+\tau) + X(t-T_o)\,X(t+\tau-T_o)]$

$R_y(\tau) = [R_x(\tau) - R_x(\tau - T_o) - R_x(\tau + T_o) + R_x(\tau)]$

$R_y(\tau) = 2\,R_x(\tau) - R_x(\tau - T_o) - R_x(\tau + T_o)$

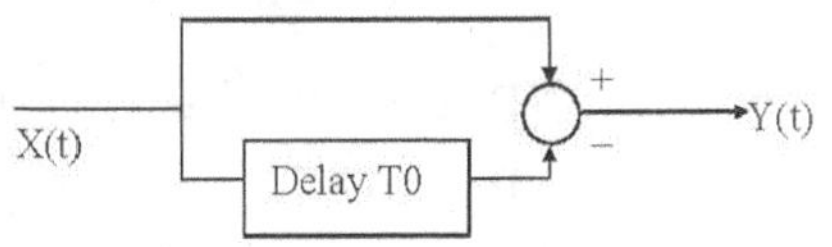

So, the autocorrection function of $R_y(\tau)$ of the output $Y(t)$ is $2R_x(\tau) - R_x(\tau - T_o) - R_x(\tau + T_o)$.

61. PSD of noise input,

$S_N(f) = 0.50$ W/Hz.

Power of y(t),

$$P_y = \int_{-\infty}^{\infty} S_N(f)|H(f)^2|df = 0.5\int_{-\infty}^{\infty} |H|(f)^2\,dt$$

$$\because h(t) = \frac{1}{2}.e^{\frac{-t^2}{2}}$$

$$p_y = \frac{1}{2}\int_{-\infty}^{\infty}\left(\frac{1}{2}e^{\frac{-t^2}{2}}\right)^2 dt$$

$$= \frac{1}{8}\int_{-\infty}^{\infty} e^{-t^2}\,dt$$

$$= \frac{\sqrt{\pi}}{8}$$

$p_y = 0.2215$ W.

62. Power in y(t) $= \left[\begin{matrix}\text{Power in} \\ h(t) \times X(t)\end{matrix}\right] + [\text{Power in } Z(t)]$

Power in $h(t) \times X(t) = \int_{-\infty}^{\infty} |H(f)|^2\,S_{XX}(f)df$

$$= \int_{-\infty}^{\infty} |H(f)|^2\,(5)df = \int_{-5}^{5} (025)(.5)df$$

$= (10)(1.25) = 12.5$ W

Power in $S_2(f)$ = Area under power spectral density

$$P_{z(t)} = \left(\frac{1}{2}\right)(10)(1) = 5\text{ W}$$

$\therefore$ Power in y(t) = (12.5 + 5) W = 17.5 W

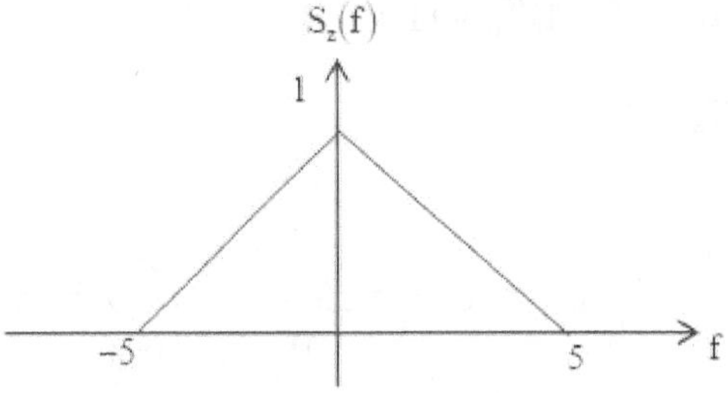

63. Narrow band representation of noise is

$n(t) = n_c(t).\cos(\omega_c t) - n_s(t).\sin(\omega_c t)$.

Its a = envelope is $R(t) = \sqrt{(n_c)^2 + (n_s)^2}$.

where, $n_c(t)$ and $n_s(t)$ are two independent, zero mean Gaussian processes, with same. The resulting envelope is Rayleigh variable.

64. A pre-amplifier is a very large gain amplifier with low noise figure. Noise figure of cascade amplifier is

$$F = F_1 + \frac{F_2 - 1}{a_1} + \frac{F_3 - 1}{a_1 a_2} + ...$$

$$+ \frac{F_n - 1}{a_1 a_2 a_3 ... G_{n-1}}$$

Therefore placing the pre-amplifier on the antenna side of the waveguide will result in the reduction of overall noise figure of the system.

65. Given : $T_e = 21$ K, $T_a = 300$ K,

Gain (G) in dB = 13 G = 19.95

Cable loss = 3 dB = $10 \log_{10} (\lambda)$

$\lambda = 1.955$.

For a cable, Noise figure (F_2) = Cable loss (λ) = 1.995.

Noise figure of amplifier $= F_1 = 1 + \frac{T_e}{T_a}$

$$1 + \frac{21}{300} = 1.07$$

Noise figure of cascade amplifier

$$F = F_1 + \frac{F_2 - 1}{a}$$

$$F = 1.07 + \frac{1.995 - 1}{19.95} = 1.12$$

F = 0.49 dB.

Effective noise temperature of the cascade amplifier,

$$T_e^1 = (F - 1) T_e = (1.12 - 1) \times 300$$

$$T_e^1 = 36K$$

66.

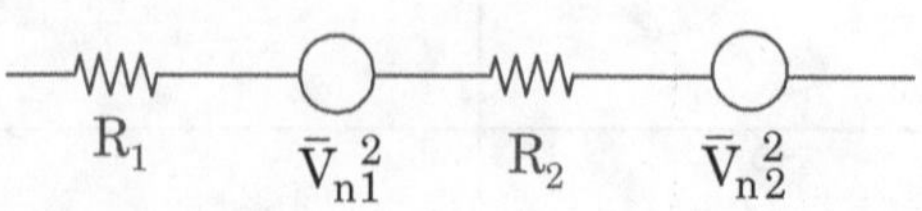

$$\overline{V}_n^2 = \overline{V}_{n_1}^2 + \overline{V}_{n_2}^2$$

$= 4 \text{ K } T_e \text{ BR} = 4 \text{ KT}_e \text{ B} (R_1 + R_2)$

$\because \overline{v}_{n_1}^2 = 4KT_1BR_1$, and $\overline{V}_{n_2}^2 = 4KT_2BR_2$

$4K\,Te\,B\,(R_1 + R_2) = 4KT_1BR_1 + 4KT_2\,BR_2F$

$T_e (R_1 + R_2) = R_1 T_1 + R_2 T_2$

$$T_e = \frac{R_1T_1 + R_2T_2}{R_1 + R_2}.$$

67. In cascade, overall noise figure

= 3 + 3 = 6 dB.

68. $X \in [-a, a]$ and $P(x = -a) = P(x = a) = \frac{1}{2}$

$\gamma = X + Z \rightarrow$ Received signal

$Q(a) = 1 \times 10^{-8}$

$Q(a) \approx e^{-v^2/2}$

$Z \sim N(\beta, X, 1)$

$$f_z(z) = \frac{1}{\sqrt{2\pi}} e^{-\frac{1}{2}(Z - \beta x)}$$

$$\gamma = \begin{cases} -a + z & \text{if } x = -a \\ a + z & \text{if } x = +a \end{cases}$$

$H_1 : x = +a$

$H_0 : x = -a$

and Threshold = 0

$$f_y(y/H_1) = \frac{1}{\sqrt{2\pi}} e^{-\frac{1}{2}(y - a(1+\beta))^2}$$

$$f_y(y/H_0) = \frac{1}{\sqrt{2\pi}} e^{-\frac{1}{2}(y + a(1+\beta))^2}$$

BER :

$P_e = P(H_1)P(e/H_1) + P(H_0)P(e/H_0)$

$$= \frac{1}{2}\int_{-\infty}^{0} \frac{1}{\sqrt{2\pi}} e^{-\frac{1}{2}(y - a(1+\beta))^2} dy + \frac{1}{2}\int_{0}^{\infty} \frac{1}{\sqrt{2\pi}} e^{-\frac{1}{2}(y + a(1+\beta))^2}$$

$dy = Q(a(1 + \beta))$

$\beta = 0$

$P_e = Q(a) = 1 \times 10^{-8} = e^{-a^2/2}$

$\Rightarrow \quad a = 6.07$

$\beta = -0.3$

$P_e = Q(6.07\,(1 - 0.3))$

$= Q(4.249)$

$P_e = e^{-(4.249)^2/2} = 1.2 \times 10^{-4}$

$P_e \simeq 10^{-4}$.

69.

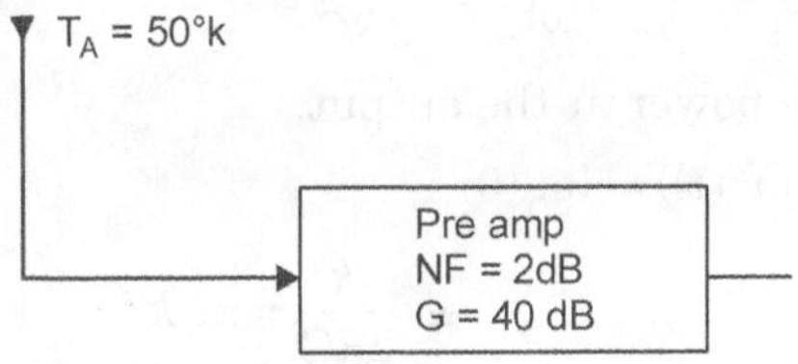

$$10 \log_{10} NF = 2dB$$

$$\log_{10} NF = 0.2$$

$$NF = 10^{0.2}$$

Noise temperature $= (NF - 1)\, T_o$

$$= (10^{0.2} - 1)\, 290°$$

$$= 169.36 \text{ K}$$

Noise power (at i/p)

$$= k\, T_e B$$

$$= 1.38 \times 10^{-23} \times (169.36 + 50) \times 12 \times 10^{6}$$

Noise power (at o/p) $= (3.632 \times 10^{-14}) \times 10^{4}$

$$= 3.73 \times 10^{-10} \text{ Watts}$$

70. **Case 1 :**

$P(t) = \phi_1(t),\; q(t) = -\phi_1(t)$

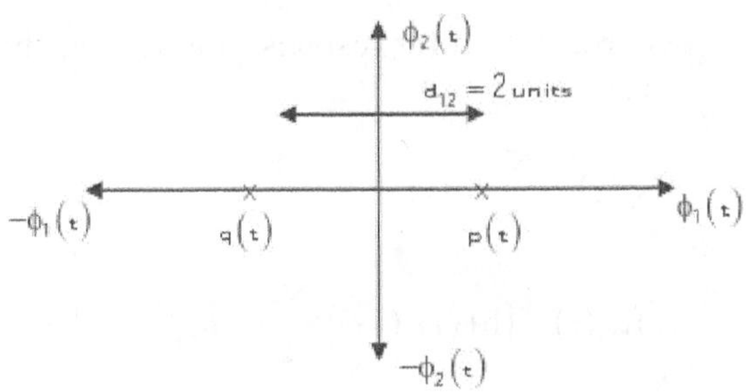

Case 2:

$P(t) = \phi_1(t)\; q(t) = \sqrt{E}\,\phi_2(t)$

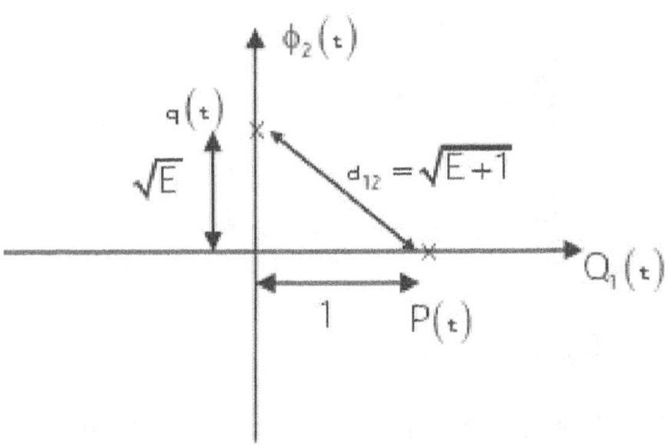

For same probability of error distance between points should be same for both cases

$$\therefore \sqrt{E+1} = 2$$

$$\Rightarrow E = 3$$

71.

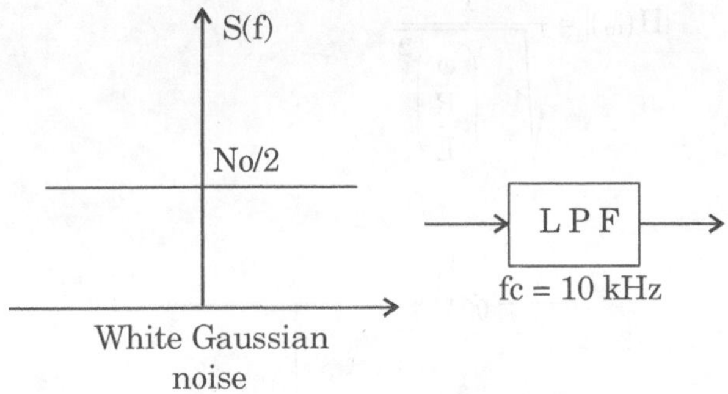

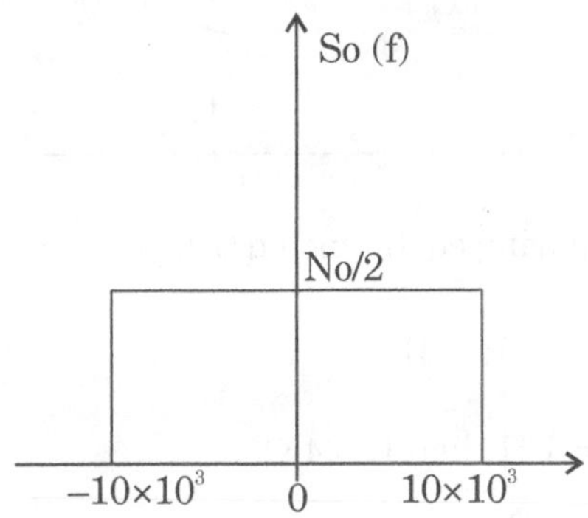

Inverse Fourier transform of $S_o(f)$ is

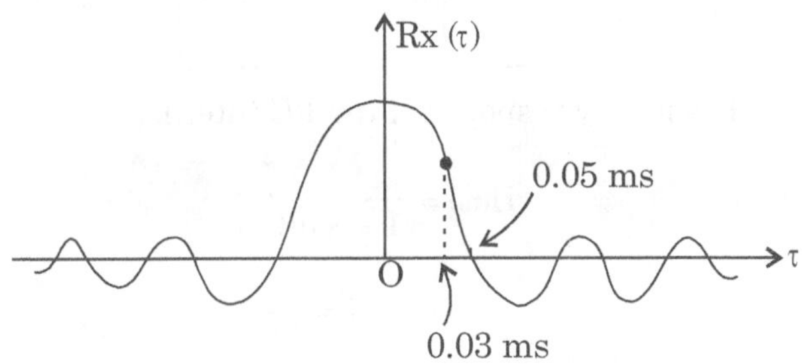

$\Rightarrow R_x(\tau) \neq 0$ at $\tau = 0.03$ ms.

Hence, samples obtained are correlated.

72. Power spectral density, $S_0(\omega) = \dfrac{16}{16+\omega^2}$

$$S_i(\omega) \longrightarrow \boxed{H(\omega)} \longrightarrow S_0(\omega)$$

$$\Rightarrow s_o(\omega) = |H|(\omega)^2 . s_i(\omega)$$

$$\frac{16}{16+\omega^2} = |H(\omega)^2 . 1|$$

$$|H(\omega)| = \frac{4}{\sqrt{16+\omega^2}} = \frac{1}{\sqrt{1+\left(\frac{\omega}{4}\right)^2}} \quad \text{...(i)}$$

It is a first - order low-pass R-L filter.

73.

$$|H(\omega)| = \frac{1}{\sqrt{1+\left(\dfrac{\omega}{\frac{R}{L}}\right)^2}}$$

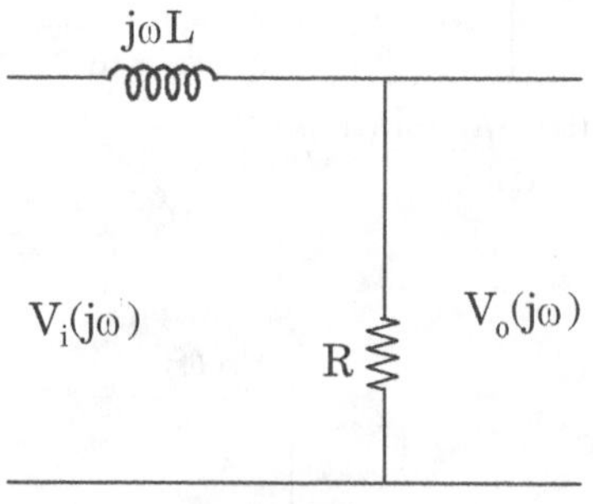

Comparing eq. (i) and eq (ii),

$$\frac{R}{L} = 4 \Rightarrow R = 4L$$

If L = 1 H then R = 4 Ω.

74.

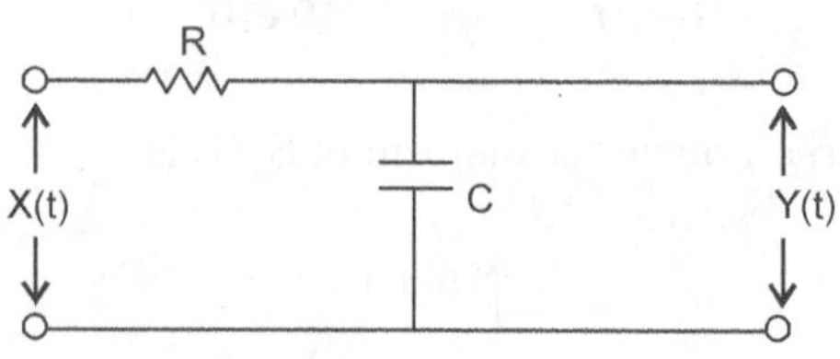

Frequency response of the RC filter is,

$$H(\omega) = \frac{1}{1+j\omega RC}$$

$$\omega = 2\pi fc = \frac{1}{RC}$$

$$S_{xx}(\omega) = k \qquad \text{[While noise process]}$$

$$S_{yy}(\omega) = |H(\omega)|^2 \cdot S_{xx}(\omega)$$

$$= \frac{1}{1+(\omega RC)^2}\cdot k$$

$$S_{yy}(\omega) = k\frac{\left(1/RC\right)^2}{\omega^2+\left(\dfrac{1}{RC}\right)^2}$$

$$= \frac{k}{2(RC)}\cdot\frac{2/RC}{\omega^2+\left(\dfrac{1}{RC}\right)^2}$$

Inverse fourier transform gives

$$R_{yy}(J) = \frac{k}{2RC}e^{-|J|/RC}$$

Noise power at the output,

$$E(Y^2(t)) = R_{YY}(0)$$

$$= \frac{k}{2RC} = \pi f_c k$$

75. Power of the output process Y(t),

$$Y(t) = |H(t)|^2 \cdot X(t)$$

$$X(t) = \frac{N_0}{2} = 1\times 10^{-10}\ \text{W/HZ}$$

$$|H(t)|^2 = \text{area under curve } |u(t)|^2$$

$$= 2\left(\frac{1}{2}\times 10\times 10^3\times 1\right) = 10^4\,\text{Hz}$$

then, $\quad V(t) = 10^{-6}\ \text{W}$

77. Variance of process. = $R_x(0)$.

Let Variance of output = $R_Y(0)$.

∵ It is zero mean.

∴ $R_y(\tau) = h(\tau) * h(-\tau) * R_N(\tau)$

where, $h(\tau)$ = filter response and $R_N(\tau)$is the input noise.

$$R_N(\tau) = \frac{N_0}{2}.\delta(\tau)$$

$$\therefore R_y(\tau) = [h(\tau)\times(-\tau)].\frac{N_0}{2}$$

$$= \frac{N_0}{2}\int_{-\infty}^{\infty} h(\tau).h(\tau+z)d\tau$$

$$= \frac{N_0}{2}\int_{-\infty}^{\infty} h^2(\tau).d\tau$$

$$= \frac{N_0}{2}\times \text{Energy} = \frac{N_0}{2}\times 3A^2$$

$$R_y(\tau) = \frac{3}{2}N_0A^2.$$

78. The given input power spectral density is as follows:

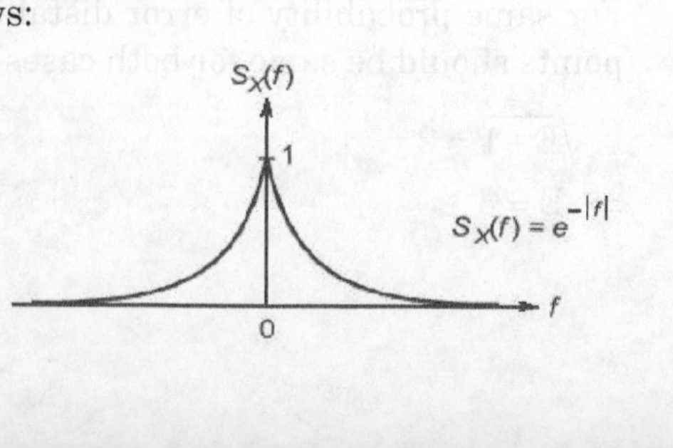

Now frequency response of the low pass filter is as follows:

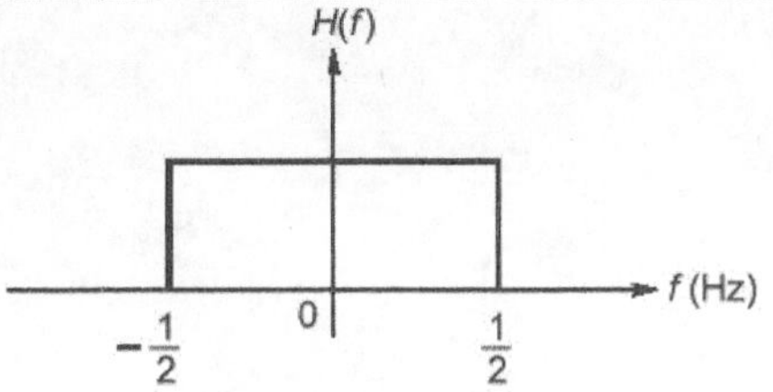

Now, $E[Y(t)] = H(0)\, E[X(t)]$

and $H(0) = 1$

$\therefore$ $E[Y(t)] = E[X(t)]$

Now $E[Y^2(t)] \neq E[X^2(t)]$

Since, LPF does not allow total power from input to output.

$$E[X^2(t)] = \int_0^\infty S_x(f)\,df = 2\text{ W}$$

As $E[Y^2(t)] \neq E[X^2(t)]$.

$E[Y^2(t)] \neq 2$

Hence only statement - I is correct.

79. When X = −1 is transmitted

$P(X = -1) = 0.2$

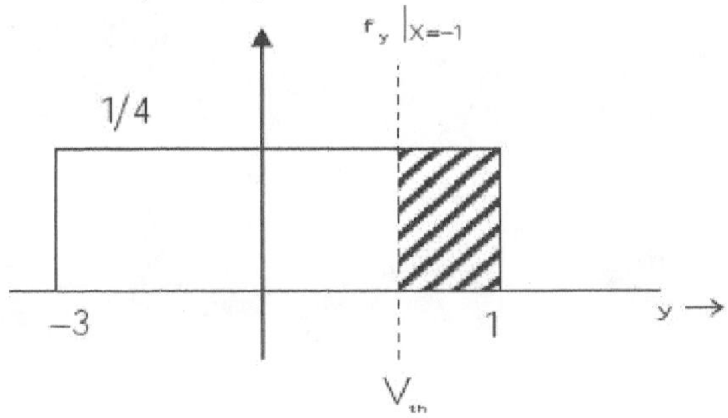

$$P_{e-1} = (0.2)\left(\frac{1}{4}\right)(1 - V_{th})$$

When X = 1 is transmitted P[x = 1] = 0.8

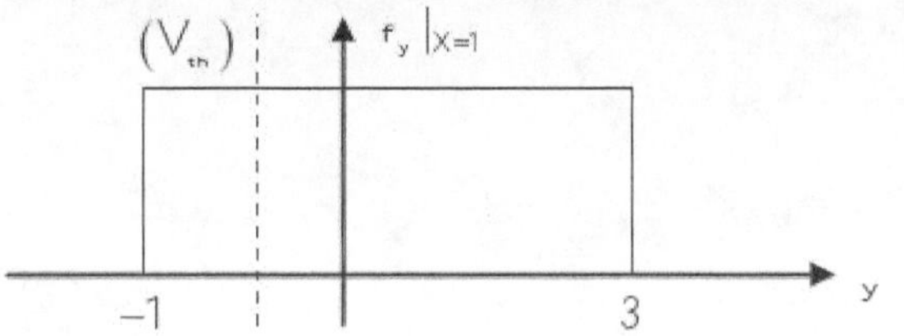

$$P_{e_1} = (0.8)\left(\frac{1}{4}\right)(V_{th} + 1)$$

$$P_{e_1} = \frac{(0.2)(1)(1 - V_{th}) + (0.8)(1)(V_{th} + 1)}{4}$$

$$P_e = \frac{0.2(1 - V_{th}) + 0.8(V_{th} + 1)}{4}$$

$$= \frac{(0.2 + 0.8) + 0.6V_{th}}{4} = \frac{1 + 0.6V_{th}}{4}$$

Now, $-1 \le V_{th} \le 1$

$\therefore P_{e_{min}}$ when $V_{th} = -1$

$$\therefore P_{e_{min}} = \frac{1 + (0.6)(-1)}{4} = \frac{0.4}{4} = 0.1$$

2

CHAPTER

Analog Communication

Analysis of Previous GATE Papers

Topics ↓	Year →		2019	2018	2017 Set 1	2017 Set 2	2016 Set 1	2016 Set 2	2016 Set 3	2015 Set 1	2015 Set 2	2015 Set 3
AMPLITUDE MODULATION	1 Mark	MCQ Type										
		Numerical Type		1			1					
	2 Marks	MCQ Type										
		Numerical Type				1						
	Total			**1**		**2**	**1**					
FREQUENCY MODULATION	1 Mark	MCQ Type										
		Numerical Type										
	2 Marks	MCQ Type										
		Numerical Type										
	Total											
PHASE MODULATION	1 Mark	MCQ Type								1		1
		Numerical Type	1									
	2 Marks	MCQ Type										
		Numerical Type				1						
	Total		**1**			**2**				**1**		**1**
SPECTRA OF AM & FM	1 Mark	MCQ Type										
		Numerical Type										
	2 Marks	MCQ Type		1								
		Numerical Type						1				
	Total			**2**				**2**				
SUPERHETERODYNE RECEIVER	1 Mark	MCQ Type										
		Numerical Type					1		1			
	2 Marks	MCQ Type										
		Numerical Type										
	Total						**1**		**1**			
CIRCUITS FOR ANALOG MODULATION	1 Mark	MCQ Type					1					
		Numerical Type										
	2 Marks	MCQ Type										
		Numerical Type										
	Total						**1**					

AMPLITUDE MODULATION

1. A 4 GHz carrier is DSB-SC modulated by a lowpass message signal with maximum frequency of 2 MHz. The resultant signal is to be ideally sampled. The minimum frequency of the sampling impulse train should be

(a) 4 MHz

(b) 8 MHz

(c) 8 GHz

(d) 3.004 GHz

[1990 : 2 Marks]

2. A PAM signal can be detected by using

(a) an ADC (b) an integrator

(c) a band pas filter (d) a high pass filter

[1995 : 1 Mark]

3. A DSB-SC signal is generated using the carrier $\cos(\omega_c t + \theta)$ and modulating signal x(t). The envelop of the DSB-SC signal is

(a) x(t)

(b) $|x(t)|$

(c) only positive portion of x(t)

(d) $x(t)\cos\theta$

[1998 : 1 Mark]

4. A modulated signal is given by,

$s(t) = m_1(t)\cos(2\pi f_c t) + m_2(t)\sin(2\pi f_c t)$;

where the baseband signal $m_1(t)$ and $m_2(t)$ have bandwidths of 10 kHz and 15 kHz, respectively. The bandwidth of the modulated signal, in kHz, is

(a) 10 (b) 15

(c) 25 (d) 30

[1999 : 1 Mark]

5. A modulated signal is given by

$s(t) = e^{-at}\cos[(\omega_c + \Delta\omega)t]u(t)$;

where a, ω_c and $\Delta\omega$ are positive constants, and $\omega_c >> \Delta\omega$. The complex envelope of s(t) is given by

(a) $\exp(-at)\exp[j(\omega_c + \Delta\omega)t]u(t)$.

(b) $\exp(-at)\exp(j\Delta\omega t)u(t)$.

(c) $\exp(j\Delta\omega t)u(t)$.

(d) $\exp[(j\omega_c + \Delta\omega)t]$.

[1999 : 1 Mark]

6. The amplitude modulated wave form $s(t) = A_c[1 + K_a m(t)]\cos\omega_c t$ is fed to an ideal envelope detector. The maximum magnitude of $K_a m(t)$ is greater than 1. Which of the following could be the detector output?

(a) $A_c m(t)$ (b) $A_c^2[1 + K_a m(t)]^2$

(c) $[A_c|1 + K_a m(t))]$ (d) $A_c|1 + K_a m(t)|^2$

[2000 : 1 Mark]

7. A message m(t) bandlimited to the frequency f_m has a power of P_m. The power of the output signal in the figure is

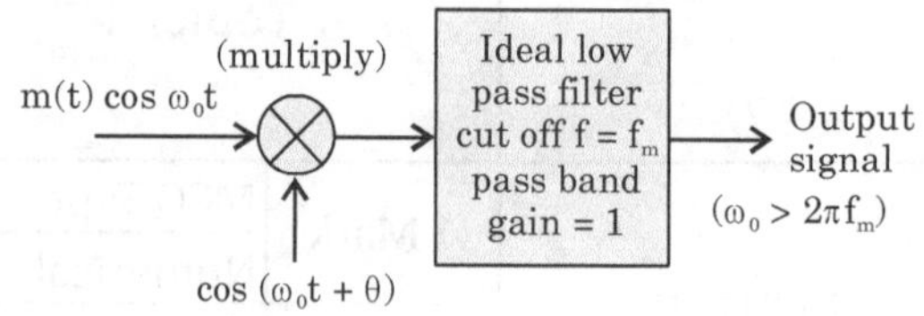

(a) $\frac{P_m \cos\theta}{2}$ (b) $\frac{P_m}{4}$

(c) $\frac{P_m \sin^2\theta}{4}$ (d) $\frac{P_m \cos^2\theta}{4}$

[2000 : 2 Marks]

8. A DSB-SC signal is to be generated with a carrier frequency $f_c = 1$ MHz using a non-linear device with the input-output characteristic

$v_0 = a_0 v_i + a_1 v_i^3$;

where, a_0 and a_i are constants. The output of the non-linear device can be filtered by an appropriate band-pass filter.

Let $v_i = A_c^i \cos(2\pi f_c^i t) + m(t)$ where m(t) is the message signal. Then the value of f_c^i (in MHz) is

(a) 1.0 (b) 0.333

(c) 0.5 (d) 3.0

[2003 : 2 Marks]

Common Data for Questions 9 and 10 :

Let $g(t) = p(t) * p(t)$, where '*' denotes convolution and $p(t) = u(t) - u(t - 1)$ with u(t) being the unit step function.

9. The impulse response of filter matched to the signal $s(t) = g(t) - \delta(t - 2) * g(t)$ is given as

(a) $s(1 - t)$ (b) $-s(1 - t)$

(c) $-s(t)$ (d) $s(t)$

[2006 : 2 Marks]

10. An Amplitude Modulated signal is given as $x_{AM}(t) = 100(p(t) + 0.5g(t))\cos \omega_c t$ in the interval $0 \leq t \leq 1$. One set of possible values of the modulating signal and modulation index would be

(a) t, 0.5 (b) t, 1.0
(c) t, 2.0 (d) t^2, 0.5

[2006 : 2 Marks]

11. Consider the amplitude modulated (AM) signal $A_c\cos \omega_c t + 2 \cos \omega_m t \cos \omega_c t$. For demodulating the signal using envelope detector, the minimum value of A_c should be

(a) 2 (b) 1
(c) 0.5 (d) 0

[2008 : 1 Mark]

12. A message signal given by

$$m(t) = \left(\frac{1}{2}\right)\cos\omega_1 t - \left(\frac{1}{2}\right)\sin\omega_2 t$$

is amplitude-modulated with a carrier of frequency ω_c to generate $s(t) = [1 + m(t)]\cos\omega_c t$. What is the power efficiency achieved by this modulation scheme?

(a) 8.33% (b) 11.11%
(c) 20 % (d) 25 %

[2009 : 2 Marks]

13. Suppose that the modulating signal is $m(t) = 2\cos(2\pi f_m t)$ and the carrier signal is $x_C(t) = A_C \cos(2\pi f_c t)$. Which one of the following is a conventional AM signal without over-modulation?

(a) $x(t) = A_C m(t)\cos(2\pi f_c t)$
(b) $x(t) = A_C[1 + m(t)]\cos(2\pi f_c t)$
(c) $x(t) = A_C\cos(2\pi f_c f) + \frac{A_C}{4} m(t)\cos(2\pi f_c t)$
(d) $x(t) = A_C\cos(2\pi f_m t)\cos(2\pi f_c t) + A_C\sin(2\pi f_m t)\sin(2\pi f_c t)$

[2010 : 1 Mark]

14. Consider sinusoidal modulation in an AM system. Assuming no overmodulation, the modulation index (m) when the maximum and minimum values of the envelope, respectively, are 3 V and 1 V, is ______.

[2014 : 1 Mark, Set-2]

15. A modulated signal is $y(t) = m(t)\cos(40000\,\pi\, t)$, where the baseband signal m(t) has frequency components less than 5 kHz only. The minimum required rate (in kHz) at which y(t) should be sampled to recover m(t) is ______.

[2014 : 1 Mark, Set-3]

16. In a double side-band (DSB) full carrier AM transmission system, if the modulation index is doubled, then the ratio of total sideband power to the carrier power increases by a factor of ______.

[2014 : 1 Mark, Set-4]

17. The amplitude of a sinusoidal carrier is modulated by a single sinusoid to obtain the amplitude modulated signal

$s(t) = 5\cos1600\pi t + 20\cos1800\pi t + 5\cos2000\pi t$

The value of the modulation index is ______.

[2016 : 1 Mark, Set-1]

18. The unmodulated carrier power in an AM transmitter is 5 kW. This carrier is modulated by a sinusoidal modulating signal. The maximum percentage of modulation is 50%. If it is reduced to 40%, then the maximum unmodulated carrier power (in kW) that can be used without overloading the transmitter is ______.

[2017 : 2 Marks, Set-2]

19. Consider the following amplitude modulated signal:

$s(t) = \cos(2000\pi t) + 4\cos(2400\pi t) + \cos(2800\pi t)$

The ratio (accurate to three decimal places) of the power of the message signal to the power of the carrier signal is ______.

[2018 : 1 Mark]

FREQUENCY MODULATION

20. A carrier $A_C \cos \omega_c t$ is frequency modulated by a signal $E_m \cos \omega_m t$. The modulation index is mf. The expression for the resulting FM signal is

(a) $A_c\cos[\omega_c t + m_f\sin \omega_m t]$
(b) $A_c\cos[\omega_c t + m_f\cos \omega_m t]$
(c) $A_c\cos[\omega_c t + 2\pi m t\sin \omega_m t]$
(d) $A_c \cos\left[\omega_c t + \frac{2\pi m_1 E_m}{\omega_m}\cos\omega_m t\right]$

[1989 : 2 Marks]

21. Which of the following schemes suffer(s) from the threshold effect?

(a) AM detection using envelope detection.
(b) AM detection using synchronous detection.
(c) FM detection using a discriminator.
(d) SSB detection with synchronous detection.

[1989 : 2 Marks]

22. A signal $x(t) = 2\cos(\pi 10^4 t)$ Volts is applied to an FM modulator with the sensitivity constant of 10 kHz/volt. Then the modulation index of the FM wave is

(a) 4 (b) 2
(c) $4/\pi$ (d) $2/\pi$ **[1989 : 2 Marks]**

23. In commercial TV transmission in India, picture and speech signals are modulated respectively as

	(Picture)		**(Speech)**
(a)	VSB	and	VSB
(b)	VSB	and	SSB
(c)	VSB	and	FM
(d)	FM	and	VSB

[1990 : 2 Marks]

24. $v(t) = 5[\cos(10^6\pi t) - \sin(10^3\pi t) \times \sin(10^6\pi t)]$ represents

(a) DSB suppressed carrier signal.
(b) AM signal.
(c) SSB upper sideband signal.
(d) Narrow band FM signal.

[1994 : 1 Mark]

25. A 10 MHz carrier is frequency modulated by a sinusoidal signal of 500 Hz, the maximum frequency deviation being 50 kHz. The bandwidth required, as given by the Carsons' rule is ______.

[1994 : 1 Mark]

26. An FM signal with a modulation index 9 is applied to a frequency tripler. The modulation index in the output signal will be

(a) 0 (b) 3
(c) 9 (d) 27 **[1996 : 2 Marks]**

Common Data for Questions 27 and 28 :

Let $m(t) = \cos[(4\pi \times 10^3)t]$ be the message signal & $c(t) = 5\cos[(2\pi \times 10^6)t]$ be the carrier.

27. c(t) and m(t) are used to generate an AM signal. The modulation index of the generated AM signal is 0.5. Then the quantity $\frac{\text{Total sideband power}}{\text{Carrier power}}$ is

(a) $\frac{1}{2}$ (b) $\frac{1}{4}$
(c) $\frac{1}{3}$ (d) $\frac{1}{8}$ **[2003 : 2 Marks]**

28. c(t) and m(t) are used to generate an FM signal. If the peak frequency deviation of the generated FM signal is three times the transmission bandwidth of the AM signal, then the coefficient of the term cos $\cos[2\pi(1008 \times 10^3 t)]$ in the FM signal (in terms of the Bessel coefficients) is

(a) $5\,J_4(3)$ (b) $\frac{5}{2}J_8(3)$
(c) $\frac{5}{2}J_8(4)$ (d) $5\,j_4(6)$

[2003 : 2 Marks]

29. A device with input x(t) and output y(t) is characterized by: $y(t) = x^2(t)$.
An FM signal with frequency deviation of 90 kHz and modulating signal bandwidth of 5 kHz is applied to this device. The bandwidth of the output signal is

(a) 370 kHz (b) 190 kHz
(c) 380 kHz (d) 95 kHz

[2005 : 2 Marks]

30. Consider the frequency modulated signal
$10\cos[2\pi \times 10^5 t + 5\sin(2\pi \times 1500t) + 7.5\sin(2\pi \times 1000t)]$ with carrier frequency of 10^5 Hz. The modulation index is

(a) 12.5 (b) 10
(c) 7.5 (d) 5

[2008 : 2 Marks]

31. Consider an FM signal
$f(t) = \cos[2\pi f_c t + \beta_1 \sin 2\pi f_2 t + \beta_2 2\pi f_2 t]$
The maximum deviation of the instantaneous frequency from the carrier frequency f_c is

(a) $\beta_1 f_1 + \beta_2 f_2$ (b) $\beta_1 f_2 + \beta_2 f_1$
(c) $\beta_1 + \beta_2$ (d) $f_1 + f_2$

[2014 : 1 Mark, Set-3]

PHASE MODULATION

32. An angle-modulated signal is given by
$s(t) = \cos 2\pi(2 \times 10^6 t + 30\sin 150t + 40\cos 150t)$.
The maximum frequency and phase deviations of s(t) are

(a) 10.5 kHz, 140π rad
(b) 6 kHz, 80π rad
(c) 10.5 kHz, 100π rad
(d) 7.5 kHz, 100π rad

33. Find the correct match between group 1 and group 2.

Group 1

P. $\{1 + km(t)\}A\sin(\omega_c t)$
Q. $km(t)A\sin(\omega_c t)$
R. $A\sin\{\omega_c t + km(t)\}$
S. $A\sin\left[\omega_c t + k\int_{-\infty}^{t} m(t)\,dt\right]$

Group 2

W. Phase modulation
X. Frequency modulation
Y. Amplitude modulation
Z. DSB-SC modulation

(a) P-Z, Q-Y, R-X, S-W
(b) P-W, Q-X, R-Y, S-Z
(c) P-X, Q-W, R-Z, S-Y
(d) P-Y, Q-Z, R-W, S-X

[2005 : 1 Mark]

34. Consider an angle modulated signal

$x(t) = 6\cos[2\pi \times 10^6 t + 2\sin(8000\pi t) + 4\cos(8000\pi t)]$ V.

The average power of x(t) is

(a) 10 W (b) 18 W

(c) 20 W (d) 28 W

[2010 : 1 Mark]

35. The signal m(t) as shown is applied both to a phase modulator (with k_p as the phase constant) and a frequency modulator with (k_f as the frequency constant) having the same carrier frequency

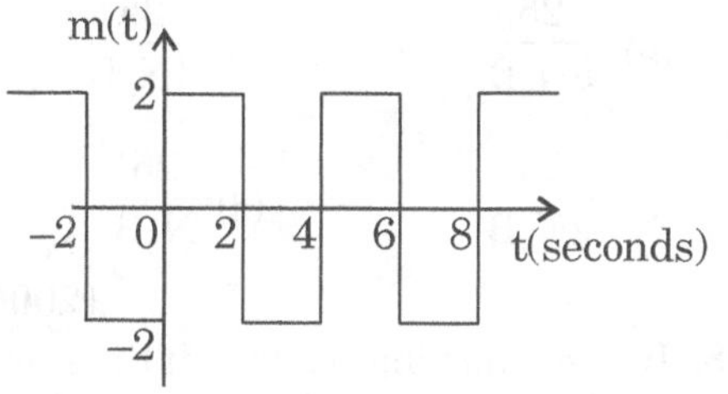

The ratio k_p/k_f (in rad/Hz) for the same maximum phase deviation is

(a) 8π (b) 4π

(c) 2π (d) π

[2012 : 2 Marks]

36. Consider the signal

$s(t) = m(t)\cos(2\pi f_c t) + \hat{m}(t)\sin(2\pi f_c t)$

where $\hat{m}(t)$ denotes the Hilber transform of m(t) and the bandwidth of m(t) is very small compared to f_c. The signal s(t) is a

(a) high-pass signal

(b) low-pass signal

(c) band-pass signal

(d) double sideband suppressed carrier signal

[2015 : 1 Mark, Set-1]

37. A message signal $m(t) = A_m\sin(2\pi f_m t)$ is used to modulate the phase of a carrier $A_c\cos(2\pi f_c t)$ to get the modulated signal $y(t) = A_c\cos(2\pi f_c t + m(t))$. The bandwidth of y(t)

(a) depends on A_m but not on f_m

(b) depends on f_m but not on A_m

(c) depends on both A_m and f_m

(d) does not depends on A_m or f_m

[2015 : 1 Mark, Set-3]

38. A modulating signal given by

$x(t) = 5\sin(4\pi 10^3 t - 10\pi\cos 2\pi 10^3 t)$ V

is fed to a phase modulator with phase deviation constant k_p = 5 rad/V. If the carrier frequency is 20 kHz, the instantaneous frequency (in kHz) at t = 0.5 ms is ______.

[2017 : 2 Marks, Set-2]

39. The baseband signal m(t) shown in the figure is phase-modulated to generate the PM signal $\varphi(t) = \cos(2\pi f_c t + km(t))$.

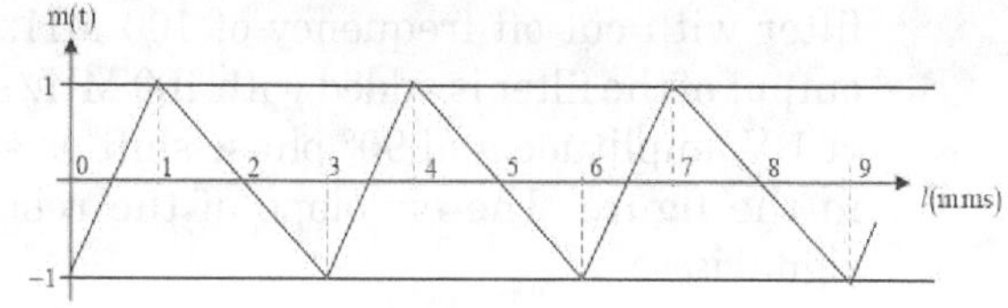

The time t on the x-axis in the figure is in milliseconds. If the carrier frequency is f_c = 50 kHz and $k = 10\pi$, the ratio of the minimum instantaneous frequency (in kHz) to the maximum instantaneous frequency (in kHz) is ______ (rounded off to 2 decimal places).

[2019 : 1 Marks]

SPECTRA OF AM & FM

40. In a FM system, a carrier of 100 MHz is modulated by a sinusoidal signal of 5 kHz. The bandwidth by Carson's approximation is 1 MHz. If y(t) = (modulated waveform)3, then by using Carson's approximation, the bandwidth of y(t) and the spacing of spectral components are, respectively

(a) 3 MHz, 5 kHz

(b) 1 MHz, 15 kHz

(c) 3 MHz, 15 kHz

(d) 1 MHz, 5 kHz

[2000 : 2 Marks]

41. A 1 MHz sinusoidal carrier is amplitude modulated by a symmetrical square wave of period 100 μsec. Which of the following frequencies will NOT be present in the modulated signal?

(a) 990 kHz

(b) 1010 kHz

(c) 1020 kHz

(d) 1030 kHz

[2002 : 1 Mark]

42. An AM signal and a narrow-band FM signal with identical carriers, modulating signals and modulation indices of 0.1 are added together. The resultant signal can be closely approximated by

(a) broadband FM

(b) SSB with carrier

(c) DSB-SC

(d) SSB without carrier

[2004 : 1 Mark]

43. A 100 MHz carrier of 1 V amplitude and a 1 MHz modulating signal of 1 V amplitude are fed to a balanced modulator. The output of the modulator is passed through an ideal high-pass filter with cut-off frequency of 100 MHz. The output of the filter is added with 100 MHz signal of 1 V amplitude and 90° phase shift as shown in the figure. The envelope of the resultant signal is

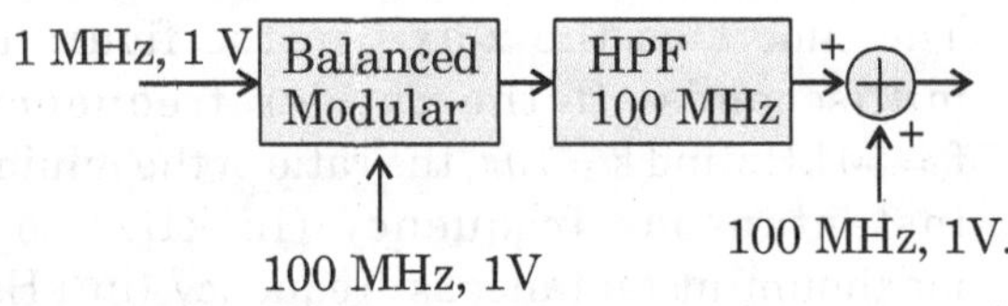

(a) constant

(b) $\sqrt{1+\sin(2\pi\times10^6 t)}$

(c) $\sqrt{\frac{5}{4}-\sin(2\pi\times10^6 t)}$

(d) $\sqrt{\frac{5}{4}+\cos(2\pi\times10^6 t)}$

[2004 : 2 Marks]

44. Which of the following analog modulation scheme requires the minimum transmitted power and minimum channel band-width?

(a) VSB (b) DSB-SC

(c) SSB (d) AM

[2005 : 1 Mark]

45. A message signal with bandwidth 10 kHz is Lower- Side Band SSB modulated with carrier frequency $f_{c1} = 10^6$ Hz. The resulting signal is then passed through a Narrow-Band Frequency Modulator with carrier frequency $f_{c2} = 10^9$ Hz. The bandwidth of the output would be

(a) 4×10^4 Hz

(b) 2×10^6 Hz

(c) 2×10^9 Hz

(d) 2×10^{10} Hz

[2006 : 2 Marks]

Common Data for Questions 46 & 47 :

Consider the following Amplitude Modulated (AM) signal, where $f_m < B$

$$x_{AM}(t) = 10(1 + 0.5\sin 2\pi f_m t)\cos 2\pi f_c t$$

46. The average side-band power for the AM signal given above is

(a) 25 (b) 12.5

(c) 6.25 (d) 3.125

[2006 : 2 Marks]

47. The AM signal gets added to a noise with Power Spectral Density $S_n(f)$ given in the figure below. The ratio of average sideband power to mean noise power would be

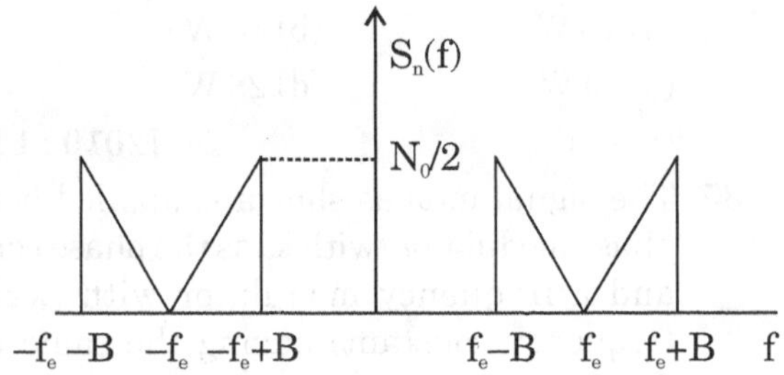

(a) $\frac{25}{8N_0B}$ (b) $\frac{25}{4N_0B}$

(c) $\frac{25}{2N_0B}$ (d) $\frac{25}{N_0B}$

[2006 : 2 Marks]

48. In the following scheme, if the spectrum M(f) of m(t) is as shown, then the spectrum Y(f) of y(t) will be

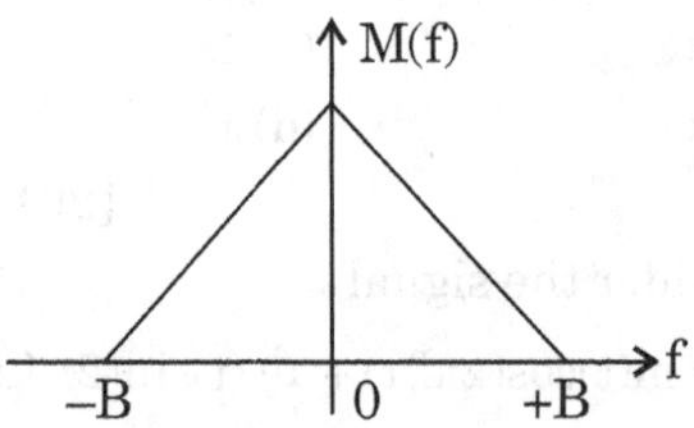

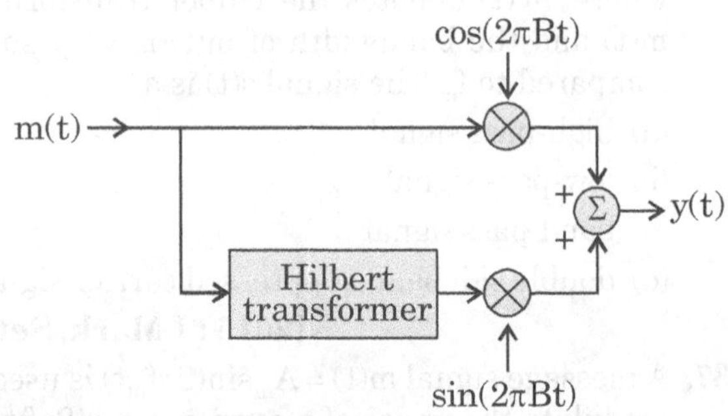

(a)

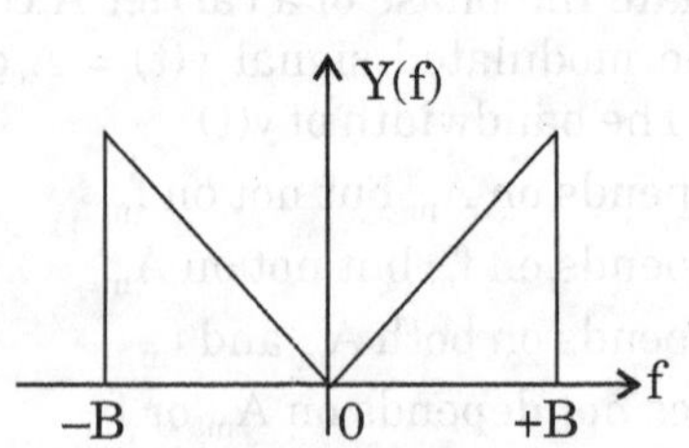

(b)

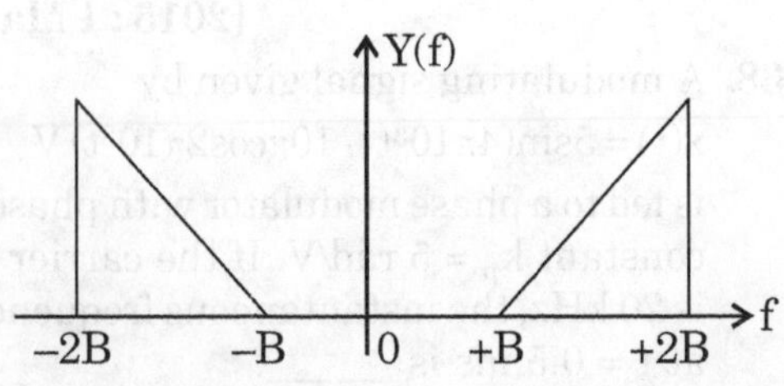

(c) 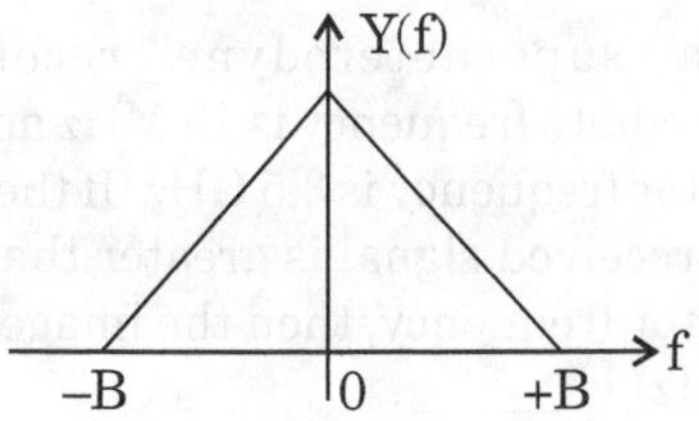

(d) 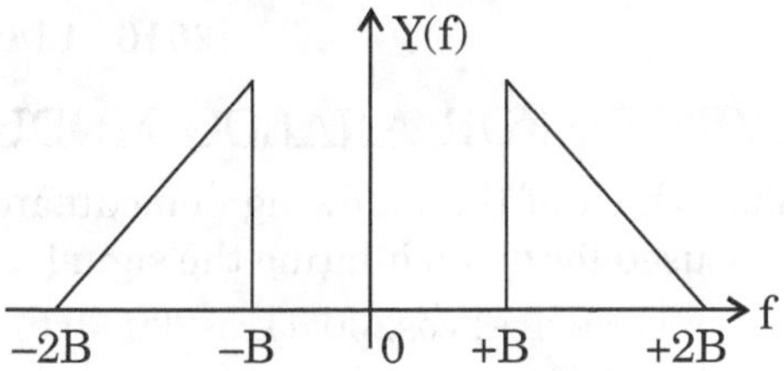

[2007 : 2 Marks]

49. The signal $\cos\omega_c t + 0.5\cos\pi_m t \sin\pi_c t$ is

(a) F Monly.

(b) AM only.

(c) both AM and FM.

(d) neither AM nor FM.

[2008 : 2 Marks]

50. For a message signal $m(t) = \cos(2\pi f_m t)$ and carrier of frequency f_c, which of the following represents a single side-band (SSB) signal?

(a) $\cos(2\pi f_m t)\cos(2\pi f_c t)$.

(b) $\cos(2\pi f_c t)$.

(c) $\cos[2\pi(f_c + f_m)t]$.

(d) $[1 + \cos(2\pi f_m t]\cos(2\pi f_c t)$.

[2009 : 1 Mark]

51. The List-I (lists the attributes) and the List-ll (lists of the modulation systems). Match the attribute to the modulation system that best meets it.

List-I

A. Power efficient transmission of signals

B. Most bandwidth efficient transmission of voice signals

C. Simplest receiver structure

D. Bandwidth efficient transmission of signals with significant dc component

List-II

1. Conventional AM
2. FM
3. VSB
4. SSB-SC

	A	B	C	D
(a)	4	2	1	3
(b)	2	4	1	3
(c)	3	2	1	4
(d)	2	4	3	1

[2011 : 1 Mark]

52. In the figure, M(f) is the Fourier transform of the message signal m(t) where A = 100 Hz and B = 40 Hz. Given $v(t) = \cos(2\pi f_c t)$ and $w(t) = \cos(2\pi(f_c + A)t)$, where $f_c > A$. The cut-off frequencies of both the filters are f_c.

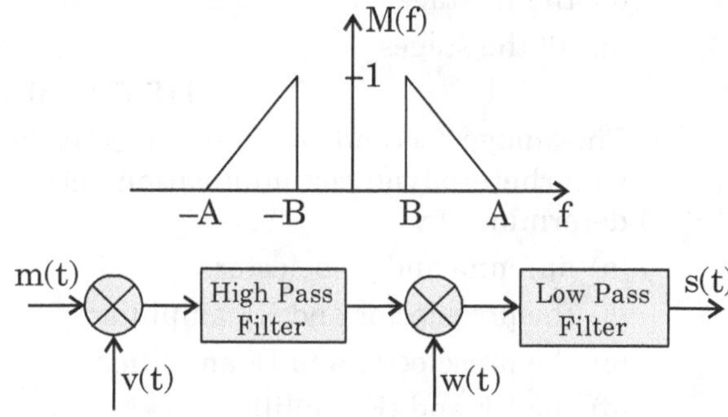

The bandwidth of the signal at the output of the modulator (in Hz) is ______.

[2014 : 2 Marks, Set-2]

53. In the system shown in Figure (a), m(t) is a low–pass signal with bandwidth W Hz. The frequency response of the band-pass filter H(f) is shown in Figure (b). If it is desired that the output signal z(t) = 10 x(t), the maximum value of W(in Hz) should be strictly less than ____.

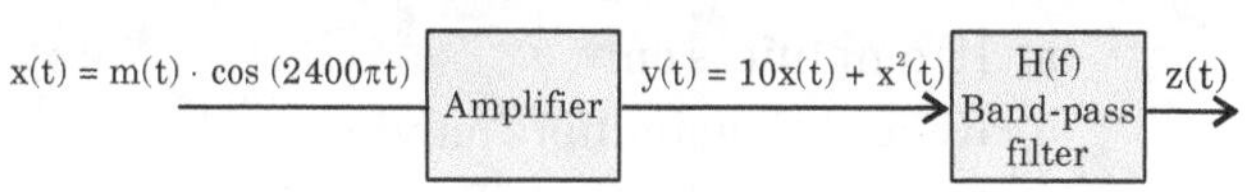

Fig. (a)

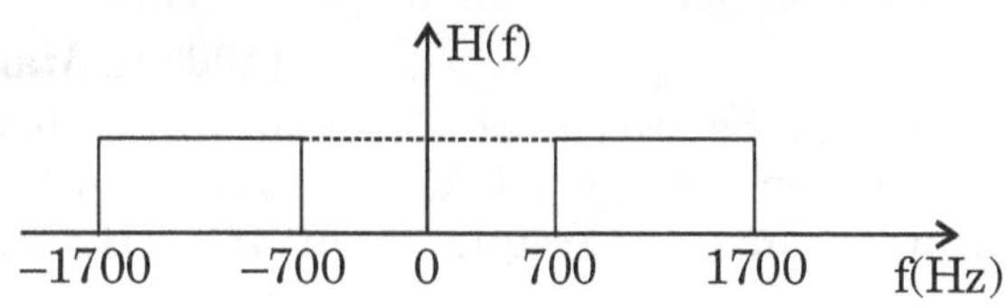

Fig. (b)

[2015 : 2 Marks, Set-1]

54. Let $c(t) = A_c\cos(2\pi f_c t)$ and $m(t) = \cos(2\pi f_m t)$. It is given that $f_c >> 5f_m$. The signal c(t) + m(t) is applied to the input of a non-linear device, whose output $v_0(t)$ is related to the input $v_i(f)$ as $v_0(t) = av_i(t) + bv_i^2(t)$, where a and b are positive constants. The output of the non-linear device is passed through an ideal band-pass filter with center frequency f_c and bandwidth 3 f_m, to produce an amplitude modulated (AM) wave. If it is desired to have the sideband power of the AM wave to be half of the carrier power, then a/b is

(a) 0.25 (b) 0.5

(c) 1 (d) 2

[2018 : 2 Marks]

SUPERHETERODYNE RECEIVER

55. In a superheterodyne AM receiver, the image channel selectivity is determined by
(a) the preselector and RF stages
(b) the preselector, RF and IF stages
(c) the IF stages
(d) all the stages

[1987 : 2 Marks]

56. The image (second) channel selectivity of a superheterodyne communication receiver is determined by
(a) antenna and preselector.
(b) the preselector and RF amplifier.
(c) the preselector and IF amplifier.
(d) the RF and IF amplifier.

[1995 : 1 Mark]

57. The image channel rejection in a superheterodyne receiver comes from
(a) IF stages only.
(b) RF stages only.
(c) detector and RF stages only.
(d) detector, RF and IF stages.

[1996 : 1 Mark]

58. The image channel selectivity of superheterodyne receiver depends upon
(a) IF amplifiers only
(b) RF and IF amplifiers only
(c) Preselector, RF and IF amplifiers
(d) Preselector and RF amplifiers only

[1998 : 1 Mark]

59. A superheterodyne receiver is to operate in the frequency range 550 kHz-1650 kHz, with the intermediate frequency of 450 kHz. Let, $R = \frac{C_{max}}{C_{min}}$ denote the required capacitance ratio of the local oscillator and I denote the image frequency (in kHz) of the incoming signal. If the receiver is tuned to 700 kHz, then
(a) R = 4.41, I = 1600
(b) R = 2.10, I = 1150
(c) R = 3.0, I = 1600
(d) R = 9.0, I = 1150

[2003 : 2 Marks]

60. A superheterodyne receiver operates in the frequency range of 58 MHz–68 MHz. The intermediate frequency f_{IF} and local oscillator frequency f_{LO} are chosen such that $f_{IF} < f_{LO}$. It is required that the image frequencies fall outside the 58 MHz-68 MHz band. The minimum required f_{IF} (in MHz) is ______.

[2016 : 1 Mark, Set-1]

61. For a superheterodyne receiver, the intermediate frequency is 15 MHz and the local oscillator frequency is 3.5 GHz. If the frequency of the received signal is greater than the local oscillator frequency, then the image frequency (in MHz) is ______.

[2016 : 1 Mark, Set-3]

CIRCUITS FOR ANALOG MODULATION

62. Which of the following demodulator(s) can be used for demodulating the signal
$x(t) = 5(1 + 2\cos 200\pi t)\cos 20000\pi t$?
(a) Envelope demodulator
(b) Square-law demodulator
(c) Synchronous demodulator
(d) None of the above

[1993 : 2 Marks]

63. Match List-I with List-II and select the correct answer using the code given below the Lists:

List-I	List-II
A. SSB	1. Envelope detector
B. AM	2. Integrate and dump
C. BPSK	3. Hi lbert transform
	4. Ratio detector
	5. PLL

Codes :

	A	B	C
(a)	3	1	2
(b)	3	2	1
(c)	2	1	3
(d)	1	2	3

[1994 : 2 Marks]

64. A PLL can be used to demodulate
(a) PAM signals (b) PCM signals
(c) FM signals (d) DSB–SC signals

[1995 : 1 Mark]

65. The Hilbert transform of $\cos \omega_1 t + \sin \omega_2 t$ is
(a) $\sin\omega_1 t - \cos\omega_2 t$ (b) $\sin\omega_1 1 + \cos\omega_2 t$
(c) $\cos\omega_1 t - \sin\omega_2 t$ (d) $\sin\omega_1 t + \sin\omega_2 t$

[2000 : 2 Marks]

66. A bandlimited signal is sampled at the Nyquist rate. The signal can be recovered by passing the samples through
(a) an RC filter.
(b) an envelope detector.
(c) a PLL.
(d) an ideal low-pass filter with the appropriate bandwidth.

[2001 : 1 Mark]

67. In the figure $m(t) = \frac{2\sin 2\pi t}{t}$, $s(t) = \cos 200\pi t$ and $n(t) = \frac{\sin 199\pi t}{t}$. The output y(t) will be

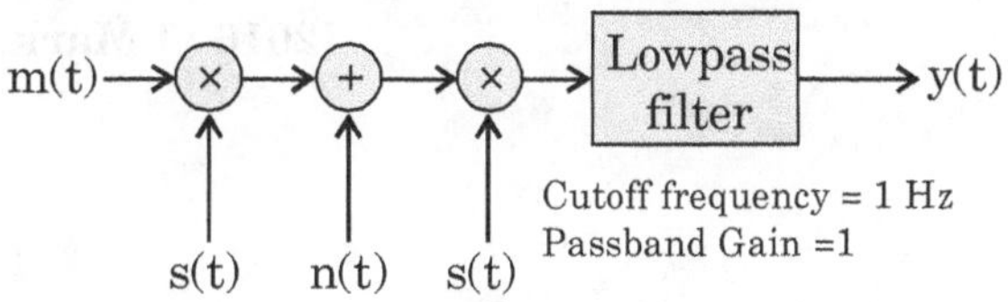

(a) $\frac{\sin 2\pi t}{t}$

(b) $\frac{\sin 2\pi t}{t} + \frac{\sin \pi t}{t}\cos 3\pi t$

(c) $\frac{\sin 2\pi t}{t} + \frac{\sin 0.5\pi t}{t}\cos 1.5\pi t$

(d) $\frac{\sin 2\pi t}{t} + \frac{\sin \pi t}{t}\cos 0.75\pi t$

[2000 : 2 Marks]

68. The input to a coherent detector is DSB-SC signal plus noise. The noise at the detector output is

(a) the in-phase component.

(b) the quadrature-component.

(c) zero.

(d) the envelope.

[2003 : 1 Mark]

69. Choose the correct one from among the alternative a, b, c, d after matching an item in Group 1 with the most appropriate item in Group 2.

Group 1

P. Ring modulator

Q. VCO

R. Foster-Seely discriminator

S. Mixer

Group 2

1. Clock recovery
2. Demodulation of FM
3. Frequency conversion
4. Summing the two inputs
5. Generation of FM
6. Generation of DSB-SC

(a) P - 1; Q - 3; R - 2; S - 4

(b) P - 6; Q - 5; R - 2; S - 3

(c) P - 6; Q - 1; R - 3; S - 2

(d) P - 5; Q - 6; R - 1; S - 3

[2003 : 2 Marks]

70. An AM signal is detected using an envelope detector. The carrier frequency and modulating signal frequency are 1 MHz and 2 kHz respectively. An appropriate value for the time constant of the envelope detector is

(a) 500 μ sec

(b) 20 μ sec

(c) 0.2 μ sec

(d) 1 μ sec

[2004 : 1 Mark]

71. Two sinusoidal signals of same amplitude and frequencies 10 kHz and 10.1 kHz are added together. The combined signal is given to an ideal frequency detector. The output of the detector is

(a) 0.1 kHz sinusoid.

(b) 20.1 kHz sinusoid.

(c) a linear function of time.

(d) a constant.

[2004 : 2 Marks]

72. The diagonal clipping in Amplitude Demodulation (using envelope detector) can be avoided if RC time-constant of the envelope detector satisfies the following condition, (here W is message bandwidth and ω is carrier frequency both in rad/sec)

(a) $RC < \frac{1}{\omega}$ (b) $RC > \frac{1}{\omega}$

(c) $RC < \frac{1}{\omega}$ (d) $RC > \frac{1}{\omega}$

[2006 : 2 Marks]

73. A message signal

$m(t) = \cos 2000\pi t + 4\cos 4000\pi t$

modulates the carrier $c(t) = \cos 2\pi f_c t$ where $f_c = 1$ MHz to produce an AM signal. For demodulating the generated AM signal using an envelope detector, the time constant RC of the detector circuit should satisfy

(a) 0.5 ms < RC < 1 ms

(b) 1 ms << RC < 0.5 ms

(c) RC << 1 μs

(d) RC >> 0.5 ms

[2011 : 2 Marks]

74. The block diagram of a frequency synthesizer consisting of a Phase Locked Loop (PLL) and a divide- by-N counter (comprising ÷ 2, ÷ 4, ÷ 8, ÷16 outputs) is sketched below. The synthesizer is excited with a 5 kHz signal (Input 1). The free-running frequency of the PLL is set to 20 kHz. Assume that the commutator switch makes contacts repeatedly in the order 1-2-3-4.

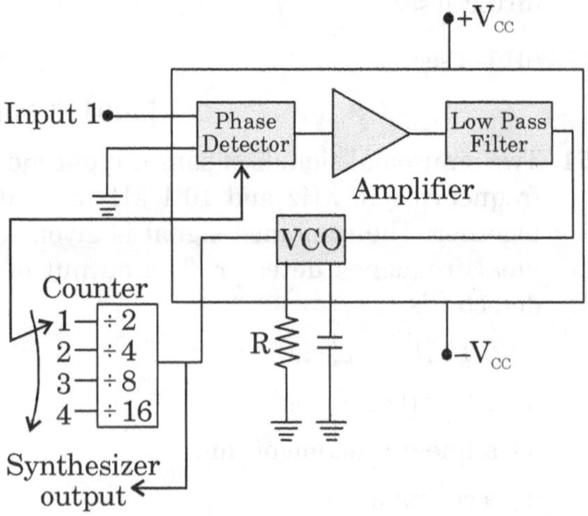

The corresponding frequency synthesized are:

(a) 10 kHz, 20 kHz, 40 kHz, 80 kHz.

(b) 20 kHz, 40 kHz, 80 kHz, 160 kHz.

(c) 80 kHz, 40 kHz, 20 kHz, 10 kHz.

(d) 160 kHz, 80 kHz, 40 kHz, 20 kHz.

[2016 : 1 Mark, Set-1]

ANSWERS

1. (b)	2. (b)	3. (b)	4. (c)	5. (c)	6. (a)	7. (d)	8. (c)	9. (c)	10. (a)
11. (a)	12. (c)	13. (c)	14. (0.5)	15. (10 kHz)	16. (4)	17. (0·5)	18. (5.208)	19. (0·25)	
20. (a)	21. (a, c)	22. (a)	23. (c)	24. (d)	25. (5)	26. (d)	27. (c)	28. (d)	29. (a)
30. (d)	31. (a)	32. (d)	33. (d)	34. (b)	35. (b)	36. (c)	37. (c)	38. (70)	39. (0.75)
40. (a)	41. (c)	42. (b)	43. (c)	44. (c)	45. (b)	46. (c)	47. (b)	48. (a)	49. (a)
50. (c)	51. (b)	52. (60)	53. (a)	54. (d)	55. (a)	56. (b)	57. (c)	58. (b)	59. (a)
60. (5)	61. (3485)	62. (a,b,c)	63. (b)	64. (c)	65. (a)	66. (d)	67. (c)	68. (c)	69. (b)
70. (b)	71. (a)	72. (a)	73. (b)	74. (a)					

EXPLANATIONS

1. Given : f_c = 4 GHz = 4000 MHz, f_M = 2 MHz.

$F_H = F_C + F_M$ = 4000 + 2 = 4002 MHz.

$F_L = F_C - F_M$ = 4000 – 2 = 3998 MHz.

Minimum sampling frequency, f_s =

$$\frac{2f_H}{K}; K_H = \frac{f_H}{f_H - f_L}$$

$$= \frac{4002}{4} \simeq 1000.$$

$$f_s = \frac{2 \times 4002}{1000} = 8.004\text{MHz}$$

$f_s \simeq 8$ MHz.

2. The correct answer is Low Pass Filter and so an integrator.

3. DSB-SC signal is s(t) = x(t) cos(ω_ct + θ)

x(t) [cos (ω_ct) cosθ – sinw_ct sinθ]

s(t) = [x(t) cosθ] cosω_ct – [x (t) sinθ]sinω_ct.

Envelope of s(t) $= \sqrt{[x(t)\cos\theta]^2 + [x(t)\sin\theta]^2}$

$= \sqrt{x^2(t)} = |x(t)|.$

4. Given : Highest frequency component, f_{max} =15 kHz.

Bandwidth of quadrature carrier multiplexing is

B.W. = $2f_{max}$ = 2 × 15 = 30 kHz.

5. Given : s(t) = e^{-at}. cos [(ω_c + Δω)t].u(t).

s(t) = e^{-at}. [cos (ω_ct). cos(Δωt) –sin (ω_ct). sin (Δωt] u(t)

Let x(t) = a cos (2πf_ct) – bsin (2πf_ct).

Then complec envelope of x(t) is x_{cE} = (a + jb).

Here, a = cos (Δωt), b = sin (Δcot)

S_{cE} = [cos (Δωt) + j sin (Δωt)] e^{-at}. u(t).

S_{cE} = $e^{j\Delta\omega t}.e^{-at}$. u(t).

6. The AM signal, when not over modulated allows the recovery of *m*(*t*) from its envelope. By using envelope detector whose operation is depicted below:

Consider the circuit show in figure (A)

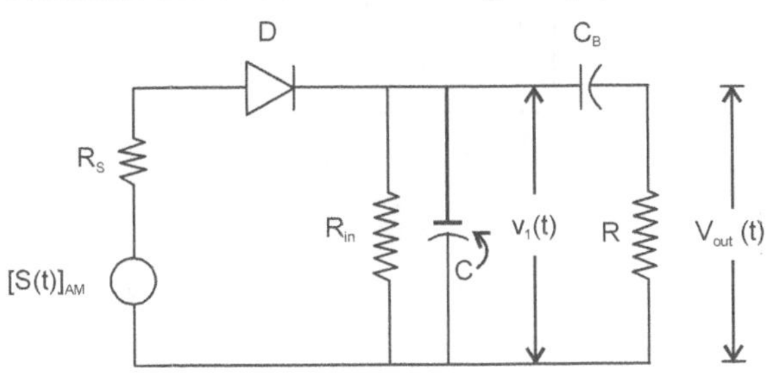

Fig. (A) The envelope detector circuit

We assume the diode *D* to be ideal. When it is forward biased, it acts as a short circuit and thereby, making the capacitor C charge through the source resistance R_s. When *D* is reverse biased, it acts as an open circuit and *C* discharges through the load resistance R_L.

As the operation of the detector circuit depends on the charge and discharge of the capacitor *C*, we shall explain this operation with the help of figure (B)

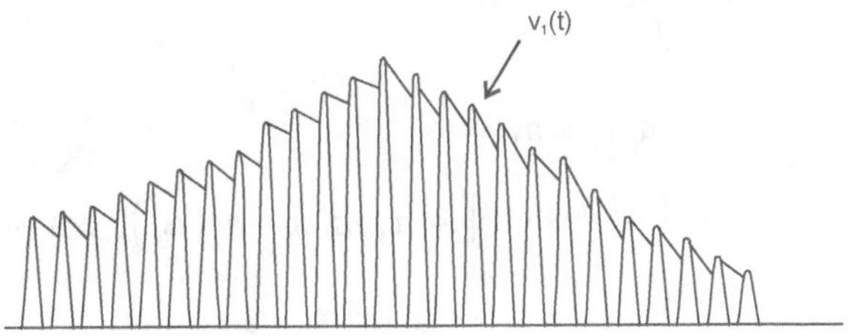

Fig.(B)

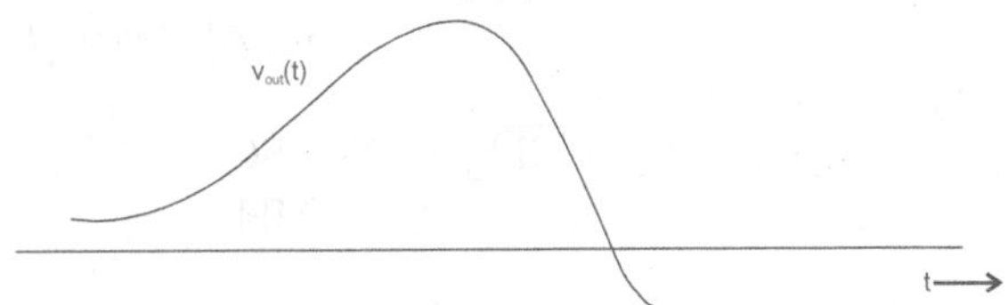

Fig.(C)

If the time constants R_sC and R_LC are properly chosen, $v_1(t)$ follows the envelope os *s*(*t*) fairly closely. During the conduction cycle of *D*, *C* quickly charges to the peak value of the carrier at that time instant. It will discharge a little during the next off cycle of the diode. The time constants of the circuit will control the ripple about the actual envelope. C_B is a blocking capacitor and the final $v_{out}(t)$ will be proportional to *m*(*t*), as shown in figure 2(*b*).

Hence, the output $V_{out}(t) = A_C m(t)$

7.

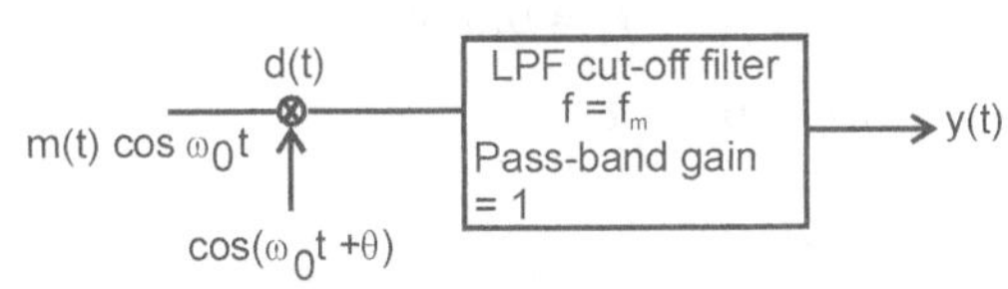

Output at product modulation

$$d(t) = m(t)\cos\omega_0 t * \cos(\omega_0 t + \theta)$$

$$= \frac{m(t)}{2}\left[\cos(2\omega_0 t + \theta) + \cos\theta\right]$$

since $\omega_0 > 2\pi f_m$

output at LPF, $y(t) = \dfrac{m(t)}{2}\cos\theta$ [higher frequency component rejected

[power in the output signal

$$P = E\left[y^2(0)\right]$$

$$= \frac{\cos^2\theta}{4} E\left[m^2(0)\right]$$

$$= \frac{\cos^2\theta}{4} \cdot P_m$$

8.

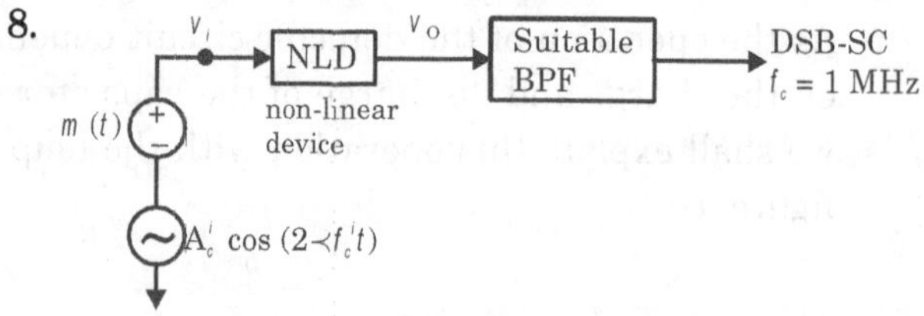

$$v_o = a_0 v_i + a_1 v_i^3$$

$$= a_0\left[A_c^i \cos\omega_c^i t\right] + a_0\, m(t) + a_1\left(A_c^i\right)^3 \cos^3\omega_c^i t$$

$$+ a_1 m^3(t) + 3a_i\, m^2(t)\, A_c^i \cos\omega_c^i t$$

$$+ 3a_1 A_c^i \cos^2\omega_c^i t \,.\, m(t)$$

$$v_o \rightarrow \boxed{\text{BPF}} \rightarrow \text{DSB-SC}$$
$$f_c = 1\,\text{MHz}$$

The useful term for DSB - SC is

$$3a_1 A_c^i \cos^2\omega_c^i t \,.\, m(t)$$

$$\Rightarrow 3a_1 A_c^i \left[\frac{1}{2} m(t) + \frac{\cos 2\omega_c^i t . m(t)}{2}\right]$$

For DSB - SC

$$\Rightarrow \frac{3a_1}{2} A_c^i \cos 2\omega_c^i t \,.\, m(t)$$

But f_c for DSB - SC is given which is 1 MHz.

Here compare $2f_c^i$ with $f_c = 1$ MHz

$$f_c^i = 0.5\,\text{MHz}$$

9. p(t) = u(t) – u(t – 1).

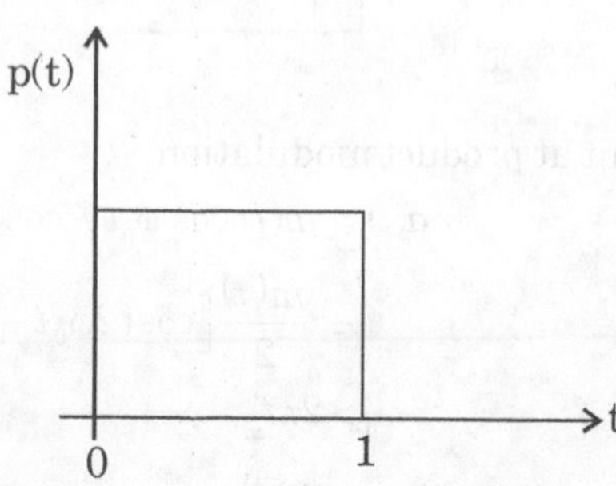

g(t) = p(t) * p(t).

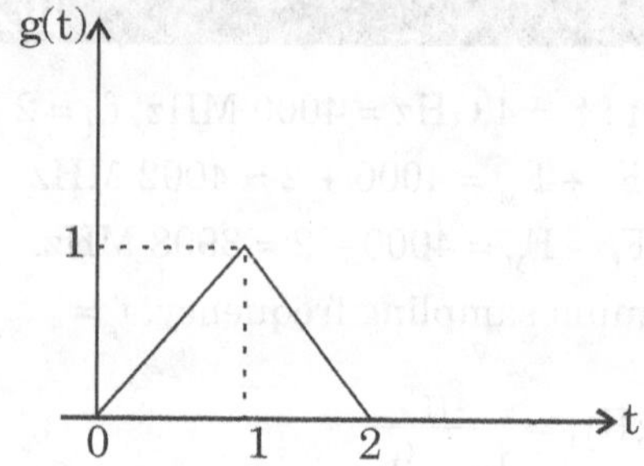

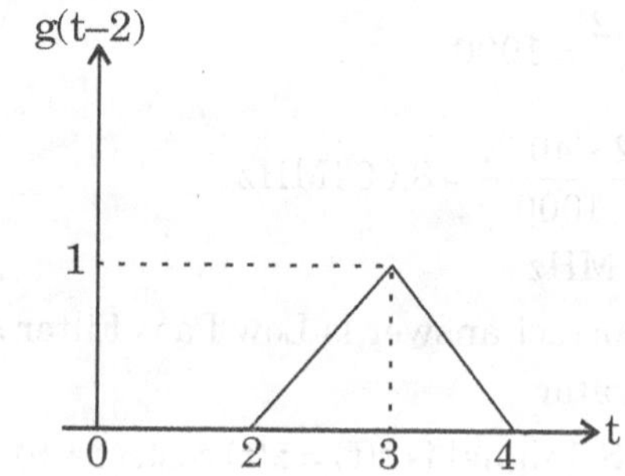

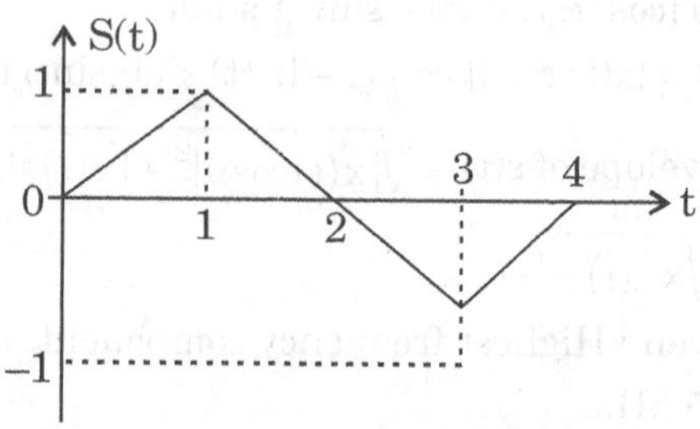

$\Rightarrow$ s(t) = g(t) – δ (t – 2) * g(t) = g(t) – g(t – 2)

$\Rightarrow$ **Impulse response of match filter is**

n(t) = s(t – T) = – s(t).

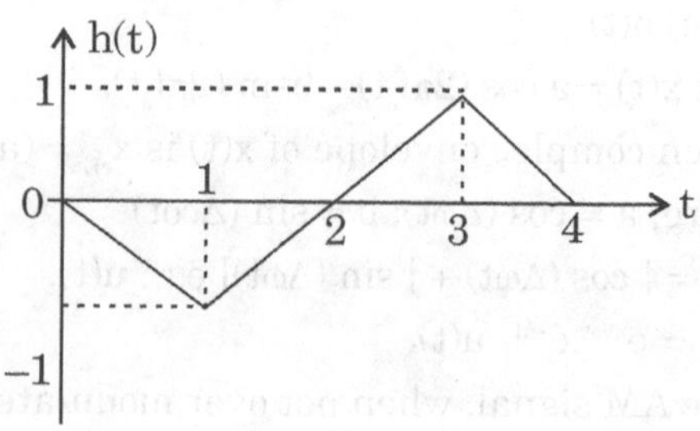

10.

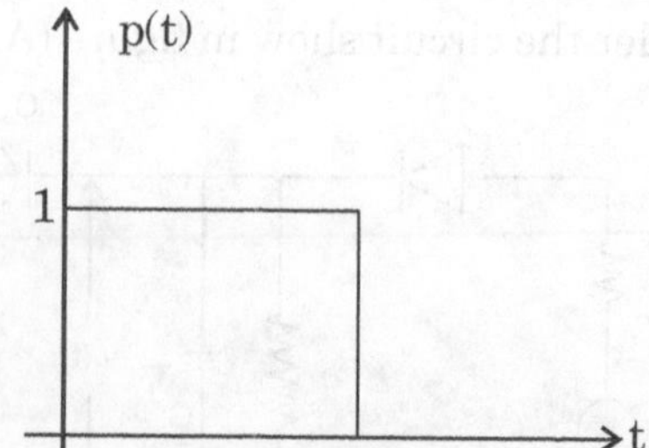

$x_{AM}(t) = 100\,[p(t) + 0.5\,g(t)]\cos w_c t$

$\Rightarrow s(t) = 100\,[1 + m(t)]\cos w_c t,\ p(t) = 1.$

$m(t) = 0.5\,g(t)$

$\Rightarrow$ modulation signal = 0.5.

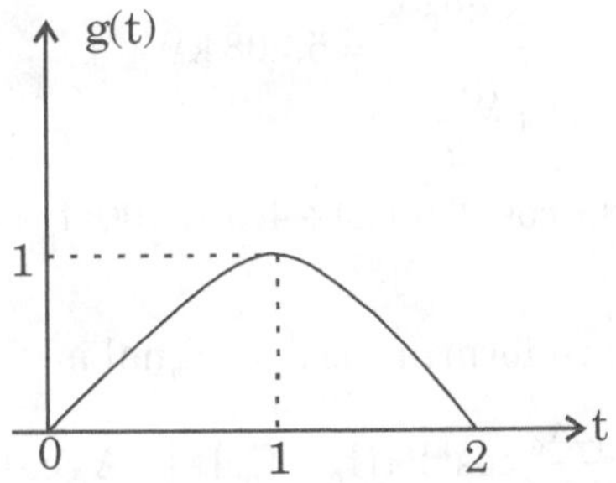

11. $x(t) = A_c \cos\omega_c t + 2\cos(\omega_m t)\cos\omega_c t$

$$= A_c \cos\omega_c t\left[1 + \frac{2}{A_c}\cos\omega_m t\right]$$

Comparing with standard equation for AM wave,

$$x(t) = A_c \cos\omega_c t\,[1 + k_a m(t)]$$

(k_a is the amplitude sensitivity)

We get that to avoid over modulation,

$$|k_a m(t)|_{max} \le 1$$

$$\Rightarrow \quad \left|\frac{2}{AC}\right| \le 1$$

or, minimum value of $A_C = 2$.

12. Power efficiency, $\eta = \dfrac{P_{SB}}{P_{TOT}}$

where, P_{SB} = power in side bands

P_{TOT} = total power transmitted

$$s(t) = \left[1 + \frac{1}{2}\cos\omega_1 t - \frac{1}{2}\sin\omega_2 t\right]\cos\omega_c t$$

$$= \cos\omega_c t + \frac{1}{4}\cos[(\omega_1 + \omega_c)t]$$

$$+ \frac{1}{4}\cos[(\omega_1 - \omega_c)t]$$

$$+ \left\{\frac{1}{4}\sin[(\omega_2 + \omega_c)t] + \frac{1}{4}\sin[(\omega_2 + \omega_c)t]\right\}$$

Power in side bands, $P_{SB} = \dfrac{1}{2}\left[\dfrac{1}{16} + \dfrac{1}{16} + \dfrac{1}{16} + \dfrac{1}{16}\right]$

$$= \frac{1}{8}$$

$$P_{TOT} = \frac{1}{2} + \frac{1}{8} = \frac{5}{8}$$

$$\therefore \quad \eta = \frac{1/8}{5/8} = \frac{1}{5}$$

$$= 0.20 = 20\%$$

Alternately

$$m_1 = \frac{1}{2}$$

$$m_2 = \frac{1}{2}$$

$$m = \sqrt{m_1^2 + m_2^2} = \frac{1}{\sqrt{2}}$$

Power efficiency $\eta = \dfrac{m^2}{2 + m^2}$

$$= \frac{\frac{1}{2}}{2 + \frac{1}{2}} = \frac{1}{5} = 20\%$$

13. Conventional AM wave is represented as,

$$x(t) = A_C[1 + k_a m(t)]\cos(2\pi f_c t) \quad ...(A)$$

For over modulation, $|k_a m(t)|_{max} > 1$

Hence, only signals given in option (*b*) and (*c*) are conventional signal.

For, $\quad x(t) = A_c\cos(2\pi f_c t) + \dfrac{A_c}{4} m(t)\cos(2\pi f_c t)$

$$|k_a m(t)|_{max} = \left|\frac{1}{4}m(t)\right|_{max} = \left|\frac{1}{4}\cdot 2\right| = \frac{1}{2} < 1$$

(Under modulated)

For, $\quad x(t) = A_c[1 + m(t)]\cos(2\pi f_c t)$

$$|k_a m(t)|_{max} = |1\cdot m(t)|_{max} = 2 > 1$$

(Over modulated)

14. $$\mu = \frac{A(t)_{max} - A(t)\min}{A(t)_{max} + A(t)\min}$$

$$\mu = \frac{3-1}{3+1} = \frac{1}{2} = 0.5$$

15. Since $m(t)$ is a base band signal with maximum frequency 5 KHz, assumed spreads as follows :

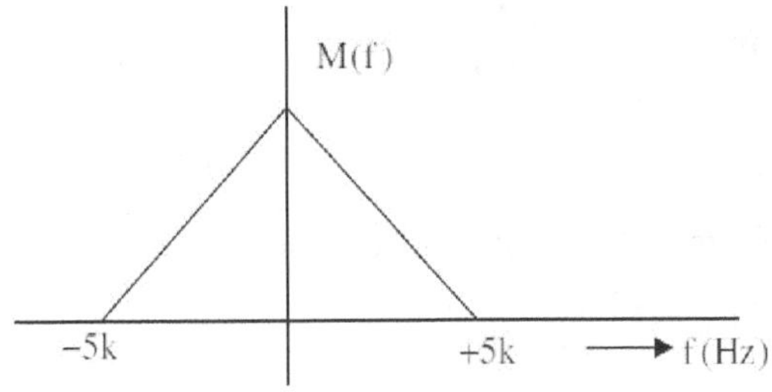

$m(t) \xrightarrow{\ f\ } M(f)$

$\because \quad y(t) = m(t)\cos(40000\pi t) \xrightarrow{7}$

$$m(f) \overset{*}{-} \frac{1}{2}[\delta(f-20k)+\delta(f+20k)]$$

$\therefore \quad y(f) = \frac{1}{2}\left[M(f-20k)+M(f+20k)\right]$

Thus the spectrum of the modulated signal is as follows :

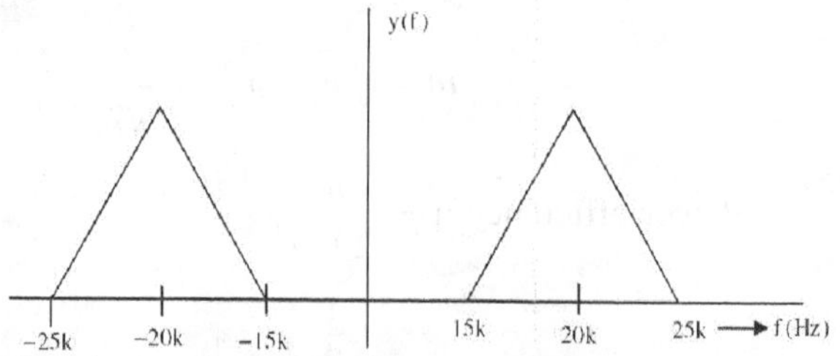

If $y(t)$ is sampled with a sampling frequency 'f_s' then the resultant signal is a periodic extension of successive replica of $y(f)$ with a period 'fs'.

It is observed that 10 KHz and 20 KHz are the two sampling frequencies which causes a replica of $M(f)$ which can be filtered out by a LPF.

Thus the minimum sampling frequency (f_s) which extracts $m(t)$ from $g(f)$ is 10 KHz.

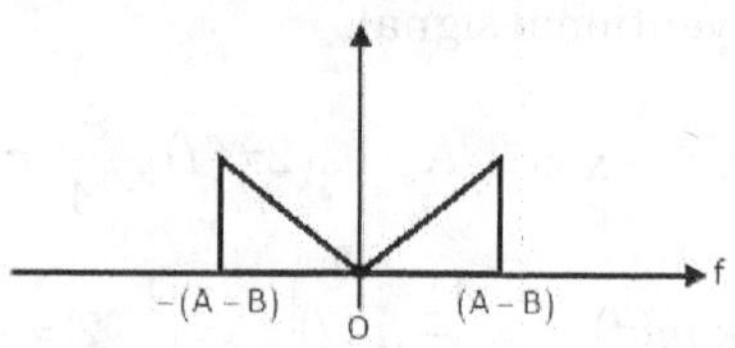

Bandwidth = A − B

= 100 − 40 = 60

16. $\dfrac{\text{Ratio of total side band power}}{\text{Carrier power}} \alpha\ \mu^2$

If it in doubled, this ratio will be come 4 times

17. Here the amplitude modulated signal

$s(t) = 5\cos 1600\pi t + 20\cos 1800\pi t + 5\cos 2200\pi t$

$$s(t) = \frac{A_c\mu}{2}\cos 2\pi(f_c - f_m)t + A_c\cos 2\pi f_c t + \frac{A_c\mu}{2}\cos 2\pi(f_c + f_m)t$$

So, $\quad A_c = 20$ and $A_c\mu = 10$

or $\quad \dfrac{A_c\mu}{2} = 5$

$\therefore \quad \mu = \dfrac{10}{20} = \dfrac{1}{2} = 0.5$

18. Given : $P_c = 5$ kW for $\mu_{max} = 0.5$

$P_{t_{max}} = P_c\left[1+\frac{\mu_{max}^2}{2}\right]$

$= 5\left[1+\frac{0.25}{2}\right]$

$P_{t_{max}} = 5.625\text{kW}$

For $\mu = 0.4$

$P_{c_{max}}\left(1+\frac{\mu^2}{2}\right) = P_{t_{max}}$

$P_{c_{max}} = \dfrac{5.625}{1+\dfrac{(0.4)^2}{2}} = 5.208\text{ kW}$

19. $s(t) = \cos(2000\pi t) + 4\cos(2400\pi t) + \cos(2800\pi t)$ ...(*i*)

Standard form of the AM signal is given below,

$$s(t) = \frac{\mu A_c}{2}\cos\left[2\pi(f_c - f_m)t\right] + A_c\cos(2\pi f_c t) + \frac{\mu A_c}{2}\cos\left[2\pi(f_c + f_m)t\right]$$

Now comparing equation (*i*) with standard equation we get

$A_c = 4$

and $\quad \dfrac{\mu A_c}{2} = 1$

$\therefore \quad \mu = \dfrac{2}{A_c}$

Now $\quad \dfrac{A_m}{A_c} = \dfrac{2}{A_c} \Rightarrow A_m = 2$

Now

$$\frac{P_m}{P_c} = \frac{\frac{1}{2}A_m^2}{\frac{1}{2}A_c^2} = \frac{A_m^2}{A_c^2} = \frac{(2)^2}{(4)^2} = \frac{1}{4} = 0.25$$

20. The expression of FM signal is given by

$$s(t) = A_c.\cos\left[\omega_c t + 2\pi k_f \int_{-\infty}^{t} m(\tau)d\tau\right].$$

for $m(t) = E_m \cos\omega_m t$.

$$s(t) = A_c\cos\left[\omega_c t + 2\pi k_f.E_m \int_{-\infty}^{t} \cos\omega_m \tau d\tau\right]$$

$$= A_c.\cos\left[\omega_c t + 2\pi k_f.\frac{E_m}{\omega_m}.\sin\omega_m t\right]$$

$$= A_c.\cos\left[\omega_c t + \frac{k_f}{E_m}.\sin\omega_m t\right]$$

$$s(t) = A_c.\cos[\omega_c t + m_f.\sin\omega_m t].$$

where $m_f = \dfrac{k_f E_m}{f_m}$ = modulation index.

21. To analyze the performance of a communication system in the presence of noise, Figure of merit, is defined as

$$\text{FOM} = \left(\frac{S}{N}\right)_{\frac{O}{P}} \Big/ \left(\frac{S}{N}\right)_{\frac{I}{P}}.$$

where $\left(\frac{S}{N}\right)_{\frac{I}{p}}$ falls below a particular value (usually 10 dB), then $\left(\frac{S}{N}\right)_{\frac{o}{p}}$ decreases rapidly and Hence makes FOM < 1 which in turn. deferiorates the system performace. It is called threshold effect.

It occurs in AM detection using envelope detector and FM detection using a discriminator.

22. Modulation index, $m_f = \frac{K_f A_m}{\omega_m}$

$$= \frac{2 \times 2\pi \times 10 \times 10^3}{\pi \times 10^4} = 4.$$

23. In commercial TV transmission in India, picture signal is modulated using VSB modulation and speech or audio signal is modulated using FM modulation.

24. For **NBFM** with $m << 1$, the modulated carrier

$c_m = E_k \, | \cos \omega_k t - m_f \sin \omega_k t \sin \omega_m t \, |$

The given equation resembles this form, the carrier and sideband are in quadrature.

25. From given data,

the frequency range of superheterodyne receiver

$(f_s) = 58 \text{ MHz} - 68\text{MHz}$

When $f_s = 58$ MHz

$f_{si} = f_s + 21\text{F} > 68$ MHz

$2\text{IF} > 10$ MHz

$\therefore$ IF ≥ 5 MHz

So, the minimum required F_{1F} is 5 MHz.

26. In a frequency multiplier circuit, the modulation index is multiplied by n. So, β'_{fm}

$\beta'_{fm} = 3 \times 9 = 27.$

27. $$P_t = P_c\left(1 + \frac{m^2}{2}\right)$$

$$\Rightarrow \frac{P_t}{P_c} = 1 + \frac{m^2}{2} = 1 + \frac{0.25}{2} = 1 + 0.125 = 1.125$$

$$\Rightarrow \frac{P_t - P_c}{P_c} = \frac{1.125 - 1}{1} = \frac{0.125}{1} = \frac{1}{8}$$

28. The FM signal is represented in terms of Bessel function as,

$$x_{FM}(t) = A \sum_{n=-\infty}^{\infty} J_n(\beta).\cos(\omega_c \pm n\omega_m)t$$

where A is the amplitude of carrier signal.

then, $\Delta\omega = 6\omega_n$ and $\beta = \frac{\Delta\omega}{\omega_m} = 6$

Given $\omega_c + n\omega_m = (1008 \times 10^3).\, 2\pi$

or $2\pi \times 10^6 + n.4\pi \times 10^3 = (1008 \times 10^3).\, 2\pi$

or $n = 4$

thus, Bessel coefficient will be $5J_4(6)$.

29. when FM signal is applied to doubler, frequency deviation doubles but f_m remains the same.

B.W.$=2\,(\Delta f + f_m) = 2(108 + 5) = 370$ kHz

30. $s(t) = 10 \cos[2\pi 10^5 t + 5 \sin(2\pi \times 500t) + 7.5 \sin(2\pi \times 1000t)]$

Let equation FM signal of type :

$$A \cos\left[\omega_c t + \frac{K_f V_m}{\omega_m} \sin\omega_{mt}\right]$$

then $\frac{K_f V_{m1}}{\omega_{m1}} = 5$; $\frac{K_f V_{m2}}{\omega_{m2}} = 7.5$

$\Rightarrow K_f V_{m1} = 5 \times 2\pi \times 1500$

and $K_f V_{m2} = 7.5 \times 2\pi \times 1000$

we have $\Delta\omega$ (frequency deviation) same in both, so

modulation index $m_f = \frac{\Delta\omega}{[\omega]_{max}}$

$$= \frac{5 \times 2\pi \times 1500}{2\pi \times 1500} = 5$$

31. Instantaneous phase

$\phi_1(t) = 2\pi f_c t + \beta_1 \sin 2\pi f_1 + \beta_2 \sin 2\pi f_2 t$

Instantaneous frequency

$$f_i(t) = \frac{d}{dt}\phi_1(t) \times \frac{1}{2\pi}$$

$= f_c + \beta_1 f_1 \cos 2\pi f_1 t + \beta_2 f_2 \cos 2\pi f_2 t$

Instantaneous frequency deviation

$= \beta_1 f_1 \cos 2\pi f_1 t + \beta_2 f_2 \cos 2\pi f_2 t$

Maximum $\Delta f = \beta_1 f_1 + \beta_2 f_2$

32. Instantaneous frequency, ω_i, is given by,

$$\omega_i = \frac{d}{dt}[\theta(t)]$$

$$= \frac{d}{dt}\left[2\pi\left(2 \times 10^6 t + 30 \sin 150t + 40 \cos 150t\right)\right]$$

and $\Delta\omega = \omega_i - \omega_c$

$= 3000\pi[3\cos 150t - 4\sin 150t]$

$= 15000\,\pi\,[\cos(150\,t + \alpha)]$,

where $\alpha = \tan^{-1}\dfrac{4}{3}$

Then maximum frequency deviation,

$$\Delta_f = \frac{|\Delta\omega|_{max}}{2\pi}$$

$$= \frac{15000\pi}{2\pi} = 7.5 \text{ kHz}$$

and $\phi(t) = 2\pi\,(30\sin 150\,t + 40\cos 150\,t)$

$$= 100\pi\left(\frac{3}{5}\sin 150\,t + \frac{4}{5}.\cos 150t\right)$$

$$= 100\pi\,.\,\sin(150\,t + \alpha')$$

where $\alpha' = \tan^{-1}\dfrac{4}{3}$.

Then, maximum phase deviation,

$$|\,\phi(t)\,|_{max} = 100\pi \text{ radian}$$

33. Standard form of amplitude modulated signal,

$s(t) = A_c[1 + K_a m(t)\cos(\omega_c t)$

Standard form of DSB-SC signal,

$s(t) = Km(t).A_c\cos(\omega_c t)$.

Standard form of phase modulated signal,

$s(t) = Ac.\cos[\omega_c t + K_p m(t)]$.

Standard form of frequency modulated signal,

$$s(t) = A_c\cos\left[\omega_c t + k_f\int_{-\infty}^{t} m(\tau)d\tau\right]$$

34. The terms $[2\sin(8000\pi t) + 4\cos(8000\pi t)]$ can be expressed as,

$$[\cos\alpha\sin(8000\pi t) + \sin\alpha\cdot\cos(8000\pi t)]$$

Hence $\cos\alpha = \dfrac{2}{\sqrt{20}} = \dfrac{1}{\sqrt{5}}, \sin\alpha = \dfrac{2}{\sqrt{5}}$

or, $\sqrt{20}\left[\sin(8000\pi t + \phi)\right]$,

where $\tan^{-1}\phi = 2$

Hence, $x(t) = 6\cos\left[2\pi\times 10^6\,t + \sqrt{20}\cdot\sin(8000\pi t + \phi)\right]$

For angle-modulated signal hainvg constant amplitude A_c.

The average power is given by,

$$P_{av} = \frac{A_c^2}{2} = \frac{6^2}{2} = 18 \text{ Watts}$$

35. In phase modulation,

Maximum Phase deviation $= K_p|m(t)|_{max} = K_p.2$

In Frequency modulation,

Maximum phase deviation $= 2\pi\,k_f\int_a^2 2\,dt = 2\pi\,K_f\times 4$

Now, $K_p.2 = 2\pi\,K_f\times 4$

$\Rightarrow \dfrac{K_p}{K_f} = 4\pi$

36. Given signal s(t) is an SSB modulated signal. Also, it is the canonical representation of a bandpass signal.

37. W_B doubled (increased) → early effect is still present but its effect less severe relative to previous W_B. Slope I_C vs V_{CE} decreases.

38. Given : modulating signal,

$x(t) = 5\sin(4000\pi t - 10\pi\cos 2000\pi t)V$.

$k_p = 5$ rad/V, $f_c = 20$ kHz

The standard phase modulated signal is

$s(t) = A_c\cos(\omega_c t + k_p x(t))$.

Instantaneous angle of the modulated signal is

$\theta(t) = \omega_c t + k_p x(t)$.

Instantaneous frequency

$$\omega_i(t) = \frac{d\theta(t)}{dt} = \omega_c + k_p\frac{dx(t)}{dt}$$

$$f_i(t) f_c + \frac{25}{2\pi}[\cos(4000\pi t - 10\pi 2000\pi t)$$

$$\times\left(4000\pi + 20000\pi^2.\sin 2000\pi t\right)].$$

At t = 0.5 ms,

$$f_i = f_c + \frac{25}{2\pi}[\cos(2\pi - 10\pi\cos(\pi)].$$

$$(4000\pi + 20000\pi^2.\sin\pi)$$

$$= f_c + \frac{25}{2\pi}(\cos 12\pi).4000\pi$$

$= f_c + 50$ kHz

($\therefore$ fc = 20 kHz)

$f_i = 70$ kHz

39. Instantaneous frequency $(f_1) = \dfrac{1}{2\pi}\dfrac{d\theta(t)}{dt}$

where $\theta(t) = 2\pi f_c t + km(t)$

$$\therefore f_1 = \frac{1}{2\pi}\frac{d}{dt}\left[2\pi f_c t + km(t)\right] = f_c + \frac{K}{2\pi}\frac{dm}{dt}$$

$$f_{i_{max}} = f_c + \frac{K}{2\pi}\frac{dm}{dt}\bigg|_{max}^{1} \text{ and } f_{i_{min}} = f_c + \frac{K}{2\pi}\frac{dm}{dt}\bigg|_{min} \quad \text{...(ii)}$$

$$\text{Now} \quad \left.\frac{dm}{dt}\right|_{max} = 2\left.\frac{dm}{dt}\right|_{min} = -1$$

Replacing the value of

$$\left.\frac{dm}{dt}\right|_{max} \text{ and } \left.\frac{dm}{dt}\right|_{min}$$

In eq. (i) and (ii) we get.

$$f_{i_{max}} = 50\text{ k} + (5)(2)\text{k} = 60\text{k}$$

$$f_{i_{min}} = (50\text{k}) + (5)(-1)\text{k} = 45\text{k}$$

$$\text{Now Required Ratio} = \frac{f_{i_{min}}}{f_{i_{max}}}$$

$$= \frac{45\text{K}}{60\text{K}}$$

$$= \frac{3}{4} = 0.75$$

40. Since $m(t)$ is sinusoidal

$$y_{FM}(t) = A_c \cos[\omega_c t + \beta \sin \omega_m t]$$

$$y(t) = [y_{FM}(t)]^3 = A_c^3 \cos^3[\omega_c t + \beta \sin \omega_m t]$$

Result. Modulation index β get multiplied by factor 3 and f_m remain constant.

$$f'_m = f'_m \; \beta' = 3\beta$$

Bandwidth, $B'_T = 2 f_m (1 + \beta') \approx 2 f_m (3\beta) \because \; 3\beta \gg 1$

hence Band width

$$B'_T = 3\text{ BT} = 3 \times 1\text{ M} = 3\text{ MHz}$$

$$f'_m = f_m = 5\text{ KHz}$$

41.

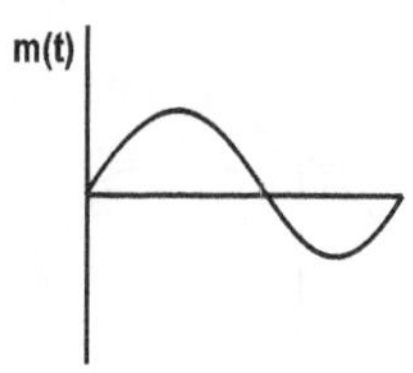

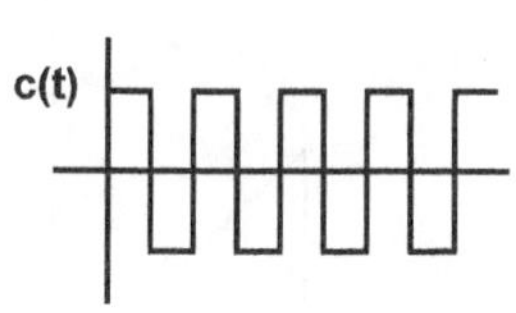

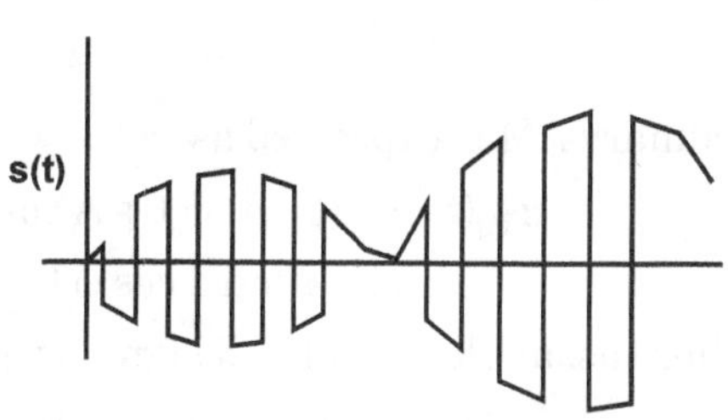

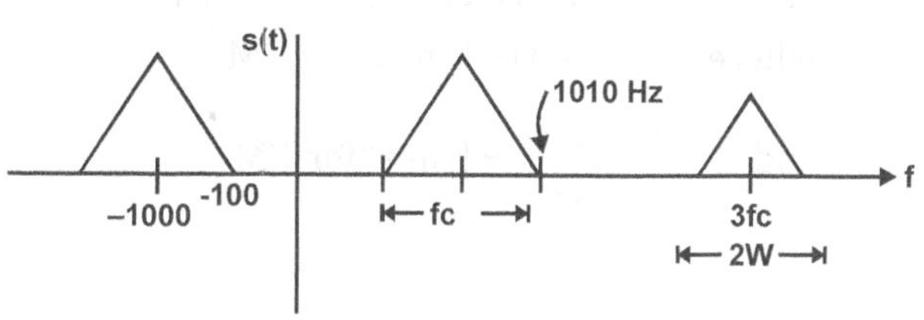

$$s(t) = c(t) \, . \, m(t)$$

Expressing square wave as,

$$c(t) = \frac{4}{\pi}\sum_{n=1}^{\infty}\frac{(-1)^{n-1}}{2n-1}\cos[2\pi f_c (2n-1)].$$

then, modulated output will be

$$s(t) = c(t) \, . \, m(t)$$

$$= \frac{4}{\pi} . \sum_{n=1}^{\infty}\frac{(-1)^{n-1}}{2n-1}\cos[2\pi f_c (2n-1)].m(t)$$

From, the spectrum it is clear than 1020 Hz signal will be absent.

42. Narrow-band FM signal is given as,

$$x_{NBFM}(t) = A_C \cos(2\pi f_c t) - \beta A_C \sin(2\pi f_m t) \sin(2\pi f_c t) \quad ...(A)$$

where $m(t) = A_m \cos(2\pi fmt)$

Carrier wave, $s(t) = A_C \cos(2\pi f_c t)$

AM signal is given as,

$$s_{AM}(t) = A_C[1 + \;\cos(2\pi f_m t)]\cos(2\pi f_c t) \quad ...(B)$$

Given $m(t)$ is same per both the signals and β =

Adding equation. (A) and equation (B), we have

$x_{NBFM}(t) + S_{AM}(t)$

$$= \underbrace{A_C \cos(2\pi f_c t)}_{\text{carrier wave}} + \underbrace{\frac{A_C\mu}{2}\cos 2\pi(f_m + f_c)t}_{\text{SSB wave}}$$

43. Output of the balanced modulator,

$$s(t) = \cos(2\pi \, 10^6 t) \, . \, \cos(2\pi \, . \, 10^8 t)$$

$$= \frac{1}{2}\left\{\cos\left[2\pi(10^8 + 10^6)t\right] + \cos\left[2\pi(10^8 - 10^6)t\right]\right\}$$

After high-pass filtering,

$$g(t) = \frac{1}{2}\cos\left[2\pi(10^8 + 10^6)t\right]$$

After adding another signal at the output, we have

$$y(t) = \frac{1}{2}\cos\left\{(10^8 + 10^6)2\pi t\right\} + \cos(10^8 \, . \, 2\pi t)\frac{1}{2}$$

Its Hilbert transform,

$$\hat{y}(t) = \frac{1}{2}\sin\left\{(10^8 + 10^6).2\pi t\right\} - \cos(10^8 . 2\pi t)$$

Then, pre-envelope,

$$y_+(t) = y(t) + j\,\hat{y}(t)$$

or,

$$y_+(t) = \frac{1}{2}e^{j2\pi(10^8+10^6)t} - j.e^{j2\pi.10^8.t}$$

and, complex-envelope

$$\tilde{y}(t) = y_+(t)\,.\,\exp(-j2\pi\,.\,10^8.\,t)$$
$$= \frac{1}{2}e^{j2\pi 10^6.t} - 1$$

$$\tilde{y}(t) = \frac{1}{2}[\cos(2\pi.10^6 t)] + j\left[\frac{1}{2}\sin 2\pi\,.\,10^6\,t - 1\right]$$

Hence envelope of $y(t)$ would be,

$$a(t) = \left[y_I^2(t) + y_Q^2(t)\right]^{1/2}$$
$$= \sqrt{\frac{5}{4} - \sin(2\pi \times 10^6 t)}$$

44. SSB has $BW = \omega_m$ and power is also minimum

45.

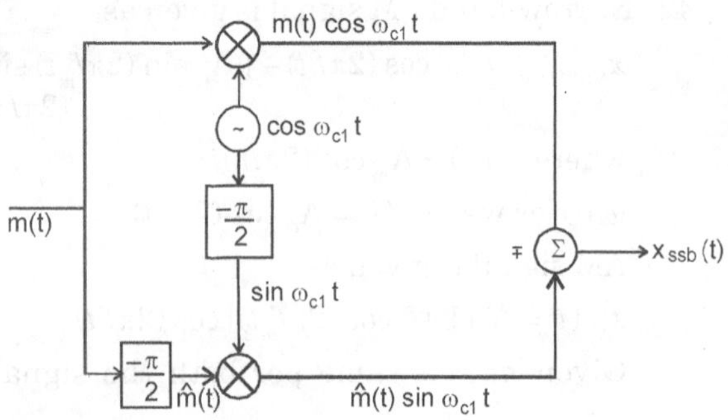

$\hat{m}(t)$ is the Hilbert transform of $m(t)$

$\omega_{c1} = 2\pi fc_1$

$m(t) = A_m \sin(2\pi f_m t)$

SSB-modulator

$x_{SSB}(t) = m(t)\cos\omega_{c_1} t \mp \hat{m}(t)\sin\omega_{c_1} t$

Lower-side-band SSB signal is represented as,

$x_{SSB}(t) = m(t)\cos\omega_{c_1} t + \hat{m}(t)\sin\omega_{c_1} t$ ...(i)

$$= A_m \cos\omega_{c_1} t\sin(2\pi f_m t) - A_m \sin\omega_{c_1} t\cdot\cos(2\pi f_m t)$$

Now, Narrow band FM modulator is shown below to which SSB signal is being applied–

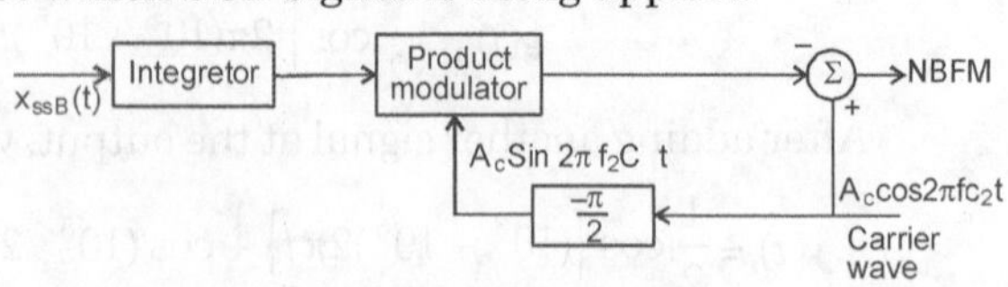

NBFM wave is represented as

$x_{NBFM}(t)$

$$= A_c\cos(2\pi f_{c_2} t) - \beta A_c\sin(2\pi f_{c_2} t)\sin(\omega_{c_1} - \omega_m)t$$

$$= A_c\cos\omega_{c_2} t - \frac{\beta A_c}{2}\left[\cos(\omega_{c_2} - \omega_{c_1} + \omega_m)t - \cos(\omega_{c_2} + \omega_{c_1} - \omega_m)t\right]$$

BW of x_{NBFM} is

$$= 2(\omega_{c_1} - \omega_m)$$
$$\simeq 2\omega_{c_1} \approx 2f_{c_1} \approx 2\times 10^6\,\text{Hz}$$

46. Given : $m_a = 0.5$, $A_c = 10$.

Side-band power,

$$P_s = P_c.\frac{m_a^2}{2} = \frac{A_c^2}{2}.\frac{m_a^2}{2} = \frac{100}{2}\times\frac{0.25}{2}$$

$p_s = 6.25$ W.

47. Side band power, $p_s = 6.25$ W.

Area under noise power spectral density is eqaul to the noise power. Mean noise power,

$$P_n = \int_{-\infty}^{\infty} S_n(t).dt$$

$$= 4\times\frac{1}{2}\times B\times\frac{N_o}{2} = N_o.B$$

The ratio of average sideband power to mean noise power

$$\frac{P_{SB}}{P_N} = \frac{6.25}{N_oB} = \frac{25}{4N_oB}.$$

48. $y(t) = m(t)\cos(2\pi\omega t) + m_h(t)\sin(2\pi\omega t)$

$$m(t) = a\cos\omega_m t$$
$$m_h(t) = a\sin\omega_m t$$

[Since $\frac{\pi}{2}$ shift by Hilbert transformer]

$$y(t) = a[\cos\omega_m t\cos 2\pi\omega t + \sin\omega_m t\sin 2\pi\omega t]$$
$$= a\cos[(\omega_m - 2\pi\omega t)]$$

This is the equation of LSB.

$$\omega_m = 2\pi f$$

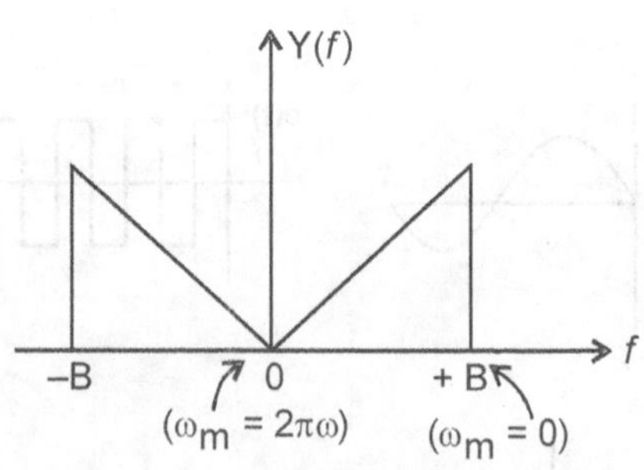

49. Ordinary AM is expressed as

$$x_{AM}(t) = m(t)\cos\omega_c t + A\cos\omega_c(t)$$
$$= [A + m(t)]\cos\omega_c t$$

Where as, angle-modulated signal is expressed

or $x_c(t) = A\cos[\omega_c(t) + \phi(t)]$

where $\phi(t) = k_p m(t)$ for PM

and $\frac{d\phi(t)}{at} = k_f m(t)$ for FM.

50. Modulated signal,

$$s(t) = m(t) \cdot \cos(2\pi f_c t)$$
$$= \cos(2\pi f_m t) \cdot \cos(2\pi f_c t)$$
$$= \frac{1}{2}[\cos 2\pi(f_c + f_m)t + \cos 2\pi(f_c - f_m)t]$$

First term represents upper side-band signal and second one represents lower side-band signal.

51. Power efficient transmission → FM

Most bandwidth efficient → SSB-SC

Transmission of voice signal

Simplest receives structure → conventional AM

Bandwidth efficient transmission of → VSB

Signals with significant DC component

52. $m(t) \leftrightarrow M(f)$

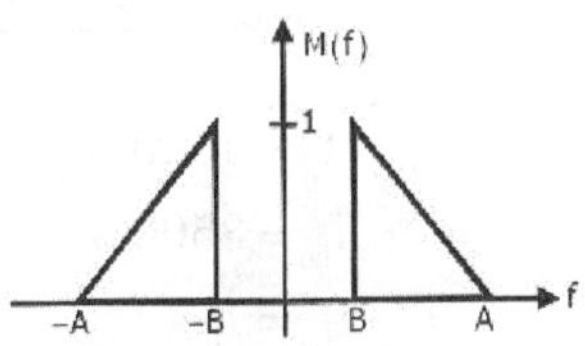

After multiplication with

$$V(t) = \cos(2\pi f_c t)$$

Let $\quad w^1(t) = m(t).V(t)$

$\Rightarrow$ $W^1(f)$ (spectrum of $w^1(t)$) is

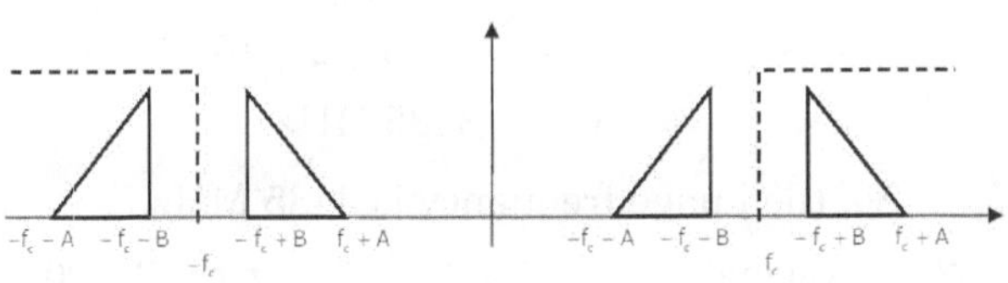

After his pass filter

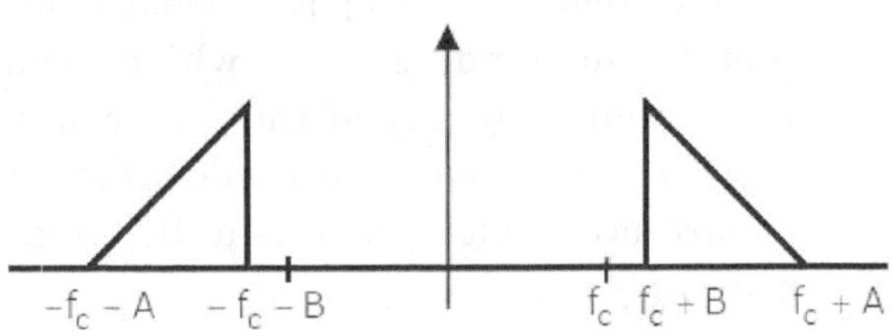

After multiplication with $\cos(2\pi(f_c + A)t)$ and low pass filter of cut off f_c

53. Given signal is $x(t) = \cos\left(10\pi t + \frac{\pi}{4}\right)$

Neglect the phase-shift $\frac{\pi}{4}$ and it can be inserted at the end result.

$\therefore$ If $\quad x_1(t) = \cos 10\pi t \xrightarrow{L} X_1(f)$

$$= \frac{1}{2}[\delta(f-5) + \delta(f+5)]$$

Given filter impulse response is,

$$h(t) = \left(\frac{\sin \pi t}{\pi t}\right)\cos\left(40\pi t - \frac{\pi}{2}\right)$$
$$= (\sin ct) \sin(40\pi t)$$

$$\therefore \quad H(f) = \text{rect } f * \frac{1}{2j}[\delta(f-20) - \delta(f+20)]$$
$$= \frac{1}{2j}[\text{rect}(f-20) - \text{rect}(f+20)]$$

$X_1(f)$ repeats with a value $f_o = 15\text{Hz}$ and each impulse value is $\frac{15}{2}$

Thus the sampled signal spectrum and the spectrum of the filter are as follows:

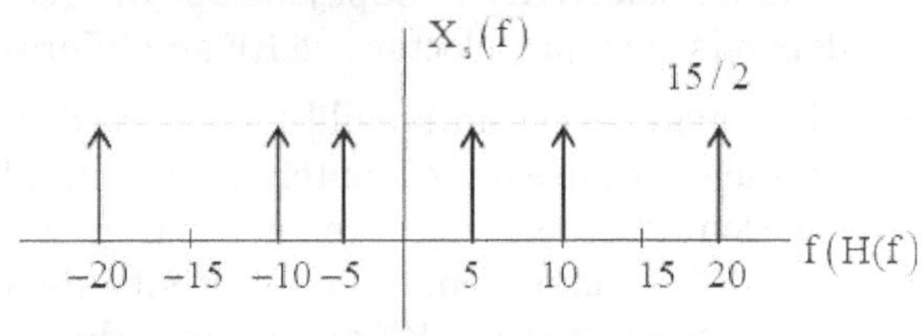

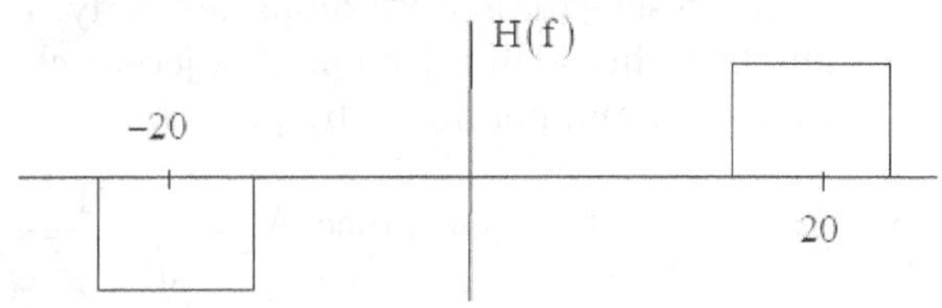

$$\therefore \quad X_s(f)\,H(f) = \frac{15}{4j}[\delta(f-20) - \delta(f+20)]$$
$$x_r(t) = \frac{15}{2}\sin(40\pi t) \rightarrow \text{recovered signal}$$
$$= \frac{15}{2}\cos\left(40\pi t - \frac{\pi}{2}\right)$$

Insert the neglected phase shift $\frac{\pi}{4}$

$$\therefore \quad x_r(t) = \frac{15}{2}\cos\left(40\pi t - \frac{\pi}{2} + \frac{\pi}{4}\right)$$
$$= \frac{15}{2}\cos\left(40\pi t - \frac{\pi}{4}\right)$$

54. $v_i(t) = A_c\cos(2\pi f_c t) + \cos(2\pi f_m t)$

$$v_o(t) = av_i(t) + bv_i^2(t)$$
$$\left[aA_c\cos(2\pi f_c t) + a\cos(2\pi f_m t)\right]$$
$$+ b[A_c^2\cos^2(2\pi f_c t) + \cos^2(2\pi f_m t)$$
$$+ 2A_c\cos(2\pi f_c t)\cos(2\pi f_m t)]$$

Now passing above signal through band pass filter we get,

$$y(t) = aA_c\cos 2\pi f_c t + 2bA_c\cos(2\pi f_c t)\cos(2\pi f_m t)$$
$$= aA_c\left[1 + \frac{2b}{a}\cos(2\pi f_m t)\right]\cos(2\pi f_c t)$$

Now, modulation index, $(\mu) = \frac{2b}{a}$

Power of side band $(P_{SB}) = \frac{\mu^2}{2} P_c = \frac{1}{2} P_c$

$\therefore \quad \mu^2 = 1 \Rightarrow \mu = 1$

$\therefore \quad \frac{2b}{a} = 1$

$\therefore \quad \frac{a}{b} = 2$

55. The image signal rejection must be achieved before it enters the IF amplifier stage because IF amplifier is not able to distinguish between the desired signal and the image signal. The image channel selectivity of supey hetrodyne receiver depends upon preselector and RF amplifiers only.

56. The image rejectinn should be achieved before IF stage because ones it enters into IF amplifire it becomes impossible to remove it from wanted signal. So, image channel selectivity depends upon preselector and RF amplifiers only. The IF amplifier helps in rejection of adjecent channel frequency and not image frequency.

57. Response cut-off resonance, $A_R = \frac{1}{\sqrt{1 + y^2 Q^2}}$

where $y = \frac{\omega}{\omega_R} - \frac{\omega_R}{\omega}$

ω = image frequency.

So the first detector gives some image rejection which may be sufficient at low tuned frequencies. As this may be insufficient at high frequencies, RF stages may be added.

58. The image rejection should be achieved before ones it enters into IF amplifier, it becomes impossible to remove it from wanted signal. So, image channel selectivity depends upon preselector and RF amplifiers only. The IF amplifier helps in rejcetion of adjacent channel frequency and not image frequency.

59.

$$I = 700 + 450 \times 2$$

$$= 1600 \text{ kHz}$$

$$f_{max} = 1650 + 450 = 2100 \text{ kHz}$$

$$f_{min} = 550 + 450 = 1000 \text{ kHz}$$

$$f = \frac{1}{2\pi} \frac{1}{\sqrt{LC}}$$

$$\Rightarrow \quad \frac{2100}{1000} = \sqrt{\frac{LC_{max}}{LC_{min}}}$$

$$\Rightarrow \quad \frac{LC_{max}}{LC_{min}} = (2.1)^2$$

$$\Rightarrow \quad \frac{C_{max}}{C_{min}} = 4.41$$

60. From given data

The frequency range of superheterodyne receiver

$(f_s) = 58 \text{ MHz} - 68\text{MHz}$

When $f_s = 58$ MHz

$f_{si} = f_s + 21F > 68$ MHz

$21F > 10$ MHz

$\therefore \quad 1F \geq 5$ MHz

So, the minimum required F_{1F} is 5 MHz.

61.

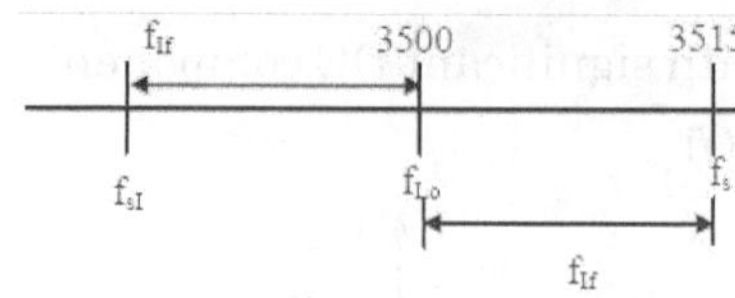

$f_{If} = 15$ MHz

$f_{Lo} = 3500$ MHz

$f_s - f_{Lo} = f_{If}$

$f_s = f_{Lo} + f_{If}$

$= 3515$ MHz

f_{si} = image frequency

$= f_s - 2 f_{If}$

$= 3515 - 2 \times 15$

$= 3485$ MHz

So, the image frequency is 3485 MHz.

62. The signal, $x(t) = 5(1 + 2 \cos 200\,\pi t) \cos 20000\,\pi t$

This is essentially an Amplitude Modulation (AM) double-sideband signal which can be demodulated by any of the 3 demodulators (*a*) (*b*) or (*c*), though envelope detection using a square law device (such as a diode) is most commonly used.

63. Both the teacher and student are wrong

$$\overline{X} = (\text{Mean}) = \sum_{i=1}^{n} p_i k_i$$

$$= 0.1 + 0.4 + 1.2 + 0.8 + 0.5$$

$$= 3.0$$

$$\overline{X}^2 = \sum_{i=1}^{n} p_i k_i^{-2}$$

$= 0.1 \times 1 + 0.2 \times 4 + 0.4 \times 9 + 0.2 \times 16 + 0.1 \times 25$

$= 0.1 + 0.8 + 3.6 + 3.2 + 2.5$

$= 10.2$

$\sigma^2 (\text{Variance}) = x^2 - x^{-2} = (-1.2)$

64. By definition and through known principle of working (of PLI)

65. Hilbert transform does change the phase by an angle $\frac{-\pi}{2}$ or 90°

If $\qquad g(t) = \cos\omega_1 t + \sin\omega_2 t$

then, Hilbert transform,

$$\hat{g}(t) = \cos\left(\omega_1 t - \frac{\pi}{2}\right) + \sin\left(\omega_2 t - \frac{\pi}{2}\right)$$

$$= \sin\omega_1 t - \cos\omega_2 t$$

66. Spectum of bandlimited signal m(t) is shown in figure.

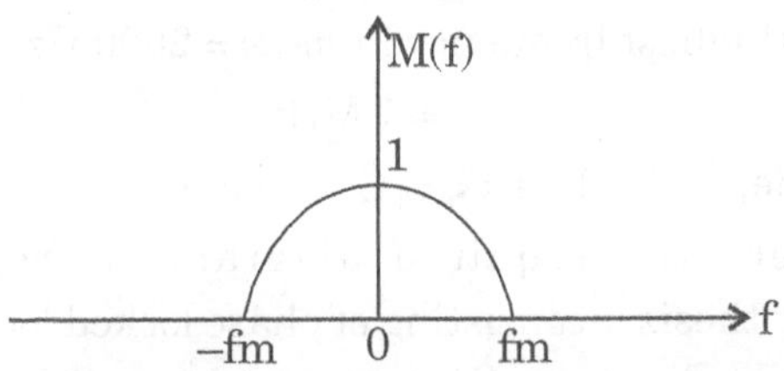

The frequency domain representation of sampled version of message signal m(t) is

$$M_s(f) = f_s \sum_{n=-\infty}^{\infty} M(f - nf_s).$$

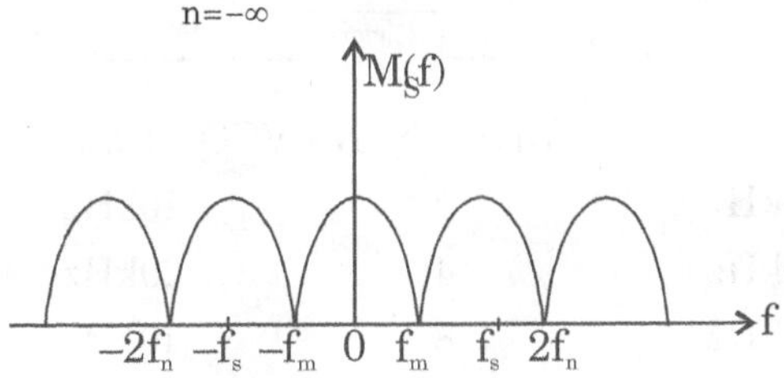

spectrum of sampled version of message signal is shows in figure.

A bandlimited signal is sampled at Nyquist rate, i.e., 2 f_m, so spectrum of sampled signal is very close to each other but they are not over-lapped.Ideal low pass filter with appropriate bandwidth must be used to recover the message signal.

67.

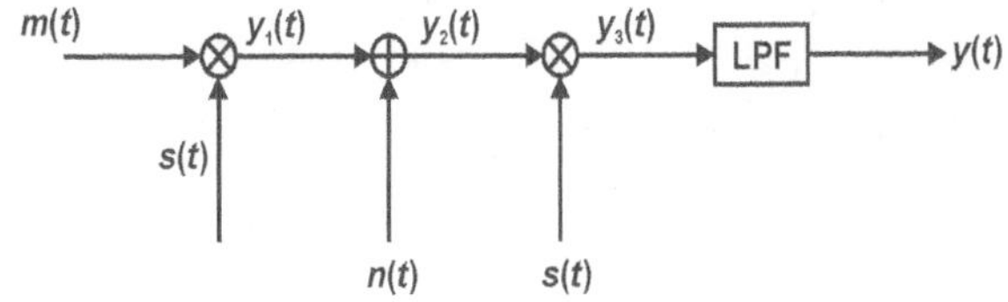

$$y_1(t) = m(t)\, s(t) = \frac{2\sin(2\pi t).\cos(200\pi t)}{t}$$

$$= \frac{\sin(202\pi t) - \sin(198\pi t)}{t} \qquad y_2(t)$$

$$= y_1(t) + n(t)$$

$$= \frac{\sin(202\pi t) - \sin(198\pi t) + \sin(199\pi t)}{t}$$

$$y_3(t) = y_2(t)\,.\,s(t)$$

$$= \frac{\sin(202\pi t).\cos(200\pi t) - \sin 198\pi t.\cos 200\pi t}{t} + \frac{\sin 199\pi t.\cos 200\pi t}{t}$$

$$= \frac{\sin 402\pi t + \sin 2\pi t - \sin 398\pi t + \sin 399\pi t + \sin(0.5\pi t).\cos(1.5\pi t)}{t}$$

After low-pass filterings, $y(t)$

$$= \frac{\sin 2\pi t + (\sin 0.5\pi t).(\cos 1.5\pi t)}{t}$$

68.

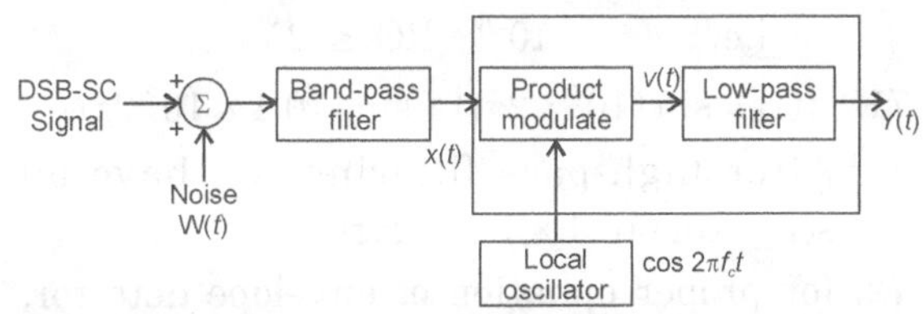

DSB – SC signal is expressed as $S(t) = CA_c \cos(2\pi f_c t).m(t)$ (here C is contant)

Average power of DSB-SC signal, $P_s = \frac{CA_c^2 P}{2}$

Hence P is the average power of message signal.

For noise spectral density of $N_0/2$, average noise power in the message bandwidth is WN_0.

then, $\qquad (SNR)_{C,DSB} = \frac{CA_C^2 P}{2WN_0}$

Using narrow band representation of the filtered noise $u(t)$, total signal at the coverent detector input,

$$x(t) = s(t) + n(t)$$

$$= CA_C\cos(2\pi f_c t)m(t) + n_I\cos(2\pi f_c t) - n_2(t)\sin(2\pi f_c t)$$

h_I = In phase component of $n(t)$

u_Q = quadrative component of $u(t)$

Output of product modulator,

$$v(t) = x(t)\cos(2\pi f_c t)$$

$$= \frac{1}{2}CA_C m(t) + \frac{1}{2}n_I(t) + \frac{1}{2}[CA_C m(t) + n_I t]\cos(4\pi f_c t) - \frac{1}{2}n_Q(t)\sin(4\pi f_c t)$$

Low-pass filter given

$$y(t) = \frac{1}{2}CA_C m(t) + \frac{1}{2}n_I(t)$$

[Only in-phase component]

69. ⇒ Ring modulator : is used to generate DSB – SC signal.

⇒ VCO : is used to generate FM in parameter variation method.

⇒ Foster-Seely discriminator : comes under phase discriminator method of FM modulation.

⇒ Mixer : is used to translate frequency from one to another.

70. $$\frac{1}{f_c} \le RC \le \frac{1}{\omega_m}$$

where, ω_m = message bandwidth

$= 2 \times 2 \times 10^3 = 4$ kHz

$f_c = 1 \times 10^6$ Hz

$\therefore \quad 10^{-6} \le RC \le \frac{1}{4} \times 10^{-3}$

i.e. $\quad 10^{-6} \le RC \le 25\ \mu$s.

71. $y(t) = \sin(10^4 \times 2\pi t) + \sin(10.1 \times 10^3\, 2\pi t)$

After high-pass filtering, we have output,

$s(t) = \sin(10.1 \times 10^3 \times 2\pi t)$

72. for proper opuation of envelope detector, time contant RC should Die between T_c and T_m, i.e., $T_c < RC < T_m 1$

$$\frac{i}{\omega_c} < RC < \frac{1}{\omega}.$$

73.

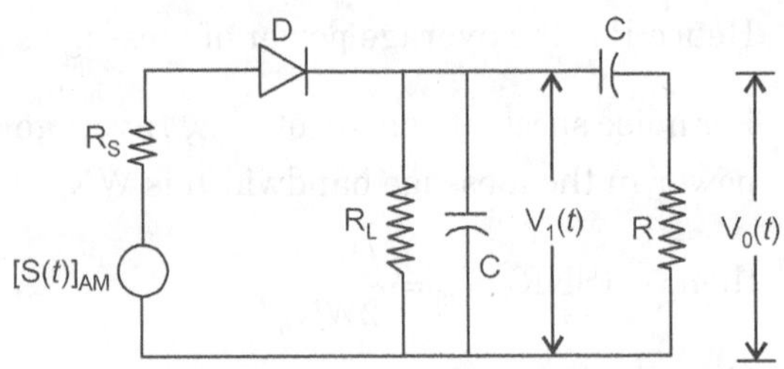

Circuit diagram of Envelope detection

Modulating signals are normaly restricted to a specific frequency range which limits the maximum rate of fall of AM's have amplitude. This rate of range of $m(t)$ depends upon the triphent frequency in $M(f)$.

During discharging f capacitor C, time constant should be large enough so that C does not discharge too much between the positive peaks of the carrier, it should be rather small enough to follow the maximum rate of charge of $m(t)$.

that is, $R_L C << \frac{1}{W}$

Now, during charging, capacitor should be charged in a smallest time possible, for this to be hold,

$$R_s C << \frac{1}{f_C}$$

Hence, the condition for satisfactory operation

$$\frac{1}{f_c} << R_L C << \frac{1}{W}$$

W(highest frequency in $m(t)$) = 2000 Hz

$f_c = 1$ MHz

The, $\quad 1\ \mu s << R_L C << 6.5$ ms

74. Here the required diagram of frequency synthesizer consisting of phase locked loop (PPL) and a divide by-N counter is shown below.

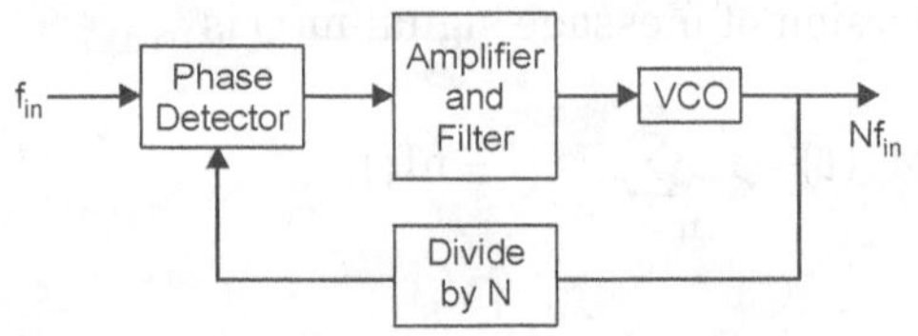

f_{in}	Divide by N	VCO output (Nf_{in})
5kHz	2	10kHz
5kHz	4	20kHz
5kHz	8	40kHz
5kHz	16	80kHz

So, the corresponding frequencies synthesizer are respectively 10 kHz, 20kHz, 40kHz, 80kHz.

3
CHAPTER

Digital Communication

Analysis of Previous GATE Papers

Topics ↓ / Year →			2019	2018	2017 Set 1	2017 Set 2	2016 Set 1	2016 Set 2	2016 Set 3	2015 Set 1	2015 Set 2	2015 Set 3
DIGITAL MOLDULATION SCHEMES	1 Mark	MCQ Type			1					1		
		Numerical Type									1	
	2 Marks	MCQ Type										
		Numerical Type										
	Total				1					1	1	
ASK, FSK, PSK, QAM	1 Mark	MCQ Type										1
		Numerical Type										
	2 Marks	MCQ Type										
		Numerical Type										
	Total											1
MATCHED FILTER RECIEVER, MAP & NL DECODING	1 Mark	MCQ Type							1			
		Numerical Type										
	2 Marks	MCQ Type			1							
		Numerical Type	1	1								
	Total		2	2	2				1			
BANDWIDTH, SNR & BER FOR DIGITAL MODLULATION	1 Mark	MCQ Type			1							
		Numerical Type		1		1	1	1				
	2 Marks	MCQ Type								1	1	
		Numerical Type						1	1	1		
	Total			1	1	1	1	3	2	4	2	

DIGITAL MODULATION SCHEMES

1. Companding in PCM systems lead to improved signal to quantization noise ratio. This improvement is for

(a) lower frequency components only.

(b) higher frequency components only.

(c) lower amplitudes only.

(d) higher amplitudes only.

[1987 : 2 Marks]

2. A signal having uniformly distributed amplitude in the interval (–V to + V), is to be encoded using PCM with uniform quantization. The signal to quantizing noise ratio is determined by the

(a) dynamic range of the siqnal.

(b) sampling rate.

(c) number of quantizing levels.

(d) power spectrum of signal.

[1988 : 2 Marks]

3. Increased pulse width in the flat top sampling, leads to

(a) attenuation of high frequencies in reproduction.

(b) attenuation of low frequencies in reproduction.

(c) greater aliasing errors in reproduction.

(d) no harmful effects in reproduction.

[1994 : 1 Mark]

4. The bandwidth required for the transmission of a PCM signal increases by a factor of______ when the number of quantization levels is increased from 4 to 64.

[1994 : 1 Mark]

5. Flat top sampling of low pass signals

(a) gives rise to aperture effect.

(b) implies oversampling.

(c) leads to aliasing.

(d) introducing delay distortion.

[1998 : 1 Mark]

6. Compression in PCM refers to relative compression of

(a) higher signal amplitudes.

(b) lower signal amplitudes.

(c) lower signal frequencies.

(d) higher signal frequencies.

[1998 : 1 Mark]

7. A sinusoidal signal with peak-to-peak amplitude of 1.536 V is quantized into 128 levels using a mid-rise uniform quantizer. The quantization-noise power is

(a) 0.768 V (b) 48×10^{-6} V^2

(c) 12×10^{-6} V^2 (d) 3.072 V

[2003 : 2 Marks]

8. The input to a linear delta modulator having a step-size $\Delta = 0.628$ is a sine wave with frequency f_m and peak amplitude E_m. If the sampling frequency f_s = 40 kHz, the combination of the sine-wave frequency and the peak amplitude, where slope overload will take place is

	E_m	f_m
(a)	0.3 V	8 kHz
(b)	1.5 V	4 kHz
(c)	1.5 V	2 kHz
(d)	3.0 V	1 kHz

[2003 : 2 Marks]

9. In a PCM system, if the code word length is increased from 6 to 8 bits, the signal to quantization noise ratio improves by the factor

(a) 8/6 (b) 12

(c) 16 (d) 8

[2004 : 1 Mark]

10. In the output of a DM speech encoder, the consecutive pulses are of opposite polarity during time interval $t_1 \le t \le t_2$. This indicates that during this interval

(a) the input to the modulator is essentially constant.

(b) the modulator is going through slope overload.

(c) the accumulator is in saturation.

(d) the speech signal is being sampled at the Nyquist rate.

[2004 : 1 Mark]

11. The minimum step-size required for a Delta-Modulator operating at 32 K samples/sec to track the signal (here u(t) is the unit-step function)

$x(t) = 125t(u(t) - u(t-1)) + (250 - 125t)(u(t-1) - u(t - 2))$ so that slope-overload is avoided, would be

(a) 2^{-10} (b) 2^{-8}

(c) 2^{-6} (d) 2^{-4}

[2006 : 2 Marks]

12. In the following figure the minimum value of the constant "C", which is to be added to $y_1(t)$ such that $y_1(t)$ and $y_2(t)$ are different, is

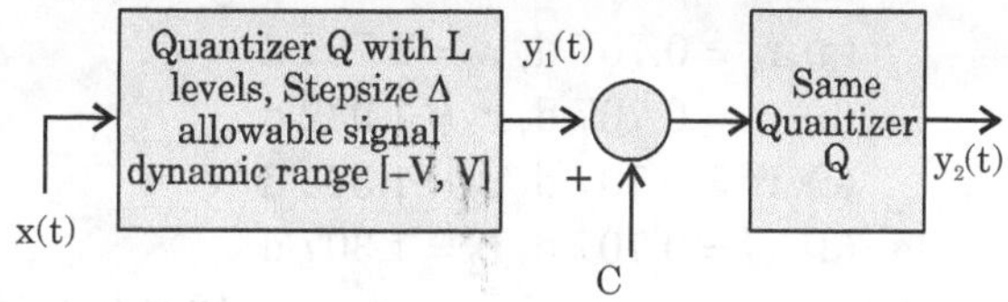

(a) Δ (b) $\frac{\Delta}{2}$

(c) $\frac{\Delta^2}{12}$ (d) $\frac{\Delta}{L}$

[2006 : 2 Marks]

13. In delta modulation, the slope overload distortion can be reduced by
(a) decreasing the step size.
(b) decreasing the granular noise.
(c) decreasing the sampling rate.
(d) increasing the step size.

[2007 : 2 Marks]

14. An analog voltage in the range 0 to 8 V is divided in 16 equal intervals for conversion to 4-bit digital output. The maximum quantization error (in V) is _____.

[2014 : 1 Mark, Set-3]

15. In a PCM system, the signal
$m(t) = \sin(100\pi t) + \cos(100\pi t)$
is sampled at the Nyquist rate. The samples are processed by a uniform quantizer with step size 0.75 V. The minimum data rate of the PCM system in bits per second is _____.

[2014 : 2 Marks, Set-3]

16. A sinusoidal signal of 2 kHz frequency is applied to a delta modulator. The sampling rate and step-size Δ of the delta modulator are 20,000 samples per second and 0.1 V, respectively. To prevent slope overload, the maximum amplitude of the sinusoidal signal (in Volts) is

(a) $\frac{1}{2\pi}$ (b) $\frac{1}{\pi}$

(c) $\frac{1}{\pi}$ (d) π

[2015 : 1 Mark, Set-1]

17. A sinusoidal signal of amplitude A is quantized by a uniform quantizer. Assume that the signal utilizes all the representation levels of the quantizer. If the signal to quantization noise ratio is 31.8 dB, the number of levels in the quantizer is _____.

[2015 : 1 Mark, Set-2]

18. Which one of the following statements about differential pulse code modulation (DPCM) is true?
(a) The sum of message signal sample with its prediction is quantized.
(b) The message signal sample is directly quantized, and its prediction is not used.
(c) The difference of message signal sample and a random signal is quantized.
(d) The difference of message signal sample with its prediction is quantized.

[2017 : 1 Mark, Set-1]

ASK, FSK, PSK, QAM

19. The message bit sequence to a DPSK modulator is 1, 1, 0, 0, 1, 1. The carrier phase during the reception of the first two message bits is π, π. The carrier phase for the remaining four message bits is

(a) $\pi, \pi, 0, \pi$ (b) $0, 0, \pi, \pi$

(c) $0, \pi, \pi, \pi$ (d) $\pi, \pi, 0, 0$

[1988 : 2 Marks]

20. In binary data transmission DPSK is preferred to PSK because
(a) a coherent carrier is not required to be generated at the receiver
(b) for a given energy per bit, the probability of error is less
(c) the 180° phase shifts of the carrier are unimportant
(d) more protection is provided against impulse noise

[1989 : 2 Marks]

21. For the signal constellation shown in the figure, the type of modulation is _____.

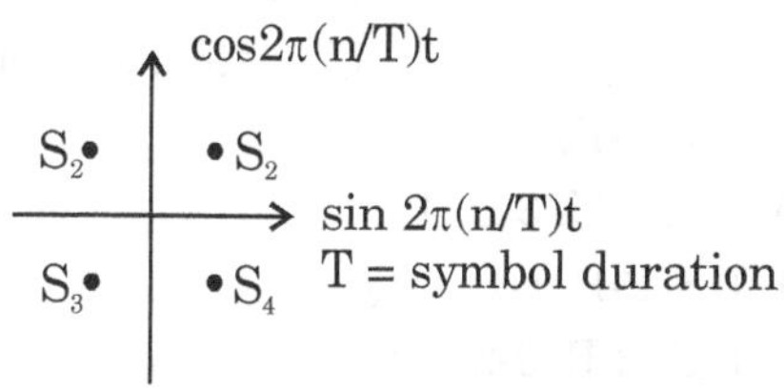

[1991 : 2 Marks]

22. Coherent demodulation of FSK signal can be detected using
(a) correlation receiver
(b) bandpass filters and envelope detectors
(c) matched filter
(d) discriminator detection

[1992 : 2 Marks]

23. For a given data rate, the bandwidth B_p of a BPSK signal and the bandwidth B_0 of the OOK signal are related as

(a) $B_p = \frac{B_0}{4}$ (b) $B_p = \frac{B_0}{4}$

(c) $B_p = B_0$ (d) $B_p = 2B_0$

[1995 : 1 Mark]

24. In a digital communication system employing Frequency Shift Keying (FSK), the 0 and 1 bit are represented by sine waves of 10 kHz and 25 kHz respectively. These waveforms will be orthogonal for a bit interval of

(a) 45 μsec (b) 200 μsec

(c) 50 μsec (d) 250 μsec

[2000 : 2 Marks]

25. If S represents the carrier synchronization at the receiver and p represents the bandwidth efficiency, then the correct statement for the coherent binary PSK is

(a) ρ = 0.5, S is required

(b) ρ = 1.0, S is required

(c) ρ = 0.5, S is not required

(d) ρ = 1.0, S is not required

[2003 : 2 Marks]

26. Choose the correct one from among the alternatives a, b, c, dafter matching an item from Group 1 with the most appropriate item in Group 2.

Group 1	Group 2
1. FM	P. Slope overload
2. DM	Q. H-law
3. PSK	R. Envelope detector
4. PCM	S. Capture effect
	T. Hilbert transform
	U. Matched filter

(a) 1-T, 2-P, 3-U, 4-S

(b) 1-S, 2-U, 3-P, 4-T

(c) 1-S, 2-P, 3-U, 4-Q

(d) 1-U, 2-R, 3-S, 4-Q

[2004 : 2 Marks]

Linked Data Questions 27 & 28 :

A four-phase and an eight-phase signal constellation are shown in the figure below.

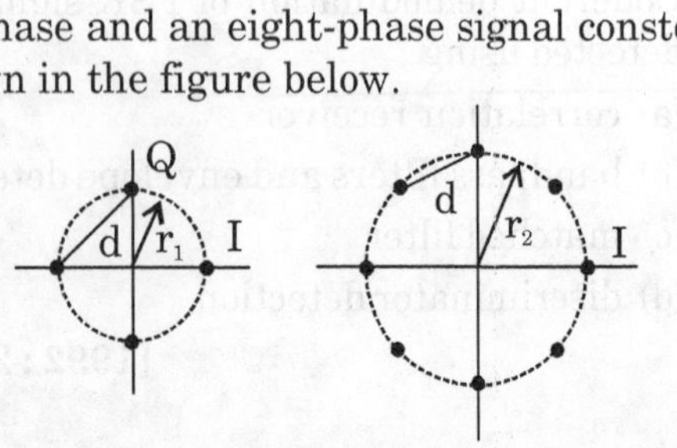

27. For the constraint that the minimum distance between pairs of signal points be d for both constellations, the radii r_1 and r_2 of the circles are

(a) r_1 = 0.707 d, r_2 = 2.782 d

(b) r_1 = 0.707 d, r_2 = 1.932 d

(c) r_1 = 0.707 d, r_2 = 1.545 d

(d) r_1 = 0.707 d, r_2 = 1.307 d

[2011 : 2 Marks]

28. Assuming high SNR and that all signals are equally probable, the additional average transmitted signal energy required by the 8-PSK signal to achieve the same error probability as the 4-PSK signal is

(a) 11.90 dB (b) 8.73 dB

(c) 6.79 dB (d) 5.33 dB

[2011 : 2 Marks]

29. The modulation scheme commonly used for transmission from GSM mobile terminals is

(a) 4-QAM

(b) 16-PSK

(c) Walsh-Hadamard orthogonal codes

(d) Gaussian Minimum Shift Keying (GMSK)

[2015 : 1 Mark, Set-3]

MATCHED FILTER RECIEVER, MAP & NL DECODING

30. The bit stream 01001 is differentially encoded using 'Delay and Ex–OR' scheme for DPSK transmission. Assuming the reference bit as a '1' and assigning phases of '0' and 'π' for 1's and 0's respectively, in the encoded sequence, the transmitted phase sequence becomes

(a) π 0 π π 0 (b) 0 π π 0 0

(c) 0 π π π 0 (d) π π 0 π π

[1992 : 2 Marks]

31. Source encoding in a data communication system is done in order to

(a) enhance the information transmission.

(b) bandpass filters and envelope rate detectors.

(c) conserve the transmitted power.

(d) discriminator detection.

[1992 : 2 Marks]

32. A 1.0 kHz signal is flat top sampled at the rate of 1800 samples/sec and the samples are applied to an ideal rectangular LPF with cut-off frequency of 1100 Hz, then the output of the filter contains

(a) only 800 Hz component.

(b) 800 Hz and 900 Hz components.

(c) 800 Hz and 1000 Hz components.

(d) 800 Hz, 900 Hz and 100 Hz components.

[1995 : 1 Mark]

33. The line code that has zero dc component for pulse transmission of random binary data is

(a) non-return to zero (NRZ).

(b) return to zero (RZ).

(c) alternate mark inversion (AMI).

(d) None of the above.

[1997 : 1 Mark]

34. The input to a matched filter is given by

$$s(t)=\begin{cases}10\sin(2\pi\times10^{6}) & 0<t<10^{-4}\text{ sec}\\ 0 & \text{Otherwise}\end{cases}$$

The peak amplitude of the filter output is

(a) 10 Volts

(b) 5 Volts

(c) 10 milliVolts

(d) 5 milliVolts

[1999 : 2 Marks]

35. Consider a sample signal

$$y(t)=5\times10^{-6}x(t)\Sigma_{n=-\infty}^{+\infty}\delta(t-nT_s)$$

where $x(t) = 10\cos(8\pi \times 10^3)t$ and $T_s = 100$ μsec. When y(t) is passed through an ideal lowpass filter with a cutoff frequency of 5 kHz, the output of the filter is

(a) $5 \times 10^{-6}\cos(8\pi \times 103)t$.

(b) $5 \times 10^{-5}\cos(8\pi \times 10^3)t$.

(c) $5 \times 10^{-1}\cos(8\pi \times 10^3)t$.

(d) $10\cos(8\pi \times 10^3)t$.

[2002 : 1 Mark]

36. A signal $x(t) = 100\cos(24\pi \times 10^3)t$ is ideally sampled with a sampling period of 50 psec and then passed through an ideal lowpass filter with cutoff frequency of 15 kHz. Which of the following frequencies is/are present at the filter output?

(a) 12 kHz only

(b) 8 kHz only

(c) 12 kHz and 9 kHz

(d) 12 kHz and 8 kHz

[2002 : 2 Marks]

37. If E_b, the energy per bit of a binary digital signal, is 10^{-5} Watt-sec and the one-sided power spectral density of the white noise, $N_0 = 10^{-6}$ W/Hz, then the output SNR of the matched filter is

(a) 26 dB

(b) 10 dB

(c) 20 dB

(d) 13 dB

[2003 : 2 Marks]

38. Consider the signal x(t) shown in the figure. Let h(t) denote the impulse response of the filter matched to x(t), with h(t) being non-zero only in the interval 0 to 4 sec. The slope of h(t) in the interval 3 < t < 4 sec is

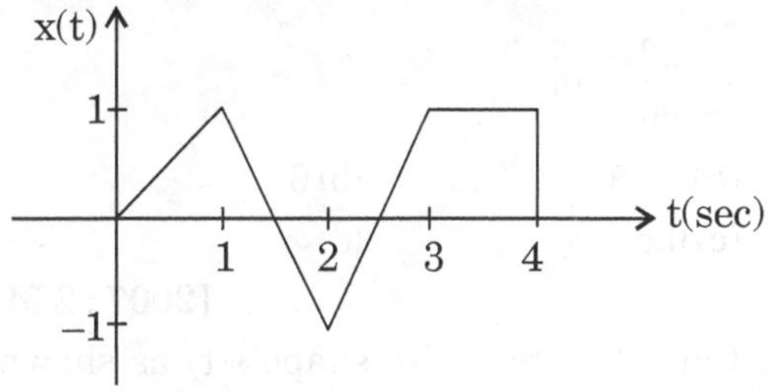

(a) $\frac{1}{2}$ sec^{-1} (b) –1 sec^{-1}

(c) $-\frac{1}{2}$ sec^{-1} (d) 1 sec^{-1}

[2004 : 2 Marks]

39. A signal as shown in the figure is applied to a matched filter. Which of the following does represent the output of this matched filter?

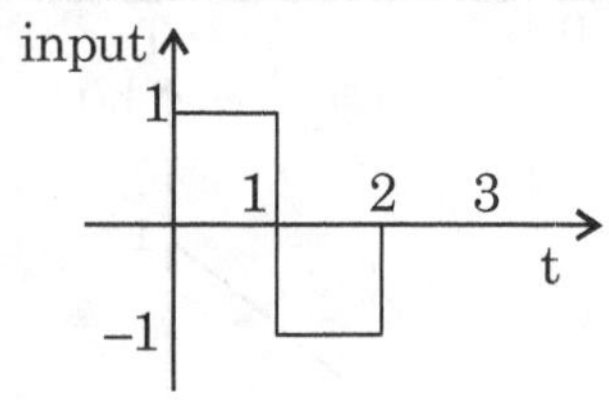

(a)
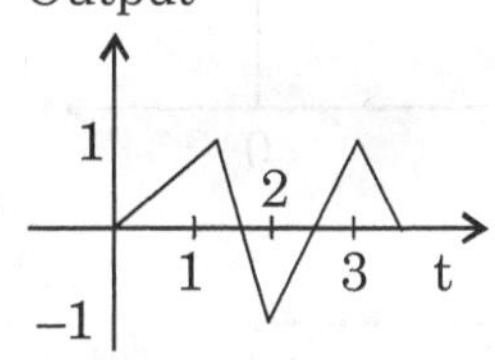

(b)
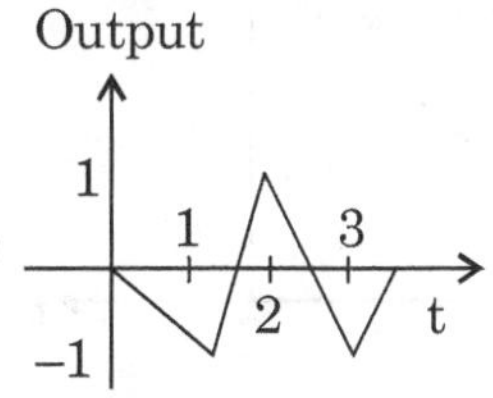

(c)
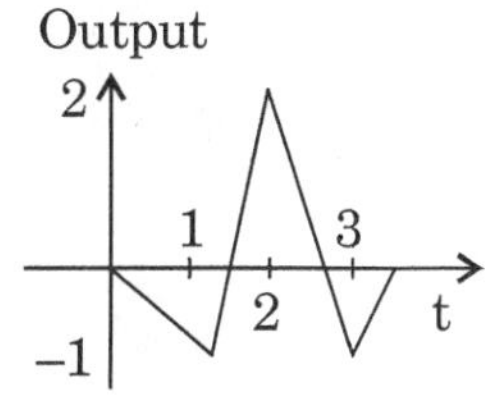

(d)
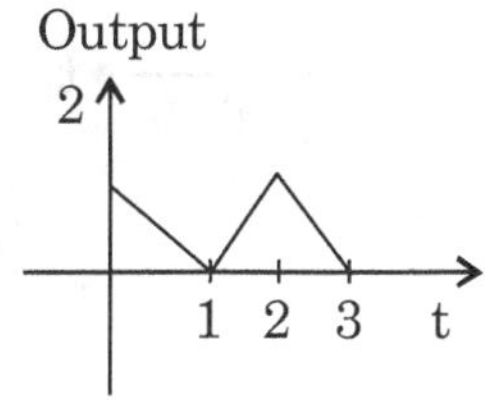

[2005 : 2 Marks]

40. The raised cosine pulse p(t) is used for zero ISI in digital communications. The expression for p(t) with unity roll-off factor is given by

$$p(t) = \frac{\sin 4\pi\omega t}{4\pi\omega t(1-16\omega^2t^2)}$$. The value of p(t) at

$t = \frac{1}{4\omega}$ is

(a) –0.5 (b) 0
(c) 0.5 (d) ∞

[2007 : 2 Marks]

41. Consider the pulse shape s(t) as shown. The impulse response h(t) of the filter matched to this pulse is

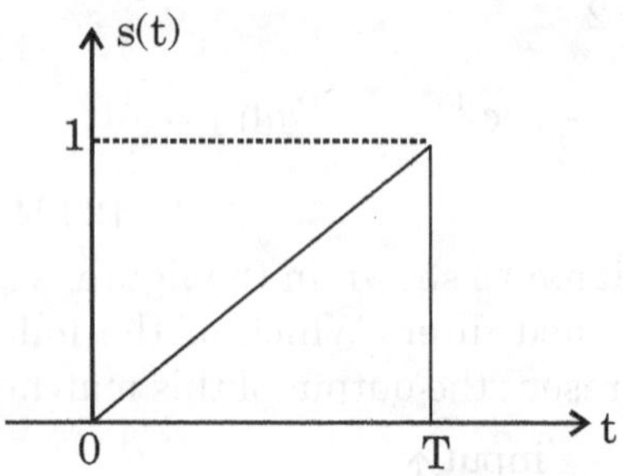

(a)

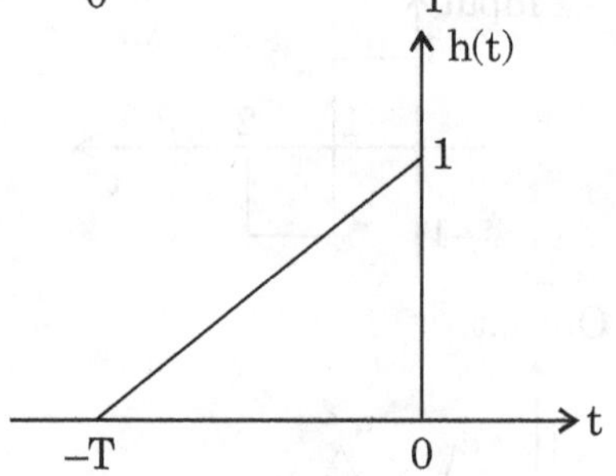

(b)

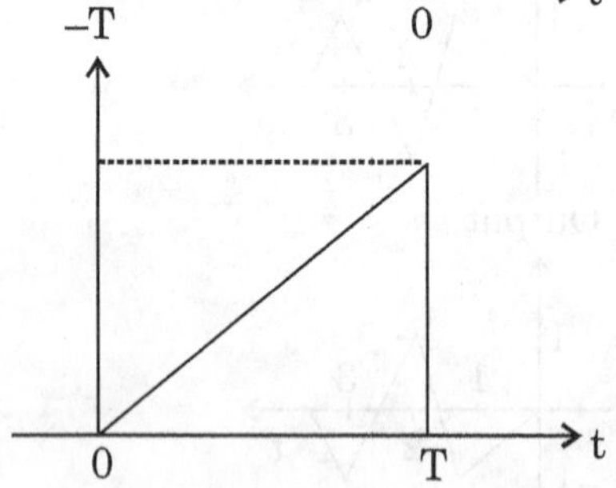

(c)

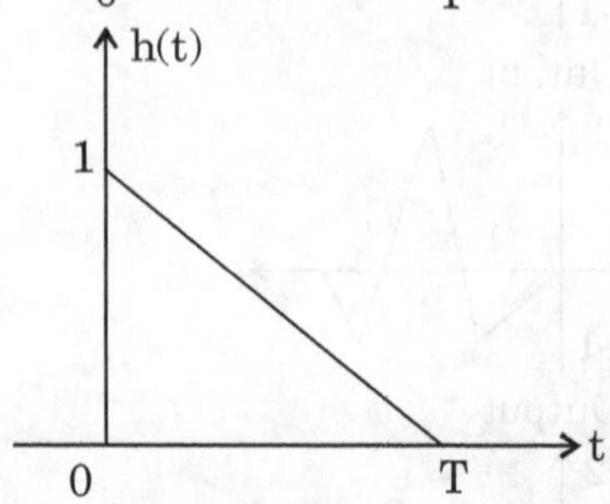

(d)

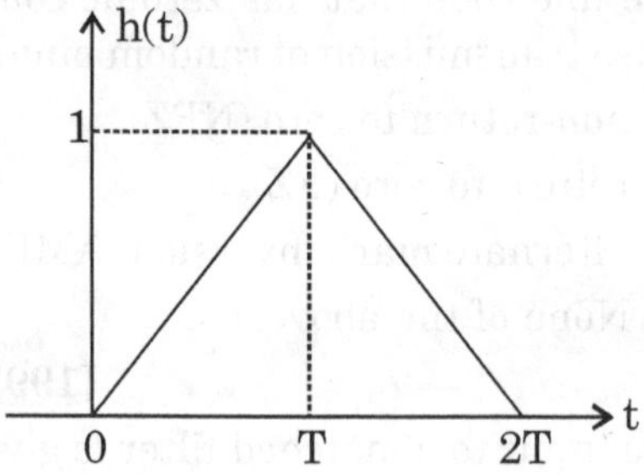

[2010 : 1 Mark]

42. A binary baseband digital communication system employs the signal

$$p(t) = \begin{cases} \frac{1}{\sqrt{T_s}}, & 0 \le t \le T_s \\ 0, & \text{otherwise;} \end{cases}$$

for transmission of bits. The graphical representation of the matched filter output y(t) for this signal will be

(a)

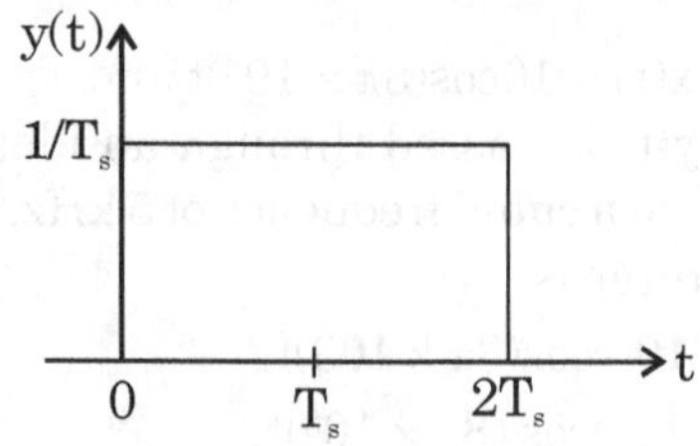

(b)

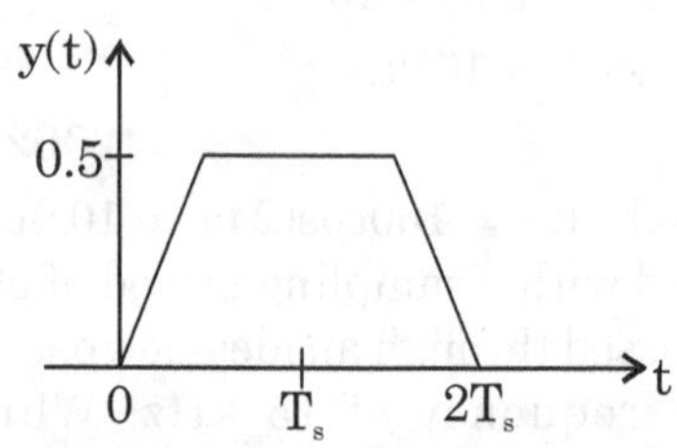

(c)

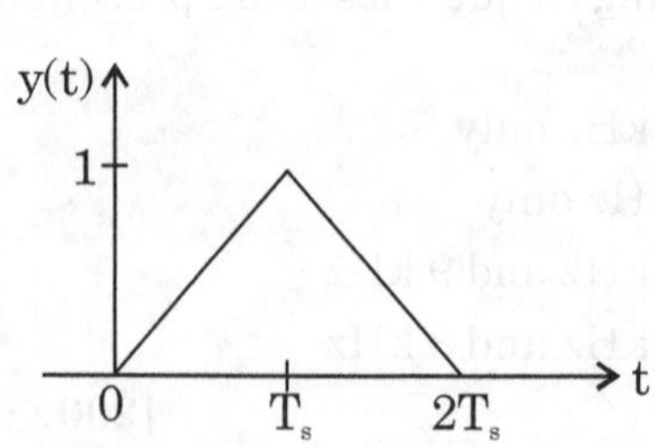

(d)

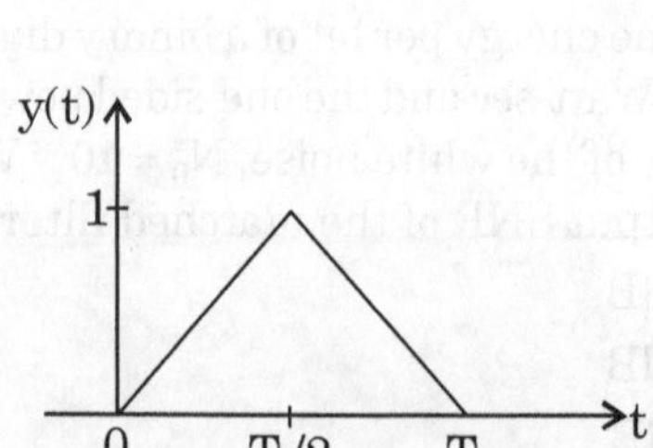

[2016 : 1 Mark, Set-3]

43. In binary frequency shift keying (FSK), the given signal waveforms are

$u_0(t) = 5\cos(20000\pi t);\ 0 \le t \le 7$ and

$u_1(t) = 5\cos(22000\pi t);\ 0 \le t < T$,

where T is the bit-duration interval and t is in seconds. Both $u_0(t)$ and $u_1(t)$ are zero outside the interval $0 \le t \le T$. With a matched filter (correlator) based receiver, the smallest positive value of T(in milliseconds) required to have $u_0(t)$ and $u_1(t)$ uncorrelated is

(a) 0.25 ms (b) 0.5 ms

(c) 0.75 ms (d) 1.0 ms

[2017 : 2 Marks, Set-1]

44. A random variable X takes values –0.5 and 0.5 with probabilities $\frac{1}{4}$ and $\frac{3}{4}$, respectively. The noisy observation of X is Y = X + Z, where Z has uniform probability density over the interval (–1, 1). X and Z are independent. If the MAP rule baaed detector outputs $\hat{X}$ as

$$\hat{X} = \begin{cases} -0.5, & Y < \alpha \\ 0.5, & Y \ge \alpha \end{cases}$$

then the value of α (accurate to two decimal places) is _______.

[2018 : 2 Marks]

45. A voice signal m (t) is in the frequency range 5 kHz to 15 kHz. The signal is amplitude modulated to generate an AM signal f(t) = A (1 + m (t)) cos $2\pi f_c t$, where f_c = 600kHz.

The Am signal f(t) is to be digitized and archived. This is done by first sampling f(t) at 1.2 times the Nyquist frequency, and then quantizing each sample using a 256-level quantizer. Finally, each quantized sample is binary coded using K bits, where K is the minimum number of bits required for the encoding. The rate, in Megabits per second (rounded off to 2 decimal places), of the resulting stream of coded bits is _____ Mbps.

[2019 : 2 Marks]

BANDWIDTH, SNR & BER FOR DIGITAL MODULATION

46. In a digital communication system, transmissions of successive bits through a noisy channel are assumed to be independent events with error probability p. The probability of at most one error in the transmission of an 8-bit sequence is

(a) $\frac{7(1-p)}{8}\frac{7(1-p)}{8+\frac{p}{8}}$

(b) $(1-p)8 + 8p(t-p)^7$

(c) $(1-p)^8 + (1-p)^7$

(d) $(1-p)^8 + p(1-p)^7$

[1988 : 2 Marks]

47. A 4 GHz carrier is DSB-SC modulated by a low pass message signal with maximum frequency of 2 MHz. The resultant signal is to be ideally sampled. The minimum frequency of the sampling impulse train should be

(a) 4 MHz (b) 8 MHz

(c) 8 GHz (d) 8.004 GHz

[1990 : 2 Marks]

48. A signal has frequency components from 300 Hz to 1.8 kHz. The minimum possible rate at which the signal has to be sampled is _____.

[1991 : 2 Marks]

49. A signal has frequency components from 300 Hz to 1.8 kHz. The minimum possible rate at which the signal has to be sampled is ______ (fill in the blank)

[1991 : 2 Marks]

50. If the number of bits per sample in a PCM system is increased from a n to n + 1, the improvement in signal to quantization nose ratio will be

(a) 3 dB (b) 6 dB

(c) 2n dB (d) n dB

[1995 : 1 Mark]

51. The signal to quantization noise ratio in an n-bit PCM system

(a) depends upon the sampling frequency employed

(b) is independent of the value of 'n'

(c) increasing with increasing value of 'n'

(d) decreases with the increasing value of 'n'

[1995 : 1 Mark]

52. The number of bits in a binary PCM system is increased from n to n + 1. As a result, the signal to quantization noise ratio will improve by a factor

(a) $\frac{(n+1)}{n}$

(b) $2^{\frac{(n+1)}{n}}$

(c) $2^{\frac{2(n+1)}{n}}$

(d) which is independent of n.

[1996 : 2 Marks]

53. A deterministic signal has the power spectrum given in figure. The minimum sampling rate needed to completely represent signal is

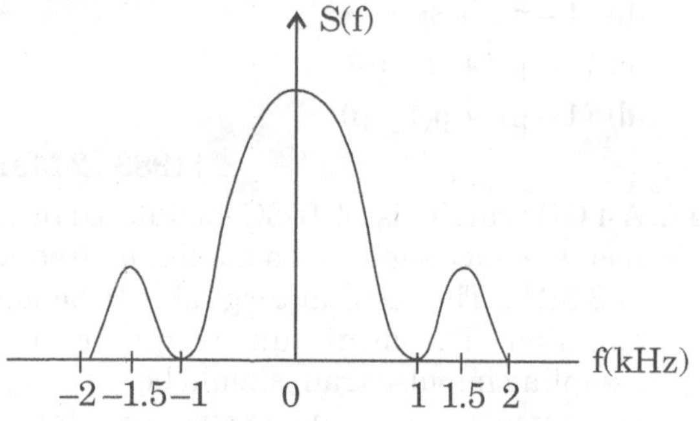

(a) 1 kHz (b) 2 kHz
(c) 3 kHz (d) None of these

54. In a PCM system with uniform quantization, increasing the number of bits from 8 to 9 will reduce the quantization noise power by a factor of

(a) 9 (b) 8
(c) 4 (d) 2

[1998 : 1 Mark]

55. The Nyquist sampling frequency (in Hz) of a signal given by $6 \times 10^4 \text{sinC}^3(400t) * 10^6 \text{sinC}^3(100t)$ is

(a) 200 (b) 300
(c) 1500 (d) 1000

[1999 : 2 Marks]

56. The peak-to-peak input to an 8-bit PCM coder is 2 Volts. The signal power-to-quantization noise power ratio (in dB) for an input of $0.5\cos(\omega_m t)$ is

(a) 47.8 (b) 43.8
(c) 95.6 (d) 99.6

[1999 : 2 Marks]

57. Four independent messages have bandwidths of 100 Hz, 100 Hz, 200 Hz, and 400 Hz, respectively. Each is sampled at the Nyquist rate, and the samples are Time Division Multiplexed (TDM) and transmitted. The transmitted sample rate (in Hz) is

(a) 1600 (b) 800
(c) 400 (d) 200

[1999 : 2 Marks]

58. A video transmission system transmits 625 picture frames per second. Each frame consists of a 400 × 400 pixel grid with 64 intensity levels per pixel. The data rate of the system is

(a) 16 Mbps (b) 100 Mbps
(c) 600 Mbps (d) 6.4 Gbps

[2001 : 2 Marks]

59. The Nyquist sampling interval, for the signal sinc(700t) + sin c(500t) is

(a) $\frac{1}{350}$ sec (b) $\frac{\pi}{350}$ sec
(c) $\frac{1}{700}$ sec (d) $\frac{\pi}{175}$ sec

[2001 : 2 Marks]

60. During transmission over a communication channel, bit errors occur independently with probability p. If a block of n bits is transmitted, the probability of at most one bit error is equal to

(a) $1-(1-p)^n$ (b) $p + (n-1)(1-p)$
(c) $np(1-p)^{n-1}$ (d) $(1-p)^n + np(1-p)^{n-1}$

[2001 : 2 Marks]

61. For a bit-rate of 8 Kbps, the best possible values of the transmitted frequencies in a coherent binary FSK system are

(a) 16 kHz and 20 kHz.
(b) 20 kHz and 32 kHz.
(c) 20 kHz and 40 kHz.
(d) 32 kHz and 40 kHz.

62. At a given probability of error, binary coherent FSK is inferior to binary coherent PSK by

(a) 6 dB (b) 3 dB
(c) 2 dB (d) 0 dB

[2003 : 1 Mark]

63. Let $x(t) = 2\cos(800\pi t) + \cos(1400\pi t)$. x(t) is sampled with the rectangular pulse train shown in the figure. The only spectral components (in kHz) present in the sampled signal in the frequency range 2.5 kHz to 3.5 kHz are

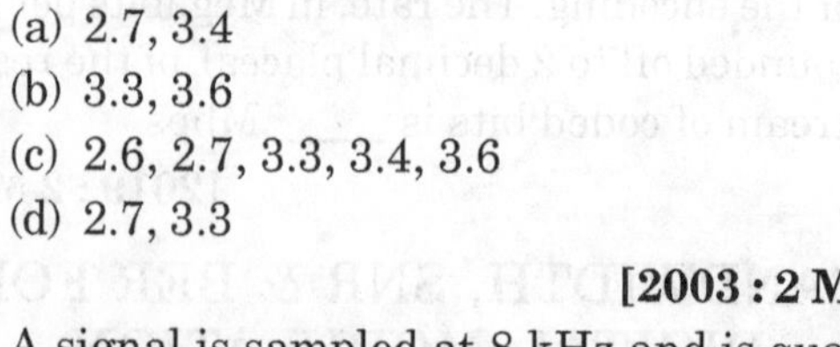

(a) 2.7, 3.4
(b) 3.3, 3.6
(c) 2.6, 2.7, 3.3, 3.4, 3.6
(d) 2.7, 3.3

[2003 : 2 Marks]

64. A signal is sampled at 8 kHz and is quantized using 8-bit uniform quantizer. Assuming SNR_q for a sinusoidal signal, the correct statement for PCM signal with a bit rate of R is

(a) R = 32 kbps, SNR_q = 25.8 dB
(b) R = 64 kbps, SNR_q = 49.8 dB
(c) R = 64 kbps, SNR_q = 55.8 dB
(d) R= 32 kbps, SNR_q = 49.8 dB

[2003 : 2 Marks]

65. A source produces binary data at the rate of 10 kbps. The binary symbols are represented as shown in the figure.

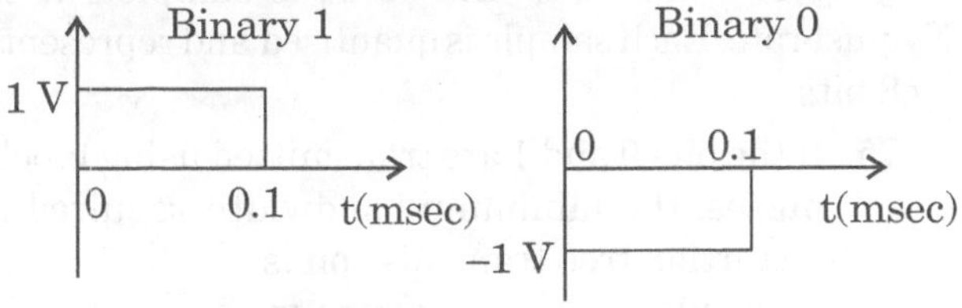

The source output is transmitted using two modulation schemes, namely Binary PSK (BPSK) and Quadrature PSK (QPSK). Let and B2 be the bandwidth requirements of BPSK and QPSK respectively. Assuming that the bandwidth of the above rectangular pulses is 10 kHz, B_1 and B_2 are

(a) $B_1 = 20$ kHz, $B_2 = 20$ kHz.

(b) $B_1 = 10$ kHz, $B_2 = 20$ kHz.

(c) $B_1 = 20$ kHz, $B_2 = 10$ kHz.

(d) $B_1 = 10$ kHz, $B_2 = 10$ kHz.

[2004 : 2 Marks]

66. Consider a binary digital communication system with equally likely 0's and 1's. When binary 0 is transmitted the detector input can lie between the levels –0.25V and +0.25V with equal probability: when binary 1 is transmitted, the voltage at the detector can have any value between 0 and 1 V with equal probability. If the detector has a threshold of 0.2 V (i.e., if the received signal is greater than 0.2 V, the bit is taken as 1), the average bit error probability is

(a) 0.15 (b) 0.2

(c) 0.05 (d) 0.5

[2004 : 2 Marks]

67. Three analog signals, having bandwidths 1200 Hz, 600 Hz and 600 Hz, are sampled at their respective Nyquist rates, encoded with 12-bit words, and time division multiplexed. The bit rate for the multiplexed signal is

(a) 115.2 kbps (b) 28.8 kbps

(c) 57.6 kpbs (d) 38.4 kbps

[2004 : 2 Marks]

68. The minimum sampling frequency (in samples/sec) required to reconstruct the following signal from its samples without distortion

$$x(t) = 5\left(\frac{\sin 2\pi 1000t}{\pi t}\right)^3 + 7\left(\frac{\sin 2\pi 1000t}{\pi t}\right)^2$$

would be

(a) 2×10^3 (b) 4×10^3

(c) 6×10^3 (d) 8×10^3

[2006 : 2 Marks]

69. During transmission over a certain binary communication channel, bit errors occur independently with probability p. The probability of AT MOST one bit in error in a block of n bits is given by

(a) p^n

(b) $1 - p^n$

(c) $np(1-p)^{n-1} + (1-p)^n$

(d) $1 - (1-p)^n$

[2007 : 2 Marks]

Common Data for Questions 70 & 71 :

Two 4-ary signal constellations are shown. It is given that ϕ_1 and ϕ_2 constitute an orthonormal basis for the two constellations. Assume that the four symbols in both the constellations are equiprobable. Let $\frac{N_0}{2}$ denote the power spectral density of white Gaussian noise.

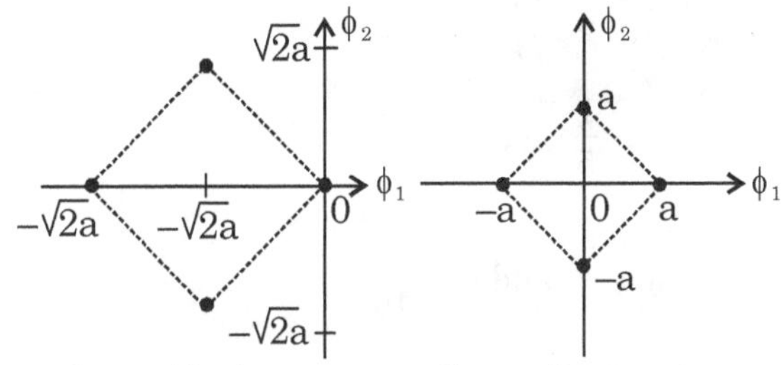

Constellation 1 Constellation 2

70. The ratio of the average energy of Constellation 1 to the average energy of Constellation 2 is

(a) $4a^2$

(b) 4

(c) 2

(d) 8

[2007 : 2 Marks]

71. If these constellations are used for digital communications over an AWGN channel, then which of the following statements is true?

(a) Probability of symbol error for Constellation 1 is lower

(b) Probability of symbol error for Constellation 1 is higher

(c) Probability of symbol error is equal for both the constellations

(d) The value of N_0 will determine which of the two constellations has a lower probability of symbol error

[2007 : 2 Marks]

Statement for Linked Answer Questions 72 and 73 :

An input to a 6-level quantizer has the probability density function f(x) as shown in the figure. Decision boundaries of the quantizer are chosen so as to maximize the entropy of the quantizer output. It is given that 3 consecutive decision boundaries are '–1', '0' and '1'.

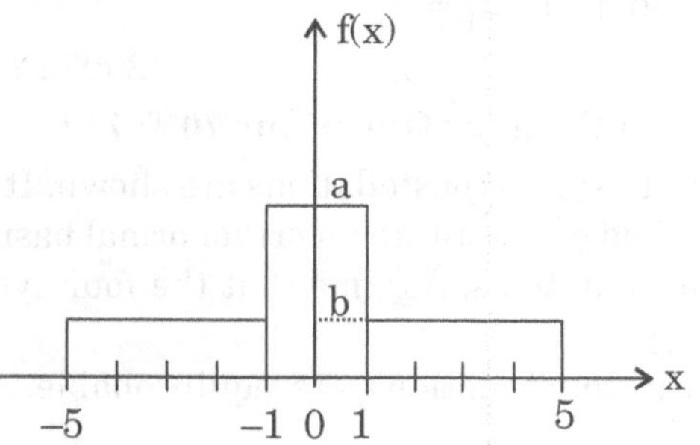

72. The values of a and b are

(a) $a = \frac{1}{6}$ and $b = \frac{1}{12}$

(b) $a = \frac{1}{5}$ and $b = \frac{3}{40}$

(c) $a = \frac{1}{4}$ and $b = \frac{1}{16}$

(d) $a = \frac{1}{3}$ and $b = \frac{1}{24}$

[2007 : 2 Marks]

73. Assuming that the reconstruction levels of the quantizer are the mid-points of the decision boundaries, the ratio of signal powr to quantization noise power is

(a) $\frac{152}{9}$ (b) $\frac{64}{3}$

(c) $\frac{76}{3}$ (d) 28

[2007 : 2 Marks]

74. Consider a Binary Symmetric Channel (BSC) with probability of error being p. To transmit a bit, say 1, we transmit a sequence of three 1 s. The receiver will interpret the received sequence to represent 1 if at least two bits are 1. The probability that the transmitted bit will be received in error is

(a) $p^3 + 3p^2(1 - p)$ (b) p^3

(c) $(1 - p)^3$ (d) $p^3 + p^2(1 - p)$

[2008 : 2 Marks]

Common Data for Questions 75, 76 and 77 :

A speech signal, band limited to 4 kHz and peak voltage varying between +5 V and –5 V, is sampled at the Nyquist rate. Each sample is quantized and represented by 8 bits.

75. If the bits 0 and 1 are transmitted using bipolar pulses, the minimum bandwidth required for distortion free transmission is

(a) 64 kHz (b) 32 kHz

(c) 8 kHz (d) 4 kHz

[2008 : 2 Marks]

76. Assuming the signal to be uniformly distributed between its peak to peak value, the signal to noise ratio at the quantizer output is

(a) 16 dB (b) 32 dB

(c) 48 dB (d) 64 dB

[2008 : 2 Marks]

77. The number of quantization levels required to reduce the quantization noise by a factor of 4 would be

(a) 1024 (b) 512

(c) 256 (d) 64

[2008 : 2 Marks]

Common data for Questions 78 and 79 :

The amplitude of a random signal is uniformly distributed between –5 V and 5 V.

78. If the signal to quantization noise ratio required in uniformly quantizing the signal is 43.5 dB, the step size of the quantization is approximately

(a) 0.0333 V (b) 0.05 V

(c) 0.0667 V (d) 0.10 V

[2009 : 2 Marks]

79. If the positive values of the signal are uniformly quantized with a step size of 0.05 V, and the negative values are uniformly quantized with a step size of 0.1 V, the resulting signal to quantization noise ratio is approximately

(a) 46 dB (b) 43.8 dB

(c) 42 dB (d) 40 dB

[2009 : 2 Marks]

80. The Nyquist sampling rate for the signal

$$s(t) = \frac{\sin(500\pi t)}{\pi t} \times \frac{\sin(700\pi t)}{\pi t}$$ is given by

(a) 400 Hz

(b) 600 Hz

(c) 1200 Hz

(d) 1400 Hz

[2010 : 2 Marks]

81. An analog signal is band-limited to 4 kHz, sampled at the Nyquist rate and the samples are quantized into 4 levels. The quantized levels are assumed to be independent and equally probable. If we transmit two quantized samples per second, the information rate is

(a) 1 bit/sec (b) 2 bits/sec

(c) 3 bits/sec (d) 4 bits/sec

[2011 : 1 Mark]

82. In a baseband communications link, frequencies upto 3500 Hz are used for signaling. Using a raised cosine pulse with 75% excess bandwidth and for no inter-symbol interference, the maximum possible signaling rate in symbols per second is

(a) 1750 (b) 2625

(c) 4000 (d) 5250

[2012 : 1 Mark]

83. A binary symmetric channel (BSC) has a transition probability of 1/8. If the binary transmit symbol X is such that P(X = 0) = 9/10, then the probability of error for an optimum receiver will be

(a) $\frac{7}{80}$ (b) $\frac{63}{80}$

(c) $\frac{9}{10}$ (d) $\frac{1}{10}$

[2012 : 2 Marks]

84. A BPSK scheme operating over an AWGN channel with noise power spectral density of $\frac{N_0}{2}$, uses equiprobable signals

$$s_1(t) = \sqrt{\frac{2E}{T}}\sin(\omega_c t) \text{ and } s_2(t) = \sqrt{\frac{2E}{T}}\sin(\omega_c t)$$

over the symbol internal (0, T). If the local oscillator in a coherent receiver is ahead in phase by 45° with respect to the received signal, the probability of error in the resulting system is

(a) $Q\left(\sqrt{\frac{2E}{N_0}}\right)$ (b) $Q\left(\sqrt{\frac{E}{N_0}}\right)$

(c) $Q\left(\sqrt{\frac{E}{2N_0}}\right)$ (d) $Q\left(\sqrt{\frac{E}{4N_0}}\right)$

[2012 : 2 Marks]

85. The bit rate of digital communication system is R kbits/s. The modulation used is 32-QAM. The minimum bandwidth required for |S| free transmission is

(a) $\frac{R}{10}$ Hz (b) $\frac{R}{10}$ kHz

(c) $\frac{R}{5}$ Hz (d) $\frac{R}{5}$ kHz

[2013 : 1 Mark]

Common Data For Questions 86 and 87 :

Bits 1 and 0 are transmitted with equal probability. At the receiver, the pdf of the respective received signals for both bits are as shown below:

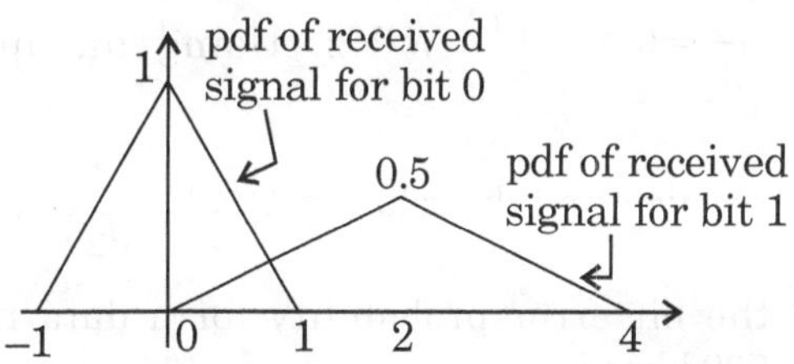

86. If the detection threshold is 1, the BER will be

(a) $\frac{1}{2}$ (b) $\frac{1}{4}$

(d) $\frac{1}{8}$ (d) $\frac{1}{16}$

[2013 : 2 Marks]

87. The optimum threshold to achieve minimum bit error rate (BER) is

(a) $\frac{1}{2}$ (b) $\frac{4}{5}$

(c) 1 (d) $\frac{3}{2}$

[2013 : 2 Marks]

88. Let $Q(\sqrt{\gamma})$ be the BER of a BPSK system over an AWGN channel with two-sided noise power spectral density $\frac{N_0}{2}$. The parameter γ is a function of bit energy and noise power spectral density. A system with two independent and identical AWGN channels with noise power spectral density $\frac{N_0}{2}$ is shown in the figure. The BPSK demodulator receives the sum of outputs of both the channels.

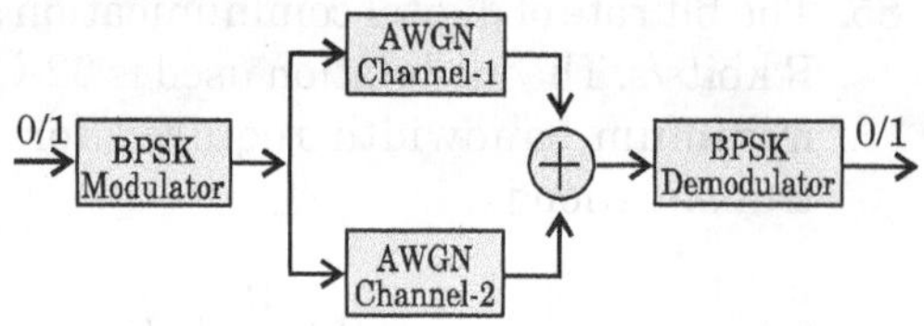

If the BER of this system is $Q(b\sqrt{\gamma})$, then the value of b is ______.

[2014 : 2 Marks, Set-1]

89. Coherent orthogonal binary FSK modulation is used to transmit two equiprobable symbol waveforms $s_1(t) = \alpha\cos 2\pi f_1 t$ & $s_2(t) = \alpha\cos 2\pi f_2 t_1$ where α = 4 mV. Assume an AWGN channel with two-sided noise power spectral density $\frac{N_0}{2} = 0.5 \times 10^{-12}$ W/Hz. Using an optimal receiver and the relation $Q(v) = \frac{1}{\sqrt{2\pi}}\int_v^{\infty} \theta^{-\frac{u^2}{2}} du$, the bit error probability for a data rate of 500 kbps is

(a) Q(2) (b) $Q(2\sqrt{2})$

(c) Q(4) (d) $Q(4\sqrt{2})$

[2014 : 2 Marks, Set-2]

90. An M-level PSK modulation scheme is used to transmit independent binary digits over a band-pass channel with bandwidth 100 kHz. The bit rate is 200 kbps and the system characteristic is a raised-cosine spectrum with 100% excess bandwidth. The minimum value of M is ______.

[2014 : 2 Marks, Set-4]

91. The input Xto Binary Symmetric Channel (BSC) shown in the figure is '1' with probability of 0.8. The cross-over probability is $\frac{1}{7}$. If the received bit Y = 0, the conditional probability that '1' was transmitted is ______.

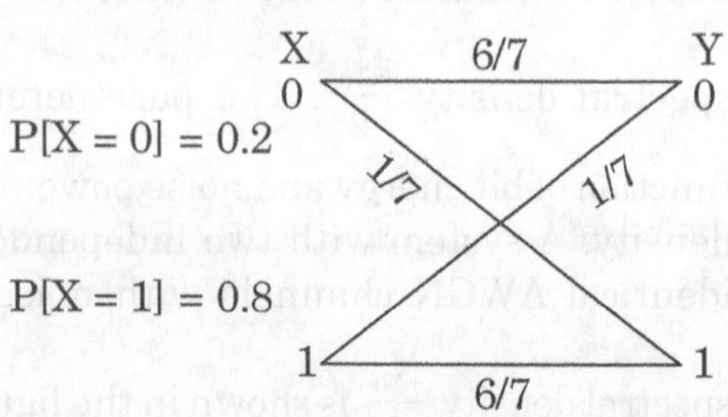

[2015 : 2 Marks, Set-1]

92. A source emits bit 0 with probability $\frac{1}{3}$ and bit 1 with probability $\frac{2}{3}$. The emitted bits are communicated to the receiver. The receiver decides for either 0 or 1 based on the received value R. It is given that the conditional density functions of R as

$$f_{R|0}(r) = \begin{cases} \frac{1}{4}, & -3 \le r \le 1 \\ 0, & \text{otherwise;} \end{cases}$$

$$f_{R|1}(r) = \begin{cases} \frac{1}{6}, & -1 \le r \le 5 \\ 0, & \text{otherwise;} \end{cases}$$

The minimum decision error probability is

(a) 0 (b) $\frac{1}{12}$

(c) $\frac{1}{9}$ (d) $\frac{1}{6}$

[2015 : 2 Marks, Set-1]

93. Consider a binary, digital communication system which uses pulses g(t) and –g(t) for transmitting bits over an AWGN channel. If the receiver uses a matched filter, which one of the following pulses will give the minimum probability of bit error?

(a)

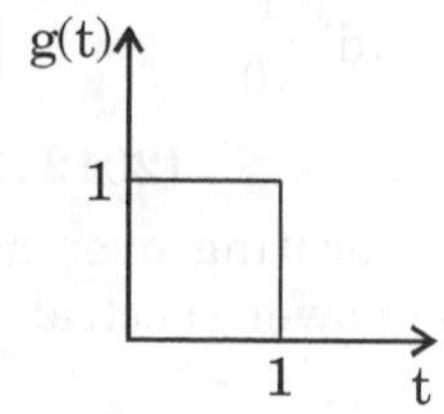

(b)

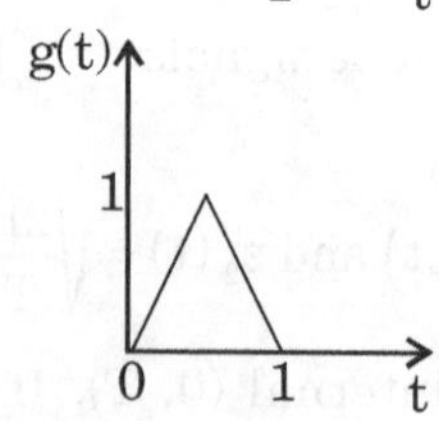

(c)

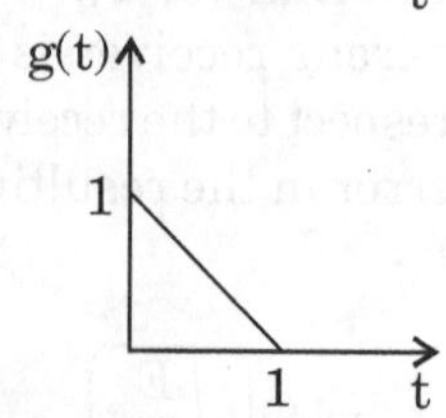

(d)

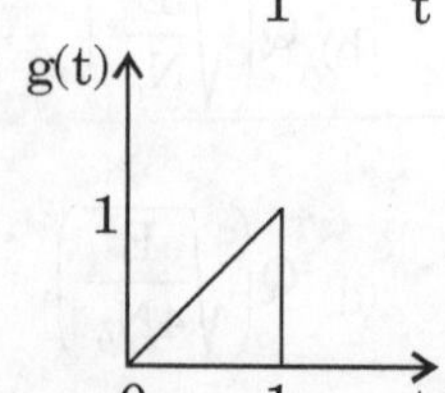

[2015 : 2 Marks, Set-2]

94. Consider a binary data transmission at a rate of 56 kbps using baseband binary pulse amplitude modulation (PAM) that is designed to have a raised-cosine spectrum. The transmission bandwidth (in kHz) required of a roll-off factor of 0.25 is ______.

[2016 : 1 Mark, Set-1]

95. An ideal band-pass channel 500 Hz- 2000 Hz is deployed for communication. A modem is designed to transmit bits at the rate of 4800 bits/s using 16-QAM. The roll-off factor of a pulse with a raised cosine spectrum that utilizes the entire frequency band is ______.

[2016 : 2 Marks, Set-2]

96. A speech signal is sampled at 8 kHz and encoded into PCM format using 8 bits/sample. The PCM data is transmitted through a baseband channel via 4-level PAM. The minimum bandwidth (in kHz) required for transmission is ______.

[2016 : 1 Mark, Set-2]

97. The bit error probability of a memoryless binary symmetric channel is 10^{-5}. If 10^5 bits are sent over this channel, then the probability that not more than one bit will be in error is ______.

[2016 : 2 Marks, Set-3]

98. In a digital communication system, the overall pulse shape p(t) at the receiver before the sampler has the Fourier transform P(f). If the symbols are transmitted at the rate of 2000 symbols per second, for which of the following cases is the inter symbol interference zero?

(a)

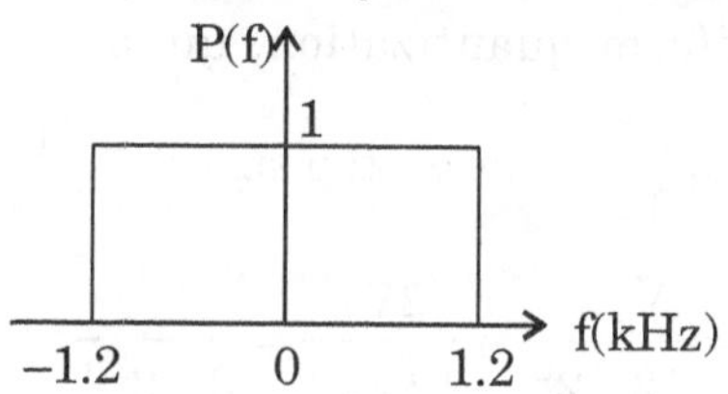

(b)

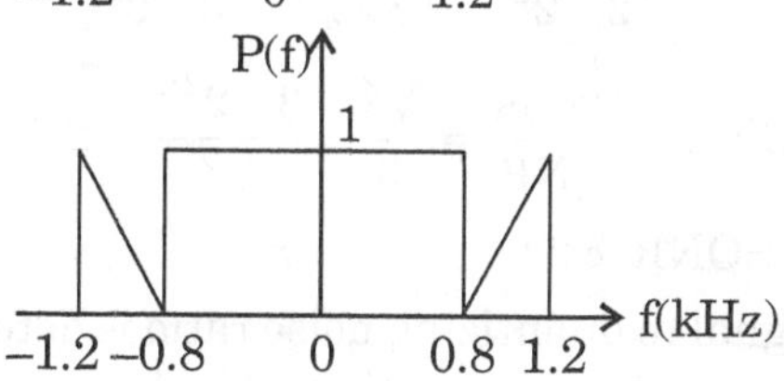

(c)

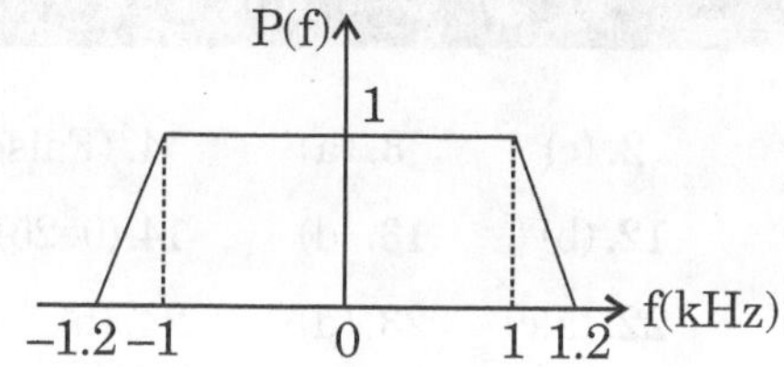

(d)

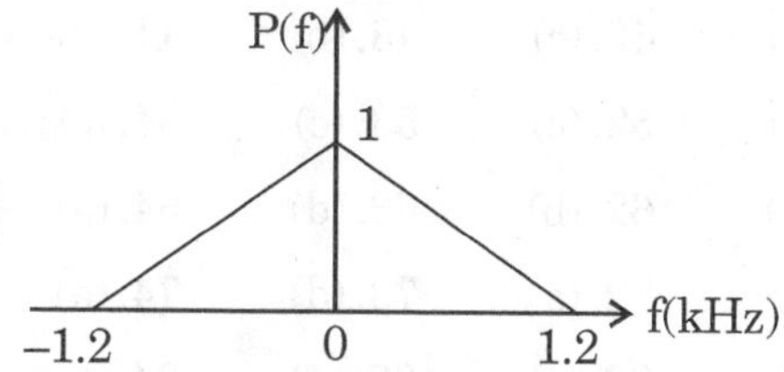

[2017 : 1 Mark, Set-1]

99. A sinusoidal message signal is converted to a PCM signal using a uniform quantizer. The required signal to quantization noise ratio (SQNR) at the output of the quantizer is 40 dB. The minimum number of bits per sample needed to achieve the desired SQNR is ______.

[2017 : 1 Mark, Set-2]

100. A binary source generates symbols $X \in (-1, 1)$ which are transmitted over a noisy channel. The probability of transmitting X = 1 is 0.5. Input to the threshold detector is R = X + N. The probability density function $f_N(n)$ of the noise N is shown below.

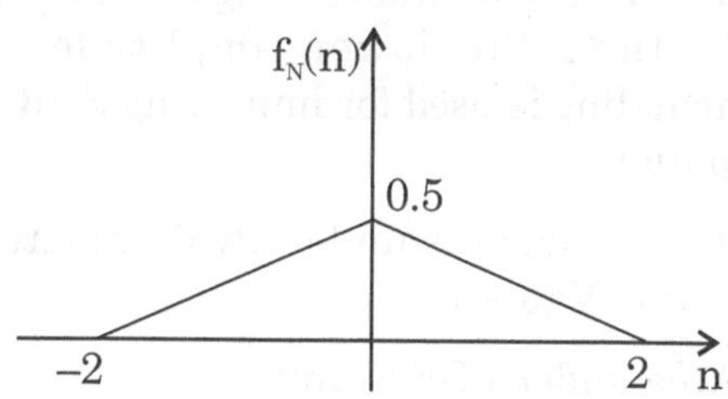

If the detection threshold is zero, then the probability of error (correct to two decimal places) is ______.

[2018 : 1 Mark]

ANSWERS

1. (c)	2. (c)	3. (a)	4. (False)	5. (a)	6. (a)	7. (c)	8. (b)	9. (b)	10. (a)
11. (b)	12. (b)	13. (d)	14. (0·25)	15. (200)	16. (a)	17. (32)	18. (d)	19. (c)	20. (a)
21. (*)	22. (a,c)	23. (d)	24. (b)	25. (a)	26. (c)	27. (d)	28. (d)	29. (d)	30. (d)
31. (a,b)	32. (b)	33. (c)	34. (a)	35. (c)	36. (d)	37. (b)	38. (b)	39. (c)	40. (c)
41. (c)	42. (c)	43. (b)	44. (–0·50)	45. (0.192)	46. (b)	47. (b)	48. (3600)	49. (3600)	50. (c)
51. (c)	52. (c)	53. (c)	54. (c)	55. (d)	56. (c)	57. (b)	58. (d)	59. (c)	60. (d)
61. (d)	62. (b)	63. (d)	64. (a)	65. (c)	66. (a)	67. (c)	68. (c)	69. (c)	70. (b)
71. (a)	72. (a)	73. (d)	74. (a)	75. (a)	76. (c)	77. (b)	78. (c)	79. (c)	80. (c)
81. (d)	82. (c)	83. (a)	84. (b)	85. (b)	86. (d)	87. (b)	88. (c)	89. (c)	90. (16)
91. (0·4)	92. (d)	93. (a)	94. (35)	95. (0·25)	96. (16)	97. (0·735)	98. (b)	99. (7)	100. (0·125)

EXPLANATIONS

1. Companding results in making SNR uniform throughout the signal irrespective of amplitude levels. Since in uniform quantiratim, step size is same, the quantization noise power is uniform throughtout the signal.

Thus, higher amplitudes of signal will have better SNR then the lower amplitudes. Hence companding is used for improving SNR at lower amplitudes.

2. Since the signal is uniformly distributed in the interval –V to + V

$\because$ Area under PDF is unity.

$K[V-(-V)] = 1$

$2KV = 1$

$K = \frac{1}{2V}$

$$\because P_x(x) = \begin{cases} 1/2V, -V \text{ to } V \\ O, \text{otherwise} \end{cases}$$

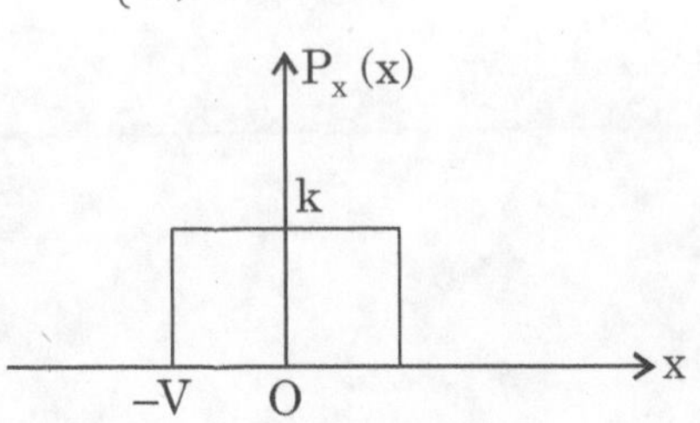

$$\text{Signal power} = \int_{-\infty}^{\infty} x^2 P_{x}(x)dx = \int_{-V}^{V} x^2 \frac{1}{2V} dx = \frac{1}{2V}\left[\frac{x^3}{3}\right]_{-V}^{V}$$

$$= \frac{1}{2V}\left[\frac{V^3}{3} - \frac{(-V^3)}{3}\right] = \frac{1}{2V} \cdot \frac{2V^3}{3}$$

$= V^2/3$

In uniform quantization, quantization noise power $= \frac{\Delta^2}{12}$ Where, step size $\Delta = \frac{V_{p-p}}{L} = \frac{V_{p-p}}{2^n}$

$$QNP = \frac{{V_{p-p}}^2}{12 \times 2^n} = \frac{(2V)^2}{12 \times 2^n} = \frac{V^2}{3 \times 2^{2n}}$$

$$SQNR = \frac{\text{Power}}{QNP} = \frac{V^2}{3} \times \frac{3 \times 2^{2n}}{V^2} = 2^{2n}$$

$\Rightarrow SQNR \propto 2^{2n}$

Signal to quantizing noise ratio is determined by the number of quantizing levels.

3.

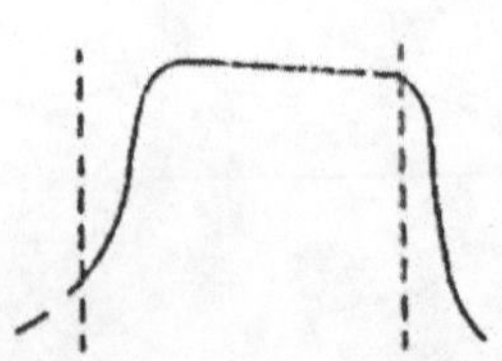

Increased rise and fall lines indicate inadequate high frequency response. Till in top indicates poor low frequency response.

4. $V_1 = \log_2 4 = 2$; $C = 2B\log_2 N$; $B = \frac{C}{4}$

$V_2 = \log_2 64 = 6 \quad B = \frac{C}{12}$

$\therefore$ Bandwidth increase by a factor $= \frac{C/12}{C/3} = \frac{1}{4}$

Multi level coding reduces the bandwidth requirement.

5. Flat top sampling of low pass signals gives rise to apatere effect

7. Step-size,________Δ $= \frac{2m_p}{L} = 0.012$

then, quantization – noise power $\langle q_e^2 \rangle = \frac{\Delta^2}{12}$

$$= \frac{(0.012)^2}{12} = 12 \times 10^{-6}\ V^2$$

$$= \frac{(0.012)^2}{12} = 12 \times 10^{-6}\ V^2$$

8. For slope overload to take place

$$|\dot{x}(t)|_{max} > f_s\Delta \quad ...(i)$$

Now $\quad x(t) = E_m \sin 2\pi f_m t$

$$\Rightarrow \quad |\dot{x}(t)|_{max} = E_m f_m \times 2\pi \quad ...(ii)$$

$$= 6.28\ E_m f_m$$

$$f_s = 40\text{ kHz}$$

$$\Delta = 0.628$$

$$\Rightarrow \quad \Delta f_s = 25.12\text{ kHz}$$

(*a*) → 6.28 × 0.3 × 8 = 15.07 < 25.12

(*b*) → 6.28 × 1.5 × 4 = 37.68 > 25.12

(*c*) → 6.28 × 1.5 × 2 = 18.44 < 25.12

(*d*) → 6.28 × 3 × 1 = 18.84 < 25.12

9. $\rho = \frac{(SNR)_2}{(SNR)_1} = \frac{2^{2n_2}}{2^{2n_1}}$

Here $n_1 = 6$, $n_2 = 8$, then

ratio, $\quad \rho = 2^{2(n_2-n_1)} = 16.$

10. When signal is constant, then in Delta Modulation consecutive pulses are of opposite polarity.

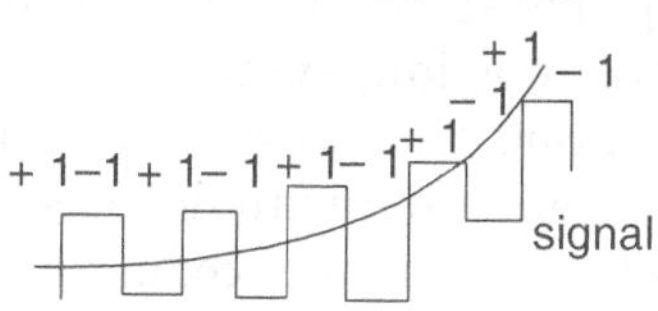

11. $x(t)$ can be sketched as

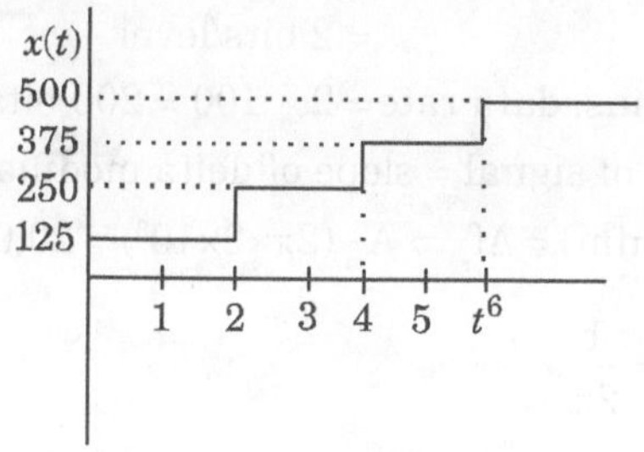

To avoid slope - overload,

$$\frac{\Delta}{T_s} \geq \frac{d_m(t)}{dt}$$

$$\Rightarrow \quad \Delta.32.1024 \geq \frac{125}{2}$$

$$\Rightarrow \quad \Delta \geq 2^{-8}$$

12. The minimum value that can make $y_1(t)$ different from $y_2(t)$ is

$$\pm\frac{LSB}{2} = \frac{\Delta}{2}$$

13. For slope overload not to occur,

$$\frac{\Delta}{T_s} \geq \frac{d}{dt}m(t)$$

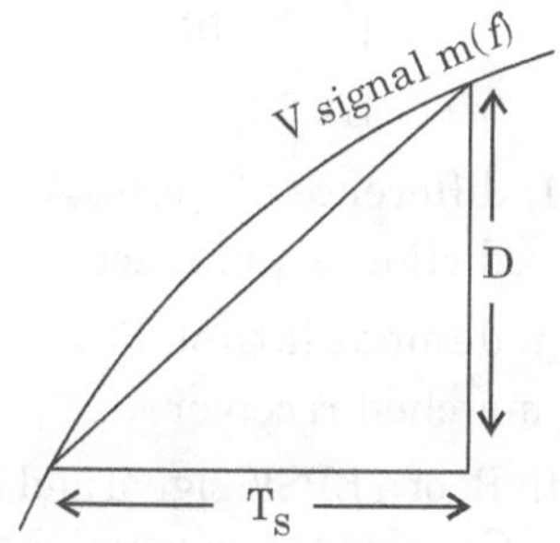

if $m(t)$ rate is high, Δ must increase.

14. Maximum quantization error is $\frac{\text{step} - \text{size}}{2}$

$$\text{step} - \text{size} = \frac{8-0}{16}$$

$$= \frac{1}{2} = 0.5V$$

Quantization error = 0.25 V

15. Nyquist rate = 2 × 50Hz

= 100 samples/sec

$$\Delta = \frac{m(t)_{max} - m(t)_{min}}{L}$$

$$\Rightarrow \quad L = \frac{\sqrt{2} - (-\sqrt{2})}{0.75}$$

$$L = \frac{2\sqrt{2}}{0.75}$$

$$= 3.77$$

$$\cong 4.$$

Number of bits required to encode '4' levels

= 2 bits/level

Thus, data rate = 2 × 100 = 200 bits/sec

16. Slope of signal = slope of delta modulator

$A_m (2\pi fm) = \Delta f_s \Rightarrow A_m (2\pi \times 2\times 10^3) = 20,000 \times 0.1$

$$A_m = \frac{1}{2\pi}$$

17. Signal power $= \frac{A^2}{2}$

Quantization step size, $\Delta = \frac{2A}{L}$

Quantization noise power $= \frac{\Delta^2}{12}$

$$= \frac{4A^2}{12L^2} = \frac{A^2}{3L^2}$$

$\Rightarrow$ Signal to quantization noise ratio $= \frac{3}{2}L^2$

Given signal to quantization noise ratio = 31.8dB or 1513.56

$\Rightarrow \quad \frac{3}{2}L^2 = 1513.56$

$\Rightarrow \quad L = 31.76$

$L \approx 32$

18. In DPCM, difference of message signal sample with its prediction is quantized.

22. Coherent demodulation FSK = correlation receiver/ matched receiver

23. Bandwidth B_p of a BPSK signal and the bandwidth B_o of the OOK signal one equal, i.e., $B_p = B_O$.

24. In a bit interval each wave should contain integral number of cycles for orthogonal to each other i.e.,

$$\Delta\omega = \frac{1}{2T}$$

25. For coherent BPSK synchronization required for detection and efficiency is $\frac{1}{2}$.

26.
FM → capture effect
DM → slope overload
PSK → matched filter
PCM → μ– law

27. For M–ary, $d = 2\sin\left(\frac{\pi}{M}\right)\sqrt{E_S}$

Distance of any point from origins is $\sqrt{E_S}$

For 4 – ary, $r_1 = \sqrt{E_{S_1}}$; For 8– ary, $r_2 = \sqrt{E_{S2}}$

For 4 – ary, M= 4, $d_1 = 2\sin\left(\frac{\pi}{4}\right) r_1$

For 8 – ary, M = 8, $d_2 = 2\sin\left(\frac{\pi}{8}\right) r_2$.

If $d_1 = d_2 = d$ then $2\sin\left(\frac{\pi}{4}\right) r_1 = d$

$$r_1 = \frac{d}{\sqrt{2}} = 0.707d.$$

$$2\sin\left(\frac{\pi}{8}\right) r_2 = \frac{d}{2\sin\left(\frac{\pi}{8}\right)} = 1.307d.$$

28. We know that $Pe \propto \sqrt{E_s}$

$$\sqrt{\frac{Es_1}{Es_2}} = \frac{r_1}{r_2} = \frac{0.707d}{1.307d} = \frac{0.707}{1.307}$$

$$\frac{Es_2}{Es_1} = \left(\frac{1.307}{0.707}\right)^2 = 3.42$$

To achieve same error, 2nd signal must have 3.42 times energy than 1st signal

In dB = 10 log (3.42) = 5.33 dB.

29. GMSK is a form of modulation used in a variety of digital ratio communication systems. It has the advantages of being able to carry digital modulation while still using the spectrum efficiently GMSK is most widely used in the GSM cellular technology.

30. bit stream 01001 DPSK

Differential encoder

ref 1 0 → 1

π → 0

1	0	1	0	0	1
	π	π	π	0	π
	1	0	1	0	0
	0	1	0	0	1
	π	π	π	0	π

31. The purpose of source encoding in a data communication system is to increase the information transmission rate and purpose of channel encoding is to decrease the probability of error. The channel coding helps in detection and correction of errors.

32. Since the sampling rate is 1800 samples/sec, the highest frequency that can be recovered is 900 Hz (sampling frequency/2). Frequencies higher than 900 Hz will be rellected as low frequency due to well known aliasing effect.

33. Alternate mask inversion (AMI) code has zero dc component for pulse transmission of random binary data

34. Maximum amplitude of matched filler output

$$= \frac{A^2 T}{2}$$

$$= \frac{10^2}{2} \times 10^{-4} = 5 \text{ mV}.$$

35. $\Rightarrow X(f) \otimes \frac{1}{T_s} \sum_{n=-\infty}^{\infty} \delta\left(f - \frac{n}{T_s}\right)$

$$\Rightarrow \frac{1}{T_s} \times 5 \times 10^{-6} \times 10 \sum_{n=-\infty}^{\infty} \left(\frac{\delta\left(f - f_m - \frac{n}{T_s}\right) + \delta\left(f + f_m - \frac{n}{T_s}\right)}{2} \right)$$

At __________ $n = 0$

$$\Rightarrow \frac{50 \times 10^{-6}}{T_s} \frac{[\delta(f - f_m) + \delta(f + f_m)]}{2}$$

Taking inverse Fourier transform

$$= \frac{50 \times 10^{-6}}{T_s} \cos \omega_m t$$

$$= \frac{50 \times 10^{-6}}{100 \times 10^{-6}} \cos \omega_m t$$

$$= 5 \times 10^{-1} \cos \omega_m t$$

36. $f_s = \frac{1}{T_s} = \frac{1}{50 \times 10^{-6}} = 20 \text{ kHz}$

$$f_c = \frac{24\pi \times 10^3}{2\pi} = 12 \text{ kHz}$$

The frequencies passed through LPF are,

$f_c, f_s - f_c$ *i.e.,* 12 kHz, 8 kHz

as LPF have cut-off frequencies is 15 kHz

37. SNR for two sided PSD $= \frac{2E_b}{N_0}$

SNR for one sided PSD $= \frac{E_b}{N_0}$

$$\therefore 10 \log_{10} \frac{10^{-5}}{10^{-6}} = 10 \text{ dB}$$

38. The impulse response of the matched filter is h (t) = x(T–t). Here T = 4 h(t) = x (4–t)

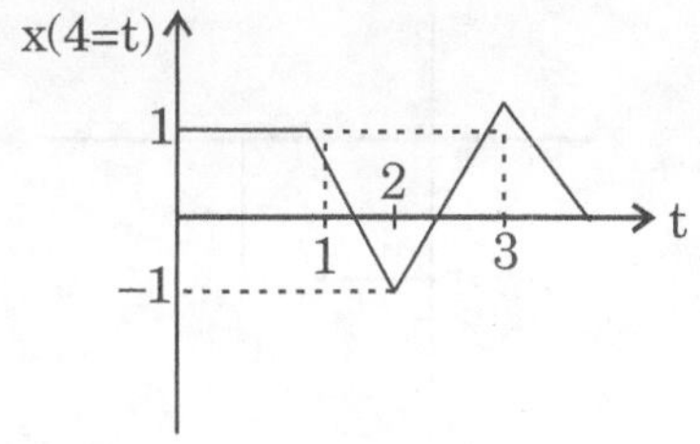

Slope in region $t \in (3, 4)$ is – 1.

39. Transfer function of matched filter

$$h(t) = s(T - t)$$

Output of matched filter having input $s(t)$

$$y(t) = h(t) * s(t) = \int_{-\infty}^{\infty} S(j)\, h(t - j)\, dj$$

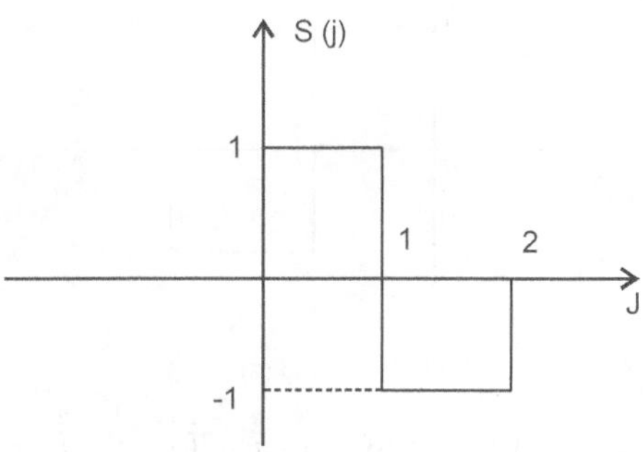

Fig (a)

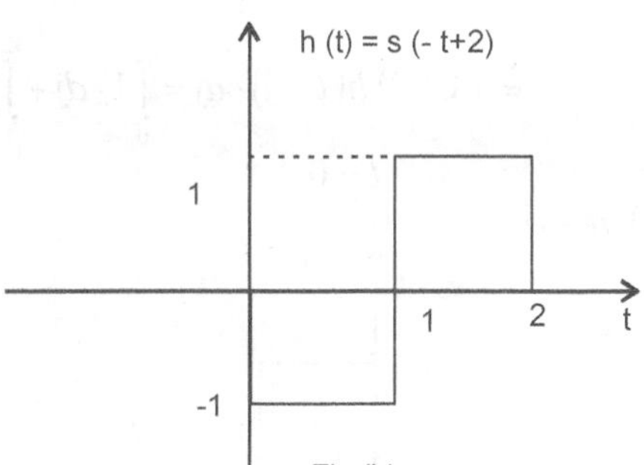

Fig (b)

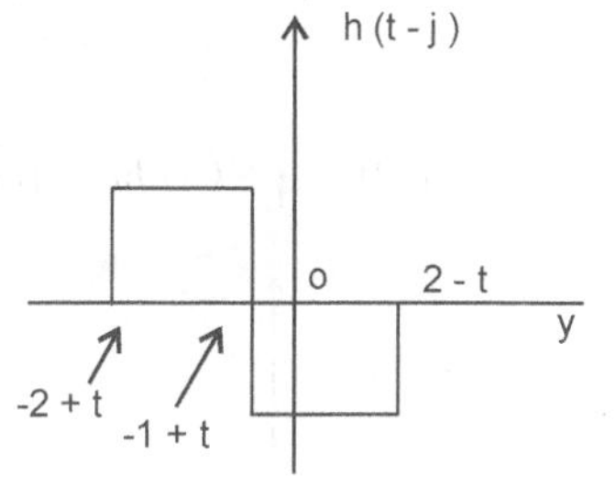

Fig (c)

For, $t = 1$,

$$P(t) = S(j)\, h(1 - j)$$

is shown in figure (*d*)

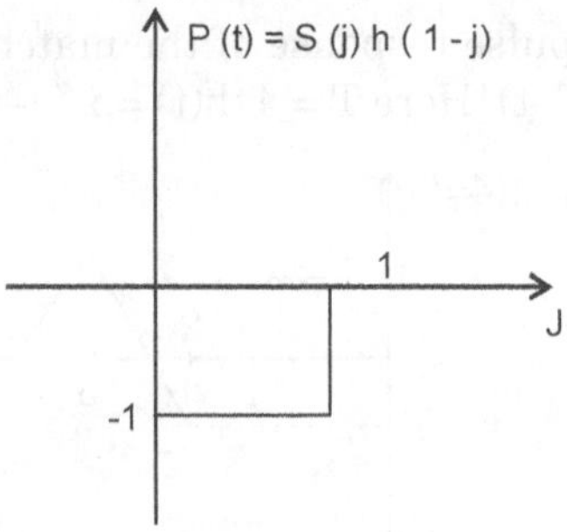

Fig (d)

$$Y(1)= \int_{-\infty}^{\infty} S(j)\, h(1-J)\, dj = \int_{0}^{1} -1 \cdot dj = -1$$

For $t = 2$

S(j) [$h(2-j)$] is shown in figure (*e*)

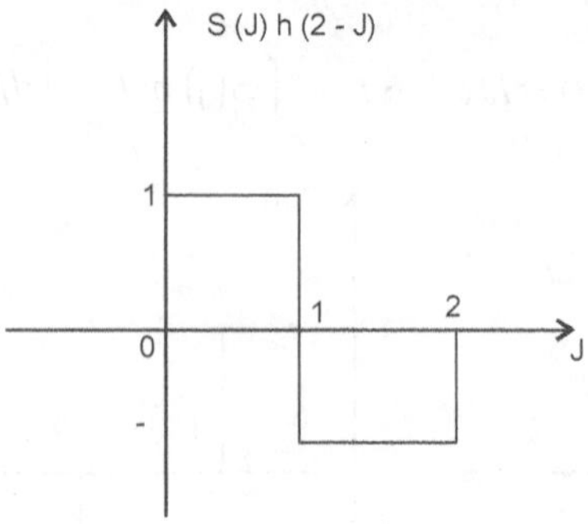

Fig (e)

$$Y(2)= \int_{-\infty}^{\infty} S(j)\, h(2-j) \cdot dj = \int_{0}^{1} 1 \cdot dj + \int_{1}^{2} -1 dj = 2$$

$$= \int_{-\infty}^{\infty} S(T)\, h(t-j) \cdot dj = \int_{0}^{1} 1 \cdot dj + \int_{1}^{2} -1 dj = 2$$

For $t = 0$

S(J)h(–J) $= 0$

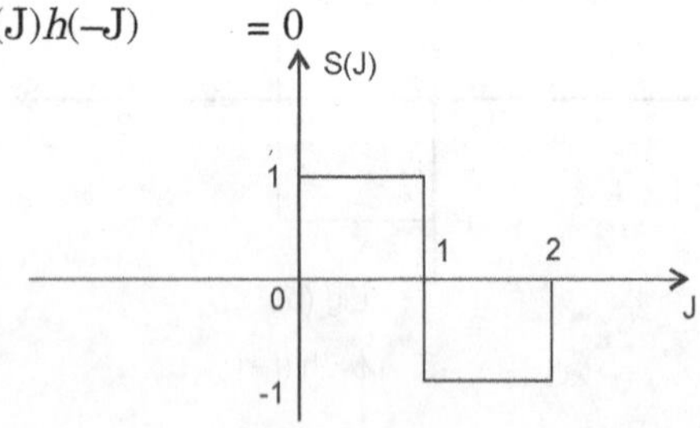

Then, $\qquad Y(0) = \int_{-\infty}^{\infty} S(J)\, h(-J) \cdot dJ = 0$

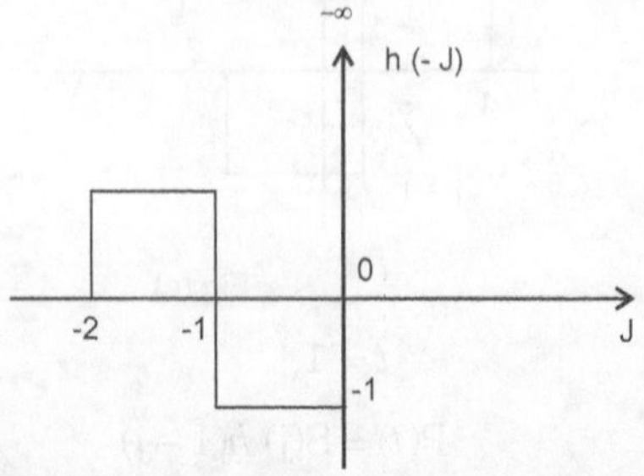

Hence, it can be inferred that right plot is in figure (*c*).

40. Given, $\quad p(t) = \dfrac{\sin 4\pi wt}{4\pi w\, t(1-16\, w^2t^2)}$

$$\lim_{t\to 1/4W} p(t) = \lim_{t\to 1/4W} \frac{\sin 4\pi wt}{4\pi w\, t(1-16\, w^2t^2)}$$

$$= \lim_{t\to 1/4W} \frac{\frac{d}{dt}(\sin 4\pi wt)}{\frac{d}{dt}[4\pi wt\,(1-16w^2t^2)]}$$

$$= \lim_{t\to 1/4W} \frac{4\pi w \cos 4\pi wt}{4\pi w\,(1-16\, w^2t^2) - 32w^2t(4\pi wt)}$$

$$= \lim_{t\to 1/4W} \frac{4\pi w \cos 4\pi wt}{4\pi w - 3\times 64\, w^3t^2}$$

$$= \frac{1}{2}$$

$$= 0.5$$

41. Impulse response of the matched filter,

$$h(t) = s(T-t)$$

Given,

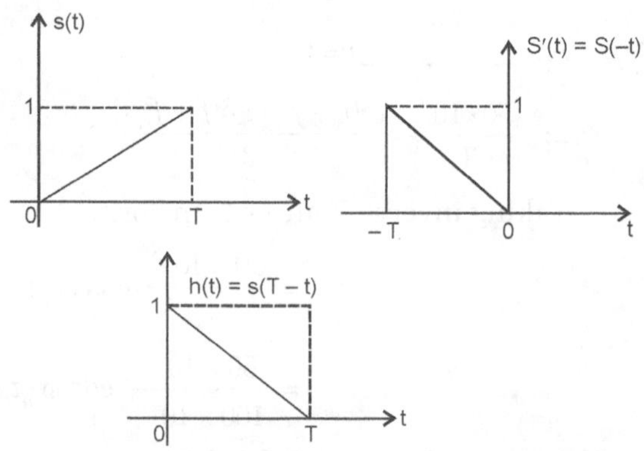

42. The graphical representation of the matched filter output $y(t)$ for this signal is given by

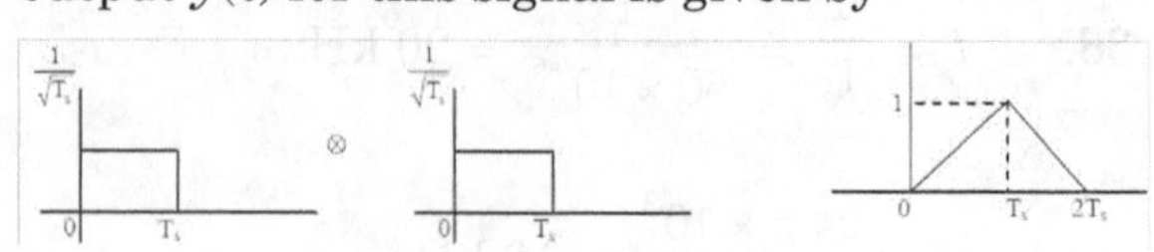

43. $\qquad u_0(t) = 5 \cos 20000\pi t$

and $\qquad u_1(t) = 5 \cos 22000\pi t$

Also $\qquad f_1 = 11000$ Hz

and $\qquad f_2 = 10000$ Hz

Now for FSK waveforms to be uncorrelated,

$$f_1 - f_2 = n\frac{R_b}{2};$$

$$n = 1, 2, 3, ...$$

$$\therefore \qquad R_b = \frac{2(f_1 - f_2)}{n} = \frac{2000}{n} \text{ bits/sec}$$

$\Rightarrow \quad R_{b(max)} = 2000$ bits/sec

($\because n = 1$ is minimum value)

Now $\quad T_{b(min)} = \dfrac{1}{R_{b(max)}} = 0.5$ ms

44.

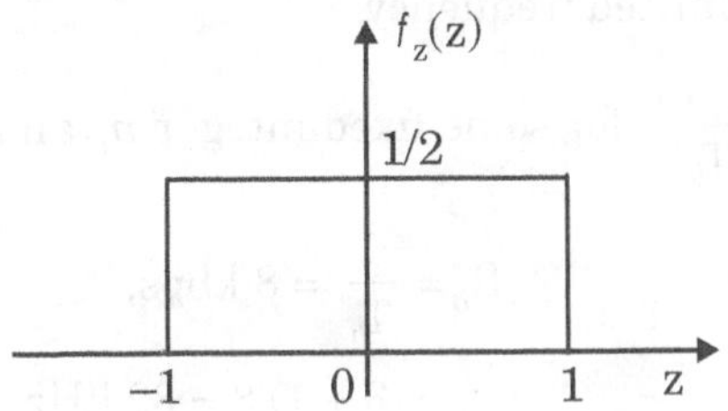

$$P(x_0) = \frac{1}{4}$$

$$P(x_1) = \frac{3}{4}$$

MAP criteria,

$$f_Y(y \mid x_0)P(x_0) \underset{x_1}{\overset{x_0}{\gtrless}} f_Y(y \mid x_1)P(x_1)$$

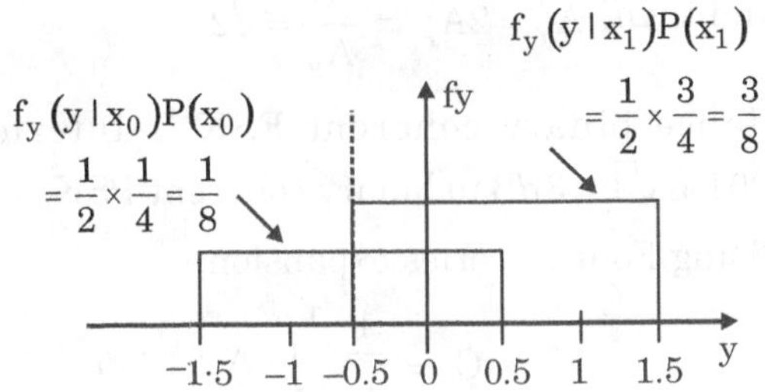

Optimum threshold exists here

So, $\quad \alpha = -0.50$

45.

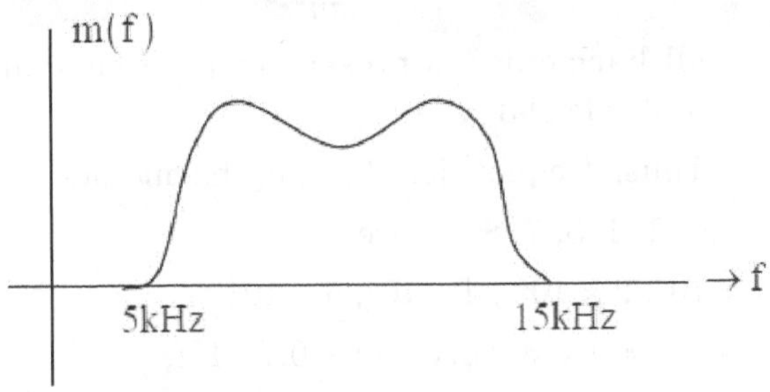

Modulation done using amplitade modulation technique

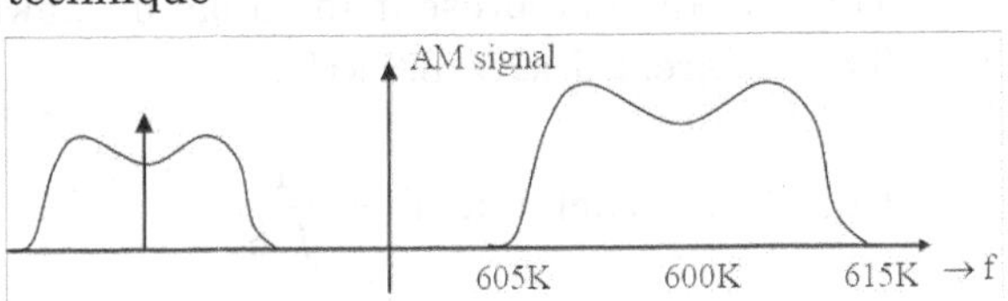

Now by sampling,

$$\frac{2f_{+1}}{n} \le f_s \le \frac{2f_1}{n-1}$$

Here $f_H = 615K$, $f_L = 605$ K.

$$\therefore \quad 1 \le n \le \frac{f_H}{B}; B = f_H - f_L$$

$\therefore \quad n \le 61.5, n = 61.$

$$\therefore \quad f_s \ge \frac{2 \times 615}{61}$$

Minimum sampling is 20 K.

$R_b = 1.2f_s \times 8 = 0.192$ M bit/s

46. Let getting an error be success. P (Success)= P, P (Failure) = l-P

P (x = at most [1]) = P (x = 0) + P (x = 1)

$= 8\,Co\,P^o x\,(1-P)^{8-0} + 8\,C_1\,(P)1\,(1-P)^{8-1}$

$= (1-P)^8 + 8\,P\,(1-P)^7$

47. Given fe = 4 GHz, fm = 2 MHz

fH = fc + fm = 4000 + 2 = 4003 MHz

F_L = fe – fm = 4000 – 2 = 3998 MHz.

B.W. = $F_H - f_L$ = 4002 – 3998

B.W. = 4 MHz

$(fs)_{min}$ = 2. (B.W.) = 2 × 4 = 8 MHz

48. Given f_H = 1800 Hz, F_L = 300 Hz

B.W. = fH–F_L = 1800 – 300 = 1500 Hz

$$K = \frac{f_H}{B.W.} = \frac{1800}{1500} = \frac{6}{5}$$

$\Rightarrow$ Minimum sampling rate, K = 1.

$$(f_s)_{min} = \frac{2f_H}{1} = \frac{2 \times 1800}{1} = 3600 \text{ samples / second}$$

49. For minimum sampling rate

$$(f_s)_{min} = \frac{2f_H}{K} = \frac{2f_H}{1} = 2f_H$$

$(f_s)_{min}$ = 2×1800 = 3600 samples/ second.

50. $R(z) = R(-z)$, for an autocorrelation function, hence, even symmetry.

51. The signal to quantization noise ratio is an n–bit PCM system is given by, SQNR $= \dfrac{3}{2}2^{2n}$

From the above equation it is clear that SQNR increases with increase is value of 'n'

52. $R(z) = R\,(-z)$, for an autocorrelation function, hence even symmetry.

53. Given f_m = 1 KHz.

Practically, 90% of the total signal strength lies in the major label so, minimum sampling rate,

$(f_s)_{min} = 2\,f_m$ = 2×1 = 2 KHz

54. Quantization noise power, QNP

$$= \frac{\Delta^2}{12}; \Delta = \frac{V_{p-p}}{2^n}$$

$$QNP = \frac{V_{p-p}^{\ 2}}{12 \times 2^{2n}} \Rightarrow QNP \propto \frac{1}{2^{2n}}$$

$$\frac{(QNP)_2}{(QNP)_1} = \frac{2^{2n}}{2^{2n_2}} = \frac{2^{2\times8}}{2^{2\times9}} = \frac{2^{16}}{2^{18}} - \frac{1}{4}.$$

$(QNP)_2 = (QNP)/_4$.

⇒ The QNP reduces by a factor of 4.

55. $x(t) = 6\times10^4 \sin C^3(400t) \times 10^6 \text{sinc}^3(100t)$.

Let $x_1(t) = 6 \times 10_4$ sinc3 (400t), $x_2(t) = 10^6$sinc3 (100t). These signals are convolved in the time – domain and hence will be multiplied in frequency domain. Hence, sampling frequency will be twice of minimum frequency of two massage signals.

Nyquist sampling frequency = $2f_m = 2 \times 150$ = 300 Hz.

56. Peak to peak of an 8 bit PCM is 2 volts.

Input signal $= 0.5\cos\omega_m t$

$$\frac{\text{Signal power}}{\text{Quantization noise power}}(db) = \frac{6n-7.2}{10}$$

$$= \frac{3}{2} \times \text{number of levels.}$$

57. Transmitted sample rate $(H_Z) = 2\,B_{max}$

$= 2 \times 400 = 800$ Hz.

58. Frame per second = 625.

Pixels per second = 400 × 400

64 intensity levels per pixel can be represented by 6–bits per pixel

∴ Data rate = 625 × 400 × 400 × 6

= 600 Mbps.

59. $g(t) = \sin c\,(700\,t) + \sin c\,(500\,t)$

$$= \frac{1}{\pi t}[\sin(\pi\,700\,t) + \sin(\pi\,500\,t)]$$

$g(t)$ is band-limited with $f_M = 350$ Hz.

Hence the Nyquist rate will be 700 Hz

then Nyquist interval is $\frac{1}{700}$ sec.

60. Probability of error = P

Probability of no error = (1 – P)

For error in at most one bit, there should be either no error or error in only one bit.

Hence the required probability

$$= {}^nC_0(1-P)^nP^0 + {}^nC_1(1-P)^{n-1}P^1$$

$$= (1-P)^n + np(1-p)^{n-1}$$

61. Transmitted frequency,

$f_i = \frac{n_c + i}{T_b}$ for some fixed integer n_C and $i = 1, 2$

Given, $R_b = \frac{1}{T_b} = 8$ kbps,

then, $f_1 = (3 + 1)\,8 = 32$ kHz

and, $f_2 = (4 + 1)\,8 = 40$ kHz

62. Probability of error, $P_e = \frac{1}{2}erfc\left(\sqrt{\frac{E_d}{2\eta}}\right)$

As P_e for FSK and PSK are same then E_d is also same for both

and then, $A_f^2 = 2A_P^2 \Rightarrow \frac{Af}{A_P} = \sqrt{2}$

Hence binary coherent FSK is interior by $20\log\sqrt{3} = 3dB$ to binary coherent PSK

63. Using Fourier series expansion,

$$C_n = \frac{1}{T_0}\int_{-T_0/6}^{T_0/6} A.e^{-Jn\omega_0 t}dt$$

$$= \frac{A}{\pi n}\sin\left(\frac{n\pi}{3}\right)$$

all harmonics is present in C_n except integer multiple of 3.

Thus, frequencies of $p(t)$ for harmonics 1, 2, 4, 5, 7, 8, ... are

10^3, 2×10^3, 4×10^3, 5×10^3 .....

∴ $p(t) \times x(t)$ gives (1 ± 0.7)kHz,

(2 ± 0.7)kHz, (4 ± 0.7)kHz ......

Thus, frequency present in range of 2.5kHz to 3.5 kHz are, 2.7 kHz, 3.3 kHz.

64. Duration of each bit, $T_b = \frac{1}{f_s.8}$

then bit rate, $R = \frac{1}{T_b} = 8f_s = 64$ kbps

In dB, $(SNR)_q = 1.76 + 6.02\,n$

$= 1.76 + (6.02) \times (8)$

$= 49.8$ dB

65. For mary PSK bandwidth required is,

$BW = \frac{2f_b}{N}$ where $M = 2^N$ and f_b is the bit rate.

For BPSK, $M = 2 = 2^N$

$\Rightarrow \quad N = 1$

Then, BW (Bandwidth) = $2f_b = 20$ KHz

$\Rightarrow \quad B_1 = 20$ KHz

For QPSK, $M = 4 = 2^N$

$\Rightarrow \quad N = 2$

then, Bandwidth, $B_2 = \frac{2f_b}{2} = 10$ KHz

66. Signal is greater than 0.2 V

Hence probability of error

That 0 is transmitted and 1 is received

$= 0.25 - 0.2 = 0.05.$

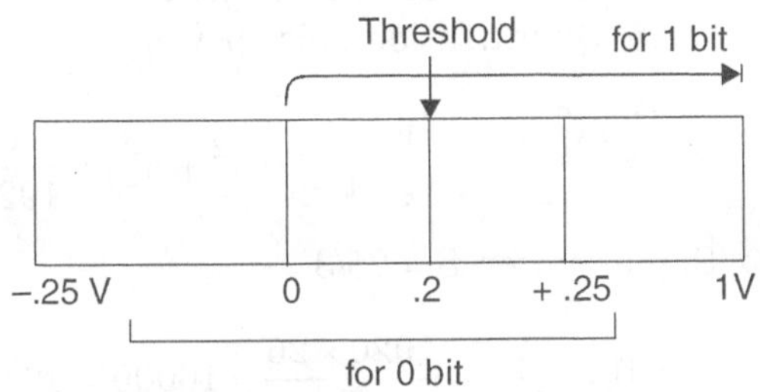

67. The three analog signals having band widths 1200 Hz, 600 Hz, 600 Hz have samples/sec of 2400, 1200 and 1200 respectively. Hence the total of 4800 samples /sec.

Then bit rate = 4800 sample/sec × 12

= 57.6 kbps

68. The Bandwidth of $x(t)$ is determined by the highest frequency component of $[\sin c\,(2000\,t)]^3$ or $[\sin c\,(2000\,t)]^2$, whichever is the largest.

Since multiplication in the time domain compounds to convolution in the frequency domain, signal $[\text{sinc}(2000t)]^3$ has a bandwidth equal to thrice to that of $[\text{sinc}(2000\,t)]$. Further signal $[\text{sinc}(2000\,t)]^2$ has a bandwidth equal to twice to that of $[\text{sinc}(2000\,t)]$.

as $\sin c(2\omega t) \rightleftharpoons \frac{1}{2\omega}\text{rect}\left(\frac{f}{2\omega}\right)$

Signal $[\text{sinc}(2000t)]$ has a bandwidth 1000 Hz and Nyquist rate = 2000 Hz

Hence, $x(t)$ has a sampling rate

$3 \times 2000 \text{ Hz} = 6 \times 10^3$

69. Atmost one-bit error

$= p(1 \text{ bit error}) + p(\text{no bit error})$

$= {}^nC_1 \times p^1 \times (1-p)^{n-1} + ({}^nC_0\ p^0\ (1-p)^n)$

$= np\,(1-p)^{n-1} + (1-p)^n$

70. Average energy of constellation 1 is

$$E_1 = \frac{0 + 4_a^2 + 4_a^2 + 4_a^2}{4} = 4_a^2$$

Average energy of constellation 2 is

$$E_2 = \frac{a^2 + a^2 + a^2 + a^2}{4} = a^2$$

$$\Rightarrow \frac{E_1}{E_2} = \frac{4_a^2}{a^2} = 4$$

71. The probability of error decreases with increase in average energy . As constellation 1 has more average energy than that of constellation 2. So, the probability of symbol error for constellation 1 is lower.

72. To maximize the entropy, all the decission boundaries should be equiprobable.

$$\int_1^5 p_X(x)dx = \frac{1}{3} \Rightarrow \int_1^5 b dx = \frac{1}{3} \Rightarrow b[5-1] = \frac{1}{3}$$

$$b = \frac{1}{12}$$

$$\Rightarrow \int_{-1}^{1} P_X(x)dx = \frac{1}{3} \Rightarrow \int_{-1}^{1} a\,dx = \frac{1}{3}$$

a [1– (–1)] = 1/3

a = 1/6

73. Signal power,

$$S = E\left[x^2\right] = \int_{-\infty}^{\infty} x^2 f_x(x)dx = 2\left[\int_0^1 ax^2dx + \int_1^5 bx^2dx\right]$$

$$= 2\left[a\left(\frac{x^3}{3}\right)_0^1 + b\left(\frac{x^3}{3}\right)_1^5\right] = 2\left[\frac{a}{3} + \frac{124b}{3}\right]$$

$$= 2\left[\frac{1}{6\times 3} + \frac{124}{12\times 3}\right] = 7$$

Quantization noise power, $N_Q = \sum_{L=1}^{2} \frac{\Delta_1^2}{12}, P(\Delta i)$

$\Delta_1 = 1$ V for $-1<x<1$.

$\Delta_2 = 2$ V for $1<|x|<S$.

$$P(\Delta_1) = P\{-1 < x < 1\} = 2a = \frac{1}{3}$$

$$P(\Delta_2) = 1 - P(\Delta_1) = 1 - \frac{1}{3} = \frac{2}{3}$$

$$\Rightarrow N_Q = \frac{1}{3}\left(\frac{1^2}{12}\right) + \frac{2}{3}\left(\frac{2^2}{12}\right) = \frac{1}{4}$$

Hence, $\text{SQNR} = \frac{S}{N_Q} = 28.$

74. p : transmitted bit will be received in error if at receiver, the bits are

$$\begin{matrix} 000 & \text{or} & 001 \\ \downarrow & & 010 \\ p^3 & & 100 \end{matrix} \Bigg\} \rightarrow 3\,p^2(1-p)$$

$$P_e = p^3 + 3\,p^2(1-p)$$

Alternately

The probability of ib being received in error is,

$$\binom{n}{i} p^i (1-p)^{n-i}$$

Hence the probability of error P_e is

$$p_e = \sum_{i=2}^{3} \binom{3}{i} p^i (1-p)^{3-i}$$

or $$p_e = \binom{3}{2} p^2 (1-p) + \binom{3}{3} p^3 (1-p)^0$$

or $$p_e = 3p^2(1-p) + p_3$$

75. Binary sequence represented by bipolar pulsed so, B.W. = R_b = nfs = 8×8K = 64 kHz.

76. Signal to noise ratio, $\left(\frac{S_o}{N_D}\right)_{dB} = 6n\,\text{dB}.$

where, n = number of bits per sample quantized

$$\left(\frac{S_o}{N_O}\right)_{dB} = 6 \times 8 = 48\,\text{dB}.$$

77. Quantization noise,

$$Nq = \frac{\Delta^2}{12}, \Delta = \text{step size} = \frac{V_{p-p}}{2^n}.$$

$$\therefore N'q = \frac{Nq}{4} \Rightarrow \frac{\Delta'^2}{12}, = \frac{\Delta^2}{12 \times 4}.$$

$$\Delta' = \frac{\Delta}{2}$$

$$\Rightarrow \frac{V_{p-p}}{2n'} = \frac{V_{p-p}}{2 \times 2^n} \Rightarrow 2n' = 2 \times 2^8 = 2^9 = 512.$$

Therefore number of quantization levels required to reduce the quantization noise by a factor would be 512.

78. Signal power,

$$S = \int_{-\infty}^{\infty} x^2 f x(x) dx = \int_{-5}^{5} \frac{1}{10} x^2 dx = \frac{25}{3}.$$

$$\text{SQNR} = 43.5\,\text{dB} = 10^{4.35}$$

$$N_Q = \frac{\Delta^2}{12} = \frac{S}{\text{SQNR}}$$

$$\Rightarrow \Delta = \sqrt{\frac{12 \times 5}{\text{SQNR}}} = \sqrt{\frac{12 \times 25}{3 \times 10^{4.35}}} = 0.0668\text{V}.$$

79. Given D_1 = 0.05V (for + Ve Values), D_2 = 0.1V (for – Ve values).

The amplitude of the input signal is symmetrically distributed about zero. Hence P(D1) = P(D2) = 0.5. Quantization noise power,

$$N_Q = \frac{\Delta_1^2 P_1}{12} + \frac{\Delta_2^2 P_2}{12} = \frac{1}{24}\left(\Delta_1^2 + \Delta_2^2\right) = \frac{1}{1920}$$

Signal power, S = 25/3

$$\text{SQNR} = \frac{S}{N_O} = \frac{1920 \times 25}{3} = 16000 = 42\text{dB}.$$

80. $$s(t) = \frac{\sin 500\pi t}{\pi t} \times \frac{\sin 700\pi t}{\pi t}$$

Assume, $$s_1(t) = \frac{\sin 500\pi t}{\pi t}$$

and $$s_2(t) = \frac{\sin 700\pi t}{\pi t}$$

$s_1(t)$ and its fourier transform is shown below:

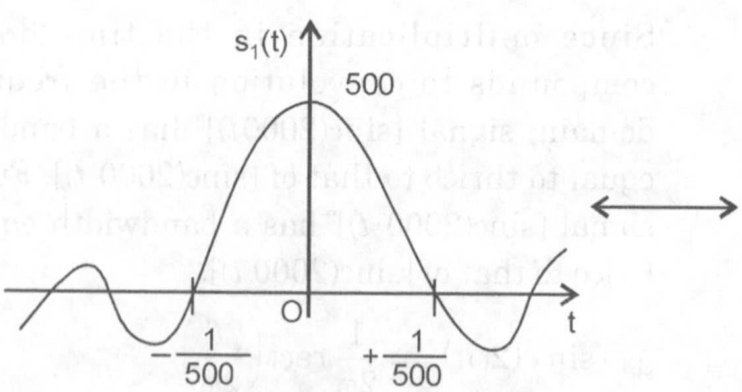

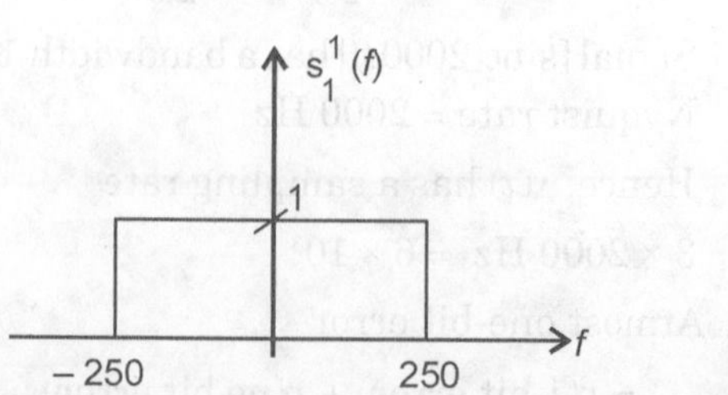

$s_2(t)$ and its fouriar transfroum is shown below :

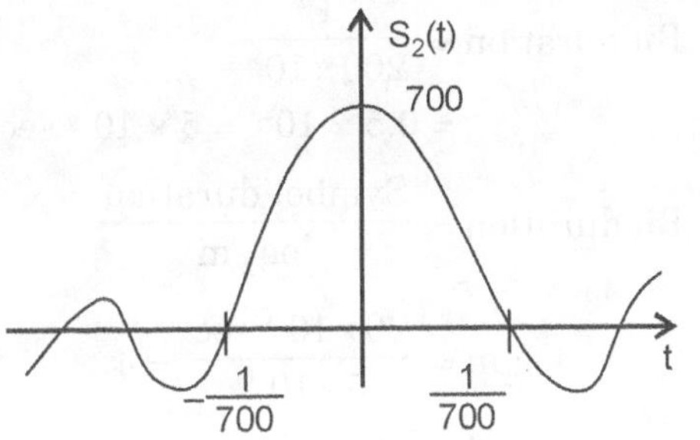

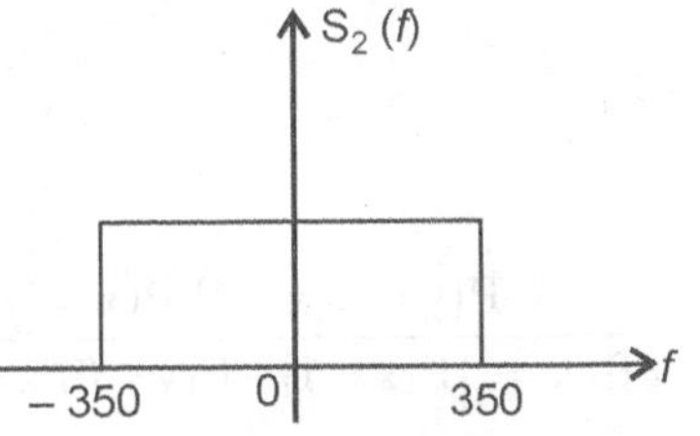

From frequency convolution theorem, $x_1(t)x_2(t) \leftrightarrow 1\, x_1(f) \cdot x_2(f)$, we find that $s(t)$ is band-limited signal and its bandwidth is equal to the sum of bandwidths of $s_1(t)$ and $s_2(t)$, i.e., 600 Hz. Thus, Nyquist rate is 1200 Hz.

81. Entropy, $4(x) = \sum_{i=1}^{4} P_i \log_2\left(\frac{1}{P_i}\right)$

As quantized level are independent and equiprobable

$$P_1 = P_2 = P_3 = P_4 = \frac{1}{4}$$

then, $H(x) = 4\left[\frac{1}{4}\log_2 4\right] = 2$ bits/sample

And, Information rate, R = Ω4

= 2 sample/sec × 2 bits/sample

= 4 bits/sec.

82. $B_T = \frac{1}{2}R_s(\beta + 1)$

$R_s \rightarrow$ Symbol rate

$\Rightarrow \quad R_s = \frac{2 \times \beta_T}{\beta + 1}$

$\Rightarrow \quad \beta = 0.75$

$\Rightarrow \quad R_s = \frac{2 \times 3500}{1 + 0.75} = 4000$ symbols/sec

83. $P[X = 0] = \frac{0}{10}$

$$P[X = 1] = 1 - P[X = 0] = \frac{1}{10}$$

Probability of error for an optimum receiver,

$P_e = P[X = 1]$ (1 – transmission probability)

$$= \frac{1}{10}\left(1 - \frac{1}{8}\right) = \frac{7}{80}$$

84. Given : $\phi = 45°$

For imperfect synchronization, probability of error in BPSK is

$$Pe = Q\left(\sqrt{\frac{2E}{No}}\text{Cos}2\phi\right) = Q\left(\sqrt{\frac{2E}{N}}.\text{Cos}^2 45°\right)$$

$$Pe = Q\sqrt{\frac{E}{N_0}}$$

85. Bit rate given = R Kbits/second

Modulation = 32-QAM

No. of bits/symbol = 5[$\log_2 32$]

Symbol rate = $\frac{R}{5}$ k symbols/second

Finally we are transmitting symbols.

$E_T \rightarrow$ transmission bandwidth

$$B_T = \frac{R(\text{symbol rate})}{(1 + \alpha)} = \frac{R}{5(1 + \alpha)}$$

For B_T to be minimum, α has to be maximum

$\Rightarrow B_T = \frac{R}{5 \times 2} = \frac{R}{10}$

Maximum value of α is '1' which is a roll off factor

86. BER is given as

Pe = P(0). P (1/0) + P (1). P (0/1)

If detection threshold is 1, then

P (0) = P (1) = 1/2.

$$P\left(\frac{y = 1}{x = 0}\right) = \int_1^{\infty} f(z/0)dz = 0$$

$$P\left(\frac{y = 0}{x = 1}\right) = \int_0^{1} f(z/1)dz = \frac{1}{4} \times \frac{1}{4} \times 1 = \frac{1}{8}$$

$$\therefore Pe = \frac{1}{2} \times 0 + \frac{1}{2} \times \frac{1}{8} = \frac{1}{16}.$$

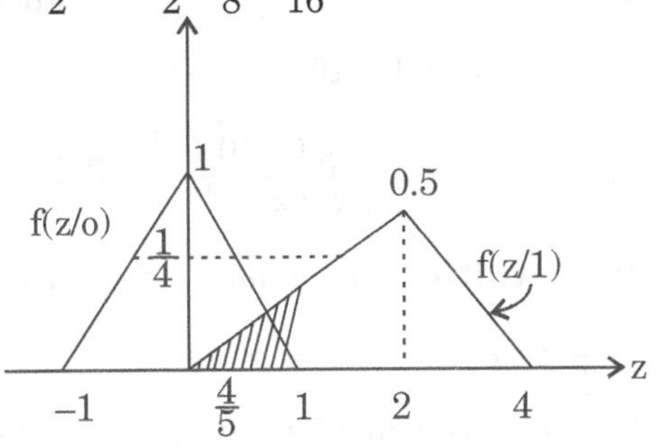

87. Optimum threshold is given by the point of intersection of two pdf curves.

$f(z/0) = 1 - |z|;\ |z| \le 1$

$f(z/1) = z/4;\ 0 < z < 2.$

The point of intersection which decides optimum threshold is

$1 - z = 1/4$

$z = 4/5$

88. $$x(t) = \sqrt{2}\left(\frac{\sin(\pi t/5)}{\pi t/5}\right)\sin\left(\pi t - \frac{\pi}{4}\right)$$

we can write above expression as

$$x(t) = -\sqrt{2}\left(\frac{\sin(\pi t/5)}{\pi t/5}\right)\left[\cos\frac{\pi}{4}\sin\pi t - \sin\frac{\pi}{4}\cos\pi t\right]$$

$$= \left(\frac{\sin(\pi t/5)}{\pi t/5}\right)\cos\pi t - \frac{\sin(\pi t/5)}{(\pi t/5)}\sin(\pi t)$$

Also

$x(t) = x_c(t)\cos 2\pi f_c t - x_s(t)\sin(2\pi f_c t)$

[Low pass representation of Bandpass signals]

$$x_c(t) = \frac{\sin(\pi t/5)}{\pi t/5}$$

$$x_s(t) = \frac{\sin(\pi t/5)}{\pi t/5}$$

$x_{ce}(t)$ is the complex envelope of $x(t)$

$$x_{ce}(t) = x_c(t) + jx_s(t)$$

$$= \frac{\sin(\pi t/5)}{\pi t/5}[1+j]$$

$$= \frac{\sqrt{2}\sin(\pi t/5)}{(\pi t/5)}e^{j\pi/4}$$

89. For Binary F_{SK}

Bit error probability $= Q\left(\sqrt{\frac{E}{N_o}}\right)$

$E \rightarrow$ Energy per bit [no. of symbols = No. of bits]

$$E = \frac{A^2T}{2},$$

$A = 4 \times 10^{-3}$,

$$T = \frac{1}{500\times10^3}[\text{inverse of data rate}]$$

$$\Rightarrow \quad E = \frac{16\times10^{-6}\times2\times10^{-6}}{2} = 16\times10^{-12}$$

$N_0 = 1\times10^{-12}$

$$\Rightarrow \quad P_e = Q\left(\sqrt{\frac{16\times10^{-12}}{1\times10^{-12}}}\right) = Q(4).$$

90. Bandwidth requirement for m-level PSK

$$= \frac{1}{T}(1+\alpha)$$

[Where T is symbol duration, α is roll of factor]

$$\Rightarrow \quad \frac{1}{T}(1+\alpha) = 100\times10^3$$

$\alpha = 1$ [100% excess bandwidth]

$$\Rightarrow \quad \frac{1}{T}(2) = 100\times10^3$$

$$\Rightarrow \quad T = \frac{2}{100\times10^3} = 20 \text{ sec}$$

$$\text{Bit duration} = \frac{1}{200\times10^3}$$

$$= 0.5\times10^{-5} = 5\times10^{-6}\text{ sec}$$

$$\text{Bit duration} = \frac{\text{Symbol duration}}{\log_2 m}$$

$$\Rightarrow \quad \log_2 m = \frac{20\times10^{-6}\text{ sec}}{5\times10^{-6}} = 4$$

$$\Rightarrow \quad M = 16$$

91. $P\left(\frac{x=1}{y=0}\right)$

$$= \frac{P(y=0/x=1).P(x=1)}{P(y=0/x=1)P(x=1)+P(y=0/x=0)P(x=0)}$$

$$= \frac{(1/7)\times0.8}{\frac{1}{7}\times0.8+\frac{6}{7}\times0.2}$$

$= 0.4$

93. Optimum receiver for AWGN channel is given by matched filter.

In case of matched filter receiver,

$$\text{Probability of error} = Q\left(\sqrt{\frac{2E}{N_u}}\right)$$

$\Rightarrow$ Probability of error is minimum for which E is maximum

Now looking at options

Energy in option (*a*) = $1^2 = 1$

Energy in option (*c*) and (*d*) is same = 1/3

$$\text{Energy in option }(b) = 2\left[\int_0^{1/2}(2t)^2\,dt\right]$$

$$= 2\left[\int_0^{1/2}4t^2\,dt\right]$$

$$= 2.4\left(\frac{t^3}{3}\right)\Big|_0^{1/2} = 1/3$$

Thus option (*a*) is correct answer.

94. From given data

binary data transmission (R_b) = 56 kbps

Roll-off factor (α) = 0.25

∴ Transmission bandwidth

$$(\text{BW}) = \frac{R_b}{2}[1+\alpha]$$

$$= \frac{56}{2}\left[1+0.25\right]\text{kHz}$$

$$= 28 \times 1.25 \text{ kHz}$$

$$= 35 \text{ kHz}$$

95. Here the range of Channel spectrum 500 Hz – 2000 Hz (Here R_b = rate of transmission)

Hence Bandwidth (BW) = 1500 Hz

$$BW = \frac{R_b}{\log_2 M}(1+\alpha)$$

$$1500 = \frac{R_b}{\log_2 16}(1+\alpha)$$

$$1500 = \frac{4800}{\log_2 16}(1+\alpha)$$

$$1500 = 1200\,(1+\alpha)$$

$$\Rightarrow \qquad \alpha = 0.25$$

Hence, the roll-off factor of a-pulse is 0.25.

96. From the given data

f_s = 8 kHz (speech signal)

n = 8 bits/sample ; M = 4

The minimum band width

$$BW_{min} = \frac{R_b}{2\log_2 M} = \frac{R_b}{2\log_2 4} = \frac{f_s \times n}{2\log_2 2^2}$$

$$= \frac{f_s \times n}{4\log_2 2} = \frac{f_s \times n}{4} = \frac{8 \times 8}{4} = 16 \text{ kHz}$$

Hence the minimum band width is 16 kHz.

97. $P = 10^{-5}$ $N = 10^5$

Given question can be solved by two methods.

Method 1 :

Binomial : ${}^nC_x\, p^x\, q^{n-x}$

$$P[x = 0] + P[x = 1] = {}^{10^5}C_0 (10^{-5})^0 (1-10^{-5})^{10^5} + {}^{10^5}C_1 (10^{-5})^1 (1-10^{-5})^{10^5-1}$$

$$= (1)\,(1) \times 0.367 + 0.367$$

$$= 0.735$$

Method 1 :

$$\text{Poisson} = \frac{e^{-\lambda}\lambda^x}{x!}$$

$$\lambda = np = 10^{-5} \times 10^5 = 1$$

(since $n = 10^5$, $p = 10^{-5}$)

$$= \frac{e^{-1}(1)^1}{1!} + e^{-1}$$

$$\Rightarrow \qquad 2 \times e^{-1} = 0.735$$

So, the probability that not more than one bit will be in error is 0.735.

98. From the given figure.

One period of signal

$$x_1(t) = u(t) - u(t-T/2)$$

$$\therefore \qquad X_1(s) = \frac{1}{s} - \frac{e^{-sT/2}}{s}$$

$$= \frac{1-e^{-sT/2}}{s}$$

$$X(s) = \frac{1}{1-e^{-sT}} X_1(s)$$

$$\left(\text{Here } X_1(s) = \frac{1-e^{-sT/2}}{s}\right)$$

$$\therefore \qquad X(s) = \frac{1}{s\left(1+e^{-sT/2}\right)}$$

100. Let s_0 and s_1 be the transmitted symbols representing the transmitted value {–1, 1} respectively and let r_0 and r_1 be the received symbols.

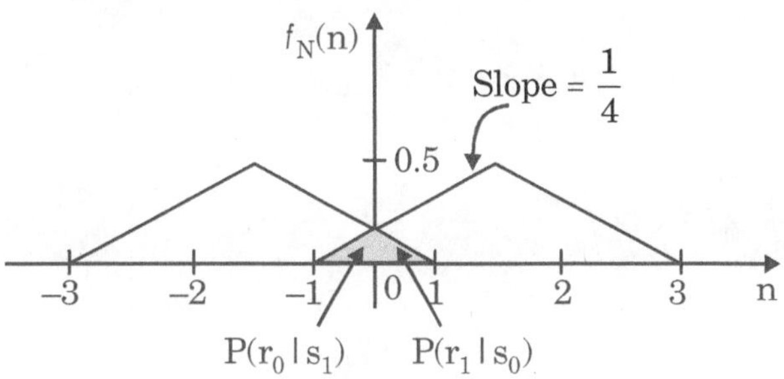

Now Probability of error, (P_e)

$$= P(s_1)P(r_0 \mid s_1) + P(s_0)P(r_1 \mid s_0)$$

$$P(r_0 \mid s_1) = P(r_1 \mid s_0)$$

$$= \frac{1}{2} \times 1 \times \frac{1}{4} = \frac{1}{8}$$

$$P(s_0) = P(s_1) = \frac{1}{2} \text{ (given)}$$

$$\therefore \quad P_e = \left(\frac{1}{2} \times \frac{1}{8} + \frac{1}{2} \times \frac{1}{8}\right)$$

$$= \frac{1}{2}\left(\frac{1}{8} + \frac{1}{8}\right)$$

4 CHAPTER

Information Theory & Coding

Analysis of Previous GATE Papers												
	Year → Topics ↓		2019	2018	2017 Set 1	2017 Set 2	2016 Set 1	2016 Set 2	2016 Set 3	2015 Set 1	2015 Set 2	2015 Set 3
FUNDAMENTALS OF ERROR CORRECTION & HAMMING CODE	**1 Mark**	MCQ Type	1			1						
		Numerical Type		1					1			
	2 Marks	MCQ Type				1						
		Numerical Type					1		1			
		Total	**1**	**1**		**3**	**2**		**3**			
TIMING & FREQUENCY SYNCHRONI ZATION	**1 Mark**	MCQ Type										
		Numerical Type										
	2 Marks	MCQ Type										
		Numerical Type										
		Total										
INTERSYMBOL INTERFERENCE & MITIGATION	**1 Mark**	MCQ Type										
		Numerical Type			1			1				
	2 Marks	MCQ Type										
		Numerical Type					1	1				
		Total			**1**		**2**	**3**				
BASICS OF TDMA, FDMA & CDMA	**1 Mark**	MCQ Type										
		Numerical Type										
	2 Marks	MCQ Type										
		Numerical Type								1		
		Total								**2**		

FUNDAMENTALS OF ERROR CORRECTION & HAMMING CODE

1. A source produces 4 symbols with probability $\frac{1}{2}, \frac{1}{4}, \frac{1}{8}$ and $\frac{1}{8}$. For this source, a practical coding scheme has an average codeword length of 2 bits/symbols. The efficiency of the code is

(a) 1 (b) $\frac{7}{8}$

(c) $\frac{1}{2}$ (d) $\frac{1}{4}$ **[1989 : 2 Marks]**

2. An image uses 512 × 512 picture elements. Each of the picture elements can take any of the 8 distinguishable intensity levels. The maximum entropy in the above image will be

(a) 2097152 bits (b) 786432 bits

(c) 648 bits (d) 144 bits

[1990 : 2 Marks]

3. A memoryless source emits n symbols each with a probability p. The entropy of the source as a function of n

(a) increases as logn.

(b) decreases as log(1/n).

(c) increases as n.

(d) increases as nlogn. **[2008 : 2 Marks]**

4. A communication channel with AWGN operating at a signal to noise ratio SNR >> 1 and bandwidth 6 has capacity C_1. If the SNR is doubled keeping B constant, the resulting capacity C_2 is given by

(a) $C_2 \approx 2C_1$ (b) $C_2 \approx C_1 + B$

(c) $C_2 \approx C_1 + 2B$ (d) $C_2 \approx C_1 + 0.3B$

[2009 : 2 Marks]

5. A source alphabet consists of N symbols with the probability of the first two symbols being the same. A source encoder increases the probability of the first symbol by a small amount ε and decreases that of the second by ε. After encoding, the entropy of the source

(a) increases.

(b) remains the same.

(c) increases only if N = 2.

(d) decreases. **[2012 : 1 Mark]**

6. Let U and V be two independent and identically distributed random variables such that $P(U = +1) = P(U = -1) = \frac{1}{2}$. The entropy H(U + V) in bits is

(a) $\frac{3}{4}$ (b) 1

(c) $\frac{3}{2}$ (d) $\log_2 3$

[2013 : 2 Marks]

7. The capacity of a Binary Symmetric Channel (BSC) with cross-over probability 0.5 is ______.

[2014 : 1 Mark, Set-1]

8. The capacity of a band-limited additive white Gaussian noise (AWGN) channel is given by $C = W \log_2 \left(1 + \frac{P}{\sigma^2 W}\right)$ bits per second (bps), where W is the channel bandwidth, P is the average power received and σ^2 is the one-sided power spectral density of the AWGN. For a fixed $\frac{P}{\sigma^2} = 1000$, the channel capacity (in kbps) with infinite bandwidth ($W \to \infty$) is approximately

(a) 1.44

(b) 1.08

(c) 0.72

(d) 0.36

[2014 : 1 Mark, Set-2]

9. Consider a discrete memoryless source with alphabet $S = (s_0, s_1, s_2, s_3, s_4, ...)$ and respective probabilities of occurrence $P = \left(\frac{1}{2}, \frac{1}{4}, \frac{1}{8}, \frac{1}{16}, \frac{1}{32}, ...\right)$. The entropy of the source (in bits) is ______.

[2016 : 2 Marks, Set-1]

10. An analog baseband signal, bandlimited to 100 Hz, is sampled at the Nyquist rate. The samples are quantized into four message symbols that occur independently with probabilities $p_1 = p_4 = 0.125$ and $p_2 = p_3$. The information rate (bits/sec) of the message source is ______.

[2016 : 1 Mark, Set-3]

11. A voice-grade AWGN (additive white Gaussian noise) telephone channel has a bandwidth of 4.0 kHz and two-sided noise power spectral density $\frac{\eta}{2} = 2.5 \times 10^{-5}$ Watt per Hz. If information at the rate of 52 kbps is to be transmitted over this channel with arbitrarily small bit error rate, then the minimum bit-energy E_h (in mJ/bit) necessary is ______.

[2016 : 2 Marks, Set-3]

12. Which one of the following graphs shows the Shannon capacity (Channel capacity) in bits of a memoryless binary symmetric channel with crossover probability p?

(a)

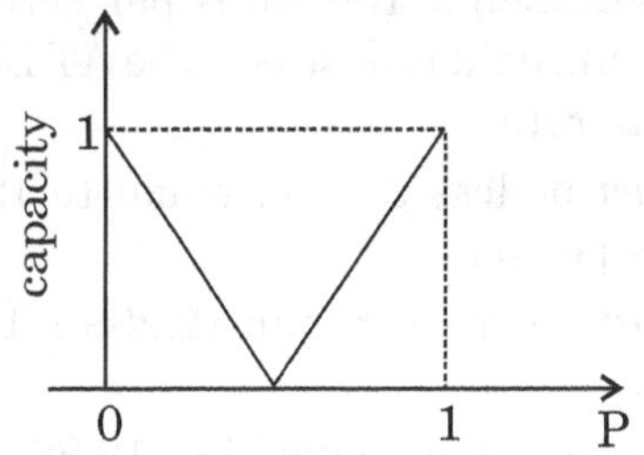

(b)

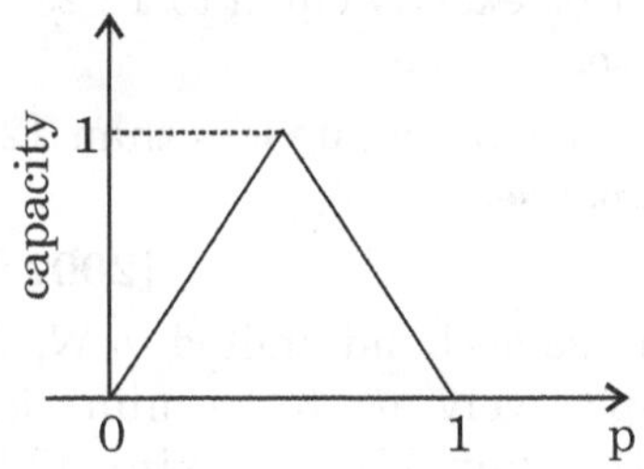

(c)

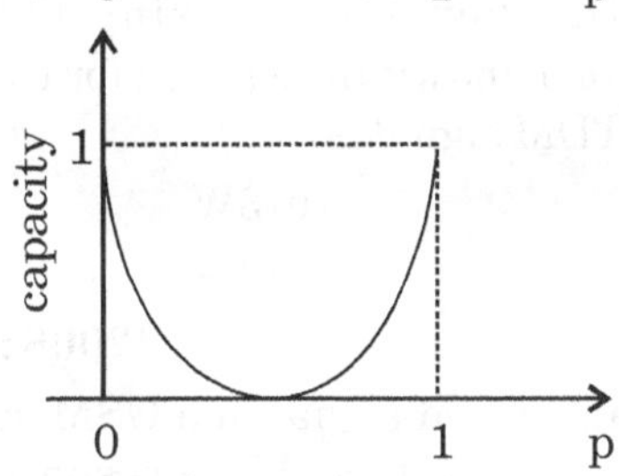

(d)

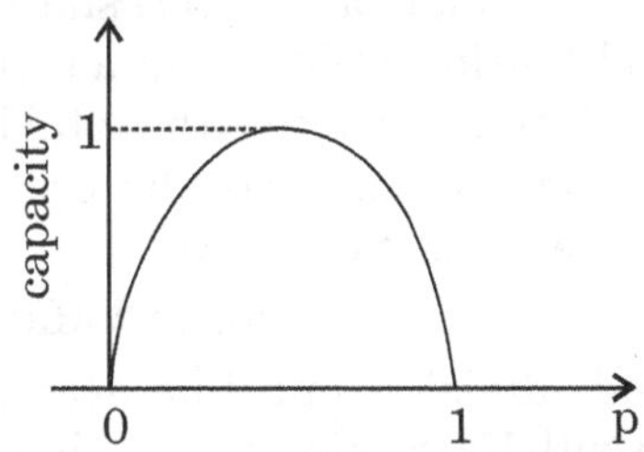

[2017 : 1 Mark, Set-2]

13. Consider a binary memoryless channel characterized by the transition probability diagram shown in the figure.

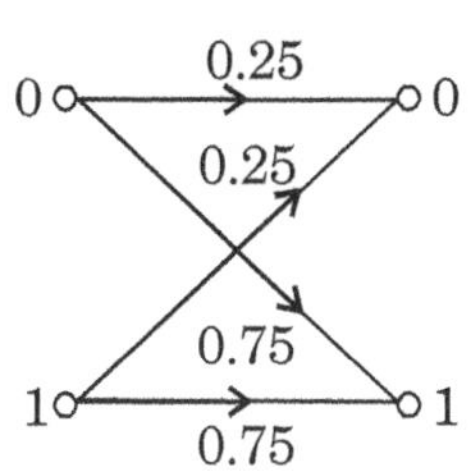

The channel is

(a) lossless (b) noiseless

(c) useless (d) deterministic

[2017 : 2 Marks, Set-2]

14. Consider a binary channel code in which each codeword has a fixed length of 5 bits. The Hamming distance between any pair of distinct codewords in this code is at least 2. The maximum number of codewords such a code can contain is ______.

[2018 : 1 Mark]

15. A linear Hamming code is used to map 4-bit messages to 7-bit codewords. The encoder mapping is linear. If the message 0001 is mapped to the codeword 0000111, and the message 0011 is mapped to the codeword 1100110, then the message 0010 is mapped to

(A) 0010011

(B) 1111111

(C) 1111000

(D) 1100001

[2019 : 1 Marks]

TIMING & FREQUENCY SYNCHRONIZATION

16. A source generates three symbols with probabilities 0.25, 0.25, 0.50 at a rate of 3000 symbols per second. Assuming independent generation of symbols, the most efficient source encoder would have average bit rate is

(a) 6000 bits/sec (b) 4500 bits/sec

(c) 3000 bits/sec (d) 1500 bits/sec

[2006 : 2 Marks]

INTERSYMBOL INTERFERENCE & MITIGATION

17. A digital communication system uses a repetition code for channel encoding/decoding. During transmission, each bit is repeated three times (0 is transmitted as 000, and 1 is transmitted as 111). It is assumed that the source puts ou: symbols independently and with equal probability. The decoder operates as follows: In a block of three received bits, if the number of zeros exceeds the number of ones, the decoder decides in favor of a 0, and if the number of ones exceeds the number of zeros, the decoder decides in favor of a 1. Assuming a binary symmetric channel with crossover probability p = 0.1. The average probability of error is ______.

[2016 : 2 Marks, Set-1]

18. A discrete memoryless source has an alphabet (a_1, a_2, a_3, a_4) with corresponding probabilities $\left(\frac{1}{2},\frac{1}{4},\frac{1}{8},\frac{1}{8}\right)$. The minimum required average codeword length in bits to represent this source for error-free reconstruction is

[2016 : 1 Mark, Set-2]

19. A binary communication system makes use of the symbols "zero" and "one". There are channel errors. Consider the following events:

x_0 : a "zero" is transmitted.

v_1 : a "one" is transmitted.

y_0 : a "zero" is received.

y_1 : a "one" is received.

The following probabilities are given:

$$P(x_0)=\frac{1}{2}, P(y_0 \mid x_0)=\frac{3}{4}, \text{and } P(y_0 \mid x_1)=\frac{1}{2}.$$

The information in bits that you obtain when you learn which symbol has been received (while you know that a "zero" has been transmitted) is ______.

[2016 : 2 Marks, Set-2]

20. Let, (X_1, X_2) be independent random variables. X, has mean 0 and variance 1, while X_2 has mean 1 and variance 4. The mutual information $/(X_1 : X2)$ between X_1 and X_2 in bits is ______.

[2017 : 1 Mark, Set-1]

BASICS OF TDMA, FDMA & CDMA

21. Quadrature multiplexing is

(a) the same as FDM.

(b) the same as TDM.

(c) a combination of FDM and TDM.

(d) quite different from FDM and TDM.

[1998 : 1 Mark]

22. In a GSM system, 8 channels can co-exist in 200 kHz bandwidth using TDMA. A GSM based cellular operator is allocated 5 MHz bandwidth. Assuming a frequency reuse factor of $\frac{1}{5}$, i.e. a five-cell repeat pattern, the maximum number of simultaneous channels that can exist in one cell is

(a) 200 (b) 40

(c) 25 (d) 5

[2007 : 2 Marks]

23. In a Direct Sequence CDMA system the chip rate is 1.2288 x 106 chips per second. If the processing gain is desired to be AT LEAST 100, the data rate

(a) must be less than or equal to 12.288×10 bits per sec.

(b) must be greater than 12.288×10^3 bits per sec.

(c) must be exactly equal to 12.288×10^3 bits per sec.

(d) can take any value less than 122.88×10^3 bits per sec.

[2007 : 2 Marks]

24. Four messages band limited to W, W, 2W and 3W respectively are to be multiplexed using Time Division Multiplexing (TDM). The minimum bandwidth required for transmission of this TDM signal is

(a) W (b) 3W

(c) 6W (d) 7W

[2008 : 2 Marks]

25. The transmitted signal in a GSM system is of 200 kHz bandwidth and 8 users share a common bandwidth using TDMA. If at a given time 12 users are talking in a cell, the total bandwidth of the signal received by the base station of the cell will be at least (in kHz) ______.

[2015 : 2 Marks, Set-1]

26. In a code-division multiple access (CDMA) system with N = 8 chips, the maximum number of users who can be assigned mutually orthogonal signature sequences is ______.

[2014 : 1 Mark, Set-1]

ANSWERS

1. (b) 2. (b) 3. (a) 4. (b) 5. (d) 6. (c) 7. (–0· 01 to 0· 01) 8. (a) 9. (1· 75)
10. (13) 11. (31· 503) 12. (c) 13. (c) 14. (16) 15. (d) 16. (b) 17. (0.028) 18. (1· 75)
19. (0· 405) 20. (0) 21. (d) 22. (b) 23. (a) 24. (d) 25. (400) 26. (7· 99 to 8· 01)

EXPLANATIONS

1. Entropy, $H = -\sum_{i=1}^{n} P_i \log_2(P_i)$

$$= -\left[\frac{1}{2}\log_2\left(\frac{1}{2}\right) + \frac{1}{4}\log_2\left(\frac{1}{4}\right) + \frac{1}{8}\log_2\left(\frac{1}{8}\right) + \frac{1}{8}\log_2\left(\frac{1}{8}\right)\right]$$

$$= \frac{1}{2} + \frac{1}{4}\times 2 + \frac{1}{8}\times 3 + \frac{1}{8}\times 3$$

$$H = 1 + \frac{3}{4} = 7/4$$

Code efficiency, $\eta = \frac{H}{L}\times 100\% = \frac{7/4}{2}\times 100\% = \frac{7}{8}.$

2. $n = \log_2(L) = \log_2(8) = 3.$

Maximum entropy = 512×512×n = 512×512×3 = 786432 bits.

3. For n symbols, $P = \frac{1}{n}$

$\therefore$ $$\text{Entropy} = \sum_{k=1}^{n} P_k \log_2 \frac{1}{P_k}$$

$$= \log_2\left(\frac{1}{P}\right)$$

$$= \log_2(n)$$

4. According to Shannon-Hartely law,

channel capacity, $C_1 = B\log_2\left(1+\frac{S}{N}\right) \approx B\log_2\left(\frac{S}{N}\right)$

$$C_2 = B\log_2\left(\frac{2S}{N}\right)$$

$$= B\log_2^2 + B\log\frac{S}{N}$$

$$C_2 = B + C_1$$

5. Entropy is maximum when all symbols are equiprobable.

If the probability of symbols are different then entropy is going to decrease.

6. $P(U+V)_{=2} = \frac{1}{2}\times\frac{1}{2} = 4.$

$$P(U+V=0) = \frac{1}{4}+\frac{1}{4} = \frac{1}{2}.$$

$$P(U+V=-2) = \frac{1}{2}\times\frac{1}{2} = \frac{1}{4}.$$

$$\therefore H(U+V) = \frac{1}{2}\log_2^2 + 2\times\frac{1}{4}\times\log_2^4$$

$$= \frac{1}{2}+\frac{1}{2}\times 2 = \frac{3}{2}.$$

U	V	(U + V)
+1	+1	+2
+1	–1	0
–1	+1	0
–1	–1	–2

7. Channel capacity of BSC is

$$C = P\log_2 P + (1-P)\log_2(1-P) + 1$$
$$= 0.5\log_2 0.5 + 0.5\log_2 0.5 + 1$$
$$= 0[\because \log_2 0.5 = 1]$$

It is the case of channel with independent input and output, hence C = 0

8. $$C = \lim_{w\to\infty} \omega\log_2\left[1+\frac{P}{\sigma^2\omega}\right]$$

$$= \lim_{w\to\infty}\frac{\omega\ln\left[1+\frac{P}{\sigma^2\omega}\right]}{\ln 2}$$

$$= \frac{1}{\ln 2}\lim_{w\to\infty}\frac{\ln\left[1+\frac{P}{\sigma^2\omega}\right]}{\frac{P}{\sigma^2\omega}}\cdot\frac{P}{\sigma^2}$$

$$= \frac{P}{\sigma^2 \ln_2}\lim_{w\to\infty}\underbrace{\frac{\ln\left[1+\frac{P}{\sigma^2\omega}\right]}{\frac{P}{\sigma^2\omega}}}_{\text{This limit is equivalent to}}$$

$$\lim_{w\to\infty}\frac{\ln[1+x]}{x} = 1$$

$$= \frac{P}{\sigma^2\cdot\ln_2}$$

$$= \ln_2 e\frac{P}{\sigma^2} = 1.44 \text{ KGpa}$$

9. The minimum required average codeword length in bits for error free reconstruction

$$L_{min} = H\,(\text{Entropy})$$

$$H = \frac{1}{2}\log_2 2+\frac{1}{4}\log_2 4 +\frac{1}{8}\log_2 8+\frac{1}{8}\log_2 8$$

$$= \frac{1}{2}\log_2 2+\frac{1}{4}\log_2 2^2+\frac{1}{8}\log_2 2^3 +\frac{1}{8}\log_2 2^3$$

$$= \frac{1}{2}\log_2 2+\frac{1}{4}\times 2\log_2 2+\frac{1}{8}\times 3\log_2 2 +\frac{1}{8}\times 3\log_2 2$$

$$= \frac{1}{2}+\frac{1}{2}+\frac{3}{8}+\frac{3}{8} = 1.75$$

$$\Rightarrow \quad L_{min} = 1.75 \text{ bits/word}$$

Hence the minimum required average codeword length is 1.75 bits/word.

10. Given, $f_m = 33$ Hz, $f_s = 46$ Hz

The frequency in sampled singal = ± 33, 13, 79, 59, 125. .... The above frequencies are passed to a LPF of cut-off frequency 23 Hz. The output frequency = 13 Hz.

11. From given data,

Channel transmission rate (C) = 52 kbps

Channel band width B = 4 kHz

$$\frac{\eta}{2} = 2.5\times 10^{-5}$$

$$N = 4\times 10^3\times 2.5\times 10^{-5}\times 2$$

$$C = B\log_2\left[1+\frac{S}{N}\right]$$

$$S = 1638.2$$

$$E_b = \frac{S}{R_b} = \frac{\text{J/sec}}{\text{bits/sec}} = 31.503$$

$$\frac{C}{B} = \log_2(1+S/N)$$

$$\Rightarrow \log_2(1+S/N) = \frac{C}{B}$$

$$\Rightarrow (1+S/N) = 2^{C/B} = 2^{52/4} = 2^{13} = 8192$$

$$\Rightarrow S/N = 8191$$

$$\therefore\ S = 8191\times N$$

$$\Rightarrow S = 8191\times 4\times 10^3\times 2.5\times 10^{-5}\times 2$$

$$= 819.1\times 2$$

$$E_b = \frac{819.1\times 2}{R_b} = 31.503$$

So the minimum bit energy (E_b) is 31.503 mJ/bit.

12. The channel capacity of a memoryless binary symmetric channel can be expressed as

$$C = 1+p\log_2^{p}+(1-p)\log_2(1-p)$$

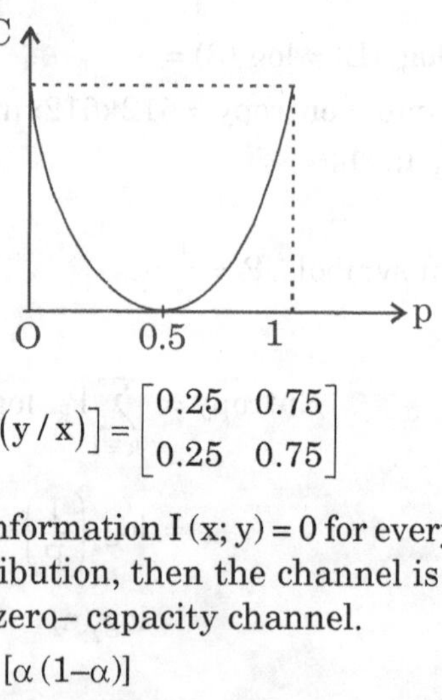

13. Given : $[P(y/x)] = \begin{bmatrix}0.25 & 0.75\\ 0.25 & 0.75\end{bmatrix}$

If mutual information I (x; y) = 0 for every possible input distribution, then the channel is called as useless or zero– capacity channel.

Let $[P(x)] = [\alpha\ (1-\alpha)]$

Then, $H(x) = -\alpha\log_2\alpha - (1-\alpha)\log_2(1-\alpha)$ bits/symbol

$$[P(y)] = [P(y)]\left[P\left(\frac{y}{x}\right)\right] = [0.25\ \ 0.75]$$

$$[P(x,y)] = \begin{bmatrix}\alpha/4 & 3\alpha/4\\ (1-\alpha)/4 & 3(1-\alpha)/4\end{bmatrix}$$

$$H\left(\frac{x}{y}\right) = -\sum_i\sum_j P(x_i,y_i)\log_2 P\left(\frac{x_i}{y_i}\right) \text{ bits/symbol}$$

$= -\alpha \log_2 \alpha - (1-\alpha)\log_2(1-\alpha)$

$$I(x;y) = H(x) - H\left(\frac{x}{y}\right) = 0.$$

So, the given binary memoryless channel is a "useless" channel.

14. For, n = 5 and d_{min} = 2 (given)

For d_{min} = 2, the codewords can be formed as follows:

0	0	0	0	1
0	0	0	1	0
0	0	1	0	0
0	0	1	1	1
0	1	0	0	0
0	1	0	1	1
0	1	1	0	1
0	1	1	1	0
1	0	0	0	0
1	0	0	1	1
1	0	1	0	1
1	0	1	1	0
1	1	0	0	1
1	1	0	1	0
1	1	1	0	0
1	1	1	1	1

⇒ Total 16 codewords are possible

15. Message (1) ⇒ 0001

Message (2) ⇒ 0011

0000111 ⇒ Codeword (1)

1100110 ⇒ Codeword (2)

Since it is a linear hamming code.

Message (1) + Message (2) results in codeword (1) + codeword (2)

⇒ Addition of binary is logical XOR

∴ Message Codeword

Message	Codeword
0001	0000111
⊕ 0011	⊕ 1100110
0010	1100001
↗ Message (3)	↗ Codeword (3)

16. Entropy, $H(x) = -\sum_{i=1}^{m} P(x_i)\log_2 P(x_i)$

For three independent sources,

$$H(X) = -\sum_{i=1}^{3} P(x_i)\log_2 P(x_i)$$

$$= -\left[P(x_1)\log_2 P(x_1) + P(x_2)\log_2 P(x_2) + P(x_3)\cdot\log_2 P(x_3)\right]$$

$$= -\left[0.25\log_2 0.25 + 0.25\log_2 0.25 + 0.50\log_2 0.50\right]$$

$$= \frac{3}{2} \text{ bits/symbol}$$

Symbol rate, r = 3000 symbols/sec

Average bit rate $R = rH(X)$

$$= \frac{3}{2} \times 3000 = 4500 \text{ bits/sec}$$

17. Probability of error

$$P_e = P^3 + 3P^2(1-P)$$

$$P = 0.1$$

$$P_e = (0.1)^3 + 3 \times (0.1)^2(1-0.1)$$

$$= 0.001 + 3 \times 0.01 \times 0.9$$

$$= 0.001 + 0.027$$

$$= 0.028$$

18. The minimum required average codeword length in bits for error free reconstruction

$$L_{min} = H \text{ (Entropy)}$$

$$H = \frac{1}{2}\log_2 2 + \frac{1}{4}\log_2 4 + \frac{1}{8}\log_2 8 + \frac{1}{8}\log_2 8$$

$$= \frac{1}{2}\log_2 2 + \frac{1}{4}\log_2 2^2 + \frac{1}{8}\log_2 2^3 + \frac{1}{8}\log_2 2^3$$

$$= \frac{1}{2}\log_2 2 + \frac{1}{4}\times 2\log_2 2 + \frac{1}{8}\times 3\log_2 2 + \frac{1}{8}\times 3\log_2 2$$

$$= \frac{1}{2} + \frac{1}{2} + \frac{3}{8} + \frac{3}{8} = 1.75$$

$$\Rightarrow \quad L_{min} = 1.75 \text{ bits/word}$$

Hence the minimum required average codeword length is 1.75 bits/word.

19.

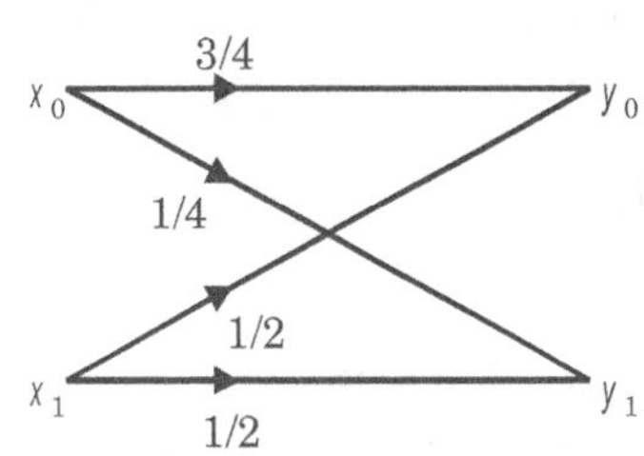

$$P(x_0) = \frac{1}{2}; \quad P(x_1) = \frac{1}{2}$$

$$P\left(\frac{y_0}{x_0}\right) = \frac{3}{4}; \quad P\left(\frac{y_0}{x_1}\right) = \frac{1}{2}$$

$$[P(x, y)] = [P(x)]_d [P(y \mid x)]$$

$$= \begin{bmatrix} 1/2 & 0 \\ 0 & 1/2 \end{bmatrix} \begin{bmatrix} 3/4 & 1/4 \\ 1/2 & 1/2 \end{bmatrix}$$

$$[P(x, y)] = \begin{bmatrix} 3/8 & 1/8 \\ 1/4 & 1/4 \end{bmatrix}$$

$$P(y \mid x_0) = -\sum_{k=0}^{1} P(x_0, y_k) \log_2 P\left(\frac{y_k}{x_0}\right)$$

$$= -\{P(x_0, y_0)\log_2 P(y_0 \mid x_0) + P(x_0, y_1)\log_2 P(y_1 \mid x_0)\}$$

$$= -\left\{\frac{3}{8}\log_2 \frac{3}{4} + \frac{1}{8}\log_2 \frac{1}{4}\right\}$$

$$= 0.405$$

20. Mutual information of two random variables is a measure of the mutual dependence of the two variables.

Given that, X and Y are independent.
Hence, I(X : Y) = 0

21. Quadrature carrier multiplexing utilizes carrier phase shifting and synchronous detection to permit two DSB signals to occupy the same frequency band. is the scheme where same carrier frequency is used for two different DSB signals. It is also known as quadrature amplitude (QAM). So, quadrature multiplexing is quite different from FDM and TDM.

22. Total available band width = 5 MHz

⇒ Frequency reuse factor $\frac{1}{5}$, so five cell repeat pattern so available band width for each cell

$$(B.W)_{cell} = \frac{(BW)_{total}}{5}$$

$$= \frac{5}{5} \text{MHz}$$

$$= 1 \text{ MHz}$$

$$\Rightarrow \quad (B.W)_{channel} = 200 \text{ kHz}$$

Total number of channel in each cell,

$$(N)_{cell} = \frac{(BW)_{cell}}{(B.W)_{channel}}$$

$$= \frac{1M}{200k} = 5$$

There is 8 channel coexist in same channel band width using TDMA.

So, total number of simultaneous channel that conexist = 5 × 8 = 40

23. In Direct Sequence CDMA system,

$$\text{Process Gain} = G_P = \frac{f_{\text{chip rate}}}{f_{\text{data rate}}}$$

Given, G_p min = 100

$G_p \geq 100$

$$\therefore \quad G_p = \frac{f_{\text{chip rate}}}{f_{\text{data rate}}} \geq 100$$

$$\Rightarrow \quad \frac{f_{\text{chip rate}}}{100} \geq f_{\text{data rate}}$$

$$\Rightarrow \quad \frac{1.2288 \times 10^6}{100} \geq f_{\text{data rate}}$$

So, $f_{\text{data rate}}$ must be less than = 12.288 × 10^3 bit/sec

25. It is given that GSM requires 200 kHz for 8 users and uses TDMA scheme to accomodate them. Thus for the next users we will require an extra of 200 kHz bandwidth. Thus, 400 kHz bandwidth is to be used.

26. $$\text{Spreading factor (SF)} = \frac{\text{chip rate}}{\text{symbol rate}}$$

This if a single symbol is represented by a code of 8 chips

Chip rate = 80 × symbol rate

$$\text{S.F. (Spreading Factor)} = \frac{8 \times \text{symbol rate}}{\text{symbol rate}} = 8$$

Spread factor (or) process gain and determine to a certain extent the upper limit of the total number of uses supported simultaneously by a station.

Unit - X

Network Theory

Syllabus

Network Solution Methods: nodal and mesh analysis; Network theorems: superposition, Thevenin and Norton's, maximum power transfer; Wye Delta transformation; Steady state sinusoidal analysis using phasors; Time domain analysis of simple linear circuits; Solution of network equations using Laplace transform; Frequency domain analysis of RLC circuits; Linear port network parameters: driving point and transfer functions; State equations for networks.

Contents

1
CHAPTER

Basics of Circuit Theory

Analysis of Previous GATE Papers

Topics ↓		Year →	2019	2018	2017 Set 1	2017 Set 2	2016 Set 1	2016 Set 2	2016 Set 3	2015 Set 1	2015 Set 2	2015 Set 3
TYPES OF CIRCUITS AND SOURCES	1 Mark	MCQ Type										
		Numerical Type				1				1		
	2 Marks	MCQ Type										
		Numerical Type										
	Total					1				1		
CIRCUIT ELEMENTS (RESISTOR, INDUCTOR, CAPACITOR)	1 Mark	MCQ Type										
		Numerical Type										
	2 Marks	MCQ Type										
		Numerical Type										
	Total											
KIRCHOFF'S LAW	1 Mark	MCQ Type										
		Numerical Type										
	2 Marks	MCQ Type		1								
		Numerical Type		1				1				
	Total			4				2				
WYE-DELTA TRANSFORMATION	1 Mark	MCQ Type										
		Numerical Type										
	2 Marks	MCQ Type						1				
		Numerical Type							1			
	Total							2	2			
NODAL AND MESH ANALYSIS	1 Mark	MCQ Type										
		Numerical Type										
	2 Marks	MCQ Type										
		Numerical Type										
	Total											

TYPES OF CIRCUITS AND SOURCES

1. The voltage V in figure is always equal to

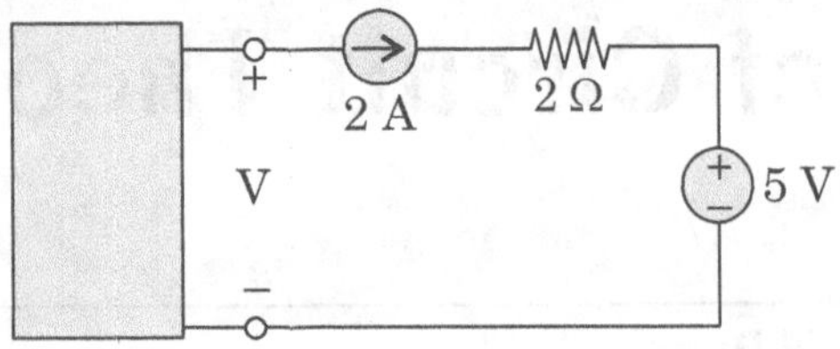

(a) 9 V (b) 5 V

(c) 1 V (d) None of these

[1997 : 1 Mark]

2. The voltage e_0 in the figure is

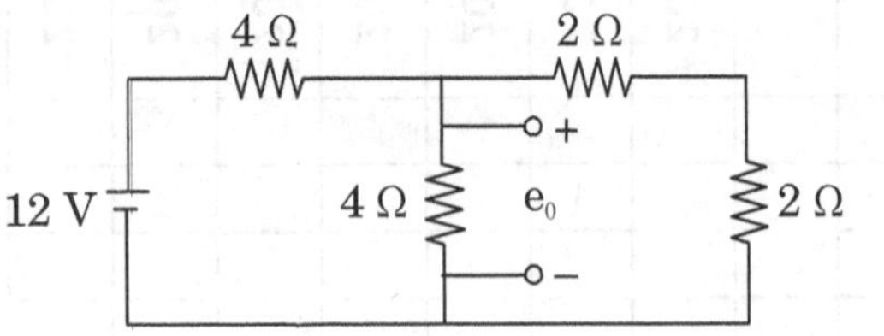

(a) 2 V (b) $\frac{4}{3}$ V

(c) 4 V (d) 8 V

[2001 : 2 Marks]

3. A fully charged mobile phone with a 12 V battery is good for a 10 minute talk-time. Assume that, during the talk-time the battery delivers a constant current of 2 A and its voltage drops linearly from 12 V to 10 V as shown in the figure. How much energy does the battery deliver during this talk-time?

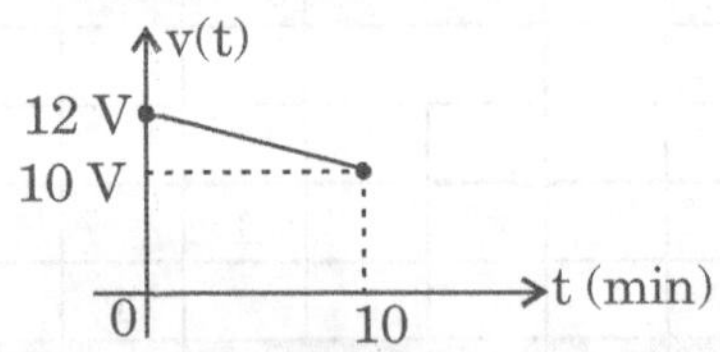

(a) 220 J (b) 12 kJ

(c) 13.2 kJ (d) 14.4 J

[2009 : 1 Mark]

4. The circuit shown in the figure represents a

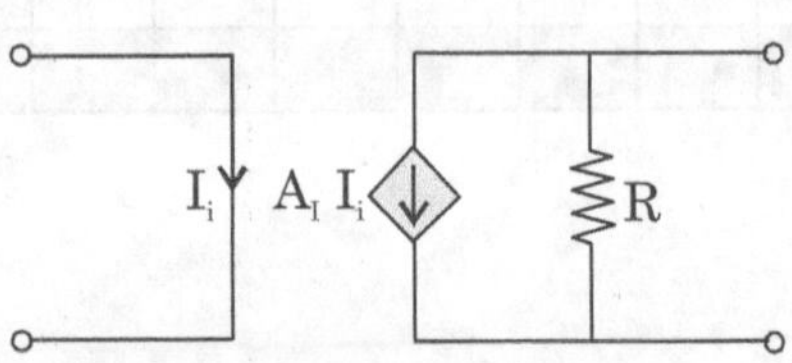

(a) voltage controlled voltage source

(b) voltage controlled current source

(c) current controlled current source

(d) current controlled voltage source

[2014 : 1 Mark, Set-4]

CIRCUIT ELEMENTS (RESISTOR, INDUCTOR, CAPACITOR)

5. A square waveform as shown in figure is applied across 1 mH ideal inductor. The current through the inductor is a____ wave of____ peak amplitude.

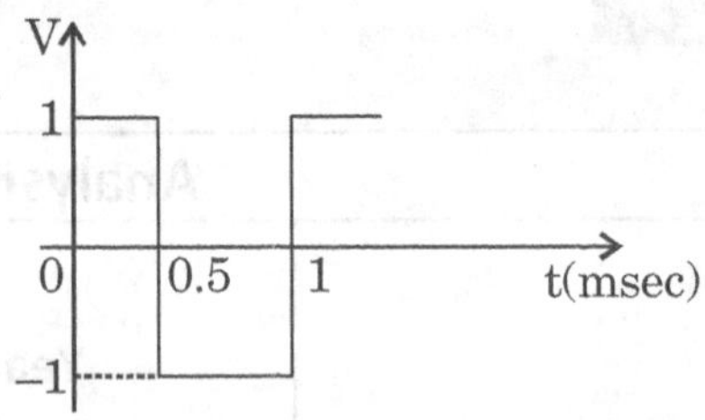

[1987 : 2 Marks]

6. Three capacitors C_1, C_2 and C_3 whose values are 10 μF, 5 μF, and 2 μF respectively, have breakdown voltages of 10 V, 5 V and 2 V respectively. For the interconnection shown below, the maximum safe voltage in volts that can be applied across the combination, and the corresponding total charge in μC stored in the effective capacitance across the terminals are, respectively

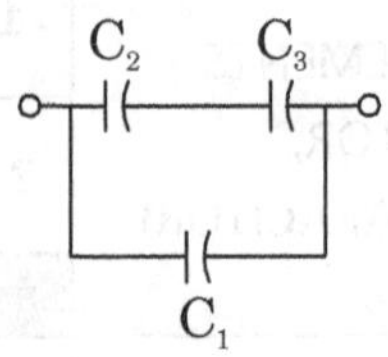

(a) 2.8 and 36 (b) 7 and 119

(c) 2.8 and 32 (d) 7 and 80

[1992 : 2 Marks]

7. The equivalent resistance in the infinite ladder network shown in the figure, is R_e.

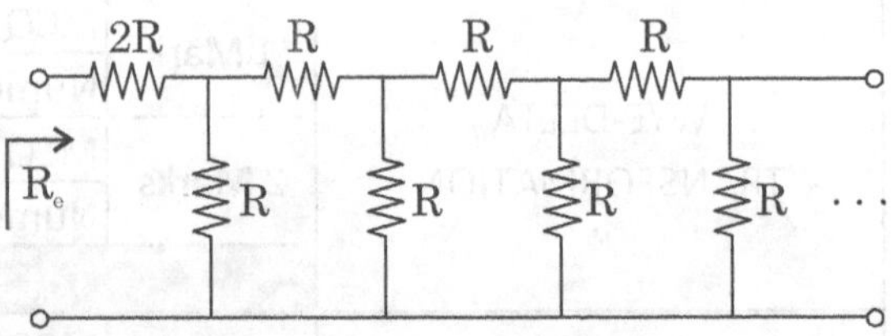

The value R_e/R is ______

[1993 : 2 Marks]

8. In the network shown in the figure, all resistors are identical with R = 300 Ω. The resistance R_{ab} (in Ω) of the network is _____.

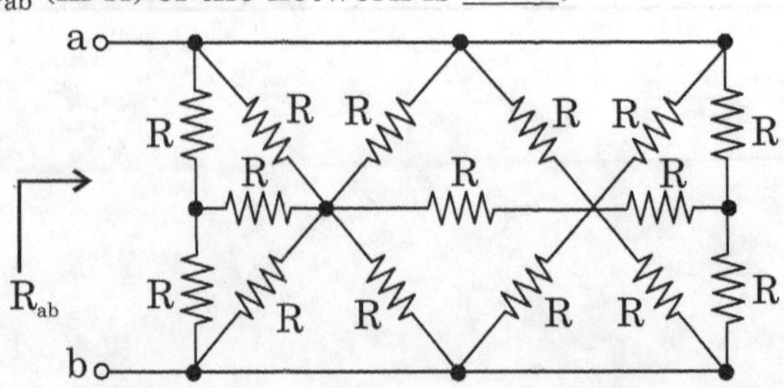

R = 300 Ω

[1994 : 1 Mark]

9. A connection is made consisting of resistance A in series with a parallel combination of resistances B and C. Three resistors of value 10 Ω, 5 Ω, 2 Ω are provided. Consider all possible permutations of the given resistors into the positions A, B, C and identify the configurations with maximum possible overall resistance, and also the ones with minimum possible overall resistance. The ratio of maximum to minimum values of the resistances (up to second decimal place) is ______.

[1995 : 1 Mark]

10. A DC voltage source is connected across a series R-L-C circuit. Under steady-state conditions, the applied DC voltage drops entirely across the

(a) R only (b) L only

(c) C only (d) R and L combination

[1995 : 1 Mark]

11. Consider a DC voltage source connected to a series R-C circuit. When the steady-state reaches, the ratio of the energy stored in the capacitor to the total energy supplied by the voltage source, is equal to

(a) 0.362 (b) 0.500

(c) 0.632 (d) 1.000

[2015 : 1 Mark, Set-1]

KIRCHOFF'S LAW

12. A network contains linear resistors and ideal voltage sources. If values of all the resistors are doubled, then the voltage across each resistor is

(a) halved.

(b) doubled.

(c) increased by four times.

(d) not changed.

[1993 : 2 Marks]

13. The two electrical subnetworks N_1 and N_2 are connected through three resistors as shown in fig. The voltages across 5 ohm resistor and 1 ohm resistor are given to be 10 V and 5 V, respectively. Then voltage across 15 ohm resistor is

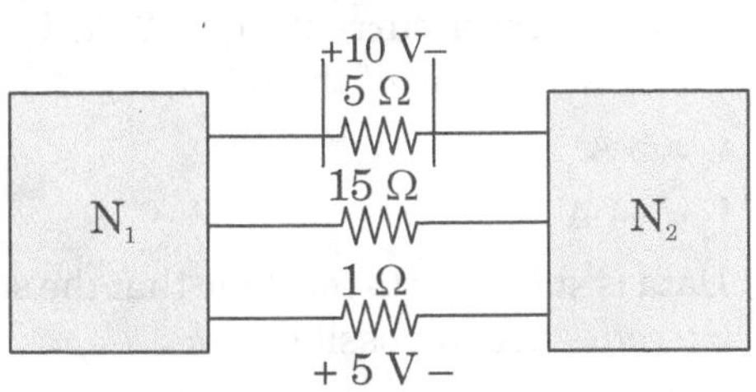

(a) −105 V (b) + 105 V

(c) −15 V (d) + 15 V

[1993 : 2 Marks]

14. The current i_4 in the circuit of figure is equal to

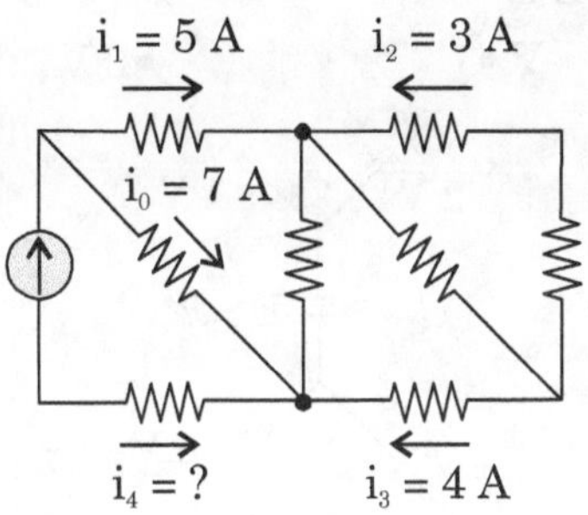

(a) 12 A (b) 4 A

(c) 4 A (d) None of these

[1997 : 3 Marks]

15. The voltage V in figure is equal to

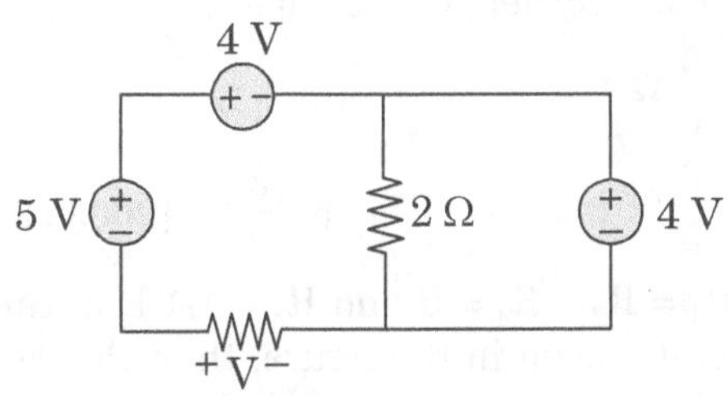

(a) 3 V (b) −3 V

(c) 5 V (d) None of these

[1997 : 1 Marks]

16. The voltage V in figure is

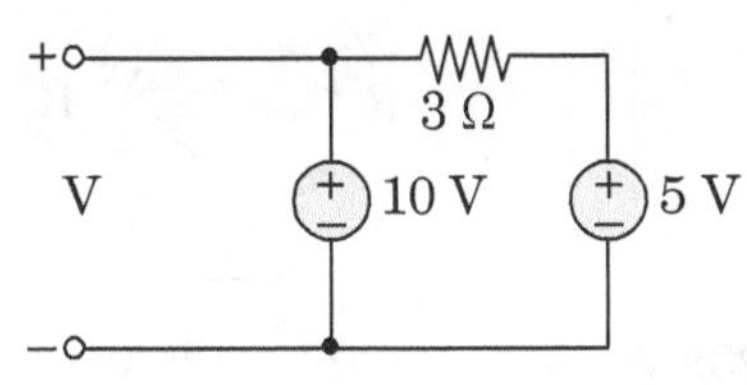

(a) 10 V (b) 15 V

(c) 5 V (d) None of these

[1997 : 1 Mark]

17. In the circuit of the figure, the voltage v(t) is

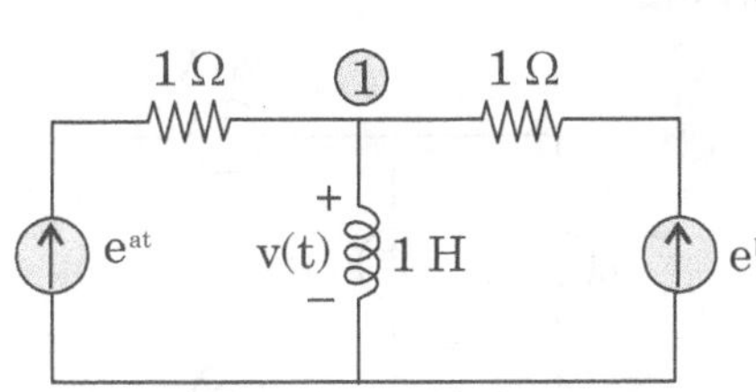

(a) $e^{at} - e^{bt}$

(b) $e^{at} + e^{bt}$

(c) $ae^{at} - be^{bt}$

(d) $ae^{at} + be^{bt}$

18. In the circuit of the figure, the value of the voltage source E is

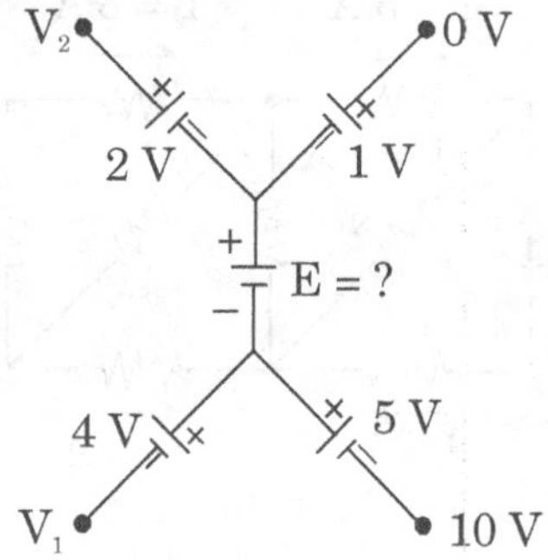

(a) –16 V (b) 4 V
(c) –6 V (d) 16 V **[2000 : 1 Mark]**

19. Twelve 1 Ω resistances are used as edges to form a cube. The resistance between two diagonally opposite corners of the cube is

(a) $\frac{5}{6}\,\Omega$ (b) 1 Ω
(c) $\frac{6}{5}\,\Omega$ (c) $\frac{3}{2}\,\Omega$ **[2003 : 2 Marks]**

20. If $R_1 = R_2 = R_4 = R$ and $R_3 = 1.1\,R$ in the bridge circuit shown in the figure, then the reading in the ideal voltmeter connected between a and b is

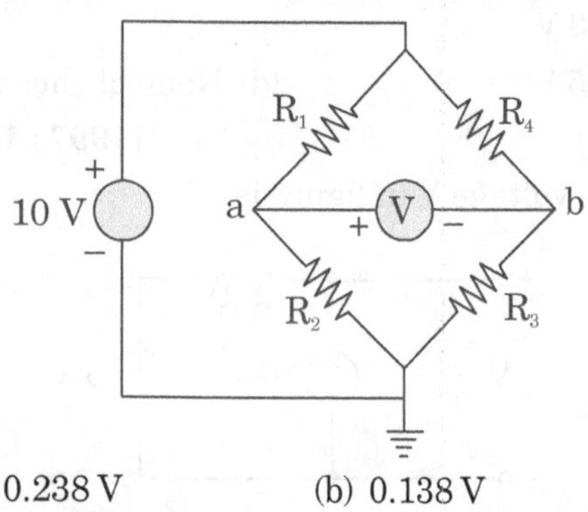

(a) 0.238 V (b) 0.138 V
(c) –0.238 V (d) 1 V

[2005 : 2 Marks]

21. In the interconnection of ideal sources shown in the figure, it is known that the 60 V source is absorbing power.

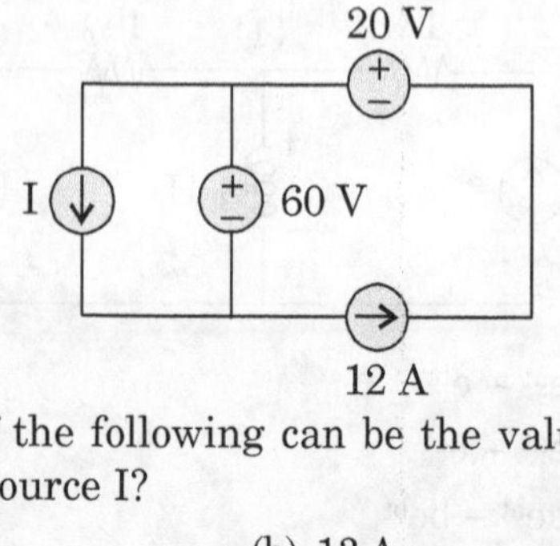

Which of the following can be the value of the current source I?

(a) 10 A (b) 13 A
(c) 15 A (d) 18 A **[2009 : 1 Mark]**

22. It $V_A - V_B = 6$ V, then $V_C - V_D$ is

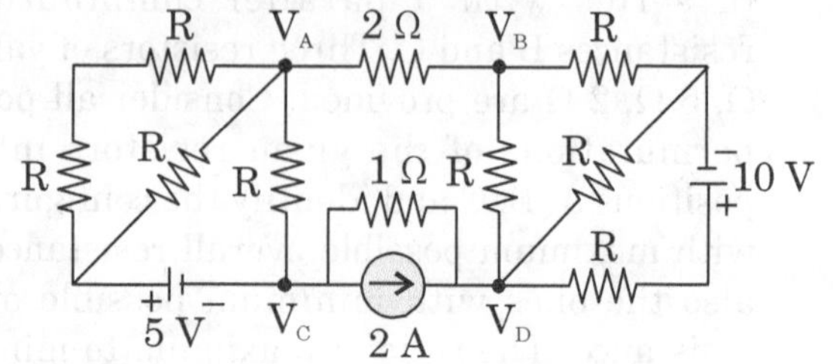

(a) –5 V (b) 2 V
(c) 3 V (d) 6 V

[2012 : 2 Marks]

Common Data For Questions 23 and 24 : Consider the following figure:

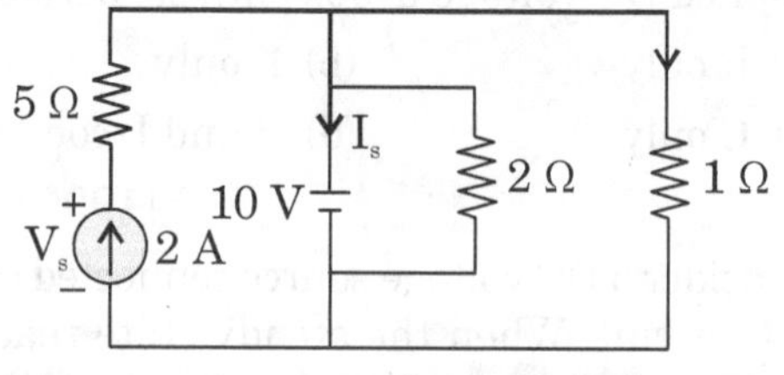

23. The current (I_s) in amps and voltage source, (V_s) in volts across the current source respectively, are

(a) 13, –20 (b) 8, –10
(c) –8, 20 (d) –13, 20

[2013 : 2 Marks]

24. The current in the 1 Ω resistor in Amps is

(a) 2 (b) 3.33
(c) 10 (d) 12

[2013 : 2 Marks]

25. Consider the configuration shown in the figure which is a portion of a larger electrical network

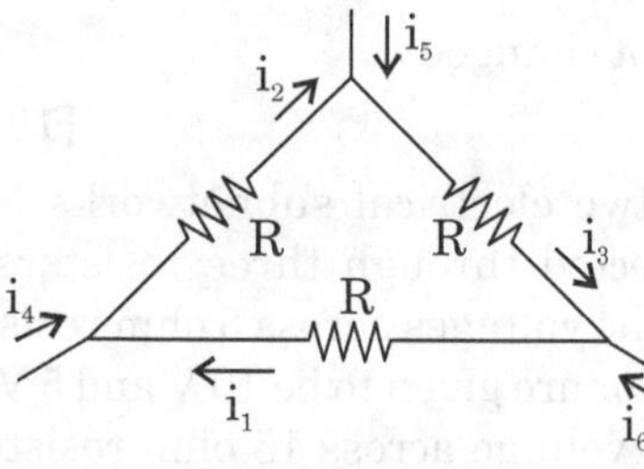

For R = 1 Ω and currents i_1 = 2 A, i_4 = –1 A, i_s = –4 A, which one of the following is TRUE?

(a) i_6 = 5 A
(b) i_3 = –4 A
(c) Data is sufficient to conclude that the supposed currents are impossible
(d) Data is insufficient to identify the currents i_1, i_2 and i_6

[2014 : 1 Mark, Set-1]

26. The magnitude of current (in mA) through the resistor R_2 in the figure shown is ______.

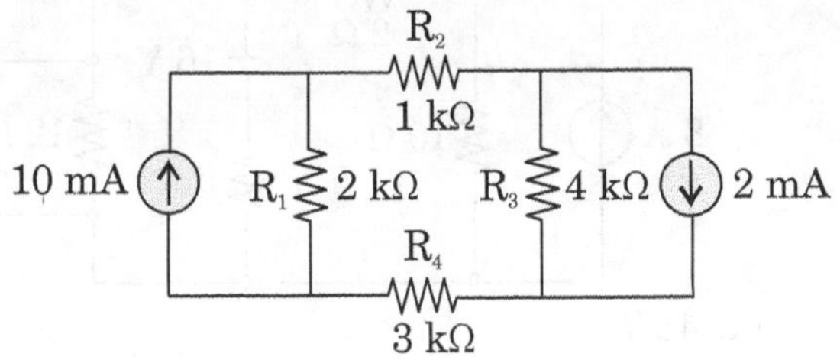

[2014 : 1 Mark, Set-4]

27. In the given circuit, the values of V_1, and V_2 respectively are

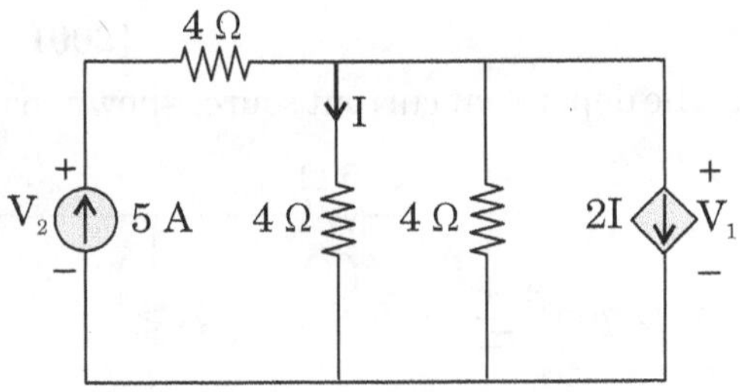

(a) 5 V, 25 V (b) 10 V, 30 V

(c) 15 V, 35 V (d) 0 V, 20 V

28. In the circuit shown in the figure, the magnitude of the current (in amperes) through R_2 is ______.

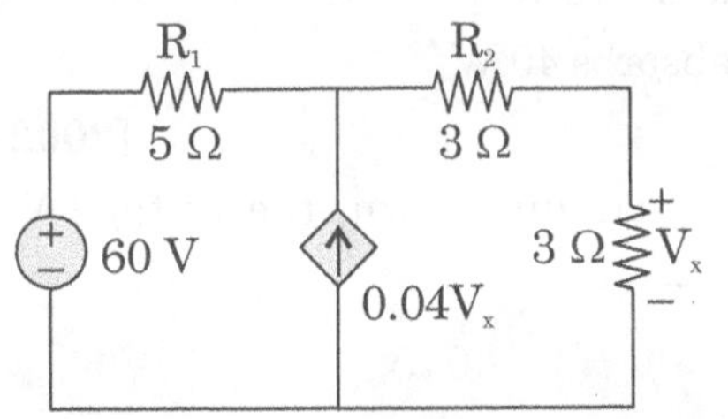

[2016 : 2 Marks, Set-2]

29. Consider the network shown below with $R_1 = 1\ \Omega$, $R_2 = 2\ \Omega$ and $R_3 = 3\ \Omega$. The network is connected to a constant voltage source of 11 V.

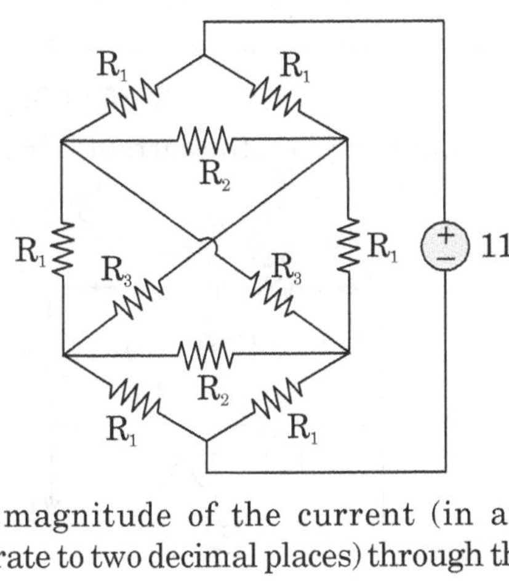

The magnitude of the current (in amperes, accurate to two decimal places) through the source is ______.

[2018 : 2 Marks]

30. The voltage V_{C1}, V_{C2} and V_{C3} across the capacitors in the circuit in fig., under steady state, are respectively.

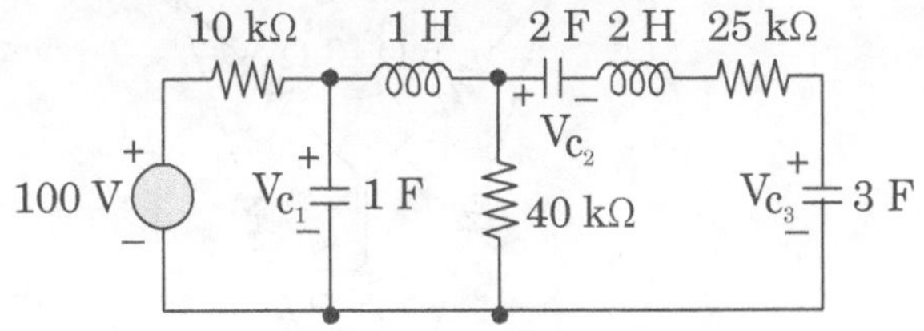

(a) 80 V, 32 V 48 V (b) 80 V, 48 V, 32 V

(c) 20 V, 8 V, 12 V (d) 20 V, 12V, 8 V

[2018 : 2 Marks]

WYE-DELTA TRANSFORMATION

31. A Delta-connected network with its Wye-equivalent is shown in the figure. The resistances R_1, R_2 and R_3 (in ohms) are respectively

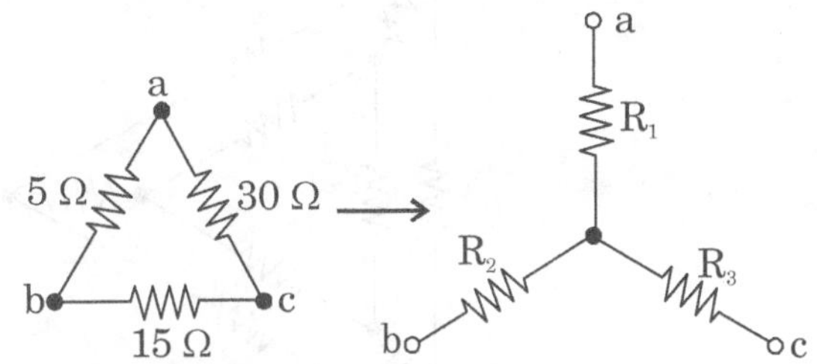

(a) 1.5, 3 and 9 (b) 3, 9 and 1.5

(c) 9, 3 and 1.5 (d) 3, 1.5 and 9

[1999 : 2 Mark]

32. Consider a delta connection of resistors and its equivalent star connection as shown below. If all elements of the delta connection are scaled by a factor k, k > 0, the elements of the corresponding star equivalent will be scaled by a factor of

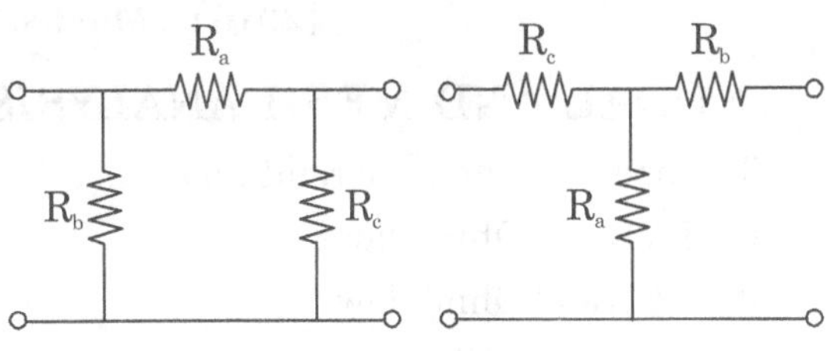

(a) K^2 (b) K

(c) $\frac{1}{K}$ (d) $\sqrt{K}$

[2013 : 1 Mark]

33. A Y-network has resistances of 10 Ω each in two of its arms, while the third arm has a resistance of 11 Ω. In the equivalent Δ-network, the lowest value (in Ω) among the three resistances is______.

[2014 : 2 Marks, Set-1]

34. For the Y-network shown in the figure, the value of R_1 (in Ω) in the equivalent Δ-network is ______.

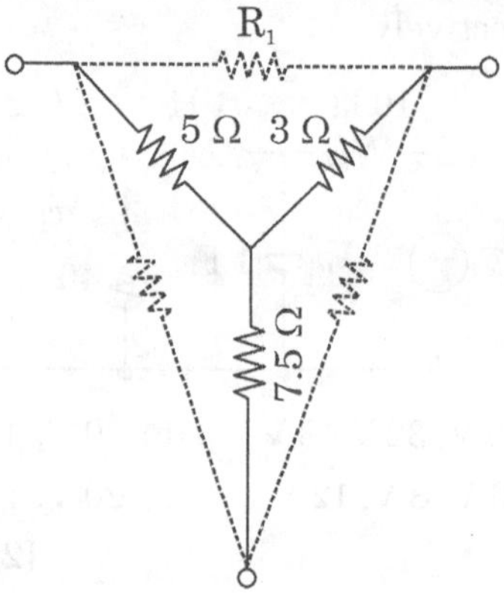

[2014 : 2 Marks, Set-3]

35. In the given circuit, each resistor has a value equal to 1 Ω.

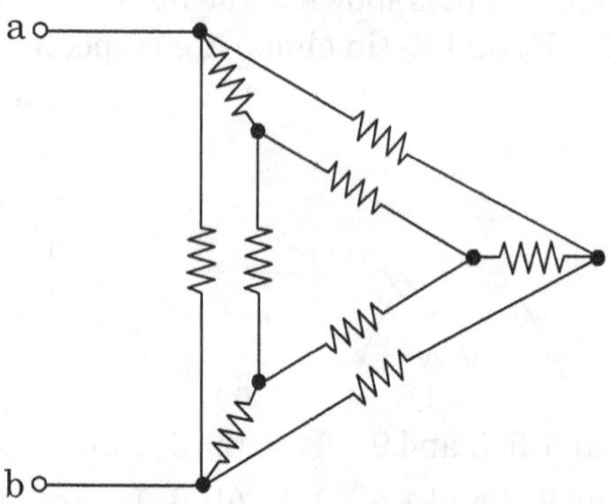

What is the equivalent resistance across the terminals a and b?

(a) $\frac{1}{6}\Omega$ (b) $\frac{1}{3}\Omega$

(c) $\frac{9}{20}\Omega$ (d) $\frac{8}{15}\Omega$

[2016 : 2 Marks, Set-2]

NODAL AND MESH ANALYSIS

36. The nodal method of circuit analysis is based on

(a) KVL and Ohm's law

(b) KCL and Ohm's law

(c) KCL and KVL

(d) KCL, KVL and Ohm's law **[1997 : 1 Mark]**

37. The voltage across the terminals a and b in fig. is

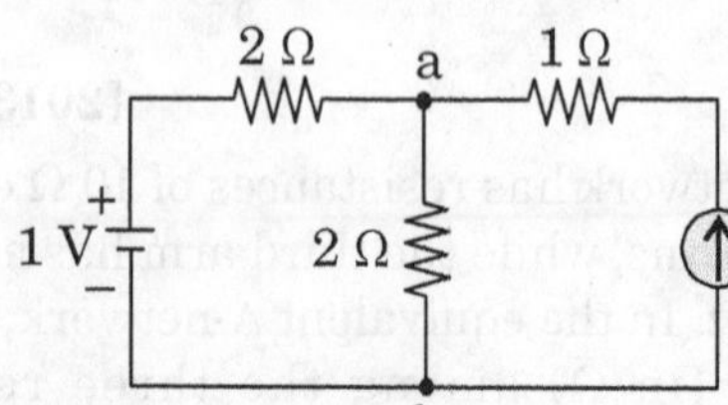

(a) 0.5 V (d) 3.0 V

(a) 3.5 V (c) 4.0 V **[1998 : 1 Mark]**

38. The voltage e_0 in the figure is

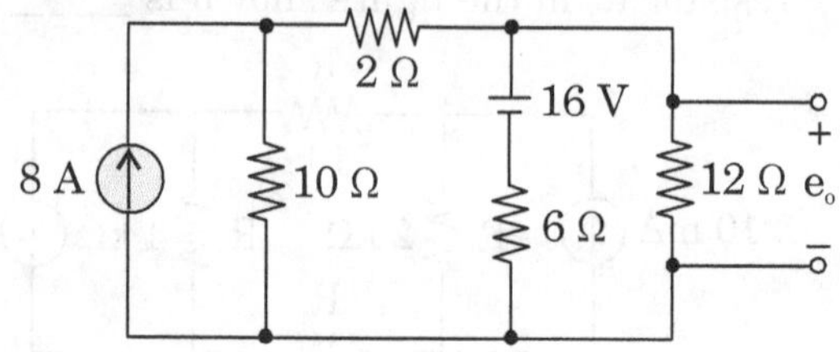

(a) 48 V

(b) 24 V

(c) 36 V

(d) 28 V

[2001 : 2 Mark]

39. The dependent current source shown in the figure

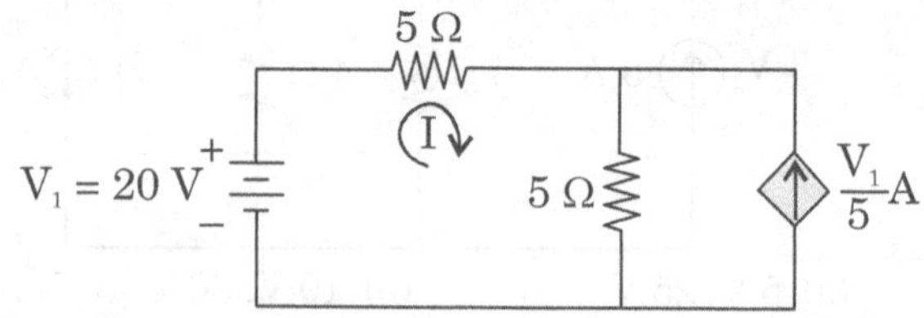

(a) delivers 80 W

(b) absorbs 80 W

(c) delivers 40 W

(d) absorbs 40 W

[2002 : 1 Mark]

40. In the circuit shown, the voltage V_s(in Volts) is _______

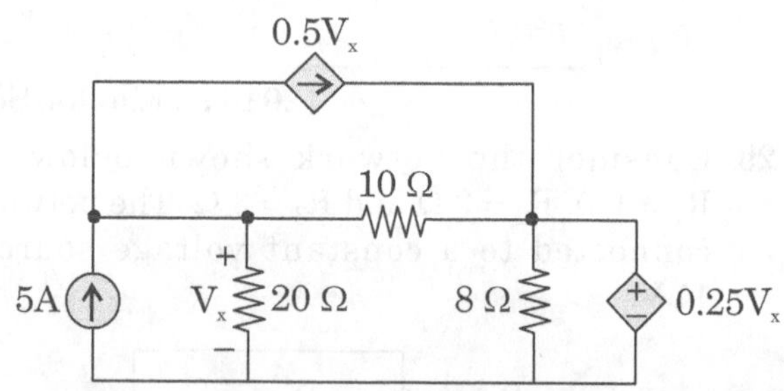

[2015 : 1 Mark, Set-3]

41. In the figure shown, the current i (in Ampere) is _______.

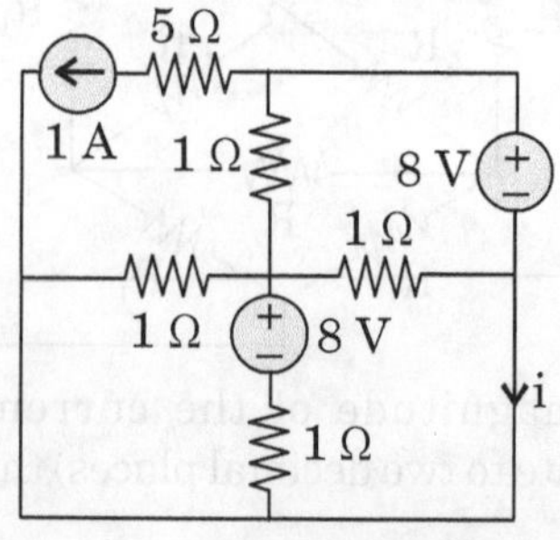

[2016 : 2 Marks, Set-3]

ANSWERS

1. (d)	**2.** (c)	**3.** (c)	**4.** (c)	**5.** (0.5)	**6.** (c)	**7.** (2.618)	**8.** (100)	**9.** (2.143)	**10.** (c)
11. (b)	**12.** (d)	**13.** (a)	**14.** (d)	**15.** (a)	**16.** (a)	**17.** (d)	**18.** (a)	**19.** (a)	**20.** (c)
21. (a)	**22.** (a)	**23.** (d)	**24.** (c)	**25.** (a)	**26.** (2.8)	**27.** (a)	**28.** (5)	**29.** (8)	**30.** (b)
31. (d)	**32.** (b)	**33.** (29.09)	**34.** (10)	**35.** (a)	**36.** (b)	**37.** (c)	**38.** (d)	**39.** (a)	**40.** (8)
41. (1)									

EXPLANATIONS

1. In given circuit the voltage across 2 A current source is not known (it can have any value of voltage). So, it is not possible to find the value of voltage V.

2. 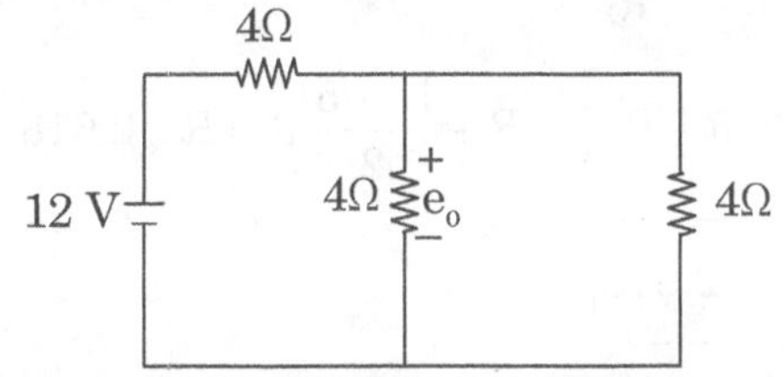

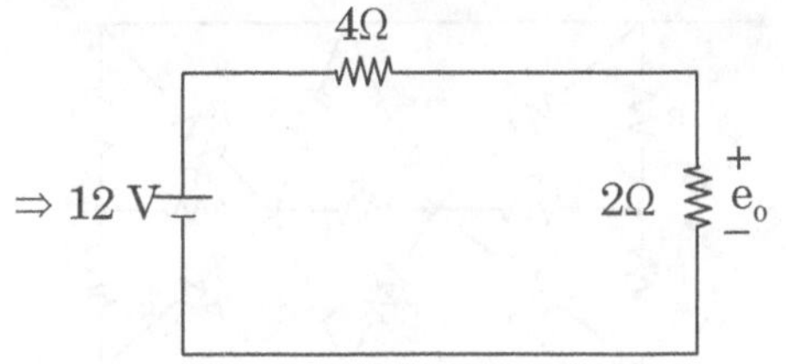

$$e_0 = 12 \times \frac{2}{2+4} = 4 \text{volt.}$$

3. Energy delivered by battery

$$E = P.t$$
$$= V.I.t = V.t \times 2 \text{ A}$$

$\therefore$ Vt = Area under (V–t) curve

$$= \frac{1}{2} \times 2 \times 600 + 10 \times 600 = 6600$$

$\Rightarrow E = 6600 \times 2 = 13.2$ kJ.

4. 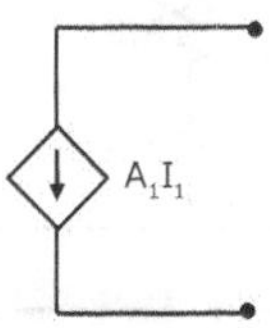

The dependent source represents a current controlled current source

5. Current through an inductor is given by

$$i_L(t) = \frac{1}{L}\int_{-\infty}^{t} v_L(t)\, dt = \frac{1}{L}\int_{-\infty}^{0} v_L(t)\, dt + \frac{1}{L}\int_{0}^{t} v_L(t)\, dt$$

$$i_L(t) = I_L(0) + \frac{1}{L}\int_{0}^{t} v_L(t)\, dt = I_L(0) + \frac{1}{L}\int_{0}^{t} V(t) dt$$

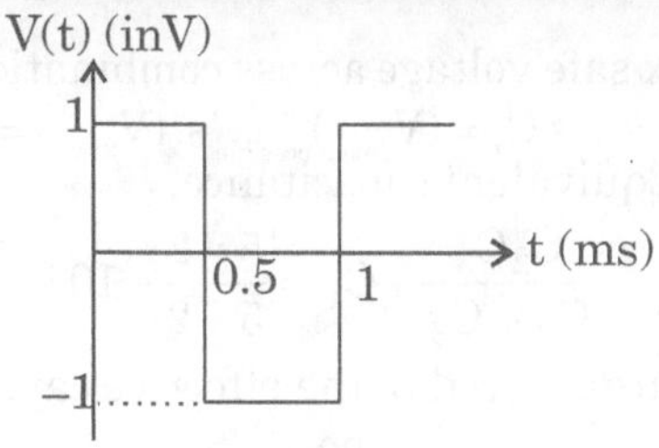

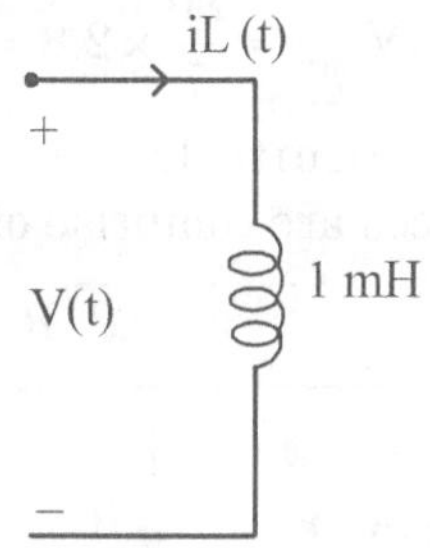

In given waveform of voltage V(t), there is no information of V(t) for t< 0. Now, ansuring $I_L(0) = 0$A.

$\Rightarrow$ V(t) can be written as

$V(+) = u(t) - 2u(t - 0.5) + 2u(t - 1) - ...$

Now, $i_L(t) = r(t) - 2r(t - 0.5) + 2r(t - 1) - ...$

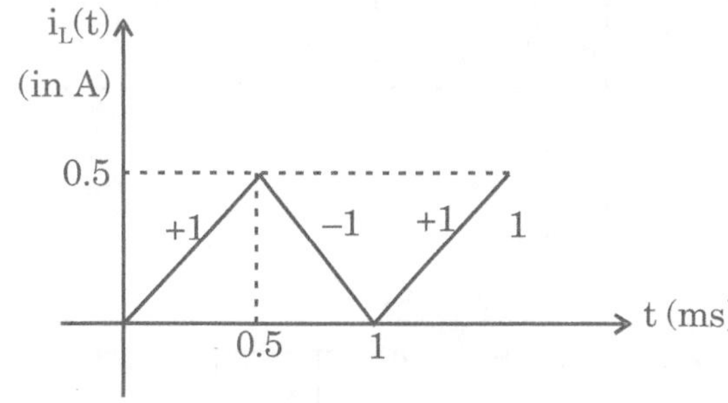

Peak amplitude of $i_L(t) = 0.5$ A

6. Given: $C_1 = 10\ \mu F$, $(V_1)_{max} = 10$ V

$C_2 = 5\ \mu F$ $(V_2)_{max} = 5$ V

$C_3 = 2\ \mu F$, $(V_3)_{max} = 2$ V

$\Rightarrow$ In series connection(s), charge (current) is same, capacitors C_2 and C_3 are in series.

$Q_2 = Q_3$

$$\frac{V_2}{V_3} = \frac{C_3}{C_2}$$

Since, voltage is inversely proportional to capacitance.

We need to check for max safe voltage for $C_2 (C_2 < C_3)$.

$$\left(V_{2\max}\right)_{\text{possible}} = \frac{C_3}{C_2}(V_3)_{\max} = \frac{2\times 2}{5} = 0.8 \text{ V} < 5 \text{ V}$$

Max safe voltage across combination of C_2 and C_3 or across $C_1 = (V_{2\max})_{\text{possible}} + (V_3)_{\max} = 2+0.8 = 2.8$ V.

$\Rightarrow$ Equivalent capacitance,

$$C_{eq} = \frac{C_2 C_3}{C_2 + C_3} + C_1 = \frac{5\times 2}{5+2} + 10 = \frac{80}{7}\ \mu\text{F}$$

Charge stored in the effective capacitance

$$= C_{eq}(V)_{\max} = \frac{80}{7} \times 2.8 = 32\ \mu\text{c}$$

7. For an infinite ladder network, if all the resistances are comprise of same value R, then

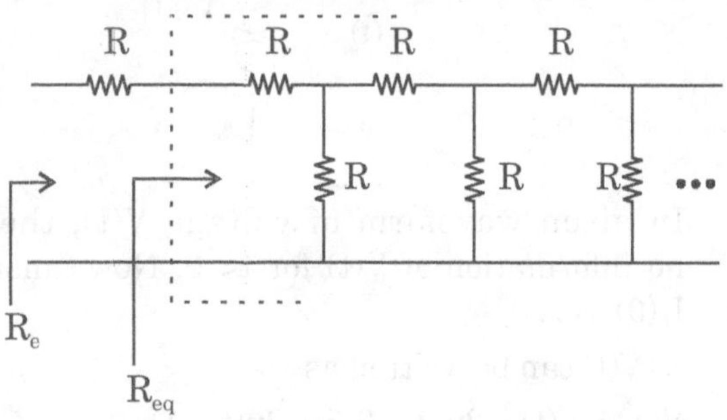

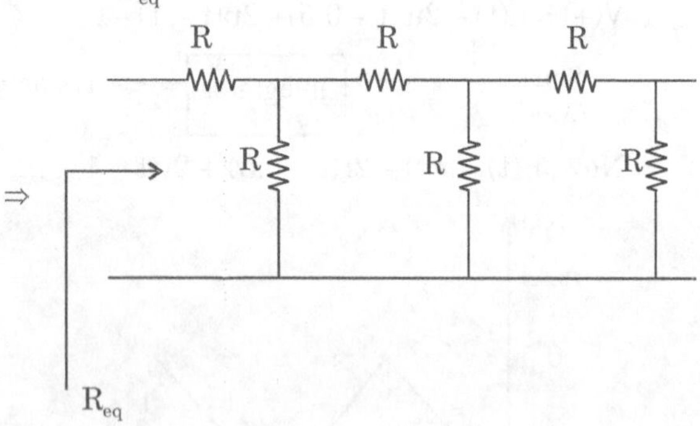

$\Downarrow$

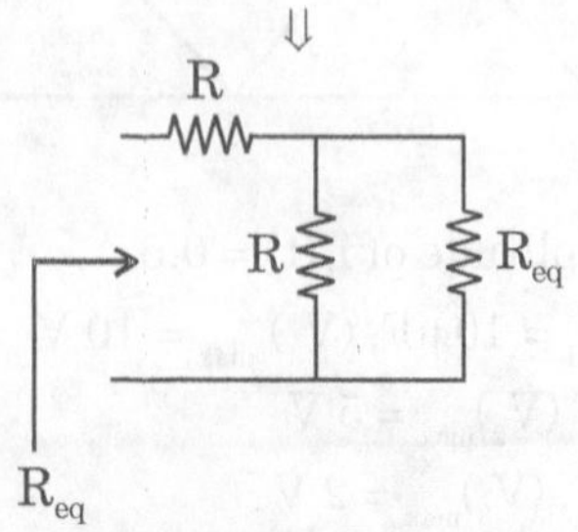

$$\Rightarrow \quad R_{eq} = R + \frac{R.R_{eq}}{R + R_{eq}}$$

$$\Rightarrow \quad R_{eq}^2 + R\frac{R}{eq} = R^2 + \frac{R.R}{eq + R.R_{eq}}$$

$$\Rightarrow \quad R_{eq}^2 - R.R_{eq} - R^2 = 0$$

$$R_{eq} = \frac{R \pm \sqrt{R^2 + 4R^2}}{2}$$

$$R_{eq} = \left(\frac{1+\sqrt{5}}{2}\right) R \text{ (valid value)}$$

Now,

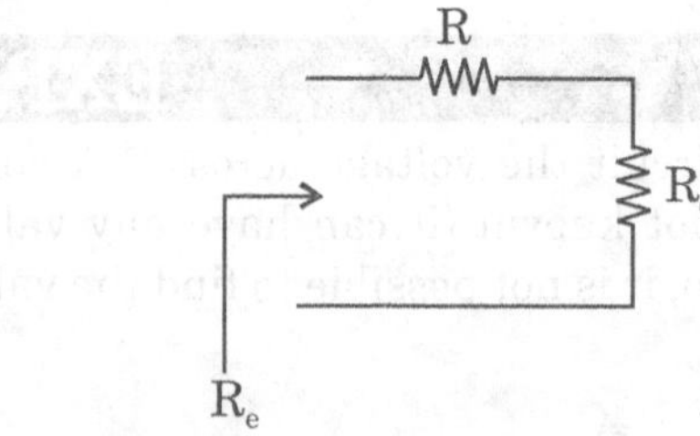

$$R_e = R + R_{eq} = R + \frac{1+\sqrt{5}}{2} R = R \times 2.618$$

$$\boxed{\frac{R_e}{R} = 2.618}$$

8.

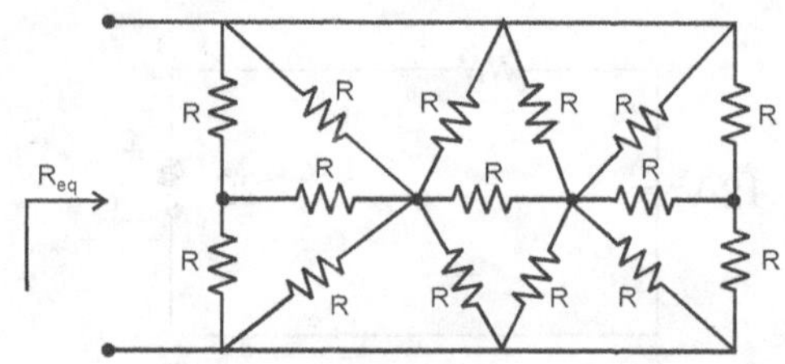

R = 300 Ω

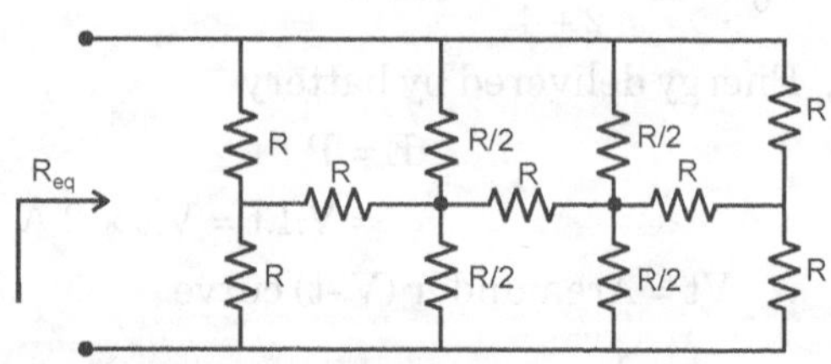

These are balanced Wheatstone bridges.

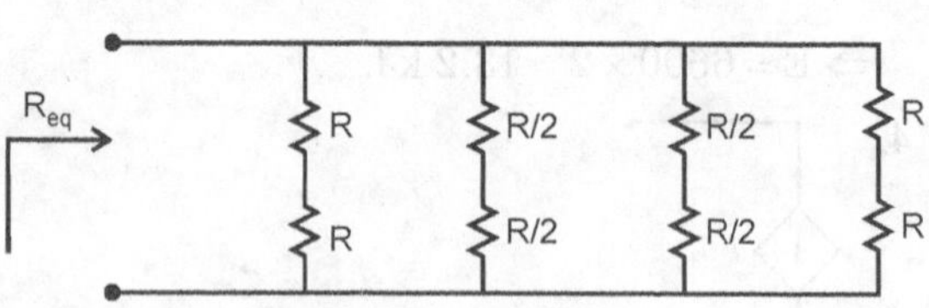

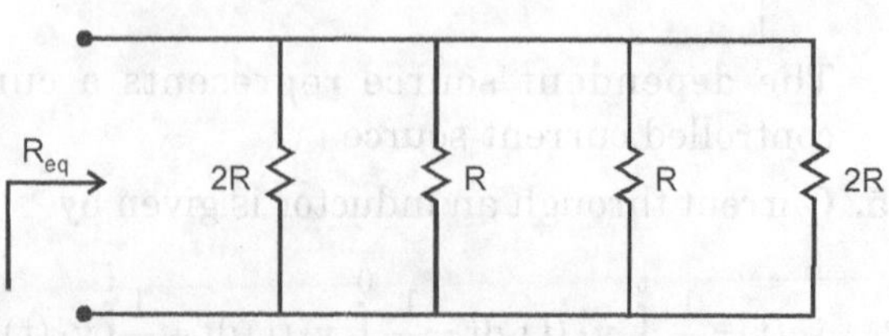

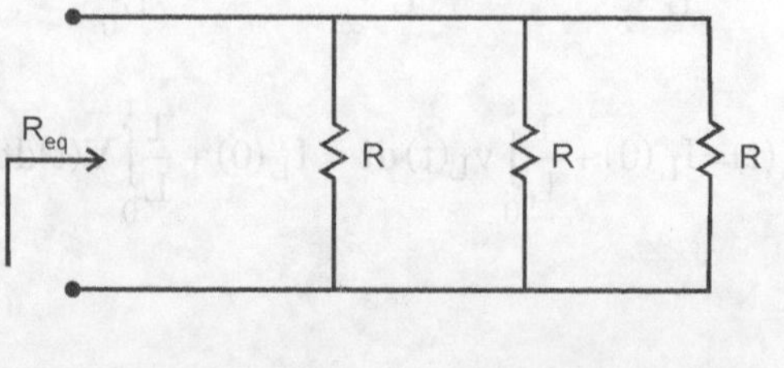

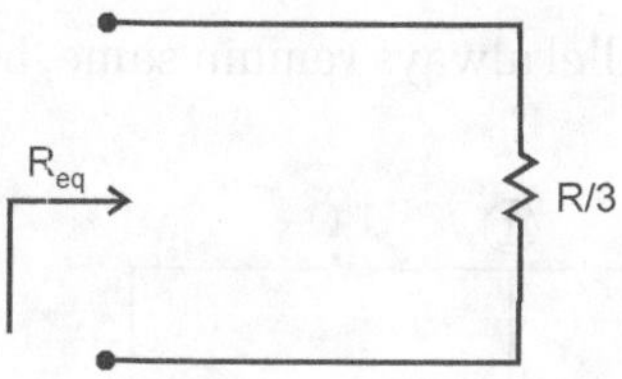

$$R_{eq} = \frac{R}{3} = \frac{300}{3} = 100\ \Omega$$

$$R_{eq} = 100\ \Omega$$

9.

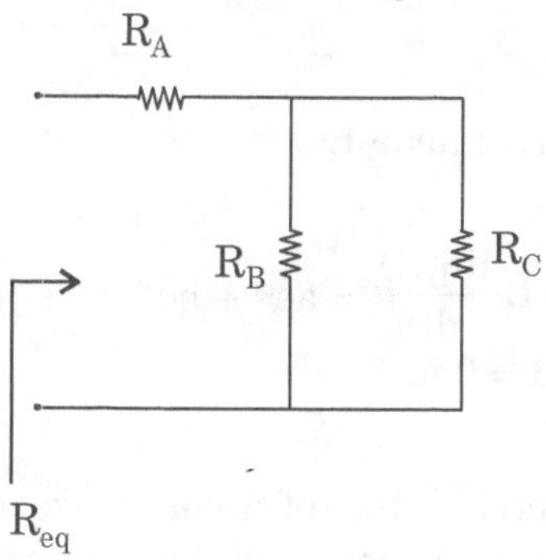

Given that.

R_A or R_B or $R_C \equiv 10\ \Omega$ or $5\ \Omega$ or $2\ \Omega$.

For R_{eq} (max) :

$R_A = 10\ \Omega$, $R_B = 5\ \Omega$ or $2\ \Omega$

$R_C = 2\ \Omega$ or $5\ \Omega$.

$$R_{eq}\,(\text{max}) = 10 + \; = 10 + (2 \,||\, 5) = \frac{80}{7}\ \Omega$$

For R_{eq} (min) :

$R_A = 2\ \Omega$, $R_B = 10\ \Omega$ or $5\ \Omega$

$R_C = 5\ \Omega$ or $10\ \Omega$

$$R_{eq}\,(\text{min}) = 2 + (5 \,||\, 10)$$

$$= \frac{16}{3}\ \Omega.$$

$$\frac{R_{eq}\,(\text{max})}{R_{eq}\,(\text{min})} = \frac{\frac{80}{7}}{\frac{16}{3}} = 2.143.$$

10. ⇒ for a dc voltage source across series RLC circuit, the inductor will act as short-cirarit ($v_L = 0$ V) and capacitor as open-circeuit and no current will to in steady-state.

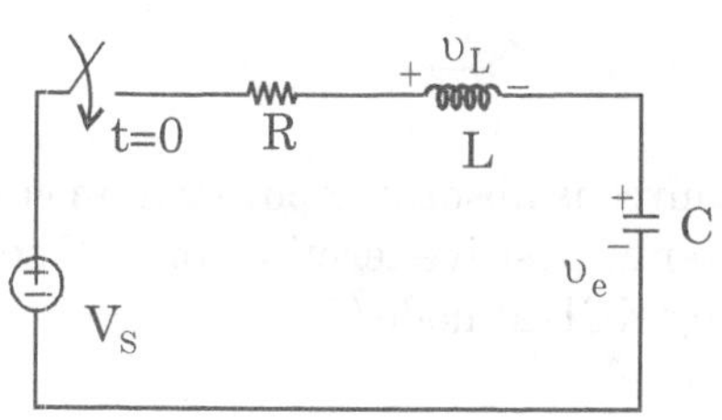

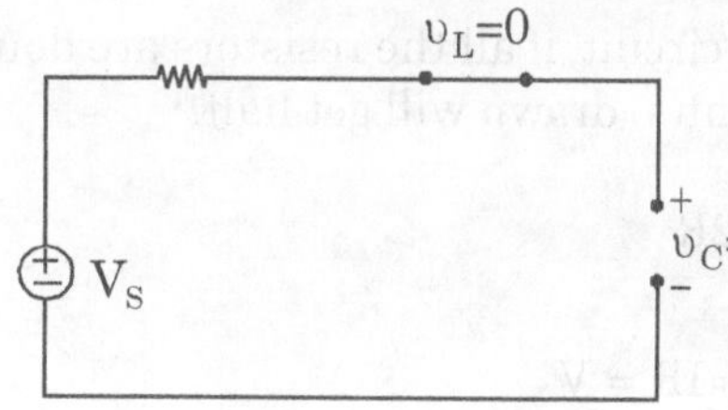

For t >> 0

Hemce, all the source voltage will appear across capacitor only.

11. Assuring the capacitor to be initially uncharged. The across the capacitor,

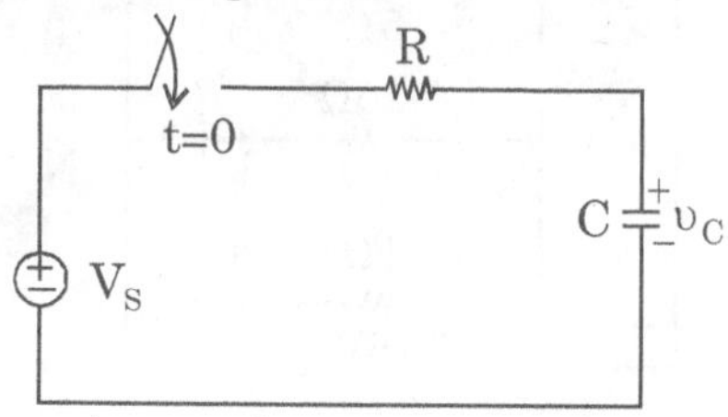

$$v_0(t) = V_s\left(1 - e^{\frac{-t}{RC}}\right) \text{ volts, } t > 0.$$

using KVL for t > 0, $-V_s + V_R + V_C = 0$

$$-V_S + V_R + V_S - V_S e^{\frac{-t}{RC}} = 0$$

$$V_R = V_S : e^{\frac{-t}{RC}}$$

$$\text{Power dissipated in the resistor} = \frac{V_R^2}{R} = \frac{V_S^2 e^{\frac{-2t}{RC}}}{R}$$

Energy dissipated on the resitor

$$= \int_0^\infty \frac{V_R^2}{R} dt = \frac{V_S^2}{R}\int_0^\infty e^{\frac{-2t}{RC}} dt = \frac{V_s^2}{R}.\frac{RC}{2}.1 = \frac{1}{2}CV_S^2$$

Power in the capactor = $v_c(t) \,.\, i_c(t)$

$$= v_c(t).\frac{cdv_c(t)}{dt}$$

$$= V_S\left[1 - e\frac{-t}{R_C}\right].c\frac{d}{dt}\left[V_S - V_S e\frac{-t}{R_C}\right]$$

$$= \frac{V_S^2}{R}\left(e\frac{-t}{R_C} - e\frac{-2t}{R_C}\right)$$

Energy stored in the capaitor

$$= \int_0^\infty \frac{V_S^2}{R}\left(e\frac{-t}{R_C} - e\frac{-2t}{R_C}\right) dt = \frac{V_S^2}{R}\left[R_C - \frac{R_C}{2}\right] = \frac{1}{2}CV_S^2$$

$$\text{Total energy supplied } \frac{1}{2}CV_S^2 + \frac{1}{2}CV_S^2 = CV_S^2$$

$$\text{Energy stored in capaitor} = \frac{1}{2}CV_S^2.$$

$$\Rightarrow \quad \frac{\text{Energy stored in capaitor}}{\text{Total energy supplied}} = 0.5.$$

12. In a linear circuit, if all the resistors are doubled then current(s) drawn will get half.

$$I' = \frac{I}{2} R' = 2R$$

$$V' = \frac{I}{2} 2R = IR = V$$

13. According to KCL (algebraic sum of all the currents associated with a closed boundary is zero;

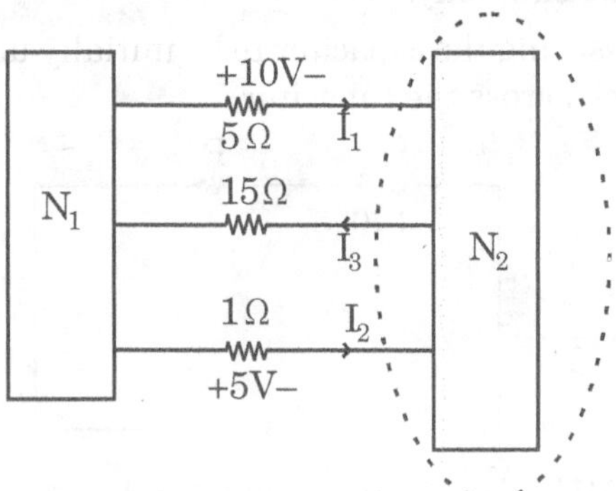

Closed boundary

$I_1 + I_2 - I_3 = 0$

$$\frac{10}{5} + \frac{5}{1} - I_3 = 0$$

$I_3 = 7A$

voltage across 15 Ω resistor

$= -7 \times 15 = -105$ V

14.

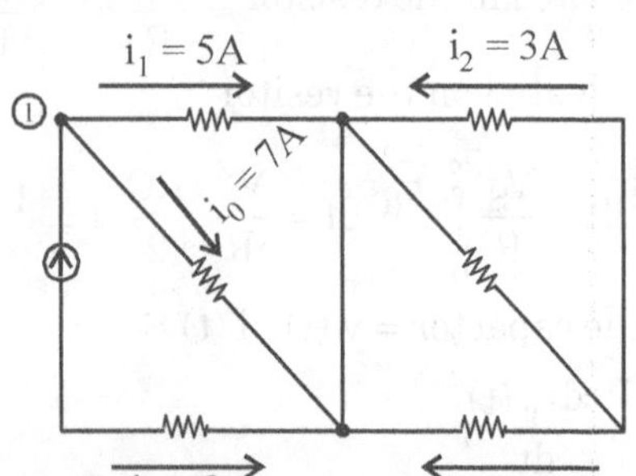

Applying KCL at node (1)

$i_o + i_4 + i_1 = 0$

$7 + i_4 + 5 = 0$

$i_4 = -12$ A.

15.

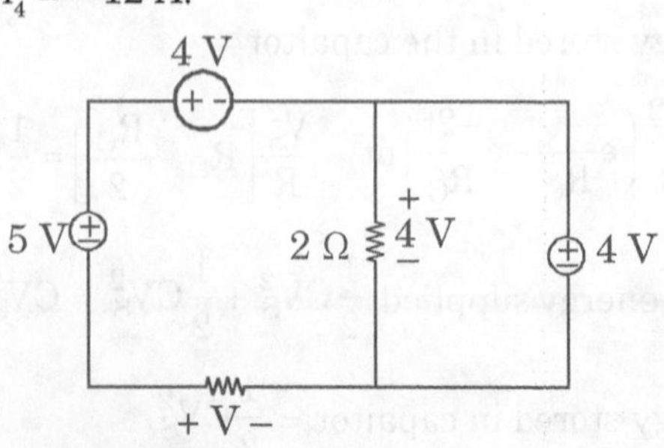

Applying KVL

$-5 + 4 + 4 - V = 0$

V = 3 volt.

16. Voltage in parallel always remain same, hence, V = 10 volt.

17.

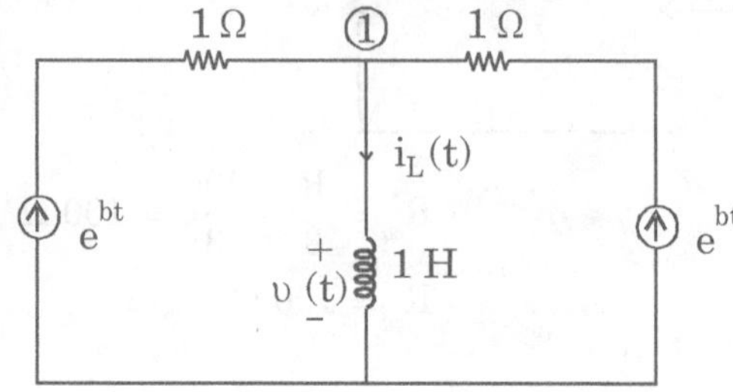

Applying KCL at node 1,

$e^{at} + e^{bt} = i_L(t)$.

$$\text{voltage } \vartheta(t) = L\frac{di_L(t)}{dt} = ae^{at} + be^{bt}$$

18. $10 + 5 + E + 1 = 0$

$\Rightarrow E = -16$ V

19. Due to symmetric nature of circuit between any two diagonals, the current will be distributed symmetrically.

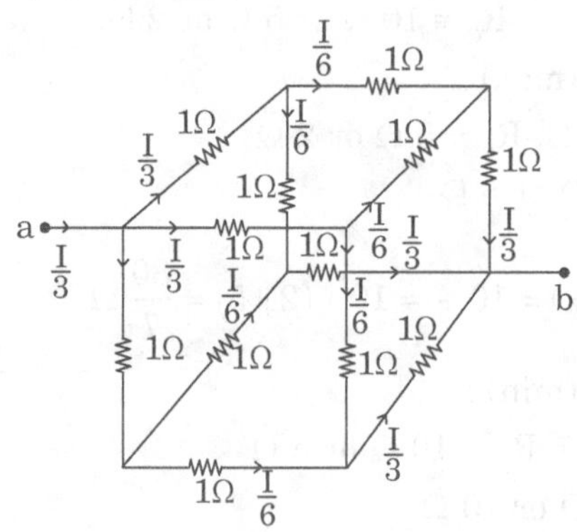

$$-V_{ab} + 1 \times \frac{I}{3} + 1 \times \frac{I}{6} + 1 \times \frac{I}{3} = 0$$

$$\frac{V_{ab}}{I} = R_{ab} = \frac{5}{6}\ \Omega.$$

20. Voltmeter reading, $V_{in} = V_{pb} - V_{ap} = 4.7519 - 5$
$= -0.238$ volts

(replace it in last segment of solution).

21.

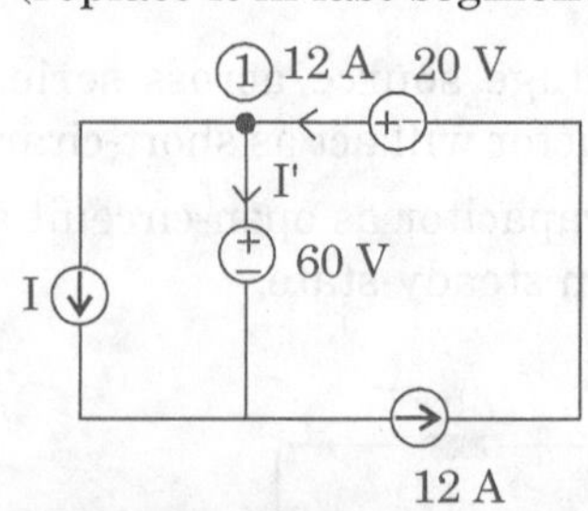

If 60 V source is absorbing power than current I will enter at positive terminal of 60 V source. Now using KCL at node (1),

$I' + I = 12$

$I' = 12 - I$

$\boxed{I < 12A}$

22.

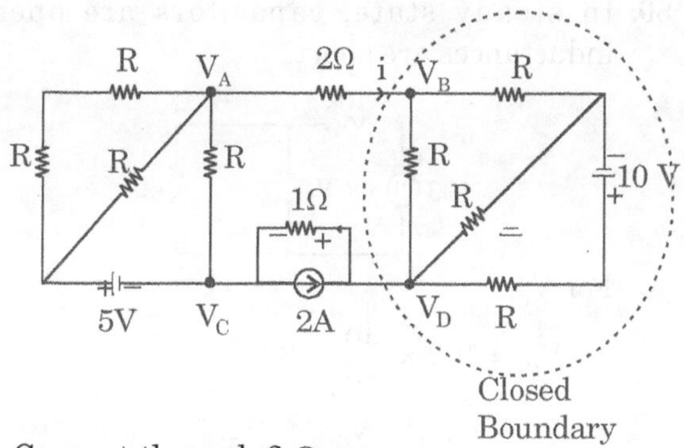

Current through 2 Ω

$$\text{resistor}, i = \frac{V_A - V_B}{2} = \frac{6}{2} = 3\text{ A}$$

According to KCL (algebraic sum of all the currents associated with a closed boundary is zero).

i + 2= i'

i' = 5A.

$V_C - V_D = -5\times 1 = -5V.$

23.

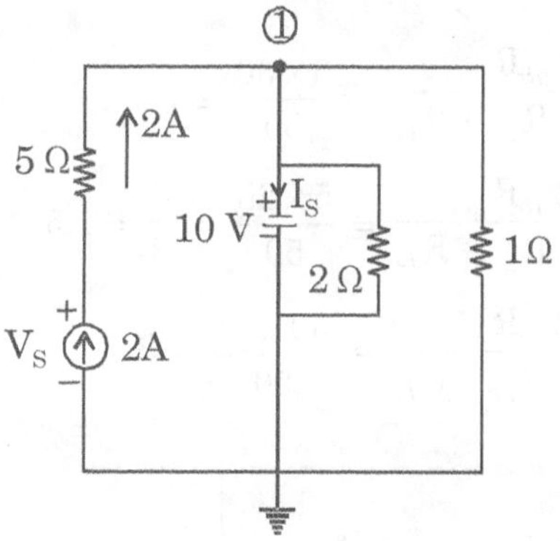

$$\text{Current through 1 Ω resistor} = \frac{10}{1} = 10\text{ A}$$

$$\text{Current through 2 Ω resistor} = \frac{10}{2} = 5\text{ A}$$

Using KCL at hode (1),

$2\text{ A} = I_S + 10 + 5$

$\boxed{I_S = -13\text{ A}}$

$\therefore V_1 = 10$ V

$V_S - V1 = 5\times2$

$V_S = 10 + 10 = 20$ V.

24. Current through 1 Ω resistor $= \frac{10}{1} = 10$ A.

25. Given: $i_1 = 2$ A, $i_4 = -1$ A, $i_5 = -4$ A.

Applying KCL at

Node 1: $i_2 + i_5 = i_3$

Node 2: $i_3 + i_6 = i_1$

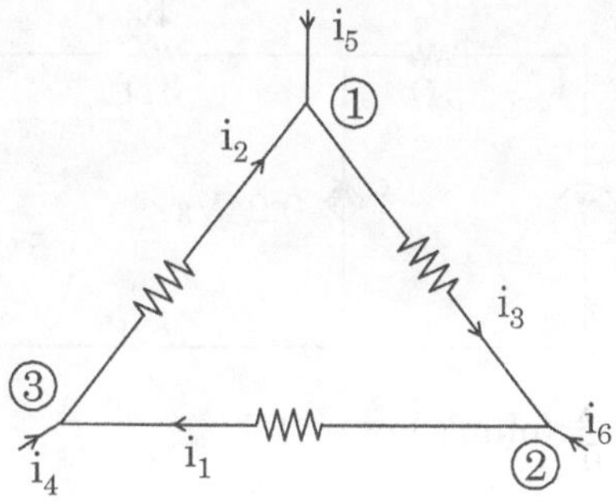

Node 3: $i_1 + i_4 = i_2$

$$\frac{i}{2} + i_5 + i_6 + i_4 = \frac{i}{2}$$

$-4 + i_6 + (-1) = 0$

$\boxed{i_6 = 5\text{A.}}$

$i_3 = i_1 - i_6 = 2 - 5 = -3$A.

26. By source transformation

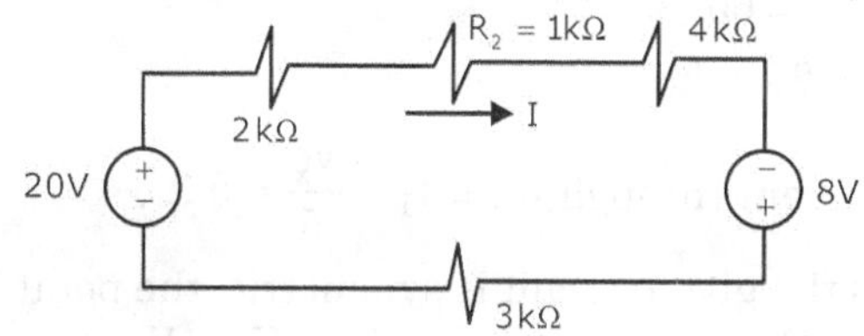

By KVL,

20 – 10k.I + 8 = 0

$$\Rightarrow I = \frac{28}{10k}$$

$\Rightarrow I = 2.8$ mA

27. Redrawing the given circuit we get

$\Rightarrow v_1 - 0 = 4I$

$\Rightarrow v_1 = 4I$ (Ohm's low)

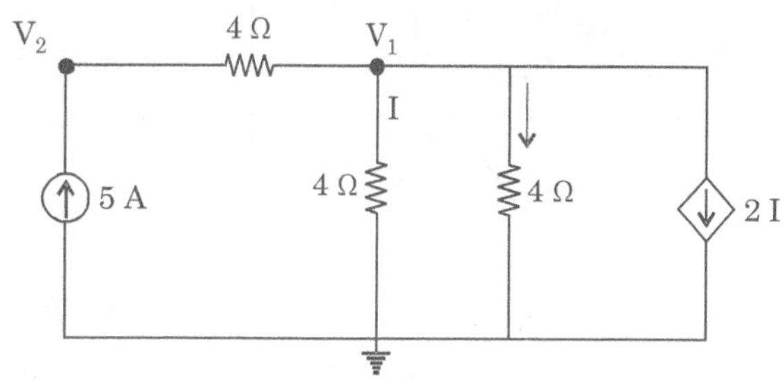

$$\Rightarrow \frac{v_1}{4} + \frac{v_1}{4} + 2I = 5 \text{ (KCL)}$$

$$\frac{v_1}{4} + \frac{v_1}{4} + \frac{2v_1}{4} = 5$$

$V_1 = 5$ volt

$\Rightarrow V_2 - V_1 = 4 \times 5$ (Ohm's law)

$V_2 = 25$ volt.

28.

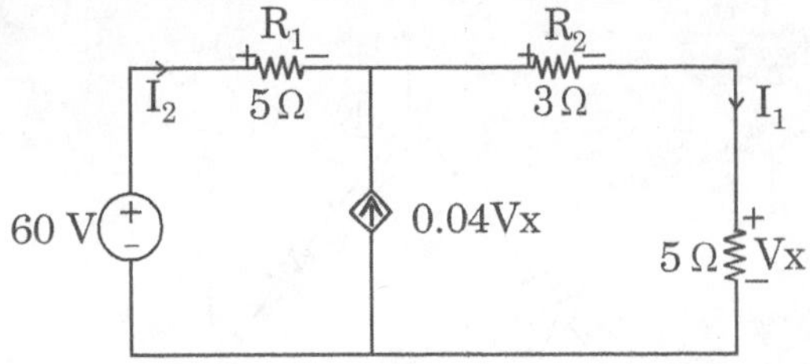

$$I_1 = \frac{v_x}{5} \text{ (ohm's law)}$$

$$\Rightarrow I_2 = -0.04v_x + \frac{v_x}{5} \text{ (KCL)}$$

$I_2 = 0.16_x$

Applying KVL,

$-60 + 5I_2 + 3I_1 + V_x = 0$

$$5 \times 0.16v_x + 3 \times \frac{v_x}{5} + v_x = 60$$

$2.4v_x = 60$

$v_x = 25$ volt

Current through R2 $= I_1 = \frac{v_x}{5} = 5$ A .

29. As the given circuit is symmetric, the point B and C are at same point, i. e., $V_B = V_C$ so current through R_2 is zero.

Points D are E are also equipotentials, hence, $V_D = V_E$.

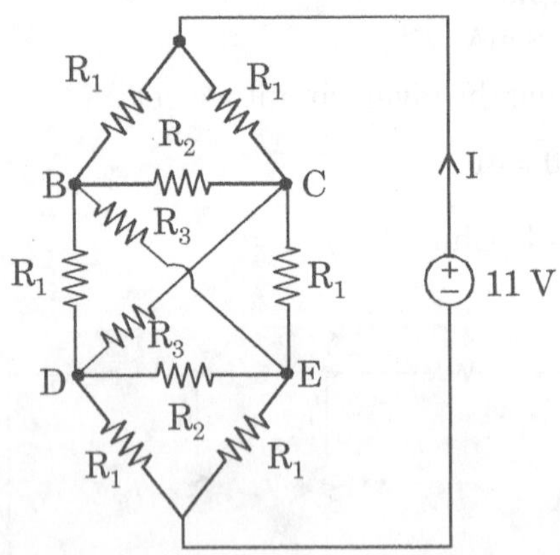

⇒

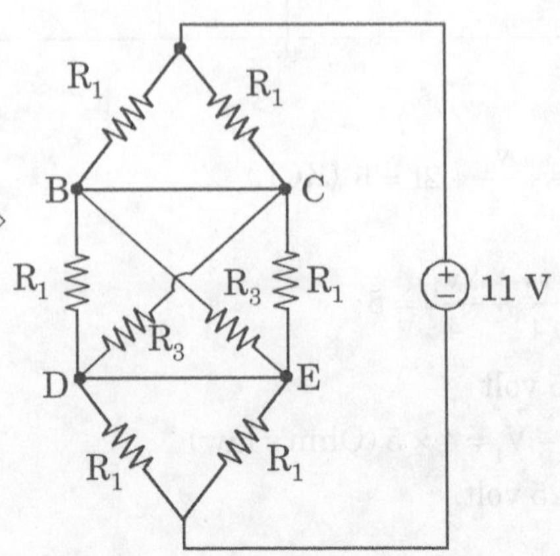

30. In steady state, capacitors are open and inductances are short.

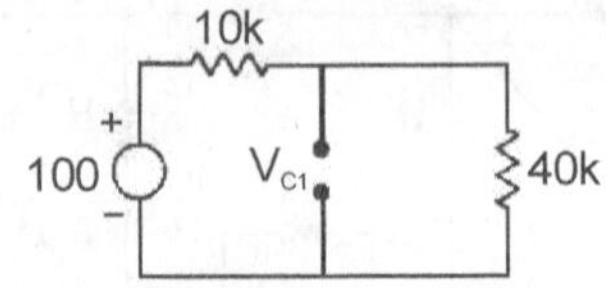

For V_{C1}

$$V_{C1} = 100 \times \frac{40}{50} = 80 \text{ V}$$

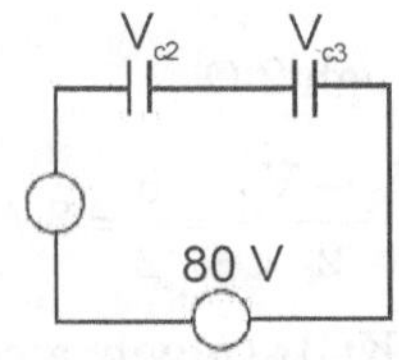

For V_{C2} and V_{C3}

$$V_{C2} = 80 \times \frac{C_3}{C_2 + C_3} = 80 \times \frac{3}{5} = 48 \text{ V}$$

$$V_{C3} = 80 \times \frac{C_2}{C_2 + C_3} = 16 \times 2 = 32 \text{ V}$$

31. $$R_1 = \frac{R_{ab}R_{ac}}{R_{ab} + R_{ac} + R_{bc}} = \frac{5 \times 30}{50} = 3$$

$$R_2 = \frac{R_{ab}R_{ac}}{R_{ab} + R_{ac} + R_{bc}} = \frac{5 \times 25}{50} = \frac{3}{2} = 1.5$$

$$R_3 = \frac{R_{ac}R_{ac}}{R_{ab} + R_{ac} + R_{bc}} = \frac{30 \times 15}{50} = 9.$$

32.

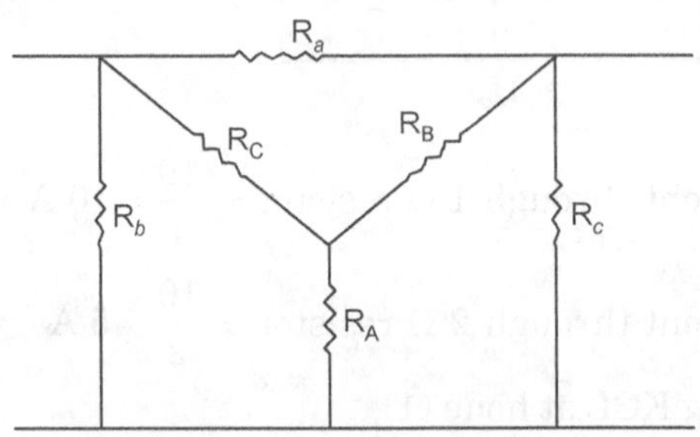

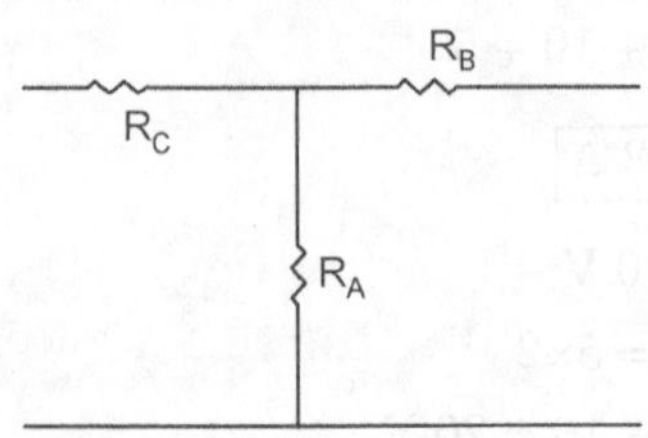

$$R_c = \frac{R_a \cdot R_b}{R_a + R_b + R_c}$$ as R_a is scaled by factor k

$$R_c' = \frac{R_a' \cdot R_b'}{R_a' + R_b' + R_c'} = \frac{k^2 R_a \cdot R_b}{k(R_a + R_b + R_c)}$$

$$= k \cdot \frac{R_a \times R_b}{R_a + R_b + c}$$

So elements corresponding to star equivalence will be scaled by factor k.

33.

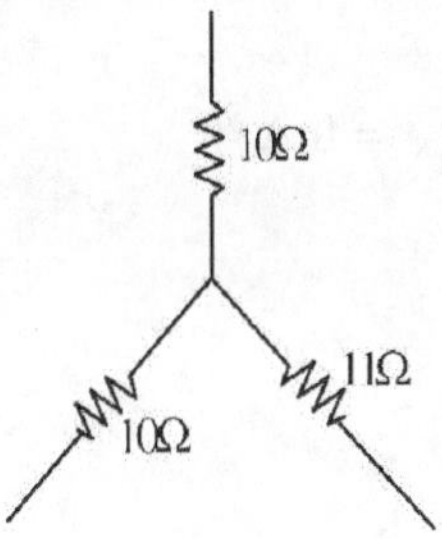

Star Connection

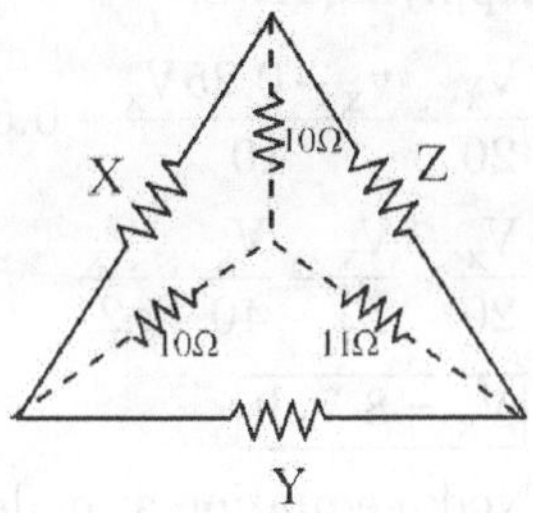

Delta Connection

$X = 29.09\ \Omega$

$y = 32\ \Omega$

$z = 32\ \Omega$

$$X = \frac{(10)(10) + (10)(11) + (10)(11)}{11}\ \Omega$$

$$Y = \frac{(10)(10) + (10)(11) + (10)(11)}{10}\ \Omega$$

$$Z = \frac{(10)(10) + (10)(11) + (10)(11)}{10}\ \Omega$$

i.e, Lowest value among three resistances is 29.09 Ω

34.

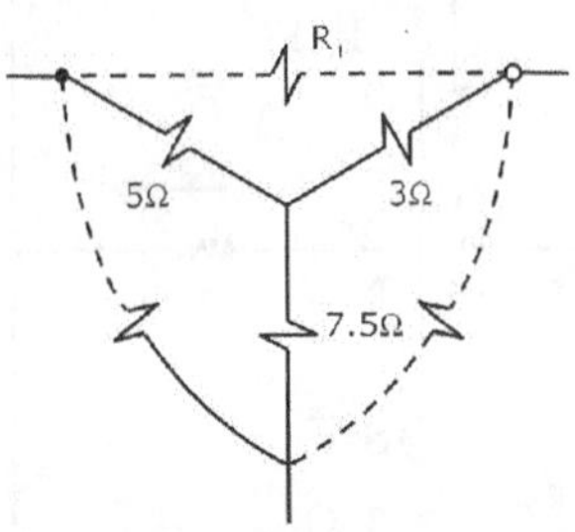

$$R_1 = \frac{(7.5)(5) + (3)(5) + (7.5)(3)}{7.5}\ \Omega$$

$R_1 = 10\ \Omega$

35. (a)

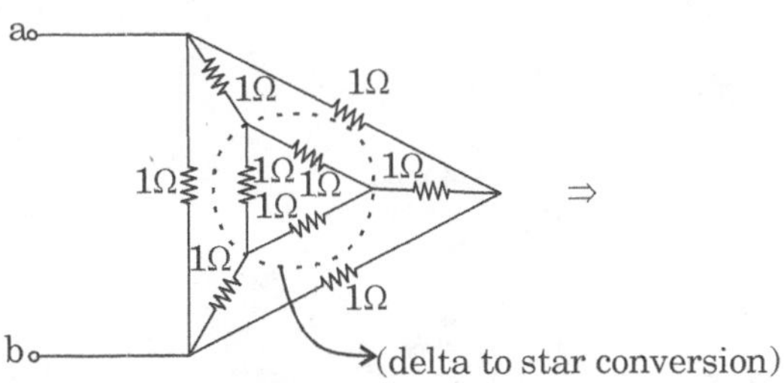

(b)

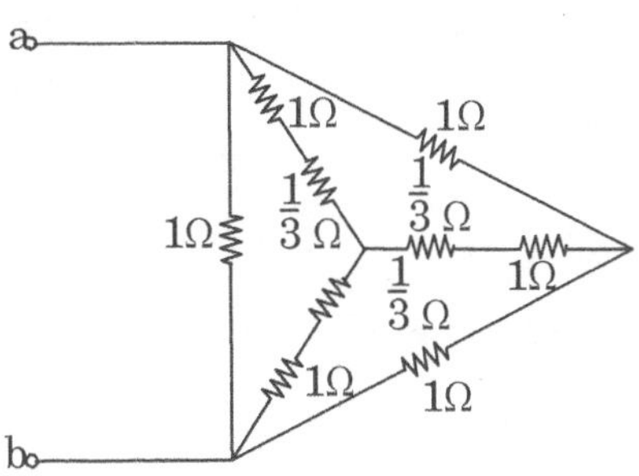

(c)

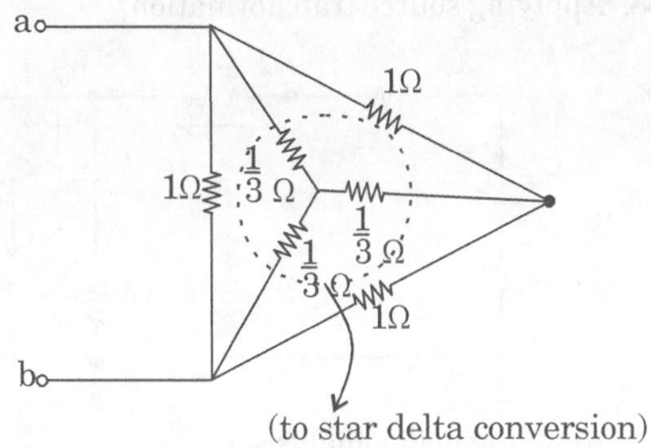

(d)

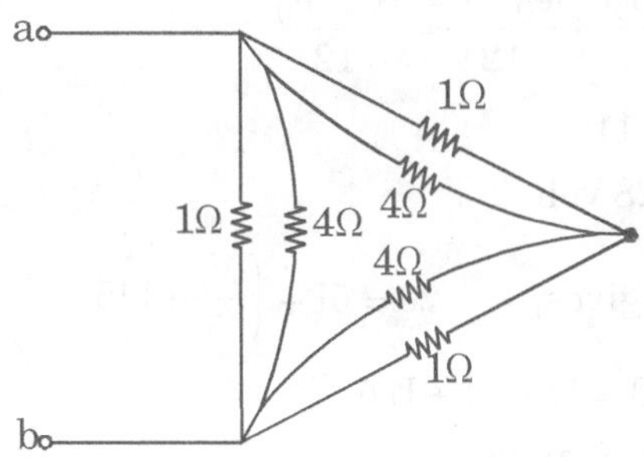

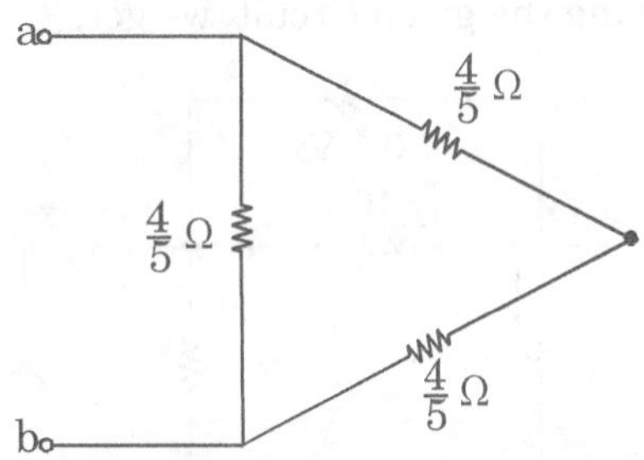

$$R_{ab} = \left(\frac{4}{5} + \frac{4}{5}\right) \| \frac{4}{5} = \frac{8}{15}\ \Omega.$$

36. The nodal method of circuit analysis is based on KCL and Ohm's law

37. It is required to find voltage across the terminals *a* and *b*.

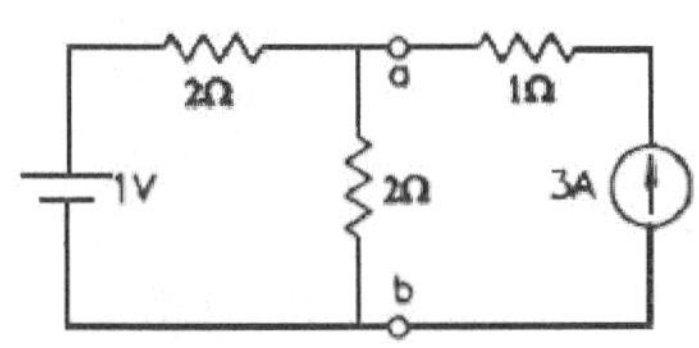

With respect to node *b* apply Nodal equation.

$$\frac{V_a - 1}{2} + \frac{V_a}{2} - 3 = 0$$

$$2V_a - 1 - 6 = 0$$

$$\therefore\ V_a = 3.5\ V$$

38. Applying source transformation,

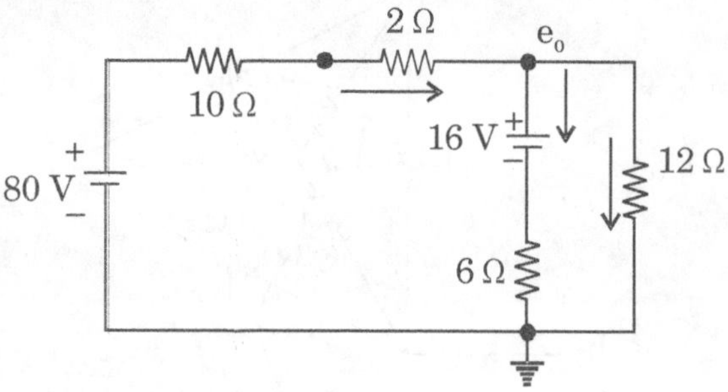

Applying nodal analysis,

$$\frac{e_0 - 16}{6} + \frac{e_0 - 0}{12} = \frac{8_0 - e_0}{12}$$

$4e_0 = 112$

$e_0 = 28$ volt

39. KVL gives, $20 = 5I + \left(\frac{V_1}{5} + I\right)5$

or $20 = 5I + (4 + I)\,5$

$I = 0$ amp.

$\therefore$ Power delivered $= 4^2 \times 5 = $ **80 W.**

40. Redrawing the given circuit, we get,

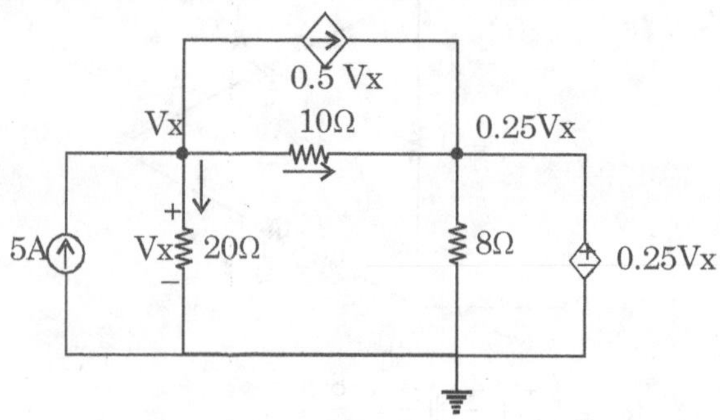

Applying KCL,

$$\frac{V_x}{20} + \frac{V_x - 0.25V_x}{10} + 0.5V_x = 5\text{ A}$$

$$\frac{V_x}{20} + \frac{V_x}{10} - \frac{V_x}{40} + \frac{V_x}{2} = 5$$

$$\boxed{V_x = 8\text{ Volt}}$$

41. Nodal equation at node (1),

$$\frac{8 - v_1}{1} + \frac{8 - v_1}{1} = \frac{v_1}{1} + \frac{v_1}{1}$$

$V_1 = 4$ volt.

Applying KCL at node (2)

$$i_1 = \frac{8 - v_1}{1} + 1 = 5\text{ A.}$$

Applying KCL at node (3)

$$i_1 = \frac{v_1 - 0}{1} - i_1 = 4 - 5 = 1\text{ A.}$$

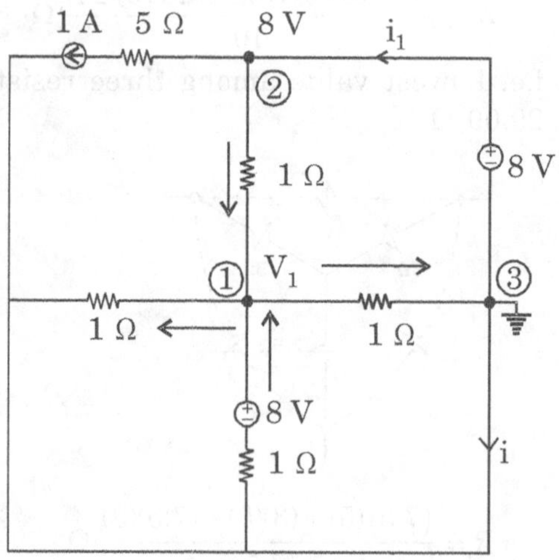

CHAPTER 2

Circuit Theorems

Analysis of Previous GATE Papers

Topics ↓	Year →		2019	2018	2017 Set 1	2017 Set 2	2016 Set 1	2016 Set 2	2015 Set 1	2015 Set 2	2015 Set 3
SUPERPOSITION	1 Mark	MCQ Type									
		Numerical Type									
	2 Marks	MCQ Type									
		Numerical Type									
	Total										
THEVENIN & NORTON THEOREM	1 Mark	MCQ Type									
		Numerical Type									1
	2 Marks	MCQ Type									
		Numerical Type				1				1	
	Total					**2**				**2**	**1**
MAXIMUM POWER TRANSFER THEOREM	1 Mark	MCQ Type						1			
		Numerical Type									
	2 Marks	MCQ Type									
		Numerical Type					1				
	Total						**2**	**1**			
RECIPROCITY THEOREM	1 Mark	MCQ Type	1								
		Numerical Type									
	2 Marks	MCQ Type									
		Numerical Type									
	Total		**1**								

SUPERPOSITION

1. A DC circuit shown in figure has a voltage source V, a current source I and several resistors. A particular resistor R dissipates a power of 4 Watts when V alone is active. The same resistor R dissipates a power of 9 Watts when I alone is active. The power dissipated by R when both sources are active will be

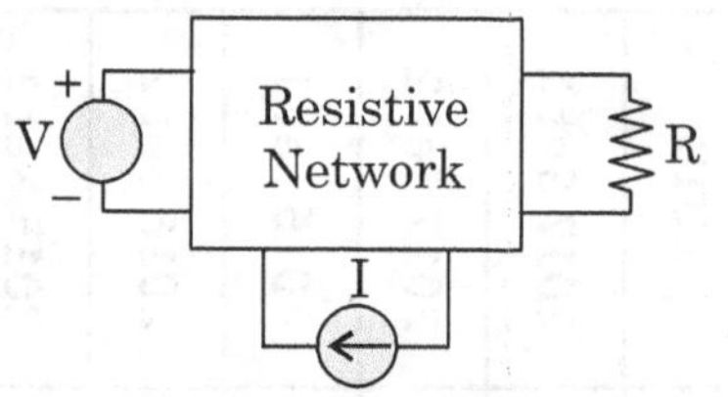

(a) 1 W (b) 5 W
(c) 13 W (d) 25 W

[1995 : 1 Mark]

2. In the circuit of figure, the power dissipated in the resistor R is 1 W when only source '1' is present and '2' is replaced by a short. The power dissipated in the same resistor R is 4 W when only source '2' is present and '1' is replaced by a short. When both the sources '1' and '2' are present, the power dissipated in R will be

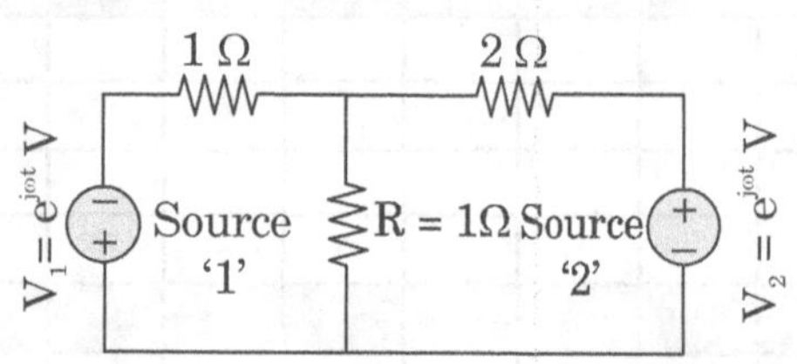

(a) 1 W (b) 3 W
(c) 3 W (d) 5 W

[1998 : 2 Marks]

3. Superposition theorem is NOT applicable to networks containing
(a) non linear elements
(b) dependent voltage sources
(c) dependent current sources
(d) transformers

[1998 : 1 Mark]

4. In the figure shown, the value of the current I (in Amperes) is ________.

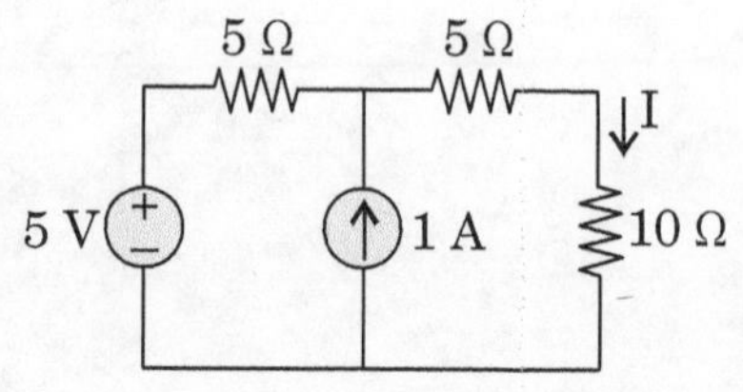

[2014 : 1 Mark]

THEVENIN & NORTON THEOREM

5. In the circuit of figure, when switch S_1 is closed, the ideal ammeter M_1 reads when S_1 is is kept open? (The value of E is not specified).

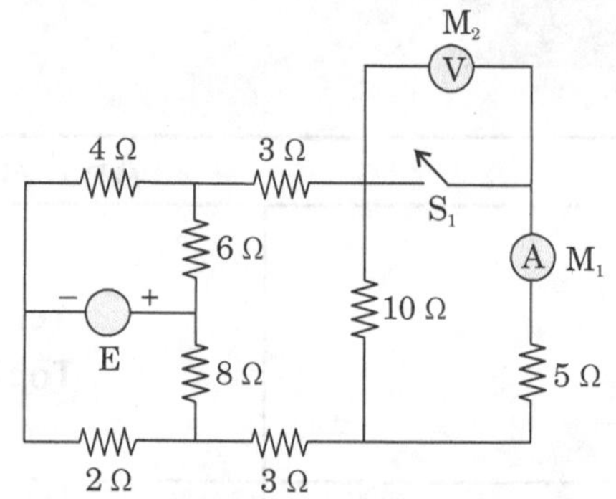

[1993 : 2 Marks]

6. For the circuit shown in the figure, Thevenin's voltage and Thevenin's equivalent resistance at terminals a–b is

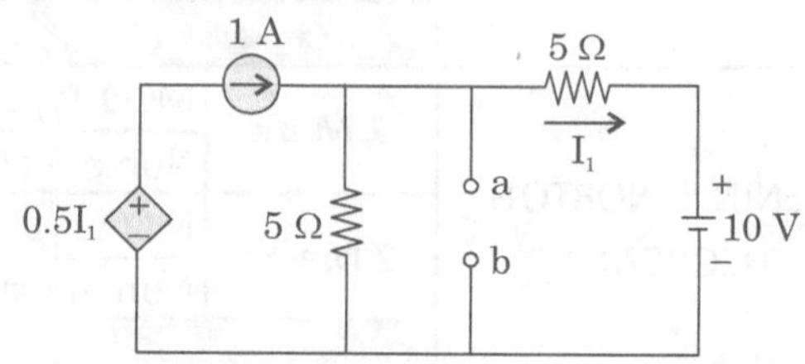

(a) 5 V and 2 Ω (b) 7.5 V and 2.5 Ω
(c) 4 V and 2 Ω (d) 3 V and 2.5 Ω

[2005 : 2 Marks]

7. For the circuit shown in the figure, the Thevenin voltage and resistance looking into X-Y are

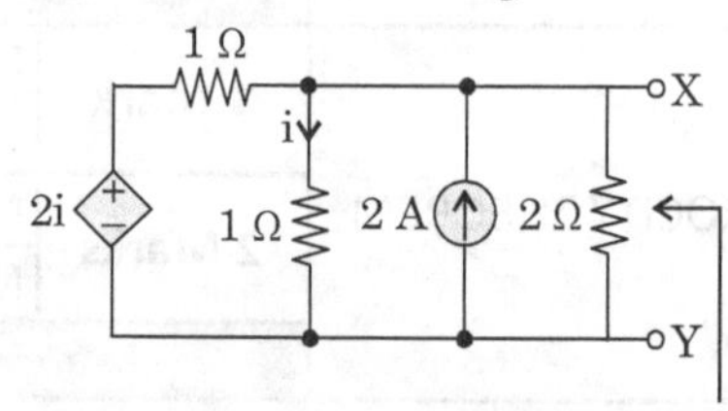

(a) 4/3 V, 2 Ω (b) 4 V, 2/3 Ω
(c) 4/3 V, 2/3 Ω (d) 4 V, 2 Ω

[2007 : 2 Marks]

8. The impedance looking into nodes 1 and 2 in the given circuit is

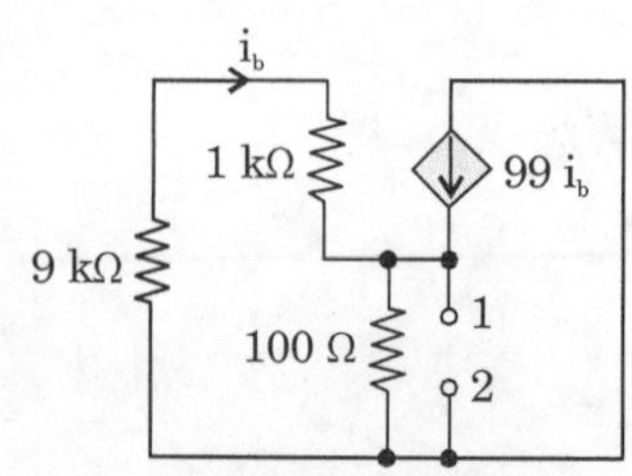

(a) 50 Ω (b) 100 Ω
(c) 5 Ω (d) 10.1 kΩ

[2012 : 1 Mark]

9. Norton's theorem states that a complex network connected to a load can be replaced with an equivalent impedance

(a) in series with a current source

(b) in parallel with a voltage source

(c) in series with a voltage source

(d) in parallel with a current source

[2014 : 1 Mark, Set-2]

10. In the circuit shown the Norton equivalent resistance (in Ω) across terminals a–b is _____.

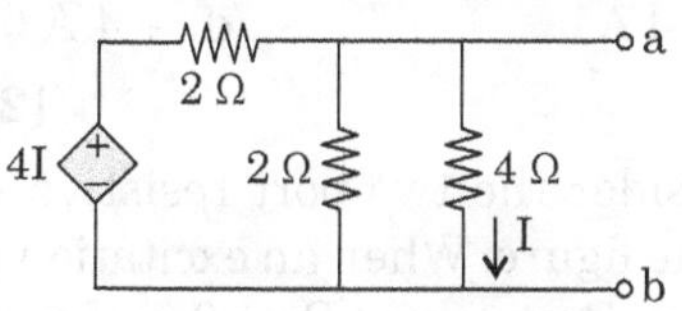

[2015 : 2 Marks, Set-2]

11. For the current shown in the figure, the Thevenin equivalent Voltage (in Volts) across terminals a–b is _______.

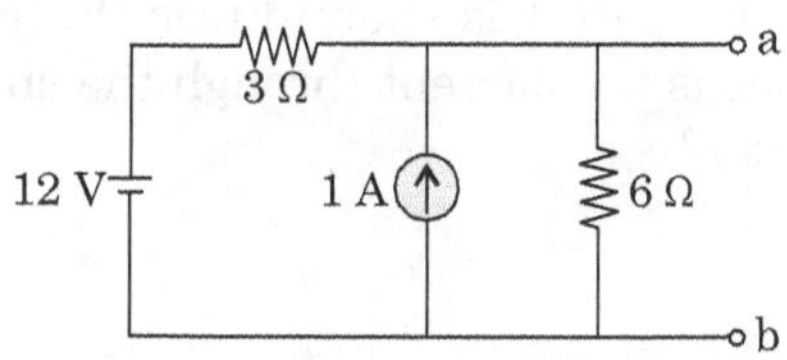

[2015 : 1 Mark, Set-3]

12. Consider the circuit shown in the figure.

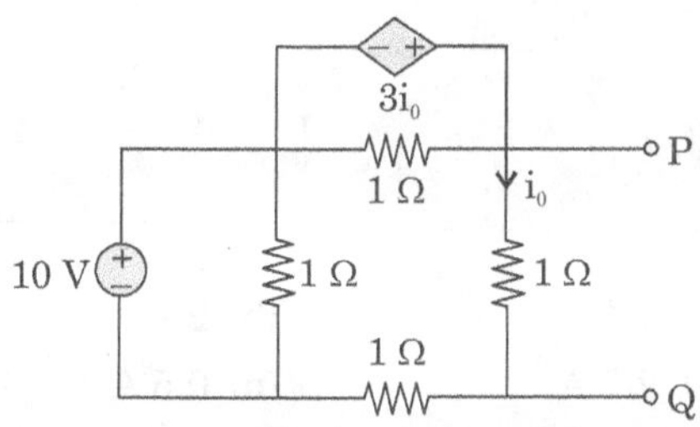

The Thevenin equivalent resistance (in Ω) across P-Q is _____. **[2017 : 2 Mark, Set-2]**

MAXIMUM POWER TRANSFER THEOREM

13. The value of the resistance, R, connected across the terminals, A and B, (ref. fig.), which will absorb the maximum power, is

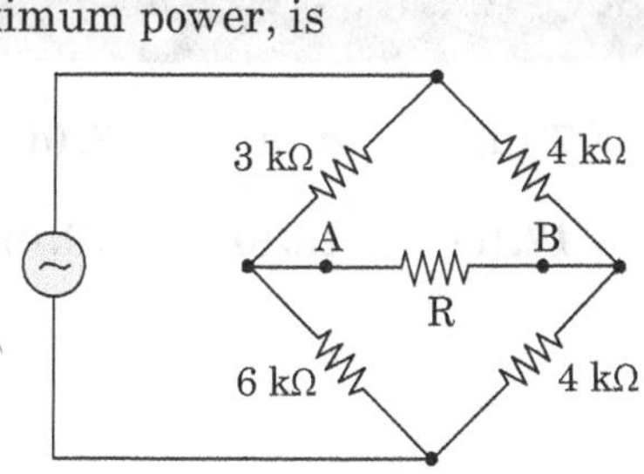

(a) 4.00 k Ω (b) 4.11 k Ω

(c) 8.00 k Ω (d) 9.00 k Ω

[1995 : 1 Marks]

14. The value of R (in ohms) required for maximum power transfer in the network shown in the figure is

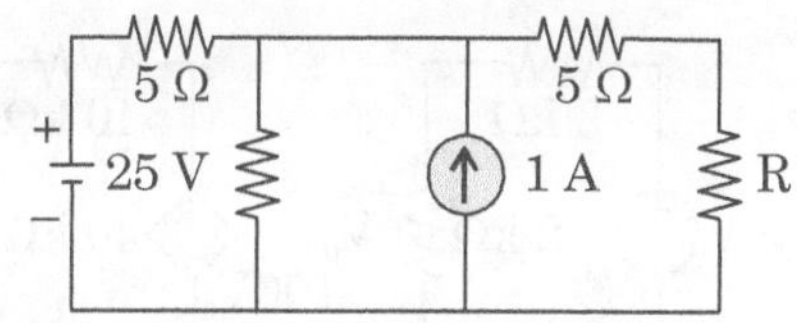

(a) 2 (b) 4

(c) 8 (d) 16

[1999 : 2 Marks]

15. In the network of the figure, the maximum power is delivered to R_L if its value is

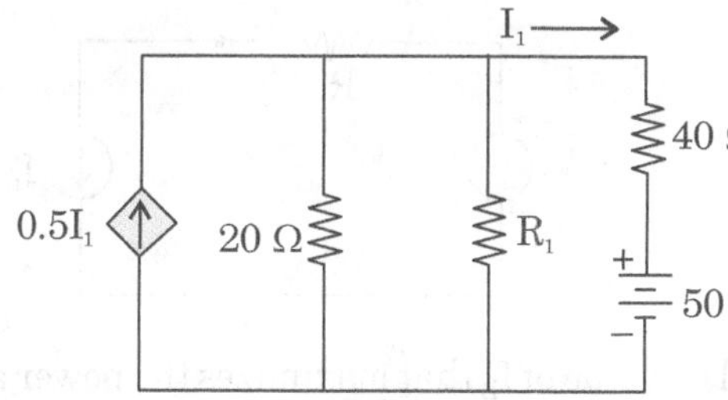

(a) 16 Ω (b) $\frac{40}{3}$ Ω

(c) 60 Ω (d) 20 Ω

[2002 : 2 Marks]

16. The maximum power that can be transferred to the load resistor R_L from the voltage source in the figure is

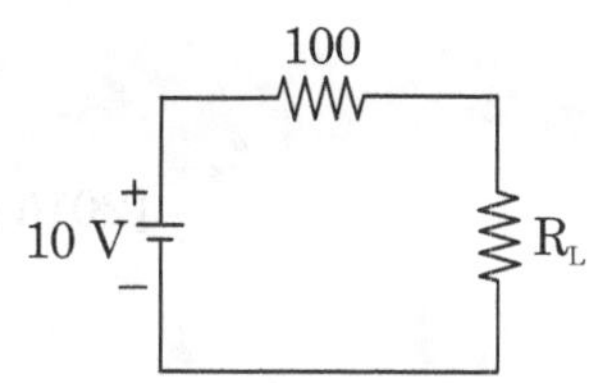

(a) 1 W (b) 10 W

(c) 0.25 W (d) 0.5 W

[2005 : 1 Mark]

17. In the circuit shown what value of R_L maximizes the power delivered to R_L?

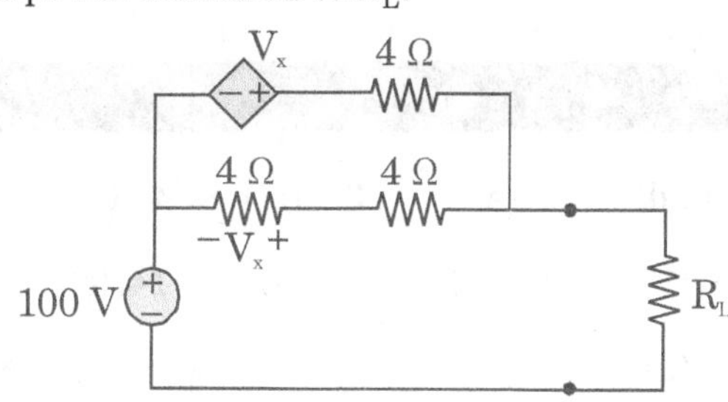

(a) 2.4 Ω (b) $\frac{8}{3}$ Ω

(c) 4 Ω (d) 6 Ω

[2009 : 2 Marks]

18. In the circuit shown in the figure, the maximum power (in watt) delivered to the resistor R is _____.

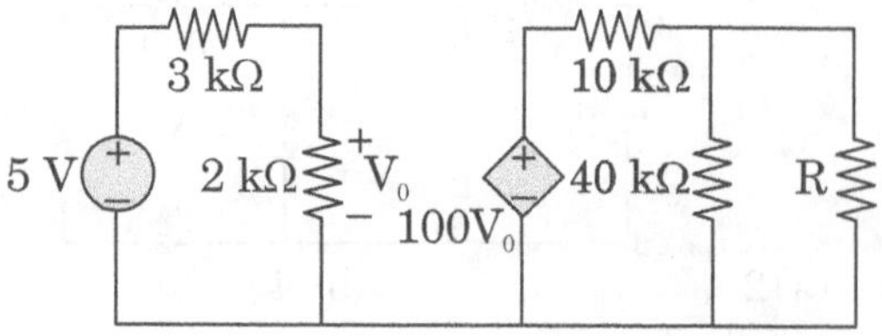

[2016 : 2 Mark, Set-1]

19. In the circuit shown below, V_S is constant voltage source and I_L is a constant current load.

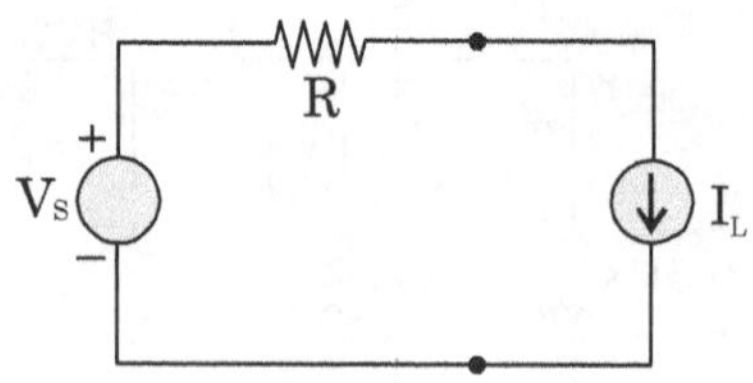

The value of I_L that maximizes the power absorbed by the constant load is

(a) $\frac{V_S}{4R}$

(b) $\frac{V_S}{2R}$

(c) $\frac{V_S}{R}$

(d) ∞

[2016 : 1 Mark, Set-2]

RECIPROCITY THEOREM

20. Use the data of the figure (a). The current i in the circuit of the figure (b) is

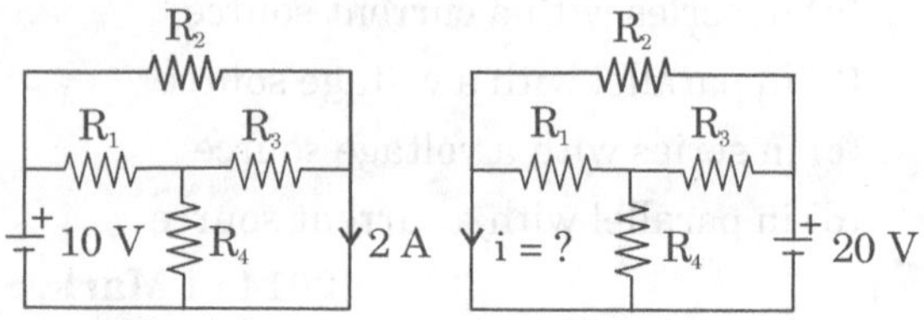

Fig. (a) Fig. (b)

(a) –2 A (b) 2 A

(c) –4 A (d) + 4 A

[2000 : 2 Marks]

21. Consider the two-port resistive network shown in the figure. When an excitation of 5 V is applied across Port 1, and Port 2 is shorted, the current through the short circuit at Port 2 is measured to be 1 A (see (a) in the figure).
Now, if an excitation of 5 V is applied across port 2, and port 1 is shorted (see (b) in the figure), what is the current through the short circuit at port 1?

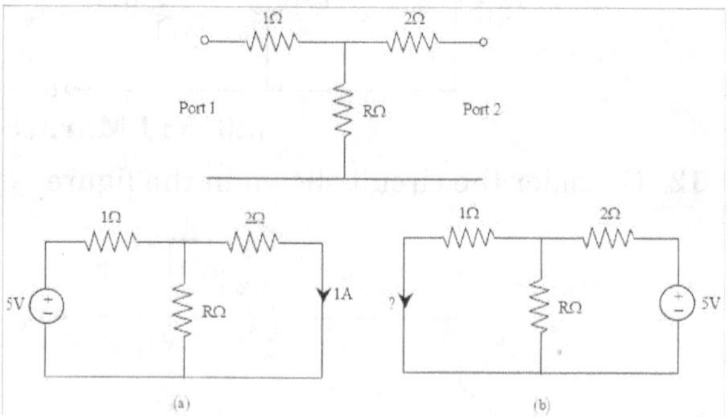

(a) 1 A (b) 2 A

(c) 2.5 A (d) 0.5 A

[2019 : 1 Mark]

ANSWERS

1. (a,d)	**2.** (a)	**3.** (a)	**4.** (0.5)	**5.** (5)	**6.** (b)	**7.** (d)	**8.** (a)	**9.** (d)	**10.** (4/3)
11. (10)	**12.** (–1)	**13.** (a)	**14.** (c)	**15.** (a)	**16.** (c)	**17.** (c)	**18.** (0.8)	**19.** (b)	**20.** (c)
21. (a)									

EXPLANATIONS

1. Option (a) and (d) both are correct.

Since the network is linear, the current through the resistor R can be expresol as a linear combination of V and I.

$\therefore \quad i = aV \pm bI$

The power dissipated ie., i^2R is given as 4W then I = D and qW when V = 0

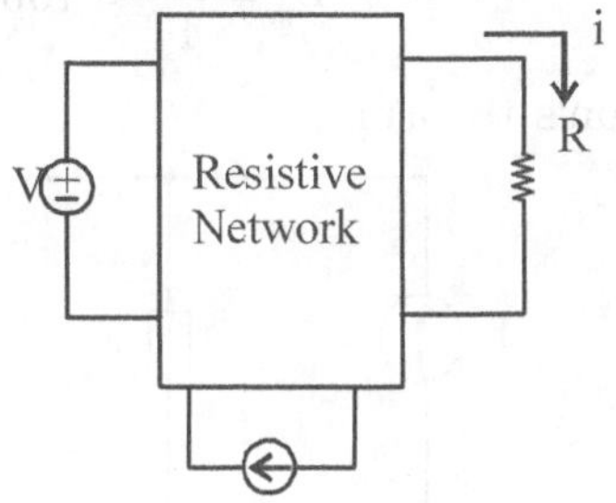

$\therefore \; P_V = (aV)2R = 4$ and $P_I = (bI)^2R = q$

$$aV = \sqrt{\frac{Pv}{R}} \text{ and } BI = \sqrt{\frac{P_I}{R}}$$

The power that will be dissipated when both source are acting simultaneously is given by,

$$P_{VI} = (aV \pm bI)^2 R = (aV)^2R + (bI)^2R \pm 2aV(bI)R$$

$$= P_V + P_I \pm 2\sqrt{\frac{P_v}{R}}\sqrt{\frac{P_I}{R}} \times R = \left(\sqrt{P_V} \pm \sqrt{P_I}\right)^2$$

$$P_{VI} = \left(\sqrt{4} \pm \sqrt{9}\right)^2 = (2+3)^2 \text{ or } (2-3)^2$$

$$P_{VI} = 25 \text{ W or } 1 \text{ W}.$$

2. Since, response (voltage or current) produced in resister R due to both sources are opposite. Using superposition principle,

Net power dissipated in

$$R = \left(\sqrt{P_1} - \sqrt{P_2}\right)^2$$

$$= \left(\sqrt{1} - \sqrt{4}\right)^2 = (1-2)^2 = 1 \text{ W}$$

3. Superposition theorem is not applicable to networks containing non-linear elements.

4.

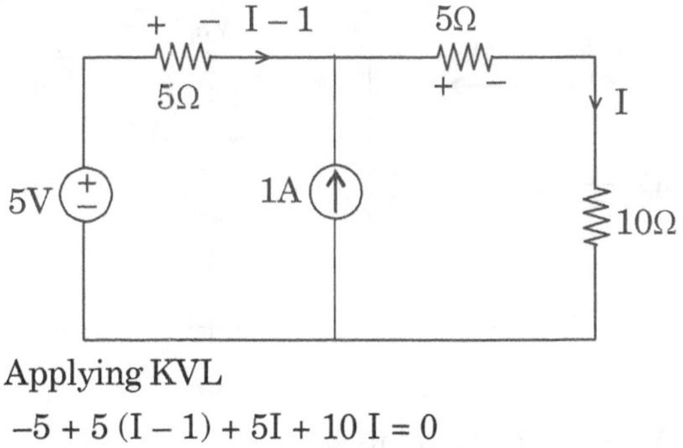

Applying KVL

$$-5 + 5(I-1) + 5I + 10I = 0$$

$$20I = 10$$

$$I = 0.5 \text{ A}$$

5. When switch S_1 is closed, the ideal ammeter M_1 reads the short-circuit or Norton's equivalent current I_N, i.e., $I_N = S_A$.

For R_N:

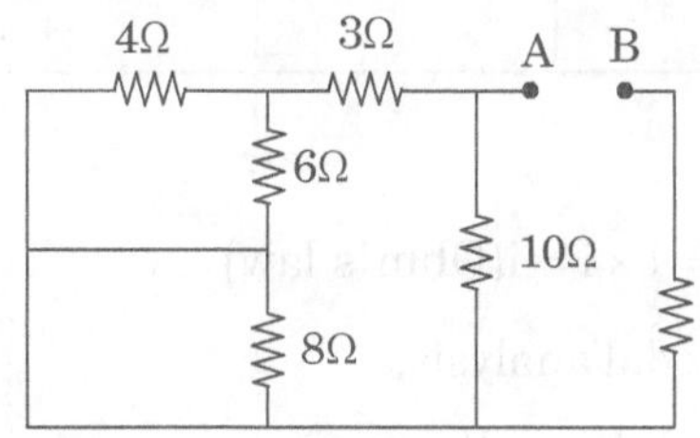

$$R_N = [(4\|6 + 2\|8) + 3 + 3]\|10 + 5$$

$$= (2.4 + 1.6 + 3 + 3)\|10 + 5 = 10\|10 + 5 = 5 + 5$$

$$R_N = 10\ \Omega$$

The equivalent circuit is given as

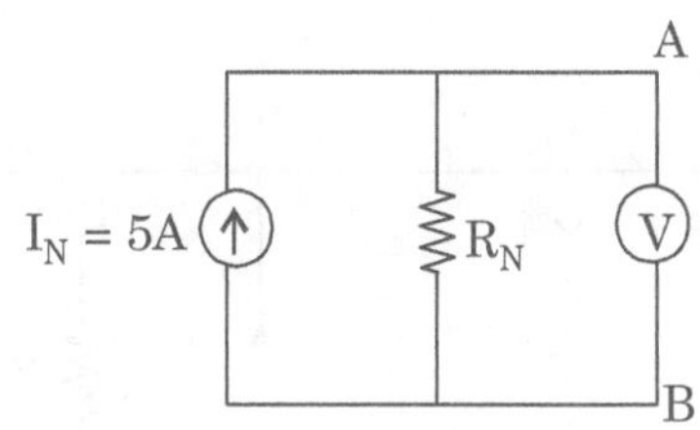

When switch S_1 is kept open, reading of voltmeter $= I_N \cdot R_N = 5 \times 10 = 50$ V.

6. For V_{th}:-

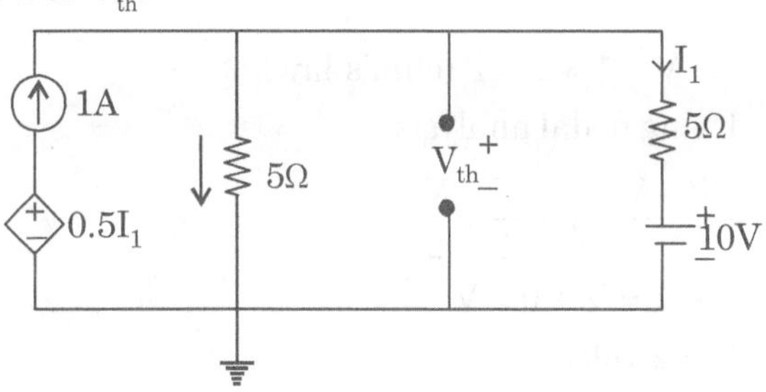

Using nodal analysis

$$\frac{V_{th} - 0}{5} + \frac{V_{th} - 10}{5} = 1$$

$$2\frac{V_{th}}{5} = 3$$

$$V_{th} = 7.5 \text{ volt.}$$

For R_{th}:-

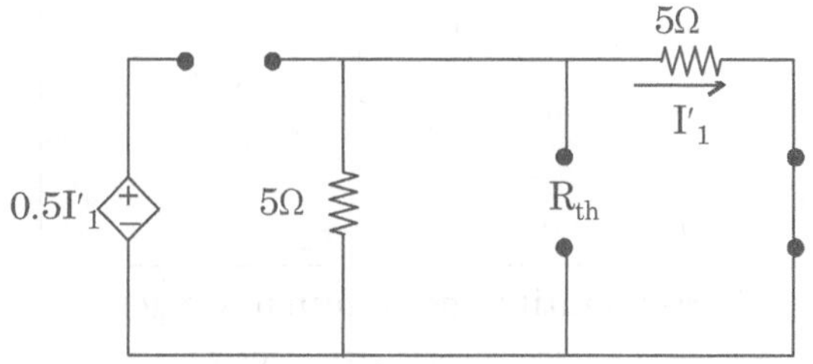

$$R_{th} = 5\,||\,5 = 2.5\ \Omega$$

7.

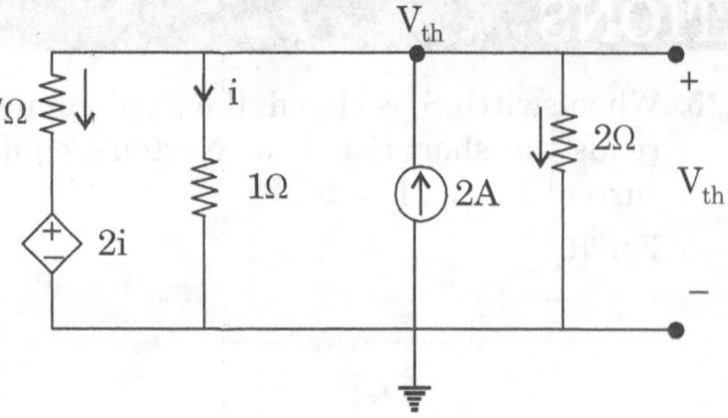

$\Rightarrow V_{th} = 1 \times i = i(\text{Ohm's law})$

Using nodal analysis,

$$\frac{V_{th} - 2i}{1} + \frac{V_{th}}{1} + \frac{V_{th}}{2} = 2$$

$$-V_{th} + V_{th} + \frac{V_{th}}{2} = 2$$

$$\boxed{V_{th} = 4V}$$

For R_{th}:-

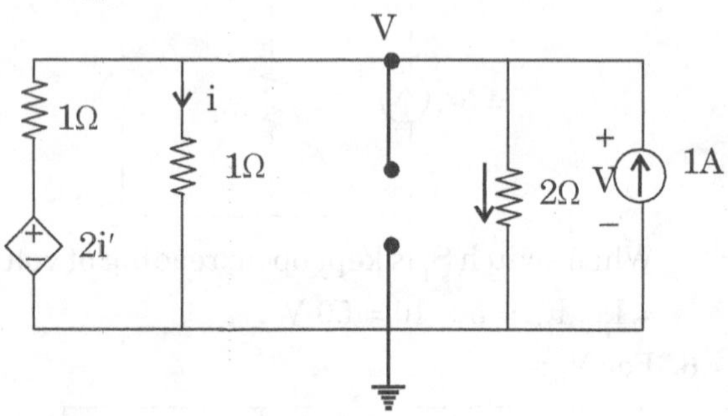

$\Rightarrow$ V = 1 × i′ = i′ (ohm's law)

Using nodal analysis,

$$\frac{2i' - V}{1} + 1 = \frac{V}{1} + \frac{V}{2}$$

$$V + 1 = V + 0.5\ V$$

V = 2 volt.

$$\Rightarrow R_{th} = \frac{V}{1\ A} = \frac{2}{1} = 2\ \Omega$$

8. Open-circuiting terminal 1-2.

$$V_{oc} = (99\ i_b + i_b) \cdot 100 = 10^4 i_b$$

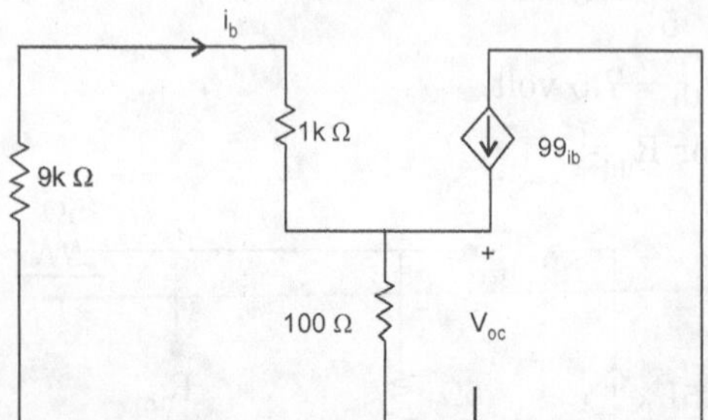

Short-circuiting the terminal, we get

$$I_{sc} = i_b + +99i_b$$

$$I_{sc} = 100i_b$$

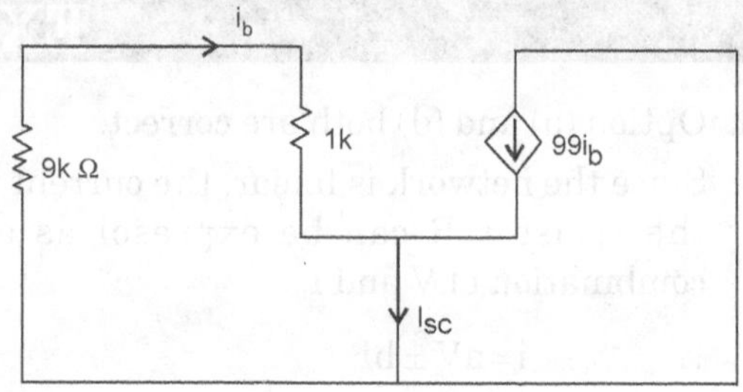

Therefore, impedance working into nodes 1 and 2,

$$Z_{th} = \frac{V_{oc}}{I_{sc}} = 100\ \Omega$$

9. Norton's theorm

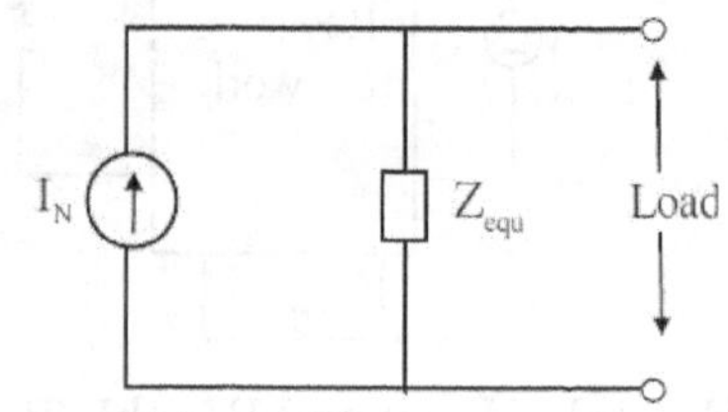

10. For R_N:-

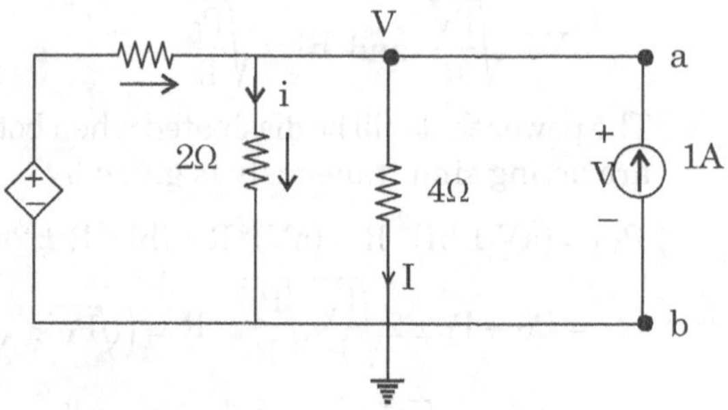

V = 4 × I = 4I (ohm's law)

Nodal analysis at node (a),

$$\frac{4I - V}{2} + 1 = \frac{V}{2} + \frac{V}{4}$$

$$0 + 1 = \frac{3V}{4}$$

$$V = \frac{4}{3}\text{Volt.}$$

$$\Rightarrow R_N = \frac{V}{1\ A} = \frac{4}{3}\ \Omega.$$

11.

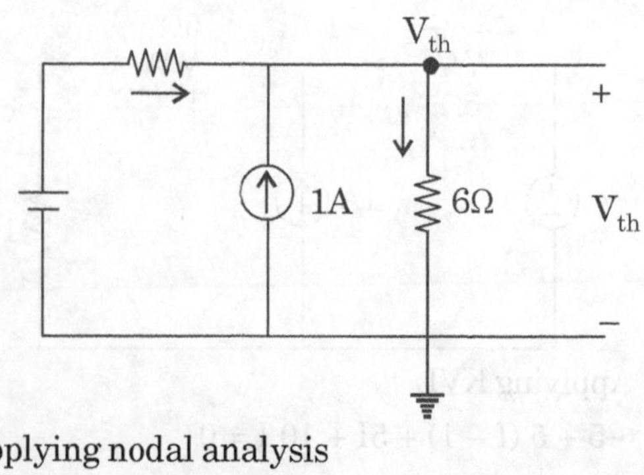

Applying nodal analysis

$$\frac{12 - V_{th}}{3} + 1 = \frac{V_{th}}{6}$$

$$4+1=\frac{V_{th}}{6}+\frac{V_{th}}{3}=\frac{V_{th}}{2}$$

V_{th} = 10 volts.

12.

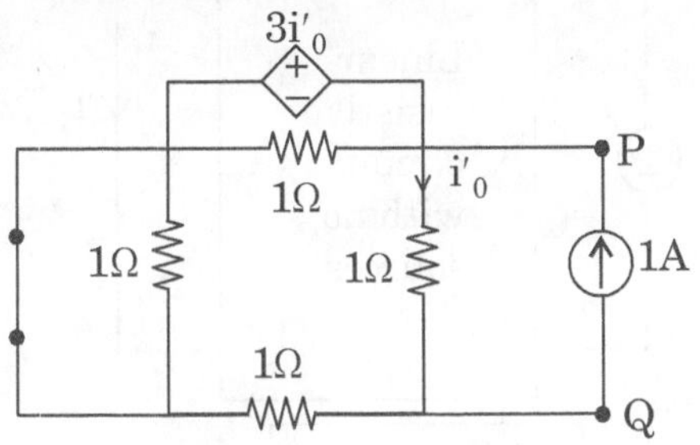

$\Rightarrow$

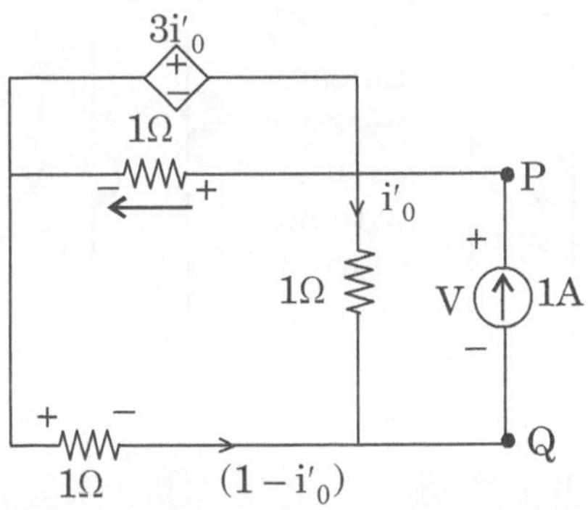

Using ohm's low

$$i=\frac{3i_0'}{1}=3i_0'$$

$$i_0'=\frac{V}{1}=V$$

Applying KVL in loop

$$-V+3i_0'+1.(1-i_0')=0$$

$$V=2i_0'+1$$

V = 2. V + 1

V = –1 volt

$$\Rightarrow R_{th}=\frac{V}{1A}=-1\ \Omega u$$

13. The circuit is drawn below by shorting the voltage source

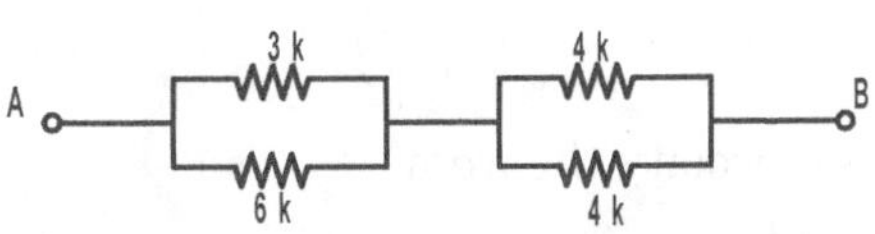

Taking the Thevenin's equivalent between points A, B

$$R_{Th}=\left(\frac{6\times 3}{6+3}+\frac{4\times 4}{4+4}\right)\ k\Omega$$

$$= 4\ k\Omega$$

For maximum power transfer,

$$R = R_{Th}$$
$$= 4\ k\Omega$$

14. For max. power transfer R must be equal to R_{th}.

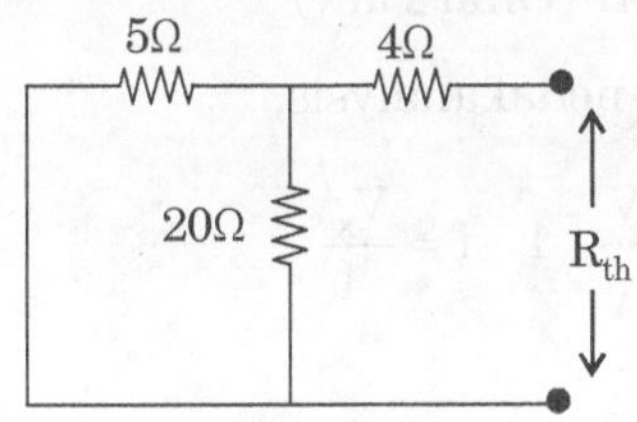

$$R_{th}=5\|20+4=4+4\ \Omega$$

$$R_{th}=8\ \Omega$$

$$R=R_{th}=8\ \Omega$$

15. To find max. power delivered to R_L, we need to calculate R_{th} across R_L.

For R_{th}:

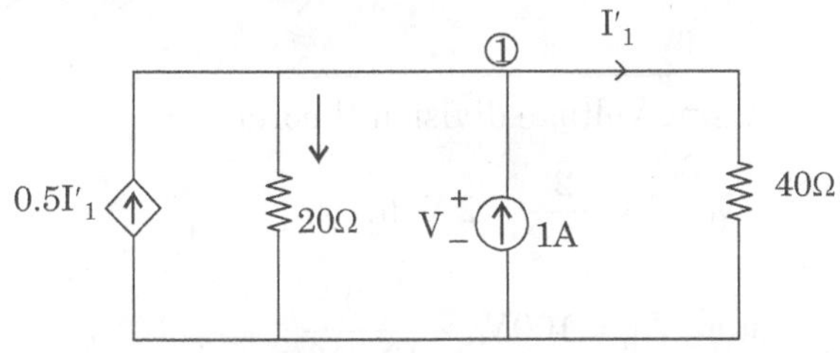

V = 40. I_1' (Ohm's law)

Using nodal analysis at node (1)

$$0.5I_1'+1=\frac{V}{20}+I_1'$$

$2.5I_1' = 1$

$$I_1'=\frac{1}{2.5}$$

$$v=40\times\frac{1}{2.5}=16\ V$$

$$\Rightarrow R_L=R_{th}=\frac{V}{1A}=16\ \Omega.$$

16. For max. power transfer, R_L = 100 Ω.

Max. power transferred,

$$P_{max}=\frac{V_{th}^2}{4R_L}=\frac{10^2}{4\times 100}=0.25\ W.$$

17. For max. power transferred to R_L; $R_L = R_{th}$.

For R_{th}:-

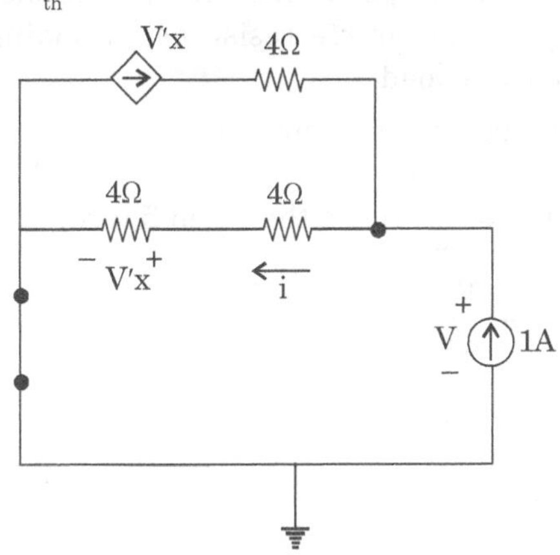

$V_x' = 4i$ (Ohm's law)

Using nodal analysis,

$$\frac{V_x' - V}{4} + 1 - i = \frac{V_x'}{4}$$

$$\frac{V}{4} = 1$$

$$\Rightarrow R_L = R_{th} = \frac{V}{1\,A} = 4\,\Omega$$

18. For V_{th}:-

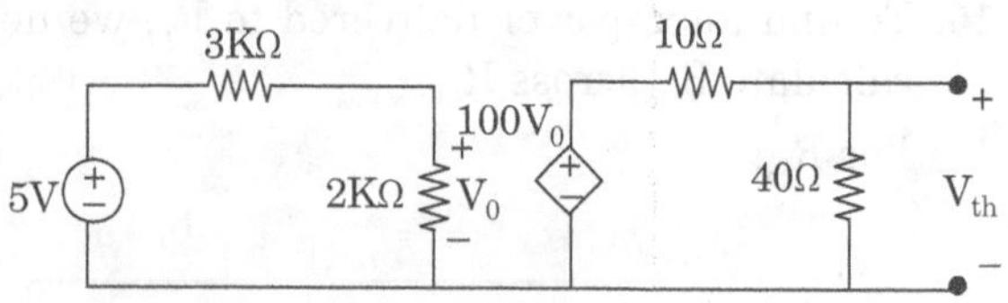

Using voltage division theorem,

$$V_0 = 5 \times \frac{2}{2+3} = 2 \text{ Volt.}$$

$$\text{and } V_{th} = 100V_0 \times \frac{40}{40+10}$$

$$= 200 \times \frac{40}{50} = 160 \text{ Volt}$$

For R_{th}:-

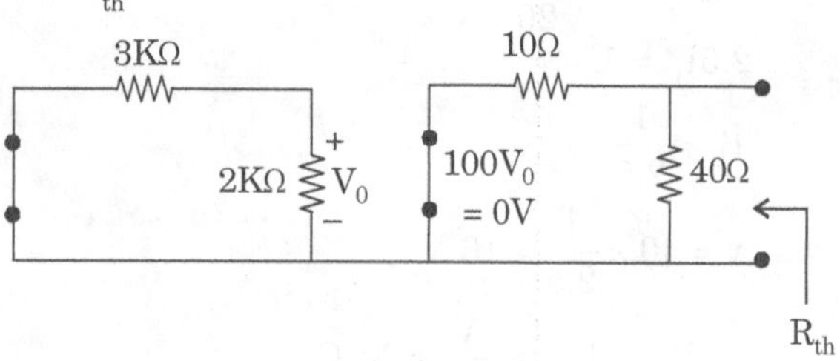

$$R_{th} = \frac{40 \times 10}{40+10}\,K\Omega$$

$R_{th} = 8\,K\Omega$.

Max. power delivered

$$= \frac{V_{th}^2}{4R_{th}} = \frac{160^2}{4 \times 8 \times 10^3} = 0.8\text{ W.}$$

19. In maximum power transfer, half of the voltage drop across source resistance, remaining half across the load.

∴ Voltage across source (R)

$$I_L R = \frac{V_s}{2} \text{ (from the given figure)}$$

$$I_L = \frac{V_s}{2R}$$

20. According to reciprocity theorem for networks shown below:

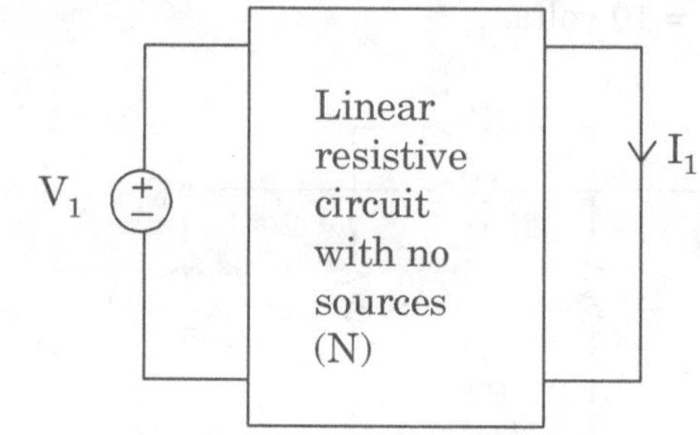

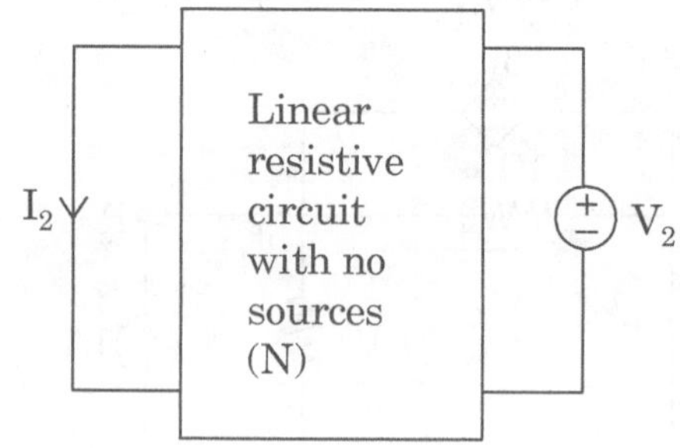

$$\boxed{\frac{V_i}{I_1} = \frac{V_2}{I_2}}$$

$$\frac{10}{2} = \frac{20}{i} \Rightarrow i = \frac{20}{5} = 4\text{ A.}$$

21.

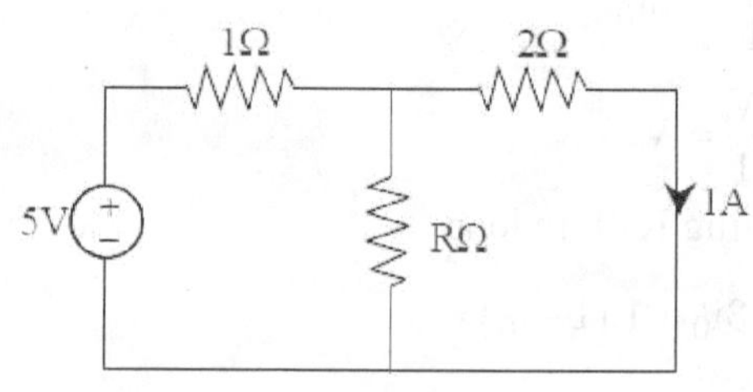

(figure a)

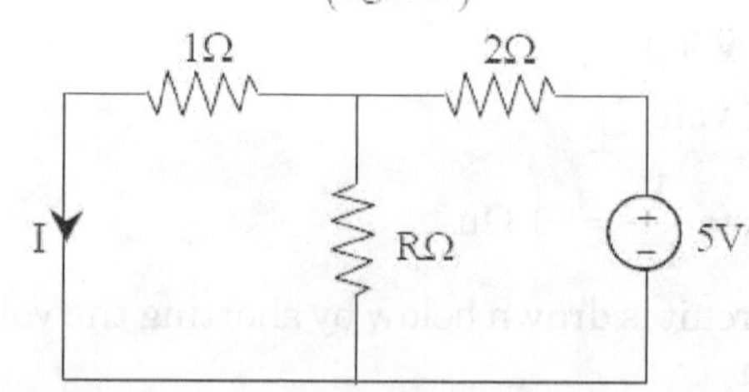

(figure b)

By reciprocity theorem,

Current I = 1 A

$$\left[\frac{V}{I} = k, \frac{V_1}{V_2} = \frac{I_1}{I_2}\right]$$

3

CHAPTER

Transient Analysis

Analysis of Previous GATE Papers												
	Year → Topics ↓		2019	2018	2017 Set 1	2017 Set 2	2016 Set 1	2016 Set 2	2016 Set 3	2015 Set 1	2015 Set 2	2015 Set 3
A.C. FUNDAMENTALS	1 Mark	MCQ Type										
		Numerical Type										
	2 Marks	MCQ Type										
		Numerical Type										
	Total											
1-PHASE & 3-PHASE CIRCUIT ANALYSIS	1 Mark	MCQ Type										
		Numerical Type			1			1			1	
	2 Marks	MCQ Type										
		Numerical Type			1							1
	Total				3			1			1	2
CIRCUIT THEOREMS	1 Mark	MCQ Type									1	
		Numerical Type							1			
	2 Marks	MCQ Type	1									
		Numerical Type								1		
	Total		2						1	2	1	
RESONANCE	1 Mark	MCQ Type										
		Numerical Type			1	1		1		1		
	2 Marks	MCQ Type								1	1	
		Numerical Type										
	Total				1	1		1		3	2	
MAGNETICALLY COUPLED CIRCUITS	1 Mark	MCQ Type										
		Numerical Type										
	2 Marks	MCQ Type										
		Numerical Type										
	Total											

A.C. FUNDAMENTALS

1. In figure, A_1, A_2 and A_3 are ideal ammeters. If A_1 reads 5 A, A_2 reads 12 A, then A_3 should read

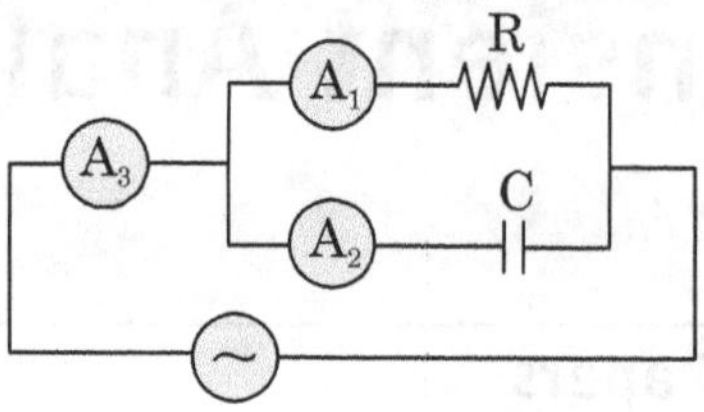

(a) 7 A (b) 12 A
(c) 13 A (d) 17 A

[1990 : 2 Marks]

2. The current, i(t), through a 10 Ω resistor in series with an inductance, is given by
$i(t) = 3 + 4\sin(100t + 45°) + 4\sin(300f + 60°)$ Amperes.
The RMS value of the current and the power dissipated in the circuit are:
(a) $\sqrt{41}$ A, 410 W respectively
(b) $\sqrt{35}$ A, 350 W respectively
(c) 5 A, 250 W respectively
(d) 11 A, 1210 W respectively

[1995 : 1 Mark]

3. In fig., A_1, A_2 and A_3 are ideal ammeters. If A_2 and A_3 read 3 A and 4 A respectively, then A_1 should read

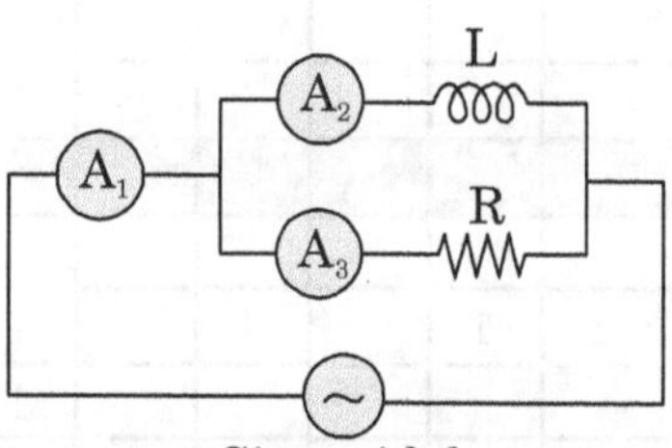

Sinusoidal voltage source

(a) 1 A (b) 5 A
(c) 7 A (d) None of the above

[1996 : 1 Mark]

4. A periodic variable x is shown in the figure as a function of time. The root-mean-square (rms) value of x is ______.

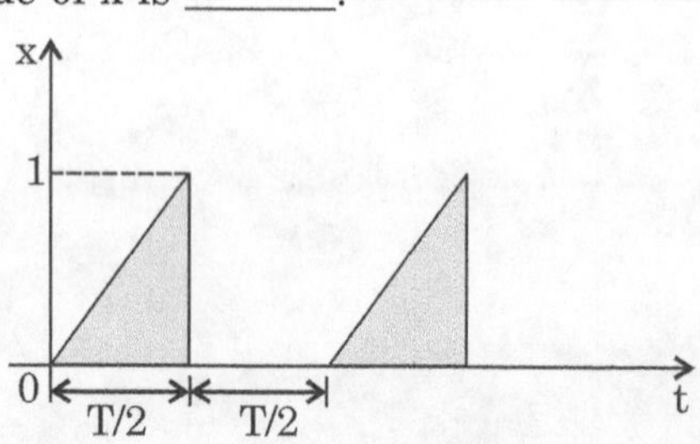

[2014 : 2 Marks, Set-1]

1-PHASE & 3-PHASE CIRCUIT ANALYSIS

5. If each branch of a Delta circuit has impedance $\sqrt{3}Z$, then each branch of the equivalent Wye circuit has impedance

(a) $\frac{Z}{\sqrt{3}}$ (b) 3Z
(c) $3\sqrt{3}Z$ (d) $\frac{Z}{3}$ **[1987 : 2 Marks]**

6. If the 3-phase balanced source in the figure delivers 1500 W at a leading power factor 0.844, then the value of Z_L (in ohm) is approximately

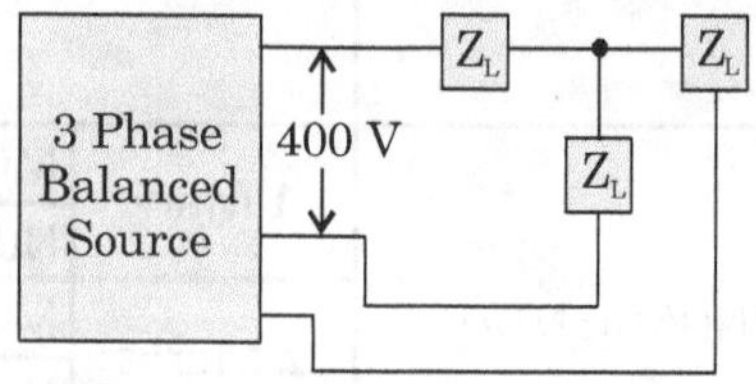

(a) $9\angle 32.44°$ (b) $80\angle 32.44°$
(c) $80\angle 32.44°$ (d) $90\angle -32.44°$

[1987 : 2 Marks]

7. In the circuit shown below, the current I is equal to
(a) $1.4\angle 0°$ A. (b) $2.0\angle 0°$ A.
(c) $2.8\angle 0°$ A. (d) $3.2\angle 0°$ A.

[1987 : 2 Marks]

8. In the circuit shown in the figure, the value of node voltage V_2 is

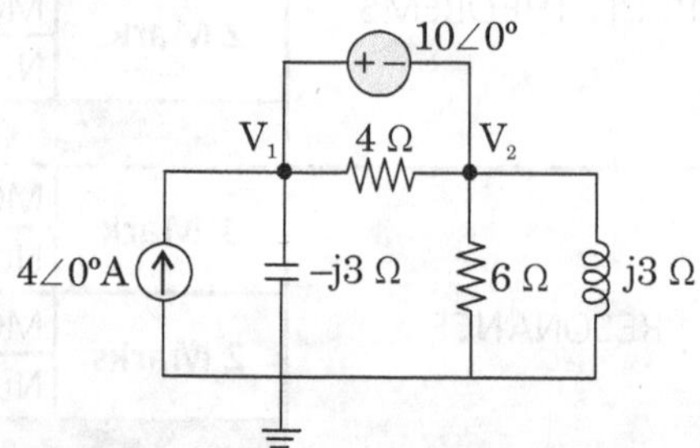

(a) 22 + j2 V. (b) 2 + j22 V.
(c) 22 – j2 V. (d) 2 – j22 V.

[1987 : 2 Marks]

9. The value of current through the 1 Farad capacitor of figure is

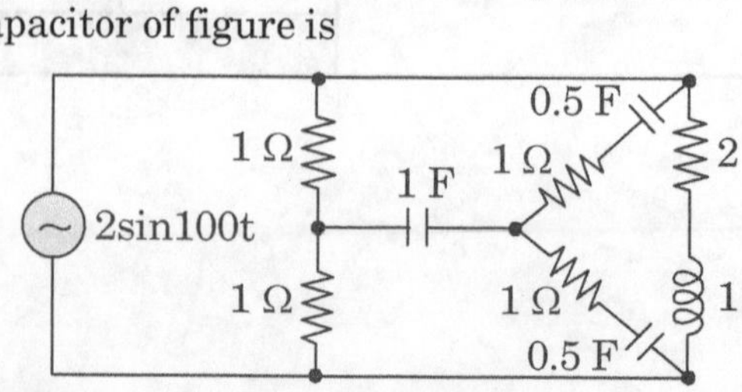

(a) zero (b) one
(c) two (d) three

[1987 : 2 Marks]

10. For the series RLC circuit of fig. 1, the partial phasor diagram at a certain frequency is a shown in fig. 2. The operating frequency of the circuit is

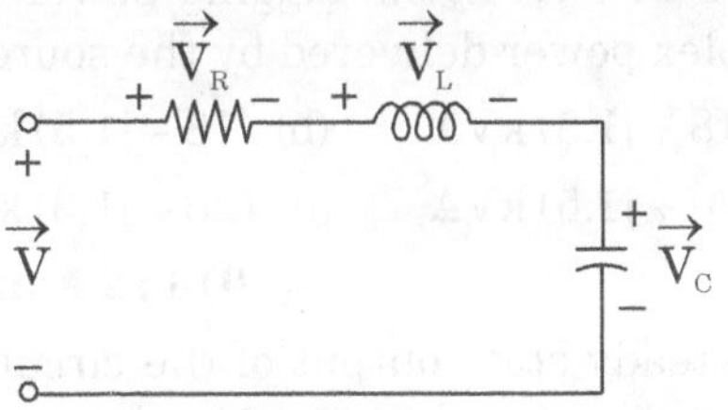

Figure-1

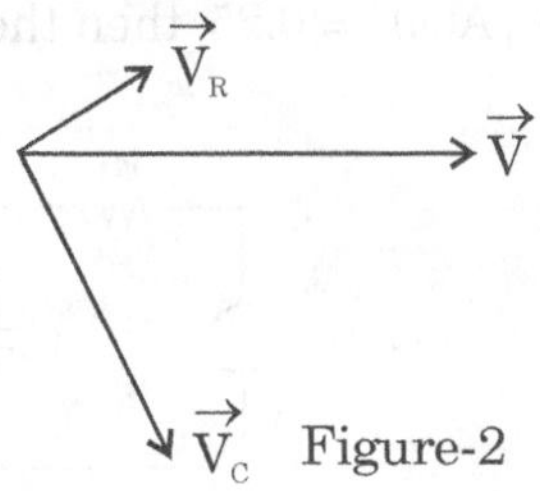

Figure-2

(a) equal to the resonance frequency
(b) less than the resonance frequency
(c) greater than the resonance frequency
(d) not zero

[1992 : 2 Marks]

11. The parallel RLC circuit shown in figure is in resonance. In this circuit

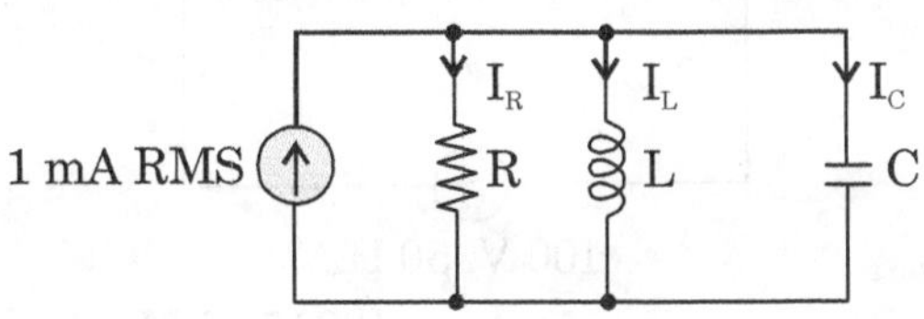

(a) $|I_R| < 1$ mA (b) $|I_R + I_L| > 1$ mA
(c) $|I_R + I_C| < 1$ mA (d) $|I_L + I_C| > 1$ mA

[1998 : 1 Mark]

12. When the angular frequency ω in the figure is varied from 0 to ∞, the locus of the current phasor I_2 is given by

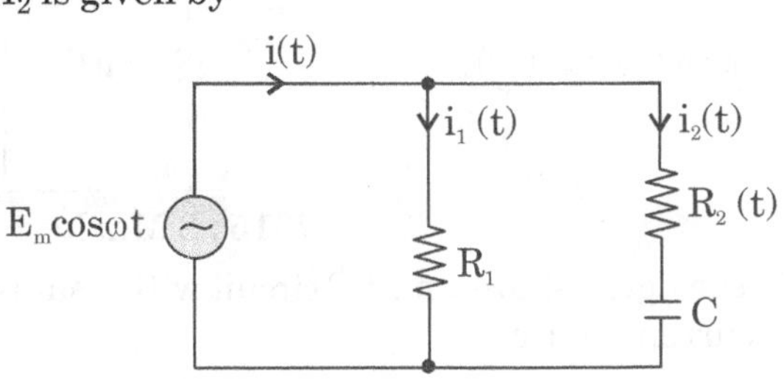

(a)

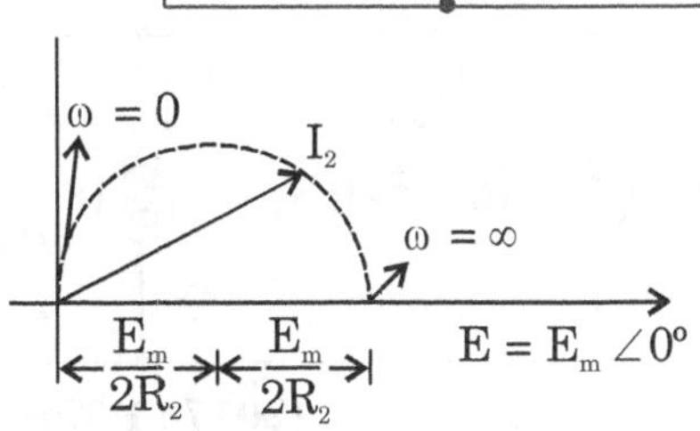

(b)

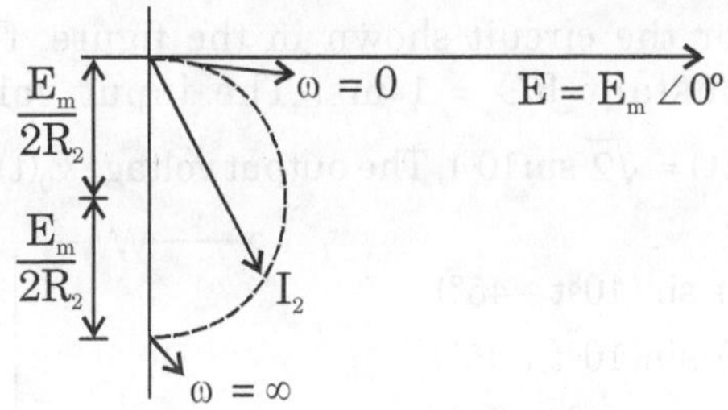

(c)

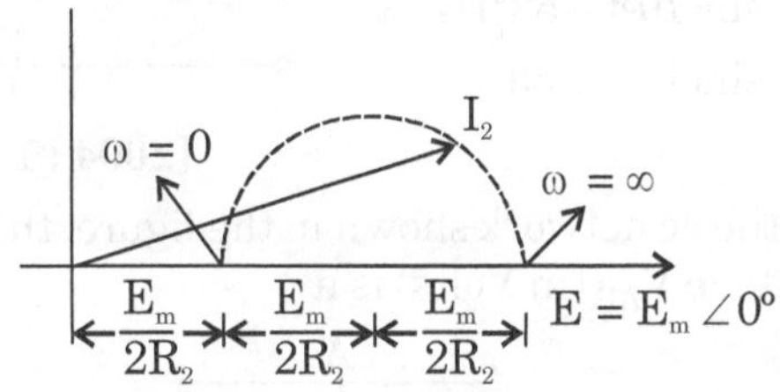

(d)

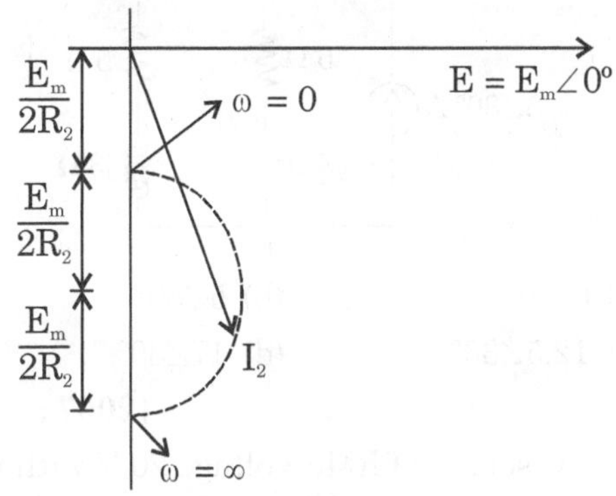

13. An input voltage $v(t) = 10\sqrt{5}\cos(t + 10°) + 10\sqrt{5}\cos(2t + 10°)$ V is applied to a series combination of resistance R = 1 Ω and an inductance L = 1 H. The resulting steady-state current i(t) in Ampere is

(a) $10\cos(t + 55°) + 10\cos(2t + 10° + \tan^{-1}2)$.

(b) $10\cos(t + 55°) + 10\sqrt{\frac{3}{2}}\cos(2t + 55)$.

(c) $10\cos(t - 35°) + 10\sqrt{\frac{3}{2}}\cos(2t + 10° - \tan^{-1}2)$

(d) $10\cos(t - 35) + 10\sqrt{\frac{3}{2}}\cos(2t - 35°)$.

[2003 : 2 Marks]

14. The circuit shown in the figure, with $R = \frac{1}{3}\,\Omega$, $L = \frac{1}{4}$ H, C = 3 F has input voltage v(t) = sin2t. The resulting current i(t) is

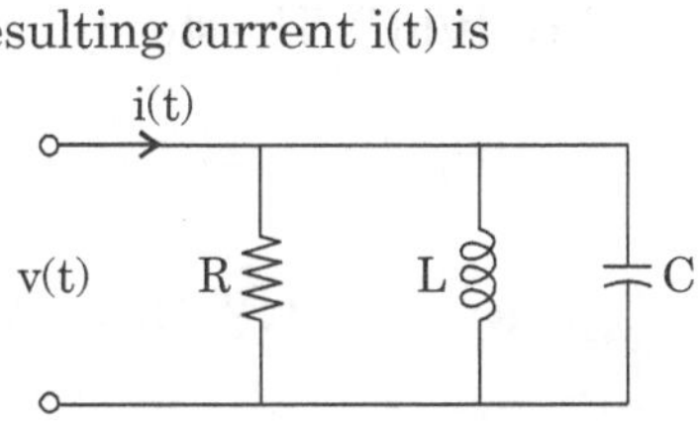

(a) 5sin(2t + 53.1°) (b) 5sin(2t – 53.1°)
(c) 25sin(2t + 53.1°) (d) 25sin(2t – 53.1°)

[2004 : 1 Mark]

15. For the circuit shown in the figure, the time constant RC = 1 ms. The input voltage is $v_i(t) = \sqrt{2}\sin 10^3 t$. The output voltage $v_0(t)$ is equal to

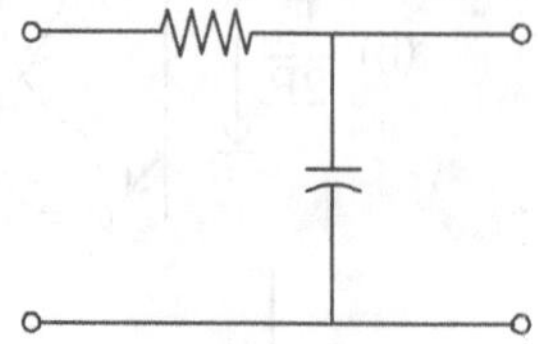

(a) $\sin(10^3t - 45°)$
(b) $\sin(10^3t + 45°)$
(c) $\sin(10^3t - 53°)$
(d) $\sin(10^3t + 53°)$

[2004 : 1 Marks]

16. In the ac network shown in the figure, the phasor voltage V_{AB} (in Volts) is a

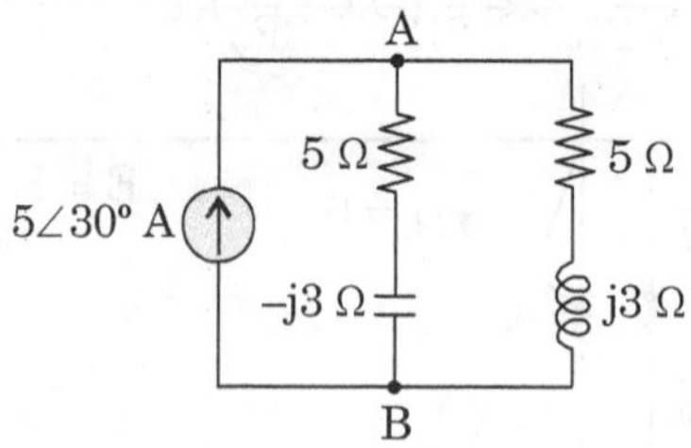

(a) 0
(b) 5∠30°
(c) 12.5∠30°
(d) 17∠30°

[2007 : 2 Marks]

17. An AC source of RMS voltage 20 V with internal impedance $Z_s = (1 + 2j)\,\Omega$ feeds a load of impedance $Z_L = (7 + 4j)\,\Omega$ in the figure below. The reactive power consumed by the load is

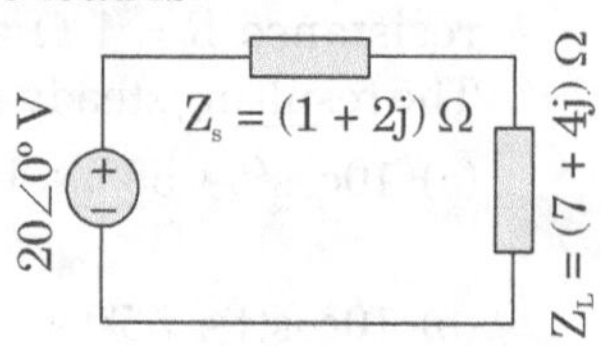

(a) 8 VAR
(b) 16 VAR
(c) 28 VAR
(d) 32 VAR

[2009 : 2 Marks]

18. The current I in the circuit shown is

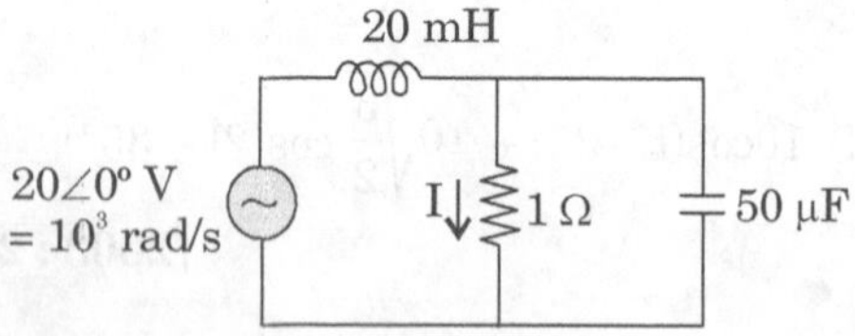

(a) –j1 A
(b) j1 A
(c) 0 A
(d) 20 A

19. The circuit shown below is driven by a sinusoidal input $v_i = V_p\cos(t/RC)$. The steady output v_0 is

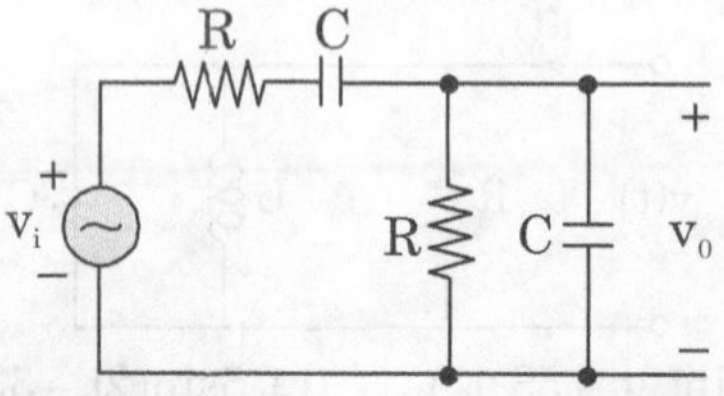

(a) $(V_p/3)\cos(t/RQ)$
(b) $(V_p/3)\cos(t/RC)$
(b) $(V_p/2)\sin(t/RC)$
(d) $(V_p/2)\sin(t/RC)$

20. A 230 V rms source supplies power to two loads connected in parallel. The first load draws 10 kW at 0.8 leading power factor and the second one draws 10 kVA at 0.8 lagging power factor. The complex power delivered by the source is

(a) (18 + j1.5) kVA
(b) (18 – j1.5) kVA
(c) (20 + j1.5) kVA
(d) (20 – j1.5) kVA

[2014 : 2 Marks, Set-1]

21. The steady state output of the circuit shown in the figure is given by $y(t) = A(\omega)\sin(\omega t + \phi(\omega))$. If the amplitude $|A(\omega)| = 0.25$, then the frequency ω is

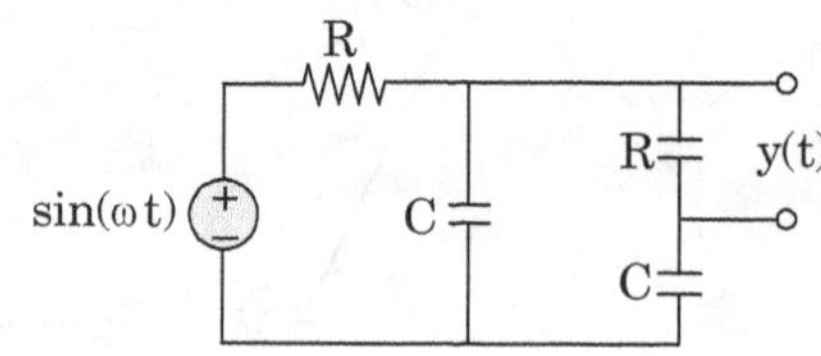

(a) $\dfrac{1}{\sqrt{3}RC}$

(b) $\dfrac{2}{\sqrt{3}RC}$

(c) $\dfrac{1}{RC}$

(d) $\dfrac{2}{RC}$

[2014 : 2 Marks, Set-4]

22. The voltage (V_c) across the capacitor (in Volts) in the network shown in ______.

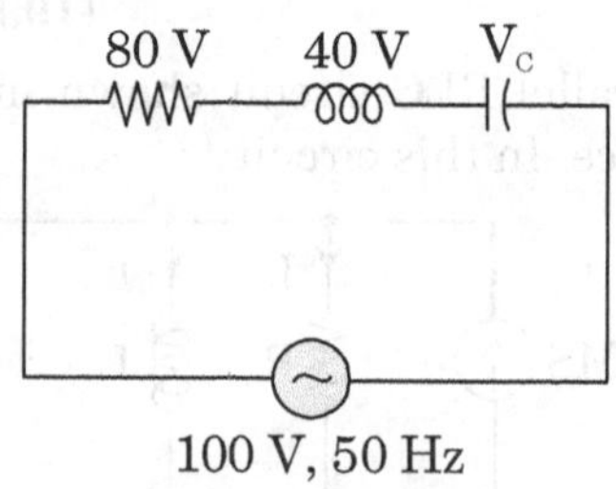

[2015 : 1 Mark, Set-2]

23. In the circuit shown, the current /flowing through the 50 Ω resistor will be zero if the value of capacitor C (in μF) is ______.

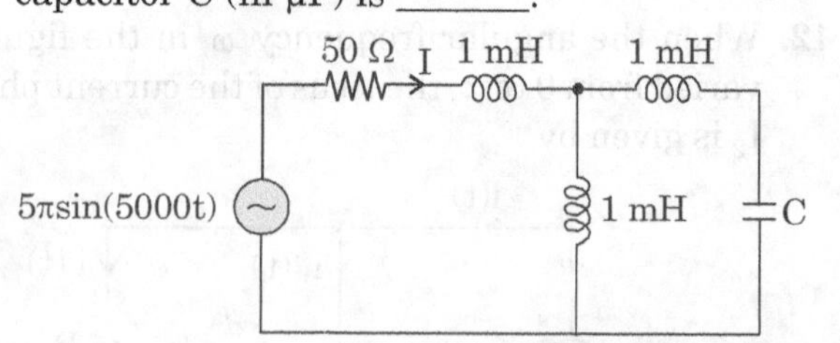

[2015 : 2 Marks, Set-3]

24. The figure shows an RLC circuit with a sinusoidal current source ______.

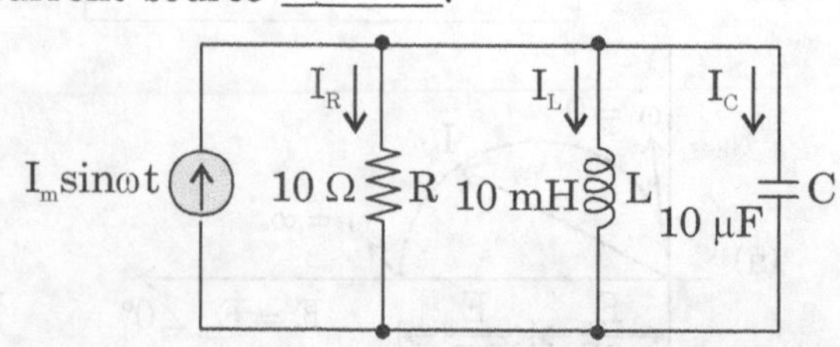

[2017 : 1 Mark, Set-1]

25. The figure shows an RLC circuit excited by the sinusoidal voltage 100cos(3t) Volts, where t is in seconds. The ratio $\frac{\text{amplitude of } V_2}{\text{amplitude of } V_1}$ is ______.

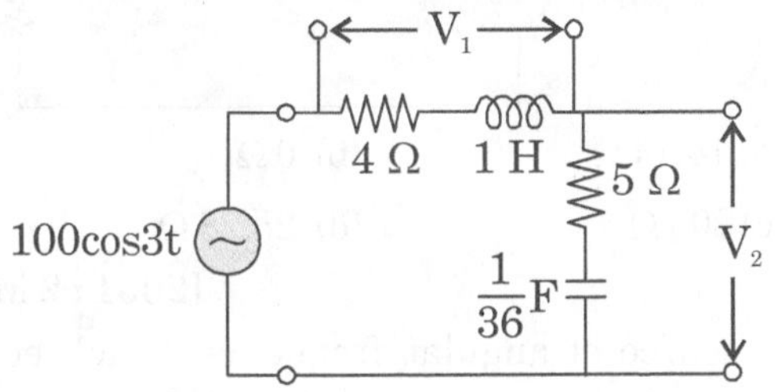

[2017 : 2 Marks, Set-1]

26. For the circuit given in the figure, the voltage V_C (in volts) across the capacitor is

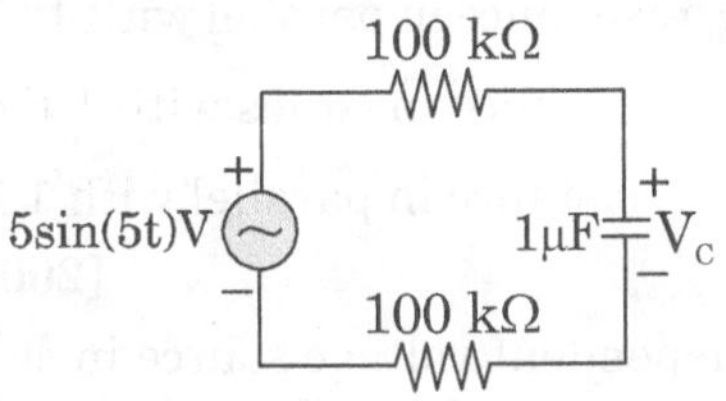

(a) $1.25\sqrt{2}\sin(5t - 0.2\pi)$

(b) $1.25\sqrt{2}\sin(5t - 0.125\pi)$

(c) $2.5\sqrt{2}\sin(5t - 0.25\pi)$

(d) $2.5\sqrt{2}\sin(5t - 0.125\pi)$

27. In the circuit of Fig. the equivalent impedance seen across terminals A – B is

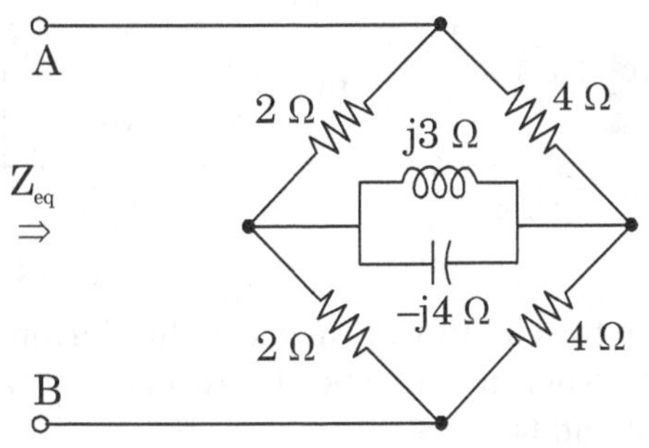

(a) (16/3) Ω

(b) (8/3) Ω

(c) (8/3+12j Ω

(d) None of the above

CIRCUIT THEOREMS

28. For the circuit shown in the figure, the instantaneous current $i_1(t)$ is

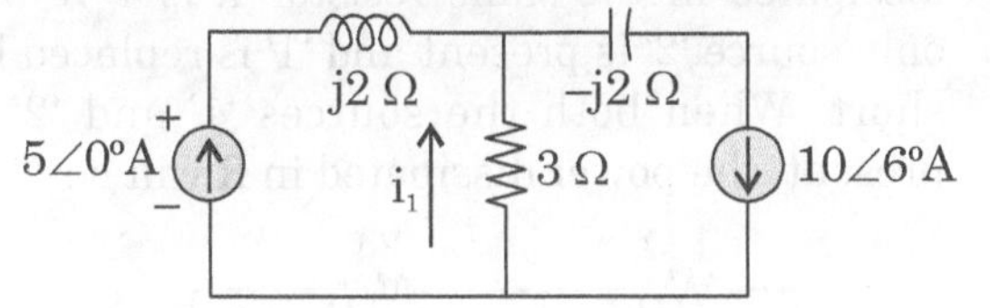

(a) $\frac{10\sqrt{3}}{2}\angle 90°$ A (b) $-\frac{10\sqrt{3}}{2}\angle 90°$ A

(c) $5\angle 60°$ A (d) $5\angle -60°$ A

[1987 : 2 Marks]

29. In the circuit shown, the average value of the voltage Vab (in Volts) in steady state condition is ______.

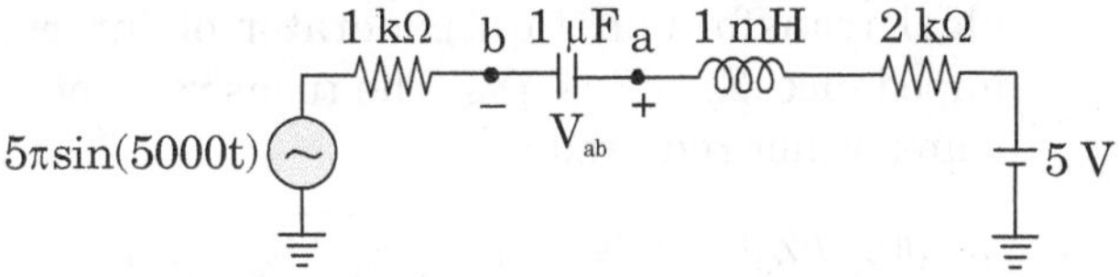

[1988 : 2 Marks]

30. In the RLC circuit shown in the figure, the input voltage is given by

$v_i(t) = 2\cos(200t) + 4\sin(500t)$

The output voltage $v_0(t)$ is

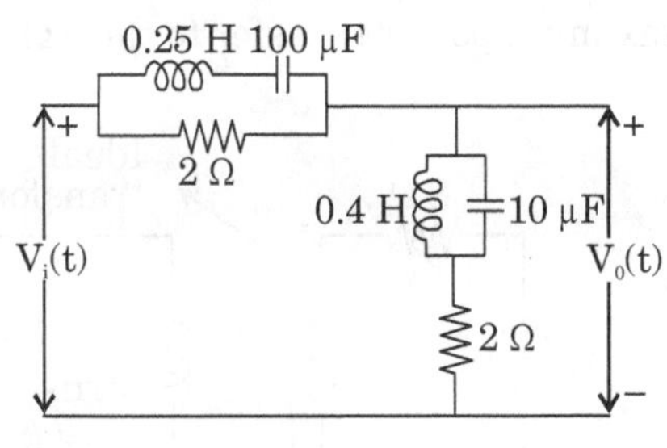

(a) cos(200t) + 2sin(500t)

(b) 2cos(200t) + 4sin(500t)

(c) sin(200t) + 2cos(500t)

(d) 2sin(200t) + 4cos(500t)

[1989 : 2 Marks]

31. If an impedance Z_L is is connected across a voltage source V with source impedance Z_s, then for maximum power transfer the load impedance must be equal to

(a) source impedance Z_S

(b) complex conjugate of Z_S

(c) real part of Z_S

(d) imaginary part of Z_S

[1989 : 2 Marks]

32. In the circuit of figure, the power dissipated in the resistor R is 1 W when only source '1' is present and '2' is replaced by a short. The power dissipated in the same resistor R is 4 W when only source '2' is present and '1' is replaced by a short. When both the sources '1' and '2' are present, the power dissipated in R will be

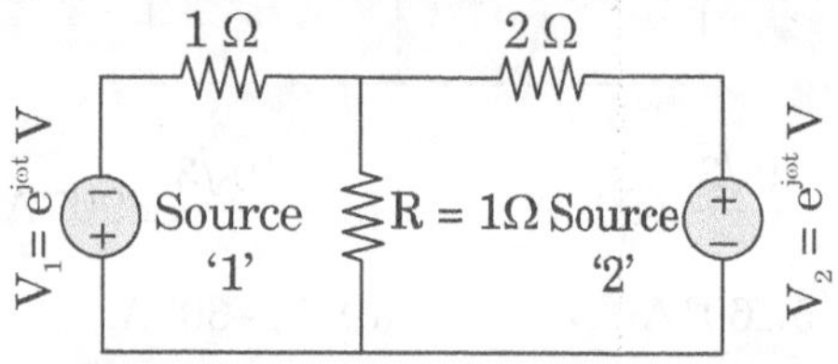

(a) 1 W (b) 3 W

(c) 3 W (d) 5 W **[1989 : 2 Marks]**

33. A load, $Z_L = R_L + jX_L$ is to be matched, using an ideal transformer, to a generator of internal impedance, $Z_S = R_S + jX_S$. The turns ratio of the transformer required is

(a) $\sqrt{|Z_L / Z_S|}$ (b) $\sqrt{|R_L / R_S|}$

(c) $\sqrt{|R_L / Z_S|}$ (d) $\sqrt{|Z_L / R_S|}$

[1992 : 2 Marks]

34. If the secondary winding of the ideal transformer shown in the circuit of figure has 40 turns, the number of turns in the primary winding for maximum power transfer to the 2 Ω resistor will be

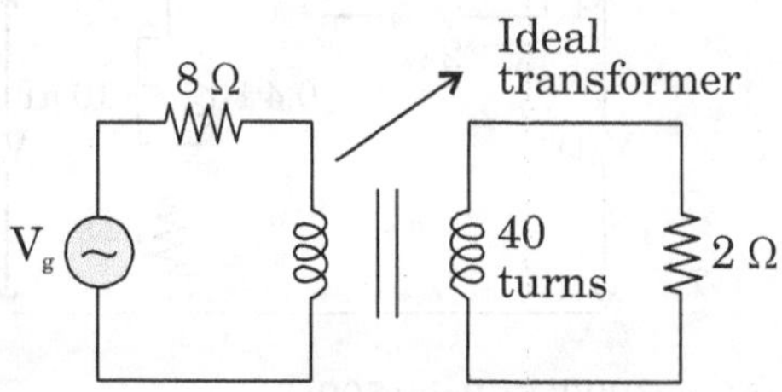

(a) 20 (b) 40

(c) 80 (d) 160 **[1993 : 1 Marks]**

35. A generator of internal impedance, Z_G, delivers maximum power to a load impedance, Z_L, only if Z_L = _____. **[1994 : 1 Marks]**

36. The Thevenin equivalent voltage V_{TH} appearing between the terminals A and B of the network shown in the figure is given by

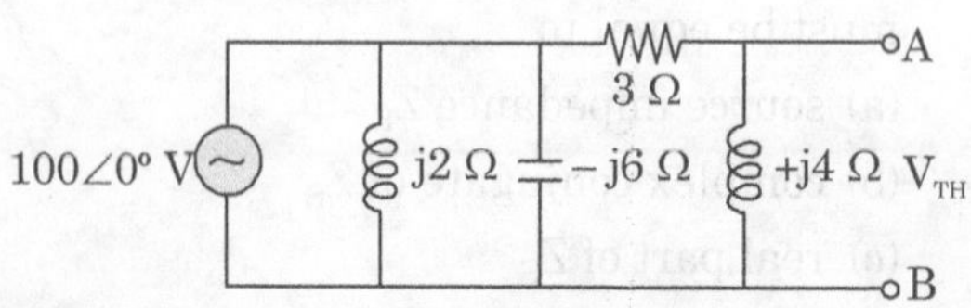

(a) j16(3–j4) (b) j16(3+j4)

(c) 16(3+j4) (d) 16(3–j4)

[1999 : 2 Marks]

37. In the figure, the value of the load resistor R which maximizes the powe delivered to it is

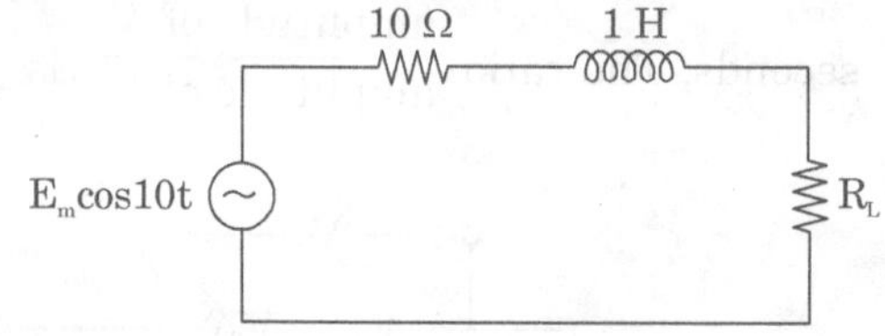

(a) 14.14 Ω (b) 0 Ω

(c) 200 Ω (d) 28.28 Ω

[2001 : 2 Marks]

38. A source of angular frequency 1 rad/sec has a source impedance consisting of 1 Ω resistance in series with 1 H inductace. The load that will obtain the maximum power transfer is

(a) 1 Ω resistance

(b) 1 Ω resistance in parallel with 1 H inductance

(c) 1 Ω resistance in series with 1 F capacitor

(d) 1 Ω resistance in parallel with 1 F capacitor

[2003 : 1 Mark]

39. An independent voltage source in series with an impedance $Z_S = R_S + jX_S$ delivers a maximum average power to a load impedance Z_L when

(a) $Z_L = R_S + jX_S$ (b) $Z_L = R_S$

(c) $Z_L = jX_S$ (d) $Z_L = R_S - jX_S$

[2007 : 1 Mark]

40. The Thevenin equivalen impedance Z_{Th} between the nodes P and Q in the following circuit is

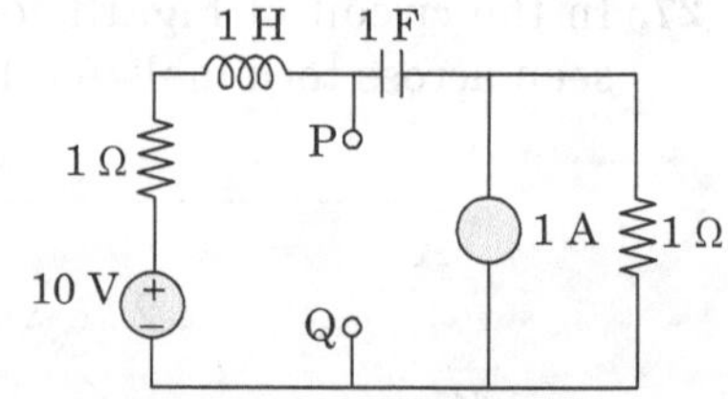

(a) 1

(b) $1 + s + \frac{1}{s}$

(c) $2 + s + \frac{1}{s}$

(d) $\frac{s^2 + s + 1}{s^2 + 2s + 1}$

[2008 : 2 Marks]

41. In the circuit shown below, the Norton equivalent currnet in amperes with respect to the terminals P and Q is

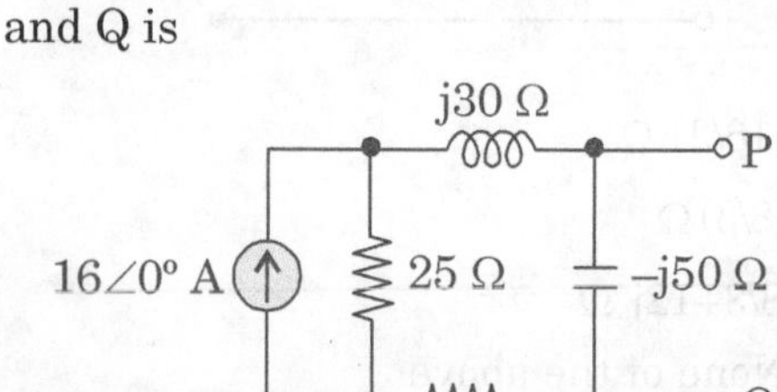

(a) 6.4 – j4.8 (b) 6.56 – j7.87

(c) 10 + j0 (d) 16 + j0

[2011 : 1 Mark]

42. Assuming both the voltage sources are in phase, the value of R for which maximum power is transferred from circuit A to circuit B is

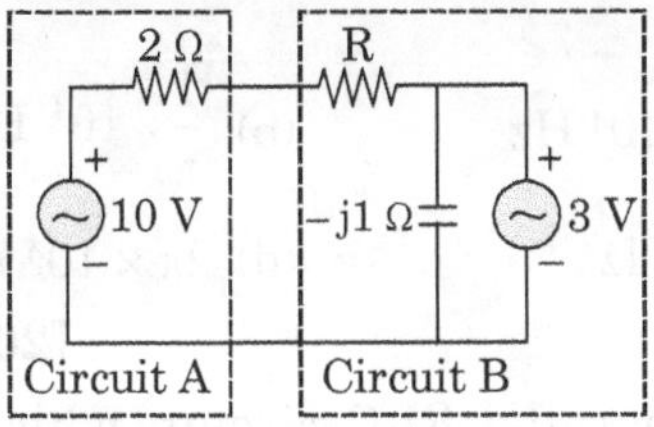

(a) 0.8 Ω (b) 1.4 Ω

(c) 2 Ω (d) 2.8 Ω

[2012 : 2 Marks]

43. A source $v_s(t) = V\cos 100\pi t$ has an internal impedance of $(4 + j3)\ \Omega$. If a purely resistive load connected to this surce has to extract the maximum power out of the source, its value in ohm should be

(a) 3 (b) 4

(c) 5 (d) 7 **[2013 : 1 Mark]**

44. In the circuit shown below, if the source voltage $V_S = 100\angle 53.13^\circ$ V then the Thevenin's equivalent voltage in Volts as seen by the load resistance R_L is

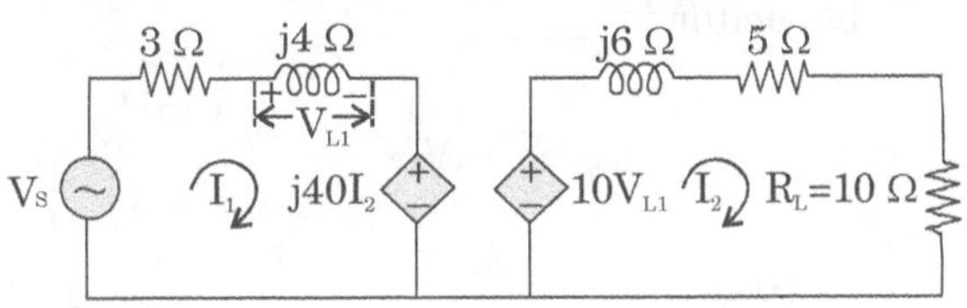

(a) $100\angle 90^\circ$ (b) $800\angle 90^\circ$

(c) $800\angle 90^\circ$ (d) $100\angle 60^\circ$

[2013 : 2 Marks]

45. In the circuit shown in the figure, the angular frequency ω (in rad/s), at which the Norton equivalent impedance as seen from terminals b–b' is purely resistive, is _______.

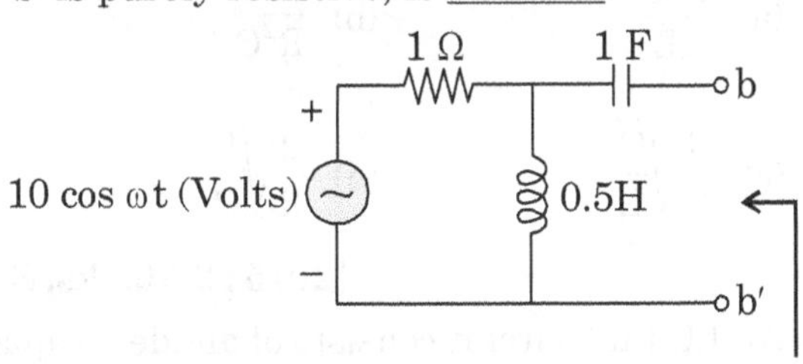

[2014 : 2 Mark, Set-3]

46. In the given circuit, the maximum power (in Watts) that can be transferred to the load R_L is

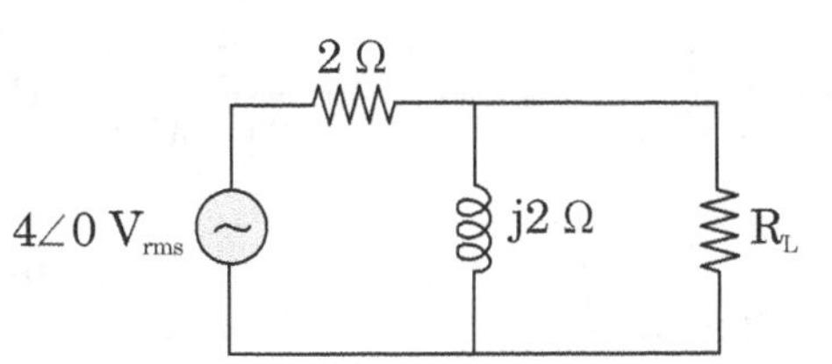

[2014 : 1 Mark, Set-3]

47. For maximum power transfer between two cascaded sections of an electrical network, the relationship betweeen the output impedance Z_1 of the first section to the input impedance Z_2 of the second section is

(a) $Z_2 = Z_1$ (b) $Z_2 = -Z_1$

(c) $Z_2 = Z_1^*$ (d) $Z_2 = -Z_1^*$

[2015 : 2 Marks, Set-1]

48. In the circuit shown, if v(t) = 2sin(1000t) volts, R = 1kΩ, and C = 1μF, then the steady-state current i(t), in milliamperes (mA) is

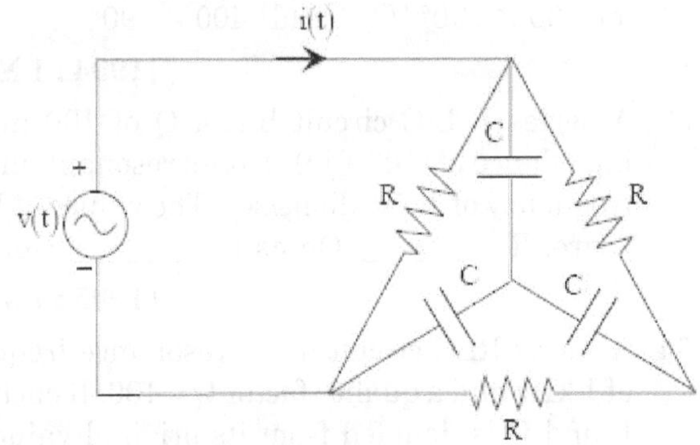

(A) sin(1000t) + cos(1000t)

(B) sin(1000t) + 3cos(1000t)

(C) 2sin(1000t) + 2cos(1000t)

(D) 3sin(1000t) + cos(1000t) **[2019 : 2 Marks]**

RESONANCE

49. The transfer function $H(s) = \dfrac{V_0(s)}{V_i(s)}$ of an R-L-C circuit is given by $H(s) = \dfrac{10^6}{s^2 + 20s + 10^6}$. The Quality factor (Q-factor) of this circuit is

(a) 25 (b) 50

(c) 100 (d) 5000

[1991 : 2 Marks]

50. For the circuit shown in the figure, the initial conditions are zero. Its transfer function

$H(s) = \dfrac{V_0(s)}{V_i(s)}$ is :

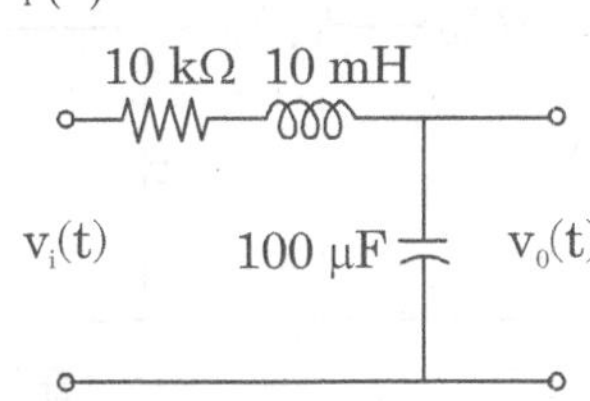

(a) $\dfrac{1}{s^2 + 10^6 s + 10^6}$ (b) $\dfrac{10^6}{s^2 + 10^3 s + 10^6}$

(c) $\dfrac{10^3}{s^2 + 10^3 s + 10^6}$ (d) $\dfrac{10^6}{s^2 + 10^6 s + 10^6}$

[1991 : 2 Marks]

51. In a series RLC high Q circuit, the current peaks at a frequency

(a) equal to the resonant frequency

(b) greater than the resonant frequency

(c) less than the resonant frequency

(d) None of the above is true **[1991 : 2 Marks]**

52. A series LCR circuit consisting of R = 10 Ω, $|X_L|$ = 20 Ω. and $|X_C|$ = 20 Ω. is connected across an ac supply of 200 V rms. The rms voltage across the capacitor is

(a) 200 ∠ –90° V. (b) 200 ∠ + 90° V.

(c) 400 ∠ +90° V. (d) 400 ∠ –90° V.

[1994 : 1 Mark]

53. A series R-L-C circuit has a Q of 100 and an impedance of (100 + j0) Ω at its resonant angular frequency of 10^7 radians/sec. The values of R and L are: R = _______ Ohms L = _______ Henries.

[1995 : 1 Mark]

54. A series RLC circuit has a resonance frequency of 1 kHz and a quality factor Q = 100. If each of R, L and C is doubled from its original value, the new Q of the circuit is

(a) 25 (b) 50

(c) 100 (d) 200 **[2003 : 1 Mark]**

55. Consider the following statements S_1 and S_2

S_1: At the resonant frequency the impedance of a series R-L-C circuit is zero.

S_2: In a parallel G-L-C circuit, increasing the conductance G results in increase in its Q factor.

Which one of the following is correct?

(a) S_1 is FALSE and S_2 is TRUE

(b) Both S_1 and S_2 are TRUE

(c) S_1 is TRUE and S_2 is FALSE

(d) Both S_1 and S_2 are FALSE

[2004 : 2 Marks]

56. The condition on R, L and C such that the step response y(t) in the figure has no oscillations, is

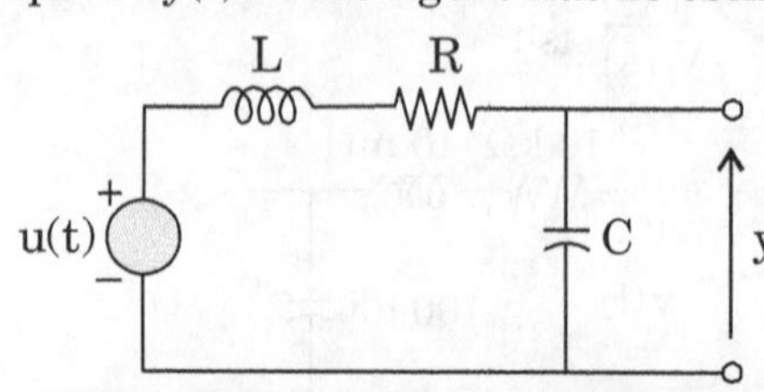

(a) $R \ge \frac{1}{2}\sqrt{\frac{L}{C}}$ (b) $R \ge \sqrt{\frac{L}{C}}$

(c) $R \ge 2\sqrt{\frac{L}{C}}$ (d) $R = \sqrt{\frac{1}{LC}}$

[2005 : 1 Mark]

57. In a series RLC circuit, R = 2 kΩ, L = 1 H, and $C = \frac{1}{400}$ μF. The resonant frequency is

(a) 2×10^4 Hz (b) $\frac{1}{\pi} \times 10^4$ Hz

(c) 10^4 Hz (d) $2\pi \times 10^4$ Hz

[2005 : 1 Mark]

58. For a parallel RLC circuit, which one of the following statements is NOT correct?

(a) The bandwidth of the circuit decreases if R is increased

(b) The bandwidth of the circuit remains same if L is increased

(c) At resonance, input impedance is a real quantity

(d) At resonance, the magnitude of input impedance attains its minimum value

[2010 : 1 Mark]

59. In the circuit shown, at resonance, the amplitude of the sinusoidal voltage (in Volts) across the capacitor is _______.

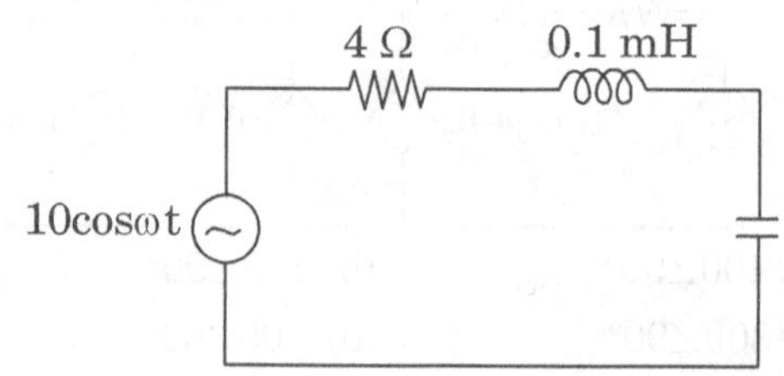

[2015 : 1 Marks, Set-1]

60. The damping ratio of a series RLC circuit can be expressed as

(a) $\frac{R^2C}{2L}$ (b) $\frac{2L}{R^2C}$

(c) $\frac{R}{2}\sqrt{\frac{C}{L}}$ (d) $\frac{2}{R}\sqrt{\frac{L}{C}}$

[2015 : 2 Marks, Set-1]

61. An LC tank circuit consists of an ideal capacitor C connected in parallel with a coil of inductance L having an internal resistance R. The resonant frequency of the tank circuit is

(a) $\frac{1}{2\pi\sqrt{LC}}$ (b) $\frac{1}{2\pi\sqrt{LC}}\sqrt{1-R^2\frac{C}{L}}$

(c) $\frac{1}{2\pi\sqrt{LC}}\sqrt{1-\frac{L}{R^2C}}$ (d) $\frac{1}{2\pi\sqrt{LC}}\left(1-R^2\frac{C}{L}\right)$

[2015 : 2 Marks, Set-2]

62. At very high frequencies, the peak output voltage V_0 (in Volts) is _______.

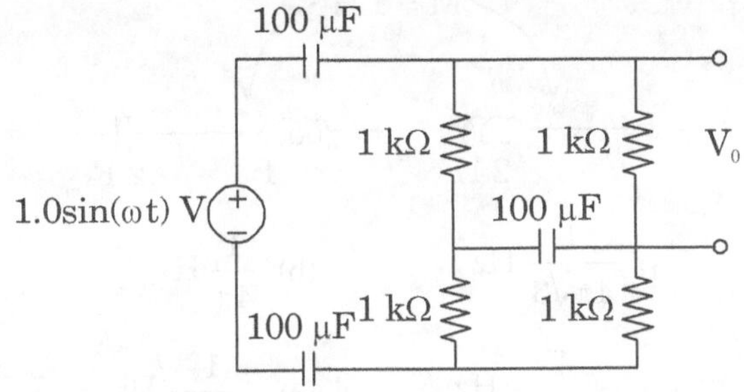

63. The figure shows an RLC circuit with a sinusoidal current source.

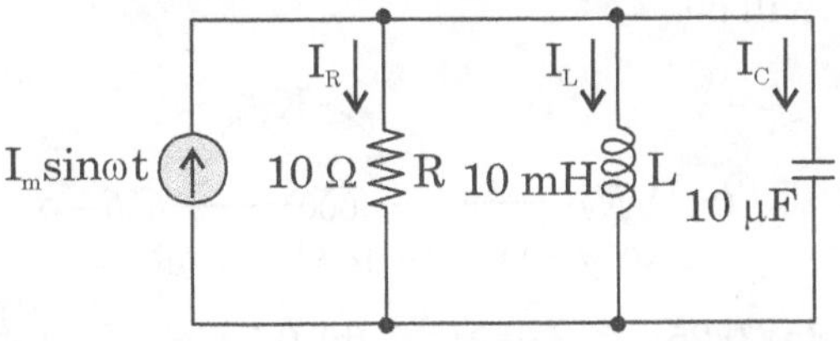

[2017 : 1 Mark, Set-1]

64. In the circuit shown, V is a sinusoidal voltage source. The current I is in phase with voltage V. The ratio

$$\frac{\text{amplitude of voltage across the capacitor}}{\text{amplitude of voltage across the capacitor}}$$

is _______.

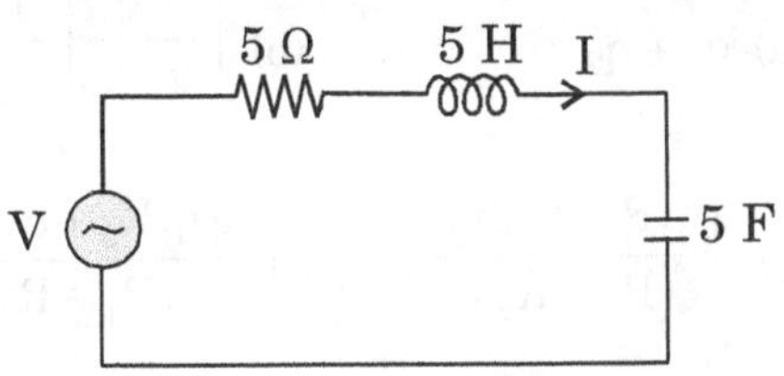

[2017 : 1 Mark, Set-2]

MAGNETICALLY COUPLED CIRCUITS

65. Two 2 H inductance coils are connected in series and are also magnetically coupled to each other the coefficient of coupling being 0.1. The total inductance of the combination can be

(a) 0.4 H

(b) 3.2 H

(c) 4.0 H

(d) 4.4 H

[1995 : 1 Mark]

66. The current flowing through the resistance R in the circuit in the figure has the form Pcos4t, where P is

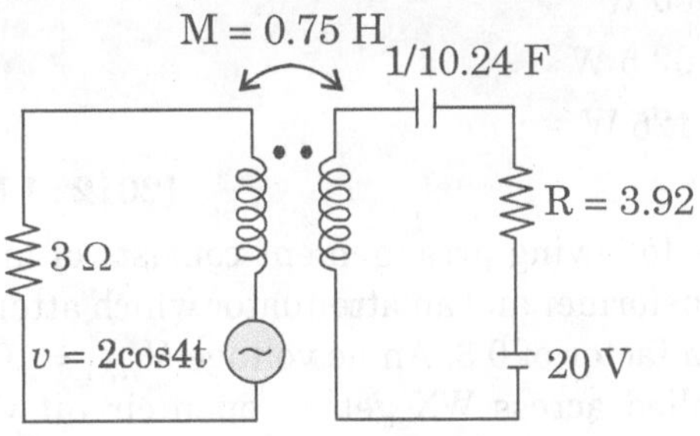

(a) (0.18 + j0.72)   (b) (0.46 + j1.90)

(c) –(0.18 + j 1.90)   (d) –(0.192 + j0.144)

[2003 : 2 Mark]

67. The equivalent inductance measured between the terminals 1 and 2 for the circuit shown in the figure is

(a) $L_1 + L_2 + M$

(b) $L_1 + L_2 - M$

(c) $L_1 + L_2 + 2M$

(d) $L_2 + L_2 - 2M$

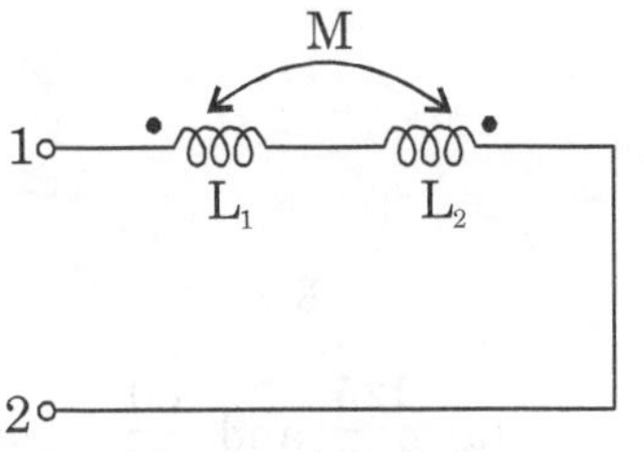

68. Impedance Z as shown in the given figure is

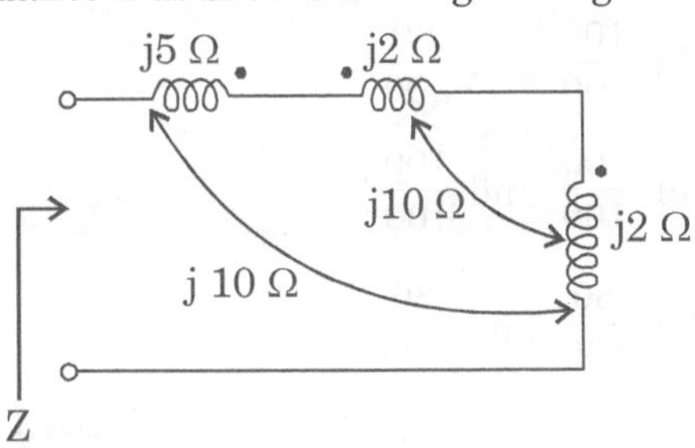

(b) j29 Ω   (b) j9 Ω

(c) j19 Ω   (d) j39 Ω

[2005 : 2 Marks]

69. In the circuit shown below, the current through the inductor is

(a) $\frac{2}{1+j}$ A

(b) $\frac{-1}{1+j}$ A.

(c) $\frac{1}{1+j}$ A.

(d) 0 A.

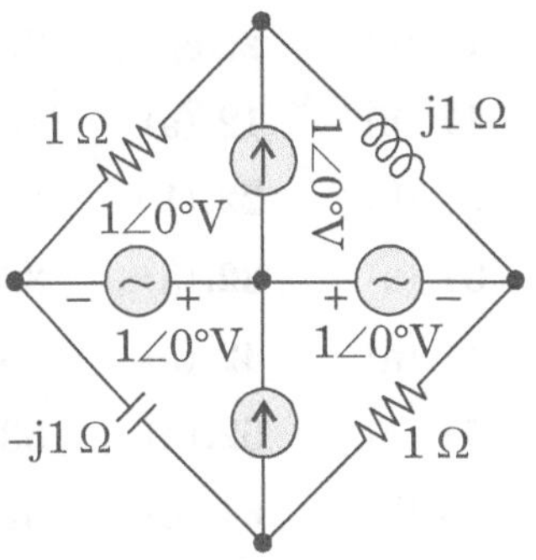

[2012 : 1 Mark]

70. The average power delivered to an impedance $(4 - j3)\ \Omega$ by a current $5\cos(100\pi t + 100)$ A is

(a) 44.2 W

(b) 50 W

(c) 62.5 W

(d) 125 W

[2012 : 1 Mark]

71. The following arrangement consists of an ideal transformer and an attenuator which attenuates by a factor of 0.8. An ac voltage $V_{WX1} = 100$ V is applied across WX get an open circuit voltage V_{YZ_1} across YZ. Next, an ac voltage $V_{YZ2} = 100$ V is applied across YZ to get an open circuit voltage V_{WX2} across WX. Then V_{YZ_1}/V_{WX_1}, V_{WX_2}/V_{YZ_2} are respectively,

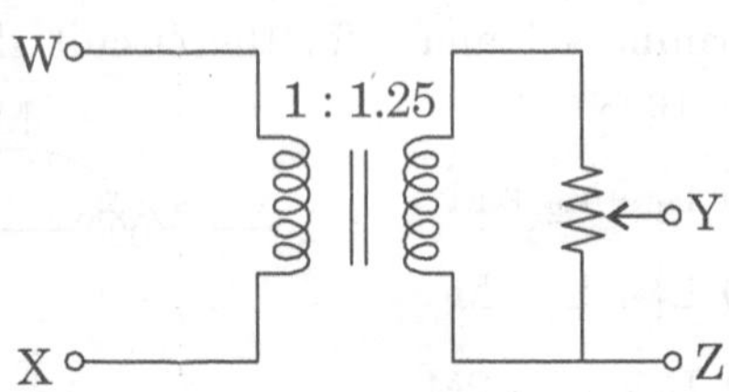

(a) $\frac{125}{100}$ and $\frac{80}{100}$

(b) $\frac{100}{100}$ and $\frac{80}{100}$

(c) $\frac{100}{100}$ and $\frac{100}{100}$

(d) $\frac{80}{100}$ and $\frac{80}{100}$

[2013 : 2 Marks]

72. The resonant frequency of the series circuit shown in figure is

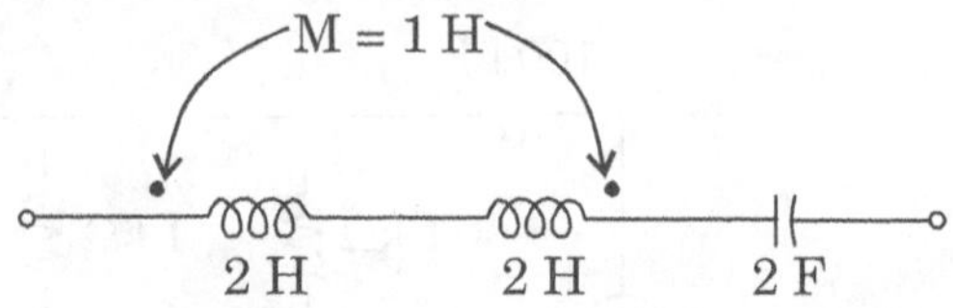

(a) $\frac{1}{4\pi\sqrt{3}}$ Hz

(b) $\frac{1}{4\pi}$ Hz

(c) $\frac{1}{2\pi\sqrt{10}}$ Hz

(d) $\frac{1}{4\pi\sqrt{2}}$ Hz

73. In the series circuit shown in figure, for series resonance, the value of the coupling coefficient k will be

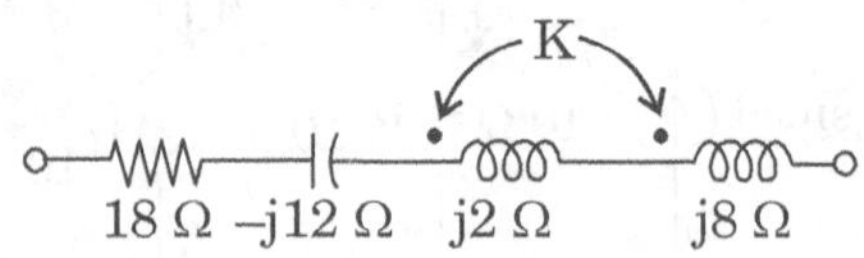

(a) 0.25

(b) 0.5

(c) 0.999

(d) 1.0

[2013 : 1 Mark]

74. Two magnetically uncoupled inductive coils have Q factors q_1 and q_2 at the chosen operating frequency. Their respective resistance are R_1 and R_2. When connected in series, their effective Q factor at the same operating frequency is

(a) $q_1 + q_2$

(b) $\left(\frac{1}{q_1}\right) + \left(\frac{1}{q_2}\right)$

(c) $\frac{(q_1R_1 + q_2R_2)}{(R_1 + R_2)}$

(d) $\frac{(q_1R_2 + q_2R_1)}{(R_1 + R_2)}$

[2013 : 2 Marks]

ANSWERS

1. (c)	**2.** (c)	**3.** (b)	**4.** (0.408)	**5.** (a)	**6.** (d)	**7.** (b)	**8.** (d)	**9.** (a)	**10.** (b)
11. (b)	**12.** (a)	**13.** (c)	**14.** (a)	**15.** (a)	**16.** (d)	**17.** (b)	**18.** (a)	**19.** (a)	**20.** (b)
21. (b)	**22.** (100)	**23.** (20)	**24.** (0.316)	**25.** (2.6)	**26.** (c)	**27.** (b)	**28.** (a)	**29.** (5)	**30.** (b)
31. (b)	**32.** (a)	**33.** (a)	**34.** (c)	**35.** (Z_G)	**36.** (d)	**37.** (a)	**38.** (c)	**39.** (d)	**40.** (a)
41. (a)	**42.** (a)	**43.** (c)	**44.** (c)	**45.** (2r)	**46.** (1.414)	**47.** (c)	**48.** (d)	**49.** (b)	**50.** (d)
51. (b)	**52.** (d)	**53.** (100, 10^3)		**54.** (b)	**55.** (d)	**56.** (c)	**57.** (b)	**58.** (d)	**59.** (25)
60. (b)	**61.** (b)	**62.** (0.5)	**63.** (0.316)	**64.** (0.2)	**65.** (d)	**66.** (*)	**67.** (d)	**68.** (b)	**69.** (c)
70. (b)	**71.** (c)	**72.** (b)	**73.** (a)	**74.** (c)					

EXPLANATIONS

1. $\tilde{I}_3 = \tilde{I}_1 + \tilde{I}_2 = \tilde{I}_1 + j\tilde{I}_2$

$$\tilde{I}_3 = \sqrt{I_1^2 + I_2^2} = \sqrt{5^2 + 12^2}$$

$$\tilde{I}_3 = \sqrt{169} = 13 \text{ A.}$$

2. $i = 3 + 4\sin(1000 + 45°) + 4\sin(300 + 60°)$ A

$$\text{RMS value} = \sqrt{3^2 + \left(\frac{4}{\sqrt{2}}\right)^2 + \left(\frac{4}{\sqrt{2}}\right)^2}$$

$$= 5 \text{ A}$$

$$\text{Power} = i^2 R$$

$$= 25 \times 10$$

$$= 250 \text{ W}$$

3. $\tilde{I}_1 = \tilde{I}_2 + \tilde{I}_3 = \tilde{I}_2 + j\tilde{I}_3$

$$I_1 = \sqrt{I_2^2 + I_3^2} = \sqrt{3^2 + 4^2} = 5 \text{ A.}$$

4. $$x(t) = \begin{cases} \frac{2}{T}.t; & 0 \le t, \frac{T}{2} \\ 0; & \frac{T}{2} \le t < T \end{cases}$$

RMS value of

$$x(t) = \sqrt{\frac{1}{T}\int_0^T x^2(t)\,dt} = \sqrt{\frac{1}{T}\int_0^{\frac{T}{2}} \left(\frac{2}{T}\right)^2 dt + 0}$$

$$= \sqrt{\frac{1}{T} \times \frac{4}{T^2} \times \left(\frac{t^3}{3}\right)_0^T}$$

$$= \sqrt{\frac{4}{T^3} \cdot \frac{T^3}{24}} = \sqrt{\frac{1}{6}} = 0.408$$

5. If the impedances are balanced then,

$$Z_\Delta = 3Z_Y$$

$$\Rightarrow Z_Y = \frac{\sqrt{3}\,Z}{3} = \frac{Z}{\sqrt{3}}.$$

6. $3\,V_p I_p \cos\theta = 1500$

$$\text{or } 3.\left(\frac{V_L}{\sqrt{3}}\right).\left(\frac{V_L}{\sqrt{3}\,Z_L}\right)\cos\theta = 1500$$

$$\text{or } Z_L = \frac{V_L^2 . \cos\theta}{1500}$$

$$= \frac{(400)^2.(0.844)}{1500} = 90\,\Omega$$

and $\theta = \angle - \cos^{-1}(0.844)$

7.

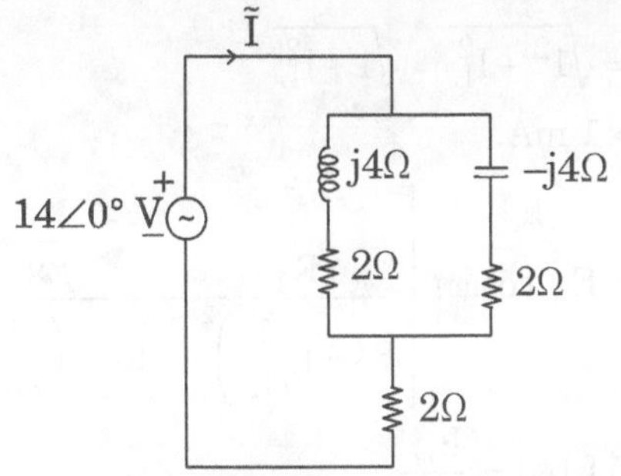

Converting delta connected resistors into star form we get the circuit shown below;

Equivalent impedance of the circuit,

$$\tilde{Z} = \frac{(2 + j4)(2 - j4)}{2 + j4 + 2 - j4} + 2 = \frac{4 + 16}{4} + 2$$

$$\tilde{Z} = 7\,\Omega$$

$$\text{Current}, \tilde{I} = \frac{14\angle 0°}{\tilde{Z}} = \frac{14\angle 0°}{7\angle 0°} = 2\angle 0° \text{A.}$$

9. Redrawing the given circuit as shown below:

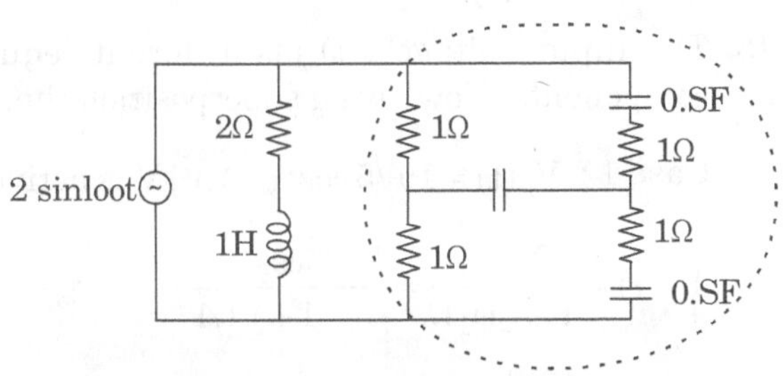

In above Circuit a balanced bridge is formed. Hence, no current will how through 1 F capacitor.

10. In series RLC circuit, voltage across rerstance, $\left(\text{hence } \tilde{V}_R\right)$ is is sa,e phase with arrent $\tilde{I}_R$ and this current is same is all the elements.

In given phasor diagram, $\left(\text{hence } \tilde{V}_R\right)$ is leading the voltage $\tilde{V}$ in the circuit. So, the given circuit is behaving as capacitive circuit. Now,

$$\tilde{V}_C > \tilde{V}_L$$

$$j\tilde{I}_R X_C > j\tilde{I}_R X_L$$

$$X_C > X_L$$

$$\frac{1}{\omega C} > \omega L$$

$$\omega^2 < \frac{1}{LC} < \omega r^2$$

$$\boxed{\omega < \omega_r \text{ and } \omega \ne 0}$$

11. At resonance in parallel RLC circuit,

$I_R = I = 1$ mA.

$|I_R + I_L| = \sqrt{1^2 + I_L^2} = \sqrt{1 + I_L^2}$

$|I_R + I_L| > 1$ mA.

12. $$I_2(\omega) = E_m \cos \omega t \left[\frac{R_2}{R_2^2 + \left(\frac{1}{\omega C}\right)^2} - \frac{\frac{1}{j\omega e}}{R_2^2 + \left(\frac{1}{\omega C}\right)^2} \right]$$

At $\omega = \infty$, $I_2(\omega) = \frac{E_m}{R_2}$.

Also, I_2 is leading with voltage phasor.

At $\omega = 0$, $I_2(0) = 0$.

Thus desired locus should be

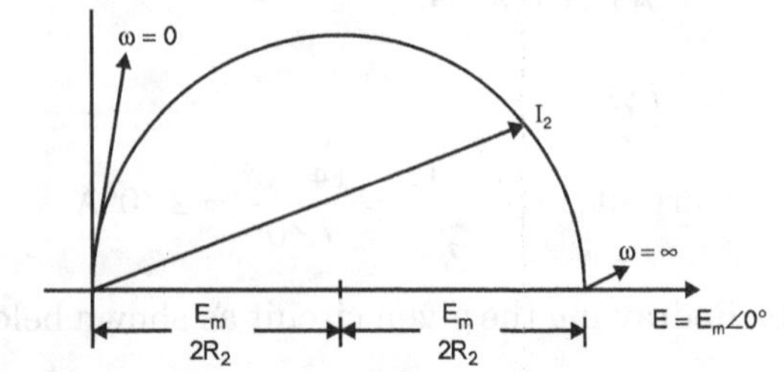

13. The input voltage v(t) jas different fequency components. Now, using superposition theorem;

Case 1 : $V_1(t) = 10\sqrt{5}\cos(t + 10°)$ V is active.

$$L_1'(t) = \frac{V_1(t)}{R + j\omega_1 L} = \frac{10\sqrt{5}\text{cas}(t + 10°)}{1 + j.1.1}$$

$$= \frac{10\sqrt{2}\text{cas}(t + 10°)}{\sqrt{2}\angle 45°}$$

$L_1'(t) = 10.\text{cas}(t - 30°)$ A.

Case 2 : $V_2(t) = 10\sqrt{5}\cos(2t + 10°)$ V is active.

$$i_2(t) = \frac{V2(t)}{R + j\omega_2 L} = \frac{10\sqrt{5}\cos(2t + 10°)}{1 + j.2.1}$$

$$= \frac{10\sqrt{5}\cos(2t + 10°)}{\sqrt{5}\angle \tan^{-12}}$$

Now, $i(t) = i_1(t) + i_2(t)$

$= 10\cos(t - 350°) + 10\cos(2t + 10° - \tan^{-12})$.

14. Not admittance,

$$Y = \left(\frac{1}{R} + \frac{1}{j\omega L} + j\omega c\right); \omega = \text{rad/s}.$$

$$Y = \left(3 + \frac{1}{j \times 2 \times \frac{1}{4}} + j \times 2 \times 3\right) = (3 - j^2 \times j^6) = 3 + j^4$$

The resulting current,

$i(t) = Y.u(t)$

$= (3 + j^4).\sin 2t$

$= 5\angle 53.13°.\sin 2t$

$= 5\sin(2t + 53.13°)$ A.

15. $$v_0(t) = \frac{v_i(t)}{R + \frac{1}{j\omega C}} \cdot \left(\frac{1}{j\omega C}\right)$$

$$= \frac{\sqrt{2}\sin 10^3 t}{j + 1} = \sin(10^3 t - 45°)$$

16. $Z_{AB} = Z_1 \parallel Z_2 = (5 - 3j) \parallel (5 + 3j)$

$Z_{AB} = \frac{34}{10}$

$V_{AB} = I.\,Z_{AB}$

$= (5\angle 30°)(\frac{34}{10}\angle 0°)$

$= 17\angle 30°$ volts

17. $$I = \frac{20}{z_{total}} = \frac{20}{\sqrt{8^2 + 6^2}} = 2\text{ A}$$

$[\because Z_{total} = Z_s + Z_c = 8 + 6j]$

∴ Reactive power

= Power on reactive element of load

$= 4 \times 2^2 = 16$ VAR

18. Applying nodal analysis at node (1),

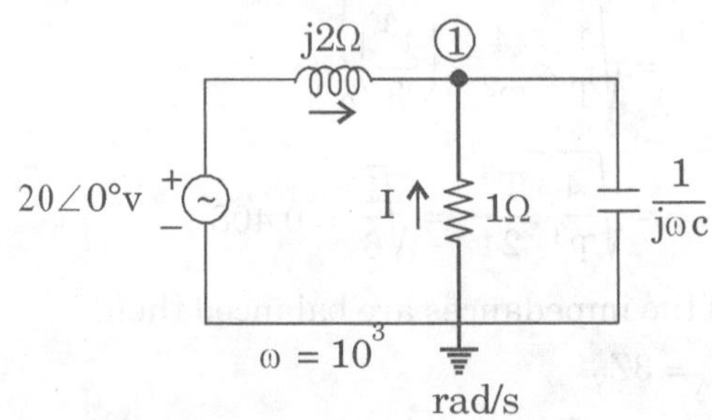

$$\frac{20\angle 0° - V_1}{j \times 10^3 \times 20 \times 10^{-3}} = \frac{V_1}{1} + V_i\left(j \times 10^3 \times 50 \times 10^{-6}\right)$$

$$\frac{20\angle 0°}{j \times 20} = \frac{V_1}{20} = V_i + j\frac{V_1}{20}$$

$$-j1 = V_1 + J\frac{V_1}{20} - j\frac{V_1}{20}$$

$V_1 = -j1$ Volt.

$$\Rightarrow I = \frac{V_1}{1} = -j1\text{ A}.$$

19. Since, $V_i = V_p \cos\left(\frac{t}{RC}\right)$; $\omega = \frac{1}{RC}$

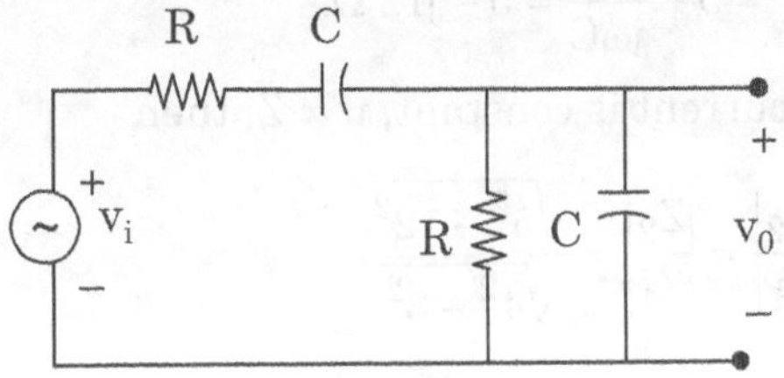

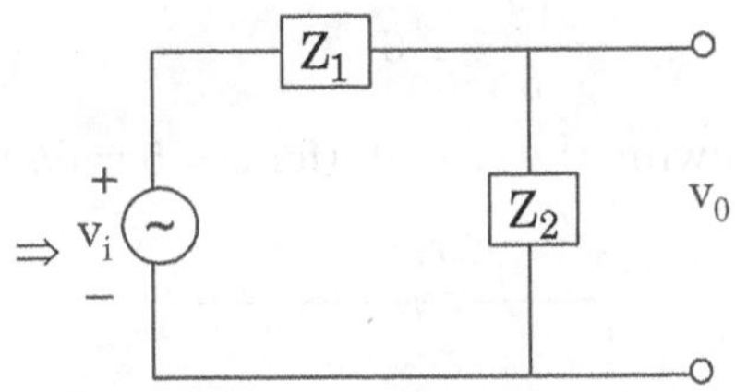

$$\Rightarrow \tilde{Z}_1 = R + \frac{1}{j\omega c}$$

$$R + \frac{1}{j.\frac{1}{RC}.C} = R + \frac{R}{j} = R(1-j)$$

$$\Rightarrow \tilde{Z}_2\left(R\,||\,\frac{1}{j\omega C}\right) = \left(R\,||\,\frac{1}{j\frac{1}{RC}.C}\right)$$

$$= \left(R\,||\,\frac{R}{j}\right) = R(1\,||\,-j1)$$

$$\tilde{Z}_2 = \frac{(-j1)1R}{1-j1} \times \frac{1+j1}{1+j1} = \frac{R(1-j)}{2}$$

Using voltage division theorem,.

$$v_0 = \frac{\tilde{Z}_2}{\tilde{Z}_1 + \tilde{Z}_2}.v_i = \frac{\frac{R(1-j)}{2}}{\frac{R}{2}(1-j) + R(1-j)}$$

$$v_i = \frac{1}{3}.v_i$$

$$v_0 = \left(\frac{v_p}{3}\right)\cos\left(\frac{t}{RC}\right).$$

20. Load I : - P_1 = 10 kW, cos ϕ_1 = 0.8 (lead)

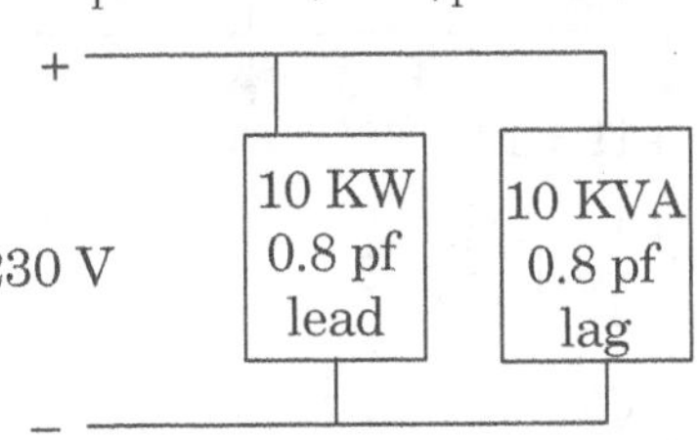

$Q_1 = -\tan \phi_1 .\, P_1.$

$$Q_1 = -10 \times \frac{3}{4} = -7.5 \text{ kVAR}$$

$$\tilde{S}_1 = P_1 + jQ_1 = 10 - j7.5 \text{ kVA}$$

Load II:- S_2 = 10 kVA, cosϕ_2 = 0.8 (lag)

$P_2 = S_2 .\cos\phi_2 = 10 \times 0.8 = 8$ kW

$Q_2 = +S_2 .\sin\phi_2 = 10 \times 0.6 = 6$ kVAR

$$\tilde{S}_2 = 8 + j6 \text{ kVA}$$

The complex power delivered by the source

$$\tilde{S} = \tilde{S}_1 + \tilde{S}_2 = 10 - 7.5 + 8 + j6$$

$$\tilde{S} = 18 - j1.5 \text{ kVA}$$

21.

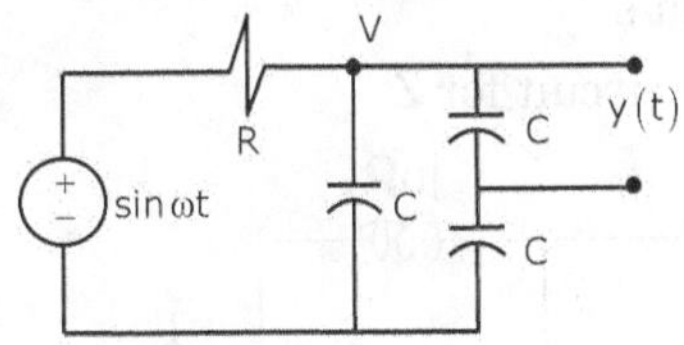

By nodal method,

$$\frac{V - 1\angle 0°}{R} + \frac{V}{\left(1/j\omega C\right)} + \frac{V}{\left(2/j\omega C\right)} = 0$$

$$V\left[\frac{1}{R} + j\omega C + \frac{j\omega C}{2}\right] = \frac{1\angle 0°}{R}$$

$$V = \frac{2}{2 + 3j\omega RC}$$

$$Y = \frac{V}{2} \Rightarrow \frac{1}{2 + j\omega 3RC}$$

$$\text{Given } |A(\omega)| = \frac{1}{4} \Rightarrow \frac{1}{\sqrt{4 + 9R^2c^2.\omega^2}}$$

$$\Rightarrow \omega = \frac{2}{\sqrt{3}RC}$$

22.

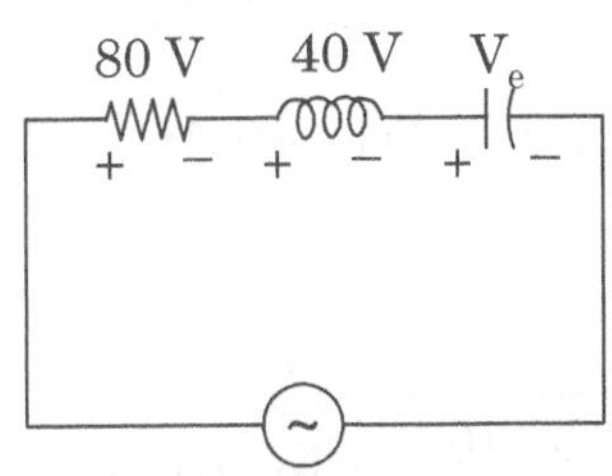

$100V_2$, 50 Hz

Applying KVL in loop

$-100 + 80 + j40 - jV_C = 0$

$100 = 80 + j(40 - V_C)$

$100^2 = 80^2 + (40 - V_C)^2.$

$$40 - V_C = \sqrt{100^2 - 80^2} = \sqrt{3600} = 60$$

V_C = 100 Volts

23.

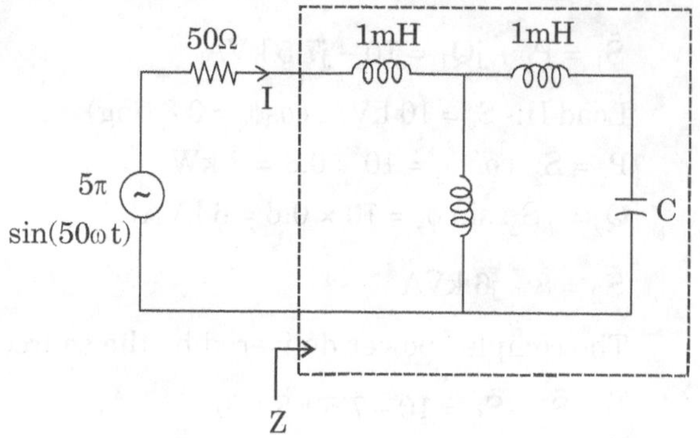

The current I is zero if impedance z is infinite.

$\Rightarrow \omega = 5000$ rad/s

Redrawing the circuit for Z

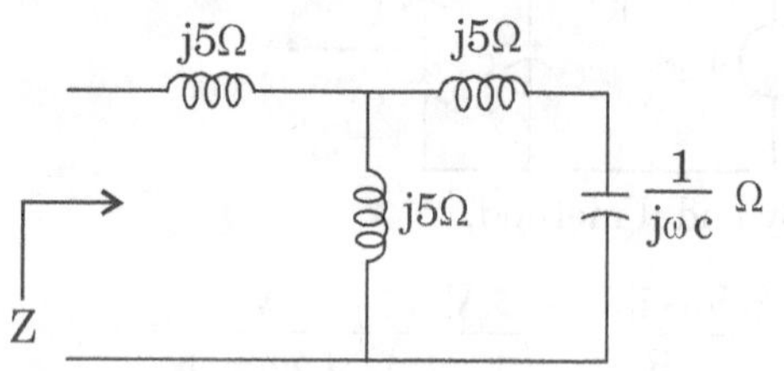

$$\Rightarrow \tilde{X}_L = j\omega L = j \times 5 \times 10^3 \times 10^{-3} = j5\ \Omega.$$

$$\Rightarrow Z = j5 + \frac{j5 \times \left(j5 + \frac{1}{j\omega C}\right)}{j5 + j5 + \frac{1}{j\omega C}}$$

For $Z \to \infty$, $j5 + j5 + \frac{1}{j\omega C} = 0$

$$j10 - \frac{j}{\omega C} = 0$$

$$C = \frac{1}{10\omega} = \frac{1}{5 \times 10^3 \times 10} = 20\ \mu F$$

24. For parallel RLC circuit in resonance condition,

$\tilde{I}_R = \tilde{I}$,

$\tilde{I}_L = Q.\tilde{I}\angle -90°$, $\tilde{I}_C = Q.\tilde{I}\angle -90°$

$$\Rightarrow \left|\frac{I_L}{I_R}\right| = \frac{QI}{I}$$

$$Q = \sqrt{\frac{C}{L}} = 10\sqrt{\frac{10 \times 10^{-6}}{10 \times 10^{-3}}} = 10 \times 0.0316 = 0.316$$

25.

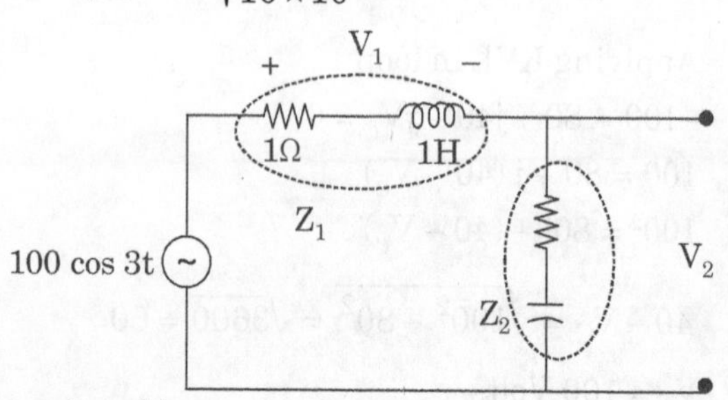

$\omega = 3$ rad/s.

$$\tilde{Z}_1 = 4 + j\omega.1 = 4 + j3\ \Omega$$

$$\tilde{Z}_2 = 5 + \frac{1}{j\omega C} = 5 - j12\ \Omega$$

If current is constant, $u \propto Z$, then

$$\frac{|V_2|}{|V_1|} = \frac{|Z_2|}{|Z_1|} = \frac{\sqrt{5^2 + 12^2}}{\sqrt{4^2 + 3^2}}$$

$$= \frac{13}{5} = 2.6$$

26. Redrawing the circuit, (for $\omega = 5$ rad/s)

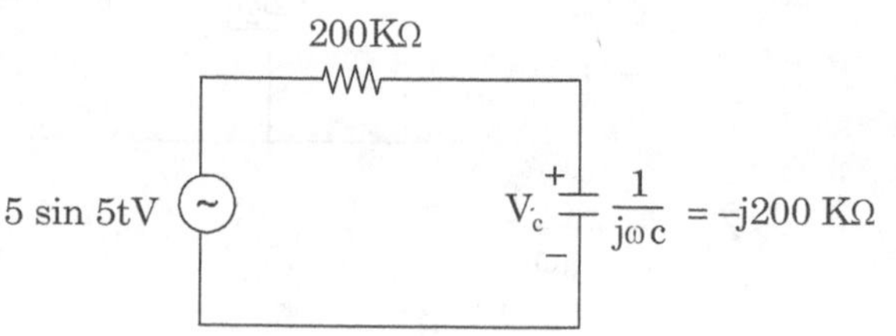

$$\tilde{V}_C = 5\angle 0° \times \frac{(-j200)}{200 - j200}$$

$$= \frac{5\angle \times 1\angle -90°}{\sqrt{2}\angle -45°}$$

$$= \frac{5}{\sqrt{2}}\angle -45° = 2.5\sqrt{2}\sin\left(5t - \frac{\pi}{4}\right)$$

$$\tilde{V}_C = 2.5\sqrt{2}\sin(5t - 0.25\pi)\ \text{volt.}$$

27. The given circuit is a balanced bridge, hence,

$$Z_{eq} = (2\,||\,4) + (2\,||\,4)$$

$$= \frac{4}{3} + \frac{4}{3} = \frac{8}{3}\ \Omega.$$

28. Applying Nodal analysis at node (1)

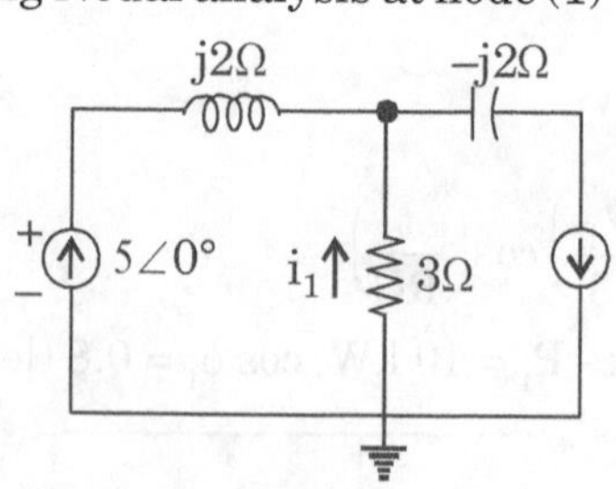

$$5\angle 0° + i1 = 10\angle 60°$$

$$i_1 = 10\left(\frac{1}{2} + j\frac{\sqrt{3}}{2}\right) - 5$$

$$i_1 = j10\frac{\sqrt{3}}{2}$$

$$i_1 = \frac{10\sqrt{3}}{2}\angle 90°\ \text{A}$$

29. Redrawing the given circuit,

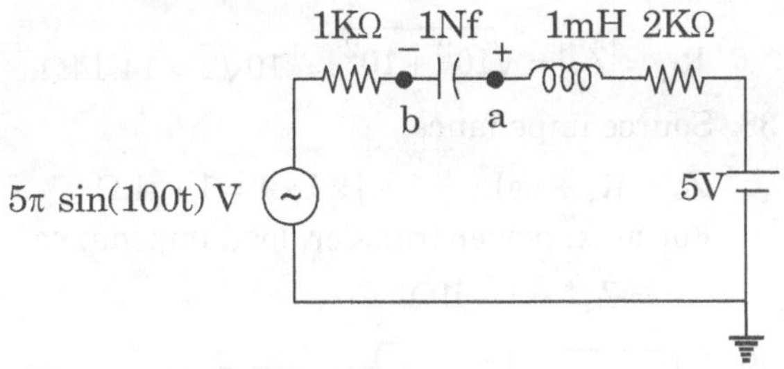

Using superpositioin theorem:

Case 1: Only 5 V source is active.

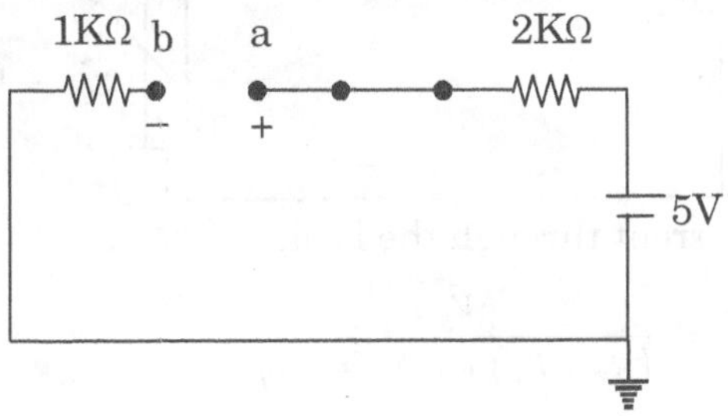

$Vab_1 = +5$ V

$V_{ab} = V_{ab} + V_{ab_2}$

$V_{ab} = 5 + V_p$

sin(5t + v)

Average value

of $V_{ab} = 5$ V.

Case 2:

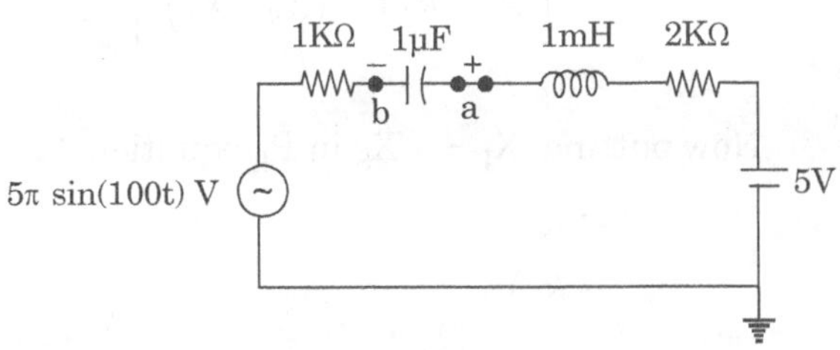

(let) $V_{ab_2} = V_P \sin(500t + \theta)$

30. The input voltage is given by

$$V_i(t) = 2\cos 200t + 4 \sin 500t$$

Let us apply Superposition theorem only consider 2cos 200*t*, then circuit becomes

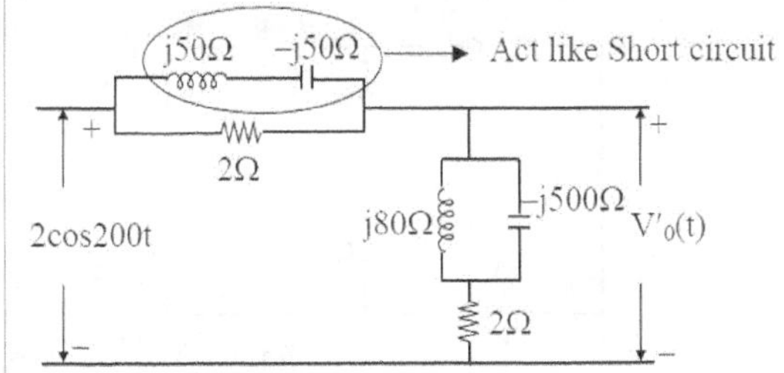

So, $\quad V_0'(t) = 2\cos 200t$

Now consider only 4sin500*t*, then circuit becomes

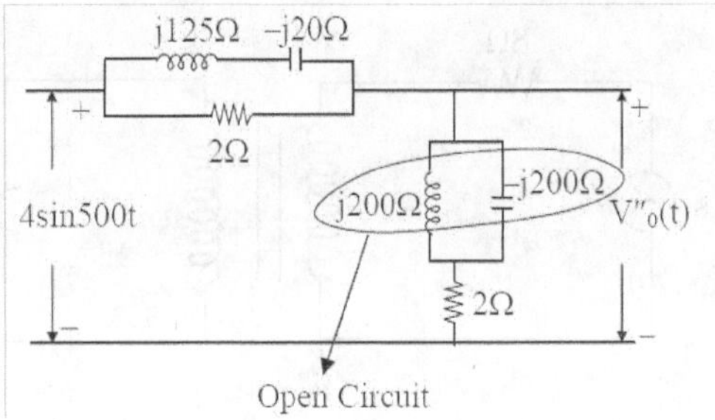

So, again $V_0^n(t) = 4 \sin 500t$

Finally (according to superposition theorem)

$$V_0(t) = V_0'(t) + V_0^n(t)$$

$$\therefore \quad V_0(t) = 2\cos(200t) + 4\sin(500\ t)$$

31. For maximum power transfer from source to load, load impedance, $Z_L = Z_S^*$ (Complex conjugate of source impedance)

32. Since, response (voltage or current) produced in resister R due to both sources are opposite. Using superposition principle,

Net power dissipated in

$$R = \left(\sqrt{P_1} - \sqrt{P_2}\right)^2$$

$$= \left(\sqrt{1} - \sqrt{4}\right)^2 = (1-2)^2 = 1 \text{ W}$$

33.

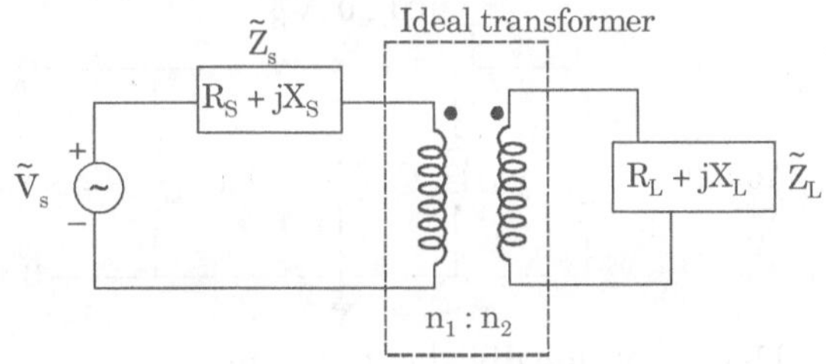

Using impedance transfer property of transformer,

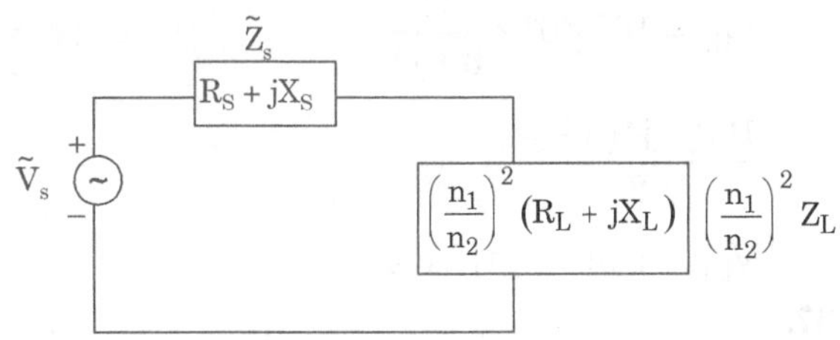

For max. power transfer or impedance matching

$$\left(\frac{n_1}{n_2}\right)^2 |Z_L| = |Z_S|$$

$$\left|\frac{|Z_L|}{|Z_S|}\right| = \left(\frac{n_2}{n_1}\right)^2.$$

$$\frac{n_2}{n_1} = \sqrt{\left|\frac{Z_L}{Z_S}\right|}$$

34.

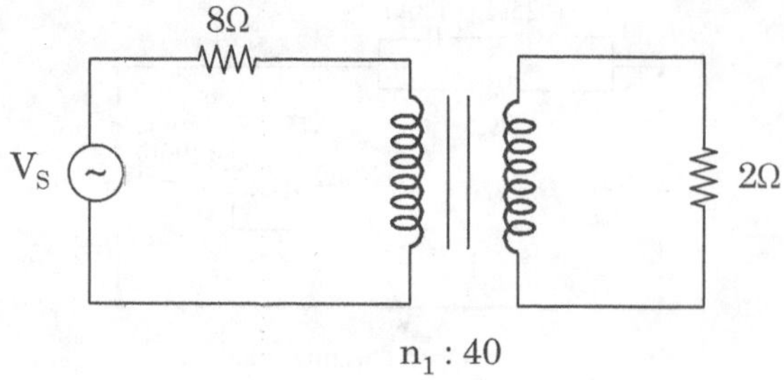

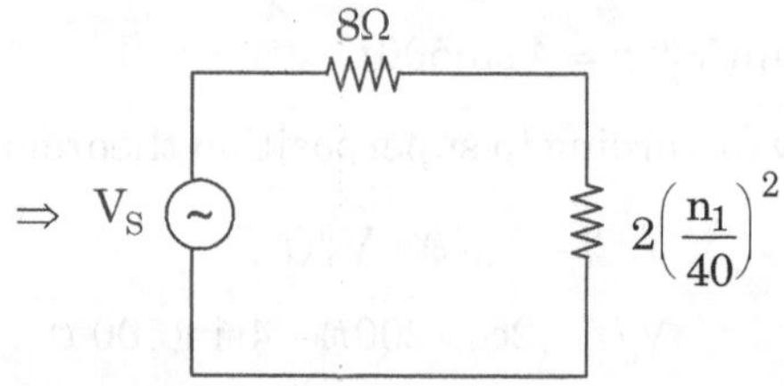

For maximum power transfer,

$$2\left(\frac{n_1}{n_0}\right)^2 = 8$$

$$\frac{n_1}{n_0} = 2$$

n_1 = 80 turns

35. For max. power transfer to load.

$Z_L = Z_G{}^*$ (Complex conjugate of Z_G).

36.

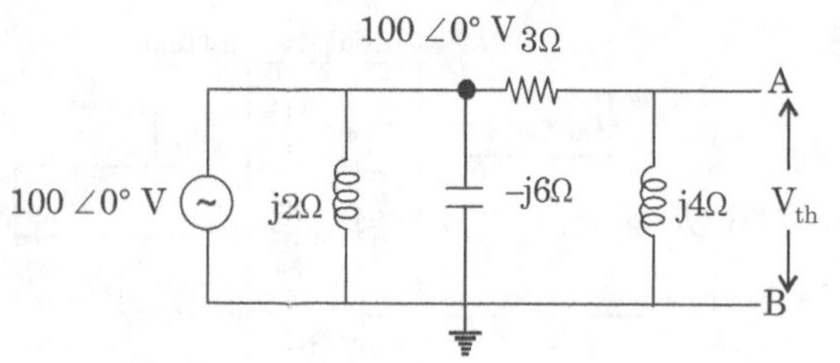

Using voltage division theorem.

$$\tilde{V}_{th} = 100\angle 0° \times \frac{j4}{3+j4}$$

$$\frac{100 \times j4(3-j4)}{25}$$

$$\tilde{V}_{th} = j16(3 - j4) \text{ Volt}$$

37.

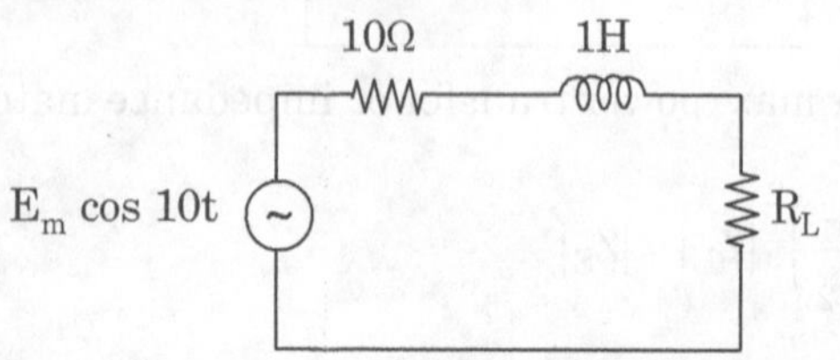

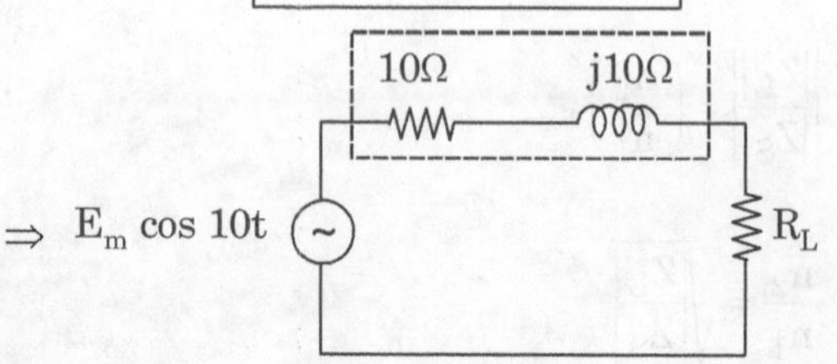

For max. power transfer to R_L,

$$R_L = |\tilde{Z}_S| = \sqrt{10^2+10^2} = 10\sqrt{2} = 14.14\Omega$$

38. Source impedance,

$$\tilde{Z}_S = R_S + j\omega L_S = 1 + j\times 1\times 1 = 1 + j1\ \Omega$$

For max. power transfer, load impedance

$$= \tilde{Z}_S{}^* = 1 - j1\Omega.$$

39.

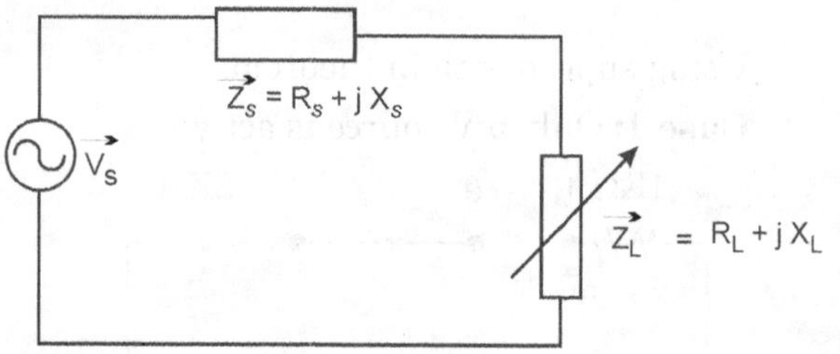

Current through the load,

$$\vec{I_L} = \frac{\vec{V_S}}{(R_S+R_L)+j(X_S+X_L)}$$

$$I_L = |\vec{I_L}| = \frac{V_S}{\sqrt{(R_S+R_L)^2+(X_S+X_L)^2}}$$

Average power to load,

$$P_L = I_L{}^2 R_L = \frac{V_S{}^2 R_L}{(R_S+R_L)^2+(X_S+X_L)^2}$$

$$\therefore \quad \frac{\partial P_L}{\partial X_L} = \frac{-2(X_S+X_C)V_S{}^2 R_L}{\left[(R_S+R_L)^2+(X_S+X_L)^2\right]^2} = 0$$

$\therefore \quad X_L = -X_S$ ..... (*i*)

Now putting $X_L = -X_S$ in P_L equation

$$P_L = \frac{V_S{}^2 R_L}{(R_S+R_L)^2}$$

But $\quad \frac{\partial P_L}{\partial R_L} = 0$

$\therefore \quad R_L = R_S$ ..... (*ii*)

From equations (*i*) and (*ii*), $Z_L = R_S - jX_S$

40. To find out Z_{th} make all the voltage sources short circuit and make all current sources open.

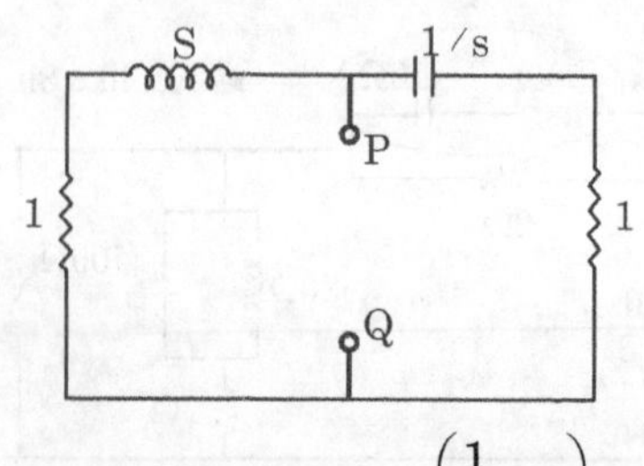

$$Z_{th} = (s+1) \parallel \left(\frac{1}{s}+1\right) = 1$$

or, $\quad Z_{TH} = \frac{(s^2+2s+1)}{s^2+2s+1} = 1$

41. When terminals P and Q are short circuit. Then circuit becomes

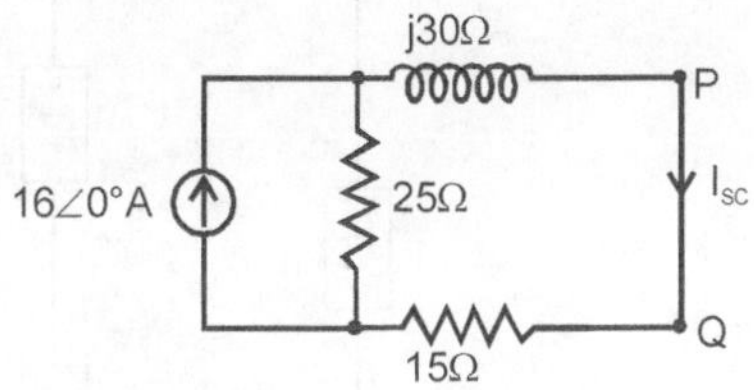

From current division rules

$$I_{SC} = \frac{16(25)}{25+15+j30} = \frac{(16)(25)}{40+j30}$$

$$= \frac{(16)25)}{10(4+j3)} = 6.4 - j4.8$$

42.

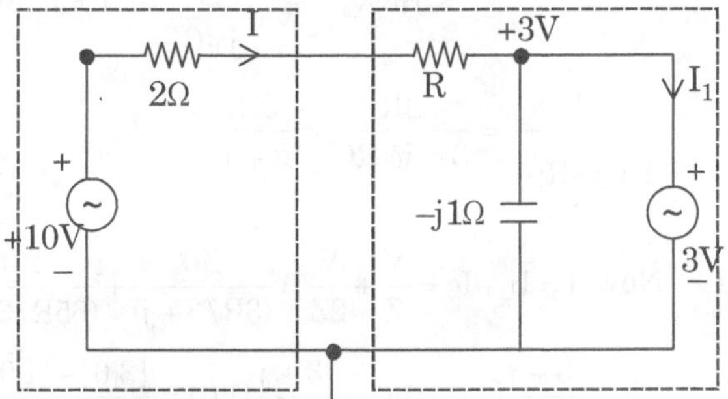

Circuit A Circuit B

Current through R,

$$I = \frac{10-3}{2+R} = \frac{7}{2+R} \text{ A.}$$

Current through 3V source,

$$I_1 = I - \frac{3}{\left(-j^1\right)} = I - j^3$$

Power delivered to circuit B from circuit A,

$$P = I^2R + 3I_1 = \left(\frac{7}{2+R}\right)^3 R + \left(\frac{7}{2+R} - j^3\right).3$$

For P to be maximum, $\frac{\partial P}{\partial R} = 0$

$$\left(\frac{7}{2+R}\right)^2 - \frac{98}{(2+R)^3}.R - \frac{21}{(2+R)^2} = 0$$

$$49\,(2 + R) - 98R - 21\,(2 + R) = 0$$

$$98 - 48R - 87R - 42 - 21 = 0$$

$$R = \frac{56}{70} = 0.8\ \Omega.$$

43. For pure resistive load to extract the max. power.

$$R_L = \left|\tilde{Z}_s\right| = \sqrt{R_s^2 + X_s^2}$$

$$= \sqrt{r^2 + 3^2} = 5\ \Omega$$

44. For V_{th}, R_L must be removed. Hence, $I_2 = 0$. & $j40I_2 = 0$

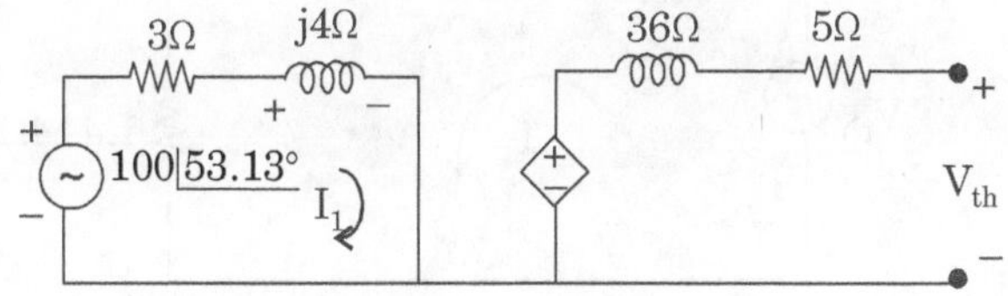

Using voltage division theorem,

$$\tilde{V}_4 = 100\underline{|53.10°} \times \frac{j^4}{3+j^4} = \frac{400\underline{|53.13+90°}}{5\underline{|53.13°}}$$

$$\tilde{V}_{L_1}\ 80\underline{|90°}$$

$$\tilde{V}_{th} = 10\tilde{V}_{L_1}\ 800\underline{|90°} \text{ Volts.}$$

45.

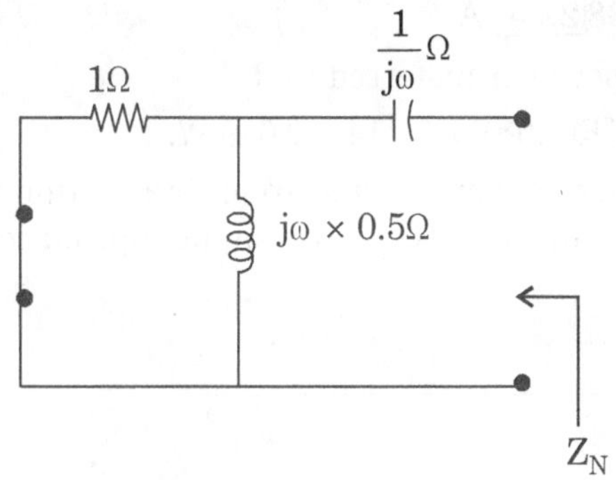

$$\tilde{Z}_N = \frac{1 \times j\frac{\omega}{2}}{1 + j\frac{\omega}{2}} + \frac{1}{j\omega}$$

$$\frac{j\omega}{2+j\omega} + \frac{1}{j\omega} = \frac{2 + j\omega - \omega^2}{-\omega^2 + j2\omega} \times \frac{\left(-\omega^2 - j^2\omega\right)}{\left(-\omega^2 - j^2\omega\right)}$$

$$= \frac{-\left(2-\omega^2\right)\omega^2 + 2\omega^2 - j\left(\omega^3\right) + j4\omega}{\omega^4 + 4\omega^2}$$

46. For max. power transfer to the load, $R_L = \left|\tilde{Z}_{th}\right|$

For Z_{th}:

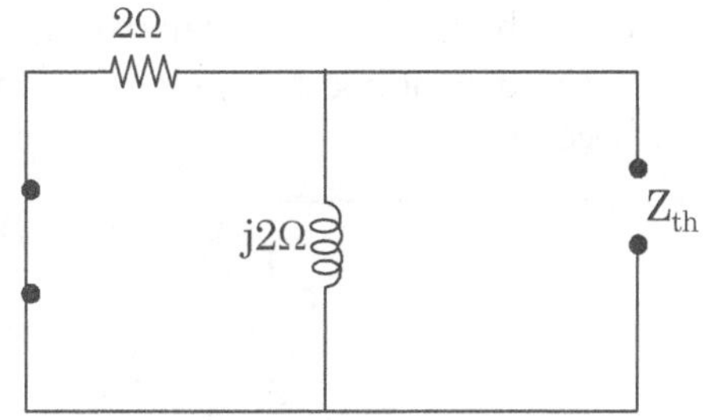

$$\tilde{Z}_{th} = \frac{2 \times j^2}{2 + j^2} = \frac{j^2}{1 + j^1} = \frac{2\angle 90°}{\sqrt{2}\angle 45°} = 1.414\underline{|45°}\ \Omega$$

$$\Rightarrow R_L = \left|\tilde{Z}_{th}\right| = 1.414\ \Omega$$

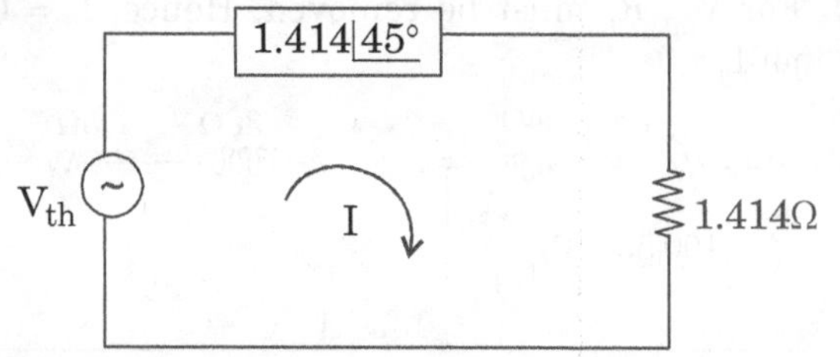

For V_{th}:-

$$\tilde{V}_{th} = \frac{40\angle 0^\circ}{2+j^2} \times j^2 = \frac{4\angle 90^\circ}{\sqrt{2}\angle 45^\circ}$$

$$\tilde{V}_{th} = 2.828\angle 45^\circ V$$

$$\tilde{I}_{th} = \frac{2.828\angle 45^\circ}{1.414\angle 45^\circ + 1.414}$$

$\tilde{I} = 1.08\angle 22.5^\circ$ A.

Max. power transferred to R_L

$= I^2R(1.08)^2 \times 1.414 = 1.649$ W.

47. For max. Power transfer from first section with output impedance (Z_1) to the input impedance (Z_2);

$$\boxed{Z_2 = Z_1}$$

48.

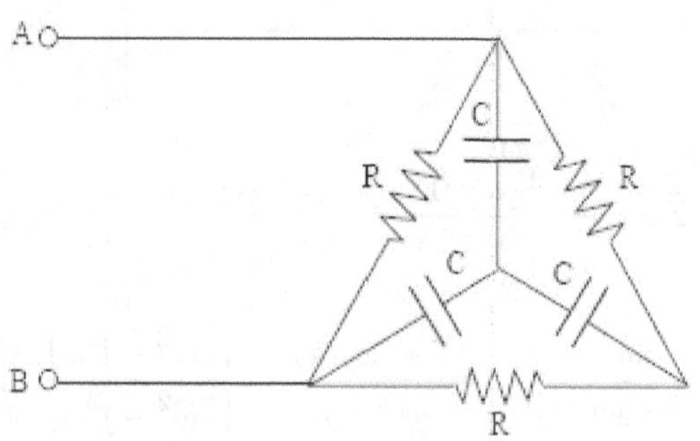

It is given that $V(t) = 2\sin 100t = \overline{V} = 2\angle 0^\circ$

$R = 1\ k\Omega$, $C = 1\ \mu F$

By observity the circuit we can say

$$\overline{I} = \frac{\overline{V}}{Z}$$

→ When each element of star network are same then its corresponding delta element are same and it becomes

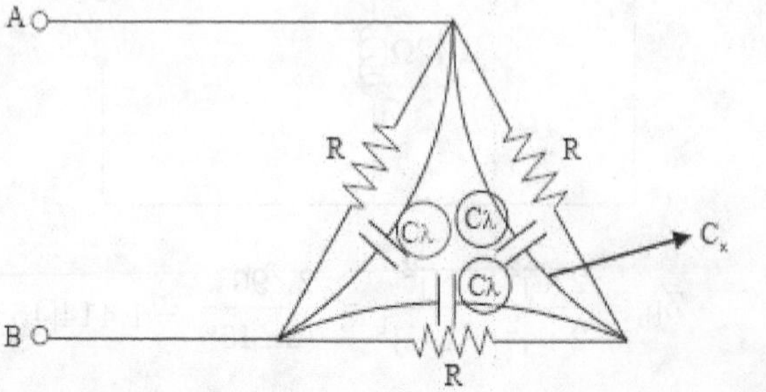

$Z_\Delta = 3Z^*$, but in capacitor case $Z = \frac{1}{j\omega C}$, So if the capacitor of star network are C each then in its delta equivalent it becomes $C/3 = C_x$

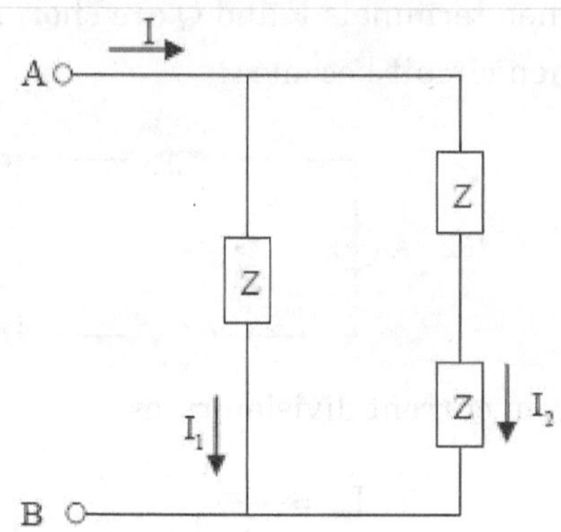

Further redrawing the network.

Where $Z = R \,||\, \frac{1}{j\omega C_x} = \frac{R/jwC_x}{R + \frac{1}{j\omega C_x}} = \frac{R}{1 + j\omega RC_x}$

$$\frac{R}{1 + j\omega R\frac{C}{3}} = \frac{3R}{3 + j\omega RC} = \frac{3R}{3+j}$$

Now $\overline{I} = \overline{I}_1 + \overline{I}_2 = \frac{\overline{V}}{Z} + \frac{\overline{V}}{2Z} = \frac{2\angle 0}{(3R/3+j)} + \frac{2\angle 0}{(65R/3+j)}$

$$= \frac{3+j}{3R}[2\angle 0 + 1\angle 0] = \frac{3+j}{R}(1\angle 0) = \frac{[3\angle 0] + [1\angle 90]}{R}$$

$= 3\angle 0 + 1\angle 90$ mA ($\because 12 - 1000\ \Omega$)

$= 3\sin 1000t + \sin(1000t + 90^\circ)$

49. $\left(R + sL + \frac{1}{Cs}\right) I(s) = V_i(s)$

$$\frac{I(s)}{V_i(s)} = \frac{s}{L\left(s^2 + \frac{Rs}{L} + \frac{1}{LC}\right)}$$

$$\frac{I(s)}{sC\,V_i(s)} = \frac{1/LC}{\left(s^2 + \frac{R}{L}s + \frac{1}{LC}\right)}$$

$$= \frac{\omega_n^2}{(s^2 + 2\omega_n s + \omega_n^2)}$$

$$\omega_n = \frac{1}{\sqrt{LC}}$$

$$\xi = \frac{R}{2L}\sqrt{LC} = \frac{R}{2}\sqrt{\frac{C}{L}}$$

Given transfer function $= \frac{10^6}{s^2 + 20s + 10^6}$

$$Q = \frac{\omega L}{R} = \frac{1}{R\omega C}$$

$$Q = \frac{1}{R}\sqrt{\frac{L}{C}} = \frac{1}{2\xi} = \frac{1}{2 \times 10^{-2}} = 50.$$

50. $V_i(s) = RI(s) + Ls.I(s) + \frac{1}{Cs}I(s)$

and $V_c(s) = \frac{1}{Cs}I(s)$

then $\frac{V_c(s)}{V_i(s)} = \frac{1}{LCs^2 + RCs + 1}$

or $\frac{V_c(s)}{V_i(s)} = \frac{10^6}{s^2 + 10^6 s + 10^6}$

51.

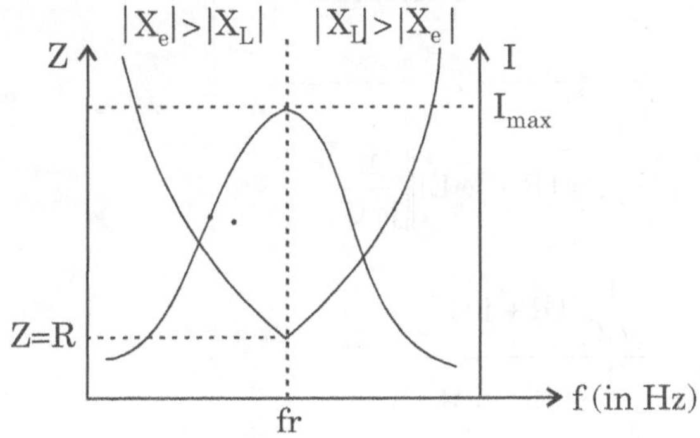

Resonance in series RLC circuit.

52. The cirarit is in resonance because

$|X_L| = |X_C| = 20\ \Omega.$

$I_{max} = \frac{V}{R} = \frac{200}{10} = 20$ A.

voltage across the capitare,

$\tilde{V}C = I_{max} \cdot (-jX_c) = 20 \cdot (-j20)$

$= -j400 = 400\underline{|-90°}$ Volt.

53. At resonance the R–L–C circuit is resistive, thus $R = 100\ \Omega$,

$$Q = \frac{\omega L}{R},$$

or $L = \frac{QR}{\omega} = \frac{10 \times 100}{10^7} = 10^3$ H

54. Resonant frequency, $\omega_0 = 2\pi f_0 = \frac{1}{\sqrt{LC}}$

and Quality factor, $Q = \frac{1}{\omega_0 RC}$

When, R, L, C all are doubled, then,

$$\omega'_0 = \frac{\omega_0}{2}$$

and, $Q' = \frac{1}{\omega_0' R'C'} = \frac{2}{4(\omega_0 RC)} = \frac{Q}{2} = 50$

55. Statement S_1: FALSE

Impedance of series RLC circuit at resonance is minimum, i. e,

$$z = R.$$

Statement S_2 : FALSE

In parallel RLC (or GLC) circuit, Q-factor

$$= R\sqrt{\frac{C}{L}} = \frac{1}{G}\sqrt{\frac{C}{L}}.$$

If C and L are comtant, $Q \propto \frac{1}{G}$

56. Transfer function is $T(s)$

Now $y(t) = \frac{\int idt}{C}$

$\therefore \quad Y(s) = \frac{I}{sC}$

$U(s) = \left(\frac{1}{sC} + R + Ls\right)$

$\therefore \quad \frac{Y(s)}{U(s)} = \frac{1/sC}{1/sC + R + Ls} = \frac{1}{LCs^{2+} + RCs + 1}$

$T(s) = \frac{1/LC}{s^2 + \frac{Rs}{L} + \frac{1}{LC}}$

$\therefore \quad \omega_n^2 = \frac{1}{LC}$

$\Rightarrow \quad \omega_n = \frac{1}{\sqrt{LC}}$

$\Rightarrow \quad 2\xi\omega_n = \frac{R}{L}$

$\therefore \quad 2\xi = \frac{R}{L}\sqrt{LC} = R\sqrt{\frac{C}{L}}$

$\Rightarrow \quad \xi = \frac{R}{2}\sqrt{\frac{C}{L}}$

For no oscillations, $\xi \geq 1$

$\therefore \quad \frac{R}{2}\sqrt{\frac{C}{L}} \geq 1$

$\Rightarrow \quad R \geq 2\sqrt{\frac{L}{C}}$

57. Resonant frequency $= \frac{1}{2\pi\sqrt{LC}}$

$$= \frac{1}{2\pi\sqrt{1 \times \frac{10^{-6}}{400}}} = \frac{1}{\pi} \times 10^4 \text{ Hz}$$

58. $I = I_m \sin \omega t$

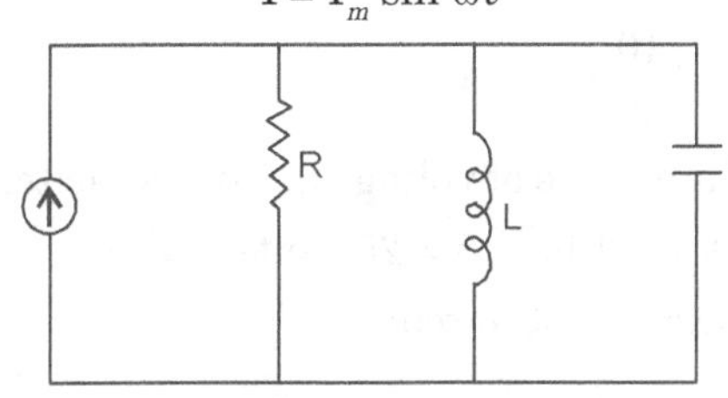

Parallel RCL circuit

Admittance,

$$Y = \frac{1}{R} + j\omega C + \frac{1}{j\omega L} = \frac{1}{R} + J\left(\omega C - \frac{1}{\omega L}\right)$$

Frequency response of voltage amplitude

$$|V| = \frac{I_m}{\sqrt{\left(\frac{1}{R}\right)^2 + \left(\omega C - \frac{1}{1L}\right)^2}}$$

At resonance, $\omega_o C - \frac{1}{\omega L} = 0, \omega = \frac{1}{\sqrt{LC}}$

Impedance, $Z = \frac{1}{Y} = R$

[Real quality and having maximum]

Half-power frequencies,

$$\omega_1, \omega_2 = -\frac{1}{2RC} \pm \sqrt{\left(\frac{1}{2RC}\right)^2 + \frac{1}{LC}}$$

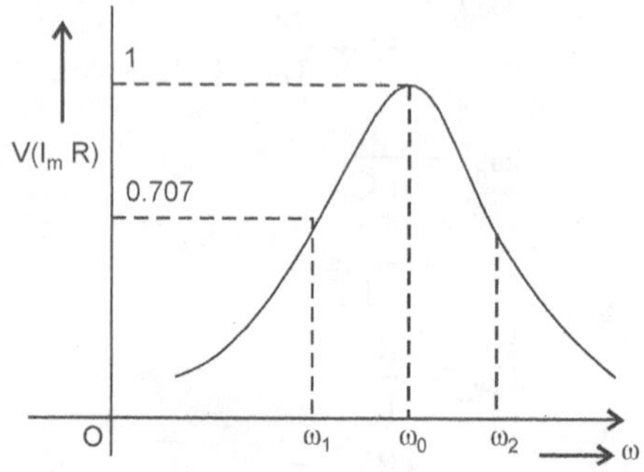

Then, BW (Bandwidth) $= \omega_2 - \omega_1 = \frac{1}{RC}$

BW decreases mean R is increased

Also, BW is independent of any variation in L.

59. Resonant frequency,

$$\omega_r = \frac{1}{\sqrt{LC}} = \frac{1}{\sqrt{0.1 \times 10^{-3} \times 1 \times 10^{-6}}}$$

$$\omega_r = 10^5 \text{ rad/s}$$

$$\Rightarrow X_C = \frac{1}{\omega C} = \frac{1}{10^5 \times 1 \times 10^{-6}} = 10\ \Omega$$

$\Rightarrow$ Amplitude of current (at resonance),

$$I_{max} = \frac{V_{max}}{R} = \frac{10}{4} = 2.5 \text{ A.}$$

$\Rightarrow$ Amplitude of voltage across capacitor

$V_c = X_c I = 10 \times 2.5 = 25$ Volts.

60. For series RLC circuit:

damping coefficient, $\alpha = \frac{R}{2L}$

resonant frequency, $\omega_r = \frac{1}{\sqrt{LC}}$

damping ratio, $\varepsilon = \frac{\alpha}{\omega_r} = \frac{R}{2}\sqrt{\frac{C}{L}}$

61.

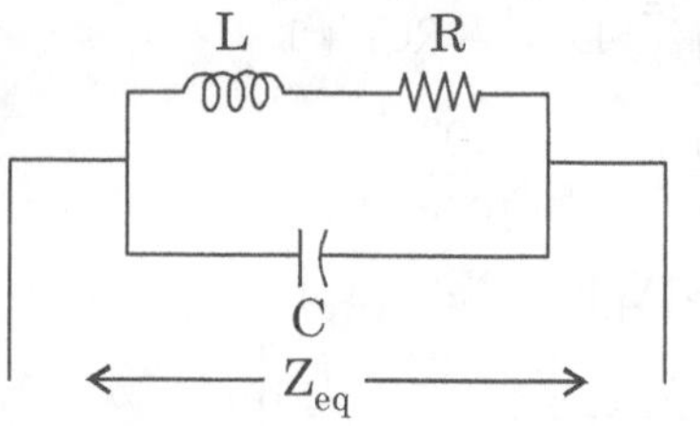

$$Z_{eq} = (R + j\omega L) \Big\| \frac{1}{j\omega C}$$

$$\frac{\frac{1}{j\omega C}.(R + j\omega L)}{\frac{1}{j\omega C} + R + j\omega L}$$

$$= \frac{\frac{R}{j\omega C} + \frac{L}{C}}{R + j\left(\omega L - \frac{1}{\omega C}\right)} \times \frac{R - j\left(\omega L - \frac{1}{\omega C}\right)}{R - j\left(\omega L - \frac{1}{\omega C}\right)}$$

$$Z_{eq} = \frac{\frac{-jR^2}{\omega C} - j\frac{L}{C}\left(\omega L - \frac{1}{\omega C}\right) - \frac{R}{\omega C}\left(\omega L - \frac{1}{\omega C}\right) + \frac{RL}{C}}{R^2 + \left(\omega L - \frac{1}{\omega C}\right)^2}$$

Equating imaginary part to be zero, i.e, Im $\{Z_{eq}\} = 0$

$$\frac{-R^2}{\omega C} - \frac{L}{C}\left(\omega L - \frac{1}{\omega C}\right) = 0$$

$$\frac{R}{\omega C} + \frac{\omega L^2}{C} - \frac{L}{\omega C^2} = 0$$

$$CR^2 + \omega^2 L^2 C - L = 0$$

$$\omega^2 = \frac{L - R^2 C}{L^2 C}$$

$$\omega = \frac{1}{\sqrt{LC}}\sqrt{1 - \frac{R^2 C}{L}}$$

$$f = \frac{1}{2\pi\sqrt{LC}}\sqrt{1 - \frac{R^2 C}{L}}$$

62. $\therefore X_C = \frac{1}{\omega C}$. If ω is very high then $X_c \to 0$.

Capacitor can be considered as short-circuited. Given circuit is balanced wheat-stone bridge.

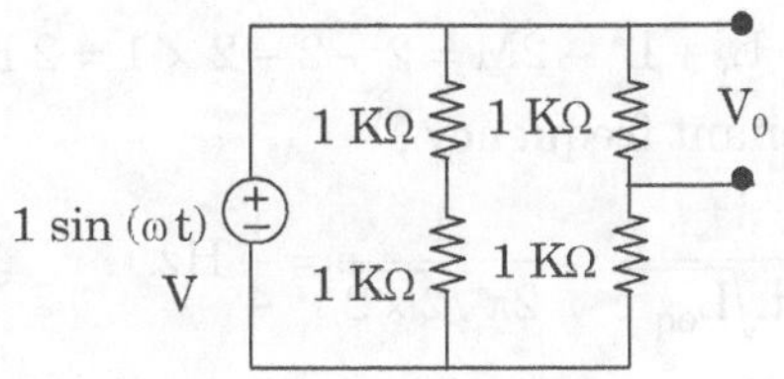

Using voltage division theorem,

$$V_0 = 1\sin\omega t \times \frac{1}{1+1} = 0.5\sin(\omega t)$$

Peak output voltage = 0.5 Volt.

63. For parallel RLC circuit in resonance condition,

$$\tilde{I}_R = \tilde{I},$$

$$\tilde{I}_L = Q.\tilde{I}\angle -90°,\quad \tilde{I}_C = Q.\tilde{I}\angle -90°$$

$$\Rightarrow \left|\frac{I_L}{I_R}\right| = \frac{QI}{I}$$

$$Q = \sqrt{\frac{C}{L}} = 10\sqrt{\frac{10\times10^{-6}}{10\times10^{-3}}} = 10\times0.0316 = 0.316$$

64. If V and I are in same phase then circuit must be in resonance. Hence, at resonance

$V_C = QV_R$.

$$\frac{\text{Amplitude of } V_C}{\text{Amplitude of } V_R}$$

$$Q = \frac{1}{R}\sqrt{\frac{L}{C}} = \frac{1}{5}\sqrt{\frac{5}{5}} = 0.2$$

65.

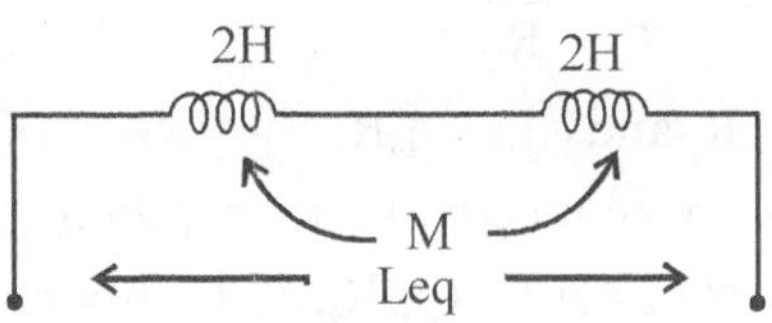

$$M = k\sqrt{L_1L_2} = 0.1\sqrt{2\times2} = 0.2\text{ H}$$

$$Leq = L_1 + L_2 \pm 2M\begin{pmatrix}+\text{aiding.}\\ -\text{oppossing}\end{pmatrix}$$

$$Leq = 2+2\pm2\times0.2$$

66. Here inductance has not been given, therefore ignoring inductance, we have, in the voltage source circuit,

$$I_1 = \frac{2\cos 4t}{3} = 0.67\cos4t$$

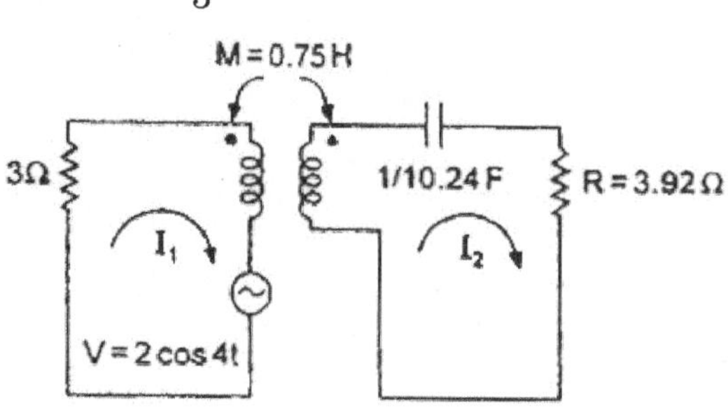

and, $$I_2 = \frac{-j\omega MI_1}{R - j/\omega C} = \frac{-j4\times0.75I_1}{3.92-2.56j}$$

$$= \frac{-j3\times0.67\cos 4t}{3.92-2.56j} = (0.23-0.35j)\cos4t$$

Now, none of options seems to be true, it can be inferred from the question that something may be missing there.

67.

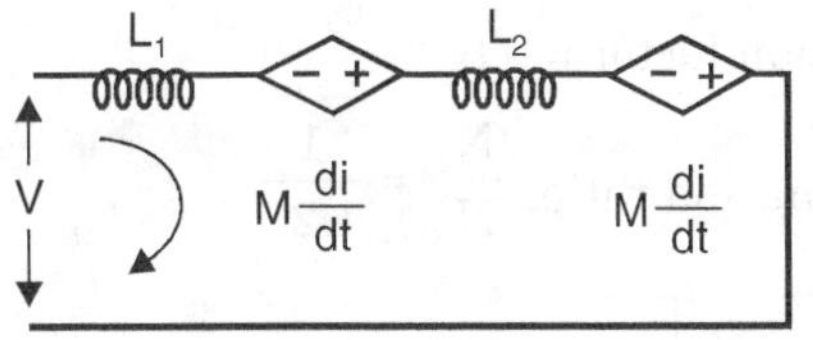

$$V = L_1\frac{di}{dt} - M\frac{di}{dt} + (L_2 - M)\frac{di}{dt}$$

$$= (L_1 + L_2 - 2M)\frac{di}{dt} = L\frac{di}{dt}$$

$$L = L_1 + L_2 - 2M$$

68.

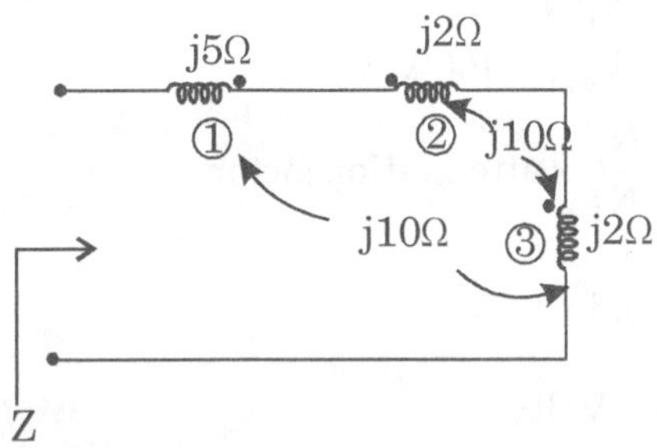

Coil 1 and 3 are om series oppossing and coil 2 and 3 are in series aiding. Now,

$$Z = (j5 - j10) + (j2 + 10)\ + (j2 + j10 - j10)\Omega$$

$$Z = ja\Omega.$$

69. Applying KCL at node (1), $i + 1 = i_L$.

Using KVL in loop,

$$1\times i + j1(i+1) - 1\angle0° + 1\angle0° = 0$$

$$i + j^i + j^1 = 0$$

$$i = -\frac{j^1}{1+j}$$

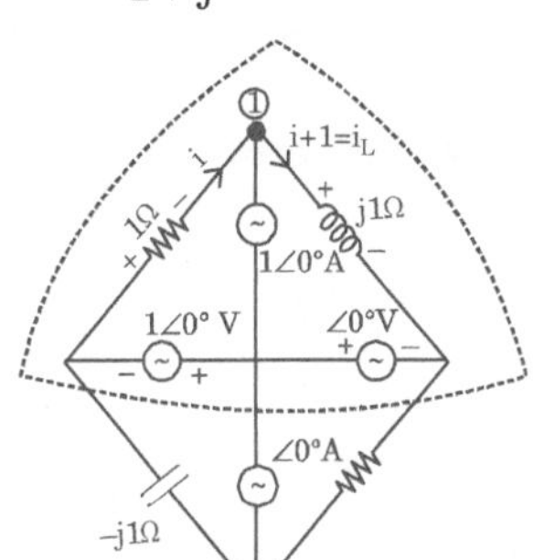

Current through inductor,

$$i_L = i + 1 = \frac{-j}{1+j} + 1 = \frac{1}{1+j}\text{ A.}$$

70. $z = 4 - j3 = R_L - jX_C;\ R_L = 4$

$I = 5\cos(100\pi t + 100) = I_m \cos(\omega t + \alpha)$

$$P = \frac{1}{2} I_m^2 R_L = \frac{1}{2} \times 5^2 \times 4 = 50 \text{ W}$$

71. For an ideal transformer,

$$\frac{V_1}{V_2} = \frac{N_1}{N_2}$$

Attenuation factor = 0.8

transformation ratio, $\frac{N_1}{N_2} = \frac{1}{1.25}$

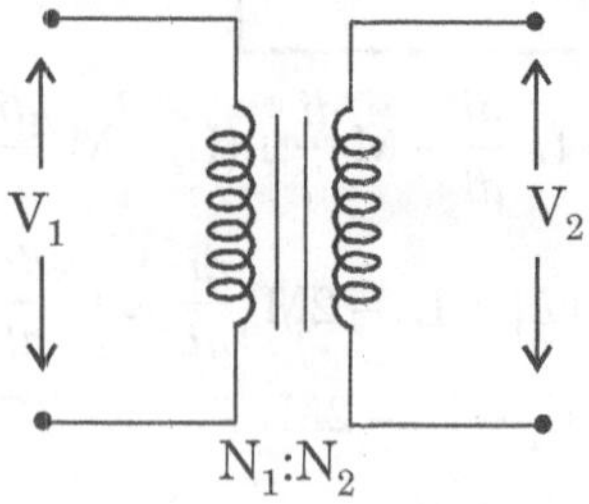

Case I : $V_{WX1} = 100$ V.

$$\frac{V_{YZ1}}{V_{Wx1}} = \frac{N_2}{N_1} \times \text{attenuation factor}$$

$$\frac{V_{YZ1}}{V_{Wx1}} = 0.8 \times 1.25 = 1.$$

$V_{YZ1} = 100$ Volt

Case II : $V_{YZ2} = 100$ V

$$\frac{V_{YZ2}}{V_{Wx2}} = 0.8 \times 1.25 = 1.$$

$V_{Wx2} = 100$ V.

Now, $\frac{V_{yz1}}{V_{Wx1}} = \frac{100}{100}$ and $\frac{V_{YZ2}}{V_{WX2}} = \frac{100}{100}$.

72. 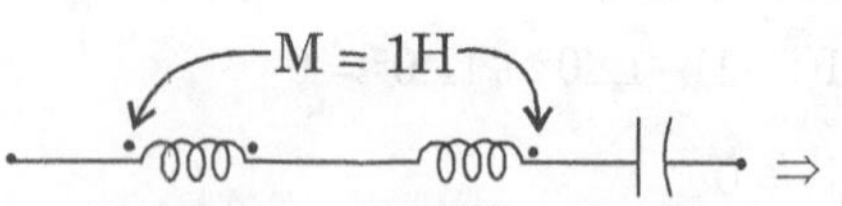

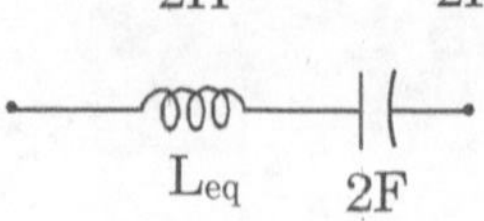

$L_{eq} = L_1 + L_2 - 2M = 2 + 2 - 2 \times 1 = 2$ H

Resonant frequency ,

$$fr\ \frac{1}{2\pi\sqrt{L_{eq}.c}} = \frac{1}{2\pi\sqrt{2 \times 2}} = \frac{1}{4} \text{ Hz.}$$

73. 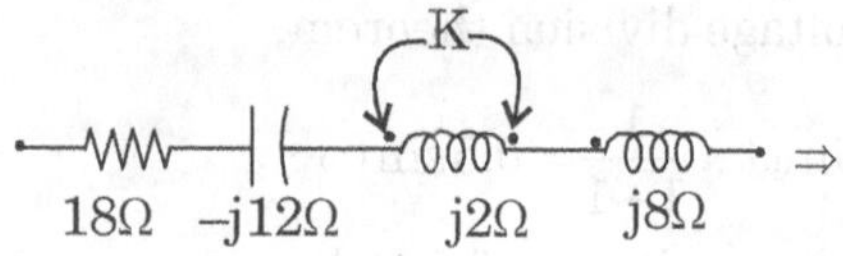

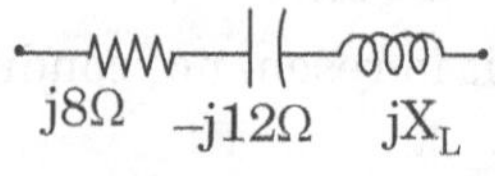

$$\Rightarrow\ L = L_1 + L_2 + 2M = L_1 + L_2 + 2K\sqrt{L_1 L_2}$$

$$j\omega L = j\omega L_1 + j\omega L_2 + 2k\sqrt{j\omega L_1 \cdot j\omega L_2}$$

$$\tilde{X}L = j^2 + j^8 + 2k\sqrt{j^2 \cdot j^8}$$

$$\tilde{X}_L = j^{10} + j^{8k}$$

At resonance, $|\tilde{X}_L| = |\tilde{X}_C|$

$10 + 8k = 12$

$$K = \frac{1}{4} = 0.25$$

74. Individual Q-factor of coils are

$$q_1 = \frac{\omega L_1}{R_1} \text{ and } q_2 = \frac{\omega L_2}{R_2}.$$

$\Rightarrow \omega L_1 = q_1 R_1$ and $\omega L_2 = q_2 R_2$.

When coils are connected in series, then

$q.R = \omega L_1 + coL_2 = q_1 R_1 + q_2 R_2$. $[\because R = R_1 + R_2]$

$$q = \frac{q_1 R_1 + q_2 R_2}{R_1 + R_2}$$

4 CHAPTER

Transient Analysis

Analysis of Previous GATE Papers

Topics ↓		Year →	2019	2018	2017 Set 1	2017 Set 2	2016 Set 1	2016 Set 2	2016 Set 3	2015 Set 1	2015 Set 2	2015 Set 3
1ST ORDER CIRCUITS	1 Mark	MCQ Type						1		1		
		Numerical Type										
	2 Marks	MCQ Type										
		Numerical Type	1	1	1	1			1	1	1	
	Total		**2**	**2**	**2**	**2**		**1**	**2**	**3**	**2**	
2ND ORDER CIRCUITS	1 Mark	MCQ Type										
		Numerical Type										
	2 Marks	MCQ Type										
		Numerical Type										
	Total											

1ST ORDER CIRCUITS

1. A series RC circuit is connected to a DC voltage source at time t = 0. The relation between the source voltage v_s, the resistance R, the capacitance C, and the current i(t) is given below:

$$v_s = Ri(t) + \frac{1}{C}\int_0^t i(t)dt$$

Which one of the following represents the current i(t)?

(a) 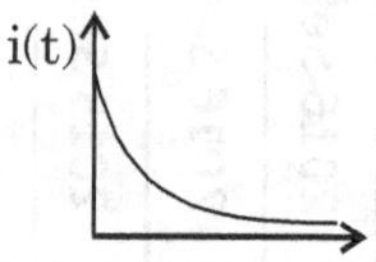

(b) 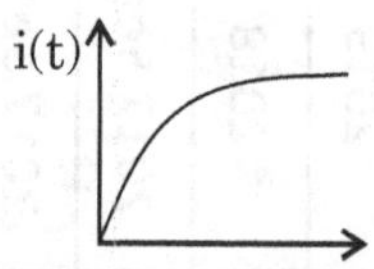

(c) 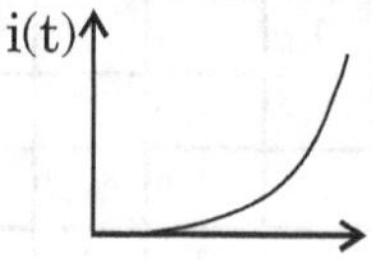

(d) 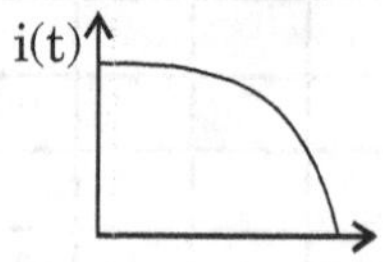

[1991 : 2 Marks]

2. For the compensated attenuator of figure, the impulse response under the condition $R_1C_1 = R_2C_2$ is

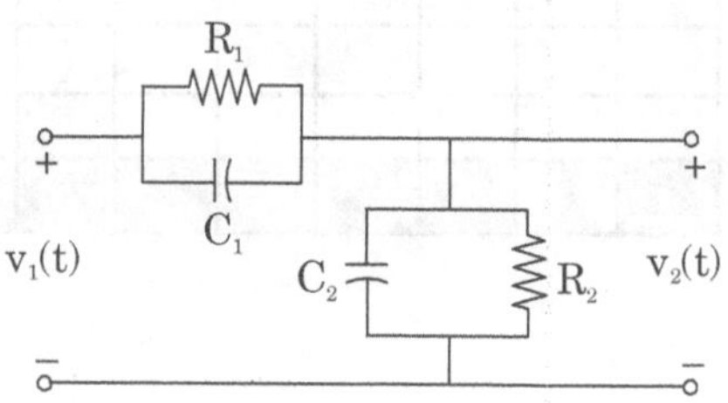

(a) $\frac{R_2}{R_1+R_2}\left[1 - e^{\frac{1}{R_1C1}}\right]u(t)$

(b) $\frac{R_2}{R_1+R_2}\delta(t)$

(c) $\frac{R_2}{R_1+R_2}u(t)$

(d) $\frac{R_2}{R_1+R_2}e^{\frac{1}{R_1C_1}}u(t)$ **[1992 : 2 Marks]**

3. A ramp voltage, v(t) = 100t Volts, is applied to an RC differentiating circuit with R = 5 kΩ and C = 4 μF. The maximum output voltage is

(a) 0.2 volt

(b) 2.0 volts

(c) 10.0 volts

(d) 50.0 volts

[1994 : 1 Marks]

4. In the circuit of fig. the energy absorbed by the 4 Ω resistor in the time interval (0, ∞) is

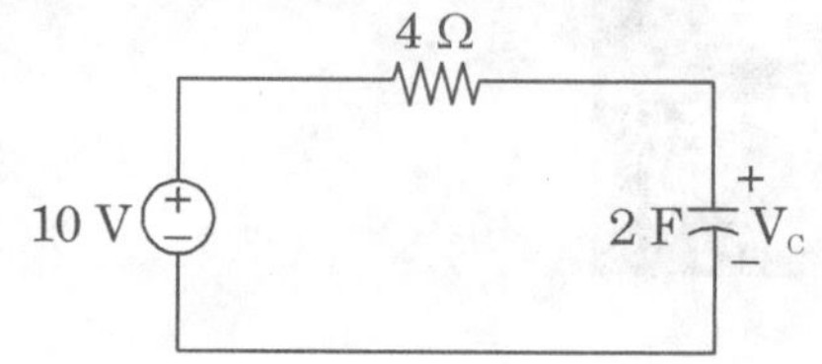

(a) 36 Joules

(b) 16 Joules

(c) 256 Joules

(d) None of the above

[1997 : 2 Marks]

5. In the figure, the switch was closed for a long time before opening at t = 0. The voltae V_x at t = 0+ is

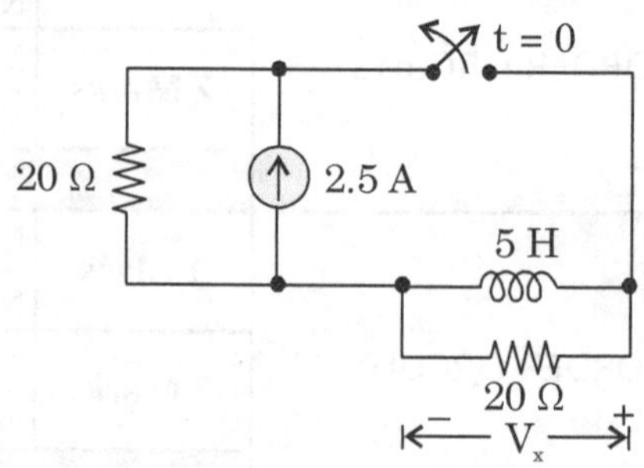

(a) 25 V

(b) 50 V

(c) –50 V

(d) 0 V **[2002 : 1 Mark]**

6. For the R-L circuit shown in the figure, the input voltage $v_i(t) = u(t)$. The current i(t) is

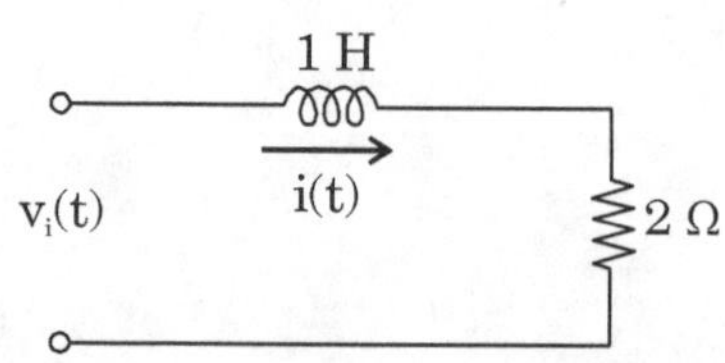

(a) 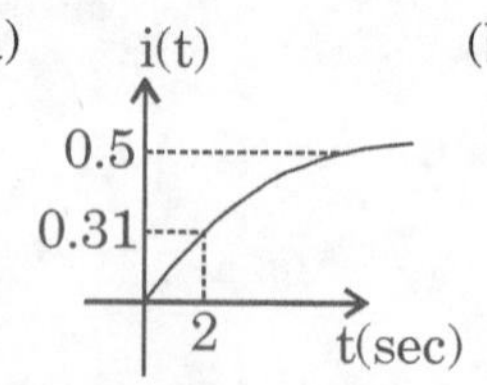

(b) 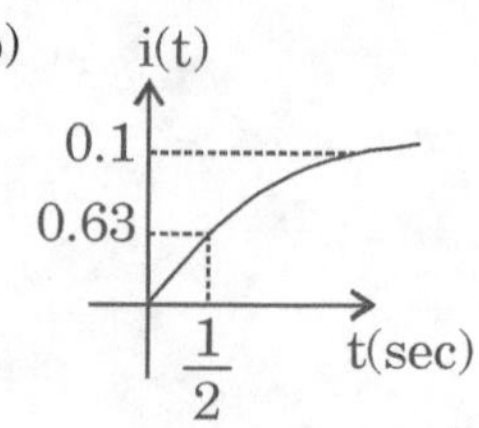

(c) 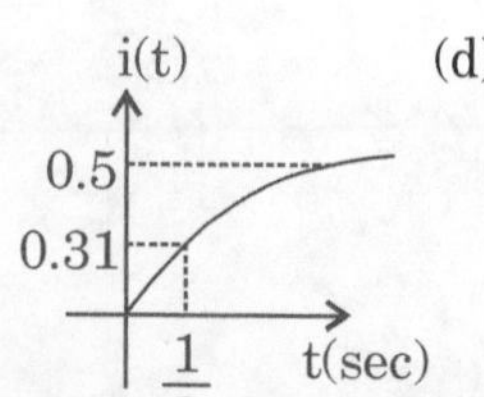

(d) 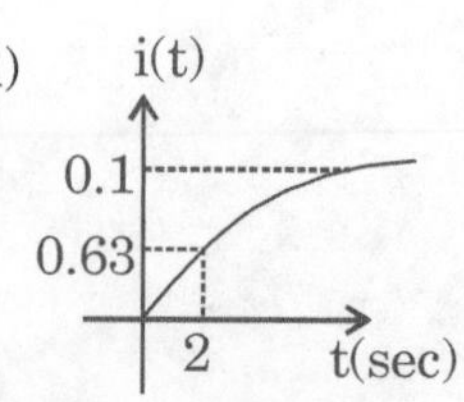

[2004: 2 Marks]

7. A square pulse of 3 volts amplitude is applied to C-R circuit shown in the figure. The capacitor is initially uncharged. The output voltage V_2 at time t = 2 sec is

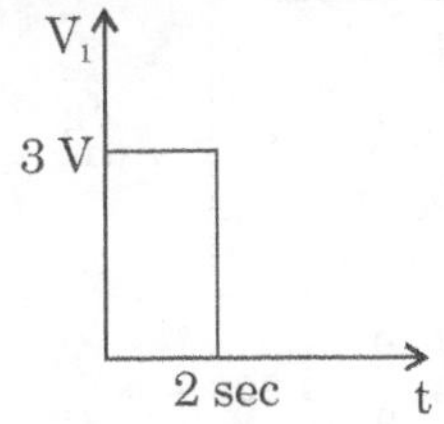

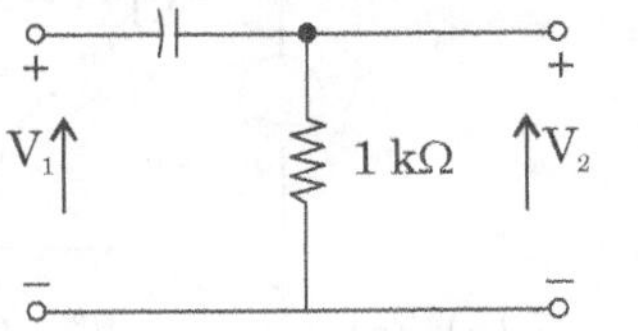

(a) 3 V (b) –3 V
(c) 4 V (d) –4 V **[2005 : 2 Marks]**

8. A 2 mH inductor with some initial current can be represented as shown below, where s is the Laplace Transform variable. The value of initial current is

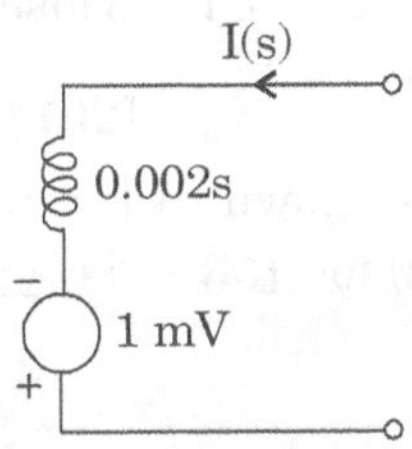

(a) 0.5 A (b) 2.0 A
(c) 1.0 A (d) 0.0 A **[2006 : 1 Mark]**

9. In the figure shown below, assume that all the capacitors are initially uncharged. If $v_i(t) = 10u(t)$ Volts, $v_0(t)$ is given by

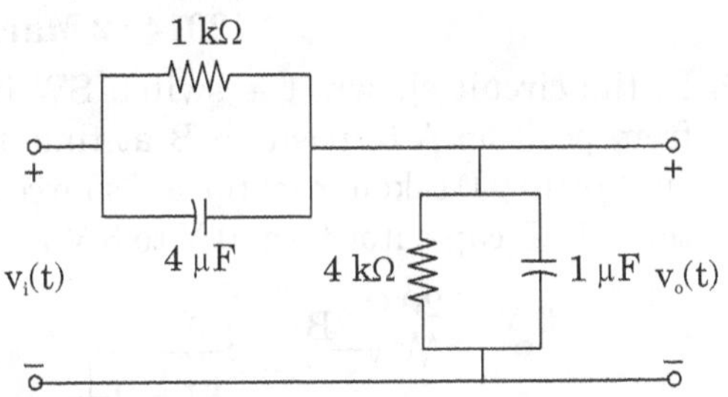

(a) $8e^{-t/0.004}$ Volts (b) $8(1 - e^{-t/0.004})$ Volts
(c) 8u(t) Volts (d) 8 Volts

[2006 : 1 Mark]

10. In the circuit shown, V_C is 0 volts at t = 0 sec. For t > 0, the capacitor current $i_c(t)$, where t is in seconds, is given by

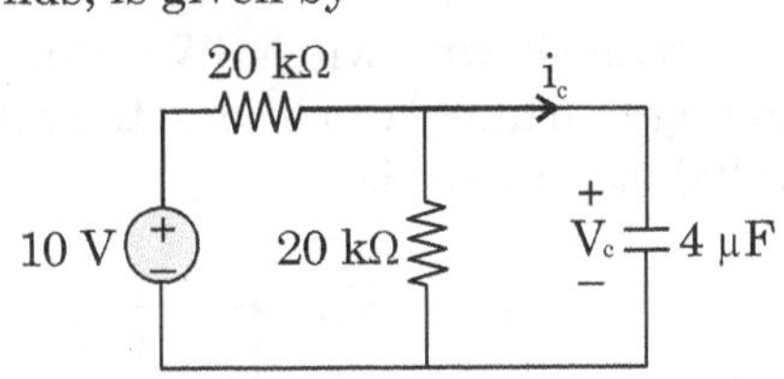

(a) 0.50exp(–25t) mA
(b) 0.25exp(–25t) mA
(c) 0.50exp(–12.5t) mA
(d) 0.25exp(–6.25t) mA **[2007 : 1 Marks]**

11. In the following circuit, the switch S is closed at t = 0. The rate of change of current $\frac{di}{dt}(0^+)$ is given by

(a) 0

(b) $\frac{R_s I_s}{L}$

(c) $\frac{(R + R_s)I_s}{L}$

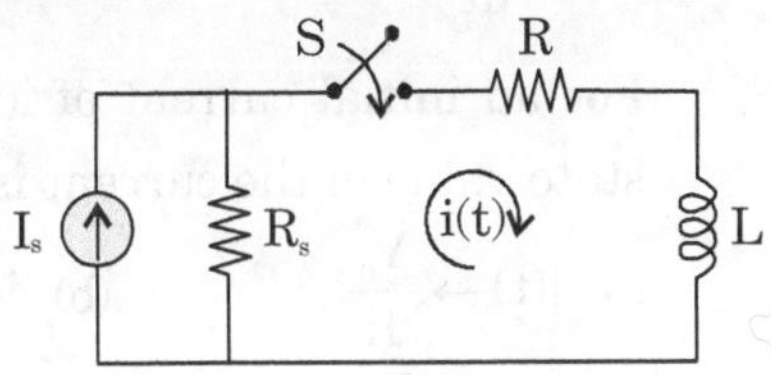

(d) ∞ **[2008 : 2 Marks]**

12. The circuit shown in the figure is used to charge the capacitor C alternately from two current sources as indicated. The switches S_1 and S_3 are mechanically coupled and connected as follows:

For $2nT < t < (2n + 1)T$, (n = 0, 1,2 ....) S_1 to P_1 and S_2 to P_2.

For $(2n + 1)T < t < (2n + 2)T$, (n = 0, 1,2, ...) S_1 to Q_1 and S_2 to Q_2.

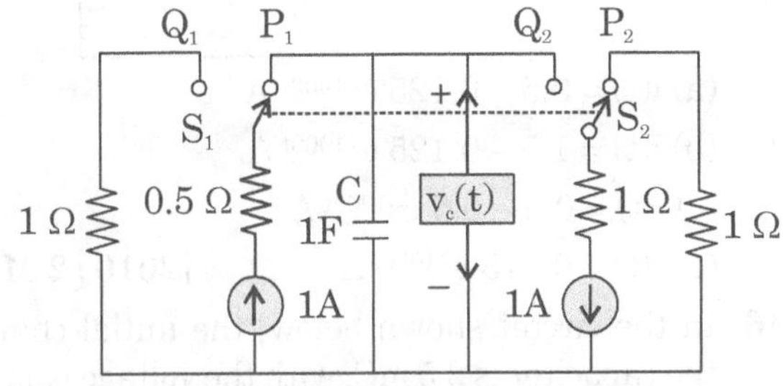

Assume that the capacitor has zero initial charge. Given that u(t) is a unit step function, the voltage $v_c(t)$ across the capacitor is given by

(a) $\sum_{n=0}^{\infty}(-1)^n tu(t - nT)$

(b) $u(t) + 2\sum_{n=1}^{\infty}(-1)^n u(t - nT)$

(c) $tu(t) + 2\sum_{n=1}^{\infty}(-1)^n (t - nT)u(t - nT)$

(d) $\sum_{n=1}^{\infty}[0.5 - e^{-(t-2nT)} + 0.5e^{-(t-2nT-T)}]$

[2008 : 2 Marks]

13. The switch in the circuit shown was on position a for a long time, and is moved to position at time t = 0. The current i(t) for t > 0 is given by

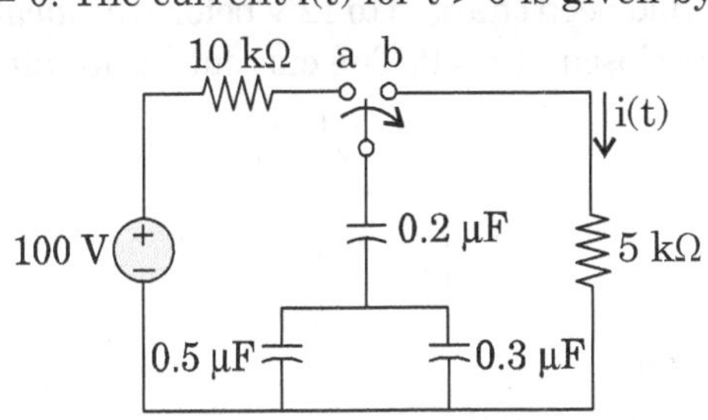

(a) $0.2e^{-125t}u(t)$ mA (b) $20e^{-1250t}u(t)$ mA
(c) $0.2e^{-125t}u(t)$ mA (d) $20e^{-1000t}u(t)$ mA

[2009 : 2 Marks]

14. The time domain behaviour of an RL circuit is represented by

$$L\frac{di}{dt} + Ri = V_0(1 + Be^{-Rt/L} \sin t)u(t)$$

For an initial current of $i(0) = \frac{V_0}{R}$, the steady state value of the current is given by

(a) $i(t) \to \frac{V_0}{R}$ (b) $i(t) \to \frac{2V_0}{R}$

(c) $i(t) \to \frac{V_0}{R}(1+B)$ (d) $i(t) \to \frac{2V_0}{R}(1+B)$

[2009 : 2 Marks]

15. In the circuit shown the switch, the switch S is open for a long time and is closed at t = 0. The current i(t) for t > 0⁺ is

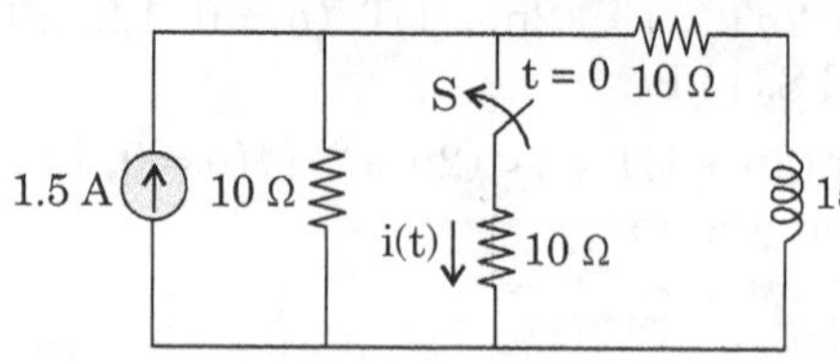

(a) $i(t) = 0.5 - 0.125\, e^{-1000t}$ A.

(b) $i(t) = 1.5 - 0.125\, e^{-1000t}$ A.

(c) $i(t) = 0.5 - 0.5\, e^{-1000t}$ A.

(d) $i(t) = 0.375\, e^{-1000t}$ A.

[2010 : 2 Marks]

16. In the circuit shown below, the initial charge on the capacitor is 2.5 mC, with the voltage polarity as indicated. The switch is closed at time t = 0. The current i(t) at a time t after the switch is closed is

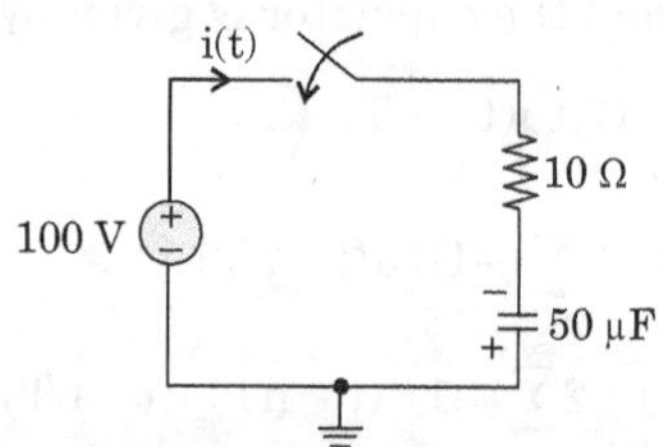

(a) $i(t) = 15\exp(-2\times10^3 t)$ A.

(b) $i(t) = 5\exp(-2\times10^3 t)$ A.

(c) $i(t) = 10\exp(-2\times10^3 t)$ A.

(d) $i(t) = -5\exp(-2\times10^3 t)$ A.

[2011 : 2 Marks]

17. In following figure, C_1 and C_2 are ideal capacitors. C_1 had been charged to 12 V before the ideal switch S is closed at t = 0. The current i(t) for all t is

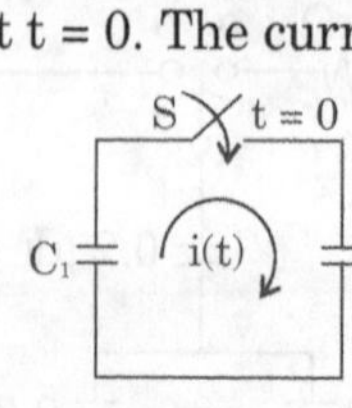

(a) zero

(b) a step function

(c) an exponentially decaying function

(d) an impulse function

[2012 : 1 Mark]

18. In the figure shown, the capacitor is initially uncharged. Which one of the following expressions describes the current i(t) (in mA) for t > 0?

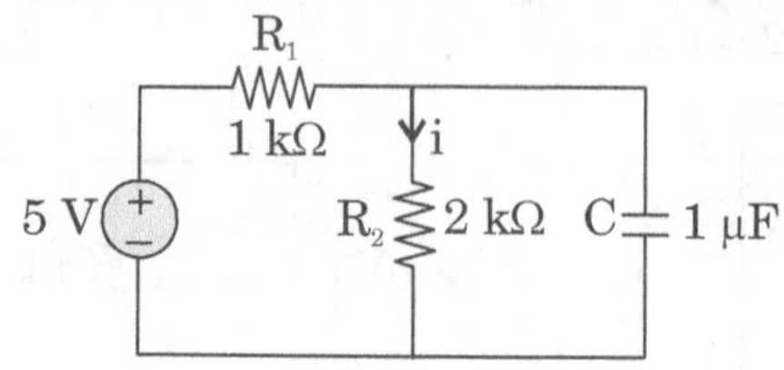

(a) $i(t) = \frac{5}{3}(1 - e^{-t/T}),\ T = \frac{2}{3}$ m sec

(b) $i(t) = \frac{5}{2}(1 - e^{-t/T}),\ T = \frac{2}{3}$ msec

(c) $i(t) = \frac{5}{3}(1 - e^{-t/T}),\ T = 3$ msec

(d) $i(t) = \frac{5}{2}(1 - e^{-t/T}),\ T = 3$ msec

[2014 : 2 Marks, Set-2]

19. In the circuit shown in the figure, the value of $v_0(t)$ (in volts) for $t \to \infty$ is ________.

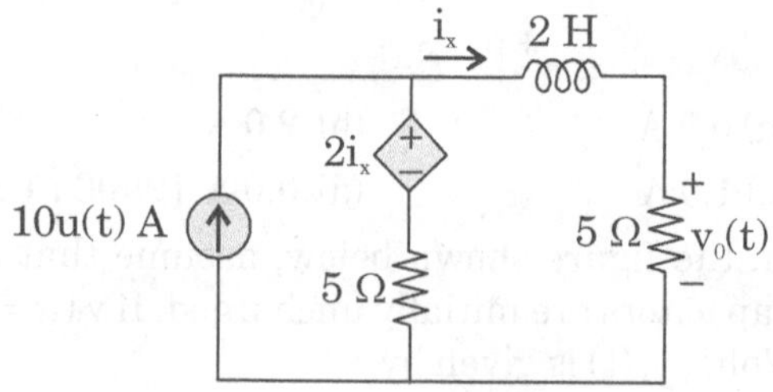

[2014 : 2 Mark, Set-4]

20. In the circuit shown, the switch SW is thrown from position A to position B at time t =0. The energy (in μJ) taken from the 3 V source to charge the 0.1 μF capacitor form 0 V to 3 V is

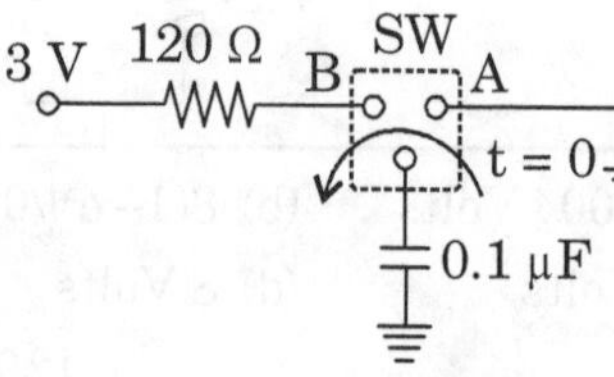

(a) 0.3 (b) 0.45

(c) 0.9 (d) 3

[2015 : 1 Mark, Set-1]

21. In the circuit shown, switch SW is closed at t = 0. Assuming zero initial conditions, the value of $v_c(t)$ (in volts) at t = 1 sec is ________.

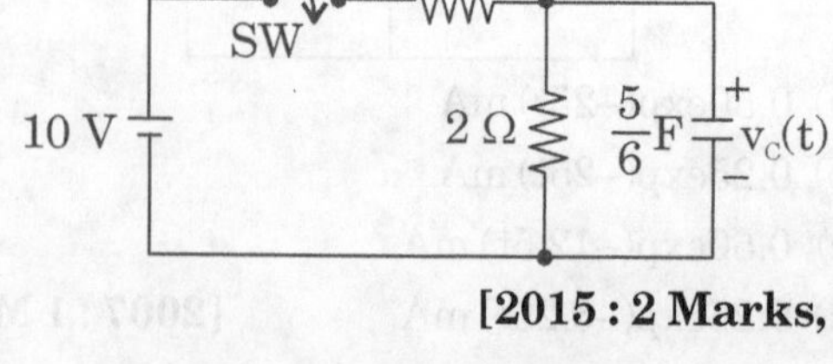

[2015 : 2 Marks, Set-1]

22. In the circuit shown, the initial voltages across the capacitors C_1 and C_2 and 1 V and 3 V, respectively. The switch is closed at time t = 0. The total energy dissipated (in Joules) in the resistor R until steady state is reached, is ____.

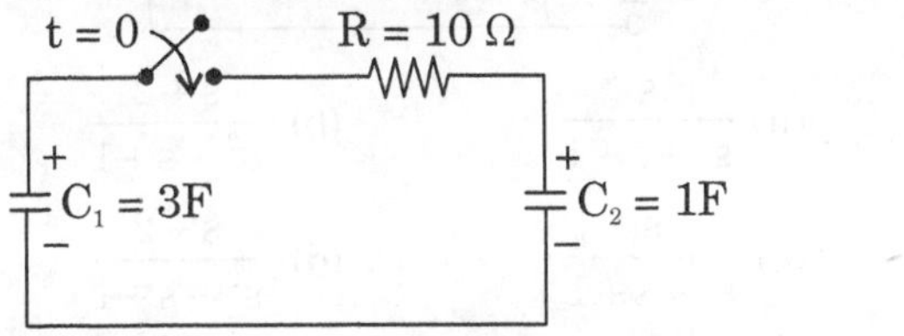

[2015 : 2 Markd, Set-2]

23. The switch has been in position 1 for a long time and abruptly changes to position 2 at t = 0.

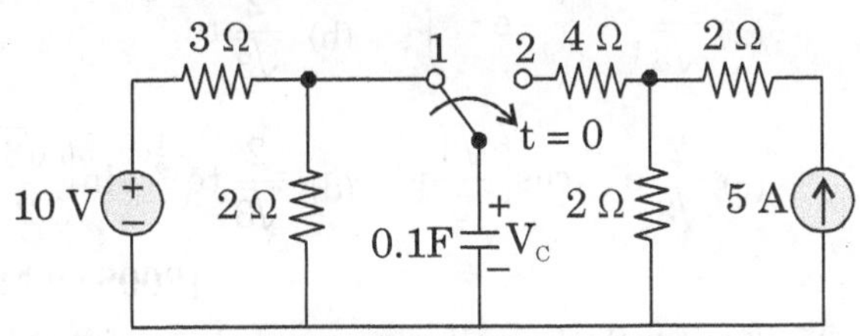

If time t is seconds the capacitor voltage V_C (in volts) for t > 0 is given by

(a) 4(1 – exp(–t/0.5))

(b) 10 – 6 exp(–t/0.5)

(c) 4(1– exp(–t/0.6))

(d) 10 – 6 exp(–t/0.6)

[2016 : 1 Mark, Set-2]

24. Assume that the circuit in the figure has reached the steady state before time t = 0 when the 3 resistor suddenly burns out, resulting in an open circuit. The current i(t) (in ampere) at t = 0^+ is _____.

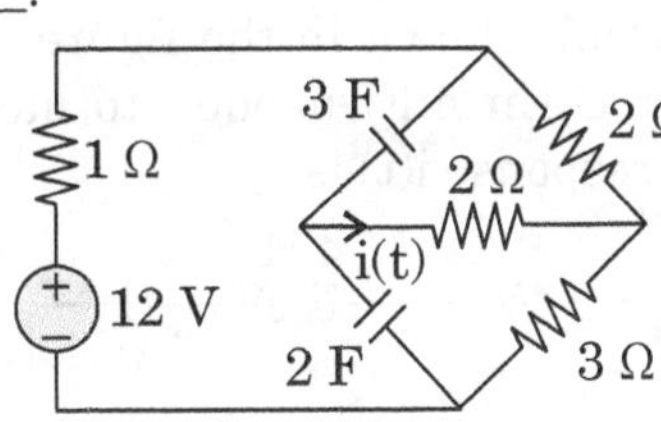

[2016 : 2 Marks, Set-3]

25. In the circuit shown, the voltage $V_{IN}(t)$ is described by:

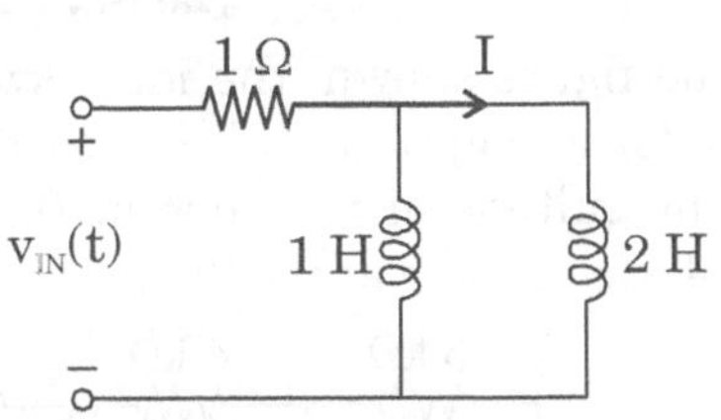

where t is in seconds. The time (in seconds) at which the current I in the circuit will reach the value 2 Amperes is _______.

[2017 : 2 Mark, Set-1]

26. The switch in the circuit, shown in the figure, was open for a long time and is closed at t = 0

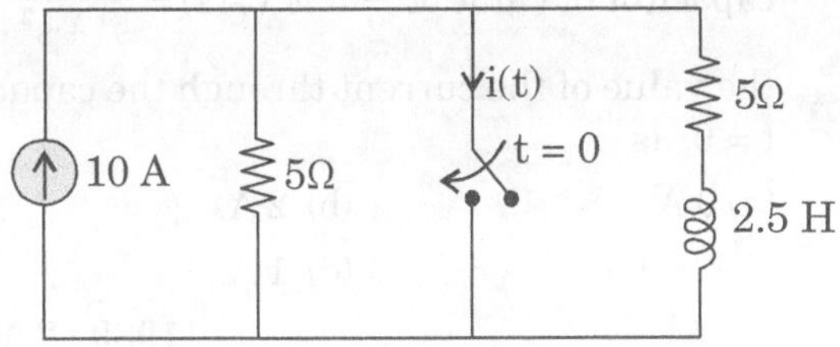

The current i(t) (in ampere) at t = 0.5 seconds is _______. **[2017 : 2 Mark, Set-2]**

27. For the circuit given in the figure, the magnitude of the loop current (in amperes, corect to three decimal places) 0.5 second after closing the switch is _______.

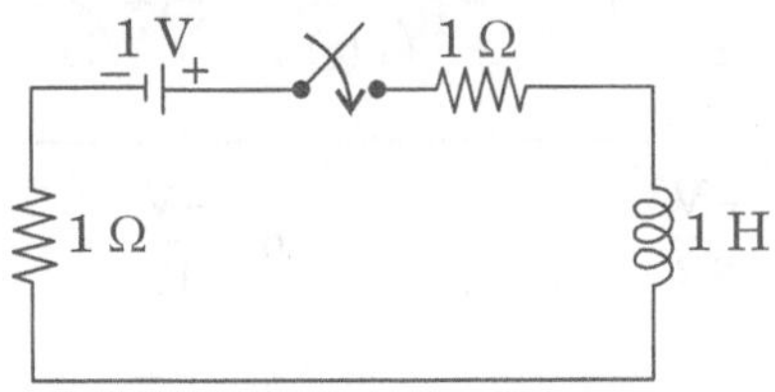

[2018 : 2 Mark]

28. The RC circuit shown below has a variable resistance R(t) given by the following expression:

$$R(t) = R_0\left(t - \frac{t}{T}\right) \text{ for } 0 \le t < T$$

Where $R_0 = 1\,\Omega$, C = 1 F. We are also given that $T = 3R_0C$ and the source voltage is V_S = 1 V. If the current at time t = 0 is 1A, then the current I(t), in amperes, at time t = T/2 is _____ (rounded off to 2 decimal places).

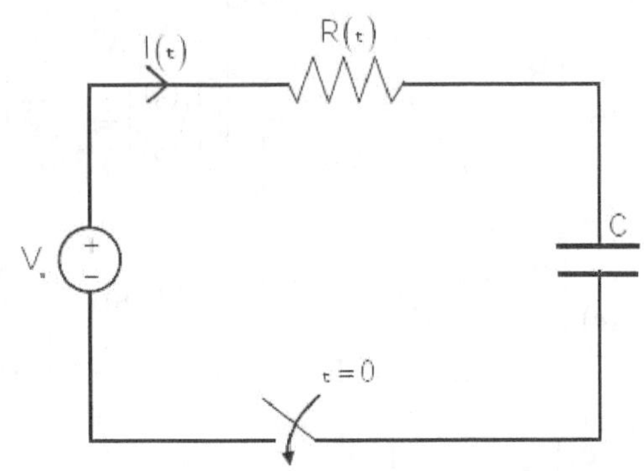

[2019 : 2 Mark]

2ND ORDER CIRCUITS

29. A 10 Ω resistor, a 1 H inductor and 1 μF capacitor are connected in parallel. The combination is driven by a unit step current. Under the steady state condition, the source current flows through

(a) the resistor (b) the inductor

(c) the capacitor only (d) All the three elements

[1989 : 2 Marks]

30. If the Laplace transform of the voltage across a capacitor of value of $\frac{1}{2}$ F is $V_C(s) = \frac{s+1}{s^3+s^2+s+1}$, the value of the current through the capacitor at $t = 0^+$ is

(a) 0 A. (b) 2 A.

(c) (1/2) A. (d) 1 A.

[1989 : 2 Marks]

31. At $t = 0^+$, the current i_1 is

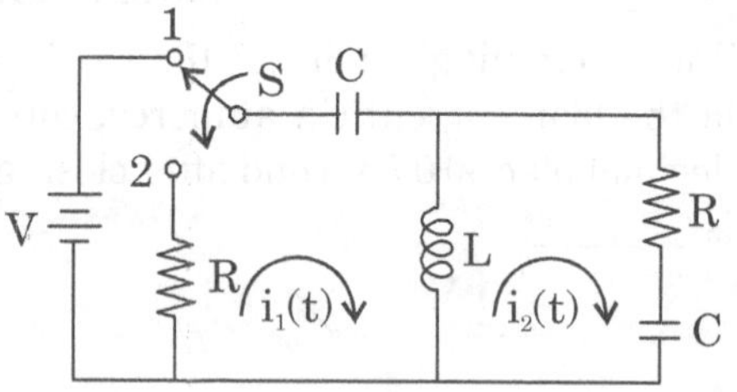

(a) $\frac{-V}{2R}$ (b) $\frac{-V}{R}$

(c) $\frac{-V}{4R}$ (d) zero

[2003 : 2 Marks]

32. $I_1(s)$ and $I_2(s)$ are the Laplace transforms of $i_1(t)$ and $i_2(t)$ respectively. The equations for the loop currents $I_1(s)$ and $I_2(s)$ for the circuit shown in the figure, after the switch is brought from position 1 to position 2 at t = 0, are

(a) $\begin{bmatrix} R+Ls+\frac{1}{Cs} & -Ls \\ -Ls & R+\frac{1}{Cs} \end{bmatrix} \begin{bmatrix} I_1(s) \\ I_2(s) \end{bmatrix} = \begin{bmatrix} \frac{V}{s} \\ 0 \end{bmatrix}$

(b) $\begin{bmatrix} R+Ls+\frac{1}{Cs} & -Ls \\ -Ls & R+\frac{1}{Cs} \end{bmatrix} \begin{bmatrix} I_1(s) \\ I_2(s) \end{bmatrix} = \begin{bmatrix} -\frac{V}{s} \\ 0 \end{bmatrix}$

(c) $\begin{bmatrix} R+Ls+\frac{1}{Cs} & -Ls \\ -Ls & R+Ls\frac{1}{Cs} \end{bmatrix} \begin{bmatrix} I_1(s) \\ I_2(s) \end{bmatrix} = \begin{bmatrix} -\frac{V}{s} \\ 0 \end{bmatrix}$

(d) $\begin{bmatrix} R+Ls+\frac{1}{Cs} & -Ls \\ -Ls & R+Ls\frac{1}{Cs} \end{bmatrix} \begin{bmatrix} I_1(s) \\ I_2(s) \end{bmatrix} = \begin{bmatrix} \frac{V}{s} \\ 0 \end{bmatrix}$

[2003 : 2 Marks]

33. The circuit shown in the figure has initial current $i_L(0^-) = 1$ A through the inductor and an initial voltage $V_C(0^-) = -1$ V across the capacitor. For input $v(t) = u(t)$, the Laplace transform of the current $i(t)$ for $t > 0$ is

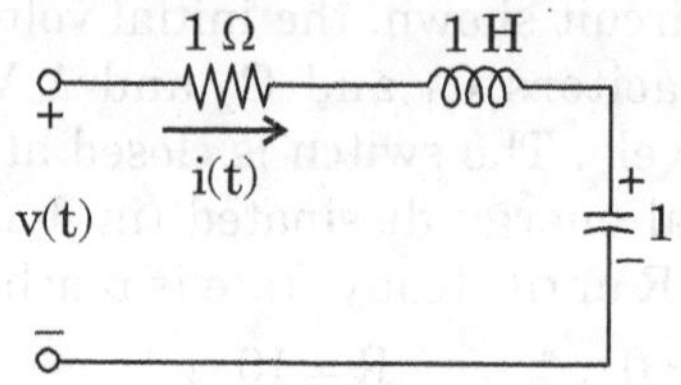

(a) $\frac{s}{s^2+s+1}$ (b) $\frac{s+2}{s^2+s+1}$

(c) $\frac{s}{s^2+s+1}$ (d) $\frac{s-2}{s^2+s+1}$

[2004 : 2 Marks]

34. For $t > 0$, the output voltage $v_c(t)$ is

(a) $\frac{2}{\sqrt{3}}\left(e^{-\frac{1}{2}t} - e^{\frac{\sqrt{3}}{2}t}\right)$ (b) $\frac{2}{\sqrt{3}}te^{-\frac{1}{2}t}$

(c) $\frac{2}{\sqrt{3}}e^{-\frac{1}{2}t}\cos\left(\frac{\sqrt{3}}{2}t\right)$ (d) $\frac{2}{\sqrt{3}}te^{-\frac{1}{2}t}\sin\left(\frac{\sqrt{3}}{2}t\right)$

[2008 : 2 Marks]

35. For $t > 0$, the voltage across the resistor is

(a) $\frac{1}{\sqrt{3}}\left(e^{\frac{\sqrt{3}}{2}t} - e^{\frac{1}{2}t}\right)$

(b) $e^{\frac{\sqrt{3}}{2}t}\left[\cos\left(\frac{\sqrt{3}t}{2}\right) - \frac{1}{\sqrt{3}}\sin\left(\frac{\sqrt{3}t}{2}\right)\right]$

(c) $\frac{2}{\sqrt{3}}e^{\frac{\sqrt{3}}{2}t}\sin\left(\frac{\sqrt{3}t}{2}\right)$

(d) $\frac{s^2+s+1}{s^2+2s+1}$

[2008 : 2 Marks]

36. In the circuit shown in the figure, the value of capacitor C (in mF) needed to have critically damped response i(t) is ______.

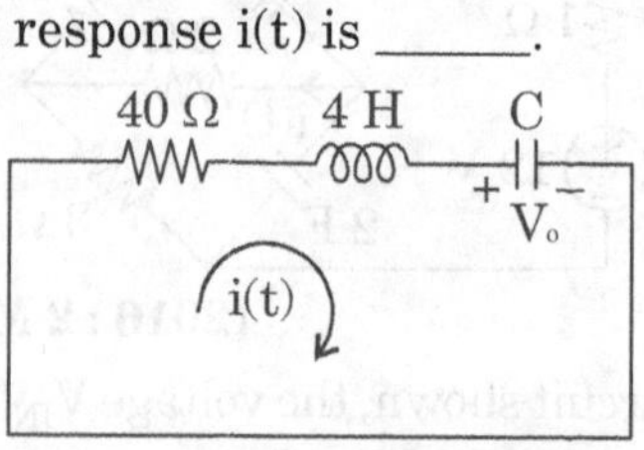

[2014 : 2 Marks, Set-1]

37. In the figure shown, the ideal switch has been open for a long time. If it is closed at t = 0, then the magnitude of the current (in mA) through the 4 kΩ resistor at $t = 0^+$ is ______.

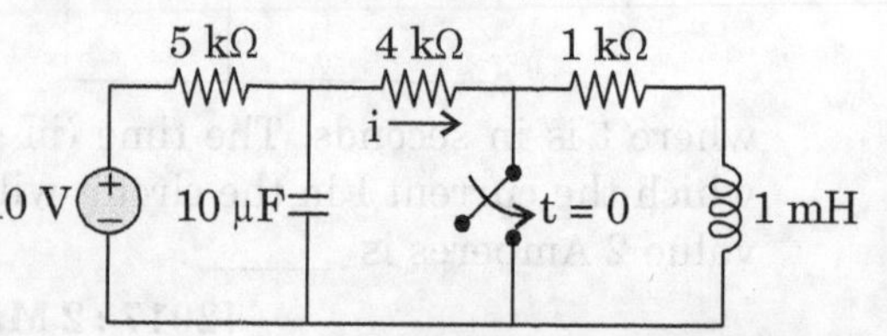

[2014 : 1 Mark, Set-2]

ANSWERS

1. (a)	**2.** (b)	**3.** (b)	**4.** (d)	**5.** (c)	**6.** (c)	**7.** (b)	**8.** (a)	**9.** (c)	**10.** (a)
11. (b)	**12.** (a)	**13.** (b)	**14.** (a)	**15.** (a)	**16.** (a)	**17.** (d)	**18.** (c)	**19.** (31.25)	**20.** (c)
21. (2.528)	**22.** (1.5)	**23.** (d)	**24.** (1)	**25.** (0.3405)		**26.** (8.16)	**27.** (0.316)	**28.** (0.10)	**29.** (b)
30. (c)	**31.** (a)	**32.** (c)	**33.** (b)	**34.** (d)	**35.** (b)	**36.** (10)	**37.** (c)		

EXPLANATIONS

1. In a series RC circuit,

→ Initially at $t = 0$, capacitor charges with a current of $\frac{V_s}{R}$ and in steady state at $t = \infty$, capacitor behaves like open circuit and no current flows through the circuit

→ So the current $i(t)$ represents an exponential decay function

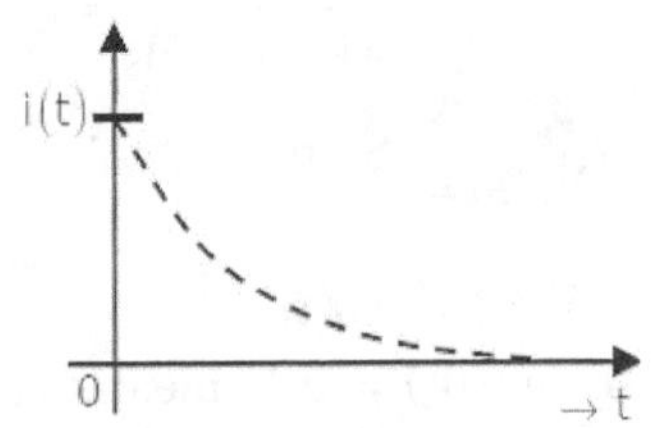

2.

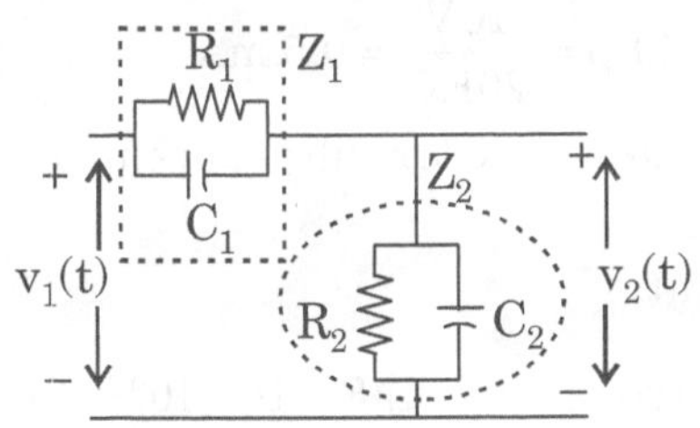

$$Z_1(s) = \frac{R_1 \cdot \frac{1}{sC}}{R_1 + \frac{1}{sC}} = \frac{R_1}{sCR_1 + 1}$$

$$Z_2(s) = \frac{R_2 \cdot \frac{1}{sC_2}}{R_2 + \frac{1}{sC_2}} = \frac{R_2}{SC_2R_2 + 1}$$

using voltage division theorem,

$$V_2(s) = V_1(s) \cdot \frac{Z_2(s)}{Z_1 + Z_2}$$

$$= V_1(s) \cdot \frac{\frac{R_2}{(SC_2R_2 + 1)}}{\frac{R_1}{sCR_1 + 1} + \frac{R_2}{SC_2R_2 + 1}}$$

$\therefore$ $R_1C = R_2C_2$ $\quad V_2(s) = V_1(s) \cdot \frac{R_2}{R_1 + R_2}$.

For impulse response, $V_1(s) = 1 \Rightarrow V_1(t) = s(t)$.

$$V_2(t) = \frac{R_2}{R_1 + R_2} \cdot \delta(t).$$

3. Output voltage of a differentiator circuit,

$$v_0(t) = RC\frac{dr.(t)}{dt} = s \times 10^3 \times 4 \times 10^{-6} \cdot \frac{d}{dt}(100t)$$

$V_0(t) = 2$ Volts

4. Given : $V_c(0) = 6$ V.

At $t = 0^+$, $V_R = 10 - 6 = 4$ V.

$$I_R(0^+) = \frac{4}{4} = 1 \text{ A}.$$

$I_R(\infty) = 0$ A.

Time comtant, $\tau = RC = 4 \times 2 = 8$ s.

$$I_R = (\infty) + \left[I_R(0^+) - I_R(\infty)e^{\frac{-t}{\tau}}\right]$$

$$I_R = 0 + (1 - 0)e^{\frac{-t}{8}} = e^{\frac{-t}{8}}$$

Energy absorbed by 4 Ω resistor in intaval $(0,\infty)$

$$= \int_0^\infty I_R^2 R\, dt = \int_0^\infty 4e^{\frac{-2t}{\theta}} dt = 4\int_0^\infty e^{\frac{-t}{4}} dt$$

$= -16.\ [0 - 1] = 16$ Joules.

5. When switch was closed, circuit was in steady-state. All the current had passed through inductor.

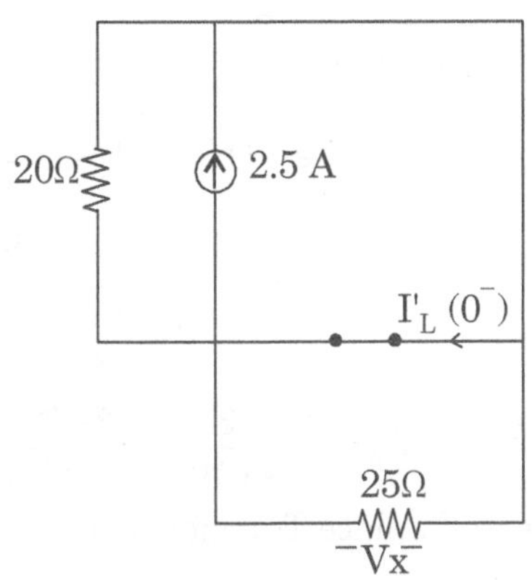

$i_L(0^-) = 2.5$ A.

$V_x(0^-) = 0$ V.

At $t = 0^+$

$\Rightarrow V_x = -R \times 2.5 = -20 \times 2.5 = -50$ Volt.

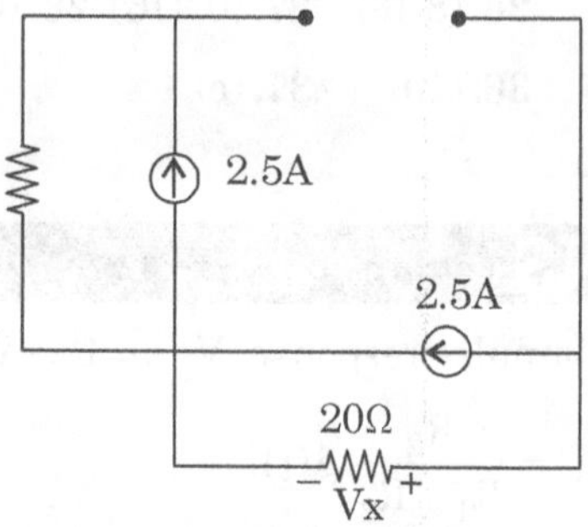

6. $$v_i(t) = Ri(t) + L\frac{di(t)}{dt}$$

With Laplace transformation, we have

$$V_i(s) = [R + Ls]\, I(s)$$

or, $$I(s) = \frac{V_i(s)}{R+Ls} = \frac{1}{s(2+s)}$$

...[since, $V_i(s) = \frac{1}{s}$]

In time-domain, $i(t) = \frac{u(t)}{2}\,[1 - e^{-2t}]$

At $t = \frac{1}{2}$, $i(t) = 0.31$

and at $t = \infty$, $i(t) = 0.5$

7. Transfer function,

$$T(s) = \frac{R}{R + \frac{1}{Cs}}$$

or, $$\frac{V_o(s)}{V_i(s)} = \frac{R}{R + \frac{1}{Cs}} \quad ...(1)$$

$$V_i(t) = 3[u(t) - u(t-2)]$$

$$V_i(s) = \frac{3}{s}(1 - e^{-2s})$$

Their from equation. (1),

$$V_o(s) = \frac{3}{s}(1 - e^{-2s}) \cdot \frac{R}{R + \frac{1}{Cs}}$$

$$= \frac{3}{s}(1 - e^{-2s}) \frac{s}{s + 10^4}$$

$$V_o(s) = \frac{3}{s+10^4} - \frac{3e^{-2s}}{s+10^4}$$

Inverse Laplace Transform gives,

$$V_o(t) = 3u(t)\, e^{-10^4 t} - 310^{-4(t-2)}\, u(t-2)$$

At $t = 2$ sec,

$$V_o(2) = 3 \times 10^{-4} \times 2 - 3 \cong -3 \text{ Volts}$$

8. Voltage across inductor, $vL(t) = L\frac{di(t)}{dt}$

Applying hapless transform

$$V(s) = sLI(s) - LI(0^+) \quad (\therefore LI(0^+) = 1 \text{ mV})$$

$$I(0^+) = \frac{1 \text{ mV}}{2 \text{ mH}} = 0.5 \text{ A.}$$

9. $$\frac{V_0(s)}{V_i(s)} = \frac{\frac{R_0}{R_0C_0s+1}}{\frac{R_0}{R_0C_0s+1} + \frac{R_i}{R_iC_is+1}}$$

$$= \frac{R_0(R_iC_is+1)}{R_0R_i(C_i+C_0)s + R_0 + R_i}$$

$$= \frac{4K(4s+1)}{4K^2 5s + 5K} = \frac{4}{5}$$

$$V_0(s) = \frac{8}{s}$$

$$\Rightarrow \quad V_0(t) = 8u(t)$$

10. At $t = 0^+$, $v_c(0^+) = 0$ V means capcitor is short-circuited; $i_c(0^+) = \frac{10\text{V}}{20\text{K}\Omega} = 0.5$ mA.

At steady-state, capacitor will get open-circuited, $i_c(\infty) = 0$ A.

Time-contant,

$$\tau = Req.c = (20 \,||\, 20\text{K}) \times 4\mu\text{F} = 10 \times 10^3 \times 4 \times 10^{-6}$$

$$= 40 \text{ ms.}$$

$$\Rightarrow i_c(t) = i_c(\infty) + [i_c(0) - i_c(\infty)]\, e^{\frac{-t}{\tau}}$$

$$= 0 + \left[0.5 \times 10^{-3} - 0\right]^{\frac{-t}{\tau}} = 0.5\, e^{-25t} \text{ mA.}$$

11. Initially $i(0^+) = 0$ due to inductor

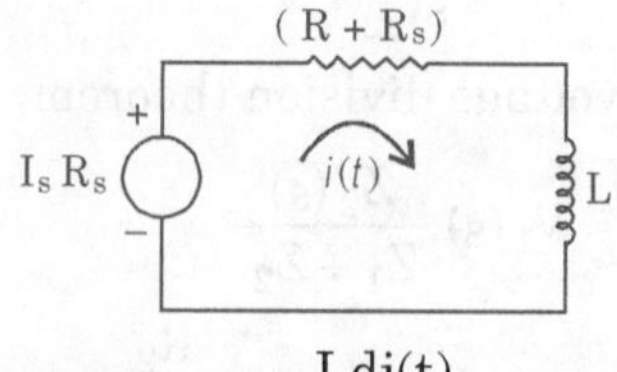

$$I_s R_s = (R + R_s)\, i(t) + \frac{Ldi(t)}{dt}$$

For $t = 0^+$, $I_s R_s = (R + R_s)\, i(0^+) + \frac{Ldi(0^+)}{dt}$

$$i(0^+) = 0$$

$$\Rightarrow \frac{di(0^+)}{dt} = \frac{I_s R_s}{L}$$

12. For $n = 0, 0 \le t \le T$, S_1P_1 and S_2P_2 are connected as shown

$$V_c(t) = \frac{1}{C}\int_0^t c\,dt = t, 0 \le t \le T$$

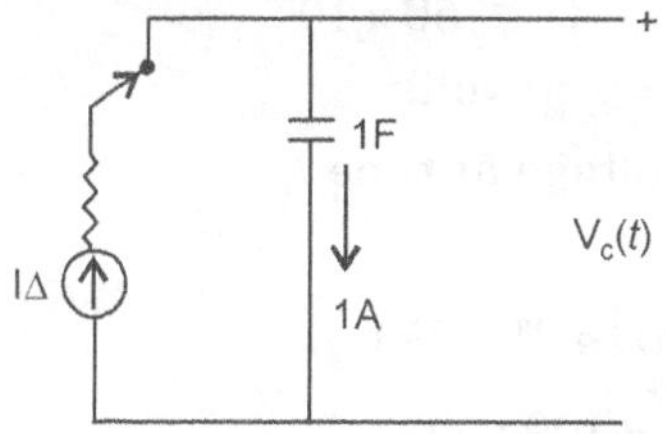

For $T \le t \le 2T$, S_1Q_1 and S_2Q_2 are connected

$$V_c(t) - V_c(T) = \frac{1}{C}\int_T^t i \cdot dt = -\int_T^t dt$$

[$i = -1A$ as flowing in opposite direction]

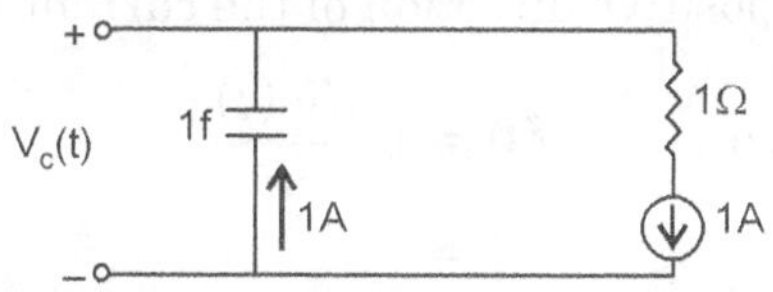

or, $V_c(t) - T = -(t - T)$

or, $V_c(t) = 2T - t$, for $T \le t \le 2T$.

Hence, $V_c(t)$ can be expressed as,

$$V_c(t) = t\,u(t) + 2[-(t - T)\,u(t - T)]$$

In generalised terms,

$$V_i(t) = t\,u(t) + 2\sum_{n=1}^{\infty}(-1)^n (t - nT)\,u(t - nT)$$

13. When the switch is in the position (*a*), the circuit will be as shown,

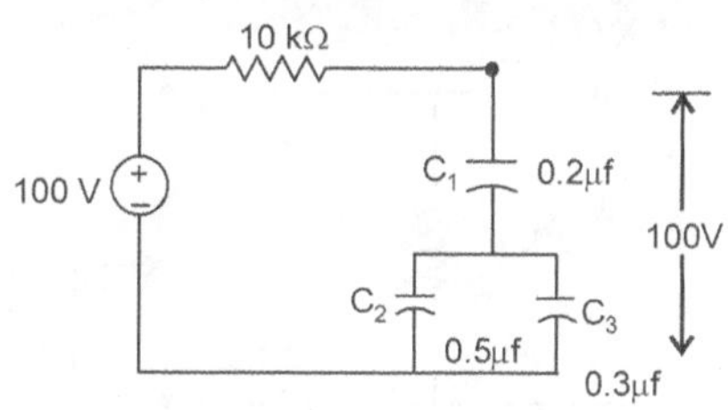

$C_1 = 0.2$ F

$C_2 = 0.5$ F

$C_3 = 0.3$ F

In steady state all the capacitors are open circuited,

⇒

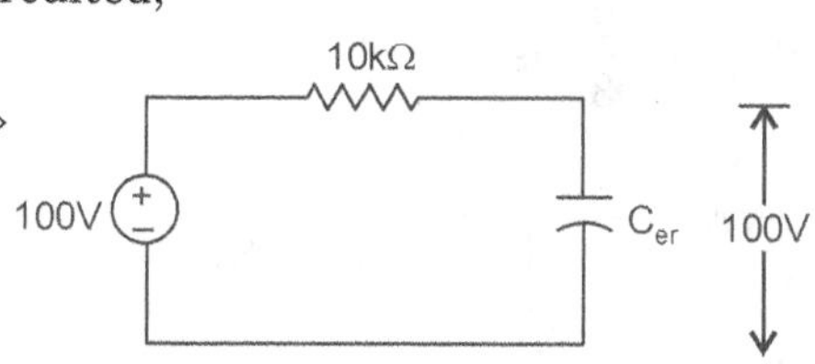

$$C_{eg} = \frac{C_1(C_2 + C_3)}{C_1 + C_2 + C_3} = 0.16 \text{ F}$$

When the switch is thrown to position (*b*)

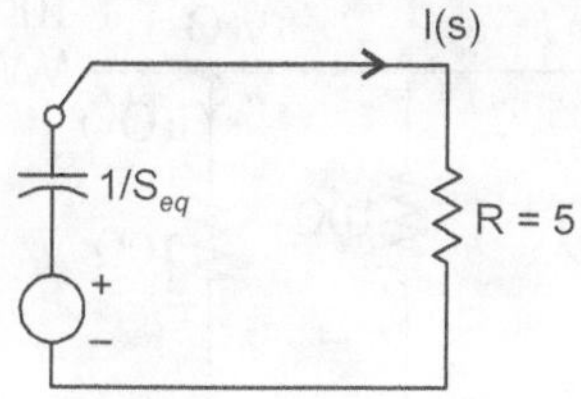

$$\frac{V(0+)}{s} = \frac{100}{s}$$

$$I(S) = \frac{V(0+)}{s} \cdot \frac{1}{R + \frac{1}{sC_{eq}}}$$

$$= \frac{V(0+)}{R} \cdot \frac{1}{a + \frac{1}{RC_{eq}}}$$

Inverse laplace given,

$$i(t) = V_F + (V_1 - V_F)\,e^{-\frac{T}{RC}}$$

$$= \frac{V(0+)}{R}.e^{-\frac{t}{RC_{eq}}}.V(t)$$

$$= \frac{100}{5k}e^{-1250t} \cdot V(t)$$

$$= 20e^{-1250t}V(t) \text{ mA}$$

14. Taking Laplace transform, we have

$$sLI(s) - Li(0) + RI(s) = V_0\left(\frac{1}{s} + B\frac{1}{\left(s + \frac{R}{L}\right)^2 + L}\right)$$

For steady state current ,

$$\lim_{t\to\infty} i(t) = \lim_{s\to 0} sI(s)$$

$$\therefore \quad I(s) = \frac{V_0}{sL + R}\left(\frac{1}{s} + \frac{B}{\left(s + \frac{R}{L}\right)^2 + 1}\right) + \frac{LV_0}{R}\frac{s}{(R + Ls)}$$

or $\lim_{s\to 0} sI(s) = \frac{V_0}{R}$

15. At t = 0– (steady-state)

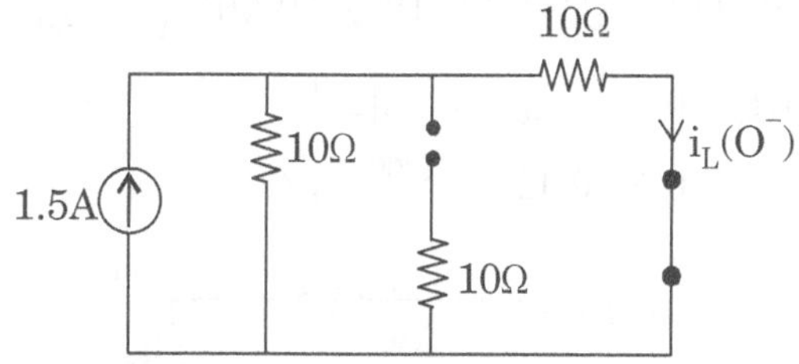

$$i_L(0^-) = 1.5 \times \frac{10}{10 + 10} = 0.75 \text{ A}$$

$i_L(0^-) = i_L(0^+) = 0.75$ A

At $t = 0^+$,

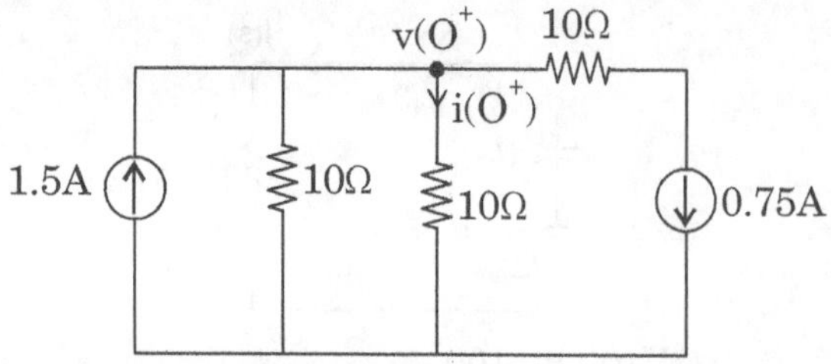

Inductor is replaced by a went source with initial value i.e;

$i_L(0^-) = i_L(0^+) = 0.75$ A.

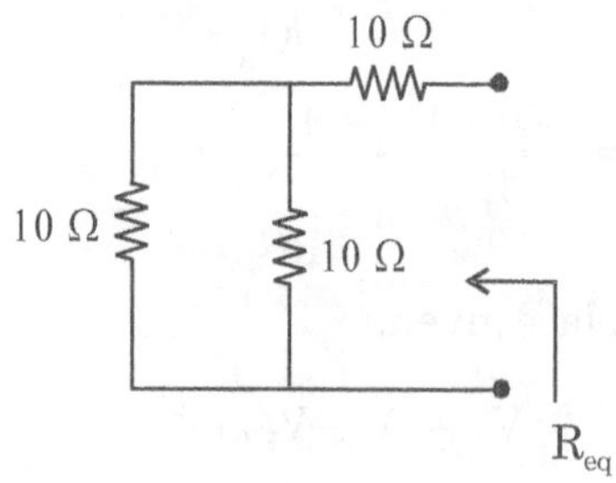

$$R_{eq} = (10 \parallel 10) + 10 = 15\ \Omega$$

$$\tau = \frac{L}{R_{eq}} = 10_5^{-3}.$$

Applying nodal analysis,

$$\frac{v(10^+)}{10} + \frac{v(0^+)}{10} - 1.5 + 0.75 = 0$$

v $(0^+) = 5 \times 0.75 = 3.75$ V.

$$i(0^+) = \frac{v(0^+)}{10} = 0.375\text{A}.$$

At $t \to \infty$ (steady-state),

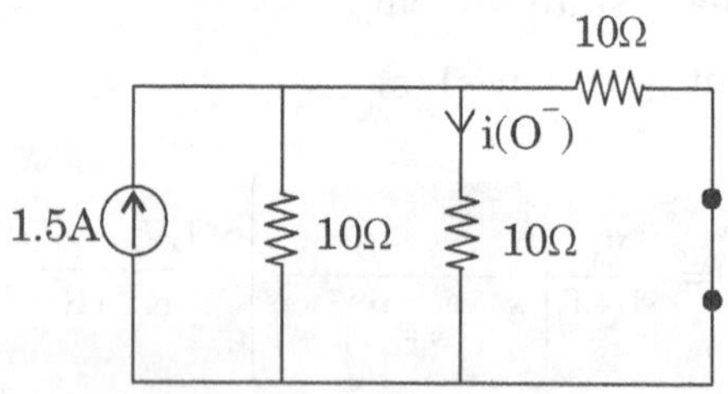

$$i(\infty) = \frac{1.5 \times 5}{10 + 5} = 0.5 \text{ A}.$$

$$\text{current } i(t) = i(\infty) + [i(0) - i(\infty)]e^{\frac{-t}{\tau}}$$

$$i(t) = 0.5 + [0.375 - 0.5]e^{-1000t}$$

$$i(t) = 0.5 - 0.125\ e^{-1000t} \text{ A}.$$

16.

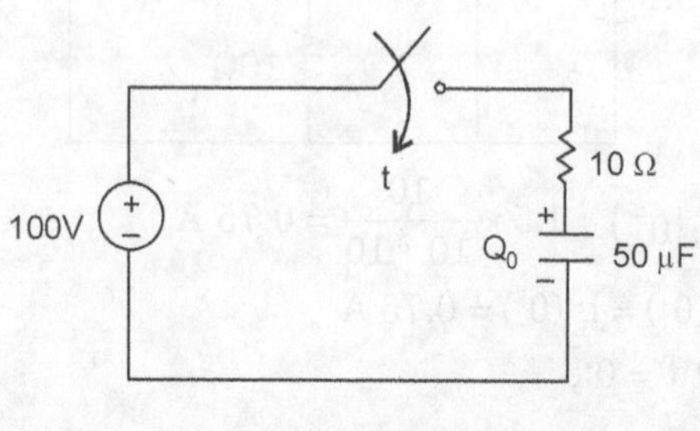

Q_0(Initial charge on capacitor)

$$= CV_c(0-) = CV_c(0+)$$

$$\Rightarrow \quad V_c(0-) = V_c(0+)$$

$$= \frac{Q_0}{C} = \frac{2.5 \times 10^{-3}}{50 \times 10^{-6}}$$

$$= -50 \text{ Volts}$$

The capacitor voltage at time t,

$$V_c(t) =$$

$$[V_c(0+) - V_c(\infty)]e^{-t/RC} + V_c(\infty)$$

$$V_c(\infty) = 100 \text{ volts}$$

$$V_C(t) = [-50 - 100]\,e^{-2\times10^3 t} + 100$$

$$= -150\,e^{-2\times10^3 t} + 100$$

[The sign is –ve because the capacitor voltage is in positive direction of the current.]

then, $\quad i(t) = C\dfrac{dv_c(t)}{dt}$

$$=$$

$$(50 \times 10^{-6}) \times (-150 \times -2 \times 10^3)\,e^{-2\times10^3 t}$$

$$= +15\,e^{-2\times10^3 t}$$

17. When switch in closed at $t = 0$, capacitor C_1 will discharge and C_2 will get charge since both C_1 and C_2 are ideal and there is no-resistance in the circuit charging and discharing time constant will be zero.

Thus current will exist like an impulse function.

18. C = 1 μF. $v_c(0^-) = 0$ V.

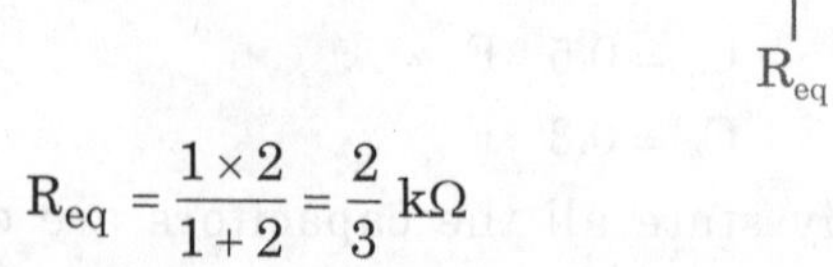

$$R_{eq} = \frac{1 \times 2}{1 + 2} = \frac{2}{3}\text{ k}\Omega$$

$$T = R_{eq}.C = \frac{2}{3}\text{ ms.}$$

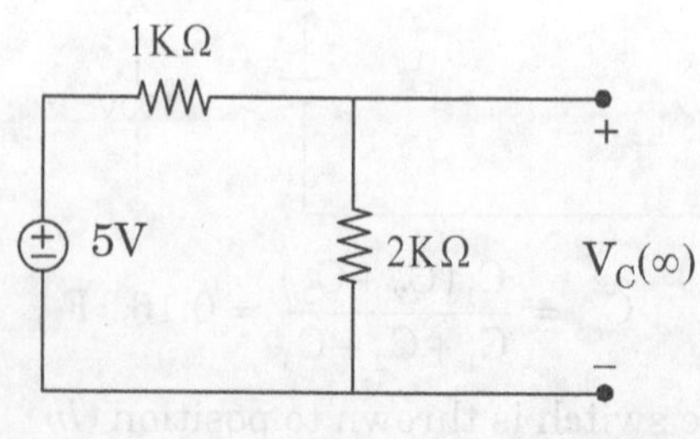

$$v_c(\infty) = 5\frac{\times 2}{2+1} = \frac{10}{3}\text{ Volt.}$$

$$v_c(t) = v_c(\infty) + \left[v_c(0^-) - v(\infty)\right]e^{\frac{-t}{T}}$$

$$= \frac{10}{3} + \left[0 - \frac{10}{3}\right]e^{\frac{-t}{T}}$$

$$v_c(t) = \frac{10}{3}\left[1 - e^{\frac{-t}{T}}\right]\text{ Volt.}$$

19.

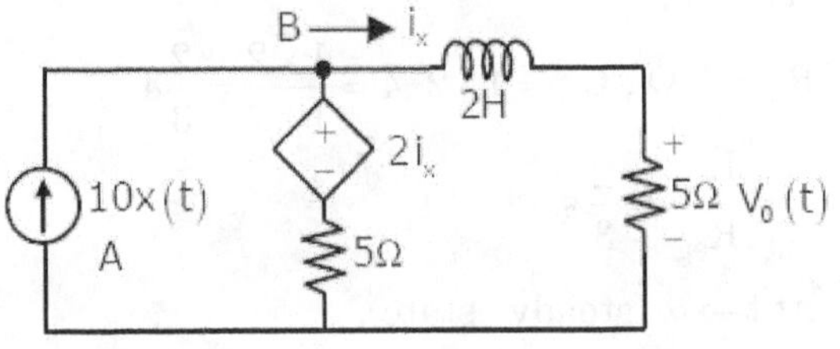

For $t \to \infty$, i.e., at steady state, inductor will behave as a shot circuit and hence $V_B = 5.i_x$

By KCL at node B, $-10 + V_B - 2i_x + i_x = 0$

$$\Rightarrow \quad i_x = \frac{50}{8}$$

$$V_0(t) = 5i_x(t)$$

$$\Rightarrow \quad V_0(t) = \frac{250}{8} = 31.25\text{ Volts}$$

20. At $t = 0^-$

$v_c(0^-) = 0$ V.

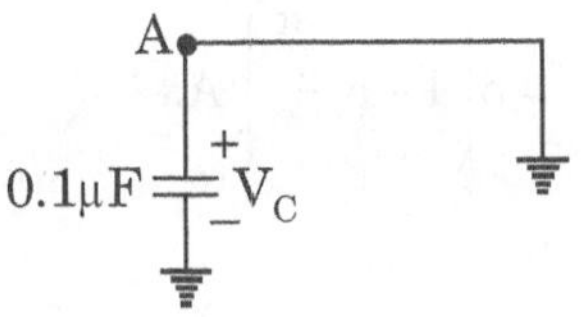

At $t = 0^+$,

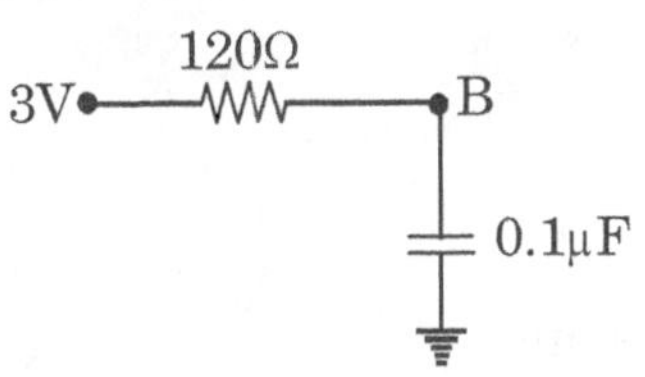

$v_c(0^-) = v_c(0^+) = 0$V.

$v_c(\infty) = 3$ V.

$\tau = R.\,C = 120 \times 0.1\times10^{-6}$

$\tau = 12 \times 10^{-6}$ s.

$$v_c(t) = v(\infty) + \left[v(0) - v(\infty)\right]e^{\frac{-t}{\tau}}$$

$$i_c(t) = c\frac{dv_c(t)}{dt} = \frac{1}{40}.e^{\frac{-t}{\tau}}$$

$$\text{Energy} = \int_0^\infty v_c i_c dt = 0 = \frac{-3}{40} \times \tau\, e^{\frac{-t}{\tau}}\Big|_0^0 = 0.9\ \mu\text{J.}$$

21. At $t = 0^-$,

$v_c(0^-) = v_c(0^+) = 0$ V

$$R_{eq} = \frac{2\times3}{2+3} = \frac{6}{5}\ \Omega.$$

$$\tau = R_{eq}.C = \frac{6}{5}\times 1\text{ s.}$$

At $t = 0^+$,

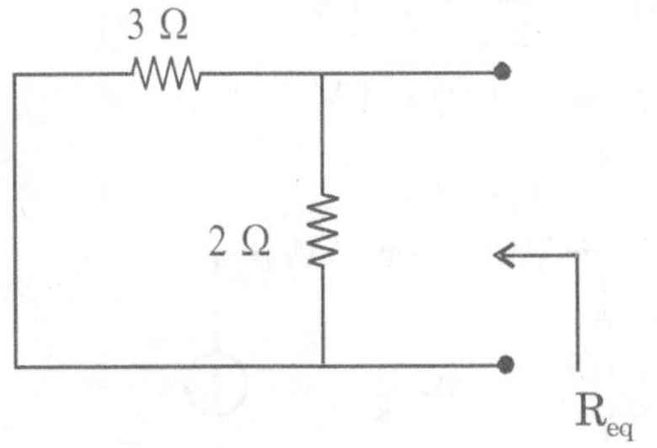

$\Rightarrow v_c(\infty)$

$$\Rightarrow v_c(\infty) = \frac{2}{2+3}\times 10 = 4\text{V.} \quad = 4\left(1 - e^{\frac{-t}{1}}\right)$$

$$\Rightarrow v_c(t) = 4\left(1 - e^{-t}\right)\text{ Volt.}$$

$$v_c(t = 1\text{ s}) = 4\left(1 - e^{-1}\right) = 2.528\text{ Volt.}$$

22.

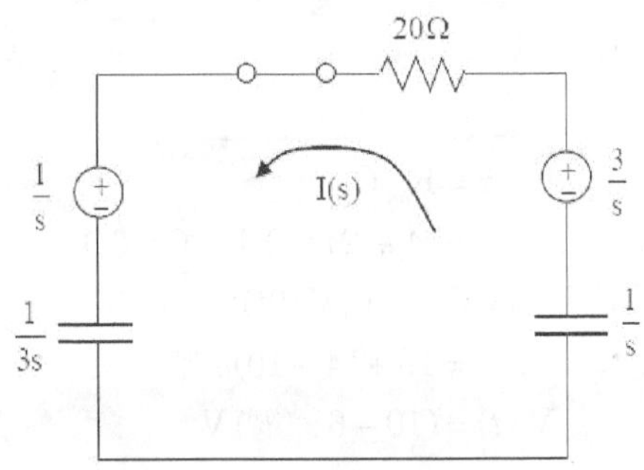

$$I(s) = \frac{\left(\frac{3}{s} - \frac{1}{s}\right)}{\left(10 + \frac{1}{3s} + \frac{3}{3s}\right)}$$

$$I(s) = \frac{2}{\left(10s + \frac{4}{3}\right)} = \frac{2}{10\left(s + \frac{4}{30}\right)}$$

$$i(t) = \frac{1}{5}e^{-\frac{4}{30}t};\ t \geq 0$$

$$E_R = \int_0^\infty i^2(t)10dt = \left(\frac{10}{25}\right)\int_0^\infty e^{\frac{-4}{15}t}dt$$

$$= \frac{10}{25}.\left[\frac{e^{\frac{4}{15}t}}{\frac{-4}{15}}\right]_0^\infty$$

$$= 0 - \frac{10}{25}\times\frac{15}{-4} = 1.5\text{ J}$$

23. At $t = 0^-$, Switch is at position-1

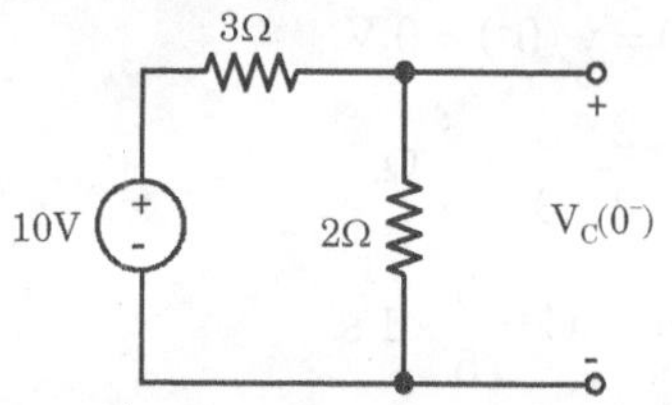

where, $V_c(0^-) = \dfrac{10 \times 2}{2+3} = 4\text{ V}$ ...(1)

$V_c(0^-) = V_c(0^+) = 4\text{ V}$

At $t = \infty$,

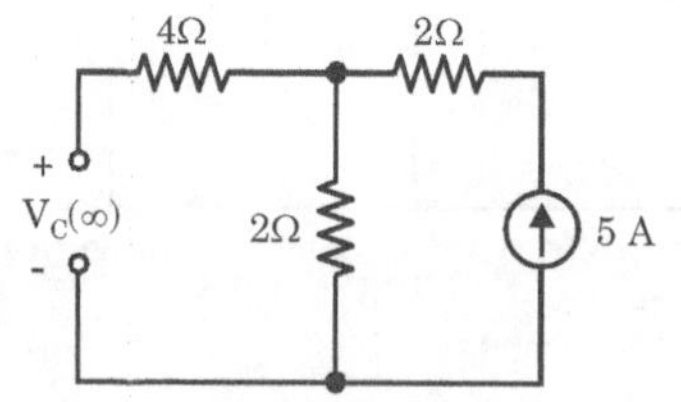

$V_c(\infty) = 5 \times 2 = 10\text{ V}$ ...(2)

The time constant of the circuit is

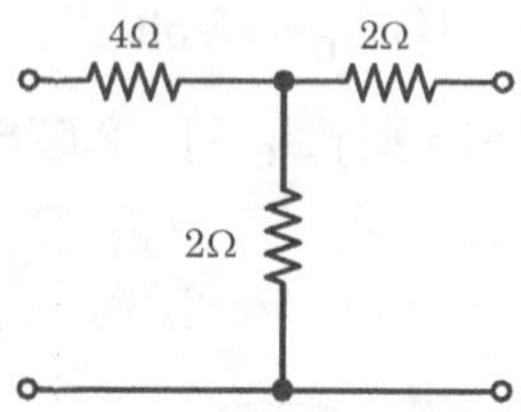

$$\tau = R_{eq}C_{eq}$$
$$= (4 + 2) \times 0.1 = 6 \times 0.1 = 0.6 \text{ sec}$$

Therefore, $V_c(t) = V_c(\infty) + [V_c(0^+) - V_c(\infty)]e^{-t/\tau}$

$$= 10 + (4 - 10)e^{-t/0.6}$$
$$V_c(t) = (10 - 6e^{-t/0.6})\text{ V}$$

24. At $t = 0^-$ (steady-state),

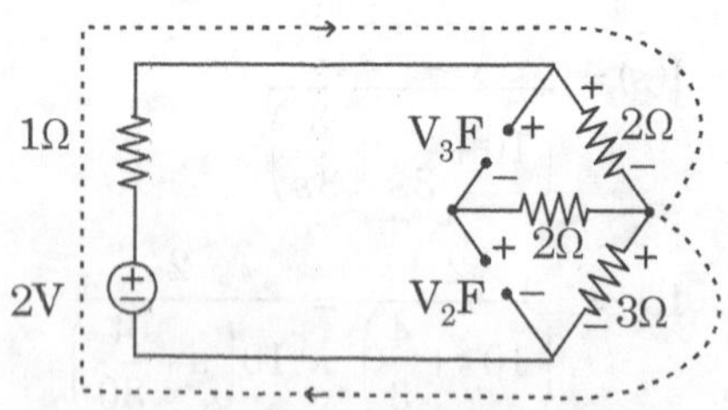

Using voltage division theorem,

$$V_{3F} = 12 \times \frac{2}{2+3+1} = 4\text{ V}$$

$$V_{2F} = 12 \times \frac{3}{2+3+1} = 6\text{ V.}$$

At $t = 0^+$,

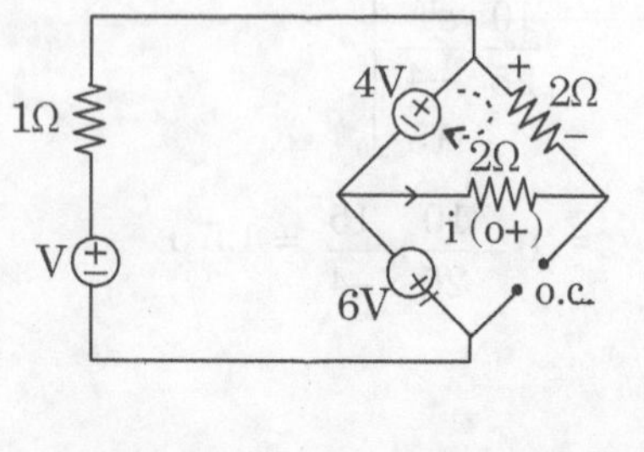

Applying KVL in loop,

$$-4V + [-2\,i(0^+)] + [-2 \times i(0^+)] = 0$$

$$i(0^+) = \frac{-4}{2+2} = -1\text{ A}$$

Note:- The direction of current is not mentioned is the question.

25. At $t = 0^-$ (steady-state),

$I_L(0^-) = I_L(0^-) = 0\text{ A}$

At $t > 0$,

$$R_{eq} = 1\,\Omega\,;\ L_{eq} = 1 \,\|\, 2\ \ 4 = \frac{1 \times 2}{1+2} = \frac{2}{3}4.$$

$$\tau = \frac{L_{eg}}{R_{eg}} = \frac{2}{3}\text{ s.}$$

At $t \to \infty$ (steady - state),

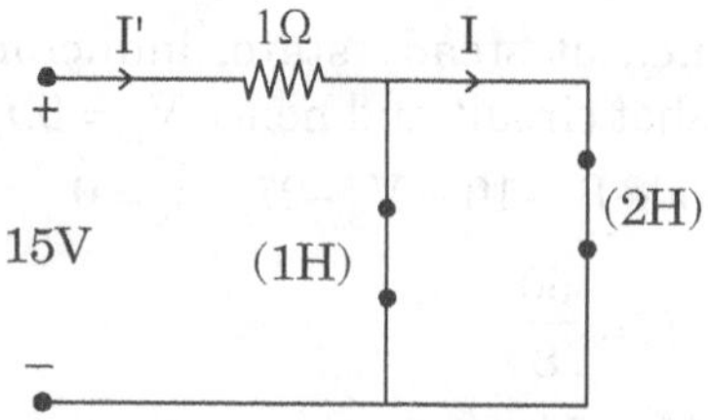

$$I' = \frac{V_m}{R} = \frac{15}{1} = 15\text{ A.}$$

Using currant division rule,

$$I(\infty) = \frac{1}{1+2} \times 15 = 5\text{ A.}$$

$$I(t) = 5 + (0-5)e^{\frac{-t}{\tau}} = 5\left(1 - e^{\frac{-3t}{2}}\right)\text{ A.}$$

Given $I'(t) = 2$ A.

$$5\left(1 - e^{\frac{-3t}{2}}\right) = 2$$

$$e^{\frac{-3t}{2}} = \frac{3}{5}$$

$t = 0.3405.$

26. At $t = 0^-$ (steady-state),

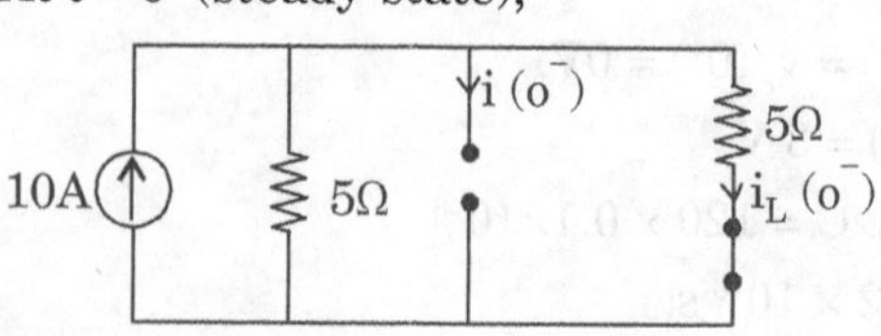

$i(0^-) = 0\text{ A}$

$$i_L(0^-) = \frac{5}{5+5} \times 10 = 5\text{ A.}$$

At $t > 0$,

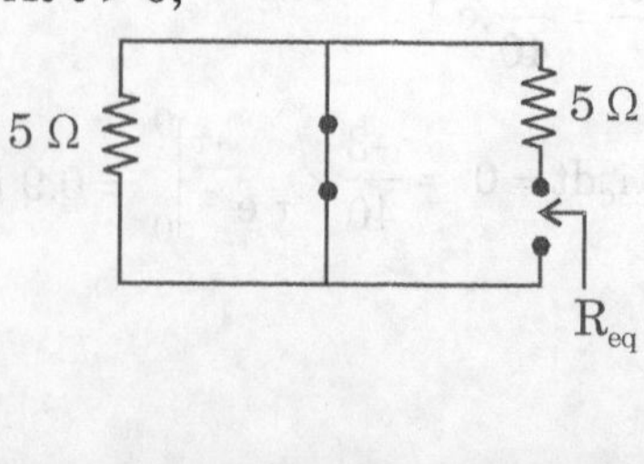

$R_{eq} = 5\ \Omega$

$$\tau = \frac{L}{R_{eq}} = \frac{2.5}{5} = \frac{1}{2}\text{ s.}$$

At $t \to \infty$ (steady - state),

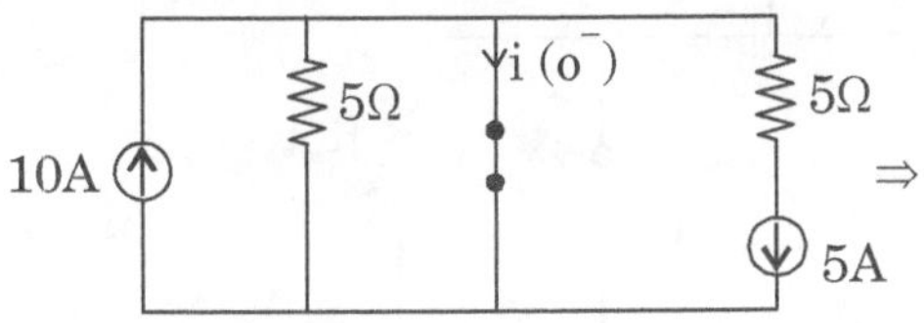

Al t + = 0^+

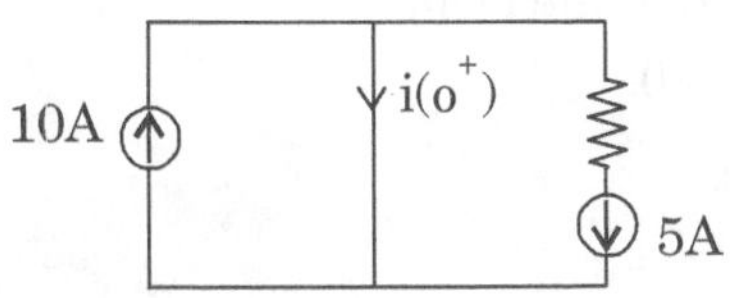

$i(0^+) = 10 - 5 = 5$ A

At $t \to \infty$ (steady - state),

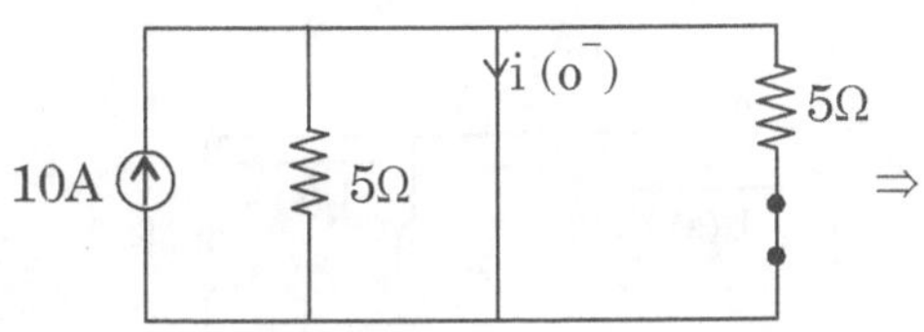

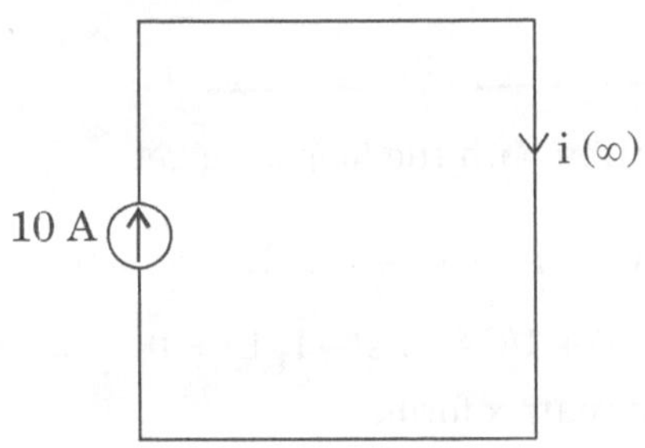

$i(\infty) = 10$ A.

$$\Rightarrow i(t) = i(\infty) + [i(0) - i(\infty)]e^{\frac{-t}{\tau}}$$

$$= 10 + (5 - 10)e^{\frac{-t}{0.5}}$$

$$i(t) = 10 - 5\,e^{-2t}$$

$$i(0.5) = 10 - 5e^{-1} = 8.16\text{ A.}$$

27. At $t = 0^-$,

$i(0^-) = 0$ A.

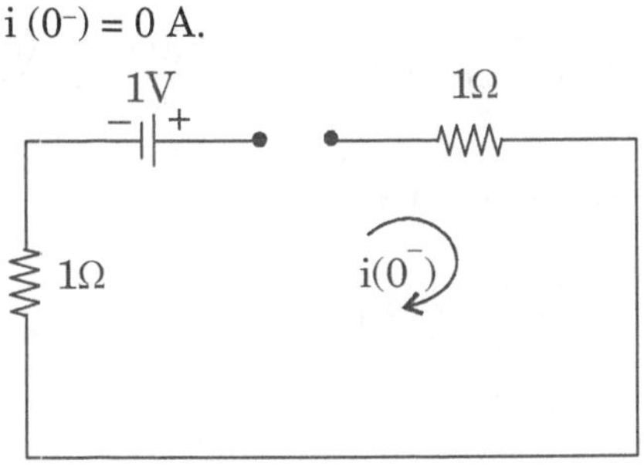

At $t \to \infty$

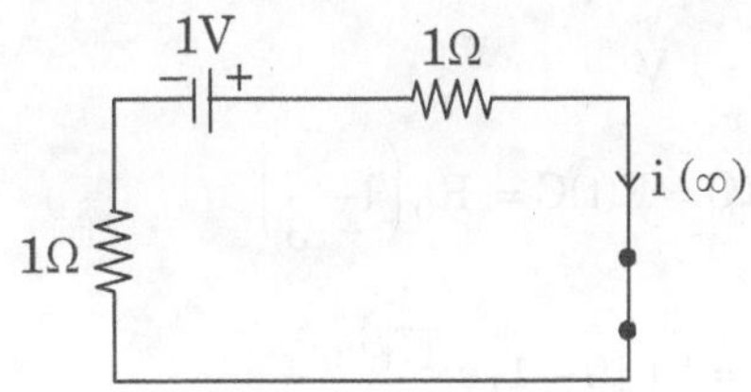

$$i(\infty) = \frac{1}{1+1} 0.5\text{ A.}$$

$$i(0^+) = i(0^-) = 0\text{ A.}$$

$$i(t) = 0.5\left(1 - e^{\frac{-t}{0.5}}\right).$$

$$i(0.5) = 0.5(1 - e^{-1}) = 0.316\text{ A.}$$

$$\tau = \frac{L}{R_{eq}} = \frac{1}{2}$$

$\tau = 0.5$

28. For $t > 0$, Redrawing the given network.

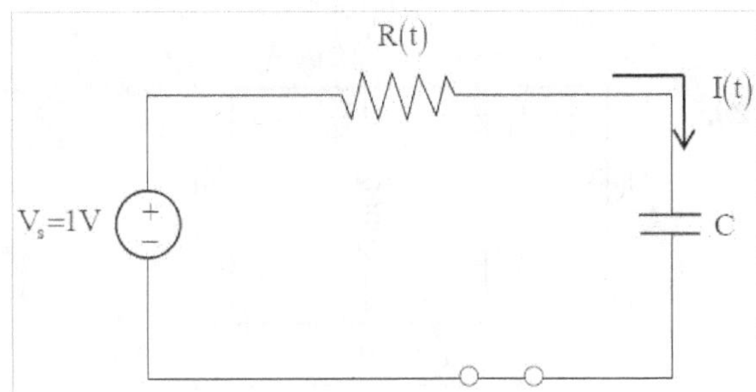

Given

$R_0 = 1\ \Omega$, $C = 1$ F, $T = 3\,R_0C$

Now $R(t) = R_0\left(1 - \frac{t}{T}\right)1,\ 0 \le t \le T$

$$R_0\left(1 - \frac{t}{3R_0C}\right) = R_0\left(1 - \frac{t}{3}\right) = (1 - t/3)$$

$\therefore\ R(0) = R_0 = 1\ \Omega$

Redrawing the network at $t = 0$,

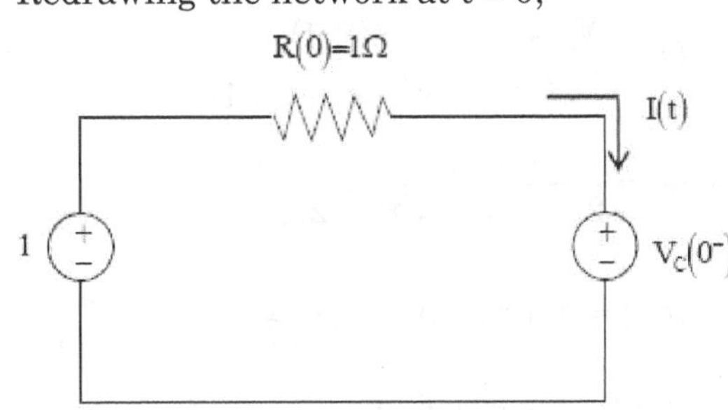

$$\text{Now } I(0) = \frac{1 - V_c(0^-)}{R(0)}$$

$$\therefore\ 1 = \frac{1 - V_C(0^-)}{1} \qquad (\because I(0) = 1\text{ A})$$

$\therefore\ 1 = 1 - V_C(0) \Rightarrow V_C(0^-) = 0$ V

Now for $t = \infty$, capacitor will be open circuit

$$V_C(\infty) = 1\text{ V}$$

$$\text{Now, } \tau = RC = R(t)C = R_0\left(1-\frac{t}{3}\right)$$

$$\text{Now } V_C(t) = 1 + [0-1]\,e^{\frac{-t}{(1-t/3)}}$$

$$\therefore\ V_C\left(\frac{T}{2}\right) = 1 - e^{-\left(\frac{T/2}{1-T/6}\right)}$$

$$= 1 - e^{-\left(\frac{3/2}{1-3/6}\right)} \quad (\because T = 3)$$

$$= 1 - e^{-\left(\frac{1.5}{0.5}\right)} = 1 - e^{-3} = 0.95$$

$$\text{Now, } I(t) = \frac{V_s(t) - V_C(t)}{R(t)}$$

$$\therefore\ I\left(\frac{T}{2}\right) = \frac{V_s\left(\frac{T}{2}\right) - V_C\left(\frac{T}{2}\right)}{R\left(\frac{T}{2}\right)} = \frac{1-0.95}{0.5} = 0.10$$

29. 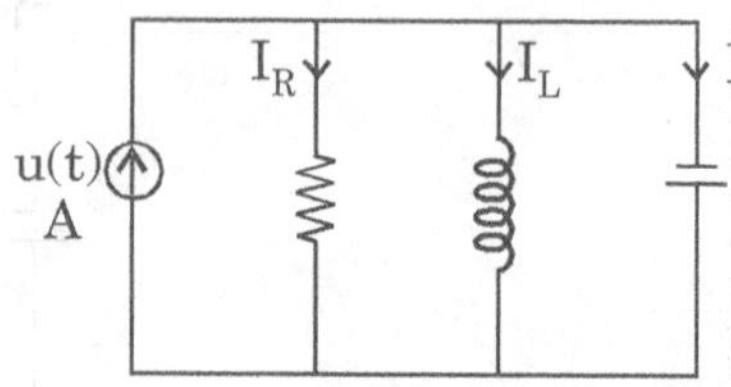

At t = 0

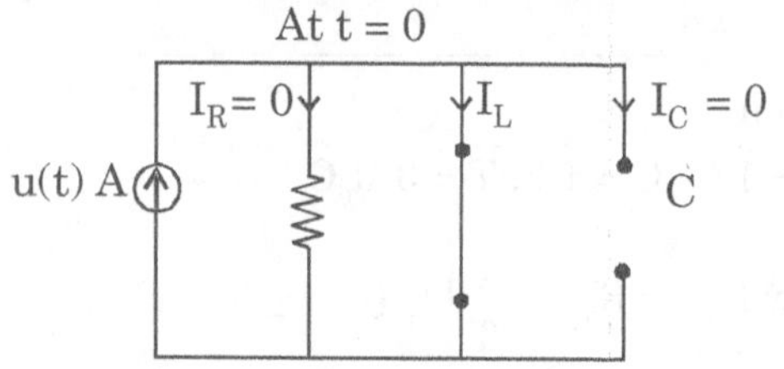

At steady state

⇒ At steady-state, source current will flow through inductor only.

30. Capacitive reactance ,

$$I_c(s) = \frac{V_c(s)}{X_c(s)} = \frac{s(SH)}{2(s^3+s^2+s+1)}$$

$$I_c(s) = \frac{s(s+1)}{2(s^2+1)(s+1)} = \frac{s}{2(s^2+1)}$$

$$i_c(0^+) = \lim_{s\to\infty} SI_c(s) = \lim_{s\to\infty}\frac{s^2}{2(s^2+1)} = \frac{1}{2+0} = \frac{1}{2}\text{ A.}$$

31. At t = 0⁻, the circuit was in steady state.

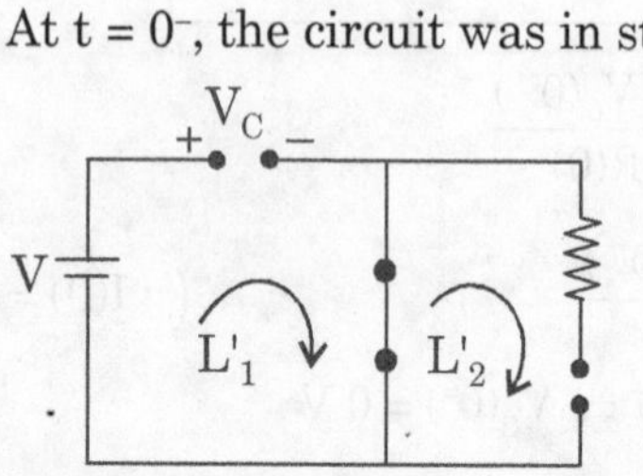

$i_1(0^-) = i_2(0^-) = 0$ A.

$v_c(0^-) = V$.

At $t = 0^+$:

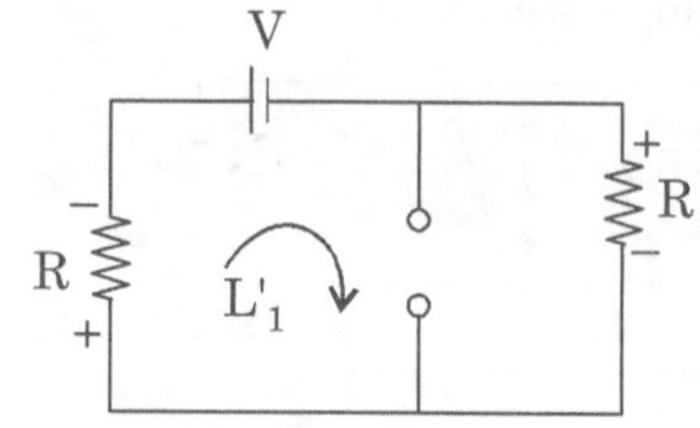

Applying KVL in the loop,

$i_1R + V + i_1R = 0$.

$$i_1 = \frac{-V}{2R}$$

32. Transformed circuit representation is as shown below

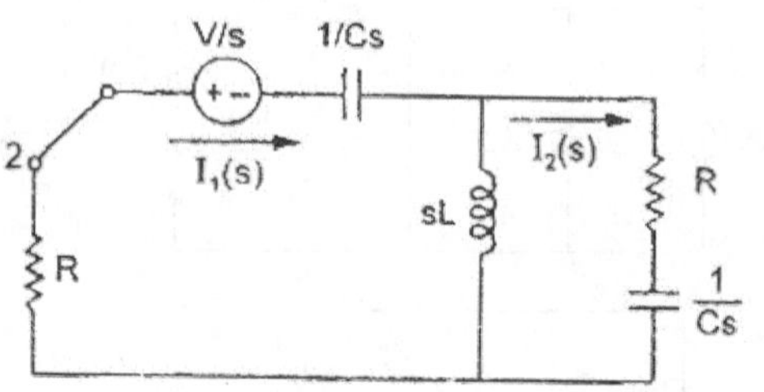

Using KVL in both the loops, we get

$$I_1(s)\,.\,[1/Cs + sL + R] - I_2\,.\,sL + \frac{V}{s} = 0$$

and $I_2(s)\,[R + 1/Cs + Ls] - I_1\,Ls = 0$

Writing in matrix form,

$$\begin{bmatrix} R+sL+\dfrac{1}{Cs} & -Ls \\ -Ls & R+Ls+\dfrac{1}{Cs}\end{bmatrix}\begin{bmatrix} I_1(s) \\ I_2(s)\end{bmatrix} = \begin{bmatrix} -V/s \\ 0\end{bmatrix}$$

33. Using KVL around the loop, we have

$$v(t) = Ri(t) + \frac{Ldi(t)}{dt} + \frac{1}{C}\int_0^\infty i(t).dt$$

Taking Laplace transformation both of sides, we have

$$V(s) = RI(s) + LsI(s) - LI(0+) + \frac{I(s)}{sC} + \frac{v_c(o+)}{s}$$

$$\Rightarrow\ \frac{1}{s} = I(s) + sI(s) - 1 + \frac{I(s)}{s} - \frac{1}{s}$$

$$\Rightarrow\ \frac{2}{s} + 1 = \frac{I(s)}{s}\left[s^2+s+1\right]$$

$$\Rightarrow\ I(s) = \frac{s+2}{s^2+s+1}.$$

34.

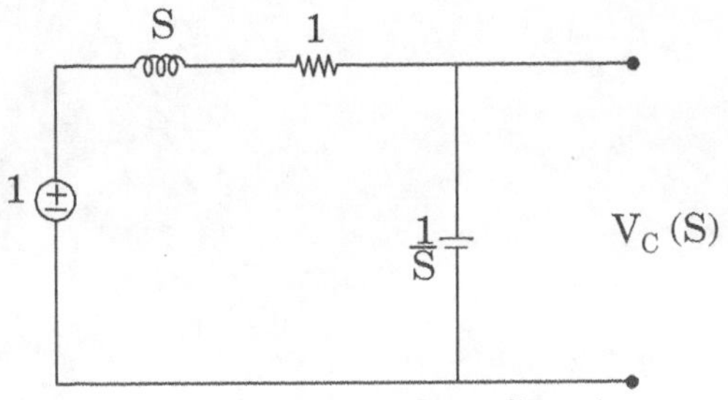

S-equivalent Circuit

$$V_c(s) = 1.\frac{\frac{1}{s}}{\frac{1}{s}+s+1}$$

$$V_c(s) = \frac{1}{s^2+s+1} = \frac{1}{\left(s+\frac{1}{2}\right)^2+\left(\frac{\sqrt{3}}{2}\right)^2}$$

$$V_c(t) = \frac{2}{3}e^{\frac{-t}{2}}.\sin\left(\frac{\sqrt{3}}{2}t\right) \text{ Volt.}$$

35. Voltage across sinister,

$$V_R(s) = 1.\frac{1}{\left(s+1+\frac{1}{s}\right)} = \frac{s}{s^2+s+1}$$

$$V_R(s) = 1.\frac{s+1\frac{1}{2}}{\left(s+\frac{1}{2}\right)^2+\left(\frac{\sqrt{3}}{2}\right)^2} - \frac{\left(\frac{1}{2}\right)\left(\frac{\sqrt{3}}{2}\right)\left(\frac{2}{\sqrt{3}}\right)}{\left(s+\frac{1}{2}\right)^2+\left(\frac{\sqrt{3}}{2}\right)^2}$$

$$v_R(t) = e^{\frac{-t}{2}}\cos\left(\frac{\sqrt{3}}{2}t\right) - \frac{1}{\sqrt{3}}e^{\frac{-t}{2}}\sin\left(\frac{\sqrt{3}}{2}t\right)$$

$$v_R(t) = e^{\frac{-t}{2}}\left[\cos\left(\frac{\sqrt{3}}{2}t\right) - \frac{1}{\sqrt{3}}\sin\left(\frac{\sqrt{3}}{2}t\right)\right] \text{ Volt}$$

36. For critical damping,

$$\xi = \frac{1}{2Q} = 1$$

where, Q = Quality factor

For series circuit,

$$Q = \frac{1}{R}\sqrt{\frac{L}{C}}$$

$$\frac{1}{\frac{2}{R}\sqrt{\frac{L}{C}}} = 1$$

$$C = \left(\frac{2}{R}\right)^2 \times L = \left(\frac{2}{40}\right)^2 \times 4 = 10 \text{ mF.}$$

37. At $t = 0^-$ (Steady-state),

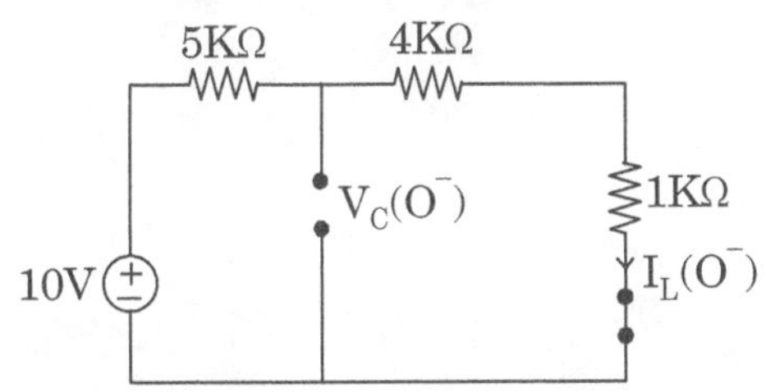

$v_c(0^-) = v_c(0^+) = SV$

$i_L(0^-) = i_L(0^+) = 1$ mA

At $t = 0^+$ (switch is closed),

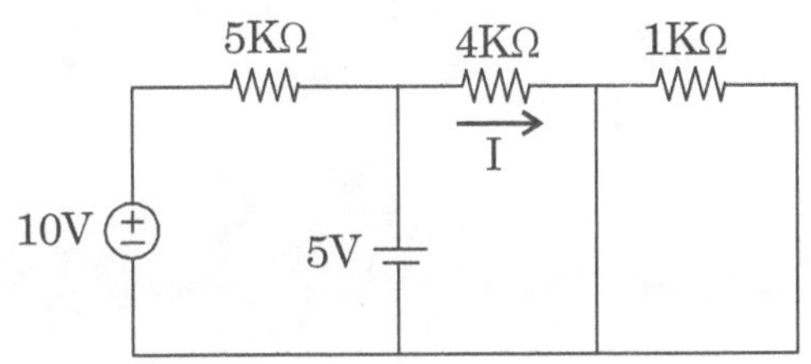

$$I = \frac{5}{4\times10^3} = 1.25 \text{ mA.}$$

5
CHAPTER

Two Port Networks

Analysis of Previous GATE Papers

	Year → Topics ↓		2019	2018	2017 Set 1	2017 Set 2	2016 Set 1	2016 Set 2	2016 Set 3	2015 Set 1	2015 Set 2	2015 Set 3
NETWORK PARAMETERS (Z, Y H AND T)	1 Mark	MCQ Type									1	
		Numerical Type		1				1				
	2 Marks	MCQ Type							1			1
		Numerical Type										
	Total			1				1	2		1	2
INTERCONNECTION OF NETWORKS	1 Mark	MCQ Type					1					
		Numerical Type										
	2 Marks	MCQ Type										
		Numerical Type										
	Total						1					

NETWORK PARAMETERS (Z, Y H AND T)

1. The condition AD – BC = 1 for a two-port network implies that the network is a

(a) reciprocal network

(b) lumped element network

(c) lossless network

(d) unilateral element network **[1989 : 2 Marks]**

2. The open circuit impedance matrix of the two port network shown in figure is

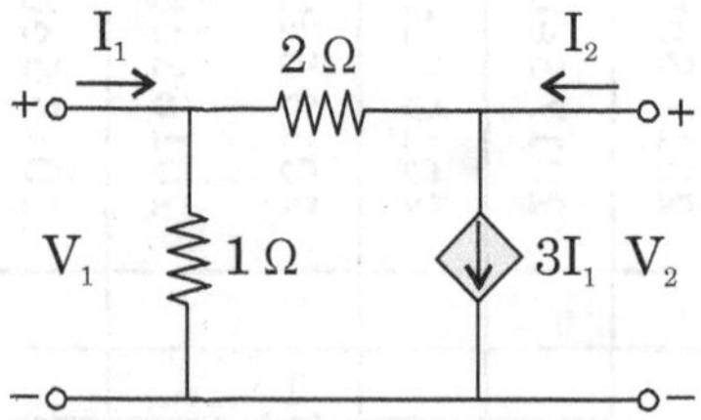

(a) $\begin{vmatrix} -2 & 1 \\ -8 & 3 \end{vmatrix}$ (b) $\begin{vmatrix} -2 & -8 \\ -8 & 3 \end{vmatrix}$

(c) $\begin{vmatrix} 0 & 1 \\ 1 & 0 \end{vmatrix}$ (d) $\begin{vmatrix} 2 & -1 \\ -1 & 3 \end{vmatrix}$

[1990 : 2 Marks]

3. For a 2-port network to be reciprocal,

(a) $Z_{11} = Z_{22}$ (b) $y_{21} = y_{12}$

(c) $h_{21} = -h_{12}$ (d) AD - BC = 0

[1992 : 2 Marks]

4. The condition, that a 2-port network is reciprocal, can be expressed in terms of its ABCD parameters as ______. **[1994 : 1 Mark]**

5. The shor-circuit admittance matrix of a two-port network is

$$\begin{bmatrix} 0 & -1/2 \\ 1/2 & 0 \end{bmatrix}$$

The two-port network is

(a) non-reciprocal and passive

(b) non-reciprocal and active

(c) reciprocal and passive

(d) reciprocal and active **[1998 : 1 Mark]**

6. A 2-port network is shown in the figure. The parameter h_{21} for this network can be given by

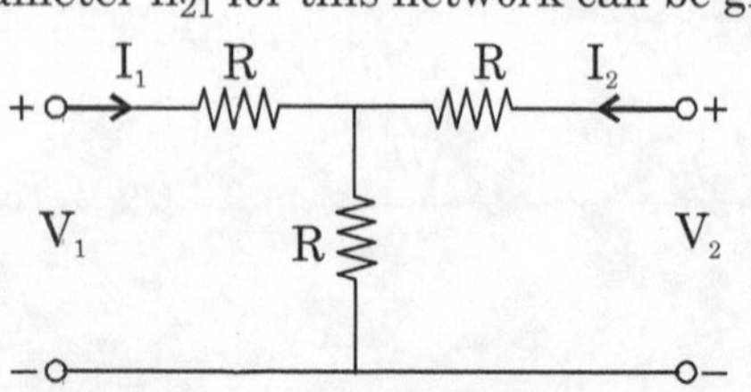

(a) –1/2 (b) +1/2

(c) –3/2 (d) +3/2

[1999 : 1 Mark]

7. The admittance parameter y_{12} in the 2-port network in figure is

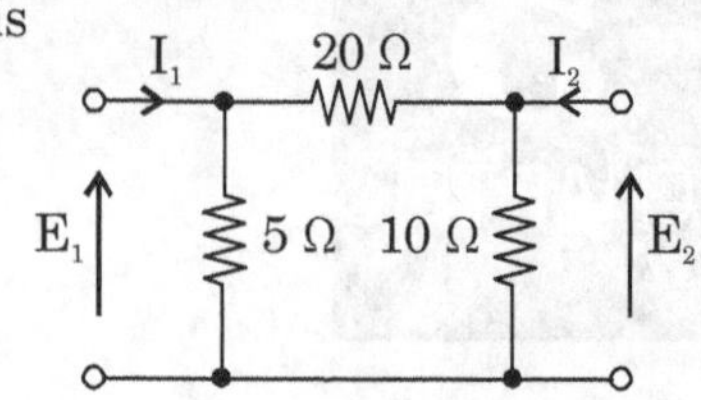

(a) -0.2 mho

(b) 0.1 mho

(c) -0.05 mho

(d) 0.05 mho

[2001 : 1 Mark]

8. The Z-parameters Z_{11} and Z_{21} for the 2-port network in the figure are

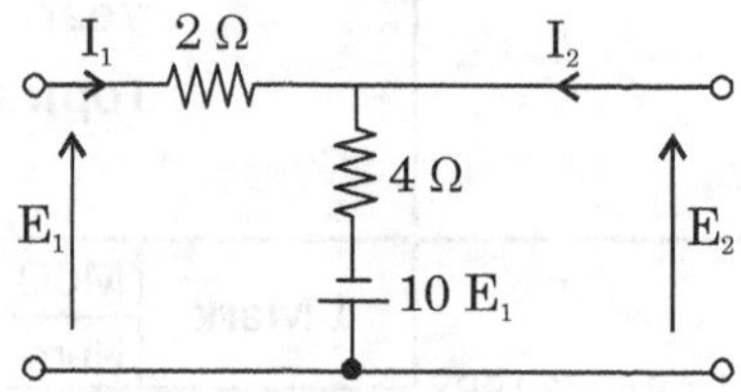

(a) $Z_{11} = \frac{-6}{11}\ \Omega;\ Z_{21} = \frac{16}{11}\ \Omega$

(b) $Z_{11} = \frac{-6}{11}\ \Omega;\ Z_{21} = \frac{4}{11}\ \Omega$

(c) $Z_{11} = \frac{-6}{11}\ \Omega;\ Z_{21} = \frac{-16}{11}\ \Omega$

(d) $Z_{11} = \frac{4}{11}\ \Omega;\ Z_{21} = \frac{4}{11}\ \Omega$ **[2001 : 2 Marks]**

9. The impedance parameters Z_{11} and Z_{12} of the two-port network in the figure are

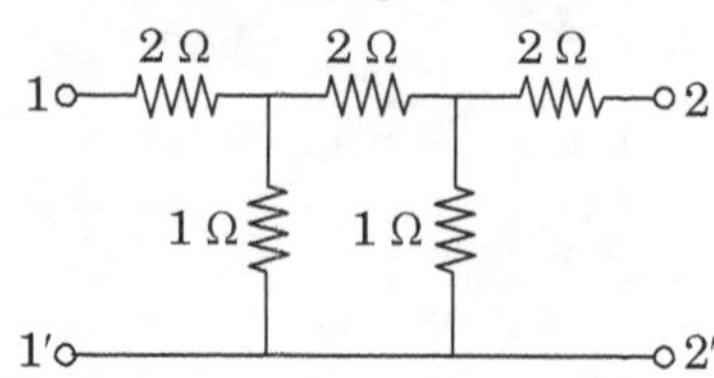

(a) $Z_{11} = 2.75\ \Omega$ and $Z_{12} = 0.25\ \Omega$

(b) $Z_{11} = 3\ \Omega$ and $Z_{12} = 0.5\ \Omega$

(c) $Z_{11} = 3\ \Omega$ and $Z_{12} = 0.25\ \Omega$

(d) $Z_{11} = 2.25\ \Omega$ and $Z_{12} = 0.5\ \Omega$

[2003 : 2 Marks]

10. For the lattice circuit shown in the figure, $Z_a = j2\,\Omega$ and $Z_b = 2\,\Omega$. The values of the open circuit impedance parameters $Z = \begin{bmatrix} Z_{11} & Z_{12} \\ Z_{21} & Z_{22} \end{bmatrix}$ are

(a) $\begin{bmatrix} 1-j & 1+j \\ 1+j & 1+j \end{bmatrix}$

(b) $\begin{bmatrix} 1-j & 1+j \\ -1+j & 1-j \end{bmatrix}$

(c) $\begin{bmatrix} 1+j & 1+j \\ 1-j & 1-j \end{bmatrix}$

(d) $\begin{bmatrix} 1+j & -1+j \\ -1+j & 1+j \end{bmatrix}$

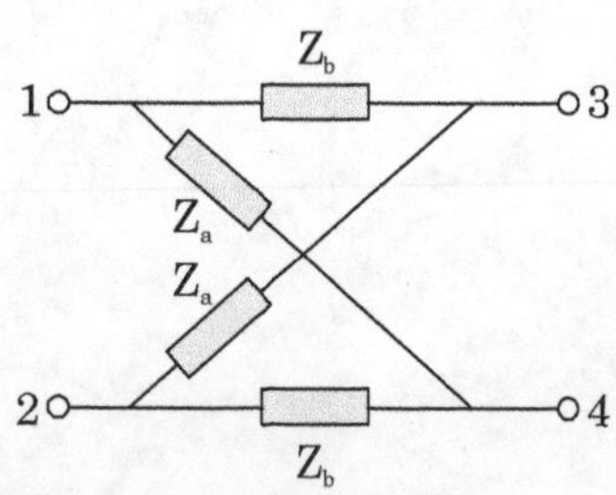

[2004 : 2 Marks]

11. The ABCD parameters of an ideal n:1 transformer shown in the figure are $\begin{bmatrix} n & 0 \\ 0 & x \end{bmatrix}$. The value of x will be

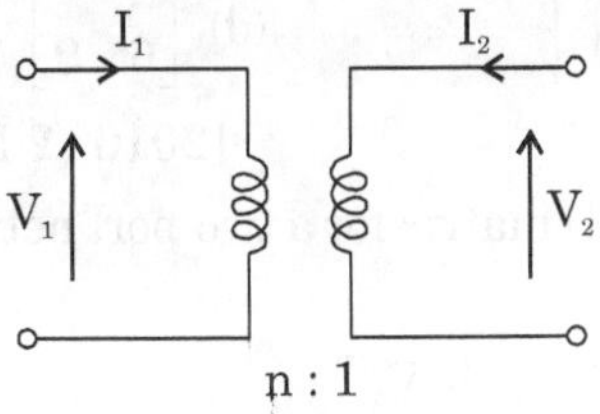

(a) n (b) $\frac{1}{n}$

(c) n^2 (d) $\frac{1}{n^2}$

[2005 : 1 Mark]

12. The h-parameters of the circuit shown in the figure are

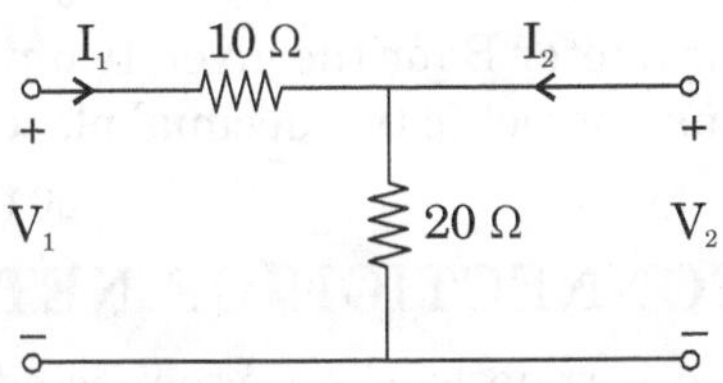

(a) $\begin{bmatrix} 0.1 & 0.1 \\ -0.1 & 0.3 \end{bmatrix}$ (b) $\begin{bmatrix} 10 & -1 \\ 1 & 0.05 \end{bmatrix}$

(c) $\begin{bmatrix} 30 & 20 \\ 20 & 20 \end{bmatrix}$ (d) $\begin{bmatrix} 10 & 1 \\ -1 & 0.05 \end{bmatrix}$

[2005 : 2 Marks]

13. In the two port network shown in the figure below, Z_{12} and Z_{21} are, respectively

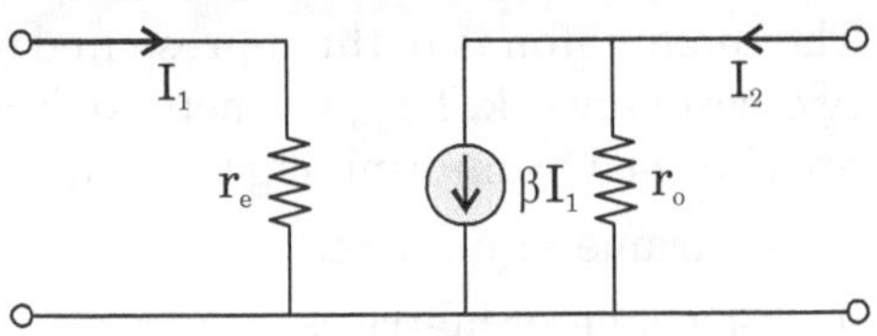

(a) r_e and βr_0 (b) 0 and $-\beta r_0$

(c) 0 and βr_0 (d) r_e and $-\beta r_0$

[2006 : 1 Mark]

14. For the two-port network shown below, the short-circuit admittance parameter matrix is

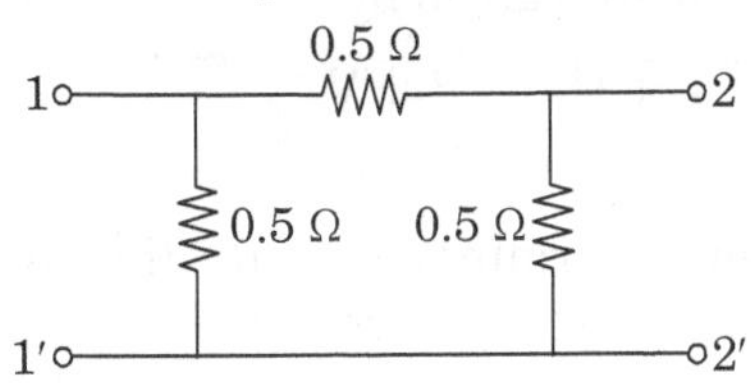

(a) $\begin{bmatrix} 4 & -2 \\ -1 & 4 \end{bmatrix}$S (b) $\begin{bmatrix} 1 & -0.5 \\ -0.5 & 1 \end{bmatrix}$S

(c) $\begin{bmatrix} 1 & 0.5 \\ 0.5 & 1 \end{bmatrix}$S (d) $\begin{bmatrix} 4 & 2 \\ 2 & 4 \end{bmatrix}$S

[2010 : 1 Mark]

15. In the circuit shown below, the network N is described by the following Y matrix:

$Y = \begin{bmatrix} 0.1\text{ S} & -0.01\text{ S} \\ 0.01\text{ S} & 0.1\text{ S} \end{bmatrix}$. The voltage gain $\frac{V_2}{V_1}$ is

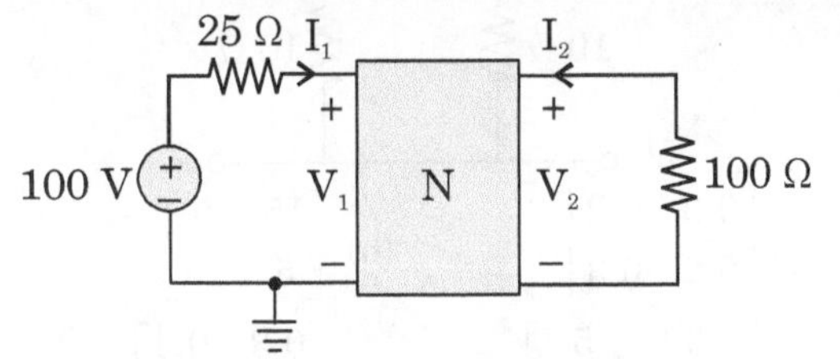

(a) 1/90 (b) – 1/90

(c) – 1/99 (d) – 1/11

[2011 : 2 Marks]

Common Data for 16 and 17:

With 10 V DC connected at port a in the linear non-reciprocal two-port network shown below, the following were observed :

(i) 1 Ω connected at port B draws a current of 3 A.

(ii) 2.5 Ω connected at port B draws a current of 2 A.

16. For the same network, with 6 V DC connected at port A, 1 Ω connected at port B draws 7/3 A. If 8 V DC is connected to port A, the open circuit voltage at port B is

(a) 6 V (b) 7 V

(c) 8 V (d) 9 V

[2012 : 2 Marks]

17. With 10 V DC connected at port A, the current drawn by 7 Ω connected at port B is

(a) 3/7 A (b) 5/7 A

(c) 1 A (d) 9/7 A

[2012 : 2 Marks]

18. For the two-port network shown in the figure, the impedance (Z) matrix (in Ω) is

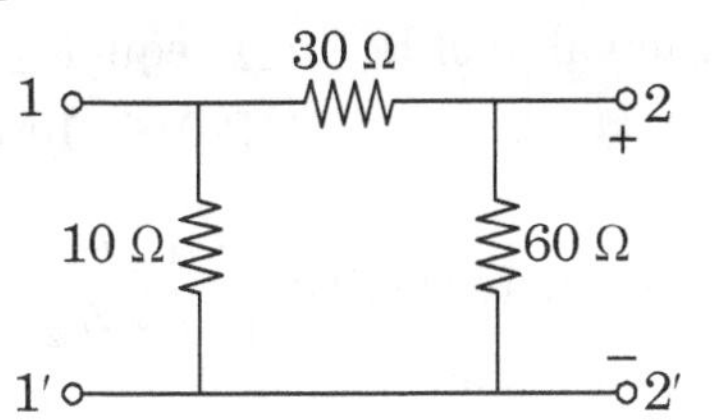

(a) $\begin{bmatrix} 6 & 24 \\ 42 & 9 \end{bmatrix}$ (b) $\begin{bmatrix} 9 & 8 \\ 8 & 24 \end{bmatrix}$

(c) $\begin{bmatrix} 9 & 6 \\ 6 & 24 \end{bmatrix}$ (d) $\begin{bmatrix} 42 & 6 \\ 6 & 60 \end{bmatrix}$

[2014 : 2 Marks, Set-4]

19. The 2-port admittance matrix of the circuit shown is given by

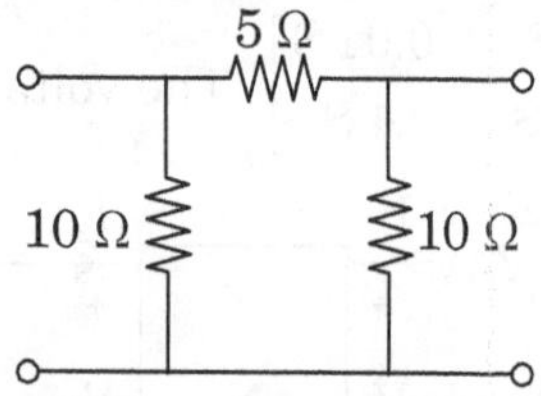

(a) $\begin{bmatrix} 0.3 & 0.2 \\ 0.2 & 0.3 \end{bmatrix}$ (b) $\begin{bmatrix} 15 & 5 \\ 5 & 15 \end{bmatrix}$

(c) $\begin{bmatrix} 3.33 & 5 \\ 5 & 3.33 \end{bmatrix}$ (d) $\begin{bmatrix} 0.3 & 0.4 \\ 0.4 & 0.3 \end{bmatrix}$

[2015 : 1 Mark, Set-2]

20. The ABCD parameters of the following 2 port network are

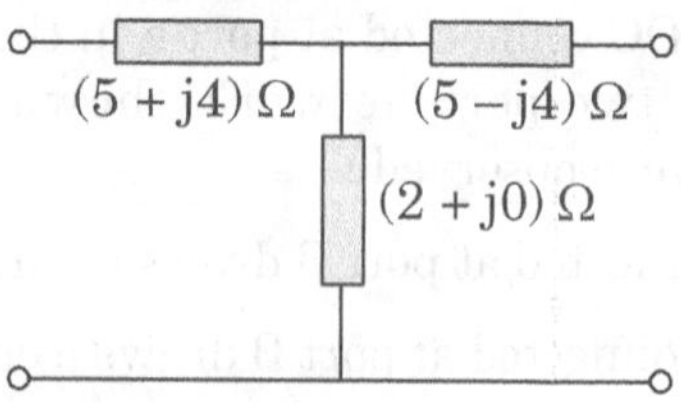

(a) $\begin{bmatrix} 3.5+j2 & 20.5 \\ 20.5 & 3.5-j2 \end{bmatrix}$ (b) $\begin{bmatrix} 3.5+j2 & 0.5 \\ 0.5 & 3.5-j2 \end{bmatrix}$

(c) $\begin{bmatrix} 10 & 2+j0 \\ 2+j0 & 10 \end{bmatrix}$ (d) $\begin{bmatrix} 7+j0 & 0.5 \\ 30.5 & 7-j4 \end{bmatrix}$

[2015 : 2 Marks, Set-3]

21. The Z-parameter matrix for the two-port network shown is

$$\begin{bmatrix} 2j\omega & j\omega \\ j\omega & 3+2j\omega \end{bmatrix}$$

Where the entries are in Ω

Suppose $Z_b(j\omega) = R_b + j\omega$

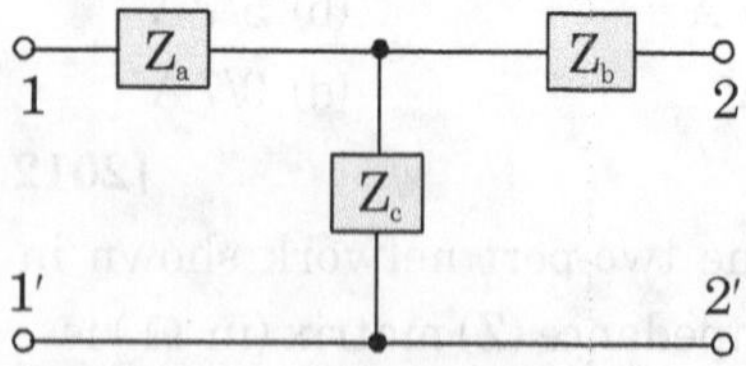

Then the value of R_b (in Ω) equal ______.

[2016 : 1 Mark, Set-2]

22. The Z-parameter matrix $\begin{bmatrix} Z_{11} & Z_{12} \\ Z_{21} & Z_{22} \end{bmatrix}$ for the two-port network shown is

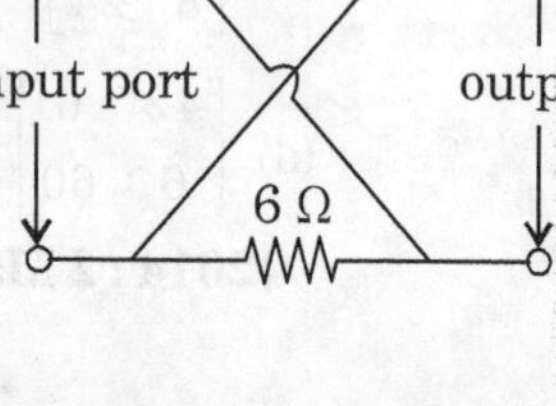

(a) $\begin{bmatrix} 2 & -2 \\ -2 & 2 \end{bmatrix}$ (b) $\begin{bmatrix} 2 & 2 \\ 2 & 2 \end{bmatrix}$

(c) $\begin{bmatrix} 9 & -3 \\ 6 & 9 \end{bmatrix}$ (d) $\begin{bmatrix} 9 & 3 \\ 6 & 9 \end{bmatrix}$

[2016 : 2 Marks, Set-3]

23. The ABCD matrix for a two-port network is defined by:

$$\begin{bmatrix} V_1 \\ I_1 \end{bmatrix} = \begin{bmatrix} A & B \\ C & D \end{bmatrix} \begin{bmatrix} V_2 \\ -I_2 \end{bmatrix}$$

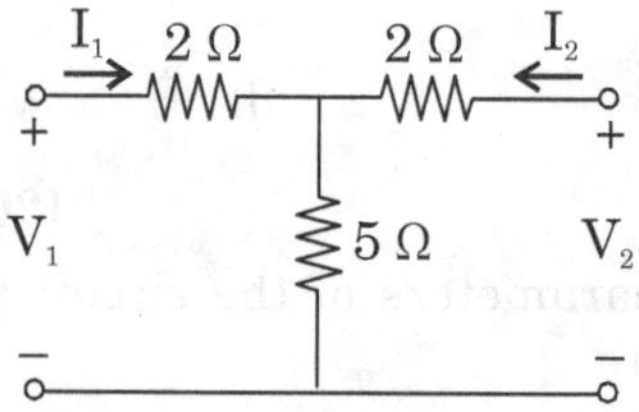

The parameter B for the given two-port network (in ohms, correct to two decimal places) is ______.

[2018 : 1 Mark]

INTERCONNECTION OF NETWORKS

24. Two two-port networks are connected in parallel. The combination is to be represented as a single two-port network. The parameters of this network are obtained by addition of the individual

(a) z parameters (b) h parameters

(c) y parameters (d) ABCD parameters

[1988 : 2 Marks]

25. Two two-port networks are connected in cascade. The combinaton is to the represented as a single two-port network. The parameters of the network are obtained by multiplying the individual

(a) z-parameter matrices

(b) h-parameter matrices

(c) y-parameter matrices

(d) ABCD-parameter matrices

[1991 : 2 Marks]

26. A two-port network is represented by ABCD parameters given by

$$\begin{bmatrix} V_1 \\ I_1 \end{bmatrix} = \begin{bmatrix} A & B \\ C & D \end{bmatrix} \begin{bmatrix} V_2 \\ -I_2 \end{bmatrix}$$

If port-2 terminated by R_L, the input impedance seen at port-1 is given by

(a) $\dfrac{A+BR_L}{C+DR_L}$ (b) $\dfrac{AR_L+C}{BR_L+D}$

(c) $\dfrac{DR_L+A}{BR_L+C}$ (d) $\dfrac{B+AR_L}{D+CR_L}$

[2006 : 1 Mark]

Linked Answer Questions 27 to 28:

A two-port nework shown below is excited by external dc sources. The voltages and currents are measured with voltmeters V_1, V_2 and ammeters A_1, A_2 (all assumed to be ideal as indicated). Under following switch conditions, the readings obtained are

(i) S_1–Open, S_2–closed $A_1 = 0$ A, $V_1 = 4.5$ V, $V_2 = 1.5$ V, $A_2 = 1$ A

(ii) S_1 –Closed, S_2 -Open $A_1 = 4$ A, $V_1 = 6$ V, $V_2 = 6$ V, $A_2 = 0$ A

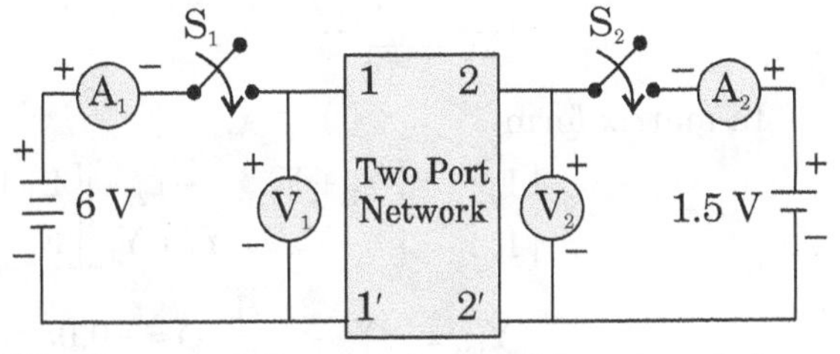

27. The Z-parameter matrix for this network is

(a) $\begin{bmatrix} 1.5 & 1.5 \\ 4.5 & 1.5 \end{bmatrix}$

(b) $\begin{bmatrix} 1.5 & 4.5 \\ 1.5 & 4.5 \end{bmatrix}$

(c) $\begin{bmatrix} 1.5 & 4.5 \\ 1.5 & 1.5 \end{bmatrix}$

(d) $\begin{bmatrix} 4.5 & 1.5 \\ 1.5 & 4.5 \end{bmatrix}$

[2008 : 2 Marks]

28. The h-parameter matrix for this network is

(a) $\begin{bmatrix} -3 & 3 \\ -1 & 0.67 \end{bmatrix}$ (b) $\begin{bmatrix} -3 & -1 \\ 3 & 0.67 \end{bmatrix}$

(c) $\begin{bmatrix} 3 & 3 \\ 1 & 0.67 \end{bmatrix}$ (d) $\begin{bmatrix} 3 & 1 \\ -3 & -0.67 \end{bmatrix}$

[2008 : 2 Marks]

29. In the h-parameter model of the 2-port network given in the figure shown, the value of h_{22} (in S) is ______.

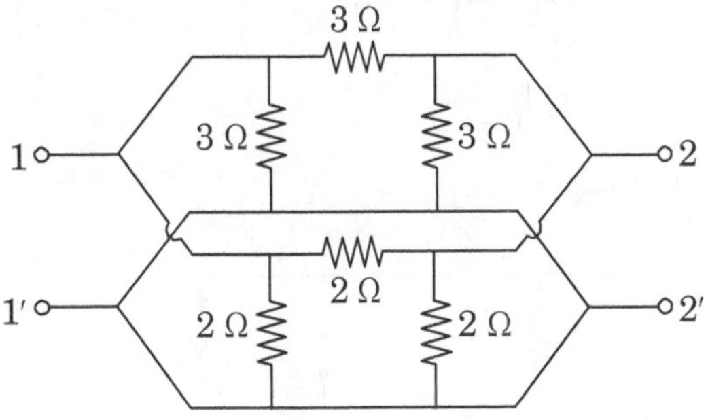

[2014 : 2 Marks, Set-2]

30. Consider a two-port network with the transmission matrix ; $T = \begin{bmatrix} A & B \\ C & D \end{bmatrix}$

If the network is reciprocal, then

(a) $T^{-1} = T$

(b) $T^2 = T$

(c) Determinant(T) = 0

(d) Determinant(T) = 1 **[2016 : 1 Mark, Set-1]**

ANSWERS

1. (a)	**2.** (a)	**3.** (b, c)	**4.** (AD – BC = 1)		**5.** (b)	**6.** (a)	**7.** (c)	**8.** (c)	**9.** (a)
10. (d)	**11.** (b)	**12.** (d)	**13.** (b)	**14.** (d)	**15.** (d)	**16.** (b)	**17.** (c)	**18.** (c)	**19.** (a)
20. (b)	**21.** (3)	**22.** (a)	**23.** (4.8)	**24.** (c)	**25.** (d)	**26.** (d)	**27.** (c)	**28.** (a)	**29.** (1.25)
30. (d)									

EXPLANATIONS

1. For a passive two-port network to be reciprocal,

AD – BC = 1

2. For Z_{11}:-

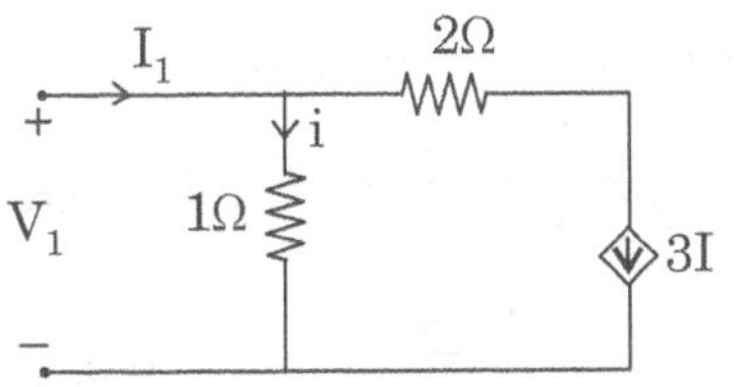

$i = I_1 - 3I_1 = -2I_1$.

$$Z_{11} = \left.\frac{V_1}{I_1}\right|_{I_1=0} = \frac{1 \times i}{I_1}$$

$Z_{11} = -2\ \Omega$.

For Z_{12}:-

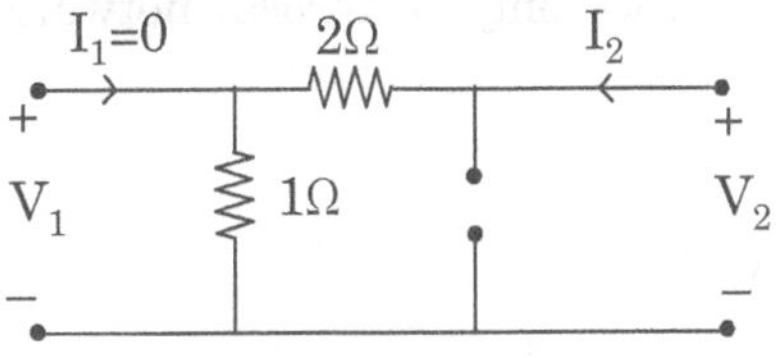

$$Z_{12} = \left.\frac{V_1}{I_2}\right|_{I_1=0}$$

$$= \frac{1.I_2}{I_2} = 1\ \Omega$$

For Z_{21}:-

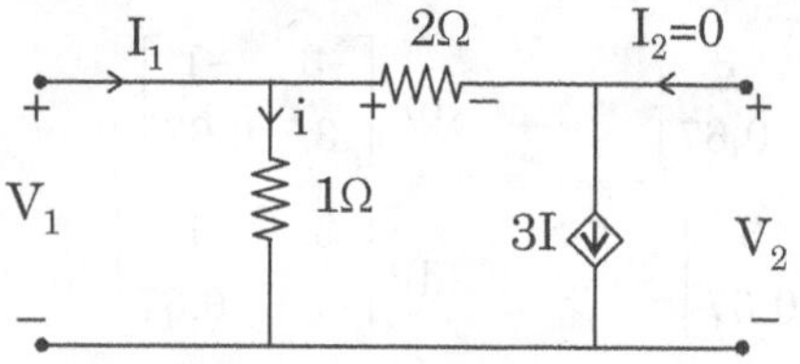

$I_1 = i + 3I_1$

$i = -2I_1$.

$-V_1 + 6I_1 + V_2 = 0$

$V_{21} = -2I_1 - 6I_1 = -8I_1$.

$$Z_{21} = \left.\frac{V_2}{I_1}\right|_{I_2=0} = -8\ \Omega$$

For Z_{22}:-

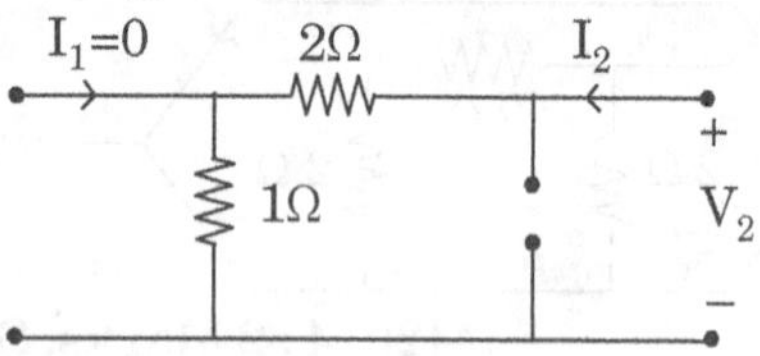

$$Z_{22} = \left.\frac{V_2}{I_2}\right|_{I_1=0} = \frac{2I_2 + I_2}{I_2}$$

$Z_{22} = 3\ \Omega$.

$$[Z] = \begin{bmatrix} Z_{11} & Z_{12} \\ Z_{21} & Z_{22} \end{bmatrix} = \begin{bmatrix} -2 & 1 \\ -8 & 3 \end{bmatrix}$$

3. For a two-port network to reciprocal,

z – parameters: $z_{12} = z_{21}$

y – parameters : $y_{12} = y_{21}$

h – parameters: $h_{12} = -h_{21}$

g – parameters: $g_{12} = -g_{21}$

T – parameters: AD – BC = 1

4. AD – BC = 1

5. Given short circuit admittance matrix of a 2 port network.

$$Y = \begin{bmatrix} 0 & -1/2 \\ 1/2 & 0 \end{bmatrix}$$

For reciprocal $y_{12} = y_{21}$, Here $\frac{1}{2} \neq -\frac{1}{2}$

It is active since only reciprocal networks are passive.

6. $h_{12} = \left.\frac{I_2}{I_1}\right|_{V_2=0}$

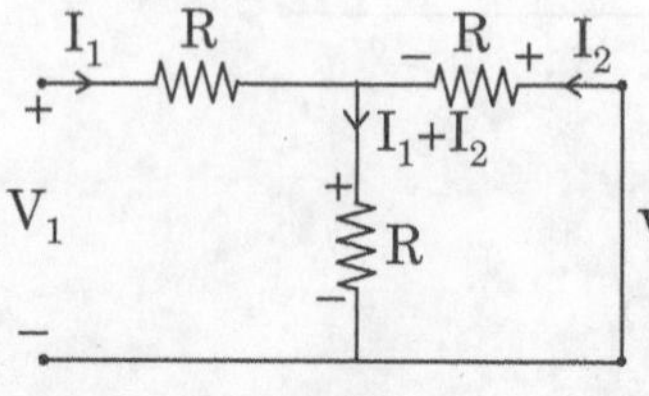

Applying KVL,

$-R(I_1 + I_2) - RI_2 = 0$

$I_2 = -\frac{I_1}{2}$

$\Rightarrow h_{12} = \frac{-1}{2}$

7. $I_1 = E_1Y_1 + (E_1 - E_2)Y_2$

$Y_1 = \frac{1}{5}\ \mho, Y_2 = \frac{1}{20}\ \mho$

$I_2 = E_2Y_3 + (-E_1 + E_2)Y_2$

$Y_3 = \frac{1}{10}\ \mho$

In matrix form,

$$\begin{bmatrix} I_1 \\ I_2 \end{bmatrix} = \begin{bmatrix} Y_1 + Y_2 & -Y_2 \\ -Y_2 & Y_2 + Y_3 \end{bmatrix} \begin{bmatrix} E_1 \\ E_2 \end{bmatrix}$$

$$Y_{12} = -Y_2 = -\frac{1}{20}\ \mho = -0.05 \text{ mho}$$

8.

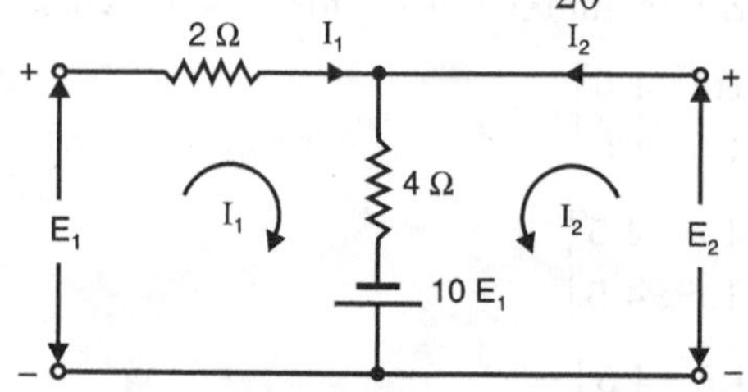

Loop law's to both the loops,

$11E_1 = 6I_1 + 4I_2$ ...(*i*)

$E_2 = (I_1 + I_2)\,4 - 10\,E_1$ ...(*ii*)

Putting the expression for E_1 from equation (*i*) in equation (*ii*), we get

$$E_2 = (I_1 + I_2)\,4 - \frac{10[2I_1 + 4(I_1 + I_2)]}{11}$$

$$= I_1\left[4 - \frac{20}{11} - \frac{40}{11}\right] + I_2\left[4 - \frac{40I_2}{11}\right]$$

Putting $I_2 = 0$ as, $Z_{21} = \left.\frac{E_2}{I_1}\right|_{I_2=0}$

$$E_2 = I_1\left[\frac{44 - 60}{11}\right] = -\frac{16}{11}I_1$$

From equation (*i*), we get

$$\therefore \quad Z_{11} = \left.\frac{E_1}{I_1}\right|_{I_2=0} = \frac{6}{11}\ \Omega.$$

9. Using ∇ – Y conversion, we have

$$\text{Here, } Z_{AP} = \frac{Z_{AB}\,Z_{AC}}{Z_{AB} + Z_{AC} + Z_{BC}} = \frac{2}{4} = 0.5$$

$$Z_{PB} = \frac{Z_{AB}\,Z_{BC}}{Z_{AB} + Z_{BC} + Z_{AC}} = \frac{2}{4} = 0.5$$

$$Z_{PC} = \frac{Z_{AC}\,Z_{BC}}{Z_{AB} + Z_{BC} + Z_{AC}} = \frac{1}{4} = 0.25$$

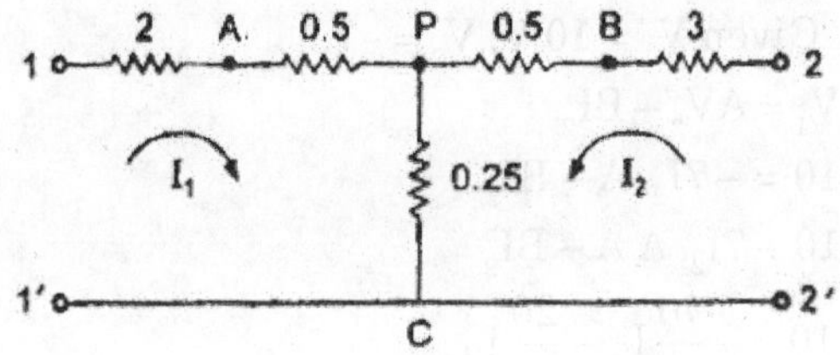

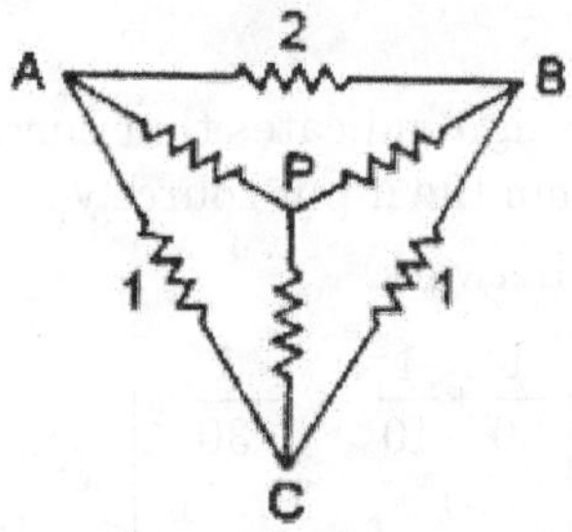

Now, from the resultant network,

$$V_1 = 2.75\, I_1 + 0.25\, I_2$$

$$= Z_{11}\, I_1 + Z_{12}\, I_2$$

Thus, $Z_{11} = 2.75\ \Omega$,

$Z_{12} = 0.25\ \Omega$

10.

$$V_1 = I_1 Z_{11} + I_2 Z_{12}$$

$$V_2 = I_1 Z_{21} + I_2 Z_{22}$$

$$Z_{11} = \left.\frac{V_1}{I_1}\right|_{I_2=0}$$

$$\frac{V_1}{(2+2j)} = i\,;\ I_1 - i = \frac{V_1}{2+2j}$$

$$I_1 = \frac{V_1}{2+2j} + \frac{V_1}{2+2j} = \frac{V_1}{1+j}$$

$$Z_{11} = 1 + j$$

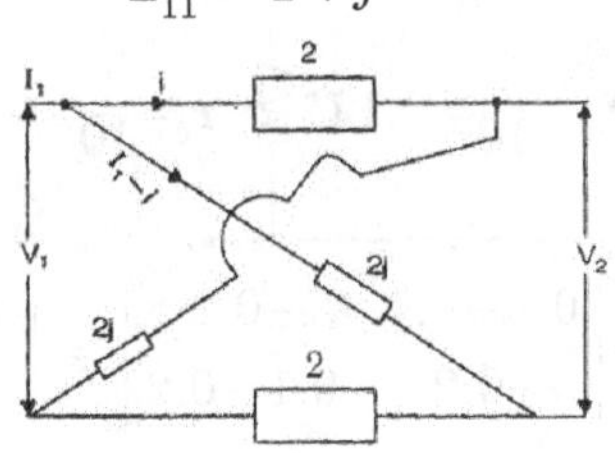

Likewise $\quad Z_{22} = \left.\frac{V_2}{I_2}\right|_{I_1=0} = 1 + j$

$$\frac{V_2}{2j} - \frac{V_2}{2} = I_1$$

$$\frac{V_2}{I_1} = -1 + j$$

$$= Z_{12}$$

Likwise, $Z_{21} = -1 + j$

$$\begin{bmatrix} Z_{11} & Z_{12} \\ Z_{21} & Z_{22} \end{bmatrix} = \begin{bmatrix} 1+j & -1+j \\ -1+j & 1+j \end{bmatrix}$$

11. For given transformer

$$\frac{V_1}{V_2} = \frac{N_1}{N_2} = n$$

or, $\quad V_1 = V_2 n$

and $\quad \frac{I_1}{I_2} = -\frac{N_2}{N_1} = \frac{-1}{n}$

or, $\quad I_1 = -\frac{I_2}{n}$

In terms of ABCD parameters,

$$\begin{bmatrix} V_1 \\ I_1 \end{bmatrix} = \begin{bmatrix} n & 0 \\ 0 & \frac{1}{n} \end{bmatrix} \begin{bmatrix} I_1 \\ -I_2 \end{bmatrix}$$

Comparing with the given matrix,

$$X = \frac{1}{n}$$

12.

$$\begin{bmatrix} V_1 \\ I_2 \end{bmatrix} = \begin{bmatrix} h_{11} & h_{12} \\ h_{21} & h_{22} \end{bmatrix} \begin{bmatrix} I_1 \\ V_2 \end{bmatrix}$$

For given circuit

$$V_1 - 10I_1 - 20\,(I_1 + I_2) = 0;$$

$$V_2 - 20\,(I_1 + I_2) = 0$$

$\Rightarrow \quad V_1 - 30\, I_1 + 20\, I_2 = 0;$

$$V_2 - 20\, I_1 + 20\, I_2 = 0$$

$$V_1 = 30\, I_1 + 20\, I_2;\ V_2 = 20\, I_1 + 20\, I_2$$

$\therefore \quad V_1 - V_2 = 10\, I_1$

$\Rightarrow \quad V_1 = V_2 + 10\, I_1$

Now, $\quad V_1 = h_{11}\, I_1 + h_{12}\, V_2$

$\therefore \quad h_{11} = 10,$

and $\quad h_{12} = 1$

$$20I_2 = V_1 - 30\, I_1$$

$$= V_2 + 10\, I_1 - 30\, I_1$$

$$= V_2 - 20\, I_1$$

$\Rightarrow \quad I_2 = \frac{1}{20} V_2 - I_1$

$\therefore \quad h_{21} = -1$ and $h_{22} = 0.05$

Hence, h-parameter are $\begin{bmatrix} 10 & 1 \\ -1 & 0.05 \end{bmatrix}$

13.

$$V_1 = r_e I_1 \qquad ...(i)$$

$$I_2 = \beta I_1 + \frac{V_2}{r_0} \qquad ...(ii)$$

Rearranging equation (*ii*), we get

$$V_2 = -\beta\, r_0 I_1 + r_0 I_2 \qquad ...(iii)$$

By equations (*i*) and (*iii*), we get

$$Z_{12} = 0,\ Z_{21} = -\beta r_0$$

14. Given circuit is

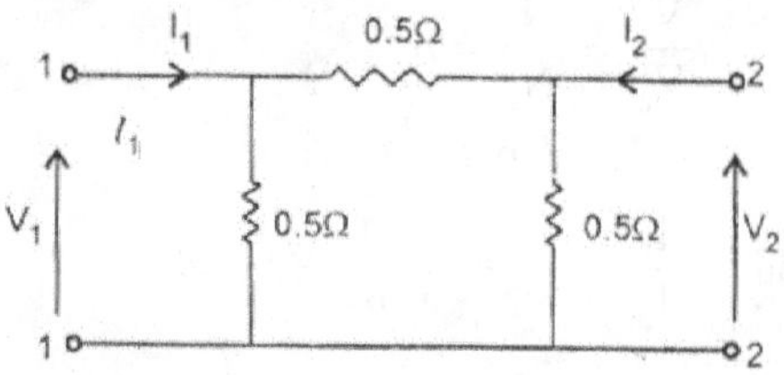

$$\begin{bmatrix} I_1 \\ I_2 \end{bmatrix} = \begin{bmatrix} Y_{11} & Y_{12} \\ Y_{21} & Y_{22} \end{bmatrix} \begin{bmatrix} V_1 \\ V_2 \end{bmatrix}$$

$$Y_{11} = \left.\frac{I_1}{V_1}\right|_{V_2=0} = \frac{1}{0.5110 \times 5} = \frac{1}{0.25} = 4$$

$$Y_{12} = \left.\frac{I_1}{V_2}\right|_{V_1=0} = \frac{1}{0.5} = 2$$

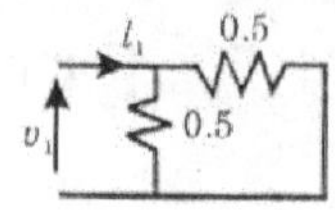

$$Y_{21} = \left.\frac{I_2}{V_1}\right|_{V_2=0} = \frac{1}{0.5} = 2$$

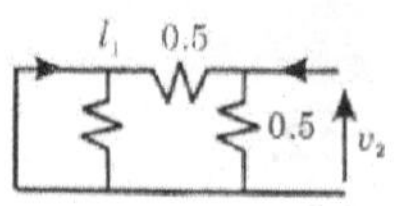

$$Y_{22} = \left.\frac{I_2}{V_2}\right|_{V_1=0} = \frac{1}{0.5110 \times 5} = 4$$

Hence short-circuit admittance parameter matrix is

$$Y = \begin{bmatrix} 4 & 2 \\ 2 & 4 \end{bmatrix} S.$$

15. $V_1 = 100V + 25I_1$, $V_2 = -I_2 R_L$

$V_2 = -100I_2$

$$I_2 = Y_{z1} V_1 + Y_{z2} V_2$$

$$\therefore \quad -0.01V_2 = 0.01 V_1 + 0.1V_2$$

$$\Rightarrow \quad \frac{V_2}{V_1} = \frac{-1}{11}$$

16. Given (i)$V_1 = 10V$, $I_1 = -3$ A.

$V_2 = 3$ V

$\Rightarrow V_1 = AV_2 - BI_2$.

$10 = 3A + 3B$ ...(1)

Given (ii) $V_2 = 5V$, $I_2 = -2$ A

$10 = 5A + 2B$ ....(2)

$$A = \frac{10}{9}, B = \frac{20}{9}$$

Now, $V_1 = 8$ V, $(V_2)_{oc} = ?$, $I_2 = 0$

$V_1 = AV_2 - BI_2$.

$8 = A(V_2)_{oc} - B \times 0$

$$(V_2)_{oc} = \frac{8}{A} = \frac{8}{\frac{10}{9}} = 7.2 \text{ V}$$

17. Given $V_1 = 10$ V, $V_2 = -7.I_2$.

$V_1 = AV_2 - BI_2$.

$10 = -7I_2. A - BI_2$

$10 - 7I_2. A A - BI_2$

$$10 = -\frac{70}{9}I_2 - \frac{20}{9}I_2$$

$I_2 = -1$ A

$\Rightarrow$ Here, negative sign indicates that current is drawn is drawn from the input source y_1.

18. For the two-part network

$$Y_{matrix} = \begin{bmatrix} \frac{1}{30} + \frac{1}{10} & -\frac{1}{30} \\ \frac{-1}{30} & \frac{1}{60} + \frac{1}{30} \end{bmatrix}$$

$$Z_{matrix} = [Y]^{-1}$$

$$Z = \begin{bmatrix} 0.1333 & -0.0333 \\ -0.0333 & 0.05 \end{bmatrix}$$

$$Z = \begin{bmatrix} 9 & 6 \\ 6 & 24 \end{bmatrix}$$

19.

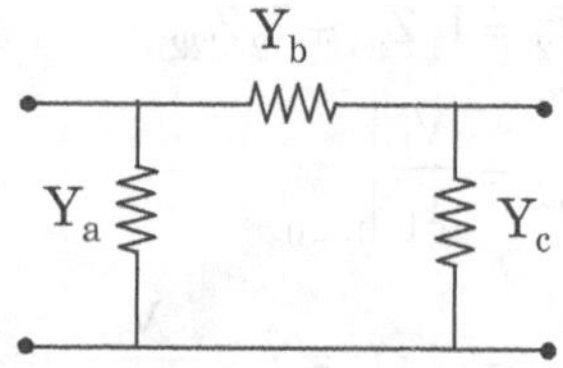

$$[Y] = \begin{bmatrix} Y_a + Y_b & -Y_b \\ -Y_b & Y_b + Y_c \end{bmatrix}$$

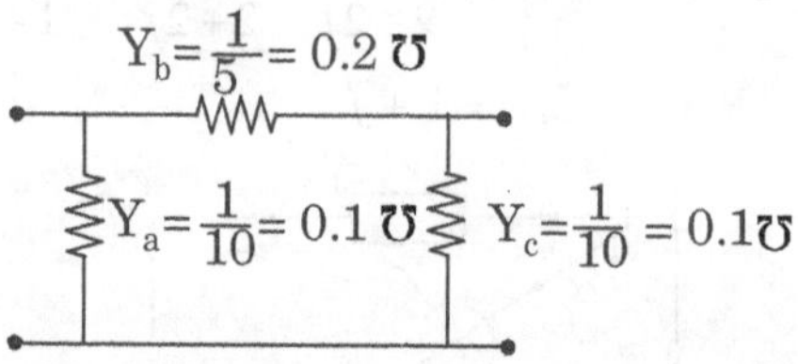

$$[Y] = \begin{bmatrix} 0.2 + 0.1 & -0.2 \\ -0.2 & 0.1 + 0.2 \end{bmatrix} = \begin{bmatrix} 0.3 & -0.2 \\ -0.2 & 0.3 \end{bmatrix}$$

$\Rightarrow$ If negative sign is ignored, option (A) is correct.

20. For the standard 'T' network, obtain the Z-matrix first and then convert it into T-matrix

$$Z = \begin{bmatrix} 7 + j4 & 2 \\ 2 & 7 - j4 \end{bmatrix}$$

$$\Delta Z = [(7 + j4)(7 - j4)] - 4$$

$$= 49 + 16 - 4 = 16$$

$$A = \frac{Z_{11}}{Z_{21}} = \frac{7 + j4}{2} = 3.5 + j2$$

$$B = \frac{\Delta Z}{Z_{21}} = \frac{61}{2} = 30.5$$

$$C = \frac{1}{Z_{21}} = \frac{1}{2} = 0.5$$

$$D = \frac{Z_{22}}{Z_{21}} = \frac{7 - j4}{2} = 3.5 - j2$$

$$T = \begin{bmatrix} 3.5 + j2 & 30.5 \\ 0.5 & 3.5 - j2 \end{bmatrix}$$

21. For T - network

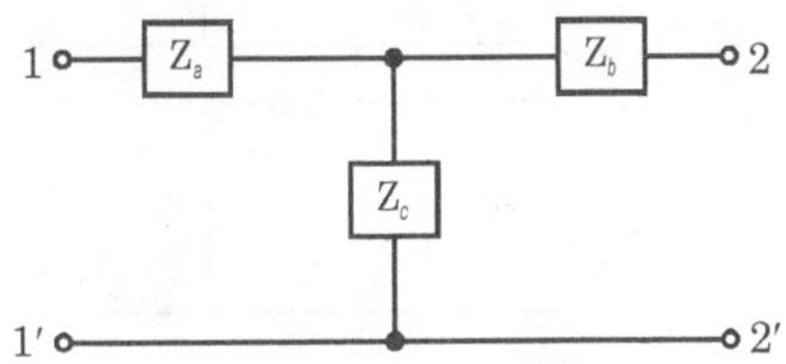

$$Z_{11} = Z_a + Z_c$$

$$Z_{22} = Z_b + Z_c$$

and $Z_{12} = Z_{21} = Z_c$

Given $[Z] = \begin{bmatrix} 2j\omega & j\omega \\ j\omega & 3 + 2j\omega \end{bmatrix}$

Therefore $Z_{12} = j\omega$

and $Z_{22} = 3 + 2j\omega$

$$= 3 + j\omega + j\omega$$

$$= Z_b + Z_c = R_b + j\omega + Z_c$$

$\therefore$ $R_b = 3\ \Omega$

Hence the value of R_b is 3Ω.

22. $I_2 = 0$:-

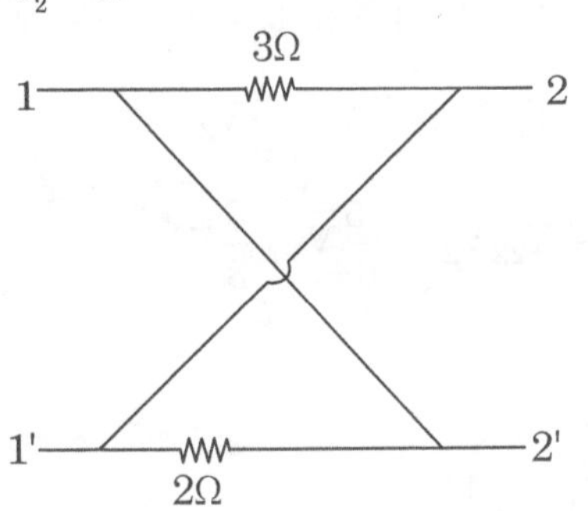

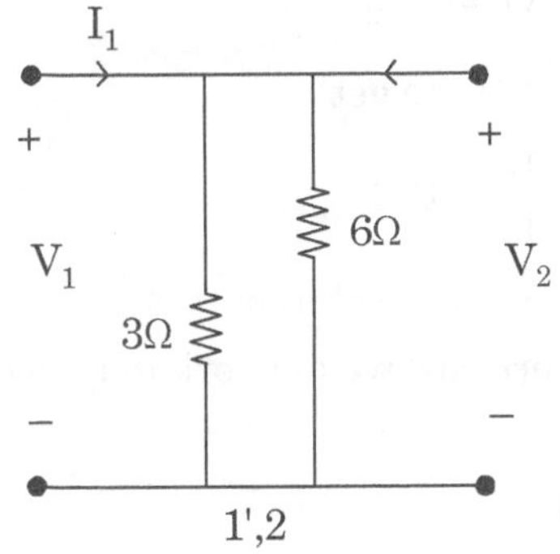

$$Z_{11} \frac{V_1}{I_1}\bigg|_{I_2=0} = \frac{3 \times 6}{3+6} = 2\ \Omega;$$

$$V_2 = -3 \times I_1 \times \frac{6}{6+3} = -2I_1$$

$$Z_{21} = \frac{V_2}{I_1}\bigg|_{I_2=0} = -2\ \Omega$$

$I_1 = 0$:-

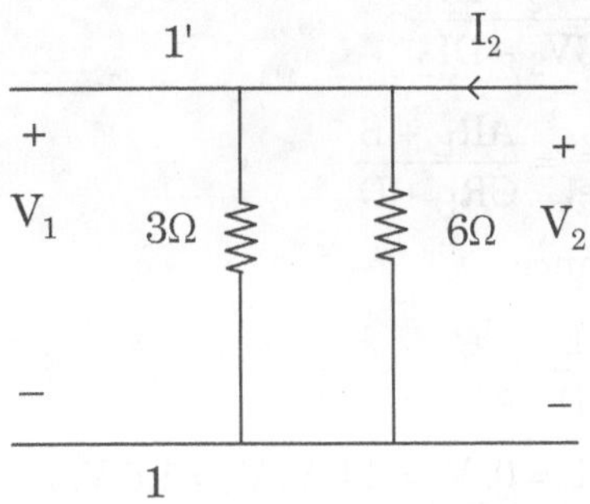

$$Z_{22} = \frac{V_2}{I_2}\bigg|_{I_1=0} = 3\,||\,6 = 2\ \Omega$$

$$Z_{12} = \frac{V_1}{I_2}\bigg|_{I_1=0} - 6 \times \frac{3}{3+6} = -2\ \Omega$$

$$[Z] = \begin{bmatrix} 2 & -2 \\ -2 & 2 \end{bmatrix}$$

23. $B = \frac{-V_1}{I_2}\bigg|_{V_2=0}$

$\Rightarrow$ Port 2 is short-circuited.

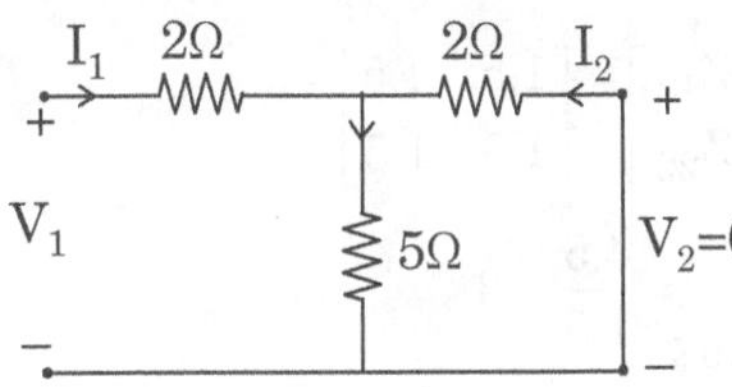

$$\Rightarrow I_1 = \frac{V_1}{2 + \frac{2 \times 5}{2+5}} = \frac{7V_1}{14+10} = \frac{7V_1}{24}$$

$$\Rightarrow I_2 = -I_1 \times \frac{5}{5+2} = -\frac{5V_1}{24}$$

$$B = \frac{-V_1}{I_2} = \frac{24}{5} = 4.8\ \Omega$$

24. Two-port network with admittance matrices (Y_A & Y_B) are connected is parallel having resultant admittance

$[Y] = [Y_A] + [Y_B]$

25. For two-port network connected is cascade.

$$\begin{bmatrix} A & B \\ C & D \end{bmatrix} = \begin{bmatrix} A_1 & B_1 \\ C_1 & D_1 \end{bmatrix}\begin{bmatrix} A_2 & B_2 \\ C_2 & D_2 \end{bmatrix}$$

26.

ABCD parameters: $V_1 = AV_2 - BI_2$

$I_1 = CV_2 - DI_2$

$\therefore$ $V_2 = -I_2 R_L$

Now, $\frac{V_1}{I_1} = \frac{AV_2 - BI_2}{CV_2 - DI_2}$

$$\frac{-A.R_L - BI_2}{C.R_LI_2 - DI_2} = \frac{AR_L + B}{CR_L + D}$$

Input impedance

$$\frac{V_1}{I_1} = \frac{AR_L + B}{CR_L + D}$$

27. Given : For $I_1 = 0, V_1 = 4.5$ V, $V_2 = 1.5$ V, $I_2 = 1$ A.

For $I_2 = 0, V_1 = 6$ V, $V_2 = 6$ V, $I_2 = 4$ A.

$$\Rightarrow Z_{11} \frac{V_1}{I_1}\bigg|_{I_2=0} = \frac{6}{4} = 1.5\ \Omega,$$

$$Z_{12} \frac{V_1}{I_2}\bigg|_{I_2=0} = \frac{4.5}{1} = 4.5\ \Omega,$$

$$Z_{21} \frac{V_2}{I_1}\bigg|_{I_2=0} = \frac{6}{4} = 1.5\ \Omega,$$

$$Z_{22} \frac{V_2}{I_2}\bigg|_{I_1=0} = \frac{1.5}{1} = 1.5\ \Omega.$$

$$[Z] = \begin{bmatrix} Z_{11} & Z_{12} \\ Z_{21} & Z_{22} \end{bmatrix} = \begin{bmatrix} 1.5 & 4.5 \\ 1.5 & 1.5 \end{bmatrix}$$

28. $h_{12} = \frac{V_1}{V_2}\bigg|_{I_1=0} = \frac{4.5}{1.5} = 3$

$$h_{22} = \frac{I_2}{V_2}\bigg|_{I_1=0} = \frac{1}{1.5} = 0.67.$$

$\Rightarrow$ When, $V_2 = 0\ Z_{21} + I_1 + Z_{22}\ I_2 = 0$

$$I_2 = -\frac{Z_{21}}{Z_{22}} I_1$$

$$V_1 = Z_{11}I_1 + Z_{12}\left(-\frac{Z_{21}}{Z_{22}}.I_1\right)$$

$$\Rightarrow h_{11} = \frac{V_1}{I_1} = Z_{11} - \frac{Z_{12}.Z_{21}}{Z_{22}}$$

$$1.5 - \frac{4.5 \times 1.5}{1.5} = -3.$$

$$\Rightarrow h_{21} = \frac{I_2}{I_1}\bigg|_{V_2=0} = -\frac{Z_{21}}{Z_{22}} = \frac{-1.5}{1.5} = -1$$

$$[h] = \begin{bmatrix} h_{11} & h_{12} \\ h_{21} & h_{22} \end{bmatrix} = \begin{bmatrix} -3 & 3 \\ -1 & 0.67 \end{bmatrix}$$

29. If two-port networks are connected in parallel, then their y-parameters are added.

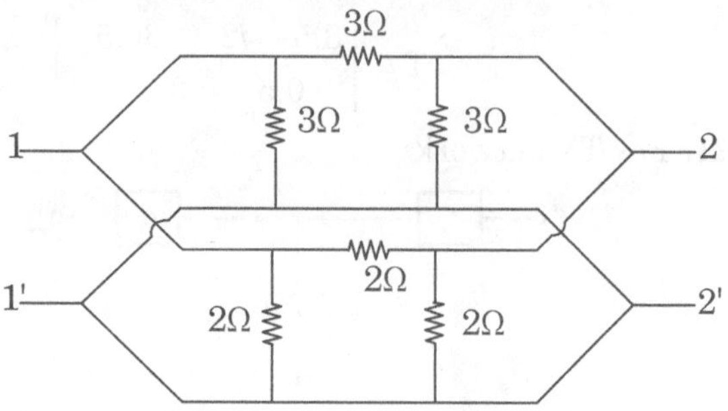

For Network:

$$[y_1] = \begin{bmatrix} \frac{1}{3}+\frac{1}{3} & -\frac{1}{3} \\ -\frac{1}{3} & \frac{1}{3}+\frac{1}{3} \end{bmatrix} = \begin{bmatrix} \frac{2}{3} & -\frac{1}{3} \\ -\frac{1}{3} & \frac{2}{3} \end{bmatrix}$$

For Network 2:

$$[y_2] = \begin{bmatrix} \frac{1}{2}+\frac{1}{2} & -\frac{1}{2} \\ -\frac{1}{2} & \frac{1}{2}+\frac{1}{2} \end{bmatrix} = \begin{bmatrix} 1 & -\frac{1}{2} \\ -\frac{1}{2} & 1 \end{bmatrix}$$

$$[y] = [y_1] + [y_2] = \begin{bmatrix} \frac{5}{3} & -\frac{5}{6} \\ -\frac{5}{6} & \frac{5}{3} \end{bmatrix}$$

$$I_1 = Y_{11}V_1 + Y_{12}V_2 = \frac{5}{3}V_1 - \frac{5}{6}V_2 \qquad ...(1)$$

$$I_2 = Y_{12}V_1 + Y_{22}V_2 = \frac{-5}{6}V_1 + \frac{5}{3}V_2 \qquad ...(2)$$

$$\Rightarrow h_{22} \frac{I_2}{V_2}\bigg|_{I_1=0}$$

from eq(1), $V_1 = \frac{1}{2}V_2$ ...(3)

from eq and (3), we get

$$h_{22} = \frac{I_2}{V_2} = \frac{15}{12} = 1.25$$

30. Consider it as new solution

For a two-port passive network to be reciprocal.

$$|T| = \begin{vmatrix} A & B \\ C & D \end{vmatrix} = 1$$

6

CHAPTER

Graph Theory

Analysis of Previous GATE Papers													
	Year → Topics ↓		2019	2018	2017 Set 1	2017 Set 2	2016 Set 1	2016 Set 2	2016 Set 3	2015 Set 1	2015 Set 2	2015 Set 3
L INEAR ORIENTED GRAPHS	1 Mark	MCQ Type										
		Numerical Type										
	2 Marks	MCQ Type										
		Numerical Type										
	Total											

LINEAR ORIENTED GRAPHS

1. The minimum number of equations required to analyze the circuit shown in the figure is

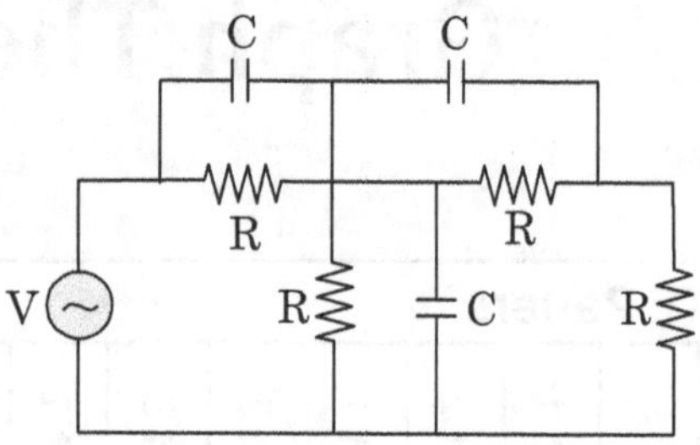

(a) 3 (b) 4

(c) 6 (d) 7 **[1991 : 1 Mark]**

2. Relative to a given fixed tree of a network,

(a) link currents form an independent set.

(b) branch currents form an independent set.

(c) link voltages form an independent set.

(d) branch voltages form an independent.

[1992 : 2 Marks]

3. The number of independent loops for a network with n nodes and b branches is

(a) $n - 1$

(b) $b - n$

(c) $b - n + 1$

(d) independent of the number of nodes

[1996 : 1 Mark]

4. A network has 7 nodes and 5 independent loops. The number of branches in the network is

(a) 12 (b) 12

(c) 11 (d) 10 **[1998 : 1 Mark]**

5. Identify which of the following is NOT a tree of the graph shown in the figure?

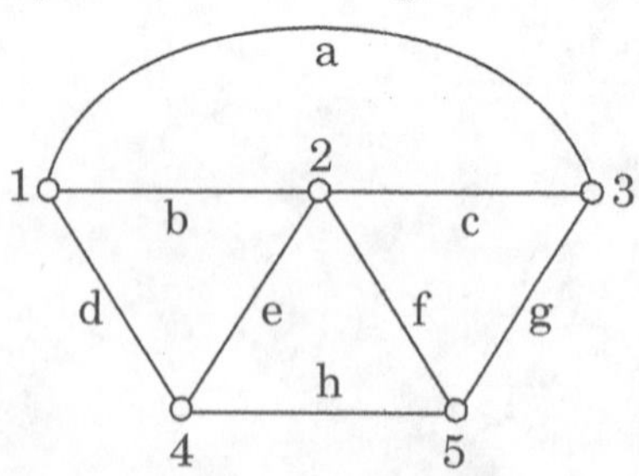

(a) begh (b) defg

(c) adhg (d) aegh

[1999 : 1 Mark]

6. The differential equation for the current i(t) in the circuit of the figure is

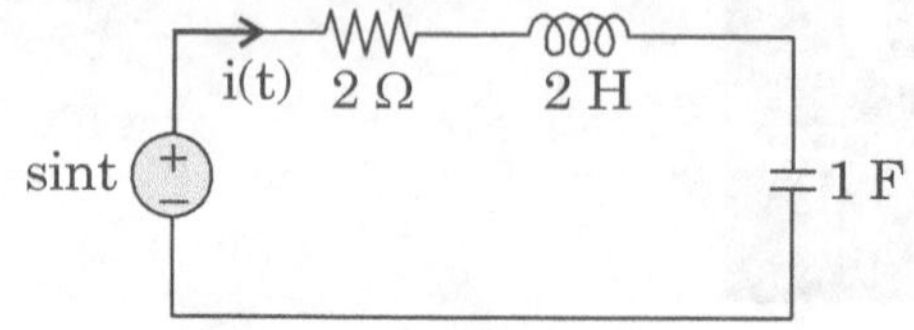

(a) $2\frac{d^2i}{d^2t} + 2\frac{di}{dt} + i(t) = \sin t$

(b) $\frac{d^2i}{d^2t} + 2\frac{di}{dt} + 2i(t) = \cos t$

(c) $2\frac{d^2i}{d^2t} + 2\frac{di}{dt} + i(t) = \cos t$

(d) $\frac{d^2i}{d^2t} + 2\frac{di}{dt} + 2i(t) = \sin t$ **[2003 : 2 Marks]**

7. Consider the network graph shown in the figure. Which one of the following is NOT a 'tree' of this graph?

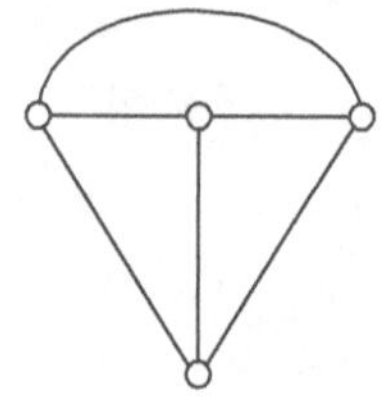

(a) 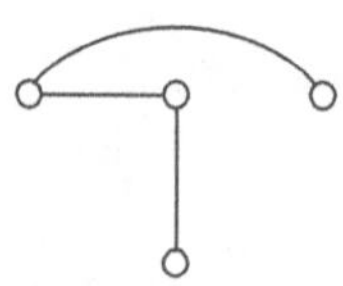

(b) 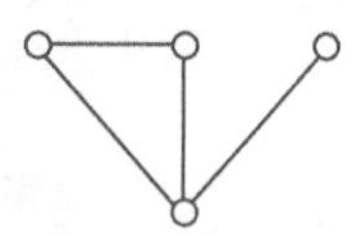

(c) 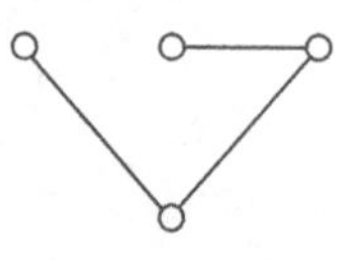

(d) 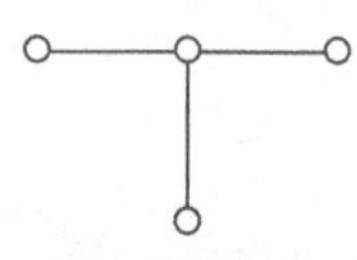

[2004 : 1 Mark]

8. In the following graph, the number of trees (P) and the number of cut-sets (Q) are

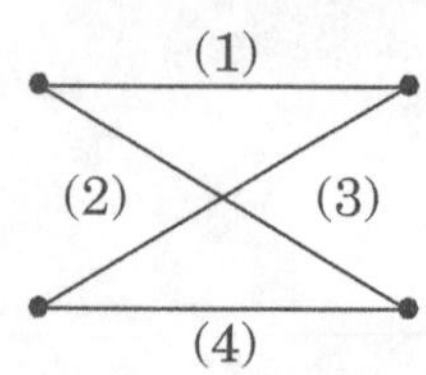

(a) P = 2, Q = 2 (b) P = 2, Q = 6

(c) P = 4, Q = 6 (d) P = 4, Q = 10

[2008 : 1 Mark]

ANSWERS

1. (a) **2.** (a,d) **3.** (c) **4.** (c) **5.** (c) **6.** (c) **7.** (b) **8.** (c)

EXPLANATIONS

1. Number of nodes (n) = 4

   Number of branches or elements (b) = 8

   Number of unique mesh equations (l) = b-n+1

   = 8 – 4 + 1 = 5.

   Number of unique nodal equations = n –1 = 4 – 1= 3.

   Since nodal equations < mesh equations; nodal method will be preferred.

2. Link currents form an independent set.

   All branch currents may be expressed in terms of link currents. Branch voltages form an independent set.

   All branch to branch voltages (node pair voltages) may be expressed in terms of branch voltages.

   Fig (*a*) shows a network graph while

   Fig (*b*) shows one of the possible trees

   Fig (*c*) shows the four link currents i_1, i_2, i_3 and i_4. All branch currents J_1 to J_8 may be expressed in terms of i_1 to i_4.

   Fig (*d*) shows the four tree branch voltages e_1, e_2, e_3 and e_4. All branch to branch voltages may be expressed in terms of e_1 to e_4.

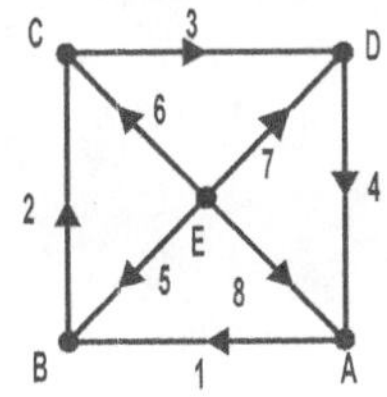

(*a*) Network graph

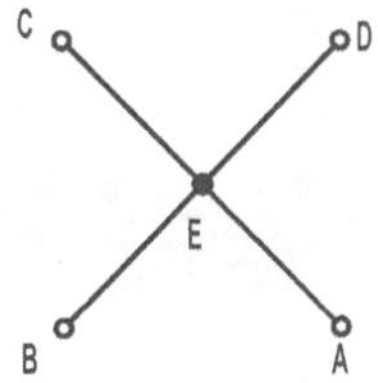

(*b*) A tree

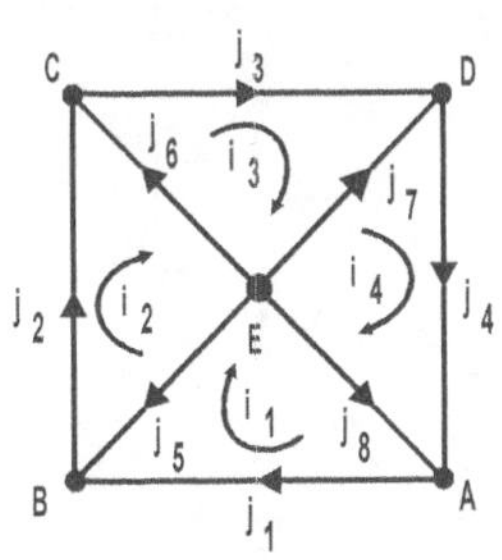

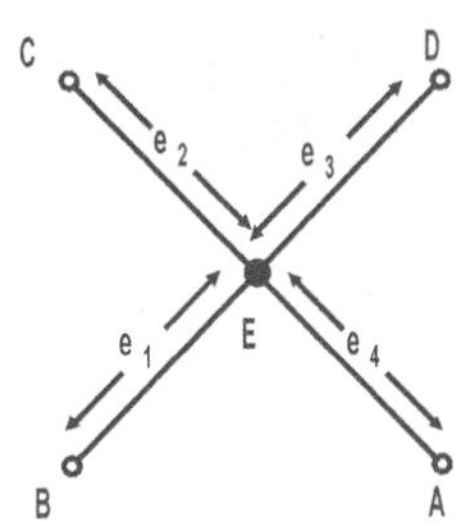

3. Number of independent loop (or mesh) equations

   l = b – n + 1

4. Number of nodes, n = 7

   Number of loops, l = 5.

   Number of branches,

   b = l + n – 1 = 5 + 7 – 1 = 11.

5. adfg is not a tree of the graph because one node (4) is not connected to other nodes.

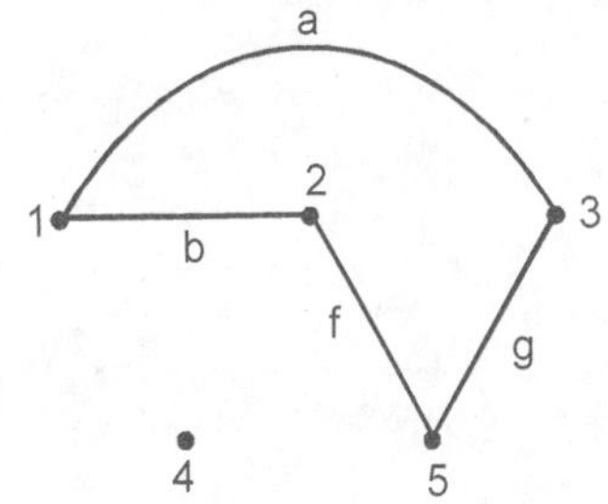

   Tree is any set of branches which does not contain any loop and connects every node to other nodes.

6. Using KVL around the loop, we have

$$\sin t = 2i(t) + 2\frac{di(t)}{dt} + \frac{1}{C}\int i(t)\,.dt$$

$$\text{Differentiating, } \cos t = \frac{2\,di(t)}{dt} + 2\frac{d^2 i(t)}{dt^2} + i(t)$$

$$\text{i.e.} \qquad 2\frac{d^2 i}{dt^2} + 2\frac{di}{dt} + i(t) = \cos t$$

7. Options (*b*) contains a loop and for a tree there is no loop exist.

8. Different trees are shown here in figure (*a*):

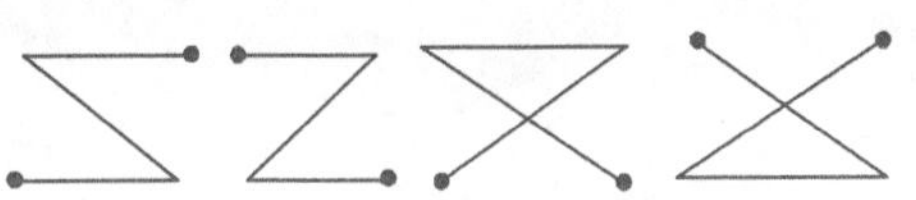

Fig. (*a*)

Different cut set are as shown in figure (*b*):

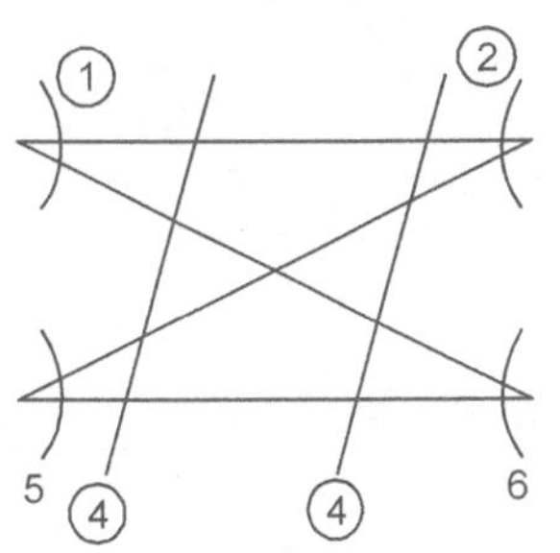

Fig. (*b*)

7

CHAPTER

Network Functions

Analysis of Previous GATE Papers												
Year → Topics ↓			2019	2018	2017 Set 1	2017 Set 2	2016 Set 1	2016 Set 2	2016 Set 3	2015 Set 1	2015 Set 2	2015 Set 3
TRANSFER FUNCTION	1 Mark	MCQ Type										
		Numerical Type										
	2 Marks	MCQ Type										
		Numerical Type										
	Total											
DRIVING POINT FUNCTION	1 Mark	MCQ Type										
		Numerical Type										
	2 Marks	MCQ Type										
		Numerical Type										
	Total											
FILTERS	1 Mark	MCQ Type										
		Numerical Type										
	2 Marks	MCQ Type										
		Numerical Type										
	Total											

TRANSFER FUNCTION

1. The equivalent inductance measured between the terminals 1 and 2 for the circuit shown in the figure is

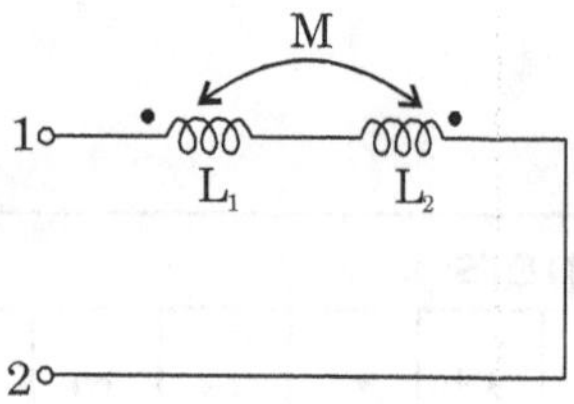

(a) $L_1 + L_2 + M$ (b) $L_1 + L_2 - M$
(c) $L_1 + L_2 + 2M$ (d) $L_2 + L_2 - 2M$

2. Consider the building block called 'Network N' shown in the figure.
Let C= 100 μF and R= 10 kΩ.

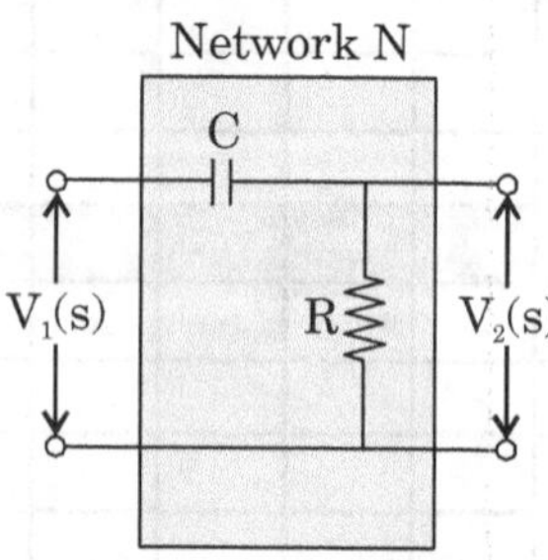

Two such blocks are connected in cascade, as shown in the figure.

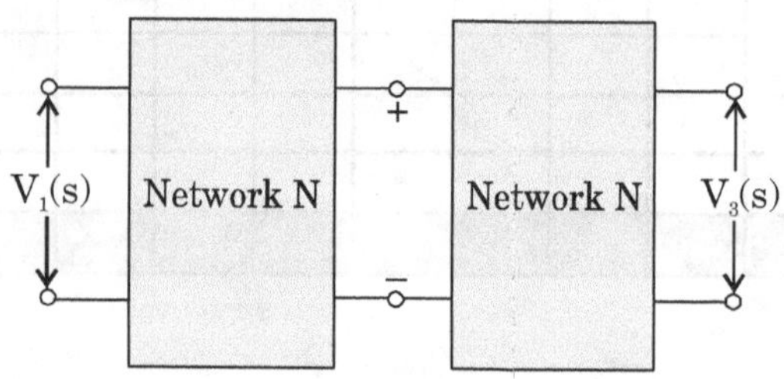

The transfer function $\frac{V_3(s)}{V_1(s)}$ of the cascaded network is

(a) $\frac{s}{1+s}$. (b) $\frac{s^2}{1+3s+s^2}$.

(c) $\left(\frac{s}{1+s}\right)^2$. (d) $\frac{s}{2+s}$.

[1988 : 2 Marks]

3. For the transfer function of a physical two-port network

(a) all the zeros must lie only in the left half of the s-plane
(b) the poles may lie anywhere in the s-plane
(c) the poles lying on the imaginary axis must be simple
(d) a pole may lie at origin

[1989 : 2 Marks]

4. If the transfer function of the following network is $\frac{V_o(s)}{V_i(s)} = \frac{1}{2+sCR}$,

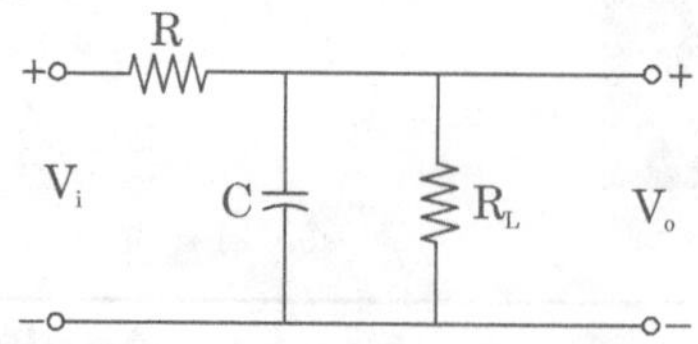

the value of the load resistance R_L is

(a) R/4 (b) R/2
(c) R (d) 2R **[2009 : 1 Mark]**

5. The transfer function $\frac{V_2(s)}{V_1(s)}$ of the circuit shown below is

(a) $\frac{0.5s+1}{s+1}$

(b) $\frac{3s+6}{s+2}$

(c) $\frac{s+2}{s+1}$

(d) $\frac{s+1}{s+2}$

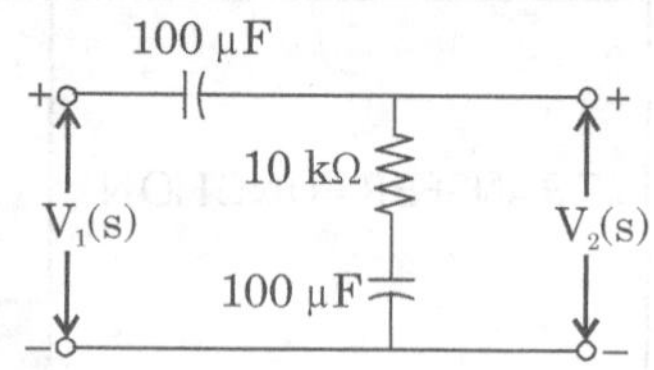

[2013 : 1 Mark]

DRIVING POINT FUNCTION

6. Of the networks, N_1, N_2, N_3 and N_4 of figure, the networks having identical driving point function are

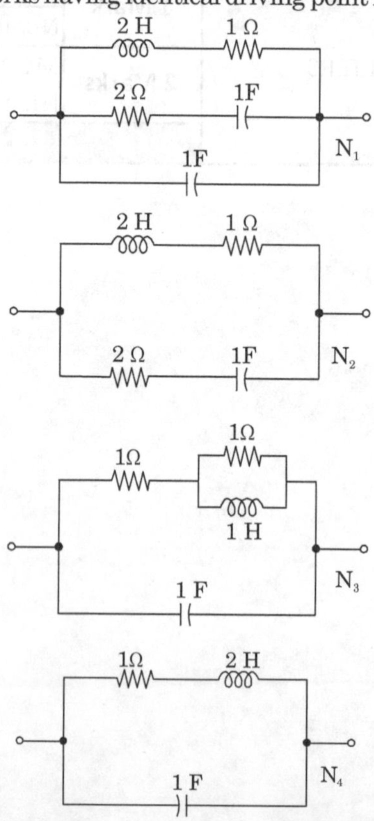

(a) N_1 and N_2 (b) N_2 and N_4
(c) N_1 and N_3 (d) N_1 and N_4

[1987 : 2 Marks]

7. A driving point admittance function has pole and zero locations as shown below. The range of s for which the function can be realized using passive elements is

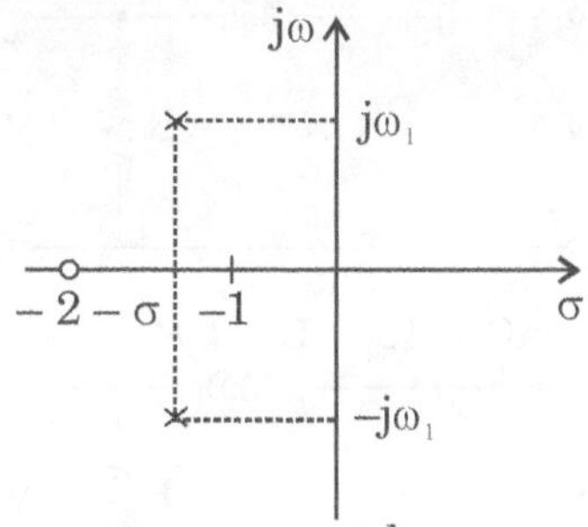

s-plane

(a) $\sigma < -1$ (b) $\sigma > 1$
(c) $\sigma < 1$ (d) $\sigma > -1$

[1988 : 2 Marks]

8. The necessary and sufficient condition for a rational function of s, T(s) to be a driving point impedance of an RC network is that all poles and zeros should be
(a) simple and lie on the negative realaxis of the s-plane
(b) complex and lie in the left half of the s-plane
(c) complex and lie in the right half of the s-plane
(d) simple and lie on the positive real axis of the s-plane **[1991 : 2 Marks]**

9. Indicate True/False and give reason (For the following question) $Z(s) = \dfrac{5}{s^2+4}$ represents the input impedance of a network. **[1994 : 2 Marks]**

10. The driving-point impedance Z(s) of a network has the pole-zero locations as shown in the figure. If Z(0) = 3, then Z(s) is

(a) $\dfrac{3(s+3)}{s^2+2s+3}$

(b) $\dfrac{2(s+3)}{s^2+2s+2}$

(c) $\dfrac{3(s+3)}{s^2-2s-2}$

(d) $\dfrac{2(s-3)}{s^2-2s-3}$

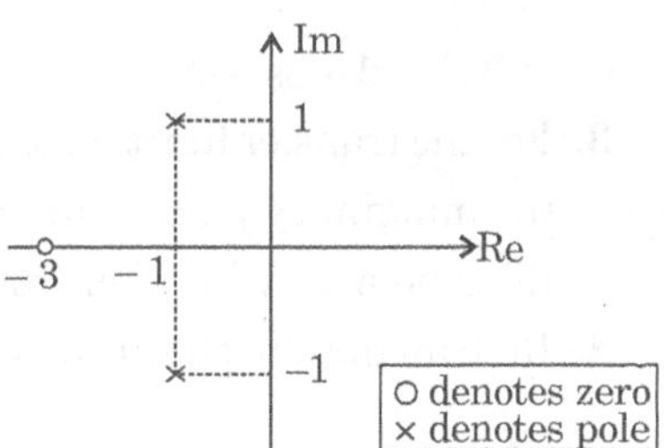

s-plane

[2003 : 2 Marks]

11. The first and the last critical frequency of an RC-driving point impedance function must respectively be
(a) a zero and a pole (b) a zero and a zero
(c) a pole and a pole (d) a pole and a zero

[2005 : 1 Mark]

12. The first and the last critical frequencies (singularities) of a driving point impedance function of a passive network having two kinds of elements, are a pole and a zero respectively. The above property will be satisfied by
(a) RL network only
(b) RC network only
(c) LC network only
(d) RC as well as RL networks **[2006 : 1 Mark]**

13. A negative resistance R_{neg} is connected to a passive network N having driving point impedance as shown below. For $Z_2(s)$ to be positive real

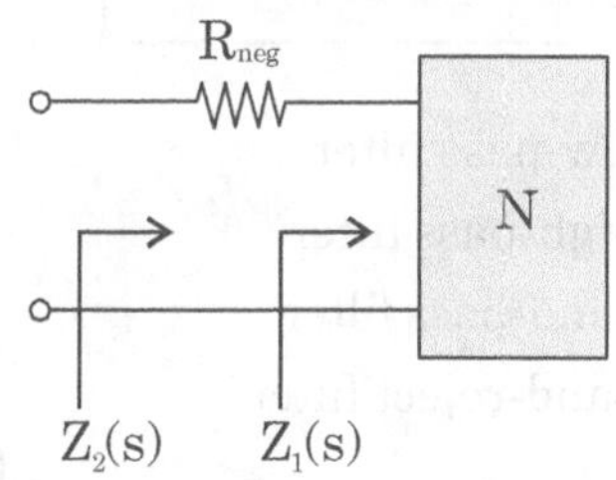

(a) $|R_{neg}| \le \text{Re}\, Z_1(j\omega), \forall \omega$

(b) $|R_{neg}| \le |Z_1(j\omega)|, \forall \omega$

(c) $|R_{neg}| \le |\text{Im}\, Z_1(j\omega)|, \forall \omega$

(d) $|R_{neg}| \le |\angle Z_1(j\omega)|, \forall \omega$ **[2006 : 1 Mark]**

14. The driving point impedance of the following network

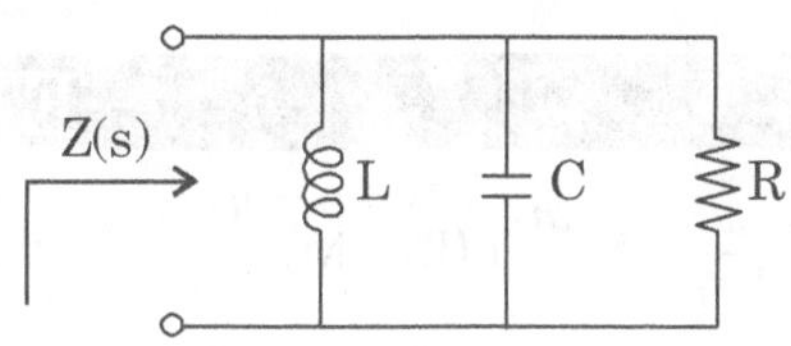

is given by $Z(s) = \dfrac{0.2s}{s^2+0.1s+2}$. The component values are

(a) L = 5 H, R = 0.5 Ω, C = 0.1 F
(b) L = 5 H, R = 0.5 Ω, C = 5 F
(c) L = 5 H, R = 2 Ω, C = 0.1 F
(d) L = 0.1 H, R = 2 Ω, C = 5 F **[2008 : 2 Marks]**

FILTERS

15. The circuit of the figure represents a

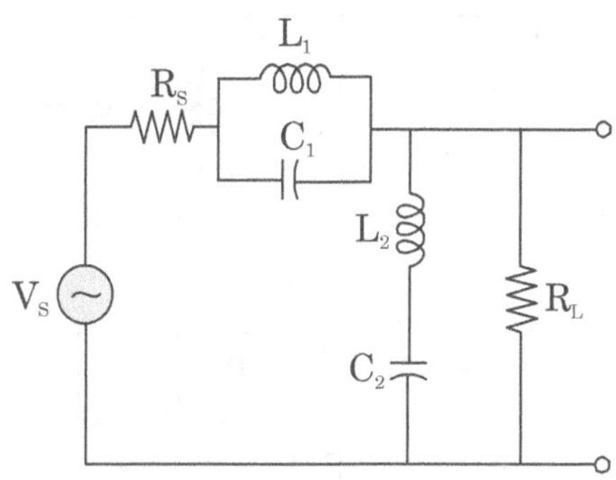

(a) low pass filter (b) high pass filter
(c) band pass filter (d) band reject filter

[2000 : 1 Mark]

16. The RC circuit shown in figure is

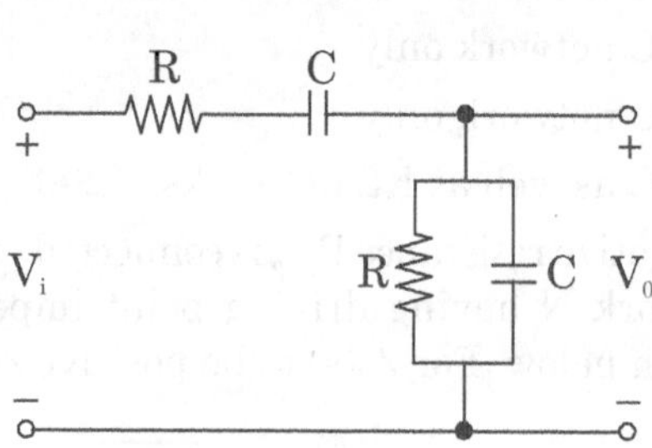

(a) a low-pass filter
(b) a high-pass filter
(c) a band-pass filter
(d) a band-reject filter

[2007 : 1 Mark]

17. Two series resonant filters are as shown in the figure. Let the 3-dB bandwidth of Filter 1 be B_1 and that of filter 2 be B_2. The value of B_1/B_2 is

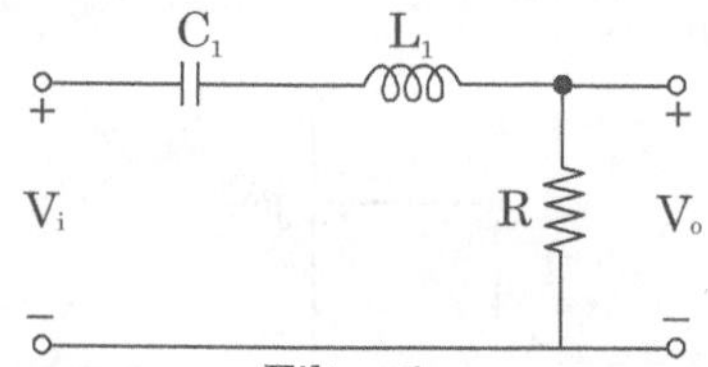

Filter-1

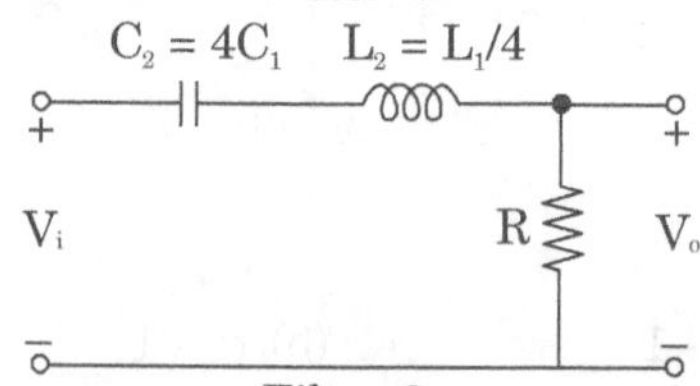

Filter-2

(a) 4
(b) 1
(c) $\frac{1}{2}$
(d) $\frac{1}{4}$

[2008: 2 Marks]

ANSWERS

1. (d)	**2.** (b)	**3.** (d)	**4.** (c)	**5.** (d)	**6.** (b)	**7.** (b)	**8.** (a)	**9.** (a)	**10.** (b)
11. (d)	**12.** (b)	**13.** (a)	**14.** (d)	**15.** (d)	**16.** (c)	**17.** (d)			

EXPLANATIONS

1. $V = L_1\frac{di}{dt} - M\frac{di}{dt} + (L_2 - M)\frac{di}{dt}$

$$= (L_1 + L_2 - 2M)\frac{di}{dt} = L\frac{di}{dt}$$

$$L = L_1 + L_2 - 2M$$

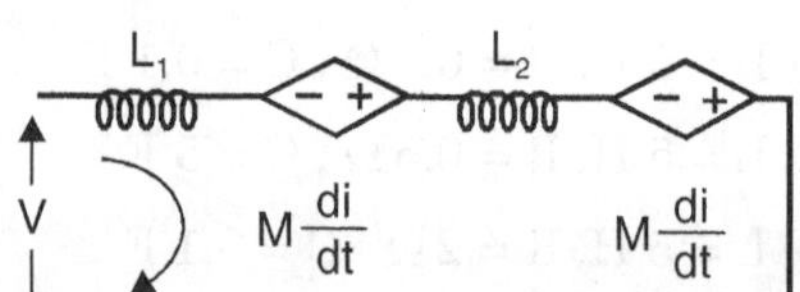

2. Two blocks are connected in cascade, Represent in s-domain,

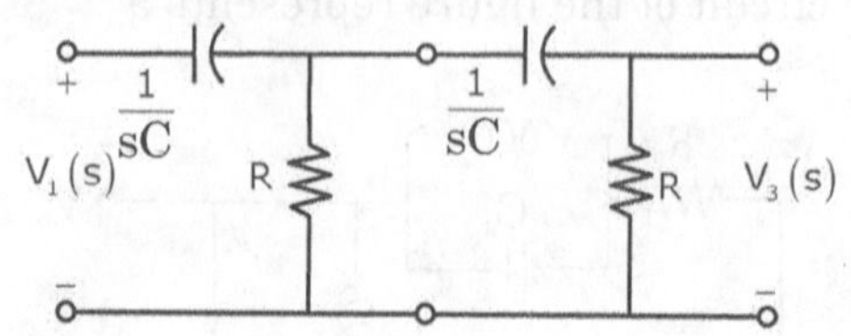

$$\frac{V_3(s)}{V_1(s)} = \frac{R\cdot R}{\frac{1}{sc}\left[R+R+\frac{1}{sC}\right]+R\left[\frac{1}{sC}+R\right]}$$

$$= \frac{R\cdot R}{\frac{1}{sC}\cdot\frac{1}{sC}\left[2R(sC)+1\right]+\frac{R}{sC}\left[1+RsC\right]}$$

$$= \frac{s^2C^2.R.R}{\left[1+2R(sC)\right]+RsC+R^2s^2C^2}$$

$$= \frac{s^2.100\times100\times10^{-6}\times10^{-6}\times10\times10\times10^3\times10^3}{S^2\times100\times10^6\times10^4\times10^{-12}+3s+100\times10^{-6}\times10^4+1}$$

$$\frac{V_3(s)}{V_1(s)} = \frac{s^2}{1+3s+s^2}$$

3. For the transfer function of physical 2-port network
(i) imaginary poles must be simple, or
(ii) a pole may lie at origin.

4. Redrawing the circuit in s-domain,

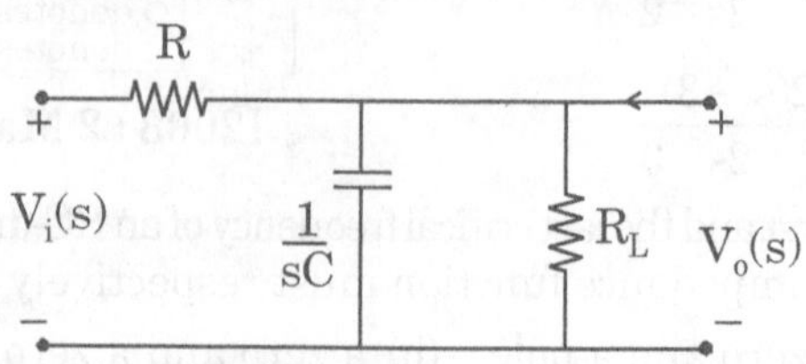

$$\frac{V_o(s)}{V_i(s)} = \frac{\dfrac{\left(\frac{R_L}{sC}\right)}{\left(R_L+\frac{1}{sC}\right)}}{R+\dfrac{\frac{R_L}{sC}}{\frac{1}{sC}}} = \frac{\dfrac{R_L}{(sCR_L+1)}}{R+\dfrac{R_L}{sCR_L+1}}$$

$$\frac{V_o(s)}{V_i(s)} = \frac{R_L}{sCR.R_L + R + R_L}$$

On comparing with $\dfrac{V_o(s)}{V_i(s)} = \dfrac{1}{2+sCR}$

$$\boxed{R = R_L}$$

5. Taking Laplace transformation of the circuit,

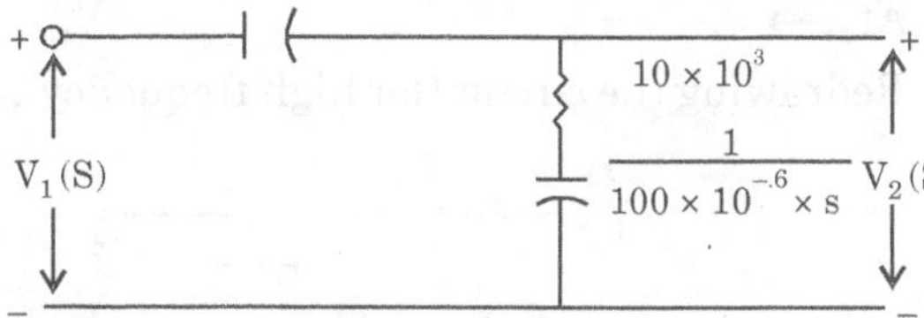

By applying voltage devider rule:

$$V_2(s) = \frac{10\times10^3 + \dfrac{10^4}{s}}{10\times10^3 + \dfrac{10^4}{s} + \dfrac{10^4}{s}} \times V_1(s)$$

$$\frac{V_2(s)}{V_1(s)} = \frac{1+\dfrac{1}{s}}{1+\dfrac{2}{s}} = \frac{s+1}{s+2}$$

7. For relatization of driving point admittance function,

$\sigma - 1 > 0$

$\sigma > 1$

8. The necessary and sufficient condition for the driving point impedance function is that its poles and zeros should be simple and alternate on the negative real axis of the s-plane.

9. For Z(s) to represent the input impedance of a passive network, the numerator and denominator degrees should not differ by more than 1.

$\Rightarrow Z(s) = \dfrac{5}{s^2+4}$ cannot represent the input impedance of network. (FALSE)

10. From the figure, $Z(s) = \dfrac{K(s+3)}{(s+1+i)(s+1-i)}$

$$= \frac{K(s+3)}{[(s+1)^2+1]}$$

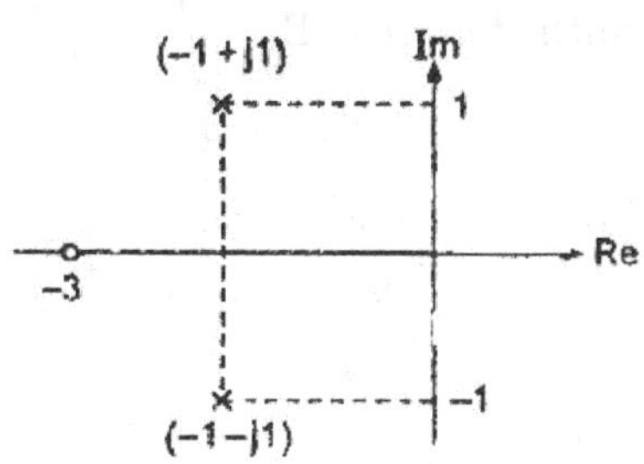

As, $Z(0) = 3$

then, $\dfrac{3K}{2} = 3$

or $K = 2$

then, $Z(s) = \dfrac{2(s+3)}{(s+1)^2+1}$

$$= \frac{2(s+3)}{s^2+2s+2}$$

11. For Series RC network

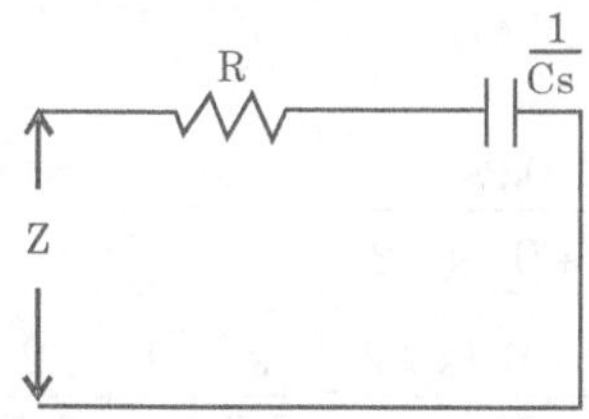

Fig - (a)

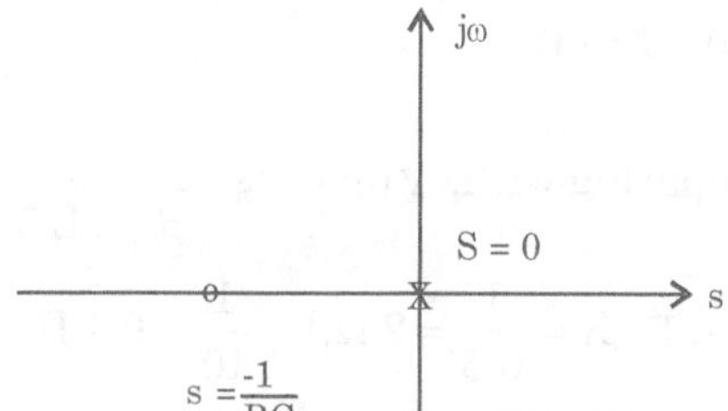

Fig - (b)

$$Z = R + \frac{1}{Cs};\ Z = \frac{RCs+1}{Cs}$$

Z has pole at origin and zero at $s = \dfrac{-1}{RC}$

From figure (*b*), first critical frequency is at s = 0 and last critical frequency is at $s = \dfrac{-1}{RC}$

For parallel RC network

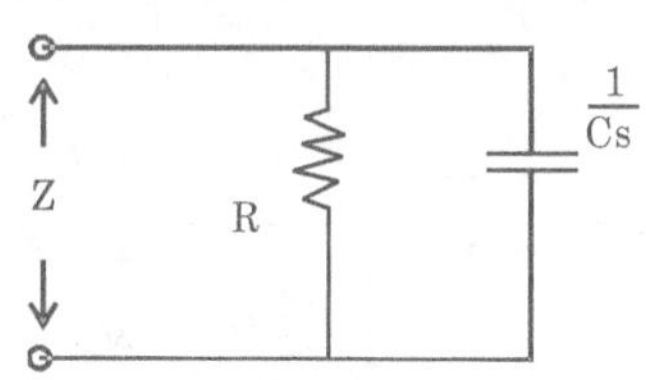

Fig - (c)

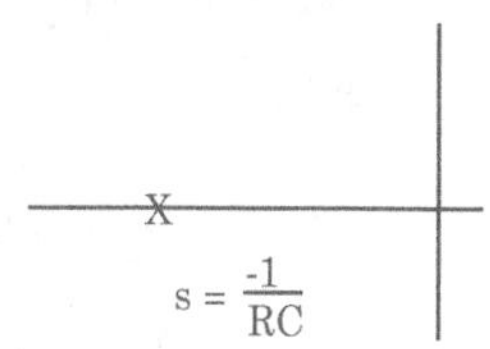

Fig - (d)

$$Z = \frac{R\cdot\dfrac{1}{Cs}}{R+\dfrac{1}{Cs}} = \frac{R}{RCs+1}$$

From figure (*d*), first critical frequency is at $s = \dfrac{-1}{RC}$ (pole) and last critical frequency is zero.

12. RC impedance function has

(i) first critical frequency due to pole,

(ii) last critical frequency due to zero.

13. For $Z_2(s)$ to be positive real

$$\text{Re}\{Z_1(s)\} \geq |R_{neg}|$$

$$\Rightarrow |R_{neg}| \leq \text{Re}\{Z_1(j\omega)\} \ \forall \text{all } \omega.$$

14. Redrawing the circuit in s-domain,

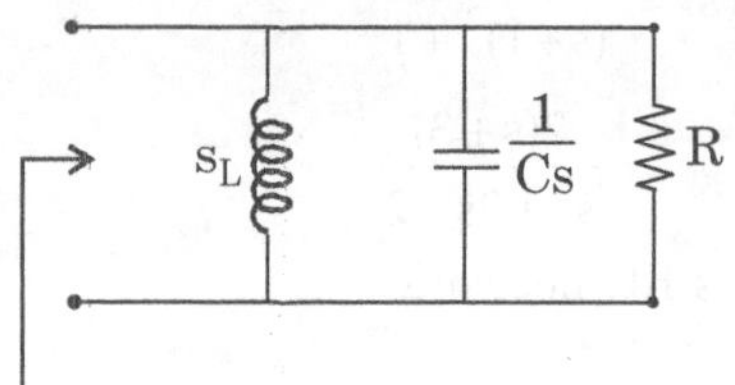

$$Z(s) = \frac{0.2s}{s^2 + 0.1s + 2}$$

$$Y(s) = \frac{s^2 + 0.1s + 2}{0.2s} = \frac{s}{0.2} + \frac{1}{2} + \frac{2}{0.2s}$$

$$Y(s) = 5s + 0.5 + \frac{10}{s}$$

Comparing with, $Y(s) = Cs + \frac{1}{R} + \frac{1}{sL}$

$$C = 5\text{ F},\ R = \frac{1}{0.5} = 2\ \Omega, l = \frac{1}{10} = 0.1\text{ H}.$$

15. Analyzing the circuit for low frequency $(\omega \cong 0)$ and high frequency $(\omega \to \infty)$

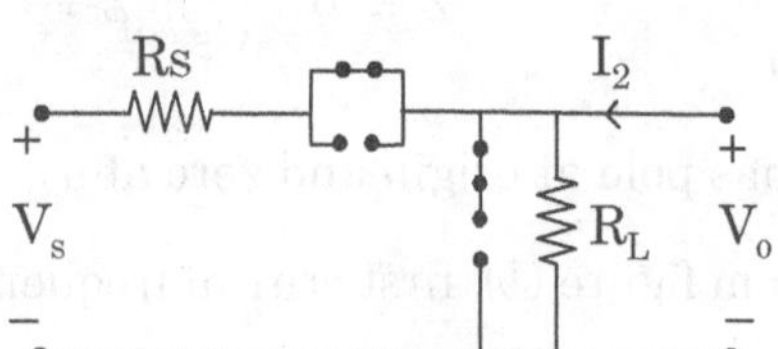

$$\frac{V_0}{V_s} = \frac{R_L}{R_S + R_L} \text{ (finite value)}$$

At $\omega \to \infty$,

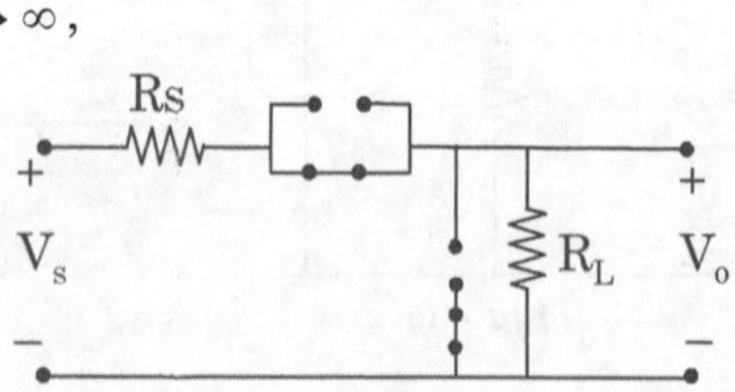

$$\frac{V_0}{V_s} = \frac{R_L}{R_L + R_S} \text{ (finite value)}$$

At $\omega = \frac{1}{\sqrt{LC}}$ (resonant frequency),

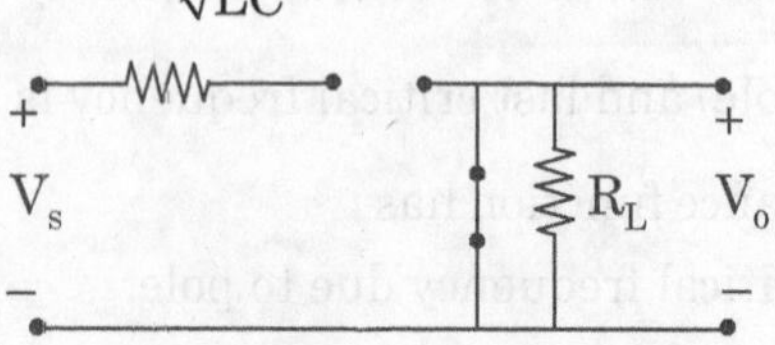

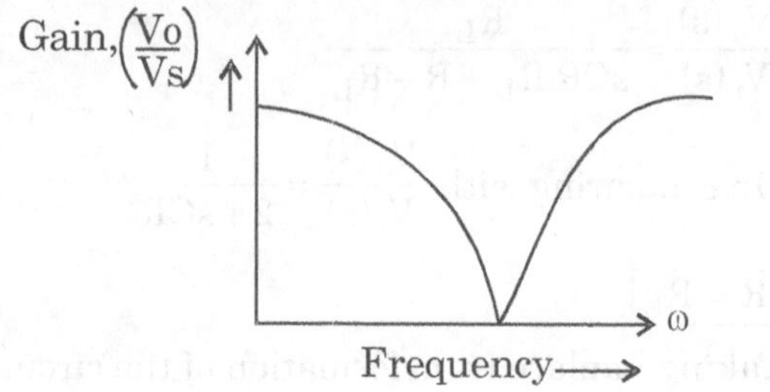

16. At $\omega \to \infty$:-

Redrawing the circuit (for high frequency),

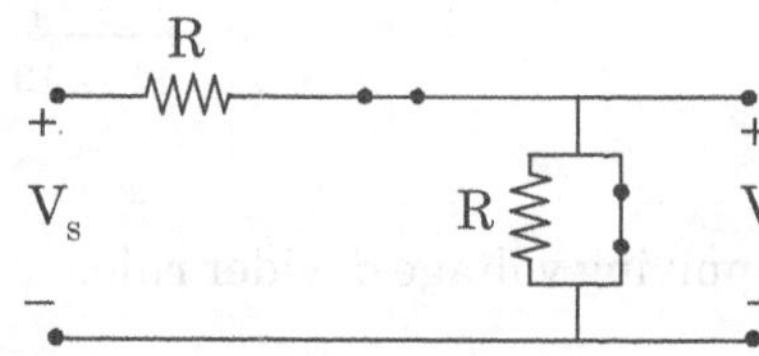

$$\frac{V_o}{V_i} = 0$$

At $\omega \to 0$:-

Redrawing the circuit (for low frequency),

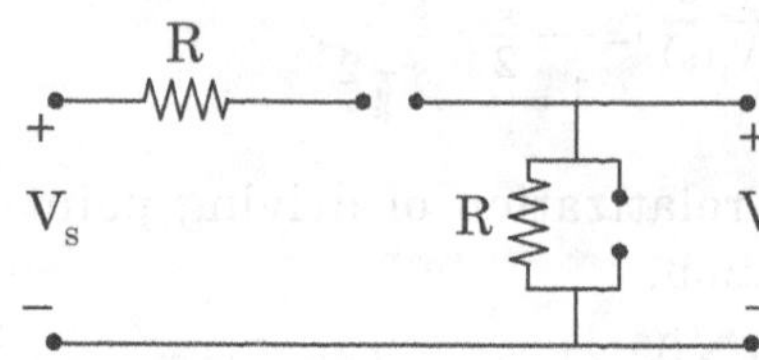

$$\frac{V_o}{V_i} = 0$$

There is no resonant frequency because of absence of So, the frequency response of the circuit is

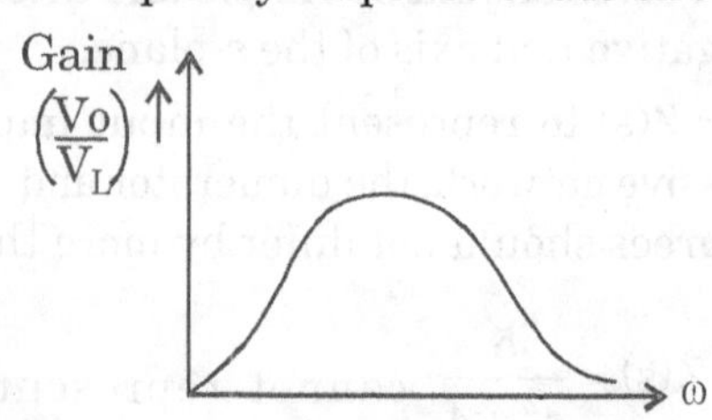

Hence, the circuit is band-pass filter.

17. In general, bandwidth of series RLC circuit = $\frac{R}{L}$

Bandwidth of filter 1, $B_1 = \frac{R}{L_1}$

Bandwidth of filter 2, $B_2 = \frac{R}{L_2} = \frac{R}{\frac{L_1}{4}} = \frac{4R}{L_1}$

$$\Rightarrow \frac{B_1}{B_2} = \frac{1}{4}.$$

Unit - XI

General Aptitude

Syllabus

Verbal Ability: English grammar, sentence completion, verbal analogies, word groups, instructions, critical reasoning and verbal deduction.

Numerical Ability: Numerical computation, numerical estimation, numerical reasoning and data interpretation.

Contents

1

CHAPTER

Verbal Ability

Analysis of Previous GATE Papers												
	Year → Topics ↓		2019	2018	2017 Set 1	2017 Set 2	2016 Set 1	2016 Set 2	2016 Set 3	2015 Set 1	2015 Set 2	2015 Set 3
Verbal Ability	1 Mark	MCQ Type	4	2	2	2	3	2	2	3	3	3
		Numerical Type										
	2 Marks	MCQ Type	1	2	1	1	1	1	1	3	3	2
		Numerical Type										
	Total		6	6	4	4	5	4	4	9	9	7

1. Which of the following options is the closest in meaning to the word below ?

(a) Cyclic (b) Indirect

(c) Confusing (d) Crooked

[2010 : 1 Mark]

2. The question below consists of a pair of related words followed by four pairs of words. Select the pair that best expresses the relation in the original pair.

Unemployed: Worker

(a) Fallow: Land (b) Unaware: Sleeper

(c) Wit : Jester (d) Renovated: House

[2010 : 1 Mark]

3. Choose the most appropriate word from the options given below to complete the following sentence:

If we manage to _______ our natural resources, we would leave a better planet for our children.

(a) uphold (b) restrain

(c) cherish (d) conserve

[2010 : 1 Mark]

4. Choose the most appropriate word from the options given below to complete the following sentence:

His rather casual remarks on politics _______ his lack of seriousness about the subject.

(a) masked (b) belied

(c) betrayed (d) suppressed

[2010 : 1 Mark]

5. Modern warfare has changed from large scale clashes of armies to suppression of civilian populations. Chemical agents that do their work silently appear to be suited to such warfare; and regretfully, there exist people in militrary establishments who think that chemical agents are useful fools for their cause.

Which of the following statements best sums up the meaning of the above passage?

(a) Modern warfare has resulted in civil strife.

(b) Chemical agents are useful in modern warfare.

(c) Use of chemical agents in warfare would be undesirable.

(d) People in military establishments like to use chemical agents in war.

[2010 : 2 Marks]

6. The question below consists of a pair of related words followed by four pairs of words. Select the pair that best expresses the relation in the original pair:

Gladiator: Arena

(a) dancer: stage (b) commuter: train

(c) teacher: classroom (d) lawyer: courtroom

[2011 : 1 Mark]

7. Choose the most appropriate word from the options given below to complete the following sentence:

It was her view that the country's problems had been_______by foreign technocrats, so that to invite them to come back would be counter-productive.

(a) identified (b) ascertained

(c) exacerbated (d) analysed

[2011 : 1 Mark]

8. Choose the word from the options given below that is most nearly opposite in meaning to the given word:

Frequency

(a) periodicity (b) rarity

(c) gradualness (d) persistency

[2011 : 1 Mark]

9. Choose the most appropriate word from the options given below to complete the following sentence:

Under ethical guidelines recently adopted by the Indian Medical Association, human genes are to be manipulated only to correct diseases for which_______treatments are unsatisfactory.

(a) similar (b) most

(c) uncommon (d) available

[2011 : 1 Mark]

10. The horse has played a little known but very important role in the field of medicine. Horses were injected with toxins of diseases until their blood built up immunities. Then a serum was made from their blood. Serums to fight with diphtheria and tetanus were developed this way.

It can be inferred from the passage, that horses were

(a) given immunity to diseases

(b) -generally quite immune to diseases

(c) given medicines to fight toxins

(d) given diphtheria and tetanus serums

[2011 : 2 Marks]

11. Which one of the following options is the closest in meaning to the word given below? Latitude

(a) Eligibility (b) Freedom
(c) Coercion (d) Meticulousness

[2012 : 1 Mark]

12. One of the parts (A, B C, D) in the sentence given below contains an ERROR. Which one of the following is INCORRECT?

I requested that he should be given the driving test today instead of tomorrow.

(a) requested that
(b) should be given
(c) the driving test
(d) instead of tomorrow **[2012 : 1 Mark]**

13. Choose the most appropriate alternative from the options given below to complete the following sentence:

If the tired soldier wanted to lie down, he _______ the mattress out on the balcony.

(a) should take (b) shall take
(c) should have taken (d) will have taken

[2012 : 1 Mark]

14. Choose the most appropriate word from the options given below to complete the following sentence:

Given the seriousness of the situation that he had to face, his_______was impressive.

(a) beggary (b) nomenclature
(c) jealousy (d) nonchalance

[2012 : 1 Mark]

15. One of the legacies of the Roman legions was discipline. In the legions, military law prevailed and discipline was brutal. Discipline on the battlefield kept units obedient, intact and fighting, even when the odds and conditions were against them.

Which one of the following statements best sums up the meaning of the above passage?

(a) Thorough regimentation was the main reason for the efficiency of the Roman legions even in adverse circumstances.
(b) The legions were treated inhumanly as if the men were animals.
(c) Discipline was the armies inheritance from their seniors.
(d) The harsh discipline to which the legions were subjected to led to the odds and conditions being against them.

[2012 : 2 Marks]

16. Choose the grammatically CORRECT sentence:

(a) Two and two add four.
(b) Two and two become four.
(c) Two and two are four.
(d) Two and two make four.

[2013 : 1 Mark]

17. Statement: You can always give me a ring whenever you need.

Which one of the following is the best inference from the above statement?

(a) Because I have a nice caller tune.
(b) Because I have a better telephone facility.
(c) Because a friend in need is a friend indeed.
(d) Because you need not pay towards the telephone bills when you give me a ring.

[2013 : 1 Mark]

18. Complete the sentence:

Dare_______mistakes.

(a) commit (b) to commit
(c) committed (d) committing

[2013 : 1 Mark]

19. They were requested not to quarrel with others. Which one of the following options is the closest in meaning to the word quarrel?

(a) make out (b) call out
(c) dig out (d) fall out

[2013 : 1 Mark]

20. Statement : There were different streams of freedom movements in colonial India carried out by the moderates, liberals, radicals, socialists, and so on.

Which one of the following is the best inference from the above statement?

(a) The emergence of nationalism in colonial India led to our Independence.
(b) Nationalism in India emerged in the context of colonialism.
(c) Nationalism in India is homogeneous.
(d) Nationalism in India is heterogeneous.

[2013 : 2 Marks]

21. Choose the most appropriate phrase from the options given below to complete the following sentence. The aircraft_______take off as soon as its flight plan was filed.

(a) is allowed to (b) will be allowed to
(c) was allowed to (d) has been allowed to

[2014 : 1 Mark, Set-1]

22. Read the statements:

All women are entrepreneurs.

Some women are doctors.

Which of the following conclusions can be logically inferred from the above statements?

(a) All women are doctors

(b) All doctors are entrepreneurs

(c) All entrepreneurs are women

(d) Some entrepreneurs are doctors

[2014 : 1 Mark, Set-1]

23. Choose the most appropriate word from the options given below to complete the following sentence.

Many ancient cultures attributed disease to supernatural causes. However, modern science has largely helped ______ such rotions.

(a) impel (b) dispel

(c) propel (d) repel

[2014 : 1 Mark, Set-1]

24. Choose the most appropriate word from the options given below to complete the following sentence.

Communication and interpersonal skills are ______ important in their own ways.

(a) each (b) both

(c) all (d) either

[2014 : 1 Mark, Set-2]

25. Which of the options given below best completes the following sentence?

She will feel much better if she ______.

(a) will get some rest

(b) gets some rest

(c) will be getting some rest

(d) is getting some rest

[2014 : 1 Mark, Set-2]

26. Choose the most appropriate pair of words from the options given below to complete the following sentence.

She could not ______ the thought of ______ the election to her bitter rival.

(a) bear, loosing (b) bare, loosing

(c) bear, losing (d) bare, losing

[2014 : 1 Mark, Set-2]

27. The value of one U.S. dollar is 65 Indian Rupees today, compared to 60 last year. The Indian Rupee has ______.

(a) depressed (b) depreciated

(c) appreciated (d) stabilized

[2014 : 1 Mark, Set-3]

28. 'Advice' is ______.

(a) a verb

(b) a noun

(c) an adjective

(d) both a verb and a noun

[2014 : 1 Mark, Set-3]

29. Which of the following options is the closest in meaning to the word underlined in the sentence below?

In a democracy, everybody has the freedom to <u>disagree</u> with the government.

(a) dissent (b) descent

(c) decent (d) decadent

[2014 : 1 Mark, Set-4]

30. After the discussion, Tom said to me, 'Please revert!'. He expects me to ______.

(a) retract (b) get back to him

(c) move in reverse (d) retreat

[2014 : 1 Mark, Set-4]

31. While receiving the award, the scientist said, "I feel vindicated". Which of the following is closest in meaning to the word vindicated'?

(a) punished (b) substantiated

(c) appreciated (d) chastened

[2014 : 1 Mark, Set-4]

32. Choose the appropriate word/phrase, out of the four options given below, to complete the following sentence:

Frogs ______.

(a) croak (b) roar

(c) hiss (d) patter

[2015 : 1 Mark, Set-1]

33. Choose the word most similar meaning to the given word:

Educe

(a) Exert (b) Educate

(c) Extract (d) Extend

[2015 : 1 Mark, Set-1]

34. Choose the most appropriate word from the options given below to complete the following sentence.

The principal presented the chief guest with a ______, as token of appreciation.

(a) momento (b) memento

(c) momentum (d) moment

[2015 : 1 Mark, Set-1]

35. The following question presents a sentence, part of which is underlined. Beneath the sentence you find four ways of phrasing the underlined part. Following the requirements of the standard written English, select the answer that produces the most effective sentence.

Tuberculosis, together with its effects, ranks one of <u>the leading causes of death</u> in India.

(a) ranks as one of the leading causes of death

(b) rank as one of the leading causes of death

(c) has the rank of one of the leading causes of death

(d) are one of the leading causes of death

[2015 : 2 Marks, Set-1]

36. Humpty Dumpty sits on a wall every day while having lunch. The wall sometimes breaks. A person sitting on the wall falls if the wall breaks.

Which one of the statements below is logically valid and can be inferred from the above sentences?

(a) Humpty Dumpty always falls white having lunch

(b) Humpty Dumpty does not fall sometimes while having lunch

(c) Humpty Dumpty never falls during dinner

(d) When Humpty Dumpty does not sit on the wall, the wall does not break

[2015 : 2 Marks, Set-1]

37. Read the following paragraph and choose the correct statement.

Climate change has reduced human security and threatened human well being. An ignored reality of human progress is that human security largely depends upon environmental security. But on the contrary, human progress seems contradictory to environmental security. To keep up both at the required level is a challenge to be addressed by one and all. One of the ways to curb the climate change may be suitable scientific innovations, while the other may be the Gandhian perspective on small scale progress with focus on sustainability.

(a) Human progress and security are positively associated with environmental security.

(b) Human progress is contradictory to environmental security.

(c) Human security is contradictory to environmental security.

(d) Human progress depends upon environmental security.

[2015 : 2 Marks, Set-1]

38. Choose the word most similar in meaning to the given word:

Awkward

(a) Inept (b) Graceful

(c) Suitable (d) Dreadful

[2015 : 1 Mark, Set-2]

39. Choose the appropriate word/phase, out of the four options given belcw, to complete the following sentence:

Dhoni, as well as the other team members of the Indian team _______ present on the occasion.

(a) were (b) was

(c) has (d) have

[2015 : 1 Mark, Set-2]

40. What is the adverb for the given word below?

Misogynous

(a) Misogynousness (b) Misogynity

(c) Misogynously (d) Misogynous

[2015 : 1 Mark, Set-2]

41. Given below are two statements followed by two conclusions. Assuming these statements to be true, decide which one logically follows.

Statements:

I. All film star are playback singers.

II. All film directors are film stars. Conclusions:

I. All film directors are playback singers.

II. Some film stars are film directors.

(a) Only conclusion I follows.

(b) Only conclusion I nor II follows.

(c) Neither conclusion I nor II follows.

(d) Both conclusions I and II follow.

[2015 : 2 Marks, Set-2]

42. In the following sentence certain parts are underlined and marked P, Q and P. One of the parts may contain certain error or may not be acceptable in standard written communication. Select the part contain ng an error. Choose D as your answer if there is no error.

The student concreted all the errors that the <u>instructor marked</u> on the <u>answer book</u>,

(a) P

(b) Q

(c) P

(d) No Error

[2015 : 2 Marks, Set-2]

43. Lamenting the gradual sidelining of the arts in school curricula, a group of prominent artists wrote to the Chief Minister last year, asking him to allocate more funds to support arts education in schools. However, no such increase has been announced in this year's Budget. The artists expressed their deep anguish at their request not being approved, but many of them remain optimistic about finding in the future,

Which of the statement(s) below is/are logically valid and can be inferred from the above statements?

(i) The artists expected funding for the arts to increase this year.

(ii) The Chief Minister was receptive to the idea of increasing funding for the arts.

(iii) The Chief Minister is a prominent artists.

(iv) Schools are giving less importance to arts education nowadays.

(a) (iii) and (iv) (b) (i) and (iv)
(c) (i), (ii) and (iv) (d) (i) and (iii)

[2015 : 2 Marks, Set-2]

44. Choose the correct verb to fill in the blank below:
Let us ______.

(a) introvert (b) alternate
(c) atheist (d) altruist

[2015 : 1 Mark, Set-3]

45. Choose the most appropriate word from the options given below to complete the following sentence.

If the athlete had wanted to come first in the race, he ______ several hours every day.

(a) should practice (b) should have practised
(c) practised (d) should be practising

[2015 : 1 Mark, Set-3]

46. Choose the most suitable one word substitute for the following expression:

Connection of a road or way

(a) Perrinacious (b) Viaticum
(c) Clandestine (d) Ravenous

[2015 : 1 Mark, Set-3]

47. Ram and Shyam shared a secret and promised to each other that it would remain between them. Ram express himself in one of the following ways as given in the choices below. Identify the correct way as per standard English.

(a) It would remain between you and me.
(b) It would remain between I and you.
(c) It would remain between you and I.
(d) It would remain with me.

[2015 : 2 Marks, Set-3]

48. In the following question, the first and the last sentence of the passage are in order and numbered 1 and 6. The rest of the passage is split into 4 parts and numbered as 2, 3, 4 and 5. These 4 parts are not arranged in proper order. Read the sentences and arrange them in a logical sequence to make a passage and choose the correct sequence from the given options.

1. On Diwali, the family rises early in the morning.
2. The whole family, including :he young and the old enjoy doing this.
3. Children let off fireworks later in the night with their friends.
4. At sunset, the lamps are lit and the family performs various rituals.
5. Father, mother and children visit relatives and exchange gifts and sweets.
6. Houses looks so pretty with lighted lamps all around.

(a) 2, 5, 3, 4 (b) 5, 2, 4, 3
(c) 3, 5, 4, 2 (d) 4, 5, 2, 3

[2015 : 2 Marks, Set-3]

49. Which of the following is CORRECT with respect to grammar and usage? Mount Everest is ______.

(a) the highest peak in the world
(b) highest peak in the world
(c) one of highest peak in the world
(d) one of the highest peak in the world

[2016 : 1 Mark, Set-1]

50. The policeman asked the victim of a theft, "What did you ______?

(a) loose (b) lose
(c) loss (d) louse

[2016 : 1 Mark, Set-1]

51. Despite the new medicine's ______ in treating diabetes, it is not ______ widely.

(a) effectiveness - prescribed
(b) availability - used
(c) prescription - available
(d) acceptance - proscribed

[2016 : 1 Mark, Set-1]

52. In a world filled with uncertainty, he was glad to have many good friends. He has always assisted them in times of need and was confident that they would reciprocate. However, the events of the last week proved him wrong.

Which of the following inference(s) is/are logically valid and can be inferred from the above passage?

I. His friends were always asking him to help them.

II. He felt that when in need of help, his friends would let him down.

III. He was sure that his friends would help him when in need.

IV. His friends did not help him last week,

(a) I and II (b) III and IV

(c) III only (d) IV only

[2016 : 2 Marks, Set-1]

53. The students ______ the teacher on teachers' day for twenty years of dedicated teaching.

(a) facilitated (b) felicitated

(c) fantasized (d) facilitated

[2016 : 1 Mark, Set-2]

54. After India's cricket world cup victory in 1985, Shrotria who was playing both tennis and cricket till then, decided to concentrate only on cricket. And the rest is history.

What does the underlined phrase mean in this context?

(a) history will rest in peace

(b) rest is recorded in history books

(c) rest is well known

(d) rest is archaic

[2016 : 1 Mark, Set-2]

55. Social science disciplines were in existence in an amorphous form until the colonial period when they were institutionalized. In varying degrees, they were intended to further the colonial interest. In the time of globalization and tne economic rise of postcolonial countries like India, conventional ways of knowledge production have become obsolete.

Which of the following can be logically inferred from the above statements?

I. Social science disciplines have become obsolete.

II. Social science disciplines had a pre-colonial origin.

III. Social science disciplines always promote colonialism.

IV. Social science must maintain disciplinary boundaries.

(a) II only (b) I and III only

(c) II and IV only (d) III and IV only

[2016 : 2 Marks, Set-2]

56. An apple costs Rs. 10. An onion costs Rs.8.

Select the most suitable sentence with respect to grammar and usage.

(a) The price of an apple is greater than an onion.

(b) The price of an apple is more than onion.

(c) The price of an apple is greater than of an onion.

(d) Apples are more costlier than onions.

[2016 : 1 Mark, Set-3]

57. The Buddha said, "Holding on to anger is Ike grasping a hot coal with the intent of throwing it at someone else; you are the one who gets burnt."

Select the word below which is closest in meaning to the word underlined above.

(a) burning (b) igniting

(c) clutching (d) flinging

[2016 : 1 Mark, Set-3]

58. The overwhelming number of people infected with rabies in India has been-flagged by the World Health Organization as a source of concern. It is estimated that inoculating 70% of pets and stray dogs against rabies can lead to a significant reduction in the number of people infected with rabies.

Which of the following can be logically inferred from the above sentences?

(a) The number of people in India infected with rabies is high.

(b) The number of people in other parts of the world who are infected with rabies is low

(c) Rabies can be eradicated in India by vaccinating 70% of stray dogs

(d) Stray dogs are the main source of rabies worldwide

[2016 : 2 Marks, Set-3]

59. I ______ made arrangements had I ______ informed earlier.

(a) could have, been (b) would have, being

(c) had, have (d) had been, been

[2017 : 1 Mark, Set-1]

60. She has a sharp tongue and it can occasionally turn ______.

(a) hurtful

(b) left

(c) methodical

(d) vital

[2017 : 1 Mark, Set-1]

61. "If you are looking for a history of India, or for an account of the rise and fall of the British Raj, or for the reason of the cleaving of the subcontinent into two mutually antagonistic parts and the effects this mutilation will have in the respective sections, and ultimately on Asia, you will not find it in these pages; for though I have spent a lifetime in the country, I lived too near the seat of events, and was too intimately associated with the actors, to get the perspective needed for the impartial recording of these matters".

Here, the word 'antagonistic' is closest in meaning to

(a) impartial (b) argumentative
(c) separated (d) hostile

[2017 : 2 Marks, Set-1]

62. It is ______ to read this year's textbook ______ the last year's.

(a) easier, than (b) most easy, than
(c) easier, from (d) easiest, from

[2017 : 1 Mark, Set-2]

63. The ninth and the tenth of this month are Monday and Tuesday ______.

(a) figuratively (b) retrospectively
(c) respectively (d) rightfully

[2017 : 1 Mark, Set-2]

64. "If you are looking for a history of India, or for an account of the rise and fall of the British Raj or for the reason of the cleaving of the subcontinent into two mutually antagonistic parts and the effects this mutilation will have in the respective sections, and ultimately on Asia, you will not find it in these pages: for though I have spent a lifetime in the country. I lived too near the seat of events, and was too intimately associated with the actors, to get the perspective needed for the impartial recording of these matters."

Which of the following statements best reflects the author's opinion?

(a) An intimate association does not allow for the necessary perspective.
(b) Matters are recorded with an impartial perspective.
(c) An intimate association offers an impartial perspective.
(d) Actors are typically associated with the impartial recording of matters.

[2017 : 2 Marks, Set-2]

65. "By giving him the last ______ of the cake, you will ensure lasting ______ in our house today."

The words that best fill the blanks in the above sentence are

(a) peas, piece (b) piece, peace
(c) peace, piece (d) peace, peas

[2018 : 1 Mark]

66. "Even though there is a vast scope for its ______, tourism has remained a/an ______ area."

The words that best fill the blanks in the above sentence are

(a) improvement, neglected
(b) rejection, approved
(c) fame, glum
(d) interest, disinterested

[2018 : 1 Mark]

67. A coastal region with unparalleled beauty is home to many species of animals. It is dotted with coral reefs and unspoilt white sandy beaches. It has remained inaccessible to tourists due to poor connectivity and lack of accommodation. A company has spotted the opportunity and is planning to develop a luxury resort with helicopter service to the nearest major city airport. Environmentalists are upset that this would lead to the region becoming crowded and polluted like any other major beach resorts.

Which one of the following statements can be logically inferred from the information given in the above paragraph?

(a) The culture and tradition of the local people will be influenced by the tourists.
(b) The region will become crowded and polluted due to tourism.
(c) The coral reefs are on the decline and could soon vanish.
(d) Helicopter connectivity would lead to an increase in tourists coming to the region.

[2018 : 2 Marks]

68. The Cricket Board has long recognized John's potential as a leader of the team. However, his on-field Temper has always been a matter of concern for them since his junior days. While this aggression has filled stadia with die-hard fans, it has taken a toll on his own batting. Until recently, it appeared that he found it difficult to convert his aggression into big scores. Over the past three seasons though, that picture of John has been replaced by a cerebral, calculative and successful batsman-captain. After many years, it appears that the team has finally found a complete captain. Which of the following statements can be logically inferred from the above paragraph?

(i) Even as a junior cricketer, John was considered a good captain.

(ii) Finding a complete captain is a challenge.

(iii) Fans and the Cricket Board have differing views on what they want in a captain.

(iv) Over the past three seasons John has accumulated big scores.

(a) (i), (ii) and (iii) only
(b) (iii) and (iv) only
(c) (ii) and (iv) only
(d) (i), (ii), (iii) and (v)

[2018 : 2 Marks]

69. Five different books (P, Q, R, S, T) are to be arranged on a shelf. The books R and S are to be arranged first and second, respectively from the right side of the shelf. The number of different order in which P, Q and T may be arranged is ________.

(a) 2 (b) 120
(c) 6 (d) 12

[2019 : 1 Mark]

70. The boat arrived ________ dawn.

(a) on (b) at
(c) under (d) in

[2019 : 1 Mark]

71. When he did not come home, she ________ him lying dead on the roadside somewhere

(a) concluded (b) pictured
(c) notice (d) looked

[2019 : 1 Mark]

72. The strategies that the company ________ to sell its products ________ house-to-house marketing.

(a) uses, include
(b) use, includes
(c) uses, including
(d) used, includes

[2019 : 1 Mark]

73. "Indian history was written by British historians– extremely well documented and researched, but not always impartial. History had to serve its purpose: Everything was made subservient to the glory of the Union Jack. Latter-day Indian scholar presented a contrary picture."

From the text above, we can infer that :

Indian history written by British historians ____.

(a) was well documented and not researched but was always biased
(b) was not well documented and researched and was sometimes biased
(c) was well documented and researched but was sometimes biased
(d) was not well documented and researched and was always biased

[2019 : 2 Marks]

ANSWERS

1. (b)	2. (d)	3. (a)	4. (c)	5. (c)	6. (d)	7. (c)	8. (b)	9. (*)	10. (b)
11. (b)	12. (b)	13. (a)	14. (d)	15. (*)	16. (d)	17. (c)	18. (b)	19. (b)	20. (d)
21. (c)	22. (d)	23. (b)	24. (b)	25. (b)	26. (c)	27. (b)	28. (c)	29. (a)	30. (b)
31. (b)	32. (a)	33. (c)	34. (b)	35. (a)	36. (b)	37. (*)	38. (a)	39. (b)	40. (c)
41. (d)	42. (*)	43. (b)	44. (b)	45. (b)	46. (b)	47. (a)	48. (*)	49. (a)	50. (b)
51. (a)	52. (b)	53. (b)	54. (c)	55. (a)	56. (c)	57. (c)	58. (a)	59. (a)	60. (a)
61. (d)	62. (*)	63. (*)	64. (*)	65. (b)	66. (a)	67. (b)	68. (c)	69. (c)	70. (b)
71. (b)	72. (a)	73. (c)							

EXPLANATIONS

1. **Circuitous** means round about or not direct
So circuitous : indirect.
4. Betrayed, **means** 'showed' or revealed.
5. Use of chemical agents in warfare would be undesirable
6. Given relationship is worker: workplace. A gladiator is
   (*i*) a person, usually a professional combatant trained to entertain the public by engaging in mortal combat with another person or a wild.
   (*ii*) A person engaged in a controversy or debate, especially in public.
7. Clues: foreign technocrats did something negatively to the problems – so it is counter-productive to invite them. All other options are non-negative. Best choice is exacerbated which means aggravated or worsened.
8. Best antonym is rarity which means shortage or scarcity.
10. From the passage it cannot be inferred that horses are given immunity as in option (*a*), since the aim is to develop medicine and in turn immunize humans. Option (*b*) is correct since it is given that horses develop immunity after some time. Refer "until their blood built up immunities". Even option(*c*) is invalid since medicine is not built till immunity is developed in the horses. Option (*d*) is incorrect since specific examples are cited to illustrate and this cannot capture the essence.
32. Frogs make 'croak' sound.
33. The word similar in meaning to Educe is Extract.
34. The principal presented the chief guest with a memento, as token of appreciation.
45. For condition regarding something which already happened, should have practiced is the correct choice.
50. Lose (verb)
51. Here 'effectiveness' is noun and 'prescribed' is verb. So these words are apt and befitting with the given word 'medicine'.
55. Social science disciplines had a pre-colonial origin.
57. The meaning of underlined word grasping means clutching (or holding something tightly).
59. Use of conditional sentence based on past participle form.
I could have made arrangements had I been informed earlier.
60. Hurtful means causing pain or suffering or something that is damaging or harmful. The expression 'sharp tongue' defines a bitter or critical manner of speaking.
61. Antagonist is a adversary or one who opposes/contends against another. Hence the closest meaning to word "antagonistic" is hostile.
67. (A) is beyond the scope of given information option (c) can also be discarded on the same grounds.
The argument deals with the coastal region becoming crowded and polluted because of the upcoming luxury resort. Option (b) precisely underlines the theme of the para.
68. Statement (i) is not true as nowhere it is mentioned that John was a captain in junior team. The introductory line emphasizes on the board recognizing John's potential (Latent quality/possibility) as leader of the team.
Statement (iii) also manipulates the facts mentioned in the argument.
The 3rd statement of the argument while this aggression has filled stadia with die-hard fans does not indicate fans expectations from John as a caption.
Statement (ii) The concluding statement of the para suggests that finding a completer captain is a tough task as it look John many years to become a successful and calculative batsman - captain.
Statement (iv) can be explicity concluded from the last 4 lines of the para.
69. ∴ The number of different orders in which P, Q and T arranged = $3! \times 2 \times 1 = 6$

2

CHAPTER

Reasoning Ability

Analysis of Previous GATE Papers												
	Year → / Topics ↓		2019	2018	2017 Set 1	2017 Set 2	2016 Set 1	2016 Set 2	2016 Set 3	2015 Set 1	2015 Set 2	2015 Set 3
Reasoning Ability	1 Mark	MCQ Type			1	2	1	1	2	1		1
		Numerical Type										
	2 Marks	MCQ Type	2		1	1	1	1	1	1		1
		Numerical Type										
	Total		4		3	4	3	3	4	3		3

1. Hari (H), Gita (G), Irfan (I) and Saira (S) are siblings (i.e. brothers and sisters). All were born on Ist January. The age difference between any two successive siblings (that is born one after another) is less than 3 years. Given the following facts:

1. Hari's age + Gita's age > Irfan's age + Saira's age.
2. The age difference between Gita and Saira is 1 year. However, Gita is not the oldest and Saira is not the youngest.
3. There are no twins.

In what order were they born (oldest first)?

(a) HSIG (b) SGHI
(c) IGSH (d) IHSG

[2010 : 2 Marks]

2. What is the next number in the series?

12 35 81 173 357 _____ .

[2014 : 1 Mark, Set-1]

3. Find the odd one from the following group:

WEKO IQWA FNTX NVBD

(a) WEKO (b) IQWA
(c) FNTX (d) NVBD

[2014 : 2 Marks, Set-1]

4. Fill in the missing number in the series.

2 3 6 15 _?_ 157.5 630

[2014 : 1 Mark, Set-2]

5. Find the odd one in the following group

QWZB, BHKM, WCGJ, MSVX,

(a) QWZB (b) BHKM
(c) WCGJ (d) MSVX

[2014 : 2 Marks, Set-2]

6. Lights of four colors (red, blue, green, yellow) are hung on a ladder. On every step of the ladder there are two lights. If one of the lights is red, the other light on that step will always be blue. If one of the lights on a step is green, the other light on that step will always be yellow. Which of the following statements is not necessarily correct?

(a) The number of red lights is equal to the number of blue lights.
(b) The number of green lights is equal to the number of yellow lights.
(c) The sum of the red and green lights is equal to the sum of the yellow and blue lights.
(d) The sum of the red and blue lights is equal to the sum of the green and yellow lights.

[2014 : 2 Marks, Set-2]

7. "India is a country of rich heritage and cultural diversity." Which one of the following facts best supports the claim made in the above sentence?

(a) India is a union of 28 states and 7 union territories.
(b) India has a population of over 1.1 billion.
(c) India is home to 22 official languages and thousands of diaiects
(d) The Indian cricket team draws players from over ten states.

[2014 : 1 Mark, Set-3]

8. The next term in the series 31, 54, 36, 24, ... is _______.

[2014 : 1 Mark, Set-3]

9. In which of the following options will the expression $P < M$ be definitely true?

(a) $M < R > P > S$ (b) $M > S < P < F$
(c) $Q < M < F = P$ (d) $P = A < R < M$

[2014 : 1 Mark, Set-3]

10. Find the next term in the sequence:

7 G, 11 K, 13 M, _______.

(a) 15 Q (b) 17 Q
(c) 15 P (d) 17 P

[2014 : 2 Marks, Set-3]

11. Find the next term in the sequence:

13M, 17Q, 19S, _______.

(a) 21 W (b) 21 V
(c) 23 W (d) 23 V

[2014 : 2 Marks, Set-4]

12. If 'KCLFTSB' stands for 'best of luck' and 'SHSWDG' stands for 'good wishes', which of the following indicates 'ace the exam'?

(a) MCHTX (b) MXHTC
(c) XMHCT (d) XMHTC

[2014 : 2 Marks, Set-4]

13. Operates $\square$, $\diamond$ and $\rightarrow$ are defined by :

$a \square b = \frac{a-b}{a+b}$; $a \diamond b = \frac{a+b}{a-b}$; $a \rightarrow b = ab$. Find the value of $(66 \square 6) \rightarrow (66 \diamond 6)$

(a) –2
(b) –1
(c) 1
(d) 2

[2015 : 1 Mark, Set-1]

14. Fill in the missing value

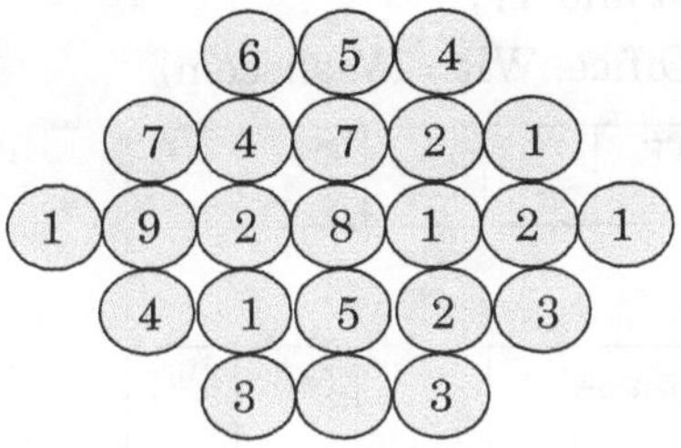

[2015 : 2 Marks, Set-1]

15. Ms. X will be Bagdogra from 01/05/2014 to 20/05/2014 and from 22/05/2014 to 31/05/2014. On the morning of 21/05/2014, she will reach Kochi via Mumbai.

Which one of the statements below is logically valid and can be inferred from the above sentences?

(a) Ms. X will be in Kochi for one day, only in May.

(b) Ms. X will be in Kochi for only one day in May.

(c) Ms. X will be only in Kochi for one day in May.

(d) Only Ms. X will be in Kochi for one day in May.

[2015 : 2 Marks, Set-3]

16. Find the missing sequence in the letter series below:

A, CD, GHI?, UVWXY

(a) LMN (b) MNO

(c) MNOP (d) NOPQ

[2015 : 1 Mark, Set-3]

17. Michael lives 10 km away from where I live. Ahmed lives 5 km away and Susan lives 7 km away from where I live. Arun is farther away than Ahmed but closer than Susan from where I live. From the information provided here, what is one possible distance (in km) at which I live from Arun's place?

(a) 3.00 (b) 4.99

(c) 6.02 (d) 7.01

[2016 : 1 Mark, Set-1]

18. Leela is older than her cousin Pavithra, Pavithra's brother Shiva is older than Leela. When Pavithra and Shiva are visiting Leela, all there like to play chess. Pavithra wins more often than Leela does. Which one of the following statements must be TRUE based on the above?

(a) When Shiva plays chess with Leela and Pavithra, he often loses.

(b) Leela is the oldest of three.

(c) Shiva is better chess player than Pavithra.

(d) Pavithra is the youngest of the three.

[2016 : 2 Marks, Set-1]

19. Based on the given statements, se ect the appropriate option with respect to grammar and usage. Statements

I. The height of Mr. X is 6 feet.

II. The height of Mr. Y is 5 feet.

(a) Mr. Xis longer than Mr. Y.

(b) Mr. X is more elongated than Mr. V.

(c) Mr. X is taller than Mr. Y.

(d) Mr. X is lengthier than Mr. Y.

[2016 : 1 Mark, Set-2]

20. M and N start from the same location. M travels 10 km East and then 10 km North-East. N travels 5 km South and then 4 km south-East. What is the shortest distance (in km)between Mand N at the end of their travel?

(a) 18.60 (b) 22.50

(c) 20.61 (d) 25.00

[2016 : 2 Marks, Set-2]

21. IM has a son O and a daughter R. He has no other children. Eis the mother of Pand daughter-in-law of M. How is P related to M?

(a) P is the son-in-law of M

(b) P is IIle grandchild of M

(c) P is the daughter-in law of M

(d) P is the grandfather of M

[2016 : 1 Mark, Set-3]

22. The number that least fits this set: (324, 441, 97 and 64)is ______.

(a) 324 (b) 441

(c) 97 (d) 64

[2016 : 1 Mark, Set-3]

23. A flat is shared by four first year undergraduate students. They agreed to allow the oldest of them to enjoy some extra space in the flat. Manu is two months older than Sravan, who is three months younger than Trideep. Pavan is one month older than Sravan. Who should occupy the extra space in the flat?

(a) Manu (b) Sravan

(c) Trideep (d) Pavan

[2016 : 2 Marks, Set-3]

24. Some tables are shelves. Some shelves are chairs. All chairs are benches. Which of the following conclusions can be deduced from the preceding sentences?

(i) At least one bench is a table.

(ii) At least one shelf is a bench.

(iii) At least one chair is a table.

(iv) All benches are chairs.

(a) only (i) (b) only (ii)
(c) only (ii) and (iii) (d) only (iv)

[2017 : 1 Mark, Set-1]

25. S, T, U, V, W, X, Y and Zare seated around a circular table. 'T's neighbours are Y and V. Z is seated third to the left of T and second to the right of S. U's neighbours are S and Y; and T and W are not seated opposite each other. Who is third to the left of V?

(a) X (b) W
(c) U (d) T

[2017 : 2 Marks, Set-1]

26. A rule states that in order to drink beer, one must be over 18 years old. In a bar there are 4 people. Pis 16years old,Q is 25 years old, R is drinking milkshake and S is drinking a beer. What must be checked to ensure that the rule is being followed?

(a) Only P's drink
(b) Only P's drink and S's age
(c) Only S's age
(d) Only P's drink. Q's drink and S's age

[2017 : 1 Mark, Set-2]

27. Fatima starts from point P, goes North for 3 km and then East for 4 km to reach point Q. She then turns to face point P and goes 15 km in that direction. She then goes North for 6 km. How far is she from point Pand in which direction should she go to reach point P?

(a) 8 km, East (b) 12 km, North
(c) 6 km, East (d) 10 km, North

[2017 : 1 Mark, Set-2]

28. Each of P, Q, R, S, W, X, Y and Z has been married at most once. Xand Yare married and have two children Pand Q. Z is the grandfather of the daughter S of P. Further, Z and W are married and are parents of P. Which one of the following must necessarily be FALSE?

(a) X is the mother-in-law of P.
(b) Pand R are net married to each other.
(c) Pisa son of X and Y.
(d) Q cannot be married to P.

[2017 : 2 Marks, Set-2]

29. Five people P, Q, R, S and T work in a bank. P and Q don't like each other but have to share an office till T gets a promotion and moves to the big office next to the garden. R, who is currently sharing an office with T wants to move to the adjacent office with S, the handsome new intern.

Given the floor plan, what is the current location of Q, R and T?

(O = Office, WR = Washroom)

(a)
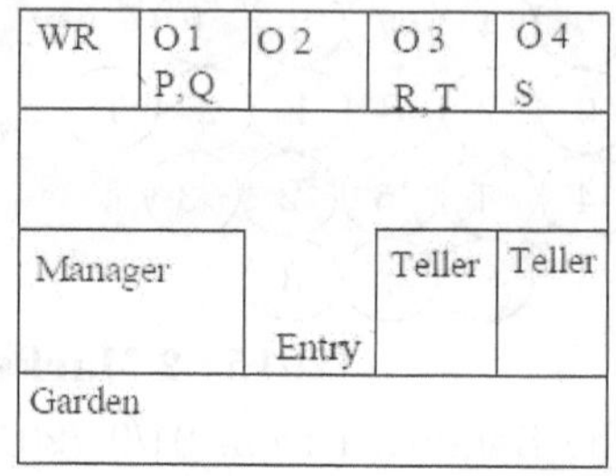

(b)
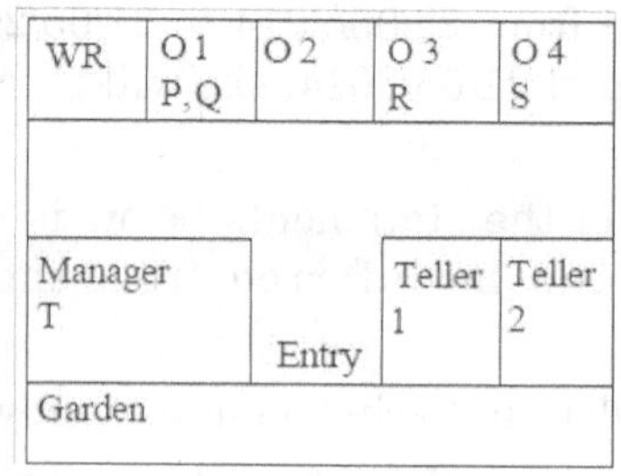

(c)
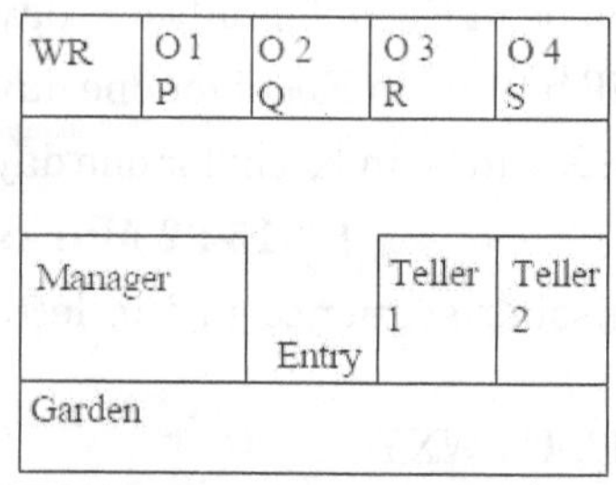

(d)
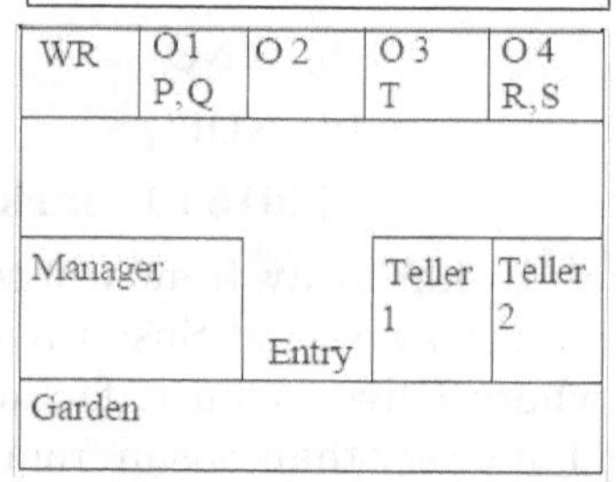

[2019 : 2 Marks]

30. Four people are standing in a line facing you. They are Rahul, Mathew, Seema and Lohit. One is an engineering, one is a doctor, one a teacher and another a dancer. You are told that:

1. Mathew is not standing next to Seema
2. There are two people standing beween Lohit and the engineer
3. Rahul is not a doctor
4. The teacher and the dancer are standing next to each other.
5. Seema is turning to her right to speak to the doctor standing next to her.

Who amongst them is an engineer?

(a) Rahul (b) Mathew
(c) Seema (d) Lohit

[2019 : 2 Marks]

ANSWERS

1. (b)	2. (725 to 725)	3. (d)	4. (45)	5. (*)	6. (d)	7. (c)	8. (16)	9. (d)
10. (b)	11. (c)	12. (b)	13. (c)	14. (3)	15. (b)	16. (c)	17. (c)	18. (d)	19. (c)
20. (c)	21. (b)	22. (c)	23. (c)	24. (b)	25. (a)	26. (*)	27. (*)	28. (*)	29. (a)
30. (b)									

EXPLANATIONS

1. $H + G > I + S$ ...(1)

and $G - S = 1$ ...(2)

G is not oldest, S is not youngest

$\therefore$ $H + 1 > I$

Irfan older than Hari

Gita older than Sarita

From given option SGHI

2. The given series is

12, 35, 81, 173, 357, ...

The given series follows the following pattern

$12 \times 2 + 11 = 35$

$35 \times 2 + 11 = 81$

$81 \times 2 + 11 = 173$

$173 \times 2 + 11 = 357$

$357 \times 2 + 11 = 725$

3. 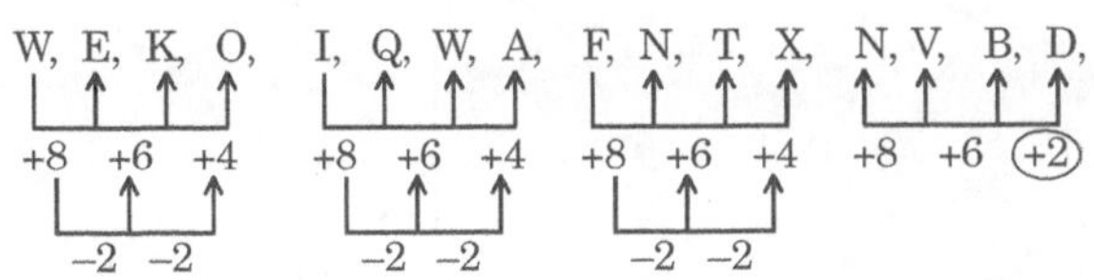

Hence the odd one from the following group is N, V, B, D.

4. 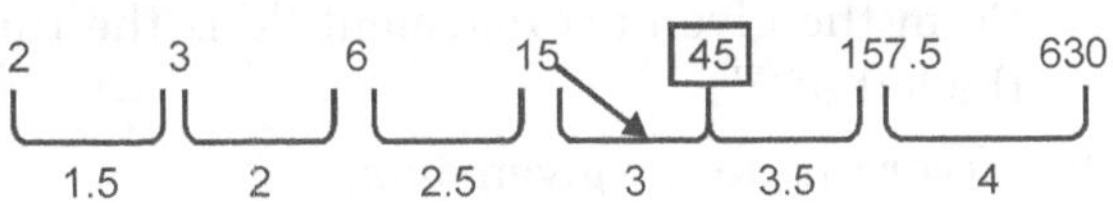

$\frac{\text{2nd number}}{\text{1st number}}$ is in increasing order as shown above

7. Diversity is shown in terms of difference language.

8. $81 - 54 = 27;\ 27 \times \frac{2}{3} = 18$

$54 - 36 = 18;\ 18 \times \frac{2}{3} = 12$

$36 - 24 = 12;\ 12 \times \frac{2}{3} = 8$

$\therefore$ $24 - 8 = 16$

12. KCLFTSB: BST-Best, F-Of, LCK-Luck (Reverse order)

SHSWDG: GD-Good, WSHS-Wishes (Reverse order)

Similarly "ace the Exam'- C-Ace, T-The, XM-Exam

13. $66\square 6 = \frac{66-6}{66+6} = \frac{60}{72} = \frac{5}{6}$

$66\diamond 6 = \frac{66+6}{66-6} = \frac{72}{60} = \frac{6}{5}$

$(66\square 6) \rightarrow (66\diamond 6) = \frac{5}{6} \times \frac{6}{5} = 1$

14. Middle number is the average of number on both sides

$\therefore$ Average of 3 and 3 is $\frac{3+3}{2} = \frac{6}{2} = 3$

15. Second sentence says that Ms. X reaches Kochi on 21/05/2014. Also she has to be in Bagdogora on 22/05/2014.

$\therefore$ She stays in Kochi for only one day in may.

16. 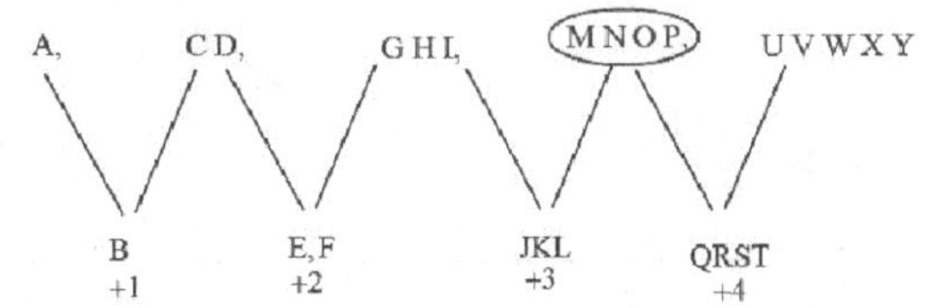

17. From given data, the following diagram can be possible.

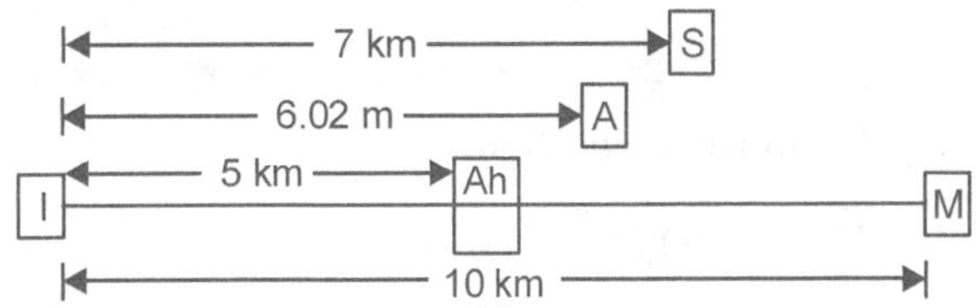

Here S → Susan

A → Arun

Ah → Ahmed

M → Michael

From the above diagram, Arun lives farthest away than Ahmed means more than 5 km but closer than Susan means less than 7 km, from the given alternatives.

So, only option 'c' is possible.

18. From the given question two statements will be followed.

For statment I

Arrange the given data according to their ages

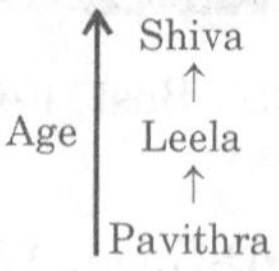

For statement II

Arrange the given data according to their winning

So, from statement I and II, it is clear that only option (*d*) is possible (i.e., statement I).

19.

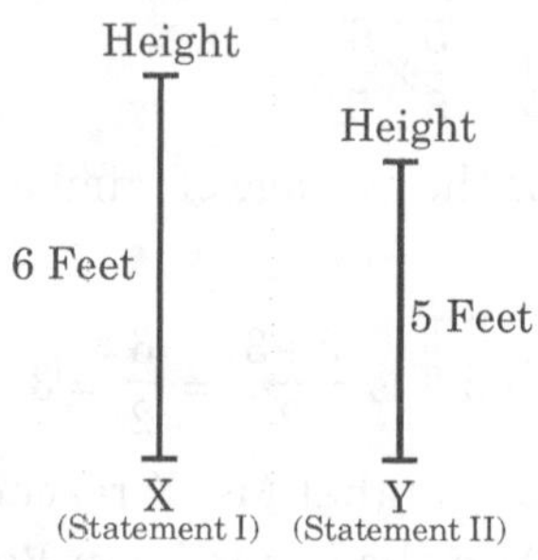

Hence from the given figure Mr. X is taller than Mr. Y by 1 foot.

20.

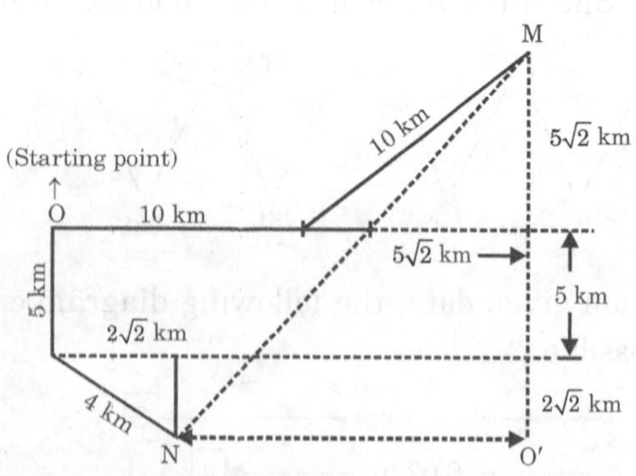

From the given figure

$$MN = \sqrt{(O'M)^2 + (O'N)^2}$$

$$O'M = 5\sqrt{2} + 5 + 2\sqrt{2} = 5 + 7\sqrt{2}$$

$$O'N = 10 + 5\sqrt{2} - 2\sqrt{2} = 10 + 3\sqrt{2}$$

$$MN = \sqrt{(5+7\sqrt{2})^2 + (10+3\sqrt{2})^2}$$

$$= \sqrt{25 + 98 + 70\sqrt{2} + 100 + 18 + 60\sqrt{2})}$$

$$\approx 20.61 \text{ km}$$

21.

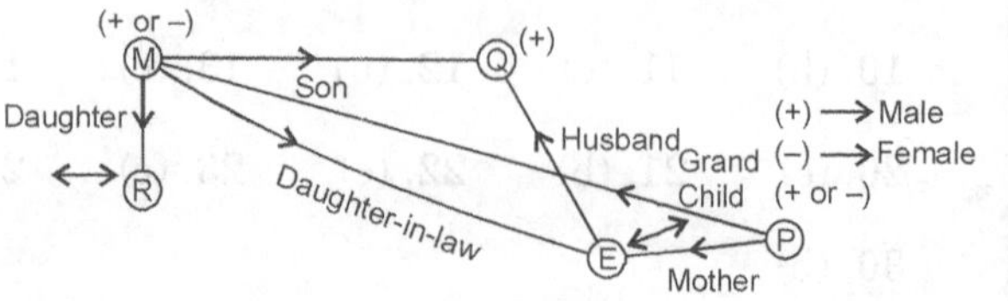

So, from the above relation diagram it is clear that P is the grandchild of M.

23. Manu age = sravan age + 2 months

Manu age = Trideep age – 3 months

Pavan age = Sravan's age + 1 month

From the above statement

Trideep age > Man > Pavan > Sravan

Hence, Trideep can occupy the extra space in the flat.

24. From given condition

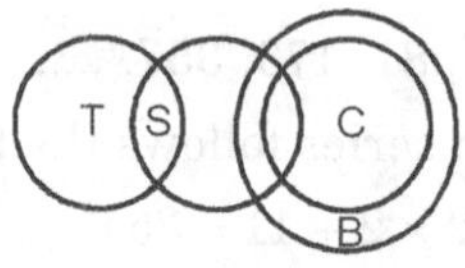

Only conclusion (*ii*) follows

25. Following circular seating arrangement can be drawn from the given data

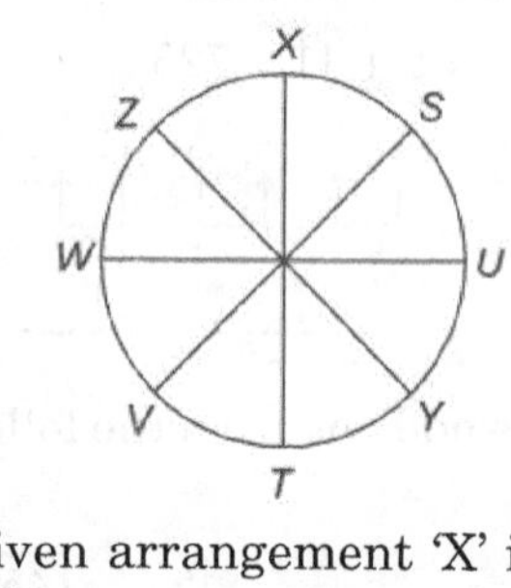

From the given arrangement 'X' is the third to the left of 'V'.

30. According to the given data;

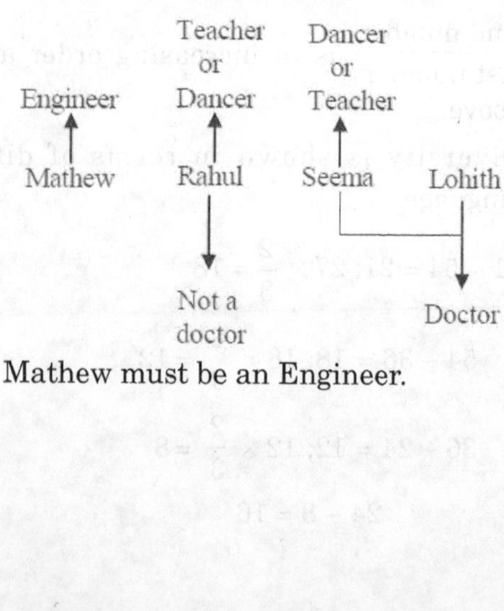

∴ Mathew must be an Engineer.

3

CHAPTER

Numerical Ability

Analysis of Previous GATE Papers												
	Year → Topics ↓		2019	2018	2017 Set 1	2017 Set 2	2016 Set 1	2016 Set 2	2016 Set 3	2015 Set 1	2015 Set 2	2015 Set 3
Numerical Ability	1 Mark	MCQ Type	1	3	2	1	1	2	1	1	2	1
		Numerical Type										
	2 Marks	MCQ Type	2	3	3	3	3	3	3	1	2	2
		Numerical Type										
	Total		5	9	8	7	7	8	7	3	6	5

1. 25 persons are in a room. 15 of them play hockey, 17 of them play football and 10 of them play both hockey and football. Then the number of persons playing neither hockey nor football is

(a) 2 (b) 17

(c) 13 (d) 3

[2010 : 1 Mark]

2. If 137 + 276 = 435, how much is 731 + 672?

(a) 534 (b) 1403

(c) 1623 (d) 1531

[2010 : 2 Marks]

3. 5 skilled workers can build a wall in 20 days; 8 semiskilled workers can build a wall in 25 days; 10 unskilled workers can build a wall in 30 days. If a team has 2 skilled, 6 semiskilled and 5 unskilled workers, how long will it take to build the wall?

(a) 20 days (b) 18 days

(c) 16 days (d) 15 days

[2010 : 2 Marks]

4. Given digits 2, 2, 3, 3, 3, 4, 4, 4, 4 how many distinct 4 digit numbers greater than 3000 can be formed?

(a) 50 (b) 51

(c) 52 (d) 54

[2010 : 2 Marks]

5. There are two candidates P and Q in an election. During the campaign 40% of the voters promised to vote for P, and rest for Q. However, on the day of election 15% of the voters went back on their promise to vote for P and instead voted for Q. 25% of the voters went back on their promise to vote for Q and instead voted for P. Suppose, P lost by 2 votes, then what was the total number of voters?

(a) 100 (b) 110

(c) 90 (d) 95

[2011 : 1 Mark]

6. The fuel consumed by a motorcycle during a journey while travelling at various speeds is indicated in the graph below

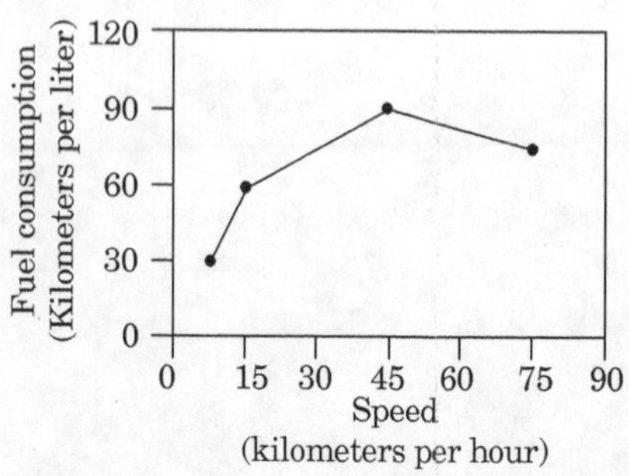

The distance covered during four laps of the journey are listed in the table below

Lap	Distance (kilometers)	Average speed (kilometers per hour)
P	15	15
Q	75	45
R	40	75
S	10	10

From the given data, we can conclude that the fuel consumed per kilometre was least during the lap

(a) P (b) Q

(c) R (d) S

[2011 : 2 Marks]

7. Three friends, R, S and T shared toffee from a bowl, R took $\frac{1}{3}$ rd of the toffees, but returned to the bowl. S took $\frac{1}{4}$ th of what was left but returned three toffees to the bowl. T took half of the remainder but returned two back into the bowl. If the bowl had 17 toffees left, how many toffees were originally there in the bowl?

(a) 38 (b) 31

(c) 48 (d) 41

[2011 : 2 Marks]

8. Given that $f(y)=\frac{|y|}{y}$, and q is any non-zero real number, the value of $|f(q)-f(-q)|$ is

(a) 0 (b) -1

(c) 1 (d) 2

[2011 : 2 Marks]

9. The sum of n terms of the series 4 + 44 + 444 + ... is

(a) $\left(\frac{4}{81}\right)[10^{n+1}-9n-1]$

(b) $\left(\frac{4}{81}\right)[10^{n-1}-9n-1]$

(c) $\left(\frac{4}{81}\right)[10^{n+1}-9n-10]$

(d) $\left(\frac{4}{81}\right)[10^{n}-9n-10]$

[2011 : 2 Marks]

10. If $(1.001)^{1259} = 5.52$ $(1.001)^{2062} = 7.85$, then $(1.001)^{3321}$

(a) 2.23 (b) 4.33

(c) 11.37 (d) 27.64

[2012 : 1 Mark]

11. The data given in the following table summarizes the monthly budget of an average household.

Category	**Amount**
Food	4000
Clothing	1200
Rent	2000
Savings	1500
Others	1800

The approximate percentage of the monthly budget NOT spent on savings is

(a) 10% (b) 14%

(c) 81% (d) 86%

[2012 : 2 Marks]

12. A and B friends. They decide to meet between 1 PM and 2 PM on a given day. There is a condition that whoever arrives first will not wait for the other for more than 15 minutes. The probability that they will meet on that day is

(a) $\frac{1}{4}$ (b) $\frac{1}{16}$

(c) $\frac{7}{16}$ (d) $\frac{9}{16}$

[2012 : 2 Marks]

13. Raju has 14 currency notes in his pocket consisting of only Rs. 20 notes and Rs. 10 notes. The total money value of the notes is Rs. 230. The number of Rs. 10 notes that Raju has is

(a) 5 (b) 6

(c) 9 (d) 10

[2012 : 2 Marks]

14. There are eight bags of rice looking alike, seven of which have equal and one is slightly heavier. The weighing balance is of unlimited capacity. Using this balance the minimum number of weighting required to identify the heavier bag is

(a) 2 (b) 3

(c) 4 (d) 8

[2012 : 2 Marks]

15. In the summer of 2012, in New Delhi, the mean temperature of Monday to Wednesday was 41°C and of Tuesday to Thursday was 43°C. If the temperature on Thursday was 15% higher than that of Monday, then the temperature in °C on Thursday was

(a) 40 (b) 43

(c) 46 (d) 49

[2013 : 1 Mark]

16. A car travels 8 km in the first quarter of an hour, 6 km in the second quarter arid 16 km in the third quarter. The average speed of the car in km per hour over the entire journey is

(a) 30 (b) 36

(c) 40 (d) 24

[2013 : 2 Marks]

17. Find the sum to n terms of the series

10 + 84 + 734 + ...

(a) $\frac{9(9^n+1)}{10}+1$ (b) $\frac{9(9^n-1)}{8}+1$

(c) $\frac{9(9^n-1)}{8}+n$ (d) $\frac{9(9^n-1)}{8}+n^2$

[2013 : 2 Marks]

18. The set of values of p for which the roots of the equation $3x^2 + 2x + p(p - 1) = 0$ are of opposite sign is

(a) $(\infty, 0)$ (b) $(0, 1)$

(c) $(1, \infty)$ (d) $(0, \infty)$

[2013 : 2 Marks]

19. What is the change that a leap year, selected at random, will contain 53 Saturdays?

(a) $\frac{2}{7}$ (b) $\frac{3}{7}$

(b) $\frac{1}{7}$ (d) $\frac{5}{7}$

[2013 : 2 Marks]

20. The statistics of runs scored in a series by four batsmen are provided in the following table. Who is the most consistent batsman of these four?

Batsman	**Average**	**Standard Deviation**
K	31.2	5.21
L	46.0	6.35
M	54.4	6.22
N	17.9	5.90

(a) K (b) L

(c) M (d) N

[2014 : 1 Mark, Set-1]

21. For submitting tax returns, all resident males with annual income below Rs.10 lakh should fill up Form P and all resident females with income below Rs. 8 lakh should fill up Form O. All people with incomes above Rs.10 lakh should fill up Form R, except non residents with income above Rs. 15 lakhs, who should fill up Form S. All others should fill Form T. An example of a person who should fill Form 7 is

(a) a resident male with annual income Rs.9 lakh.

(b) a resident female with annual income Rs.9 lakh.

(c) a non-resident male with annual income Rs. 16 lakh.

(d) a non-resident female with annual income Rs. 16 lakh.

[2014 : 2 Marks, Set-1]

22. A train that is 280 metres long, travelling at a uniform speed, crosses a platform in 60 seconds and passes a man standing on the platform in 20 seconds. What is the length of the platform in metres?

[2014 : 2 Marks, Set-1]

23. The exports and imports (in crores of Rs.) of a country from 2000 to 2007 are given in the following bar chart. If the trade deficit is defined as excess of imports over exports, in which year is the trade deficit 1/5th of the exports?

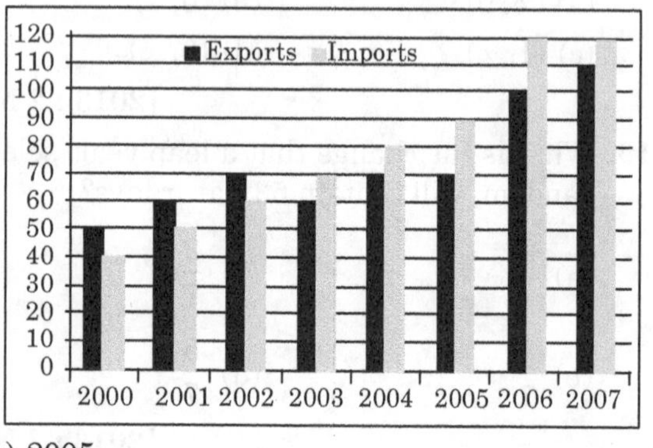

(a) 2005

(b) 2004

(c) 2007

(d) 2006

[2014 : 2 Marks, Set-1]

24. You are given three coins: one has heads on both faces, the second has tails on both faces, and the third has a head on one face and a tail on the other. You choose a coin at random and toss it, and it comes up heads. The probability that the other face is tails is

(a) $\frac{1}{4}$

(b) $\frac{1}{3}$

(c) $\frac{1}{2}$

(d) $\frac{2}{3}$

[2014 : 2 Marks, Set-1]

25. A regular die has six sides with numbers 1 to 6 marked on its sides. If a very large number of throws show the following frequencies of occurrence:

$1 \rightarrow 0.167$; $2 \rightarrow 0.167$; $3 \rightarrow 0.152$; $4 \rightarrow 0.166$; $5 \rightarrow 0.168$; $6 \rightarrow 0.180$.

We call this die

(a) irregular (b) biased

(c) Gaussian (d) insufficient

[2014 : 1 Mark, Set-2]

26. The sum of eight consecutive odd numbers is 656. The average of four consecutive even numbers is 87. What is the sum of the smallest odd number and second largest even number?

[2014 : 2 Marks, Set-2]

27. The total exports and revenues from the exports of a country are given in the two charts shown below. The pie chart for exports shows the quantity of each item exported as a percentage of the total quantity of exports. The pie chart for the revenues shows the percentage of the total revenue generated through export of each item. The total quantity of exports of all the items is 500 thousand tonnes and the total revenues are 250 crore rupees. Which item among the following has generated the maximum revenue per kg?

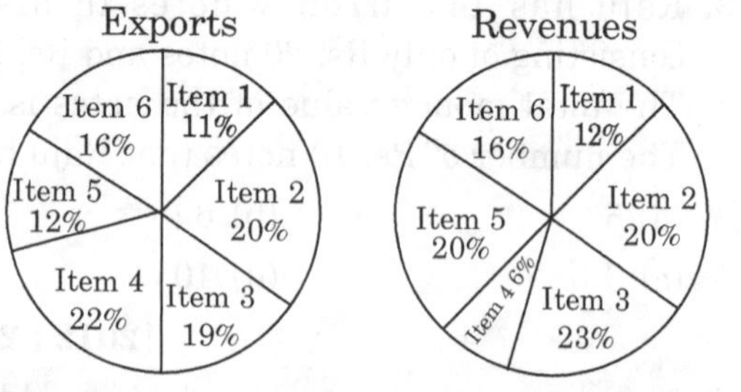

(a) Item 2

(b) Item 3

(c) Item 6

(d) Item 5

[2014 : 2 Marks, Set-2]

28. It takes 30 minutes to empty a half-full tank by draining it at a constant rate. It is decided to simultaneously pump water into the half-full tank while draining it. What is the rate at which water has to be pumped in so that it gets fully filled in 10 minutes?

(a) 4 times the draining rate

(b) 3 times the draining rate

(c) 2.5 times the draining rate

(d) 2 times the draining rate

[2014 : 2 Marks, Set-2]

29. The multi-level hierarchical pie chart shows the population of animals in a reserve forest. The correct conclusions from this information are:

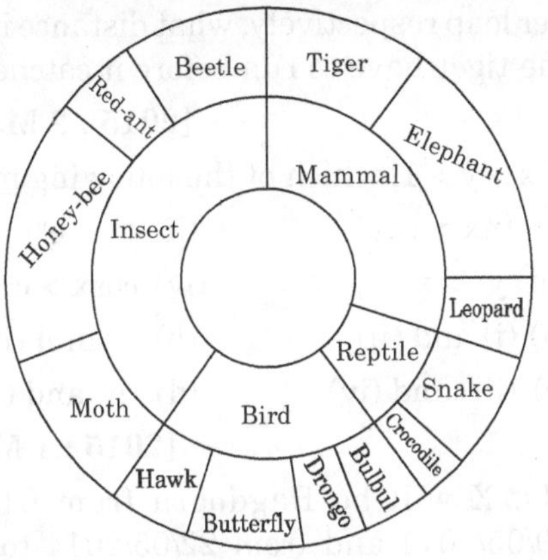

(i) Butterflies are birds

(ii) There are more tigers in this forest than red ants.

(iii) All reptiles in this forest are either snakes or crocodiles.

(iv) Elephants are the largest mammals in this forest.

(a) (i) and (ii) only

(b) (i), (ii), (iii) and (iv)

(c) (i), (iii) and (iv) only

(d) (i), (ii) and (iii) only

[2014 : 2 Marks, Set-3]

30. A man can row at 8 km per hour in still water. If it takes him thrice as long to row upstream, as to row downstream, then find the stream velocity in km per hour. **[2014 : 2 Marks, Set-3]**

31. A firm producing air purifiers sold 200 units in 2012. The following pie chart presents the share of raw material, labour, energy, plant & machinery, and transportation costs in the total manufacturing cost of the firm in 2012. The expenditure on labour in 2012 is Rs. 4,50,000. In 2013, the raw material expenses increased by 30% and all other expenses increased by 20%. If the company registered a profit of Rs. 10 lakhs in 2012, at what price (in Rs.) was each air purifier sold?

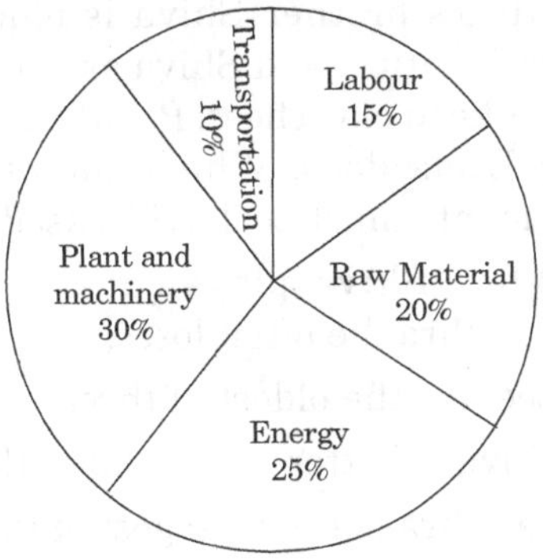

[2014 : 2 Marks, Set-3]

32. A batch of one hundred bulbs is inspected by testing four randomly chosen bulbs. The batch is rejected if even one of the bulbs is defective. A batch typically has five defective bulbs. The probability that the current batch is accepted is ______.

[2014 : 2 Marks, Set-3]

33. Let $f(x, y)= x^n y^n = P$. If x is doubled and y is halved, the new value of f is

(a) $2^{n-m}P$ (b) $2^{m-n}P$

(c) $2(n - m)P$ (d) $2(m - n)P$

[2014 : 1 Mark, Set-4]

34. In a sequence of 12 consecutive odd numbers, the sum of the first 5 numbers is 425. What is the sum of the last 5 numbers in the sequence?

[2014 : 1 Mark, Set-4]

35. Industrial consumption of power doubled from 2000- 2001 to 2010-2011. Find the annual rate of increase in percent assuming it to be uniform over the years.

(a) 5.6 (b) 7.2

(c) 10.0 (d) 12.2

[2014 : 2 Marks, Set-4]

36. A firm producing air purifiers sold 200 units in 2012. The following pie chart presents the share of raw material, labour, energy, plant & machinery, and transportation costs in the total manufacturing cost of the firm in 2012. The expenditure on labour in 2012 is Rs.4,50,000. In 2013, the raw material expenses increased by 30% and all other expenses increased by 20%. What is the percentage increase in total cost for the company in 2013?

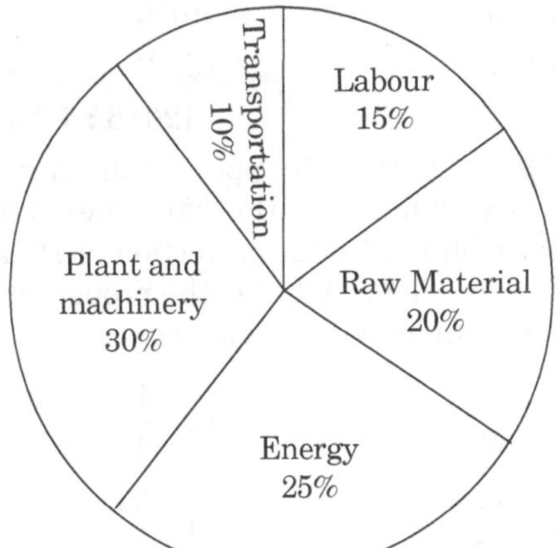

[2014 : 2 Marks, Set-4]

37. A five digit number is formed using the digits 1, 3, 5, 7 and 9 without repeating any of them. What is the sum of all such possible five digit numbers?

(a) 6666660 (b) 6666600

(c) 6666666 (d) 6666606

[2014 : 2 Marks, Set-4]

38. If $\log_x\left(\frac{5}{7}\right) = -\frac{1}{3}$, then the value of x is

(a) $\frac{343}{125}$ (b) $\frac{125}{343}$

(c) $-\frac{25}{49}$ (d) $-\frac{49}{25}$

[2015 : 1 Mark, Set-1]

39. A cube side 3 units is formed using a set of smaller cubes of side 1 unit. Find the proportion of the number of faces of the smaller cubes visible to those which are NOT visible.

(a) 1 : 4 (b) 1 : 3

(c) 1 : 2 (d) 2 : 3

[2015 : 2 Marks, Set-1]

40. An electric bus has onboard instruments that report the total electricity consumed since the start of the trip as well as the total distance covered. During a single day of operation, the bus travels on stretches M, N, O and P, in that order. The cumulative distance travelled and the corresponding electricity consumption are shown in the table below:

Stretch	Cumulative distance (km)	Electricity used (kWh)
M	20	12
N	45	25
O	75	45
P	100	57

The stretch where the electricity consumption per km is minimum is

(a) M (b) N

(c) O (d) P

[2015 : 1 Mark, Set-2]

41. Ram and Ramesh appeared in an interview for two vacancies in the same department. The probability of Ram's selection is 1/6 and that of Ramesh is 1/8. What is the probability that only one of them will be selected?

(a) $\frac{47}{48}$ (b) $\frac{1}{4}$

(c) $\frac{13}{48}$ (d) $\frac{35}{48}$

42. If $a^2 + b^2 + c = 1$, then ab + bc + ac lies in the interval

(a) $\left[1, \frac{2}{3}\right]$ (b) $\left[-\frac{1}{2}, 1\right]$

(c) $\left[-\frac{1}{2}, 1\right]$ (d) [2, –4]

[2015 : 2 Mark, Set-2]

43. A tiger is 50 leaps of its own behind a deer. The tiger takes 5 leaps per minute to the deer's 4. If the tiger and the deer cover 8 metre and 5 metre per leap respectively, what distance in metres will the tiger have to run before it catches the deer?

[2015 : 2 Marks, Set-2]

44. If x > y > 1, which of the following <u>must</u> be true?

(i) $\ln x > \ln y$ (ii) $e^x > e^y$

(iii) $y^x > x^y$ (iv) cosx > cosy

(a) (i) and (ii) (b) (i) and (iii)

(c) (iii) and (iv) (d) (ii) and (iv)

[2015 : 1 Mark, Set-3]

45. Ms. X will be Bagdogra from 01/05/2014 to 20/05/2014 and from 22/05/2014 to 31/05/2014. On the morning of 21/05/2014, she will reach Kochi via Mumbai.

Which one of the statements below is logically valid and can be inferred from the above sentences?

(a) Ms. X will be in Kochi for one day, only in May.

(b) Ms. X will be in Kochi for only one day in May.

(c) Ms. X will be only in Kochi for one day in May.

(d) Only Ms. X will be in Kochi for one day in May.

[2015 : 2 Marks, Set-3]

46. Find the missing sequence in the letter series below:

A, CD, GHI?, UVWXY

(a) LMN (b) MNO

(c) MNOP (d) NOPQ

[2015 : 1 Mark, Set-3]

47. Michael lives 10 km away from where I live. Ahmed lives 5 km away and Susan lives 7 km away from where I live. Arun is farther away than Ahmed but closer than Susan from where I live. From the information provided here, what is one possible distance (in km) at which I live from Arun's place?

(a) 3.00 (b) 4.99

(c) 6.02 (d) 7.01

[2016 : 1 Mark, Set-1]

48. Leela is older than her cousin Pavithra, Pavithra's brother Shiva is older than Leela. When Pavithra and Shiva are visiting Leela, all there like to play chess. Pavithra wins more often than Leela does. Which one of the following statements must be TRUE based on the above?

(a) When Shiva plays chess with Leela and Pavithra, he often loses.

(b) Leela is the oldest of three.

(c) Shiva is better chess player than Pavithra.

(d) Pavithra is the youngest of the three.

[2016 : 2 Marks, Set-1]

49. If $q^{-a} = \frac{1}{r}$ and $r^{-b} = \frac{1}{s}$ and $s^{-c} = \frac{1}{q}$, the value of abc is ______.

(a) $(rqs)^{-1}$ (b) 0

(c) 1 (d) r + q + s

[2016 : 2 Marks, Set-1]

50. P, Q, R and S are working on a project. Q can finish the task in 25 days, working alone for 12 hours a day. Pcan finish the task in 50 days, working alone for 12 hours per day. Q worked 12 hours a day but took sick leave in the beginning for two days. R worked 18 hours a day on all days. What is the ratio of work done by Q and R after 7 days from the start of the projects?

(a) 10:11 (b) 11:10

(c) 20:21 (d) 21:20

[2016 : 2 Marks, Set-1]

51. Given (9 inches)$^{1/2}$ = (0.25 yards)$^{1/2}$, which one of the following statements is TRUE?

(a) 3 inches = 0.5 yards

(b) 9 inches = 1.5 yards

(c) 9 inches = 0.25 yards

(d) 81 inches = 0.0625 yards

[2016 : 1 Mark, Set-2]

52. S.M.E and Fare working in shifts in a team to finish a project. M works with twice the efficiency of others but for half as many days as E worked. S and M have 6 hour shifts in a day. whereas E and Fhave 12 hours shifts. What is the ratio of contribution of M to contribution of E in the project?

(a) 1 : 1 (b) 1 : 2

(c) 1 : 4 (d) 2 : 1

[2016 : 1 Mark, Set-2]

53. The Venn diagram shows the preference of the student population for leisure activities.

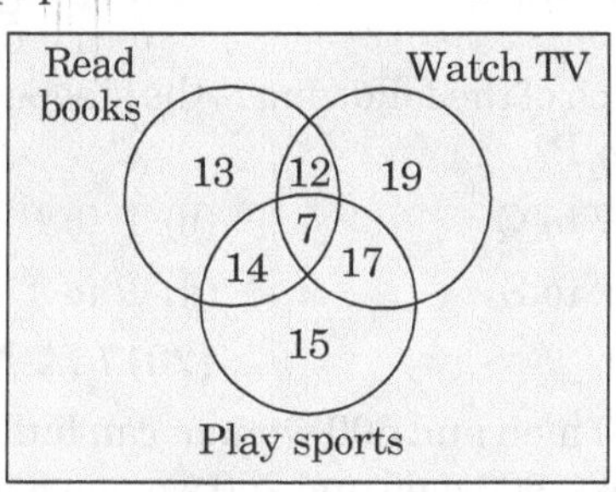

From the data given, the number of students who like to read books or play sports is ______.

(a) 44 (b) 51

(c) 79 (d) 108

[2016 : 2 Marks, Set-2]

54. Two and a quarter hours back, when seen in a mirror, the reflection of a wall clock without number markings seemed to show 1 : 30. What is the actual current time shown by the clock?

(a) 8:15 (b) 11:15

(c) 12:15 (d) 12:45

[2016 : 2 Marks, Set-2]

55. A wire of length 340 mm is to be cut into two parts. One of the parts is to be made into a square and the other into a rectangle where sides are in the ratio of 1:2. What is the length of the side of the square (in mm)such that the combined area of the square and the rectangle is a MINIMUM?

(a) 30 (b) 40

(c) 120 (d) 180

[2016 : 2 Marks, Set-2]

56. It takes 10s and 15s, respectively, for two trains travelling at different constant speeds to completely pass a telegraph post. The length of the first train is 120 m and that of the second train is 150 m. The magnitude of the difference in the speeds of the two trains (in m/s) is ______.

(a) 2.0 (b) 10.0

(c) 12.0 (d) 22.0

[2016 : 1 Mark, Set-3]

57. The velocity V of a vehicle along a straight line is measured in m/s and plotted as shown with respect to time in seconds. At the end of the 7 seconds, how much will the odometer reading increase by (in m)?

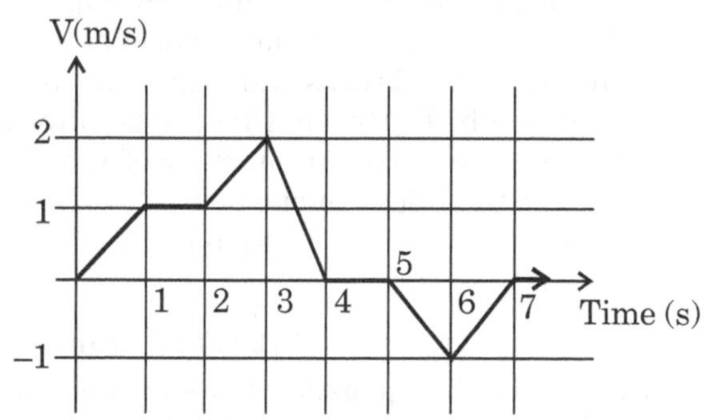

(a) 0 (b) 3

(c) 4 (d) 5

[2016 : 2 Marks, Set-3]

58. Find the area bounded by the lines 3x + 2y = 14, 2x − 3y = 5 in the first quadrant.

(a) 14.95 (b) 15.25

(c) 15.70 (d) 20.35

[2016 : 2 Marks, Set-3]

59. A straight line is fit to a data set (In x, y). This line intercepts the abscissa at In x = 0.1 and has a slope of -0.02. What is the value of y at x = 5 from the fit?

(a) –0.030 (b) –0.014
(c) 0.014 (d) 0.030

[2016 : 2 Marks, Set-3]

60. In the summer, water consumption is known to decrease overall by 25%. A Water Board official states that in the summer household consumption decreases by 20%, while other consumption increases by 70%.
Which of the following statements is correct?

(a) The ratio of household to other consumption is $\frac{8}{17}$.
(b) The ratio of household to other consumption is $\frac{1}{17}$.
(c) The ratio of household to other consumption is $\frac{17}{8}$.
(d) There are errors in the official's statement.

[2017 : 1 Mark, Set-1]

61. 40% of deaths on city roads may be attributed to drunken driving. The number of degrees needed to represent this as a slice of a pie-chart is

(a) 120 (b) 144
(c) 160 (d) 212

[2017 : 1 Mark, Set-1]

62. Trucks (10 m long)and cars (5 m long) go on a single lane bridge. There must be a gap of at least 20 m after each truck and a gap of at least 15 m after each car. Trucks and cars travel at a speed of 36 km/h. If cars and trucks go alternately. What is the maximum number of vehicles that can use the bridge in one hour?

(a) 1440 (b) 1200
(c) 720 (d) 600

[2017 : 2 Marks, Set-1]

63. A contour line joins locations having the same height above the mean sea level. The following is a contour plot of a geographical region. Contour lines are shown at 25 m intervals in this plot.

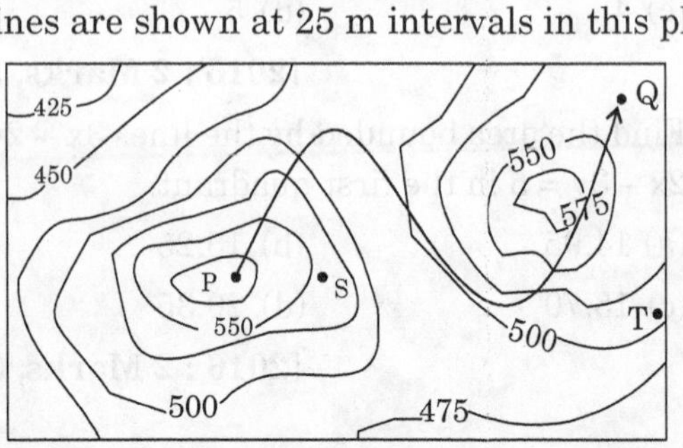

The path from P to Q is best described by

(a) P to Q (b) P to R
(c) P to S (d) P to T

[2017 : 2 Marks, Set-1]

64. There are 3 Indians and 3 Chinese in a group of 6 people. How many subgroups of this group can we choose so that every subgroup has at least one Indian?

(a) 56 (b) 52
(c) 48 (d) 44

[2017 : 2 Marks, Set-1]

65. 500 students are taking one or more courses out of Chemistry, Physics and Mathematics. Registration records indicate course enrolment as follows: Chemistry (329), Physics (186), Mathematics (295), Chemistry and Physics (83), Chemistry and Mathematics (217) and Physics and Mathematics (63). How many students are taking all 3 subjects?

(a) 37 (b) 43
(c) 47 (d) 53

[2017 : 1 Mark, Set-2]

66. A contour line joins locations having the same height above the mean sea level. The following is a contour plot of a geographical region. Contour lines are shown at 25 m intervals in this plot.

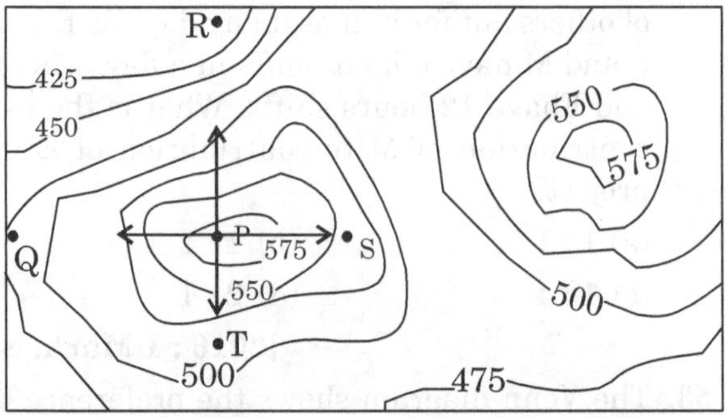

Which of the following is the steepest path leaving from P?

(a) P to Q (b) P to P
(c) P to S (d) P to T

[2017 : 2 Marks, Set-2]

67. 1200 men and 500 women can build a bridge in 2 weeks, 900 men and 250 women will take 3 weeks to build the same bridge. How many men will be needed to build ths bridge in one week?

(a) 3000 (b) 3300
(c) 3600 (d) 3900

[2017 : 2 Marks, Set-2]

68. The number of 3-digit numbers such that the digit 1 is never to the immediate right of 2 is

(a) 781 (b) 791

(c) 881 (d) 891

[2017 : 2 Marks, Set-2]

69. What is the value of $1+\frac{1}{4}+\frac{1}{16}+\frac{1}{64}+\frac{1}{256}+...?$

(a) 2 (b) $\frac{7}{4}$

(c) $\frac{3}{2}$ (d) $\frac{4}{3}$

[2018 : 1 Mark]

70. A 1.5 m in tall person is standing at a distance of 3 m from a lamp post. The light from the lamp at the top of the post casts her shadow. The length of the shadow is twice her height. What is the height of the lamp post in meters?

(a) 1.5 (b) 3

(c) 4.5 (d) 6

[2018 : 1 Mark]

71. If the number 715 ? 423 is divisible 3 (? denotes the missing digit in the thousandths place), then the smallest whole number iri [tie place of ? is ______.

(a) 0 (c) 5

(b) 2 (d) 6

[2018 : 1 Mark]

72. Two alloys A and B contain gold and copper in the ratios of 2 : 3 and 3 : 7 by mass, respectively. Equal masses of alloys A and B are melted to make an alloy C. The ratio of gold to copper in alloy C is ______.

(a) 5 : 10

(b) 7 : 13

(c) 6 : 11

(d) 9 : 13

[2018 : 2 Marks]

73. Leila aspires to buy a car worth Rs. 10,00,000 after 5 years. What is the minimum amount in Rupees that she should deposit now in a bank which offers 10% annual rate of interest, if the interest was compounded annually?

(a) 5,00,000

(c) 6,66,667

(b) 6,21,000

(d) 7,50,000

[2018 : 2 Marks]

74. A cab was involved in a hit and run accident at night You are given the following data about the cabs in the city and the accident.

(i) 85% of cabs in the city are green and the remaining cabs are blue.

(ii) A witness identified the cab involved in the accident as blue.

(iii) It is known that a witness can correctly identify the cab colour only 80% of the time. cab?

Which of the following options is closest to the probability that the accident was caused by a blue cab?

(a) 12% (b) 15%

(c) 41% (d) 80%

[2018 : 2 Marks]

75. It would take one machine 4 hours to complete a production order and another machine 2 hours to complete the same order. If both machines work simultaneously at their respective constant rates, the time taken to complete the same order is ______ hours.

(a) 2/3 (b) 7/3

(c) 4/3 (d) 3/4

[2019 : 1 Mark]

76. Two design consultants, P and Q, started working from 8 AM for a client. The client budgeted a total of USD 3000 for the consultants. P stopped working when the hour hand moved by 210 degrees on the clock. Q stopped working when the hour hand moved by 240 degrees. P took two tea breaks of 15 minutes each during her shift, but took no lunch break. Q took only one lunch break for 20 minutes, but no tea breaks. The market rate for consultants is USD 200 per hour and breaks are not paid. After paying the consultants, the client shall have USE _____ remaining in the budget.

(a) 000.00

(b) 433.33

(c) 166.67

(d) 300.00

[2019 : 2 Marks]

77. The bar graph in Panel (a) shows the proportion of male and female illiterates in 2001 and 2011. The proportions of males and females in 2001 and 2011 are given in Panel (b) and (c), respectively. The total population did not change during this period.

The percentage increase in the total number of literate from 2001 to 2011 is ________.

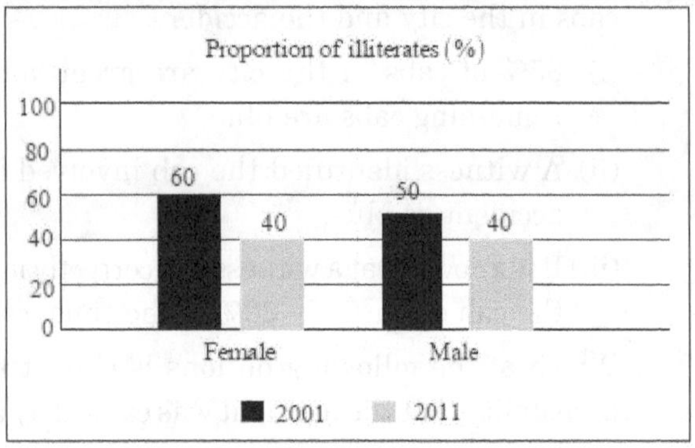

Panel (a)

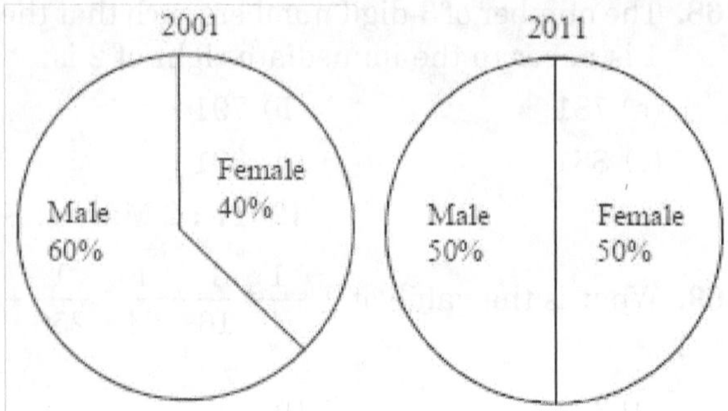

Panel (b) Panel (c)

(a) 33.43 (b) 35.43

(c) 34.43 (d) 30.43

[2019 : 2 Marks]

ANSWERS

1. (*)	2. (*)	3. (d)	4. (b)	5. (a)	6. (a)	7. (c)	8. (d)	9. (c)	10. (d)
11. (d)	12. (c)	13. (a)	14. (a)	15. (c)	16. (*)	17. (d)	18. (b)	19. (a)	20. (a)
21. (b)	22. (*)	23. (d)	24. (b)	25. (b)	26. (163)	27. (d)	28. (a)	29. (d)	30. (b)
31. (20,000)		32. (0· 8145)		33. (a)	34. (d)	35. (b)	36. (20,000)		37. (b)
38. (a)	39. (c)	40. (b)	41. (b)	42. (b)	43. (800)	44. (*)	45. (c)	46. (*)	47. (a)
48. (c)	49. (c)	50. (c)	51. (c)	52. (b)	53. (d)	54. (d)	55. (b)	56. (a)	57. (d)
58. (b)	59. (a)	60. (d)	61. (b)	62. (a)	63. (c)	64. (a)	65. (*)	66. (*)	67. (*)
68. (*)	69. (d)	70. (b)	71. (b)	72. (b)	73. (b)	74. (a)	75. (c)	76. (c)	77. (d)

EXPLANATIONS

3. 5 skilled workers build wall in 20 days

1 skilled worker build wall in 20 × 5 days

Hence in 1 day, part of work done by skilled work

$$= \frac{1}{100}$$

Similarly in 1 day part of work done by semi-skilled workers

$$= \frac{1}{25 \times 8}$$

and in 1 day part of work done by un-skilled worker

$$= \frac{1}{30 \times 10}$$

So part of work done in 1 day by 2 skilled, 6 semi-skilled and 5 unskilled

$$= \frac{2}{100} + \frac{6}{200} + \frac{5}{300} = \frac{1}{15}$$

So work done by given workers in days = 15

4. 2, 2, 3, 3, 4, 4, 4, 4 $x > 3000$

322 — 3, 4	344 — 2, 3, 4
323 — 2, 3, 4	422 — 3, 4
324 — 2, 3, 4	423 — 2, 3, 4
332 — 2, 3, 4	424 — 2, 3, 4
333 — 2, 4	432 — 2, 3, 4
334 — 2, 3, 4	433 — 2, 3, 4
342 — 2, 3, 4	434 — 2, 3, 4
343 — 2, 3, 4	442 — 2, 3, 4
	443 — 2, 3, 4
	444 — 2, 3, 4

6.

P	Q
40%	60%
–6%	+6%
+15%	–15%
49%	51%

∴ 2% = 2 ⇒ 100% = 100

7.

	Fuel consumption	Actual
P	60 km/l	$\frac{15}{60} = \frac{1}{4} l$
Q	90 km/l	$\frac{75}{90} = \frac{5}{6} l$
R	75 km/l	$\frac{40}{75} = \frac{8}{15} l$
S	30 km/l	$\frac{10}{30} = \frac{1}{3} l$

8. Let total number of toffees in bowl be x

R took $\frac{1}{3}$ of toffees and returned 4 to the bowl

∴ Number of toffees with R = $\frac{1}{3}x - 4$

Remaining of toffees in bowl = $\frac{2}{3}x + 4$

Number of toffees with S = $\frac{1}{4}\left(\frac{2}{3}x+4\right) - 3$

Remaining toffees in bowl = $\frac{3}{4}\left(\frac{2}{3}x+4\right)+4$

Number of toffees with T = $\frac{1}{2}\left(\frac{3}{4}\left(\frac{2}{3}x+4\right)+4\right)+2$

Remaining toffees in bowl

$$= \frac{1}{2}\left[\frac{3}{4}\left(\frac{2}{3}x+4\right)+4\right]+2$$

Given : $\frac{1}{2}\left[\frac{3}{4}\left(\frac{2}{3}x+4\right)+4\right]+2 = 17$

$$\Rightarrow \frac{3}{4}\left(\frac{2}{3}x+4\right) = 27$$

$$\Rightarrow x = 48$$

9. Given : $f(y) = \frac{|y|}{y}$

$$\Rightarrow f(q) = \frac{|q|}{q}$$

$$f(-q) = \frac{|-q|}{-q} = \frac{-|q|}{q}$$

$$|f(q) - f(q)| = \frac{|q|}{q} + \frac{|q|}{q} = \frac{2|q|}{q} = 2$$

10. Let S = 4 (1 + 11 + 111 + ...)

$$= \frac{4}{9}(9 + 99 + 999 + ...)$$

$$= \frac{4}{9}\{(10-1) + (10^2-1) + (10^3-1) + ...\}$$

$$= \frac{4}{9}\{(10 + 10^2 + ... + 10^n) - n\}$$

$$= \frac{4}{9}\left\{10\frac{(10^n-1)}{9} - n\right\}$$

$$= \frac{4}{81}\{10^{n+1} - 9n - 10\}$$

11. Let $1.001 = x$

$x^{1259} = 3.52$, and $x^{2062} = 7.85$

$\therefore \quad x^{3321} = x^{1259} \cdot x^{2062}$

$= 3.52 \times 7.85$

$= 27.64$

12. Total budget = 10,500

Expenditure other than savings = 9000

Hence approximate percentage of monthly budget

$$= \frac{9000}{10500} = 86\%$$

13.

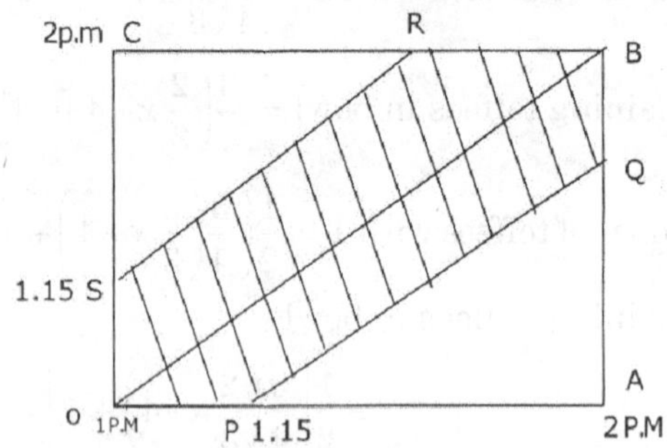

OB is the line when both A and B arrive at same time.

Total sample sapce = $60 \times 60 = 3600$

Favourable cases = Area of OABC – 2(Area of SRC)

$$= 3600 - 2 \times \left(\frac{1}{2} \times 45 \times 45\right)$$

$$= 1575$$

$\therefore$ Required probability $= \dfrac{1575}{3600} = \dfrac{7}{16}$

14. Let number of ₹ 20 notes be x and ₹ 10 notes be y

$\therefore \quad 20x + 10y = 230 \quad \ldots (1)$

and $\quad x + y = 14 \quad \ldots (2)$

Solving equations (1) and (2), we have

$x = 9$ and $y = 5$

Hence numbers of 10 rupee notes are 5.

15. Let us categorize bags in three groups as

$A_1\ A_2\ A_3 \quad B_1\ B_2\ B_3 \quad C_1\ C_2$

1st weighing A vs B

Case-1	**Case-2**
$A_1\ A_2\ A_3 = B_1\ B_2\ B_3$	$A_1\ A_2\ A_3 \neq B_1\ B_2\ B_3$
Then either C_1 or C_2 is heavier	Either A or B would be heavier (Say A > B)
2nd weighing	
C_1 vs C_2	A_1 vs A_2
If $C_1 > C_2$, then C_1	If $A_1 = A_2$, then A_3
If $C_1 < C_2$, then C_2	If $A_1 > A_2$, then A_1
	If $A_1 < A_2$, then A_2

16. $\dfrac{\text{Mon + Tues + Wed.}}{3} = 41$

Mon + Tues + Wed. = 123 ...(1)

$\dfrac{\text{Tues + Wed + Thurs.}}{3} = 43°$

Tue + Wed + Thu. = 129° ...(2)

(2) – (1)

Tues + Wed + Thu – (Mon + Tues. + Wed.)

= 129 – 123 = 6°

Thu. – Mon. = 6° $\Rightarrow \dfrac{115x}{100} - x = 6°$

Thus. = Mon $\times \dfrac{115}{100} = \dfrac{15x}{100} = 6°$

Mon = $x \quad x = 40°$

Thurs = $\dfrac{115x}{100}$

$\therefore \quad$ Thurs = $\dfrac{115 \times 40}{100} = 46°$

20. at a leap year

52 weeks, and 2 extra day they are (Mon, Tues) (Tues, Wed) (Wed Thu.) (Thus. Fri) (Fri. Sat) (Sat. Sun) (Sun Mond.)

$n(s) = 7$

$n(E) = 2 \quad P(E) = \dfrac{2}{7}$

24. By reading the instructions, it is clear that person who fills form T is a resident female with annual income of ₹ 9 Lakh.

26. For 2005,

trade deficit = (90 – 70) crores = 20 crores

Now $\dfrac{1}{5}^{\text{th}}$ of export = $\dfrac{1}{5}(70)$ crores

= 14 crores ≠ 20 crores.

Hence option (*a*) is wrong.

For 2004,

Trade deficit = (80 – 70) crores = 10 crores

Now $\dfrac{1}{5}^{\text{th}}$ of export= $\dfrac{1}{5}(70)$ crores

= 14 crores ≠ 20 crores.

Hence option (*b*) is wrong.

For 2007

Trade deficit = (120 – 110) crores

= 10 crores

Now $\dfrac{1}{5}^{\text{th}}$ of export = $\dfrac{1}{5}(110)$ crores

= 22 crores ≠ 10 crores.

Hence option (*c*) is wrong.

For 2006,

Trade deficit = (120 – 100) crores

= 20 crores

Now $\frac{1}{5}^{th}$ of export = $\frac{1}{5}$ (100) crores

= 20 crores = Trade deficts

Hence option (*d*) is correct.

27. The probability that the other face is tail is $\frac{1}{3}$.

28. For a very large number of throws, the frequency should be same for unbiased throw. As it not same, then the die is baised.

32. Eight consecutive odd number = 656

$a - 6, a - 1, a - 2, a, a + 2, a + 4, a + 6$

$a + 8 = 656$

$a = 81$

Smallest $m = 75$ ...(1)

Average consecutive even numbers

$$\Rightarrow \frac{a-2+a+a+2+a+4}{4} = 87$$

$\Rightarrow a = 86$

Second largest number = 88 ...(2)

Adding equation (1) and (2),

75 + 88 = 163

33. Item : 2

$$\frac{\frac{20}{100}\times 250\times 10^7}{\frac{20}{100}\times 500\times 10^3}$$

$0.5\times 10^4 = 5\times 10^3 \boxed{1} = \text{Item 2}$

Item : 3

$$\frac{23\times 250\times 10^7}{19\times 500\times 10^3}$$

1.2 = Item 3

Item : 6

$\frac{19}{16} = 1.18 = \text{Item 6}$

Item : 5

$\frac{20}{12} = \frac{5}{3} = 1.6 \Rightarrow \boxed{1.6 = \text{Item 5}}$

34. $V_{half} = 30(s)$ drawing rate = s

Total volume = 60 S tank

$(s^1)(10) - (s)10 = 30s$

$s^1(s) - s = 3s$

$s1 = 4s$

$s^1 = 4$ drawing rate

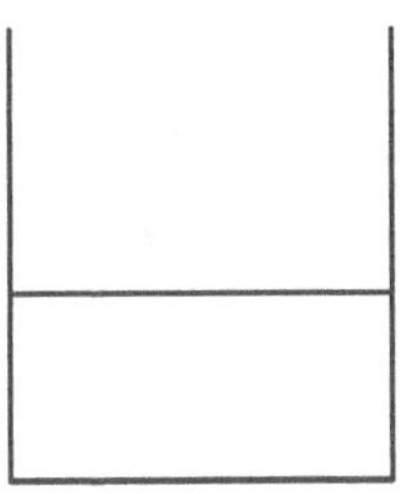

39. It is not mentioned that elephant is the largest animal.

40. Speed of man = 8;

Left distance = d

Time taken = $\frac{d}{8}$

Upstream: Speed of stream = s

$\Rightarrow$ speed upstream = S' = (8 – s)

$$t' = \left(\frac{d}{8-s}\right)$$

Downstream:

Given speed downstream = $t'' = \frac{d}{8+s}$

$$\Rightarrow 3t' = t'' \Rightarrow \frac{3d}{8-s} = \frac{d}{8+s}$$

$$\Rightarrow \frac{3d}{8-s} = \frac{d}{8+s}$$

$\Rightarrow$ s = 4 km/hr

41. Total expenditure = $\frac{15}{100}$ x = 4,50,000

$x = 3\times 10^6$

Profit = 10 lakhs

So, total selling price = 40,00,000 ...(1)

Total purifies = 200 ...(2)

S.P. of each purifier = (1)/(2) = 20,000

42. Probability for one bulb to be non-defective is $\frac{95}{100}$

$\therefore$ Probabilities that none of the bulbs is defectives $\left(\frac{95}{100}\right)^4 = 0.8145$

43.
$$P' = 2^n X^n \left(\frac{1}{2}\right)^m y^m$$

$$= 2^{n-m} X^n Y^m = 2^{n-m} P$$

44. 8th observation is 7 × 2 = 14 more than 1st observation

9th observation is 14 more than 2nd observation

10th observation is 14 more than 3rd observation

11th observation is 14 more than 4th observation

12th observation is 14 more than 5th observation

Total 14 × 5 =70

Sum of the first five numbers = 425

Sum of last five numbers = 495

48. Total expenditure = $\frac{15}{100}$ $x = 4,50,000$

$x = 3 \times 10^6$

Profit = 10 lakhs

So, total selling price = 40,00,000 ...(1)

Total purifies = 200 ...(2)

S.P. of each purifier = (1)/(2) = 20,000

49. The digit in unit place is selected in 4! Ways

The digit in tens place is selected in 4! Ways

The digit in hundreds place is selected in 4! Ways

The digit in thousands place is selected in 4! Ways

The digit in ten thousands place is selected in 4! Ways

Sum of all values for 1

$4! \times 1 \times (10^0 + 10^1 + 10^2 + 10^3 + 10^4)$

$= 4! \times 11111 \times 1$

Similarly for '3' $4! \times (11111) \times 3$

Similarly for '5' $4! \times (11111) \times 5$

Similarly for '7' $4! \times (11111) \times 7$

Similarly for '9' $4! \times (11111) \times 9$

$\therefore$ sum of all such numbers

$4! \times (11111) \times (1 + 3 + 5 + 7 + 9)$

$= 24 \times (11111) \times 25$

$= 6666600$

50. $\frac{5}{7} = x^{-1/3} \Rightarrow \frac{7}{5} = x^{1/3} \Rightarrow \left(\frac{7}{5}\right)^3 = x \Rightarrow x = \frac{343}{125}$

52.

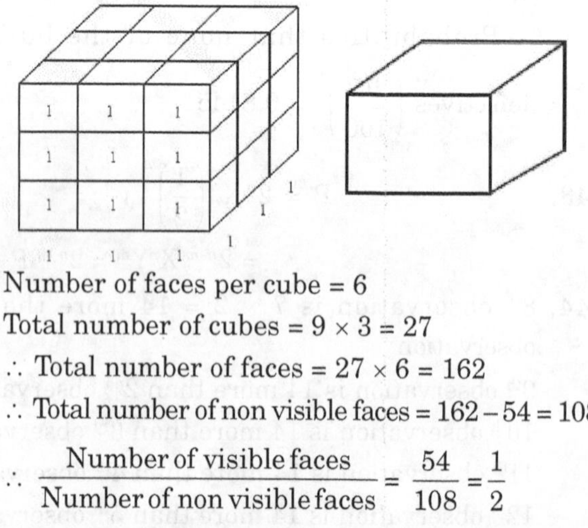

Number of faces per cube = 6

Total number of cubes = 9 × 3 = 27

$\therefore$ Total number of faces = 27 × 6 = 162

$\therefore$ Total number of non visible faces = 162 − 54 = 108

$\therefore \frac{\text{Number of visible faces}}{\text{Number of non visible faces}} = \frac{54}{108} = \frac{1}{2}$

54. For $M \Rightarrow \frac{12}{20} = 0.6$

$N \Rightarrow \frac{25}{45} = 0.55$

$O \Rightarrow \frac{45}{75} = 0.6$

$P \Rightarrow \frac{57}{100} = 0.57$

55. P (Ram) = 1/6; p(Ramesh) = 1/8

p(only at) = p (Ram) × p (not ramesh) + p(Ramesh) × $p(n_0 \times R_{am})$

$= \frac{1}{6} + \frac{7}{8} + \frac{1}{8} \times \frac{5}{6}$

$\Rightarrow \frac{12}{40} = \frac{1}{4}$

56. 1 8 M

57. Tiger – 1 leap ⇒ 8 meter

Speed = 5 leap/hr = 40m/min

Deer → 1 leap = 5 meter

speed = 4hr = 20m/min

Let at time '*t*' the tiger catches the deer.

$\therefore$ Distance travelled by deer + initial distance between them

50 × 8 ⇒ 400m = distance covered by tiger.

$\Rightarrow \quad 40 \times t = 400 + 20t$

$\Rightarrow \quad t = \frac{400}{200} = 20$ min

⇒ total distance

$400 + 20 \times t = 800$ m

60. log tan 1° + log tan 89° = log(tan 1° × tan 89°)

= log(tan 1° × cot 1°)

= log 1 = 0

Using the same logic total sum is '0'.

63. Given,

Total no. of fruits = 5692000

Unripe type of apples = 45% of 15% of 5692000

$= \frac{45}{100} \times \frac{15}{100} \times 5692000$

$= 384210$

Ripe type of apples = (100 – 66)% × (100 – 15)% × 5692000

$= \frac{34}{100} \times \frac{85}{100} \times 5692000$

$= 1644988$

$\therefore$ Total no. of apples = 384210 + 1644988

= 2029198

65. The percentage of people moving through a tuberculosis prone zone remains infected but does not show symptoms of disease.

$$= (100 - 30)\% \times 50\% = \frac{70}{100} \times \frac{50}{100} = \frac{35}{100} = 35\%$$

67. Given $q^{-a} = \frac{1}{r}$

$\Rightarrow \frac{1}{q^a} = \frac{1}{r}$

$\Rightarrow q^a = r$ ...(*i*)

$r^{-b} = \frac{1}{s}$

$\Rightarrow \frac{1}{r^b} = \frac{1}{s}$

$\Rightarrow s = r^b$ ...(*ii*)

$s^{-c} = \frac{1}{q}$

$\Rightarrow \frac{1}{s^c} = \frac{1}{q}$

$\Rightarrow s^c = q$ ...(*iii*)

From equation (*i*),

$q^a = r$

$(s^c)^a = r$ (from eq. (*iii*))

$S^{ac} = r$

$s^{ac} = (s)^{\frac{1}{b}}$ (from eq. (*ii*))

$s^{abc} = s = s^1$

$\therefore abc = 1$

So the value of *abc* is 1

68. Q can finish 1 day's 1 hour's work

$$= \frac{1}{25 \times 12} = \frac{1}{300}$$

R can finish 1 day's 1 hour's work

$$= \frac{1}{50 \times 12}$$

$$= \frac{1}{600}$$

Now Q working hours

$= (7 - 2) \times 12$

$= 60$ hr

$\therefore$ R working hours $= 7 \times 18 = 126$

After 7 days, the ratio of work done by Q & R is

Q : R

$$\frac{60}{300} : \frac{126}{600}$$

Q : R = 120 : 126

= 20 : 21

70. $(9 \text{ inches})^{1/2} = (0.25 \text{ yards})^{1/2}$,

Solving we get 9 inch = 0.25 yards

(since 1 inch = 0.028 yard)

71. M is twice as efficient as E but worked for half as many days. So in this case they will do equal work if their shifts had same timings. But M's shift is for 6 hours, while E's shift for 12 hours. Hence, E will do twice the work as M.

Ratio of contribution of M : E in work is 1 : 2.

72. Given Venn diagram is

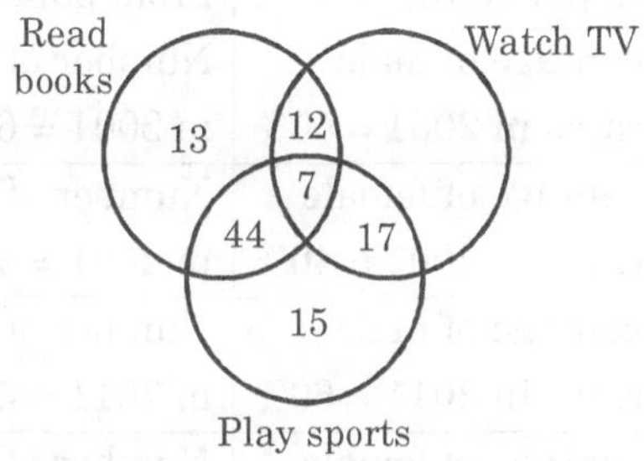

The number of students who like to read books or play sports

= 13 + 12 + 44 + 7 + 15 + 17 = 108

73.

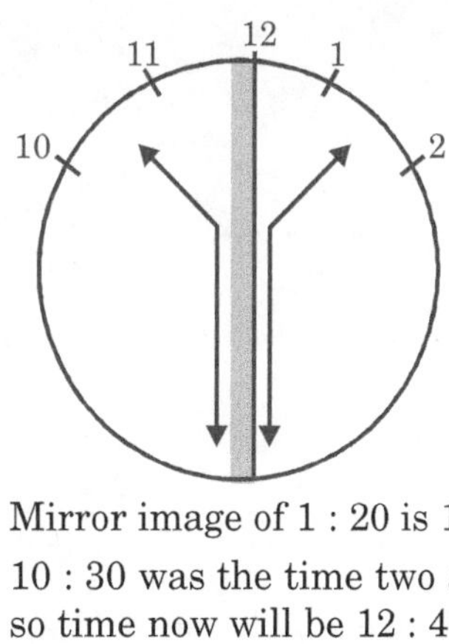

Mirror image of 1 : 20 is 10 : 30

10 : 30 was the time two and quarter hour back so time now will be 12 : 45

74. Time taken $= \left(\frac{4 \times 2}{4 + 2}\right)$ hr

$= \frac{4}{3}$ hours

75. Given, P and Q started working from 8 A.M. for a client

Total budget = USD 3000

P worked exactly 7 hours and took 30 min break in between.

$\Rightarrow$ 'P' worked number of hours = 6.5 hours

'Q' worked exactly 8 hours and took 20 min break in between.

$\Rightarrow$ 'Q' worked number of hours $\approx$ 7.67 hours

Client amount paid by for both P and Q

= USD 200/hr

$\therefore$ Total USD Paid = $(6.5 \times 200 + 7.67 \times 200)$

$= (1300 + 1534) = 2834$

Remaining amount left

$= 3000 - 2834 \approx 166$ (approx)

76. Let us assume, that population = 100 [2001 – 2011]

From panel (a);	From panel (b & c)
Percentage of male literates in 2001 = 50%	Number of males in 2001 = 60
Percentage of female literates in 2001 = 40%	Number of females in 2001 = 40
Percentage of male literates in 2011 = 60%	Number of males in 2011 = 50
Percentage of female literates in 2011 = 60%	Number of females in 2011 = 50

$\therefore$ Number of male literates in 2001

$$= 60 \times \frac{50}{100} = 30$$

Number of male literates in 2011

$$= 50 \times \frac{60}{100} = 30$$

Number of female literates in 2001

$$= 40 \times \frac{40}{100} = 16$$

Total number of literates in 2001

$$= 30 + 16 = 46$$

Number of female literates in 2011

$$= 50 \times \frac{60}{100} = 30$$

Total number of literates in 2011

$$= 30 + 30 = 60$$

$\therefore$ Percentage increase in the total number of literates from 2001–2011

$$= \left[\frac{60-46}{46} \times 100\right]\% = \left[\frac{14}{46} \times 100\right]\% = 30.43\%$$